BIOLOGY

Fifth Edition

Peter H. Raven

Director, Missouri Botanical Garden;
Engelmann Professor of Botany
Washington University

George B. Johnson

Professor of Biology
Washington University

Significant contributions
to the animal biology chapters by

Stuart Ira Fox
Pierce College

Boston Burr Ridge, IL Dubuque, IA Madison, WI New York San Francisco St. Louis
Bangkok Bogotá Caracas Lisbon London Madrid
Mexico City Milan New Delhi Seoul Singapore Sydney Taipei Toronto

WCB/McGraw-Hill

A Division of The McGraw·Hill Companies

BIOLOGY, FIFTH EDITION

2 3 4 5 6 7 8 9 0 VNH/VNH 9 3 2 1 0 9

ISBN 0–697–35353–2

Vice president and editorial director: *Kevin T. Kane*
Publisher: *Michael D. Lange*
Sponsoring editor: *Patrick E. Reidy*
Senior developmental editor: *Elizabeth M. Sievers*
On-site developmental editor: *Megan Jackman*
Project manager: *Marilyn M. Sulzer*
Senior production supervisor: *Mary E. Haas*
Design manager: *Stuart D. Paterson*
Photo research coordinator: *John C. Leland*
Art editor: *Jodi K. Banowetz*
Senior supplement coordinator: *Audrey A. Reiter*
Compositor: *Shepherd, Inc.*
Typeface: *10/12 Janson*
Printer: *Von Hoffmann Press, Inc.*

Cover design: *Christopher Reese*
Interior design: *Kathy Theis*
Cover image: *"Tree of Life" by Alfredo Arreguin*
Illustrations done by: *Art and Science Inc.; Medical & Scientific Illustration; William C. Ober and Claire W. Garrison*

The credits section for this book begins on page 1247 and is considered an extension of the copyright page.

Library of Congress Cataloging-in-Publication Data
Raven, Peter H.
    Biology / Peter H. Raven, George B. Johnson. — 5th ed.
        p.   cm.
    Includes bibliographical references and index.
    ISBN 0-697-35353-2
    1. Biology.   I. Johnson, George B.   (George Brooks),   1942-   .
II. Title.
QH308.2.R38   1999
570—dc21   97–48875
                                                        CIP

INTERNATIONAL EDITION
Copyright © 1999. Exclusive rights by The McGraw-Hill Companies, Inc. for manufacture and export. This book cannot be re-exported from the country to which it is consigned by McGraw-Hill. The International Edition is not available in North America.
When ordering this title, use ISBN 0–07–115511–2.

www.mhhe.com

Brief Contents

iii

Section A General Principles

Chemical Biology

Part I The Origin of Living Things 1

Cell Biology

Part II Biology of the Cell 79

Part III Energetics 145

Genetics

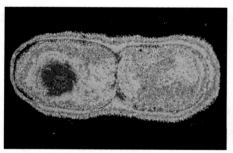

Part IV Reproduction and Heredity 205

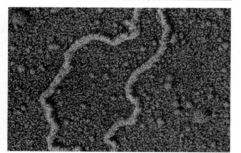

Part V Molecular Genetics 277

Evolution and Ecology

Part VI Evolution 389

Part VII Ecology 463

Contents

Chemical Biology

Part I   *The Origin of Living Things*

1

The Science of Biology   1

Science is the process of testing ideas against observation. Darwin developed his ideas about evolution by testing them against a wealth of observation. In this text science will provide the framework for your exploration of life.

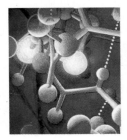

2

The Nature of Molecules   19

Organisms are chemical machines, and to understand them we must first learn a little chemistry. We first explore how atoms are linked together into molecules. The character of the water molecule in large measure determines what organisms are like.

3

The Chemical Building Blocks of Life   37

The four kinds of large macromolecules that are the building blocks of organisms are each built up of long chains of carbon atoms. In each, the macromolecule is assembled as a long chain of subunits, like pearls in a necklace or cars of a railway train.

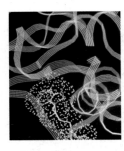

4

The Origin and Early History of Life   61

Little is known about how life originated on earth. If it originated spontaneously, as most biologists surmise, then it must have evolved very quickly, as microfossils of bacteria are found in rocks formed soon after the earth's surface cooled.

Cell Biology

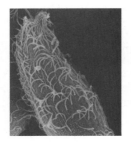

5   *Cell Structure*   79

Bacterial cells have little internal organization, while the cells of eukaryotes are subdivided by internal membranes into numerous compartments with different functions. Compartmentalization is the hallmark of the eukaryotic cell.

6   *Membranes*   105

Every cell is encased within a thin membrane that separates it from its environment. The membrane is a mosaic of proteins floating on a sheet of lipid that provide channels into the cell for both molecules and information.

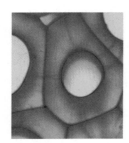

7   *Cell-Cell Interactions*   127

Cells receive molecular signals with protein receptors on or within the plasma membrane. The information passes into the cell interior as a cascade of interactions that greatly amplify the strength of the original signal.

Genetics

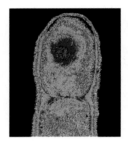

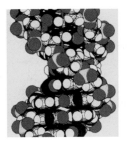
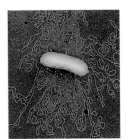

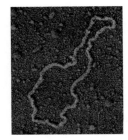

Evolution and Ecology

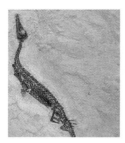

23 Population Ecology 463

The way in which a population grows depends upon how many young individuals it contains. The way in which a species reproduces represents an evolutionary trade-off between reproductive cost and investment in survival.

23.1 Populations are individuals of the same species that live together.
23.2 Population dynamics depend critically upon age distribution.
23.3 Life histories often reflect trade-offs between reproduction and survival.
23.4 Population growth is limited by the environment.

24 Community Ecology 479

Organisms make complex evolutionary adjustments to living together within communities. Some adjustments involve cooperation, others capturing prey or avoiding being captured. In general, communities with more species interacting with one another are more stable.

24.1 Competition shapes communities.
24.2 Evolution often fosters cooperation.
24.3 Predators and their prey coevolve.
24.4 Biodiversity promotes community stability.

25 Dynamics of Ecosystems 503

Ecological systems are dynamic "machines" in which chemicals cycle between organisms and the environment and energy is used as it passes through the system. Much energy is lost to heat at each step of its journey through living things.

25.1 Chemicals cycle within ecosystems.
25.2 Ecosystems are structured by who eats whom.
25.3 Energy flows through ecosystems.

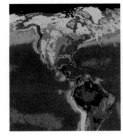

26 The Biosphere 517

The sun powers major movements of the atmosphere that circulate heat and moisture over the surface of the globe, creating different climates in different locales. Each characteristic climate favors a particular community of organisms adapted to living there.

26.1 Organisms must cope with a varied environment.
26.2 Climate shapes the character of ecosystems.
26.3 Biomes are widespread terrestrial ecosystems.
26.4 Aquatic ecosystems cover much of the earth.

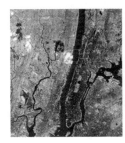

27 The Future of the Biosphere 543

The world's human population is growing at an explosive rate, placing a severe strain on the world's ecosystems. No one knows if the world can sustain the nearly 6 billion people it now contains—and the population will soon grow much larger.

27.1 The world's human population is growing explosively.
27.2 Improvements in agriculture are needed to feed a hungry world.
27.3 Human activity is placing the environment under stress.
27.4 Solving environmental problems requires individual involvement.

Section B   Organismal Biology

Simple Organisms

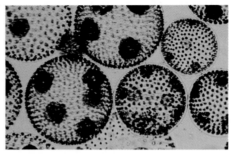

Plants

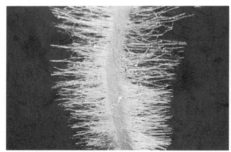

Animals I

Animals II

Simple Organisms

Part VIII *Viruses and Simple Organisms*

28 How We Classify Organisms 559

The living world is a rich tapestry of diversity, teeming with different kinds of organisms. One of the great challenges of biology is to find sensible ways to name and classify kinds of organisms, ways that tell us about them and how they relate to each other.

28.1 Biologists name organisms in a systematic way.
28.2 Taxonomy is the science of classifying organisms.
28.3 All living organisms are grouped into one of a few major categories.

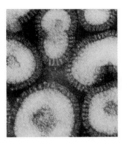

29 Viruses 575

Viruses are not considered organisms because they lack cellular structure. However, because they can replicate themselves within the cells of organisms, viruses are able to evolve and are responsible for a broad array of serious diseases.

29.1 Viruses are fragments of DNA or RNA that have detached from genomes.
29.2 Bacterial viruses exhibit two sorts of reproductive cycles.
29.3 HIV is a typical animal virus.
29.4 Viruses are responsible for many important human diseases.
29.5 Viruses are not the only nonliving infectious agents.

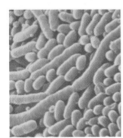

30 Bacteria 591

Bacteria are the simplest organisms, composed of single cells that lack the complex internal organization seen in all other cells. Biologists believe bacteria were the first organisms to evolve on earth, and they are easily the most numerous and successful.

30.1 Bacteria are the smallest and most numerous organisms.
30.2 Bacterial cell structure is more complex than commonly supposed.
30.3 Bacteria exhibit considerable diversity in both structure and metabolism.
30.4 Bacteria are responsible for many human diseases.

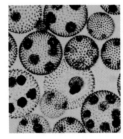

31 Protists 605

All eukaryotic organisms that are not fungi, plants, or animals are lumped together in a catch-all category called protists. Except for the marine algae, all protists are unicellular, but the diversity among protists is so great that it is difficult to compare them.

31.1 The kingdom Protista is by far the most diverse of any kingdom.
31.2 Protists are grouped into fifteen very distinctive phyla.

32 Fungi 629

Fungi are extraordinarily strange multicellular organisms whose cells share cytoplasm and nuclei. Fungi absorb their food from their surroundings, first excreting digestive enzymes, then soaking up the resulting soup of molecular fragments.

32.1 Fungi are unlike any other kind of organism.
32.2 Fungi are classified by their reproductive structures.
32.3 Fungi form two key symbiotic associations.

Plants

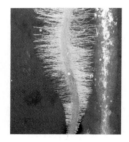

Animals I

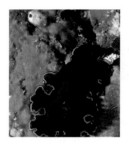
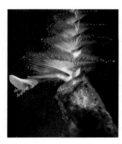

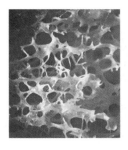

46 *Organization of the Animal Body* 903

Although they at first glance seem quite different from one another, all vertebrates are basically similar in body design, using the same array of tissues to form a similar array of organs. Vertebrates differ principally in degree of terrestrial adaptation.

47 *Locomotion* 919

Muscle tissue, because it can contract, allows animals to move about. Some animals, like fish and snakes, move their entire bodies, while others move limbs such as arms, legs, or wings. The contraction of muscle is powered by ATP and controlled by the nervous system.

48 *Fueling Body Activities: Digestion* 935

Before the cells of an animal can assimilate fatty acids, sugars, and amino acids to use in manufacturing ATP, these "food" molecules must be cleaved out of far larger fats, carbohydrates, and proteins. This process of digestion takes place in the stomach and small intestine.

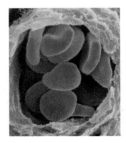

49 *Circulation* 957

The highway that carries materials from one place to another in the vertebrate body is a network of flexible pipes called arteries and veins. A muscular pump called the heart pushes a rich fluid called blood through these pipes to every organ in the body.

50 *Respiration* 973

One of the most important cargos carried by blood are two gases, oxygen and carbon dioxide. In the lungs, blood picks up oxygen from the air and dumps carbon dioxide acquired from the body's tissues as a by-product of their respiratory metabolism.

Animals II

Part XIII *Regulating the Animal Body*

51 *The Nervous System* 991

The activities of the many organs of the vertebrate body are coordinated by the nervous system, an extensive network of neuron cells connecting all organs of the body. At a central coordinating center, the brain, information is processed, associations made, and commands given.

52 *Sensory Systems* 1021

Information used by the nervous system to coordinate the body's activities comes from a system of sensors that collects information about the body's condition, its position in space, and what is going on around it. Vertebrates differ substantially in which sensors they employ.

53 *The Endocrine System* 1045

The vertebrate nervous system uses long-lasting chemical signals called hormones to coordinate many physiological and developmental processes. The cells that secrete these hormones are often regulated by a set of command hormones released by the brain.

54 *The Immune System* 1067

Vertebrates defend themselves from infection by viruses, bacteria, and other microbes with an army of circulating cells that seek out and check the identity of all cells in the body. When a stranger is identified, other "killer" cells are called in to deal with the invader.

55 *Maintaining Homeostasis* 1095

Vertebrates maintain internal conditions at constant values, often quite different from those of their environment. This is particularly true of temperature and salt levels. Vertebrates go to great pains to maintain body temperature and avoid water loss or gain.

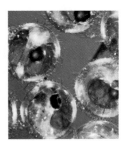

Preface

Biology was published in 1986 and, in the dozen years since its initial publication, has sold almost half a million copies. These have been exciting years for the science of biology, marked by great advances in many fields. The twin impacts of gene technology and evolutionary thinking have literally transformed the science of biology. In the last year alone, great strides were made in understanding AIDS and developing therapies for cancer, in sorting out the role of productivity in promoting ecosystem stability, and in tracing human origins in Africa. Much of the excitement inherent in scientific advance is reflected in controversial reports that have received considerable publicity. Recently, for example, we saw the reported cloning of an adult sheep, the suggestion of ancient life on a Mars rock, and the Nobel Prize recognition of a claim that mad-cow disease is caused by infectious proteins called prions.

The need to keep *Biology* current with these rapid advances is an obvious reason why we revised this text again this past year. A second and equally important reason is that the way biology is being taught has also undergone a sea change. There is far more emphasis today on the teaching of concepts than there was a dozen years ago, and this has led us to make significant changes in how we present material. Technology plays a greater role in teaching than it used to, both with interactive CD-ROMs, and more recently with the wealth of information that can be accessed via the internet.

The Approach of this Revision

Our approach to revising *Biology* has been guided by five considerations, each of which has influenced in a significant way how we have revised each chapter of this book:

1. **Focus on concepts.** Throughout the text we attempted to emphasize even more strongly the ideas of biology, the conceptual framework that is the core of what we want students to learn. Our efforts to reorganize the contents of each chapter into conceptual modules, discussed on the facing page, were in large measure driven by our desire to more strongly emphasize concepts, and particularly to bring out clearly the conceptual skeleton of the chapter.

2. **Reinforcing ideas.** The most effective way to learn biology is to frame the consideration of new material in terms of what has already been taught. Thus the idea of chemiosmosis, introduced in the discussion of membrane proteins in chapter 6, is subsequently used to explain a key aspect of how cells harvest energy in chapter 9, and form ATP in photosynthesis in chap-

ter 10. The mechanisms used by cells to control the cell cycle, outlined in chapter 11, play a central role in the discussion of molecular mechanisms of cancer in chapter 17. What you learn of viruses in chapter 29 plays a key role in understanding AIDS in chapter 56.

3. **Emphasizing relevance to students.** Because so much of what is going on in biology today directly affects the lives of students, we attempted to present very clear explanations of these key issues. The physiological nature of drug addiction, the way in which cigarette-induced mutations disable cell-cycle control mechanisms, the effects of alcohol on fetal development—these and other issues are discussed explicitly and in detail.

4. **Keeping up with new developments.** The ability of biologists to study cells in molecular detail continues to revolutionize biology. Much of what is important in cell and molecular biology today had not even been anticipated a few years ago.

5. **Careful editing.** The fifth edition has gone through a scrupulous review process employing experienced instructors as well as a large cadre of expert scientists to eliminate inaccurate information or misstatements.

Guiding Themes

Evolution is the core of the science of biology, and from the first words we wrote, has always been a central theme of *Biology*. Evolution provides a context for understanding broad biological phenomena, such as the marked differences in anatomy and physiology among the vertebrates. It also provides a context for exploring the details of quite specific processes, such as the rapid evolution of the HIV virus during the course of an AIDS infection.

Exploration—the way in which scientists investigate the unknown—is a key element of our presentation of biology. Throughout this revision we provide detailed explanations of how experiments have led to our knowledge of life's processes. We chose the cover of this edition to highlight this exploration. Hidden within the image are many elements not immediately obvious at first look. Examine the cover closely for faces!

Structure-and-function provides an orientation for examining how cells and multicellular organisms *work*. The cause-and-effect relationships that underlie anatomy and physiology are a core element in organizing how we present much of the material in this text. Understanding such mechanisms provides an important element of integration as students learn biology.

What We Did in this Revision

Learning Modules

A key element in our attempt to direct student focus toward concepts has been to reorganize the internal structure of each chapter into a series of discrete learning modules, each occupying one or two pages (a few use more), and ending with a summary. The concept outline that appears at the beginning of each chapter thus represents the conceptual skeleton of the chapter, allowing students to readily grasp how the concepts relate to one another and to the overall theme of the chapter.

This sort of modular organization has been used for several years in a variety of biology texts. It has proven a very effective way to point students toward the key ideas on which they need to focus. At first glance it might appear too confining an approach for a majors text like *Biology*, but we were surprised to find how naturally the material fell into this organization. The material was in almost all cases already arrayed into conceptual blocks, as this is the natural way to teach biology—making the individual treatments fit on one or two pages neatly was largely a matter of adjusting the sizes of figures. In practically no case was material cut to make a treatment *fit*. Far more often, we found ourselves adding new material. So the entire previous edition is still here, and even more. It is just presented in easy-to-grasp pieces. Traditionally, page layout happens late in the production process, long after the authors are finished with the manuscript. Beginning each major topic at the start of a page, the hallmark of this modular organization, would be practically impossible under that arrangement. In this revision, by contrast, each page of each chapter was layed out by the authors as we revised that chapter. Content and presentation were no longer independent, but rather part of the same creative process. We feel it has made for a much more effective teaching tool, while not in any way compromising content.

Expanded Coverage

In this revision we have greatly expanded our coverage of botany to eight full chapters. One four-chapter part, Plant Form & Function, covers elements of plant anatomy and physiology, and another, Plant Growth & Reproduction, covers plant reproduction and development, including a chapter devoted specifically to plant molecular biology. We have also greatly expanded our coverage of ecology, adding significant coverage of many current issues. Population and community ecology, weak points in previous editions, have been particularly strengthened.

In every other chapter of the text, we have added material where significant advances have taken place, from our understanding of how the cell cycle is controlled, to recent findings about the course of human evolution. Few chapters escaped significant improvement in coverage.

New Art Program

When *Biology* was first published in 1986, it was the first full-color majors text ever, although that is hard to comprehend in today's world of splashy five-color texts. In intervening years we have added to the art, improving individual pieces but inevitably introducing a variety of styles and conventions. In this revision we have gone back and started from scratch, introducing an entirely new art program. Every figure in the text has been reconsidered and redone, so that all illustrations are now consistent in style as well as content. Biological molecules and cell structures, for example, now are consistently color-coded to aid in student comprehension. Cross-sections of biological membranes look the same wherever they are encountered, and so do other elements of cell and body structure. This approach enrolls visual learning as a powerful ally in the teaching process.

Technology

In the last decade, and particularly in the three years since the last edition of *Biology*, sweeping changes in the use of technology have greatly altered how biology is taught in our classrooms. While the initial loudly-trumpeted advent of laser discs had little impact (thousands of extra images added little worth to biology teaching, just as a painting is rarely improved by giving the artist more paint), fully interactive CD-ROMs that allow students to explore freely and carry out meaningful experiments are now making real contributions to biology teaching in classrooms all over the country.

You will find a new CD-ROM tutorial available with this fifth edition of *Biology*. The *Essential Study Partner* (ESP) is available **free** with the fifth edition of *Biology* and contains high quality 3-D animations, interactive study activities, illustrated overviews of key topics in the text, and supplementary quizzing and exams that students will find extremely valuable. This is a study tool that your students must have, so they receive it free.

An even greater—in fact, explosive—impact is being had by access to the internet. Every text now has its own web site. Our web site, http://www.mhhe.com/biolink, provides a wealth of opportunities to the student and teacher. These include readings, sample tests, and other elements traditionally provided in study guides, as well as a wide range of other enrichments. The three most important, in our judgment, are monthly **updates** written by us that provide the student with current information about rapidly-changing areas of biology, **internet links** provided with each chapter that let students see the range of internet opportunities available, and **enhancement chapters** written by us that allow students and instructors to explore rapidly changing areas of biology in more depth. The three enhancement chapters included with this edition are *Conservation Biology*, *Dinosaurs*, and *The Revolution in Taxonomy*. We did not wish to lengthen an already-long text, and so have placed these chapters on our website, http://www.mhhe.com/biolink, where anyone using our text will have access to them.

Acknowledgments

We have been teaching biology together for 26 years, and are consistently surprised, each and every year, by how much fun it is. This book has been part of that fun, written in the spirit that biology ought to be engaging and interesting as well as comprehensive and authoritative. This fifth edition of *Biology* would not have been possible without the contributions of many others. Stuart Fox made major contributions to the animal biology section, James Traniello to the animal behavior chapters, and Don Briskin & Margaret Gawienowski to the chapter on plant molecular biology. Kingsley Stern provided detailed reviews of the botany section, and teachers around the country made numerous suggestions that have improved the book substantially. Every one of you has our heartfelt thanks.

A major feature of this edition has been the reformatting of the presentation into conceptual modules. This formidable task would not have been possible without the effort of Megan Jackman, our on-site developmental editor. Her intelligence and perseverance played a major role in the high quality of this book.

As any author knows, a textbook is made not by a writer but by a publishing team, a group of people that guide the raw book written by the authors through a year-long process of review, editing, fine-tuning and production. This edition was particularly fortunate in its book team, led by Elizabeth Sievers, a developmental editor who proved both a delight to work with and a reliable guide to the complicated process of getting the book out on time and in the best possible shape.

As always, we have had the support of wives and family who have seen less of us than they might like because of the pressures of getting this revision done. As this is the fifth edition of *Biology*, they have become accustomed to the many hours this book draws us away from them, a hidden price of textbook writing of which they are fully aware.

Acknowledgments would not be complete without thanking the generations of students who have used the many editions of this text. They have taught us at least as much as we have taught them.

Finally, we need to thank our reviewers. Every text owes a great deal to those faculty across the country who review it. Serving as sensitive antennae for errors and sounding boards for new approaches, reviewers are among the most valuable tools at an author's disposal. Many improvements in this edition are the direct result of their suggestions. Every one of them has our sincere thanks.

Expert Reviewers

Vernon Ahmadjian   *Clark University*
Sylvester Allred   *Northern Arizona University*
Jane Aloi   *Saddleback College*
Kenneth W. Andersen   *Gannon University*
Howard Arnott   *University of Texas at Arlington*
Robert W. Atherton   *University of Wyoming*

Bert Atsma   *Union County College*
Joyce Azevedo   *Southern Adventist University*
Milton Bailey   *Pulaski Technical College*
Frank Baker   *Golden West College*
Steve Baker   *Oxford College of Emory University*
Carla Barnwell   *University of Illinois*
Steven Bassett   *Southeast Community College*
Ruth E. Beattie   *University of Kentucky*
Robert Bergad   *Century College*
Joseph S. Bettencourt   *Marist College*
Prakash Bhuta   *Eastern Washington University*
Charles Biernbaum   *College of Charleston*
Del Blackburn   *Clark College*
Andrew R. Blaustein   *Oregon State University*
Hessel Bouma III   *Calvin College*
Bradley Bowden   *Alfred University*
Jim Brammer   *North Dakota State University*
Stan Braude   *Washington University*
Michelle A. Briggs   *Lycoming College*
Richard D. Brown   *Brunswick Community College*
Arthur L. Buikema, Jr.   *Virginia Tech*
William Bukovsan   *SUNY at Oneonta*
W. Barkley Butler   *Indiana University of Pennsylvania*
Michael A. Campbell   *Pennsylvania State University at Erie-Behrend College*
Kirit D. Chapatwala   *Selma University*
William H. Chrouser   *Warner Southern College*
Clarence W. Clark   *Morehouse College*
Robert Cleland   *University of Washington*
Richard Clements   *Chattanooga State Technical Community College*
Joe Coelho   *Western Illinois University*
William Cohen   *University of Kentucky*
Mariette Cole   *Concordia University, St. Paul*
Wade L. Collier   *Manatee Community College*
Gary Coté   *Radford University*
Neil Crenshaw   *Indiana River Community College*
Joe W. Crim   *University of Georgia*
Carol Crowder   *North Harris College*
Kenneth J. Curry   *University of Southern Mississippi*
Marya Czech   *Lourdes College*
Lisa Danko   *Mercyhurst College*
Garry Davies   *University of Alaska-Anchorage*
Keith R. Davis   *Ohio State University*
Teresa DeGolier   *Bethel College*
Jean DeSaix   *University of North Carolina*
Elizabeth A. Desy   *Southwest State University*
Walter J. Diehl   *Mississippi State University*
Ronald Dietz   *Pennsylvania State University*
Linda K. Dion   *University of Delaware*
Richard J. Driskill   *Delaware State University*
Lynn Ebersole   *Northern Kentucky University*
Phillip Eichman   *University of Rio Grande*
Donald Emmeluth   *Fulton-Montgomery Community College*
Bruce Evans   *Huntington College*
Robert Evans   *Rutgers University-Camden Campus*
David Evans   *University of Florida*
Lee Fairbanks   *Pennsylvania State University*
Guy Farish   *Adams State College*
Clarence Fouche   *Virginia Intermont College*
Stuart I. Fox   *Pierce College*
Carl Frankel   *Penn State University*
Donald French   *Oklahoma State University*

Jannon Fuchs   *University of North Texas*
Greg Garman   *Centralia College*
Dalton Gossett   *Louisiana State University in Shreveport*
Merrill Gassman   *University of Illinois at Chicago*
Robert George   *University of Wyoming*
Florence Gleason   *University of Minnesota*
David Goldstein   *Wright State University*
Glenn A. Gorelick   *Citrus College*
Joseph Goy   *Harding University*
Stephen C. Grebe   *The American University*
Leo T. Hall   *Phillips University*
W. E. Hamilton   *Penn State University, New Kensington Campus*
Denny Harris   *University of Kentucky*
Lauraine Hawkins   *Pennsylvania State University*
Danny G. Herman   *Montcalm Community College*
Deborah D. Hettinger   *Texas Lutheran University*
Theresa Hoffman-Till   *Independent Consultant*
Rebecca Holberton   *University of Mississippi*
Dale Holen   *Pennsylvania State University*
James Horwitz   *Palm Beach Community College*
Herbert W. House   *Elon College*
Lauren D. Howard   *Norwich University*
Terry L. Hufford   *George Washington University*
Allen Hunt   *Elizabethtown Community College*
Mack Ivey   *University of Arkansas*
Amiel Jarstfer   *LeTourneau University*
Rusell Johnson   *Ricks College*
Carolyn H. Jones   *Vincennes University*
Craig T. Jordan   *University of Texas at San Antonio*
Geoffrey Jowett   *Savannah College of Art and Design*
Eric Juterbock   *Ohio State University*
Carl S. Keener   *Pennsylvania State University*
Mary Evelyn B. Kelley   *St. Mary's College*
Lori Kelman   *Iona College*
Donald Kirk   *Shasta College*
William Kleinelp   *Middlesex County College*
Daniel Klionsky   *University of California-Davis*
Tim Knight   *Ouachita Baptist University*
Paul Lago   *University of Mississippi*
Allan J. Landwer   *Hardin-Simmons University*
V. A. Langman   *Louisiana State University*
Hugh Lefcort   *Gonzaga University*
Andrea LeSchack   *Florida Community College at Jacksonville*
Harvey B. Lillywhite   *University of Florida*
Roger Lloyd   *Florida Community College*
Paul Lutz   *University of North Carolina at Greensboro*
Jon Maki   *Eastern Kentucky University*
Charles Mallery   *University of Miami*
Betty McGuire   *Smith College*
J. Philip McLaren   *Eastern Nazarene College*
Mary Lou McReynolds   *Hopkinsville Community College*
Michael Meighan   *University of California-Berkeley*

John Messick   *Missouri Southern State College*
Ralph R. Meyer   *University of Cincinnati*
Melissa Michael   *University of Illinois*
Herbert Monoson   *Bradley University*
Michele Morek   *Brescia College*
Debra M. Moriarity   *University of Alabama in Huntsville*
David Morimoto   *Regis College*
Richard Mortensen   *Albion College*
Robert Muckel   *Doane College*
Colleen Nolan   *St. Mary's University*
Gregg Orloff   *Emory University*
Anne F. Pacitti   *Thomas Jefferson University*
Michael J. Patrick   *Pennsylvania State University, Altoona College*
Carol A. Paul   *Suffolk Community College*
Shelley Penrod   *North Harris College*
R. Gordon Perry   *Fairleigh Dickinson University*
John Pleasants   *Iowa State University*
Barbara Pleasants   *Iowa State University*
Donald R. Powers   *George Fox University*
Rudy Prins   *Western Kentucky University*
Maralyn Renner   *College of the Redwoods*
Quentin Reuer   *University of Alaska, Anchorage*
Barbra A. Roller   *Florida International University*
Lynette Rushton   *South Puget Sound Community College*
Mark Sanders   *University of California-Davis*
David L. Schultz   *Nicholls State University*
Peter J. Scott   *Memorial University of Newfoundland*
Robert Seagull   *Hofstra University*
Roger Seeber, Jr.   *West Liberty State College*
Marilyn L. Shaver   *Ashland Community College*
James Slock   *Kings College*
Lisa Shimeld   *Crafton Hills College*
Brian Shmaefsky   *Kingwood College*
J. Kenneth Shull   *Appalachian State University*
Paul Small   *Eureka College*
Joanna Smith   *Bullbrier Press*
Teresa Snyder-Leiby   *Marist College*
Pamela Soltis   *Washington State University*
Marshall D. Sundberg   *Emporia State University*
Kenneth W. Thomulka   *Philadelphia College of Pharmacy & Science*
James Traniello   *Boston University*
Terry Trobec   *Oakton Community College*
Linda Van Thiel   *Wayne State University*
Frank Watson   *St. Andrews College*
Harold Webster   *Pennsylvania State University, DuBois Campus*
Charles R. Wert   *Linn-Benton Community College*
Peter Wilkin   *Purdue University North Central*
Esther Wilson   *University of Wisconsin-Parkside*
Dan Wivagg   *Baylor University*
Roger J. Young   *Drury College*
Stam M. Zervanos   *Pennsylvania State University*

Significant Changes to the Fifth Edition

In keeping with the goals of this revision, you will find significant changes throughout the text. In reviewing the table of contents, you will see that we've expanded the text from 55 chapters to 60 chapters. This change was in direct response to reviewers who said that we shortchanged some key areas in the fourth edition. We expanded the coverage of the viruses and simple organisms (Part VIII) from four to five chapters. We expanded the coverage of plants from five to eight chapters. We expanded the coverage of animal diversity, devoting entire chapters to both arthropods and echinoderms. We also expanded the coverage of animal form and function, devoting entire chapters to both respiration and circulation.

We changed the format and organization of the text to follow a "learning modules" approach. The material is presented according to key concepts that are numbered for ease of identification. The student can get an overview of the chapter content by reading the key concept statements and the headings that follow each statement. A walkthrough of the learning modules is presented on the next page.

Other significant changes to the content are listed below:

How Genes Work (chapter 15)

Ribosomal RNA now includes the E site; the treatment of transcription is greatly expanded, covering a two-page spread; and the discussion and illustrations of translation have been updated to include the action of the ribosomal E site.

Cancer (chapter 17)

This chapter now includes a new section on how cancer is caused by chemicals, including a new table of chemical carcinogens in the workplace; a new table of cancer-causing genes in humans; new findings on lung cancer, including how smoking damages the gene *p 53*; and a new section with up-to-date information on curing cancer with molecular therapies targeted to different stages in the cell cycle and with a genetically-engineered virus.

Populaton Ecology (chapter 23)

Many new additions to this chapter include: the levels of ecological organization are listed and defined, a discussion of metapopulations has been added, demography has been expanded to a two-page spread that includes a new section of life tables, a new two-page spread on life history trade-offs and adaptations has been added, the discussion of *r*-selected and *K*-selected adaptations has been revised, and human population growth has been updated with new figures and a new table.

Community Ecology (chapter 24)

New additions to this chapter include: resource partitioning, including character displacement in Darwin's finches; a two-page spread on biodiversity and community structure, including species richness and species diversity discussions and trends; biogeographic patterns of species diversity; island biogeography; an expanded and revised section on succession; and preserving biodiversity.

Viruses (chapter 29)

The lytic and lysogenic cycles are discussed in the text; the new finding of the mechanism of how cholera-causing bacteria are transformed by a bacteriophage has been added; the HIV infection cycle has been updated to include the latest findings on co-receptors and the latency period; a new section has been added on the future of HIV treatment, including art and discussions of the variety of methods that are being researched; and the section on prions and viroids has been updated to include prion formation and mad-cow disease.

Plant Molecular Biology (chapter 40)

This entire chapter is new and includes up-to-date discussions, art, and photos covering the size, organization, and mapping of plant genomes; plant tissue culture methods and applications, and genetic engineering in plants.

Animals (chapter 41)

New additions to this chapter include a new phylogeny of the animals, and an illustrated table of the major animal phyla, a section discussing and illustrating four key transitions in body plan, an updated classification of the cnidarians, and a discussion with a photo of the newly discovered animal phylum Cycliophora.

The Immune System (chapter 54)

A new section on the evolution of the immune system discusses and illustrates the progression from invertebrates to vertebrates and the types of immune cells that have evolved; an illustrated table of the cells of the immune system has been added; the discussion of B cells has been revised and includes a clear layout of the humoral immune response, a discussion of how the diversity of antibodies is produced by somatic rearrangement, and the uses of the antibodies in medical diagnoses; an up-to-date AIDS epidemic chart has been added; and a new finding of how DNA vaccines may get around antigen shifting is included.

Cellular Mechanisms of Development (chapter 57)

A new section on the reversal of determination discusses the recent animal cloning experiments, and a new section on aging discusses and illustrates recent theories of aging.

A Guide to the "Learning Modules"

You will find the chapters in this edition organized in conceptual units. Each chapter presents three or four general concepts that together explain the key ideas that the chapter addresses. Each of these concepts is presented as a numbered statement, and is explored in a series of learning modules; each module is typically a one or two page spread (some use three or four pages) ending with a statement summarizing that module. The numbered general concepts explored in specific supporting modules provide the student with a clear outline of how the various topics covered in the chapter relate to one another, revealing in a direct way the conceptual skeleton of the chapter, so students don't miss the "big picture."

Here's How It Works

Each chapter opening page contains a Concept Outline that presents an overview of the conceptual units in the chapter. A portion of the Concept Outline for chapter 6 is reproduced here. The Key Concepts are numbered and the supporting topics are listed under the key concepts and provide the student with an overview of contents discussed in the chapter.

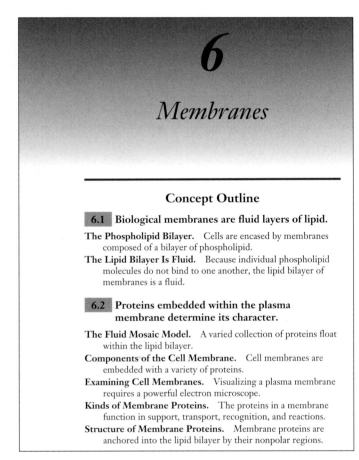

6

Membranes

Concept Outline

6.1 Biological membranes are fluid layers of lipid.

The Phospholipid Bilayer. Cells are encased by membranes composed of a bilayer of phospholipid.

The Lipid Bilayer Is Fluid. Because individual phospholipid molecules do not bind to one another, the lipid bilayer of membranes is a fluid.

6.2 Proteins embedded within the plasma membrane determine its character.

The Fluid Mosaic Model. A varied collection of proteins float within the lipid bilayer.

Components of the Cell Membrane. Cell membranes are embedded with a variety of proteins.

Examining Cell Membranes. Visualizing a plasma membrane requires a powerful electron microscope.

Kinds of Membrane Proteins. The proteins in a membrane function in support, transport, recognition, and reactions.

Structure of Membrane Proteins. Membrane proteins are anchored into the lipid bilayer by their nonpolar regions.

The numbered concepts begin the discussion of a general concept with a conceptual unit covering usually one page or a two-page spread. The first topic under the concept 6.1, "Biological membranes are fluid layers of lipid," is reproduced here.

| 6.1 | Biological membranes are fluid layers of lipid. |

The Phospholipid Bilayer

The membranes that encase all living cells are sheets of lipid only two molecules thick; more than 10,000 of these sheets piled on one another would just equal the thickness of this sheet of paper. The lipid layer that forms the foundation of a cell membrane is composed of molecules called **phospholipids** (figure 6.2).

Phospholipids

Like the fat molecules you studied in chapter 3, a phospholipid has a backbone derived from a three-carbon molecule called glycerol. Attached to this backbone are fatty acids, long chains of carbon atoms ending in a carboxyl (—COOH) group. A fat molecule has three such chains, one attached to each carbon in the backbone; because these chains are nonpolar, they do not form hydrogen bonds

The concept module "The Phospholipid Bilayer" ends with a summary statement shown below, as do all of the topics, whether they cover one page or two or more pages. This makes it very clear to the student where a topic begins and ends.

---

phospholipid molecules that make up the lipid bilayer, the membranes of every cell also contain proteins that extend through the lipid bilayer, providing passageways across the membrane.

The basic foundation of biological membranes is a lipid bilayer, which forms spontaneously. In such a layer, the nonpolar hydrophobic tails of phospholipid molecules point inward, forming a nonpolar barrier to water-soluble molecules.

The organization of the material into conceptual units is carried through to the end of the chapter. The chapter summaries at the end of the chapter are organized as a "Summary of Concepts" according to the "Concept Outline" found at the beginning of the chapter. Each numbered concept is followed by summary statements of the material found in the conceptual unit.

Summary of Concepts

6.1 Biological membranes are fluid layers of lipid.

- Every cell is encased within a fluid bilayer sheet of phospholipid molecules called the plasma membrane.

6.2 Proteins embedded within the plasma membrane determine its character.

- Proteins that are embedded within the plasma membrane have their hydrophobic regions exposed to the hydrophobic interior of the bilayer, and their hydrophilic regions exposed to the cytoplasm or the extracellular fluid.
- Membrane proteins can transport materials into or out of the cell, they can mark the identity of the cell, or they

- Osmosis is the diffusion of water. Since all organisms are composed of mostly water, maintaining osmotic balance is essential to life.

6.4 Bulk transport utilizes endocytosis.

- Materials or volumes of fluid that are too large to pass directly through the cell membrane can move into or out of cells through endocytosis or exocytosis, respectively.
- In these processes, the cell expends energy to change the shape of its plasma membrane, allowing the cell to engulf materials into a temporary vesicle (endocytosis), or eject materials by fusing a filled vesicle with the plasma membrane (exocytosis).

A section entitled "Discussing Key Terms" is far more than a list of key terms in the chapter. This section discusses the significance of the term within the context of the material covered in the chapter.

Discussing Key Terms

1. **Lipid bilayer** The foundation of the plasma membrane is a sheet composed of two layers of phospholipids, with the lipids oriented so that their hydrophobic ends face the interior of the sheet. The interior is thus very nonpolar, creating an effective barrier to the passage of polar substances.

2. **Transmembrane proteins** Numerous proteins are anchored in the plasma membrane by nonpolar segments that pass through the nonpolar interior of the bilayer. Some transmembrane proteins transport substances into and out of the cell, others act as receptors for chemical signals from other cells or the

4. **Osmosis** Net movement of water across a membrane. When two aqueous solutions are separated by a membrane that restricts the free passage of solutes, water diffuses across the membrane in the direction of higher solute concentration.

5. **Endocytosis and exocytosis** In endocytosis, portions of the plasma membrane invaginate and fuse, trapping a volume of extracellular liquid and its contents within a vesicle. Similarly, in exocytosis, a vesicle merges with the plasma membrane and discharges its contents to the exterior.

6. **Active transport** The movement of certain substances

Aids for Students and Instructors

Technology

We offer students and professors various technology products to support the fifth edition of *Biology*.

Essential Study Partner (ESP) CD-ROM This CD-ROM tutorial supports and enhances the material presented in the fifth edition of *Biology*. The ESP is offered free with the text and will enhance student learning with the use of high quality 3-D animations, activities that require the student to be active in the learning process, module and unit quizzes that take the student back into the material on the CD to review immediately, and illustrated overviews of key topics. Students will find the ESP to be a very valuable study tool.

Visual Resource Library CD-Rom

The WCB/McGraw-Hill *Visual Resource Library* CD-ROM for *Biology* contains nearly all of the illustrations found in the fifth edition. The CD-ROM contains an easy-to-use program that enables users to quickly view images, and easily import the images into PowerPoint to create multimedia presentations or use the already prepared PowerPoint presentations.

In addition to this text-specific *Visual Resource Library* CD-ROM, we offer two additional CD-ROMs, *Visual Resource Library* for Botany and for Environmental Science. Instructors can enhance classroom presentations by using images from these two CD-ROMs.

The Dynamic Human CD-ROM Version 2.0

This guide to anatomy and physiology interactively illustrates the complex relationships between anatomical structures and their functions in the human body. Realistic 3-D visuals are the premier feature of this exciting learning tool.

Explorations in Human Biology CD-ROM; Explorations in Cell Biology and Genetics CD-ROM

These interactive CDs, by George Johnson, feature 33 different interactive modules that cover key topics in biology.

Virtual Physiology Laboratory CD-ROM

This CD-ROM features ten simulations of the most common and important animal-based experiments ordinarily performed in introductory lab courses. The program contains video, audio, and text to clarify complex physiological functions.

Life Science Living Lexicon CD-ROM
by William N. Marchuk, Red Deer College

A Life Science Living Lexicon CD-ROM contains a comprehensive collection of life science terms, including definitions of their roots, prefixes, and suffixes as well as audio pronunciations and illustrations. The Lexicon is student interactive, providing quizzing and notetaking capabilities. It contains 4,500 terms, which can be broken down for study into the following categories: anatomy and physiology, botany, cell and molecular biology, genetics, ecology and evolution, and zoology.

Exploring the Internet on the Raven/Johnson Web Site
http://www.mhhe.com/biolink

This text-specific site has been developed exclusively for users of the fifth edition of *Biology*. When visiting the site, students can access additional study aids including quizzes, explore links to other relevant biology sites, catch up on current information, and pursue other activities, including content updates and enhancement chapters.

Life Science Animations 3-D Videotape

The *Life Science Animations* 3-D videotape contains 42 high-quality 3-D animations. These animations bring visual movement to biological processes that are difficult to understand. These 42 animations also appear as part of the *Essential Study Partner* CD-ROM, which provides students with a wonderful reinforcement of concepts and biological processes.

Life Science Animations Videotape Series (6 tapes)

Complex processes such as active transport and osmosis come to life in this series. Students can now review more than 65 animations (in the six-tape set) of the most difficult to learn concepts.

Biology Start Up Software (Mac Only)

This 5-set computer tutorial has complete coverage of basic biological principles such as cellular respiration and cell division.

Math Prep for Biology Software

With this computer tutorial, students are given practice with their math skills through biology-specific math applications. This tutorial allows students to master their math skills for further life science study.

Customization

Chapter-by-chapter customization allows you to customize *Biology* to fit your ideal course. Select only the chapters you use, and WCB/McGraw-Hill will bind them in full color. Contact your WCB/McGraw-Hill sales representative for more information or call 1–800–228–0634.

Additional Textbook Aids

Course Solutions. To help instructors incorporate technology and additional study aids into their course, a text-specific *Course Solutions* manual is available. Organized by chapter, the *Course Solutions* offers suggestions on how ancillaries can be used in lecture and as enhancements outside of lecture.

Student Study Guide, written by Margaret Gould Burke and Ronald M. Taylor, contains chapter overviews for review and page-referenced key terms as well as student exercises in the form of fill-in-the-blank, matching, multiple-choice, and essay questions. Complete answers and explanations for the exercises are provided at the end of each chapter.

Biology Laboratory Manual, a full-color lab manual written by Darrell S. Vodopich of Baylor University and Randy Moore of the University of Louisville, contains approximately fifty laboratory exercises. The lab manual is customizable by chapter in full color and is accompanied by a *Laboratory Resource Guide.*

Transparency Set of 300 acetates contains 300 images from the text. The text contains an entirely new art program so this set of transparencies will be completely new.

Instructor's Manual and Test Item File, written by Linda Van Thiel of Wayne State University, contains chapter synopsis, chapter objectives, key terms, chapter outline, instructional strategy, and a list of related visual resources. The manual also provides lists of biological supply companies, and media resources as well as answers to the review questions in the text. The *test item file,* prepared by L. Rao Ayyagari of Lindenwood College, is available in softcover or on disk *(Microtest)* in Mac and Windows platforms. The Test Item File contains an average of forty questions per chapter in the form of fill-in-the-blank, multiple-choice, and matching.

Additional Supplements

How Scientists Think
by George Johnson

A paperbound text describing twenty-one experiments that have shaped our understanding of genetics and molecular biology. It fosters critical thinking and reinforces the scientific method.

The AIDS Booklet
by Frank D. Cox

This booklet describes how AIDS and related diseases are commonly spread so that readers can protect themselves and their friends against this debilitating and deadly disease. This booklet is updated quarterly to give readers the most current information.

Basic Chemistry for Biology, second edition,
by Carolyn Chapman of Suffolk County Community College

A self-paced book that leads students through basic concepts of inorganic and organic chemistry.

How to Study Science, 2nd Edition
by Fred Drewes, Suffolk County Community College

This excellent workbook offers students helpful suggestions for meeting the considerable challenges of a college science course. It offers tips on how to take notes, how to get the most out of laboratories, and how to overcome science anxiety. The book's unique design helps students develop critical thinking skills while facilitating careful notetaking.

Critical Thinking Case Study Workbook
by Robert Allen

This ancillary includes 34 critical thinking case studies that are designed to immerse students in the "process of science" and challenge them to solve problems in the same way biologists do. The case studies are divided into three levels of difficulty (introductory, intermediate, and advanced) to afford instructors greater choice and flexibility. An answer key accompanies this workbook.

About the Authors

Peter Raven (left) and George Johnson

Dr. Peter H. Raven is director of the Missouri Botanical Garden and Englemann Professor of Botany at Washington University. Dr. Raven oversees the Garden's internationally recognized research program in tropical botany, one of the most active in the world in the study and conservation of imperiled habitats. A distinguished scientist, Dr. Raven is a member of the National Academy of Science and the National Research Council, and is a MacArthur and a Guggenheim fellow. He has received numerous honors and awards for his botanical research and work in tropical conservation.

In addition to coauthoring this text with Dr. George Johnson, Dr. Raven has authored fifteen other books and more than 450 articles.

Dr. George B. Johnson is professor of biology at Washington University in St. Louis, where he has taught undergraduates for 26 years. He regularly teaches general biology courses for majors and non-majors and has recently taken on two large biology courses, one focused on the biology underlying key current issues, the other on the biology of dinosaurs. Also professor of genetics at Washington University's school of medicine, Dr. Johnson is a student of population genetics and evolution, renowned for his pioneering studies of genetic variability. He has authored more than fifty scientific publications and seven texts, including *The Living World* (a very successful non-majors biology text) and two widely-used high school biology textbooks. Dr. Johnson has also pioneered the development of interactive CD-ROMs for biology teaching.

1

The Science of Biology

Concept Outline

FIGURE 1.1
A replica of the *Beagle*, off the southern coast of South America. The famous English naturalist, Charles Darwin, set forth on H.M.S. *Beagle* in 1831, at the age of 22.

You are about to embark on a journey—a journey of discovery about the nature of life. Over 150 years ago, a young English naturalist named Charles Darwin set sail on a similar journey on board H.M.S. *Beagle* (figure 1.1 shows a replica of the *Beagle*). What Darwin learned on his five-year voyage led directly to his development of the theory of evolution by natural selection, a theory that has become the core of the science of biology. Darwin's voyage seems a fitting place to begin our exploration of biology, the scientific study of living organisms and how they have evolved. Before we begin, however, let's take a moment to think about what biology is and why it's important.

Properties of Life

In its broadest sense, *biology is the study of living things—the science of life.* Living things come in an astounding variety of shapes and forms, and biologists study life in many different ways. They live with gorillas, collect fossils, and listen to whales. They isolate viruses, grow mushrooms, and examine the structure of fruit flies. They read the messages encoded in the long molecules of heredity and count how many times a hummingbird's wings beat each second.

What makes something "alive"? Anyone could deduce that a galloping horse is alive and a car is not, but *why*? We cannot say, "If it moves, it's alive," because a car can move, and gelatin can wiggle in a bowl. They certainly are not alive. What characteristics *do* define life? All living organisms share five basic characteristics:

1. **Order.** All organisms consist of one or more cells with highly ordered structures: atoms make up molecules, which construct cellular organelles, which are contained within cells. This hierarchical organization continues at higher levels in multicellular organisms and among organisms (figure 1.2).
2. **Sensitivity.** All organisms respond to stimuli. Plants grow toward a source of light, and your pupils dilate when you walk into a dark room.
3. **Growth, development, and reproduction.** All organisms are capable of growing and reproducing, and they all possess hereditary molecules that are passed to their offspring, ensuring that the offspring are of the same species. Although crystals also "grow," their growth does not involve hereditary molecules.
4. **Regulation.** All organisms have regulatory mechanisms that coordinate the organism's internal functions. These functions include supplying cells with nutrients, transporting substances through the organism, and many others.
5. **Homeostasis.** All organisms maintain relatively constant internal conditions, different from their environment, a process called homeostasis.

All living things share certain key characteristics: order, sensitivity, growth, development and reproduction, regulation, and homeostasis.

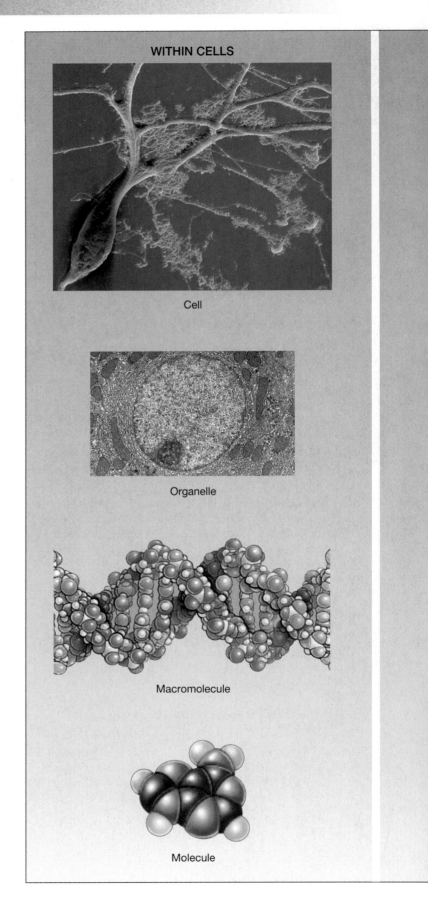

WITHIN CELLS

Cell

Organelle

Macromolecule

Molecule

FIGURE 1.2
Hierarchical organization of living things. Life is highly organized—from small and simple to large and complex, within cells, within multicellular organisms, and among organisms.

WITHIN MULTICELLULAR ORGANISMS

Organism

Organ System

Organ

Tissue

AMONG ORGANISMS

Ecosystem

Community

Species

Population

The Nature of Science

Biology is a fascinating and important subject, because it dramatically affects our daily lives and our futures. Many biologists are working on problems that critically affect our lives, such as the world's rapidly expanding population and diseases like cancer and AIDS. The knowledge these biologists gain will be fundamental to our ability to manage the world's resources in a suitable manner, to prevent or cure diseases, and to improve the quality of our lives and those of our children and grandchildren. For these reasons, an understanding of biology is increasingly necessary for any educated person.

To understand biology, you must first understand the nature of science. Biology is one of the most successful of the "natural sciences," explaining what our world is like. The basic tool a scientist uses is thought. To understand the nature of science, it is useful to focus for a moment on how scientists think. They reason in two ways: deductively and inductively.

Deductive Reasoning

Deductive reasoning applies general principles to predict specific results. Over 2200 years ago, the Greek Eratosthenes used deductive reasoning to accurately estimate the circumference of the earth. At high noon on the longest day of the year, when the sun's rays hit the bottom of a deep well in the city of Syene, Egypt, Eratosthenes measured the length of the shadow cast by a tall obelisk in Alexandria, about 800 kilometers to the north. Because he knew the distance between the two cities and the height of the obelisk, he was able to employ the principles of Euclidean geometry to correctly deduce the circumference of the earth (figure 1.3). This sort of analysis of specific cases using general principles is an example of deductive reasoning. It is the reasoning of mathematics and philosophy and is used to test the validity of general ideas in all branches of knowledge. General principles are constructed and then used as the basis for examining specific cases to which they apply.

Inductive Reasoning

Inductive reasoning uses specific observations to *construct* general scientific principles. *Webster's Dictionary* defines *science* as systematized knowledge derived from observation and experiment carried on to determine the principles underlying what is being studied. In other words, a scientist determines principles from observations, discovering general principles by careful examination of specific cases. Inductive reasoning first became important to science in the

FIGURE 1.3
Deductive reasoning: How Eratosthenes estimated the circumference of the earth using deductive reasoning. 1. On a day when sunlight shone straight down a deep well at Syene in Egypt, Eratosthenes measured the length of the shadow cast by a tall obelisk in the city of Alexandria, about 800 kilometers away. **2.** The shadow's length and the obelisk's height formed two sides of a triangle. Using the recently developed principles of Euclidean geometry, he calculated the angle, a, to be 7° and 12′, exactly 1/50 of a circle (360°). **3.** If angle a = 1/50 of a circle, then the distance between the obelisk (in Alexandria) and the well (in Syene) must equal 1/50 of the circumference of the earth. **4.** Eratosthenes had heard that it was a 50-day camel trip from Alexandria to Syene. Assuming that a camel travels about 18.5 kilometers per day, he estimated the distance between obelisk and well as 925 kilometers (using different units of measure, of course). **5.** Eratosthenes thus deduced the circumference of the earth to be 50 × 925 = 46,250 kilometers. Modern measurements put the distance from the well to the obelisk at just over 800 kilometers. Employing a distance of 800 kilometers, Eratosthenes's value would have been 50 × 800 = 40,000 kilometers. The actual circumference is 40,075 kilometers.

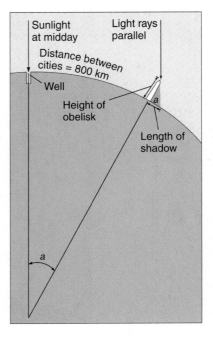

1600s in Europe, when Francis Bacon, Isaac Newton, and others began to use the results of particular experiments to infer general principles about how the world operates. If you release an apple from your hand, what happens? The apple falls to the ground. From a host of simple, specific observations like this, Newton inferred a general principle: all objects fall toward the center of the earth. What Newton did was construct a mental model of how the world works, a family of general principles consistent with what he could see and learn. Scientists do the same today. They use specific observations to build general models, and then test the models to see how well they work.

> **Science is a way of viewing the world that focuses on objective information, putting that information to work to build understanding.**

How Science Is Done

How do scientists establish which general principles are true from among the many that might be true? They do this by systematically testing alternative proposals. If these proposals prove inconsistent with experimental observations, they are rejected as untrue. After making careful observations concerning a particular area of science, scientists construct a **hypothesis,** which is a suggested explanation that accounts for those observations. A hypothesis is a proposition that might be true. Those hypotheses that have not yet been disproved are retained. They are useful because they fit the known facts, but they are always subject to future rejection if, in the light of new information, they are found to be incorrect.

Testing Hypotheses

We call the test of a hypothesis an **experiment** (figure 1.4). Suppose that a room appears dark to you. To understand why it appears dark, you propose several hypotheses. The first might be, "There is no light in the room because the light switch is turned off." An alternative hypothesis might be, "There is no light in the room because the lightbulb is burned out." And yet another alternative hypothesis might be, "I am going blind." To evaluate these hypotheses, you would conduct an experiment designed to eliminate one or more of the hypotheses. For example, you might test your hypotheses by reversing the position of the light switch. If you do so and the light does not come on, you have disproved the first hypothesis. Something other than the setting of the light switch must be the reason for the darkness. Note that a test such as this does not prove that any of the other hypotheses are true; it merely demonstrates that one of them is not. A successful experiment is one in which one or more of the alternative hypotheses is demonstrated to be inconsistent with the results and is thus rejected.

As you proceed through this text, you will encounter a great deal of information, often accompanied by explanations. These explanations are hypotheses that have withstood the test of experiment. Many will continue to do so; others will be revised as new observations are made by biologists. Biology, like all science, is in a constant state of change, with new ideas appearing and replacing old ones.

FIGURE 1.4

How science is done. This diagram illustrates the way in which scientific investigations proceed. First, scientists make observations that raise a particular question. They develop a number of potential explanations (hypotheses) to answer the question. Next, they carry out experiments in an attempt to eliminate one or more of these hypotheses. Then, predictions are made based on the remaining hypotheses, and further experiments are carried out to test these predictions. As a result of this process, the least unlikely hypothesis is selected. If it is validated by numerous experiments and stands the test of time, the hypothesis may eventually be considered a theory.

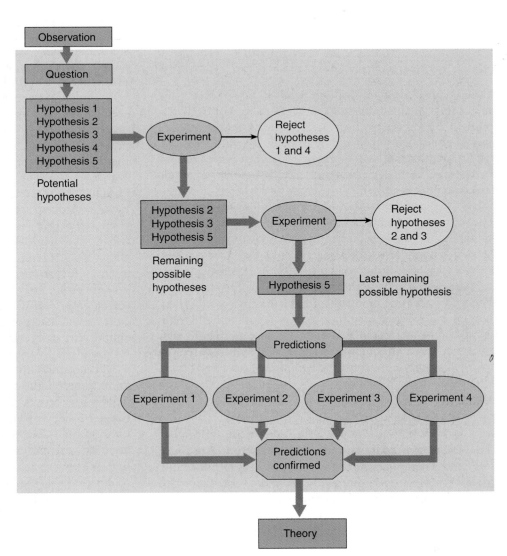

Establishing Controls

Often we are interested in learning about processes that are influenced by many factors, or **variables.** To evaluate alternative hypotheses about one variable, all other variables must be kept constant. This is done by carrying out two experiments in parallel: in the first experiment, one variable is altered in a specific way to test a particular hypothesis; in the second experiment, called the **control experiment,** that variable is left unaltered. In all other respects the two experiments are identical, so any difference in the outcomes of the two experiments must result from the influence of the variable that was changed. Much of the challenge of experimental science lies in designing control experiments that isolate a particular variable from other factors that might influence a process. For example, a horticulturist might want to know which nutrient most affects plant growth. She could take five pots containing the same type of plant, add a different nutrient to four of them, and leave the fifth pot without any added nutrients at all. All five pots would be exposed to the same amount of sunlight and receive the same amount of water. If all other conditions were identical in the five pots, any difference in the growth of the plants could be attributed to the only variable that was altered, the nutrient.

Using Predictions

A successful scientific hypothesis needs to be not only valid but useful—it needs to tell you something you want to know. A hypothesis is most useful when it makes predictions, because those predictions provide a way to test the validity of the hypothesis. If an experiment produces results inconsistent with the predictions, the hypothesis must be rejected. On the other hand, if the predictions are supported by experimental testing, the hypothesis is supported. The more experimentally supported predictions a hypothesis makes, the more valid the hypothesis is. For example, Einstein's hypothesis of relativity was at first provisionally accepted because no one could devise an experiment that invalidated it. The hypothesis made a clear prediction: that the sun would bend the path of light passing by it. When this prediction was tested in a total eclipse, the light from background stars was indeed bent. Because this result was unknown when the hypothesis was being formulated, it provided strong support for the hypothesis, which was then accepted with more confidence.

Developing Theories

Hypotheses that stand the test of time, often tested but never rejected, are called **theories.** Thus one speaks of the general principle first proposed by Newton as the theory of gravity. Scientists use the word *theory* in a very different sense than the general public does. To a scientist, theories are the solid ground of science, that of which we are most certain. In contrast, to the general public, *theory* implies just the opposite—a *lack* of knowledge, or a guess. Not surprisingly, this difference in subjective meaning often results in confusion. In this text, *theory* will always be used in its scientific sense, in reference to a generally accepted scientific principle.

However, there is no absolute truth in science, only varying degrees of uncertainty. The possibility remains that future evidence will cause a theory to be revised. Therefore, a scientist's acceptance of a theory is always provisional.

Some theories are so strongly supported that the likelihood of their being rejected in the future is small. The theory of evolution, for example, is backed by such a wealth and diversity of evidence that all but a few scientists accept it with as much confidence as they do the theory of gravity. Thus, to suggest that evolution is "just a theory" is misleading. We will examine this particular theory later in this chapter as an example of how scientists carry out their work. Evolution is a particularly important theory to biologists, since it provides the conceptual framework that unifies biology as a science.

Research and the Scientific Method

It used to be fashionable to speak of the "scientific method" as consisting of an orderly sequence of logical "either/or" steps. Each step would reject one of two mutually incompatible alternatives, as if trial-and-error testing would inevitably lead one through the maze of uncertainty that always impedes scientific progress. If this were indeed so, a computer would make a good scientist. But science is not done this way. As British philosopher Karl Popper has pointed out, successful scientists without exception design their experiments with a pretty fair idea of how the results are going to come out. They have what Popper calls an "imaginative preconception" of what the truth might be. A hypothesis that a successful scientist tests is not just any hypothesis; rather, it is an educated guess or a hunch, in which the scientist integrates all that he or she knows and allows his or her imagination full play, in an attempt to get a sense of what *might* be true (see Box: How Biologists Do Their Work). It is because insight and imagination play such a large role in scientific progress that some scientists are so much better at science than others, just as Beethoven and Mozart stand out among most other composers.

Some scientists perform what is called basic research, which is intended to extend the boundaries of what we know. These individuals typically work at universities, and their research is usually financially supported by their institutions and by external sources, such as the government, industry, and private foundations. Basic research is as diverse as its name implies. Some basic scientists attempt to find out how certain cells take up specific chemicals, while others count the number of dents in tiger teeth. The information generated by basic research contributes to the growing body of scientific knowledge, and it provides the scientific foundation utilized by applied research. Scientists who conduct applied research are often employed in

How Biologists Do Their Work

The Consent

Late in November, on a single night
Not even near to freezing, the ginkgo trees
That stand along the walk drop all their leaves
In one consent, and neither to rain nor to wind
But as though to time alone: the golden and green
Leaves litter the lawn today, that yesterday
Had spread aloft their fluttering fans of light.
What signal from the stars? What senses took it in?
What in those wooden motives so decided
To strike their leaves, to down their leaves,
Rebellion or surrender? And if this
Can happen thus, what race shall be exempt?
What use to learn the lessons taught by time,
If a star at any time may tell us: Now.

Howard Nemerov

What is bothering the poet Howard Nemerov is that life is influenced by forces he cannot control or even identify. It is the job of biologists to solve puzzles such as the one he poses, to identify and try to understand those things that influence life.

Nemerov asks why ginkgo trees (figure 1.A) drop all their leaves at once. To find an answer to questions such as this, biologists and other scientists pose *possible* answers and then try to determine which answers are false. Tests of alternative possibilities are called experiments. To learn why the ginkgo trees drop all their leaves simultaneously, a scientist would first formulate several possible answers, called hypotheses:

Hypothesis 1: Ginkgo trees possess an internal clock that times the release of leaves to match

FIGURE 1.A
A ginkgo tree.

the season. On the day Nemerov describes, this clock sends a "drop" signal (perhaps a chemical) to all the leaves at the same time.

Hypothesis 2: The individual leaves of ginkgo trees are each able to sense day length, and when the days get short enough in the fall, each leaf responds independently by falling.

Hypothesis 3: A strong wind arose the night before Nemerov made his observation, blowing all the leaves off the ginkgo trees.

Next, the scientist attempts to eliminate one or more of the hypotheses by conducting an experiment. In this case, one might cover some of the leaves so that they cannot use light to sense day length. If hypothesis 2 is true, then the covered leaves should not fall when the others do, since they are not receiving the same information. Suppose, however, that despite the covering of some of the leaves, all the leaves still fall together. This result would eliminate hypothesis 2 as a possibility. Either of the other hypotheses, and many others, remain possibilities.

This simple experiment with ginkgoes points out the essence of scientific progress: science does not prove that certain explanations are true; rather, it proves that others are not. Hypotheses that are inconsistent with experimental results are rejected, while hypotheses that are not proven false by an experiment are provisionally accepted. However, hypotheses may be rejected in the future when more information becomes available, if they are inconsistent with the new information. Just as a computer can find the correct path through a maze by trying and eliminating false paths, so scientists work to find the correct explanations of natural phenomena by eliminating false possibilities.

A hypothesis that is tested in many experiments but never rejected is called a theory. In biology, the hypothesis that mitochondria, tiny structures within cells, are the descendants of bacteria is a theory. We are not certain that this hypothesis is correct, but it is supported by an overwhelming amount of evidence, and most biologists accept it with confidence. Their acceptance is provisional, however. Should new evidence contradict the theory, it would have to be revised or rejected.

some kind of industry. Their work may involve the manufacturing of food additives, creating of new drugs, or testing the quality of the environment.

After developing a hypothesis and performing a series of experiments, a scientist writes a paper carefully describing the experiment and its results. He or she then submits the paper for publication in a scientific journal, but before it is published, it must be reviewed and accepted by other scientists who are familiar with that particular field of research. This process of careful evaluation, called peer review, lies at the heart of modern science, fostering careful work, precise description, and thoughtful analysis. When an important discovery is announced in a paper, other scientists attempt to reproduce the result, providing a check on accuracy and honesty. Non-reproducible results are not taken seriously for long.

The explosive growth in scientific research during the second half of the twentieth century is reflected in the enormous number of scientific journals now in existence. Although some, such as *Science* and *Nature*, are devoted to a wide range of scientific disciplines, most are extremely specialized: *Cell Motility and the Cytoskeleton, Glycoconjugate Journal, Mutation Research*, and *Synapse* are just a few examples.

> **The scientific process involves the rejection of hypotheses that are inconsistent with experimental results or observations. Hypotheses that are consistent with available data are conditionally accepted. The formulation of the hypothesis often involves creative insight. A theory is a hypothesis that is supported by a great deal of evidence.**

Darwin's Theory of Evolution

The theory of evolution proposes that one type of organism can gradually evolve into another. This famous theory provides a good example of how a scientist develops a hypothesis and how, after much testing, it is eventually accepted as a theory.

Charles Robert Darwin (1809–1882; figure 1.5) was an English naturalist who, after 30 years of study and observation, wrote one of the most famous and influential books of all time. This book, *On the Origin of Species by Means of Natural Selection, or The Preservation of Favoured Races in the Struggle for Life*, created a sensation when it was published, and the ideas Darwin expressed in it have played a central role in the development of human thought ever since.

In Darwin's time, most people believed that the various kinds of organisms and their individual structures resulted from direct actions of the Creator (and to this day many people still believe this to be true). Species were thought to be specially created and unchangeable, or immutable, over the course of time. In contrast to these views, a number of earlier philosophers had presented the view that living things must have changed during the history of life on earth. Darwin proposed a concept he called natural selection as a coherent, logical explanation for this process, and he brought his ideas to wide public attention. His book, as its title indicates, presented a conclusion that differed sharply from conventional wisdom. Although his theory did not challenge the existence of a Divine Creator, Darwin argued that this Creator did not simply create things and then leave them forever unchanged. Instead, Darwin's God expressed Himself through the operation of natural laws that produced change over time, or **evolution.** These views put Darwin at odds with most people of his time, who believed in a literal interpretation of the Bible and accepted the idea of a fixed and constant world. His revolutionary theory deeply troubled not only many of his contemporaries but Darwin himself.

FIGURE 1.5
Charles Darwin. This newly rediscovered photograph taken in 1881, the year before Darwin died, appears to be the last ever taken of the great biologist.

The story of Darwin and his theory begins in 1831, when he was 22 years old. On the recommendation of one of his professors at Cambridge University, he was selected to serve as naturalist on a five-year navigational mapping expedition around the coasts of South America (figure 1.6), aboard H.M.S. *Beagle* (figure 1.7). During this long voyage, Darwin had the chance to study a wide variety of plants and animals

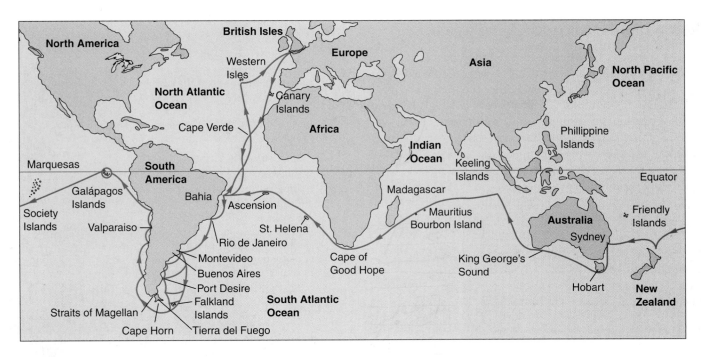

FIGURE 1.6
The five-year voyage of H.M.S. *Beagle*. Most of the time was spent exploring the coasts and coastal islands of South America, such as the Galápagos Islands. Darwin's studies of the animals of the Galápagos Islands played a key role in his eventual development of the theory of evolution by means of natural selection.

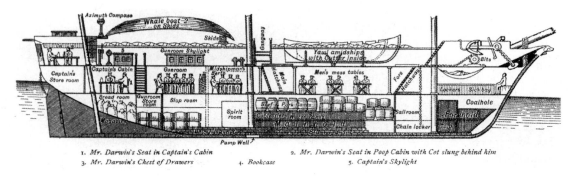

1. *Mr. Darwin's Seat in Captain's Cabin*   2. *Mr. Darwin's Seat in Poop Cabin with Cot slung behind him*
3. *Mr. Darwin's Chest of Drawers*   4. *Bookcase*   5. *Captain's Skylight*

FIGURE 1.7
Cross section of the *Beagle*. A 10-gun brig of 242 tons, only 90 feet in length, the *Beagle* had a crew of 74 people! After he first saw the ship, Darwin wrote to his college professor Henslow: "The absolute want of room is an evil that nothing can surmount."

on continents and islands and in distant seas. He was able to explore the biological richness of the tropical forests, examine the extraordinary fossils of huge extinct mammals in Patagonia at the southern tip of South America, and observe the remarkable series of related but distinct forms of life on the Galápagos Islands, off the west coast of South America. Such an opportunity clearly played an important role in the development of his thoughts about the nature of life on earth.

When Darwin returned from the voyage at the age of 27, he began a long period of study and contemplation. During the next 10 years, he published important books on several different subjects, including the formation of oceanic islands from coral reefs and the geology of South America. He also devoted eight years of study to barnacles, a group of small marine animals with shells that inhabit rocks and pilings, eventually writing a four-volume work on their classification and natural history. In 1842, Darwin and his family moved out of London to a country home at Down, in the county of Kent. In these pleasant surroundings, Darwin lived, studied, and wrote for the next 40 years.

Darwin was the first to propose natural selection as an explanation for the mechanism of evolution that produced the diversity of life on earth. His hypothesis grew from his observations on a five-year voyage around the world.

Darwin's Evidence

One of the obstacles that had blocked the acceptance of any theory of evolution in Darwin's day was the incorrect notion, widely believed at that time, that the earth was only a few thousand years old. The discovery of thick layers of rocks, evidences of extensive and prolonged erosion, and the increasing numbers of diverse and unfamiliar fossils discovered during Darwin's time made this assertion seem less and less likely. The great geologist Charles Lyell (1797–1875), whose *Principles of Geology* (1830) Darwin read eagerly as he sailed on the *Beagle*, outlined for the first time the story of an ancient world of plants and animals in flux. In this world, species were constantly becoming extinct while others were emerging. It was this world that Darwin sought to explain.

What Darwin Saw

When the *Beagle* set sail, Darwin was fully convinced that species were immutable. Indeed, it was not until two or three years after his return that he began to consider seriously the possibility that they could change. Nevertheless, during his five years on the ship, Darwin observed a number of phenomena that were of central importance to him in reaching his ultimate conclusion (table 1.1). For example, in the rich fossil beds of southern South America, he observed fossils of extinct armadillos similar in form to the armadillos that still lived in the same area (figure 1.8). Why would similar living and fossil organisms be in the same area unless the earlier form had given rise to the other?

Table 1.1   Darwin's Evidence that Evolution Occurs

FOSSILS

1. Extinct species, such as the fossil armadillo in figure 1.8, most closely resemble living ones in the same area, suggesting that one had given rise to the other.
2. In rock strata (layers), progressive changes in characteristics can be seen in fossils from earlier and earlier layers.

GEOGRAPHICAL DISTRIBUTION

3. Lands with similar climates, such as Australia, South Africa, California, and Chile, have unrelated plants and animals, indicating that diversity is not entirely influenced by climate and environment.
4. The plants and animals of each continent are distinctive; all South American rodents belong to a single group, structurally similar to the guinea pigs, for example, while most of the rodents found elsewhere belong to other groups.

OCEANIC ISLANDS

5. Although oceanic islands have few species, those they do have are often unique (endemic) and show relatedness to one another, such as the Galápagos tortoises (figure 1.9). This suggests that the tortoises and other groups of endemic species developed after their mainland ancestors reached the islands and are, therefore, more closely related to one another.
6. Species on oceanic islands show strong affinities to those on the nearest mainland. Thus, the finches of the Galápagos Islands (figure 1.10*a*) closely resemble a finch seen on the western coast of South America (figure 1.10*b*). The Galápagos finches do *not* resemble the birds on the Cape Verde Islands, islands in the Atlantic Ocean off the coast of Africa that are similar to the Galápagos. Darwin visited the Cape Verde Islands and many other island groups personally and was able to make such comparisons on the basis of his own observations.

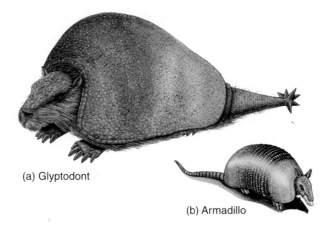

(a) Glyptodont

(b) Armadillo

FIGURE 1.8
Fossil evidence of evolution. The now-extinct glyptodont (a) was a 2000-kilogram South American armadillo, much larger than the modern armadillo (b), which weighs an average of about 4.5 kilograms. (Drawings are not to scale.) Finding the fossils of organisms such as the glyptodonts and noting their similarity to living organisms found in the same regions suggested to Charles Darwin that evolution had taken place.

Repeatedly, Darwin saw that the characteristics of similar species varied somewhat from place to place. These geographical patterns suggested to him that organismal lineages change gradually as species migrate from one area to another. On the Galápagos Islands, off the coast of Ecuador, Darwin encountered giant land tortoises. Surprisingly, these tortoises were not all identical. In fact, local residents and the sailors who captured the tortoises for food could tell which island a particular tortoise had come from just by looking at its shell (figure 1.9). This distribution of physical variation suggested that all of the tortoises were related, but that they had changed slightly in appearance after becoming isolated on different islands.

In a more general sense, Darwin was struck by the fact that the plants and animals on these relatively young volcanic islands resembled those on the nearby coast of South America (figure 1.10). If each one of these plants and animals had been created independently and simply placed on

(a)

(b)

FIGURE 1.9
Galápagos tortoises. Tortoises with large, domed shells (a) are found in relatively moist habitats. Those with lower, saddleback-type shells (b), in which the front of the shell is bent up and exposes the head and part of the neck, are found in dry habitats. Differences of this kind make it possible to identify the races of tortoises that inhabit the different islands of the Galápagos.

(a)

(b)

FIGURE 1.10
Darwin's finches. (a) One of Darwin's Galápagos finches, the medium ground finch. (b) The blue-black grassquit, found in grasslands along the Pacific Coast from Mexico to Chile. This species may be the ancestor of Darwin's finches.

the Galápagos Islands, why didn't they resemble the plants and animals of islands with similar climates, such as those off the coast of Africa, for example? Why did they resemble those of the adjacent South American coast instead?

The fossils and patterns of life that Darwin observed on the voyage of the *Beagle* eventually convinced him that evolution had taken place.

Inventing the Theory of Natural Selection

It is one thing to observe the results of evolution, but quite another to understand how it happens. Darwin's great achievement lies in his formulation of the hypothesis that evolution occurs because of natural selection.

Darwin and Malthus

Of key importance to the development of Darwin's insight was his study of Thomas Malthus's *Essay on the Principle of Population* (1798). In his book, Malthus pointed out that populations of plants and animals (including human beings) tend to increase geometrically, while the ability of humans to increase their food supply increases only arithmetically. A *geometric progression* is one in which the elements increase by a constant *factor*; for example, in the progression 2, 6, 18, 54, . . . , each number is three times the preceding one. An *arithmetic progression*, in contrast, is one in which the elements increase by a constant *difference*; in the progression 2, 6, 10, 14, . . . , each number is four greater than the preceding one (figure 1.11).

Because populations increase geometrically, virtually any kind of animal or plant, if it could reproduce unchecked, would cover the entire surface of the world within a surprisingly short time. Instead, populations of species remain fairly constant year after year, because death limits population numbers. Malthus's conclusion provided the key ingredient that was necessary for Darwin to develop the hypothesis that evolution occurs by natural selection.

Natural Selection

Sparked by Malthus's ideas, Darwin saw that although every organism has the potential to produce more offspring than can survive, only a limited number actually do survive and produce further offspring. Combining this observation with what he had seen on the voyage of the *Beagle*, as well as with his own experiences in breeding domestic animals, Darwin made an important association (figure 1.12): *Those individuals that possess superior physical, behavioral, or other attributes are more likely to survive than those that are not so well endowed.* By surviving, they gain the opportunity to pass on their favorable characteristics to their offspring. As the frequency of these characteristics increases in the population, the nature of the population as a whole will gradually change. Darwin called this process **selection.** The driving force he identified has often been referred to as *survival of the fittest.*

Darwin was thoroughly familiar with variation in domesticated animals and began *On the Origin of Species* with a detailed discussion of pigeon breeding. He knew that breeders selected certain varieties of pigeons and other animals, such as dogs, to produce certain characteristics, a process Darwin called **artificial selection.** Once this had been done, the animals would breed true for the characteristics

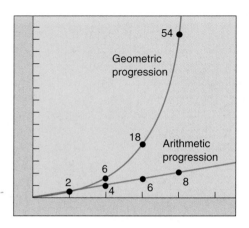

FIGURE 1.11
Geometric and arithmetic progressions. An arithmetic progression increases by a constant difference (e.g., units of 1 or 2 or 3), while a geometric progression increases by a constant factor (e.g., × 2 or × 3 or × 4). Malthus contended that the human growth curve was geometric, but the human food production curve was only arithmetic. Can you see the problems this difference would cause?

"Can we doubt . . . that individuals having any advantage, however slight, over others, would have the best chance of surviving and procreating their kind? On the other hand, we may feel sure that any variation in the least degree injurious would be rigidly destroyed. This preservation of favorable variations, I call Natural Selection."

FIGURE 1.12
An excerpt from Charles Darwin's *On the Origin of Species*.

that had been selected. Darwin had also observed that the differences purposely developed between domesticated races or breeds were often greater than those that separated wild species. Domestic pigeon breeds, for example, show much greater variety than all of the hundreds of wild species of pigeons found throughout the world. Such relationships suggested to Darwin that evolutionary change could occur in nature too. Surely if pigeon breeders could foster such variation by "artificial selection," nature could do the same, playing the breeder's role in selecting the next generation—a process Darwin called **natural selection.**

Darwin's theory provides a simple and direct explanation of biological diversity, or why animals are different in different places: because habitats differ in their requirements and opportunities, the organisms with characteristics favored locally by natural selection will tend to vary in different places.

Darwin Drafts His Argument

Darwin drafted the overall argument for evolution by natural selection in a preliminary manuscript in 1842. After showing the manuscript to a few of his closest scientific friends, however, Darwin put it in a drawer, and for 16 years turned to other research. No one knows for sure why Darwin did not publish his initial manuscript—it is very thorough and outlines his ideas in detail. Some historians have suggested that Darwin was shy of igniting public criticism of his evolutionary ideas—there could have been little doubt in his mind that his theory of evolution by natural selection would spark controversy. Others have proposed that Darwin was simply refining his theory all those years, although there is little evidence he altered his initial manuscript in all that time.

Wallace Has the Same Idea

The stimulus that finally brought Darwin's theory into print was an essay he received in 1858. A young English naturalist named Alfred Russel Wallace (1823–1913) sent the essay to Darwin from Malaysia; it concisely set forth the theory of evolution by means of natural selection, a theory Wallace had developed independently of Darwin. Like Darwin, Wallace had been greatly influenced by Malthus's 1798 essay. Colleagues of Wallace, knowing of Darwin's work, encouraged him to communicate with Darwin. After receiving Wallace's essay, Darwin arranged for a joint presentation of their ideas at a seminar in London. Darwin then completed his own book, expanding the 1842 manuscript which he had written so long ago, and submitted it for publication.

Publication of Darwin's Theory

Darwin's book appeared in November 1859 and caused an immediate sensation. Many people were deeply disturbed by the suggestion that human beings were descended from the same ancestor as apes (figure 1.13). Darwin did not actually discuss this idea in his book, but it followed directly from the principles he outlined. In a subsequent book, *The Descent of Man*, Darwin presented the argument directly, building a powerful case that humans and living apes have common ancestors. Although people had long accepted

FIGURE 1.13
Darwin greets his monkey ancestor. In his time, Darwin was often portrayed unsympathetically, as in this drawing from an 1874 publication.

that humans closely resembled apes in many characteristics, the possibility that there might be a direct evolutionary relationship was unacceptable to many. Darwin's arguments for the theory of evolution by natural selection were so compelling, however, that his views were almost completely accepted within the intellectual community of Great Britain after the 1860s.

The fact that populations do not really expand geometrically implies that nature acts to limit population numbers. The traits of organisms that survive to produce more offspring will be more common in future generations—a process Darwin called natural selection.

Evolution After Darwin: Testing the Theory

More than a century has elapsed since Darwin's death in 1882. During this period, the evidence supporting his theory has grown progressively stronger. There have also been many significant advances in our understanding of how evolution works. Although these advances have not altered the basic structure of Darwin's theory, they have taught us a great deal more about the mechanisms by which evolution occurs. We will briefly explore some of this evidence here; in chapter 20 we will return to the theory of evolution and examine the evidence in more detail.

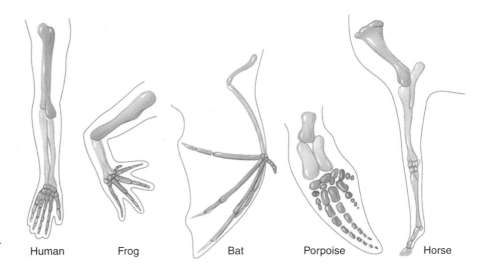

Human   Frog   Bat   Porpoise   Horse

FIGURE 1.14
Homology among vertebrate limbs. The forelimbs of four mammals and a frog show the ways in which the relative proportions of the forelimb bones have changed in relation to the particular way of life of each organism.

The Fossil Record

Darwin predicted that the fossil record would yield intermediate links between the great groups of organisms, for example, between fishes and the amphibians thought to have arisen from them, and between reptiles and birds. We now know the fossil record to a degree that was unthinkable in the nineteenth century. Recent discoveries of microscopic fossils have extended the known history of life on earth back to about 3.5 billion years ago. The discovery of other fossils has supported Darwin's predictions and has shed light on how organisms have, over this enormous time span, evolved from the simple to the complex. For vertebrate animals especially, the fossil record is rich and exhibits a graded series of changes in form, with the evolutionary parade visible for all to see (see Box: Why Study Fossils?).

The Age of the Earth

In Darwin's day, some physicists argued that the earth was only a few thousand years old. This bothered Darwin, because the evolution of all living things from some single original ancestor would have required a great deal more time. Using evidence obtained by studying the rates of radioactive decay, we now know that the physicists of Darwin's time were wrong: the earth was formed about 4.5 billion years ago.

The Mechanism of Heredity

Darwin received some of his sharpest criticism in the area of heredity. At that time, no one had any concept of genes or of how heredity works, so it was not possible for Darwin to explain completely how evolution occurs. Theories of heredity in Darwin's day seemed to rule out the possibility of genetic variation in nature, a critical requirement of Darwin's theory. Genetics was established as a science only at the start of the twentieth century, 40 years after the publication of Darwin's *On the Origin of Species*. When scientists began to understand the laws of inheritance (discussed in chapter 13), the heredity problem with Darwin's theory vanished. Genetics accounts in a neat and orderly way for the production of new variations in organisms.

Comparative Anatomy

Comparative studies of animals have provided strong evidence for Darwin's theory. In many different types of vertebrates, for example, the same bones are present, indicating their evolutionary past. Thus, the forelimbs shown in figure 1.14 are all constructed from the same basic array of bones, modified in one way in the wing of a bat, in another way in the fin of a porpoise, and in yet another way in the leg of a frog or a horse. The bones are said to be **homologous** in the different vertebrates; that is, they have the same evolutionary origin, but they now differ in structure and function. This contrasts with **analogous** structures, such as the wings of birds and butterflies, which have similar structure and function but different evolutionary origins.

Molecular Biology

Biochemical tools are now of major importance in efforts to reach a better understanding of how evolution occurs. Within the last few years, for example, evolutionary biologists have begun to "read" genes, much as you are reading this page. They have learned to recognize the order of the

Why Study Fossils?

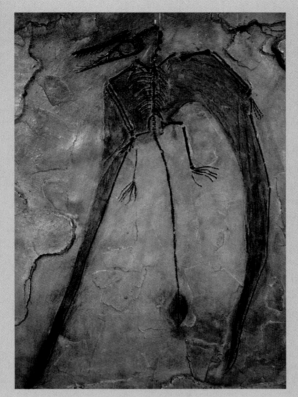

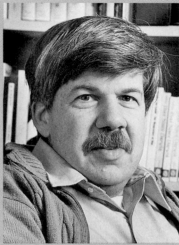

Stephen Jay Gould
Harvard University

I grew up on the streets of New York City, in a family of modest means and little formal education, but with a deep love of learning. Like many urban kids who become naturalists, my inspiration came from a great museum—in particular, from the magnificent dinosaurs on display at the American Museum of Natural History. As we all know from *Jurassic Park* and other sources, dinomania in young children (I was five when I saw my first dinosaur) is not rare—but nearly all children lose the passion, and the desire to become a paleontologist becomes a transient moment between policeman and fireman in a chronology of intended professions. But I persisted and became a professional paleontologist, a student of life's history as revealed by the evidence of fossils (though I ended up working on snails rather than dinosaurs!). Why?

I remained committed to paleontology because I discovered, still as a child, the wonder of one of the greatest transforming ideas ever discovered by science: evolution. I learned that those dinosaurs, and all creatures that have ever lived, are bound together in a grand family tree of physical relationships, and that the rich and fascinating changes of life, through billions of years in geological time, occur by a natural process of evolutionary transformation—"descent with modification," in Darwin's words. I was thrilled to learn that humans had arisen from ape-like ancestors, who had themselves evolved from the tiny mouse-like

Flight has evolved three separate times among vertebrates. Birds and bats are still with us, but pterosaurs, such as the one pictured, became extinct with the dinosaurs about 65 million years ago.

mammals that had lived in the time of dinosaurs and seemed then so inconspicuous, so unsuccessful, and so unpromising.

Now, at mid-career (I was born in 1941) I remain convinced that I made the right choice, and committed to learn and convey, as much as I can as long as I can, about evolution and the history of life. We can learn a great deal about the *process* of evolution by studying modern organisms. But history is complex and unpredictable—and principles of evolution (like natural selection) cannot specify the *pathway* that life's history has actually followed. Paleontology holds the archives of the pathway—the fossil record of past life, with its fascinating history of mass extinctions, periods of rapid change, long episodes of stability, and constantly changing patterns of dominance and diversity. Humans represent just one tiny, largely fortuitous, and late-arising twig on the enormously arborescent bush of life. Paleontology is the study of this grandest of all bushes.

"letters" of the long DNA molecules, which are present in every living cell and which provide the genetic information for that organism. By comparing the sequences of "letters" in the DNA of different groups of animals or plants, we can specify the degree of relationship among the groups more precisely than by any other means. In many cases, detailed family trees can then be constructed. The consistent pattern emerging from a growing mountain of data is one of progressive change over time, with more distantly related species showing more differences in their DNA than closely related ones, just as Darwin's theory predicts. By measuring the degree of difference in the genetic coding, and by interpreting the information available from the fossil record, we can even estimate the *rates* at which evolution is occurring in different groups of organisms.

Development

Twentieth-century knowledge about growth and development further supports Darwin's theory of evolution. Striking similarities are seen in the developmental stages of many organisms of different species. Human embryos, for example, go through a "fish-like" stage with gills, a "tadpole-like" stage with a tail, and even a stage when the embryo has fur! Thus, the development of an organism (its ontogeny) appears to go through stages that bear a resemblance to the evolutionary history of the species as a whole (its phylogeny).

Since Darwin's time, new discoveries of the fossil record, genetics, anatomy, and development all support Darwin's theory.

Core Principles of Biology

From centuries of biological observation and inquiry, one organizing principle has emerged: biological diversity reflects history, a record of success, failure, and change extending back to a period soon after the formation of the earth. The explanation for this diversity, the theory of evolution by natural selection, will form the backbone of your study of biological science, just as the theory of the covalent bond is the backbone of chemistry, or the theory of quantum mechanics is that of physics. Evolution by natural selection is a thread that runs through everything you will learn in this book.

Basic Principles

The first half of this book is devoted to a description of the basic principles of biology, introduced through a levels-of-organization framework (see figure 1.2). At the molecular, organellar, and cellular levels of organization, you will be introduced to *cell biology*. You will learn how cells are constructed and how they grow, divide, and communicate. At the organismal level, you will learn the principles of *genetics*, which deal with the way that individual traits are transmitted from one generation to the next. At the population level, you will examine *evolution*, the gradual change in populations from one generation to the next, which has led through natural selection to the biological diversity we see around us. Finally, at the community and ecosystem levels, you will study *ecology*, which deals with how organisms interact with their environments and with one another to produce the complex communities characteristic of life on earth.

Organisms

The second half of the book is devoted to an examination of organisms, the products of evolution. It is estimated that at least 5 million different kinds of plants, animals, and microorganisms exist, and their diversity is *incredible* (fig-

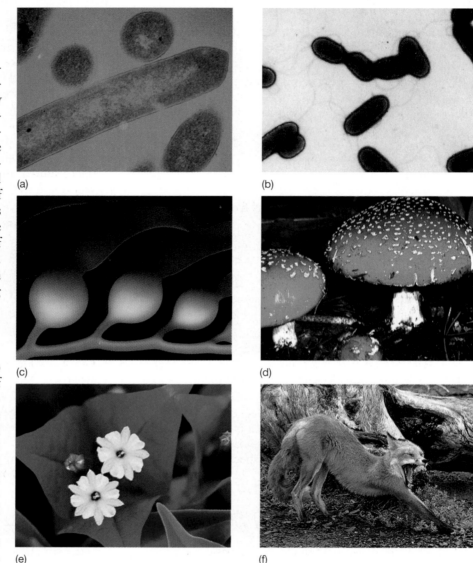

FIGURE 1.15
The diversity of life. Biologists categorize all living things into six major groups: (a) archaebacteria, including bacteria such as this methanogenic bacterium; (b) eubacteria, such as this soil bacterium; (c) protists, including a diverse array of organisms such as this algae; (d) fungi; (e) plants; and (f) animals.

ure 1.15). Later in the book, we will take a particularly detailed look at the vertebrates, the group of animals of which we are members. We will consider the vertebrate body and how it functions, since this information is of greatest interest and importance to most students.

As you proceed through this book, what you learn at one stage will give you the tools to understand the next. The core principle of biology is that biological diversity is the result of a long evolutionary journey.

Summary of Concepts

1.1 Biology is the science of life.

- Living things are highly organized, whether as single cells or as multicellular organisms with several hierarchical levels.
- Living things respond to stimuli. For example, they may flinch away from pain or grow toward sunlight.
- Living things grow, develop, and reproduce using hereditary mechanisms.
- Living things possess regulatory systems that control the living processes of the organism, often maintaining relatively constant internal conditions.

1.2 Scientists form generalizations from observations.

- Science is the determination of general principles from observation and experimentation.
- Scientists select the best hypotheses by using controlled experiments to eliminate alternative hypotheses that are inconsistent with observations.
- A group of related hypotheses supported by a large body of evidence is called a theory. In science, a theory represents what we are most sure about. However, there are no absolute truths in science, and even theories are accepted only conditionally.
- Scientists conduct basic research, designed to gain information about natural phenomena in order to contribute to our overall body of knowledge, and applied research, devoted to solving specific problems with practical applications.

1.3 Darwin's theory of evolution illustrates how science works.

- One of the central theories of biology is Darwin's theory that evolution occurs by natural selection. It states that certain individuals have heritable traits that allow them to produce more offspring in a given kind of environment than other individuals lacking those traits. Consequently, those traits will increase in frequency through time.
- Because environments differ in their requirements and opportunities, the traits favored by natural selection will vary in different environments.
- This theory is supported by a wealth of evidence acquired over more than a century of testing and questioning.

1.4 This book is organized to help you learn biology.

- Biological diversity is the result of a long history of evolutionary change. For this reason evolution is the core of the science of biology.
- Considered in terms of levels-of-organization, the science of biology can be said to consist of subdisciplines focusing on particular levels. Thus one speaks of molecular biology, cell biology, organismal biology, population biology, and community biology.
- Considered in terms of kinds of organisms, the science of biology attempts to understand over 5 million different kinds of organisms.

Discussing Key Terms

1. **Life** Living things share the following basic characteristics: a degree of orderliness; the ability to respond to stimuli; the capacity to grow, develop, and reproduce using hereditary molecules; and the possession of regulatory processes that control and coordinate life functions.
2. **The nature of science** A scientist makes observations and, from the patterns he or she sees, formulates hypotheses about why things happen the way they do.
3. **The experimental method** Scientists follow a logical progression of steps to develop and test their hypotheses. Predictions are made, and carefully controlled experiments are performed to test those predictions.
4. **Scientific theories** Hypotheses supported by a great deal of evidence are called theories. A scientific theory is the best possible explanation scientists are able to develop to account for natural phenomena.
5. **Natural selection** Individuals with characteristics more suitable for survival and reproduction will tend to leave more offspring, and so become more common in future generations.
6. **Evolution** The theory that natural selection directs the formation of new kinds of organisms forms the core of modern biology.

Review Questions

1. What are the characteristics of living things?

2. What is the difference between deductive and inductive reasoning? What is a hypothesis?

3. What are variables? How are control experiments used in testing hypotheses?

4. How does a hypothesis become a theory? At what point does a theory become accepted as an absolute truth, no longer subject to any uncertainty?

5. What is the difference between basic and applied research?

6. What are the major points of Darwin's theory of evolution by natural selection? How did the writings of Lyell and Malthus help Darwin develop his theory?

7. Describe the evidence that led Darwin to propose that evolution occurs by means of natural selection. What evidence gathered since the publication of Darwin's theory has lent further support to the theory?

8. What is the difference between homologous and analogous structures? Give an example of each.

Thought Questions

1. Imagine that you are a scientist asked to test the following hypothesis: the disappearance of a particular species of fish from a lake in the northeastern United States is due to acid rain resulting from industrial air pollution. What alternative hypotheses could you formulate? What experiments would you conduct to test these hypotheses? How would you use control experiments to isolate the influence of acid rain from that of other variables?

2. It is sometimes argued that Darwin's reasoning is circular, because he first *defined* the "fittest" individuals as those that leave the most offspring, and then observed that the fittest leave the most offspring. Do you think this is a fair criticism of Darwin's theory of evolution as described in this chapter?

3. On the Galápagos Islands, Darwin saw a variety of finches, but few varieties of other small birds. Imagine that you are visiting another group of islands about as far away from the South American mainland as are the Galápagos, but far enough from the Galápagos that no birds travel between the two island groups. Would you expect to find a variety of finches on this second island group as well? Explain your reasoning.

Internet Links

Science Resources
http://www.scicentral.com
SCIENCE CENTRAL presents a meta-directory of science on the web, providing links to resources in all fields, and lists of college and university web pages.

Web Software Useful to Biologists
http://lib-www.ucr.edu/
INFOMINE is a shortcut to more than 11,000 scholarly internet links, many in the natural sciences.

Read Darwin's Books
http://www.literature.org/Works/Charles-Darwin/
The complete text of Darwin's three key books on evolution: Voyage of the Beagle, The Origin of Species, & The Descent of Man.

Web Software Useful to Biologists
http:www.geocities.com/CapeCanaveral/1957/software.html
This site presents a list of software available over the internet that would be useful to biologists and biology teachers, each keyed to research or instructional use, nature of application employed, and type of computer (Windows or Macintosh).

Daily Science Updates
http://www.sciencedaily.com/
SCIENCE DAILY is an on-line magazine that publishes a daily digest of the science stories making the news. A great way to keep up with exciting developments.

For Further Reading

Browne, J.: *Charles Darwin: Voyaging. Volume 1 of a Biography*, Knopf, New York, 1995. A brilliantly penetrating retelling of Darwin's life up to his decision to write *Origin of Species*. Written by a former editor of Darwin's vast correspondence, this is one of the fullest and most revealing accounts of Darwin yet written.

Desmond, A., and J. Moore: *Darwin*, Warner Books, New York, 1992. A marvelously detailed account of how Darwin developed his theory of natural selection, from a modern "social" approach that emphasizes Darwin the person.

Moore, J. A.: *Science as a Way of Knowing—The Foundations of Modern Biology*, Harvard University Press, Cambridge, Mass, 1993. An outstanding exposition of evolution and the whole field of biology.

Smith, J. M.: *The Problems of Biology*, Oxford University Press, London, 1986. A succinct and witty guide to what biology is all about and the deep problems on which all research elaborates.

Weiner, J.: *The Beak of the Finch*, Knopf, New York, 1994. A remarkable account of how biologists study evolution in action among the Galápagos finches today.

2

The Nature of Molecules

Concept Outline

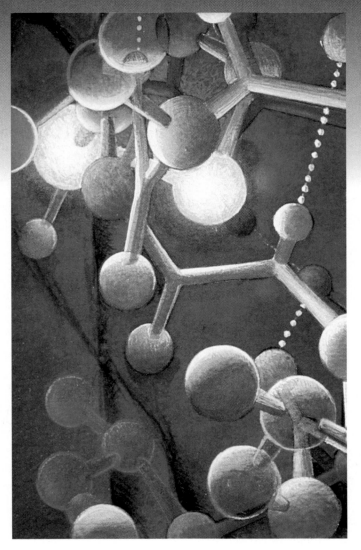

FIGURE 2.1
Cells are made of molecules. Specific, often simple, combinations of atoms yield an astonishing diversity of molecules within the cell, each with unique functional characteristics.

About 10 to 20 billion years ago, an enormous explosion likely marked the beginning of the universe. With this explosion began the process of evolution, which eventually led to the origin and diversification of life on earth. When viewed from the perspective of 20 billion years, life within our solar system is a recent development, but to understand the origin of life, we need to consider events that took place much earlier. The same processes that led to the evolution of life were responsible for the evolution of molecules (figure 2.1). Thus, our study of life on earth begins with physics and chemistry. As chemical machines ourselves, we must understand chemistry to begin to understand our origins.

Atoms

Any substance in the universe that has mass (see below) and occupies space is defined as **matter.** All matter is composed of extremely small particles called **atoms** (figure 2.2). Because of their size, atoms are difficult to study. Not until early in this century did scientists carry out the first experiments suggesting what an atom is like.

The Structure of Atoms

Objects as small as atoms can be "seen" only indirectly, by using very complex technology such as tunneling microcopy. We now know a great deal about the complexities of atomic structure, but the simple view put forth in 1913 by the Danish physicist Niels Bohr provides a good starting point. Bohr proposed that every atom possesses an orbiting cloud of tiny subatomic particles called **electrons** whizzing around a core like the planets of a miniature solar system. At the center of each atom is a small, very dense nucleus formed of two other kinds of subatomic particles, **protons** and **neutrons** (figure 2.3).

Within the nucleus, the cluster of protons and neutrons is held together by a force that works only over short subatomic distances. Each proton carries a positive (+) charge, and each electron carries a negative (–) charge. Typically an atom has one electron for each proton. The number of protons (the atom's **atomic number**) determines the chemical character of the atom, because it dictates the number of electrons orbiting the nucleus and available for chemical activity. Neutrons, as their name implies, possess no charge.

Atomic Mass

The terms *mass* and *weight* are often used interchangeably, but they have slightly different meanings. *Mass* refers to the amount of a substance, while *weight* refers to the force gravity exerts on a substance. Hence, an object has the same *mass* whether it is on the earth or the moon, but its *weight* will be greater on the earth, since the earth's gravitational force is greater than the moon's. The **atomic mass** of an atom is equal to the sum of the masses of its protons and neutrons. Atoms that occur naturally on earth contain from 1 to 92 protons and up to 146 neutrons.

The mass of atoms and subatomic particles is measured in units called *daltons*. To give you an idea of just how small these units are, note that it takes 602 million million billion (6.02×10^{23}) daltons to make 1 gram! A proton weighs approximately 1 dalton (actually 1.009 daltons), as does a neutron (1.007 daltons). In contrast, electrons weigh only 1/1840 of a dalton, so their contribution to the overall mass of an atom is negligible.

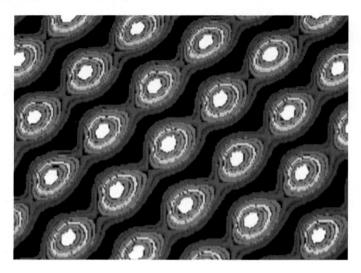

FIGURE 2.2
Individual atoms on the surface of a silicon crystal. This photograph was taken by means of the newly developed technique of tunneling microscopy.

Isotopes

Atoms with the same atomic number (that is, the same number of protons) have the same chemical properties and are said to belong to the same **element.** Formally speaking, an element is any substance that cannot be broken down to any other substance by ordinary chemical means. However, while all atoms of an element have the same number of protons, they may not all have the same number of neutrons. Atoms of an element that possess different numbers of neutrons are called **isotopes** of that element. Most elements in nature exist as mixtures of different isotopes. Carbon (C), for example, has three isotopes, all containing six protons (figure 2.4). Over 99% of the carbon found in

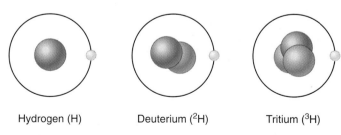

Hydrogen (H)   Deuterium (^{2}H)   Tritium (^{3}H)

FIGURE 2.3
Atoms. The smallest atom is hydrogen (atomic mass 1), whose nucleus consists of a single proton. Hydrogen also exists as two other naturally occurring isotopes, whose nuclei contain neutrons as well as the single proton: deuterium (one neutron), and tritium (two neutrons).

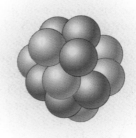

Carbon–12
6 Protons
6 Neutrons
6 Electrons

Carbon–13
6 Protons
7 Neutrons
6 Electrons

Carbon–14
6 Protons
8 Neutrons
6 Electrons

FIGURE 2.4
The three most abundant isotopes of carbon.

nature exists as an isotope with six neutrons. Because its total mass is 12 daltons (6 from protons plus 6 from neutrons), this isotope is referred to as carbon-12, and symbolized ^{12}C. Most of the rest of the naturally occurring carbon is carbon-13, an isotope with seven neutrons. The rarest carbon isotope is carbon-14, with eight neutrons. Unlike the other two isotopes, carbon-14 is unstable: its nucleus tends to break up into elements with lower atomic numbers. This nuclear breakup, which emits a significant amount of energy, is called radioactive decay, and isotopes that decay in this fashion are **radioactive isotopes.**

Some radioactive isotopes are more unstable than others and therefore decay more readily. For any given isotope, however, the rate of decay is constant. This rate is usually expressed as the **half-life,** the time it takes for one half of the atoms in a sample to decay. Carbon-14, for example, has a half-life of about 5600 years. A sample of carbon containing 1 gram of carbon-14 today would contain 0.5 gram of carbon-14 after 5600 years, 0.25 gram 11,200 years from now, 0.125 gram 16,800 years from now, and so on. By determining the ratios of the different isotopes of carbon and other elements in biological samples and in rocks, scientists are able to accurately determine when these materials formed.

While there are many useful applications of radioactivity, there are also harmful side effects that must be considered in any planned use of radioactive substances. Radioactive substances emit energetic subatomic particles that have the potential to severely damage living cells, producing mutations in their genes, and, at high doses, cell death. Consequently, exposure to radiation is now very carefully controlled and regulated. Scientists who work with radioactivity (basic researchers as well as applied scientists such as X-ray technologists) wear radiation-sensitive badges to monitor the total amount of radioactivity to which they are exposed. Each month the badges are collected and scrutinized. Thus, employees whose work places them in danger of excessive radioactive exposure are equipped with an "early warning system."

Electrons

The positive charges in the nucleus of an atom are counterbalanced by negatively charged electrons orbiting at varying distances around the nucleus. Thus, atoms with the same number of protons and electrons are electrically neutral, having no net charge.

Electrons are maintained in their orbits by their attraction to the positively charged nucleus. Sometimes other forces overcome this attraction and an atom loses one or more electrons. In other cases, atoms may gain additional electrons. Atoms in which the number of electrons does not equal the number of protons are known as **ions,** and they do carry a net electrical charge. An atom that has more protons than electrons has a net positive charge and is called a **cation.** For example, an atom of sodium (Na) that has lost one electron becomes a sodium ion (Na⁺), with a charge of +1. An atom that has fewer protons than electrons carries a net negative charge and is called an **anion.** A chlorine atom (Cl) that has gained one electron becomes a chloride ion (Cl⁻), with a charge of –1.

An atom consists of a nucleus of protons and neutrons surrounded by a cloud of electrons. The number of its electrons largely determines the chemical properties of an atom. Atoms that have the same number of protons but different numbers of neutrons are called isotopes. Isotopes of an atom differ in atomic mass but have similar chemical properties.

Electrons Determine the Chemical Behavior of Atoms

The key to the chemical behavior of an atom lies in the arrangement of its electrons in their orbits. It is convenient to visualize individual electrons as following discrete circular orbits around a central nucleus, as in the Bohr model of the atom. However, such a simple picture is not realistic. It is not possible to precisely locate the position of any individual electron precisely at any given time. In fact, a particular electron can be anywhere at a given instant, from close to the nucleus to infinitely far away from it.

However, a particular electron is more likely to be located in some positions than in others. The area around a nucleus where an electron is most likely to be found is called the **orbital** of that electron (figure 2.5). Some electron orbitals near the nucleus are spherical (*s* orbitals), while others are dumbbell-shaped (*p* orbitals). Still other orbitals, more distant from the nucleus, may have different shapes. Regardless of its shape, no orbital may contain more than two electrons.

Almost all of the volume of an atom is empty space, because the electrons are quite far from the nucleus relative to its size. If the nucleus of an atom were the size of an apple, the orbit of the nearest electron would be more than 1600 meters away. Consequently, the nuclei of two atoms never come close enough in nature to interact with each other. It is for this reason that an atom's electrons, not its protons or neutrons, determine its chemical behavior. This also explains why the isotopes of an element, all of which have the same arrangement of electrons, behave the same way chemically.

Energy within the Atom

All atoms possess energy, defined as the ability to do work. Because electrons are attracted to the positively charged nucleus, it takes work to keep them in orbit, just as it takes work to hold an apple in your hand against the pull of gravity. The apple is said to possess potential energy, the ability to do work, because of its position; if you were to release it, the apple would fall and its energy would be reduced. Conversely, if you were to move the apple to the top of a building, you would increase its potential energy. Similarly, electrons have potential energy of position. To oppose the attraction of the nucleus and move the electron to a more distant orbital requires an input of energy and results in an electron with greater potential energy. This is how chlorophyll captures energy from light during photosynthesis (chapter 10)—the light excites electrons in the chlorophyll. Moving an electron closer to the nucleus has the opposite effect: energy is released, usually as heat, and the electron ends up with less potential energy (figure 2.6).

A given atom can possess only certain discrete amounts of energy. Like the potential energy of an apple on a step of a staircase, the potential energy contributed by the position of an electron in an atom can have only certain values. Every atom exhibits a ladder of potential energy values, rather than a continuous spectrum of possibilities, a discrete set of orbits at particular distances from the nucleus.

During some chemical reactions, electrons are transferred from one atom to another. In such reactions, the loss of an electron is called **oxidation,** and the gain of an electron is called **reduction** (figure 2.7). It is important to realize that when an electron is transferred in this way, it keeps its en-

FIGURE 2.5
Electron orbitals. The lowest energy level or electron shell, which is nearest the nucleus, is level *K.* It is occupied by a single *s* orbital, referred to as 1*s.* The next highest energy level, *L*, is occupied by four orbitals: one *s* orbital (referred to as the 2*s* orbital) and three *p* orbitals (each referred to as a 2*p* orbital). The four *L*-level orbitals compactly fill the space around the nucleus, like two pyramids set base-to-base.

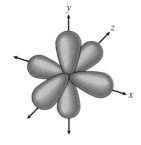

2*s* Orbital

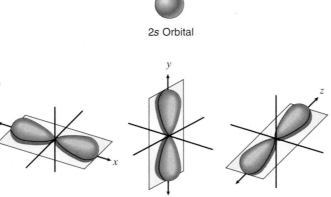

1*s* Orbital

2*p* Orbitals

Orbital for energy level *K*:
one spherical orbital (1*s*)

Orbitals for energy level *L*:
one spherical orbital (2*s*) and
three dumbbell-shaped orbitals (2*p*)

Composite of
all *p* orbitals

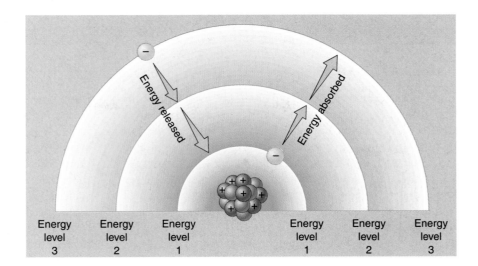

FIGURE 2.6
Atomic energy levels. When an electron absorbs energy, it moves to higher energy levels farther from the nucleus. When an electron releases energy, it falls to lower energy levels closer to the nucleus.

Energy level 3 · Energy level 2 · Energy level 1 · Energy level 1 · Energy level 2 · Energy level 3

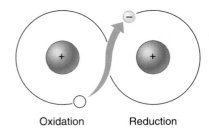

Oxidation · Reduction

FIGURE 2.7
Oxidation and reduction. Oxidation is the loss of an electron; reduction is the gain of an electron.

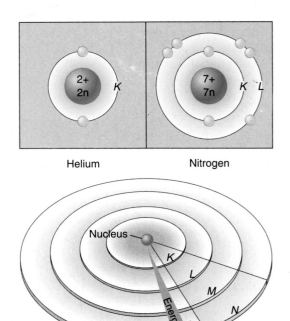

Helium · Nitrogen

Nucleus · K · L · M · N · Energy level

FIGURE 2.8
Electron energy levels for helium and nitrogen. Gold balls represent the electrons. Each concentric circle represents a different distance from the nucleus and, thus, a different electron energy level.

ergy of position. In organisms, chemical energy is stored in high-energy electrons that are transferred from one atom to another in reactions involving oxidation and reduction.

Since the amount of energy an electron possesses is related to its distance from the nucleus, electrons that are the same distance from the nucleus have the same energy, even if they occupy different orbitals. Such electrons are said to occupy the same **energy level.** In a schematic diagram of an atom (figure 2.8), the nucleus is represented as a small circle and the electron energy levels are drawn as concentric rings, with the energy level increasing with distance from the nucleus. Be careful not to confuse energy levels, which are drawn as rings to indicate an electron's *energy*, with orbitals, which have a variety of three-dimensional shapes and indicate an electron's most likely *location*.

Electrons orbit a nucleus in paths called orbitals. No orbital can contain more than two electrons, but many orbitals may be the same distance from the nucleus and, thus, contain electrons of the same energy.

Kinds of Atoms

There are 92 naturally occurring elements, each with a different number of protons and a different arrangement of electrons. When the nineteenth-century Russian chemist Dmitri Mendeleev arranged the known elements in a table according to their atomic mass (figure 2.9), he discovered one of the great generalizations in all of science. Mendeleev found that the elements in the table exhibited a pattern of chemical properties that repeated itself in groups of eight elements. This periodically repeating pattern lent the table its name: the periodic table of elements.

The Periodic Table

The eight-element periodicity that Mendeleev found is based on the interactions of the electrons in the outer energy levels of the different elements. These electrons are called **valence electrons** and their interactions are the basis for the differing chemical properties of the elements. For most of the atoms important to life, an outer energy level can contain no more than eight electrons; the chemical behavior of an element reflects how many of the eight positions are filled. Elements possessing all eight electrons in their outer energy level are **inert,** or nonreactive; they include helium (He), neon (Ne), argon (Ar), krypton (Kr), xenon (Xe), and radon (Rn). In sharp contrast, elements with seven electrons (one fewer than the maximum number of eight) in their outer energy level, such as fluorine (F), chlorine (Cl), and bromine (Br), are highly reactive. They tend to gain the extra electron needed to fill the energy level. Elements with only one electron in their outer energy level, such as lithium (Li), sodium (Na), and potassium (K), are also very reactive; they tend to lose the single electron in their outer level.

Mendeleev's periodic table thus leads to a useful generalization, the **octet rule** (Latin *octo,* "eight") or **rule of eight:** atoms tend to establish completely full outer energy levels. Most chemical behavior can be predicted quite accurately from this simple rule, combined with the tendency of atoms to balance positive and negative charges.

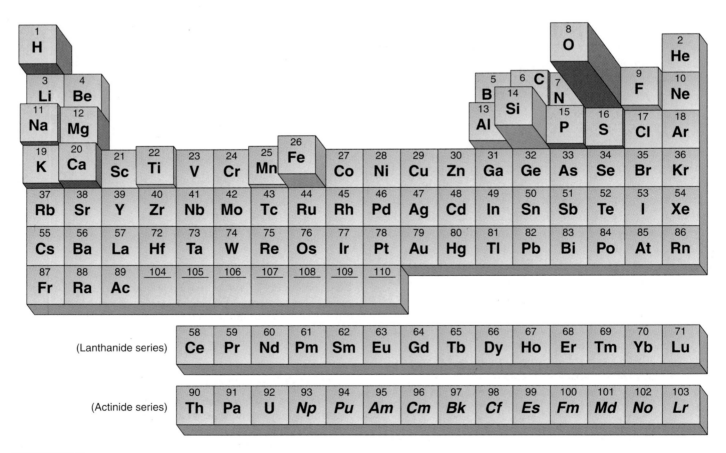

FIGURE 2.9
Periodic table of the elements. In this representation, the frequency of elements that occur in the earth's crust is indicated by the height of the block. Elements found in significant amounts in living organisms are shaded in blue.

Table 2.1 The Most Common Elements on Earth and Their Distribution in the Human Body

Element	Symbol	Atomic Number	Approximate Percent of Earth's Crust by Weight	Percent of Human Body by Weight	Importance or Function
Oxygen	O	8	46.6	65.0	Required for cellular respiration; component of water
Silicon	Si	14	27.7	Trace	
Aluminum	Al	13	6.5	Trace	
Iron	Fe	26	5.0	Trace	Critical component of hemoglobin in the blood
Calcium	Ca	20	3.6	1.5	Component of bones and teeth; triggers muscle contraction
Sodium	Na	11	2.8	0.2	Principal positive ion outside cells; important in nerve function
Potassium	K	19	2.6	0.4	Principal positive ion inside cells; important in nerve function
Magnesium	Mg	12	2.1	0.1	Critical component of many energy-transferring enzymes
Hydrogen	H	1	0.14	9.5	Electron carrier; component of water and most organic molecules
Manganese	Mn	25	0.1	Trace	
Fluorine	F	9	0.07	Trace	
Phosphorus	P	15	0.07	1.0	Backbone of nucleic acids; important in energy transfer
Carbon	C	6	0.03	18.5	Backbone of organic molecules
Sulfur	S	16	0.03	0.3	Component of most proteins
Chlorine	Cl	17	0.01	0.2	Principal negative ion outside cells
Vanadium	V	23	0.01	Trace	
Chromium	Cr	24	0.01	Trace	
Copper	Cu	29	0.01	Trace	Key component of many enzymes
Nitrogen	N	7	Trace	3.3	Component of all proteins and nucleic acids
Boron	B	5	Trace	Trace	
Cobalt	Co	27	Trace	Trace	
Zinc	Zn	30	Trace	Trace	Key component of some enzymes
Selenium	Se	34	Trace	Trace	
Molybdenum	Mo	42	Trace	Trace	Key component of many enzymes
Tin	Sn	50	Trace	Trace	
Iodine	I	53	Trace	Trace	Component of thyroid hormone

Distribution of the Elements

Of the 92 naturally occurring elements on earth, only 11 are found in organisms in more than trace amounts (0.01% or higher). These 11 elements have atomic numbers less than 21 and, thus, have low atomic masses. Table 2.1 lists the levels of various elements in the human body; their levels in other organisms are similar. Inspection of this table suggests that the distribution of elements in living systems is by no means accidental. The most common elements inside organisms are not the elements that are most abundant in the earth's crust. For example, silicon, aluminum, and iron constitute 39.2% of the earth's crust, but they exist in trace amounts in the human body. On the other hand, carbon atoms make up 18.5% of the human body but only 0.03% of the earth's crust.

92 elements occur naturally on earth; only 11 of them are found in significant amounts in living organisms. Four of them—oxygen, hydrogen, carbon, nitrogen—constitute 96.3% of the weight of your body.

Ionic Bonds Form Crystals

A group of atoms held together by energy in a stable association is called a **molecule.** When a molecule contains atoms of more than one element, it is called a **compound.** The atoms in a molecule are joined by **chemical bonds;** these bonds can result when atoms with opposite charges attract (ionic bonds), when two atoms share one or more pairs of electrons (covalent bonds), or when atoms interact in other ways. We will start by examining **ionic bonds,** which form when atoms with opposite electrical charges (ions) attract.

A Closer Look at Table Salt

Common table salt, sodium chloride (NaCl), is a lattice of ions in which the atoms are held together by ionic bonds (figure 2.10). Sodium has 11 electrons: 2 in the inner energy level, 8 in the next level, and 1 in the outer (valence) level. The valence electron is unpaired (free) and has a strong tendency to join with another electron. A stable configuration can be achieved if the valence electron is lost to another atom that also has an unpaired electron. The loss of this electron results in the formation of a positively charged sodium ion, Na^+.

The chlorine atom has 17 electrons: 2 in the inner energy level, 8 in the next level, and 7 in the outer level. Hence, one of the orbitals in the outer energy level has an unpaired electron. The addition of another electron to the outer level fills that level and causes a negatively charged chloride ion, Cl^-, to form.

When placed together, metallic sodium and gaseous chlorine react swiftly and explosively, as the sodium atoms donate electrons to chlorine, forming Na^+ and Cl^- ions. Because opposite charges attract, the Na^+ and Cl^- remain associated in an **ionic compound,** NaCl, which is electrically neutral. However, the electrical attractive force holding NaCl together is not directed specifically between particular Na^+ and Cl^- ions, and no discrete sodium chloride molecules form. Instead, the force exists between any one ion and all neighboring ions of the opposite charge, and the ions aggregate in a crystal matrix with a precise geometry. Such aggregations are what we know as salt crystals. If a salt such as NaCl is placed in water, the electrical attraction of the water molecules, for reasons we will point out later in this chapter, disrupts the forces holding the ions in their crystal matrix, causing the salt to dissolve into a roughly equal mixture of free Na^+ and Cl^- ions.

> **An ionic bond is an attraction between ions of opposite charge in an ionic compound. Such bonds are not formed between particular ions in the compound; rather, they exist between an ion and all of the oppositely charged ions in its immediate vicinity.**

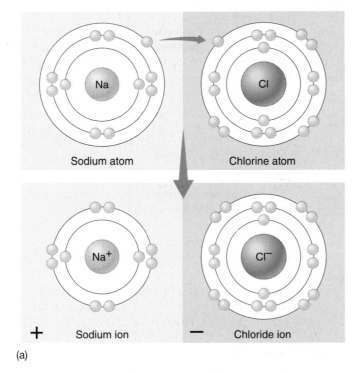

Sodium atom   Chlorine atom

Na⁺ Sodium ion   Cl⁻ Chloride ion

(a)

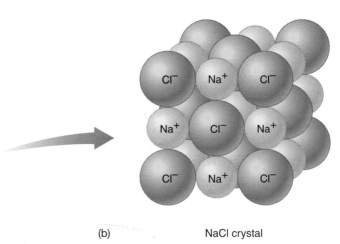

(b)   NaCl crystal

FIGURE 2.10
The formation of ionic bonds by sodium chloride. (a) When a sodium atom donates an electron to a chlorine atom, the sodium atom becomes a positively charged sodium ion, and the chlorine atom becomes a negatively charged chloride ion. (b) Sodium chloride forms a highly regular lattice of alternating sodium ions and chloride ions.

Covalent Bonds Build Stable Molecules

Covalent bonds form when two atoms share one or more pairs of valence electrons. Consider hydrogen (H) as an example. Each hydrogen atom has an unpaired electron and an unfilled outer energy level; for these reasons the hydrogen atom is chemically unstable. When two hydrogen atoms are close to each other, however, each atom's electron can orbit both nuclei. In effect, the nuclei are able to share their electrons. The result is a diatomic (two-atom) molecule of hydrogen gas (figure 2.11).

The molecule formed by the two hydrogen atoms is stable for three reasons:

1. **It has no net charge.** The diatomic molecule formed as a result of this sharing of electrons is not charged, because it still contains two protons and two electrons.
2. **The octet rule is satisfied.** Each of the two hydrogen atoms can be considered to have two orbiting electrons in its outer energy level. This satisfies the octet rule, because each shared electron orbits both nuclei and is included in the outer energy level of *both* atoms.
3. **It has no free electrons.** The bonds between the two atoms also pair the two free electrons.

Unlike ionic bonds, covalent bonds are formed between two specific atoms, giving rise to true, discrete molecules. While ionic bonds can form regular crystals, the more specific associations made possible by covalent bonds allow the formation of complex molecular structures.

Covalent Bonds Can Be Very Strong

The strength of a covalent bond depends on the number of shared electrons. Thus **double bonds,** which satisfy the octet rule by allowing two atoms to share *two* pairs of electrons, are stronger than **single bonds,** in which only one electron pair is shared. This means more chemical energy is required to break a double bond than a single bond. The strongest covalent bonds are **triple bonds,** such as those that link the two nitrogen atoms of nitrogen gas molecules. Covalent bonds are represented in chemical formulations as lines connecting atomic symbols, where each line between two bonded atoms represents the sharing of one pair of electrons. The **structural formulas** of hydrogen gas and oxygen gas are H–H and O=O, respectively, while their **molecular formulas** are H_2 and O_2.

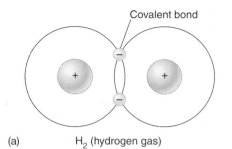

(a) H_2 (hydrogen gas)

(b)

FIGURE 2.11

Hydrogen gas. (a) Hydrogen gas is a diatomic molecule composed of two hydrogen atoms, each sharing its electron with the other. (b) The flash of fire that consumed the *Hindenburg* occurred when the hydrogen gas that was used to inflate the dirigible combined explosively with oxygen gas in the air to form water.

Molecules with Several Covalent Bonds

Molecules often consist of more than two atoms. One reason that larger molecules may be formed is that a given atom is able to share electrons with more than one other atom. An atom that requires two, three, or four additional electrons to fill its outer energy level completely may acquire them by sharing its electrons with two or more other atoms.

For example, the carbon atom (C) contains six electrons, four of which are in its outer energy level. To satisfy the octet rule, a carbon atom must gain access to four additional electrons; that is, it must form four covalent bonds. Because four covalent bonds may form in many ways, carbon atoms are found in many different kinds of molecules.

Chemical Reactions

The formation and breaking of chemical bonds, the essence of chemistry, is called a **chemical reaction.** All chemical reactions involve the shifting of atoms from one molecule or ionic compound to another, without any change in the number or identity of the atoms. For convenience, we refer to the original molecules before the reaction starts as **reactants,** and the molecules resulting from the chemical reaction as **products.** For example:

$$A - B + C - D \;\rightarrow\; A - C + B + D$$
$$\text{reactants} \qquad\qquad \text{products}$$

The extent to which chemical reactions occur is influenced by several important factors:

1. **Temperature.** Heating up the reactants increases the rate of a reaction (as long as the temperature isn't so high as to destroy the molecules).
2. **Concentration of reactants and products.** Reactions proceed more quickly when more reactants are available. An accumulation of products typically speeds reactions in the reverse direction.
3. **Catalysts.** A catalyst is a substance that increases the rate of a reaction. It doesn't alter the reaction's equilibrium between reactants and products, but it does shorten the time needed to reach equilibrium, often dramatically. In organisms, proteins called enzymes catalyze almost every chemical reaction.

A covalent bond is a stable chemical bond formed when two atoms share one or more pairs of electrons.

(a)   (b)   (c)

FIGURE 2.12
Water takes many forms. As a liquid, water fills our rivers and runs down over the land to the sea. (a) The iceberg on which the penguins are holding their meeting was formed in Antarctica from a huge block of ice that broke away into the ocean water. (b) When water cools below 0° C, it forms beautiful crystals, familiar to us as snow and ice. However, water is not always plentiful. (c) At Badwater, in Death Valley, California, there is no hint of water except for the broken patterns of dried mud.

Chemistry of Water

Of all the molecules that are common on earth, only **water** exists as a liquid at the relatively low temperatures that prevail on the earth's surface, three-fourths of which is covered by liquid water (figure 2.12). When life was originating, water provided a medium in which other molecules could move around and interact without being held in place by strong covalent or ionic bonds. Life evolved as a result of these interactions, and it is still inextricably tied to water. Life began in water and evolved there for 3 billion years before spreading to land. About two-thirds of any organism's body is composed of water, and no organism can grow or reproduce in any but a water-rich environment. It is no accident that tropical rain forests are bursting with life, while dry deserts appear almost lifeless except when water becomes temporarily plentiful, such as after a rainstorm.

The Atomic Structure of Water

Water has a simple atomic structure. It consists of an oxygen atom bound to two hydrogen atoms by two single covalent bonds (figure 2.13*a*). The resulting molecule is stable: it satisfies the octet rule, has no unpaired electrons, and carries no net electrical charge.

The single most outstanding chemical property of water is its ability to form weak chemical associations with only 5–10% of the strength of covalent bonds. This

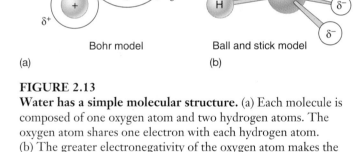

Bohr model   Ball and stick model
(a)   (b)

FIGURE 2.13
Water has a simple molecular structure. (a) Each molecule is composed of one oxygen atom and two hydrogen atoms. The oxygen atom shares one electron with each hydrogen atom. (b) The greater electronegativity of the oxygen atom makes the water molecule polar: water carries a partial negative charge (δ^-) near the oxygen atom and a partial positive charge (δ^+) near the hydrogen atoms.

property, which derives directly from the structure of water, is responsible for much of the organization of living chemistry.

The chemistry of life is water chemistry. The way in which life first evolved was determined in large part by the chemical properties of the liquid water in which that evolution occurred.

Water Atoms Act Like Tiny Magnets

Both the oxygen and the hydrogen atoms attract the electrons they share in the covalent bonds of a water molecule; this attraction is called **electronegativity.** However, the oxygen atom is more electronegative than the hydrogen atoms, so it attracts the electrons more strongly than do the hydrogen atoms. As a result, the shared electrons in a water molecule are far more likely to be found near the oxygen nucleus than near the hydrogen nuclei. This stronger attraction for electrons gives the oxygen atom a partial negative charge (δ^-), as though the electron cloud were denser near the oxygen atom than around the hydrogen atoms. Because the water molecule as a whole is electrically neutral, each hydrogen atom carries a partial positive charge (δ^+). The Greek letter delta (δ) signifies a partial charge, much weaker than the full unit charge of an ion.

What would you expect the shape of a water molecule to be? Each of water's two covalent bonds has a partial charge at each end, δ^- at the oxygen end and δ^+ at the hydrogen end. The most stable arrangement of these charges is a *tetrahedron*, in which the two negative and two positive charges are approximately equidistant from one another (figure 2.13*b*). The oxygen atom lies at the center of the tetrahedron, the hydrogen atoms occupy two of the apexes, and the partial negative charges occupy the other two apexes. This results in a bond angle of 104.5° between the two covalent oxygen-hydrogen bonds. (In a regular tetrahedron, the bond angles would be 109.5°; in water, the partial negative charges occupy more space than the hydrogen atoms, and, therefore, they compress the oxygen-hydrogen bond angle slightly.)

The water molecule, thus, has distinct "ends," each with a partial charge, like the two poles of a magnet.

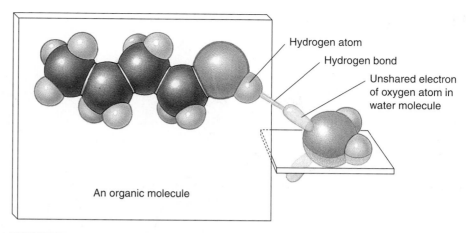

Hydrogen atom

Hydrogen bond

Unshared electron of oxygen atom in water molecule

An organic molecule

FIGURE 2.14
Structure of a hydrogen bond.

(These partial charges are much less than the unit charges of ions, however.) Molecules that exhibit charge separation are called **polar molecules** because of their magnet-like poles, and water is one of the most polar molecules known. *The polarity of water underlies its chemistry and the chemistry of life.*

Polar molecules interact with one another, as the δ^- of one molecule is attracted to the δ^+ of another. Since many of these interactions involve hydrogen atoms, they are called **hydrogen bonds** (figure 2.14). Each hydrogen bond is individually very weak and transient, lasting on average only 1/100,000,000,000 second (10^{-11} sec). However, the cumulative effects of large numbers of these bonds can be enormous. Water forms an abundance of hydrogen bonds, which are responsible for many of its important physical properties (table 2.2).

The water molecule is very polar, with ends that exhibit partial positive and negative charges. Opposite charges attract, forming weak linkages called hydrogen bonds.

Table 2.2 The Properties of Water		
Property	**Explanation**	**Example of Benefit to Life**
Cohesion	Hydrogen bonds hold water molecules together	Leaves pull water upward from the roots; seeds swell and germinate
High specific heat	Hydrogen bonds absorb heat when they break, and release heat when they form, minimizing temperature changes	Water stabilizes the temperature of organisms and the environment
High heat of vaporization	Many hydrogen bonds must be broken for water to evaporate	Evaporation of water cools body surfaces
Lower density of ice	Water molecules in an ice crystal are spaced relatively far apart because of hydrogen bonding	Because ice is less dense than water, lakes do not freeze solid, and they overturn in spring
High polarity	Polar water molecules are attracted to ions and polar compounds, making them soluble	Many kinds of molecules can move freely in cells, permitting a diverse array of chemical reactions

Water Clings to Polar Molecules

The polarity of water causes it to be attracted to other polar molecules. When the other molecules are also water, the attraction is referred to as **cohesion.** When the other molecules are of a different substance, the attraction is called **adhesion.** It is because water is cohesive that it is a liquid, and not a gas, at moderate temperatures.

The cohesion of liquid water is also responsible for its **surface tension.** Small insects can walk on water (figure 2.15) because at the air-water interface all of the hydrogen bonds in water face downward, causing the molecules of the water surface to cling together. Water is adhesive to any substance with which it can form hydrogen bonds. That is why substances containing polar molecules get "wet" when they are immersed in water, while those that are composed of nonpolar molecules (such as oils) do not.

The attraction of water to substances like glass with surface electrical charges is responsible for capillary action: if a glass tube with a narrow diameter is lowered into a beaker of water, the water will rise in the tube above the level of the water in the beaker, because the adhesion of water to the glass surface, drawing it upward, is stronger than the force of gravity, drawing it down (figure 2.16). The narrower the tube, the greater the electrostatic forces between the water and the glass, and the higher the water rises.

Water Stores Heat

Water moderates temperature through two properties: its high specific heat and its high heat of vaporization. The temperature of any substance is a measure of how rapidly its individual molecules are moving. Because of the many hydrogen bonds that water molecules form with one another, a large input of thermal energy is required to break these bonds before the individual water molecules can begin moving about more freely and so have a higher temperature. Therefore, water is said to have a high **specific heat,** which is defined as the amount of heat that must be absorbed or lost by one gram of a substance to change its temperature by one degree Celsius (°C). Specific heat measures the extent to which a substance resists changing its temperature when it absorbs or loses heat. Since polar substances tend to form hydrogen bonds, and energy is needed to break these bonds, the more polar a substance is, the higher is its specific heat. The specific heat of water

FIGURE 2.15
Cohesion. Some insects, such as this water strider, literally walk on water. In this photograph you can see the dimpling the insect's feet make on the water as its weight bears down on the surface. Because the surface tension of the water is greater than the force that one foot brings to bear, the strider glides atop the surface of the water rather than sinking.

(1 calorie/gram/°C) is twice that of most carbon compounds and nine times that of iron. Only ammonia, which is more polar than water and forms very strong hydrogen bonds, has a higher specific heat than water (1.23 calories/gram/°C). Still, only 20% of the hydrogen bonds are broken as water heats from 0 to 100°C.

Because of its high specific heat, water heats up more slowly than almost any other compound and holds its temperature longer when heat is no longer applied. This characteristic enables organisms, which have a high water content, to maintain a relatively constant internal temperature. The heat generated by the chemical reactions inside cells would destroy the cells, if it were not for the high specific heat of the water within them.

FIGURE 2.16
Capillary action. Capillary action causes the water within a narrow tube to rise above the surrounding water; the adhesion of the water to the glass surface, which draws water upward, is stronger than the force of gravity, which tends to draw it down. The narrower the tube, the greater the surface area available for adhesion for a given volume of water, and the higher the water rises in the tube.

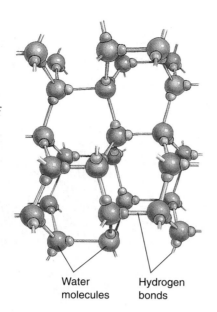

FIGURE 2.17
The role of hydrogen bonds in an ice crystal. When water cools below 0° C, it forms a regular crystal structure in which the four partial charges of one water molecule interact with the opposite charges of other water molecules. Because water forms a crystal latticework, ice is less dense than liquid water and floats. If it did not, inland bodies of water far from the earth's equator might never thaw fully.

Water molecules   Hydrogen bonds

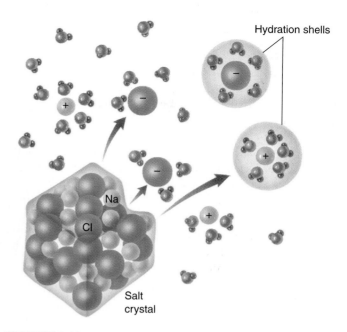

Hydration shells

Na
Cl
Salt crystal

FIGURE 2.18
Why salt dissolves in water. When a crystal of table salt dissolves in water, individual Na+ and Cl− ions break away from the salt lattice and become surrounded by water molecules. Water molecules orient around Cl− ions so that their partial positive poles face toward the negative Cl− ion; water molecules surrounding Na+ ions orient in the opposite way, with their partial negative poles facing the positive Na+ ion. Surrounded by hydration shells, Na+ and Cl− ions never reenter the salt lattice.

A considerable amount of heat energy (586 calories) is required to change one gram of liquid water into a gas. Hence, water also has a high **heat of vaporization.** Since the transition of water from a liquid to a gas requires the input of energy to break its many hydrogen bonds, the evaporation of water from a surface causes cooling of that surface. Many organisms dispose of excess body heat by evaporative cooling; for example, humans and many other vertebrates sweat.

At low temperatures, water molecules are locked into a crystal-like lattice of hydrogen bonds, forming the solid we call ice (figure 2.17). Interestingly, ice is less dense than liquid water because the hydrogen bonds in ice space the water molecules relatively far apart. This unusual feature enables icebergs to float. Were it otherwise, ice would cover nearly all bodies of water, with only shallow surface melting annually.

Water Is a Powerful Solvent

Water is an effective solvent because of its ability to form hydrogen bonds. Water molecules gather closely around any substance that bears an electrical charge, whether that substance carries a full charge (ion) or a charge separation (polar molecule). For example, sucrose (table sugar) is composed of molecules that contain slightly polar hydroxyl (OH) groups. A sugar crystal dissolves rapidly in water because water molecules can form hydrogen bonds with individual hydroxyl groups of the sucrose molecules. Therefore, sucrose is said to be *soluble* in water. Every time a sucrose molecule dissociates or breaks away from the crystal, water molecules surround it in a cloud, forming a **hydration shell** and preventing it from associating with other sucrose molecules. Hydration shells also form around ions such as Na+ and Cl− (figure 2.18).

Water Organizes Nonpolar Molecules

Water molecules always tend to form the maximum possible number of hydrogen bonds. When nonpolar molecules such as oils, which do not form hydrogen bonds, are placed in water, the water molecules act to exclude them. The nonpolar molecules are forced into association with one another, thus minimizing their disruption of the hydrogen bonding of water. In effect, they shrink from contact with water and for this reason they are referred to as **hydrophobic** (Greek *hydros,* "water" and *phobos,* "fearing"). In contrast, polar molecules, which readily form hydrogen bonds with water, are said to be **hydrophilic** ("water-loving").

The tendency of nonpolar molecules to aggregate in water is known as **hydrophobic exclusion.** By forcing the hydrophobic portions of molecules together, water causes these molecules to assume particular shapes. Different molecular shapes have evolved by alteration of the location and strength of nonpolar regions. As you will see, much of the evolution of life reflects changes in molecular shape that can be induced in just this way.

Water molecules, which are very polar, cling to one another, so that it takes considerable energy to separate them. Water also clings to other polar molecules, causing them to be soluble in water solution, but water tends to exclude nonpolar molecules.

Water Ionizes

The covalent bonds within a water molecule sometimes break spontaneously. In pure water at 25° C, only 1 out of every 550 million water molecules undergoes this process. When it happens, one of the protons (hydrogen atom nuclei) dissociates from the molecule. Because the dissociated proton lacks the negatively charged electron it was sharing in the covalent bond with oxygen, its own positive charge is no longer counterbalanced, and it becomes a positively charged ion, H^+. The rest of the dissociated water molecule, which has retained the shared electron from the covalent bond, is negatively charged and forms a **hydroxide ion** (OH^-). This process of spontaneous ion formation is called **ionization**:

$$H_2O \rightarrow OH^- + H^+$$
$$\text{water} \qquad \text{hydroxide ion} \qquad \text{hydrogen ion}$$

At 25° C, a liter of water contains 1/10,000,000 (or 10^{-7}) mole of H^+ ions. (A **mole** is defined as the weight in grams that corresponds to the summed atomic masses of all of the atoms in a molecule. In the case of H^+, the atomic mass is 1, and a mole of H^+ ions would weigh 1 gram. One mole of any substance always contains 6.02×10^{23} molecules of the substance.) Therefore, the **molar concentration** of hydrogen ions (represented as $[H^+]$) in pure water is 10^{-7} mole/liter. Actually the hydrogen ion usually associates with another water molecule to form a hydronium (H_3O^+) ion.

pH

A more convenient way to express the hydrogen ion concentration of a solution is to use the **pH scale** (figure 2.19). This scale defines pH as the negative logarithm of the hydrogen ion concentration in the solution:

$$pH = -\log [H^+]$$

Since the logarithm of the hydrogen ion concentration is simply the exponent of the molar concentration of H^+, the pH equals the exponent times –1. Thus, pure water, with an $[H^+]$ of 10^{-7} mole/liter, has a pH of 7. Recall that for every H^+ ion formed when water dissociates, an OH^- ion is also formed, meaning that the dissociation of water produces H^+ and OH^- in equal amounts. Therefore, a pH value of 7 indicates neutrality—a balance between H^+ and OH^-—on the pH scale.

Note that the pH scale is *logarithmic*, which means that a difference of 1 on the scale represents a tenfold change in hydrogen ion concentration. This means that a solution with a pH of 4 has *ten times* the concentration of H^+ than is present in one with a pH of 5.

Acids. Any substance that dissociates in water to increase the concentration of H^+ ions is called an acid. Acidic solutions have pH values below 7. The stronger an acid is, the

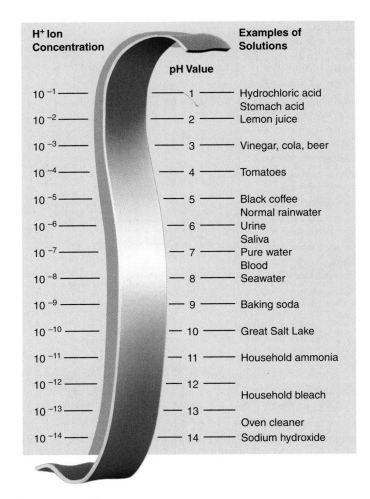

FIGURE 2.19
The pH scale. The pH value of a solution indicates its concentration of hydrogen ions. Solutions with a pH less than 7 are acidic, while those with a pH greater than 7 are basic. The scale is logarithmic, so that a pH change of 1 means a tenfold change in the concentration of hydrogen ions. Thus, lemon juice is 100 times more acidic than tomato juice, and seawater is 10 times more basic than pure water, which has a pH of 7.

more H^+ ions it produces and the lower its pH. For example, hydrochloric acid (HCl), which is abundant in your stomach, ionizes completely in water. This means that 10^{-1} mole per liter of HCl will dissociate to form 10^{-1} mole per liter of H^+ ions, giving the solution a pH of 1. The pH of champagne, which bubbles because of the carbonic acid dissolved in it, is about 2.

Bases. A substance that combines with H^+ ions when dissolved in water is called a base. By combining with H^+ ions, a base lowers the H^+ ion concentration in the solution. Basic (or alkaline) solutions, therefore, have pH values above 7. Very strong bases, such as sodium hydroxide (NaOH), have pH values of 12 or more.

Buffers

The pH inside almost all living cells, and in the fluid surrounding cells in multicellular organisms, is fairly close to 7. Most of the biological catalysts (enzymes) in living systems are extremely sensitive to pH; often even a small change in pH will alter their shape, thereby disrupting their activities and rendering them useless. For this reason it is important that a cell maintain a constant pH level.

Yet the chemical reactions of life constantly produce acids and bases within cells. Furthermore, many animals eat substances that are acidic or basic; Coca Cola, for example, is a strong (although dilute) acidic solution. Despite such variations in the concentrations of H^+ and OH^-, the pH of an organism is kept at a relatively constant level by buffers (figure 2.20).

A **buffer** is a substance that acts as a reservoir for hydrogen ions, donating them to the solution when their concentration falls and taking them from the solution when their concentration rises. What sort of substance will act in this way? Within organisms, most buffers consist of pairs of substances, one an acid and the other a base. The key buffer in human blood is an acid-base pair consisting of carbonic acid (acid) and bicarbonate (base). These two substances interact in a pair of reversible reactions. First, carbon dioxide (CO_2) and H_2O join to form carbonic acid (H_2CO_3), which in a second reaction dissociates to yield bicarbonate ion (HCO_3^-) and H^+ (figure 2.21). If some acid or other substance adds H^+ ions to the blood, the HCO_3^- ions act as a base and remove the excess H^+ ions by forming H_2CO_3. Similarly, if a basic substance removes H^+ ions from the blood, H_2CO_3 dissociates, releasing more H^+ ions into the blood. The forward and reverse reactions that interconvert H_2CO_3 and HCO_3^- thus stabilize the blood's pH.

The reaction of carbon dioxide and water to form carbonic acid is important because it permits significant

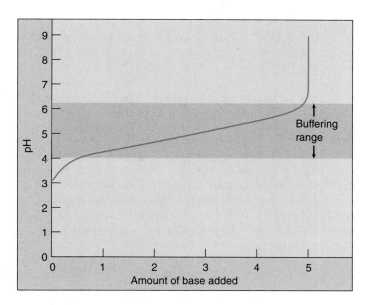

FIGURE 2.20
Buffers minimize changes in pH. Adding a base to a solution neutralizes some of the acid present, and so raises the pH. Thus, as the curve moves to the right, reflecting more and more base, it also rises to higher pH values. What a buffer does is to make the curve rise or fall very slowly over a portion of the pH scale, called the "buffering range" of that buffer.

amounts of carbon, essential to life, to enter water from the air. As we will discuss in chapter 4, biologists believe that life first evolved in the early oceans. These oceans were rich in carbon because of the reaction of carbon dioxide with water.

In a condition called blood acidosis, human blood, which normally has a pH of about 7.4, drops 0.2–0.4 points on the pH scale. This condition is fatal if not treated immediately. The reverse condition, blood alkalosis, involves an increase in blood pH of a similar magnitude and is just as serious.

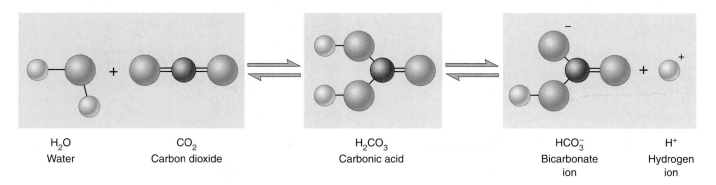

H_2O	CO_2	H_2CO_3	HCO_3^-	H^+
Water	Carbon dioxide	Carbonic acid	Bicarbonate ion	Hydrogen ion

FIGURE 2.21
Buffer formation. Carbon dioxide and water combine chemically to form carbonic acid (H_2CO_3). The acid then dissociates in water, freeing H^+ ions. This reaction makes carbonated beverages acidic, and produced the carbon-rich early oceans that cradled life.

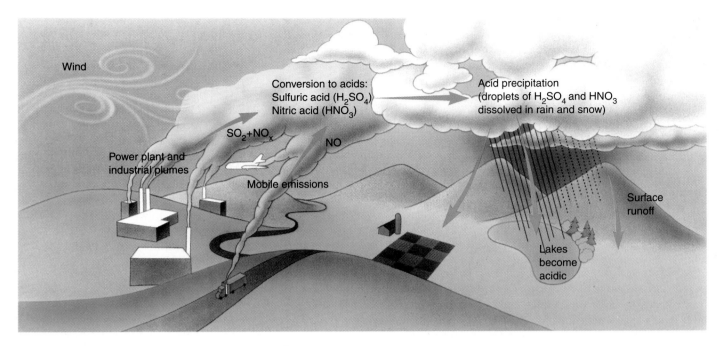

FIGURE 2.22
The formation of acid precipitation. Sulfur and nitrogen compounds released by the combustion of fossil fuels combine with water in the atmosphere to form sulfuric and nitric acids, which then fall back to the earth in the form of acid precipitation, often with devastating effects.

Acid Rain

Because changes in pH can harm living organisms, changes in the pH of the environment can have drastic effects. Acid precipitation (rain, fog, or snow; figure 2.22) can kill all of the fish in a lake or trees in a forest, cause developmental defects in plants and animals, seriously affect human health, and even erode monuments and buildings made out of limestone and marble (figure 2.23). Acid precipitation is a global concern, since excess production of industrial pollutants in one country can easily affect other countries, even on other continents. The United States implemented the Clean Air Act in 1990, which has contributed to decreased levels of sulfur dioxide and other air pollutants in this country.

> **The pH of a solution is the negative logarithm of the H^+ ion concentration in the solution. Thus, low pH values indicate high H^+ concentrations (acidic solutions), and high pH values indicate low H^+ concentrations (basic solutions). Even small changes in pH can be harmful to life.**

FIGURE 2.23
Effects of acid precipitation on limestone. Limestone is rich in calcium carbonate, which is easily degraded by even weak acids. This tombstone has been badly eroded, its original message obliterated for all time, because of its exposure to the elements—including acid rain.

Summary of Concepts

2.1 Atoms are nature's building material.

- The smallest stable particles of matter are protons, neutrons, and electrons, which associate to form atoms.
- The core, or nucleus, of an atom consists of protons and neutrons; the electrons orbit around the nucleus in a cloud. The farther an electron is from the nucleus, the faster it moves and the more energy it possesses.
- The chemical behavior of an atom is largely determined by the distribution of its electrons and in particular by the number of electrons in its outermost (highest) energy level. There is a strong tendency for atoms to have a completely filled outer level; electrons are lost, gained, or shared until this condition is reached.

2.2 The atoms of living things are among the smallest.

- More than 95% of the weight of an organism consists of oxygen, hydrogen, carbon, and nitrogen, all of which form strong covalent bonds with one another.

2.3 Chemical bonds hold molecules together.

- A molecule is a stable collection of atoms held together by chemical bonds.
- Ionic bonds form when electrons transfer from one atom to another, and the resulting oppositely charged ions attract one another.
- Covalent bonds form when two atoms share electrons. They are responsible for the formation of most biologically important molecules.

2.4 Water is the cradle of life.

- The chemistry of life is the chemistry of water (H_2O). The central oxygen atom in water attracts the electrons it shares with the two hydrogen atoms. As a result, the oxygen atom is electron-rich (bearing a partial negative charge) and the hydrogen atoms are electron-poor (bearing a partial positive charge). This charge separation gives water positive and negative poles, making it a polar molecule.
- A hydrogen bond is formed between the partial positive charge of a hydrogen atom in one molecule and the partial negative charge of another atom, either in another molecule or in a different portion of the same molecule.
- Water is cohesive and adhesive, has a great capacity for storing heat, is a good solvent for other polar molecules, and tends to exclude nonpolar molecules.
- Water ionizes spontaneously to a slight degree, producing H^+ and OH^- ions. At 25° C, the concentration of H^+ that results from this dissociation is 10^{-7} mole/liter. The H^+ concentration in a solution is expressed by the pH scale, in which pH equals the negative logarithm of the H^+ concentration. Thus, the pH of pure water is 7.
- Maintenance of a relatively constant pH is critical in living systems. Since the pH scale is logarithmic, even a small deviation in pH represents a substantial change in H^+ concentration. Organisms regulate pH through a system of buffers that add H^+ when solutions are too basic and remove H^+ when they are too acidic.

Discussing Key Terms

1. **Atom** An atom is the smallest particle into which a chemical can be divided and still retain its characteristic properties. It is composed of a nucleus of protons and neutrons surrounded by a cloud of electrons.
2. **Molecule** A molecule is an aggregate of atoms held together by covalent chemical bonds.
3. **Covalent bond** A covalent bond is a chemical bond formed by the sharing of one or more pairs of electrons by two atoms.
4. **Hydrogen bond** Hydrogen bonds form when hydrogen atoms that are covalently bound to one atom are also attracted to a neighboring electronegative atom, either in the same molecule or in a separate molecule. Hydrogen bonds are not covalent bonds and are only 1/10 as strong, but in great numbers they can stabilize interactions between large organic molecules such as proteins.
5. **pH** The pH scale measures the concentration of hydrogen ions (H^+) in water. Solutions containing a high H^+ concentration have a low pH and are said to be acidic, while those containing a low H^+ concentration have a high pH and are called basic or alkaline.
6. **Buffer** A buffer is a substance that acts as a reservoir for H^+ ions, donating them to a solution when their concentration falls and taking them from the solution when their concentration rises. One of the most biologically important buffers is formed when carbon dioxide combines with water to form carbonic acid, which dissociates into H^+ and bicarbonate ions.

Review Questions

1. An atom of nitrogen has 7 protons and 7 neutrons. What is its atomic number? What is its atomic mass? How many electrons does it have?

2. How do the isotopes of a single element differ from each other? How do their chemical properties compare?

3. The half-life of radium-226 is 1620 years. If a sample of material contains 16 milligrams of radium-226, how much will it contain in 1620 years? How much will it contain in 3240 years? How long will it take for the sample to contain 1 milligram of radium-226?

4. What is a cation? What is an anion?

5. Which has more potential energy, an electron that is closer to an atom's nucleus, or an electron that is farther from the nucleus? What happens to this energy of position when the electron is transferred to another atom in a chemical reaction? What is the process in which an atom loses an electron, and what is the process in which an atom gains an electron?

6. What is the octet rule, and how does it affect the chemical behavior of atoms?

7. What is the difference between an ionic bond and a covalent bond? Give an example of each.

8. What types of atoms participate in the formation of hydrogen bonds? How does the strength of a hydrogen bond compare with that of a covalent bond? How do hydrogen bonds contribute to water's high specific heat?

9. What types of molecules are hydrophobic? What types are hydrophilic? Why do these two types of molecules behave differently in water?

10. Define pH. What is the pH of a solution that has a hydrogen ion concentration of 10^{-3} mole/liter? Would such a solution be acidic or basic? What is the pH of a solution that has a hydrogen ion concentration of 10^{-9} mole/liter? Would such a solution be acidic or basic?

11. How does the carbonic acid-bicarbonate buffer system respond to the addition of excess hydrogen ions to a solution? How does this buffer system respond when another substance removes hydrogen ions from the solution? Explain how this buffer system helps an organism maintain a constant pH.

Thought Questions

1. Carbon (atomic number 6) and silicon (atomic number 14) both have just four electrons in their outer energy levels. Ammonia (NH_3) is even more polar than water. Why do you suppose life evolved into organisms composed of carbon chains in water solution rather than organisms composed of silicon in ammonia?

2. Carbon atoms can share four electron pairs when forming molecules. Why doesn't carbon form a bimolecular gas, as hydrogen (one pair of shared electrons), oxygen (two pairs of shared electrons), and nitrogen (three pairs of shared electrons) do?

Internet Links

Water
http://www.webdirectory.com/Water_Resources/
A directory of sites on the web related to water, including both chemistry and environmental matters.

Interactive Chemistry Tutorial
http://tqd.advanced.org/2923/html/home3.html
CHEMISTRY TUTOR provides an excellent review of basic chemistry which a student may review at his or her own pace.

Scientific Method
http://www.xnet.com/~blatura/skep_1.html
A discussion of the scientific method that answers ten key questions about how science works.

Another Interactive Periodic Table
http://www.shef.ac.uk/chemistry/web-elements/
A wonderful interactive periodic table. Click on any element and you get three-dimensional images, what compounds it forms, melting points, isotopes, as well as the substance's history and where it can be found.

For Further Reading

Atkins, P. W.: *Molecules*, Scientific American Library, New York, 1987. A delightful journey among the molecules most familiar to us.

Cox, T.: "Origin of the Chemical Elements," *New Scientist*, February 3, 1990, pages 1–4. The heavier elements making up most of the earth—and us—were created by the birth and death of generations of stars.

Gorham, E.: "Atmospheric Chemistry: Neutralizing Acid Rain," *Nature*, vol. 367, January 27, 1994, page 321. A brief discussion of the unanticipated "extras" that contribute to acid rain.

Mertz, W.: "The Essential Trace Elements," *Science*, vol. 213, 1981, pages 1332–38. An account of the roles of rare elements in human metabolism and the effects of their absences.

Raven, P. R., L. R. Berg, and G. B. Johnson: *Environment*, ed. 2, Saunders College Publishing, Fort Worth, 1995. A user-friendly text on environmental biology, with a particularly lucid discussion of acid precipitation.

Tyson, J. L.: "Scientists to Turn High Beam on Molecules," *Christian Science Monitor*, March 2, 1994, page 14. A very readable discussion of the Advanced Photon Source project underway at the Argonne National Laboratory.

Weinberg, S.: *The Discovery of Subatomic Particles*, Scientific American Library, New York, 1983. Documents the discovery of subatomic particles and introduces the nontechnical reader to the fundamentals of classic physics.

3

The Chemical Building Blocks of Life

Concept Outline

3.1 Molecules are the building blocks of life.

The Chemistry of Carbon. Because individual carbon atoms can form multiple covalent bonds, organic molecules can be quite complex.

3.2 Carbohydrates contain many CH bonds.

Kinds of Carbohydrates. Sugars are simple carbohydrates, often consisting of six-carbon rings.

Linking Sugars Together. Sugars can be linked together to form long polymers, or polysaccharides.

Structural Carbohydrates. Structural carbohydrates like cellulose are chains of sugars linked in a way that enzymes cannot easily attack.

3.3 Lipids are not soluble in water.

Fats. Fats are clusters of long carbon chains that are very nonpolar and so are quite insoluble in water.

Fatty Acids. A component of fats, fatty acids determine whether a fat is saturated or unsaturated.

Other Kinds of Lipids. Many other kinds of molecules within cells are insoluble in water, including many in biological membranes.

3.4 Proteins perform the chemistry of the cell.

The Many Functions of Proteins. Proteins can be catalysts, transporters, supporters, and regulators.

Amino Acids Are the Building Blocks of Proteins. Proteins are long chains of various combinations of amino acids.

The Shape of Globular Proteins. A protein's shape is determined by its amino acid sequence.

How Proteins Fold. The distribution of nonpolar amino acids along a protein chain largely determines how the protein folds.

How Proteins Unfold. When conditions such as pH or temperature fluctuate, proteins may denature or unfold.

3.5 Nucleic acids store the genetic information.

Information Molecules. Nucleic acids, long chains of nucleotides, store information in cells.

The Structure of Nucleic Acids. RNA is a single-chain polymer of nucleotides, while DNA possesses two chains twisted around each other.

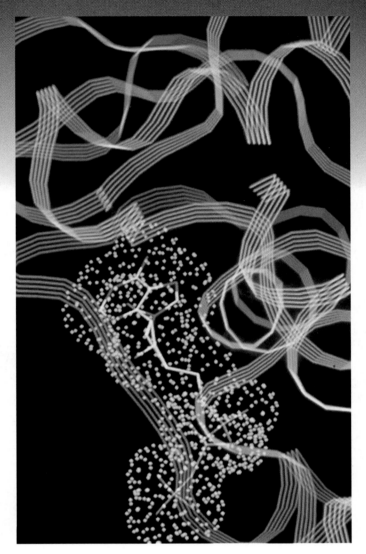

FIGURE 3.1
Computer-generated model of a macromolecule. Pictured is the enzyme responsible for releasing energy from sugar. This complex molecule consists of hundreds of different amino acids linked into chains that form the characteristic coils and folds seen here.

Molecules are extremely small compared with the familiar world we see about us. Imagine: there are more water molecules in a cup than there are stars in the sky. Many other molecules are gigantic, compared with water, consisting of thousands of atoms. These atoms are organized into hundreds of smaller molecules that are linked together into long chains (figure 3.1). These enormous molecules, almost always synthesized by living things, are called macromolecules. As we shall see, there are four general types of macromolecules, the basic chemical building blocks from which all organisms are assembled.

The Chemistry of Carbon

In chapter 2 we discussed how atoms combine to form molecules. In this chapter, we will focus on **organic molecules,** those chemical compounds that contain carbon. Because every carbon atom possesses four valence electrons in its outer shell—a shell that holds eight—carbon can form up to four covalent bonds. Because carbon can readily form single, double, and even triple bonds with itself, it can readily form chains of carbon atoms.

Carbon chains, in which carbon atoms are linked together, form the framework of biological molecules. These chains may be straight, branched, or closed into rings, with atoms of hydrogen, oxygen, nitrogen or other elements bonded to individual carbon atoms at many positions on the chain. As you can imagine, all of these possibilities generate an incredible variety of molecules.

Organic molecules consisting only of carbon and hydrogen are called **hydrocarbons.** Propane gas, for example, is a hydrocarbon consisting of a chain of three carbon atoms, with eight hydrogen atoms bound to it. Carbon contributes a total of 12 valence electrons (three atoms with four electrons each), four of which are involved in the two carbon-carbon bonds. That leaves eight available electrons, each of which forms a covalent bond with a hydrogen atom. Since carbon-hydrogen covalent bonds store considerable energy, hydrocarbons make good fuels. Petroleum, for example, is rich in hydrocarbons.

Functional Groups

Most biologically important organic molecules are not hydrocarbons. Instead of hydrogen atoms, other elements are attached to their carbon framework. In general, any organic molecule can be thought of as a carbon-based core to which specific groups of atoms with definite chemical properties are attached. We refer to these attached groups of atoms as **functional groups.** For example, a hydrogen atom bonded to an oxygen atom (—OH) is called a *hydroxyl group.*

Functional groups have definite chemical properties that they retain no matter where they occur. The hydroxyl group, for example, is polar, because its oxygen atom, being very electronegative, draws electrons toward itself (as we saw in chapter 2). Figure 3.2 illustrates the hydroxyl group and other biologically important functional groups. Most chemical reactions that occur within organisms involve the transfer of a functional group as an intact unit from one molecule to another.

Biological Macromolecules

Some of the organic molecules in organisms are small and simple, containing only one or a few functional groups. Others, especially those that play structural roles or store genetic information, are large complex assemblies called **macromolecules.** In many cases, these macromolecules are polymers, molecules built by linking together a large number of small, similar chemical subunits, like railroad cars coupled to form a train. For example, complex carbohydrates like starch are polymers of simple ring-shaped sugars, proteins are polymers of amino acids, and nucleic acids (DNA and RNA) are polymers of nucleotides.

Biological macromolecules are traditionally grouped into four major categories: carbohydrates, lipids, proteins, and nucleic acids (figure 3.3).

Group	Chemical Formula	Structural Formula	Ball-and-Stick Model	Found In:
Hydroxyl	—OH	—OH	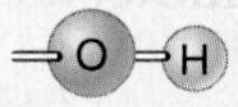	Alcohols
Carbonyl	C=O	C=O	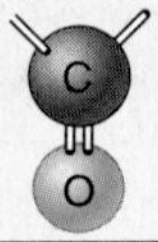	Formaldehyde
Carboxyl	—COOH	—C(=O)OH	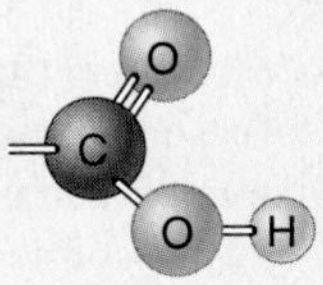	Vinegar
Amino	—NH₂	—N(H)(H)	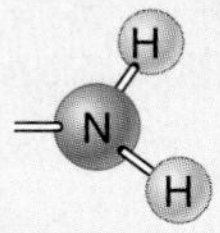	Ammonia
Sulfhydryl	—SH	—S—H	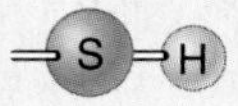	Rubber
Phosphate	—PO₄	—O—P(=O)(O⁻)—O⁻	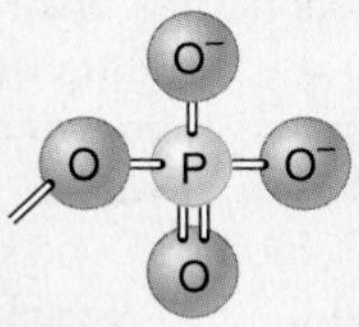	ATP
Methyl	—CH₃	—C(H)(H)—H	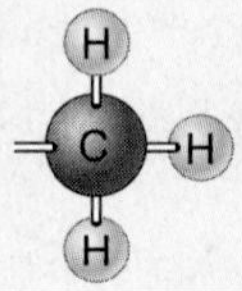	Methane gas

FIGURE 3.2
The primary functional chemical groups. These groups tend to act as units during chemical reactions and confer specific chemical properties on the molecules that possess them. Hydroxyl groups, for example, make a molecule more basic, while carboxyl groups make a molecule more acidic.

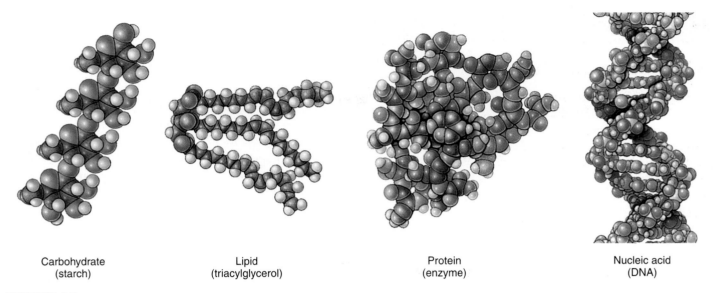

Carbohydrate
(starch)

Lipid
(triacylglycerol)

Protein
(enzyme)

Nucleic acid
(DNA)

FIGURE 3.3
Biological macromolecules. Representatives of the four fundamental kinds of biological macromolecules: carbohydrates, lipids, proteins, and nucleic acids.

Building Macromolecules

Although the four categories of macromolecules contain different kinds of subunits, they are all assembled in the same fundamental way: to form a covalent bond between two subunit molecules, an —OH group is removed from one subunit and a hydrogen atom (H) is removed from the other (figure 3.4a). This process is called a dehydration (water-losing) reaction, or **dehydration synthesis,** because the removal of the —OH group and H during the synthesis of a new molecule in effect constitutes the removal of a molecule of water (H_2O). For every subunit that is added to a macromolecule, one water molecule is removed. Energy is required to break the chemical bonds when water is extracted from the subunits, so cells must supply energy to assemble macromolecules. Reactions in which macromolecules are built from smaller subunits are called *anabolic reactions.* These and other biochemical reactions require that the reacting substances be held close together and that the correct chemical bonds be stressed and broken. This process of positioning and stressing, termed catalysis, is carried out in cells by a special class of proteins known as enzymes.

While cells synthesize some macromolecules, they disassemble others into their constituent subunits by performing *catabolic reactions.* These reactions are essentially the reverse of dehydration—a molecule of water is added instead of removed (figure 3.4b). In this process, which is called **hydrolysis** (Greek *hydro,* "water" + *lyse,* "break"), a hydrogen atom is attached to one subunit and a hydroxyl group to the other, breaking a specific covalent bond in the macromolecule. Hydrolytic reactions release the energy that was stored in the bonds that were broken.

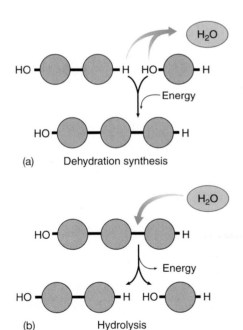

(a)   Dehydration synthesis

(b)   Hydrolysis

FIGURE 3.4
Making and breaking macromolecules. (a) Biological macromolecules are polymers formed by linking subunits together. The covalent bond between the subunits is formed by dehydration synthesis, an energy-requiring process that creates a water molecule for every bond formed. (b) Breaking the bond between subunits requires the returning of a water molecule with a subsequent release of energy, a process called hydrolysis.

Polymers are large molecules consisting of long chains of similar subunits joined by dehydration reactions, in which a hydroxyl (—OH) group is removed from one subunit and a hydrogen atom (H) is removed from the other.

3.2 Carbohydrates contain many CH bonds.

Kinds of Carbohydrates

We will begin our discussion of macromolecules that make up the bodies of organisms (table 3.1) with carbohydrates. Carbohydrates function as energy-storage molecules as well as structural elements. Some are small, simple molecules, while others form long polymers.

Sugars Are Simple Carbohydrates

The **carbohydrates** are a loosely defined group of molecules that contain carbon, hydrogen, and oxygen in the molar ratio 1:2:1. Their empirical formula (which lists the atoms in the molecule with subscripts to indicate how many there are of each) is $(CH_2O)_n$, where n is the number of carbon atoms. Because they contain many carbon-hydrogen (C—H) bonds, which release energy when they are broken, carbohydrates are well suited for energy storage.

Monosaccharides. The simplest of the carbohydrates are the simple sugars, or **monosaccharides** (Greek *mono*, "single" + Latin *saccharum*, "sugar"). Simple sugars may contain as few as three carbon atoms, but those that play the central role in energy storage have six. Their empirical formula is:

$$C_6H_{12}O_6, \text{ or } (CH_2O)_6$$

Six-carbon sugars can exist in a straight-chain form, but in water solution they almost always form rings. The most important of these for energy storage is *glucose* (figure 3.5), a six-carbon sugar which has seven energy-storing C—H bonds.

Disaccharides. Many familiar sugars like sucrose are "double sugars," two monosaccharides joined by a covalent bond. Called **disaccharides,** they often play a role in the transport of sugars, as we will discuss shortly.

Polysaccharides. **Polysaccharides** are macromolecules made up of monosaccharide subunits. Starch is a polysaccharide used by plants to store energy. It consists entirely of glucose molecules, linked one after another in long chains. Cellulose is a polysaccharide that serves as a structural building material in plants. It too consists entirely of glucose molecules linked together into chains, but the links are difficult for enzymes to break.

Table 3.1 Macromolecules			
Macromolecule	**Subunit**	**Function**	**Example**
CARBOHYDRATES			
Starch, glycogen	Glucose	Energy storage	Potatoes
Cellulose	Glucose	Cell walls	Paper
Chitin	Modified glucose	Structural support	Crab shells
LIPIDS			
Fats	Glycerol and three fatty acids	Energy storage	Butter; soap
Phospholipids	Glycerol, two fatty acids, phosphate, and polar R groups	Cell membranes	Lecithin
Prostaglandins	Five-carbon rings with two nonpolar tails	Chemical messengers	Prostaglandin E (PGE)
Steroids	Four fused carbon rings	Membranes; hormones	Cholesterol; estrogen
Terpenes	Long carbon chains	Pigments; structural	Carotene; rubber
PROTEINS			
Globular	Amino acids	Catalysis; transport	Hemoglobin
Structural	Amino acids	Support	Hair; silk
NUCLEIC ACIDS			
DNA	Nucleotides	Encodes genes	Chromosomes
RNA	Nucleotides	Needed for gene expression	Messenger RNA

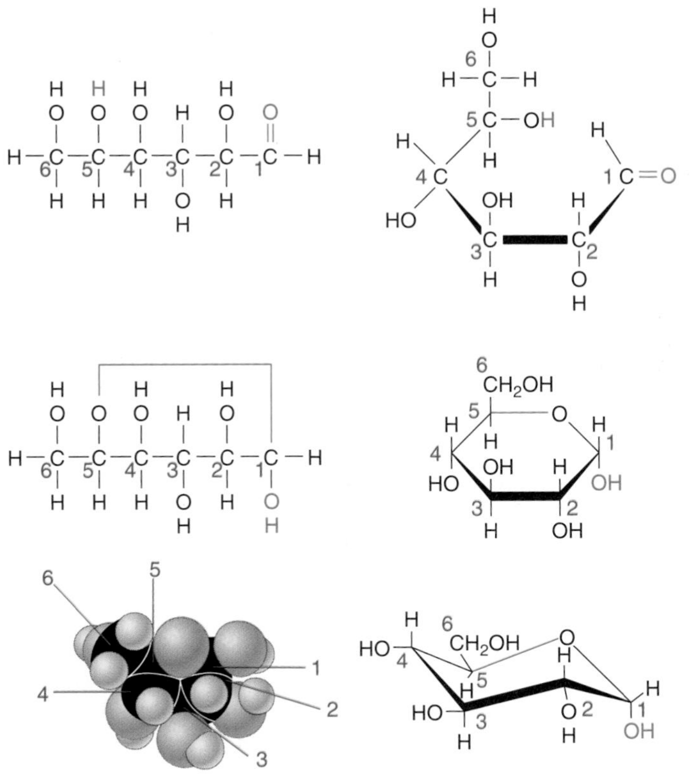

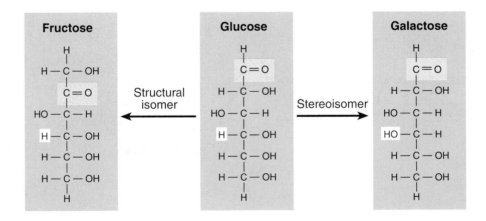

Sugar Isomers

Glucose is not the only sugar with the formula $C_6H_{12}O_6$. Other common six-carbon sugars such as fructose and galactose also have this same empirical formula (figure 3.6). These sugars are **isomers,** or alternative forms, of glucose. Even though isomers have the same empirical formula, their atoms are arranged in different ways; that is, their three-dimensional structures are different. These structural differences often account for substantial functional differences between the isomers. Glucose and fructose, for example, are *structural isomers*. In fructose, the double-bonded oxygen is attached to an internal carbon rather than to a terminal one. Your taste buds can tell the difference, as fructose tastes much sweeter than glucose, despite the fact that both sugars have the same chemical composition. This structural difference also has an important chemical consequence: the two sugars form different polymers.

Unlike fructose, galactose has the same bond structure as glucose; the only difference between galactose and glucose is the orientation of one hydroxyl group. Because the hydroxyl group positions are mirror images of each other, galactose and glucose are called *stereoisomers*. Again, this seemingly slight difference has important consequences, since this hydroxyl group is often involved in creating polymers with distinct functions, such as starch (energy storage) and cellulose (structural support).

Sugars are among the most important energy-storage molecules in organisms, containing many energy-storing C—H bonds. The structural differences among sugar isomers can confer substantial functional differences upon the molecules.

FIGURE 3.5
Structure of the glucose molecule. Glucose is a linear six-carbon molecule that forms a ring shape in solution. The structure of the ring can be represented in many ways; the ones shown here are the most common, with the carbons conventionally numbered (in green) so that the forms can be compared easily. The bold, darker lines represent portions of the molecule that are projecting out of the page toward you—remember, these are three-dimensional molecules!

FIGURE 3.6
Isomers and stereoisomers. Glucose, fructose, and galactose are isomers with the empirical formula $C_6H_{12}O_6$. A structural isomer of glucose, such as fructose, has identical chemical groups bonded to different carbon atoms, while a stereoisomer of glucose, such as galactose, has identical chemical groups bonded to the same carbon atoms but in different orientations.

Linking Sugars Together

Transport Disaccharides

Most organisms transport sugars within their bodies. In humans, the glucose that circulates in the blood does so as a simple monosaccharide. In plants and many other organisms, however, glucose is converted into a transport form before it is moved from place to place within the organism. In such a form it is less readily metabolized (used for energy) during transport. Transport forms of sugars are commonly made by linking two monosaccharides together to form a disaccharide (Greek *di*, "two"). Disaccharides serve as effective reservoirs of glucose because the normal glucose-utilizing enzymes of the organism cannot break the bond linking the two monosaccharide subunits. Enzymes that can do so are typically present only in the tissue where the glucose is to be used.

Transport forms differ depending on which monosaccharides link to form the disaccharide. Glucose forms transport disaccharides with itself and many other monosaccharides, including fructose and galactose. When glucose forms a disaccharide with its structural isomer, fructose, the resulting disaccharide is *sucrose*, or table sugar (figure 3.7*a*). Sucrose is the form in which most plants transport glucose and the sugar that most humans (and other animals) eat. Sugarcane is rich in sucrose, and so are sugar beets.

When glucose is linked to its stereoisomer, galactose, the resulting disaccharide is *lactose*, or milk sugar. Many mammals supply energy to their young in the form of lactose. Adults have greatly reduced levels of lactase, the enzyme required to cleave lactose into its two monosaccharide components, and thus cannot metabolize lactose as efficiently. Most of the energy that is channeled into lactose production is therefore reserved for their offspring.

Storage Polysaccharides

Organisms store the metabolic energy contained in monosaccharides by converting them into disaccharides, such as *maltose* (figure 3.7*b*), which are then linked together into insoluble forms that are deposited in specific storage areas in their bodies. These insoluble polysaccharides are long polymers of monosaccharides formed by dehydration synthesis. Plant polysaccharides formed from glucose are called **starches.** Plants store starch as granules within chloroplasts and other organelles. Because glucose is a key metabolic fuel, the stored starch provides a reservoir of energy available for future needs. Food energy can be retrieved by hydrolyzing the links that bind the glucose subunits together.

The starch with the simplest structure is *amylose*, which is composed of many hundreds of glucose molecules linked together in long, unbranched chains. Each linkage occurs between the number 1 carbon of one glucose molecule and the number 4 carbon of another, so that amylose is, in effect, a longer form of maltose. The long chains of amylose tend to coil up in water (figure 3.8*a*), a property that renders amylose insoluble. Potato starch is about 20% amylose. When amylose is digested by a sprouting potato plant (or by an animal that eats a potato), enzymes first break it into fragments of random length, which are more soluble because they are shorter. Baking or boiling potatoes has the same effect, breaking the chains into fragments. Another enzyme then cuts these fragments into molecules of maltose. Finally, the maltose is cleaved into two glucose molecules, which cells are able to metabolize.

Most plant starch, including the remaining 80% of potato starch, is a somewhat more complicated variant of amylose called *amylopectin* (figure 3.8*b*). Pectins are branched polysaccharides. Amylopectin has short, linear amylose

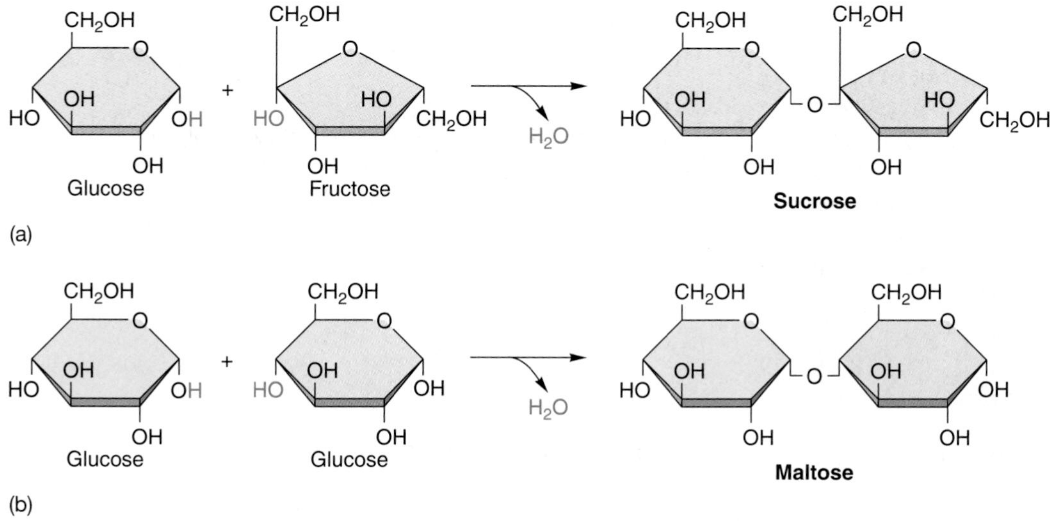

FIGURE 3.7
Disaccharides. Some disaccharides are used to transport glucose from one part of an organism's body to another; one example is sucrose (a), which is found in sugarcane. Other disaccharides, such as maltose in grain (b), are used for storage.

branches consisting of 20 to 30 glucose subunits. In some plants these chains are cross-linked. The cross-links create an insoluble mesh of glucose, which can be degraded only in the presence of yet another kind of enzyme. The size of the mesh differs from plant to plant; in rice about 100 amylose chains, each with one or two cross-links, forms the mesh.

The animal version of starch is *glycogen*. Like amylopectin, glycogen is an insoluble polysaccharide containing branched amylose chains. In glycogen, the average chain length is much greater and there are more branches than in plant starch (figure 3.8*c*). Humans and other vertebrates store excess food energy as glycogen in the liver and in muscle cells; when the demand for energy in a tissue increases, glycogen is hydrolyzed to release glucose.

Nonfattening Sweets

Imagine a kind of table sugar that looks, tastes, and cooks like the real thing, but has no calories or harmful side effects. You could eat mountains of candy made from such sweeteners without gaining weight. As Louis Pasteur discovered in the late 1800s, most sugars are "right-handed" molecules, in that the hydroxyl group that binds a critical carbon atom is on the right side. However, "left-handed" sugars, in which the hydroxyl group is on the left side, can be made readily in the laboratory. These synthetic sugars are mirror-image chemical twins of the natural form, but the enzymes that break down sugars in the human digestive system can tell the difference. To digest a sugar molecule, an enzyme must first grasp it, much like a shoe fitting onto a foot, and all of the body's enzymes are right-handed! A left-handed sugar just doesn't fit, any more than a shoe for the right foot fits onto a left foot.

The Latin word for "left" is *levo*, and left-handed sugars are called **levo-**, or **1-sugars.** They do not occur in nature except for trace amounts in red algae, snail eggs, and seaweed. Because they pass through the body without being used, they can let diet-conscious sweet-lovers have their cake and eat it, too. Nor will they contribute to tooth decay, since bacteria cannot metabolize them, either.

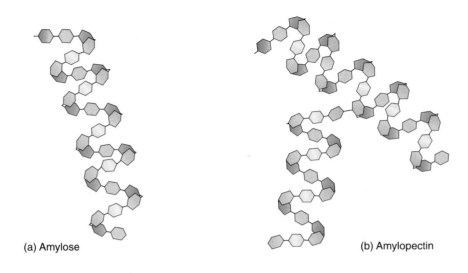

(a) Amylose

(b) Amylopectin

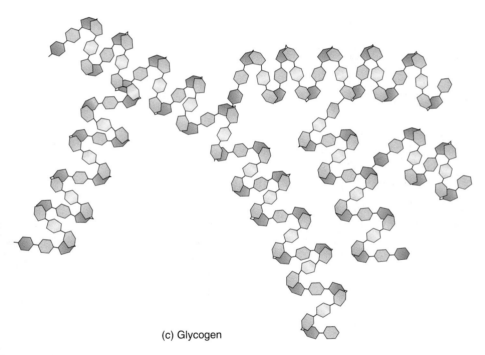

(c) Glycogen

FIGURE 3.8
Storage polysaccharides. Starches are long glucose polymers that store energy in plants. (a) The simplest starches are long chains of maltose called amylose, which tend to coil up in water. (b) Most plants contain more complex starches called amylopectins, which are branched. (c) Animals store glucose in glycogen, which is more extensively branched than amylopectin and contains longer chains of amylose.

Starches are glucose polymers. Most starches are branched and some are cross-linked. The branching and cross-linking render the polymer insoluble and protect it from degradation.

Structural Carbohydrates

While some chains of sugars store energy, others serve as structural material for cells.

Cellulose

For two glucose molecules to link together, the glucose subunits must be the same form. Glucose can form a ring in two ways, with the hydroxyl group attached to the carbon where the ring closes being locked into place either below or above the plane of the ring. If below, it is called the **alpha form,** and if above, the **beta form.** All of the glucose subunits of the starch chain are alpha-glucose. When a chain of glucose molecules consists of all beta-glucose subunits, a polysaccharide with very different properties results. This **structural polysaccharide** is *cellulose*, the chief component of plant cell walls (figure 3.9). Cellulose is chemically similar to amylose, with one important difference: the starch-degrading enzymes that occur in most organisms cannot break the bond between two beta-glucose sugars. This is not because the bond is stronger, but rather because its cleavage requires an enzyme most organisms lack. Because cellulose cannot be broken down readily, it works well as a biological structural material and occurs widely in this role in plants. Those few animals able to break down cellulose find it a rich source of energy. Certain vertebrates, such as cows, can digest cellulose by means of bacteria and protists they harbor in their intestines which provide the necessary enzymes.

Chitin

The structural building material in insects, many fungi, and certain other organisms is called chitin (figure 3.10). *Chitin* is a modified form of cellulose with a nitrogen group added to the glucose units. When cross-linked by proteins, it forms a tough, resistant surface material that serves as the hard exoskeleton of arthropods such as insects and crustaceans (see chapter 43). Few organisms are able to digest chitin.

Structural carbohydrates are chains of sugars that are not easily digested. They include cellulose in plants and chitin in arthropods and fungi.

FIGURE 3.9
A journey into wood.
This jumble of cellulose fibers (a) is from a yellow pine (*Pinus ponderosa*) (20×). (b) While starch chains consist of alpha-glucose subunits, (c) cellulose chains consist of beta-glucose subunits. Cellulose fibers can be very strong and are quite resistant to metabolic breakdown, which is one reason why wood is such a good building material.

(a)

α form of glucose
(b)

β form of glucose
(c)

Starch: chain of α glucose subunits

Cellulose: chain of β glucose subunits

FIGURE 3.10
Chitin. Chitin, which might be considered to be a modified form of cellulose, is the principal structural element in the external skeletons of many invertebrates, such as this lobster.

Lipids are a loosely defined group of molecules with one main characteristic: they are insoluble in water. The most familiar lipids are fats and oils.

Fats

When organisms store the energy of glucose molecules for long periods, they usually convert most of the glucose into **fats,** molecules which contain even more C—H bonds than do carbohydrates. The ratio of H to O in carbohydrates is 2:1, but in fat molecules it is much higher. Unlike the O—H bonds of water, the C—H bonds of carbohydrates and fats are nonpolar and cannot form hydrogen bonds. Because fat molecules contain a large number of nonpolar C—H bonds, they are hydrophobically excluded by water molecules, which tend to form hydrogen bonds with other polar molecules. The result is that the fat molecules cluster together. Therefore, like starches, fats are insoluble and can be deposited at specific locations within the organism. However, while starches are insoluble because they are long polymers, fats are insoluble because they are nonpolar.

Fats are one kind of lipid. Oils such as olive oil, corn oil, and coconut oil are also lipids, as are waxes such as beeswax and earwax (see table 3.1). Fats are composite molecules, each made up of two kinds of subunits:

1. *Glycerol*, a three-carbon alcohol, with each carbon bearing a hydroxyl group. Glycerol forms the backbone of a fat molecule.
2. *Fatty acids*, long hydrocarbon chains ending in a carboxyl (—COOH) group. Three fatty acids are attached to the glycerol backbone in a fat molecule.

The structure of an individual fat molecule, such as the one diagrammed in figure 3.11, consists simply of a glycerol molecule with a fatty acid joined to each of its three carbon atoms:

$$
\begin{array}{c}
H \\
| \\
H-C-\text{Fatty acid} \\
| \\
H-C-\text{Fatty acid} \\
| \\
H-C-\text{Fatty acid} \\
| \\
H
\end{array}
$$

Because it contains three fatty acids, a fat molecule is called a triglyceride, or, more properly, a **triacylglycerol.** The three fatty acids of a triacylglycerol need not be identical, and often they differ markedly from one another.

A diet rich in fats is one of several factors that are thought to contribute to heart disease, particularly to atherosclerosis, a condition in which deposits of fatty tissue called plaque adhere to the lining of blood vessels, blocking the flow of blood. Fragments of plaque, breaking off from a deposit, are a major cause of strokes.

Individual fat molecules are triacylglycerols, with three long fatty acid chains linked to a glycerol molecule. Because the C—H bonds in fats are very nonpolar, fats are not water-soluble and thus are classified as lipids.

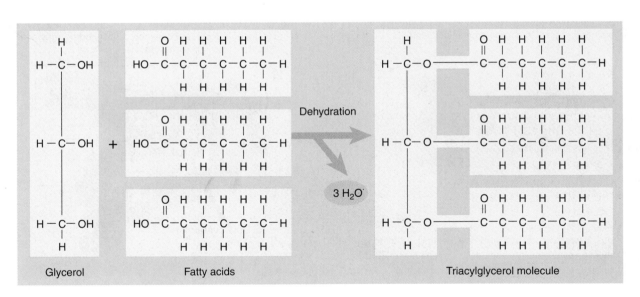

FIGURE 3.11
Triacylglycerols are made from simple pieces. Triacylglycerols are composite molecules, made up of three fatty acid molecules coupled by dehydration synthesis to a single glycerol backbone.

Fatty Acids

The hydrocarbon chains of fatty acids vary in length; the most common are even-numbered chains of 14 to 20 carbons. If all of the internal carbon atoms in the fatty acid chains are bonded to at least two carbon atoms, the fatty acid is said to be **saturated,** because it contains the maximum possible number of hydrogen atoms (figure 3.12). If a fatty acid has double bonds between one or more pairs of successive carbon atoms, the fatty acid is said to be **unsaturated,** because the double bonds replace some of the hydrogen atoms, causing the fatty acids to contain fewer than the maximum number of hydrogen atoms. If a given fatty acid has more than one double bond, it is said to be **polyunsaturated.** Fats made from polyunsaturated fatty acids have low melting points because their fatty acid chains bend at the double bonds, preventing the fat molecules from aligning closely with one another. Consequently, a polyunsaturated fat such as corn oil is usually liquid at room temperature and is called an oil. In contrast, most saturated fats such as those in butter are solid at room temperature.

Fats as Food

Most fats contain over 40 carbon atoms. The ratio of energy-storing C—H bonds to carbon atoms in fats is more than twice that of carbohydrates, making fats much more efficient molecules for storing chemical energy. On the average, fats yield about 9 kilocalories (kcal) of chemical energy per gram, as compared with somewhat less than 4 kcal per gram for carbohydrates. As you might expect, more highly saturated fats are richer in energy than less saturated fats, as they have more C—H bonds. Consequently, saturated animal fats usually contain more calories than unsaturated vegetable fats.

All fats produced by animals are saturated (except some fish oils), while most plant fats are unsaturated. The exceptions are the tropical oils (palm oil and coconut oil), which are saturated despite their fluidity at room temperature. Because of the risks of a diet rich in saturated fats, many manufacturers are now producing foods without using tropical oils or animal fats. It is possible to convert an oil into a solid fat by adding hydrogen. Peanut butter sold in stores is usually artificially hydrogenated to make the peanut fats solidify, preventing them from separating out as oils while the jar sits on the store shelf. However, artificially hydrogenating unsaturated fats seems to eliminate the health advantage they have over saturated fats, as it makes both equally rich in C—H bonds. Therefore, it now appears that margarine made from hydrogenated corn oil is no better for your health than butter.

When an organism consumes excess carbohydrate, it is converted into starch, glycogen, or fats and reserved for future use. The reason that many humans gain weight as they grow older is that the amount of energy they need decreases with age, while their intake of food does not. Thus, an increasing proportion of the carbohydrate they ingest is available to be converted into fat.

Fats are efficient energy-storage molecules because of their high concentration of C—H bonds.

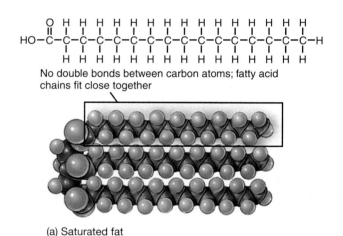

No double bonds between carbon atoms; fatty acid chains fit close together

(a) Saturated fat

Double bonds present between carbon atoms; fatty acid chains do not fit close together

(b) Unsaturated fat

FIGURE 3.12
Saturated and unsaturated fats. (a) Palmitic acid, with no double bonds and, thus, a maximum number of hydrogen atoms bonded to the carbon chain, is a saturated fatty acid. Many animal triacylglycerols (fats) are saturated. Because their fatty acid chains can fit closely together, these triacylglycerols form immobile arrays called hard fat. (b) Linoleic acid, with three double bonds and, thus, fewer than the maximum number of hydrogen atoms bonded to the carbon chain, is an unsaturated fatty acid. Plant fats are typically unsaturated. The many kinks the double bonds introduce into the fatty acid chains prevent the triacylglycerols from closely aligning and produce oils, which are liquid at room temperature.

FIGURE 3.13
Not all lipids are triacylglycerols. These structures represent the four other major classes of biologically important lipids: (a) phospholipids, (b) steroids, (c) terpenes, and (d) prostaglandins.

(a) Phospholipid (phosphatidyl choline)

(b) Steroid (cholesterol)

(c) Terpene (citronellol)

(d) Prostaglandin (PGE)

Other Kinds of Lipids

Organisms contain many other kinds of lipids besides fats (figure 3.13). For example, the membranes of cells are composed of modified fats called *phospholipids*. Phospholipids have two fatty acid chains rather than three; one of the fatty acid chains of fat is here replaced by a phosphate group. The phospholipid molecule can be thought of as having a polar "head" at one end (the phosphate group) and two long, nonpolar "tails" at the other. In water, the nonpolar tails of nearby phospholipids aggregate away from the water, forming two layers of tails pointed toward each other—a lipid bilayer (figure 3.14). Lipid bilayers are the basic framework of biological membranes, discussed in detail in chapter 6. Membranes often also contain *steroids*, a type of lipid composed of four carbon rings; most animal cell membranes contain the steroid cholesterol. Other steroids, such as testosterone and estrogen, function in multicellular organisms as hormones.

Terpenes constitute a third category of lipids. These long-chain lipids are components of many biologically important pigments, such as the photosynthetic pigment chlorophyll in plants and the visual pigment retinal in vertebrate eyes. Rubber is also a terpene.

Prostaglandins are a group of about 20 lipids that are modified fatty acids, with two nonpolar "tails" attached to a five-carbon ring. Prostaglandins occur in many vertebrate tissues, where they appear to act as local chemical messengers.

Cells contain many kinds of molecules that are not soluble in water.

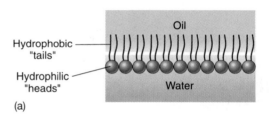

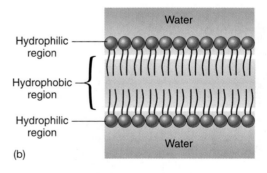

FIGURE 3.14
Phospholipids. (a) At an oil-water interface, phospholipid molecules will orient so that their polar (hydrophilic) heads are in the polar medium, water, and their nonpolar (hydrophobic) tails are in the nonpolar medium, oil. (b) When surrounded by water, phospholipid molecules arrange themselves into two layers with their heads extending outward and their tails inward.

The Many Functions of Proteins

Proteins are the third major group of macromolecules that make up the bodies of organisms (see table 3.1). Proteins perform many functions (table 3.2 and figure 3.15).

1. **Enzyme catalysis.** We have already encountered one class of proteins, **enzymes,** which are biological catalysts that facilitate specific chemical reactions. Because of this property, the appearance of enzymes was one of the most important events in the evolution of life. Enzymes are globular proteins, with a three-dimensional shape that fits snugly around the chemicals they work on, facilitating chemical reactions by stressing particular chemical bonds.

2. **Defense.** Other globular proteins use their shapes to "recognize" foreign microbes and cancer cells. These cell surface receptors form the core of the body's hormone and immune systems.

3. **Transport.** A variety of globular proteins transport specific small molecules and ions. The transport protein hemoglobin, for example, transports oxygen in the blood, and myoglobin, a similar protein, transports oxygen in muscle. Iron is transported in blood by the protein transferrin.

Table 3.2 The Many Functions of Proteins			
Function	**Class of Protein**	**Examples**	**Use**
Metabolism (Catalysis)	Enzymes	Hydrolytic enzymes	Cleave polysaccharides
		Proteases	Break down proteins
		Polymerases	Produce nucleic acids
		Kinases	Phosphorylate sugars and proteins
Defense	Immunoglobulins	Antibodies	Mark foreign proteins for elimination
	Toxins	Snake venom	Block nerve function
Cell recognition	Cell surface antigens	MHC proteins	"Self" recognition
Transport throughout body	Globins	Hemoglobin	Carries O_2 and CO_2 in blood
		Myoglobin	Carries O_2 and CO_2 in muscle
		Cytochromes	Electron transport
Membrane transport	Transporters	Sodium-potassium pump	Excitable membranes
		Proton pump	Chemiosmosis
		Anion channels	Transport Cl^- ions
Structure/Support	Fibers	Collagen	Cartilage
		Keratin	Hair, nails
		Fibrin	Blood clot
Motion	Muscle	Actin	Contraction of muscle fibers
		Myosin	Contraction of muscle fibers
Osmotic regulation	Albumin	Serum albumin	Maintains osmotic concentration of blood
Regulation of gene action	Repressors	*lac* repressor	Regulates transcription
Regulation of body functions	Hormones	Insulin	Controls blood glucose levels
		Vasopressin	Increases water retention by kidneys
		Oxytocin	Regulates uterine contractions and milk production
Storage	Ion binding	Ferritin	Stores iron, especially in spleen
		Casein	Stores ions in milk
		Calmodulin	Binds calcium ions

FIGURE 3.15
Some of the more common structural proteins. (a) Collagen: strings of a tennis racket from gut tissue; (b) fibrin: scanning electron micrograph of a blood clot (3000×); (c) keratin: a peacock feather; (d) silk: a spider's web; (e) keratin: human hair.

4. **Support.** Fibrous proteins play structural roles; these structural proteins include keratin in hair, fibrin in blood clots, and collagen, which forms the matrix of skin, ligaments, tendons, and bones and is the most abundant protein in a vertebrate body.

5. **Motion.** Muscles contract through the sliding motion of two kinds of protein filament: actin and myosin. **Contractile proteins** also play key roles in the cell's cytoskeleton and in moving materials within cells.

6. **Regulation.** Small proteins called hormones serve as **intercellular messengers** in animals. Proteins also play many regulatory roles within the cell, turning on and shutting off genes during development, for example. In addition, proteins also receive information, acting as cell surface receptors.

Proteins carry out a diverse array of functions, including catalysis, defense, transport of substances, motion, and regulation of cell and body functions.

Amino Acids Are the Building Blocks of Proteins

Although proteins are complex and versatile molecules, they are all polymers of only 20 amino acids. Many scientists believe amino acids were among the first molecules formed in the early earth. It seems highly likely that the oceans that existed early in the history of the earth contained a wide variety of amino acids.

Amino Acid Structure

An **amino acid** is a molecule containing an amino group (—NH₂), a carboxyl group (—COOH), and a hydrogen atom, all bonded to a central carbon atom:

$$H_2N - \underset{\underset{\textstyle H}{|}}{\overset{\overset{\textstyle R}{|}}{C}} - COOH$$

Each amino acid has unique chemical properties determined by the nature of the side group (indicated by R) covalently bonded to the central carbon atom. For example, when the side group is —CH₂OH, the amino acid (serine) is polar, but when the side group is —CH₃, the amino acid (alanine) is nonpolar. The 20 common amino acids are grouped into five chemical classes, based on their side groups:

1. Nonpolar amino acids, such as leucine, often have R groups that contain —CH₂ or —CH₃.
2. Polar uncharged amino acids, such as threonine, have R groups that contain oxygen (or only —H).
3. Ionizable amino acids, such as glutamic acid, have R groups that contain acids or bases.
4. Aromatic amino acids, such as phenylalanine, have R groups that contain an organic (carbon) ring with alternating single and double bonds.
5. Special-function amino acids have unique individual properties; methionine often is the first amino acid in a chain of amino acids, proline causes kinks in chains, and cysteine links chains together.

Each amino acid affects the shape of a protein differently depending on the chemical nature of its side group. Portions of a protein chain with numerous nonpolar amino acids, for example, tend to fold into the interior of the protein by hydrophobic exclusion.

In addition to its R group, each amino acid, when ionized, has a positive amino (NH₃⁺) group at one end and a negative carboxyl (COO⁻) group at the other end. The amino and carboxyl groups on a pair of amino acids can undergo a dehydration reaction, losing a molecule of water and forming a covalent bond. A covalent bond that links two amino acids is called a **peptide bond** (figure 3.16). The two amino acids linked by such a bond are not free to rotate around the

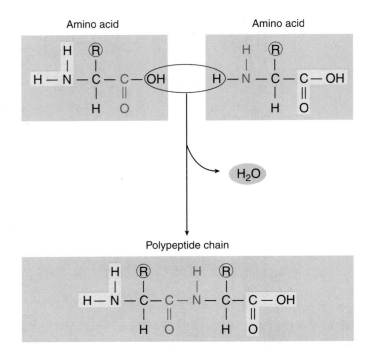

FIGURE 3.16
The peptide bond. A peptide bond forms when the —NH₂ end of one amino acid joins to the —COOH end of another. Because of the partial double-bond nature of peptide bonds, the resulting peptide chain cannot rotate freely around these bonds.

N—C linkage because the peptide bond has a partial double-bond character. The stiffness of the peptide bond is one factor that makes it possible for chains of amino acids to form coils and other regular shapes.

Proteins Are Chains of Amino Acids

A **protein** is composed of one or more long chains, or **polypeptides,** composed of amino acids linked by peptide bonds. It was not until the pioneering work of Frederick Sanger in the early 1950s that it became clear that each kind of protein had a specific amino acid sequence. Sanger succeeded in determining the amino acid sequence of insulin and in so doing demonstrated clearly that this protein had a defined sequence, the same for all insulin molecules in the solution. Although many different amino acids occur in nature, only 20 commonly occur in proteins. Figure 3.17 illustrates these 20 "common" amino acids and their side groups.

A protein is a polymer containing up to 20 different kinds of amino acids. The amino acids fall into five chemical classes, each with different properties. These properties determine the nature of the resulting protein.

FIGURE 3.17

The 20 common amino acids. Each amino acid has the same chemical backbone, but differs in the side, or R, group it possesses. Six of the amino acids are nonpolar because they have —CH₂ or —CH₃ in their R groups. Two of the six are bulkier because they contain ring structures, which classifies them also as aromatic. Another six are polar because they have oxygen or just hydrogen in their R groups; these amino acids, which are uncharged, differ from one another in how polar they are. Five other amino acids are polar and, because they have a terminal acid or base in their R group, are capable of ionizing to a charged form. The remaining three have special chemical properties that allow them to help form links between protein chains or kinks in proteins.

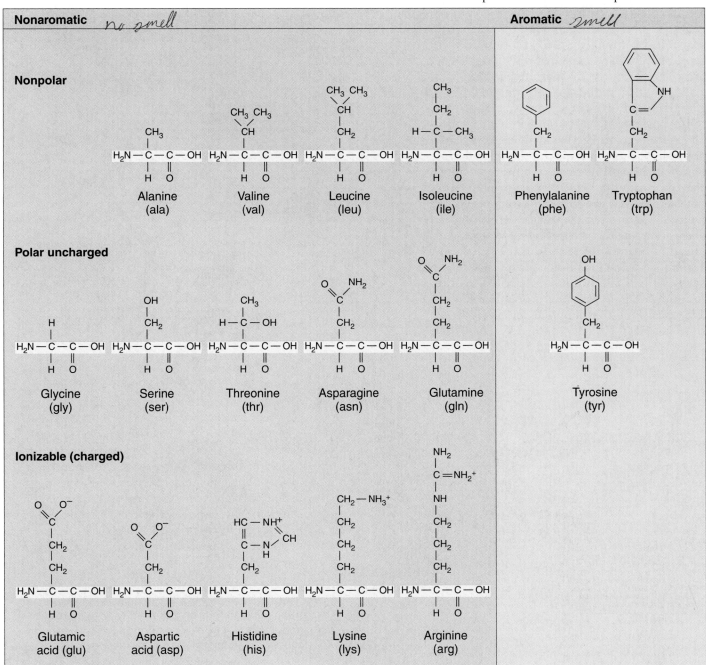

Nonaromatic *no smell* Aromatic *smell*

Nonpolar: Alanine (ala), Valine (val), Leucine (leu), Isoleucine (ile), Phenylalanine (phe), Tryptophan (trp)

Polar uncharged: Glycine (gly), Serine (ser), Threonine (thr), Asparagine (asn), Glutamine (gln), Tyrosine (tyr)

Ionizable (charged): Glutamic acid (glu), Aspartic acid (asp), Histidine (his), Lysine (lys), Arginine (arg)

Special structural property: Proline (pro), Methionine (met), Cysteine (cys)

The Shape of Globular Proteins

The shape of a protein is very important because it determines the protein's function. If we picture a polypeptide as a long strand similar to a reed, a protein might be the basket woven from it.

Overview of Protein Structure

Many proteins consist of long amino acid chains folded into complex shapes. What do we know about the shape of these so-called globular proteins? One way to study the shape of something as small as a protein is to look at it with very short wavelength energy—with X rays. X-ray diffraction is a painstaking procedure that allows the investigator to build up a three-dimensional picture of the position of each atom. The first globular protein to be analyzed in this way was myoglobin, soon followed by the related protein hemoglobin. As more and more proteins were added to the list, a general principle became evident: in every protein studied, essentially all the internal amino acids are nonpolar ones, amino acids such as leucine, valine, and phenylalanine. Water's tendency to hydrophobically exclude nonpolar molecules literally shoves the nonpolar portions of the amino acid chain into the protein's interior. This positions the nonpolar amino acids in close contact with one another, leaving little empty space inside. Polar and charged amino acids are restricted to the surface of the protein except for the few that play key functional roles.

Levels of Protein Structure

The structure of globular proteins is traditionally discussed in terms of four levels of structure, as *primary*, *secondary*, *tertiary*, and *quaternary* (figure 3.18). Because of progress in our knowledge of protein structure, two additional levels of structure are increasingly distinguished by molecular biologists: *motifs* and *domains*. Because these latter two elements play important roles in coming chapters, we introduce them here.

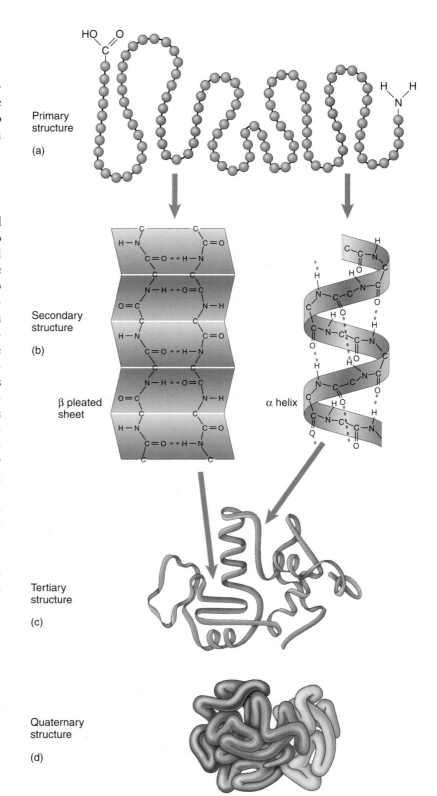

FIGURE 3.18
Levels of protein structure. The amino acid sequence of a protein, called its (a) primary structure, encourages hydrogen bonds to form between nearby amino acids. This produces coils (b) called alpha helices and fold-backs called beta-pleated sheets; these coils and fold-backs constitute the protein's secondary structure. The protein folds up on itself further to assume a three-dimensional (c) tertiary structure. Many proteins aggregate with other polypeptide chains in clusters; this clustering is called the (d) quaternary structure of the protein.

Primary Structure. The specific amino acid sequence of a protein is its primary structure. This sequence is determined by the nucleotide sequence of the gene that encodes the protein. Because the R groups that distinguish the various amino acids play no role in the peptide backbone of proteins, a protein can consist of any sequence of amino acids. Thus a protein containing 100 amino acids could form any of 20^{100} different amino acid sequences (that's the same as 10^{130}, or 1 followed by 130 zeros—more than the number of atoms known in the universe). This is an important property of proteins, since it permits such great diversity.

Secondary Structure. The amino acid side groups are not the only portions of proteins that form hydrogen bonds. The —COOH and —NH$_2$ groups of the main chain also form quite good hydrogen bonds—so good that their interactions with water might be expected to offset the tendency of nonpolar sidegroups to be forced into the protein interior. Inspection of the protein structures determined by X-ray diffraction reveals why they don't—the polar groups of the main chain form hydrogen bonds with each other! Two patterns of H bonding occur. In one, hydrogen bonds form along a single chain, linking one amino acid to another farther down the chain. This tends to pull the chain into a coil called an alpha (α) helix. In the other pattern, hydrogen bonds occur across two chains, linking the amino acids in one chain to those in the other. Often, many parallel chains are linked, forming a pleated, sheet-like structure called a β-pleated sheet. The folding of the amino acid chain by hydrogen bonding into these characteristic coils and pleats is called a protein's secondary structure.

Motifs. The elements of secondary structure can combine in proteins in characteristic ways called motifs, or sometimes "supersecondary structure." One very common motif is the $\beta \alpha \beta$ motif, which creates a fold or crease; the so-called "Russmann fold" at the core of nucleotide binding sites in a wide variety of proteins is an $\alpha \beta \alpha \beta \alpha \beta$ motif. A second motif that occurs in many proteins is the β barrel, a β sheet folded around to form a tube. A third type of motif, the α turn α motif, is important because many proteins use it to bind the DNA double helix.

Tertiary Structure. The final folded shape of a protein, which positions the various motifs and folds nonpolar side groups into the interior, is called a protein's tertiary structure. A protein is driven into its tertiary structure by hydrophobic interactions with water. The final folding of a protein is determined by its primary structure—by the chemical nature of its side groups. Many proteins can be fully unfolded ("denatured") and will spontaneously refold back into their characteristic shape.

The stability of a globular protein, once it has folded into its 3-D shape, is strongly influenced by how well its interior fits together. When two nonpolar chains in the interior are in very close proximity, they experience a form of molecular attraction called van der Waal's forces. Individually quite weak, these forces can add up to a strong attraction when many of them come into play, like the combined strength of hundreds of hooks and loops on a strip of Velcro. They are effective forces only over short distances, however; there are no "holes" or cavities in the interior of proteins. That is why there are so many different nonpolar amino acids (alanine, valine, leucine, isoleucine). Each has a different sized R group, allowing very precise fitting of nonpolar chains within the protein interior. Now you can understand why a mutation that converts one nonpolar amino acid within the protein interior (alanine) into another (leucine) very often disrupts the protein's stability; leucine is a lot bigger than alanine and disrupts the precise way the chains fit together within the protein interior. A change in even a single amino acid can have profound effects on protein shape.

Domains. Many proteins in your body are encoded within your genes in functional sections called exons (exons will be discussed in detail in chapter 15). Each exon-encoded section of a protein, typically 100 to 200 amino acids long, folds into a structurally independent globular unit called a **domain.** A single polypeptide chain connects the domains of a protein, like a rope tied into several adjacent knots. Often the domains of a protein have quite separate functions—one domain of an enzyme might bind a cofactor, for example, and another the enzyme's substrate.

Quaternary Structure. When two or more polypeptide chains associate to form a functional protein, the individual chains are referred to as subunits of the protein. The subunits need not be the same. Hemoglobin, for example, is a protein composed of two α-chain subunits and two β-chain subunits. A protein's subunit arrangement is called its quaternary structure. In proteins composed of subunits, the interfaces where the subunits contact one another are often nonpolar, and play a key role in transmitting information between the subunits about individual subunit activities.

A change in the identity of one of these amino acids can have profound effects. Sickle cell hemoglobin is a mutation that alters the identity of a single amino acid at the corner of the β subunit from polar glutamate to nonpolar valine. Putting a nonpolar amino acid on the surface creates a "sticky patch" that causes one hemoglobin molecule to stick to another, forming long nonfunctional chains and leading to the cell sickling characteristic of this hereditary disorder.

Globular protein structure can be viewed at six levels:
1. the amino acid sequence, or primary structure,
2. coils and sheets, called secondary structure, 3. folds or creases, called motifs, 4. the three-dimensional shape called tertiary structure, 5. functional units called domains, 6. individual polypeptide subunits associated in a quaternary structure.

How Proteins Fold

How does a protein fold into a specific shape? Nonpolar amino acids play a key role. Until recently, investigators thought that newly made proteins fold spontaneously as hydrophobic interactions with water shove nonpolar amino acids into the protein interior. We now know this is too simple a view. Protein chains can fold in so many different ways that trial-and-error would simply take too long. In addition, as the open chain folds its way toward its final form, nonpolar "sticky" interior portions are exposed during intermediate stages. If these intermediate forms are placed in a test tube in the same protein environment that occurs in a cell, they stick to other unwanted protein partners, forming a gluey mess.

Chaperone Proteins

How do cells avoid this? A vital clue came in studies of unusual mutations in *E. coli* bacteria that prevented bacterial viruses from replicating in *E. coli* cells—it turned out the virus proteins could not fold properly! Further study revealed that normal cells contain special **chaperone proteins** that help new proteins fold correctly (figure 3.19). When the *E. coli* gene encoding the chaperone protein is disabled by mutation, the bacteria die, clogged with lumps of incorrectly folded proteins. In the absence of chaperone proteins, fully 30% of the bacteria's proteins fail to fold to the right shape.

Molecular biologists have now identified more than 17 kinds of proteins that act as **molecular chaperones.** Many are heat shock proteins, produced in greatly elevated amounts if a cell is exposed to elevated temperature; high temperatures cause proteins to unfold, and heat shock chaperones help the cell's proteins refold.

There is considerable controversy about how chaperones work. It was first thought that they provided a protected environment within which folding could take place unhindered by other proteins, but it now seems more likely that chaperones rescue proteins that are caught in a wrongly folded state, giving them another chance to fold correctly. When investigators "fed" chaperones to a deliberately misfolded protein called malate dehydrogenase, the protein was rescued, refolding to its active shape.

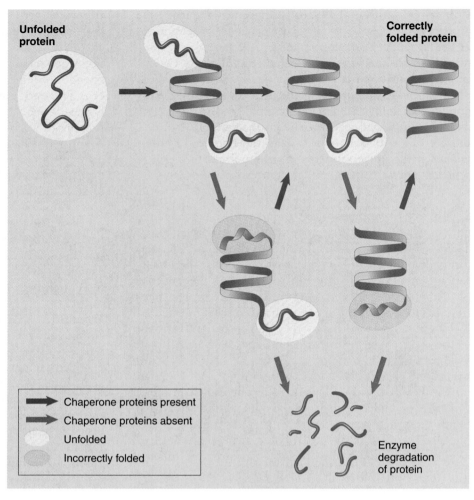

FIGURE 3.19
A current model of protein folding. A newly synthesized protein rapidly folds into characteristic motifs composed of α helices and β sheets, but these elements of structure are only roughly aligned in an open conformation. Subsequent folding occurs more slowly, by trial and error. This process is aided by chaperone proteins, which appear to recognize improperly folded proteins and unfold them, giving them another chance to fold properly. Eventually, if proper folding is not achieved, the misfolded protein is degraded by proteolytic enzymes.

Protein Folding and Disease

There are tantalizing suggestions that chaperone protein deficiencies may play a role in certain diseases by failing to facilitate the intricate folding of key proteins. Cystic fibrosis is a hereditary disorder in which a mutation disables a protein that plays a vital part in moving ions across cell membranes. In at least some cases, the vital membrane protein appears to have the correct amino acid sequence, but fails to fold to its final form. It has also been speculated that chaperone deficiency may be a cause of the protein clumping in brain cells that produces the amyloid plaques characteristic of Alzheimer's disease. Many researchers are testing this idea.

Proteins called chaperones aid newly produced proteins to fold properly.

How Proteins Unfold

If a protein's environment is altered, the protein may change its shape or even unfold. This process is called **denaturation**. Proteins can be denatured when the pH, temperature, or ionic concentration of the surrounding solution is changed. When proteins are denatured, they are usually rendered biologically inactive. This is particularly significant in the case of enzymes. Since practically every chemical reaction in a living organism is catalyzed by a specific enzyme, it is vital that a cell's enzymes remain functional. That is the rationale behind traditional methods of salt-curing and pickling: prior to the ready availability of refrigerators and freezers, the only practical way to keep microorganisms from growing in food was to keep the food in a solution containing high salt or vinegar concentrations, which denatured the enzymes of microorganisms and kept them from growing on the food.

Most enzymes function within a very narrow range of physical parameters. Blood-borne enzymes that course through a human body at a pH of about 7.4 would rapidly become denatured in the highly acidic environment of the stomach. On the other hand, the protein-degrading enzymes that function at a pH of 2 or less in the stomach would be denatured in the basic pH of the blood. Similarly, organisms that live near oceanic hydrothermal vents have enzymes that work well at the temperature of this extreme environment (over 100° C). They cannot survive in cooler waters, because their enzymes would denature at lower temperatures. Any given organism usually has a "tolerance range" of pH, temperature, and salt concentration. Within that range, its enzymes maintain the proper shape to carry out their biological functions.

When a protein's normal environment is reestablished after denaturation, a small protein may spontaneously refold into its natural shape, driven by the interactions between its nonpolar amino acids and water (figure 3.20). Larger proteins can rarely refold spontaneously because of the complex nature of their final shape. It's important to distinguish denaturation from **dissociation**. The four subunits of hemoglobin (figure 3.21) may dissociate into four individual molecules (two α-globin and two β-globin) without denaturation of the folded globin proteins, and will readily reassume their four-subunit quaternary structure.

Every globular protein has a narrow range of conditions in which it folds properly; outside that range, proteins tend to unfold.

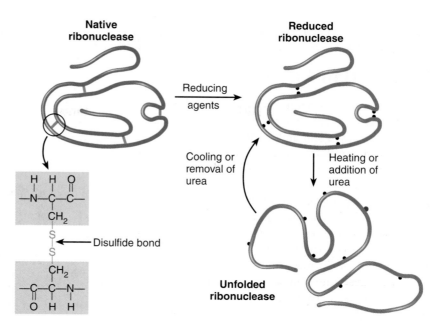

FIGURE 3.20
Primary structure determines tertiary structure. When the protein ribonuclease is treated with reducing agents to break the covalent disulfide bonds that cross-link its chains, and then placed in urea or heated, the protein denatures (unfolds) and loses its enzymatic activity. Upon cooling or removal of urea, it refolds and regains its enzymatic activity. This demonstrates that no information but the amino acid sequence of the protein is required for proper folding: the primary structure of the protein determines its tertiary structure.

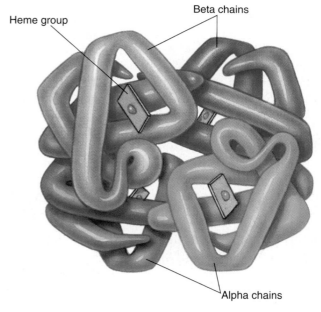

FIGURE 3.21
The four subunits of hemoglobin. The hemoglobin molecule is made of four globin protein subunits, informally referred to as polypeptide chains. The two lower α chains, identical α-globin proteins, are shaded pink; the two upper β chains, identical β-globin proteins, are shaded blue.

Information Molecules

Nucleic acids are the fourth major kind of macromolecule within cells (see table 3.1). Nucleic acids are the information storage devices of cells, just as disks or tapes store the information that computers use, blueprints store the information that builders use, and road maps store the information that tourists use. There are two varieties of nucleic acids: *deoxyribonucleic acid* (**DNA**; see figure 3.3) and *ribonucleic acid* (**RNA**). The way in which DNA encodes the information used to assemble proteins is similar to the way in which the letters on a page encode information (see chapter 14). Unique among macromolecules, nucleic acids are able to serve as templates to produce precise copies of themselves, so that the information that specifies what an organism is can be copied and passed down to its descendants. For this reason, DNA is often referred to as the hereditary material. Cells use the alternative form of nucleic acid, RNA, to read the cell's DNA-encoded information and direct the synthesis of proteins. RNA is similar to DNA in structure and is made as a transcript copy of portions of the DNA. This transcript passes out into the rest of the cell, where it serves as a blueprint specifying a protein's amino acid sequence. This process will be described in detail in chapter 15.

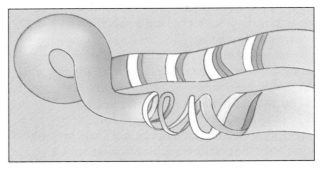

FIGURE 3.22
The first photograph of a DNA molecule. This micrograph, with sketch below, shows the two strands magnified a million times! The molecule is so slender that it would take 50,000 of them to equal the diameter of a human hair.

"Seeing" DNA

DNA molecules cannot be seen with an optical microscope, which is incapable of resolving anything smaller than 1000 atoms across. An electron microscope can image structures as small as a few dozen atoms across, but still cannot resolve the individual atoms of a DNA strand. This limitation was finally overcome in the last decade with the introduction of the scanning-tunneling microscope (figure 3.22).

How do these microscopes work? Imagine you are in a dark room with a chair. To determine the shape of the chair, you could shine a flashlight on it, so that the light bounces off the chair and forms an image on your eye. That's what optical and electron microscopes do; in the latter, the "flashlight" emits a beam of electrons instead of light. You could, however, also reach out and feel the chair's surface with your hand. In effect, you would be putting a probe (your hand) near the chair and measuring how far away the surface is. In a scanning-tunneling microscope, computers advance a probe over the surface of a molecule in steps smaller than the diameter of an atom.

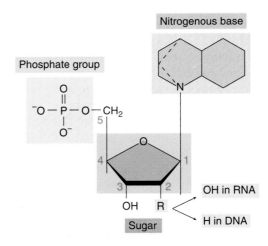

FIGURE 3.23
Structure of a nucleotide. The nucleotide subunits of DNA and RNA are made up of three elements: a five-carbon sugar, an organic nitrogenous base, and a phosphate group.

DNA encodes information, which is copied off as RNA molecules that direct the assembly of specific sequences of amino acids in proteins.

The Structure of Nucleic Acids

Nucleic acids are long polymers of repeating subunits called **nucleotides.** Each nucleotide consists of three components: a five-carbon sugar (ribose in RNA and deoxyribose in DNA); a phosphate (—PO$_4$) group; and an organic nitrogen-containing base (figure 3.23). When a nucleic acid polymer forms, the phosphate group of one nucleotide binds to the hydroxyl group of another, releasing water and forming a phosphodiester bond. A **nucleic acid,** then, is simply a chain of five-carbon sugars linked together by phosphodiester bonds with an organic base protruding from each sugar (figure 3.24).

Two types of organic bases occur in nucleotides. The first type, *purines*, are large, double-ring molecules found in both DNA and RNA; they are adenine (A) and guanine (G). The second type, *pyrimidines*, are smaller, single-ring molecules; they include cytosine (C, in both DNA and RNA), thymine (T, in DNA only), and uracil (U, in RNA only).

In addition to serving as subunits of DNA and RNA, nucleotide bases play other critical roles in the life of a cell. For example, adenine is a key component of the molecule *adenosine triphosphate* (ATP; figure 3.25), the energy currency of the cell. It also occurs in the molecules *nicotinamide adenine dinucleotide* (NAD$^+$) and *flavin adenine dinucleotide* (FAD$^+$), which carry electrons whose energy is used to make ATP.

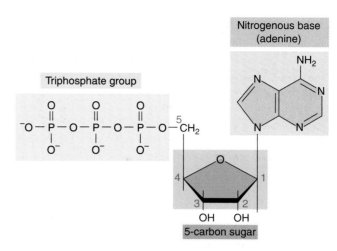

FIGURE 3.25
ATP. Adenosine triphosphate (ATP) contains adenine, a five-carbon sugar, and three phosphate groups. This molecule serves to transfer energy rather than store genetic information.

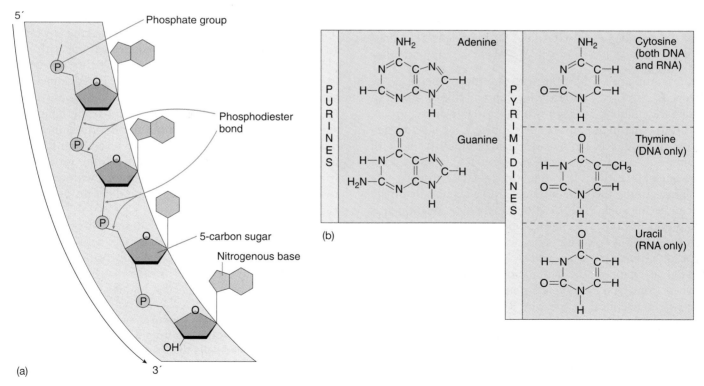

FIGURE 3.24
The structure of a nucleic acid and the organic nitrogen-containing bases. (a) In a nucleic acid, nucleotides are linked to one another via phosphodiester bonds, with organic bases protruding from the chain. (b) The nucleotides contain two types of bases: purines and pyrimidines. In DNA, thymine replaces the uracil found in RNA.

DNA

Organisms encode the information specifying the amino acid sequences of their proteins as sequences of nucleotides in the DNA. This method of encoding information is very similar to that by which the sequences of letters encode information in a sentence. While a sentence written in English consists of a combination of the 26 different letters of the alphabet in a specific order, the code of a DNA molecule consists of different combinations of the four types of nucleotides in specific sequences such as CGCTTACG. The information encoded in DNA is used in the everyday metabolism of the organism and is passed on to the organism's descendants.

DNA molecules in organisms exist not as single chains folded into complex shapes, like proteins, but rather as double chains. Two DNA polymers wind around each other like the outside and inside rails of a circular staircase. Such a winding shape is called a helix, and a helix composed of two chains winding about one another, as in DNA, is called a **double helix.** Each step of DNA's helical staircase is a base-pair, consisting of a base in one chain attracted by hydrogen bonds to a base opposite it on the other chain. These hydrogen bonds hold the two chains together as a duplex (figure 3.26). The base-pairing rules are rigid: adenine can pair only with thymine (in DNA) or with uracil (in RNA), and cytosine can pair only with guanine. The bases that participate in base-pairing are said to be **complementary** to each other. Additional details of the structure of DNA and how it interacts with RNA in the production of proteins are presented in chapters 14 and 15.

RNA

RNA is similar to DNA, but with two major chemical differences. First, RNA molecules contain ribose sugars in which the number 2 carbon is bonded to a hydroxyl group. In DNA, this hydroxyl group is replaced by a hydrogen atom. Second, RNA molecules utilize uracil in place of thymine. Uracil has the same structure as thymine, except that one of its carbons lacks a methyl (—CH₃) group.

Transcribing the DNA message into a chemically different molecule such as RNA allows the cell to tell which is the original information storage molecule and which is the transcript. DNA molecules are always double-stranded (except for a few single-stranded DNA viruses that will be discussed in chapter 30), while the RNA molecules transcribed from DNA are typically single-stranded. Although there is no chemical reason why RNA cannot form double helices as DNA does, cells do not possess the enzymes necessary to assemble double strands of RNA, as they do for DNA. Using two different molecules, one single-stranded and the other double-stranded, separates the role of DNA in storing hereditary information from the role of RNA in using this information to specify protein structure.

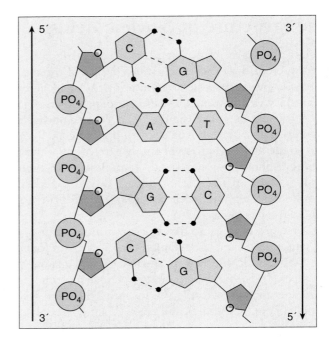

FIGURE 3.26
Hydrogen bonds hold the DNA double helix together.
Hydrogen bond formation (dashed lines) between the organic bases, called base-pairing, causes the two chains of a DNA duplex to bind to each other.

Which Came First, DNA or RNA?

The information necessary for the synthesis of proteins is stored in the cell's double-stranded DNA base sequences. The cell uses this information by first making a working RNA transcript of it, RNA nucleotides pairing with complementary DNA nucleotides. By storing the information in DNA while using a complementary RNA sequence to actually direct protein synthesis, the cell does not expose the information-encoding DNA chain to the dangers of single-strand cleavage every time the information is used. Therefore, DNA is thought to have evolved from RNA as a means of preserving the genetic information, protecting it from the ongoing wear and tear associated with cellular activity. This genetic system has come down to us from the very beginnings of life.

The cell uses the single-stranded, short-lived RNA transcript to direct the synthesis of a protein with a specific sequence of amino acids. Thus, the information flows from DNA to RNA to protein, a process that has been termed the "central dogma" of molecular biology.

A nucleic acid is a long chain of five-carbon sugars with an organic base protruding from each sugar. DNA is a double-stranded helix that stores hereditary information as a specific sequence of nucleotide bases. RNA is a single-stranded molecule that transcribes this information to direct protein synthesis.

3.1 Molecules are the building blocks of life.

- The chemistry of living systems is the chemistry of carbon-containing compounds.
- Carbon's unique chemical properties allow it to polymerize into chains by dehydration synthesis, forming the four key biological macromolecules: carbohydrates, lipids, proteins, and nucleic acids.

3.2 Carbohydrates contain many CH bonds.

- Carbohydrates consist of monosaccharides or polymers of monosaccharides, and are used principally for energy, which is stored in their carbon-hydrogen (C—H) bonds.
- The most metabolically important carbohydrate is glucose, a six-carbon sugar. Organisms often transport sugars as disaccharides, two monosaccharides that are linked together and, therefore, cannot be utilized while they are being transported.
- Excess energy resources may be stored in complex sugar polymers called starches (in plants) and glycogen (in animals and fungi).

3.3 Lipids are not soluble in water.

- Fats are one type of water-insoluble molecules called lipids.
- Fats are molecules that contain many more C—H bonds than carbohydrates and, thus, provide a more efficient form of long-term energy storage.
- Other lipids include phospholipids, steroids, terpenes, and prostaglandins.

3.4 Proteins perform the chemistry of the cell.

- Proteins are linear polymers of amino acids.
- Because the 20 amino acids that occur in proteins have side groups with different chemical properties, the function and shape of a protein are critically affected by its particular sequence of amino acids.
- Proteins are typically characterized as fibrous or globular.

3.5 Nucleic acids store the genetic information.

- Hereditary information is stored as a sequence of nucleotides in a linear polymer called DNA, which exists in cells as a double helix.
- Cells use the information in DNA by producing a complementary single strand of RNA from the DNA template. Elsewhere in the cell, this RNA directs the synthesis of a protein whose amino acid sequence corresponds to the nucleotide sequence of the DNA from which the RNA was transcribed.

Discussing Key Terms

1. **Macromolecule**   A very large molecule assembled by linking subunits together into a chain. Chains of subunits form the backbone of the four major groups of organic macromolecules: carbohydrates, lipids, proteins, and nucleic acids.

2. **Dehydration synthesis**   The formation of a chemical bond through the elimination of water. In this process, a covalent bond forms between two subunits when one subunit loses —OH and the other loses —H to form H_2O.

3. **Isomer**   One of a group of molecules identical to one another in atomic composition but differing in arrangement.

4. **Polymer**   A macromolecule composed of a long chain of similar or identical subunits; starch is a polymer of glucose.

5. **Hydrophobic folding**   The tendency of protein chains to fold into three-dimensional shapes as their nonpolar segments aggregate to the inside of the molecule, away from water.

6. **Complementarity**   In a DNA double helix, each nucleotide has a specific partner, A with T and G with C; thus, knowledge of the sequence of one strand of the helix fully determines the sequence of the other.

Review Questions

1. What types of molecules are formed by dehydration reactions? What types of molecules are formed by hydrolysis? Describe the basic features of each type of reaction.

2. What three elements are present in a carbohydrate, and what is their molar ratio? What chemical feature of carbohydrates makes them well suited for energy storage?

3. What does it mean to say that glucose, fructose, and galactose are isomers? Which two are structural isomers, and how do they differ from each other? Which two are stereoisomers, and how do they differ from each other?

4. What is the chemical difference between starch and cellulose? Which of these two types of molecules cannot be digested by most organisms, and why can't it be digested?

5. What are the two kinds of subunits that make up a fat molecule, and how are they arranged in the molecule?

6. Describe the difference between a saturated and an unsaturated fat. Which type of fat is usually a liquid at room temperature, and why?

7. What is the basic structure of an amino acid? What part of an amino acid determines its particular chemical properties? How are amino acids linked to form proteins?

8. Explain what is meant by the primary, secondary, tertiary, and quaternary structure of a protein.

9. What are the three components of a nucleotide? How are nucleotides linked to form nucleic acids?

10. Which of the purines and pyrimidines are capable of forming base-pairs with each other? How are these base-pairs involved in the structure of DNA?

11. How do DNA and RNA differ from each other? What are the primary functions of these molecules inside cells?

Thought Questions

1. Why do you suppose humans circulate the monosaccharide, glucose, in their blood, rather than employing a disaccharide like sucrose as a transport sugar, as plants do?

2. A bird such as a gull may walk on hot sand during the summer and ice during the winter. Likewise, the needles of a pine tree may be exposed to searing heat in the summer and snow in the winter. How would you expect this variation in temperature to affect the fats and proteins in the bird's feet and the pine tree's needles? What mechanisms might enable organisms like these to ensure that their fats and proteins continue to function normally over such wide temperature ranges?

3. Of all possible DNA nucleotide sequences, what sequence of base pairs would dissociate most easily into single strands if the DNA duplex were heated gently? Explain your reasoning.

4. Imagine you had made an RNA copy of each strand of a DNA double helix. If you were now to mix these two RNA single-strand copies together, do you think they would form a double-stranded RNA molecule? Why do you suppose DNA is double-stranded, while RNA is single-stranded?

Internet Links

Molecules on Display
http://www.csc.fi/lul/chem/graphics.html
The CHEMIST'S ART GALLERY offers both still and dynamic images of proteins, and animated chemical reactions.

Library of Molecules in 3D
http://cwis.nyu.edu/pages/mathmol/library/index.html
NYU maintains this library of 3D images of molecules, including many discussed in this chapter.

Image Library of Biological Macromolecules
http://www.imb-jena.de/IMAGE.html
A very fine library of images of macromolecules from Jena, Germany.

Chance and Scientific Discovery
http://www.gene.com/ae/AE/AEC/CC/chance.html
A concise essay in the roles that chance and luck play in scientific discovery. A powerful argument that chance favors the prepared mind.

For Further Reading

Cohen, J. S., and M. E. Hogan: "The New Genetic Medicines," *Scientific American*, December 1994, pages 76–82. Using synthetic DNA molecules as "magic bullets" to target specific cancers and viruses without harming other tissues!

Doolittle, R.: "Proteins," *Scientific American*, October 1985, pages 88–99. Proteins, by virtue of their flexible structure, are able to bind to many other molecules, making the machinery of life possible.

Hartl, F. U.: "Molecular chaperones in cellular protein folding," *Nature*, June 1996, v. 381, pages 571–80. A current account of the rapidly changing ideas in this "hot" research area.

Holzman, D.: "Protein Folding," *American Scientist*, May–June 1994, pages 267–74. An insightful discussion of protein folding mechanisms and the impact they may have on biology in general.

Maddox, J.: "Does Folding Determine Protein Configuration?" *Nature*, July 7, 1994, page 13. More discussion of the eternal question, "Why do proteins fold?"

Various Authors: "The Molecules of Life," *Scientific American*, October 1985. An issue devoted entirely to presenting a comprehensive view of what is known about biologically important molecules, with individual articles on DNA, RNA, and many kinds of proteins.

4

The Origin and Early History of Life

Concept Outline

4.1 All living things share key characteristics.

What Is Life? All known organisms share certain general properties, and to a large degree these properties define what we mean by life.

4.2 There are many ideas about the origin of life.

Ideas About the Origin of Life. There are both religious and scientific views about the origin of life. This text treats only the latter—the only one scientifically testable.

What the Early Earth Was Like. The atmosphere of the early earth was rich in hydrogen, providing a ready supply of energetic electrons with which to build organic molecules.

Testing the Spontaneous Origin Hypothesis. Experiments attempting to duplicate the conditions of early earth produce many of the key molecules of living organisms.

4.3 The first cells had little internal structure.

Theories About the Origin of Cells. The first cells are thought to have arisen spontaneously, but there is little agreement as to the mechanism.

The Earliest Cells. The earliest fossils are of bacteria too small to see with the unaided eye.

Do Bacteria Exist on Other Worlds? It seems probable that life has evolved on other worlds besides our own. The possible presence of ancient bacteria on Mars is a source of current speculation.

4.4 The first eukaryotic cells were larger and more complex than bacteria.

The First Eukaryotic Cells. Fossils of the first eukaryotic cells do not appear in rocks until 1.5 billion years ago, over 2 billion years after bacteria.

4.5 Biologists classify organisms into six kingdoms.

The Kingdoms of Life. Multicellular life is restricted to the four eukaryotic kingdoms of life.

FIGURE 4.1
The origin of life. The fortuitous mix of physical events and chemical elements at the right place and time created the first living cells on earth.

Although recent photographs taken by the *Hubble Space Telescope* have revived controversy about the age of the universe, it seems clear the earth itself was formed about 4.5 billion years ago. The oldest surviving terrestrial rocks, 4.3 billion years old, contain no definite traces of life—at least none we can recognize with our current level of technology. Some rocks 3.9 billion years old contain carbonates, which geologists believe resulted from life processes. The oldest microfossils are 3.5 billion years old. We can only speculate what happened in that early time. It seems to have taken just the right combination of physical events and chemical processes to give rise to life on earth (figure 4.1).

The earth formed as a hot mass of molten rock. As the earth cooled, much of the water vapor present in its atmosphere condensed into liquid water, which accumulated on the surface in chemically rich oceans. The primitive oceans must not have been pleasant places. It is odd to think of life originating from a dilute, hot, smelly soup of ammonia, formaldehyde, formic acid, cyanide, methane, hydrogen sulfide, and organic hydrocarbons. Yet from such an ocean emerged the organisms from which all subsequent life-forms are derived. The way in which this happened is a puzzle and may forever remain so. Nevertheless, one cannot escape a certain curiosity about the earliest steps that eventually led to the origin of all living things on earth, including ourselves. How did organisms evolve from the complex molecules that swirled in the early oceans?

What Is Life?

Before we can address this question, we must first consider what qualifies something as "living." What *is* life? This is a difficult question to answer, largely because life itself is not a simple concept. If you try to write a definition of "life," you will find that it is not an easy task, because of the loose manner in which the term is used. Imagine a situation in which two astronauts encounter a large, amorphous blob on the surface of a planet. How would they determine whether it is alive?

Movement.   One of the first things the astronauts might do is observe the blob to see if it moves. Most animals move about (figure 4.2), but movement from one place to another in itself is not diagnostic of life. Most plants and even some animals do not move about, while numerous nonliving objects, such as clouds, do move. The criterion of movement is thus neither *necessary* (possessed by all life) nor *sufficient* (possessed only by life).

Sensitivity.   The astronauts might prod the blob to see if it responds. Almost all living things respond to stimuli (figure 4.3). Plants grow toward light, and animals retreat from fire. Not all stimuli produce responses, however. Imagine kicking a redwood tree or singing to a hibernating bear. This criterion, although superior to the first, is still inadequate to define life.

Death.   The astronauts might attempt to kill the blob. All living things die, while inanimate objects do not. Death is not easily distinguished from disorder, however; a car that breaks down has not died, since it was never alive. Death is simply the loss of life, so this is a circular definition at best. Unless one can detect life, death is a meaningless concept, and hence a very inadequate criterion for defining life.

Complexity.   Finally, the astronauts might cut up the blob, to see if it is complexly organized. All living things are complex. Even the simplest bacteria contain a bewildering array of molecules, organized into many complex structures. However a computer is also complex, but not alive. Complexity is a *necessary* criterion of life, but it is not *sufficient* in itself to identify living things, since many complex things are not alive.

FIGURE 4.2
Movement. Animals have evolved mechanisms that allow them to move about in their environment. While some animals, like this giraffe, move on land, others move through water or air.

FIGURE 4.3
Sensitivity. This father lion is responding to a stimulus: he has just been bitten on the rump by his cub. As far as we know, all organisms respond to stimuli, although not always to the same ones or in the same way. Had the cub bitten a tree instead of its father, the response would not have been as dramatic.

FIGURE 4.4
Cellular organization (150×). These *Paramecia*, which are complex, single-celled organisms called protists, have just ingested several yeast cells. The yeasts, stained red in this photograph, are enclosed within membrane-bounded sacs called digestive vacuoles. A variety of other organelles are also visible.

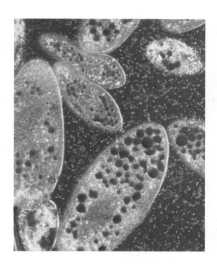

FIGURE 4.5
Proteinoid microspheres (1250×). These protocells, hollow microspheres formed from protein molecules, possess many of the characteristics of living cells. If you look carefully, for example, you can see that each is bounded by a bilayer membrane. The first cells are thought by some to have evolved from protein microspheres such as these.

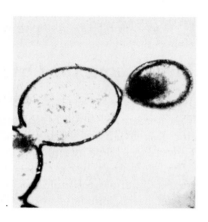

To determine whether the blob is alive, the astronauts would have to learn more about it. Probably the best thing they could do would be to examine it more carefully and determine whether it resembles the organisms we are familiar with, and if so, how.

Fundamental Properties of Life

As we discussed in chapter 1, all known organisms share certain general properties. To a large degree, these properties define what we mean by life. The following fundamental properties are shared by all organisms on earth.

Cellular organization. All organisms consist of one or more *cells*—complex, organized assemblages of molecules enclosed within membranes (figure 4.4).

Sensitivity. All organisms respond to stimuli—though not always to the same stimuli in the same ways.

Growth. All living things assimilate energy and use it to grow, a process called *metabolism*. Plants, algae, and some bacteria use sunlight to create covalent carbon-carbon bonds from CO_2 and H_2O through photosynthesis. This transfer of the energy in covalent bonds is essential to all life on earth.

Development. Multicellular organisms undergo systematic gene-directed changes as they grow and mature.

Reproduction. All living things reproduce, passing on traits from one generation to the next. Although some organisms live for a very long time, no organism lives forever, as far as we know. Because all organisms die, ongoing life is impossible without *reproduction*.

Regulation. All organisms have regulatory mechanisms that coordinate internal processes.

Homeostasis. All living things maintain relatively constant internal conditions, different from their environment.

The Key Role of Heredity

Are these properties adequate to define life? Is a membrane-enclosed entity that grows and reproduces alive? Not necessarily. Soap bubbles and proteinoid microspheres (figure 4.5) spontaneously form hollow bubbles that enclose a small volume of water. These spheres can enclose energy-processing molecules, and they may also grow and subdivide. Despite these features, they are certainly not alive. Therefore, the criteria just listed, although necessary for life, are not sufficient to define life. One ingredient is missing—a mechanism for the preservation of improvement.

Heredity. All organisms on earth possess a *genetic system* that is based on the replication of a long, complex molecule called DNA. This mechanism allows for adaptation and evolution over time, also distinguishing characteristics of living things.

To understand the role of heredity in our definition of life, let us return for a moment to proteinoid microspheres. When we examine an individual microsphere, we see it at that precise moment in time but learn nothing of its predecessors. It is likewise impossible to guess what future droplets will be like. The droplets are the passive prisoners of a changing environment, and it is in this sense that they are not alive. The essence of being alive is the ability to encompass change and to reproduce the results of change permanently. Heredity, therefore, provides the basis for the great division between the living and the nonliving. Change does not become evolution unless it is passed on to a new generation. A genetic system is the sufficient condition of life. Some changes are preserved because they increase the chances of survival in a hostile world, while others are lost. Not only did life evolve—evolution is the very essence of life.

All living things on earth are characterized by cellular organization, heredity, and a handful of other characteristics that serve to define the term life.

Ideas About the Origin of Life

The question of how life originated is not easy to answer, since it is impossible to go back in time and observe life's beginnings; nor are there any witnesses. There is testimony in the rocks of the earth, but it is not easily read, and often it is silent on issues crying out for answers. There are, in principle, at least three possibilities:

1. **Special creation.** Life-forms may have been put on earth by supernatural or divine forces.
2. **Extraterrestrial origin.** Life may not have originated on earth at all; instead, life may have infected earth from some other planet.
3. **Spontaneous origin.** Life may have evolved from inanimate matter, as associations among molecules became more and more complex.

Special Creation. The hypothesis of special creation, that a divine God created life is at the core of most major religions. The oldest hypothesis about life's origins, it is also the most widely accepted. Far more Americans, for example, believe that God created life on earth than believe in the other two hypotheses. Many take a more extreme position, accepting the biblical account of life's creation as factually correct. This viewpoint forms the basis for the very unscientific "scientific creationism" viewpoint discussed in chapter 21.

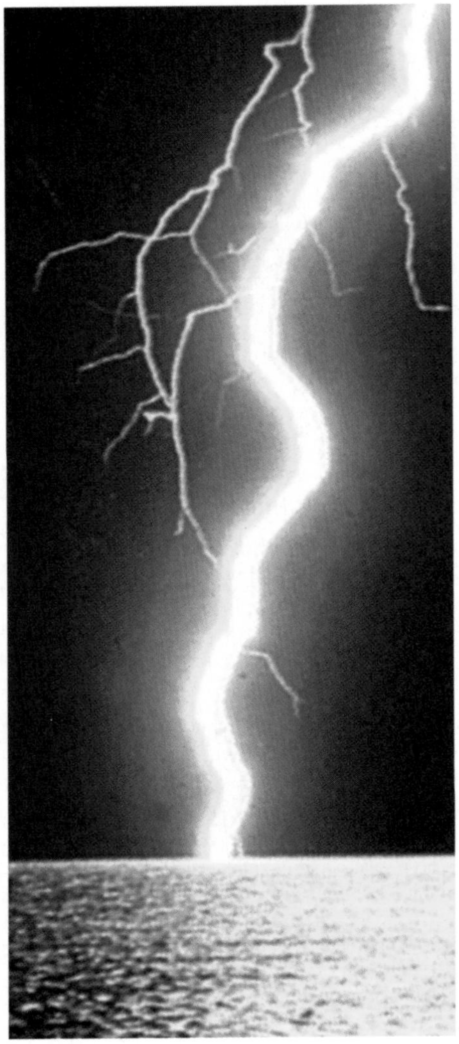

Extraterrestrial Origin. The hypothesis of **panspermia** proposes that meteors or cosmic dust may have carried life to earth, perhaps as an extraterrestrial infection of spores originating on a planet of a distant star. This hypothesis, which has a long and colorful history, cannot be rejected based on evidence currently available to science. As the planets of our solar system become better known, this hypothesis becomes more attractive. For example, recently discovered suggestions of fossils in rocks from Mars, and the discovery of liquid water under the surface

FIGURE 4.6
Lightning. Before life evolved, the simple molecules in the earth's atmosphere combined to form more complex molecules. The energy that drove these chemical reactions may have come from lightning and forms of geothermal energy.

of Jupiter's ice-shrouded moon Europa, might lend credence to this idea. However, because this hypothesis can never even in principle be disproven, it is no more scientific than the biblical account of creation.

Spontaneous Origin. Most scientists tentatively accept the hypothesis of spontaneous origin, that life evolved from inanimate matter. In this view, the force leading to life was selection. As changes in molecules increased their stability and caused them to persist longer, these molecules could initiate more and more complex associations, culminating in the evolution of cells.

Taking a Scientific Viewpoint

In this book we will focus on the third possibility, attempting to understand whether the forces of evolution could have led to the origin of life and, if so, how the process might have occurred. This is not to say that the third possibility is definitely the correct one. Any one of the three possibilities might be true. Nor does the third possibility preclude religion (a divine agency might have acted via evolution) or extraterrestrial origin (life may have infected earth from some other world, and then evolved). However, we are limiting the scope of our inquiry to scientific matters. Since only the third possibility permits testable hypotheses to be constructed, it provides the only *scientific* explanation—that is, an explanation that can be tested and potentially disproved.

In our search for this understanding, we must look back to the early times. There are fossils of simple living things, bacteria, in rocks 3.5 billion years old. They tell us that life originated during the first billion years of the history of our planet. As we attempt to determine how this process took place, we will first focus on how organic molecules may have originated (figure 4.6), and then we will consider how those molecules might have become organized into living cells.

Spontaneous origin is the only testable hypothesis of life's origin currently available.

What the Early Earth Was Like

Geochemists who study the conditions of the primitive earth believe that as the primitive earth cooled and its rocky crust formed, volcanoes blasted enormous amounts of material and gases from the earth's molten core skyward. These gases formed a cloud around the earth and were held as an atmosphere by the earth's gravity.

A Reducing Atmosphere

Not all geochemists agree on the exact composition of this early atmosphere, but they do agree that it contained principally carbon dioxide and nitrogen gas, along with significant amounts of water. It is probable, although not certain, that the early atmosphere also contained compounds in which hydrogen atoms were bonded to the other light elements (sulfur, nitrogen, and carbon) also present in the atmosphere of early earth. These compounds would have been hydrogen sulfide (H_2S), ammonia (NH_3), and methane (CH_4). The atmosphere is also thought by most investigators to have been rich in hydrogen gas (H_2), although there is debate on this point.

We refer to such an atmosphere as a *reducing atmosphere* because of the ample availability of hydrogen atoms and their associated electrons. Little, if any, oxygen gas was present. In such a reducing atmosphere, it would not take as much energy as it would today to form the carbon-rich molecules from which life evolved. Our atmosphere changed once living organisms began to harness the energy in sunlight to split water molecules and form complex carbon molecules, giving off gaseous oxygen molecules in the process. The earth's atmosphere is now approximately 21% oxygen. In the oxidizing atmosphere that exists today, the spontaneous formation of complex carbon molecules cannot occur.

High Temperatures

The more we learn about earth's early history, the more likely it seems that earth's first organisms emerged and lived at very high temperatures. Rubble from the forming solar system slammed into early earth from 4.6 to 3.8 billion years ago, keeping the surface molten hot. As the bombardment slowed down, temperatures dropped. By about 3.8 billion years ago, ocean temperatures are thought to have dropped to a hot 49–88°C (120–190°F). Between 3.8 and 3.5 billion years ago, life first appeared on earth. Recent studies suggest that life may have begun closer to 3.8 billion years ago, promptly after the earth was inhabitable—rocks that date back to 3.8 billion years ago contain traces of carbon-13 and carbon-12 isotopes in ratios that are characteristic of biological life. Thus, as intolerable as early earth's infernal temperatures seem to us today, they gave birth to life.

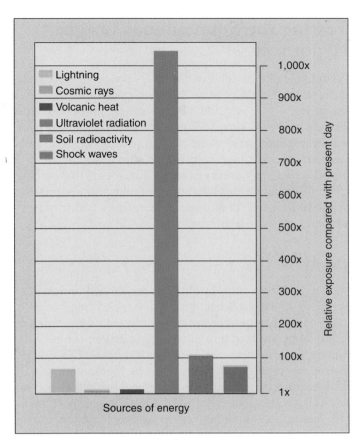

FIGURE 4.7

Sources of energy for the synthesis of complex molecules in the atmosphere of the primitive earth. Today's ambient exposure to the six forms of energy depicted in the histogram above would be at a level of one on the *y*-axis. At the time of synthesis of organic molecules, however, ultraviolet exposure was over 1000 times what it is today.

The first step in the origin of life, therefore, probably took place in a hot reducing atmosphere, devoid of gaseous oxygen, and different from the atmosphere that exists now. On earth today, we are shielded from the effects of solar ultraviolet radiation by a layer of ozone gas (O_3) in the upper atmosphere, and it is particularly difficult to imagine the enormous flux of ultraviolet energy to which the early earth's surface must have been exposed. Subjected to ultraviolet energy and to lightning in intense electrical storms, heat from violent volcanic eruptions and radioactive decay, and other sources of energy (figure 4.7), the gases of the early earth's atmosphere underwent chemical reactions with one another, forming a complex assemblage of molecules.

When life first appeared on earth, the atmosphere was devoid of oxygen gas and very hot. It is believed that these conditions fueled chemical reactions that gave rise to life.

Testing the Spontaneous Origin Hypothesis

What kinds of molecules might have been produced on the early earth? One way to answer this question is to repeat the process: (1) assemble an atmosphere similar to that thought to have existed; (2) exclude gaseous oxygen from the atmosphere, since none was present in the earth's atmosphere at that time; (3) place this atmosphere over liquid water, which was present on the surface of the cooling earth; (4) maintain this mixture at a temperature somewhat below 100°C; and (5) simulate lightening by bombarding it with energy in the form of sparks.

The Miller-Urey Experiment

Stanley L. Miller and Harold C. Urey carried out this experiment in 1953 (figure 4.8). They found that within a week, 15% of the carbon originally present as methane gas (CH_4) had converted into other simple carbon compounds. Among these compounds were formaldehyde (CH_2O) and hydrogen cyanide (HCN; figure 4.9). These compounds then combined to form simple molecules, such as formic acid (HCOOH) and urea (NH_2CONH_2), and more complex molecules containing carbon-carbon bonds, including the amino acids glycine and alanine.

As we saw in chapter 3, amino acids are the basic building blocks of proteins, and proteins are one of the major kinds of molecules of which organisms are composed. In similar experiments performed later by other scientists, more than 30 different carbon compounds were identified, including the amino acids glycine, alanine, glutamic acid, valine, proline, and aspartic acid. Other biologically important molecules were also formed in these experiments. For example, hydrogen cyanide contributed to the production of a complex ring-shaped molecule called adenine—one of the bases found in DNA and RNA. Thus, the key molecules of life could have formed in the atmosphere of the early earth.

The Path of Chemical Evolution

A raging debate among biologists who study the origin of life concerns which organic molecules came first, RNA or proteins. Scientists are divided into two camps: the "RNA world" group feels that without a hereditary molecule, other molecules could not have formed consistently; the "protein-first" group argues that without enzymes (which are proteins), nothing could replicate at all, heritable or not. Both arguments have their strong points. The "RNA world" argument earned support when Thomas Cech at the University of Colorado discovered **ribozymes,** RNA molecules that can behave as enzymes, catalyzing their own assembly. Others have shown that rRNA, the RNA contained in ribosomes (discussed in chapter 5), is actually responsible for polymerizing amino acids to form proteins.

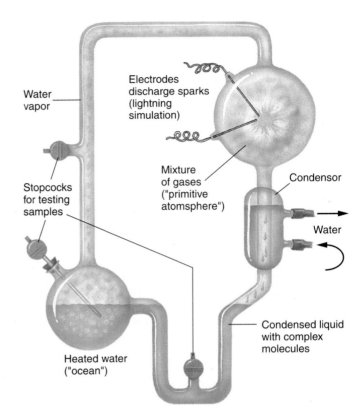

FIGURE 4.8

The Miller-Urey experiment. The apparatus consisted of a closed tube connecting two chambers. The upper chamber contained a mixture of gases thought to resemble the primitive earth's atmosphere. Electrodes discharged sparks through this mixture, simulating lightning. Condensers then cooled the gases, causing water droplets to form, which passed into the second heated chamber, the "ocean." Any complex molecules formed in the atmosphere chamber would be dissolved in these droplets and carried to the ocean chamber, from which samples were withdrawn for analysis.

Therefore, the RNA in ribosomes also functions as an enzyme, in a sense. If RNA has the ability to pass on inherited information and the capacity to act like an enzyme, were proteins really needed?

The "protein-first" proponents argue that nucleotides, the individual units of nucleic acids such as RNA, are too complex to have formed spontaneously, and certainly too complex to form spontaneously again and again. In an effort to shed light on this problem, Julius Rebek and a number of other chemists have created synthetic nucleotide-like molecules in the laboratory that are able to replicate. Moving even further, Rebek and his colleagues have created synthetic molecules that could replicate and "make mistakes." This simulates mutation, a necessary ingredient for the process of evolution.

While there is no doubt that simple proteins are easier to synthesize from abiotic components than nucleotides, both can form in the laboratory under the right conditions. Deciding which came first is a chicken-and-egg paradox. Un-

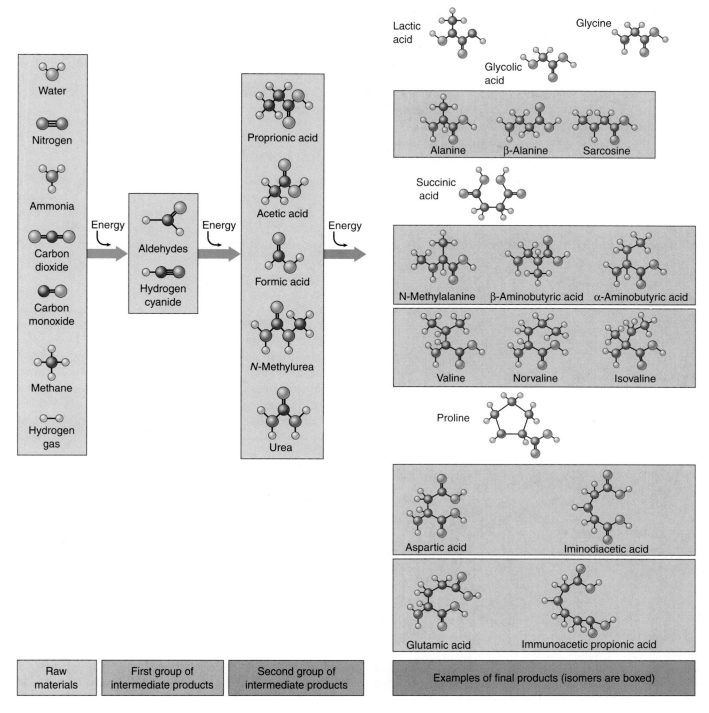

FIGURE 4.9
Results of the Miller-Urey experiment. Seven simple molecules, all gases, were included in the original mixture. Note that oxygen was not among them; instead, the atmosphere was rich in hydrogen. At each stage of the experiment, more complex molecules were formed: first aldehydes, then simple acids, then more complex acids. Among the final products, the molecules that are structural isomers of one another are grouped together in boxes. In most cases only one isomer of a compound is found in living systems today, although many may have been produced in the Miller-Urey experiment.

fortunately, we are unable to travel back in time and watch the "warm little pond" Darwin spoke of. All we can do is try to recreate the process in the laboratory, as Miller and Urey and a generation of subsequent scientists have done.

Molecules that are the building blocks of living organisms form spontaneously under conditions designed to simulate those of the primitive earth.

Theories About the Origin of Cells

The evolution of cells required early organic molecules to assemble into a functional, interdependent unit. Cells, discussed in the next chapter, are essentially little bags of fluid. What the fluid contains depends on the individual cell, but every cell's contents differ from the environment outside the cell. Therefore, an early cell may have floated along in a dilute "primordial soup," but its interior would have had a higher concentration of specific organic molecules.

Cell Origins: The Importance of Bubbles

How did these "bags of fluid" evolve from simple organic molecules? Although the answer to this question is a matter for debate, most scientists believe bubbles were involved in this evolutionary step. A bubble, such as those produced by soap solutions, is a hollow spherical structure. Certain molecules, particularly those with hydrophobic regions, will spontaneously form bubbles in water. The structure of the bubble shields the hydrophobic regions of the molecules from contact with water. If you have ever watched the ocean surge upon the shore, you may have noticed the foamy froth created by the agitated water. The primitive oceans were more than likely very frothy places bombarded by ultraviolet and other ionizing radiation, and exposed to an atmosphere that probably contained methane and other simple organic molecules.

Oparin's Bubble Theory

The first bubble theory is attributed to Alexander Oparin, a Russian chemist with extraordinary insight. In the mid-1930s, Oparin suggested that the present-day atmosphere was incompatible with the creation of life. He proposed that life must have arisen from nonliving matter under a set of very different environmental circumstances some time in the distant history of the earth. His was the theory of **primary abiogenesis** (primary because all living cells are now known to come from previously living cells, except in that first case). At the same time, J. B. S. Haldane, a British geneticist, was also independently espousing the same views. Oparin decided that in order for cells to evolve, they must have had some means of developing chemical complexity, separating their contents from their environment by means of a cell membrane, and concentrating materials within themselves. He termed these early, chemical-concentrating bubble-like structures **protobionts.**

Oparin's theories were published in English in 1938, and for awhile most scientists ignored them. However, Harold Urey, an astronomer at the University of Chicago, was quite taken with Oparin's ideas. He convinced one of his graduate students, Stanley Miller, to follow Oparin's rationale and see if he could "create" life. The Urey-Miller experiment has proven to be one of the most significant experiments in the history of science. As a result Oparin's ideas became better known and more widely accepted.

A Host of Bubble Theories

Different versions of "bubble theories" have been championed by numerous scientists since Oparin. The bubbles they propose go by a variety of names; they may be called *microspheres, protocells, protobionts, micelles, liposomes,* or *coacervates,* depending on the composition of the bubbles (lipid or protein) and how they form. In all cases, the bubbles are hollow spheres, and they exhibit a variety of cell-like properties. For example, the lipid bubbles called **coacervates** form an outer boundary with two layers that resembles a biological membrane. They grow by accumulating more subunit lipid molecules from the surrounding medium, and they can form bud-like projections and divide by pinching in two, like bacteria. They also can contain amino acids and use them to facilitate various acid-base reactions, including the decomposition of glucose. Although they are not alive, they obviously have many of the characteristics of cells.

A Bubble Scenario

It is not difficult to imagine that a process of chemical evolution involving bubbles or microdrops preceded the origin of life (figure 4.10). The early oceans must have contained untold numbers of these microdrops, billions in a spoonful, each one forming spontaneously, persisting for a while, and then dispersing. Some would, by chance, have contained amino acids with side groups able to catalyze growth-promoting reactions. Those microdrops would have survived longer than ones that lacked those amino acids, because the persistence of both proteinoid microspheres and lipid coacervates is greatly increased when they carry out metabolic reactions such as glucose degradation and when they are actively growing.

Over millions of years, then, the complex bubbles that were better able to incorporate molecules and energy from the lifeless oceans of the early earth would have tended to persist longer than the others. Also favored would have been the microdrops that could use these molecules to expand in size, growing large enough to divide into "daughter" microdrops with features similar to those of their "parent" microdrop. The daughter microdrops have the same favorable combination of characteristics as their parent, and would have grown and divided, too. When a way to facilitate the reliable transfer of new ability from parent to offspring developed, heredity—and life—began.

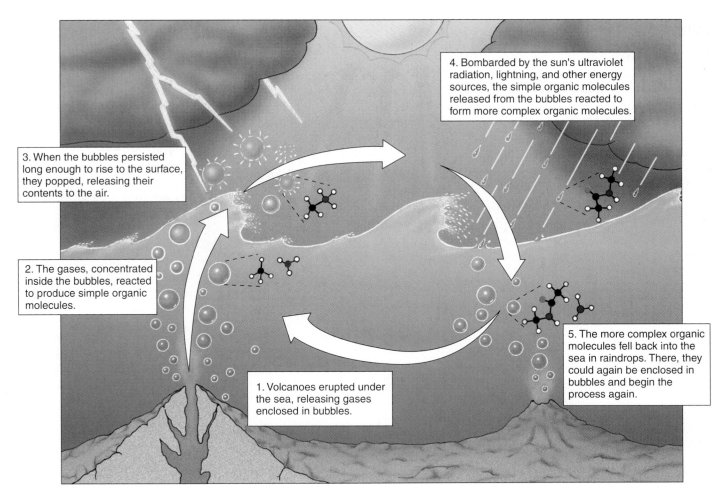

FIGURE 4.10
A current bubble hypothesis. In 1986 geophysicist Louis Lerman proposed that the chemical processes leading to the evolution of life took place within bubbles on the ocean's surface.

Current Thinking

Whether the early bubbles that gave rise to cells were lipid or protein remains an unresolved argument. While it is true that lipid microspheres (coacervates) will form readily in water, there appears to be no mechanism for their heritable replication. On the other hand, one *can* imagine a heritable mechanism for protein microspheres. Although protein microspheres do not form readily in water, Sidney Fox and his colleagues at the University of Miami have shown that they can form under dry conditions.

The discovery that RNA can act as an enzyme to assemble new RNA molecules on an RNA template has raised the interesting possibility that neither coacervates nor protein microspheres were the first step in the evolution of life. Perhaps the first components were RNA molecules, and the initial steps on the evolutionary journey led to increasingly complex and stable RNA molecules. Later, stability might have improved further when a lipid (or possibly protein) microsphere surrounded the RNA. At present, those studying this problem have not arrived at a consensus about whether RNA evolved before or after a bubble-like structure that likely preceded cells.

Little is known about how the first cells originated. Current hypotheses involve chemical evolution within bubbles, but there is no general agreement about their composition, or about how the process occurred.

The Earliest Cells

What do we know about the earliest life-forms? The fossils found in ancient rocks show an obvious progression from simple to complex organisms, beginning about 3.5 billion years ago. Life may have been present earlier, but rocks of such great antiquity are rare, and fossils have not yet been found in them.

Microfossils

The earliest evidence of life appears in **microfossils,** fossilized forms of microscopic life (figure 4.11). What do we know about these earliest life-forms? Microfossils were small (1–2 micrometers in diameter) and single-celled, lacked external appendages, and had little evidence of internal structure. Thus, they physically resemble present-day **bacteria** (figure 4.12), although some ancient forms cannot be matched exactly. We call organisms with this simple body plan **prokaryotes,** from the Greek words meaning "before" and "kernel," or "nucleus." The name reflects their lack of a **nucleus,** a spherical organelle characteristic of the more complex cells of **eukaryotes.** Judging from the fossil record, eukaryotes did not appear until about 1.5 billion years ago. Therefore, for at least 2 billion years—nearly half the age of the earth—bacteria were the only organisms that existed.

Ancient Bacteria: Archaebacteria

Most organisms living today are adapted to the relatively mild conditions of present-day earth. However, if we look in unusual environments, we encounter organisms that are quite remarkable, differing in form and metabolism from other living things. Sheltered from evolutionary alteration in unchanging habitats that resemble earth's early environment, these living relics are the surviving representatives of the first ages of life on earth. In places such as the oxygen-free depths of the Black Sea or the boiling waters of hot springs and deep sea vents, we can find bacteria living at very high temperatures without oxygen.

These unusual bacteria are called **archaebacteria,** from the Greek word for "ancient ones." Among the first to be studied in detail have been the **methanogens,** or methane-producing bacteria, among the most primitive bacteria that exist today. These organisms are typically simple in form and are able to grow only in an oxygen-free environment; in fact, oxygen poisons them. For this reason they are said to grow "without air," or **anaerobically** (Greek *an,* "without" + *aer,* "air" + *bios,* "life"). The methane-producing bacteria convert CO_2 and H_2 into methane gas (CH_4). They resemble all other bacteria in that they possess hereditary machinery based on DNA, a cell membrane composed of lipid molecules, an exterior cell wall, and a metabolism based on an energy-carrying molecule called ATP. However, the resemblance ends at that point.

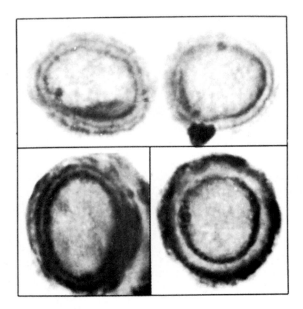

FIGURE 4.11
Cross-section of fossil bacteria. These microfossils from the Bitter Springs formation of Australia are of ancient cyanobacteria, far too small to be seen with the unaided eye. In this electron micrograph, the cell walls are clearly evident.

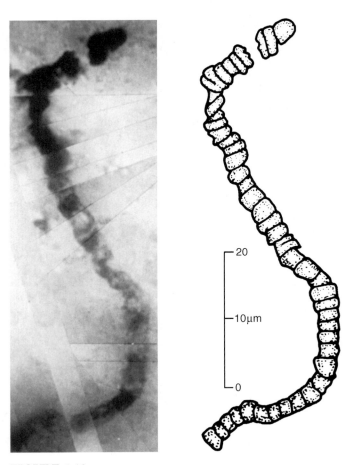

FIGURE 4.12
The oldest microfossil. This ancient bacterial fossil, discovered by J. William Schopf of UCLA in 3.5-billion-year-old rocks in western Australia, is similar to present-day cyanobacteria, as you can see by comparing it to figure 4.13.

Unusual Cell Structures

When the details of cell wall and membrane structure of the methane-producing bacteria were examined, they proved to be different from those of all other bacteria. Archaebacteria are characterized by a conspicuous lack of a protein cross-linked carbohydrate material called **peptidoglycan** in their cell walls, a key compound in the cell walls of most modern bacteria. Archaebacteria also have unusual lipids in their cell membranes that are not found in any other group of organisms. There are also major differences in some of the fundamental biochemical processes of metabolism, different from those of all other bacteria. The methane-producing bacteria are survivors from an earlier time when oxygen gas was absent.

Earth's First Organisms?

Other archaebacteria that fall into this classification are some of those that live in very salty environments like the Dead Sea (**extreme halophiles**—"salt lovers") or very hot environments like hydrothermal volcanic vents under the ocean (**extreme thermophiles**—"heat lovers"). Thermophiles have been found living comfortably in boiling water. Indeed, many kinds of thermophilic archaebacteria thrive at temperatures of 110°C (230°F). Because these thermophiles live at high temperatures similar to those that may have existed when life first evolved, microbiologists speculate that thermophilic archaebacteria may be relics of earth's first organisms.

Just how different are extreme thermophiles from other organisms? A methane-producing archaebacteria called *Methanococcus* isolated from deep sea vents provides a startling picture. These bacteria thrive at temperatures of 88°C (185°F) and crushing pressures 245 times greater than at sea level. In 1996 molecular biologists announced that they had succeeded in determining the full nucleotide sequence of *Methanococcus*. This was possible because archaebacterial DNA is relatively small—it has only 1700 genes, coded in a DNA molecule only 1,739,933 nucleotides long (a human cell has 2000 times more!). The thermophile nucleotide sequence proved to be astonishingly different from the DNA sequence of any other organism ever studied; fully two thirds of its genes are unlike any ever known to science before! Clearly these archaebacteria separated from other life on earth a long time ago. Preliminary comparisons to the gene sequences of other bacteria suggest that archaebacteria split from other types of bacteria over 3 billion years ago, soon after life began.

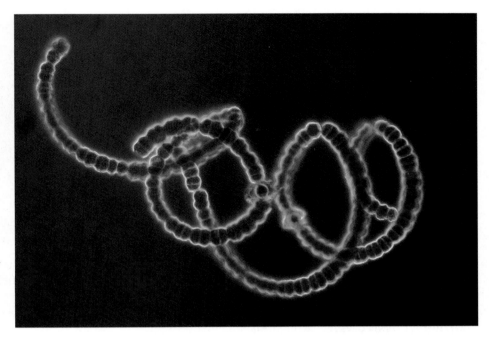

FIGURE 4.13
Living cyanobacteria. Although not multicellular, these bacteria often aggregate into chains such as those seen here.

Eubacteria

The second major group of bacteria, the **eubacteria,** have very strong cell walls and a simpler gene architecture. Most bacteria living today are eubacteria. Included in this group are bacteria that have evolved the ability to capture the energy of light and transform it into the energy of chemical bonds within cells. These organisms are *photosynthetic*, as are plants and algae.

One type of photosynthetic eubacteria that has been important in the history of life on earth is the cyanobacteria, sometimes called "blue-green algae" (figure 4.13). They have the same kind of chlorophyll pigment that is most abundant in plants and algae, as well as other pigments that are blue or red. Cyanobacteria produce oxygen as a result of their photosynthetic activities, and when they appeared at least 3 billion years ago, they played a decisive role in increasing the concentration of free oxygen in the earth's atmosphere from below 1% to the current level of 21%. As the concentration of oxygen increased, so did the amount of ozone in the upper layers of the atmosphere. The thickening ozone layer afforded protection from most of the ultraviolet radiation from the sun, radiation that is highly destructive to proteins and nucleic acids. Certain cyanobacteria are also responsible for the accumulation of massive limestone deposits.

All bacteria now living are members of either Archaebacteria or Eubacteria.

Do Bacteria Exist on Other Worlds?

The life-forms that evolved on earth mirror the nature of this planet and its history. The evolution of carbon-based life-forms appears to be possible only within the narrow range of temperatures that exist on earth, which depends on the earth's distance from the sun. If the earth were farther from the sun, it would be colder, and chemical processes would be greatly slowed down. Water would be a solid, and many carbon compounds would be brittle. On the other hand, if the earth were closer to the sun, it would be far hotter, chemical bonds would be less stable, and few carbon compounds would be chemically stable enough to persist.

The size of the earth also has played an important role in the evolution of life because it has permitted the earth to have a gaseous atmosphere. If the earth were smaller, it would not have sufficient gravitational pull to hold an atmosphere. If it were larger, it might have such a dense atmosphere that all solar radiation would be absorbed before it reached the surface of the planet.

Thus if we wish to look for life on other planets, it seems reasonable first to explore those with conditions similar to earth's.

Ancient Bacteria on Mars?

A dull grey chunk of rock collected in 1984 in Antarctica has recently excited science with the possibility that it contains evidence of life on Mars. Analysis of gases trapped within small pockets of the rock indicate it is a meteorite from Mars. It is, in fact, the oldest rock known to science—fully 4.5 billion years old. Back then, when this rock formed on Mars, that cold, arid planet was much warmer, flowed with water, and had a carbon dioxide atmosphere—conditions not remarkably different from those that spawned life on earth.

Intrigued by the parallel, a group of NASA scientists set out to investigate the meteorite for evidence of biological activity. The search did not yield conclusive results, but several features of the meteorite seem to point in the direction of bacterial life!

Three aspects of the Mars meteorite suggested the presence of ancient bacteria:

1. **Biotic combinations of minerals.** In acid-pitted portions of the meteorite, the iron oxide called magnetite and the iron sulfide called pyrrhotite are found together, associated with microfossil-like elements. Magnetite is, as its name suggests, magnetic. Many earthly bacteria secrete particles of magnetite within themselves, probably to help them navigate by orienting to the earth's magnetic field. Iron sulfide is also a common product of bacteria that do not need oxygen to respire. These two minerals do not precipitate together under acid

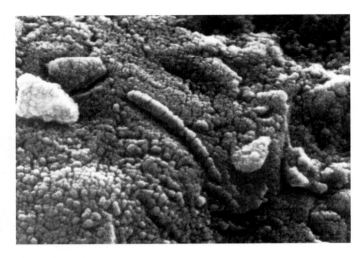

FIGURE 4.14
Evidence of life? In a Martian meteorite, patches of carbonate some 50 micrometers across contain microfossils like this half-micrometer "tube-like structural form."

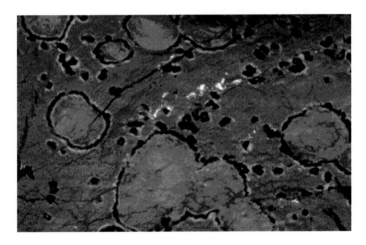

FIGURE 4.15
A Martian "microfossil" in cross-section. It is not yet known if these structures have true cell walls, as all bacteria do.

conditions, but they often do when associated with bacterial growth. However, independent work published on the same meteorite a year earlier concluded that the mineral deposits formed within the meteorite at temperatures much too hot for life.

2. **Organic residues.** Polycyclic aromatic hydrocarbons (PAHs) are found in the interior of the meteorite, where they could not be the result of contamination. These large complex organic molecules have previously been found in meteorites, the residue of nonbiological reactions. However, the PAHs in the Martian meteorite are unusual. They resemble neither common terres-

trial nor meteoritic ones. Instead, they look surprisingly like the sort of PAHs which bacteria dissolve into when subjected to rock-forming processes.

3. **Bacteria-like structures.** When examined with powerful electron microscopes, carbonate patches within the meteorite exhibit what look like microfossils, some 20–100 nanometers in length (figure 4.14). 100 times smaller than any known bacteria, it is not clear they actually are fossils, but the resemblance to bacteria is striking. While **carbonate spheres** tend to grow on rocks brought out of the Antarctic cold, the spheres within this meteorite must have formed before the meteorite arrived on earth, because some of them were cracked by whatever shock blasted the rock from the Martian surface. Also, the iron oxide and iron sulfide deposits lie *within* the spheres. Closer examination of cross-sections (figure 4.15) will reveal whether the possible microfossils have cell walls, as all bacteria do, and whether they contain amino acids.

FIGURE 4.16
Is there life elsewhere? Currently the most likely candidate for life elsewhere within the solar system is Europa, one of the many moons of the large planet Jupiter. Europa is covered with ice, and satellite photos taken in 1996 suggest that under the ice lies a water ocean. What looks like crisscrossing lines may actually be icy debris pushed up by volcanic activity. And, if volcanoes are generating heat, a subsurface ocean containing life may exist. The conditions on Europa now are far less hostile to life than the conditions that existed in the oceans of the primitive earth.

Has life evolved on other worlds within our solar system? There are planets other than ancient Mars with conditions not unlike those on earth. Europa, a large ice-covered moon of Jupiter, is a promising candidate (figure 4.16).

Life in Other Solar Systems

There are undoubtedly many other worlds in the universe with physical characteristics that resemble those of our planet. The universe contains some 10^{20} (100,000,000,000,000,000,000) stars similar to our sun. We don't know how many of these stars have planets, but it seems increasingly likely that many do. In 1996 astronomers succeeded in actually detecting several planets orbiting distant stars. At least 10% of stars are thought to have planetary systems. If only 1 in 10,000 of these planets is the right size and at the right distance from its star to duplicate the conditions in which life originated on earth, the "life experiment" will have been repeated 10^{15} times (that is, a million billion times). Consequently, it does not seem likely that we are alone.

Because other meteorites that supposedly harbored evidence of life have proven to be contaminated, the report of possible bacterial fossils on the Martian meteorite has met with considerable skepticism. But contamination does not seem to be a problem in this case; the PAHs, present in the interior of the meteorite, are absent from the outer rind, where the meteorite melted during atmospheric entry.

Viewed as a whole, the evidence of bacterial life associated with the Mars meteorite is at best indirect and suggestive, rather than compelling. Clearly, more painstaking research remains to be done before the discovery can claim a scientific consensus.

Bacteria in Alien Earth Environments

The possibility that life on earth actually originated in the vicinity of deep-sea hydrothermal vents is gaining popularity. At the bottom of the ocean, where these vents spewed out a rich froth of molecules, the geological turbulence and radioactive energy battering the land was absent, and things were comparatively calm. The thermophilic archaebacteria found near these vents today are the most ancient group of organisms living on earth. Perhaps the gentler environment of the ocean depths was the actual cradle of life.

Nor should we overlook the possibility that life processes might have evolved in different ways on other planets. A functional genetic system, capable of accumulating and replicating changes and thus of adaptation and evolution, could theoretically evolve from molecules other than carbon, hydrogen, nitrogen, and oxygen in a different environment. Silicon, like carbon, needs four electrons to fill its outer energy level, and ammonia is even more polar than water. Perhaps under radically different temperatures and pressures, these elements might form molecules as diverse and flexible as those carbon has formed on earth.

The final question about life, of course, is whether it has an end. Does evolution tend toward any fixed form or state? Some have suggested that if life might have emerged on any of 10^{15} worlds, we should have heard from some other life-forms by now—unless the life experiment does not work, with life always tending to destroy itself. We can only hope that this is not so and that peace and progress are perpetual possibilities.

While there is no conclusive evidence of bacterial life associated with meteorites, it seems very possible that life has evolved on other worlds in addition to our own.

All fossils more than 1.5 billion years old are generally similar to one another structurally. They are small, simple cells; most measure 0.5 to 2 micrometers in diameter, and none are more than about 6 micrometers in diameter. These simple cells eventually evolved into larger, more complex forms—the first eukaryotic cells (figure 4.17).

The First Eukaryotic Cells

In rocks about 1.5 billion years old, we begin to see the first microfossils that are noticeably different in appearance from the earlier, simpler forms (figure 4.18). These cells are much larger than bacteria and have internal membranes and thicker walls. Cells more than 10 micrometers in diameter rapidly increased in abundance. Some fossilized cells 1.4 billion years old are as much as 60 micrometers in diameter; others, 1.5 billion years old, contain what appear to be small, membrane-bound structures. Many of these fossils have elaborate shapes—some exhibit highly branched filaments, tetrahedral configurations, or spines.

These early fossils mark a major event in the evolution of life: A new kind of organism had appeared. These new cells are called **eukaryotes,** from the Greek words for "true" and "nucleus," because they possess an internal structure called a nucleus. All organisms other than the bacteria are eukaryotes. The early eukaryotes rapidly evolved to produce all of the diverse organisms that inhabit the earth today, including humans (figure 4.19).

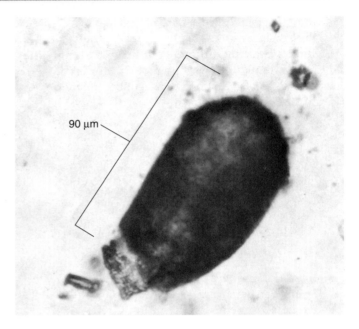

90 µm

FIGURE 4.17
Microfossil of a primitive eukaryote. This single-celled microfossil is about 800 million years old. All life was unicellular until about 700 million years ago.

		Geological evidence	Millions of years ago	Life forms
CAMBRIAN	PHANEROZOIC			
PRECAMBRIAN	PROTEROZOIC	Oldest multicellular fossils	570 600	Appearance of first multicellular organisms
				Appearance of first eukaryotes
		Oldest compartmentalized fossil cells	1500	Appearance of aerobic (oxygen-using) respiration
		Disappearance of iron from oceans and formation of iron oxides	2500	Appearance of oxygen-forming photosynthesis (cyanobacteria)
	ARCHEAN	Oldest definite fossils	3500	Appearance of chemoautotrophs (sulfate respiration)
				Appearance of life (prokaryotes): anaerobic (methane-producing) bacteria and anaerobic (hydrogen sulfide-forming) photosynthesis
		Oldest dated rocks	4500	Formation of the earth

FIGURE 4.18
The geological timescale. The periods refer to different stages in the evolution of life on earth. Certain fossils date back to the Archean Period, and very different ones are found in samples from the Proterozoic Era. The time scale is calibrated by examining rocks containing particular kinds of fossils; the fossils are dated by determining the degree of spontaneous decay of radioactive isotopes locked within rock when it was formed.

A primitive eukaryote called *Pelomyxa palustris* offers a glimpse of what the earliest eukaryotes may have been like, since it seems to represent a transition between the prokaryotes and the eukaryotes. *Pelomyxa* is a single-celled, amoeboid protist that lives at the bottom of ponds. Although it has a nucleus, its nucleus divides more like a prokaryotic than a eukaryotic cell, without a complex of microtubules to separate the daughter chromosomes in a dividing cell. *Pelomyxa* possesses no mitochondria as other eukaryotes do, but living within it are bacteria that serve a similar function.

Origin of Mitochondria and Chloroplasts

Bacteria that live within other cells and perform specific functions for their host cells are called *endosymbiotic bacteria*. Their widespread presence led Lynn Margulis of the University of Massachusetts to champion the **endosymbiotic theory** in the early 1970s. This theory, now widely accepted, suggests that a critical stage in the evolution of eukaryotic cells involved endosymbiotic relationships with prokaryotic organisms. According to this theory, energy-producing bacteria may have come to reside within larger bacteria, eventually evolving into what we now know as mitochondria. Similarly, photosynthetic bacteria may have come to live within other larger bacteria, leading to the evolution of chloroplasts, the photosynthetic organelles of plants and algae. Bacteria with flagella, long whip-like cellular appendages used for propulsion, may have become symbiotically involved with nonflagellated bacteria to produce larger, motile cells. The fact that we now witness so many symbiotic relationships lends general support to this theory. Even stronger support comes from the observation that present-day organelles such as mitochondria, chloroplasts, and centrioles contain their own DNA, which is remarkably similar to the DNA of bacteria in size and character.

Sexual Reproduction

Eukaryotic cells also possess the ability to reproduce sexually, something prokaryotes cannot do effectively. **Sexual reproduction** is the process of producing offspring, with two copies of each chromosome, by fertilization, the union of two cells that each have one copy of each chromosome. The great advantage of sexual reproduction is that it allows for frequent genetic recombination, which generates the variation that is the raw material for evolution. Not all eukaryotes reproduce sexually, but most have the capacity to do so. The evolution of meiosis and sexual reproduction (discussed in chapter 12) led to the tremendous explosion of diversity among the eukaryotes.

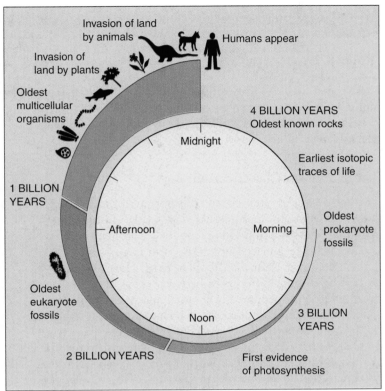

FIGURE 4.19
A clock of biological time. A billion seconds ago, it was 1963, and most students using this text had not yet been born. A billion minutes ago, Jesus was alive and walking in Galilee. A billion hours ago, the first human had not been born. A billion days ago, no biped walked on earth. A billion months ago, the first dinosaurs had not yet been born. A billion years ago, no creature had ever walked on the surface of the earth.

Multicellularity

Diversity was also promoted by the development of **multicellularity.** Some single eukaryotic cells began living in association with others, in colonies. Eventually individual members of the colony began to assume different duties, and the colony began to take on the characteristics of a single individual. Multicellularity has arisen many times among the eukaryotes. Practically every organism big enough to see with the unaided eye is multicellular, including all animals and plants. The great advantage of multicellularity is that it fosters specialization; some cells devote all of their energies to one task, other cells to another. Few innovations have had as great an impact on the history of life as the specialization made possible by multicellularity.

For at least the first 2 billion years of life on earth, all organisms were bacteria. About 1.5 billion years ago, the first eukaryotes appeared.

The Kingdoms of Life

Confronted with the great diversity of life on earth today, biologists have attempted to categorize similar organisms in order to better understand them, giving rise to the science of taxonomy.

The Kingdom Classification

The early Greeks often classified things according to size, color, or whether they possessed a "soul." The earliest **kingdom classification** is attributed to Carolus Linnaeus in 1758. Linnaeus placed organisms into one of two kingdoms: Plantae or Animalia. In 1866, Ernst Haeckel, recognizing the distinct differences between some of the more primitive organisms and the familiar plants and animals, added a third kingdom, Protista, which included the bacteria, protists, and sponges. Copeland separated the bacteria in a fourth kingdom, Monera, in 1956. (He also replaced the name Protista with Protoctista, but scientists tend to recognize the earliest names in taxonomy, so subsequent naming schemes have gone back to "Protista.") All of these classification systems include the fungi with the plants; in 1969, Whittaker put them in a separate fifth kingdom, Fungi, because unlike plants they obtain their organic molecules by consuming other organisms. Margulis came out with a similar scheme of five kingdoms in 1974, although she placed the *motile* fungi in the kingdom Protista. Recent molecular studies have led to the addition of a sixth kingdom, Archaebacteria, containing primitive bacteria with unusual cell walls and membranes, and eukaryote-like genetic mechanisms. Additional molecular evidence suggests that the ciliates, protists such as the laboratory favorite *Paramecium*, may belong in their own kingdom, since they use the genetic code differently than all other organisms.

The Six Kingdoms

In later chapters, we will discuss taxonomy and classification in detail, but for now we can generalize that all living things fall into one of six kingdoms (figure 4.20):

Kingdom Archaebacteria: Prokaryotes that lack a peptidoglycan cell wall, including the methanogens and extreme halophiles and thermophiles.

Kingdom Eubacteria: Prokaryotic organisms with a peptidoglycan cell wall, including cyanobacteria, soil bacteria, nitrogen-fixing bacteria, and pathogenic (disease-causing) bacteria.

Kingdom Protista: Eukaryotic, primarily unicellular (although algae are multicellular), photosynthetic or heterotrophic organisms, such as amoebas and paramecia.

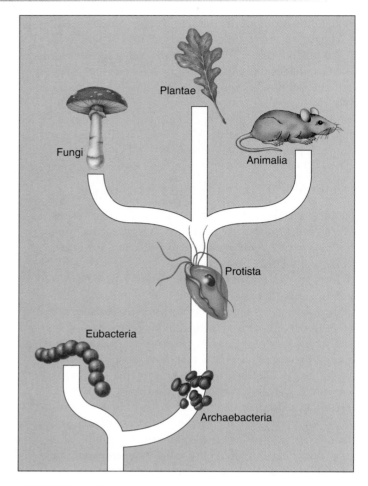

FIGURE 4.20
The kingdoms of life.

Kingdom Fungi: Eukaryotic, mostly multicellular (although yeasts are unicellular), heterotrophic, usually nonmotile organisms, with cell walls of chitin, such as mushrooms.

Kingdom Plantae: Eukaryotic, multicellular, nonmotile, usually terrestrial, photosynthetic organisms, such as trees, grasses, and mosses.

Kingdom Animalia: Eukaryotic, multicellular, motile, heterotrophic organisms, such as sponges, spiders, newts, penguins, and humans.

As more is learned about living things, particularly from the newer evidence that DNA studies provide, scientists will continue to reevaluate the relationships among the kingdoms of life.

Biologists place living organisms into six general categories called kingdoms. The two most ancient kingdoms contain bacteria.

Summary of Concepts

4.1 All living things share key characteristics.

- All living things are characterized by cellular organization, growth, reproduction, and heredity.
- Other properties commonly exhibited by living things include movement and sensitivity to stimuli.

4.2 There are many ideas about the origin of life.

- Of the many explanations of how life might have originated, only the theory of evolution provides a scientifically testable explanation.
- Experiments re-creating the atmosphere of primitive earth, with the energy sources and temperatures thought to be prevalent at that time, have led to the spontaneous formation of amino acids and other biologically significant molecules.

4.3 The first cells had little internal structure.

- The first cells are thought to have arisen from aggregations of molecules that were more stable and, therefore, persisted longer.
- It has been suggested that RNA may have arisen before cells did, and subsequently became packaged within a membrane.

- Microfossils of bacteria are found continuously in the fossil record as far back as 3.5 billion years ago in the oldest rocks suitable for the preservation of organisms.
- Bacteria were the only life-forms on earth for 2 billion years or more. At least three kinds of bacteria were present in ancient times: methane utilizers, anaerobic photosynthesizers, and eventually O_2-forming photosynthesizers.
- There are approximately 10^{20} stars in the universe similar to our sun. It is almost certain that life has evolved on planets circling some of them.

4.4 The first eukaryotic cells were larger and more complex than bacteria.

- The first eukaryotes can be seen in the fossil record 1.3 to 1.5 billion years ago. All organisms other than bacteria are their descendants.

4.5 Biologists classify organisms into six kingdoms.

- Biologists group all living organisms into six "kingdoms," each profoundly different from the others.
- The two most ancient kingdoms contain prokaryotes (bacteria); the other four contain eukaryotes.

Discussing Key Terms

1. **Spontaneous origin**   This theory argues that life arose spontaneously from nonliving matter. It is currently the only scientifically testable explanation, as it makes predictions that can be tested and potentially shown to be false.

2. **Bubble theories**   All living things consist of one or more compartments called cells. So-called bubble theories suggest that cells evolved from prebiotic molecules that aggregated in stable clusters resembling tiny bubbles.

3. **Microfossils**   Microfossils are fossils too small to see without a powerful microscope. Microfossils of bacteria have been found in earth rocks 3.5 billion years old. Hence, these bacteria appeared only a billion years after the earth formed.

4. **Prokaryotes (bacteria)**   For the first 2 billion years of life, simple single-celled organisms called prokaryotes, or, more informally, bacteria were the only living things on the earth.

5. **Eukaryotes**   Only 1.3 to 1.5 billion years ago, prokaryotes gave rise to eukaryotes, more complex single-celled organisms with membrane-bounded organelles. These organisms, which eventually developed the ability to reproduce sexually, ultimately gave rise to all other living things.

6. **Multicellularity**   Biologists believe single-cell eukaryotes associated in colonies, cells of which eventually began to perform different functions. This specialization led to the evolution of multicellular organisms.

Review Questions

1. What molecules are thought to have been present in the atmosphere of the early earth? Which molecule that was notably absent then is now a major component of the atmosphere, and what process was responsible for increasing its concentration in the atmosphere?

2. How did Urey and Miller simulate the process in which organic compounds are thought to have been produced on the early earth's surface? What types of compounds were produced in their simulation and in similar experiments performed by other scientists?

3. What evidence supports the argument that RNA evolved first on the early earth? What evidence supports the argument that proteins evolved first?

4. What are the characteristics of living things? Of these, which are *necessary* characteristics (possessed by all living things), and which are *sufficient* characteristics (possessed only by living things)?

5. What are coacervates, and what characteristics do they have in common with organisms? Are they alive? Why or why not?

6. What were the earliest known organisms like, and when did they appear? What present-day organisms do they resemble?

7. How do archaebacteria differ from eubacteria? Which group contains bacteria that are capable of photosynthesis?

8. From a scientific viewpoint, how likely is it that life has evolved on other planets? What characteristics must a planet have to be conducive to the evolution of carbon-based life-forms?

9. When did the first eukaryotes appear? How did they differ from the earlier prokaryotes? By what mechanism are they thought to have evolved from the earlier prokaryotes?

10. What are the six kingdoms of life currently recognized by biologists, and what sorts of organisms are contained in each?

Thought Questions

1. In Fred Hoyle's science fiction novel, *The Black Cloud*, the earth is approached by a large interstellar cloud of gas that orients itself around the sun. Scientists soon discover that the cloud is feeding on the sun, absorbing the sun's energy through the excitation of electrons in the outer energy levels of cloud molecules, in a process similar to the photosynthesis that occurs on earth. Different portions of the cloud are isolated from each other by associations of ions created by this excitation. Electron currents pass between these portions, much as they do on the surface of the human brain, endowing the cloud with self-awareness, memory, and the ability to think. Using electrical discharges, the cloud is able to communicate with humans and describe its history. It tells human scientists that it originated as a small extrusion from an ancestral cloud, and that since then it has grown by the absorption of molecules and energy from stars like our sun, on which it has been feeding. Soon the cloud moves off in search of other stars. Is it alive?

Internet Links

The Grand Canyon: Window Through Time
http://www.gorp.com/gcjunkies/strata2.htm
A delightful interactive that allows you to click on each layer of the virtual Grand Canyon and learn about the time during which that layer was deposited.

Microbes Deep Inside the Earth
http://www.sciam.com/1096issue/1096onstott.html
A 1996 Scientific American article about recently-discovered microorganisms that dwell deep within the earth's crust and may reveal clues to the origin of life.

Fossil Cyanobacteria
http://www.ucmp.berkeley.edu/bacteria/cyanofr.html
The cyanobacteria have an extensive fossil record, dating back 3.5 billion years. This site from Cal Berkeley explores that record, with many beautiful micrographs.

Observed Instances of Speciation
http://www.talkorigins.org/gaqs/faq-speciation.html
Responding to creationists claims that "no one has seen species form" is this detailed compilation of observed instances of speciation.

For Further Reading

Cairns-Smith, A. G.: "The First Organisms," *Scientific American*, June 1985, pages 90–101. An interesting article arguing that clay, not primordial soup, provided the fundamental materials from which life arose.

Dawkins, R.: *The Blind Watchmaker*, W. W. Norton, New York, 1986. Modern evolutionary theory is put into clear and vivid language by this enthusiastic popular paperback writer.

Gibson, E. K. and others: "The Case for Relic Life on Mars," *Scientific American*, December 1997, pages 58–65. A meteorite found in Antarctica offers controversial hints of microbial life on Mars. This article reviews the controverisal clams.

Horgan, J.: "In the Beginning . . . ," *Scientific American*, February 1991, pages 116–25. A review of the very active controversy among scientists about how life first emerged on earth.

Joyce, G.: "RNA Evolution and the Origin of Life," *Nature*, vol. 338, March 1989, pages 217–24. A review of what scientists now think about the role of RNA in early life-forms, in light of new findings from molecular biology.

Monastersky, R.: "The Rise of Life on Earth," *National Geographic*, March 1998, pages 54–81. A comprehensive, beautifully illustrated review of what scientists currently believe about life's origins on earth.

5

Cell Structure

Concept Outline

5.1 All organisms are composed of cells.

Cells. A cell is a membrane-bound unit that contains DNA and cytoplasm.

The Cell Theory. All organisms are cells or aggregates of cells, descendants of the first cells.

Cells Are Small. The greater relative surface area of small cells enables more rapid communication between the cell interior and the environment.

5.2 Eukaryotic cells are far more complex than bacterial cells.

Bacteria Are Simple Cells. Bacterial cells are small and lack interior organization.

Eukaryotic Cells Have Complex Interiors. Eukaryotic cells are compartmentalized by endomembranes.

5.3 Take a tour of a eukaryotic cell.

The Nucleus: Information Center for the Cell. The nucleus of a eukaryotic cell isolates the cell's DNA.

The Endoplasmic Reticulum: Compartmentalizing the Cell. An extensive system of membranes subdivides the cell interior.

The Golgi Apparatus: Delivery System of the Cell. A system of membrane channels collects, modifies, packages, and distributes molecules within the cell.

Vesicles: Enzyme Storehouses. Sacs that contain enzymes digest or modify particles in the cell.

Ribosomes: Sites of Protein Synthesis. An RNA-protein complex directs the production of proteins.

Organelles That Contain DNA. Some organelles with very different functions contain their own DNA.

The Cytoskeleton: Interior Framework of the Cell. A network of protein fibers supports the shape of the cell and anchors organelles to fixed locations.

Cell Movement. Eukaryotic cells can move by moving cytoskeleton fibers.

5.4 Symbiosis played a key role in the origin of eukaryotes.

Endosymbiosis. Mitochondria and chloroplasts may have arisen from prokaryotes engulfed by eukaryotes.

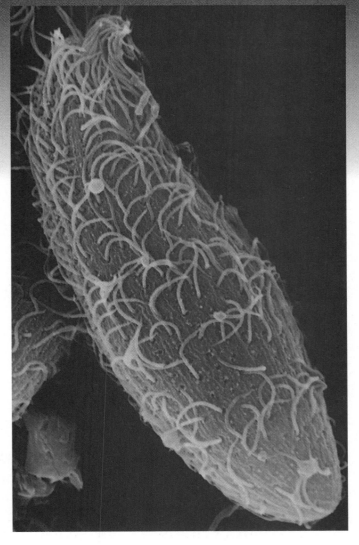

FIGURE. 5.1
The single-celled protist *Dileptus*. The hair-like projections that cover its surface are cilia, which it undulates to propel itself through the water (1000×).

All organisms are composed of cells. The gossamer wing of a butterfly is a thin sheet of cells, and so is the glistening corneal layer covering your eyes. The hamburger or tomato you eat is composed of cells, and its contents soon become part of your cells. Some organisms consist of a single cell too small to see with the unaided eye (figure 5.1), while others, like us, are composed of many cells. Cells are so much a part of life as we know it that we cannot imagine an organism that is not cellular in nature. In this chapter we will take a close look at the internal structure of cells. In the following chapters, we will focus on cells in action—on how they communicate with their environment, grow, and reproduce.

Cells

What is a typical cell like, and what would we find inside it? The general plan of cellular organization varies in the cells of different organisms, but despite these modifications, all cells resemble each other in certain fundamental ways (figure 5.2). Before we begin our detailed examination of cell structure, let's first summarize three major features all cells have in common: a plasma membrane, a nucleoid or nucleus, and cytoplasm.

The Plasma Membrane Surrounds the Cell

The **plasma membrane** encloses a cell and separating its contents from its surroundings. The plasma membrane is a phospholipid bilayer about 5 to 10 nanometers (5–10 billionths of a meter) thick with proteins embedded in it. Viewed in cross-section with the electron microscope, such membranes appear as two dark lines separated by a lighter area. This distinctive appearance arises from the tail-to-tail packing of the phospholipid molecules that make up the membrane (see figure 3.14). The proteins of a membrane have large hydrophobic domains, which associate with and become embedded in the phospholipid bilayer.

The proteins of the plasma membrane are in large part responsible for a cell's ability to interact with its environment. *Transport proteins* help molecules and ions move across the plasma membrane, either from the environment to the interior of the cell or vice versa. *Receptor proteins* induce changes within the cell when they come in contact with specific molecules in the environment, such as hormones. *Markers* identify the cell as a particular type. This is especially important in multicellular organisms, whose cells must be able to recognize each other as they form tissues.

We'll examine the structure and function of cell membranes more thoroughly in chapter 6.

The Central Portion of the Cell Contains the Genetic Material

Every cell contains DNA, the hereditary molecule. In **prokaryotes** (bacteria), most of the genetic material lies in a single circular molecule of DNA. It typically resides near the center of the cell in an area called the **nucleoid**, but this area is not segregated from the rest of the cell's interior by membranes. By contrast, the DNA of **eukaryotes** is contained in the **nucleus**, which is surrounded by two membranes. In both types of organisms, the DNA contains the genes that code for the proteins synthesized by the cell.

The Cytoplasm Comprises the Rest of the Cell's Interior

A semifluid matrix called the **cytoplasm** fills the interior of the cell, exclusive of the nucleus (nucleoid in prokaryotes) lying within it. The cytoplasm contains the chemical wealth of the cell: the sugars, amino acids, and proteins the cell uses to carry out its everyday activities. In eukaryotic cells, the cytoplasm also contains specialized membrane-bounded compartments called **organelles.**

A cell is a membrane-bounded unit that contains the DNA hereditary machinery and cytoplasm.

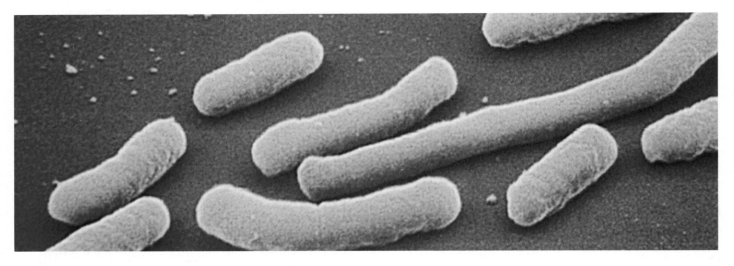

FIGURE 5.2
Bacterial cells. Some cells are very simple, such as these bacterial cells, while others are very specialized and complex. But all cells have three characteristics in common: a plasma membrane that surrounds the cell, a portion of the cell that contains the genetic material, and an interior matrix called the cytoplasm.

The Cell Theory

A general characteristic of cells is their microscopic size. While there are a few exceptions—the marine alga *Acetabularia* can be up to 5 centimeters long—a typical eukaryotic cell is 10 to 100 micrometers (10–100 millionths of a meter) in diameter (figure 5.3); most bacterial cells are only 1 to 10 micrometers in diameter.

Because cells are so small, no one observed them until microscopes were invented in the midseventeenth century. Robert Hooke first described cells in 1665, when he used a microscope he had built to examine a thin slice of cork, a nonliving tissue found in the bark of certain trees. Hooke observed a honeycomb of tiny, empty (since the cells were dead) compartments. He called the compartments in the cork *cellulae* (Latin, "small rooms"), and the term has come down to us as *cells*. The first living cells were observed a few years later by the Dutch naturalist Antonie van Leeuwenhoek, who called the tiny organisms that he observed "animalcules," meaning little animals. For another century and a half, however, biologists failed to recognize the importance of cells. In 1838, botanist Matthias Schleiden made a careful study of plant tissues and developed the first statement of the cell theory. He stated that all plants "are aggregates of fully individualized, independent, separate beings, namely the cells themselves." In 1839, Theodor Schwann reported that all animal tissues also consist of individual cells.

The **cell theory**, in its modern form, includes the following three principles:

1. All organisms are composed of one or more cells, and the life processes of metabolism and heredity occur within these cells.
2. Cells are the smallest living things, the basic units of organization of all organisms.
3. Cells arise only by division of a previously existing cell. Although life likely evolved spontaneously in the environment of the early earth, biologists have concluded that no additional cells are originating spontaneously at present. Rather, life on earth represents a continuous line of descent from those early cells.

All organisms are cells or aggregates of cells.

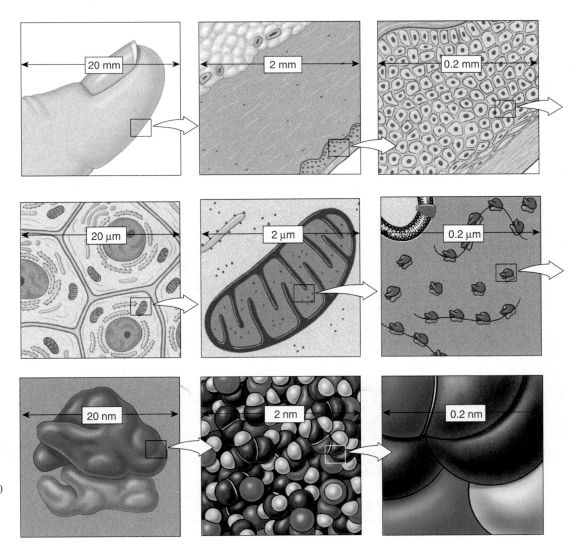

FIGURE 5.3
The size of cells and their contents. This diagram shows the size of human skin cells, organelles, and molecules. In general, the diameter of a human skin cell is 20 micrometers (μm), of a mitochondrion is 2 μm, of a ribosome is 20 nanometers (nm), of a protein molecule is 2 nm, and of an atom is 0.2 nm.

Cells Are Small

How many cells are big enough to see with the unaided eye? Other than egg cells, not many. Most are less than 50 micrometers in diameter, far smaller than the period at the end of this sentence. It would take a string of two thousand red blood cells to circle a pencil!

The Resolution Problem

How do we study cells if they are too small to see? The key is to understand *why* we can't see them. The reason we can't see such small objects is the limited resolution of the human eye. **Resolution** is defined as the minimum distance two points can be apart and still be distinguished as two separated points. When two objects are closer together than about 100 micrometers, the light reflected from each strikes the same "detector" cell at the rear of the eye. Only when the objects are farther than 100 micrometers apart will the light from each strike different cells, allowing your eye to resolve them as two objects rather than one.

Microscopes

One way to increase resolution is to increase magnification, so that small objects appear larger. Robert Hooke and Antonie van Leeuwenhoek were able to see small cells by magnifying their size, so that the cells appeared larger than the 100-micrometer limit imposed by the structure of the human eye. Hooke and van Leeuwenhoek accomplished this feat with **microscopes** that magnified images of cells by bending light through a glass lens. The size of the image that falls on the sheet of detector cells lining the back of your eye depends on how close the object is to your eye—the closer the object, the bigger the image. Your eye, however, is incapable of focusing comfortably on an object closer than about 25 centimeters, because the eye is limited by the size and thickness of its lens. Hooke and van Leeuwenhoek assisted the eye by interposing a glass lens between object and eye. The glass lens adds additional focusing power. Because the glass lens makes the object appear closer, the image on the back of the eye is bigger than it would be without the lens.

Modern light microscopes use two magnifying lenses (and a variety of correcting lenses) that act like back-to-back eyes. The first lens focuses the image of the object on the second lens, which magnifies it again and focuses it on the back of the eye. Microscopes that magnify in stages using several lenses are called **compound microscopes.** The finest structures they can resolve are about 200 nanometers thick. An image from a compound microscope is shown in figure 5.4*a*.

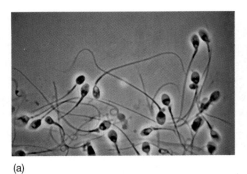
(a)

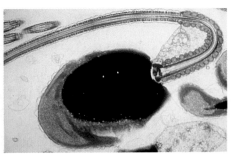
(b)

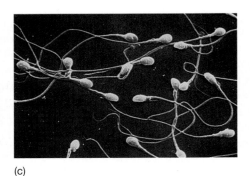
(c)

FIGURE 5.4
Human sperm cells viewed with three different microscopes. (a) Image of sperm taken with a light microscope. (b) Transmission electron micrograph of a sperm cell. (c) Scanning electron micrograph of sperm cells.

Increasing Resolution

Light microscopes, even compound ones, are not powerful enough to resolve many structures within cells. For example, a membrane is only 5 nanometers thick. Why not just add another magnifying stage to the microscope and so increase its resolving power? This approach doesn't work, because when two objects are closer than a few hundred nanometers, the light beams reflecting from the two images start to overlap. The only way two light beams can get closer together and still be resolved is if their "wavelengths" are shorter.

One way to avoid overlap is by using a beam of electrons rather than a beam of light. Electrons have a much shorter wavelength, and a microscope employing electron beams has 1000 times the resolving power of a light microscope. **Transmission electron microscopes,** so called because the electrons used to visualize the specimens are transmitted through the material, are capable of resolving objects only 0.2 nanometer apart—just twice the diameter of a hydrogen atom! Figure 5.4*b* shows a transmission electron micrograph.

A second kind of electron microscope, the **scanning electron microscope,** beams the electrons onto the surface of the specimen from a fine probe that passes rapidly back and forth. The electrons reflected back from the surface of the specimen, together with other electrons that the specimen itself emits as a result of the bombardment, are amplified and transmitted to a television screen, where the image can be viewed and photographed. Scanning electron microscopy yields striking three-dimensional images and has improved our understanding of many biological and physical phenomena (figure 5.4*c*).

Why Aren't Cells Larger?

Most cells are not large for practical reasons. The most important of these is communication. The different regions of a cell need to communicate with one another in order for the cell as a whole to function effectively. Proteins and organelles are being synthesized, and materials are continually entering and leaving the cell. All of these processes involve the diffusion of substances at some point, and the larger a cell is, the longer it takes for substances to diffuse from the plasma membrane to the center of the cell. For this reason, an organism made up of many relatively small cells has an advantage over one composed of fewer, larger cells.

The advantage of small cell size is readily visualized in terms of the **surface area-to-volume ratio** (figure 5.5). As a cell's size increases, its volume increases much more rapidly than its surface area. For a spherical cell, the increase in surface area is equal to the square of the increase in diameter, while the increase in volume is equal to the cube of the increase in diameter. Thus, if two cells differ by a factor of 10 cm in diameter, the larger cell will have 10^2, or 100 times, the surface area, but 10^3, or 1000 times, the volume, of the smaller cell (figure 5.6). A cell's surface provides its only opportunity for interaction with the environment, since all substances enter and exit a cell via the plasma membrane. This membrane plays a key role in controlling cell function, and because small cells have more surface area per unit of volume than large ones, the control is more effective when cells are relatively small.

Multicellular organisms usually consist of many small cells rather than a few large ones because small cells function more effectively. They have a greater relative surface area, enabling more rapid communication between the center of the cell and the environment.

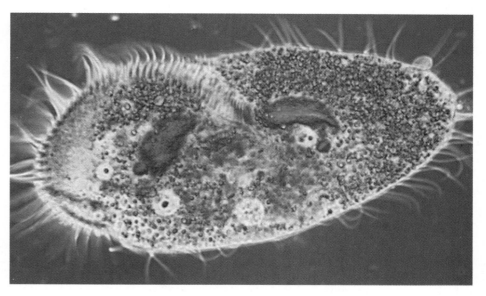

FIGURE 5.5
The advantages of small cell size. This *Paramecium* is smaller than the dot at the end of this sentence. This single-celled organism does not grow as big as a watermelon because of the limitations imposed by the surface-to-volume ratio; the smaller the cell, the more surface area per unit of volume, and the more efficient the cellular processes.

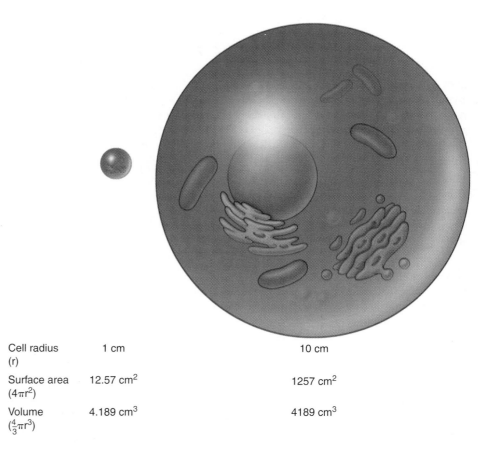

Cell radius (r)	1 cm	10 cm
Surface area ($4\pi r^2$)	12.57 cm^2	1257 cm^2
Volume ($\frac{4}{3}\pi r^3$)	4.189 cm^3	4189 cm^3

FIGURE 5.6
Surface area-to-volume ratio. As a cell gets larger, its volume increases at a faster rate than its surface area. If the cell radius increases by 10 times, the surface area increases by 100 times, but the volume increases by 1000 times. A cell's surface area must be large enough to meet the needs of its volume.

Bacteria Are Simple Cells

Prokaryotes, the bacteria, are the simplest organisms. Scientists have identified over 2500 species, but doubtless many times that number actually exist. Although these species are diverse in form (figure 5.7), their organization is fundamentally similar: they are small cells surrounded by a membrane and encased within a rigid cell wall, with no distinct interior compartments (figure 5.8). Sometimes bacterial cells adhere in chains or masses, but the individual cells function independently of one another.

Strong Cell Walls

Most bacteria are encased by a strong **cell wall** composed of *peptidoglycan*, which consists of a carbohydrate matrix (polymers of sugars) that is cross-linked by short polypeptide units. No eukaryotes possess cell walls with this type of chemical composition. Bacteria may be classified into two types based on differences in their cell walls detected by the Gram staining procedure. The name refers to the Danish microbiologist Hans Christian Gram, who developed the procedure to detect the presence of certain disease-causing bacteria. **Gram-positive** bacteria have a thick, single-layered cell wall that retains the Gram stain within the cell, causing the stained cells to appear purple under a microscope. More complex cell walls have evolved in other groups of bacteria. In them, the wall is multilayered and does not retain the Gram stain; such bacteria are characterized as **gram-negative.** The susceptibility of bacteria to antibiotics often depends on the structure of their cell walls.

Long chains of sugars called polysaccharides cover the cell walls of many bacteria. They enable a bacterium to adhere to teeth, skin, food—practically any surface that will support their growth. Many disease-causing bacteria secrete a jelly-like protective capsule of polysaccharide around the cell.

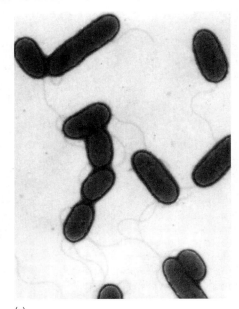

(a)

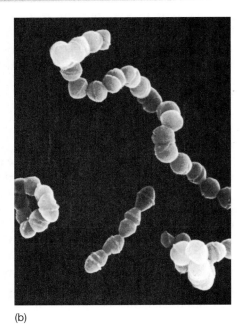

(b)

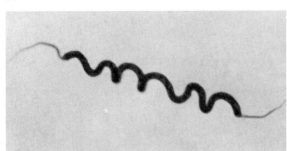

(c)

FIGURE 5.7
An assortment of bacteria.
The cells of *Pseudomonas* (a) are peanut-shaped, with daughter cells often adhering in short clusters. *Streptococcus* (b) has spherical cells that link together in long chains. *Spirilla* (34,000×) (c) has long, twisted shapes with terminal flagella (500×).

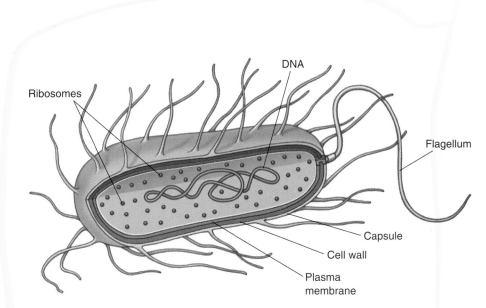

FIGURE 5.8
Structure of a bacterial cell. Generalized cell organization of a bacterium.

Simple Interior Organization

If you were to look at an electron micrograph of a bacterial cell, you would be struck by the cell's simple organization. There are few, if any, internal compartments, and while they contain simple organelles like ribosomes, most have no membrane-bounded organelles, the kinds of distinct structures so characteristic of eukaryotic cells. Nor do bacteria have a true nucleus. The entire cytoplasm of a bacterial cell is one unit with no internal support structure. Consequently, the strength of the cell comes primarily from its rigid wall (see figure 5.8).

The plasma membrane of a bacterial cell carries out some of the functions organelles perform in eukaryotic cells. When a bacterial cell divides, for example, the bacterial chromosome, a simple circle of DNA, replicates before the cell divides. The two DNA molecules that result from the replication attach to the plasma membrane at different points, ensuring that each daughter cell will contain one of the identical units of DNA. Moreover, some photosynthetic bacteria (e.g., *Cyanobacteria* or *Prochloron*, figure 5.9) have an extensively folded plasma membrane, with the folds extending into the cell's interior. These membrane folds contain the bacterial pigments connected with photosynthesis.

Because a bacterial cell contains no membrane-bounded organelles, the DNA, enzymes, and other cytoplasmic constituents have access to all parts of the cell. Reactions are not compartmentalized as they are in eukaryotic cells, and the whole bacterium operates as a single unit.

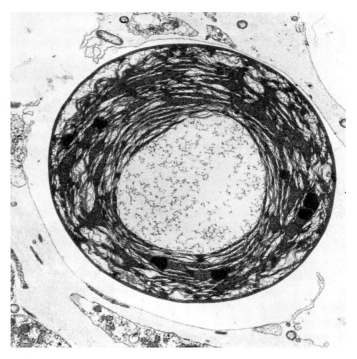

FIGURE 5.9
Electron micrograph of a photosynthetic bacterial cell. Extensive folded photosynthetic membranes are visible in this *Prochloron* cell. The single, circular DNA molecule is located in the clear area in the central region of the cell.

Rotating Flagella

Some bacteria use a flagellum (plural, flagella) to move about. **Flagella** are long, thread-like structures protruding from the surface of a cell that are used in locomotion and feeding. Bacterial flagella are protein fibers that extend out from a bacterial cell. There may be one or more per cell, or none, depending on the species. Bacteria can swim at speeds up to 20 cell diameters per second by rotating their flagella like screws (figure 5.10). A "motor" unique to bacteria that is embedded within their cell walls and membranes powers the rotation. Only a few eukaryotic cells have structures that truly rotate.

Bacteria are small cells that lack interior organization. They are encased by an exterior wall composed of carbohydrates cross-linked by short polypeptides, and some are propelled by rotating flagella.

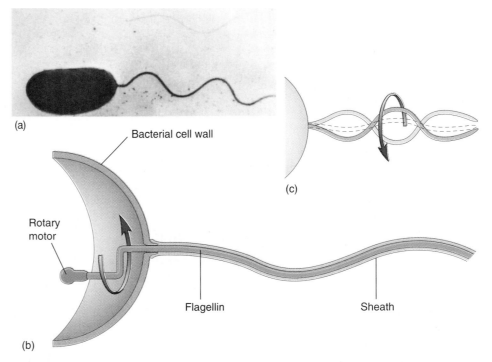

FIGURE 5.10
Bacteria swim by rotating their flagella. (a) The photograph is of *Vibrio cholerae*, the microbe that causes the serious disease cholera. The unsheathed core visible at the top of the photograph is composed of a single crystal of the protein flagellin. (b) In intact flagella, the core is surrounded by a flexible sheath. Imagine that you are standing inside the *Vibrio* cell, turning the flagellum like a crank. (c) You would create a spiral wave that travels down the flagellum, just as if you were turning a wire within a flexible tube. The bacterium creates this kind of rotary motion when it swims.

Eukaryotic Cells Have Complex Interiors

Eukaryotic cells (figures 5.11 and 5.12) are far more complex than prokaryotic cells. The hallmark of the eukaryotic cell is compartmentalization. The interiors of eukaryotic cells contain numerous **organelles,** membrane-bounded structures that close off compartments within which multiple biochemical processes can proceed simultaneously and independently. Plant cells often have a large membrane-bounded sac called a **central vacuole,** which stores proteins, pigments, and waste materials. Both plant and animal cells contain **vesicles,** smaller sacs that store and transport a variety of materials. Inside the nucleus, the DNA is wound tightly around pro-

FIGURE 5.11

Structure of an animal cell. (a) A generalized diagram of an animal cell. (b) Micrograph of a rat pancreas cell with drawings detailing organelles.

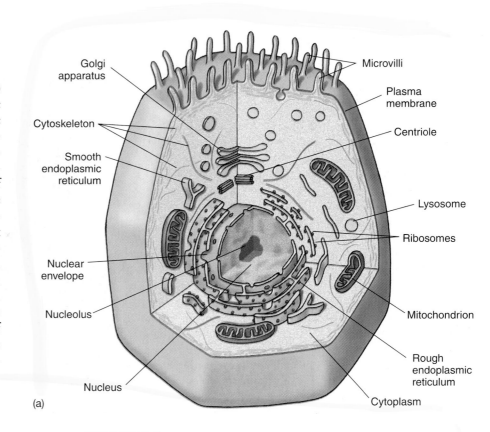

(a)

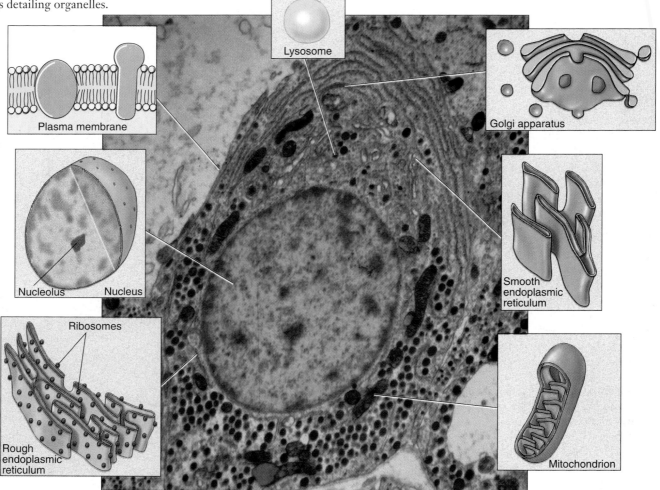

(b)

teins and packaged into compact units called **chromosomes.** All eukaryotic cells are supported by an internal protein scaffold, the **cytoskeleton.** While the cells of animals and some protists lack cell walls, the cells of fungi, plants, and many protists have strong **cell walls** composed of cellulose or chitin fibers embedded in a matrix of other polysaccharides and proteins. This composition is very different from the peptidoglycan that makes up bacterial cell walls. Let's now examine the structure and function of the internal components of eukaryotic cells in more detail.

Eukaryotic cells contain membrane-bounded organelles that carry out specialized functions.

FIGURE 5.12
Structure of a plant cell. A generalized illustration (a) and micrograph (b) of a plant cell. Most mature plant cells contain large central vacuoles, which occupy a major portion of the internal volume of the cell.

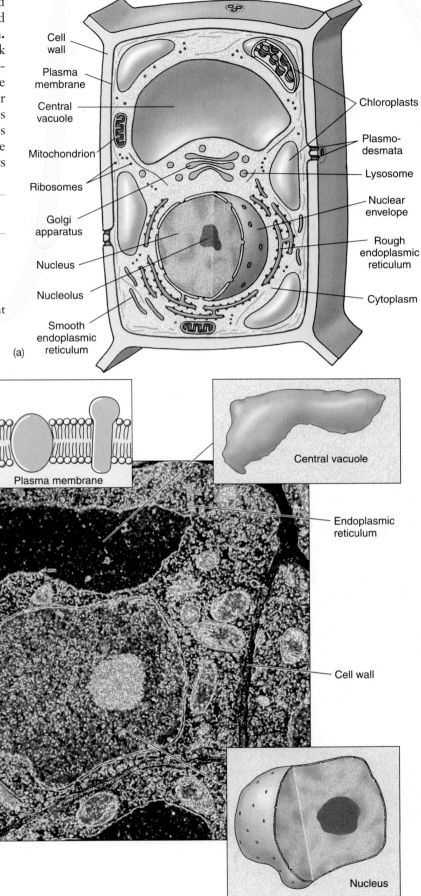

The Nucleus: Information Center for the Cell

The largest and most easily seen organelle within a eukaryotic cell is the **nucleus** (Latin, for kernel or nut), first described by the English botanist Robert Brown in 1831. Nuclei are roughly spherical in shape and, in animal cells, they are typically located in the central region of the cell (figure 5.13). In some cells, a network of fine cytoplasmic filaments seems to cradle the nucleus in this position. The nucleus is the repository of the genetic information that directs all of the activities of a living eukaryotic cell. Most eukaryotic cells possess a single nucleus, although the cells of fungi and some other groups may have several to many nuclei. Mammalian erythrocytes (red blood cells) lose their nuclei when they mature. Many nuclei exhibit a dark-staining zone called the nucleolus, which is a region where intensive synthesis of ribosomal RNA is taking place.

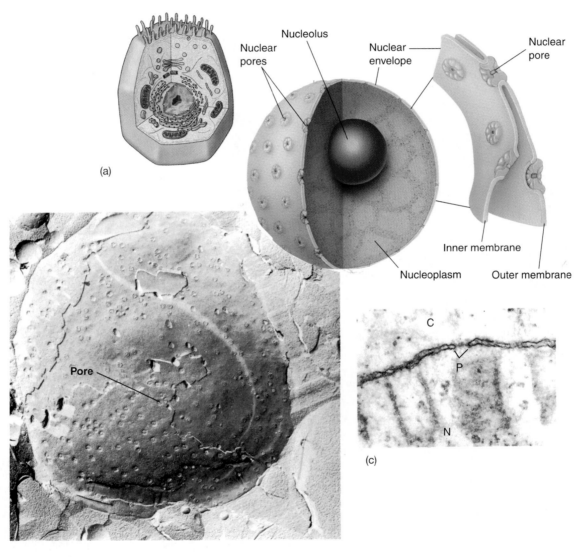

FIGURE 5.13

The nucleus. (a) The nucleus is composed of a double membrane, called a nuclear envelope, enclosing a fluid-filled interior containing the chromosomes. In cross-section, the individual nuclear pores are seen to extend through the two membrane layers of the envelope; the dark material within the pore is protein, which acts to control access through the pore. (b) A freeze-fracture scanning electron micrograph of a cell nucleus showing nuclear pores (9500×). (c) A transmission electron micrograph of the nuclear membrane showing nuclear pores (c = cytoplasm; p = pore; n = nucleus).

The Nuclear Envelope: Getting In and Out

The surface of the nucleus is bounded by *two* phospholipid bilayer membranes, which together make up the **nuclear envelope** (see figure 5.13). The outer membrane of the nuclear envelope is continuous with the cytoplasm's interior membrane system, called the endoplasmic reticulum. Scattered over the surface of the nuclear envelope, like craters on the moon, are shallow depressions called **nuclear pores.** These pores form 50 to 80 nanometers apart at locations where the two membrane layers of the nuclear envelope pinch together. Rather than being empty, nuclear pores are filled with proteins that act as molecular channels, permitting certain molecules to pass into and out of the nucleus. Passage is restricted primarily to two kinds of molecules: (1) proteins moving into the nucleus to be incorporated into nuclear structures or to catalyze nuclear activities; and (2) RNA and protein-RNA complexes formed in the nucleus and exported to the cytoplasm.

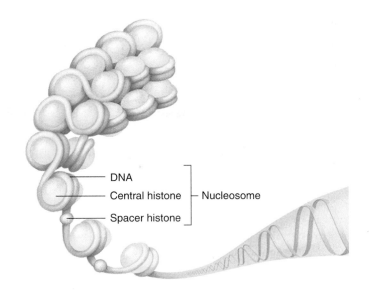

FIGURE 5.14
Nucleosomes. Each nucleosome is a region in which the DNA is wrapped tightly around a cluster of histone proteins.

The Chromosomes: Packaging the DNA

In both bacteria and eukaryotes, DNA contains the hereditary information specifying cell structure and function. However, unlike the DNA of bacteria, the DNA of eukaryotes is divided into several linear **chromosomes.** Except when a cell is dividing, its chromosomes are fully extended into thread-like strands, called **chromatin,** of DNA complexed with protein. This open arrangement allows proteins to attach to specific nucleotide sequences along the DNA. Without this access, DNA could not direct the day-to-day activities of the cell. The chromosomes are associated with packaging proteins called **histones.** When a cell prepares to divide, the DNA coils up around the histones into a highly condensed form. In the initial stages of this condensation, units of histone can be seen with DNA wrapped around like a sash. Called **nucleosomes,** these initial aggregations resemble beads on a string (figure 5.14). Coiling continues until the DNA is in a compact mass. Under a light microscope, these fully condensed chromosomes are readily seen in dividing cells as densely staining rods (figure 5.15). After cell division, eukaryotic chromosomes uncoil and can no longer be individually distinguished with a light microscope. Uncoiling the chromosomes into a more extended form permits RNA polymerase, an enzyme that makes RNA copies of DNA, to gain access to the DNA molecule. Only by means of these RNA copies can the hereditary information in the DNA be used to direct the synthesis of proteins.

The nucleus of a eukaryotic cell contains the cell's hereditary apparatus and isolates it from the rest of the cell. A distinctive feature of eukaryotes is the organization of their DNA into complex chromosomes.

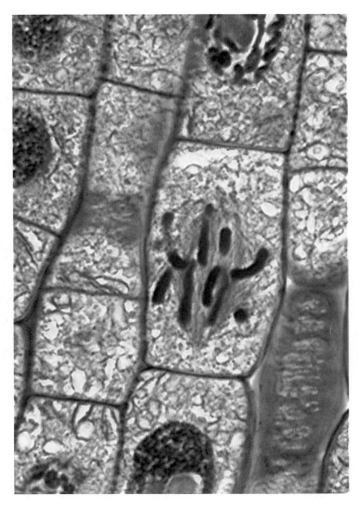

FIGURE 5.15
Eukaryotic chromosomes. These condensed chromosomes within an onion root tip are visible under the light microscope (500×).

The Endoplasmic Reticulum: Compartmentalizing the Cell

As seen through a light microscope, the interiors of eukaryotic cells exhibit various organelles (table 5.1) embedded within a relatively featureless matrix. An electron microscope, however, reveals a striking difference—the cell interiors appear to be packed with membranes. So thin that they are invisible under the lower resolving power of the light microscope, this **endomembrane system** fills the cell, dividing it into compartments, channeling the passage of molecules through the interior of the cell, and providing surfaces for protein and lipid synthesis. The presence of these membranes in eukaryotic cells constitutes one of the most fundamental distinctions between eukaryotes and prokaryotes.

The largest of the internal membranes is called the **endoplasmic reticulum (ER).** The term *endoplasmic* means "within the cytoplasm," and the term *reticulum* is Latin for "a little net." Like the plasma membrane, the ER is composed of a lipid bilayer embedded with proteins. It weaves in sheets through the interior of the cell, creating a series of channels and interconnections between its folds (figure 5.16).

Rough ER: Manufacturing Proteins for Export

The ER surface regions that are devoted to protein synthesis are heavily studded with **ribosomes,** large molecular aggregates of protein and ribonucleic acid (RNA) that translate RNA copies of genes into protein (We will examine ribosomes in detail later in this chapter). Through the electron microscope, these ribosome-rich regions of the ER appear pebbly, like the surface of sandpaper, and they are therefore called **rough ER** (see figure 5.16).

The proteins synthesized on the surface of the rough ER may be used in the cell or exported from the cell. These proteins contain special amino acid sequences called **signal sequences.** As a new protein is made by a free ribosome (one not attached to a membrane), the signal sequence of the growing polypeptide attaches to a recognition factor that carries the ribosome and its partially completed protein to a "docking site" on the surface of the ER. As the protein is assembled it passes through the ER membrane into the interior ER compartment and moves to a vesicle-forming system called the Golgi apparatus (figure 5.17). The protein then travels within vesicles to the inner surface of the plasma membrane, where it is released to the outside of the cell.

Table 5.1	Eukaryotic Cell Structures and Their Functions	
Structure	**Description**	**Function**
Cell wall	Outer layer of cellulose or chitin; or absent	Protection; support
Cytoskeleton	Network of protein filaments	Structural support; cell movement
Flagella (cilia)	Cellular extensions with 9 + 2 arrangement of pairs of microtubules	Motility or moving fluids over surfaces
Plasma membrane	Lipid bilayer with embedded proteins	Regulates what passes into and out of cell; cell-to-cell recognition
Endoplasmic reticulum	Network of internal membranes	Forms compartments and vesicles; participates in protein and lipid synthesis
Nucleus	Structure (usually spherical) surrounded by double membrane that contains chromosomes	Control center of cell; directs protein synthesis and cell reproduction
Golgi apparatus	Stacks of flattened vesicles	Packages proteins for export from cell; forms secretory vesicles
Lysosomes	Vesicles derived from Golgi apparatus that contain hydrolytic digestive enzymes	Digest worn-out organelles and cell debris; play role in cell death
Microbodies	Vesicles formed from incorporation of lipids and proteins containing oxidative and other enzymes	Isolate particular chemical activities from rest of cell
Mitochondria	Bacteria-like elements with double membrane	"Power plants" of the cell; sites of oxidative metabolism
Chloroplasts	Bacteria-like elements with membranes containing chlorophyll, a photosynthetic pigment	Sites of photosynthesis in plant cells
Chromosomes	Long threads of DNA that form a complex with protein	Contain hereditary information
Nucleolus	Site of genes for rRNA synthesis	Assembles ribosomes
Ribosomes	Small, complex assemblies of protein and RNA, often bound to endoplasmic reticulum	Sites of protein synthesis

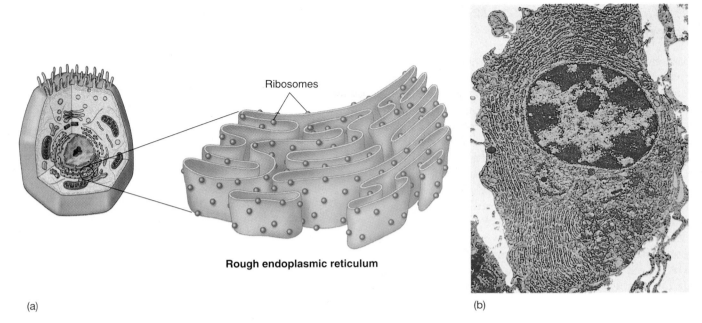

Rough endoplasmic reticulum

(a)

(b)

FIGURE 5.16

Rough endoplasmic reticulum. (a) In the drawing you can see that the ribosomes (often grouped into "polysomes") are associated with only one side of the rough ER; the other side is the boundary of a separate compartment within the cell into which the ribosomes extrude newly made proteins destined for secretion. (b) The electron micrograph (10,000×) is of a rat liver cell, rich in ER-associated ribosomes.

Smooth ER: Organizing Internal Activities

Regions of the ER with relatively few bound ribosomes are referred to as **smooth ER.** The membranes of the smooth ER contain many embedded enzymes, most of them active only when associated with a membrane. Enzymes anchored within the ER, for example, catalyze the synthesis of a variety of carbohydrates and lipids. In cells that carry out extensive lipid synthesis, such as those in the testes, intestine, and brain, smooth ER is particularly abundant. In the liver, the enzymes of the smooth ER are involved in the detoxification of drugs including amphetamines, morphine, codeine, and phenobarbital.

Some vesicles form at the plasma membrane by budding inward, a process called endocytosis, then move into the cytoplasm and fuse with the smooth endoplasmic reticulum.

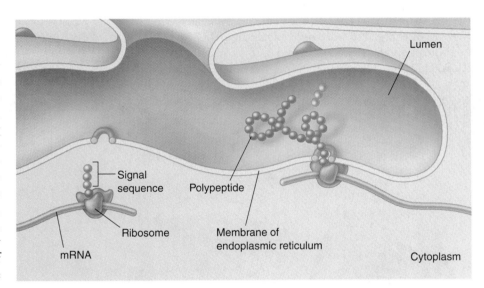

FIGURE 5.17

Signal sequences direct proteins to their destinations in the cell. In this example, a sequence of hydrophobic amino acids (the signal sequence) on a secretory protein attaches them (and the ribosomes making them) to the membrane of the ER. As the protein is synthesized, it passes into the lumen (internal chamber) of the ER. The signal sequence is clipped off after the leading edge of the protein enters the lumen.

The endoplasmic reticulum (ER) is an extensive system of folded membranes that divides the interior of eukaryotic cells into compartments and channels. Rough ER synthesizes proteins for export, while smooth ER organizes the synthesis of lipids and other biosynthetic activities.

The Golgi Apparatus: Delivery System of the Cell

At various locations within the endomembrane system, flattened stacks of membranes called **Golgi bodies** occur, often interconnected with one another. These structures are named for Camillo Golgi, the nineteenth-century Italian physician who first called attention to them. The numbers of Golgi bodies a cell contains ranges from 1 or a few in protists, to 20 or more in animal cells and several hundred in plant cells. They are especially abundant in glandular cells, which manufacture and secrete substances. Collectively the Golgi bodies are referred to as the **Golgi apparatus** (figure 5.18).

The Golgi apparatus functions in the collection, packaging, and distribution of molecules synthesized at one place in the cell and utilized at another location in the cell. A Golgi body has a front and a back, with distinctly different membrane compositions at the opposite ends. The front, or receiving end, is called the *cis* face, and is usually located near ER. Materials move to the *cis* face in transport vesicles that bud off of the ER. These vesicles fuse with the *cis* face, emptying their contents into the interior, or lumen, of the Golgi apparatus. These ER-synthesized molecules then pass through the channels of the Golgi apparatus until they reach the back, or discharging end, called the *trans* face, where they are discharged in secretory vesicles (figure 5.19).

Proteins and lipids manufactured on the rough and smooth ER membranes are transported into the Golgi apparatus and modified as they pass through it. The most common alteration is the addition or modification of short sugar chains, forming a *glycoprotein* when a polysaccharide is complexed to a protein, and *glycolipids*, consisting of a polysaccharide bound to a lipid. In many instances, enzymes in the Golgi apparatus modify existing glycoproteins and glycolipids made in the ER by cleaving a sugar from their sugar chain or modifying one or more of the sugars.

The newly formed or altered glycoproteins and glycolipids collect at the ends of the Golgi bodies, in flattened stacked membrane folds called **cisternae** (Latin, "collecting vessels"). Periodically, the membranes of the cisternae push together, pinching off small, membrane-bounded secretory vesicles containing the glycoprotein and glycolipid molecules. These vesicles then move to other locations in the cell, distributing the newly synthesized molecules to their appropriate destinations. **Liposomes** are synthetically manufactured vesicles that contain any variety of desirable substances (such as drugs), and can be injected into the body. Since the membrane of liposomes is similar to plasma and organellar membranes, these liposomes serve as an effective and natural delivery system to cells and may prove to be of great therapeutic value.

> The Golgi apparatus is the delivery system of the eukaryotic cell. It collects, packages, modifies, and distributes molecules that are synthesized at one location within the cell and used at another.

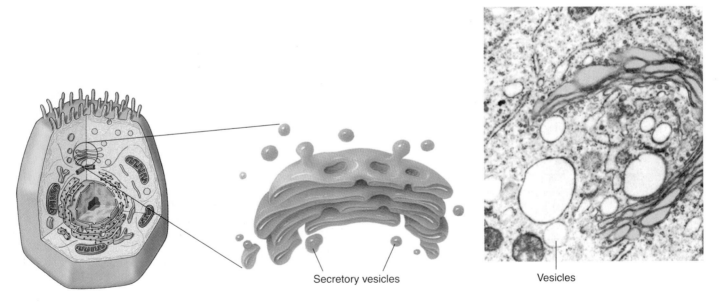

Secretory vesicles

Vesicles

FIGURE 5.18

The Golgi apparatus. The Golgi apparatus is a smooth, concave membranous structure located near the middle of the cell. It receives material for processing on one surface and sends the material packaged in vesicles off the other. The substance in a vesicle could be for export out of the cell or for distribution to another region within the same cell.

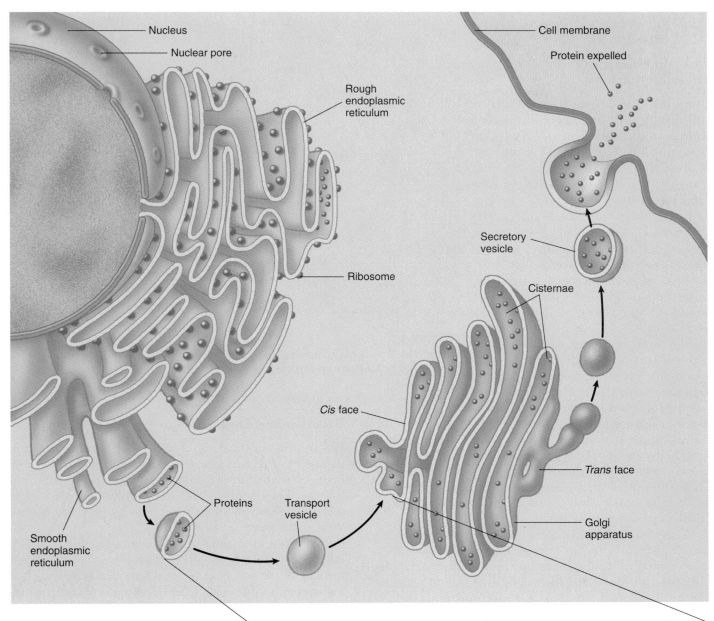

FIGURE 5.19

How proteins are transported within the cell. Proteins are manufactured at the ribosome and then released into the internal compartments of the rough ER. If the newly synthesized proteins are to be used at a distant location in or outside of the cell, they are transported within vesicles that bud off the rough ER and travel to the *cis* face, or receiving end, of the Golgi apparatus. There they are modified and packaged into secretory vesicles. The secretory vesicles then migrate from the *trans* face, or discharging end, of the Golgi apparatus to other locations in the cell, or they fuse with the cell membrane, releasing their contents to the external cellular environment.

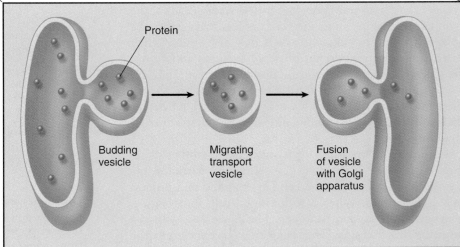

Vesicles: Enzyme Storehouses

Lysosomes: Intracellular Digestion Centers

Lysosomes, membrane-bounded digestive vesicles, are also components of the endomembrane system and seem to arise from the Golgi apparatus. They contain in a concentrated mix the digestive enzymes of the cell, which catalyze the rapid breakdown of proteins, nucleic acids, lipids, and carbohydrates. Throughout the lives of eukaryotic cells, lysosomal enzymes break down old organelles, recycling their component molecules and making room for newly formed organelles. For example, mitochondria are replaced in some tissues every 10 days.

The digestive enzymes in lysosomes function best in an acidic environment. Lysosomes actively engaged in digestion keep their battery of hydrolytic enzymes (enzymes that catalyze the hydrolysis of molecules) fully active by pumping protons into their interiors and thereby maintaining a low internal pH. Lysosomes that are not functioning actively do not maintain an acidic internal pH and are called *primary lysosomes.* When a primary lysosome fuses with a food vacuole or other organelle, its pH falls and its arsenal of hydrolytic enzymes is activated; it is then called a *secondary lysosome.*

What prevents lysosomes from digesting *themselves?* Although the answer is not entirely clear, the process must require energy, because eukaryotic cells that are metabolically inactive die as the hydrolytic enzymes of primary lysosomes digest the lysosomal membranes from within. When these membranes disintegrate, the digestive enzymes of the lysosomes pour out into the cytoplasm of the cell and destroy it. Thus, the very process that repairs the ravages of time in eukaryotic cells may also lead to their destruction. Bacteria do not possess lysosomes and do not die when they are metabolically inactive; instead, they are able to remain quiescent until altered conditions restore their metabolic activity. This property greatly heightens their ability to persist under unfavorable environmental conditions.

In addition to breaking down organelles and other structures within cells, lysosomes also eliminate particles (including other cells) that the cell has engulfed in a process called *phagocytosis.* When a white blood cell, for example, phagocytizes a passing pathogen, lysosomes fuse with the resulting "food vesicle," releasing their enzymes into the vesicle and degrading the material within (figure 5.20).

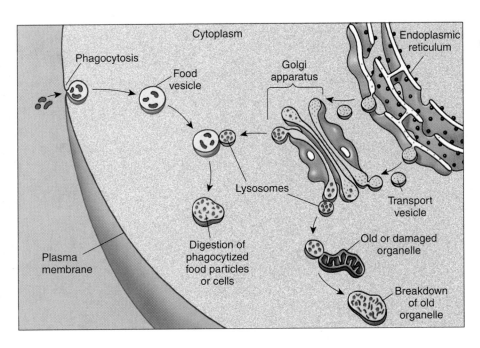

FIGURE 5.20

Lysosomes. Lysosomes contain hydrolytic enzymes that digest particles or cells taken into the cell by phagocytosis and break down old organelles.

Peroxisomes: Detoxifiers of Hydrogen Peroxide

Eukaryotic cells contain a variety of enzyme-bearing, membrane-enclosed vesicles called **microbodies.** Microbodies are found in the cells of plants, animals, fungi, and protists. The distribution of enzymes into microbodies is one of the principal ways in which eukaryotic cells organize their metabolism.

While lysosomes bud from the endomembrane system, microbodies grow by incorporating lipids and protein, then dividing. Plant cells have a special type of microbody called a **glyoxysome** that contains enzymes that convert fats into carbohydrates. Another type of microbody, a **peroxisome,** contains enzymes that catalyze the removal of electrons and associated hydrogen atoms. If these oxidative enzymes were not isolated within microbodies, they would tend to short-circuit the metabolism of the cytoplasm, which often involves adding hydrogen atoms to oxygen. The name *peroxisome* refers to the hydrogen peroxide produced as a by-product of the activities of the oxidative enzymes in the microbody. Hydrogen peroxide is dangerous to cells because of its violent chemical reactivity. However, peroxisomes also contain the enzyme catalase, which breaks down hydrogen peroxide into harmless water and oxygen.

Lysosomes and peroxisomes are vesicles that contain digestive and detoxifying enzymes. The isolation of these enzymes in vesicles protects the rest of the cell from inappropriate digestive activity.

Ribosomes: Sites of Protein Synthesis

Although the DNA in a cell's nucleus encodes the amino acid sequence of each protein in the cell, the proteins are not assembled there. A simple experiment demonstrates this: if a brief pulse of radioactive amino acid is administered to a cell, the radioactivity shows up associated with newly made protein, not in the nucleus, but in the cytoplasm. When investigators first carried out these experiments, they found that protein synthesis was associated with large RNA-protein complexes they called **ribosomes**. To make proteins, the ribosome attaches to the messenger RNA (mRNA) transcribed from a gene and uses the information to direct the synthesis of a protein.

Ribosomes are made up of several molecules of a special form of RNA called ribosomal RNA, or rRNA, bound within a complex of several dozen different proteins. Ribosomes are among the most complex molecular assemblies found in cells. Each ribosome is composed of two subunits (figure 5.21). The subunits join to form a functional ribosome only when they attach to messenger RNA in the cytoplasm. Bacterial ribosomes are smaller than eukaryotic ribosomes.

A bacterial cell typically has a few thousand ribosomes, while a metabolically active eukaryotic cell, such as a human liver cell, contains several million. Proteins that function in the cytoplasm are made by free ribosomes suspended in the cytoplasm, while proteins bound within membranes or destined for export from the cell are assembled by ribosomes bound to rough ER.

The Nucleolus Manufactures Ribosomal Subunits

When cells are synthesizing a large number of proteins, they must first make a large number of ribosomes. To facilitate this, many hundreds of copies of the portion of the DNA encoding the rRNA are clustered together on the chromosome. By transcribing RNA molecules from this cluster, the cell rapidly generates large numbers of the molecules needed to produce ribosomes.

At any given moment, many rRNA molecules dangle from the chromosome at the sites of these clusters of genes that encode rRNA. Proteins that will later form part of the ribosome complex bind to the dangling rRNA molecules. These areas where ribosomes are being assembled are easily visible within the nucleus as one or more dark-staining regions, called **nucleoli** (singular, **nucleolus;** figure 5.22). Nucleoli can be seen under the light microscope even when the chromosomes are extended, unlike the rest of the chromosomes, which are visible only when condensed.

Ribosomes provide a framework for protein synthesis in the cytoplasm.

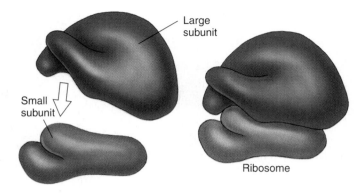

FIGURE 5.21
A ribosome. Ribosomes consist of a large and a small subunit composed of rRNA and protein. The individual subunits are synthesized in the nucleolus and then move through the nuclear pores to the cytoplasm, where they assemble. Ribosomes serve as sites of protein synthesis.

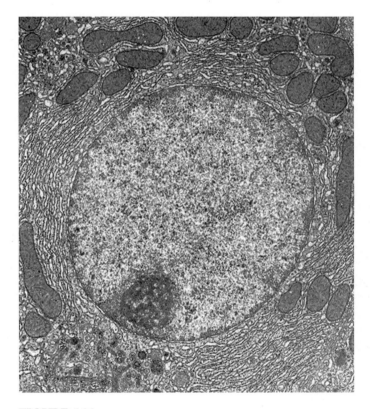

FIGURE 5.22
The nucleolus. This is the interior of a rat liver cell, magnified about 6000 times. A single large nucleus occupies the center of the micrograph. The electron-dense area in the lower left of the nucleus is the nucleolus, the area where the major components of the ribosomes are produced. Partly formed ribosomes can be seen around the nucleolus.

Organelles That Contain DNA

Among the most interesting cell organelles are those in addition to the nucleus that contain DNA.

Mitochondria: The Cell's Chemical Furnaces

Mitochondria (singular, *mitochondrion*) are typically tubular or sausage-shaped organelles about the size of bacteria and found in all types of eukaryotic cells (figure 5.23). Mitochondria are bounded by two membranes: a smooth outer membrane and an inner one folded into numerous contiguous layers called **cristae** (singular, *crista*). The cristae partition the mitochondrion into two compartments: a **matrix,** lying inside the inner membrane; and an outer compartment, or **intermembrane space,** lying between the two mitochondrial membranes. On the surface of the inner membrane, and also submerged within it, are proteins that carry out oxidative metabolism, the oxygen-requiring process by which energy in macromolecules is stored in ATP.

Mitochondria have their own DNA; this DNA contains several genes that produce proteins essential to mitochondria's role as centers of oxidative metabolism. All of these genes are copied into RNA and used to make proteins within the mitochondrion. In this process, the mitochondria employ small RNA molecules and ribosomal components that the mitochondrial DNA also encodes. However, most of the genes that produce the enzymes used in oxidative metabolism are located in the nucleus.

A eukaryotic cell does not produce brand new mitochondria each time the cell divides. Instead, the mitochondria themselves divide in two, doubling in number, and these are partitioned between the new cells. Most of the components required for mitochondrial division are encoded by genes in the nucleus and translated into proteins by cytoplasmic ribosomes. Mitochondrial replication is, therefore, impossible without nuclear participation, and mitochondria thus cannot be grown in a cell-free culture.

Chloroplasts: Where Photosynthesis Takes Place

The photosynthetic cells of plants and other organisms that carry out photosynthesis (using light energy to manufacture organic molecules) typically contain from one to several hundred **chloroplasts.** Chloroplasts bestow an obvious advantage on the organisms that possess them: they can manufacture their own food. The number of chloroplasts in a cell depends on the organism or, in the case of multicellular photosynthetic organ-

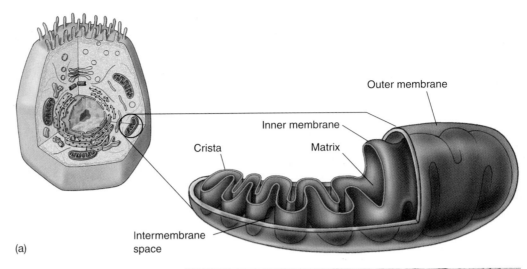

(a)

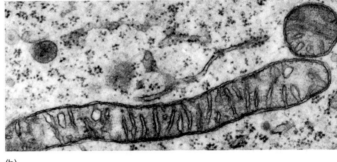

(b)

FIGURE 5.23
Mitochondria. (a) The inner membrane of a mitochondrion is shaped into folds called cristae, which greatly increase the surface area for oxidative metabolism. (b) Mitochondria in cross-section and cut lengthwise (70,000×).

isms, the kind of cell. Chloroplasts contain the pigment chlorophyll that gives most plants their green color.

The chloroplast body is enclosed, like the mitochondrion, within two membranes that resemble those of mitochondria (figure 5.24). However, chloroplasts are larger and more complex than mitochondria. In addition to the outer and inner membranes, which lie in close association with each other, chloroplasts have a closed compartment of stacked membranes called **grana** (singular, *granum*), which lie internal to the inner membrane. A chloroplast may contain a hundred or more grana, and each granum may contain from a few to several dozen disk-shaped structures called **thylakoids.** On the surface of the thylakoids are the light-capturing photosynthetic pigments, to be discussed in depth in chapter 10. Surrounding the thylakoid is a fluid matrix called the *stroma*.

Like mitochondria, chloroplasts contain DNA, but many of the genes that specify chloroplast components are also located in the nucleus. Some of the elements used in the photosynthetic process, including the specific protein components necessary to accomplish the reaction, are synthesized entirely within the chloroplast.

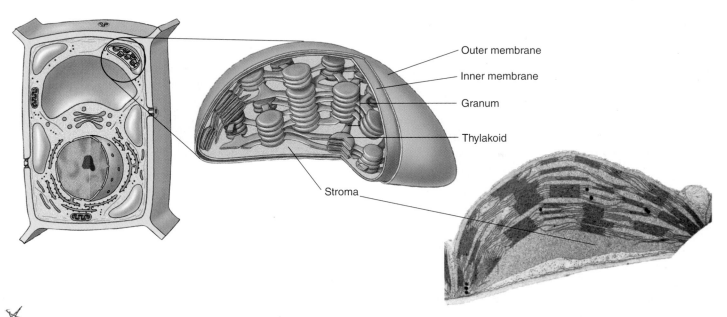

FIGURE 5.24

Chloroplast structure. The inner membrane of a chloroplast is fused to form stacks of closed vesicles called thylakoids. Within these thylakoids, photosynthesis takes place. Thylakoids are typically stacked one on top of the other in columns called grana.

When deprived of light for long periods, chloroplasts lose much of their internal structure and become **leucoplasts.** In root cells and some other plant cells, leucoplasts may serve as starch storage sites. A leucoplast that stores starch (amylose) is sometimes termed an **amyloplast.** These organelles—chloroplasts, leucoplasts, and amyloplasts—are collectively called **plastids.** All plastids come from the division of existing plastids.

Centrioles: Microtubule Assembly Centers

Centrioles are barrel-shaped organelles found in the cells of animals and most protists. They occur in pairs, usually located at right angles to each other near the nuclear membranes (figure 5.25): the pair is referred to as a **centrosome.** At least some centrioles contain DNA, which apparently is involved in producing their structural proteins. Centrioles help to assemble **microtubules,** long, hollow cylinders of the protein tubulin. Microtubules influence cell shape, move the chromosomes in cell division, and provide the functional internal structure of flagella and cilia, as we will discuss later. Centrioles are the best known examples of **microtubule-organizing centers (MTOCs).** The cells of plants and fungi lack centrioles, and cell biologists are still in the process of characterizing their MTOCs.

Both mitochondria and chloroplasts contain specific genes related to some of their functions, but both depend on nuclear genes for other functions. Some centrioles also contain DNA, which apparently helps control the synthesis of their structural proteins.

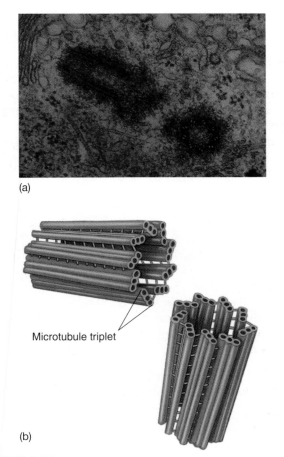

(a)

(b)

FIGURE 5.25

Centrioles. (a) This electron micrograph shows a pair of centrioles (180,000×). The round shape is a centriole in cross-section; the rectangular shape is a centriole in longitudinal section. (b) Each centriole is composed of nine triplets of microtubules.

The Cytoskeleton: Interior Framework of the Cell

The cytoplasm of all eukaryotic cells is crisscrossed by a network of protein fibers that supports the shape of the cell and anchors organelles to fixed locations. This network, called the **cytoskeleton** (figure 5.26), is a dynamic system, constantly forming and disassembling. Individual fibers form by **polymerization,** as identical protein subunits attract one another chemically and spontaneously assemble into long chains. Fibers disassemble in the same way, as one subunit after another breaks away from one end of the chain.

Cells from plants and animals contain three types of cytoskeletal fibers, each formed from a different kind of subunit (figure 5.27):

1. **Actin filaments.** Actin filaments are long fibers about 7 nanometers in diameter. Each filament is composed of two protein chains loosely twined together like two strands of pearls (figure 5.28a). Each "pearl," or subunit, on the chains is the globular protein **actin.** Actin molecules spontaneously form these filaments, even in a test tube; a cell regulates the rate of their formation through other proteins that act as switches, turning on polymerization when appropriate. Actin filaments are responsible for cellular movements such as contraction, crawling, "pinching" during division, and formation of cellular extensions.

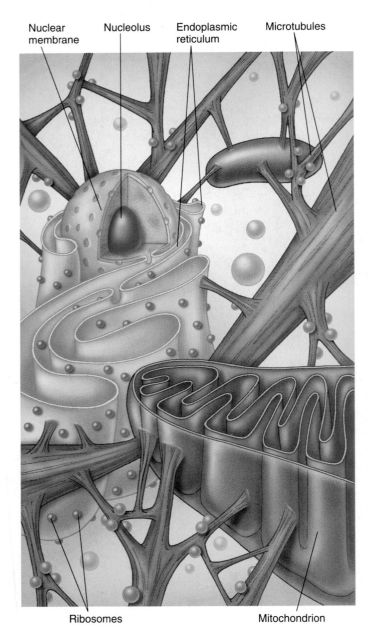

Nuclear membrane · Nucleolus · Endoplasmic reticulum · Microtubules · Ribosomes · Mitochondrion

FIGURE 5.26
The cytoskeleton. In this diagrammatic cross-section of a eukaryotic cell, the cytoskeleton, a network of microtubules, support organelles such as mitochondria.

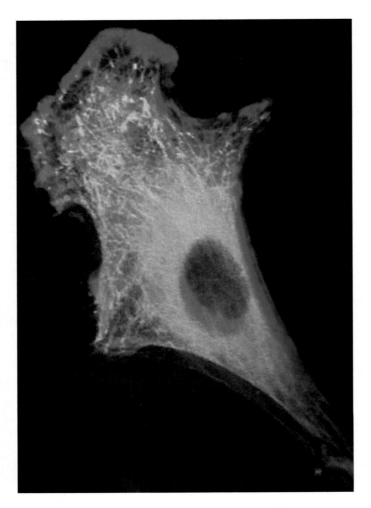

FIGURE 5.27
Cytoskeletal fibers. A complex web of proteins runs through the cytoplasm of the eukaryotic cell. The different proteins are made visible by fluorescence microscopy; actin appears blue, microtubules appear green, and intermediate fibers appear red.

2. Microtubules. Microtubules are hollow tubes about 25 nanometers in diameter, each composed of a ring of 13 protein protofilaments (figure 5.28*b*). Globular proteins consisting of heterodimers of alpha and beta *tubulin* subunits polymerize to form the 13 protofilaments. The protofilaments are arrayed side by side around a central core, giving the microtubule its characteristic tube shape. Microtubules form from MTOC nucleation centers near the center of the cell and radiate toward the periphery. They are in a constant state of flux, continually polymerizing and depolymerizing (the average half-life of a cytoskeleton microtubule is about 10 minutes, and that of a spindle microtubule as short as 20 seconds), unless stabilized by the binding of guanosine triphosphate (GTP) to the ends, which inhibits depolymerization. The ends of the microtubule are designated as "1" (away from the nucleation center) or "2" (toward the nucleation center). Along with allowing for cellular movement, microtubules are responsible for moving materials within the cell itself. Special motor proteins, discussed later in this chapter, move cellular organelles around the cell on microtubular "tracks." *Kinesin* proteins move organelles toward the "1" end (toward the cell periphery), and *dyneins* move them toward the "2" end (toward the center of the cell).

3. Intermediate filaments. The most durable element of the cytoskeleton is a system of tough, fibrous protein molecules twined together in an overlapping arrangement (figure 5.28*c*). These fibers are characteristically 8 to 10 nanometers in diameter, intermediate in size between actin filaments and microtubules (which is why they are called intermediate filaments). Once formed, intermediate filaments are stable and usually do not break down. Intermediate filaments constitute a heterogeneous group of cytoskeletal fibers. The most common type, composed of protein subunits called *vimentin*, provides structural stability for many kinds of cells. *Keratin*, another class of intermediate filament, is found in epithelial cells (cells that line organs and body cavities) and associated structures such as hair and fingernails. The intermediate filaments of nerve cells are called *neurofilaments*.

> Elements of the cytoskeleton crisscross the cytoplasm, supporting the cell shape and anchoring organelles in place.

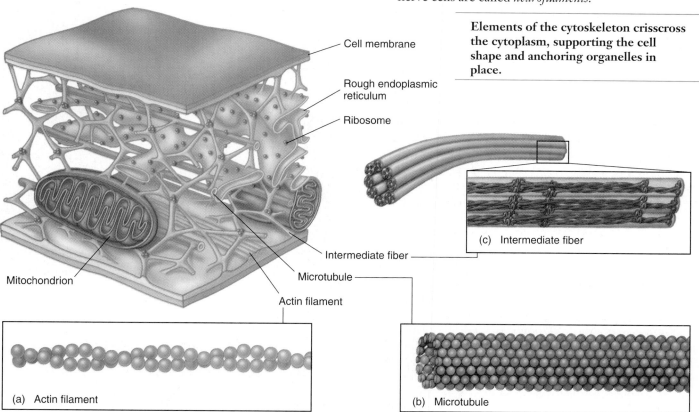

FIGURE 5.28
Molecules that make up the cytoskeleton. (a) Actin filaments. Actin filaments are made of two strands of the fibrous protein actin twisted together and usually occur in bundles. Actin filaments are ubiquitous, although they are concentrated below the plasma membrane in bundles known as stress fibers, which may have a contractile function. (b) Microtubules. Microtubules are composed of 13 stacks of tubulin protein subunits arranged side by side to form a tube. Microtubules are comparatively stiff cytoskeletal elements that serve to organize metabolism and intracellular transport in the nondividing cell. (c) Intermediate filaments. Intermediate filaments are composed of overlapping staggered tetramers of protein. This molecular arrangement allows for a rope-like structure that imparts tremendous mechanical strength to the cell.

Cell Movement

Essentially all cell motion is tied to the movement of actin filaments, microtubules, or both. Intermediate filaments act as intracellular tendons, preventing excessive stretching of cells, and actin filaments play a major role in determining the shape of cells. Because actin filaments can form and dissolve so readily, they enable some cells to change shape quickly. If you look at the surfaces of such cells under a microscope, you will find them alive with motion, as projections, called **microvilli** in animal cells, shoot outward from the surface and then retract, only to shoot out elsewhere moments later (figure 5.29).

Some Cells Crawl

It is the arrangement of actin filaments within the cell cytoplasm that allows cells to "crawl," *literally!* Crawling is a significant cellular phenomenon, essential to inflammation, clotting, wound healing, and the spread of cancer. White blood cells in particular exhibit this ability. Produced in the bone marrow, these cells are released into the circulatory system and then eventually crawl out of capillaries and into the tissues to destroy potential pathogens. The crawling mechanism is an exquisite example of cellular coordination.

Cells exist in a *gel-sol* state; that is, at any given time, some regions of the cell are rigid (*gel*) and some are more fluid (*sol*). The cell is typically more sol-like in its interior, and more gel-like at its perimeter. To crawl, the cell creates a weak area in the gel perimeter, and then forces the fluid (sol) interior through the weak area, forming a **pseudopod** ("false foot"). As a result a large section of cytoplasm oozes off in a different direction, but still remains within the plasma membrane. Once extended, the pseudopod stabilizes into a gel state, assembling actin filaments. Specific membrane proteins in the pseudopod stick to the surface the cell is crawling on, and the rest of the cell is dragged in that direction. The pressure required to force out a developing pseudopod is created when actin filaments in the trailing end of the cell contract, just as squeezing a water balloon at one end forces the balloon to bulge out at the other end.

Actin filaments play a role in other types of cell movement. For example, during cell reproduction (see chapter 11), chromosomes move to opposite sides of a dividing cell because they are attached to shortening microtubules. The cell then pinches in two when a belt of actin filaments contracts like a purse string. Muscle cells also use actin filaments to contract their cytoskeletons. The fluttering of an eyelash, the flight of an eagle, and the awkward crawling of a baby all depend on these cytoskeletal movements within muscle cells.

Not only is the cytoskeleton responsible for the cell's shape and movement, but it also provides a scaffold that holds certain enzymes and other macromolecules in defined areas of the cytoplasm. Many of the enzymes involved in cell metabolism, for example, bind to actin filaments; so do ribosomes. By moving and anchoring particular enzymes near one another, the cytoskeleton, like the endoplasmic reticulum, organizes the cell's activities.

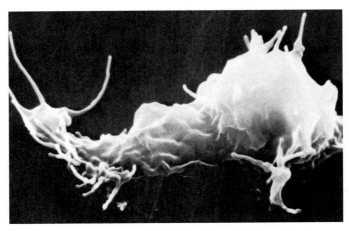

(a)

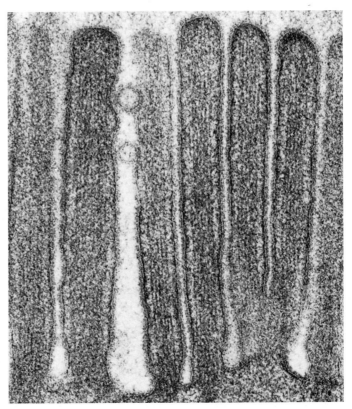

(b)

FIGURE 5.29
The surfaces of some cells are in constant motion. (a) This amoeba, a single-celled protist, is advancing toward you, its advancing edges extending projections outward. The moving edges have been said to resemble the ruffled edges of a skirt (15 μm). (b) Animal cells, like this one lining the human intestine, often have projections called microvilli. Microvilli can change their length quickly. They often appear to pop up almost instantaneously and to disappear just as quickly.

Intracellular Molecular Motors

Certain eukaryotic cells must move materials from one place to another in the cytoplasm. Most cells use the endomembrane system as an intracellular highway; the Golgi apparatus packages materials into vesicles that move through the channels of the endoplasmic reticulum to the far reaches of the cell. However, this highway is only effective over short distances. When a cell has to transport materials through long extensions like the axon of a nerve cell, the ER highways are too slow. For these situations, eukaryotic cells have developed high-speed locomotives that run along microtubular tracks.

Embedded within the membrane of long-distance transport vesicles are linking proteins that bind the vesicles specifically to one of two types of motor proteins. As nature's tiniest motors, these motor proteins literally pull the transport vesicles along the microtubular tracks. One vesicle-linking protein, *kinectin*, is an integral membrane protein of the endoplasmic reticulum. Kinectin binds the vesicles that display it on their membranes to the motor protein *kinesin*. Kinesin then uses ATP to power its movement toward the cell periphery, dragging the vesicle with it as it travels along the microtubule. Another vesicle protein (or perhaps a slight modification of kinesin—further research is needed to determine which) binds vesicles to the motor protein *dynein*, which directs movement in the opposite direction, inward toward the cell's center. The destination of a particular transport vesicle and its contents is thus determined by the nature of the linking protein embedded within the vesicle's membrane.

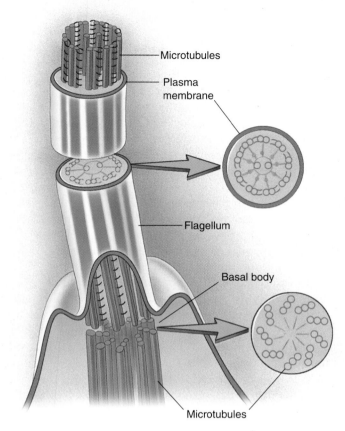

FIGURE 5.30
Structure of a eukaryotic flagellum. A eukaryotic flagellum originates directly from a basal body. The flagellum has two microtubules in its core connected by radial spokes to an outer ring of nine paired microtubules.

Swimming with Flagella and Cilia

Earlier in this chapter, we described the structure of bacterial flagella. Eukaryotic cells have a *completely different* kind of flagellum, consisting of a circle of nine microtubule pairs surrounding two central microtubules; this arrangement is referred to as the **9 + 2 structure** (figure 5.30). As pairs of microtubules move past one another, the eukaryotic flagellum undulates rather than rotates. When examined carefully, each flagellum proves to be an outward projection of the cell's interior, containing cytoplasm and enclosed by the plasma membrane. The microtubules of the flagellum are derived from a **basal body,** situated just below the point where the flagellum protrudes from the surface of the cell.

The flagellum's complex microtubular apparatus evolved early in the history of eukaryotes. Although the cells of many multicellular and some unicellular eukaryotes today no longer possess flagella and are nonmotile, an organization similar to the 9 + 2 arrangement of microtubules can still be found within them, in structures called **cilia** (singular, *cilium*). Cilia are short cellular projections that are often organized in rows (see figure 5.1). They are more numerous than flagella on the cell surface, but have the same internal structure. In many multicellular organisms, cilia carry out tasks far removed from their original function of propelling cells through water. In several kinds of vertebrate tissues, for example, the beating of rows of cilia moves water over the tissue surface. The sensory cells of the vertebrate ear also contain cilia; sound waves bend these cilia, the initial sensory input of hearing. Thus, the 9 + 2 structure of flagella and cilia appears to be a fundamental component of eukaryotic cells.

Some eukaryotic cells use pseudopodia to crawl about within multicellular organisms, while many protists swim using flagella and cilia. Materials are transported within cells by special motor proteins.

5.4 Symbiosis played a key role in the origin of eukaryotes.

Endosymbiosis

Symbiosis is a close relationship between organisms of different species that live together. The theory of **endosymbiosis** proposes that today's eukaryotic cells evolved by a symbiosis in which one species of prokaryote was engulfed by and lived inside another species of prokaryote (figure 5.31). According to the endosymbiont theory, the engulfed prokaryotes provided their hosts with certain advantages associated with their special metabolic abilities. Two key eukaryotic organelles are believed to be the descendants of these endosymbiotic prokaryotes: mitochondria, which are thought to have originated as bacteria capable of carrying out oxidative metabolism; and chloroplasts, which apparently arose from photosynthetic bacteria.

The endosymbiont theory is supported by a wealth of evidence. Both mitochondria and chloroplasts are surrounded by two membranes; the inner membrane probably evolved from the plasma membrane of the engulfed bacterium, while the outer membrane is probably derived from the plasma membrane or endoplasmic reticulum of the host cell. Mitochondria are about the same size as most bacteria, and the cristae formed by their inner membranes resemble the folded membranes in various groups of bacteria. Mitochondrial ribosomes are also similar to bacterial ribosomes in size and structure. Both mitochondria and chloroplasts contain circular molecules

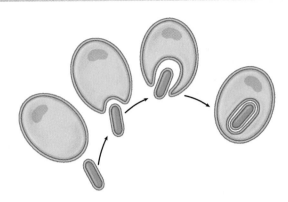

FIGURE 5.31
Endosymbiosis. This figure shows how a double membrane may have been created during the symbiotic origin of mitochondria.

of DNA similar to those in bacteria. Finally, mitochondria divide by simple fission, splitting in two just as bacterial cells do, and they apparently replicate and partition their DNA in much the same way as bacteria. Table 5.2 compares and reviews the features of three types of cells.

Eukaryotic cells are thought to have arisen by endosymbiosis.

Table 5.2	A Comparison of Bacterial, Animal, and Plant Cells		
	Bacterium	**Animal**	**Plant**
EXTERIOR STRUCTURES			
Cell wall	Present (protein-polysaccharide)	Absent	Present (cellulose)
Cell membrane	Present	Present	Present
Flagella	May be present (single strand)	May be present	Absent except in sperm of a few species
INTERIOR STRUCTURES			
ER	Absent	Usually present	Usually present
Ribosomes	Present	Present	Present
Microtubules	Absent	Present	Present
Centrioles	Absent	Present	Absent
Golgi apparatus	Absent	Present	Present
OTHER ORGANELLES			
Nucleus	Absent	Present	Present
Mitochondria	Absent	Present	Present
Chloroplasts	Absent	Absent	Present
Chromosomes	A single circle of naked DNA	Multiple; DNA-protein complex	Multiple; DNA-protein complex
Lysosomes	Absent	Usually present	Present as "spherosomes"
Vacuoles	Absent	Absent or small	Usually a large single vacuole

Summary of Concepts

5.1 All organisms are composed of cells.

- The cell is the smallest unit of life. All living things are made of cells.
- The cell is composed of a nuclear region, which holds the hereditary apparatus, enclosed within the cytoplasm.
- In all cells, the cytoplasm is bounded by a membrane of phospholipid and protein.
- Cells must be small to increase their surface area-to-volume ratio. The smaller the cell, the greater the surface area-to-volume ratio and the more efficiently the cell can function.

5.2 Eukaryotic cells are far more complex than bacterial cells.

- Bacteria, which have prokaryotic cell structure, do not have membrane-bounded organelles within their cells. Their DNA molecule is circular.
- The eukaryotic cell is larger and more complex, with many internal compartments.

5.3 Take a tour of a eukaryotic cell.

- A eukaryotic cell is organized into three principal zones: the nucleus, the cytoplasm, and the plasma membrane. Located in the cytoplasm are numerous organelles, which perform specific functions for the cell.
- Many of these organelles, such as the endoplasmic reticulum, Golgi apparatus (which gives rise to lysosomes), and nucleus, are part of a complex endomembrane system.
- Mitochondria and chloroplasts are part of the energy-processing system of the cell.
- The cytoskeleton encompasses a variety of fibrous proteins that provide structural support and perform other functions for the cell.
- Many eukaryotic cells possess flagella or cilia having a 9 + 2 arrangement of microtubules; sliding of the microtubules past one another bends these cellular appendages.
- Eukaryotic cells move by adjusting their cytoskeleton, and, in some cases, by waving flagella.
- Cells transport materials long distances within the cytoplasm by packaging them into vesicles that are pulled by motor proteins along microtubule tracks.

5.4 Symbiosis played a key role in the origin of eukaryotes.

- Present-day mitochondria and chloroplasts probably evolved as a consequence of early endosymbiosis: the ancestor of the eukaryotic cell engulfed a bacterium, and the bacterium continued to function within the host cell.
- Modern mitochondria and chloroplasts are now required for eukaryotic cell function. Although these organelles still contain DNA, they cannot function without the genetic information located in the cell's nucleus.

Discussing Key Terms

1. **Cell theory**   The cell theory states that every living organism is composed of cells. Each cell is a membrane-bounded volume of cytoplasm containing DNA that carries the hereditary instructions dictating how the cell will grow and develop. All cells arise from existing cells.

2. **Surface area-to-volume ratio**   The amount of surface area per the amount of volume. Large cells have less surface area per unit volume than small cells. A greater surface area-to-volume ratio allows the cell more interaction with its environment and more precise and efficient intra- and intercellular communication.

3. **Compartmentalization**   Prokaryotes differ from eukaryotes in many ways, but perhaps the most fundamental difference is that their cells lack interior organization. Eukaryotic cells contain an endomembrane system, which organizes their interiors into specialized compartments where diverse functions are carried out.

4. **Endosymbiotic origin of eukaryotes**   The theory by which mitochondria and chloroplasts, the ATP-providing organelles of eukaryotic cells, are thought to have arisen when ancient eukaryotes engulfed bacterial cells that eventually evolved into mitochondria or chloroplasts.

Review Questions

1. What are the three principles of the cell theory?

2. How does the surface area-to-volume ratio of cells limit the size that cells can attain?

3. How are prokaryotes different from eukaryotes in terms of their cell walls, interior organization, and flagella?

4. What is the endoplasmic reticulum? What is its function? How does rough ER differ from smooth ER?

5. What is the nuclear envelope? What types of molecules move through nuclear pores?

6. What are ribosomes? Where are they assembled?

7. What is the function of the Golgi apparatus? How do the substances released by the Golgi apparatus make their way to other locations in the cell?

8. How do some types of microbodies protect cells from the effects of hydrogen peroxide?

9. Describe the basic structure of a mitochondrion. What types of eukaryotic cells contain mitochondria? What function do mitochondria perform?

10. What types of eukaryotic cells contain chloroplasts? What unique metabolic activity occurs in chloroplasts?

11. What cellular functions do centrioles participate in?

12. What three kinds of fibers make up the cytoskeleton? Which of the fibers are stable and which are changeable?

13. How do cilia compare with eukaryotic flagella? What does the term "9 + 2 structure" refer to?

14. What is the endosymbiont theory? What is the evidence supporting this theory?

Thought Questions

1. Some cells are much larger than others. Given the constraints imposed by the surface area-to-volume ratio, how would you expect the level of activity in large cells to compare with that in small cells? What assumptions must you make to answer this question?

2. What mechanism could account for the appearance of the 9 + 2 and other similar structures in three different cellular components (centrioles, flagella, and cilia)?

3. Mitochondria and chloroplasts are thought to be the evolutionary descendants of living cells that were engulfed by other cells. Are mitochondria and chloroplasts alive?

Internet Links

Cells Alive!
http://www.cellsalive.com/
Pictures and movies of living cells, many of them quite striking.

Virtual Cell
http://ampere.scale.uiuc.edu/~m-lexa/cell/cell.html
An interactive tour of a plant cell that lets you zoom in in 3D to see chloroplasts and mitochondria and even explore their interiors.

For Further Reading

Alberts, B., D. Bray, J. Lewis, M. Raff, K. Roberts, and J. Watson: *Molecular Biology of the Cell*, ed. 3, Garland Press, New York, 1994. A comprehensive treatment, this is probably the best text ever written on modern cell biology.

DeDuve, C: "The Birth of Complex Cells," *Scientific American*, April 1996, pages 50–57. All animals, plants, and fungi owe their existence to the remnants of tiny primitive bacteria.

Glover, D., C. Gonzalez, and J. Raff: "The Centrosome," *Scientific American*, June 1993, pages 62–68. Recent information concerning the structure and function of centrioles.

Hall, J., Z. Ramanis, and D. Luck: "Basal Body/Centriolar DNA," *Cell*, vol. 59, 1989, pages 121–32. The exciting discovery that centrioles contain DNA, a key prediction of the Margulis endosymbiont theory.

Heidemann, S.: "A New Twist on Integrins and the Cytoskeleton," *Science*, vol. 260, May 1993, pages 1080–81. How internal and external cellular proteins cooperate to allow cellular movement.

Lasic, D.: "Liposomes," *American Scientist*, vol. 80, 1992, pages 20–31. A very readable account of how liposomes are manufactured and how they can be used as medical delivery systems.

McDermott, J.: "A Biologist Whose Heresy Redraws Earth's Tree of Life," *Smithsonian*, August 1989, pages 71–81. Engaging account of the scientific approach taken by Lynn Margulis, the leading contemporary advocate of the endosymbiont theory of the origin of some organelles.

Rothman, J., and L. Orci: "Budding Vesicles in Living Cells," *Scientific American*, March 1996, pages 70–75. Complex machinery forms the tiny containers that store proteins and shuttle them to and fro in cells.

Stossel, T.: "The Machinery of Cell Crawling," *Scientific American*, vol. 271, September 1994, pages 54–63. The role of protein scaffolding in the mechanism of cell crawling is discussed.

6

Membranes

Concept Outline

6.1 Biological membranes are fluid layers of lipid.

The Phospholipid Bilayer. Cells are encased by membranes composed of a bilayer of phospholipid.

The Lipid Bilayer Is Fluid. Because individual phospholipid molecules do not bind to one another, the lipid bilayer of membranes is a fluid.

6.2 Proteins embedded within the plasma membrane determine its character.

The Fluid Mosaic Model. A varied collection of proteins float within the lipid bilayer.

Components of the Cell Membrane. Cell membranes are embedded with a variety of proteins.

Examining Cell Membranes. Visualizing a plasma membrane requires a powerful electron microscope.

Kinds of Membrane Proteins. The proteins in a membrane function in support, transport, recognition, and reactions.

Structure of Membrane Proteins. Membrane proteins are anchored into the lipid bilayer by their nonpolar regions.

6.3 Passive transport across membranes moves down the concentration gradient.

Diffusion. Random molecular motion results in a net movement of molecules to regions of lower concentration.

Facilitated Diffusion. Passive movement across a membrane is often through specific channels.

Osmosis. Polar solutes attract a cloud of water molecules, which are then not free to diffuse away.

6.4 Bulk transport utilizes endocytosis.

Bulk Passage Into and Out of the Cell. To transport large particles, membranes form vesicles.

6.5 Active transport across membranes is powered by energy from ATP.

Active Transport. Cells transport molecules up a concentration gradient using ATP-powered channels.

Coupled Channels. Active transport of ions drives coupled uptake of other molecules up their concentration gradients.

Chloride Channels and Cystic Fibrosis. Individuals with cystic fibrosis have defective chloride channels.

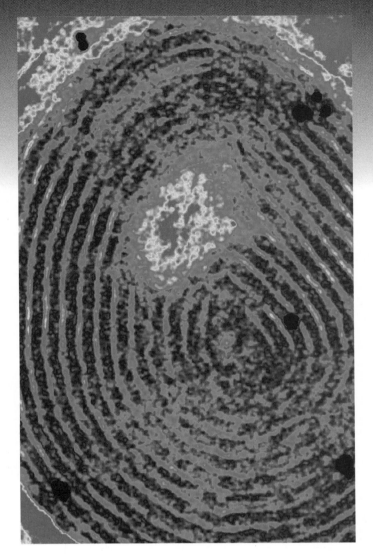

FIGURE 6.1
Membranes within an immune system cell. This cell is not only encased by a lipid membrane, but also contains interior compartments with extensive internal membranes.

Among a cell's most important activities are its transactions with the environment, a give and take that never ceases. Without it, life could not persist. While living cells are encased within a lipid membrane through which few water-soluble substances can pass, the membrane contains protein passageways that permit specific substances to move in and out of the cell and allow the cell to exchange information with its environment. We call this delicate skin of protein molecules embedded in a thin sheet of lipid (figure 6.1) a **plasma membrane**. This chapter will examine the structure and function of this remarkable membrane.

The Phospholipid Bilayer

The membranes that encase all living cells are sheets of lipid only two molecules thick; more than 10,000 of these sheets piled on one another would just equal the thickness of this sheet of paper. The lipid layer that forms the foundation of a cell membrane is composed of molecules called **phospholipids** (figure 6.2).

Phospholipids

Like the fat molecules you studied in chapter 3, a phospholipid has a backbone derived from a three-carbon molecule called glycerol. Attached to this backbone are fatty acids, long chains of carbon atoms ending in a carboxyl (—COOH) group. A fat molecule has three such chains, one attached to each carbon in the backbone; because these chains are nonpolar, they do not form hydrogen bonds with water, and the fat molecule is not water-soluble. A phospholipid, by contrast, has only two fatty acid chains attached to its backbone. The third carbon on the backbone is attached instead to a highly polar organic alcohol that readily forms hydrogen bonds with water. Since this alcohol is attached by a phosphate group, the molecule is called a *phospho*lipid.

One end of a phospholipid molecule is, therefore, strongly nonpolar (water-insoluble), while the other end is strongly polar (water-soluble). The two nonpolar fatty acids extend in one direction, roughly parallel to each other, and the polar alcohol group points in the other direction. Because of this structure, phospholipids are often diagrammed as a polar head with two dangling nonpolar tails (as in figure 6.2*b*).

Phospholipids Form Bilayer Sheets

What happens when a collection of phospholipid molecules is placed in water? The polar water molecules repel the long nonpolar tails of the phospholipids as the water molecules seek partners for hydrogen bonding. Due to the polar nature of the water molecules, the nonpolar tails of the phospholipids end up packed closely together, sequestered as far as possible from water. Every phospholipid molecule orients to face its polar head toward water and its nonpolar tails away. When *two* layers form with the tails facing each other, no tails ever come in contact with water. The resulting structure is called a **lipid bilayer** (figure 6.3). Lipid bilayers form spontaneously, driven by the tendency of water molecules to form the maximum number of hydrogen bonds.

The nonpolar interior of a lipid bilayer impedes the passage of any water-soluble substances through the bilayer, just as a layer of oil impedes the passage of a drop of water ("oil and water do not mix"). This barrier to the passage of water-soluble substances is the key biological property of the lipid

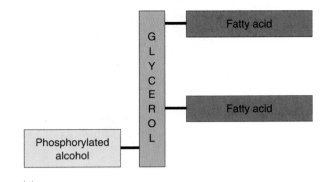

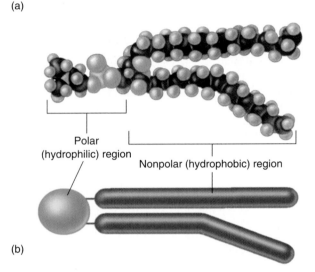

FIGURE 6.2
Phospholipid structure. (a) A phospholipid is a composite molecule similar to a triacylglycerol, except that only two fatty acids are bound to the glycerol backbone; a phosphorylated alcohol occupies the third position on the backbone. (b) Because the phosphorylated alcohol usually extends from one end of the molecule and the two fatty acid chains extend from the other, phospholipids are often diagrammed as a polar head with two nonpolar hydrophobic tails.

bilayer. It means that a cell, if fully encased within a pure lipid bilayer, would be completely impermeable to water-soluble molecules such as sugars, polar amino acids, and proteins. No cell, however, is so encased. In addition to the phospholipid molecules that make up the lipid bilayer, the membranes of every cell also contain proteins that extend through the lipid bilayer, providing passageways across the membrane.

The basic foundation of biological membranes is a lipid bilayer, which forms spontaneously. In such a layer, the nonpolar hydrophobic tails of phospholipid molecules point inward, forming a nonpolar barrier to water-soluble molecules.

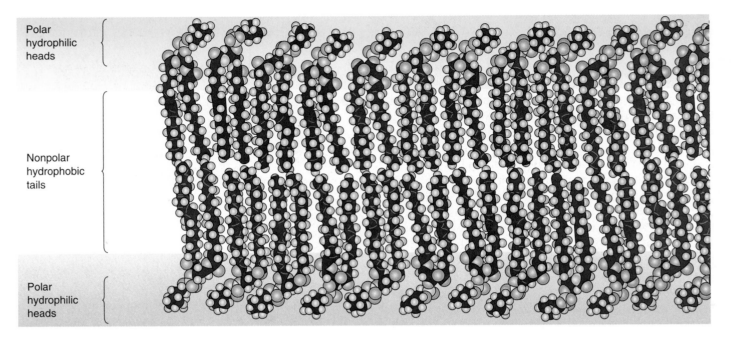

Polar
hydrophilic
heads

Nonpolar
hydrophobic
tails

Polar
hydrophilic
heads

FIGURE 6.3
A phospholipid bilayer. The basic structure of every plasma membrane is a double layer of lipid, in which phospholipids aggregate to form a bilayer with a nonpolar interior. Because the temperature is above the phase transition temperature, the tails do not align perfectly and the membrane is "fluid." Individual phospholipid molecules can move from one place to another in the membrane.

The Lipid Bilayer Is Fluid

A lipid bilayer is stable because water's affinity for hydrogen bonding never stops. Just as surface tension holds a soap bubble together, even though it is made of a liquid, so the hydrogen bonding of water holds a membrane together. But while water continually drives phospholipid molecules into this configuration, it does not locate specific phospholipid molecules relative to their neighbors in the bilayer. As a result, individual phospholipids and unanchored proteins are free to move about within the membrane (figure 6.4).

Phospholipid bilayers are fluid, with the viscosity of olive oil (and like oil, their viscosity increases as the temperature decreases). Some membranes are more fluid than others, however. The tails of individual phospholipid molecules are attracted to one another when they line up close together. This causes the membrane to become less fluid, because aligned molecules must pull apart from one another before they can move about in the membrane. The greater the degree of alignment, the less fluid the membrane. Some phospholipid tails do not align well because they contain one or more double bonds between carbon atoms, introducing kinks in the tail. Membranes containing these types of phospholipids are more fluid than membranes that lack them. Most membranes also contain steroid

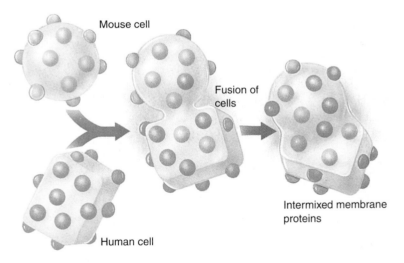

Mouse cell

Fusion of
cells

Intermixed membrane
proteins

Human cell

FIGURE 6.4
Proteins move about in membranes. Protein movement within membranes can be demonstrated easily by labeling the plasma membrane proteins of a mouse cell with fluorescent antibodies and then fusing that cell with a human cell. Within an hour, the labeled and unlabeled proteins are intermixed throughout the hybrid cell's plasma membrane.

lipids like cholesterol, which can either increase or decrease membrane fluidity, depending on temperature.

The lipid bilayer is liquid like a soap bubble, rather than solid like a rubber balloon.

The Fluid Mosaic Model

A plasma membrane (figure 6.5) is composed of both lipids and globular proteins. For many years, biologists thought the protein covered the inner and outer surfaces of the phospholipid bilayer like a coat of paint. The widely accepted Davson-Danielli model, proposed in 1935, portrayed the membrane as a sandwich: a phospholipid bilayer between two layers of globular protein. This model, however, was not consistent with what researchers were learning in the 1960s about the structure of membrane proteins. Unlike most proteins found within cells, membrane proteins are not very soluble in water—they possess long stretches of nonpolar hydrophobic amino acids. If such proteins indeed coated the surface of the lipid bilayer, as the Davson-Danielli model suggests, then their nonpolar portions would separate the polar portions of the phospholipids from water, causing the bilayer to dissolve! Since this doesn't happen, there is clearly something wrong with the model.

In 1972, S. Singer and G. Nicolson revised the model in a simple but profound way: they proposed that the globular proteins are *inserted* into the lipid bilayer, with their nonpo-

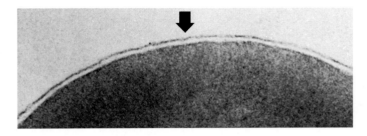

FIGURE 6.5
The plasma membrane. This electron micrograph of a red blood cell clearly shows the plasma membrane (*arrow*), which consists of two layers of phospholipid molecules.

lar segments in contact with the nonpolar interior of the bilayer and their polar portions protruding out from the membrane surface. In this model, called **the fluid mosaic model,** a mosaic of proteins float in the fluid lipid bilayer like boats on a pond (figure 6.6).

The fluid mosaic model proposes that membrane proteins are embedded within the lipid bilayer.

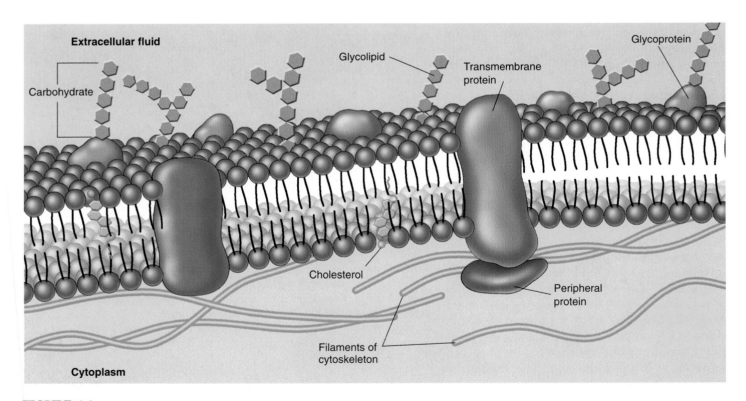

FIGURE 6.6
The fluid mosaic model of the plasma membrane. A variety of proteins protrude through the plasma membrane of animal cells, and nonpolar regions of the proteins tether them to the membrane's nonpolar interior. The three principal classes of membrane proteins are transport proteins, receptors, and cell surface markers. Carbohydrate chains are often bound to the extracellular portion of these proteins, as well as to the membrane phospholipids. These chains serve as distinctive identification tags, unique to particular cells.

Components of the Cell Membrane

A eukaryotic cell contains many membranes. While they are not all identical, they share the same fundamental architecture. Cell membranes are assembled from four components (table 6.1):

1. **Lipid bilayer.** Every cell membrane is composed of a phospholipid bilayer. The other components of the membrane are enmeshed within the bilayer, which provides a flexible matrix and, at the same time, imposes a barrier to permeability.

2. **Transmembrane proteins.** A major component of every membrane is a collection of proteins that float on or in the lipid bilayer. These proteins provide passageways that allow substances and information to cross the membrane. Many membrane proteins are not fixed in position; they can move about, as the phospholipid molecules do. Some membranes are crowded with proteins, while in others, the proteins are more sparsely distributed.

3. **Network of supporting fibers.** Membranes are structurally supported by intracellular proteins that reinforce the membrane's shape. For example, a red blood cell has a characteristic biconcave shape because a scaffold of proteins called spectrin links proteins in the plasma membrane with actin filaments in the cell's cytoskeleton. Membranes use networks of other proteins to control the lateral movements of some key membrane proteins, anchoring them to specific sites.

4. **Exterior proteins and glycolipids.** Membrane sections assemble in the endoplasmic reticulum, transfer to the Golgi complex, and then are transported to the plasma membrane. The endoplasmic reticulum adds chains of sugar molecules to membrane proteins and lipids, creating a "sugar coating" called the glycocalyx that extends from the membrane on the outside of the cell only. Different cell types exhibit different varieties of these glycoproteins and glycolipids on their surfaces, which act as cell identity markers.

Membranes are composed of a lipid bilayer within which proteins are anchored. Plasma membranes are supported by a network of fibers and coated on the exterior with cell identity markers.

Table 6.1 Components of the Cell Membrane

Component	Composition	Function	How It Works	Example
Phospholipid molecules	Phospholipid bilayer	Provides permeability barrier, matrix for proteins	Excludes water-soluble molecules from nonpolar interior of bilayer	Bilayer of cell is impermeable to water-soluble molecules, like glucose
Transmembrane proteins	Carriers	Transport molecules across membrane against gradient	"Escort" molecules through the membrane in a series of conformational changes	Glycophorin channel for sugar transport
	Channels	Passively transport molecules across membrane	Create a tunnel that acts as a passage through membrane	Sodium and potassium channels in nerve cells
	Receptors	Transmit information into cell	Signal molecules bind to cell-surface portion of the receptor protein; this alters the portion of the receptor protein within the cell, inducing activity	Specific receptors bind peptide hormones and neurotransmitters
Interior protein network	Spectrins	Determine shape of cell	Form supporting scaffold beneath membrane, anchored to both membrane and cytoskeleton	Red blood cell
	Clathrins	Anchor certain proteins to specific sites, especially on the exterior cell membrane in receptor-mediated endocytosis	Proteins line coated pits and facilitate binding to specific molecules	Localization of low-density lipoprotein receptor within coated pits
Cell surface markers	Glycoproteins	"Self"-recognition	Create a protein/carbohydrate chain shape characteristic of individual	Major histocompatibility complex protein recognized by immune system
	Glycolipid	Tissue recognition	Create a lipid/carbohydrate chain shape characteristic of tissue	A, B, O blood group markers

Examining Cell Membranes

Biologists examine the delicate, filmy structure of a cell membrane using electron microscopes that provide clear magnification to several thousand times.

Transmission Electron Microscopy

In **transmission electron microscopy (TEM)** the tissue of choice is embedded in a hard matrix, usually some sort of epoxy (figure 6.7), and cut with a microtome, a machine with a very sharp blade that makes incredibly thin slices. The knife moves up and down as the specimen advances toward it, causing transparent "epoxy shavings" less than 1 millimeter thick to peel away from the block of tissue. These shavings are placed on a grid and a beam of electrons is directed through the grid in the TEM. At the high magnification an electron microscope provides, resolution is good enough to reveal the double layers of a membrane.

Scanning Electron Microscopy

Scanning electron microscopy (SEM) does not shoot electrons through the tissue, but rather bounces electrons off it. To prepare a cell membrane for SEM examination, tissue is often "freeze-fractured"— embedded in a medium and quick-frozen with liquid nitrogen. The frozen tissue is then "tapped" with a knife, causing a crack between the phospholipid layers of membranes. Proteins, carbohydrates, pits, pores, channels, or any other structure affiliated with the membrane will pull apart (whole, usually) and stick with one side of the split membrane. To examine the membrane halves with SEM, the newly exposed surfaces must be coated with an electron-attracting substance, usually platinum, gold, or some other metal or alloy. Once the topography of the membrane has been preserved by the metals and carbon, the actual tissue is removed, and the carbon/metal "cast" is examined with the SEM, creating a strikingly different view of the membrane.

> **Visualizing a plasma membrane requires a very powerful electron microscope. Electrons can either be passed through a sample or bounced off it.**

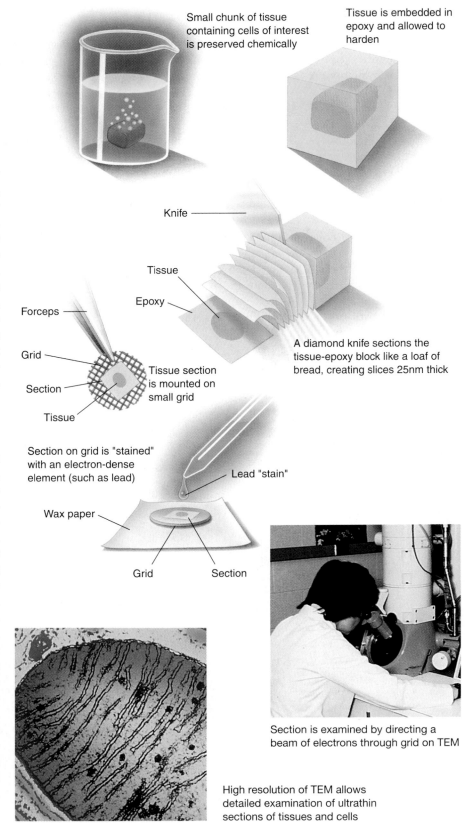

Small chunk of tissue containing cells of interest is preserved chemically

Tissue is embedded in epoxy and allowed to harden

Knife

Tissue

Epoxy

A diamond knife sections the tissue-epoxy block like a loaf of bread, creating slices 25nm thick

Forceps

Grid

Section

Tissue

Tissue section is mounted on small grid

Section on grid is "stained" with an electron-dense element (such as lead)

Lead "stain"

Wax paper

Grid

Section

Section is examined by directing a beam of electrons through grid on TEM

High resolution of TEM allows detailed examination of ultrathin sections of tissues and cells

FIGURE 6.7
Examining membranes with transmission electron microscopy (TEM).

Kinds of Membrane Proteins

As we've seen, the plasma membrane is a complex assembly of proteins enmeshed in a fluid array of phospholipid molecules. This enormously flexible design permits a broad range of interactions with the environment, some directly involving membrane proteins (figure 6.8). Though cells interact with their environment through their plasma membranes in many ways, we will focus on six key classes of membrane protein in this and the following chapter (chapter 7).

1. **Transport channels.** Membranes are very selective, allowing only certain substances to enter or leave the cell. In some instances, they take up molecules already present in the cell in high concentration.
2. **Enzymes.** Cells carry out many chemical reactions on the interior surface of the plasma membrane, using enzymes attached to the membrane.
3. **Cell surface receptors.** Membranes are exquisitely sensitive to chemical messages, detecting them with receptor proteins on their surfaces that act as antennae.
4. **Cell surface identity markers.** Membranes carry cell surface markers that identify them to other cells. Every cell type carries its own ID tags, a specific combination of cell surface proteins characteristic of that cell.
5. **Cell adhesion proteins.** Cells use specific proteins to glue themselves to one another. Some act like Velcro, while others form a more permanent bond.
6. **Attachments to the cytoskeleton.** Surface proteins that interact with other cells are often anchored to the cytoskeleton by linking proteins.

The many proteins embedded within a membrane carry out a host of functions, many of which are associated with transport of materials or information across the membrane.

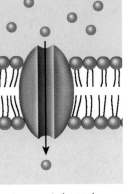 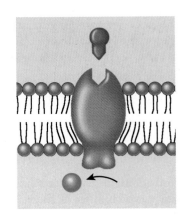 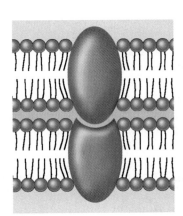

Outside

Plasma membrane

Inside

Selective transport channel

Enzyme

Cell surface receptor

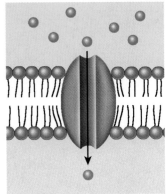 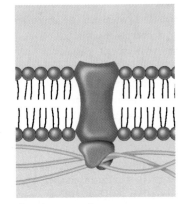

Cell surface identity marker

Cell adhesion

Attachment to the cytoskeleton

FIGURE 6.8
Functions of plasma membrane proteins. Membrane proteins act as transport channels, enzymes, cell surface receptors, and cell surface markers, as well as aiding in cell-to-cell adhesion and securing the cytoskeleton.

Structure of Membrane Proteins

If proteins float on lipid bilayers like ships on the sea, how do they manage to extend through the membrane to create channels, and how can certain proteins be anchored into particular positions on the cell membrane?

Anchoring Proteins in the Bilayer

Many membrane proteins are attached to the external surface of the membrane by a special "anchor" affiliated with a phospholipid molecule called *phosphatidylinositol.* Like a ship tied up to a floating dock, these proteins are free to move about on the surface of the membrane tethered to the phospholipid.

In contrast, other proteins actually traverse the lipid bilayer. The part of the protein that extends through the lipid bilayer, in contact with the nonpolar interior, consists of one or more nonpolar helices or several β-pleated sheets of nonpolar amino acids (figure 6.9). Because water avoids nonpolar amino acids much as it does nonpolar lipid chains, the nonpolar portions of the protein are held within the interior of the lipid bilayer. Although the polar ends of the protein protrude from both sides of the membrane, the protein itself is locked into the membrane by its nonpolar segments. Any movement of the protein out of the membrane, in either direction, brings the nonpolar regions of the protein into contact with water, which "shoves" the protein back into the interior.

Extending Proteins across the Bilayer

Cells contain a variety of different **transmembrane proteins,** which differ in the way they traverse the bilayer, depending on their functions

Anchors. A single nonpolar segment is adequate to anchor a protein in the membrane. Anchoring proteins of this sort attach the spectrin network of the cytoskeleton to the interior of the plasma membrane (figure 6.10). Many proteins that function as receptors for extracellular signals are also "single-pass" anchors that pass through the membrane only once. The portion of the receptor that extends out from the cell surface binds to specific hormones or other molecules when the cell encounters them; the binding induces changes at the other end of the protein, in the cell's interior. In this way, information outside the cell is translated into action within the cell. The mechanisms of cell signaling will be addressed in detail in chapter 7.

Channels. Other proteins have several helical segments that thread their way back and forth through the membrane, creating a passage through the bilayer like the hole in a doughnut. Some of these proteins serve as channels. For example, one of the key transmembrane proteins that carries out photosynthesis contains a total of 10 nonpolar

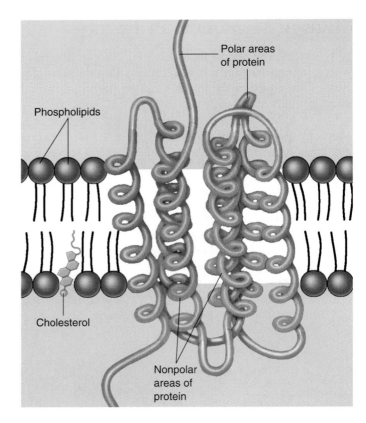

FIGURE 6.9
How nonpolar regions lock proteins into membranes. A spiral helix of nonpolar amino acids (red) extends across the nonpolar lipid interior, while polar (purple) portions of the protein protrude out from the bilayer. The protein cannot move in or out because such a movement would drag nonpolar segments of the protein into contact with water.

helical segments. The protein thus passes across the membrane 10 times, creating a crescent-shaped channel through the membrane (figure 6.11). All water-soluble molecules or ions that enter or leave the cell are either transported by carriers or pass through channels. Each carrier or channel admits only certain substances.

Pores. Some transmembrane proteins have extensive nonpolar regions with secondary configurations of β-pleated sheets instead of α-helices. The β-sheets form a characteristic motif, folding back over themselves to form a pore in the membrane called a β-barrel. Scientists have identified and studied many proteins of this type, appropriately named *porins,* in the outer membrane of some bacteria (figure 6.12).

Transmembrane proteins are anchored into the bilayer by their nonpolar segments. While anchor proteins may pass through the bilayer only once, many channels and pores are created by proteins that pass back and forth through the bilayer repeatedly, creating a circular hole in the bilayer.

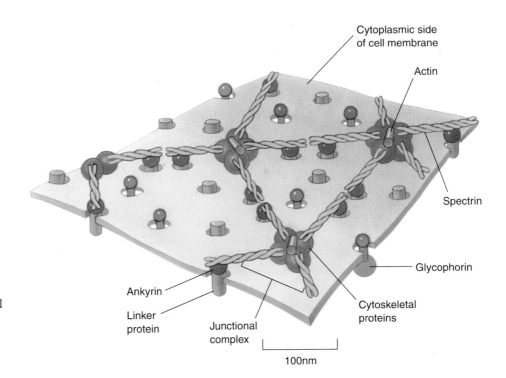

FIGURE 6.10

Anchoring proteins. Spectrin extends as a mesh anchored to the cytoplasmic side of a red blood cell plasma membrane. The spectrin protein is represented as a twisted dimer, attached to the membrane by special proteins such as junctional complexes, ankyrin, and glycophorins. This cytoskeletal protein network confers resiliency to the red blood cell.

Cytoplasmic side of cell membrane

Actin

Spectrin

Glycophorin

Cytoskeletal proteins

Junctional complex

Linker protein

Ankyrin

100nm

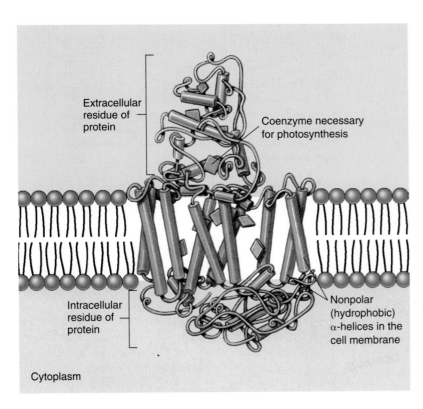

Extracellular residue of protein

Coenzyme necessary for photosynthesis

Intracellular residue of protein

Nonpolar (hydrophobic) α-helices in the cell membrane

Cytoplasm

FIGURE 6.11

A channel protein. This transmembrane protein mediates photosynthesis in the bacterium *Rhodopseudomonas viridis.* Its hydrophobic regions are within the hydrophobic center of the bilayer. The protein traverses the membrane many times, creating a hollow channel through the bilayer.

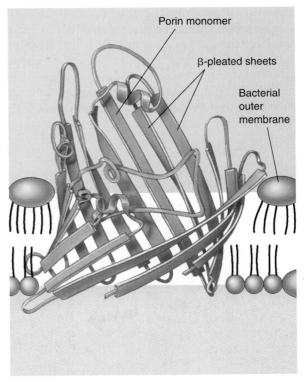

Porin monomer

β-pleated sheets

Bacterial outer membrane

FIGURE 6.12

A pore protein. The bacterial transmembrane protein porin creates large open tunnels called pores in the outer membrane of a bacterium. Sixteen strands of β-pleated sheets run antiparallel to each other, creating a β-barrel in the bacterial outer cell membrane. The tunnel allows water and other materials to pass through the membrane.

Diffusion

Molecules and ions dissolved in water are in constant motion, moving about randomly. This random motion causes a net movement of these substances from regions where their concentration is high to regions where their concentration is lower, a process called **diffusion** (figure 6.13). Net movement driven by diffusion will continue until the concentrations in all regions are the same. You can demonstrate diffusion by filling a jar to the brim with ink, capping it, placing it at the bottom of a bucket of water, and then carefully removing the cap. The ink molecules will slowly diffuse out from the jar until there is a uniform concentration in the bucket and the jar. This uniformity in the concentration of molecules is a type of equilibrium.

Selective Permeability

No cell, however, is limited to these bulk-phase mechanisms of exchange with the environment. As we have seen, cell membranes are studded with proteins, some of which act as channels. Because a given channel will allow only certain kinds of molecules or ions to pass through, the plasma membrane is **selectively permeable**—that is, it is permeable to some substances but not others. If a cell possesses many different kinds of membrane transport proteins, it is able to control the entry and exit of a wide variety of substances. Selective permeability makes it possible for cells to function as isolated compartments that concentrate specific substances together in particular combinations.

Diffusion of Ions through Channels

One of the simplest ways for a substance to diffuse across a cell membrane is through a channel, as ions do. Ions are solutes (substances dissolved in water) with an unequal number of protons and electrons. Those with an excess of protons are positively charged and called *cations*. Ions with more electrons are negatively charged and called *anions*. Because they are charged, ions interact well with polar molecules like water but are repelled by the nonpolar interior of a phospholipid bilayer. Therefore, ions cannot move between the cytoplasm of a cell and the extracellular fluid without the assistance of membrane transport proteins. **Ion channels** have a water-filled pore that spans the membrane. Ions can diffuse through the pore in either direction without coming into contact with the hydrophobic tails of the phospholipids in the membrane, and the transported ions do not bind to or otherwise interact with the channel proteins. Two conditions determine the direction of net movement of the ions: their relative concentrations on either side of the membrane, and the voltage across the membrane (a topic we'll explore in chapter 51). Each type of channel is specific for a particular ion, such as calcium (Ca^{++}) or chloride (Cl^-), or in some cases for a few kinds of ions. Ion channels play an essential role in signaling by the nervous system.

Diffusion is the net movement of substances to regions of lower concentration as a result of random spontaneous motion. It tends to distribute substances uniformly. Membrane transport proteins allow only certain molecules and ions to diffuse through the plasma membrane.

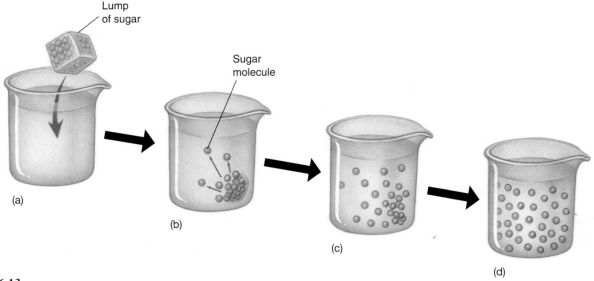

FIGURE 6.13
Diffusion. If a lump of sugar is dropped into a beaker of water (a), its molecules dissolve (b) and diffuse (c). Eventually, diffusion results in an even distribution of sugar molecules throughout the water (d).

Facilitated Diffusion

Carriers, another class of membrane proteins, transport ions as well as other solutes like sugars and amino acids across the membrane. Like channels, carriers are specific for a certain type of solute and can transport substances in either direction across the membrane. Unlike channels, however, they facilitate the movement of solutes across the membrane by physically binding to them on one side of the membrane and releasing them on the other. Again, the direction of the solute's net movement simply depends on its *concentration gradient* across the membrane. If the concentration is greater in the cytoplasm, the solute is more likely to bind to the carrier on the cytoplasmic side of the membrane and be released on the extracellular side. This will cause a net movement from inside to outside. If the concentration is greater in the extracellular fluid, the net movement will be from outside to inside. Thus, the net movement always occurs from areas of high concentration to low, just as it does in simple diffusion, but carriers facilitate the process. For this reason, this mechanism of transport is sometimes called **facilitated diffusion** (figure 6.14).

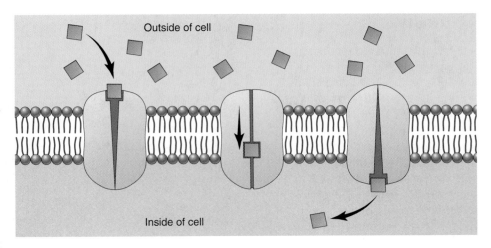

FIGURE 6.14
Facilitated diffusion is a carrier-mediated transport process. Molecules bind to a receptor on the extracellular side of the cell and are conducted through the plasma membrane by a membrane protein.

Facilitated Diffusion in Red Blood Cells

Several examples of facilitated diffusion by carrier proteins can be found in the membranes of vertebrate red blood cells (RBCs). One RBC carrier protein, for example, transports a different molecule in each direction: Cl^- in one direction and bicarbonate ion (HCO_3^-) in the opposite direction. As you will learn in chapter 50, this carrier is important in transporting carbon dioxide in the blood.

A second important facilitated diffusion carrier in RBCs is the glucose transporter. Red blood cells keep their internal concentration of glucose low through a chemical trick: they immediately add a phosphate group to any entering glucose molecule, converting it to a highly-charged glucose phosphate that cannot pass back across the membrane. This maintains a steep concentration gradient for glucose, favoring its entry into the cell. The glucose transporter that carries glucose into the cell does not appear to form a channel in the membrane for the glucose to pass through. Instead, the transmembrane protein appears to bind the glucose and then flip its shape, dragging the glucose through the bilayer and releasing it on the inside of the plasma membrane. Once it releases the glucose, the glucose transporter reverts to its original shape. It is then available to bind the next glucose molecule that approaches the outside of the cell.

Carrier-Mediated Processes Saturate

A characteristic feature of carrier-mediated transport is that its rate is saturable. In other words, if the concentration gradient of a substance is progressively increased, its rate of transport will also increase to a certain point and then level off. Further increases in the gradient will produce no additional increase in rate. The explanation for this observation is that there are a limited number of carriers in the membrane. When the concentration of the transported substance rises high enough, all of the carriers will be in use and the capacity of the transport system will be saturated. In contrast, substances that move across the membrane by simple diffusion (diffusion through channels in the bilayer without the assistance of carriers) do not show saturation.

Facilitated diffusion provides the cell with a ready way to prevent the buildup of unwanted molecules within the cell or to take up needed molecules, such as sugars, that may be present outside the cell in high concentrations. Facilitated diffusion has three essential characteristics:

1. **It is *specific*.** Any given carrier transports only certain molecules or ions.
2. **It is *passive*.** The direction of net movement is determined by the relative concentrations of the transported substance inside and outside the cell.
3. **It *saturates*.** If all relevant protein carriers are in use, increases in the concentration gradient do not increase the transport rate.

Facilitated diffusion is the transport of molecules and ions across a membrane by specific carriers in the direction of lower concentration of those molecules or ions.

Osmosis

The cytoplasm of a cell contains ions and molecules, such as sugars and amino acids, dissolved in water. The mixture of these substances and water is called an **aqueous solution.** Water, the most common of the molecules in the mixture, is the **solvent,** and the substances dissolved in the water are **solutes.** The ability of water and solutes to diffuse across membranes has important consequences.

Molecules Diffuse down a Concentration Gradient

Both water and solutes diffuse from regions of high concentration to regions of low concentration; that is, they diffuse down their concentration gradients. When two regions are separated by a membrane, what happens depends on whether or not the solutes can pass freely through that membrane. Most solutes, including ions and sugars, are not lipid-soluble and, therefore, are unable to cross the lipid bilayer of the membrane. Water molecules, in contrast, are small enough to pass between lipid molecules, so they can diffuse across the membrane. This form of net water movement across a membrane by diffusion is called **osmosis** (figure 6.15).

The concentration of *all* solutes in a solution determines the **osmotic concentration** of the solution. If two solutions have unequal osmotic concentrations, the solution with the higher concentration is **hyperosmotic** (Greek *hyper,* "more than"), and the solution with the lower concentration is **hypoosmotic** (Greek *hypo,* "less than"). If the osmotic concentrations of two solutions are equal, the solutions are **isosmotic** (Greek *iso,* "the same").

In cells, a plasma membrane separates two aqueous solutions, one inside the cell (the cytoplasm) and one outside (the extracellular fluid). The direction of the net diffusion of water across this membrane is determined by the osmotic concentrations of the solutions on either side (figure 6.16). For example, if the cytoplasm of a cell were *hypo*osmotic to the extracellular fluid, water would diffuse out of the cell, toward the solution with the higher concentration of solutes (and, therefore, the lower concentration of unbound water molecules). This loss of water from the cytoplasm would cause the cell to shrink until the osmotic concentrations of the cytoplasm and the extracellular fluid become equal.

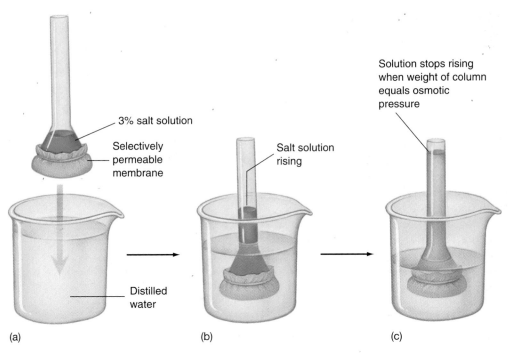

FIGURE 6.15
An experiment demonstrating osmosis. (a) The end of a tube containing a salt solution is closed by stretching a selectively permeable membrane across its face; the membrane allows the passage of water molecules but not salt ions. (b) When this tube is immersed in a beaker of distilled water, the salt cannot cross the membrane, but water can. The water entering the tube causes the salt solution to rise in the tube. (c) Water will continue to enter the tube from the beaker until the weight of the column of water in the tube exerts a downward force equal to the force drawing water molecules upward into the tube. This force is referred to as osmotic pressure.

Osmotic Pressure

What would happen if the cell's cytoplasm were *hyper*osmotic to the extracellular fluid? In this situation, water would diffuse into the cell from the extracellular fluid, causing the cell to swell. The pressure of the cytoplasm pushing out against the cell membrane, or **hydrostatic pressure,** would increase. On the other hand, the **osmotic pressure** (see figure 6.15 and figure 6.17), defined as the pressure that must be applied to stop the osmotic movement of water across a membrane, would also be at work. If the membrane were strong enough, the cell would reach an equilibrium, at which the osmotic pressure, which tends to drive water into the cell, is exactly counterbalanced by the hydrostatic pressure, which tends to drive water back out of the cell. However, a plasma membrane by itself cannot withstand large internal pressures, and an isolated cell under such conditions would burst like an overinflated balloon. Accordingly, it is important for animal cells to maintain isosmotic conditions. The cells of bacteria, fungi, plants, and many protists, in contrast, are surrounded by strong cell walls. The cells of these organisms can withstand high internal pressures without bursting.

(a)

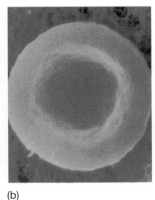
(b)

(c)

FIGURE 6.16
Osmosis. These scanning electron micrographs show what happens when a human red blood cell is placed in hyperosmotic, isosmotic, and hypoosmotic solutions. (a) In a hyperosmotic solution, such as very salty water, water moves out of the cell toward the higher concentration of solutes, causing the cell to shrivel. (b) In an isosmotic solution, the concentration of solutes on either side of the membrane is the same. Osmosis still occurs, but water diffuses into and out of the cell at the same rate, and the cell doesn't change size. (c) In a hypoosmotic solution, such as distilled water, the concentration of solutes is higher within the cell than without, so the net movement of water is into the cell, and the cell swells and explodes.

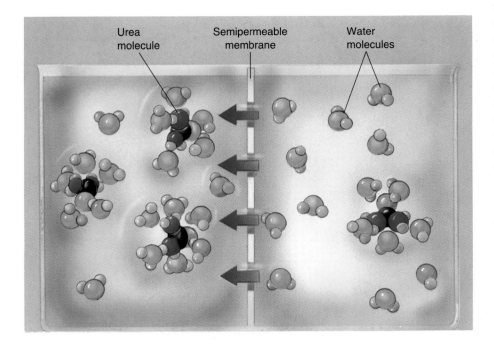

FIGURE 6.17
How solutes create osmotic pressure. Charged or polar substances are soluble in water because they form hydrogen bonds with water molecules clustered around them. When a polar solute (illustrated here with urea) is added to the solution on one side of a membrane, the water molecules that gather around each urea molecule are no longer free to diffuse across the membrane; in effect, the polar solute has reduced the number of free water molecules on that side of the membrane increasing the osmotic pressure. Because the hypoosmotic side of the membrane (on the right, with less solute) has more unbound water molecules than the hyperosmotic side (on the left, with more solute), water moves by diffusion from the right to the left.

Maintaining Osmotic Balance

Organisms have developed many solutions to the osmotic dilemma posed by being hyperosmotic to their environment.

Extrusion. Some single-celled eukaryotes like the protist *Paramecium* use organelles called contractile vacuoles to remove water. Each vacuole collects water from various parts of the cytoplasm and transports it to the central part of the vacuole, near the cell surface. The vacuole possesses a small pore that opens to the outside of the cell. By contracting rhythmically, the vacuole pumps the water out of the cell through the pore.

Isosmotic Solutions. Some organisms that live in the ocean adjust their internal concentration of solutes to match that of the surrounding seawater. Isosmotic with respect to their environment, there is no net flow of water into or out of these cells. Many terrestrial animals solve the problem in a similar way, by circulating a fluid through their bodies that bathes cells in an isosmotic solution. The solute concentration of the fluid must be monitored constantly to ensure that it matches the concentration inside the cells. The blood in your body, for example, contains a high concentration of the protein albumin, which elevates the solute concentration of the blood to match your cells.

Turgor. Most plant cells are hyperosmotic to their immediate environment, containing a high concentration of solutes in their central vacuoles. The resulting internal hydrostatic pressure, known as **turgor pressure,** presses the plasma membrane firmly against the interior of the cell wall, making the cell rigid. The newer, softer portions of trees and shrubs and of plants that lack wood altogether depend to a large extent on turgor pressure to maintain their shape. These plants wilt when they lack sufficient water.

Osmosis is the diffusion of water, but not solutes, across a membrane.

Bulk Passage Into and Out of the Cell

Endocytosis

The lipid nature of their biological membranes raises a second problem for cells. The substances cells use as fuel are for the most part large, polar molecules that cannot cross the hydrophobic barrier a lipid bilayer creates. How do organisms get these substances into their cells? One process many single-celled eukaryotes employ is **endocytosis** (figure 6.18). In this process the plasma membrane extends outward and envelops food particles. Cells use three major types of endocytosis: phagocytosis, pinocytosis, and receptor-mediated endocytosis.

Phagocytosis and Pinocytosis. If the material the cell takes in is particulate (made up of discrete particles), such as an organism or some other fragment of organic matter (figure 6.18, *a*, *c*, and *d*), the process is called **phagocytosis** (Greek *phagein*, "to eat" + *cytos*, "cell"). If the material the cell takes in is liquid (figure 6.18*b*), it is called **pinocytosis** (Greek *pinein*, "to drink"). Pinocytosis is common among animal cells. Mammalian egg cells, for example, "nurse" from surrounding cells; the nearby cells secrete nutrients that the maturing egg cell takes up by pinocytosis. Virtually all eukaryotic cells constantly carry out these kinds of endocytosis, trapping particles and extracellular fluid in vesicles and ingesting them. Endocytosis rates vary from one cell type to another. They can be surprisingly high: some types of white blood cells ingest 25% of their cell volume each hour!

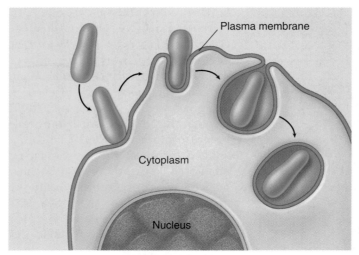

Phagocytosis

(a)

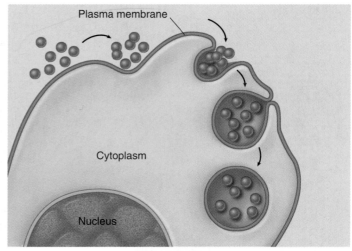

Pinocytosis

(b)

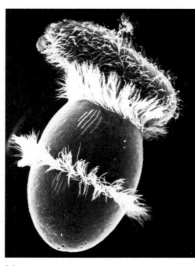

(c)

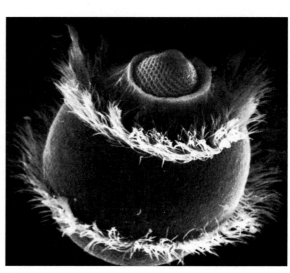

(d)

FIGURE 6.18
Endocytosis. Both phagocytosis (a) and pinocytosis (b) are forms of endocytosis. (c) The large egg-shaped protist *Didinium nasutum* has just begun eating the smaller protist *Paramecium;* (d) its meal is practically over.

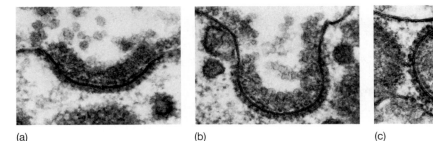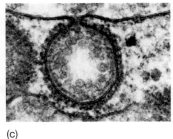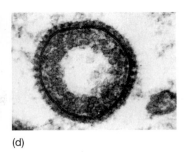

(a) (b) (c) (d)

FIGURE 6.19
Receptor-mediated endocytosis (80,000×). A coated pit appears in the plasma membrane of a developing egg cell, covered with a layer of proteins (a). A coating of protein molecules can be seen just beneath the pit, on the cytoplasmic side of the membrane. When an appropriate collection of molecules gathers in the coated pit, the pit deepens (b) as the outer membrane of the cell closes in behind the pit (c), and the pit seals off to form a coated vesicle, which carries the molecules into the cell (d).

Receptor-Mediated Endocytosis. Specific molecules are often transported into eukaryotic cells through **receptor-mediated endocytosis.** Cells that perform this process have indented pits on their outer plasma membrane, coated with the protein **clathrin.** The pits act like molecular mousetraps, closing over to form an internal vesicle when the right molecule enters the pit (figure 6.19). The trigger that releases the trap is a receptor protein embedded in the membrane of the pit, which detects the presence of a particular target molecule and reacts by initiating endocytosis. The process is highly specific and very fast. **Fluid-phase endocytosis** is the receptor-mediated pinocytosis of fluids. It is important to understand that endocytosis in itself does not bring substances directly into the cytoplasm of a cell. The material taken in is still separated from the cytoplasm by the membrane of the vesicle, as figure 6.18*a* and *b* clearly show.

Exocytosis

The reverse of endocytosis is **exocytosis,** the discharge of material from vesicles at the cell surface (figure 6.20). In plant cells, exocytosis is an important means of exporting the materials needed to construct the cell wall through the plasma membrane. Among protists, contractile vacuole discharge is a form of exocytosis. In animal cells, exocytosis provides a mechanism for secreting many hormones, neurotransmitters, digestive enzymes, and other substances.

Cells import bulk materials by engulfing them with their plasma membranes in a process called endocytosis; similarly, they extrude or secrete material through exocytosis.

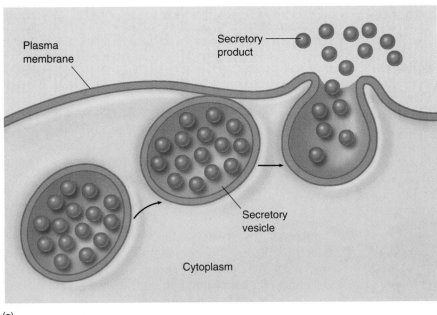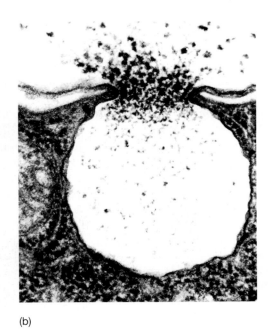

(a)

(b)

FIGURE 6.20
Exocytosis. (a) Proteins and other molecules are secreted from cells in small packets called vesicles, whose membranes fuse with the plasma membrane, releasing their contents to the cell surface. (b) A transmission electron micrograph showing exocytosis.

Active Transport

While diffusion, facilitated diffusion, and osmosis are passive transport processes that move materials down their concentration gradients, cells can also move substances across the membrane *up* their concentration gradients. This process requires the expenditure of energy and is therefore called **active transport.** Like facilitated diffusion, active transport involves highly selective protein carriers within the membrane. These carriers bind to the transported substance, which could be an ion or a simple molecule like a sugar (figure 6.21), an amino acid, or a nucleotide to be used in the synthesis of DNA.

Active transport is one of the most important functions of any cell. It enables a cell to take up additional molecules of a substance that is already present in its cytoplasm in concentrations higher than in the extracellular fluid. Without active transport, for example, liver cells would be unable to accumulate glucose molecules from the blood plasma, since the glucose concentration is often higher inside the liver cells than it is in the plasma. Active transport also enables a cell to move substances from its cytoplasm to the extracellular fluid despite higher external concentrations.

The Sodium-Potassium Pump

More than one-third of all of the energy expended by an animal cell that is not actively dividing is used in the active transport of sodium (Na^+) and potassium (K^+) ions. Most animal cells have a low internal concentration of Na^+, relative to their surroundings, and a high internal concentration of K^+. They maintain these concentration differences by actively pumping Na^+ out of the cell and K^+ in. The remarkable protein that transports these two ions across the cell membrane is known as the **sodium-potassium pump** (figure 6.22). The cell obtains the energy it needs to operate the pump from adenosine triphosphate (ATP), a molecule we'll learn more about in chapter 8.

The important characteristic of the sodium-potassium pump is that it is an *active* transport process, transporting Na^+ and K^+ from areas of low concentration to areas of high concentration. This transport up their concentration gradients is the opposite of the passive transport in diffusion; it is achieved only by the constant expenditure of metabolic energy. The sodium-potassium pump works through a series of conformational changes in the transmembrane protein:

Step 1. Three sodium ions bind to the cytoplasmic side of the protein, causing the protein to change its conformation.

Step 2. In its new conformation, the protein binds a molecule of ATP and cleaves it into adenosine diphosphate and phosphate (ADP + P_i). ADP is released, but

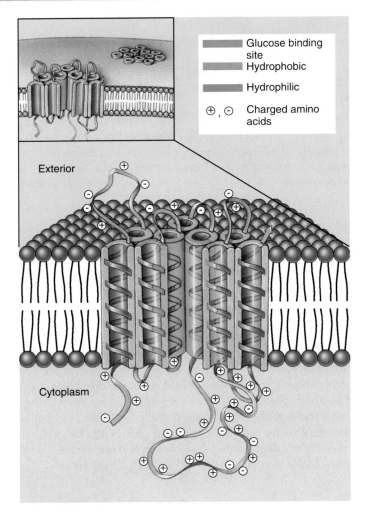

FIGURE 6.21
A glucose transport channel. The molecular structure of this particular glucose transport carrier is known in considerable detail. The protein's 492 amino acids form a folded chain that traverses the lipid membrane 12 times. Amino acids with charged groups are less stable in lipid and are thus exposed to the cytoplasm or the extracellular fluid. Researchers think the center of the protein consists of five helical segments with glucose-binding sites (in red) facing inward. A conformational change in the protein transports glucose across the membrane by shifting the position of the glucose-binding sites.

the phosphate group remains bound to the protein. The protein is now phosphorylated.

Step 3. The phosphorylation of the protein induces a second conformational change in the protein. This change translocates the three Na^+ across the membrane, so they now face the exterior. In this new conformation, the protein has a low affinity for Na^+, and the three bound Na^+ dissociate from the protein and diffuse into the extracellular fluid.

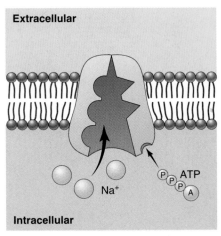

Extracellular

Intracellular

Na+

ATP

1. Protein in membrane binds intracellular sodium.

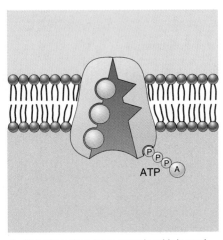

ATP

2. ATP phosphorylates protein with bound sodium.

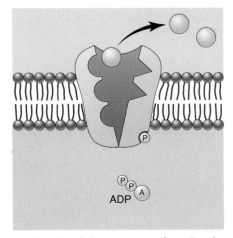

ADP

3. Phosphorylation causes conformational change in protein, allowing sodium to leave.

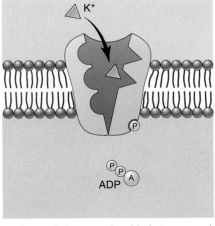

K+

ADP

4. Extracellular potassium binds to exposed sites.

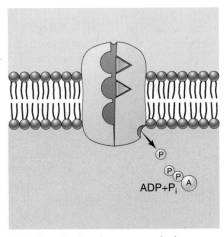

ADP+P_i

5. Binding of potassium causes dephosphorylation of protein.

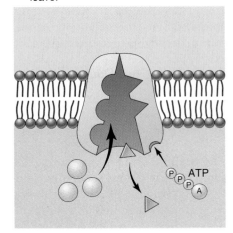

ATP

6. Dephosphorylation of protein triggers change back to original conformation, potassium moves into cell, and the cycle repeats.

FIGURE 6.22

The sodium-potassium pump. The protein channel known as the sodium-potassium pump transports sodium (Na+) and potassium (K+) ions across the cell membrane. For every three Na+ that are transported out of the cell, two K+ are transported into the cell. The sodium-potassium pump is fueled by ATP.

Step 4. The new conformation has a high affinity for K+, two of which bind to the extracellular side of the protein as soon as it is free of the Na+.

Step 5. The binding of the K+ causes another conformational change in the protein, this time resulting in the dissociation of the bound phosphate group.

Step 6. Freed of the phosphate group, the protein reverts to its original conformation, exposing the two K+ to the cytoplasm. This conformation has a low affinity for K+, so the two bound K+ dissociate from the protein and diffuse into the interior of the cell. The original conformation has a high affinity for Na+; when these ions bind, they initiate another cycle.

Three Na+ leave the cell and two K+ enter in every cycle. The changes in protein conformation that occur during the cycle are rapid, enabling each carrier to transport as many as 300 Na+ per second. The sodium-potassium pump appears to be ubiquitous in animal cells, although cells vary widely in the number of pump proteins they contain. This active transport process is responsible for maintaining the voltage difference that exists across nerve cell membranes, as we'll see in chapter 51. Moreover, the Na+ gradient established by the sodium-potassium pump drives many other transport processes as well.

Active transport moves a solute across a membrane up its concentration gradient, using protein carriers driven by the expenditure of chemical energy.

Coupled Channels

Many molecules are transported into cells up a concentration gradient through channels that do not utilize ATP. Instead, the molecules move hand-in-hand with sodium ions or protons that are moving *down* their concentration gradients. This type of active transport, called **cotransport,** has two components:

1. **Establishing the down gradient.** ATP is used to establish the sodium ion or proton *down* gradient, which is greater than the *up* gradient of the molecule to be transported.
2. **Traversing the up gradient.** Cotransport channels (also called coupled channels) carry the molecule and either a sodium ion or a proton together across the membrane.

Because the *down* gradient of the sodium ion or proton is greater than the *up* gradient of the molecule to be transported, the net movement across the channel is in the direction of the down gradient, typically into the cell.

Establishing the *Down* Gradient

Either the sodium-potassium pump or the proton pump establishes the down gradient that powers most active transport processes of the cell.

The Sodium-Potassium Pump. The sodium-potassium pump actively pumps sodium ions out of the cell, powered by energy from ATP. This establishes a sodium ion concentration gradient that is lower inside the cell.

The Proton Pump. The **proton pump** pumps protons (H⁺ ions) across a membrane using energy derived from energy-rich molecules or from photosynthesis. This creates a proton gradient, in which the concentration of protons is higher on one side of the membrane than the other. Membranes are impermeable to protons, so the only way protons can diffuse down their concentration gradient back through the membrane is through a second cotransport channel.

Traversing the *Up* Gradient

Animal cells accumulate many amino acids and sugars against a concentration gradient: the molecules are transported into the cell from the extracellular fluid, even though their concentrations are higher inside the cell. These molecules couple with sodium ions to enter the cell down the Na⁺ concentration gradient established by the sodium-potassium pump. In this **cotransport process,** Na⁺ and a specific sugar or amino acid simultaneously bind to the same transmembrane protein on the outside of the cell (figure 6.23). Both are then translocated to the inside of the cell, but in the process Na⁺ moves *down* its concentration gradient while the sugar or amino acid moves *up* its concentration gradient. In effect, the cell uses some of the energy stored in the Na⁺ concentration gradient to accumulate sugars and amino acids. Since the Na⁺ gradient is established and maintained by the sodium-potassium

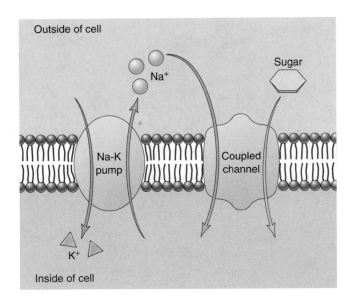

FIGURE 6.23
Cotransport through a coupled channel. A membrane protein transports sodium ions into the cell, down their concentration gradient, at the same time it transports a sugar molecule into the cell. The gradient driving the Na⁺ entry is so great that sugar molecules can be brought in *against* their concentration gradient.

pump, the inward transport of sugars and amino acids occurs as a secondary consequence of the primary active transport of Na⁺ by the sodium-potassium pump.

In a related process, called **countertransport,** the inward movement of Na⁺ is coupled with the outward movement of another substance, such as Ca⁺⁺ or H⁺. As in cotransport, both Na⁺ and the other substance bind to the same transport protein, but in this case they bind on opposite sides of the membrane and are moved in opposite directions. In countertransport, the cell uses the energy released as Na⁺ moves down its concentration gradient into the cell to extrude a substance up its concentration gradient.

The cell uses the proton *down* gradient established by the proton pump (figure 6.24) in ATP production. The movement of protons through their cotransport channel is coupled to the production of ATP, the energy-storing molecule we mentioned earlier. Thus, the cell expends energy to produce ATP, which provides it with a convenient energy storage form that it can employ in its many activities. The coupling of the proton pump to ATP synthesis, called **chemiosmosis,** is responsible for almost all of the ATP produced from food (chapter 9) and all of the ATP produced by photosynthesis (chapter 10). We know that proton pump channels are ancient, since they are present in bacteria as well as in eukaryotes. The mechanisms for transport across plasma membranes that we have considered in this chapter are summarized in table 6.2.

Many molecules are cotransported into cells up their concentration gradients by coupling their movement to that of sodium ions or protons moving down their concentration gradients.

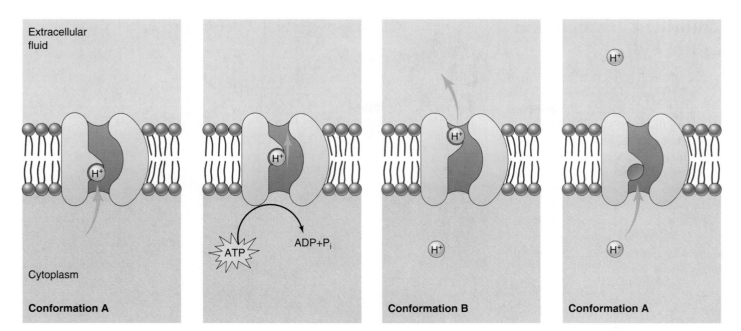

FIGURE 6.24
The proton pump. In this general model of energy-driven proton pumping, the transmembrane protein that acts as a proton pump is driven through a cycle of two conformations: A and B. The cycle A→B→A goes only one way, causing protons to be pumped from the inside to the outside of the membrane. ATP powers the pump.

	Table 6.2	Mechanisms for Transport across Cell Membranes	
Process	**Passage through Membrane**	**How It Works**	**Example**
PASSIVE PROCESSES			
Diffusion	Direct	Random molecular motion produces net migration of molecules toward region of lower concentration	Movement of oxygen into cells
Facilitated diffusion	Protein channel	Molecule binds to carrier protein in membrane and is transported across; net movement is toward region of lower concentration	Movement of glucose into cells
Osmosis	Direct	Diffusion of water across differentially permeable membrane	Movement of water into cells placed in a hypotonic solution
ACTIVE PROCESSES			
Endocytosis			
Phagocytosis	Membrane vesicle	Particle is engulfed by membrane, which folds around it and forms a vesicle	Ingestion of bacteria by white blood cells
Pinocytosis	Membrane vesicle	Fluid droplets are engulfed by membrane, which forms vesicles around them	"Nursing" of human egg cells
Carrier-mediated endocytosis	Membrane vesicle	Endocytosis triggered by a specific receptor	Cholesterol uptake
Exocytosis	Membrane vesicle	Vesicles fuse with plasma membrane and eject contents	Secretion of mucus
Active transport			
Na^+/K^+ pump	Protein carrier	Carrier expends energy to export Na^+ against a concentration gradient	Coupled uptake of glucose into cells against its concentration gradient
Proton pump	Protein carrier	Carrier expends energy to export protons against a concentration gradient	Chemiosmotic generation of ATP

Chloride Channels and Cystic Fibrosis

Cystic fibrosis (CF) is a fatal human disease in which the mucus of affected individuals is far thicker than normal. The buildup of this thickened mucus causes chronic tissue inflammation and eventually to the replacement of injured cells with scar tissue, blocking the airways of the lungs and the ducts of the pancreas and liver. Patients typically die of pneumonia (or related lung disorders) or cirrhosis of the liver. Many also experience pancreatic failure: when no pancreatic enzymes reach the intestines, fats are not digested properly, and patients experience a deficiency in fat-soluble vitamins that leads to rickets and other serious vitamin deficiency disorders. Cystic fibrosis is usually thought of as a children's disease because affected individuals, if untreated, usually die by the age of 4 or 5. Today, with therapy (antibiotics to combat pneumonia, fat-soluble vitamins, and low-fat diets), affected individuals survive an average of 27 years.

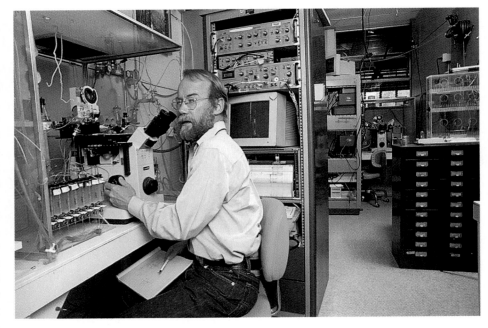

FIGURE 6.25

Dr. Paul Quinton in his laboratory. Dr. Quinton is the discoverer of the key link between cystic fibrosis and chloride ion channels.

Cystic fibrosis is a hereditary disorder, resulting from a defect in a single gene passed down from parent to child. It is the most common fatal genetic disease among Caucasians. About 1 in 25 individuals possesses at least one copy of the defective gene. Most carriers are not affected with the disorder; only those children who inherit two copies of the defective gene, one from each parent, succumb to cystic fibrosis—about 1 in 2500 Caucasian children.

Cystic fibrosis has proved to be difficult to study. Many organs are affected, and until recently it was impossible to identify the nature of the defective gene responsible for the disorder. In 1985, Paul Quinton obtained the first clue (figure 6.25). Quinton seized on a commonly observed characteristic of cystic fibrosis patients—their sweat is abnormally salty—and performed the following experiment. He isolated a sweat duct from a small piece of skin of a person with cystic fibrosis and placed it in a solution of salt (NaCl) that was three times as concentrated as the NaCl inside the duct. He then monitored the movement of ions. Diffusion is expected to drive both the sodium and chloride ions into the duct because of the higher outer ion concentrations. In skin isolated from normal individuals, both Na^+ and Cl^- indeed entered the duct, crossing the membrane easily. In skin isolated from cystic fibrosis patients, however, only Na^+ entered the ducts—no Cl^- crossed the membrane! Chloride channels did not appear to be functioning in these individuals.

The physical basis of the disease now became clear: affected individuals were not able to export chloride ions from their epithelial cells (cells lining body cavities and organs). Its buildup caused water to be drawn into these cells by osmosis, dehydrating and thickening the surrounding mucus. The defective gene was isolated in 1987, and its position on a particular human chromosome pinpointed in 1989. The gene encodes the protein *cystic fibrosis transmembrane conductance regulator* (CFTR). One of a family of transmembrane proteins that use ATP to transport certain ions and other molecules into and out of cells, the CFTR protein has four domains: two 6-pass channel elements and two ATP-binding elements. It functions to adjust the relative charge gradient across the cell membrane, pumping Cl^- out as required. In individuals with CF, the defective gene prevents this.

Now scientists are searching for ways to transfer working copies of the CFTR gene into cystic fibrosis patients. Initial attempts in 1992, using common cold viruses as carriers and animals as patients, were successful, and in 1994 CF mice with transplanted genes were completely cured. Human trials are beginning. For the first time, we can hope for a cure for cystic fibrosis.

Individuals with the inherited disorder cystic fibrosis produce a defective chloride channel. Their cells, unable to export chloride ions, take up water by osmosis, dangerously thickening the body's mucus.

Summary of Concepts

6.1 Biological membranes are fluid layers of lipid.

- Every cell is encased within a fluid bilayer sheet of phospholipid molecules called the plasma membrane.

6.2 Proteins embedded within the plasma membrane determine its character.

- Proteins that are embedded within the plasma membrane have their hydrophobic regions exposed to the hydrophobic interior of the bilayer, and their hydrophilic regions exposed to the cytoplasm or the extracellular fluid.

- Membrane proteins can transport materials into or out of the cell, they can mark the identity of the cell, or they can behave as receptors for extracellular information.

6.3 Passive transport across membranes moves down the concentration gradient.

- Diffusion is the kinetic movement of molecules or ions from an area of high concentration to an area of low concentration.

- Osmosis is the diffusion of water. Since all organisms are composed of mostly water, maintaining osmotic balance is essential to life.

6.4 Bulk transport utilizes endocytosis.

- Materials or volumes of fluid that are too large to pass directly through the cell membrane can move into or out of cells through endocytosis or exocytosis, respectively.

- In these processes, the cell expends energy to change the shape of its plasma membrane, allowing the cell to engulf materials into a temporary vesicle (endocytosis), or eject materials by fusing a filled vesicle with the plasma membrane (exocytosis).

6.5 Active transport across membranes is powered by energy from ATP.

- Cells use active transport to move substances across the plasma membrane against their concentration gradients, either accumulating them within the cell or extruding them from the cell. Active transport requires energy from ATP, either directly or indirectly.

Discussing Key Terms

1. **Lipid bilayer**   The foundation of the plasma membrane is a sheet composed of two layers of phospholipids, with the lipids oriented so that their hydrophobic ends face the interior of the sheet. The interior is thus very nonpolar, creating an effective barrier to the passage of polar substances.

2. **Transmembrane proteins**   Numerous proteins are anchored in the plasma membrane by nonpolar segments that pass through the nonpolar interior of the bilayer. Some transmembrane proteins transport substances into and out of the cell, others act as receptors for chemical signals from other cells or the environment, and still others serve as markers that identify particular cell types.

3. **Diffusion**   Diffusion is a form of passive transport, the net movement of molecules and ions toward regions where their concentrations are lower.

4. **Osmosis**   Net movement of water across a membrane. When two aqueous solutions are separated by a membrane that restricts the free passage of solutes, water diffuses across the membrane in the direction of higher solute concentration.

5. **Endocytosis and exocytosis**   In endocytosis, portions of the plasma membrane invaginate and fuse, trapping a volume of extracellular liquid and its contents within a vesicle. Similarly, in exocytosis, a vesicle merges with the plasma membrane and discharges its contents to the exterior.

6. **Active transport**   The movement of certain substances across the plasma membrane by ATP-powered transport proteins. These proteins can transport substances against their concentration gradients.

Review Questions

1. Describe the structure of a phospholipid molecule. How do phospholipids and fats differ in structure and in their interactions with water?

2. What type of physical arrangement do phospholipid molecules spontaneously assume when they are placed in water? How does this arrangement impede the passage of water-soluble substances?

3. How would increasing the number of phospholipids with double bonds between carbon atoms in their tails affect the fluidity of a membrane?

4. Describe the two basic types of structures that are characteristic of proteins that span membranes.

5. Define diffusion. What condition or state does diffusion ultimately lead to?

6. What is osmosis? What type of structure must a system have in order for osmosis to occur?

7. If a cell's cytoplasm were hyperosmotic to the extracellular fluid, how would the concentration of solutes in the cytoplasm compare with that in the extracellular fluid? How would these concentrations compare if the cytoplasm were hypoosmotic to the extracellular fluid?

8. How do phagocytosis and pinocytosis differ?

9. Describe the mechanism of receptor-mediated endocytosis.

10. What two factors determine the direction of net movement of ions through ion channels?

11. In what two ways does facilitated diffusion differ from simple diffusion across a membrane?

12. How does active transport differ from facilitated diffusion? How is it similar to facilitated diffusion?

13. How many sodium and potassium ions are moved across the membrane by the sodium-potassium pump in each pump cycle, and in which directions are they moved? What molecule provides the energy that drives the pump?

14. What types of substances are moved across the plasma membrane by cotransport, and in which directions are they moved? What types of substances are moved across the plasma membrane by countertransport, and in which directions are they moved?

15. What is the proton pump? What biochemical process is it coupled to?

Thought Questions

1. If the viscosity of membranes increases at low temperatures, how do cells keep their plasma membranes fluid in cold environments (e.g., in the water beneath the ice on a frozen lake)?

2. Why is a lipid bilayer membrane freely permeable to water, which is quite polar, but not freely permeable to ammonia, which is also polar and has molecules of about the same size?

3. Cells can concentrate many internal molecules used for fuel by coupling their movement into the cell to the sodium gradient established by the sodium-potassium pump. What happens to the potassium ions that are also transported by this pump?

Internet Links

Electron Microscopy
http://www.unl.edu/CMRAcfem/em.htm
An excellent tour of the electron microscope, maintained by the University of Nebraska.

How a Scanning Electron Microscope Works
http://hypatia.dartmouth.edu/levey/ssml/equipment/SEM/SEM_intro.html
This Dartmouth College web site explains how SEM provides surface images, with detailed treatment of imaging and sample preparation.

For Further Reading

Gabriel, S. E., K. N. Brigman, B. H. Koller, R. C. Boucher, and M. J. Stutts: "Cystic Fibrosis Heterozygote Resistance to Cholera Toxin in the Cystic Fibrosis Mouse Model," *Science*, October 7, 1994. How the gene for cystic fibrosis may confer resistance to cholera.

Jan, L. Y., and Y. N. Jan: "Potassium Channels and Their Evolving Gates," *Nature*, September 8, 1994. How potassium channels preferentially transport potassium, how they are related to other channels, and how their activity is controlled.

Lienhard, G., and others: "How Cells Absorb Glucose," *Scientific American*, January 1992, pages 86–91. A description of research on the structure of the glucose transport protein.

McNeil, P.: "Cell Wounding and Healing," *American Scientist*, May 1991, pages 222–35. A discussion of the role the cell membrane plays in tissue repair.

Rennie, J.: "Leaky Channels," *Scientific American*, January 1991. A devastating inherited disease may be linked to a faulty channel protein in the cell membrane.

Simon, S.: "Enter the 'Swinging Gate,'" *Nature*, September 8, 1994, pages 103–4. A brief discussion on how conformational changes in channel proteins affect ion transport.

7

Cell-Cell Interactions

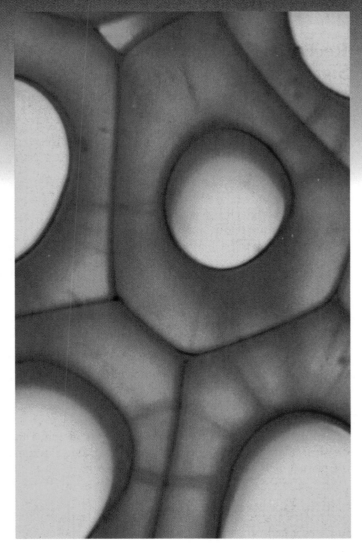

FIGURE 7.1
Persimmon cells in close contact with one another. These plant cells and all cells, no matter what their function, interact with their environment, including the cells around them.

Concept Outline

7.1 Cells signal one another with chemicals.

Receptor Proteins and Signaling between Cells. Receptor proteins embedded in the plasma membrane change shape when they bind specific signal molecules, triggering a chain of events within the cell.

Types of Cell Signaling. Cell signaling can occur between adjacent cells, although chemical signals called hormones act over long distances.

7.2 Proteins in the cell and on its surface receive signals from other cells.

Intracellular Receptors. Some receptors are located within the cell cytoplasm. These receptors respond to lipid-soluble signals, such as steroid hormones.

Cell Surface Receptors. Most cell-to-cell signals are water-soluble and cannot penetrate membranes. Instead, the signals are received by transmembrane proteins protruding out from the cell surface.

7.3 Follow the journey of information into the cell.

Initiating the Intracellular Signal. Cell surface receptors often use "second messengers" to transmit a signal to the cytoplasm.

Amplifying the Signal: Protein Kinase Cascades. Surface receptors and second messengers amplify signals as they travel into the cell, often toward the cell nucleus.

7.4 Cell surface proteins mediate cell-cell interactions.

The Expression of Cell Identity. Cells possess on their surfaces a variety of tissue-specific identity markers that identify both the tissue and the individual.

Intercellular Adhesion. Cells attach themselves to one another with protein links. Some of the links are very strong, others more transient.

Communicating between Cells. Many adjacent cells have direct passages that link their cytoplasms, permitting the passage of ions and small molecules.

Did you know that each of the 100 trillion cells of your body shares one key feature with the cells of tigers, bumblebees, and persimmons (figure 7.1)—a feature that most bacteria and protists lack? Your cells touch and communicate with one another. Sending and receiving a variety of chemical signals, they coordinate their behavior so that your body functions as an integrated whole, rather than as a massive collection of individual cells acting independently. The ability of cells to communicate with one another is the hallmark of multicellular organisms. In this chapter we will look in detail at how the cells of multicellular organisms interact with one another, first exploring how they signal one another with chemicals and then examining the ways in which their cell surfaces interact to organize tissues and body structures.

Receptor Proteins and Signaling between Cells

Communication between cells is common in nature. Single-celled eukaryotes such as yeast, for example, initiate sex by secreting small peptides (protein chains only a few amino acids long) called *mating factors*. A mating factor alerts cells of the opposite mating type that the secreting cell is ready to mate. This communication path is quite elaborate, involving a peptide signal, specific membrane receptor proteins that recognize the signal, and a variety of proteins that transmit the signal into the receiving cell's interior.

Cell signaling occurs in all multicellular organisms, providing an indispensable mechanism for cells to influence one another. The cells of multicellular organisms use a variety of molecules as signals, including not only peptides, but also large proteins, individual amino acids, nucleotides, steroids and other lipids, and even dissolved gases like nitric oxide. Some of these molecules are attached to the surface of the signaling cell; others are secreted through the plasma membrane or released by exocytosis.

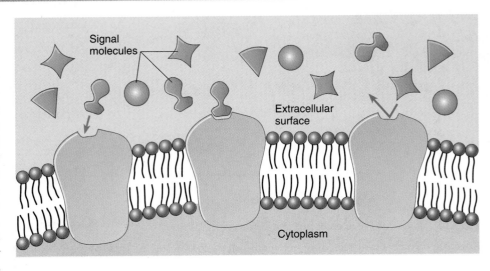

FIGURE 7.2
Cell surface receptors recognize only specific molecules. Signal molecules will bind only to those cells displaying receptor proteins with a shape into which they can fit snugly.

Cell Surface Receptors

Any given cell of a multicellular organism is exposed to a constant stream of signals. At any time, hundreds of different chemical signals may be in the environment surrounding the cell. However, each cell responds only to certain signals and ignores the rest (figure 7.2), like a person following the conversation of one or two individuals in a noisy, crowded room. How does a cell "choose" which signals to respond to? Located on or within the cell are **receptor proteins,** each with a three-dimensional shape that fits the shape of a specific signal molecule. When a signal molecule approaches a receptor protein of the right shape, the two can bind. This binding induces a change in the receptor protein's shape, ultimately producing a response in the cell. Hence, a given cell responds to the signal molecules that fit the particular set of receptor proteins it possesses, and ignores those for which it lacks receptors.

The Hunt for Receptor Proteins

The characterization of receptor proteins has presented a very difficult technical problem, because of their relative scarcity in the cell. Since these proteins may constitute less than 0.01% of the total mass of protein in a cell, purifying them is analogous to searching for a particular grain of sand in a sand dune! However, two recent techniques have enabled cell biologists to make rapid progress in this area.

Monoclonal antibodies. The first method uses *monoclonal antibodies*. An antibody is an immune system protein that, like a receptor, binds specifically to another molecule. Each individual immune system cell can make only one particular type of antibody, which can bind to only one specific target molecule. Thus, a cell-line derived from a single immune system cell (a clone) makes one specific antibody (a *monoclonal* antibody). Monoclonal antibodies that bind to particular receptor proteins can be used to isolate those proteins from the thousands of others in the cell.

Gene isolation. The second technique applies *genetic engineering*, which we will consider in detail in chapter 18. In this approach, the genes that encode receptor proteins are identified, and the structure of the proteins is deduced from the sequence of nucleotides in the genes.

Remarkably, these techniques have revealed that the enormous number of receptor proteins can be grouped into just a handful of "families" containing many related receptors. Later in this chapter we will meet some of the members of these receptor families.

Cells in a multicellular organism communicate with others by releasing signal molecules that bind to receptor proteins on the surface of the other cells. Recent advances in protein isolation have yielded a wealth of information about the structure and function of these proteins.

Types of Cell Signaling

Cells communicate through any of four basic mechanisms, depending primarily on the distance between the signaling and responding cells (figure 7.3). In addition to using these four basic mechanisms, some cells actually send signals to themselves, secreting signals that bind to specific receptors on their own plasma membranes. This process, called **autocrine signaling,** is thought to play an important role in reinforcing developmental changes.

Direct Contact

As we saw in chapter 6, the surface of a eukaryotic cell is a thicket of proteins, carbohydrates, and lipids attached to and extending outward from the plasma membrane. When cells are very close to one another, some of the molecules on the cells' plasma membranes may bind together in specific ways. Many of the important interactions between cells in early development occur by means of direct contact between cell surfaces (figure 7.3a). We'll examine contact-dependent interactions more closely later in this chapter.

Paracrine Signaling

Signal molecules released by cells can diffuse through the extracellular fluid to other cells. If those molecules are taken up by neighboring cells, destroyed by extracellular enzymes, or quickly removed from the extracellular fluid in some other way, their influence is restricted to cells in the immediate vicinity of the releasing cell. Signals with such short-lived, local effects are called **paracrine** signals (figure 7.3b). Like direct contact, paracrine signaling plays an important role in early development, coordinating the activities of clusters of neighboring cells.

Endocrine Signaling

If a released signal molecule remains in the extracellular fluid, it may enter the organism's circulatory system and travel widely throughout the body. These longer lived signal molecules, which may affect cells very distant from the releasing cell, are called **hormones,** and this type of intercellular communication is known as **endocrine** signaling (figure 7.3c). Chapter 53 discusses endocrine signaling in detail. Both animals and plants use this signaling mechanism extensively.

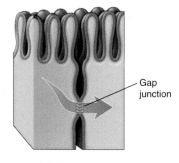

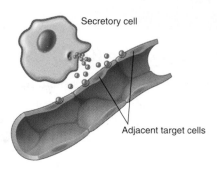

(a) Direct contact

(b) Paracrine signaling

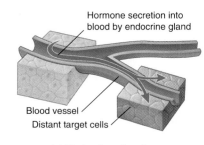

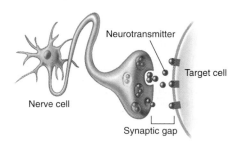

(c) Endocrine signaling

(d) Synaptic signaling

FIGURE 7.3

Four kinds of cell signaling. Cells communicate in several ways. (a) Two cells in *direct contact* with each other may send signals across gap junctions. (b) In *paracrine signaling,* secretions from one cell have an effect only on cells in the immediate area. (c) In *endocrine signaling,* hormones are released into the circulatory system, which carries them to the target cells. (d) Chemical *synaptic signaling* involves transmission of signal molecules, called neurotransmitters, from a neuron over a small synaptic gap to the target cell.

Synaptic Signaling

In animals, the cells of the nervous system provide rapid communication with distant cells. Their signal molecules, **neurotransmitters,** do not travel to the distant cells through the circulatory system like hormones do. Rather, the long, fiberlike extensions of nerve cells release neurotransmitters from their tips very close to the target cells (figure 7.3d). The narrow gap between the two cells is called a **chemical synapse.** While paracrine signals move through the fluid between cells, neurotransmitters cross the synapse and persist only briefly. We will examine synaptic signaling more fully in chapter 51.

Adjacent cells can signal others by direct contact, while nearby cells that are not touching can communicate through paracrine signals. Two other systems mediate communication over longer distances: in endocrine signaling the blood carries hormones to distant cells, and in synaptic signaling nerve cells secrete neurotransmitters from long cellular extensions close to the responding cells.

Intracellular Receptors

All cell signaling pathways share certain common elements, including a chemical signal that passes from one cell to another and a receptor that receives the signal in or on the target cell. We've looked at the sorts of signals that pass from one cell to another. Now let's consider the nature of the receptors that receive the signals. Table 7.1 summarizes the types of receptors we will discuss in this chapter, as well as a variety of ways cells make physical contact with one another.

Many cell signals are lipid-soluble or very small molecules that can readily pass across the plasma membrane of the target cell and into the cell, where they interact with a receptor. Some bind to protein receptors located in the cytoplasm; others pass across the nuclear membrane as well and bind to receptors within the nucleus. These **intracellular receptors** (figure 7.4) may trigger a variety of responses in the cell, depending on the function of the receptor.

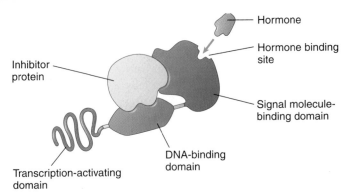

FIGURE 7.4
Basic structure of a gene-regulating intracellular receptor. These receptors are located within the cell and function in the reception of signals such as steroid hormones, vitamin D, and thyroid hormone.

Table 7.1   Cell Communicating Mechanisms

Mechanism	Structure	Function	Example
INTRACELLULAR RECEPTORS	No extracellular signal-binding site	Receives signals from lipid-soluble or noncharged, nonpolar small molecules	Receptors for NO, steroid hormone, vitamin D, and thyroid hormone
CELL SURFACE RECEPTORS			
Chemically gated ion channels	Multipass transmembrane protein forming a central pore	Molecular "gates" triggered chemically to open or close	Neurons
Enzymic receptors	Single-pass transmembrane protein	Binds signal extracellularly, catalyzes response intracellularly	Phosphorylation of protein kinases
G protein-linked receptors	Seven-pass transmembrane protein with cytoplasmic binding site for G protein	Binding of signal to receptor effects a conformational change in protein, which activates GTP; GTP detaches to deliver signal inside the cell	Peptide hormones, rod cells in the eyes
PHYSICAL CONTACT WITH OTHER CELLS			
Surface markers	Variable; integral proteins or glycolipids in cell membrane	Identify the cell	MHC complexes, blood groups, antibodies
Tight junctions	Tightly bound, leakproof, fibrous protein "belt" that surrounds cell	Organizing junction: holds cells together, allowing materials to pass *through* but not *between* the cells	Junctions between epithelial cells in the gut
Adherens junctions	Transmembrane fibrous proteins	Anchoring junction: "roots" extracellular matrix to cytoskeleton	Tissues with high mechanical stress, such as the skin
Desmosomes	Intermediate filaments of cytoskeleton linked to adjoining cells through cadherins	Anchoring junction: "buttons" cells together	Epithelium
Gap junctions	Six transmembrane connexon proteins creating a "pipe" that connects cells	Communicating junction: allows passage of small molecules from cell to cell in a tissue	Excitable tissue such as heart muscle
Plasmodesmata	Cytoplasmic connections between gaps in adjoining plant cell walls	Communicating junction between plant cells	Plant tissues

Receptors that Act as Gene Regulators

Some intracellular receptors act as regulators of gene transcription. Among them are the receptors for steroid hormones, such as cortisol, estrogen, and progesterone, as well as the receptors for a number of other small, lipid-soluble signal molecules, such as vitamin D and thyroid hormone. All of these receptors have similar structures; the genes that code for them may well be the evolutionary descendants of a single ancestral gene. Because of their structural similarities, they are all part of the **intracellular receptor superfamily.**

Each of these receptors has a binding site for DNA. In its inactive state, the receptor typically cannot bind DNA because an inhibitor protein occupies the binding site. When the signal molecule binds to another site on the receptor, the inhibitor is released and the DNA binding site is exposed (figure 7.5). The receptor then binds to a specific nucleotide sequence on the DNA, which activates (or, in a few instances, suppresses) a particular gene, usually located adjacent to the regulatory site.

The lipid-soluble signal molecules that intracellular receptors recognize tend to persist in the blood far longer than water-soluble signals. Most water-soluble hormones break down within minutes, and neurotransmitters within seconds or even milliseconds. A steroid hormone like cortisol or estrogen, on the other hand, persists for hours.

The target cell's response to a lipid-soluble cell signal can vary enormously, depending on the nature of the cell. This is true even when different target cells have the same intracellular receptor, for two reasons: First, the binding site for the receptor on the target DNA differs from one cell type to another, so that different genes are affected when the signal-receptor complex binds to the DNA, and second, most eukaryotic genes have complex controls. We will discuss them in detail in chapter 16, but for now it is sufficient to note that several different regulatory proteins are usually involved in reading a eukaryotic gene. Thus the intracellular receptor interacts with different signals in different tissues. Depending on the cell-specific controls operating in different tissues, the effect the intracellular receptor produces when it binds with DNA will vary.

Receptors that Act as Enzymes

Other intracellular receptors act as enzymes. A very interesting example is the receptor for the signal molecule, **nitric oxide (NO).** A small gas molecule, NO diffuses readily out of the cells where it is produced and passes directly into neighboring cells, where it binds to the enzyme guanylyl cyclase. Binding of NO activates the enzyme, enabling it to catalyze the synthesis of cyclic guanosine monophosphate

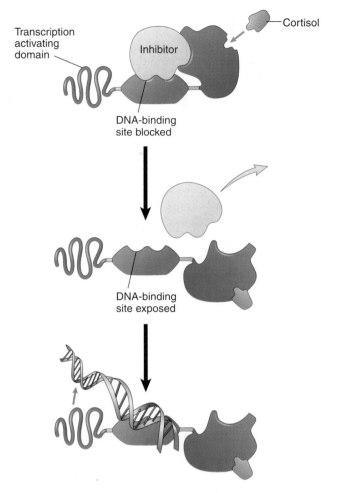

FIGURE 7.5
How intracellular receptors regulate gene transcription. In this model, the binding of the steroid hormone cortisol to a DNA regulatory protein causes it to alter its shape. The inhibitor is released, exposing the DNA-binding site of the regulatory protein. The DNA binds to the site, positioning a specific nucleotide sequence over the transcription activating domain of the receptor and initiating transcription.

(GMP), an intracellular messenger molecule that produces cell-specific responses such as the relaxation of smooth muscle cells.

NO has only recently been recognized as a signal molecule in vertebrates. Already, however, a wide variety of roles have been documented. For example, when the brain sends a nerve signal relaxing the smooth muscle cells lining the walls of vertebrate blood vessels, the signal molecule acetylcholine released by the nerve near the muscle does not interact with the muscle cell directly. Instead, it causes nearby epithelial cells to produce NO, which then causes the smooth muscle to relax, allowing the vessel to expand and thereby increase blood flow.

Many target cells possess intracellular receptors, which are activated by substances that pass through the plasma membrane.

Cell Surface Receptors

Most signal molecules are water-soluble, including neurotransmitters, peptide hormones, and the many proteins that multicellular organisms employ as "growth factors" during development. Water-soluble signals cannot diffuse through cell membranes. Therefore, to trigger responses in cells, they must bind to receptor proteins on the surface of the cell. These **cell surface receptors** (figure 7.6) convert the extracellular signal to an intracellular one, responding to the binding of the signal molecule by producing a change within the cell's cytoplasm. Most of a cell's receptors are cell surface receptors, and almost all of them belong to one of three receptor superfamilies: chemically gated ion channels, enzymic receptors, and G protein-linked receptors.

Chemically Gated Ion Channels

Chemically gated ion channels are receptor proteins that ions pass through. The receptor proteins that bind many neurotransmitters have the same basic structure (figure 7.6*a*). Each is a "multipass" transmembrane protein, meaning that the chain of amino acids threads back and forth across the plasma membrane several times. In the center of the protein is a pore that connects the extracellular fluid with the cytoplasm. The pore is big enough for ions to pass through, so the protein functions as an **ion channel.** The channel is said to be chemically gated because it opens when a chemical (the neurotransmitter) binds to it. The type of ion (sodium, potassium, calcium, chloride, for example) that flows across the membrane when a chemically gated ion channel opens depends on the specific three-dimensional structure of the channel.

Enzymic Receptors

Many cell surface receptors either act as enzymes or are directly linked to enzymes (figure 7.6*b*). When a signal molecule binds to the receptor, it activates the enzyme. In almost all cases, these enzymes are **protein kinases,** enzymes that add phosphate groups to proteins. Most enzymic receptors have the same general structure. Each is a single-pass transmembrane protein (the amino acid chain passes through the plasma membrane only once); the portion that binds the signal molecule lies outside the cell, and the portion that carries out the enzyme activity is exposed to the cytoplasm.

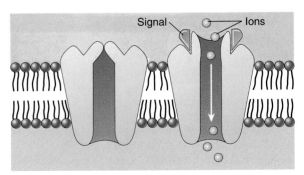

(a) Chemically gated ion channel

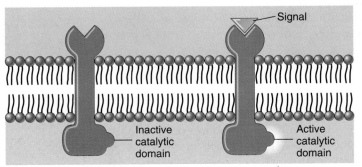

(b) Enzymic receptor

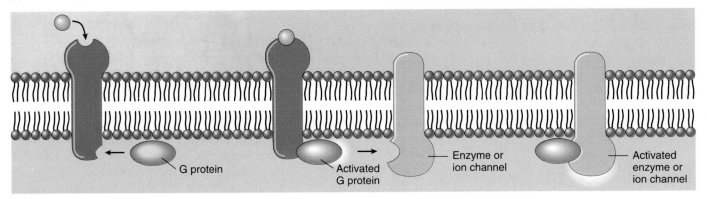

(c) G protein-linked receptor

FIGURE 7.6

Cell surface receptors. (a) Chemically gated ion channels are multipass transmembrane proteins that form a pore in the cell membrane. This pore is opened or closed by chemical signals. (b) Enzymic receptors are single-pass transmembrane proteins that bind the signal on the extracellular surface. A catalytic region on their cytoplasmic portion then initiates enzymatic activity inside the cell. (c) G protein-linked receptors bind to the signal outside the cell and to G proteins inside the cell. The G protein then activates an enzyme or ion channel, mediating the passage of a signal from the cell's surface to its interior.

G Protein-Linked Receptors

A third class of cell surface receptors acts indirectly on enzymes or ion channels in the plasma membrane with the aid of an assisting protein, called a *guanosine triphosphate* (GTP)-*binding protein*, or **G protein** (figure 7.6c). Receptors in this category use G proteins to mediate passage of the signal from the membrane surface into the cell interior.

Discovery of G Proteins. Martin Rodbell of the National Institute of Environmental Health Sciences and Alfred Gilman of the University of Texas Southwestern Medical Center received the 1994 Nobel Prize for Medicine or Physiology for their work on G proteins. Rodbell, a biochemist, had theorized about the existence of G proteins for several years. His views often drew skepticism from his colleagues, as they contradicted the prevailing thinking on how receptors work. In 1980, Gilman's group finally managed to isolate and purify a G protein, thus proving their existence.

Rodbell and Gilman's work has proven to have significant ramifications. G proteins are involved in the mechanism employed by over half of all medicines in use today. Studying G proteins will vastly expand our understanding of how these medicines work. Furthermore, the investigation of G proteins should help elucidate how cells communicate in general and how they contribute to the overall physiology of organisms. As Gilman says, G proteins are "involved in everything from sex in yeast to cognition in humans."

The Largest Family of Cell Surface Receptors. Scientists have identified more than 100 different G protein-linked receptors, more than any other kind of cell surface receptor. They mediate an incredible range of cell signals, including peptide hormones, neurotransmitters, fatty acids, and amino acids. Despite this great variation in specificity, however, all G protein-linked receptors whose amino acid sequences are known have a similar structure. They are almost certainly closely related in an evolutionary sense, arising from a single ancestral sequence. Each of these G protein-linked receptors is a seven-pass transmembrane protein (figure 7.7)—a single polypeptide chain that threads back and forth across the lipid bilayer seven times, creating a channel through the membrane.

Evolutionary Origin of G Protein-Linked Receptors. As research revealed the structure of G protein-linked receptors, an interesting pattern emerged: the same seven-pass structural motif is seen again and again, in sensory receptors such as the light-activated rhodopsin protein in the vertebrate eye, in the light-activated bacteriorhodopsin proton pump that plays a key role in bacterial photosynthesis, in the receptor that recognizes the yeast mating factor protein discussed earlier, and in many other sensory receptors. Vertebrate rhodopsin is in fact a G protein-linked receptor

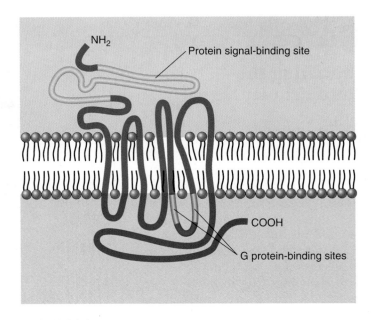

FIGURE 7.7
The G protein-linked receptor is a seven-pass transmembrane protein.

and utilizes a G protein. Bacteriorhodopsin is not. The occurrence of the seven-pass structural motif in both, and in so many other G protein-linked receptors, suggests that this motif is a very ancient one, and that G protein-linked receptors may have evolved from sensory receptors of single-celled ancestors.

How G Protein-Linked Receptors Work. G proteins are mediators that initiate a diffusible signal in the cytoplasm. They forge the key link between the receptor on the cell surface and signal pathways within the cytoplasm. When a signal arrives, it finds the G protein nestled into the G protein-linked receptor on the cytoplasmic side of the plasma membrane. Once the signal molecule binds to the receptor, the G protein-linked receptor changes shape. This change in receptor shape twists the G protein, causing it to bind GTP, so that it is now "activated." The activated G protein then diffuses away from the receptor. This starts a chain of events (described in the next section) that ultimately brings about the cell's response.

Most receptors are located on the surface of the plasma membrane. Chemically gated ion channels open or close when signal molecules bind to the channel, allowing specific ions to diffuse through. Enzyme receptors typically activate intracellular proteins by phosphorylation. G protein-linked receptors activate an intermediary protein, which then effects the intracellular change.

Initiating the Intracellular Signal

Some enzymic receptors and most G protein-linked receptors carry the signal molecule's message into the target cell by utilizing other substances to relay the message within the cytoplasm. These other substances, small molecules or ions commonly called **second messengers,** or intracellular mediators, alter the behavior of particular proteins by binding to them and changing their shape. The two most widely used second messengers are cyclic adenosine monophosphate (cAMP) and calcium.

cAMP

All animal cells studied thus far use **cAMP** as a second messenger (chapter 53 discusses cAMP in detail). To see how cAMP typically works as a messenger, let's examine what happens when the hormone epinephrine binds to a particular type of G protein-linked receptor called the β-adrenergic receptor (figure 7.8). When epinephrine binds with this receptor, it activates a G protein, which then stimulates the enzyme **adenylyl cyclase** to produce large amounts of cAMP within the cell (figure 7.9a). The cAMP then binds to and activates the enzyme α-kinase, which adds phosphates to specific proteins in the cell. The effect this phosphorylation has on cell function depends on the identity of the cell and the proteins that are phosphorylated. In muscle cells, for example, the α-kinase phosphorylates and thereby activates enzymes that stimulate the breakdown of glycogen into glucose and inhibit the synthesis of glycogen from glucose. Glucose is then more available to the muscle cells for metabolism.

Calcium

Calcium (Ca^{++}) ions serve even more widely than cAMP as second messengers. Ca^{++} levels inside the cytoplasm of a cell are normally very low (less than 10^{-7} M), while outside the cell and in the endoplasmic reticulum Ca^{++} levels are quite high (about 10^{-3} M). Chemically gated calcium channels in the endoplasmic reticulum membrane act as switches; when they open, Ca^{++} rushes into the cytoplasm and triggers proteins sensitive to Ca^{++} to initiate a variety of

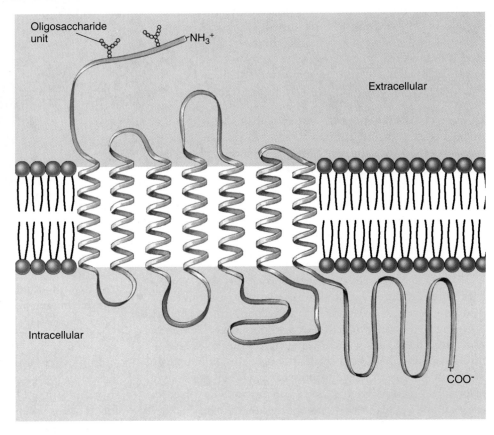

FIGURE 7.8
Structure of the β-adrenergic receptor. The receptor is a G protein-linked molecule which, when it binds to an extracellular signal molecule, stimulates voluminous production of cAMP inside the cell, which then effects the cellular change.

activities. For example, the efflux of Ca^{++} from the endoplasmic reticulum causes skeletal muscle cells to contract and some endocrine cells to secrete hormones.

The gated Ca^{++} channels are opened by a G protein-linked receptor. In response to signals from other cells, the receptor activates its G protein, which in turn activates the enzyme, *phospholipase C.* This enzyme catalyzes the production of *inositol trisphosphate* (IP_3) from phospholipids in the plasma membrane. The IP_3 molecules diffuse through the cytoplasm to the endoplasmic reticulum and bind to the Ca^{++} channels. This opens the channels and allows Ca^{++} to flow from the endoplasmic reticulum into the cytoplasm (figure 7.9b).

Ca^{++} initiates some cellular responses by binding to *calmodulin,* a 148-amino acid cytoplasmic protein that contains four binding sites for Ca^{++} (figure 7.10). When four Ca^{++} ions are bound to calmodulin, the calmodulin/Ca^{++} complex binds to other proteins, and activates them.

Cyclic AMP and Ca^{++} often behave as second messengers, intracellular substances that relay messages from receptors to target proteins.

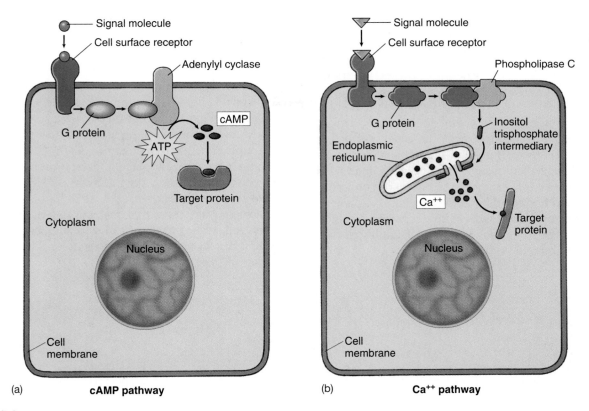

(a) **cAMP pathway**

(b) **Ca++ pathway**

FIGURE 7.9

How second messengers work. (a) The cyclic AMP (cAMP) pathway. An extracellular receptor binds to a signal molecule and, through a G protein, activates the membrane-bound enzyme, adenylyl cyclase. This enzyme catalyzes the synthesis of cAMP, which binds to the target protein to intiate the cellular change. (b) The calcium (Ca++) pathway. An extracellular receptor binds to another signal molecule and, through another G protein, activates the enzyme phospholipase C. This enzyme stimulates the production of inositol trisphosphate, which binds to and opens calcium channels in the membrane of the endoplasmic reticulum. Ca++ is released into the cytoplasm, effecting a change in the cell.

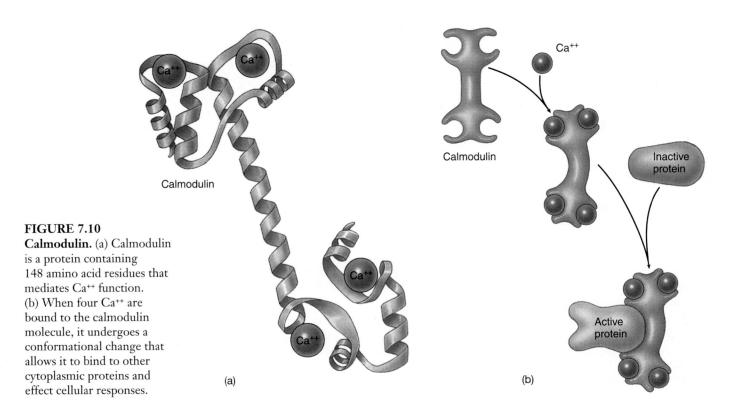

FIGURE 7.10

Calmodulin. (a) Calmodulin is a protein containing 148 amino acid residues that mediates Ca++ function. (b) When four Ca++ are bound to the calmodulin molecule, it undergoes a conformational change that allows it to bind to other cytoplasmic proteins and effect cellular responses.

Amplifying the Signal: Protein Kinase Cascades

Both enzyme-linked and G protein-linked receptors receive signals at the surface of the cell, but as we've seen, the target cell's response rarely takes place there. In most cases the signals are relayed to the cytoplasm or the nucleus by second messengers, which influence the activity of one or more enzymes or genes and so alter the behavior of the cell. But a messenger's diffusion across the cytoplasm tends to blunt the impact of a signal, since diffusing molecules travel slowly and become diluted in the cytoplasm. Therefore, most enzyme-linked and G protein-linked receptors use a chain of other protein messengers to amplify the signal as it is being relayed to the nucleus.

How is the signal amplified? Imagine a relay race where, at the end of each stage, the finishing runner tags five new runners to start the next stage. The number of runners would increase dramatically as the race progresses: 1, then 5, 25, 125, and so on. The same sort of process takes place as a signal is passed from the cell surface to the cytoplasm or nucleus. First the receptor activates a stage-one protein, almost always by phosphorylating it. The receptor either adds a phosphate group directly, or adds GTP, which contains an activating phosphate. Once activated, each of these stage-one proteins in turn activates a large number of stage-two proteins; then each of them activates a large number of stage-three proteins, and so on (figure 7.11). A single cell surface receptor can thus stimulate a cascade of protein kinases to amplify the signal.

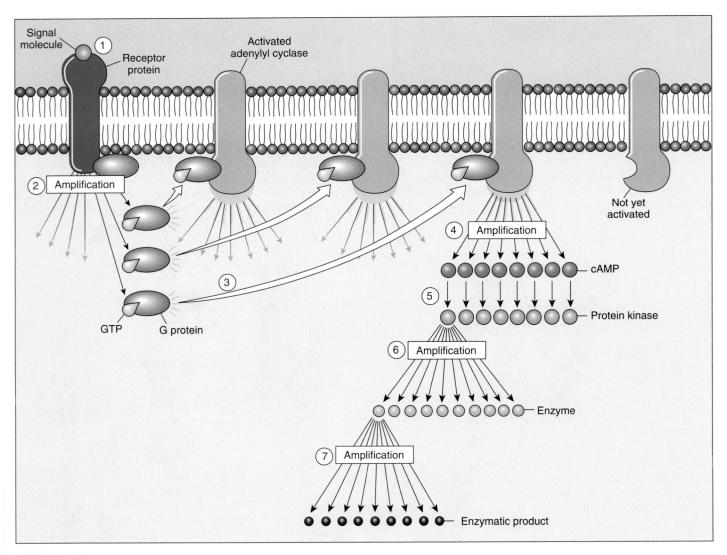

FIGURE 7.11

Signal amplification. There is opportunity for amplification at many steps of the cell-signaling process, which can ultimately produce a large response by the cell. One cell surface receptor (1), for example, may activate many G protein molecules (2), each of which activates a molecule of adenylyl cyclase (3), yielding an enormous production of cAMP (4). Each cAMP molecule in turn will activate a protein kinase (5), which can phosphorylate and thereby activate several copies of a specific enzyme (6). Each of *those* enzymes can then catalyze many chemical reactions (7). Thus, one cell surface receptor can trigger the production of many thousands of molecules of product.

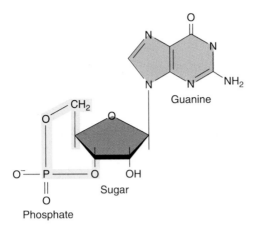

FIGURE 7.12

Cyclic-GMP. Cyclic-GMP is a guanosine monophosphate nucleotide molecule with the single phosphate group attached to a sugar residue in two places (this cyclic part is shown in yellow). Cyclic-GMP is an important second messenger linking G proteins to signal transduction pathways within the cytoplasm.

The Vision Amplification Cascade

Let's trace a protein amplification cascade to see exactly how one works. In vision, a single light-activated rhodopsin (a G protein-linked receptor) activates hundreds of molecules of the G protein transducin in the first stage of the relay. In the second stage, each transducin causes an enzyme to modify thousands of molecules of a special inside-the-cell messenger called cyclic-GMP (figure 7.12). (We will discuss cyclic GMP in more detail later.) In about 1 second, a single rhodopsin signal passing through this two-step cascade produces more than 10^5 (100,000) cyclic-GMP molecules (figure 7.13)!

The Cell Division Amplification Cascade

The amplification of signals traveling from the plasma membrane to the nucleus can be even more complex than the process we've just described. Cell division, for example, is controlled by a receptor that acts as a protein kinase. The receptor responds to growth-promoting signals by phosphorylating an intracellular protein called ras, which then activates a series of interacting phosphorylation cascades, some with five or more stages. If the ras protein becomes hyperactive for any reason, the cell acts as if it is being constantly stimulated to divide. Ras proteins were first discovered in cancer cells. A mutation of the gene that encodes ras had caused it to become hyperactive, resulting in unrestrained cell proliferation. Almost one-third of all human cancers have such a mutation in a *ras* gene.

A small number of surface receptors can generate a vast intracellular response, as each stage of the pathway amplifies the next.

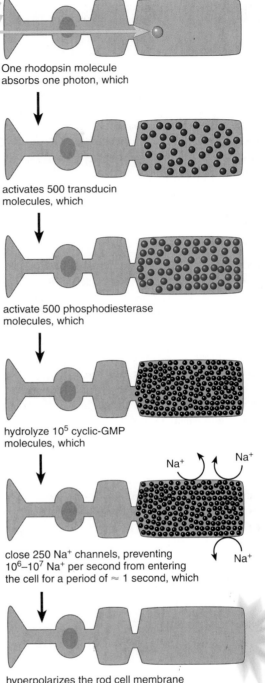

One rhodopsin molecule absorbs one photon, which

activates 500 transducin molecules, which

activate 500 phosphodiesterase molecules, which

hydrolyze 10^5 cyclic-GMP molecules, which

close 250 Na⁺ channels, preventing 10^6–10^7 Na⁺ per second from entering the cell for a period of ≈ 1 second, which

hyperpolarizes the rod cell membrane by 1 mV, sending a visual signal to the brain.

FIGURE 7.13

The role of signal amplification in vision. In this vertebrate rod cell (the cells of the eye responsible for interpreting light and dark), *one* single rhodopsin pigment molecule, when excited by a photon, ultimately yields *100,000* split cGMP molecules, which will then effect a change in the membrane of the rod cell, which will be interpreted by the organism as a visual event.

The Expression of Cell Identity

With the exception of a few primitive types of organisms, the hallmark of multicellular life is the development of highly specialized groups of cells called **tissues,** such as skin, blood, and muscle. Remarkably, each cell within a tissue performs the functions of that tissue and no other, even though all cells of the body are derived from a single fertilized cell and contain the same genetic information. How do cells sense where they are, and how do they "know" which type of tissue they belong to?

Tissue-Specific Identity Markers

As it develops, each cell type acquires a unique set of cell surface molecules. These molecules serve as markers proclaiming the cells' tissue-specific identity. Other cells that make direct physical contact with them "read" the markers.

Glycolipids.   Most tissue-specific cell surface markers are glycolipids (figure 7.14), lipids with carbohydrate heads. The glycolipids on the surface of red blood cells are also responsible for the differences among A, B, and O blood types. As the cells in a tissue divide and differentiate, the population of cell surface glycolipids changes dramatically.

MHC Proteins.   The immune system uses other cell surface markers to distinguish between "self" and "nonself" cells. All of the cells of a given individual, for example, have the same "self" markers, called *major histocompatibility complex* (MHC) *proteins.* Because practically every individual makes a different set of MHC proteins, they serve as distinctive identity tags for each individual. The MHC proteins and other self-identifying markers are single-pass proteins anchored in the plasma membrane,

and many of them are members of a large superfamily of receptors, the immunoglobulins (figure 7.15). Cells of the immune system continually inspect the other cells they encounter in the body, triggering the destruction of cells that display foreign or "nonself" identity markers.

> **Every cell contains a specific array of marker proteins on its surface. These markers identify each type of cell in a very precise way.**

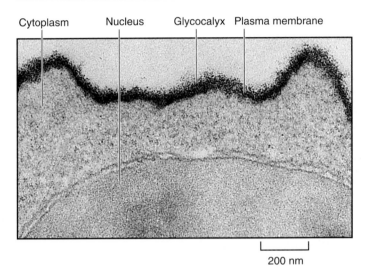

Cytoplasm   Nucleus   Glycocalyx   Plasma membrane

200 nm

FIGURE 7.14
The glycocalyx (70,000×). In this transmission electron micrograph of a lymphocyte, the glycocalyx, a coating of glycolipids, is visible as an electron-dense fuzzy layer immediately overlying the plasma membrane.

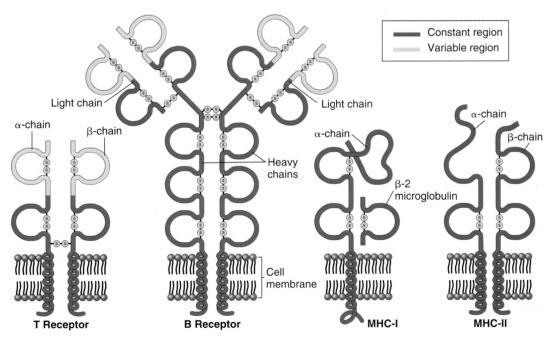

FIGURE 7.15
Structure of the immunoglobulin family of cell surface marker proteins. T and B cell receptors help mediate the immune response in organisms by recognizing and binding to foreign cell markers. MHC antigens label cells as "self," so that the immune system attacks only invading entities, such as bacteria, viruses, and usually even the cells of transplanted organs!

Light chain   Light chain
α-chain   β-chain   α-chain   α-chain   β-chain
Heavy chains
β-2 microglobulin
Cell membrane
Constant region
Variable region
T Receptor   **B Receptor**   **MHC-I**   **MHC-II**

Intercellular Adhesion

Not all physical contacts between cells in a multicellular organism are fleeting touches. In fact, most cells are in physical contact with other cells at all times, usually as members of organized tissues such as those in the lungs, heart, or gut. These cells and the mass of other cells clustered around them form long-lasting or permanent connections with each other called **cell junctions** (figure 7.16). The nature of the physical connections between the cells of a tissue in large measure determines what the tissue is like. Indeed, a tissue's proper functioning often depends critically upon how the individual cells are arranged within it. Just as a house cannot maintain its structure without nails and cement, so a tissue cannot maintain its characteristic architecture without the appropriate cell junctions.

Cell junctions are divided into three categories, based upon the functions they serve (figure 7.17): tight junctions, anchoring junctions, and communicating junctions.

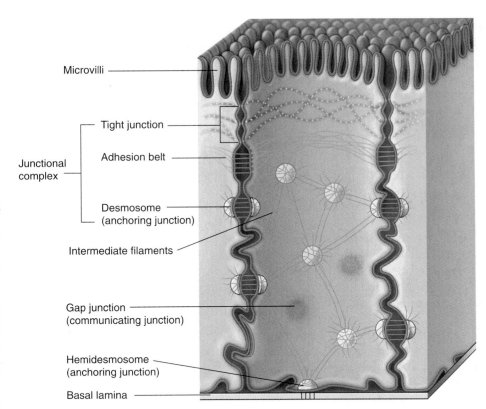

FIGURE 7.16
A summary of cell junction types. Gut epithelial cells are used here to illustrate the comparative structures and locations of common cell junctions.

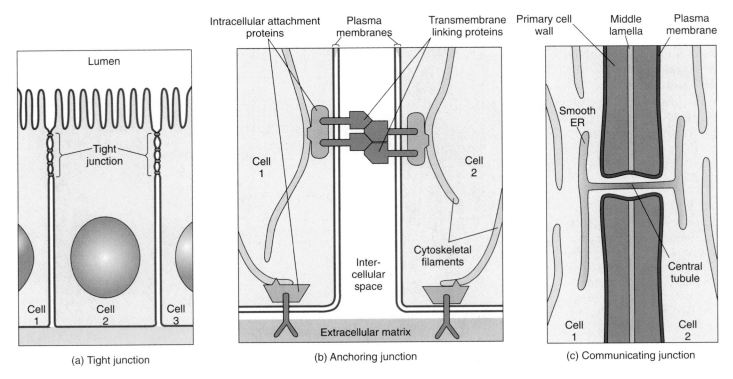

(a) Tight junction

(b) Anchoring junction

(c) Communicating junction

FIGURE 7.17
The three types of cell junctions. These three models represent current thinking on how the structures of the three major types of cell junctions facilitate their function: (a) tight junction; (b) anchoring junction; (c) communicating junction.

Tight Junctions

Sometimes called occluding junctions, **tight junctions** connect the plasma membranes of adjacent cells in a sheet, preventing small molecules from leaking between the cells and through the sheet (figure 7.18). This allows the sheet of cells to act as a wall within the organ, keeping molecules on one side or the other.

Creating Sheets of Cells. The cells that line an animal's digestive tract are organized in a sheet only one cell thick. One surface of the sheet faces the inside of the tract and the other faces the extracellular space where blood vessels are located. Tight junctions encircle each cell in the sheet, like a belt cinched around a pair of pants. The junctions between neighboring cells are so securely attached that there is no space between them for leakage. Hence, nutrients absorbed from the food in the digestive tract must pass directly through the cells in the sheet to enter the blood.

Partitioning the Sheet. The tight junctions between the cells lining the digestive tract also partition the plasma membranes of these cells into separate compartments. Transport proteins in the membrane facing the inside of the tract carry nutrients from that side to the cytoplasm of the cells. Other proteins, located in the membrane on the opposite side of the cells, transport those nutrients from the cytoplasm to the extracellular fluid, where they can enter the blood. For the sheet to absorb nutrients properly, these proteins must remain in the correct locations within the fluid membrane. Tight junctions effectively segregate the proteins on opposite sides of the sheet, preventing them from drifting within the membrane from one side of the sheet to the other. When tight junctions are experimentally disrupted, just this sort of migration occurs.

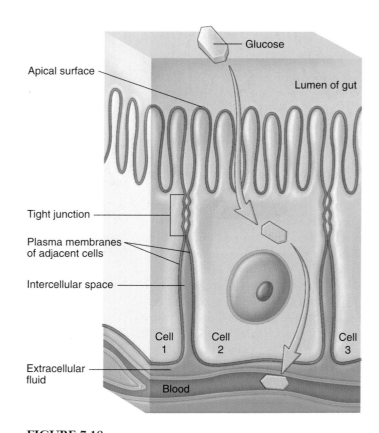

FIGURE 7.18
Tight junctions. Encircling the cell like a tight belt, these intercellular contacts ensure that materials move through the cells rather than between them.

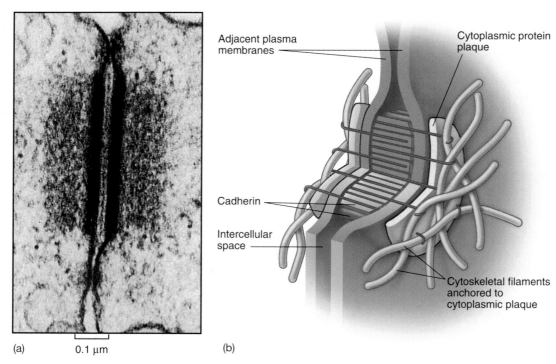

(a)   0.1 μm   (b)

FIGURE 7.19
Desmosomes. (a) Cytoskeletal filaments join cells together in several places. (b) Cadherin proteins cause adjoining cells to adhere.

Anchoring Junctions

Anchoring junctions mechanically attach the cytoskeleton of a cell to the cytoskeletons of other cells or to the extracellular matrix. They are commonest in tissues subject to mechanical stress, such as muscle and skin epithelium.

Cadherin-Mediated Links. Anchoring junctions called desmosomes connect the cytoskeletons of adjacent cells (figure 7.19), while hemidesmosomes anchor epithelial cells to a basement membrane. Proteins called **cadherins,** most of which are single-pass transmembrane glycoproteins, create the critical link. A variety of attachment proteins link the short cytoplasmic end of a cadherin to the intermediate filaments in the cytoskeleton. The other end of the cadherin molecule projects outward from the plasma membrane, joining directly with a cadherin protruding from an adjacent cell in a firm handshake binding the cells together.

Connections between proteins tethered to the cytoskeleton are much more secure than connections between freefloating membrane proteins. Proteins are suspended within the membrane by relatively weak interactions between the nonpolar portions of the protein and the membrane lipids. It would not take much force to pull an untethered protein completely out of the membrane, as if pulling an unanchored raft out of the water.

Cadherins can also connect the actin frameworks of cells in cadherin-mediated junctions (figure 7.20). When they do, they form less stable links between cells than when they connect intermediate filaments. Many kinds of actin-linking cadherins occur in different tissues, as well as in the same tissue at different times. During vertebrate development, the migration of neurons in the embryo is associated with changes in the type of cadherin expressed on their plasma membranes. This suggests that gene-controlled changes in cadherin expression may provide the migrating cells with a "roadmap" to their destination.

Integrin-Mediated Links. Anchoring junctions called adherens junctions are another type of junction that connects the actin filaments of one cell with those of neighboring cells or with the extracellular matrix (figure 7.21). The linking proteins in these junctions are members of a large superfamily of cell surface receptors called **integrins.** Each integrin is a transmembrane protein composed of two different glycoprotein subunits that extend outward from the plasma membrane. Together, these subunits bind a protein component of the extracellular matrix, like two hands clasping a pole. There appear to be many different kinds of integrin (cell biologists have identified 20), each with a slightly different shaped "hand." The exact component of the matrix that a given cell binds to depends on which combination of integrins that cell has in its plasma membrane.

Tight junctions and anchoring junctions cause cells to adhere tightly to one another.

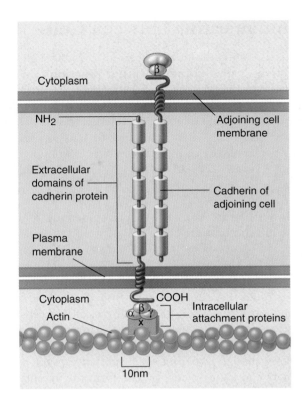

FIGURE 7.20
A cadherin-mediated junction. The cadherin molecule is anchored to actin in the cytoskeleton and passes through the membrane to interact with the cadherin of an adjoining cell.

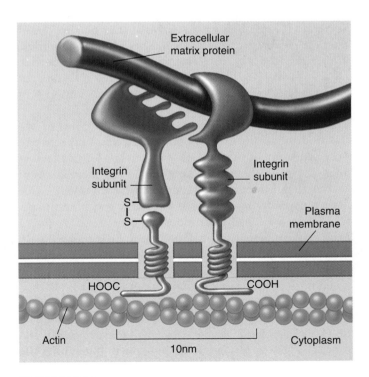

FIGURE 7.21
An integrin-mediated junction. These adherens junctions link the action filaments inside cells to their neighbors and to the extracellular matrix.

Communicating between Cells

Many cells communicate with adjacent cells through direct connections, called **communicating junctions.** In these junctions a chemical signal passes directly from one cell to an adjacent one. Communicating junctions establish direct physical connections that link the cytoplasms of two cells together, permitting small molecules or ions to pass from one to the other. In animals, these direct communication channels between cells are called gap junctions. In plants, they are called plasmodesmata.

Gap Junctions in Animals.   Communicating junctions called **gap junctions** are composed of structures called *connexons*, complexes of six identical transmembrane proteins (figure 7.22). The proteins in a connexon are arranged in a circle to create a channel through the plasma membrane that protrudes several nanometers from the cell surface. A gap junction forms when the connexons of two cells align perfectly, creating an open channel spanning the plasma membranes of both cells. Gap junctions provide passageways large enough to permit small substances, such as simple sugars and amino acids, to pass from the cytoplasm of one cell to that of the next, yet small enough to prevent the passage of larger molecules such as proteins. The connexons hold the plasma membranes of the paired cells about 4 nanometers apart; the term gap junction contrasts this separation with the more-or-less direct contact between the lipid bilayers in a tight junction.

Gap junction channels are dynamic structures that can open or close in response to a variety of factors, including Ca^{++} and H^+ ions. This gating serves at least one important function. When a cell is damaged, its plasma membrane often becomes leaky. Ions in high concentrations outside the cell, such as Ca^{++}, flow into the damaged cell and shut its gap junction channels. This isolates the cell and so prevents the damage from spreading to other cells.

Plasmodesmata in Plants.   In plants, cell walls separate every cell from all others. Cell-cell junctions occur only at holes or gaps in the walls, where the plasma membranes of adjacent cells can come into contact with each other. Cytoplasmic connections that form across the touching plasma membranes are called **plasmodesmata** (figure 7.23). Every living cell of a higher plant is thought to be connected with its neighbors by these junctions. Plasmodesmata function much like gap junctions in animal cells, although their structure is more complex. Unlike gap junctions, plasmodesmata are lined with plasma membrane and contain a central tubule that connects the endoplasmic reticulum of the two cells. These connections linking every cell of the plant are likely to play an important role in integrating the activities of the plant body.

Communicating junctions permit the controlled passage of small molecules or ions between cells.

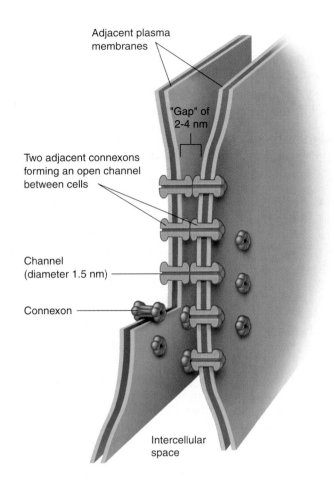

FIGURE 7.22
Gap junctions. Connexons in gap junctions create passageways that connect the cytoplasms of adjoining cells. Gap junctions readily allow the passage of small molecules and ions required for rapid communication (such as in heart tissue), but do not allow the passage of larger molecules like proteins.

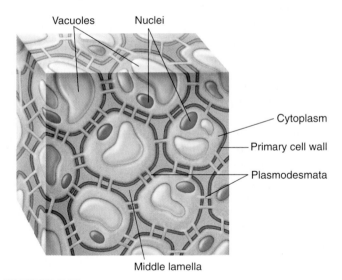

FIGURE 7.23
Plasmodesmata. Plant cells can communicate through specialized openings in their cell walls, called plasmodesmata, where the cytoplasms of adjoining cells are connected.

Summary of Concepts

7.1 Cells signal one another with chemicals.

- Cell signaling is accomplished through the recognition of signal molecules by target cells.

7.2 Proteins in the cell and on its surface receive signals from other cells.

- The binding of a signal molecule to an intracellular receptor usually initiates transcription of specific regions of DNA, ultimately resulting in the production of specific proteins.
- Cell surface receptors bind to specific molecules in the extracellular fluid. In some cases, this binding causes the receptor to enzymatically alter other (usually internal) proteins, typically through phosphorylation.
- G proteins behave as intracellular shuttles, moving from an activated receptor to other areas in the cell.

7.3 Follow the journey of information into the cell.

- There are usually several amplifying steps between the binding of a signal molecule to a cell surface receptor and the response of the cell. These steps often involve phosphorylation by protein kinases.

7.4 Cell surface proteins mediate cell-cell interactions.

- Tight junctions and desmosomes enable cells to adhere in tight, leakproof sheets, holding the cells together such that materials cannot pass between them.
- Gap junctions (in animals) and plasmodesmata (in plants) permit small substances to pass directly from cell to cell through special passageways.

Discussing Key Terms

1. **Cell signaling**   Cells send chemical signals to other cells; these signal molecules bind to specific protein receptors inside or on the plasma membrane of the other cells. Intracellular receptors typically act by binding to DNA and so affecting the transcription of particular genes. Other receptors are transmembrane proteins that act at the cell surface.

2. **Cell surface receptors**   Protein receptors on plasma membrane act as enzymes when they bind to signal molecules, often activating other proteins by phosphorylating them.

3. **G Proteins**   These important membrane proteins act as relay switches between the cell surface and the interior. They are activated by cell surface receptors, initiating a cytoplasmic signal that passes into the interior.

4. **Signal amplification**   The message relayed by enzymic and G protein receptors may be amplified as it passes into the cell interior by a cascading series of enzymatic reactions, each stage of which produces many outputs for every input.

5. **Adhering cell junctions**   Junctions that lock cells together in particular orientations, thereby stabilizing cell-to-cell contact. These junctions typically connect the cytoskeletons of the adjacent cells rather than their more fluid plasma membranes.

6. **Communicating cell junctions**   Cell junctions that permit small molecules and ions to pass from one cell to another.

Review Questions

1. What determines which signal molecules in the extracellular environment a cell will respond to?

2. How do paracrine, endocrine, and synaptic signaling differ?

3. Describe two of the ways in which intracellular receptors control cell activities.

4. What structural features are characteristic of chemically gated ion channels, and how are these features related to the function of the channels?

5. What structural features are characteristic of enzymic cell surface receptors? What intracellular effect do these receptors usually produce when they bind to a signal molecule?

6. What are G proteins? How do they participate in cellular responses mediated by G protein-linked receptors?

7. What are second messengers? Describe the basic mechanism of signaling systems that use cAMP as a second messenger.

Describe the basic mechanism of signaling systems that use calcium as a second messenger.

8. How does the binding of a single signal molecule to a cell surface receptor result in an amplified response within the target cell?

9. What are two functions of cell surface markers? Describe the structures of the cell surface molecules that serve these functions.

10. What are the functions of tight junctions? What are the functions of desmosomes and adherens junctions, and what proteins are involved in these junctions?

11. What are the molecular components that make up gap junctions? What sorts of substances can pass through gap junctions?

12. Where are plasmodesmata found? What cellular constituents are found in plasmodesmata?

Thought Questions

1. At first glance, the signaling systems that involve cell surface receptors may appear rather complex and indirect, with their use of G proteins, second messengers, and often multiple stages of enzymes. What are the advantages of such seemingly complex response systems?

2. *Shigella flexneri* is one of several species of bacteria that cause shigellosis, or bacillary dysentery. Recent evidence has shown that *Shigella flexneri* can't spread between the epithelial cells of the intestines without the expression of cadherin by those cells. Why do you suppose it can't?

Internet Links

Nanoworld
http://www.uq.oz.au/nanoworld/images_1.html
A spectacular collection of electron microscope images, a tour of the world of the very small.

A Course in Cell Biology
http://gened.emc.maricopa.edu/bio/bio181/BIOBK/
BioBookCELL2.html
A general introduction to cell biology, particularly for students with little background.

For Further Reading

Berridge, M.: "Cell Signaling: A Tale of Two Messengers," *Nature*, September 1993, pages 388–89. A discussion of second messengers and their functions in cell signaling.

Citi, S.: "The Molecular Organization of Tight Junctions," *The Journal of Cell Biology*, April 1993, pages 485–89. An in-depth, technical look at how tight junctions are organized in the cell.

Darnell, J., H. Lodish, and D. Baltimore: *Molecular Cell Biology*, ed. 2, Scientific American Books, New York, 1990. A good cell biology text, with particular strengths in cell-to-cell interactions.

Horwitz, A. F.: "Integrins and Health," *Scientific American*, May 1997, pages 68–75. A fascinating discussion of the role of adhesive cell surface molecules called integrins in a variety of cell functions.

Linder, M., and A. Gilman: "G Proteins," *Scientific American*, July 1992, pages 56–61. A good, basic discussion of what G proteins are and how they work. Gilman, the second author, won the 1994 Nobel Prize for this research.

Marx, J.: "Medicine: A Signal Award for Discovering G Proteins," in "Eight Get the Call to Stockholm," *Science*, October 1994, pages 368–69. An account of how Rodbell and Gilman received the 1994 Nobel Prize for Medicine or Physiology for their work on G proteins.

Putney, J.: "Excitement about Calcium Signaling in Inexcitable Cells," *Science*, October 1993, pages 676–78. A discussion about calcium in the body, its role as a second messenger, and signal transduction.

8

Energy and Metabolism

Concept Outline

8.1 The laws of thermodynamics describe how energy changes.

The Flow of Energy in Living Things. Potential energy is present in the electrons of atoms, and so can be transferred from one molecule to another.

The Laws of Thermodynamics. Energy is never lost but as it is transferred, more and more of it dissipates as heat.

Free Energy. In a chemical reaction, the energy released or supplied is the difference in bond energies between reactants and products, corrected for disorder.

Activation Energy. To start a chemical reaction, a small input of energy is required to destabilize existing chemical bonds.

8.2 Enzymes are biological catalysts.

Enzymes. Globular proteins called enzymes catalyze chemical reactions within cells.

How Enzymes Work. Enzymes are shaped to fit their substrates snugly, forcing chemically reactive groups close enough to facilitate a reaction.

Factors Affecting Enzyme Activity. Enzymes work most efficiently at an optimal temperature and pH.

Enzyme Cofactors. Metal ions or other substances often help enzymes carry out their catalysis.

Enzymes Take Many Forms. Some enzymes are associated in complex groups; other catalysts are not even proteins.

8.3 ATP is the energy currency of life.

What Is ATP? Cells store and release energy from the phosphate bonds of ATP, the energy currency of the cell.

8.4 Metabolism is the chemical life of a cell.

Biochemical Pathways: The Organizational Units of Metabolism. Biochemical pathways are the organizational units of metabolism.

The Evolution of Metabolism. The major metabolic processes evolved over a long period, building on what came before.

FIGURE 8.1
Lion at lunch. Energy that this lion extracts from its meal of giraffe will be used to power its roar, fuel its running, and build a bigger lion.

Life can be viewed as a constant flow of energy, channeled by organisms to do the work of living. Each of the significant properties by which we define life—order, growth, reproduction, responsiveness, and internal regulation—requires a constant supply of energy (figure 8.1). Deprived of a source of energy, life stops. Therefore, a comprehensive study of life would be impossible without discussing **bioenergetics,** the analysis of how energy powers the activities of living systems. In this chapter, we will focus on energy—on what it is and how organisms capture, store, and use it.

The Flow of Energy in Living Things

Energy is defined as the capacity to do work. It can be considered to exist in two states. **Kinetic energy** is the energy of motion. Moving objects perform work by causing other matter to move. **Potential energy** is stored energy. Objects that are not actively moving but have the capacity to do so possess potential energy. A boulder perched on a hilltop has potential energy; as it begins to roll downhill, some of its potential energy is converted into kinetic energy. Much of the work that living organisms carry out involves transforming potential energy to kinetic energy.

Energy can take many forms: mechanical energy, heat, sound, electric current, light, or radioactive radiation. Because it can exist in so many forms, there are many ways to measure energy. The most convenient is in terms of heat, because all other forms of energy can be converted into heat. In fact, the study of energy is called **thermodynamics,** meaning heat changes. The unit of heat most commonly employed in biology is the **kilocalorie** (kcal). One kilocalorie is equal to 1000 calories (cal), and one calorie is the heat required to raise the temperature of one gram of water one degree Celsius (°C). (It is important not to confuse calories with a term related to diets and nutrition, the Calorie with a capital C, which is actually another term for kilocalorie.) Another energy unit, often used in physics, is the **joule;** one joule equals 0.239 cal.

Oxidation-Reduction

Energy flows into the biological world from the sun, which shines a constant beam of light on the earth. It is estimated that the sun provides the earth with more than 13×10^{23} calories per year, or 40 million billion calories per second! Plants, algae, and certain kinds of bacteria capture a fraction of this energy through photosynthesis. In photosynthesis, energy garnered from sunlight is used to combine small molecules (water and carbon dioxide) into more complex molecules (sugars). The energy is stored as potential energy in the covalent bonds between atoms in the sugar molecules. Recall from chapter 2 that an atom consists of a central nucleus surrounded by one or more orbiting electrons, and a covalent bond forms when two atomic nuclei share electrons. Breaking such a bond requires energy to pull the nuclei apart. Indeed, the strength of a covalent bond is measured by the amount of energy required to break it. For example, it takes 98.8 kcal to break one mole (6.023×10^{23}) of carbon-hydrogen (C—H) bonds.

During a chemical reaction, the energy stored in chemical bonds may transfer to new bonds. In some of these reactions, electrons actually pass from one atom or molecule to another. When an atom or molecule loses an electron, it is

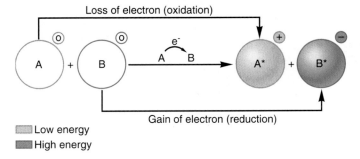

FIGURE 8.2
Redox reactions. Oxidation is the loss of an electron; reduction is the gain of an electron. In this example, the charges of molecules A and B are shown in small circles to the upper right of each molecule. Molecule A loses energy as it loses an electron, while molecule B gains energy as it gains an electron.

said to be oxidized, and the process by which this occurs is called **oxidation.** The name reflects the fact that in biological systems oxygen, which attracts electrons strongly, is the most common electron acceptor. Conversely, when an atom or molecule gains an electron, it is said to be reduced, and the process is called **reduction.** Oxidation and reduction always take place together, because every electron that is lost by an atom through oxidation is gained by some other atom through reduction. Therefore, chemical reactions of this sort are called **oxidation-reduction (redox) reactions** (figure 8.2). Energy is transferred from one molecule to another via redox reactions. The reduced form of a molecule thus has a higher level of energy than the oxidized form.

Oxidation-reduction reactions play a key role in the flow of energy through biological systems because the electrons that pass from one atom to another carry energy with them. The amount of energy an electron possesses depends on how far it is from the nucleus and how strongly the nucleus attracts it. Light (and other forms of energy) can add energy to an electron and boost it to a higher energy level. When this electron departs from one atom (oxidation) and moves to another (reduction), the electron's added energy is transferred with it, and the electron orbits the second atom's nucleus at the higher energy level. The added energy is stored as potential chemical energy that the atom can later release when the electron returns to its original energy level.

Energy is the capacity to do work, either actively (kinetic energy) or stored for later use (potential energy). Energy is often transferred with electrons. Oxidation is the loss of an electron; reduction is the gain of one.

The Laws of Thermodynamics

Running, thinking, singing, reading these words—all activities of living organisms involve changes in energy. A set of universal laws we call the Laws of Thermodynamics govern all energy changes in the universe, from nuclear reactions to the buzzing of a bee.

The First Law of Thermodynamics

The first of these universal laws, the **First Law of Thermodynamics,** concerns the amount of energy in the universe. It states that energy can change from one form to another (from potential to kinetic, for example) but it can never be destroyed, nor can new energy be made. The total amount of energy in the universe remains constant.

The lion eating a giraffe in figure 8.1 is in the process of acquiring energy. Rather than creating new energy or capturing the energy in sunlight, the lion is merely transferring some of the potential energy stored in the giraffe's tissues to its own body (just as the giraffe obtained the potential energy stored in the plants it ate while it was alive). Within any living organism, this chemical potential energy can be shifted to other molecules and stored in different chemical bonds, or it can convert into other forms, such as kinetic energy, light, or electricity. During each conversion, some of the energy dissipates into the environment as **heat,** a measure of the random motions of molecules (and, hence, a measure of one form of kinetic energy). Energy continuously flows through the biological world in one direction, with new energy from the sun constantly entering the system to replace the energy dissipated as heat.

Heat can be harnessed to do work only when there is a heat gradient, that is, a temperature difference between two areas (this is how a steam engine functions). Cells are too small to maintain significant internal temperature differences, so heat energy is incapable of doing the work of cells. Thus, although the total amount of energy in the universe remains constant, the energy available to do work decreases, as progressively more of it dissipates as heat.

The Second Law of Thermodynamics

The **Second Law of Thermodynamics** concerns this transformation of potential energy into heat, or random molecular motion. It states that the disorder (more formally called *entropy*) in the universe is continuously

Disorder happens "spontaneously"

Organization requires energy

FIGURE 8.3
Entropy in action. As time elaspes, a child's room becomes more disorganized. It takes effort to clean it up.

increasing. Put simply, disorder is more likely than order. For example, it is much more likely that a column of bricks will tumble over than that a pile of bricks will arrange themselves spontaneously to form a column. In general, energy transformations proceed spontaneously to convert matter from a more ordered, less stable form, to a less ordered, more stable form (figure 8.3).

Entropy

Entropy is a measure of the disorder of a system, so the Second Law of Thermodynamics can also be stated simply as "entropy increases." When the universe formed 10 to 20 billion years ago, it held all the potential energy it will ever have. It has become progressively more disordered ever since, with every energy exchange increasing the amount of entropy in the universe.

The First Law of Thermodynamics states that energy cannot be created or destroyed; it can only undergo conversion from one form to another. The Second Law of Thermodynamics states that disorder (entropy) in the universe is increasing. Life converts energy from the sun to other forms of energy that drive life processes; the energy is never lost, but as it is used, more and more of it is converted to heat, the energy of random molecular motion.

Free Energy

Because it takes energy to break chemical bonds, those bonds tend to hold the atoms in a molecule together. Heat energy, acting in the opposite way, increases atomic motion and makes it easier for the atoms to pull apart. Both chemical bonding and heat have a significant influence on a molecule, the former reducing disorder and the latter increasing it. The net effect, the amount of energy actually available to break and subsequently form other chemical bonds, is called the **free energy** of that molecule. In a more general sense, free energy is defined as the energy available to do work in any system. In a molecule within a cell, where pressure and volume usually do not change, the free energy is denoted by the symbol G (for "Gibbs' free energy," which limits the system being considered to the cell). G is equal to the energy contained in a molecule's chemical bonds (called *enthalpy* and designated **H**) minus the energy unavailable because of disorder (called *entropy* and given the symbol **S**) times the absolute temperature, **T**, in degrees Kelvin (K = °C + 273):

$$G = H - TS$$

Chemical reactions break some bonds in the reactants and form new bonds in the products. Consequently, reactions can produce changes in free energy. When a chemical reaction occurs under conditions of constant temperature, pressure, and volume—as do most biological reactions—the change in free energy (ΔG) is simply:

$$\Delta G = \Delta H - T \Delta S$$

The change in free energy, or ΔG, is a fundamental property of chemical reactions. In some reactions, the ΔG is positive. This means that the products of the reaction contain *more* free energy than the reactants; the bond energy (H) is higher or the disorder (S) in the system is lower. Such reactions do not proceed spontaneously. They require an input of energy and are thus **endergonic** ("inward energy"). In other reactions, the ΔG is negative. The products of the reaction contain less free energy than the reactants; either the bond energy is lower or the disorder is higher, or both. Such reactions tend to proceed spontaneously. Any chemical reaction will tend to proceed spontaneously if the difference in disorder (T ΔS) is *greater* than the difference in bond energies between reactants and products (ΔH). These reactions release the excess free energy as heat and are thus **exergonic** ("outward energy"). Figure 8.4 sums up these reactions.

Free energy is the energy available to do work. In the chemical reactions carried out within cells, the change in free energy (ΔG) is the difference in bond energies between reactants and products (ΔH), minus any change in the degree of disorder of the system (T ΔS). Any reaction whose products contain less free energy than the reactants (ΔG is negative) will tend to proceed spontaneously.

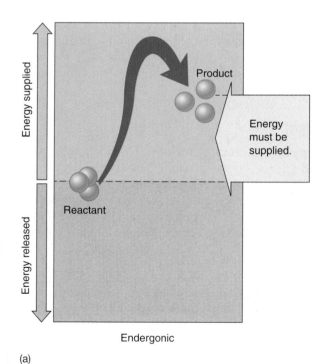

(a) Endergonic

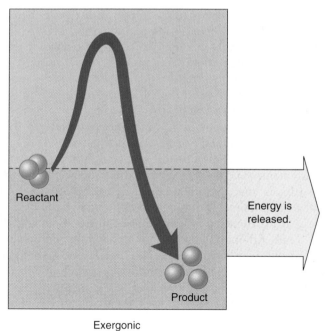

(b) Exergonic

FIGURE 8.4

Energy in chemical reactions. (a) In an endergonic reaction, the products of the reaction contain more energy than the reactants, and the extra energy must be supplied for the reaction to proceed. (b) In an exergonic reaction, the products contain less energy than the reactants, and the excess energy is released.

Activation Energy

If all chemical reactions that release free energy tend to occur spontaneously, why haven't all such reactions already occurred? One reason they haven't is that most reactions require an input of energy to get started. Before it is possible to form new chemical bonds, even bonds that contain less energy, it is first necessary to break the existing bonds, and that takes energy. The extra energy required to destabilize existing chemical bonds and initiate a chemical reaction is called **activation energy** (figure 8.5*a*).

The rate of an exergonic reaction depends on the activation energy required for the reaction to begin. Reactions with larger activation energies tend to proceed more slowly, since fewer molecules succeed in overcoming the initial energy hurdle. Activation energies are not constant, however. Stressing particular chemical bonds can make them easier to break. The process of influencing chemical bonds in a way that lowers the activation energy needed to initiate a reaction is called **catalysis**, and substances that accomplish this are known as catalysts (figure 8.5*b*).

Catalysts cannot violate the basic laws of thermodynamics; they cannot, for example, make an endergonic reaction proceed spontaneously. By reducing the activation energy, a catalyst accelerates both the forward and the reverse reactions by exactly the same amount. Hence, it does not alter the proportion of reactant ultimately converted into product.

To grasp this, imagine a bowling ball resting in a shallow depression on the side of a hill. Only a narrow rim of dirt below the ball prevents it from rolling down the hill. Now imagine digging away that rim of dirt. If you remove enough dirt from below the ball, it will start to roll down the hill—but removing dirt from below the ball will *never* cause the ball to roll UP the hill! Removing the lip of dirt simply allows the ball to move freely; gravity determines the direction it then travels. Lowering the resistance to the ball's movement will promote the movement dictated by its position on the hill.

Similarly, the direction in which a chemical reaction proceeds is determined solely by the difference in free energy. Like digging away the soil below the bowling ball on the hill, catalysts reduce the energy barrier preventing the reaction from proceeding. Catalysts don't favor endergonic reactions any more than digging makes the hypothetical bowling ball roll uphill. Only exergonic reactions proceed spontaneously, and catalysts cannot change that. What catalysts *can* do is make a reaction proceed much faster.

The rate of a reaction depends on the activation energy necessary to initiate it. Catalysts reduce the activation energy and so increase the rates of reactions, although they do not change the final proportions of reactants and products.

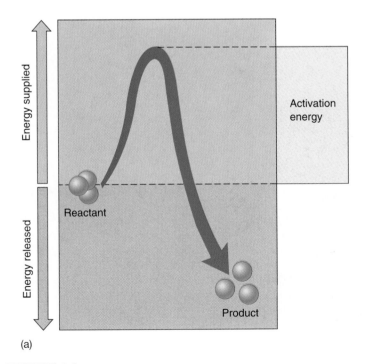

(a)

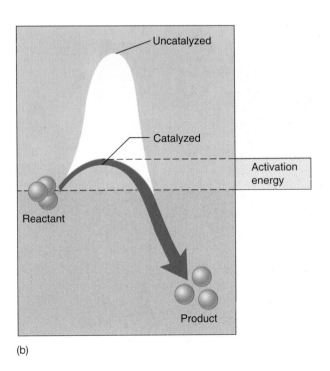

(b)

FIGURE 8.5
Activation energy and catalysis. (a) Exergonic reactions do not necessarily proceed rapidly because energy must be supplied to destabilize existing chemical bonds. This extra energy is the activation energy for the reaction. (b) Catalysts accelerate particular reactions by lowering the amount of activation energy required to initiate the reaction.

Enzymes are biological catalysts.

Enzymes

The chemical reactions within living organisms are regulated by controlling the points at which catalysis takes place. Life itself is, therefore, regulated by catalysts. The agents that carry out most of the catalysis in living organisms are proteins called **enzymes.** (There is increasing evidence that some types of biological catalysis are carried out by RNA molecules.) The unique three-dimensional shape of an enzyme enables it to stabilize a temporary association between substrates, the molecules that will undergo the reaction. By bringing two substrates together in the correct orientation, or by stressing particular chemical bonds of a substrate, an enzyme lowers the activation energy required for new bonds to form. The reaction thus proceeds much more quickly than it would without the enzyme. Because the enzyme itself is not changed or consumed in the reaction, only a small amount of an enzyme is needed, and it can be used over and over.

As an example of how an enzyme works, let's consider the reaction of carbon dioxide and water to form carbonic acid. This important enzyme-catalyzed reaction occurs in vertebrate red blood cells:

$$CO_2 + H_2O \longrightarrow H_2CO_3$$
carbon dioxide water   carbonic acid

This reaction may proceed in either direction, but because it has a large activation energy, the reaction is very slow in the absence of an enzyme: perhaps 200 molecules of carbonic acid form in an hour in a cell. Reactions that proceed this slowly are of little use to a cell. Cells overcome this problem by employing an enzyme within their cytoplasm called *carbonic anhydrase* (enzyme names usually end in "–ase"). Under the same conditions, but in the presence of carbonic anhydrase, an estimated 600,000 molecules of carbonic acid form every *second!* Thus, the enzyme increases the reaction rate more than 10 million times.

Thousands of different kinds of enzymes are known, each catalyzing one or a few specific chemical reactions. By facilitating particular chemical reactions, the enzymes in a cell determine the course of metabolism—the collection of all chemical reactions—in that cell. Different types of cells contain different sets of enzymes, and this difference contributes to structural and functional variations among cell types. The chemical reactions taking place within a red blood cell differ from those that occur within a nerve cell, in part because the cytoplasm and membranes of red blood cells and nerve cells contain different arrays of enzymes.

Cells use proteins called enzymes as catalysts to lower activation energies.

Catalysis: A Closer Look at Carbonic Anhydrase

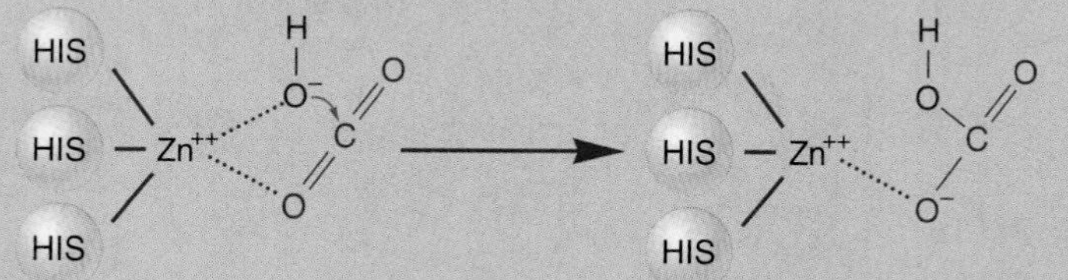

FIGURE 8.A

One of the most rapidly acting enzymes in the human body is carbonic anhydrase, which plays a key role in blood by converting dissolved CO_2 into carbonic acid, which dissociates into bicarbonate and hydrogen ions:

$$CO_2 + H_2O \rightarrow H_2CO_3 \rightarrow HCO_3^- + H^+$$

Fully 70% of the CO_2 transported by the blood is transported as bicarbonate ion. This reaction is exergonic, but its energy of activation is significant, so that little conversion to bicarbonate occurs spontaneously. In the presence of the enzyme carbonic anhydrase, however, the rate of the reaction accelerates by a factor of more than 10 million!

How does carbonic anhydrase catalyze this reaction so effectively? The active site of the enzyme is a deep cleft traversing the enzyme, as if it had been cut with the blade of an ax. Deep within the cleft, some 1.5 nm from the surface, are located three histidines, their imidazole (nitrogen ring) groups all pointed at the same place in the center of the cleft. Together they hold a zinc ion firmly in position. This zinc ion will be the cutting blade of the catalytic process.

Here is how the zinc catalyzes the reaction. Immediately adjacent to the position of the zinc atom in the cleft are a group of amino acids that recognize and bind carbon dioxide. The zinc atom interacts with this carbon dioxide molecule, orienting it in the plane of the cleft. Meanwhile, water bound to the zinc is rapidly converted to hydroxide ion. This hydroxide ion is now precisely positioned to attack the carbon dioxide. When it does so, HCO_3^- is formed—and the enzyme is unchanged (figure 8.A).

Carbonic anhydrase is an effective catalyst because it brings its two substrates into close proximity and optimizes their orientation for reaction. Other enzymes use other mechanisms. Many, for example, use charged amino acids to polarize substrates or electronegative amino acids to stress particular bonds. Whatever the details of the reaction, however, the precise positioning of substrates achieved by the particular shape of the enzyme always plays a key role.

How Enzymes Work

Most enzymes are globular proteins with one or more pockets or clefts on their surface called **active sites** (figure 8.6). Substrates bind to the enzyme at these active sites, forming an **enzyme-substrate complex.** For catalysis to occur within the complex, a substrate molecule must fit precisely into an active site. When that happens, amino acid side groups of the enzyme end up in close proximity to certain bonds of the substrate. These side groups interact chemically with the substrate, usually stressing or distorting a particular bond and consequently lowering the activation energy needed to break the bond. The substrate, now a product, then dissociates from the enzyme.

Proteins are not rigid. The binding of a substrate induces the enzyme to adjust its shape slightly, leading to a better *induced fit* between enzyme and substrate (figure 8.7). This interaction may also facilitate the binding of other substrates; in such cases, the substrate itself "activates" the enzyme to receive other substrates.

Enzymes typically catalyze only one or a few similar chemical reactions because they are specific in their choice of substrates. This specificity is due to the active site of the enzyme, which is shaped so that only a certain substrate molecule will fit into it.

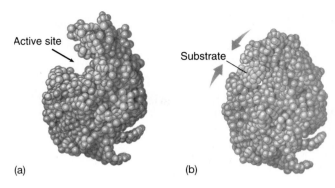

(a) (b)

FIGURE 8.6
How the enzyme lysozyme works. (a) A groove runs through lysozyme that fits the shape of the polysaccharide (a chain of sugars) that makes up bacterial cell walls. (b) When such a chain of sugars, indicated in yellow, slides into the groove, its entry induces the protein to alter its shape slightly and embrace the substrate more intimately. This induced fit positions a glutamic acid residue in the protein next to the bond between two adjacent sugars, and the glutamic acid "steals" an electron from the bond, causing it to break.

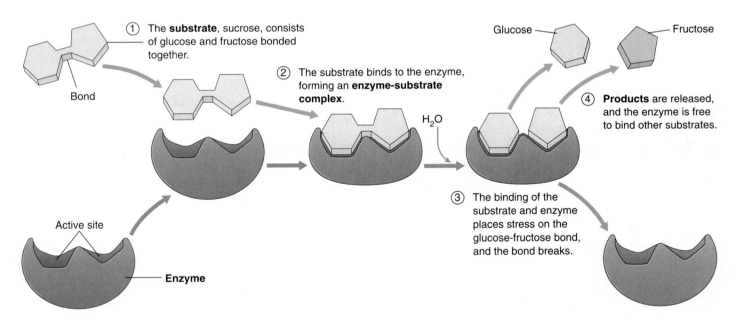

① The **substrate**, sucrose, consists of glucose and fructose bonded together.

② The substrate binds to the enzyme, forming an **enzyme-substrate complex**.

③ The binding of the substrate and enzyme places stress on the glucose-fructose bond, and the bond breaks.

④ **Products** are released, and the enzyme is free to bind other substrates.

Bond

Active site

Enzyme

H_2O

Glucose

Fructose

FIGURE 8.7
The catalytic cycle of an enzyme. Enzymes increase the speed with which chemical reactions occur, but they are not altered themselves as they do so. In the reaction illustrated here, the enzyme sucrase is splitting the sugar sucrose (present in most candy) into two simpler sugars: glucose and fructose. (1) First, the sucrose substrate binds to the active site of the enzyme, fitting into a depression in the enzyme surface. (2) The binding of sucrose to the active site forms an enzyme-substrate complex and induces the sucrase molecule to alter its shape, fitting more tightly around the sucrose. (3) Amino acid residues in the active site, now in close proximity to the bond between the glucose and fructose components of sucrose, break the bond. (4) The enzyme releases the resulting glucose and fructose fragments, the products of the reaction, and is then ready to bind another molecule of sucrose and run through the catalytic cycle once again. This cycle is often summarized by the equation: **E + S ↔ [ES] ↔ E + P,** where E = enzyme, S = substrate, ES = enzyme-substrate complex, and P = products.

Factors Affecting Enzyme Activity

Any chemical or physical factor that alters an enzyme's three-dimensional shape—such as temperature, pH, salt concentration, and the binding of specific regulatory molecules—can affect the enzyme's ability to catalyze a reaction.

Temperature

Increasing the temperature of an uncatalyzed reaction will increase its rate, since the additional heat represents an increase in random molecular movement. The rate of an enzyme-catalyzed reaction also increases with temperature, but only up to a point called the *temperature optimum* (figure 8.8*a*). Below this temperature, the hydrogen bonds and hydrophobic interactions that determine the enzyme's shape are not flexible enough to permit the induced fit that is optimum for catalysis. Above the temperature optimum, these forces are too weak to maintain the enzyme's shape against the increased random movement of the atoms in the enzyme. At these higher temperatures, the enzyme denatures, as we described in chapter 3. Most human enzymes have temperature optima between 35°C and 40°C, a range that includes normal body temperature. Bacteria that live in hot springs have more stable enzymes (that is, enzymes held together more strongly), so the temperature optima for those enzymes can be 70°C or higher.

pH

Ionic interactions between oppositely charged amino acid residues, such as glutamic acid (–) and lysine (+), also hold enzymes together. These interactions are sensitive to the hydrogen ion concentration of the fluid the enzyme is dissolved in, because changing that concentration shifts the balance between positively and negatively charged amino acid residues. For this reason, most enzymes have a **pH optimum** that usually ranges from pH 6 to 8. Those enzymes able to function in very acid environments are proteins that maintain their three-dimensional shape even in the presence of high levels of hydrogen ion. The enzyme pepsin, for example, digests proteins in the stomach at pH 2, a very acidic level (figure 8.8*b*).

Inhibitors and Activators

Enzyme activity is sensitive to the presence of specific substances that bind to the enzyme and cause changes in its shape. Through these substances, a cell is able to regulate which enzymes are active and which are inactive at a particular time. This allows the cell to increase its efficiency and to control changes in its characteristics during development. A substance that binds to an enzyme and *decreases* its activity is called an **inhibitor.** Very often, the end product of a biochemical pathway acts as an inhibitor of an early re-

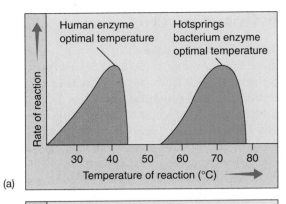
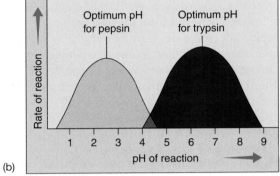

FIGURE 8.8
Enzymes are sensitive to their environment. The activity of an enzyme is influenced by both (a) temperature and (b) pH. Most human enzymes, such as the protein-degrading enzyme trypsin, work best at temperatures of about 40°C and within a pH range of 6 to 8.

action in the pathway, a process called *feedback inhibition* (to be discussed later).

Enzyme inhibition occurs in two ways: **competitive inhibitors** compete with the substrate for the same binding site, displacing a percentage of substrate molecules from the enzymes; **noncompetitive inhibitors** bind to the enzyme in a location other than the active site, changing the shape of the enzyme and making it unable to bind to the substrate (figure 8.9). Most noncompetitive inhibitors bind to a specific portion of the enzyme called an **allosteric site** (Greek *allos*, "other" + *steros*, "form"). These sites serve as chemical on/off switches; the binding of a substance to the site can switch the enzyme between its active and inactive configurations. A substance that binds to an allosteric site and reduces enzyme activity is called an **allosteric inhibitor** (figure 8.9*b*). Alternatively, **activators** bind to allosteric sites and keep the enzymes in their active configurations, thereby *increasing* enzyme activity.

Enzymes have an optimum temperature and pH, at which the enzyme functions most effectively. Inhibitors decrease enzyme activity, while activators increase it.

Enzyme Cofactors

Enzyme function is often assisted by additional chemical components known as **cofactors.** For example, the active sites of many enzymes contain metal ions that help draw electrons away from substrate molecules. The enzyme carboxypeptidase digests proteins by employing a zinc ion (Zn^{++}) in its active site to remove electrons from the bonds joining amino acids. Other elements, such as molybdenum and manganese, are also used as cofactors. Like zinc, these substances are required in the diet in small amounts. When the cofactor is a nonprotein organic molecule, it is called a **coenzyme.** Many vitamins are parts of coenzymes.

In numerous oxidation-reduction reactions that are catalyzed by enzymes, the electrons pass in pairs from the active site of the enzyme to a coenzyme that serves as the electron acceptor. The coenzyme then transfers the electrons to a different enzyme, which releases them (and the energy they bear) to the substrates in another reaction. Often, the electrons pair with protons (H^+) as hydrogen atoms. In this way, coenzymes shuttle energy in the form of hydrogen atoms from one enzyme to another in a cell.

One of the most important coenzymes is the hydrogen acceptor **nicotinamide adenine dinucleotide (NAD$^+$)** (figure 8.10). The NAD$^+$ molecule is composed of two nucleotides bound together. As you may recall from chapter 3, a nucleotide is a five-carbon sugar with one or more phosphate groups attached to one end and an organic base attached to the other end. The two nucleotides that make up NAD$^+$, nicotinamide monophosphate (NMP) and adenine monophosphate (AMP), are joined head-to-head by their phosphate groups. The two nucleotides serve different functions in the NAD$^+$ molecule: AMP acts as the core, providing a shape recognized by many enzymes; NMP is the active part of the molecule, contributing a site that is readily reduced (that is, easily accepts electrons).

When NAD$^+$ acquires an electron and a hydrogen atom (actually, two electrons and a proton) from the active site of an enzyme, it is reduced to NADH. The NADH molecule now carries the two energetic electrons and the proton. The oxidation of energy-containing molecules, which provides energy to cells, involves stripping electrons from those molecules and donating them to NAD$^+$. As we'll see, much of the energy of NADH is transferred to another molecule.

The activity of enzymes is often facilitated by cofactors, which can be metal ions or other substances. Cofactors that are nonprotein organic molecules are called coenzymes.

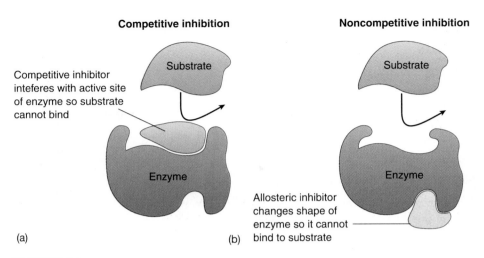

Competitive inhibition

Competitive inhibitor inteferes with active site of enzyme so substrate cannot bind

Substrate

Enzyme

(a)

Noncompetitive inhibition

Substrate

Enzyme

Allosteric inhibitor changes shape of enzyme so it cannot bind to substrate

(b)

FIGURE 8.9
How enzymes can be inhibited. (a) In competitive inhibition, the inhibitor interferes with the active site of the enzyme. (b) In noncompetitive inhibition, the inhibitor binds to the enzyme at a place away from the active site, effecting a conformational change in the enzyme so that it can no longer bind to its substrate.

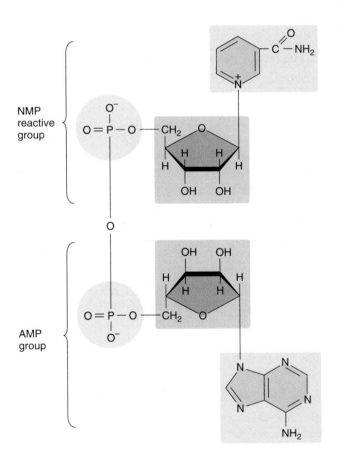

FIGURE 8.10
The chemical structure of nicotinamide adenine dinucleotide (NAD$^+$). This key cofactor is composed of two nucleotides, NMP and AMP, attached head-to-head.

Enzymes Take Many Forms

While many enzymes are suspended in the cytoplasm of cells, free to move about and not attached to any structure, other enzymes function as integral parts of cell structures and organelles.

Multienzyme Complexes

Often in cells the several enzymes catalyzing the different steps of a sequence of reactions are loosely associated with one another in noncovalently-bonded assemblies called *multienzyme complexes.* The bacterial pyruvate dehydrogenase multienzyme complex seen in figure 8.11 contains enzymes that carry out three sequential reactions in oxidative metabolism. Each complex has multiple copies of each of the three enzymes—60 protein subunits in all. The many subunits work in concert, like a tiny factory.

Multienzyme complexes offer significant advantages in catalytic efficiency:

1. The rate of any enzyme reaction is limited by the frequency with which the enzyme collides with its substrate. If a series of sequential reactions occur within a multienzyme complex, the product of one reaction can be delivered to the next enzyme without releasing it to diffuse away.
2. Because the reacting substrate never leaves the complex during its passage through the series of reactions, the possibility of unwanted side reactions is eliminated.
3. All of the reactions that take place within the multienzyme complex can be controlled as a unit.

In addition to pyruvate dehydrogenase, which controls entry to the Krebs cycle, several other key processes in the cell are catalyzed by multienzyme complexes. One well-studied system is the fatty acid synthetase complex that catalyzes the synthesis of fatty acids from two-carbon precursors. There are seven different enzymes in this multienzyme complex, and the reaction intermediates remain associated with the complex for the entire series of reactions.

Not All Biological Catalysts Are Proteins

Until a few years ago, most biology textbooks contained statements such as "Enzymes are the catalysts of biological systems." We can no longer make that statement without qualification. As discussed in chapter 4, Tom Cech and his colleagues at the University of Colorado reported in 1981 that certain reactions involving RNA molecules appear to be catalyzed in cells by RNA itself, rather than by enzymes.

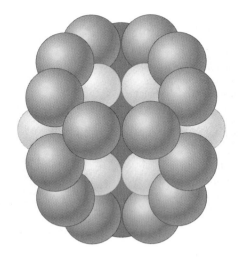

(a)

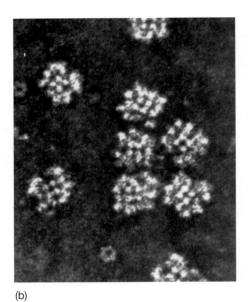

(b)

FIGURE 8.11

The enzyme pyruvate dehydrogenase. The enzyme (model, a) that carries out the oxidation of pyruvate is one of the most complex enzymes known—it has 60 subunits, many of which can be seen in the electron micrograph (b) (200,000×).

This initial observation has been corroborated by additional examples of RNA catalysis in the last few years. Like enzymes, these RNA catalysts, which are loosely called "ribozymes," greatly accelerate the rate of particular biochemical reactions and show extraordinary specificity with respect to the substrates on which they act.

There appear to be at least two sorts of ribozymes. Those that carry out *intra*molecular catalysis have folded structures and catalyze reactions on themselves. Those that carry out *inter*molecular catalysis act on other molecules without themselves being changed in the process. Many important cellular reactions involve small RNA molecules, including reactions that chip out unnecessary sections from RNA copies of genes, that prepare ribosomes for protein synthesis, and that facilitate the replication of DNA within mitochondria. In all of these cases, the possibility of RNA catalysis is being actively investigated. It seems likely, particularly in the complex process of photosynthesis, that both enzymes and RNA play important catalytic roles.

The ability of RNA, an informational molecule, to act as a catalyst has stirred great excitement among biologists, as it appears to provide a potential answer to the "chicken-and-egg" riddle posed by the spontaneous origin of life hypothesis discussed in chapter 3. Which came first, the protein or the nucleic acid? It now seems at least possible that RNA may have evolved first and catalyzed the formation of the first proteins.

Not all biological catalysts float free in the cytoplasm. Some are part of other structures, and some are not even proteins.

What Is ATP?

The chief energy currency all cells use is a molecule called **adenosine triphosphate (ATP).** The bulk of the energy that plants harvest during photosynthesis is channeled into ATP production, as is most of the energy stored in fat and starch. Cells use their supply of ATP to power almost every energy-requiring process they carry out, from supplying activation energy for chemical reactions and actively transporting substances across membranes to moving through their environment and growing.

Structure of the ATP Molecule

Each ATP molecule is composed of three smaller components (figure 8.12). The first component is a five-carbon sugar, ribose, which serves as the backbone to which the other two subunits are attached. The second component is adenine, an organic molecule composed of two carbon-nitrogen rings. Each of the nitrogen atoms in the ring has an unshared pair of electrons and weakly attracts hydrogen ions. Adenine, therefore, acts chemically as a base and is usually referred to as a nitrogenous base (it is one of the four nitrogenous bases found in DNA). The third component of ATP is a triphosphate group (three phosphates linked in a chain).

How ATP Stores Energy

The key to how ATP stores energy lies in its triphosphate group. Phosphate groups are highly negatively charged, so they repel one another strongly. Because of the electrostatic repulsion between the charged phosphate groups, the two covalent bonds joining the phosphates are unstable. The ATP molecule is often referred to as a "coiled spring," the phosphates straining away from one another.

The unstable bonds holding the phosphates together in the ATP molecule have a low activation energy and are easily broken. When they break, they can transfer a considerable amount of energy. In most reactions involving ATP, only the outermost high-energy phosphate bond is hydrolyzed, cleaving off the phosphate group on the end. When this happens, ATP becomes **adenosine diphosphate (ADP),** and energy equal to 7.3 kcal/mole is released under standard conditions. The liberated phosphate group usually attaches temporarily to some intermediate molecule. When that molecule is dephosphorylated, the phosphate group is released as inorganic phosphate (P_i).

How ATP Powers Energy-Requiring Reactions

Cells use ATP to drive endergonic reactions. Such reactions do not proceed spontaneously, because their products possess more free energy than their reactants. However, if

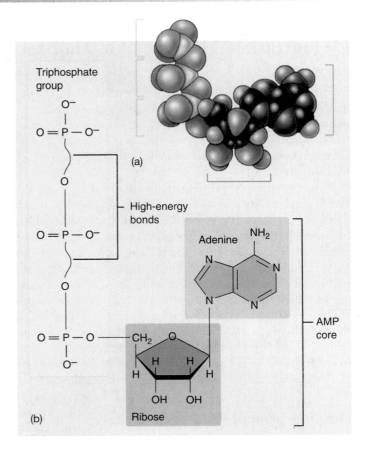

FIGURE 8.12
The ATP molecule. (a) The model and (b) structural diagram both show that like NAD^+, ATP has a core of AMP. In this case, however, the reactive group added to the end of the AMP phosphate group is not another nucleotide but rather a chain of two additional phosphate groups. The bonds (~) connecting these two phosphate groups to each other and to AMP are energy-storing bonds.

the cleavage of ATP's terminal high-energy bond releases more energy than the other reaction consumes, the overall energy change of the two coupled reactions will be exergonic (energy releasing) and they will both proceed. Because almost all endergonic reactions require less energy than is released by the cleavage of ATP, ATP can provide most of the energy a cell needs.

The same feature that makes ATP an effective energy donor—the instability of its phosphate bonds—precludes it from being a good long-term energy storage molecule. Fats and carbohydrates serve that function better. Most cells do not maintain large stockpiles of ATP. Instead, they typically have only a few seconds' supply of ATP at any given time, and they continually produce more from ADP and P_i.

The instability of its phosphate bonds makes ATP an excellent energy donor.

Biochemical Pathways: The Organizational Units of Metabolism

This living chemistry, the total of all chemical reactions carried out by an organism, is called **metabolism** (Greek *metabole*, "change"). Those reactions that expend energy to make or transform chemical bonds are called *anabolic* reactions, or **anabolism**. Reactions that harvest energy when chemical bonds are broken are called *catabolic* reactions, or **catabolism.**

Organisms contain thousands of different kinds of enzymes that catalyze a bewildering variety of reactions. Many of these reactions in a cell occur in sequences called **biochemical pathways.** In such pathways, the product of one reaction becomes the substrate for the next (figure 8.13). Biochemical pathways are the organizational units of metabolism, the elements an organism controls to achieve coherent metabolic activity. Most sequential enzyme steps in biochemical pathways take place in specific compartments of the cell; the steps of the citric acid cycle (chapter 9), for example, occur inside mitochondria. By determining where many of the enzymes that catalyze these steps are located, we can "map out" a model of metabolic processes in the cell.

How Biochemical Pathways Evolved

In the earliest cells, the first biochemical processes probably involved energy-rich molecules scavenged from the environment. Most of the molecules necessary for these processes are thought to have existed in the "organic soup" of the early oceans. The first catalyzed reactions are thought to have been simple, one-step reactions that brought these molecules together in various combinations. Eventually, the energy-rich molecules became depleted in the external environment, and only organisms that had evolved some means of making those molecules from other substances in the environment could survive. Thus, a hypothetical reaction,

$$\begin{array}{c} F \\ + \longrightarrow H \\ G \end{array}$$

where two energy-rich molecules (F and G) react to produce compound H and release energy, became more complex when the supply of F in the environment ran out. A new reaction was added in which the depleted molecule, F, is made from another molecule, E, which was also present in the environment:

$$\begin{array}{c} E \longrightarrow F \\ + \longrightarrow H \\ G \end{array}$$

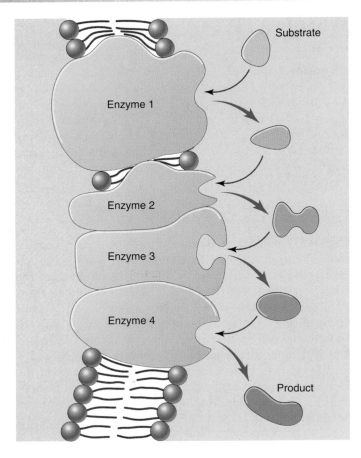

FIGURE 8.13
A biochemical pathway. The original substrate is acted on by enzyme 1, changing the substrate to a new form reocgnized by enzyme 2. Each enzyme in the pathway acts on the product of the previous stage.

When the supply of E in turn became depleted, organisms that were able to make it from some other available precursor, D, survived. When D became depleted, those organisms in turn were replaced by ones able to synthesize D from another molecule, C:

$$\begin{array}{c} C \longrightarrow D \longrightarrow E \longrightarrow F \\ + \longrightarrow H \\ G \end{array}$$

This hypothetical biochemical pathway would have evolved slowly through time, with the final reactions in the pathway evolving first and earlier reactions evolving later. Looking at the pathway now, we would say that the organism, starting with compound C, is able to synthesize H by means of a series of steps. This is how the biochemical pathways within organisms are thought to have evolved— not all at once, but one step at a time, backward.

How Biochemical Pathways Are Regulated

For a biochemical pathway to operate efficiently, its activity must be coordinated and regulated by the cell. Not only is it unnecessary to synthesize a compound when plenty is already present, doing so would waste energy and raw materials that could be put to use elsewhere. It is, therefore, advantageous for a cell to temporarily shut down biochemical pathways when their products are not needed.

The regulation of simple biochemical pathways often depends on an elegant feedback mechanism: the end product of the pathway binds to an allosteric site on the enzyme that catalyzes the first reaction in the pathway. In the hypothetical pathway we just described, the enzyme catalyzing the reaction C ——→ D would possess an allosteric site for H, the end product of the pathway. As the pathway churned out its product and the amount of H in the cell increased, it would become increasingly likely that one of the H molecules would encounter the allosteric site on the C ——→ D enzyme. If the product H functioned as an allosteric inhibitor of the enzyme, its binding to the enzyme would essentially shut down the reaction C ——→ D. Shutting down this reaction, the first reaction in the pathway, effectively shuts down the whole pathway. Hence, as the cell produces increasing quantities of the product H, it automatically inhibits its ability to produce more. This mode of regulation is called **feedback inhibition** (figure 8.14).

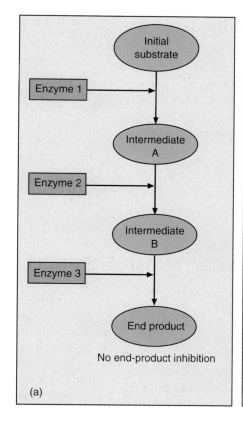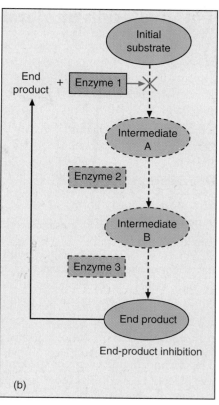

No end-product inhibition

(a)

End-product inhibition

(b)

FIGURE 8.14

Feedback inhibition. (a) A biochemical pathway with no feedback inhibition. (b) A biochemical pathway in which the final end product becomes the allosteric effector for the first enzyme in the pathway. In other words, the formation of the pathway's final end product stops the pathway.

> A biochemical pathway is an organized series of reactions, often regulated as a unit.

A Vocabulary of Metabolism

activation energy The energy required to destabilize chemical bonds and to initiate a chemical reaction.

catalysis Acceleration of the rate of a chemical reaction by lowering the activation energy.

coenzyme A nonprotein organic molecule that plays an accessory role in enzyme-catalyzed reactions, often by acting as a donor or acceptor of electrons. NAD+ is a coenzyme.

endergonic reaction A chemical reaction to which energy from an outside source must be added before the reaction proceeds; the opposite of an exergonic reaction.

entropy A measure of the randomness or disorder of a system. In cells, it is a measure of how much energy has become so dispersed (usually as evenly distributed heat) that it is no longer available to do work.

exergonic reaction. An energy-yielding chemical reaction. Exergonic reactions tend to proceed spontaneously, although activation energy is required to initiate them.

free energy Energy available to do work.

kilocalorie 1000 calories. A calorie is the heat required to raise the temperature of 1 gram of water by 1°C.

metabolism The sum of all chemical processes occurring within a living cell or organism.

oxidation The loss of an electron by an atom or molecule. It occurs simultaneously with reduction of some other atom or molecule because an electron that is lost by one is gained by another.

reduction The gain of an electron by an atom or molecule. Oxidation-reduction reactions are an important means of energy transfer within living systems.

substrate A molecule on which an enzyme acts; the initial reactant in an enzyme-catalyzed reaction.

The Evolution of Metabolism

Metabolism has changed a great deal as life on earth has evolved. This has been particularly true of the reactions organisms use to capture energy from the sun to build organic molecules (anabolism), and then break down organic molecules to obtain energy (catabolism). These processes, the subject of the next two chapters, evolved in concert with each other.

Degradation

The most primitive forms of life are thought to have obtained chemical energy by degrading, or breaking down, organic molecules that were abiotically produced.

The first major event in the evolution of metabolism was the origin of the ability to harness chemical bond energy. At an early stage, organisms began to store this energy in the bonds of ATP, an energy carrier used by all organisms today.

Glycolysis

The second major event in the evolution of metabolism was glycolysis, the initial breakdown of glucose. As proteins evolved diverse catalytic functions, it became possible to capture a larger fraction of the chemical bond energy in organic molecules by breaking chemical bonds in a series of steps. For example, the progressive breakdown of the six-carbon sugar glucose into three-carbon molecules is performed in a series of 10 steps that results in the net production of two ATP molecules. The energy for the synthesis of ATP is obtained by breaking chemical bonds and forming new ones with less bond energy, the energy difference being channeled into ATP production. This biochemical pathway is called glycolysis.

Glycolysis undoubtedly evolved early in the history of life on earth, since this biochemical pathway has been retained by all living organisms. It is a chemical process that does not appear to have changed for well over 3 billion years.

Anaerobic Photosynthesis

The third major event in the evolution of metabolism was anaerobic photosynthesis. Early in the history of life, some organisms evolved a different way of generating ATP, called photosynthesis. Instead of obtaining energy for ATP synthesis by reshuffling chemical bonds, as in glycolysis, these organisms developed the ability to use light to pump protons out of their cells, and to use the resulting proton gradient to power the production of ATP, a process called chemiosmosis.

Photosynthesis evolved in the absence of oxygen and works well without it. Dissolved H_2S, present in the oceans beneath an atmosphere free of oxygen gas, served as a ready source of hydrogen atoms for building organic molecules. Free sulfur was produced as a by-product of this reaction.

Nitrogen Fixation

Nitrogen fixation was the fourth major step in the evolution of metabolism. Proteins and nucleic acids cannot be synthesized from the products of photosynthesis, since both of these biologically critical molecules contain nitrogen. Obtaining nitrogen atoms from N_2 gas, a process called *nitrogen fixation*, requires the breaking of an $N \equiv N$ triple bond. This important reaction evolved in the hydrogen-rich atmosphere of the early earth, an atmosphere in which no oxygen was present. Oxygen acts as a poison to nitrogen fixation, which today occurs only in oxygen-free environments, or in oxygen-free compartments within certain bacteria.

Oxygen-Forming Photosynthesis

The substitution of H_2O for H_2S in photosynthesis was the fifth major event in the history of metabolism. Oxygen-forming photosynthesis employs H_2O rather than H_2S as a source of hydrogen atoms and their associated electrons. Because it garners its hydrogen atoms from reduced oxygen rather than from reduced sulfur, it generates oxygen gas rather then free sulfur.

More than 2 billion years ago, small cells capable of carrying out this oxygen-forming photosynthesis, such as cyanobacteria, became the dominant forms of life on earth. Oxygen gas began to accumulate in the atmosphere. This was the beginning of a great transition that changed conditions on earth permanently. Our atmosphere is now 20.9% oxygen, every molecule of which is derived from an oxygen-forming photosynthetic reaction.

Aerobic Respiration

Aerobic respiration is the sixth and final event in the history of metabolism. This cellular process harvests energy by stripping energetic electrons from organic molecules. Aerobic respiration employs the same kind of proton pumps as photosynthesis, and is thought to have evolved as a modification of the basic photosynthetic machinery. However, the hydrogens and their associated electrons are not obtained from H_2S or H_2O, as in photosynthesis, but rather from the breakdown of organic molecules.

Biologists think that the ability to carry out photosynthesis without H_2S first evolved among purple nonsulfur bacteria, which obtain their hydrogens from organic compounds instead. It was perhaps inevitable that among the descendants of these respiring photosynthetic bacteria, some would eventually do without photosynthesis entirely, subsisting only on the energy and hydrogens derived from the breakdown of organic molecules. The mitochondria within all eukaryotic cells are thought to be their descendants.

Six major innovations highlight the evolution of metabolism as we know it today.

8.1 **The laws of thermodynamics describe how energy changes.**

- Energy is the capacity to bring about change, to provide motion against a force, or to do work.

- Kinetic energy is actively engaged in doing work, while potential energy has the capacity to do so. Many energy transformations of living things involve transformations of potential energy into kinetic energy.

- An oxidation-reduction (redox) reaction is one in which an electron is taken from one atom or molecule (oxidation) and donated to another (reduction).

- The First Law of Thermodynamics states that the amount of energy in the universe is constant; energy is neither lost nor created.

- The Second Law of Thermodynamics states that disorder in the universe (entropy) tends to increase. As a result, energy spontaneously converts to less-ordered forms.

- Any chemical reaction whose products contain less free energy than the original reactants will tend to proceed spontaneously. However, the difference in free energy does not determine the rate of the reaction.

- The rate of a reaction depends on the amount of activation energy required to break existing bonds.

- Catalysis is the process of lowering activation energies by stressing chemical bonds.

8.2 **Enzymes are biological catalysts.**

- Enzymes are the major catalysts of cells; they affect the rate of a reaction but not the ultimate balance between reactants and products.

- Cells contain many different enzymes, each of which catalyzes a specific reaction.

- The specificity of an enzyme is due to its active site, which fits only one or a few types of substrate molecules.

8.3 **ATP is the energy currency of life.**

- Cells obtain energy from photosynthesis and the oxidation of organic molecules and use it to manufacture ATP from ADP and phosphate.

- The energy stored in ATP is then used to drive endergonic reactions.

8.4 **Metabolism is the chemical life of a cell.**

- Generally, the final reactions of a biochemical pathway evolved first; preceding reactions in the pathway were added later, one step at a time.

1. **Energy**   All of the activities of a cell are driven by energy, either supplied by the sun or contained in chemical bonds.

2. **Oxidation-reduction**   When an electron (often associated with a proton as a hydrogen atom) is transferred from one molecule to another, the molecule that loses the electron is said to be oxidized, and the molecule that gains the electron is said to be reduced. Electrons that move in this fashion often carry considerable energy with them.

3. **Entropy**   Energy spontaneously converts to less-ordered forms. Consequently, relationships among molecules tend to become more disordered (entropy increases) as energy moves toward its least organized form, the random molecular motion known as heat.

4. **Free energy**   The energy present in the chemical bonds of a molecule, minus the energy that is unavailable because of disorder, is that molecule's free energy. More generally, free energy is the energy available to do work.

5. **Exergonic**   Any chemical reaction whose products contain less free energy than the original reactants will tend to proceed spontaneously.

6. **Activation energy**   The energy that must be supplied to break the existing chemical bonds before a reaction can proceed. Even spontaneous reactions require this initial activation energy.

7. **Catalysis**   The process of lowering the activation energy required for specific chemical reactions to proceed. Enzymes stress particular chemical bonds so that they are more likely to break.

8. **Enzyme specificity**   An enzyme catalyzes only one or a few particular reactions because only certain potential substrate molecules can fit snugly into the critical cavities (active sites) on the enzyme's surface.

9. **Metabolism**   Metabolism encompasses all the chemical reactions in a living cell, allowing organisms to capture and harvest energy.

Review Questions

1. What is the difference between anabolism and catabolism?

2. What are the two basic states in which energy can exist? How are they different?

3. Define *oxidation* and *reduction*. Why must these two reactions always occur in concert? What is usually exchanged between reactants and products in a redox reaction?

4. State the First and Second Laws of Thermodynamics.

5. What is heat? What is entropy? What is free energy?

6. What is the difference between an exergonic and an endergonic reaction? Which type of reaction tends to proceed spontaneously?

7. Define *activation energy*. How is it related to reaction rate? How does a catalyst affect activation energy and reaction rate? How does a catalyst affect the final proportion of reactant converted into product?

8. How are the rates of enzyme-catalyzed reactions affected by temperature? How are they affected by pH? What is the molecular basis for the effects of these conditions on reaction rate?

9. What is the difference between the active site and an allosteric site on an enzyme? How does a competitive inhibitor of enzyme activity differ from a noncompetitive inhibitor?

10. How do cofactors and coenzymes participate in the chemical reactions that occur inside cells?

11. What part of the ATP molecule contains the bond that is employed to provide energy for most of the endergonic reactions in cells?

12. What is a biochemical pathway? How does feedback inhibition regulate the activity of a biochemical pathway?

Thought Questions

1. Almost no sunlight penetrates into the deep ocean. However, many fish that live there attract prey and potential mates by producing their own light. Where does that light come from? Does its generation require energy?

2. Oxidation-reduction reactions can involve a wide variety of molecules. Why do you suppose those involving hydrogen and oxygen are the ones of paramount importance in biological systems?

3. Some people argue that evolution, which is generally associated with progressive increases in the complexity (order) of organisms, cannot occur because entropy (disorder) is increasing in the universe. Is this argument valid?

Internet Links

Metabolism
http://expasy.hcuge.ch/
This site, hosted by the University of Geneva, provides an electronic version of the well-known "metabolic pathways" map of Buehringer Mannheim Corp., as well as access to many protein and nucleic acid databases.

Metabolism Problem Set
http://www.biology.arizona.edu/biochemistry/problem_sets/metabolism/metabolism.html
A selection of interactive multiple-choice questions covering many aspects of metabolism, part of the excellent Biology Project of the University of Arizona.

For Further Reading

Diamond, J.: "Quantitative Design of Life," *Nature*, December 1993, pages 405–6. A discussion of the evolution of physiology.

Harold, F.: *The Vital Force: A Study of Bioenergetics*, San Francisco, W. H. Freeman and Company, 1986. A comprehensive review of the advances that have been made in the last 10 years in our understanding of how organisms process energy.

Hinkle, P.C., and R.E. McCarty: "How Cells Make ATP," *Scientific American*, March 1978, pages 104–23. A description of how cells use electrons stripped from foodstuffs to make ATP.

Kraut, J.: "How Do Enzymes Work?" *Science*, October 28, 1988, vol. 242, pages 533–40. A clear presentation of the idea that enzymes work by stabilizing transitory transition states.

Lehninger, A., D. Nelson, and M. Cox: *Principles of Biochemistry*, ed. 2, 1993, Worth Publishers, New York. A standard comprehensive text for biochemistry.

Riddihough, G.: "Picture an Enzyme at Work," *Nature*, April 1993, page 793. A discussion of how conformational changes in enzymes enable them to function.

Wachtershauser, G.: "Evolution of the First Metabolic Cycles," *Proceedings of the National Academy of Science USA*, 1990, vol. 87, pages 200–4. A suggestion that metabolic cycles played a key role in the evolution of life.

Westheimer, F.: "Why Nature Chose Phosphates," *Science*, March 6, 1987, vol. 235, pages 1173–77. A wonderfully thoughtful analysis of the appropriateness of phosphates for biological but not laboratory chemistry.

9

How Cells Harvest Energy

Concept Outline

9.1 Cells harvest the energy in chemical bonds.

Using Chemical Energy to Drive Metabolism. The energy in C—H, C—O, and other chemical bonds can be captured and used to fuel the synthesis of ATP.

9.2 Cellular respiration oxidizes food molecules.

An Overview of Glucose Catabolism. The chemical energy in sugar is harvested by both substrate-level phosphorylation and by aerobic respiration.

Stage One: Glycolysis. The 10 reactions of glycolysis capture energy from glucose by reshuffling the bonds.

Stage Two: The Oxidation of Pyruvate. Pyruvate, the product of glycolysis, is oxidized to acetyl-CoA.

Stage Three: The Krebs Cycle. In a series of reactions, electrons are stripped from acetyl-CoA.

Harvesting Energy by Extracting Electrons. The respiration of glucose is a series of oxidation-reduction reactions that release energy by repositioning electrons closer to oxygen atoms.

Stage Four: The Electron Transport Chain. The electrons harvested from glucose pass through a chain of membrane proteins that use the energy to pump protons, driving the synthesis of ATP.

Summarizing Aerobic Respiration. Aerobic respiration harvests over half the energy in the chemical bonds of glucose.

Regulating Aerobic Respiration. High levels of ATP tend to shut down cellular respiration by inhibiting key reactions.

9.3 Catabolism of proteins and fats can yield considerable energy.

Glucose Is Not the Only Food. Proteins and fats are dismantled and the products fed into cellular respiration.

9.4 Cells can metabolize food without oxygen.

Fermentation. Fermentation allows continued metabolism in the absence of oxygen by donating the electrons harvested from glycolysis to organic molecules.

FIGURE 9.1
Harvesting chemical energy. Organisms such as these harvest mice depend on the energy stored in the chemical bonds of the food they eat to power their life processes.

Life is driven by energy. All the activities organisms carry out—the swimming of bacteria, the purring of a cat, your reading of these words—use energy. In this chapter, we will discuss the processes all cells use to derive chemical energy from organic molecules and to convert that energy to ATP. We will consider photosynthesis, which uses light energy rather than chemical energy, in detail in chapter 10. We examine the conversion of chemical energy to ATP first because all organisms, both photosynthesizers and the organisms that feed on them (like the field mice in figure 9.1), are capable of harvesting energy from chemical bonds. As you will see, though, this process and photosynthesis have much in common.

Using Chemical Energy to Drive Metabolism

Plants, algae, and some bacteria harvest the energy of sunlight through photosynthesis, converting radiant energy into chemical energy. These organisms, along with a few others that use chemical energy in a similar way, are called **autotrophs** ("self-feeders"). All other organisms live on the energy autotrophs produce and are called **heterotrophs** ("fed by others"). At least 95% of the kinds of organisms on earth—all animals and fungi, and most protists and bacteria—are heterotrophs.

Where is the chemical energy in food, and how do heterotrophs harvest it to carry out the many tasks of living (figure 9.2)? Most foods contain a variety of carbohydrates, proteins, and fats, all rich in energy-laden chemical bonds. Carbohydrates and fats, for example, possess many carbon-hydrogen (C—H), as well as carbon-oxygen (C—O) bonds. The job of extracting energy from this complex organic mixture is tackled in stages. First, enzymes break the large molecules down into smaller ones, a process called **digestion.** Then, other enzymes dismantle these fragments a little at a time, harvesting energy from C—H and other chemical bonds at each stage. This process is called **catabolism.**

While you obtain energy from many of the constituents of food, it is traditional to focus first on the catabolism of carbohydrates. We will follow the six-carbon sugar, glucose ($C_6H_{12}O_6$), as its chemical bonds are progressively harvested for energy. Later, we will come back and examine the catabolism of proteins and fats.

Cellular Respiration

The energy in a chemical bond can be visualized as potential energy borne by the electrons that make up the covalent bond. Cells harvest this energy by putting the electrons to work, often to produce ATP, the energy currency of the cell. Afterward, the energy-depleted electron (associated with a proton as a hydrogen atom) is donated to some other molecule. When oxygen gas (O_2) accepts the hydrogen atom, water forms, and the process is called **aerobic respiration.** When an inorganic molecule other than oxygen accepts the hydrogen, the process is called **anaerobic respiration.** When an organic molecule accepts the hydrogen atom, the process is called **fermentation.**

FIGURE 9.2
Start every day with a good breakfast. The carbohydrates, proteins, and fats in this fish contain energy that the bear's cells can use to power their daily activities.

Chemically, there is little difference between the catabolism of carbohydrates in a cell and the burning of wood in a fireplace. In both instances, the reactants are carbohydrates and oxygen, and the products are carbon dioxide, water, and energy:

$$C_6H_{12}O_6 + 6\ O_2 \longrightarrow 6\ CO_2 + 6\ H_2O + \text{energy (heat or ATP)}$$

The change in free energy in this reaction is –720 kilocalories (–3012 kilojoules) per mole of glucose under the conditions found within a cell (the traditional value of – 686 kilocalories, or – 2870 kJ, per mole refers to standard conditions—room temperature, one atmosphere of pressure, etc.). This change in free energy results largely from the breaking of the six C—H bonds in the glucose molecule. The negative sign indicates that the products possess *less* free energy than the reactants. The same amount of energy is released whether glucose is catabolized or burned, but when it is burned most of the energy is released as heat. This heat cannot be used to perform work in cells. The key to a cell's ability to harvest useful energy from the catabolism of food molecules such as glucose is its conversion of a portion of the energy into a more useful form. Cells do this by using some of the energy to drive the production of ATP, a molecule that can power cellular activities.

The ATP Molecule

Adenosine triphosphate (ATP) is the energy currency of the cell, the molecule that transfers the energy captured during respiration to the many sites that use energy in the cell. How is ATP able to transfer energy so readily? Recall from chapter 8 that ATP is composed of a sugar (ribose) bound to an organic base (adenine) and a chain of three phosphate groups. As shown in figure 9.3, each phosphate group is negatively charged. Because like charges repel each other, the linked phosphate groups push against the bond that holds them together. Like a cocked mousetrap, the linked phosphates store the energy of their electrostatic repulsion. Transferring a phosphate group to another molecule relaxes the electrostatic spring of ATP, at the same time cocking the spring of the molecule that is phosphorylated. This molecule can then use the energy to undergo some change that requires work.

How Cells Use ATP

Cells use ATP to do most of those activities that require work. One of the most obvious is movement. Some bacteria swim about, propelling themselves through the water by rapidly spinning a long, tail-like flagellum, much as a ship moves by spinning a propeller. During your development as an embryo, many of your cells moved about, crawling over one another to reach new positions. Movement also occurs within cells. Tiny fibers within muscle cells pull against one another when muscles contract. Mitochondria pass a meter or more along the narrow nerve cells that connect your feet with your spine. Chromosomes are pulled by microtubules during cell division. All of these movements by cells require the expenditure of ATP energy.

A second major way cells use ATP is to *drive endergonic reactions*. Many of the synthetic activities of the cell are endergonic, because building molecules takes energy. The chemical bonds of the products of these reactions contain more energy, or are more organized, than the reactants. The reaction can't proceed until that extra energy is supplied to the reaction. It is ATP that provides this needed energy.

How ATP Drives Endergonic Reactions

How does ATP drive an endergonic reaction? The enzyme that catalyzes the endergonic reaction has *two* binding sites on its surface, one for the reactant and another for ATP. The ATP site splits the ATP molecule, liberating over 7 kcal (30 kJ) of chemical energy. This energy pushes the reactant at the second site "uphill," driving the endergonic reaction. (In a similar way, you can make water in a swimming pool leap straight up in the air, despite the fact that gravity prevents water from rising spontaneously—just jump in the pool! The energy you add going in more than compensates for the force of gravity holding the water back.)

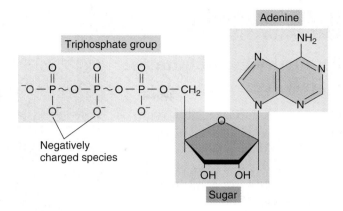

FIGURE 9.3
Structure of the ATP molecule. ATP is composed of an organic base and a chain of phosphates attached to opposite ends of a five-carbon sugar. Notice that the charged regions of the phosphate chain are close to one another. These like charges tend to repel one another, giving the bonds that hold them together a particularly high energy transfer potential.

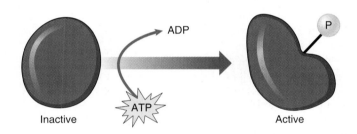

FIGURE 9.4
How ATP drives an endergonic reaction. In many cases, a phosphate group split from ATP activates a protein, catalyzing an endergonic process.

When the splitting of ATP molecules drives an energy-requiring reaction in a cell, the two parts of the reaction—ATP-splitting and endergonic—take place in concert. In some cases, the two parts both occur on the surface of the same enzyme; they are physically linked, or "coupled," like two legs walking. In other cases, a high-energy phosphate from ATP attaches to the protein catalyzing the endergonic process, activating it (figure 9.4). Coupling energy-requiring reactions to the splitting of ATP in this way is one of the key tools cells use to manage energy.

The catabolism of glucose into carbon dioxide and water in living organisms releases about 720 kcal (3012 kJ) of energy per mole of glucose. This energy is captured in ATP, which stores the energy by linking charged phosphate groups near one another. When the phosphate bonds in ATP are hydrolyzed, energy is released and available to do work.

An Overview of Glucose Catabolism

Cells are able to make ATP from the catabolism of organic molecules in two different ways.

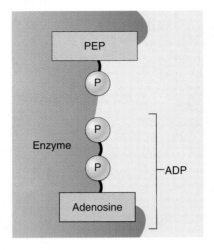
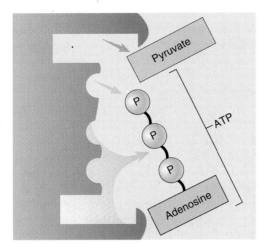

1. **Substrate-level phosphorylation.** In the first, called **substrate-level phosphorylation,** ATP is formed by transferring a phosphate group directly to ADP from a phosphate-bearing intermediate (figure 9.5). During glycolysis, discussed below, the chemical bonds of glucose are shifted around in reactions that provide the energy required to form ATP.

2. **Aerobic respiration.** In the second, called **aerobic respiration,** ATP forms as electrons are harvested, transferred along the electron transport chain, and eventually donated to oxygen gas. Eukaryotes produce the majority of their ATP from glucose in this way.

FIGURE 9.5

Substrate-level phosphorylation. Some molecules, such as phosphoenolpyruvate (PEP), possess a high-energy phosphate bond similar to the bonds in ATP. When PEP's phosphate group is transferred enzymatically to ADP, the energy in the bond is conserved and ATP is created.

In most organisms, these two processes are combined. To harvest energy to make ATP from the sugar glucose in the presence of oxygen, the cell carries out a complex series of enzyme-catalyzed reactions that occur in four stages: the first stage captures energy by substrate-level phosphorylation through glycolysis, the following three stages carry out aerobic respiration by oxidizing the end product of glycolysis.

Glycolysis

Stage One: Glycolysis. The first stage of extracting energy from glucose is a 10-reaction biochemical pathway called **glycolysis** that produces ATP by substrate-level phosphorylation. The enzymes that catalyze the glycolytic reactions are in the cytoplasm of the cell, not bound to any membrane or organelle. Two ATP molecules are used up early in the pathway, and four ATP molecules are formed by substrate-level phosphorylation. This yields a net of two ATP molecules for each molecule of glucose catabolized. In addition, four electrons are harvested as NADH that can be used to form ATP by aerobic respiration. Still, the total yield of ATP is small. When the glycolytic process is completed, the two molecules of pyruvate that are formed still contain most of the energy the original glucose molecule held.

Glycolysis occurs in all organisms, and the glycolytic reactions that form ATP by substrate-level phosphorylation can occur with or without oxygen. In animals, however, the harvesting of energetic electrons by glycolysis cannot take place indefinitely in the absence of oxygen. As a result, animal cells without oxygen soon die.

Aerobic Respiration

Stage Two: Pyruvate Oxidation. In the second stage, pyruvate, the end product from glycolysis, is converted into carbon dioxide and a two-carbon molecule called acetyl-CoA. For each molecule of pyruvate converted, one molecule of NADH is synthesized.

Stage Three: The Krebs Cycle. The third stage introduces this acetyl-CoA into a cycle of nine reactions called the **Krebs cycle,** named after the British biochemist, Sir Hans Krebs, who discovered it. (The Krebs cycle is also called the citric acid cycle, for the citric acid, or citrate, formed in its first step, and less commonly, the tricarboxylic acid cycle, because citrate has three carboxyl groups.) In the Krebs cycle, two more ATP molecules are extracted by substrate-level phosphorylation, and a large number of electrons are removed by NADH.

Stage Four: Electron Transport Chain. In the fourth stage, the energetic electrons carried by NADH are employed to drive the synthesis of a large amount of ATP by the electron transport chain.

Pyruvate oxidation, the reactions of the Krebs cycle, and ATP production by electron transport chains occur inside the mitochondria of all eukaryotes, as well as within many forms of bacteria. Recall from chapter 5 that mitochondria are thought to have evolved from bacteria. Although plants and algae can produce ATP by photosynthesis, they also produce ATP by aerobic respiration, just as animals and other nonphotosynthetic eukaryotes do. Figure 9.6 provides an overview of aerobic respiration.

Anaerobic Respiration

In the presence of oxygen, cells can respire aerobically. In the absence of oxygen, some organisms respire anaerobically, using different inorganic electron acceptors than oxygen. For example, many bacteria use sulfur, nitrate, or other inorganic compounds as the electron acceptor in place of oxygen.

Methanogens.   Among the heterotrophs that practice anaerobic respiration are primitive archaebacteria such as the thermophiles discussed in chapter 4. Some of these, called methanogens, use CO_2 as the electron acceptor, reducing CO_2 to CH_4 (methane) with the hydrogens derived from organic molecules produced by other organisms.

Sulfur Bacteria.   Evidence of a second anaerobic respiratory process among primitive bacteria is seen in a group of rocks about 2.7 billion years old, known as the Woman River iron formation. Organic material in these rocks is enriched for the light isotope of sulfur, ^{32}S, relative to the heavier isotope ^{34}S. No known geochemical process produces such enrichment, but biological sulfur reduction does, in a process still carried out today by certain primitive bacteria. In this sulfate respiration, the bacteria derive energy from the reduction of inorganic sulfates (SO_4) to H_2S. The hydrogen atoms are obtained from organic molecules other organisms produce. These bacteria thus do the same thing methanogens do, but they use SO_4 as the oxidizing (that is, electron-accepting) agent in place of CO_2.

The sulfate reducers set the stage for the evolution of photosynthesis, creating an environment rich in H_2S. As discussed in chapter 8, the first form of photosynthesis obtained hydrogens from H_2S using the energy of sunlight.

In aerobic respiration, the cell harvests energy from glucose molecules in a sequence of four major pathways: glycolysis, pyruvate oxidation, the Krebs cycle, and the electron transport chain. Oxygen is the final electron acceptor. Anaerobic respiration donates the harvested electrons to other inorganic compounds.

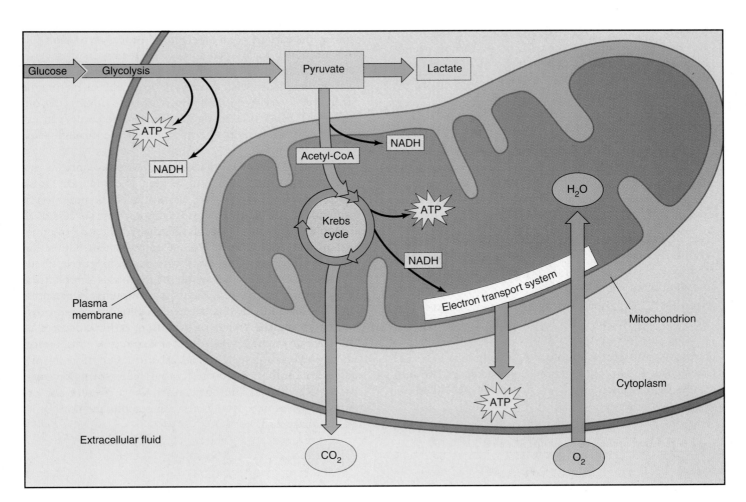

FIGURE 9.6
An overview of aerobic respiration.

Stage One: Glycolysis

The metabolism of primitive organisms focused on glucose. Glucose molecules can be dismantled in many ways, but primitive organisms evolved a glucose-catabolizing process that releases enough free energy to drive the synthesis of ATP in coupled reactions. This process, called glycolysis, occurs in the cytoplasm and involves a sequence of 10 reactions (figure 9.7) that convert glucose into 2 three-carbon molecules of pyruvate. For each molecule of glucose that passes through this transformation, the cell nets two ATP molecules by substrate-level phosphorylation.

Priming

The first half of glycolysis consists of five sequential reactions that convert one molecule of glucose into two molecules of the three-carbon compound, glyceraldehyde 3-phosphate (G3P). These reactions demand the expenditure of ATP, so they are an energy-requiring process.

Step A: Glucose priming. Three reactions "prime" glucose by changing it into a compound that can be cleaved readily into 2 three-carbon phosphorylated molecules. Two of these reactions require the cleavage of ATP, so this step requires the cell to use two ATP molecules.

Step B: Cleavage and rearrangement. In the first of the remaining pair of reactions, the six-carbon product of step A is split into 2 three-carbon molecules. One is G3P, and the other is then converted to G3P by the second reaction.

Substrate-Level Phosphorylation

In the second half of glycolysis, five more reactions convert G3P into pyruvate in an energy-yielding process that generates ATP. Overall, then, glycolysis is a series of 10 enzyme-catalyzed reactions in which some ATP is invested in order to produce more.

Step C: Oxidation. Two electrons and one proton are transferred from G3P to NAD^+, forming NADH. Note that NAD^+ is an ion, and that both electrons in the new covalent bond come from G3P.

Step D: ATP generation. Four reactions convert G3P into another three-carbon molecule, pyruvate. This process generates two ATP molecules (see figure 9.5).

Because each glucose molecule is split into two G3P molecules, the overall reaction sequence yields two molecules of ATP, as well as two molecules of NADH and two of pyruvate:

4 ATP (2 ATP for each of the 2 G3P molecules in step D)
− 2 ATP (used in the two reactions in step A)
_____
2 ATP

Under the nonstandard conditions within a cell, each ATP molecule produced represents the capture of about 12 kcal (50 kJ) of energy per mole of glucose, rather than the 7.3 traditionally quoted for standard conditions. This means glycolysis harvests about 24 kcal/mole (100 kJ/mole). This is not a great deal of energy. The total energy content of the chemical bonds of glucose is 686 kcal (2870 kJ) per mole, so glycolysis harvests only 3.5% of the chemical energy of glucose.

Although far from ideal in terms of the amount of energy it releases, glycolysis does generate ATP. For more than a billion years during the anaerobic first stages of life on earth, it was the primary way heterotrophic organisms generated ATP from organic molecules. Like many biochemical pathways, glycolysis is believed to have evolved backward, with the last steps in the process being the most ancient. Thus, the second half of glycolysis, the ATP-yielding breakdown of G3P, may have been the original process early heterotrophs used to generate ATP. The synthesis of G3P from glucose would have appeared later, perhaps when alternative sources of G3P were depleted.

All Cells Use Glycolysis

The glycolytic reaction sequence is thought to have been among the earliest of all biochemical processes to evolve. It uses no molecular oxygen and occurs readily in an anaerobic environment. All of its reactions occur free in the cytoplasm; none is associated with any organelle or membrane structure. Every living creature is capable of carrying out glycolysis. Most present-day organisms, however, can extract considerably more energy from glucose through aerobic respiration.

Why does glycolysis take place even now, since its energy yield in the absence of oxygen is comparatively so paltry? The answer is that evolution is an incremental process: change occurs by improving on past successes. In catabolic metabolism, glycolysis satisfied the one essential evolutionary criterion: it was an improvement. Cells that could not carry out glycolysis were at a competitive disadvantage, and only cells capable of glycolysis survived the early competition of life. Later improvements in catabolic metabolism built on this success. Glycolysis was not discarded during the course of evolution; rather, it served as the starting point for the further extraction of chemical energy. Metabolism evolved as one layer of reactions added to another, just as successive layers of paint cover the walls of an old building. Nearly every present-day organism carries out glycolysis as a metabolic memory of its evolutionary past.

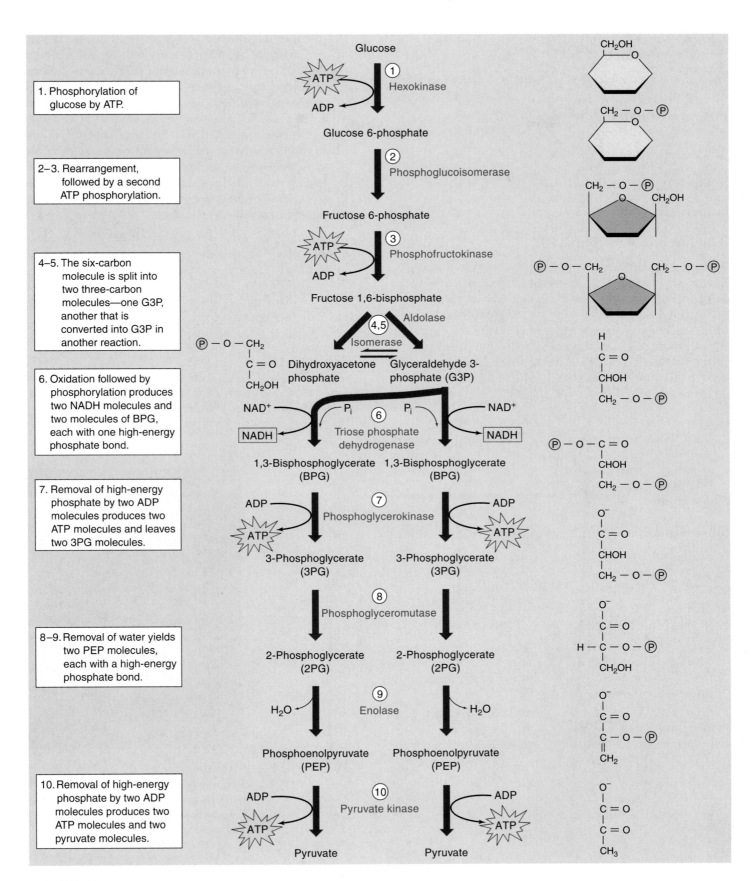

FIGURE 9.7
The glycolytic pathway. The first five reactions convert a molecule of glucose into two molecules of G3P. The second five reactions convert G3P into pyruvate.

Closing the Metabolic Circle: The Regeneration of NAD⁺

Inspect for a moment the net reaction of the glycolytic sequence:

Glucose + 2 ADP + 2 P_i + 2 NAD⁺ $\longrightarrow$
2 Pyruvate + 2 ATP + 2 NADH + 2 H⁺ + 2 H_2O

You can see that three changes occur in glycolysis: (1) glucose is converted into two molecules of pyruvate; (2) two molecules of ADP are converted into ATP; and (3) two molecules of NAD⁺ are converted into NADH.

The Need to Recycle NADH

As long as food molecules that can be converted into glucose are available, a cell can continually churn out ATP to drive its activities. In doing so, however, it accumulates NADH and depletes the pool of NAD⁺ molecules. A cell does not contain a large amount of NAD⁺, and for glycolysis to continue, NADH must be recycled into NAD⁺. Some other molecule than NAD⁺ must ultimately accept the hydrogen atom taken from G3P and be reduced. Two molecules can carry out this key task (figure 9.8):

1. **Aerobic respiration.** Oxygen is an excellent electron acceptor. Through a series of electron transfers, the hydrogen atom taken from G3P can be donated to oxygen, forming water. This is what happens in the cells of eukaryotes in the presence of oxygen. Because air is rich in oxygen, this process is also referred to as aerobic metabolism.

2. **Fermentation.** When oxygen is unavailable, an organic molecule can accept the hydrogen atom instead. Such fermentation plays an important role in the metabolism of most organisms (figure 9.9), even those capable of aerobic respiration.

The fate of the pyruvate that is produced by glycolysis depends upon which of these two processes takes place. The aerobic respiration path starts with the oxidation of pyruvate to a molecule called acetyl-CoA, which is then further oxidized in a series of reactions called the Krebs cycle. The fermentation path, by contrast, involves the reduction of all or part of pyruvate. We will start by examining aerobic respiration, then look briefly at fermentation.

Glycolysis generates a small amount of ATP by reshuffling the bonds of glucose molecules. In glycolysis, two molecules of NAD⁺ are reduced to NADH. NAD⁺ must be regenerated for glycolysis to continue unabated.

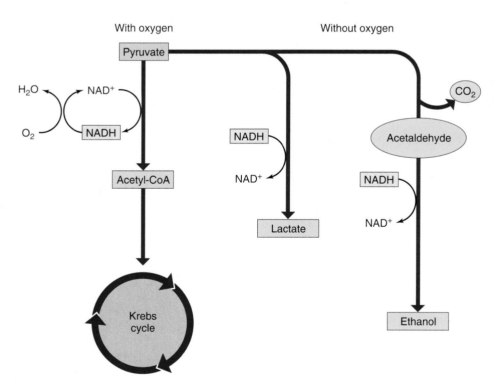

FIGURE 9.8

What happens to pyruvate, the product of glycolysis? In the presence of oxygen, pyruvate is oxidized to acetyl-CoA, which enters the Krebs cycle. In the absence of oxygen, pyruvate is instead reduced, accepting the electrons extracted during glycolysis and carried by NADH. When pyruvate is reduced directly, as in muscle cells, the product is lactate. When CO_2 is first removed from pyruvate and the product, acetaldehyde, is then reduced, as in yeast cells, the product is ethanol.

FIGURE 9.9

Fermentation. The conversion of pyruvate to ethanol takes place naturally in grapes left to ferment on vines, as well as in fermentation vats of crushed grapes. Yeasts carry out the process, but when their conversion increases the ethanol concentration to about 12%, the toxic effects of the alcohol kill the yeast cells. What is left is wine.

Stage Two: The Oxidation of Pyruvate

In the presence of oxygen, the oxidation of glucose that begins in glycolysis continues where glycolysis leaves off—with pyruvate. In eukaryotic organisms, the extraction of additional energy from pyruvate takes place exclusively inside mitochondria. The cell harvests pyruvate's considerable energy in two steps: first, by oxidizing pyruvate to form acetyl-CoA, and then by oxidizing acetyl-CoA in the Krebs cycle.

Producing Acetyl-CoA

Pyruvate is oxidized in a single "decarboxylation" reaction that cleaves off one of pyruvate's three carbons. This carbon then departs as CO_2 (figure 9.10a). This reaction produces a two-carbon fragment called an acetyl group, as well as a pair of electrons and their associated hydrogen, which reduce NAD^+ to NADH. The reaction is complex, involving three intermediate stages, and is catalyzed within mitochondria by a *multienzyme complex*. As chapter 8 noted, such a complex organizes a series of enzymatic steps so that the chemical intermediates do not diffuse away or undergo other reactions. Within the complex, component polypeptides pass the substrates from one enzyme to the next, without releasing them. *Pyruvate dehydrogenase*, the complex of enzymes that removes CO_2 from pyruvate, is one of the largest enzymes known: it contains 60 subunits! In the course of the reaction, the acetyl group removed from pyruvate combines with a cofactor called coenzyme A (CoA), forming a compound known as **acetyl-CoA:**

$$\text{Pyruvate} + NAD^+ + \text{CoA} \longrightarrow \text{Acetyl-CoA} + \text{NADH} + CO_2$$

This reaction produces a molecule of NADH, which is later used to produce ATP. Of far greater significance than the reduction of NAD^+ to NADH, however, is the production of acetyl-CoA (figure 9.10b). Acetyl-CoA is important because so many different metabolic processes generate it. Not only does the oxidation of pyruvate, an intermediate in carbohydrate catabolism, produce it, but the metabolic breakdown of proteins, fats, and other lipids also generate acetyl-CoA. Indeed, almost all molecules catabolized for energy are converted into acetyl-CoA. Acetyl-CoA is then channeled into fat synthesis or into ATP production, depending on the organism's energy requirements. Acetyl-CoA is a key point of focus for the many catabolic processes of the eukaryotic cell.

Using Acetyl-CoA

Although the cell forms acetyl-CoA in many ways, only a limited number of processes use acetyl-CoA. Most of it is either directed toward energy storage (lipid synthesis, for example) or oxidized in the Krebs cycle to produce ATP. Which of these two options is taken depends on the level of ATP in the cell. When ATP levels are high, the oxida-

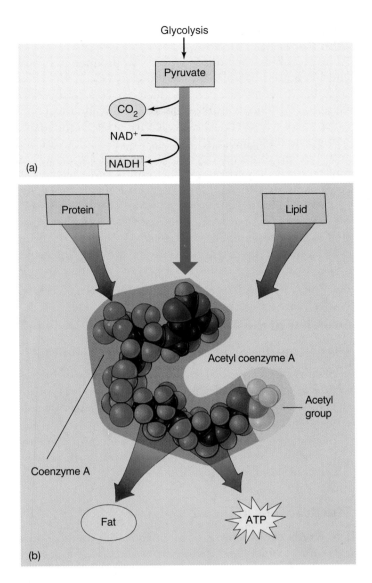

FIGURE 9.10
The oxidation of pyruvate. (a) This complex reaction involves the reduction of NAD^+ to NADH and is thus a significant source of metabolic energy. (b) Its product, acetyl-CoA, is the starting material for the Krebs cycle. Almost all molecules that are catabolized for energy are converted into acetyl-CoA, which is then channeled into fat synthesis or into ATP production.

tive pathway is inhibited, and acetyl-CoA is channeled into fatty acid synthesis. This explains why many animals (humans included) develop fat reserves when they consume more food than their bodies require. Alternatively, when ATP levels are low, the oxidative pathway is stimulated, and acetyl-CoA flows into energy-producing oxidative metabolism.

In the second energy-harvesting stage of glucose catabolism, pyruvate is decarboxylated, yielding acetyl-CoA, NADH, and CO_2. This process occurs within the mitochondrion.

Stage Three: The Krebs Cycle

After glycolysis catabolizes glucose to produce pyruvate, and pyruvate is oxidized to form acetyl-CoA, the third stage of extracting energy from glucose begins. In this third stage, acetyl-CoA is oxidized in a series of nine reactions called the Krebs cycle. These reactions occur in the matrix of mitochondria (figure 9.11). In this cycle, the two-carbon acetyl group of acetyl-CoA combines with a four-carbon molecule called oxaloacetate. The resulting six-carbon molecule then goes through a sequence of electron-yielding oxidation reactions, during which two CO_2 molecules split off, restoring oxaloacetate. The oxaloacetate is then recycled to bind to another acetyl group. In each turn of the cycle, a new acetyl group replaces the two CO_2 molecules lost, and more electrons are extracted to drive proton pumps that generate ATP.

Overview of the Krebs Cycle

The nine reactions of the Krebs cycle occur in two steps:

Step A: Priming. Three reactions prepare the six-carbon molecule for energy extraction. First, acetyl-CoA joins the cycle, and then chemical groups are rearranged.

Step B: Energy extraction. Four of the six reactions in this step are oxidations in which electrons are removed, and one generates an ATP equivalent directly by substrate-level phosphorylation.

The Reactions of the Krebs Cycle

The **Krebs cycle** consists of nine sequential reactions that cells use to extract energetic electrons and drive the synthesis of ATP. A two-carbon group from acetyl-CoA enters the cycle at the beginning, and two CO_2 molecules and several electrons are given off during the cycle.

Reaction 1: Condensation. The two-carbon group from acetyl-CoA joins with a four-carbon molecule, oxaloacetate, to form a six-carbon molecule, citrate. This condensation reaction is irreversible, committing the two-carbon acetyl group to the Krebs cycle. The reaction is inhibited when the cell's ATP concentration is high and stimulated when it is low. Hence, when the cell possesses ample amounts of ATP, the Krebs cycle shuts down and acetyl-CoA is channeled into fat synthesis.

Reactions 2 and 3: Isomerization. Before the oxidation reactions can begin, the hydroxyl (—OH) group of citrate must be repositioned. This is done in two steps: first, a water molecule is removed from one carbon; then, water is added to a different carbon. As a result, an —H group and an —OH group change positions. The product is an isomer of citrate called isocitrate.

Reaction 4: The First Oxidation. In the first energy-yielding step of the cycle, isocitrate undergoes an oxidative decarboxylation reaction. First, isocitrate is oxidized, yielding a pair of electrons that reduce a molecule of NAD^+ to NADH. Then the oxidized intermediate is decarboxylated; the central carbon atom splits off to form CO_2, yielding a five-carbon molecule called α-ketoglutarate.

Reaction 5: The Second Oxidation. Next, α-ketoglutarate is decarboxylated by a multienzyme complex similar to pyruvate dehydrogenase. The succinyl group left after the removal of CO_2 joins to coenzyme A, forming succinyl-CoA. In the process, two electrons are extracted, and they reduce another molecule of NAD^+ to NADH.

Reaction 6: Substrate-Level Phosphorylation. The linkage between the four-carbon succinyl group and CoA is a high-energy bond. In a coupled reaction similar to those that take place in glycolysis, this bond is cleaved, and the energy released drives the phosphorylation of guanosine diphosphate (GDP), forming guanosine triphosphate (GTP). GTP is readily converted into ATP, and the four-carbon fragment that remains is called succinate.

Reaction 7: The Third Oxidation. Next, succinate is oxidized to fumarate. The free energy change in this reaction is not large enough to reduce NAD^+. Instead, **flavin adenine dinucleotide (FAD^+)** is the electron acceptor. Unlike NAD^+, FAD^+ is not free to diffuse within the mitochondrion; it is an integral part of the inner mitochondrial membrane. Its reduced form, $FADH_2$, contributes electrons to the electron transport chain in the membrane.

Reactions 8 and 9: Regeneration of Oxaloacetate. In the final two reactions of the cycle, a water molecule is added to fumarate, forming malate. Malate is then oxidized, yielding a four-carbon molecule of oxaloacetate and two electrons that reduce a molecule of NAD^+ to NADH. Oxaloacetate, the molecule that began the cycle, is now free to combine with another two-carbon acetyl group from acetyl-CoA and reinitiate the cycle.

The Products of the Krebs Cycle

In the process of aerobic respiration, glucose is entirely consumed. The six-carbon glucose molecule is first cleaved into a pair of three-carbon pyruvate molecules during glycolysis. One of the carbons of each pyruvate is then lost as CO_2 in the conversion of pyruvate to acetyl-CoA, and the other two carbons are lost as CO_2 during the oxidations of the Krebs cycle. All that is left to mark the passing of the glucose molecule into six CO_2 molecules is its energy, some of which is preserved in four ATP molecules and in the reduced state of 12 electron carriers. Ten of these carriers are NADH molecules; the other two are $FADH_2$.

The Krebs cycle generates two ATP molecules per molecule of glucose, the same number generated by glycolysis. More importantly, the Krebs cycle and the oxidation of pyruvate harvest many energized electrons, which can be directed to the electron transport chain to drive the synthesis of much more ATP.

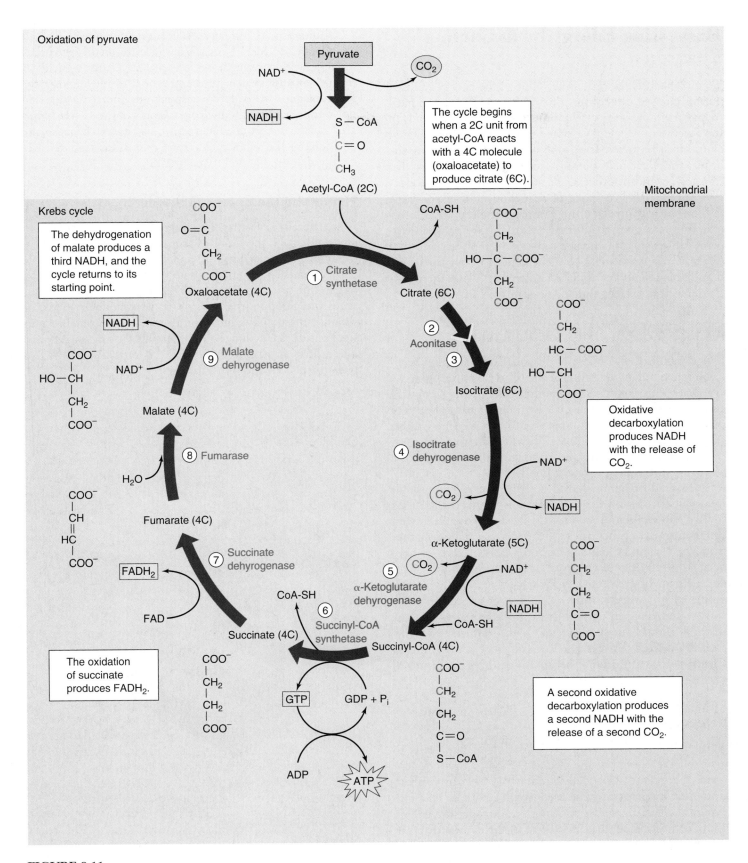

FIGURE 9.11

The Krebs cycle. This series of reactions takes place within the matrix of the mitochondrion. For the complete breakdown of a molecule of glucose, the two molecules of acetyl-CoA produced by glycolysis and pyruvate oxidation will each have to make a trip around the Krebs cycle. Notice the changes that occur in the carbon skeletons of the molecules as they proceed through the cycle.

Harvesting Energy by Extracting Electrons

To understand how cells direct some of the energy released during glucose catabolism into ATP production, we need to take a closer look at the electrons in the C—H bonds of the glucose molecule. We stated in chapter 8 that when an electron is removed from one atom and donated to another, the electron's potential energy of position is also transferred. In this process, the atom that receives the electron is reduced. We spoke of reduction in an all-or-none fashion, as if it involved the complete transfer of an electron from one atom to another. Often this is just what happens. However, sometimes a reduction simply changes the *degree of sharing* within a covalent bond. Let us now revisit that discussion and consider what happens when the transfer of electrons is incomplete.

A Closer Look at Oxidation-Reduction

The catabolism of glucose is an oxidation-reduction reaction. The covalent electrons in the C—H bonds of glucose are shared approximately equally between the C and H atoms because carbon and hydrogen nuclei have about the same affinity for valence electrons (that is, they exhibit similar *electronegativity*). However, when the carbon atoms of glucose react with oxygen to form carbon dioxide, the electrons in the new covalent bonds take a different position. Instead of being shared equally, the electrons that were associated with the carbon atoms in glucose shift far toward the oxygen atom in CO_2 because oxygen is very electronegative. Since these electrons are pulled farther from the carbon atoms, the carbon atoms of glucose have been oxidized (loss of electrons) and the oxygen atoms reduced (gain of electrons). Similarly, when the hydrogen atoms of glucose combine with oxygen atoms to form water, the oxygen atoms draw the shared electrons strongly toward them; again, oxygen is reduced and glucose is oxidized. In this reaction, oxygen is an oxidizing (electron-attracting) agent because it oxidizes the atoms of glucose.

Releasing Energy

The key to understanding the oxidation of glucose is to focus on the energy of the shared electrons. In a covalent bond, energy must be added to remove an electron from an atom, just as energy must be used to roll a boulder up a hill. The more electronegative the atom, the steeper the energy hill that must be climbed to pull an electron away from it. However, energy is released when an electron is shifted away from a less electronegative atom and *closer* to a more electronegative atom, just as energy is released when a boulder is allowed to roll down a hill. In the catabolism of glucose, energy is released when glucose is oxidized, as electrons relocate closer to oxygen (figure 9.12).

Glucose is an energy-rich food because it has an abundance of C—H bonds. Viewed in terms of oxidation-reduction, glucose possesses a wealth of electrons held far from their atoms, all with the potential to move closer toward oxygen. In oxidative respiration, energy is released not simply because the hydrogen atoms of the C—H bonds are transferred from glucose to oxygen, but because the positions of the valence electrons shift. This shift releases energy that can be used to make ATP.

Harvesting the Energy in Stages

It is generally true that the larger the release of energy in any single step, the more of that energy is released as heat (random molecular motion) and the less there is available to be channeled into more useful paths. In the combustion of gasoline, the same amount of energy is released whether all of the gasoline in a car's gas tank explodes at once, or whether the gasoline burns in a series of very small explosions inside the cylinders. By releasing the energy in gasoline a little at a time, more of the energy can be used to push the pistons and move the car.

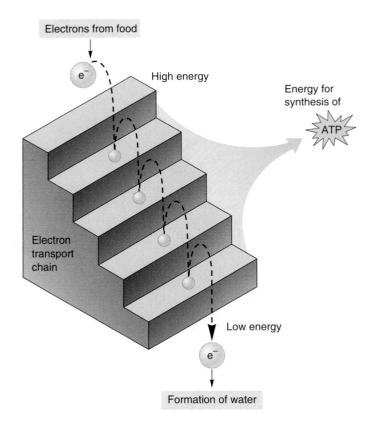

FIGURE 9.12
How electron transport works. This diagram shows how ATP is generated when electrons transfer from one energy level to another. Rather than releasing a single explosive burst of energy, electrons "fall" to lower and lower energy levels in steps, releasing stored energy with each fall as they tumble to the lowest (most electronegative) electron acceptor.

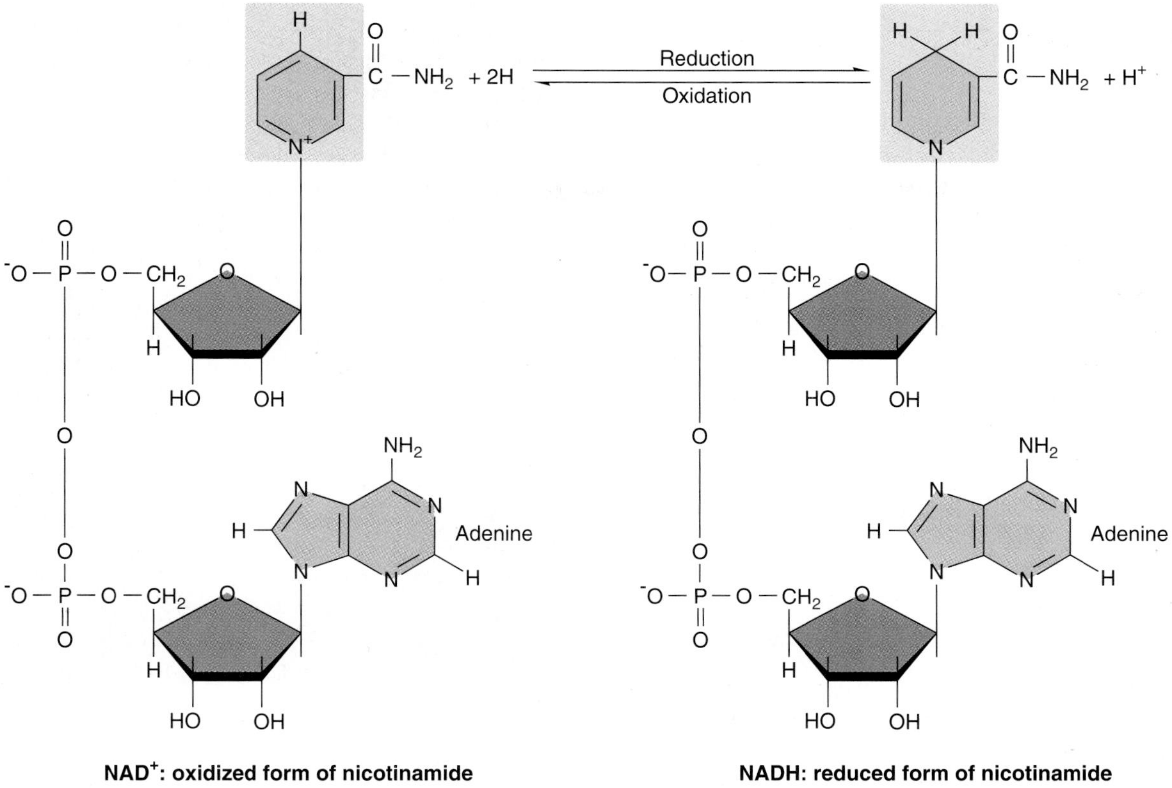

NAD⁺: oxidized form of nicotinamide | **NADH: reduced form of nicotinamide**

FIGURE 9.13

NAD⁺ and NADH. This dinucleotide serves as an "electron shuttle" during cellular respiration. NAD⁺ accepts electrons from catabolized macromolecules and is reduced to NADH.

The same principle applies to the oxidation of glucose inside a cell. If all of the hydrogen were transferred to oxygen in one explosive step, releasing all of the free energy at once, the cell would recover very little of that energy in a useful form. Instead, cells burn their fuel much as a car does, a little at a time. The six hydrogens in the C—H bonds of glucose are stripped off in stages in the series of enzyme-catalyzed reactions collectively referred to as glycolysis and the Krebs cycle. We have had a great deal to say about these reactions already in this chapter. Recall that the hydrogens are removed by transferring them to a coenzyme carrier, NAD⁺ (figure 9.13). Discussed in chapter 8, NAD⁺ is a very versatile electron acceptor, shuttling energy-bearing electrons throughout the cell. In harvesting the energy of glucose, NAD⁺ acts as the primary electron acceptor.

Following the Electrons

As you examine these reactions, try not to become confused by the changes in electrical charge. Always *follow the electrons*. Enzymes extract two hydrogens—that is, two electrons and two protons—from glucose and transfer both electrons and one of the protons to NAD⁺. The other proton is released as a hydrogen ion, H⁺, into the surrounding solution. This transfer converts NAD⁺ into NADH; that is,

two negative electrons and one positive proton are added to one positively charged NAD⁺ to form NADH, which is electrically neutral.

Energy captured by NADH is *not* harvested all at once. Instead of being transferred directly to oxygen, the two electrons carried by NADH are passed along the **electron transport chain** if oxygen is present. This chain consists of a series of molecules, mostly proteins, embedded within the inner membranes of mitochondria. NADH delivers electrons to the top of the electron transport chain and oxygen captures them at the bottom. The oxygen then joins with hydrogen ions to form water. At each step in the chain, the electrons move to a slightly more electronegative carrier, and their positions shift slightly. Thus, the electrons move *down* an energy gradient. The entire process releases a total of 53 kcal/mole (222 kJ/mole) under standard conditions. The transfer of electrons along this chain allows the energy to be extracted gradually. In the next section, we will discuss how this energy is put to work to drive the production of ATP.

The catabolism of glucose involves a series of oxidation-reduction reactions that release energy by repositioning electrons closer to oxygen atoms. Energy is thus harvested from glucose molecules in gradual steps, using NAD⁺ as an electron carrier.

Stage Four: The Electron Transport Chain

The NADH and FADH$_2$ molecules formed during the first three stages of aerobic respiration each contain a pair of electrons that were gained when NAD$^+$ and FAD$^+$ were reduced. The NADH molecules carry their electrons to the inner mitochondrial membrane, where they transfer the electrons to a series of membrane-associated proteins collectively called the electron transport chain.

Moving Electrons through the Electron Transport Chain

The first of the proteins to receive the electrons is a complex, membrane-embedded enzyme called **NADH dehydrogenase.** A carrier called ubiquinone then passes the electrons to a protein-cytochrome complex called the bc$_1$ complex. This complex, along with others in the chain, operates as a proton pump, driving a proton out across the membrane. Cytochromes are respiratory proteins that contain heme groups, complex carbon rings with many alternating single and double bonds and an iron atom in the center.

The electron is then carried by another carrier, *cytochrome c,* to the cytochrome oxidase complex (figure 9.14). This complex uses four such electrons to reduce a molecule of oxygen, each oxygen then combines with 2 hydrogen ions to form water:

$$O_2 + 4\,H^+ + 4\,e^- \longrightarrow 2\,H_2O$$

This series of membrane-associated electron carriers is collectively called the electron transport chain (figure 9.15).

NADH contributes its electrons to the first protein of the electron transport chain, NADH dehydrogenase. FADH$_2$, which is always attached to the inner mitochondrial membrane, feeds its electrons into the electron transport chain later, to ubiquinone.

It is the availability of a plentiful electron acceptor (often oxygen) that makes oxidative respiration possible. As we'll see in chapter 10, the electron transport chain used in aerobic respiration is similar to, and may well have evolved from, the chain employed in aerobic photosynthesis.

> The electron transport chain is a series of five membrane-associated proteins. Electrons delivered by NADH and FADH$_2$ are passed from protein to protein along the chain, like a baton in a relay race.

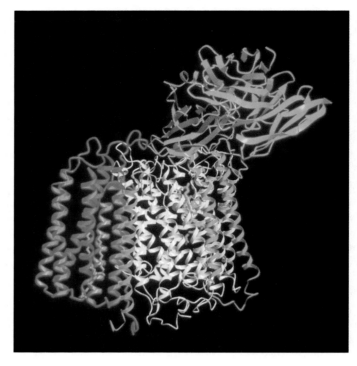

FIGURE 9.14

The cytochrome oxidase complex. The largest membrane protein with a fully known structure, cytochrome oxidase is composed of 11 protein subunits that form a channel through the membrane for protons and electrons. Electrons are sent down a pair of electronic "bucket brigades," each consisting of a pair of iron-containing groups in the interior of the channel.

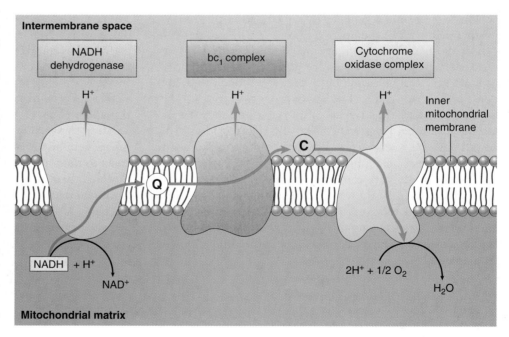

FIGURE 9.15

The electron transport chain. High-energy electrons harvested from catabolized molecules are transported (*red arrows*) by mobile electron carriers (ubiquinone, marked Q, and cytochrome c, marked C) along a chain of membrane proteins. Three proteins use portions of the electrons' energy to pump protons (*blue arrows*) out of the matrix and into the intermembrane space. The electrons are finally donated to oxygen to form water.

Building an Electrochemical Gradient

In eukaryotes, aerobic metabolism takes place within the mitochondria present in virtually all cells. The internal compartment, or matrix, of a mitochondrion contains the enzymes that carry out the reactions of the Krebs cycle. As the electrons harvested by oxidative respiration are passed along the electron transport chain, the energy they release transports protons out of the matrix and into the outer compartment, sometimes called the intermembrane space. Three transmembrane proteins in the inner mitochondrial membrane (see figure 9.15) actually accomplish the transport. The flow of excited electrons induces a change in the shape of these pump proteins, which causes them to transport protons across the membrane. The electrons contributed by NADH activate all three of these proton pumps, while those contributed by $FADH_2$ activate only two.

Producing ATP: Chemiosmosis

As the proton concentration in the outer compartment rises above that in the matrix, the matrix becomes slightly negative in charge. This internal negativity attracts the positively charged protons and induces them to reenter the matrix. The higher outer concentration tends to drive protons back in by diffusion; since membranes are relatively impermeable to ions, most of the protons that reenter the matrix pass through special proton channels in the inner mitochondrial membrane. When the protons pass through, these channels synthesize ATP from ADP + P_i within the matrix. The ATP is then transported by facilitated diffusion out of the mitochondrion and into the cell's cytoplasm. Because the chemical formation of ATP is driven by a diffusion force similar to osmosis, this process is referred to as **chemiosmosis** (figure 9.16).

Thus, the electron transport chain uses electrons harvested in aerobic respiration to pump a large number of protons across the inner mitochondrial membrane. Their subsequent reentry into the mitochondrial matrix drives the synthesis of ATP by chemiosmosis. Figure 9.17 summarizes the overall process.

The electrons harvested from glucose are pumped out of the mitochondrial matrix by the electron transport chain. The return of the protons into the matrix by diffusion through special channels is coupled to the generation of ATP.

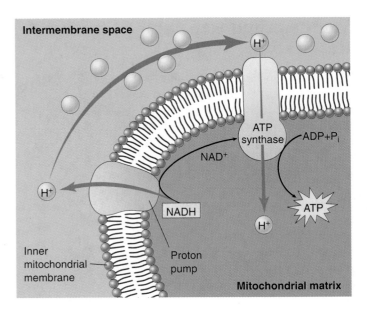

FIGURE 9.16
Chemiomosis. NADH transports high-energy electrons harvested from the catabolism of macromolecules to "proton pumps" that use the energy to pump protons out of the mitochondrial matrix. As a result, the concentration of protons outside the inner mitochondrial membrane rises, inducing protons to diffuse back into the matrix. Many of the protons pass through special channels that couple the reentry of protons to the production of ATP.

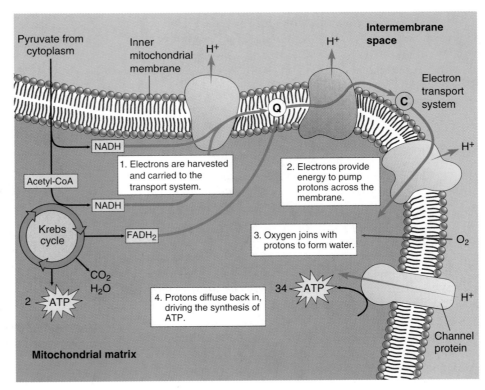

FIGURE 9.17
An overview of ATP generation. This process begins with pyruvate, the product of glycolysis, and ends with the synthesis of ATP.

Summarizing Aerobic Respiration

How much metabolic energy does a cell actually gain from the electrons harvested from a molecule of glucose, using the electron transport chain to produce ATP by chemiosmosis?

Theoretical Yield

The chemiosmotic model suggests that one ATP molecule is generated for each proton pump activated by the electron transport chain. Since the electrons from NADH activate three pumps and those from $FADH_2$ activate two, we would expect each molecule of NADH and $FADH_2$ to generate three and two ATP molecules, respectively. However, because eukaryotic cells carry out glycolysis in their cytoplasm and the Krebs cycle within their mitochondria, they must transport the two molecules of NADH produced during glycolysis across the mitochondrial membranes, which requires one ATP per molecule of NADH. Thus, the net ATP production is decreased by two. Therefore, the overall ATP production resulting from aerobic respiration *theoretically* should be 4 (from substrate-level phosphorylation during glycolysis) + 30 (3 from each of 10 molecules of NADH) + 4 (2 from each of 2 molecules of $FADH_2$) − 2 (for transport of glycolytic NADH) = 36 molecules of ATP (figure 9.18).

Actual Yield

The amount of ATP *actually* produced in a eukaryotic cell during aerobic respiration is somewhat lower than 36, for two reasons. First, the inner mitochondrial membrane is somewhat "leaky" to protons, allowing some of them to re-enter the matrix without passing through ATP-generating channels. Second, mitochondria often use the proton gradient generated by chemiosmosis for purposes other than ATP synthesis (such as transporting pyruvate into the matrix). Consequently, the actual measured values of ATP generated by NADH and $FADH_2$ are closer to 2.5 for each NADH and 1.5 for each $FADH_2$. With these corrections, the overall harvest of ATP from a molecule of glucose in a eukaryotic cell is closer to 4 (from substrate-level phosphorylation) + 25 (2.5 from each of 10 molecules of NADH) + 3 (1.5 from each of 2 molecules of $FADH_2$) − 2 (transport of glycolytic NADH) = 30 molecules of ATP.

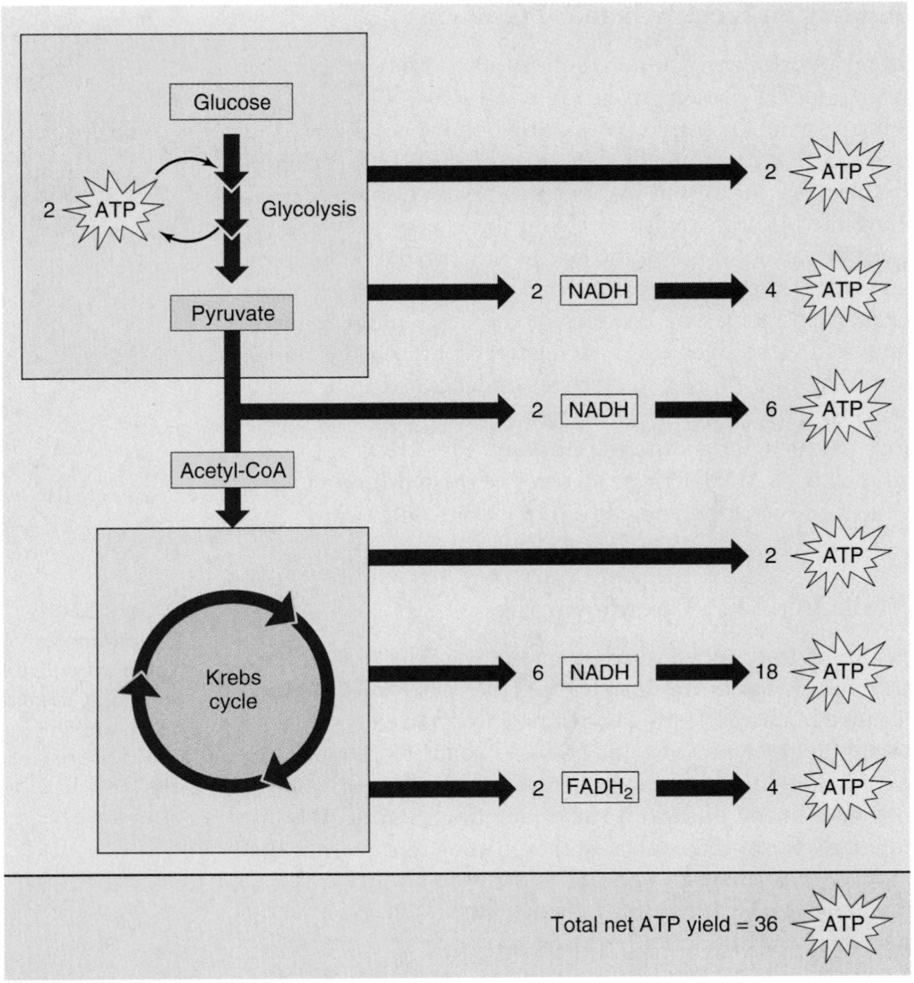

FIGURE 9.18
Theoretical ATP yield. The theoretical yield of ATP harvested from glucose by aerobic respiration totals 36 molecules.

The catabolism of glucose by aerobic respiration, in contrast to glycolysis, is quite efficient. Aerobic respiration in a eukaryotic cell harvests about $12 \times 30 \div 686 = 52\%$ of the energy available in glucose (By comparison, a typical car converts only about 25% of the energy in gasoline into useful energy). The efficiency of oxidative respiration at harvesting energy establishes a natural limit on the maximum length of food chains.

The high efficiency of aerobic respiration was one of the key factors that fostered the evolution of heterotrophs. With this mechanism for producing ATP, it became feasible for nonphotosynthetic organisms to derive metabolic energy exclusively from the oxidative breakdown of other organisms. As long as some organisms captured energy by photosynthesis, others could exist solely by feeding on them.

Oxidative respiration produces approximately 30 molecules of ATP from each molecule of glucose in eukaryotic cells. This represents more than half of the energy in the chemical bonds of glucose.

Regulating Aerobic Respiration

When cells possess plentiful amounts of ATP, the key reactions of glycolysis, the Krebs cycle, and fatty acid breakdown are inhibited, slowing ATP production. The regulation of these biochemical pathways by the level of ATP is an example of feedback inhibition. Conversely, when ATP levels in the cell are low, ADP levels are high; and ADP activates enzymes in the pathways of carbohydrate catabolism to stimulate the production of more ATP.

Control of glucose catabolism occurs at two key points of the catabolic pathway (figure 9.19). The control point in glycolysis is the enzyme phosphofructokinase, which catalyzes reaction 3, the conversion of fructose phosphate to fructose bisphosphate. This is the first reaction of glycolysis that is not readily reversible, committing the substrate to the glycolytic sequence. High levels of ADP relative to ATP (implying a need to convert more ADP to ATP) stimulate phosphofructokinase, committing more sugar to the catabolic pathway; so do low levels of citrate (implying the Krebs cycle is not running at full tilt and needs more input). The main control point in the oxidation of pyruvate occurs at the committing step in the Krebs cycle with the enzyme pyruvate decarboxylase. It is inhibited by high levels of NADH (implying no more is needed).

Another control point in the Krebs cycle is the enzyme citrate synthetase, which catalyzes the first reaction, the conversion of oxaloacetate and acetyl-CoA into citrate. High levels of ATP inhibit citrate synthetase (as well as pyruvate decarboxylase and two other Krebs cycle enzymes), shutting down the catabolic pathway.

Relative levels of ADP and ATP regulate the catabolism of glucose at key committing reactions.

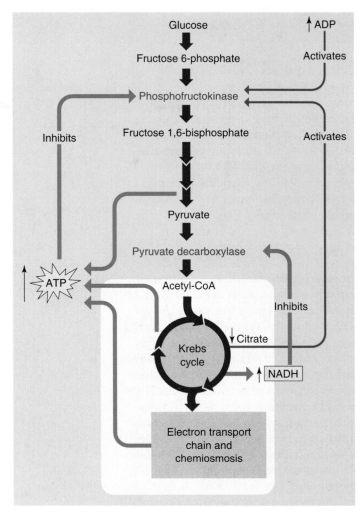

FIGURE 9.19
Control of glucose catabolism. The relative levels of ADP and ATP control the catabolic pathway at two key points: the committing reactions of glycolysis and the Krebs cycle.

A Vocabulary of ATP Generation

aerobic respiration The portion of cellular respiration that requires oxygen as an electron acceptor; it includes pyruvate oxidation and the Krebs cycle.

anaerobic respiration Cellular respiration in which inorganic electron acceptors other than oxygen are used; it includes glycolysis.

cellular respiration The oxidation of organic molecules to produce ATP in which the final electron acceptor is organic; it includes aerobic and anaerboic respiration.

chemiosmosis The passage of high-energy electrons along the electron transport chain, which is coupled to the pumping of protons across a membrane and the return of protons to the original side of the membrane through ATP-generating channels.

fermentation Alternative ATP-producing pathway performed by some cells in the absence of oxygen, in which the final electron acceptor is an organic molecule.

maximum efficiency The maximum number of ATP molecules generated by oxidizing a substance, relative to the free energy of that substance; in organisms, the actual efficiency is usually less than the maximum.

oxidation The loss of an electron. In cellular respiration, high-energy electrons are stripped from food molecules, oxidizing them.

photosynthesis The chemiosmotic generation of ATP and complex organic molecules powered by the energy derived from light.

substrate-level phosphorylation The generation of ATP by the direct transfer of a phosphate group to ADP from another phosphorylated molecule.

Glucose Is Not the Only Food

Thus far we have discussed oxidative respiration of glucose, which organisms obtain from the digestion of carbohydrates or from photosynthesis. Other organic molecules, particularly proteins and fats, are also important sources of energy (figure 9.20).

Cellular Respiration of Protein

Proteins are first broken down into their individual amino acids. The nitrogen-containing side group (the amino group) is then removed from each amino acid in a process called **deamination.** A series of reactions convert the carbon chain that remains into a molecule that takes part in glycolysis or the Krebs cycle. For example, alanine is converted into pyruvate, glutamate into α-ketoglutarate, and aspartate into oxaloacetate. The reactions of glycolysis and the Krebs cycle then extract the high-energy electrons from these molecules and put them to work making ATP.

Cellular Respiration of Fat

Fats are broken down into fatty acids plus glycerol. The tails of fatty acids typically have 16 or more —CH_2 links, and the many hydrogen atoms in these long tails provide a rich harvest of energy. Fatty acids are oxidized in the matrix of the mitochondrion. Enzymes there remove the two-carbon acetyl groups from the end of each fatty acid tail until the entire fatty acid is converted into acetyl groups. Each acetyl group then combines with coenzyme A to form acetyl-CoA. This process is known as **β-oxidation.**

How much ATP does the catabolism of fatty acids produce? Let's compare a hypothetical six-carbon fatty acid with the six-carbon glucose molecule, which we've said yields about 30 molecules of ATP in a eukaryotic cell. Two rounds of β-oxidation would convert the fatty acid into three molecules of acetyl-CoA. Each round requires one molecule of ATP to prime the process, but it also produces one molecule of NADH and one of $FADH_2$. These molecules together yield four molecules of ATP (assuming 2.5 ATPs per NADH and 1.5 ATPs per $FADH_2$). The oxidation of each acetyl-CoA in the Krebs cycle ultimately produces an additional 10 mol-

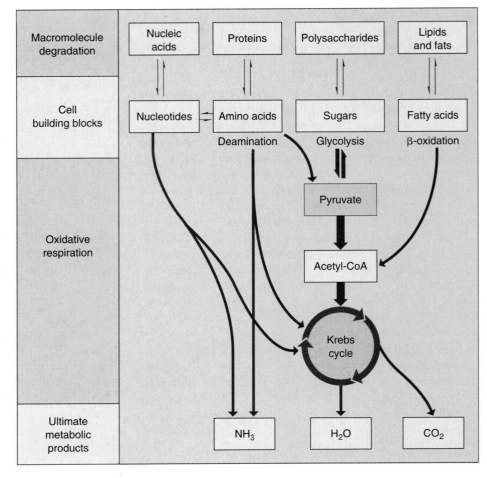

FIGURE 9.20

How cells extract chemical energy. All eukaryotes and many prokaryotes extract energy from organic molecules by oxidizing them. The first stage of this process, breaking down macromolecules into their constituent parts, yields little energy. The second stage, oxidative or aerobic respiration, extracts energy, primarily in the form of high-energy electrons, and produces water and carbon dioxide.

ecules of ATP. Overall, then, the ATP yield of a six-carbon fatty acid would be approximately 8 (from two rounds of β-oxidation) – 2 (for priming those two rounds) + 30 (from oxidizing the three acetyl-CoAs) = 36 molecules of ATP. Therefore, the respiration of a six-carbon fatty acid yields 20% more ATP than the respiration of glucose. Moreover, a fatty acid of that size would weigh less than two-thirds as much as glucose, so a gram of fatty acid contains more than twice as many kilocalories as a gram of glucose. That is why fat is a storage molecule for excess energy in many types of animals. If excess energy were stored instead as carbohydrate, as it is in plants, animal bodies would be much bulkier.

Proteins, fats, and other organic molecules are also metabolized for energy. The amino acids of proteins are first deaminated, while fats undergo a process called β-oxidation.

Metabolic Efficiency and the Length of Food Chains

It has been estimated that a heterotroph limited to glycolysis captures only 3.5% of the energy in the food it consumes. Hence, if such a heterotroph preserves 3.5% of the energy in the autotrophs it consumes, then any other heterotrophs that consume the first hetertroph will capture through glycolysis 3.5% of the energy in it, or 0.12% of the energy available in the original autotrophs. A very large base of autotrophs would thus be needed to support a small number of heterotrophs.

When organisms became able to extract energy from organic molecules by oxidative metabolism, this constraint became far less severe, because the efficiency of oxidative respiration is estimated to be about 52–63%. This increased efficiency results in the transmission of much more energy from one trophic level to another than does glycolysis. (A trophic level is a step in the movement of energy through an ecosystem.) The efficiency of oxidative metabolism has made possible the evolution of food chains, in which autotrophs are consumed by heterotrophs, which are consumed by other heterotrophs, and so on. You will read more about food chains in chapter 25.

Even with oxidative metabolism, approximately two-thirds of the available energy is lost at each trophic level, and that puts a limit on how long a food chain can be. Most food chains, like the one illustrated in figure 9.A, involve only three or rarely four trophic levels. Too much energy is lost at each transfer to allow chains to be much longer than that. For example, it would be impossible for a large human population to subsist by eating lions captured from the Serengeti Plain of Africa; the amount of grass available there would not support enough zebras and other herbivores to maintain the number of lions needed to feed the human population. Thus, the ecological complexity of our world is fixed in a fundamental way by the chemistry of oxidative respiration.

Stage 1: Photosynthesizers

Stage 2:Herbivores

Stage 3: Carnivore

Stage 4: Scavengers

Stage 5: Refuse utilizers

FIGURE 9.A

A food chain in the savannas, or open grasslands, of East Africa. Stage 1: *Photosynthesizer*. The grass under these palm trees grows actively during the hot, rainy season, capturing the energy of the sun and storing it in molecules of glucose, which are then converted into starch and stored in the grass. Stage 2: *Herbivore*. These large antelopes, known as wildebeests, consume the grass and transfer some of its stored energy into their own bodies. Stage 3: *Carnivore*. The lion feeds on wildebeests and other animals, capturing part of their stored energy and storing it in it's own body. Stage 4: *Scavenger*. This hyena and the vultures occupy the same stage in the food chain as the lion. They are also consuming the body of the dead wildebeest, which has been abandoned by the lion. Stage 5: *Refuse utilizer*. These butterflies, mostly *Precis octavia*, are feeding on the material left in the hyena's dung after the food the hyena consumed had passed through its digestive tract. At each of these four levels, only about a third or less of the energy present is used by the recipient.

Fermentation

In the absence of oxygen, aerobic metabolism cannot occur, and cells must rely exclusively on glycolysis to produce ATP. Under these conditions, the hydrogen atoms generated by glycolysis are donated to organic molecules in a process called **fermentation.**

Bacteria carry out more than a dozen kinds of fermentations, all using some form of organic molecule to accept the hydrogen atom from NADH and thus recycle NAD^+:

$$\text{Organic molecule} + \text{NADH} \longrightarrow \text{Reduced organic molecule} + NAD^+$$

Often the reduced organic compound is an organic acid—such as acetic acid, butyric acid, propionic acid, or lactic acid—or an alcohol.

Ethanol Fermentation

Eukaryotic cells are capable of only a few types of fermentation. In one type, which occurs in single-celled fungi called yeast, the molecule that accepts hydrogen from NADH is pyruvate, the end product of glycolysis itself. Yeast enzymes remove a terminal CO_2 group from pyruvate through decarboxylation, producing a two-carbon molecule called acetaldehyde. The CO_2 released causes bread made with yeast to rise, while bread made without yeast (unleavened bread) does not. The acetaldehyde accepts a hydrogen atom from NADH, producing NAD^+ and **ethanol** (ethyl alcohol). This particular type of fermentation is of great interest to humans, since it is the source of the ethanol in wine and beer (figure 9.21). Ethanol is a by-product of fermentation that is actually toxic to yeast; as it approaches a concentration of about 12%, it begins to kill the yeast. That explains why naturally fermented wine contains only about 12% ethanol.

Lactic Acid Fermentation

Most animal cells regenerate NAD^+ without decarboxylation. Muscle cells, for example, use an enzyme called lactate dehydrogenase to transfer a hydrogen atom from NADH back to the pyruvate that is produced by glycolysis. This reaction converts pyruvate into lactic acid and regenerates NAD^+ from NADH. It therefore closes the metabolic circle, allowing glycolysis to continue as long as glucose is available. Circulating blood removes excess lactate (the ionized form of lactic acid) from muscles, but when removal cannot keep pace with production, the accumulating lactic acid interferes with muscle function and contributes to muscle fatigue.

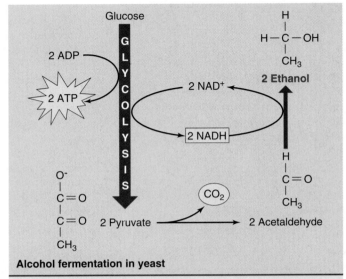

Alcohol fermentation in yeast

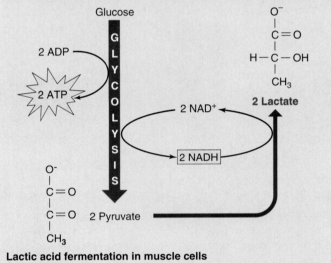

Lactic acid fermentation in muscle cells

FIGURE 9.21

How wine is made. Yeasts carry out the conversion of pyruvate to ethanol. This takes place naturally in grapes left to ferment on vines, as well as in fermentation vats containing crushed grapes. When the ethanol concentration reaches about 12%, its toxic effects kill the yeast; what remains is wine. Muscle cells convert pyruvate into lactate, which is less toxic than ethanol. However, lactate is still toxic enough to produce a painful sensation in muscles during heavy exercise, when oxygen in the muscles is depleted.

In fermentation, which occurs in the absence of oxygen, the electrons that result from the glycolytic breakdown of glucose are donated to an organic molecule, regenerating NAD^+ from NADH.

Summary of Concepts

9.1 Cells harvest the energy in chemical bonds.

- Organisms acquire ATP from photosynthesis or cellular respiration, a process in which the chemical energy in the bonds of organic molecules is harvested.

- The reactions of cellular respiration are oxidation-reduction (redox) reactions. Those that require a net input of free energy are coupled to the cleavage of ATP, which releases free energy.

- The mechanics of cellular respiration are often dictated by electron behavior, which is in turn influenced by the presence of electron acceptors. Some atoms, such as oxygen, are very electronegative and thus behave as good oxidizing agents.

9.2 Cellular respiration oxidizes food molecules.

- In eukaryotic cells, the oxidative respiration of pyruvate takes place within the matrix of mitochondria.

- The electrons generated in the process are passed along the electron transport chain, a sequence of electron carriers in the inner mitochondrial membrane.

- Some of the energy released by passage of electrons along the electron transport chain is used to pump protons out of the mitochondrial matrix. The reentry of protons into the matrix is coupled to the production of ATP. This process is called chemiosmosis. The ATP then leaves the mitochondrion by facilitated diffusion.

9.3 Catabolism of proteins and fats can yield considerable energy.

- The catabolism of fatty acids begins with β-oxidation and provides more energy than the catabolism of carbohydrates.

9.4 Cells can metabolize food without oxygen.

- Fermentation is an anaerobic process that uses an organic molecule instead of oxygen as a final electron acceptor.

- It occurs in bacteria as well as eukaryotic cells, including yeast and the muscle cells of animals.

Discussing Key Terms

1. **Cellular respiration**   The oxidation of organic molecules within cells that is accomplished by harvesting electrons, which are used to synthesize ATP and then donated to an inorganic molecule; in aerobic respiration, the final acceptor of the electrons is oxygen.

2. **Redox reaction**   The donation and acceptance of an electron from one atom to another. The energy of that electron is reduced if it moves to a lower energy level in the recipient atom.

3. **Electronegativity**   The attraction for valence electrons that atoms have. Some atoms have a much stronger electronegativity than others; highly electronegative atoms like oxygen attract electrons strongly and are good oxidizing agents.

4. **Electron transport chain**   The series of electron carriers within the inner mitochondrial membrane through which electrons harvested from organic molecules are passed. Each carrier is slightly more electronegative than the previous one. Three of the carriers respond to the passage of an electron by pumping protons across the membrane, from the matrix of the mitochondrion to the cytoplasm.

5. **Chemiosmosis**   When protons are pumped out of the mitochondrial matrix, the matrix becomes slightly negative in charge, and so creates a driving force for the positively charged protons to reenter the matrix. Most of the protons that reenter do so through ATP-producing channels. Chemiosmosis is the synthesis of ATP driven by the movement of protons toward the negatively charged matrix.

6. **Fermentation**   The process by which cells, in the absence of oxygen, donate the hydrogens generated by glycolysis to an organic molecule.

7. **β-oxidation**   Fatty acids, the basic units of fat, are oxidized by removing two-carbon fragments that are directed into the cell's pathways for aerobic respiration.

Review Questions

1. What is the difference between an autotroph and a heterotroph? How does each obtain energy?

2. What is the difference between digestion and catabolism? Which provides more energy?

3. What is substrate-level phosphorylation? At what points during the oxidation of glucose in a eukaryotic cell does it occur?

4. Where in a eukaryotic cell does glycolysis occur? What molecules are produced from the carbon atoms in glucose as a result of glycolysis? What is the *net* production of ATP during glycolysis, and why is it different from the number of ATP molecules *synthesized* during glycolysis?

5. How is glycolysis affected by the absence of oxygen? Given an unlimited supply of glucose, what factor ultimately limits glycolysis?

6. By what two mechanisms can the NADH that results from glycolysis be converted back into NAD$^+$? How do these mechanisms differ in terms of electron acceptors and ATP production? Which mechanism is aerobic, and which is anaerobic?

7. Where in a eukaryotic cell do the oxidation of pyruvate, the Krebs cycle, and the passage of electrons along the electron transport chain occur? What is the *theoretical maximum* number of ATP molecules produced during the oxidation of a glucose molecule by these processes? Why is the *actual* number of ATP molecules produced usually lower than the theoretical maximum?

8. How is the passage of electrons along the electron transport chain coupled to the production of ATP? How is this coupling affected by the absence of oxygen?

9. How is acetyl-CoA produced during the aerobic oxidation of carbohydrates, and what happens to it? How is it produced during the aerobic oxidation of fatty acids, and what happens to it then?

10. How do the amounts of ATP produced by the aerobic oxidation of glucose and fatty acids compare? Which type of substance contains more energy on a per-weight basis?

Thought Questions

1. As explained in chapter 5, mitochondria are thought to have evolved from bacteria that were engulfed by and lived symbiotically within early eukaryotic cells. Why haven't present-day eukaryotic cells dispensed with mitochondria, placing all of the mitochondrial genes in the nucleus and carrying out all of the metabolic functions of mitochondria within the cytoplasm?

2. Why do plants typically store their excess energy as carbohydrate rather than fat?

Internet Links

Glycolysis
http://bmbwww.leeds.ac.uk/designs/glyintro/home.html
A good general overview of glycolysis, from the University of Leeds, England.

Glycolysis & the Krebs Cycle
http://esg~www.mit.edu:8001/esgbio/glycolysis/dir.html
A collection of links to university sites teaching various aspects of energy metabolism.

Krebs Cycle Animation
http://tidepool.st.usm.edu/crswr/pagest/krebsmov.html
An animated movie of the Krebs cycle, allowing you to follow the path of energy and carbon from one reaction to the next.

On Being a Scientist
http://www.nap.edu/readingroom/books/obas/
A booklet on the process of science published by the National Academy of Science that includes discussions of experimental design, the role of values in scientific investigation, and misconduct in science. Recommended.

For Further Reading

Babcock, G., and M. Wikstrom: "Oxygen Activation and the Conservation of Energy in Cell Respiration," *Nature*, March 1992, pages 301–8. Proteins that carry out proton pumping and oxygen reduction now appear to share a similar but very unusual bimetallic iron-copper reaction center.

Dickerson, R.: "Cytochrome *c* and the Evolution of Energy Metabolism," *Scientific American*, March 1980, pages 136–54. A superb description of how the metabolism of modern organisms evolved.

Harold, F.: *The Vital Force: A Study Of Bioenergetics*, 1986. Freeman Press, New York. A detailed account of chemiosmosis, focusing on the physical basis of the proton motive force.

Hinkle, P., and R. McCarty: "How Cells Make ATP," *Scientific American*, March 1978, pages 104–25. A classic summary of oxidative respiration, with a clear account of the events that happen at the mitochondrial membrane.

Levine, M., and others: "Structure of Pyruvate Kinase and Similarities with Other Enzymes: Possible Implications for Protein Taxonomy and Evolution," *Nature*, 1978, vol. 271, pages 626–30. An advanced article, well worth the effort, that explains how the enzymes of oxidative metabolism may have evolved.

Wachtershauser, G.: "Evolution of the First Metabolic Cycles," *Proceedings of the National Academy of Science USA*, 1990, vol. 87, pages 200–4. A suggestion that metabolic cycles played a key role in the evolution of life.

10

Photosynthesis

FIGURE 10.1
Capturing energy. These sunflower plants, growing vigorously in the August sun, are capturing light energy for conversion into chemical energy through photosynthesis.

Concept Outline

In the previous chapter we described how cells extract chemical energy from food molecules and use that energy to power their activities. In this chapter, we will examine photosynthesis, the process by which certain organisms capture energy from sunlight and use it to build food molecules rich in chemical energy (figure 10.1). We will first look at an overview of photosynthesis and then review the long path of investigation that led to our current understanding of it. Subsequently, we will briefly consider how photosynthesis might have evolved among the photosynthetic bacteria. Finally, we will conduct a more detailed examination of the particular pattern of photosynthesis characteristic of the photosynthetic eukaryotes, plants and algae.

The Chloroplast as a Photosynthetic Machine

Life is powered by sunshine. The energy used by most living cells comes ultimately from the sun, captured by plants, algae and bacteria through the process of photosynthesis. The diversity of life is only possible because our planet is awash in energy streaming earthward from the sun. Each day, the radiant energy that reaches the earth equals about 1 million Hiroshima-sized atomic bombs. Photosynthesis captures about 1% of this huge supply of energy, using it to provide the energy that drives all life.

The Photosynthetic Process

Photosynthesis occurs in many kinds of bacteria and algae, and in the leaves and sometimes the stems of green plants. Figure 10.2 describes the levels of organization in a plant leaf. Recall from chapter 5 that the cells of plant leaves contain organelles called chloroplasts that actually carry out the photosynthetic process. No other structure in a

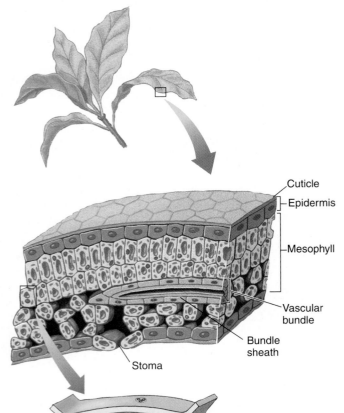

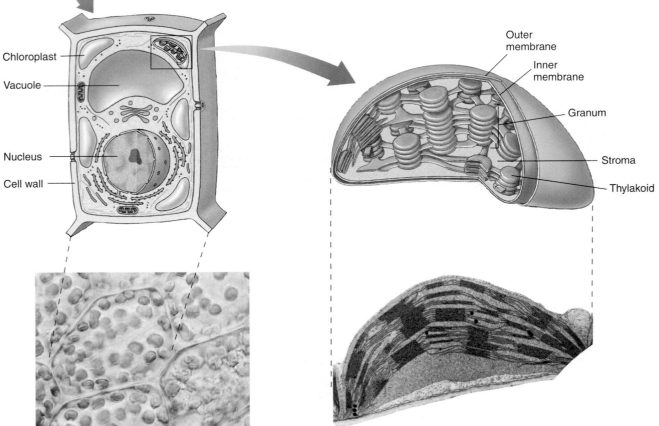

FIGURE 10.2
Journey into a leaf. A plant leaf possesses a thick layer of cells (the mesophyll) rich in chloroplasts. The flattened thylakoids in the chloroplast are stacked into columns called grana (singular, granum). The light-dependent reactions take place on the thylakoid

plant is able to carry out photosynthesis. Photosynthesis takes place in three stages: (1) capturing energy from sunlight; (2) using the energy to make ATP and reducing power in the form of a compound called NADPH; and (3) using the ATP and NADPH to power the synthesis of organic molecules from CO_2 in the air (carbon fixation).

The first two processes take place in the presence of light and are commonly called the **light-dependent reactions.** The third process (formation of organic molecules from atmospheric CO_2) can take place with or without light and involves the **light-independent reactions** of photosynthesis. As long as ATP and NADPH are available, the light-independent reactions occur as readily in the absence of light as in its presence.

The following simple equation summarizes the overall process of photosynthesis:

$$6\ CO_2 + 12\ H_2O + light \longrightarrow C_6H_{12}O_6 + 6\ H_2O + 6\ O_2$$

carbon dioxide   water   glucose   water   oxygen

Inside the Chloroplast

The internal membranes of chloroplasts are organized into sacs called *thylakoids*, and often numerous thylakoids are stacked on one another in columns called *grana*. Surrounding the thylakoid membrane system is a semiliquid substance called *stroma*. In the membranes of thylakoids,

photosynthetic pigments are grouped together in a network called a **photosystem.**

Each pigment molecule within the photosystem network is capable of capturing photons. A lattice of proteins holds the pigments in close contact with one another. When light of a proper wavelength strikes a pigment molecule in the photosystem, the resulting excitation passes from one chlorophyll molecule to another. The excited electron does not transfer physically—it is the *energy* that passes from one molecule to another. A crude analogy to this form of energy transfer is the initial "break" in a game of pool. If the cue ball squarely hits the point of the triangular array of 15 pool balls, the two balls at the far corners of the triangle fly off, but none of the central balls move. The energy passes through the central balls to the most distant ones.

Eventually the energy arrives at a key chlorophyll molecule within the reaction center, a chlorophyll touching a membrane-bound protein. The energy is transferred as an excited electron to that protein, which passes it on to a series of other membrane proteins that put the energy to work making ATP and NADPH and building organic molecules. The photosystem thus acts as a large antenna, amplifying the power of individual pigment molecules to gather light.

The reactions of photosynthesis take place within thylakoid membranes within chloroplasts in leaf cells.

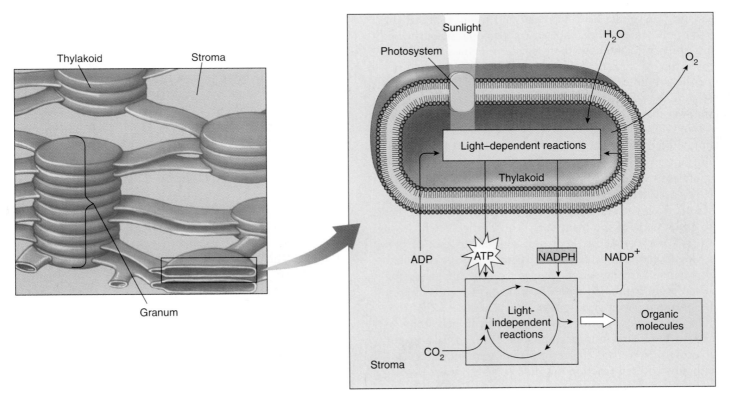

FIGURE 10.2 (continued)
membrane and generate the ATP and NADPH that fuel the light-independent reactions. The fluid interior matrix of a chloroplast, the stroma, contains the enzymes that carry out the light-independent reactions.

The Role of Soil and Water

The story of how we learned about photosynthesis is one of the most interesting in science and serves as a good introduction to this complex process. The story starts over 300 years ago, with a simple but carefully designed experiment by a Belgian doctor, Jan Baptista van Helmont (1577–1644). From the time of the Greeks, plants were thought to obtain their food from the soil, literally sucking it up with their roots; van Helmont thought of a simple way to test the idea. He planted a small willow tree in a pot of soil after weighing the tree and the soil. The tree grew in the pot for several years, during which time van Helmont added only water. At the end of five years, the tree was much larger: its weight had increased by 74.4 kilograms. However, *all of this added mass could not have come from the soil*, because the soil in the pot weighed only 57 grams less than it had five years earlier! With this experiment, van Helmont demonstrated that the substance of the plant was not produced only from the soil. He incorrectly concluded that mainly the water he had been adding accounted for the plant's increased mass.

A hundred years passed before the story became clearer. The key clue was provided by the English scientist Joseph Priestly, in his pioneering studies of the properties of air. On the 17th of August, 1771, Priestly "accidentally hit upon a method of restoring air that had been injured by the burning of candles." He "put a [living] sprig of mint into air in which a wax candle had burnt out and found that, on the 27th of the same month, another candle could be burned in this same air." Somehow, the vegetation seemed to have restored the air! Priestly found that while a mouse could not breathe candle-exhausted air, air "restored" by vegetation was not "at all inconvenient to a mouse." The key clue, then, was that living vegetation *adds something to the air.*

How does vegetation "restore" air? Twenty-five years later, Dutch physician Jan Ingenhousz solved the puzzle. Working over several years, Ingenhousz reproduced and significantly extended Priestly's results, demonstrating that air was restored only in the presence of sunlight, and only by a plant's green leaves, not by its roots. He proposed that the green parts of the plant carry out a process (which we now call photosynthesis) that uses sunlight to split carbon dioxide (CO_2) into carbon and oxygen. He suggested that the oxygen was released as O_2 gas into the air, while the carbon atom combined with water to form carbohydrates. His proposal was a good guess; chemists later found that the proportions of carbon, oxygen, and hydrogen atoms in carbohydrates are indeed about one atom of carbon per molecule of water (as the term *carbohydrate* indicates). Ingenhousz's overall reaction for photosynthesis was:

$$CO_2 + H_2O + \text{light energy} \longrightarrow (CH_2O) + O_2$$

This simple and perfectly reasonable hypothesis was accepted as true for over a hundred years, but, as often happens in science, photosynthesis turned out not to be so simple after all.

The true meaning of Ingenhousz's deceptively simple equation was sorted out in the early 1930s by a graduate student at Stanford University, C. B. van Niel, who was studying photosynthesis in bacteria. One of the types of bacteria he was studying, the purple sulfur bacteria, does not release oxygen during photosynthesis; instead, they convert hydrogen sulfide (H_2S) into globules of pure elemental sulfur that accumulate inside themselves. The process that van Niel observed was

$$CO_2 + 2\,H_2S + \text{light energy} \longrightarrow (CH_2O) + H_2O + 2\,S$$

The striking parallel between this equation and Ingenhousz's equation led van Niel to propose that the generalized process of photosynthesis is in fact

$$CO_2 + 2\,H_2A + \text{light energy} \longrightarrow (CH_2O) + H_2O + 2\,A$$

In this equation, the substance H_2A serves as an electron donor. In photosynthesis performed by green plants, H_2A is water, while among purple sulfur bacteria, H_2A is hydrogen sulfide. The product, A, comes from the splitting of H_2A. Therefore, the O_2 produced during green plant photosynthesis results from splitting water, not carbon dioxide.

When isotopes came into common use in biology in the early 1950s, it became possible to test van Niel's revolutionary proposal. Investigators examined photosynthesis in green plants supplied with ^{18}O water; they found that the ^{18}O label ended up in oxygen gas rather than in carbohydrate, just as van Niel had predicted:

$$CO_2 + 2\,H_2^{18}O + \text{light energy} \longrightarrow (CH_2O) + H_2O + {}^{18}O_2$$

In algae and green plants, the carbohydrate typically produced by photosynthesis is the sugar, glucose, which has six carbons. The complete balanced equation for photosynthesis in these organisms thus becomes

$$6\,CO_2 + 12\,H_2O + \text{light energy} \longrightarrow C_6H_{12}O_6 + 6\,O_2 + 6\,H_2O.$$

Initially, researchers thought soil added mass to a growing plant. Van Helmont showed this was not true. Van Niel discovered that photosynthesis splits water molecules, incorporating the carbon atoms of carbon dioxide gas and the hydrogen atoms of water into organic molecules and leaving oxygen gas.

The Role of Light

Both Ingenhousz's early equation for photosynthesis and van Niel's currently accepted equation include one factor we have not discussed: light energy. What role does light play in photosynthesis? At the beginning of this century, the English plant physiologist F. F. Blackman began to address this question. In 1905, he came to the startling conclusion that photosynthesis is in fact a two-stage process, only one of which requires light.

Blackman measured the effects of different light intensities and temperatures on photosynthesis. As long as light intensity was relatively low, he found that photosynthesis could be accelerated by increasing the amount of light, but not by increasing the temperature (figure 10.3). At high light intensities, however, an increase in temperature greatly accelerated photosynthesis. Based on these data, Blackman concluded that photosynthesis consists of an initial set of what he called "light" reactions, that are largely independent of temperature, and a second set of "dark" reactions, that are independent of light. Do not be confused by Blackman's labels—the so-called "dark" reactions occur in the light (in fact, they require the products of the light reactions); their name simply indicates that light is not directly involved in those reactions. Most modern plant physiologists now call them "light-independent reactions" to emphasize this point.

Blackman found that increased temperature increases the rate of the dark reactions, but only up to about 35°C. Higher temperatures caused the rate to fall off rapidly. Because 35°C is the temperature at which many plant enzymes begin to be denatured (the hydrogen bonds that hold an enzyme in its particular catalytic shape begin to be disrupted), Blackman concluded that enzymes must carry out the dark reactions. We now know that the first stage of photosynthesis, the light-dependent or "light" reactions, uses the energy of light to reduce NADP (an electron carrier molecule) to NADPH and to manufacture ATP. The NADPH and ATP from the first stage of photosynthesis are then used in the second stage, the light-independent or "dark" reactions, to reduce the carbon in carbon dioxide and form a simple sugar whose carbon skeleton can be used to synthesize other organic molecules. The incorporation of the carbon atoms of CO_2 into organic molecules in the dark reactions is called **carbon fixation.**

From Blackman's work and that of many other investigators, it became clear that the energy of sunlight is used in the light reactions of photosynthesis to drive the reduction

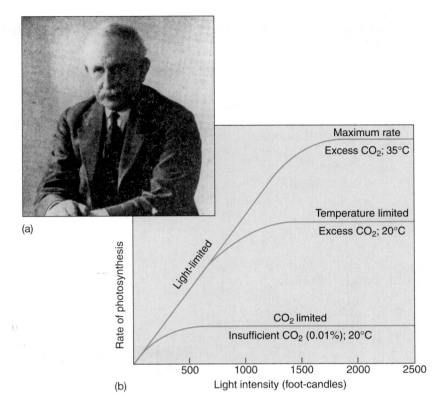

(a)

(b)

FIGURE 10.3
Discovery of the dark reactions. (a) Blackman measured photosynthesis rates under differing light intensities, CO_2 concentrations, and temperatures. (b) As this graph shows, light is the limiting factor at low light intensities, while temperature and CO_2 concentration are the limiting factors at higher light intensities. Foot-candles are a standard measure of light brightness (another is the *lux*; 1 foot-candle $\cong$ 10 *lux*). The actual energy of light is measured in joules/sec, or Watts. Bright sunlight is $\approx$ 10,000 foot-candles, 108,000 *lux*, or 500 W m^{-2}.

of carrier molecules; in other words, photosynthesis is a redox process. Recall from chapter 9 that all redox processes involve the transfer of energy. During oxidative respiration, energy is released from sugar molecules when electrons associated with hydrogen atoms in the sugar are transported by carriers like NADH to oxygen, forming water as a final by-product. The electrons extracted from sugar lose potential energy as they are drawn down the electron transport chain by strongly electronegative oxygen atoms, and mitochondria use the released energy to manufacture ATP. In photosynthesis, the path of the electrons' journey is reversed. Water is split and electrons (with associated hydrogen atoms) are extracted and transferred to the carbon atoms in carbon dioxide, reducing them and forming a sugar. Flowing in this direction, the electrons *gain* potential energy as they pass from water to sugar. The role of light in photosynthesis is to supply this energy.

Blackman showed that capturing photosynthetic energy requires sunlight, while building organic molecules does not.

The Biophysics of Light

Where is the energy in light? What is there in sunlight that a plant can use to reduce carbon dioxide? This is the mystery of photosynthesis, the one factor fundamentally different from processes such as respiration. To answer these questions, we will need to consider the physical nature of light itself. Perhaps the best place to start is with a curious experiment carried out in a laboratory in Germany in 1887. A young physicist, Heinrich Hertz, was attempting to verify a highly mathematical theory that predicted the existence of electromagnetic waves. To see whether such waves existed, Hertz designed a clever experiment. On one side of a room he constructed a powerful spark generator that consisted of two large, shiny metal spheres standing near each other on tall, slender rods. When a very high static electrical charge was built up on one sphere, sparks would jump across to the other sphere.

After constructing this device, Hertz set out to investigate whether the sparking would create invisible electromagnetic waves, so-called radio waves, as predicted by the mathematical theory. On the other side of the room, he placed the world's first radio receiver, a thin metal hoop on an insulating stand. There was a small gap at the bottom of the hoop, so that the hoop did not quite form a complete circle. When Hertz turned on the spark generator across the room, he saw tiny sparks passing across the gap in the hoop! This was the first demonstration of radio waves. But Hertz noted another curious phenomenon. When light was shining across the gap on the hoop, the sparks were produced more readily. This unexpected facilitation, called the photoelectric effect, puzzled investigators for many years.

The photoelectric effect was finally explained by Albert Einstein in 1905. Light consists of units of energy called photons, and some of these photons were being absorbed by the metal atoms of the hoop. In this process, some of the electrons in the metal atoms were boosted into higher energy levels, which caused them to be ejected from the atoms. In effect, light was "blasting" electrons from the metal surface at the ends of the hoop, creating ions and thus facilitating the passage of the electric spark induced by the radio waves.

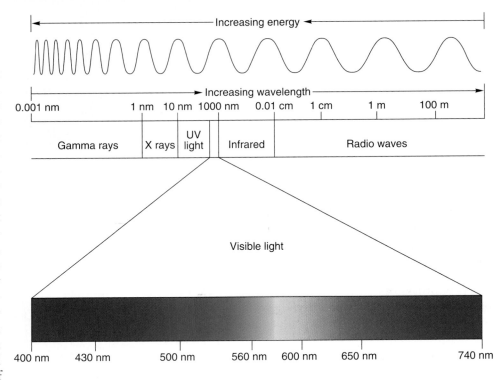

FIGURE 10.4
The electromagnetic spectrum. Light is a form of electromagnetic energy conveniently thought of as a wave. The shorter the wavelength of light, the greater its energy. Visible light represents only a small part of the electromagnetic spectrum between 400 and 740 nanometers.

The Energy in Photons

Photons do not all possess the same amount of energy. Instead, the energy content of a photon is inversely proportional to the wavelength of the light: short-wavelength light contains photons of higher energy than long-wavelength light. X rays, which contain a great deal of energy, have very short wavelengths—much shorter than visible light, making them ideal for high-resolution microscopes.

Hertz had noted that the strength of the photoelectric effect depends on the wavelength of light; short wavelengths are much more effective than long ones in producing the photoelectric effect. Einstein's theory of the photoelectric effect provides an explanation: sunlight contains photons of many different energy levels, only some of which our eyes perceive as visible light. The highest energy photons, at the short-wavelength end of the electromagnetic spectrum (figure 10.4), are gamma rays, with wavelengths of less than 1 nanometer; the lowest energy photons, with wavelengths of up to thousands of meters, are radio waves. Within the visible portion of the spectrum, violet light has the shortest wavelength and the most energetic photons, and red light has the longest wavelength and the least energetic photons.

Ultraviolet Light

Sunlight contains a significant amount of ultraviolet (UV) light, which, because of its shorter wavelength, possesses considerably more energy than visible light. UV light is thought to have been an important source of energy on the primitive earth when life originated. Today's atmosphere contains ozone (derived from oxygen gas), which absorbs most of the UV photons in sunlight, but anyone who has been sunburned knows that a considerable amount of UV light still manages to penetrate the atmosphere. As we will describe in a later chapter, mysterious "holes" have been recently discovered in the ozone layer. These holes appear to be due to human activities and threaten to cause an enormous jump in the incidence of human skin cancers throughout the world.

Absorption Spectra and Pigments

How does a molecule "capture" the energy of light? A photon can be envisioned as a very fast-moving packet of energy. When it strikes a molecule, its energy is either lost as heat or absorbed by the electrons of the molecule, boosting those electrons into higher energy levels. Whether or not the photon's energy is absorbed depends on how much energy it carries (defined by its wavelength) and on the chemical nature of the molecule it hits. As we saw in chapter 2, electrons occupy discrete energy levels in their orbits around atomic nuclei. To boost an electron into a different energy level requires just the right amount of energy, just as reaching the next rung on a ladder requires you to raise your foot just the right distance. A specific atom can, therefore, absorb only certain photons of light—namely, those that correspond to the atom's available electron energy levels. As a result, each molecule has a characteristic **absorption spectrum,** a range of photons it is capable of absorbing.

Molecules that absorb light well are called **pigments.** Organisms have evolved a variety of different pigments, but there are only two general types used in green plant photosynthesis: carotenoids and chlorophylls. Chlorophylls absorb photons within narrow energy ranges. Two kinds of chlorophyll in plants, chlorophylls *a* and *b*, preferentially absorb violet-blue and red light (figure 10.5). Neither of these pigments absorbs photons with wavelengths between about 500 and 600 nanometers, and light of these wavelengths is, therefore, reflected by plants. When these pho-

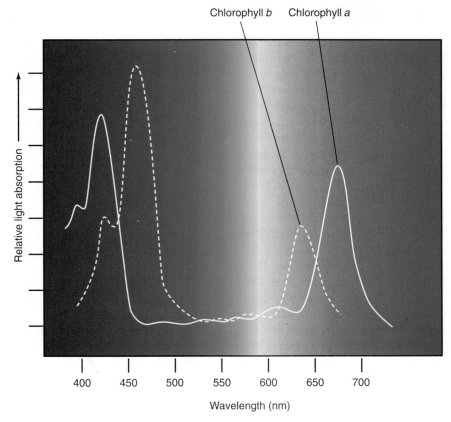

FIGURE 10.5
The absorption spectrum of chlorophyll. The peaks represent wavelengths of sunlight that the two common forms of photosynthetic pigment, chlorophyll *a* (*solid line*) and chlorophyll *b* (*dashed line*), strongly absorb. These pigments absorb predominately violet-blue and red light in two narrow bands of the spectrum and reflect the green light in the middle of the spectrum. Carotenoids (not shown here) absorb mostly blue and green light and reflect orange and yellow light.

tons are subsequently absorbed by retinal in our eyes, we perceive them as green.

Chlorophyll *a* is the main photosynthetic pigment and is the only pigment that can act directly to convert light energy to chemical energy. However, chlorophyll *b*, acting as an **accessory** or secondary light-absorbing pigment, complements and adds to the light absorption of the primary pigment, chlorophyll *a*. Chlorophyll *b* has an absorption spectrum shifted toward the green wavelengths. Therefore, chlorophyll *b* can absorb photons chlorophyll *a* cannot. Chlorophyll *b* therefore greatly increases the proportion of the photons in sunlight that plants can harvest. An important group of accessory pigments, the carotenoids, assist in photosynthesis by capturing energy from light of wavelengths that are not efficiently absorbed by chlorophyll.

In photosynthesis, photons of light are absorbed by pigments; the wavelength of light absorbed depends upon the specific pigment.

Chlorophylls and Carotenoids

Chlorophylls absorb photons by means of an excitation process analogous to the photoelectric effect. These pigments contain a complex ring structure, called a porphyrin ring, with alternating single and double bonds. At the center of the ring is a magnesium atom. Photons absorbed by the pigment molecule excite electrons in the ring, which are then channeled away through the carbon-bond system. Several small side groups attached to the outside of the ring alter the absorption properties of the molecule in different kinds of chlorophyll (figure 10.6).

Once Ingenhousz demonstrated that only the green parts of plants can "restore" air, researchers suspected chlorophyll was the primary pigment that plants employ to absorb light in photosynthesis. Experiments conducted in the 1800s clearly verified this suspicion. One such experiment, performed by T. W. Englemann in 1882 (figure 10.7), serves as a particularly elegant example, simple in design and clear in outcome. Englemann set out to characterize the **action spectrum** of photosynthesis, that is, the relative effectiveness of different wavelengths of light in promoting photosynthesis. He carried out the entire experiment utilizing a single slide mounted on a microscope. To obtain different wavelengths of light, he placed a prism under his microscope,

splitting the light that illuminated the slide into a spectrum of colors. He then arranged a filament of algal cells across the spectrum, so that different parts of the filament were illuminated with different wavelengths, and allowed the algae to carry out photosynthesis. To assess how fast photosynthesis was proceeding, Englemann chose to monitor the rate of oxygen production. Lacking a mass spectrometer and other modern instruments, he added aerotactic (oxygen-seeking) bacteria to the slide; he knew they would gather along the filament at locations where oxygen was being produced. He found that the bacteria accumulated in areas illuminated by red and violet light, the two colors most strongly absorbed by chlorophyll.

All plants and algae, and all but one primitive group of photosynthetic bacteria, use chlorophyll *a* as their primary pigments. It is reasonable to ask why these photosynthetic organisms do not use a pigment like retinal, which has a broad absorption spectrum that covers the range of 500 to 600 nanometers. The most likely hypothesis involves *photoefficiency*. Although retinal absorbs a broad range of wavelengths, it does so with relatively low efficiency. Chlorophyll, in contrast, absorbs in only two narrow bands, but does so with high efficiency. Therefore, plants and most other photosynthetic organisms achieve far higher overall photon capture rates with chlorophyll than with other pigments.

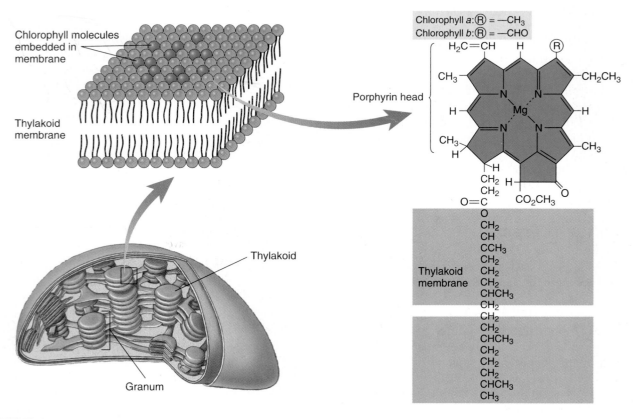

FIGURE 10.6
Chlorophyll. Chlorophyll molecules consist of a porphyrin head and a hydrocarbon tail that anchors the pigment molecule to hydrophobic regions of proteins embedded within the membranes of thylakoids. The only difference between the two chlorophyll molecules is the substitution of a —CHO (aldehyde) group in chlorophyll *b* for a —CH₃ (methyl) group in chlorophyll *a*.

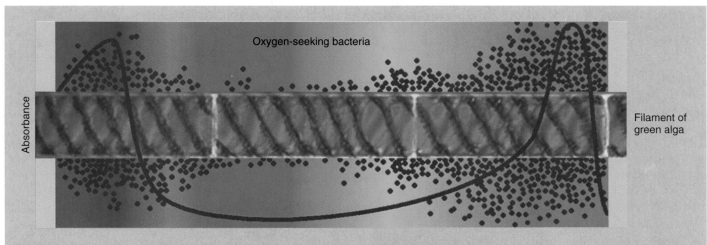

T. W. Englemann revealed the action spectrum of photosynthesis in the filamentous alga *Spirogyra* in 1882. Englemann used the rate of oxygen production to measure the rate of photosynthesis. As his oxygen indicator, he chose bacteria that are attracted by oxygen. In place of the mirror and diaphragm usually used to illuminate objects under view in his microscope, he substituted a "microspectral apparatus," which, as its name implies, produced a tiny spectrum of colors that it projected upon the slide under the microscope. Then he arranged a filament of algal cells parallel to the spread of the spectrum. The oxygen-seeking bacteria congregated mostly in the areas where the violet and red wavelengths fell upon the algal filament.

FIGURE 10.7
Constructing an action spectrum for photosynthesis. As you can see, the action spectrum for photosynthesis that Englemann revealed in his experiment parallels the absorption spectrum of chlorophyll (see figure 10.5).

Oak leaf in summer

Oak leaf in autumn

FIGURE 10.8
Fall colors are produced by carotenoids and other accessory pigments. During the spring and summer, chlorophyll in leaves masks the presence of carotenoids and other accessory pigments. When cool fall temperatures cause leaves to cease manufacturing chlorophyll, the chlorophyll is no longer present to reflect green light, and the leaves reflect the orange and yellow light that carotenoids and other pigments do not absorb.

Carotenoids consist of carbon rings linked to chains with alternating single and double bonds. They can absorb photons with a wide range of energies, although they are not always highly efficient in transferring this energy. Carotenoids assist in photosynthesis by capturing energy from light of wavelengths that are not efficiently absorbed by chlorophylls (figure 10.8).

A typical carotenoid is β-carotene, whose two carbon rings are connected by a chain of 18 carbon atoms with alternating single and double bonds. Splitting a mole-cule of β-carotene into equal halves produces two mole-cules of vitamin A. Oxidation of vitamin A produces retinal, the pigment used in vertebrate vision. This ex-plains why carrots, which are rich in β-carotene, en-hance vision.

A pigment is a molecule that absorbs light. The wavelengths absorbed by a particular pigment depend on the available energy levels to which light-excited electrons can be boosted in the pigment.

Organizing Pigments into Photosystems

The light-dependent reactions of photosynthesis occur on membranes (figure 10.9). In bacteria like those studied by van Niel, the plasma membrane itself is the photosynthetic membrane. In plants and algae, by contrast, photosynthesis is carried out by organelles that are the evolutionary descendants of photosynthetic bacteria, chloroplasts—the photosynthetic membranes exist *within* the chloroplasts. The light-dependent reactions take place in three stages:

1. **Primary photoevent.** A photon of light is captured by a pigment. The result of this primary photoevent is the excitation of an electron within the pigment.
2. **Electron transport.** The excited electron is shuttled along a series of electron-carrier molecules embedded within the photosynthetic membrane until it arrives at a transmembrane proton-pumping channel (sometimes referred to as a "redox pump"). Its arrival at the pump induces the transport of a proton across the membrane. The electron is then passed to an acceptor.
3. **Chemiosmosis.** The transport of protons back across drives the chemiosmotic synthesis of ATP, just as it does in aerobic respiration.

Discovery of Photosystems

One way to study how pigments absorb light is to measure the dependence of the output of photosynthesis on the intensity of illumination—that is, how much photosynthesis is produced by how much light. When experiments of this sort are done on plants, they show that the output of photosynthesis increases linearly at low intensities but lessens at higher intensities, finally saturating at high-intensity light. Saturation occurs because all the light-absorbing capacity of the plant is in use; adding more light doesn't increase the output because there is nothing to absorb the added photons.

It is tempting to think that at saturation, all of a plant's pigment molecules are in use. In 1932 plant physiologists Robert Emerson and William Arnold set out to test this hypothesis in an organism where they could measure both the number of chlorophyll molecules and the output of photosynthesis. In their experiment, they measure the oxygen yield of photosynthesis when *Chlorella* (unicellular green algae) were exposed to very brief light flashes lasting only a few microseconds. Assuming the hypothesis of pigment saturation to be correct, they expected to find that as they increased the intensity of the flashes, the yield per flash would increase, until each chlorophyll molecule

 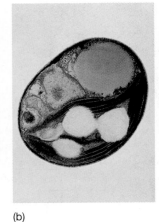 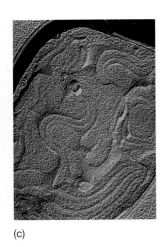

(a)   (b)   (c)

FIGURE 10.9

A sampling of photosynthetic membranes in autotrophic organisms. (a) Green plant *Coleus blumei* (5,000×), (b) unicellular green algae *Chlorella*, and (c) nitrogen-fixing cyanobacterium *Nostoc museorum* (60,000×).

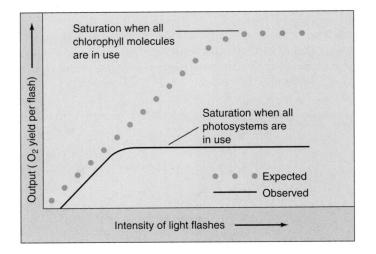

FIGURE 10.10

Emerson and Arnold's experiment. When photosynthetic saturation is achieved, further increases in intensity cause no increase in output.

absorbed a photon, which would then be used in the light-dependent reactions, producing a molecule of O_2.

Unexpectedly, this is not what happened (figure 10.10). Instead, saturation was achieved much earlier, with only one molecule of O_2 per 2500 chlorophyll molecules! This led Emerson and Arnold to conclude that light is absorbed not by independent pigment molecules, but rather by clusters of chlorophyll and accessory pigment molecules which have come to be called *photosystems*. In a photosystem such as those of *Chlorella*, light is absorbed by any one of the hundreds of pigment molecules in a photosystem, which transfer their excitation energy to one with a lower energy level than the others. This **reaction center** of the photosystem acts as an energy sink, trapping the excitation energy. It was the saturation of these reaction centers, not individual molecules, that was observed by Emerson and Arnold.

Architecture of a Photosystem

In chloroplasts and all but the most primitive bacteria, light is captured by such photosystems. Each photosystem is a network of chlorophyll *a* molecules and accessory pigments held within a protein matrix on the surface of the photosynthetic membrane. Like a magnifying glass focusing light on a precise point, a photosystem channels the excitation energy gathered by any one of its pigment molecules to a specific molecule, the reaction center chlorophyll. This molecule then passes the energy out of the photosystem so it can be put to work driving the synthesis of ATP and organic molecules.

A photosystem thus consists of two closely linked components: (1) *an antenna complex* of hundreds of pigment molecules that gather photons and feed the captured light energy to the reaction center; and (2) *a reaction center*, consisting of one or more chlorophyll *a* molecules in a matrix of protein, that passes the energy out of the photosystem.

The Antenna Complex. The antenna complex captures photons from sunlight (figure 10.11). In chloroplasts, the antenna complex is a web of chlorophyll molecules linked together and held tightly on the thylakoid membrane by a matrix of proteins. Varying amounts of carotenoid accessory pigments may also be present. The protein matrix serves as a sort of scaffold, holding individual pigment molecules in orientations that are optimal for energy transfer.

The excitation energy resulting from the absorption of a photon passes from one pigment molecule to an adjacent molecule on its way to the reaction center. After the transfer, the excited electron in each molecule returns to the low-energy level it had before the photon was absorbed. Consequently, it is energy, not the excited electrons themselves, that passes from one pigment molecule to the next. The antenna complex funnels the energy from many electrons to the reaction center.

The Reaction Center. The reaction center is a transmembrane protein-pigment complex. In the reaction center of purple photosynthetic bacteria, which is simpler than in chloroplasts but better understood, a pair of chlorophyll *a* molecules acts as a trap for photon energy, passing an excited electron to an acceptor precisely positioned as its neighbor. This moves the photon excitation away from the chlorophylls and is the key conversion of light to chemical energy.

Figure 10.12 shows the transfer of energy from the reaction center to the primary electron acceptor. By energizing an electron of the reaction center chlorophyll, light creates a strong electron donor where none existed before. The chlorophyll transfers the energized electron to the primary acceptor, a molecule of quinone, reducing the quinone and converting it to a strong electron donor. A weak electron donor then donates a low-energy electron to the chlorophyll, restoring it to its original condition. In the purple bacterium, this donor is a cytochrome, and the journey of the electron follows a circular path, as we shall discuss later. In plant chloroplasts, water serves as the electron donor.

> Photon energy is captured from light in photosystems by pigments that focus the energy on reaction centers that employ it to excite electrons that are channeled away to do chemical work.

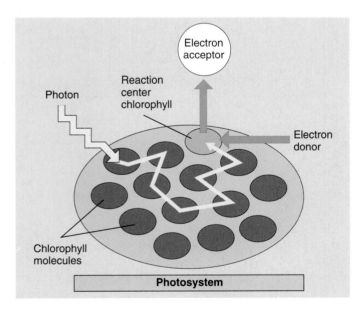

FIGURE 10.11
How the antenna complex works. When light of the proper wavelength strikes any pigment molecule within a photosystem, the light is absorbed by that pigment molecule. The excitation energy is then transferred from one molecule to another within the cluster of pigment molecules until it encounters the reaction center chlorophyll *a*. The reaction center chlorophyll channels the energy in the form of high-energy electrons out of the photosystem to the energy-harvesting machinery.

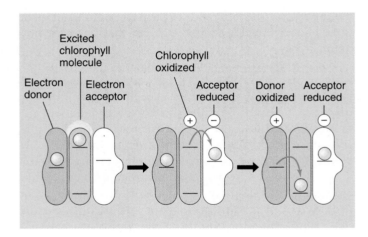

FIGURE 10.12
Converting light to chemical energy. The reaction center chlorophyll donates a light-energized electron to the primary electron acceptor, reducing it. The oxidized chlorophyll then fills its electron "hole" by oxidizing a donor molecule.

How Photosystems Convert Light to Chemical Energy

Bacteria Use a Single Photosystem

Photosynthetic units are thought to have evolved more than 3 billion years ago in bacteria similar to the sulfur bacteria studied by van Niel. In these bacteria, the absorption of a photon of light at a peak absorption of 870 nanometers by the photosystem results in the transmission of an energetic electron along an electron transport chain, eventually combining with a proton to form a hydrogen atom. In the sulfur bacteria (figure 10.13), the proton is extracted from hydrogen sulfide, leaving elemental sulfur as a by-product. In bacteria that evolved later, as well as in plants and algae, the proton comes from water, producing oxygen as a by-product.

The ejection of an electron from the bacterial reaction center leaves it short one electron. Before the photosystem of the sulfur bacteria can function again, an electron must be returned. These bacteria channel the electron back to the pigment through an electron transport system similar to the one described in chapter 9; the electron's passage drives a proton pump that promotes the chemiosmotic synthesis of ATP. One molecule of ATP is produced for every three electrons that follow this path. Viewed overall (figure 10.14), the path of the electron originally extracted from the photosystem is thus a circle. Chemists therefore call this process **cyclic photophosphorylation.** Note, however, that the electron that left the P_{870} reaction center was a high-energy electron, boosted by the absorption of a photon of light, while the electron that returns has only as much energy as it had before the photon was absorbed. The difference in the energy of that electron is the photosynthetic payoff, the energy that drives the proton pump.

For more than a billion years, cyclic photophosphorylation was the only form of photosynthetic light reaction that organisms used. However, its major limitation is that it is geared only toward energy production, not toward biosynthesis. Most photosynthetic organisms incorporate atmospheric carbon dioxide into carbohydrates. Because the carbohydrate molecules are more reduced (have more hydrogen atoms) than carbon dioxide, a source of reducing power (that is, hydrogens) must be provided. Cyclic photophosphorylation does not do this, and bacteria that are restricted to this process must scavenge hydrogens from other sources, an inefficient undertaking.

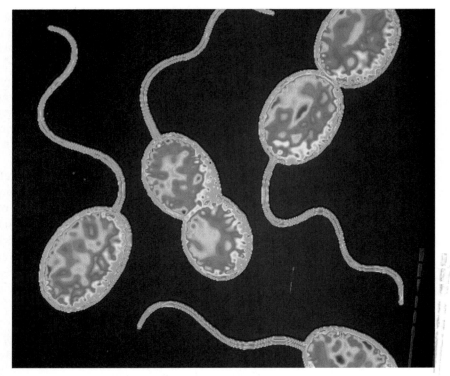

FIGURE 10.13
Purple sulfur bacteria. Because their photosynthetic system is relatively simple, sulfur bacteria have been the subjects of intensive study.

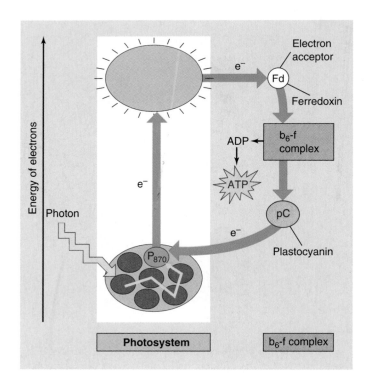

FIGURE 10.14
The path of an electron in purple sulfur bacteria. When a light-energized electron is ejected from the photosystem reaction center (P_{870}), it passes in a circle, eventually returning to the photosystem from which it was ejected.

Why Plants Use Two Photosystems

After the sulfur bacteria appeared, other kinds of bacteria evolved an improved version of the photosystem that overcame the limitation of cyclic photophosphorylation in a neat and simple way: a second, more powerful photosystem using another arrangement of chlorophyll *a* was grafted onto the original. This great evolutionary advance took place when the cyanobacteria originated, no less than 2.8 billion years ago.

In this second photosystem, called **photosystem II,** molecules of chlorophyll *a* are arranged with a different geometry, so that more shorter wavelength, higher energy photons are absorbed than in the ancestral photosystem, which is called **photosystem I.** As in the ancestral photosystem, energy is transmitted from one pigment molecule to another within the antenna complex of these photosystems until it reaches the reaction center, a particular pigment molecule positioned near a strong membrane-bound electron acceptor. In photosystem II, the absorption peak (that is, the wavelength of light most strongly absorbed) of the pigments is approximately 680 nanometers; therefore, the reaction center pigment is called P_{680}. The absorption peak of photosystem I pigments in plants is 700 nanometers, so its reaction center pigment is called P_{700}. Working together, the two photosystems carry out a noncyclic light-dependent reaction.

The use of two photosystems solves the problem of obtaining reducing power in a simple and direct way, by harnessing the energy of two photosystems. The scheme shown in figure 10.15, called a *Z diagram*, illustrates the two electron-energizing steps, one catalyzed by each photosystem. The electrons originate from water, which holds onto its electrons very tightly (redox potential = +820 mV), and end up in NADPH, which holds its electrons much more loosely (redox potential = –320 mV). The use of two separate photosystems in series provides more than enough energy to energize an electron all the way from the bottom of photosystem II to the top of photosystem I, with enough left over to enable the electron transport chain that links the two photosystems to pump protons across the thylakoid membrane and so power chemiosmotic ATP production.

In sulfur bacteria, excited electrons ejected from the reaction center travel a circular path, driving a proton pump and then returning to their original photosystem. Plants employ two photosystems in series, which generates power to reduce NADP+ to NADPH with enough left over to make ATP.

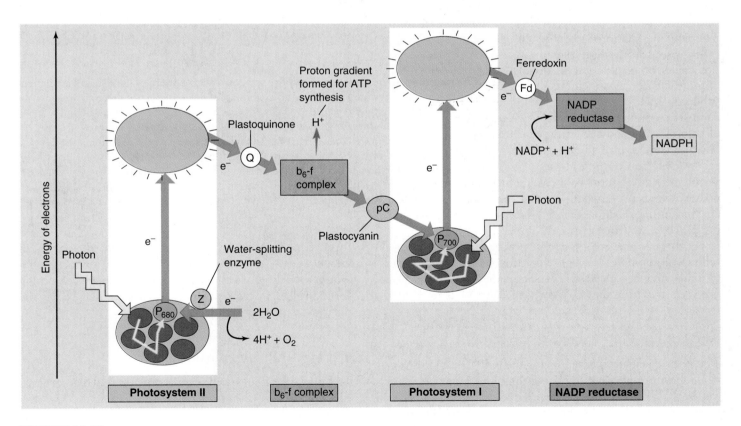

FIGURE 10.15
A Z diagram of photosystems I and II. Two photosystems work sequentially. First, a photon of light ejects a high-energy electron from photosystem II; that electron is used to pump a proton across the membrane, contributing chemiosmotically to the production of a molecule of ATP. The ejected electron then passes along a chain of cytochromes to photosystem I. When photosystem I absorbs a photon of light, it ejects a high-energy electron used to drive the formation of NADPH.

How the Two Photosystems of Plants Work Together

Plants use the two photosystems discussed earlier in series, first one and then the other, to produce both ATP and NADPH. This two-stage process is called **noncyclic photophosphorylation,** because the path of the electrons is not a circle—the electrons ejected from the photosystems do not return to it, but rather end up in NADPH. The photosystems are replenished instead with electrons obtained by splitting water. Photosystem II acts first. High-energy electrons generated by photosystem II are used to synthesize ATP and then passed to photosystem I to drive the production of NADPH. For every pair of electrons obtained from water, one molecule of NADPH and slightly more than one molecule of ATP are produced.

Photosystem II

The reaction center of photosystem II, called P_{680}, closely resembles the reaction center of purple bacteria. It consists of more than 10 transmembrane protein subunits. The light-harvesting antenna complex consists of some 250 molecules of chlorophyll *a* and accessory pigments bound to several protein chains. In photosystem II the oxygen atoms of two water molecules bind to a cluster of manganese atoms embedded within an enzyme in the reaction center symbolized Z. In a way that is poorly understood, this enzyme splits water, removing electrons one at a time to fill the holes left in the reaction center by departure of light-energized electrons. As soon as four electrons have been removed from the two water molecules, O_2 is released.

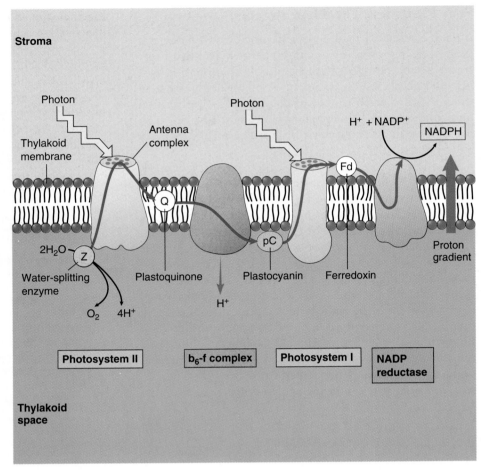

FIGURE 10.16

The photosynthetic electron transport system. When a photo of light strikes a pigment molecule in photosystem II, it excites an electron. This electron is coupled to a proton stripped from water by a Z protein and is passed along a chain of membrane-bound cytochrome electron carriers (*red arrow*). When water is split, oxygen is released from the cell, and the hydrogen ions remain in the thylakoid space. At the proton pump (b_6-f complex), the energy supplied by the photon is used to transport a proton across the membrane into the thylakoid. The concentration of hydrogen ions within the thylakoid thus increases further. When photosystem I absorbs another photon of light, its pigment passes a second high-energy electron to a reduction complex, which generates NADPH.

The Path to Photosystem I

The primary electron acceptor for the light-energized electrons leaving photosystem II is a quinone molecule, as it was in the bacterial photosystem described earlier. The reduced quinone which results (*plastoquinone*, symbolized Q) is a strong electron donor; it passes the excited electron to a proton pump called the *b_6-f complex* embedded within the thylakoid membrane (figure 10.16). This complex closely resembles the bc_1 complex in the respiratory electron trans-port chain of mitochondria discussed in chapter 9. Arrival of the energetic electron causes the b_6-f complex to pump a proton into the thylakoid space. A small copper-containing protein called *plastocyanin* (symbolized PC) then carries the electron to photosystem I.

Making ATP: Chemiosmosis

Each thylakoid is a closed compartment into which protons are pumped from the stroma by the b_6-f complex. The thylakoid membrane is impermeable to protons, so protons cross back out almost exclusively via the channels provided by *ATP synthases*. These channels protrude like knobs on the external surface of the thylakoid membrane. As protons pass out of the thylakoid through the ATP synthase channel, ADP is phosphorylated to ATP and released into the stroma, the fluid matrix inside the chloroplast. The stroma contains the enzymes that catalyze the light-independent reactions of carbon fixation (figure 10.17).

Photosystem I

The reaction center of photosystem I, called P_{700}, is a transmembrane complex consisting of at least 13 protein subunits. Energy is fed to it by an antenna complex consisting of 130 chlorophyll *a* and accessory pigment molecules. Photosystem I accepts an electron from plastocyanin into the hole created by the exit of a light-energized electron. This arriving electron has by no means lost all of its light-excited energy; almost half remains. Thus, the absorption of a photon of light energy by photosystem I boosts the electron leaving the reaction center to a very high energy level. Unlike photosystem II and the bacterial photosystem, photosystem I does not rely on quinones as electron acceptors. Instead, it passes electrons to an iron-sulfur protein called *ferredoxin* (Fd).

Making NADPH

Photosystem I passes electrons to ferredoxin on the stromal side of the membrane (outside the thylakoid). The reduced ferredoxin carries a very-high-potential electron. Two of them, from two molecules of reduced ferredoxin, are then donated to a molecule of NADP+ to form NADPH. The reaction is catalyzed by the membrane-bound enzyme *NADP reductase*. Because the reaction occurs on the stromal side of the membrane and involves the uptake of a proton in forming NADPH, it contributes further to the proton gradient established during photosynthetic electron transport.

Making More ATP

The passage of an electron from water to NADPH in the noncyclic photophosphorylation described previously generates one molecule of NADPH and slightly more than one molecule of ATP. However, as you will learn later in this chapter, building organic molecules takes more energy than that—it takes one-and-a-half ATP

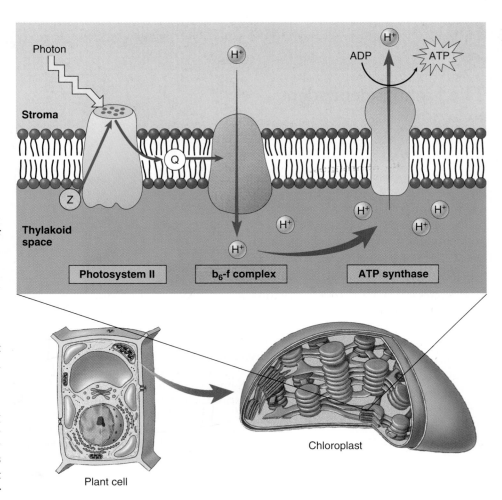

FIGURE 10.17

Chemoiosmosis in a chloroplast. The b_6-f complex embedded in the thylakoid membrane pumps protons into the interior of the thylakoid. ATP is produced on the outside surface of the membrane (stroma side), as protons diffuse back out of the thylakoid through ATP synthase channels.

molecules per NADPH molecule to fix carbon. To produce the extra ATP, many plant species are capable of short-circuiting photosystem I, switching photosynthesis into a *cyclic photophosphorylation* mode, so that the light-excited electron leaving photosystem I is used to make ATP instead of NADPH. The energetic electron is simply passed back to the b_6-f complex rather than passing on to NADP+. The b_6-f complex pumps out a proton, adding to the proton gradient driving the chemiosmotic synthesis of ATP. The relative proportions of cyclic and noncyclic photophosphorylation in these plants determines the relative amounts of ATP and NADPH available for building organic molecules.

The electrons that photosynthesis employs to form energy-rich reduced organic molecules are obtained from water; the residual oxygen atoms of the water molecules combine to form oxygen gas.

10.4 Cells use the energy and reducing power captured by the light reactions to make organic molecules.

The Light-Independent Reactions

Stated very simply, photosynthesis is a way of making organic molecules from carbon dioxide (CO_2). Organic molecules are highly reduced compared with CO_2, since they contain many C—H bonds. To build organic molecules, cells use raw materials provided by the light-dependent reactions:

1. **Energy.** ATP (provided by photosystem II) drives the endergonic reactions.
2. **Reducing power.** NADPH (provided by photosystem I) provides a source of hydrogens and the energetic electrons needed to bind them to carbon atoms. Much of the light energy captured in photosynthesis ends up invested in the energy-rich C—H bonds of sugars.

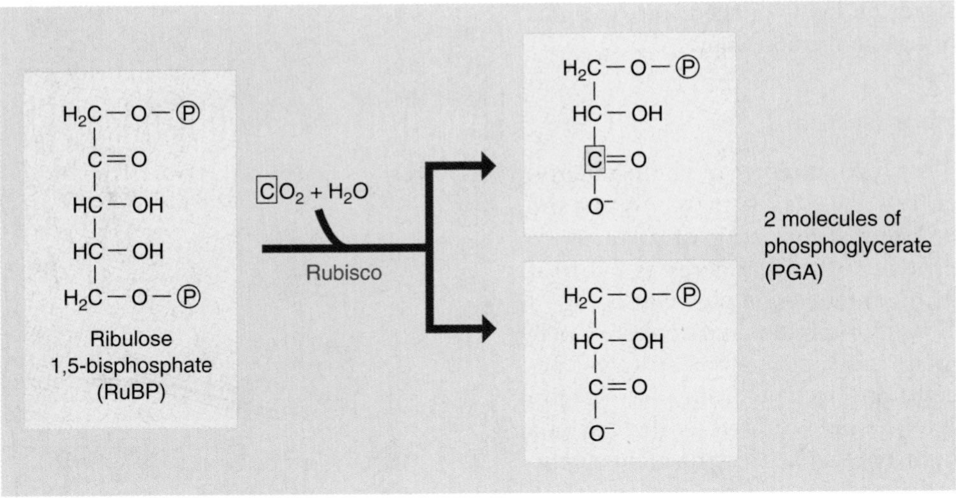

FIGURE 10.18
The key step in carbon fixation. Melvin Calvin and his coworkers at the University of California worked out the first step of what later became known as the Calvin cycle. They exposed photosynthesizing algae to radioactive carbon dioxide ($^{14}CO_2$). By following the fate of a radioactive carbon atom, they found that it first binds to a molecule of ribulose 1,5-bisphosphate (RuBP), then immediately splits, forming two molecules of phosphoglycerate (PGA). One of these PGAs contains the radioactive carbon atom. In 1948, workers isolated the enzyme responsible for this remarkable carbon-fixing reaction: rubisco.

Carbon Fixation

The key stage in the light-independent reactions—the event that makes the reduction of CO_2 possible—is the attachment of CO_2 to an organic molecule. A very special molecule is needed. Photosynthetic cells produce this molecule by reassembling the bonds of two intermediates in glycolysis, fructose 6-phosphate and glyceraldehyde 3-phosphate, to form the energy-rich five-carbon sugar, *ribulose1,5-bisphosphate (RuBP)*.

CO_2 binds to RuBP in the key process called **carbon fixation**, forming 2 three-carbon molecules of *phosphoglycerate (PGA)* (figure 10.18). The enzyme that carries out this reaction, *ribulose bisphosphate carboxylase* (usually called *rubisco* for short) is a very large enzyme present in the chloroplast stroma. It works very sluggishly, processing only about three molecules of RuBP per second (a typical enzyme processes about 1000 substrate molecules per second). Because it works so slowly, many copies of rubisco are needed. In a typical leaf, rubisco constitutes over 50% of all the protein. It is thought to be the most abundant protein on earth.

The Calvin Cycle

The light-independent reactions of photosynthesis can take place readily in the dark. The carbon-fixing reaction proceeds readily because RuBP is energy-rich. However, producing a supply of RuBP takes a series of reactions that consume large amounts of ATP and NADPH. They form a cycle of enzyme-catalyzed steps similar to the Krebs cycle. Because the cycle begins when CO_2 binds RuBP to form PGA, and PGA contains three carbon atoms, this process is called **C_3 photosynthesis.**

In a series of reactions (figure 10.19), 3 molecules of CO_2 are fixed by rubisco to produce 6 molecules of PGA (containing $6 \times 3 = 18$ carbon atoms in all, 3 from CO_2 and 15 from RuBP). The 18 carbon atoms then undergo a cycle of reactions that regenerates the 3 molecules of RuBP used in the initial step (containing $3 \times 5 = 15$ carbon atoms). This leaves one molecule of glyceraldehyde 3-phosphate (three carbon atoms) as the net gain (see figure 10.18).

The net equation of the light-independent reactions is:

$$3\ CO_2 + 9\ ATP + 6\ NADPH + water \longrightarrow$$
$$glyceraldehyde\ 3\text{-}phosphate + 8\ P_i + 9\ ADP + 6\ NADP^+$$

This cycle of reactions is called the **Calvin cycle,** after its discoverer, Melvin Calvin of the University of California, Berkeley. With three full turns of the cycle, three molecules of carbon dioxide enter, a molecule of glyceraldehyde 3-phosphate (G3P) is produced, and three molecules of RuBP are regenerated.

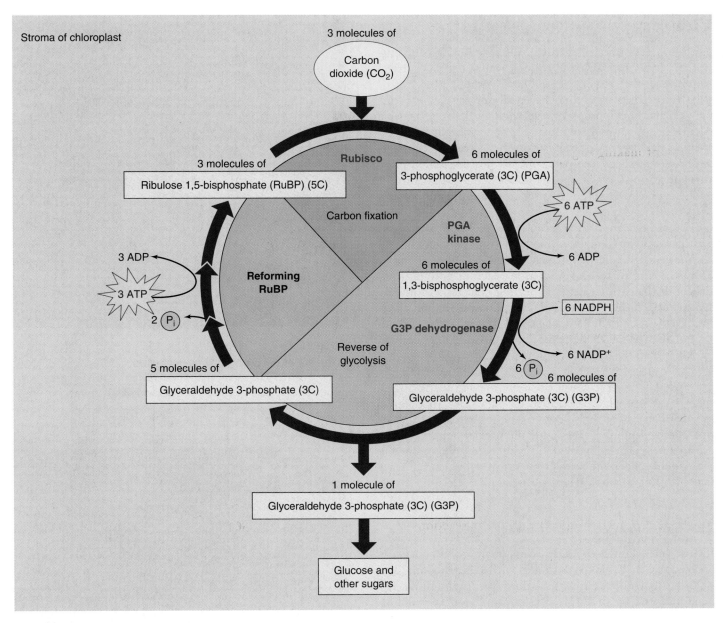

FIGURE 10.19
The Calvin cycle. For every three molecules of CO_2 that enter the cycle, one molecule of the three-carbon compound, glyceraldehyde 3-phosphate (G3P), is produced. Notice that the process requires energy stored in ATP and NADPH, which are generated by the light-dependent reactions. This process occurs in the stroma of the chloroplast.

Output of the Calvin Cycle

The glyceraldehyde 3-phosphate that is the product of the Calvin cycle is a three-carbon sugar that is a key intermediate in glycolysis. Much of it is exported from the chloroplast to the cytoplasm of the cell, where the reversal of several reactions in glycolysis allows it to be converted to fructose 6-phosphate and glucose 1-phosphate, and from that to sucrose, a major transport sugar in plants (sucrose, common table sugar, is a disaccharide made of fructose and glucose).

In times of intensive photosynthesis, glyceraldehyde 3-phosphate levels in the stroma of the chloroplast rise. As a consequence, some glyceraldehyde 3-phosphate in the chloroplast is converted to glucose 1-phosphate, in an analogous set of reactions to those done in the cytoplasm, by reversing several reactions of glycolysis. The glucose 1-phosphate is then combined into an insoluble polymer, forming long chains of starch stored as bulky starch grains in chloroplasts.

Most plants incorporate carbon dioxide into sugars by means of a cycle of reactions called the Calvin cycle, which is driven by the ATP and NADPH produced in the light-dependent reactions but which can itself take place in the dark.

Photorespiration

Evolution does not necessarily result in optimum solutions. Rather, it favors workable solutions that can be derived from others that already exist. Photosynthesis is no exception. Rubisco, the enzyme that catalyzes the key carbon-fixing reaction of photosynthesis (figure 10.20), provides a decidedly suboptimal solution. This enzyme has a second enzyme activity that interferes with the Calvin cycle, *oxidizing* ribulose 1,5-bisphosphate. In this process, called **photorespiration**, CO_2 is released *without* the production of ATP or NADPH. Hence, photorespiration undoes the work of photosynthesis.

The carboxylation and oxidation of ribulose 1,5-bisphosphate are catalyzed at the same active site on rubisco, and compete with each other. Under normal conditions at 25°C the rate of the carboxylation reaction is four times that of the oxidation reaction, meaning that 20% of photosynthetically fixed carbon is lost to photorespiration. This loss rises substantially as temperature increases, because the rate of the oxidation reaction increases with temperature far faster than the carboxylation reaction.

The loss of fixed carbon as a result of photorespiration is not trivial. Plants that use the Calvin cycle to fix atmospheric carbon, which are called C_3 plants, lose between a fourth and a half of their photosynthetically fixed carbon in this way, depending largely upon the temperature. In tropical climates, especially those in which the temperature is often above 28°C, the problem is a severe one, and it has a major impact on tropical agriculture.

The Crassulacean Acid Pathway

Plants that adapted to these warmer environments have evolved two principal ways to deal with this problem. One strategy to facilitate photosynthesis in hot regions has been adopted by many succulent (water-storing) plants such as cacti, pineapples, and some members of about two dozen other plant groups (figure 10.21). This mode of carbon fixation is called **crassulacean acid metabolism (CAM),** after the plant family Crassulaceae (the stonecrops or hens-and-chicks), in which it was first discovered. In these plants, the stomata (singular, stoma), specialized openings in the leaves of all plants through which CO_2 enters and water vapor is lost, open during the night and close during the day. This pattern of stomatal opening and closing is the reverse of that in most plants. Closing stomata during the day prevents water loss and reduces photorespiration by preventing CO_2 from leaving the leaves, thereby maintaining a high ratio of CO_2 to O_2 within the leaves. The CO_2 necessary for producing sugars is provided from organic molecules made during the night by the Calvin cycle.

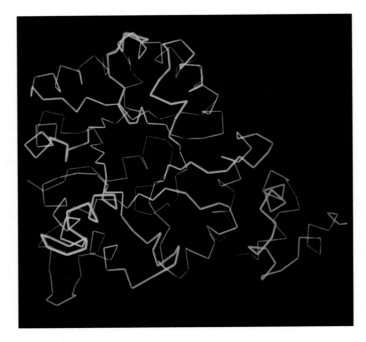

FIGURE 10.20
Structure of the catalytic domain of rubisco. The structural motif is an αβ barrel. The eight parallel β strands are shown in yellow, the eight α strands in pink, and the rest of the molecule in blue.

FIGURE 10.21
A familiar CAM plant. Pineapples and many other tropical succulents are CAM plants.

The C₄ Pathway

A second approach, taken by a number of grasses, including corn, sugarcane, and sorghum, is called **C₄ photosynthesis.** These plants create high local levels of CO_2 in order to favor the carboxylation reaction of rubisco (figure 10.22). They first carboxylate a three-carbon metabolite called phosphoenolpyruvate, producing oxaloacetate. Oxaloacetate is in turn converted into the intermediate malate, which is transported to an adjacent bundle-sheath cell. Inside the bundle-sheath cell, malate is decarboxylated to produce pyruvate, releasing CO_2. Because bundle-sheath cells are impermeable to CO_2, the CO_2 is retained within them in high concentrations. Pyruvate returns to the mesophyll cell, where *two* of the high-energy bonds in an ATP molecule are split to convert the pyruvate back into phosphoenolpyruvate, thus completing the cycle.

The enzymes that carry out the Calvin cycle in a C₄ plant are located within the bundle-sheath cells (figure 10.23), where the increased CO_2 concentration inhibits photorespiration. Because each CO_2 molecule is transported into the bundle-sheath cells at a cost of two high-energy ATP bonds, and because six carbons must be fixed to form a molecule of glucose, 12 additional molecules of ATP are required to form a molecule of glucose. In C₄ photosynthesis, the energetic cost of forming glucose is almost twice that of C₃ photosynthesis: 30 molecules of ATP versus 18. Nevertheless, C₄ photosynthesis is advantageous in a hot climate: photorespiration would otherwise remove more than half of the carbon fixed.

Photorespiration short-circuits photosynthesis. CAM and C₄ plants circumvent this problem through modifications of leaf architecture and photosynthetic chemistry that locally increase CO_2 concentrations. CAM plants isolate CO_2 production temporally, C₄ plants spatially.

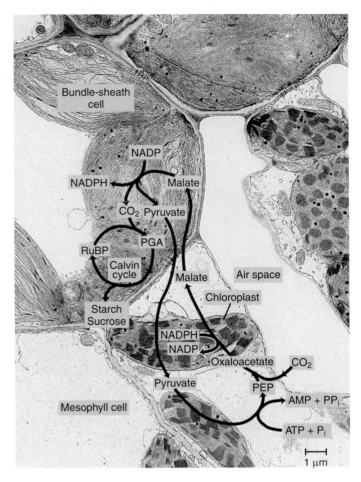

FIGURE 10.22
Carbon fixation in C₄ plants. This process is called the C₄ pathway because the starting material, oxaloacetate, is a molecule containing four carbons.

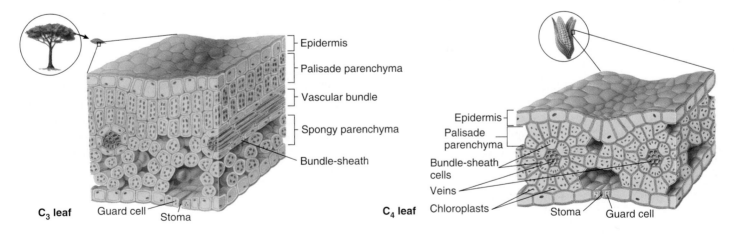

FIGURE 10.23
Comparing C₃ and C₄ plant leaves. In a C₃ leaf, CO_2 is taken up by widely spaced mesophyll cells (the spongy parenchyma). In a C₄ leaf, CO_2 is released from C₄ compounds within bundle-sheath cells; these cells, relatively impermeable to CO_2, build up high concentrations of CO_2.

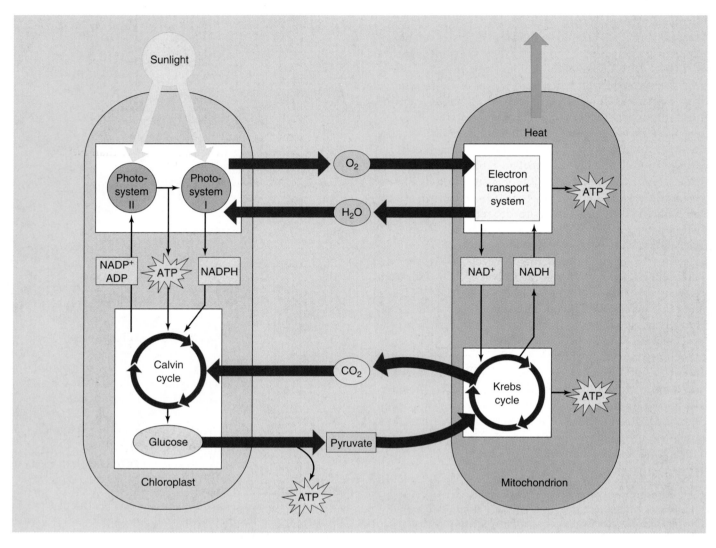

FIGURE 10.24

Chloroplasts and mitochondria: Completing an energy cycle. Water and oxygen gas cycle between chloroplasts and mitochondria within a plant cell, as do glucose and CO_2. Cells with chloroplasts require an outside source of CO_2 and water and generate glucose and oxygen. Cells without chloroplasts, such as animal cells, require an outside source of glucose and oxygen and generate CO_2 and water.

The Energy Cycle

The energy-capturing metabolisms of the chloroplasts studied in this chapter and the mitochondria studied in the previous chapter are intimately related. Photosynthesis uses the products of respiration as starting substrates, and respiration uses the products of photosynthesis as its starting substrates (figure 10.24). The Calvin cycle even uses part of the ancient glycolytic pathway, run in reverse, to produce glucose. And, the principal proteins involved in electron transport in plants are related to those in mitochondria, and in many cases are actually the same.

Photosynthesis is but one aspect of plant biology, although it is an important one. In chapters 33 through 40, we will examine plants in more detail. We have treated photosynthesis here, in a section devoted to cell biology, because photosynthesis arose long before plants did, and all organisms depend directly or indirectly on photosynthesis for the energy that powers their lives.

Summary of Concepts

10.1 Photosynthesis takes place in chloroplasts.

• Light is used by plants, algae, and some bacteria in a process called photosynthesis to convert atmospheric carbon (CO_2) into carbohydrate.

10.2 Learning about photosynthesis: An experimental journey.

• A series of simple experiments demonstrated that plants capture energy from light and use it to convert the carbon atoms of CO_2 and the hydrogen atoms of water into organic molecules.

10.3 Pigments capture energy from sunlight.

• Light consists of energy packets called photons; the shorter the wavelength of light, the more its energy.

• When photons are absorbed by a pigment, electrons in the pigment are boosted to a higher energy level.

• Photosynthesis seems to have evolved in organisms similar to the purple sulfur bacteria, which use an array of chlorophyll and accessory pigment molecules (a photosystem) to channel photon excitation energy into a single pigment molecule.

• That molecule then donates an electron to an electron transport chain, which drives a proton pump and ultimately returns the electron to the pigment, in a process called cyclic photophosphorylation.

• Descendants of the original photosynthetic bacteria employ two photosystems which can generate enough energy to use H_2O rather than H_2S as a hydrogen source.

• In organisms with two photosystems, light is first absorbed by photosystem II, removing an electron from one chlorophyll *a* molecule, P_{680}. This process has two effects: (1) the absence of the high-energy electron causes photosystem II to seek another electron, which it obtains from the splitting of water, leaving O_2 as a by-product; and (2) the high-energy electron is passed to

photosystem I, driving a proton pump in the process and bringing about the chemiosmotic synthesis of ATP.

• When the electron from P_{680} arrives at photosystem I, P_{700} absorbs a photon of light, energizing one of its electrons. Electrons from photosystem I are channeled to a primary electron acceptor, which reduces NADP$^+$ to NADPH. Use of NADPH rather than NADH allows plants and algae to keep the processes of photosynthesis and oxidative respiration separate from each other.

10.4 Cells use the energy and reducing power captured by the light reactions to make organic molecules.

• The ATP and reducing power produced by the light-dependent reactions are used to fix carbon in a series of light-independent reactions called the Calvin cycle.

• In this process, ribulose 1,5-bisphosphate (RuBP) is carboxylated, and the product is split into two molecules of phosphoglycerate (PGA). The PGA molecules are run through a series of reactions to form molecules of glyceraldehyde phosphate, some of which are used to reconstitute RuBP. The remainder of the glyceraldehyde phosphate molecules enter the cell's metabolism as newly-fixed carbon.

• RuBP carboxylase, the enzyme that fixes carbon in the Calvin cycle, also carries out an oxidative reaction that uses the products of photosynthesis, a process called photorespiration.

• Many tropical plants inhibit photorespiration by expending ATP to increase the intracellular concentration of CO_2. This process, called the C_4 pathway, nearly doubles the energetic cost of synthesizing glucose. In warm climates, though, the cost is less than that of the glucose molecules that would otherwise be converted back into carbon dioxide by photorespiration and lost.

Discussing Key Terms

1. **Photosystem**   Chlorophyll pigments are organized into integrated units called photosystems that by sharing excitation greatly amplify the light-gathering capacity of the individual pigment molecules.

2. **Reducing power**   A source of hydrogens that is needed to reduce CO_2 to form organic molecules. In organisms with two photosystems, one of them drives the synthesis of NADPH, thus providing reducing power. Organisms with one photosystem must scavenge a source of hydrogens and are less efficient.

3. **Carbon fixation**   Photosynthesis fixes carbon; that is, carbon from the environment (usually in the form of atmospheric CO_2) is converted into carbohydrates

through the so-called light-independent reactions of photosynthesis.

4. **Photorespiration**   The process by which plants and algae sometimes convert the carbon fixed by photosynthesis back into CO_2 and heat, undoing the efforts of photosynthesis. This occurs because the carboxylase enzyme binds oxygen as well as CO_2.

5. **C_4 photosynthesis**   An alternative photosynthetic process used by plants in warmer climates to combat the loss of CO_2 through photorespiration. C_4 plants concentrate CO_2 within special cells to outcompete O_2 at the carboxylase enzyme.

Review Questions

1. How did van Helmont determine that plants do not obtain their food from the soil?

2. What is the overall equation for photosynthesis? Where do the oxygen atoms in the O_2 produced during photosynthesis come from? How was the answer to that question ascertained?

3. What are the basic units of light energy? Which wavelengths of visible light are most energetic, and which are least energetic?

4. How is the energy of light captured by a pigment molecule? Why does light reflected by the pigment chlorophyll appear green?

5. How did Englemann show that chlorophyll is likely to be involved in photosynthesis?

6. What are the three main stages in the light-dependent reactions of photosynthesis? What is the function of the reaction center chlorophyll? What is the function of the primary electron acceptor?

7. Explain how photosynthesis in the sulfur bacteria is a cyclic process. What is its energy yield in terms of ATP molecules synthesized per electron?

8. How do the two photosystems in plants and algae work? Which stage generates ATP and which generates NADPH?

9. What occurs in the light-independent reactions of photosynthesis? How do the light-independent reactions use the ATP and NADPH produced in the light-dependent reactions?

10. In the Calvin cycle, what happens to ribulose 1,5-bisphosphate? What happens to glyceraldehyde 3-phosphate (G3P)?

11. In a C_3 plant, where do the light-dependent reactions occur? Where do the light-independent reactions occur?

12. What is photorespiration? What advantage do C_4 plants have over C_3 plants with respect to photorespiration? What disadvantage do C_4 plants have that limits their distribution primarily to warm regions of the earth?

Thought Questions

1. What is the advantage of having many pigment molecules in each photosystem but only one reaction center chlorophyll? In other words, why not couple *every* pigment molecule directly to an electron acceptor?

2. The two photosystems, P_{680} and P_{700}, of cyanobacteria, algae, and plants yield an oxidant capable of cleaving water. How might the subsequent evolution of cellular respiration have been different if the two photosystems had not evolved, and all photosynthetic organisms were restricted to the cyclic photophosphorylation used by the sulfur bacteria?

3. In theory, a plant kept in total darkness could still manufacture glucose—if it were supplied with *which* molecules?

Internet Links

An Illustrated Review of Photosynthesis
http://gened.emc.maricopa.edu/bio/bio181/BIOBK/BioBookPS.html
A short, well-illustrated review of photosynthesis, with simple clear diagrams.

The Photosynthetic Process
http://www.life.uiuc.edu/govindjee/paper/gov.html#51
A good illustrated review of the current state of knowledge of photosynthesis by John Whitmarsh and Govindjee, two leading photosynthesis researchers.

Photosynthesis Problem Sets
http://www.biology.arizona.edu/biochemistry/problem_sets/photosynthesis_1/
http://www.biology.arizona.edu/biochemistry/problem_sets/photosynthesis_2/
Sets of interactive multiple-choice questions covering the light (1) and dark (2) reactions of photosynthesis, part of the Biology Project of the University of Arizona.

For Further Reading

Barber, J., and B. Andersson: "Revealing the Blueprint of Photosynthesis," *Nature*, July 7, 1994, pages 31–34. With increased understanding of how the photosynthetic mechanism works, how might we be able to put it to technological use?

Deisenhofer, J., and H. Michel: "The Photosynthetic Reaction Center of the Purple Bacterium *Rhodopseudomonas*," *Science*, 1988, vol. 245, pages 1463–73. The Nobel Prize address for the first description at the molecular level of a photosynthetic reaction center.

Govindjee, and W. Coleman: "How Plants Make Oxygen," *Scientific American*, February 1990, pages 50–58. How plants and some bacteria exploit solar energy to split water molecules into oxygen gas. (The first author, an expert in primary photochemistry, uses only one name; his last name, Astana, was dropped by his father as a protest against the caste system in India.)

Nash, L.: "Case No. 5: Plants and the Atmosphere," in *Harvard Case Histories in Experimental Science*, vol. 2, J. Conant, (Ed.), Harvard University Press, Cambridge, MA, 1964. An engaging description of the early work on photosynthesis, often presented in the words of the investigators themselves.

Pakrasi, H.: Genetic Analysis of the Form and Function of Photosystem I and Photosystem II. *Annual Reviews of Genetics*, 1995, vol. 29, pages 755–76. A very clear review article discussing recent key findings about photosynthetic photosystems. Technical but highly recommended.

Raven, P., R. Evert, and S. Eichhorn: *Biology of Plants*, 6th ed., Worth, New York, 1999. The definitive botany text, by one of this text's authors, comprehensive and easy to read.

11

How Cells Divide

Concept Outline

11.1 Bacteria divide far more simply than do eukaryotes.

Cell Division in Prokaryotes. Bacterial cells divide by splitting in two.

11.2 Chromosomes are highly ordered structures.

Discovery of Chromosomes. All eukaryotic cells contain chromosomes, but different organisms possess differing numbers of chromosomes.

The Structure of Eukaryotic Chromosomes. Proteins play an important role in packaging DNA in chromosomes.

11.3 Mitosis is a key phase of the cell cycle.

Phases of the Cell Cycle. The cell cycle consists of three growth phases, a nuclear division phase, and a cytoplasmic division stage.

Interphase: Preparing for Mitosis. In interphase, the cell grows, replicates its DNA, and prepares for cell division.

Prophase: Formation of the Mitotic Apparatus. In prophase, the chromosomes condense and microtubules attach sister chromosomes to opposite poles of the cell.

Metaphase: Division of the Centromeres. In metaphase, chromosomes align along the center of the cell.

Anaphase and Telophase: Separation of the Chromatids and Reformation of the Nuclei. In anaphase, the chromosomes separate; in telophase the spindle dissipates and the nuclear envelope re-forms.

Cytokinesis. In cytokinesis, the cytoplasm separates into two roughly equal halves.

11.4 The cell cycle is carefully controlled.

General Strategy of Cell Cycle Control. At three points in the cell cycle, feedback from the cell determines whether the cycle will continue.

Molecular Mechanisms of Cell Cycle Control. Special proteins regulate the "checkpoints" of the cell cycle.

Cancer and the Control of Cell Proliferation. Cancer results from damage to genes encoding proteins that regulate the cell division cycle.

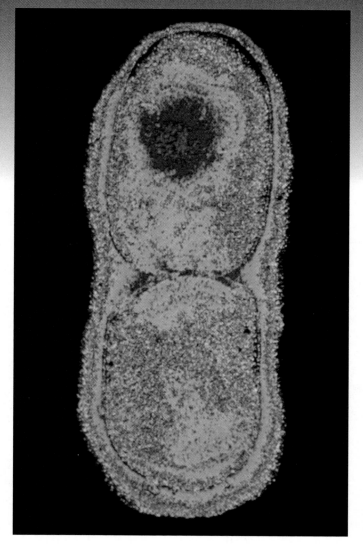

FIGURE 11.1
Cell division in bacteria. It's hard to imagine fecal coliform bacteria as beautiful, but here is *Escherichia coli*, inhabitant of the large intestine and the biotechnology lab, spectacularly caught in the act of fission.

All species of organisms—bacteria, alligators, the weeds in a lawn—grow and reproduce. From the smallest of creatures to the largest, all species produce offspring like themselves and pass on the hereditary information that makes them what they are. In this chapter, we begin our consideration of heredity with an examination of how cells reproduce (figure 11.1). The mechanism of cell reproduction and its biological consequences have changed significantly during the evolution of life on earth.

Cell Division in Prokaryotes

In bacteria, which are prokaryotes and lack a nucleus, cell division consists of a simple procedure called **binary fission** (literally, "splitting in half"), in which the cell divides into two equal or nearly equal halves. The genetic information, or *genome*, replicates early in the life of the cell. It exists as a single, circular, double-stranded DNA molecule. Fitting this DNA circle into the bacterial cell is a remarkable feat of packaging—fully stretched out, the DNA of a bacterium like *Escherichia coli* is about 500 times longer than the cell itself.

The DNA circle is attached at one point to the cytoplasmic surface of the bacterial cell's plasma membrane. At a specific site on the DNA molecule called the *replication origin*, a battery of more than 22 different proteins begins the process of copying the DNA (figure 11.2). When these enzymes have proceeded all the way around the circle of DNA, the cell possesses two copies of the genome. These "daughter" genomes are attached side-by-side to the plasma membrane.

The growth of a bacterial cell to about twice its initial size induces the onset of cell division. First, the cell lays down new plasma membrane and cell wall materials in the zone between the attachment sites of the two daughter genomes. A new plasma membrane grows between the genomes; eventually, it reaches all the way into the center of the cell, dividing it in two (figure 11.3). Because the membrane forms between the two genomes, each new cell is assured of retaining one of the genomes. Finally, a new cell wall forms around the new membrane.

The evolution of the eukaryotes introduced several additional factors into the process of cell division. Eukaryotic cells are much larger than bacteria, and their genomes con-

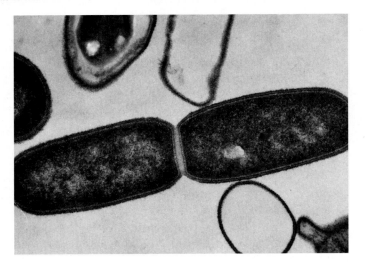

FIGURE 11.3

Fission (40,000×). Bacteria divide by a process of simple cell fission. Note the newly formed plasma membrane between the two daughter cells.

tain much more DNA. Eukaryotic DNA is contained in a number of linear chromosomes, whose organization is much more complex than that of the single, circular DNA molecules in bacteria. In chromosomes, DNA forms a complex with packaging proteins called histones and is wound into tightly condensed coils.

Bacteria divide by binary fission. Fission begins between the points where the two genomes are bound to the plasma membrane, ensuring that one genome will end up in each daughter cell.

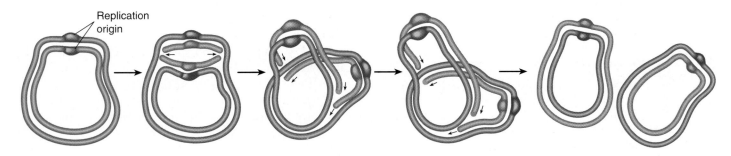

FIGURE 11.2

How bacterial DNA replicates. The replication of the circular DNA molecule that constitutes the genome of a bacterium begins at a single site, called the replication origin. The replication enzymes move out in both directions from that site. When they meet on the far side of the molecule, replication is complete.

Discovery of Chromosomes

Chromosomes were first observed by the German embryologist Walther Fleming in 1882, while he was examining the rapidly dividing cells of salamander larvae. When Fleming looked at the cells through what would now be a rather primitive light microscope, he saw minute threads within their nuclei that appeared to be dividing lengthwise. Fleming called their division **mitosis,** based on the Greek word *mitos,* meaning "thread."

Chromosome Number

Since their initial discovery, chromosomes have been found in the cells of all eukaryotes examined. Their number may vary enormously from one species to another. A few kinds of organisms—such as the Australian ant *Myrmecia* spp.; the plant *Haplopappus gracilis,* a relative of the sunflower that grows in North American deserts; and the fungus *Penicillium*—have only 1 pair of chromosomes, while some ferns have more than 500 pairs (table 11.1). Most eukaryotes have between 10 and 50 chromosomes in their body cells.

Human cells each have 46 chromosomes, consisting of 23 nearly identical pairs (figure 11.4). Each of these 46 chromosomes contains thousands of genes that play important roles in determining how a person's body develops and functions. For this reason, possession of all the chromosomes is essential to survival. Humans missing even one chromosome, a condition called monosomy, do not survive embryonic development. Nor does the human embryo develop properly with an extra copy of any one chromosome, a condition called trisomy. For all but a few of the smallest chromosomes, trisomy is fatal, and even in those few cases,

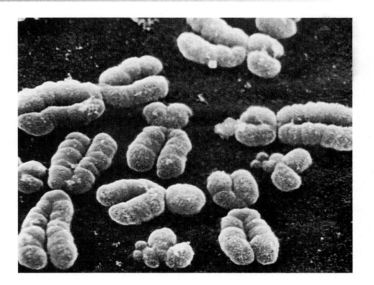

FIGURE 11.4
Human chromosomes. This photograph (950×) shows human chromosomes as they appear immediately before nuclear division. Each DNA strand has already replicated, forming identical sister chromatids held together by a constriction called the centromere.

serious problems result. Individuals with an extra copy of the very small chromosome 21, for example, develop more slowly than normal and are mentally retarded, a condition called Down syndrome.

> All eukaryotic cells store their hereditary information in chromosomes, but different kinds of organisms utilize very different numbers of chromosomes to store this information.

Table 11.1 Chromosome Number in Selected Eukaryotes

Group	Total Number of Chromosomes	Group	Total Number of Chromosomes
FUNGI		Bread wheat	42
Neurospora (haploid)	7	Sugarcane	80
Saccharomyces (a yeast)	16	Horsetail	216
		Adder's tongue fern	1262
INSECTS			
Mosquito	6	**VERTEBRATES**	
Drosophila	8	Opossum	22
Honeybee	32	Frog	26
Silkworm	56	Mouse	40
		Human	46
PLANTS		Chimpanzee	48
Haplopappus gracilis	2	Horse	64
Garden pea	14	Chicken	78
Corn	20	Dog	78

The Structure of Eukaryotic Chromosomes

In the century since the discovery of chromosomes, we have learned a great deal about their structure and composition.

Composition of Chromatin

Chromosomes are composed of **chromatin,** a complex of DNA and protein; most are about 40% DNA and 60% protein. A significant amount of RNA is also associated with chromosomes because chromosomes are the sites of RNA synthesis. The DNA of a chromosome is one very long, double-stranded fiber that extends unbroken through the entire length of the chromosome. A typical human chromosome contains about 140 million (1.4×10^8) nucleotides in its DNA. The amount of information one chromosome contains would fill about 280 printed books of 1000 pages each, if each nucleotide corresponded to a "word" and each page had about 500 words on it. Furthermore, if the strand of DNA from a single chromosome were laid out in a straight line, it would be about 5 centimeters (2 inches) long. Fitting such a strand into a nucleus is like cramming a string the length of a football field into a baseball—and that's only 1 of 46 chromosomes! In the cell, however, the DNA is coiled, allowing it to fit into a much smaller space than would otherwise be possible.

Chromosome Coiling

How can this long DNA fiber coil so tightly? If we gently disrupt a eukaryotic nucleus and examine the DNA with an electron microscope, we find that it resembles a string of beads (figure 11.5). Every 200 nucleotides, the DNA duplex is coiled around a core of eight histone proteins, forming a complex known as a **nucleosome.** Unlike most proteins, which have an overall negative charge, histones are positively charged, due to an abundance of the basic

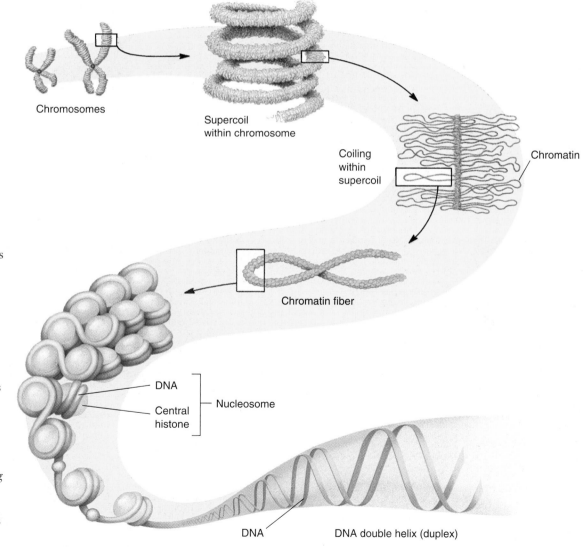

FIGURE 11.5
Levels of eukaryotic chromosomal organization.
Nucleotides assemble into long double strands of DNA molecules. These strands require further packaging to fit into the cell nucleus. The DNA duplex is tightly bound to and wound around proteins called *histones.* The DNA-wrapped histones then aggregate into structures called *nucleosomes.* The nucleosomes then coalesce into *chromatin* fibers, ultimately coiling around into *supercoils* that make up the form of DNA recognized as a *chromosome.*

Chromosomes

Supercoil within chromosome

Coiling within supercoil

Chromatin

Chromatin fiber

DNA

Central histone

Nucleosome

DNA

DNA double helix (duplex)

amino acids arginine and lysine. They are thus strongly attracted to the negatively charged phosphate groups of the DNA. The histone cores thus act as "magnetic forms" that promote and guide the coiling of the DNA. Further coiling occurs when the string of nucleosomes wraps up into higher order coils called supercoils.

Highly condensed portions of the chromatin are called **heterochromatin.** Some of these portions remain permanently condensed, so that their DNA is never expressed. The remainder of the chromosome, called **euchromatin,** is condensed only during cell division, when compact packaging facilitates the movement of the chromosomes. At all other times, euchromatin is present in an open configuration, and its genes can be expressed. The way chromatin is packaged when the cell is not dividing is not well understood beyond the level of nucleosomes and is a topic of intensive research.

Chromosome Karyotypes

Chromosomes may differ widely in appearance. They vary in size, staining properties, the location of the *centromere* (a constriction found on all chromosomes), the relative length of the two arms on either side of the centromere, and the positions of constricted regions along the arms. The particular array of chromosomes that an individual possesses is called its **karyotype** (figure 11.6). Karyotypes show marked differences among species and sometimes even among individuals of the same species.

To examine a human karyotype, investigators collect a blood sample and add chemicals that induce the white blood cells in the sample to divide. Later, they add other chemicals to stop cell division at a stage when the chromosomes are most condensed and thus most easily distinguished from one another. The cells are then broken open and their contents, including the chromosomes, spread out and stained. To facilitate the examination of the karyotype, the chromosomes are usually photographed, and the outlines of the chromosomes are cut out of the photograph and arranged in order (see figure 11.6).

How Many Chromosomes Are in a Cell?

With the exception of the **gametes** (eggs or sperm) and a few specialized tissues, every cell in a human body is **diploid (2n).** This means that the cell contains two nearly identical copies of each of the 23 types of chromosomes, for a total of 46 chromosomes. The **haploid (1n)** gametes contain only one copy of each of the 23 chromosome types, while certain tissues have unusual numbers of chromosomes—many liver cells, for example, have two nuclei, while mature red blood cells have no nuclei at all. The two copies of each chromosome in body cells are called **homologous chromosomes,** or **homologues** (Greek *homologia,* "agreement"). Before cell division, each homologue replicates, producing two identical **sister chromatids** joined at the **centromere,** a condensed area found

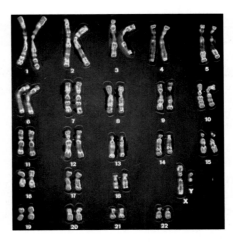

FIGURE 11.6
A human karyotype. The individual chromosomes that make up the 23 pairs differ widely in size and in centromere position. In this preparation, the chromosomes have been specifically stained to indicate further differences in their composition and to distinguish them clearly from one another.

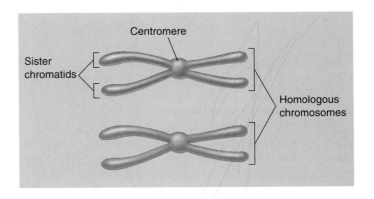

FIGURE 11.7
The difference between homologous chromosomes and sister chromatids. Homologous chromosomes are a pair of the same chromosome—say, chromosome number 16. Sister chromatids are the two replicas of a single chromosome held together by the centromeres after DNA replication.

on all eukaryotic chromosomes (figure 11.7). Hence, as cell division begins, a human body cell contains a total of 46 replicated chromosomes, each composed of two sister chromatids joined by one centromere. The cell thus contains 46 centromeres and 92 chromatids (2 sister chromatids for each of 2 homologues for each of 23 chromosomes). The cell is said to contain 46 chromosomes rather than 92 because, by convention, the number of chromosomes is obtained by counting centromeres.

Eukaryotic genomes are larger and more complex than those of bacteria. Eukaryotic DNA is packaged tightly into chromosomes, enabling it to fit inside cells. Haploid cells contain one set of chromosomes, while diploid cells contain two sets.

Phases of the Cell Cycle

The increased size and more complex organization of eukaryotic genomes over those of bacteria required radical changes in the process by which the two replicas of the genome are partitioned into the daughter cells during cell division. This division process is diagrammed as a **cell cycle,** consisting of five phases (figure 11.8).

The Five Phases

G₁ is the primary growth phase of the cell. For many organisms, this encompasses the major portion of the cell's life span.

S is the phase in which the cell synthesizes a replica of the genome.

G₂ is the second growth phase, in which preparations are made for genomic separation. During this phase, mitochondria and other organelles replicate, chromosomes condense, and microtubules begin to assemble at a spindle.

M is the phase of the cell cycle in which the microtubular apparatus assembles, binds to the chromosomes, and moves the sister chromatids apart. Called **mitosis,** this process is the essential step in the separation of the two daughter genomes. We will discuss mitosis as it occurs in animals and plants, where the process does not vary much (it is somewhat different among fungi and some protists). Although mitosis is a continuous process, it is traditionally subdivided into four stages: prophase, metaphase, anaphase, and telophase (figure 11.9).

C is the phase of the cell cycle when the cytoplasm divides, creating two daughter cells. This phase is called **cytokinesis.** In animal cells, the microtubule spindle helps position a contracting ring of actin that constricts like a drawstring to pinch the cell in two. In cells with a cell wall, such as plant cells, a plate forms between the dividing cells.

Duration of the Cell Cycle

The time it takes to complete a cell cycle varies greatly among organisms. Cells in growing embryos can complete their cell cycle in under 20 minutes; the briefest known animal nuclear division cycles occur in fruit fly embryos (8 minutes). Cells such as these simply divide their nuclei as quickly as they can replicate their DNA, without cell growth. Half of the cycle is taken up by S, half by M, and essentially none by G₁ or G₂. Because mature cells require time to grow, most of their cycles are much longer than

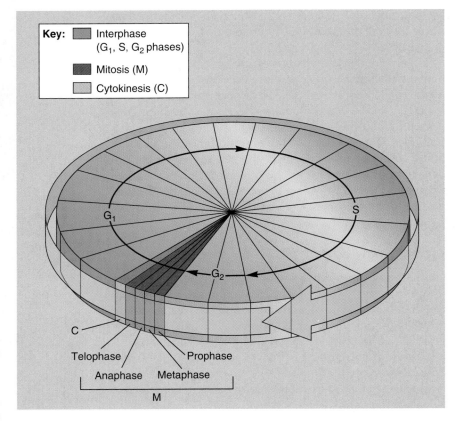

FIGURE 11.8
The cell cycle. Each wedge represents one hour of the 22-hour cell cycle in human cells growing in culture. G₁ represents the primary growth phase of the cell cycle, S the phase during which a replica of the genome is synthesized, and G₂ the second growth phase.

those of embryonic tissue. Typically, a rapidly dividing mammalian cell completes its cell cycle in about 24 hours, but some cells, like certain cells in the human liver, have cell cycles lasting more than a year. During the cycle, growth occurs throughout the G₁ and G₂ phases (sometimes referred to as "gap" phases, as they separate S from M), as well as during the S phase. The M phase takes only about an hour, a small fraction of the entire cycle.

Most of the variation in the length of the cell cycle from one organism or tissue to the next occurs in the G₁ phase. Cells often pause in G₁ before DNA replication and enter a resting state called **G₀ phase;** they may remain in this phase for days to years before resuming cell division. At any given time, most of the cells in an animal's body are in G₀ phase. Some, such as muscle and nerve cells, remain there permanently; others, such as liver cells, can resume G₁ phase in response to factors released during injury.

Most eukaryotic cells repeat a process of growth and division referred to as the cell cycle. The cycle can vary in length from a few minutes to several years.

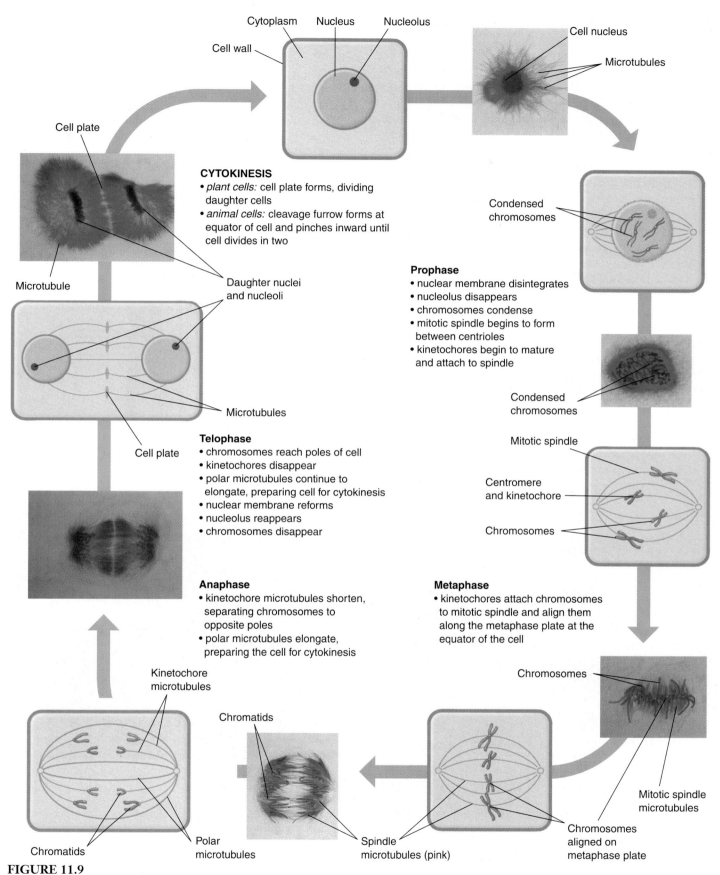

CYTOKINESIS
- *plant cells:* cell plate forms, dividing daughter cells
- *animal cells:* cleavage furrow forms at equator of cell and pinches inward until cell divides in two

Prophase
- nuclear membrane disintegrates
- nucleolus disappears
- chromosomes condense
- mitotic spindle begins to form between centrioles
- kinetochores begin to mature and attach to spindle

Telophase
- chromosomes reach poles of cell
- kinetochores disappear
- polar microtubules continue to elongate, preparing cell for cytokinesis
- nuclear membrane reforms
- nucleolus reappears
- chromosomes disappear

Anaphase
- kinetochore microtubules shorten, separating chromosomes to opposite poles
- polar microtubules elongate, preparing the cell for cytokinesis

Metaphase
- kinetochores attach chromosomes to mitotic spindle and align them along the metaphase plate at the equator of the cell

Labels: Cytoplasm, Nucleus, Nucleolus, Cell wall, Cell nucleus, Microtubules, Cell plate, Condensed chromosomes, Microtubule, Daughter nuclei and nucleoli, Microtubules, Cell plate, Condensed chromosomes, Mitotic spindle, Centromere and kinetochore, Chromosomes, Chromosomes, Kinetochore microtubules, Chromatids, Chromatids, Polar microtubules, Spindle microtubules (pink), Mitotic spindle microtubules, Chromosomes aligned on metaphase plate

FIGURE 11.9

Mitosis and cytokinesis. Mitosis (separation of the two genomes) occurs in four stages—prophase, metaphase, anaphase, and telophase—and is followed by cytokinesis (division into two separate cells). In this depiction, the chromosomes of the African blood lily, *Haemanthus katharinae*, are stained blue, and microtubules are stained red.

Interphase: Preparing for Mitosis

The events that occur during interphase, made up of the G_1, S, and G_2 phases, are very important for the successful completion of mitosis. During G_1, cells undergo the major portion of their growth. During the S phase, each chromosome replicates to produce two sister chromatids, which remain attached to each other at the **centromere.** The centromere is a point of constriction on the chromosome, containing a specific DNA sequence to which is bound a disk of protein called a **kinetochore.** This disk functions as an attachment site for fibers that assist in cell division (figures 11.10 and 11.11). Each chromosome's centromere is located at a characteristic site.

The cell grows throughout interphase. The G_1 and G_2 segments of interphase are periods of active growth, when proteins are synthesized and cell organelles produced. The cell's DNA replicates only during the S phase of the cell cycle.

After the chromosomes have replicated in S phase, they remain fully extended and uncoiled. This makes them invisible under the light microscope. In G_2 phase, they begin the long process of **condensation,** coiling ever more tightly. Special *motor proteins* are involved in the rapid final condensation of the chromosomes that occurs early in mitosis. Also during G_2 phase, the cells begin to assemble the machinery they will later use to move the chromosomes to opposite poles of the cell. In animal cells, a pair of microtubule-organizing centers called **centrioles** replicate. All eukaryotic cells undertake an extensive synthesis of *tubulin,* the protein of which microtubules are formed.

Interphase is that portion of the cell cycle in which the chromosomes are invisible under the light microscope because they are not yet condensed. It includes the G_1, S, and G_2 phases. In the G_2 phase, the cell mobilizes its resources for cell division.

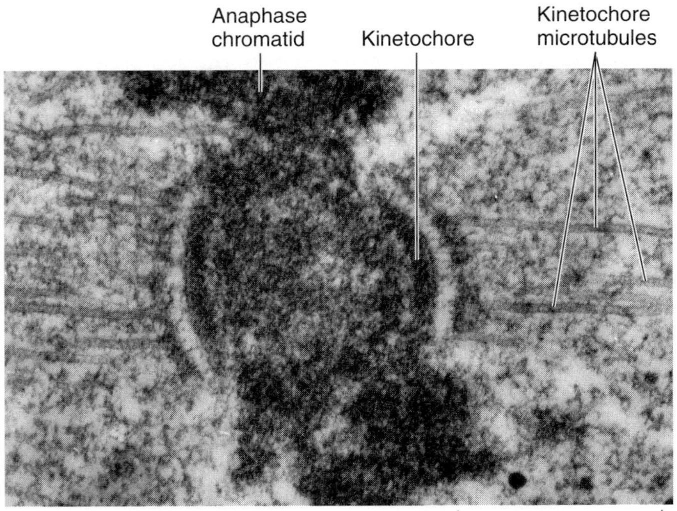

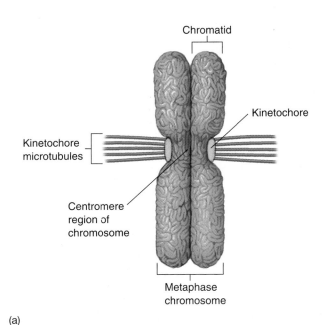

(a)

(b)

FIGURE 11.10
How microtubules attach to the kinetochore. In this electron micrograph of a green alga chromosome, the kinetochore plaques can be clearly seen, their inner edges contiguous with the centromere. The kinetochore consists of several (typically three) layers of protein, with microtubules embedded in them.

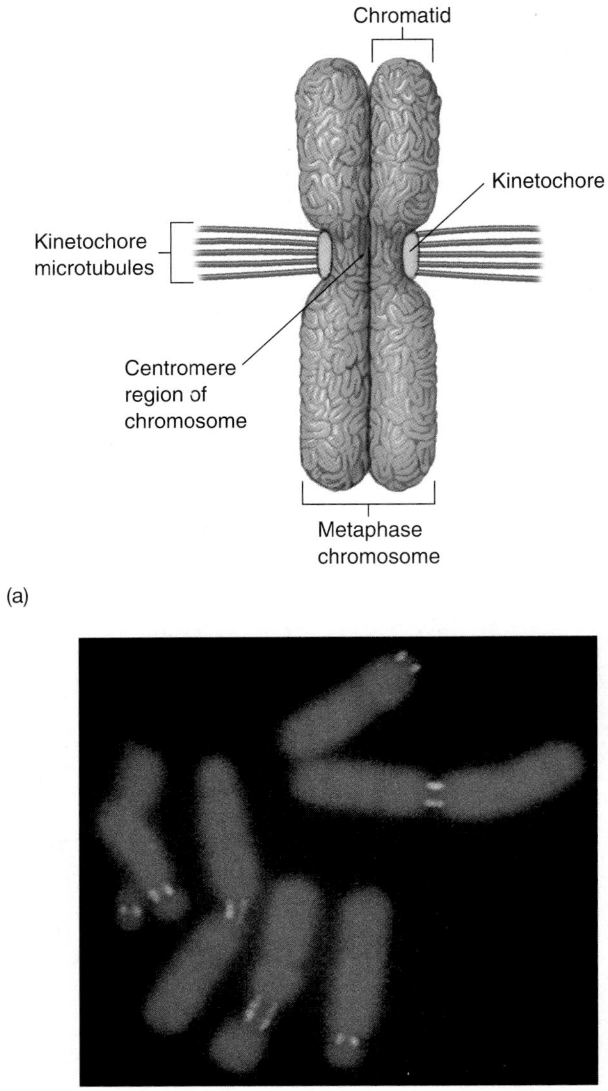

FIGURE 11.11
Kinetochores. (a) In a metaphase chromosome, kinetochore microtubules are anchored to proteins at the centromere. (b) Labeled with antibodies directed against them, the kinetochore proteins can be seen at the centromeres of metaphase chromosomes in these cultured marsupial cells.

Prophase: Formation of the Mitotic Apparatus

When the chromosome condensation initiated in G_2 phase reaches the point at which individual condensed chromosomes first become visible with the light microscope, the first stage of mitosis, **prophase,** has begun (figure 11.12). The condensation process continues throughout prophase; consequently, some chromosomes that start prophase as minute threads appear quite bulky before its conclusion. Ribosomal RNA synthesis ceases when the portion of the chromosome bearing the rRNA genes is condensed.

Assembling the Spindle Apparatus

The assembly of the microtubular apparatus that will later separate the sister chromatids also continues during prophase. In animal cells, the two centriole pairs formed during G_2 phase begin to move apart early in prophase, forming between them an axis of microtubules referred to as **spindle fibers.** By the time the centrioles reach the opposite poles of the cell, they have established a bridge of microtubules called the **spindle apparatus** between them. In plant cells, a similar bridge of microtubular fibers forms between opposite poles of the cell, although no microtubule-organizing center is visible with the light microscope, and centrioles are absent.

During the formation of the spindle apparatus, the nuclear envelope breaks down and the endoplasmic reticulum reabsorbs its components. At this point, then, the microtubular spindle fibers extend completely across the cell, from one pole to the other. Their orientation determines the plane in which the cell will subsequently divide, through the center of the cell at right angles to the spindle apparatus.

In animal cell mitosis, the centrioles extend a radial array of microtubules toward the plasma membrane when they reach the poles of the cell. This arrangement of microtubules is called an **aster.** Although the aster's function is not fully understood, it probably braces the centrioles against the membrane and stiffens the point of microtubular attachment during the retraction of the spindle. Plant cells, which have rigid cell walls, do not form asters.

Linking Sister Chromatids to Opposite Poles

Each chromosome possesses two kinetochores, one attached to the centromere region of each sister chromatid (see figure 11.11a). As prophase continues, a second group of microtubules appears to grow from the poles of the cell toward the centromeres. These microtubules connect the kinetochores on each pair of sister chromatids to the two poles of the spindle. Because microtubules extending from the two poles attach to opposite sides of the centromere, they attach one sister chromatid to one pole and the other sister chromatid to the other pole. This arrangement is absolutely critical to the process of mitosis; any mistakes in microtubule positioning can be disastrous. The attachment of the two sides of a centromere to the same pole, for example, leads to a failure of the sister chromatids to separate, so that they end up in the same daughter cell.

Prophase is the stage of mitosis characterized by chromosome condensation. During prophase, microtubules attach the centromeres joining pairs of sister chromatids to opposite poles of the spindle apparatus.

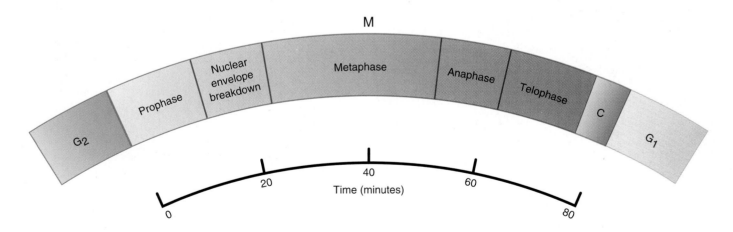

FIGURE 11.12
Time course of mitosis in a typical mammalian cell. The duration of prophase varies considerably in different cell types. All times are shorter for embryonic cells, which divide rapidly.

Metaphase: Division of the Centromeres

The second stage of mitosis, **metaphase,** begins when the chromosomes align in the center of the cell. When viewed with a light microscope, the chromosomes appear to array themselves in a circle along the inner circumference of the cell, as the equator girdles the earth (figure 11.13). An imaginary plane perpendicular to the axis of the spindle that passes through this circle is called the *metaphase plate.* The metaphase plate is not an actual structure, but rather an indication of the future axis of cell division. Positioned by the microtubules attached to the kinetochores of their centromeres, all of the chromosomes line up on the metaphase plate, their centromeres neatly arrayed in a circle, equidistant from the two poles of the cell.

As metaphase ends, the centromeres divide. Each centromere splits in two, freeing the two sister chromatids from each other. The centromeres of all the chromosomes separate simultaneously, but the mechanism that achieves this synchrony is not known.

Metaphase is the stage of mitosis characterized by the alignment of the chromosomes in a ring along the inner circumference of the cell. Each chromosome is drawn to that position by the microtubules extending from its centromere to the two poles of the spindle apparatus. The centromeres divide at the end of metaphase, thus freeing the sister chromatids.

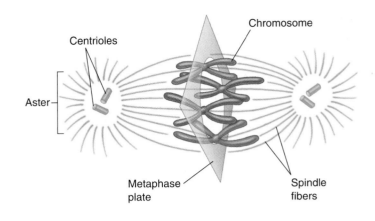

FIGURE 11.13
Metaphase. In metaphase, the chromosomes array themselves in a circle around the spindle midpoint.

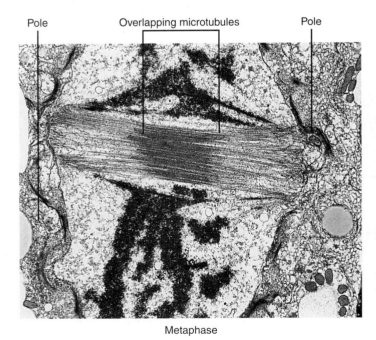

Metaphase

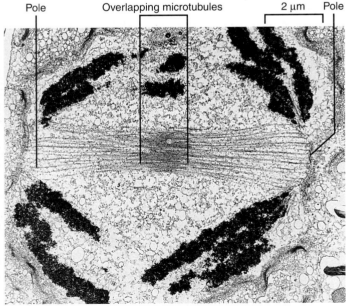

Late anaphase

FIGURE 11.14
Microtubules slide past each other as the chromosomes separate. In these electron micrographs of dividing diatoms, the overlap of the microtubules lessens markedly during spindle elongation as the cell passes from metaphase to anaphase.

Anaphase and Telophase: Separation of the Chromatids and Reformation of the Nuclei

Of all the stages of mitosis, **anaphase** is the shortest and the most beautiful to watch. Freed from each other, the sister chromatids are pulled rapidly toward the poles to which their kinetochores are attached. In the process, two forms of movement take place simultaneously, each driven by microtubules.

First, *the poles move apart* as microtubular spindle fibers physically anchored to opposite poles slide past each other, away from the center of the cell (figure 11.14). Because another group of microtubules attach the chromosomes to the poles, the chromosomes move apart, too. If a flexible membrane surrounds the cell, it becomes visibly elongated.

Second, *the centromeres move toward the poles* as the microtubules that connect them to the poles shorten. This shortening process is not a contraction; the microtubules do not get any thicker. Instead, tubulin subunits are removed from the kinetochore ends of the microtubules by the organizing center. As more subunits are removed, the chromatid-bearing microtubules are progressively disassembled, and the chromatids are pulled ever closer to the poles of the cell.

When the sister chromatids separate in anaphase, the accurate partitioning of the replicated genome—the essential element of mitosis—is complete. In **telophase**, the spindle apparatus disassembles, as the microtubules are broken down into tubulin monomers that can be used to construct the cytoskeletons of the daughter cells. A nuclear envelope forms around each set of sister chromatids, which can now be called chromosomes, since each has its own centromere. The chromosomes soon begin to uncoil into the more extended form that permits gene expression. One of the early group of genes expressed are the rRNA genes, resulting in the reappearance of the nucleolus. Figure 11.15 highlights the sequence of phases in mitosis.

> **Anaphase is the stage of mitosis characterized by the physical separation of sister chromatids. The poles of the cell are pushed apart by microtubular sliding, and the sister chromatids are drawn to opposite poles by the shortening of the microtubules attached to them. Telophase is the stage of mitosis during which the spindle apparatus assembled during prophase is disassembled, nuclear envelopes are reestablished, and the normal expression of genes present in the chromosomes is reinitiated.**

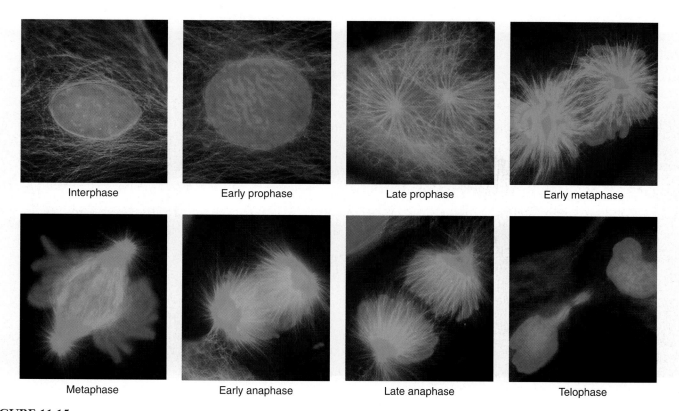

Interphase   Early prophase   Late prophase   Early metaphase

Metaphase   Early anaphase   Late anaphase   Telophase

FIGURE 11.15
Mitosis in an animal cell. The chromosomes in these cultured newt lung cells were stained with a blue fluorescent dye, and the microtubules with a green fluorescent antibody.

Cytokinesis

Mitosis is complete at the end of telophase. The eukaryotic cell has partitioned its replicated genome into two nuclei positioned at opposite ends of the cell. While mitosis was going on, the cytoplasmic organelles, including mitochondria and chloroplasts (if present), were reassorted to areas that will separate and become the daughter cells. The replication of organelles takes place before cytokinesis, often in the S or G_2 phase. Cell division is still not complete at the end of mitosis, however, because the division of the cell proper has not yet begun. The phase of the cell cycle when the cell actually divides is called **cytokinesis.** It generally involves the cleavage of the cell into roughly equal halves.

Cytokinesis in Animal Cells

In animal cells and the cells of all other eukaryotes that lack cell walls, cytokinesis is achieved by means of a constricting belt of actin filaments. As these filaments slide past one another, the diameter of the belt decreases, pinching the cell and creating a *cleavage furrow* around the cell's circumference (figure 11.16*a*). As constriction proceeds, the furrow deepens until it eventually slices all the way into the center of the cell. At this point, the cell is divided in two (figure 11.16*b*).

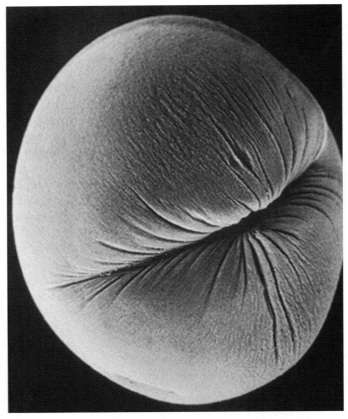

(a)

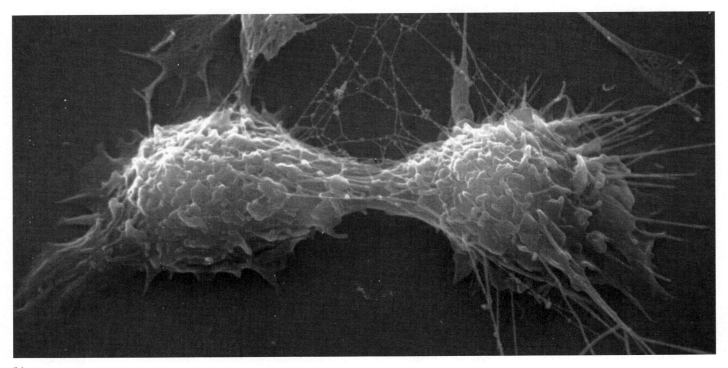

(b)

FIGURE 11.16
Cytokinesis in animal cells. (a) A cleavage furrow forms around a dividing sea urchin egg (30×). (b) The completion of cytokinesis in an animal cell. The two daughter cells are still joined by a thin band of cytoplasm occupied largely by microtubules.

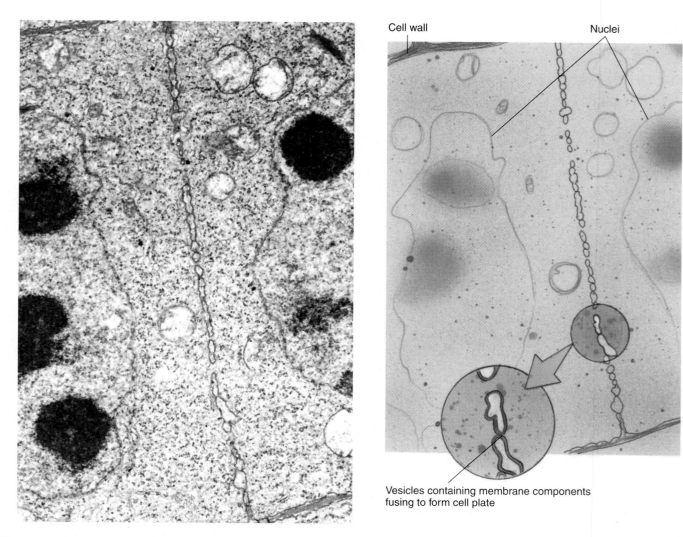

Cell wall

Nuclei

Vesicles containing membrane components fusing to form cell plate

FIGURE 11.17

Cytokinesis in plant cells. In this photograph and companion drawing, a cell plate is forming between daughter nuclei. Once the plate is complete, there will be two cells.

Cytokinesis in Plant Cells

Plant cells possess a cell wall far too rigid to be squeezed in two by actin filament. Instead, these cells assemble membrane components in their interior, at right angles to the spindle apparatus (figure 11.17). This expanding membrane partition, called a **cell plate,** continues to grow outward until it reaches the interior surface of the plasma membrane and fuses with it, effectively dividing the cell in two. Cellulose is then laid down on the new membranes, creating two new cell walls. The space between the daughter cells becomes impregnated with pectins and is called a **middle lamella.**

Cytokinesis in Fungi and Protists

In fungi and some groups of protists, the nuclear membrane does not dissolve and, as a result, all the events of mitosis occurs entirely *within* the nucleus. Only after mitosis is complete in these organisms does the nucleus then divide into two daughter nuclei, and one nucleus goes to each daughter cell during cytokinesis. This separate nuclear division phase of the cell cycle does not occur in plants, animals, or most protists.

After cytokinesis in any eukaryotic cell, the two daughter cells contain all of the components of a complete cell. While mitosis ensures that both daughter cells contain a full complement of chromosomes, no similar mechanism ensures that organelles such as mitochondria and chloroplasts are distributed equally between the daughter cells. However, as long as some of each organelle are present in each cell, the organelles can replicate to reach the number appropriate for that cell.

Cytokinesis is the physical division of the cytoplasm of a eukaryotic cell into two daughter cells.

General Strategy of Cell Cycle Control

The events of the cell cycle are coordinated in much the same way in all eukaryotes. The control system human cells utilize first evolved among the protists over a billion years ago; today, it operates in essentially the same way in fungi as it does in humans.

The goal of controlling any cyclic process is to adjust the duration of the cycle to allow sufficient time for all events to occur. In principle, a variety of methods can achieve this goal. For example, an internal "clock" can be employed to allow adequate time for each phase of the cycle to be completed. This is how many organisms control their daily activity cycles. The disadvantage of using such a clock to control the cell cycle is that it is not very flexible. One way to achieve a more flexible and sensitive regulation of a cycle is simply to let the completion of each phase of the cycle trigger the beginning of the next phase, as a runner passing a baton starts the next leg in a relay race. Until recently, biologists thought this type of mechanism controlled the cell division cycle. However, we now know that eukaryotic cells employ a separate, centralized controller to regulate the process: at critical points in the cell cycle, further progress depends upon a central set of "go/no-go" switches that are regulated by feedback from the cell.

This mechanism is the same one engineers use to control many processes. For example, the furnace that heats a home in the winter typically goes through a daily heating cycle. When the daily cycle reaches the morning "turn on" checkpoint, sensors report whether the house temperature is below the set point (for example, 70° F). If it is, the thermostat triggers the furnace, which warms the house. If the house is already at least that warm, the thermostat does not start up the furnace. Similarly, the cell cycle has key checkpoints where feedback signals from the cell about its size and the condition of its chromosomes can either trigger subsequent phases of the cycle, or delay them to allow more time for the current phase to be completed.

Architecture of the Control System

Three principal checkpoints control the cell cycle in eukaryotes (figure 11.18):

Cell growth is assessed at the G_1 checkpoint. Located near the end of G_1, just before entry into S phase, this checkpoint makes the key decision of whether the cell should divide, delay division, or enter a resting stage (figure 11.19). In yeasts, where researchers first studied this checkpoint, it is called START. If conditions are favorable for division, the cell begins to copy its DNA, initiating S phase. The G_1 checkpoint is where the more complex

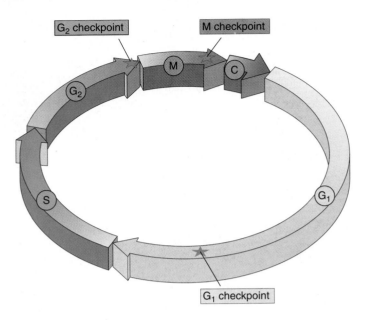

FIGURE 11.18
Control of the cell cycle. Cells use a centralized control system to check whether proper conditions have been achieved before passing three key "checkpoints" in the cell cycle.

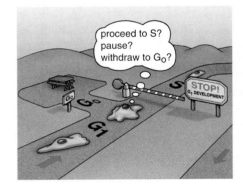

**FIGURE 11.19
The G_1 checkpoint.** Feedback from the cell determines whether the cell cycle will proceed to the S phase, pause, or withdraw into G_0 for an extended rest period.

eukaryotes typically arrest the cell cycle if environmental conditions make cell division impossible, or if the cell passes into G_0 for an extended period.

The success of DNA replication is assessed at the G_2 checkpoint. The second checkpoint, which occurs at the end of G_2, triggers the start of M phase. If this checkpoint is passed, the cell initiates the many molecular processes that signal the beginning of mitosis.

Mitosis is assessed at the M checkpoint. Occurring at metaphase, the third checkpoint triggers the exit from mitosis and cytokinesis and the beginning of G_1.

The cell cycle is controlled at three checkpoints.

Molecular Mechanisms of Cell Cycle Control

Exactly how does a cell achieve central control of the division cycle? The basic mechanism is quite simple. A set of proteins sensitive to the condition of the cell interact at the checkpoints to trigger the next events in the cycle. Two key types of proteins participate in this interaction: cyclin-dependent protein kinases and cyclins (figure 11.20).

The Cyclin Control System

Cyclin-dependent protein kinases (Cdks) are enzymes that phosphorylate (add phosphate groups to) the serine and threonine amino acids of important cellular enzymes and other proteins. At the G_2 checkpoint, for example, Cdks phosphorylate histones, nuclear membrane filaments, and the microtubule-associated proteins that form the mitotic spindle. Phosphorylation of these components of the cell division machinery initiates activities that carry the cycle past the checkpoint into mitosis.

   Cyclins are proteins that bind to Cdks, enabling the Cdks to function as enzymes. Cyclins are so named because they are destroyed and resynthesized during each turn of the cell cycle (figure 11.21). Different cyclins regulate the interphase two cell cycle checkpoints.

The G_2 Checkpoint.   During G_2, the cell gradually accumulates G_2 cyclin (also called mitotic cyclin). This cyclin binds to Cdk to form a complex called MPF (mitosis-promoting factor). At first, MPF is not active in carrying the cycle past the G_2 checkpoint. But eventually, other cellular enzymes phosphorylate and so activate a few molecules of MPF. These activated MPFs in turn increase the activity of the enzymes that phosphorylate MPF, setting up a positive feedback that leads to a very rapid increase in the cellular concentration of activated MPF. When the level of activated MPF exceeds the threshold necessary to trigger mitosis, G_2 phase ends.

   MPF sows the seeds of its own destruction. The length of time the cell spends in M phase is determined by the activity of MPF, for one of its many functions is to activate proteins that destroy cyclin. As mitosis proceeds to the end of metaphase, Cdk levels stay relatively constant, but increasing amounts of G_2 cyclin are degraded, causing progressively less MPF to be available and so initiating the events that end mitosis. After mitosis, the gradual accumulation of new cyclin starts the next turn of the cell cycle.

The G_1 Checkpoint.   The G_1 checkpoint is thought to be regulated in a similar fashion. In unicellular eukaryotes such as yeasts, the main factor triggering DNA replication is cell size. Yeast cells grow and divide as rapidly as possible, and they make the START decision by comparing the volume of cytoplasm to the size of the genome. As a cell grows, its cytoplasm increases in size, while the amount of DNA remains constant. Eventually a threshold ratio is reached that promotes the production of cyclins and thus triggers the next round of DNA replication and cell division.

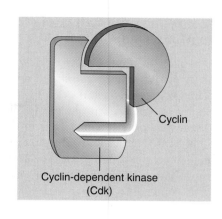

FIGURE 11.20
A complex of two proteins triggers passage through cell cycle checkpoints. Cdk is a protein kinase that activates numerous cell proteins by phosphorylating them. Cyclin is a regulatory protein required to activate Cdk; in other words, Cdk does not function unless cyclin is bound to it.

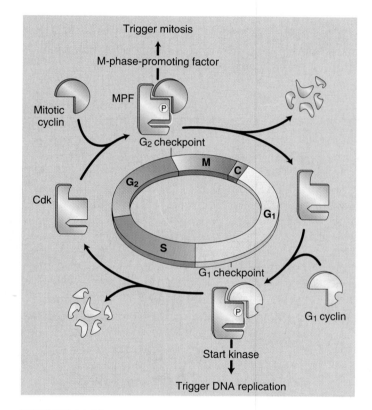

FIGURE 11.21
How cell cycle control works. As the cell cycle passes through the G_1 and G_2 checkpoints, Cdk becomes associated with different cyclins and, as a result, activates different cellular processes. At the completion of each phase, the cyclins are degraded, bringing Cdk activity to a halt until the next set of cyclins appears.

Controlling the Cell Cycle in Multicellular Eukaryotes

The cells of multicellular eukaryotes are not free to make individual decisions about cell division, as yeast cells are. The body's organization cannot be maintained without severely limiting cell proliferation, so that only certain cells divide, and only at appropriate times. The way that cells inhibit individual growth of other cells is apparent in mammalian cells growing in tissue culture: a single layer of cells expands over a culture plate until the growing border of cells comes into contact with neighboring cells, and then the cells stop dividing. If a sector of cells is cleared away, neighboring cells rapidly refill that sector and then stop dividing again. How are cells able to sense the density of the cell culture around them? Each growing cell apparently takes up minute amounts of positive regulatory signals called **growth factors,** proteins that stimulate cell division (such as MPF; figure 11.22). When neighboring cells have taken up what little growth factor is present, not enough is left to trigger cell division in any one cell.

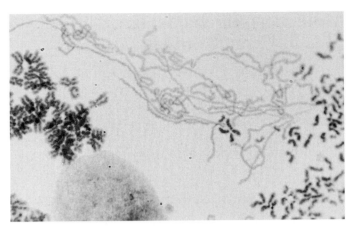

FIGURE 11.22
Evidence that MPF controls the entrance to M phase. When a mitotic cell is fused with a G_1 interphase cell whose chromosomes are still single unreplicated chromatids, the unreplicated chromatids condense anyway, driven into mitosis by a regulatory protein called mitosis-promoting factor (MPF) contributed by the mitotic cell.

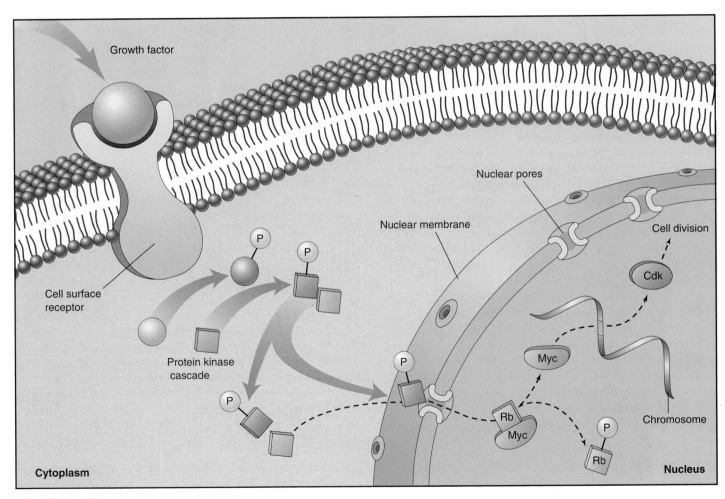

FIGURE 11.23
The cell proliferation-signaling pathway. Binding of a growth factor sets in motion a cascading intracellular signaling pathway (described in chapter 7), which activates nuclear regulatory proteins that trigger cell division. In this example, when the nuclear protein Rb is phosphorylated, another nuclear protein (Myc) is released and is then able to stimulate the production of Cdk proteins.

Growth Factors and the Cell Cycle

As you may recall from chapter 7 (cell-cell interactions), growth factors work by triggering intracellular signaling systems. Fibroblasts, for example, possess numerous receptors on their plasma membranes for one of the first growth factors to be identified: platelet-derived growth factor (PDGF). When PDGF binds to a membrane receptor, it initiates an amplifying chain of internal cell signals that stimulates cell division. PDGF was discovered when investigators found that fibroblasts would grow and divide in tissue culture only if the growth medium contained blood serum (the liquid that remains after blood clots); blood plasma (blood from which the cells have been removed without clotting) would not work. The researchers hypothesized that platelets in the blood clots were releasing into the serum one or more factors required for fibroblast growth. Eventually, they isolated such a factor and named it PDGF. Growth factors such as PDGF override cellular controls that otherwise inhibit cell division. When a tissue is injured, a blood clot forms and the release of PDGF triggers neighboring cells to divide, helping to heal the wound. Only a tiny amount of PDGF (approximately 10^{-10} M) is required to stimulate cell division.

Characteristics of Growth Factors.

Over 50 different proteins that function as growth factors have been isolated (table 11.2 lists a few), and more undoubtedly exist. A specific cell surface receptor "recognizes" each growth factor, its shape fitting that growth factor precisely. When the growth factor binds with its receptor, the receptor reacts by triggering events within the cell (figure 11.23). The cellular selectivity of a particular growth factor depends upon which target cells bear its unique receptor. Some growth factors, like PDGF and epidermal growth factor (EGF), affect a broad range of cell types, while others affect only specific types. For example, nerve growth factor (NGF) promotes the growth of certain classes of neurons, and erythropoietin triggers cell division in red blood cell precursors. Most animal cells need a combination of several different growth factors to overcome the various controls that inhibit cell division.

The G_0 Phase.

If cells are deprived of appropriate growth factors, they stop at the G_1 checkpoint of the cell cycle. With their growth and division arrested, they remain in the G_0 phase, as we discussed earlier. This nongrowing state is distinct from the interphase stages of the cell cycle, G_1, S, and G_2.

It is the ability to enter G_0 that accounts for the incredible diversity seen in the length of the cell cycle among different tissues. Epithelial cells lining the gut divide more than twice a day, constantly renewing the lining of the digestive tract. By contrast, liver cells divide only once every year or two, spending most of their time in G_0 phase. Mature neurons and muscle cells usually never leave G_0.

Two groups of proteins, cyclins and Cdks, interact to regulate the cell cycle. Cells also receive protein signals called growth factors that affect cell division.

Table 11.2	Growth Factors of Mammalian Cells	
Growth Factor	**Range of Specificity**	**Effects**
Epidermal growth factor (EGF)	Broad	Stimulates cell proliferation in many tissues; plays a key role in regulating embryonic development
Erythropoietin	Narrow	Required for proliferation of red blood cell precursors and their maturation into erythrocytes (red blood cells)
Fibroblast growth factor (FGF)	Broad	Initiates the proliferation of many cell types; inhibits maturation of many types of stem cells; acts as a signal in embryonic development
Insulin-like growth factor	Broad	Stimulates metabolism of many cell types; potentiates the effects of other growth factors in promoting cell proliferation
Interleukin-2	Narrow	Triggers the division of activated T lymphocytes during the immune response
Mitosis-promoting factor (MPF)	Broad	Regulates entrance of the cell cycle into the M phase
Nerve growth factor (NGF)	Narrow	Stimulates the growth of neuron processes during neural development
Platelet-derived growth factor (PDGF)	Broad	Promotes the proliferation of many connective tissues and some neuroglial cells
Transforming growth factor β (TGF-β)	Broad	Accentuates or inhibits the responses of many cell types to other growth factors; often plays an important role in cell differentiation

Cancer and the Control of Cell Proliferation

The unrestrained, uncontrolled growth of cells, called cancer, is addressed more fully in chapter 17. However, cancer certainly deserves mention in a chapter on cell division, as it is essentially a disease of cell division—a failure of cell division *control*. Recent work has identified one of the culprits. Working independently, scientists researching such diverse fields as genetics, molecular biology, cell biology, and cancer have repeatedly identified what has proven to be the same gene! Officially dubbed *p53* (researchers italicize the gene symbol to differentiate it from the protein) and popularly referred to as the "Guardian Angel gene," this gene plays a key role in the G_1 checkpoint of cell division. The gene's product, the p53 protein, monitors the integrity of DNA, checking that it has been successfully replicated and is undamaged. If the p53 protein detects damaged DNA, it halts cell division and stimulates the activity of special enzymes to repair the damage. Once the DNA has been repaired, *p53* allows cell division to continue. In cases where the DNA is irreparable, *p53* then directs the cell to kill itself, activating an apoptosis (cell suicide) program (see chapter 57 for a discussion of apoptosis).

By halting division in damaged cells, *p53* prevents the development of many mutated cells, and it is therefore considered a tumor-suppressor gene (even though its activities are not limited to cancer prevention). Scientists have found that *p53* is entirely absent or damaged beyond use in the majority of cancerous cells they have examined! It is precisely because *p53* is nonfunctional that these cancer cells are able to repeatedly undergo cell division without being halted at the G_1 checkpoint (figure 11.24). To test this, scientists administered healthy p53 protein to rapidly dividing cancer cells in a petri dish: the cells soon ceased dividing and died.

Scientists at Johns Hopkins University School of Medicine have further reported that cigarette smoke causes mutations in the *p53* gene. This study, published in 1995, reinforced the strong link between smoking and cancer described in chapter 17.

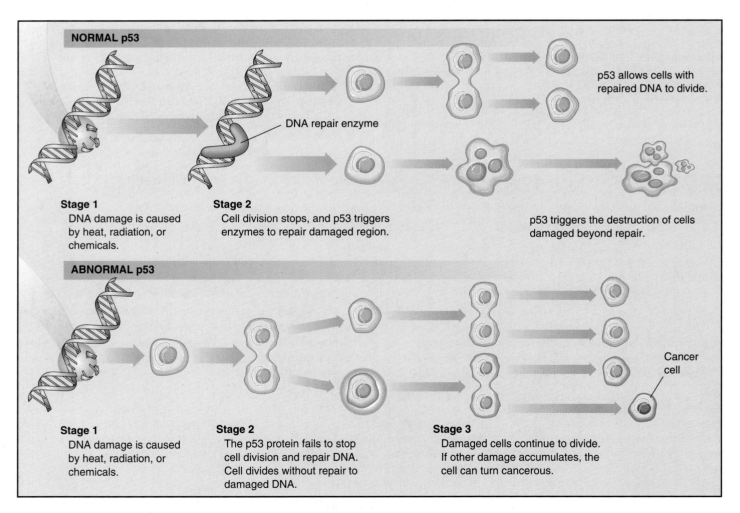

NORMAL p53

Stage 1
DNA damage is caused by heat, radiation, or chemicals.

DNA repair enzyme

Stage 2
Cell division stops, and p53 triggers enzymes to repair damaged region.

p53 allows cells with repaired DNA to divide.

p53 triggers the destruction of cells damaged beyond repair.

ABNORMAL p53

Stage 1
DNA damage is caused by heat, radiation, or chemicals.

Stage 2
The p53 protein fails to stop cell division and repair DNA. Cell divides without repair to damaged DNA.

Stage 3
Damaged cells continue to divide. If other damage accumulates, the cell can turn cancerous.

Cancer cell

FIGURE 11.24
Cell division and p53 protein. Normal p53 protein monitors DNA, destroying cells with irreparable damage to their DNA. Abnormal p53 protein fails to stop cell division and repair DNA. As damaged cells proliferate, cancer develops.

Growth Factors and Cancer

How do growth factors influence the cell cycle? As you have seen, there are two different approaches, one positive and the other negative.

Proto-oncogenes. PDGF and many other growth factors utilize the positive approach, stimulating cell division. They trigger passage through the G_1 checkpoint by aiding the formation of cyclins and so activating genes that promote cell division. Genes that normally stimulate cell division are sometimes called *proto-oncogenes* because mutations that cause them to be over-expressed or hyperactive convert them into oncogenes (Greek *onco*, "cancer"), leading to the excessive cell proliferation that is characteristic of cancer (figure 11.25). Even a single mutation (creating a heterozygote) can lead to cancer if the other cancer-preventing genes are nonfunctional. Geneticists, using Mendel's terms, call such mutations of proto-oncogenes *dominant*.

Some 30 different proto-oncogenes are known. Some act very quickly after stimulation by growth factors. Among the most intensively studied of these are *myc*, *fos*, and *jun*, all of which cause unrestrained cell growth and division when they are overexpressed. In a normal cell, the *myc* proto-oncogene appears to be important in regulating the G_1 checkpoint. Cells in which *myc* expression is prevented will not divide, even in the presence of growth factors. A critical activity of *myc* and other genes in this group of immediately responding proto-oncogenes is to stimulate a second group of "delayed response" genes, including those that produce cyclins and Cdk proteins (figure 11.26).

Tumor-suppressor Genes. Other growth factors utilize a negative approach to cell cycle control. They block passage through the G_1 checkpoint by preventing cyclins from binding to Cdk, thus inhibiting cell division. Genes that normally inhibit cell division are called *tumor-suppressor genes*. When mutated, they can also lead to unrestrained cell division, but only if both copies of the gene are mutant. Hence, these cancer-causing mutations are *recessive*.

The most thoroughly understood of the tumor-suppressor genes is the retinoblastoma (*Rb*) gene. This gene was originally cloned from children with a rare form of eye cancer inherited as a recessive trait, implying that the normal gene product was a cancer suppressor that helped keep cell division in check. The *Rb* gene encodes a protein present in ample amounts within the nucleus. This protein interacts with many key regulatory proteins of the cell cycle, but how it does so depends upon its state of phosphorylation. In G_0 phase, the Rb protein is dephosphorylated. In this

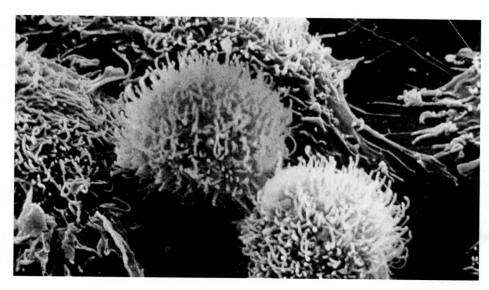

FIGURE 11.25

A cancerous fibroblast. "Cancer" takes its name from the crab-like appearance of cancer cells. Instead of dividing smoothly, these cells extend numerous projections, allowing them to adhere and metastasize (spread) easily.

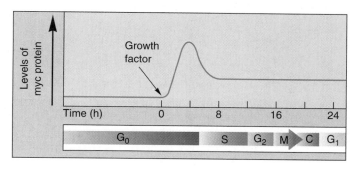

FIGURE 11.26

The role of *myc* in triggering cell division. The addition of a growth factor leads to transcription of the *myc* gene and rapidly increasing levels of the myc protein. This causes G_0 cells to enter the S phase and begin proliferating.

state, it binds to and ties up a set of regulatory proteins, like myc and fos, needed for cell proliferation, blocking their action and so inhibiting cell division (see figure 11.23). When phosphorylated, the Rb protein releases its captive regulatory proteins, freeing them to act and so promoting cell division. Growth factors lessen the inhibition the Rb protein imposes by activating kinases that phosphorylate it. Free of Rb protein inhibition, cells begin to produce cyclins and Cdk, pass the G_1 checkpoint, and proceed through the cell cycle. Figure 11.27 summarizes the types of genes that can cause cancer when mutated.

The progress of mitosis is regulated by the interaction of two key classes of proteins, cyclin-dependent protein kinases and cyclins. Some growth factors accelerate the cell cycle by promoting cyclins and Cdks, others suppress it by inhibiting their action.

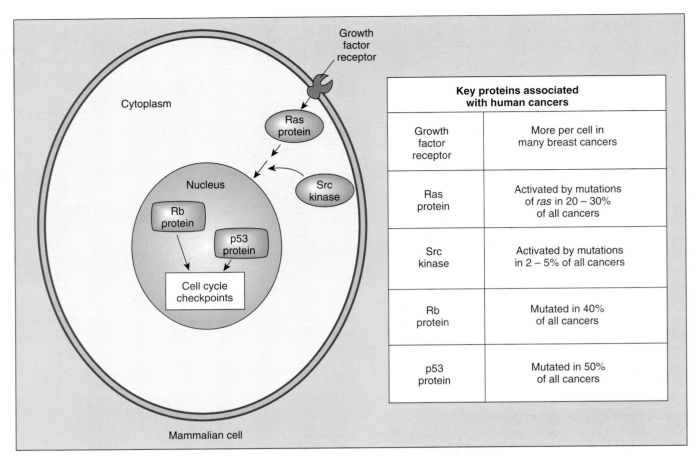

FIGURE 11.27

Mutations cause cancer. Mutations in genes encoding key components of the cell division-signaling pathway are responsible for many cancers. Among them are proto-oncogenes encoding growth factor receptors, such as ras protein, and kinase enzymes, such as src, that aid ras function. Mutations that disrupt tumor-suppressor proteins, such as Rb and p53, also foster cancer development.

A Vocabulary of Cell Division

binary fission Asexual reproduction of a cell by division into two equal or nearly equal parts. Bacteria divide by binary fission.

centromere A constricted region of a chromosome about 220 nucleotides in length, composed of highly repeated DNA sequences (satellite DNA). During mitosis, the centromere joins the two sister chromatids and is the site to which the kinetochores are attached.

chromatid One of the two strands of a replicated chromosome, joined by a single centromere to the other strand.

chromatin The complex of DNA and proteins of which eukaryotic chromosomes are composed.

chromosome The structure within cells that contains the genes. In eukaryotes, it consists of a single linear DNA molecule associated with proteins. The DNA is replicated during S phase, and the replicas separated during M phase.

cytokinesis Division of the cytoplasm of a cell after nuclear division.

euchromatin The portion of a chromosome that is extended except during cell division, and from which RNA is transcribed.

heterochromatin The portion of a chromosome that remains permanently condensed and, therefore, is not transcribed into RNA. Most centromere regions are heterochromatic.

homologues Homologous chromosomes; in diploid cells, one of a pair of chromosomes that carry equivalent genes.

kinetochore A disk of protein bound to the centromere and attached to microtubules during mitosis, linking each chromatid to the spindle apparatus.

microtubule A hollow cylinder, about 25 nanometers in diameter, composed of subunits of the protein tubulin. Microtubules lengthen by the addition of tubulin subunits to their end(s) and shorten by the removal of subunits.

mitosis Nuclear division in which replicated chromosomes separate to form two genetically identical daughter nuclei. When accompanied by cytokinesis, it produces two identical daughter cells.

nucleosome The basic packaging unit of eukaryotic chromosomes, in which the DNA molecule is wound around a cluster of histone proteins. Chromatin is composed of long strings of nucleosomes that resemble beads on a string.

11.1 Bacteria divide far more simply than do eukaryotes.

- Bacterial cells divide by simple binary fission.

- The two replicated circular DNA molecules attach to the plasma membrane at different points, and fission is initiated between those points.

11.2 Chromosomes are highly ordered structures.

- Eukaryotic DNA forms a complex with histones and other proteins and is packaged into chromosomes. Some of the DNA is permanently condensed into heterochromatin, while the rest is condensed only during cell division.

- In eukaryotic cells, DNA replication is completed during the S phase of the cell cycle, and during the G_2 phase the cell makes its final preparation for mitosis.

- Along with G_1, these two phases constitute the portion of the cell cycle called interphase, which alternates with mitosis and cytokinesis.

11.3 Mitosis is a key phase of the cell cycle.

- The first stage of mitosis is prophase, during which the mitotic spindle apparatus forms. At the end of prophase, the nuclear envelope disassembles and microtubules attach each pair of sister chromatids to the two poles of the cell.

- In the second stage of mitosis, metaphase, the chromosomes are arranged in a circle around the periphery of the cell; this circle lies in a plane through the center of the cell at right angles to the spindle axis.

- At the end of metaphase, the centromeres joining each pair of sister chromatids separate, freeing the sister chromatids from each other.

- The third stage of mitosis is anaphase, during which the chromatids physically separate and are pulled to opposite poles of the cell by the microtubules attached to their centromeres.

- In the fourth and final stage of mitosis, telophase, the mitotic apparatus is disassembled, the nuclear envelope re-forms, and the chromosomes uncoil.

- When mitosis is complete, the cell divides in two, so that the two sets of chromosomes separated by mitosis end up in different daughter cells. In animal cells, a central, drawstring-like constriction furrow driven by actin contraction pinches the cell in two. In plant cells, membrane vesicles accumulate in the center of the cell and then fuse to form a membrane plate.

- In plant cells and other cells with a cell wall, an expanding cell plate forms perpendicular to the spindle apparatus along the midline of the cell.

11.4 The cell cycle is carefully controlled.

- The cell cycle is regulated by two types of proteins, cyclins and cyclin-dependent protein kinases, which permit progress past key "checkpoints" in the cell cycle only if the cell is ready to proceed further.

- Failures of cell cycle regulation can lead to uncontrolled cell growth and lie at the root of cancer.

Discussing Key Terms

1. **Binary fission**   Method by which bacteria reproduce simply by duplicating their circular DNA and dividing in half. Bacteria have no complex cell cycle like that shown by eukaryotic cells.

2. **The cell cycle**   Eukaryotic cells go through a controlled cycle of growth (G_1 phase, in which cells spend most of their time), DNA replication (S phase), preparation for cell division (G_2 phase), mitosis (M phase), and cytokinesis (C phase).

3. **Mitosis**   Eukaryotic cells contain much more DNA than bacterial cells. Eukaryotic DNA is packaged into chromosomes, which are replicated before cell division.

The replicated chromosomes, called sister chromatids, are separated in a process called mitosis, and one replica is passed to each end of the cell.

4. **Cytokinesis**   Following mitosis, most cells undergo cytoplasmic cleavage, or cytokinesis. In cells without a cell wall, the cell body is pinched in two by a belt of actin filaments that draw inward around the cell's midsection.

5. **Cell cycle "checkpoints"**   At several "checkpoints" in the cell cycle, further progress depends upon successfully passing critical tests that measure the condition of the cell. This regulation is carried out by a variety of cyclins and cyclin-dependent protein kinases.

Review Questions

1. How is the genome replicated prior to binary fission in a bacterial cell? What mechanism ensures that each daughter cell receives one of the replicated genomes?

2. What are nucleosomes composed of, and how do they participate in the coiling of DNA?

3. What is the difference between heterochromatin and euchromatin in terms of their appearance under the light microscope? What is the difference in terms of gene expression?

4. What is a karyotype? How are chromosomes distinguished from one another in a karyotype?

5. Which phases of the cell cycle are specifically associated with the process of cell division? Which phases are the longest in the cells of embryos? Which phase is generally the longest in the cells of a mature eukaryote?

6. What happens to the chromosomes during S phase? What happens to them during G_2 phase? What other event necessary for cell division also occurs during G_2 phase?

7. What changes with respect to ribosomal RNA occur during prophase? What characteristic structure of the nucleus do these changes affect?

8. What event signals the initiation of metaphase? What event indicates the end of metaphase?

9. What molecular mechanism seems to be responsible for the movement of the poles during anaphase? What molecular mechanism seems to be responsible for the movement of the centromeres during anaphase?

10. Describe three events that occur during telophase.

11. How is cytokinesis in animal cells different from that in plant cells?

12. What aspects of the cell cycle are controlled by the G_1, G_2, and M checkpoints? How are cyclins and cyclin-dependent protein kinases involved in cell cycle regulation at checkpoints?

13. How do proto-oncogenes differ from tumor-suppressor genes in the mechanisms by which they lead to excessive cell proliferation and cancer?

Thought Questions

1. The plant *Haplopappus gracilis* has only two chromosomes, while the adder's tongue fern (*Ophioglossum* spp.) has 1262. There is much less variation in chromosome number among mammals. Can you suggest a reason for the wide variation in chromosome number among plants but not mammals?

2. Colchicine is a poison that binds to tubulin and prevents its assembly into microtubules; cytochalasins are compounds that bind to the ends of actin filaments and prevent their elongation. What effects do you think these two substances would have on cell division in animal cells?

3. If you could construct an artificial chromosome, what elements would you introduce into it, at a minimum, so that it could function normally in mitosis?

Internet Links

Quick Time Mitosis
http://www.botany.utexas.edu/facstaff/facpages/ksata/ecpf96/9/index.html
A quick time movie of mitosis from the University of Texas.

Virtual Cell Division
http://www.biology.uc.edu/ugenetic/
Prepared by the biology staff of the University of Cincinnati, these online lessons provide virtual tutorials on mitosis and meiosis.

Animated Meiosis
http://www.biology.yale.edu/animatedMeiosis.nclk
A movie of meiosis, employing simple diagrams to illustrate the process.

Cancer and the Cell Cycle
http://wsrv.clas.virginia.edu/~rjh9u/cdk.html
A brief, accurate, and up-to-date summary from the University of Virginia of the ways eukaryotic cells control the cell cycle, with detailed information on cyclins and checkpoints.

For Further Reading

Greider, C., and E. Blackburn: "Telomeres, Telomerase, and Cancer," *Scientific American*, February 1996, pages 92–97. Telomerase enzymes act on the ends of chromosomes, shortening them with each cell division.

Hackney, D.: "Polar Explorations," *Nature*, vol. 376, July 20, 1995, pages 215–16. This and three other articles in this issue discuss "motor proteins" in the kinesin superfamily of fibrous proteins. Kinesins are responsible for intracellular movement of organelles and the movements of chromosomes during mitosis.

Koshland, D.: "Mitosis: Back to the Basics," *Cell*, vol. 77, July 1, 1994, pages 951–54. A detailed overview of the entire process.

Murray, A., and M. Kirschner: "What Controls the Cell Cycle?" *Scientific American*, March 1991, pages 56–63. How cell division is controlled. One protein plays a key role in virtually all organisms.

Peters, G.: "The Cell Cycle: Stifled by Inhibitions," *Nature*, vol. 371, September 15, 1994, pages 204–5. A discussion of Cdks and cyclins and their effects on the cell cycle.

Pluta, A., and others: "The Centromere: Hub of Chromosomal Activities," *Science*, vol. 270, December 8, 1995, pages 1591–94. A current review of our rapidly expanding knowledge of centromeres.

"What You Need to Know About Cancer," *Scientific American*, September 1996. An entire issue devoted to what is currently known of cancer's cause, detection, treatment, and prevention.

12

Sexual Reproduction and Meiosis

Concept Outline

12.1 Meiosis produces sexual gametes.

Discovery of Reduction Division. Sexual reproduction does not increase chromosome number because gamete production by meiosis involves a decrease in chromosome number.

The Sexual Life Cycle. Sexual individuals inherit chromosomes from two parents.

12.2 Meiosis involves two nuclear divisions.

Unique Features of Meiosis. Two unique features of meiosis are synapsis and reduction division.

Prophase I. Homologous chromosomes pair intimately, and undergo crossing over that locks them together.

Metaphase I. Spindle microtubules align the chromosomes in the central plane of the cell.

Completing Meiosis. The second meiotic division is like a mitosis, but has a very different outcome.

12.3 The evolution of sex led to increased genetic variability.

Why Sex? Sex appears to have evolved as a mechanism to repair DNA.

The Evolutionary Consequences of Sex. Sexual reproduction increases genetic variability by shuffling combinations of genes.

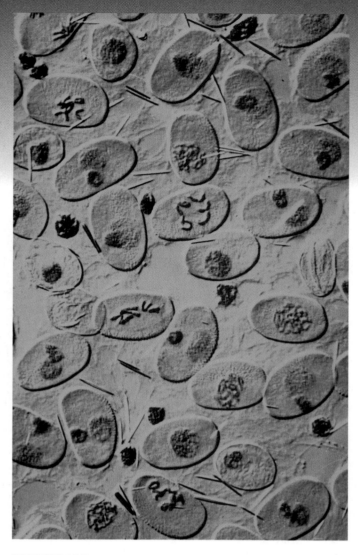

FIGURE 12.1
Plant cells undergoing meiosis (600×). This preparation of pollen cells of a spiderwort, *Tradescantia*, was made by freezing the cells and then fracturing them. It shows several stages of meiosis.

Most animals and plants reproduce sexually. Gametes of opposite sex unite to form a cell that, dividing repeatedly by mitosis, eventually gives rise to an adult body with some 100 trillion cells. The gametes that give rise to the initial cell are the products of a special form of cell division called meiosis (figure 12.1), the subject of this chapter. Far more intricate than mitosis, the details of meiosis are not as well understood. The basic process, however, is clear. Also clear are the profound consequences of sexual reproduction: it plays a key role in generating the tremendous genetic diversity that is the raw material of evolution.

Discovery of Reduction Division

Only a few years after Walther Fleming's discovery of chromosomes in 1882, Belgian cytologist Pierre-Joseph van Beneden was surprised to find different numbers of chromosomes in different types of cells in the roundworm *Ascaris*. Specifically, he observed that the **gametes** (eggs and sperm) each contained two chromosomes, while the **somatic** (nonreproductive) cells of embryos and mature individuals each contained four.

Fertilization

From his observations, van Beneden proposed in 1887 that an egg and a sperm, each containing half the complement of chromosomes found in other cells, fuse to produce a single cell called a **zygote**. The zygote, like all of the somatic cells ultimately derived from it, contains two copies of each chromosome. The fusion of gametes to form a new cell is called **fertilization,** or **syngamy**.

Reduction Division

It was clear even to early investigators that gamete formation must involve some mechanism that reduces the number of chromosomes to half the number found in other cells. If it did not, the chromosome number would double with each fertilization, and after only a few generations, the number of chromosomes in each cell would become impossibly large. For example, in just 10 generations, the 46 chromosomes present in human cells would increase to over 47,000 (46×2^{10}).

The number of chromosomes does not explode in this way because of a special reduction division that occurs during gamete formation, producing cells with half the normal number of chromosomes. The subsequent fusion of two of these cells ensures a consistent chromosome number from one generation to the next. This reduction division process, known as **meiosis,** is the subject of this chapter.

> **Meiosis is a process of cell division in which the number of chromosomes in certain cells is halved during gamete formation.**

FIGURE 12.2

Sexual *and* asexual reproduction. Not all organisms reproduce exclusively asexually or sexually; some do both. The strawberry reproduces both asexually (runners) and sexually (flowers).

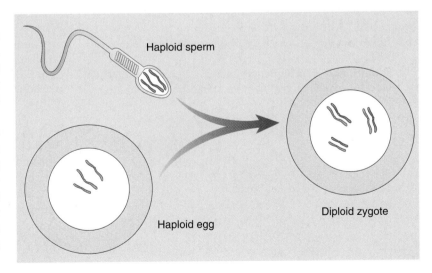

FIGURE 12.3

Diploid cells carry chromosomes from two parents. A diploid cell contains two versions of each chromosome, one contributed by the haploid egg of the mother, the other by the haploid sperm of the father.

The Sexual Life Cycle

Meiosis and fertilization together constitute a cycle of reproduction. Two sets of chromosomes are present in the somatic cells of adult individuals, making them **diploid** cells (Greek *di*, "two"), but only one set is present in the gametes, which are thus **haploid** (Greek *haploos*, "one"). Reproduction that involves this alternation of meiosis and fertilization is called **sexual reproduction** (figure 12.2). Its outstanding characteristic is that offspring inherit chromosomes from *two* parents (figure 12.3). You, for example, inherited 23 chromosomes from your mother, contributed by the egg fertilized at your conception, and 23 from your father, contributed by the sperm that fertilized that egg.

Somatic Tissues

The life cycles of all sexually reproducing organisms follow the same basic pattern of alternation between the diploid and haploid chromosome numbers (figures 12.4 and 12.5). After fertilization, the resulting zygote begins to divide by mitosis. This single diploid cell eventually gives rise to all of the cells in the adult. These cells are called **somatic** cells, from the Latin word for "body." Except when rare accidents occur, or in special variation-creating situations such as occur in the immune system, every one of the adult's somatic cells is genetically identical to the zygote.

In unicellular eukaryotic organisms, including most protists, individual cells function as gametes, fusing with other gamete cells. The zygote may undergo mitosis, or it may divide immediately by meiosis to give rise to haploid individuals. In plants, the haploid cells that meiosis produces divide by mitosis, forming a multicellular haploid phase. Certain cells of this haploid phase eventually differentiate into eggs or sperm.

Germ-Line Tissues

In animals, the cells that will eventually undergo meiosis to produce gametes are set aside from somatic cells early in the course of development. These cells are often referred to as **germ-line** cells. Both the somatic cells and the gamete-producing germ-line cells are diploid, but while *somatic cells undergo mitosis* to form genetically identical, diploid daughter cells, gamete-producing *germ-line cells undergo meiosis*, producing haploid gametes.

In the sexual life cycle, there is an alternation of diploid and haploid generations.

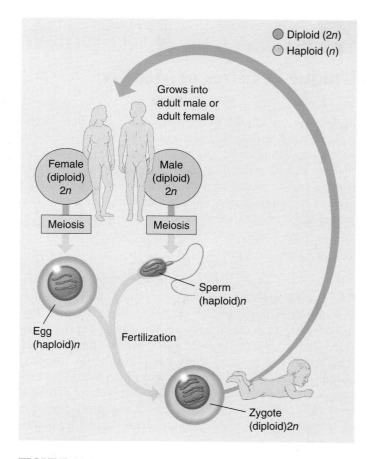

FIGURE 12.4
The sexual life cycle. In animals, the completion of meiosis is followed soon by fertilization. Thus, the vast majority of the life cycle is spent in the diploid stage.

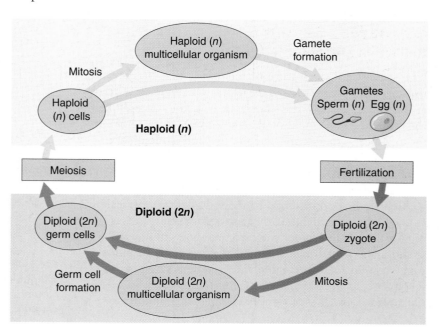

FIGURE 12.5
Alternation of generations. In sexual reproduction, haploid cells or organisms alternate with diploid cells or organisms.

Unique Features of Meiosis

The mechanism of cell division varies in important details in different organisms. This is particularly true of chromosomal separation mechanisms, which differ substantially in protists and fungi from the process in plants and animals that we will describe here. Meiosis in a diploid organism consists of two rounds of division, mitosis of one. Although meiosis and mitosis have much in common, meiosis has two unique features: synapsis and reduction division.

Synapsis

The first of these two features happens early during the first nuclear division. Following chromosome replication, *homologous chromosomes*, or *homologues* (see chapter 11), *pair all along their length, and genetic exchange occurs between them* while they are thus physically joined (figure 12.6*a*). The process of forming these complexes of homologous chromosomes is called **synapsis**, and the exchange process that occurs between paired chromosomes is called **crossing over**. Chromosomes are then drawn together along the equatorial plane of the dividing cell; subsequently, homologues are pulled by microtubules toward opposite poles of the cell. When this process is complete, the cluster of chromosomes at each pole contains one of the two homologues of each chromosome. Each pole is haploid, containing half the number of chromosomes present in the original diploid cell. Sister chromatids do not separate from each other in the first nuclear division, so each homologue is still composed of two chromatids.

Reduction Division

The second unique feature of meiosis is that *the chromosomes do not replicate between the two nuclear divisions*, so that at the end of the two divisions each cell contains only half the original complement of chromosomes (figure 12.6*b*). In most respects, the second meiotic division is identical to a normal mitotic division. However, because of the crossing over that occurred during the first division, the sister chromatids in meiosis II are not identical to each other.

Meiosis is a continuous process, but it is most easily studied when we divide it into arbitrary stages, just as we did for mitosis in chapter 11. The two stages of meiosis are traditionally called meiosis I and meiosis II. Like mitosis, each stage of meiosis is subdivided further into prophase, metaphase, anaphase, and telophase (figure 12.7). In meiosis, however, prophase I is more complex than in mitosis.

In meiosis, homologous chromosomes become intimately associated and do not replicate between the two nuclear divisions.

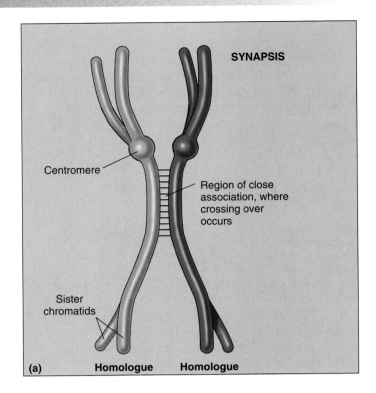

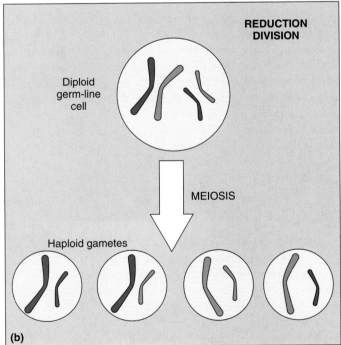

FIGURE 12.6
Unique features of meiosis. (a) Synapsis draws homologous chromosomes together, creating a situation where the two chromosomes can physically exchange arms, a process called crossing over. (b) Reduction division, by omitting a chromosome duplication before meiosis II, produces haploid gametes, thus ensuring that chromosome number remains stable during the reproduction cycle.

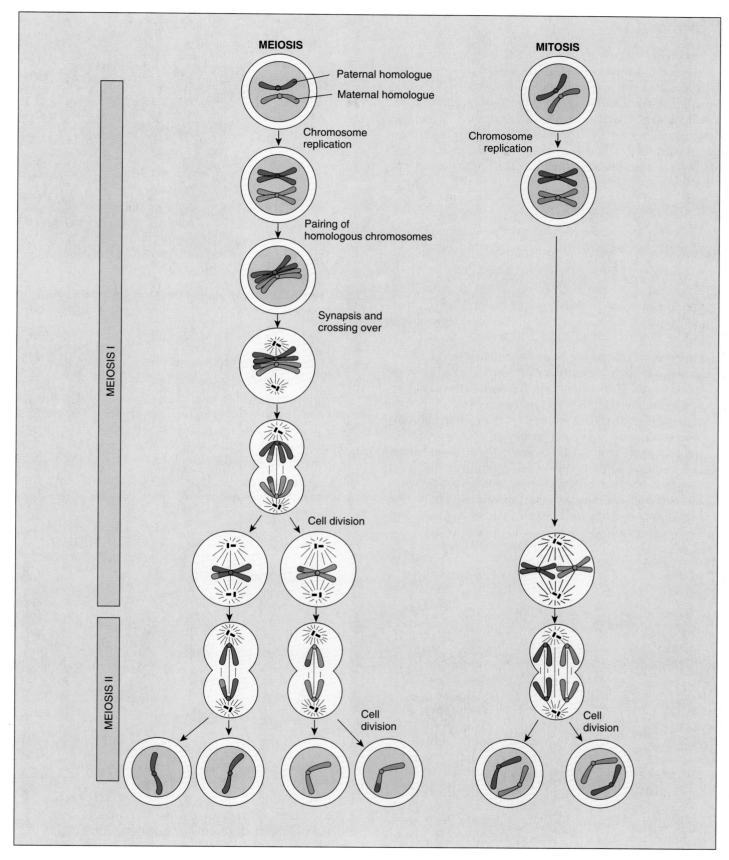

FIGURE 12.7

A comparison of meiosis and mitosis. Meiosis involves two nuclear divisions with no DNA replication between them. It thus produces four daughter cells, each with half the original number of chromosomes. Crossing over occurs in prophase I of meiosis. Mitosis involves a single nuclear division after DNA replication. It thus produces two daughter cells, each containing the original number of chromosomes.

Prophase I

In prophase I of meiosis, the DNA coils tighter, and individual chromosomes first become visible under the light microscope as a matrix of fine threads. Since the DNA has already replicated before the onset of meiosis, each of these threads actually consists of two sister chromatids joined at their centromeres. In prophase I, homologous chromosomes become closely associated in synapsis, exchange segments by crossing over, and then separate.

An Overview

Prophase I is traditionally divided into five sequential stages: leptotene, zygotene, pachytene, diplotene, and diakinesis.

Leptotene. Chromosomes condense tightly.

Zygotene. A lattice of protein is laid down between the homologous chromosomes in the process of synapsis, forming a structure called a *synaptonemal complex* (figure 12.8).

Pachytene. Pachytene begins when synapsis is complete (just after the synaptonemal complex form; figure 12.9), and lasts for days. This complex, about 100 nm across, holds the two replicated chromosomes in precise register, keeping each gene directly across from its partner on the homologous chromosome, like the teeth of a zipper. Within the synaptonemal complex, the DNA duplexes unwind at certain sites, and single strands of DNA form base-pairs with complementary strands *on the other homologue*. The synaptonemal complex thus provides the structural framework that enables crossing over between the homologous chro-

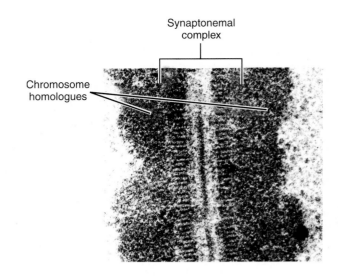

FIGURE 12.8
Structure of the synaptonemal complex. A portion of the synaptonemal complex of the ascomycete *Neotiella rutilans*, a cup fungus.

mosomes. As you will see, this has a key impact on how the homologues separate later in meiosis.

Diplotene. At the beginning of diplotene, the protein lattice of the synaptonemal complex disassembles. Diplotene is a period of intense cell growth. During this period the chromosomes decondense and become very active in transcription.

Diakinesis. At the beginning of diakinesis, the transition into metaphase, transcription ceases and the chromosomes recondense.

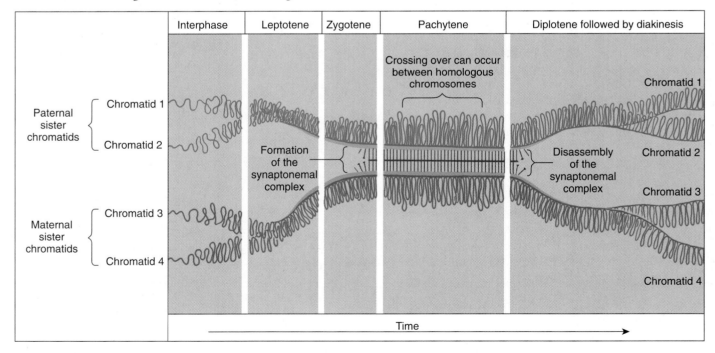

FIGURE 12.9
Time course of prophase I. The five stages of prophase I represent stages in the formation and subsequent disassembly of the synaptonemal complex, the protein lattice that holds homologous chromosomes together during synapsis.

Synapsis

During prophase, the ends of the chromatids attach to the nuclear envelope at specific sites. The sites the homologues attach to are adjacent, so that the members of each homologous pair of chromosomes are brought close together. They then line up side by side, apparently guided by heterochromatin sequences, in the process called synapsis.

Crossing Over

Within the synaptonemal complex recombination is thought to be carried out during pachytene by very large protein assemblies called **recombination nodules**. A nodule's diameter is about 90 nm, spanning the central element of the synaptonemal complex. Spaced along the synaptonemal complex, these recombination nodules act as large multienzyme "recombination machines," each nodule bringing about a recombination event. The details of the crossing over process are not well understood, but involve a complex series of events in which DNA segments are exchanged between nonsister or sister chromatids. In humans, an average of two or three such crossover events occur per chromosome pair.

When crossing over is complete, the synaptonemal complex breaks down, and the homologous chromosomes are released from the nuclear envelope and begin to move away from each other. At this point, there are four chromatids for each type of chromosome (two homologous chromosomes, each of which consists of two sister chromatids). The four chromatids do not separate completely, however, because they are held together in two ways: (1) the two sister chromatids of each homologue, recently created by DNA replication, are held near by their common centromeres; and (2) the paired homologues are held together at the points where crossing over occurred within the synaptonemal complex.

Chiasma Formation

Evidence of crossing over can often be seen under the light microscope as an X-shaped structure known as a **chiasma** (Greek, "cross;" plural, **chiasmata;** figure 12.10). The presence of a chiasma indicates that two chromatids (one from each homologue) have exchanged parts (figure 12.11). Like small rings moving down two strands of rope, the chiasmata move to the end of the chromosome arm as the homologous chromosomes separate.

FIGURE 12.10
Chiasmata. This micrograph shows two distinct crossovers, or chiasmata.

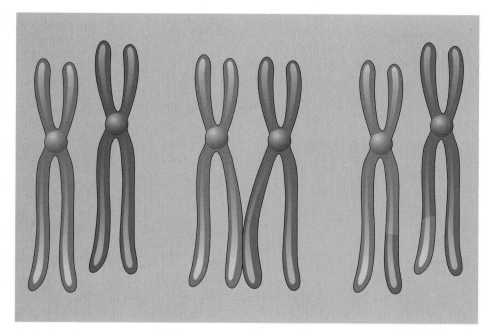

FIGURE 12.11
The results of crossing over. During crossing over, nonsister (shown above) or sister chromatids may exchange segments.

> Synapsis is the close pairing of homologous chromosomes that takes place early in prophase I of meiosis. Crossing over occurs between the paired DNA strands, creating the chromosomal configurations known as chiasmata. The two homologues are locked together by these exchanges and they do not disengage readily.

Metaphase I

By the second stage of meiosis I, the nuclear envelope has dispersed and the microtubules form a spindle, just as in mitosis. During diakinesis of prophase I, the chiasmata move down the paired chromosomes from their original points of crossing over, eventually reaching the ends of the chromosomes. At this point, they are called terminal chiasmata. Terminal chiasmata hold the homologous chromosomes together in metaphase I, so that only one side of each centromere faces outward from the complex; the other side is turned inward toward the other homologue (figure 12.12). Consequently, spindle microtubules are able to attach to kinetochore proteins only on the outside of each centromere, and the centromeres of the two homologues attach to microtubules originating from opposite poles. This one-sided attachment is in marked contrast to the attachment in mitosis, when kinetochores on *both* sides of a centromere bind to microtubules.

Each joined pair of homologues then lines up on the metaphase plate. The orientation of each pair on the spindle axis is random: either the maternal or the paternal homologue may orient toward a given pole (figure 12.13). Figure 12.14 illustrates the alignment of chromosomes during metaphase I.

Chiasmata play an important role in aligning the chromosomes on the metaphase plate.

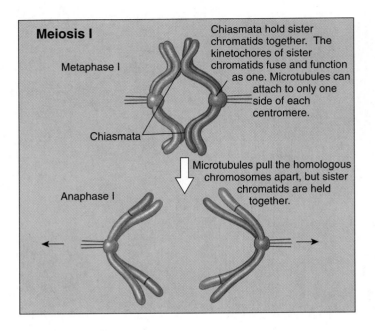

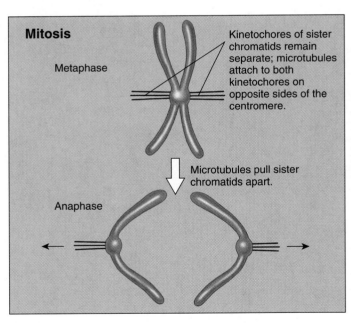

FIGURE 12.12
Chiasmata created by crossing over have a key impact on how chromosomes align in metaphase I. In the first meiotic division, the chiasmata hold one sister chromatid to the other sister chromatid; consequently, the spindle microtubules can bind to only one side of each centromere, and the homologous chromosomes are drawn to opposite poles. In mitosis, microtubules attach to *both* sides of each centromere; when the microtubules shorten, the sister chromatids are split and drawn to opposite poles.

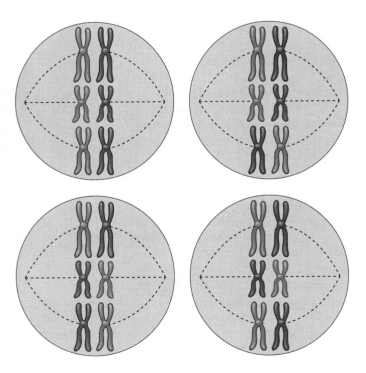

FIGURE 12.13
Random orientation of chromosomes on the metaphase plate. The number of possible chromosome orientations equals 2 raised to the power of the number of chromosome pairs. In this hypothetical cell with three chromosome pairs, eight (2^3) possible orientations exist, four of them illustrated here. Each orientation produces gametes with different combinations of parental chromosomes.

FIGURE 12.14
The stages of meiosis in a lily. Note the arrangement of chromosomes in metaphase I.

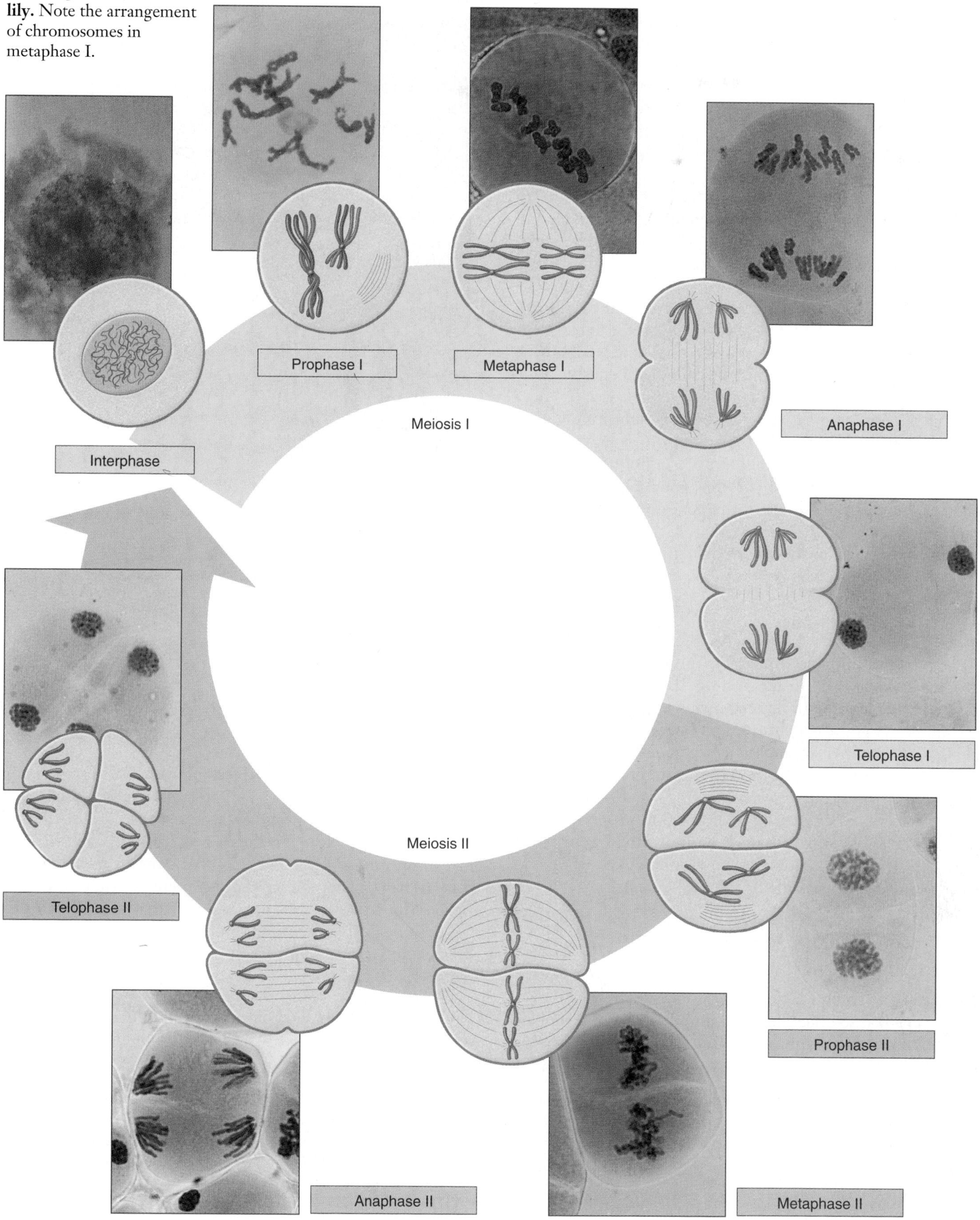

Interphase

Prophase I

Metaphase I

Meiosis I

Anaphase I

Telophase I

Prophase II

Meiosis II

Metaphase II

Anaphase II

Telophase II

Completing Meiosis

After the long duration of prophase and metaphase, which together make up 90% or more of the time meiosis I takes, meiosis I rapidly concludes. Anaphase I and telophase I proceed quickly, followed—without an intervening period of DNA synthesis—by the second meiotic division.

Anaphase I

In anaphase I, the microtubules of the spindle fibers begin to shorten. As they shorten, they break the chiasmata and pull the centromeres toward the poles, dragging the chromosomes along with them. Because the microtubules are attached to kinetochores on only one side of each centromere, the individual centromeres are not pulled apart to form two daughter centromeres, as they are in mitosis. Instead, the entire centromere moves to one pole, taking both sister chromatids with it. When the spindle fibers have fully contracted, each pole has a complete haploid set of chromosomes consisting of one member of each homologous pair. Because of the random orientation of homologous chromosomes on the metaphase plate, a pole may receive either the maternal or the paternal homologue from each chromosome pair. As a result, the genes on different chromosomes assort independently; that is, meiosis I results in the **independent assortment** of maternal and paternal chromosomes into the gametes.

Telophase I

By the beginning of telophase I, the chromosomes have segregated into two clusters, one at each pole of the cell. Now the nuclear membrane re-forms around each daughter nucleus. Since each chromosome within a daughter nucleus replicated before meiosis I began, each now contains two sister chromatids attached by a common centromere. Importantly, *the sister chromatids are no longer identical*, because of the crossing over that occurred in prophase I (figure 12.15). Cytokinesis may or may not occur after telophase I. The second meiotic division, meiosis II, occurs after an interval of variable length.

The Second Meiotic Division

After a typically brief interphase, in which no DNA synthesis occurs, the second meiotic division begins.

Meiosis II resembles a normal mitotic division. Prophase II, metaphase II, anaphase II, and telophase II follow in quick succession.

Prophase II.   At the two poles of the cell the clusters of chromosomes enter a brief prophase II, each nuclear envelope breaking down as a new spindle forms.

Metaphase II.   In metaphase II, spindle fibers bind to both sides of the centromeres.

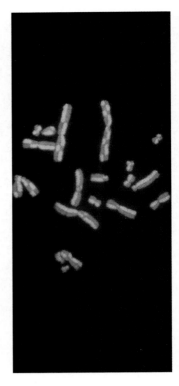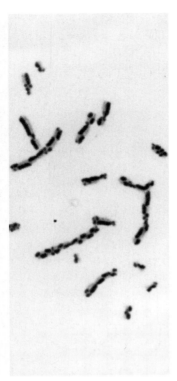

FIGURE 12.15
After meiosis I, sister chromatids are not identical. So-called "harlequin" chromosomes, each containing one fluorescent DNA strand, provide vivid testimony of the reciprocal exchange of genetic material during meiosis I between sister chromatids.

Anaphase II.   The spindle fibers contract, splitting the centromeres and moving the sister chromatids to opposite poles.

Telophase II.   Finally, the nuclear envelope re-forms around the four sets of daughter chromosomes.

The final result of this division is four cells containing haploid sets of chromosomes. No two are alike, because of the crossing over in prophase I. The nuclei are then reorganized, and nuclear envelopes form around each haploid set of chromosomes. The cells that contain these haploid nuclei may develop directly into gametes, as they do in animals. Alternatively, they may themselves divide mitotically, as they do in plants, fungi, and many protists, eventually producing greater numbers of gametes or, as in the case of some plants and insects, adult individuals of varying ploidy.

During meiosis I, homologous chromosomes move toward opposite poles in anaphase I, and individual chromosomes cluster at the two poles in telophase I. At the end of meiosis II, each of the four haploid cells contains one copy of every chromosome in the set, rather than two. Because of crossing over, no two cells are the same. These haploid cells may develop directly into gametes, as in animals, or they may divide by mitosis, as in plants, fungi, and many protists.

Why Sex?

Not all reproduction is sexual. In **asexual reproduction,** an individual inherits all of its chromosomes from a single parent and is, therefore, genetically identical to its parent. Bacterial cells reproduce asexually, undergoing binary fission to produce two daughter cells containing the same genetic information. Most protists reproduce asexually except under conditions of stress; then they switch to sexual reproduction. Among plants, asexual reproduction is common, and many other multicellular organisms are also capable of reproducing asexually. In animals, asexual reproduction often involves the budding off of a localized mass of cells, which grows by mitosis to form a new individual.

Even when meiosis and the production of gametes occur, there may still be reproduction without sex. The development of an adult from an unfertilized egg, called **parthenogenesis,** is a common form of reproduction in arthropods. Among bees, for example, fertilized eggs develop into diploid females, but unfertilized eggs develop into haploid males. Parthenogenesis even occurs among the vertebrates (figure 12.16). Some lizards, fishes, and amphibians are capable of reproducing in this way; their unfertilized eggs undergo a mitotic nuclear division without cell cleavage to produce a diploid cell, which then develops into an adult.

Recombination Can Be Destructive

If reproduction can occur without sex, why does sex occur at all? This question has generated considerable discussion, particularly among evolutionary biologists. Sex is of great evolutionary advantage for populations or species, which benefit from the variability generated by meiotic recombination (crossing over) and gene segregation. However, evolution occurs because of changes at the level of *individual* survival and reproduction, rather than at the population level, and no obvious advantage accrues to the progeny of an individual that engages in sexual reproduction. In fact, recombination is a destructive as well as a constructive process in evolution. The segregation of chromosomes during meiosis tends to disrupt advantageous combinations of genes more often than it creates new, better adapted combinations; as a result, some of the diverse progeny produced by sexual reproduction will not be as well adapted as their parents were. In fact, the more complex the adaptation of an individual organism, the less likely that recombination will improve it, and the more likely that recombination will disrupt it. It is, therefore, a puzzle to know what a well-adapted individual gains from participating in sexual reproduction, since *all* of its progeny could maintain its successful gene combinations if that individual simply reproduced asexually.

FIGURE 12.16
Development from an unfertilized egg. *Cnemidophorus exsanguis,* a parthenogenetic lizard.

Synapsis Evolved to Repair DNA

If recombination is often detrimental to an individual's progeny, then what benefit promoted the evolution of sexual reproduction? Although the answer to this question is unknown, we can gain some insight by examining the protists. Meiotic recombination is often absent among the protists, which typically undergo sexual reproduction only occasionally. Often the fusion of two haploid cells occurs only under stress, creating a diploid zygote.

Why do some protists form a diploid cell in response to stress? Biologists think this occurs because only a diploid cell can effectively repair certain kinds of chromosome damage, particularly double-strand breaks in DNA. Both radiation and chemical events within cells can induce such breaks. As organisms became larger and longer-lived, it must have become increasingly important for them to be able to repair such damage. The synaptonemal complex, which in early stages of meiosis precisely aligns pairs of homologous chromosomes, may well have evolved originally as a mechanism for repairing double-strand damage to DNA, using the homologous chromosome as a template. A transient diploid phase would have provided an opportunity for such repair. In yeast, mutations that inactivate the repair system for double-strand breaks of the chromosomes also prevent crossing over, suggesting a common mechanism for both synapsis and repair processes.

The close association between homologous chromosomes that occurs during meiosis probably evolved as mechanisms to repair chromosomal damage, in which homologous chromosomes were used as templates.

The Evolutionary Consequences of Sex

While our knowledge of how sex evolved is sketchy, it is abundantly clear that sexual reproduction has an enormous impact on how species evolve today, because of its ability to rapidly generate new genetic combinations. Three mechanisms each make key contributions: independent assortment, crossing over, and random fertilization.

Independent Assortment.   The reassortment of genetic material that takes place during meiosis is the principal factor that has made possible the evolution of eukaryotic organisms, in all their bewildering diversity, over the past 1.5 billion years. Sexual reproduction represents an enormous advance in the ability of organisms to generate genetic variability. To understand, recall that most organisms have more than one chromosome. In human beings, for example, each gamete receives one homologue of each of the 23 chromosomes, but which homologue of a particular chromosome it receives is determined randomly (figure 12.17). Each of the 23 pairs of chromosomes segregates independently, so 2^{23} (more than 8 million) different possible kinds of gametes can be produced.

Crossing Over.   The DNA exchange that occurs when the arms of nonsister chromatids cross over adds even more recombination to the independent assortment of chromosomes that occurs later in meiosis. Thus, the number of possible genetic combinations that can occur among gametes is virtually unlimited.

Random Fertilization.   Furthermore, because the zygote that forms a new individual is created by the fusion of *two* gametes, each produced independently, fertilization squares the number of possible outcomes ($2^{23} \times 2^{23} = 70$ trillion).

Importance of Generating Diversity

Whatever the forces that led to sexual reproduction, its evolutionary consequences have been profound. No genetic process generates diversity more quickly; and, as you will see in later chapters, genetic diversity is the raw material of evolution, the fuel that drives it and determines its potential directions. In many cases, the pace of evolution appears to increase as the level of genetic diversity increases. Programs for selecting larger stature in domesticated animals such as cattle and sheep, for example, proceed rapidly at first, but then slow as the existing genetic combinations are exhausted; further progress must then await the generation of new gene combinations. Racehorse breeding provides a graphic example: thoroughbred racehorses are all descendants of a small initial number of individuals, and selection for speed has accomplished all it can with this limited amount of genetic variability—the winning times in major races ceased to improve decades ago.

Paradoxically, the evolutionary process is thus both revolutionary and conservative. It is revolutionary in that the pace of evolutionary change is quickened by genetic recombination, much of which results from sexual reproduction. It is conservative in that evolutionary change is not always favored by selection, which may instead preserve existing combinations of genes. These conservative pressures appear to be greatest in some asexually reproducing organisms that do not move around freely and that live in especially demanding habitats. In vertebrates, on the other hand, the evolutionary premium appears to have been on versatility, and sexual reproduction is the predominant mode of reproduction by an overwhelming margin.

Sexual reproduction increases genetic variability through (1) crossing over in prophase I of meiosis; (2) independent assortment in metaphase I of meiosis; and (3) random fertilization.

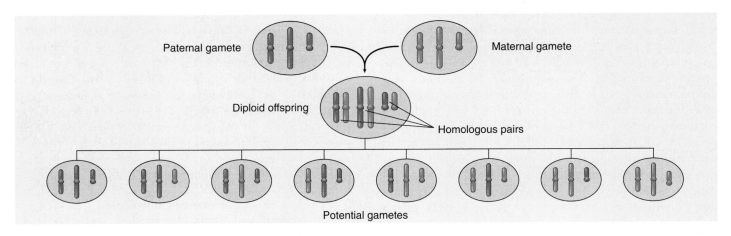

FIGURE 12.17

Independent assortment increases genetic variability. Independent assortment contributes new gene combinations to the next generation because the orientation of chromosomes on the metaphase plate is random. In the cells shown above with three chromosome pairs, eight different gametes can result, each with different combinations of parental chromosomes.

12.1 Meiosis produces sexual gametes.

- Meiosis is a special form of nuclear division that produces the gametes of the sexual cycle. It involves two chromosome separations but only one chromosome replication.

- In sexual reproduction, meiosis and syngamy alternate, producing haploid and diploid stages, respectively. Each diploid individual is produced by the union of two haploid gametes originating from different parents.

- In meiosis, diploid cells proceed through two division cycles where homologous chromosomes are separated from each other, but the DNA is replicated only once. The products of meiosis are thus haploid cells, which contain only half the number of chromosomes found in the diploid somatic cells of the organism.

12.2 Meiosis involves two nuclear divisions.

- The crossing over that occurs between homologues during synapsis is an essential element of meiosis.

- Because crossing over binds the homologues together, only one side of each homologue is accessible to the spindle fibers. Hence, the spindle fibers separate the paired homologues rather than the sister chromatids.

- Ultimately, the shortening of the spindle fibers breaks the terminal chiasmata that link the homologues.

- At the end of meiosis I, one homologue of each chromosome type is present at each of the two poles of the dividing nucleus.

- The homologues still consist of two chromatids, which may differ from each other as a result of the crossing over that occurred during synapsis.

- No further DNA replication occurs before the second nuclear division, which is essentially a mitotic division occurring at each of the two poles.

- The sister chromatids of each chromosome are separated, resulting in the formation of four daughter nuclei, each with half the number of chromosomes that were present before meiosis.

- Cytokinesis may or may not occur at this point. Because each daughter nucleus has one copy of every chromosome, any or all of the nuclei can give rise to gametes.

12.3 The evolution of sex led to increased genetic variability.

- In asexual reproduction, mitosis produces offspring genetically identical to the parent.

- Meiosis is thought to have evolved initially as a mechanism to repair double-strand breaks in DNA, in which the broken chromosome is paired with its homologue while it is being repaired.

- The evolutionary significance of meiosis is that it generates large amounts of recombination, rapidly reshuffling gene combinations and so producing variability upon which evolutionary processes can act.

Discussing Key Terms

1. **Meiosis**   Meiosis consists of a pair of serial nuclear divisions. It includes three unique features: synapsis (the intimate pairing of homologous chromosomes), crossing over, and a lack of chromosome replication before the second nuclear division.

2. **Sexual reproduction**   Reproduction involving an alternation of fertilization and meiosis. Haploid gamete cells produced by meiosis unite in sexual reproduction to produce a diploid organism.

3. **Crossing over**   In prophase of meiosis I, homologous chromosomes pair along their length, a process called synapsis; within the region of close pairing, the paired chromatids often exchange sections. This phenomenon, called "crossing over," is an important source of genetic variation. As a result of crossing over with their homologues, the sister chromatids at the end of the first meiotic division are no longer identical.

4. **Independent assortment**   The independent assortment that results from random alignment of chromosomes on the metaphase plate provides for a tremendous amount of variability among offspring, important for the process of evolution.

5. **Asexual reproduction**   In some organisms, haploid or diploid cells undergo mitotic divisions to yield new individuals via processes such as budding or fission. Without meiosis and sexual reproduction, cellular nuclei divide without the opportunity for synapsis, crossing over, and recombination. Asexually produced organisms are, except for chance mutations, identical to their parents, offering no improved "varieties" for evolution to select.

Review Questions

1. What are the cellular products of meiosis called, and are they haploid or diploid? What is the cellular product of syngamy called, and is it haploid or diploid?

2. What three unique features distinguish meiosis from mitosis?

3. What are synaptonemal complexes? How do they participate in crossing over? At what stage during meiosis are they formed?

4. How many chromatids are present for each type of chromosome at the completion of crossing over? What two structures hold the chromatids together at this stage?

5. How is the attachment of spindle microtubules to centromeres in metaphase I of meiosis different from that which occurs in metaphase of mitosis? What effect does this difference have on the movement of chromosomes during anaphase I?

6. What mechanism is responsible for the independent assortment of chromosomes?

7. How do the chromosomes present at either pole of the cell during telophase I compare with the chromosomes present in the cell before the start of meiosis? What is responsible for the differences between a given chromosome's chromatids during telophase I?

8. How does meiosis II compare with mitosis? What is the chromosomal complement of the nuclei resulting from meiosis II?

9. What types of organisms are capable of asexual reproduction? How do the progeny of asexual reproduction differ from those produced via sexual reproduction with respect to their genetic similarity to their parents?

10. What is one of the current scientific explanations for the evolution of sexual reproduction?

11. By what three mechanisms does sexual reproduction increase genetic variability? How does this increase in genetic variability affect the evolution of species?

Thought Questions

1. Humans have 23 pairs of chromosomes. Ignoring the effects of crossing over, what proportion of a woman's eggs contain only chromosomes she received from her mother?

2. Occasionally, the homologues of a particular chromosome fail to separate during meiosis I, or the sister chromatids of a particular homologue fail to separate during meiosis II. Using the hypothetical example of an organism with two pairs of chromosomes in the adult, describe the chromosomal makeup of

the eggs that would result from each of these errors in meiosis. If these eggs were fertilized by normal sperm, what would the chromosomal makeup of the resulting zygotes be like?

3. Many sexually reproducing lizard species are able to generate local populations that reproduce by parthenogenesis. What do you imagine the sex of these local parthenogenetic populations would be: male, female, or neuter? Explain your reasoning.

Internet Links

Cell Cycle & Mitosis Tutorial
http://www.biology.arizona.edu:80/cell_bio/tutorials/cell_cycle/main.html
A good tutorial on mitosis and the cell cycle, prepared by the University of Arizona.

Virtual Cell Division
http://www.biology.uc.edu/vgenetic/
Prepared by the biology staff of the University of Cincinnati, these on-line lessons provide virtual tutorials on mitosis and meiosis.

Cancer Today
http://oncolink.upenn.edu/upcc/
The University of Pennsylvania Cancer Center's ONCOLINK site provides a rich source of information on current cancer research.

Cells Alive
http://www.cellsalive.com/ecoli.htm
Watching bacteria grow and multiply is only one aspect of this rich and rewarding site, a great place to browse for micrographs, animations, and short movies of many cell activities.

For Further Reading

Kondrashov, A. S.: "The Asexual Ploidy Cycle and the Origin of Sex," *Nature*, vol. 370, July 23, 1994, pages 213–16. Why would asexual organisms have an alternation of haploid and diploid life stages? The author supplies some suggestions and speculates on the origin of sex.

Marx, J.: "Chromosomes Yield New Clue to Pairing in Meiosis," *Science*, vol. 273, July 5, 1996, pages 35–36. Matching heterochromatin sequences appear to play a crucial role in holding paired chromosomes together in a synapsis.

McKim, K., and S. Hawley: "Chromosomal Control of Meiotic Cell Division," *Science*, vol. 270, December 8, 1995, pages 1595–1600. Chromosomes play a crucial role in control of the meiotic cell cycle.

Murray, A. W.: *The Cell Cycle: An Introduction*, Oxford University Press, New York, 1993. An in-depth discussion of the events occurring in the life of a cell, including a detailed section on meiosis.

Pickett-Heaps, J., D. Tippit, and K. Porter: "Rethinking Mitosis," *Cell*, vol. 29, 1982, pages 729–44. An advanced but very rewarding assessment of the evidence concerning the evolution of mitosis and meiosis.

13

Patterns of Inheritance

Concept Outline

13.1 Mendel solved the mystery of heredity.

Early Ideas about Heredity: The Road to Mendel. Before Mendel, biologists believed in the direct transmission of traits.

Mendel and the Garden Pea. Mendel, a monk, experimented with heredity in edible peas, as had many others, but he counted his results.

What Mendel Found. Mendel found that contrasting traits segregated among second-generation progeny in the ratio 3:1.

Mendel's Model of Heredity. Mendel proposed that information rather than the trait itself is inherited, with each parent contributing one copy.

How Mendel Interpreted His Results. Mendel found that one alternative of a trait could mask the other in heterozygotes, but both could subsequently be expressed in homozygotes of future generations.

Mendelian Inheritance Is Not Always Easy to Analyze. A variety of factors can disguise the Mendelian segregation of alleles.

13.2 Genes are on chromosomes.

Chromosomes: The Vehicles of Mendelian Inheritance. Mendelian segregation reflects the random assortment of chromosomes in meiosis.

Genetic Recombination. Crossover frequency indicates the physical distance between genes and is used to construct genetic maps.

13.3 Human genetics follows Mendelian principles.

Multiple Alleles: The ABO Blood Groups. The human ABO blood groups are determined by three *I* gene alleles.

Human Chromosomes. Humans possess 23 pairs of chromosomes, one of them determining the sex.

Human Abnormalities Due to Alterations in Chromosome Number. Loss or addition of chromosomes has serious consequences.

Human Genetic Disorders. Many heritable human disorders are the result of recessive mutations in genes.

Genetic Counseling. Some gene defects can be detected early in pregnancy.

FIGURE 13.1
Human beings are extremely diverse in appearance. The differences between us are partly inherited and partly the result of environmental factors we encounter in our lives.

Every living creature is a product of the long evolutionary history of life on earth. While all organisms share this history, only humans wonder about the processes that led to their origin. We are still far from understanding everything about our origins, but we have learned a great deal. Like a partially completed jigsaw puzzle, the boundaries have fallen into place, and much of the internal structure is becoming apparent. In this chapter, we will discuss one piece of the puzzle—the enigma of heredity. Why do groups of people from different parts of the world often differ in appearance (figure 13.1)? Why do the members of a family tend to resemble one another more than they resemble members of other families?

Early Ideas about Heredity: The Road to Mendel

As far back as written records go, patterns of resemblance among the members of particular families have been noted and commented on (figure 13.2). Some familial features are unusual, such as the protruding lower lip of the European royal family Hapsburg, evident in pictures and descriptions of family members from the thirteenth century onward. Other characteristics, like the occurrence of redheaded children within families of redheaded parents, are more common (figure 13.3). Inherited features, the building blocks of evolution, will be our concern in this chapter.

Like many great puzzles, the riddle of heredity seems simple now that it has been solved. The solution was not an easy one to find, however. Our present understanding is the culmination of a long history of thought, surmise, and investigation. At every stage we have learned more, and as we have done so, the models we use to describe the mechanisms of heredity have changed to encompass new facts.

Classical Assumption 1: Constancy of Species

Two concepts provided the basis for most of the thinking about heredity before the twentieth century. The first is that *heredity occurs within species.* For a very long time people believed that it was possible to obtain bizarre composite animals by breeding (crossing) widely different species. The minotaur of Cretan mythology, a creature with the body of a bull and the torso and head of a man, is one example. The giraffe was thought to be another; its scientific name, *Giraffa camelopardalis,* suggests the belief that it was the result of a cross between a camel and a leopard. From the Middle Ages onward, however, people discovered that such extreme crosses were not possible and that variation and heredity occur mainly within the boundaries of a particular species. Species were thought to have been maintained without significant change from the time of their creation.

Classical Assumption 2: Direct Transmission of Traits

The second early concept related to heredity is that *traits are transmitted directly.* When variation is inherited by offspring from their parents, *what* is transmitted? The ancient Greeks suggested that the parents' body parts were transmitted directly to their offspring. Hippocrates called this type of reproductive material *gonos,* meaning "seed." Hence, a characteristic such as a misshapen limb was the result of material that came from the misshapen limb of a parent. Information from each part of the body was supposedly passed along independently of the information

FIGURE 13.2
Heredity is responsible for family resemblance. Family resemblances are often strong—a visual manifestation of the mechanism of heredity. This is the Johnson family, the wife and daughters of one of the authors. While each daughter is different, all clearly resemble their mother.

FIGURE 13.3
Red hair is inherited. Many different traits are inherited in human families. This redhead is exhibiting one of these traits.

from the other parts, and the child was formed after the hereditary material from all parts of the parents' bodies had come together.

This idea was predominant until fairly recently. For example, in 1868, Charles Darwin proposed that all cells and tissues excrete microscopic granules, or "gemmules," that are passed to offspring, guiding the growth of the corresponding part in the developing embryo. Most similar theories of the direct transmission of hereditary material assumed that the male and female contributions *blend* in the offspring. Thus, parents with red and brown hair would produce children with reddish brown hair, and tall and short parents would produce children of intermediate height.

Koelreuter Demonstrates Hybridization between Species

Taken together, however, these two concepts lead to a paradox. If no variation enters a species from outside, and if the variation within each species blends in every generation, then all members of a species should soon resemble one another exactly. Obviously, this does not happen. Individuals within most species differ widely from each other, and they differ in characteristics that are transmitted from generation to generation.

How could this paradox be resolved? Actually, the resolution had been provided long before Darwin, in the work of the German botanist Josef Koelreuter. In 1760, Koelreuter carried out the first successful **hybridizations** of plant species, crossing different strains of tobacco and obtaining fertile offspring. The hybrids differed in appearance from both parent strains. When individuals within the hybrid generation were crossed, the offspring were highly variable. Some of these offspring resembled plants of the hybrid generation (their parents), but a few resembled the original strains (their grandparents).

The Classical Assumptions Fail

Koelreuter's work represents the beginning of modern genetics, the first clues pointing to the modern theory of heredity. Koelreuter's experiments provided an important clue about how heredity works: the traits he was studying could be masked in one generation, only to reappear in the next. This pattern contradicts the theory of direct transmission. How could a trait that is transmitted directly be latent and then reappear? Nor were the traits of Koelreuter's plants blended. A contemporary account stated that the traits reappeared in the third generation "fully restored to all their original powers and properties."

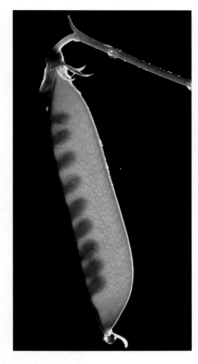

FIGURE 13.4
The garden pea, *Pisum sativum*. Easy to cultivate and able to produce many distinctive varieties, the garden pea was a popular experimental subject in investigations of heredity as long as a century before Gregor Mendel's experiments.

It is worth repeating that the offspring of Koelreuter's crosses were not identical to one another. Some resembled the hybrid generation, while others did not. The alternative forms of the traits Koelreuter was studying were distributed among the offspring. A modern geneticist would say the alternative forms of each trait were **segregating** among the progeny of a mating, meaning that some offspring exhibited one alternative form of a trait (for example, hairy leaves), while other offspring from the same mating exhibited a different alternative (smooth leaves). This segregation of alternative forms of a trait provided the clue that led Gregor Mendel to his understanding of the nature of heredity.

Knight Studies Heredity in Peas

Over the next hundred years, other investigators elaborated on Koelreuter's work. Prominent among them were English gentleman farmers trying to improve varieties of agricultural plants. In one such series of experiments, carried out in the 1790s, T. A. Knight crossed two true-breeding varieties (varieties that remain uniform from one generation to the next) of the garden pea, *Pisum sativum* (figure 13.4). One of these varieties had purple flowers, and the other had white flowers. All of the progeny of the cross had purple flowers. Among the offspring of these hybrids, however, were some plants with purple flowers and others, less common, with white flowers. Just as in Koelreuter's earlier studies, a trait from one of the parents disappeared in one generation only to reappear in the next.

In these deceptively simple results were the makings of a scientific revolution. Nevertheless, another century passed before the process of gene segregation was fully appreciated. Why did it take so long? One reason was that early workers did not quantify their results. A numerical record of results proved to be crucial to understanding the process. Knight and later experimenters who carried out other crosses with pea plants noted that some traits had a "stronger tendency" to appear than others, but they did not record the numbers of the different classes of progeny. Science was young then, and it was not obvious that the numbers were important.

Early geneticists demonstrated that some forms of an inherited trait (1) can disappear in one generation only to reappear unchanged in future generations; (2) segregate among the offspring of a cross; and (3) are more likely to be represented than their alternatives.

Mendel and the Garden Pea

The first quantitative studies of inheritance were carried out by Gregor Mendel, an Austrian monk (figure 13.5). Born in 1822 to peasant parents, Mendel was educated in a monastery and went on to study science and mathematics at the University of Vienna, where he failed his examinations for a teaching certificate. He returned to the monastery and spent the rest of his life there, eventually becoming abbot. In the garden of the monastery, Mendel initiated a series of experiments on plant hybridization (figure 13.6). The results of these experiments would ultimately change our views of heredity irrevocably.

Why Mendel Chose the Garden Pea

For his experiments, Mendel chose the garden pea, the same plant Knight and many others had studied earlier. The choice was a good one for several reasons. First, many earlier investigators had produced hybrid peas by crossing different varieties. Mendel knew that he could expect to observe segregation of traits among the offspring. Second, a large number of true-breeding varieties of peas were available. Mendel initially examined 32. Then, for further study, he selected lines that differed with respect to seven easily distinguishable traits, such as round versus wrinkled seeds and purple versus white flowers, a characteristic Knight had studied. Third, pea plants are small and easy to grow, and they have a short generation time. Thus, one can conduct experiments involving numerous plants, grow several generations in a single year, and obtain results relatively quickly.

A fourth advantage of studying peas is that the sexual organs of the pea are enclosed within the flower (figure 13.7). The flowers of peas, like those of most flowering plants, contain both male and female sex organs. Furthermore, the gametes produced by the male and female parts of the same flower, unlike those of many flowering plants, can fuse to form viable offspring. Fertilization takes place automatically within an individual flower if it is not disturbed, resulting in offspring that are the progeny from a single individual. Therefore, one can either let individual flowers engage in **self-fertilization,** or remove the flower's male parts before fertilization and introduce pollen from a strain with alternative characteristics, thus performing *cross-pollination* which results in **cross-fertilization.**

FIGURE 13.5
Gregor Johann Mendel. Cultivating his plants in the garden of a monastery in Brunn, Austria (now Brno, Czech Republic), Mendel studied how differences among varieties of peas were inherited when the varieties were crossed. Similar experiments had been done before, but Mendel was the first to appreciate the significance of the results.

FIGURE 13.6
The garden where Mendel carried out his plant-breeding experiments. Gregor Mendel did his most important scientific experiments in this small garden in a monastery.

Mendel's Experimental Design

Mendel was careful to focus on only a few specific differences between the plants he was using and to ignore the countless other differences he must have seen. He also had the insight to realize that the differences he selected to analyze must be comparable. For example, he appreciated that trying to study the inheritance of round seeds versus tall height would be useless; the traits, like apples and oranges, are not comparable.

Mendel usually conducted his experiments in three stages:

1. First, he allowed pea plants of a given variety to produce progeny by self-fertilization for several generations. Mendel was thus able to assure himself that the forms of traits he was studying were indeed constant, transmitted unchanged from generation to generation. Pea plants with white flowers, for example, when crossed with each other, produced only offspring with white flowers, regardless of the number of generations.

2. Mendel then performed crosses between varieties exhibiting alternative forms of traits. For example, he removed the male parts from the flower of a plant that produced white flowers and fertilized it with pollen from a purple-flowered plant. He also carried out the reciprocal cross, using pollen from a white-flowered individual to fertilize a flower on a pea plant that produced purple flowers (figure 13.8).

3. Finally, Mendel permitted the hybrid offspring produced by these crosses to self-pollinate for several generations. By doing so, he allowed the alternative forms of a trait to segregate among the progeny. This was the same experimental design that Knight and others had used much earlier. But Mendel went an important step farther: he counted the numbers of offspring of each type in each succeeding generation. No one had ever done that before. The quantitative results Mendel obtained proved to be of supreme importance in revealing the process of heredity.

Mendel's experiments with the garden pea involved crosses between true-breeding varieties, followed by a generation or more of inbreeding.

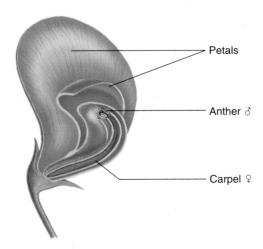

FIGURE 13.7
Structure of the pea flower (longitudinal section). In a pea plant flower, the petals enclose the male anther (containing pollen grains, which give rise to haploid sperm) and the female carpel (containing ovules, which give rise to haploid eggs). This ensures that self-fertilization will take place unless the flower is disturbed.

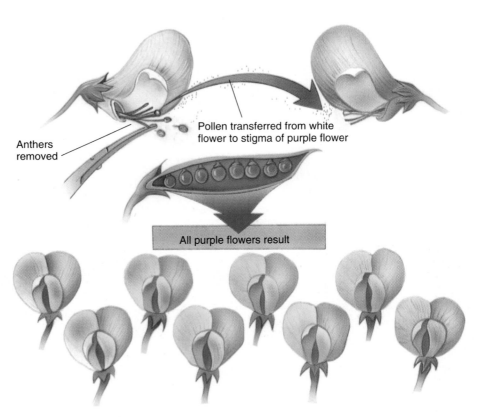

FIGURE 13.8
How Mendel conducted his experiments. Mendel pushed aside the petals of a white flower and cut off the anthers, the source of the pollen. He then placed that pollen onto the stigma (part of the carpel) of a similarly castrated purple flower, causing cross-fertilization to take place. All the seeds in the pod that resulted from this pollination were hybrids of the white-flowered male parent and the purple-flowered female parent. After planting these seeds, Mendel observed what kinds of plants they produced. All of the progeny of this cross had purple flowers.

What Mendel Found

The seven traits Mendel studied in his experiments possessed several variants that differed from one another in ways that were easy to recognize and score (figure 13.9). We will examine in detail Mendel's crosses with flower color. His experiments with other traits were similar, and they produced similar results.

The F_1 Generation

When Mendel crossed two contrasting varieties of peas, such as white-flowered and purple-flowered plants, the hybrid offspring he obtained did not have flowers of intermediate color, as the theory of blending inheritance would predict. Instead, in every case the flower color of the offspring resembled one of their parents. It is customary to refer to these offspring as the **first filial** (*filius* is Latin for "son"), or F_1, generation. Thus, in a cross of white-flowered with purple-flowered plants, the F_1 offspring all had purple flowers, just as Knight and others had reported earlier.

Mendel referred to the trait expressed in the F_1 plants as **dominant** and to the alternative form that was not expressed in the F_1 plants as **recessive.** For each of the seven pairs of contrasting forms of traits that Mendel examined, one of the pair proved to be dominant and the other recessive.

Trait	Dominant vs. recessive		F_2 generation		Ratio
			Dominant form	Recessive form	
Flower color	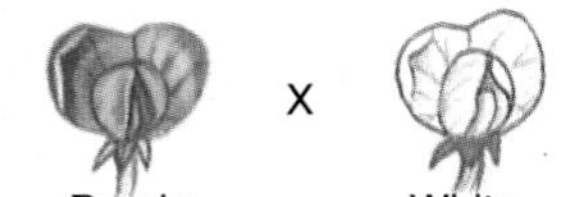 Purple X White		705	224	3.15:1
Seed color	 Yellow X Green		6022	2001	3.01:1
Seed shape	 Round X Wrinkled		5474	1850	2.96:1
Pod color	 Green X Yellow		428	152	2.82:1
Pod shape	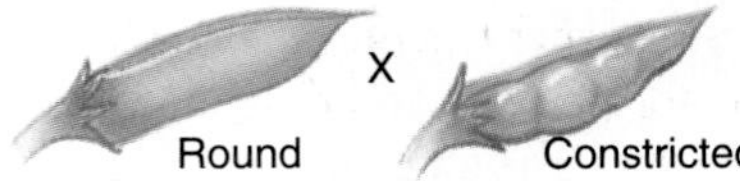 Round X Constricted		882	299	2.95:1
Flower position	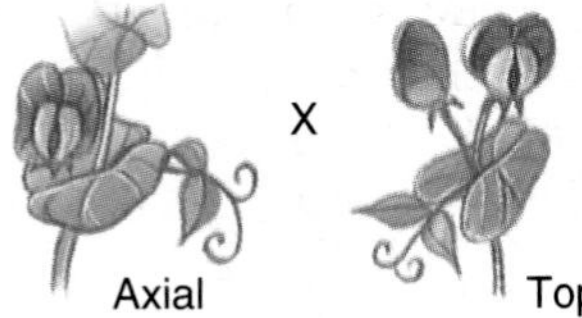 Axial X Top		651	207	3.14:1
Plant height	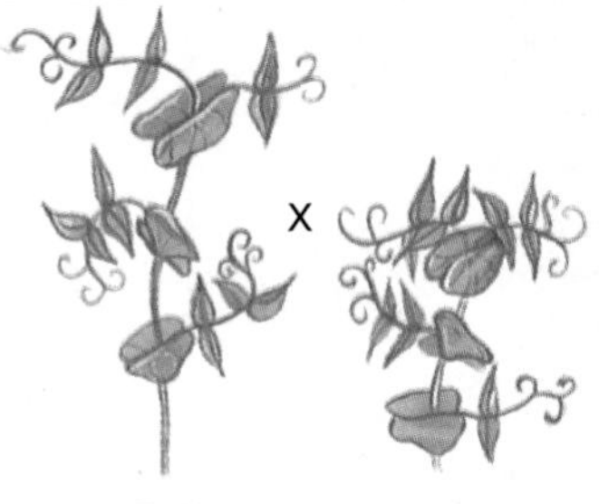 Tall X Dwarf		787	277	2.84:1

FIGURE 13.9

Mendel's experimental results. This table illustrates the seven pairs of contrasting traits Mendel studied in his crosses of the garden pea and presents the data he obtained for these crosses. Each pair of traits appeared in the F_2 generation in very close to a 3:1 ratio.

The F₂ Generation

After allowing individual F₁ plants to mature and self-pollinate, Mendel collected and planted the seeds from each plant to see what the offspring in the **second filial**, or **F₂**, generation would look like. He found, just as Knight had earlier, that some F₂ plants exhibited white flowers, the recessive form of the trait. Latent in the F₁ generation, the recessive form reappeared among some F₂ individuals.

Believing the proportions of the F₂ types would provide some clue about the mechanism of heredity, Mendel counted the numbers of each type among the F₂ progeny. In the cross between the purple-flowered F₁ plants, he counted a total of 929 F₂ individuals (see figure 13.9). Of these, 705 (75.9%) had purple flowers and 224 (24.1%) had white flowers. Approximately ¼ of the F₂ individuals exhibited the recessive form of the trait. Mendel obtained the same numerical result with the other six traits he examined: ¾ of the F₂ individuals exhibited the dominant form of the trait, and ¼ displayed the recessive form. In other words, the dominant: recessive ratio among the F₂ plants was always close to 3:1. Mendel carried out similar experiments with other traits, such as wrinkled versus round seeds (figure 13.10), and obtained the same result.

FIGURE 13.10
Seed shape: a Mendelian trait. One of the differences Mendel studied affected the shape of pea plant seeds. In some varieties, the seeds were round, while in others, they were wrinkled. As you can see, the wrinkled seeds look like dried-out versions of the round ones.

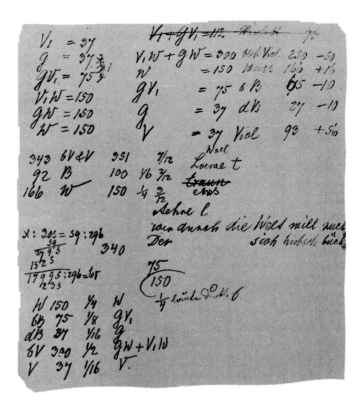

FIGURE 13.11
A page from Mendel's notebook. In these notes, Mendel is trying various ratios in an unsuccessful attempt to explain a segregation ratio disguised by phenotypes that are so similar he cannot distinguish them from one another.

A Disguised 1:2:1 Ratio

Mendel went on to examine how the F₂ plants passed traits on to subsequent generations. He found that the recessive ¼ were always true-breeding. In the cross of white-flowered with purple-flowered plants, for example, the white-flowered F₂ individuals reliably produced white-flowered offspring when they were allowed to self-fertilize. By contrast, only ⅓ of the dominant purple-flowered F₂ individuals (¼ of all F₂ offspring) proved true-breeding, while ⅔ were not. This last class of plants produced dominant and recessive individuals in the third filial (F₃) generation in a 3:1 ratio. This result suggested that, for the entire sample, the 3:1 ratio that Mendel observed in the F₂ generation was really a disguised 1:2:1 ratio: ¼ pure-breeding dominant individuals, ½ not-pure-breeding dominant individuals, and ¼ pure-breeding recessive individuals (figure 13.11).

When Mendel crossed two contrasting varieties and counted the offspring in the subsequent generations, he found all of the offspring in the first generation exhibited one (dominant) trait, and none exhibited the other (recessive) trait. In the following generation, 25% were pure-breeding for the dominant trait, 50% were hybrid for the two traits and appeared dominant, and 25% were pure-breeding for the recessive trait.

Mendel's Model of Heredity

From his experiments, Mendel was able to understand four things about the nature of heredity. *First*, the plants he crossed did not produce progeny of intermediate appearance, as a theory of blending inheritance would have predicted. Instead, different plants inherited each alternative intact, as a discrete characteristic that either was or was not visible in a particular generation. *Second*, Mendel learned that for each pair of alternative forms of a trait, one alternative was not expressed in the F_1 hybrids, although it reappeared in some F_2 individuals. **The "invisible" trait must therefore be latent (present but not expressed) in the F_1 individuals.** *Third*, the pairs of alternative forms of the traits examined segregated among the progeny of a particular cross, some individuals exhibiting one form of a trait, some the other. *Fourth*, pairs of alternatives were expressed in the F_2 generation in the ratio of ¾ dominant to ¼ recessive. This characteristic 3:1 segregation is often referred to as the **Mendelian ratio.**

To explain these results, Mendel proposed a simple model. It has become one of the most famous models in the history of science, containing simple assumptions and making clear predictions. The model has five elements:

1. Parents do not transmit physiological traits directly to their offspring. Rather, they transmit discrete information about the traits, what Mendel called "factors." These factors later act in the offspring to produce the trait. In modern terms, we would say that information about the alternative forms of traits that an individual expresses is *encoded* by the factors that it receives from its parents.

2. Each individual receives two factors that may code for the same form or for two alternative forms of the trait. We now know that there are two factors for each trait present in each individual because these factors are carried on chromosomes, and each adult individual is *diploid*. When the individual forms gametes (eggs or sperm), they contain only one of each kind of chromosome; the gametes are *haploid*. Therefore, only one factor for each trait of the adult organism is contained in the gamete. Which of the two factors for each trait ends up in a particular gamete is randomly determined.

3. Not all copies of a factor are identical. In modern terms, the alternative forms of a factor, leading to alternative forms of a trait, are called **alleles.** When two haploid gametes containing exactly the same allele of a factor fuse during fertilization to form a zygote, the offspring that develops from that zygote is said to be **homozygous;** when the two haploid gametes contain different alleles, the individual offspring is **heterozygous.**

In modern terminology, Mendel's factors are called **genes.** We now know that each gene is composed of a particular DNA nucleotide sequence (chapter 3). The particular location of a gene on a chromosome is referred to as the gene's **locus** (plural, loci).

FIGURE 13.12

A recessive trait. Blue eyes are considered a recessive trait in humans, although many genes influence eye color.

4. The two alleles, one contributed by the male gamete and one by the female, do not influence each other in any way. In the cells that develop within the new individual, these alleles remain discrete. They neither blend with nor alter each other. (Mendel referred to them as "uncontaminated.") Thus, when the individual matures and produces its own gametes, the alleles for each gene segregate randomly into these gametes, as described in point 2.

5. The presence of a particular allele does not ensure that the form of the trait encoded by it will be expressed in an individual carrying that allele. In heterozygous individuals, only one allele (the dominant one) is expressed, while the other (recessive) allele is present but unexpressed. To distinguish between the presence of an allele and its expression, modern geneticists refer to the totality of alleles that an individual contains as the individual's **genotype** and to the physical appearance of that individual as its **phenotype.** The phenotype of an individual is the observable outward manifestation of its genotype, the result of the functioning of the enzymes and proteins encoded by the genes it carries. In other words, the genotype is the blueprint, and the phenotype is the visible outcome.

These five elements, taken together, constitute Mendel's model of the hereditary process. Many traits in humans also exhibit dominant or recessive inheritance, similar to the traits Mendel studied in peas (figure 13.12, table 13.1).

The genes that an individual has are referred to as its genotype; the outward appearance of the individual is referred to as its phenotype.

How Mendel Interpreted His Results

Does Mendel's model predict the results he actually obtained? To test his model, Mendel first expressed it in terms of a simple set of symbols, and then used the symbols to interpret his results. It is very instructive to do the same. Consider again Mendel's cross of purple-flowered with white-flowered plants. We will assign the symbol P to the dominant allele, associated with the production of purple flowers, and the symbol p to the recessive allele, associated with the production of white flowers. By convention, genetic traits are usually assigned a letter symbol referring to their more common forms, in this case "P" for purple

flower color. The dominant allele is written in upper case, as P; the recessive allele (white flower color) is assigned the same symbol in lower case, p.

In this system, the genotype of an individual that is true-breeding for the recessive white-flowered trait would be designated pp. In such an individual, both copies of the allele specify the white-flowered phenotype. Similarly, the genotype of a true-breeding purple-flowered individual would be designated PP, and a heterozygote would be designated Pp (dominant allele first). Using these conventions, and denoting a cross between two strains with ×, we can symbolize Mendel's original cross as $pp \times PP$ (figure 13.13).

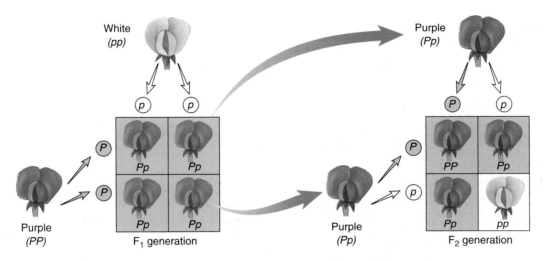

FIGURE 13.13

Mendel's cross of pea plants differing in flower color. All of the offspring of the first cross (the F_1 generation) are Pp heterozygotes with purple flowers. When two heterozygous F_1 individuals are crossed, three kinds of F_2 offspring are possible: PP homozygotes (purple flowers); Pp heterozygotes (also purple flowers); and pp homozygotes (white flowers). Therefore, in the F_2 generation, the ratio of dominant to recessive type is 3:1.

Table 13.1	Some Dominant and Recessive Traits in Humans		
Recessive Traits	**Phenotypes**	**Dominant Traits**	**Phenotypes**
Common baldness	M-shaped hairline receding with age	Middigital hair	Presence of hair on middle segment of fingers
Albinism	Lack of melanin pigmentation	Brachydactyly	Short fingers
Alkaptonuria	Inability to metabolize homogenistic acid	Huntington's disease	Degeneration of nervous system, starting in middle age
Red-green color blindness	Inability to distinguish red or green wavelengths of light	Phenylthiocarbamide (PTC) sensitivity	Ability to taste PTC as bitter
Cystic fibrosis	Abnormal gland secretion, leading to liver degeneration and lung failure	Camptodactyly	Inability to straighten the little finger
Duchenne muscular dystrophy	Wasting away of muscles during childhood	Hypercholesterolemia (the most common human Mendelian disorder—1:500)	Elevated levels of blood cholesterol and risk of heart attack
Hemophilia	Inability to form blood clots	Polydactyly	Extra fingers and toes
Sickle cell anemia	Defective hemoglobin that causes red blood cells to curve and stick together		

The F₁ Generation

Using these simple symbols, we can now go back and reexamine the crosses Mendel carried out. Since a white-flowered parent (*pp*) can produce only *p* gametes, and a pure purple-flowered (homozygous dominant) parent (*PP*) can produce only *P* gametes, the union of an egg and a sperm from these parents can produce only heterozygous *Pp* offspring in the F₁ generation (see figure 13.13). Because the *P* allele is dominant, all of these F₁ individuals are expected to have purple flowers. The *p* allele is present in these heterozygous individuals, but it is not phenotypically expressed. This is the basis for the latency Mendel saw in recessive traits.

The F₂ Generation

When F₁ individuals are allowed to self-fertilize, the *P* and *p* alleles segregate randomly during gamete formation. Their subsequent union at fertilization to form F₂ individuals is also random, not being influenced by which alternative alleles the individual gametes carry. What will the F₂ individuals look like? The possibilities may be visualized in a simple diagram called a **Punnett square,** named after its originator, the English geneticist Reginald Crundall Punnett (figure 13.14). Mendel's model, analyzed in terms of a Punnett square, clearly predicts that the F₂ generation should consist of ¾ purple-flowered plants and ¼ white-flowered plants, a phenotypic ratio of 3:1.

The Laws of Probability Can Predict Mendel's Results

A different way to express Mendel's result is to say that there are three chances in four (¾) that any particular F₂ individual will exhibit the dominant trait, and one chance in four (¼) that an F₂ individual will express the recessive trait. Stating the results in terms of probabilities allows simple predictions to be made about the outcomes of crosses. If both F₁ parents are *Pp* (heterozygotes), the probability that a particular F₂ individual will be *pp* (homozygous recessive) is the probability of receiving a *p* gamete from the male (½) times the probability of receiving a *p* gamete from the female (½), or ¼. This is the same operation we perform in the Punnett square illustrated in figure 13.14. The ways probability theory can be used to analyze Mendel's results is discussed in detail on page 272.

Further Generations

As you can see in figure 13.13, there are really three kinds of F₂ individuals: ¼ are pure-breeding, white-flowered individuals (*pp*); ½ are heterozygous, purple-flowered individuals (*Pp*); and ¼ are pure-breeding, purple-flowered individuals (*PP*). The 3:1 phenotypic ratio is really a disguised 1:2:1 genotypic ratio.

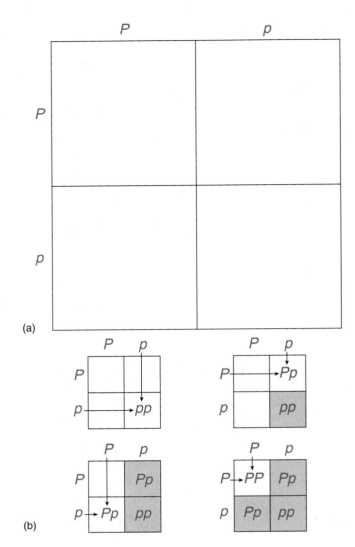

FIGURE 13.14
A Punnett square. (a) To make a Punnett square, place the different possible types of female gametes along one side of a square and the different possible types of male gametes along the other. (b) Each potential zygote can then be represented as the intersection of a vertical line and a horizontal line.

Mendel's First Law of Heredity: Segregation

Mendel's model thus accounts in a neat and satisfying way for the segregation ratios he observed. Its central assumption—that alternative alleles of a trait segregate from each other in heterozygous individuals and remain distinct—has since been verified in many other organisms. It is commonly referred to as **Mendel's First Law of Heredity,** or the **Law of Segregation.** As you saw in chapter 12, the segregational behavior of alternative alleles has a simple physical basis, the alignment of chromosomes at random on the metaphase plate. It is a tribute to the intellect of Mendel's analysis that he arrived at the correct scheme with no knowledge of the cellular mechanisms of inheritance; neither chromosomes nor meiosis had yet been described.

The Testcross

To test his model further, Mendel devised a simple and powerful procedure called the **testcross.** Consider a purple-flowered plant. It is impossible to tell whether such a plant is homozygous or heterozygous simply by looking at its phenotype. To learn its genotype, you must cross it with some other plant. What kind of cross would provide the answer? If you cross it with a homozygous dominant individual, all of the progeny will show the dominant phenotype whether the test plant is homozygous or heterozygous. It is also difficult (but not impossible) to distinguish between the two possible test plant genotypes by crossing with a heterozygous individual. However, if you cross the test plant with a homozygous recessive individual, the two possible test plant genotypes will give totally different results (figure 13.15):

Alternative 1: unknown individual homozygous (*PP*). *PP* × *pp*: all offspring have purple flowers (*Pp*)

Alternative 2: unknown individual heterozygous (*Pp*). *Pp* × *pp*: ½ of offspring have white flowers (*pp*) and ½ have purple flowers (*Pp*)

To perform his testcross, Mendel crossed heterozygous F_1 individuals back to the parent homozygous for the recessive trait. He predicted that the dominant and recessive traits would appear in a 1:1 ratio, and that is what he observed.

For each pair of alleles he investigated, Mendel observed phenotypic F_2 ratios of 3:1 (see figure 13.13) and testcross ratios very close to 1:1, just as his model predicted.

Testcrosses can also be used to determine the genotype of an individual when two genes are involved. Mendel carried out many two-gene crosses, some of which we will soon discuss. He often used testcrosses to verify the genotypes of particular dominant-appearing F_2 individuals. Thus an F_2 individual showing both dominant traits (*A_ B_*) might have any of the following genotypes: *AABB, AaBB, AABb* or *AaBb*. By crossing dominant-appearing F_2 individuals with homozygous recessive individuals (that is, *A_ B_* × *aabb*), Mendel was able to determine if either or both of the traits bred true among the progeny, and so to determine the genotype of the F_2 parent:

AABB	trait A breeds true	trait B breeds true
AaBB	————	trait B breeds true
AAbb	trait A breeds true	————
AaBb	————	————

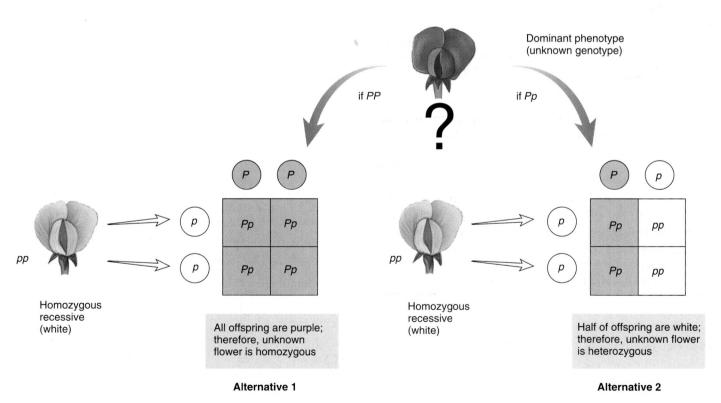

FIGURE 13.15
A testcross. To determine whether an individual exhibiting a dominant phenotype, such as purple flowers, is homozygous or heterozygous for the dominant allele, Mendel crossed the individual in question with a plant that he knew to be homozygous recessive, in this case a plant with white flowers.

Mendel's Second Law of Heredity: Independent Assortment

After Mendel had demonstrated that different alleles of a given gene segregate independently of each other in crosses, he asked whether different genes also segregate independently. Mendel set out to answer this question in a straightforward way. He first established a series of pure-breeding lines of peas that differed in just two of the seven pairs of characteristics he had studied. He then crossed contrasting pairs of the pure-breeding lines to create heterozygotes. In a cross involving different seed shape alleles (round, *R*, and wrinkled, *r*) and different seed color alleles (yellow, *Y*, and green, *y*), all the F₁ individuals were identical, each one heterozygous for both seed shape (*Rr*) and seed color (*Yy*). The F₁ individuals of such a cross are **dihybrids,** individuals heterozygous for each of two genes.

The third step in Mendel's analysis was to allow the dihybrids to self-fertilize. If the alleles affecting seed shape and seed color were segregating independently, then the probability that a particular pair of seed shape alleles would occur together with a particular pair of seed color alleles would be simply the product of the individual probabilities that each pair would occur separately. Thus, the probability that an individual with wrinkled green seeds (*rryy*) would appear in the F₂ generation would be equal to the probability of observing an individual with wrinkled seeds (¼) times the probability of observing one with green seeds (¼), or ¹⁄₁₆.

Since the genes concerned with seed shape and those concerned with seed color are each represented by a pair of alternative alleles in the dihybrid individuals, four types of gametes are expected: *RY*, *Ry*, *rY*, and *ry*. Therefore, in the F₂ generation there are 16 possible combinations of alleles, each of them equally probable (figure 13.16). Of these, 9 possess at least one dominant allele for each gene (signified *R__Y__*, where the dash indicates the presence of either allele) and, thus, should have round, yellow seeds. Of the rest, 3 possess at least one dominant *R* allele but are homozygous recessive for color (*R__yy*); 3 others possess at least one dominant *Y* allele but are homozygous recessive for shape (*rrY__*); and 1 combination among the 16 is homozygous recessive for both genes (*rryy*). The hypothesis that color and shape genes assort independently thus predicts that the F₂ generation of this dihybrid cross will display a 9:3:3:1 ratio: nine individuals with round, yellow seeds, three with round, green seeds, three with wrinkled, yellow seeds, and one with wrinkled, green seeds (see figure 13.16).

What did Mendel actually observe? From a total of 556 seeds from dihybrid plants he had allowed to self-fertilize, he observed:

315 round yellow (*R__Y__*)

108 round green (*R__yy*)

101 wrinkled yellow (*rrY__*)

 32 wrinkled green (*rryy*)

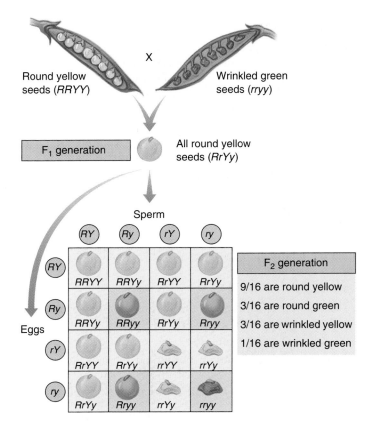

FIGURE 13.16
Analyzing a dihybrid cross. This Punnett square analyzes the results of Mendel's dihybrid cross between plants with round yellow seeds and plants with wrinkled green seeds. The ratio of the four possible combinations of phenotypes is predicted to be 9:3:3:1, the ratio that Mendel found.

These results are very close to a 9:3:3:1 ratio (which would be 313:104:104:35). Consequently, the two genes appeared to assort completely independently of each other. Note that this independent assortment of different genes in no way alters the independent segregation of individual pairs of alleles. Round versus wrinkled seeds occur in a ratio of approximately 3:1 (423:133); so do yellow versus green seeds (416:140). Mendel obtained similar results for other pairs.

Mendel's discovery is often referred to as **Mendel's Second Law of Heredity,** or the **Law of Independent Assortment.** Genes that assort independently of one another, like the seven genes Mendel studied, usually do so because they are located on different chromosomes, which segregate independently during the meiotic process of gamete formation. A modern restatement of Mendel's Second Law would be that *genes that are located on different chromosomes assort independently during meiosis.*

> Mendel summed up his discoveries about heredity in two laws. Mendel's First Law of Heredity states that alternative alleles of a trait segregate independently; his Second Law of Heredity states that genes located on different chromosomes assort independently.

Mendelian Inheritance Is Not Always Easy to Analyze

Mendel's original paper describing his experiments, published in 1866, is charming and interesting to read. His explanations are clear, and the logic of his arguments is presented lucidly. Although Mendel's results did not receive much notice during his lifetime, three different investigators independently rediscovered his pioneering paper in 1900, 16 years after his death. They came across it while searching the literature in preparation for publishing their own findings, which closely resembled those Mendel had presented more than three decades earlier.

Modified Mendelian Ratios

In the decades following the rediscovery of Mendel in 1900, many investigators set out to test Mendel's ideas. Initial work was carried out primarily in agricultural animals and plants, since techniques for breeding these organisms were well established. However, scientists attempting to confirm Mendel's theory often had trouble obtaining the same simple ratios he had reported. This was particularly true for dihybrid crosses. Recall that when individuals heterozygous for two different genes mate (a dihybrid cross), four different phenotypes are possible among the progeny: offspring may display the dominant phenotype for both genes, either one of the genes, or for neither gene. Sometimes, however, it is not possible for an investigator to identify successfully each of the four phenotypic classes, because two or more of the classes look alike. Such situations proved confusing to investigators following Mendel.

Epistasis

One example of such difficulty in identification is seen in the analysis of particular varieties of corn, *Zea mays*. Some commercial varieties exhibit a purple pigment called anthocyanin in their seed coats, while others do not. In 1918, geneticist R. A. Emerson crossed two pure-breeding corn varieties, neither exhibiting anthocyanin pigment. Surprisingly, all of the F_1 plants produced purple seeds.

When two of these pigment-producing F_1 plants were crossed to produce an F_2 generation, 56% were pigment producers and 44% were not. What was happening? Emerson correctly deduced that two genes were involved in producing pigment, and that the second cross had thus been a dihybrid cross like those performed by Mendel. Mendel had predicted 16 equally possible ways gametes could combine with each other, resulting in genotypes with a phenotypic ratio of 9:3:3:1 (9 + 3 + 3 + 1 = 16). How many of these were in each of the two types Emerson obtained? He multiplied the fraction that were pigment producers (0.56) by 16 to obtain 9, and multiplied the fraction that were not (0.44) by 16 to obtain 7. Thus, Emerson had a **modified ratio** of 9:7 instead of the usual 9:3:3:1 ratio.

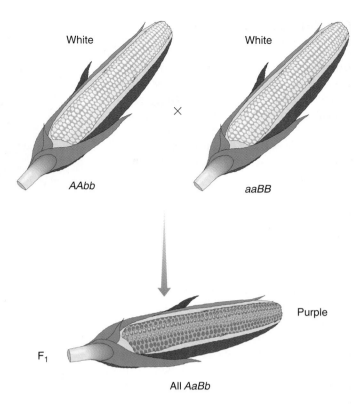

FIGURE 13.17
How epistasis affects grain color. The purple pigment found in some varieties of corn is the product of a two-step biochemical pathway. Unless both enzymes are active (the plant has a dominant allele for each of the two genes, *A* and *B*), no pigment is expressed.

Why Was Emerson's Ratio Modified? When genes act sequentially, as in a biochemical pathway, an allele expressed as a defective enzyme early in the pathway blocks the flow of material through the rest of the pathway. This makes it impossible to judge whether the later steps of the pathway are functioning properly. Such gene interaction, where one gene can interfere with the expression of another gene, is the basis of the phenomenon called **epistasis.**

The pigment anthocyanin is the product of a two-step biochemical pathway:

$$\text{Starting molecule} \xrightarrow{\text{Enzyme 1}} \text{Intermediate} \xrightarrow{\text{Enzyme 2}} \text{Anthocyanin}$$
(Colorless)   (Colorless)   (Purple)

To produce pigment, a plant must possess at least one good copy of each enzyme gene (figure 13.17). The dominant alleles encode functional enzymes, but the recessive alleles encode nonfunctional enzymes. Of the 16 genotypes predicted by random assortment, 9 contain at least one dominant allele of both genes; they produce purple progeny. The remaining 7 genotypes lack dominant alleles at either or both loci (3 + 3 + 1 = 7) and so are phenotypically the same (nonpigmented), giving the phenotypic ratio of 9:7 that Emerson observed. The inability to score enzyme 2 when enzyme 1 is nonfunctional is an example of epistasis.

Continuous Variation

Few phenotypes are the result of the action of only one gene. Instead, most traits reflect the action of **polygenes,** many genes that act sequentially or jointly. When multiple genes act jointly to influence a trait such as height or weight, the trait often shows a range of small differences. Because all of the genes that play a role in determining phenotypes such as height or weight segregate independently of one another, one sees a gradation in the degree of difference when many individuals are examined (figure 13.18). We call this graduation **continuous variation.** The greater the number of genes that influence a trait, the more continuous the expected distribution of the versions of that trait.

How can one describe the variation in a trait such as the height of the individuals in figure 13.18*a*? Individuals range from quite short to very tall, with average heights more common than either extreme. What one often does is to group the variation into categories—in this case, by measuring the heights of the individuals in inches, rounding fractions of an inch to the nearest whole number. Each height, in inches, is a separate phenotypic category. Plotting the numbers in each height category produces a histogram, such as that in figure 13.18*b*. The histogram approximates an idealized bell-shaped curve, and the variation can be characterized by the mean and spread of that curve.

Pleiotropic Effects

Often, an individual allele will have more than one effect on the phenotype. Such an allele is said to be **pleiotropic.** When the pioneering French geneticist Lucien Cuenot studied yellow fur in mice, a dominant trait, he was unable to obtain a true-breeding yellow strain by crossing individual yellow mice with each other. Individuals homozygous for the yellow allele died, because the yellow allele was pleiotropic: one effect was yellow color, but another was a lethal developmental defect. A pleiotropic gene alteration may be dominant with respect to one phenotypic consequence (yellow fur) and recessive with respect to another (lethal developmental defect). In pleiotropy, one gene affects many traits, in marked contrast to polygeny, where many genes affect one trait. Pleiotropic effects are difficult to predict, because the genes that affect a trait often perform other functions we may know nothing about.

(a)

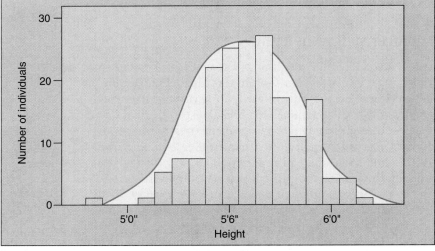

(b)

FIGURE 13.18

Height is a continuously varying trait. (a) This photograph shows the variation in height among students of the 1914 class of the Connecticut Agricultural College. Because many genes contribute to height and tend to segregate independently of one another, there are many possible combinations of those genes. (b) The cumulative contribution of different combinations of alleles to height forms a *continuous* spectrum of possible heights—a random distribution, in which the extremes are much rarer than the intermediate values.

Pleiotropic effects are characteristic of many inherited disorders, such as cystic fibrosis and sickle cell anemia, both discussed later in this chapter. In these disorders, multiple symptoms can be traced back to a single gene defect. In cystic fibrosis, patients exhibit clogged blood vessels, overly sticky mucus, salty sweat, liver and pancreas failure, and a battery of other symptoms. All are pleiotropic effects of a single defect, a mutation in a gene that encodes a chloride ion transmembrane channel. In sickle cell anemia, a defect in the oxygen-carrying hemoglobin molecule causes anemia, heart failure, increased susceptibility to pneumonia, kidney failure, enlargement of the spleen, and many other symptoms. It is usually difficult to deduce the nature of the primary defect from the range of its pleiotropic effects.

Lack of Complete Dominance

Not all alternative alleles are fully dominant or fully recessive in heterozygotes. Some pairs of alleles instead produce a heterozygous phenotype that is either intermediate between those of the parents (incomplete dominance), or representative of both parental phenotypes (codominance). For example, in the cross of red and white flowering Japanese four o'clocks described in figure 13.19, all the F$_1$ offspring had pink flowers—indicating that neither red nor white flower color was dominant. Does this example of incomplete dominance argue that Mendel was wrong? Not at all. When two of the F$_1$ pink flowers were crossed, they produced red-, pink-, and white-flowered plants in a 1:2:1 ratio. Heterozygotes are simply intermediate in color.

Environmental Effects

The degree to which an allele is expressed may depend on the environment. Some alleles are heat-sensitive, for example. Traits influenced by such alleles are more sensitive to temperature or light than are the products of other alleles. The arctic foxes in figure 13.20, for example, make fur pigment only when the weather is warm. Similarly, the *ch* allele in Himalayan rabbits and Siamese cats encodes a heat-sensitive version of tyrosinase, one of the enzymes mediating the production of melanin, a dark pigment. The ch version of the enzyme is inactivated at temperatures above about 33°C. At the surface of the main body and head, the temperature is above 33°C and the tyrosinase enzyme is inactive, while it is more active at body extremities such as the tips of the ears and tail, where the temperature is below 33°C. The dark melanin pigment this enzyme produces causes the ears, snout, feet, and tail of Himalayan rabbits and Siamese cats to be black.

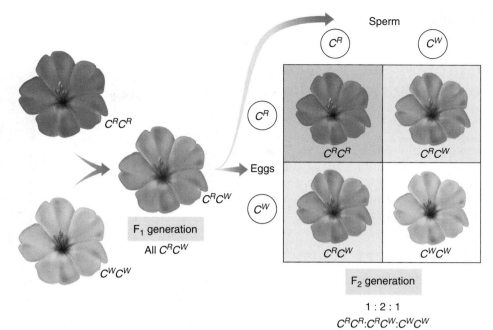

FIGURE 13.19

Incomplete dominance. In a cross between a red-flowered Japanese four o'clock, genotype C^RC^R, and a white-flowered one (C^WC^W), neither allele is dominant. The heterozygous progeny have pink flowers and the genotype C^RC^W. If two of these heterozygotes are crossed, the phenotypes of their progeny occur in a ratio of 1:2:1 (red:pink:white).

FIGURE 13.20

Environmental effects on an allele. An arctic fox in winter has a coat that is almost white, so it is difficult to see the fox against a snowy background. In summer, the same fox's fur darkens to a reddish brown, so that it resembles the color of the surrounding tundra. Heat-sensitive alleles control this color change.

A variety of factors can disguise the Mendelian segregation of alleles. Among them are gene interactions that produce epistasis, the continuous variation that results when many genes contribute to a trait, incomplete dominance that produces heterozygotes unlike either parent, and environmental influences on the expression of phenotypes.

Chromosomes: The Vehicles of Mendelian Inheritance

Chromosomes are not the only kinds of structures that segregate regularly when eukaryotic cells divide. Centrioles also divide and segregate in a regular fashion, as do the mitochondria and chloroplasts (when present) in the cytoplasm. Therefore, in the early twentieth century it was by no means obvious that chromosomes were the vehicles of hereditary information.

The Chromosomal Theory of Inheritance

A central role for chromosomes in heredity was first suggested in 1900 by the German geneticist Karl Correns, in one of the papers announcing the rediscovery of Mendel's work. Soon after, observations that similar chromosomes paired with one another during meiosis led directly to **the chromosomal theory of inheritance,** first formulated by the American Walter Sutton in 1902.

Several pieces of evidence supported Sutton's theory. One was that reproduction involves the initial union of only two cells, egg and sperm. If Mendel's model were correct, then these two gametes must make equal hereditary contributions. Sperm, however, contain little cytoplasm, suggesting that the hereditary material must reside within the nuclei of the gametes. Furthermore, while diploid individuals have two copies of each pair of homologous chromosomes, gametes have only one. This observation was consistent with Mendel's model, in which diploid individuals have two copies of each heritable gene and gametes have one. Finally, chromosomes segregate during meiosis, and each pair of homologues orients on the metaphase plate independently of every other pair. Segregation and independent assortment were two characteristics of the genes in Mendel's model.

A Problem with the Chromosomal Theory

However, investigators soon pointed out one problem with this theory. If Mendelian traits are determined by genes located on the chromosomes, and if the independent assortment of Mendelian traits reflects the independent assortment of chromosomes in meiosis, why does the number of traits that assort independently in a given kind of organism often greatly exceed the number of chromosome pairs the organism possesses? This seemed a fatal objection, and it led many early researchers to have serious reservations about Sutton's theory.

Morgan's White-Eyed Fly

The essential correctness of the chromosomal theory of heredity was demonstrated long before this paradox was

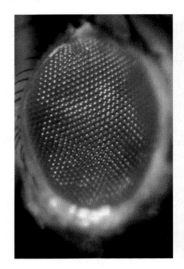

FIGURE 13.21
Red-eyed (wild type) and white-eyed (mutant) *Drosophila*.
The white-eyed defect is hereditary, the result of a mutation in a gene located on the X chromosome. By studying this mutation, Morgan first demonstrated that genes are on chromosomes.

resolved. A single small fly provided the proof. In 1910 Thomas Hunt Morgan, studying the fruit fly *Drosophila melanogaster*, detected a **mutant** male fly, one that differed strikingly from normal flies of the same species: its eyes were white instead of red (figure 13.21).

Morgan immediately set out to determine if this new trait would be inherited in a Mendelian fashion. He first crossed the mutant male to a normal female to see if red or white eyes were dominant. All of the F_1 progeny had red eyes, so Morgan concluded that red eye color was dominant over white. Following the experimental procedure that Mendel had established long ago, Morgan then crossed the red-eyed flies from the F_1 generation with each other. Of the 4252 F_2 progeny Morgan examined, 782 (18%) had white eyes. Although the ratio of red eyes to white eyes in the F_2 progeny was greater than 3:1, the results of the cross nevertheless provided clear evidence that eye color segregates. However, there was something about the outcome that was strange and totally unpredicted by Mendel's theory—*all of the white-eyed F_2 flies were males!*

How could this result be explained? Perhaps it was impossible for a white-eyed female fly to exist; such individuals might not be viable for some unknown reason. To test this idea, Morgan testcrossed the female F_1 progeny with the original white-eyed male. He obtained both white-eyed and red-eyed males and females in a 1:1:1:1 ratio, just as Mendelian theory predicted. Hence, a female could have white eyes. Why, then, were there no white-eyed females among the progeny of the original cross?

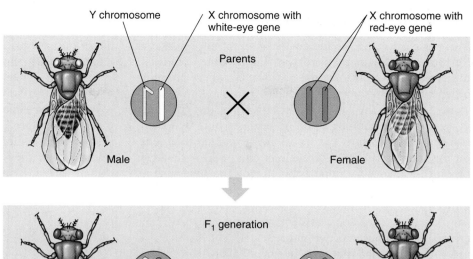

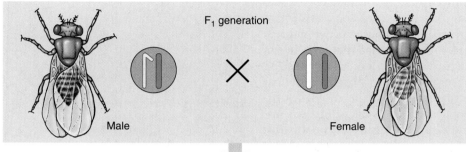

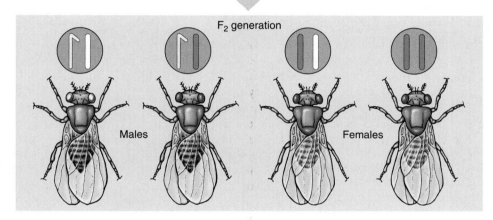

FIGURE 13.22
Morgan's experiment demonstrating the chromosomal basis of sex linkage in *Drosophila*. The white-eyed mutant male fly was crossed with a normal female. The F_1 generation flies all exhibited red eyes, as expected for flies heterozygous for a recessive white-eye allele. In the F_2 generation, all of the white-eyed flies were male.

Sex Linkage

The solution to this puzzle involved sex. In *Drosophila*, the sex of an individual is determined by the number of copies of a particular chromosome, the **X chromosome**, that an individual possesses. A fly with two X chromosomes is a female, and a fly with only one X chromosome is a male. In males, the single X chromosome pairs in meiosis with a large, dissimilar partner called the **Y chromosome**. The female thus produces only X gametes, while the male produces both X and Y gametes. When fertilization involves an X sperm, the result is an XX zygote, which develops into a female; when fertilization involves a Y sperm, the result is an XY zygote, which develops into a male.

The solution to Morgan's puzzle is that the gene causing the white-eye trait in *Drosophila* resides only on the X chromosome—it is absent from the Y chromosome. (We now know that the Y chromosome in flies carries almost no functional genes.) A trait determined by a gene on the sex chromosome is said to be **sex-linked.** Knowing the white-eye trait is recessive to the red-eye trait, we can now see that Morgan's result was a natural consequence of the Mendelian assortment of chromosomes (figure 13.22).

Morgan's experiment was one of the most important in the history of genetics because it presented the first clear evidence that the genes determining Mendelian traits do indeed reside on the chromosomes, as Sutton had proposed. The segregation of the white-eye trait has a one-to-one correspondence with the segregation of the X chromosome. In other words, Mendelian traits such as eye color in *Drosophila* assort independently because chromosomes do. When Mendel observed the segregation of alternative traits in pea plants, he was observing a reflection of the meiotic segregation of chromosomes.

Mendelian traits assort independently because they are determined by genes located on chromosomes that assort independently in meiosis.

Genetic Recombination

Morgan's experiments led to the general acceptance of Sutton's chromosomal theory of inheritance. Scientists then attempted to resolve the paradox that there are more independently assorting Mendelian genes than chromosomes. In 1903 the Dutch geneticist Hugo de Vries suggested that this paradox could be resolved only by assuming that homologous chromosomes exchange elements during meiosis. In 1909, French cytologist F. A. Janssens provided evidence to support this suggestion. Investigating chiasmata produced during amphibian meiosis, Janssens noticed that of the four chromatids involved in each chiasma, two crossed each other and two did not. He suggested that this crossing of chromatids reflected a switch in chromosomal arms between the paternal and maternal homologues, involving one chromatid in each homologue. His suggestion was not accepted widely, primarily because it was difficult to see how two chromatids could break and rejoin at exactly the same position.

Crossing Over

Later experiments clearly established that Janssens was indeed correct. One of these experiments, performed in 1931 by American geneticist Curt Stern, is described in figure 13.23. Stern studied two sex-linked eye traits in *Drosophila* strains whose X chromosomes were visibly abnormal at both ends. He first examined many flies and identified those in which an exchange had occurred with respect to the two eye traits. He then studied the chromosomes of those flies to see if their X chromosomes had exchanged arms. Stern found that all of the individuals that had exchanged eye traits also possessed chromosomes that had exchanged abnormal ends. The conclusion was inescapable: genetic exchanges of traits such as eye color involve the physical exchange of chromosome arms, a phenomenon called **crossing over.** Crossing over creates new combinations of genes, and is thus a form of **genetic recombination.**

The chromosomal exchanges Stern demonstrated provide the solution to the paradox, because crossing over can occur between homologues anywhere along the length of the chromosome, in locations that seem to be randomly determined. Thus, if two different genes are located relatively far apart on a chromosome, crossing over is more likely to occur somewhere between them than if they are located close together. Two genes can be on the same chromosome and still show independent assortment if they are located so far apart on the chromosome that crossing over occurs regularly between them (figure 13.24).

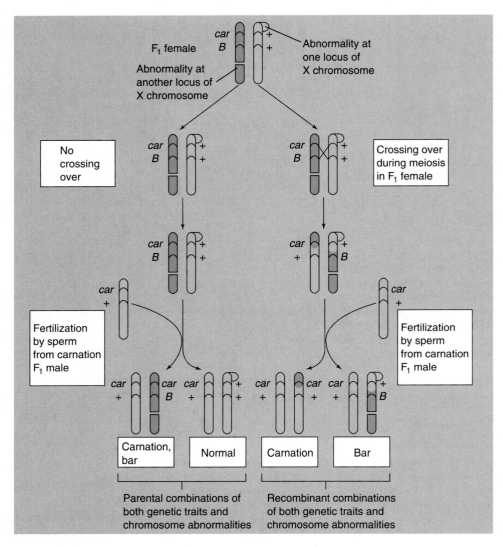

FIGURE 13.23
Stern's experiment demonstrating the physical exchange of chromosomal arms during crossing over. Stern monitored crossing over between two genes, the recessive carnation eye color (*car*) and the dominant bar-shaped eye (*B*), on chromosomes with physical peculiarities visible under a microscope. Whenever these genes recombined through crossing over, the chromosomes recombined as well. Therefore, the recombination of genes reflects a physical exchange of chromosome arms. The "+" notation on the chromosomes refers to the wild-type allele, the most common allele for a particular gene.

Using Recombination to Make Genetic Maps

Because crossing over is more frequent between two genes that are relatively far apart than between two that are close together, the frequency of crossing over can be used to map the relative positions of genes on chromosomes. In a cross, the proportion of progeny exhibiting an exchange between two genes is a measure of the frequency of crossover events between them, and thus indicates the relative distance separating them. The results of such crosses can be used to construct a **genetic map** that measures distance between genes in terms of the frequency of recombination. One "map unit" is defined as the distance within which a crossover event is expected to occur in an average of 1% of gametes. A map unit is now called a **centimorgan,** after Thomas Hunt Morgan.

In recent times new technologies have allowed geneticists to create gene maps based on the relative positions of specific gene sequences called *restriction sequences* because they are recognized by DNA-cleaving enzymes called restriction endonucleases. Restriction maps, discussed in chapter 18, have largely supplanted genetic recombination maps for detailed gene analysis because they are far easier to produce. Recombination maps remain the method of choice for genes widely separated on a chromosome.

The Three-Point Cross.   In constructing a genetic map, one simultaneously monitors recombination among three or more genes located on the same chromosome, referred to as **syntenic** genes. When genes are close enough together on a chromosome that they do not assort independently, they are said to be **linked** to one another. A cross involving three linked genes is called a **three-point cross.** Data obtained by Morgan on traits encoded by genes on the X chromosome of *Drosophila* were used by his student A. H. Sturtevant, to draw the first genetic map (figure 13.25). By convention, the most common allele of a gene is often denoted on a map with the symbol "+" and is designated as **wild type.** All other alleles are assigned specific symbols.

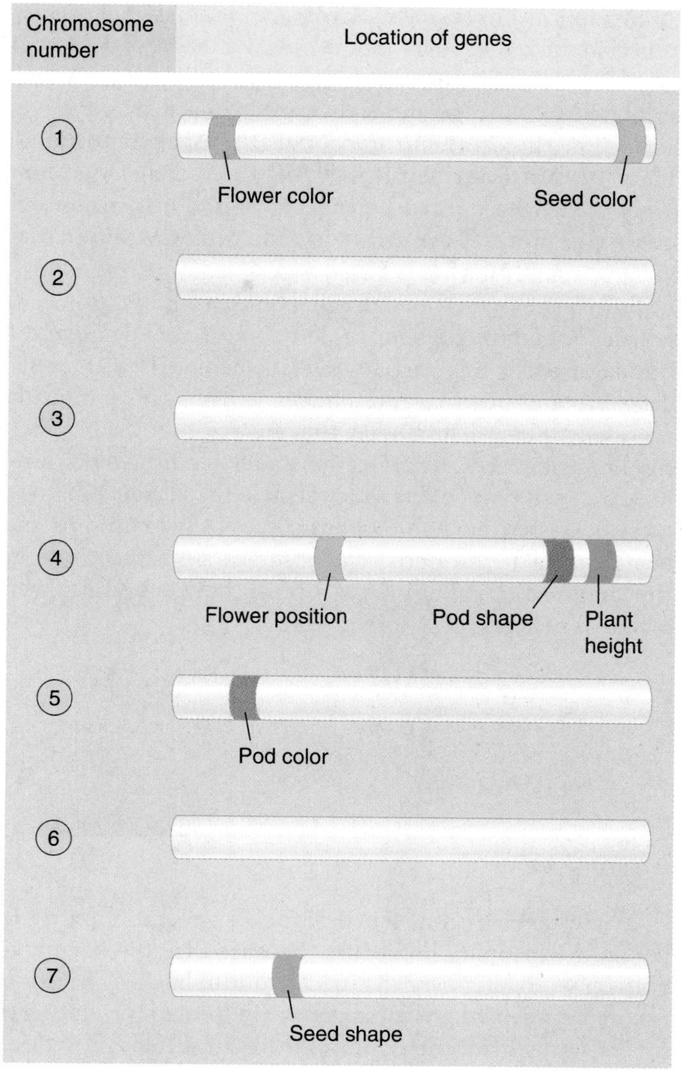

FIGURE 13.24
The chromosomal locations of the seven genes studied by Mendel in the garden pea. The genes for plant height and pod shape are very close to each other and rarely recombine. Plant height and pod shape were not among the pairs of traits Mendel examined in dihybrid crosses. One wonders what he would have made of the linkage he surely would have detected had he tested this pair of traits.

FIGURE 13.25
The first genetic map. This map of the X chromosome of *Drosophila* was prepared in 1913 by A. H. Sturtevant, a student of Morgan. On it he located the relative positions of five recessive traits that exhibited sex linkage by estimating their relative recombination frequencies in genetic crosses. Sturtevant arbitrarily chose the position of the *yellow* gene as zero on his map to provide a frame of reference. The higher the recombination frequency, the farther apart the two genes.

Five traits		Recombination frequencies	Genetic map
y	Yellow body color	*y* and *w* 0.010	.58 *r*
w	White eye color	*v* and *m* 0.030	
v	Vermilion eye color	*v* and *r* 0.269	
m	Miniature wing	*v* and *w* 0.300	
r	Rudimentary wing	*v* and *y* 0.322	.34 *m*
		w and *m* 0.327	.31 *v*
		y and *m* 0.355	
		w and *r* 0.450	.01 *w* 0 *y*

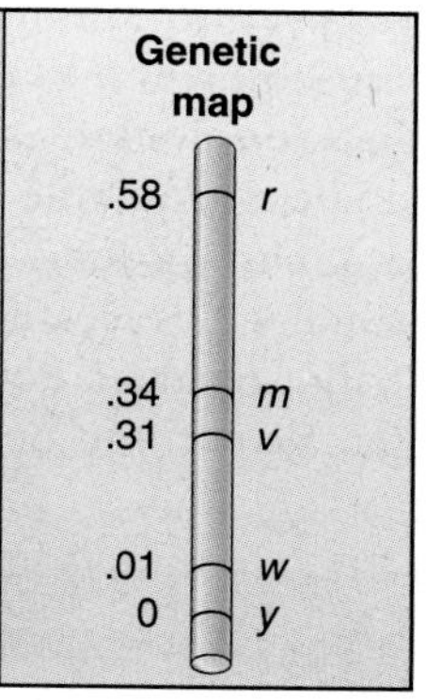

Analyzing a Three-Point Cross. The first genetic map was constructed by A. H. Sturtevant, a student of Morgan's in 1913. He studied several traits of *Drosophila*, all of which exhibited sex linkage and thus were encoded by genes residing on the same chromosome (the X chromosome). Here we will describe his study of three traits: *y*, yellow body color (the normal body color is grey), *w*, white eye color (the normal eye color is red), and *min*, miniature wing (the normal wing is 50% longer).

Sturtevant carried out the mapping cross by crossing a female fly homozygous for the three recessive alleles with a normal male fly that carried none of them. All of the progeny were thus heterozygotes. Such a cross is conventionally represented by a diagram like the one that follows, in which the lines represent gene locations and + indicates the normal, or "wild type" allele. Each female fly participating in a cross possesses two homologous copies of the chromosome being mapped, and both chromosomes are represented in the diagram. Crossing over occurs between these two copies in meiosis.

$$\text{P generation} \quad \frac{y\ w\ min}{y\ w\ min} \quad \times \quad \frac{+++}{\text{(Y chromosome)}}$$

$$\downarrow$$

$$\text{F}_1 \text{ generation} \atop \text{females} \quad \frac{y\ w\ min}{+\ +\ +}$$

These heterozygous females, the F_1 generation, are the key to the mapping procedure. Because they are heterozygous, any crossing over that occurs during meiosis will, if it occurs between where these genes are located, produce gametes with different combinations of alleles for these genes—in other words, recombinant chromosomes. Thus, a crossover between the homologous X chromosomes of such a female in the interval between the *y* and *w* genes will yield recombinant [*y* +] and [+ *w*] chromosomes, which are different combinations than we started with. (In the parental chromosomes, *w* is always linked with *y* and + linked with +.)

$$\frac{y\ w}{+\ +} \overset{\times}{} \quad \rightarrow \quad \frac{y\ +}{+\ w}$$

In order to see all the recombinant types that might be present among the gametes of these heterozygous flies, Sturtevant conducted a testcross. He crossed female heterozygous flies to males recessive for all three traits and examined the progeny. Since males contribute either a Y chromosome with no genes on it or an X chromosome with recessive alleles at all three loci, the male contribution does not disguise the potentially recombinant female chromosomes.

Table 13.2 summarizes the results Sturtevant obtained. The parentals are represented by the highest number of progeny and the double crossovers by the lowest number. To analyze his data, Sturtevant considered the traits in pairs and determined which involved a crossover event.

1. For the body trait (*y*) and the eye trait (*w*), the first two classes, [+ +] and [*y w*], involve no crossovers (they are parental combinations). In table 13.2, no progeny numbers are tabulated for these two classes on the "body-eye" column (a dash appears instead).
2. The next two classes have the same body-eye combination as the parents, [+ +] and [*y w*], so again no numbers are entered as recombinants under body-eye crossover type.
3. The next two classes, [+ *w*] and [*y* +], do *not* have the same body-eye combinations as the parent chromosomes), so the observed numbers of progeny are recorded, 16 and 12, respectively.
4. The last two classes also differ from parental chromosomes in body-eye combination, so again the observed numbers of each class are recorded, 1 and 0.
5. The sum of the numbers of observed progeny that are recombinant for body (*y*) and eye (*w*) is 16 + 12 + 1, or 29. Since the total number of progeny is 2205, this represents 29/2205, or 0.01315. The percentage of recombination between *y* and *w* is thus 1.315%, or 1.3 centimorgans.

To estimate the percentage of recombination between eye (*w*) and wing (*min*), one proceeds in the same manner, obtaining a value of 32.608%, or 32.6 centimorgans. Similarly, body (*y*) and wing (*min*) are separated by a recombination distance of 33.832%, or 33.8 centimorgans.

From this, then, we can construct our genetic map. The biggest distance, 33.8 centimorgans, separates the two outside genes, which are evidently *y* and *min*. The gene *w* is between them, near *y*.

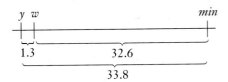

The two distances 1.3 and 32.6 do not add up to 33.8 but rather to 33.9. The difference, 0.1, represents chromosomes in which two crossovers occurred, one between *y* and *w* and another between *w* and *min*. These chromosomes do not exhibit recombination between *y* and *min*.

Genetic maps such as this are the key tools in genetic analysis, permitting an investigator reliably to predict how a newly discovered trait, once it has been located on the chromosome map, will recombine with many others.

Table 13.2 Sturtevant's Results

	Phenotypes			Number of Progeny	Crossover Types		
	Body	Eye	Wing		Body-Eye	Eye-Wing	Body-Wing
Parental	+	+	+	758	—	—	—
	y	*w*	*min*	700	—	—	—
Single crossover	+	+	*min*	401	—	401	401
	y	*w*	+	317	—	317	317
	+	*w*	*min*	16	16	—	16
	y	+	+	12	12	—	12
Double crossover	+	*w*	+	1	1	1	—
	y	+	*min*	0	0	0	—
TOTAL				2205	29	719	746
Recombination frequency (%)					1.315	32.608	33.832

The Human Genetic Map

Genetic maps of human chromosomes (figure 13.26) are of great importance. Knowing where particular genes are located on human chromosomes can often be used to tell whether a fetus at risk of inheriting a genetic disorder actually has the disorder. The genetic-engineering techniques described in chapter 18 have begun to permit investigators to isolate specific genes and determine their nucleotide sequences. It is hoped that knowledge of differences at the gene level may suggest successful therapies for particular genetic disorders and that knowledge of a gene's location on a chromosome will soon permit the substitution of normal genes for dysfunctional ones. Because of the great potential of this approach, investigators are working hard to assemble a detailed map of the entire human genome, the so-called **human genome project,** described in chapter 18. Initially, this map will consist of a "library" of thousands of small fragments of DNA whose relative positions are known. Investigators wishing to study a particular gene will first use techniques described in chapter 18 to screen this library and determine which fragment carries the gene of interest. They will then be able to analyze that fragment in detail. In parallel with this mammoth undertaking, the entire genomes of other, smaller genomes have already been sequenced, including yeasts and several bacteria. Progress on the human genome is rapid, and the full map is expected within the decade.

Gene maps locate the relative positions of different genes on the chromosomes of an organism. Traditionally produced by analyzing the relative amounts of recombination in genetic crosses, gene maps are increasingly being made by analyzing the sizes of fragments made by restriction enzymes.

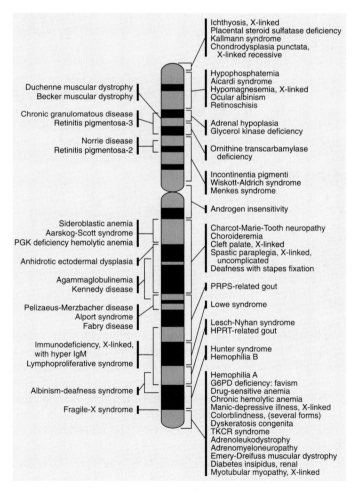

FIGURE 13.26
The human X chromosome gene map. Over 59 diseases have been traced to specific segments of the X chromosome. Many of these disorders are also influenced by genes on other chromosomes.

Multiple Alleles: The ABO Blood Groups

A gene may have more than two alleles in a population, and most genes possess several different alleles. Often, no single allele is dominant; instead, each allele has its own effect, and the alleles are considered **codominant.**

A human gene that exhibits more than one codominant allele is the gene that determines ABO blood type. This gene encodes an enzyme that adds sugar molecules to lipids on the surface of red blood cells. These sugars act as recognition markers for cells in the immune system and are called *cell surface antigens.* The gene that encodes the enzyme, designated *I*, has three common alleles: I^B, whose product adds the sugar galactose; I^A, whose product adds galactosamine; and *i*, which codes for a protein that does not add a sugar.

Different combinations of the three *I* gene alleles occur in different individuals because each person possesses two copies of the chromosome bearing the *I* gene and may be homozygous for any allele or heterozygous for any two. An individual heterozygous for the I^A and I^B alleles produces both forms of the enzyme and adds both galactose and galactosamine to the surfaces of red blood cells. Because both alleles are expressed simultaneously in heterozygotes, the I^A and I^B alleles are codominant. Both I^A and I^B are dominant over the *i* allele because both I^A or I^B alleles lead to sugar addition and the *i* allele does not. The different combinations of the three alleles produce four different phenotypes (figure 13.27):

1. Type A individuals add only galactosamine. They are either I^AI^A homozygotes or I^Ai heterozygotes.
2. Type B individuals add only galactose. They are either I^BI^B homozygotes or I^Bi heterozygotes.
3. Type AB individuals add both sugars and are I^AI^B heterozygotes.
4. Type O individuals add neither sugar and are *ii* homozygotes.

These four different cell surface phenotypes are called the **ABO blood groups** or, less commonly, the Landsteiner blood groups, after the man who first described them. As Landsteiner noted, a person's immune system can distinguish between these four phenotypes. If a type A individual receives a transfusion of type B blood, the recipient's immune system recognizes that the type B blood cells possess a "foreign" antigen (galactose) and attacks the donated blood cells, causing the cells to clump, or agglutinate. This also happens if the donated blood is type AB. However, if the donated blood is type O, no immune attack will occur, as there are no galactose antigens on the surfaces of blood cells produced by the type O donor. In general, any individual's immune system will tolerate a transfusion of type O blood. Because neither galactose nor galactosamine is foreign to type AB individuals (whose red blood cells have both sugars), those individuals may receive any type of blood.

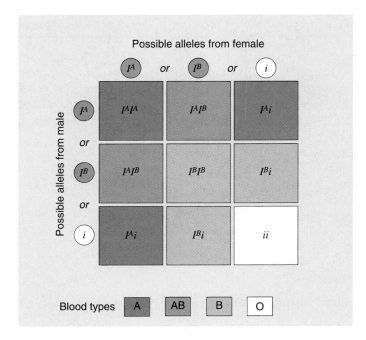

FIGURE 13.27
Multiple alleles control the ABO blood groups. Different combinations of the three *I* gene alleles result in four different blood type phenotypes: type A (either I^AI^A homozygotes or I^Ai heterozygotes), type B (either I^BI^B homozygotes or I^Bi heterozygotes), type AB (I^AI^B heterozygotes), and type O (*ii* homozygotes).

The Rh Blood Group

Another set of cell surface markers on human red blood cells are the **Rh blood group** antigens, named for the rhesus monkey in which they were first described. About 85% of adult humans have the Rh cell surface marker on their red blood cells, and are called Rh-positive. Rh-negative persons lack this cell surface marker because they are homozygous recessive for the gene encoding it.

If an Rh-negative person is exposed to Rh-positive blood, the Rh surface antigens of that blood are treated like foreign invaders by the Rh-negative person's immune system, which proceeds to make antibodies directed against the Rh antigens. This most commonly happens when an Rh-negative woman gives birth to an Rh-positive child (whose father is Rh-positive). Some fetal red blood cells cross the placental barrier and enter the mother's bloodstream, where they induce the production of "anti-Rh" antibodies. In subsequent pregnancies, the mother's antibodies can cross back to the new fetus and cause its red blood cells to clump, leading to a potentially fatal condition called erythroblastosis fetalis.

Many blood group genes possess multiple alleles, several of which may be common within populations.

Human Chromosomes

It wasn't until 1956 that techniques were developed that allowed investigators to determine the exact number of chromosomes in human cells. We now know that each human somatic cell normally has 46 chromosomes, which in meiosis form 23 pairs. By convention, the chromosomes are divided into seven groups (designated A through G), each characterized by a different size, shape, and appearance. The differences among the chromosomes are most clearly visible when the chromosomes are arranged in order in a karyotype (figure 13.28). Techniques that stain individual segments of chromosomes with different-colored dyes make the identification of chromosomes unambiguous. Like a fingerprint, each chromosome always exhibits the same pattern of colored bands.

Sex Chromosomes

Of the 23 pairs of human chromosomes, 22 are perfectly matched in both males and females and are called **autosomes.** The remaining pair, the **sex chromosomes,** consist of two similar chromosomes in females and two dissimilar chromosomes in males. In humans, as in *Drosophila* (but by no means in all diploid species), females are designated XX and males XY. One of the pair of sex chromosomes in the male (the Y chromosome) is highly condensed and bears few functional genes in most organisms. Because few of the genes on the Y chromosome are expressed, recessive alleles on a male's single X chromosome have no *active* counterpart on the Y chromosome. Some of the active genes the Y chromosome does possess are responsible for the features associated with "maleness" in humans. Consequently, any individual with *at least* one Y chromosome is a male, and any individual without a Y chromosome is a female.

Barr Bodies

Although males have only one copy of the X chromosome and females have two, female cells do not produce twice as much of the proteins encoded by genes on the X chromosome. Instead, one of the X chromosomes in females is inactivated early in embryonic development, shortly after the embryo's sex is determined. Which X chromosome is inactivated varies randomly from cell to cell. If a woman is heterozygous for a sex-linked trait, some of her cells will express one allele and some the other. The inactivated and highly condensed X chromosome is visible as a deeply staining **Barr body** attached to the nuclear membrane (figure 13.29).

One of the 23 pairs of human chromosomes carries the genes that determine sex. The gene determining maleness is located on a version of the sex chromosome called Y, which has few other transcribed genes.

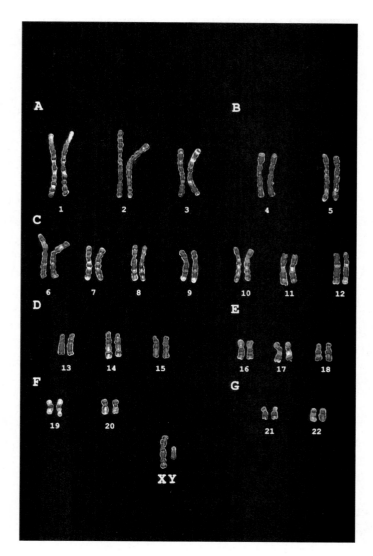

FIGURE 13.28
A human karyotype. This karyotype shows the colored banding patterns, arranged by class A–G.

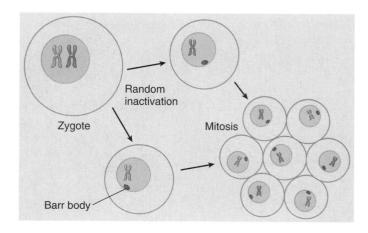

FIGURE 13.29
Barr bodies. In the developing female embryo, one of the X chromosomes (determined randomly) condenses and becomes inactivated. These condensed X chromosomes, called Barr bodies, then attach to the nuclear membrane.

Human Abnormalities Due to Alterations in Chromosome Number

Occasionally, homologues or sister chromatids fail to separate properly in meiosis, leading to the acquisition or loss of a chromosome in a gamete. This condition, called **primary nondisjunction,** can result in individuals with severe abnormalities if the affected gamete forms a zygote.

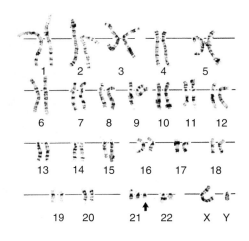

FIGURE 13.30
Down syndrome. As shown in this male karyotype, Down syndrome is associated with trisomy of chromosome 21. A child with Down syndrome sitting on his father's knee.

Nondisjunction Involving Autosomes

Almost all humans of the same sex have the same karyotype, for the same reason that all automobiles have engines, transmissions, and wheels: other arrangements don't work well. Humans who have lost even one copy of an autosome (called **monosomics**) do not survive development. In all but a few cases, humans who have gained an extra autosome (called **trisomics**) also do not survive. However, five of the smallest autosomes—those numbered 13, 15, 18, 21, and 22—can be present in humans as three copies and still allow the individual to survive for a time. The presence of an extra chromosome 13, 15, or 18 causes severe developmental defects, and infants with such a genetic makeup die within a few months. In contrast, individuals who have an extra copy of chromosome 21 or, more rarely, chromosome 22, usually survive to adulthood. In such individuals, the maturation of the skeletal system is delayed, so they generally are short and have poor muscle tone. Their mental development is also affected, and children with trisomy 21 or trisomy 22 are always mentally retarded.

Down Syndrome. The developmental defect produced by trisomy 21 (figure 13.30) was first described in 1866 by J. Langdon Down; for this reason, it is called **Down syndrome** (formerly "Down's syndrome"). About 1 in every 750 children exhibits Down syndrome, and the frequency is similar in all racial groups. Similar conditions also occur in chimpanzees and other related primates. In humans, the defect is associated with a particular small portion of chromosome 21. When this chromosomal segment is present in three copies instead of two, Down syndrome results. In 97% of the human cases examined, all of chromosome 21 is present in three copies. In the other 3%, a small portion of chromosome 21 containing the critical segment has been added to another chromosome by translocation; it exists along with the normal two copies of chromosome 21. This condition is known as *translocation Down syndrome.*

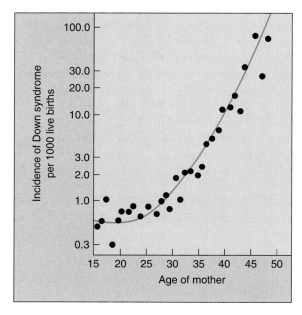

FIGURE 13.31
Correlation between maternal age and the incidence of Down syndrome. As women age, the chances they will bear a child with Down syndrome increase. After a woman reaches 35, the frequency of Down syndrome increases rapidly.

Not much is known about the developmental role of the genes whose duplication produces Down syndrome, although clues are beginning to emerge from current research. Some researchers suspect that the gene or genes that produce Down syndrome are similar or identical to some of the genes associated with cancer and with Alzheimer's disease. The reason for this suspicion is that one of the human cancer-causing genes (to be described in chapter 17) and the gene causing

Alzheimer's disease are located on the segment of chromosome 21 associated with Down syndrome. Moreover, cancer is more common in children with Down syndrome. The incidence of leukemia, for example, is 11 times higher in children with Down syndrome than in unaffected children of the same age.

How does Down syndrome arise? In humans, it comes about almost exclusively as a result of primary nondisjunction of chromosome 21 during egg formation. The cause of these primary nondisjunctions is not known, but their incidence, like that of cancer, increases with age (figure 13.31). In mothers younger than 20 years of age, the risk of giving birth to a child with Down syndrome is about 1 in 1700; in mothers 20 to 30 years old, the risk is only about 1 in 1400. In mothers 30 to 35 years old, however, the risk rises to 1 in 750, and by age 45, the risk is as high as 1 in 16!

Primary nondisjunctions are far more common in women than in men because all of the eggs a woman will ever produce have developed to the point of prophase in meiosis I by the time she is born. By the time she has children, her eggs are as old as she is. In men, by contrast, new sperm develop daily. Therefore, there is a much greater chance for problems of various kinds, including those that cause primary nondisjunction, to accumulate over time in the gametes of women than in those of men. For this reason, the age of the mother is more critical than that of the father in couples contemplating childbearing.

Nondisjunction Involving the Sex Chromosomes

Individuals that gain or lose a sex chromosome do not generally experience the severe developmental abnormalities caused by similar changes in autosomes. Such individuals may reach maturity, but they have somewhat abnormal features.

The X Chromosome. When X chromosomes fail to separate during meiosis, some of the gametes that are produced possess both X chromosomes and so are XX gametes; the other gametes that result from such an event have no sex chromosome and are designated "O" (figure 13.32).

If an XX gamete combines with an X gamete, the resulting XXX zygote develops into a female with one functional X chromosome and two Barr bodies. She is sterile but usually normal in other respects. If an XX gamete instead combines with a Y gamete, the effects are more serious. The resulting XXY zygote develops into a sterile male who has many female body characteristics and, in some cases, diminished mental capacity. This condition, called *Klinefelter syndrome*, occurs in about 1 out of every 500 male births.

If an O gamete fuses with a Y gamete, the resulting OY zygote is nonviable and fails to develop further because humans cannot survive when they lack the genes on the X chromosome. If, on the other hand, an O gamete fuses with an X gamete, the XO zygote develops into a sterile female of

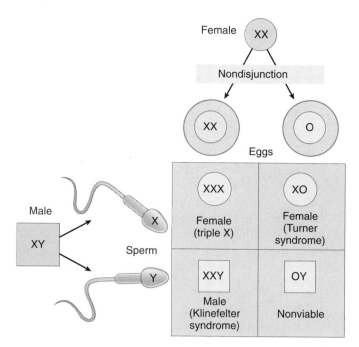

FIGURE 13.32
How nondisjunction can produce abnormalities in the number of sex chromosomes. When nondisjunction occurs in the production of female gametes, the gamete with two X chromosomes (XX) produces Klinefelter males (XXY) and XXX females. The gamete with no X chromosome (O) produces Turner females (XO) and nonviable OY males lacking any X chromosome.

short stature, with a webbed neck and immature sex organs that do not undergo changes during puberty. The mental abilities of an XO individual are in the low-normal range. This condition, called *Turner syndrome*, occurs roughly once in every 5000 female births.

The Y Chromosome. The Y chromosome can also fail to separate in meiosis, leading to the formation of YY gametes. When these gametes combine with X gametes, the XYY zygotes develop into fertile males of normal appearance. The frequency of the XYY genotype is about 1 per 1000 newborn males, but it is approximately 20 times higher among males in penal and mental institutions. This observation has led to the highly controversial suggestion that XYY males are inherently antisocial, a suggestion supported by some studies but not by others. In any case, most XYY males do not develop patterns of antisocial behavior.

Gene dosage plays a crucial role in development, so humans do not tolerate the loss or addition of chromosomes well. Autosome loss is always lethal, and an extra autosome is with few exceptions lethal too. Additional sex chromosomes have less serious consequences, although they can lead to sterility.

Human Genetic Disorders

In chapter 17 we will discuss the process of mutation, which produces variant alleles. At this point, we will note only that mutation involves random changes in genes, and that such changes rarely improve the functioning of the proteins those genes encode, just as randomly changing a wire in a computer rarely improves the computer's functioning. Therefore, variant alleles arising from mutation are rare in populations of organisms.

Nevertheless, some alternative alleles with detrimental effects are present in populations. Usually, they are recessive to other alleles. When two seemingly normal individuals who are heterozygous for such an allele produce offspring homozygous for the allele, the offspring suffer the detrimental effects of the mutant allele. When a detrimental allele occurs at a significant frequency in a population, the harmful effect it produces is called a **genetic disorder.** Table 13.3 lists some of the most prevalent genetic disorders in humans. We know a great deal about some of them, and much less about many others. Learning how to prevent them is one of the principal goals of human genetics.

Most Gene Defects Are Rare: Tay-Sachs Disease

Tay-Sachs disease is an incurable hereditary disorder in which the brain deteriorates. Affected children appear normal at birth and usually do not develop symptoms until about the eighth month, when signs of mental deterioration appear. The children are blind within a year after birth, and they rarely live past five years of age.

Tay-Sachs disease is rare in most human populations, occurring in only 1 of 300,000 births in the United States. However, the disease has a high incidence among Jews of Eastern and Central Europe (Ashkenazi), and among American Jews, 90% of whom trace their ancestry to East-

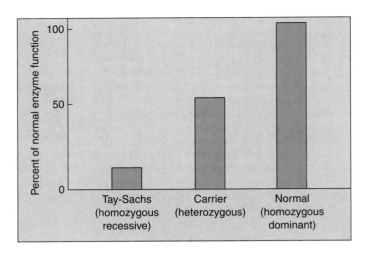

FIGURE 13.33
Tay-Sachs disease. Homozygous individuals (*left bar*) typically have less than 10% of the normal level of hexosaminidase A (*right bar*), while heterozygous individuals (*middle bar*) have about 50% of the normal level—enough to prevent deterioration of the central nervous system.

ern and Central Europe. In these populations, it is estimated that 1 in 28 individuals is a heterozygous carrier of the disease, and approximately 1 in 3500 infants has the disease. Because the disease is caused by a recessive allele, most of the people who carry the defective allele do not themselves develop symptoms of the disease.

The Tay-Sachs allele produces the disease by encoding a nonfunctional form of the enzyme hexosaminidase A. This enzyme breaks down *gangliosides*, a class of lipids occurring within the lysosomes of brain cells (figure 13.33). As a result, the lysosomes fill with gangliosides, swell, and eventually burst, releasing oxidative enzymes that kill the cells. There is no known cure for this disorder.

Table 13.3 Some Important Genetic Disorders				
Disorder	**Symptom**	**Defect**	**Dominant/ Recessive**	**Frequency among Human Births**
Cystic fibrosis	Mucus clogs lungs, liver, and pancreas	Failure of chloride ion transport mechanism	Recessive	1/2500 (Caucasians)
Sickle cell anemia	Poor blood circulation	Abnormal hemoglobin molecules	Recessive	1/625 (African Americans)
Tay-Sachs disease	Deterioration of central nervous system in infancy	Defective enzyme (hexosaminidase A)	Recessive	1/3500 (Ashkenazi Jews)
Phenylketonuria	Brain fails to develop in infancy	Defective enzyme (phenylalanine hydroxylase)	Recessive	1/12,000
Hemophilia	Blood fails to clot	Defective blood clotting factor VIII	Sex-linked recessive	1/10,000 (Caucasian males)
Huntington's disease	Brain tissue gradually deteriorates in middle age	Production of an inhibitor of brain cell metabolism	Dominant	1/24,000
Muscular dystrophy (Duchenne)	Muscles waste away	Degradation of myelin coating of nerves stimulating muscles	Sex-linked recessive	1/3700 (males)
Hypercholesterolemia	Excessive cholesterol levels in blood, leading to heart disease	Abnormal form of cholesterol cell surface receptor	Dominant	1/500

Gene Defects Are Inherited in Families: Hemophilia

When a blood vessel ruptures, the blood in the immediate area of the rupture forms a solid gel called a clot. The clot forms as a result of the polymerization of protein fibers circulating in the blood. A dozen proteins are involved in this process, and all must function properly for a blood clot to form. A mutation causing any of these proteins to loose their activity leads to a form of **hemophilia,** a hereditary condition in which the blood is slow to clot or does not clot at all.

Hemophilias are recessive disorders, expressed only when an individual does not possess any copy of the normal allele and so cannot produce one of the proteins necessary for clotting. Most of the genes that encode the blood-clotting proteins are on autosomes, but two (designated VIII and IX) are on the X chromosome. These two genes are sex-linked: any male who inherits a mutant allele of either of the two genes will develop hemophilia because his other sex chromosome is a Y chromosome that lacks any alleles of those genes.

The most famous instance of hemophilia, often called the Royal hemophilia, is a sex-linked form that arose in the royal family of England. This hemophilia was caused by a mutation in gene IX that occurred in one of the parents of Queen Victoria of England (1819–1901; figure 13.34). In the five generations since Queen Victoria, 10 of her male descendants have had hemophilia. The present British royal family has escaped the disorder because Queen Victoria's son, King Edward VII, did not inherit the defective allele, and all the subsequent rulers of England are his descendants. Three of Victoria's nine children did receive the defective allele, however, and they carried it by marriage into many of the other royal families of Europe (figure 13.35). It is still being transmitted to future generations through these family lines—except in Russia, where all of the five children of Alexandra, Victoria's granddaughter, were killed soon after the Russian revolution in 1917. (Speculation that one daughter, Anastasia, might have survived ended in 1996 when DNA analysis confirmed the identity of her remains.)

FIGURE 13.34

Queen Victoria of England, surrounded by some of her descendants in 1894. Of Victoria's four daughters who lived to bear children, two, Alice and Beatrice, were carriers of Royal hemophilia. Two of Alice's daughters are standing behind Victoria (wearing feathered boas): Princess Irene of Prussia (right), and Alexandra (left), who would soon become Czarina of Russia. Both Irene and Alexandra were also carriers of hemophilia.

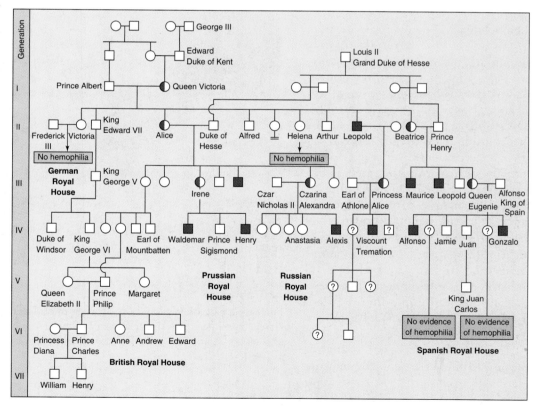

FIGURE 13.35

The Royal hemophilia pedigree. Queen Victoria's daughter Alice introduced hemophilia into the Russian and Austrian royal houses, and Victoria's daughter Beatrice introduced it into the Spanish royal house. Victoria's son Leopold, himself a victim, also transmitted the disorder in a third line of descent. Half-shaded symbols represent carriers with one normal allele and one defective allele; fully shaded symbols represent affected individuals.

Gene Defects Often Affect Specific Proteins: Sickle Cell Anemia

Sickle cell anemia is a heritable disorder first noted in Chicago in 1904. Afflicted individuals have defective molecules of hemoglobin, the protein within red blood cells that carries oxygen. Consequently, these individuals are unable to properly transport oxygen to their tissues. The defective hemoglobin molecules stick to one another, forming stiff, rod-like structures and resulting in the formation of sickle-shaped red blood cells (figure 13.36). As a result of their stiffness and irregular shape, these cells have difficulty moving through the smallest blood vessels; they tend to accumulate in those vessels and form clots. People who have large proportions of sickle-shaped red blood cells tend to have intermittent illness and a shortened life span.

The hemoglobin in the defective red blood cells differs from that in normal red blood cells in only one of hemoglobin's 574 amino acid subunits. In the defective hemoglobin, the amino acid valine replaces a glutamic acid at a single position in the protein. Interestingly, the position of the change is far from the active site of hemoglobin where the iron-bearing heme group binds oxygen. Instead, the change occurs on the outer edge of the protein. Why then is the result so catastrophic? The sickle cell mutation puts a very nonpolar amino acid on the surface of the hemoglobin protein, creating a "sticky patch" that sticks to other such patches—nonpolar amino acids tend to associate with one another in polar environments like water. As one hemoglobin adheres to another, ever-longer chains of hemoglobin molecules form.

Individuals heterozygous for the sickle cell allele are generally indistinguishable from normal persons. However, some of their red blood cells show the sickling characteristic when they are exposed to low levels of oxygen. The allele responsible for sickle cell anemia is particularly common among people of African descent; about 9% of African Americans are heterozygous for this allele, and about 0.2% are homozygous and therefore have the disorder. In some groups of people in Africa, up to 45% of all individuals are

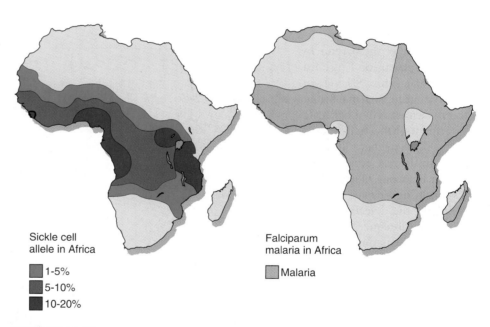

FIGURE 13.36
Sickle cell anemia. In individuals homozygous for the sickle cell trait, many of the red blood cells have sickle or irregular shapes, such as the cell on the far right.

Sickle cell allele in Africa

■ 1-5%
■ 5-10%
■ 10-20%

Falciparum malaria in Africa

□ Malaria

FIGURE 13.37
The sickle cell allele confers resistance to malaria. The distribution of sickle cell anemia closely matches the occurrence of malaria in central Africa. This is not a coincidence. The sickle cell allele, when heterozygous, confers resistance to malaria, a very serious disease.

heterozygous for this allele, and fully 6% are homozygous and express the disorder. What factors determine the high frequency of sickle cell anemia in Africa? It turns out that heterozygosity for the sickle cell anemia allele increases resistance to malaria, a common and serious disease in central Africa (figure 13.37). We will discuss this situation in more detail in chapter 19.

Not All Gene Defects Are Recessive: Huntington's Disease

Not all hereditary disorders are recessive. **Huntington's disease** is a hereditary condition caused by a dominant allele that causes the progressive deterioration of brain cells (figure 13.38). Perhaps 1 in 24,000 individuals develops the disorder. Since the allele is dominant, every individual that carries the allele expresses the disorder. Nevertheless, the disorder persists in human populations because its symptoms usually do not develop until the affected individuals are more than 30 years old, and by that time most of those individuals have already had children. Consequently, the allele is often transmitted before the lethal condition develops. A person who is heterozygous for Huntington's disease has a 50% chance of passing the *disease* to his or her children (even though the other parent does not have the disorder). In contrast, the carrier of a recessive disorder such as cystic fibrosis has a 50% chance of passing the *allele* to offspring and must mate with another carrier to risk bearing a child with the disease.

Some Gene Defects May Soon Be Curable: Cystic Fibrosis

Some of the most common and serious gene defects result from single recessive mutations, including many of the defects listed in table 13.3. Recent developments in gene technology have raised the hope that this class of disorders may be curable. Perhaps the best example is **cystic fibrosis,** the most common fatal genetic disorder among Caucasians (figure 13.39). As we learned in chapter 6, affected individuals secrete a thick mucus that clogs their lungs, and the passages of their pancreas and liver. About 1 in 20 Caucasians has a copy of the defective gene but shows no symptoms; homozygous recessive individuals make up about 1 in 2500 Caucasian children. These individuals inevitably die from complications that result from their disease.

We know that the cause of cystic fibrosis is a defect in the way certain cells regulate the transport of chloride ions across their membranes. Cystic fibrosis occurs when an individual is homozygous for an allele that encodes a defective version of the protein that regulates the chloride transport channel. This allele is recessive to the gene that codes for the normal version of the regulatory protein, so the chloride channels of heterozygous individuals function normally and the individuals do not develop cystic fibrosis.

The gene responsible for cystic fibrosis, dubbed CFTR for *cystic fibrosis transmembrane regulator*, was isolated in 1989, and attempts are underway to introduce healthy copies of the gene into cystic fibrosis patients. This **gene transfer therapy** was carried out successfully with mice in 1994, but initial attempts to introduce the gene into humans (using cold viruses to carry the gene) have not yet succeeded. These procedures are discussed in detail in chapter 18.

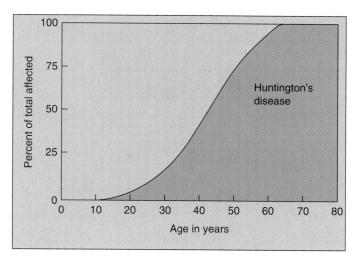

FIGURE 13.38
Huntington's disease is a dominant genetic disorder. It is because of the late age of onset of this disease that it persists despite the fact that it is dominant and fatal.

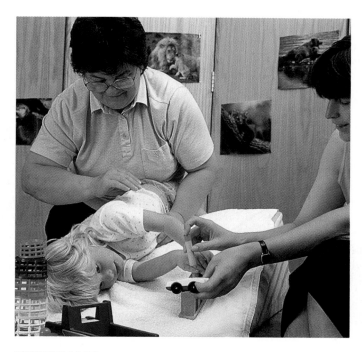

FIGURE 13.39
A child with cystic fibrosis. In cystic fibrosis patients, the mucus that normally lines the insides of the lungs thickens, making breathing difficult. Affected children are expected to live into their late twenties.

Many heritable disorders are the result of recessive mutations in genes encoding critical proteins such as those that clot blood, carry oxygen, or transport chloride ions into and out of cells. All such disorders are potentially curable if ways can be found to successfully introduce undamaged copies of the genes into affected individuals.

Genetic Counseling

Although most genetic disorders cannot yet be cured, we are learning a great deal about them, and progress toward successful therapy is being made in many cases. In the absence of a cure, however, the only recourse is to try to avoid producing children with these conditions. The process of identifying parents at risk of producing children with genetic defects and of assessing the genetic state of early embryos is called *genetic counseling.*

If a genetic defect is caused by a recessive allele, how can potential parents determine the likelihood that they carry the allele? One way is through pedigree analysis, often employed as an aid in genetic counseling. By analyzing a person's pedigree, it is sometimes possible to estimate the likelihood that the person is a carrier for certain disorders. For example, if one of your relatives has been afflicted with a recessive genetic disorder such as cystic fibrosis, it is possible that you are a heterozygous carrier of the recessive allele for that disorder. When a couple is expecting a child, and pedigree analysis indicates that both of them have a significant probability of being heterozygous carriers of a recessive allele responsible for a serious genetic disorder, the pregnancy is said to be a **high-risk pregnancy.** In such cases, there is a significant probability that the child will exhibit the clinical disorder.

Another class of high-risk pregnancies is that in which the mothers are more than 35 years old. As we have seen, the frequency of birth of infants with Down syndrome increases dramatically in the pregnancies of older women (see figure 13.31).

When a pregnancy is diagnosed as being high-risk, many women elect to undergo *amniocentesis,* a procedure that permits the prenatal diagnosis of many genetic disorders (figure 13.40). In the fourth month of pregnancy, a sterile hypodermic needle is inserted into the expanded uterus of the mother, removing a small sample of the amniotic fluid bathing the fetus. Within the fluid are free-floating cells derived from the fetus; once removed, these cells can be grown in cultures in the laboratory. During amniocentesis, the position of the needle and that of the fetus are usually observed by means of *ultrasound* (figure 13.41). The sound waves used in ultrasound are not harmful to mother or fetus, and they permit the person withdrawing the amniotic fluid to do so without damaging the fetus. In addition, ultrasound can be used to examine the fetus for signs of major abnormalities.

In recent years, physicians have increasingly turned to a new, less invasive procedure for genetic screening called **chorionic villi sampling.** In this procedure, the physician

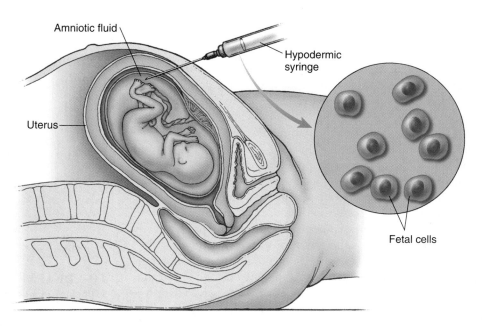

FIGURE 13.40
Amniocentesis. A needle is inserted into the amniotic cavity, and a sample of amniotic fluid, containing some free cells derived from the fetus, is withdrawn into a syringe. The fetal cells are then grown in culture and their karyotype and many of their metabolic functions are examined.

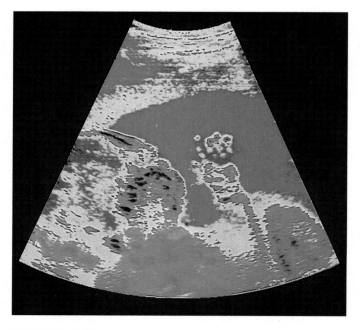

FIGURE 13.41
An ultrasound view of a fetus. During the fourth month of pregnancy, when amniocentesis is normally performed, the fetus usually moves about actively. The head of the fetus above is to the left.

removes cells from the chorion, a membranous part of the placenta that nourishes the fetus. This procedure can be used earlier in pregnancy (by the eighth week) and yields results much more rapidly than does amniocentesis.

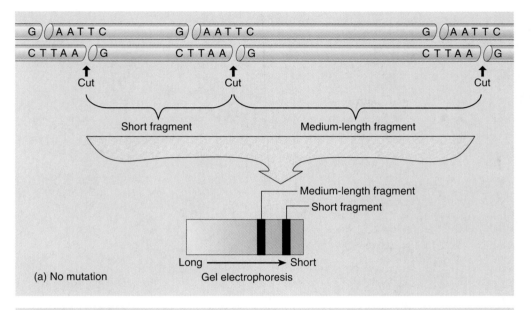

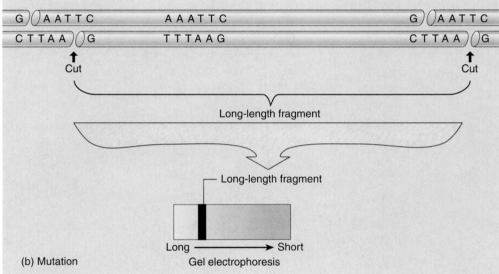

(a) No mutation

(b) Mutation

FIGURE 13.42
RFLPs. Restriction fragment length polymorphisms (RFLPs) are playing an increasingly important role in genetic identification. In (a), the restriction endonuclease cuts the DNA molecule in three places, producing two fragments. In (b), the mutation of a single nucleotide from G to A (see top fragment) alters a restriction endonuclease cutting site. Now the enzyme no longer cuts the DNA molecule at that site. As a result, a single long fragment is obtained, rather than two shorter ones. Such a change is easy to detect when the fragments are subjected to a technique called gel electrophoresis.

To test for certain genetic disorders, genetic counselors can look for three things in the cultures of cells obtained from amniocentesis or chorionic villi sampling. First, analysis of the karyotype can reveal aneuploidy and gross chromosomal alterations. Second, in many cases it is possible to test directly for the *proper functioning of enzymes* involved in genetic disorders. The lack of normal enzymatic activity signals the presence of the disorder. Thus, the lack of the enzyme responsible for breaking down phenylalanine signals PKU, the absence of the enzyme responsible for the breakdown of gangliosides indicates Tay-Sachs disease, and so forth.

Third, genetic counselors can look for an *association with known genetic markers*. For sickle cell anemia, Huntington's disease, and one form of muscular dystrophy (a genetic disorder characterized by weakened muscles), investigators have found other mutations on the same chromosomes that, by chance, occur at about the same place as the mutations that cause those disorders. By testing for the presence of these other mutations, a genetic counselor can identify individuals with a high probability of possessing the disorder-causing mutations. Finding such mutations in the first place is a little like searching for a needle in a haystack, but persistent efforts have proved successful in these three disorders. The associated mutations are detectable because they alter the length of the DNA segments that **restriction enzymes** produce when they cut strands of DNA at particular places (see chapter 18). Therefore, these mutations produce what are called **restriction fragment length polymorphisms,** or **RFLPs** (figure 13.42).

Many gene defects can be detected early in pregnancy, allowing for appropriate planning by the prospective parents.

Probability and Allele Distribution

Many, although not all, alternative alleles produce discretely different phenotypes. Mendel's pea plants were tall or dwarf, had purple or white flowers, and produced smooth or shriveled seeds. The eye color of a fruit fly may be red or white, and the skin color of a human may be pigmented or albino. When only two alternative alleles exist for a given trait, the distribution of phenotypes among the progeny of a cross is referred to as a **binomial distribution.**

As an example, consider the distribution of sexes in humans. Imagine that a couple has chosen to have three children. How likely is it that two of the children will be boys and one will be a girl? The frequency of any particular possibility is referred to as its **probability** of occurrence. Let p symbolize the probability of having a boy at any given birth and q symbolize the probability of having a girl. Since any birth is equally likely to produce a girl or boy:

$$p = q = \tfrac{1}{2}$$

Table 13.A shows eight possible gender combinations among the three children. The sum of the probabilities of the eight different ways must equal one. Thus:

$$p^3 + 3p^2q + 3pq^2 + q^3 = 1$$

The probability that the three children will be two boys and one girl is:

$$3p^2q = 3 \times (\tfrac{1}{2})^2 \times (\tfrac{1}{2}) = \tfrac{3}{8}$$

To test your understanding, try to estimate the probability that two parents heterozygous for the recessive allele producing albinism (a) will have one albino child in a family of three. First, set up a Punnett square:

		Father's Gametes	
		A	**a**
Mother's	**A**	AA	Aa
Gametes	**a**	Aa	aa

You can see that one-fourth of the children are expected to be albino (aa). Thus, for any given birth the probability of an albino child is $\tfrac{1}{4}$. This probability can be symbolized by q. The probability of a nonalbino child is $\tfrac{3}{4}$, symbolized by p. Therefore, the probability that there will be one albino child among the three children is:

$$3p^2q = 3 \times (\tfrac{3}{4})^2 \times (\tfrac{1}{4}) = \tfrac{27}{64}, \text{ or } 42\%$$

This means that the chance of producing one albino child in the three is 42%.

Table 13.A Binomial Distribution of the Sexes of Children in Human Families

Composition of Family	Order of Birth	Calculation	Probability	
3 boys	bbb	$p \times p \times p$	p^3	
2 boys and 1 girl	bbg	$p \times p \times q$	p^2q	
	bgb	$p \times q \times p$	p^2q	$3p^2q$
	gbb	$q \times p \times p$	p^2q	
1 boy and 2 girls	ggb	$q \times q \times p$	pq^2	
	gbg	$q \times p \times q$	pq^2	$3pq^2$
	bgg	$p \times q \times q$	pq^2	
3 girls	ggg	$q \times q \times q$	q^3	

A Vocabulary of Genetics

allele One of two or more alternative forms of a gene.

diploid Having two sets of chromosomes, which are referred to as *homologues*. Animals are diploid as well as plants in the dominant phase of their life cycle as are some protists.

dominant allele An allele that dictates the appearance of heterozygotes. One allele is said to be dominant over another if a heterozygous individual with one copy of that allele has the same appearance as a homozygous individual with two copies of it.

gene The basic unit of heredity; a sequence of DNA nucleotides on a chromosome that encodes a polypeptide or RNA molecule and so determines the nature of an individual's inherited traits.

genotype The total set of genes present in the cells of an organism. This term is often also used to refer to the set of alleles at a single gene locus.

haploid Having only one set of chromosomes. Gametes, certain animals, protists and fungi, and certain stages in the life cycle of plants are haploid.

heterozygote A diploid individual carrying two different alleles of a gene on its two homologous chromosomes. Most human beings are heterozygous for many genes.

homozygote A diploid individual whose two copies of a gene are the same. An individual carrying identical alleles of a gene on both homologous chromosomes is said to be *homozygous* for that gene.

locus The location of a gene on a chromosome.

phenotype The realized expression of the genotype; the observable manifestation of a trait (affecting an individual's structure, physiology, or behavior) that results from the biological activity of the DNA molecules.

recessive allele An allele whose phenotypic effect is masked in heterozygotes by the presence of a dominant allele.

Summary of Concepts

13.1 Mendel solved the mystery of heredity.

- Koelreuter noted the basic facts of heredity a century before Mendel. He found that alternative traits segregate in crosses and may mask each other's appearance. Mendel, however, was the first to quantify his data, counting the numbers of each alternative type among the progeny of crosses.

- By counting progeny types, Mendel learned that the alternatives that were masked in hybrids (the F_1 generation) appeared only 25% of the time in the F_2 generation. This finding, which led directly to Mendel's model of heredity, is usually referred to as the Mendelian ratio of 3:1 dominant-to-recessive traits.

- Mendel deduced from the 3:1 ratio that traits are specified by discrete "factors" that do not blend. He deduced that pea plants contain two factors for each trait that he studied (we now know this is because the plants are diploid). When a plant is heterozygous for a trait, the two factors for that trait are not the same, and one factor, which Mendel described as dominant, determines the appearance, or phenotype, of the individual. We now refer to Mendel's factors as genes and to alternative forms of genes as alleles.

- When two heterozygous individuals mate, an individual offspring has a 50% (that is, random) chance of obtaining the dominant allele from the father and a 50% chance of obtaining the dominant allele from the mother; therefore, the probability of being homozygous recessive is 25%. The progeny thus appear as ¾ dominant and ¼ recessive, a dominant-to-recessive ratio of 3:1.

- When two genes are located on different chromosomes, the alleles included in an individual gamete are distributed at random. The allele for one gene included in the gamete has no influence on which allele of the other gene is included in the gamete. Such genes are said to assort independently.

- Because phenotypes are often influenced by more than one gene, the ratios of alternative phenotypes observed in crosses sometimes deviate from the simple ratios predicted by Mendel. This is particularly true in epistatic situations, where the product of one gene masks another.

13.2 Genes are on chromosomes.

- The first clear evidence that genes reside on chromosomes was provided by Thomas Hunt Morgan, who demonstrated that the segregation of the white-eye trait in *Drosophila* is associated with the segregation of the X chromosome, which is involved in sex determination.

- The first genetic evidence that crossing over occurs between chromosomes was provided by Curt Stern, who showed that when two Mendelian traits exchange during a cross, so do visible abnormalities on the ends of the chromosomes bearing those traits.

- The frequency of crossing over between genes can be used to construct genetic maps, which are representations of the physical locations of genes on chromosomes, inferred from the frequency of crossing over between particular pairs of genes.

13.3 Human genetics follows Mendelian principles.

- Primary nondisjunction results when chromosomes do not separate during meiosis, leading to gametes with missing or extra chromosomes. In humans, the loss of an autosome is invariably fatal. Gaining an extra autosome, which leads to a condition called trisomy, is also fatal, with only two exceptions: trisomy of chromosomes 21 and 22.

- Some genetic disorders are relatively common in human populations; others are rare. Many of the most important genetic disorders are associated with recessive alleles, which may lead to the production of defective versions of enzymes that normally perform critical functions. Because such traits are determined by recessive alleles and, therefore, are not expressed in heterozygotes, the alleles are not eliminated from the human population, even though their effects in homozygotes may be lethal.

- While there are no cures for any genetic disorder at present, many of these conditions can be identified through genetic therapy.

Discussing Key Terms

1. **Homozygous/heterozygous**   Different versions of the same gene are called alleles; each diploid offspring receives one allele from each parent. A diploid individual containing two copies of the same allele is homozygous; an individual containing two different alleles is heterozygous.

2. **Dominance**   In many cases, one (dominant) allele will "mask" the presence of another (recessive) allele in a heterozygous individual, so that the individual shows the dominant trait.

3. **The chromosomal theory of inheritance**   Mendelian traits are usually the result of the expression of particular genes located on chromosomes. Each gene is a segment of DNA encoding a protein.

4. **Nondisjunction**   Errors during meiosis may yield gametes with either extra or missing chromosomes. In a few instances, these gametes can produce zygotes that develop into individuals with mild to profound abnormalities.

Review Questions

1. How did Koelreuter's experiments on tobacco plants conflict with the ideas regarding heredity that were prevalent at that time? Why weren't the implications of his results recognized for a century?

2. What characteristics of the garden pea made this organism a good choice for Mendel's experiments on heredity?

3. How did Mendel produce self-fertilization in the garden pea? How did he produce cross-fertilization?

4. To determine whether a purple-flowered pea plant of unknown genotype is homozygous or heterozygous, what type of plant should it be crossed with? What would the offspring of this cross be like if the plant with unknown genotype were homozygous? What would the offspring be like if it were heterozygous?

5. In a dihybrid cross, what is the ratio of expected phenotypes? What fraction of the offspring should be homozygous recessive for both traits? What fraction of the offspring should be homozygous dominant for both traits?

6. What is primary nondisjunction? How is it related to Down syndrome? In humans, how are the age and sex of an individual related to the likelihood of producing gametes affected by nondisjunction?

7. What is the sex chromosome genotype of an individual with Klinefelter syndrome? Is such an individual genetically male or female? Why? Would such an individual usually have male or female body characteristics?

8. Is Huntington's disease a dominant or a recessive genetic disorder? Why is it maintained at its current frequency in human populations?

Thought Questions

1. Why did Mendel observe only two alleles of any given trait in the crosses that he carried out?

2. How might Mendel's results and the model he formulated have been different if the traits he chose to study were governed by alleles exhibiting incomplete dominance or codominance?

3. If a heterozygous, Rh-positive woman and a homozygous, Rh-negative man produce an Rh-negative fetus, will the fetus develop antibodies against the Rh antigens and kill the mother? Explain your reasoning.

Internet Links

Mendel
http://ww.stg.brown.edu/MendelWeb/
An outstanding site from Brown University for those interested in Mendel and his experiments, with excellent supplementary materials.

Classic Papers in Mendelian Genetics
http://www.esp.org/foundations/genetics/classical/
A wonderful collection of original historically-significant papers, including key papers by Mendel, Bateson, Morgan, and many others.

Virtual Drosophila Crosses
http://uflylab.calstatela.edu/edesktop/VirtApps/Vflylab
IntroVflylab.html
The VIRTUAL FLY LAB allows you to learn the principles of genetic inheritance by mating virtual fruit flies and analyzing the offspring.

Online Mendelian Inheritance in Man
http://www3.ncbi.nlm.nih.gov/omim
The single largest resource on human genetics, this site, maintained by the National Center of Biotechnology Information of the NIH, provides an extensive data base of genetic disorders and information on individual human genes. Over 9,300 genes are included, with roughly 50 new genes added each month.

Searching For a Cure
http://www.hhmi.org/GeneticTrail/
BLAZING A GENETIC TRAIL provides a wonderful account of human hereditary disorders, their causes, and the on-going search for cures. Prepared by the Howard Hughes Medical Institute. Highly Recommended.

For Further Reading

Blixt, S.: "Why Didn't Gregor Mendel Find Linkage?" *Nature*, vol. 256, 1975, page 206. Modern information on the chromosomal location of the genes Mendel studied.

Corcos, A., and F. Monaghan: "Mendel's Work and Its Rediscovery: A New Perspective," *Critical Reviews in Plant Sciences*, vol. 9, May 1990, pages 197–212. An evaluation of the myths surrounding Mendel's work.

Diamond, J.: "Blood, Genes, and Malaria," *Natural History*, February 1989. An account of the evolutionary history of sickle cell anemia.

Mendel, G.: "Experiments on Plant Hybridization," (1866), translated and reprinted in *The Origins of Genetics: A Mendel Source Book*, C. Stern and E. Sherwood (eds.), W. H. Freeman, San Francisco, 1966. Mendel's original research, largely ignored for over 30 years.

Morgan, T. H.: "Sex-Limited Inheritance in *Drosophila*," *Science*, vol. 32, 1910, pages 120–22. Morgan's original account of his famous analysis of the inheritance of the white-eye trait.

Mulligan, R.: "The Basic Science of Gene Therapy," *Science*, vol. 260, May 14, 1993, pages 926–32. An overview of gene therapy, how far we've come, and how it works.

Patterson, D.: "The Causes of Down Syndrome," *Scientific American*, August 1987, pages 52–60. A cluster of genes on chromosome 21 associated with Down syndrome are being identified and studied.

Verma, I.: "Gene Therapy," *Scientific American*, November 1990, pages 68–84. Treatment of genetic disorders by introducing healthy genes into the body of an affected individual is producing exciting results.

Mendelian Genetics Problems

1. The illustration describes Mendel's cross of *wrinkled* and *round* seed characters. (Hint: Do you expect all the seeds in a pod to be the same?) What is wrong with this diagram?

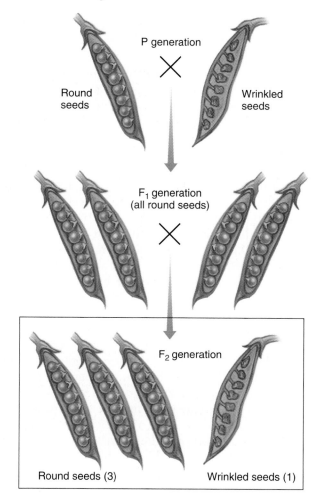

P generation

Round seeds ✕ Wrinkled seeds

F₁ generation (all round seeds)

✕

F₂ generation

Round seeds (3) Wrinkled seeds (1)

2. The annual plant *Haplopappus gracilis* has two pairs of chromosomes 1 and 2. In this species, the probability that two traits *a* and *b* selected at random will be on the same chromosome is equal to the probability that they will both be on chromosome 1 ($\frac{1}{2} \times \frac{1}{2} = \frac{1}{4}$, or 0.25), plus the probability that they will both be on chromosome 2 (also $\frac{1}{2} \times \frac{1}{2} = \frac{1}{4}$, or 0.25), for an overall probability of $\frac{1}{2}$, or 0.5. In general, the probability that two randomly selected traits will be on the same chromosome is equal to $\frac{1}{n}$ where *n* is the number of chromosome pairs. Humans have 23 pairs of chromosomes. What is the probability that any two human traits selected at random will be on the same chromosome?

3. Among Hereford cattle there is a dominant allele called *polled;* the individuals that have this allele lack horns. Suppose you acquire a herd consisting entirely of polled cattle, and you carefully determine that no cow in the herd has horns. Some of the calves born that year, however, grow horns. You remove them from the herd and make certain that no horned adult has gotten into your pasture. Despite your efforts, more horned calves are born the next year. What is the reason for the appearance of the horned calves? If your goal is to maintain a herd consisting entirely of polled cattle, what should you do?

4. An inherited trait among humans in Norway causes affected individuals to have very wavy hair, not unlike that of a sheep. The trait, called *woolly,* is very evident when it occurs in families; no child possesses woolly hair unless at least one parent does. Imagine you are a Norwegian judge, and you have before you a woolly-haired man suing his normal-haired wife for divorce because their first child has woolly hair but their second child has normal hair. The husband claims this constitutes evidence of his wife's infidelity. Do you accept his claim? Justify your decision.

5. In human beings, Down syndrome, a serious developmental abnormality, results from the presence of three copies of chromosome 21 rather than the usual two copies. If a female exhibiting Down syndrome mates with a normal male, what proportion of her offspring would you expect to be affected?

6. Many animals and plants bear recessive alleles for *albinism,* a condition in which homozygous individuals lack certain pigments. An albino plant, for example, lacks chlorophyll and is white, and an albino human lacks melanin. If two normally pigmented persons heterozygous for the same albinism allele marry, what proportion of their children would you expect to be albino?

7. You inherit a racehorse and decide to put him out to stud. In looking over the stud book, however, you discover that the horse's grandfather exhibited a rare disorder that causes brittle bones. The disorder is hereditary and results from homozygosity for a recessive allele. If your horse is heterozygous for the allele, it will not be possible to use him for stud, since the genetic defect may be passed on. How would you determine whether your horse carries this allele?

8. In the fly *Drosophila,* the allele for dumpy wings (*d*) is recessive to the normal long-wing allele (*d⁺*), and the allele for white eye (*w*) is recessive to the normal red-eye allele (*w⁺*). In a cross of $d^+d^+w^+w \times d^+dww$, what proportion of the offspring are expected to be "normal" (long wings, red eyes)? What proportion are expected to have dumpy wings and white eyes?

9. Your instructor presents you with a *Drosophila* with red eyes, as well as a stock of white-eyed flies and another stock of flies homozygous for the red-eye allele. You know that the presence of white eyes in *Drosophila* is caused by homozygosity for a recessive allele. How would you determine whether the single red-eyed fly was heterozygous for the white-eye allele?

10. Some children are born with recessive traits (and, therefore, must be homozygous for the recessive allele specifying the trait), even though neither of the parents exhibits the trait. What can account for this?

11. You collect two individuals of *Drosophila*, one a young male and the other a young, unmated female. Both are normal in appearance, with the red eyes typical of *Drosophila*. You keep the two flies in the same bottle, where they mate. Two weeks later, the offspring they have produced all have red eyes. From among the offspring, you select 100 individuals, some male and some female. You cross each individually with a fly you know to be homozygous for the recessive allele *sepia*, which produces black eyes when homozygous. Examining the results of your 100 crosses, you observe that in about half of the crosses, only red-eyed flies were produced. In the other half, however, the progeny of each cross consists of about 50% red-eyed flies and 50% black-eyed flies. What were the genotypes of your original two flies?

12. Hemophilia is a recessive sex-linked human blood disease that leads to failure of blood to clot normally. One form of hemophilia has been traced to the royal family of England, from which it spread throughout the royal families of Europe. For the purposes of this problem, assume that it originated as a mutation either in Prince Albert or in his wife, Queen Victoria.
    a. Prince Albert did not have hemophilia. If the disease is a sex-linked recessive abnormality, how could it have originated in Prince Albert, a male, who would have been expected to exhibit sex-linked recessive traits?
    b. Alexis, the son of Czar Nicholas II of Russia and Empress Alexandra (a granddaughter of Victoria), had hemophilia, but their daughter Anastasia did not. Anastasia died, a victim of the Russian revolution, before she had any children. Can we assume that Anastasia would have been a carrier of the disease? Would your answer be different if the disease had originated in Nicholas II or in Alexandra?

13. In 1986, *National Geographic* magazine conducted a survey of its readers' abilities to detect odors. About 7% of Caucasians in the United States could not smell the odor of musk. If neither parent could smell musk, none of their children were able to smell it. On the other hand, if the two parents could smell musk, their children generally could smell it, too, but a few of the children in those families were unable to smell it. Assuming that a single pair of alleles governs this trait, is the ability to smell musk best explained as an example of dominant or recessive inheritance?

14. A couple with a newborn baby is troubled that the child does not resemble either of them. Suspecting that a mix-up occurred at the hospital, they check the blood type of the infant. It is type O. As the father is type A and the mother type B, they conclude a mix-up must have occurred. Are they correct?

15. Mabel's sister died of cystic fibrosis as a child. Mabel does not have the disease, and neither do her parents. Mabel is pregnant with her first child. If you were a genetic counselor, what would you tell her about the probability that her child will have cystic fibrosis?

16. How many chromosomes would you expect to find in the karyotype of a person with Turner syndrome?

17. A woman is married for the second time. Her first husband has blood type A and her child by that marriage has type O. Her new husband has type B blood, and when they have a child its blood type is AB. What is the woman's blood genotype and blood type?

18. Two intensely freckled parents have five children. Three eventually become intensely freckled and two do not. Assuming this trait is governed by a single pair of alleles, is the expression of intense freckles best explained as an example of dominant or recessive inheritance?

19. Total color blindness is a rare hereditary disorder among humans. Affected individuals can see no colors, only shades of gray. It occurs in individuals homozygous for a recessive allele, and it is not sex-linked. A man whose father is totally color blind intends to marry a woman whose mother is totally color blind. What are the chances they will produce offspring who are totally color blind?

20. A normally pigmented man marries an albino woman. They have three children, one of whom is an albino. What is the genotype of the father?

21. Four babies are born in a hospital, and each has a different blood type: A, B, AB, and O. The parents of these babies have the following pairs of blood groups: A and B, O and O, AB and O, and B and B. Which baby belongs to which parents?

22. A couple both work in an atomic energy plant, and both are exposed daily to low-level background radiation. After several years, they have a child who has Duchenne muscular dystrophy, a recessive genetic defect caused by a mutation on the X chromosome. Neither the parents nor the grandparents have the disease. The couple sue the plant, claiming that the abnormality in their child is the direct result of radiation-induced mutation of their gametes, and that the company should have protected them from this radiation. Before reaching a decision, the judge hearing the case insists on knowing the sex of the child. Which sex would be more likely to result in an award of damages, and why?

14

DNA: The Genetic Material

Concept Outline

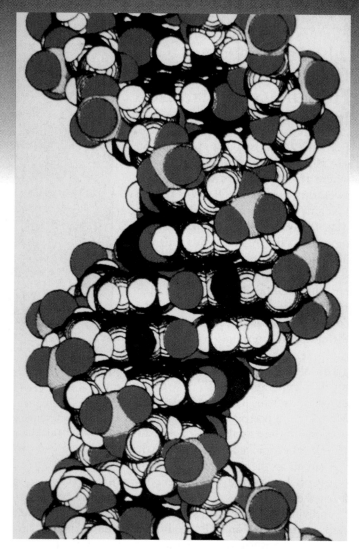

FIGURE 14.1
DNA. The hereditary blueprint in each cell of all living organisms is a very long, slender molecule called deoxyribonucleic acid (DNA).

The realization that patterns of heredity can be explained by the segregation of chromosomes in meiosis raised a question that occupied biologists for over 50 years: What is the exact nature of the connection between hereditary traits and chromosomes? This chapter describes the chain of experiments that have led to our current understanding of the molecular mechanisms of heredity (figure 14.1). The experiments are among the most elegant in science. Just as in a good detective story, each conclusion has led to new questions. The intellectual path taken has not always been a straight one, the best questions not always obvious. But however erratic and lurching the course of the experimental journey, our picture of heredity has become progressively clearer, the image more sharply defined.

The Hammerling Experiment: Cells Store Hereditary Information in the Nucleus

Perhaps the most basic question one can ask about hereditary information is where it is stored in the cell. To answer this question, Danish biologist Joachim Hammerling, working at the Max Plank Institute for Marine Biology in Berlin in the 1930s, cut cells into pieces and observed the pieces to see which were able to express hereditary information. For this experiment, Hammerling needed cells large enough to operate on conveniently and differentiated enough to distinguish the pieces. He chose the unicellular green alga *Acetabularia*, which grows up to 5 cm (figure 14.2), as a **model organism** for his investigations. Just as Mendel used pea plants and Sturtevant used fruit flies as model organisms, Hammerling picked an organism that was suited to the specific experimental question he wanted to answer, assuming that what he learned could then be applied to other organisms.

Individuals of the genus *Acetabularia* have distinct foot, stalk, and cap regions; all are differentiated parts of a single cell. The nucleus is located in the foot. As a preliminary experiment, Hammerling amputated the caps of some cells and the feet of others. He found that when he amputated the cap, a new cap regenerated from the remaining portions of the cell (foot and stalk). When he amputated the foot, however, no new foot regenerated from the cap and stalk. Hammerling, therefore, hypothesized that the hereditary information resided within the foot of *Acetabularia*.

Surgery on Single Cells

To test his hypothesis, Hammerling selected individuals from two species of the genus *Acetabularia* in which the caps look very different from one another: *A. mediterranea* has a disk-shaped cap, and *A. crenulata* has a branched, flower-like cap. Hammerling grafted a stalk from *A. crenulata* to a foot from *A. mediterranea* (figure 14.3). The cap that regenerated looked somewhat like the cap of *A. crenulata*, though not exactly the same.

FIGURE 14.2
The marine green alga *Acetabularia*. Although *Acetabularia* is a large organism with clearly differentiated parts, such as the stalks and elaborate caps visible here, each individual is actually a single cell. *Acetabularia* has been the subject of many experiments in developmental biology.

Hammerling then cut off this regenerated cap and found that a disk-shaped cap exactly like that of *A. mediterranea* formed in the second regeneration and in every regeneration thereafter. This experiment supported Hammerling's hypothesis that the instructions specifying the kind of cap are stored in the foot of the cell, and that these instructions must pass from the foot through the stalk to the cap.

In his regeneration experiment, the initial flower-shaped cap was somewhat intermediate in shape, unlike the disk-shaped caps of subsequent generations. Hammerling speculated that this initial cap, which resembled that of *A. crenulata*, was formed from instructions already present in the transplanted stalk when it was excised from the original *A. crenulata* cell. In contrast, all of the caps that regenerated subsequently used new information derived from the foot of the *A. mediterranea* cell the stalk had been grafted onto. In some unknown way, the original instructions that had been present in the stalk were eventually "used up." We now understand that genetic instructions (in the form of messenger RNA, discussed in chapter 15) pass from the nucleus in the foot upward *through the stalk* to the developing cap.

Hereditary information in *Acetabularia* is stored in the foot of the cell, where the nucleus resides.

Transplantation Experiments: Each Cell Contains a Full Set of Genetic Instructions

Since the nucleus is contained in the foot of *Acetabularia*, Hammerling's experiments suggested that the nucleus is the repository of hereditary information in a cell. A direct test of this hypothesis was carried out in 1952 by American embryologists Robert Briggs and Thomas King. Using a glass pipette drawn to a fine tip and working with a microscope, Briggs and King removed the nucleus from a frog egg. Without the nucleus, the egg did not develop. However, when they replaced the nucleus with one removed from a frog embryo cell, the egg developed into an adult frog. Clearly, the nucleus was directing the egg's development.

Successfully Transplanting Nuclei

Can every nucleus in an organism direct the development of an entire adult individual? The experiment of Briggs and King did not answer this question definitively, because the nuclei they transplanted from frog embryos into eggs often caused the eggs to develop abnormally. Two experiments performed soon afterward gave a clearer answer to the question. In the first, John Gurdon, working with another species of frog at Oxford and Yale, transplanted nuclei from tadpole cells into eggs from which the nuclei had been removed. The experiments were difficult—it was necessary to synchronize the division cycles of donor and host. However, in many experiments, the eggs went on to develop normally, indicating that the nuclei of cells in later stages of development retain the genetic information necessary to direct the development of all other cells in an individual.

Totipotency in Plants

In the second experiment, F. C. Steward at Cornell University in 1958 placed small fragments of fully developed carrot tissue (isolated from a part of the vascular system called the phloem) in a flask containing liquid growth medium. Steward observed that when individual cells broke away from the fragments, they often divided and developed into multicellular roots. When he immobilized the roots by placing them in a solid growth medium, they went on to develop normally into entire, mature plants. (This experiment is discussed more completely in chapter 39.) Steward's experiment makes it clear that, even in adult tissues, the nuclei of individual plant cells are "totipotent"—each contains a full set of hereditary instructions and can generate an entire adult individual. As you will learn in chapter 18, animal cells, like plant cells, are totipotent, and a single adult animal cell can generate an entire adult animal.

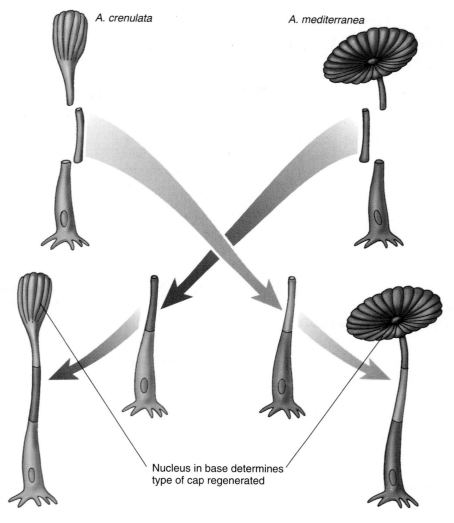

A. crenulata   A. mediterranea

Nucleus in base determines
type of cap regenerated

FIGURE 14.3

Hammerling's *Acetabularia* reciprocal graft experiment. Hammerling grafted a stalk of each species of *Acetabularia* onto the foot of the other species. In each case, the cap that eventually developed was dictated by the nucleus-containing foot rather than by the stalk.

Hereditary information is stored in the nucleus of eukaryotic cells. With rare exceptions, each nucleus in any eukaryotic cell contains a full set of genetic instructions.

The Griffith Experiment: Hereditary Information Can Pass between Organisms

The identification of the nucleus as the repository of hereditary information focused attention on the chromosomes, which were already suspected to be the vehicles of Mendelian inheritance. Specifically, biologists wondered how the **genes,** the units of hereditary information studied by Mendel, were actually arranged in the chromosomes. They knew that chromosomes contained both protein and deoxyribonucleic acid (DNA). Which of these held the genes? Starting in the late 1920s and continuing for about 30 years, a series of investigations addressed this question. We will describe three different kinds of experiments, each of which yielded a clear answer in a simple and elegant manner.

Discovery of Transformation

In 1928, British microbiologist Frederick Griffith made a series of unexpected observations while experimenting with pathogenic (disease-causing) bacteria. When he infected mice with a virulent strain of *Streptococcus pneumoniae* bacteria (then known as *Pneumococcus*), the mice died of blood poisoning. However, when he infected similar mice with a mutant strain of *S. pneumoniae* that lacked the virulent strain's polysaccharide coat, the mice showed no ill effects. The coat was apparently necessary for infection. The normal pathogenic form of this bacterium is referred to as the S form because it forms smooth colonies on a culture dish. The mutant form, which lacks an enzyme needed to manufacture the polysaccharide capsule, is called the R form because it forms rough colonies.

To determine whether the polysaccharide coat itself had a toxic effect, Griffith injected dead bacteria of the virulent S strain into mice; the mice remained perfectly healthy. As a control, he injected mice with a mixture containing dead S bacteria of the virulent strain and live coatless R bacteria, each of which by itself did not harm the mice (figure 14.4). Unexpectedly, the mice developed disease symptoms and many of them died. The blood of the dead mice was found to contain high levels of live, virulent *Streptococcus* type S bacteria, which had surface proteins characteristic of the live (previously R) strain. Somehow, the information specifying the polysaccharide coat had passed from the dead, virulent S bacteria to the live, coatless R bacteria in the mixture, permanently transforming the coatless R bacteria into the virulent S variety.

Hereditary information can pass from dead cells to living ones, and transform them.

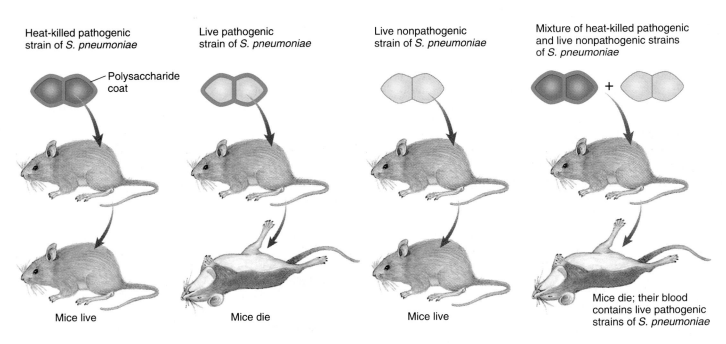

Heat-killed pathogenic strain of *S. pneumoniae*

Polysaccharide coat

Mice live

Live pathogenic strain of *S. pneumoniae*

Mice die

Live nonpathogenic strain of *S. pneumoniae*

Mice live

Mixture of heat-killed pathogenic and live nonpathogenic strains of *S. pneumoniae*

Mice die; their blood contains live pathogenic strains of *S. pneumoniae*

FIGURE 14.4
Griffith's discovery of transformation. The pathogenic bacterium *Streptococcus pneumoniae* kills many of the mice it is injected into, but only if the bacterial cells are covered with a polysaccharide coat, which the bacteria themselves synthesize. However, the coat itself is not the agent of disease. When Griffith injected mice with dead bacteria that possessed polysaccharide coats, the mice were unharmed. Similarly, an injection of live, coatless bacteria produced no ill effects. But when Griffith injected a mixture of dead bacteria with polysaccharide coats and live bacteria without such coats, many of the mice died, and virulent bacteria with coats were recovered. Griffith concluded that the live cells had been "transformed" by the dead ones; that is, genetic information specifying the polysaccharide coat had passed from the dead cells to the living ones.

The Avery Experiments: The Transforming Principle Is DNA

The agent responsible for transforming *Streptococcus* went undiscovered until 1944. In a classic series of experiments, Oswald Avery and his coworkers Colin MacLeod and Maclyn McCarty characterized what they referred to as the "transforming principle." They first prepared the mixture of dead S *Streptococcus* and live R *Streptococcus* that Griffith had used. Then Avery and his colleagues removed as much of the protein as they could from their preparation, eventually achieving 99.98% purity. Despite the removal of nearly all protein, the transforming activity was not reduced (figure 14.5). Moreover, the properties of the transforming principle resembled those of DNA in several ways:

1. When the purified principle was analyzed chemically, the array of elements agreed closely with DNA.
2. In an ultracentrifuge, the transforming principle migrated like DNA; in electrophoresis and other chemical and physical procedures, it also acted like DNA.
3. Extracting the lipid and protein from the purified transforming principle did not reduce its activity.
4. Protein-digesting enzymes did not affect the principle's activity; nor did RNA-digesting enzymes.
5. The DNA-digesting enzyme DNase destroyed all transforming activity.

The evidence was overwhelming. They concluded that "a nucleic acid of the deoxyribose type is the fundamental unit of the transforming principle of *Pneumococcus* Type III"—in essence, that DNA is the hereditary material.

> Avery's experiments demonstrate conclusively that DNA is the hereditary material.

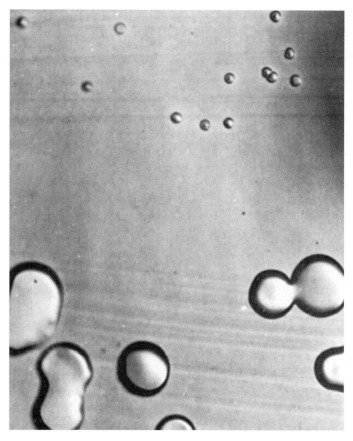

FIGURE 14.5
Transformation of *Streptococcus*. This photo, from the original publication of Avery and coworkers, shows the transformation of nonpathogenic R *Streptococcus* (the small colonies) to pathogenic S *Streptococcus* (the large colonies) in an extract prepared from heat-killed S *Streptococcus*.

The Identification of the Transforming Principle

In May 1943, Oswald Avery described his efforts to identify the "transforming principle" in a letter to his brother Roy:

For the past 2 years, first with MacLeod and now with Dr. McCarty, I have been trying to find out what is the chemical nature of the substance in the bacterial extract which induces this specific change. The crude extract of Type III is full of capsular polysaccharide, C (somatic) carbohydrate, nucleoproteins, free nucleic acids of both the yeast and thymus type, lipids, and other cell constituents. Try to find in the complex mixtures the active principle! Try to isolate and chemically identify the particular substance that will by itself, when brought into contact with the R cell derived from Type II, cause it to elaborate Type III capsular polysaccharide and to acquire all the aristocratic distinctions of the same specific type of cells as that from which the extract was prepared! Some job, full of headaches and heartbreaks. But at last perhaps we have it.

. . . if we prove to be right—and of course that is a big if—then it means that both the chemical nature of the inducing stimulus is known and the chemical structure of the substance produced is also known, the former being thymus nucleic acid, the latter Type III polysaccharide, and both are thereafter reduplicated in the daughter cells and after innumerable transfers without further addition of the inducing agent and the same active and specific transforming substance can be recovered far in excess of the amount originally used to induce the re-action. Sounds like a virus—may be a gene. But with mechanisms I am not now concerned. One step at a time and the first step is what is the chemical nature of the transforming principle? Someone else can work out the rest. Of course the problem bristles with implications. It touches the biochemistry of the thymus type of nucleic acids which are known to constitute the major part of chromosomes but have been thought to be alike regardless of origin and species. It touches genetics, enzyme chemistry, cell metabolism and carbohydrate synthesis. But today it takes a lot of well documented evidence to convince anyone that the sodium salt of deoxyribose nucleic acid, protein free, could possibly be endowed with such biologically active and specific properties and that is the evidence we are now trying to get. It is lots of fun to blow bubbles but it is wiser to prick them yourself before someone else tries to.

The Hershey–Chase Experiment: Some Viruses Direct Their Heredity with DNA

Avery's results were not widely accepted at first, as many biologists preferred to believe that proteins were the repository of hereditary information. Additional evidence supporting Avery's conclusion was provided in 1952 by Alfred Hershey and Martha Chase, who experimented with **bacteriophages,** viruses that attack bacteria. Viruses, described in more detail in chapter 29, consist of either DNA or RNA (ribonucleic acid) surrounded by a protein coat. When a *lytic* (potentially cell-rupturing) bacteriophage infects a bacterial cell, it first binds to the cell's outer surface and then injects its hereditary information into the cell. There, the hereditary information directs the production of thousands of new viruses within the bacterium. The bacterial cell eventually ruptures, or lyses, releasing the newly made viruses.

To identify the hereditary material injected into bacterial cells at the start of an infection, Hershey and Chase used the bacteriophage T2, which contains DNA rather than RNA. They labeled the viral DNA with a radioactive isotope of phosphorus, ^{32}P, and labeled the protein coats with a radioactive isotope of sulfur, ^{35}S. These isotopes are easily distinguished from each other, since they emit particles with different energies when they decay. After the labeled viruses were permitted to infect bacteria, the bacterial cells were agitated violently to remove the protein coats of the infecting viruses from the surfaces of the bacteria. This procedure removed nearly all of the ^{35}S label (and thus nearly all of the viral protein) from the bacteria. However, the ^{32}P label (and thus the viral DNA) had transferred to the interior of the bacteria, and the viruses subsequently released from the infected bacteria contained the ^{32}P label (figure 14.6). Hence, the hereditary information injected into the bacteria that specified the new generation of viruses was DNA and not protein.

Considerable evidence accumulated over the past 40 years indicates that DNA serves as the hereditary material in eukaryotes as well as in bacteria and most viruses. For example, purified DNA has been used to change the genetic nature of eukaryotic cells in tissue culture and has been injected into the fertilized eggs of *Drosophila* and mice, thereby altering the characteristics of the adults that develop from those eggs.

The hereditary material of bacteriophage is DNA and not protein.

**FIGURE 14.6
The Hershey and Chase experiment.**
(a) Viral DNA and protein coats were labeled with radioactive isotopes. (b) As this electron micrograph shows, individual DNA strands enter a bacterial cell from viruses bound to its surface. (c) Hershey and Chase then agitated the bacteria to remove the viruses and empty protein coats from the cell surfaces. (d, e) Within the cell, the labeled DNA directs the synthesis and assembly of the parts needed to make new viruses. Hershey and Chase found ^{35}S radioactivity in the medium, and ^{32}P radioactivity inside the bacterial cells. Thus, they concluded that the viral DNA, not the viral protein, was responsible for directing the production of new viruses.

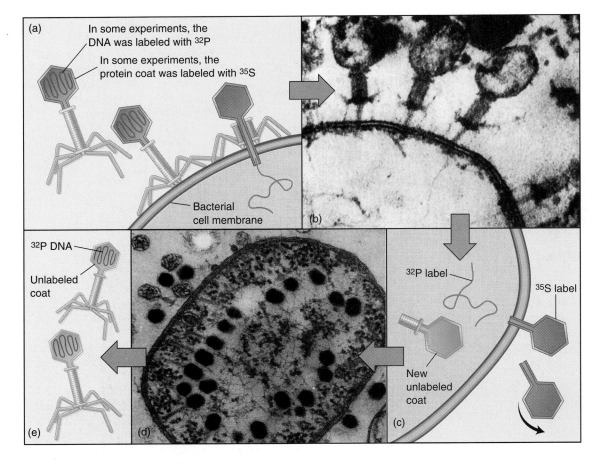

The Fraenkel-Conrat Experiment: Other Viruses Direct Their Heredity with RNA

Some viruses contain RNA instead of DNA, and yet they manage to reproduce quite satisfactorily. What genetic material do *they* use?

In 1957, Heinz Fraenkel-Conrat and coworkers answered this question for two RNA-containing viruses: tobacco mosaic virus (TMV), which infects the leaves of tobacco leaves, and Holmes ribgrass virus (HRV) which infects grass. TMV, the better studied, consists of a single strand of RNA 6390 nucleotides long, surrounded by a protein coat of 2130 identical subunits. The protein can be separated from the RNA by a simple chemical treatment. When this is done, the isolated RNA is infective, while the protein is not, suggesting that RNA is the hereditary material of these viruses. If the dissociated RNA and protein subunits are mixed together in solution, they recombine to form fully active virus particles.

Fraenkel-Conrat and his coworkers further investigated this conclusion with a simple but compelling exchange experiment. First they chemically dissociated each virus, separating its protein coat from its RNA. They then manufactured hybrid viruses by combining the protein of one with the RNA of the other. When they infected healthy tobacco plants with a hybrid virus composed of HRV RNA and TMV protein, the tobacco leaves developed lesions characteristic of HRV (figure 14.7). Clearly, the hereditary properties of the virus were determined by the nucleic acid in its core, not the protein in its coat.

Retroviruses

Later studies have shown that many other viruses contain RNA rather than DNA. When DNA viruses infect a cell, their DNA is often inserted into the host cell's DNA as if they were the cell's own genes. Viruses containing RNA use a more indirect method. They first make an intermediate double-stranded form of DNA from the RNA, using a special kind of polymerase enzyme called reverse transcriptase. This DNA copy may then insert into the cell's DNA. Because the path of information flows from RNA to DNA rather than from DNA to RNA, these RNA viruses are called **retroviruses.** The human immunodeficiency virus (HIV) that causes acquired immunodeficiency syndrome (AIDS) is a retrovirus, as are many tumor-forming viruses. Transcription of the retrovirus RNA, necessary to produce new virus particles, only takes place after its DNA copy has been integrated into the host's DNA. Thus, integration is an obligatory step in the life cycle of a retrovirus.

DNA is the genetic material for all cellular organisms and most viruses, although some viruses use RNA.

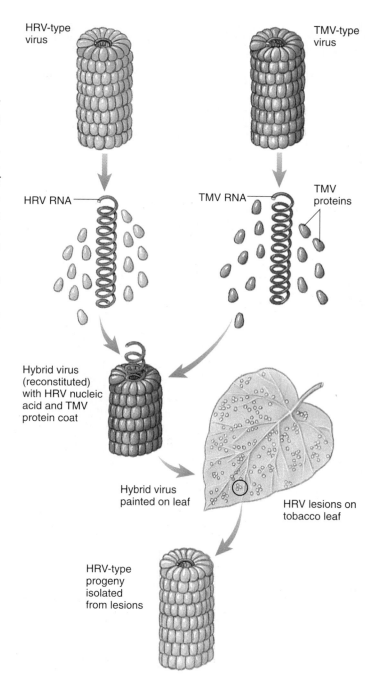

FIGURE 14.7
Fraenkel-Conrat's virus reconstitution experiment. Both tobacco mosaic virus (TMV) and Holmes ribgrass virus (HRV) are RNA viruses that infect plants, causing lesions on the leaves. Because the two viruses produce different kinds of lesions, the source of any particular infection can be identified. In this experiment, the protein and RNA in each virus were separated. Then, hybrid viruses consisting of HRV RNA and TMV protein were produced. When tobacco leaves were exposed to the hybrid HRV viruses, they developed HRV-type lesions, from which normal HRV viruses could be isolated in great numbers. No TMV viruses could be isolated from the lesions. Thus, the RNA rather than the protein contained the information necessary to specify the production of the viruses.

The Chemical Nature of Nucleic Acids

A German chemist, Friedrich Miescher, discovered DNA in 1869, only four years after Mendel's work was published. Miescher extracted a white substance from the nuclei of human cells and fish sperm. The proportion of nitrogen and phosphorus in the substance was different from that in any other known constituent of cells, which convinced Miescher that he had discovered a new biological substance. He called this substance "nuclein," since it seemed to be specifically associated with the nucleus.

Levene's Analysis: DNA Is a Polymer

Because Miescher's nuclein was slightly acidic, it came to be called **nucleic acid.** For 50 years biologists did little research on the substance, because nothing was known of its function in cells. In the 1920s, the primary structure of nucleic acids was determined by the biochemist P. A. Levene, who found that DNA contains three main components (figure 14.8): (1) phosphate (PO_4) groups; (2) five-carbon sugars; and (3) nitrogen-containing bases called **purines** (adenine, A, and guanine, G) and **pyrimidines** (thymine, T, and cytosine, C; RNA contains uracil, U, instead of T). From the roughly equal proportions of these components, Levene concluded correctly that DNA and RNA molecules are made of repeating units of the three components, strung one after another in a long chain. Each unit, consisting of a sugar attached to a phosphate group and a base, is called a **nucleotide.** The identity of the base distinguishes one nucleotide from another.

To identify the various chemical groups in DNA and RNA, it is customary to number the carbon atoms of the base and the sugar and then refer to any chemical group attached to a carbon atom by that number. In the sugar, four of the carbon atoms together with an oxygen atom form a five-membered ring. As illustrated in figure 14.9, the carbon atoms are numbered 1′ to 5′, proceeding clockwise from the oxygen atom; the prime symbol (′) indicates that the number refers to a carbon in a sugar rather than a base. Under this numbering scheme, the phosphate group is attached to the 5′ carbon atom of the sugar, and the base is attached to the 1′ carbon atom. In addition, a free hydroxyl (—OH) group is attached to the 3′ carbon atom.

The 5′ phosphate and 3′ hydroxyl groups allow DNA and RNA to form long chains of nucleotides, because these two groups can react chemically with each other. The reaction between the phosphate group of one nucleotide and the hydroxyl group of another is a dehydration synthesis, eliminating a water molecule and forming a covalent bond that links the two groups (figure 14.10). The linkage is called a **phosphodiester bond** because the phosphate group is now linked to the two sugars by means of a pair of ester (P—O—C) bonds. The two-unit polymer resulting from this reaction still has a free 5′ phosphate group at one end and a free 3′ hydroxyl group at the other, so it can link to other nucleotides. In this way, many thousands of nucleotides can join together in long chains.

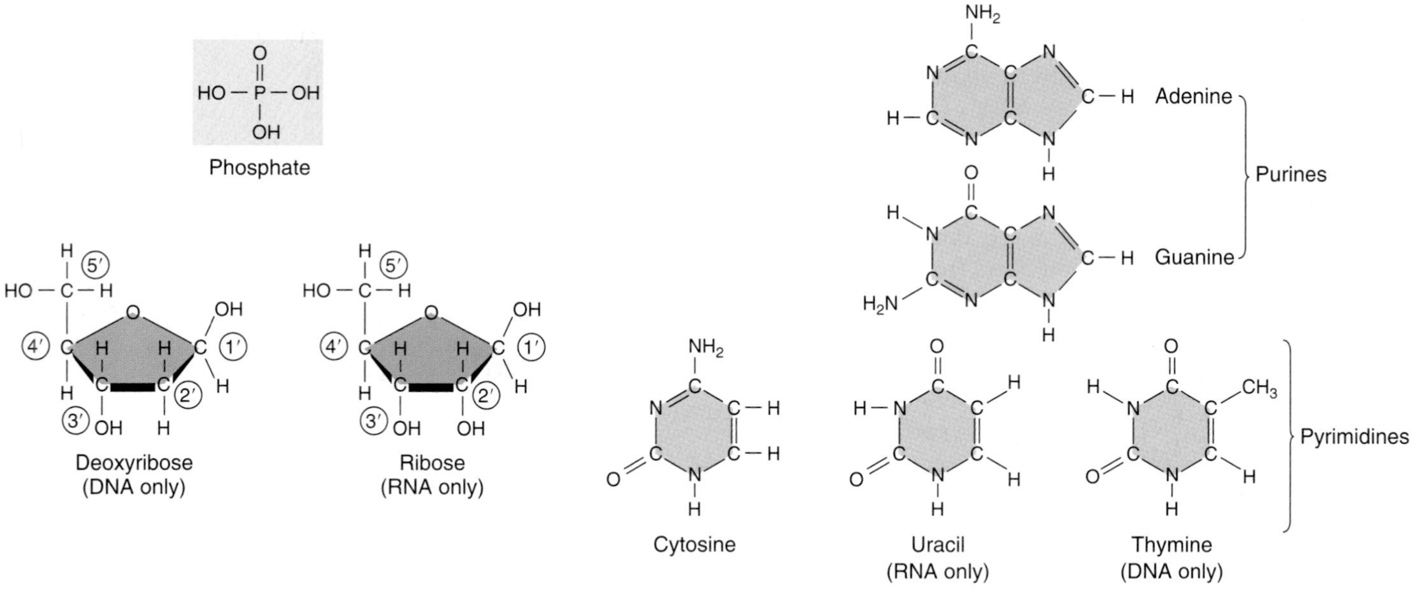

FIGURE 14.8
Nucleotide subunits of DNA and RNA. The nucleotide subunits of DNA and RNA are composed of three elements: a nitrogenous base (either a purine or a pyrimidine), a phosphate group, and a five-carbon sugar (deoxyribose in DNA and ribose in RNA).

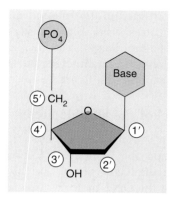

FIGURE 14.9
Numbering the carbon atoms in a nucleotide. The carbon atoms in the sugar of the nucleotide are numbered 1′ to 5′, proceeding clockwise from the oxygen atom. The "prime" symbol (′) indicates that the carbon belongs to the sugar rather than the base.

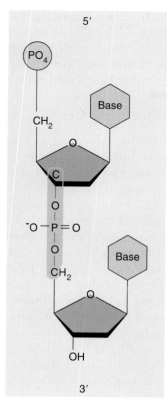

FIGURE 14.10
A phosphodiester bond.

Any linear strand of DNA or RNA, no matter how long, will always have a free 5′ phosphate group at one end and a free 3′ hydroxyl group at the other. Therefore, every DNA and RNA molecule has an intrinsic directionality, and we can refer unambiguously to each end of the molecule. By convention, the sequence of bases is usually expressed in the 5′-to-3′ direction. Thus, the base sequence "GTC-CAT" refers to the sequence,

$$5′ \text{ pGpTpCpCpApT—OH } 3′$$

where the phosphates are indicated by "p." Note that this is not the same molecule as that represented by the reverse sequence:

$$5′ \text{ pTpApCpCpTpG—OH } 3′$$

Levene's early studies indicated that all four types of DNA nucleotides were present in roughly equal amounts. This result, which later proved to be erroneous, led to the mistaken idea that DNA was a simple polymer in which the four nucleotides merely repeated (for instance, GCAT . . . GCAT . . . GCAT . . . GCAT . . .). If the sequence never varied, it was difficult to see how DNA might contain the hereditary information; this was why Avery's conclusion that DNA is the transforming principle was not readily accepted at first. It seemed more plausible that DNA was simply a structural element of the chromosomes, with proteins playing the central genetic role.

Chargaff's Analysis: DNA Is Not a Simple Repeating Polymer

When Levene's chemical analysis of DNA was repeated using more sensitive techniques that became available after World War II, quite a different result was obtained. The four nucleotides were *not* present in equal proportions in DNA molecules after all. A careful study carried out by Erwin Chargaff showed that the nucleotide composition of DNA molecules varied in complex ways, depending on the source of the DNA (table 14.1). This strongly suggested that DNA was not a simple repeating polymer and might have the information-encoding properties genetic material must have. Despite DNA's complexity, however, Chargaff observed an important underlying regularity: *the amount of adenine present in DNA always equals the amount of thymine, and the amount of guanine always equals the amount of cytosine.* These findings are commonly referred to as **Chargaff's rules:**

1. The proportion of A always equals that of T, and the proportion of G always equals that of C:

$$A = T, \text{ and } G = C.$$

2. It follows that there is always an equal proportion of purines (A and G) and pyrimidines (C and T).

A single strand of DNA or RNA consists of a series of nucleotides joined together in a long chain. In all natural DNA molecules, the proportion of A equals that of T, and the proportion of G equals that of C.

	Table 14.1 Chargaff's Analysis of DNA Nucleotide Base Compositions			
	Base Composition (Mole Percent)			
Organism	**A**	**T**	**G**	**C**
Escherichia coli (K12)	26.0	23.9	24.9	25.2
Mycobacterium tuberculosis	15.1	14.6	34.9	35.4
Yeast	31.3	32.9	18.7	17.1
Herring	27.8	27.5	22.2	22.6
Rat	28.6	28.4	21.4	21.5
Human	30.9	29.4	19.9	19.8

Source: Data from E. Chargaff and J. Davidson (editors), *The Nucleic Acides*, 1955, Academic Press, New York, NY.

The Three-Dimensional Structure of DNA

As it became clear that DNA was the molecule that stored the hereditary information, investigators began to puzzle over how such a seemingly simple molecule could carry out such a complex function.

Franklin: X-ray Diffraction Patterns of DNA

The significance of the regularities pointed out by Chargaff were not immediately obvious, but they became clear when a British chemist, Rosalind Franklin (figure 14.11), carried out an X-ray crystallographic analysis of DNA. In X-ray crystallography, a molecule is bombarded with a beam of X rays. When individual rays encounter atoms, their path is bent or diffracted, and the diffraction pattern is recorded on photographic film. The patterns resemble the ripples created by tossing a rock into a smooth lake. When carefully analyzed, they yield information about the three-dimensional structure of a molecule.

X-ray crystallography works best on substances that can be prepared as perfectly regular crystalline arrays. However, it was impossible to obtain true crystals of natural DNA at the time Franklin conducted her analysis, so she had to use DNA in the form of fibers. Franklin worked in the laboratory of British biochemist Maurice Wilkins, who was able to prepare more uniformly oriented DNA fibers than anyone had previously. Using these fibers, Franklin succeeded in obtaining crude diffraction information on natural DNA. The diffraction patterns she obtained suggested that the DNA molecule had the shape of a helix, or corkscrew, with a diameter of about 2 nanometers and a complete helical turn every 3.4 nanometers (figure 14.12).

Watson and Crick: A Model of the Double Helix

Learning informally of Franklin's results before they were published in 1953, James Watson and Francis Crick, two young investigators at Cambridge University, quickly worked out a likely structure for the DNA molecule (figure 14.13), which we now know was substantially correct. They analyzed the problem deductively, first building models of the nucleotides, and then trying to assemble the nucleotides into a molecule that matched what was known about the structure of DNA. They tried various possibilities before they finally hit on the idea that the molecule might be a simple **double helix** (figure 14.14), with the bases of two strands pointed inward toward each other, forming **base-pairs.** In their model, base-pairs always consist of purines, which are large, pointing toward pyrimidines, which are small, keeping the diameter of the molecule a constant 2 nanometers. Because hydrogen bonds can form between the bases in a base-pair, the double helix is stabilized as a duplex DNA molecule composed of two **antiparallel strands,** one chain running 3′ to 5′ and the other 5′ to 3′. The base-pairs are planar (flat) and stack 0.34 nm apart hydrophobically, contributing to the overall stability of the molecule.

FIGURE 14.11
The researcher. Rosalind Franklin developed techniques for taking X-ray diffraction pictures of fibers of DNA.

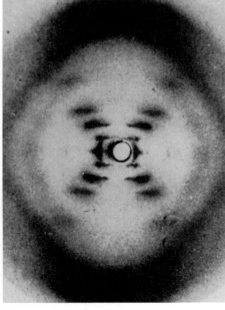

FIGURE 14.12
The essential clue. This is the telltale X-ray diffraction photograph of DNA fibers made in 1953 by Rosalind Franklin in the laboratory of Maurice Wilkins.

FIGURE 14.13
The discoverers. In 1953 James Watson (*left*), and Francis Crick (*right*) deduced the structure of DNA from Chargaff's rules and Franklin's diffraction studies.

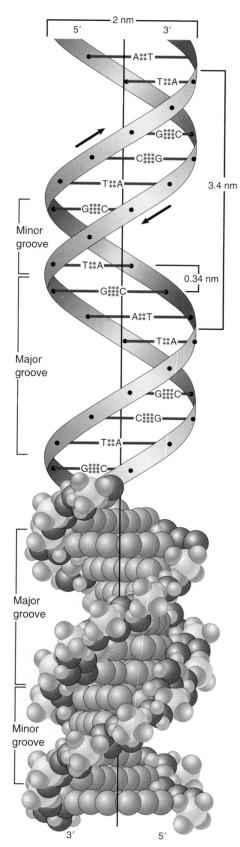

FIGURE 14.14
A model of the DNA double helix. The X-ray diffraction studies of Rosalind Franklin suggested the dimensions of the double helix.

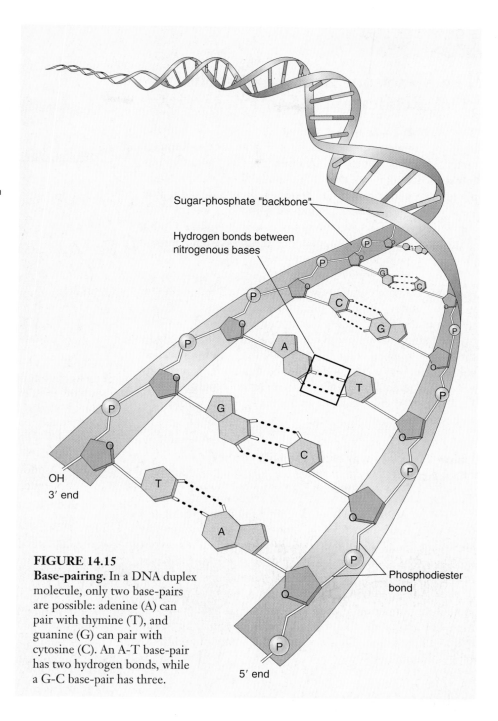

FIGURE 14.15
Base-pairing. In a DNA duplex molecule, only two base-pairs are possible: adenine (A) can pair with thymine (T), and guanine (G) can pair with cytosine (C). An A-T base-pair has two hydrogen bonds, while a G-C base-pair has three.

The Watson–Crick model explained why Chargaff had obtained the results he had: in a double helix, adenine forms two hydrogen bonds with thymine, but it will not form hydrogen bonds properly with cytosine. Similarly, guanine forms three hydrogen bonds with cytosine, but it will not form hydrogen bonds properly with thymine. Consequently, adenine and thymine will always occur in the same proportions in any DNA molecule, as will guanine and cytosine, because of this base-pairing (figure 14.15).

The DNA molecule is a double helix, the strands held together by base-pairing.

The Meselson–Stahl Experiment: DNA Replication Is Semiconservative

The Watson–Crick model immediately suggested that the basis for copying the genetic information is **complementarity.** One chain of the DNA molecule may have any conceivable base sequence, but this sequence completely determines the sequence of its partner in the duplex. For example, if the sequence of one chain is 5'-ATTGCAT-3', the sequence of its partner *must* be 3'-TAACGTA-5'. Thus, each chain in the duplex is a complementary mirror image of the other.

The complementarity of the DNA duplex provides a ready means of accurately duplicating the molecule. If one were to "unzip" the molecule, one would need only to assemble the appropriate complementary nucleotides on the exposed single strands to form two daughter duplexes with the same sequence. This form of DNA replication is called **semiconservative,** because while the sequence of the original duplex is conserved after one round of replication, the duplex itself is not. Instead, each strand of the duplex becomes part of another duplex.

Using Heavy Isotopes to Density-Label DNA Strands

The hypothesis of semiconservative replication was tested in 1958 by Matthew Meselson and Franklin Stahl of the California Institute of Technology. These two scientists grew bacteria in a medium containing the heavy isotope of nitrogen, ^{15}N, which became incorporated into the bases of the bacterial DNA. After several generations, the DNA of these bacteria was denser than that of bacteria grown in a medium containing the lighter isotope of nitrogen, ^{14}N. Meselson and Stahl then transferred the bacteria from the ^{15}N medium to the ^{14}N medium and collected the DNA at various intervals.

Separating DNA Strands by Density

By dissolving the DNA they had collected in a heavy salt called cesium chloride and then spinning the solution at very high speeds in an ultracentrifuge, Meselson and Stahl were able to separate DNA strands of different densities. The enormous centrifugal forces generated by the ultracentrifuge caused the cesium ions to migrate toward the bottom of the centrifuge tube, creating a gradient of cesium concentration, and thus of density. Each DNA strand floats or sinks in the gradient until it reaches the position where its density exactly matches the density of the cesium there. Because ^{15}N strands are denser than ^{14}N strands, they migrate farther down the tube to a denser region of the cesium gradient.

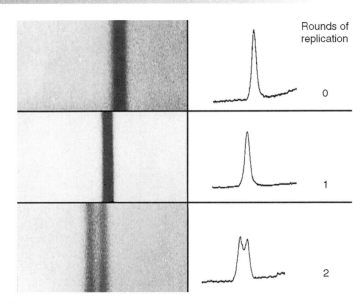

FIGURE 14.16
The key result of the Meselson and Stahl experiment. These bands of DNA, photographed on the left and scanned on the right, are from the density-gradient centrifugation experiment of Meselson and Stahl. At 0 generation, all DNA is heavy; after one replication all DNA has a hybrid density; after two replications, all DNA is hybrid or light.

The Key Result: Replication Alters DNA Density

The DNA collected immediately after the transfer was all dense. However, after the bacteria completed their first round of DNA replication in the ^{14}N medium, the density of their DNA had decreased to a value intermediate between ^{14}N-DNA and ^{15}N-DNA. After the second round of replication, two density classes of DNA were observed, one intermediate and one equal to that of ^{14}N-DNA (figure 14.16).

Interpreting the Result

Meselson and Stahl interpreted their results as follows: after the first round of replication, each daughter DNA duplex was a hybrid possessing one of the heavy strands of the parent molecule and one light strand; when this hybrid duplex replicated, it contributed one heavy strand to form another hybrid duplex and one light strand to form a light duplex (figure 14.17). Thus, this experiment clearly confirmed the prediction of the Watson-Crick model that DNA replicates in a semiconservative manner.

The basis for the great accuracy of DNA replication is complementarity. A DNA molecule is a duplex, containing two strands that are complementary mirror images of each other, so either one can be used as a template to reconstruct the other.

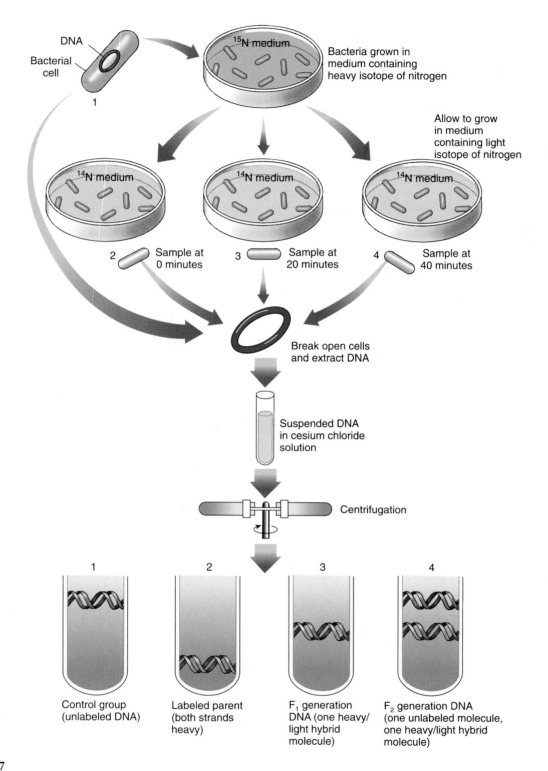

FIGURE 14.17
The Meselson and Stahl experiment: evidence demonstrating semiconservative replication. Bacterial cells were grown for several generations in a medium containing a heavy isotope of nitrogen (^{15}N) and then were transferred to a new medium containing the normal lighter isotope (^{14}N). At various times thereafter, samples of the bacteria were collected, and their DNA was dissolved in a solution of cesium chloride, which was spun rapidly in a centrifuge. Because the cesium ion is so massive, it tends to settle toward the bottom of the spinning tube, establishing a gradient of cesium density. DNA molecules sink in the gradient until they reach a place where their density equals that of the cesium; they then "float" at that position. DNA containing ^{15}N is denser than that containing ^{14}N, so it sinks to a lower position in the cesium gradient. After one generation in ^{14}N medium, the bacteria yielded a single band of DNA with a density between that of ^{14}N-DNA and ^{15}N-DNA, indicating that only one strand of each duplex contained ^{15}N. After two generations in ^{14}N medium, two bands were obtained; one of intermediate density (in which one of the strands contained ^{15}N) and one of low density (in which neither strand contained ^{15}N). Meselson and Stahl concluded that replication of the DNA duplex involves building new molecules by separating strands and assembling new partners on each of these templates.

The Replication Complex

To be effective, DNA replication must be fast and accurate. The machinery responsible has been the subject of intensive study for 40 years, and we now know a great deal about it. The replication of DNA begins at one or more sites on the DNA molecule where there is a specific sequence of nucleotides called a **replication origin**. There the DNA replicating enzyme **DNA polymerase III** and other enzymes begin a complex process that catalyzes the addition of nucleotides to the growing complementary strands of DNA (table 14.2 lists the proteins involved in bacteria). Before considering the replication process in detail, let's take a closer look at DNA polymerase III.

DNA Polymerase III

The first DNA polymerase enzyme to be characterized, DNA polymerase I, is a relatively small enzyme that plays a key supporting role in DNA replication. The true replicating enzyme, dubbed DNA polymerase III, is some 10 times larger and far more complex in structure. It contains 10 different kinds of polypeptide chains, as illustrated in figure 14.18. The enzyme is a dimer, with two similar multisubunit complexes. Each complex catalyzes the replication of one DNA strand.

A variety of different proteins play key roles within each complex. The subunits include a single large catalytic α subunit that catalyzes 5′ to 3′ addition of nucleotides to a growing chain, a smaller ε subunit that proofreads 3′ to 5′ for mistakes, and a ring-shaped β₂ dimer subunit that clamps the polymerase III complex around the DNA double helix. Polymerase III progressively threads the DNA through the enzyme complex, moving it at a rapid rate, some 1000 nucleotides per second (100 full turns of the helix, 0.34 micrometers).

Table 14.2	DNA Replication Proteins of *E. coli*		
Protein	**Role**	**Size (kd)**	**Molecules per Cell**
Helicase	Unwinds the double helix	300	20
Primase	Synthesizes RNA primers	60	50
Single-strand binding protein	Stabilizes single-stranded regions	74	300
DNA gyrase	Relieves torque	400	250
DNA polymerase III	Synthesizes DNA	≈900	20
DNA polymerase I	Erases primer and fills gaps	103	300
DNA ligase	Joins the ends of DNA segments	74	300

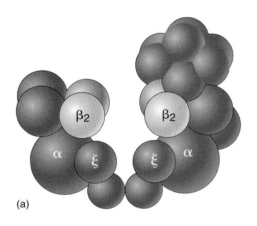

(a)

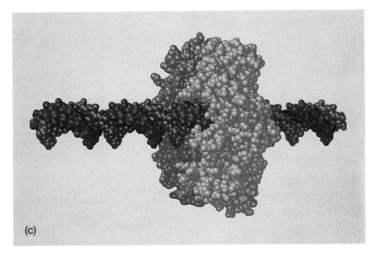

(b)

(c)

FIGURE 14.18
The DNA polymerase III molecule. (a) The molecule contains 10 kinds of protein chains. The protein is a dimer because both strands of the DNA duplex must be replicated simultaneously. The catalytic (α) subunits, the proofreading (ξ) subunits, and the "sliding clamp" (β₂) subunits (*yellow* and *blue*) are labeled. (b) The "sliding clamp" units encircle the DNA template and (c) move it through the catalytic subunit like a rope drawn through a ring.

The Need for a Primer

One of the features of DNA polymerase III is that it can add nucleotides only to a chain of nucleotides that is already paired with the parent strands. Hence, DNA polymerase cannot link the first nucleotides in a newly synthesized strand. Instead, another enzyme, an RNA polymerase called **primase,** constructs an **RNA primer,** a sequence of about 10 RNA nucleotides complementary to the parent DNA template. DNA polymerase III recognizes the primer and adds DNA nucleotides to it to construct the new DNA strands. The RNA nucleotides in the primers are then replaced by DNA nucleotides.

The Two Strands of DNA Are Assembled in Different Ways

Another feature of DNA polymerase III is that it can add nucleotides only to the 3′ end of a DNA strand (the end with an —OH group attached to a 3′ carbon atom). This means that replication always proceeds in the 5′ → 3′ direction on a growing DNA strand. Since the two parent strands of a DNA molecule are antiparallel, *the new strands are oriented in opposite directions* along the parent templates at each replication fork (figure 14.19). Therefore, the new strands must be elongated by different mechanisms! The **leading strand,** which elongates *toward* the replication fork, is built up simply by adding nucleotides continuously

to its growing 3′ end. In contrast, the **lagging strand,** which elongates *away from* the replication fork, is synthesized discontinuously as a series of short segments that are later connected. These segments, called **Okazaki fragments,** are about 100 to 200 nucleotides long in eukaryotes and 1000 to 2000 nucleotides long in prokaryotes. Each Okazaki fragment is synthesized by DNA polymerase III in the 5′ → 3′ direction, beginning at the replication fork and moving away from it. When the polymerase reaches the 5′ end of the lagging strand, another enzyme, **DNA ligase,** attaches the fragment to the lagging strand. The DNA is further unwound, new RNA primers are constructed, and DNA polymerase III then jumps ahead 1000 to 2000 nucleotides (toward the replication fork) to begin constructing another Okazaki fragment. If one looks carefully at electron micrographs showing DNA replication in progress, one can sometimes see that one of the parent strands near the replication fork appears single-stranded over a distance of about 1000 nucleotides. Since the synthesis of the leading strand is continuous, while that of the lagging strand is discontinuous, the overall replication of DNA is said to be **semidiscontinuous.**

DNA is replicated by the complex multisubunit enzyme DNA polymerase III, which replicates one strand continuously and the other discontinuously using RNA primers to initiate each segment.

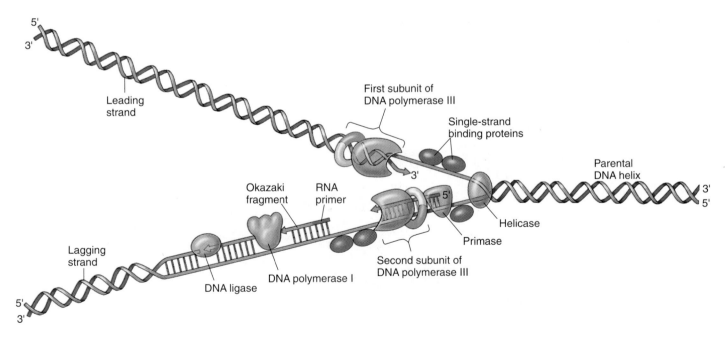

FIGURE 14.19
A DNA replication fork. Helicase enzymes separate the strands of the double helix, and single-strand binding proteins stabilize the single-stranded regions. Replication occurs by two mechanisms. (1) *Continuous synthesis:* After primase adds a short RNA primer, DNA polymerase III adds nucleotides to the 3′ end of the leading strand. DNA polymerase I then replaces the RNA primer with DNA nucleotides. (2) *Discontinuous synthesis:* Primase adds a short RNA primer (*green*) ahead of the 5′ end of the lagging strand. DNA polymerase III then adds nucleotides to the primer until the gap is filled in. DNA polymerase I replaces the primer with DNA nucleotides, and DNA ligase attaches the short segment of nucleotides to the lagging strand.

The Replication Process

The replication of the DNA double helix is a complex process that has taken decades of research to understand. It takes place in five interlocking steps:

1. **Opening up the DNA double helix.** The very stable DNA double helix must be opened up and its strands separated from each other for semiconservative replication to occur.

   Step one: Initiating replication. The binding of **initiator proteins** to the replication origin starts an intricate series of interactions that opens the helix.

   Step two: Unwinding the duplex. After initiation, "unwinding" enzymes called **helicases** bind to and move along one strand, shouldering aside the other strand as they go.

   Step three: Stabilizing the single strands. The unwound portion of the DNA double helix is stabilized by **single-strand binding protein,** which binds to the exposed single strands, protecting them from cleavage and preventing them from rewinding.

   Step four: Relieving the torque generated by unwinding. For replication to proceed at 1000 nucleotides per second, the parental helix ahead of the replication fork must rotate 100 revolutions per second! To relieve the resulting twisting, called torque, enzymes known as topisomerases—or, more informally, **gyrases**—cleave a strand of the helix, allow it to swivel around the intact strand, and then reseal the broken strand.

2. **Building a primer.** New DNA cannot be synthesized on the exposed templates until a primer is constructed, as DNA polymerases require 3′ primers to initiate replication. The necessary primer is a short stretch of RNA, added by a specialized RNA polymerase called *primase* in a multisubunit complex informally called a *primasome.* Why an RNA primer, rather than DNA? Starting chains on exposed templates introduces many errors; RNA marks this initial stretch as "temporary," making this error-prone stretch easy to excise later.

3. **Assembling complementary strands.** Next, the dimeric DNA polymerase III then binds to the replication fork. While the leading strand complexes with one half of the polymerase dimer, the lagging strand is thought to loop around and complex with the other half of the polymerase dimer (figure 14.20). Moving in concert down the parental double helix, DNA polymerase III catalyzes the formation of complementary sequences on each of the two single strands at the same time.

4. **Removing the primer.** The enzyme DNA polymerase I now removes the RNA primer and fills in the gap, as well as any gaps between Okazaki fragments.

5. **Joining the Okazaki fragments.** After any gaps between Okazaki fragments are filled in, the enzyme DNA ligase joins the fragments to the lagging strand.

DNA replication involves many different proteins that open and unwind the DNA double helix, stabilize the single strands, synthesize RNA primers, assemble new complementary strands on each exposed parental strand—one of them discontinuously—remove the RNA primer, and join new discontinuous segments on the lagging strand.

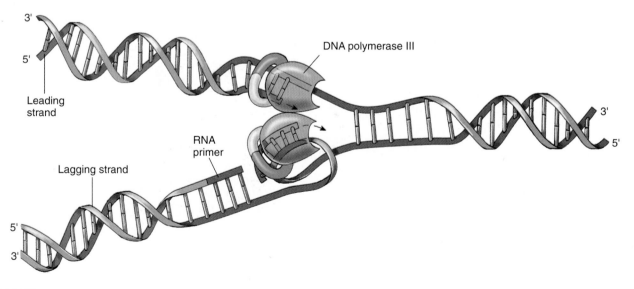

FIGURE 14.20
How DNA polymerase III works. This diagram presents a current view of how DNA polymerase III works. Note that the DNA on the lagging strand is folded to allow the dimeric DNA polymerase III molecule to replicate both strands of the parental DNA duplex simultaneously. This brings the 3′ end of each completed Okazaki fragment close to the start site for the next fragment.

Eukaryotic DNA Replication

In eukaryotic cells, the DNA is packaged in nucleosomes (figure 14.21) within chromosomes (figure 14.22). Each individual zone of a chromosome replicates as a discrete section called a **replication unit,** or **replicon,** which can vary in length from 10,000 to 1 million base-pairs; most are about 100,000 base-pairs long. Each replication unit has its own origin of replication, and multiple units may be undergoing replication at any given time, as can be seen in electron micrographs of replicating chromosomes (figure 14.23). Each unit replicates in a way fundamentally similar to prokaryotic DNA replication, using similar enzymes.

The advantage of having multiple origins of replication in eukaryotes is speed: replication takes approximately eight hours in humans cells, but if there were only one origin, it would take 100 times longer. Regulation of the replication process ensures that only one copy of the DNA is ultimately produced. How a cell achieves this regulation is not yet completely clear. It may involve periodic inhibitor or initiator proteins on the DNA molecule itself.

> To copy (replicate) a DNA molecule, the two strands of the molecule are separated, and each strand is used as a template for the assembly of a new complementary strand, thus forming two daughter duplexes.

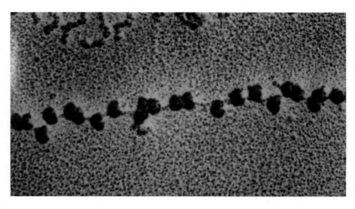

FIGURE 14.21
Nucleosomes. In this electron micrograph of rat liver DNA, nucleosomes look like beads on a string. Each nucleosome is a region in which the DNA duplex is wound tightly around a cluster of histone proteins.

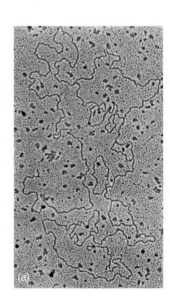
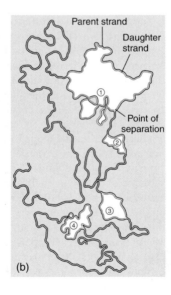

FIGURE 14.22
DNA of a single human chromosome. This chromosome has been "exploded," or relieved, of most of its packaging proteins. The residual protein scaffolding appears as the dark material in the lower part of the micrograph.

FIGURE 14.23
Eukaryotic chromosomes possess numerous replication forks spaced along their length. Four replication units (each with two replication forks) are producing daughter strands (a) in this electron micrograph, as indicated in *red* in the (b) corresponding drawing.

The One-Gene/One-Polypeptide Hypothesis

As the structure of DNA was being solved, other biologists continued to puzzle over how the genes of Mendel were related to DNA.

Garrod: Inherited Disorders Can Involve Specific Enzymes

In 1902, a British physician, Archibald Garrod, was working with one of the early Mendelian geneticists, his countryman William Bateson, when he noted that certain diseases he encountered among his patients seemed to be more prevalent in particular families. By examining several generations of these families, he found that some of the diseases behaved as if they were the product of simple recessive alleles. Garrod concluded that these disorders were Mendelian traits and that they had resulted from changes in the hereditary information in an ancestor of the affected families.

Garrod investigated several of these disorders in detail. In alkaptonuria the patients produced urine that contained homogentisic acid (alkapton). This substance oxidized rapidly when exposed to air, turning the urine black. In normal individuals, homogentisic acid is broken down into simpler substances. With considerable insight, Garrod concluded that patients suffering from alkaptonuria lacked the enzyme necessary to catalyze this breakdown. He speculated that many other inherited diseases might also reflect enzyme deficiencies.

Beadle and Tatum: Genes Specify Enzymes

From Garrod's finding, it took but a short leap of intuition to surmise that the information encoded within the DNA of chromosomes acts to specify particular enzymes. This point was not actually established, however, until 1941, when a series of experiments by Stanford University geneticists George Beadle and Edward Tatum provided definitive evidence on this point. Beadle and Tatum deliberately set out to create Mendelian mutations in chromosomes and then studied the effects of these mutations on the organism.

FIGURE 14.24
Beadle and Tatum's procedure for isolating nutritional mutants in *Neurospora.* This fungus grows easily on an artificial medium in test tubes. In this experiment, spores were irradiated to increase the frequency of mutation; they were then placed on a "complete" medium that contained all of the nutrients necessary for growth. Once the fungal colonies were established on the complete medium, individual spores were transferred to a "minimal" medium that lacked various substances the fungus could normally manufacture. Any spore that would not grow on the minimal medium but would grow on the complete medium contained one or more mutations in genes needed to produce the missing nutrients. To determine which gene had mutated, the minimal medium was supplemented with particular substances. The mutation illustrated here produced an arginine mutant, a collection of cells that lost the ability to manufacture arginine. These cells will not grow on minimal medium but will grow on minimal medium with only arginine added.

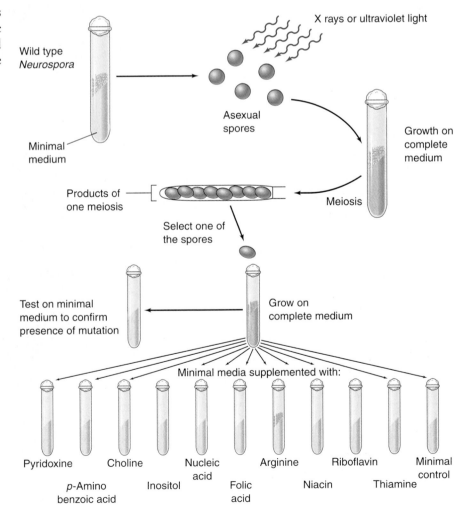

A Defined System. One of the reasons Beadle and Tatum's experiments produced clear-cut results is that the researchers made an excellent choice of experimental organism. They chose the bread mold *Neurospora*, a fungus that can be grown readily in the laboratory on a defined medium (a medium that contains only known substances such as glucose and sodium chloride, rather than some uncharacterized mixture of substances such as ground-up yeasts). Beadle and Tatum exposed *Neurospora* spores to X rays, expecting that the DNA in some of the spores would experience damage in regions encoding the ability to make compounds needed for normal growth (figure 14.24). DNA changes of this kind are called mutations, and organisms that have undergone such changes (in this case losing the ability to synthesize one or more compounds) are called mutants. Initially, they allowed the progeny of the irradiated spores to grow on a defined medium containing all of the nutrients necessary for growth, so that any growth-deficient mutants resulting from the irradiation would be kept alive.

Isolating Growth-Deficient Mutants. To determine whether any of the progeny of the irradiated spores had mutations causing metabolic deficiencies, Beadle and Tatum placed subcultures of individual fungal cells on a "minimal" medium that contained only sugar, ammonia, salts, a few vitamins, and water. Cells that had lost the ability to make other compounds necessary for growth would not survive on such a medium. Using this approach, Beadle and Tatum succeeded in identifying and isolating many growth-deficient mutants.

Identifying the Deficiencies. Next the researchers added various chemicals to the minimal medium in an attempt to find one that would enable a given mutant strain to grow. This procedure allowed them to pinpoint the nature of the biochemical deficiency that strain had. The addition of arginine, for example, permitted several mutant strains, dubbed *arg* mutants, to grow. When their chromosomal positions were located, the *arg* mutations were found to cluster in three areas (figure 14.25).

One-Gene/One-Polypeptide

For each enzyme in the arginine biosynthetic pathway, Beadle and Tatum were able to isolate a mutant strain with a defective form of that enzyme, and the mutation was always located at one of a few specific chromosomal sites. Most importantly, they found there was a different site for each enzyme. Thus, each of the mutants they examined had a defect in a single enzyme, caused by a mutation at a single site on one chromosome. Beadle and Tatum concluded that genes produce their effects by specifying the structure of enzymes and that each gene encodes the structure of one enzyme. They called this relationship the **one-gene/one-enzyme hypothesis.** Because many enzymes contain multiple protein or polypeptide subunits, each encoded by a separate gene, the relationship is today more commonly referred to as **"one-gene/one-polypeptide."**

Enzymes are responsible for catalyzing the synthesis of all the parts of an organism. They mediate the assembly of nucleic acids, proteins, carbohydrates, and lipids. Therefore, by encoding the structure of enzymes and other proteins, DNA specifies the structure of the organism itself.

Genetic traits are expressed largely as a result of the activities of enzymes. Organisms store hereditary information by encoding the structures of enzymes and other proteins in their DNA.

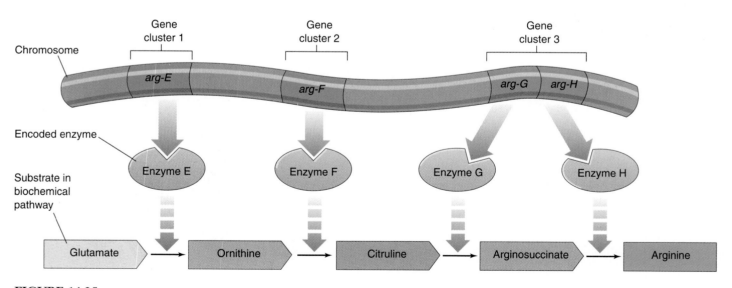

FIGURE 14.25
Evidence for the "one-gene/one-polypeptide" hypothesis. The chromosomal locations of the many arginine mutants isolated by Beadle and Tatum cluster around three locations. These locations correspond to the locations of the genes encoding the enzymes that carry out arginine biosynthesis.

How DNA Encodes Protein Structure

What kind of information must a gene encode to specify a protein? For some time, the answer to that question was not clear, since protein structure seemed to be impossibly complex. It was not even known whether a particular kind of protein always had the same sequence of amino acids.

Sanger: Proteins Consist of Defined Sequences of Amino Acids

The picture changed in 1953, the same year in which Watson and Crick unraveled the structure of DNA. That year, the English biochemist Frederick Sanger, after many years of work, announced the complete sequence of amino acids in the protein insulin. Insulin, a small protein hormone, was the first protein for which the amino acid sequence was determined. Sanger's achievement was extremely significant because it demonstrated for the first time that proteins consisted of definable sequences of amino acids—for any given form of insulin, every molecule has the same amino acid sequence. Sanger's work soon led to the sequencing of many other proteins, and it became clear that all enzymes and other proteins are strings of amino acids arranged in a certain definite order. The information needed to specify a protein such as an enzyme, therefore, is an ordered list of amino acids.

Ingram: Single Amino Acid Changes in a Protein Can Have Profound Effects

Following Sanger's pioneering work, Vernon Ingram in 1956 discovered the molecular basis of sickle cell anemia, a protein defect inherited as a Mendelian disorder. By analyzing the structures of normal and sickle cell hemoglobin, Ingram, working at Cambridge University, showed that sickle cell anemia is caused by a change from glutamic acid to valine at a single position in the protein (figure 14.26). The alleles of the gene encoding hemoglobin differed only in their specification of this one amino acid in the hemoglobin amino acid chain.

These experiments and other related ones have finally brought us to a clear understanding of the unit of heredity. The sequence of nucleotides that encodes a particular protein is called a gene. Although most genes encode proteins, some genes are devoted to the production of special forms of RNA, many of which play important roles in protein synthesis.

In general, the path of information is from DNA to RNA to protein, an observation often referred to as the Central Dogma of molecular genetics. In the next chapter, we will further explore this information pathway.

The amino acid sequence of a particular protein is specified by a corresponding sequence of nucleotides in the DNA. This nucleotide sequence is called a gene.

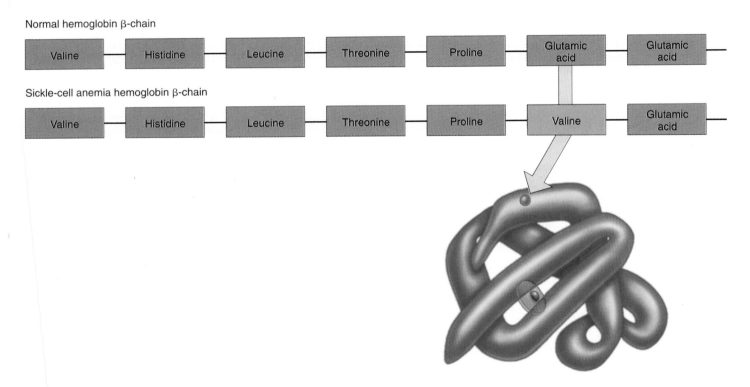

FIGURE 14.26
The molecular basis of a hereditary disease. Sickle cell anemia is produced by a recessive allele of the gene that encodes the hemoglobin β-chains. It represents a change in a single amino acid, from glutamic acid to valine at the sixth position in the chains, which consequently alters the tertiary structure of the hemoglobin molecule, reducing its ability to carry oxygen.

Summary of Concepts

14.1 What is the genetic material?

- Eukaryotic cells store hereditary information within the nucleus.

- When the nucleus is transplanted into another cell, the hereditary specifications of the organism are also transplanted.

- In viruses, bacteria, and eukaryotes, the hereditary information resides in nucleic acids. The transfer of nucleic acids can lead to the transfer of hereditary traits.

- The hereditary material is DNA in all cellular organisms and some viruses; it is RNA in other viruses.

- When radioactively labeled DNA viruses infect bacteria, the DNA but not the protein coat of the viruses enters the bacterial cells, indicating that the hereditary material is DNA rather than protein.

14.2 What is the structure of DNA?

- Chargaff showed that the proportion of adenine in DNA always equals that of thymine, and the proportion of guanine always equals that of cytosine.

- DNA has the structure of a double helix, consisting of two chains of nucleotides held together by hydrogen bonds between adenines and thymines, and between guanines and cytosines.

- The sequence of these nucleotides determines the structures of the proteins they encode.

14.3 How does DNA replicate?

- During the S phase of the cell cycle, the hereditary message in DNA is replicated with great accuracy.

- Copies of the DNA double helix are made by a complex of enzymes that unravels the double helix and assembles a new complementary partner on each of the single strands.

- During replication, the DNA duplex is unwound, and two new strands are assembled in opposite directions along the original strands. One strand elongates by the continuous addition of nucleotides to its growing end; the other is constructed by the addition of segments containing 100 to 2000 nucleotides, which are then joined to the end of that strand.

- A variety of enzymes participate in the replication process, including helicase, DNA polymerase, and DNA ligase.

14.4 What is a gene?

- Most hereditary traits reflect the actions of enzymes.

- The traits are hereditary because the information necessary to specify the structure of the enzymes is stored within the DNA.

- Each enzyme is encoded by a specific region of the DNA called a gene.

Discussing Key Terms

1. **Complementarity**   DNA is composed of two strands of nucleotides wrapped around each other in a double helix. Hydrogen bonding holds the two strands together, A pairing with T, and G with C; none of the other potential pairings form proper hydrogen bonds.

2. **Semiconservative replication**   Because DNA consists of two strands that are complementary to each other, each strand can serve as a template for the assembly of a new copy of the other.

3. **Gene**   A gene is the entire DNA nucleotide sequence necessary to specify a functional protein or RNA molecule.

4. **Transformation**   The transfer of DNA from one cell to another can alter the genetic makeup of the recipient cell. The transformation of bacteria observed by Griffith was the result of the passage of DNA from dead pathogenic strains to live nonpathogenic ones, transforming them to pathogens.

5. **Discontinuous DNA synthesis**   While one strand of DNA is synthesized by adding onto the growing end, the other strand is synthesized in the opposite direction, first assembling a short segment inward toward the growing end of the strand, and then joined to it.

Review Questions

1. In Hammerling's experiments on *Acetabularia*, what happened when a stalk from *A. crenulata* was grafted to a foot from *A. mediterranea?* What did these experiments suggest about the location of hereditary information within *Acetabularia* and cells in general?

2. How did Briggs, King, and Gurdon use frog eggs to determine whether the nucleus of a single cell contains a full complement of hereditary information in an organism? How did Steward use carrot tissue to address the same question?

3. What did Avery conclude was the primary component of the transforming principle in Griffith's experiments? What was the evidence supporting his conclusion?

4. How did Hershey and Chase determine which component of bacterial viruses contains the viruses' hereditary information? How did Fraenkel-Conrat answer this question? Did the two sets of experimenters obtain the same answer?

5. What are the three main components of a DNA molecule? What is the three-dimensional shape of the molecule? What mechanism is responsible for holding the strands of the molecule together, and how does this mechanism fit with Chargaff's observations on the proportions of purines and pyrimidines in DNA?

6. What does it mean to say that the two strands in a DNA molecule exhibit complementarity? How did Meselson and Stahl show that DNA replication is semiconservative?

7. How is the leading strand of a DNA duplex replicated? How is the lagging strand replicated? What is the basis for the requirement that the leading and lagging strands be replicated by different mechanisms?

8. What hypothesis did Beadle and Tatum test in their experiments on *Neurospora?* What did they do to change the DNA in individuals of this organism? How did they determine whether any of these changes affected enzymes in biosynthetic pathways?

Thought Questions

1. From an extract of human cells growing in tissue culture, you obtain a white, fibrous substance. How would you distinguish whether it was DNA, RNA, or protein?

2. The human genome contains approximately 3 billion (3×10^9) nucleotide base-pairs, and each nucleotide in a strand of DNA takes up about 0.34 nanometer (0.34×10^{-9} meter). How long would the human genome be if it were fully extended?

3. Cells were obtained from a patient with a viral infection. The DNA extracted from these cells consisted of two forms: double-stranded human DNA and single-stranded viral DNA. The base compositions of these two forms of DNA were as follows:

	A	C	G	T
Form 1	22.1%	27.9%	27.9%	22.1%
Form 2	31.3%	31.3%	18.7%	18.7%

Which form consisted of single-stranded viral DNA? Explain your reasoning.

Internet Links

A Molecular Biology Tutorial

http://www.biology.arizona.edu/molecular_bio/molecular_bio.html

This excellent tutorial from the University of Arizona features tutorials and problem sets on the molecular genetics of bacteria, and explores the basics of nucleic acid structure and function.

Rotating a DNA Molecule

http://www.sirius.com/~johnkyrk/DNAfast.html

A beautiful animation of a revolving DNA double helix that lets you gain a fuller appreciation of DNA in three dimensions.

For Further Reading

Crick, F. H. C.: "The Discovery of the Double Helix Was a Matter of Selecting the Right Problem and Sticking to It," *Chronicle of Higher Education*, October 5, 1988. Francis Crick's own recollections of the hectic days when he and James Watson deduced that the structure of DNA is a double helix.

Greider, C., and E. Blackburn: "Telomeres, Telomerase, and Cancer," *Scientific American*, February 1996, pages 92–97. Telomerase enzymes act on chromosome ends, shortening them with each cell division.

Judson, H. F.: *The Eighth Day of Creation*, Simon and Schuster, New York, 1979. The definitive historical account of the experimental unraveling of the mechanisms of heredity, based on personal interviews with the participants. This book provides a feel of how science is really conducted.

Marx, J.: "How DNA Replication Originates," *Science*, December 8, 1995, pages 1585–88. Researchers are beginning to understand the cellular machinery that ensures that DNA copies itself once—and only once—during each cell division cycle.

Watson, J. D.: *The Double Helix*, Athenaeum Publishing Company, New York, 1968. A lively, often irreverent account of what it was like to discover the structure of DNA, recounted by someone in a position to know.

Watson, J. D., and F. H. C. Crick: "A Structure for Deoxyribose Nucleic Acid," *Nature*, 1953, vol. 171, page 737. The original report of the double helical structure of DNA. Only one page long, this paper marks the birth of molecular genetics.

15

Genes and How They Work

Concept Outline

15.1 The Central Dogma traces the flow of gene-encoded information.

Cells Use RNA to Make Protein.   The information in genes is expressed in two steps, first being transcribed into RNA, and the RNA then being translated into protein.

15.2 Genes encode information in three-nucleotide code words.

The Genetic Code.   The sequence of amino acids in a protein is encoded in the sequence of nucleotides in DNA, three nucleotides encoding an amino acid.

15.3 Genes are first transcribed, then translated.

Transcription.   The enzyme RNA polymerase unwinds the DNA helix and synthesizes an RNA copy of one strand.
Translation.   mRNA is translated by activating enzymes that select tRNAs to match amino acids.
A Closer Look at the Mechanism of Protein Synthesis.   Proteins are synthesized on ribosomes, which provide a framework for the interaction of tRNA and mRNA.

15.4 Eukaryotic gene transcripts are edited.

The Discovery of Introns.   Eukaryotic genes contain extensive material that is not translated.
Differences between Bacterial and Eukaryotic Protein Synthesis.   Protein synthesis is broadly similar in bacteria and eukaryotes, although it differs in some respects.

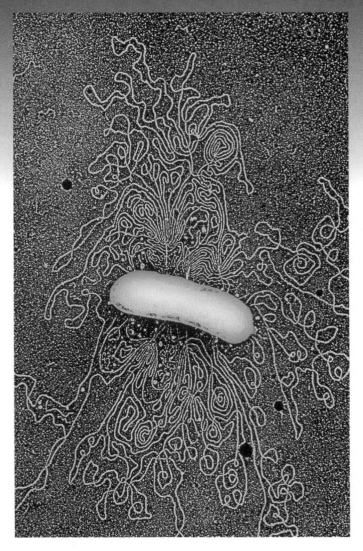

FIGURE 15.1
The unraveled chromosome of an *E. coli* bacterium. This complex tangle of DNA represents the full set of assembly instructions for the living organism *E. coli*.

Every cell in your body contains more information than is stored in this book. This information consists of the hereditary instructions specifying that you will have arms rather than fins, hair rather than feathers, and two eyes rather than one. The color of your eyes, the texture of your fingernails, and all of the other traits you receive from your parents are recorded in the cells of your body. As we have seen, this information is contained in long molecules of DNA (figure 15.1). The essence of heredity is the ability of cells to use the information in their DNA to produce particular proteins, thereby affecting what the cells will be like. In that sense, proteins are the tools of heredity. In this chapter, we will examine how proteins are synthesized from the information in DNA, using both prokaryotes and eukaryotes as models.

Cells Use RNA to Make Protein

To find out how a eukaryotic cell uses its DNA to direct the production of particular proteins, you must first ask where in the cell the proteins are made. We can answer this question by placing cells in a medium containing radioactively labeled amino acids for a short time. The cells will take up the labeled amino acids and incorporate them into polypeptide chains, which fold into proteins. If we then look to see where in the cells radioactive proteins first appear, we will find that it is not in the nucleus, where the DNA is, but rather in the cytoplasm, on large protein aggregates called **ribosomes** (figure 15.2). These polypeptide-making factories are very complex, composed of several RNA molecules and over 50 different proteins. Protein synthesis involves three different sites on the ribosome surface, called the P, A, and E sites, discussed later in this chapter. As you will recall from chapter 3, RNA is similar to DNA (figure 15.3). Its presence in ribosomes hints that RNA molecules play an important role in polypeptide synthesis.

Kinds of RNA

The class of RNA found in ribosomes is called **ribosomal RNA (rRNA)**. During polypeptide synthesis, rRNA and the ribosomal proteins provide the site where polypeptides are assembled. In addition to rRNA, there are two other major classes of RNA in cells. **Transfer RNA (tRNA)** molecules both transport the amino acids to the ribosome for use in building the polypeptides and position each amino acid at the correct place on the elongating polypep-

tide chain. Human cells contain about 45 different kinds of tRNA molecules. **Messenger RNA (mRNA)** molecules are long strands of RNA that are transcribed from DNA and that travel to the ribosomes to direct precisely *which* amino acids are assembled into polypeptides.

These RNA molecules, together with ribosomal proteins and certain enzymes, constitute a system that reads the genetic messages encoded by nucleotide sequences in the DNA and produces the polypeptides that those sequences specify. As we will see, biologists have also learned to read these messages. In so doing, they have learned a great deal about what genes are and how they are able to dictate what a protein will be like and when it will be made.

The Central Dogma

All organisms, from the simplest bacteria to ourselves, use the same basic mechanism of gene expression, so fundamental to life as we know it that it is often referred to as the "Central Dogma": Information passes from the genes (DNA) to an RNA copy of the gene, and the RNA copy directs the sequential assembly of a chain of amino acids (figure 15.4). Said briefly,

$$DNA \rightarrow RNA \rightarrow protein$$

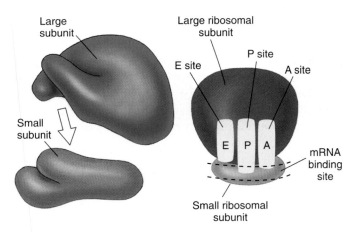

FIGURE 15.2
A ribosome is composed of two subunits. The smaller subunit fits into a depression on the surface of the larger one. The A, P, and E sites on the ribosome, discussed later in this chapter, play key roles in protein synthesis.

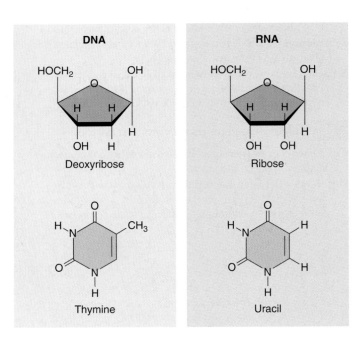

FIGURE 15.3
How RNA differs from DNA. There are two important differences. First, in place of the sugar deoxyribose, RNA contains ribose, which has an additional oxygen atom. Second, RNA contains the pyrimidine uracil (U) instead of thymine (T). Other differences are that RNA does not have a regular helical structure and is usually single-stranded.

Transcription: An Overview

The first step of the Central Dogma is the transfer of information from DNA to RNA, which occurs when an mRNA copy of the gene is produced. Like all classes of RNA, mRNA is formed on a DNA template. Because the DNA sequence in the gene is transcribed into an RNA sequence, this stage is called **transcription.** Transcription is initiated when the enzyme **RNA polymerase** binds to a particular binding site called a **promoter** located at the beginning of a gene. Starting there, the RNA polymerase moves along the strand into the gene. As it encounters each DNA nucleotide, it adds the corresponding complementary RNA nucleotide to a growing mRNA strand. Thus, guanine (G), cytosine (C), thymine (T), and adenine (A) in the DNA would signal the addition of C, G, A, and uracil (U), respectively, to the mRNA.

When the RNA polymerase arrives at a transcriptional "stop" signal at the opposite end of the gene, it disengages from the DNA and releases the newly assembled RNA chain. This chain is a complementary transcript of the gene from which it was copied.

Translation: An Overview

The second step of the Central Dogma is the transfer of information from RNA to protein, which occurs when the information contained in the mRNA transcript is used to direct the sequence of amino acids during the synthesis of polypeptides by ribosomes. This process is called **translation** because the nucleotide sequence of the mRNA transcript is translated into an amino acid sequence in the polypeptide. Translation begins when an rRNA molecule within the ribosome recognizes and binds to a "start" sequence on the mRNA. The ribosome then moves along the mRNA molecule, three nucleotides at a time. Each group of three nucleotides is a code that specifies which amino acid will be added to the growing polypeptide chain. The ribosome continues in this fashion until it encounters a translational "stop" signal; then it disengages from the mRNA and releases the completed polypeptide.

The two steps of the Central Dogma, taken together, are a concise summary of the events involved in the expression of an active gene. Biologists refer to this process as **gene expression.**

The information encoded in genes is expressed in two phases: transcription, in which an RNA polymerase enzyme assembles an mRNA molecule whose nucleotide sequence is complementary to the DNA nucleotide sequence of the gene; and translation, in which a ribosome assembles a polypeptide, whose amino acid sequence is specified by the nucleotide sequence in the mRNA.

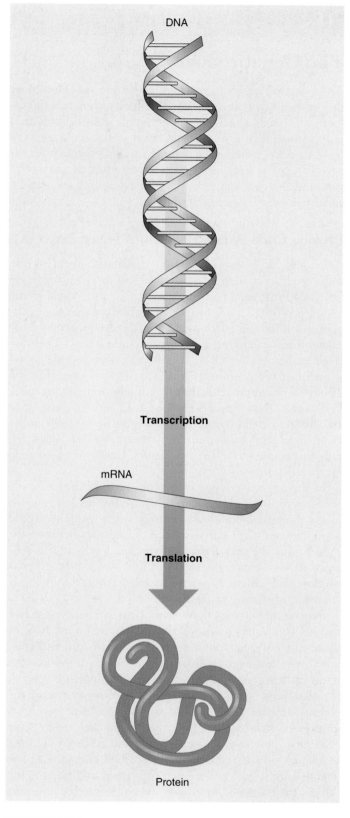

FIGURE 15.4
The Central Dogma of gene expression. DNA is transcribed to make mRNA, which is translated to make a protein.

The Genetic Code

The essential question of gene expression is, "How does the *order* of nucleotides in a DNA molecule encode the information that specifies the order of amino acids in a polypeptide?" In other words, what is the nature of the "genetic code"? The answer came in 1961, through an experiment led by Francis Crick. That experiment was so elegant and the result so critical to understanding the genetic code that we will describe it in detail.

Proving Code Words Have Only Three Letters

Crick and his colleagues reasoned that the genetic code most likely consisted of a series of blocks of information called **codons,** each corresponding to an amino acid in the encoded protein. They further hypothesized that the information within one codon was probably a sequence of three nucleotides specifying a particular amino acid. They arrived at the number three, because a two-nucleotide codon would not yield enough combinations to code for the 20 different amino acids that commonly occur in proteins. With four DNA nucleotides (G, C, T, and A), only 4^2, or 16, different pairs of nucleotides could be formed. However, these same nucleotides can be arranged in 4^3, or 64, different combinations of three, more than enough to code for the 20 amino acids.

In theory, the codons in a gene could lie immediately adjacent to each other, forming a continuous sequence of transcribed nucleotides. Alternatively, the sequence could be punctuated with untranscribed nucleotides between the codons, like the spaces that separate the words in this sentence. It was important to determine which method cells employ, since these two ways of transcribing DNA imply different translating processes.

To choose between these alternative mechanisms, Crick and his colleagues used a chemical to delete one, two, or three nucleotides from a viral DNA molecule and then asked whether a gene downstream of the deletions was transcribed correctly. When they made a single deletion or two deletions near each other, the **reading frame** of the genetic message shifted, and the downstream gene was transcribed as nonsense. However, when they made three deletions, the correct reading frame was restored, and the sequences downstream were transcribed correctly. They obtained the same results when they made additions to the DNA consisting of one, two, or three nucleotides. As shown in figure 15.5, these results could not have been obtained if the codons were punctuated by untranscribed nucleotides. Thus, Crick and his colleagues concluded that the genetic code is read in increments consisting of three nucleotides (in other words, it is a **triplet code**) and that reading occurs continuously without punctuation between the three-nucleotide units.

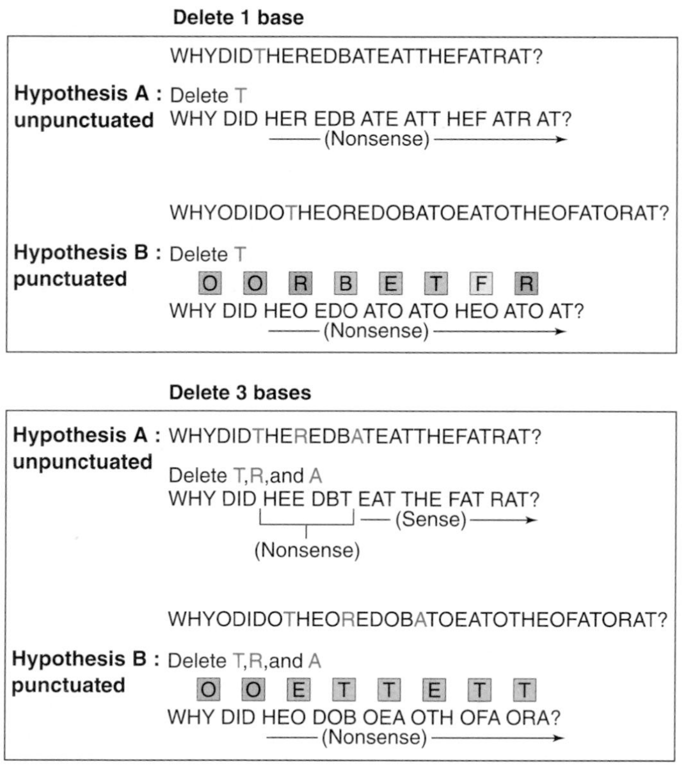

FIGURE 15.5
Using frame-shift alterations of DNA to determine if the genetic code is punctuated. The hypothetical genetic message presented here is "Why did the red bat eat the fat rat?" Under hypothesis B, which proposes that the message is punctuated, the three-letter words are separated by nucleotides that are not read (indicated by the letter "O").

Breaking the Genetic Code

Within a year of Crick's experiment, other researchers succeeded in determining the amino acids specified by particular three-nucleotide units. Marshall Nierenberg discovered in 1961 that adding the synthetic mRNA molecule polyU (an RNA molecule consisting of a string of uracil nucleotides) to cell-free systems consisting of mixtures of RNA and protein isolated from ruptured cells resulted in the production of the polypeptide polyphenylalanine (a string of phenylalanine amino acids). Therefore, one of the three-nucleotide sequences specifying phenylalanine is UUU. In 1964, Nierenberg and Philip Leder developed a powerful **triplet binding assay** in which a specific triplet was tested to see which radioactive amino acid (complexed to tRNA) it would bind. Some 47 of the 64 possible triplets gave unambiguous results. Har Gobind Khorana decoded the remaining 17 triplets by constructing artificial mRNA molecules of defined sequence and examining what polypeptides they directed. In these ways, all 64 possible three-nucleotide sequences were tested, and the full genetic code was determined (table 15.1).

Table 15.1 The Genetic Code

First Letter	Second Letter								Third Letter
	U		**C**		**A**		**G**		
U	UUU	Phenylalanine	UCU	Serine	UAU	Tyrosine	UGU	Cysteine	U
	UUC		UCC		UAC		UGC		C
	UUA	Leucine	UCA		UAA	Stop	UGA	Stop	A
	UUG		UCG		UAG	Stop	UGG	Tryptophan	G
C	CUU	Leucine	CCU	Proline	CAU	Histidine	CGU	Arginine	U
	CUC		CCC		CAC		CGC		C
	CUA		CCA		CAA	Glutamine	CGA		A
	CUG		CCG		CAG		CGG		G
A	AUU	Isoleucine	ACU	Threonine	AAU	Asparagine	AGU	Serine	U
	AUC		ACC		AAC		AGC		C
	AUA	Methionine; Start	ACA		AAA	Lysine	AGA	Arginine	A
	AUG		ACG		AAG		AGG		G
G	GUU	Valine	GCU	Alanine	GAU	Aspartate	GGU	Glycine	U
	GUC		GCC		GAC		GGC		C
	GUA		GCA		GAA	Glutamate	GGA		A
	GUG		GCG		GAG		GGG		G

A codon consists of three nucleotides read in the sequence shown. For example, ACU codes for threonine. The first letter, A, is in the First Letter column; the second letter, C, is in the Second Letter column; and the third letter, U, is in the Third Letter column. Each of the mRNA codons is recognized by a corresponding anticodon sequence on a tRNA molecule. Some tRNA molecules recognize more than one codon in mRNA, but they always code for the same amino acid. In fact, most amino acids are specified by more than one codon. For example, threonine is specified by four codons, which differ only in the third nucleotide (ACU, ACC, ACA, and ACG).

The Code Is Practically Universal

The genetic code is the same in almost all organisms. For example, the codon AGA specifies the amino acid arginine in bacteria, in humans, and in all other organisms whose genetic code has been studied. The universality of the genetic code is among the strongest evidence that all living things share a common evolutionary heritage. Because the code is universal, genes transcribed from one organism can be translated in another; the mRNA is fully able to dictate a functionally active protein. Similarly, genes can be transferred from one organism to another and be successfully transcribed and translated in their new host. This universality of gene expression is central to many of the advances of genetic engineering. Many commercial products such as the insulin used to treat diabetes are now manufactured by placing human genes into bacteria. The bacteria then serve as tiny factories to turn out prodigious quantities of the protein encoded by the transferred human gene.

But Not Quite

In 1979, investigators began to determine the complete nucleotide sequences of the mitochondrial genomes in humans, cattle, and mice. It came as something of a shock when these investigators learned that the genetic code used by these mammalian mitochondria was not quite the same as the "universal code" that has become so familiar to biologists. Although most of the code was the same, there were some differences. In the mitochondrial genomes, what should have been a "stop" codon, UGA, was instead read as the amino acid tryptophan; AUA was read as methionine rather than isoleucine; and AGA and AGG were read as "stop" rather than arginine. Furthermore, minor differences from the universal code have also been found in the genomes of chloroplasts and ciliates (certain types of protists).

Thus, it appears that the genetic code is not quite universal. Some time ago, presumably after they began their endosymbiotic existence, mitochondria and chloroplasts began to read the code differently, particularly the portion of the code associated with "stop" signals. Under most conditions, any change involving termination signals would be expected to be lethal. How such changes could have arisen in organelles, much less in eukaryotes like ciliates, is a puzzle for which we currently have no answer.

Within genes that encode proteins, the nucleotide sequence of DNA is read in blocks of three consecutive nucleotides, without punctuation between the blocks. Each block, or codon, codes for one amino acid.

Transcription

The first step in gene expression is the production of an RNA copy of the DNA sequence encoding the gene, a process called **transcription.** To understand the mechanism behind the transcription process, it is useful to focus first on RNA polymerase, the remarkable enzyme responsible for carrying it out (figure 15.6).

RNA Polymerase

RNA polymerase is best understood in bacteria. Bacterial RNA polymerase is very large and complex, consisting of five subunits: two α subunits bind regulatory proteins, a β' subunit binds the DNA template, a β subunit binds RNA nucleoside subunits, and a σ subunit recognizes the promoter and initiates synthesis. Only one of the two strands of DNA, called the **template strand,** is transcribed. The RNA transcript's sequence is complementary to the template strand. The strand of DNA that is not transcribed is called the **coding strand.** It has the same sequence as the RNA transcript, except T takes the place of U. The coding strand is also known as the sense (+) strand, and the template strand as the antisense (−) strand.

In both bacteria and eukaryotes, the polymerase adds ribonucleotides to the growing 3′ end of an RNA chain. No primer is needed, and synthesis proceeds in the 5′ → 3′ direction. Bacteria contain only one RNA polymerase enzyme, while eukaryotes have three different RNA polymerases: RNA polymerase I synthesizes rRNA in the nucleolus; RNA polymerase II synthesizes mRNA; and RNA polymerase III synthesizes tRNA.

Promoter Sites

Transcription starts at RNA polymerase binding sites called **promoter sites** on the DNA template strand. A promoter site is about 60 base-pairs long and is not itself transcribed by the polymerase that binds to it. Striking similarities are evident in the sequences of different promoters. For example, two six-base sequences are common to almost all bacterial promoter sites, a TTGACA sequence called the **−35 sequence,** located 35 nucleotides upstream of the position where transcription actually starts, and a TATAAT sequence called the **−10 sequence,** located 10 nucleotides upstream of the start site. In eukaryotic DNA, the sequence TATAAA, called the **TATA box,** is located at −25 and is very similar to the prokaryotic −10 sequence but is farther from the start site.

Promoters differ widely in efficiency. Strong promoters cause frequent initiations of transcription, as often as every 2 seconds in some bacteria. Weak promoters may tran-

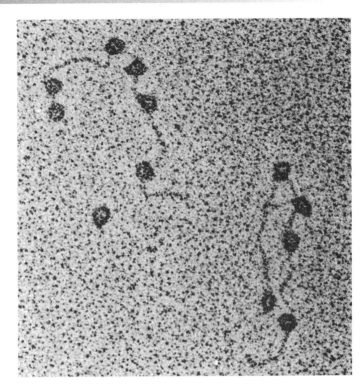

FIGURE 15.6
RNA polymerase. In this electron micrograph, the dark circles are RNA polymerase molecules bound to several promoter sites on bacterial virus DNA.

scribe only once every 10 minutes. Most strong promoters have unaltered −35 and −10 sequences, while weak promoters often have substitutions within these sites.

Initiation

The binding of RNA polymerase to the promoter is the first step in gene transcription. In bacteria, a subunit of RNA polymerase called σ **(sigma)** recognizes the −10 sequence in the promoter and binds RNA polymerase there. Importantly, this subunit can detect the −10 sequence without unwinding the DNA double helix. In eukaryotes, the −25 sequence plays a similar role in initiating transcription, as it is the binding site for a key protein factor. Other eukaryotic factors then bind one after another, assembling a large and complicated **transcription complex.** The eukaryotic transcription complex is described in detail in the following chapter.

Once bound to the promoter, the RNA polymerase begins to unwind the DNA helix. Measurements indicate that bacterial RNA polymerase unwinds a 17-base-pair segment, nearly two turns of the DNA double helix. This sets the stage for the assembly of the RNA chain.

Elongation

The transcription of the RNA chain starts with ATP or GTP. One of these forms the 5′ end of the chain, which grows in the 5′ → 3′ direction as ribonucleotides are added. Unlike DNA synthesis, a primer is not required. The region containing the RNA polymerase, DNA, and growing RNA transcript is called the **transcription bubble** because it contains a locally unwound "bubble" of DNA (figure 15.7). Within the bubble, the first 12 bases of the newly synthesized RNA strand temporarily form a helix with the template DNA strand. Corresponding to not quite one turn of the helix, this stabilizes the positioning of the 3′ end of the RNA so it can interact with an incoming ribonucleotide. The RNA-DNA hybrid helix rotates each time a nucleotide is added so that the 3′ end of the RNA stays at the catalytic site.

The transcription bubble moves down the DNA at a constant rate, about 50 nucleotides per second, leaving the growing RNA strand protruding from the bubble. After the transcription bubble passes, the nowtranscribed DNA is rewound as it leaves the bubble.

Unlike DNA polymerase, RNA polymerase has no proofreading capability. Transcription thus produces many more copying errors than replication. These mistakes, however, are not transmitted to progeny. Most genes are transcribed many times, so a few faulty copies are not harmful.

Termination

At the end of a gene are "stop" sequences that cause the formation of phosphodiester bonds to cease, the RNA-DNA hybrid within the transcription bubble to dissociate, the RNA polymerase to release the DNA, and the DNA within the transcription bubble to rewind. The simplest stop signal is a series of GC base-pairs followed by a series of AT base-pairs. The RNA transcript of this stop region forms a GC hairpin (figure 15.8), followed by four or more U ribonucleotides. How does this structure terminate transcription? The hairpin causes the RNA polymerase to pause immediately after the polymerase has synthesized it, placing the polymerase directly over the run of four uracils. The pairing of U with DNA's A is the weakest of the four hybrid base-pairs and is not strong enough to hold the hybrid strands together during the long pause. Instead, the RNA strand dissociates from the DNA within the transcription bubble, and transcription stops. A variety of protein factors aid hairpin loops in terminating transcription of particular genes.

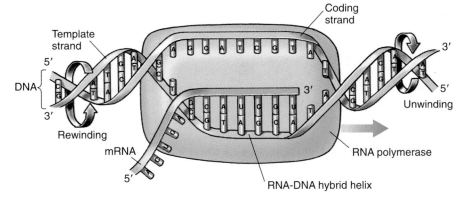

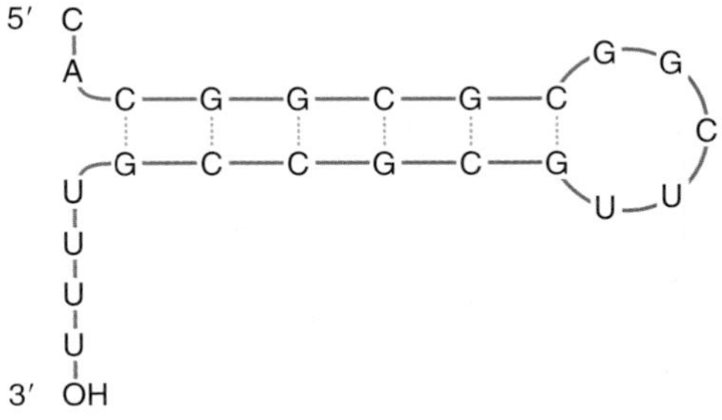

FIGURE 15.7
Model of a transcription bubble. The DNA duplex unwinds as it enters the RNA polymerase complex and rewinds as it leaves. One of the strands of DNA functions as a template, and nucleotide building blocks are assembled into RNA from this template.

FIGURE 15.8
A GC hairpin. This structure stops gene transcription.

Posttranscriptional Modifications

In eukaryotes, every mRNA transcript must travel a long journey out from the nucleus into the cytoplasm before it can be translated. Eukaryotic mRNA transcripts are modified in several ways to aid this journey:

5′ caps. Transcripts usually begin with A or G. However, in eukaryotes, this terminal ribonucleotide is modified in a highly unusual way: a GTP is added backwards. What happens is that the terminal phosphate of the 5′ A or G is removed, and then a very unusual 5′-5′ linkage forms with GTP. Called a **5′ cap,** this structure protects the 5′ end of the RNA template from nucleases and phosphatases during its long journey through the cytoplasm. Without these caps, RNA transcripts are rapidly degraded.

3′ poly-A tails. After a eukaryotic transcript is released from the transcription bubble, the 3′ end is cleaved off at a specific site containing the sequence AAUAAA. A special poly-A polymerase enzyme then adds about 250 A ribonucleotides to the 3′ end of the transcript. Called a **3′ poly-A tail,** this long string of A protects the transcript from degradation by nucleases. It also appears to make the transcript a better template for protein synthesis.

Transcription is carried out by the enzyme RNA polymerase, aided in eukaryotes by many other proteins.

Translation

Working out the genetic code was a great step forward, removing the mystery from the process of gene expression. However, it did not explain how the information stored in a sequence of nucleotides such as UUU actually specifies a particular amino acid such as phenylalanine. To understand how this happens, we must look more carefully at the first events in translation.

In prokaryotes, translation begins when the initial portion of an mRNA molecule binds to an rRNA molecule in a ribosome. The mRNA lies on the ribosome in such a way that only one of its codons is exposed at the polypeptide-making site at any time. A tRNA molecule possessing the complementary three-nucleotide sequence, or **anticodon,** binds to the exposed codon on the mRNA (figure 15.9). Because this tRNA molecule carries a particular amino acid, that amino acid and no other is added to the polypeptide in that position. As the mRNA molecule moves through the ribosome, successive codons on the mRNA message are exposed, and a series of tRNA molecules bind one after another to the exposed codons. Each of these tRNA molecules carries an attached amino acid, which it adds to the end of the growing polypeptide chain (figure 15.10).

There are about 45 different kinds of tRNA molecules. Why are there 45 and not 64 tRNAs (one for each codon)? Because the third base-pair of a tRNA anticodon allows some "wobble," some tRNAs recognize more than one codon.

If you think about it, we have not yet answered the key question of how the genetic code is translated. While it is true that anticodon sequences on tRNA molecules bind to specific codons on the mRNA, this does not solve the puzzle, but merely substitutes a new question: How do particular amino acids become associated with particular tRNA molecules? The key translation step, which pairs the three-nucleotide sequences with appropriate amino acids, is carried out by a remarkable set of enzymes called activating enzymes.

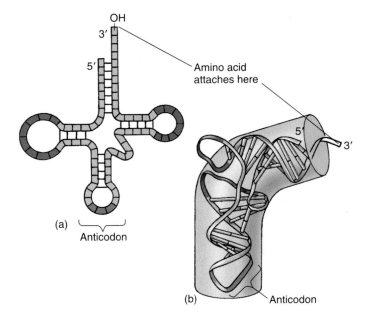

FIGURE 15.9
The structure of tRNA. (a) In the two-dimensional schematic, the three loops of tRNA are unfolded. Two of the loops bind to the ribosome during polypeptide synthesis, and the third loop contains the anticodon sequence. Amino acids attach to the free, single-stranded —OH end. (b) In the three-dimensional structure, the loops of tRNA are folded.

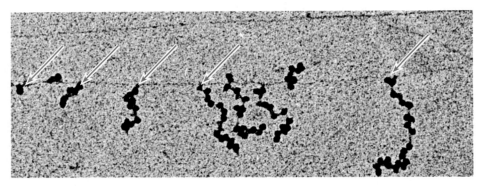

FIGURE 15.10
Translation in action. Bacteria have no nucleus and hence no membrane barrier between the DNA and the cytoplasm. In this electron micrograph of genes being transcribed in the bacterium *Escherichia coli,* you can see every stage of the process. The arrows point to RNA polymerase molecules. From each mRNA molecule dangling from the DNA, a series of ribosomes is assembling polypeptides. These clumps of ribosomes are sometimes called "polyribosomes."

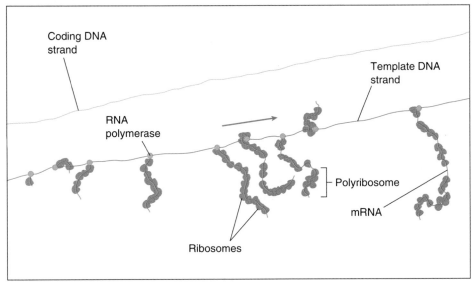

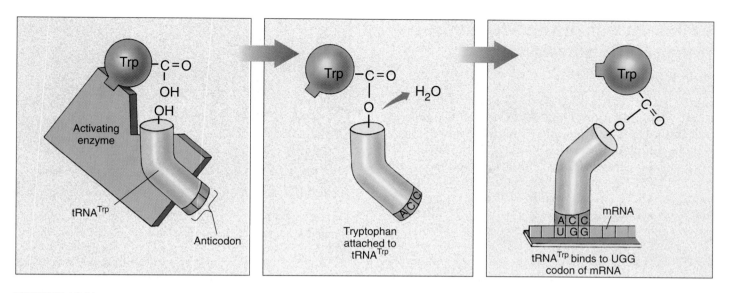

FIGURE 15.11
Activating enzymes "read" the genetic code. Each kind of activating enzyme recognizes and binds to a specific amino acid, such as tryptophan; it also recognizes and binds to the tRNA molecules with anticodons specifying that amino acid, such as ACC for tryptophan. In this way, activating enzymes link the tRNA molecules to specific amino acids.

Activating Enzymes

Particular tRNA molecules become attached to specific amino acids through the action of activating enzymes called **aminoacyl-tRNA synthetases,** one of which exists for each of the 20 common amino acids (figure 15.11). Therefore, each of these enzymes must correspond to both a specific anticodon sequence on a tRNA molecule and a particular amino acid. Some activating enzymes correspond to only one anticodon and thus only one tRNA molecule. Others recognize two, three, four, or six different tRNA molecules, each with a different anticodon but coding for the same amino acid (see table 15.1). If one considers the nucleotide sequence of mRNA a coded message, then the 20 activating enzymes are responsible for decoding that message.

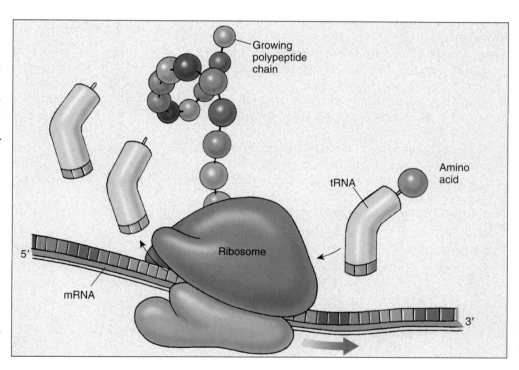

FIGURE 15.12
Translation. The mRNA strand acts as a template for tRNA molecules. The appropriate tRNA brings its particular amino acid into place as the ribosome moves along the mRNA in three-nucleotide steps, and the polypeptide chain elongates.

"Start" and "Stop" Signals

There is no tRNA with an anticodon complementary to three of the 64 codons: UAA, UAG, and UGA. These codons, called **nonsense codons,** serve as "stop" signals in the mRNA message, marking the end of a polypeptide. The "start" signal that marks the beginning of a polypeptide within an mRNA message is the codon AUG, which also encodes the amino acid methionine. The ribosome uses the first AUG that it encounters in the mRNA message to signal the start of translation. Figure 15.12 summarizes translation.

Activating enzymes match tRNAs to their appropriate amino acids, thereby interpreting the mRNA codon message.

A Closer Look at the Mechanism of Protein Synthesis

Initiation

In prokaryotes, polypeptide synthesis begins with the formation of an **initiation complex.** First, a tRNA molecule carrying a chemically modified methionine called *N*-formylmethionine (fMet-tRNA) binds to the small ribosomal subunit. Proteins called **initiation factors** position the fMet-tRNA on the ribosomal surface at the *P site* (for peptidyl), where peptide bonds will form. Nearby, two other sites will form: the *A site* (for aminoacyl), where successive amino acid-bearing tRNAs will bind, and the *E site* (for exit), where empty tRNAs will exit the ribosome (figure 15.13). This initiation complex, guided by another initiation factor, then binds to the anticodon AUG on the mRNA. Proper positioning of the mRNA is critical because it determines the reading frame—that is, which groups of three nucleotides will be read as codons. Moreover, the complex must bind to the beginning of the mRNA molecule, so that all of the transcribed gene will be translated. In bacteria, the beginning of each mRNA molecule is marked by a *leader sequence* complementary to one of the rRNA molecules on the ribosome. This complementarity allows base-pairs to form between the mRNA and rRNA molecules, ensuring that the mRNA message is read from the beginning. Bacteria often include several genes within a single mRNA transcript (polycistronic mRNA), while each eukaryotic gene is transcribed on a separate mRNA (monocistronic mRNA).

Initiation in eukaryotes is similar, although it differs in two important ways. First, in eukaryotes, the initiating amino acid is methionine rather than *N*-formylmethionine. Second, the initiation complex is far more complicated than in bacteria, containing nine or more protein factors, many consisting of several subunits. Eukaryotic initiation complexes are discussed in detail in the following chapter.

Elongation

After the initiation complex has formed, the large ribosome subunit binds, exposing the mRNA codon adjacent to the initiating AUG codon, and so positioning it for interaction with another amino acid-bearing tRNA molecule. When a tRNA molecule with the appropriate anticodon appears, proteins called elongation factors assist in binding it to the exposed mRNA codon at the A site (figure 15.14). When the second tRNA binds to the ribosome, it places its amino acid directly adjacent to the initial methionine, which is still attached to its tRNA molecule, which in turn is still bound to the ribosome. The two amino acids undergo a

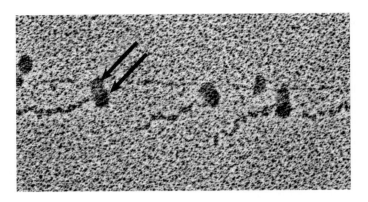

FIGURE 15.14
Elongation in action. These ribosomes are reading along an mRNA molecule from the fly *Chironomus tentans*. They read from left to right, each assembling polypeptides that dangle behind like the tail of a tadpole. Clearly visible are the two subunits (*arrows*) of each ribosome translating the mRNA.

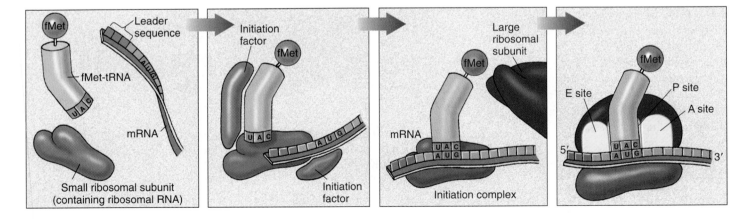

FIGURE 15.13
Formation of the initiation complex. In prokaryotes, proteins called initiation factors play key roles in positioning the small ribosomal subunit and the *N*-formylmethionine, or fmet-tRNA, molecule at the beginning of the mRNA message. When the fmet-tRNA is positioned over the first AUG codon of the mRNA, the large ribosomal subunit binds, forming the P, A, and E sites where successive tRNA molecules bind to the ribosomes, and polypeptide synthesis begins.

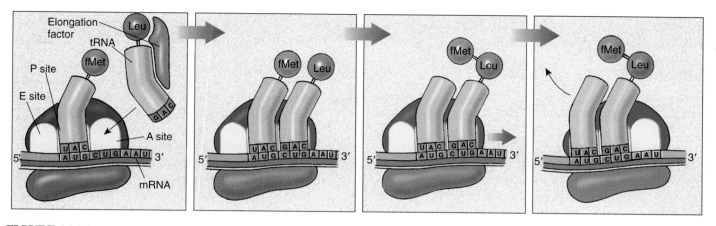

FIGURE 15.15

Translocation. The initiating fMet-tRNA in prokaryotes (Met-tRNA in eukaryotes) occupies the P site, and a tRNA molecule with an anticodon complementary to the exposed mRNA codon binds at the A site. fMet is transferred to the incoming amino acid (Leu), as the ribosome moves three nucleotides to the right along the mRNA. The empty fMet-tRNA moves to the E site to exit the ribosome, the growing polypeptide chain moves to the P site, and the A site is again exposed and ready to bind the next amino acid-laden tRNA.

chemical reaction, catalyzed by *peptidyl transferase,* which releases the initial methionine from its tRNA and attaches it instead by a peptide bond to the second amino acid.

In a process called **translocation** (figure 15.15), the ribosome now moves (translocates) three more nucleotides along the mRNA molecule in the 5' → 3' direction, guided by other elongation factors. This movement relocates the initial tRNA to the E site and ejects it from the ribosome, repositions the growing polypeptide chain (at this point containing two amino acids) to the P site, and exposes the next codon on the mRNA at the A site. When a tRNA molecule recognizing that codon appears, it binds to the codon at the A site, placing its amino acid adjacent to the growing chain. The chain then transfers to the new amino acid, and the entire process is repeated.

Termination

Elongation continues in this fashion until a chain-terminating nonsense codon is exposed (for example, UAA in figure 15.16). Nonsense codons do not bind to tRNA, but they are recognized by **release factors,** proteins that release the newly made polypeptide from the ribosome.

The first step in protein synthesis is the formation of an initiation complex. Each step of the ribosome's progress exposes a codon, to which a tRNA molecule with the complementary anticodon binds. The amino acid carried by each tRNA molecule is added to the end of the growing polypeptide chain.

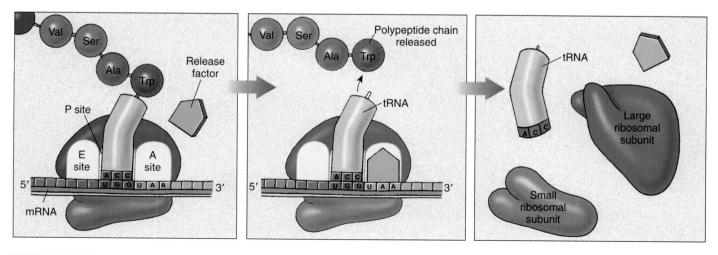

FIGURE 15.16

Termination of protein synthesis. There is no tRNA with an anticodon complementary to any of the three termination signal codons, such as the UAA nonsense codon illustrated here. When a ribosome encounters a termination codon, it therefore stops translocating. A specific release factor facilitates the release of the polypeptide chain by breaking the covalent bond that links the polypeptide to the P-site tRNA.

The Discovery of Introns

While the mechanisms of protein synthesis are similar in bacteria and eukaryotes, they are not identical. One difference is of particular importance. Unlike bacterial genes, most eukaryotic genes are much larger than they need to be to produce the polypeptides they code for. Such genes contain long sequences of nucleotides, known as **introns,** that do not code for any portion of the polypeptide specified by the gene. Introns are inserted between **exons,** much shorter sequences in the gene that do code for portions of the polypeptide.

Virtually every nucleotide within the transcribed portion of a bacterial gene is part of an amino acid-specifying codon, and the sequence of amino acids in the encoded protein matches the sequence of codons in the gene. Scientists assumed for many years that these features of gene expression were the same in all organisms. In the late 1970s, however, biologists were amazed to discover that many of the characteristics of prokaryotic gene expression did not apply to eukaryotes. In particular, they found that eukaryotic proteins are encoded by RNA segments that are excised from several locations along what is called the **primary RNA transcript** (or **primary transcript**) and then spliced together to form the mRNA that is eventually translated in the cytoplasm. The experiment that revealed this unexpected mode of gene expression consisted of several steps:

1. The mRNA transcribed from a particular gene was isolated and purified. For example, ovalbumin mRNA could be obtained fairly easily from unfertilized eggs.

2. Molecules of DNA complementary to the isolated mRNA were synthesized with the enzyme **reverse transcriptase.** These DNA molecules, which are called "copy" DNA (cDNA), had the same nucleotide sequence as the template strand of the gene that produced the mRNA.

3. With genetic engineering techniques (chapter 18), the portion of the nuclear DNA containing the gene that produced the mRNA was isolated. This procedure is referred to as *cloning* the gene in question.

4. Single-stranded forms of the cDNA and the nuclear DNA were mixed and allowed to pair with each other (to *hybridize*).

When the researchers examined the resulting hybrid DNA molecules with an electron microscope, they found that the DNA did not appear as a single duplex. Instead, they observed unpaired loops. In the case of the ovalbumin gene, they discovered seven loops, corresponding to sites where the nuclear DNA contained long nucleotide sequences not present in the cDNA. The conclusion was inescapable: nucleotide sequences must have been removed from the gene transcript before the cytoplasmic mRNA was translated into protein. These removed sequences are introns, and the remaining sequences are exons (figure 15.17). Because introns are excised from the RNA transcript before it is translated into protein, they do not affect the structure of the protein encoded by the gene in which they occur.

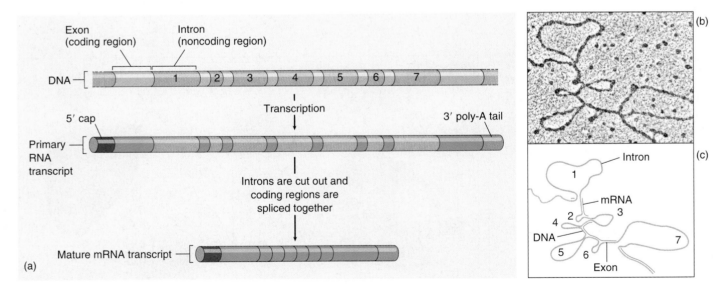

FIGURE 15.17

The eukaryotic gene that codes for ovalbumin in eggs containing introns. (a) The ovalbumin gene and its primary RNA transcript contain seven segments not present in the mRNA the ribosomes use to direct protein synthesis. Enzymes cut these segments (introns) out and splice together the remaining segments (exons). (b) The seven loops are the seven introns represented in the schematic drawing (c) of the mature mRNA transcript hybridized to DNA.

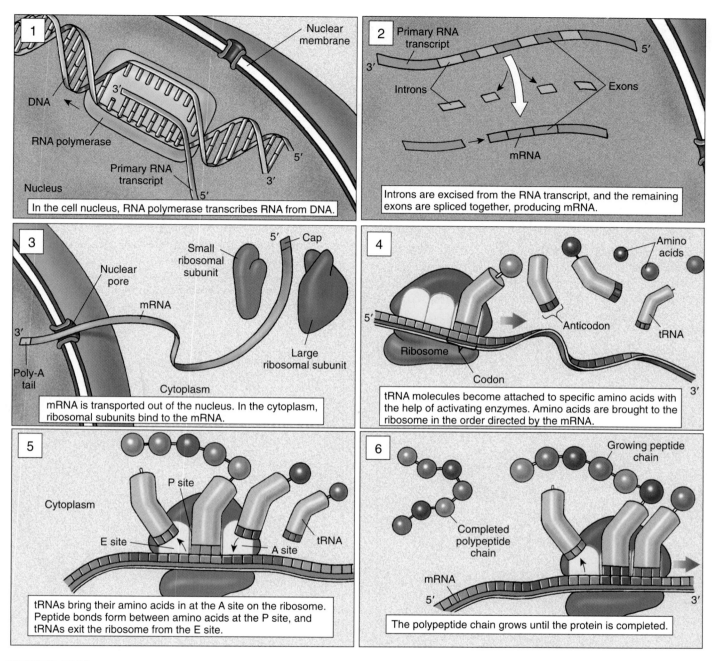

FIGURE 15.18
An overview of protein synthesis in eukaryotes.

1 In the cell nucleus, RNA polymerase transcribes RNA from DNA.

2 Introns are excised from the RNA transcript, and the remaining exons are spliced together, producing mRNA.

3 mRNA is transported out of the nucleus. In the cytoplasm, ribosomal subunits bind to the mRNA.

4 tRNA molecules become attached to specific amino acids with the help of activating enzymes. Amino acids are brought to the ribosome in the order directed by the mRNA.

5 tRNAs bring their amino acids in at the A site on the ribosome. Peptide bonds form between amino acids at the P site, and tRNAs exit the ribosome from the E site.

6 The polypeptide chain grows until the protein is completed.

RNA Splicing

When a gene is transcribed, the primary RNA transcript (that is, the gene copy as it is made by RNA polymerase, before any modification occurs) contains sequences complementary to the entire gene, including introns as well as exons. However, in a process called **RNA processing,** or **splicing,** the intron sequences are cut out of the primary transcript before it is used in polypeptide synthesis; therefore, those sequences are not translated. The remaining sequences, which correspond to the exons, are spliced together to form the final, "processed" mRNA molecule that is translated. In a typical human gene, the introns can be 10 to 30 times larger than the exons. For example, even though only 432 nucleotides are required to encode the 144 amino acids of hemoglobin, there are actually 1356 nucleotides in the primary mRNA transcript of the hemoglobin gene. Figure 15.18 summarizes eukaryotic protein synthesis.

Much of a eukaryotic gene is not translated. Noncoding segments scattered throughout the gene are removed from the primary transcript before the mRNA is translated.

Differences between Bacterial and Eukaryotic Protein Synthesis

1. Most eukaryotic genes possess introns. With the exception of a few genes in the Archaebacteria, prokaryotic genes lack introns (figure 15.19).

2. Individual bacterial mRNA molecules often contain transcripts of several genes. By placing genes with related functions on the same mRNA, bacteria coordinate the regulation of those functions. Eukaryotic mRNA molecules rarely contain transcripts of more than one gene. Regulation of eukaryotic gene expression is achieved in other ways.

3. Because eukaryotes possess a nucleus, their mRNA molecules must be completely formed and must pass across the nuclear membrane before they are translated. Bacteria, which lack nuclei, often begin translation of an mRNA molecule before its transcription is completed.

4. In bacteria, translation begins at an AUG codon preceded by a special nucleotide sequence. In eukaryotic cells, mRNA molecules are modified at the 5′ leading end after transcription, adding a 5′ cap, a methylated guanosine triphosphate. The cap initiates translation by binding the mRNA, usually at the first AUG, to the small ribosomal subunit.

5. Bacterial mRNA molecules are translated by ribosomes as they are transcribed. Eukaryotic mRNA molecules are modified before they are translated: in addition to the 5′ cap, they add a 3′ poly-A tail consisting of some 200 adenine (A) nucleotides. These modifications appear to delay the destruction of the mRNA by cellular enzymes.

6. The ribosomes of eukaryotes are a little larger than those of bacteria.

Protein synthesis is broadly similar in bacteria and eukaryotes, although it differs in some details.

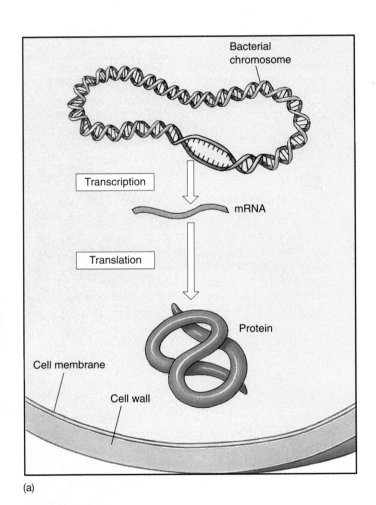

(a)

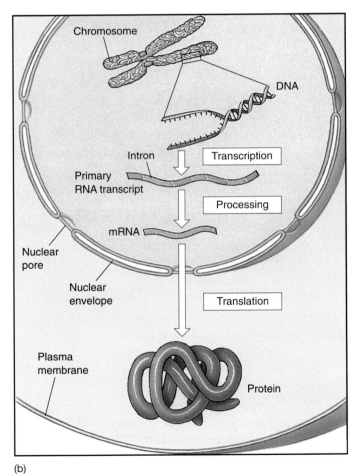

(b)

FIGURE 15.19

Gene information is processed differently in prokaryotes and eukaryotes. (a) Bacterial genes are transcribed into mRNA, which is translated immediately. Hence, the sequence of DNA nucleotides corresponds exactly to the sequence of amino acids in the encoded polypeptide. (b) Eukaryotic genes are typically different, containing long stretches of nucleotides called introns that do not correspond to amino acids within the encoded polypeptide. Introns are removed from the primary RNA transcript of the gene before the mRNA directs the synthesis of the polypeptide.

Summary of Concepts

15.1 The Central Dogma traces the flow of gene-encoded information.

- There are three principal kinds of RNA: messenger RNA (mRNA), transcripts of genes used to direct the assembly of amino acids into proteins; ribosomal RNA (rRNA), which combines with proteins to make up the ribosomes that carry out the assembly process; and transfer RNA (tRNA), molecules that transport the amino acids to the ribosome for assembly into proteins.

15.2 Genes encode information in three-nucleotide code words.

- The sequence of nucleotides in DNA encodes the sequence of amino acids in proteins. The mRNA transcribed from the DNA is read by ribosomes in increments of three nucleotides called codons.

15.3 Genes are first transcribed, then translated.

- During transcription, the enzyme RNA polymerase manufactures mRNA molecules with nucleotide sequences complementary to particular segments of the DNA.

- During translation, the mRNA sequences direct the assembly of amino acids into proteins on cytoplasmic ribosomes.

- The information in a gene and in an mRNA molecule is read in three-nucleotide blocks called codons.

- On the ribosome, the mRNA molecule is positioned so that only one of its codons is exposed at any time.

- This exposure permits a tRNA molecule with the complementary base sequence (anticodon) to bind to it.

- Attached to the other end of the tRNA is an amino acid, which is added to the end of the growing polypeptide chain.

15.4 Eukaryotic gene transcripts are edited.

- Most eukaryotic genes contain noncoding sequences (introns) interspersed randomly between coding sequences (exons).

- The portions of an mRNA molecule corresponding to the introns are removed from the primary RNA transcript before the remainder is translated.

Discussing Key Terms

1. **Transcription** The enzyme RNA polymerase binds to a particular site called a promoter at the beginning of a gene and then moves along one DNA strand, assembling a complementary RNA version of the gene called an mRNA transcript.

2. **Translation** In bacteria, ribosomes bind to the mRNA transcript as it is being transcribed and use it to direct the synthesis of the encoded protein, a process called translation. In eukaryotes, the mRNA must first be transported out of the nucleus before ribosomes located in the cytoplasm can translate it.

3. **The genetic code** DNA encodes proteins by specifying each amino acid with a three-nucleotide codon.

4. **Activating enzymes read the code** Each mRNA codon binds a particular tRNA, which inserts the amino acid it bears at the growing end of the nascent polypeptide chain. The choice of which amino acid to attach to which tRNA is made by activating enzymes, the components of the cell that "read" the genetic code.

5. **Eukaryotic genes are fragmented** In eukaryotic genes, the encoded information is typically fragmented into multiple segments called exons that are embedded within a background of noncoding sequences called introns. Before the eukaryotic mRNA leaves the nucleus, the introns are cut away from the primary transcript, and the exons spliced together to form the processed mRNA message that is translated.

Review Questions

1. What are the three major classes of RNA? What is the function of each type?

2. What is the function of RNA polymerase in transcription? What determines where RNA polymerase begins and ends its function?

3. How did Crick and his colleagues determine whether the nucleotides in a gene are read in a simple or a punctuated sequence? What did they find?

4. How many nucleotides are used to specify each amino acid? How was the nucleotide sequence that codes for each amino acid determined?

5. What is a codon? What is an anticodon? What is a nonsense codon?

6. During protein synthesis, what mechanism ensures that only one amino acid is added to the growing polypeptide at a time? What mechanism ensures the correct amino acid is added at each position in the polypeptide?

7. How does an mRNA molecule specify where the polypeptide it encodes should begin? How does it specify where the polypeptide should end?

8. What roles do elongation factors play in translation?

9. What is an intron? What is an exon? How is each involved in the mRNA molecule that is ultimately translated?

Thought Questions

1. The nucleotide sequence of a hypothetical eukaryotic nuclear gene is:

TAC ATA CTA GTT AC<u>G</u> TCG CCC
GGA AAT ATC

If a mutation in this gene were to change the fifteenth nucleotide (underlined) from guanine to thymine, what effect do you think it might have on the expression of this gene?

2. Investigators have found that 96% of the amino acids in yeast proteins are specified by only 25 of the 61 codons that code for amino acids; the other 36 codons are rarely used, and their corresponding tRNAs exist at much lower concentrations inside the cells. Can you suggest a reason why so many tRNAs are rarely used?

Internet Links

Exploring Protein Structure
http://www.seqnet.dl.ac.uk/PPS/course/index.html
A virtual biology class from Birkbeck College, London, on the basics of protein structure and synthesis.

From DNA to Protein
http://ampere.scale.uiuc.edu/pb102/get.cgi?lect+5
A lecture from the superb online Biology class of the University of Illinois, with clear explanations, colorful graphics, and an instructive quiz.

For Further Reading

Berg, P.: *Dealing with Genes: The Language of Heredity*, University Science Books, Mill Valley, CA, 1992. A book by a leader in the field; Paul Berg won the 1980 Nobel Prize in chemistry for his work on the development of recombinant DNA.

Doolittle, W. F.: "The Origin and Function of Intervening Sequences in DNA—A Review," *American Naturalist*, vol. 130, December 1987, pages 915–28. A thoughtful and comprehensive review of the many ideas about why introns exist and what they do.

Dorit, R., L. Schoenbach, and W. Gilbert: "How Big Is the Universe of Exons?" *Science*, vol. 250, December 1990, pages 1377–82. A controversial attempt to estimate the number of exons needed to construct all known proteins. A very important first step in addressing this key evolutionary question.

Nierhaus, K.: "The Three-Site Elongation Model for the Ribosome Elongation Cycle," *Biochemistry*, vol. 29, May 1990, pages 4997–5007. An exciting suggestion that the ribosome contains an E (exit) site as well as A and P sites.

Nowak, R.: "Mining Treasures from 'Junk DNA'," *Science*, vol. 263, February 4, 1994, pages 608–10. If "junk DNA" really is junk, why do we have so much of it?

Ross, J.: "The Turnover of Messenger RNA," *Scientific American*, April 1989, pages 48–55. The level of many proteins in the body is determined by how fast the messenger RNA encoding them is broken down, rather than by how speedily new transcripts are churned out.

Steitz, J.: "Snurps," *Scientific American*, June 1988, pages 56–63. Small nuclear ribonuclear proteins (snurps) help remove introns from mRNA transcripts and are an active subject of research.

Todorov, I.: "How Cells Maintain Stability," *Scientific American*, December 1990, pages 66–75. An account of how cells shut down their protein-making factories in times of stress and start them back up afterward.

16

Control of Gene Expression

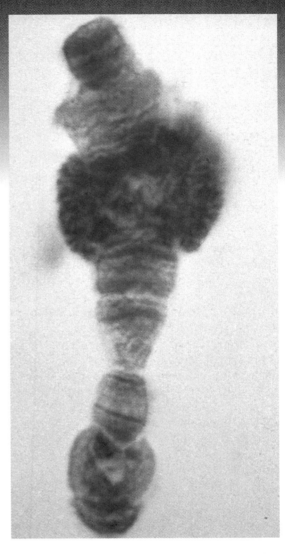

FIGURE 16.1
Chromosome puffs. In this chromosome of the fly *Drosophila melanogaster*, individual active genes can be visualized as "puffs" on the chromosomes. The RNA being transcribed from the DNA template has been radioactively labeled, and the dark specks indicate its position on the chromosome.

Concept Outline

I n an orchestra, all of the instruments do not play all the time; if they did, all they would produce is noise. Instead, a musical score determines which instruments in the orchestra play when. Similarly, all of the genes in an organism are not expressed at the same time, each gene producing the protein it encodes full tilt. Instead, different genes are expressed at different times, with a genetic score written in regulatory regions of the DNA determining which genes are active when (figure 16.1).

315

An Overview of Transcriptional Control

Control of gene expression is essential to all organisms. In bacteria, it allows the cell to take advantage of changing environmental conditions. In multicellular organisms, it is critical for directing development and maintaining homeostasis.

Regulating Promoter Access

One way to control transcription is to regulate the initiation of transcription. In order for a gene to be transcribed, RNA polymerase must have access to the DNA helix and must be capable of binding to the gene's **promoter,** a specific sequence of nucleotides at one end of the gene that tells the polymerase where to begin transcribing. How is the initiation of transcription regulated? Protein-binding nucleotide sequences on the DNA regulate the initiation of transcription by modulating the ability of RNA polymerase to bind to the promoter. These protein-binding sites are usually only 10 to 15 nucleotides in length (even a large protein has a "footprint," or binding area, of only about 20 nucleotides). Hundreds of these regulatory sequences have been characterized, and each provides a binding site for a specific protein able to recognize the sequence. Binding the protein to the regulatory sequence either *blocks* transcription by getting in the way of RNA polymerase, or *stimulates* transcription by facilitating the binding of RNA polymerase to the promoter.

Transcriptional Control in Prokaryotes

Control of gene expression is accomplished very differently in bacteria than in the cells of complex multicellular organisms. Bacterial cells have been shaped by evolution to grow and divide as rapidly as possible, enabling them to exploit transient resources. In bacteria, the primary function of gene control is to adjust the cell's activities to its immediate environment. Changes in gene expression alter which enzymes are present in the cell in response to the quantity and type of available nutrients and the amount of oxygen present. Almost all of these changes are fully reversible, allowing the cell to adjust its enzyme levels up or down as the environment changes.

Transcriptional Control in Eukaryotes

The cells of multicellular organisms, on the other hand, have been shaped by evolution to be protected from transient changes in their immediate environment. Most of them experience fairly constant conditions. Indeed, **homeostasis**—the maintenance of a constant internal environment—is considered by many to be the hallmark of multicellular organisms. Although cells in such organisms still respond to signals in their immediate environment (such as growth factors and hormones) by altering gene expression, in doing so they participate in regulating the body as a whole. In multicellular organisms with relatively constant internal environments, the primary function of gene control in a cell is not to respond to that cell's immediate environment, but rather to participate in regulating the body as a whole.

Some of these changes in gene expression compensate for changes in the physiological condition of the body. Others mediate the decisions that *produce* the body, ensuring that the right genes are expressed in the right cells at the right time during development. The growth and development of multicellular organisms entail a long series of biochemical reactions, each catalyzed by a specific enzyme. Once a particular developmental change has occurred, these enzymes cease to be active, lest they disrupt the events that must follow. To produce these enzymes, genes are transcribed in a carefully prescribed order, each for a specified period of time. In fact, many genes are activated only once, producing irreversible effects. In many animals, for example, **stem cells** develop into differentiated tissues like skin cells or red blood cells, following a fixed genetic program that often leads to programmed cell death. The one-time expression of the genes that guide this program is fundamentally different from the reversible metabolic adjustments bacterial cells make to the environment. In all multicellular organisms, changes in gene expression within particular cells serve the needs of the whole organism, rather than the survival of individual cells.

Posttranscriptional Control

Gene expression can be regulated at many levels. By far the most common form of regulation in both bacteria and eukaryotes is **transcriptional control,** that is, control of the transcription of particular genes by RNA polymerase. Other less common forms of control occur after transcription, influencing the mRNA that is produced from the genes or the activity of the proteins encoded by the mRNA. These controls, collectively referred to as **posttranscriptional controls,** will be discussed briefly later in this chapter.

Gene expression is controlled at the transcriptional and posttranscriptional levels. Transcriptional control, more common, is effected by the binding of proteins to regulatory sequences within the DNA.

How to Read a Helix without Unwinding It

It is the ability of certain proteins to bind to *specific* DNA regulatory sequences that provides the basic tool of gene regulation, the key ability that makes transcriptional control possible. To understand how cells control gene expression, it is first necessary to gain a clear picture of this molecular recognition process.

Looking into the Major Groove

Molecular biologists used to think that the DNA helix had to unwind before proteins could distinguish one DNA sequence from another; only in this way, they reasoned, could regulatory proteins gain access to the hydrogen bonds between base-pairs. We now know it is unnecessary for the helix to unwind because proteins can bind to its outside surface, where the edges of the base-pairs are exposed. Careful inspection of a DNA molecule reveals two helical grooves winding round the molecule, one deeper than the other. Within the deeper groove, called the **major groove**, the nucleotides' hydrophobic methyl groups, hydrogen atoms, and hydrogen bond donors and acceptors are accessible to proteins. The pattern created by these chemical groups is unique for each of the four possible base-pair arrangements, providing a ready way for a protein nestled in the groove to read the sequence of bases (figure 16.2).

DNA Binding Motifs

Protein-DNA recognition is an area of active research; so far, the structures of over 30 regulatory proteins have been analyzed. Although each protein is unique in its fine details, the part of the protein that actually binds to the DNA is much less variable. Almost all of these proteins employ one of a small set of **structural,** or **DNA-binding, motifs,** particular bends of the protein chain that permit it to interlock with the major groove of the DNA helix.

Regulatory proteins identify specific sequences on the DNA double helix, without unwinding it, by inserting DNA-binding motifs into the major groove of the double helix where the edges of the bases protrude.

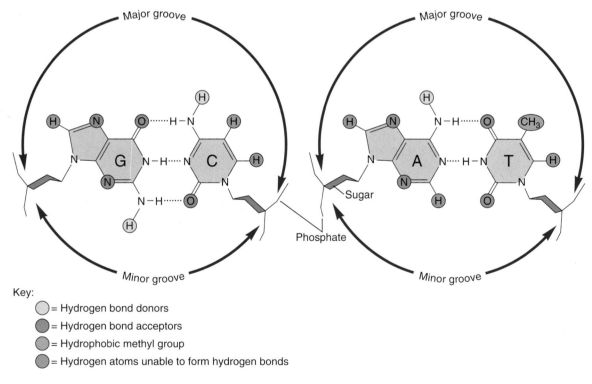

Key:
- ◯ = Hydrogen bond donors
- ● = Hydrogen bond acceptors
- ◕ = Hydrophobic methyl group
- ◑ = Hydrogen atoms unable to form hydrogen bonds

FIGURE 16.2
Reading the major groove of DNA. Looking down into the major groove of a DNA helix, we can see the edges of the bases protruding into the groove. Each of the four possible base-pair arrangements (two are shown here) extends a unique set of chemical groups into the groove, indicated in this diagram by differently colored balls. A regulatory protein can identify the base-pair arrangement by this characteristic signature.

Four Important Motifs

The Helix-Turn-Helix Motif

The most common DNA-binding motif is the **helix-turn-helix,** constructed from two α-helical segments of the protein linked by a short nonhelical segment (figure 16.3). The first DNA-binding motif recognized, the helix-turn-helix motif has since been identified in hundreds of DNA-binding proteins.

A close look at the structure of a helix-turn-helix motif reveals how proteins containing such motifs are able to interact with the major groove of DNA. Interactions between the helical segments of the motif hold them at roughly right angles to each other. When this motif is pressed against DNA, one of the helical segments (called the recognition helix) fits snugly in the major groove of the DNA molecule, while the other butts up against the outside of the DNA molecule, helping to ensure the proper positioning of the recognition helix. Most DNA regulatory sequences recognized by helix-turn-helix motifs occur in symmetrical pairs. Such sequences are bound by proteins containing two helix-turn-helix motifs separated by 3.4 nm, the distance required for one turn of the DNA helix (figure 16.4). Having *two* protein/DNA-binding sites doubles the zone of contact between protein and DNA and so greatly strengthens the bond that forms between them.

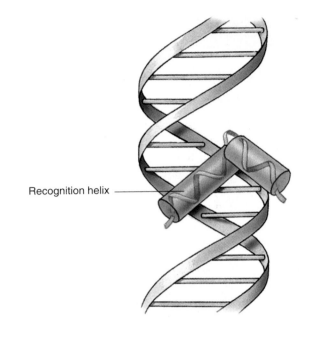

Recognition helix

FIGURE 16.3
The helix-turn-helix motif. One helical region, called the recognition helix, actually fits into the major groove of DNA. There it contacts the edges of base-pairs, enabling it to recognize specific sequences of DNA bases.

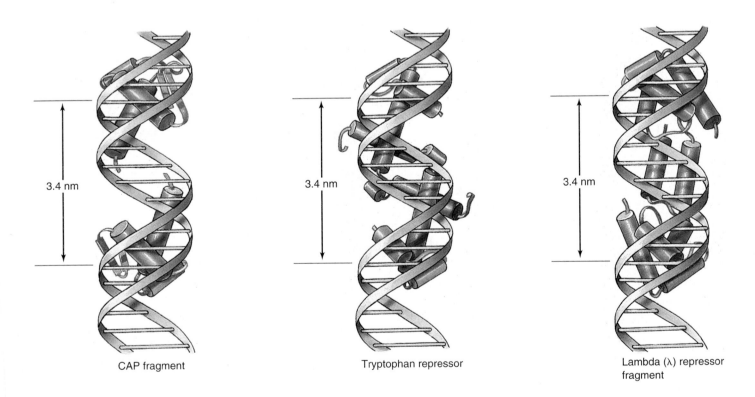

CAP fragment

Tryptophan repressor

Lambda (λ) repressor fragment

FIGURE 16.4
How the helix-turn-helix binding motif works. The three regulatory proteins illustrated here all bind to DNA as dimers. In each case, the two copies of the recognition helix (*red*) are separated by 3.4 nm, precisely the spacing of one turn of the DNA helix. This allows the regulatory proteins to slip into two adjacent portions of the major groove in DNA, providing a strong attachment.

The Homeodomain Motif

A special class of helix-turn-helix motifs plays a critical role in development in a wide variety of eukaryotic organisms, including humans. These motifs were discovered when researchers began to characterize a set of homeotic mutations in *Drosophila* (mutations that alter how the parts of the body are assembled). They found that the mutant genes encoded regulatory proteins whose normal function was to initiate key stages of development by binding to developmental switch-point genes. More than 50 of these regulatory proteins have been analyzed, and they all contain a nearly identical sequence of 60 amino acids, the **homeodomain** (figure 16.5*b*). The center of the homeodomain is occupied by a helix-turn-helix motif that binds to the DNA. Surrounding this motif within the homeodomain is a region that always presents the motif to the DNA in the same way.

The Zinc Finger Motif

A different kind of DNA-binding motif uses one or more zinc atoms to coordinate its binding to DNA. Called **zinc fingers** (figure 16.5*c*), these motifs exist in several forms. In one form, a zinc atom links an α-helical segment to a β-sheet segment so that the helical segment fits into the major groove of DNA. This sort of motif often occurs in clusters, the β-sheets spacing the helical segments so that each helix contacts the major groove. The more zinc fingers in the cluster, the stronger the protein binds to the DNA. In other forms of the zinc finger motif, the β-sheet's place is taken by another helical segment.

The Leucine Zipper Motif

In yet another DNA-binding motif, two different protein subunits cooperate to create a single DNA-binding site. This motif is created where a region on one of the subunits containing several hydrophobic amino acids (usually leucines) interacts with a similar region on the other subunit. This interaction holds the two subunits together at those regions, while the rest of the subunits are separated. Called a **leucine zipper**, this structure has the shape of a "Y," with the two arms of the Y being helical regions that fit into the major groove of DNA (figure 16.5*d*). Because the two subunits can contribute quite different helical regions to the motif, leucine zippers allow for great flexibility in controlling gene expression.

Regulatory proteins bind to the edges of base-pairs exposed in the major groove of DNA. Most contain structural motifs such as the helix-turn-helix, homeodomain, zinc finger, or leucine zipper.

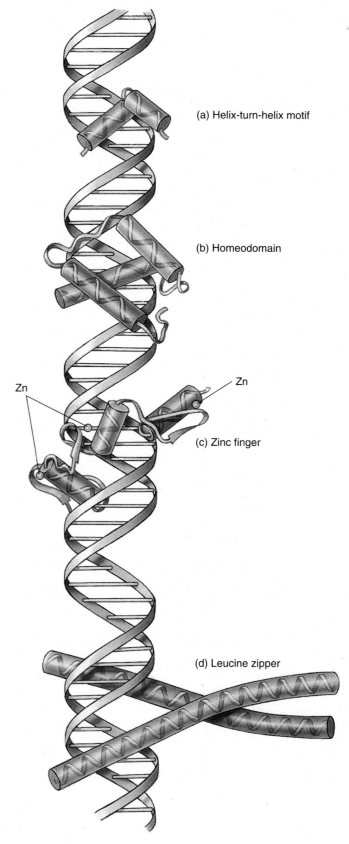

(a) Helix-turn-helix motif

(b) Homeodomain

Zn   Zn

(c) Zinc finger

(d) Leucine zipper

FIGURE 16.5
Major DNA-binding motifs.

Controlling Transcription Initiation

How do organisms use regulatory DNA sequences and the proteins that bind them to control when genes are transcribed? The same basic controls are used in bacteria and eukaryotes, but eukaryotes employ several additional elements that reflect their more elaborate chromosomal structure. We will begin by discussing the relatively simple controls found in bacteria.

Repressors Are *OFF* Switches

A typical bacterium possesses genes encoding several thousand proteins, but only a few are transcribed at any one time; the others are held in reserve until needed. When the cell encounters a potential food source, for example, it begins to manufacture the enzymes necessary to metabolize that food. Perhaps the best-understood example of this type of transcriptional control is the regulation of tryptophan-producing genes (*trp* genes), which was investigated in the pioneering work of Charles Yanofsky and his students at Stanford University.

Operons. The bacterium *Escherichia coli* uses proteins encoded by a cluster of five genes to manufacture the amino acid tryptophan. All five genes are transcribed together as a unit called an operon, producing a single, long piece of mRNA. RNA polymerase binds to a promoter located at the beginning of the first gene, and then proceeds down the DNA, transcribing the genes one after another. Regulatory proteins shut off transcription by binding to an operator site immediately in front of the promoter and often overlapping it.

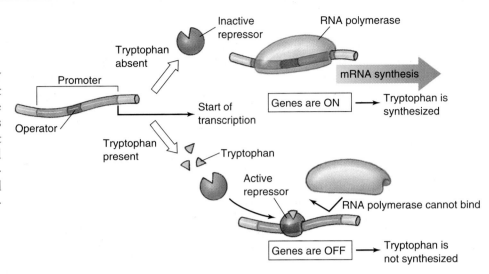

FIGURE 16.6
How the *trp* operon is controlled. The tryptophan repressor cannot bind the operator (which is located *within* the promoter) unless tryptophan first binds to the repressor. Therefore, in the absence of tryptophan, the promoter is free to function and RNA polymerase transcribes the operon. In the presence of tryptophan, the tryptophan-repressor complex binds tightly to the operator, preventing RNA polymerase from initiating transcription.

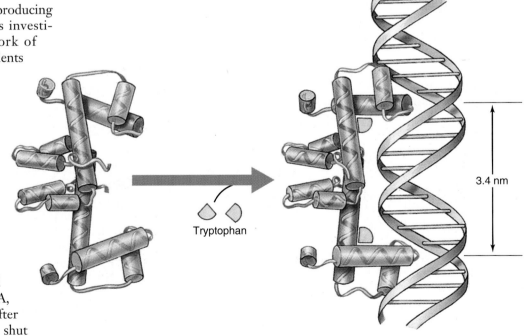

FIGURE 16.7
How the tryptophan repressor works. The binding of tryptophan to the repressor increases the distance between the two recognition helices in the repressor, allowing the repressor to fit snugly into two adjacent portions of the major groove in DNA.

FIGURE 16.8

How regulatory proteins bind to DNA. Studies of the repressor protein that blocks transcription of the genes of the bacterial virus called lambda (λ) have played a key role in our understanding of OFF switches. (a) the λ virus repressor protein. Protein dimers reach into adjacent grooves of the λ DNA helix. Binding of the λ repressor blocks the binding of RNA polymerase. (b) The CRO protein. Binding of a protein called CRO actually bends the λ DNA helix.

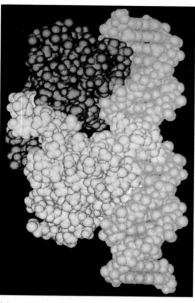

(a)

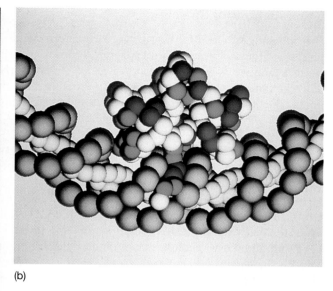

(b)

When tryptophan is present in the medium surrounding the bacterium, the cell shuts off transcription of the *trp* genes by means of a tryptophan **repressor,** a helix-turn-helix regulatory protein that binds to the operator site located within the *trp* promoter (figure 16.6). Binding of the repressor to the operator prevents RNA polymerase from binding to the promoter. The key to the functioning of this control mechanism is that the tryptophan repressor cannot bind to DNA unless it has first bound to two molecules of tryptophan. The binding of tryptophan to the repressor alters the orientation of a pair of helix-turn-helix motifs in the repressor, causing their recognition helices to fit into adjacent major grooves of the DNA (figures 16.7 and 16.8).

Thus, the bacterial cell's synthesis of tryptophan depends upon the absence of tryptophan in the environment. When the environment lacks tryptophan, there is nothing to activate the repressor, so the repressor cannot prevent RNA polymerase from binding to the *trp* promoter. The *trp* genes are transcribed, and the cell proceeds to manufacture tryptophan from other molecules. On the other hand, when tryptophan is present in the environment, it binds to the repressor, which is then able to bind to the *trp* promoter. This blocks transcription of the *trp* genes, and the cell's synthesis of tryptophan halts.

Activators Are *ON* Switches

Not all regulatory switches shut genes off—some turn them on. In these instances, bacterial promoters are deliberately constructed to be poor binding sites for RNA polymerase, and the genes these promoters govern are thus rarely transcribed—unless something happens to improve the promoter's ability to bind RNA polymerase. This can happen if a regulatory protein called a **transcriptional activator** binds to the DNA nearby. By contacting the poly-

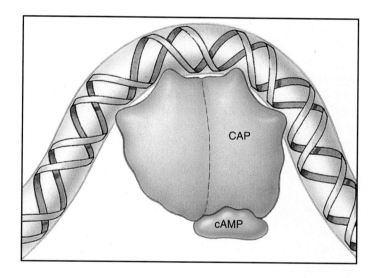

FIGURE 16.9

How CAP works. Binding of the catabolite activator protein (CAP) to DNA causes the DNA to bend around it. This increases the exposure of the promoter to RNA polymerase.

merase protein itself, the activator protein helps hold the polymerase against the DNA promoter site so that transcription can begin.

A well-understood transcriptional activator is the catabolite activator protein (CAP) of *E. coli*, which initiates the transcription of genes that allow *E. coli* to use other molecules as food when glucose is not present. Falling levels of glucose lead to higher intracellular levels of the signaling molecule, cyclic-AMP (cAMP), which binds to the CAP protein. When cAMP binds to it, the CAP protein changes shape, enabling its helix-turn-helix motif to bind to the DNA near any of several promoters. Consequently, those promoters are activated and their genes can be transcribed (figure 16.9).

Combinations of Switches

By combining ON and OFF switches, bacteria can create sophisticated transcriptional control systems. A particularly well-studied example is the *lac* operon of *E. coli* (figure 16.10). This operon is responsible for producing three proteins that import the disaccharide lactose into the cell and break it down into two monosaccharides: glucose and galactose.

The Activator Switch. The *lac* operon possesses two regulatory sites. One is a CAP site located adjacent to the *lac* promoter. It ensures that the *lac* genes are not transcribed when ample amounts of glucose are already present. In the absence of glucose, a high level of cAMP builds up in the cell. Consequently, cAMP is available to bind to CAP and allow it to change shape, bind to the DNA, and activate the *lac* promoter (figure 16.11). In the presence of glucose, cAMP levels are low, CAP is unable to bind to the DNA, and the *lac* promoter is not functional.

The Repressor Switch. Whether the *lac* genes are actually transcribed in the absence of glucose is determined by the second regulatory site, the **operator,** which is located adjacent to the promoter. A protein called the *lac* repressor is capable of binding to the operator, but only when lactose is absent. Because the operator and the promoter are close together, the repressor covers part of the promoter when it binds to the operator, preventing RNA polymerase from binding and so blocking transcription of the *lac* genes. These genes are then said to be "repressed" (figure 16.12). As a

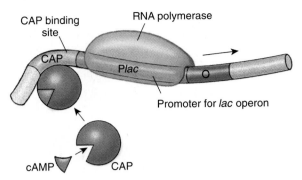

(a) Glucose low, promoter activated

(b) Glucose high, promoter not activated

FIGURE 16.11
How the CAP site works. The CAP molecule can attach to the CAP binding site only when the molecule is bound to cAMP.
(a) When glucose levels are low, cAMP is abundant and binds to CAP. The cAMP-CAP complex binds to the CAP site, bends in the DNA, and gives RNA polymerase access to the promoter.
(b) When glucose levels are high, cAMP is scarce, and CAP is unable to activate the promoter.

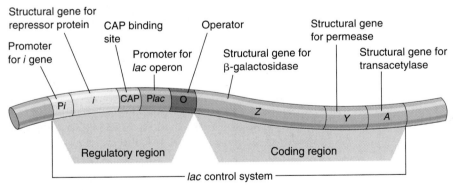

FIGURE 16.10
The *lac* region of the *Escherichia coli* chromosome. The *lac* operon consists of a promoter, an operator, and three structural genes. In addition, there is a binding site for the catabolite activator protein (CAP), which affects whether or not RNA polymerase will bind to the promoter. Gene *i* codes for a repressor protein, which will bind to the operator and block transcription of the *lac* genes. The three structural genes, *Z*, *Y*, and *A* encode the two enzymes and the permease involved in the metabolism of lactose. Because they encode the primary structure (amino acid sequence) of enzymes, they are called structural genes.

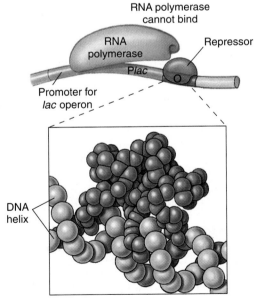

FIGURE 16.12
The *lac* repressor. Because the repressor fills the major groove of the DNA helix, RNA polymerase cannot attach to the promoter, and transcription is blocked.

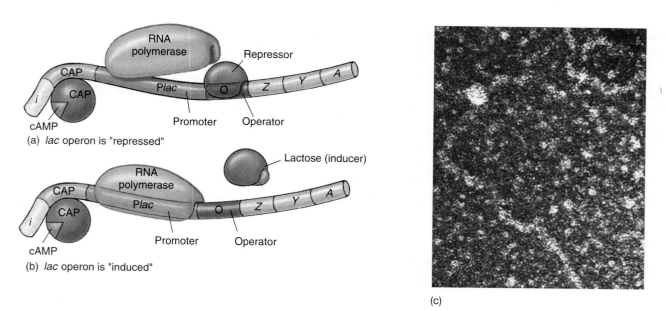

(a) *lac* operon is "repressed"

(b) *lac* operon is "induced"

(c)

FIGURE 16.13

How the *lac* repressor works. (a) The *lac* operon is shut down ("repressed") when the repressor protein is bound to the operator site. Because promoter and operator sites overlap, RNA polymerase and the repressor cannot bind at the same time, any more than two people can sit in the same chair at once. (b) The *lac* operon is transcribed ("induced") when CAP is bound and when lactose binding to the repressor changes its shape so that it can no longer sit on the operator site and block polymerase binding. (c) Electron micrograph of the *lac* repressor (the large white sphere).

result, the cell does not transcribe genes whose products it has no use for. However, when lactose is present, a lactose isomer binds to the repressor, twisting its binding motif away from the major groove of the DNA. This prevents the repressor from binding to the operator and so allows RNA polymerase to bind to the promoter, initiating transcription of the *lac* genes. Transcription of the *lac* operon is said to have been "induced" by lactose (figure 16.13).

This two-switch control mechanism thus causes the cell to produce lactose-utilizing proteins whenever lactose is present but glucose is not, enabling it to make a metabolic decision to produce only what the cell needs, conserving its resources (figure 16.14).

Bacteria regulate gene expression transcriptionally through the use of repressor and activator "switches," such as the *trp* repressor and the CAP activator. The transcription of some clusters of genes, such as the *lac* operon, is regulated by both repressors and activators.

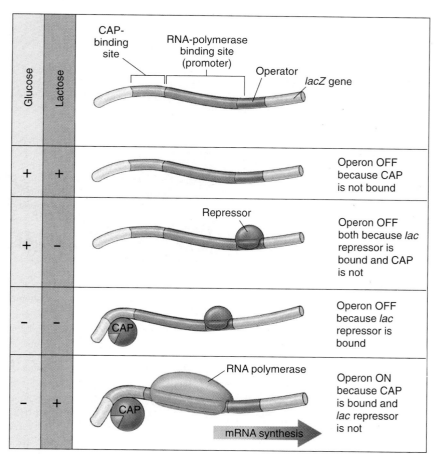

FIGURE 16.14

Two regulatory proteins control the *lac* operon. Together, the *lac* repressor and CAP provide a very sensitive response to the cell's need to utilize lactose-metabolizing enzymes.

Designing a Complex Gene Control System

As we have seen, combinations of ON and OFF control switches allow bacteria to regulate the transcription of particular genes in response to the immediate metabolic demands of their environment. All of these switches work by interacting directly with RNA polymerase, either blocking or enhancing its binding to specific promoters. There is a limit to the complexity of this sort of regulation, however, because only a small number of switches can be squeezed into and around one promoter. In a eukaryotic organism that undergoes a complex development, many genes must interact with one another, requiring many more interacting elements than can fit around a single promoter (table 16.1).

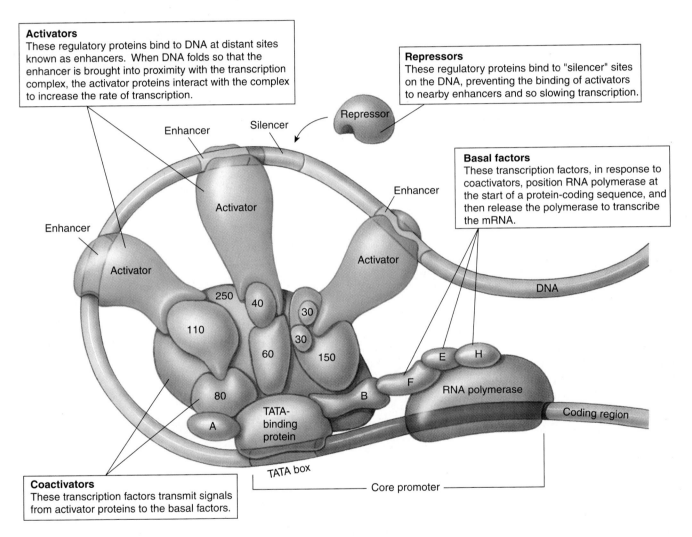

Activators
These regulatory proteins bind to DNA at distant sites known as enhancers. When DNA folds so that the enhancer is brought into proximity with the transcription complex, the activator proteins interact with the complex to increase the rate of transcription.

Repressors
These regulatory proteins bind to "silencer" sites on the DNA, preventing the binding of activators to nearby enhancers and so slowing transcription.

Basal factors
These transcription factors, in response to coactivators, position RNA polymerase at the start of a protein-coding sequence, and then release the polymerase to transcribe the mRNA.

Coactivators
These transcription factors transmit signals from activator proteins to the basal factors.

FIGURE 16.15
The structure of a human transcription complex. The transcription complex that positions RNA polymerase at the beginning of a human gene consists of four kinds of proteins. Basal factors (the green shapes at bottom of complex with letter names) are transcription factors that are essential for transcription but cannot by themselves increase or decrease its rate. They include the TATA-binding protein, the first of the basal factors to bind to the core promoter sequence. Coactivators (the tan shapes that form the bulk of the transcription complex, named according to their molecular weights) are transcription factors that link the basal factors with regulatory proteins called activators (the red shapes). The activators bind to enhancer sequences at other locations on the DNA. The interaction of individual basal factors with particular activator proteins is necessary for proper positioning of the polymerase, and the rate of transcription is regulated by the availability of these activators. When a second kind of regulatory protein called a repressor (the purple shape) binds to a so-called "silencer" sequence located adjacent to or overlapping an enhancer sequence, the corresponding activator that would normally have bound that enhancer is no longer able to do so. The activator is thus unavailable to interact with the transcription complex and initiate transcription.

Table 16.1 Some Gene Regulatory Proteins and the DNA Sequences They Recognize

Species	Regulatory Protein	DNA Sequence Recognized*
Escherichia coli	*lac* repressor	AATTGTGAGCGGATAACAATT TTAACACTCGCCTATTGTTAA
	CAP	TGTGAGTTAGCTCACT ACACTCAATCGAGTGA
	λ repressor	TATCACCGCCAGAGGTA ATAGTGGCGGTCTCCAT
Yeast	GAL4	CGGAGGACTGTCCTCCG GCCTCCTGACAGGAGGC
	MAT α2	CATGTAATT GTACATTAA
	GCN4	ATGACTCAT TACTGAGTA
Drosophila melanogaster	Krüppel	AACGGGTTAA TTGCCCAATT
	bicoid	GGGATTAGA CCCTAATCT
Human	Spl	GGGCGG CCCGCC
	Oct-1	ATGCAAAT TACGTTTA
	GATA-1	TGATAG ACTATC

*Each regulatory protein is able to recognize a family of closely related DNA sequences; only one member of each family is listed here.

In eukaryotes, this physical limitation is overcome by having distant sites on the chromosome exert control over the transcription of a gene. In this way, many regulatory sequences scattered around the chromosomes can influence a particular gene's transcription. This "control-at-a-distance" mechanism includes two features not found in bacteria: a set of proteins that help bind RNA polymerase to the promoter, and modular regulatory proteins that bind to distant sites. These two features produce a truly flexible control system.

Eukaryotic Transcription Factors

For RNA polymerase to successfully bind to a eukaryotic promoter and initiate transcription, a set of proteins called **transcription factors** (see table 16.1) must first assemble on the promoter, forming a complex that guides and stabilizes the binding of the polymerase (figure 16.15). The assembly process begins some 25 nucleotides upstream from the transcription start site, where a transcription factor composed of many subunits binds to a short TATA sequence. Other transcription factors then bind, eventually forming a full transcription factor complex able to capture RNA polymerase. In many instances, the transcription factor complex then phosphorylates the bound polymerase, disengaging it from the complex so that it is free to begin transcription.

The binding of several different transcription factors provides numerous points where control over transcription

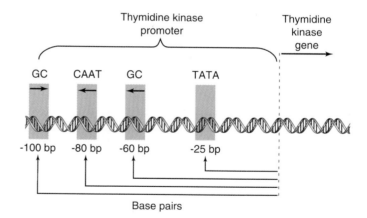

FIGURE 16.16
A eukaryotic promoter. This promoter for the gene encoding the enzyme thymidine kinase contains the TATA box that the initiation factor binds to, as well as three other DNA sequences that direct the binding of other elements of the transcription complex.

may be exerted (figure 16.16). Anything that reduces the availability of a particular factor (for example, by regulating the promoter that governs the transcription of that factor) or limits its ease of assembly into the transcription factor complex will inhibit transcription.

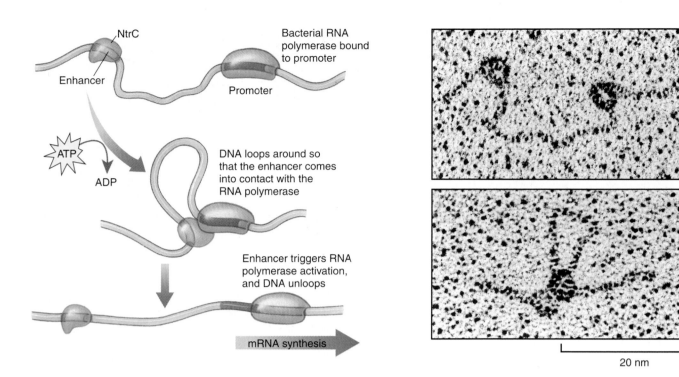

FIGURE 16.17

An enhancer in action. When the bacterial activator NtrC binds to an enhancer, it causes the DNA to loop over to a distant site where RNA polymerase is bound, activating transcription. While such enhancers are rare in bacteria, they are common in eukaryotes.

Enhancers

A key advance in the evolution of eukaryotic gene transcription was the advent of regulatory proteins composed of two distinct modules, or domains. The **DNA-binding domain** physically attaches the protein to the DNA at a specific site, using one of the structural motifs discussed earlier, while the **regulatory domain** interacts with other regulatory proteins.

The great advantage of this modular design is that it uncouples regulation from DNA binding, allowing a regulatory protein to bind to a specific DNA sequence at one site on a chromosome and exert its regulation over a promoter at another site, which may be thousands of nucleotides away. The distant sites where these regulatory proteins bind are called **enhancers.** Although enhancers also occur in exceptional instances in bacteria (figure 16.17), they are the rule rather than the exception in eukaryotes.

How can regulatory proteins affect a promoter when they bind to the DNA at enhancer sites located far from the promoter? Apparently the DNA loops around so that the enhancer is positioned near the promoter. This brings the regulatory domain of the protein attached to the enhancer into direct contact with the transcription factor complex attached to the promoter (figure 16.18).

The enhancer mode of transcriptional control that has evolved in eukaryotes adds a great deal of flexibility to the control process. The positioning of regulatory sites at a distance permits a large number of different regulatory sequences scattered about the DNA to influence a particular gene.

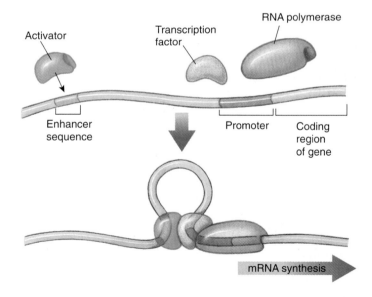

FIGURE 16.18

How enhancers work. The enhancer site is located far away from the gene being regulated. Binding of an activator (*red*) to the enhancer allows the activator to interact with the transcription factors (*green*) associated with RNA polymerase, activating transcription.

Transcription factors and enhancers confer great flexibility on the control of gene expression in eukaryotes.

The Effect of Chromosome Structure on Gene Regulation

The way DNA is packaged into chromosomes can have a profound effect on gene expression. As we saw in chapter 11, the DNA of eukaryotes is packaged in a highly compact form that enables it to fit into the cell nucleus. DNA is wrapped tightly around histone proteins to form nucleosomes (figure 16.19) and then the strand of nucleosomes is twisted into 30-nm filaments.

Promoter Blocking by Nucleosomes

Intensive study of eukaryotic chromosomes has shown that histones positioned over promoters block the assembly of transcription factor complexes. Therefore, transcription factors appear unable to bind to a promoter packaged in a nucleosome. In this way, nucleosomes may prevent continuous transcription initiation. On the other hand, nucleosomes do *not* inhibit activators and RNA polymerase. The regulatory domains of activators attached to enhancers apparently are able to displace the histones that block a promoter. In fact, this displacement of histones and the binding of activator to promoter are required for the assembly of the transcription factor complex. Once transcription has begun, RNA polymerase seems to push the histones aside as it traverses the nucleosome.

DNA Methylation

Chemical **methylation** of the DNA was once thought to play a major role in gene regulation in vertebrate cells. The addition of a methyl group to cytosine creates

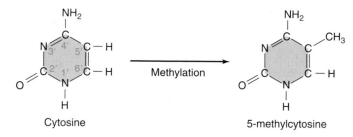

FIGURE 16.20
DNA methylation. Cytosine is methylated at the 5′ position, creating 5-methylcytosine. Because the methyl group is positioned to the side, it does not interfere with the hydrogen bonds of a GC base-pair.

5-methylcytosine but has no effect on base-pairing with guanine (figure 16.20), just as the addition of a methyl group to thymine produces uracil without affecting base-pairing with adenine. Many inactive mammalian genes are methylated, and it was tempting to conclude that methylation caused the inactivation. However, methylation is now viewed as having a less direct role, blocking accidental transcription of "turned-off" genes. Vertebrate cells apparently possess a protein that binds to clusters of 5-methylcytosine, preventing transcriptional activators from gaining access to the DNA. DNA methylation in vertebrates thus ensures that once a gene is turned off, it stays off.

Transcriptional control of gene expression occurs in eukaryotes despite the tight packaging of DNA into nucleosomes.

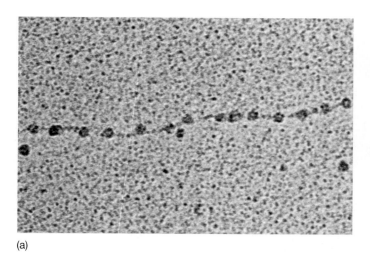

(a)

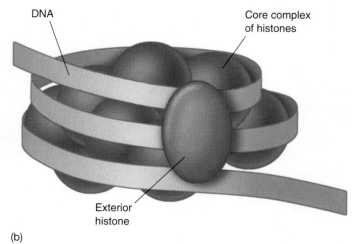

(b)

FIGURE 16.19
Nucleosomes. (a) In the electron micrograph, the individual nucleosomes have diameters of about 10 nm. (b) In the diagram of a nucleosome, the DNA double helix is wound around a core complex of eight histones; one additional histone binds to the outside of the nucleosome, exterior to the DNA.

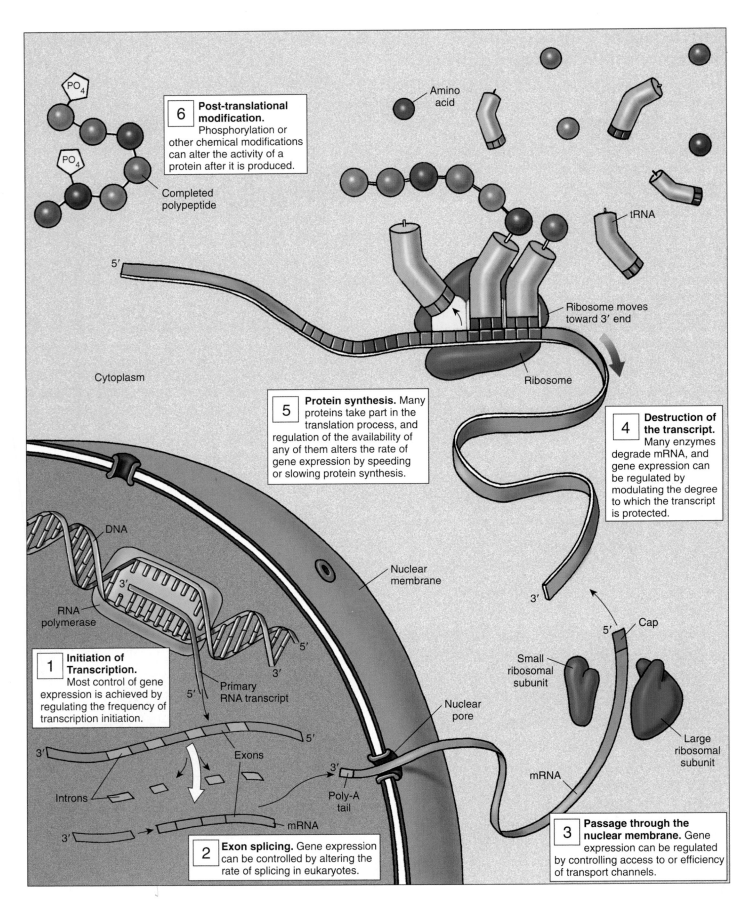

6 **Post-translational modification.** Phosphorylation or other chemical modifications can alter the activity of a protein after it is produced.

Completed polypeptide

Amino acid

tRNA

Ribosome moves toward 3' end

Ribosome

Cytoplasm

5 **Protein synthesis.** Many proteins take part in the translation process, and regulation of the availability of any of them alters the rate of gene expression by speeding or slowing protein synthesis.

4 **Destruction of the transcript.** Many enzymes degrade mRNA, and gene expression can be regulated by modulating the degree to which the transcript is protected.

DNA

Nuclear membrane

RNA polymerase

1 **Initiation of Transcription.** Most control of gene expression is achieved by regulating the frequency of transcription initiation.

Primary RNA transcript

Exons

Introns

mRNA

Cap

Small ribosomal subunit

Large ribosomal subunit

mRNA

Nuclear pore

Poly-A tail

2 **Exon splicing.** Gene expression can be controlled by altering the rate of splicing in eukaryotes.

3 **Passage through the nuclear membrane.** Gene expression can be regulated by controlling access to or efficiency of transport channels.

FIGURE 16.21
Six levels where gene expression can be controlled in eukaryotes.

Posttranscriptional Control in Eukaryotes

Thus far we have discussed gene regulation entirely in terms of transcription initiation, that is, when and how often RNA polymerase starts "reading" a particular gene. Most gene regulation appears to occur at this point. However, there are many other points after transcription where gene expression could be regulated in principle (figure 16.21), and all of them serve as control points for at least some eukaryotic genes. In general, these posttranscriptional control processes involve the recognition of specific sequences on the primary RNA transcript by regulatory proteins or other RNA molecules.

Processing of the Primary Transcript

As we learned in chapter 15, most eukaryotic genes have a patchwork structure, being composed of numerous short coding sequences (exons) embedded within long stretches of noncoding sequences (introns). The initial mRNA molecule copied from a gene by RNA polymerase, the **primary transcript,** is a faithful copy of the entire gene, including introns as well as exons. Before the primary transcript is translated, the introns, which comprise on average 90% of the transcript, are removed in a process called **RNA processing,** or **RNA splicing.** Particles called *small nuclear ribonucleoproteins,* or *snRNPs,* are thought to play a role in RNA splicing. These particles reside in the nucleus of a cell and are composed of proteins and a special type of RNA called *small nuclear RNA,* or *snRNA* (more informally, **snurps**). The ends of introns are marked by short sequences of nucleotides complementary to sequences on the snRNA. When multiple snRNPs combine to form a larger complex called a **spliceosome,** the intron loops out and is excised (figures 16.22 and 16.23).

RNA splicing provides a potential point where the expression of a gene can be controlled, because exons can be spliced together in different ways, allowing a variety of different polypeptides to be assembled from the same gene! Alternative splicing is common in insects and vertebrates, with two or three different proteins typically produced from one gene. In many cases, gene expression is regulated by changing which splicing event occurs during different stages of development or in different tissues.

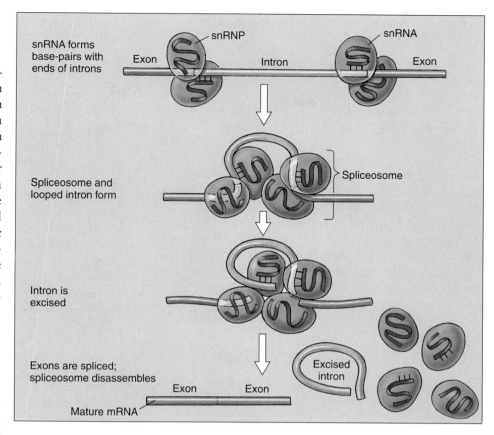

FIGURE 16.22
How spliceosomes process RNA. Particles called snRNPs contain snRNA that interacts with the ends of introns. Several snRNPs come together and form a spliceosome. As the intron forms a loop, the ends are cut, and the exons are spliced together. The spliceosome then disassembles and releases the mature mRNA.

FIGURE 16.23
RNA splicing in action. In this electron micrograph, a spliceosome is processing a β-globin primary RNA transcript, cutting introns out and splicing exons together.

Transport of the Processed Transcript Out of the Nucleus

Processed mRNA transcripts exit the nucleus through the nuclear pores described in chapter 5. The passage of a transcript across the nuclear membrane is an active process that requires that the transcript be recognized by receptors lining the interior of the pores. Specific portions of the transcript, such as the poly-A tail, appear to play a role in this recognition. The transcript cannot move through a pore as long as any of the splicing enzymes remain associated with the transcript, ensuring that partially processed transcripts are not exported into the cytoplasm.

There is little hard evidence that gene expression is regulated at this point, although it could be. On average, about 10% of transcribed genes are exon sequences, but only about 5% of the total mRNA produced as primary transcript ever reaches the cytoplasm. This suggests that about half of the exon primary transcripts never leave the nucleus, but it is not clear whether the disappearance of this mRNA is selective.

Selecting Which mRNAs Are Translated

The translation of a processed mRNA transcript by the ribosomes in the cytoplasm involves a complex of proteins called translation factors. In at least some cases, gene expression is regulated by modification of one or more of these factors. In other instances, **translation repressor proteins** shut down translation by binding to the beginning of the transcript, so that it cannot attach to the ribosome. In humans, the production of ferritin (an iron-storing protein) is normally shut off by a translation repressor protein called aconitase. Aconitase binds to a 30-nucleotide sequence at the beginning of the ferritin mRNA, forming a stable loop to which ribosomes cannot bind. When the cell encounters iron, the binding of iron to aconitase causes the aconitase to dissociate from the ferritin mRNA, freeing the mRNA to be translated and increasing ferritin production 100-fold.

Selectively Degrading mRNA Transcripts

Unlike bacterial mRNA transcripts, which typically have a half-life of about 3 minutes, eukaryotic mRNA transcripts are very stable. For example, β-globin gene transcripts have a half-life of over 10 hours, an eternity in the fast-moving metabolic life of a cell. The transcripts encoding regulatory proteins and growth factors, however, are usually much less stable, with half-lives of less than 1 hour. What makes these particular transcripts so unstable? In many cases, they contain specific sequences near their 3′ ends that make them attractive targets for enzymes that degrade mRNA. A sequence of A and U nucleotides near the 3′ poly-A tail of a transcript promotes removal of the tail, which destabilizes the mRNA. Histone transcripts, for example, have a half-life of about 1 hour in cells that are actively synthesizing DNA; at other times during the cell cycle, the poly-A tail is lost and the transcripts are degraded within minutes. Other mRNA transcripts contain sequences near their 3′ ends that are recognition sites for endonucleases, which causes these transcripts to be digested quickly. The short half-lives of the mRNA transcripts of many regulatory genes are critical to the function of those genes, as they enable the levels of regulatory proteins in the cell to be altered rapidly.

Although less common than transcriptional control, posttranscriptional control of gene expression occurs in eukaryotes via RNA splicing, translation repression, and selective degradation of mRNA transcripts.

A Vocabulary of Gene Expression

anticodon   The three-nucleotide sequence on one end of a tRNA molecule that is complementary to and base-pairs with an amino acid-specifying codon in mRNA.

codon   The basic unit of the genetic code; a sequence of three adjacent nucleotides in DNA or mRNA that codes for one amino acid or for polypeptide termination.

exon   A segment of eukaryotic DNA that is both transcribed into mRNA and translated into protein. Exons are typically scattered within much longer stretches of non-translated intron sequences.

intron   A segment of eukaryotic DNA that is transcribed into mRNA but removed before translation.

nonsense codon   A codon (UAA, UAG, or UGA) for which there is no tRNA with a complementary anticodon; a chain-terminating codon often called a "stop" codon.

operator   A site of negative gene regulation; a sequence of nucleotides near or within the promoter that is recognized by a repressor. Binding of the repressor to the operator prevents binding of RNA polymerase to the promoter and so blocks transcription.

operon   A cluster of functionally related genes transcribed into a single mRNA molecule. A common mode of gene regulation in prokaryotes, it is rare in eukaryotes other than fungi.

promoter   A site upstream from a gene to which RNA polymerase attaches to initiate transcription.

repressor   A protein that regulates transcription by binding to the operator and so preventing RNA polymerase from attaching to the promoter.

RNA polymerase   The enzyme that transcribes DNA into RNA.

transcription   The RNA polymerase-catalyzed assembly of an RNA molecule complementary to a strand of DNA.

translation   The assembly of a polypeptide on the ribosomes, using mRNA to direct the sequence of amino acids.

Summary of Concepts

16.1 Gene expression is controlled by regulating transcription.

- When the product of a specific gene is required by the cell, the gene is transcribed by RNA polymerase; when the gene product is not required, other mechanisms act to inhibit the gene's transcription.
- Regulatory sequences are short stretches of DNA that function in transcriptional control but are not transcribed themselves.
- Regulatory proteins recognize and bind to specific regulatory sequences on the DNA.
- The binding of a regulatory protein can either stimulate or inhibit the transcription of a specific gene.

16.2 Regulatory proteins read DNA without unwinding it.

- Regulatory proteins possess structural motifs that allow them to fit snugly into the major groove of DNA, where the sides of the base-pairs are exposed.
- Common structural motifs include the helix-turn-helix, homeodomain, zinc finger, and leucine zipper.

16.3 Bacteria limit transcription by blocking the polymerase.

- Many genes are transcriptionally regulated through repressors, proteins that bind to the DNA at or near the promoter and thereby inhibit transcription of the gene.

- Genes may also be transcriptionally regulated through activators, proteins that bind to the DNA and thereby stimulate the binding of RNA polymerase to the promoter.
- Transcription is often controlled by a *combination* of repressors and activators.

16.4 Transcriptional control in eukaryotes operates at a distance.

- In eukaryotes, RNA polymerase cannot bind to the promoter unless aided by a family of transcription factors.
- Anything that interferes with the activity of the transcription factors can block gene expression.
- Eukaryotic DNA is packaged tightly in nucleosomes within chromosomes. This packaging appears to provide some inhibition of transcription, although regulatory proteins and RNA polymerase can still activate specific genes even when they are so packaged.
- Gene expression can also be regulated at the posttranscriptional level, through RNA splicing, translation repressor proteins, and the selective degradation of mRNA transcripts.

Discussing Key Terms

1. **Regulatory sequences** Transcriptional controls are exerted by means of specific, short DNA sequences, which are not themselves translated. In prokaryotes, such regulatory sequences are located near the promoter, but in eukaryotes they are often far from the promoter.

2. **Structural motifs** The DNA-binding portions of almost all regulatory proteins employ one of a small set of structural motifs, shapes that enable the proteins to fit into the major groove of DNA.

3. **Repressors** Bacteria inhibit the transcription of genes through the use of repressors, proteins that bind to or near the promoter site, preventing RNA polymerase from occupying it and so blocking transcription.

4. **Activators** Bacteria also possess regulatory proteins that activate transcription, typically by improving the promoter's ability to bind to RNA polymerase.

5. **Enhancers** Eukaryotes have distant DNA-binding sites for regulatory proteins called enhancers that interact with a gene's transcription complex to modulate its rate of transcription.

6. **Promoters** The binding site of RNA polymerase, located upstream from a gene, is called a promoter.

Review Questions

1. What features of the major groove allow regulatory proteins to use it to locate DNA-binding sites?

2. What is a helix-turn-helix motif? What sort of developmental events are homeodomain motifs involved in? What are zinc fingers? What is a leucine zipper?

3. Define the following terms: promoter, operon, operator.

4. Compare the effects of repressors and activators on the activity of RNA polymerase and on transcription.

5. Describe the mechanism by which the transcription of *trp* genes is regulated in *Escherichia coli* when tryptophan is present in the environment.

6. What intracellular messenger participates in the transcriptional control the CAP protein exerts in *E. coli*? Describe how this messenger is involved.

7. Describe the mechanism by which the transcription of *lac* genes is regulated in *E. coli* when glucose is absent but lactose is present in the environment.

8. How do transcription factors promote transcription in eukaryotic cells? How do the enhancers of eukaryotic cells differ from most regulatory sites on bacterial DNA?

9. What role does the methylation of DNA likely play in transcriptional control?

10. How does the *primary* RNA transcript of a eukaryotic gene differ from the mRNA transcript of that gene as it is translated in the cytoplasm?

11. How can a eukaryotic cell control the translation of mRNA transcripts after they have been transported from the nucleus to the cytoplasm?

Thought Questions

1. The life-span of a typical bacterium (for instance, *E. coli*) is typically very short—anywhere from 30 minutes to an hour or two. If the bacterial culture were to grow for 50 years, do you think bacterial genes would be regulated as they are now? Why or why not?

2. All human beings have a rich growth of *E. coli* bacteria in their large intestine. Will the bacteria in a lactose-intolerant individual who is careful never to consume anything with lactose (milk sugar) in it ever have access to lactose? Why or why not?

Internet Links

Molecular Genetics
http://www.gene.com/ae/AE
This comprehensive ACCESS EXCELLENCE site, created by the 1994 Woodrow Wilson National Leadership Program in Biology, contains, among many other resources, a fine curriculum on molecular genetics. This excellent site also provides news updates on scientific discoveries, notice of science TV programs, and links to information about educational technology.

Regulating Gene Expression in Bacteria
http://esg-www.mit.edu:8001/esgbio/pge/pgedir.html
A chapter from the MIT BIOLOGY HYPERTEXTBOOK, comprehensive and up-to-date, with clear graphics and instructive problem sets.

For Further Reading

Lewis, M., and others: "Crystal Structure of the Lactose Operon Repressor and Its Complexes with DNA and Inducer," *Science*, March 1996, vol. 271, pages 1247–54. A helix-turn-helix motif governs the binding of this regulatory protein to DNA.

Tijan, R.: "Molecular Machines That Control Genes," *Scientific American*, February 1995, pages 54–61. A stunning discussion of the regulatory proteins that assemble on DNA to direct transcription, and what happens when they malfunction.

Tijan, R., and T. Maniatis: "Transcriptional Activation: A Complex Puzzle with Few Easy Pieces," *Cell*, vol. 77, April 8, 1994, pages 5–8. RNA, RNA polymerase, and gene expression are discussed. This particular volume of *Cell* has many detailed papers on transcription.

Welch, W.: "How Cells Respond to Stress," *Scientific American*, vol. 268, May 1993, page 56. Cells produce stress proteins that repair damage and help proteins fold properly.

Wingender, E.: *Gene Regulation in Eukaryotes*, VCH Publishers, New York, 1993. An up-to-date, in-depth look at eukaryotic gene regulation, including discussions of transcription factors, RNA polymerases, and signal transduction.

Wolffe, A.: "Transcription: In Tune with the Histones," *Cell*, vol. 77, April 8, 1994, pages 13–16. A look at what is known about how transcription is affected by histones. This particular volume of *Cell* has many detailed papers on transcription.

17

Altering the Genetic Message

FIGURE 17.1
Cancer. A scanning electron micrograph of deadly cancer cells (8000×).

In general, the genetic message can be altered in two broad ways: mutation and recombination. A change in the content of the genetic message—the base sequence of one or more genes—is referred to as a *mutation*. Some mutations alter the identity of a particular nucleotide, while others remove or add nucleotides to a gene. A change in the position of a portion of the genetic message is referred to as *recombination*. Some recombination events move a gene to a different chromosome; others alter the location of only part of a gene. In this chapter, we will first consider gene mutation, using cancer as a focus for our inquiry (figure 17.1). Then we will turn to recombination, focusing on how it has affected the organization of the eukaryotic genome.

Mutations Are Rare but Important

The cells of eukaryotes contain an enormous amount of DNA. If the DNA in all of the cells of an adult human were lined up end-to-end, it would stretch nearly 100 billion kilometers—60 times the distance from Earth to Jupiter! The DNA in any multicellular organism is the final result of a long series of replications, starting with the DNA of a single cell, the fertilized egg. Organisms have evolved many different mechanisms to avoid errors during DNA replication and to preserve the DNA from damage. Some of these mechanisms "proofread" the replicated DNA strands for accuracy and correct any mistakes. The proofreading is not perfect, however. If it were, no variation in the nucleotide sequences of genes would be generated.

Mistakes Happen

In fact, cells do make mistakes during replication, and damage to the genetic message also occurs, causing mutation (figure 17.2). However, change is rare. Typically, a particular gene is altered in only one of a million gametes. If changes were common, the genetic instructions encoded in DNA would soon degrade into meaningless gibberish. Limited as it might seem, the steady trickle of change that does occur is the very stuff of evolution. Every difference in the genetic messages that specify different organisms arose as the result of genetic change.

The Importance of Genetic Change

All evolution begins with alterations in the genetic message: mutation creates new alleles, gene transfer and transposition alter gene location, reciprocal recombination shuffles and sorts these changes, and chromosomal rearrangement alters the organization of entire chromosomes. Some changes in germ-line tissue produce alterations that enable an organism to leave more offspring, and those changes tend to be preserved as the genetic endowment of future generations. Other changes reduce the ability of an organism to leave offspring. Those changes tend to be lost, as the organisms that carry them contribute fewer members to future generations.

FIGURE 17.2
Mutation. Normal fruit flies have one pair of wings extending from the thorax. This fly is a mutant because of changes in *ultrabithorax*, a gene regulating a critical stage of development; it possesses two thoracic segments and thus two sets of wings.

Evolution can be viewed as the selection of particular combinations of alleles from a pool of alternatives. The rate of evolution is ultimately limited by the rate at which these alternatives are generated. Genetic change through mutation and recombination provides the raw material for evolution.

Genetic changes in somatic cells do not pass on to offspring, and so have less evolutionary consequence than germ-line change. However, changes in the genes of somatic cells can have an important immediate impact, particularly if the gene affects development or is involved with regulation of cell proliferation.

Rare changes in genes, called mutations, can have significant effects on the individual when they occur in somatic tissue, but are only inherited if they occur in germ-line tissue. Inherited changes provide the raw material for evolution.

Kinds of Mutation

Because mutations can occur randomly anywhere in a cell's DNA, most mutations are detrimental, just as making a random change in a computer program or a musical score usually worsens performance. The consequences of a detrimental mutation may be minor or catastrophic, depending on the function of the altered gene.

Mutations in Germ-Line Tissues

The effect of a mutation depends critically on the identity of the cell in which the mutation occurs. During the embryonic development of all multicellular organisms, there comes a point when cells destined to form gametes (**germ-line cells**) are segregated from those that will form the other cells of the body (**somatic cells**). In plants and fungi, this decision happens very late in development; this means that a mutation in any cell can pass to the next generation, since any cell can potentially develop into an adult individual. In most animals, by contrast, the decision is made early in development. Only when a mutation occurs within a germ-line cell is it passed to subsequent generations as part of the hereditary endowment of the gametes derived from that cell.

Mutations in Somatic Tissues

Mutations in germ-line tissue are of enormous biological importance because they provide the raw material from which natural selection produces evolutionary change. Change can occur only if there are new, different allele combinations available to replace the old. Mutation produces new alleles, and recombination puts the alleles together in different combinations. In animals, it is the occurrence of these two processes in germ-line tissue that is important to evolution, since mutations in somatic cells (**somatic mutations**) are not passed from one generation to the next. However, a somatic mutation may have drastic effects on the individual organism in which it occurs, since it *is* passed on to all of the cells that are descended from the original mutant cell. Thus, if a mutant lung cell divides, all cells derived from it will carry the mutation. Somatic mutations of lung cells are, as we shall see, the principal cause of lung cancer in humans.

Point Mutations

One category of mutational changes affects the message itself, producing alterations in the sequence of DNA nucleotides (Table 17.1 summarizes the sources and types of mutations.). If alterations involve only one or a few base-pairs in the coding sequence, they are called **point mutations.** While some point mutations arise due to spontaneous pairing errors that occur during DNA replication, others result from damage to the DNA caused by **mutagens,** usually radiation or chemicals. The latter class of mutations is of particular practical importance because modern industrial societies often release many chemical mutagens into the environment.

Changes in Gene Position

Another category of mutations affects the way the genetic message is organized. In both bacteria and eukaryotes, individual genes may move from one place in the genome to another by **transposition.** When a particular gene moves to a different location, its expression or the expression of neighboring genes may be altered. In addition, large segments of chromosomes in eukaryotes may change their relative locations or undergo duplication. Such **chromosomal rearrangements** often have drastic effects on the expression of the genetic message.

Point mutations are changes in the hereditary message of an organism. They may result from spontaneous errors during DNA replication or from damage to the DNA due to radiation or chemicals.

Table 17.1	Sources and Types of Mutation	
Source	**Primary Effect**	**Type of Mutation**
MUTATIONAL		
Ionizing radiation	Two-strand breaks in DNA	Deletions, translocations
Ultraviolet radiation	Pyrimidine dimers	Errors in nucleotide choice during repair
Chemical mutagens	Modification of a base, leading to mispairing	Single nucleotide substitution
Spontaneous error	Isomerization of a base	Single nucleotide substitution
	Slipped mispairing	Frame-shift changes, short deletion
RECOMBINATIONAL		
Transposition	Insertion of transposon into gene	Insertional inactivation
Mispairing of repeated sequences	Unequal crossing over	Deletions, additions, inversions
Homologue pairing	Gene conversion	Single or multiple nucleotide substitution

Point Mutations

Physical Damage to DNA

Ionizing Radiation. High-energy forms of radiation, such as X rays and gamma rays, are highly mutagenic. When such radiation reaches a cell, it is absorbed by the atoms it encounters, imparting energy to the electrons in their outer shells. These energized electrons are ejected from the atoms, leaving behind free radicals, ionized atoms with unpaired electrons. Because most of a cell's atoms reside in water molecules, the great majority of free radicals created by ionizing radiation are produced from water. Free radicals react violently with other molecules, including DNA. Hence, most of the damage ionizing radiation inflicts on DNA is indirect.

When a free radical breaks *both* phosphodiester bonds of a DNA helix, causing a **double-strand break,** the cell's usual mutational repair enzymes cannot fix the damage. The two fragments created by the break must be aligned while the phosphodiester bonds between them form again. Bacteria have no mechanism to achieve this alignment, and double-strand breaks are lethal to their descendants. In eukaryotes, which almost all possess multiple copies of their chromosomes, the synaptonemal complex assembled in meiosis is used to pair the fragmented chromosome with its homologue. In fact, it is speculated that meiosis may have evolved initially as a mechanism to repair double-strand breaks in DNA.

Ultraviolet Radiation. Ultraviolet (UV) radiation, the component of sunlight that tans (and burns), contains much less energy than ionizing radiation. It does not induce atoms to eject electrons, and thus it does not produce free radicals. The only molecules capable of absorbing UV radiation are certain organic ring compounds, whose outer-shell electrons become reactive when they absorb UV energy.

DNA strongly absorbs UV radiation in the pyrimidine bases, thymine and cytosine. If one of the nucleotides on either side of the absorbing pyrimidine is also a pyrimidine, a double covalent bond forms between them. The resulting cross-link between adjacent pyrimidines is called a **pyrimidine dimer** (figure 17.3). In most cases, cellular UV repair systems either cleave the bonds that link the adjacent pyrimidines or excise the entire pyrimidine dimer from the strand and fill in the gap, using the other strand as a template (figure 17.4). In those rare instances in which a pyrimidine dimer goes unrepaired, DNA polymerase may fail to replicate the portion of the strand that includes the dimer, skipping ahead and leaving the problem area to be filled in later. This filling-in process is often error-prone, however, and it may create mutational changes in the base sequence of the

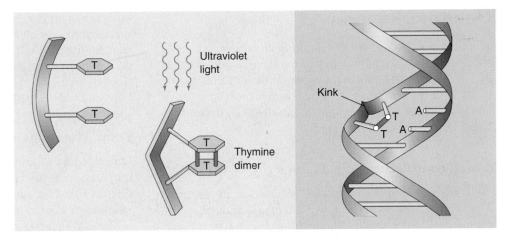

FIGURE 17.3

Making a pyrimidine dimer. When two pyrimidines, such as two thymines, are adjacent to each other in a DNA strand, the absorption of UV radiation can cause covalent bonds to form between them—creating a pyrimidine dimer. The dimer introduces a "kink" into the double helix that prevents replication of the duplex by DNA polymerase.

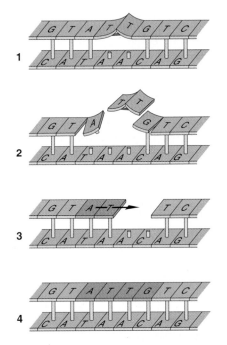

FIGURE 17.4
Repair of a pyrimidine dimer. Some pyrimidine dimers are repaired by excising the dimer, as well as a short run of nucleotides on either side of it, and then filling in the gap using the other strand as a template.

gap region. Some unrepaired pyrimidine dimers block DNA replication altogether, which is lethal to the cell.

Sunlight can wreak havoc on the cells of the skin because its UV light causes mutations. Indeed, a strong and direct association exists between exposure to bright sunlight, UV-induced DNA damage, and skin cancer. A deep tan is *not* healthy! A rare hereditary disorder among humans called **xeroderma pigmentosum** causes these problems after a lesser exposure to UV. Individuals with this disorder develop extensive skin tumors after exposure to sunlight because they lack a mechanism for repairing the DNA damage UV radiation causes. Because of the many different proteins involved in excision and repair of pyrimidine dimers, mutations in as many as eight different genes cause the disease.

Chemical Modification of DNA

Many mutations result from direct chemical modification of the DNA. The chemicals that act on DNA fall into three classes: (1) chemicals that resemble DNA nucleotides but pair incorrectly when they are incorporated into DNA (figure 17.5). Some of the new AIDS chemotherapeutic drugs are analogues of nitrogenous bases that are inserted into the viral or infected cell DNA. This DNA cannot be properly transcribed, so viral growth slows; (2) chemicals that remove the amino group from adenine or cytosine, causing them to mispair; and (3) chemicals that add hydrocarbon groups to nucleotide bases, also causing them to mispair. This last group includes many particularly potent mutagens commonly used in laboratories, as well as compounds sometimes released into the environment, such as mustard gas.

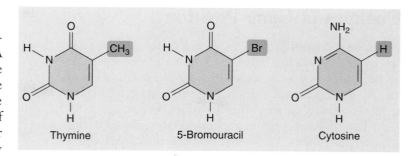

FIGURE 17.5
Chemicals that resemble DNA bases can cause mutations. For example, DNA polymerase cannot distinguish between thymine and 5-bromouracil, which are similar in shape. Once incorporated into a DNA molecule, however, 5-bromouracil tends to rearrange to a form that resembles cytosine and pairs with guanine. When this happens, what was originally an A-T base-pair becomes a G-C base-pair.

Spontaneous Mutations

Many point mutations occur spontaneously, without exposure to radiation or mutagenic chemicals. Sometimes nucleotide bases spontaneously shift to alternative conformations, or isomers, which form different kinds of hydrogen bonds than the normal conformations. During replication, DNA polymerase pairs a different nucleotide with the isomer than it would have otherwise selected. Unrepaired spontaneous errors occur in fewer than one in a billion nucleotides per generation, but they are still an important source of mutation.

Sequences sometimes misalign when homologous chromosomes pair, causing a portion of one strand to loop out. These misalignments, called **slipped mispairing,** are usually only transitory, and the chromosomes quickly revert to the normal arrangement (figure 17.6). If the error-correcting system of the cell encounters a slipped mispairing before it reverts, however, the system will attempt to "correct" it, usually by excising the loop. This may result in a **deletion** of several hundred nucleotides from one of the chromosomes. Many of these deletions start or end in the middle of a codon, thereby shifting the reading frame by one or two bases. These so-called **frame-shift mutations** cause the gene to be read in the wrong three-base groupings, distorting the genetic message, just as the deletion of the letter F from the sentence, THE FAT CAT ATE THE RAT shifts the reading frame of the sentence, producing the meaningless message, THE ATC ATA TET HER AT. Some chemicals specifically promote deletions and frameshift mutations by stabilizing the loops produced during slipped mispairing, thus increasing the time the loops are vulnerable to excision.

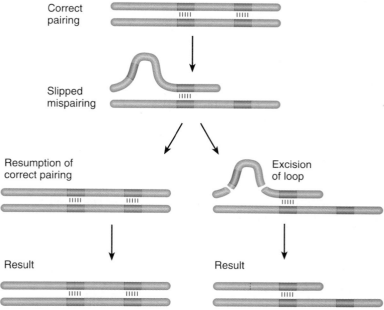

FIGURE 17.6
Slipped mispairing. Slipped mispairing occurs when a sequence is present in more than one copy on a chromosome and the copies on homologous chromosomes pair out of register, like a shirt buttoned wrong. The loop this mistake produces is sometimes excised by the cell's repair enzymes, producing a short deletion and often altering the reading frame. Any chemical that stabilizes the loop increases the chance it will be excised.

The major sources of physical damage to DNA are ionizing radiation, which breaks the DNA strands; ultraviolet radiation, which creates nucleotide cross-links whose removal often leads to errors in base selection; and chemicals that modify DNA bases and alter their base-pairing behavior. Unrepaired spontaneous errors in DNA replication occur only rarely.

Changes in Gene Position

Chromosome location is an important factor in determining whether genes are transcribed. Some genes cannot be transcribed if they are adjacent to a tightly coiled region of the chromosome, even though the same gene can be transcribed normally in any other location. Transcription of many chromosomal regions appears to be regulated in this manner; the binding of specific proteins regulates the degree of coiling in local regions of the chromosome, determining the accessibility RNA polymerase has to genes located within those regions.

Chromosomal Rearrangements

Chromosomes undergo several different kinds of gross physical alterations that have significant effects on the locations of their genes. The two most important are **translocations,** in which a segment of one chromosome becomes part of another chromosome, and **inversions,** in which the orientation of a portion of a chromosome is reversed. Translocations often have significant effects on gene expression. Inversions, on the other hand, usually do not alter gene expression, but they are nonetheless important. Recombination within a region that is inverted on one homologue but not the other (figure 17.7) leads to serious problems: none of the gametes that contain chromatids produced following such a crossover event will have a complete set of genes.

Other chromosomal alterations change the number of gene copies an individual possesses. Particular genes or segments of chromosomes may be deleted or duplicated, whole chromosomes may be lost or gained *(aneuploidy),* and entire sets of chromosomes may be added *(polyploidy).* Most deletions are harmful, since they halve the number of gene copies within a diploid genome and thus seriously affect the level of transcription. Duplications cause gene imbalance and are also usually harmful.

Insertional Inactivation

Many small segments of DNA are capable of moving from one location to another in the genome, using an enzyme to cut and paste themselves into new genetic neighborhoods. We call these mobile bits of DNA transposable elements, or **transposons.** Transposons select their new locations at random, and are as likely to enter one segment of a chromosome as another. Inevitably, some transposons end up inserted into genes, and this almost always inactivates the gene. The encoded protein now has a large meaningless chunk inserted within it, disrupting its structure. This form of mutation, called **insertional inactivation,** is common in nature. Indeed, it seems to be one of the most significant causes of mutation. The original white-eye mutant of *Drosophila* discovered by Morgan (chapter 13) is the result of a transposition event, a transposon nested within a gene encoding a pigment-producing enzyme.

As you might expect, a variety of human gene disorders are the result of transposition. The human transposon called *Alu,* for example, is responsible for an X-linked hemophilia, inserting into clotting factor IX and placing a premature stop codon there. It also causes inherited high levels of cholesterol (hypercholesterolemia), *Alu* elements inserting into the gene encoding the low density lipoprotein (LDL) receptor. In one very interesting case, a *Drosophila* transposon called *Mariner* proves responsible for a rare human neurological disorder called Charcot-Marie-Tooth disease, in which the muscles and nerves of the legs and feet gradually wither away. The Mariner transposon is inserted into a key gene called *CMT* on chromosome 17, creating a weak site where the chromosome can break. No one knows what the *Drosophila* transposon is doing in the human genome.

Many mutations result from changes in gene location or from insertional inactivation.

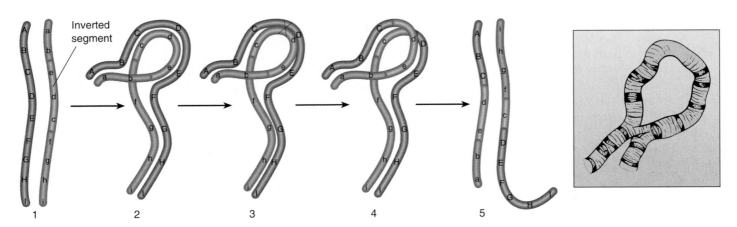

FIGURE 17.7

The consequence of inversion. (1) When a segment of a chromosome is inverted, (2) it can pair in meiosis only by forming an internal loop. (3) Any crossing over that occurs within the inverted segment during meiosis will result in nonviable gametes; some genes are lost from each chromosome, while others are duplicated (4 and 5). For clarity, only two strands are shown, although crossing over occurs in the four-strand stage. The pairing that occurs between inverted segments is sometimes visible under the microscope as a characteristic loop (*inset*).

What Is Cancer?

Cancer is a growth disorder of cells. It starts when an apparently normal cell begins to grow in an uncontrolled and invasive way (figure 17.8). The result is a cluster of cells, called a **tumor,** that constantly expands in size. Cells that leave the tumor and spread throughout the body, forming new tumors at distant sites, are called **metastases** (figure 17.9). Cancer is perhaps the most pernicious disease. Of the children born in 1999, one-third will contract cancer at some time during their lives; one-fourth of the male children and one-third of the female children will someday die of cancer. Most of us have had family or friends affected by the disease. In 1997, 560,000 Americans died of cancer.

Not surprisingly, researchers are expending a great deal of effort to learn the cause of this disease. Scientists have made a great deal of progress in the last 20 years using molecular biological techniques, and the rough outlines of understanding are now emerging. We now know that cancer is a gene disorder of somatic tissue, in which damaged genes fail to properly control cell proliferation. The cell division cycle is regulated by a sophisticated group of proteins described in chapter 11. Cancer results from the mutation of the genes encoding these proteins.

Cancer can be caused by chemicals that mutate DNA or in some instances by viruses that circumvent the cell's normal proliferation controls. Whatever the immediate cause, however, all cancers are characterized by unrestrained growth and division. Cell division never stops in a cancerous line of cells. Cancer cells are virtually immortal—until the body in which they reside dies.

Cancer is unrestrained cell proliferation caused by damage to genes regulating the cell division cycle.

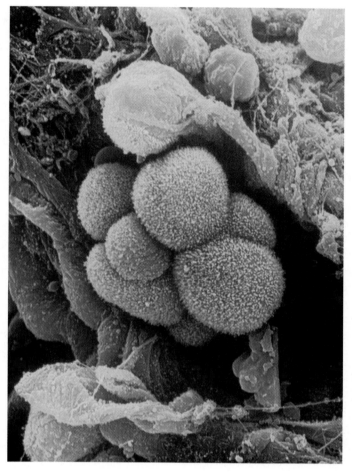

FIGURE 17.8
Lung cancer cells (530×). These cells are from a tumor located in the alveolus (air sac) of a lung.

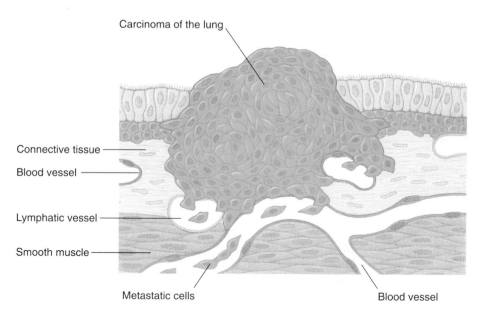

FIGURE 17.9
Portrait of a cancer. This ball of cells is a carcinoma (cancer tumor) developing from epithelial cells that line the interior surface of a human lung. As the mass of cells grows, it invades surrounding tissues, eventually penetrating lymphatic and blood vessels, both plentiful within the lung. These vessels carry metastatic cancer cells throughout the body, where they lodge and grow, forming new masses of cancerous tissue.

Carcinoma of the lung

Connective tissue

Blood vessel

Lymphatic vessel

Smooth muscle

Metastatic cells

Blood vessel

Kinds of Cancer

Cancer can occur in almost any tissue, so a bewildering number of different cancers occur. Tumors arising from cells in connective tissue, bone, or muscle are known as **sarcomas,** while those that originate in epithelial tissue such as skin are called **carcinomas.** The three deadliest human cancers are lung cancer, cancer of the colon and rectum, and breast cancer (table 17.2). Lung cancer, responsible for the most cancer deaths, is largely preventable; most cases result from smoking cigarettes. Colorectal cancers appear to be fostered by the high-meat diets so favored in the United States. The cause of breast cancer (figure 17.10) is still a mystery, although in 1994 and 1995 researchers isolated two genes responsible for hereditary susceptibility to breast cancer, *BRCA1* and *BRCA2* (Breast Cancer genes #1 and #2 located on human chromosomes 17 and 23); their discovery offers hope that researchers will soon be able to unravel the fundamental mechanism leading to breast cancer.

The search for the cause of cancer has focused in part on environmental factors, including ionizing radiation such as X rays and a variety of chemicals (figure 17.11). Agents thought to cause cancer are called **carcinogens.** The association of particular chemicals with cancer, particularly chemicals that are potent mutagens, led researchers early on to the suspicion that cancer might be caused, at least in part, by chemicals, the so-called **chemical carcinogenesis theory.**

Cancers occur in all tissues, but are more common in some than others.

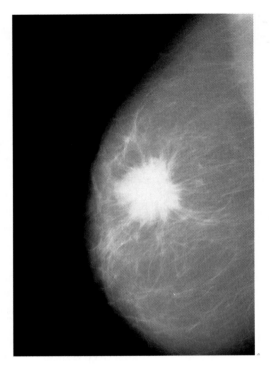

FIGURE 17.10
Mammogram of a breast with late-stage breast cancer.
A cancerous tumor (the white mass) appears as a dense area on X-ray film. Regular mammograms, recommended for women aged 35 and older, allow physicians to detect and intercept the potential development of cancer.

Table 17.2	Incidence of Cancer in the United States in 1997		
Type of Cancer	**New Cases**	**Deaths**	**% of Cancer Deaths**
Lung	178,100	160,400	29
Colon and rectum	131,200	54,900	10
Breast	181,600	44,190	8
Leukemia/lymphoma	89,400	46,590	8
Prostate	209,900	41,800	7
Pancreas	27,600	28,100	5
Stomach	22,400	14,000	3
Ovary	26,800	14,200	3
Liver	13,600	12,400	3
Nervous system/eye	19,700	13,450	2
Bladder	54,500	11,700	2
Oral cavity	30,750	8,440	1
Malignant melanoma	40,300	7,300	1
Kidney	28,800	11,300	1
Cervix/uterus	49,400	10,800	1
Sarcoma (connective tissue)	9,100	5,510	1
All other cancers	269,250	74,920	15

In the United States in 1997 there were 1,382,400 reported cases of new cancers and 560,000 cancer deaths, indicating that roughly half the people who develop cancer die from it.
Source: Data from the American Cancer Society, Inc., 1997.

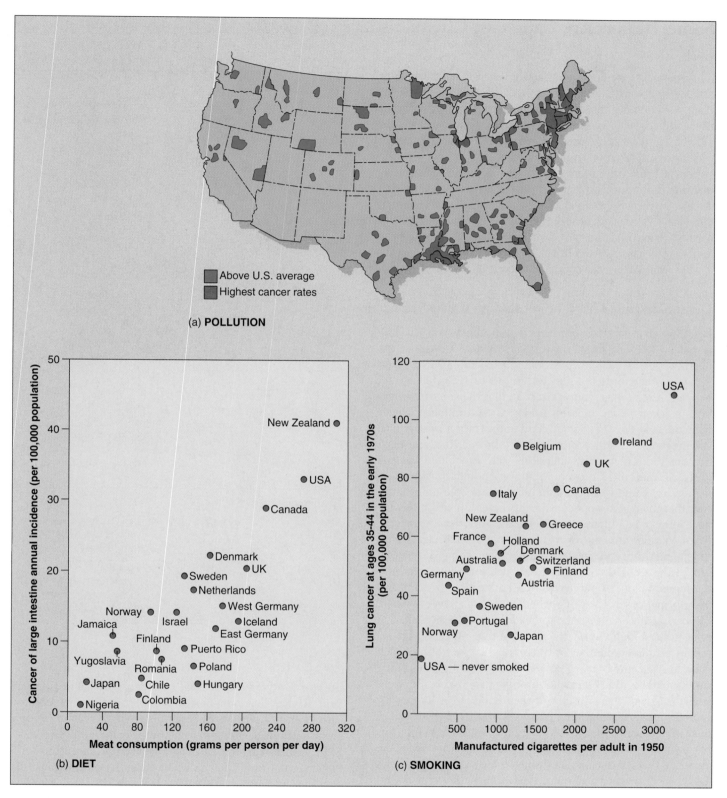

FIGURE 17.11

Potential cancer-causing agents. (a) The incidence of cancer per 1000 people is not uniform throughout the United States. The incidence is higher in cities and in the Mississippi Delta, suggesting that pollution and pesticide runoff may contribute to the development of cancer. (b) One of the deadliest cancers in the United States, cancer of the large intestine, is uncommon in many other countries. Its incidence appears to be related to the amount of meat a person consumes: a high-meat diet slows the passage of food through the intestine, prolonging exposure of the intestinal wall to digestive waste. (c) The biggest killer among cancers is lung cancer, and the most deadly environmental agent producing lung cancer is cigarette smoke. The incidence of lung cancer among men 35 to 44 years of age in various countries strongly correlates with the cigarette consumption in that country 20 years earlier.

Some Tumors Are Caused by Chemicals

Early Ideas

The chemical carcinogenesis theory was first advanced over 200 years ago in 1761 by Dr. John Hill, an English physician, who noted unusual tumors of the nose in heavy snuff users and suggested tobacco had produced these cancers. In 1775, a London surgeon, Sir Percivall Pott, made a similar observation, noting that men who had been chimney sweeps exhibited frequent cancer of the scrotum, and suggesting that soot and tars might be responsible. British sweeps washed themselves infrequently and always seemed covered with soot. Chimney sweeps on the continent, who washed daily, had much less of this scrotal cancer. These and many other observations led to the hypothesis that cancer results from the action of chemicals on the body.

Demonstrating That Chemicals Can Cause Cancer

It was over a century before this hypothesis was directly tested. In 1915, Japanese doctor Katsusaburo Yamagiwa applied extracts of coal tar to the skin of 137 rabbits every 2 or 3 days for 3 months. Then he waited to see what would happen. After a year, cancers appeared at the site of application in seven of the rabbits. Yamagiwa had induced cancer with the coal tar, the first direct demonstration of chemical carcinogenesis. In the decades that followed, this approach demonstrated that many chemicals were capable of causing cancer. Importantly, most of them were potent mutagens.

Because these were lab studies, many people did not accept that the results applied to real people. Do tars in fact induce cancer in humans? In 1949, the American physician Ernst Winder and the British epidemiologist Richard Doll independently reported that lung cancer showed a strong link to the smoking of cigarettes, which introduces tars into the lungs. Winder interviewed 684 lung cancer patients and 600 normal controls, asking whether each had ever smoked. Cancer rates were 40 times higher in heavy smokers than in nonsmokers. Doll's study was even more convincing. He interviewed a large number of British physicians, noting which ones smoked, then waited to see which would develop lung cancer. Many did. Overwhelmingly, those who did were smokers. From these studies, it seemed likely as long as 50 years ago that tars and other chemicals in cigarette smoke induce cancer in the lungs of persistent smokers. While this suggestion was (and is) resisted by the tobacco industry, the evidence that has accumulated since these pioneering studies makes a clear case, and there is no longer any real doubt. Chemicals in cigarette smoke cause cancer.

Carcinogens Are Common

In ongoing investigations over the last 50 years, many hundreds of synthetic chemicals have been shown capable of causing cancer in laboratory animals. Among them are

Table 17.3	Chemical Carcinogens in the Workplace	
Chemical	**Cancer**	**Workers at Risk for Exposure**
COMMON EXPOSURE		
Benzene	Myelogenous leukemia	Painters; dye users; furniture finishers
Diesel exhaust	Lung	Railroad and bus-garage workers; truckers; miners
Mineral oils	Skin	Metal machining
Pesticides	Lung	Sprayers
Cigarette tar	Lung	Smokers
UNCOMMON EXPOSURE		
Asbestos	Mesothelioma, lung	Brake-lining, insulation workers
Synthetic mineral fibers	Lung	Wall and pipe insulation; duct wrapping
Hair dyes	Bladder	Hairdressers and barbers
Paint	Lung	Painters
Polychlorinated biphenyl's	Liver, skin	Hydraulic fluids and lubricants; inks; adhesives; insecticides
Soot	Skin	Chimney sweeps; bricklayers; firefighters; heating-unit service workers
RARE EXPOSURE		
Arsenic	Lung, skin	Insecticide/herbicide sprayers; tanners; oil refiners
Formaldehyde	Nose	Hospital and lab workers; wood product, paper, textiles, and metal product workers

trichloroethylene, asbestos, benzene, vinyl chloride, arsenic, arylamide, and a host of complex petroleum products with chemical structures resembling chicken wire. People in the workplace encounter chemicals daily (table 17.3).

In addition to identifying potentially dangerous substances, what have the studies of potential carcinogens told us about the nature of cancer? What do these cancer-causing chemicals have in common? *They are all mutagens, each capable of inducing changes in DNA.*

Chemicals that produce mutations in DNA are often potent carcinogens. Tars in cigarette smoke, for example, are the direct cause of most lung cancers.

Other Tumors Result from Viral Infection

Chemical mutagens are not the only carcinogens, however. Some tumors seem almost certainly to result from viral infection. Viruses can be isolated from certain tumors, and these viruses cause virus-containing tumors to develop in other individuals. About 15% of human cancers are associated with viruses.

A Virus That Causes Cancer

In 1911, American medical researcher Peyton Rous reported that a virus, subsequently named **Rous avian sarcoma virus (RSV),** was associated with chicken sarcomas. He found that RSV could infect and initiate cancer in chicken fibroblast (connective tissue) cells growing in culture; from those cancerous cells, more viruses could be isolated. Rous was awarded the 1966 Nobel Prize in Physiology or Medicine for this discovery.

RSV proved to be an RNA virus, or **retrovirus.** When retroviruses infect a cell, they make a DNA copy of their RNA genome and may insert that copy into the host cell's DNA.

How RSV Causes Cancer

How does RSV initiate cancer? When RSV was compared to a closely related virus, RAV-O, which is unable to transform normal chicken cells into cancerous cells, the two viruses proved to be identical except for one gene that was present in RSV but absent from RAV-O. That gene was called the *src* gene, short for sarcoma.

How do viral genes cause cancer? An essential clue came in 1970, when temperature-sensitive RSV mutants were isolated. These mutants would transform tissue culture cells into cancer cells at 35°C, but not at 41°C. Temperature sensitivity of this kind is almost always associated with proteins. It seemed likely, therefore, that the *src* gene was actively transcribed by the cell, rather than serving as a recognition site for some sort of regulatory protein. This was an exciting result because it suggested that the protein specified by this cancer-causing gene, or **oncogene,** could be isolated and its properties studied.

The *src* protein was first isolated in 1977. It turned out to be an enzyme of moderate size that phosphorylates (adds a phosphate group to) the tyrosine amino acids of proteins. Such enzymes, called **tyrosine kinases,** are not common in animal cells. One of the few that is known is an enzyme that also serves as a plasma membrane receptor for **epidermal growth factor,** a protein that signals the initiation of cell division. This finding raised the exciting possibility that RSV may cause cancer by introducing into cells an active form of a normally quiescent growth-promoting enzyme. Later experiments showed this is indeed the case.

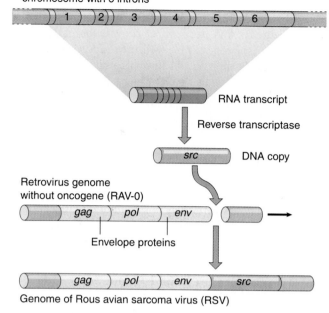

FIGURE 17.12

How a chicken gene got into the RSV genome. RSV contains only a few genes that encode the viral protein coat, envelope proteins (encoded by the *gag* and *env* genes) and reverse transcriptase (encoded by *pol*). It also contains the *src* gene that causes sarcomas, which the RAV-O virus lacks. RSV originally obtained its *src* gene from chickens, where a copy of the gene occurs normally and is controlled by the chicken's regulatory genes.

Origin of the *src* Gene

Does the *src* gene actually integrate into the host cell's chromosome along with the rest of the RSV genome? One way to answer this question is to prepare a radioactive version of the gene, allow it to bind to complementary sequences on the chicken chromosomes, and examine where the chromosomes become radioactive. The result of this experiment is that radioactive *src* DNA does in fact bind to the site where RSV DNA is inserted into the chicken genome—but it also binds to a second site where there is no part of the RSV genome!

The explanation for the second binding site is that the *src* gene is not exclusively a viral gene. It is also a growth-promoting gene that evolved in and occurs normally in chickens. This normal chicken gene is the second site where *src* binds to chicken DNA. Somehow, an ancestor of RSV picked up a copy of the normal chicken gene in some past infection. Now part of the virus, the gene is transcribed under the control of viral promoters rather than under the regulatory system of the chicken genome (figure 17.12).

Studies of RSV reveal that cancer results from the inappropriate activity of growth-promoting genes that are less active or completely inactive in normal cells.

Cancer and the Cell Cycle

An important technique used to study tumors is called **transfection.** In this procedure, the nuclear DNA from tumor cells is isolated and cleaved into random fragments. Each fragment is then tested individually for its ability to induce cancer in the cells that assimilate it.

Using transfection, researchers have discovered that most human tumors appear to result from the mutation of genes that regulate the cell cycle. Sometimes the mutation of a single gene is all that is needed to transform normally dividing cells into cancerous cells in tissue culture. In many cases, the gene that mutates to cause cancer is a gene previously associated with a cancer-causing virus. By comparing cancer-causing genes with their normal, nonmutated counterparts, investigators have gained a clearer picture of how mutation of certain genes can lead to cancer (table 17.4).

Point Mutations Can Lead to Cancer

The difference between a normal gene encoding a protein that regulates the cell cycle and a cancer-inducing version can be a single point mutation in the DNA. In one case of *ras*-induced bladder cancer, for example, a single DNA base change from guanine to thymine converts a glycine in the normal *ras* protein into a valine in the cancer-causing version. Several other *ras*-induced human carcinomas have been shown to also involve single nucleotide substitutions.

Telomerase and Cancer

Telomeres are short sequences of nucleotides repeated thousands of times on the ends of chromosomes. They are covered by a tight cap of proteins that prevent an enzyme called telomerase from lengthening the sequence. Because DNA polymerase is unable to copy chromosomes all the way to the tip (there is no place for the primer necessary to copy the last Okazaki fragment), telomeric segments are lost every time a cell divides. The telomere sequence eventually becomes so short that it does not bind enough cap proteins to prevent telomerase from lengthening the sequence. At that point, the telomerase makes the telomere sequence grow long again. When it has grown longer, enough cap proteins are able to bind again and prevent telomerase from acting. Telomeres thus alternate between growing and shrinking in a dynamic equilibrium.

The **telomere cycle,** only recently understood, refutes a popular hypothesis about telomerase and cancer. Researchers found telomerase in human ovarian tumor cells. These cells contain mutations that inactivate the cell control that blocks the transcription of the telomerase gene. Believing that telomeres shortened as cells aged, researchers hypothesized that telomerase produced in these cells was preventing telomere shortening and allowing the cells to continue to proliferate and gain the immortality of cancer cells.

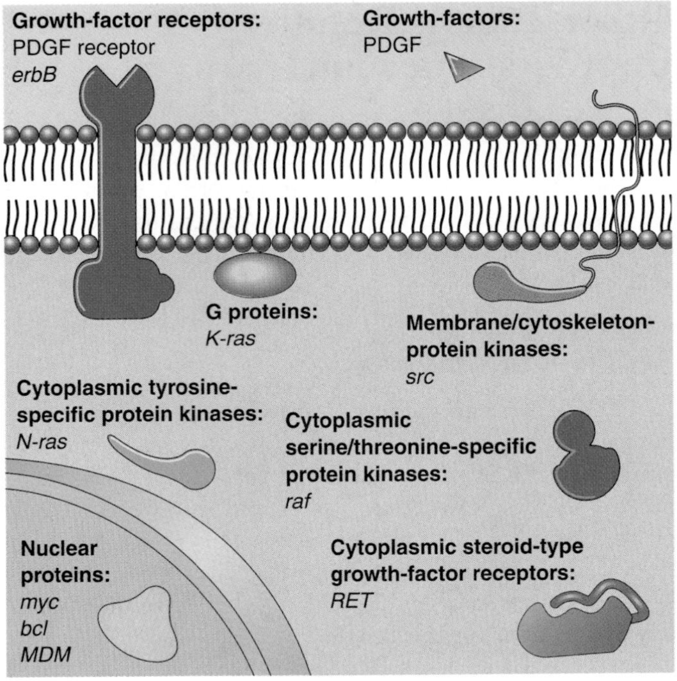

FIGURE 17.13
The main classes of oncogenes. Before they are altered by mutation to their cancer-causing condition, oncogenes are called proto-oncogenes (that is, genes able to become oncogenes). Illustrated here are the principal classes of proto-oncogenes, with some typical representatives indicated.

Mutations in Proto-Oncogenes: Accelerating the Cell Cycle

Most cancers are the direct result of mutations in growth-regulating genes. There are two general classes of cancer-inducing mutations: mutations of proto-oncogenes and mutations of tumor-suppressor genes.

Genes known as **proto-oncogenes** encode proteins that stimulate cell division. Mutations that overactivate these stimulatory proteins cause the cells that contain them to proliferate excessively. Mutated proto-oncogenes become cancer-causing genes called **oncogenes** (Greek *onco-,* "tumor") (figure 17.13) Often the induction of these cancers involves changes in the activity of a receptor on the surface of the plasma membrane. In a normal cell, these receptors activate intracellular signaling pathways that trigger passage of the G_1 checkpoint of cell proliferation (see figure 11.20).

All of these oncogenes are genetically dominant. Among the most widely studied are *myc* and *ras*. Expression of *myc* stimulates the production of cyclins and cyclin-dependent protein kinases (Cdks), key elements in regulating the checkpoints of cell division.

The *ras* gene product is involved in the cellular response to epidermal growth factor (EGF), an intercellular signal that normally initiates cell proliferation. When EGF binds to a specific receptor protein on the plasma membrane of epithelial cells, the portion of the receptor that protrudes into the cytoplasm stimulates the *ras* protein to bind to GTP. The *ras* protein/GTP complex phosphorylates, activating two cytoplasmic kinases and triggering an intracellular signaling system (see chapter 7). The final step in the pathway is the activation of transcription factors that trigger cell proliferation. Cancer-causing mutations in *ras* greatly reduce the amount of EGF necessary to initiate cell proliferation.

Mutations in Tumor-Suppressor Genes: Inactivating the Cell's Inhibitors of Proliferation

If the first class of cancer-inducing mutations "steps on the accelerator" of cell division, the second class of cancer-inducing mutations "removes the brakes." Cell division is normally turned off in healthy cells by proteins that prevent cyclins from binding to Cdks. The genes that encode these proteins are called **tumor-suppressor genes.** Their mutant alleles are genetically recessive.

Among the most widely studied tumor-suppressor genes are *Rb*, *p16*, *p21*, and *p53*. The unphosphorylated product of the *Rb* gene ties up transcription factor E2F, which transcribes several genes required for passage through the G_1 checkpoint into S phase of the cell cycle (figure 17.14). The proteins encoded by *p16* and *p21* reinforce the tumor-suppressing role of the Rb protein, preventing its phosphorylation by binding to the appropriate Cdk/cyclin complex and inhibiting its kinase

activity. The p53 protein senses the integrity of the DNA and is activated if the DNA is damaged (figure 17.15). It appears to act by inducing the transcription of *p21*, which binds to cyclins and Cdk and prevents them from interacting. One of the reasons repeated smoking leads inexorably to lung cancer is that it induces *p53* mutations. Indeed, almost half of all cancers involve mutations of the *p53* gene.

A Family of Cancer-Causing Genes

Mutation of a particular proto-oncogene or tumor-suppressor gene can lead to a variety of different clinical forms of cancer, such as bladder cancer, lung cancer, etc., depending upon the tissue in which the mutation occurs. Thus while there are numerous clinical forms of cancer, there are no more than a few dozen different genes whose mutation can lead to cancer.

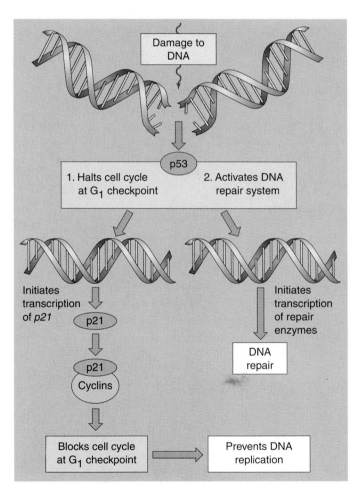

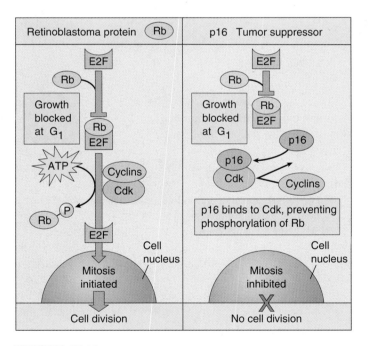

FIGURE 17.14

How the tumor-supressor genes *Rb* and *p16* interact to block cell devision. The retinoblastoma protein (Rb) binds to the transcription factor (E2F) that activates genes in the nucleus, preventing this factor from initiating mitosis. The G_1 checkpoint is passed when Cdk interacts with cyclins to phosphorylate Rb, releasing E2F. The p16 tumor-supressor protein reinforces Rb's inhibitory action by binding to Cdk so that Cdk is not available to phosphorylate Rb.

FIGURE 17.15

The role of tumor-supressor *p53* in regulating the cell cycle. The p53 protein works at the G_1 checkpoint to check for DNA damage. If the DNA is damaged, p53 activates the DNA repair system and stops the cell cycle at the G_1 checkpoint (before DNA replication). This allows time for the damage to be repaired. p53 stops the cell cycle by inducing the transcription of *p21*. The p21 protein then binds to cyclins and prevents them from complexing with Cdk.

Table 17.4 Some Genes Implicated in Human Cancers

Gene	Product	Cancer
ONCOGENES		
Genes Encoding Growth Factors or Their Receptors		
erb-B	Receptor for epidermal growth factor	Glioblastoma (a brain cancer); breast cancer
erb-B2	A growth factor receptor (gene also called *neu*)	Breast cancer; ovarian cancer; salivary gland cancer
PDGF	Platelet-derived growth factor	Glioma (a brain cancer)
RET	A growth factor receptor	Thyroid cancer
Genes Encoding Cytoplasmic Relays in Intracellular Signaling Pathways		
K-ras	Protein kinase	Lung cancer; colon cancer; ovarian cancer; pancreatic cancer
N-ras	Protein kinase	Leukemias
Genes Encoding Transcription Factors That Activate Transcription of Growth-Promoting Genes		
c-myc	Transcription factor	Lung cancer; breast cancer; stomach cancer; leukemias
L-myc	Transcription factor	Lung cancer
N-myc	Transcription factor	Neuroblastoma (a nerve cell cancer)
Genes Encoding Other Kinds of Proteins		
bcl-2	Protein that blocks cell suicide	Follicular B cell lymphoma
bcl-1	Cyclin D1, which stimulates the cell cycle clock (gene also called *PRAD1*)	Breast cancer; head and neck cancers
MDM2	Protein antagonist of p53 tumor-supressor protein	Wide variety of sarcomas (connective tissue cancers)
TUMOR-SUPRESSOR GENES		
Genes Encoding Cytoplasmic Proteins		
APC	Step in a signaling pathway	Colon cancer; stomach cancer
DPC4	A relay in signaling pathway that inhibits cell division	Pancreatic cancer
NF-1	Inhibitor of ras, a protein that stimulates cell division	Neurofibroma; myeloid leukemia
NF-2	Inhibitor of ras	Meningioma (brain cancer); schwannoma (cancer of cells supporting peripheral nerves)
Genes Encoding Nuclear Proteins		
MTS1	p16 protein, which slows the cell cycle clock	A wide range of cancers
p53	p53 protein, which halts cell division at the G_1 checkpoint	A wide range of cancers
Rb	Rb protein, which acts as a master brake of the cell cycle	Retinoblastoma; breast cancer; bone cancer; bladder cancer
Genes Encoding Proteins of Unknown Cellular Locations		
BRCA1	?	Breast cancer; ovarian cancer
BRCA2	?	Breast cancer
VHL	?	Renal cell cancer

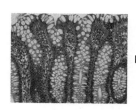
Normal tissue

Loss of *APC*

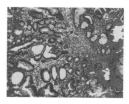
Small benign polyp

Mutation of *K-ras* and *DCC*

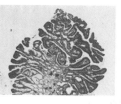
Large benign polyp

Mutation of *p53*

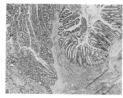
Cancer

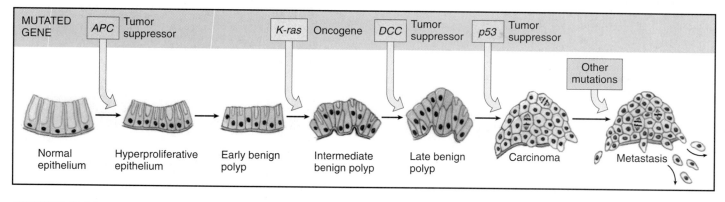

MUTATED GENE — *APC* Tumor suppressor — *K-ras* Oncogene — *DCC* Tumor suppressor — *p53* Tumor suppressor — Other mutations

Normal epithelium → Hyperproliferative epithelium → Early benign polyp → Intermediate benign polyp → Late benign polyp → Carcinoma → Metastasis

FIGURE 17.16

The progression of mutations that commonly lead to colorectal cancer. The fatal metastasis is the last of six serial changes that the epithelial cells lining the rectum undergo. One of these changes is brought about by mutation of a proto-oncogene, and three of them involve mutations that inactivate tumor-suppressor genes.

Cancer-Causing Mutations Accumulate over Time

Cells control proliferation at several checkpoints, and all of these controls must be inactivated for cancer to be initiated. Therefore, the induction of most cancers involves the mutation of multiple genes; four is a typical number (figure 17.16). In many of the tissue culture cell lines used to study cancer, most of the controls are already inactivated, so that mutations in only one or a few genes transform the line into cancerous growth. The need to inactivate several regulatory genes almost certainly explains why most cancers occur in people over 40 years old (figure 17.17); in older persons, there has been more time for individual cells to accumulate multiple mutations. It is now clear that mutations, including those in potentially cancer-causing genes, do accumulate over time. Using the polymerase chain reaction (PCR), researchers in 1994 searched for a certain cancer-associated gene mutation in the blood cells of 63 cancer-free people. They found that the mutation occurred 13 times more often in people over 60 years old than in people under 20.

> **Cancer is a disease in which the controls that normally restrict cell proliferation do not operate. In some cases, cancerous growth is initiated by the inappropriate activation of proteins that regulate the cell cycle; in other cases, it is initiated by the inactivation of proteins that normally suppress cell division.**

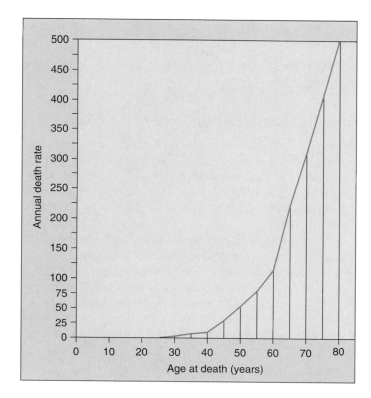

FIGURE 17.17

The annual death rate from cancer climbs with age. The rate of cancer deaths increases steeply after age 40 and even more steeply after age 60, suggesting that several independent mutations must accumulate to give rise to cancer.

Smoking and Cancer

How can we prevent cancer? The most obvious strategy is to minimize mutational insult. Anything that decreases exposure to mutagens can decrease the incidence of cancer, since exposure has the potential to mutate a normal gene into an oncogene. It is no accident that the most reliable tests for the carcinogenicity of a substance are tests that measure the substance's mutagenicity.

The Association between Smoking and Cancer

About a third of all cases of cancer in the United States are directly attributable to cigarette smoking. The association between smoking and cancer is particularly striking for lung cancer (figure 17.18). Studies of male smokers show a highly positive correlation between the number of cigarettes smoked per day and the incidence of lung cancer (figure 17.19). For individuals who smoke two or more packs a day, the risk of contracting lung cancer is at least 40 times greater than it is for nonsmokers, whose risk level approaches zero. Clearly, an effective way to avoid lung cancer is not to smoke. Other studies have shown a clear relationship between cigarette smoking and reduced life expectancy (figure 17.20). Life insurance companies have calculated that smoking a single cigarette lowers one's life expectancy by 10.7 minutes (longer than it takes to smoke the cigarette)! Every pack of 20 cigarettes bears an unwritten label:

> "The price of smoking this pack of cigarettes is 3½ hours of your life."

Smoking Introduces Mutagens to the Lungs

Over half a million people died of cancer in the United States in 1995; about 29% of them died of lung cancer. About 140,000 persons were diagnosed with lung cancer each year in the 1980s. Around 90% of them died within three years after diagnosis; 96% of them were cigarette smokers.

Smoking is a popular pastime. In the United States, 29% of the population smokes, and U. S. smokers consumed 520 billion cigarettes in 1989. The smoke emitted from these cigarettes contains some 3000 chemical components, including vinyl chloride, benzo[a]pyrenes, and nitroso-*nor*-nicotine, all potent mutagens. Smoking places these mutagens into direct contact with the tissues of the lungs.

Mutagens in the Lung Cause Cancer

Introducing powerful mutagens to the lungs causes considerable damage to the genes of the epithelial cells that line the lungs and are directly exposed to the chemicals. Among the genes that are mutated as a result are some whose normal function is to regulate cell proliferation. When these genes are damaged, lung cancer results.

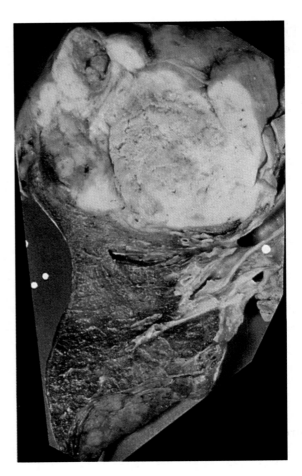

FIGURE 17.18
Photo of a cancerous lung in an adult human. The bottom half of the lung is normal, while a cancerous tumor has completely taken over the top half. The cancer cells will eventually break through into the lymph and blood vessels and spread through the body.

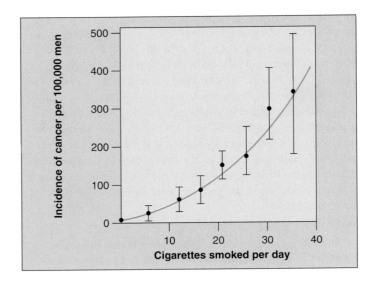

FIGURE 17.19
Smoking causes cancer. The annual incidence of lung cancer per 100,000 men clearly increases with the number of cigarettes smoked per day.

This process has been clearly demonstrated for benzo[a]pyrene (BP), one of the potent mutagens released into cigarette smoke from tars in the tobacco. The epithelial cells of the lung absorb BP from tobacco smoke and chemically alter it to a derivative form, benzo[a]pyrene-diolepoxide (BPDE), that binds directly to the tumor-suppressor gene *p53* and mutates it to an inactive form. The protein encoded by *p53* oversees the G_1 cell cycle checkpoint described in chapter 11 and is one of the body's key mechanisms for preventing uncontrolled cell proliferation. The destruction of *p53* in lung epithelial cells greatly hastens the onset of lung cancer—*p53* is mutated to an inactive form in over 70% of lung cancers. When examined, the *p53* mutations in cancer cells almost all occur at one of three "hot spots." The key evidence linking smoking and cancer is that when the mutations of *p53* caused by BPDE from cigarettes are examined, they occur at the same three specific "hot spots!"

The Incidence of Cancer Reflects Smoking

Cigarette manufacturers argue that the causal connection between smoking and cancer has not been proved, and that somehow the relationship is coincidental. Look carefully at the data presented in figure 17.21 and see if you agree. The upper graph, compiled from data on American men, shows the incidence of smoking from 1900 to 1990 and the incidence of lung cancer over the same period. Note that as late as 1920, lung cancer was a rare disease. About 20 years after the incidence of smoking began to increase among men, lung cancer also started to become more common.

Now look at the lower graph, which presents data on American women. Because of social mores, significant numbers of American women did not smoke until after World War II, when many social conventions changed. As late as 1963, when lung cancer among males was near current levels, this disease was still rare in women. In the United States that year, only 6588 women died of lung cancer. But as more women smoked, more developed lung cancer, again with a lag of about 20 years. American women today have achieved equality with men in the numbers of cigarettes they smoke, and their lung cancer death rates are today approaching those for men. In 1990, more than 49,000 women died of lung cancer in the United States. The current annual rate of deaths from lung cancer in male and female smokers is 180 per 100,000, or about 2 out of every 1000 smokers *each year*.

The easiest way to avoid cancer is to avoid exposure to mutagens. The single greatest contribution one can make to a longer life is not to smoke.

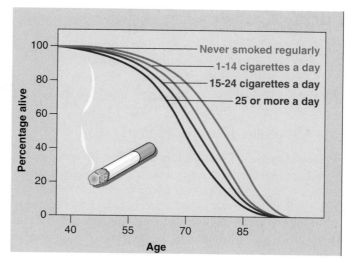

FIGURE 17.20
Tobacco kills one smoker in two. The world's longest-running survey of smoking, begun in 1951 in Britain, revealed that by 1994 the death rate for smokers had climbed to three times the rate for nonsmokers among men 35 to 69 years of age.
Source: Data from *New Scientist*, October 15, 1994.

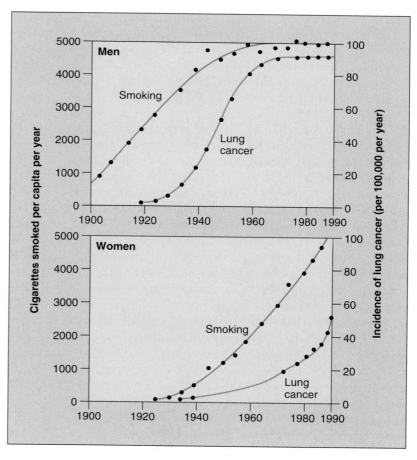

FIGURE 17.21
The incidence of lung cancer in men and women. What do these graphs indicate about the connection between smoking and lung cancer?

Curing Cancer

Potential cancer therapies are being developed on many fronts. Many of the most exciting possibilities, which we can loosely call molecular therapies, focus on different stages of the cell's decision-making process (figure 17.22).

New Molecular Therapies

1. Receiving the Signal to Divide. The first step in the decision process is the reception of a "divide" signal, usually a small protein called a growth factor released from a neighboring cell. The growth factor is received by a special receptor on the cell surface whose shape fits the growth factor like a hand fits a glove. Mutations that increase the number of receptors on the cell surface amplify the division signal and so lead to cancer. Over 20% of breast cancers prove to overproduce the receptor for epidermal growth factor (EGF).

Therapies directed at this stage of the decision process utilize the human immune system to attack cancer cells. Special protein molecules called "monoclonal antibodies," created by genetic engineering, are the therapeutic agents. These monoclonal antibodies are designed to seek out and stick to EGF. Like waving a red flag, the presence of the monoclonal antibody then calls down attack by the immune system on the cell with EGF. Because the cancer cells of breast cancer patients overproduce EGF, they are killed preferentially. Genentech's monoclonal antibody, called "anti-HER2," has given promising results in initial tests: of 43 advanced breast cancer patients in whom chemotherapy had failed, 14 stabilized and 5 went into remission when treated with anti-HER2.

2. The Relay Switch. The second step in the decision process is the passage of the signal into the cell's interior, the cytoplasm. This is carried out in normal cells by a protein called Ras that acts as a relay switch. When growth factor binds to a receptor like EGF, the adjacent Ras protein acts like it has been "goosed," contorting into a new shape. This new shape is chemically active, and initiates a chain of reactions that passes the "divide" signal inward toward the nucleus. Mutated forms of the Ras protein behave like a relay switch stuck in the "ON" position, continually instructing the cell to divide when it should not. 30% of all cancers have a mutant form of Ras.

Therapies directed at this stage of the decision process take advantage of the fact that normal Ras proteins are inactive when made. Only after it has been modified by a special enzyme with the jaw-breaking name *farnesyl transferase* does Ras protein become able to function as a relay switch. Drugs that inhibit farnesyl transferase offer great promise as anticancer therapies. In tests on animals, farnesyl transferase inhibitors induce the regression of tumors and prevent the formation of new ones.

3. Amplifying the Signal. The third step in the decision process is the amplification of the signal within the cytoplasm. Just as a TV signal needs to be amplified in order to be received at a distance, so a "divide" signal must be amplified if it is to reach the nucleus at the interior of the cell, a very long journey at a molecular scale. Cells use an ingenious trick to amplify the signal. Ras, when "ON," acts as an enzyme, a protein kinase. It activates other protein kinases that in their turn activate still others. The trick is that once a protein kinase enzyme is activated, it goes to work like a demon, activating hoards of others every second! And each and every one it activates behaves the same way too, activating still more, in a cascade of ever-widening effect. At each stage of the relay, the signal is amplified a thousand-fold. Mutations stimulating any of the protein kinases can dangerously increase the already amplified signal and lead to cancer. 5% of all cancers, for example, have a mutant hyperactive form of the protein kinase Src.

Therapies directed at this stage of the decision process initially employed protein kinase inhibitors. However, the cell uses nearly 1000 protein kinases for other important jobs in the cell, all with highly similar structures, so that generalized protein kinase inhibitors disrupted the activity of many of them, leading to undesirable side effects. A new approach is to prepare so-called "anti-sense RNA" directed specifically against Src or other cancer-inducing kinase mutations. The idea is that the *src* gene uses a complementary copy of itself to manufacture the Src protein (the "sense" RNA or messenger RNA), and a mirror image complementary copy of the sense RNA ("anti-sense RNA") will stick to it, gumming it up so it can't be used to make Src protein. The approach appears promising. In tissue culture, anti-sense RNAs inhibit the growth of cancer cells, and some also appear to block the growth of human tumors implanted in laboratory animals. Human clinical trials are underway.

4. Releasing the Brake. The fourth step in the decision process is the removal of the "brake" the cell uses to restrain cell division. In healthy cells this brake, a tumor suppressor protein called Rb, blocks the activity of another protein called E2F. When free, E2F directs the cell to copy its DNA. Normal cell division is triggered to begin when Rb is inhibited, unleashing E2F. Mutations which destroy Rb release E2F from its control completely, leading to ceaseless cell division. 40% of all cancers have a defective form of Rb.

Therapies directed at this stage of the decision process are only now being attempted. They focus on drugs able to inhibit E2F, which should halt the growth of tumors arising from inactive Rb. Experiments in mice in which the E2F genes have been destroyed provide a model system to study such drugs, which are being actively investigated.

5. Checking That Everything Is Ready. The most exciting of the new therapies focuses on the final step in the decision process, the mechanism used by the cell to ensure that its DNA is undamaged and ready to divide. This job is carried out in healthy cells by the tumor-suppressor protein p53. Sometimes called the "Guardian Angel" of the cell, p53 inspects the DNA, and when it detects damaged or foreign DNA it stops cell division and activates the cell's DNA repair systems. If the damage doesn't get repaired in a reasonable time, p53 pulls the plug, triggering events that kill the cell. In this way, mutations such as those that cause cancer are either repaired or the cells containing them eliminated. If p53 is itself destroyed by mutation, future damage accumulates unrepaired. Among this damage are mutations that lead to cancer, mutations that would have been repaired by a healthy p53. 50% of all cancers have a disabled p53. Fully 70 to 80% of lung cancers have a mutant inactive p53—the chemical benzo-[*a*]pyrene in cigarette smoke is a potent mutagen of the *p53* gene.

In 1996 a therapy was suggested for cancers with a mutant p53. The story of its development spans 20 years. In 1977, virologists discovered that adenovirus (responsible for mild colds) could not reproduce within human cells without a working copy of a virus gene dubbed *E1B*. What does the protein encoded by *E1B* do that is so crucial? Inside human cells, the virus E1B protein binds to the human p53 protein! For over 10 years this finding wasn't appreciated, as little was known about p53. Only 5 years ago did researchers learn that p53 prevents cells from replicating damaged or foreign DNA. The infected human cell will not replicate the foreign adenovirus DNA unless p53 is inactive. That is why adenovirus needs a working version of E1B—to bind to p53 in the human cell and prevent this watchdog from blocking DNA replication.

In 1992 Frank McCormick, a biochemist from ONYX Pharmaceuticals in Richmond, California, realized that because adenovirus must turn off p53 to replicate, an adenovirus without E1B could not disable p53 and would be unable to grow in healthy cells—but it *would* grow just fine in cells lacking p53. What human cells lack p53? Cancer cells! If McCormick was right, adenovirus with disabled E1B would not grow in normal human cells, but would grow in, and destroy, a wide range of cancer cells (those with defective p53).

Initial studies reported in October 1996 have been very promising. In tissue culture the E1B-negative adenovirus does not grow in healthy skin cells, but does grow in a wide variety of tumor cells, including colon and lung cancer cells. When human tumor cells are introduced into mice lacking an immune system and allowed to produce substantial tumors, 60% of the tumors simply disappear when treated with E1B-deficient adenovirus, and do not reappear later. Initial human trials have been started.

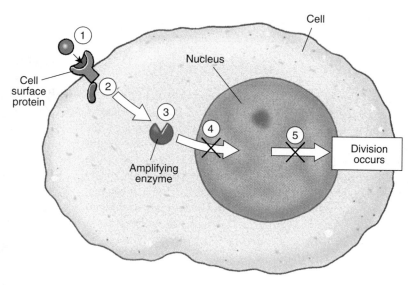

FIGURE 17.22
New molecular therapies for cancer target five differet stages in the cell's decision-making process. Scientists are actively researching ways to combat cancer at five different stages in cells: (*1*) on the cell surface, where the cell receives the signal, a growth factor, to divide; (*2*) just inside the cell, where a protein relay switch passes on the divide signal; (*3*) in the cytoplasm, where enzymes amplify the signal; (*4*) inside the nucleus, where a "brake," a particular protein that prevents DNA replication, is released; and (*5*) inside the nucleus, where certain proteins check that everything is ready, more specifically, that the replicated DNA is not damaged.

While E1B-deficient adenovirus offers great promise as a therapy for a wide range of cancers, including most lung cancers, a significant technical hurdle remains. Recall that the initial animal tests were done on mice with no immune system. Humans have active immune systems. In cancer patients the adenovirus therapy may be neutralized by the patient's immune system before the virus has a chance to do any good, simply because most people have had adenovirus colds in the past and so can be expected to carry antibodies directed against adenoviruses. In these people, such antibodies might attack any E1B-deficient adenovirus introduced to fight cancer. Anticipating this problem, investigators are exploring alternative viruses that would not provoke an immune response.

Molecular therapies such as those described here are only part of a wave of potential treatments under development and clinical trial. The clinical trials will take years to complete, but by the turn of the century we can expect to greet a new millennium in which cancer is becoming a curable disease.

Understanding of how mutations produce cancer has progressed to the point where promising potential therapies can be tested.

An Overview of Recombination

Mutation is a change in the *content* of an organism's genetic message, but it is not the only source of genetic diversity. Diversity is also generated when existing elements of the genetic message move around within the genome. As an analogy, consider the pages of this book. A point mutation would correspond to a change in one or more of the letters on the pages. For example, " . . . in one or more of the letters *of* the pages" is a mutation of the previous sentence, in which an "n" is changed to an "f." A significant alteration is also achieved, however, when we move the position of words, as in " . . . in one or more of the pages on the letters." The change alters (and destroys) the meaning of the sentence by exchanging the position of the words "letters" and "pages." This second kind of change, which represents an alteration in the genomic *location* of a gene or a fragment of a gene, demonstrates **genetic recombination.**

Gene Transfer

Viewed broadly, genetic recombination can occur by two mechanisms (table 17.5). In **gene transfer,** one chromosome or genome donates a segment to another chromosome or genome. The transfer of genes from the human immunodeficiency virus (HIV) to a human chromosome is an example of gene transfer. Because gene transfer occurs in both prokaryotes and eukaryotes, it is thought to be the more primitive of the two mechanisms.

Reciprocal Recombination

In contrast, **reciprocal recombination,** in which two chromosomes trade segments, is strictly a eukaryotic phenomenon. It is exemplified by the crossing over that occurs between homologous chromosomes during meiosis. **Chromosome assortment** during meiosis is another form of reciprocal recombination. Discussed in chapters 12 and 13,

FIGURE 17.23
A Nobel Prize for discovering gene transfer by transposition. Barbara McClintock receiving her Nobel Prize in 1983.

it is responsible for the 9:3:3:1 ratio of phenotypes in a dihybrid cross and occurs only in eukaryotes.

Genetic recombination is a change in the genomic association among genes. It often involves a change in the position of a gene or portion of a gene. Recombination of this sort may result from one-way gene transfer or reciprocal gene exchange.

Table 17.5	Classes of Genetic Recombination
Class	**Occurrence**
GENE TRANSFERS	
Conjugation	Occurs predominantly but not exclusively in bacteria and is targeted to specific locations in the genome
Transposition	Common in both bacteria and eukaryotes; genes move to new genomic locations, apparently at random
RECIPROCAL RECOMBINATIONS	
Crossing over	Requires the pairing of homologous chromosomes and may occur anywhere along their length
Unequal crossing over	The result of crossing over between mismatched segments; leads to gene duplication and deletion
Gene conversion	Occurs when homologous chromosomes pair and one is "corrected" to resemble the other
Independent assortment	Haploid cells produced by meiosis contain only one randomly selected member of each pair of homologous chromosomes

Gene Transfer

Genes are not fixed in their locations on chromosomes or the circular DNA molecules of bacteria; they can move around. Some genes move because they are part of small, circular, extrachromosomal DNA segments called **plasmids.** Plasmids enter and leave the main genome at specific places where a nucleotide sequence matches one present on the plasmid. Plasmids occur primarily in bacteria, in which the main genomic DNA can interact readily with other DNA fragments. About 5% of the DNA that occurs in a bacterium is plasmid DNA. Some plasmids are very small, containing only one or a few genes, while others are quite complex and contain many genes. Other genes move within **transposons,** which jump from one genomic position to another at random in both bacteria and eukaryotes. Together, the movement of plasmids and transposons is responsible for much of the genetic recombination in organisms.

Gene transfer by means of plasmid movement was discovered by Joshua Lederberg and Edward Tatum in 1947. Three years later, transposons were discovered by Barbara McClintock. However, her work implied that the position of genes in a genome need not be constant. Researchers accustomed to viewing genes as fixed entities, like beads on a string, did not readily accept the idea of transposons. Therefore, while Lederberg and Tatum were awarded a Nobel Prize for their discovery in 1958, McClintock did not receive the same recognition for hers until 1983 (figure 17.23).

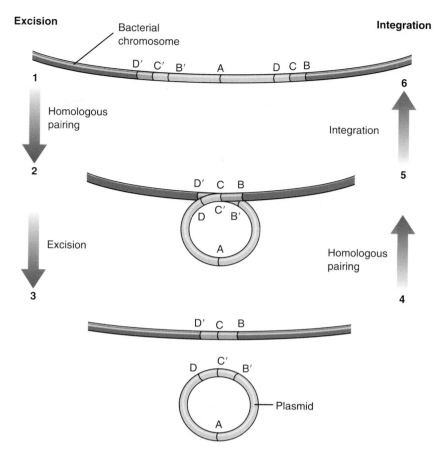

FIGURE 17.24

Integration and excision of a plasmid. Because the ends of the two sequences in the bacterial genome are the same (D′, C′, B′, and D, C, B), it is possible for the two ends to pair. Steps 1–3 show the sequence of events if the strands exchange during the pairing. The result is excision of the loop and a free circle of DNA—a plasmid. Steps 4–6 show the sequence when a plasmid integrates itself into a bacterial genome.

Plasmid Creation

To understand how plasmids arise, consider a hypothetical stretch of bacterial DNA that contains two copies of the same nucleotide sequence. It is possible for the two copies to base-pair with each other and create a transient "loop," or double duplex. All cells have recombination enzymes that can cause such double duplexes to undergo a **reciprocal exchange,** in which they exchange strands. As a result of the exchange, the loop is freed from the rest of the DNA molecule and becomes a plasmid (figure 17.24, steps 1–3). Any genes between the duplicated sequences (such as gene A in figure 17.24) are transferred to the plasmid.

Once a plasmid has been created by reciprocal exchange, DNA polymerase will replicate it if it contains a replication origin, often without the controls that restrict the main genome to one replication per cell division. Consequently, some plasmids may be present in multiple copies, others in just a few copies, in a given cell.

Integration

A plasmid created by recombination can reenter the main genome the same way it left. Sometimes the region of the plasmid DNA that was involved in the original exchange, called the **recognition site,** aligns with a matching sequence on the main genome. If a recombination event occurs anywhere in the region of alignment, the plasmid will integrate into the genome (figure 17.24, steps 4–6). Integration can occur wherever any shared sequences exist, so plasmids may be integrated into the main genome at positions other than the one from which they arose. If a plasmid is integrated at a new position, it transfers its genes to that new position.

Transposons and plasmids transfer genes to new locations on chromosomes. Plasmids can arise from and integrate back into a genome wherever DNA sequences in the genome and in the plasmid match.

Gene Transfer by Conjugation

One of the startling discoveries Lederberg and Tatum made was that plasmids can pass from one bacterium to another. The plasmid they studied was part of the genome of *Escherichia coli*. It was given the name F for fertility factor, since only cells that had that plasmid integrated into their DNA could act as plasmid donors. These cells are called Hfr cells (for "high-frequency recombination"). The F plasmid contains a DNA replication origin and several genes that promote its transfer to other cells. These genes encode protein subunits that assemble on the surface of the bacterial cell, forming a hollow tube called a **pilus.**

When the pilus of one cell (F+) contacts the surface of another cell that lacks a pilus, and therefore does not contain an F plasmid (F−), the pilus forms a **conjugation bridge** between the two cells and mobilizes the plasmid within the first cell for transfer. First, the F plasmid binds to a site on the interior of the F+ cell just beneath the pilus. Then, by a process called **rolling-circle replication**, the F plasmid begins to copy its DNA at the binding point. As it is replicated, the single-stranded copy of the plasmid passes through the pilus and into the other cell. There a complementary strand is added, creating a new, stable F plasmid (figure 17.25). In this way, genes are passed from one bacterium to another. This transfer of genes between bacteria is called **conjugation.**

In an Hfr cell, with the F plasmid integrated into the main bacterial genome rather than free in the cytoplasm, the F plasmid can still organize the transfer of genes. In this case, the integrated F region binds beneath the pilus and initiates the *replication of the bacterial genome*, transferring the newly replicated portion to the recipient cell. Transfer proceeds as if the bacterial genome were simply a part of the F plasmid. By studying this phenomenon, researchers have been able to locate the positions of different genes in bacterial genomes (figure 17.26).

Plasmids transfer copies of bacterial genes (and even entire genomes) from one bacterium to another.

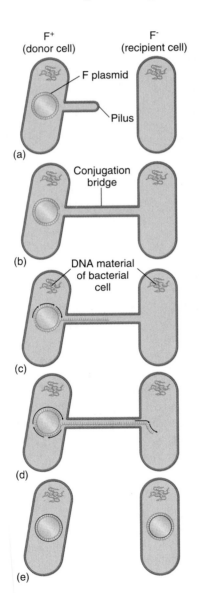

(a)

(b)

(c)

(d)

(e)

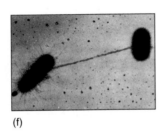

(f)

FIGURE 17.25
Gene transfer between bacteria. (a) Donor cells (F+) contain an F plasmid that recipient cells (F−) lack. (b) The long pilus connecting the two cells is called a conjugation bridge. (c) The F plasmid replicates itself and transfers the copy across the bridge. The remaining strand of the plasmid serves as a template to build a replacement. (d) When the single strand enters the recipient cell, it serves as a template to assemble a double-stranded plasmid. (e) When the process is complete, both cells contain a complete copy of the plasmid. (f) The electron micrograph shows two bacteria undergoing conjugation.

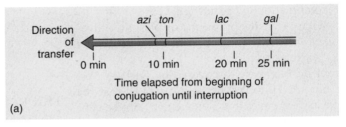

(a)

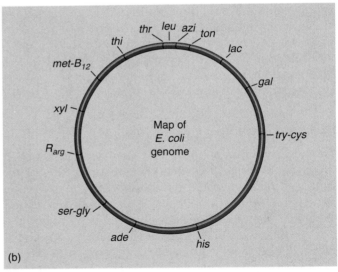

(b)

FIGURE 17.26
A conjugation map of the *E. coli* chromosome. Scientists have been able to break the *E. coli* conjugation bridges by agitating the cell suspension rapidly in a blender at different intervals after the start of conjugation. In this way, investigators have been able to locate the positions of various genes along the bacterial genome. (a) The closer the genes are to the origin of replication, the sooner one has to turn on the blender to block their transfer. (b) A map of the *Escherichia coli* genome was developed using this method.

Gene Transfer by Transposition

Like plasmids, transposons (figure 17.27) move from one genomic location to another. After spending many generations in one position, a transposon may abruptly move to a new position in the genome, carrying various genes along with it. Transposons encode an enzyme called **transposase,** that inserts the transposon into the genome (figure 17.28). Because this enzyme usually does not recognize any particular sequence on the genome, transposons appear to move to random destinations.

The movement of any given transposon is relatively rare: it may occur perhaps once in 100,000 cell generations. Although low, this rate is still about 10 times as frequent as the rate at which random mutational changes occur. Furthermore, there are many transposons in most cells. Hence, over long periods of time, transposition can have an enormous evolutionary impact.

One way this impact can be felt is through mutation. The insertion of a transposon into a gene often destroys the gene's function, resulting in what is termed **insertional inactivation.** This phenomenon is thought to be the cause of a significant number of the spontaneous mutations observed in nature.

Transposition can also facilitate **gene mobilization,** the bringing together in one place of genes that are usually located at different positions in the genome. In bacteria, for example, a number of genes encode enzymes that make the bacteria resistant to antibiotics such as penicillin, and many of these genes are located on plasmids. The simultaneous exposure of bacteria to multiple antibiotics, a common medical practice some years ago, favors the persistence of plasmids that have managed to acquire several resistance genes. Transposition can rapidly generate such composite plasmids, called **resistance transfer factors** (RTFs), by moving antibiotic resistance genes from several plasmids to one. Bacteria possessing RTFs are thus able to survive treatment with a wide variety of antibiotics. RTFs are thought to be responsible for much of the recent difficulty in treating hospital-engendered *Staphylococcus aureus* infections and the new drug-resistant strains of tuberculosis.

Transposition is the one-way transfer of genes to a randomly selected location in the genome. The genes move because they are associated with mobile genetic elements called transposons.

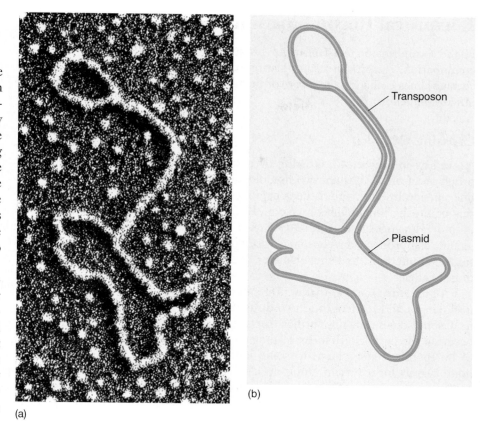

(a)

(b)

FIGURE 17.27

Transposon. Transposons form characteristic stem-and-loop structures called "lollipops" because their two ends have the same nucleotide sequence as inverted repeats. These ends pair together to form the stem of the lollipop. (a) Electron micrograph (33,800×), (b) Schematic sketch of the micrograph.

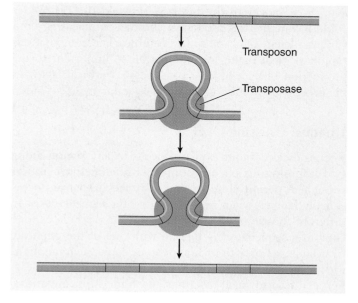

FIGURE 17.28

Transposition. Transposase does not recognize any particular DNA sequence; rather, it selects one at random, moving the transposon to a random location. Some transposons leave a copy of themselves behind when they move.

Reciprocal Recombination

In the second major mechanism for producing genetic recombination, reciprocal recombination, two homologous chromosomes exchange all or part of themselves during the process of meiosis.

Crossing Over

As we saw in chapter 12, crossing over occurs in the first prophase of meiosis, when two homologous chromosomes line up side by side within the synaptonemal complex. At this point, the homologues exchange DNA strands at one or more locations (see figure 12.12). This exchange of strands can produce chromosomes with different combinations of mutations, and thus gametes with new combinations of alleles.

Imagine, for example, that a giraffe has genes encoding neck length and leg length at two different loci on one of its chromosomes. Imagine further that a recessive mutation occurs at the neck length locus, leading after several rounds of independent assortment to some individuals that are homozygous for a variant "long-neck" allele. Similarly, a recessive mutation at the leg length locus leads to homozygous "long-leg" individuals.

It is very unlikely that these two mutations would arise at the same time in the same individual, since the probability of two independent events occurring together is the product of their individual probabilities. If the spontaneous occurrence of both mutations in a single individual were the only way to produce a giraffe with both a long neck and long legs, it would be extremely unlikely that such an individual would ever occur. Because of recombination, however, a crossover in the interval between the two genes could in one meiosis produce a chromosome bearing both variant alleles. This ability to reshuffle gene combinations rapidly is what makes recombination so important to the production of natural variation and so a critical factor in the process of evolution by natural selection in eukaryotes.

Unequal Crossing Over

Reciprocal recombination can occur in any region along two homologous chromosomes with sequences similar enough to permit close pairing. Mistakes in pairing occasionally happen when several copies of a sequence exist in different locations on a chromosome. In such cases, one copy of a sequence may line up with one of the duplicate copies instead of with its homologous copy. Such misalignment causes slipped mispairing, which, as we discussed earlier, can lead to small deletions and frame-shift mutations. If a crossover occurs in the pairing region, it will result in unequal crossing over, since the two homologues will exchange segments of unequal length.

In unequal crossing over, one chromosome gains extra copies of the multicopy sequences, while the other

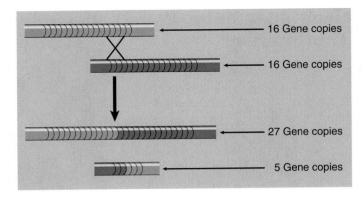

FIGURE 17.29
Unequal crossing over. When a repeated sequence pairs out of register, a crossover within the region will produce one chromosome with fewer gene copies and one with more. Much of the gene duplication that has occurred in eukaryotic evolution may well be the result of unequal crossing over.

chromosome loses them (figure 17.29). This process can generate a chromosome with hundreds of copies of a particular gene, lined up side by side in tandem array.

Because the genomes of most eukaryotes possess multiple copies of transposons scattered throughout the chromosomes, unequal crossing over between copies of transposons located in different positions has had a profound influence on gene organization in eukaryotes. As we shall see later, most of the genes of eukaryotes appear to have been duplicated one or more times during their evolution.

Gene Conversion

Because the two homologues that pair within a synaptonemal complex are not identical, some nucleotides in one homologue are not complementary to their counterpart in the other homologue with which it is paired. These occasional nonmatching pairs of nucleotides are called **mismatch pairs.**

As you might expect, the cell's error-correcting machinery is able to detect mismatch pairs. If a mismatch is detected during meiosis, the enzymes that "proofread" new DNA strands during DNA replication correct it. The mismatched nucleotide in one of the homologues is excised and replaced with a nucleotide complementary to the one in the other homologue. Its base-pairing partner in the first homologue is then replaced, producing two chromosomes with the same sequence. This error correction causes one of the mismatched sequences to convert into the other, a process called **gene conversion.**

Unequal crossing over is a crossover between chromosomal regions that are similar in nucleotide sequence but are not homologous. Gene conversion is the alteration of one homologue by the cell's error-detection and repair system to make it resemble the other homologue.

Trinucleotide Repeats

In 1991, a new kind of change in the genetic material was reported, one that involved neither changes in the identity of nucleotides (mutation) nor changes in the position of nucleotide sequences (recombination), but rather an increase in the number of copies of repeated trinucleotide sequences. Called **trinucleotide repeats,** these changes appear to be the root cause of a surprisingly large number of inherited human disorders.

The first examples of disorders resulting from the expansion of trinucleotide repeat sequences were reported in individuals with *fragile X syndrome* (the most common form of mental retardation) and *spinal muscular atrophy*. In both disorders, genes containing runs of repeated nucleotide triplets (CGG in fragile X syndrome and CAG in spinal muscular atrophy) exhibit large increases in copy number. In individuals with fragile X syndrome, for example, the CGG sequence is repeated hundreds and sometimes thousands of times (figure 17.30), whereas in normal individuals it repeats only about 30 times.

Ten additional human genes are now known to have alleles with expanded trinucleotide repeats (figure 17.31). Many (but not all) of these alleles are GC-rich. A few of the alleles appear benign, but most are associated with heritable disorders, including Huntington's disease, myotonic dystrophy, and a variety of neurological ataxias. In each case, the expansion transmits as a dominant trait. Often the repeats are found within the exons of their genes, but sometimes, as in the case of fragile X syndrome, they are located outside the coding segment. Furthermore, although the repeat number is stably transmitted in normal families, it shows marked instability once it has abnormally expanded. Siblings often exhibit unique repeat lengths, and the repeat number tends to increase in subsequent generations.

As the repeat number increases, disease severity tends to increase in step. In fragile X syndrome, the CGG triplet number first increases from the normal stable range of 5 to 55 times (the most common allele has 29 repeats) to an unstable number of repeats ranging from 50 to 200, with no

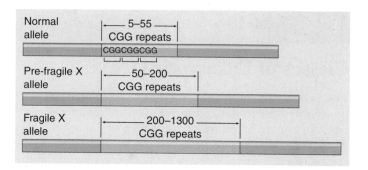

FIGURE 17.30
CGG repeats in fragile X alleles. The CGG triplet is repeated approximately 30 times in normal alleles. Individuals with pre-fragile X alleles show no detectable signs of the syndrome but do have increased numbers of CGG repeats. In fragile X alleles, the CGG triplet repeats hundreds of times.

detectable effect. In offspring, the number increases markedly, with copy numbers ranging from 200 to 1300, with significant mental retardation (see figure 17.30). Similarly, the normal allele for myotonic dystrophy has 5 GTC repeats. Mildly affected individuals have about 50, and severely affected individuals have up to 1000.

Trinucleotide repeats appear common in human genes, but their function is unknown. Nor do we know the mechanism behind trinucleotide repeat expansion. It may involve unequal crossing over, which can readily produce copy-number expansion, or perhaps some sort of stutter in the DNA polymerase when it encounters a run of triplets. The fact that di- and tetranucleotide repeat expansions are not found seems an important clue. Undoubtedly, further examples of this remarkable class of genetic change will be reported in the future. Considerable research is currently focused on this extremely interesting area.

Many human genes contain runs of a trinucleotide sequence. Their function is unknown, but if the copy number expands, hereditary disorders often result.

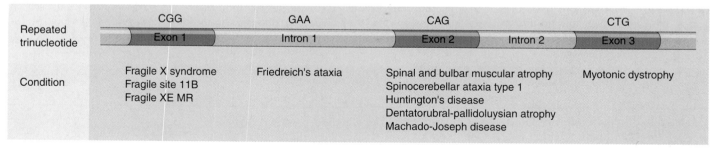

FIGURE 17.31
A hypothetical gene showing the locations and types of trinucleotide repeats associated with various human diseases. The CGG repeats of fragile X syndrome, fragile XE mental retardation (MR), and fragile site 11B occur in the first exon of their respective genes. GAA repeats characteristic of Friedreich's ataxia exist in the first intron of its gene. The genes for five different diseases, including Huntington's disease, have CAG repeats within their second exons. Lastly, the myotonic dystrophy gene contains CTG repeats within the third exon.

Classes of Eukaryotic DNA

The two main mechanisms of genetic recombination, gene transfer and reciprocal recombination, are directly responsible for the architecture of the eukaryotic chromosome. They determine where genes are located and how many copies of each exist. To understand how recombination shapes the genome, it is instructive to compare the effects of recombination in bacteria and eukaryotes.

Comparing Bacterial and Eukaryotic DNA Sequences

Bacterial genomes are relatively simple, containing genes that almost always occur as single copies. Unequal crossing over between repeated transposition elements in their circular DNA molecules tends to *delete* material, fostering the maintenance of a minimum genome size (figure 17.32a). For this reason, these genomes are very tightly packed, with few or no noncoding nucleotides. Recall the efficient use of space in the organization of the *lac* genes described in chapter 16.

In eukaryotes, by contrast, the introduction of *pairs* of homologous chromosomes (presumably because of their importance in repairing breaks in double-stranded DNA) has led to a radically different situation. Unequal crossing over between homologous chromosomes tends to promote the *duplication* of material rather than its reduction (figure 17.32b). Consequently, eukaryotic genomes have been in a constant state of flux during the course of their evolution. Multiple copies of genes have evolved, some of them subsequently diverging in sequence to become different genes, which in turn have duplicated and diverged.

Six different classes of eukaryotic DNA sequences are commonly recognized, based on the number of copies of each (table 17.6).

Transposons

Transposons exist in multiple copies scattered about the genome. In *Drosophila*, for example, more than 30 different transposons are known, most of them present at 20 to 40 different sites throughout the genome. In all, the known transposons of *Drosophila* account for perhaps 5% of its DNA. Mammalian genomes contain fewer kinds of transposons than the genomes of many other organisms, although the transposons in mammals are repeated more often. The family of human transposons called *ALU* elements, for example, typically occurs about 300,000 times in each cell. Transposons are transcribed but appear to play

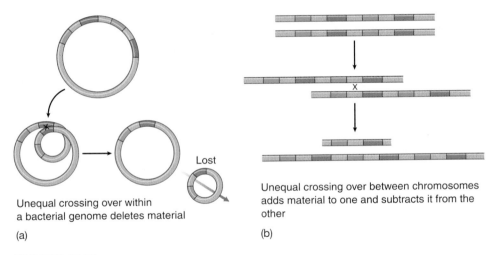

Unequal crossing over within a bacterial genome deletes material

(a)

Unequal crossing over between chromosomes adds material to one and subtracts it from the other

(b)

FIGURE 17.32
Unequal crossing over has different consequences in bacteria and eukaryotes.
(a) Bacteria have a circular DNA molecule, and a crossover between duplicate regions within the molecule deletes the intervening material. (b) In eukaryotes, with two versions of each chromosome, crossing over adds material to one chromosome; thus, gene amplification occurs in that chromosome.

Table 17.6 Classes of DNA Sequences Found in Eukaryotes	
Class	**Description**
Transposons	Thousands of copies scattered around the genome
Tandem clusters	Clusters containing hundreds of nearly identical copies of a gene
Multigene families	Clusters of a few to several hundred copies of related but distinctly different genes
Satellite DNA	Short sequences present in millions of copies per genome
Dispersed pseudogenes	Inactive members of a multigene family separated from other members of the family
Single-copy genes	Genes that exist in only one copy in the genome

no functional role in the life of the cell. As noted earlier in this chapter, many transposition events carry transposons into the exon portions of genes, disrupting the function of the protein specified by the gene transcript. These insertional inactivations are thought to be responsible for many naturally occurring mutations.

Tandem Clusters

A second class consists of DNA sequences that are repeated many times, one copy following another in tandem array. By transcribing all of the copies in these **tandem clusters** simultaneously, a cell can rapidly obtain large amounts of the product they encode. For example, the genes encoding rRNA are present in several hundred copies in most eukaryotic cells. Because these clusters are active sites of rRNA synthesis, they are readily visible in cytological preparations, where they are called **nucleolar organizer regions.** When transcription of the rRNA gene clusters ceases during cell division, the nucleolus disappears from view under the microscope, but it reappears when transcription begins again.

The genes present in a tandem cluster are very similar in sequence but not always identical; some may differ by one or a few nucleotides. Each gene in the cluster is separated from its neighboring copies by a short "spacer" sequence that is not transcribed. Unlike the genes, the spacers in a cluster vary considerably in sequence and in length.

Multigene Families

As we have learned more about the nucleotide sequences of eukaryotic genomes, it has become apparent that many genes exist as parts of **multigene families,** groups of related but distinctly different genes that often occur together in a cluster. Multigene families differ from tandem clusters in that they contain far fewer genes (from three to several hundred), and those genes differ much more from one another than the genes in tandem clusters. Despite their differences, the genes in a multigene family are clearly related in their sequences, making it likely that they arose from a single ancestral sequence through a series of unequal crossing over events. For example, studies of the evolution of the hemoglobin multigene family indicate that the ancestral globin gene is at least 800 million years old. By the time modern fishes evolved, this ancestral gene had already duplicated, forming the α and β forms (figure 17.33). Later, after the evolutionary divergence of amphibians and reptiles, these two globin gene forms moved apart on the chromosome; the mechanism of this movement is not known, but it may have involved transposition. In mammals, two more waves of duplication occurred to produce the array of 11 globin genes found in the human genome. Three of these genes are silent, encoding nonfunctional proteins. Other genes are expressed only during embryonic (ζ and ϵ) or fetal (γ) development. Only four (δ, β, α_1, and α_2) encode the polypeptides that make up adult human hemoglobin.

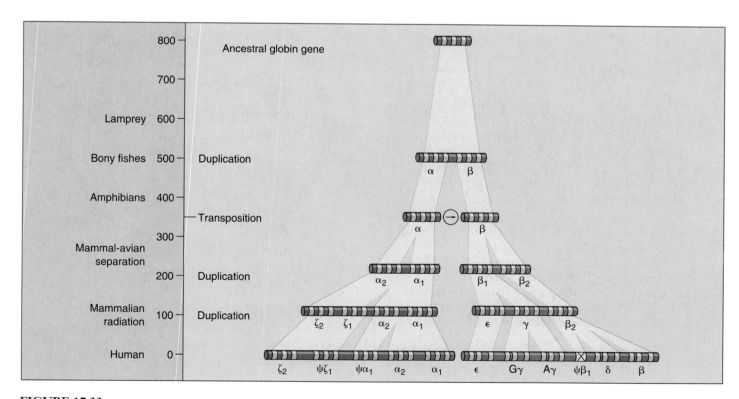

FIGURE 17.33
The family of globin genes. One ancestral form eventually diverged into the 11 forms found in the human genome.

Satellite DNA

Some short nucleotide sequences are repeated several million times in eukaryotic genomes. These sequences are collectively called **satellite DNA** and occur outside the main body of DNA. Almost all satellite DNA is either clustered around the centromere or located near the ends of the chromosomes, at the telomeres (figure 17.34). These regions of the chromosomes remain highly condensed, tightly coiled, and untranscribed throughout the cell cycle; this suggests that satellite DNA may serve some sort of structural function, such as initiating the pairing of homologous chromosomes in meiosis. About 4% of the human genome consists of satellite DNA.

Dispersed Pseudogenes

Silent copies of a gene, inactivated by mutation, are called **pseudogenes.** Such mutations may affect the gene's promoter (see chapter 16), shift the reading frame of the gene, or produce a small deletion. While some pseudogenes occur within a multigene family cluster, others are widely separated. The latter are called **dispersed pseudogenes** because they are believed to have been dispersed from their original position within a multigene family cluster. No one suspected the existence of dispersed pseudogenes until a few years ago, but they are now thought to be of major evolutionary significance in eukaryotes.

Single-Copy Genes

Ever since eukaryotes appeared, processes such as unequal crossing over between different copies of transposons have repeatedly caused segments of chromosomes to duplicate, and it appears that no portion of the genome has escaped this phenomenon. The duplication of genes, followed by the conversion of some of the copies into pseudogenes, has probably been the major source of "new" genes during the evolution of eukaryotes. As pseudogenes accumulate mutational changes, a fortuitous combination of changes may eventually result in an active gene encoding a protein with different properties. When that new gene first arises, it is a **single-copy gene,** but in time it, too, will be duplicated. Thus, a single-copy gene is but one stage in the cycle of duplication and divergence that has characterized the evolution of the eukaryotic genome (figure 17.35).

Gene sequences in eukaryotes vary greatly in copy number, some occurring many thousands of times, others only once. Many protein-encoding eukaryotic genes occur in several nonidentical copies, some of them not transcribed.

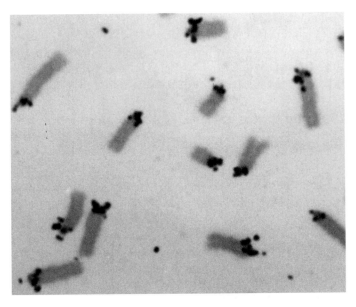

FIGURE 17.34
Satellite DNA. These mouse chromsomes were exposed to a solution containing radioactive RNA complementary to mouse satellite DNA. The labeled RNA (*dark spots*) has bound to its complementary sequences on the DNA, showing that the satellite sequences are localized near the ends of chromosomes.

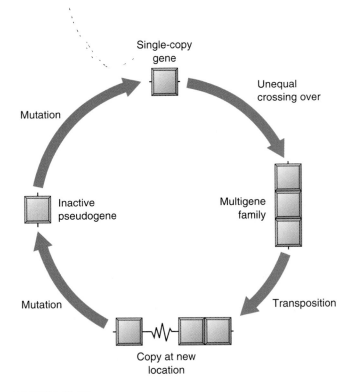

FIGURE 17.35
The cycle of gene duplication and divergence in eukaryotes.
Copies of genes continually duplicate and diverge into new genes.

Summary of Concepts

17.1 Mutations are changes in the genetic message.

- A mutation is any change in the hereditary message.
- Mutations that change one or a few nucleotides are called point mutations. They may arise as a result of damage from ionizing or ultraviolet radiation, chemical mutagens, or errors in pairing during DNA replication.
- Slipped mispairing, in which nonhomologous sequences on homologous chromosomes pair up during meiosis, can lead to deletions. When a deletion interrupts a codon, the result is a frame-shift mutation, in which the downstream portion of the gene is transcribed out of register.

17.2 Cancer results from mutation of growth-regulating genes.

- Cancer is a disease in which the regulatory controls that normally restrain cell division are disrupted.
- A variety of environmental factors, including ionizing radiation, chemical mutagens, and viruses, have been implicated in causing cancer.
- The testing of DNA fragments from tumor cells for their ability to induce cancer in other cells has led to the identification of a number of cancer-causing genes. Each of these genes has a function in normal cell proliferation; when modified by mutation, they escape the cell's normal regulatory mechanisms and lead to uncontrolled proliferation.
- The best way to avoid getting cancer is to avoid exposure to mutagens, especially those in cigarette smoke.

17.3 Recombination alters gene location.

- Recombination is the creation of new gene combinations. It includes changes in the position of genes or fragments of genes as well as the exchange of entire chromosomes during meiosis.
- Genes may be transferred between bacteria when they are included within small circles of DNA called plasmids. Plasmids are excised from or inserted into a bacterial cell's main genome by a process called reciprocal exchange.
- Transposition is the random movement of genes within transposons to new locations in the genome. It is responsible for many naturally occurring mutations, as the insertion of a transposon into a gene often inactivates the gene.
- Crossing over involves a physical exchange of genetic material between homologous chromosomes during the close pairing that occurs in meiosis. It may produce chromosomes that have different combinations of alleles.

17.4 Genomes are continually evolving.

- Satellite sequences are short sequences of nucleotides repeated millions of times. They are commonly associated with nontranscribed regions of chromosomes and probably play a structural role.
- Tandem clusters are genes that occur in thousands of copies grouped together at one or a few sites on a chromosome. These genes encode products that are required by the cell in large amounts.
- Multigene families consist of tens of copies of genes clustered at one site on a chromosome. The genes in a multigene family diverge in sequence more than the genes in a tandem cluster.

Discussing Key Terms

1. **Mutation** Any alteration of the content of a genetic message is a mutation. The changing of just one or a few nucleotides (as a result of replication error, chemical alteration, or physical damage from radiation) is called a point mutation.

2. **Slipped mispairing** Because some DNA sequences exist in multiple copies, homologous chromosomes occasionally misalign during synapses, causing some of the DNA to be looped out and creating deletions and frame-shift mutations.

3. **Recombination** Any process that alters the location of a gene within the genome is a form of recombination. The combining and mixing of nucleotide sequences from two separate genomes in meiosis contributes to variation in eukaryotic genomes.

4. **Crossing over** During early prophase of meiosis, homologous chromosomes line up and pair in close register. Strand exchange at this time can lead to the exchange of chromosomal arms, an important source of recombination among eukaryotes.

5. **Trinucleotide repeats** When trinucleotides in genes become repeated hundreds of times, points of instability result that can lead to hereditary disorders.

6. **Transposons** In eukaryotic genomes, genes may move randomly from one location to another within transposons. Insertion of a transposon into a gene can be a significant source of mutation.

7. **Multigene families** Related but distinct eukaryotic genes that often occur together in a cluster and arose from a single ancestral sequence.

Review Questions

1. What is the relationship between ionizing radiation and free radicals? Why are free radicals so damaging to cells?

2. What are pyrimidine dimers? How do they form? How are they repaired? What may happen if they are not repaired?

3. What is slipped mispairing? Explain how it can cause deletions and frame-shift mutations.

4. How do the consequences of mutations differ in germ-line cells and somatic cells ?

5. What is transfection? What has it revealed about the genetic basis of cancer?

6. About how many genes can be mutated to cause cancer? Why do most cancers require mutations in multiple genes?

7. What is genetic recombination? What mechanisms produce it? Which of these mechanisms occurs in prokaryotes, and which occurs in eukaryotes?

8. What is a plasmid? What is a transposon? How are plasmids and transposons similar, and how are they different?

9. What are mismatched pairs? How are they corrected? What effect does this correction have on the genetic message?

10. What is unequal crossing over? Under what circumstances is it most likely to occur? What is its effect on homologous chromosomes?

11. What kinds of genes exist in multigene families? How are these families thought to have evolved?

12. What are pseudogenes? How might they have been involved in the evolution of single-copy genes?

Thought Questions

1. In a colony of mice maintained for medical research, a hairless mouse is born. What evidence would you need to accept that this variant represents a genetic mutation?

2. While most bacterial genes exist as single copies, many eukaryotic genes exist multiple copies. Therefore, while a newly arisen mutation usually alters the phentype of a bacterium, it alters only one member of a multigene family in a eukaryote. Does this mean that mutations in eukaryotes do not usually alter the phenotype? If so, how does evolution occur in eukaryotes?

3. Analysis of the structure of dispersed pseudogenes suggests that RNA transcripts, after being processed in the cytoplasm, may gain reentry into the genome as DNA copies of the transcripts. Since many processes within the cytoplasm alter RNA molecules, does this passage of RNA sequences (via DNA copies) back into the genome imply that the cytoplasmic condition may modify the genetic message?

Internet Links

Primer on Molecular Genetics
http://www.bis.med.jhmi.edu/Dan/DOE/intro.html
This Department of Energy site provides a basic primer of the concepts of molecular genetics, complete with a very useful hypertext glossary.

Access Gene Technology
http://www.gene.com/ae/AE/AEPC/WWC/1994/
Genentech's superb ACCESS EXCELLENCE site provides a wealth of activities and resouces. The section on biotechnology is particularly fine, with many real examples of current research.

For Further Reading

Bartecchi, C. E., and others: "The Global Tobacco Epidemic," *Scientific American*, May 1995, pages 44–51. Cigarette smoking has stopped declining in the United States and is rising elsewhere in the world.

"Cancer," *Scientific American*, vol. 273, September 1996. An entire issue devoted to reviewing the current status of the war against cancer.

Cavenee, W. K., and R. L. White: "The Genetic Basis of Cancer," *Scientific American*, March 1995, pages 72–79. An accumulation of gene mutations causes normal cells to become cancerous.

Federoff, N.: "Transposable Genetic Elements in Maize," *Scientific American*, June 1984, pages 85–98. An excellent introduction to the studies for which Barbara McClintock was awarded the 1983 Nobel Prize in Physiology or Medicine, rephrased in molecular terms that were not available when she made her important discoveries.

Greider, C., and E. Blackburn: "Telomeres, Telomerase, and Cancer," *Scientific American*, vol. 274, February 1996, pages 92–97. An unusual enzyme called telomerase counteracts the telomere shortening associated with cessation of cell division.

Hartwell, L., and M. Kastan: "Cell Cycle Control and Cancer," *Science*, December 1994, pages 1721–28. Recent advances in our understanding of the cell cycle reveal how mutations can alter control of the cycle, leading to unrestrained growth.

Liotta, L.: "Cancer Cell Invasion and Metastasis," *Scientific American*, February 1992, pages 54–63. An up-to-date account of what is known about what makes a tumor cell metastatic.

Miki, Y., and others: "A Strong Candidate for the Breast and Ovarian Cancer Susceptibility Gene *BRCA1*," *Science*, October 1994, pages 66–71. A landmark research report that represents the first major success in solving the mystery of breast cancer.

Varmus, H., and R. A. Weinberg: *Genes and the Biology of Cancer*, Scientific American Library, New York, 1993. Two famous cancer researchers present a first-hand view of cancer. Highly recommended.

Weinberg, R.: *Racing to the Beginning of the Road—The Search for the Origin of Cancer*. Harmony Books, New York, 1996. An extremely well-written account of the history of cancer research, by a key investigator. Highly recommended.

18

Gene Technology

Concept Outline

18.1 **The ability to manipulate DNA has led to a new genetics.**

Restriction Endonucleases. Enzymes that cleave DNA at specific sites allow DNA segments from different sources to be spliced together.

Using Restriction Endonucleases to Manipulate Genes. Fragments produced by cleaving DNA with restriction endonucleases can be spliced into plasmids, which can be used to insert the DNA into host cells.

18.2 **Genetic engineering involves easily understood procedures.**

The Four Stages of a Genetic Engineering Experiment. Gene engineers cut DNA into fragments that they splice into vectors that carry the fragments into cells.

Working with Gene Clones. Gene technology is used in a variety of procedures involving DNA manipulation.

18.3 **Biotechnology is producing a scientific revolution.**

DNA Sequence Technology. The complete nucleotide sequence of the genomes of many organisms are now known. The unique DNA of every individual can be used to identify sperm, blood, or other tissues.

Medical Applications. Many drugs and vaccines are now produced with gene technology.

Agricultural Applications. Gene engineers have developed crops resistant to pesticides and pests, as well as superior animals.

Cloning. Recent experiments show it is possible to clone agricultural animals, a result with many implications for both agriculture and society.

Ethics and Regulation. Genetic engineering raises important questions about danger and privacy.

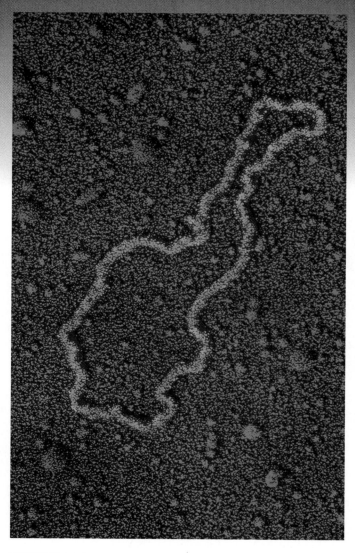

FIGURE 18.1
A famous plasmid. The circular molecule in this electron micrograph is pSC101, the first plasmid used successfully to clone a vertebrate gene. Its name comes from the fact that it was the one-hundred-and-first plasmid isolated by Stanley Cohen.

Over the past decade, the development of new and powerful techniques for studying and manipulating DNA has revolutionized genetics (figure 18.1). These techniques have allowed biologists to intervene directly in the genetic fate of organisms for the first time. In this chapter, we will explore these technologies and consider how they apply to specific problems of great practical importance. Few areas of biology will have as great an impact on our future lives.

Restriction Endonucleases

In 1980, geneticists used the relatively new technique of gene splicing, which we will describe in this chapter, to introduce the human gene that encodes **interferon** into a bacterial cell's genome. Interferon is a rare blood protein that increases human resistance to viral infection, and medical scientists have been interested in its possible usefulness in cancer therapy. This possibility was difficult to investigate before 1980, however, because purification of the large amounts of interferon required for clinical testing would have been prohibitively expensive, given interferon's scarcity in the blood. An inexpensive way to produce interferon was needed, and introducing the gene responsible for its production into a bacterial cell made that possible. The cell that had acquired the human interferon gene proceeded to produce interferon at a rapid rate, and to grow and divide. Soon there were millions of interferon-producing bacteria in the culture, all of them descendants of the cell that had originally received the human interferon gene.

The Advent of Genetic Engineering

This procedure of producing a line of genetically identical cells from a single altered cell, called **cloning,** made every cell in the culture a miniature factory for producing interferon. The human insulin gene has also been cloned in bacteria, and now large amounts of insulin, a hormone essential for treating some forms of diabetes, can be manufactured at relatively little expense. Beyond these clinical applications, cloning and related molecular techniques are used to obtain basic information about how genes are put together and regulated. The interferon experiment and others like it marked the beginning of a new genetics, **genetic engineering.**

The essence of genetic engineering is the ability to cut DNA into recognizable pieces and rearrange those pieces in different ways. In the interferon experiment, a piece of DNA carrying the interferon gene was inserted into a plasmid, which then carried the gene into a bacterial cell. Most other genetic engineering approaches have used the same general strategy, bringing the gene of interest into the target cell by first incorporating it into a plasmid or an infective virus. To make these experiments work, one must be able to cut the source DNA (human DNA in the interferon experiment, for example) and the plasmid DNA in such a way that the desired fragment of source DNA can be spliced permanently into the plasmid. This cutting is performed by enzymes that recognize and cleave specific sequences of nucleotides in DNA. These enzymes are the basic tools of genetic engineering.

Discovery of Restriction Endonucleases

Scientific discoveries often have their origins in seemingly unimportant observations that receive little attention by researchers before their general significance is appreciated. In the case of genetic engineering, the original observation was that bacteria use enzymes to defend themselves against viruses.

Most organisms eventually evolve means of defending themselves from predators and parasites, and bacteria are no exception. Among the natural enemies of bacteria are bacteriophages, viruses that infect bacteria and multiply within them. At some point, they cause the bacterial cells to burst, releasing thousands more viruses. Through natural selection, some types of bacteria have acquired powerful weapons against these viruses: they contain enzymes called **restriction endonucleases** that fragment the viral DNA as soon as it enters the bacterial cell. Many restriction endonucleases recognize specific nucleotide sequences in a DNA strand, bind to the DNA at those sequences, and cleave the DNA at a particular place within the recognition sequence.

Why don't restriction endonucleases cleave the bacterial cells' own DNA as well as that of the viruses? The answer to this question is that bacteria modify their own DNA, using other enzymes known as **methylases** to add methyl ($-CH_3$) groups to some of the nucleotides in the bacterial DNA. When nucleotides within a restriction endonuclease's recognition sequence have been methylated, the endonuclease cannot bind to that sequence. Consequently, the bacterial DNA is protected from being degraded at that site. Viral DNA, on the other hand, has not been methylated and therefore is not protected from enzymatic cleavage.

How Restriction Endonucleases Cut DNA

The sequences recognized by restriction endonucleases are typically four to six nucleotides long, and they are often palindromes. This means the nucleotides at one end of the recognition sequence are complementary to those at the other end, so that the two strands of the DNA duplex have the same nucleotide sequence running in opposite directions for the length of the recognition sequence. Two important consequences arise from this arrangement of nucleotides.

First, because the same recognition sequence occurs on both strands of the DNA duplex, the restriction endonuclease can bind to and cleave both strands, effectively cutting the DNA in half. This ability to cut across both strands is almost certainly the reason that restriction endonucleases have evolved to recognize nucleotide sequences with twofold rotational symmetry.

Second, because the bond cleaved by a restriction endonuclease is typically not positioned in the center of the recognition sequence to which it binds, and because the DNA strands are antiparallel, the cut sites for the two strands of a duplex are offset from each other (figure 18.2). After cleavage, each DNA fragment has a single-stranded end a few nucleotides long. The single-stranded ends of the two fragments are complementary to each other.

Some restriction endonucleases cleave the center of a four- or six-nucleotide sequence, producing fragments without single-stranded ends. Such fragments do not spontaneously reassociate, so the endonucleases that produce them are often used in genetic engineering procedures where it is important to prevent spontaneous reassociation.

Why Restriction Endonucleases Are So Useful

There are hundreds of bacterial restriction endonucleases, and each one has a specific recognition sequence. By chance, a particular endonuclease's recognition sequence is likely to occur somewhere in any given sample of DNA; the shorter the sequence, the more often it will arise by chance within a sample. Therefore, a given restriction endonuclease can probably cut DNA from any source into fragments. Each fragment will have complementary single-stranded ends characteristic of that endonuclease. Because of their complementarity, these single-stranded ends can pair with each other (consequently, they are sometimes called "sticky ends"). Once their ends have paired, two fragments can then be joined together with the aid of the enzyme **DNA ligase,** which re-forms the phosphodiester bonds of DNA. What makes restriction endonucleases so valuable for genetic engineering is the fact that *any* two fragments produced by the same restriction endonuclease can be joined together. Fragments of elephant and ostrich DNA cleaved by the same endonuclease can be joined to one another as readily as two bacterial DNA fragments.

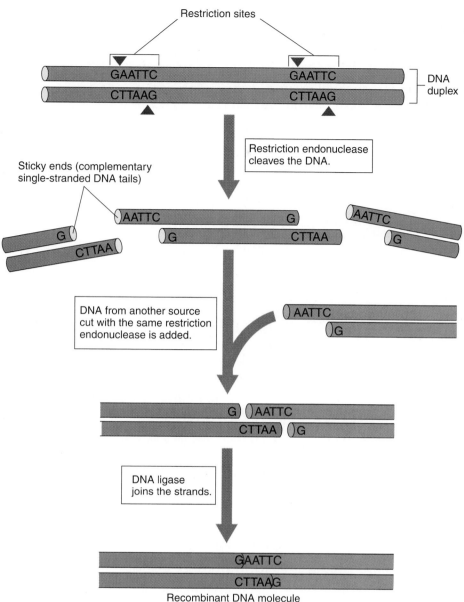

FIGURE 18.2
Many restriction endonucleases produce DNA fragments with "sticky ends." The restriction endonuclease *Eco*RI always cleaves the sequence GAATTC between G and A. Because the same sequence occurs on both strands, both are cut. However, the two sequences run in opposite directions on the two strands. As a result, single-stranded tails are produced that are complementary to each other, or "sticky."

Genetic engineering involves manipulating specific genes by cutting and rearranging DNA. A restriction endonuclease cleaves DNA at a specific site, generating in most cases two fragments with short single-stranded ends. Because these ends are complementary to each other, any pair of fragments produced by the same endonuclease, from any DNA source, can be joined together.

Using Restriction Endonucleases to Manipulate Genes

A **chimera** is a mythical creature with the head of a lion, body of a goat, and tail of a serpent. Although no such creatures existed in nature, biologists have made chimeras of a more modest kind through genetic engineering.

Constructing pSC101

One of the first chimeras was manufactured from a bacterial plasmid called a resistance transfer factor by American geneticists Stanley Cohen and Herbert Boyer in 1973. Cohen and Boyer used a restriction endonuclease (called *Escherichia coli* restriction endonuclease I, or *Eco*RI) to cut the plasmid into fragments. One fragment, 9000 nucleotides in length, contained both the origin of replication necessary for replicating the plasmid and a gene that conferred resistance to the antibiotic tetracycline (*tet^r*). Because both ends of this fragment were cut by the same restriction endonuclease, they could be ligated to form a circle, a smaller plasmid Cohen dubbed pSC101 (figure 18.3).

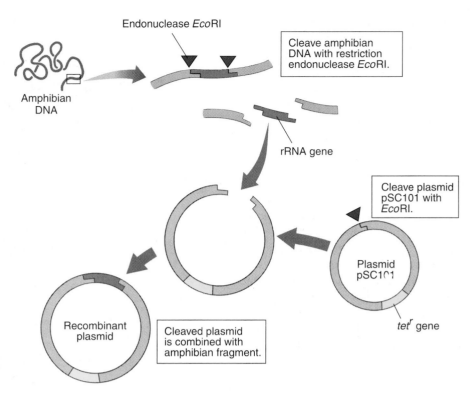

FIGURE 18.3

One of the first genetic engineering experiments. This diagram illustrates how Cohen and Boyer inserted an amphibian gene encoding rRNA into pSC101. The plasmid contains a single site cleaved by the restriction endonuclease *Eco*RI; it also contains *tet^r*, a gene which confers resistance to the antibiotic tetracycline. The rRNA-encoding gene was inserted into pSC101 by cleaving the amphibian DNA and the plasmid with *Eco*RI and allowing the complementary sequences to pair.

Using pSC101 to Make Recombinant DNA

Cohen and Boyer also used *Eco*RI to cleave DNA that coded for rRNA that they had isolated from an adult amphibian, the African clawed toad, *Xenopus laevis*. They then mixed the fragments of *Xenopus* DNA with pSC101 plasmids that had been "reopened" by *Eco*RI and allowed bacterial cells to take up DNA from the mixture. Some of the bacterial cells immediately became resistant to tetracycline, indicating that they had incorporated the pSC101 plasmid with its antibiotic-resistance gene. Furthermore, some of these pSC101-containing bacteria also began to produce toad ribosomal RNA! Cohen and Boyer concluded that the toad rRNA gene must have been inserted into the pSC101 plasmids in those bacteria. In other words, the two ends of the pSC101 plasmid, produced by cleavage with *Eco*RI, had joined to the two ends of a toad DNA fragment that contained the rRNA gene, also cleaved with *Eco*RI.

The pSC101 plasmid containing the toad rRNA gene is a true chimera, an entirely new genome that never existed in nature and never would have evolved by natural means. It is a form of **recombinant DNA,** that is, DNA created in the laboratory by joining together pieces of different genomes to form a novel combination.

Other Vectors

The introduction of foreign DNA fragments into host cells has become common in molecular genetics. The genome that carries the foreign DNA into the host cell is called a **vector.** Newer-model plasmids, with names like pUC18, pGEM®, or pBluescript®, can be induced to make hundreds of copies of themselves and thus of the foreign genes they contain. Entry into bacterial cells can also be achieved by using a bacterial virus, such as λ virus, as a vector instead of a plasmid. Not all vectors have bacterial targets, however. Animal viruses, for example, have served as vectors to carry bacterial genes into monkey cells, and animal genes have even been introduced into plant cells.

One of the first recombinant genomes produced by genetic engineering was a bacterial plasmid into which an amphibian ribosomal RNA gene was inserted. Viruses can also be used as vectors to insert foreign DNA into host cells and create recombinant genomes.

Examples of Gene Manipulation

HERMAN THE WONDER BULL

GenPharm, a California biotechnology company, engineered Herman, a bull that possesses the gene for human lactoferrin (HLF). HLF confers antibacterial and iron transport properties to humans. Many of Herman's female offspring now produce milk containing HLF, and GenPharm intends to build a herd of transgenic cows for the large-scale commercial production of HLF.

WILT-PROOF FLOWERS

Ethylene, the plant hormone that causes fruit to ripen, also causes flowers to wilt. Researchers at Purdue have found the gene that makes flower petals respond to ethylene by wilting and replaced it with a gene insensitive to ethylene. The transgenic carnations they produced lasted for 3 weeks after cutting, while normal carnations last only 3 days.

WEEVIL-PROOF PEAS

Not only has gene technology afforded agriculture viral and pest control in the field, it has also provided a pest control technique for the storage bin. A team of U.S. and Australian scientists have engineered a gene that is expressed only in the seed of the pea plant. The enzyme inhibitor encoded by this gene inhibits feeding by weevils, one of the most notorious pests affecting stored crops. The worldwide ramifications are significant since up to 40% of stored grains are lost to pests.

SUPER SALMON!

Canadian fisheries scientists have inserted recombinant growth hormone genes into developing salmon embryos, creating the first transgenic salmon. Not only do these transgenic fish have shortened production cycles, they are, on an average, *11 times heavier* than nontransgenic salmon! The implications for the fisheries industry and for worldwide food production are obvious.

The Four Stages of a Genetic Engineering Experiment

Like the experiment of Cohen and Boyer, most genetic engineering experiments consist of four stages: DNA cleavage, production of recombinant DNA, cloning, and screening.

Stage 1: DNA Cleavage

A restriction endonuclease is used to cleave the source DNA into fragments. Because the endonuclease's recognition sequence is likely to occur many times within the source DNA, cleavage will produce a large number of different fragments.

A different set of fragments may be obtained by employing endonucleases that recognize different sequences. The fragments can be separated from one another according to their size by electrophoresis (figure 18.4).

Stage 2: Production of Recombinant DNA

The fragments of DNA are inserted into plasmids or viral DNA, which have been cleaved with the same restriction endonuclease as the source DNA.

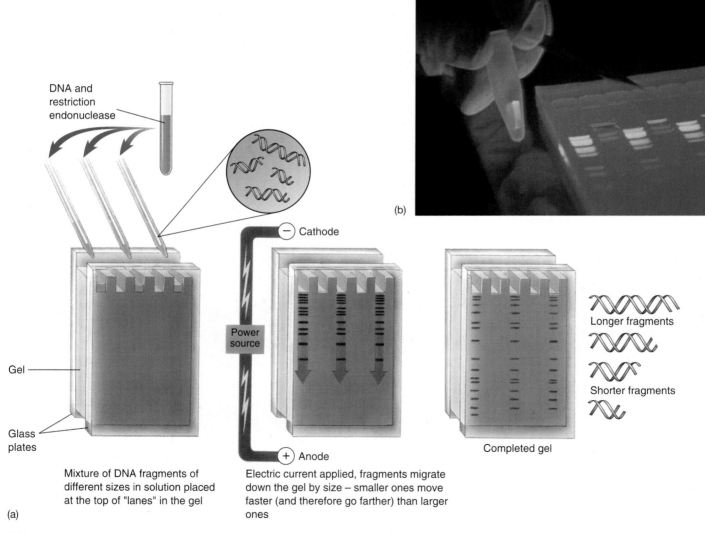

FIGURE 18.4
Stage 1: Using restriction endonucleases to cleave DNA and electrophoresis to resolve the fragment. (a) After restriction endonucleases have cleaved the DNA, the fragments are loaded on a gel, and an electric current is applied. The DNA fragments migrate through the gel, with bigger ones moving more slowly. The fragments can be visualized easily, as the migrating bands fluoresce in UV light. (b) In the photograph, one band of DNA has been excised from the gel for further analysis and can be seen glowing in the tube the technician holds.

Stage 3: Cloning

The plasmids or viruses serve as vectors that can introduce the DNA fragments into cells—usually, but not always, bacteria (figure 18.5). As each cell reproduces, it forms a clone of cells that all contain the fragment-bearing vector. Each clone is maintained separately, and all of them together constitute a clone library of the original source DNA.

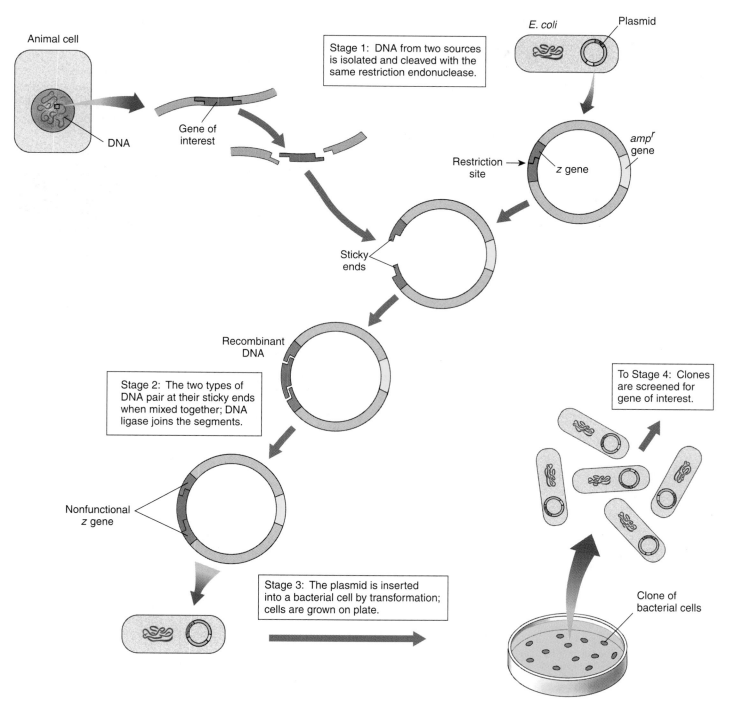

FIGURE 18.5

Stages 2 and 3: Using plasmids as vectors to clone the restriction fragments in bacteria. In stage 1, DNA containing the gene of interest (in this case, from an animal cell) and DNA from a plasmid are cleaved with the same restriction endonuclease. The genes *amp*^{r} and *z* are contained within the plasmid and used for screening a clone (stage 4). In stage 2, the two cleaved sources of DNA are mixed together and pair at their sticky ends. In stage 3, the recombinant DNA is inserted into a bacterial cell, which reproduces and forms clones. In Stage 4, the bacterial clones will be screened for the gene of interest.

Stage 4: Screening

The clones containing a specific DNA fragment of interest, often a fragment that includes a particular gene, are identified from the clone library. Let's examine this stage in more detail, as it is generally the most challenging in any genetic engineering experiment.

4–I: The Preliminary Screening of Clones. To make the screening of clones easier, investigators initially try to eliminate from the library any clones that do not contain vectors, as well as clones whose vectors do not contain fragments of the source DNA. The first category of clones can be eliminated by employing a vector with a gene that confers resistance to a specific antibiotic, such as tetracycline, penicillin, or ampicillin. In Figure 18.6a, the gene *amp^r* is incorporated into the plasmid and confers resistance to the antibiotic ampicillin. When the clones are exposed to a medium containing that antibiotic, only clones that contain the vector will be resistant to the antibiotic and able to grow.

One way to eliminate clones with vectors that do not have an inserted DNA fragment is to use a vector that, in addition to containing antibiotic resistance genes, contains the z gene which produces β-galactosidase, an enzyme that enables the cells to metabolize the sugar, X-gal. Metabolism of X-gal results in the formation of a blue reaction product, so any cells whose vectors contain a functional version of this gene will turn blue in the presence of X-gal (figure 18.6b). However, if one uses a restriction endonuclease whose recognition sequence lies within the z gene, the gene will be cleaved when recombinants are formed. If a fragment of the source DNA inserts into the vector at the cleavage site, the gene will be inactivated and the cell will be unable to metabolize X-gal. Therefore, cells with vectors that contain a fragment of source DNA should remain colorless in the presence of X-gal.

Any cells that are able to grow in a medium containing the antibiotic but don't turn blue in the medium with X-gal must have incorporated a vector with a fragment of source DNA. Identifying cells that have a *specific* fragment of the source DNA is the next step in screening clones.

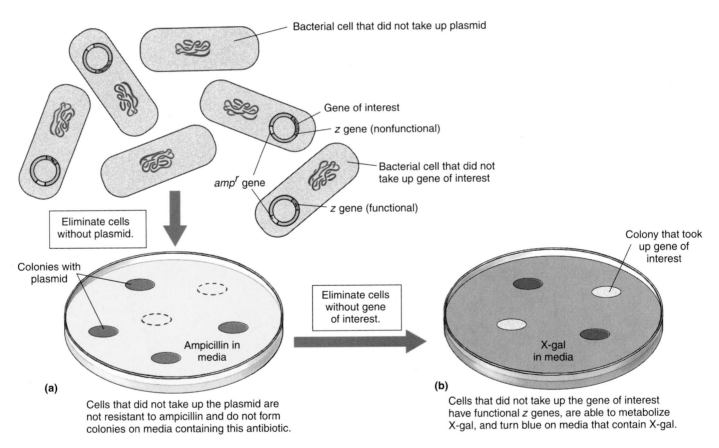

FIGURE 18.6
Stage 4-I: Using antibiotic resistance and X-gal as preliminary screens of restriction fragment clones. Bacteria are transformed with recombinant plasmids that contain a gene (*amp^r*) that confers resistance to the antibiotic ampicillin and a gene (z) that produces β-galactosidase, the enzyme which enables the cells to metabolize the sugar X-gal. (a) Only those bacteria that have incorporated a plasmid will be resistant to ampicillin and will grow on a medium that contains the antibiotic. (b) Ampicillin-resistant bacteria will be able to metabolize X-gal if their plasmid does *not* contain a DNA fragment inserted in the z gene; such bacteria will turn blue when grown on a medium containing X-gal. Bacteria with a plasmid that has a DNA fragment inserted within the z gene will not be able to metabolize X-gal and, therefore, will remain colorless in the presence of X-gal.

4–II: Finding the Gene of Interest.

A clone library may contain anywhere from a few dozen to many thousand individual fragments of source DNA. Many of those fragments will be identical, so to assemble a complete library of the entire source genome, several hundred thousand clones could be required. A complete *Drosophila* (fruit fly) library, for example, contains more than 40,000 different clones; a complete human library consisting of fragments 20 kilobases long would require close to a million clones. To search such an immense library for a clone that contains a fragment corresponding to a particular gene requires ingenuity, but many different approaches have been successful.

The most general procedure for screening clone libraries to find a particular gene is **hybridization** (figure 18.7). In this method, the cloned genes form base-pairs with complementary sequences on another nucleic acid. The complementary nucleic acid is called a **probe** because it is used to probe for the presence of the gene of interest. At least part of the nucleotide sequence of the gene of interest must be known to be able to construct the probe.

In this method of screening, bacterial colonies containing an inserted gene are grown on agar. Some cells are transferred to a filter pressed onto the colonies, forming a replica of the plate. The filter is then treated with a solution that denatures the bacterial DNA and that contains a radioactively labeled probe. The probe hybridizes with complementary single-stranded sequences on the bacterial DNA.

When the filter is laid over photographic film, areas that contain radioactivity will expose the film (autoradiography). Only colonies which contain the gene of interest hybridize with the radioactive probe and emit radioactivity onto the film. The pattern on the film is then compared to the original master plate, and the gene-containing colonies may be identified.

> Genetic engineering generally involves four stages: cleaving the source DNA; making recombinants; cloning copies of the recombinants; and screening the cloned copies for the desired gene. Screening can be achieved by making the desired clones resistant to certain antibiotics and giving them other properties that make them readily identifiable.

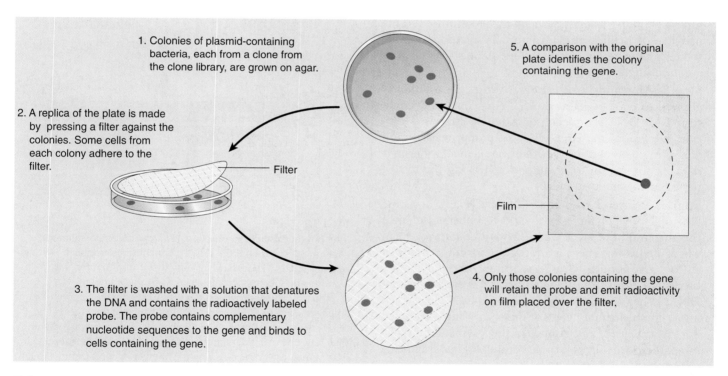

1. Colonies of plasmid-containing bacteria, each from a clone from the clone library, are grown on agar.

2. A replica of the plate is made by pressing a filter against the colonies. Some cells from each colony adhere to the filter.

Filter

3. The filter is washed with a solution that denatures the DNA and contains the radioactively labeled probe. The probe contains complementary nucleotide sequences to the gene and binds to cells containing the gene.

4. Only those colonies containing the gene will retain the probe and emit radioactivity on film placed over the filter.

5. A comparison with the original plate identifies the colony containing the gene.

Film

FIGURE 18.7

Stage 4-II: Using hybridization to identify the gene of interest. (1) Each of the colonies on these bacterial culture plates represents millions of clones descended from a single cell. To test whether a certain gene is present in any particular clone, it is necessary to identify colonies whose cells contain DNA that hybridizes with a probe containing DNA sequences complementary to the gene. (2) Pressing a filter against the master plate causes some cells from each colony to adhere to the filter. (3) The filter is then washed with a solution that denatures the DNA and contains the radioactively labeled probe. (4) Only those colonies that contain DNA that hybridizes with the probe, and thus contains the gene of interest, will expose film in autoradiography. (5) The film is then compared to the master plate to identify the gene-containing colony.

Working with Gene Clones

Once a gene has been successfully cloned, a variety of procedures are available to characterize it.

Getting Enough DNA to Work with: The Polymerase Chain Reaction

Once a particular gene is identified within the library of DNA fragments, the final requirement is to make multiple copies of it. One way to do this is to insert the identified fragment into a bacterium; after repeated cell divisions, millions of cells will contain copies of the fragment. A far more direct approach, however, is to use DNA polymerase to copy the gene sequence of interest through the **polymerase chain reaction (PCR;** figure 18.8). Kary Mullis developed PCR in 1983 while he was a staff chemist at the Cetus Corporation; in 1993, it won him the Nobel Prize for chemistry. PCR can amplify specific sequences or add sequences (such as endonuclease recognition sequences) as primers to cloned DNA. There are three steps in PCR:

Step 1: Denaturation. First, an excess of primer (typically a synthetic sequence of 20 to 30 nucleotides) is mixed with the DNA fragment to be amplified. This mixture of primer and fragment is heated to about 98° C. At this temperature, the DNA fragment dissociates into single strands.

Step 2: Annealing of Primers. Next, the solution is allowed to cool to about 60° C. As it cools, the single strands of DNA reassociate into double strands. However, because of the large excess of primer, each strand of the fragment base-pairs with a complementary primer flanking the region to be amplified, leaving the rest of the fragment single-stranded.

Step 3: Primer Extension. Now a very heat-stable type of DNA polymerase, called Taq polymerase (after the thermophilic bacterium *Thermus aquaticus,* from which Taq is extracted) is added, along with a supply of all four nucleotides. Using the primer, the polymerase copies the rest of the fragment as if it were replicating DNA. When it is done, the primer has been lengthened into a complementary copy of the entire single-stranded fragment. Because *both* strands of the fragment are replicated, there are now two copies of the original fragment.

Steps 1–3 are now repeated, and the two copies become four. It is not necessary to add any more polymerase, as the heating step does not harm this particular enzyme. Each heating and cooling cycle, which can be as short as 1 or 2 minutes, doubles the number of DNA molecules. After 20 cycles, a single fragment produces more than one million (2^{20}) copies! In a few hours, 100 billion copies of the fragment can be manufactured.

PCR, now fully automated, has revolutionized many aspects of science and medicine because it allows the investigation of minute samples of DNA. In criminal investiga-

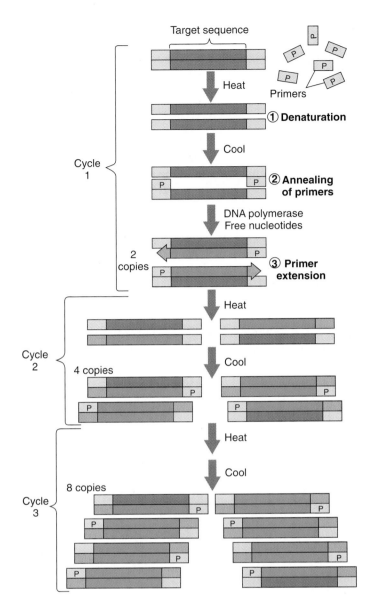

FIGURE 18.8

The polymerase chain reaction. (1) *Denaturation.* A solution containing primers and the DNA fragment to be amplified is heated so that the DNA dissociates into single strands. (2) *Annealing of primers.* The solution is cooled, and the primers bind to complementary sequences on the DNA flanking the region to be amplified. (3) *Primer extension.* DNA polymerase then copies the remainder of each strand, beginning at the primer. Steps 1–3 are then repeated with the replicated strands. This process is repeated many times, each time doubling the number of copies, until enough copies of the DNA fragment exist for analysis.

tions, "DNA fingerprints" are prepared from the cells in a tiny speck of dried blood or at the base of a single human hair. Physicians can detect genetic defects in very early embryos by collecting a few sloughed-off cells and amplifying their DNA. PCR could also be used to examine the DNA of historical figures such as Abraham Lincoln and of now-extinct species, as long as even a minuscule amount of their DNA remains intact.

Identifying DNA: Southern Blotting

Once a gene has been cloned, it may be used as a probe to identify the same or a similar gene in another sample (figure 18.9). In this procedure, called a **Southern blot,** DNA from the sample is cleaved into restriction fragments with a restriction endonuclease, and the fragments are spread apart by gel electrophoresis. The double-stranded helix of each DNA fragment is then denatured into single strands by making the pH of the gel basic, and the gel is "blotted" with a sheet of nitrocellulose, transferring some of the DNA strands to the sheet. Next, a probe consisting of purified, single-stranded DNA corresponding to a specific gene (or mRNA transcribed from that gene) is poured over the sheet. Any fragment that has a nucleotide sequence complementary to the probe's sequence will hybridize (base-pair) with the probe. If the probe has been labeled with ^{32}P, it will be radioactive, and the sheet will show a band of radioactivity where the probe hybridized with the complementary fragment.

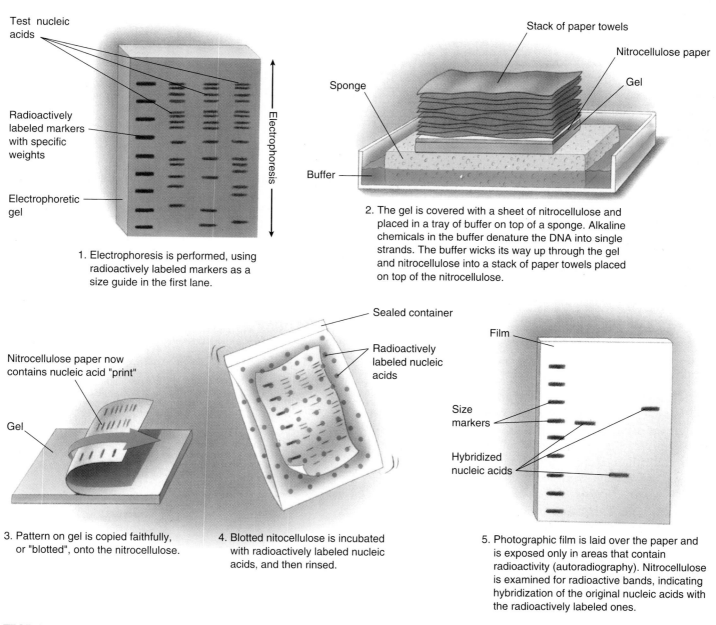

1. Electrophoresis is performed, using radioactively labeled markers as a size guide in the first lane.

2. The gel is covered with a sheet of nitrocellulose and placed in a tray of buffer on top of a sponge. Alkaline chemicals in the buffer denature the DNA into single strands. The buffer wicks its way up through the gel and nitrocellulose into a stack of paper towels placed on top of the nitrocellulose.

3. Pattern on gel is copied faithfully, or "blotted", onto the nitrocellulose.

4. Blotted nitocellulose is incubated with radioactively labeled nucleic acids, and then rinsed.

5. Photographic film is laid over the paper and is exposed only in areas that contain radioactivity (autoradiography). Nitrocellulose is examined for radioactive bands, indicating hybridization of the original nucleic acids with the radioactively labeled ones.

FIGURE 18.9

The Southern blot procedure. E. M. Southern developed this procedure in 1975 to enable DNA fragments of interest to be visualized in a complex sample containing many other fragments of similar size. The DNA is separated on a gel, then transferred ("blotted") onto a solid support medium such as nitrocellulose paper or a nylon membrane. It is then incubated with a radioactive single-strand copy of the gene of interest, which hybridizes to the blot at the location(s) where there is a fragment with a complementary sequence. The positions of radioactive bands on the blot identify the fragments of interest.

Distinguishing Differences in DNA: RFLP Analysis

Often a researcher wishes not to find a specific gene, but rather to identify a particular *individual* using a specific gene as a marker. One powerful way to do this is to analyze **restriction fragment length polymorphisms,** or **RFLPs** (figure 18.10). Point mutations, sequence repetitions, and transposons (chapter 17) that occur within or between the restriction endonuclease recognition sites will alter the length of the DNA fragments (restriction fragments) the restriction endonucleases produce. DNA from different individuals rarely has exactly the same array of restriction sites and distances between sites, so the population is said to be polymorphic (having many forms) for their restriction fragment patterns. By cutting a DNA sample with a particular restriction endonuclease, separating the fragments according to length on an electrophoretic gel, and then using a radioactive probe to identify the fragments on the gel, one can obtain a pattern of bands often unique for each strand of DNA analyzed. These "DNA fingerprints" are used in forensic analysis during criminal investigations. RFLPs are also useful as markers to identify particular groups of people at risk for some genetic disorders.

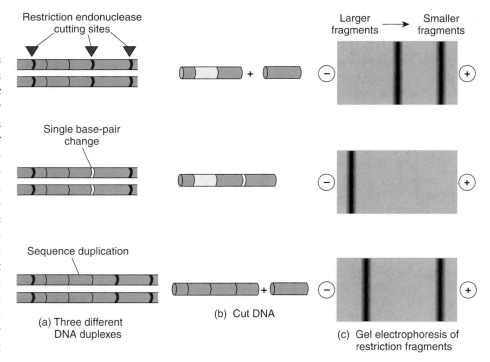

FIGURE 18.10

Restriction fragment length polymorphism (RFLP) analysis. (a) Three strands of DNA differ in their restriction sites due to a single base-pair substitution in one case and a sequence duplication in another case. (b) When the strands are cut with a restriction endonuclease, different numbers and sizes of fragments are produced. (c) Gel electrophoresis separates the fragments, and different banding patterns result.

Making an Intron-Free Copy of a Eukaryotic Gene

Recall from chapter 15 that eukaryotic genes are encoded in exons separated by numerous nontranslated introns. When the gene is transcribed to produce the primary transcript, the introns are cut out during RNA processing to produce the mature mRNA transcript. When transferring eukaryotic genes into bacteria, it is desirable to transfer DNA already processed this way, instead of the raw eukaryotic DNA, because bacteria lack the enzymes to carry out the processing. To do this, genetic engineers first isolate from the cytoplasm the mature mRNA corresponding to a particular gene. They then use an enzyme called reverse transcriptase to make a DNA version of the mature mRNA transcript (figure 18.11). The single strand of DNA can then serve as a template for the synthesis of a complementary strand. In this way, one can produce a double-stranded molecule of DNA that contains a gene lacking introns. This molecule is called **complementary DNA,** or **cDNA.**

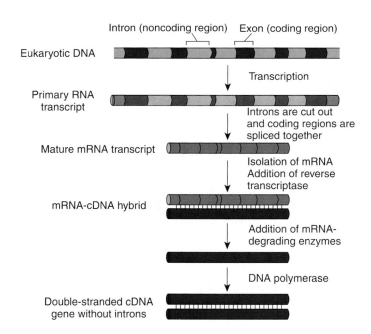

FIGURE 18.11

The formation of cDNA. A mature mRNA transcript is isolated from the cytoplasm of a cell. The enzyme reverse transcriptase is then used to make a DNA strand complementary to the processed mRNA. That newly made strand of DNA is the template for the enzyme DNA polymerase, which assembles a complementary DNA strand along it, producing cDNA, a double-stranded DNA version of the intron-free mRNA.

Sequencing DNA: The Sanger Method

Most DNA sequencing is currently done using the "chain termination" technique developed initially by Frederick Sanger, for which he earned his *second* Nobel Prize (figure 18.12). (1) A short single-stranded primer is added to the end of a single-stranded DNA fragment of unknown sequence. The primer provides a 3′ end for DNA polymerase. (2) The primed fragment is added, along with DNA polymerase and a supply of all four deoxynucleotides (d-nucleotides), to four synthesis tubes. Each contains a different *di*deoxynucleotide (dd-nucleotide); such nucleotides lack both the 2′ and the 3′ —OH groups and are thus chain-terminating. The first tube, for example, contains ddATP and stops synthesis whenever ddA is incorporated into DNA instead of dATP. Because of the relatively low concentration of ddATP compared to dATP, ddA will not necessarily be added to the first A site; this tube will contain a series of fragments of different lengths, corresponding to the different distances the polymerase traveled from the primer before a ddA was incorporated. (3) These fragments can be separated according to size by electrophoresis. (4) A radioactive label (here dATP*) allows the fragments to be visualized on X-ray film, and the newly made sequence can be read directly from the film. Try it. (5) The original fragment has the complementary sequence.

Techniques such as Southern blotting and PCR enable investigators to identify specific genes and produce them in large quantities, while RFLP analysis and the Sanger method identify individuals and unknown gene sequences.

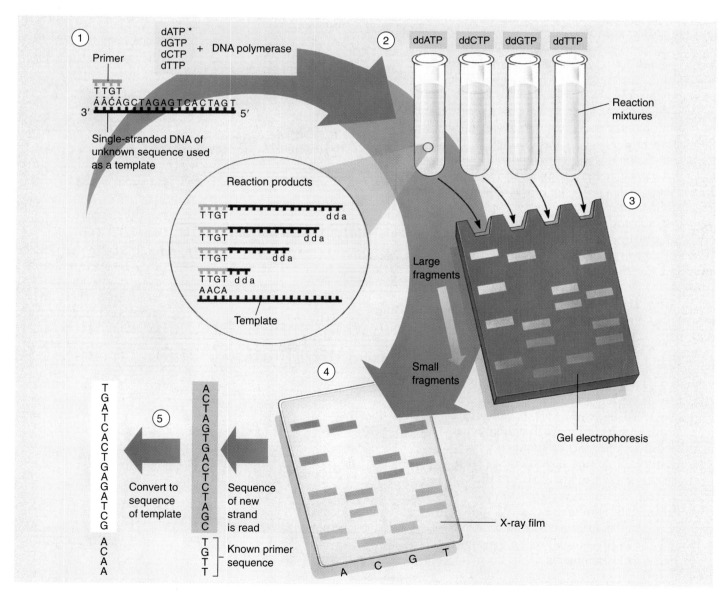

FIGURE 18.12
The Sanger dideoxynucleotide sequencing method.

DNA Sequence Technology

The 1980s saw an explosion of interest in **biotechnology,** the application of genetic engineering to practical human problems. Let us examine some of the major areas where these techniques have been put to use.

Genome Sequencing

The international scientific community has mounted a major effort to sequence the entire human genome, one restriction fragment at a time. Because the human genome contains some 3000 Mb (million nucleotide base-pairs), this task presents no small challenge.

However, genetic engineering techniques are enabling us to learn a great deal more about the human genome. Several clonal libraries of the human genome have been assembled, using large-size restriction fragments. Any cloned gene can now be localized to a specific chromosomal site by using radioactive probes to detect *in situ* hybridization (that is, binding between the probe and a complementary sequence on the chromosome). Genes are now being mapped at an astonishing rate: genes that contribute to dyslexia, obesity, and cholesterol-proof blood are some of the important ones that were mapped in 1994 and 1995 alone! With an understanding of where specific genes are located in the human genome and how they work, it is not difficult to imagine a future in which virtually any genetic disease could be treated or perhaps even cured with **gene therapy.** As we mentioned in chapter 13, some success has already been reported in treating patients who have cystic fibrosis with a genetically corrected version of the cystic fibrosis gene.

An exciting scientific by-product of the human genome project has been the complete genome sequencing of many microorganisms with smaller genomes, on the order of a few Mb (table 18.1). In general, about half of the genes prove to have a known function; what the other half of the genes are doing is a complete mystery. Only one eukaryotic genome has been sequenced in its entirety, that of brewer's yeast *Saccharomyces cerevisiae;* many of its approximately 6000 genes have a similar structure to some human genes. The complete sequences of many much larger genomes are expected to be completed soon, including the malarial *Plasmodium* parasite (30 Mb), the nematode (100 Mb), the plant *Arabidopsis* (100 Mb) (figure 18.13), the fruit fly *Drosophila* (120 Mb), and the mouse (300 Mb).

Table 18.1	Genome Sequencing Projects	
Organism	**Genome Size (Mb)**	**Description**
ARCHAEBACTERIA		
Archaeoglobus fulgidus	2.2	Sulfate-reducing chemoautotroph
Methanobacterium thermoautotrophicum	1.8	Reduces CO_2 to methane
Methanococcus jannaschii	1.7	Extreme thermophile
EUBACTERIA		
Escherichia coli	4.6	Laboratory standard
Mycobacterium tuberculosis	4.4	Tuberculosis
Bacillus subtilis	4.2	Forms endospores
Synechocystis spp.	3.6	Gliding cyanobacterium
Haemophilus influenzae	1.8	Meningitis; head infections
Helicobacter pylori	1.7	Ulcers
Borrelia burgdorferi	1.4	Lyme disease
Treponema pallidum	1.1	Syphilis
Mycoplasma pneumoniae	0.8	Pneumonia
Mycoplasma genitalium	0.6	Smallest known genome
FUNGI		
Saccharomyces cerevisiae	13	Baker's yeast

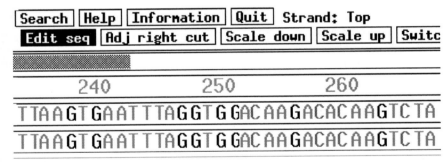

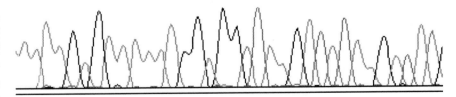

FIGURE 18.13

Part of the genome sequence of the plant *Arabidopsis.* Data from an automated DNA-sequencing run shows the nucleotide sequence for a small section of the *Arabidopsis* genome. Automated DNA sequencing has greatly increased the speed at which genomes can be sequenced.

DNA Fingerprinting

Figure 18.14 shows the DNA fingerprints a prosecuting attorney presented in a rape trial in 1987. They consisted of autoradiographs, parallel bars on X-ray film resembling the line patterns of the universal price code found on groceries. Each bar represents the position of a DNA restriction endonuclease fragment produced by techniques similar to those described in figures 18.4 and 18.10. The lane with many bars represents a standardized control. Two different probes were used to identify the restriction fragments. A vaginal swab had been taken from the victim within hours of her attack; from it semen was collected and the semen DNA analyzed for its restriction endonuclease patterns.

Compare the restriction endonuclease patterns of the semen to that of the suspect Andrews. You can see that the suspect's two patterns match that of the rapist (and are not at all like those of the victim). Clearly the semen collected from the rape victim and the blood sample from the suspect came from the same person. The suspect was Tommie Lee Andrews, and on November 6, 1987, the jury returned a verdict of guilty. Andrews became the first person in the United States to be convicted of a crime based on DNA evidence.

Since the Andrews verdict, DNA fingerprinting has been admitted as evidence in more than 2000 court cases (figure 18.15). While some probes highlight profiles shared by many people, others are quite rare. Using several probes, identity can be clearly established or ruled out.

Just as fingerprinting revolutionized forensic evidence in the early 1900s, so DNA fingerprinting is revolutionizing it today. A hair, a minute speck of blood, a drop of semen can all serve as sources of DNA to damn or clear a suspect. As the man who analyzed Andrews' DNA says: "It's like leaving your name, address, and social security number at the scene of the crime. It's that precise." Of course, laboratory analyses of DNA samples must be carried out properly—sloppy procedures could lead to a wrongful conviction. After widely publicized instances of questionable lab procedures, national standards are being developed.

The genomes of several organisms have been completely sequenced. When DNA is digested with restriction endonucleases, distinctive profiles on electrophoresis gels can be used to identify tissue from a specific individual.

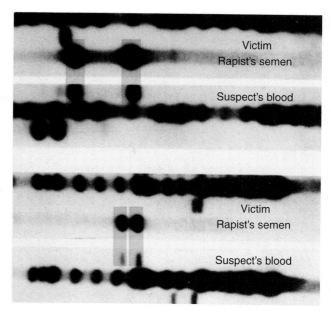

FIGURE 18.14
Two of the DNA profiles that led to the conviction of Tommie Lee Andrews for rape in 1987. The two DNA probes seen here were used to characterize DNA isolated from the victim, the semen left by the rapist, and the suspect. The dark channels are multiband controls. There is a clear match between the suspect's DNA and the DNA of the rapist's semen in these.

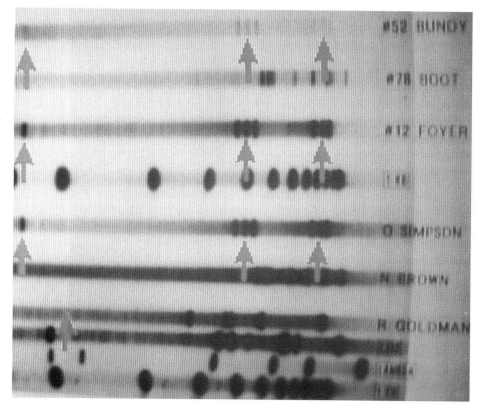

FIGURE 18.15
The DNA profiles of O. J. Simpson and blood samples from the murder scene of his former wife from his highly publicized and controversial murder trial in 1995.

Medical Applications

Pharmaceuticals

The first and perhaps most obvious commercial application of genetic engineering was the introduction of genes that encode clinically important proteins into bacteria. Because bacterial cells can be grown cheaply in bulk (fermented in giant vats, like the yeasts that make beer), bacteria that incorporate recombinant genes can synthesize large amounts of the proteins those genes specify. This method has been used to produce several forms of human insulin and interferon, as well as commercially valuable proteins such as growth hormone (figure 18.16) and erythropoietin, which stimulates red blood cell production.

Among the medically important proteins now manufactured by these approaches are **atrial peptides,** small proteins that may provide a new way to treat high blood pressure and kidney failure. Another is **tissue plasminogen activator,** a human protein synthesized in minute amounts that causes blood clots to dissolve and may be effective in preventing heart attacks and strokes.

A problem with this general approach has been the difficulty of separating the desired protein from the others the bacteria make. The purification of proteins from such complex mixtures is both time-consuming and expensive, but it is still easier than isolating the proteins from the tissues of animals (for example, insulin from dog pancreases), which is how such proteins used to be obtained. Recently, however, researchers have succeeded in producing RNA transcripts of cloned genes; they can then use the transcripts to produce only these proteins in a test tube containing the transcribed RNA, ribosomes, cofactors, amino acids, tRNA, and ATP.

Gene Therapy

In 1990, researchers first attempted to combat genetic defects by the transfer of human genes. When a hereditary disorder is the result of a single defective gene, an obvious way to cure the disorder is to add a working copy of the gene. This approach is being use in an attempt to combat cystic fibrosis, and it offers potential for treating muscular dystrophy and a variety of other disorders (table 18.2). One of the first successful attempts was the transfer of a gene encoding the enzyme adenosine deaminase into the bone marrow of two girls suffering from a rare blood disorder caused by the lack of this enzyme. However, while many clinical trials are underway, no others have yet proven successful. This extremely promising approach will require a lot of additional effort.

FIGURE 18.16
Genetically engineered human growth hormone. These two mice are genetically identical, but the large one has one extra gene: the gene encoding human growth hormone. The gene was added to the mouse's genome by human genetic engineers and is now a stable part of the mouse's genetic endowment.

Table 18.2   Diseases Being Treated in Clinical Trials of Gene Therapy
Disease
Cancer (melanoma, renal, cell, ovarian, neuroblastoma, brain, head and neck, lung, liver, breast, colon, prostate, mesothelioma, leukemia, lymphoma, multiple myeloma)
SCID (severe combined immunodeficiency)
Cystic fibrosis
Gaucher's disease
Familial hypercholesterolemia
Hemophilia
Purine nucleoside phosphorylase deficiency
Alpha-1 antitrypsin deficiency
Fanconi's anemia
Hunter's syndrome
Chronic granulomatous disease
Rheumatoid arthritis
Peripheral vascular disease
AIDS

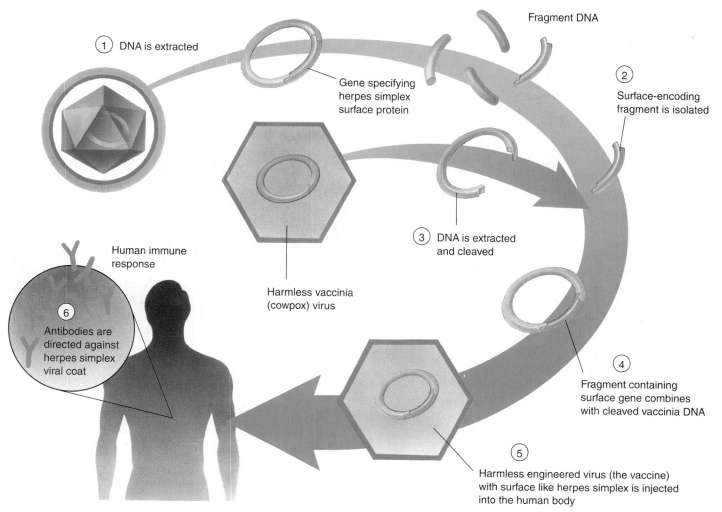

① DNA is extracted

Gene specifying
herpes simplex
surface protein

Fragment DNA

**② Surface-encoding
fragment is isolated**

**③ DNA is extracted
and cleaved**

Harmless vaccinia
(cowpox) virus

Human immune
response

**④ Fragment containing
surface gene combines
with cleaved vaccinia DNA**

**⑤ Harmless engineered virus (the vaccine)
with surface like herpes simplex is injected
into the human body**

**⑥ Antibodies are
directed against
herpes simplex
viral coat**

FIGURE 18.17
Constructing a subunit vaccine for herpes simplex.

Piggyback Vaccines

Another area of potential significance involves the use of genetic engineering to produce **subunit vaccines** against viruses such as those that cause herpes and hepatitis. Genes encoding part of the protein-polysaccharide coat of the herpes simplex virus or hepatitis B virus are spliced into a fragment of the vaccinia (cowpox) virus genome (figure 18.17). The vaccinia virus, which British physician Edward Jenner used almost 200 years ago in his pioneering vaccinations against smallpox, is now used as a vector to carry the herpes or hepatitis viral coat gene into cultured mammalian cells. These cells produce many copies of the recombinant virus, which has the outside coat of a herpes or hepatitis virus. When this recombinant virus is injected into a mouse or rabbit, the immune system of the infected animal produces antibodies directed against the coat of the recombinant virus. It therefore develops an immunity to herpes or hepatitis virus. Vaccines produced in this way are harmless, since the vaccinia virus is benign and only a small fragment of the DNA from the disease-causing virus is introduced via the recombinant virus.

The great attraction of this approach is that it does not depend upon the nature of the viral disease. In the future, similar recombinant viruses may be injected into humans to confer resistance to a wide variety of viral diseases.

In 1995, the first clinical trials began of a novel new kind of **DNA vaccine,** one that depends not on antibodies but rather on the second arm of the body's immune defense, the so-called cellular immune response, in which blood cells known as killer T cells attack infected cells. The infected cells are attacked and destroyed when they stick fragments of foreign proteins onto their outer surfaces that the T cells detect (the discovery by Peter Doherty and Rolf Zinkernagel that infected cells do so led to their receiving the Nobel Prize in medicine in 1996). The first DNA vaccines spliced an influenza virus gene encoding an internal nucleoprotein into a plasmid, which was then injected into mice. The mice developed strong cellular immune responses to influenza. New and controversial, the approach offers great promise.

Genetic engineering has produced commercially valuable proteins, gene therapies, and, possibly, new and powerful vaccines.

Agricultural Applications

Another major area of genetic engineering activity is manipulation of the genes of key crop plants. In plants the primary experimental difficulty has been identifying a suitable vector for introducing recombinant DNA. Plant cells do not possess the many plasmids that bacteria do, so the choice of potential vectors is limited. The most successful results thus far have been obtained with the **Ti** (tumor-inducing) **plasmid** of the plant bacterium *Agrobacterium tumefaciens*, which infects broadleaf plants such as tomato, tobacco, and soybean. Part of the Ti plasmid integrates into the plant DNA, and researchers have succeeded in attaching other genes to this portion of the plasmid (figure 18.18). The characteristics of a number of plants have been altered using this technique, which should be valuable in improving crops and forests. Among the features scientists would like to affect are resistance to disease, frost, and other forms of stress; nutritional balance and protein content; and herbicide resistance. Unfortunately, *Agrobacterium* generally does not infect cereals such as corn, rice, and wheat, but alternative methods are being developed to introduce new genes into them.

A recent advance in genetically manipulated fruit is Calgene's "Flavr Savr" tomato, which has been approved for sale by the USDA (figure 18.19). The tomato has been engineered to inhibit genes that cause cells to produce ethylene. In tomatoes and other plants, ethylene acts as a hormone to speed fruit ripening. In Flavr Savr tomatoes, inhibition of ethylene production delays ripening. The result is a tomato that can stay on the vine longer and that resists overripening and rotting during transport to market.

FIGURE 18.19
Flavr Savr tomatoes. These genetically engineered tomatoes stay fresh longer.

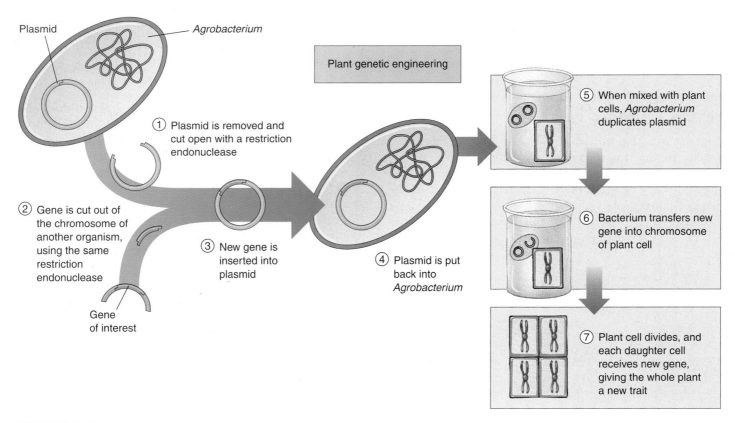

FIGURE 18.18
The Ti plasmid. This *Agrobacterium tumefaciens* plasmid is used in plant genetic engineering.

Herbicide Resistance

Recently, broadleaf plants have been genetically engineered to be resistant to **glyphosate,** the active ingredient in Roundup, a powerful, biodegradable herbicide that kills most actively growing plants (figure 18.20). Glyphosate works by inhibiting an enzyme called EPSP synthetase, which plants require to produce aromatic amino acids. Humans do not make aromatic amino acids; they get them from their diet, so they are unaffected by glyphosate. To make glyphosate-resistant plants, agricultural scientists used a Ti plasmid to insert extra copies of the EPSP synthetase genes into plants. These engineered plants produce 20 times the normal level of EPSP synthetase, enabling them to synthesize proteins and grow despite glyphosate's suppression of the enzyme. In later experiments, a bacterial form of the EPSP synthetase gene that differs from the plant form by a single nucleotide was introduced into plants via Ti plasmids; the bacterial enzyme in these plants is not inhibited by glyphosate.

These advances are of great interest to farmers, since a crop resistant to Roundup would never have to be weeded if the field were simply treated with the herbicide. Because Roundup is a broad-spectrum herbicide, farmers would no longer need to employ a variety of different herbicides, most of which kill only a few kinds of weeds. Furthermore, glyphosate breaks down readily in the environment, unlike many other herbicides commonly used in agriculture. A plasmid is actively being sought for the introduction of the EPSP synthetase gene into cereal plants, making them also glyphosate-resistant.

Nitrogen Fixation

A long-range goal of agricultural genetic engineering is to introduce the genes that allow soybeans and other legume plants to "fix" nitrogen into key crop plants. These so-called *nif* **genes** are found in certain symbiotic root-colonizing bacteria. Living in the root nodules of legumes (figure 18.21), these bacteria break the powerful triple bond of atmospheric nitrogen gas, converting N_2 into NH_3 (ammonia). The plants then use the ammonia to make amino acids and other nitrogen-containing molecules. Other plants lack these bacteria and cannot fix nitrogen, so they must obtain their nitrogen from the soil. Farmland where these crops are grown soon becomes depleted of nitrogen, unless nitrogenous fertilizers are applied. Worldwide, farmers applied over 60 million metric tons of such fertilizers in 1987, an expensive undertaking. Farming costs would be much lower if major crops like wheat and corn could be engineered to carry out biological nitrogen fixation. However, introducing the nitrogen-fixing genes from bacteria into plants has proved difficult because these genes do not seem to function properly in eukaryotic cells. Researchers are actively experimenting with other species of nitrogen-fixing bacteria whose genes might function better in plant cells.

FIGURE 18.20
Genetically engineered herbicide resistance. All four of these petunia plants were exposed to equal doses of the herbicide Roundup. The two on top were genetically engineered to be resistant to glyphosate, the active ingredient of Roundup, while the two on the bottom were not.

FIGURE 18.21
Root nodules of soybeans. The ball-shaped nodules on these soybean roots contain *Rhizobium* bacteria that carry out nitrogen fixation.

Insect Resistance

Many commercially important plants are attacked by insects, and the traditional defense against such attacks is to apply insecticides. Over 40% of the chemical insecticides used today are targeted against boll weevils, bollworms, and other insects that eat cotton plants. Genetic engineers are now attempting to produce plants that are resistant to insect pests, removing the need to use many externally applied insecticides.

The approach is to insert into crop plants genes encoding proteins that are harmful to the insects that feed on the plants but harmless to other organisms. One such insecticidal protein has been identified in *Bacillus thuringiensis*, a soil bacterium (figure 18.22). When the tomato hornworm caterpillar ingests this protein, enzymes in the caterpillar's stomach convert it into an insect-specific toxin, causing paralysis and death. Because these enzymes are not found in other animals, the protein is harmless to them. Using the Ti plasmid, scientists have transferred the gene encoding this protein into tomato and tobacco plants. They have found that these **transgenic** plants are indeed protected from attack by the insects that would normally feed on them (figure 18.23). In 1995, the EPA approved altered forms of potato, cotton, and corn. The genetically altered potato can kill the Colorado potato beetle, a common pest. The altered cotton is resistant to cotton bollworm, budworm, and pink bollworm. The corn has been altered to resist the European corn borer and other mothlike insects.

Monsanto scientists screening natural compounds extracted from plant and soil samples have recently isolated a new insect-killing compound from a fungus, the enzyme cholesterol oxidase. Apparently, the enzyme disrupts membranes in the insect gut. The fungus gene, called the Bollgard gene after its discoverer, has been successfully inserted into a variety of crops. It kills a wide range of insects, including the cotton boll weevil and the Colorado potato beetle, both serious agricultural pests. Field tests began in 1996.

Some insect pests attack plant roots, and *B. thuringiensis* is being employed to counter that threat as well. This bacterium does not normally colonize plant roots, so biologists have introduced the *B. thuringiensis* insecticidal protein gene into root-colonizing bacteria, especially strains of *Pseudomonas*. Field testing of this promising procedure has been approved by the Environmental Protection Agency.

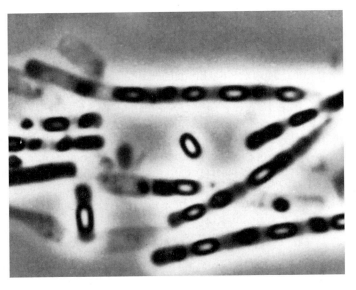

FIGURE 18.22
Bacillus thuringiensis, **a bacterium that produces a protein toxic to many caterpillars.** The white spots within the bacteria are protein crystals that are transformed into a toxin by enzymes in the caterpillar's stomach.

FIGURE 18.23
A successful experiment in pest resistance. Both of these tomato plants were exposed to destructive caterpillars under laboratory conditions. The nonengineered plant (*left*) has been almost completely eaten, while the engineered plant (*right*) shows virtually no signs of damage.

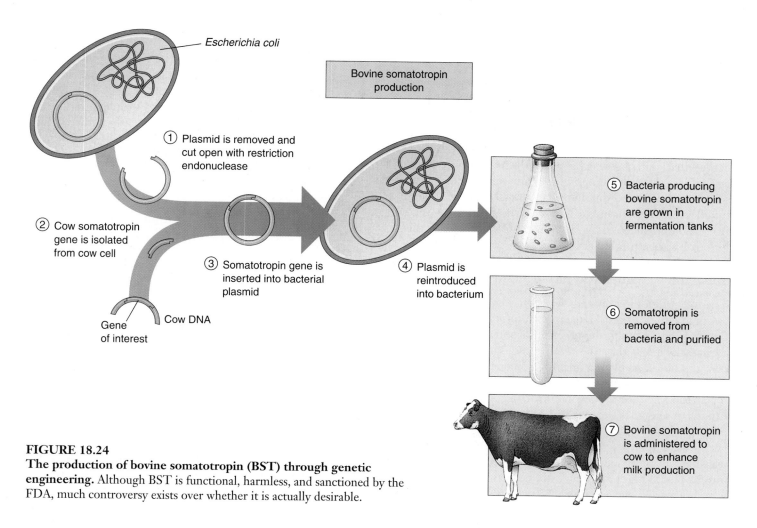

① Plasmid is removed and cut open with restriction endonuclease

② Cow somatotropin gene is isolated from cow cell

③ Somatotropin gene is inserted into bacterial plasmid

④ Plasmid is reintroduced into bacterium

Gene of interest

Cow DNA

Escherichia coli

Bovine somatotropin production

⑤ Bacteria producing bovine somatotropin are grown in fermentation tanks

⑥ Somatotropin is removed from bacteria and purified

⑦ Bovine somatotropin is administered to cow to enhance milk production

FIGURE 18.24
The production of bovine somatotropin (BST) through genetic engineering. Although BST is functional, harmless, and sanctioned by the FDA, much controversy exists over whether it is actually desirable.

Farm Animals

The gene encoding the growth hormone somatotropin was one of the first to be cloned successfully. In 1994 Monsanto received federal approval to make its recombinant bovine somatotropin (BST) commercially available, and dairy farmers worldwide began to add the hormone as a supplement to their cows' diets, increasing the animals' milk production (figure 18.24). Genetically engineered somatotropin is also being tested to see if it increases the muscle weight of cattle and pigs (figure 18.25), and as a treatment for human disorders in which the pituitary gland fails to make adequate levels of somatotropin, producing dwarfism. BST ingested in milk or meat has no effect on humans, because it is a protein and is digested in the stomach. Nevertheless, BST has met with some public resistance, due primarily to generalized fears of gene technology. Some people mistrust milk produced through genetic engineering, even though the milk itself is identical to other milk. Problems concerning public perception are not uncommon as gene technology makes an even greater impact on our lives.

Transgenic animals engineered to have specific desirable genes are becoming increasingly available to breeders. Now, instead of selectively breeding for several generations to produce a racehorse or a stud bull with desirable quali-

FIGURE 18.25
A meatier pig. Researchers are trying to increase the ratio of muscle mass to fat in pigs by introducing pig growth hormone.

ties, the process can be shortened by simply engineering such an animal right at the start.

Gene technology is revolutionizing agriculture, increasing yields and resistance to pests, and producing animals with desirable traits.

Cloning

The difficulty in using transgenic animals to improve livestock is in getting enough of them. Breeding produces offspring only slowly, and recombination acts to undo the painstaking work of the genetic engineer. Ideally, one would like to "Xerox" many exact genetic copies of the transgenic strain—but until 1997 it was commonly accepted that animals can't be cloned. Now the holy grail of agricultural genetic engineers seems within reach. In 1997, scientists announced the first successful cloning of differentiated vertebrate tissue, a lamb grown from a cell taken from an adult sheep. This startling result promises to revolutionize agricultural science.

Spemann's "Fantastical Experiment"

The idea of cloning animals was first suggested in 1938 by German embryologist Hans Spemann (often called the "father of modern embryology") who proposed what he called a "fantastical experiment": remove the nucleus from an egg cell, and put in its place a nucleus from another cell.

It was 14 years before technology advanced far enough for anyone to take up Spemann's challenge. In 1952, two American scientists, Robert Briggs and T. J. King used very fine pipettes to suck the nucleus from a frog egg (frog eggs are unusually large, making the experiment feasible) and transfer a nucleus sucked from a body cell of an adult frog into its place. The experiment did not work when done this way, but partial success was achieved 18 years later by the British developmental biologist John Gurdon, who in 1970 inserted nuclei from advanced toad embryos rather than adult tissue. The toad eggs developed into tadpoles, but died before becoming adults.

The Path to Success

For 14 years, nuclear transplant experiments were attempted without success. Technology continued to advance however, until finally in 1984, Steen Willadsen, a Danish embryologist working in Texas, succeeded in cloning a sheep using a nucleus from a cell of an early embryo. This exciting result was soon replicated by others in a host of other organisms, including cattle, pigs, and monkeys.

Only early embryo cells seemed to work, however. Researchers became convinced that animal cells become irreversibly "committed" after the first few cell divisions of the developing embryo, and after that nuclei from differentiated animal cells cannot be used to clone entire organisms.

We now know this conclusion to have been unwarranted. The key advance for unraveling this puzzle was made in Scotland by geneticist Keith Campbell, a specialist in studying the cell cycle of agricultural animals. By the early 1990s, knowledge of how the cell cycle is controlled, advanced by cancer research, had led to an understanding that cells don't divide until conditions are appropriate. Just as a washing machine checks that the water has completely emptied before initiating the spin cycle, so the cell checks that everything needed is on hand before initiating cell division. Campbell reasoned "Maybe the egg and the donated nucleus need to be at the same stage in the cell cycle."

This proved to be a key insight. In 1994 researcher Neil First, and in 1995 Campbell himself working with reproductive biologist Ian Wilmut, succeeded in cloning farm animals from advanced embryos by first starving the cells, so that they paused at the beginning of the cell cycle at the G_2 checkpoint. Two starved cells are thus held in resting phase at the same point in the cell cycle. The stage was now set for success.

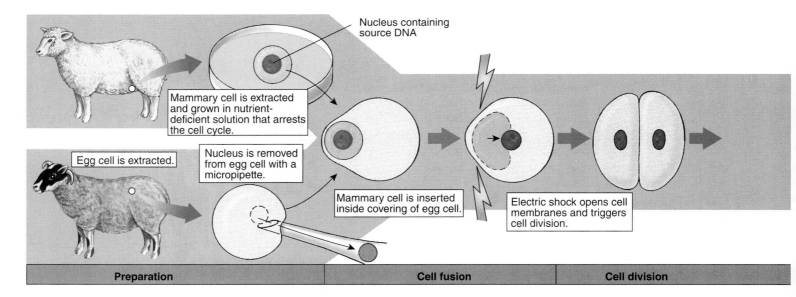

Nucleus containing source DNA

Mammary cell is extracted and grown in nutrient-deficient solution that arrests the cell cycle.

Egg cell is extracted.

Nucleus is removed from egg cell with a micropipette.

Mammary cell is inserted inside covering of egg cell.

Electric shock opens cell membranes and triggers cell division.

| Preparation | Cell fusion | Cell division |

FIGURE 18.26
Wilmut's animal cloning experiment. Wilmut combined a nucleus from a mammary cell and an egg cell (with its nucleus removed) to successfully clone a sheep.

Wilmut's Lamb

Wilmut then set out to attempt the key breakthrough, the experiment that had eluded researchers since Spemann proposed it 59 years before: to transfer the nucleus from an adult differentiated cell into an enucleated egg, and allow the resulting embryo to grow and develop in a surrogate mother, hopefully producing a healthy animal.

Wilmut removed cells from the udder of a 6-year-old sheep (figure 18.26). Mammary cells, the origin of these cells, gave the clone its name, "Dolly" after the country singer Dolly Parton. The cells were grown in tissue culture, and some frozen so that in the future it would be possible with genetic fingerprinting to prove that a clone was indeed genetically identical with the cells from the 6-year-old sheep.

In preparation for cloning, Wilmut's team greatly reduced for 5 days the concentration of serum on which the sheep mammary cells were subsisting. In parallel preparation, eggs obtained from a ewe were enucleated, the nucleus of each egg carefully removed with a micropipette.

Mammary cells and egg cells were then surgically combined in January of 1996, the mammary cells inserted inside the covering around the egg cell. Wilmut then applied a brief electrical shock. A neat trick, this causes the plasma membranes surrounding the two cells to become leaky, so that the contents of the mammary cell passes into the egg cell. The shock also kick-starts the cell cycle, causing the cell to begin to divide.

After 6 days, in 30 of 277 tries, the dividing embryo reached the hollow-ball "blastula" stage, and 29 of these were transplanted into surrogate mother sheep. A little over 5 months later, on July 5, 1997, one sheep gave birth to a lamb. This lamb, "Dolly," was the first successful clone generated from a differentiated animal cell.

The Future of Cloning

Wilmut's successful cloning of fully differentiated sheep cells is a milestone event in gene technology. Even though his procedure proved inefficient (only one of 277 trials succeeded), it established the point beyond all doubt that cloning of adult animal cells *can* be done. In 1998 researchers succeeded in greatly improving the efficiency of cloning. Seizing upon the key idea in Wilmut's experiment, to clone a resting-stage cell, they returned to the nuclear transplant procedure pioneered by Briggs and King, but used 800 resting-stage cumulus cells from a female mouse. From them they produced 10 healthy mouse clones, all identical to the female source, and, soonafter, 7 clones of the clones.

Transgenic cloning can be expected to have a major impact on medicine as well as agriculture. Animals with human genes can be used to produce rare hormones. For example, sheep that have recently been genetically engineered to secrete a protein called alpha-1 antitrypsin (helpful in relieving the symptoms of cystic fibrosis) into their milk may be cloned, greatly cheapening the production of this expensive drug.

It is impossible not to speculate on the possibility of cloning a human. There is no reason to believe such an experiment would not work, but many reasons to question whether it should be done. The ethics of human cloning have not been seriously addressed—no one thought it possible until Wilmut's lamb pointed the way. As much of Western thought is based on the concept of human individuality, we can expect the possibility of human cloning to engender considerable controversy.

Recent experiments have demonstrated the possibility of cloning differentiated mammalian tissue, opening the door for the first time to practical transgenic cloning of farm animals. To date, no human has been cloned, although the experiment seems feasible.

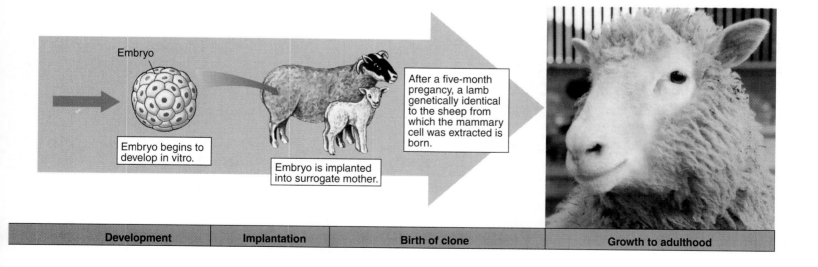

Embryo

Embryo begins to develop in vitro.

Embryo is implanted into surrogate mother.

After a five-month pregancy, a lamb genetically identical to the sheep from which the mammary cell was extracted is born.

| Development | Implantation | Birth of clone | Growth to adulthood |

Ethics and Regulation

The advantages afforded by genetic engineering are revolutionizing our lives. But what are the disadvantages, the potential costs and dangers of genetic engineering? Many people, including influential activists and members of the scientific community, have expressed concern that genetic engineers are "playing God" by tampering with genetic material. This is more than an ethical or moral problem, with ramifications that could be quite serious (figure 18.27). For instance, what would happen if one fragmented the DNA of a cancer cell, and then incorporated the fragments at random into vectors that were propagated within bacterial cells? Might there not be a danger that some of the resulting bacteria would transmit an infective form of cancer? What would happen if a killer virus was engineered? Could genetically engineered products administered to plants or animals turn out to be dangerous for consumers after several generations? What kind of unforeseen impact on the ecosystem might "improved" crops have? Suppose a genetically superior plant was released into the wild, where it out-competed all other existing plants. Is it ethical to create "genetically superior" organisms, including humans?

Many of the public's concerns about genetic engineering are not well founded. Less than 1% of known bacteria are pathogenic to humans, and almost *all* of the bacteria used in genetic engineering are not. In fact, most of those used in genetic engineering could not live in or on a human host at all. Nor does a rot-resistant tomato appear to be threatening. In most cases, genetically altered crops differ by just one or a few genes, which are often obtained from other lines of the same species. Recombinant technology achieves the same kind of gene reassortment as the genetic crosses of Mendel, only faster. In that sense, science isn't doing anything that nature isn't already doing. Mutations occur naturally on a random basis, and organisms with favorable mutations are selected for; that is the essence of evolution.

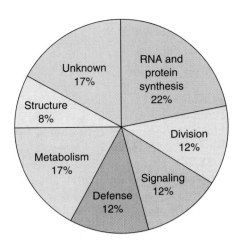

FIGURE 18.27
We are learning an incredible amount about ourselves.
Comparing partial sequences of human genes revealed by the human genome project with the vast array of genes known from other organisms, researchers have been able to make a fair guess as to the functions of many of the human genes. This chart presents percentages of genes that appear to be devoted to each of the major activities of a typical human cell. How ethically may we use this information?
From W. A. Haseltine, "Discovering Genes for New Medicines."
Copyright © 1997 by Scientific American, Inc. All rights reserved.

The genetic "dabbling" performed by humans is minuscule when compared with that performed by nature.

This does not mean that mistakes cannot occur, or that Nature cannot take something synthetically engineered and "improve" on it, for better or worse. Still, the benefits of genetic engineering seem to far outweigh the risks.

> Genetic engineering affords great opportunities for progress in medicine, food production, manufacturing, and forensics. Although many are concerned about the possible risks of genetic engineering, the risks are in fact slight, and the potential benefits are substantial.

Summary of Concepts

18.1 The ability to manipulate DNA has led to a new genetics.

- Genetic engineering involves the isolation of specific genes and their transfer to new genomes.
- The key to genetic engineering technology is a special class of enzymes called restriction endonucleases, which cleave DNA molecules into fragments. These fragments usually have short, single-stranded ends that are complementary to each other.
- Because the ends of each fragment cleaved by a given restriction endonuclease are complementary, such fragments can pair with each other and be spliced together, whatever their source. Therefore, DNA fragments from very different genomes can be combined.
- The first such recombinant DNA was made by Cohen and Boyer in 1973, when they inserted a toad ribosomal RNA gene into a bacterial plasmid.

18.2 Genetic engineering involves easily understood procedures.

- Genetic engineering experiments consist of four stages: cleavage of DNA, production of recombinant DNA, cloning, and screening for the gene(s) of interest.
- Preliminary screening can be accomplished by making the desired clones resistant to an antibiotic; hybridization can then be employed to identify the gene of interest.
- Gene technologies, including PCR, Southern blotting, RFLP analysis, and the Sanger method, enable researchers to isolate genes and produce them in large quantities.

18.3 Biotechnology is producing a scientific revolution.

- Extensive research on the human genome has yielded important information about the location of genes, such as those that may be involved in dyslexia, obesity, and resistance to high blood cholesterol levels.
- Genetic engineering has fostered many advances in the production of drugs, including interferon, insulin, atrial peptides, and tissue plasminogen activator.
- Gene splicing holds great promise as a clinical tool, particularly in the prevention of disease. For example, a gene from a pathogenic virus can be spliced into the DNA of a harmless virus; the recombinant virus, which is still harmless, can then be injected into animals as a vaccine to induce the formation of antibodies against the pathogenic virus.
- A major focus of genetic engineering activity has been agriculture, where genes conferring resistance to herbicides or insect pests have been incorporated into crop plants.
- Recent experiments open the way for cloning of genetically altered animals and suggest that human cloning is feasible.
- The impact of genetic engineering has skyrocketed over the past decade, providing many useful innovations for society. Although genetic engineering is rigorously controlled and constantly monitored by the scientific community and federal agencies, its moral and ethical aspects still provide a topic for heated debates.

Discussing Key Terms

1. **Cloning**   Specific genes or segments of DNA can be inserted into living cells. The cells are then grown to produce multiple copies, or clones, of the gene. When a nucleus is placed into an enucleated egg, an entire organism may be cloned.

2. **Restriction endonucleases**   The key tool of genetic engineers is a special group of enzymes that recognize specific short sequences of DNA and cleave the DNA at those sites.

3. **Recombinant DNA**   Many restriction endonucleases cleave their DNA recognition sequences asymmetrically, leaving DNA fragments with "sticky ends." This allows DNA fragments from different sources—even from different organisms—to be spliced together. In 1973, the first recombinant genome was created by inserting a frog gene into a bacterial plasmid.

4. **PCR**   Polymerase chain reaction is a procedure for amplifying tiny samples of DNA into amounts sufficient for detailed analysis. Using a temperature cycle, single-stranded DNA is copied by a polymerase. The resulting double helices are made single-stranded, and each is copied by a polymerase; as the cycle repeats, the amount of the sequence increases exponentially.

5. **RFLP**   Restriction fragment length polymorphisms are a form of genetic variation. Individuals differ in the location of restriction sites; this results in different-sized fragments when the DNA is cleaved with restriction endonucleases that recognize those sites.

6. **DNA fingerprinting**   Restriction endonuclease fragments of highly variable DNA regions can be used to identify individuals, much like fingerprints.

7. **Transgenic organisms**   Scientists have been able to move genes from one species to another, introducing desired characteristics such as disease resistance or increased size into plants and animals.

8. **Bioethics**   It is important for scientists and society in general to assess whether the benefits of particular applications of genetic engineering outweigh the costs.

Review Questions

1. What features do most restriction endonuclease recognition sequences have in common? What are the ends of the DNA fragments created by most restriction endonucleases like? Why do these ends enable fragments from different genomes to be spliced together?

2. Describe the procedure used to eliminate clones that have not incorporated a vector in a genetic engineering experiment. Describe the procedure used to eliminate clones with vectors that do not have an inserted DNA fragment.

3. What is used as a probe in a Southern blot? With what does the probe hybridize? How are the regions of hybridization visualized?

4. What are restriction fragment length polymorphisms? How can they be used to identify individuals in forensic analyses?

5. What is the primary vector used to introduce genes into plant cells? What types of plants are generally infected by this vector? Describe three examples of how this vector has been used for genetic engineering, and explain the agricultural significance of each example.

6. How is the genetic engineering of bovine somatotropin (BST) used to increase milk production in the dairy industry? What effect would BST in milk have on persons who drink it?

Thought Questions

1. A major focus of genetic engineering has been the attempt to produce large quantities of scarce human metabolites by placing the appropriate human genes into bacteria. Prodigious numbers of bacteria are produced readily and cheaply, and large amounts of the metabolite can be made. Human insulin is now manufactured in this way. However, if one attempts to use this approach to produce human hemoglobin (β-globin), the experiment does not work. Even if the proper clone from a clone library is identified with an appropriate radioactive probe, the fragment containing the β-globin gene is successfully incorporated into a plasmid, and *E. coli* are successfully infected with the chimeric plasmid, no β-globin is produced by the infected cells. Why doesn't this experiment work?

2. In the novel *Jurassic Park*, genetic engineers recreate dinosaurs from samples of fossilized dinosaur DNA. The story relies in part on discoveries and technology that already exist. For example, mosquitoes and other blood-sucking insects dating from the age of the dinosaurs have been found preserved in amber; researchers have been successful in obtaining and analyzing DNA contained in 18-million-year-old fossils of leaves; and transgenic animals have been produced after foreign genes were inserted into the eggs from which they developed. Given these successes, why hasn't the *Jurassic Park* experiment been completed in reality? What stumbling blocks must still be overcome? What ethical issues should be considered before scientists are permitted to attempt such an experiment?

Internet Links

Genes From a Student's Prospective
http://vector.cshl.org/
The GENE ALMANAC site of the Cold Spring Harbor Laboratory provides timely information about genes, in a format students will appreciate.

Genomes That Have Been Sequenced So Far
http://www.tigr.org/tdb/mdb/mdb.html
A listing of microbial genomes whose sequences have been published, with references to the publications, and an update on those in the progress of being sequence.

Biotechnology "In The News"
http://schmidel.com/bionet/biotech.htm
BioChemNet's section on Biotechnology provides a rich exploration of what's going on in biotechnology, with a fine collection of links to tutorials, references, research tools, and virtual laboratory learning opportunities.

Exploring the Human Genome Project
http://www.kumc.edu/gec/
A useful site that provides a wealth of links to web resources and curricular material concerning the human genome project.

For Further Reading

Anderson, W. F.: "Gene Therapy," *Scientific American*, September 1995, pages 123–28. An update on the many clinical trials of gene therapy, involving several hundred patients.

Capecchi, M. R.: "Targeted Gene Replacement," *Scientific American*, vol. 270, March 1994, pages 52–59. Researchers at the University of Utah can change the nucleotide sequences of every single cell in a mouse. Can all genes be changed, or fixed?

Cohen, J. S., and M. E. Hogan: "The New Genetic Medicines," *Scientific American*, December 1994, pages 76–82. Can antisense and triplex agent DNA be used as "magic bullets" in combating cancer and deadly viral diseases?

Haseltine, W. A.: "Discovering Genes for New Medicines," *Scientific American*, March 1997, pages 92–97. An up-to-date look at some of the progress in gene technology.

Mullis, K.: "The Unusual Origin of the Polymerase Chain Reaction," *Scientific American*, April 1990, pages 56–65. How a critical advance in gene engineering technology was made.

Neufeld, P. J., and N. Colman: "When Science Takes the Stand," *Scientific American*, May 1990, pages 46–53. DNA and other evidence is increasingly applied to the solution of criminal cases but must be used with caution.

Stix, G.: "A Recombinant Feast: New Bioengineered Crops Move Toward Market," *Scientific American*, March 1995, pages 38–40. A progress report on the growing role of genetic engineering in modern agriculture.

Various authors: "Making Gene Therapy Work," *Scientific American*, June 1997, pages 95–123. A special report on current gene therapy research, including its role in fighting cancer and AIDS.

19

Genes within Populations

Concept Outline

FIGURE 19.1
Genetic variation. The range of genetic material in a population is expressed in a variety of ways—including color.

No other human being is exactly like you (unless you have an identical twin). Often the particular characteristics of an individual have an important bearing on its survival, on its chances to reproduce, and on the success of its offspring. Evolution is driven by such consequences. Genetic variation that influences these characteristics provides the raw material for natural selection, and natural populations contain a wealth of such variation. In plants (figure 19.1), insects, and vertebrates, practically every gene exhibits some level of variation. In this chapter, we will explore genetic variation in natural populations and consider the evolutionary forces that cause allele frequencies in natural populations to change. These deceptively simple matters lie at the core of evolutionary biology.

Gene Variation Is the Raw Material of Evolution

Microevolution Leads to Macroevolution

When the word *"evolution"* is mentioned, it is difficult not to conjure up images of dinosaurs, woolly mammoths frozen in blocks of ice, or Darwin observing a monkey. Traces of ancient life forms, now extinct, survive as fossils that help us piece together the evolutionary story. With such a background, we usually think of evolution in terms of changes that take place over long periods of time, changes in the kinds of animals and plants on earth as new forms replace old ones. This kind of evolution, **macroevolution,** is evolutionary change on a grand scale, encompassing the origins of novel designs, evolutionary trends, new kinds of organisms penetrating new habitats, and major episodes of extinction.

Much of the focus of Darwin's theory of natural selection, however, is directed not at the way in which new species are formed from old ones, but rather at the way that changes occur within species. Natural selection is the process whereby some individuals in a population, which possess certain inherited characteristics, produce more surviving offspring than individuals lacking these characteristics. As a result, the population will gradually come to include more and more individuals with the advantageous characteristics. In this way the population evolves. Changes of this sort within populations—changes in gene frequencies—represent **microevolution.**

Adaptation results from microevolutionary changes that increase the likelihood of survival and reproduction by an organism in a particular environment. In essence, Darwin's explanation of evolution is that adaptation by natural selection is responsible for evolutionary changes *within* a species (microevolution), and that the accumulation of these changes leads to the development of new species (macroevolution).

The Key Is the Source of the Variation

Darwin agreed with many earlier philosophers and naturalists who deduced that the many kinds of organisms around us were produced by a process of evolution. Unlike his predecessors, however, Darwin proposed natural selection as the mechanism of evolution. He proposed that new kinds of organisms evolve from existing ones because some individuals have traits that allow them to produce more offspring that, in turn, carry those traits.

Natural selection was by no means the only mechanism proposed. A rival theory, championed by the prominent biologist Jean-Baptiste Lamarck, was that evolution occurred by **the inheritance of acquired characteristics.** According to Lamarck, individuals passed on to offspring body

FIGURE 19.2
How did giraffes evolve a long neck? The giraffe's near relative, the okapi, has a short neck. Lamarck believed the giraffe lengthened its neck by stretching to reach tree leaves, then passed the change to offspring. Darwin proposed instead that giraffes who happened to be born with longer necks were more successful and consequently produced more surviving offspring. The key difference in the two proposals is whether the variation is acquired or genetic.

and behavior changes acquired during their lives. Thus, Lamarck proposed that ancestral giraffes with short necks tended to stretch their necks to feed on tree leaves, and this extension of the neck was passed on to subsequent generations, leading to the long-necked giraffe (figure 19.2). In Darwin's theory, by contrast, the variation is not created by experience but already exists when selection acts on it.

> **Darwin proposed that natural selection on variants within populations leads to the evolution of different species.**

Gene Variation in Nature

According to Darwin's theory, evolution is a series of adaptive changes brought about by natural selection, resulting in allele frequency changes. In considering this theory, it is best to start by looking at the genetic variation present among individuals within a species. This is the raw material available for the selective process.

Measuring Levels of Genetic Variation

As we saw in chapter 13, a natural population can contain a great deal of genetic variation. This is true not only of humans but of all organisms. How much variation usually occurs? Biologists have looked at many different genes in an effort to answer this question:

1. **Blood groups.** Chemical analysis has revealed the existence of more than 30 blood group genes in humans, in addition to the ABO locus. At least a third of these genes are routinely found in several alternative allelic forms in human populations. In addition to these, there are more than 45 variable genes encoding other proteins in human blood cells and plasma which are not considered blood groups. Thus, there are more than 75 genetically variable genes in this one system alone.
2. **Enzymes.** Alternative alleles of genes specifying particular enzymes are easy to distinguish by measuring how fast the alternative proteins migrate in an electric field (a process called **electrophoresis**). A great deal of variation exists at enzyme-specifying loci. About 5% of the enzyme loci of a typical human are heterozygous: if you picked an individual at random, and in turn selected one of the enzyme-encoding genes of that individual at random, the chances are 1 in 20 (5%) that the gene you selected would be heterozygous in that individual.

Considering the entire human genome, it is fair to say that almost all people are different from one another. This is also true of other organisms, except for those that reproduce asexually. In nature, genetic variation is the rule.

Enzyme Polymorphism

Many loci in a given population have more than one allele at frequencies significantly greater than would occur from mutation alone. Researchers refer to a locus with more variation than can be explained by mutation as **polymorphic** (*poly*, "many," *morphic*, "forms") (figure 19.3). The extent of such variation within natural populations was not even suspected a few decades ago, until modern techniques such as gel electrophoresis made it possible to examine enzymes and other proteins directly. We now know that most populations of insects and plants are polymorphic (that is, have more than one allele occurring at a frequency greater than 5%) at more than half of their enzyme-encoding loci,

FIGURE 19.3
Polymorphic variation. These Australian snails, all of the species *Bankivia fasciata*, exhibit considerable variation in pattern and color. Individual variations are heritable and passed on to offspring.

although vertebrates are somewhat less polymorphic. **Heterozygosity** (that is, the probability that a randomly selected gene will be heterozygous) is about 15% in *Drosophila* and other invertebrates, between 5% and 8% in vertebrates, and around 8% in outcrossing plants. These high levels of genetic variability provide ample supplies of raw material for evolution.

DNA Sequence Polymorphism

With the advent of gene technology, it has become possible to assess genetic variation even more directly by sequencing the DNA itself. In a pioneering study in 1989, Martin Kreitman sequenced ADH genes isolated from 11 individuals of the fruit fly *Drosophila melanogaster*. He found 43 variable sites, only one of which had been detected by protein electrophoresis! In the following decade, numerous other studies of variation at the DNA level have confirmed these findings: abundant variation exists in both the coding regions of genes and in their nontranslated introns—considerably more variation than we can detect examining enzymes with electrophoresis.

Natural populations contain considerable amounts of genetic variation—more than can be accounted for by mutation alone.

Population genetics is the study of the properties of genes in populations. Genetic variation within natural populations was a puzzle to Darwin and his contemporaries. The way in which meiosis produces genetic segregation among the progeny of a hybrid had not yet been discovered. Selection, scientists then thought, should always favor an optimal form, and so tend to eliminate variation.

The Hardy–Weinberg Principle

Two people independently solved and almost simultaneously published the solution to the puzzle of why genetic variation persists in 1908—G. H. Hardy, an English mathematician, and G. Weinberg, a German physician. They pointed out that the original proportions of the genotypes in a population will remain constant from generation to generation, as long as the following assumptions are met:

1. The population size is very large.
2. Random mating is occurring.
3. No mutation takes place.
4. No genes are input from other sources (no migration takes place).
5. No selection occurs.

Dominant alleles thus do not, in fact, replace recessive ones. Because their proportions do not change, the genotypes are said to be in **Hardy–Weinberg equilibrium.**

In algebraic terms, the Hardy–Weinberg principle is written as an equation. Its form is what is known as a binomial expansion. For a gene with two alternative alleles, which we will call B and b, the equation looks like this:

$$(p + q)^2 = \underset{\substack{\text{(Individuals}\\ \text{homozygous}\\ \text{for allele } B)}}{p^2} + \underset{\substack{\text{(Individuals}\\ \text{heterozygous}\\ \text{with alleles } B + b)}}{2pq} + \underset{\substack{\text{(Individuals}\\ \text{homozygous}\\ \text{for allele } b)}}{q^2}$$

In statistics, **frequency** is defined as the proportion of individuals falling within a certain category in relation to the total number of individuals under consideration. Thus, in a population of 100 cats, with 84 black and 16 white cats, the respective frequencies would be 0.84 (or 84%) and 0.16 (or 16%). In the algebraic terms of this equation, the letter p designates the frequency of one allele, the letter q the frequency of the alternative allele. By convention, the more common of the two alleles is designated p, the rarer q. Because there are only two alleles, p plus q must always equal 1.

If we assume that the white cats are homozygous recessive for a gene we designate b, and the black cats are therefore either homozygous dominant BB or heterozygous Bb, we can calculate the **allele frequencies** of the two alleles in the population from the proportion of black and white individuals. If $q^2 = 0.16$ (the frequency of white cats), then $q = 0.4$. Therefore, p, the frequency of allele B, would be 0.6 ($1.0 - 0.4 = 0.6$). We can now easily calculate the **genotype frequencies:** there are $p^2 = (0.6)^2 \times 100$ (the number of individuals in the total population), or 36 homozygous dominant BB individuals. The heterozygous individuals have the Bb genotype, and there would be $2pq$, or ($2 \times 0.6 \times 0.4$) $\times$ 100, or 48 heterozygous Bb individuals.

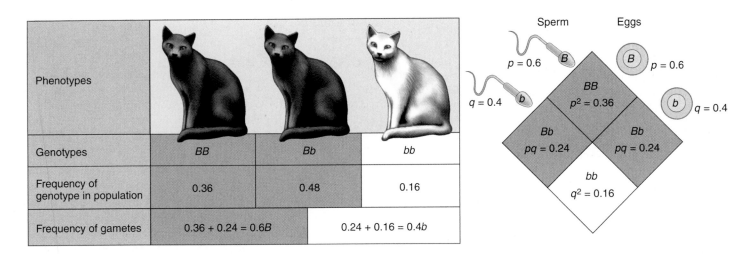

FIGURE 19.4
The Hardy–Weinberg equilibrium. In the absence of factors that alter them, the frequencies of gametes, genotypes, and phenotypes remain constant generation after generation.

Using the Hardy–Weinberg Equation

The Hardy–Weinberg equation is a simple extension of the Punnett square described in chapter 13, with two alleles assigned frequencies p and q. Figure 19.4 allows you to trace genetic reassortment during sexual reproduction and see how it affects the frequencies of the B and b alleles during the next generation. In constructing this diagram, we have assumed that the union of sperm and egg in these cats is random, so that all combinations of b and B alleles are equally likely. For this reason, the alleles are mixed randomly and represented in the next generation in proportion to their original representation. Each individual in each generation has a 0.6 chance of receiving a B allele ($p = 0.6$) and a 0.4 chance of receiving a b allele ($q = 0.4$).

In the next generation, therefore, the chance of combining two B alleles is p^2, or 0.36 (that is, 0.6×0.6), and approximately 36% of the individuals in the population will continue to have the BB genotype. The frequency of bb individuals is q^2 (0.4×0.4) and so will continue to be about 16%, and the frequency of Bb individuals will be $2pq$ ($2 \times 0.6 \times 0.4$), or approximately 48%. Phenotypically, we will still see approximately 84 black individuals (with either BB or Bb genotypes) and 16 white individuals (with the bb genotype) in the population.

This simple relationship has proved extraordinarily useful in assessing actual situations. Consider the recessive allele responsible for the serious human disease cystic fibrosis. This allele is present in North Americans of Caucasian descent at a frequency q of about 22 per 1000 individuals, or 0.022. What proportion of North American Caucasians, therefore, is expected to express this trait? The frequency of double recessive individuals (q^2) is expected to be 0.022×0.022, or 1 in every 2000 individuals. What proportion is expected to be heterozygous carriers? If the frequency of the recessive allele q is 0.022, then the frequency of the dominant allele p must be $1 - 0.022$, or 0.978. The frequency of heterozygous individuals ($2pq$) is thus expected to be $2 \times 0.978 \times 0.022$, or 43 in every 1000 individuals.

How valid are these calculated predictions? For many genes, they prove to be very accurate. Most human populations, for example, are large and effectively random-mating and are, therefore, similar to the "ideal" population envisioned by Hardy and Weinberg. As we will see, for some genes the calculated predictions do *not* match the actual values. The reasons they do not tell us a great deal about evolution.

Why Do Allele Frequencies Change?

According to the Hardy–Weinberg principle, both the allele and genotype frequencies in a large, random-mating population will remain constant from generation to generation if no mutation, no migration, and no selection occurs. The stipulations tacked onto the end of the statement are

Table 19.1	Agents of Evolutionary Change
Factor	**Description**
Mutation	The ultimate source of variation. Individual mutations occur so rarely that mutation alone does not change allele frequency much.
Migration	A very potent agent of change. Populations exchange members.
Nonrandom mating	Inbreeding is the most common form. It does not alter allele frequency but decreases the proportion of heterozygotes.
Genetic drift	Statistical accidents. Usually occurs only in very small populations.
Selection	The only form that produces *adaptive* evolutionary changes. Only rapid for allele frequencies greater than 0.01.

important. In fact, they are the key to the importance of the Hardy–Weinberg principle, because individual allele frequencies *are* changing all the time in natural populations, with some alleles becoming more common than others. The Hardy–Weinberg principle establishes a convenient baseline against which to measure such changes. By looking at how various factors alter the proportions of homozygotes and heterozygotes in populations (usually expressed as the **heterozygosity,** the likelihood that a randomly selected individual will be heterozygous at the gene locus), we can identify the forces affecting particular situations we observe.

Many factors can alter allele frequencies. Only five, however, alter the proportions of homozygotes and heterozygotes enough to produce significant deviations from the proportions predicted by the Hardy–Weinberg principle: mutation, migration (including both immigration into and emigration out of a given population), genetic drift (random loss of alleles, which is more likely in small populations), nonrandom mating, and selection (table 19.1). Of these, only selection produces adaptive evolutionary change because only in selection does the result depend on the nature of the environment. The other factors operate relatively independently of the environment, so the changes they produce are not shaped by environmental demands.

The Hardy–Weinberg principle states that in a large population mating at random and in the absence of other forces that would change the proportions of the different alleles at a given locus, the process of sexual reproduction (meiosis and fertilization) alone will not change these proportions.

Five Agents of Evolutionary Change

1. Mutation

Mutation from one allele to another can obviously change the proportions of particular alleles in a population. Mutation rates are generally so low that they have little effect on the Hardy–Weinberg proportions of common alleles. A single gene may mutate about 1 to 10 times per 100,000 cell divisions (although *some* genes mutate much more frequently than that). Because most environments are constantly changing, it is rare for a population to be stable enough to accumulate changes in allele frequency produced by a process this slow. Nonetheless, mutation is the ultimate source of genetic variation and thus makes evolution possible.

2. Migration

Migration is the movement of individuals from one population into another. It can be a powerful agent of change because members of two different populations may exchange genetic material. Sometimes migration is obvious, as when an animal moves from one place to another. If the characteristics of the newly arrived animal differ from those of the animals already there, and if the newcomer is adapted well enough to the new area to survive and mate successfully, the genetic composition of the receiving population may be altered. Other important kinds of migration are not as obvious. These subtler movements include the drifting of gametes or immature stages of plants or marine animals from one place to another (figure 19.5). Male gametes of flowering plants are often carried great distances by insects and other animals that visit their flowers. Seeds may also blow in the wind or be carried by animals or other agents to new populations far from their place of origin. However it occurs, migration can alter the genetic characteristics of populations and prevent them from maintaining Hardy–Weinberg equilibrium. Even low levels of migration tend to homogenize allele frequencies among populations and keep them from diverging. The evolutionary role of migration in particular populations is often difficult to assess, as the degree of divergence is also strongly influenced by the prevailing selective forces.

Gene flow is the movement of genes from one population to another. Such movement can take place by migration, but it can also come about by hybridization between individuals belonging to adjacent populations. The degree to which gene flow takes place depends on the size of the unoccupied area that separates two populations of a particular species, the degree to which individuals tend to move about, and the distance two individuals usually travel to mate.

FIGURE 19.5

Migration. The yellowish green cloud around these Monterey pines is pollen being dispersed by the wind. The male gametes within the pollen reach the egg cells of the pine passively in this way. In genetic terms, such dispersal is a form of migration.

3. Nonrandom Mating

Individuals with certain genotypes sometimes mate with one another more commonly than would be expected on a random basis, a phenomenon known as nonrandom mating. **Inbreeding** (mating with relatives) is a type of nonrandom mating that causes the frequencies of particular genotypes to differ greatly from those predicted by the Hardy–Weinberg principle. Inbreeding does not change the frequency of the alleles, but rather increases the proportion of homozygous individuals. This is why populations of self-fertilizing plants consist primarily of homozygous individuals, while **outcrossing** plants, which interbreed with individuals different from themselves, have a higher proportion of heterozygous individuals.

Because inbreeding increases the proportion of homozygous individuals in a population, it tends to promote double recessive combinations. It is for this reason that marriage between close relatives is discouraged and to some degree outlawed—it increases the possibility of producing children homozygous for an allele associated with one or more of the recessive genetic disorders discussed in chapter 13.

4. Genetic Drift

In small populations, frequencies of particular alleles may change drastically by chance alone. Such changes in allele frequency occur randomly, as if the frequencies were drifting, and are thus known as **genetic drift.** For this reason, a population must be large to be in Hardy–Weinberg equilibrium. If the gametes of only a few individuals form the next generation, the alleles they carry may by chance not be representative of the parent population from which they were drawn, as illustrated in figure 19.6, where a small number of individuals are removed from a bottle containing many. By chance, most of the individuals removed are blue, so the new population has a much higher population of blue individuals than the parent one had.

A series of small populations that are isolated from one another may come to differ strongly as a result of genetic drift, even when other evolutionary forces are pushing allele frequencies in different directions. Because of genetic drift, harmful alleles may increase in frequency in small populations, despite selective disadvantage, and favorable alleles may be lost even though selectively advantageous. It is interesting to realize that humans have lived in small groups for much of the course of their evolution; consequently, genetic drift may have been a particularly important factor in the evolution of our species. Two important causes of genetic drift are founder effects and bottlenecks.

Founder Effects. Sometimes one or a few individuals are dispersed and become the founders of a new, isolated population at some distance from their place of origin. These pioneers are not likely to have all the alleles present in the source population. Rare alleles in the source population may be a significant fraction of the new population's genetic endowment. This phenomenon, in which rare alleles may be more common in new populations, is called the **founder effect.** Founder effects are not rare in nature. Many self-pollinating plants start new populations from a single seed.

Founder effects have been particularly important in the evolution of organisms on distant oceanic islands, such as the Hawaiian Islands and the Galápagos Islands visited by Darwin. Most of the organisms in such areas probably derive from one or a few initial "founders." In a similar way, isolated human populations are often dominated by genetic features characteristic of their particular founders.

The Bottleneck Effect. Even if organisms do not move from place to place, occasionally their populations may be drastically reduced in size. This may result from flooding, drought, earthquakes, and other natural forces, or from progressive changes in the environment. The few surviving individuals constitute a random genetic sample of the original population. Such a restriction in genetic variability has been termed the **bottleneck effect.**

Some living species appear to be severely depleted genetically and have probably suffered from a bottleneck ef-

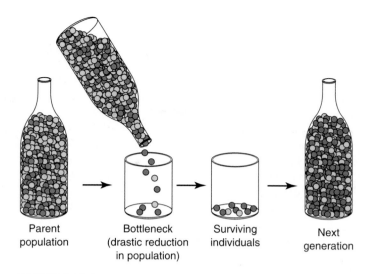

FIGURE 19.6
Genetic drift: The bottleneck effect. The parent population contains roughly equal numbers of blue and yellow individuals. By chance, the few individuals selected for the next generation are mostly blue. The bottleneck occurs because so few individuals form the next generation.

fect in the past. As shown by Steve O'Brien of the National Institutes of Health, all living cheetahs are practically identical genetically. Lacking the genetic diversity that might lend some of them resistance, they are therefore very susceptible to disease. The simplest explanation for this similarity is that there was a drastic reduction in the size of cheetah populations in the recent past and that all cheetahs living today have descended from a very few individuals. They have passed through a genetic bottleneck.

Bottleneck effects can occur within large populations when only a few individuals breed. Imagine a population containing only two alleles of a gene, B and b, in equal frequency (that is, $p = q = 0.5$). In a large Hardy–Weinberg population, the genotype frequencies are expected to be 0.25 BB, 0.50 Bb, and 0.25 bb. If only a small sample produces the next generation, large deviations in these genotype frequencies can occur by chance. Imagine, for example, that four gametes form the next generation, and that by chance they are two Bb heterozygotes and two BB homozygotes—the allele frequencies in the next generation are $p = 0.75$ and $q = 0.25$! If you were to replicate this experiment 1000 times, one of the two alleles would be missing entirely from about 8 of the 1000 populations. This leads to an important conclusion: genetic drift leads to the loss of alleles in isolated populations.

Genetic drift reduces genetic variation within isolated populations, but individual isolates may differ from one another substantially. Thus a researcher may conclude that a population is diverse if the isolating barriers are not apparent, when in fact the population consists of numerous genetically uniform isolated subpopulations.

5. Selection

As Darwin pointed out, some individuals leave behind more progeny than others, and the rate at which they do so is affected by their inherited characteristics. We describe the results of this process as **selection** and speak of both **artificial selection** and **natural selection.** In artificial selection, the breeder selects for the desired characteristics, as in the pigeons shown in figure 19.7. In natural selection, environmental conditions determine which individuals in a population produce the most offspring and thereby affect the proportions of genes among individuals of future populations. This is the key point in Darwin's proposal: that the environment imposes the conditions that determine the results of selection and thus the direction of evolution.

Selection to Match Climatic Conditions. Many studies of selection in nature have focused on genes encoding enzymes because in such cases the investigator can directly assess the consequences to the organisms of changes in the frequency of alternative enzyme alleles. Often investigators find that enzyme allele frequencies vary latitudinally, with one allele more common in northern populations but progressively less common at more southern locations (figure 19.8). The implication is that the more northern allele encodes a form of the enzyme that functions better in colder climates. Such **clines** in allele frequency are reported, for example, in a classic study by the Danish biologists Christiansen and Frydenberg of hemoglobin variation in the ocean fish *Zoarces viviparus.* One hemoglobin allele is predominant in the Atlantic Ocean northwest of Denmark but becomes progressively rarer in samples taken along a line southeast into the Baltic Sea, until populations are reached in which the allele is not present at all. Another protein, the enzyme phosphoglucomutase, shows no such cline in its variation, although sampled from the same fish.

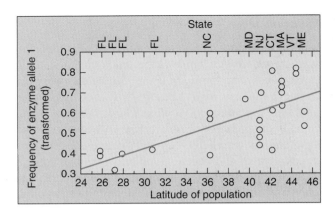

FIGURE 19.8
A latitudinal cline. The frequency of one allele of the enzyme alcohol dehydrogenase is greater in populations of *Drosophila* at more northern latitudes. (After Berry and Kreitman, 1993)

Similar clines have been reported for many other proteins. However, the presence of a cline in an allele's frequency is not in itself good evidence of selection at work on that enzyme locus. The cline may instead be the result of selection acting on some other gene located nearby on the chromosome, a phenomenon called **linkage disequilibrium.** To argue for selection, the enzyme locus must be examined more closely. A superb example is seen in studies of latitudinal clines in allele frequencies of the enzyme alcohol dehydrogenase (Adh) in *Drosophila melanogaster* along the east coast of the United States (see figure 19.8). Biochemical studies show that the allelic form of the enzyme (called an **allozyme**) that is more frequent in the north is more active at low temperatures than the allozyme that is more frequent in the south. When examined at the DNA level at 20 variable sites within the *Adh* gene, only the two allozyme alleles and one other site show a cline, a result that rules out linkage disequilibrium. Selection is clearly maintaining this cline in allele frequency.

(a)   (b)   (c)

FIGURE 19.7
Darwin's pigeons. In *On the Origin of Species*, Charles Darwin wrote, "Believing that it is always best to study some special group, I have, after deliberation, taken up domestic pigeons . . . if shown to an ornithologist, and he were told that they were wild birds, would certainly, I think, be ranked by him as well-defined species." The differences obtained by artificial selection of (a) the wild European rock pigeon and such domestic races as (b) the red fantail and (c) the fairy swallow, with its fantastic tufts of feathers around its feet, are indeed so great that the birds probably would, if wild, be classified in entirely different major groups.

Selection to Avoid Predators. Many of the most dramatic documented instances of adaptation involve genetic changes which decrease the probability of capture by a predator. The caterpillar larvae of the common sulphur butterfly *Colias eurytheme* usually exhibit a dull Kelly green color, providing excellent camouflage on the alfalfa plants on which they feed. An alternative bright blue color morph is kept at very low frequency because this color renders the larvae highly visible on the food plant, making it easier for bird predators to see them.

The way the shell markings in the land snail *Cepaea nemoralis* match its background habitat reflects the same pattern of avoiding predation by camouflage (figure 19.9). If one examines the broken shells of bird-predated snails and compares them with the snail population of the area, it is always the most conspicuous snails that birds preferentially take. The snail populations are polymorphic for many patterns of banding and color because the snail's environment is diverse; different habitats conceal different morphs.

Selection in Laboratory Environments. One way to assess the action of selection is to carry out artificial selection in the laboratory. Strains that are genetically identical except for the gene subject to selection can be crossed so that the possibility of linkage disequilibrium does not confound the analysis. Populations of bacteria provide a particularly powerful tool for studying selection in the laboratory because bacteria have a short generation time (less than an hour) and can be grown in huge numbers in growth vats called chemostats. In pioneering studies, Dan Hartl and coworkers backcrossed bacteria with different alleles of the enzyme 6-phosphogluconate dehydrogenase (6-PGD) into a homogeneous genetic background, and then compared the growth of the different strains when they were fed only gluconate, the enzyme's substrate. Hartl found that all of the alleles grew at the same rate! The different alleles were thus **selectively neutral** in a normal genetic background. However, when Hartl disabled an alternative biochemical pathway for the metabolism of gluconate, so that only 6-PGD mediated the utilization of this sole source of carbon, he obtained very different results: several alleles were markedly superior to others. Selection was clearly able to operate on these alleles, but only under certain conditions.

Selection for Pesticide Resistance. A particularly clear example of selection in action in natural populations is provided by the recent investigation by Taylor, Shen, and Kreitman of pesticide resistance in the tobacco budworm, *Heliothis virescens*. The widespread use of insecticides in agriculture has led to the rapid evolution of resistance in many pest species. One widely used type of insecticide, pyrethroids, acts to disable voltage-gated sodium channels in insect neurons, paralyzing the insect; pyrethroids have no effect on vertebrates. 660 tobacco budworms were collected from field populations in Georgia, Texas, Arizona, and Louisiana and tested for paralysis by pyrethroids. The budworms were also characterized genetically by examining

FIGURE 19.9
Selection in action. The snail *Cepaea nemoralis* is highly polymorphic for shell color and banding pattern. It appears that natural selection maintains this polymorphism by favoring different combinations in different microhabitats, each favored where it offers superior camouflage.

DNA variation at the sodium channel locus and also at an unrelated locus. The sensitivity of budworms to the insecticide was seen to closely parallel genetic variation at the sodium channel locus, while not reflecting changes at the unrelated locus.

A single gene also confers resistance to the herbicide triazine in the pigweed *Amaranthus hybridus*, one of about 28 agricultural weeds that have evolved resistance to this chemical. Triazine inhibits photosynthesis by binding to a protein in the chloroplast membrane. The gene encoding the protein has been sequenced. When the nucleotide sequence of the many resistant and susceptible strains were compared, they differed at only three nucleotide positions, all producing the same single amino acid substitution in the chloroplast protein, from serine (susceptible) to glycine (resistant).

Five factors can bring about a deviation from the proportions of homozygotes and heterozygotes predicted by the Hardy–Weinberg principle. Only selection regularly produces adaptive evolutionary change, but the genetic constitution of individual populations, and thus the course of evolution, can also be affected by mutation, migration, nonrandom mating, and genetic drift.

Identifying the Evolutionary Forces Maintaining Polymorphism

The Adaptive Selection Theory

As evidence began to accumulate in the 1970s that natural populations exhibit a great deal of genetic polymorphism (that is, many alleles of a gene exist in the population), the question arose: What evolutionary force is maintaining the polymorphism? As we have seen, there are in principle five processes that act on allele frequencies: mutation, migration, nonrandom mating, genetic drift, and selection. Because migration and nonrandom mating are not major influences in most natural populations, attention focused on the other three forces.

The first suggestion, advanced by R. C. Lewontin (one of the discoverers of enzyme polymorphism) and many others, was that selection was the force acting to maintain the polymorphism. Natural environments are often quite heterogeneous, so selection might reasonably be expected to pull gene frequencies in different directions within different microhabitats, generating a condition in which many alleles persist. This proposal is called the **adaptive selection theory.**

The Neutral Theory

A second possibility, championed by the great Japanese geneticist Moto Kimura, was that a balance between mutation and genetic drift is responsible for maintaining polymorphism. Kimura used elegant mathematics to demonstrate that, even in the absence of selection, natural populations could be expected to contain considerable polymorphism if mutation rates (generating the variation) were high enough and population sizes (promoting genetic drift) were small enough. In this proposal, selection is not acting, differences between alleles being "neutral to selection." The proposal is thus called the **neutral theory.**

Kimura's theory, while complex, can be stated simply:

$$\bar{H} = 1/(4N_e \mu + 1)$$

$\bar{H}$, the mean heterozygosity, is the likelihood that a randomly selected member of the population will be heterozygous at a randomly selected locus. In a population without selection, this value is influenced by two variables, the effective population size (N_e) and the mutation rate (μ).

The peculiar difficulty of the neutral theory is that the level of polymorphism, as measured by $\bar{H}$, is determined by the *product* of a very large number, N_e, and a very small number, μ, both very difficult to measure with precision. As a result, the theory can account for almost any value of $\bar{H}$, making it very difficult to prove or disprove. As you might expect, a great deal of controversy has resulted.

Testing the Neutral Theory

Choosing between the adaptive selection theory and the neutral theory is not simple, for they both appear to account equally well for much of the data on gene polymorphism in natural populations. A few well-characterized instances where selection acts on enzyme alleles do not settle the more general issue. An attempt to test the neutral theory by examining large-scale patterns of polymorphism sheds light on the difficulty of choosing between the two theories:

Population size: According to the neutral theory, polymorphism as measured by $\bar{H}$ should be proportional to the effective population size N_e, assuming the mutation rate among neutral alleles μ is constant. Thus, $\bar{H}$ should be much greater for insects than humans, as there are far more individuals in an insect population than in a human one. When DNA sequence variation is examined, the fruit fly *Drosophila melanogaster* indeed exhibits sixfold higher levels of variation, as the theory predicts; but when enzyme polymorphisms are examined, levels of variation in fruit flies and humans are similar. If the level of DNA variation correctly mirrors the predictions of the neutral theory, then something (selection?) is increasing variation at the enzyme level in humans. These sorts of patterns argue for rejection of the neutral theory.

The nearly neutral model: One way to rescue the neutral theory from these sorts of difficulties is to retreat from the assumption of strict neutrality, modifying the theory to assume that many of the variants are slightly deleterious rather than strictly neutral to selection. With this adjustment, it is possible to explain many of the population-size-dependent/large-scale patterns. However, little evidence exists that the wealth of enzyme polymorphism in natural populations is in fact slightly deleterious.

As increasing amounts of DNA sequence data become available, a detailed picture of variation at the DNA level is emerging. It seems clear that most nucleotide substitutions that change amino acids are disadvantageous and are eliminated by selection. But what about the many protein alleles that are seen in natural populations? Are they nearly neutral or advantageous? No simple answer is yet available, although the question is being actively investigated. Levels of polymorphism at enzyme-encoding genes may depend on both the action of selection on the gene (the adaptive selection theory) and on the population dynamics of the species (the nearly neutral theory), with the relative contribution varying from one gene to the next.

Adaptive selection clearly maintains some enzyme polymorphisms in natural populations. Genetic drift seems to play a major role in producing the variation we see at the DNA level. For most enzyme-level polymorphism, investigators cannot yet choose between the selection theory and the nearly neutral theory.

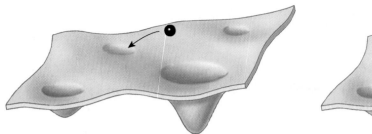

(a) Mild selection, represented by a slight incline of the adaptive topography, causes marble to roll.

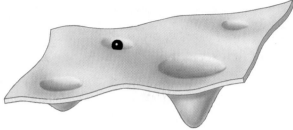

(b) Marble finds a shallow depression.

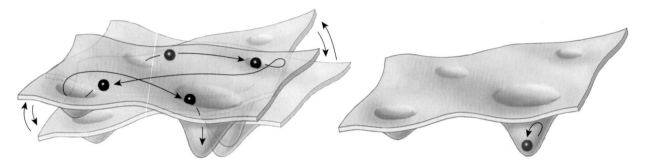

(c) Genetic drift, represented by a rocking adaptive topography, causes marble to roll around randomly.

(d) Marble, by chance, may find a deeper hole where its position is more stable.

FIGURE 19.10
The shifting balance theory. In this inverted adaptive topography, the marble rolls to different adaptive optima, genetic drift facilitating its exploration of the possibilities.

The Shifting Balance Theory

The **shifting balance theory** of evolution, developed by the great population geneticist Sewell Wright, provides a framework within which both adaptive selection and genetic drift play major roles. Wright's theory focuses on the way in which genes interact in determining phenotypes, with the appropriate balance of alleles at different loci determining fitness. While its details are beyond the scope of an introductory text, some sense of it can be gained by viewing a simple model. Imagine two genes *A* and *B*, each with two alleles, which interact in their effect on fitness. For every combination of allele frequencies at the two gene loci, a certain fitness results. If you plot the frequency of *A* versus *B* and connect the points with the same fitness, you obtain a family of curves that looks like a topographical map. The "mountains" represent particularly advantageous combinations of alleles. Wright called such displays **adaptive topographies.**

Now picture the adaptive topography upside down (figure 19.10). Adaptive peaks are now represented as "holes", depressions in the adaptive surface. Imagine dropping a marble onto the surface and lifting one edge; gravity plays the role of selection. The marble will role downhill along the surface, eventually falling to the bottom of the first hole it encounters. The steeper the incline you create by lifting the edge (the greater the selection), the faster the marble rolls (the more rapid the adaptation). The marble stays in its hole, held by gravity (selection), even though a far deeper hole may exist nearby. The optimal solution may never appear—selection drives the process to the first workable solution that presents itself.

So what is the basis for adaptive improvement, the hallmark of evolution? Strangely enough, it is random genetic drift. Start gently rocking your topography. The marble will start to roll around in its depression, to climb out of its hole. This is the analogue of genetic drift, shifting the balance between the two gene loci in random directions. Now rock it even more. This is the critical stage of the process: random movements may cause your marble to roll all the way out of its depression. Back on the adaptive surface, it is free to roll about and may encounter a deeper hole. It is by random perturbations induced by genetic drift that the landscape of adaptive possibilities is explored. Because deeper depressions are more difficult to leave, over time the random exploration of the topography will take the marble to the deepest hole. Random exploration eventually leads to adaptive improvement.

> The shifting balance theory suggests that genetic drift continually perturbs the balance of alleles contributing to fitness, disturbances which allow different selective combinations to be tested by selection.

Forms of Selection

Selection operates in natural populations of a species as skill does in a football game. In any individual game, it can be difficult to predict the winner, since chance can play an important role in the outcome; but over a long season, the teams with the most skillful players usually win the most games. In nature, those individuals best suited to their environments tend to win the evolutionary game by leaving the most offspring, although chance can play a major role in the life of any one individual. Selection is a statistical concept, just as betting is. Although you cannot predict the fate of any one individual or any one coin toss, it *is* possible to predict which *kind* of individual will tend to become more common in populations of a species, as it is possible to predict the proportion of heads after many coin tosses.

In nature many traits, perhaps most, are affected by more than one gene. The interactions between genes are typically complex, as you saw in chapter 13. For example, alleles of many different genes play a role in determining human height

(see figure 13.18). In such cases, selection operates on all the genes, influencing most strongly those that make the greatest contribution to the phenotype. How selection changes the population depends on which genotypes are favored.

Disruptive Selection

In some situations, selection acts to eliminate rather than to favor intermediate types. A clear example is the different color patterns of the African butterfly *Papilio dardanus*. In different parts of Africa, the color pattern of this butterfly is dramatically different, in each instance closely mimicking some other species that birds do not like to eat (*Papilio dardanus* is said to be a "mimic"). Birds detect and eat any patterns that do not look like distasteful butterflies, so any intermediate patterns that occur are selected against. In this case, selection is acting to eliminate the intermediate phenotypes, in effect partitioning the population into homozygous groups. This form of selection is called **disruptive selection** (figure 19.11*a*).

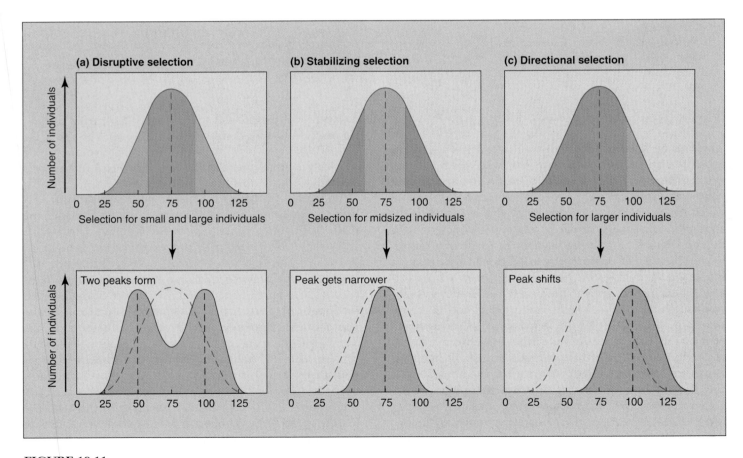

FIGURE 19.11
Three kinds of natural selection. (a) In *disruptive selection*, individuals in the middle of the range of phenotypes of a certain trait are selected against, and the extreme forms of the trait are favored. (b) In *stabilizing selection*, individuals with midrange phenotypes are favored, with selection acting against both ends of the range of phenotypes. (c) In *directional selection*, individuals concentrated toward one extreme of the array of phenotypes are favored.

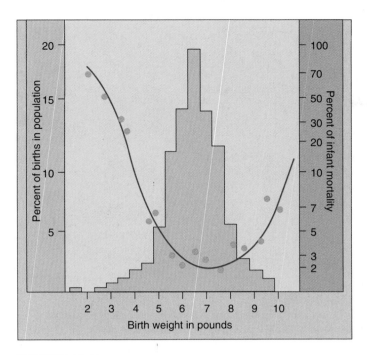

FIGURE 19.12
Stabilizing selection for birth weight in human beings. The death rate among babies (red curve; right *y*-axis) is lowest at an intermediate birth weight; both smaller and larger babies have a greater tendency to die than those around the optimum weight (blue area; left *y*-axis) of between 7 and 8 pounds.

Stabilizing Selection

When selection acts to eliminate *both* extremes from an array of phenotypes (figure 19.11*b*), the result is to increase the frequency of the already common intermediate type. In effect, selection is operating to prevent change away from this middle range of values. In a classic study carried out after an "uncommonly severe storm of snow, rain, and sleet" on February 1, 1898, 136 starving English sparrows were collected and brought to the laboratory of H. C. Bumpus at Brown University in Providence, Rhode Island. Of these, 64 died and 72 survived. Bumpus took standard measurements on all the birds. He found that among males, the surviving birds tended to be bigger, as one might expect from the action of directional selection. However, among females, the birds that perished were not smaller, on the average, than those that survived. But among them were many more individuals that had extreme measurements—measurements unusual for the population as a whole. Selection had acted most strongly against these individuals. When selection acts in this way, the population contains fewer individuals with alleles promoting extreme types. Selection has not changed the most common phenotype of the population, but rather made it even more common by eliminating extremes. Many examples similar to Bumpus's female sparrows are known. In humans, infants with intermediate weight at birth have the highest survival rate (figure 19.12). In ducks and chickens, eggs of intermediate

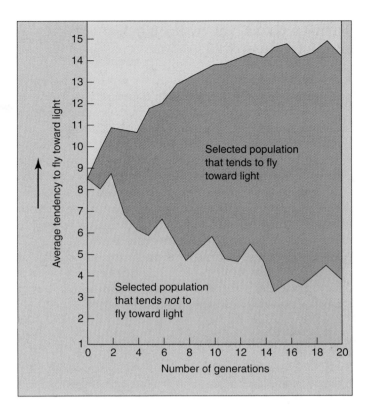

FIGURE 19.13
Directional selection for phototropism in *Drosophila*. In generation after generation, individuals of the fly *Drosophila* were selectively bred to obtain two populations. When flies with a strong tendency to fly toward light were used as parents for the next generation, their offspring had a greater tendency to fly toward light (top curve). When flies that tended *not* to fly toward light were used as parents for the next generation, their offspring had an even greater tendency not to fly toward light (bottom curve).

weight have the highest hatching success. This form of selection is called **stabilizing selection.**

Directional Selection

When selection acts to eliminate one extreme from an array of phenotypes (figure 19.10*c*), the genes promoting this extreme become less frequent in the population. Thus, in the *Drosophila* population illustrated in figure 19.13, the elimination of flies that move toward light causes the population to contain fewer individuals with alleles promoting such behavior. If you were to pick an individual at random from the new fly population, there is a smaller chance it would spontaneously move toward light than if you had selected a fly from the old population. Selection has changed the population in the direction of lower light attraction. This form of selection is called **directional selection.**

Selection on traits affected by many genes can favor both extremes of the trait, or intermediate values, or only one extreme.

Limits to What Selection Can Accomplish

Although selection is perhaps the most powerful of the five principal agents of genetic change, there are limits to what it can accomplish. These limits arise because alternative alleles may interact in different ways with other genes, and these interactions tend to set limits on how much a phenotype can be altered. For example, selecting for large clutch size in barnyard chickens eventually leads to eggs with thinner shells that break more easily. Because of limits imposed by gene interactions, strong selection is apt to result in rapid change initially, but the change soon comes to a halt as the interactions between genes increase. For this reason, we do not have gigantic cattle that yield twice as much meat as our leading strains, chickens that lay twice as many eggs as the best layers do now, or corn with an ear at the base of every leaf, instead of just a few leaves.

Selecting for Desirable Traits

In 1988, two investigators at Trinity College in Dublin, Ireland, carried out a fascinating analysis of the performances of thoroughbred horses that underscores this point. Over 80% of the gene pool of the thoroughbred horses racing today goes back to 31 known ancestors from the late eighteenth century. Despite intense directional selection on thoroughbreds, their performance times have not improved for the last 50 years (figure 19.14). Although the investigators concluded that the lack of improvement was probably not caused by a depletion of genetic variation in the thoroughbred stock, it is difficult to devise alternative explanations.

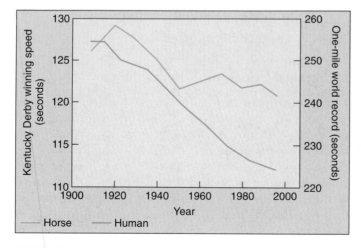

FIGURE 19.14
Selection for increased speed in racehorses is no longer effective. Kentucky Derby winning speeds have not improved significantly since 1950. By contrast, human running speed, as measured by the world's record for the mile, continues to improve, presumably reflecting not changes in genes but improvements in training.

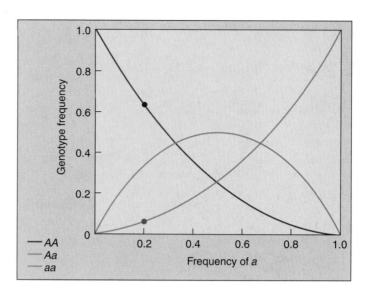

FIGURE 19.15
The relationship between allele frequency and genotype frequency. If allele *a* is present at a frequency of 0.2, the double recessive genotype *aa* is only present at a frequency of 0.04. In other words, only 4 in 100 individuals will have a homozygous recessive genotype, while 64 in 100 will have a homozygous dominant genotype.

Selection Against Rare Alleles

A second factor limits what selection can accomplish: selection acts only on phenotypes. Only those characteristics that are expressed in an organism's phenotype can affect the ability of that organism to produce progeny. For this reason, selection does not operate efficiently on rare recessive alleles, simply because there is no way to select them unless they come together as homozygotes. For example, when a recessive allele *a* is present at a frequency *q* equal to 0.2, 20% of the alleles for that particular gene will be *a*, but only four out of a hundred individuals (q^2) will be double recessive and display the phenotype associated with this allele (figure 19.15). For lower allele frequencies, the effect is even more dramatic: if the frequency in the population of the recessive allele *q* = 0.01, the frequency of recessive homozygotes in that population will be only 1 in 10,000.

The fact that selection acts on phenotypes rather than genotypes means that selection against undesirable genetic traits in humans or domesticated animals is difficult unless the heterozygotes can also be detected. For example, if a particular recessive allele *r* (*q* = 0.01) was considered undesirable, and none of the homozygotes for this allele were allowed to breed, it would take 1000 generations, or about 25,000 years in humans, to lower the allele frequency by half to 0.005. At this point, after 25,000 years of work, the frequency of homozygotes would still be 1 in 40,000, or 25% of what it was initially.

Selection cannot alter a trait with little or no genetic variation and will not eliminate recessive alleles.

Summary of Concepts

19.1 Genes vary in natural populations.

- Macroevolution describes evolution above the species level, the sweeping changes that evolution produces over time.

- Microevolution refers to evolutionary changes that occur within species. It includes both adaptation and random changes. Adaptation may lead to species formation, and, thus, ultimately to macroevolution.

- By the 1860s, natural selection was widely accepted as the correct explanation for the process of evolution. The field of evolution did not progress much further, however, until the 1920s because of the lack of a suitable explanation of how hereditary traits are transmitted.

- Invertebrates and outcrossing plants are often heterozygous at about 12% to 15% of their loci; the corresponding value for vertebrates is about 4% to 8%. These levels of genetic variation are much higher than any imagined before methods suitable to detect them were introduced in the 1960s.

19.2 Why do allele frequencies change in populations?

- Studies of how allele frequencies shift within populations allow investigators to study evolution in action.

- Meiosis does not alter allele frequencies within populations. Unless selection or some other force acts on the genes, the frequencies of their alleles remain unchanged from one generation to the next.

- Those individuals in a population that have combinations of alleles that better adapt them to their environment are said to be more "fit" because they tend to leave more offspring than those having other combinations of alleles.

- Selection acts only indirectly, on the phenotype, rather than directly on the genotype; it also acts on entire individuals, and, thus, assemblages of genes, rather than on individual genes. Both of these points are important in determining the consequences of selection.

19.3 Selection can act on traits affected by many genes.

- There are three principal kinds of selection. Directional selection acts to eliminate one extreme from an array of phenotypes; stabilizing selection acts to eliminate *both* extremes; and disruptive selection acts to eliminate rather than to favor the intermediate type.

Discussing Key Terms

1. **Adaptation**   A change in structure, physiology, or behavior caused by natural selection that promotes the likelihood of an organism's survival and reproduction in a particular environment.

2. **Natural selection**   The survival and differential reproductive success of individuals bearing certain allele combinations over others, producing shifts in those frequencies from one generation to the next. Natural selection can only operate on those traits that nature can "see," that is, phenotype. By acting on phenotype, selection ultimately alters the genotype by shifting allele frequencies.

3. **Heterozygosity**   The likelihood a randomly selected locus will display more than one allele in an individual. The extent to which a population contains genetic variability is often characterized as its heterozygosity.

4. **Population genetics**   The branch of genetics that deals with the behavior of genes in populations. Understanding population genetics is essential to understanding evolution.

5. **Hardy–Weinberg distribution**   The Hardy–Weinberg principle provides the baseline for all population genetic theory. It illustrates the fact that in large populations with random mating, allele and genotype frequencies—and, consequently, phenotype frequencies—will remain constant indefinitely, provided that selection, mutation, migration, nonrandom mating, and genetic drift do not occur.

6. **Fitness**   Fitness is a measure of the tendency of some organisms to leave more offspring than competing members of the same population. The genetic traits possessed by the fit individuals will appear in greater proportions among members of succeeding generations.

7. **Gene polymorphism**   The presence in a population of more than one allele of a gene at a significant frequency.

8. **Genetic drift**   Random fluctuations in allele frequencies over time. Most frequent in small populations, genetic drift can result in the permanent loss of alleles from a population.

Review Questions

1. Define *macroevolution* and *microevolution*. How are they related?

2. What is adaptation? How does it fit into Darwin's concept of evolution?

3. What is genetic polymorphism? Which are more polymorphic, invertebrates or vertebrates? What has polymorphism to do with evolution?

4. State the Hardy–Weinberg principle in mathematical terms and define all variables.

5. Given that allele *A* is present in a large random-mating population at a frequency of 54 per 100 individuals, what is the proportion of individuals in that population expected to be heterozygous for the allele? homozygous dominant? homozygous recessive?

6. What are the five factors that can alter the proportions of homozygotes and heterozygotes from the predicted Hardy–Weinberg values? Which produce adaptive evolutionary change? Why?

7. What is gene flow? How does it take place? What factors determine the degree to which gene flow takes place?

8. What is genetic drift? Why is it dependent on the size of the population?

9. Why does the founder effect have such a profound influence on a population's genetic makeup? How does the bottleneck effect differ from the founder effect?

10. Why does nonrandom mating adversely affect the Hardy–Weinberg prediction? What effect does inbreeding have on allele frequency? Why is marriage between close relatives discouraged?

11. Define *selection*. How does it alter allele frequencies? What are the three types of selection? Give an example of each.

12. Why are there limitations to the success of selection?

Thought Questions

1. The North American human population is similar to the ideal Hardy–Weinberg population in that it is very large (over 270 million people in the United States and Canada alone) and generally random-mating. Although mutation occurs, it alone does not lead to great changes in allele frequencies. However, migration from Latin American and Asian countries occurs at relatively high levels—perhaps 1% per year. The following data were obtained in 1976 by geneticist A. E. Mourant about relative numbers of individuals bearing the two alleles of the *MN* blood group:

	MM	**MN**	**NN**	**Total**
Observed individuals	1787	3037	1305	6129

Do these data suggest that migration, selection, or some other factor is disrupting the Hardy–Weinberg proportions of the three genotypes? What are the allele frequencies of *M* and *N*?

Internet Links

View From the Smithsonian
http://www.mnh.si.edu/
The NATIONAL MUSEUM OF NATURAL HISTORY site allows you on-line visits to many collections providing important evidence about evolution.

How Evolution Works
http://nitro.biosci.arizona.edu/courses/EEB182/Lecture02/lect2.html
A series of useful lectures from the University of Arizona on Natural Selection, Hardy Weinberg, and Genetic Drift.

For Further Reading

Berry, A., and M. Kreitman: "Molecular Analysis of an Allozyme Cline: Alcohol Dehydrogenase in *Drosophila melanogaster* on the East Coast of North America," *Genetics*, vol. 134, July 1993, pages 869–93. An instant classic, this paper is a superb illustration of how difficult it is to conclusively demonstrate the operation of natural selection; this is science done as it should be.

Caro, T., and others: "Ecological and Genetic Factors in Conservation—A Cautionary Tale," *Science*, January 1994, pages 485–87. Inbreeding is clearly important in the management of captive populations, but its relevance to small groups of animals living in the wild may have been overstated.

Cohn, J.: "Genetics for Wildlife Conservation," *BioScience*, March 1990, pages 167–71. DNA analysis provides valuable information for managing both natural and captive populations of endangered species.

Cunningham, P.: "The Genetics of Thoroughbred Horses," *Scientific American*, May 1991, pages 92–98. An enjoyable account of how breeding has influenced thoroughbred racehorses.

Gillis, A.: "Getting a Picture of Human Diversity," *BioScience*, vol. 44, January 1994, pages 8–11. A report on how population geneticists are using variation in human genes to track the history of *Homo sapiens*.

Lemonick, M.: "A Terrible Beauty," *Time*, December 12, 1994, pages 64–70. By selecting for show-ring looks, animal breeders are crippling America's purebred dogs.

Morin, P., and others: "Kin Selection, Social Structure, Gene Flow, and the Evolution of Chimpanzees," *Science*, August 1994, pages 1193–1201. By extracting DNA from chimp hairs, researchers have for the first time been able to study kinship patterns among wild chimps—and possibly identify a new species.

Weiner, J.: *The Beak of the Finch: Evolution in Real Time*, New York: Alfred A. Knopf, 1994. A highly recommended account of ongoing studies of evolution in action among Darwin's Galápagos finches, discussed in detail in chapter 20.

20

The Evidence for Evolution

Concept Outline

20.1 Natural selection explains adaptive microevolution.

Sickle Cell Anemia. Natural selection favors a mutation in the gene encoding hemoglobin in areas where malaria is prevalent, as it increases resistance to the disease.

Peppered Moths and Industrial Melanism. Natural selection favors dark-colored moths in areas of heavy pollution, while light-colored moths survive better in unpolluted areas.

The Beaks of Darwin's Finches. Natural selection favors stouter bills in dry years, when large tough-to-crush seeds are the only food available to finches.

20.2 The evidence for macroevolution is extensive.

The Fossil Record. When fossils are arranged in the order of their age, a continual series of change is seen, new changes being added at each stage.

The Molecular Record. When gene or protein sequences from organisms are arranged in the order that the organisms diverged, the sequences show an accumulation of increasing numbers of changes over time.

The Anatomical Record. When anatomical features of living animals are examined, evidence of shared ancestry is often apparent.

Convergent Evolution. Evolution favors similar forms under similar circumstances.

20.3 The theory of evolution has proven controversial.

Scientific Creationism. Disputing the theory of evolution, some believe instead that God created all species much as they are now, a belief that is untestable and thus not science.

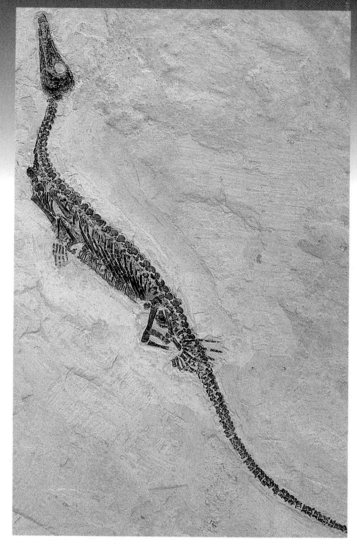

FIGURE 20.1
A window into the past. The fossil remains of the reptile *Mesosaurus* found in Permian sediments in Africa and South America provided one of the earliest clues to a former connection between the two continents. *Mesosaurus* was a freshwater species measuring less than half a meter long and so clearly incapable of a transatlantic swim. Therefore, it must have lived in the lakes and rivers of a formerly contiguous landmass that later became divided as Africa and South America drifted apart in the Cretaceous.

Of all the major ideas of biology, the theory that today's organisms evolved from now-extinct ancestors (figure 20.1) is perhaps the best known to the general public. This is not because the average person truly understands the basic facts of evolution, but rather because many people mistakenly believe that it represents a challenge to their religious beliefs. Similar highly publicized criticisms of evolution have occurred ever since Darwin's time. For this reason, it is important that, during the course of your study of biology, you address the issue squarely: Just what is the evidence for evolution?

Sickle Cell Anemia

The best evidence for evolution is that it can be seen in action. We will review three classic cases, starting with mutations in the blood protein hemoglobin. Sickle cell anemia, a hereditary disease affecting hemoglobin molecules, was first detected in 1904 by a Chicago physician, Ernest Iron (figure 20.2). A West Indian black student exhibited symptoms of severe anemia that appeared related to abnormal red blood cells. Dr. Iron noted, "The shape of the red cells was very irregular, but what especially attracted attention was the large number of thin, elongated, sickle-shaped and crescent-shaped forms." The disease was found to be common among African Americans. In chapter 13, we noted that this disorder, which affects roughly 3 African Americans out of every 1000, is associated with a particular recessive allele. Using the Hardy–Weinberg equation, you can calculate the frequency of the sickle cell allele in the African-American population; this frequency is the square root of 0.003, or approximately 0.054. In contrast, the frequency of the allele among white Americans is only about 0.001.

Molecular Basis of the Disease

Sickle cell anemia is often fatal. Until therapies were developed to more effectively treat its symptoms, almost all affected individuals died as children. Even today, 31% of patients in the United States die by the age of 15. The disease occurs because of a single amino acid change, repeated in the two β-chains of the hemoglobin molecule. In this change, a valine replaces the usual glutamic acid at a location on the surface of the protein near the oxygen-binding site. Unlike glutamic acid, valine is nonpolar (hydrophobic). Its presence on the surface of the molecule creates a "sticky" patch that attempts to escape from the polar water environment by binding to another similar patch. As long as oxygen is bound to the hemoglobin molecule there is no problem, since the oxygen atoms shield the critical area of the surface. When oxygen levels fall, such as after exercise or at high altitudes, oxygen is not so readily bound to hemoglobin, and the exposed sticky patch then binds to similar patches on other molecules, eventually producing long, fibrous clumps (figure 20.3). The result is a deformed, "sickle-shaped" red blood cell.

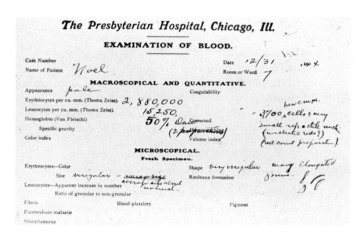

FIGURE 20.2
The first known sickle cell anemia patient. Dr. Ernest Iron's blood examination report on his patient Walter Clement Noel, December 31, 1904, described his oddly shaped red blood cells.

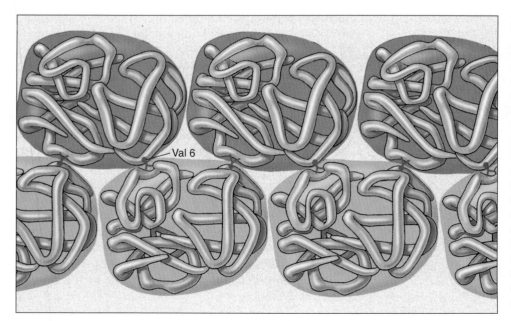

FIGURE 20.3
Why the sickle cell mutation causes hemoglobin to clump. The sickle cell mutation changes the sixth amino acid in the hemoglobin β-chain (position B6) from glutamic acid (very polar) to valine (nonpolar). The unhappy result is that the nonpolar valine at position B6, protruding from a corner of the hemoglobin molecule, fits into a nonpolar pocket on the opposite side of another hemoglobin molecule, causing the two molecules to clump together. As each molecule has both a B6 valine and an opposite nonpolar pocket, long chains form. When polar glutamic acid (the normal allele) occurs at position B6, it is not attracted to the nonpolar pocket, and no clumping occurs.
Copyright © Irving Geis.

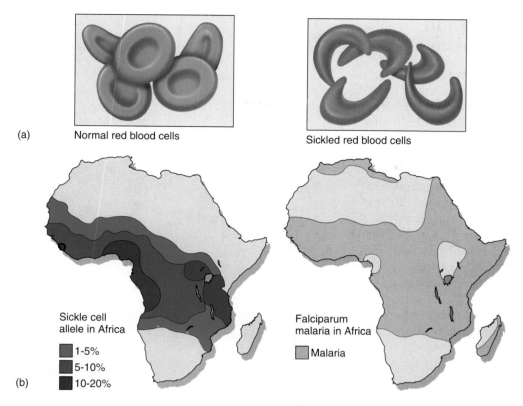

(a) Normal red blood cells

Sickled red blood cells

Sickle cell allele in Africa

- 1-5%
- 5-10%
- 10-20%

(b)

Falciparum malaria in Africa

- Malaria

FIGURE 20.4

Frequency of sickle cell allele and distribution of falciparum malaria. (a) The red blood cells of people homozygous for the sickle cell allele collapse into sickled shapes when the oxygen level in the blood is low. (b) The distribution of the sickle cell allele in Africa coincides closely with that of falciparum malaria.

Malaria and Balancing Selection

Individuals who are heterozygous for the valine-specifying allele (designated allele *S*) are said to possess the sickle cell trait. They produce some sickle-shaped red blood cells, but only 2% of the number seen in homozygous individuals.

The average incidence of the *S* allele in the central African population is about 0.12, far higher than that found among African Americans. From the Hardy–Weinberg principle, you can calculate that 1 in 5 Central African individuals are heterozygous at the *S* allele, and 1 in 100 develops the fatal form of the disorder. People who are homozygous for the sickle cell allele almost never reproduce because they usually die before they reach reproductive age. Why is the *S* allele not eliminated from the central African population by selection, rather than being maintained at such high levels? People who are heterozygous for the sickle cell allele are much less susceptible to malaria— one of the leading causes of illness and death, especially among young children—in the areas where the allele is common. In addition, for reasons not yet understood, women who are heterozygous are more fertile than are those who lack the allele. Consequently, even though most homozygous recessive individuals die before they have children, the sickle cell allele is maintained at high levels in these populations (it is selected for) because of its association with resistance to malaria in heterozygotes and with increased fertility in female heterozygotes.

As Darwin's theory predicts, it is the environment that acts to maintain the sickle cell allele at high frequency. In this case the environmental characteristic that is exercising selection is the presence of malaria. For people living in areas where malaria is common, having the sickle cell allele in the heterozygous condition has adaptive value (figure 20.4). Among African Americans, however, many of whom have lived for some 15 generations in a country where malaria has been relatively rare and is now essentially absent, the environment does not place a premium on resistance to malaria. Consequently, no adaptive value counterbalances the ill effects of the disease; in this non-malarial environment, selection is acting to eliminate the *S* allele. Only 1 in 375 African Americans develop sickle cell anemia, far fewer than in Central Africa.

The hemoglobin allele *S*, responsible for sickle cell anemia in homozygotes, is maintained by balancing selection in Central Africa, where heterozygotes for the *S* allele are resistant to malaria.

Peppered Moths and Industrial Melanism

The peppered moth, *Biston betularia*, is a European moth that rests on tree trunks during the day. Until the mid-nineteenth century, almost every captured individual of this species had light-colored wings. From that time on, individuals with dark-colored wings increased in frequency in the moth populations near industrialized centers until they made up almost 100% of these populations. Black individuals had a dominant allele that was present but very rare in populations before 1850. Biologists soon noticed that in industrialized regions where the dark moths were common, the tree trunks were darkened almost black by the soot of pollution. Dark moths were much less conspicuous resting on them than were light moths. In addition, the air pollution that was spreading in the industrialized regions had killed many of the light-colored lichens on tree trunks, making the trunks darker.

Selection for Melanism

Can Darwin's theory explain the increase in the frequency of the dark allele? Why did dark moths gain a survival advantage around 1850? An amateur moth collector named J. W. Tutt proposed what became the most commonly accepted hypothesis explaining the decline of the light-colored moths. He suggested that peppered forms were more visible to predators on sooty trees that have lost their lichens. Consequently, birds ate the peppered moths resting on the trunks of trees during the day. The black forms, in contrast, were at an advantage because they were camouflaged (figure 20.5). Although Tutt initially had no evidence, British ecologist Bernard Kettlewell tested the hypothesis in the 1950s by rearing populations of peppered moths with equal numbers of dark and light individuals. Kettlewell then released these populations into two sets of woods: one, near heavily polluted Birmingham, the other, in unpolluted Dorset. Kettlewell set up rings of traps around the woods to see how many of both kinds of moths survived. To evaluate his results, he had marked the released moths with a dot of paint on the underside of their wings, where birds could not see it.

In the polluted area near Birmingham, Kettlewell trapped 19% of the light moths, but 40% of the dark ones.

FIGURE 20.5
Tutt's hypothesis explaining industrial melanism. These photographs show color variants of the peppered moth, *Biston betularia*. Tutt proposed that the dark moth is more visible to predators on unpolluted trees (*top*), while the light moth is more visible to predators on bark blackened by industrial pollution (*bottom*).

This indicated that dark moths had a far better chance of surviving in these polluted woods, where the tree trunks were dark. In the relatively unpolluted Dorset woods, Kettlewell recovered 12.5% of the light moths but only 6% of the dark ones. This indicated that where the tree trunks were still light-colored, light moths had a much better chance of survival. Kettlewell later solidified his argument by placing hidden blinds in the woods and actually filming birds eating the moths. Sometimes the birds Kettlewell observed actually passed right over a moth that was the same color as its background.

Industrial Melanism

Industrial melanism is a term used to describe the evolutionary process in which darker individuals come to predominate over lighter individuals since the industrial revolution as a result of natural selection. The process is widely believed to have taken place because the dark organisms are better concealed from their predators in habitats that have been darkened by soot and other forms of industrial pollution, as suggested by Kettlewell's research.

Dozens of other species of moths have changed in the same way as the peppered moth in industrialized areas throughout Eurasia and North America, with dark forms becoming more common from the mid-nineteenth century onward as industrialization spread.

Selection Against Melanism

In the second half of the twentieth century, with the widespread implementation of pollution controls, these trends are reversing, not only for the peppered moth in many areas in England, but also for many other species of moths throughout the northern continents. These examples provide some of the best documented instances of changes in allelic frequencies of natural populations as a result of natural selection due to specific factors in the environment.

In England, the pollution promoting industrial melanism began to reverse following enactment of Clean Air legislation in 1956. Beginning in 1959, the *Biston* population at Caldy Common outside Liverpool has been sampled each year. The frequency of the melanic (dark) form has dropped from a high of 94% in 1960 to its current (1994) low of 19% (figure 20.6). Similar reversals have been documented at numerous other locations throughout England. The drop correlates well with a drop in air pollution, particularly with tree-darkening sulfur dioxide and suspended particulates.

Interestingly, the same reversal of industrial melanism appears to have occurred in America during the same time that it was happening in England. Industrial melanism in the American subspecies of the peppered moth was not as widespread as in England, but it has been well-documented at a rural field station near Detroit. Of 576 peppered moths collected there from 1959 to 1961, 515 were melanic, a frequency of 89%. The American Clean Air act, passed in 1963, led to significant reductions in air pollution. Resampled in 1994, the Detroit field station peppered moth population had only 15% melanic moths (see figure 20.6)! The moths in Liverpool and Detroit, both part of the same natural experiment, exhibit strong evidence of natural selection.

Reconsidering the Target of Natural Selection

Tutt's hypothesis, widely accepted in the light of Kettlewell's studies, is currently being reevaluated. The problem is that the recent selection against melanism does not appear to correlate with changes in tree lichens. At Caldy Common, the light form of the peppered moth began its increase in frequency long before lichens began to reappear on the trees. At the Detroit field station, the lichens never changed significantly as the dark moths first became dominant and then declined over the last 30 years. In fact, investigators have not been able to find peppered moths on Detroit trees at all, whether covered with lichens or not. Wherever the moths rest during the day, it does not appear to be on tree bark. Some evidence suggests they rest on leaves on the treetops, but no one is sure.

The action of selection may depend on other differences between light and dark forms of the peppered moth besides their wing coloration. Researchers report, for example, a clear difference in their ability to survive as caterpillars under a variety of conditions. Perhaps natural selection is targeting the caterpillars rather than the adults. While we can't say exactly what is going on yet, researchers are actively investigating one of the best documented instances of natural selection in action.

> **Natural selection has favored the dark form of the peppered moth in areas subject to severe air pollution, perhaps because on darkened trees they are less easily seen by moth-eating birds. Selection has in turn favored the light form as pollution has abated.**

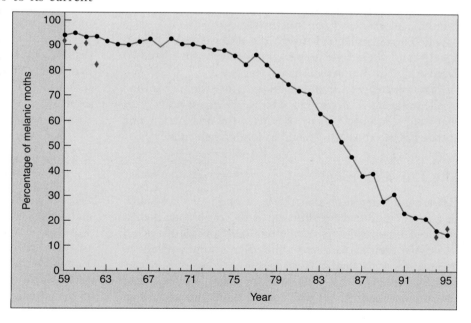

FIGURE 20.6

Selection against melanism. The circles indicate the frequency of melanic *Biston* moths at Caldy Common in England, sampled continuously from 1959 to 1995. Diamonds indicate frequencies in Michigan from 1959 to 1962 and from 1994 to 1995.

Source: Data from Grant, et al., "Parallel Rise and Fall of Melanic Peppered Moths" in *Journal of Heredity*, vol. 87, 1996, Oxford Unversity Press.

The Beaks of Darwin's Finches

Darwin's finches are a classic example of evolution by natural selection. Darwin collected 31 specimens of finch from three islands when he visited the Galápagos Islands off the coast of Ecuador in 1835. Darwin, not an expert on birds, had trouble identifying the specimens, believing by examining their bills that his collection contained wrens, "gross-beaks," and blackbirds. You can see Darwin's sketches of four of these birds in figure 20.7.

The Importance of the Beak

Upon Darwin's return to England, ornithologist John Gould examined the finches. Gould recognized that Darwin's collection was in fact a closely related group of distinct species, all similar to one another except for their bills. In all, there were 13 species. The two ground finches with the larger bills in figure 20.7 feed on seeds which they crush in their beaks, while the two with narrower bills eat insects. One species is a fruit eater, another a cactus eater, yet another a "vampire" that creeps up on seabirds and uses its sharp beak to drink their blood. Perhaps most remarkable are the tool users, woodpecker finches that pick up a twig, cactus thorn, or leaf stalk, trim it into shape with their bills, and then poke it into dead branches to pry out grubs.

The correspondence between the beaks of the 13 finch species and their food source immediately suggested to Darwin that evolution had shaped them:

"Seeing this gradation and diversity of structure in one small, intimately related group of birds, one might really fancy that from an original paucity of birds in this archipelago, one species has been taken and modified for different ends."

Was Darwin Wrong?

If Darwin's suggestion that the beak of an ancestral finch had been "modified for different ends" is correct, then it ought to be possible to see the different species of finches acting out their evolutionary roles, each using their bills to acquire their particular food specialty. The four species that crush seeds within their bills, for example, should feed on different seeds, those with stouter beaks specializing on harder-to-crush seeds.

Many biologists visited the Galápagos after Darwin, but it was 100 years before any tried this key test of his hypothesis. When the great naturalist David Lack finally set out to do this in 1938, observing the birds closely for a full 5 months, his observations seemed to contradict Darwin's proposal! Lack often observed many different species of finch feeding together on the same seeds. His data indicated that the stout-beaked species and the slender-beaked species were feeding on the very same array of seeds.

We now know that it was Lack's misfortune to study the birds during a wet year, when food was plentiful. The finch's beak is of little importance in such flush times; slender and stout beaks work equally well to gather the abundant tender small seeds. Later work has revealed a very different picture during leaner, dry years, when few seeds are available and the difference between survival and starvation depends on being able to eat them.

A Closer Look

The key to successfully testing Darwin's proposal that the beaks of Galápagos finches are adaptations to different food sources proved to be patience. Starting in 1973, Peter and Rosemary Grant of Princeton University and generations of their students have studied the medium ground finch *Geospiza fortis* (figure 20.8) on a tiny island in the center of the Galápagos called Daphne Major. These finches feed preferentially on small tender seeds, produced in abundance by plants in wet years. The birds resort to larger, drier seeds, which are harder to crush, only when small seeds are hard to find. Such lean times come during long periods of dry weather, when plants produce few seeds, large or small.

FIGURE 20.7
Darwin's own sketches of Galápagos finches. From Darwin's *Journal of Researches:* (1) large ground finch *Geospiza magnirostris;* (2) medium ground finch *Geospiza fortis;* (3) small tree finch *Camarhynchus parvulus;* (4) warbler finch *Certhidea olivacea.*

The Grants quantified beak shape among the medium ground finches of Daphne Major by carefully measuring beak depth (width of beak, from top to bottom, at its base) on individual birds. Measuring many birds every year, they were able to assemble for the first time a detailed portrait of evolution in action. The Grants found that beak depth changed from one year to the next in a predictable fashion. During droughts, beak depth increased, only to decrease again when wet seasons returned (figure 20.9).

Could these changes in beak dimension reflect the action of natural selection? An alternative possibility might be that the changes in beak depth do not reflect changes in gene frequencies, but rather are simply a response to diet, with poorly fed birds having stouter beaks. To rule out this possibility, the Grants measured the relation of parent bill size to offspring bill size, examining many broods over several years. The depth of the bill was passed down faithfully from one generation to the next, suggesting the differences in bill size indeed reflected gene differences. Switching eggs between the nests of stout-billed and slender-billed birds would provide even stronger evidence, but these experiments are not practical because they would seriously disrupt the small *G. fortis* population of Daphne Major, the subject of so many years of study.

Darwin Was Right after All

If the year-to-year changes in beak depth indeed reflect genetic changes, as now seems likely, and these changes can be predicted by the pattern of dry years, then Darwin was right after all—natural selection does seem to be operating to adjust the beak to its food supply. Birds with stout beaks have an advantage during dry periods, for they can break the large, dry seeds that are the only food available. When small seeds become plentiful once again with the return of wet weather, a smaller beak proves a more efficient tool for harvesting smaller seeds.

Among Darwin's finches, natural selection adjusts the shape of the beak in response to the nature of the available food supply, adjustments which can be seen to be occurring even today.

FIGURE 20.8
The subject of the Grants' study. The medium ground finch, *Geospiza fortis*, feeds on seeds that it crushes in its bill.

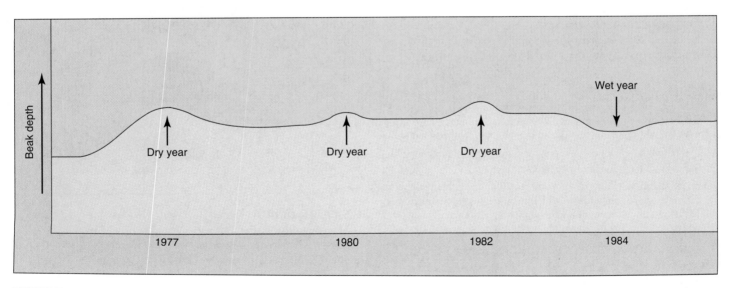

FIGURE 20.9
Evidence that natural selection alters beak size in *Geospiza fortis*. In dry years, when only large, tough seeds are available, the mean beak size increases. In wet years, when many small seeds are available, smaller beaks become more common.

The Fossil Record

Adaptation within natural populations constitutes strong evidence that Darwin was right in arguing that natural selection can bring about genetic change within populations. The examples we reviewed offer direct and compelling evidence of microevolutionary change. What of macroevolution, however? What is the evidence that macroevolution has led to the diversity of life on earth (table 20.1)?

The most direct evidence of macroevolution is found in the fossil record. Today we have a far more complete understanding of this record than was available in Darwin's time. Fossils are the preserved remains, tracks, or traces of once-living organisms. Fossils are created when organisms become buried in sediment, the calcium in bone or other hard tissue mineralizes, and the surrounding sediment eventually hardens to form rock. The fossils contained in layers of sedimentary rock reveal a history of life on earth.

Dating Fossils

By dating the rocks in which fossils occur, we can get an accurate idea of how old the fossils are. In Darwin's day, rocks were dated by their position with respect to one another (*relative dating*); rocks in deeper strata are generally older. Knowing the relative positions of sedimentary rocks and the rates of erosion of different kinds of sedimentary rocks in different environments, geologists of the nineteenth century derived a fairly accurate idea of the relative ages of rocks.

Today, rocks are dated by measuring the degree of decay of certain radioisotopes contained in the rock (*absolute dating*); the older the rock, the more its isotopes have decayed. Because radioactive isotopes decay at a constant rate unaltered by temperature or pressure, the isotopes in a rock act as an internal clock, measuring the time since the rock was formed. This is a more accurate way of dating rocks and provides dates stated in millions of years, rather than relative dates.

Table 20.1 Examples of the Evidence for Evolution

THE FOSSIL RECORD

When fossils are arrayed in the order of their age, a continual series of changes is seen.

THE MOLECULAR RECORD

The longer organisms have been separated according to the fossil record, the more differences are seen in their DNA and proteins.

HOMOLOGY

Many organisms exhibit organs that are similar in structure to those in a recent common ancestor. This is evidence of evolutionary relatedness.

DEVELOPMENT

The early developmental stages of an organism often exhibit characteristics seen in the early stages of other organisms, suggesting that the two groups are related.

VESTIGIAL STRUCTURES

Many vertebrates contain structures with reduced or no function that resemble functional structures of other vertebrates, suggesting that the structures are inherited from a common ancestor.

CONVERGENCE

The marsupials in Australia closely resemble the placental mammals of the rest of the world, which argues that parallel selection has occurred. Similarly, euphorbs and cacti, two unrelated desert plants, resemble each other.

PATTERNS OF DISTRIBUTION

Inhabitants of oceanic islands resemble forms of the nearest mainland organisms but show some differences, which suggests that they evolved from mainland migrants.

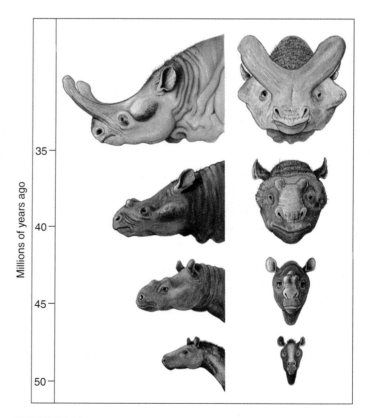

FIGURE 20.10

Macroevolution. Evolution in a group of hoofed mammals known as titanotheres between the Early Eocene Epoch (about 50 million years ago) and the Early Oligocene Epoch (about 36 million years ago). The small, bony protuberances that began to appear by the Middle Eocene (about 45 million years ago) evolved into relatively large, blunt horns.

A History of Evolutionary Change

When fossils are arrayed according to their age, from oldest to youngest, they often provide evidence of successive evolutionary change. Among the hoofed mammals illustrated in figure 20.10, small, bony bumps on the nose can be seen to change continuously, until they become large, blunt horns. In the evolution of horses, the number of toes on the front foot gradually diminishes from four to one. About 200 million years ago, oysters underwent a change from small curved shells to larger, flatter ones, with progressively flatter fossils being seen in the fossil record over a period of 12 million years (figure 20.11). A host of other examples all illustrate a record of successive change. The demonstration of this successive change is one of the strongest lines of evidence that evolution has occurred.

Gaps in the Fossil Record

While many gaps interrupted the fossil record in Darwin's era, even then, scientists knew of the *Archaeopteryx* fossil transitional between reptiles and birds. Today, the fossil record is far more complete, particularly among the vertebrates; fossils have been found linking all the major groups. The forms linking mammals to reptiles is particularly well known. A series of extinct marine mammals linking whales to terrestrial, four-legged hoofed ancestors filled in one of the last significant gaps (figure 20.12).

The fossil record provides a clear record of continuous evolutionary change.

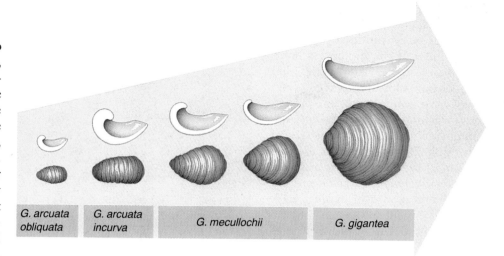

| *G. arcuata obliquata* | *G. arcuata incurva* | *G. mecullochii* | *G. gigantea* |

FIGURE 20.11

Evolution of shell shape in oysters. Over 12 million years of the Early Jurassic Period, the shells of this group of coiled oysters became larger, thinner, and flatter. These animals rested on the ocean floor in a special position called the "life position," and it may be that the larger, flatter shells were more stable against potentially disruptive water movements.

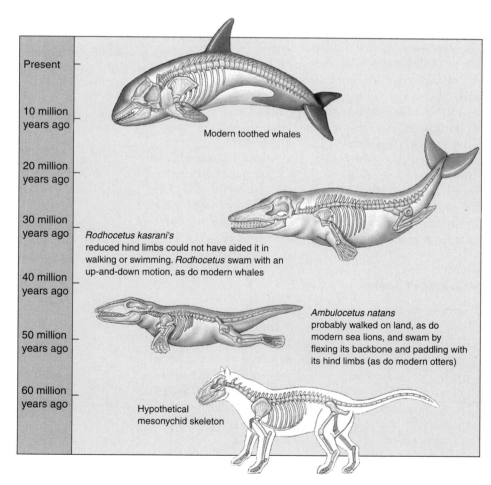

Present

10 million years ago

Modern toothed whales

20 million years ago

30 million years ago

Rodhocetus kasrani's reduced hind limbs could not have aided it in walking or swimming. *Rodhocetus* swam with an up-and-down motion, as do modern whales

40 million years ago

Ambulocetus natans probably walked on land, as do modern sea lions, and swam by flexing its backbone and paddling with its hind limbs (as do modern otters)

50 million years ago

60 million years ago

Hypothetical mesonychid skeleton

FIGURE 20.12

Whale "missing links." The recent discoveries of *Ambulocetus* and *Rodhocetus* have filled in the gaps between the mesonychids, the hypothetical ancestral link between the whales and the hoofed mammals, and present-day whales.

The Molecular Record

Traces of our evolutionary past are also evident at the molecular level. We possess color vision genes that have become more complex as vertebrates have evolved, and employ pattern formation genes during early development that all animals share. If you think about it, the fact that organisms have evolved successively from relatively simple ancestors implies that a record of evolutionary change is present in the cells of each of us, in our DNA. According to evolutionary theory, every evolutionary change involves the substitution of new versions of genes for old ones, the new alleles arising from the old by mutation and coming to predominance through favorable selection. Thus, a series of evolutionary changes involves a continual accumulation of genetic changes in the DNA. Organisms that are more distantly related will have accumulated a greater number of evolutionary differences, while two species that are more closely related will share a greater portion of their DNA. This pattern of divergence is clearly seen in a human hemoglobin polypeptide (figure 20.13). Chimpanzees, gorillas, orangutans, and macaques, vertebrates more closely related to humans, have fewer differences from humans in the 146-amino-acid hemoglobin β-chain than do more distantly related mammals, like dogs. Nonmammalian vertebrates differ even more, and nonvertebrate hemoglobins are the most different of all.

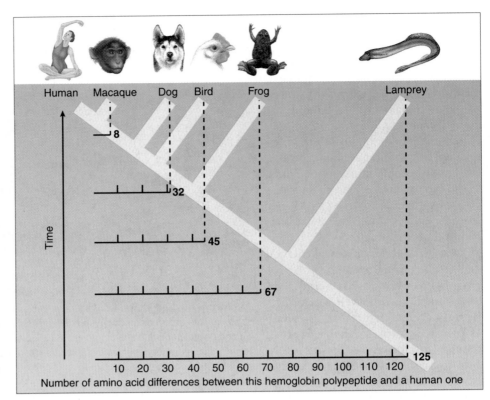

FIGURE 20.13
Molecules reflect evolutionary divergence. You can see that the greater the evolutionary distance from humans (yellow cladogram), the greater the number of amino acid differences in the vertebrate hemoglobin polypeptide.

Molecular Clocks

This same pattern is seen when DNA sequences from various organisms are compared. For example, the longer the time since the organisms diverged, the greater the number of differences in the nucleotide sequence of the gene for cytochrome *c*—a protein, as we learned in chapter 9, that plays a key role in oxidative metabolism (figure 20.14). The changes appear to accumulate in cytochrome *c* at a constant rate, a phenomenon sometimes referred to as a "molecular clock."

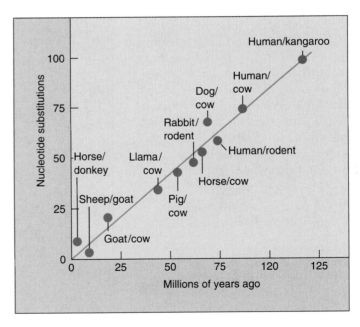

FIGURE 20.14
The evolution of cytochrome *c*. When the time since each pair of organisms presumably diverged is plotted against the number of nucleotide differences, the result is a straight line, suggesting that the cytochrome *c* gene is evolving at a constant rate.

Proteins Evolve at Different Rates

The constant rate at which cytochrome *c* has evolved raises the interesting question of whether other proteins also evolve at constant rates. As a rule of thumb, highly conserved proteins like hemoglobin or cytochrome *c* provide the best molecular clocks, but all proteins for which data are available appear to accumulate changes over time. However, different proteins evolve at very different rates, as can be seen in figure 20.15, which presents estimates of the number of "acceptable" point mutations (that is, changes successfully incorporated into the protein) per 100 amino acids per million years. The fastest rate of change appears to be in fibrinopeptides, while the most highly conserved protein is histone H4. Even faster rates of change are seen for pseudogenes, which are not transcribed, suggesting that molecular evolution proceeds more quickly when less constrained by selection.

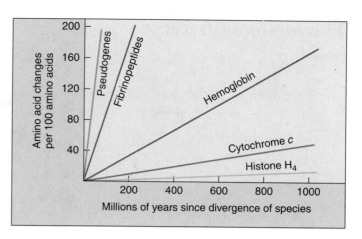

FIGURE 20.15
Different molecular clocks tick at different speeds.
Pseudogenes and fibrinopeptides evolve more rapidly than hemoglobin, which evolves more rapidly than cytochrome *c* or histone H4.

Phylogenetic Trees

The same regular pattern of change is seen in the globins and many other proteins (figure 20.16). Some genes, such as the ones specifying the protein hemoglobin, have been well studied, and the entire time course of their evolution can be laid out with confidence by tracing the origin of particular substitutions in their nucleotide sequences. The pattern of descent obtained is called a **phylogenetic tree.** It represents the evolutionary history of the gene. Note that the successive changes in the hemoglobin molecule produce a tree that closely reflects the evolutionary relationships predicted by a study of anatomy. Whales, dolphins, and porpoises cluster together, as do the primates and the hoofed animals. The pattern of accumulating changes seen in the molecular record constitutes strong direct evidence for macroevolution.

> **The genes encoding proteins have undergone continual evolution, accumulating increasing numbers of changes over time.**

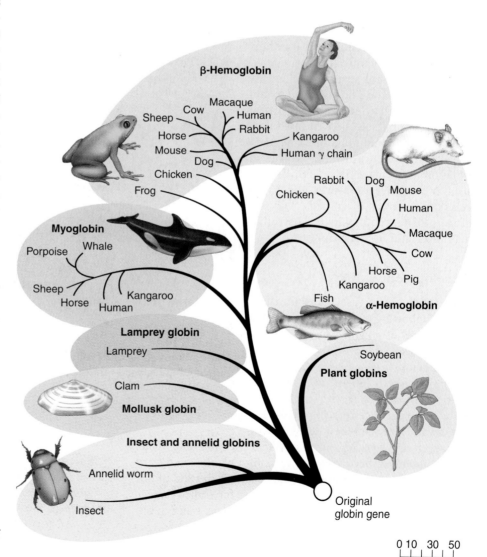

FIGURE 20.16
Evolution of the globin gene. The length of the various lines corresponds to the number of nucleotide substitutions in the gene.

The Anatomical Record

Homology

As vertebrates evolved, the same bones were sometimes put to different uses. Yet the bones are still seen, their presence betraying their evolutionary past. For example, the forelimbs of vertebrates are all **homologous structures,** that is, structures with different appearances and functions that all derived from the same body part in a common ancestor. You can see in figure 20.17 how the bones of the forelimb have been modified in one way in the wings of bats, in another way in the fins of porpoises, and in yet other ways in the forelimbs of frogs, horses, and humans.

Not all similar features are homologous. **Analogous structures** come to resemble each other and have similar functions as the result of parallel evolution in separate lineages. For example, the flippers of penguins and dolphins are analogous structures that originated from very different structures in ancestors of two separate lineages (birds and mammals) and then were modified through natural selection to look alike and serve the same function. The marsupial mammals of Australia evolved in isolation from placental mammals, but similar selective pressures generated very similar kinds of animals. We will examine the evolution of Australian mammals later in this chapter.

Development

Some of the strongest anatomical evidence supporting evolution comes from comparisons of how organisms develop. In many cases, the evolutionary history of an organism can be seen to unfold during its development, with the embryo exhibiting characteristics of the embryos of its ancestors (figure 20.18). For example, early in their development, human embryos possess gill slits, like a fish; at a later stage, every human embryo has a long bony tail, the vestige of which we carry to adulthood as the coccyx at the end of our spine. Human fetuses even possess a fine fur (called *lanugo*) during the fifth month of development. These relict developmental forms suggest strongly that our development has evolved, with new instructions layered on top of old ones.

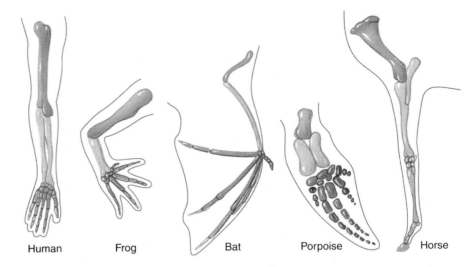

Human   Frog   Bat   Porpoise   Horse

FIGURE 20.17
Homology among the bones of the forelimb. Although these structures show considerable differences in form and function, the same basic bones are present in the forelimbs of humans, frogs, bats, porpoises, and horses.

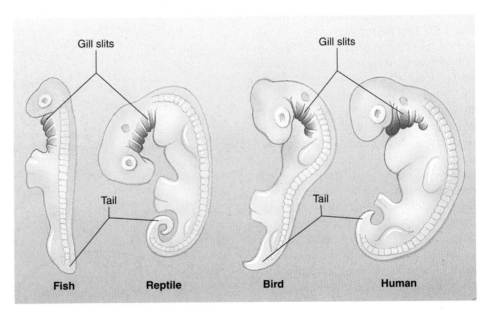

Fish   Reptile   Bird   Human

FIGURE 20.18
Our embryos show our evolutionary history. The embryos of various groups of vertebrate animals show the features they all share early in development, such as gill slits (*in purple*) and a tail.

The fact that seemingly different organisms exhibit similar embryological forms provides direct evidence of an evolutionary relationship. Slugs and giant ocean squids, for example, do not bear much superficial resemblance to each other, but the similarity of their embryological forms provides convincing evidence that they are both mollusks.

Vestigial Structures

Many organisms possess *vestigial structures* that have no apparent function, but that resemble structures their presumed ancestors had. Humans, for example, possess a complete set of muscles for wiggling their ears, just as a coyote does (table 20.2). Boa constrictors have hip bones and rudimentary hind legs. Manatees have fingernails on their fins (which evolved from legs). Figure 20.19 illustrates the skeleton of a baleen whale, which contains pelvic bones, as other mammal skeletons do, even though such bones serve no known function in the whale. The human vermiform appendix is apparently vestigial; it represents the degenerate terminal part of the cecum, the blind pouch or sac in which the large intestine begins. In other mammals such as mice, the cecum is the largest part of the large intestine and functions in storage—usually of bulk cellulose in herbivores. Although some suggestions have been made, it is difficult to assign any current function to the vermiform appendix. In many respects, it is a dangerous organ: quite often it becomes infected, leading to an inflammation called appendicitis; without surgical removal, the appendix may burst, allowing the contents of

the gut to come in contact with the lining of the body cavity, a potentially fatal event. It is difficult to understand vestigial structures such as these as anything other than evolutionary relics, holdovers from the evolutionary past. They argue strongly for the common ancestry of the members of the groups that share them, regardless of how different they have subsequently become.

> **Comparisons of the anatomy of different living animals often reveal evidence of shared ancestry. In some instances, the same organ has evolved to carry out different functions, in others, an organ loses its function altogether. Sometimes, different organs evolve in similar ways when exposed to the same selective pressures.**

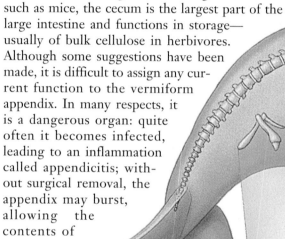

FIGURE 20.19
Vestigial features. The skeleton of a baleen whale, a representative of the group of mammals that contains the largest living species, contains pelvic bones. These bones resemble those of other mammals, but are only weakly developed in the whale and have no apparent function.

Table 20.2	Some Vestigial Traits in Humans
Trait	**Description**
Ear-wiggling muscles	Three small muscles around each ear that are large and important in some mammals, such as dogs, turning the ears toward a source of sound. Few people can wiggle their ears, and none can turn them toward sound.
Muscles that make body hairs stand on end	Muscles that lift hairs, creating a thick layer of trapped air that provides insulation in hairier mammals. In humans, these muscles just make goosebumps.
Tail	Present in human and all vertebrate embryos. In humans, the tail is reduced; most adults only have three to five tiny tail bones and, occasionally, a trace of a tail-extending muscle.
Appendix	Structure which presumably had a digestive function in some of our ancestors, like the cecum of some herbivores. In humans, it varies in length from 5–15 cm, and some people are born without one.
Wisdom teeth	Molars that are often useless and sometimes even trapped in the jawbone. Some people never develop wisdom teeth.

Based on a suggestion by Dr. Leslie Dendy, Department of Science and Technology, University of New Mexico, Los Alamos.

Convergent Evolution

Different geographical areas sometimes exhibit groups of plants and animals of strikingly similar appearance, even though the organisms may be only distantly related. It is difficult to explain so many similarities as the result of coincidence. Instead, natural selection appears to have favored parallel evolutionary adaptations in similar environments. Because selection in these instances has tended to favor changes that made the two groups more alike, their phenotypes have converged. This form of evolutionary change is referred to as **convergent evolution,** or sometimes, **parallel evolution.** As discussed earlier, similarity due to convergent evolution is analogy, not homology.

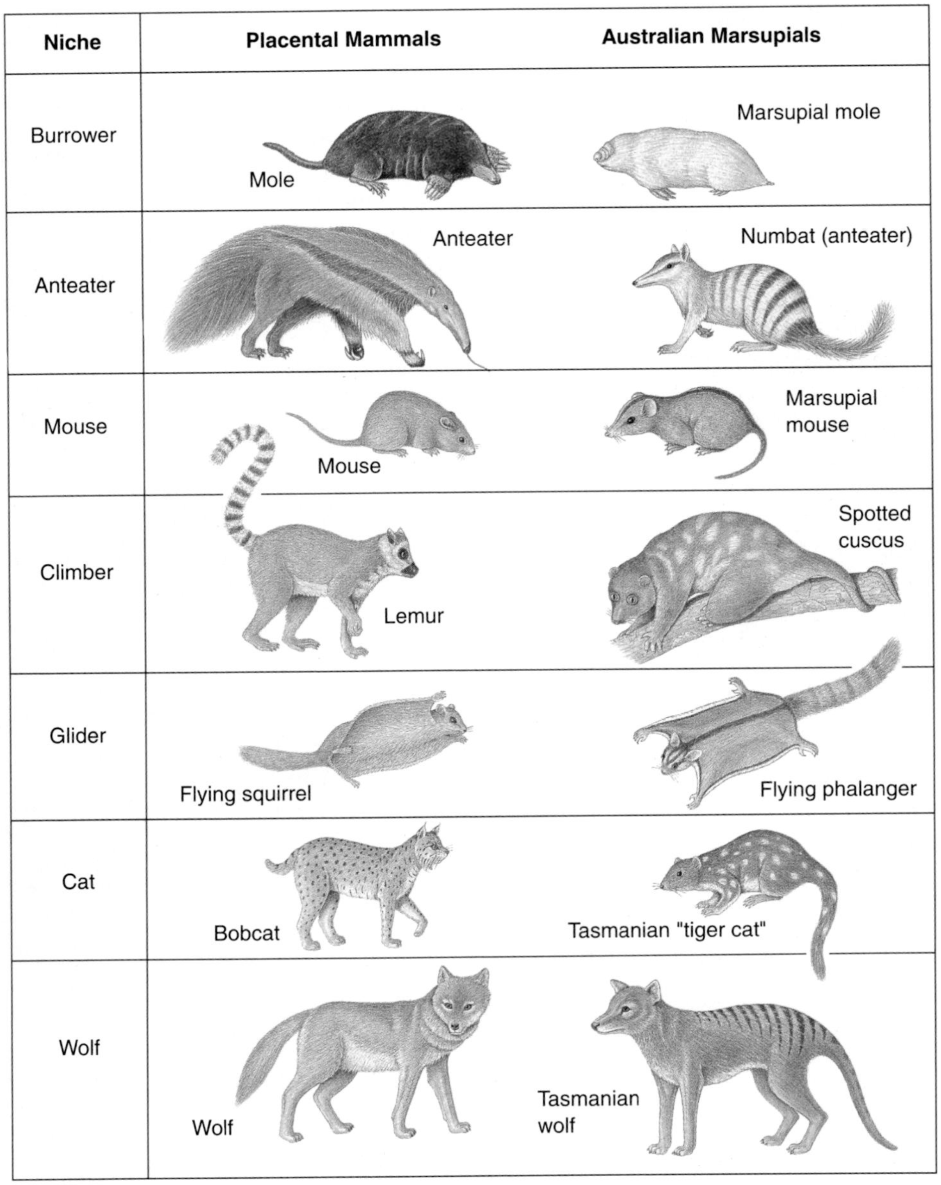

The Marsupial-Placental Convergence

In the best known case of convergent evolution, two major groups of mammals, marsupials and placentals, have evolved in a very similar way, even though the two lineages have been living independently on separate continents. Australia, the home of the Australian marsupials, separated from the other continents more than 50 million years ago. For these 50 million years, the only mammals in Australia have been marsupials, members of a group in which the young are born in a very immature condition and held in a pouch until they are ready to emerge into the outside world. Marsupials, which evolved earlier than placental mammals, most likely arrived in Australia before its separation from Antarctica. Placental mammals evolved later and are the dominant mammalian group throughout most of the other continents. Only recently, as a result of human activity, have a few placental mammals arrived in Australia.

What are the Australian marsupials like? To an astonishing degree, they resemble the placental mammals living today on the other continents (figure 20.20). The similarity between some individual members of these two sets of mammals argues strongly that they are the result of convergent evolution, similar forms having evolved in different, isolated areas because of similar selective pressures in similar environments.

Niche	Placental Mammals	Australian Marsupials
Burrower	Mole	Marsupial mole
Anteater	Anteater	Numbat (anteater)
Mouse	Mouse	Marsupial mouse
Climber	Lemur	Spotted cuscus
Glider	Flying squirrel	Flying phalanger
Cat	Bobcat	Tasmanian "tiger cat"
Wolf	Wolf	Tasmanian wolf

FIGURE 20.20

Convergent evolution. Marsupials in Australia resemble placental mammals in the rest of the world. They evolved in isolation after Australia separated from other continents.

Homology versus Analogy

How do we know when two similar characters are homologous and when they are analogous? As we have seen, adaptation favoring different functions can obscure homologies, while convergent evolution can create analogues that appear as similar as homologues. There is no hard and fast answer to this question, and the determination of true homologues is often a thorny issue in biological classification. As we have seen in comparing vertebrate embryos, and again in comparing slugs and squids, studies of embryonic development often reveal features not apparent when studying adult organisms. In general, the more complex two structures are, the less likely they evolved independently.

FIGURE 20.21
A Galápagos tortoise most closely resembles South American tortoises. Isolated on these remote islands, the tortoise have evolved distinctive forms. This natural experiment is being terminated, however. Since Darwin's time, much of the natural habitat of the larger islands has been destroyed by human intrusion. Goats introduced by settlers, for example, have drastically altered the vegetation.

Darwin and Patterns of Recent Divergence

Darwin was the first to present evidence that animals and plants living on oceanic islands resemble most closely the forms on the nearest continent—a relationship that only makes sense as reflecting common ancestry. The Galápagos tortoise in figure 20.21 is more similar to South American tortoises than to those of any other continent. This kind of relationship strongly suggests that the island forms evolved from individuals that came from the adjacent mainland at some time in the past. Thus, the Galápagos finches of figure 20.7 have beaks different from their South American relatives. In the absence of evolution, there seems to be no logical explanation of why individual kinds of island plants and animals would be clearly related to others on the nearest mainland, but still have some divergent features. As Darwin pointed out, this relationship provides strong evidence that macroevolution has occurred.

A similar resemblance to mainland birds can be seen in an island finch Darwin never saw—a solitary finch species living on Cocos Island, a tiny, remote volcanic island located 630 kilometers to the northeast of the Galápagos. This finch does not resemble European finches, or the finches of Australia, Africa, or North America. Instead, it resembles the finches of Costa Rica, 500 kilometers to the east.

Of course, because of adaptation to localized habitats, island forms are not *identical* to those on the nearby continents. The tortoises have evolved different shell shapes, for example; those living in moist habitats have dome-shaped shells while others living in dry places have low, saddle-backed shells with the front of the shell bent up to expose the head and neck. Similarly, the Galápagos finches have evolved from a single presumptive ancestor into 13 species, each specialized in a different way. These Galápagos tortoises and finches have evolved in concert with the continental forms, from the same ancestors, but the two lineages have diverged rather than converged.

It is fair to ask how Darwin knew that the Galápagos tortoises and finches do not represent the convergence of unrelated island and continental forms (analogues) rather than the divergence of recently isolated groups (homologues). While either hypothesis would argue for natural selection, Darwin chose divergence of homologues as by far the simplest explanation, since the tortoises and finches differ by only a few traits, and are similar in many.

In sum, the evidence for macroevolution is overwhelming. In the next chapter, we will consider Darwin's proposal that microevolutionary changes have led directly to macroevolutionary changes, the key argument in his theory that evolution occurs by natural selection.

Evolution favors similar forms under similar circumstances. Convergence is the evolution of similar forms in different lineages when exposed to the same selective pressures. Divergence is the evolution of different forms in the same lineage when exposed to different selective pressures.

Scientific Creationism

In the century since he proposed it, Darwin's theory of evolution by natural selection has become nearly universally accepted by biologists as the best available explanation for biological diversity, its predictions supported by the experiments and observations of generations of scientists. There is ongoing controversy among serious students as to the details of how evolution has occurred, just as there is controversy in every active scientific field. But there is no controversy about Darwin's basic finding that natural selection has played and is continuing to play the central role in the process of evolution.

The Creationism Movement

Evolution is not the only way in which the diversity of life on earth has been explained. The clear distinction between science and religion sometimes gets muddled. Thus, a number of individuals, mainly in the United States and starting largely in the 1970s, have put forward a view they title **"scientific creationism."** This view holds that the Biblical account of the origin of the earth is literally true, that the earth is much younger than most scientists believe, and that all species of organisms were individually created and appeared at their creation essentially the same as they appear today. Scientific creationists are arguing in the courts that their views should be taught alongside evolution in classrooms. They argue that if both evolution and scientific creationism provide scientific explanations of biological diversity, then teachers have an obligation to present *both* views, taught side by side, so that students can choose knowledgeably between them.

This does not seem to be a bad argument if you accept the premise, which is that the view of the "scientific creationists" is indeed scientific. The confusion is not in the beliefs of the scientific creationists, which are religious beliefs that many people hold, but rather in their labeling of these beliefs as "scientific."

Creationism Is Not Science

Scientific creationism should not be labeled science for three reasons:

1. It is not supported by any empirical observations.
2. It does not infer its principles from observation, as does all science.
3. Its assumptions lead to no testable, falsifiable hypotheses.

The hypothesis that species of organisms were created separately by a supernatural agency is untestable, and as such, lies outside the realm of science.

The Creationist Objections to Evolution

Creationists Raise Six Principal Objections to Evolution:

1. **There Are No Intermediates.** *"No one ever saw a fin on the way to becoming a leg,"* creationists claim, pointing to the many gaps in the fossil record in Darwin's day. Since then, most vertebrate intermediates *have* been found.

2. **Earth Is Not Old Enough.** *"The radioactive dating indicating that the world is billions of years old is not reliable,"* they argue, pointing out that living oysters yield carbon-14 dates indicating an age of 3000 years. In fact, there is no doubt about the validity of radiodating. Oysters make their shells from calcium carbonate in seawater, which is dissolved limestone, and limestone is made of old seashells from oysters that lived thousands of years ago. To reject the validity of radiodating is to abandon all we know of modern physics.

3. **The Clockmaker Argument.** *"The organs of living creatures are too complex for a random process to have produced—the existence of a clock is evidence of the existence of a clockmaker."* Biologists do not agree. The intermediates in the evolution of the mammalian ear can be seen in fossils, and many intermediate "eyes" are known in various invertebrates. It is only because soft tissues are rarely preserved in fossils that we don't have an even fuller record.

4. **Evolution Violates the Second Law of Thermodynamics.** *"A bunch of soda cans doesn't jump neatly into a stack—things become more disorganized due to random events, not more organized."* Biologists point out that this argument ignores what the second law really says: disorder increases *in a closed system,* which the earth most certainly is not. Energy continually enters the biosphere from the sun, fueling life and all the processes that organize it.

5. **Proteins Cannot Assemble Randomly.** *"Amino acids form peptide bonds by dehydration synthesis, with water as a product; because this is a reversible reaction, it cannot take place in water."* True; but biologists do not argue that it did. Instead, they propose that the initial macromolecules were RNA, which *has* been shown to form chains spontaneously in water, and that RNA, acting as an enzyme, then catalyzed the formation of proteins.

6. **Proteins Are Too Improbable.** *"Hemoglobin has 141 amino acids in its chain. The probability that the first one would be leucine is 1/20, and that all 141 would be the ones they are by chance is $(1/20)^{141}$, an impossibly rare event!"* This is statistical foolishness—you cannot use probability to argue backwards. The probability that a student in a classroom has a particular birthday is 1/365; arguing this way, the probability that everyone in a class of 50 would have the birthdays they do is $(1/365)^{50}$, and yet there the class sits.

20.1 Natural selection explains adaptive microevolution.

- Natural populations provide clear evidence of microevolutionary change.

- Sickle cell anemia occurs because of an altered hemoglobin molecule associated with a particular allele. If this allele is present in homozygous form, it is often but not invariably lethal; if it is present in heterozygous form, it not only does not produce anemia, but it also confers resistance to malaria and increases female fertility. For these reasons, the allele has maintained high frequencies in areas with a high incidence of malaria.

- The British populations of the peppered moth, *Biston betularia*, consisted mostly of light-colored individuals before the Industrial Revolution. Over the last two centuries, populations that occur in heavily polluted areas, where the tree trunks are darkened with soot, have come to consist mainly of dark-colored (melanic) individuals—a result of rapid natural selection.

- More recently, efforts to clean up the air have reversed the direction of natural selection, and light-colored moths are favored.

20.2 The evidence for macroevolution is extensive.

- Two direct lines of evidence argue that macroevolution has occurred: (1) the fossil record, which exhibits successive anatomical changes correlated with age; and (2) the molecular record, which exhibits accumulated gene and protein changes, the amount of change correlated with age as determined in the fossil record.

- Several indirect lines of evidence argue that macroevolution has occurred, including successive changes in homologous structures, developmental patterns, vestigial structures, parallel patterns of evolution, and patterns of distribution.

20.3 The theory of evolution has proven controversial.

- A nonscientific view is one whose principles are not derived from observation or supported by observation. "Scientific creationism" is an example of a nonscientific view.

Discussing Key Terms

1. **Adaptation**   Natural selection favors changes that increase an organism's ability to survive in its environment.

2. **Microevolution**   Changes in allele frequencies within populations of a species are microevolutionary changes that can lead to adaptive change.

3. **Macroevolution**   Extinction and species formation interact to bring about changes in the array of species alive at any time, a process called macroevolution.

4. **Homology**   The similarity between two structures that arose because of a common evolutionary origin is called homology.

5. **Convergent evolution**   Similar structures sometimes evolve in organisms that are not directly related, usually reflecting similar patterns of selection.

6. **Scientific creationism**   Creationists have attempted to gain equal time in public education, arguing that creation is just another scientific theory. Actually, the term *scientific creationism* is an oxymoron: "creation" in a biblical or divine sense cannot be examined or explained scientifically.

Review Questions

1. Why is the sickle cell anemia allele maintained at high levels in some Central African populations? What are the advantages or disadvantages to being homozygous (*AA* or *SS*) or heterozygous (*AS*)?

2. Why is the frequency of the sickle cell anemia allele lower among African Americans than among natives of Central Africa?

3. Why did the frequency of light-colored moths decrease and that of dark-colored moths increase with the advent of industrialism? What is industrial melanism?

4. What direct and indirect evidence supports macroevolution?

5. How did scientists date fossils in Darwin's day? Why are scientists today able to date rocks more accurately?

6. How does the molecular record indicate evolutionary change?

7. What is homology? How does it support evolutionary theory?

8. Why do whales still have pelvic bones even though they seem to serve no useful function? What term is used to describe these apparently useless structures?

9. What is convergent evolution? Give examples.

10. How did Darwin's studies of island populations provide evidence for evolution?

11. Based on the explanation of the scientific process outlined in chapter 1, is "scientific creationism" truly scientific? Why or why not?

Thought Questions

1. In Central Africa there is a low frequency of a third hemoglobin allele, called *C*, in addition to the *A* and *S* alleles discussed in this chapter. Individuals heterozygous for *C* and the normal allele *A* are susceptible to malaria, just as *AA* homozygotes are, but *CC* individuals are resistant to malaria—and, unlike *SS* homozygotes, do not develop anemia! Assuming that the Bantu people entered Central Africa relatively recently from a land where malaria is not common, and that the *C* and *S* alleles were both rare among the original settlers, can you suggest a reason that *CC* individuals have not become predominant?

2. Imagine you were sitting on the Supreme Court in the fall of 1986, hearing a case in which it was argued that creation science should be taught in public schools as a legitimate alternative scientific explanation of biological diversity. What is the best case lawyers might have made for and against this proposition? The Supreme Court announced a decision in June 1987. How would you have voted and why?

Internet Links

A Virtual Library on Evolution
http://golgi.harvard.edu/biopages/evolution.html
The WORLD WIDE WEB VIRTUAL LIBRARY ON EVOLUTION contains links to a wide array of evolution sources, including museums and many sites of individual research scientists. Strong emphasis on molecular evolution and taxonomy.

Transitional Vertebrate Fossils
http://cns-web.bu.edu/pub/dorman/trans_faq.html
A University of Washington site that answers the challenge "there aren't any transitional fossils" with a wealth of examples from the vertebrate fossil record.

For Further Reading

Behe, M.: *Darwin's Black Box*, The Free Press, New York, 1996. A creationist recounts the argument for intelligent design from the point of view of molecular biology.

Dawkins, R.: *The Blind Watchmaker*, W. W. Norton & Company, New York, 1986. A brilliant exposition of the factors involved in evolution by natural selection and of contemporary reasoning concerning them.

Futuyma, D.: *Science on Trial: The Case for Evolution*, Sinauer Associates, Sunderland, Mass., 1983. An excellent exposition of the serious errors in the creationist argument, still timely after 15 years.

Gilkey, L.: *Creationism on Trial: Evolution and God at Little Rock*, Winston Press, Minneapolis, MN, 1985. A book, written by a theologian, outlining the case for creationism as argued in the courts.

Gordon, M., and E. Olson: *Invasions of the Land*, Columbia University Press, New York, 1995. A fine overview of the role of adaptation in macroevolution.

Grant, B., D. Owen, and C. Clarke: "Parallel Rise and Fall of Melanic Peppered Moths in America and Britain," *Journal of Heredity*, September 1996, pages 351–57. A suggestion that the role of lichens has been overemphasized in studying industrial melanism.

Raup, D.: "The Role of Extinction in Evolution," *Proc. Nat. Acad. Sci. USA*, Vol. 91, July 1994, pages 6758–63. A very interesting reevaluation of the role extinction should play in evolutionary theory.

Webb, G.: *The Evolution Controversy in America*, University of Kentucky Press, Lexington, 1994. An update on the ongoing battle between creationists and evolutionists.

Weiner, J.: *The Beak of the Finch*, Knopf, New York, 1994. For over 20 generations of Galápagos finches, biologists Peter and Rosemary Grant have closely monitored evolution in action. A delightful, engaging, and very informative book that won the Pulitzer Prize for nonfiction in 1995. Highly recommended to anyone interested in evolution.

21

The Origin of Species

Concept Outline

FIGURE 21.1
A group of Galápagos iguanas bask in the sun on their isolated island. How does isolation contribute to the formation of new species?

The kinds of changes that have occurred over the last few hundred years in populations of peppered moths and finches (chapter 20) occur continuously in nature. Not surprisingly, many of the best documented changes in natural populations are those that have occurred in response to human alterations in the environment. Human activities have dominated habitats all over the world for centuries, and populations of moths, bacteria, and other organisms have had to evolve in response to them or become extinct. Similar changes occur in populations of organisms undisturbed by human activities (figure 21.1). As a result of selection, organisms have come to occupy diverse habitats, invade new areas, or live successfully under new conditions. Such microevolutionary changes may result in the formation of new species.

The Pace of Evolution

Different kinds of organisms evolve at different rates. Mammals, for example, evolve relatively slowly. On the basis of a relatively complete fossil record, it has been estimated that a good average value for the duration of a "typical" mammal species, from formation of the species to its extinction, might be about 200,000 years. American paleontologist George Gaylord Simpson has pointed out that certain groups of animals, such as lungfishes, are apparently evolving even more slowly than mammals. In fact, Simpson estimated that there has been little evolutionary change among lungfishes over the past 150 million years, and even slower rates of evolution occur in other groups.

Evolution in Spurts?

Not only does the rate of evolution differ greatly from group to group, but evolution within a group apparently proceeds rapidly during some periods and relatively slowly during others. The fossil record provides evidence for such variability in evolutionary rates, and evolutionists are very interested in understanding the factors that account for it. In 1972, paleontologists Niles Eldredge of the American Museum of Natural History in New York and Stephen Jay Gould of Harvard University proposed that evolution normally proceeds in spurts. They claimed that the evolutionary process is a series of **punctuated equilibria.** Evolutionary innovations would occur and give rise to new lines; then these lines might persist unchanged for a long time, in "equilibrium." Eventually there would be a new spurt of evolution, creating a "punctuation" in the fossil record. Eldredge and Gould contrast their theory of punctuated equilibrium with that of **gradualism,** or gradual evolutionary change, which they claimed was what Darwin and most earlier students of evolution had considered normal (figure 21.2).

Eldredge and Gould proposed that **stasis,** or lack of evolutionary change, would be expected in large populations under diverse and conflicting selective pressures. In contrast, rapid evolution of new species would usually occur when populations were small, isolated, and possibly already differing from their parental population as a result of the founder

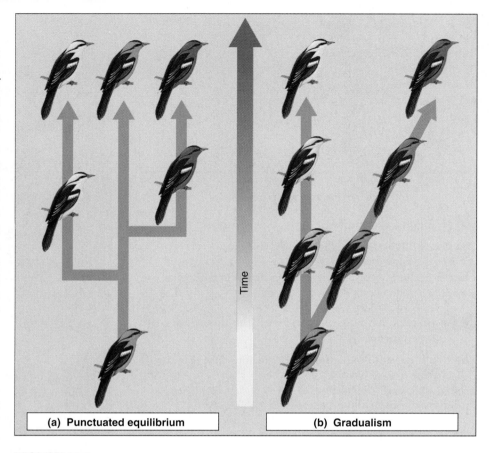

(a) Punctuated equilibrium   **(b) Gradualism**

FIGURE 21.2
Two views of the pace of macroevolution. (a) Punctuated equilibrium surmises that species formation occurs in bursts, separated by long periods of quiet, while (b) gradualism surmises that species formation is constantly occurring.

effect. This, combined with selective pressures from a new environment, could bring about rapid change.

Unfortunately, the distinctions are not as clear-cut as implied by this discussion. Some well-documented groups such as African mammals clearly have evolved gradually, and not in spurts. Other groups, like marine bryozoa, seem to show the irregular pattern of evolutionary change the punctuated equilibrium model predicts. One could argue, however, that the fossil record in these instances is incomplete because of changes in the conditions under which fossils are deposited, making the interpretation of many of the "gaps" problematic. Despite these difficulties, the punctuated equilibrium model has provided a useful perspective for considering the mode and pace of evolution.

The punctuated equilibrium model assumes that evolution occurs in spurts, between which there are long periods in which there is little evolutionary change. The gradualism model assumes that evolution proceeds gradually, with successive change in a given evolutionary line.

The Nature of Species

Whether it occurs gradually or in spurts, evolution introduces new species. How might adaptive changes in natural populations lead to the origin of species? Darwin was extremely interested in this question, which lies at the core of his proposal that microevolution leads to macroevolution. As we begin to consider the problem, we must first examine what *species* means and how the concept has changed through the years.

The Taxonomic Species Concept

John Ray (1627–1705), an English clergyman and scientist, was one of the first to propose a definition of **species.** In about 1700, he indicated how a species could be recognized: all individuals that belonged to a species could breed with one another and produce progeny still of that species. Even if two different-looking individuals appeared among the progeny of a single mating, they were still considered to belong to the same species. All dogs were one species, all pigeons, and so on; carp, however, were not the same species as goldfish, nor mallards the same species as teal, and so forth.

In an informal way, people had always recognized species. The word *species* is simply Latin for "kind." With John Ray's interpretation, species began to be regarded as an important biological unit that could be catalogued and understood. Along with other scientists of his time, Ray believed that species were individually created by the Supreme Being and did not change, a view widely held until challenged by Darwin in 1859. Darwin explained the relative constancy of species by saying that each had its own distinct role in nature, what biologist G. Evelyn Hutchinson was much later to call its **niche.** Thus, each species exploits different resources in different ways, occurs in a particular kind of place, is active at certain times of the day, and eats certain foods—all attributes of its unique niche in the ecosystem. Natural selection favored changes that improved the "fit" of an organism to its particular environment. In the taxonomic species concept, the adaptation of different populations to different environments is the key event in species formation.

The Biological Species Concept

With the emergence of population genetics in the 1920s, a desire developed to define the species in genetic terms. Much of the early experimental work involved *Drosophila*, a small fruit fly widely distributed around the world, and the concept of species that began to emerge was shaped by the experiences of biologists working with this organism. The **biological species concept** they developed was summed up by evolutionist Ernst Mayr as follows: species are:

"... groups of actually or potentially interbreeding natural populations which are reproductively isolated from other such groups."

In other words, the biological species concept says that hybrids between species occur rarely in nature. On the other hand, individuals that belong to the same species are able to interbreed freely. The key is hybridization. Under the biological species concept, the erection of breeding barriers is the key event in species formation.

Problems with the Biological Species Concept

The biological species concept works very well for most terrestrial animals. In fishes and plants, by contrast, scientists often have little or no experimental evidence on which to base their decisions about what constitutes a species. They recognize species primarily based on the *taxonomic species concept*, which defines species based on the differences in their features. In some groups of plants there are essentially no barriers to hybridization between the species, while other groups have strong barriers to hybridization, and still others reproduce asexually. That is why botanists have not so readily accepted definitions of species that claim that all individuals of a species can interbreed with one another, but not with individuals of other species. Botanists tend instead to use a taxonomic species concept.

Patterns of Variation within Species

Within the units classified as species, populations that occur in different areas may be more or less distinct from one another. Such groups of distinctive individuals are informally called **races** and may be classified taxonomically as **subspecies** or **varieties.** In areas where these populations approach one another, many individuals may have a combination of distinctive features characteristic of each population. In other words, distinct-appearing populations within a species usually intergrade or merge characteristics with one another when they occur together, and many of these individuals may not match either race in their characteristics. In contrast, when species occur together, they usually do not intergrade, although they may hybridize occasionally.

In some groups of organisms, even local races are not capable of interbreeding with one another. This pattern occurs in many annual plants. In contrast, tree species, some groups of mammals, and fishes generally are able to form fertile hybrids with one another, even though they may not do so in nature. For other plants and animals, it is not known whether the species can form hybrids.

Thus, you can see that the biological species concept, while appropriate for some kinds of organisms, is not a useful way of looking at others. In this text, we thus define a species as a group of organisms that is unlike other such groups of organisms and does not hybridize extensively with them in nature.

Species are groups of organisms that differ in one or more characteristics and do not hybridize extensively if they occur together in nature.

The Divergence of Populations

The rate at which populations evolve to suit their environment depends on the strength of the selective forces at work. If these forces are strong, the populations will change rapidly. But how do different populations diverge to become different species? What prevents them from sharing any changes that occur? How do populations come to differ enough that breeding barriers have a chance to evolve? One obvious way for populations to diverge is physical separation. In **allopatric speciation** geographically isolated populations diverge into distinct species. In **sympatric speciation,** by contrast, populations within a common area split into species.

The practical problem in using these terms to describe real speciation events is that we rarely know enough to discern between them. Local populations are often able to adjust individually and effectively to the demands of their particular environments in subtle ways that may not be obvious to an observer. The checkerspot butterflies described in figure 21.3, for example, occupy a single open grassland that seems at first glance uniform, but the butterflies effectively subdivide this habitat into discrete zones, forming distinct populations that only rarely exchange individuals. Invisible to an investigator, the boundaries separating the zones are clearly obvious to the butterflies. Only because this butterfly population has been carefully studied do we know of its subdivision—a more cursory examination would have suggested a single butterfly population. If these checkerspot butterfly populations were to diverge and eventually form different species, the event would seem on cursory examination to be sympatric, when in fact the speciation would actually be allopatric.

(a) (b) (c)

FIGURE 21.3
The boundaries of populations are not always obvious. (a) The butterfly *Euphydryas editha bayensis* occurs on (b) Jasper Ridge, a biological preserve in the foothills above Stanford University, just south of San Francisco, California. The butterflies appear to constitute one continuous population throughout the open grassland, which is an "island" surrounded by oak forest and chaparral. Extensive studies by Paul Ehrlich and his colleagues over more than 30 years involved marking the butterflies with dots of ink on their wings, releasing them, and recapturing them. These studies revealed that this butterfly species actually exists as a series of discontinuous populations between which there is little movement of individuals. (c) Each of these local populations, separated by dashed lines on the maps, is free to respond to the selective forces characteristic of its particular part of the ridge. Changes in the density of these populations, which have remained essentially constant in overall distribution since the late 1950s at least, are shown on the maps.

Reducing Dispersal Can Increase Divergence

Geographic isolation promotes divergence by allowing natural selection to differentiate between populations. Anything that reduces dispersal tends to increase isolation and thus to favor increased divergence. In a particularly lucid study, Martin Cody and Jacob Overton of the University of California, Los Angeles, explored how natural selection has favored the loss of dispersal ability in plants confined to islands. They examined weedy plants of the daisy family (Asteraceae) with wind-dispersed seeds—wild lettuce, groundsel, and false dandelion. The wind-dispersed seeds of these plants have two parts: the tiny seed proper (the *achene*), surrounded by an enormous ball of fluff (the *pappus*). The two-part seed forms a parachute well adapted to being carried by the wind.

The bigger the fluff ball and the lighter the achene, the longer the seed remains in the air and the farther it is likely to be carried. This dispersal ability can be quantified as the ratio of pappus volume to achene volume (V_p/V_a), which reaches values as high as 17,000.

Cody and Overton sampled 240 tiny islands in Barkley Sound off the coast of Vancouver Island in British Columbia, Canada, varying in size from around one square kilometer to a few square meters. For each island, they compared the dispersal ability of the island plants with plants of the same species from the nearby mainland. In their first study, they found that the island weeds had reduced dispersal ability, compared to the mainland weeds. On such small islands the weed populations are often temporary, disappearing after only a few years, only to be recolonized again from the mainland in a continual cycle. For a reduction in dispersal to have evolved on the islands, selection must have operated very quickly. This prediction was confirmed by resampling the island plants over a period of 10 years. Because populations often became extinct and new ones established during this period, Cody and Overton were able to measure parachute and payload sizes in populations of known age, from 1 year to 10 years or more. They found achene size (V_a) increases with population age, while parachute size (V_p) decreases (figure 21.4). Thus the ratio V_p/V_a decreases sharply, as do time aloft and dispersal distance.

Natural selection has clearly acted on these island weeds to reduce dispersal—probably because seeds with larger parachutes, lighter achenes, and longer times aloft get blown out to sea and so do not contribute to the next generation.

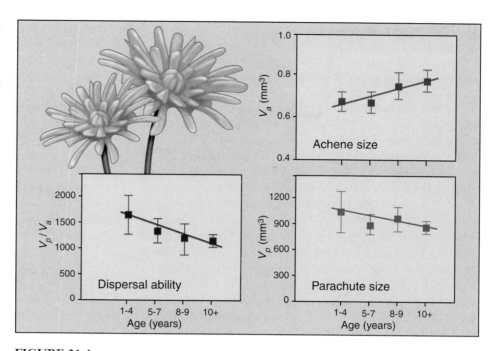

FIGURE 21.4

Natural selection for decreased dispersal in island weeds. As weeds colonize tiny islands off the west coast of Canada, natural selection favors heavier achene seed cores (V_a increases) and smaller parachutes of fluff (V_p decreases). As a result, dispersal ability drops (V_p/V_a falls). Lighter seeds with bigger parachutes are more likely to be carried out to sea and not to contribute to the next generation.

How Much Divergence Does It Take to Make a Species?

How much divergence does it take to create a new species? How many gene changes does it take? Since Darwin, the traditional view has been that new species arise by the accumulation of many small genetic differences. While there is little doubt that many species have formed in this gradual way, new techniques of molecular biology suggest that in at least some cases, the evolution of a new species may involve very few genes. Studying two species of monkey flower found in the western United States, researchers found only a few genes separate the two species, even though at first glance the two species appear to be very different. One, *Mimulus lewisii*, has the pale pink flowers and concentrated nectar optimal for attracting bumblebees, while *M. cardinalis* has the red flowers and copious dilute nectar typical of hummingbird-pollinated flowers. Using gene technologies like those described in chapter 18, the researchers found that all of the major differences in the flowers were attributable to what appear to be single genes! Because individual genes have such powerful effects, species as different as these two can evolve in relatively few steps.

Natural selection promotes divergence by favoring adaptation to local habitats and in some instances by decreasing dispersal. The number of genes involved need not be large.

Ecological Races

What kinds of patterns can be expected to result from the differentiation of populations? One consequence is that the individuals of a species in one part of its range often look different from those that occur elsewhere. The existence of such *races* fascinated Darwin because he considered them an intermediate stage in the evolution of species. The ecologically defined races we will first discuss may change over time to become the clusters of species that we will consider next. Both provide important examples of the evolution of populations in nature.

Ecological races were first studied in detail in plants. As every gardener knows, the same species of plant may differ greatly in appearance, depending on the place where it is grown. This is true even for genetically identical divisions of the same plant, called *clones*. For example, a plant that is usually in the sun often produces leaves unlike the leaves it produces in the shade; shade leaves are usually thinner and broader and have more internal air spaces than sun leaves. Thus, it seemed to many botanists in the nineteenth century that environmental factors, rather than genetic differences, might account for many of the differences between races and even between species of plants. Future studies were to prove them wrong.

Ecotypes in Plants

In the 1920s and 1930s, Swedish botanist Göte Turesson performed a series of experiments designed to test whether differences between races of plants were largely genetically determined or mainly caused by environmental factors. Turesson observed that many plants have distinctive races that grow in different habitats. These races differ from one another in characteristics such as height, leaf size and shape, degree of hairiness, flowering time, and branching pattern. Turesson dug up individuals representing these races and cultivated them together in his experimental garden at Lund, Sweden. In nearly every case he found that the unique features of individual races were maintained when the plants were grown in a common environment. Most of the characteristics he observed, therefore, had a genetic basis; a few were environmental. Turesson called the ecological races that he studied and that proved to have a genetic basis **ecotypes.**

The important studies Turesson initiated were continued in California by a group of scientists from the Department of Plant Biology of the Carnegie Institution of Washington, located on the campus of Stanford University. The Carnegie investigators used three major transplant stations for their investigations, one near Stanford at sea level; the second at 1400 meters elevation on the forested, western slopes of the Sierra Nevada; and the third at 3050 meters elevation near the crest of that range. Planting genetically identical individuals—divisions of the same plant—at the three stations, the scientists were able to demonstrate the existence of ecotypes in a number of different plant species and also to confirm the physiological differences between them. These differences made each ecotype better suited to grow in a particular area; for example, those that occurred naturally in areas of summer drought exhibited dormancy during that period.

Ultimately, these studies and others led to the conclusion that most of the differences between individuals, populations, races, and species of plants have a genetic basis. Despite the fact that plants change certain characteristics in relation to the environments in which they grow, the differences between them are usually fixed genetically in the course of their evolution.

Ecological Races in Animals

Similar patterns of variation also appear in animals (figure 21.5). The differences may be morphological or physiological, and the differences between subspecies may be striking. For example, the larger races of some species of birds may consist of individuals that weigh three or four times as much as individuals of the smaller races. Races may differ from one another in their tolerance to different temperatures, in the speed of their larval development, in their behavioral characteristics—in short, in any feature that can be measured and studied. Their features are almost always genetically determined. Breeds of dogs, races maintained by artificial selection, serve as an example of how races may differ in appearance. A Russian wolfhound weighing 30 or 40 kilograms and a Mexican chihuahua weighing less than 1 kilogram are indisputably considered members of the same species: *Canis familiaris*, the domesticated dog. Yet if you were a scientist visiting from another planet, would you think they were in the same genus, much less the same species?

Human Races

Human beings, like all other species, have differentiated in their characteristics as they have spread throughout the world. Local populations in one area often appear significantly different from those that live elsewhere. For example, northern Europeans often have blond hair, fair skin, and blue eyes, while Africans often have black hair, dark skin, and brown eyes. In addition, other features, such as the ABO blood groups discussed in chapter 13, differ in proportion from area to area. These traits may play a role in adapting the particular populations to their environments. Blood groups may be associated with immunity to diseases more common in certain geographical areas, and dark skin shields the body from the damaging effects of ultraviolet radiation, which is much stronger in the tropics than in temperate regions.

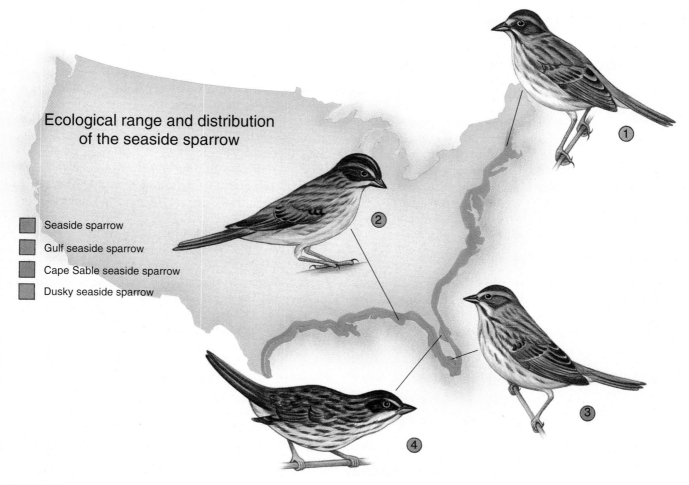

FIGURE 21.5

Ecological races or subspecies in animal populations. Subspecies of the seaside sparrow, *Ammodramus maritimus*, are local in distribution; some are in danger of extinction because of the alteration of their habitats. The widespread subspecies, *Ammodramus maritimus maritimus* (1) is the most common; *A. m. fisheri* (2) occurs along the Gulf Coast. The Cape Sable seaside sparrow, *A. m. mirabilis* (3), occurs in a small area of southwestern Florida. The last subspecies, the dusky seaside sparrow (4), *A. m. nigrescens*, occurred only near Titusville, Florida. The last individual, a male, died in captivity in 1987.

All human beings are capable of mating with one another and producing fertile offspring. The reasons that they do or do not choose to associate with one another are purely psychological and behavioral (cultural). The number of groups into which the human species might logically be divided has long been a point of contention. Some contemporary anthropologists divide people into as many as 30 "races," others as few as three: Caucasoid, Negroid, and Oriental. American Indians, Bushmen, and Aborigines are examples of particularly distinctive subunits that are sometimes regarded as distinct groups.

The problem with classifying people or other organisms into races in this fashion is that the characteristics used to define the races are usually not well correlated with one another, and so the determination of race is always somewhat arbitrary. In human beings, it is simply not possible to delimit clearly defined races that can be recognized by a particular combination of characteristics. Variation patterns are somewhat correlated with geographical distribution, but they do not form clearly defined units that we can classify. Different groups of people have constantly intermingled and interbred with one another during the entire course of history. Today, the differences between human "races" are breaking down rapidly as large numbers of individuals constantly move over the face of the globe, mixing with one another and recombining their characteristics.

In any event, individual differences within what might be considered races are greater than the differences between such units. This is a sound biological basis for dealing with each human being on his or her own merits and not as a member of a particular "race."

> **Populations of organisms tend to become increasingly different from one another in all of their characteristics. If the process continues long enough, or the selective forces are strong enough, the populations may become so different that they are considered distinct species.**

Prezygotic Isolating Mechanisms

Once species have formed, how do they keep their separate identity? The reasons they retain their new identity fall into two categories: **prezygotic isolating mechanisms,** which prevent the formation of zygotes; and **postzygotic isolating mechanisms,** which prevent the proper functioning of zygotes after they form. In the following sections we will discuss various isolating mechanisms in these two categories and offer examples that illustrate how the isolating mechanisms operate to help species retain their identities.

Geographical Isolation

Most species do not exist together in the same places. Species that occur in different places are generally adapted to local conditions, and they may differ in many ways associated with their different habitats. Because they are physically separated from one another, they have no opportunity to hybridize. They may, however, hybridize if they are brought together in zoos, parks, or botanical gardens.

One example of a geographically isolated species, the English oak, *Quercus robur*, occurs throughout areas of Europe that have a relatively mild, oceanic climate. Its characteristics are similar to those of the valley oak, *Quercus lobata*, of California, and quite different from the scrub oak, *Quercus dumosa*, also of California and the adjacent Baja peninsula (figure 21.6). All of these species can hybridize with one another and form fertile hybrids. The English oak does not hybridize with the others in nature simply because its geographical range does not overlap with theirs.

Similarly, although today African lions, *Panthera leo*, and Asian tigers, *P. tigris*, do not naturally occur together, they do mate and produce hybrids in zoos. The hybrids in which the tiger is the father, called "tigons," are viable and fertile; less is known about "ligers," hybrids in which the lion is the father. No barrier to hybridization between lions and tigers has evolved during their long isolation from each other. Natural selection has not favored a reduction in hybridization for the simple reason that no hybridization has been possible.

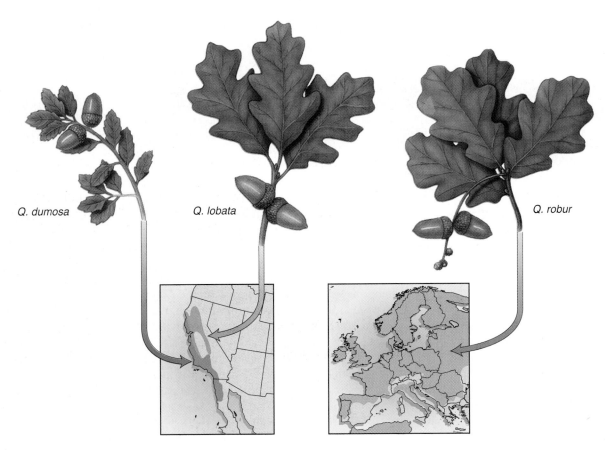

Q. dumosa   Q. lobata   Q. robur

FIGURE 21.6
Distance is an effective barrier for keeping species apart. Leaf and acorn characteristics differ substantially in three species of oaks, *Quercus dumosa, Q. lobata,* and *Q. robur.* The two California species occur in different habitats and so remain different.

Ecological Isolation

Even if two species occur in the same area, they may utilize different portions of the habitat and thus not hybridize because they do not encounter each other. For example, in India, the ranges of lions and tigers overlapped until about 150 years ago. Even when they did, however, there were no records of natural hybrids. Lions stayed mainly in the open grassland and hunted in groups called prides; tigers tended to be solitary creatures of the forest (figure 21.7). Because of their ecological and behavioral differences, lions and tigers rarely came into direct contact with each other, even though their ranges overlapped over thousands of square kilometers.

In another example, the ranges of two toads, *Bufo woodhousei* and *B. americanus*, overlap in some areas. Although these two species can produce viable hybrids, they usually do not interbreed because they utilize different portions of the habitat for breeding. While *B. woodhousei* prefers to breed in streams, *B. americanus* breeds in rainwater puddles. Similarly, the ranges of two species of dragonflies overlap in Florida. However, the dragonfly *Progomphus obscurus* lives near rivers and streams, and *P. alachuenis* lives near lakes.

Similar situations occur among plants. We have already mentioned two species of oaks that occur in California: the valley oak, *Quercus lobata*, and the scrub oak, *Q. dumosa*. The valley oak, a graceful deciduous tree which can be as tall as 35 meters, occurs in the fertile soils of open grassland on gentle slopes and valley floors in central and southern California. In contrast, the scrub oak is an evergreen shrub, usually only 1 to 3 meters tall, which often forms the kind of dense scrub known as chaparral. The scrub oak is found on steep slopes in less fertile soils. Hybrids between these different oaks do occur and are fully fertile, but they are rare. The sharply distinct habitats of their parents limit their occurrence together, and there is no intermediate habitat where the hybrids might flourish.

(a)

(b)

(c)

FIGURE 21.7
Lions and tigers are ecologically isolated. The ranges of lions and tigers used to overlap in India. However, lions and tigers do not hybridize in the wild because they utilize different portions of the habitat. (a) Lions live in open grassland. (b) Tigers are solitary animals that live in the forest. (c) Hybrids, such as this tiglon, have been successfully produced in captivity, but hybridization does not occur in the wild.

Behavioral Isolation

In chapter 60, we will consider the often elaborate courtship and mating rituals of some groups of animals. Related species of organisms such as birds often differ in their courtship rituals, which tends to keep these species distinct in nature even if they inhabit the same places (figure 21.8).

More than 500 species of flies of the genus *Drosophila* live in the Hawaiian Islands. This is one of the most remarkable concentrations of species in a single animal genus found anywhere. The genus occurs throughout the world, but nowhere are the flies more diverse in external appearance or behavior than in Hawaii. Many of these flies differ greatly from other species of *Drosophila*, exhibiting characteristics that can only be described as bizarre.

The Hawaiian species of *Drosophila* are long-lived and often very large compared with their relatives on the mainland. The females are more uniform than the males, which are often bizarrely distinctive. The males display complex territorial behavior and elaborate courtship rituals. Some of these are shown in figure 21.9.

The mating behavior patterns among Hawaiian species of *Drosophila* are of great importance in maintaining the distinctiveness of the individual species. For example, despite the great differences between them, *D. heteroneura* and *D. silvestris* are very closely related. Hybrids between them are fully fertile. The two species occur together over a wide area on the island of Hawaii, yet hybridization has been observed at only one locality. The very different and complex behavioral characteristics of these flies obviously play a major role in maintaining their distinctiveness. Another example of behavioral isolation is seen in four species of leopard frogs (genus *Rana*) that are similar in appearance but differ in their mating calls (figure 21.10).

FIGURE 21.8
Differences in courtship rituals can isolate related bird species. These Galápagos blue-footed boobies select their mates only after an elaborate courtship display. This male is lifting his feet in a ritualized high-step that shows off his bright blue feet.

(a)   (b)   (c)

FIGURE 21.9
Behavioral isolation among the Hawaiian *Drosophila*. Closely related species of Hawaiian flies of the genus *Drosophila* engage in different territorial defense and courtship behaviors. (a) *Drosophila silvestris*. After approaching the female from the rear, the male has lunged forward, vibrating his wings, with his head under the wings of the female. He has raised his forelegs up and over the female's abdomen; specialized hairs on the dorsal surface of one of the leg segments then "drum" over the dorsal surface of the female's abdomen. (b) *Drosophila heteroneura*. A male with extended wings is approaching and displaying toward a female in typical courtship posture. (c) *Drosophila heteroneura*. Two males have locked antennae, aggressively defending their territories. (d) *Drosophila clavisetae*. In this species, the males spray a chemical signal over the female.

(d)

Other Prezygotic Isolating Mechanisms

Temporal Isolation. *Lactuca graminifolia* and *L. canadensis*, two species of wild lettuce, grow together along roadsides throughout the southeastern United States. Hybrids between these two species are easily made experimentally and are completely fertile. But such hybrids are rare in nature because *L. graminifolia* flowers in early spring and *L. canadensis* flowers in summer. When their blooming periods overlap, as they do occasionally, the two species do form hybrids, which may become locally abundant.

Many species of closely related amphibians have different breeding seasons that prevent hybridization between the species. For example, five species of frogs of the genus *Rana* occur together in most of the eastern United States. But hybrids are rare because the peak breeding time is different for each of them.

Mechanical Isolation. Structural differences prevent mating between some related species of animals. Aside from such obvious features as size, the structure of the male and female copulatory organs may be incompatible. In many insect and other arthropod groups, the sexual organs, particularly those of the male, are so diverse that they are used as a primary basis for classification.

Similarly, flowers of related species of plants often differ significantly in their proportions and structures. Some of these differences limit the transfer of pollen from one plant species to another. For example, bees may pick up the pollen of one species on a certain place on their bodies; if this area does not come into contact with the receptive structures of the flowers of another plant species, the pollen is not transferred.

Prevention of Gamete Fusion. In animals that shed their gametes directly into water, eggs and sperm derived from different species may not attract one another. Many land animals may not hybridize successfully because the sperm of one species may function so poorly within the reproductive tract of another that fertilization never takes place. In plants, the growth of pollen tubes may be impeded in hybrids between different species. In both plants and animals the operation of such isolating mechanisms prevents the union of gametes even following successful mating.

Prezygotic isolating mechanisms lead to reproductive isolation by preventing the formation of hybrid zygotes.

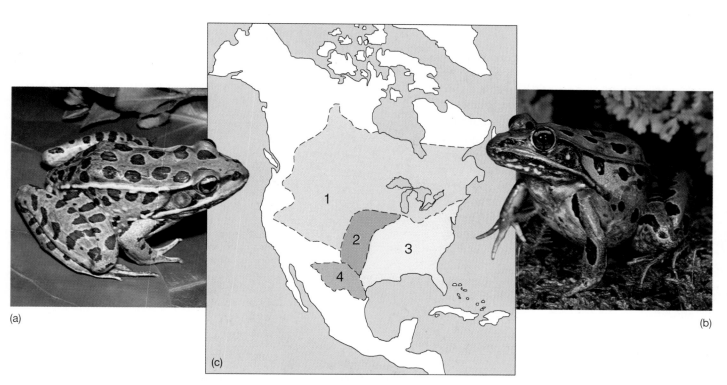

(a) (b) (c)

FIGURE 21.10
Behavioral isolation in leopard frogs. (a) The Rio Grande leopard frog, *Rana berlandieri*, and (b) the southern leopard frog, *Rana utricularia*, look alike but are separate species. (c) Numbers indicate the following species in the geographical ranges shown: 1, *Rana pipiens*; 2, *Rana blairi*; 3, *Rana utricularia*; 4, *Rana berlandieri*. These four species resemble one another closely in their external features. Their status as separate species was first suspected when hybrids between them produced defective embryos in some combinations. At first, biologists thought that this indicated differentiation within a species along a geographical gradient, but when the mating calls of the four species proved to differ substantially, their distinctiveness became evident.

Postzygotic Isolating Mechanisms

All of the factors we have discussed up to this point tend to prevent hybridization. If hybridization does occur and zygotes are produced, many factors may still prevent those zygotes from developing into normal, functional, fertile F_1 individuals (figure 21.11). Development in any species is a complex process. In hybrids, the genetic complements of two species may be so different that they cannot function together normally in embryonic development. For example, hybridization between sheep and goats usually produces embryos that die in the earliest developmental stages.

Leopard frogs (*Rana pipiens* complex) of the eastern United States are a group of similar species, assumed for a long time to constitute a single species. Previously, it was assumed that hybrids between some of these frogs were scarce because they came from different regions. But careful examination revealed that although the species appear similar, successful mating between them is rare. Many of the hybrid combinations cannot be produced even in the laboratory.

Many examples of this kind, in which similar species have been distinguished only as a result of hybridization experiments, are known in plants. Sometimes the hybrid embryos can be removed at an early stage and grown in an artificial medium. When these hybrids are supplied with extra nutrients or other supplements that compensate for their weakness or inviability, they may complete their development normally.

Even if the hybrids survive the embryo stage, they may not develop normally. If the hybrids are weaker than their parents, they will almost certainly be eliminated in nature. Even if they are vigorous and strong, as in the case of the mule, a hybrid between a horse and a donkey (figure 21.12), they may still be sterile and thus incapable of contributing to succeeding generations. The development of sex organs in hybrids may be abnormal, the chromosomes derived from the respective parents may not pair properly, or their fertility may simply be lower than normal for other reasons.

Postzygotic isolating mechanisms are those in which hybrid zygotes fail to develop or develop abnormally, or in which hybrids cannot become established in nature.

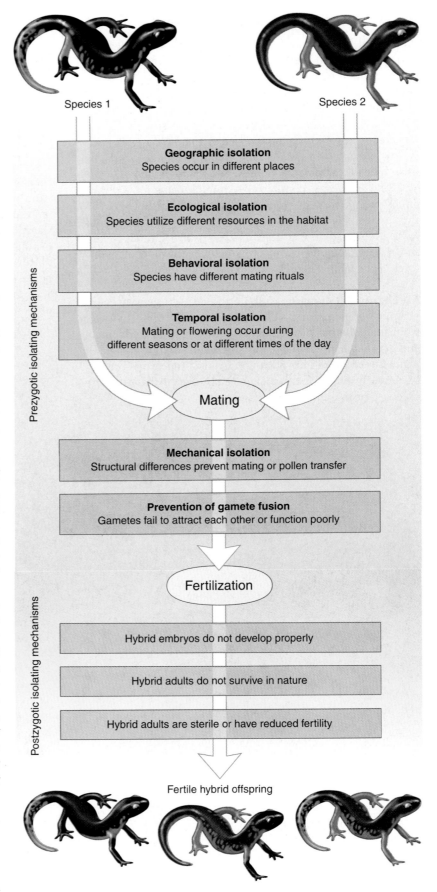

FIGURE 21.11
Reproductive isolating mechanisms. A variety of different mechanisms can prevent successful reproduction between individuals of different species.

FIGURE 21.12
The mule, a sterile hybrid between a female horse and a male donkey. By any standard, the horse and donkey are distinct species; mules, of course, cannot interbreed with either.

Reproductive Isolation: An Overview

All of the kinds of reproductive isolation we have discussed can arise in populations that are diverging from one another. They can also occur occasionally within species, as would be expected since such differences first evolved in this way. Some populations of particular species, therefore, cannot hybridize with one another, just as if they were distinct species as judged by this criterion.

The formation of species is a continuous process, one that we can understand because of the existence of intermediate stages at all levels of differentiation. If populations that are partly differentiated come into contact with one another, they may still be able to interbreed freely, and the differences between them may disappear over the course of time. If their hybrids are partly sterile, or not as well-adapted to the existing habitats as their parents, these hybrids will be at a disadvantage. As a result, selection may favor factors that limit the ability of the differentiated populations to hybridize. Individual plants or animals that do not hybridize may be more fit than those that do.

Most species are separated by combinations of the isolating mechanisms we have discussed. For example, two related species may occur in different habitats, produce their gametes at different times of the year, have different behav-

ioral patterns, and produce inviable embryos even if they do mate. Patterns in which more than one factor limits the frequency of hybrids between two species presumably arise for two reasons:

1. The factors that limit hybridization arise primarily as by-products of adaptive change in populations. Consequently, several different kinds of factors that limit hybridization often emerge simultaneously and may characterize the differentiated populations.
2. If differentiated populations do come into contact with one another, natural selection may strengthen the isolating mechanisms already present. If, for example, the hybrids do not complete development beyond the embryo stage, any factors that limit the hybridization that produces them (prezygotic isolating mechanisms) would provide an advantage. Individuals that form hybrids that do not function well in nature waste reproductive energy and are less fit than individuals that do not form such hybrids.

Both kinds of forces that limit hybridization build up as a part of the overall process of change in populations that are isolated from one another, although they may sometimes be strengthened when and if the differentiated populations migrate into contact with one another.

The Role of Polyploidy in Species Formation

Among plants, fertile individuals often arise from sterile ones through **polyploidy,** which doubles the chromosome number of the original sterile hybrid individual. A polyploid cell, tissue, or individual has more than two sets of chromosomes. Polyploid cells and tissues occur spontaneously and reasonably often in all organisms, although in many they are soon eliminated. A hybrid may be sterile simply because its sets of chromosomes, derived from male and female parents of different species, do not pair with one another. If the chromosome number of such a hybrid doubles, the hybrid, as a result of the doubling, will have a duplicate of each chromosome. In that case, the chromosomes will pair, and the fertility of the polyploid hybrid individual may be restored. It is estimated that about half of the approximately 260,000 species of plants have a polyploid episode in their history, including many of great commercial importance, such as bread wheat, cotton, tobacco, sugarcane, bananas, and potatoes. As you might imagine, the advantages a polyploid plant offers for natural selection can be substantial; hence the significance of polyploidy in the evolution of plants.

As a result of the way in which they originate, species are often separated by more than one factor. Some of these factors prevent the species from hybridizing; others limit the success of the interspecific hybrids once they are formed.

Darwin's Finches

One of the most visible manifestations of evolution is the existence of groups of closely related species that have recently evolved from a common ancestor by occupying different habitats. This type of **adaptive radiation** occurred among the 13 species of Darwin's finches on the Galápagos Islands. Presumably, the ancestor of Darwin's finches reached these islands before other land birds, so that when it arrived, all of the types of habitats where birds occur on the mainland were unoccupied. As the new arrivals moved into these vacant niches and adopted new lifestyles, they were subjected to diverse sets of selective pressures. Under these circumstances, the ancestral finches rapidly split into a series of diverse populations, some of which evolved into separate species.

The descendants of the original finches that reached the Galápagos Islands now occupy many different kinds of habitats on the islands (figure 21.13). These habitats encompass a variety of niches comparable to those several distinct groups of birds occupy on the mainland. The 13 species that inhabit the Galápagos comprise four groups:

1. **Ground finches.** There are six species of seed-eating *Geospiza* ground finches. Most of the ground finches feed on seeds. The size of their bills is related to the size of the seeds they eat. One of the ground finches, the large cactus ground finch *G. conirostris*, feeds primarily on cactus flowers and fruits and has a longer, larger, more pointed bill than the others.

2. **Tree finches.** There are five species of insect-eating *Camarhynchus* tree finches. Four species have bills that are suitable for feeding on insects. One has a parrot-like beak and feeds on buds and fruit in trees. The woodpecker finch has a chisel-like beak. This unusual bird carries around a twig or a cactus spine, which it uses to probe for insects in deep crevices.

3. **Warbler finch.** This unusual bird plays the same ecological role in the Galápagos woods that warblers play on the mainland, searching continually over the leaves and branches for insects. It has a slender, warbler-like beak.

4. **Vegetarian finch.** The very heavy bill of this bud-eating bird is used to wrench buds from branches.

Darwin's finches, all derived from one similar mainland species, have radiated widely on the Galápagos Islands in the absence of competition.

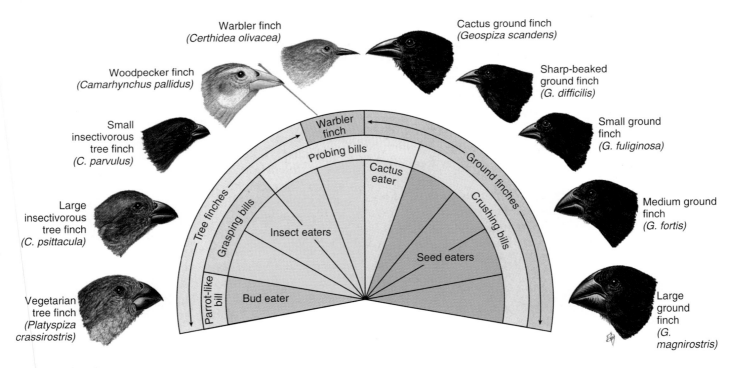

FIGURE 21.13
Darwin's finches. Ten species of Darwin's finches from Isla Santa Cruz, one of the Galápagos Islands, show differences in bills and feeding habits. The bills of several of these species resemble those of distinct families of birds on the mainland. This condition presumably arose when the finches evolved new species in habitats lacking small birds. The woodpecker finch uses cactus spines to probe in crevices of bark and rotten wood for food. Scientists believe all of these birds derived from a single common ancestor, a finch like the one in figure 1.10*a*.

Hawaiian *Drosophila*

Our second example of a cluster of species is the fly genus *Drosophila* on the Hawaiian Islands, which we mentioned earlier as an example of behavioral isolation. There are at least 1250 species of this genus throughout the world, and more than a quarter are found only in the Hawaiian Islands (figure 21.14). New species of *Drosophila* are still being discovered in Hawaii, although the rapid destruction of the native vegetation is making the search more difficult. Aside from their sheer number, Hawaiian *Drosophila* species are unusual because of the morphological and behavioral traits discussed earlier (see figure 21.9). No comparable species of *Drosophila* are found anywhere else in the world.

A second, closely related genus of flies, *Scaptomyza*, also forms a species cluster in Hawaii, where it is represented by as many as 300 species. A few species of *Scaptomyza* are found outside of Hawaii, but the genus is better represented there than elsewhere. In addition, species intermediate between *Scaptomyza* and *Drosophila* exist in Hawaii, but nowhere else. The genera are so closely related that scientists have suggested that all of the estimated 800 species of these two genera that occur in Hawaii may have derived from a single common ancestor.

The native Hawaiian flies are closely associated with the remarkable native plants of the islands and are often abundant in the native vegetation. Evidently, when their ancestors first reached these islands, they encountered many "empty" habitats that other kinds of insects and other animals occupied elsewhere. The evolutionary opportunities the ancestral *Drosophila* flies found were similar to those the ancestors of Darwin's finches in the Galápagos Islands encountered, and both groups evolved in a similar way. Many of the Hawaiian *Drosophila* species are highly selective in their choice of host plants for their larvae and in the part of the plant they use. The larvae of various species live in rotting stems, fruits, bark, leaves, or roots, or feed on sap.

New islands have continually arisen from the sea in the region of the Hawaiian Islands. As they have done so, they appear to have been invaded successively by the various *Drosophila* groups present on the older islands. New species have evolved as new islands have been colonized. The Hawaiian species of *Drosophila* have had even greater evolutionary opportunities than Darwin's finches because of their restricted ecological niches and the variable ages of the islands. They clearly tell one of the most unusual evolutionary stories found anywhere in the world.

The adaptive radiation of about 800 species of the flies *Drosophila* and *Scaptomyza* on the Hawaiian Islands, probably from a single common ancestor, is one of the most remarkable examples of intensive species formation found anywhere on earth.

(a) *Drosophila mulli*

(b) *Drosophila primaeva*

(c) *Drosophila digressa*

FIGURE 21.14
Hawaiian *Drosophila*. The hundreds of species that have evolved on the Hawaiian Islands are extremely variable in appearance, although genetically almost identical.

Lake Victoria Cichlid Fishes

Lake Victoria is an immense shallow freshwater sea about the size of Switzerland in the heart of equatorial East Africa, until recently home to an incredibly diverse collection of over 200 species of cichlid fishes.

Recent Radiation

This cluster of species appears to have evolved recently and quite rapidly. By sequencing the cytochrome *b* gene in many of the lake's fish species and using variation among the species as a molecular clock (see figure 20.15), researchers like Axel Meyer at the State University of New York, Stony Brook, have been able to estimate that the first cichlids entered Lake Victoria only 200,000 years ago, colonizing from the Nile. Dramatic changes in water level encouraged species formation. As the lake rose, it flooded new areas and opened up new habitat. Many of the species may have originated after the lake dried down 14,000 years ago, isolating local populations in small lakes until the water level rose again.

FIGURE 21.15
Cichlid fishes of Lake Victoria. These four species of the genus *Haplochromis* occupy different ecological niches, although they are similar in appearance. *H. acidens* (*upper right*) feeds on dense stands of plants; *H. prognathus* (*upper left*) is a fish predator; *H. microdon* (*lower left*) feeds on fish eggs and larvae; *H. saxicola* (*lower right*) feeds on insect larvae on the lake bottom.

Cichlid Diversity

These small, perchlike fishes range from 2 to 10 inches in length, and the males come in endless varieties of colors. In initial surveys, far from complete, over 300 closely related species were described! The most diverse assembly of vertebrates known to science, the Lake Victoria cichlids defy simple description. We can gain some sense of the vast range of types by looking at how different species eat (figure 21.15). There are mud biters, algae scrapers, leaf chewers, snail crushers, snail shellers (who pounce on slow-crawling snails and spear their soft parts with long curved teeth before the snail can retreat into its shell), zooplankton eaters, insect eaters, prawn eaters, and fish eaters. Scale scraping cichlids rasp slices of scales off of other fish.

There are even cichlid species that are "pedophages," eating the young of other cichlids. All Lake Victoria cichlids are mouthbrooders, the females keeping their young inside their mouths to protect them. Some pedophage species operate as suckers, others as rammers. Some rammers shoot toward the mother from below and behind, ramming into her throat, and then eating the ejected brood before the surprised mother can recover. Another rammer crashes down from above, kamikaze-like, onto the nose of the mother.

Abrupt Extinction

Much of this diversity is gone. In the 1950s, the Nile perch, a commercial fish with a voracious appetite, was introduced on the Ugandan shore of Lake Victoria. Since then it has spread through the lake, eating its way through the cichlids. By 1990 all the open-water cichlid species were extinct, as well as many living in rocky shallow regions. Over 70% of all the named Lake Victoria cichlid species had disappeared, as well as untold numbers of species that had yet to be described. The Nile perch, in the meantime, has become a superb source of food for people living around the lake. The isolation of Lake Victoria from other kinds of fishes played a primary role in the explosive radiation of cichlid fishes, and when that isolation broke down with the introduction of the Nile perch, the bloom of speciation ended.

Very rapid speciation occurred among cichlid fishes isolated in Lake Victoria, but widespread extinction followed when the isolation ended.

New Zealand Alpine Buttercups

Adaptive radiations as we have described in Galápagos finches, Hawaiian *Drosophila*, and cichlid fishes seem to be favored by *periodic isolation*. Finches and *Drosophila* invade new islands, local species evolve, and they in turn reinvade the home island, in a cycle of expanding diversity. Similarly, cichlids become isolated by falling water levels, evolving separate species in isolated populations that later are merged when the lake's water level rises again.

A clear example of the role periodic isolation plays in species formation can be seen in the alpine buttercups (genus *Ranunculus*) of New Zealand (figure 21.16). More species of alpine buttercup grow on the two islands of New Zealand than in all of North and South America combined. Detailed studies by the Canadian taxonomist Fulton Fisher revealed that the evolutionary mechanism responsible for inducing this diversity is recurrent isolation associated with the recession of glaciers. The 14 species of alpine *Ranunculus* occupy five distinctive habitats: *snowfields* (rocky crevices among outcrops in permanent snowfields at 7000 to 9000 feet elevation); *snowline fringe* (rocks at lower margin of snowfields between 4000 and 7000 ft); *stony debris* (scree slopes of exposed loose rocks at 2000 to 6000 ft); *sheltered situations* (shaded by rock or shrubs at 1000 to 6000 ft); and *boggy habitats* (sheltered slopes and hollows, poorly drained tussocks at elevations between 2500 and 5000 ft).

Ranunculus species have invaded these five habitats repeatedly as glaciers have formed, joining mountains together, receded to isolate the species, and then re-formed, joining the alpine chain once again (figure 21.17). Parallel sets of "ecospecies" have evolved independently to occupy similar habitats, so that pairs of distantly related species now occur in close proximity on glacial snow, scree slopes, and bogs.

Recurrent isolation promotes species formation.

FIGURE 21.16
A New Zealand alpine buttercup. Fourteen species of alpine *Ranunculus* grow among the glaciers and mountains of New Zealand, including this *R. lyallii*, the giant buttercup.

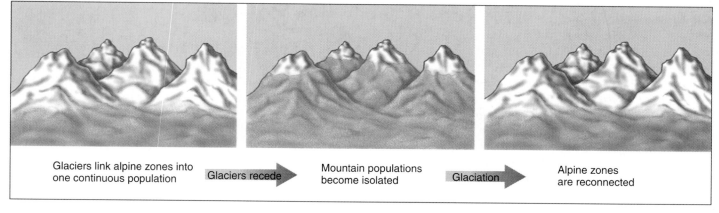

Glaciers link alpine zones into one continuous population → Glaciers recede → Mountain populations become isolated → Glaciation → Alpine zones are reconnected

FIGURE 21.17
Periodic glaciation encouraged species formation among alpine buttercups in New Zealand. The formation of extensive glaciers during the Pleistocene linked the alpine zones of many mountains together. When the glaciers receded, these alpine zones were isolated from one another, only to become reconnected with the advent of the next glacial period. During periods of isolation, populations of alpine buttercups diverged in the isolated habitats.

Sexual Selection and the Origin of Species

In addition to natural selection, Charles Darwin considered sexual selection a major feature of evolution. He devoted an entire book, *The Descent of Man, and Selection in Relation to Sex*, to this topic. Many subsequent writers on evolution, however, have denied both the major significance and the distinctiveness of sexual selection, a judgment that is now outdated.

Sexual selection is defined as differential reproduction that results from variable success in obtaining mates due to combat or courtship. This contrasts with natural selection, which Darwin defined as differential reproduction related to the "struggle for existence." Sexual selection results from a contest or struggle between rivals of the same sex for mates.

The theory of sexual selection argues that evolutionary change can originate from the divergence of characteristics used in sexual displays; social factors may sometimes outweigh ecological factors in determining the evolution of distinctive features. They may lead to incompatibility of populations because of social and courtship incompatibility. Even a slight divergence in key aspects of courtship behavior may be sufficient to put "misfit" males at a severe disadvantage, with female discrimination against them a strong barrier to interbreeding. Sexual selection is clearly a distinctive factor in the evolution of life, and it is presently an important subject of research.

Males and females of many animal species exhibit marked differences in secondary sexual characteristics, a phenomenon known as *sexual dimorphism*. The most common form of sexual dimorphism is difference in size, with the male usually larger. A second widespread form involves male adornments, such as antlers on male deer (figure 21.18*a*), elk, and moose; manes on male lions; and colorful plumage on male ducks and other birds.

In some types of sexual dimorphism, the secondary sexual characteristic is used in direct competition with other males. Male mountain sheep, for example, use their horns in butting contests as they compete for a female. Competition among males is particularly common in species where a single male maintains a harem of females.

(a)

(b)

FIGURE 21.18
Sexual selection. Sexual selection acts to exaggerate sexual differences. (a) The male white-tailed deer has antlers that the female lacks. (b) This male widow bird has a tail longer than its body and head—which attracts females of the species.

In other cases of sexual dimorphism, the male with the most impressive features attracts the most females. A classic example of this form of sexual selection is the evolution of the astonishing tail of the African long-tailed widow bird—their tails are longer than their bodies and heads combined (figure 21.18*b*)! Females strongly prefer long-tailed individuals, the longer the better. When investigator Malte Andersson in 1982 cut short the tails of some males and lengthened the tails of others by gluing on extra feathers, the changes in tail length had no effect on male–male competition; each successfully defended its display site. However, females were attracted four times as often by the long-tailed males.

Sexual selection leads to sexual dimorphism

Summary of Concepts

21.1 Species are the basic units of evolution.

- The pace of evolution is not constant among all organisms. Some scientists believe it occurs in spurts, others argue that it proceeds gradually.

- Species are groups of organisms that differ from one another in one or more characteristics and do not hybridize freely when they come into contact in their natural environment. Many species cannot hybridize with one another at all.

21.2 Species formation begins with the divergence of populations.

- Populations change as they adapt to the demands of their environments. Even populations of a given species that are close to one another geographically are often effectively isolated. Such populations are free to diverge in response to the demands of their particular environments.

- In general, if the selective forces on populations differ greatly, the populations will diverge rapidly; if the selective forces are similar, the populations will diverge slowly.

- Ecological races and subspecies differentiate within species but often still intergrade with one another. The differences between them, in both plants and animals, are mostly genetically fixed.

21.3 Species formation is completed by the evolution of reproductive barriers.

- Among the factors that separate populations and species are geographical, ecological, temporal, behavioral, and mechanical isolation, as well as factors that inhibit the fusion of gametes or the normal development of the hybrid organisms.

- Reproductive isolation between species arises as a normal by-product of the continual differentiation of populations.

- Some isolating mechanisms (prezygotic) prevent hybrid formation; others (postzygotic) prevent hybrids from surviving and reproducing.

- Hybridization between different species, especially in trees and shrubs, creates a source of new genotypes that selection can act upon. If the chromosomes of the parents cannot pair in a hybrid, spontaneous doubling may produce a polyploid individual.

21.4 Clusters of species reflect rapid evolution.

- Clusters of species arise when populations differentiate to fill several niches. On islands, differentiation is often rapid because of numerous open habitats. In many continental areas, differentiation is not as rapid; in local situations, as when many different kinds of plants are developing close to one another, differentiation may occur rapidly.

- Sexual selection favors sexual dimorphism, causing certain individuals of one sex to inherit traits that afford them greater success in breeding.

Discussing Key Terms

1. **Species**   The basic unit of evolution and taxonomy is the species. In animals, species tend to be groups of interbreeding individuals that do not breed with other groups; but in plants, fungi, and many protists and bacteria, this distinction is less clear.

2. **Dispersal**   The ability of individuals or their seeds to move away from their home populations has a major impact on how rapidly a species colonizes new habitats.

3. **Divergence**   As separated populations adapt to different circumstances, they become increasingly different genetically.

4. **Ecotypes**   Local populations of a species become adapted to their local habitat. Such ecotypes or ecological races are the first step in species formation.

5. **Subspecies**   When ecological races begin to differ in obvious features, they are referred to as subspecies.

6. **Reproductive isolation**   A key event in species formation is the erection of a barrier to breeding between two isolated groups. Barriers to reproduction may operate prior to mating and zygote formation or after the zygote has been formed.

7. **Prezygotic isolating mechanisms**   Reproductive isolation is favored by mechanisms that prevent the formation of zygotes.

8. **Postzygotic isolating mechanisms**   Reproductive isolation is also favored by mechanisms that prevent the proper functioning of zygotes after they are formed.

9. **Sexual selection**   The favoring of characteristics in one sex, as a result of competition among individuals of that sex for mates, is called sexual selection.

Review Questions

1. How does the concept of punctuated equilibrium differ from gradualism as they both relate to evolution?

2. What is the modern definition of a species?

3. Define the term *niche*. How many species can occupy a niche?

4. Why do populations of organisms change?

5. How does selection relate to population divergence?

6. If two populations of the same species are geographically close and are under similar selective forces, will they retain similar characteristics over time? Why or why not?

7. What barriers exist to hybrid formation and success? Which are prezygotic and which are postzygotic isolating mechanisms?

8. When are hybrids at a fitness disadvantage? What can be the result of this disadvantage?

9. What is adaptive radiation? What types of habitats encourage it? Why?

10. What is the primary mechanism that maintains the integrity of the species clusters of *Drosophila* in the Hawaiian Islands?

11. How does sexual selection differ from natural selection?

12. What environmental conditions foster rapid evolution?

13. Define the term *polyploidy*.

Thought Questions

1. What circumstances select against hybridization between species? When is it selected for?

2. Polyploid animals are far less common than polyploid plants, particularly perennial plants. Why do you think this might be so?

3. When two partly differentiated populations of a single species come into contact with one another after a period of isolation their differences may increase or decrease. What would be the result if their differences increased? Decreased? What conditions would favor each outcome?

Internet Links

Read Darwin's Books
http://www.literature.org/Works/Charles-Darwin/
Charles Darwin wrote 22 books during his long life. This site presents the complete text of Darwin's three key books on evolution: Voyage of the Beagle, The Origin of Species, & The Descent of Man. *An excellent writer, Darwin makes biology come alive. More than a century later, you will discover that his books are a delight to read.*

A History of Ideas About Evolution
http://www.ucmp.berkeley.edu/history/evolution.html
A wonderful tour of the history of evolutionary thinking before and after Darwin, featuring the contributions of 28 significant scientists and thinkers.

For Further Reading

Avise, J.: *Molecular Markers, Natural History and Evolution.* Chapman and Hall, New York, 1994. An up-to-date account of how molecular biology is revolutionizing our ability to study evolution.

Browne, J.: *Charles Darwin: Voyaging, Volume I of a Biography.* Knopf/Cape, 1995. A delightful book that takes a detailed look at Darwin's life in light of the then-understood status of evolutionary biology. First of two volumes, highly recommended.

Cody, M. and J. Overton: "Short-term Evolution of Reduced Dispersal in Island Plant Populations," *Journal of Ecology,* 1996, vol. 84, pages 53–61. An unusually clear and simple example of evolution by natural selection.

Futuyma, D.: *Evolutionary Biology,* 2d. ed., Sinauer Associates, Sunderland, MA, 1986. This text presents a good discussion of the development of evolutionary thought, combined with a process-oriented treatment of the whole field.

Goldschmidt, T.: *Darwin's Dreampond,* The MIT Press, Cambridge, Mass., 1996. An award-winning account by a Dutch fish taxonomist of how the colorful and incredibly diverse cichlids he was classifying were driven to extinction when Nile perch were introduced into Lake Victoria.

Gould, S. J.: *Ever Since Darwin,* W. W. Norton, New York, 1977. An entertaining and insightful collection of essays on evolution and Darwinism.

Grant, P.: "Natural Selection and Darwin's Finches," *Scientific American,* October 1991, pages 82–87. In the Galápagos, finches evolve very rapidly—populations are altered significantly by a single season of drought.

Kerr, R.: "Did Darwin Get It All Right?" *Science,* March 1995, vol. 267, pages 1421–22. A thorough study of species formation in bryozoans indicates that new species appear abruptly in the fossil record after long periods of stability, providing evidence supporting the punctuated equilibrium model.

Little, C.: *Terrestrial Invasion: An Ecophysiological Approach to the Origins of Land Animals,* Cambridge University Press, Cambridge, England, 1990. An excellent treatment of the factors involved in the evolution of land animals.

May, R.: "How Many Species Inhabit the Earth?" *Scientific American,* October 1992, pages 42–48. A good guess at how many species exist now, how they got here, and the implications for conservation.

22

How Humans Evolved

Concept Outline

22.1 The evolutionary path to humans starts with the advent of primates.

The Evolutionary Path to Apes. Primates first evolved 65 million years ago, giving rise first to prosimians and then to monkeys.

How the Apes Evolved. Apes, including our closest relatives, the chimpanzees, arose from an ancestor common to Old World monkeys.

The Origins of Bipedalism. The ability to walk upright on two legs marks the beginning of hominid evolution.

22.2 The first hominids to evolve were australopithecines.

An Evolutionary Tree with Many Branches. The first hominids were australopithecines, of which there were several different kinds.

The Root of the Hominid Tree. One can draw the hominid family tree in two very different ways, either lumping variants together or splitting them into separate species.

22.3 The genus *Homo* evolved in Africa.

African Origin: Early *Homo*. There may have been several species of early *Homo*, with brains significantly larger than those of australopithecines.

Out of Africa: *Homo erectus*. The first hominid species to leave Africa was the relatively large-brained *H. erectus*, the longest lived species of *Homo*.

22.4 Modern humans evolved quite recently.

The Last Stage of Hominid Evolution. Modern humans evolved within the last 600,000 years, our own species within the last 200,000 years.

Out of Africa—Again? Though hotly debated, it seems most likely that *H. sapiens* evolved in Africa and then migrated out of Africa, replacing *H. erectus* in Europe and later in Asia.

The Emergence of Modern Humans. Our species is unique in evolving culturally.

FIGURE 22.1
The trail of our ancestors. These fossil footprints, made in Africa 3.7 million years ago, look as if they might have been left by a mother and child walking on the beach. But these tracks, preserved in volcanic ash, are not human. They record the passage of two individuals of the genus *Australopithecus*, the group from which our genus, *Homo*, evolved.

In his book *The Descent of Man* (1871), Darwin used arguments from comparative anatomy, development, and behavior to support the idea that humans evolved from now-extinct African apes—the ancestors of today's gorilla and chimpanzee. Although little fossil evidence existed in 1871 to support Darwin's case, numerous fossil discoveries since Darwin's death strongly support his hypothesis (figure 22.1). Human evolution is the part of evolution that usually interests people most. This chapter describes the evolutionary journey that has led to humans. It is an exciting story, replete with controversy.

The Evolutionary Path to Apes

The story of human evolution begins around 65 million years ago, with the explosive radiation of a group of small, arboreal mammals called the Archonta. These primarily insectivorous mammals had large eyes and were most likely **nocturnal** (active at night). Their radiation gave rise to different types of mammals, including bats, tree shrews, and **primates.**

The Earliest Primates

Two distinct features of primates allowed them to succeed in the arboreal, insect-eating environment:

1. **Grasping fingers and toes.** Unlike the clawed feet of tree shrews and squirrels, primates have prehensile (grasping) hands and feet that let them grip limbs, hang from branches, seize food, and, in some primates, use tools. The first digit in many primates is **opposable** and at least some, if not all, of the digits have nails.
2. **Binocular vision.** Unlike the eyes of shrews and squirrels, which sit on each side of the head so that the two fields of vision do not overlap, the eyes of primates are shifted forward to the front of the face. This produces overlapping **binocular vision** that lets the brain judge distance precisely—important to an animal moving through the trees.

Other mammals have binocular vision, but only primates have both binocular vision and grasping hands, making them particularly well adapted to their environment. While early primates were mostly insectivorous, their dentition began to change from the shearing, triangular-shaped molars specialized for insect eating to the more flattened, square-shaped molars and rodent-like incisors specialized for plant eating. Primates that evolved later show also a continuous reduction in snout length and number of teeth.

The Evolution of Prosimians

About 40 million years ago, the earliest primates split into two groups: the prosimians and the anthropoids. The **prosimians** ("before monkeys") looked something like a cross between a squirrel and a cat and were common in North America, Europe, Asia, and Africa. Only a few prosimians survive today, representing two lineages that split off and evolved from the earliest primates at roughly the same time as the anthropoids. One lineage gave rise to lemurs and lorises and another to tarsiers. In addition to having grasping digits and binocular vision, prosimians have large eyes with increased visual acuity (figure 22.2). They feed on fruits, leaves, and flowers among slender tree branches, often at night. Most prosimians are nocturnal, and many lemurs have long tails for balancing.

(a)

(b)

FIGURE 22.2
Prosimians. (a) The tarsier, a prosimian native to tropical Asia. (b) A ring-tailed lemur. All living lemurs are found in Madagascar.

Origin of the Anthropoids

The anthropoids, or higher primates, include monkeys, apes, and humans. Anthropoids are almost all diurnal—that is, active during the day—feeding mainly on fruits and leaves. Evolution favored many changes in eye design, including color vision, that were adaptations to daytime foraging. An expanded cortex governs the improved senses, with the braincase forming a larger portion of the head. Anthropoids, like the relatively few diurnal prosimians, live in groups with complex social interactions. In addition, the anthropoids tend to care for their young for prolonged periods, allowing for a long childhood of learning and the development of the brain.

One of the most contentious issues in primate biology is the identity of the first anthropoid, the common ancestor of monkeys, apes, and humans. One candidate is a 45-million-year-old Chinese fossil called *Eosimias* ("dawn ape") known only from two jaws. If it is a primate, it is very old, living just 25 million years after the dinosaurs became extinct. Another candidate, known from several skulls with jaws and teeth, is a 37-million-year-old Egyptian lemur-like primate about the size of a squirrel, called *Catopithecus.* The skulls have accepted anthropoid characteristics, such as a complete bony cone around the eye socket, shovel-shaped incisor teeth, and forehead bones that are fused together rather than separate. While no clear determination of the earliest anthropoid is possible without more fossils, current evidence favors *Catopithecus.*

The early anthropoids, now extinct, are thought to have evolved in Africa. Their direct descendants are a very successful group of primates, the monkeys.

New World Monkeys. About 30 million years ago, some anthropoids migrated to South America, where they developed in isolation. Their South American descendants, known as the **New World monkeys,** are easy to identify: all are arboreal, they have flat spreading noses, and many of them grasp objects with long prehensile tails (figure 22.3*a*).

Old World Monkeys. Around 25 million years ago, anthropoids that remained in Africa split into two lineages: one gave rise to the **Old World monkeys** and one gave rise to the hominoids, or human line. Old World monkeys have been evolving separately from New World monkeys for at least 30 million years and include ground-dwelling as well as arboreal species. None of the Old World monkeys have prehensile tails. Their nostrils are close together, their noses point downward, and some have toughened pads of skin for prolonged sitting (figure 22.3*b*).

The earliest primates arose from small, tree-dwelling, insect-eaters and gave rise to prosimians and then anthropoids. Early anthropoids gave rise to New World monkeys and Old World monkeys.

(a)

(b)

FIGURE 22.3
New and Old World monkeys. (a) All New World monkeys are arboreal, and many have prehensile tails. (b) Old World monkeys lack prehensile tails, and many are ground dwellers.

How the Apes Evolved

From anthropoid ancestors evolved **hominoids**, including the **apes** and the **hominids** (humans and their direct ancestors). The living apes consist of the gibbon (genus *Hylobates*), orangutan (*Pongo*), gorilla (*Gorilla*), and chimpanzee (*Pan*) (figure 22.4). Apes have larger brains than monkeys, and they lack tails. With the exception of the gibbon, which is small, all living apes are larger than any monkey. Apes exhibit the most adaptable behavior of any mammal except human beings. Once widespread in Africa and Asia, apes are rare today, living in relatively small areas. No apes ever occurred in North or South America.

The First Hominoid

Considerable controversy exists about the identity of the first hominoid. During the 1980s it was commonly believed that the common ancestor of apes and hominids was a late Miocene ape living 5 to 10 million years ago. In 1932, a candidate fossil, an 8-million-year-old jaw with teeth, had been unearthed in India. It was called *Ramapithecus* (after the Hindu deity Rama). However, these fossils have never been found in Africa, and more complete fossils discovered in 1981 made it clear that *Ramapithecus* is in fact closely related to the orangutan. Attention has now shifted to an earlier Miocene ape, *Proconsul*, which has many of the characteristics of Old World monkeys but lacks a tail and has ape-like hands, feet, and pelvis. However, because very few fossils have been recovered from the period 5 to 10 million years ago, it is not yet possible to identify with certainty the first hominoid ancestor.

(b)

(c)

(a)

(d)

FIGURE 22.4
Apes. (a) Mueller gibbon, *Hylobates muelleri*. (b) Orangutan, *Pongo pygmaeus*. (c) Gorilla, *Gorilla gorilla*. (d) Chimpanzee, *Pan troglodytes*.

Which Ape Is Our Closest Relative?

Studies of ape DNA have explained a great deal about how the living apes evolved. The Asian apes evolved first. The line of apes leading to gibbons diverged from other apes about 15 million years ago, while orangutans split off about 10 million years ago (see figure 22.8). Neither are closely related to humans.

The African apes evolved more recently, between 6 and 10 million years ago. These apes are the closest living relatives to humans; some taxonomists have even advocated placing humans and the African apes in the same zoological family, the Hominidae. Chimpanzees are more closely related to humans than gorillas are, diverging from the ape line less than 6 million years ago. Because this split was so recent, the genes of humans and chimpanzees have not had time to evolve many differences—humans and chimpanzees share 98.4% of their nuclear DNA, a level of genetic similarity normally found between sibling species of the same genus! A human hemoglobin molecule differs from its chimpanzee counterpart in only a single amino acid.

Gorilla DNA differs from human DNA by about 2.3%. This somewhat greater genetic difference reflects the greater time since gorillas split off from the ape line, some 8 million years ago. Sometime after the gorilla lineage diverged, the common ancestor of all hominids split off from the ape line to begin the evolutionary journey leading to hominids. Fossils of the earliest hominids, described later in the chapter, suggest that the common ancestor of the hominids was more like a chimpanzee than a gorilla.

Comparing Apes to Hominids

The common ancestor of apes and hominids is thought to have been an arboreal climber. Much of the subsequent evolution of the hominoids reflected different approaches to locomotion. Hominids became **bipedal**, walking upright, while the apes evolved knuckle-walking, supporting their weight on the back sides of their fingers (monkeys, by contrast, use the palms of their hands).

Humans depart from apes in several areas of anatomy related to bipedal locomotion (figure 22.5). Because humans walk on two legs, their vertebral column is more curved than an ape's, and the human spinal cord exits from the bottom rather than the back of the skull. The human pelvis has become broader and more bowl-shaped, with the bones curving forward to center the weight of the body over the legs. The hip, knee, and foot (in which the human big toe no longer splays sideways) have all changed proportions.

Being bipedal, humans carry much of the body's weight on the lower limbs, which comprise 32% to 38% of the body's weight and are longer than the upper limbs; human upper limbs do not bear the body's weight and make up only 7% to 9% of human body weight. African apes walk on all fours, with the upper and lower limbs both bearing the body's weight; in gorillas, the longer upper limbs account for 14% to 16% of body weight, the somewhat shorter lower limbs for about 18%.

Hominoids, the apes and hominids, arose from Old World monkeys. Among living apes, chimpanzees seem the most closely related to humans.

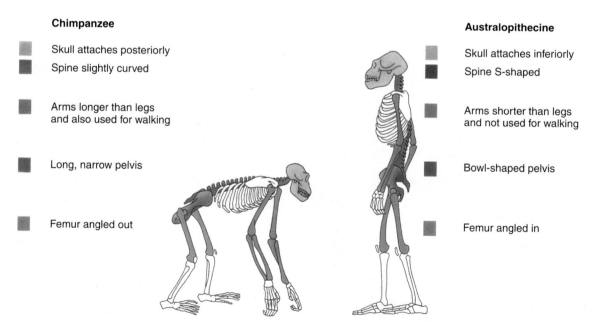

Chimpanzee

- Skull attaches posteriorly
- Spine slightly curved

- Arms longer than legs and also used for walking

- Long, narrow pelvis

- Femur angled out

Australopithecine

- Skull attaches inferiorly
- Spine S-shaped

- Arms shorter than legs and not used for walking

- Bowl-shaped pelvis

- Femur angled in

FIGURE 22.5

A comparison of ape and hominid skeletons. Early humans, such as australopithecines, were able to walk upright because their arms were shorter, their spinal cord exited from the bottom of the skull, their pelvis was bowl-shaped and centered the body weight over the legs, and their femurs angled inward, directly below the body, to carry its weight.

The Origins of Bipedalism

For much of this century, biologists have debated the sequence of events that led to the evolution of hominids. A key element may have been bipedalism. Bipedalism seems to have evolved as our ancestors left dense forests for grasslands and open woodland (figure 22.6). One school of thought proposes that hominid brains enlarged first, and then hominids became bipedal. Another school sees bipedalism as a precursor to bigger brains. Those who favor the brain-first hypothesis speculate that human intelligence was necessary to make the decision to walk upright and move out of the forests and onto the grassland. Those who favor the bipedalism-first hypothesis argue that bipedalism freed the forelimbs to manufacture and use tools, favoring the subsequent evolution of bigger brains.

A treasure trove of fossils unearthed in Africa has settled the debate. These fossils demonstrate that bipedalism extended back 4 million years ago; knee joints, pelvis, and leg bones all exhibit the hallmarks of an upright stance. Substantial brain expansion, on the other hand, did not appear until roughly 2 million years ago. In hominid evolution, upright walking clearly preceded large brains.

Remarkable evidence that early hominids were bipedal is a set of some 69 hominid footprints found at Laetoli, East Africa. Two individuals, one larger than the other, walked side-by-side for 27 meters, their footprints preserved in

FIGURE 22.7
A Laetoli footprint. This footprint, almost certainly of *Australopithecus afarensis*, is 3.7 million years old.

3.7-million-year-old volcanic ash! These footprints came from an early hominid, walking upright. The impression in the ash (figure 22.7) reveals a strong heelstrike and a deep indentation made by the big toe, much as you might make in sand when pushing off to take a step. Importantly, the big toe is not splayed out to the side as in a monkey or ape—the footprints were clearly made by a hominid.

The evolution of bipedalism marks the beginning of the hominids. Why bipedalism evolved in hominids remains a matter of controversy. No tools appeared until 2.5 million years ago, so toolmaking seems an unlikely cause. Alternative ideas suggest that walking upright is faster and uses less energy than walking on four legs; that an upright posture permits hominids to pick fruit from trees and see over tall grass; that being upright reduces the body surface exposed to the sun's rays; that an upright stance aided the wading of aquatic hominids, and that bipedalism frees the forelimbs of males to carry food back to females, encouraging pair-bonding. All of these suggestions have their proponents, and none are universally accepted. The origin of bipedalism, the key event in the evolution of hominids (figure 22.8), seems destined to remain a mystery for now.

The evolution of bipedalism—walking upright—marks the beginning of hominid evolution, although no one is quite sure why bipedalism evolved.

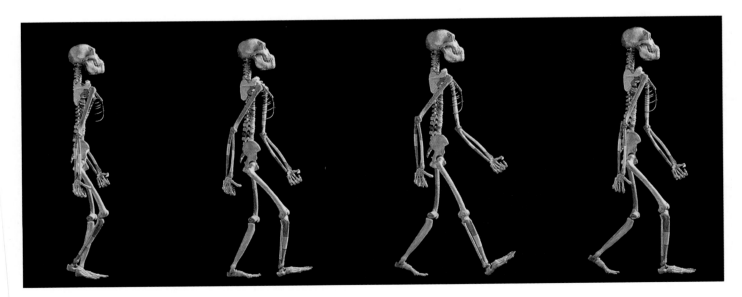

FIGURE 22.6
A reconstruction of an early hominid walking upright. These articulated plaster skeletons, made by Owen Lovejoy and his students at Kent State University, depict an early hominid (*Australopithecus afarensis*) walking upright.

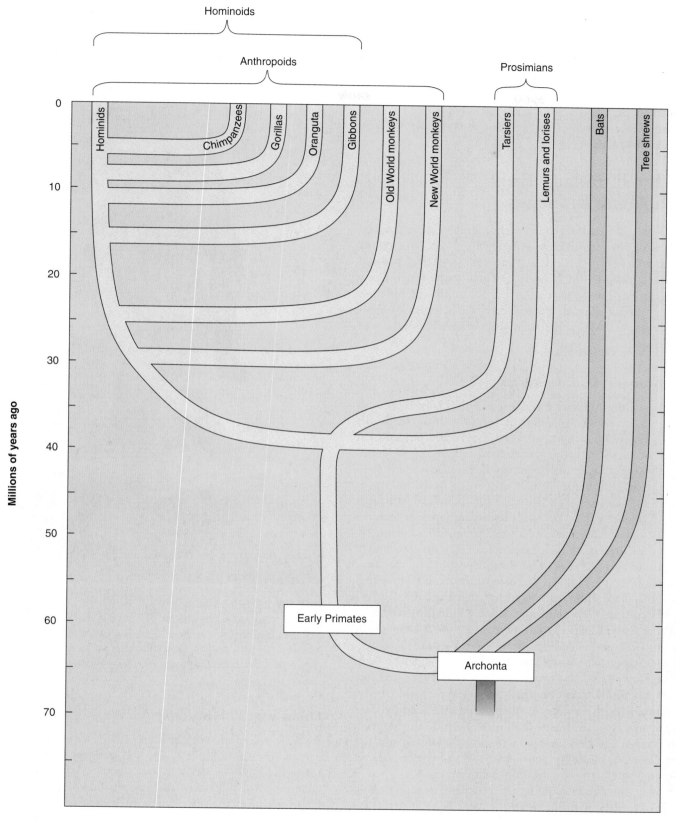

FIGURE 22.8
A primate evolutionary tree. The most ancient of the primates are the prosimians, while the hominids were the most recent to evolve.

Five to 10 million years ago, the world's climate began to get cooler, and the great forests of Africa were largely replaced with savannas and open woodland. In response to these changes, a new kind of ape was evolving, one that was bipedal. These new apes are classified as hominids—that is, of the human line.

An Evolutionary Tree with Many Branches

There are two major groups of hominids: three to seven species of the genus *Homo* (depending how you count them) and seven species of the older, smaller-brained genus *Australopithecus*. In every case where the fossils allow a determination to be made, the hominids are bipedal, walking upright. Bipedal locomotion is the hallmark of hominid evolution. We will first discuss *Australopithecus*, and then *Homo*.

Discovery of *Australopithecus*

The first hominid was discovered in 1924 by Raymond Dart, an anatomy professor at Johannesburg in South Africa. One day, a mine worker brought him an unusual chunk of rock—actually, a rock-hard mixture of sand and soil. Picking away at it, Professor Dart uncovered a skull unlike that of any ape he had ever seen. Beautifully preserved, the skull was of a five-year-old individual, still with its milk teeth. While the skull had many apelike features such as a projecting face and a small brain, it had distinctly human features as well—for example, a rounded jaw unlike the pointed jaw of apes. The ventral position of the foramen magnum (the hole at the base of the skull from which the spinal cord emerges) suggested that the creature had walked upright. Dart concluded it was a human ancestor.

What riveted Dart's attention was that the rock in which the skull was embedded had been collected near other fossils that suggested that the rocks and their fossils were several million years old! At that time, the oldest reported fossils of hominids were less than 500,000 years old, so the ancientness of this skull was unexpected and exciting. Scientists now estimate Dart's skull to be 2.8 million years old. Dart called his find *Australopithecus africanus* (from the Latin *australo*, meaning "southern," and the Greek *pithecus*, meaning "ape"), the ape from the south of Africa.

Today, fossils are dated by the relatively new process of single-crystal/laser-fusion dating. A laser beam melts a single potassium feldspar crystal, releasing argon gas, which is measured in a gas mass spectrometer. Since the argon in the crystal has accumulated by radioactive decay from potassium at a known rate, the amount released reveals the age of the rock and thus of nearby fossils. The margin of error is less than 1%.

FIGURE 22.9
"Lucy." This reconstruction, at The Living World education center of the St. Louis Zoo, was based on a careful study of muscle attachments to the skull and skeleton. Lucy is the most complete skeleton of *Australopithecus* discovered so far.

Other Kinds of *Australopithecus*

In 1938, a second, stockier kind of *Australopithecus* was unearthed in South Africa. Called *A. robustus*, it had massive teeth and jaws. In 1959, in east Africa, Mary Leakey discovered a third kind of *Australopithecus*—*A. boisei* (after Charles Boise, an American-born businessman who contributed to the Leakeys' projects)—who was even more stockily built. Like the other australopithecines, *A. boisei* was very old—almost 2 million years. Nicknamed "Nutcracker man," *A. boisei* had a great bony ridge on the crest of the head to anchor its immense jaw muscles, like a Mohawk haircut of bone.

In 1974, anthropologist Don Johanson went to the remote Afar Desert of Ethiopia in search of early human fossils and hit the jackpot. He found the most complete, best preserved australopithecine skeleton known. Nicknamed "Lucy," the skeleton was 40% complete and over 3 million years old (figure 22.9). The skeleton and other similar fossils have been assigned the scientific name *Australopithecus afarensis* (from the Afar Desert). The shape of the pelvis indicated that Lucy was a female, and her leg bones proved she walked upright. Her teeth were distinctly hominid, but her head was shaped like an ape's, and her brain was no larger than that of a chimpanzee, about 400 cubic centimeters. (For comparison, your brain is about 1350 cubic centimeters.)

More than 300 specimens of *A. afarensis* have since been discovered, making it one of the best known australopithecines. Overall the skull is ape-like in appearance, dominated by a strongly projecting face. Brain size averages 430 cubic centimeters (cm^3), about the size of an orange.

In the last 10 years, three additional kinds of australopithecine have been reported: two massively boned ancestors of *A. boisei* called *A. aethiopicus* and *A. bahrelghazali*, and fragmentary evidence of a very old species called *A. anamensis*. These seven species provide ample evidence that australopithecines were a diverse group, and additional species will undoubtedly be described by future investigators. The evolution of hominids seems to have begun with an initial radiation of numerous species.

Early Australopithecines Were Bipedal

We now know australopithecines from hundreds of fossils. The structure of these fossils clearly indicate that australopithecines walked upright. These early hominids weighed about 18 kilograms and were about 1 meter tall. Their dentition was distinctly hominid, but their brains were not any larger than those of apes, generally 500 cm^3 or less. *Homo* brains, by comparison, are usually larger than 600 cm^3 (figure 22.10); modern *Homo sapiens* brains average 1350 cm^3.

Australopithecine fossils have been found only in Africa. Although all the fossils to date come from sites in South and East Africa (except for one specimen from Chad), it is probable that they lived over a much broader area of Africa. Only in South and East Africa, however, are sediments of the proper age exposed to fossil hunters.

The australopithecines were hominids that walked upright and lived in Africa over 3 million years ago.

FIGURE 22.10
Two hominids who lived side-by-side in Koobi Fora, Kenya, 1.7 million years ago. On the left is *Australopithecus boisei*, on the right *Homo ergaster*. Both fossils were recovered from sediments of the same age at the same site.

The Root of the Hominid Tree

The Oldest Known Hominid

In 1994, a remarkable, nearly complete fossil skeleton was unearthed in Ethiopia (the country where Lucy was found). The skeleton is still being painstakingly assembled, but it seems almost certainly to have been bipedal; the foramen magnum, for example, is situated far forward, as in other bipedal hominids. Some 4.4 million years old, it is the most ancient hominid yet discovered. It is significantly more ape-like than any australopithecine and so has been assigned to a new genus, *Ardipithecus ramidus* (figure 22.11). In the local Afar language, *ardi* means "ground," while *ramid* means "root"; *pithecus* is the Greek word for "ape."

The First Australopithecine

In 1995, hominid fossils of nearly the same age, 4.2 million years old, were found in the Rift Valley in Kenya. The fossils are fragmentary, but they include a partial tibia (leg bone) with clear bipedal characteristics, confirming that the fossils are hominid. They have been assigned to the genus *Australopithecus* rather than *Ardipithecus* because the fossils are much less apelike than *A. ramidus*. While clearly australopithecine, the fossils are intermediate in many ways between apes and *A. afarensis*. Numerous fragmentary specimens of the species, named *Australopithecus anamensis*, have since been found.

Most researchers agree that these slightly built *A. anamensis* individuals represent the true base of our family tree, the first members of the genus *Australopithecus*, and thus ancestor to *A. afarensis* and all other australopithecines.

Differing Views of the Hominid Family Tree

The traditional view of human evolution pictures the process as progressing from simple to complex, one form replacing another in an unbroken linear sequence. This clearly has little relevance to the australopithecines. The seven species are not related in any simple linear way. Their evolutionary relationships resemble a bush more than a single stem.

A Cladistic Approach. How ought we organize what we know of the relationships between the australopithecines? One approach employs a system known as cladistics (figure 22.12). Described in detail in chapter 28, cladistics attempts to array a collection of organisms in an objective way that accurately portrays ancestral relationships. In cladistics, evolutionary relationships are characterized by innovations: individuals are grouped into "clades" when they share a characteristic inherited from a recent common ancestor, an innovation unseen in more ancient ancestors.

FIGURE 22.11
Our earliest known ancestor. A tooth from *Ardipithecus ramidus*, discovered in 1994. The name *ramidus* is from the Latin word for "root," as this is thought to be the root of the hominoid family tree. The earliest known hominid, at 4.4 million years old, *A. ramidus* was about the size of a chimpanzee and apparently could walk upright.

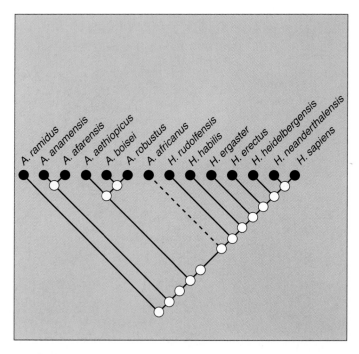

FIGURE 22.12
A hominid cladogram. In this cladogram, the three "robust" species (*A. aethiopicus, A. boisei, A. robustus*) are grouped on one branch of the cladogram, while the two most primitive species (*A. anamensis, A. afarensis*) group on another. *A. africanus* is the australopithecine most closely associated with genus *Homo*, although most anthropologists would now assign that honor to *A. afarensis*.
From Johanson and Edgar, *From Lucy to Language*, 1996. Copyright © 1996 Simon and Schuster. Reprinted by permission.

Lumpers and Splitters. Investigators take two different philosophical approaches to characterizing the diverse group of African hominid fossils. One group focuses on common elements in different fossils and tends to lump together fossils that share key characters. Differences between the fossils are attributed to diversity within the group. The hominid phylogenetic tree in figure 22.13a illustrates such a lumper's view.

Other investigators focus more pointedly on the differences between hominid fossils. They are more inclined to assign fossils that exhibit differences to different species. The hominid phylogenetic tree in figure 22.13b presents a splitter's view. Where the "lumpers" tree presents three species of *Homo*, for example, the "splitters" tree presents no fewer than seven! The splitters approach puts more of the data to work in attempting to ferret out evolutionary relationships and more accurately represents the view of most human paleontologists.

The root of the hominid evolutionary tree is only imperfectly known. The earliest australopithecine yet described is *A. anamensis*, over 4 million years old.

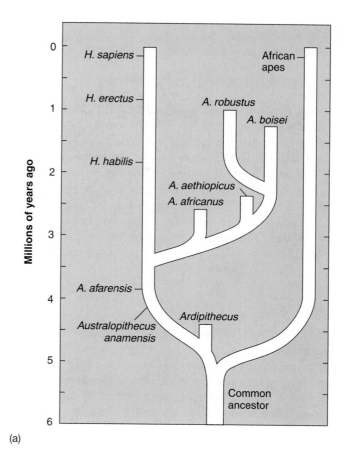

(a)

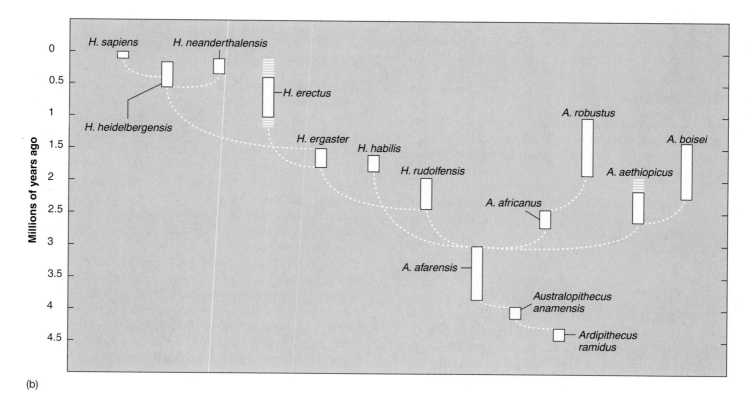

(b)

FIGURE 22.13
Two views of the hominid phylogenetic tree. (a) A lumper's approach. Six species of *Australopithecus* and only three species of *Homo* are included. (b) A splitter's approach. In this tree, the most widely accepted, the vertical bars show the known dates of the first and last appearances of proposed species. Six species of *Australopithecus* and seven of *Homo* are included.

African Origin: Early *Homo*

The first humans evolved from australopithecine ancestors about 2 million years ago. The exact ancestor has not been clearly defined, but is commonly thought to be *A. afarensis.* Only within the last 30 years have a significant number of fossils of early *Homo* been uncovered. An explosion of interest has fueled intensive field exploration in the last few years, and new finds are announced yearly, so every year our picture of the base of the human evolutionary tree grows clearer. The account given here will undoubtedly be outdated by future discoveries, but it provides a good example of science at work.

Homo habilis

In the early 1960s, stone tools were found scattered among hominid bones close to the site where *A. boisei* had been unearthed. Although the fossils were badly crushed, painstaking reconstruction of the many pieces suggested a skull with a brain volume of about 680 cubic centimeters, larger than the australopithecine range of 400 to 550 cubic centimeters. Because of its association with tools, this early human was called *Homo habilis*, meaning "handy man." Partial skeletons discovered in 1986 indicate that *H. habilis* was small in stature, with arms longer than legs and a skeleton much like *Australopithecus.* Because of its general similarity to australopithecines, many researchers at first questioned whether this fossil was human.

Homo rudolfensis

In 1972, Richard Leakey, working east of Lake Rudolf in northern Kenya, discovered a virtually complete skull about the same age as *H. habilis.* The skull, 1.9 million years old, had a brain volume of 750 cubic centimeters and many of the characteristics of human skulls—it was clearly human and not australopithecine. Some anthropologists assign this skull to *H. habilis*, arguing it is a large male. Other anthropologists assign it to a separate species, *H. rudolfensis*, because of its substantial brain expansion (figure 22.14).

Homo ergaster

Of the early *Homo* fossils being discovered, some do not easily fit into either of these species (figure 22.15). They tend to have even larger brains than *H. rudolfensis*, with skeletons less like an australopithecine and more like a modern human in both size and proportion. Interestingly, they also have small cheek teeth, as modern humans do. Some anthropologists have placed these specimens in a third species of early *Homo*, *H. ergaster* (*ergaster* is from the Greek for "workman").

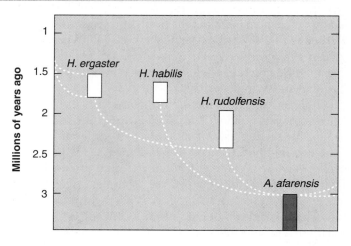

FIGURE 22.14
A closeup of the hominid evolutionary tree showing early *Homo*. Some scientists recognize three species of early *Homo*: *H. habilis* (fossils that are clearly human and not australopithecine), *H. rudolfensis* (fossils with a largely expanded brain compared to *H. habilis*), and *H. ergaster* (fossils that have more modern dentition and even larger brains). Other scientists call all early *Homo* species *Homo habilis*.

How Diverse Was Early *Homo?*

Because so few fossils have been found of early *Homo*, there is lively debate about whether they should all be lumped into *H. habilis* or split into the three species *H. rudolfensis*, *H. habilis*, and *H. ergaster*. If the three species designations are accepted, as increasing numbers of researchers are doing, then it would appear that *Homo* underwent an adaptive radiation (as described in chapter 21) with *H. rudolfensis* the most ancient species, followed by *H. habilis* and then *H. ergaster*. Because of its modern skeleton, *H. ergaster* is thought the most likely ancestor to later species of *Homo*.

The great diversity among early *Homo* fossils serves to stress the complexity of the early evolution of the genus. Increasingly, anthropologists are taking the view that human evolution has not been a simple linear process, as has been traditionally believed. Rather, as with other mammals, the early portion of the human evolutionary tree appears to be a "bush" exhibiting diverse branches.

Early species of *Homo*, the oldest members of our genus, had a distinctly larger brain than australopithecines and most likely used tools. There may have been several different species.

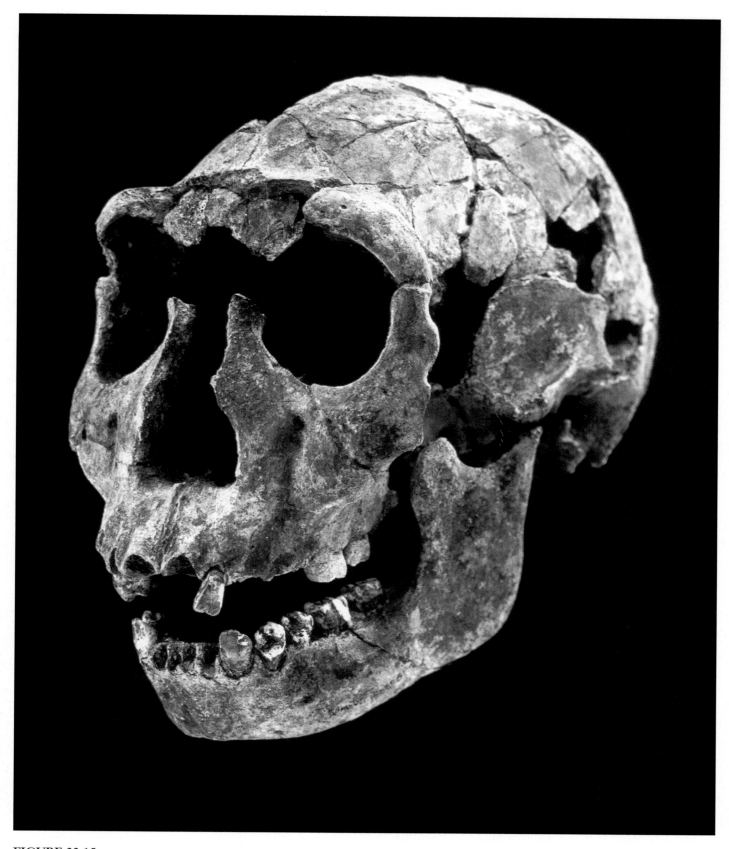

FIGURE 22.15
Early *Homo*. This skull (actual size) of a boy who apparently died in early adolescence is 1.6 million years old and has been assigned to the species *Homo ergaster*. Much larger than earlier hominids, he was about 1.5 meters in height and weighed approximately 47 kilograms.

Out of Africa: *Homo erectus*

Details about early *Homo* are sketchy, since only a few specimens are available for study. In fact, some scientists still dispute whether early *Homo* fossils qualify as truly human—they had not moved far from their australopithecine roots. There are no such doubts about **Homo erectus,** the species that appears in the fossil record after early *Homo* (figure 22.16). The many specimens that have been found show that *H. erectus* was a true human. From the neck down, *H. erectus* is very similar to modern humans, and the *H. erectus* brain is far larger than that of any early *Homo*. With the appearance of *H. erectus*, hominid evolution enters a new phase. For the first time, hominids will leave Africa.

Java Man

After the publication of Darwin's book *On the Origin of Species* in 1859, there was much public discussion about "the missing link," the fossil ancestor common to both humans and apes. Darwin was convinced our origins lay in Africa. Dutch doctor and anatomist Eugene Dubois was swayed by Alfred Russell Wallace's conviction that our origins lay instead in southeast Asia. Dubois was particularly intrigued by the orangutans, the "old men of the forest" from Java and Borneo. Many of the orangutan's anatomical features seemed to fit the picture of a "missing link." So Dubois closed his practice and sought out fossil evidence of the missing link in the orangutan's home country of Java.

Dubois enrolled as an army surgeon in the Royal Dutch East Indies Army and set up practice, first in Sumatra, then in a village on the Solo river in eastern Java. In 1891, while digging into a hill that villagers claimed held "dragon bones," his team unearthed a skullcap and, upstream, a thigh bone. Dubois was very excited by his find for three reasons:

1. The structure of the thigh bone indicated that the individual had long, straight legs and was an excellent walker.
2. The size of the skullcap suggested a *very* large brain, about 1000 cubic centimeters—much larger than any ape's.
3. Most surprising, the bones seemed very old—perhaps 500,000 years old, judged by other fossils Dubois unearthed along with them.

Dubois's fossil hominid was far older than any fossil hominid discovered up to that time, and at first few scientists were willing to accept it as an ancient species of human. Finally, after years of arguing to an unconvinced audience that the bones were indeed human, Dubois in disgust buried the "Java man" skullcap and thigh bone under the floorboards of his dining room. For 30 years he refused to let anyone see them. In the years since Dubois' died, about

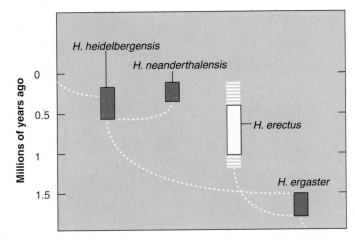

FIGURE 22.16
A closeup of the hominid evolutionary tree showing *Homo erectus.* Some scientists recognize *Homo erectus* as a separate branch that arose from early *Homo*. *H. ergaster* was likely the direct ancestor to modern humans.

40 individuals similar in character and age to his fossil have been found in Java, including in 1969 the nearly complete skull of an adult male.

Peking Man

A generation after his death, scientists were forced to admit that Dubois had been right all along. In the 1920s, a skull was discovered in a cave on "Dragon Bone Hill" some 40 kilometers south of Peking (now Beijing), China, that closely resembled Java man. Continued excavation at the site initially revealed 14 skulls, many excellently preserved, together with lower jaws and other bones. Crude tools were also found, and most important of all, the ashes of campfires. In anticipation of the Japanese invasion of China at the onset of World War II, researchers made careful casts of "Peking man" and the other fossils and distributed them for study to laboratories around the world. The originals were loaded onto a truck and evacuated from Peking in December 1941 at the beginning of the Japanese invasion, only to disappear into the confusion of history. No one knows what happened to the truck or its priceless cargo.

In the years since, continued investigation of Dragon Bone Hill has revealed a few additional specimens. Fortunately, the farsighted preservation of the high-quality casts, and the molds from which they were made, has enabled scientists to gain a clear view of what we now know to be Chinese *H. erectus*. In all, 40 individuals have been described, the largest collection of *H. erectus* fossils from any one place and a third of the known fossils of this species. The five complete skullcaps have a mean cranial capacity of 1043 cubic centimeters, much larger than any earlier *Homo*.

Homo erectus

Java man and Peking man are now recognized as belonging to the same species—*Homo erectus*. At a height of 5 feet, *H. erectus* was much taller than *H. habilis*. *Homo erectus* had a large brain, about 1000 cubic centimeters (figure 22.17), and walked erect. Its skull had prominent brow ridges and, like modern man, a rounded jaw. The shape of the skull interior suggests that *H. erectus* was able to talk.

Traditionally, *H. erectus* had been considered the direct ancestor of modern humans, but today investigators reserve that honor for *H. ergaster*, believing it to be the common ancestor of both *H. erectus* and modern humans. Because there are so few fossils, however, the matter is not clearcut.

Homo Migrates out of Africa

Homo erectus has been shown to have come out of Africa. In 1976, a complete *H. erectus* skull discovered in East Africa was dated at 1.5 million years old, a million years older than the Java and Peking finds. The appearance of *H. erectus* in Africa 1.5 million years ago marked the beginning of the great human expansion. Far more successful than *H. habilis*, *H. erectus* quickly spread and became abundant in Africa and within a million years had migrated into Asia and Europe.

The Success of *Homo erectus*

A social species, *H. erectus* lived in tribes of 20 to 50 people, often dwelling in caves, although evidence also indicates that they built crude wooden shelters. They successfully hunted large animals, butchered them using flint and bone tools, and cooked them over fires. The site in China where Peking man was found contains the remains of horses, bears, elephants, deer, and rhinoceroses.

Homo erectus survived longer than any other species of human, for nearly a million years. These very adaptable humans only disappeared in Africa about 500,000 years ago, as modern humans were emerging. Interestingly, *H. erectus* survived much longer in Asia, until about 250,000 years ago.

Of the species of humans that have evolved, *Homo erectus* has been the most successful, lasting for nearly a million years and spreading out of Africa across Europe and Asia.

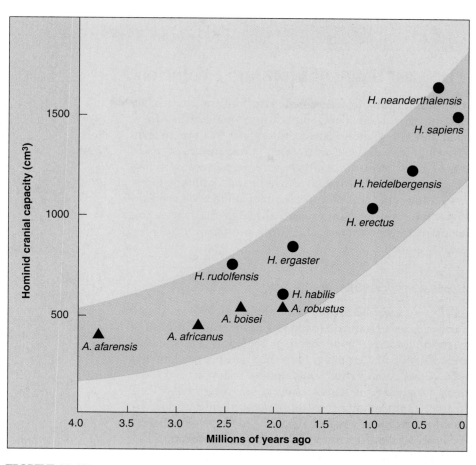

FIGURE 22.17

Brain size increased as hominids evolved. *Homo erectus* had a larger brain than early *Homo*, which were in turn larger than those of the australopithecines they shared East African grasslands with. Maximum brain size (and apparently body size) was attained by *H. neanderthalensis*. Both brain and body size appear to have declined some 10% in recent millennia.

The Last Stage of Hominid Evolution

The evolutionary journey to modern humans entered its final phase when modern humans first appeared in Africa about 600,000 years ago. Investigators who focus on human diversity consider there to have been three species of modern humans: *Homo heidelbergensis*, *H. neanderthalensis* and *H. sapiens* (figure 22.18). Other investigators lump the three species into one, *H. sapiens* ("wise man"). Here we will follow the three-species approach, although the matter remains quite controversial and either position can be reasonably argued.

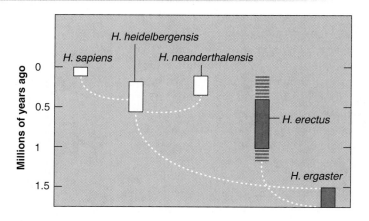

FIGURE 22.18

A closeup of the hominid evolutionary tree showing modern *Homo*. Some scientists recognize three species of modern *Homo*: *H. neanderthalensis*, *H. heidelbergensis*, and *H. sapiens*. Other scientists recognize only one species, *H. sapiens*.

Homo neanderthalensis

As *H. erectus* was becoming rarer, about 130,000 years ago, a new species of human arrived in Europe from Africa. Because the first fossil to be found, in 1856, was from the Neander Valley of Germany, these early European humans were called Neanderthals (*thal* means "valley" in old German). Originating in Africa, *Homo neanderthalensis* appeared to have branched off from the ancestral line leading to modern humans as long as 500,000 years ago—mitochondrial DNA recovered from fossils is that different from living human mDNA. Compared with modern humans, European Neanderthals were short, stocky, and powerfully built. Their skulls were massive, with protruding faces and heavy, bony ridges over the brows. Their brains were larger than those of modern humans.

The Neanderthals made diverse tools, including scrapers, spearheads, and hand axes. They lived in huts or caves. Rare at first outside of Africa, they became progressively more abundant in Europe and Asia, and by 70,000 years ago had become common. Neanderthals took care of their injured and sick and commonly buried their dead, often placing food, weapons, and even flowers with the bodies. Such attention to the dead strongly suggests that they believed in a life after death. This is the first evidence of the symbolic thinking characteristic of modern humans.

Homo heidelbergensis

Both the great age of Neanderthals and their high degree of specialization argue that they are not the direct ancestor of *H. sapiens*, and suggest instead that *H. sapiens* evolved from a somewhat older species, *H. heidelbergensis*.

The fossil evidence for *H. heidelbergensis* is fragmentary. The oldest fossil is from Ethiopia and is around 600,000 years old, making it a contemporary of *H. erectus* in Africa, but features of its anatomy are significantly advanced. Like Neanderthals, *H. heidelbergensis* has a bony keel running along the midline of the skull, a thick ridge over the eye sockets, and a large brain. But, its forehead and nasal bones are very like those of *H. sapiens*.

Homo sapiens

The oldest fossil known of *Homo sapiens*, our own species, is from Ethiopia and is about 130,000 years old. Other fossils from Israel that appear to be between 100,000 and 120,000 years old confirm that our species arose less than 200,000 years ago. *H. sapiens* is the only surviving species of *Homo*, and indeed is the only surviving hominid.

Among the best evidence for the recent origin of *H. sapiens* are 20 well-preserved skeletons with skulls found in a cave near Nazareth in Israel. Modern dating techniques date these humans to between 90,000 and 100,000 years old. The skulls are modern in appearance, with high, short braincases, vertical foreheads with only slight brow ridges, and a cranial capacity of roughly 1550 cubic centimeters, well within the range of modern humans.

Outside of Africa and the Middle East, there are no clearly dated *H. sapiens* fossils older than roughly 40,000 years of age. The implication is that *H. sapiens* evolved in Africa, then migrated to Europe and Asia. Acceptance of the relatively recent African origin of *H. sapiens* (the "out-of-Africa" model) implies that modern-day Asian *H. sapiens* evolved from African ancestors; an opposing view (the "multiregional" model) argues that the human races independently evolved from *H. erectus* in different parts of the world. This matter is still widely debated among students of human evolution.

Modern humans evolved within the last 600,000 years. Our own species, *H. sapiens*, appears to have evolved in Africa less than 200,000 years ago.

Out of Africa—Again?

The origin of human races is a much-debated point among scientists studying human evolution. Investigators favoring the multiregional model argue that each of the races of *H. sapiens* evolved *in situ* from *H. erectus* and that each adapted to a different environment. Others favoring the out-of-Africa model believe that the same species would be unlikely to evolve more than once and argue that human races appeared after *H. sapiens* evolved from *H. erectus* in Africa.

The Mitochondrial DNA Controversy

In 1987, scientists at Berkeley, California, studying mitochondrial DNA from humans all over the world made the startling claim that all human races originated from one *H. sapiens* ancestor in Africa.

They chose mitochondrial DNA to study hominid evolution because the DNA within mitochondria is transmitted only by females. Females' eggs carry many mitochondria that become part of a new baby, while sperm contribute no mitochondria to the new baby (sperm carry their mitochondria wrapped around their tails and do not inject them into the egg during fertilization). For that reason, particular versions of a mitochondrial gene can be traced back through a family tree, from mother to grandmother to great-grandmother.

However, after much initial excitement about the "mitochondrial Eve," the result is no longer widely accepted. The computer program the Berkeley scientists used to reach their conclusion, called PAUP (for phylogenetic analysis using parsimony), gives different results depending upon how the data are entered! Thus, the exact human tree cannot be reliably traced using this approach.

However, the greatest number of different mitochondrial DNA sequences occur among modern Africans. Since DNA accumulates mutations over time, the oldest populations should show the greatest genetic diversity. This result is consistent with the hypothesis that humans have been living in Africa longer than on any other continent, and from there spread to all parts of the world, retracing the path taken by *H. erectus* half a million years before (figure 22.19).

A Clearer Picture from Nuclear DNA

A clearer analysis is possible using chromosomal DNA, some highly variable segments of which are far more variable than mitochondrial DNA, so providing more "markers" to compare. In 1996 a large team of geneticists from six countries reported the results of a comprehensive nuclear DNA study. The team analyzed, in human populations all over the world, two variable parts of the CD4 gene (the cell surface receptor on white blood cells that recognizes the AIDS-causing HIV virus) located on human chromosome 12. Sixteen hundred individuals were examined in 42 populations. A total of 24 different versions of the two segments were found. Fully 21 of them were present in human populations in Africa, while 3 were found in Europeans and only 2 in Asians and in Americans. This argues forcefully that chromosome 12 has existed in African humans far longer than among non-African humans. Together with fossils of early *H. sapiens* from Africa and Israel, these results strongly support an out-of-Africa model of human origins.

> While still hotly debated, it appears likely that *H. sapiens* evolved once in Africa, then migrated out of Africa to Europe and Asia.

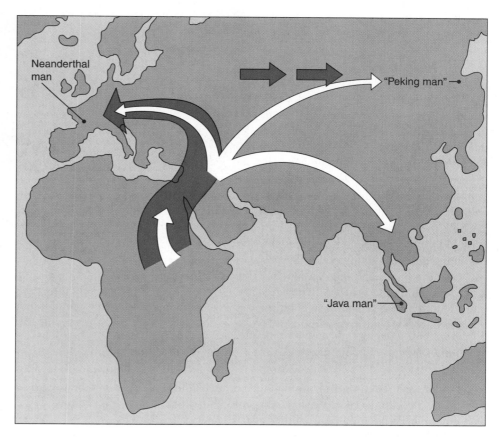

FIGURE 22.19

Out of Africa—twice? A still-controversial theory suggests that *Homo* spread from Africa to Europe and Asia twice. First, *Homo erectus* (white arrow) spread as far as Java and China. Later, *H. erectus* was replaced by *Homo sapiens* (red arrow) in a second wave of migration.

The Emergence of Modern Humans

About 34,000 years ago, the European Neanderthals were abruptly replaced by people of essentially modern character, the **Cro-Magnons** (named after the valley in France where they were discovered). Scientists can only speculate on the reasons for this sudden replacement, but it was complete all over Europe in a short period. The available evidence indicates that the Cro-Magnons came from Africa: *H. sapiens* fossils of essentially modern aspect but as old as 100,000 years have been found there. Cro-Magnons appear to have replaced the Neanderthals completely in the Middle East by 40,000 years ago and then spread across Europe, coexisting and possibly even interbreeding with Neanderthals for several thousand years.

Cro-Magnons used sophisticated stone tools and also made tools out of bones and horns. They had a complex social organization and are thought to have had fully modern language capabilities. They lived by hunting. The world was cooler than it is now—it was the time of the last great Ice Age—and Europe was covered with grasslands inhabited by large herds of grazing animals. These animals are pictured in elaborate and often beautiful cave paintings throughout Europe (figure 22.20).

Humans of modern appearance eventually spread across Siberia to North America, which they reached at least 13,000 years ago, after the ice had begun to retreat and while a land bridge still connected Siberia and Alaska. By about 12,500 years ago, humans had reached all the way down to southern Chile. At that time, fewer than 5 million people populated the entire world (compared with nearly 6 billion now).

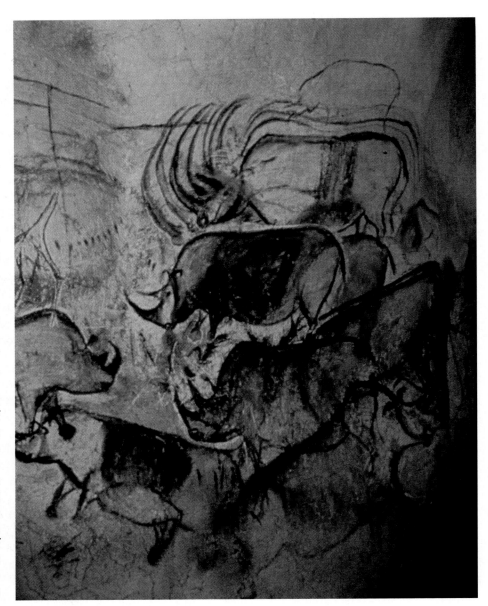

FIGURE 22.20
Cro-Magnon art. Rhinoceroses are among the animals depicted in this remarkable cave painting found in 1995 near Vallon-Pont d'Arc, France.

Homo sapiens Are Unique

We humans are animals and the product of an evolution marked by a continual increase in brain size. While not the only animal capable of conceptual thought, we have refined and extended this ability until it has become the hallmark of our species. We use symbolic language, and with words can shape concepts out of experience. This power has allowed us to transmit the accumulation of experience from one generation to the next. Thus, we have what few, if any, other animals have ever experienced: cultural evolution. Through culture, humans more than any other animals have found ways to change and mold our environment to our needs, rather than evolving in response to environmental demands. We control our biological future in a way never before possible—an exciting potential and a frightening responsibility.

Our species migrated out of Africa less than 50,000 years ago, replacing *Homo erectus* first in Europe, then in Asia.

22.1 The evolutionary path to humans starts with the advent of primates.

- Primates first appeared about 65 million years ago, evolving from arboreal insect eaters.

- Prehensile (grasping) fingers and toes and binocular vision were distinct adaptations that allowed early primates to be successful in their particular environments.

- Mainly diurnal (day-active) anthropoids and mainly nocturnal (night-active) prosimians diverged about 40 million years ago. Anthropoids include monkeys, apes, and humans, and all exhibit complex social interactions and enlarged brains.

- New World monkeys are anthropoids that migrated to South America about 30 million years ago. Today they are all arboreal, and many have prehensile tails.

- Old World monkeys evolved in Africa about 25 million years ago. Today they are both arboreal and ground-dwelling, and none have prehensile tails.

- The hominoids evolved from anthropoid ancestors about 25 million years ago. Hominoids consist of the apes (gibbons, orangutans, gorillas, and chimpanzees) and the hominids (human beings and their direct ancestors).

- Hominids are distinguished from apes by their ability to walk upright.

22.2 The first hominids to evolve were australopithecines.

- Early hominids belonging to the genus *Australopithecus* were ancestral to humans. They exhibited bipedalism (walking upright on two feet) and lived in Africa over 4 million years ago.

- *Ardipithecus ramidus* is the oldest known hominid. It seems to have walked upright but is more ape-like than *Australopithecus*.

- *Australopithecus anamensis* is over 4.2 million years old and is the oldest known australopithecine.

22.3 The genus *Homo* evolved in Africa.

- Hominids with an enlarged brain and the ability to use tools belong to the genus *Homo*.

- Species of early *Homo* appeared in Africa about 2 million years ago and became extinct about 1.5 million years ago.

- *Homo erectus* appeared in Africa at least 1.5 million years ago and had a much larger brain than early species of *Homo*.

- Within a million years, *Homo erectus* migrated from Africa to Europe and Asia, as evidenced by the Java and Peking fossils.

- *Homo erectus* walked erect, had a large brain, (1000 cubic centimeters), and presumably was able to talk.

22.4 Modern humans evolved quite recently.

- The modern species of *Homo* appeared about 600,000 years ago in Africa and about 350,000 years ago in Eurasia.

- Some scientists recognize *H. heidelbergensis* as the oldest modern human, believing that it gave rise to *H. neanderthalensis* and *H. sapiens*.

- The Neanderthals appeared in Europe about 130,000 years ago. They made diverse tools and showed evidence of symbolic thinking.

- Studies of mitochondrial DNA suggest (but do not yet prove) that all of today's human races originated from *H. sapiens* in Africa.

- The Cro-Magnons (essentially modern humans) began to replace the Neanderthals about 40,000 years ago and depicted herds of animals in their cave paintings.

Discussing Key Terms

1. **Primates** These mammals have grasping fingers and toes and binocular vision. The order contains prosimians (tarsiers and lemurs), monkeys, apes, and humans.

2. **Hominids** Humans and their ape-like ancestors are members of two genera: *Homo* and *Australopithecus*. The earliest hominids arose less than 5 million years ago.

3. **Australopithecines** The immediate ancestors of humans, australopithecines first appeared in the fossil record some 4.4 million years ago. They walked upright and had teeth with many human characteristics.

4. **The genus *Homo*** Of the human species that have evolved, only *Homo sapiens* still survives today. All *Homo* species are bipedal and possess large brains relative to body size.

5. **African origins** *Homo habilis* and *H. erectus* originated in Africa. Whether *H. sapiens* also originated in Africa is currently a matter of controversy.

Review Questions

1. Which characteristics were selected for in the earliest primates to allow them to become successful in their environment?

2. How do monkeys differ from prosimians?

3. How are apes distinguished from monkeys?

4. What is the best explanation for why the nucleotide sequence in the DNA of humans and chimpanzees is more than 98% similar?

6. When did the first hominids appear? What were they called? What distinguished them from the apes?

7. Who was Raymond Dart, and what information did he provide about human evolution?

8. Why is there some doubt in the scientific community that *Homo habilis* was a true human?

9. How did *Homo erectus* differ from *Homo habilis?*

10. The greatest number of different mitochondrial DNA sequences in humans occurs in Africa. What does this tell us about human evolution?

11. Where did Neanderthals originate, and where did they migrate? What evidence is there that Neanderthals may have believed in life after death?

12. How did Cro-Magnons differ from Neanderthals? Is there any evidence that they coexisted with Neanderthals? If so, where and when?

Thought Questions

1. Form a hypothesis for why all living lemurs on earth are on the island of Madagascar.

2. Modern humans, *Homo sapiens*, evolved from *Homo* ancestors less than 1 million years ago. Do you think evolution of the genus *Homo* is over, or might another species of humans evolve within the next million years? Do you think this would involve the extinction of *H. sapiens?*

3. Studies of primate DNA have revealed that humans differ from gorillas in only 2.3% of DNA nucleotide sequences, and from chimpanzees in only 1.6%. This degree of genetic similarity is the same usually seen among "sibling" species (that is, species that have only recently evolved from a common ancestor). Yet humans have been assigned not only to a different genus but to a different family! Do you think this is legitimate, or are humans just a rather unusual kind of African ape?

Internet Links

The Primate Gallery
http://www.selu.com/~bio/PrimateGallery/index.html
A beautiful site providing a wealth of images and information on humanity's nearest relatives.

Creationists & Human Evolution
http://www.talkorigins.org/origins/faqs-qa.html
A thoughtful and carefully prepared FAQ intended to answer creationist claims that there is no evidence for human evolution.

The Dawn of Humans: The First Steps
http://www.nationalgeographic.com/media/ngm/9702/
    hilights_splash003.html#d
Highlights of one of a brilliant series of articles on human evolution in the National Geographic appearing over the last several years.

For Further Reading

Gore, R.: "Neanderthals," *National Geographic*, January 1996, pages 2–35. Early humans in Europe were quite different from people today.

Johanson, D. and B. Edgar.: *From Lucy to Language*, Simon & Schuster, New York, 1996. An excellent overview of the current ferment within human anthropology, with superb photographs of the key fossils.

Johanson, D.: "Face-to-Face with Lucy's Family," *National Geographic*, March 1996, pages 96–115. The discovery of *Australopithecus afarensis*, the immediate ancestor of humans.

Kahn, P. and A. Gibbons: "DNA from an Extinct Human," *Science*, vol. 277, July 1997, pages 176–78. A careful study of DNA from Neanderthals reveals them to be far older than previously thought.

Leakey, M.: "The Dawn of Humans—The Farthest Horizon," *National Geographic*, September 1995, pages 38–51. The base of the hominid tree is revealed by fossils of ape-like hominids that walked upright 4 million years ago.

Leakey, M. and A. Walker: "Early Hominid Fossils from Africa," *Scientific American*, June 1997, pages 74–79. A new species of bipedal Australopithecus, *A. anamensis*, is the earliest member of the genus yet discovered.

Tattersall, I.: *The Fossil Trail: How We Know What We Think We Know About Human Evolution*, Oxford University Press, New York, 1995. A lovely account of how what we expect to find influences our interpretation of what we do find in the fossil record.

Tattersall, I.: "Out of Africa Again . . . and Again," *Scientific American*, April 1997, pages 60–67. Recent results support a single and comparatively recent origin for *H. sapiens*, very likely in Africa.

Thorne, A. and M. Wolpoff: "The Multiregional Evolution of Humans," *Scientific American*, April 1992, pages 76–83. The argument against the African origin of modern *H. sapiens*.

23

Population Ecology

Concept Outline

23.1 Populations are individuals of the same species that live together.

Population Ecology. All populations have characteristic features such as size, density, degree of dispersion, and age structure that determine if they grow and how fast.

23.2 Population dynamics depend critically upon age distribution.

Demography. The growth rate of a population is a sensitive function of its age structure; populations with many young individuals grow rapidly as these individuals enter reproductive age.

23.3 Life histories often reflect trade-offs between reproduction and survival.

The Cost of Reproduction. Evolutionary success is a trade-off between investment in current reproduction and in growth that promotes future reproduction.

Life History Adaptations. Life history characteristics such as clutch size, number of reproductive events per lifetime, and age at first reproduction represent trade-offs between reproductive cost and investment in survival.

23.4 Population growth is limited by the environment.

Biotic Potential. Populations grow if the birth rate exceeds the death rate, until they reach the carrying capacity of their environment, which ultimately determines how large the population will be.

Applying Growth Models to Real Populations. Some populations exhibit rapid opportunistic growth, while others, specialized to survive in highly competitive situations, grow more slowly.

The Influence of Population Density. Some of the factors that regulate a population's growth depend upon the size of the population; others do not.

Human Populations. The human population has been growing rapidly for 300 years, since technological innovations dramatically reduced the death rate.

FIGURE 23.1
Life takes place in populations. This population of gannets is subject to the rigorous effects of reproductive strategy, competition, predation, and other limiting factors.

Ecology, the study of how organisms relate to one another and to their environments, is a complex and fascinating area of biology that has important implications for each of us. In our exploration of ecological principles, we will first consider the properties of populations, emphasizing population dynamics (figure 23.1). In chapter 24, we will discuss communities and the interactions that occur in them. Chapter 25 moves on to the dynamics of ecosystems, and chapter 26 considers the properties of major aquatic and terrestrial biological communities. Chapter 27 deals with the future of the biosphere, a future that will depend to an ever-increasing extent on how we understand and use the ecological principles presented in this section.

Population Ecology

Organisms live as members of **populations,** groups of individuals of a species that live together. The huge cities that now pepper the earth (figure 23.2) are local clusters of people, portions of the 6 billion humans that make up the world's human population. In this chapter, we will consider the properties of populations, focusing on elements that influence whether a population will grow, and if so, how fast. The explosive growth of the world's human population in the last few centuries provides a focus for our inquiry.

The Science of Ecology

Ecology is the study of how organisms interact with their environment and one another. The study of ecology helps us understand why particular kinds of organisms live in one place and not another. Ecologists try to determine the physical and biological variables that govern distributions and numbers of organisms. They also try to determine principles that predict the future consequences of interactions between organisms and their environment that might have an impact on the world's uncertain future.

Levels of Ecological Organization

Ecologists consider groups of organisms at four progressively more encompassing levels of organization.

Populations.   Individuals of the same species that live together are members of a population. They potentially interbreed with one another, share the same habitat, and use the same pool of resources the habitat provides.

Communities.   Populations of different species that live together in the same place are called communities. Different species typically utilize different resources within the habitat they share.

Ecosystems.   A community and the nonliving factors with which it interacts is called an ecosystem. An ecosystem regulates the flow of energy, ultimately derived from the sun, and the cycling of the essential elements on which the lives of its constituent organisms depend.

Biomes.   Biomes are major terrestrial assemblages of plants, animals, and microorganisms that occur over wide geographical areas and that have distinct characteristics. Examples include deserts, tropical forests, and grasslands; similar groupings occur in marine and freshwater habitats.

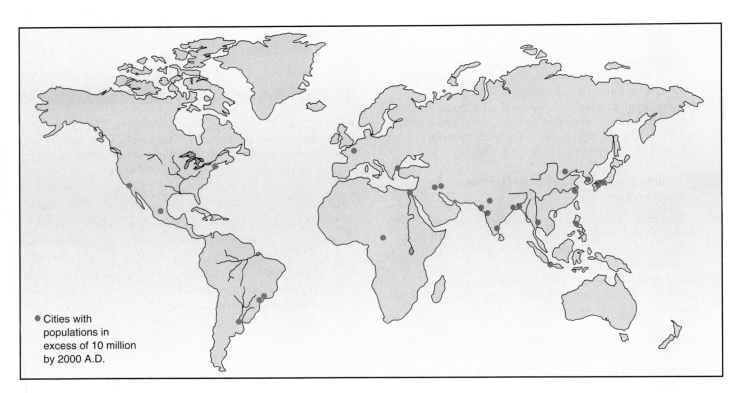

● Cities with populations in excess of 10 million by 2000 A.D.

FIGURE 23.2
Megacities. An increasing portion of the world's population is crowding into cities. In 1920, only 360 million people lived in cities; by the end of the century, nearly 3 billion people will be city dwellers, almost half the world's population. In 1950, only seven megacities had populations larger than 5 million; by 2000, there will be 57 megacities, 42 of them in the Third World. Twenty-six cities are likely to have a population greater than 10 million by 2000 A.D.

Population Structure

A population consists of the individuals of a given species that occur together at one place and time. This flexible definition allows us to speak in similar terms of the world's human population, the population of protozoa in the gut of an individual termite, or the population of deer that inhabit a forest.

Population Size. One of the important features of any population is its size. Population size has a direct bearing on the ability of a given population to survive. Very small populations are the most likely to become extinct; random events and natural disturbances are more likely to endanger a population if it contains only a few individuals. Inbreeding can also be a negative factor in the survival of a small population. Along with lowering vigor by its direct genetic effects, inbreeding also reduces the level of genetic variability, detracting from the population's ability to adjust to changing conditions. If an entire species consists of only one or a few small populations, that species is likely to become extinct, especially if it lives in areas undergoing radical changes.

Population Density. Population density is also extremely important to survival. If the individuals that make up a population are widely spaced, they may rarely encounter one another. This can limit their reproductive capabilities, and therefore their survival, even if the absolute number of individuals over a wide area is relatively high.

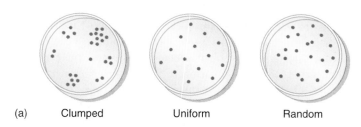

(a)   Clumped   Uniform   Random

(b)

Population Dispersion. Another key characteristic of population structure is dispersion, the way in which individuals of a population are arranged. They may be randomly spaced, uniformly spaced, or clumped (figure 23.3).

Randomly spaced. Individuals are randomly spaced within populations when they do not interact strongly with one another or with nonuniform aspects of their microenvironment. Random distributions are not common in nature.

Uniformly spaced. Individuals are uniformly spaced within a population when they interact directly with one another. Regular spacing of animal populations often results from social interaction, while in plants it may reflect competition for sunlight or water.

Clumped. Individuals clump into groups or clusters in response to uneven distribution of resources in their immediate environments. Clumped distributions are common in nature because individual animals, plants, and microorganisms tend to prefer microhabitats defined by soil type, moisture, or certain kinds of host trees.

Metapopulations

Clumped populations with clumps or patches that undergo periodic extinction and recolonization are called **metapopulations**. Typically, a metapopulation occupies a fragmented habitat, a patchwork of suitable sites scattered within large stretches of unsuitable habitat. In northern California, for example, the larval foodplant of checkerspot butterflies (*Euphydryas editha*) is a plant, *Plantago erecta*, that grows only on rock outcroppings scattered about within grassy meadows; only on these outcroppings can local subpopulations persist. A metapopulation will persist as long as its local recolonization rate equals or exceeds its local extinction rate—that is, if enough migration takes place between subpopulations to recolonize areas of extinction. Local subpopulations of *E. editha* became completely extinct between 1975 and 1977, but were repopulated in 1988 by individuals from other sites. The study of metapopulations has become very important in conservation biology as natural habitats become increasingly fragmented.

> A population is a group of individuals of the same species living together at the same place and time. Although the individuals of a population may be evenly spaced or randomly dispersed, clumped distribution patterns are the most frequent in nature.

FIGURE 23.3
Population dispersion. (a) Different arrangements of bacterial colonies, and (b) starlings distributed uniformly along telephone wires.

Demography

Demography is the statistical study of populations. The term comes from two Greek words: *demos*, "the people" (the same root we see in the word *democracy*), and *graphos*, "measurement." Demography therefore means measurement of people, or, by extension, of the characteristics of populations. Demography is the science that helps predict how population sizes will change in the future. Populations grow if births outnumber deaths and shrink if deaths outnumber births. Because birth and death rates depend significantly on age and sex, the future size of a population depends on its present age structure and sex ratio.

Age Structure

Many annual plants and insects time their reproduction to particular seasons of the year and then die. All members of these populations are the same age. Perennial plants and longer-lived animals contain individuals of more than one generation, so that in any given year individuals of different ages are reproducing within the population. A group of individuals of the same age is referred to as a **cohort.**

Within a population, every cohort has a characteristic birth rate, or **fecundity,** defined as the number of offspring produced in a standard time (for example, per year), and a characteristic death rate, or **mortality,** the number of individuals that die in that period. The rate of a population's growth depends directly upon the difference between these two rates.

The relative number of individuals in each cohort defines a population's **age structure.** Because individuals of different ages have different fecundity and death rates, age structure has a critical impact on a population's growth rate. Populations with a large proportion of young individuals, for example, tend to grow rapidly because an increasing proportion of their individuals are reproductive.

Sex Ratio

The proportion of males and females in a population is its **sex ratio.** The number of births is usually directly related to the number of females, but may not be as closely related to the number of males in species where a single male can mate with several females. In deer, elk, lions, and many other animals, a reproductive male guards a "harem" of females with which he mates, while preventing other males from mating with them. In such species, a reduction in the number of males simply changes the identities of the reproductive males without reducing the number of births. Among monogamous species like many birds, by contrast, in which pairs form long-lasting reproductive relationships, a reduction in the number of males can directly reduce the number of births.

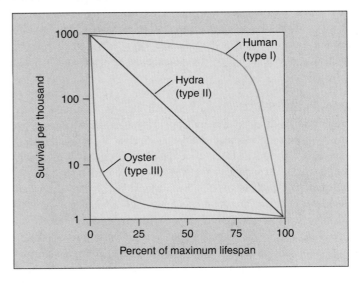

FIGURE 23.4
Survivorship curves. By convention, survival (the vertical axis) is plotted on a log scale. Humans have a type I life cycle, the hydra (an animal related to jellyfish) type II, and oysters type III.

Mortality and Survivorship Curves

A population's intrinsic rate of increase depends on the ages of the organisms in it and the reproductive performance of the individuals in the various age groups. When a population lives in a constant environment for a few generations, its **age distribution**—the proportion of individuals in different age categories—tends to stabilize. This distribution differs greatly from species to species and even, to some extent, from population to population within a given species. Depending on the mating system of the species, sex ratio and generation time can also have a significant effect on population growth. A population whose size remains fairly constant through time is called a stable population. In such a population, births plus immigration must balance deaths plus emigration.

One way to express the age distribution characteristics of populations is through a **survivorship curve.** Survivorship is defined as the percentage of an original population that survives to a given age. Examples of different kinds of survivorship curves are shown in figure 23.4. In hydra, animals related to jellyfish, individuals are equally likely to die at any age, as indicated by the straight survivorship curve (type II). Oysters, like plants, produce vast numbers of offspring, only a few of which live to reproduce. However, once they become established and grow into reproductive individuals, their mortality rate is extremely low (type III survivorship curve). Finally, even though human babies are susceptible to death at relatively high rates, mortality rates in humans, as in many animals and protists, rise steeply in the postreproductive years (type I survivorship curve).

Life Tables

Human survivorship curves tell us that different age cohorts die at very different rates. To estimate how much longer, on average, an individual of a given age can be expected to live, life insurance companies have developed mortality summaries called **life tables.** These tables indicate the chance of survival at any given age. Ecologists use similar life tables to assess how populations in nature are changing.

Life tables can be constructed by following the fate of a cohort from birth until death, noting the number of individuals that die each year. Constructing such a cohort life table is not an easy task. A cohort must be identified and followed for many years, even though its individuals mingle freely with individuals of other cohorts. A very nice example of such a study was performed by V. Lowe on red deer (*Cervus elaphus*) on the small island of Rhum, Scotland, from 1957 to 1966. The deer live for up to 16 years, and females become capable of breeding when 4 years old. In 1957, Lowe and his coworkers made a careful count of all the deer on the island, including the number of calves (deer less than one year old). Lowe chose the female calves as the cohort he would follow in subsequent years. Each year from 1957 to 1966, every deer that had died from natural causes or been shot was examined and aged by examining its teeth. This let Lowe determine which dead deer had been calves in 1957. The life table for this cohort of female red deer appears in table 23.1, and the indicated survivorship curve is shown in figure 23.5. The convex shape of the survival curve (type I) indicates a fairly consistent increase in the risk of mortality with age.

In table 23.1, the first column indicates the age of the cohort (that is, the number of years since 1957). The second column indicates the proportion of the original 1957 cohort still alive at the beginning of that year. The third column indicates the proportion of the original cohort that died during that year. The fourth column presents the

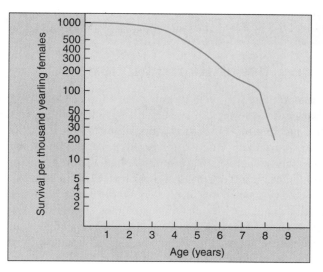

FIGURE 23.5
Survivorship curve for a cohort of female red deer on the island of Rhum. The cohort studied consisted of calves in the year 1957. Reproductive maturity typically begins at age four years. Only then does mortality begin to be significant.

mortality rate, the proportion of individuals that started that year alive but died by the end of it. Little mortality occurs among the young; most deer in this female cohort die only after reaching reproductive age (four years old).

Cohort life tables are difficult to obtain for most species of animals, so a less satisfactory but still useful alternative approach must be taken. In these instances, ecologists estimate mortality from the age structure of a population at one moment in time, a so-called *static*, or *cross-sectional, life table.*

The growth rate of a population is a sensitive function of its age structure. In some species mortality is focused among the young, in others among the old; in only a few is mortality independent of age.

Age (years) x	**Table 23.1** Life Table for a Cohort of Female Red Deer (After Lowe, 1969)		
	Proportion of Original Cohort Surviving to the Beginning of Age-Class X l_x	**Proportion of Original Cohort Dying During Age-Class X** d_x	**Mortality Rate** a_x
1	1.000	0	0
2	1.000	0.061	0.061
3	0.939	0.185	0.197
4	0.754	0.249	0.330
5	0.505	0.200	0.396
6	0.305	0.119	0.390
7	0.186	0.054	0.290
8	0.132	0.107	0.810
9	0.025	0.025	1.000

The Cost of Reproduction

The complete life cycle of an organism constitutes its **life history.** Life histories are very diverse, with different organisms making different choices about how to live. Some organisms reproduce only rarely, others frequently. Some lavish parental care on their young, others do not. Some have many offspring, others few. Darwin's theory of evolution suggests that evolution should favor high reproductive value, so why do organisms take so many different approaches? Why doesn't every organism reproduce immediately after its own birth, produce large families of large offspring, care for them intensively, and do this repeatedly throughout a long life, while outcompeting others, escaping predators, and capturing food with ease? The answer is that no one organism can do all of this—there are simply not enough resources available.

As a general rule, natural selection favors the life history that produces the highest **total reproductive value (TRV).** There are many ways for organisms to balance the conflicting opportunities that nature provides to maximize TRV, which is why there are so many different life histories.

Reproductive Trade-Offs

All life histories involve significant trade-offs. Because resources are limited, a change that increases reproduction may decrease survival and reduce future reproduction (figure 23.6). Thus, a Douglas fir tree that produces more cones increases its **current reproductive value (CRV),** but it also grows more slowly, diminishing its future **residual reproductive value (RRV).** Birds that have more offspring each year (increasing CRV) have a higher probability of dying during that year or producing smaller clutches the following year (both decreasing RRV). Conversely, individuals that delay reproduction may grow faster and larger, enhancing future reproduction. Ecologists refer to the reduction in future reproductive potential embodied in RRV as the **cost of reproduction (CR).** Natural selection will favor the life history with the greatest sum of CRV and RRV, with CRV tending to go up as RRV goes down. The "cost" incurred by contemporary reproduction contributes to a decrease in RRV. This trade-off between reproduction and its cost crucially influences how life histories evolve.

One key trade-off in life histories involves getting the most "bang for the buck" for a given reproductive investment. In simplest terms, this trade-off is between the number of offspring and their individual size. In general, the larger the offspring, the fewer produced. Thus in commercial poultry there is a negative correlation between egg size and number. A related investment concerns the degree of parental care, with a trade-off between the number of offspring and the amount of food each receives.

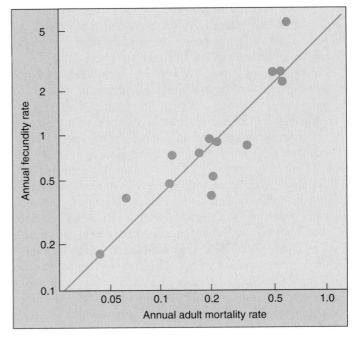

FIGURE 23.6
Reproduction has a price. Increased fecundity in birds correlates with higher mortality. Birds that raise more offspring per year have a higher probability of dying during that year.

The Role of Habitat

Does natural selection favor different life histories in different environments? Do particular life histories tend to occur in particular kinds of habitats? We can evaluate how present reproduction (contributing to CRV) and growth (contributing to RRV) combine to determine total reproductive value in different habitats by looking at two contrasting habitat types:

High CR. In high cost-of-reproduction habitats, any reduced growth that results from present reproduction has a significant negative impact on RRV. Thus in habitats where there is intense competition between individuals, reducing present reproduction may, by allowing increased growth, result in improved competitive ability and hence increased RRV.

Low CR. In low cost-of-reproduction habitats, RRV is not affected much by present reproduction. Thus when mortality is unavoidable (when temporary ponds dry out, for example) increased size is likely to be worthless in the future. In such habitats, total reproductive value is much the same whatever the level of growth.

Total reproductive value is a trade-off between investment in current reproduction and in growth that promotes future reproduction.

Life History Adaptations

Optimum Clutch Size

The number of offspring produced by a particular reproductive event is referred to as the **clutch size.** The cost-of-reproduction trade-off is between greater numbers of small offspring or fewer large ones. The ecologist David Lack proposed in 1947 that natural selection will favor a compromise clutch size which allows the maximum number to survive to maturity. This has come to be known as the "Lack clutch size." Many attempts have been made to test Lack's proposal in birds and insects by adding or removing eggs from natural clutches and seeing which is ultimately the most productive. These tests suggest that Lack's proposal is wrong: the clutch size observed naturally is *not* the most productive. Lack's proposal ignores the cost of reproduction. A large clutch may extract too high a price in terms of RRV. The clutch size favored by natural selection will thus be smaller than what appears to be the most productive clutch size in field experiments (figure 23.7).

Reproductive Events per Lifetime

The trade-off between age and fecundity plays a key role in many life histories. Annual plants and most insects focus all of their reproductive resources on a single large event and then die. This life history adaptation is called **semelparity** (from the Latin *semel,* "once," and *parito,* "to beget"). Organisms that produce offspring several times over many seasons exhibit a life history adaptation called **iteroparity** (from the Latin *itero,* "to repeat"). Species that reproduce yearly must avoid overtaxing themselves in any one reproductive episode (figure 23.8). Semelparity, or "big bang" reproduction, is usually found in short-lived species in which the cost of staying alive between broods is great, such as plants growing in harsh climates. Semelparity is also favored when fecundity entails large reproductive cost, as when Pacific salmon migrate upriver to their spawning grounds.

Age at First Reproduction

Among mammals and many other animals, longer lived species reproduce later (figure 23.9). Birds, for example, gain experience as juveniles before expending the high costs of reproduction. In long-lived animals, the relative advantage of juvenile experience outweighs the energy investment in survival and growth. In shorter lived animals, on the other hand, quick reproduction is more critical than juvenile training, and reproduction tends to occur earlier.

Life history adaptations involve many trade-offs between reproductive cost and investment in survival. Different kinds of animals and plants employ quite different approaches.

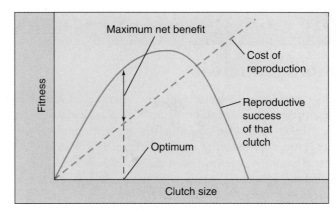

FIGURE 23.7
Optimum clutch size. Theory predicts that optimum clutch size is where the net benefit is greatest, where the distance between the cost line and benefit curve (reproductive success of the clutch) is greatest.

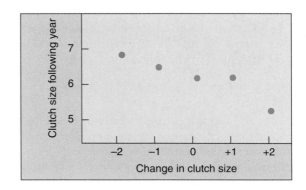

FIGURE 23.8
Reproductive events per lifetime. Increasing current reproductive effort (CRV) decreases residual reproductive value (RRV). Adding eggs to nests of collared flycatchers (which increases the reproductive efforts of the female rearing the young) decreases clutch size the following year; removing eggs from the nest increases the next year's clutch size.

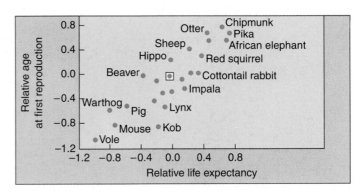

FIGURE 23.9
Age at first reproduction. Among mammals, compensating for the effects of size, age at first reproduction increases with life expectancy at birth. Each dot represents a species. Values are given relative to the species symbolized □. (After Begon et al., 1996.)

Biotic Potential

Most populations tend to remain a relatively constant size, regardless of how many offspring they produce. As you saw in chapter 1, Darwin based his theory of natural selection partly on this seeming contradiction. He reasoned that even the slow-breeding elephants would soon cover the world if their reproduction were unchecked. Natural selection occurs because of checks on reproduction, with some individuals reproducing less often than others. To understand populations, we must consider how they grow and what factors in nature limit population growth.

The Exponential Growth Model

The simplest model of population growth assumes a population growing without limits at its maximal rate. This rate, symbolized r and called the **biotic potential,** is the rate at which a population of a given species will increase when no limits are placed on its rate of growth. In mathematical terms, this is defined by the following formula:

$$\frac{dN}{dt} = r_i N$$

where N is the number of individuals in the population, dN/dt is the rate of change in its numbers over time, and r_i is the intrinsic rate of natural increase for that population—its innate capacity for growth.

The actual rate of population increase, r, is defined as the difference between the birth rate and the death rate corrected for any movement of individuals in or out of the population, whether net emigration (movement out of the area) or net immigration (movement into the area). Thus,

$$r = (b - d) + (i - e)$$

Movements of individuals can have a major impact on population growth rates. For example, the increase in human population in the United States during the closing decades of the twentieth century is mostly due to immigrants. Less than half of the increase came from the reproduction of the people already living there.

The innate capacity for growth of any population is exponential. Even when the *rate* of increase remains constant, the actual increase in the *number* of individuals accelerates rapidly as the size of the population grows. This sort of growth pattern is similar to that obtained by compounding interest on an investment. In practice, such patterns prevail only for short periods, usually when an organism reaches a new habitat with abundant resources (figure 23.10). Natural examples include dandelions reaching the fields, lawns, and meadows of North America from Europe for the first time; algae colonizing a newly-formed pond; or the first terrestrial immigrants arriving on an island recently thrust up from the sea.

FIGURE 23.10
An example of a rapidly increasing population. European purple loosestrife, *Lythrum salicaria*, became naturalized over thousands of square miles of marshes and other wetlands in North America. It was introduced sometime before 1860 and has had a negative impact on many native plants and animals.

Carrying Capacity

No matter how rapidly populations grow, they eventually reach a limit imposed by shortages of important environmental factors, such as space, light, water, or nutrients. A population ultimately stabilizes at a certain size, called the **carrying capacity** of the particular place where it lives. The carrying capacity, symbolized by K, is the number of individuals that place can support, a dynamic rather than static measure as the characteristics of the place change.

The Logistic Growth Model

As a population approaches its carrying capacity, its rate of growth slows greatly, because fewer resources remain for each new individual to use. The growth curve of such a population, which is always limited by one or more factors in the environment, can be approximated by the following **logistic growth equation:**

$$\frac{dN}{dt} = rN\left(\frac{K-N}{K}\right)$$

In this logistic model of population growth, the growth rate of the population (dN/dt) equals its rate of increase (r multiplied by N, the number of individuals present at any one time), adjusted for the amount of resources available. The adjustment is made by multiplying rN by the fraction of K still unused (K minus N, divided by K). As N increases (the population grows in size), the fraction by which r is multiplied (the remaining resources) becomes smaller and smaller, and the rate of increase of the population declines.

In mathematical terms, as N approaches K, the *rate* of population growth (dN/dt) begins to slow, until it reaches 0 when $N = K$ (figure 23.11). In practical terms, factors such as increasing competition among more individuals for a given set of resources, the buildup of waste, or an increased rate of predation causes the decline in the rate of population growth.

Graphically, if you plot N versus t (time) you obtain an **S**-shaped **sigmoid growth curve** characteristic of most biological populations. The curve is called "sigmoid" because its shape has a double curve like the letter S. As the size of a population stabilizes at the carrying capacity, its rate of growth slows down, eventually coming to a halt (figure 23.12).

Processes such as competition for resources, emigration, and the accumulation of toxic wastes all tend to increase as a population approaches its carrying capacity for a particular habitat. The resources for which the members of the population are competing may be food, shelter, light, mating sites, mates, or any other factor the species needs to carry out its life cycle and reproduce.

The size at which a population stabilizes in a particular place is defined as the carrying capacity of that place for that species. Populations grow to the carrying capacity of their environment.

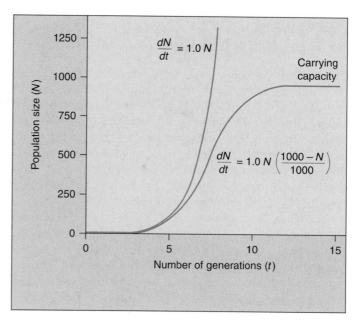

FIGURE 23.11
Two models of population growth. The red line illustrates the exponential growth model for a population with an r of 1.0. The blue line illustrates the logistic growth model in a population with $r = 1.0$ and $K = 1000$ individuals. At first, logistic growth accelerates exponentially, then, as resources become limiting, the death rate increases and growth slows. Growth ceases when the death rate equals the birth rate. The carrying capacity (K) ultimately depends on the resources available in the environment.

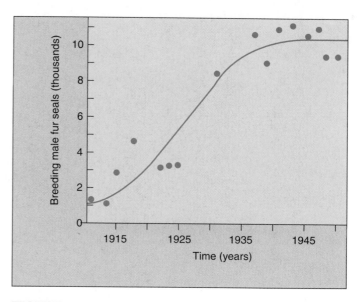

FIGURE 23.12
Most natural populations exhibit logistic growth. These data present the history of a fur seal (*Callorhinus ursinus*) population on St. Paul Island, Alaska. Driven almost to extinction by hunting at the turn of the century, the fur seal made a comeback after hunting was banned in 1911. Today the number of breeding males with "harems" oscillates around 10,000 individuals, presumably the carrying capacity of the island for fur seals.

Applying Growth Models to Real Populations

Many species, such as annual plants, some insects, and bacteria, have very fast rates of population growth. Their growth is not yet limited by dwindling environmental resources. For example, the growth rate of some organisms cannot be controlled effectively by reducing their population sizes or limiting their resources. In such species, small surviving populations soon enter an exponential pattern of growth and regain their original sizes. The growth of these populations is best described by an exponential growth model.

The populations of most animals have much slower rates of growth, best described by the logistic growth model. Populations of organisms that have sigmoid growth curves become limited in number as available resources become limiting. The number of individuals supported at this limit is the carrying capacity of the environment, or K.

Some life history adaptations of a population favor very rapid growth, among them reproducing early, producing

Adaptation	r-Selected Populations	K-Selected Populations
Age at first reproduction	Early	Late
Homeostatic capability	Limited	Often extensive
Lifespan	Short	Long
Maturation time	Short	Long
Mortality rate	Often high	Usually low
Number of offspring produced per reproductive episode	Many	Few
Number of reproductions per lifetime	Usually one	Often several
Parental care	None	Often extensive
Size of offspring or eggs	Small	Large

Table 23.2 *r*-Selected and *K*-Selected Life History Adaptations

Source: Data from E. R. Pianka, *Evolutionary Ecology,* 4th edition, 1987, Harper & Row, New York, NY.

many small offspring that mature quickly, and engaging in other aspects of "big bang" reproduction. Using the terms of the exponential model, these adaptations, all favoring a high rate of increase *r*, are called **r-selected adaptations.** Examples of organisms displaying *r*-selected life history adaptations include dandelions, aphids, mice, and cockroaches (figure 23.13).

Other life history adaptations favor survival in an environment where individuals are competing for limited resources, among them reproducing late, having small numbers of large offspring that mature slowly and receive intensive parental care, and other aspects of iteroparous reproduction. In terms of the logistic model, these adaptations, all favoring reproduction near the carrying capacity of the environment *K*, are called **K-selected adaptations.** Examples of organisms displaying *K*-selected life history adaptations include coconut palms, whooping cranes, and whales.

Most natural populations show life history adaptations that exist along a continuum ranging from completely *r*-selected traits to completely *K*-selected traits. Table 23.2 outlines the adaptations at the extreme ends of the continuum.

Some life history adaptations favor near-exponential growth, others the more competitive logistic growth. Most natural populations exhibit a combination of the two.

FIGURE 23.13
The consequences of exponential growth. All organisms have the potential to produce populations larger than those that actually occur in nature. The German cockroach (*Blatella germanica*), a major household pest, produces 80 young every six months. If every cockroach that hatched survived for three generations, kitchens might look like this theoretical culinary nightmare concocted by the Smithsonian Museum of Natural History.

The Influence of Population Density

Many factors act to regulate the growth of populations in nature. Some of these factors act independently of the size of the population; others do not.

Density-Independent Effects

Effects that are independent of the size of a population and act to regulate its growth are called **density-independent effects.** Density-independent effects, such as weather and physical disruption of the habitat, operate regardless of population size.

Density-Dependent Effects

Effects that are dependent on the size of the population and act to regulate its growth are called **density-dependent effects.** Among animals, these effects may be accompanied by hormonal changes that can alter behavior that will directly affect the ultimate size of the population. One striking example occurs in migratory locusts ("short-horned" grasshoppers). When they become crowded, the locusts produce hormones that cause them to enter a migratory phase; the locusts take off as a swarm and fly long distances to new habitats (figure 23.14). Density-dependent effects, in general, have an increasing effect as population size increases. As the population grows, the individuals in the population compete with increasing intensity for limited resources. Charles Darwin proposed that these effects resulted in natural selection and improved adaptation as individuals compete for the limiting factors.

Maximizing Population Productivity

In natural systems that are exploited by humans, such as agricultural systems and fisheries, the aim is to maximize productivity by exploiting the population early in the rising portion of its sigmoid growth curve. At such times, populations and individuals are growing rapidly, and net productivity—in terms of the amount of material incorporated into the bodies of these organisms—is highest.

Commercial fisheries attempt to operate so that they are always harvesting populations in the steep, rapidly growing parts of the curve. The point of **maximal sustainable yield** lies partway up the sigmoid curve. Harvesting the population of an economically desirable species near this point will result in the best sustained yields. Overharvesting a population that is smaller than this critical size can destroy its productivity for many years or even drive it to extinc-

FIGURE 23.14
Density-dependent effects. Migratory locusts, *Locusta migratoria*, are a legendary plague of large areas of Africa and Eurasia. At high population densities, the locusts have different hormonal and physical characteristics and take off as a swarm. The most serious infestation of locusts in 30 years occurred in North Africa in 1988.

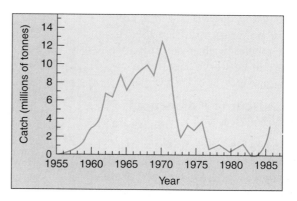

FIGURE 23.15
Catch history of the Peruvian anchovy fishery. Catches fell precipitously after the 1972 El Niño lowered ocean productivity at the same time that fishing intensity increased. The result was overharvesting and a precipitous drop in fish populations.

tion. This evidently happened in the Peruvian anchovy fishery after the populations had been depressed by the 1972 El Niño (figure 23.15). It is often difficult to determine population levels of commercially valuable species, and without this information it is hard to determine the yield most suitable for long-term, productive harvesting.

Density-dependent effects are caused by factors that come into play particularly when the population size is larger; density-independent effects are controlled by factors that operate regardless of population size.

Human Populations

Humans exhibit many *K*-selected life history traits, including small brood size, late reproduction, and a high degree of parental care. These life history traits evolved during the early history of hominids, when the limited resources available from the environment controlled population size. Throughout most of human history, our populations have been regulated by food availability, disease, and predators. While unusual disturbances, including floods, plagues, and droughts no doubt affected the pattern of human population growth, the overall size of the human population grew only slowly during our early history. Two thousand years ago, perhaps 130 million people populated the earth. It took a thousand years for that number to double, and it was 1650 before it had doubled again, to about 500 million. For over 16 centuries, the human population was characterized by very slow growth. In this respect, human populations resembled many other species with predominantly *K*-selected life history adaptations.

The Advent of Exponential Growth

Starting in the early 1700s, changes in technology have given humans more control over their food supply, enabled them to develop superior weapons to ward off predators, and led to the development of cures for many diseases. At the same time, improvements in shelter and storage capabilities have made humans less vulnerable to climatic uncertainties. These changes allowed humans to expand the carrying capacity of the habitats in which they lived, and thus to escape the confines of logistic growth and reenter the exponential phase of the sigmoidal growth curve.

Responding to the lack of environmental constraints, the human population has grown explosively over the last 300 years. While the birth rate has remained essentially unchanged at about 30 per 1000 per year over this period, the death rate has fallen dramatically, from 29 per 1000 per year to its present level of about 13 per 1000 per year. This difference between birth and death

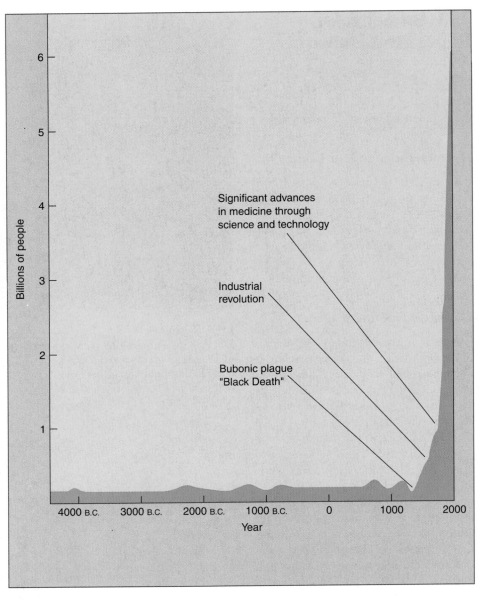

FIGURE 23.16
History of human population size. Temporary increases in death rate, even severe ones like the Black Death of the 1400s, have little lasting impact. Explosive growth began with the industrial revolution in the 1700s, which produced a significant long-term lowering of the death rate. The current population is 6 billion, and at the current rate will double in 39 years.

rates (17 per 1000) means that the population is growing at the rate of 1.7% per year.

A 1.7% annual growth rate may not seem large, but it has produced a current human population of 6 billion people (figure 23.16)! At this growth rate, 100 million people are added to the world population annually, and the human population will double in 40 years. As we will discuss in chapter 27, both the current human population level and the projected growth rate have potential consequences for our future that are extremely grave.

Population Pyramids

While the human population as a whole continues to grow rapidly at the close of the twentieth century, this growth is not occurring uniformly over the planet. Some countries, like Mexico, are growing rapidly, their birth rate greatly exceeding their death rate (figure 23.17). Other countries are growing much more slowly. The rate at which a population can be expected to grow in the future can be assessed graphically by means of a **population pyramid**—a bar graph displaying the numbers of people in each age category. Males are conventionally shown to the left of the vertical age axis, females to the right. A human population pyramid thus displays the age composition of a population by sex. In most human population pyramids, the number of older females is disproportionately large compared to the number of older males, because females in most regions have a longer life expectancy than males.

Viewing such a pyramid, one can predict demographic trends in births and deaths. In general, rectangular "pyramids" are characteristic of countries whose populations are stable, their numbers neither growing nor shrinking. A triangular pyramid is characteristic of a country that will exhibit rapid future growth, as most of its population has not yet entered the child-bearing years. Inverted triangles are characteristic of populations that are shrinking.

Examples of population pyramids for the United States and Kenya in 1990 are shown in figure 23.18. In the nearly rectangular population pyramid for the United States, the cohort (group of individuals) 55 to 59 years old represents people born during the Depression and is smaller in size than the cohorts in the preceding and following years. The

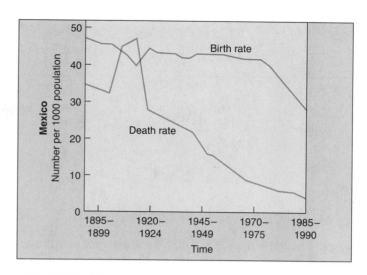

FIGURE 23.17
Why the population of Mexico is growing. The death rate (*red line*) in Mexico has been falling steadily throughout this century, while the birth rate (*blue line*) remained fairly steady until 1970. The difference between birth and death rates has fueled a high growth rate. Efforts begun in 1970 to reduce the birth rate have been quite successful, although the growth rate remains rapid.

cohorts 25 to 44 years old represent the "baby boom." The rectangular shape of the population pyramid indicates that the population of the United States is not expanding rapidly. The very triangular pyramid of Kenya, by contrast, predicts explosive future growth. The population of Kenya is predicted to double in less than 20 years.

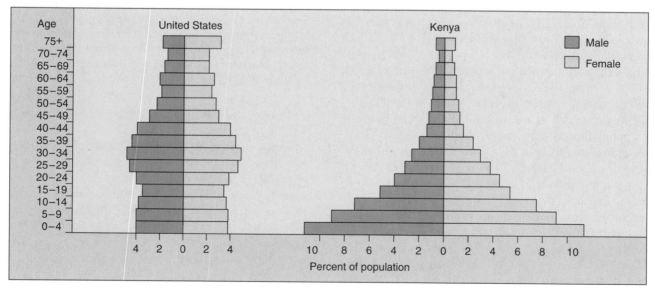

FIGURE 23.18
Population pyramids from 1990. Population pyramids are graphed according to a population's age distribution. Kenya's pyramid has a broad base because of the great number of individuals below child-bearing age. When all of the young people begin to bear children, the population will experience rapid growth. The U.S. pyramid demonstrates a larger number of individuals in the "baby boom" cohort—the pyramid bulges because of an increase in births between 1945 and 1964.

Table 23.3 A Comparison of 1994 Population Data in Developed and Developing Countries			
	United States *(highly developed)*	**Brazil** *(moderately developed)*	**Ethiopia** *(poorly developed)*
Fertility rate	2.1	3.0	6.9
Doubling time at current rate (yr)	98	40	22
Infant mortality rate	8.3	66	110
Life expectancy at birth (yrs)	76	67	52
Per capita GNP (U.S. $; 1992)	$23,120	$2,770	$110

An Uncertain Future

The earth's rapidly growing human population constitutes perhaps the greatest challenge to the future of the biosphere, the world's interacting community of living things. Humanity is adding 100 million people a year to the earth's population—a million people every three days, 250 every minute! In more rapidly growing countries, the resulting population increase is staggering (table 23.3). India, for example, had a population of 853 million in 1996; by 2020 its population will exceed 1.4 billion!

A key element in the world's population growth is its uneven distribution among countries. Of the billion people that will be added to the world's population in the 1990s, 90% will live in developing countries (figure 23.19). This is leading to a major reduction in the fraction of the world's population that lives in industrialized countries. In 1950, fully one third of the world's population lived in industrialized countries; by 1996 that proportion had fallen to one quarter; in 2020 the proportion will have fallen to one sixth. Thus the world's population growth will be centered in the parts of the world least equipped to deal with the pressures of rapid growth.

Rapid population growth in developing countries has the harsh consequence of increasing the gap between rich and poor. Today 23% of the world's population lives in the industrialized world with a per capita income of $17,900, while 77% of the world's population lives in developing countries with a per capita income of only $810. The disproportionate wealth of the industrialized quarter of the world's population is evidenced by the fact that 85% of the world's capital wealth is in the industrial world, only 15% in developing countries. 80% of all the energy used today is consumed by the industrial world, only 20% by developing countries. Perhaps most worrisome for the future, fully 94% of all scientists and engineers reside in the industrialized world, only 6% in developing countries. Thus the problems created by the future's explosive population growth will be faced by countries with little of the world's scientific or technological expertise.

No one knows whether the world can sustain today's population of 6 billion people, much less the far greater populations expected in the future. As chapter 27 (*The Future of the Biosphere*) outlines, the world ecosystem is already under considerable stress. We cannot reasonably expect to

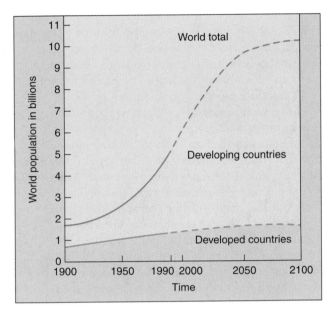

FIGURE 23.19

Most of the worldwide increase in population since 1950 has occurred in developing countries. The age structures of developing countries indicate that this trend will increase in the near future. The stabilizing of the world's population at about 10 billion is an optimistic World Bank/United Nations prediction that assumes significant worldwide reductions in growth rate. If the world's population continues to increase at its 1996 rate, there will be over 30 billion humans by 2100!

continue to expand its carrying capacity indefinitely, and indeed we already seem to be stretching the limits. It seems unavoidable that to restrain the world's future population growth, birth and death rates must be equalized. If we are to avoid catastrophic increases in the death rate, the birth rates must fall dramatically. Faced with this grim dichotomy, significant efforts are underway worldwide to lower birth rates. These efforts are already having a noticeable effect in countries like Mexico (see figure 23.17), although there is far to go.

The human population has been growing rapidly for 300 years, since technological innovations dramatically reduced the death rate.

Summary of Concepts

23.1 Populations are individuals of the same species that live together.

- Populations are the same species in one place; communities are populations of different species that live together in a particular place. A community and the nonliving components of its environment combine to form an ecosystem. Biomes are major terrestrial assemblages of plants, animals, and microorganisms in wide geographical areas that have definite characteristics.

- Populations may be dispersed in a clumped, uniform, or random manner. Clumped patterns are the most common.

23.2 Population dynamics depend critically upon age distribution.

- The growth rate of a population depends on its age structure, and to a lesser degree, sex ratio.

- Survivorship curves describe the characteristics of mortality in different kinds of populations. In type I populations, a large proportion of individuals live out a full lifespan. Type II populations have a constant mortality rate throughout their lifespan. In type III populations, many individuals die in the early stages of growth, but an individual surviving beyond that point is likely to live a long time.

23.3 Life histories often reflect trade-offs between reproduction and survival.

- Organisms balance investment in current reproduction with investment in growth and future reproduction.

23.4 Population growth is limited by the environment.

- Population size will change if birth and death rates differ, or if there is net migration into or out of the population. The intrinsic rate of increase of a population is defined as its biotic potential.

- Many populations exhibit a sigmoid growth curve, with a relatively slow start in growth, a rapid increase, and then a leveling off when the carrying capacity of the environment is reached. Populations can be harvested most effectively while in the rapid growth phase.

- Large broods and rapid rates of population growth characterize r-strategists. K-strategists are limited in population size by the carrying capacity of their environments; they tend to have fewer offspring and slower rates of population growth.

- Density-independent factors have the same impact on a population no matter what its density; a sudden storm flattening a field of wheat is density-independent. Density-dependent effects depend upon the density of the population; behavioral responses to overcrowding are density-dependent.

Discussing Key Terms

1. **Populations**   A group of potentially interbreeding individuals of a single species living together in one place at one time.

2. **Population growth rate**   The rate of growth of any population is the difference between the birth rate and the death rate per individual per unit of time, in the absence of immigration or emigration.

3. **Demography**   The statistical study of populations determines sex ratio and age structure, a key element in predicting the future of population growth.

4. **Life history**   Every species has its own reproductive strategy, dictating how often it reproduces and how many offspring it has.

5. **Carrying capacity**   As a population approaches the maximum population size its habitat can support, or its carrying capacity, its rate of growth slows.

6. **Reproductive strategies**   Some organisms (such as cockroaches) produce hundreds or thousands of offspring at a time to ensure that at least a few survive (r-strategists), while other organisms (such as whales) produce only a few offspring but invest considerable time in caring for them (K-strategists).

7. **Density-dependent effects**   Density-dependent effects have an increasing effect on population growth as population size increases.

8. **Density-independent effects**   Density-independent effects influence population growth no matter what the population size.

Review Questions

1. Which is more likely to become extinct: a small population or a large one? Why?

2. What are the three types of dispersion in a population? Which type is most frequently seen in nature? Why?

3. What is survivorship? Describe the three types of survivorship curves and give examples of each.

4. What is demography? What two factors are taken into account in demographic studies? What are the characteristics of a stable population?

5. Define the biotic potential of a population. What is the definition for the actual rate of population increase? What other two factors affect it?

6. What is an exponential capacity for growth? When does this type of growth naturally occur? Give an example.

7. What is carrying capacity? Is this a static or dynamic measure? Why?

8. In general, what types of growth-regulating effects are density-dependent? What types of effects are density-independent?

9. Why is it best to harvest individuals of a productive population partway up the sigmoid growth curve rather than when the population is at its carrying capacity? What is the result of harvesting small populations?

Thought Questions

1. Why does the net productivity of an ecosystem decrease as it becomes mature?

2. Many of our most serious pests, including rats, cockroaches, and mosquitoes, are *r*-strategists. What features of their life histories make them particularly difficult to control?

3. The sigmoid growth curve described by the logistic growth equation assumes that the growing population does not destroy the ability of the environment to support the population. Some environmentalists argue that the rapidly growing human population is doing just that. Would you expect this to produce a crash in population numbers, or a gradual fall to a new equilibrium?

Internet Links

Modeling Population Growth
http://www.geom.umn.edu/education/calc-init/population/
A very high-quality interactive module from the University of Minnesota on population growth, interactive, well presented and documented.

Amazing Kudzu
http://www.sa.ua.edu/brent/kudzu.htm
An introduced plant, Kudzu now covers over 7 million acres of the Southern United States. This site allows you to investigate how it was introduced, and the ongoing efforts to control its spread.

Population Biology
http://www.psc.lsa.umich.edu/
The POPULATION STUDIES CENTER of the University of Michigan carries out interdisciplinary research on populations.

On-line Population Ecology
http://www.gypsymoth.ento.vt.edu/~sharov/PopEcol/popecol.html
A full-blown on-line course in population ecology from Virginia Tech, with interactive population models and many links and references. An excellent site.

A Growing Human Population
http://www.prb.org/prb/
The home page of the POPULATION REFERENCE BUREAU is a great source of statistics, with data sheets on 198 countries and a monthly magazine devoted to population issues.

How Ecologists Use Life Tables
http://viner.ento.vt.edu/~sharov/PopEcol/lec6/lifetab.html
An excellent lecture from Virginia Tech on how life tables are used in ecology, with many helpful links.

For Further Reading

Begon, M., J. L. Harper, and C. R. Townsend: *Ecology: Individuals, Populations, and Communities*, 3d. ed., Blackwell Science, Cambridge, Mass., 1996. Perhaps the best undergraduate general ecology text, crammed with data and analysis.

Caro, T. M. and M. K. Laurenson: "Ecological and Genetic Factors in Conservation: A Cautionary Tale," *Science*, vol. 263, January 28, 1994, pages 485–86. An interesting discussion of how data gained in captivity may not reflect results observed in the wild.

Daily, G. C. and P. R. Erlich: "Population, Sustainability, and the Earth's Carrying Capacity," *BioScience*, September 1992, pages 761–70. An argument for sustaining the growing human population without undermining the planet's potential for supporting future generations.

Durning, A.: *How Much Is Enough?* Norton, New York, 1995. An insightful and disturbing look at our high-consumption society and at where this behavior seems to be leading us.

Raven, P., L. Berg, and G. Johnson. *Environment*, Saunders, Philadelphia, 1995. An environmental science text by the authors of this text. Perhaps not surprisingly, we think it quite a good one.

Saunders, D., R. Hobbs, and C. Margules: "Biological Consequences of Ecosystem Fragmentation: A Review," *Conservation Biology*, vol. 5, 1991, pages 18–32. A good statement of how fragmentation, by reducing effective population size, can increase the dangers of extinction.

24

Community Ecology

Concept Outline

FIGURE 24.1
Communities involve interactions between disparate groups. This clownfish is one of the few species that can nestle safely among the stinging tentacles of the sea anemone—a classic example of a symbiotic relationship.

All the organisms that live together in a place are called a community. The myriad species that inhabit a tropical rainforest are a community. Indeed, every inhabited place on earth supports its own particular array of organisms, biological communities most of which have existed for a long time. Over time, the different species have made many complex adjustments to community living (figure 24.1), evolving together and forging relationships that give the community its character and stability. Both competition and cooperation have played key roles; in this chapter, we will look at these and other factors in community ecology.

Communities

The magnificent redwood forest that extends along the coast of central and northern California and into the southwestern corner of Oregon is an example of a **community**. Within it, the most obvious organisms are the redwood trees, *Sequoia sempervirens* (figure 24.2). These trees are the sole survivors of a genus that was once distributed throughout much of the Northern Hemisphere. A number of other plants and animals are regularly associated with redwood trees, including the sword fern and beetle illustrated in figure 24.3. Their coexistence is in part made possible by the special conditions the redwood trees themselves create, providing shade, water (dripping from the branches), and relatively cool temperatures. This particular distinctive assemblage of organisms is called the redwood community. The organisms characteristic of this community have each had a complex and unique evolutionary history. They evolved at different times in the past and then came to be associated with the redwoods.

FIGURE 24.2
The redwood community. The redwood forest of coastal California and southwestern Oregon is dominated by the redwoods (*Sequoia sempervirens*) themselves.

We recognize this community mainly because of the redwood trees, and its boundaries are determined by the redwood's distribution. The distributions of the other organisms in the redwood community may differ a good deal. Some organisms may not be distributed as widely as the redwoods, and some may be distributed over a broader range. In the redwood community or any other community, the ranges of the different organisms overlap; that is why they occur together.

Many communities are very similar in species composition and appearance over wide areas. For example, the open savanna that stretches across much of Africa includes many plant and animal species that coexist over thousands of square kilometers. Interactions among these organisms occur in a similar manner throughout these grassland communities, and some interactions have evolved over millions of years.

We recognize a community largely because of the presence of its dominant species, but many other kinds of organisms are also characteristic of each community. A community exists in a place because the ranges of its species overlap there.

FIGURE 24.3
Plants and animals in the redwood community. (a) Sword fern (*Polystichum munitum*). (b) Redwood sorrel (*Oxalis oregana*). (c) A ground beetle (*Scaphinotus velutinus*) feeding on a slug on a sword fern leaf. The ecological requirements of each of these organisms differ, but they overlap enough that the organisms occur together in the redwood community.

The Niche and Competition

Each organism in an ecosystem confronts the challenge of survival in a different way. The **niche** an organism occupies is the sum total of all the ways it utilizes the resources of its environment. The niche of a species may be thought of as its biological role in the community. A niche may be described in terms of space utilization, food consumption, temperature range, appropriate conditions for mating, requirements for moisture, and other factors. A full portrait of an organism's niche would also take into account its behavior and the ways in which this behavior changes at different seasons and different times of the day. *Niche* is not synonymous with **habitat,** the place where an organism lives. *Habitat* is a place, *niche* a pattern of living.

Sometimes organisms are not able to occupy their entire niche because somebody else is using it. We call such situations when two organisms attempt to utilize the same resource **competition.** Competition is the struggle of two organisms to utilize the same resource when there is not enough of the resource to satisfy both. Fighting over resources is referred to as **interference competition;** consuming shared resources is called **exploitative competition.**

Interspecific competition refers to the interactions between individuals of different species when both require the same scarce resource. Interspecific competition is often greatest between organisms that obtain their food in similar ways; thus, green plants compete mainly with other green plants, herbivores with other herbivores, and carnivores with other carnivores. In addition, competition is more acute between similar organisms than between those that are less similar. Interspecific competition occurs between members of different species and is to be distinguished from **intraspecific competition,** which occurs between individuals of a single species.

The Realized Niche

Because of competition, organisms may not be able to occupy the entire niche they are theoretically capable of using, called the **fundamental niche** (or theoretical niche). The actual niche the organism is able to occupy in the presence of competitors is called its **realized niche.**

In a classic study, J. H. Connell of the University of California, Santa Barbara, investigated competitive interactions between two species of barnacles that grow together on rocks along the coast of Scotland. Barnacles are marine animals (crustaceans) that have free-swimming larvae. The larvae eventually settle down, cementing themselves to rocks and remaining attached for the rest of their lives. Of the two species Connell studied, *Chthamalus stellatus* lives in shallower water, where tidal action often exposed it to air, and *Semibalanus balanoides* (called *Balanus balanoides* prior to 1995) lives lower down, where it is rarely exposed to the atmosphere (figure 24.4). In the deeper zone, *Semibalanus* could always outcompete *Chthamalus* by crowding it off the rocks, undercutting it, and replacing it even where it had

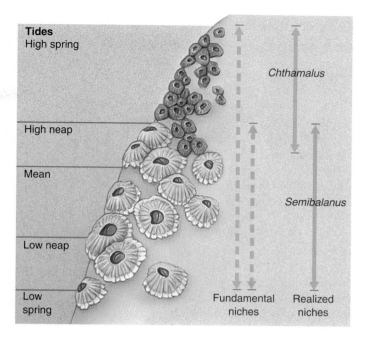

FIGURE 24.4
Competition among two species of barnacles limits niche use. The neap tide is the extreme high or low tide of each lunar month, while the spring tide is produced when the sun, earth, and moon are aligned. *Chthamalus* can live in both deep and shallow zones (its fundamental niche), but *Semibalanus* forces *Chthamalus* out of the part of its fundamental niche that overlaps the realized niche of *Semibalanus*.

begun to grow. When Connell removed *Semibalanus* from the area, however, *Chthamalus* was easily able to occupy the deeper zone, indicating that no physiological or other general obstacles prevented it from becoming established there. In contrast, *Semibalanus* could not survive in the shallow-water habitats where *Chthamalus* normally occurs; it evidently does not have the special physiological and morphological adaptations that allow *Chthamalus* to occupy this zone. Thus, the fundamental niche of the barnacle *Chthamalus* in Connell's experiments in Scotland included that of *Semibalanus*, but its realized niche was much narrower because *Chthamalus* was outcompeted by *Semibalanus* in its theoretical niche.

Another interesting example illustrating the properties of a niche involves flour beetles of the genus *Tribolium*. If *Tribolium* is grown in pure flour along with beetles of a second genus, *Oryzaephilus*, it will drive *Oryzaephilus* to extinction by means that are only partly understood. If, however, small pieces of glass tubing are placed in the flour, providing refuges for *Oryzaephilus*, then both kinds of flour beetle will coexist indefinitely. This experiment suggests why so many kinds of organisms can coexist in a structurally complex ecosystem, such as a tropical rainforest or coral reef.

A niche may be defined as the way in which an organism utilizes its environment.

Gause and the Principle of Competitive Exclusion

In classic experiments carried out between 1934 and 1935, Russian ecologist G. F. Gause studied competition among three species of *Paramecium*, a tiny protist. All three species grew well alone in culture tubes, preying on bacteria and yeasts that fed on oatmeal suspended in the culture fluid. However, when Gause grew *P. aurelia* together with *P. caudatum* in the same culture tube, the numbers of *P. caudatum* always declined to extinction, leaving *P. aurelia* the only survivor. Why? Gause found *P. aurelia* was able to grow six times faster than its competitor *P. caudatum* because it was able to better utilize the limited available resources.

From experiments such as this, Gause formulated what is now called the **principle of competitive exclusion**. This principle states that if two species are competing for a resource, the species that uses the resource more efficiently will eventually eliminate the other locally—no two species with the same niche can coexist.

Is competitive exclusion the inevitable outcome of competition for limited resources, as Gause's principle states? No. The outcome depends on the fierceness of the competition and on the degree of similarity between the fundamental niches of the competing species. If the species can avoid competing, they may coexist.

Niche Overlap

In a revealing experiment, Gause challenged *Paramecium caudatum*—the defeated species in his earlier experiments—with a third species, *P. bursaria*. Because he expected these two species to also compete for the limited bacterial food supply, Gause thought one would win out, as had happened in his previous experiments. But that's not what happened. Instead, both species survived in the culture tubes; the paramecia found a way to divide the food resources. How did they do it? In the upper part of the culture tubes, where the oxygen concentration and bacterial density were high, *P. caudatum* dominated because it was better able to feed on bacteria. However, in the lower part of the tubes, the lower oxygen concentration favored the growth of a different potential food, yeast, and *P. bursaria* was better able to eat this food. The fundamental niche of each species was the whole culture tube, but the realized niche of each species was only a portion of the tube. Because the niches of the two species did not overlap too much, both species were able to survive. Figure 24.5 summarizes Gause's experiments. The graph also demonstrates the negative effect competition had on the participants. Both species reach about twice the density when grown without a competitor.

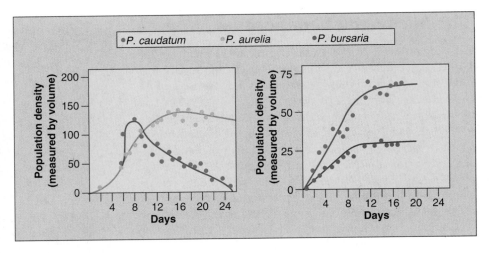

FIGURE 24.5
Competitive exclusion among three species of *Paramecium*. In the microscopic world, *Paramecium* is a ferocious predator. Paramecia eat by ingesting their prey; their cell membranes surround bacterial or yeast cells, forming a food vacuole containing the prey cell. In his experiments, Gause found that *Paramecium caudatum* would decline to extinction when grown with *P. aurelia* because they shared the same realized niche, and *P. aurelia* outcompeted *P. caudatum* for food resources. However, *P. caudatum* and *P. bursaria* were able to coexist because the two have different realized niches and thus avoided competition.

Competitive Exclusion

Gause's principle of competitive exclusion can be restated to say that *no two species can occupy the same niche indefinitely.* Certainly species can and do coexist while competing for the same resources; we have seen many examples of such relationships. Nevertheless, Gause's theory predicts that when two species coexist on a long-term basis, their niches will always differ in one or more features; otherwise, one species will outcompete the other and the extinction of the second species will inevitably result, a process referred to as **competitive exclusion.**

Niche is, of course, a complex concept, one that involves all facets of the environment that are important to individual species. Within the past decade a vigorous debate has arisen concerning the role of competitive exclusion, not only in determining the structure of communities but also in setting the course of evolution. When one or more resources are obviously limited, as in periods of drought, the role of competition becomes much more obvious than when they are not. On the other hand, especially for plants, the factors important in defining a niche are often difficult to demonstrate, and alternative explanations are being sought for the coexistence of large numbers of species.

No two species can occupy the same niche indefinitely without competition driving one to extinction.

Resource Partitioning

Gause's exclusion principle has a very important consequence: persistent competition between two species is rare in natural communities. Either one species drives the other to extinction, or natural selection reduces the competition between them. When the late Princeton ecologist Robert MacArthur studied five species of warblers, small insect-eating forest songbirds, he found that they all appeared to be competing for the same resources. However, when he studied them more carefully, he found that each species actually fed in a different part of spruce trees and so ate different subsets of insects. One species fed on insects near the tips of branches, a second within the dense foliage, a third on the lower branches, a fourth high on the trees and a fifth at the very apex of the trees. Thus, each species of warbler had evolved so as to utilize a different portion of the spruce tree resource. They *subdivided the niche*, partitioning the available resource so as to avoid direct competition with one another.

Resource partitioning can often be seen in similar species that occupy the same geographical area. Called **sympatric species** (Greek, *syn*, "same" and *patria*, "country"), these species avoid competition by living in different portions of the habitat or by utilizing different food or other resources (figure 24.6). Species that do not live in the same geographical area, called **allopatric species** (Greek, *allos*, "other" and *patria*, "country"), often utilize the same habitat locations and food resources—since they are not in competition, natural selection does not favor evolutionary changes that subdivide their niche.

When a pair of species occupy the same habitat (that is, when they are sympatric), they tend to exhibit greater differences in morphology and behavior than the same two species do when living in different habitats (that is, when they are allopatric). Called **character displacement,** the differences evident between sympatric species are thought to have been favored by natural selection as a mechanism to facilitate habitat partitioning and thus reduce competition. Thus, the two Darwin's finches in figure 24.7 have bills of similar size where the finches are allopatric, each living on an island where the other does not occur. On islands where they are sympatric, the two species have evolved beaks of different sizes, one adapted to larger seeds, the other to smaller ones.

Sympatric species partition available resources, reducing competition between them.

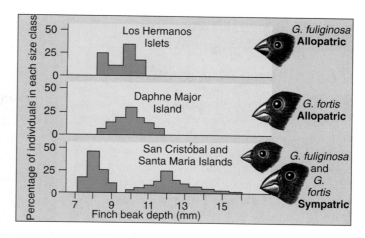

FIGURE 24.7
Character displacement in Darwin's finches. These two species of finches (genus *Geospiza*) have bills of similar size when allopatric, but different size when sympatric.

FIGURE 24.6
Resource partitioning among sympatric lizard species. Species of *Anolis* lizards in the Caribbean studied by Washington University ecologist Jonathan Losos partition their tree habitats in a variety of ways. Some species of anoles occupy the canopy of trees (a), others use twigs on the periphery (b), and still others are found at the base of the trunk (c). In addition, some use grassy areas in the open (d). When two species occupy the same part of the tree, they either utilize different-sized insects as food or partition the thermal microhabitat; for example, one might only be found in the shade, whereas the other would only bask in the sun. Most interestingly, the same pattern of resource partitioning has evolved independently on different Caribbean islands.

Coevolution and Symbiosis

The plants, animals, protists, fungi, and bacteria that live together in communities have changed and adjusted to one another continually over a period of millions of years. For example, many features of flowering plants have evolved in relation to the dispersal of the plant's gametes by animals (figure 24.8). These animals, in turn, have evolved a number of special traits that enable them to obtain food or other resources efficiently from the plants they visit, often from their flowers. In addition, the seeds of many flowering plants have features that make them more likely to be dispersed to new areas of favorable habitat.

Such interactions, which involve the long-term, mutual evolutionary adjustment of the characteristics of the members of biological communities, are examples of **coevolution.** In this chapter, we will consider some examples of coevolution, including symbiotic relationships and predator-prey interactions.

Symbiosis Is Widespread

In **symbiotic relationships** two or more kinds of organisms live together in often elaborate and more-or-less permanent relationships. All symbiotic relationships carry the potential for coevolution between the organisms involved, and in many instances the results of this coevolution are fascinating. Examples of symbiosis include *lichens*, which are associations of certain fungi with green algae or cyanobacteria. Lichens are discussed in more detail in chapter 32. Another important example are *mycorrhizae*, the association between fungi and the roots of most kinds of plants. The fungi expedite the plant's absorption of certain nutrients, and the plants in turn provide the fungi with carbohydrates. Similarly, root nodules that occur in legumes and certain other kinds of plants contain bacteria that fix atmospheric nitrogen and make it available to their host plants.

In the tropics, leafcutter ants are often so abundant that they can remove a quarter or more of the total leaf surface of the plants in a given area. They do not eat these leaves directly; rather, they take them to underground nests, where they chew them up and inoculate them with the spores of particular fungi. These fungi are cultivated by the ants and brought from one specially prepared bed to another, where they grow and reproduce. In turn, the fungi constitute the primary food of the ants and their larvae. The relationship between leafcutter ants and these fungi is an excellent example of symbiosis.

FIGURE 24.8
Pollination by bat. Many flowers have coevolved with other species to facilitate pollen transfer. Insects are widely known as pollinators, but they're not the only ones. Notice the cargo of pollen on the bat's snout.

Kinds of Symbiosis

The major kinds of symbiotic relationships include (1) **commensalism,** in which one species benefits while the other neither benefits nor is harmed; (2) **mutualism,** in which both participating species benefit; and (3) **parasitism,** in which one species benefits but the other is harmed. Parasitism can also be viewed as a form of predation, although the organism that is preyed upon does not necessarily die.

Coevolution is a term that describes the long-term evolutionary adjustments of species to one another. In symbiosis two or more species live together.

Commensalism

Commensalism is a symbiotic relationship that benefits one species and neither hurts nor helps the other. In nature, individuals of one species are often physically attached to members of another. For example, epiphytes are plants that grow on the branches of other plants. In general, the host plant is unharmed, while the epiphyte that grows on it benefits. Similarly, various marine animals, such as barnacles, grow on other, often actively moving sea animals like whales and thus are carried passively from place to place. These "passengers" presumably gain more protection from predation than they would if they were fixed in one place, and they also reach new sources of food. The increased water circulation that such animals receive as their host moves around may be of great importance, particularly if the passengers are filter feeders. The gametes of the passenger are also more widely dispersed than would be the case otherwise.

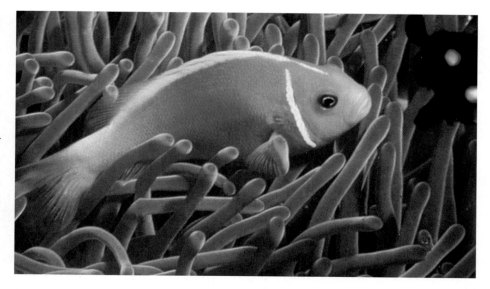

FIGURE 24.9
Commensalism in the sea. Clownfishes, such as this *Amphiprion perideraion* in Guam, often form symbiotic associations with sea anemones, gaining protection by remaining among their tentacles and gleaning scraps from their food. Different species of anemones secrete different chemical mediators; these attract particular species of fishes and may be toxic to the fish species that occur symbiotically with other species of anemones in the same habitat. There are 26 species of clownfishes, all found only in association with sea anemones; 10 species of anemones are involved in such associations, so that some of the anemone species are host to more than one species of clownfish.

Examples of Commensalism

The best known examples of commensalism involve the relationships between certain small tropical fishes and sea anemones, marine animals that have stinging tentacles (see chapter 41). These fish have evolved the ability to live among the tentacles of sea anemones, even though these tentacles would quickly paralyze other fishes that touched them (figure 24.9). The anemone fishes feed on the detritus left from the meals of the host anemone, remaining uninjured under remarkable circumstances.

On land, an analogous relationship exists between birds called oxpeckers and grazing animals such as cattle or rhinoceros. The birds spend most of their time clinging to the animals, picking off parasites and other insects, carrying out their entire life cycles in close association with the host animals. Cattle egrets, which have extended their range greatly during the past few decades, engage in a loosely coupled relationship of this kind (figure 24.10).

FIGURE 24.10
Commensalism between cattle egrets and an African cape buffalo. The egrets eat insects off the buffalo.

When Is Commensalism Commensalism?

In each of these instances, it is difficult to be certain whether the second partner receives a benefit or not; there is no clear-cut boundary between commensalism and mutualism. For instance, it may be advantageous to the sea anemone to have particles of food removed from its tentacles; it may then be better able to catch other prey. Similarly, while often thought of as commensalism, the association of grazing mammals and gleaning birds is actually an example of mutualism. The mammal benefits by having parasites and other insects removed from its body, but the birds also benefit by gaining a dependable source of food.

Commensalism is the benign use of one organism by another.

Mutualism

Mutualism is a symbiotic relationship among organisms in which both species benefit. Examples of mutualism are of fundamental importance in determining the structure of biological communities. Some of the most spectacular examples of mutualism occur among flowering plants and their animal visitors, including insects, birds, and bats. As we will see in chapter 33, during the course of their evolution, the characteristics of flowers have evolved in large part in relation to the characteristics of the animals that visit them for food and, in doing so, spread their pollen from individual to individual. At the same time, characteristics of the animals have changed, increasing their specialization for obtaining food or other substances from particular kinds of flowers.

FIGURE 24.11
Mutualism: ants and aphids. These ants are tending to willow aphids, feeding on the "honeydew" that the aphids excrete continuously, moving the aphids from place to place, and protecting them from potential predators.

Another example of mutualism involves ants and aphids. Aphids, also called greenflies, are small insects that suck fluids from the phloem of living plants with their piercing mouthparts. They extract a certain amount of the sucrose and other nutrients from this fluid, but they excrete much of it in an altered form through their anus. Certain ants have taken advantage of this—in effect, domesticating the aphids (figure 24.11). The ants carry the aphids to new plants, where they come into contact with new sources of food, and then consume as food the "honeydew" that the aphids excrete.

Ants and Acacias

A particularly striking example of mutualism involves ants and certain Latin American species of the plant genus *Acacia*. In these species, certain leaf parts, called stipules, are modified as paired, hollow thorns; these particular species are called "bull's horn acacias." The thorns are inhabited by stinging ants of the genus *Pseudomyrmex*, which do not nest anywhere else. Like all thorns that occur on plants, the acacia horns serve to deter herbivores.

At the tip of the leaflets of these acacias are unique, protein-rich bodies called Beltian bodies, named after Thomas Belt, a nineteenth-century British naturalist who first wrote about them after seeing them in Nicaragua. Beltian bodies do not occur in species of *Acacia* that are not inhabited by ants, and their role is clear: they serve as a primary food for the ants. In addition, the plants secrete nectar from glands near the bases of their leaves. The ants consume this nectar as well, feeding it and the Beltian bodies to their larvae.

Obviously, this association is beneficial to the ants, and one can readily see why they inhabit acacias of this group. The ants and their larvae are protected within the swollen thorns, and the trees provide a balanced diet, including the sugar-rich nectar and the protein-rich Beltian bodies. What, if anything, do the ants do for the plants? This question had fascinated observers for nearly a century until it was answered by Daniel Janzen, then a graduate student at the University of California, Berkeley, in a beautifully conceived and executed series of field experiments.

Whenever any herbivore lands on the branches or leaves of an acacia inhabited by ants, the ants immediately attack and devour the herbivore. Thus, the ants protect the acacias from being eaten, and the herbivore also provides additional food for the ants, which continually patrol the acacia's branches. Related species of acacias that do not have the special features of the bull's horn acacias and are not protected by ants have bitter-tasting substances in their leaves that the bull's horn acacias lack. Evidently, these bitter-tasting substances protect the acacias in which they occur from herbivores in a different way.

The ants that live in the bull's horn acacias also help their hosts to compete with other plants. The ants cut away any branches of other plants that touch the bull's horn acacia in which they are living. They create, in effect, a tunnel of light through which the acacia can grow, even in the lush deciduous forests of lowland Central America. Without the ants, as Janzen showed experimentally by poisoning the ant colonies that inhabited individual plants, the acacia is unable to compete successfully in this habitat. Finally, the ants bring organic material into their nests. The parts they do not consume, together with their excretions, provide the acacias with an abundant source of nitrogen.

Mutualism involves cooperation between species, to the mutual benefit of both.

Parasitism

Parasitism may be regarded as a special form of symbiosis in which the predator, or parasite, is much smaller than the prey and remains closely associated with it. Parasitism is harmful to the prey organism and beneficial to the parasite. The concept of parasitism seems obvious, but individual instances are often surprisingly difficult to distinguish from predation and from other kinds of symbiosis.

External Parasites

Parasites that feed on the exterior surface of an organism are external parasites, or **ectoparasites.** Many instances of external parasitism are known (figure 24.12). Lice, which live on the bodies of vertebrates—mainly birds and mammals—are normally considered parasites. Mosquitoes are not considered parasites, even though they draw food from birds and mammals in a similar manner to lice, because their interaction with their host is so brief.

Parasitoids are insects that lay eggs on living hosts. This behavior is common among wasps, whose larvae feed on the body of the unfortunate host, often killing it.

Internal Parasites

Vertebrates are parasitized internally by **endoparasites,** members of many different phyla of animals and protists. Invertebrates also have many kinds of parasites that live within their bodies. Bacteria and viruses are not usually considered parasites, even though they fit our definition precisely.

Internal parasitism is generally marked by much more extreme specialization than external parasitism, as shown by the many protist and invertebrate parasites that infect humans. The more closely the life of the parasite is linked with that of its host, the more its morphology and behavior are likely to have been modified during the course of its evolution. The same is true of symbiotic relationships of all sorts. Conditions within the body of an organism are different from those encountered outside and are apt to be much more constant. Consequently, the structure of an internal parasite is often simplified, and unnecessary armaments and structures are lost as it evolves.

(a)

(b)

FIGURE 24.12
A parasitic flowering plant and a close, nonparasitic relative. Both are members of the plant family Convolvulaceae, the morning glory family. (a) Dodder (*Cuscuta*) is a parasite, while (b) morning glory (*Convolvulus*) is not. Dodder has lost its chlorophyll and its leaves in the course of its evolution and is heterotrophic—unable to manufacture its own food. Instead, it obtains its food from the host plants it grows on.

Brood Parasitism

Not all parasites consume the body of their host. In brood parasitism, birds like cowbirds and European cuckoos lay their eggs in the nests of other species. The host parents raise the brood parasite as if it were one of their own clutch, in many cases investing more in feeding the imposter than in feeding their own offspring. The brood parasite reduces the reproductive success of the foster parent hosts, so it is not surprising that evolution has fostered the hosts' ability to detect parasite eggs and reject them.

In parasitism, one organism serves as a host to another organism, usually to the host's disadvantage.

24.3 | Predators and their prey coevolve.

Plant Defenses against Herbivores

Predator-prey interactions are inter-actions between organisms in which one organism uses the other for food. Plants have evolved many mechanisms to defend themselves from herbivores. The most obvious are **morphological defenses**: thorns, spines, and prickles play an important role in discouraging browsers, and plant hairs, especially those that have a glandular, sticky tip, deter insect herbivores. Some plants, such as grasses, deposit silica in their leaves, both strengthening and protect-ing themselves. If enough silica is pres-ent in their cells, these plants are sim-ply too tough to eat.

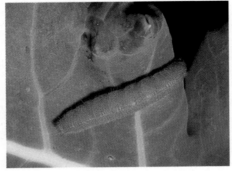
(a)

(b)

FIGURE 24.13

Insect herbivores are well suited to their hosts. (a) The green caterpillars of the cabbage butterfly, *Pieris rapae*, are camouflaged on the leaves of cabbage and other plants on which they feed. Although mustard oils protect these plants against most herbivores, the cabbage butterfly caterpillars are able to break down the mustard oil compounds. (b) An adult cabbage butterfly.

Chemical Defenses

Significant as these morphological adaptations are, the chemical defenses that occur so widely in plants are even more crucial. Best known and perhaps most important in the defenses of plants against herbivores are **secondary chemical compounds.** These are distinguished from pri-mary compounds, which are regular components of the major metabolic pathways, such as respiration. Virtually all plants, and apparently many algae as well, contain very structurally diverse secondary compounds that are either toxic to most herbivores or disturb their metabolism so greatly that they are unable to complete normal develop-ment. Consequently, most herbivores tend to avoid the plants that possess these compounds.

The mustard family (Brassicaceae) is characterized by a group of chemicals known as mustard oils. These are the substances that give the pungent aromas and tastes to such plants as mustard, cabbage, watercress, radish, and horse-radish. The same tastes we enjoy signal the presence of toxic chemicals to many groups of insects. Similarly, plants of the milkweed family (Asclepiadaceae) and the related dogbane family (Apocynaceae) produce a milky sap that de-ters herbivores from eating them. In addition, these plants usually contain cardiac glycosides, molecules named for their drastic effect on heart function in vertebrates.

The Evolutionary Response of Herbivores

Certain groups of herbivores are associated with each fam-ily or group of plants protected by a particular kind of sec-ondary compound. These herbivores are able to feed on these plants without harm, often as their exclusive food source. For example, cabbage butterfly caterpillars (sub-family Pierinae) feed almost exclusively on plants of the mustard and caper families, as well as on a few other small families of plants that also contain mustard oils (figure 24.13). Similarly, caterpillars of monarch butterflies and their relatives (subfamily Danainae) feed on plants of the milkweed and dogbane families. How do these animals manage to avoid the chemical defenses of the plants, and what are the evolutionary precursors and ecological conse-quences of such patterns of specialization?

We can offer a potential explanation for the evolution of these particular patterns. Once the ability to manufacture mustard oils evolved in the ancestors of the caper and mus-tard families, the plants were protected for a time against most or all herbivores that were feeding on other plants in their area. At some point, certain groups of insects—for ex-ample, the cabbage butterflies—developed the ability to break down mustard oils and thus feed on these plants with-out harming themselves. Having evolved this ability, the butterflies were able to use a new resource without compet-ing with other herbivores for it. Often, in groups of insects such as cabbage butterflies, sense organs have evolved that are able to detect the secondary compounds that their food plants produce. Clearly, the relationship that has formed between cabbage butterflies and the plants of the mustard and caper families is an example of coevolution.

The members of many groups of plants are protected from most herbivores by their secondary compounds. Once the members of a particular herbivore group evolve the ability to feed on them, these herbivores gain access to a new resource, which they can exploit without competition from other herbivores.

Animal Defenses against Predators

Some animals that feed on plants rich in secondary compounds receive an extra benefit. When the caterpillars of monarch butterflies feed on plants of the milkweed family, they do not break down the cardiac glycosides that protect these plants from herbivores. Instead, the caterpillars concentrate and store the cardiac glycosides in fat bodies; they then pass them through the chrysalis stage to the adult and even to the eggs of the next generation. The incorporation of cardiac glycosides thus protects all stages of the monarch life cycle from predators. A bird that eats a monarch butterfly quickly regurgitates it (figure 24.14) and in the future avoids the conspicuous orange-and-black pattern that characterizes the adult monarch. Some birds, however, appear to have acquired the ability to tolerate the protective chemicals. These birds eat the monarchs.

Defensive Coloration

Many insects that feed on milkweed plants are brightly colored; they advertise their poisonous nature using an ecological strategy known as **warning coloration**, or **aposematic coloration.** Showy coloration is characteristic of animals that use poisons and stings to repel predators, while organisms that lack specific chemical defenses are seldom brightly colored. In fact, many have **cryptic coloration**—color that blends with the surroundings and thus hides the individual from predators (figure 24.15). Camouflaged animals usually do not live together in groups because a predator that discovers one individual gains a valuable clue to the presence of others.

Chemical Defenses

Animals also manufacture and use a startling array of substances to perform a variety of defensive functions. Bees, wasps, predatory bugs, scorpions, spiders, and many other arthropods use chemicals to defend themselves and to kill their prey. In addition, various chemical defenses have evolved among marine animals and the vertebrates, including venomous snakes, lizards, fishes, and some birds. The poison-dart frogs of the family Dendrobatidae produce toxic alkaloids in the mucus that covers their brightly colored skin (figure 24.16). Some of these toxins are so powerful that a few micrograms will kill a person if injected into the bloodstream. More than 200 different alkaloids have been isolated from these frogs, and some are playing important roles in neuromuscular research. There is an intensive investigation of marine animals, algae, and flowering plants for new drugs to fight cancer and other diseases, or as sources of antibiotics.

Animals defend themselves against predators with warning coloration, camouflage, and chemical defenses such as poisons and stings.

(a)   (b)

FIGURE 24.14
A blue jay learns that monarch butterflies taste bad. (a) This cage-reared jay that had never seen a monarch butterfly before tried eating one. (b) The same jay regurgitated the butterfly a few minutes later. This bird is not likely to attempt to eat an orange-and-black insect again.

FIGURE 24.15
Cryptic coloration. An inchworm caterpillar (*Necophora quernaria*) (hanging from the upper twig) closely resembles a twig.

FIGURE 24.16
Vertebrate chemical defenses. Frogs of the family Dendrobatidae, abundant in the forests of Latin America, are extremely poisonous to vertebrates. Dendrobatids advertise their toxicity with aposematic coloration, as shown here.

Predator-Prey Cycles

Predation is the consuming of one organism by another. In this sense, predation includes everything from a leopard capturing and eating an antelope, to a deer grazing on spring grass.

The relationships between large carnivores and grazing mammals have a major impact on biological communities in many parts of the world. Appearances, however, are sometimes deceiving. On Isle Royale in Lake Superior, moose reached the island by crossing over ice in an unusually cold winter and multiplied freely there in isolation. When wolves later reached the island by crossing over the ice, naturalists widely assumed that the wolves were playing a key role in controlling the moose population. More careful studies have demonstrated that this is not in fact the case. The moose that the wolves eat are, for the most part, old or diseased animals that would not survive long anyway. In general, the moose are controlled by food availability, disease, and other factors rather than by the wolves (figure 24.17).

FIGURE 24.17
Wolves chasing a moose—what will the outcome be? On Isle Royale, Michigan, a large pack of wolves pursue a moose. They chased this moose for almost 2 kilometers; it then turned and faced the wolves, who by that time were exhausted from running through chest-deep snow. The wolves lay down and the moose walked away.

Refuges Promote Cycles

When experimental populations are set up under simple laboratory conditions, the predator often exterminates its prey and then becomes extinct itself, having nothing left to eat (figure 24.18). However, if refuges are provided for the prey, its population will drop to low levels but not to extinction. Low prey population levels will then provide inadequate food for the predators, causing the predator population to decrease. When this occurs, the prey population can recover. In this situation the predators and prey populations may continue in this cyclical pattern for some time.

Cycles in Hare Populations: A Case Study

Population cycles are characteristic of some species of small mammals, such as lemmings, and they appear to be stimulated, at least in some situations, by their predators. Ecologists have studied cycles in hare populations since the 1920s. They have found that the North American snowshoe hare (*Lepus americanus*) follows a "10-year cycle" (in reality, it varies from 8 to 11 years). Its numbers fall tenfold to 30-fold in a typical cycle, and 100-fold changes can occur. Two factors appear to be generating the cycle: food plants and predators.

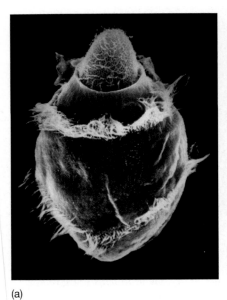

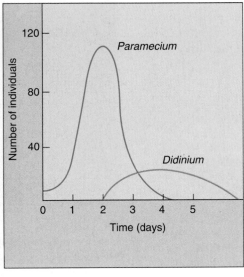

FIGURE 24.18
Predator-prey in the microscopic world. (a) Egg-shaped *Didinium*, the predator, has almost completely ingested its prey, the smaller protist *Paramecium*. (b) The graph demonstrates that when *Didinium* is added to a *Paramecium* population, the numbers of *Didinium* initially rise, while the numbers of *Paramecium* steadily fall. When the *Paramecium* population is depleted, however, the *Didinium* individuals also die.

(a)

(b)

(a)

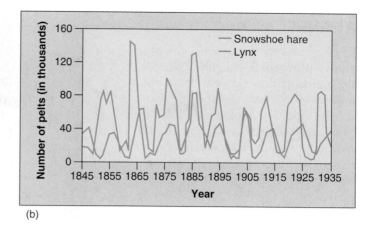

(b)

FIGURE 24.19

A predator-prey cycle. (a) A snowshoe hare being chased by a lynx. (b) The numbers of lynxes and snowshoe hares oscillate in tune with each other in northern Canada. The data are based on numbers of animal pelts from 1845 to 1935. As the number of hares grows, so does the number of lynxes, with the cycle repeating about every nine years. Both predators (lynxes) and available food resources control the number of hares. The number of lynxes is controlled by the availability of prey (snowshoe hares).

Food plants. The preferred foods of snowshoe hares are willow and birch twigs. As hare density increases, the quantity of these twigs decreases, forcing the hares to feed on high-fiber (low-quality) food. Lower birth rates, low juvenile survivorship, and low growth rates follow. The hares also spend more time searching for food, exposing them more to predation. The result is a precipitous decline in willow and birch twig abundance, and a corresponding fall in hare abundance. It takes two to three years for the quantity of mature twigs to recover.

Predators. A key predator of the snowshoe hare is the Canada lynx (*Lynx canadensis*). The Canada lynx shows a "10-year" cycle of abundance that seems remarkably entrained to the hare abundance cycle (figure 24.19). As hare numbers increase, lynx numbers do, too, rising in response to the increased availability of lynx food. When hare numbers fall, so do lynx numbers, their food supply depleted.

Which factor is responsible for the predator-prey oscillations? Do increasing numbers of hares lead to overharvesting of plants (a hare-plant cycle) or do increasing numbers of lynx lead to overharvesting of hares (a hare-lynx cycle)? Field experiments carried out by C. Krebs and coworkers in 1992 provide an answer. Krebs set up experimental plots in Canada's Yukon containing hare populations. If food is added (no food effect) and predators excluded (no predator effect) from an experimental area, hare numbers increase tenfold and stay there—the cycle is lost. However, the cycle is retained if either of the factors is allowed to operate alone: exclude predators but don't add food (food effect alone), or add food in presence of predators (predator effect alone). Thus, both factors can affect the cycle, which, in practice, seems to be generated by the interaction between the two factors.

Predation Reduces Competition

Predator-prey interactions are an essential factor in the maintenance of communities that are rich and diverse in species. By controlling the levels of some species, predators make possible the continued survival of other species in the same community. The predators prevent or greatly reduce competitive exclusion by reducing the numbers of individuals of competing species. Such patterns are particularly characteristic of biological communities in marine intertidal habitats. For example, in preying selectively on bivalves, sea stars prevent bivalves from monopolizing such habitats, opening up space for many other organisms. When sea stars are removed from a habitat, species diversity falls precipitously, the seafloor community coming to be dominated by a few species of bivalves. Because predation tends to reduce competition in natural communities, it is usually a mistake to attempt to eliminate a major predator such as wolves or mountain lions from a community. The result is to decrease rather than increase the biological diversity of the community, the opposite of what is intended.

A given predator may often feed on two, three, or more kinds of plants or animals in a given community. The predator's choice depends partly on the relative abundance of the prey options. In other words, a predator may feed on species *A* when it is abundant and then switch to species *B* when *A* is rare. Similarly, a given prey species may be a primary source of food for increasing numbers of species as it becomes more abundant, which tends to limit the size of its population automatically. Such feedback is a key factor in structuring many natural communities.

Predators and their prey often show similar cyclic oscillations, often promoted by refuges that prevent prey populations from being driven to extinction.

Mimicry

During the course of their evolution, many unprotected (nonpoisonous) species have come to resemble distasteful ones that exhibit aposematic coloration. The unprotected mimic gains an advantage by looking like the distasteful model. Two types of mimicry have been identified: Batesian mimicry and Müllerian mimicry.

Batesian Mimicry

Batesian mimicry is named for Henry Bates, the nineteenth-century British naturalist who first brought this type of mimicry to general attention in 1857. In his journeys to the Amazon region of South America, Bates discovered many instances of palatable insects that resembled brightly colored, distasteful species. He reasoned that if the unprotected animals (mimics) are present in numbers that are low relative to those of the distasteful species that they resemble (the model), the mimics will be avoided by predators, who are fooled by the disguise into thinking the mimic actually is the distasteful model.

It is important that mimics be rare relative to models. If the unprotected mimic animals are too common, or if they do not live among their models, many will be eaten by predators that have not yet learned to avoid model individuals with those particular characteristics. As a result, the predator learns that animals with that coloration are good rather than bad to eat.

Many of the best known examples of Batesian mimicry occur among butterflies and moths. Obviously, predators in systems of this kind must use visual cues to hunt for their prey; otherwise, similar color patterns would not matter to potential predators. There is also increasing evidence indicating that Batesian mimicry can also involve nonvisual cues, such as olfaction, although such examples are less obvious to humans.

The kinds of butterflies that provide the models in Batesian mimicry are, not surprisingly, members of groups whose caterpillars feed on only one or a few closely related plant families. The plant families on which they feed are strongly protected by toxic chemicals. The model butterflies incorporate the poisonous molecules from these plants into their bodies. The mimic butterflies, in contrast, belong to groups in which the feeding habits of the caterpillars are not so

(a) **Model**

(b) **Batesian Mimic**

FIGURE 24.20

A Batesian mimic. (a) The model. Monarch butterflies (*Danaus plexippus*) are protected from birds and other predators by the cardiac glycosides they incorporate from the milkweeds and dogbanes they feed on as larvae. Adult monarch butterflies advertise their poisonous nature with warning coloration. (b) The mimic. Viceroy butterflies, (*Limenitis archippus*) are Batesian mimics of the poisonous monarch. Although the viceroy is not related to the monarch, it looks a lot like it, so predators that have learned not to eat distasteful monarchs avoid viceroys, too.

restricted. As caterpillars, these butterflies feed on a number of different plant families unprotected by toxic chemicals.

One often-studied mimic among North American butterflies is the viceroy, *Limenitis archippus* (figure 24.20). This butterfly, which resembles the poisonous monarch, ranges from central Canada through much of the United States and into Mexico. The caterpillars feed on willows and cottonwoods, and neither caterpillars nor adults were thought to be distasteful to birds, although recent findings may dispute this. Interestingly, the Batesian mimicry seen in the adult viceroy butterfly does not extend to the caterpillars: viceroy caterpillars are camouflaged on leaves, resembling bird droppings, while the monarch's distasteful caterpillars are very conspicuous.

(a)

(b)

(c)

(d)

Müllerian Mimics

FIGURE 24.21

Müllerian mimics. Because the color patterns of these insects are very similar, and because they all sting, they are Müllerian mimics. The yellow jacket (a) the masarid wasp (b), the sand wasp (c), and anthidiine bee (d) all act as models for each other, strongly reinforcing the color pattern that they all share.

Müllerian Mimicry

Another kind of mimicry, **Müllerian mimicry,** was named for German biologist Fritz Müller, who first described it in 1878. In Müllerian mimicry, several unrelated but protected animal species come to resemble one another (figure 24.21). Thus, different kinds of stinging wasps have yellow-and-black-striped abdomens, but they may not all be descended from a common yellow-and-black-striped ancestor. In general, yellow-and-black and bright red tend to be common color patterns that warn away predators relying on vision. If animals that resemble one another are all poisonous or dangerous, they gain an advantage because a predator will learn more quickly to avoid them.

In both Batesian and Müllerian mimicry, mimic and model must not only look alike but also act alike if predators are to be deceived. For example, the members of several families of insects that resemble wasps behave surprisingly like the wasps they mimic, flying often and actively from place to place.

In Batesian mimicry, unprotected species resemble others that are distasteful. Both species exhibit aposematic coloration. If they are relatively scarce, the unprotected mimics will be avoided by predators. In Müllerian mimicry, two or more unrelated but protected species resemble one another, thus achieving a kind of group defense.

Community Structure

Ecologists have long sought to understand why some ecosystems are more stable than others—better able to avoid permanent change and return to normal after disturbances like land clearing, fire, invasion by plagues of insects, or severe storm damage. Most ecologists now agree that biologically diverse ecosystems are generally more stable than simple ones. Ecosystems with many different kinds of organisms support a more complex web of interactions, and an alternative niche is thus more likely to exist to compensate for the effect of a disruption.

Species Richness

The complexity of a biological community depends both on the number of different species it contains and on the relative numbers of individuals of each species. The number of species an ecosystem's biological community contains is referred to as its **species richness.** Communities that are more species-rich are, other things being equal, more complex.

Different ecosystems vary greatly in the species richness of their biological communities. Forests typically support more species than grasslands, and tropical communities more species than temperate ones. Fewer species inhabit harsh habitats like deserts and polar tundra than inhabit more temperate places.

Species Diversity

However, species richness alone is not an adequate measure of community complexity, because communities also differ greatly in the relative abundance of the species they contain. Species richness misses the information that some species are rare and others common. A community with 10 species, each represented by the same number of individuals, is much more diverse than a community with 10 species, 1 of which contains 91% of all the individuals present, and the other 9 species containing 1% each. To account for differences in relative species abundance, ecologists use a weighted measure of species richness, **species diversity.**

The most commonly used measure of species diversity is the **Shannon diversity index, H.** H is calculated by determining, for each species, the proportion of individuals (P) that it contributes to the total in the sample. Each value of P is then multiplied by its natural log, and the products for the different species are added together.

$$H = - \text{[the sum of } P\,(\ln P) \text{ for all the species]}$$

Table 24.1 presents data on the species composition of two deciduous forests in Indiana. While they have the same species richness, one is more diverse ($H = 1.8$) than the other ($H = 1.2$), as more species are common within it.

	Forest A		Forest B	
Species	**P**	**$P\,(\ln P)$**	**P**	**$P(\ln P)$**
Sugar maple (*Acer saccharum*)	0.17	−0.30	0.01	−0.05
Shagbark hickory (*Carya ouata*)	0.08	−0.20	0.07	−0.19
American beech (*Fagus grandifolia*)	0.28	−0.36	0.05	−0.15
White ash (*Fraxinus americanus*)	0.03	−0.10	0.02	−0.08
Black walnut (*Juglans nigra*)	0.00	0.00	0.01	−0.05
Yellow poplar (*Liriodendron tulipifera*)	0.20	−0.32	0.02	−0.08
White oak (*Quercus alba*)	0.01	−0.05	0.63	−0.29
Red oak (*Quercus rubra*)	0.03	−0.11	0.18	−0.31
Basswood (*Tilia americana*)	0.18	−0.31	0.00	0.00
Elm (*Ulmus americana*)	0.02	−0.08	0.01	−0.05
	1.00	$H = 1.8$	1.00	$H = 1.2$

Table 24.1 Species Diversity of Two Forests in Turkey Run State Park, Indiana

At maximum diversity, these forests would have each of the nine species present in equal frequency; this would yield an H value of 2.2. A low-diversity forest, say one in which eight of the species were present at only 1% and one dominant species at 92%, would yield an H value of just below 0.5. These two forests are both diverse, although forest A ($H = 1.8$) is significantly more diverse than forest B ($H = 1.2$).

Factors Promoting Species Richness

A number of factors might, in theory, influence the species richness of a community. It is often extremely difficult to sort out the relative contributions of different factors. Among the many that may be important, we will discuss three: ecosystem productivity, spatial heterogeneity, and climate. Two additional factors that may play an important role, the evolutionary age of the community and the degree to which the community has been disturbed, will be examined later in this chapter.

Ecosystem Productivity. When natural communities are examined, species richness often correlates with productivity (figure 24.22). In a key experimental test of this hypothesis, David Tilman of the University of Minnesota and some 50 coworkers burned, plowed, planted, and hand-tended a large number of 100-square-foot experimental plots in a Minnesota prairie. They tended 147 plots in all, each containing from 1 to 24 native prairie plant species. Each of the plots was monitored to estimate how much growth was occurring and how much nitrogen the growing plants were taking up from the soil. Tilman found that the more species a plot had, the more nitrogen they had taken up. In his study, increased biodiversity clearly leads to greater productivity.

Spatial Heterogeneity. Environments that are more spatially heterogeneous—that contain more soil types, topographies, and other habitat variations—can be expected to accommodate more species because they provide a greater variety of microhabitats, microclimates, places to hide from predators, and so on. In general, the species richness of animals tends to reflect the species richness of the plants in their community, while plant species richness reflects the spatial heterogeneity of the ecosystem. The plants provide a biologically derived spatial heterogeneity of microhabitats to the animals. Thus, the number of lizard species in the American southwest mirrors the structural diversity of the plants (figure 24.23).

Climate. The role of climate is more difficult to assess. On the one hand, more species might be expected to coexist in a seasonal environment than in a constant one, because a predictably changing seasonal environment provides a changing climate that may favor different species at different times of the year. On the other hand, unpredictable climatic variation may have just the opposite effect: stable environments are able to support specialized species that would be unable to survive where conditions fluctuate dramatically. Thus, the number of mammal species along the west coast of North America increases as the temperature range decreases (figure 24.24).

Species richness promotes ecosystem productivity and is fostered by spatial heterogeneity and stable climate.

(a)

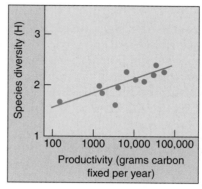
(b)

FIGURE 24.22
Ecosystem productivity.
(a) In a study of prairie ecosystems, plots with less species diversity (*foreground*) had a lower productivity than plots with more species (*background*). (b) The species diversity (*H*) of cladocerans, tiny pond animals, is higher in lakes with higher productivity.

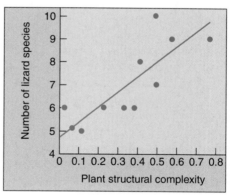

FIGURE 24.23
Spatial heterogeneity.
The species richness of desert lizards is positively correlated with the structural complexity of the plant cover in desert sites in the American southwest.

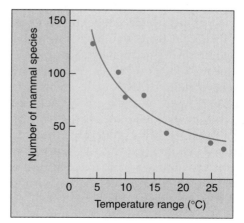

FIGURE 24.24
Climate. The species richness of mammals is inversely correlated with monthly mean temperature range along the west coast of North America.

Biogeographic Patterns of Species Diversity

Since before Darwin, biologists have recognized that there are more different kinds of animals and plants in the tropics than in temperate regions. For many species, there is a steady increase in species richness from the arctic to the tropics. Called a **species diversity cline,** such a biogeographic gradient in numbers of species correlated with latitude has been reported for plants and animals, including birds (figure 24.25), mammals, reptiles.

Why Are There More Species in the Tropics?

For the better part of a century, ecologists have puzzled over the cline in species diversity from the arctic to the tropics. The difficulty has not been in forming a reasonable hypothesis of why there are more species in the tropics, but rather in sorting through the many reasonable hypotheses that suggest themselves. Here we will consider five of the most commonly discussed suggestions:

Evolutionary age. It has often been proposed that the tropics have more species than temperate regions because the tropics have existed over long and uninterrupted periods of evolutionary time, while temperate regions have been subject to repeated glaciations. The greater age of tropical communities would have allowed complex population interactions to coevolve within them, fostering a greater variety of plants and animals in the tropics.

However, recent work suggests that the long-term stability of tropical communities has been greatly exaggerated. An examination of pollen within undisturbed soil cores reveals that during glaciations the tropical forests contracted to a few small refuges surrounded by grassland. This suggests that the tropics have not had a continuous record of species richness over long periods of evolutionary time; unfortunately, the fossil record is too sparse to assess the past species richness of the tropics.

Higher productivity. A second often-advanced hypothesis is that the tropics contain more species because this part of the earth receives more solar radiation than temperate regions do. The argument is that more solar energy, coupled to a year-round growing season, greatly increases the overall photosynthetic activity of plants in the tropics. If we visualize the tropical forest as a pie (total resources) being cut into slices (species niches), we can see that a larger pie accommodates more slices. However, many field studies have indicated that species richness is highest at intermediate levels of productivity. Accordingly, increasing productivity would be expected to lead to lower, not higher, species richness. Perhaps the long column of vegetation down through which light passes in a tropical forest produces a wide range of frequencies and intensities, creating a greater variety of light environments and in this way promoting species diversity.

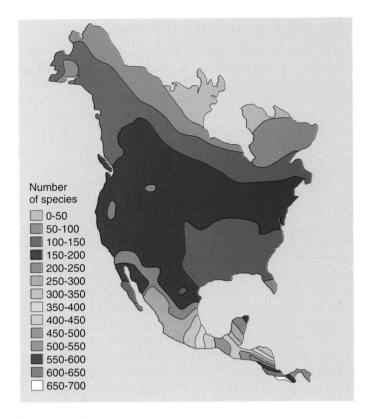

FIGURE 24.25
A latitudinal cline in species richness. Among North and Central American birds, a marked increase in the number of species occurs as one moves toward the tropics. Fewer than 100 species are found at arctic latitudes, while more than 600 species live in southern Central America.

Number of species

- 0–50
- 50–100
- 100–150
- 150–200
- 200–250
- 250–300
- 300–350
- 350–400
- 400–450
- 450–500
- 500–550
- 550–600
- 600–650
- 650–700

Predictability. There are no seasons in the tropics. Tropical climates are stable and predictable, one day much like the next. These unchanging environments might encourage specialization, with niches subdivided to partition resources and so avoid competition. The expected result would be a larger number of more specialized species in the tropics, which is what we see. Many field tests of this hypothesis have been carried out, and almost all support it, reporting larger numbers of narrower niches in tropical communities than in temperate areas.

Predation. Many reports indicate that predation may be more intense in the tropics. In theory, more intense predation could reduce the importance of competition, permitting greater niche overlap and thus promoting greater species richness.

Spatial heterogeneity. As noted earlier, spatial heterogeneity promotes species richness. Tropical forests, by virtue of their complexity, create a variety of microhabitats and so may foster larger numbers of species.

No one really knows why there are more species in the tropics, but there are plenty of suggestions.

Island Biogeography

Oceanic islands provide a natural laboratory for studying the factors that promote species richness. In 1967, Robert MacArthur of Princeton University and Edward O. Wilson of Harvard University proposed that the number of species on such islands is related to the size of the island.

The Equilibrium Model

MacArthur and Wilson reasoned that species are constantly being dispersed to islands, so islands have a tendency to accumulate more and more species. At the same time that new species are added, however, other species are lost by extinction. Once the number of species fills the capacity of that island, no more species can be established unless one of the species already there becomes extinct. Every island of a given size, then, has a characteristic equilibrium number of species that tends to persist through time (the intersection point in figure 24.26*a*), although the individual species may change.

MacArthur and Wilson's equilibrium theory proposes that island species richness is a dynamic equilibrium between colonization and extinction. Both island size and distance from the mainland would play important roles. We would expect smaller islands to have higher rates of extinction because their population sizes would, on average, be smaller. Also, we would expect fewer colonizers to reach islands that lie farther from the mainland. Thus, small islands far from the mainland have the fewest species; large islands near the mainland have the most (figure 24.26*b*).

The predictions of this simple model bear out well in field data. Asian Pacific bird species (figure 24.26*c*) exhibit a positive correlation of species richness with island size, but a negative correlation of species richness with distance from the mainland (New Guinea).

FIGURE 24.27
A test of the equilibrium model. Scientists erected a scaffold to entirely cover this small mangrove island in the Florida Keys. Covered with a plastic sheet, the island became an enclosed experimental system.

Testing the Equilibrium Model

An interesting test of the equilibrium model was carried out on small mangrove islands in the Florida Keys. Islands about 12 meters across were inhabited by 25 to 40 species of arthropods; larger islands had more species, smaller islands fewer. After a careful census, six of the islands were covered with plastic sheets and fumigated with an insecticide to kill all arthropods (figure 24.27). These islands were recolonized within a year, and rapidly attained a steady-state number of species. For each island, the capacity of that island set the equilibrium number of species, while chance determined the exact identity of the species.

Species richness on islands is a dynamic equilibrium between colonization and extinction.

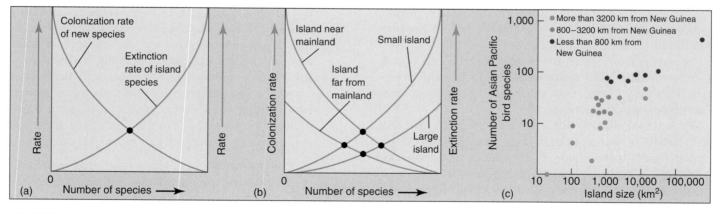

FIGURE 24.26
The equilibrium model of island biogeography. (a) Island species richness reaches an equilibrium (*black dot*) when the colonization rate of new species equals the extinction rate of species on the island. (b) The equilibrium shifts when the colonization rate is affected by distance from the mainland and when the extinction rate is affected by size of the island. Species richness is positively correlated with island size and inversely correlated with distance from the mainland. (c) Small distant Asian islands have fewer bird species, bearing out the principles of the equilibrium model.

Ecological Succession

Even when the climate of an area remains stable year after year, ecosystems have a tendency to change from simple to complex in a process known as **succession.** This process is familiar to anyone who has seen a vacant lot or cleared woods slowly become occupied by an increasing number of plants, or a pond become dry land as it is filled with vegetation encroaching from the sides.

Secondary Succession

If a wooded area is cleared and left alone, plants will slowly reclaim the area. Eventually, traces of the clearing will disappear and the area will again be woods. This kind of succession, which occurs in areas where an existing community has been disturbed, is called secondary succession. Humans are often responsible for initiating secondary succession, but it may also take place when a fire has burned off an area, or in abandoned agricultural fields.

Primary Succession

In contrast, primary succession occurs on bare, lifeless substrate, such as rocks, or in open water, where organisms gradually move into an area and change its nature. Primary succession occurs in lakes left behind after the retreat of glaciers, on volcanic islands that rise above the sea, and on land exposed by retreating glaciers (figure 24.28). Primary succession that occurs on land is called **xerarch** succession, to distinguish it from the **hydrarch** primary succession that occurs in open water. Primary succession on glacial moraines provides an example (figure 24.29). On bare, mineral-poor soil, lichens grow first, forming small pockets of soil. Acidic secretions from the lichens help to break down the substrate and add to the accumulation of soil. Mosses then colonize these pockets of soil, eventually building up enough nutrients in the soil for alder shrubs to take hold. Over a hundred years, the alders build up the soil nitrogen levels until spruce are able to thrive, eventually crowding out the alder and forming a dense spruce forest.

In a similar example involving hydrarch succession, an **oligotrophic** lake—one poor in nutrients—may gradually, by the accumulation of organic matter, become **eutrophic**—rich in nutrients.

Primary succession in different habitats may over the long term arrive at the same kinds of vegetation—vegetation characteristic of the region as a whole. This relationship led American ecologist F. E. Clements, at about the turn of the century, to propose the concept of a final **climax vegetation** (and the related term *climax community*). With an increasing realization that (1) the climate keeps changing, (2) the process of succession is often very slow, and (3) the nature of a region's vegetation is being determined to an increasing extent by human activities, ecologists do not consider the concept of "climax vegetation" to be as useful as they once did.

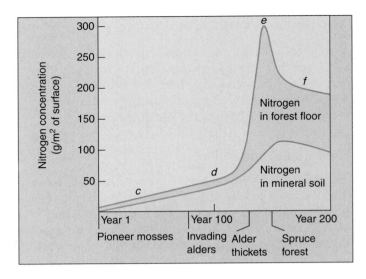

FIGURE 24.28
Plant succession produces progressive changes in the soil. Initially the glacial moraine at Glacier Bay, Alaska, portrayed in figure 24.29, had little soil nitrogen, but nitrogen-fixing alders led to a buildup of nitrogen in the soil, encouraging the subsequent growth of the conifer forest. Letters in the graph correspond to photographs in parts *c, d, e,* and *f* of figure 24.29.

Why Succession Happens

Succession happens because species alter the habitat and the resources available in it, often in ways that favor other species. Three dynamic concepts are of critical importance in the process: tolerance, inhibition, and facilitation.

1. **Tolerance.** Early successional stages are characterized by weedy *r*-selected species that do not compete well in established communities but are tolerant of the harsh, abiotic conditions in barren areas.
2. **Facilitation.** The weedy early successional stages introduce local changes in the habitat that favor other, less weedy species. Thus, the mosses in the Glacier Bay succession of figure 24.29 fix nitrogen that allows alders to invade. The alders in turn lower soil pH as their fallen leaves decompose, and spruce and hemlock, which require acidic soil, are able to invade.
3. **Inhibition.** Sometimes the changes in the habitat caused by one species, while favoring other species, inhibit the growth of the species that caused them. Alders, for example, do not grow as well in acidic soil as the spruce and hemlock that replace them.

As ecosystems mature, and more *K*-selected species replace *r*-selected ones, species richness and total biomass increase but net productivity decreases. Because earlier successional stages are more productive than later ones, agricultural systems are intentionally maintained in early successional stages to keep net productivity high.

Communities evolve to have greater total biomass and species richness in a process called succession.

FIGURE 24.29

Primary succession at Alaska's Glacier Bay. (a) The sides of the glacier have been retreating at a rate of some 8 meters a year, leaving behind (b) exposed soil from which nitrogen and other minerals have been leached out. The first invaders of these exposed sites are (c) pioneer moss species with nitrogen-fixing mutualistic microbes. Within 20 years, (d) young alder shrubs take hold. Rapidly fixing nitrogen, they soon form (e) dense thickets. As soil nitrogen levels rise, (f) spruce crowd out the mature alders, forming a forest.

Preserving Biodiversity

The Role of Disturbance

Disturbances often interrupt the succession of plant communities to stages marked by greater species richness and more complex interactions between species. Disturbances severe enough to disrupt succession include calamities such as forest fires, drought, and floods. Animals may also cause severe disruptions. Gypsy moths can devastate a forest by consuming its trees. Unregulated deer populations may grow explosively, the deer overgrazing and so destroying the forest they live in, in the same way too many cattle overgraze a pasture by eating all available grass down to the ground.

Human activity has the greatest disruptive impact of all. All over the world, humans are reducing extensive forests to isolated patches by clear-cut logging and land clearance for farming and development. Clear-cut slopes and abandoned fields quickly become dominated by early secondary succession growth; weedy and shrubby vegetation now covers much of the United States where mature communities once grew. In chapter 27, we will examine the impact of human disruption of the earth's environment more closely.

Keystone Species

Some communities are more resistant to disturbance than others. The very complexity of highly diverse ecosystems buffers them from everyday insult, although it also makes them more likely to have particular points of vulnerability which, if damaged, can have far-reaching consequences. A species that interacts in critical ways with many other elements of an ecosystem is called a **keystone species** (figure 24.30). Because the loss of a keystone species may affect many other organisms in a diverse community, such species represent points of particular sensitivity in the ecosystem—places where the complex ecological machine can be easily broken. Think of how you would design a computer to protect it from "crashing"—like a complex ecosystem, you would make it with a diverse array of redundant (repeated) circuits which could take over the functions of others if the need arose. However, this added complexity comes with a price: the advanced chip is uniquely vulnerable to damage at those points where functional crossover takes place. A single flaw can destroy the operation of the entire computer, while a simple chip would have lost only one circuit.

The Importance of Biodiversity

If we wish to preserve natural ecosystems, we should do all we can to preserve their biological diversity, while at the same time realizing that diverse ecosystems can be damaged too. What biologists can do in the face of this crisis is to help design intelligent plans for locating those organisms most likely to be of use and saving them from extinction. They must also participate in sound, globally based

(a)

(b)

FIGURE 24.30
A keystone species. (a) In a controlled experiment in a coastal ecosystem, an investigator removed a key predator (*Pisaster*). (b) In response, fiercely competitive mussels exploded in growth, effectively crowding out seven other indigenous species.

schemes to preserve as much as possible of the biological diversity of life on earth. With the loss of tropical forest and other biological communities throughout the world, we are permanently losing many opportunities not only to gain knowledge, but to increase human prosperity. It is biologists who must understand this message and inform their fellow citizens of its importance.

Succession is often disrupted by natural or human causes, stopping the progression toward more diverse communities. Sometimes removal of a key species is all that is needed to seriously disrupt a biological community.

Summary of Concepts

24.1 Competition shapes communities.

- Communities are sets of organisms that share a habitat; they exist because the ecological requirements and environmental tolerances of the organisms that constitute them overlap in the areas where they occur together.

- Each species plays a specific role in its ecosystem; this role is called its niche.

- An organism's fundamental niche is the total niche that the organism would occupy in the absence of competition. Its realized niche is the actual niche it occupies in nature.

- Two species cannot occupy the same niche for long; one will outcompete the other, driving it to extinction.

24.2 Evolution often fosters cooperation.

- Coevolution occurs when different kinds of organisms adjust to one another by genetic change over long periods of time.

- Many organisms have coevolved to a point of dependence. In some instances the relationship is mutually beneficial (mutualism), in others only one organism benefits while the other is unharmed (commensalism), and in still others one organism serves as a host to another, usually to the host's disadvantage (parasitism).

24.3 Predators and their prey coevolve.

- Plants are often protected from herbivores by chemicals they manufacture. Such chemicals, which are not part of the primary metabolism of the plant, are called secondary compounds.

- Warning, or aposematic, coloration is characteristic of organisms that are poisonous, sting, or are otherwise harmful. In contrast, cryptic coloration, or camouflage, is characteristic of nonpoisonous organisms.

- Predator-prey relationships are of crucial importance in limiting population sizes in nature.

- Some palatable organisms benefit from mimicry, bearing a striking resemblance to a local noxious species. Unrelated species that are noxious sometimes resemble each other, achieving a kind of group defense.

24.4 Biodiversity promotes community stability.

- Species richness is fostered by spatial heterogeneity and stable climates.

- Many more species live in the tropics than in temperate regions, but the reason is not well understood.

- Primary succession takes place in barren areas, like rocks or open water. Secondary succession takes place in areas where the original communities of organisms have been disturbed.

Discussing Key Terms

1. **Niche**   The specific role a species plays in its environment is its niche. In the absence of any competition, an organism can occupy its full fundamental niche, but the real-life interactions in dynamic ecosystems often limit the niche organisms to a realized niche.

2. **Competitive exclusion**   If two species compete for the same limited resources, one will be able to use them more efficiently and will drive the other to extinction. This principle can be restated in terms of the niche concept: no two species can occupy the same realized niche indefinitely.

3. **Coevolution**   Coevolution is the long-term development of adaptations in organisms that interact so closely that each acts as a selective force on the other.

4. **Symbiosis**   Symbiotic relationships are those in which two kinds of organisms live together. There are three principal sorts of symbiotic relationships: in commensalism, one benefits and the other is unaffected; in mutualism, each organism benefits; and in parasitism, one benefits and the other is harmed.

5. **Mimicry**   In Batesian mimicry, a palatable or nontoxic organism resembles another organism that is distasteful or toxic. In Müllerian mimicry, several toxic or dangerous kinds of organisms resemble one another.

6. **Species diversity**   The number of species in a community (species richness), weighted for differences in relative abundance, is a measure of species diversity.

7. **Succession**   The progressive development of a community is referred to as succession. The first colonization of a lifeless area is primary succession; recolonizing a disrupted area secondary succession.

Review Questions

1. What is the difference between interspecific competition and intraspecific competition? What is Gause's principle of competitive exclusion?

2. Is the term *niche* synonymous with the term *habitat*? Why or why not? How does an organism's fundamental niche differ from its realized niche?

3. Classify the following symbiotic relationships as commensalism, mutualism, or parasitism: ants on the acacia tree; dodder and host plant; sea anemones and clownfish; ants and aphids; lice and birds; lichens; legume root nodules; and fleas and humans.

4. What morphological defenses do plants use to defend themselves against herbivores?

5. What is the difference between primary and secondary plant chemical compounds? How do secondary chemical compounds protect plants from being overgrazed?

6. Consider aposematic coloration, cryptic coloration, and Batesian mimicry. Which would be associated with an adult viceroy butterfly? Which would be associated with a larval monarch butterfly? Which would be associated with a larval viceroy butterfly?

7. Why have scientists altered the concept of a final, climax vegetation in a given ecosystem? What types of organisms are often associated with early stages of succession?

Thought Questions

1. If an island contained no carnivores of any type, only herbivores, what would happen if a generalist carnivore arrived (one that would eat a variety of different organisms)? Would the number of species on the island increase or decrease? What conditions would favor each answer? Why? If the carnivore was a specialist and fed only on a certain type of bird, how would this change your predictions?

2. Most chemically protected prey species produce toxins that make them taste bad or make the predator sick. If the prey is tasted by a predator, the prey usually dies. If it dies, how have the chemicals protected it?

Internet Links

Investigating Predator Prey Models
http://fisher.teorekol.lu.se/simulation_server/
The SIMULATION SERVER allows you to interactively investigate the behavior of the Lotka-Volterra ecological models of predator prey relationships.

Biological Pest Control
http://www.nysaes.cornell.edu/ent/biocontrol/
From Cornell, a site that illustrates how agricultural scientists are using natural interactions between organisms to protect crops without chemical pesticides, insecticides, or herbicides.

Microbial Ecology
http://commtechlab.msu.edu/sites/dlc-me/
The DIGITAL LEARNING CENTER FOR MICROBIAL ECOLOGY provides an interesting look at the soil microbes so critial for recycling in natural ecosystems.

Keystone Species
http://www.bagheera.com/CLASROOM/spotlite/spkey.htm
A lecture from the ENDANGERED SPECIES CLASSROOM with links to case studies.

For Further Reading

Case, T. J. and M. L. Cody: "Testing Theories of Island Biogeography," *American Scientist*, 1987, vol. 75, pages 402–11. On islands in the Gulf of California, a natural laboratory for testing theories of island biogeography exists.

Chapin, F. Stuart, III, et. al.: "Biotic Control over the Functioning of Ecosystems," *Science*, July 1997, vol. 277, pages 500–504. A seminal review by David Tilman and his students describing the role of biodiversity in stabilizing ecosystems.

Connor, E. F. and D. Simberloff: "Competition, Scientific Method, and Null Models in Ecology," *American Scientist*, 1986, vol. 74, pages 155–62. A central article in the continuing debate about the role of competition in structuring natural communities.

Devries, P.: "Stinging Caterpillars, Ants, and Symbiosis," *Scientific American*, October 1992, pages 76–82. An article about the intricate symbiosis between ants and caterpillars, a symbiosis that says a lot about interdependence within ecosystems and how it arises.

Handel, S. and A. Beattie: "Seed Dispersal by Ants," *Scientific American*, August 1990, pages 76–83. Many plant species induce ants to spread their seeds with special food lures and other adaptations.

Nadkarni, N. M.: "Diversity of Species and Interactions in the Upper Tree Canopy of Forest Ecosystems," *American Zoologist*, 1994, vol. 34 (1), 1994, pages 70–78. How important is diversity to nutrient cycling?

Quammen, D.: *The Song of the Dodo: Island Biogeography in an Age of Extinction*, Scribner, New York, 1996. A wonderful account of how biologists have come to understand the special biology of islands. Look particularly to the treatment of Wallace.

Rennie, J.: "Living Together," *Scientific American*, January 1992, pages 120–33. An excellent overview of how coevolution and symbiosis have shaped life today.

Tilman, D., D. Wedin, J. Knops: "Productivity and Sustainability Influenced by Biodiversity in Grassland Ecosystems," *Nature*, February 22, 1996, vol. 379, pages 718–20. An important experiment demonstrating that biodiversity promotes productivity.

Worster, D.: *Nature's Economy: A History of Ecological Ideas*, Cambridge University Press, New York, 1994. An insightful look at the evolution of ecological theory.

25

Dynamics of Ecosystems

Concept Outline

25.1 Chemicals cycle within ecosystems.

The Water Cycle. Water cycles between the atmosphere and the oceans, although deforestation has broken the cycle in some ecosystems.

The Carbon Cycle. Photosynthesis captures carbon from the atmosphere; respiration returns it.

The Nitrogen Cycle. Nitrogen is captured from the atmosphere by the metabolic activities of bacteria; other bacteria degrade organic nitrogen, returning it to the atmosphere.

The Phosphorus Cycle. Of all nutrients that plants require, phosphorus tends to be the most limiting.

Biogeochemical Cycles Illustrated: Recycling in a Forest Ecosystem. In a classic experiment, the role of forests in retaining nutrients is assessed.

25.2 Ecosystems are structured by who eats whom.

Trophic Levels. Energy passes through ecosystems in a limited number of steps, typically three or four.

25.3 Energy flows through ecosystems.

Primary Productivity. Plants produce biomass by photosynthesis, while animals produce biomass by consuming plants or other animals.

The Energy in Food Chains. As energy passes through an ecosystem, a good deal is lost at each step.

Ecological Pyramids. The biomass of a trophic level is less, the further it is from the primary production of photosynthesizers.

FIGURE 25.1
Mushrooms serve a greater function than haute cuisine.
Mushrooms and other organisms are crucial recyclers in ecosystems, breaking down dead and decaying material and releasing critical elements such as carbon and nitrogen back into nutrient cycles.

The ecosystem is the most complex level of biological organization. In every ecosystem there is a regulated transfer of energy—ultimately derived from the sun—and a controlled cycling of nutrients. The earth is a closed system with respect to chemicals, but an open system in terms of energy. Collectively, the organisms in ecosystems regulate the capture and expenditure of energy and the cycling of chemicals (figure 25.1). As we will see in this chapter, all organisms, including humans, depend on the ability of other organisms—plants, algae, and some bacteria—to recycle the basic components of life.

All of the substances that occur in organisms cycle through ecosystems in **biogeochemical cycles,** cyclical paths involving both biological and chemical processes. On a global scale, only a very small portion of these substances is contained within the bodies of organisms; almost all of it exists in nonliving reservoirs: the atmosphere, water, or rocks. Carbon (in the form of carbon dioxide), nitrogen, and oxygen enter the bodies of organisms primarily from the atmosphere, while phosphorus, potassium, sulfur, magnesium, calcium, sodium, iron, and cobalt, all required for plant and animal growth, come from rocks. All organisms require carbon, hydrogen, oxygen, nitrogen, phosphorus, and sulfur in relatively large quantities; they require other elements in smaller amounts.

The cycling of materials in ecosystems begins when these chemicals are incorporated into the bodies of organisms from nonliving reservoirs such as the atmosphere or the waters of oceans or rivers. Many minerals, for example, first enter water from weathered rock, then pass into organisms when they drink the water. Materials pass from the organisms that first acquire them into the bodies of other organisms that eat them, until ultimately, through decomposition, they complete the cycle and return to the nonliving world.

The Water Cycle

The water cycle (figure 25.2) is the most familiar of all biogeochemical cycles. All life depends directly on the presence of water; the bodies of most organisms consist mainly of this substance. Water is the source of hydrogen ions, whose movements generate ATP in organisms. For that reason alone, it is indispensable to their functioning.

The Path of Free Water

The oceans cover three-fourths of the earth's surface. From the oceans, water evaporates into the atmosphere, a process powered by energy from the sun. Over land approximately 90% of the water that reaches the atmosphere is moisture that evaporates from the surface of plants through a process called transpiration (chapter 35). Most precipitation falls directly into the oceans, but some falls on land, where it passes into surface and subsurface bodies of fresh water. Only about 2% of all the water on earth is captured in any form—frozen, held in the soil, or incorporated into the bodies of organisms. All of the rest is free water, circulating between the atmosphere and the oceans.

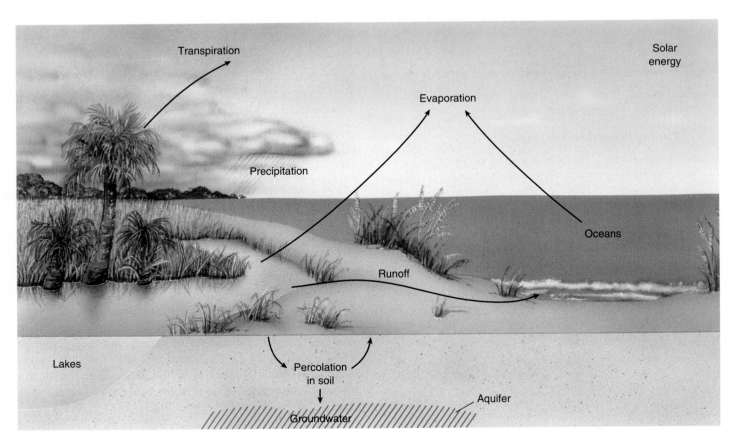

FIGURE 25.2
The water cycle. Water circulates from atmosphere to earth and back again.

The Importance of Water to Organisms

Organisms live or die on the basis of their ability to capture water and incorporate it into their bodies. Plants take up water from the earth in a continuous stream. Crop plants require about 1000 kilograms of water to produce one kilogram of food, and the ratio in natural communities is similar. Animals obtain water directly or from the plants or other animals they eat. The amount of free water available at a particular place often determines the nature and abundance of the living organisms present there.

Groundwater

Much less obvious than surface water, which we see in streams, lakes, and ponds, is groundwater, which occurs in aquifers—permeable, saturated, underground layers of rock, sand, and gravel. In many areas, groundwater is the most important reservoir of water. It amounts to more than 96% of all fresh water in the United States. The upper, unconfined portion of the groundwater constitutes the water table, which flows into streams and is partly accessible to plants; the lower confined layers are generally out of reach, although they can be "mined" by humans. The water table is recharged by water that percolates through the soil from precipitation as well as by water that seeps downward from ponds, lakes, and streams. The deep aquifers are recharged very slowly from the water table.

Groundwater flows much more slowly than surface water, anywhere from a few millimeters to a meter or so per day. In the United States, groundwater provides about 25% of the water used for all purposes and provides about 50% of the population with drinking water. Rural areas tend to depend almost exclusively on wells to access groundwater, and its use is growing at about twice the rate of surface water use. In the Great Plains of the central United States, the extensive use of the Ogallala Aquifer as a source of water for agricultural needs as well as for drinking water is depleting it faster than it can be naturally recharged. This seriously threatens the agricultural production of the area and similar problems are appearing throughout the drier portions of the globe.

Because of the greater rate of groundwater use, and because it flows so slowly, the increasing chemical pollution of groundwater is also a very serious problem. It is estimated that about 2% of the groundwater in the United States is already polluted, and the situation is worsening. Pesticides, herbicides, and fertilizers have become a serious threat to water purity. Another key source of groundwater pollution consists of the roughly 200,000 surface pits, ponds, and lagoons that are actively used for the disposal of chemical wastes in the United States alone. Because of the large volume of water, its slow rate of turnover, and its inaccessibility, removing pollutants from aquifers is virtually impossible.

FIGURE 25.3
Deforestation breaks the water cycle. As time goes by, the consequences of tropical deforestation may become even more severe, as the extensive erosion in this deforested area of Madagascar shows.

Breaking the Water Cycle

In dense forest ecosystems such as tropical rainforests, more than 90% of the moisture in the ecosystem is taken up by plants and then transpired back into the air. Because so many plants in a rainforest are doing this, the vegetation is the primary source of local rainfall. In a very real sense, these plants create their own rain: the moisture that travels up from the plants into the atmosphere falls back to earth as rain.

Where forests are cut down, the organismic water cycle is broken, and moisture is not returned to the atmosphere. Water drains away from the area to the sea instead of rising to the clouds and falling again on the forest. As early as the late 1700s, the great German explorer Alexander von Humbolt reported that stripping the trees from a tropical rainforest in Colombia prevented water from returning to the atmosphere and created a semiarid desert. It is a tragedy of our time that just such a transformation is occurring in many tropical areas, as tropical and temperate rainforests are being clear-cut or burned in the name of "development" (figure 25.3). Much of Madagascar, a California-sized island off the east coast of Africa, has been transformed in this century from lush tropical forest into semiarid desert by deforestation. Because the rain no longer falls, there is no practical way to reforest this land. The water cycle, once broken, cannot be easily reestablished.

Water cycles between oceans and atmosphere. Some 96% of the fresh water in the United States consists of groundwater, which provides 25% of all the water used in this country.

The Carbon Cycle

The **carbon cycle** is based on carbon dioxide, which makes up only about 0.03% of the atmosphere. Worldwide, the synthesis of organic compounds from carbon dioxide and water through photosynthesis (see chapter 10) results in the fixation of about 10% of the roughly 700 billion metric tons of carbon dioxide in the atmosphere each year (figure 25.4). This enormous amount of biological activity takes place as a result of the combined activities of photosynthetic bacteria, protists, and plants. All heterotrophic organisms, including nonphotosynthetic bacteria and protists, fungi, animals, and a few plants that have lost the ability to photosynthesize, obtain their carbon indirectly from the organisms that fix it. When their bodies decompose, microorganisms release carbon dioxide back to the atmosphere. From there, it can be reincorporated into the bodies of other organisms.

Most of the organic compounds formed as a result of carbon dioxide fixation in the bodies of photosynthetic organisms are ultimately broken down and released back into the atmosphere or water. Certain carbon-containing compounds, such as cellulose, are more resistant to breakdown than others, but certain bacteria and fungi, as well as a few kinds of insects, are able to accomplish this feat. Some cellulose, however, accumulates as undecomposed organic matter such as peat. The carbon in this cellulose may eventually be incorporated into fossil fuels such as oil or coal.

In addition to the roughly 700 billion metric tons of carbon dioxide in the atmosphere, approximately 1 trillion metric tons are dissolved in the ocean. More than half of this quantity is in the upper layers, where photosynthesis takes place. The fossil fuels, primarily oil and coal, contain more than 5 trillion additional metric tons of carbon, and between 600 million and 1 trillion metric tons are locked up in living organisms at any one time. In global terms, photosynthesis and respiration (see chapter 9) are approximately balanced, but the balance has been shifted recently because of the consumption of fossil fuels. The combustion of coal, oil, and natural gas has released large stores of carbon into the atmosphere as carbon dioxide. The increase of carbon dioxide in the atmosphere appears to be changing global climates, and may do so even more rapidly in the future, as we will discuss in more detail in chapter 27.

About 10% of the estimated 700 billion metric tons of carbon dioxide in the atmosphere is fixed annually by the process of photosynthesis.

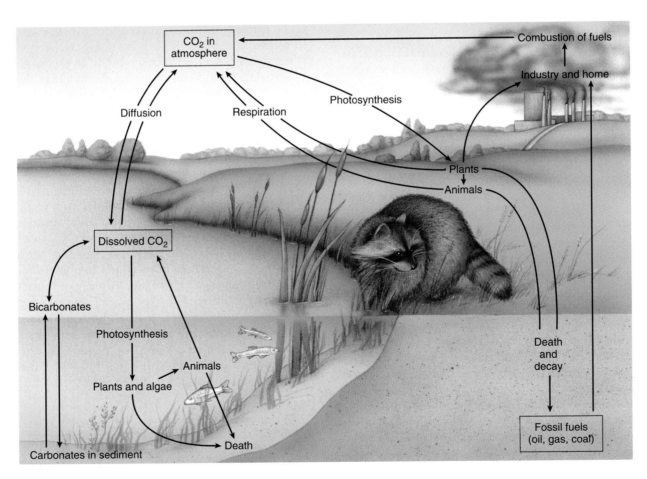

FIGURE 25.4
The carbon cycle. Photosynthesis captures carbon; respiration returns it to the atmosphere.

The Nitrogen Cycle

All living organisms depend on the results of nitrogen fixation to synthesize proteins, nucleic acids, and other necessary nitrogen-containing compounds. Nitrogen gas constitutes 78% of the earth's atmosphere, but the biologically available fixed nitrogen in the soil, oceans, and the bodies of organisms is only about 0.03% of that amount. Nitrogen cycles between organisms and nitrogen reservoirs via the **nitrogen cycle** (figure 25.5).

Relatively few kinds of organisms—all of them bacteria—can convert, or fix, atmospheric nitrogen into forms that can be used for biological processes. The triple bond that links together the two atoms that make up diatomic atmospheric nitrogen (N_2) makes it a very stable molecule. In living systems the cleavage of atmospheric nitrogen is catalyzed by a complex of three proteins—ferredoxin, nitrogen reductase, and nitrogenase. This process uses ATP as a source of energy, electrons derived from photosynthesis or respiration, and a powerful reducing agent. The overall reaction of nitrogen fixation is written:

$$N_2 + 3H_2 \rightarrow 2NH_3$$

Some genera of bacteria have the ability to fix atmospheric nitrogen. Most are free-living, but some form symbiotic relationships with the roots of legumes (plants of the pea family, Fabaceae) and other plants. Only the symbiotic bacteria fix enough nitrogen to be of major significance in nitrogen production. Because of the activities of such organisms in the past, a large reservoir of ammonia and nitrates now exists in most ecosystems. This reservoir is the immediate source of much of the nitrogen used by organisms.

Nitrogen-containing compounds, such as proteins in plant and animal bodies, are decomposed rapidly by certain bacteria and fungi. These bacteria and fungi use the amino acids they obtain through decomposition to synthesize their own proteins and to release excess nitrogen in the form of ammonium ions (NH_4^+), a process known as **ammonification.** The ammonium ions can be converted to soil nitrites and nitrates by certain kinds of organisms and then be absorbed by plants.

A certain proportion of the fixed nitrogen in the soil is steadily lost. Under anaerobic conditions, nitrate is often converted to nitrogen gas (N_2) and nitrous oxide (N_2O), both of which return to the atmosphere. This process, which several genera of bacteria carry out, is called **denitrification.**

Nitrogen becomes available to organisms almost entirely through the metabolic activities of bacteria, some free-living and others which live symbiotically in the roots of legumes and other plants.

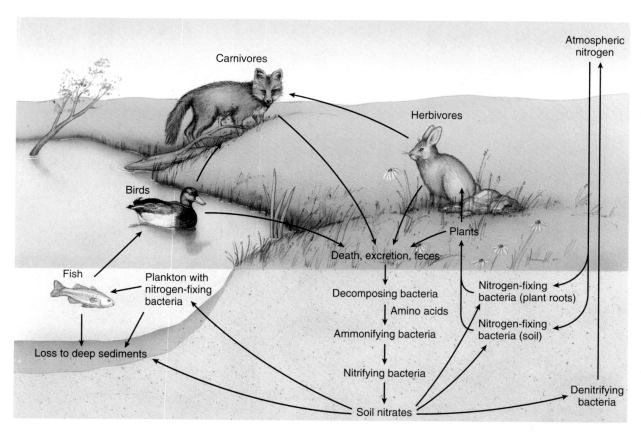

FIGURE 25.5
The nitrogen cycle. Certain bacteria fix atmospheric nitrogen, converting it to a form living organisms can use. Other bacteria decompose nitrogen-containing compounds from plant and animal materials, returning it to the atmosphere.

The Phosphorus Cycle

In all biogeochemical cycles other than those involving water, carbon, oxygen, and nitrogen, the reservoir of the nutrient exists in mineral form, rather than in the atmosphere. The **phosphorus cycle** (figure 25.6) is presented as a representative example of all other mineral cycles because of the critical role phosphorus plays in plant nutrition worldwide.

Of all the required plant nutrients other than nitrogen, phosphorus is the most likely to be scarce enough to limit plant growth. Phosphates, in the form of phosphorus anions, exist in soil only in small amounts. This is because they are relatively insoluble and are present only in certain kinds of rocks. As phosphates weather out of soils, they are transported by rivers and streams to the oceans, where they accumulate in sediments. They are naturally brought back up again only by the uplift of lands, such as occurs along the Pacific coast of North and South America, or by marine animals. Such animals are often consumed by seabirds, which deposit enormous amounts of guano (feces) rich in phosphorus along certain coasts. Guano deposits have tra-ditionally been used for fertilizer. Crushed phosphate-rich rocks, found in certain regions, are also used for fertilizer. The seas are the only inexhaustible source of phosphorus, one of the reasons that deep-seabed mining now looks so commercially attractive.

Every year, millions of tons of phosphate are added to agricultural lands in the belief that it becomes fixed to and enriches the soil. In general, four times as much phosphate as a crop requires is added each year. This is usually in the form of **superphosphate,** which is soluble calcium dihydrogen phosphate, $Ca(H_2PO_4)_2$, derived by treating bones or apatite, the mineral form of calcium phosphate, with sulfuric acid. But the enormous quantities of phosphates that are being added annually to the world's agricultural lands are not leading to proportionate gains in crops. Plants can apparently use only so much of the phosphorus that is added to the soil. Agricultural scientists are actively seeking new ways to approach the problem of phosphate supply.

Phosphates are relatively insoluble and are present in most soils only in small amounts. They often are so scarce that their absence limits plant growth.

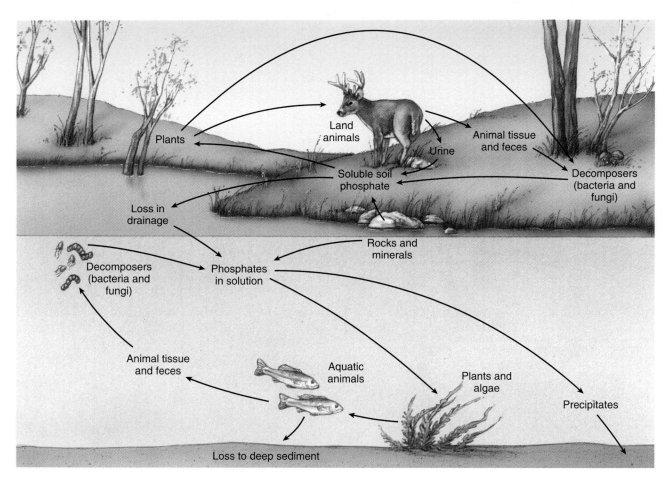

FIGURE 25.6
The phosphorus cycle. Phosphates weather from soils into water, enter plants and animals, and are redeposited in the soil when plants and animals decompose.

Biogeochemical Cycles Illustrated: Recycling in a Forest Ecosystem

An ongoing series of studies conducted at the Hubbard Brook Experimental Forest in New Hampshire has revealed in impressive detail the overall recycling pattern of nutrients in an ecosystem. The way this particular ecosystem functions, and especially the way nutrients cycle within it, has been studied since 1963 by Herbert Bormann of the Yale School of Forestry and Environmental Studies, Gene Likens of The Institute of Ecosystem Studies, and their colleagues. These studies have yielded much of the available information about the cycling of nutrients in forest ecosystems. They have also provided the basis for the development of much of the experimental methodology that is being applied successfully to the study of other ecosystems.

Hubbard Brook is the central stream of a large watershed that drains a region of temperate deciduous forest. To measure the flow of water and nutrients within the Hubbard Brook ecosystem, concrete weirs with V-shaped notches were built across six tributary streams. All of the water that flowed out of the valleys had to pass through the notches, since the weirs were anchored in bedrock. The researchers measured the precipitation that fell in the six valleys, and determined the amounts of nutrients that were present in the water flowing in the six streams. By these methods, they demonstrated that the undisturbed forests in this area were very efficient at retaining nutrients; the small amounts of nutrients that precipitated from the atmosphere with rain and snow were approximately equal to the amounts of nutrients that ran out of the valleys. These quantities were very low in relation to the total amount of nutrients in the system. There was a small net loss of calcium—about 0.3% of the total calcium in the system per year—and small net gains of nitrogen and potassium.

In 1965 and 1966, the investigators felled all the trees and shrubs in one of the six watersheds and then prevented regrowth by spraying the area with herbicides. The effects were dramatic. The amount of water running out of that valley increased by 40%. This indicated that water that previously would have been taken up by vegetation and ultimately evaporated into the atmosphere was now running off. For the four-month period from June to September 1966, the runoff was four times higher than it had been during comparable periods in the preceding years. The amounts of nutrients running out of the system also greatly increased; for example, the loss of calcium was 10 times higher than it had been previously. Phosphorus, on the other hand, did not increase in the stream water; it apparently was locked up in the soil.

The change in the status of nitrogen in the disturbed valley was especially striking (figure 25.7). The undisturbed ecosystem in this valley had been accumulating nitrogen at a rate of about 2 kilograms per hectare per year, but the deforested ecosystem *lost* nitrogen at a rate of about 120 kilo-

(a)

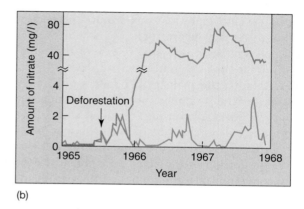

(b)

FIGURE 25.7

The Hubbard Brook experiment. (a) A 38-acre watershed was completely deforested, and the runoff monitored for several years. (b) Deforestation greatly increased the loss of minerals in runoff water from the ecosystem. The red curve represents nitrate in the runoff water from the deforested watershed; the blue curve nitrate in runoff water from an undisturbed neighboring watershed.

grams per hectare per year. The nitrate level of the water rapidly increased to a level exceeding that judged safe for human consumption, and the stream that drained the area generated massive blooms of cyanobacteria and algae. In other words, the fertility of this logged-over valley decreased rapidly, while at the same time the danger of flooding greatly increased. This experiment is particularly instructive in the 1990s, as large areas of tropical rainforest are being destroyed to make way for cropland, a topic that will be discussed further in chapter 27.

> When the trees and shrubs in one of the valleys in the Hubbard Brook watershed were cut down and the area was sprayed with herbicide, water runoff and the loss of nutrients from that valley increased. Nitrogen, which had been accumulating at a rate of about 2 kilograms per hectare per year, was lost at a rate of 120 kilograms per hectare per year.

Trophic Levels

An ecosystem includes autotrophs and heterotrophs. **Autotrophs** are plants, algae, and some bacteria that are able to capture light energy and manufacture their own food. To support themselves, **heterotrophs,** which include animals, fungi, most protists and bacteria, and nongreen plants, must obtain organic molecules that have been synthesized by autotrophs. Autotrophs are also called **primary producers,** and heterotrophs are also called **consumers.**

Once energy enters an ecosystem, usually as the result of photosynthesis, it is slowly released as metabolic processes proceed. The autotrophs that first acquire this energy provide all of the energy heterotrophs use. The organisms that make up an ecosystem delay the release of the energy obtained from the sun back into space.

Green plants, the primary producers of a terrestrial ecosystem, generally capture about 1% of the energy that falls on their leaves, converting it to food energy. In especially productive systems, this percentage may be a little higher. When these plants are consumed by other organisms, only a portion of the plant's accumulated energy is actually converted into the bodies of the organisms that consume them.

Several different levels of consumers exist. The **primary consumers,** or herbivores, feed directly on the green plants. **Secondary consumers,** carnivores and the parasites of animals, feed in turn on the herbivores. **Decomposers** break down the organic matter accumulated in the bodies of other organisms. Another more general term that includes decomposers is **detritivores.** Detritivores live on the refuse of an ecosystem. They include large scavengers, such as crabs, vultures, and jackals, as well as decomposers.

All of these categories occur in any ecosystem. They represent different **trophic levels,** from the Greek word *trophos,* which means "feeder." Organisms from each trophic levels, feeding on one another, make up a series called a **food chain** (figure 25.8). The length and complexity of food chains vary greatly. In real life, it is rather rare for a given kind of organism to feed only on one other type of organism. Usually, each organism feeds on two or more kinds and in turn is eaten by several other kinds of organisms. When diagrammed, the relationship appears as a series of branching lines, rather than a straight line; it is called a **food web** (figure 25.9).

A certain amount of the energy ingested and retained by the organisms at a given trophic level goes toward heat production, and a great deal more of it is used for digestion and work. Usually 40% or less of the energy ingested goes toward growth and reproduction. An invertebrate typically uses about a quarter of this 40% for growth; in other words, about 10% of the food an invertebrate eats is turned into its own body and thus into potential food for its predators. Although the comparable figure varies from approxi-

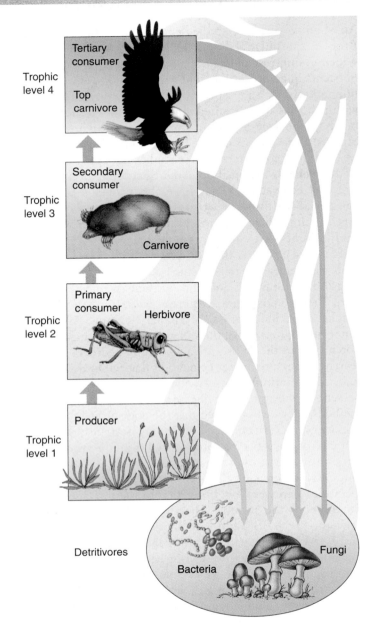

FIGURE 25.8
Trophic levels within a food chain. Plants obtain their energy directly from the sun, placing them at trophic level 1. Animals that eat plants (primary consumers or herbivores) are at trophic level 2. Animals that eat plant-eating animals (carnivores) are at trophic level 3 (secondary consumers) or higher. Detritivores use all trophic levels for food.

mately 5% in carnivores to nearly 20% for herbivores, 10% is a good average value for the amount of organic matter that reaches the next trophic level.

Energy passes through ecosystems, a good deal being lost at each step.

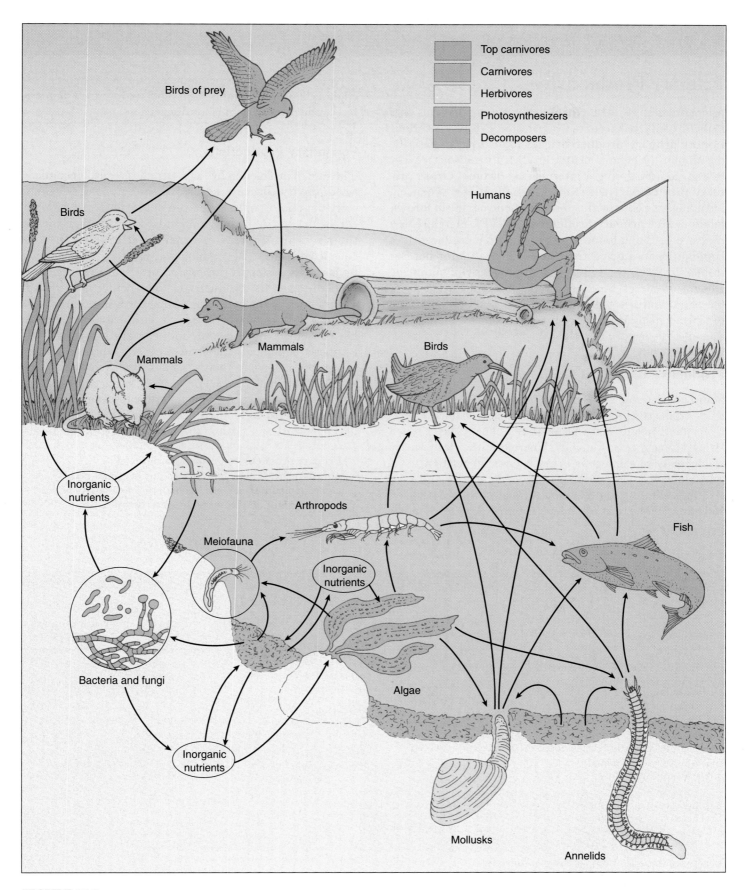

Top carnivores
Carnivores
Herbivores
Photosynthesizers
Decomposers

Birds of prey

Birds

Mammals

Mammals

Inorganic nutrients

Bacteria and fungi

Meiofauna

Arthropods

Inorganic nutrients

Algae

Inorganic nutrients

Mollusks

Humans

Birds

Fish

Annelids

FIGURE 25.9
The food web in a salt marsh shows the complex interrelationships among organisms. The meiofauna are very small animals that live between the grains of sand.

Primary Productivity

Approximately 1% to 5% of the solar energy that falls on a plant is converted to organic material. **Primary production** or **primary productivity** are terms used to describe the amount of organic matter produced from solar energy in a given area during a given period of time. **Gross primary productivity** is the total organic matter produced, including that used by the photosynthetic organism for respiration. **Net primary productivity (NPP),** therefore, is a measure of the amount of organic matter produced in a community in a given time that is available for heterotrophs. It equals the gross primary productivity minus the amount of energy expended by the metabolic activities of the photosynthetic organisms. The net weight of all of the organisms living in an ecosystem, its **biomass,** increases as a result of its net production.

Productive Biological Communities

Some ecosystems, such as wetlands, have a high net primary productivity (table 25.1). A rainforest also has a relatively high net primary productivity, but it has a much larger biomass than wetlands. Therefore, the net primary productivity of rainforests is much lower in relation to its total biomass.

Tropical forests and wetlands normally produce between 1500 and 3000 grams of organic material per square meter per year. Corresponding figures for other communities include 1200 to 1300 grams for temperate forests, 900 grams for savanna, and 90 grams for deserts.

Secondary Productivity

The rate of production by heterotrophs is called **secondary productivity.** Because herbivores and carnivores cannot carry out photosynthesis, they do not manufacture biomolecules or energy-rich compounds. Instead, they obtain them by eating plants or other heterotrophs. In both aquatic and terrestrial communities, secondary productivity by herbivores is approximately an order of magnitude less than the primary productivity upon which it is based. Where does all the energy in plants that is not captured by herbivores go (figure 25.10)? First, much of the biomass is not consumed by herbivores and instead supports the decomposer community (bacteria, fungi and detritivorous animals). Second, some energy is not assimilated by the herbivore's body but is passed on as feces to the decomposers. Third, not all the energy which herbivores assimilate is actually converted to biomass; some of it is lost as heat produced by work and random thermodynamic motion.

Primary productivity occurs as a result of photosynthesis, which is carried out by green plants, algae, and some bacteria. Secondary productivity is the production of new biomass by heterotrophs.

		Table 25.1 Terrestrial Ecosystem Productivity Per Year			
		Net Primary Productivity (NPP)		**Biomass**	
Ecosystem Type	**Area** (10^6 km^2)	**NPP Per Unit Area** (g/m^2)	**World NPP** (10^9 t)	**Biomass** (kg/m^2)	**World Biomass** (10^9 t)
Extreme desert, rock, sand, and ice	24.0	3	0.07	0.02	0.5
Desert and semidesert shrub	18.0	90	1.6	0.7	13
Tropical rainforest	17.0	2200	37.4	45	765
Savanna	15.0	900	13.5	4	60
Cultivated land	14.0	650	9.1	1	14
Boreal forest	12.0	800	9.6	20	240
Temperate grassland	9.0	600	5.4	1.6	14
Woodland and shrubland	8.5	700	6.0	6	50
Tundra and alpine	8.0	140	1.1	0.6	5
Tropical seasonal forest	7.5	1600	12.0	35	260
Temperate deciduous forest	7.0	1200	8.4	30	210
Temperate evergreen forest	5.0	1300	6.5	35	175
Wetlands	2.0	2000	4.0	15	30

Source: After Whittaker, 1975.

The Energy in Food Chains

Food chains generally consist of only three or four steps. So much energy is lost at each step that very little usable energy remains in the system after it has been incorporated into the bodies of organisms at four successive trophic levels.

Community Energy Budgets

Lamont Cole of Cornell University studied the flow of energy in a freshwater ecosystem in Cayuga Lake in upstate New York. He calculated that about 150 of each 1000 calories of potential energy fixed by algae and cyanobacteria are transferred into the bodies of small heterotrophs (figure 25.11). Of these, about 30 calories are incorporated into the bodies of smelt, the principal secondary consumers of the system. If humans eat the smelt, they gain about 6 of the 1000 calories that originally entered the system. If trout eat the smelt and humans eat the trout, humans gain only about 1.2 calories of the 1000.

Relationships of this kind make it clear that eating meat is an inefficient way to harvest primary productivity. Organisms, including people, that subsist on an all-plant diet have more of the energy fixed by photosynthesis available to them than carnivores do. Such considerations will become increasingly important in future efforts to maximize the yield of food for a hungry and increasingly overcrowded world.

Factors Limiting Community Productivity

Communities with higher productivity can in theory support longer food chains. The limit on a community's productivity is determined ultimately by the amount of sunlight it receives, for this determines how much photosynthesis can occur. This is why in the deciduous forests of North America the net primary productivity increases as the growing season lengthens. NPP is higher in warm climates than cold ones not only because of the longer growing seasons, but also because more nitrogen tends to be available in warm climates, where nitrogen-fixing bacteria are more active.

> **Considerable energy is lost at each step of food chains, which limits their length. In general, more productive food chains can support longer food chains.**

FIGURE 25.11
The food web in Cayuga Lake. In Cayuga Lake, one of New York's Finger Lakes, autotrophic plankton (algae and cyanobacteria) fix the energy of the sun, heterotrophic plankton feed on them, and are both consumed by smelt. The smelt are eaten by trout, with about a fivefold loss in fixed energy; for humans, the amount of biomass available in smelt is at least five times greater than that available in trout, although humans prefer to eat trout.

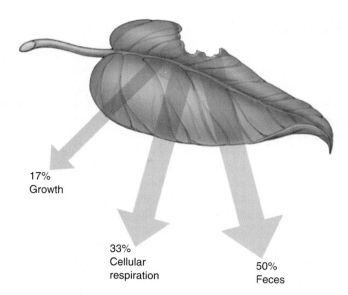

17%
Growth

33%
Cellular
respiration

50%
Feces

FIGURE 25.10
How heterotrophs utilize food energy. A heterotroph assimilates only a fraction of the energy it consumes. For example, if a "bite" comprises 500 Joules of energy (one Joule = 0.239 calories), about 50%, 250 J, is lost in feces, about 33%, 165 J, is used to fuel cellular respiration, and about 17%, 85 J, is converted into insect biomass. Only this 85 J is available to the next trophic level.

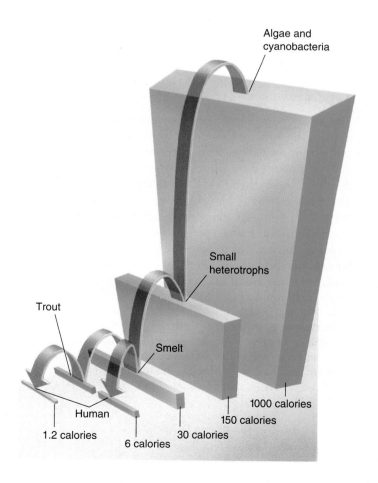

Algae and
cyanobacteria

Small
heterotrophs

Trout

Smelt

Human

1.2 calories

6 calories

30 calories

150 calories

1000 calories

Ecological Pyramids

A plant fixes about 1% of the sun's energy that falls on its green parts. The successive members of a food chain, in turn, process into their own bodies about 10% of the energy available in the organisms on which they feed. For this reason, there are generally far more individuals at the lower trophic levels of any ecosystem than at the higher levels. Similarly, the biomass of the primary producers present in a given ecosystem is greater than the biomass of the primary consumers, with successive trophic levels having a lower and lower biomass and correspondingly less potential energy. Larger animals are characteristically members of the higher trophic levels. To some extent, they *must* be larger to be able to capture enough prey in the lower trophic levels.

These relationships, if shown diagrammatically, appear as pyramids (figure 25.12). We can speak of "pyramids of biomass," "pyramids of energy," "pyramids of number," and so forth, as characteristic of ecosystems.

Inverted Pyramids

Some aquatic ecosystems have inverted biomass pyramids. For example, in a planktonic ecosystem—dominated by small organisms floating in water—the turnover of photosynthetic phytoplankton at the lowest level is very rapid, with zooplankton consuming phytoplankton so quickly that the phytoplankton (the producers at the base of the food chain) can never develop a large population size. Because the phytoplankton reproduce very rapidly, the community can support a population of heterotrophs that is larger in biomass and more numerous than the phytoplankton (see figure 25.12b).

Top Carnivores

The loss of energy that occurs at each trophic level places a limit on how many top-level carnivores a community can support. As we have seen, only about one-thousandth of the energy captured by photosynthesis passes all the way through a three-stage food chain to a tertiary consumer such as a snake or hawk. This explains why there are no predators that subsist on lions or eagles—the biomass of these animals is simply insufficient to support another trophic level.

In the pyramid of numbers, top-level predators tend to be fairly large animals. Thus, the small residual biomass available at the top of the pyramid is concentrated in a relatively small number of individuals.

> Because energy is lost at every step of a food chain, the biomass of primary producers (photosynthesizers) tends to be greater than that of the herbivores that consume them, and herbivore biomass greater than the biomass of the predators that consume them.

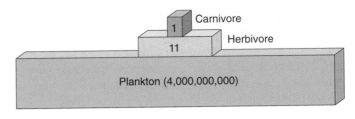

Pyramid of numbers

(a)

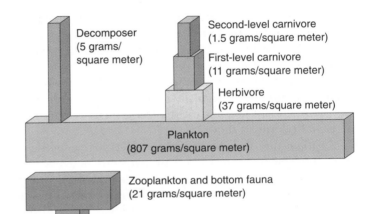

Pyramid of biomass

(b)

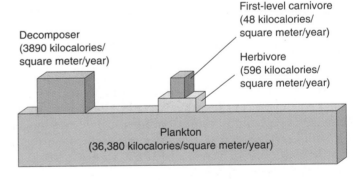

Pyramid of energy

(c)

FIGURE 25.12
Ecological pyramids. Ecological pyramids measure different characteristics of each trophic level. (a) Pyramid of numbers. (b) Pyramids of biomass, both normal (*top*) and inverted (*bottom*). (c) Pyramid of energy.

25.1 **Chemicals cycle within ecosystems.**

- Ecosystems regulate the flow of energy derived from the sun and the cycling of nutrients.

- Fully 98% of the water on earth cycles through the atmosphere. In the United States, 96% of the fresh water is groundwater. The contamination of our limited water supplies is a serious problem.

- About 10% of the roughly 700 billion metric tons of free carbon dioxide in the atmosphere is fixed each year through photosynthesis. An additional trillion metric tons of carbon dioxide is dissolved in the ocean, and five times that amount is locked up in coal, oil, and gas. About as much carbon exists in living organisms at any one time as is present in the atmosphere.

- Carbon, nitrogen, and oxygen have gaseous or liquid reservoirs, as does water. All of the other nutrients, such as phosphorus, are contained in solid mineral reservoirs.

- Several genera of symbiotic and free-living bacteria convert atmospheric nitrogen into ammonia. Other bacteria, in turn, convert the ammonia into nitrites and then to nitrates by other bacteria. Nitrates are incorporated into the bodies of plants and converted back into ammonium ions, which are used in the manufacture of many kinds of molecules in living organisms. The breakdown of these nitrogenous molecules either converts them to recyclable forms or releases atmospheric nitrogen.

- Phosphorus is a key component of many biological molecules; it weathers out of soils and is transported to the world's oceans. Phosphorus is relatively scarce in rocks; this scarcity often limits or excludes the growth of certain kinds of plants.

- The Hubbard Brook experiments vividly illustrate the role a forested ecosystem plays in regulating nutrient cycles. When a particular tributary valley was deforested, 4 times as much water and 10 times as much calcium ran off as had previously. Nitrogen, which had been accumulating in the area at a rate of about 2 kilograms per hectare per year, was now lost at a rate of about 120 kilograms per hectare per year.

25.2 **Ecosystems are structured by who eats whom.**

- Plants convert about 1% to 5% of the light energy that falls on their leaves to food energy. These producers constitute one trophic level.

- The herbivores that eat the plants and the carnivores that eat the herbivores constitute two more trophic levels.

- At each level, only about 10% of the energy available in the food is fixed in the body of the consumer. For this reason, food chains are always relatively short.

25.3 **Energy flows through ecosystems.**

- The primary productivity of a community is a measure of the biomass photosynthesis produces within it.

- As energy passes through the trophic levels of an ecosystem, much is lost at each step.

- Ecological pyramids reflect this energy loss.

1. **Ecosystems** All of the organisms that live in a place, together with the nonliving characteristics of the place, interact dynamically as an ecosystem (ecological system).

2. **Biogeochemical cycles** Elements critical to organisms (oxygen, water, carbon, nitrogen, and phosphorus) cycle between organisms and reservoirs.

3. **Water cycle** Water cycles continuously from the atmosphere to the earth, to the oceans, and back to the atmosphere again.

4. **Carbon cycle** Carbon cycles from CO_2 in the atmosphere via photosynthesis to the organic molecules within organisms. About 10% of atmospheric carbon dioxide is fixed each year through photosynthesis.

5. **Nitrogen cycle** The primary nitrogen reservoir is the atmosphere. Atmospheric nitrogen is "fixed" as ammonia and then converted by organisms into other forms.

6. **Phosphorus cycle** Phosphorus is an important component of nucleic acids and other biological molecules. Phosphorus enters ecosystems through the weathering of rocks.

7. **Energy flow** Energy flows into ecosystems from the sun and passes from one organism to another, most of it lost at each step.

8. **Food webs** A branching food chain, a series of organisms that feed on each other, is a food web.

Review Questions

1. What are biogeochemical cycles? What are the primary reservoirs for the chemicals in these cycles? Are more of the life-sustaining chemicals found in these reservoirs or in the earth's living organisms?

2. Where is over 90% of the fresh water in the United States located? Why is the pollution of this water such a serious problem?

3. What are the earth's repositories for nitrogen? What is the process of nitrogen fixation? What types of organisms fix nitrogen?

4. What is denitrification? Which organisms carry it out? What would happen if all the organisms that can carry out this process died?

5. How is the phosphorus cycle different from the water, carbon, nitrogen, and oxygen cycles? What are the natural sources for phosphorus?

6. What effect does deforestation have on the water cycle and flood control? What is its effect on nutrient cycles and overall fertility of the land?

7. What is the difference between primary productivity, gross primary productivity, and net primary productivity?

8. What type of biological community is most productive? Rank intertidal zones, deserts, tropical rainforests, and temperate forests from highest to lowest in terms of primary productivity.

9. How efficient is the energy transfer from one trophic level to the next? Which type of diet, carnivorous or herbivorous, provides more food value to any given living organism? Why? Is it more efficient to feed the starving peoples of the world corn or hamburgers?

Thought Questions

1. The net carbon dioxide in the atmosphere is increasing. Where is this carbon dioxide coming from? How could we stop this increase?

2. Cultivated land has the same mean net primary productivity (NPP) as temperate grassland, about 600 grams per square meter.

Cultivated land is typically far less diverse than grassland, however. Can you think of a reason why the two NPPs are so similar?

3. Why are pyramids of energy never inverted, even in aquatic systems?

Internet Links

Virtual Biomes
http://www.mobot.org/MBGnet/sets/index.htm
A graphics-rich introduction to the six major biomes presented by the Missouri Botanical Garden.

Photo-Journey Through a Rainforest
http://www.ecofuture.org/ecofuture/pk/pkar9512.html
From the PLANET KEEPERS site, take a real journey through a Costa Rican rainforest.

Learning About Coral Reefs
http://www.reef.edu.au
The REEF EDUCATION NETWORK provides a basic guide to reef communities, as well as descriptions and links to the people and laboratories involved in reef research.

The Weather Visualizer
http://www.atmos.uiuc.edu/
From the University of Illinois, a current weather map for any region of the United States — you choose satellite, radar, etc.

For Further Reading

Baskin, Y.: "Losing a Lake," *Discover*, March 1994, pages 72–81. The saga of Lake Victoria, the largest on the African continent, is discussed. Introduced species of opportunistic fish as well as centuries of pollution are making a huge negative impact.

Bormann, F. H. and G. E. Likens: *Pattern and Process in a Forested Ecosystem*, Springer-Verlag, New York, 1994. A fascinating and well-written description of the Hubbard Brook experiments.

Colvinaux, P. A.: "The Past and Future Amazon," *Scientific American*, May 1989, pages 102–8. A very interesting article, based on current research, about historical changes in the Amazon rainforest ecosystem.

Houghton, R. A. and G. M. Woodwell: "Global Climatic Change," *Scientific American*, April 1989, pages 36–44. A good summary of research on global warming.

Kusler, J. A., W. J. Mitsch, and J. S. Larson: "Wetlands," *Scientific American*, vol. 210, January 1994, pages 64–70. The complexity of and vital role played by wetlands are discussed, along with the environmental policy that may or may not protect them.

Paul, W. M., and others: "The Global Carbon Cycle," *American Scientist*, vol. 78, pages 310–26. 1990. The dynamic responses of natural systems to carbon dioxide may determine the future of the earth's climate.

Pimm, S., and others: "Food Web Patterns and Their Consequences," *Nature*, vol. 350, April 1991, pages 669–74. A review of current ecological knowledge of food webs.

Sanderson, S. L. and R. Wassersug: "Suspension-Feeding Vertebrates," *Scientific American*, March 1990, pages 96–101. Animals that filter their food out of the water can reap the abundance of plankton and grow in huge numbers or to enormous size.

Sisson, R.: "Tide Pools: Windows Between Land and Sea," *National Geographic*, vol. 169 (2), 1986, pages 252–59. A beautifully illustrated tour of a California tide pool, alive with organisms.

Spencer, C. N., B. R. McClelland, and J. A. Stanford: "Shrimp Stocking, Salmon Collapse, and Eagle Displacement," *BioScience*, vol. 41, 1991, pages 14–21. Cascading interactions in the food web of a large aquatic ecosystem.

26

The Biosphere

FIGURE 26.1
Life in the biosphere. In this satellite image, orange zones are largely arid. Almost every environment on earth can be described in terms of temperature and moisture. These physical parameters have great bearing on the forms of life that are able to inhabit a particular region.

Concept Outline

26.1 Organisms must cope with a varied environment.

The Environmental Challenge. Habitats vary in ways important to survival.

Adaptations to Environmental Change. Organisms cope with environmental variation with physiological, morphological, and behavioral adaptations.

26.2 Climate shapes the character of ecosystems.

The Sun and Atmospheric Circulation. The sun powers major movements in atmospheric circulation.

Atmospheric Circulation, Precipitation, and Climate. Latitude and elevation have important effects on climate, although other factors affect regional climate.

26.3 Biomes are widespread terrestrial ecosystems.

The Major Biomes. Characteristic communities called biomes occur in different climatic regions. Variations in temperature and precipitation are good predictors of what biomes will occur where. Major biomes include tropical rainforest, savanna, desert, grassland, temperate deciduous forest, temperate evergreen forest, taiga, and tundra.

Other Biomes. Minor biomes include polar ice, mountain zone, chaparral, warm moist evergreen forest, tropical monsoon forest, and semidesert.

26.4 Aquatic ecosystems cover much of the earth.

Patterns of Circulation in the Oceans. The world's oceans circulate in huge circles deflected by the continents.

Life in the Oceans. Most of the major groups of organisms originated and are still represented in the sea.

Marine Ecosystems. The communities of the ocean are delineated primarily by depth.

Freshwater Habitats. Like miniature oceans, ponds and lakes support different communities at different depths.

Productivity of Freshwater Ecosystems. Freshwater ecosystems are often highly productive.

The biosphere includes all living communities on earth, from the profusion of life in the tropical rainforests to the photosynthetic phytoplankton in the world's oceans. In a very general sense, the distribution of life on earth reflects variations in the world's environments, principally in temperature and the availability of water. Figure 26.1 is a satellite image of North and South America, collected over eight years, the colors keyed to the relative abundance of chlorophyll, a good indicator of rich biological communities. Phytoplankton and algae produce the dark red zones in the oceans and along the seacoasts. Green and dark green areas on land are dense forests, while orange areas like the deserts of western South America are largely barren of life.

517

The Environmental Challenge

How Environments Vary

The nature of the physical environment in large measure determines what organisms live in a place. Key elements include:

Temperature. Most organisms are adapted to live within a relatively narrow range of temperatures and will not thrive if temperatures are colder or warmer. The growing season of plants, for example, is importantly influenced by temperature.

Water. Plants and all other organisms require water. On land, water is often scarce, so patterns of rainfall have a major influence on life.

Sunlight. Almost all ecosystems rely on energy captured by photosynthesis; the availability of sunlight influences the amount of life an ecosystem can support, particularly below the surface in marine communities.

Soil. The physical consistency, pH, and mineral composition of soil often severely limit plant growth, particularly the availability of nitrogen and phosphorus.

FIGURE 26.2

Meeting the challenge of obtaining moisture in a desert. On the dry sand dunes of the Namib Desert in southwestern Africa, the beetle *Onymacris unguicularis* collects moisture from the fog by holding its abdomen up at the crest of a dune to gather condensed water.

Range and Grain of Environmental Variation

Variations in temperature, rainfall, sunlight, soil composition, and other factors have a major influence on what organisms occur in different places. A key element is the spatial size of the variation relative to the size and mobility of the organism, what ecologists refer to as **environmental grain.** In a "coarse-grained" environment, the patches are large relative to the size and activity of the organism, so individual organisms can select among patches. The plants in a field represent a coarse-grained environment to a bee, who pollinates clover while ignoring the many other kinds of plants present. In a "fine-grained" environment, the varying patches are small relative to the way the organism uses them, so small that the organism may ignore them. A cow may ignore the difference between clover and grass plants, eating them indiscriminately.

Temporal variation may also be coarse-grained or fine-grained. Daily variation in temperature or moisture is usually fine-grained for long-lived animals and plants, but it can be coarse-grained for short-lived insects whose adult lifespans may be compressed into a few hours. Seasonal variation in climate, on the other hand, is almost always coarse-grained.

Active and Passive Approaches to Coping with Environmental Variation

An individual encountering coarse-grained environmental variation may choose to maintain a "steady-state" internal environment in the face of the variation, an approach known as maintaining **homeostasis.** Many animals and plants actively employ physiological, morphological, or behavioral mechanisms to maintain homeostasis. The beetle in figure 26.2 is using a behavioral mechanism to cope with drastic changes in water availability. Other animals and plants simply conform to the environment in which they find themselves, their bodies adopting the temperature, salinity, and other aspects of their surroundings.

Resource Allocation

In chapter 23, we noted that no one organism can do everything optimally, for there are simply not enough available resources. Faced with limited resources, organisms are forced to make compromises, allocating the available energy among the key tasks of maintaining the body, growing, and reproducing. After the body's basics needs are met, homeotherms that maintain their own constant body temperature expend some 80% of assimilated energy on generating heat, an expenditure that a conforming organism need not make.

Much of the variation among biological communities reflects variation in the physical environment.

Adaptations to Environmental Change

Physiology

Many organisms are able to adapt to coarse-grained variation by making physiological adjustments. Thus, your body constricts the blood vessels on the surface of your face on a cold day, reducing heat loss and giving your face a "flush." The frogs in figure 26.3, exposed for several days to relatively low temperatures (5 °C) shift their temperature response downward along the temperature scale. The rate of their metabolism at a given temperature, as measured by oxygen consumption, depends upon the temperature to which the frog is acclimated.

Some insects avoid freezing by adding glycerol "antifreeze" to their blood; others tolerate freezing by converting much of their glycogen reserves into alcohols that protect their cell membranes from freeze damage.

Morphology

Animals that maintain a constant internal temperature (endotherms) in a cold environment have adaptations that tend to minimize energy expenditure. Thus, mammals from colder climates often have shorter ears and limbs (Allen's rule) and larger bodies (Bergmann's rule) to limit heat loss. Both mechanisms reduce the surface area across which the animals lose heat.

Some mammals escape some of the costs of maintaining a constant body temperature during winter by hibernating during the coldest season, behaving, in effect, like conformers. Many other mammals utilize their fur as insulation to retain body heat during the winter. In general, the thicker the fur, the greater the insulation (figure 26.4). Thus, a wolf's fur is some three times as thick in winter as summer and insulates more than twice as well.

Behavior

Many animals deal with coarse-grained variation in the environment by simply avoiding it. The tropical lizards in Figure 26.5 manage to maintain a fairly uniform body temperature in an open habitat by basking in patches of sun. The same lizards living in shaded forests with no opportunity for basking are temperature conformers, their bodies adopting the temperature of their surroundings.

Behavioral adaptations can be extreme. The spadefoot toad *Scaphiophus*, which lives in the deserts of North America, can burrow nearly a meter below the surface and remain there for as long as nine months of each year, its metabolic rate greatly reduced, living on fat reserves. When moist cool conditions return, the toads emerge and breed. The young toads mature rapidly and burrow back underground.

Organisms use a variety of physiological, morphological, and behavioral mechanisms to avoid conforming to coarse-grained variation in the environment.

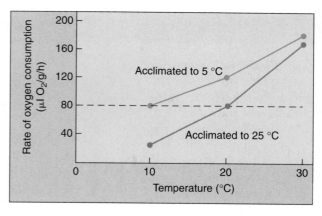

FIGURE 26.3
Physiological adaptation. Frogs (*Rana pipiens*) adjust their metabolism to prevailing temperatures. Cold-acclimated frogs consume as much oxygen at 10 °C as frogs acclimated to warmer temperatures consume at 20 °C.

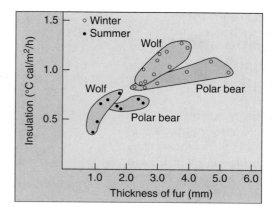

FIGURE 26.4
Morphological adaptation. Fur thickness in North American mammals has a major impact on the degree of insulation the fur provides.

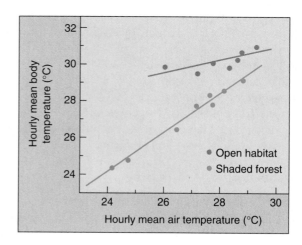

FIGURE 26.5
Behavioral adaptation. The Puerto Rican lizard *Anolis cristatellus* maintains a relatively constant temperature by seeking out and basking in patches of sunlight; in shaded forests, this behavior is not possible and body temperature conforms to the surroundings.

The distribution of biomes (see discussion later in this chapter) results from the interaction of the features of the earth itself, such as different soil types or the occurrence of mountains and valleys, with two key physical factors: (1) the amount of solar heat that reaches different parts of the earth and seasonal variations in that heat; and (2) global atmospheric circulation and the resulting patterns of oceanic circulation. Together these factors dictate local climate, and so determine the amounts and distribution of precipitation.

The Sun and Atmospheric Circulation

The earth receives an enormous quantity of heat from the sun in the form of shortwave radiation, and it radiates an equal amount of heat back to space in the form of longwave radiation. About 10^{24} calories arrive at the upper surface of the earth's atmosphere each year, or about 1.94 calories per square centimeter per minute. About half of this energy reaches the earth's surface. The wavelengths that reach the earth's surface are not identical to those that reach the outer atmosphere. Most of the ultraviolet radiation is absorbed by the oxygen (O_2) and ozone (O_3) in the atmosphere. As we will see in chapter 27, the depletion of the ozone layer, apparently as a result of human activities, poses serious ecological problems.

Why the Tropics Are Warmer

The world contains a great diversity of biomes because its climate varies so much from place to place. On a given day, Miami, Florida, and Bangor, Maine, often have very different weather. There is no mystery about this. Because the earth is a sphere, some parts of it receive more energy from the sun than others. This variation is responsible for many of the major climatic differences that occur over the earth's surface, and, indirectly, for much of the diversity of biomes.

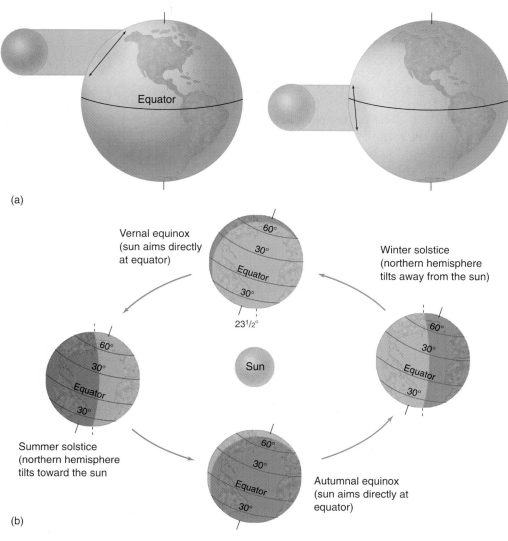

(a)

(b)

FIGURE 26.6
Relationships between the earth and the sun are critical in determining the nature and distribution of life on earth. (a) A beam of solar energy striking the earth in the middle latitudes spreads over a wider area of the earth's surface than a similar beam striking the earth near the equator. (b) The rotation of the earth around the sun has a profound effect on climate. In the northern and southern hemispheres, temperatures change in an annual cycle because the earth tilts slightly on its axis in relation to the path around the sun.

The tropics are warmer than temperate regions because the sun's rays arrive almost perpendicular to regions near the equator. Near the poles the angle of incidence of the sun's rays spreads them out over a much greater area, providing less energy per unit area (figure 26.6a).

The earth's annual orbit around the sun and its daily rotation on its own axis are both important in determining world climate (figure 26.6b). Because of the annual cycle, and the inclination of the earth's axis at approximately 23.5° from its plane of revolution around the sun, there is a progression of seasons in all parts of the earth away from the equator. One pole or the other is tilted closer to the sun at all times except during the spring and autumn equinoxes.

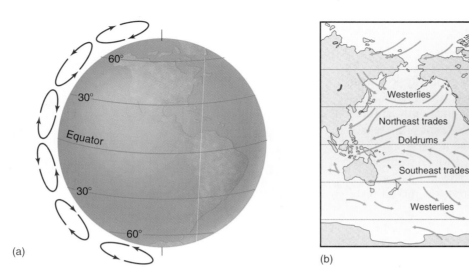

(a)

(b)

FIGURE 26.7
General patterns of atmospheric circulation. (a) The pattern of air movement toward and away from the earth's surface. (b) The major wind currents across the face of the earth.

Major Atmospheric Circulation Patterns

The moisture-holding capacity of air increases when it warms and decreases when it cools. High temperatures near the equator encourage evaporation and create warm, moist air. As this air rises and flows toward the poles, it cools and loses most of its moisture (figure 26.7). Consequently, the greatest amounts of precipitation on earth fall near the equator. This equatorial region of rising air is one of low pressure, called the **doldrums,** which draws air from both north and south of the equator. When the air masses that have risen reach about 30° north and south latitude, the dry air, now cooler, sinks and becomes reheated. As the air reheats, its evaporative capacity increases, creating a zone of decreased precipitation. The air, still warmer than in the polar regions, continues to flow toward the poles. It rises again at about 60° north and south latitude, producing another zone of high precipitation. At this latitude there is another low-pressure area, the polar front. Some of this rising air flows back to the equator and some continues north and south, descending near the poles and producing another zone of low precipitation before it returns to the equator.

Air Currents Generated by the Earth's Rotation

Related to these bands of north-south circulation are three major air currents generated mainly by the interaction of the earth's rotation with patterns of worldwide heat gain. Between about 30° north latitude and 30° south latitude, the trade winds blow, from the east-southeast in the southern hemisphere and from the east-northeast in the northern hemisphere. The trade winds blow all year long and are the steadiest winds found anywhere on earth. They are stronger in winter and weaker in summer. Between 30° and 60° north and south latitude, strong prevailing westerlies

FIGURE 26.8
Winds may dominate the climate. In India during the winter, the trade winds blow off the land to the sea. In the summer, however, moisture-laden winds blow from the ocean to the land and stir up great dust storms and then bring heavy rains called monsoons.

blow from west to east and dominate climatic patterns in these latitudes (figure 26.8), particularly along the western edges of the continents. Weaker winds, blowing from east to west, occur farther north and south in their respective hemispheres.

Warm air rises near the equator, descends and produces arid zones at about 30° north and south latitude, flows toward the poles, then rises again at about 60° north and south latitude, and moves back toward the equator. Part of this air, however, moves toward the poles, where it produces zones of low precipitation.

Atmospheric Circulation, Precipitation, and Climate

As we have discussed, precipitation is generally low near 30° north and south latitude, where air is falling and warming, and relatively high near 60° north and south latitude, where it is rising and cooling. Partly as a result of these factors, all the great deserts of the world lie near 30° north or south latitude. Other major deserts are formed in the interiors of large continents. These areas have limited precipitation because of their distance from the sea, the ultimate source of most moisture.

Rain Shadows

Other deserts occur because mountain ranges intercept moisture-laden winds from the sea. When this occurs, the air rises and the moisture-holding capacity of the air decreases, resulting in increased precipitation on the windward side of the mountains—the side from which the wind is blowing. As the air descends the other side of the mountains, the leeward side, it is warmed, and its moisture-holding capacity increases, tending to block precipitation. In California, for example, the eastern sides of the Sierra Nevada Mountains are much drier than the western sides, and the vegetation is often very different. This phenomenon is called the **rain shadow effect** (figure 26.9).

Regional Climates

Four relatively small areas, each located on a different continent, share a climate that resembles that of the Mediterranean region. So-called Mediterranean climates are found in portions of Baja, California, and Oregon; in central Chile; in southwestern Australia; and in the Cape region of South Africa (figure 26.10). In all of these areas, the prevailing westerlies blow during the summer from a cool ocean onto warm land. As a result, the air's moisture-holding capacity increases, the air absorbing moisture and creating hot rainless summers. Such climates are unusual on a world scale. In the five regions where they occur, many unique kinds of plants and animals, often local in distribution, have evolved. Because of the prevailing westerlies, the great deserts of the world (other than those in the interiors of continents) and the areas of Mediterranean climate lie on the western sides of the continents.

Another kind of regional climate occurs in southern Asia. The monsoon climatic conditions characteristic of India and southern Asia occur during the summer months. During the winter, the trade winds blow from the east-northeast off the cool land onto the warm sea. From June to October, though, when the land is heated, the direction of the air flow reverses, and the winds veer around to blow

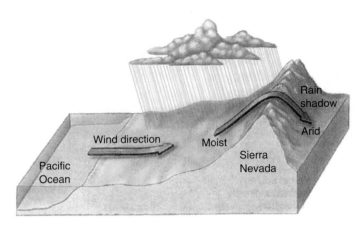

FIGURE 26.9
The rain shadow effect. Moisture-laden winds from the Pacific Ocean rise and are cooled when they encounter the Sierra Nevada Mountains. As their moisture-holding capacity decreases, precipitation occurs, making the middle elevation of the range one of the snowiest regions on earth; it supports tall forests, including those that include the famous giant sequoias (*Sequoiadendron giganteum*). As the air descends on the east side of the range, its moisture-holding capacity increases again, and the air picks up rather than releases moisture from its surroundings. As a result, desert conditions prevail on the east side of the mountains.

FIGURE 26.10
A South African hillside. Very similar scenes occur in California and Italy, which have very much the same climate.

onto the Indian subcontinent and adjacent areas from the southwest bringing rain. The duration and strength of the monsoon winds spell the difference between food sufficiency and starvation for hundreds of millions of people in this region each year.

Latitude

Temperatures are higher in tropical ecosystems for a simple reason: more sunlight per unit area falls on tropical latitudes. Solar radiation is most intense when the sun is directly overhead, and this only occurs in the tropics, where sunlight strikes the equator perpendicularly. As figure 26.11 shows, the highest mean global temperatures occur near the equator (that is, 0 latitude). Because there are no seasons in the tropics, there is little variation in mean monthly temperature in tropical ecosystems. As you move from the equator into temperate latitudes, sunlight strikes the earth at a more oblique angle, so that less falls on a given area. As a result, mean temperatures are lower. At temperate latitudes, temperature variation increases because of the increasingly marked seasons.

Seasonal changes in wind circulation produce corresponding changes in ocean currents, sometimes causing nutrient-rich cold water to well up from ocean depths. This produces "blooms" among the plankton and other organisms living near the surface. Similar turnover occurs seasonally in freshwater lakes and ponds, bringing nutrients from the bottom to the surface in the fall and again in the spring.

Elevation

Temperature also varies with elevation, with higher altitudes becoming progressively colder. At any given latitude, air temperature falls about 6 °C for every 1000-meter increase in elevation. The ecological consequences of temperature varying with elevation are the same as temperature varying with latitude (figure 26.12). Thus, in North America a 1000-meter increase in elevation results in a temperature drop equal to that of an 880-kilometer increase in latitude. This is one reason "timberline" (the elevation above which trees do not grow) occurs at progressively lower elevations as one moves farther from the equator.

Microclimate

Climate also varies on a very fine scale within ecosystems. Within the litter on a forest floor, there is considerable variation in shading, local temperatures, and rates of evaporation from the soil. Called **microclimate**, these very localized climatic conditions can be very different from those of the overhead atmosphere. Gardeners spread straw over newly seeded lawns to create such a moisture-retaining microclimate.

The great deserts and associated arid areas of the world mostly lie along the western sides of continents at about 30° north and south latitude. Mountain ranges tend to intercept rain, creating deserts in their shadow. In general, temperatures are warmer in the tropics and at lower elevations.

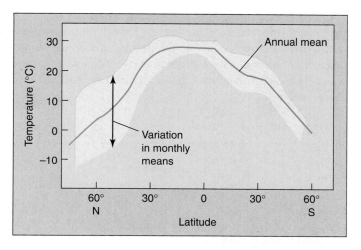

FIGURE 26.11
Temperature varies with latitude. The blue line represents the annual mean temperature at latitudes from the North Pole to Antarctica.

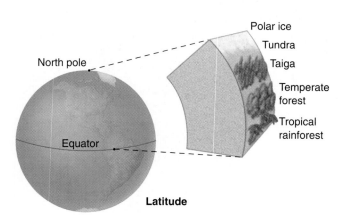

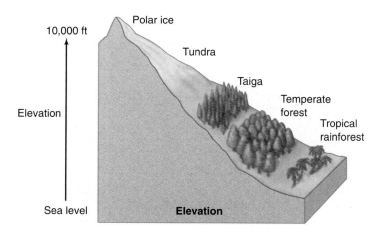

FIGURE 26.12
Elevation affects the distribution of biomes much as latitude does. Biomes that normally occur far north and far south of the equator at sea level also occur in the tropics at high mountain elevations. Thus, on a tall mountain in southern Mexico or Guatemala, one might see a sequence of biomes like the one illustrated here.

The Major Biomes

Biomes are major communities of organisms that have a characteristic appearance and that are distributed over a wide land area defined largely by regional variations in climate. As you might imagine from such a broad definition, there are many ways to classify biomes, and different ecologists may assign the same community to different biomes. There is little disagreement, however, about the reality of biomes as major biological communities—only about how to best describe them.

Distribution of the Major Biomes

Eight major biome categories are presented in this text: tropical rainforest, savanna, desert, temperate grassland, temperate deciduous forest, temperate evergreen forest, taiga, and tundra. These biomes occur worldwide, occupying large regions that can be defined by rainfall and temperature.

Six additional biomes are presented, although some ecologists regard them as subsets of the eight major ones: polar ice, mountain zone, chaparral, warm moist evergreen forest, tropical monsoon forest, and semidesert. They vary remarkably from one another because they have evolved in regions with very different climates.

Distributions of the 14 biomes are mapped in figure 26.13. Although each is by convention named for the dominant vegetation (deciduous forest, evergreen forest, grassland, and so on) each biome is also characterized by particular animals, fungi, and microorganisms adapted to live as members of that community. Wolves, caribou or reindeer, polar bears, hares, lynx, snowy owls, deer flies, and mosquitoes inhabit the tundra all over the world and are as much a defining characteristic of the tundra biomes as the low, shrubby, matlike vegetation.

Biomes and Climate

Many different environmental factors play a role in determining which biomes are found where. Two key parameters are

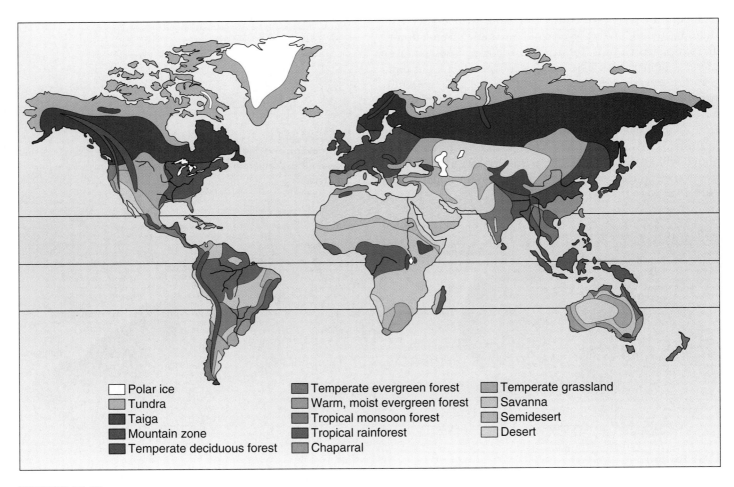

FIGURE 26.13
The distribution of biomes. Each biome is similar in structure and appearance wherever it occurs on earth.

available moisture and temperature. Figure 26.14 presents data on ecosystem productivity as a function of annual precipitation and of annual mean temperature: ecosystem productivity is strongly influenced by both. This is not to say that other factors such as soil structure and its mineral composition (discussed in detail in chapter 36), or seasonal versus constant climate, are not also important. Different places with the same annual precipitation and temperature sometimes support different biomes, so other factors must also be important. Nevertheless, these two variables do a fine job of predicting what biomes will occur in most places, as figure 26.15 illustrates.

If there were no mountains and no climatic effects caused by the irregular outlines of continents and by different sea temperatures, each biome would form an even belt around the globe, defined largely by latitude. In truth, these other factors also greatly affect the distribution of biomes. Distance from the ocean has a major impact on rainfall, and elevation affects temperature—the summits of the Rocky Mountains are covered with a vegetation type that resembles the tundra which normally occurs at a much higher latitude.

Major biological communities called biomes can be distinguished in different climatic regions. These communities, which occur in regions of similar climate, are much the same wherever they are found. Variation in annual mean temperature and precipitation are good predictors of what biome will occur where.

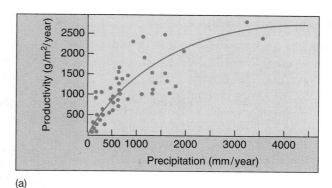

(a)

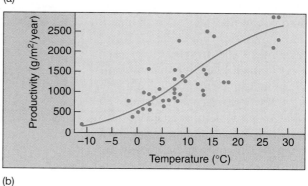

(b)

FIGURE 26.14
The effects of precipitation and temperature on primary productivity. The net primary productivity of ecosystems at 52 locations around the globe depends significantly upon (a) mean annual precipitation and (b) mean annual temperature.

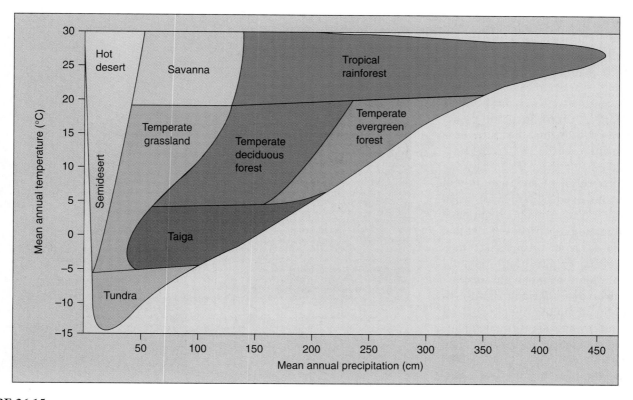

FIGURE 26.15
Temperature and precipitation are excellent predictors of biome distribution. At mean annual precipitations between 50 and 150 cm, other factors such as seasonal drought, fire, and grazing also have a major influence on biome distribution.

Tropical Rainforests

Rainforests, which receive 140 to 450 centimeters of rain a year, are the richest ecosystems on earth (figure 26.16). They contain at least half of the earth's species of terrestrial plants and animals—more than 2 million species! In a single square mile of tropical forest in Rondonia, Brazil, there are 1200 species of butterflies—twice the total number found in the United States and Canada combined. The communities that make up tropical rainforests are diverse in that each kind of animal, plant, or microorganism is often represented in a given area by very few individuals. There are extensive tropical rainforests in South America, Africa, and Southeast Asia. But the world's rainforests are being destroyed, and countless species, many of them never seen by humans, are disappearing with them. A quarter of the world's species will disappear with the rainforests during the lifetime of many of us.

FIGURE 26.16
Tropical rainforest.

Savannas

In the dry climates that border the tropics are the world's great grasslands, called **savannas** (figure 26.17). Savanna landscapes are open, often with widely spaced trees, and rainfall (75–125 centimeters annually) is seasonal. Many of the animals and plants are active only during the rainy season. The huge herds of grazing animals that inhabit the African savanna are familiar to all of us. Such animal communities lived in North America during the Pleistocene epoch but have persisted mainly in Africa. On a global scale, the savanna biome is transitional between tropical rainforest and desert. As these savannas are increasingly converted to agricultural use to feed rapidly expanding human populations in subtropical areas, their inhabitants are struggling to survive. The elephant and rhino are now endangered species; lion, giraffe, and cheetah will soon follow.

FIGURE 26.17
Savanna.

FIGURE 26.18
Desert.

Deserts

In the interior of continents are the world's great deserts, especially in Africa (the Sahara), Asia (the Gobi) and Australia (the Great Sandy Desert). **Deserts** are dry places where less than 25 centimeters of rain falls in a year—an amount so low that vegetation is sparse and survival depends on water conservation (figure 26.18). Plants and animals may restrict their activity to favorable times of the year, when water is present. To avoid high temperatures, most desert vertebrates live in deep, cool, and sometimes even somewhat moist burrows. Those that are active over a greater portion of the year emerge only at night, when temperatures are relatively cool. Some, such as camels, can drink large quantities of water when it is available and then survive long, dry periods. Many animals simply migrate to or through the desert, where they exploit food that may be abundant seasonally.

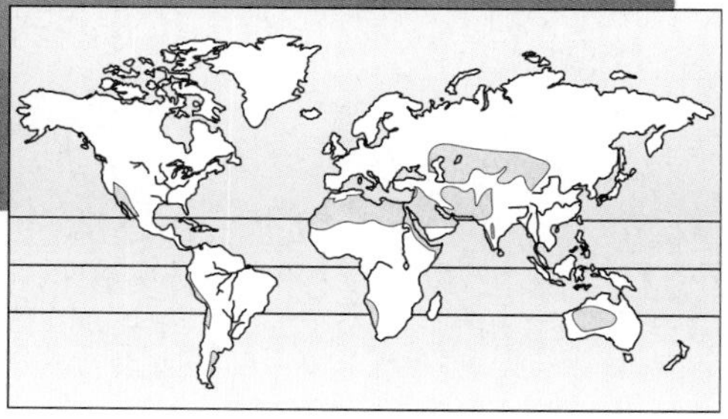

Temperate Grasslands

Halfway between the equator and the poles are temperate regions where rich **grasslands** grow (figure 26.19). These grasslands once covered much of the interior of North America, and they were widespread in Eurasia and South America as well. Such grasslands are often highly productive when converted to agricultural use. Many of the rich agricultural lands in the United States and southern Canada were originally occupied by **prairies,** another name for temperate grasslands. The roots of perennial grasses characteristically penetrate far into the soil, and grassland soils tend to be deep and fertile. Temperate grasslands are often populated by herds of grazing mammals. In North America, huge herds of bison and pronghorns once inhabited the prairies. The herds are almost all gone now, with most of the prairies having been converted to the richest agricultural region on earth.

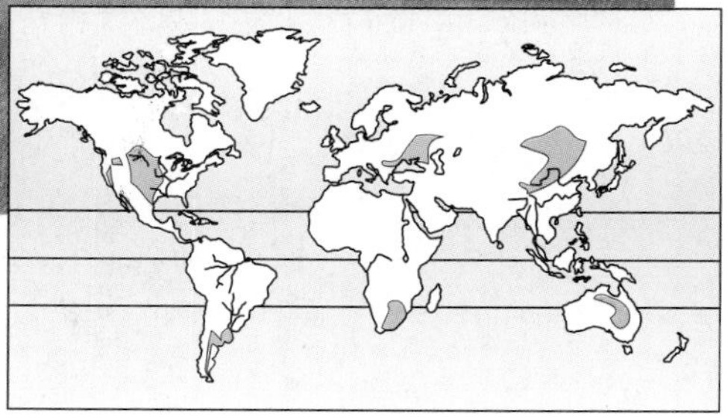

FIGURE 26.19
Temperate grassland.

Temperate Deciduous Forests

Mild climates (warm summers and cool winters) and plentiful rains promote the growth of **deciduous** (hardwood) **forests** in Eurasia, the northeastern United States, and eastern Canada (figure 26.20). A deciduous tree is one that drops its leaves in the winter. Deer, bears, beavers, and raccoons are the familiar animals of the temperate regions. Because the temperate deciduous forests represent the remnants of more extensive forests that stretched across North America and Eurasia several million years ago, these remaining areas—especially those in eastern Asia and eastern North America—share animals and plants that were once more widespread. Alligators, for example, are found today only in China and in the southeastern United States. The deciduous forest in eastern Asia is rich in species because climatic conditions have historically remained constant. Many perennial herbs live in areas of temperate deciduous forest.

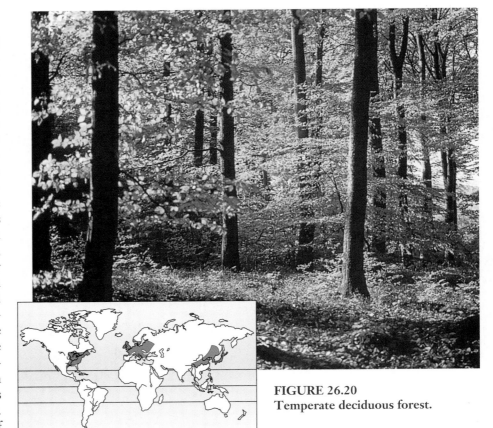

FIGURE 26.20
Temperate deciduous forest.

Temperate Evergreen Forests

Temperate evergreen forests (figure 26.21) occur in regions where winters are cold and there is a strong, seasonal dry period. The pine forests of the western United States, the California oak woodlands, and the Australian eucalyptus forests are typical temperate evergreen forests. Temperate evergreen forests are often mixed with deciduous trees in Europe and the southeastern United States, and are characteristic of regions with more nutrient-poor soils. Such temperate-mixed evergreen forests represent a broad transitional zone between temperate deciduous forests to the south and taiga to the north. Many of these forests are endangered by overlogging, particularly in the western United States.

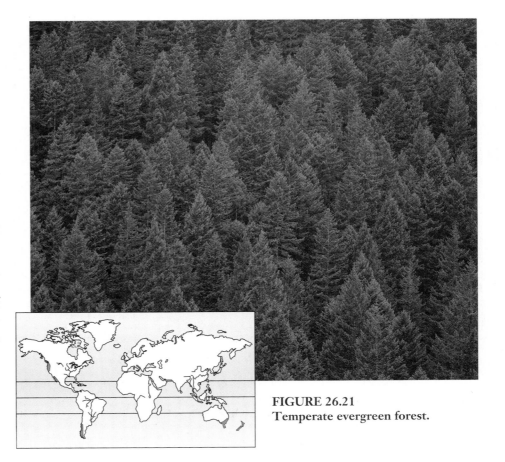

FIGURE 26.21
Temperate evergreen forest.

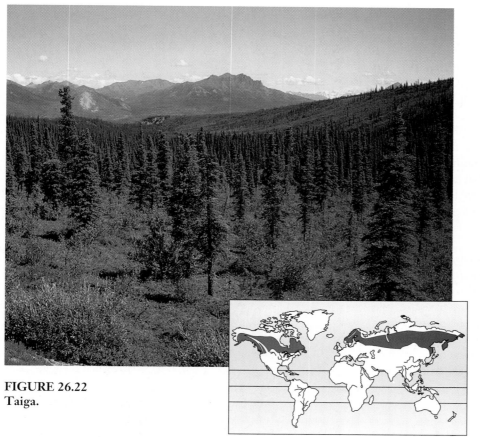

Taiga

A great ring of northern forests of coniferous trees (spruce, hemlock, and fir) extends across vast areas of Asia and North America. Coniferous trees are ones with leaves like needles that are kept all year long. This ecosystem, called **taiga,** is one of the largest on earth (figure 26.22). Here, the winters are long and cold, and most of the limited precipitation falls in the summer. Because the taiga has too short a growing season for farming, few people live there. Many large mammals, including elk, moose, deer, and such carnivores as wolves, bears, lynx, and wolverines, live in the taiga. Traditionally, fur trapping has been extensive in this region, as has lumber production. Marshes, lakes, and ponds are common and are often fringed by willows or birches. Most of the trees occur in dense stands of one or a few species.

FIGURE 26.22
Taiga.

Tundra

In the far north, above the great coniferous forests and south of the polar ice, few trees grow. There the grassland, called **tundra,** is open, windswept, and often boggy (figure 26.23). Enormous in extent, this ecosystem covers one-fifth of the earth's land surface. Very little rain or snow falls. When rain does fall during the brief arctic summer, it sits on frozen ground, creating a sea of boggy ground. **Permafrost,** or permanent ice, usually exists within a meter of the surface. Trees are small and are mostly confined to the margins of streams and lakes. As in taiga, herbs of the tundra are perennials that grow rapidly during the brief summers. Large grazing mammals, including musk-oxen, caribou, reindeer, and carnivores, such as wolves, foxes, and lynx, live in the tundra. Lemming populations rise and fall on a long-term cycle, with important effects on the animals that prey on them.

FIGURE 26.23
Tundra.

Other Biomes

Polar Ice

Ice caps lie over the Arctic Ocean and Greenland in the north and Antarctica in the south (figure 26.24). The poles receive almost no precipitation, so although ice is abundant, fresh water is scarce. The sun barely rises in the winter months. Life in Antarctica is largely limited to the coasts. Because the Antarctic ice cap lies over a landmass, it is not warmed by the latent heat of circulating ocean water and becomes very cold. As a result, only bacteria, algae, and some small insects inhabit the vast Antarctic interior.

Mountain Zone (Alpine)

Because increasing altitude produces many of the same changes in temperature and moisture as increasing latitude, the tops of mountains have a typical windswept vegetation similar in many respects to tundra. Few if any trees are able to grow in this alpine zone, which, like the polar tundra regions, is alive with life in the warm summer months. During the harsh winter, little grows. The alpine zone extends down to lower elevations on mountains located at higher latitudes (figure 26.25).

Chaparral

The **chaparral** of California and adjacent regions is historically derived from deciduous forests. Chaparral occurs in the five areas of the world with a Mediterranean, or summer-dry, climate: the Mediterranean area itself, California, central Chile, the Cape region of South Africa, and southwestern Australia. Chaparral consists of evergreen, sometimes spiny shrubs and low trees that form extensive communities adapted to periodic fires (figure 26.26).

FIGURE 26.25
Mountain zone.

FIGURE 26.24
Polar ice.

FIGURE 26.26
Chaparral.

Warm, Moist Evergreen Forest

Massive evergreen forests occur in temperate regions where winters are mild and moisture is plentiful (figure 26.27). These can be seen in central China, in the pine forests covering much of the southeastern United States, and in the coastal redwood forests of northern California.

Tropical Monsoon Forest

Tropical monsoon forests (figure 26.28), also called tropical upland forests, occur in the tropics and semitropics at slightly higher latitudes than rainforests or where local climates are drier. Most trees in these forests are deciduous, losing many of their leaves during the dry season. Leaves are smaller, and photosynthesis is depressed because of lower temperatures. Rainfall is typically very seasonal, measuring several inches daily in the monsoon season and approaching drought conditions in the dry season, particularly in locations far from oceans, such as in central India.

Semidesert (Tropical Dry Forest)

Semidesert (figure 26.29), or tropical dry forests, occur in tropical regions with less rain than monsoon forests but more rain than savannas. Vegetation is dominated by bushes and trees with thorns and spikes, which is why these regions are also known as thornwood forests. Little or no rain falls for eight or nine months during the winter. Plants survive on one or a few short heavy rains received during the summer wet season, growing intensively in response to the moisture. The brief rain is followed by the long dry season, when leaves fall from plants and little or no growth occurs.

FIGURE 26.27
Warm, moist evergreen forest.

FIGURE 26.28
Tropical monsoon forest.

FIGURE 26.29
Semidesert.

Patterns of Circulation in the Oceans

Patterns of ocean circulation are determined by the patterns of atmospheric circulation, but they are modified by the locations of landmasses. Oceanic circulation is dominated by huge surface gyres (figure 26.30), which move around the subtropical zones of high pressure between approximately 30° north and 30° south latitude. These gyres move clockwise in the northern hemisphere and counterclockwise in the southern hemisphere. The ways they redistribute heat profoundly affects life not only in the oceans but also on coastal lands. For example, the Gulf Stream, in the North Atlantic, swings away from North America near Cape Hatteras, North Carolina, and reaches Europe near the southern British Isles. Because of the Gulf Stream, western Europe is much warmer and more temperate than eastern North America at similar latitudes.

As a general principle, western sides of continents in temperate zones of the northern hemisphere are warmer than their eastern sides; the opposite is true of the southern hemisphere. In addition, winds passing over cold water onto warm land increase their moisture-holding capacity, limiting precipitation.

In South America, the Humboldt Current carries phosphorus-rich cold water northward up the west coast. Phosphorus is brought up from the ocean depths by the upwelling of cool water that occurs as offshore winds blow from the mountainous slopes that border the Pacific Ocean. This nutrient-rich current helps make possible the abundance of marine life that supports the fisheries of Peru and northern Chile. Marine birds, which feed on these organisms, are responsible for the commercially important, phosphorus-rich, guano deposits on the seacoasts of these countries.

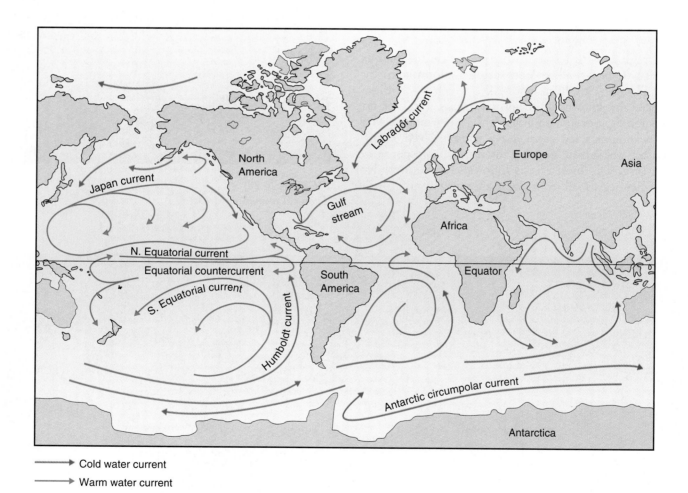

Cold water current
Warm water current

FIGURE 26.30
Ocean circulation. Water moves in the oceans in great surface spiral patterns called gyres; they profoundly affect the climate on adjacent lands.

El Niño Southern Oscillations and Ocean Ecology

Every Christmas a tepid current sweeps down the coast of Peru and Ecuador from the tropics, reducing the fish population slightly and giving local fishermen some time off. The local fishermen named this Christmas current *El Niño* ("The Christ Child"). Now, though, the term is reserved for a catastrophic version of the same phenomenon, one that occurs every two to seven years and is felt not only locally but on a global scale.

Scientists now have a pretty good idea of what goes on in an El Niño. Normally the Pacific Ocean is fanned by constantly blowing east-to-west trade winds that push warm surface water away from the ocean's eastern side (Peru, Ecuador, and Chile) and allow cold water to well up from the depths in its place, carrying nutrients that feed plankton and hence fish. This surface water piles up in the west, around Australia and the Philippines, making it several degrees warmer and a meter or so higher than the eastern side of the ocean. But if the winds slacken briefly, warm water begins to slosh back across the ocean.

Once this happens, ocean and atmosphere conspire to ensure it keeps happening. The warmer the eastern ocean gets, the warmer and lighter the air above it becomes, and hence more similar to the air on the western side. This reduces the difference in pressure across the ocean. Because a pressure difference is what makes winds blow, the easterly trades weaken further, letting the warm water continue its eastward advance.

The end result is to shift the weather systems of the western Pacific Ocean 6000 km eastward. The tropical rainstorms that usually drench Indonesia and the Philippines are caused when warm seawater abutting these islands causes the air above it to rise, cool, and condense its moisture into clouds. When the warm water moves east, so do the clouds, leaving the previously rainy areas in drought. Conversely, the western edge of South America, its coastal waters usually too cold to trigger much rain, gets a soaking, while the upwelling slows down. During an El Niño, commercial fish stocks virtually disappear from the waters of Peru and northern Chile, and plankton drop to a twentieth of their normal abundance. The commercially valuable anchovy fisheries of Peru were essentially destroyed by the 1972 El Niño.

That is just the beginning. El Niño's effects are propagated across the world's weather systems (figure 26.31). Violent winter storms lash the coast of California, accompanied by flooding, and El Niño produces colder and wetter winters than normal in Florida and along the Gulf Coast. The American midwest experiences heavier-then-normal rains, as do Israel and its neighbors. A particularly severe El Niño, beginning in October 1986 and extending through early 1988 brought unusual drought in the states of Washington and Oregon in the summer of 1987.

Though the effects of El Niños are now fairly clear, what triggers them still remains a mystery. Models of these weather disturbances suggest that the climatic change that triggers El Niño is "chaotic." Wind and ocean currents return again and again to the same condition, but never in a regular pattern, and small nudges can send them off in many different directions—including an El Niño.

The world's oceans circulate in huge gyres deflected by continental landmasses. Circulation of ocean water redistributes heat, warming the western side of continents. Disturbances in ocean currents like El Niño can have profound influences on world climate.

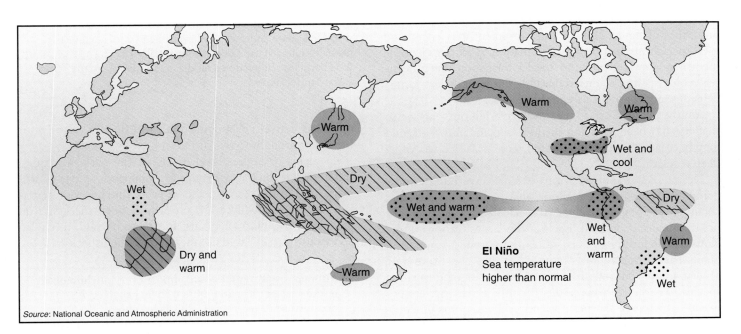

Source: National Oceanic and Atmospheric Administration

FIGURE 26.31
An El Niño winter. El Niño currents produce unusual weather patterns all over the world as warm waters from the western Pacific move eastward.

Life in the Oceans

Nearly three-quarters of the earth's surface is covered by ocean. Oceans have an *average* depth of more than 3 kilometers, and they are, for the most part, cold and dark. Heterotrophic organisms inhabit even the greatest ocean depths, which reach nearly 11 kilometers in the Marianas Trench of the western Pacific Ocean. Photosynthetic organisms are confined to the upper few hundred meters of water. Organisms that live below this level obtain almost all of their food indirectly, as a result of photosynthetic activities that occur above.

The supply of oxygen can often be critical in the ocean, and as water temperatures become warmer, the water holds less oxygen. For this reason, the amount of available oxygen becomes an important limiting factor for organisms in warmer marine regions of the globe. Carbon dioxide, in contrast, is almost never limited in the oceans. The distribution of minerals is much more uniform in the ocean than it is on land, where individual soils reflect the composition of the parent rocks from which they have weathered.

Frigid and bare, the floors of the deep sea have long been considered a biological desert. Recent close-up looks taken by marine biologists, however, paint a different picture (figure 26.32). The ocean floor is teeming with life. Often miles deep, thriving in pitch darkness under enormous pressure, crowds of marine invertebrates have been found in hundreds of deep samples from the Atlantic and Pacific. Rough estimates of deep-sea diversity have soared to millions of species. Many appear endemic (local). The diversity of species is so high it may rival that of tropical rainforests! This profusion is unexpected. New species usually require some kind of barrier in order to diverge (see chapter 21), and the ocean floor seems boringly uniform. However, little migration occurs among deep populations. Like the *Euphydryas* butterflies of figure 21.3, lack of movement may encourage local specialization and species formation. A patchy environment may also contribute to species formation there; deep-sea ecologists find evidence that fine but nonetheless formidable resource barriers arise in the deep sea.

Another conjecture is that the extra billion years or so that life has been evolving in the sea compared with land may be a factor in the unexpected biological richness of its deep recesses.

Despite the many new forms of small invertebrates now being discovered on the seafloor, and the huge biomass that occurs in the sea, more than 90% of all *described* species of organisms occur on land. Each of the largest groups of organisms, including insects, mites, nematodes, fungi, and plants has marine representatives, but they constitute only a very small fraction of the total number of described species. There are two reasons for this. First, barriers between habitats are sharper on land, and variations in elevation, parent rock, degree of exposure, and other factors have been crucial to the evolution of the millions of species of terrestrial organisms. Second, there are simply few tax-

FIGURE 26.32
Food comes to the ocean floor from above. Looking for all the world like some undersea sunflower, the two sea anemones (actually animals) use a glass-sponge stalk to catch "marine snow," food particles raining down on the ocean floor from the ocean surface miles above.

onomists actively classifying the profusion of ocean floor life being brought to the surface.

In terms of higher level diversity, the pattern is quite different. Of the major groups of organisms—phyla—most originated in the sea, and almost every one has representatives in the sea. Only a few phyla have been successful on land or in freshwater habitats, but these have given rise to an extraordinarily large number of described species.

Although representatives of almost every phylum occur in the sea, an estimated 90% of living species of organisms are terrestrial. This is because of the enormous evolutionary success of a few phyla on land, where the boundaries between different habitats are sharper than they are in the sea.

Marine Ecosystems

The marine environment consists of three major habitats: (1) the **neritic zone,** the zone of shallow waters along the coasts of continents; (2) the **pelagic zone,** the area of water above the ocean floor; and (3) the **benthic zone,** the actual ocean floor (figure 26.33). The part of the ocean floor that drops to depths where light does not penetrate is called the **abyssal zone.**

The Neritic Zone

The **neritic zone** of the ocean is the area less than 300 meters below the surface along the coasts of continents and islands. The zone is small in area, but it is inhabited by large numbers of species (figure 26.34). The intense and sometimes violent interaction between sea and land in this zone gives a selective advantage to well-secured organisms that can withstand being washed away by the continual beating of the waves. Part of this zone, the **intertidal region,** sometimes called the **littoral region,** is exposed to the air whenever the tides recede.

Because of its accessibility to land, the intertidal region must have been home to the ancestors of the first organisms that colonized terrestrial habitats. Perhaps the greater complexity needed to anchor and fasten the animals and plants that dwell in this zone constituted a kind of preadaptation to life on land, where environmental stresses are even more extensive. The organisms that live successfully in habitats regularly exposed to air must have some kind of waterproof covering or habits that protect them from the air's drying action when the tide is out. Such adaptations are of central importance for terrestrial organisms.

The world's great fisheries are in shallow waters over continental shelves, either near the continents themselves or in the open ocean, where huge banks come near the surface. Nutrients, derived from land, are much more abundant in coastal and other shallow regions, where upwelling from the depths occurs, than in the open ocean. This accounts for the great productivity of the continental shelf fisheries. The preservation of these fisheries, a source of high-quality protein exploited throughout the world, has become a growing concern. In Chesapeake Bay, where complex systems of rivers enter the ocean from heavily populated areas, environmental stresses have become so severe that they not only threaten the continued existence of formerly highly productive fisheries, but also diminish the quality of human life in these regions (see figure 26.35). Increased runoff from farms and sewage effluent in areas like Chesapeake Bay add large amounts of nutrients to the water. This increased nutrient supply allows an increase in the numbers of some marine organisms. The increased populations then use up more and more of the oxygen in the water and thus may disturb established populations of organisms such as oysters. Climatic shifts may magnify these effects, and large numbers of marine animals die suddenly as a result.

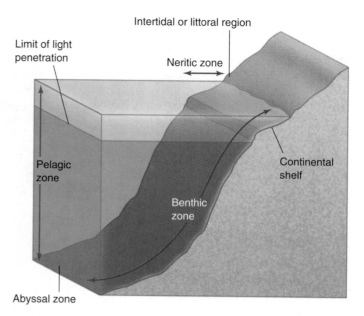

FIGURE 26.33
Marine ecosystems. Ecologists classify marine communities into neritic, pelagic, benthic, and abyssal zones, according to depth (which affects how much light penetrates) and distance from shore.

FIGURE 26.34
Diversity is great in coastal regions. Fishes and many other kinds of animals find food and shelter among the kelp beds in the coastal waters of temperate regions.

About three-fourths of the surface area of the world's oceans are located in the tropics. In these waters, where the water temperature remains about 21 °C, coral reefs can grow. These highly productive ecosystems can successfully concentrate nutrients, even from the relatively nutrient-poor waters characteristic of the tropics.

FIGURE 26.35
Chesapeake Bay. Chesapeake Bay has more than 11,300 kilometers of shoreline and drains more than 166,000 square kilometers in one of the most densely populated and heavily industrialized areas in North America. The body of open water is about 320 kilometers long and, at some points, nearly 50 kilometers wide. (a) Large metropolitan areas and shipping facilities make the bay one of the busiest natural harbors anywhere. (b) One of the most biologically productive bodies of water in the world, the bay yielded an annual average of about 275,000 kilograms of fish in the 1960s but only a tenth as much in the 1980s. (c) This grebe is coated from an oil spill off the mouth of the Potomac River. (d) Uncontrolled erosion and increases in nutrients cloud the water and block the light needed for photosynthesis, upsetting the delicate ecological balance on which the productivity of the bay depends.

The Pelagic Zone

Drifting freely in the upper, well-illuminated waters of the pelagic zone, a diverse biological community exists, primarily consisting of microscopic organisms called **plankton.** Fish and other larger organisms that swim in these waters constitute the *nekton,* whose members feed on plankton and one another. Together, the organisms that make up the plankton and the nekton provide all the food for those that live below. Some members of the plankton, including protists and some bacteria, are photosynthetic. Collectively, these organisms account for about 40% of all photosynthesis that takes place on earth. Most plankton live in the top 100 meters of the sea, the zone into which light from the surface penetrates freely. Perhaps half of the total photosynthesis in this zone is carried out by organisms less than 10 micrometers in diameter—at the lower limits of size for organisms—including cyanobacteria and algae, organisms so small that their abundance and ecological importance have been unappreciated until relatively recently.

Many heterotrophic protists and animals live in the plankton and feed directly on photosynthetic organisms and on one another. Gelatinous animals, especially jellyfish and ctenophores, are abundant in the plankton but are relatively poorly known because their fragility makes them difficult to collect and study. Their bodies may consist of as much as 95% water. The largest animals that have ever existed on earth, baleen whales, graze on plankton and nekton as their only source of food. These whales are heavier than the largest dinosaurs known. A number of heterotrophic, free-swimming organisms, such as fishes and crustaceans, move up into the plankton at times—in some instances, on a regular cycle—to feed on the organisms there.

Populations of organisms that make up plankton can increase rapidly, and the turnover of nutrients in the sea is so great that the productivity in these systems, although still low, was seriously underestimated in the past. Since nitrogen and phosphorus are often present in only small amounts and organisms may be relatively scarce, this productivity reflects rapid use and recycling rather than an abundance of these nutrients. The smallest organisms turn over phosphorus much more rapidly than larger ones, and their role in these complex and productive ecosystems is just starting to be understood properly.

The Benthic Zone

The seafloor at depths below 1000 meters, the **abyssal zone,** has about twice the area of all the land on earth. The sea floor itself, sometimes called the **benthic zone,** is a thick blanket of mud, consisting of fine particles that have settled from the overlying water and accumulated over millions of years. Because of high pressures (an additional atmosphere of pressure for every 10 meters of depth), cold temperatures (2°–3 °C), darkness, and lack of food, the first deep-sea biologists thought that nothing could live on the seafloor. In fact, as we discussed, recent work has shown that the number of species that live at great depth is quite high. Most of these animals are only a few millimeters in size, although larger ones also occur in these regions. Some of the larger ones are bioluminescent (figure 26.36*a*) and thus are able to communicate with one another or attract their prey.

Many deep-sea animals are known from only a few samples. Today, they are sampled quantitatively by removing cubes of mud from the seafloor. Such samples have revealed a great diversity of animal species, indicating that the deep sea represents a major reservoir of largely unknown biological diversity.

Most organic matter in plankton is recycled in the surface layers of the ocean. Animals on the sea bottom depend on the meager leftovers from organisms living kilometers overhead. The low densities and small size of most deep-sea bottom animals is in part a consequence of this limited food supply. In 1977, oceanographers diving in a research submarine were surprised to find dense clusters of large animals living on geothermal energy at a depth of 2500 meters. These deep-sea oases occur where seawater circulates through porous rock at sites where molten material from beneath the earth's crust comes close to the rocky surface. A series of these areas form the Mid-Ocean Ridge, a worldwide feature where basalt erupts through the ocean floor.

This water is heated to temperatures in excess of 350 °C and, in the process, becomes rich in reduced compounds. These compounds, such as hydrogen sulfide, provide energy for bacterial primary production through chemosynthesis instead of photosynthesis. Mussels, clams, and large red-plumed worms in a phylum unrelated to any shallow-water invertebrates cluster around the vents (figure 26.36*b*). Bacteria live symbiotically within the tissues of these animals. The animal supplies a place for the bacteria to live and transfers CO_2, H_2S, and O_2 to them for their growth; the bacteria supply the animal with organic compounds to use as food. Polychaete worms (chapter 42), anemones, and limpets live on free-living chemosynthetic bacteria. Crabs act as scavengers and predators, and some of the fish are also predators. This is one of the few ecosystems on earth that does not depend on the sun's energy.

About 40% of the world's photosynthetic productivity is estimated to occur in the oceans. Of this, perhaps half is carried out by organisms less than 10 micrometers long. The turnover of nutrients in the plankton is much more rapid than in most other ecosystems, and the total amounts of nutrients are very low.

(a)

(b)

FIGURE 26.36

Life in the abyssal and benthic zones. (a) The luminous spot below the eye of this deep-sea fish results from the presence of a symbiotic colony of luminous bacteria. Similar luminous signals are a common feature of deep-sea animals that move about. (b) These giant beardworms live along vents where water jets from fissures at 350 °C and then cools to the 2 °C of the surrounding water.

Freshwater Habitats

Freshwater habitats are distinct from both marine and terrestrial ones, but they are limited in area. Inland lakes cover about 1.8% of the earth's surface, and running water (streams and rivers) covers about 0.3%. All freshwater habitats are strongly connected with terrestrial ones, with marshes and swamps constituting intermediate habitats. In addition, a large amount of organic and inorganic material continuously enters bodies of fresh water from communities growing on the land nearby (figure 26.37). Many kinds of organisms are restricted to freshwater habitats (figure 26.38). When organisms live in rivers and streams, they must be able to swim against the current or attach themselves in such a way as to resist the effects of current, or risk being swept away.

Ponds and Lakes

Small bodies of fresh water are called ponds, and larger ones lakes. Because water absorbs light passing through it at wavelengths critical to photosynthesis (every meter absorbs 40% of the red and about 2% of the blue), the distribution of photosynthetic organisms is limited to the upper **photic zone;** heterotrophic organisms occur in the lower **aphotic zone** where very little light penetrates.

Ponds and lakes, like the ocean, have three zones where organisms occur, distributed according to the depth of the water and its distance from shore (figure 26.39). The *littoral zone* is the shallow area along the shore. The *limnetic zone* is the well-illuminated surface water away from the shore, inhabited by plankton and other organisms that live in open water. The *profundal zone* is the area below the limits where light can effectively penetrate.

FIGURE 26.37
A nutrient-rich stream. Much organic material falls or seeps into streams from communities along the edges. This input increases the stream's biological productivity.

(a)

(b)

FIGURE 26.38
Freshwater organisms. (a) This speckled darter and (b) this giant waterbug with eggs on its back can only live in freshwater habitats.

Thermal Stratification

Thermal stratification is characteristic of larger lakes in temperate regions (figure 26.40). In summer, warmer water forms a layer at the surface known as the *epilimnion*. Cooler water, called the *hypolimnion* (about 4 °C), lies below. An abrupt change in temperature, the thermocline, separates these two layers. Depending on the climate of the particular area, the epilimnion may become as much as 20 meters thick during the summer.

In autumn the temperature of the epilimnion drops until it is the same as that of the hypolimnion, 4 °C. When this occurs, epilimnion and hypolimnion mix—a process called fall overturn. Since water is densest at about 4 °C, further cooling of the water as winter progresses creates a layer of cooler, lighter water, which freezes to form a layer of ice at the surface. Below the ice, the water temperature remains between 0° and 4 °C, and plants and animals can survive. In spring, the ice melts, and the surface water warms up. When it warms back to 4 °C, it again mixes with the water below. This process is known as spring overturn. When lake waters mix in the spring and fall, nutrients formerly held in the depths of the lake are returned to the surface, and oxygen from surface waters is carried to the depths.

> Freshwater habitats include several distinct life zones. These zones shift seasonally in temperate lakes and ponds. In the spring and fall, when their temperatures are equal, shallower and deeper waters of the lake mix, with oxygen being carried to the depths and nutrients being brought to the surface.

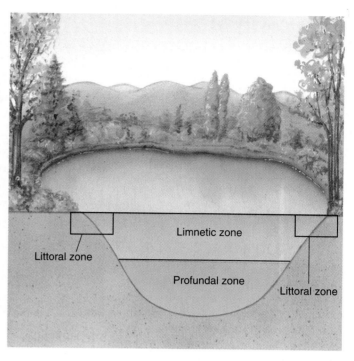

FIGURE 26.39
The three zones in ponds and lakes. A shallow "edge" (littoral) zone lines the periphery of the lake, where attached algae and their insect herbivores live. An open-water surface (limnetic) zone lies across the entire lake and is inhabited by floating algae, zooplankton, and fish. A dark, deep-water (profundal) zone overlies the sediments at the bottom of the lake and contains numerous bacteria and worm-like organisms that consume dead debris settling at the bottom of the lake.

FIGURE 26.40
Stratification in fresh water.
The pattern stratification in a large pond or lake in temperate regions is upset in the spring and fall overturns. Of the three layers of water shown, the hypolimnion consists of the densest water, at 4 °C; the epilimnion consists of warmer water that is less dense; and the thermocline is the zone of abrupt change in temperature that lies between them. If you have dived into a pond in temperate regions in the summer, you have experienced the existence of these layers.

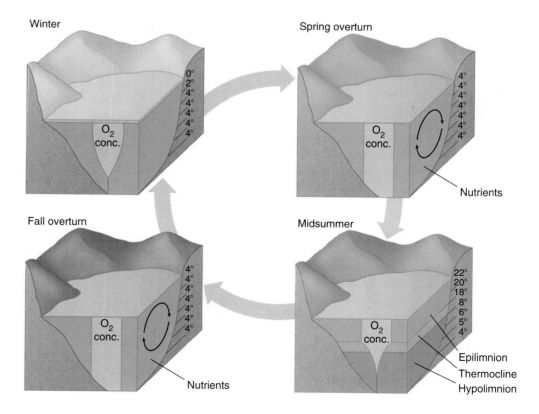

Productivity of Freshwater Ecosystems

Some aquatic communities, such as fast-moving streams, are not highly productive. Since the moving water washes away plankton, the photosynthesis that supports the community is limited to algae attached to the surface and to rooted plants.

The Productivity of Lakes

Lakes can be divided into two categories based on their production of organic matter. **Eutrophic lakes** contain an abundant supply of minerals and organic matter. As the plentiful organic material drifts below the thermocline from the well-illuminated surface waters of the lake, it provides a source of energy for other organisms. Most of these are oxygen-requiring organisms that can easily deplete the oxygen supply below the thermocline during the summer months. The oxygen supply of the deeper waters cannot be replenished until the layers mix in the fall. This lack of oxygen in the deeper waters of some lakes may have profound effects, such as allowing relatively harmless materials such as sulfates and nitrates to convert into toxic materials such as hydrogen sulfide and ammonia.

In **oligotrophic lakes,** organic matter and nutrients are relatively scarce. Such lakes are often deeper than eutrophic lakes and have very clear blue water. Their hypolimnetic water is always rich in oxygen. Oligotrophic lakes are highly susceptible to nutrient pollution because they have such limited quantities of nutrients (figure 26.41). Excess phosphorus from sources such as fertilizer runoff, sewage, and detergents can quickly lead to harmful effects such as runaway weed growth or the promotion of algae "blooms." These can rapidly deplete the lake's oxygen supply, killing the natural fish population.

The Productivity of Wetlands

Swamps, marshes, bogs, and other **wetlands** covered with water support a wide variety of water-tolerant plants, called hydrophytes ("water plants"), and a rich diversity of invertebrates, birds, and other animals. Wetlands are among the most productive ecosystems on earth (table 26.1). They also play a key ecological role by providing water storage basins that moderate flooding. Many wetlands are being disrupted by human "development" of what is sometimes perceived as otherwise useless land, but government efforts are now underway to protect the remaining wetlands.

FIGURE 26.41
Oligotrophic lakes are highly susceptible to pollution. Lake Washington is an oligotrophic lake near Seattle, Washington. The drainage from fertilizers applied to the plantings around residences, business concerns, and recreational facilities bordering the lake poses an ever-present threat to the organisms that inhabit its deep blue water.

Table 26.1	The Most Productive Ecosystems
Ecosystem	**Net Primary Productivity per unit area (g/m²)**
Coral reefs	2500
Tropical rainforest	2200
Wetlands	2000
Tropical seasonal forest	1600
Estuaries	1500
Temperate evergreen forest	1300
Temperate deciduous forest	1200
Savanna	900
Boreal forest	800
Cultivated land	650
Continental shelf	360
Lake and stream	250
Open ocean	125
Extreme desert, rock, sand, and ice	3

Source: Whitaker, 1975.

The most productive freshwater ecosystems are wetlands. Most lakes are far less productive, limited by lack of nutrients.

26.1 Organisms must cope with a varied environment.

- Organisms employ physiological, morphological, and behavioral mechanisms to cope with variations in the environment.

26.2 Climate shapes the character of ecosystems.

- Warm air rises near the equator and flows toward the poles, descending at about 30° north and south latitude. Because the air falls in these regions, it is warmed, and its moisture-holding capacity increases. The great deserts of the world are formed in these drier latitudes.

26.3 Biomes are widespread terrestrial ecosystems.

- We have grouped our discussion of biomes in 14 major categories. These are (1) tropical rainforest; (2) savanna; (3) desert; (4) temperate grassland; (5) temperate deciduous forest; (6) temperate evergreen forest; (7) taiga; (8) tundra; (9) polar ice; (10) mountain zone; (11) chaparral; (12) warm, moist evergreen forest; (13) tropical monsoon forest; and (14) semidesert.

- Tropical rainforests contain at least half of all species of plants, animals, fungi, and microorganisms. These organisms are the most directly threatened by extinction in the near future.

- Savannas, which were much more widespread during the Pleistocene Epoch, are highly productive ecosystems. They are often inhabited by large herds of grazing mammals and their predators, including humans.

- Deserts are the hottest and driest habitats on earth. They are of great biological interest because of the extreme behavioral, morphological, and physiological adaptations of the plants and animals that live in them.

- Temperate grasslands cover large areas in North America, Eurasia, and South America. When they receive sufficient precipitation, they are often well-suited to agriculture.

- Temperate deciduous forests once dominated huge areas of the northern hemisphere. Today, they are confined largely to eastern North America and parts of China.

- Temperate evergreen forests occur in western North America, Chile, and parts of Asia and Australia.

- Taiga is a vast coniferous forest that stretches across Eurasia and North America. Its winters are long and cold, but plants grow rapidly during the long summer days.

- Even farther north is the tundra, which covers about 20% of the earth's land surface. It consists largely of open grassland, often boggy in summer, over a layer of permafrost.

26.4 Aquatic ecosystems cover much of the earth.

- The ocean contains three major environments: the neritic zone, the pelagic zone, and the benthic zone.

- The neritic zone, which lies along the coasts, is small in area but very productive and rich in species.

- The surface layers of the pelagic zone are home to plankton (drifting organisms) and nekton (actively swimming ones). The productivity of this zone has been underestimated because of the very small size (less than 10 μm) of many of its key organisms and because of its rapid turnover of nutrients.

- The benthic zone is home to a surprising number of species.

- Freshwater habitats constitute only about 2.1% of the earth's surface; most are ponds and lakes. These possess a littoral zone, a limnetic zone, and a profundal zone. The waters in these zones mix seasonally, delivering oxygen to the bottom and nutrients to the surface.

Discussing Key Terms

1. **Climate** The major terrestrial communities of organisms are structured largely by climate, particularly by temperature and rainfall.

2. **Biomes** The same forms of terrestrial communities tend to occur in different places with the same climates. The world's surface can be divided into about 14 types of terrestrial communities called biomes.

3. **Ocean communities** A great wealth of nutrients and biomass are present in the world's oceans, particularly in warm coastal regions and the planktonic stratum of the surface zone.

4. **Freshwater communities** Very little of the earth's water is stored as fresh water in ponds, rivers, and lakes. Nutrient cycling occurs in temperate lakes in the spring and fall. At these times, the lake water mixes, bringing nutrients to surface waters and oxygen to the depths.

Review Questions

1. What is a biome? What are the two key physical factors that affect the distribution of biomes across the earth?

2. Why are the majority of great deserts located near 30° north and south latitude? Is it more likely that a desert will form in the interior or at the edge of a continent? Explain why. Why is the windward side of a mountain generally more moist than the leeward side?

3. What is the difference between plankton and nekton in the ocean's pelagic zone? How important are the photosynthetic plankton to the survival of the earth? Is the turnover of nutrients in the surface zone slow or fast?

4. What conditions of the abyssal zone led early deep-sea biologists to believe nothing lived there? What provides the energy for the deep-sea communities found around thermal vents? What kind of organisms live there?

5. How do eutrophic lakes differ from oligotrophic lakes? Which is more susceptible to the effects of pollution? Why?

6. List the eight most prominent biomes in the world according to their distance from the equator. How is this stratification imitated in changes in elevation?

7. What are the key characteristics of tropical rainforests in terms of number of species, level of specialization, and amount of rainfall?

Thought Questions

1. The largest animal that has ever existed on earth, the blue whale, has no teeth. Instead it has a horny sieve, or baleen, that can filter huge amounts of seawater. What do you think is the major part of this whale's diet? Would it be more energy-efficient for it to eat larger animals, such as seals? What would be the advantages and disadvantages?

2. What kinds of biological communities would you expect to find on the windward and leeward sides of a mountain range in an area where the annual precipitation ranged between 20 and 100 centimeters per year and fell mainly in one rainy season? How would the height of the mountain range affect the situation?

3. Near the coast in southern California are two major plant communities. Right along the ocean is the coastal sage community, which is dominated by low shrubs that often wither or lose their leaves in the summer. Higher up, on the ridges, is the chaparral, a community dominated by tall evergreen shrubs. Which of these communities would you think receives more rainfall? Which grows on better soil? Which is the more productive on an annual basis? Where would you expect to find more annual plants?

Internet Links

The Ocean from Space
http://www.athena.ivv.nasa.gov/curric/oceans/ocolor/index.html
This NASA site illustrates how satellite images of ocean color are used to estimate the productivity of the world's oceans.

Ocean Planet
http://seawifs.gsfc.nasa.gov/ocean_planet.html
An electronic version of a very popular Smithsonian exhibit on the world's largest ecosystem.

How Many Species Are There on Earth?
http://www.ciesin.org/docs/002-253/002-253.html
A classic paper by Robert May on the many factors that interact to determine the number of species in a place and on earth.

Saving Biodiversity
http://www.wri.org/wri/biodiv/index.html
The focus of the WORLD RESOURCES INSTITUTE is on conservation of biodiversity. This site has many resources focused on loss of biodiversity and ways to lessen the loss (also available in Spanish).

For Further Reading

Doherty, J.: "Alaska's Arctic Refuge," *Smithsonian*, March 1996, pages 32–43. A visit to a largely untouched ecosystem of great beauty and fragility.

Dybas, C.: "The Deep-Sea Floor Rivals Rainforests in Diversity of Life," *Smithsonian*, January 1996, pages 96–106. The deep-sea floor is an ecosystem where millions of species of worms live their lives.

Gore, R.: "Between Monterey Tides," *National Geographic*, February 1990, pages 2–43. Beautifully illustrated account of life along the California coast.

Holloway, M.: "Nurturing Nature," *Scientific American*, vol. 270, April 1994, pages 98–104. A considerable undertaking—the rescue of the Florida Everglades—is discussed.

McNaughton, S. J.: "Grazing Lawns: Animals in Herds, Plant Form, and Coevolution," *American Naturalist*, vol. 124, 1984, pages 863–86. The fascinating story of how differences in grasses and in the populations of animals that depend on grasses help determine the structure of the savanna biome.

Rützler, K. and I. Feller: "Caribbean Mangrove Swamps," *Scientific American*, March 1996, pages 94–99. These complex tropical ecosystems occur worldwide.

Science, vol. 269, July 1995: "Frontiers in Biology: Ecology." An entire issue devoted to recent progress in ecology.

27

The Future of the Biosphere

Concept Outline

27.1 The world's human population is growing explosively.

A Growing Population. The world's population of 6 billion people is growing rapidly and at current rates will double in 43 years.

27.2 Improvements in agriculture are needed to feed a hungry world.

The Future of Agriculture. Much of the effort in searching for new sources of food has focused on improving the productivity of existing crops.

27.3 Human activity is placing the environment under increasing stress.

Nuclear Power. Nuclear power, a plentiful source of energy, is neither cheap nor safe.

Carbon Dioxide and Global Warming. The world's industrialization has led to a marked increase in the atmosphere's level of CO_2, with resulting warming of climates.

Pollution. Human industrial and agricultural activity introduces significant levels of many harmful chemicals into ecosystems.

Acid Precipitation. Burning of cheap high-sulfur coal has introduced sulfur to the upper atmosphere, where it combines with water to form sulfuric acid that falls back to earth, harming ecosystems.

The Ozone Hole. Industrial chemicals called CFCs are destroying the atmosphere's ozone layer, removing an essential shield from the sun's UV radiation.

Destruction of the Tropical Forests. Much of the world's tropical forest is being destroyed by human activity.

27.4 Solving environmental problems requires individual involvement.

Environmental Science. The commitment of one person often makes a key difference in solving environmental problems.

Preserving Nonreplaceable Resources. Three key nonreplaceable resources are topsoil, groundwater, and biodiversity.

FIGURE 27.1
New York City by satellite.

The view of New York City in figure 27.1 was photographed from a satellite in the spring of 1985. At the moment this picture was taken, millions of people within its view were talking, hundreds of thousands of cars were struggling through traffic, hearts were being broken, babies born, and dead people buried. Our futures and those of everyone on the planet are linked to the unseen millions in this photograph, for we share the earth with them. A lot of people consume a lot of food and water, use a great deal of energy and raw materials, and produce a great deal of waste. They also have the potential to solve the problems that arise in an increasingly crowded world. In this chapter, we will study how human life affects the environment and how the efforts being mounted can lessen the adverse impact and increase the potential benefits of our burgeoning population.

A Growing Population

The current world population of 6 billion people is placing severe strains on the biosphere. How did it grow so large? For the past 300 years, the human birth rate (as a global average) has remained nearly constant, at about 30 births per year per 1000 people. Today it is about 25 births per year per 1000 people. However, at the same time, better sanitation and improved medical techniques have caused the death rate to fall steadily, from about 29 deaths per 1000 people per year to 13 per 1000 per year. Thus, while the birth rate has remained fairly constant and may have even decreased slightly, the tremendous fall in the death rate has produced today's enormous population (figure 27.2). The difference between the birth and death rates amounts to an annual worldwide increase of approximately 1.6%. This rate of increase may seem relatively small, but it would double the world's population in only 43 years!

The *annual* increase in world population today is nearly 100 million people, about equal to the current population of Mexico. Nearly 260,000 people are added to the world each day, or more than 180 every minute! The world population is expected to continue to rise to over 6 billion people by the end of the century, and then perhaps stabilize at a figure between 8.5 billion and 20 billion during the next century.

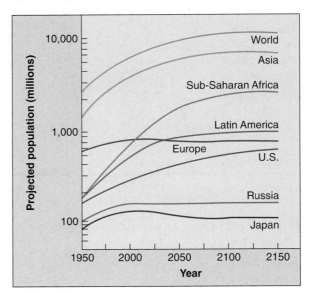

FIGURE 27.2

Anticipated growth of the global human population. Despite considerable progress in lowering birth rates, the human population will continue to grow for another century (data are presented above on a log scale). Much of the growth will center in sub-Saharan Africa, the poorest region on the globe, where the population could reach over 2 billion. Fertility rates there currently range from 3 to more than 5 children per woman, compared to fewer than 2.1 in Europe and the United States.

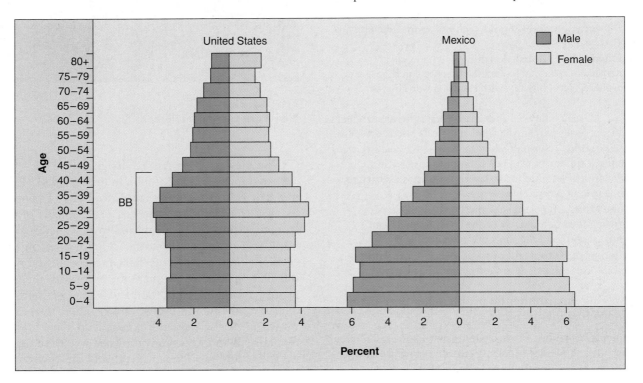

FIGURE 27.3

The age structure of the United States and Mexico in 1990. The population growth rate in Mexico, already rapid, will increase considerably in the future because so much of its population will soon be entering their reproductive years. The "bulge" in the population structure of the United States (BB) indicates the "baby boom" generation.

The Future Situation

By the year 2000, about 60% of the people in the world will be living in tropical or subtropical regions. An additional 20% will be living in China, and the remaining 20% in the developed or industrialized countries: Europe, the successor states of the Soviet Union, Japan, United States, Canada, Australia, and New Zealand. Although populations of industrialized countries are growing at an annual rate of about 0.3%, those of the developing, mostly tropical countries (excluding China) are growing at an annual rate estimated in 1995 to be about 2.2%. For every person living in an industrialized country like the United States in 1950, there were two people living elsewhere; in 2020, just 70 years later, there will be five.

As you learned in chapter 23, the age structure of a population determines how fast the population will grow. To predict the future growth patterns of a population, it is essential to know what proportion of its individuals have not yet reached childbearing age. In industrialized countries such as the United States, about a fifth of the population is under 15 years of age; in developing countries such as Mexico, the proportion is typically about twice as high (figure 27.3). Even if most tropical and subtropical countries consistently carry out the policies they have established to limit population growth, their populations will continue to grow well into the twenty-first century (figure 27.4), and industrialized countries will constitute a smaller and smaller proportion of the world's population. If India, with a 1995 population level of about 930 million people (36% under 15 years old), managed to reach a simple replacement reproductive rate by the year 2000, its population would still not stop growing until the middle of the twenty-first century. At present rates of growth, India will have a population of nearly 1.4 billion people by 2025 and will still be growing rapidly.

Population Growth Rate Starting to Decline

The United Nations has announced that the world population growth rate and world fertility were less than expected for 1995. In 1994, the world population growth rate was predicted to be 1.57 percent a year, and the actual growth rate from 1990 to 1995 was 1.48 percent. Also, they projected the average number of children per woman would be 3.1, while it was actually 2.96 from 1990 to 1995.

The U.N. attributes the decline to increased family planning efforts and the increased economic power and social status of women. While the U.N. applauds the United States for leading the world in funding family planning programs abroad, some oppose spending money on international family planning. The opposition states that money is better spent on improving education and the economy in other countries, leading to an increased awareness and lowered fertility rates. The U.N. certainly supports the improvement of education programs in developing countries, but, interestingly, it has reported increased education levels *following* a decrease in family size as a result of family planning.

FIGURE 27.4
Population growth is highest in tropical and subtropical countries. Mexico City, the world's largest city, has well over 20 million inhabitants.

Most countries are devoting considerable attention to slowing the growth rate of their populations, and there are genuine signs of progress. If these efforts are maintained, the world population may stabilize sometime in the next century. No one knows how many people the planet can support, but our use of the world's resources has not been sustainable since World War II. We clearly already have more people than can be supported with current technologies. Building a sustainable world is the most important task facing humanity's future. The quality of life available to our children in the next century will depend to a large extent on our success.

In 1998, the global human population of 6 billion people was growing at a rate of approximately 1.6% annually. At that rate, it will reach a level of more than 7 billion people by 2010, and nearly 8.4 billion by 2025.

The Future of Agriculture

One of the greatest and most immediate challenges facing today's world is producing enough food to feed our expanding population. This problem is often not appreciated by economists, who estimate that world food production has expanded 2.6 times since 1950, more rapidly than the human population. However, virtually all land that can be cultivated is already in use, and much of the world is populated by large numbers of hungry people who are rapidly destroying the sustainable productivity of the lands they inhabit. Well over 20% of the world's topsoil has been lost from agricultural lands since 1950. In the face of these massive problems, we need to consider what the prospects are for increased agricultural productivity in the future.

Finding New Food Plants

How many food plants do we use at present? Just three species—rice, wheat, and corn—supply more than half of all human energy requirements. Just over 100 kinds of plants supply over 90% of the calories we consume. Only about 5000 have ever been used for food. There may be tens of thousands of additional kinds of plants, among the 250,000 known species, that could be used for human food if their properties were fully explored and they were brought into cultivation (figure 27.5).

Agricultural scientists are attempting to identify such new crops, especially ones that will grow well in the tropics and subtropics, where the world's population is expanding most rapidly. Nearly all major crops now grown in the world have been cultivated for hundreds or even thousands of years. Only a few, including rubber and oil palms, have entered widespread cultivation since 1800.

One key feature for which nearly all of our important crops were first selected was ease of growth by relatively simple methods. Today, however, techniques of cultivation are far more sophisticated and are able to improve soil fertility and combat pests. This enables us to consider many more plants as potential crops. We are searching systematically for new crops that fit the multiple needs of modern society, in ways that would not have been considered earlier.

(a)

(b)

FIGURE 27.5
New food plants. (a) Grain amaranths (*Amaranthus* spp.) were important crops in the Latin American highlands during the days of the Incas and Aztecs. Grain amaranths are fast-growing plants that produce abundant grain rich in lysine, an amino acid rare in plant proteins but essential for animal nutrition. (b) The winged bean (*Psophocarpus tetragonolobus*) is a nitrogen-fixing tropical vine that produces highly nutritious leaves and tubers whose seeds produce large quantities of edible oil. First cultivated in New Guinea and Southeast Asia, the winged bean has spread since the 1970s throughout the tropics.

Improving the Productivity of Today's Crops

Searching for new crops is not a quick process. While the search proceeds, the most promising strategy to quickly expand the world food supply is to improve the productivity of crops that are already being grown. Much of the improvement in food production must take place in the tropics and subtropics, where the rapidly growing majority of the world's population lives, including most of those enduring a life of extreme poverty. These people cannot be fed by exports from industrial nations, which contribute only about 8% of their total food at present and whose agricultural lands are already heavily exploited. During the 1950s and 1960s, the so-called Green Revolution introduced new, improved strains of wheat and rice. The production of wheat in Mexico increased nearly tenfold between 1950 and 1970, and Mexico temporarily became an exporter of wheat rather than an importer. During the same decades, food production in India was largely able to outstrip even a population growth of approximately 2.3% annually, and China became self-sufficient in food.

Despite the apparent success of the Green Revolution, improvements were limited. Raising the new agricultural strains of plants requires the expenditure of large amounts of energy and abundant supplies of fertilizers, pesticides, and herbicides, as well as adequate machinery. For example, in the United States it requires about 1000 times as much energy to produce the same amount of wheat produced from traditional farming methods in India.

Biologists are playing a crucial role in improving existing crops and in developing new ones by applying traditional methods of plant breeding and selection to many new, nontraditional crops in the tropics and subtropics (see figure 27.5). A recent triumph in this field by scientists at the International Rice Research Institute in the Philippines announced in 1994 was the development of a new "super rice." This new strain can produce 25% higher yields than strains currently in use. If widely planted, this could lead to the production of enough rice to feed an additional 450 million people, mostly in Asia where about 90% of the world's rice is grown and consumed. The new strain also requires less water and fertilizer than the traditional strains. Fertilizer is becoming increasingly difficult for poor nations to afford.

Genetic Engineering to Improve Crops

Genetic engineering techniques (see chapter 18) will make it possible to produce plants resistant to specific herbicides. These herbicides can then control weeds much more effectively, without damaging crop plants. Genetic engineers are also developing new strains of plants that will grow successfully in areas where they previously could not grow. Eventually, desirable characteristics could be introduced into important crop plants, such as the ability to tolerate irrigation with salt water, fix nitrogen, carry out C_4 photosynthesis, or produce substances that deter pests and diseases.

The ability to transfer genes between organisms, first accomplished in a laboratory in 1973, has tremendous potential for the improvement of crop plants as the twentieth century closes.

New Approaches to Cultivation

Several new approaches may improve crop production. "No-till" agriculture, spreading widely in the United States and elsewhere in the 1990s, conserves topsoil and so is a desirable agricultural practice for many areas. On the other hand, **hydroponics,** the cultivation of plants in water containing an appropriate mixture of nutrients, holds less promise, as it does not differ remarkably in its requirements and challenges from growing plants on land, since it requires as much fertilizer and other chemicals, as well as the water itself.

The oceans were once regarded as an inexhaustible source of food, but overexploitation of their resources is limiting the world catch more each year, and these catches are costing more in terms of energy. Mismanagement of fisheries, mainly through overfishing, local pollution, and the destruction of fish breeding and feeding grounds, has lowered the catch of fish in the sea by about 20% from maximum levels. Many fishing areas that were until recently important sources of food have been depleted or closed. For example, the Grand Banks in the North Atlantic Ocean off Newfoundland, a major source of cod and other fish, are now nearly depleted. In 1994, the Canadian government prohibited all cod fishing there indefinitely, throwing 27,000 fishermen out of work, and the United States government banned all fishing on Georges Bank and other defined New England waters. Populations of Atlantic bluefin tuna have dropped 90% since 1975. The United Nations Food and Agriculture organization estimated in 1993 that 13 of 17 major ocean fisheries are in trouble, with the annual marine fish catch dropping from 86 million metric tons in 1986 to 82.5 million tons by 1992 and continuing to fall each year as the intensity of the fishing increases.

The development of new kinds of food, such as microorganisms cultured in nutrient solutions, should definitely be pursued. For example, the photosynthetic, nitrogen-fixing cyanobacterium *Spirulina* is being investigated in several countries as a possible commercial food source. It is a traditional food in Africa, Mexico, and other regions. *Spirulina* thrives in very alkaline water, and it has a higher protein content than soybeans. Ponds in which it grows are 10 times more productive than wheat fields. Such protein-rich concentrates of microorganisms could provide important nutritional supplements. However, psychological barriers must be overcome to persuade people to eat such foods, and the processing required tends to be energy-expensive.

Just over 100 kinds of plants, out of the roughly 250,000 known, supply more than 90% of all the calories we consume. Many more could be developed by a careful search for new crops.

The simplest way to gain a feeling for the dimensions of the global environmental problem we face is simply to scan the front pages of any newspaper or news magazine or to watch television. Although they are only a sampling, features selected by these media teach us a great deal about the scale and complexity of the challenge we face. We will discuss a few of the most important issues here.

Nuclear Power

At 1:24 A.M. on April 26, 1986, one of the four reactors of the Chernobyl nuclear power plant blew up. Located in Ukraine 100 kilometers north of Kiev, Chernobyl was one of the largest nuclear power plants in Europe, producing 1000 megawatts of electricity, enough to light a medium-sized city. Before dawn on April 26, workers at the plant hurried to complete a series of tests of how Reactor Number 4 performed during a power reduction and took a foolish short-cut: they shut off all the safety systems. Reactors at Chernobyl were graphite reactors designed with a series of emergency systems that shut the reactors down at low power, because the core is unstable then—and these are the emergency systems the workers turned off. A power surge occurred during the test, and there was nothing to dampen it. Power zoomed to hundreds of times the maximum, and a white-hot blast with the force of a ton of dynamite partially melted the fuel rods and heated a vast head of steam that blew the reactor apart.

The explosion and heat sent up a plume 5 kilometers high, carrying several tons of uranium dioxide fuel and fission products. The blast released over 100 megacuries of radioactivity, making it the largest nuclear accident ever reported; by comparison, the Three Mile Island accident in Pennsylvania in 1979 released 17 curies, millions of times less. This cloud traveled first northwest, then southeast, spreading the radioactivity in a band across central Europe from Scandinavia to Greece. Within a 30-kilometer radius of the reactor, at least one-fifth of the population, some 24,000 people, received serious radiation doses (greater than 45 rem). Thirty-one individuals died as a direct result of radiation poisoning, most of them firefighters who succeeded in preventing the fire from spreading to nearby reactors.

The rest of Europe received a much lower but still significant radiation dose. Data indicate that, because of the large numbers of people exposed, radiation outside of the immediate Chernobyl area can be expected to cause from 5000 to 75,000 cancer deaths.

The Promise of Nuclear Power

Our industrial society has grown for over 200 years on a diet of cheap energy. Until recently, much of this energy has been derived from burning wood and fossil fuels: coal, gas, and oil. However, as these sources of fuel become increasingly scarce and the cost of locating and extracting new deposits becomes more expensive, modern society is being forced to look elsewhere for energy. The great promise of nuclear power is that it provides an alternative source of plentiful energy. Although nuclear power is not cheap—power plants are expensive to build and operate—its raw material, uranium ore, is so common in the earth's crust that it is unlikely we will ever run out of it.

Burning coal and oil to obtain energy produces two undesirable chemical by-products: sulfur and carbon dioxide. As we will see in this chapter, the sulfur emitted from burning coal is a principal cause of acid rain, while the CO_2 produced from burning all fossil fuels is a major greenhouse gas (see the discussion of global warming in the next section). For these reasons, we need to find replacements for fossil fuels.

For all of its promise of plentiful energy, nuclear power presents several new problems that must be mastered before its full potential can be realized. You have encountered one serious challenge already in this chapter: the need to ensure safe operation of the world's approximately 390 nuclear reactors. A second challenge is the need to safely dispose of the radioactive wastes produced by the plants and to safely decommission plants that have reached the end of their useful lives (about 25 years). In 1997, over 35 plants were more than 25 years old, and not one has been safely decommissioned, its nuclear wastes disposed of. A third challenge is the need to guard against terrorism and sabotage, because the technology of nuclear power generation is closely linked to that of nuclear weapons.

For these reasons, it is important to continue to investigate and develop other promising alternatives to fossil fuels, such as solar energy and wind energy. The generation of electricity by burning fossil fuels accounts for up to 15% of global warming gas emissions in the United States. As much as 75% of the electricity produced in the United States and Canada currently is wasted through the use of inefficient appliances, according to scientists at Lawrence Berkeley Laboratory. Using highly efficient motors, lights, heaters, air conditioners, refrigerators, and other technologies already available could save huge amounts of energy and greatly reduce global warming gas emission. For example, a new, compact fluorescent lightbulb uses only 20% of the amount of electricity a conventional lightbulb uses, provides equal or better lighting, lasts up to 13 times longer, and provides substantial cost savings.

Nuclear power offers plentiful energy for the world's future, but its use involves significant problems and dangers.

Carbon Dioxide and Global Warming

By studying earth's history and making comparisons with other planets, scientists have determined that concentrations of gases in the atmosphere, particularly carbon dioxide, maintain the average temperature on earth about 25 °C higher than it would be if these gases were absent. Carbon dioxide and other gases trap the longer wavelengths of infrared light, or heat, radiating from the surface of the earth, creating what is known as a **greenhouse effect** (figure 27.6). The atmosphere acts like the glass of a gigantic greenhouse surrounding the earth.

Roughly seven times as much carbon dioxide is locked up in fossil fuels, approximately 5 trillion metric tons, as exists in the atmosphere today. Before widespread industrialization, the concentration of carbon dioxide in the atmosphere was approximately 260 to 280 parts per million (ppm). Since the extensive use of fossil fuels, the amount of carbon dioxide in the atmosphere has been increasing rapidly. Another major contribution of carbon dioxide added to the atmosphere comes from the destruction of forests, which releases large amounts of carbon dioxide when the trees and other vegetation are burned. During the 25-year period starting in 1958, the concentration of carbon dioxide increased from 315 ppm to more than 340 ppm. Climatologists have calculated that the actual mean global temperature has increased about 1 °C since 1900, a phenomenon known as **global warming.**

In a recent study, the U.S. National Research Council estimated that the concentration of carbon dioxide in the atmosphere would pass 600 ppm (roughly double the current level) by the third quarter of the next century, and might exceed that level as soon as 2035. These concentrations of carbon dioxide, if actually reached, would warm global surface air by between 1.5° and 4.5 °C. The actual increase might be considerably greater, however, because a number of trace gases, such as nitrous oxide, methane, ozone, and chlorofluorocarbons, are also increasing rapidly in the atmosphere as a result of human activities. These gases have warming, or "greenhouse," effects similar to those of carbon dioxide. One, methane, increased from 1.14 ppm in the atmosphere in 1951 to 1.68 ppm in 1986—nearly a 50% increase.

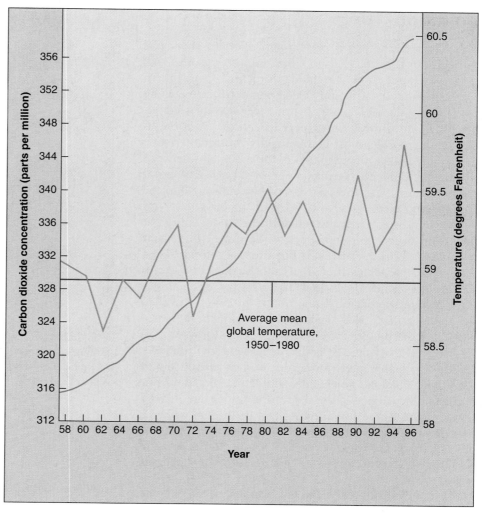

FIGURE 27.6
The greenhouse effect. The concentration of carbon dioxide in the atmosphere has steadily increased since the 1950s (*blue line*). The red line shows the general increase in average global temperature for the same period of time.
Source: Data from Geophysical Monograph, American Geophysical Union, National Academy of Sciences, and National Center for Atmospheric Research.

Major problems associated with climatic warming include rising sea levels. Sea levels may have already risen 2 to 5 centimeters from global warming. If the climate becomes so warm that the polar ice caps melt, sea levels would rise by more than 150 meters, flooding the entire Atlantic coast of North America for an average distance of several hundred kilometers inland.

Changes in the distribution of precipitation are difficult to model, although of great importance to local productivity. Certainly, changing climatic patterns are likely to make some of the best farmlands much drier than they are at present. If the climate warms as rapidly as many scientists project, the next 50 years may see greatly altered weather patterns, a rising sea level, and major shifts of deserts and fertile regions.

As the global concentration of carbon dioxide increases, the world's temperature is rising, with great potential impact on the world's climate.

Pollution

The River Rhine is a broad ribbon of water that runs through the heart of Europe. From high in the Alps that separate Italy and Switzerland, the Rhine flows north across the industrial regions of Germany before reaching Holland and the sea. Judging by the sheer amount of goods produced and shipped on or near its shores, the Rhine is one of the world's most commercially important rivers. The Rhine is also, where it crosses the mountains between Mainz and Coblenz, Germany, one of the most beautiful rivers on earth. On the first day of November 1986, the Rhine almost died.

The blow that struck at the life of the Rhine did not at first seem deadly. Firefighters were battling a blaze that morning in Basel, Switzerland. The fire was gutting a huge warehouse, into which firefighters shot streams of water to dampen the flames. The warehouse belonged to Sandoz, a giant chemical company. In the rush to contain the fire, no one thought to ask what chemicals were stored in the warehouse. By the time the fire was out, streams of water had washed 30 tons of mercury and pesticides into the Rhine.

Flowing downriver, the deadly wall of poison killed everything it passed. For hundreds of kilometers, dead fish blanketed the surface of the river. Many cities that use the water of the Rhine for drinking had little time to make other arrangements. Even the plants in the river began to die. All across Germany, from Switzerland to the sea, the river reeked of rotting fish, and not one drop of water was safe to drink.

Six months later, Swiss and German environmental scientists monitoring the effects of the accident were able to report that the blow to the Rhine was not mortal. Enough small aquatic invertebrates and plants had survived to provide a basis for the eventual return of fish and other water life, and the river was rapidly washing out the remaining residues from the spill. A lesson difficult to ignore, the spill on the Rhine has caused the governments of Germany and Switzerland to intensify efforts to protect the river from future industrial accidents and to regulate the growth of chemical and industrial plants on its shores.

The Threat of Pollution

The pollution of the Rhine is a story that can be told countless times in different places in the industrial world, from Love Canal in New York to the James River in Virginia to the town of Times Beach in Missouri. Nor are all pollutants that threaten the sustainability of life immediately toxic. Many forms of pollution arise as by-products of industry. For example, the polymers known as plastics, which we produce in abundance, break down slowly, if at all, in nature. Scientists are attempting to develop strains of bacteria that can decompose plastics, but their efforts have been largely unsuccessful. Consequently, virtually all of the plastic items that have ever been produced are still with us. Collectively, they constitute a new form of pollution.

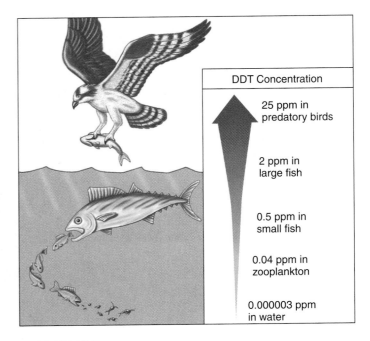

FIGURE 27.7
Biological magnification of DDT. Because DDT accumulates in animal fat, the compound becomes increasingly concentrated in higher levels of the food chain. Before DDT was banned in the United States, predatory bird populations drastically declined because DDT made their eggshells thin and fragile enough to break during incubation.

Widespread agriculture, carried out increasingly by modern methods, introduces large amounts of many new kinds of chemicals into the global ecosystem, including pesticides, herbicides, and fertilizers. Industrialized countries like the United States now attempt to carefully monitor side effects of these chemicals. Unfortunately, large quantities of many toxic chemicals no longer manufactured still circulate in the ecosystem.

For example, the chlorinated hydrocarbons, a class of compounds that includes DDT, chlordane, lindane, and dieldrin, have all been banned for normal use in the United States, where they were once widely used. They are still manufactured in the United States, however, and exported to other countries, where their use continues. Chlorinated hydrocarbon molecules break down slowly and accumulate in animal fat. Furthermore, as they pass through a food chain, they become increasingly concentrated in a process called **biological magnification** (figure 27.7). DDT caused serious problems by leading to the production of thin, fragile eggshells in many predatory bird species in the United States and elsewhere until the late 1960s, when it was banned in time to save the birds from extinction. Chlorinated compounds have other undesirable side effects and exhibit hormone-like activities in the bodies of animals.

> Chemical pollution is causing ecosystems to accumulate many harmful substances, as the result of spills and runoff from agricultural or urban use.

Acid Precipitation

The four smokestacks you see in figure 27.8 are part of the Four Corners power plant in New Mexico. This facility burns coal, sending smoke up high into the atmosphere through these stacks, each over 65 meters tall. The smoke the stacks belch out contains high concentrations of sulfur dioxide and other sulfates, which produce acid when they combine with water vapor in the air. The intent of those who designed the plant was to release the sulfur-rich smoke high in the atmosphere, where winds would disperse and dilute it, carrying the acids far away.

Environmental effects of this acidity are serious. Sulfur introduced into the upper atmosphere combines with water vapor to produce sulfuric acid, and when the water later falls as rain or snow, the precipitation is acid. Natural rainwater rarely has a pH lower than 5.6; in the northeastern United States, however, rain and snow now have a pH of about 3.8, roughly 100 times as acid.

Acid precipitation destroys life. Thousands of lakes in southern Sweden and Norway no longer support fish; these lakes are now eerily clear. In the northeastern United States and eastern Canada, tens of thousands of lakes are dying biologically as a result of acid precipitation (figure 27.9). At pH levels below 5.0, many fish species and other aquatic animals die, unable to reproduce. In southern Sweden and elsewhere, groundwater now has a pH between 4.0 and 6.0, as acid precipitation slowly filters down into the underground reservoirs.

There has been enormous forest damage in the Black Forest in Germany and in the forests of the eastern United States and Canada. It has been estimated that at least 3.5 million hectares of forest in the northern hemisphere are being affected by acid precipitation (figure 27.10), and the problem is clearly growing.

Its solution at first seems obvious: capture and remove the emissions instead of releasing them into the atmosphere. However, there are serious difficulties in executing this solution. First, it is expensive. The costs of installing and maintaining the necessary "scrubbers" in the United States are estimated to be 4 to 5 billion dollars per year. An additional difficulty is that the polluter and the recipient of the pollution are far from each other, and neither wants to pay for what they view as someone else's problem. The Clean Air Act revisions of 1990 addressed this problem in the United States significantly for the first time, and substantial worldwide progress has been made in implementing a solution.

Industrial pollutants such as nitric and sulfuric acids, introduced into the upper atmosphere by factory smokestacks, are spread over wide areas by the prevailing winds and fall to earth with precipitation called "acid rain," lowering the pH of water on the ground and killing life.

FIGURE 27.8
Four Corners power plant in New Mexico. In August 1991, the plant agreed to install "scrubbers" on its stacks, one of the first benefits of new federal clean air legislation.

FIGURE 27.9
Acid precipitation kills. In Twin Pond in the Adirondacks of upstate New York, as in many lakes in the region, high acidity levels have killed the fishes, amphibians, and most other kinds of animals and plants.

FIGURE 27.10
Damage to trees at Camels Hump Mountain in Vermont. Acid precipitation weakens trees and makes them more susceptible to pests and predators.

The Ozone Hole

The swirling colors of the satellite photos in figure 27.11 represent different concentrations of **ozone** (O_3), a different form of oxygen gas than O_2. As you can see, over Antarctica there is an "ozone hole" about the size of the United States, an area within which the ozone concentration is much less than elsewhere. This ozone hole was first reported in 1985 by British environmental scientists. Reviewing satellite data, we now know the ozone thinning appeared for the first time in 1975. The hole is not a permanent feature, but rather becomes evident each year for a few months at the onset of the Antarctic spring. Every September from 1975 onward, the ozone "hole" has reappeared. Each year the layer of ozone is thinner and the hole is larger, sometimes reaching southern New Zealand, Australia, and southern South America.

The major cause of the ozone depletion had already been suggested in 1974 by Sherwood Roland and Mario Molina, who were awarded the Nobel Prize for their work in 1995. They proposed that chlorofluorocarbons (CFCs), relatively inert chemicals used in cooling systems, fire extinguishers, and Styrofoam containers, were percolating up through the atmosphere and reducing O_3 molecules to O_2. One chlorine atom from a CFC molecule could destroy 100,000 ozone molecules in the following mechanism:

UV radiation causes CFCs to release Cl atoms:

$$CCl_3F \xrightarrow{UV} Cl + CCl_2F$$

UV creates oxygen free radicals:

$$O_2 \longrightarrow 2O$$

Cl atoms and O free radicals interact with ozone:

$$2Cl + 2O_3 \longrightarrow 2ClO + 2O_2$$
$$2ClO + 2O \longrightarrow 2Cl + 2O_2$$

Net reaction: $2O_3 \longrightarrow 3O_2$

Although other factors have also been implicated in ozone depletion, the role of CFCs is so predominant that worldwide agreements have been signed to phase out their production by the year 2000. The United States banned the production of CFCs and other ozone-destroying chemicals after 1995. Nonetheless, the CFCs that were manufactured earlier are moving slowly upward through the atmosphere. The ozone layer will be further depleted before it begins to form again.

Thinning of the ozone layer in the stratosphere, 25 to 40 kilometers above the surface of the earth, is a matter of serious concern. This layer protects key biological molecules, especially proteins and nucleic acids, from the harmful ultraviolet rays that bombard the earth continuously from the sun. Life on land may have become possible only when the oxygen layer was sufficiently thick to generate enough ozone to shield the surface of the earth from these destructive rays. This may account for the billions of years in which all life was aquatic. There may have been too much opportunity for damage to biologically significant molecules for organisms to exist on dry land.

Ultraviolet radiation is a serious human health concern. Every 1% drop in atmospheric ozone is estimated to lead to a 6% increase in the incidence of skin cancers. At middle latitudes, the approximately 3% drop that has already occurred worldwide is estimated to have increased skin cancers by as much as 20%. Skin cancers (melanomas) are one of the more lethal diseases afflicting humans.

> **Industrial CFCs released into the atmosphere react at very cold temperatures with ozone, converting it to oxygen gas. This has the effect of destroying the earth's ozone shield and exposing the earth's surface to increased levels of harmful UV radiation.**

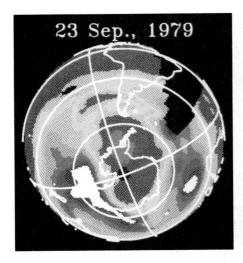 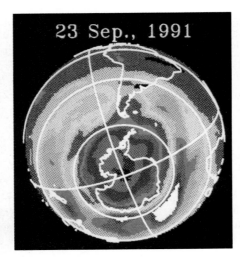 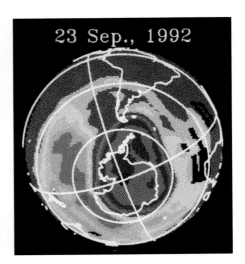

FIGURE 27.11
Satellite photos of the ozone hole over Antarctica. In these satellite views of the South Pole, the swirling colors represent ozone levels in the atmosphere. Pink and purple indicate the areas with the least amount of ozone. As you can see, there is an ozone hole over Antarctica that grew between 1979 and 1992 to about the size of the United States.

Destruction of the Tropical Forests

More than half of the world's human population lives in the tropics, and this percentage is increasing rapidly. For global stability, and for the sustainable management of the world ecosystem, it will be necessary to solve the problems of food production and regional stability in these areas. World trade, political and economic stability, and the future of most species of plants, animals, fungi, and microorganisms depends on our addressing these problems.

Rainforests Are Rapidly Disappearing

Tropical rainforests are biologically the richest of the world's biomes. Most other kinds of tropical forest, such as seasonally dry forests and savanna forests, have already been largely destroyed—because they tend to grow on more fertile soils, they were exploited by humans a long time ago. Now the rainforests, which grow on poor soils, are being destroyed. In the mid-1990s, it is estimated that only about 5.5 million square kilometers of tropical rainforest still exist in a relatively undisturbed form. This area, about two-thirds of the size of the United States (excluding Alaska), represents about half of the original extent of the rainforest. From it, about 160,000 square kilometers are being clear-cut every year, with perhaps an equivalent amount severely disturbed by shifting cultivation, firewood gathering, and the clearing of land for cattle ranching. The total area of tropical rainforest destroyed—and therefore permanently removed from the world total—amounts to an area greater than the size of Indiana each year. At this rate, all of the tropical rainforest in the world will be gone in about 30 years; but in many regions, the rate of destruction is much more rapid. As a result of this overexploitation, experts predict there will be little undisturbed tropical forest left anywhere in the world by early in the next century. Many areas now occupied by dense, species-rich forests may still be tree-covered, but the stands will be sparse and species-poor.

A Serious Matter

Not only does the disappearance of tropical forests represent a tragic loss of largely unknown biodiversity, but the loss of the forests themselves is ecologically a serious

(a)　　　　　　　　　　　　　　　　(b)

FIGURE 27.12
Destroying the tropical forests. (a) When tropical forests are cleared, the ecological consequences can be disastrous. These fires are destroying rainforest in Brazil and clearing it for cattle pasture. (b) The consequences of deforestation can be seen on these middle-elevation slopes in Ecuador, which now support only low-grade pastures and permit topsoil to erode into the rivers. These areas used to support highly productive forest, which protected the watersheds of the area, in the 1970s.

matter. Tropical forests are complex, productive ecosystems that function well in the areas where they have evolved. When people cut a forest or open a prairie in the north temperate zone, they provide farmland that we know can be worked for generations. In most areas of the tropics, people do not know how to engage in continuous agriculture. When they clear a tropical forest, they engage in a one-time consumption of natural resources that will never be available again (figure 27.12). The complex ecosystems built up over billions of years are now being dismantled, in almost complete ignorance, by humans.

What biologists must do is to learn more about the construction of sustainable agricultural ecosystems that will meet human needs in tropical and subtropical regions. The ecological concepts we have been reviewing in the last three chapters are universal principles. The undisturbed tropical rainforest has one of the highest rates of net primary productivity of any plant community on earth, and it is therefore imperative to develop ways that it can be harvested for human purposes in a sustainable, intelligent way.

More than half of the tropical rainforests have been destroyed by human activity, and the rate of loss is accelerating.

Environmental Science

Environmental scientists attempt to find solutions to environmental problems, considering them in a broad context. Unlike biology or ecology, sciences that seek to learn general principles about how life functions, environmental science is an applied science dedicated to solving practical problems. Its basic tools are derived from ecology, geology, meteorology, social sciences, and many other areas of knowledge that bear on the functioning of the environment and our management of it. Environmental science addresses the problems created by rapid human population growth: an increasing need for energy, a depletion of resources, and a growing level of pollution.

Solving Environmental Problems

The problems our severely stressed planet faces are not insurmountable. A combination of scientific investigation and public action, when brought to bear effectively, can solve environmental problems that seem intractable. Viewed simply, there are five components to solving any environmental problem:

1. **Assessment.** The first stage in addressing any environmental problem is scientific analysis, the gathering of information. Data must be collected and experiments performed to construct a model that describes the situation. This model can be used to make predictions about the future course of events.
2. **Risk analysis.** Using the results of scientific analysis as a tool, it is possible to analyze what could be expected to happen if a particular course of action were followed. It is necessary to evaluate not only the potential for solving the environmental problem, but also any adverse effects a plan of action might create.
3. **Public education.** When a clear choice can be made among alternative courses of action, the public must be informed. This involves explaining the problem in terms the public can understand, presenting the alternatives available, and explaining the probable costs and results of the different choices.
4. **Political action.** The public, through its elected officials selects a course of action and implements it. Choices are particularly difficult to implement when environmental problems transcend national boundaries.
5. **Follow-through.** The results of any action should be carefully monitored to see whether the environmental problem is being solved as well as to evaluate and improve the initial modeling of the problem. Every environmental intervention is an experiment, and we need the knowledge gained from each one to better address future problems.

Individuals Can Make the Difference

The development of appropriate solutions to the world's environmental problems must rest partly on the shoulders of politicians, economists, bankers, engineers—many kinds of public and commercial activity will be required. However, it is important not to loose sight of the key role often played by informed individuals in solving environmental problems. Often one person has made the difference; two examples serve to illustrate the point.

The Nashua River. Running through the heart of New England, the Nashua River was severely polluted by mills established in Massachusetts in the early 1900s. By the 1960s, the river was clogged with pollution and declared ecologically dead. When Marion Stoddart moved to a town along the river in 1962, she was appalled. She approached the state about setting aside a "greenway" (trees running the length of the river on both sides), but the state wasn't interested in buying land along a filthy river. So Stoddart organized the Nashua River Cleanup Committee and began a campaign to ban the dumping of chemicals and wastes into the river. The committee presented bottles of dirty river water to politicians, spoke at town meetings, recruited businesspeople to help finance a waste treatment plant, and began to clean garbage from the Nashua's banks. This citizen's campaign, coordinated by Marion Stoddart, led in large measure to the passage of the Massachusetts Clean Water Act of 1966. Industrial dumping into the river is now banned, and the river has largely recovered.

Lake Washington. A large, 86 km² freshwater lake east of Seattle, Lake Washington became surrounded by Seattle suburbs in the building boom following the Second World War. Between 1940 and 1953, a ring of 10 municipal sewage plants discharged their treated effluent into the lake. Safe enough to drink, the effluent was believed "harmless." By the mid-1950s a great deal of effluent had been dumped into the lake (try multiplying 80 million liters/day × 365 days/year × 10 years). In 1954, an ecology professor at the University of Washington in Seattle, W. T. Edmondson, noted that his research students were reporting filamentous blue-green algae growing in the lake. Such algae require plentiful nutrients, which deep freshwater lakes usually lack—the sewage had been fertilizing the lake! Edmondson, alarmed, began a campaign in 1956 to educate public officials to the danger: bacteria decomposing dead algae would soon so deplete the lake's oxygen that the lake would die. After five years, joint municipal taxes financed the building of a trunk sewer to carry the effluent out to sea. The lake is now clean.

In solving environmental problems, the commitment of one person can make a critical difference.

Preserving Nonreplaceable Resources

Among the many ways ecosystems are suffering damage, one class of problem stands out as more serious than the rest: consuming or destroying resources that we cannot replace in the future. While a polluted stream can be cleaned up, no one can restore an extinct species. In the United States, we are consuming three nonreplaceable resources at alarming rates: topsoil, groundwater, and biodiversity.

Topsoil

The United States is one of the most productive agricultural countries on earth, largely because much of it is covered with particularly fertile soils. Our Midwestern farm belt sits astride what was once a great prairie. The **topsoil** of that ecosystem accumulated bit by bit from countless generations of animals and plants until, by the time humans began to plow it, the rich soil extended down several feet.

We can never replace this rich topsoil, the capital upon which our country's greatness is built, yet we are allowing it to be lost at a rate of centimeters every decade. By repeatedly tilling (turning the soil over) to eliminate weeds,
we permit rain to wash more and more of the topsoil away, into rivers, and eventually out to sea. Our country has lost one-quarter of its topsoil since 1950! New approaches are desperately needed to lessen our reliance on intensive cultivation. Some possible solutions include using genetic engineering to make crops resistant to weed-killing herbicides and terracing to recapture lost topsoil (figure 27.13).

Groundwater

A second resource we cannot replace is **groundwater,** water trapped beneath the soil within porous rock reservoirs called aquifers. This water seeped into its underground reservoir very slowly during the last ice age over 12,000 years ago. We should not waste this treasure, for we cannot replace it.

In most areas of the United States, local governments exert relatively little control over the use of groundwater. As a result, a large portion is wasted watering lawns, washing cars, and running fountains. A great deal more is inadvertently polluted by poor disposal of chemical wastes—and once pollution enters the groundwater, there is no effective means of removing it.

FIGURE 27.13
Contour farming, or terracing, preserves topsoil. Loss of topsoil is one of the world's most pressing environmental problems. Contour farming is one way to minimize soil runoff and improve agricultural success.

Loss of Biodiversity

The most serious and rapidly accelerating of all global environmental problems is the loss of **biodiversity.** Over the past 300 years, many species of organisms, including mammals, birds, and plants, have been lost. In addition, habitat is vanishing rapidly, especially in the tropics. Scientists such as E. O. Wilson of Harvard University have calculated that as much as 20% of the world's biodiversity may be lost during the next 30 years. No more than 15% of the world's eukaryotic organisms have been described, and a much smaller proportion of tropical organisms have been named. Thus, we may never even know of the existence of many of the organisms we are driving to extinction.

As many as 50,000 species of the world total of 250,000 species of plants, 4000 of the world's 20,000 species of butterflies, and nearly 2000 of the world's 9000 species of birds could be lost during the next few decades. This is an astonishing rate of loss.

Why Preserve Biodiversity?

This loss is tragic for several reasons. First, many feel that on moral, ethical, and aesthetic grounds, we do not have the right to drive to extinction such a high proportion of what are, as far as we know, our only living companions in the universe. Many believe we should take positive, international actions such as those prescribed by the International Biodiversity Convention, introduced at the Earth Summit in Rio de Janeiro in June 1993.

Second, organisms are our only means of sustainability. If we want to solve the problem of how to occupy the world on a continuing basis, it will be the properties of organisms that makes it possible. Organisms are the only sustainable sources of food, medicine, clothing, biomass (for energy and other purposes), and shelter (figure 27.14). We have examined only a minute proportion of the existing kinds of organisms to see whether their properties are of interest. We are just beginning to be able to enumerate genetic differences between organisms and to use their genes for our advantage. We are killing species off at a rate not approached for the past 65 million years. Far more species are in danger of extinction within our lifetimes than the total number that became extinct at the end of the Cretaceous period. One can scarcely think of anything we could passively allow to happen that would be more detrimental to the human future.

Third, organisms function in communities to preserve soils, regulate water and nutrient cycles essential to plants, modulate characteristics of the atmosphere, and absorb pollution. By destroying them in the relentless drive toward "development," we are creating conditions of instability and unproductivity, including desertification, waterlogging, mineralization, and many other undesirable outcomes throughout the world.

FIGURE 27.14
Plants such as this rosy periwinkle offer useful products we have barely tapped. The rosy periwinkle (*Catharanthus roseus*), native only to Madagascar, is a garden plant now widespread in cultivation. Two drugs developed from periwinkle by Eli Lilly & Company, vinblastine and vincristine, effectively treat certain forms of leukemia. With these drugs, a child with leukemia has a 95% chance of survival past the age of 5. In 1950, the child would have had only a 20% chance to live.

Conservation Biology

Concerned about the dramatic losses in biological diversity occurring worldwide, biologists have begun concerted efforts to deal with the biodiversity crisis. A new discipline, **conservation biology,** focuses on better understanding the causes of extinction, and on discovering better ways to manage habitats and ecosystems. As it has become clear that isolated patches of habitat lose species far more rapidly than large preserves do, biologists have promoted the creation in the tropics of so-called **megareserves,** large areas of land containing a core of one or more undisturbed habitats. The key to preserving such large tracts of land is to preserve the ecological reserve in a way compatible with local land use. Thus, while no economic activity is allowed in the core regions of the megareserve, the remainder of the reserve may be used for nondestructive harvesting of resources. Linking preserved areas to carefully managed land zones creates a much larger total "patch" of habitat than would otherwise be economically practical. Pioneering these efforts, a series of eight such megareserves have been created in Costa Rica to jointly manage biodiversity and economic activity.

The biosphere faces a biodiversity crisis as the tropical rainforests are destroyed by human activity. Only enlightened land management will foster the preservation of what biodiversity remains.

27.1 The world's human population is growing explosively.

- When the college class of 1998 started school in 1980, 4 billion people inhabited the world. There are 6 billion people today, and the world's population is increasing by 100 million people per year. At this rate, over 8 billion people will be living when their children attend college, double the number in 1980.

- An explosively growing human population is placing considerable stress on the environment, consuming ever-increasing quantities of food and water, using a great deal of energy and raw materials, and producing enormous amounts of waste and pollution.

27.2 Improvements in agriculture are needed to feed a hungry world.

- Much current effort is focused on improving the productivity of existing crops, although the search for new crops continues.

27.3 Human activity is placing the environment under increasing stress.

- Human activities present many challenges to the environment, including the release of harmful materials into the environment.

- Nuclear power is a promising energy source, but the release of radioactive materials into the ecosystem may increase the incidence of cancer.

- Burning fossil fuels releases carbon dioxide, which may increase the world's temperature and alter weather and ocean levels.

- Release of pollutants into rivers may make the water unfit for aquatic life and human consumption.

- Release of industrial smoke into the upper atmosphere leads to acid precipitation that kills forests and lakes.

- Release of chemicals such as chlorofluorocarbons may destroy the atmosphere's ozone and expose the world to dangerous levels of ultraviolet radiation.

- Other challenges arise from our attempts to "develop" natural resources; for example, cutting and burning the tropical rainforests of the world to make pasture and cropland is producing a massive wave of extinction.

27.4 Solving environmental problems requires individual involvement.

- All of these challenges to our future can and must be addressed. Today, environmental scientists and concerned citizens are actively searching for constructive solutions to these problems.

- The application of the principles of biology to human problems has never been more necessary than it is now. Only the full attention of society and many talented individuals to the solution of these grave problems will make it possible for our children and grandchildren to enjoy the same benefits that we now enjoy.

1. **Population growth** The human population has grown from an estimated 130 million people 2000 years ago to 2.5 billion in 1950 to 6 billion just 48 years later. Recent rapid increases in growth threaten the planet's ability to sustain life in many regions. Although a stable population is likely to be attained in the next century, it could range from 8.5 billion to nearly 20 billion. The difference is of deep significance for the quality of life on earth.

2. **Resource consumption** Not only population size but levels of consumption and the use of technology affect the human impact on the planet. At present, humans are consuming, wasting, or diverting a large portion of the total net photosynthetic productivity on land.

3. **Crop improvement** The most promising strategy for increasing the food supply is to improve the productivity of known crops. Just over 100 kinds of crops provide more than 90% of human caloric intake. Potential opportunities also exist for developing plants as sources of food, medicine, chemicals, and shelter materials.

4. **Tropics** Population growth is most rapid in the tropics and subtropics, where fragile ecosystems must be managed with special care to be sustainable.

5. **Environmental impacts** Among the environmental challenges we face are developing appropriate energy sources; avoiding pollution, global warming, and acid precipitation; and preserving the stratospheric ozone layer.

6. **Biodiversity** The irreversible loss of biodiversity, which may lead to extinction of a fifth or more of all species on earth during the next 30 years, is the most challenging of all environmental problems and is completely irreversible. By altering the quantity and quality of life on earth so profoundly, we are greatly limiting our options for the future.

Review Questions

1. What biological event fostered the rapid growth of human populations? How did this event affect the location in which humans lived? What major cultural event eventually took place?

2. What three species supply more than half of the human energy requirements on earth? How many plants supply over 90%?

3. What are the benefits of nuclear power? What problems must we master before its full potential can be realized?

4. How has the amount of carbon dioxide in the atmosphere changed since the advent of industrialization? Why? What effect does this have on the world as a whole?

5. Why were chlorinated hydrocarbons banned in the United States? Why can you still find them as contaminants on fruits and vegetables?

6. How does acid precipitation form? Why has it been difficult to implement solutions to this problem?

7. What is the ozone layer? How is it formed? What are the harmful effects of decreasing the earth's ozone layer? What may be the primary cause of this damage?

8. Explain three reasons why the loss of biodiversity is important.

Thought Questions

1. Can we ever produce enough food and other materials to ensure that population growth will not be a matter of concern? How?

2. Explain what occurred during the Green Revolution of the 1950s and 1960s. Why don't most people consider the Green Revolution an unqualified success?

3. Some have argued that attempts by the United States to promote lower birth rates in underdeveloped tropical countries is nothing more than economic imperialism, and that it is in the best interest of these countries to allow their populations to grow as rapidly as possible. Give the reasons why you agree or disagree with this assessment.

Internet Links

Conservation Biology
http://darwin.bio.uci.edu/~sustain/bio65/
A web book, BIODIVERSITY AND CONSERVATION, created by Peter Bryant for his classes at UC Irvine, contains chapters on extinction, deforestation, and many other topics, packed with links.

Pluging In to Global Change
http://jasper.stanford.edu/GCTE/hotlink.html
An extensive collection of hot links from Stanford University that connect you to sites concerned with global change.

A Map of Acid Rain in the United States
http://nadp.nrel.colostate.edu/isopleths/
Very illuminating on-line color maps from Colorado State University showing the average pH levels for precipitation (current levels of acid rain) in the United States.

Grappling With the Greenhouse Effect
http://www.erin.gov.au/air
From Australia, a detailed look at many aspects of the environment, including the greenhouse effect, its enhancement by emissions of CO_2 and other gases, and potential effects on global climate.

Deforestation of the Amazon Viewed From Space
http://pathfinder-www.sr.unh.edu/pathfinder1/information/papers/index.html
A collection of papers published in popular journels that use satellite imagery to document destruction of the tropical rainforest.

Eye on the Planet
www.usgs.gov/Earthshots
Environmental stories told with Landsat Satellite Images.

For Further Reading

Blaustein, H. R. and D. B. Wake: "The Puzzle of Declining Amphibian Populations," *Scientific American*, April 1995, pages 52–57. The plight of numerous frogs, toads, and salamanders is discussed in light of habitat and ozone destruction. A thorough review of the species in particular danger, with beautiful photographs.

Malle, K.: "Cleaning Up the River Rhine," *Scientific American*, January 1996, pages 70–75. Intensive international efforts are reclaiming the devastated river.

Rice, R. E., R. E. Gullison, and J. W. Reid: "Can Sustainable Management Save Tropical Forests," *Scientific American*, April 1997, pages 44–49. Scientists assess the problems of preserving tropical rainforests and propose strategies for balancing the harvest of timber with the conservation of it.

Toon, O. and R. Turco: "Polar Stratospheric Clouds and Ozone Depletion," *Scientific American*, June 1991, pages 68–74. The story of how the ozone hole forms every spring in the skies over Antarctica.

Vietmeyer, N. D.: "Lesser-Known Plants of Potential Use in Agriculture and Forestry," *Science*, vol. 232, 1986, pages 1379–84. An excellent, brief survey of little-known but extremely useful plants that could improve food and firewood supplies in the future.

White, R. M.: "The Great Climate Debate," *Scientific American*, July 1990, pages 36–43. Outstanding analysis of actions we might take now in the face of impending climatic change.

28

How We Classify Organisms

Concept Outline

28.1 Biologists name organisms in a systematic way.

The Classification of Organisms. Biologists name organisms using a binomial system.

Species Names. Every kind of organism is assigned a unique name.

The Taxonomic Hierarchy. The higher groups into which an organism is placed reveal a great deal about the organism.

What Is a Species? Species are groups of similar organisms that tend not to interbreed with individuals of other groups.

28.2 Taxonomy is the science of classifying organisms.

Evolutionary Taxonomy. Traditional and cladistic interpretations of evolution differ in the emphasis they place on particular traits.

28.3 All living organisms are grouped into one of a few major categories.

The Kingdoms of Life. Living organisms are grouped into three great groups called domains, and within domains into kingdoms.

Domain Archaea (Archaebacteria). The oldest domain consists of primitive bacteria that often live in extreme environments.

Domain Bacteria (Eubacteria). Too small to see with the unaided eye, there are more individual eubacteria than any other organism.

Domain Eukarya (Eukaryotes). There are four kingdoms of eukaryotes, three of them entirely or predominantly multicellular.

Key Characteristics of Eukaryotes. Two of the most important characteristics to have evolved among the eukaryotes are multicellularity and sexuality.

Viruses: A Special Case. Viruses are not organisms, and thus do not belong to any kingdom.

FIGURE 28.1
Biological diversity. All living things are assigned to particular classifications based on characteristics such as their anatomy, development, mode of nutrition, level of organization, and biochemical composition.

All organisms share many biological characteristics. They are composed of one or more cells, carry out metabolism and transfer energy with ATP, and encode hereditary information in DNA. All species have evolved from simpler forms and continue to evolve. Individuals live in populations. These populations make up communities and ecosystems, which provide the overall structure of life on earth. So far, we have stressed these common themes, considering the general principles that apply to all organisms. Now we will consider the diversity of the biological world and focus on the differences among groups of organisms (figure 28.1). For the rest of the text, we will examine the different kinds of life on earth, from bacteria and amoebas to blue whales and sequoia trees.

The Classification of Organisms

Millions of different kinds of organisms exist; to talk about them and study them, it is necessary that they have names. Going back as early as we can trace, organisms have been grouped in basic units, such as oaks, cats, and horses. Eventually, these units were called **genera** (singular, **genus**), and they often acquired the names the Greeks and Romans had given them. Starting in the Middle Ages, these names were written in Latin, the language of scholars at that time, or given a Latin form. Oaks were assigned to the genus *Quercus*, cats to *Felis*, and horses to *Equus*—all names that the Romans applied to these groups of organisms. For genera that were not known in antiquity, new names had to be invented.

Quercus phellos
(Willow oak)

Quercus rubra
(Red oak)

(a)

(b)

FIGURE 28.2
Two species of oaks. (a) Willow oak, *Quercus phellos.* (b) Red oak, *Quercus rubra.* Although they are both oaks (*Quercus*), these two species differ sharply in leaf shape and size and in many other features, including geographical distributions.

About 300 years ago, scientists first began to develop an overall classification system for organisms and to produce comprehensive works about them. Scientists who studied and contributed to the classification of organisms became known as **taxonomists** or **systematists,** and their subject is called **taxonomy** or **systematics.**

The Polynomial System

Before the 1750s, scholars who wanted to designate a particular species usually added a series of additional descriptive terms to the name of the genus. The resulting string of Latin words was called a **polynomial.** One name for the European honeybee, for example, was *Apis pubescens, thorace subgriseo, abdomine fusco, pedibus posticis glabris utrinque margine ciliatis.* Because such polynomials were obviously cumbersome and often changed from one scholar to another, single species often did not have an easily recognized single name. A better alternative was needed.

The Binomial System

The Swedish biologist Carl Linnaeus (1707–1778) developed a simpler system while preparing encyclopedic works on plants, animals, and minerals. In the 1750s, he produced several major works that employed the cumbersome polynomial system. However, using a kind of shorthand of reference, Linnaeus included a two-part name (binomial) for each species. For example, the honeybee became *Apis mellifera.*

A Closer Look at Linnaeus

To illustrate Linnaeus's work further, lets consider how he treated two species of oaks from North America. He grouped all oaks in the genus *Quercus*, as had been the practice since Roman times. The willow oak of the southeastern United States (figure 28.2*a*) had by 1753, when Linnaeus was writing his important book on the kinds of plants, *Species Plantarum*, been described by a number of travelers and noted by several scientists. Linnaeus took the name that he adopted for it, *Quercus foliis lanceolatis integerrimis glabris* ("oak with spear-shaped, smooth leaves with absolutely no teeth along the margins"), from an account of the plants of Virginia published from 1739 to 1743 by his friend, the Dutch botanist Jan Fredrick Gronovius. For the common red oak of eastern temperate North America (figure 28.2*b*), Linnaeus devised a new name, *Quercus foliis obtuse-sinuatis setaceo-mucronatis* ("oak with leaves with deep blunt lobes bearing hairlike bristles"). For each of these species, he also presented a shorthand designation, the binomial names *Quercus phellos* and *Quercus rubra*. These have remained the official names for these species since 1753, even though Linnaeus did not intend this when he first used them in his book. He considered the polynomials the true names of the species.

The names of genera were the basic point of reference in classification systems, and these names eventually came to be written in Latin.

Species Names

Species epithets, the second part of the name of each species, are meaningless when used by themselves. But the name of the genus may be abbreviated to a single letter (for example, *Q. phellos*) if the reference to the abbreviated generic name is clear from the context.

International associations of taxonomists regularly establish the principles that govern the selection of the correct scientific names for organisms. There are different associations that are concerned with animals, plants, fungi, algae, and bacteria. Committees rule on the appropriate names in cases of doubt. For the most part, the resulting names are the same throughout the world and provide a uniform way of communicating about organisms. A Chinese-speaking biologist can be sure of the identity of *Quercus rubra* whether he or she is reading a paper written in English, Japanese, or Russian. In contrast, common names often differ greatly from place to place (figure 28.3). A "robin" in Europe is *Erithacus rubicula*, but in North America a robin is a larger bird, very different in appearance—*Turdus migratorius*. The clay-colored robin of Mexico and Central America, *Turdus grayi*, is a member of the same genus as the North American robin, but a different species. The hill robin of Asia is another very different kind of bird, *Leiothrix lutea*. The scientific names of these birds, in the context of their overall classification, make clear their degrees of relationship in a way that their common names do not. They also provide a constant point of reference for discussion anywhere in the world.

Even within one country, common names can confuse. The cat flea, sand flea, snow flea, and water flea have little in common. Despite sharing part of their common names, each of these organisms is in a different major group.

> For scientific communication, it is important to have one standard set of names. Biologists use the binomial system devised by Carl Linnaeus nearly 250 years ago.

(a)

(b)

(c)

FIGURE 28.3
Common names make poor labels. The common names corn (a), bear (b), and robin (c) bring clear images to our minds (photos on *left*), but the images would be very different to someone living in Europe or Australia (photos on *right*). There, the same common names are used to label very different species.

(a)

(b)

(c)

FIGURE 28.4
Three genera of the squirrel family, Sciuridae. These animals differ greatly in some characteristics and are adapted for different modes of life. (a) The red squirrel, *Tamiasciurus hudsonicus*, an agile dweller in the trees. (b) Townsend's chipmunk, *Eutamias townsendii*, a member of a genus of diurnal, brightly colored, very active ground-dwelling squirrels. (c) Yellow-bellied marmot, *Marmota flaviventris*, a large sciurid that lives in burrows. Yellow-bellied marmots are social, tolerant, and playful; they live in harems that consist of a single territorial male together with numerous females and their offspring.

The Taxonomic Hierarchy

In the decades following Linnaeus, taxonomists began to group genera into larger, more inclusive categories known as families. The membership of these families was intended to reflect perceived relationships between the genera included. The oaks (*Quercus*), beeches (*Fagus*), and chestnuts (*Castanea*) are included in the beech family, Fagaceae, because they share the characteristics of that family. Similarly, tree squirrels (*Tamiasciurus*), Siberian and western North American chipmunks (*Eutamias*), and marmots (*Marmota*) are included, along with other genera, in the family Sciuridae (figure 28.4). All have four front toes, five hind toes, a hairy tail and other features characteristic of the family. If one knows that a genus belongs to a particular family, one immediately knows a great many of its features.

The **taxonomic system** was eventually extended to include several more inclusive units. **Families** are grouped into **orders,** orders into **classes,** and classes into **phyla** (singular **phylum**). For historical reasons, phyla may also be called **divisions** among plants, fungi, and algae. Phyla are in turn grouped into **kingdoms,** and then **domains,** the most inclusive units of biological classification (figure 28.5).

Table 28.1 shows how the red squirrel, honeybee, and red oak are placed in taxonomic categories at the eight hierarchical levels. The categories at the different levels may include many, a few, or only one **taxon** (the general name given a taxonomic unit at any level). For example, there is only one living genus of the family Hominidae, but several living genera of Fagaceae. To someone familiar with classification or with access to the appropriate reference books, each taxon implies both a set of characteristics and a group of organisms belonging to the taxon.

Convention governs how scientific names are printed. The genus is always capitalized, but the species is not, and both are italicized, or written in distinctive print: for example, *Homo sapiens.* The scientific names of the other taxonomic units are capitalized but not printed distinctively, italicized, or underlined.

Species are grouped into genera, genera into families, families into orders, orders into classes, and classes into phyla. Phyla are the basic units within kingdoms; such a system is hierarchical.

Table 28.1	Sample Classifications of Three Organisms			
	Red Squirrel		**Honeybee**	**Red Oak**
Domain	Eukarya		Eukarya	Eukarya
Kingdom	Animalia		Animalia	Plantae
Phylum	Chordata		Arthropoda	Anthophyta
Class	Mammalia		Insecta	Dicotyledones
Order	Rodentia		Hymenoptera	Fagales
Family	Sciuridae		Apidae	Fagaceae
Genus	*Tamiasciuris*		*Apis*	*Quercus*
Species	*Tamiasciurus hudsonicus*		*Apis mellifera*	*Quercus rubra*

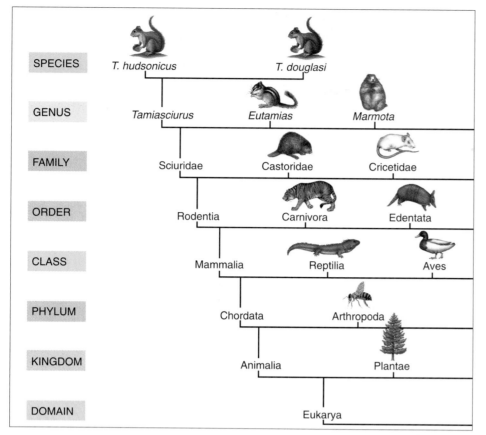

SPECIES			
	T. hudsonicus	T. douglasi	
GENUS	Tamiasciurus	Eutamias	Marmota
FAMILY	Sciuridae	Castoridae	Cricetidae
ORDER	Rodentia	Carnivora	Edentata
CLASS	Mammalia	Reptilia	Aves
PHYLUM	Chordata	Arthropoda	
KINGDOM	Animalia	Plantae	
DOMAIN		Eukarya	

FIGURE 28.5
The hierarchical system used in classifying organisms. A red squirrel is first recognized as a eukaryote (domain: Eukarya). Second, within this domain, it is an animal (kingdom: Animalia). Among the different phyla of animals, it is a vertebrate (phylum: Chordata). The squirrel's fur characterizes it as a mammal (class: Mammalia). Within this class, it is distinguished by its gnawing teeth (order: Rodentia). Next, because it has four front toes and five back toes, it is a squirrel (family: Sciuridae). Within this family, it is a tree squirrel with a black line along the side of its belly (genus: *Tamiasciurus*). Lastly, its fur is reddish (species: *Tamiasciurus hudsonicus*).

What Is a Species?

In chapter 21, we reviewed the nature of species and saw there are no absolute criteria for the definition of this category. Looking different, for example, is not a useful criterion: different individuals that belong to the same species (for example, dogs) may look very unlike one another, as different as a chihuahua and a St. Bernard. These very different-appearing individuals are fully capable of hybridizing with one another.

Among sexual species that regularly **outcross**—interbreed with individuals other than themselves—barriers to hybridization often serve as criteria for species. However, in many groups of organisms, including bacteria and many eukaryotes, **asexual reproduction**—reproduction without sex—predominates. Within species that reproduce only asexually, hybridization cannot be used as a criterion for species recognition.

Defining Species

Despite such difficulties, biologists generally agree on the organisms they classify as species based on the similarity of morphological features and ecology. As a practical definition, we can say that species are groups of organisms that remain relatively constant in their characteristics, can be distinguished from other species, and do not normally interbreed with other species in nature.

How Many Species Are There?

Scientists have described and named a total of 1.4 million species, but doubtless many more actually exist. Some groups of organisms, such as flowering plants, vertebrate animals, and butterflies, are relatively well known. Perhaps more than 90% of the total estimated number of species have already been named in these groups. Many other groups are very poorly known. It is generally accepted that only about 5% of all species have been recognized for bacteria, nematodes (roundworms), fungi, and mites (a group of organisms related to spiders).

By taking comprehensive samples of organisms from different habitats, such as the upper branches of tropical trees, soil, or the deep ocean, scientists have been able to estimate the total numbers of species that may actually exist. Ten million is a reasonable estimate, with about 15% of them marine organisms.

Most Species Live in the Tropics

Most species, perhaps 6 or 7 million, are tropical. Presently only 400,000 species have been named in tropical Asia, Africa, and Latin America combined, well under 10% of all species that occur in the tropics. This is an incredible gap in our knowledge concerning biological diversity in a world that depends on biodiversity for its sustainability.

These estimates apply to the number of eukaryotic organisms only. There is no functional way of estimating the numbers of species of prokaryotic organisms, although it is clear that only a very small fraction of all species have been discovered and characterized so far.

Species are groups of organisms that differ from one another in recognizable ways and generally do not interbreed with one another in nature.

Evolutionary Taxonomy

Taxonomy is often thought of as the gray science, old men in musty rooms arguing about names. In fact, it is one of the more lively branches of biology, often controversial and rarely dull. Perhaps the oldest and most fundamental disagreement within taxonomy concerns what its role in biology should be, what it should be trying to accomplish. There are two fundamental viewpoints, one the Linnaean approach of classifying and naming, and the other the Darwinian approach of tracing evolutionary history. Each viewpoint has led to extreme schools within taxonomy, and in practice both viewpoints have an important influence on how taxonomy is done today.

Classifying by Morphological Similarity

For many centuries, biologists have been distinguishing and naming new species by carefully noting how different kinds of creatures differ from one another. The essence of this process lies in making a judgment about which differences between species are most important; for this reason, taxonomy always involves a degree of subjectivity. In the 1950s some taxonomists began to apply numerical methods to the evaluation of similarities and differences between species. As many characteristics as possible were used for these comparisons, without extra emphasis being given to any particular one. Such comparisons became possible at this time because of the availability first of sophisticated adding machines and, later, computers. This approach, called **numerical taxonomy,** or **phenetics,** was widely applied and discussed and made possible a clearer understanding of the decisions the process of classification entails. It also enabled taxonomists to avoid confusion arising from parallel evolution (see chapter 20) because it treated the similar characteristics of unrelated organisms as only a small subset of their features as a whole. Such similar characteristics are said to be **analogous,** while those that have arisen as a result of common evolutionary descent are said to be **homologous.** In subsequent and more sophisticated applications of numerical taxonomy, specific characters were weighted (given more emphasis than others). The criteria for weighting were statistical ones, objectively defined.

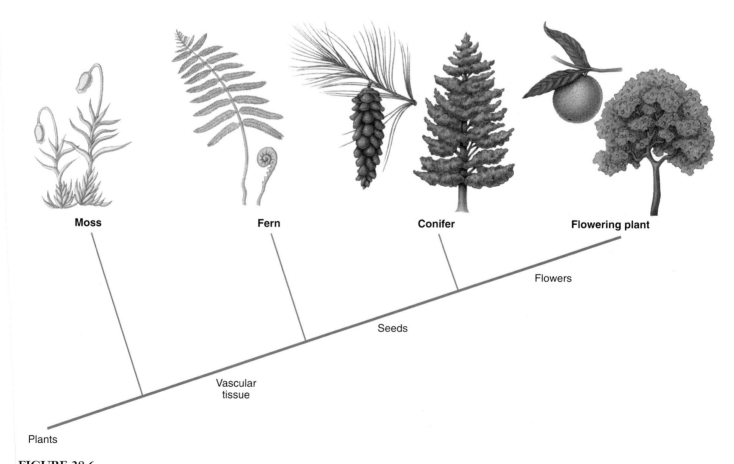

FIGURE 28.6
A cladogram. The derived characters between the cladogram branch points are shared by all organisms above the branch point and are not present in any below it.

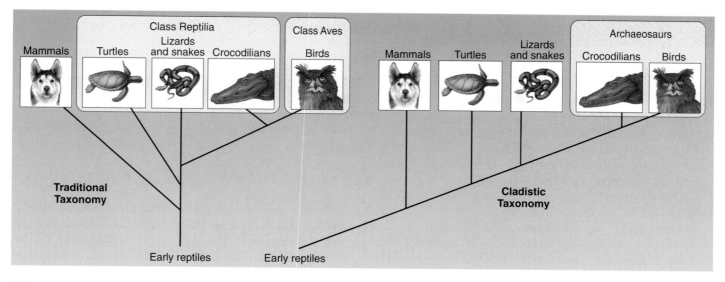

FIGURE 28.7

Traditional and cladistic interpretations of vertebrate evolution. Traditional and cladistic taxonomic analyses of the same set of data often produce different results: in these two classifications of vertebrates, notice particularly the placement of the birds. In the traditional analysis, key characteristics such as feathers and hollow bones are weighted more heavily than others, placing the birds in their own group. Cladistic analysis gives equal weight to these and many other characters and places birds in the same grouping with crocodiles, reflecting the close evolutionary relationship between the two.

Classifying by Evolutionary Relationships

At the opposite end of the taxonomic spectrum are biologists who consider only evolutionary relatedness in assigning taxonomic affinity, ignoring their degree of morphological similarity or difference. This school of taxonomy is called **cladistics.** Cladistics (from the Greek word *clados*, "branch") classifies organisms according to the historical order in which evolutionary branches arose during the history of the group. Those employing cladistic analyses often use biochemical characteristics, such as the DNA sequence divergence discussed in chapter 20, as well as the morphological features that have been used traditionally, to arrive at the best possible analyses of evolutionary history.

In cladistic analysis, the object is to ascertain which characteristics reveal common ancestry between two or more groups of organisms. Hypotheses are constructed about the ancestral condition in the group for each characteristic used in the analysis and about the number of times the **derived characters** arose in the course of its history. Derived characters are characters shared by all members of a branch but not present before the branch. On the phylogenetic tree of plants, for example, vascular tissue is a derived characteristic shared by all vascular plants, but not present in others. Among vascular plants, seeds originated only once, so all seed plants are placed on a common branch of the **cladogram,** or evolutionary tree (figure 28.6). Among the seed plants, flowers are a unique characteristic of the angiosperms. The original angiosperms had seedlings with two seedling leaves, while the monocots, a specialized group of angiosperms, have seedlings with only one leaf, a derived characteristic. The possession of flowers or seeds groups a number of different kinds of organisms and distinguishes them from all others.

The correct interpretation of particular features is of fundamental importance for the construction of accurate cladograms, although the available knowledge often makes interpretation difficult. Nonetheless, cladistic analysis is a powerful method for unraveling the evolutionary history of organisms and thus their historical relationships. It is important to keep in mind that a cladogram shows the *order* of evolutionary descent, not the *extent* of divergence.

Taxonomy Today

In practice, taxonomy today utilizes information from both phenetics and cladistics—in other words, it takes into account both the degree of difference between different groups of organisms and their evolutionary history. A clear example of the conflict that can arise between order of divergence and the magnitude of divergence is provided by the assignment of birds to a separate class, Aves, while retaining crocodiles in the class Reptilia—even though crocodiles are more closely related to birds than to other reptiles. Birds and crocodiles share many derived features, such as a four-chambered heart, that indicate their common evolutionary descent (figure 28.7). Despite this undoubted relationship, the great majority of taxonomists agree that birds are best assigned to their own class because they have diverged so fundamentally since they separated from their common ancestor with the crocodiles, evolving feathers, hollow bones, and a host of other adaptations to flight.

A cladogram presents the evolutionary history of a group of organisms, without making any judgments about the relative importance of different characteristics. Taxonomists often choose to emphasize certain characters in classifying organisms.

The Kingdoms of Life

The earliest classification systems recognized only two kingdoms of living things: animals and plants (figure 28.8*a*). But as biologists discovered microorganisms and learned more about other organisms, they added kingdoms in recognition of fundamental differences discovered among organisms (figure 28.8*b*). Most biologists now use a six-kingdom system first proposed by Carl Woese of the University of Illinois (figure 28.8*c*).

In this system, four kingdoms consist of eukaryotic organisms. The two most familiar kingdoms, **Animalia** and **Plantae,** contain only organisms that are multicellular during most of their life cycle. The kingdom **Fungi** contains multicellular forms and single-celled yeasts, which are thought to have multicellular ancestors. Fundamental differences divide these three kingdoms. Plants are mainly stationary, but some have motile sperm; fungi have no motile cells; animals are mainly motile. Animals ingest their food, plants manufacture it, and fungi digest it by means of secreted extracellular enzymes. Each of these kingdoms probably evolved from a different single-celled ancestor.

The large number of unicellular eukaryotes are arbitrarily grouped into a single kingdom called **Protista** (see chapter 31). This kingdom includes the algae, all of which are unicellular during important parts of their life cycle.

The remaining two kingdoms, **Archaebacteria** and **Eubacteria,** consist of prokaryotic organisms, which are vastly different from all other living things (see chapter 30). Archaebacteria are a diverse group including the methanogens and extreme thermophiles, and differ from the other bacteria, members of the kingdom Eubacteria.

Domains

As biologists have learned more about the archaebacteria, it has become increasingly clear that this ancient group is very different from all other organisms. When the full genomic DNA sequences of an archaebacterium and a eubacterium were first compared in 1996, the differences proved striking. Archaebacteria are as different from eubacteria as eubacteria are from eukaryotes. Recognizing this, biologists are increasingly adopting a classification of living organisms that recognizes three **domains,** a taxonomic level higher than kingdom (figure 28.8*d*). Archaebacteria are in one domain, eubacteria in a second, and eukaryotes in the third.

Living organisms are grouped into three general categories called domains. One of the domains, the eukaryotes, is subdivided into four kingdoms: protists, fungi, plants, and animals.

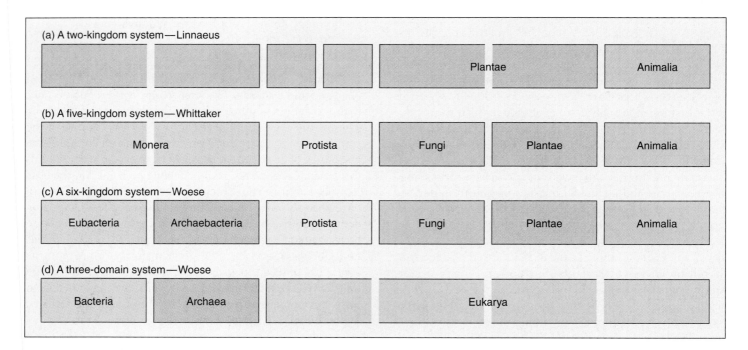

FIGURE 28.8
Different approaches to classifying living organisms. (a) Linnaeus popularized a two-kingdom approach, in which the fungi and the photosynthetic protists were classified as plants, and the nonphotosynthetic protists as animals; when bacteria were described, they too were considered plants. (b) Whittaker in 1969 proposed a five-kingdom system that soon became widely accepted. (c) Woese has championed splitting the bacteria into two kingdoms for a total of six kingdoms, or even assigning them separate domains (d).

Domain Archaea (Archaebacteria)

The term *archaebacteria* (Greek, *archaio*, ancient) refers to the ancient origin of this group of bacteria, which most likely diverged very early from the eubacteria (figure 28.9). Today, archaebacteria inhabit some of the most extreme environments on earth. Though a diverse group, all archaebacteria share certain key characteristics (table 28.2). Their cell walls lack the peptidoglycan characteristic of other bacteria. They possess very unusual lipids, and characteristic ribosomal RNA sequences. Some of their genes possess introns, unlike those of other bacteria.

The archaebacteria are grouped into three general categories: methanogens, extremophiles, and nonextreme archaebacteria.

Methanogens obtain their energy by using hydrogen gas (H_2) to reduce carbon dioxide (CO_2) to methane gas (CH_4). They are strict anaerobes, poisoned by even traces of oxygen. They live in swamps, marshes, and the intestines of mammals. Methanogens release about 2 billion tons of methane gas into the atmosphere each year.

Extremophiles are able to grow under conditions that seem extreme to us.

Thermophiles ("heat lovers") live in very hot places, typically from 60°C to 80°C. Many thermophiles have metabolisms based on sulfur. Thus the *Sulfolobus* inhabiting the hot sulfur springs of Yellowstone National Park at 70–75°C obtain their energy by oxidizing elemental sulfur to sulfuric acid. The recently-described *Pyrolobus fumarii* holds the current record for heat stability, with a 106°C temperature optimum and 113°C maximum—it is so heat tolerant that it is not killed by one-hour treatment in an autoclave (121°C)!

Halophiles ("salt lovers") live in very salty places like the Great Salt Lake in Utah, Mono Lake in California, and the Dead Sea in Israel. Whereas the salinity of seawater is around 3%, these bacteria thrive in, and indeed require, water with a salinity of 15 to 20%.

pH-tolerant archaebacteria grow in highly acid (pH = 0.7) and very basic (pH = 11) environments.

Pressure-tolerant archaebacteria have been isolated from ocean depths that require at least 300 atmospheres of pressure to survive, and tolerate up to 800 atmospheres!

Nonextreme archaebacteria grow in the same environments eubacteria do. As the genomes of archaebacteria have become better known, microbiologists have been able to identify **signature sequences** of DNA present in all archaebacteria and in no other organisms. When samples from soil or sea water are tested for genes matching these signal sequences, many of the bacteria living there prove to be archaebacteria. Clearly, archaebacteria are not restricted to extreme habitats, as microbiologists used to think.

Archaebacteria are poorly-understood bacteria that inhabit diverse environments, some of them extreme.

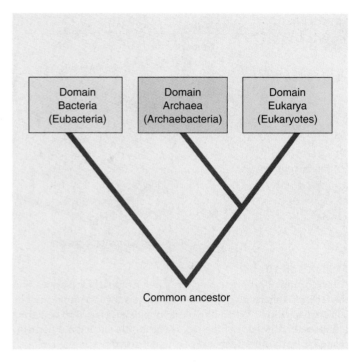

FIGURE 28.9
An evolutionary relationship among the three domains. Eubacteria are thought to have diverged early from the evolutionary line that gave rise to the archaebacteria and eukaryotes.

Table 28.2	Features of the Domains of Life		
	Domain		
Feature	**Archaea**	**Bacteria**	**Eukarya**
Amino acid that initiates protein synthesis	Methionine	Formyl-methionine	Methionine
Introns	Present in some genes	Absent	Present
Membrane-bounded organelles	Absent	Absent	Present
Membrane lipid structure	Branched	Unbranched	Unbranched
Nuclear envelope	Absent	Absent	Present
Number of different RNA polymerases	Several	One	Several
Peptidoglycan in cell wall	Absent	Present	Absent
Response to the antibiotics streptomycin and chloramphenicol	Growth not inhibited	Growth inhibited	Growth not inhibited

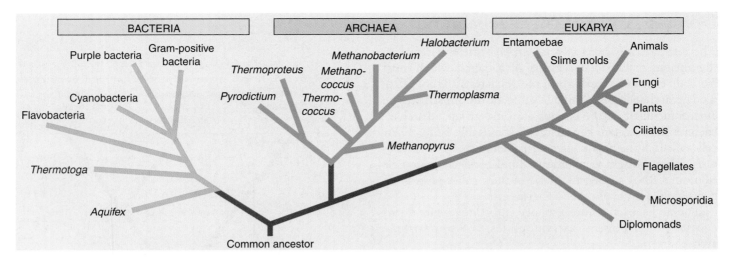

FIGURE 28.10
A tree of life. This phylogeny, prepared from rRNA analyses, shows the evolutionary relationships among the three domains. The base of the tree was determined by examining genes that are duplicated in all three domains, the duplication presumably having occurred in the common ancestor. When one of the duplicates is used to construct the tree, the other can be used to root it. This approach clearly indicates that the root of the tree is within the eubacterial domain. Archaebacteria and eukaryotes diverged later and are more closely related to each other than either is to eubacteria.

Domain Bacteria (Eubacteria)

The eubacteria are the most abundant organisms on earth. There are more living eubacteria in your mouth than there are mammals living on earth. Although too tiny to see with the unaided eye, eubacteria play critical roles throughout the biosphere. They extract from the air all the nitrogen used by organisms, and play key roles in cycling carbon and sulfur. Much of the world's photosynthesis is carried out by eubacteria.

There are many different kinds of eubacteria, and the evolutionary links between them are not well understood. While there is considerable disagreement among taxonomists about the details of bacterial classification, most recog-nize 12 to 15 major groups of eubacteria. Comparisons of the nucleotide sequences of ribosomal RNA (rRNA) molecules are beginning to reveal how these groups are related to one another and to the other two domains. One view of our current understanding of the "Tree of Life" is presented in figure 28.10. The oldest divergences represent the deepest rooted branches in the tree. The root of the tree is within the eubacterial domain. The archaebacteria and eukaryotes are more closely related to each other than to eubacteria and are on a separate evolutionary branch of the tree, even though archaebacteria and eubacteria are both prokaryotes.

Eubacteria are as different from archaebacteria as from eukaryotes.

Kingdom	Cell Type	Nuclear Envelope	Mitochondria	Chloroplasts	Cell Wall
Archaebacteria and Eubacteria	Prokaryotic	Absent	Absent	None (photosynthetic membranes in some types)	Noncellulose (polysaccharide plus amino acids)
Protista	Eukaryotic	Present	Present or absent	Present (some forms)	Present in some forms, various types
Fungi	Eukaryotic	Present	Present or absent	Absent	Chitin and other noncellulose polysaccharides
Plantae	Eukaryotic	Present	Present	Present	Cellulose and other polysaccharides
Animalia	Eukaryotic	Present	Present	Absent	Absent

Table 28.3 Characteristics of the Six Kingdoms

Domain Eukarya (Eukaryotes)

For at least 2 billion years, bacteria ruled the earth. No other organisms existed to eat them or compete with them, and their tiny cells formed the world's oldest fossils. The third great domain of life, the eukaryotes, appear in the fossil record much later, only about 1.5 billion years ago. Metabolically, eukaryotes are more uniform than bacteria. Each of the two domains of prokaryotic organisms has far more metabolic diversity than all eukaryotic organisms taken together.

Three Largely Multicellular Kingdoms

Fungi, plants, and animals are well-defined evolutionary groups, each of them clearly stemming from a different single-celled, eukaryotic ancestor. They are largely multicellular, each a distinct evolutionary line from an ancestor that would be classified in the kingdom Protista.

The amount of diversity among the protists, however, is much greater than that within or between the three largely multicellular kingdoms derived from the protists. Because of the size and ecological dominance of plants, animals, and fungi, and because they are predominantly multicellular, we recognize them as kingdoms distinct from Protista.

A Fourth Very Diverse Kingdom

When multicellularity evolved, the diverse kinds of single-celled organisms that existed at that time did not simply become extinct. A wide variety of unicellular eukaryotes and their relatives exists today, grouped together in the kingdom Protista solely because they are not fungi, plants, or animals. Protists are a fascinating group containing many organisms of intense interest and great biological significance.

The characteristics of the six kingdoms are outlined in table 28.3.

Symbiosis and the Origin of Eukaryotes

The hallmark of eukaryotes is complex cellular organization, highlighted by an extensive endomembrane system that subdivides the eukaryotic cell into functional compartments. Not all of these compartments, however, are derived from the endomembrane system. With few exceptions, all modern eukaryotic cells possess energy-producing organelles, the mitochondria. Mitochondria are about the size of bacteria and contain DNA. Comparison of the nucleotide sequence of this DNA with that of a variety of organisms indicates clearly that mitochondria are the descendants of purple bacteria that were incorporated into eukaryotic cells early in the history of the group. As we will see in chapter 31, some protist phyla have in addition acquired chloroplasts during the course of their evolution and thus are photosynthetic. These chloroplasts are derived from cyanobacteria that became symbiotic in several groups of protists early in their history.

Mitochondria and chloroplasts are both believed to have entered early eukaryotic cells by a process called **endosymbiosis** (*endo*, inside). We discussed the theory of the endosymbiotic origin of mitochondria and chloroplasts in chapter 5. Both organelles contain their own ribosomes, which are more similar to bacterial ribosomes than to eukaryotic cytoplasmic ribosomes. They manufacture their own inner membranes. They divide independently of the cell and contain chromosomes similar to those in bacteria.

Some biologists suggest that basal bodies, centrioles, flagella, and cilia may have arisen from endosymbiotic spirochaete-like bacteria. Even today, so many bacteria and unicellular protists form symbiotic alliances that the incorporation of smaller organisms with desirable features into eukaryotic cells appears to be a relatively common process.

Eukaryotic cells acquired mitochondria and chloroplasts by endosymbiosis, mitochondria being derived from purple bacteria and chloroplasts from cyanobacteria.

Means of Genetic Recombination, if Present	Mode of Nutrition	Motility	Multicellularity	Nervous System
Conjugation, transduction, transformation	Autotrophic (chemosynthetic, photosynthetic) or heterotrophic	Bacterial flagella, gliding or nonmotile	Absent	None
Fertilization and meiosis	Photosynthetic or heterotrophic, or combination of both	9 + 2 cilia and flagella; ameboid, contractile fibrils	Absent in most forms	Primitive mechanisms for conducting stimuli in some forms
Fertilization and meiosis	Absorption	Nonmotile	Present in most forms	None
Fertilization and meiosis	Photosynthetic chlorophylls *a* and *b*	None in most forms, 9 + 2 cilia and flagella in gametes of some forms	Present in all forms	None
Fertilization and meiosis	Digestion	9 + 2 cilia and flagella, contractile fibrils	Present in all forms	Present, often complex

Key Characteristics of Eukaryotes

Multicellularity

The unicellular body plan has been tremendously successful, with unicellular prokaryotes and eukaryotes constituting about half of the biomass on earth. Bacteria have continued a slow evolution for billions of years, with different groups using nearly every energy source available and occurring in every habitat imaginable. Similarly, modern protists are not only numerous, but they are extraordinarily diverse both in their form and in their biochemistry.

Yet a single cell has limits. The evolution of multicellularity allowed organisms to deal with their environments in novel ways. Distinct types of cells, tissues, and organs can be differentiated within the complex bodies of multicellular organisms. With such a functional division within its body, a multicellular organism can protect itself, resist drought efficiently, regulate its internal conditions, move about, seek mates and prey, and carry out other activities on a scale and with a complexity that would be impossible for its unicellular ancestors. With all these advantages, it is not surprising that multicellularity has arisen independently so many times.

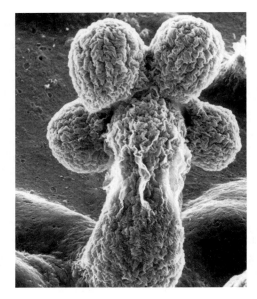

FIGURE 28.11
Colonial bacteria. No bacteria are truly multicellular. These gliding bacteria, *Stigmatella aurantiaca*, have aggregated into a structure called a fruiting body; within, some cells transform into spores.

True multicellularity, in which the activities of individual cells are coordinated and the cells themselves are in contact, occurs only in eukaryotes and is one of their major characteristics. The cell walls of bacteria occasionally adhere to one another, and bacterial cells may also be held together within a common sheath. Some bacteria form filaments, sheets, or three-dimensional aggregates (figure 28.11), but little or no functional coordination occurs. Such bacteria may be considered colonial, but none are truly multicellular. Many protists also form colonial aggregates of many cells with little differentiation or integration.

Other protists—the red, brown, and green algae, for example—have independently attained multicellularity (figure 28.12). Multicellular green algae were ancestors of the plants (see chapters 31 and 33), and, like the other photosynthetic protists, are considered plants in some classification schemes. In the system adopted here, the plant kingdom includes only multicellular land plants, a group that arose from a single ancestor in terrestrial habitats and that has a unique set of characteristics. Aquatic plants are recent derivatives.

Fungi and animals arose from unicellular protist ancestors with different characteristics. As we will see in subsequent chapters, the groups that seem to have given rise to each of these kingdoms are still in existence.

(a)

(b)

(c)

FIGURE 28.12
Multicellular protists. Three phyla of algae have become multicellular independently. They differ greatly in their photosynthetic pigments and structure. (a) The brown algae (phylum Phaeophyta) are large, complex algae that often dominate northern temperate seacoasts. (b) *Ulva* is a green alga (phylum Chlorophyta), a group that includes many unicellular as well as a number of multicellular organisms. (c) Some red algae (phylum Rhodophyta) are shown here growing on sponges. Members of this phylum occur at greater depths than the members of either of the other phyla.

Sexuality

Another major characteristic of eukaryotic organisms as a group is sexuality. Although some interchange of genetic material occurs in bacteria (see chapter 30), it is certainly not a regular, predictable mechanism in the same sense that sex is in eukaryotes. The sexual cycle characteristic of eukaryotes alternates between **syngamy,** the union of male and female gametes producing a cell with two sets of chromosomes, and meiosis, cell division producing daughter cells with one set of chromosomes. This cycle differs sharply from any exchange of genetic material found in bacteria.

Except for gametes, the cells of most animals and plants are diploid, containing two sets of chromosomes, during some part of their life cycle. A few eukaryotes complete their life cycle in the haploid condition, with only one set of chromosomes in each cell. As we have seen, in diploid cells, one set of chromosomes comes from the male parent and one from the female parent. These chromosomes segregate during meiosis. Because crossing over frequently occurs during meiosis (chapter 12), no two products of a single meiotic event are ever identical. As a result, the offspring of sexual, eukaryotic organisms vary widely, thus providing the raw material for evolution.

Sexual reproduction, with its regular alternation between syngamy and meiosis, produces genetic variation. Sexual organisms can adapt to the demands of their environments because they produce a variety of progeny.

In many of the unicellular phyla of protists, sexual reproduction occurs only occasionally. Meiosis may have originally evolved as a means of repairing damage to DNA, producing an organism better adapted to survive changing environmental conditions. The first eukaryotes were probably haploid. Diploids seem to have arisen on a number of separate occasions by the fusion of haploid cells, which then eventually divided by meiosis.

Life Cycles

Eukaryotes are characterized by three major types of life cycles (figure 28.13):

1. In the simplest cycle, found in algae, the zygote is the only diploid cell. Such a life cycle is said to be characterized by **zygotic meiosis,** since the zygote immediately undergoes meiosis.
2. In most animals, the gametes are the only haploid cells. Animals exhibit **gametic meiosis,** meiosis producing gametes which fuse, giving rise to a zygote.
3. Plants show a regular *alternation of generations* between a multicellular haploid phase and a multicellular diploid phase. The diploid phase undergoes meiosis producing haploid spores that give rise to the haploid phase, and the haploid phase produces gametes that fuse to form the zygote. The zygote is the first cell of the multicellular diploid phase. This kind of life cycle is characterized by **alternation of generations** and has **sporic meiosis.**

The complex differentiation that we associate with advanced life-forms depends on multicellularity and sexuality, which must have been highly advantageous to have evolved independently so often.

Key: ☐ Haploid ☐ Diploid

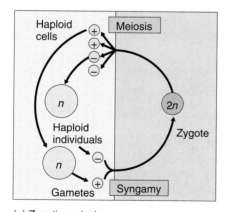

(a) Zygotic meiosis

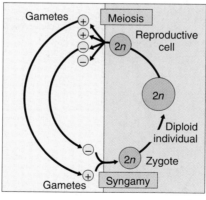

(b) Gametic meiosis

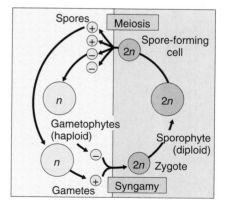

(c) Sporic meiosis

FIGURE 28.13
Diagrams of the three major kinds of life cycles in eukaryotes. (a) Zygotic meiosis, (b) gametic meiosis, and (c) sporic meiosis.

Viruses: A Special Case

Viruses are not living organisms because they do not satisfy the basic criteria of life. Unlike all living organisms, viruses are acellular—that is, they are not cells and do not consist of cells. Unlike all living organisms, viruses do not metabolize energy. Viruses also do not carry out photosynthesis, cellular respiration, or fermentation.

Viruses thus present a special classification problem. Because they are not organisms, we cannot logically place them in any of the kingdoms. Despite their simplicity—merely bits of nucleic acid usually surrounded by a protein coat—viruses are able to invade cells and direct the genetic machinery of these cells to manufacture more virus material (figure 28.14). Viruses appear to be fragments of nucleic acids originally derived from the genome of a living cell. Viruses infect organisms at all taxonomic levels (figure 28.15).

Viruses are not organisms and are not classified in the kingdoms of life.

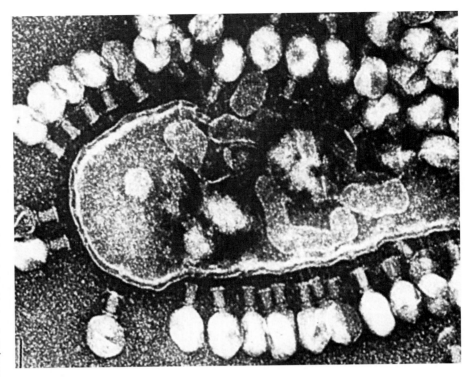

FIGURE 28.14
Viruses are cell parasites. In this micrograph, several T4 bacteriophages (viruses) are attacking an *Escherichia coli* bacterium. Some of the viruses have already entered the cell and are reproducing within it.

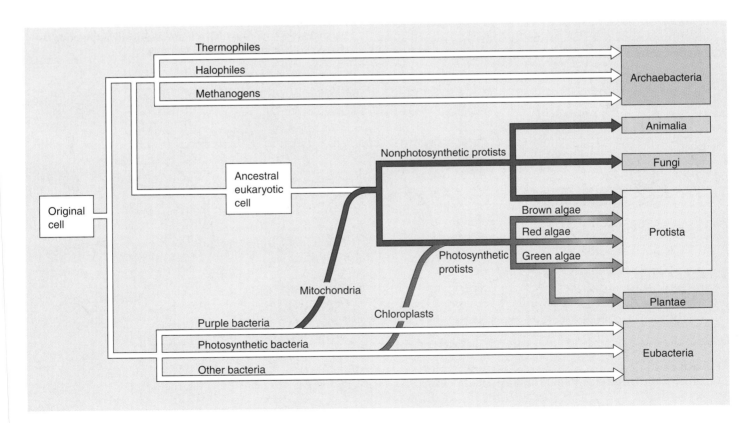

FIGURE 28.15
Diagram of the evolutionary relationships among the six kingdoms of organisms. The colored lines indicate symbiotic events. Viruses infect organisms in all taxonomic levels.

Summary of Concepts

28.1 Biologists name organisms in a systematic way.

- Carl Linnaeus, an eighteenth-century Swedish scientist, developed the binomial system of naming organisms that remains in use today. Biologists give every species a two-part (binomial) name that consists of the name of its genus plus a distinctive specific epithet.

- In the hierarchical system of classification used in biology, genera are grouped into families, families into orders, orders into classes, classes into phyla, and phyla into kingdoms. (Many students have used this time-honored mnemonic to memorize taxon order from the top down: Kings Play Chess On Fine-Grained Sand.)

- There are perhaps 10 million species of plants, animals, fungi, and eukaryotic microorganisms, but only about 1.4 million of them have been assigned names. About 15% of the total number of species are marine; the remainder are mostly terrestrial.

28.2 Taxonomy is the science of classifying organisms.

- Taxonomists may use different approaches to classify organisms.

- Phenetic classification systems are based on comparisons of either particular features of organisms or comprehensive samples of their characteristics.

- Cladistic systems of classification arrange organisms according to evolutionary relatedness, regardless of morphological similarities or differences.

28.3 All living organisms are grouped into one of a few major categories.

- The fundamental division among organisms is between prokaryotes, all of which lack a true nucleus, and eukaryotes, which have a true nucleus and several membrane-bound organelles.

- Prokaryotes, or bacteria, are assigned to two fundamentally different kingdoms, Archaebacteria and Eubacteria.

- Eukaryotes are more closely related to one another than are the two kingdoms of prokaryotes. Many distinctive evolutionary lines that consist mainly of unicellular organisms exist. Most of these are included in a very diverse kingdom, Protista.

- Three of the major evolutionary lines of eukaryotic organisms that consist principally or entirely of multicellular organisms and have particular sets of distinctive characteristics are recognized as separate kingdoms: Plantae, Animalia, and Fungi.

- True multicellularity and sexuality are found only among eukaryotes. Multicellularity confers a degree of protection from the environment and the ability to carry out a wider range of activities than is available to unicellular organisms. Sexuality permits genetic variation among descendants.

- Viruses are not organisms and are not included in the classification of organisms. They are self-replicating portions of the genomes of organisms.

Discussing Key Terms

1. **Species** Species are groups of interbreeding or potentially interbreeding individuals that are distinctly different from other groups.

2. **Kingdoms and domains** Organisms belong to six distinctive evolutionary lines called kingdoms. The two kingdoms that contain different kinds of bacteria are so different that they are sometimes considered to be in separate domains. The four that contain eukaryotes are placed in a third domain.

3. **Phenetic classification** Phenetic approaches to classification use either particular characteristics of organisms or summed characteristics to estimate the degree of difference or similarity between them.

4. **Cladistic classification** Cladistic approaches use shared, derived characteristics as a means of constructing testable hypotheses about organisms' evolutionary descent, which then serve as the basis for classification.

5. **Multicellularity** In true multicellularity, different cells of an organism carry out different functions. This allows the organism to respond to the environment and to specialize to a much greater degree than any one cell can achieve.

6. **Sexuality** Sexuality, a cyclical process involving an alternation between meiosis and syngamy, occurs only in eukaryotic organisms.

Review Questions

1. What was the polynomial system? Why didn't this system become the standard for naming particular species?

2. From the most specific to the most general, what are the names of the groups in the hierarchical taxonomic system? Which two are given special consideration in the way in which they are printed? What are these distinctions?

3. What types of features are emphasized in a phenetic classification system? What does it mean when characters are weighted?

4. What does it mean when characteristics are said to be analogous? What does it mean when they are homologous?

5. What types of features are emphasized in a cladistic classification system? What is the resulting relationship of organisms that are classified in this manner? Are most present classification schemes phenetic or cladistic?

6. How do prokaryotes differ from eukaryotes? Is there a greater fundamental difference between plants and animals or between prokaryotes and eukaryotes? Into which kingdoms are prokaryotes placed?

7. From which of the four eukaryotic kingdoms have the other three evolved? What is the primary characteristic that unifies that kingdom? In terms of nutrition and locomotion, distinguish among the four multicellular kingdoms.

8. Of the four eukaryotic kingdoms, which is the most diverse? What are some of the ways in which it is diverse? What common name is given to the autotrophic members of this kingdom? Name three of the phyla in this kingdom that are at least partly multicellular.

9. What is the apparent origin of at least two organelles found in eukaryotes? Of the two, which one is possessed by nearly all eukaryotic organisms?

10. What defines whether or not a collection of individuals of a single type is truly multicellular? Did multicellularity arise once or many times in the evolutionary process? What advantages do multicellular organisms have over unicellular ones?

11. What are the three major types of life cycles in eukaryotes? Describe the major events of each.

Thought Questions

1. What are some of the advantages of the binomial system of naming organisms?

2. Do you think a better system of classification would be to group all photosynthetic eukaryotes, regardless of whether or not they are single-celled, as plants, and to group all nonphotosynthetic ones as animals? Present arguments in favor of doing so and other arguments in favor of the system of classification used in this book.

3. What are the advantages of sexual reproduction?

Internet Links

The Tree of Life
http://phylogeny.arizona.edu/tree/phylogeny.html
This site, from the University of Arizona, lets you retrace phylogenies from life's origins out through the branches of THE TREE OF LIFE. Highly recommended.

What Critter Is That?
http://www.ucmp.berkeley.edu/help/taxaform.html
This site, from U.C. Berkeley's Museum of Paleontology, allows you to explore the kingdoms of life and find detailed information about any organism if you know its scientific or common names.

The Beast of Bodmin Moor
http://www.nhm.ac.uk/sc/bm/bm_01.htm
A wonderful interactive from the British Museum in which you must use information about comparative anatomy to determine the identity of the "beast."

Nanoworld
http://www.uq.oz.au/nanoworld/gallery.html
This marvelous site from the University of Queensland in Australia features hundreds of electron micrographs images— a fly's foot, a mollusk's spiny teeth—a garden of wonderful visual delights.

For Further Reading

May, R.: "How Many Species Inhabit the Earth?" *Scientific American*, October 1992, pages 42–48. A good guess at how many species exist now, how they got here, and the implications for conservation.

McDermott, J.: "A Biologist Whose Heresy Redraws Earth's Tree of Life," *Smithsonian*, August 1989, pages 72–80. This article explores Lynn Margulis's innovative approach to endosymbiosis which has revolutionized the way we think about organisms.

Wayne, R. and J. Gittleman: "The Problematic Red Wolf," *Scientific American*, July 1995, pages 36–39. Is the red wolf a species or a long-established hybrid of the grey wolf and the coyote?

Woese, C.: "Default Taxonomy: Ernst Mayer's View of the Microbial World," *Proc. Natl. Acad. Sci. US*, vol 95, 1998, pages 11043–11046. A spirited defense of the domain system of classification, contrasting fundamentally different views of the basic nature of biological classification, and in doing so revealing so much about the current excitement among taxonomists.

29

Viruses

Concept Outline

29.1 Viruses are fragments of DNA or RNA that have detached from genomes.

The Discovery of Viruses. The first virus to be isolated proved to consist of two chemicals, one a protein and the other a nucleic acid.

The Nature of Viruses. Viruses occur in all organisms. Able to reproduce only within living cells, viruses are not themselves alive.

29.2 Bacterial viruses exhibit two sorts of reproductive cycles.

Bacteriophages. Some bacterial viruses, called bacteriophages, rupture the cells they infect, while others integrate themselves into the bacterial chromosome to become a stable part of the bacterial genome.

Cell Transformation. Integrated bacteriophages sometimes modify the host bacterium they infect.

29.3 HIV is a typical animal virus.

AIDS. The animal virus HIV infects certain key cells of the immune system, destroying the ability of the body to defend itself from cancer and disease.

The HIV Infection Cycle. The HIV infection cycle is typically a lytic cycle, in which the HIV RNA first directs the production of a corresponding DNA, and this DNA then directs the production of progeny virus particles.

The Future of HIV Treatment. Combination therapies and chemokines offer promising avenues of AIDS therapy.

29.4 Viruses are responsible for many important human diseases.

Disease Viruses. Some of the most serious viral diseases have only recently emerged, the result of transfer from other hosts.

29.5 Viruses are not the only nonliving infectious agents.

Prions and Viroids. In some instances, proteins and "naked" RNA molecules can also transmit diseases.

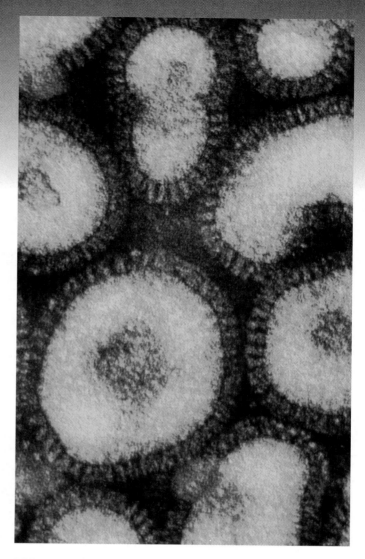

FIGURE 29.1
Influenza viruses. A virus has been referred to as "a piece of bad news wrapped up in a protein." How can something as "simple" as a virus have such a profound effect on living organisms? (30,000×)

We start our exploration of the diversity of life with viruses. Viruses are not considered to be organisms, as they cannot reproduce independently and can only replicate themselves by gaining entry into a cell and using that cell's machinery. A virus is more accurately considered a detached fragment of a genome. Because of their disease-producing potential, viruses are important biological entities. The virus particles you see in figure 29.1 produce the important disease influenza. Other viruses cause AIDS, polio, flu, and many other important human diseases. However, viruses are not restricted to humans. It is thought that the genomes of all organisms contain viruses and that new viruses are continuously "escaping" from bacterial and eukaryotic genomes.

The Discovery of Viruses

Viruses are strands of nucleic acid encased within a protein coat. They do not have the ability to grow or replicate on their own. Viruses can only reproduce when they enter cells and use the cellular machinery of their hosts. Individual kinds of viruses contain only a single type of nucleic acid, either DNA or RNA. All true organisms contain both DNA and RNA, and both are essential components of their genetic machinery. Earlier theories that viruses represent a kind of halfway point between life and nonlife have largely been abandoned. Instead, viruses are now viewed as detached fragments of the genomes of organisms. They could not have existed independently of preexisting organisms.

Stanley Isolates the First Virus

Scientists made the earliest indirect observations of viruses, other than simple observations of their effects, near the end of the nineteenth century. At that time, several groups of European scientists independently concluded that the infectious agents associated with tobacco mosaic (a plant disease) and those associated with hoof-and-mouth disease in cattle were not bacteria. They reached this conclusion because these infectious units could not be filtered out of solutions with the kinds of fine-pored porcelain filters that were routinely used to remove bacteria from various media. In 1933, Wendell Stanley of the Rockefeller Institute prepared an extract of tobacco mosaic virus and purified it. Surprisingly, the purified virus precipitated in the form of crystals (figure 29.2). Stanley was able to show that viruses can better be regarded as chemical matter than as living organisms, at least in any normal sense of the word *living*. The purified crystals still retained the ability to infect healthy tobacco plants and so they clearly were the virus, not merely a chemical derived from it. In fact,

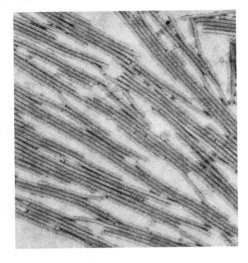

FIGURE 29.2
Tobacco mosaic virus (103,000×). This electron micrograph of tobacco mosaic virus shows virus particles in a crystalline array taken from infected tobacco leaves.

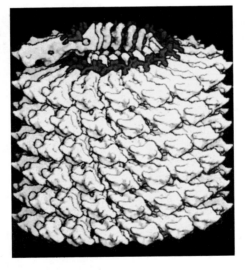

FIGURE 29.3
Computer-generated model of a portion of tobacco mosaic virus. An entire virus consists of 2130 identical protein molecules—the yellow knobs forming a cylindrical coat around the red-colored single strand of RNA in this model. The RNA molecule is the backbone of the virus and determines its shape; along it, identical protein molecules are packed tightly together, protecting the RNA. This model is based on radiographic analyses of the virus structure.

viruses can be kept indefinitely in a crystalline form and still maintain the ability to reinfect their hosts. With Stanley's experiments, scientists began to understand the nature of viruses for the first time.

Viruses Are Made of Nucleic Acid and Protein

Within a few years, other scientists followed up on Stanley's discovery and demonstrated that tobacco mosaic virus consisted of a protein in combination with a nucleic acid. They discovered that the particles that constituted this virus were rods about 300 nanometers long. Subsequently, it was shown that tobacco mosaic virus consists of RNA surrounded by a protein coat (figure 29.3). Many plant viruses have a similar composition, but other viruses have DNA in place of RNA. Nearly all viruses form a protein sheath, or **capsid**, around their nucleic acid core. Many animal viruses form an **envelope** around the capsid rich in proteins, lipids, and glycoprotein molecules.

The simple structure of viruses, the large numbers that are produced in an infection of a cell, and the fact that their genes are related to those of their host, have led many scientists to study viruses in attempts to unravel the nature of genes and how they work. For more than three decades, the history of virology has been thoroughly intertwined with those of genetics and molecular biology. In the future, it is expected that viruses will be one of the principal tools used to experimentally carry genes from one organism to another. Already, viruses are being employed in the treatment of human genetic diseases.

> Viruses are protein-coated fragments of DNA or RNA that have become detached from the genomes of cells but retain the ability to replicate themselves within cells. Because they cannot replicate on their own, they are not organisms.

The Nature of Viruses

Viruses occur in virtually every kind of organism that has been investigated for their presence. Some wreak havoc on the cells they infect, while others produce no disease or other outward sign of their presence. Viruses are often highly specific in the hosts they infect and do not reproduce anywhere else. A given organism often has more than one kind of virus. This means that there may be many more kinds of viruses than there are kinds of organisms—perhaps millions of them. Only a few thousand viruses have been described at this point.

Viral Replication

An infecting virus can be thought of as a set of instructions, not unlike a computer program. A computer's operation is directed by the instructions in its operating program, just as a cell is directed by DNA-encoded instructions. A new program can be introduced into the computer that will cause the computer to cease what it is doing and devote all of its energies to making copies of the introduced program. The new program is not itself a computer and cannot make copies of itself when it is outside the computer, lying on the desk. The introduced program, like a virus, is simply a set of instructions.

Viruses can reproduce only when they enter cells and utilize the cellular machinery of their hosts. Individual kinds of viruses contain only a single type of nucleic acid, either DNA or RNA. They are able to reproduce because they carry genes that are translated into proteins by the cell's genetic machinery, proteins that lead to the production of more viruses. Viruses also lack ribosomes and all of the enzymes necessary for protein synthesis and energy production.

Viruses Consist of a Genome in a Protein Shell

Viruses vary greatly in appearance. The smallest are only about 17 nanometers in diameter, and the largest are up to 1000 nanometers (1 micrometer) in their greatest dimension (figure 29.4). The largest viruses are barely visible with a light microscope. Most viruses can be detected only by using the higher resolution of an electron microscope. Viruses are directly comparable to molecules in size; a hydrogen atom is about 0.1 nanometer in diameter, and a large protein molecule is several hundred nanometers in its greatest dimension.

The simplest viruses consist of a single molecule of a nucleic acid surrounded by a capsid, which is made up of one to a few different protein molecules repeated many times

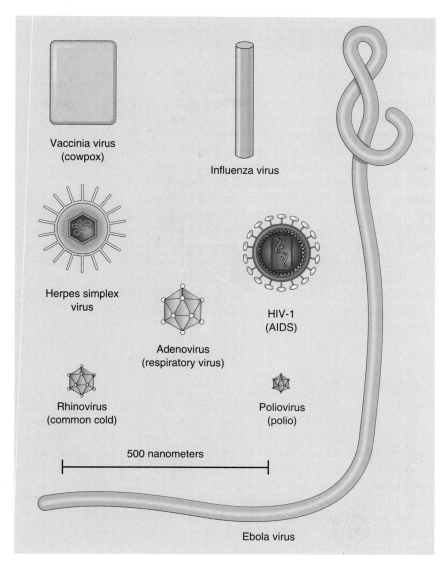

FIGURE 29.4
Virus sizes. At the scale shown here, a human hair would be 8 meters thick.

(see figure 29.3). In more complex viruses, there may be several different kinds of molecules of either DNA or RNA in each virus particle and many different kinds of proteins. Most viruses have an overall structure that is either **helical** or **isometric.** Helical viruses, such as the tobacco mosaic virus, have a rod-like or thread-like appearance. Isometric viruses have a roughly spherical shape.

The only structural pattern found so far among isometric viruses is the **icosahedron,** a figure with 20 equilateral triangular facets, like the adenovirus shown in figure 29.4. Most viruses are icosahedral in basic structure. The icosahedron is the basic design of the geodesic dome. It is the most efficient symmetrical arrangement that linear subunits can take to form a shell with maximum internal capacity.

> Viruses occur in all organisms and can only reproduce within living cells. Most are icosahedral in basic structure.

Bacteriophages

Bacteriophages are viruses that infect bacteria. They are diverse both structurally and functionally, and are united solely by their occurrence in bacterial hosts. Double-stranded DNA bacteriophages have played a key role in molecular biology. Many of these bacteriophages are large and complex, with relatively large amounts of DNA and proteins. Some of them have been named as members of a "T" series (T1, T2, and so forth); others have been given different kinds of names. To illustrate the diversity of these viruses, T3 and T7 bacteriophages are icosahedral and have short tails. In contrast, the so-called T-even bacteriophages (T2, T4, and T6) have an icosahedral head, a capsid that consists primarily of three proteins, a connecting neck with a collar and long "whiskers," a long tail, and a complex base plate (figure 29.5).

The Lytic Cycle

During the process of bacterial infection by T4 phage, at least one of the tail fibers of the phage—they are normally held near the phage head by the "whiskers"—contacts the lipoproteins of the host bacterial cell wall. The other tail fibers set the phage perpendicular to the surface of the bacterium and bring the base plate into contact with the cell surface. The tail contracts, and the tail tube passes through an opening that appears in the base plate, piercing the bacterial cell wall. The contents of the head, mostly DNA, are then injected into the host cytoplasm.

The T-series bacteriophages are all **virulent viruses,** multiplying within infected cells and eventually lysing (rupturing) them. When a virus kills the infected host cell in which it is replicating, the reproductive cycle is referred to as a **lytic cycle** (figure 29.6).

The Lysogenic Cycle

Many bacteriophages do not immediately kill the cells they infect, instead integrating their nucleic acid into the genome of the infected host cell. While residing there, it is called a **prophage.** Among the bacteriophages that do this is the lambda (λ) phage of *Escherichia coli*. We know as much about this bacteriophage as we do about virtually any other biological particle; the complete sequence of its 48,502 bases has been determined. At least 23 proteins are associated with the development and maturation of lambda phage, and many other enzymes are involved in the integration of these viruses into the host genome.

The integration of a virus into a cellular genome is called **lysogeny.** At a later time, the prophage may exit the genome and initiate virus replication. This sort of reproductive cycle, involving a period of genome integration, is called a **lysogenic cycle.** Viruses that become stably integrated within the genome of their host cells are called **lysogenic viruses** or **temperate viruses.**

Bacteriophages are a diverse group of viruses that attack bacteria. Some kill their host in a lytic cycle; others integrate into the host's genome, initiating a lysogenic cycle.

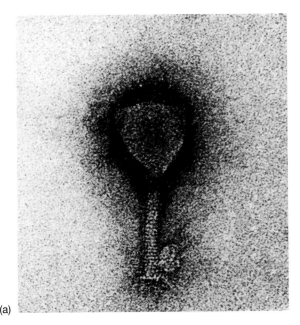

(a)

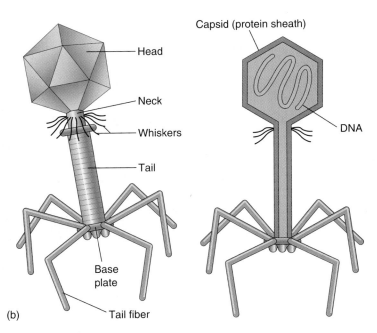

(b)

FIGURE 29.5
A bacterial virus. (a) Electron micrograph and (b) diagram of the structure of a T4 bacteriophage.

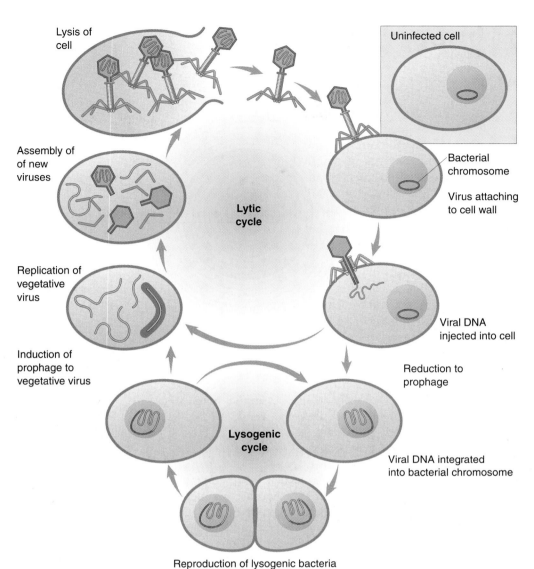

FIGURE 29.6
Lytic and lysogenic cycles of a bacteriophage. In the lytic cycle, the bacteriophage exists as viral DNA free in the bacterial host cell's cytoplasm; the viral DNA directs the production of new viral particles by the host cell until the virus kills the cell by lysis. In the lysogenic cycle, the bacteriophage DNA is integrated into the large, circular DNA molecule of the host bacterium and reproduces. It may continue to replicate and produce lysogenic bacteria or enter the lytic cycle and kill the cell. Bacteriophages are much smaller relative to their hosts than illustrated in this diagram.

Labels in figure:
Lysis of cell
Uninfected cell
Assembly of of new viruses
Bacterial chromosome
Virus attaching to cell wall
Lytic cycle
Replication of vegetative virus
Viral DNA injected into cell
Induction of prophage to vegetative virus
Reduction to prophage
Lysogenic cycle
Viral DNA integrated into bacterial chromosome
Reproduction of lysogenic bacteria

Cell Transformation

During the integrated portion of a lysogenic reproductive cycle, virus genes are often expressed. RNA polymerase reads them just as if they were host genes, and sometimes expression of these genes has an important effect on the host cell, altering it in novel ways. The genetic alteration of a cell's genome by the introduction of foreign DNA is called **transformation.**

Transforming the Cholera-Causing Bacterium

An important example of this sort of cell transformation directed by viral genes is provided by the bacterium responsible for an often-fatal human disease. The disease-causing bacteria, *Vibrio cholerae*, usually exists in a harmless form, but a second disease-causing, virulent form also occurs. In

this latter form, the bacterium causes the deadly disease cholera, but how the bacteria changed from harmless to deadly was not known until recently. Research now shows that a bacteriophage that infects *V. cholerae* introduces into the host bacterial cell a gene that codes for the cholera toxin. This gene becomes incorporated into the bacterial chromosome, where it is translated along with the other host genes, thereby transforming the benign bacterium to a disease-causing agent. The transfer occurs through bacterial pili (see chapter 30); in further experiments, mutant bacteria that did not have pili were resistant to infection by the bacteriophage. This discovery has important implications in the development of vaccines against cholera, which has been unsuccessful up to this point.

Bacteriophage transform *Vibrio cholerae* bacteria into disease-causing agents.

AIDS

A diverse array of viruses occur among animals. A good way to gain a general idea of what they are like is to look at one animal virus in detail. Here we will look at the virus responsible for a comparatively new and fatal viral disease, acquired immunodeficiency syndrome (AIDS). AIDS was first reported in the United States in 1981. It was not long before the infectious agent, human immunodeficiency virus (HIV), was identified by laboratories in France and the United States (figure 29.7). Study of HIV revealed it to be closely related to a chimpanzee virus, suggesting that it might have been introduced to humans in central Africa from chimpanzees.

Infected individuals have little resistance to infection, and nearly all of them eventually die of diseases that noninfected individuals easily ward off. Few who contract AIDS survive more than a few years untreated. The risk of HIV transmission from an infected individual to a healthy one in the course of day-to-day contact is essentially nonexistent. However, the transfer of body fluids, such as blood, semen, or vaginal fluid, or the use of nonsterile needles, between infected and healthy individuals poses a severe risk.

The incidence of AIDS is growing very rapidly in the United States. Over 1 million people were estimated to have been infected by HIV by the mid-1990s. Many—perhaps all of them—will eventually come down with AIDS. AIDS incidence is already very high in many African countries. The AIDS epidemic is discussed further in chapter 54.

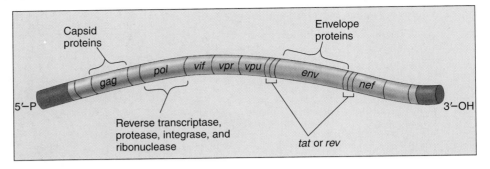

FIGURE 29.7
Gene map of human immunodeficiency virus (HIV). The HIV genome contains three main genes, *gag*, *pol*, and *env*. Proteases cleave their protein products to yield the envelope, capsid, and replication proteins. Six additional genes, unique to HIV, regulate expression of the HIV genome. Two of these, *tat* and *rev*, are transcribed to alternative reading frames.

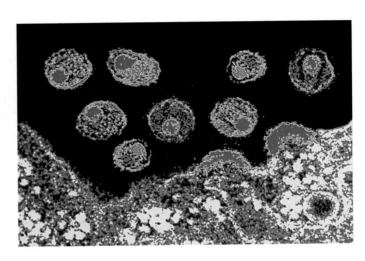

FIGURE 29.8
The AIDS virus. HIV particles exit an infected CD4+ T cell. Soon, they will spread and infect neighboring CD4+ T cells.

How HIV Compromises the Immune System

In normal individuals, an army of specialized cells patrols the bloodstream, attacking and destroying any invading bacteria or viruses. In AIDS patients, this army of defenders is vanquished. One special kind of white blood cell, called a CD4+ T cell (discussed further in chapter 54) is required to rouse the defending cells to action. In AIDS patients, the virus homes in on CD4+ T cells, infecting and killing them until none are left (figure 29.8). Without these crucial immune system cells, the body cannot mount a defense against invading bacteria or viruses. AIDS patients die of infections that a healthy person could fight off.

Clinical symptoms typically do not begin to develop until after a long latency period, generally 8 to 10 years after the initial infection with HIV. During this long interval, carriers of HIV have no clinical symptoms but are apparently fully infectious, which makes the spread of HIV very difficult to control. The reason for the long latency period puzzled researchers for some time. In some viral infections, such as herpes, the DNA inserts itself into the host cell's chromosome as a provirus, much as a bacteriophage does, where it remains inactive until some future event causes it to be removed from the chromosome and resume activity. This does not appear to be the case with HIV. Evidence indicates that the HIV infection cycle continues throughout the latent period, the immune system suppressing the ongoing infection. Eventually, however, a random mutational event in the virus allows it to quickly overcome the immune defense.

The HIV virus infects and eliminates key cells of the immune system, destroying the body's ability to defend itself from cancer and infection.

The HIV Infection Cycle

The way HIV infects humans (figure 29.9) provides a good example of how animal viruses replicate. Most other viral infections follow a similar course, although the details of entry and replication differ in individual cases.

Attachment

When HIV is introduced into the human bloodstream, the virus particle circulates throughout the entire body but will only infect CD4+ cells. Most other animal viruses are similarly narrow in their requirements. Polio goes only to certain spinal nerves, hepatitis to the liver, and rabies to the brain.

How does a virus such as HIV recognize a specific kind of target cell? Recall from chapter 7 that every kind of cell in the human body has a specific array of cell-surface glycoprotein markers that serve to identify them to other, similar cells. Each HIV particle possesses a glycoprotein (called **gp120**) on its surface that precisely fits a cell-surface marker protein called **CD4**

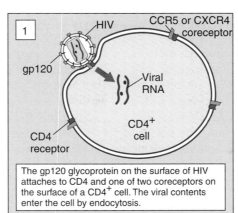
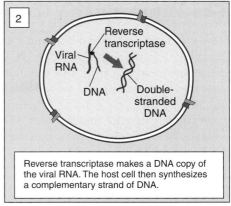

The gp120 glycoprotein on the surface of HIV attaches to CD4 and one of two coreceptors on the surface of a CD4+ cell. The viral contents enter the cell by endocytosis.

Reverse transcriptase makes a DNA copy of the viral RNA. The host cell then synthesizes a complementary strand of DNA.

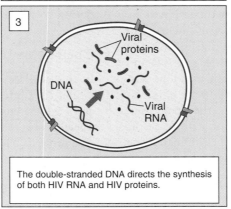
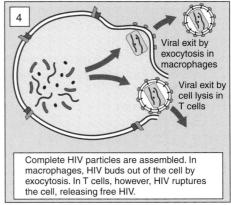

The double-stranded DNA directs the synthesis of both HIV RNA and HIV proteins.

Complete HIV particles are assembled. In macrophages, HIV buds out of the cell by exocytosis. In T cells, however, HIV ruptures the cell, releasing free HIV.

FIGURE 29.9
The HIV infection cycle.

on the surfaces of immune system cells called macrophages and T cells. Macrophages are infected first.

Entry into Macrophages

After docking onto the CD4 receptor of a macrophage, HIV requires a second macrophage receptor, called CCR5, to pull itself across the cell membrane. After gp120 binds to CD4, it goes through a conformational change that allows it to bind to CCR5. Investigators speculate that after the conformational change, the second receptor passes the gp120-CD4 complex through the cell membrane, triggering passage of the contents of the HIV virus into the cell by endocytosis, with the cell membrane folding inward to form a deep cavity around the virus.

Replication

Once inside the macrophage, the HIV particle sheds its protective coat. This leaves virus RNA floating in the cytoplasm, along with a virus enzyme that was also within the virus shell. This enzyme, called **reverse transcriptase**, synthesizes a double strand of DNA complementary to the virus RNA, often making mistakes and so creating new mutations. This double-stranded DNA directs the host cell machinery to produce many copies of the virus. HIV does not rupture and kill the macrophage cells it infects. Instead,

the new viruses are released from the cell by exocytosis. HIV synthesizes large numbers of viruses in this way, challenging the immune system over a period of years.

Entry into T Cells

During this time, HIV is constantly replicating and mutating. Eventually, by chance, HIV alters the gene for gp120 in a way that causes the gp120 protein to change its second-receptor allegiance. This new form of gp120 protein prefers to bind instead to a different second receptor, CXCR4, a receptor that occurs on the surface of T lymphocyte CD4+ cells. Soon the body's T lymphocytes become infected with HIV. This has deadly consequences, as new viruses exit the cell by rupturing the cell membrane, effectively killing the infected T cell. Thus, the shift to the CXCR4 second receptor is followed swiftly by a steep drop in the number of T cells. This destruction of the body's T cells blocks the immune response and leads directly to the onset of AIDS, with cancers and opportunistic infections free to invade the defenseless body.

HIV, the virus that causes AIDS, is an RNA virus that replicates inside human cells by first making a DNA copy of itself. It is only able to gain entrance to those cells possessing a particular cell surface marker recognized by a glycoprotein on its own surface.

The Future of HIV Treatment

New discoveries of how HIV works continue to fuel research on devising ways to counter HIV. For example, scientists are testing drugs and vaccines that act on HIV receptors, researching the possibility of blocking CCR5, and looking for defects in the structures of HIV receptors in individuals that are infected with HIV but have not developed AIDS. Figure 29.10 summarizes some of the recent developments and discoveries.

Combination Drug Therapy

A variety of drugs inhibit HIV in the test tube. These include AZT and its analogs (which inhibit virus nucleic acid replication) and protease inhibitors (which inhibit the cleavage of the large polyproteins encoded by *gag*, *poll*, and *env* genes into functional capsid, enzyme, and envelope segments). When combinations of these drugs were administered to people with HIV in a recent study, their condition was markedly improved. A combination of a protease inhibitor and two AZT analog drugs entirely eliminated the HIV virus from many of the patients' bloodstreams. Importantly, all of these patients began to receive the drug therapy within three months of contracting the virus, before their bodies had an opportunity to develop tolerance to any one of them. Widespread use of this **combination therapy** has cut the U.S. AIDS death rate by almost two-thirds since its introduction in the mid-1990s, from 43,000 AIDS deaths in 1995 to 31,000 in 1996, and just under 17,000 in 1997.

Unfortunately, this sort of combination therapy does not appear to actually succeed in eliminating HIV from the body. While the virus disappears from the bloodstream, traces of it can still be detected in lymph tissue of the patients. When combination therapy is discontinued, virus levels in the bloodstream once again rise. Because of demanding therapy schedules and many side effects, long-term combination theraphy does not seem a promising approach.

Using a Defective HIV Gene to Develop Vaccines and Drug Therapy

Recently, five people in Australia who are HIV-positive but have not developed AIDS in 14 years were found to have all received a blood transfusion from the same HIV-positive person, who also has not developed AIDS. This led scientists to believe that the strain of virus transmitted to these people has some sort of genetic defect that prevents it from effectively disabling the human immune system. In subsequent research, a defect was found in one of the nine genes present in the AIDS virus. This gene is called *nef*, named for "negative factor," and the defective version of *nef* in the six Australians seems to be missing some pieces. Viruses with the defective gene may not be able to reproduce as much, allowing the immune system to keep the virus in check.

This finding has exciting implications for developing a vaccine against AIDS. Before this, scientists have been unsuccessful in trying to produce a harmless strain of AIDS that can elicit an effective immune response. The Australian strain with the defective *nef* gene has the potential to be used in a vaccine that would arm the immune system against this and other strains of HIV.

Another potential application of this discovery is its use in developing drugs that inhibit HIV proteins that speed virus replication. It seems that the protein produced from the *nef* gene is one of these critical HIV proteins, because viruses with defective forms of *nef* do not reproduce, as seen in the cases of the six Australians. Research is currently underway to develop a drug that targets the *nef* protein.

Chemokines and CAF

In the laboratory, chemicals called **chemokines** appear to inhibit HIV infection by binding to and blocking the CCR5 and CXCR4 coreceptors. As you might expect, people long infected with the HIV virus who have not developed the disease prove to have high levels of chemokines in their blood.

The search for HIV-inhibiting chemokines is intense. Not all results are promising. Researchers report that in their tests, the levels of chemokines were not different between patients in which the disease was not progressing and those in which it was rapidly progressing. More promising, levels of another factor called **CAF** (CD8+ cell antiviral factor) *are* different between these two groups. Researchers have not yet succeeded in isolating CAF, which seems not to block receptors that HIV uses to gain entry to cells, but, instead, to prevent replication of the virus once it has infected the cells. Research continues on the use of chemokines in treatments for HIV infection, either increasing the amount of chemokines or disabling the CCR5 receptor. However, promising research on CAF suggests that it may be an even better target for treatment and prevention of AIDS.

One problem with using chemokines as drugs is that they are also involved in the inflammatory response of the immune system. The function of chemokines is to attract white blood cells to areas of infection. Chemokines work beautifully in small amounts and in local areas, but chemokines in mass numbers can cause an inflammatory response that is worse than the original infection. Injections of chemokines may hinder the immune system's ability to respond to local chemokines, or they may even trigger an out-of-control inflammatory response. Thus, scientists caution that injection of chemokines could make patients *more* susceptible to infections, and they continue to research other methods of using chemokines to treat AIDS.

Disabling Receptors

Scientists have found that a mutation in the gene that codes for the second receptor CCR5, consisting of a 32-base-pair deletion, appears to block or inhibit HIV infection. Indi-

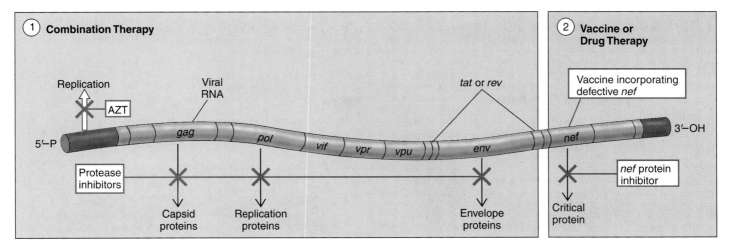

(1) Combination Therapy

Replication

AZT

Viral RNA

tat or *rev*

5′–P

gag   pol   vif   vpr   vpu   env   nef

3′–OH

Protease inhibitors

Capsid proteins

Replication proteins

Envelope proteins

(2) Vaccine or Drug Therapy

Vaccine incorporating defective *nef*

nef protein inhibitor

Critical protein

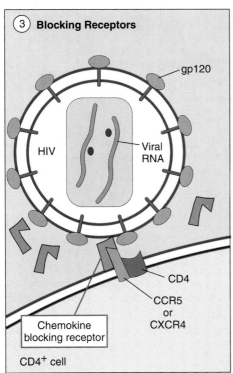

(3) Blocking Receptors

gp120

HIV

Viral RNA

CD4

CCR5 or CXCR4

Chemokine blocking receptor

CD4⁺ cell

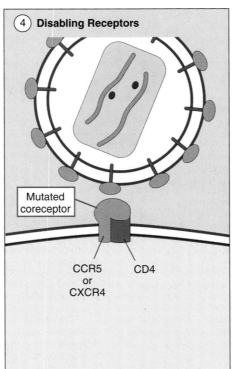

(4) Disabling Receptors

Mutated coreceptor

CCR5 or CXCR4

CD4

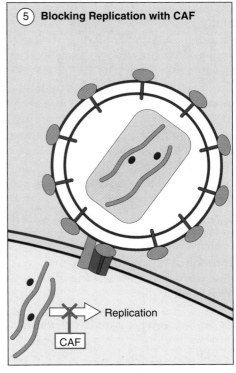

(5) Blocking Replication with CAF

Replication

CAF

FIGURE 29.10

Five ways of treating HIV. Research is currently underway to develop new treatments for HIV. Among them are these five: (1) *Combination therapy* involves using two drugs, AZT to block replication of the virus and protease inhibitors to block the normal production of critical viral proteins. (2) Using a defective form of the viral gene *nef,* scientists may be able to construct an HIV *vaccine.* Also, *drug therapy* that inhibits *nef's* protein product is being tested. (3) Other research focuses on the use of chemokine chemicals to *block receptors* (CXCR4 and CCR5), thereby disabling the mechanism HIV uses to enter CD4⁺ T cells. (4) Producing mutations that will *disable receptors* may also be possible. (5) Lastly, CAF, an antiviral factor which acts inside the CD4⁺ T cell, may be able to *block replication* of HIV.

viduals homozygous for the mutated gene allele who have been exposed to HIV or are at high risk of HIV infection have not developed AIDS. Furthermore, individuals heterozygous for the mutated allele may have some protection and may develop the disease more slowly. In one study of 1955 people, scientists found no individuals who were infected and homozygous for the mutated allele. The allele seems to be more common in Caucasian populations (10% to 11%) than in African-American populations (2%), and absent in African and Asian populations.

Treatment for AIDS involving disruption of CCR5 looks promising, as research indicates that people live perfectly well without CCR5. Attempts to block or disable CCR5 are being sought in numerous laboratories.

A cure for AIDS is not yet in hand, but many new approaches look promising.

Disease Viruses

Humans have known and feared diseases caused by viruses for thousands of years. Among the diseases that viruses cause (table 29.1) are influenza, smallpox, infectious hepatitis, yellow fever, polio, rabies, and AIDS, as well as many other diseases not as well known. In addition, viruses have been implicated in some cancers and leukemias. For many autoimmune diseases, such as multiple sclerosis and rheumatoid arthritis, and for diabetes, specific viruses have been found associated with certain cases. In view of their effects, it is easy to see why the late Sir Peter Medawar, Nobel laureate in physiology or medicine, wrote, "A virus is a piece of bad news wrapped in protein." Viruses not only cause many human diseases, but also cause major losses in agriculture, forestry, and in the productivity of natural ecosystems.

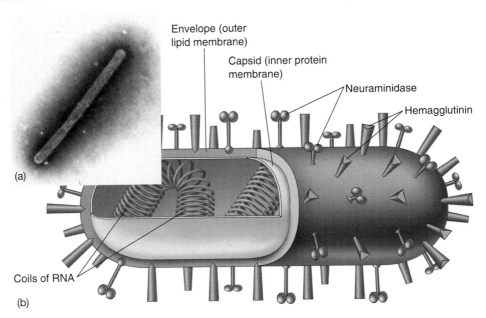

FIGURE 29.11
The influenza virus. (a) TEM of the so-called "bird flu" influenza virus, A(H5N1), which first infected humans in Hong Kong in 1997. (b) Diagram of an influenza virus. The coiled RNA has been revealed by cutting through the outer lipid-rich envelope, with its two kinds of projecting spikes, and the inner protein capsid.

Influenza

Perhaps the most lethal virus in human history has been the influenza virus. Some 21 million Americans and Europeans died of flu within 18 months in 1918 and 1919, an astonishing number.

Types. Flu viruses are RNA animal viruses. An individual flu virus resembles a rod studded with spikes composed of two kinds of protein (figure 29.11). There are three general "types" of flu virus, distinguished by their capsid (inner membrane) protein, which is different for each type: Type A flu virus causes most of the serious flu epidemics in humans, and also occurs in mammals and birds. Type B and Type C viruses are restricted to humans and rarely cause serious health problems.

Subtypes. Different strains of flu virus, called subtypes, differ in their protein spikes. One of these proteins, **hemagglutinin (H)** aids the protein in gaining access to the cell interior. The other, **neuraminidase (N)** helps the daughter virus break free of the host cell once virus replication has been completed. The structures of both the H and N molecules are known in detail. The H molecule, for example, is made of three parts, and stands on the surface of the virus somewhat like a tripod, with clublike projections on top. Each "leg" of the H molecule contains "hot spots" that display an unusual tendency to change as a result of mutation of the virus RNA during imprecise replication.

Point mutations cause changes in these spike proteins in 1 of 100,000 viruses during the course of each generation. These very variable segments of the H molecule function as targets against which the body's antibodies are directed. Because of accumulating changes in the H and N molecules, different flu vaccines are required to protect against different subtypes. Type A flu viruses are currently classified into 13 distinct H subtypes and 9 distinct N subtypes, each of which requires a different vaccine to protect against infection. Thus the type A virus that caused the Hong Kong flu epidemic of 1968 has type 3 H molecules and type 2 N molecules, and is called A(H3N2).

Importance of recombination. The greatest problem in combating flu viruses arises not through mutation, but through recombination. Viral genes are readily reassorted by genetic recombination, sometimes putting together novel combinations of H and N spikes unrecognizable by human antibodies specific for the old configuration. Viral recombination of this kind seems to have been responsible for the three major flu pandemics (that is, worldwide epidemics) that have occurred in this century, by producing drastic shifts in H N combinations. The "killer flu" of 1918, A(H1N1), killed 21 million people. The Asian flu of 1957, A(H2N2), killed over 100,000 Americans. The Hong Kong flu of 1968, A(H3N2), infected 50 million people in the United States alone, of which 70,000 died.

Table 29.1 Important Human Viral Diseases

Disease	Pathogen	Reservoir	Vector/Epidemiology
AIDS	HIV	STD	Destroys immune defenses, resulting in death by infection or cancer. Four million cases worldwide by 1995.
Chicken pox	Human herpes virus 3 (zoster varicella)	Humans	Spread through contact with infected individuals. No cure. Rarely fatal. Vaccine approved in U.S. in early 1995.
Ebola	Filoviruses	Unknown	Acute hemorrhagic fever; virus attacks connective tissue, leading to massive hemorrhaging and death. Peak mortality is 50–90% if the disease goes untreated. Outbreaks confined to local regions of central Africa.
Hepatitus B (viral)	Hepatitis B virus (HBV)	Humans	Highly infectious through contact with infected body fluids. Approximately 1% of U.S. population infected. Vaccine available, no cure. Can be fatal.
Herpes	Herpes simplex virus (HSV)	Humans	Fever blisters; spread primarily through contact with infected saliva. Very prevalent worldwide. No cure. Exhibits latency—the disease can be dormant for several years.
Influenza	Influenza viruses	Humans, ducks	Historically a major killer (22 million died in 18 months in 1918–19); wild Asian ducks, chickens, and pigs are major reservoirs. The ducks are not affected by the flu virus, which shuffles its antigen genes while multiplying within them, leading to new flu strains.
Measles	Paramyxoviruses	Humans	Extremely contagious through contact with infected individuals. Vaccine available. Usually contracted in childhood, when it is not serious; more dangerous to adults.
Mononucleosis	Epstein-Barr virus (EBV)	Humans	Spread through contact with infected saliva. May last several weeks; common in young adults. No cure. Rarely fatal.
Mumps	Paramyxovirus	Humans	Spread through contact with infected saliva. Vaccine available; rarely fatal. No cure.
Pneumonia	Influenza virus	Humans	Acute infection of the lungs, often fatal without treatment.
Polio	Poliovirus	Humans	Acute viral infection of the CNS that can lead to paralysis and is often fatal. Prior to the development of Salk's vaccine in 1954, 60,000 people a year contracted the disease in the U.S. alone.
Rabies	Rhabdovirus	Wild and domestic Canidae (dogs, foxes, wolves, coyotes, etc.)	An acute viral encephalomyelitis transmitted by the bite of an infected animal. Fatal if untreated.
Smallpox	Variola virus	Formerly humans, now only exists in two research labs—may be eliminated entirely soon	Historically a major killer; the last recorded case of smallpox was in 1977. A worldwide vaccination campaign wiped out the disease completely. There is current debate as to whether the virus should be maintained even for research purposes.
Yellow fever	Flavivirus	Humans, mosquitoes	Spread from individual to individual by mosquito bites; a notable cause of death during the construction of the Panama Canal. If untreated, this disease has a peak mortality rate of 60%.

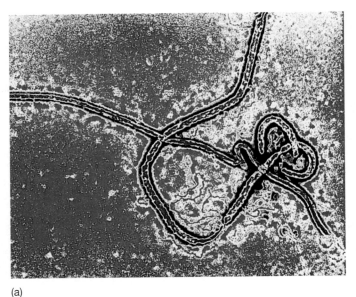
(a)

(b)

FIGURE 29.12
The Ebola virus. (a) A virus with a fatality rate that can exceed 90% in humans has reappeared in West Africa after a 20-year hiatus. Health professionals are scrambling to identify natural hosts of the virus and to devise strategies to combat transmission of the disease. (b) Two children wait in May 1995 outside the hospital in the Zaire town of Kikwit, where those infected with Ebola were isolated. Seventy percent of the infected died.

It is no accident that new strains of flu usually originate in the far east. The most common hosts of influenza virus are ducks, chickens, and pigs, which in Asia often live in close proximity to each other and to humans. Pigs are subject to infection by both bird and human strains of the virus, and individual animals are often simultaneously infected with multiple strains. This creates conditions favoring genetic recombination between strains, producing new combinations of H and N subtypes. The Hong Kong flu, for example, arose from recombination between A(H3N8) [from ducks] and A(H2N2) [from humans]. The new strain of influenza, in this case A(H3N2), then passed back to humans, creating an epidemic because the human population has never experienced that H N combination before.

A potentially deadly new strain of flu virus emerged in Hong Kong in 1997, A(H5N1). Unlike all previous instances of new flu strains, A(H5N1) passed to humans directly from birds, in this case chickens. A(H5N1) was first identified in chickens in 1961, and in the spring of 1997 devastated flocks of chickens in Hong Kong. The first human case of "bird flu" occurred in May, 1997, in a 3-year-old boy who died of the infection. The number of human infections by A(H5N1) remains small, with 17 confirmed cases by the end of the year. Five have died, and three are in intensive care units, surviving with the aid of mechanical respirators. Fortunately this strain of flu virus does not appear to spread easily from person to person. Public health officals remain concerned that the genes of A(H5N1) could yet mix with those of a human strain to create a new strain that could spread widely in the human population, and to prevent this ordered the killing of all 1.2 million chickens in Hong Kong in 1997.

Emerging Viruses

Sometimes viruses that originate in one organism pass to another, causing a disease in the new host. HIV, for example, arose in chimpanzees and relatively recently passed to humans. Influenza is fundamentally a bird virus. New pathogens arising in this way, called **emerging viruses,** represent a considerable threat in an age when airplane travel potentially allows infected individuals to move about the world quickly, spreading an infection.

Among the most lethal of emerging viruses are a collection of filamentous viruses arising in central Africa that attack human connective tissue. With lethality rates in excess of 50%, these so-called filoviruses are among the most lethal infectious diseases known. One, Ebola virus (figure 29.12), has exhibited lethality rates in excess of 90% in isolated outbreaks in central Africa. The outbreak of Ebola virus in the summer of 1995 in Zaire killed 245 people out of 316 infected—a mortality rate of 78%. The latest outbreak occurred in Gabon, West Africa, in February 1996. The natural host of Ebola is unknown.

Often carried by monkeys, emerging viruses represent a real potential threat to worldwide human health if they ever achieve widespread dissemination.

Viruses are responsible for some of the most lethal diseases of humans. Some of the most serious examples are viruses that have transferred to humans from some other host. Influenza, a bird virus, has been responsible for the most devastating epidemics in human history. Newly emerging viruses such as Ebola have received considerable public attention.

Prions and Viroids

For decades scientists have been fascinated by a peculiar group of fatal brain diseases. These diseases have the unusual property that they are transmissible from one individual to another, but it is years and often decades before the disease is detected in infected individuals. The brains of infected individuals develop numerous small cavities as neurons die, producing a marked spongy appearance. Called **transmissible spongiform encephalopathies (TSEs)**, these diseases include scrapie in sheep, "mad cow" disease in cattle, and kuru and Creutzfeldt-Jakob disease in humans.

TSEs can be transmitted between individuals of a species by injecting infected brain tissue into a recipient animal's brain. TSEs can also spread via tissue transplants and, apparently, food. Kuru was common in the Fore people of Papua New Guinea when they practiced ritual cannibalism, literally eating the brains of infected individuals. Mad cow disease spread widely among the cattle herds of England in the 1990s because cows were fed bone meal prepared from cattle carcasses to increase the protein content of their diet. Like the Fore, the British cattle were literally eating the tissue of cattle that had died of the disease.

A Heretical Suggestion

In the 1960s, British researchers T. Alper and J. Griffith noted that infectious TSE preparations remained infectious even after exposed to radiation that would destroy DNA or RNA. They suggested that the infectious agent was a protein. Perhaps, they speculated, the protein usually preferred one folding pattern, but could sometimes misfold, and then catalyze other proteins to do the same, the misfolding spreading like a chain reaction. This heretical suggestion was not accepted by the scientific community, as it violates a key tenant of molecular biology: only DNA or RNA act as hereditary material, transmitting information from one generation to the next.

Prusiner's Prions

In the early 1970s, physician Stanley Prusiner, moved by the death of a patient from Creutzfeldt-Jakob disease, began to study TSEs. Prusiner became fascinated with Alper and Griffith's hypothesis. Try as he might, Prusiner could find no evidence of nucleic acids or viruses in the infectious TSE preparations, and concluded, as Alper and Griffith had, that the infectious agent was a *protein*, which in a 1982 paper he named a **prion**, for "proteinaceous infectious particle."

Prusiner went on to isolate a distinctive prion protein, and for two decades continued to amass evidence that prions

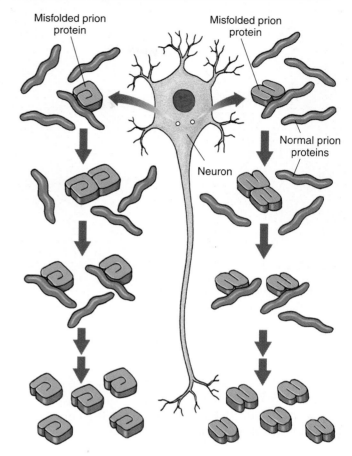

FIGURE 29.13

How prions may be maintained. Prions may cause normal prion protein to misfold simply by contacting them. When prions misfolded in different ways (blue) contact normal prion protein (purple), the normal prion protein misfolds in the same way.

play a key role in triggering TSEs. The scientific community resisted Prusiner's renegade conclusions, but eventually experiments done in Prusiner's and other laboratories began to convince many. For example, when Prusiner injected prions of a different abnormal conformations into several different hosts, these hosts developed prions with the same abnormal conformations as the parent prions (figure 29.13). In another important experiment, Charles Weissmann showed that mice genetically engineered to lack Prusiner's prion protein are immune to TSE infection. However, if brain tissue with the prion protein is grafted into the mice, the grafted tissue—but not the rest of the brain—can then be infected with TSE. In 1997, Prusiner was awarded the Nobel Prize in physiology or medicine for his work on prions.

FIGURE 29.14
Dr. Stanley B. Prusiner. For his investigation of prions, physician Stanley B. Prusiner won the 1997 Nobel Prize in Physiology or Medicine.

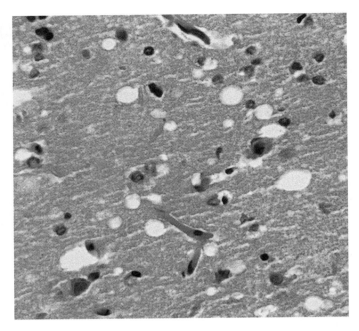

FIGURE 29.15
Damaged brain tissue. The lesions (holes) in this brain tissue are the result of a TSE called Creutzfeldt-Jakob disease (CJD) that affects humans. Prions play a key role in causing this fatal deterioration of brain tissue.

Is Prusiner Right?

Despite Prusiner's Nobel Prize (figure 29.14), not all scientists are convinced that he is right. So far no one has been able to inject a prion protein synthesized in a test tube (and thus sure to be free of any contaminating virus or nucleic acid) into a healthy animal and make it sick with TSE. Until this experiment has been done successfully, or enough becomes known about the process to understand why it cannot be, these scientists argue that the prion theory is not firmly established.

Can Humans Catch "Mad Cow Disease" From Eating Infected Meat?

Many scientists are becoming worried that prions may be transmitting such diseases to humans (figure 29.15). Specifically, they worry that prion-caused bovine spongiform encephalopathy (BSE), a brain disease in cows commonly known as **mad-cow disease,** may infect humans and produce a similar fatal disorder called Creutzfeldt-Jakob disease (CJD). In March 1996, an outbreak of mad-cow disease in Britain, with many thousands of cattle apparently affected, created widespread concern. BSE, a degeneration of the brain caused by prions, appears to have entered the British cattle herds from sheep! Sheep are subject to a prion disease called scrapie, and the disease is thought to have passed from sheep to cows through protein supplemented feed pellets containing ground-up sheep brains. The passage of prions from one species to another has British sci-

entists worried: the death of four dairy farmers in Britain from CJD in the last four years suggests that prions may be able to pass from cows to people! The case for a connection between eating British beef and CJD is strongly supported by the finding that tissue from the brains of the dead farmers and from BSE cows induces the same brain lesions in mice, while classic CJD produces quite different lesions—clearly their form of CJD was caused by the same agent that caused BSE. Thus, there appears to be legitimate cause for caution. Because the incubation period for CJD can vary from 15 to 45 years, the number of people infected by eating BSE-contaminated meat in Great Britain may not become apparent for some time.

Viroids

Viroids are tiny, naked molecules of RNA, only a few hundred nucleotides long, that are important infectious disease agents in plants. A recent viroid outbreak killed over ten million coconut palms in the Philippines. It is not clear how viroids cause disease. One clue is that viroid nucleotide sequences resemble the sequences of introns within ribosomal RNA genes. These sequences are capable of catalyzing excision from DNA—perhaps the viroids are catalyzing the destruction of chromosomal integrity.

Prions are infectious proteins that many scientists—but not all—believe are responsible for serious brain diseases. In plants, naked RNA molecules called viroids can also transmit disease.

Summary of Concepts

29.1 Viruses are fragments of DNA or RNA that have detached from genomes.

- Viruses are not alive and are not organisms; they cannot reproduce outside of living cells, since they lack the machinery to do so.
- Viruses are detached fragments of bacterial or eukaryotic genomes that are able to replicate within cells by using the genetic machinery of those cells.
- Viruses are basically either helical or isometric. Most isometric viruses are icosahedral in shape.
- The simplest viruses use the enzymes of the host cell for both protein synthesis and gene replication; the more complex ones contain up to 200 genes and are capable of synthesizing many structural proteins and enzymes.

29.2 Bacterial viruses exhibit two sorts of reproductive cycles.

- Viruses are either virulent or temperate.
- Virulent bacteriophages infect bacterial cells by injecting their viral DNA or RNA into the cell, where it directs the production of new virus particles, ultimately lysing the cell.
- Temperate bacteriophages, upon entering a bacterial cell, insert their DNA into the cell genome, where they may remain stably integrated into the bacterial genome as a prophage for many generations. Such a prophage is said to be lysogenic, because it retains the capability of exiting the bacterial genome and resuming a lytic reproductive cycle.
- Sometimes the genes of a prophage are expressed, altering the character of the infected bacterial cell, a process called transformation.

29.3 HIV is a typical animal virus.

- AIDS, a viral infection that destroys the immune system, is caused by HIV (human immunodeficiency virus), which possesses a glycoprotein on its exterior that precisely fits a specific protein called CD4 on the cell surface of certain key immune system cells. After docking on CD4, HIV enters the cell and replicates, destroying the cell.
- Considerable progress has been made in the treatment of AIDS, particularly with drugs such as protease inhibitors that block cleavage of HIV polyproteins into functional segments. Research into chemokines also appears promising.

29.4 Viruses are responsible for many important human diseases.

- Viruses are responsible for many serious human diseases. Some of the most serious, like AIDS and Ebola, have only recently transferred to humans from some other animal host.

29.5 Viruses are not the only nonliving infectious agents.

- Proteins called prions may transmit serious brain diseases from one individual to another, although the theory is controversial.
- RNA molecules can in some instances be infective disease agents.

Discussing Key Terms

1. **Emerging viruses** All organisms contain viruses. When a virus from an ape or some other animal transfers to a human, it may cause a serious or even lethal disease. Such viruses, though they may be ancient, are called "emerging" because the human disease they cause seems to appear abruptly.

2. **Prions** A prion is a protein that can infect cells and transmit its shape differences to the prion proteins in these cells without the involvement of protein synthesis or DNA.

3. **Viroids** Viroids are small naked fragments of RNA that infect plant cells, sometimes to devastating effect. With nucleotide sequences similar to introns, they may catalyze excision events that destroy the integrity of the nuclear DNA.

4. **Chemokine** Chemokines are a family of small proteins involved in different aspects of the immune system. Particular chemokines prevent HIV replication by binding to the CD4 receptor; others, interacting with CD4, trigger its entry into CD4+ cells and subsequent replication.

5. **Coreceptor** When HIV binds to CD4 on the surface of a white blood cell, it cannot enter the cell by endocytosis without the active intervention of a cell surface protein that acts as a coreceptor.

6. **Prophage** When a eukaryotic virus is integrated into a cellular genome it is called a prophage, because it retains the potential to exit the genome as a functional virus.

Review Questions

1. Why are viruses not considered to be living organisms? What is the most general structure of a virus? When do viruses act like living organisms?

2. How did early scientists come to the conclusion that the infectious agents associated with tobacco mosaic and hoof-and-mouth disease in cattle were not bacteria?

3. Are viruses specific or nonspecific in terms of the hosts that they infect? What could one assume, therefore, about the actual number of different kinds of viruses?

4. What is a virulent virus? What are temperate viruses? Why is it so difficult to treat a viral infection of either type? How is this different from treating bacterial infections?

5. What type of microscope is generally required to visualize an actual viral particle? What, therefore, is the approximate size range of viruses? What are the two types of viral structures? What shape does each impart?

6. What is a bacteriophage? How does a T4 phage infect a host cell?

7. How does a virus recognize a specific kind of target cell? What specific type of human cell does the AIDS virus infect? What does this cell possess that other human cells do not?

8. How do many animal viruses penetrate the host cell? What viral components remain in the host cell? How does a plant virus infect its host? How does a bacterial virus infect its host?

Thought Questions

1. In what ways might the early, self-replicating particles that gave rise to the first organisms have resembled, or differed from, viruses?

2. If a viral disease like influenza could kill 22 million people over 18 months in 1918-1919, why do you suppose the killing ever stopped? What prevented flu from continuing its lethal assault on humanity?

Internet Links

All You Ever Wanted to Know About Viruses.
http://www.tulane.edu/~dmsander/garryfavweb.thml
ALL THE VIROLOGY ON THE WEB is a catalog from Tulane University bursting with links and information on viruses and the diseases they cause. Among the site's many features is The Big Picture Book of Viruses, packed with virus micrographs and information searchable by name, host, structure, or disease. The site also holds online courses and provides more than 1,000 links to other virology resources.

Latest AIDS Information
http://hivinsite.ucsf.edu
IN SITE, *maintained by AIDS researchers at UC San Francisco, presents the AIDS Knowledge Base (a text on HIV treatment), the latest HIV statistics by U.S. city, and many news for features and interviews.*

An AIDS Tutorial
http://www.msichicago.org/exhibit/AIDS/
An on-line tutorial that is part of an exhibit at The Museum of Science and Industry, Chicago, with up-to-date information on the HIV lifecycle; presents the daily CDC AIDS summary.

Keeping Track of Emerging Diseases
http://www.outbreak.org
This very popular OUTBREAK site covers everything from anthrax to ebola, chronicling outbreaks with dispatches from the CDC, media clippings, and interviews with infectuous disease experts.

For Further Reading

Caldwell, J. C. and P. Caldwell: "The African AIDS Epidemic," *Scientific American*, vol. 274, March 1996, pages 62–68. Nearly 25% of the population in parts of Africa is HIV-positive. Could lack of circumcision in men in this region contribute to the high incidence of AIDS?

Fincher, J.: "America's Deadly Rendezvous with the 'Spanish Lady'," *Smithsonian*, January 1989, pages 131–45. In 1918, before the end of World War I, the world found itself fighting a killer flu for which there was no cure.

Garrett, L: *The Coming Plague*, Farrar, Straus, and Giroux, New York, 1994. Contains a frightening account of the development and spread of the Ebola virus.

Le Guenno, B: "Emerging Viruses," *Scientific American*, October 1995, pages 56–64. A frightening discussion of hemorrhagic fever viruses, about how new ones continually arise, and how their spread is facilitated.

O'Brien, S. and M. Dean: "In Search of AIDS-Resistance Genes," *Scientific American*, September 1997. A genetic allele that prevents HIV infection has been found. It encodes a defective cell surface receptor, and offers promising avenues of AIDS treatment.

Preston, R.: "Crisis in the Hot Zone," *The New Yorker*, vol. 68, October 26, 1992, pages 58–62. A very well written account of the frightening Ebola virus outbreak, a deadly disease imported accidentally by monkeys. Much media has been devoted to this outbreak, including a movie (*Outbreak*) and a best-selling book (*The Hot Zone*).

Prusiner, S.: "The Prion Diseases," *Scientific American*, January 1995, pages 48–57. Prions are infectious proteins. Once dismissed as an impossibility, they are now thought to cause a number of infectious diseases. This article, by their discoverer, recounts his investigation of these novel agents, for which he was awarded the 1997 Nobel Prize in Medicine.

30

Bacteria

Concept Outline

30.1 Bacteria are the smallest and most numerous organisms.

The Prevalence of Bacteria. The simplest of organisms, bacteria are thought to be the most ancient. They are the most abundant living organisms.

Prokaryotes versus Eukaryotes. Bacteria lack the high degree of internal compartmentalization characteristic of eukaryotes.

30.2 Bacterial cell structure is more complex than commonly supposed.

The Bacterial Surface. Some bacteria have a secondary membrane-like covering outside of their cell wall.

The Cell Interior. While bacteria lack extensive internal compartments, they may have complex internal membranes.

30.3 Bacteria exhibit considerable diversity in both structure and metabolism.

Bacterial Diversity. There are at least 16 phyla of bacteria, although many more doubtless await discovery.

Bacterial Variation. Mutation and recombination generate enormous variation within bacterial populations.

Bacterial Metabolism. Bacteria obtain carbon atoms and energy from a wide array of sources. Some can thrive in the absence of other organisms, while others must obtain their energy and carbon atoms from other organisms.

30.4 Bacteria are responsible for many human diseases.

Human Bacterial Diseases. Many serious human diseases are caused by bacteria, some of them responsible for millions of deaths each year.

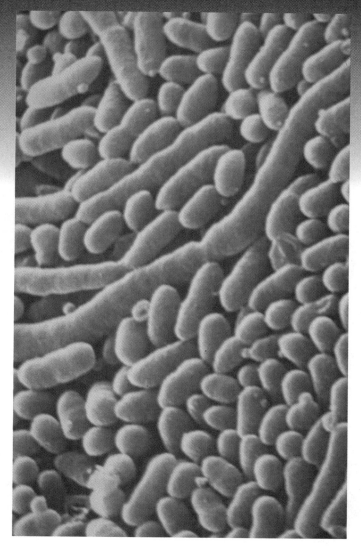

FIGURE 30.1
A colony of bacteria. With their enormous adaptability and metabolic versatility, bacteria are found in every habitat on earth, carrying out many of the vital processes of ecosystems, including photosynthesis, nitrogen fixation, and decomposition.

The simplest organisms living on earth today are bacteria, and biologists think they closely resemble the first living organisms to evolve on earth. Too small to see with the unaided eye, bacteria are the most abundant of all organisms (figure 30.1) and are the only ones characterized by prokaryotic cellular organization. Life on earth could not exist without bacteria. Bacteria make possible many of the essential functions of ecosystems, including the capture of nitrogen from the atmosphere and, in many aquatic communities, photosynthesis. Indeed, bacterial photosynthesis is thought to have been responsible for generating much of the oxygen in the earth's atmosphere. Bacteria are also responsible for many serious human diseases. An understanding of bacteria is thus essential for both scientific and practical reasons.

The Prevalence of Bacteria

Bacteria are the oldest, structurally simplest, and the most abundant forms of life on earth. They are also the only organisms with prokaryotic cellular organization. Represented in the oldest rocks from which fossils have been obtained, 3.5 to 3.8 billion years old, bacteria were abundant for over 2 billion years before eukaryotes appeared in the world (see figure 4.11).

About 4800 different kinds of bacteria are currently recognized, but there are doubtless many thousands more awaiting proper description (figure 30.2). The structural differences among various bacteria, even viewed with an electron microscope, are not great, so species are identified largely by their metabolic characteristics. Bacteria can be characterized properly only when they are grown on a defined medium because the characteristics of these organisms often change, depending on their growth conditions and the substances with which they are in contact.

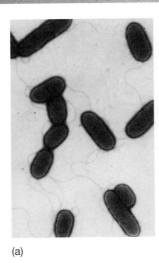
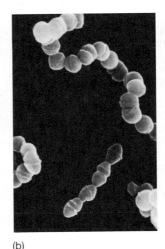
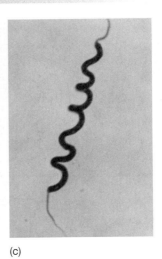

(a)   (b)   (c)

FIGURE 30.2
The diversity of bacteria. (a) *Pseudomonas aeruginosa*, a rod-shaped, flagellated bacterium (bacillus). *Pseudomonas* includes the bacteria that cause many of the most serious plant diseases. (b) *Streptococcus*. The spherical individual bacteria (cocci) adhere in chains in the members of this genus (34,000×). (c) *Spirillum volutans*, one of the spirilla. This large bacterium, which occurs in stagnant fresh water, has a tuft of flagella at each end (500×).

Importance of Bacteria

Bacteria were largely responsible for creating the properties of the atmosphere and the soil over billions of years. They are metabolically much more diverse than eukaryotes, which is why they are able to exist in such a wide range of habitats. Many bacteria are autotrophic—either photosynthetic or chemoautotrophic—and make major contributions to the world carbon balance in terrestrial, freshwater, and marine habitats. Others are heterotrophic and play a key role in world ecology by breaking down organic compounds. Some of these heterotrophic bacteria cause major diseases of plants and animals, including humans. One of the most important roles of bacteria in the world ecosystem relates to the fact that only a few genera of bacteria—and no other organisms—have the ability to fix atmospheric nitrogen and thus make it available for use by other organisms (see chapter 25).

Bacteria are very important in many industrial processes. Bacteria are now used in the production of acetic acid and vinegar, various amino acids and enzymes, and especially in the fermentation of lactose into lactic acid, which coagulates milk proteins and is used in the production of almost all cheeses, yogurt, and similar products. In the production of bread and other foods, the addition of certain strains of bacteria can lead to the enrichment of the final product with respect to its mix of amino acids, a key factor in its nutritive value. Many products traditionally manufactured using yeasts, such as ethanol, can also be made using bacteria. The comparative economics of these processes will determine which group of organisms is used in the future. Many of the most widely used antibiotics, including streptomycin, aureomycin, erythromycin, and chloromycetin, are derived from bacteria. Most antibiotics seem to be substances used by bacteria to compete with one another and fungi in nature, allowing one species to exclude others from a favored habitat.

Bacteria and Genetic Engineering

Applying genetic engineering methods to produce improved strains of bacteria for commercial use, as discussed in chapter 18, holds enormous promise for the future. Bacteria are under intense investigation, for example, as nonpolluting insect control agents. *Bacillus thuringiensis* attacks insects in nature, and improved, highly specific strains of *B. thuringiensis* have greatly increased its usefulness as a biological control agent. Bacteria have also been extraordinarily useful in our attempts to understand genetics.

Bacteria are the oldest and most abundant organisms on earth. They play a vital role both in productivity and in cycling the substances essential to all other life-forms. Certain bacteria are also the only organisms capable of fixing atmospheric nitrogen.

Prokaryotes versus Eukaryotes

Prokaryotes, or bacteria, differ from eukaryotes in numerous important features. These profound differences began to be understood properly only about 50 years ago. At that time, some scientists began to realize that these differences were far greater than the traditional ones used to separate plants and animals. These differences have been discussed in several places in this text, since a consideration of the simpler features of bacteria illustrates the evolutionary origins of many of the structures and capabilities of eukaryotes. We will review the differences briefly, referring back to earlier discussions. These differences represent some of the most fundamental distinctions that separate any groups of organisms.

1. **Multicellularity.** All bacteria are fundamentally single-celled. Some types may have cells that adhere within a matrix and form filaments, especially some cyanobacteria, but their cytoplasm is not directly interconnected, as often is the case in multicellular eukaryotes. The activities of a bacterial colony are less integrated and coordinated than those in multicellular eukaryotes. A primitive form of colonial organization occurs in gliding bacteria, which move together and form spore-bearing structures (figure 30.3). Such approaches to multicellularity are rare among bacteria.

2. **Cell size.** Most bacterial cells are only 1 micrometer or less in diameter. Most eukaryotic cells are well over 10 times that size.

3. **Chromosomes.** Eukaryotic cells have a membrane-bound nucleus containing chromosomes made up of both nucleic acids and proteins. Bacteria do not have membrane-bound nuclei, nor do they have chromosomes of the kind present in eukaryotes, in which DNA forms a structural complex with proteins. Instead, their naked circular DNA is localized in a zone of the cytoplasm called the **nucleoid**.

4. **Cell division and genetic recombination.** Cell division in eukaryotes takes place by mitosis and involves

FIGURE 30.3
Approaches to multicellularity in bacteria. *Chondromyces crocatus*, one of the gliding bacteria. The rod-shaped individuals move together, forming the composite spore-bearing structures shown here. Millions of spores, which are basically individual bacteria, are released from these structures.

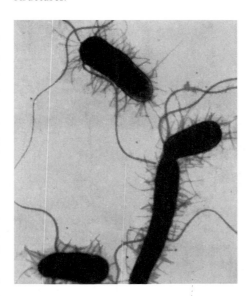

FIGURE 30.4
Flagella in the common intestinal bacterium, *Escherichia coli.* The long strands are flagella, while the shorter hair-like outgrowths are called pili.

spindles made up of microtubules. Cell division in bacteria takes place mainly by binary fission (see chapter 11). True sexual reproduction occurs only in eukaryotes and involves syngamy and meiosis, with an alternation of diploid and haploid forms. Despite their lack of sexual reproduction, bacteria do have mechanisms that lead to the transfer of genetic material. These mechanisms are far less regular than those of eukaryotes and do not involve the equal participation of the individuals between which the genetic material is transferred.

5. **Internal compartmentalization.** In eukaryotes, the enzymes for cellular respiration are packaged in mitochondria. In bacteria, the corresponding enzymes are not packaged separately but are bound to the cell membranes (see chapters 5 and 9). The cytoplasm of bacteria, unlike that of eukaryotes, contains no internal compartments or cytoskeleton and no organelles except ribosomes.

6. **Flagella.** Bacterial flagella are simple, composed of a single fiber of the protein flagellin (figure 30.4; see also chapter 5). Eukaryotic flagella and cilia are complex and have a 9 + 2 structure of microtubules (see figure 5.30). Bacterial flagella also function differently, spinning like propellers, while eukaryotic flagella have a whiplike motion.

7. **Metabolic diversity.** In photosynthetic eukaryotes, the enzymes for photosynthesis are packaged in membrane-bound organelles. Only one kind of photosynthesis occurs in eukaryotes, and it involves the release of oxygen. In photosynthetic bacteria, the enzymes for photosynthesis are bound to the cell membrane. These bacteria have several different patterns of anaerobic and aerobic photosynthesis, involving the formation of end products such as sulfur, sulfate, and oxygen (see chapter 10). This photosynthetic diversity again underscores the metabolic diversity of the bacteria.

Bacteria differ from eukaryotes in a wide variety of characteristics, a degree of difference as great as any that separates any groups of organisms.

The Bacterial Surface

Bacterial cell walls usually consist of **peptidoglycan,** a network of polysaccharide molecules connected by polypeptide cross-links. In **gram-positive** bacteria, this network makes up the basic structure of the bacterial cell wall, which is about 15 to 80 nanometers thick (figure 30.5). In the more common **gram-negative** bacteria, an outer membrane containing large molecules of lipopolysaccharide—a polysaccharide chain with lipids attached to it—covers a thin layer of peptidoglycan. This outer membrane layer makes gram-negative bacteria resistant to many antibiotics that interfere with cell wall synthesis in gram-positive bacteria. In some kinds of bacteria, an additional gelatinous layer, the capsule, surrounds the cell wall.

Bacteria are mostly simple in form, varying mainly from straight and rod-shaped **(bacilli)** or spherical **(cocci)** to long and spirally coiled **(spirilla).** Some bacteria change into stalked structures, grow long, branched filaments, or form erect structures that release **spores,** single-celled bodies that grow into new bacterial individuals. Some rod-shaped and spherical bacteria adhere end-to-end after they have divided, forming chains. Some filamentous bacteria are capable of gliding motion, often combined with rotation around a longitudinal axis. Biologists have not yet determined the mechanism by which they move.

Many kinds of bacteria have slender, rigid, helical flagella composed of the protein flagellin (figure 30.6). These flagella range from 3 to 12 micrometers in length and are very thin—only 10 to 20 nanometers thick. Flagella may be distributed all over the surface of the bacterium in which they occur, or they may be confined to one or both ends of the cell.

Pili (singular, **pilus**) are other kinds of hair-like outgrowths that occur on the cells of some bacteria (see figure 30.4). They are shorter than bacterial flagella, up to several micrometers long, and about 7.5 to 10 nanometers thick. Pili help the bacterial cells attach to appropriate substrates and exchange genetic information. The number and arrangement of flagella and pili are useful aids in bacterial identification.

Some bacteria form thick-walled **endospores** around their chromosome and a small portion of the surrounding cytoplasm when they are exposed to nutrient-poor conditions. These endospores are highly resistant to environmental stress, especially heat, and can germinate to form new individuals after decades or even centuries.

The cytoplasm of a bacterium is bounded by a plasma membrane and is often encased within a cell wall composed of one or more polysaccharide layers. There are two structurally different kinds of bacterial cell walls, gram-negative and gram-positive.

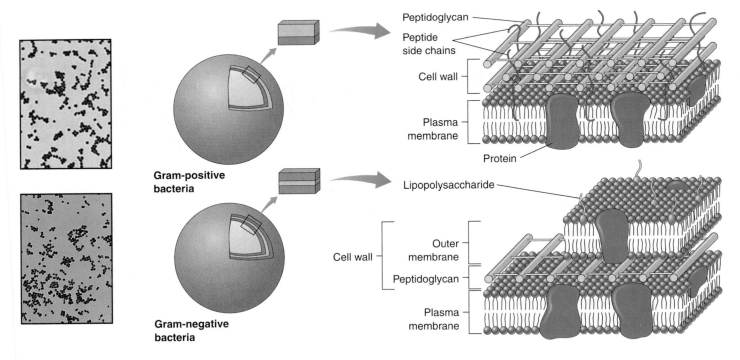

FIGURE 30.5
The Gram stain. The peptidoglycan layer encasing gram-positive bacteria traps gentian violet dye, so the bacteria appear purple in a Gram-stained smear (named after Hans Christian Gram, who developed the technique). Because gram-negative bacteria have much less peptidoglycan (located between the plasma membrane and an outer membrane), they do not retain the violet dye and so exhibit the red background stain.

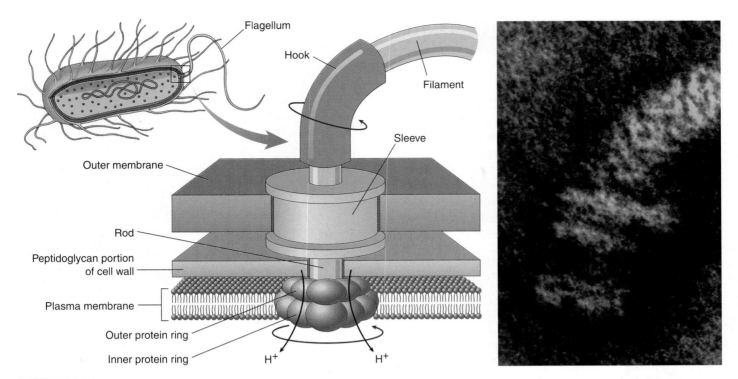

FIGURE 30.6
The flagellar motor of a gram-negative bacterium. A protein filament, composed of the protein flagellin, is attached to a protein shaft that passes through a sleeve in the outer membrane and through a hole in the peptidoglycan layer to rings of protein anchored in the cell wall and plasma membrane, like rings of ballbearings. The shaft rotates when the inner protein ring attached to the shaft turns with respect to the outer ring fixed to the cell wall. The inner ring is an H^+ ion channel, a proton pump that uses the passage of protons into the cell to power the movement of the inner ring past the outer one.

The Cell Interior

The most fundamental characteristic of bacterial cells is their prokaryotic organization. Bacterial cells lack nuclei and do not carry out mitosis, instead dividing by transverse binary fusion, as described in chapter 11. Bacterial cells lack the extensive functional compartmentalization seen within eukaryotic cells.

Internal membranes. Many bacteria possess invaginated regions of the plasma membrane, sometimes extensive, that function in respiration or photosynthesis (figure 30.7).

Nucleoid region. Bacteria do not possess the complex chromosomes characteristic of eukaryotes. Instead, their genes are encoded within a single double-stranded ring of DNA that is crammed into one region of the cell known as the **nucleoid region.** Many bacterial cells also possess small, independently-replicating circles of DNA called plasmids. Plasmids contain only a few genes, usually not essential for the cell's survival. They are best thought of as an excised portion of the bacterial chromosome.

Ribosomes. Bacterial ribosomes are smaller than those of eukaryotes and differ in protein and RNA content. Antibiotics such as tetracycline and chloramphenicol can tell the difference—they bind to bacterial ribosomes and block protein synthesis, but do not bind to eukaryotic ribosomes.

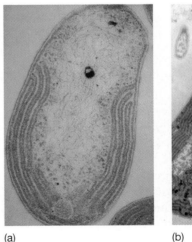
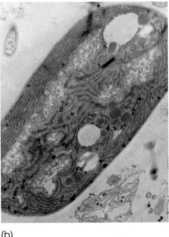

(a)   (b)

FIGURE 30.7
Bacterial cells often have complex internal membranes. This aerobic bacterium (a) exhibits extensive respiratory membranes within its cytoplasm not unlike those seen in mitochondria. This cyanobacterium (b) has thylakoid-like membranes that provide a site for photosynthesis.

The interior of a bacterial cell lacks a nucleus and is not extensively subdivided into functional compartments, although it may possess internal membranes and a nucleoid region.

Bacterial Diversity

There are over 4800 named species of bacteria, and undoubtedly many thousands more await discovery. Bacteria are not easily classified according to their forms, and only recently has enough been learned about their metabolic characteristics to develop a satisfactory overall classification comparable to that used for other organisms. Early systems for classifying bacteria relied on differential stains such as the Gram stain. Key bacterial characteristics used in classifying bacteria were:

1. Photosynthetic or nonphotosynthetic
2. Motile or nonmotile
3. Unicellular or multicellular
4. Formation of spores or dividing by transverse binary fission

With the development of genetic and molecular approaches, bacterial classifications can at last reflect true evolutionary relatedness. Lynn Margulis and Karlene Schwartz proposed a useful classification system that divides bacteria into 16 phyla, according to their most significant features. Table 30.1 outlines some of the major features of the phyla we describe.

Kinds of Bacteria

Although small and internally simple, bacteria are diverse in their internal chemistry and in many of the details of how they are assembled (figure 30.8). During the 2 billion years they evolved alone on earth, and in the 1.5 billion years since eukaryotes appeared, bacteria have adapted to many kinds of environments, including some you might consider harsh. They have successfully invaded waters that are very salty, very acidic or alkaline, and very hot or cold. They are found in hot springs where the temperatures exceed 78 °C (172 °F) and have been recovered living beneath 430 meters of ice in Antarctica!

Much of what we know of bacteria we have learned from studies in the laboratory. It is important to understand the limits this has placed on our knowledge: we have only been able to study those bacteria that can be cultured in laboratories. Field studies suggest that these represent but a small fraction of the kinds of bacteria that occur in soil, most of which cannot be cultured with existing techniques. We clearly have only scraped the surface of bacterial diversity.

As we learned in chapter 28, bacteria split into two lines early in the history of life, so different in structure and metabolism that they are as different from each other as either is from eukaryotes. The differences are so fundamental that biologists assign the two groups of bacteria to separate domains. One domain consists of the archaebacteria ("ancient bacteria"—although they are actually not as ancient as the

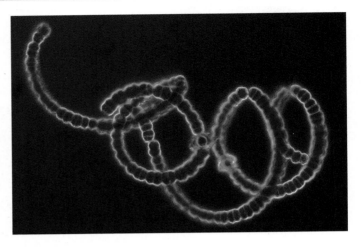

FIGURE 30.8
The cyanobacterium *Anabaena*. Individual cells adhere in filaments. The larger cells (bulging areas on the filament) are heretocysts, specialized cells in which nitrogen fixation occurs.

other bacterial domain). The only survivors today are confined to extreme environments that may resemble habitats on the early earth. The other more ancient domain consists of the eubacteria ("true bacteria"). It includes nearly all of the named species of bacteria.

Comparing Archaebacteria and Eubacteria

Archaebacteria and eubacteria differ in four ways:

1. **Cell wall.** Both kinds of bacteria typically have cell walls covering the plasma membrane that strengthen the cell. The cell walls of eubacteria are constructed of carbohydrate-protein complexes called peptidoglycan, which link together to create a strong mesh that gives the eubacterial cell wall great strength. The cell walls of archaebacteria lack peptidoglycan.
2. **Plasma membranes.** All bacteria have plasma membranes with a lipid-bilayer architecture (see chapter 6). The plasma membranes of eubacteria and archaebacteria, however, are made of very different kinds of lipids.
3. **Gene translation machinery.** Eubacteria possess ribosomal proteins and an RNA polymerase that are distinctly different from those of eukaryotes. However, the ribosomal proteins and RNA of archaebacteria are very similar to those of eukaryotes.
4. **Gene architecture.** The genes of eubacteria are not interrupted by introns, while at least some of the genes of archaebacteria do possess introns.

While superficially similar, bacteria differ from one another in a wide variety of characteristics.

Table 30.1   Bacteria

Major Group	Typical Examples		Key Characteristics
			ARCHAEBACTERIA
Archaebacteria	Methanogens, thermophiles, halophiles		Bacteria that are not members of the kingdom Eubacteria. Mostly anaerobic with unusual cell walls. Some produce methane. Others reduce sulfur.
			EUBACTERIA
Actinomycetes	*Streptomyces, Actinomyces*		Gram-positive bacteria. Form branching filaments and produce spores; often mistaken for fungi. Produce many commonly used antibiotics, including streptomycin and tetracycline. One of the most common types of soil bacteria; also common in dental plaque.
Chemoautotrophs	Sulfur bacteria, *Nitrobacter, Nitrosomonas*		Bacteria able to obtain their energy from inorganic chemicals. Most extract chemical energy from reduced gases such as H_2S (hydrogen sulfide), NH_3 (ammonia), and CH_4 (methane). Play a key role in the nitrogen cycle.
Cyanobacteria	*Anabaena, Nostoc*		A form of photosynthetic bacteria common in both marine and freshwater environments. Deeply pigmented; often responsible for "blooms" in polluted waters.
Enterobacteria	*Escherichia coli, Salmonella, Vibrio*		Gram-negative, rod-shaped bacteria. Do not form spores; usually aerobic heterotrophs; cause many important diseases, including bubonic plague and cholera.
Gliding and budding bacteria	Myxobacteria, *Chondromyces*		Gram-negative bacteria. Exhibit gliding motility by secreting slimy polysaccharides over which masses of cells glide; some groups form upright multicelluar structures carrying spores called fruiting bodies.
Pseudomonads	*Pseudomonas*		Gram-negative heterotrophic rods with polar flagella. Very common form of soil bacteria; also contain many important plant pathogens.
Rickettsias and chlamydias	*Rickettsia, Chlamydia*		Small, gram-negative intracelluar parasites. *Rickettsia* life cycle involves both mammals and arthropods such as fleas and ticks; *Rickettsia* are responsible for many fatal human diseases, including typhus (*Rickettsia prowazekii*) and Rocky Mountain spotted fever. Chlamydial infections are one of the most common sexually transmitted diseases.
Spirochaetes	*Treponema*		Long, coil-shaped cells. Common in aquatic environments; a parasitic form is responsible for the disease syphilis.

Bacterial Variation

When bacteria undergo fission, the two resulting cells are identical. After many divisions—some bacteria divide every few minutes—a large clone of cells is formed. But not all of these cells will be identical. Two processes create variation among them: mutation and genetic recombination.

Mutation

In a bacterium such as *Escherichia coli* there are about 5000 genes, judging from the amount of DNA that an individual cell contains. When considering any one of these genes, we may conservatively expect one mutation to have occurred by chance among every million copies. Since 200 individual bacteria together contain a total of a million genes, we can expect 1 of every 200 bacteria to have a mutant characteristic (figure 30.9). A spoonful of soil typically contains over a billion bacteria and therefore should contain something on the order of 5 million mutant individuals!

Because bacteria multiply so rapidly, favorable mutations can spread rapidly in a population and can change the characteristics of that population. For example, when adequate food and nutrients are available, a population of *Escherichia coli* growing under optimal conditions will double in under 20 minutes. The favorable mutation will soon be represented by large numbers of individuals better suited for growth than individuals that lack it.

The ability of bacteria to change rapidly in response to new challenges often has adverse effects on humans. Recently a number of strains of *Staphylococcus aureus* associated with serious infections in hospitalized patients have appeared, some of them with alarming frequency. Unfortunately, these strains have acquired resistance to penicillin and a wide variety of other antibiotics, so that infections are very difficult to treat. *Staphylococcus* infections provide an excellent example of the way in which mutation and intensive selection can bring about rapid change in bacterial populations. Such changes have serious medical implications when, as in the case of *Staphylococcus*, strains of bacteria emerge that are resistant to a variety of antibiotics. In the mid-1980s the routine administration of small doses of antibiotics to livestock to improve their weight gains came into question because of the danger to humans posed by the antibiotic-resistant strains of bacteria that inevitably result from this practice.

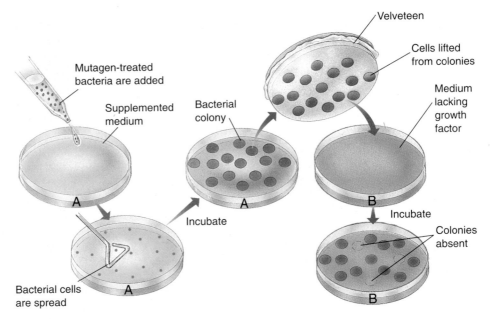

FIGURE 30.9

A mutant hunt in bacteria. Mutations in bacteria can be detected by a technique called replica plating, which allows the genetic characteristics of the colonies to be investigated without destroying them. The bacterial colonies, growing on a semisolid agar medium, are transferred from A to B using a sterile velveteen disc pressed on the plate. Plate A has a medium that includes special growth factors, while B has a medium that lacks some of these growth factors. Bacteria that are not mutated can produce their own growth factors and do not require them to be added to the medium. The colonies absent in B were unable to grow on the deficient medium and were thus mutant colonies; they were already present but undetected in A.

Genetic Recombination

Another source of genetic variation in populations of bacteria is recombination, which was discussed in detail in chapter 17. Bacterial recombination occurs by the transfer of genes from one cell to another as portions of viruses, plasmids, or other DNA fragments. The rapid transfer of newly produced, antibiotic-resistant genes by plasmids has been an important factor in the appearance of the resistant strains of *Staphylococcus aureus* discussed earlier. An even more important example in terms of human health involves the Enterobacteriaceae, the family of bacteria to which the common intestinal bacterium, *Escherichia coli*, belongs. In this family, there are many important pathogenic bacteria, including the organisms that cause dysentery, typhoid, and other major diseases. At times, some of their genetic material is exchanged with or transferred to *E. coli* by plasmids. Because of its abundance in the human digestive tract, *E. coli* poses a special threat if it acquires harmful traits.

Because of the *short* generation time of bacteria, mutation and recombination play an important role in generating genetic diversity.

Bacterial Metabolism

Bacteria have evolved many ways to acquire the carbon atoms and energy necessary for growth and reproduction. Many are **autotrophs,** organisms that obtain their carbon from inorganic CO_2. Autotrophs that obtain their energy from sunlight are called *photoautotrophs*, while those that harvest energy from inorganic chemicals are called *chemoautotrophs*. Other bacteria are **heterotrophs,** organisms that obtain at least some of their carbon from organic molecules like glucose. Heterotrophs that obtain their energy from sunlight are called *photoheterotrophs*, while those that harvest energy from organic molecules are called *chemoheterotrophs*.

Photoautotrophs. Many bacteria carry out photosynthesis, using the energy of sunlight to build organic molecules from carbon dioxide. The cyanobacteria use chlorophyll *a* as the key light-capturing pigment and use H_2O as an electron donor, leaving oxygen gas as a by-product. Other bacteria use bacteriochlorophyll as their pigment and H_2S as an electron donor, leaving elemental sulfur as the by-product.

Chemoautotrophs. Some bacteria obtain their energy by oxidizing inorganic substances. Nitrifiers, for example, oxidize ammonia or nitrite to form the nitrate that is taken up by plants. Other bacteria oxidize sulfur, hydrogen gas, and other inorganic molecules. On the dark ocean floor at depths of 2500 meters, entire ecosystems subsist on bacteria that oxidize hydrogen sulfide as it escapes from volcanic vents.

Photoheterotrophs. The so-called purple nonsulfur bacteria use light as their source of energy but obtain carbon from organic molecules such as carbohydrates or alcohols that have been produced by other organisms.

Chemoheterotrophs. Most bacteria obtain both carbon atoms and energy from organic molecules. These include decomposers and most pathogens.

How Heterotrophs Infect Host Organisms

In the 1980s, researchers studying the disease-causing species of *Yersinia*, a group of gram-negative bacteria, found that large amounts of proteins were being produced and secreted by the bacteria. These proteins lacked a signal-sequence that two known secretion mechanisms require for transport across the double membrane of gram-negative bacteria. The proteins must therefore have been secreted by a third type of system, which researchers called the *type III system.*

Recently, genes coding for the type III system are turning up in other gram-negative animal pathogens, and even in more distantly related plant pathogens. The genes seem to be more closely related to one another than do the bacteria. Furthermore, the genes are similar to those that code for bacterial flagella.

The function of these proteins is still under investigation, but it seems that some of the proteins are used to transfer other virulence proteins into nearby eukaryotic cells. Given the similarity of the type III genes to the genes that code for flagella, the transfer proteins may form a flagellum-like structure that shoots virulence proteins into the host cells. Once in the eukaryotic cells, the virulence proteins may determine the host's response to the pathogens. In *Yersinia*, proteins secreted by the type III system are injected into macrophages; they disrupt signals that tell the macrophages to engulf bacteria. *Salmonella* and *Shigella* use their type III proteins to enter the cytoplasm of eukaryotic cells and thus be protected from the immune system. The proteins secreted by *E. coli* alter the cytoskeleton of nearby intestinal eukaryotic cells, resulting in a bulge onto which the bacterial cells can tightly bind.

Currently, researchers are looking for a way to disarm the bacteria using knowledge of their internal machinery, possibly by causing the bacteria to release the virulence proteins before they are near eukaryotic cells. Others are studying the eukaryotic target proteins and the process by which they are affected.

Bacteria as Plant Pathogens

Many costly diseases of plants are associated with particular heterotrophic bacteria. Almost every kind of plant is susceptible to one or more kinds of bacterial disease. The symptoms of these plant diseases vary, but they are commonly manifested as spots of various sizes on the stems, leaves, flowers, or fruits. Other common and destructive diseases of plants, including blights, soft rots, and wilts, also are associated with bacteria. Fire blight, which destroys pears, apple trees, and related plants, is a well-known example of bacterial disease. Most bacteria that cause plant diseases are members of the group of rod-shaped bacteria known as pseudomonads (see figure 30.2*a*).

In August 1984, a bacterial disease known as citrus canker broke out in Florida, causing farmers to destroy more than 4 million citrus seedlings in four months in an effort to halt the spread of the disease. Citrus canker, which had not been detected in Florida for more than 40 years, probably was introduced on fruits that were brought from abroad by a traveler—fruits that escaped detection at a customs station. The disease is caused by one of more than 100 distinct varieties of the pseudomonad *Xanthomonas campestris*. Other varieties of this bacterium cause diseases with similar symptoms in beans, cabbages, peaches, and other plants. Citrus canker now threatens the continued existence of the Florida citrus industry, a $2.5 billion business that was already weakened by the worst freezes of the century in 1983 to 1985.

While bacteria obtain carbon and energy in many ways, most are chemoheterotrophs. Some heterotrophs have evolved sophisticated ways to infect their hosts.

Human Bacterial Diseases

Bacteria cause many diseases in humans, including cholera, leprosy, tetanus, bacterial pneumonia, whooping cough, and diphtheria (table 30.2). Members of the genus *Streptococcus* (see figure 30.2*b*) are associated with scarlet fever, rheumatic fever, pneumonia, and other infections. Tuberculosis, another bacterial disease, is still a leading cause of death in humans. These diseases are mostly spread through the air, as are some of those caused by the pathogenic genus *Staphylococcus*. Other bacterial diseases are dispersed primarily in food or water, including typhoid fever, paratyphoid fever, and bacillary dysentery. Typhus is spread among rodents and humans by insect vectors. To gain some appreciation of bacterial diseases, let's examine two more closely.

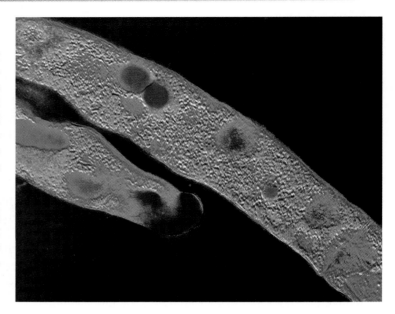

FIGURE 30.10

Mycobacterium tuberculosis. This rod-shaped bacterium is responsible for tuberculosis in humans.

Tuberculosis

Tuberculosis (TB) has been one of the great killer diseases for thousands of years. Currently, about one-third of all people worldwide are infected with *Mycobacterium tuberculosis*, the tuberculosis bacterium (figure 30.10). Eight million new cases crop up each year, with about 3 million people dying from the disease annually (the World Health Organization predicts 4 million deaths a year by 2005). In fact, in 1997, TB was the leading cause of death from a single *infectious agent* worldwide. Despite these facts, in the 1970s and 1980s there was a mistaken assumption that TB was no longer a problem in the United States, and as a result the programs to control it were dismantled. Since the mid-1980s, the United States has been experiencing a dramatic resurgence of tuberculosis. The causes of this resurgence include social factors such as poverty, crowding, homelessness, and incarceration (which have always promoted the spread of TB); immigration from high-prevalence countries (the foreign-born account for nearly one-third of the nation's TB cases); declining public health infrastructure, and the increasing prevalence of HIV infections (patients with AIDS are much more likely to develop TB than people with healthy immune systems).

In addition to the increased numbers of cases—more than 25,000 nationally as of March 1995—there have been alarming outbreaks of multidrug-resistant strains of tuberculosis—strains resistant to the best available anti-TB medications. Multidrug-resistant TB is particularly concerning because it requires much more time to treat, is more expensive to treat, and may prove to be fatal.

The basic principles of TB control are to make sure all patients complete a full course of treatment so they do not develop drug-resistant strains of the disease, and to ensure that high-risk individuals who are infected but not yet sick receive preventative therapy, which is 90% effective in reducing the likelihood of developing active TB.

Dental Caries

One human disease we do not usually consider bacterial in origin arises in the film on our teeth. This film, or plaque, consists largely of bacterial cells surrounded by a polysaccharide matrix. Most of the bacteria in plaque are filaments of rod-shaped cells classified as *Leptotricha buccalis*, which extend out perpendicular to the surface of the tooth. Many other bacterial species are also present in plaque. Tooth decay, or **dental caries,** is caused by the bacteria present in the plaque, which persists especially in places that are difficult to reach with a toothbrush. Diets that are high in sugars are especially harmful to teeth because lactic acid bacteria (especially *Streptococcus sanguis* and *S. mutans)* ferment the sugars to lactic acid, a substance that causes the local loss of calcium from the teeth. Frequent eating of sugary snacks or sucking on candy over a period of time keeps the pH level of the mouth low and actively promotes tooth decay and the formation of cavities. Bacteria will not cause decay in the absence of such a diet. Once the hard tissue has started to break down, the lysis of proteins in the matrix of the tooth enamel starts, and tooth decay begins in earnest. Fluoride makes the teeth more resistant to decay because it retards the loss of calcium. It was first realized that bacteria cause tooth decay when germ-free animals were raised. Their teeth do not decay even if they are fed sugary diets.

Table 30.2 Important Human Bacterial Diseases

Disease	Pathogen	Vector/Reservoir	Epidemiology
Anthrax	*Bacillus anthracis*	Animals, including processed skins	Bacterial infection that can be transmitted through contact or ingested. Rare except in sporadic outbreaks. May be fatal.
Botulism	*Clostridium botulinum*	Improperly prepared food	Contracted through ingestion or contact with wound. Produces acute toxic poison; can be fatal.
Chlamydia	*Chlamydia trachomatis*	Humans, STD	Urogenital infections with possible spread to eyes and respiratory tract. Occurs worldwide; increasingly common over past 20 years.
Cholera	*Vibrio cholerae*	Human feces, plankton	Causes severe diarrhea that can lead to death by dehydration; 50% peak mortality if the disease goes untreated. A major killer in times of crowding and poor sanitation; over 100,000 died in Rwanda in 1994 during a cholera outbreak.
Dental caries	*Streptococcus*	Humans	A dense collection of this bacteria on the surface of teeth leads to secretion of acids that destroy minerals in tooth enamel—sugar alone will not cause caries.
Diphtheria	*Corynebacterium diphtheriae*	Humans	Acute inflammation and lesions of mucous membranes. Spread through contact with infected individual. Vaccine available.
Gonorrhea	*Neisseria gonorrhoeae*	Humans only	STD, on the increase worldwide. Usually not fatal.
Hansen's disease (leprosy)	*Mycobacterium leprae*	Humans, feral armadillos	Chronic infection of the skin; worldwide incidence about 10–12 million, especially in southeast Asia. Spread through contact with infected individuals.
Lyme disease	*Borrelia bergdorferi*	Ticks, deer, small rodents	Spread through bite of infected tick. Lesion followed by malaise, fever, fatigue, pain, stiff neck, and headache.
Peptic ulcers	*Helicobacter pylori*	Humans	Originally thought to be caused by stress or diet, most peptic ulcers now appear to be caused by this bacterium; good news for ulcer sufferers as it can be treated with antibiotics.
Plague	*Yersinia pestis*	Fleas of wild rodents: rats and squirrels	Killed 1/4 of the population of Europe in the 14th century; endemic in wild rodent populations of the western U.S. today.
Pneumonia	*Streptococcus, Mycoplasma, Chlamydia*	Humans	Acute infection of the lungs, often fatal without treatment
Tuberculosis	*Mycobacterium tuberculosis*	Humans	An acute bacterial infection of the lungs, lymph, and meninges. Its incidence is on the rise, complicated by the development of new strains of the bacteria that are resistant to antibiotics.
Typhoid fever	*Salmonella typhi*	Humans	A systemic bacterial disease of worldwide incidence. Less than 500 cases a year are reported in the U.S. The disease is spread through contaminated water or foods (such as improperly washed fruits and vegetables). Vaccines are available for travelers.
Typhus	*Rickettsia typhi*	Lice, rat fleas, humans	Historically a major killer in times of crowding and poor sanitation; transmitted from human to human through the bite of infected lice and fleas. Typhus has a peak untreated mortality rate of 70%.

Sexually Transmitted Diseases

In addition to HIV (discussed in chapters 29 and 54), a number of other microbes cause sexually transmitted diseases (STDs). Some of these STDs are caused by bacteria; others by viruses. Viral STDs have no cures, and in some bacterial STDs, the bacteria can mutate into drug-resistant strains. With all the media coverage given to AIDS, people sometimes forget that other STDs also are hazardous to health. All sexually active persons should learn about the nature and symptoms of STDs and take the appropriate measures to prevent their transmission.

Gonorrhea. Gonorrhea is one of the most prevalent communicable diseases in North America. Caused by the bacterium *Neisseria gonorrhoeae*, gonorrhea can be transmitted through sexual intercourse or any other sexual contacts in which body fluids are exchanged, such as oral or anal intercourse. Gonorrhea can infect the throat, urethra, cervix, or rectum and can spread to the eyes and internal organs, causing conjunctivitis (a severe infection of the eyes) and arthritic meningitis (an infection of the joints). Left untreated in women, gonorrhea can cause pelvic inflammatory disease (PID), a condition in which the Fallopian tubes become scarred and blocked. PID can eventually lead to sterility.

Syphilis. Syphilis, a very destructive STD, was once prevalent but is now less common due to the advent of blood-screening procedures and the development of antibiotics. Syphilis is caused by a spirochaete bacterium, *Treponema pallidum*, that is transmitted during sexual intercourse or through direct contact with an open syphilis sore. The bacterium can also be transmitted from a mother to her fetus.

Once inside the body, the disease progresses in four distinct stages. The first, or primary stage, is characterized by the appearance of a small, painless, often unnoticed sore called a chancre. The chancre resembles a blister and occurs at the location where the bacterium entered the body about three weeks following exposure. This stage of the disease is highly infectious, and an infected person may unwittingly transmit the disease to others.

The second stage of syphilis is marked by a rash that may cover the entire body, characteristically including the palms and the soles; an infected person may have a sore throat and sores in the mouth. The bacteria can be transmitted at this stage through kissing or contact with an open sore.

The third stage of syphilis is symptomless. This stage may last for several years, and at this point, the person is no longer infectious. The final stage of syphilis is the most debilitating, however, as the damage done by the bacteria in the third stage becomes evident. Sufferers at this stage of syphilis experience heart disease, mental deficiency, and nerve damage, which may include a loss of motor functions or blindness.

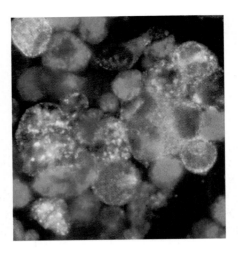

FIGURE 30.11
Chlamydia trachomatis. In this micrograph, infected cells appear green, and *C. trachomatis* (*yellow*) can be seen within them.

Chlamydia. Sometimes called the "silent STD," chlamydia is caused by an unusual bacterium, *Chlamydia trachomatis*, that has both bacterial and viral characteristics. Like a bacterium, it is susceptible to antibiotics, and, like a virus, it depends on its host to replicate its genetic material; it is an obligate internal parasite (figure 30.11). The bacterium is transmitted through vaginal, anal, or oral intercourse with an infected person.

Chlamydia is called the "silent STD" because women usually experience no symptoms until after the infection has become established. The effects of an established chlamydia infection on the female body are extremely serious. Chlamydia can cause pelvic inflammatory disease (PID), which can lead to sterility. Symptoms of PID include painful urination, painful intercourse, vaginal discharge, and irregular vaginal bleeding (such as between menstrual periods or after intercourse). Chlamydial infection of the cervix has been shown to increase the risk of acquiring HIV as well. In men, the symptoms are a watery discharge from the penis and burning or itching around the urethra.

Within the last few years, two types of tests for chlamydia have been developed that look for the presence of the bacteria in the discharge from men and women. The treatment for chlamydia is antibiotics, usually tetracycline (penicillin is not effective against chlamydia). Any woman who experiences the symptoms associated with this STD should be tested for the presence of the chlamydia bacterium; otherwise, her fertility may be at risk.

This discussion of STDs may give the impression that sexual activity is fraught with danger, and in a way, it is. It is folly not to take precautions to avoid STDs. The best way to do this is to know one's sexual partners well enough to discuss the possible presence of an STD. Condom use can also prevent transmission of most of the diseases. Responsibility for protection lies with each individual.

Bacterial diseases have a major impact worldwide. Sexually transmitted diseases (STDs) are becoming increasingly widespread among Americans as sexual activity increases.

Summary of Concepts

30.1 Bacteria are the smallest and most numerous organisms.

- Bacteria are the oldest and simplest organisms, but they are metabolically much more diverse than all other life forms combined.
- Bacteria play vital roles in cycling nutrients within ecosystems. Certain bacteria are the only organisms able to fix atmospheric nitrogen into organic molecules, a process on which all life depends.
- Bacteria differ from eukaryotes in many ways, the most important of which concern the degree of internal organization within the cell.

30.2 Bacterial cell structure is more complex than commonly supposed.

- Most bacteria have cell walls that consist of a network of polysaccharide molecules connected by polypeptide cross-links. In gram-negative bacteria, an outer membrane containing large molecules of lipopolysaccharide lies over a thin peptidoglycan layer. Gram-positive bacteria lack this outer membrane but have a thicker peptidoglycan layer.
- Bacteria are rod-shaped (bacilli), spherical (cocci), or spiral (spirilla) in form. Bacilli or cocci may adhere in small groups or chains.
- A bacterial cell does not possess a membrane-bounded nucleus, but it may exhibit a nucleoid region where the bacterial DNA is located.

30.3 Bacteria exhibit considerable diversity in both structure and metabolism.

- The two kingdoms, Archaebacteria and Eubacteria, are made up of the prokaryotes, or bacteria, with about 4800 species named so far.
- The Archaebacteria, which consist of the methanogenic bacteria and a few related groups, differ markedly from the Eubacteria and from the eukaryotes in their ribosomal sequences and in other respects.
- Mutation and genetic recombination are important sources of variability in bacteria.
- Many bacteria are autotrophic, both photosynthetic and chemoautotrophic, and make major contributions to the world carbon balance. Others are heterotrophic and play a key role in world ecology by breaking down organic compounds.
- Some heterotrophic bacteria cause major diseases in plants and animals.

30.4 Bacteria are responsible for many human diseases.

- Human diseases caused by heterotrophic bacteria include many fatal diseases that have had major impacts on human history, including cholera, plague, and typhus.
- The most lethal infectious disease today is tuberculosis, caused by a bacterium. Over 3 million people die each year from this disease.

Discussing Key Terms

1. **Prokaryote**   All bacteria are prokaryotic cells, without nuclei or other subcellular compartments. All other organisms are eukaryotes, with nuclei and an elaborate endomembrane system that subdivides the cell interior into functional compartments.

2. **Cell walls**   Most bacteria have a protective polysaccharide cell wall surrounding the cell membrane. The cell wall helps the bacterium maintain its shape in changing osmotic environments. Some bacteria also have a lipopolysaccharide layer surrounding the cell wall; they are said to be gram-negative, as Gram stain does not adhere to them.

3. **Autotroph**   Many bacteria are autotrophic, obtaining energy from sunlight (photosynthesizers) or from inorganic chemicals (chemoautotrophs).

4. **Heterotroph**   Other bacteria are heterotrophs, obtaining their energy from organic molecules. Bacteria play an important role with fungi as decomposers in food webs. Some heterotrophic bacteria are pathogens, able to infect plants or animals that cause disease.

5. **STD**   Both viruses and bacteria cause sexually transmitted diseases; unlike viral STDs, most bacterial STDs can be effectively treated with antibiotics.

Review Questions

1. Structural differences among bacteria are not great. How are different species of bacteria recognized? How is this determined?

2. In what seven ways do prokaryotes differ substantially from eukaryotes?

3. What is the structure of the bacterial cell wall? How does the cell wall differ between gram-positive and gram-negative bacteria? In general, which type of bacteria is more resistant to the action of most antibiotics? Why?

4. How do the Archaebacteria differ from the Eubacteria? What group of bacteria is the most prominent of the Archaebacteria? What unique metabolism do they exhibit? What are their oxygen requirements? How do they control this themselves?

5. Why does mutation play such an important role in creating genetic diversity in bacteria? How does bacterial recombination result in genetic diversity in bacteria? What is an example of the effects of such variability?

6. How do heterotrophic bacteria that are successful pathogens overcome the many defenses the human body uses to ward off disease? Are all infective heterotrophs pathogens?

7. What are STDs? How are they transmitted? Which STDs are caused by viruses and which are caused by bacteria? Why is the cause of chlamydia unusual?

Thought Questions

1. Justify the assertion that life on earth could not exist without bacteria.

2. What do you think the functions of antibiotics are in the bacteria that produce them?

3. What are endospores? Why would they be an advantage? Would a bacterium with the ability to form endospores have a greater chance of extinction? Why or why not?

Internet Links

Bugs in the News
http://falcon.cc.ukans.edu/~jbrown/bugs.html
From the University of Kansas, a fun newsy way to keep up with what is going on in microbiology and biotechnology.

The Microbe Zoo
http://commtechlab.msu.edu/CTLProjects/dlc-me/zoo/
A trip to the MICROBIAL ZOO visits microbes living in the ground ("dirtland"), within animals, in the water, and even in outer space. Fun and interesting.

Microbes Weird and Wonderful
http://www.wsu.edu:8080/~hurlbert/pages/micnews.html
A collection of interesting tidbits about microbes "in the news."

Plague Fever
http://www.outbreak.org
A successor to The Ebola Page of 1995, this site aims to raise public awareness of new diseases, covering everything from anthrax and emerging tropical viruses to prions and yellow fever. Outbreaks are chronicled with up-to-date dispatches, news clippings, and interviews.

For Further Reading

Blaser, M.: "The Bacteria behind Ulcers," *Scientific American*, vol. 274, February 1996, pages 104–7. A bacterial culprit, *Helicobacter pylori*, has been shown to cause many forms of peptic ulcers, gastric cancers, and gastritis—diseases previously attributed to stress, diet, smoking, and drinking. Over one-third of the world's population is thought to be infected with *H. pylori*.

Fenchel, T. and B. J. Finlay: "The Evolution of Life Without Oxygen," *American Scientist*, vol. 82, January/February 1994, pages 22–29. The role of the prokaryote in eukaryotic evolution is examined.

Iseman, M. D.: "Evolution of Drug-Resistant Tuberculosis: A Tale of Two Species," *Proc. Nat. Acad. Sci. USA*, vol. 91, March 29, 1994, pages 2428–29. Drug-resistant diseases may be the biggest threat facing modern medicine. This article discusses the mechanisms by which the tuberculosis pathogen is becoming significantly dangerous.

Kiester, E.: "A Curiosity Turned into the First Silver Bullet Against Death," *Smithsonian*, vol. 21, November 1990, page 173. A very engaging account of the discovery of penicillin.

Losick, R. and D. Kaiser: "Why and How Bacteria Communicate," *Scientific American*, February 1997, pages 68–73. Bacteria exhibit surprisingly sophisticated chemical communications.

Mee, C. L., Jr.: "How a Mysterious Disease Laid Low Europe's Masses," *Smithsonian*, February 1990, pages 67–79. In the 1300s, a third of the population of Europe died of plague brought on by fleas.

Monmaney, T.: "Marshall's Hunch," *The New Yorker*, September 20, 1993, pages 64–72. The unprecedented and unorthodox findings of an unknown doctor point toward a cure for stomach ulcers—he proved it to be an infectious disease! Highly recommended.

31

Protists

Concept Outline

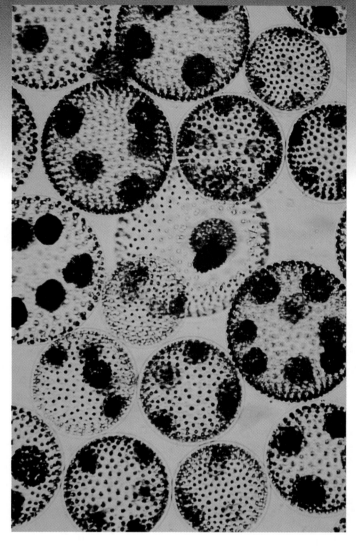

FIGURE 31.1
Volvox, **a colonial protist.** The protists are a large, diverse group of primarily single-celled organisms, a group from which the other three eukaryotic kingdoms each evolved.

Protists are the most diverse of the four kingdoms in the domain Eukaryota. The kingdom Protista contains many unicellular, colonial and multicellular groups (figure 31.1). Probably the most important statement we can make about the kingdom Protista is that it is an artificial group; as a matter of convenience, single-celled eukaryotic organisms have typically been grouped together into this kingdom. This lumps many very different and only distantly related forms together. The "single-kingdom" classification of the protista is not representative of any evolutionary relationships, in contrast to the classifications of the animals, plants, fungi, eubacteria, and archaebacteria which *do* represent evolutionary relationships. The phyla of protists are, with very few exceptions, only distantly related to one another.

The Challenge of Classifying the Protists

In a sense, placing all the protists in a single kingdom simply reflects our ignorance about how the many different types are related to one another—an ignorance that is rapidly waning due to excellent new research in the area. Of all the groups of organisms biologists study, protists are probably in the greatest state of flux when it comes to classification. There is little consensus, even among experts, as to how the different kinds of protists should be classified. Are they a single, very diverse kingdom, or are they better considered as several different kingdoms, each of equal rank with animals, plants, and fungi?

Before the introduction of the five-kingdom system of classification by Whittaker in 1969, the green, photosynthetic protists were considered plants, as were all protists that absorbed their food. Those protists that ingested their food like animals were considered small, very simple animals. These schemes of classification do not reflect the complex relationships between the different groups of protists and have generally been replaced by the classification scheme used in this book. To consider some groups of protists to be like plants, simply because they have independently acquired chloroplasts in the course of their evolution, or to consider others as "resembling" animals because they move and have not acquired chloroplasts, is highly misleading.

Since the protista are still predominantly considered part of one diverse, nonunified group, that is how we will treat them in this chapter, bearing in mind that biologists are rapidly gaining a better understanding of the evolutionary relationships among members of the kingdom Protista (figure 31.2). It seems likely that within a few years, the traditional kingdom Protista will be replaced by another more illuminating arrangement.

> The taxonomy of the protists is in a state of flux, with no general agreement.

FIGURE 31.2
The challenge of Protistan classification. Three different suggestions for protistan classification are presented, each adapted from the work of an authority in the field. Their great differences attest to the wide divergence of opinion within the field itself. The classification on the top is based on molecular variation in ribosomal subunits. The classification in the middle presents a cladistic analysis of a broad range of characters (including ribosomal subunits). The classification on the bottom outlines a more revolutionary reevaluation of the protists. Comparison of the three schemes reveals that some groups are commonly recognized as related (like ciliates and dinoflagellates), while the classification of others (like *Giardia*) is clearly in a state of flux.

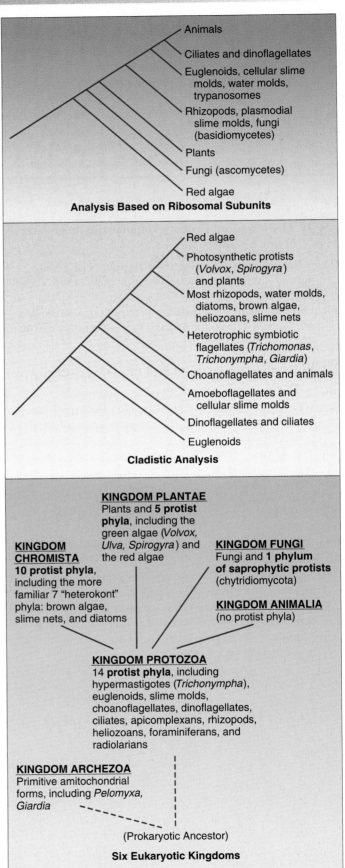

Analysis Based on Ribosomal Subunits

Cladistic Analysis

KINGDOM PLANTAE
Plants and **5 protist phyla**, including the green algae (*Volvox, Ulva, Spirogyra*) and the red algae

KINGDOM CHROMISTA
10 protist phyla, including the more familiar 7 "heterokont" phyla: brown algae, slime nets, and diatoms

KINGDOM FUNGI
Fungi and **1 phylum of saprophytic protists** (chytridiomycota)

KINGDOM ANIMALIA
(no protist phyla)

KINGDOM PROTOZOA
14 protist phyla, including hypermastigotes (*Trichonympha*), euglenoids, slime molds, choanoflagellates, dinoflagellates, ciliates, apicomplexans, rhizopods, heliozoans, foraminiferans, and radiolarians

KINGDOM ARCHEZOA
Primitive amitochondrial forms, including *Pelomyxa, Giardia*

(Prokaryotic Ancestor)

Six Eukaryotic Kingdoms

General Biology of the Protists

Protists are united on the basis of a single negative characteristic: they are not fungi, plants, or animals. In all other respects they are diverse. Many are unicellular (figure 31.3), but there are numerous colonial and multicellular groups. Most are microscopic, but some are as large as trees. They represent all symmetries, and exhibit all types of nutrition.

The Cell Surface

Protists possess a varied array of cell surfaces. Some protists, like amoebas, are surrounded only by their plasma membranes. Other protists, like algae and molds, are encased within strong cell walls. Still others, like diatoms and forams, secrete glassy shells of silica.

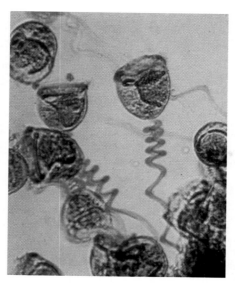

FIGURE 31.3
A unicellular protist. The protist kingdom is a catch-all kingdom for many different groups of unicellular organisms, such as this *Vorticella* (phylum Ciliophora), which is heterotrophic, feeds on bacteria, and has a retractable stalk.

Locomotor Organelles

Protists move chiefly by either flagellar rotation or pseudopodial movement. Many protists wave one or more flagella to propel themselves through the water, while others use banks of flagella called cilia to create water currents for their feeding. Pseudopodia are the chief means of locomotion among amoeba, whose pseudopods are large, blunt extensions of the cell body called lobopodia. Other related protists extend thin, branching protrusions called filopodia. Still other protists extend long, thin pseudopodia called axopodia supported by axial rods of microtubules. Axopodia can be extended or retracted. Since the tips can adhere to adjacent surfaces, the cell can move by a rolling motion, shortening the axopodia in front and extending those in the rear.

Cyst Formation

Many protists with delicate surfaces are successful in quite harsh habitats. How do they manage to survive so well? They survive inhospitable conditions by forming **cysts.** A cyst is a dormant form of a cell with a resistant outer covering in which cell metabolism is more or less completely shut down. Cyst formation is also important to many parasitic protists that must survive a harsh environment between hosts. Cysts of the soil ciliate *Colpoda* survive for up to 38 years, and at temperatures as hot as boiling water and as cold as liquid air. Not all cysts are so sturdy. Vertebrate parasitic amoebae, for example, form cysts that are quite resistant to gastric acidity, but will not tolerate desiccation or high temperature.

Nutrition

Some protists are photosynthetic autotrophs and are called **phototrophs.** Others are heterotrophs that obtain energy from organic molecules synthesized by other organisms. Among heterotrophic protists, those that ingest visible particles of food are called **phagotrophs,** or **holozoic feeders.** Those ingesting food in soluble form are called **osmotrophs,** or **saprozoic feeders.**

Phagotrophs ingest food particles into intracellular vesicles called **food vacuoles** or **phagosomes.** Lysosomes fuse with the food vacuoles, introducing enzymes that digest the food particles within. As the digested molecules are absorbed across the vacuole membrane, the food vacuole becomes progressively smaller. Any indigestible material is ultimately released to the outside by exocytosis, the food vacuole again fusing with the plasma membrane.

Reproduction

Protists typically reproduce asexually, reproducing sexually only in times of stress. Asexual reproduction involves mitosis, but the process is often somewhat different from the mitosis that occurs in multicellular animals. The nuclear membrane, for example, often persists throughout mitosis, with the microtubular spindle forming within it. In some groups, asexual reproduction involves spore formation, in others fission. The most common type of fission is **binary,** in which a cell simply splits into nearly equal halves. When the progeny cell is considerably smaller than its parent, and then grows to adult size, the fission is called **budding.** In multiple fission, or **schizogony,** common among some protists, fission is preceded by several nuclear divisions, so that fission produces several individuals almost simultaneously.

Sexual reproduction also takes place in many forms among the protists. In ciliates and some flagellates, **gametic meiosis** occurs just before gamete formation, as it does in metazoans. In the sporozoans, **zygotic meiosis** occurs directly *after* fertilization, and all the individuals that are produced are haploid until the next zygote is formed. In algae, there is **intermediary meiosis,** producing an alternation of generations similar to that seen in plants, with significant portions of the life cycle spent as haploid and as diploid.

Protists exhibit a wide range of forms, locomotion, nutrition and reproduction.

31.2 Protists are grouped into fifteen very distinctive phyla.

Seven Groups of Protists

There are some 15 major phyla of protists. It is difficult to encompass their great diversity with any simple scheme. Traditionally, texts have grouped them artificially (as was done in the nineteenth century) into photosynthesizers (algae), heterotrophs (protozoa), and absorbers (fungus-like protists).

In this text, we will group the protists into seven general groups according to major shared characteristics (table 31.1). These are characteristics that taxonomists are using today in broad attempts to classify the kingdom Protista and include presence or absence and type of cilia or flagella; presence and kinds of pigments; type of mitosis; kinds of cristae present in the mitochondria; molecular genetics of the ribosomal "S" subunit; what kind of inclusions the protist may have; overall body form (amoeboid, coccoid, and so forth.);

whether the protist has any kind of shell or other body "armor"; and modes of nutrition and movement. These represent only some of the characters used to define phylogenetic relationships.

The seven criteria we have chosen to define groups are not the only ones that might be chosen, and there is no broad agreement among biologists as to which set of criteria is preferable. As molecular analysis gives us a clearer picture of the phylogenetic relationships among the protists, more evolutionarily suitable groupings will without a doubt replace the one represented here. Table 31.2 summarizes some of the general characteristics and groupings of the 15 major phyla of protists.

The 15 major protist phyla can be conveniently grouped into seven categories according to major shared characteristics.

Table 31.1	Some Features of the Protista			
Name*	Number of Species	Chlorophylls	Wall/Shell	Flagella
SARCODINA				
Rhizopoda	Hundreds	None	Shells in some	0
Foraminifera	Hundreds	None	Shells, tests	0 (podia)
Actinopoda	Hundreds	None	Skeletons	0
ALGAE				
Rhodophyta	4000	*a*	Cellulose†	0
Phaeophyta	1500	*a + c*	Cellulose†	2
Chlorophyta	7000	*a + b*	Cellulose†	2
DIATOMS				
Chrysophyta	11,500	*a + c*	Opaline silica	0
FLAGELLATES				
Pyrrhophyta	2100	*a + b*	Mostly cellulose plates	2; Un‡
Euglenophyta	1000	*a + b* or none	Flexible pellicle	2; Un‡
Zoomastigophora	Thousands	None	None	Many
SPOROZOANS				
Apicomplexa	3900	None	Spores	Variable
CILIATES				
Ciliophora	8000	None	None	Many
MOLDS				
Acrasiomycota	70	None	Cellulose (spores)	0
Myxomycota	500	None	None	0 or 2
Oomycota	580	None	Cellulose†	2

*Phyla whose names are printed in italics consist of multicellular organisms; Chlorophyta also includes many unicellular ones, and Rhodophyta includes a few unicellular ones. Myxomycota moves about in a mass, called a plasmodium, in which there are many nuclei.

†These phyla have cellulose as the primary constituent in their cell walls, but other substances are present in some members of these phyla.

‡Un = unequal in length.

Table 31.2 Kinds of Protists

Group	Phylum	Typical Examples		Key Characteristics
Sarcodina—*Heterotrophs with no permanent locomotor apparatus*				
Amoebas	Rhizopoda	*Amoeba*		Move by pseudopodia
Forams	Foraminifera	Forams		Rigid shells; move by protoplasmic streaming
Radiolarians	Actinopoda	Radiolarians		Glassy skeletons; needle-like pseudopods
Algae—*Photosynthetic protists that are multicellular or largely multicellular*				
Red	Rhodophyta	Coralline algae		Chlorophyll *a* + red pigment
Brown	Phaeophyta	Kelp		Chlorophyll *a* + chlorophyll *c*
Green	Chlorophyta	*Chlamydomonas*		Chlorophyll *a* + chlorophyll *b*
Diatoms—*Photosynthetic protists that are unicellular, many with a double shell of silica*				
	Chrysophyta	*Diatoma*		Manufacture the carbohydrate chrysolaminarin
		Golden algae		
Flagellates—*Protists with locomotor flagella*				
Dinoflagellates	Pyrrhophyta	Red tides		Photosynthetic; unicellular; two flagella
Euglenoids	Euglenophyta	*Euglena*		Some photosynthetic; others heterotrophic; unicellular
Zoomastigotes	Zoomastigophora	Trypanosomes		Heterotrophic; unicellular
Sporozoans—*Nonmotile, spore-forming unicellular parasites*				
	Apicomplexa	*Plasmodium*		The apical end of the spores contains a complex mass of organelles
Ciliates—*Heterotrophic unicellular protists with cells of fixed shape possessing two nuclei and many cilia*				
	Ciliophora	*Paramecium*		Many cells also contain highly complex and specialized organelles
Molds—*Heterotrophs with restricted mobility and cell walls made of carbohydrate*				
Cellular slime molds	Acrasiomycota	*Dictyostelium*		Colonial aggregations of individual cells; most closely related to amoebas
Plasmodial slime molds	Myxomycota	*Fuligo*		Stream along as a multinucleate mass of cytoplasm
Water molds	Oomycota	Water molds, rusts, and mildew		Terrestrial and freshwater

The Sarcodines

Rhizopoda: The Amoebas

Hundreds of species of amoebas are found throughout the world in both fresh and salt waters. They are also abundant in soil. Many kinds of amoebas are parasites of animals. Reproduction in amoebas occurs by fission, or the direct division into two cells of equal volume. Amoebas of the phylum Rhizopoda lack cell walls, flagella, meiosis, and any form of sexuality. They do undergo mitosis, with a spindle apparatus that resembles that of other eukaryotes.

Amoebas move from place to place by means of their **pseudopods**, from the Greek words for "false" and "foot" (figure 31.4). Pseudopods are flowing projections of cytoplasm that extend and pull the amoeba forward or engulf food particles. An amoeba puts a pseudopod forward and then flows into it. Microfilaments of actin and myosin similar to those found in muscles are associated with these movements.

Some kinds of amoebas form resistant cysts. In parasitic species such as *Entamoeba histolytica*, which causes amoebic dysentery (a protist-caused disease), cysts enable the amoebas to resist digestion by their animal hosts. Mitotic division takes place within the cysts, which ultimately break and release four, eight, or even more amoebas within the digestive tracts of their host animals. The primary infection takes place in the intestine, but it often moves into the liver and other parts of the body. The cysts are dispersed in feces and may be transmitted directly from person to person, in infected food or water, or by flies. It is estimated that up to 10 million people in the United States have infections of parasitic amoebas, and some 2 million show symptoms of the disease, ranging from abdominal discomfort with slight diarrhea to much more serious conditions. In some tropical areas, more than half of the people may be infected. The spread of amoebic dysentery can be limited by proper sanitation and hygiene.

Actinosphaerium (figure 31.5) is an unusual kind of amoeba-like protist that belongs to another phylum, Actinopoda, whose members have glassy skeletons with many needle-like pseudopods.

A Possible Missing Link

Pelomyxa palustris (figure 31.6) may represent an early stage in the evolution of eukaryotic cells, a stage before they had acquired mitochondria and before mitosis had evolved. Pelomyxa lacks mitochondria and does not undergo mitosis. Its nuclei divide somewhat like bacterial cells do, by simply pinching apart into two nuclei, with new membranes forming around the daughter nuclei. The cells of *Pelomyxa* are much larger than bacteria and are visible to the naked eye. Although *Pelomyxa* lacks mitochondria, its cells do have two kinds of bacterial symbionts within them, and these may play the same role that mitochondria do in all other eukaryotes. This organism is so distinct that it is assigned to a phylum of its own, Caryoblastea.

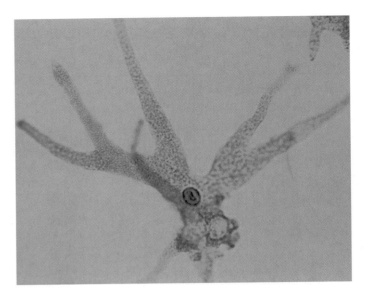

FIGURE 31.4
Amoeba proteus. This relatively large amoeba is commonly used in teaching and for research in cell biology. The projections are pseudopods; an amoeba moves by flowing into them. The nucleus of the amoeba is plainly visible.

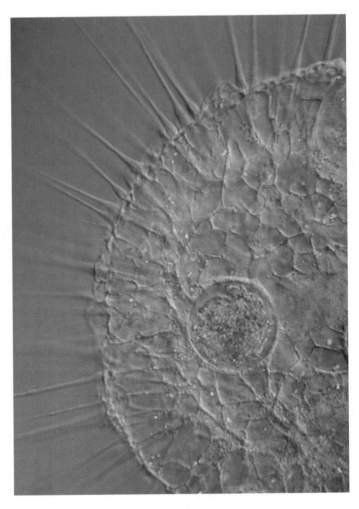

FIGURE 31.5
Actinosphaerium, **a protist of the phylum Actinopoda (300×).** This amoeba-like radiolarian has striking needle-like pseudopods.

FIGURE 31.6
FIGURE 31.6
Pelomyxa palustris. This unique, amoeba-like protist lacks mitochondria and does not undergo mitosis. *Pelomyxa* may represent a very early stage in the evolution of eukaryotic cells. This species is the only member of the phylum Caryoblastea.

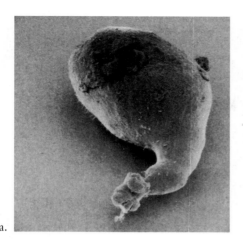

Foraminifera: Forams

Members of the phylum Foraminifera are heterotrophic marine protists. They range in diameter from about 20 micrometers to several centimeters. Characteristic of the group are pore-studded shells (called **tests**) composed of organic materials usually reinforced with grains of inorganic matter. These grains may be calcium carbonate, sand, or even plates from the shells of echinoderms or spicules (minute needles of calcium carbonate) from sponge skeletons. Depending on the building materials they use, foraminifera—often informally called "forams"—may have shells of very different appearance. Some of them are brilliantly colored red, salmon, or yellow-brown.

Most foraminifera live in sand or are attached to other organisms, but two families consist of free-floating planktonic organisms. Their tests may be single-chambered but more often are multichambered, and they sometimes have a spiral shape resembling that of a tiny snail. Thin cytoplasmic projections called **podia** emerge through openings in the tests (figure 31.7). Podia are used for swimming, gathering materials for the tests, and feeding. Forams eat a wide variety of small organisms.

The life cycles of foraminifera are extremely complex, involving an alternation between haploid and diploid generations (sporic meiosis). Forams have contributed massive accumulations of their tests to the fossil record for more than 200 million years. Because of the excellent preservation of their tests and the often striking differences among them, forams are very important as geological markers. The pattern of occurrence of different forams is often used as a guide in searching for oil-bearing strata. Limestones all over the world, including the famous white cliffs of Dover in southern England, are often rich in forams (figure 31.8).

> Amoebas and forams are unicellular, heterotrophic protists that lack cell walls, flagella, meiosis, and sexuality. Amoebas move from place to place by means of extensions called pseudopods. The pore-studded tests, or shells, of the forams have openings through which podia extend that are used for locomotion.

FIGURE 31.7
A representative of the Foraminifera (90×). A living foram with podia, thin cytoplasmic projections that extend through pores in the calcareous test, or shell, of the organism.

FIGURE 31.8
White cliffs of Dover. The limestone that forms these cliffs is composed almost entirely of fossil shells of protists, including coccolithophores (a type of algae) and foraminifera.

The Algae

Chlorophyta: The Green Algae

Green algae are of special interest, both because of their unusual diversity and because the ancestors of the plant kingdom were clearly multicellular green algae. Their chloroplasts are biochemically similar to those of the plants. They contain chlorophylls *a* and *b*, as well as carotenoids. Green algae include a very wide array of both unicellular and multicellular organisms.

Green algae are an extremely varied group of more than 7000 species. They are mostly aquatic, but some are semi-terrestrial in moist places, such as on tree trunks or in soil. Many are microscopic and unicellular, but some, such as sea lettuce, *Ulva* (see figure 28.12*b*), are tens of centimeters across and easily visible on rocks and pilings around the coasts.

Among the unicellular green algae, *Chlamydomonas* (figure 31.9) is a well-known genus. Individuals are microscopic (usually less than 25 micrometers long), green, rounded, and have two flagella at the anterior end. They move rapidly in water by beating their flagella in opposite directions. Each individual has an eyespot, which contains about 100,000 molecules of rhodopsin, the same pigment employed in vertebrate eyes. Light received by this eyespot is used by the alga to help direct its swimming. Most individuals of *Chlamydomonas* are haploid. *Chlamydomonas* reproduces asexually (by cell division) as well as sexually. In sexual reproduction, two haploid individuals fuse to form a four-flagellated zygote. The zygote ultimately enters a resting phase, called the **zygospore,** in which the flagella disappear. Meiosis occurs at the end of this resting period and results in the production of four haploid cells.

Several lines of evolutionary specialization have been derived from organisms like *Chlamydomonas*. The first is the evolution of nonmotile, unicellular green algae. *Chlamydomonas* is capable of retracting its flagella and settling down as an immobile unicellular organism if the ponds in which it lives dry out. Some common algae of soil and bark, such as *Chlorella*, are essentially like *Chlamydomonas* in this trait, but do not have the ability to form flagella. *Chlorella* is widespread in both fresh and salt water as well as soil and is only known to reproduce asexually. Recently, *Chlorella* has been widely investigated as a possible food source for hu-

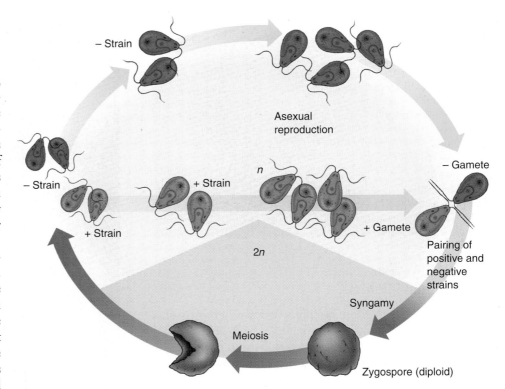

FIGURE 31.9

Life cycle of *Chlamydomonas* (Chlorophyta). Individual cells of this microscopic, biflagellated alga, which are haploid, divide asexually, producing identical copies of themselves. At times, such haploid cells act as gametes—fusing, as shown in the lower right-hand side of the diagram, to produce a zygote. The zygote develops a thick, resistant wall, becoming a zygospore; this is the only diploid cell in the entire life cycle. Within this diploid zygospore, meiosis takes place, ultimately resulting in the release of four haploid individuals. Because of the segregation during meiosis, two of these individuals are called the (+) strain, the other two the (−) strain. Only + and − individuals are capable of mating with each other when syngamy does take place, although both may divide asexually to reproduce themselves.

mans and other animals, and pilot farms have been established in Israel, the United States, Germany, and Japan.

The second major line of specialization from cells like *Chlamydomonas* concerns the formation of motile, colonial organisms. In these genera of green algae, the *Chlamydomonas*-like cells retain some of their individuality. The most elaborate of these organisms is *Volvox* (see figure 31.1), a hollow sphere made up of a single layer of 500 to 60,000 individual cells, each cell with two flagella. Only a small number of the cells are reproductive. The colony has definite anterior and posterior ends, and the flagella of all of the cells beat in such a way as to rotate the colony in a clockwise direction as it moves forward through the water. The reproductive cells of *Volvox* are located mainly at the posterior end of the colony. Some may divide asexually, bulge inward, and give rise to new colonies that initially remain within the parent colony. Others produce gametes. In some species of *Volvox*, there is a true division of labor among the different types of cells, which are specialized in relation to their ultimate function throughout the development of the organism.

FIGURE 31.10
**Stoneworts
(Chlorophyta).**
(a) Stoneworts, as seen in
this view of *Chara*, have a
complex growth pattern,
with nodal regions from
which whorls of branches
arise, and hollow,
multinucleate internodal
regions. (b) Stoneworts also
have multicellular and
complex egg-bearing
structures (*bottom*) and
round sperm-bearing
structures (*top*).

(a)

(b)

In addition to these two lines of specialization from *Chlamydomonas*-like cells, there are many other kinds of green algae of less certain derivation. Many filamentous genera, such as *Spirogyra*, with its ribbon-like chloroplasts, differ substantially from the remainder of the green algae in their modes of cell division and reproduction. Some of these genera have even been placed in separate phyla. The study of the green algae, involving modern methods of electron microscopy and biochemistry, is beginning to reveal unexpected new relationships within this phylum.

Ulva, or sea lettuce (see figure 28.12*b*), is a genus of marine green algae that is extremely widespread. The glistening individuals of this genus, often more than 10 centimeters across, consist of undulating sheets only two cells thick. Sea lettuce attaches by protuberances of the basal cells to rocks or other substrates. *Ulva* has an alternation of generations (sporic meiosis; see figure 28.13*c*) in which the **gametophytes** (haploid phase) and **sporophytes** (diploid phase) resemble one another closely.

The stoneworts, a group of about 250 living species of green algae, many of them in the genera *Chara* and *Nitella*, have complex structures (figure 31.10). Whorls of short branches arise regularly at their nodes, and the gametangia are complex and multicellular. Stoneworts are often abundant in fresh to brackish water and are common as fossils.

Rhodophyta: The Red Algae

Along with green algae and brown algae, red algae are the seaweeds we see cast up along shores and on beaches. Their characteristic colors result from phycoerythrin, a type of phycobilin pigment. Phycobilins are responsible for the colors of the cyanobacteria. Chlorophyll *a* also occurs with the phycobilins in red algae, just as it does in cyanobacteria. Cyanobacteria likely became symbiotic in the cells of the heterotrophic ancestors of the red algae and gave rise to their chloroplasts.

Almost all red algae are multicellular. The great majority of the estimated 4000 species occur in the sea (see figure 28.12*c*), and in warm waters red algae are more common than brown algae. Phycobilins are especially efficient in absorbing the green, violet, and blue light that penetrates into the deepest waters; for this reason, red algae are able to grow at greater depths than are brown or green algae. In 1985, a red alga was reported growing attached to rocks 268 meters below the surface of the sea in the Bahamas, a record for any photosynthetic organism.

Red algae have complex bodies made up of interwoven filaments of cells. The coralline algae (figure 31.11) deposit calcium carbonate in their cell walls, which otherwise are mostly cellulose. In the cell walls of other red algae there is

FIGURE 31.11
A coralline alga (Rhodophyta). In the coralline algae, the cellulose cell walls are heavily impregnated with calcium carbonate. Some species of coralline algae contribute greatly to overall food production in coral reefs, while others, such as the one shown here, occur widely elsewhere.

a mucilaginous outer component usually composed of sulfated polysaccharides such as agar and carrageenan, which make these algae important economically. Agar is used to make gel capsules, as a material for making dental impressions, and as a base for cosmetics. It is also the basis of the laboratory media on which bacteria, fungi, and other organisms are often grown. In addition, agar is used to prevent baked goods from drying out, for rapid-setting jellies, and as a temporary preservative for meat and fish in warm regions. Carrageenan is used mainly to stabilize emulsions such as paints, cosmetics, and dairy products such as ice cream. In addition to these uses, red algae such as *Porphyra*, called "nori," are eaten and, in Japan, are even cultivated as a food crop for human consumption.

The life cycles of red algae are complex but usually involve an alternation of generations (sporic meiosis). None of the red algae have flagella or cilia at any stage in their life cycle, and they may have descended directly from ancestors that never had them, especially since the red algae also lack centrioles. Together with the fungi, which also lack flagella and centrioles, red algae may be one of the most ancient groups of eukaryotes.

Phaeophyta: The Brown Algae

While the kingdom Protista contains the smallest eukaryotes, it also contains some of the longest, fastest growing and most photosynthetically productive living things. The phaeophyta, or brown algae, consist of about 1500 species of multicellular protists, almost exclusively marine. They are the most conspicuous seaweeds in many northern regions, dominating rocky shores almost everywhere in temperate North America and Eurasia. In habitats where the larger brown algae—and in particular a group known as the **kelps** (order Laminariales)—occur abundantly, they are responsible for most of the food production through photosynthesis. Many kelps are conspicuously differentiated into flattened blades, stalks, and grasping basal portions that anchor them to the rocks. The organic matter that kelp produces supports the myriad of invertebrates, fishes, and marine mammals and birds that live among these kelp forests (figure 31.12).

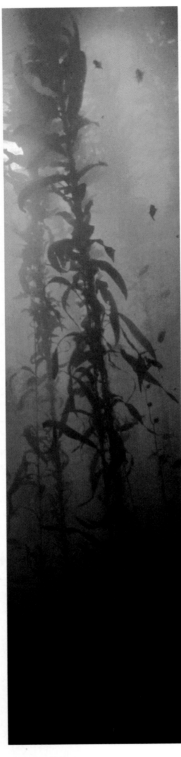

FIGURE 31.12
Brown algae (Phaeophyta).
The massive "groves" of giant kelp that occur in relatively shallow water along the coasts of the world provide food and shelter for many different kinds of organisms.

Among the larger brown algae are genera such as *Macrocystis*, in which some individuals may reach 100 meters in length. The flattened blades of this kelp float out on the surface of the water, while the base is anchored tens of meters below the surface. Another ecologically important member of this phylum is sargasso weed, *Sargassum*, which forms huge floating masses that dominate the vast Sargasso Sea, an area of the Atlantic Ocean northeast of the Caribbean. The stalks of the larger brown algae often exhibit a complex internal differentiation of conducting tissues analogous to that of plants. The chloroplasts of brown algae resemble those of diatoms and dinoflagellates in having chlorophylls *a* and *c*. How chloroplasts with these characteristics came to be shared between these groups is presently under investigation.

The life cycle of the brown algae is marked by an alternation of generations between a sporophyte and a gametophyte. The large individuals we recognize, such as the kelps, are sporophytes. The gametophytes are often much smaller, filamentous individuals, perhaps a few centimeters across. Sporangia, which produce haploid, swimming spores after meiosis, are formed on the sporophytes. These spores divide by mitosis, giving rise to individual gametophytes. There are two kinds of gametophytes in the kelps; one produces sperm, and the other produces eggs. If sperm and eggs fuse, the resulting zygotes grow into the mature kelp sporophytes, provided that they reach a favorable site.

Some of the larger kelps form extensive beds that are harvested commercially for sodium and potassium salts, iodine, and alginates (carbohydrates that are used in the formation of gels). Scientists are also investigating the possibility of using kelp, which grows continuously and produces large amounts of organic material, as an inexpensive source of fuel.

Nonmotile, unicellular algae and multicellular, flagellated colonies have been derived from green algae like *Chlamydomonas*—a biflagellated, unicellular organism. The life cycle of brown algae is marked by an alternation of generations between the diploid phase, or sporophyte, and the haploid phase, or gametophyte.

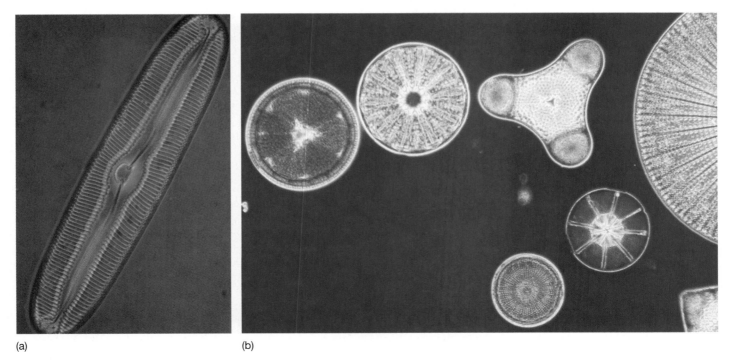

(a)
(b)

FIGURE 31.13
Diatoms (Chrysophyta). (a) A pennate (bilaterally symmetrical) diatom. (b) Several different kinds of centric (radially symmetrical) diatoms.

The Diatoms

Chrysophyta: The Diatoms

Members of the phylum Chrysophyta produce a unique carbohydrate called chrysolaminarin. The two principle groups are diatoms and golden algae. Diatoms are photosynthetic, unicellular organisms with unique double shells made of opaline silica, which are often strikingly and characteristically marked. The shells of diatoms are like small boxes with lids, one half of the shell fitting inside the other. Their chloroplasts, with chlorophylls *a* and *c*, as well as carotenoids, resemble those of the brown algae and dinoflagellates. In other respects, however, there are few similarities between these groups, and they probably do not share an immediate common ancestor.

There are more than 11,500 living species of diatoms, with many more known in the fossil record. The shells of fossil diatoms often form very thick deposits, which are sometimes mined commercially. The resulting "diatomaceous earth" is used as an abrasive or to add the sparkling quality to the paint used on roads, among other purposes. Living diatoms are often abundant both in the sea and in fresh water, where they are important food producers. Diatoms occur in the plankton and are attached to submerged objects in relatively shallow water. Many species are able to move by means of a secretion that is produced from a fine groove along each shell. The diatoms exude and perhaps also retract this secretion as they move.

There are two major groups of diatoms, one with radial symmetry (like a wheel) and the other with bilateral (two-sided) symmetry (figure 31.13). Diatom shells are rigid, and

the organisms reproduce asexually by separating the two halves of the shell, each half then regenerating another half shell within it. Because of this mode of reproduction, there is a tendency for the shells, and consequently the individual diatoms, to get smaller and smaller with each asexual reproduction. When the resulting individuals have diminished to about 30% of their original size, one may slip out of its shell, grow to full size, and regenerate a full-sized pair of new shells.

Individual diatoms are diploid. Meiosis occurs more frequently under conditions of starvation. Some marine diatoms produce numerous sperm and others a single egg. If fusion occurs, the resulting zygote regenerates a full-sized individual. In some freshwater diatoms, the gametes are amoeboid and similar in appearance.

The Golden Algae

Also included within the Chrysophyta are the golden algae, named for the yellow and brown carotenoid and xanthophyll accessory pigments in their chloroplasts, which give them a golden color. Unicellular but often colonial, these freshwater protists typically have two flagella, both attached near the same end of the cell. When ponds and lakes dry out in summer, golden algae form resistant cysts. Viable cells emerge from these cysts when wetter conditions recur in the fall.

Diatoms and golden algae are unicellular, photosynthetic organisms that produce a unique carbohydrate. Diatoms have double shells made of opaline silica.

The Flagellates

Pyrrhophyta: The Dinoflagellates

The dinoflagellates consist of about 2100 known species of unicellular, photosynthetic organisms, most of which have two flagella. A majority of the dinoflagellates are marine, and they are often abundant in the plankton, but some occur in fresh water. Some planktonic dinoflagellates are luminous and contribute to the twinkling or flashing effects that we sometimes see in the sea at night, especially in the tropics.

The flagella, protective coats, and biochemistry of dinoflagellates are distinctive, and they do not appear to be directly related to any other phylum. Their flagella beat in two grooves, one encircling the body like a belt, and the other perpendicular to it. By beating in their respective grooves, these flagella cause the dinoflagellate to rotate like a top as it moves. Many dinoflagellates are clad in stiff cellulose plates, often encrusted with silica, which give them a very unusual appearance (figure 31.14). Most have chlorophylls *a* and *c*, in addition to carotenoids, so that in the biochemistry of their chloroplasts, they resemble the diatoms and the brown algae, possibly acquiring such chloroplasts by ingesting members of those groups.

Some dinoflagellates occur as symbionts in many other groups of organisms, including jellyfish, sea anemones, mollusks, and corals (figure 31.15*a*). When dinoflagellates grow as symbionts within other cells, they lack their characteristic cellulose plates and flagella, appearing as spherical, golden-brown globules in their host cells. In such a state they are called **zooxanthellae** (figure 31.15*b* and *c*). Zooxanthellae are the primary factor responsible for the productivity of corals and their ability to grow in tropical waters, which are often extremely low in nutrients. Most of the carbon that the zooxanthellae fix is translocated to the host corals.

The poisonous and destructive "red tides" that occur frequently in coastal areas are often associated with great population explosions, or "blooms," of dinoflagellates. The pigments in the individual, microscopic cells of the dinoflagellates are responsible for the color of the water. Red tides have a profound, detrimental effect on the fishing industry in the United States. Approximately 20 species of dinoflagellates are known to produce powerful toxins that inhibit

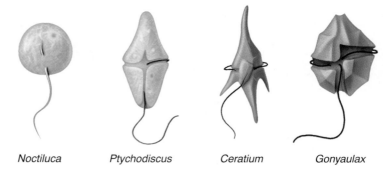

FIGURE 31.14
Some dinoflagellates: *Noctiluca, Ptychodiscus, Ceratium,* **and** *Gonyaulax.*
Noctiluca, which lacks the heavy cellulose armor characteristic of most dinoflagellates, is one of the bioluminescent organisms that cause the waves to sparkle in warm seas. In the other three genera, the shorter, encircling flagellum is seen in its groove, with the longer one projecting away from the body of the dinoflagellate.

FIGURE 31.15
Zooxanthellae, the primary factor in the productivity of coral reefs. (a) The coral animals in this reef are packed with golden-brown zooxanthellae. (b) Individuals of the sea anemone *Anthopleura elegantissima.* The brown-green pigmentation of these animals is caused almost entirely by pigments in the zooxanthellae; animals of the same species that lack zooxanthellae are milky pink. (c) Densely packed zooxanthellae in tissue of the anemone shown in (b). In this photograph, the zooxanthellae fluoresce red, while the tissue of the anemone fluoresces yellowish or bluish green.

the diaphragm and cause respiratory failure in many vertebrates. When the toxic dinoflagellates are abundant, fishes, birds, and marine mammals may die in large numbers. In addition, toxic dinoflagellates are accumulated by shellfish (mollusks), which strain them out of the water. The mollusks themselves do not die, but they become poisonous to humans and other animals that consume them.

Dinoflagellates reproduce primarily by longitudinal cell division, but sexual reproduction has also been shown in more than 10 genera. Their form of mitosis is unique: the chromosomes, which are almost entirely DNA and are permanently condensed, remain within the nucleus but are distributed along the sides of channels containing bundles of microtubules that run through the nucleus.

Euglenophyta: The Euglenoids

Most of the approximately 1000 known species of euglenoids live in fresh water. The members of this phylum clearly illustrate the impossibility of distinguishing "plants" from "animals" among the protists. About a third of the approximately 40 genera of euglenoids have chloroplasts and are fully autotrophic; the others lack chloroplasts, ingest their food, and are heterotrophic. These organisms are not significantly different from some groups of zoomastigotes (see next section), and many biologists believe that the two phyla should be merged into one.

Some euglenoids with chloroplasts may become heterotrophic if the organisms are kept in the dark; the chloroplasts become small and nonfunctional. If they are put back in the light, they may become green within a few hours. Normally photosynthetic euglenoids may sometimes feed on dissolved or particulate food.

Individual euglenoids range from 10 to 500 micrometers long and are highly variable in form. Interlocking proteinaceous strips arranged in a helical pattern form a flexible structure called the **pellicle,** which lies within the cell membrane of the euglenoids. Since its pellicle is flexible, a euglenoid is able to change its shape. Reproduction in this phylum occurs by mitotic cell division. The nuclear envelope remains intact throughout the process of mitosis. No sexual reproduction is known to occur in this group.

In *Euglena* (figure 31.16), the genus for which the phylum is named, two flagella are attached at the base of a flask-shaped opening called the **reservoir,** which is located at the anterior end of the cell. One of the flagella is long and has a row of very fine, short, hair-like projections along one side. A second, shorter flagellum is located within the reservoir but does not emerge from it. Contractile vacuoles collect excess water from all parts of the organism and empty it into the reservoir, which apparently helps regulate the osmotic pressure within the organism. The **stigma,** an organ that also occurs in the green algae (phylum Chlorophyta), is light-sensitive and aids these photosynthetic organisms to move toward light.

Cells of *Euglena* contain numerous small chloroplasts. These chloroplasts, like those of the green algae and plants, contain chlorophylls *a* and *b*, together with carotenoids. Although the chloroplasts of euglenoids differ somewhat in structure from those of green algae, they probably had a common origin. It seems likely that euglenoids acquired their chloroplasts by ingesting green algae.

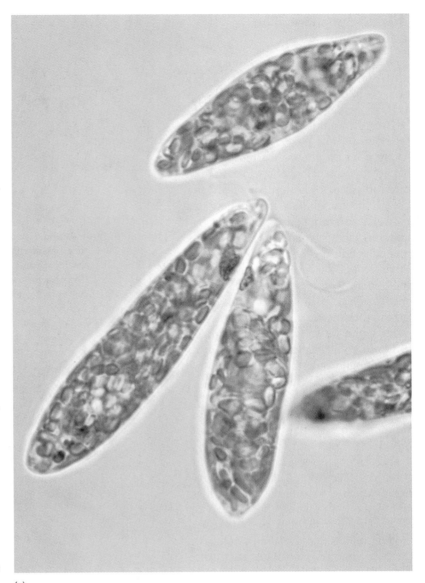

(a)

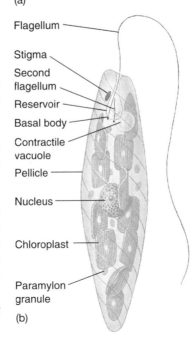

Flagellum

Stigma

Second flagellum

Reservoir

Basal body

Contractile vacuole

Pellicle

Nucleus

Chloroplast

Paramylon granule

(b)

FIGURE 31.16
Euglenoids. (a) Micrograph of individuals of the genus *Euglena* (Euglenophyta). (b) Diagram of *Euglena*. Paramylon granules are areas where food reserves are stored.

Zoomastigophora: The Zoomastigotes

The zoomastigotes are unicellular, heterotrophic organisms that are highly variable in form (figure 31.17). Each has at least one flagellum, with some species having thousands. They include both free-living and parasitic organisms. Many zoomastigotes apparently reproduce only asexually, but sexual reproduction occurs in some species. One group, the amoeboflagellates, alternates between an amoeboid stage and a flagellated stage, depending on environmental conditions. The members of another group, the trypanosomes, include the genera *Trypanosoma* (figure 31.17c) and *Crithidia*, pathogens of human beings and domestic animals. The euglenoids could be viewed as a specialized group of zoomastigotes, some of which acquired chloroplasts during the course of evolution.

Among the diseases for which the trypanosomes are responsible are sleeping sickness, or trypanosomiasis (figure 31.18), East Coast fever, and Chagas' disease, all of great importance in tropical areas. Another tropical disease, leishmaniasis, which is transmitted by sand flies, afflicts about 4 million people a year. Its effects range from skin sores to deep, eroding lesions that can almost obliterate the face. The trypanosomes that cause these diseases are spread by biting insects, including tsetse flies and assassin bugs.

A serious effort is now under way to produce a vaccine for trypanosome-caused diseases. These diseases make it impossible to raise domestic cattle for meat or milk in a large portion of Africa. Control is especially difficult because of the unique attributes of these organisms. For example, tsetse fly-transmitted trypanosomes have evolved an elaborate genetic mechanism for repeatedly changing the antigenic nature of their protective glycoprotein coat, thus dodging the antibodies their hosts produce against them (see chapter 54). Only a single one out of some 1000 to 2000 variable antigen genes is expressed at a time. Rearrangements of these genes during the asexual cycle of the organism allow for the expression of a seemingly endless variety of different antigen genes that maintain infectivity by the trypanosomes.

When the trypanosomes are ingested by a tsetse fly, they embark on a complicated cycle of development and multi-

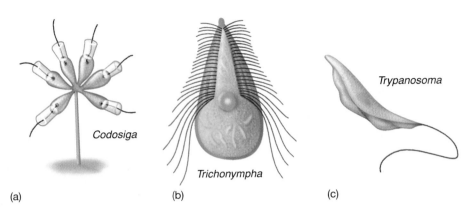

(a)   (b)   (c)

FIGURE 31.17

Three genera of zoomastigotes (Zoomastigophora), a highly diverse group.
(a) *Codosiga*, a colonial choanoflagellate that remains attached to its substrate; other colonial choanoflagellates swim around as a colony, resembling the green alga *Volvox* in this respect. (b) *Trichonympha*, one of the zoomastigotes that inhabits the guts of termites and wood-feeding cockroaches and digests cellulose there. *Trichonympha* has rows of flagella in its anterior regions. (c) *Trypanosoma*, which causes sleeping sickness, an important tropical disease. It has a single, anterior flagellum.

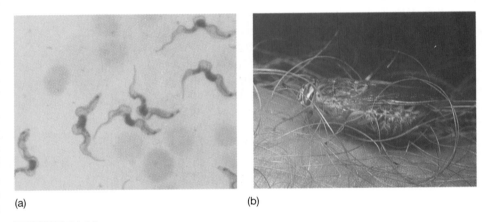

(a)   (b)

FIGURE 31.18

***Trypanosoma* is the zoomastigote that causes sleeping sickness.** (a) *Trypanosoma* among red blood cells. The nuclei (dark-staining bodies), anterior flagella, and undulating, changeable shape of the trypanosomes are visible in this photograph (500×). (b) The tsetse fly, shown here sucking blood from a human arm, can carry trypanosomes.

plication, first in the fly's gut and later in its salivary glands. Recombination has been observed between different strains of trypanosomes introduced into a single fly, thus suggesting that mating, syngamy, and meiosis occur, even though they have not been observed directly. Although most trypanosome reproduction is asexual, this sexual cycle, reported for the first time in 1986, affords still further possibilities for recombination in these organisms.

In the guts of the flies that spread them, trypanosomes are noninfective. When they are prepared for transfer to the mammalian skin or bloodstream, trypanosomes acquire the thick coat of glycoprotein antigens that protect them from the host's antibodies. When they are taken up by a fly, the trypanosomes again shed their coats. The production of vaccines against such a system is complex, but tests are underway. Releasing sterilized flies to impede the reproduction of populations is another technique used to try to control the fly population. Traps made of dark cloth and scented like cows, but poisoned with insecticides, have likewise proved effective. Research is proceeding rapidly because the presence of tsetse flies with their associated trypanosomes blocks the use of some 11 million square kilometers of potential grazing land in Africa.

Some zoomastigotes occur in the guts of termites and other wood-eating insects (see figure 31.17b). They possess enzymes that allow them to digest the wood and thus make the components of the wood available to their hosts. The relationship is similar to that between certain bacteria and protozoa that function in the rumens of cattle and related mammals (see chapter 48).

Another group of zoomastigotes, the choanoflagellates (see figure 31.17a), is most likely the group from which the sponges (phylum Porifera) and probably all animals arose (see chapter 41). Choanoflagellates have a single emergent flagellum surrounded by a funnel-shaped, contractile collar composed of closely placed filaments, a unique structure that is exactly matched in the sponges. These protists feed on bacteria strained out of the water by the collar.

Hiker's Diarrhea

Giardia lamblia is a flagellate protist (belonging to a small group called metamonads) found throughout the world, including all parts of the United States and Canada (figure 31.19). It occurs in water, including the clear water of mountain streams and the water supplies of some cities. It infects at least 40 species of wild and domesticated animals in addition to humans. In 1984 in Pittsburgh, 175,000 people had to boil their drinking water for several days following the appearance of *Giardia* in the city's water system. Although most individuals exhibit no symptoms if they drink water infested with *Giardia*, many suffer nausea, cramps, bloating, vomiting, and diarrhea. Only 35 years ago, *Giardia* was thought to be harmless; today, it is estimated that at least 16 million residents of the United States are infected by it.

Giardia lives in the upper small intestine of its host. It occurs there in a motile form that cannot survive outside the host's body. It is spread in the feces of infected individuals in the form of dormant, football-shaped cysts—sometimes at levels as high as 300 million individuals per

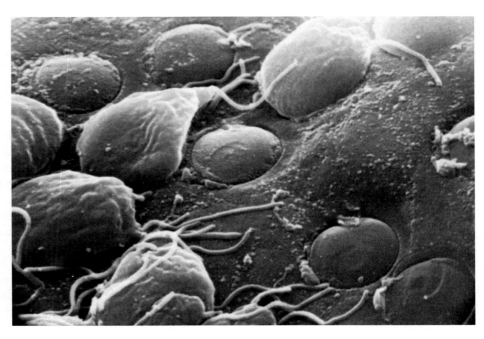

FIGURE 31.19

Giardia lamblia. Giardia are flagellated unicellular parasites that infect the human intestine. *Giardia* are very primitive, having only a rudimentary cytoskeleton and lacking mitochondria and chloroplasts. Sequencing of ribosomal RNA suggests that *Giardia* and *Pelomyxa*, the eukaryotes most closely related to prokaryotes, should be grouped together. The name Archezoa (Greek *arkhaios*, "ancient") has been suggested for the group, stressing its early divergence from bacteria as long as 2 billion years ago.

gram of feces. These cysts can survive at least two months in cool water, such as that of mountain streams. They are relatively resistant to the usual water-treatment agents such as chlorine and iodine but are killed at temperatures greater than about 65 °C. Apparently, pollution by humans seems to be the main way *Giardia* is released into stream water. There are at least three species of *Giardia* and many distinct strains; how many of them attack humans and under what circumstances are not known with certainty.

In the wilderness, good sanitation is important in preventing the spread of *Giardia*. Dogs, which readily contract and spread the disease, should not be taken into pristine wilderness areas. Drinking water should be filtered—the filter must be capable of eliminating particles as small as 1 micrometer in diameter—or boiled for at least one minute. Water from natural streams or lakes should never be consumed directly, regardless of how clean it looks. In other regions, good sanitation methods are important to prevent not only *Giardia* infection but also other diseases.

Dinoflagellates are unicellular, photosynthetic, and flagellated. Euglenoids (phylum Euglenophyta) consist of about 40 genera, about a third of which have chloroplasts similar biochemically to those of green algae and plants. The zoomastigotes are a highly diverse group of flagellated unicellular heterotrophs, containing among their members the ancestors of animals as well as the very primitive *Giardia*.

The Sporozoans

Apicomplexa: The Sporozoans

All sporozoans are nonmotile, spore-forming parasites of animals. Their spores are small, infective bodies that are transmitted from host to host. These organisms are distinguished by a unique arrangement of fibrils, microtubules, vacuoles, and other cell organelles at one end of the cell. There are 3900 described species of this phylum; best known among them is the malarial parasite, *Plasmodium*.

Sporozoans have complex life cycles that involve both asexual and sexual phases. Sexual reproduction involves an alternation of haploid and diploid generations. Both haploid and diploid individuals can also divide rapidly by mitosis, thus producing a large number of small infective individuals. Sexual reproduction involves the fertilization of a large female gamete by a small, flagellated male gamete. The zygote that results soon becomes an **oocyst.** Within the oocyst, meiotic divisions produce infective haploid spores called **sporozoites.**

An alternation between different hosts often occurs in the life cycles of sporozoans. Sporozoans of the genus *Plasmodium* are spread from person to person by mosquitoes of the genus *Anopheles* (figure 31.20); at least 65 different species of this genus are involved. When an *Anopheles* mosquito penetrates human skin to obtain blood, it injects saliva mixed with an anticoagulant. If the mosquito is infected with *Plasmodium*, it will also inject the elongated sporozoites into the bloodstream of its victim. The parasite makes its way through the bloodstream to the liver, where it rapidly divides asexually. After this division

phase, **merozoites,** the next stage of the life cycle, form, either reinvading other liver cells or entering the host's bloodstream. In the bloodstream, they invade the red blood cells, dividing rapidly within them and causing them to become enlarged and ultimately to rupture. This event releases toxic substances throughout the body of the host, bringing about the well-known cycle of fever and chills that is characteristic of malaria. The cycle repeats itself regularly every 48 hours, 72 hours, or longer.

Plasmodium enters a sexual phase when some merozoites develop into **gametocytes,** cells capable of producing gametes. There are two types of gametocytes: male and female. Gametocytes are incapable of producing gametes within their human hosts and do so only when they are extracted from an infected human by a mosquito. Within the gut of the mosquito, the male and female gametocytes form sperm and eggs, respectively. Zygotes develop within the mosquito's intestinal walls and ultimately differentiate into oocysts. Within the oocysts, repeated mitotic divisions take place, producing large numbers of sporozoites. These sporozoites migrate to the salivary glands of the mosquito, and from there they are injected by the mosquito into the bloodstream of a human, thus starting the life cycle of the parasite again.

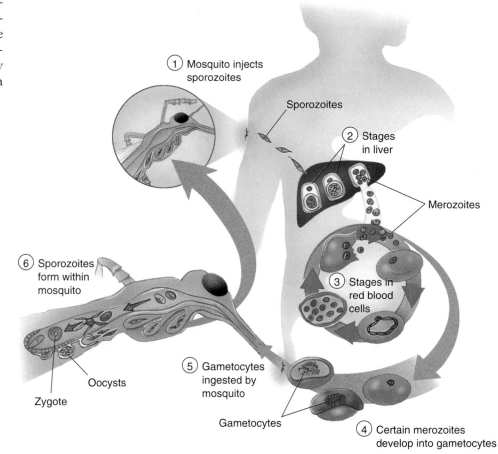

1 Mosquito injects sporozoites

Sporozoites

2 Stages in liver

Merozoites

6 Sporozoites form within mosquito

3 Stages in red blood cells

Oocysts

Zygote

5 Gametocytes ingested by mosquito

Gametocytes

4 Certain merozoites develop into gametocytes

FIGURE 31.20
The life cycle of *Plasmodium*, the sporozoan that causes malaria. *Plasmodium* has a complex life cycle that alternates between mosquitoes and mammals.

Malaria

Malaria, caused by infections by the sporozoan *Plasmodium*, is one of the most serious diseases in the world. According to the World Health Organization, about 500 million people are affected by it at any one time, and approximately 2 million of them, mostly children, die each year. Malaria kills most children under five years old who contract it. In areas where malaria is prevalent, most survivors more than five or six years old do not become seriously ill again from malaria infections. The symptoms, familiar throughout the tropics, include severe chills, fever, and sweating, an enlarged and tender spleen, confusion, and great thirst. Ultimately, a victim of malaria may die of anemia, kidney failure, or brain damage. The disease may be brought under control by the person's immune system or by drugs. As discussed in chapter 20, some individuals are genetically resistant to malaria. Other persons develop immunity to it.

Efforts to eradicate malaria have focused on (1) the elimination of the mosquito vectors; (2) the development of drugs to poison the parasites once they have entered the human body; and (3) the development of vaccines. The widescale applications of DDT from the 1940s to the 1960s led to the elimination of the mosquito vectors in the United States, Italy, Greece, and certain areas of Latin America. For a time, the worldwide elimination of malaria appeared possible, but this hope was soon crushed by the development of DDT-resistant strains of malaria-carrying mosquitoes in many regions; no fewer than 64 resistant strains were identified in a 1980 survey. Even though the worldwide use of DDT, long banned in the United States, nearly doubled from its 1974 level to more than 30,000 metric tons in 1984, its effectiveness in controlling mosquitoes is dropping. Further, there are serious environmental concerns about the use of this long-lasting chemical anywhere in the world. In addition to the problems with resistant strains of mosquitoes, strains of *Plasmodium* have appeared that are resistant to the drugs that have historically been used to kill them.

As a result of these problems, the number of new cases of malaria per year roughly doubled from the mid-1970s to the mid-1980s, largely because of the spread of resistant strains of the mosquito and the parasite. In many tropical regions, malaria is blocking permanent settlement. Scientists have therefore redoubled their efforts to produce an effective vaccine. Antibodies to the parasites have been isolated and produced by genetic engineering techniques, and they are starting to produce promising results.

Vaccines Against Malaria

The three different stages of the *Plasmodium* life cycle each produce different antigens, and they are sensitive to different antibodies. The gene encoding the sporozoite antigen was cloned in 1984, but it is not certain how effective a vaccine against sporozoites might be. When a mosquito inserts its proboscis into a human blood vessel, it injects about a thousand sporozoites. They travel to the liver within a few minutes, where they are no longer exposed to antibodies circulating in the blood. If even one sporozoite reaches the liver, it will multiply rapidly there and cause malaria. The number of malaria parasites increases roughly eightfold every 24 hours after they enter the host's body. A compound vaccination against sporozoites, merozoites, and gametocytes would probably be the most effective preventive measure, but such a compound vaccine has proven difficult to develop.

However, research completed in 1997 brings a glimmer of hope. An experimental vaccine containing one of the surface proteins of the disease-causing parasite, *P. falciparum*, seems to induce the immune system to produce defenses that are able to destroy the parasite in future infections (figure 31.21). In tests, six out of seven vaccinated people did not get malaria after being bitten by mosquitoes that carried *P. falciparum*. Although research is still underway, many are hopeful that this new vaccine may be able to fight malaria, especially in Africa, where it takes a devastating toll.

The best known of the sporozoans is the malarial parasite *Plasmodium*. Like other sporozoans, *Plasmodium* has a complex life cycle involving sexual and asexual phases and alternation between different hosts, in this case mosquitoes and humans. Malaria kills about 2 million people each year.

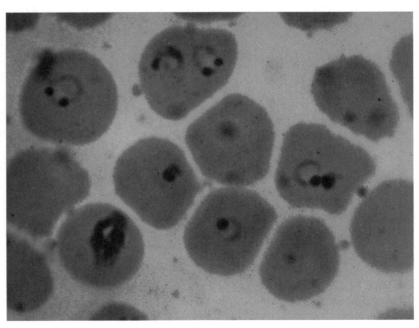

FIGURE 31.21
Vaccine against *Plasmodium falciparum*. The vaccine is visible here as dark dots within blood cells.

The Ciliates

Ciliophora: The Ciliates

As the name indicates, most members of the Ciliophora feature large numbers of cilia. These heterotrophic, unicellular protists range in size from 10 to 3000 micrometers long. About 8000 species have been named. Despite their unicellularity, ciliates are extremely complex organisms, inspiring some biologists to consider them organisms without cell boundaries rather than single cells.

Their most characteristic feature, cilia, are usually arranged either in longitudinal rows or in spirals around the body of the organism (figure 31.22a). Cilia are anchored to microtubules beneath the cell membrane, and they beat in a coordinated fashion. In some groups, the cilia have specialized locomotory and feeding functions, becoming fused into sheets, spikes, and rods which may then function as mouths, paddles, teeth, or feet. The cell wall of ciliates is a tough but flexible outer covering called the pellicle that enables the organism to squeeze through or move around many kinds of obstacles.

All ciliates that have been studied have two very different types of nuclei within their cells, small **micronuclei** and larger **macronuclei** (figure 31.22b). The micronuclei, which contain apparently normal diploid chromosomes, divide by meiosis and are able to undergo genetic recombination. Macronuclei are derived from certain micronuclei in a complex series of steps. Within the macronuclei are multiple copies of the genome, and the DNA is divided into small pieces—smaller than individual chromosomes. In one group of ciliates, these are equivalent to single genes. Macronuclei divide by elongating and constricting and play an essential role in routine cellular functions, such as the production of mRNA to direct protein synthesis for growth and regeneration.

Ciliates form vacuoles for ingesting food and regulating their water balance. Food first enters the gullet, which in the well-known ciliate *Paramecium* is lined with cilia fused into a membrane (figure 31.23). From the gullet, the food passes into food vacuoles, where enzymes and hydrochloric acid aid in its digestion. After the digested material has been completely absorbed, the vacuole empties its waste contents through a special pore in the pellicle known as the **cytoproct.** The cytoproct is essentially an exocytotic vesicle that appears periodically when solid particles are ready to be expelled. The contractile vacuoles, which function in the regulation of water balance, periodically expand and contract as they empty their contents to the outside of the organism.

Ciliates usually reproduce by transverse fission of the parent cell across its short axis, thus forming two equal offspring (figure 31.24a). In this process of cell division, the mitosis of the micronuclei proceeds normally, and the macronuclei divide as just described.

In *Paramecium*, the cells divide asexually for about 700 generations and then die if sexual reproduction has not occurred. Like most ciliates, *Paramecium* has a sexual process

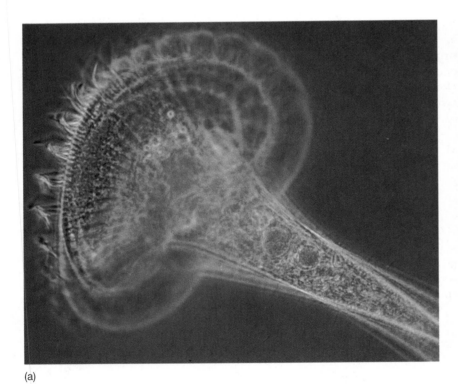

(a)

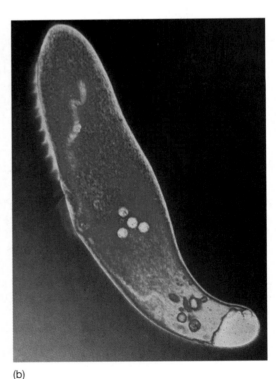

(b)

FIGURE 31.22
Ciliates (Ciliophora). (a) *Stentor*, a funnel-shaped ciliate, showing spirally arranged cilia (120×). (b) *Blepharisma japonica*, showing micronuclei (small round objects) and a macronucleus (elongated structure within cell) (400×).

called **conjugation,** in which two individual cells remain attached to each other for up to several hours (figure 31.24*b*). Only cells of two different genetically determined mating types, *odd* and *even*, are able to conjugate. Meiosis in the micronuclei of each individual produces several haploid micronuclei, and two partners exchange a pair of these micronuclei through a cytoplasmic bridge that appears between the two partners.

In each conjugating individual, the new micronucleus fuses with one of the micronuclei already present in that individual, resulting in the production of a new diploid micronucleus in each individual. After conjugation, the macronucleus in each cell disintegrates, while the new diploid micronucleus undergoes mitosis, thus giving rise to two new identical diploid micronuclei within each individual. One of these micronuclei becomes the precursor of the future micronuclei of that cell, while the other micronucleus undergoes multiple rounds of DNA replication, becoming the new macronucleus. This kind of complete segregation of the genetic material is a unique feature of the ciliates and makes them ideal organisms for the study of certain aspects of genetics.

Progeny from a sexual division in *Paramecium* must go through about 50 asexual divisions before they are able to conjugate. When they do so, their biological clocks are restarted, and they can conjugate again. After about 600 asexual divisions, however, *Paramecium* loses the protein molecules around the gullet that enable it to recognize an appropriate mating partner. As a result, the individuals are unable to mate, and death follows about 100 generations later. The exact mechanisms producing these unusual events are unknown, but they involve the accumulation of a protein, which is now being studied.

Ciliates possess characteristic cilia, and have two types of nuclei. The macronuclei contain multiple copies of certain genes, while the micronuclei contain multigene chromosomes.

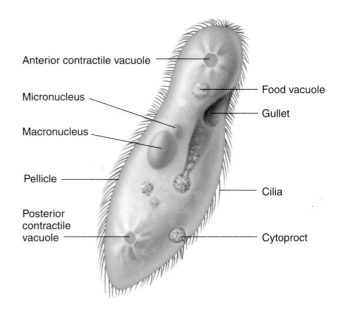

FIGURE 31.23
Paramecium. The main features of this familiar ciliate are shown.

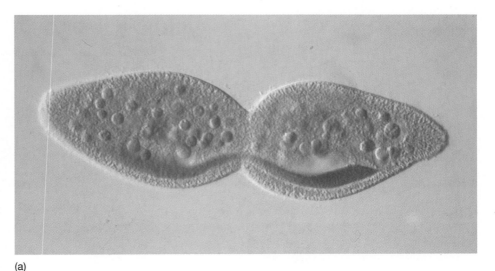

(a)

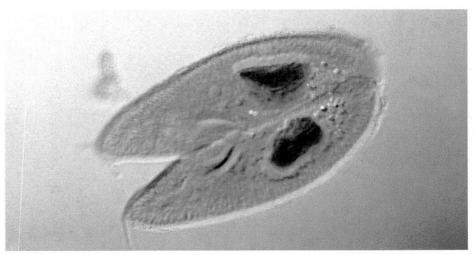

(b)

FIGURE 31.24
Life cycle of *Paramecium.* (a) When *Paramecium* reproduces asexually, a mature individual divides, and two complete individuals result. (b) In sexual reproduction, two mature cells fuse in a process called conjugation (100×).

The "Molds"

Acrasiomycota: The Cellular Slime Molds

There are about 70 species of cellular slime molds. This phylum has extraordinarily interesting features and was once thought to be related to fungi, "mold" being a general term for fungus-like organisms. In fact, the cellular slime molds are probably more closely related to amoebas (phylum Rhizopoda) than to any other group, but they have many special features that mark them as distinct. Cellular slime molds are common in fresh water, damp soil, and on rotting vegetation, especially fallen logs. They have become one of the most important groups of organisms for studies of differentiation because of their relatively simple developmental systems and the ease of analyzing them (figure 31.25).

The individual organisms of this group behave as separate amoebas, moving through the soil or other substrate and in-gesting bacteria and other smaller organisms. At a certain phase of their life cycle, the individual organisms aggregate and form a moving mass, the "slug," that eventually transforms itself into a spore-containing mass, the **sorocarp.** In the sorocarp the amoebas become encysted as spores. Some of the amoebas fuse sexually to form **macrocysts,** which have diploid nuclei; meiosis occurs in them after a short period (zygotic meiosis). Other amoebas are released directly, eventually aggregating again to form a new slug.

The development of *Dictyostelium discoideum*, a cellular slime mold, has been studied extensively because of the implication its unusual life cycle has for understanding the developmental process in general. When the individual amoebas of this species exhaust the supply of bacteria in a given area and are near starvation, they aggregate and form a compound, motile mass. The aggregation of the individual amoebas is induced by pulses of cyclic adenosine monophosphate (cAMP), which the cells begin to secrete when they

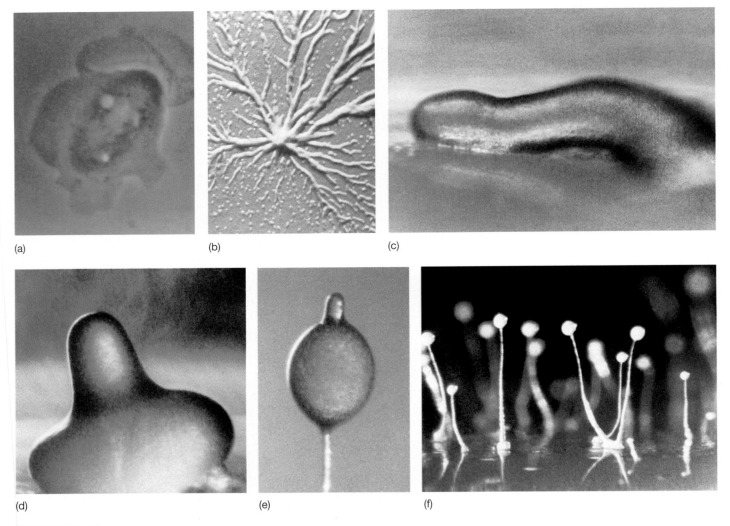

(a)   (b)   (c)

(d)   (e)   (f)

FIGURE 31.25

Development in *Dictyostelium discoideum*, a cellular slime mold. (a) First, a spore germinates, forming amoebas. These amoebas feed and reproduce until the food runs out. (b) The amoebas aggregate and move toward a fixed center. (c) Next, they form a multicellular "slug" 2 to 3 mm long that migrates toward light. (d) The slug stops moving and begins to differentiate into a sorocarp. (e) The differentiated head of a sorocarp. (f) Many sorocarps together.

are starving. In the new habitat, the colony differentiates into a multicellular sorocarp that includes a basal portion, a stalk, and a terminal swollen portion, within which spores differentiate. Each of these spores, if it falls into a suitably moist habitat, releases a new amoeba, which begins to feed, and the cycle is started again.

Myxomycota: The Plasmodial Slime Molds

Plasmodial slime molds are a group of about 500 species. These bizarre organisms stream along as a **plasmodium,** a nonwalled, multinucleate mass of cytoplasm, that resembles a moving mass of slime (figure 31.26). This is called the feeding phase, and the plasmodia may be orange, yellow, or another color. Plasmodia show a back-and-forth streaming of cytoplasm that is very conspicuous, especially under a microscope. They are able to pass through the mesh in cloth or simply flow around or through other obstacles. As they move, they engulf and digest bacteria, yeasts, and other small particles of organic matter. Plasmodia contain many nuclei, but these are not separated by cell walls. The nuclei undergo mitosis synchronously, with the nuclear envelope breaking down, but only at late anaphase or telophase. Centrioles are lacking in cellular slime molds. Although they have similar common names, there is no evidence that the plasmodial slime molds are closely related to the cellular slime molds; they differ in most features of their structure and life cycles.

When either food or moisture is in short supply, the plasmodium migrates relatively rapidly to a new area. Here it stops moving and either forms a mass in which spores differentiate or divides into a large number of small mounds, each of which produces a single, mature **sporangium,** the structure in which spores are produced.

FIGURE 31.26
A plasmodial protist. This multinucleate plasmodium moves about in search of the bacteria and other organic particles that it ingests.

These sporangia are often extremely complex in form and beautiful (figure 31.27). These spores are either diploid or haploid, depending on the condition found in the nuclei within the plasmodium. In most species of plasmodial slime molds with a diploid plasmodium, meiosis occurs in the spores within 24 hours of their formation. Three of the four nuclei in each spore disintegrate, leaving each spore with a single haploid nucleus.

The spores are highly resistant to unfavorable environmental influences and may last for years if kept dry. When conditions are favorable, they split open and release their protoplast, the contents of the individual spore. The protoplast may be amoeboid or bear two flagella. These two stages appear to be interchangeable, and conversions in either direction occur readily. Later, after the fusion of haploid protoplasts (gametes), a usually diploid plasmodium may be reconstituted by repeated mitotic divisions.

(a) (b) (c)

FIGURE 31.27
Sporangia of three genera of plasmodial slime molds (phylum Myxomycota). (a) *Arcyria.* (b) *Fuligo.* (c) Developing sporangia of *Tubifera.*

Oomycota

The oomycetes comprise about 580 species, among them the water molds, white rusts, and downy mildews. All of the members of this group are either parasites or **saprobes** (organisms that live by feeding on dead organic matter). The cell walls of the oomycetes are composed of cellulose or polymers that resemble cellulose. They differ remarkably from the chitin cell walls of fungi, with which the oomycetes have at times been grouped. Oomycete life cycles are characterized by gametic meiosis and a diploid phase; this also differs from fungi. Mitosis in the oomycetes resembles that in most other organisms, while mitosis in fungi has a number of unusual features, as you will see in chapter 32. Filamentous structures of fungi and, by convention, those of oomycetes, are called **hyphae.** Most oomycetes live in fresh or salt water or in soil, but some are plant parasites that depend on the wind to spread their spores. A few aquatic oomycetes are animal parasites.

Oomycetes are distinguished from other protists by the structure of their motile spores, or **zoospores,** which bear two unequal flagella, one of which is directed forward, the other backward. Such zoospores are produced asexually in a sporangium. Sexual reproduction in the group involves **gametangia** (singular, **gametangium**)—gamete-producing structures—of two different kinds. The female gametangium is called an **oogonium,** and the male gametangium is called an **antheridium.** The antheridia contain numerous male nuclei, which are the functional male gametes; the oogonia contain from one to eight eggs, which are the female gametes. When the contents of an antheridium flow into an oogonium, it leads to the individual fusion of male nuclei with eggs. This is followed by the thickening of the cell wall around the resulting zygote or zygotes. This produces a special kind of thick-walled cell called an **oospore,** the structure that gives the phylum its name. Details from the life cycle of one of the oomycetes, *Saprolegnia,* are shown in figure 31.28.

Aquatic oomycetes, or water molds, are common and easily cultured. Some water molds cause fish diseases, producing a kind of white fuzz on aquarium fishes. Among their terrestrial relatives are oomycetes of great importance as plant pathogens, including *Plasmopara viticola,* which causes downy mildew of grapes, and *Phytophthora infestans,* which causes the late blight of potatoes. This oomycete was responsible for the Irish potato famine of 1845 and 1847, during which about 400,000 people starved to death or died of diseases complicated by starvation. Millions of Irish people emigrated to the United States and elsewhere as a result of this disaster.

Molds are heterotrophic protists, many of which are capable of amoeba-like streaming. The feeding phase of plasmodial slime molds consists of a multinucleate mass of protoplasm; a plasmodium can flow through a cloth mesh and around obstacles. If the plasmodium begins to dry out or is starving, it forms often elaborate sporangia. Meiosis occurs in the spores once they have been cleaved within the sporangium.

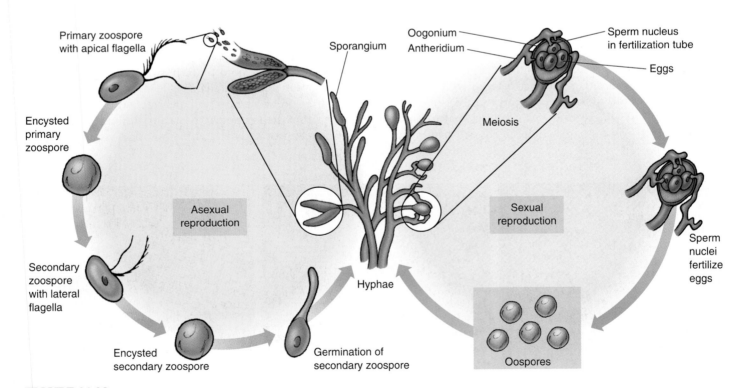

FIGURE 31.28
Life cycle of *Saprolegnia,* an oomycete. Asexual reproduction by means of flagellated zoospores is shown at left, sexual reproduction at right. Hyphae with diploid nuclei are produced by germination of both zoospores and oospores.

31.1 **The kingdom Protista is by far the most diverse of any kingdom.**

- The kingdom Protista consists of the exclusively or predominantly unicellular phyla of eukaryotes, together with three phyla that include large numbers of multicellular organisms: the red algae, brown algae, and green algae. The phyla of protists are highly diverse.

- The catch-all kingdom Protista includes all eukaryotic organisms except animals, plants, and fungi. Mitochondria originated symbiotically early in the history of the group, and chloroplasts evolved in several of its phyla subsequently.

- The three major multicellular groups of eukaryotes—plants, animals, and fungi—all originated from protists. They are not related directly to one another. The plants originated from green algae. The sponges, and probably all animals, originated from the choanoflagellates, one of the groups of zoomastigotes. The specific protistan ancestors of the fungi are unknown.

31.2 **Protists are grouped into fifteen very distinctive phyla.**

- The sarcodines are heterotrophs with no permanent locomotor apparatus.

- Green algae (phylum Chlorophyta) are a highly diverse group of organisms that are abundant in the sea, fresh water, and damp terrestrial habitats, such as on tree trunks and in soil.

- Red algae (phylum Rhodophyta), which have chlorophyll *a* and phycobilins, occur at greater depths, down to more than 250 meters. They have very complex life cycles involving an alternation of generations. Red algae lack flagellated cells and centrioles and may be descendants of one of the most ancient lines of eukaryotes.

- Brown algae (phylum Phaeophyta) are multicellular, marine protists, some reaching 100 meters in length. The kelps contribute greatly to the productivity of the sea, especially along the coasts in relatively shallow areas. Kelps have an alternation of generations in which the gametophytes are very small and filamentous and the sporophytes are large and evident.

- Diatoms (phylum Chrysophyta) are unicellular, photosynthetic protists with opaline silica shells. They include the golden algae.

- Dinoflagellates (phylum Dinoflagellata) are a major phylum of unicellular organisms that have unique chromosomes and a very unusual form of mitosis that takes place entirely within the nucleus. They are responsible for many of the poisonous red tides. In the symbiotic form known as zooxanthellae, dinoflagellates are widespread and are responsible for most of the productivity of coral reefs and of many marine organisms.

- Euglenoids (phylum Euglenophyta) have chloroplasts that share the biochemical features of those found in green algae and plants. With their flexible proteinaceous pellicles, euglenoids might better be considered one of the groups of zoomastigotes (phylum Zoomastigina), a group of heterotrophic, mostly unicellular protists that includes the organism responsible for sleeping sickness, rather than as an independent phylum.

- The malarial parasite, *Plasmodium*, is a member of the phylum Apicomplexa. Carried by mosquitoes, it multiplies rapidly in the liver of humans and other primates and brings about the cyclical fevers characteristic of malaria by releasing toxins into the bloodstream of its host.

- There are about 8000 named species of ciliates (phylum Ciliophora); these protists have a very complex morphology with numerous cilia. They also have a life cycle that involves micronuclei and macronuclei.

- Cellular slime molds (phylum Acrasiomycota) exist as single cells that aggregate, migrate as a slug, and form spores from encysted amoebas that secrete a cellulose wall.

- Plasmodial slime molds (phylum Myxomycota) move about as a plasmodium, containing numerous nuclei. On drying, starving, or being subjected to other environmental cues, plasmodial slime molds form sporangia, which cleave to form spores that may undergo meiosis.

- The oomycetes (phylum Oomycota) occur mainly in water or as plant parasites, including the organism responsible for the great Irish potato famine of the mid-nineteenth century. They have a filamentous structure and exhibit an alternation of generations.

1. **Protista**   This kingdom represents a phylogenetically puzzling group of eukaryotic organisms. The construction of this kingdom is largely one of convenience, containing all the eukaryotes that are not animals, plants, or fungi. Most protists are unicellular, although most phyla include multicellular representatives.

2. **Phyla**   Phyla are the major groups within a kingdom. There are 15 major phyla of protists, most only distantly related to one another.

Review Questions

1. Why is the kingdom Protista said to be an artificial group? How is this different from the other kingdoms?

2. What is the evolutionary significance of the Chlorophyta? Why is *Chlamydomonas* an important member of this phylum? What determines whether a collection of individuals is truly multicellular?

3. What are the structural characteristics of the members of Dinoflagellata? Why is mitosis in this group unique? What are red tides? What are zooxanthellae?

4. Why has it been so difficult to produce a vaccine for trypanosome-caused diseases? Why is it important to control these diseases?

5. What unique characteristic differentiates the members of Ciliophora from other protists? Why are these one-celled organisms not as simple as one might expect? What is the function of two vacuoles exhibited by most members of Ciliophora?

6. What differentiates the oomycetes from the kingdom Fungi, in which they were previously placed? What is the feeding strategy of this phylum? What distinguishes this group from the other protists? Why are these organisms generally considered harmful?

Thought Questions

1. If plants were derived from green algae, why don't we classify green algae as plants in this book?

2. If mitochondria and chloroplasts originated as symbiotic bacteria, what characteristics would the organism that engulfed them have?

3. Coral reefs are one of the most productive communities on earth even though tropical waters are often poor in nutrients. Why? Why do you think coral reefs are only found in tropical and subtropical waters?

Internet Links

Protist Information Gallery
http://mtlab.biol.tsukuba.ac.jp/
This server, from Tsukuba, a "science city" in Japan, offers a site loaded with interesting information about protists.

A Protist Image Bank
http://megasun.bch.umontreal.ca/protists/protists.html
A rich source of images of protists from the University of Montreal, Canada.

Protists Quicktime
http://www.nyhallsci.org/movies_nof.html
Quicktime movies of Paramecium and Amoeba in action, from the NY Hall of Science.

Malaria and Drug Resistance
http://www-micro.msb.le.ac.uk/224/Bradley/Bradley.html
A comprehensive report about malaria: the history of the disease, the biology of the protist that causes it, how it is treated, and the problems posed by rising resistance to these treatments.

The Fuss Over Pfiesteria
http://www.unc.edu/depts/cmse/science/pfiesteria.html
The center for Math and Science Education at the University of North Carolina provides this informative page devoted to a major new protist threat to commercial fisheries.

Protists as Parasites
http://www.cellsalive.com/parasit.htm
This well-illustrated selection from CELLS ALIVE focuses on protozoan parasites infecting humans.

Meet the Algae
http://www.botany.uwc.ac.za/algae/
From the University of British Columbia in Canada comes WELCOME TO THE WORLD OF ALGAE, a comprehensive site which provides a complete introduction to the world of algae.

For Further Reading

Brusca, R. and G. Brusca: *Invertebrates*, Sinauer Press, Sunderland, MA, 1990. A very thoughtful book on animal diversity with an excellent chapter on heterotrophic protists.

Lambrecht, F.: "Trypanosomes and Hominid Evolution," *BioScience*, vol. 35, 1985, pages 640–46. A fascinating article that charts the probable effects of sleeping sickness on the course of human history.

Saffo, M. B.: "New Light on Seaweeds," *BioScience*, vol. 37, 1987, pages 654–64. Discusses recent studies of the role of light-harvesting pigments in depth zonation of seaweeds.

Waters, A., and others: "*Plasmodium falciparum* Appears to Have Arisen as a Result of Lateral Transfer Between Avian and Human Hosts," *Proceedings of the National Academy of Science USA*, vol. 88, April 1991, pages 3140–44. Humans may have caught malaria from chickens.

32

Fungi

FIGURE 32.1
Spores exploding from the surface of a puffball fungus. The fungi constitute a unique kingdom of heterotrophic organisms. Along with bacteria, they are important decomposers and disease-causing organisms.

Concept Outline

32.1 Fungi are unlike any other kind of organism.

A Fungus Is Not a Plant. Unlike any plant, all fungi are filamentous heterotrophs with cell walls made of chitin.

The Body of a Fungus. Cytoplasm flows from one cell to another within the filamentous body of a fungus.

How Fungi Reproduce. Fungi reproduce sexually when filaments of different fungi encounter one another and fuse.

How Fungi Obtain Nutrients. Fungi secrete digestive enzymes and then absorb the products of the digestion.

Ecology of Fungi. Fungi are among the most important decomposers in terrestrial ecosystems.

32.2 Fungi are classified by their reproductive structures.

The Three Phyla of Fungi. There are three phyla of fungi, distinguished by their reproductive structures.

Phylum Zygomycota. In zygomycetes, the fusion of hyphae leads directly to the formation of a zygote.

Phylum Ascomycota. In ascomycetes, hyphal fusion leads to stable dikaryons that grow into massive webs of hyphae that form zygotes within a characteristic sac-like structure, the ascus. Yeasts are unicellular fungi, mostly ascomycetes, that play many important commercial and medical roles.

Phylum Basidiomycota. In basidiomycetes, dikaryons also form, but zygotes are produced within reproductive structures called basidia.

The Imperfect Fungi. Fungi that have not been observed to reproduce sexually cannot be classified into one of the three phyla.

32.3 Fungi form two key symbiotic associations.

Lichens. A lichen is a symbiotic association between a fungus and a photosynthetic alga or cyanobacterium.

Mycorrhizae. Mycorrhizae are symbiotic associations between fungi and the roots of plants.

Of all the bewildering variety of organisms that live on earth, perhaps the most unusual, the most peculiarly different from ourselves, are the fungi (figure 32.1). Mushrooms and toadstools are fungi, multicellular creatures that grow so rapidly in size that they seem to appear overnight on our lawns. At first glance, a mushroom looks like a funny kind of plant growing up out of the soil. However, when you look more closely, fungi turn out to have nothing in common with plants except that they are multicellular and grow in the ground. As you will see, the more you examine fungi, the more unusual they are.

A Fungus Is Not a Plant

The fungi are a distinct kingdom of organisms, comprising about 77,000 named species (figure 32.2). **Mycologists,** scientists who study fungi, believe there may be many more species in existence. Although fungi have traditionally been included in the plant kingdom, they lack chlorophyll and resemble plants only in their general appearance and lack of mobility. Significant differences between fungi and plants include the following:

1. **Fungi are heterotrophs.** Perhaps most obviously, a mushroom is not green. Virtually all plants are photosynthesizers, while no fungi carry out photosynthesis. Instead, fungi obtain their food by secreting digestive enzymes onto whatever they are attached, and then absorbing the organic molecules that are released by the enzymes.

2. **Fungi have filamentous bodies.** Fungi are basically filamentous in their growth form (that is, their bodies consist of long slender filaments), even though these filaments may be packed together to form complex structures like the mushroom. Plants, in contrast, are made of many box-like cells.

3. **Fungi have nonmotile sperm.** Some plants have motile sperm with flagella. No fungi do. Most fungi reproduce sexually with nuclear exchange rather than gametes.

4. **Fungi have cell walls made of chitin.** The cell walls of fungi are built of polysaccharides (chains of sugars) and chitin, the same tough material a crab shell is made of. The cell walls of plants are made of cellulose, also a strong building material.

5. **Fungi have nuclear mitosis.** Mitosis in fungi is different from that in plants or most other eukaryotes in one key respect: the nuclear envelope does not break down and re-form. Instead, mitosis takes place *within* the nucleus. A spindle apparatus forms there, dragging chromosomes to opposite poles of the *nucleus* (not the cell, as in most other eukaryotes).

You could build a much longer list, but already the take-home lesson is clear: fungi are not like plants at all! Their many unique features are strong evidence that fungi are not closely related to any other group of organisms. DNA studies confirm vast differences from other eukaryotes.

Fungi absorb their food after digesting it with secreted enzymes. This mode of nutrition, combined with a filamentous growth form, nuclear mitosis, and other traits, makes the members of this kingdom highly distinctive.

(a)

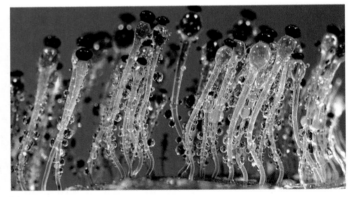
(c)

(b)

FIGURE 32.2
Representatives of the three phyla of fungi. (a) A cup fungus, *Cookeina tricholoma*, a kind of ascomycete, from the rainforest of Costa Rica. (b) *Amanita muscaria*, the fly agaric, a highly toxic basidiomycete. In the cup fungi, the spore-producing structures line the cup; in basidiomycetes that form mushrooms, like *Amanita*, they line the gills beneath the cap of the mushroom. All visible structures of fleshy fungi, such as the ones shown here, arise from an extensive network of filamentous hyphae that penetrates and is interwoven with the substrate on which they grow. (c) *Pilobolus*, a zygomycete that grows on animal feces. Stalks about 10 millimeters long contain dark spore-bearing sacs.

The Body of a Fungus

Fungi exist mainly in the form of slender filaments, barely visible to the naked eye, which are called **hyphae** (singular, **hypha**). These hyphae may be divided into cells by cross-walls called **septa** (singular, **septum**). The septa rarely form a complete barrier, except when they separate the reproductive cells. Cytoplasm characteristically flows or streams freely throughout the hyphae, passing right through major pores in the septa (figure 32.3). Because of this streaming, proteins synthesized throughout the hyphae may be carried to their actively growing tips. As a result, fungal hyphae may grow very rapidly when food and water are abundant and the temperature is optimum.

A mass of hyphae is called a **mycelium** (plural, **mycelia**). This word and the term *mycologist* are both derived from the Greek word for fungus, *myketos*. The mycelium of a fungus (figure 32.4) constitutes a system that may, in the aggregate, be many meters long. This mycelium grows through and penetrates its substrate, resulting in a unique relationship between the fungus and its environment. All parts of such a fungus are metabolically active, continually interacting with the soil, wood, or other material in which the mycelium is growing.

In two of the three phyla of fungi, reproductive structures formed of interwoven hyphae, such as mushrooms, puffballs, and morels, are produced at certain stages of the life cycle. These structures expand rapidly because of rapid elongation of the hyphae. For this reason, mushrooms can appear suddenly on your lawn.

The cell walls of fungi are formed of polysaccharides and chitin, not cellulose like those of plants and many groups of protists. Chitin is the same material that makes up the major portion of the hard shells, or exoskeletons, of arthropods, a group of animals that includes insects and crustaceans (see chapter 43).

Mitosis in fungi differs from that in most other organisms. The nuclear envelope does not break down and re-form; instead, the spindle apparatus is formed *within* it. Centrioles are lacking in all fungi; instead, fungi regulate the formation of microtubules during mitosis with small, relatively amorphous structures called **spindle plaques.** This unique combination of features strongly suggests that fungi originated from some unknown group of single-celled eukaryotes with these characteristics.

Fungi exist primarily in the form of filamentous hyphae, with usually incomplete division into individual cells by septa. These hyphae surround and penetrate the substrate on which the fungi are growing. The cell walls of fungi are composed of chitin. Mitosis occurs within the nuclear envelope, which remains intact at all times. These and other unique features indicate that fungi are not closely related to any other group of organisms.

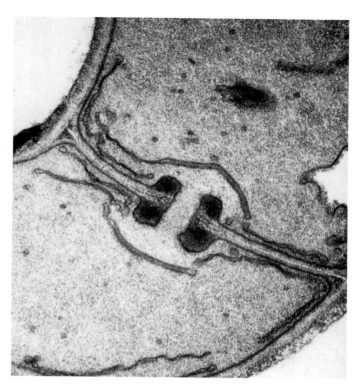

FIGURE 32.3
A septum (45,000×). This transmission electron micrograph of a section through a hypha of the basidiomycete *Inonotus tomentosus* shows a pore through which the cytoplasm streams.

FIGURE 32.4
Fungal mycelium. This mycelium is growing through leaves on the forest floor in Maryland.

How Fungi Reproduce

All fungal nuclei except for the zygote are haploid, and there are many haploid nuclei in the common cytoplasm of a fungal mycelium. When fungi reproduce sexually, hyphae of two genetically different mating types come together and fuse. In two of the three phyla of fungi, the genetically different nuclei that are associated in a common cytoplasm after fusion do not combine immediately. Instead, the two types of nuclei coexist for most of the life of the fungus. A fungal hypha containing nuclei derived from two genetically distinct individuals is called a **heterokaryotic** hypha. If all of the nuclei are genetically similar to one another, the hypha is said to be **homokaryotic**. If there are two distinct nuclei within each compartment of the hyphae, they are **dikaryotic**. If each compartment has only a single nucleus, it is **monokaryotic**. Dikaryotic hyphae have some of the genetic properties of diploids, because both genomes are transcribed. These distinctions are important in understanding the life cycles of the individual groups.

Cytoplasm in fungal hyphae normally flows through perforated septa or moves freely in their absence. Reproductive structures are an important exception to this general pattern. When reproductive structures form, they are cut off by complete septa that lack perforations or have perforations that soon become blocked. Three kinds of reproductive structures occur in fungi: (1) **sporangia**, which are involved in the formation of spores; (2) **gametangia**, structures within which gametes form; and (3) **conidiophores**, structures that produce **conidia**, multinucleate asexual spores.

Spores, always nonmotile, are a common means of reproduction among fungi. They may form as a result of either asexual or sexual processes. When spores land in a suitable place, they germinate, giving rise to a new fungal hypha. Since the spores are very small, they can remain suspended in the air for long periods of time. Because of this, fungal spores may be blown great distances from their place of origin, a factor in the extremely wide distributions of many kinds of fungi. Unfortunately, many of the fungi that cause diseases in plants and animals are spread rapidly and widely by such means. The spores of other fungi are routinely dispersed by insects and other small animals.

> Fungi reproduce sexually after two hyphae of opposite mating type fuse. Asexual reproduction by spores is a second common means of reproduction.

FIGURE 32.5
A carnivorous fungus. The oyster mushroom, *Pleurotus ostreatus*, immobilizes nematodes, which the fungus uses as a source of food.

How Fungi Obtain Nutrients

All fungi obtain their food by secreting digestive enzymes into their surroundings and then absorbing back into the fungus the organic molecules produced by this **external digestion**. The significance of the fungal body plan reflects this approach, the extensive network of hyphae providing an enormous surface area for absorption. Many fungi are able to break down the cellulose in wood, cleaving the linkages between glucose subunits and then absorbing the glucose molecules as food. That is why fungi so often grow on dead trees.

It might surprise you to know that some fungi are predatory (figure 32.5). For example, the mycelium of the edible oyster fungus, *Pleurotus ostreatus*, excretes a substance that anesthetizes tiny roundworms known as nematodes (chapter 41) that feed on the fungus. When the worms become sluggish and inactive, the fungal hyphae envelop and penetrate their bodies and absorb their nutritious contents. The fungus usually grows within living trees or on old stumps, obtaining the bulk of its glucose through the enzymatic digestion of cellulose from the wood, so that the nematodes it consumes apparently serve mainly as a source of nitrogen—a substance almost always in short supply in biological systems. Other fungi are even more active predators than *Pleurotus*, snaring, trapping, or firing projectiles into nematodes, rotifers, and other small animals on which they prey.

> Fungi secrete digestive enzymes onto organic matter and then absorb the products of the digestion.

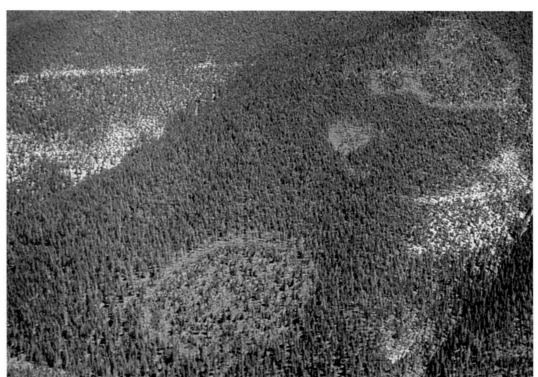

FIGURE 32.6
World's largest organism?
Armillaria, a pathogenic fungus shown here afflicting three discrete regions of coniferous forest in Montana, grows out from a central focus as a single clone. The large patch at the bottom of the picture is almost 8 hectares in diameter. The largest clone measured so far has been 15 hectares in diameter—pretty impressive for a single individual!

Ecology of Fungi

Fungi, together with bacteria, are the principal decomposers in the biosphere. They break down organic materials and return the substances locked in those molecules to circulation in the ecosystem. Fungi are virtually the only organisms capable of breaking down lignin, one of the major constituents of wood. By breaking down such substances, fungi release critical building blocks, such as carbon, nitrogen, and phosphorus, from the bodies of dead organisms and make them available to other organisms.

In breaking down organic matter, some fungi attack living plants and animals as a source of organic molecules, while others attack dead ones. Fungi often act as disease-causing organisms for both plants (figure 32.6) and animals, and they are responsible for billions of dollars in agricultural losses every year. Not only are fungi the most harmful pests of living plants, but they also attack food products once they have been harvested and stored. In addition, fungi often secrete substances into the foods that they are attacking that make these foods unpalatable or poisonous.

The same aggressive metabolism that makes fungi ecologically important has been put to commercial use in many ways. The manufacture of both bread and beer depends on the biochemical activities of **yeasts,** single-celled fungi that produce abundant quantities of ethanol and carbon dioxide. Cheese and wine achieve their delicate flavors because of the metabolic processes of certain fungi, and others make possible the manufacture of soy sauce. Vast industries depend on the biochemical manufacture of organic substances such as citric acid by fungi in culture, and yeasts are now used on a large scale to produce protein for the enrichment of animal food. Many antibiotics, including the first one that was used on a wide scale, penicillin, are derived from fungi.

Some fungi are used to convert one complex organic molecule into another, cleaning up toxic substances in the environment. For example, at least three species of fungi have been isolated that combine selenium, accumulated at the San Luis National Wildlife Refuge in California's San Joaquin Valley, with harmless volatile chemicals—thus removing excess selenium from the soil.

Two kinds of mutualistic associations between fungi and autotrophic organisms are ecologically important. **Lichens** are symbiotic associations between fungi and either green algae or cyanobacteria. They are prominent nearly everywhere in the world, especially in unusually harsh habitats such as bare rock. **Mycorrhizae,** specialized symbiotic associations between the roots of plants and fungi, are characteristic of about 80% of all plants. In each of them, the photosynthetic organisms fix atmospheric carbon dioxide and thus make organic material available to the fungi. The metabolic activities of the fungi, in turn, enhance the overall ability of the symbiotic association to exist in a particular habitat. In the case of mycorrhizae, the fungal partner expedites the plant's absorption of essential nutrients such as phosphorus. Both of these associations will be discussed further in this chapter.

Fungi are key decomposers within almost all terrestrial ecosystems and play many other important ecological and commercial roles.

The Three Phyla of Fungi

There are four groups of fungi: phylum Zygomycota, the **zygomycetes;** phylum Ascomycota, the **ascomycetes;** phylum Basidiomycota, the **basidiomycetes,** and the imperfect fungi (table 32.1). Several other groups that historically have been associated with fungi, such as the slime molds and water molds (phylum Oomycota; see chapter 31), now are considered to be protists, not fungi. Oomycetes are sharply distinct from fungi in their (1) motile spores; (2) cellulose-rich cell walls; (3) pattern of mitosis; and (4) diploid hyphae.

The three phyla of fungi are distinguished primarily by their sexual reproductive structures. In the zygomycetes, the fusion of hyphae leads directly to the formation of a zygote, which divides by meiosis when it germinates. In the other two phyla, an extensive growth of dikaryotic hyphae may lead to the formation of structures of interwoven hyphae within which are formed the distinctive kind of reproductive cell characteristic of that particular group. Nuclear fusion, followed by meiosis, occurs within these cells.

Sexual reproductive structures distinguish the three phyla of fungi.

Table 32.1 Fungi

Phylum	Typical Examples	Key Characteristics	Approximate Number of Living Species
Ascomycota	Yeasts, truffles, morels	Develop by sexual means; ascospores are formed inside a sac called an ascus; asexual reproduction is also common	30,000
Imperfect fungi	*Aspergillus, Penicillium*	Sexual reproduction has not been observed; most are thought to be ascomycetes that have lost the ability to reproduce sexually	17,000
Basidiomycota	Mushrooms, toadstools, rusts	Develop by sexual means; basidiospores are borne on club-shaped structures called basidia; the terminal hyphal cell that produces spores is called a basidium; asexual reproduction occurs occasionally	16,000
Zygomycota	*Rhizopus* (black bread mold)	Develop sexually and asexually; multinucleate hyphae lack septa, except for reproductive structures; fusion of hyphae leads directly to formation of a zygote, in which meiosis occurs just before it germinates	665

Phylum Zygomycota

The zygomycetes (phylum Zygomycota) lack septa in their hyphae except when they form sporangia or gametangia. Zygomycetes are by far the smallest of the three phyla of fungi, with only about 665 named species. Included among them are some of the more common bread molds (figure 32.7), as well as a variety of other microscopic fungi found on decaying organic material. The group is named after a characteristic feature of the life cycle of its members, the production of temporarily dormant structures called **zygosporangia.**

In the life cycle of the zygomycetes (figure 32.8), sexual reproduction occurs by the fusion of gametangia, which contain numerous nuclei. The gametangia are cut off from the hyphae by complete septa. These gametangia may be formed on hyphae of different mating types or on a single hypha. If both + and − mating strains are present in a colony, they may grow together and their nuclei may fuse. Once the haploid nuclei have fused, forming diploid zygote nuclei, the area where the fusion has taken place develops into an often massive and elaborate zygosporangium. A zygosporangium may contain one or more diploid nuclei and acquires a thick coat. Meiosis occurs during the germination of the zygosporangium. Normal, haploid hyphae grow from the haploid cells that result from this process. Except for the zygote nuclei, all nuclei of the zygomycetes are haploid.

Asexual reproduction occurs much more frequently than sexual reproduction in the zygomycetes. During asexual reproduction, hyphae grow over the surface of the bread or other material on which the fungus feeds and produce erect stalks, called **sporangiophores,** in clumps. The tips of the sporangiophores form **sporangia,** which are separated by septa. Haploid spores are produced within the sporangia. Their spores are thus shed above the substrate, in a position where they may be picked up by the wind and blown about.

> Zygomycetes form characteristic resting structures, called zygosporangia, which contain one or more zygotes. The hyphae of zygomycetes are multinucleate, with septa only where gametangia or sporangia are separated.

FIGURE 32.7
Rhizopus, **a zygomycete that grows on moist bread and other similar substrates.** The spore-bearing structures are about a centimeter tall.

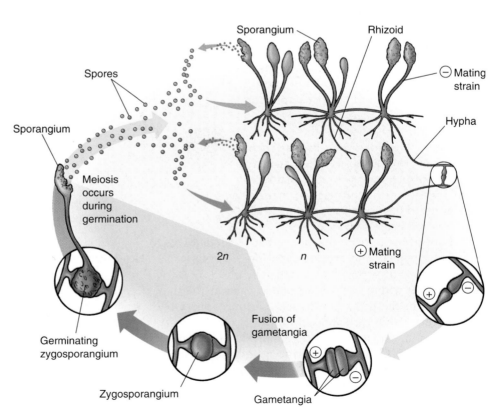

FIGURE 32.8
Life cycle of *Rhizopus,* a zygomycete. This phylum is named for its characteristic zygosporangia.

Phylum Ascomycota

The second phylum of fungi, the ascomycetes (phylum Ascomycota), is a very large group of about 30,000 named species, with more being discovered each year. Among the ascomycetes are such familiar and economically important fungi as yeasts, common molds, morels (figure 32.9), and truffles. Also included in this phylum are many serious plant pathogens, including the chestnut blight, *Cryphonectria parasitica*, and Dutch elm disease, *Ceratocystis ulmi*.

The ascomycetes are named for their characteristic reproductive structure, the microscopic, sac-like **ascus** (plural, **asci**). The zygote, which is the only diploid nucleus of the ascomycete life cycle (figure 32.10), is formed within the ascus. The asci are differentiated within a structure made up of densely interwoven hyphae, corresponding to the visible portions of a morel or cup fungus, called the **ascocarp.**

Asexual reproduction is very common in the ascomycetes. It takes place by means of **conidia** (singular, **conidium**), spores cut off by septa at the ends of modified hyphae called **conidiophores.** Many conidia are multinucleate. The hyphae of ascomycetes are divided by septa, but the septa are perforated and the cytoplasm flows along the length of each hypha. The septa that cut off the asci and conidia are initially perforated, but later become blocked.

The cells of ascomycete hyphae may contain from several to many nuclei. The hyphae may be either homokaryotic or heterokaryotic. Female gametangia, called **ascogonia,** each have a beak-like outgrowth called a **trichogyne.** When the **antheridium,** or male gametangium, forms, it fuses with the trichogyne of an adjacent ascogonium. Initially, both kinds of gametangia contain a number of nuclei. Nuclei from the antheridium then migrate through the trichogyne into the ascogonium and pair with nuclei of the opposite mating type. Heterokaryotic hyphae then arise from the area of the fusion. Throughout such hyphae, nuclei that represent the two different original mating types occur. These hyphae are thus both dikaryotic and heterokaryotic.

Asci are formed at the tips of the heterokaryotic hyphae and are separated by the formation of septa. There are two haploid nuclei within each ascus, one of each mating type represented in the dikaryotic hypha. Fusion of these two nuclei occurs within each ascus, forming a zygote. Each zygote divides immediately by meiosis, forming four haploid daughter nuclei. These usually divide again by mitosis, producing eight haploid nuclei that become walled **ascospores.** In many ascomycetes, the ascus becomes highly turgid at maturity and ultimately bursts, often at a preformed area. When this occurs, the ascospores may be thrown as far as 30 centimeters, an amazing distance considering that most ascospores are only about 10 micrometers long. This would be equivalent to throwing a baseball (diameter 7.5 centimeters) 1.25 kilometers—about 10 times the length of a home run!

(a)   (b)

FIGURE 32.9
Ascomycetes. (a) This morel, *Morchella esculenta*, is a delicious edible ascomycete that appears in early spring. (b) A cup fungus.

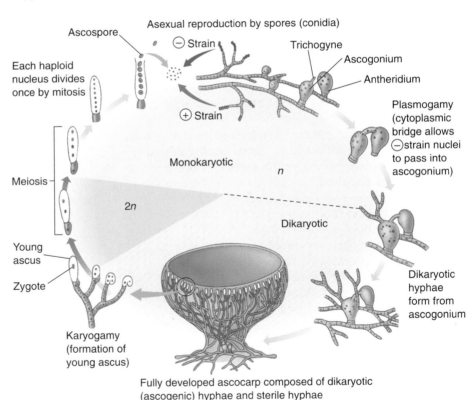

FIGURE 32.10
Life cycle of an ascomycete. The zygote forms within the ascus.

Yeasts

Yeasts, which are unicellular, are one of the most interesting and economically important groups of microscopic fungi, usually ascomycetes. There are about 40 genera of yeasts, with about 350 species. Most of their reproduction is asexual and takes place by cell fission or budding, when a smaller cell forms from a larger one (figure 32.11).

Sometimes two yeast cells will fuse, forming one cell containing two nuclei. This cell may then function as an ascus, with syngamy followed immediately by meiosis. The resulting ascospores function directly as new yeast cells.

Because they are single-celled, yeasts might be considered primitive fungi. However, it appears that they are actually reduced in structure and were originally derived from multicellular ancestors. The word *yeast* actually signifies only that these fungi are single-celled. Some yeasts have been derived from each of the three phyla of fungi, although ascomycetes are best represented. Even yeasts that were derived from ascomycetes are not necessarily directly related to one another, but instead seem to have been derived from different groups of ascomycetes.

Putting Yeasts to Work. The ability of yeasts to ferment carbohydrates, breaking down glucose to produce ethanol and carbon dioxide, is fundamental in the production of bread, beer, and wine (figure 32.12). Many different strains of yeast have been domesticated and selected for these processes. Wild yeasts—ones that occur naturally in the areas where wine is made—were important in wine making historically, but domesticated yeasts are normally used now. The most important yeast in all these processes is *Saccharomyces cerevisiae*. This yeast has been used by humans throughout recorded history. Other yeasts are important pathogens and cause diseases such as thrush and cryptococcosis; one of them, *Candida*, causes a common vaginal infection.

Over the past few decades, yeasts have become increasingly important in genetic research. They were the first eukaryotes to be manipulated extensively by the techniques of genetic engineering, and they still play the leading role as models for research in eukaryotic cells. In 1983, investigators synthesized a functional artificial chromosome in *Saccharomyces cerevisiae* by assembling the appropriate DNA molecule chemically; this has not yet been possible in any other eukaryote. In 1996, the genome sequence of *S. cerevisiae*, the first eukaryote to be sequenced entirely, was completed. With their rapid generation time and a rapidly increasing pool of genetic and biochemical information, the yeasts in general and *S. cerevisiae* in particular are becoming the eukaryotic cells of choice for many types of experiments in molecular and cellular biology. Yeasts have become, in this respect, comparable to *Escherichia coli* among the bacteria, and they are continuing to provide significant insights into the functioning of eukaryotic systems.

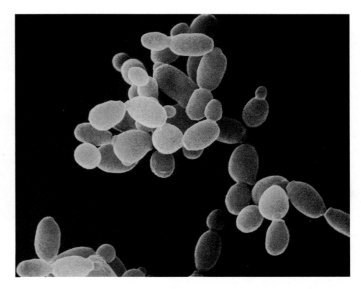

FIGURE 32.11
Scanning electron micrograph of a yeast, showing the characteristic cell division method of budding (19,000×). The cells tend to hang together in chains, a feature that calls to mind the derivation of single-celled yeasts from multicellular ancestors.

FIGURE 32.12
The use of yeasts. Baking bread involves the metabolic activities of the yeasts, single-celled ascomycetes. Wine making also uses the metabolic properties of yeasts; the same species, *Saccharomyces cerevisiae*, is usually used for both processes. In baking bread, yeasts generate carbon dioxide; in making beer or wine, yeasts are used to produce alcohol.

Ascomycetes form their zygotes within a characteristic sac-like structure, the ascus. Meiosis follows, resulting in the production of ascospores. Yeasts are unicellular fungi, mainly ascomycetes, that have evolved from hypha-forming ancestors; not all yeasts are directly related to one another. Long useful for baking, brewing, and wine making, yeasts are now becoming very important in genetic research.

Phylum Basidiomycota

The third phylum of fungi, the basidiomycetes (phylum Basidiomycota), has about 16,000 named species. More is known about some members of this group than about any other fungi. Among the basidiomycetes are not only the mushrooms, toadstools, puffballs, jelly fungi, and shelf fungi, but also many important plant pathogens among the groups called rusts and smuts (figure 32.13). Many mushrooms are used as food, but others are deadly poisonous.

Basidiomycetes are named for their characteristic sexual reproductive structure, the **basidium** (plural, **basidia**). A basidium is club-shaped. Syngamy occurs within the basidium, giving rise to the zygote, the only diploid cell of the life cycle (figure 32.14). As in all fungi, meiosis occurs immediately after the formation of the zygote. In the basidiomycetes, the four haploid products of meiosis are incorporated into **basidiospores**. In most members of this phylum, the basidiospores are borne at the end of the basidia on slender projections called **sterigmata** (singular, **sterigma**). Thus the structure of a basidium differs from that of an ascus, although functionally the two are identical. The septum that cuts off the young basidium is initially perforated but often becomes blocked.

The life cycle of a basidiomycete continues with the production of homokaryotic hyphae after spore germination. These hyphae lack septa at first. Eventually, septa form between the nuclei of the monokaryotic hyphae. A basidiomycete mycelium made up of monokaryotic hyphae is called a **primary mycelium.** Different mating types of monokaryotic hyphae may fuse, forming a dikaryotic or **secondary mycelium.** Such a mycelium is heterokaryotic, with two nuclei, representing the two different mating types, between each pair of septa. The **basidiocarps,** or mushrooms, are formed entirely of secondary (dikaryotic) mycelium. Gills on the undersurface of the cap of a mushroom form vast numbers of minute spores. It has been estimated that a mushroom with a cap that is 7.5 centimeters across produces as many as 40 million spores per hour!

(a)

(b)

FIGURE 32.13
Representative basidiomycetes. (a) Death cap mushroom, *Amanita phalloides.* When eaten, these mushrooms are usually fatal. (b) A shelf fungus, *Trametes versicolor.* Basidia line the inner surfaces of tubes on the lower surface of many shelf fungi, although some have basidia-lined gills.

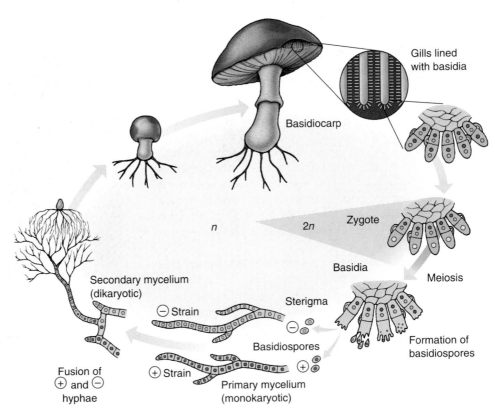

FIGURE 32.14
Life cycle of a basidiomycete. The basidium is the reproductive structure where syngamy occurs.

> Most basidiomycete hyphae are dikaryotic. Ultimately, the hyphae fuse to form basidiocarps, with basidia lining the gills on the underside. Meiosis immediately follows syngamy in these basidia.

The Imperfect Fungi

Most of the so-called imperfect fungi, a group also called deuteromycetes, are those in which the sexual reproductive stages have not been observed. This group includes some basidiomycetes and zygomycetes in addition to the well-represented ascomycetes. The group of fungi from which a particular nonsexual strain has been derived usually can be determined by the features of its hyphae and asexual reproduction. It cannot, however, be classified by the standards of that group because the classification systems are based on the features related to sexual reproduction.

There are some 17,000 described species of imperfect fungi (figure 32.15). Even though sexual reproduction is absent among imperfect fungi, a certain amount of genetic recombination occurs. This becomes possible when hyphae of different genetic types fuse, as sometimes happens spontaneously. Within the heterokaryotic hyphae that arise from such fusion, a special kind of genetic recombination called **parasexuality** may occur. In parasexuality, genetically distinct nuclei within a common hypha exchange portions of chromosomes. Recombination of this sort also occurs in other groups of fungi and seems to be responsible for some of the new pathogenic strains of wheat rust.

Economic Importance

Among the economically important genera of the imperfect fungi are *Penicillium* and *Aspergillus*. Some species of *Penicillium* are sources of the well-known antibiotic penicillin, and other species of the genus give the characteristic flavors and aromas to cheeses such as Roquefort and Camembert. Species of *Aspergillus* are used to ferment soy sauce and soy paste, processes in which certain bacteria and yeasts also play important roles. Citric acid is produced commercially with members of this genus under highly acidic conditions. Some species of both *Penicillium* and *Aspergillus* form ascocarps, but the genera are still classified primarily as imperfect fungi because the ascocarps are found rarely in only a few species. Most of the fungi that cause skin diseases in humans, including athlete's foot and ringworm, are also imperfect fungi.

(a)

(b)

(c)

FIGURE 32.15
The imperfect fungi. (a) *Verticillium alboatrum* (1350×), an important pathogen of alfalfa, has whorled conidia. The single-celled conidia of this member of the imperfect fungi are borne at the ends of the conidiophores. (b) In *Tolypocladium inflatum*, the conidia arise along the branches. This soil fungus is one of the sources of cyclosporin, a drug that suppresses immune reactions and thus assists in making human organ grafts possible; the drug was put on the market in 1979. (c) This scanning electron micrograph of *Aspergillus* shows spores, the round balls at the end of the hyphae.

Fusarium, a member of the imperfect fungi that occurs widely on food, produces highly toxic substances such as trichothecenes, which have been claimed to be agents of chemical warfare (The phenomenon of "yellow rain" in southeast Asia, once claimed to be dispersing trichothecenes, has been shown to consist of honeybee feces sprayed by swarms of bees flying at great heights). *Fusarium*, however, is very important as a dangerous agent of food spoilage.

> Imperfect fungi are fungi in which no sexual reproduction has been observed. For this reason, they cannot be classified by the standards applied to the three phyla of fungi. The great majority of the imperfect fungi are clearly ascomycetes.

Lichens

Lichens (figures 32.16 and 32.17) are symbiotic associations between a fungus and a photosynthetic partner. They provide an outstanding example of mutualism, the kind of symbiotic association that benefits both partners. Ascomycetes (including some imperfect fungi) are the fungal partners in all but about 20 of the approximately 15,000 species of lichens estimated to exist; the exceptions, mostly tropical, are basidiomycetes. Most of the visible body of a lichen consists of its fungus, but within the tissues of that fungus are found cyanobacteria, green algae, or sometimes both (figure 32.18). Specialized fungal hyphae penetrate or envelop the photosynthetic cells within them and transfer nutrients directly to the fungal partner. Biochemical "signals" sent out by the fungus apparently direct its cyanobacterial or green algal component to produce metabolic substances that it does not produce when growing independently of the fungus. The photosynthetic member of the association is normally held between thick layers of interwoven fungal hyphae and is not directly exposed to the light, but enough light penetrates the translucent layers of fungal hyphae to make photosynthesis possible. The fungi found in lichens are unable to grow normally without their photosynthetic partners.

The reproduction of lichens takes place by a combination of normal sexual processes: the formation of ascospores in the fungal component and the reproduction, usually asexual, of the photosynthetic component. These two components may come together under favorable circumstances and produce a lichen body. Alternatively, lichens can simply break up into fragments, each containing both the fungal and photosynthetic components, and can give rise to a new individual.

The durable construction of the fungus, combined with the photosynthetic properties of its partner, has enabled lichens to invade the harshest habitats at the tops of mountains, in the farthest northern and southern latitudes, and on dry, bare rock faces in the desert. In such harsh, exposed areas, lichens are often the first colonists, breaking down the rocks and setting the stage for the invasion of other organisms. Those lichens in which the primary or secondary photosynthetic partner is a cyanobacterium are able to fix atmospheric nitrogen, which is leached to the environment, where other pioneering organisms can use it.

Lichens are able to survive in inhospitable habitats partly by being able to dry or freeze to a condition that we might call suspended animation. Once the drought or cold has passed, the lichens recover quickly and resume their normal metabolic activities, including photosynthesis. The growth of lichens may be extremely slow in harsh environments: many relatively small organisms actually appear to be thousands of years old and, therefore, are among the oldest living things on earth.

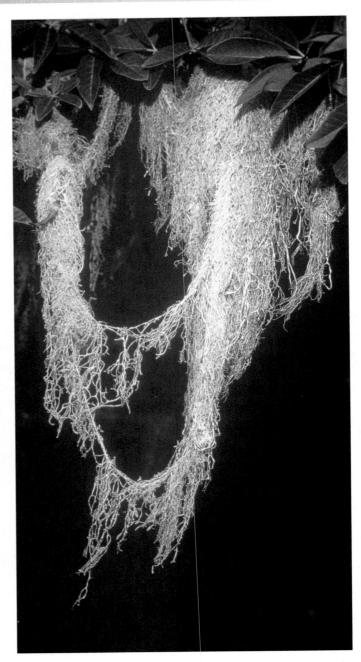

FIGURE 32.16
A fruticose lichen, *Ramalina menziesii*, growing on the twigs of a tree. The Spanish moss of the southern United States, although superficially similar to this lichen, is actually a flowering plant, *Tillandsia*, belonging to the pineapple family.

Lichens are often strikingly colored because of the presence of pigments that probably play a role in protecting the photosynthetic partner from the destructive action of the sun's rays. These same pigments may be extracted from the lichens and used as natural dyes. The traditional method of manufacturing Scotland's famous Harris tweed used fungal dies; however, synthetic dyes are now used.

FIGURE 32.17
Lichens are found in a variety of habitats. (a) Crustose (encrusting) lichens growing on a rock in California. (b) A fruticose lichen, *Cladina evansii*, growing on the ground in Florida. (c) A foliose ("leafy") lichen, *Parmotrema gardneri*, growing on the bark of a tree in a mountain forest in Panama.

Lichens and Pollution

Lichens are extremely sensitive to pollutants in the atmosphere, and thus they can be used as bioindicators of air quality. Their sensitivity results from their ability to absorb substances dissolved in rain and dew, a property on which their existence depends but one that can also prove fatal in a polluted environment. Lichens are generally absent in and around cities because of automobile traffic and industrial activity, even though suitable substrates exist. They are acutely sensitive to sulfur dioxide and are disappearing from national parks and other relatively remote areas reached increasingly by industrial pollution. The pollutants are absorbed by lichens and cause the destruction of chlorophyll, decreases in photosynthesis, and alterations in membrane permeability, among other problems. Similarly, pollutants upset the physiological balance between the fungus and the alga or cyanobacterium, and degradation or destruction of the lichen results.

Lichens are used to assess the pollution by radionuclides that occur in the vicinity of uranium mines, crashed satellites, and other similar sites, just as they were used in the days of atmospheric nuclear weapon testing to estimate the dispersal of radioactive fallout from the explosions. The absorption of radioactive dust by arctic lichens following the nuclear disaster at Chernobyl in 1985 rendered the meat of the reindeer that fed on the lichens unsuitable for human consumption for many months.

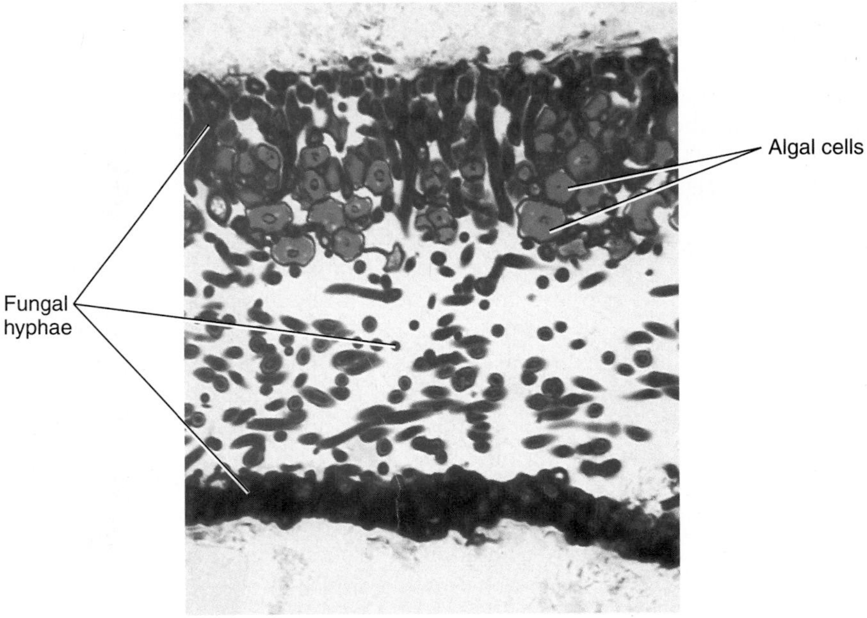

FIGURE 32.18
Stained section of a lichen (250×). This section shows fungal hyphae (*purple*) more densely packed into a protective layer on the top and, especially, the bottom layer of the lichen. The blue cells near the upper surface of the lichen are those of a green alga. These cells supply carbohydrate to the fungus.

Fungal hyphae

Algal cells

Lichens are symbiotic associations between a fungus—an ascomycete in all but a very few instances—and a photosynthetic partner, which may be a green alga or a cyanobacterium or both.

Mycorrhizae

The roots of about 90% of all kinds of vascular plants normally are involved in symbiotic relationships with certain kinds of fungi. It has been estimated that these fungi probably amount to 15% of the total weight of the world's plant roots. Associations of this kind are termed **mycorrhizae** (from the Greek words for "fungus" and "roots"). To a certain extent, the fungi in mycorrhizal associations replace and perform the same function as root hairs, the fine projections from the epidermis, or outermost cell layer, of the terminal portions of the roots. When mycorrhizae are present, they aid in the direct transfer of phosphorus, zinc, copper, and other nutrients from the soil into the roots. The plant, on the other hand, supplies organic carbon to the fungus, so the system is an example of mutualism.

There are two principal types of mycorrhizae: **endomycorrhizae,** in which the fungal hyphae penetrate the outer cells of the plant root, forming coils, swellings, and minute branches, and also extend out into the surrounding soil; and **ectomycorrhizae,** in which the hyphae surround but do not penetrate the cell walls of the roots. In both kinds of mycorrhizae, the mycelium extends far out into the soil.

Endomycorrhizae

Endomycorrhizae are by far the more common of these two types. The fungal component in them is a zygomycete. Only about 100 species of zygomycetes are known to be involved in such relationships throughout the world. These few species of zygomycetes are associated with more than 200,000 species of plants. Endomycorrhizal fungi are being studied intensively because they are potentially capable of increasing crop yields with lower phosphate and energy inputs.

The earliest fossil plants often have endomycorrhizal roots. Such associations, common in fossils of plants, may have played an important role in allowing plants to invade land. The soils available at such times would have been sterile and completely lacking in organic matter. Plants that form mycorrhizal associations are particularly successful in infertile soils; considering the fossil evidence, the suggestion that mycorrhizal associations found in the earliest plants helped them succeed on such soils seems reasonable. In addition, the most primitive vascular plants surviving today continue to depend strongly on mycorrhizae.

Ectomycorrhizae

Ectomycorrhizae (figure 32.19) involve far fewer kinds of plants than do endomycorrhizae, perhaps a few thousand. They are characteristic of certain groups of trees and shrubs, particularly those of temperate regions, including pines, firs, oaks, beeches, and willows. The fungal components in most ectomycorrhizae are basidiomycetes, but some are ascomycetes. Several different kinds of basidiomycetes may form mycorrhizal associations with one plant. Different combinations have different effects on the

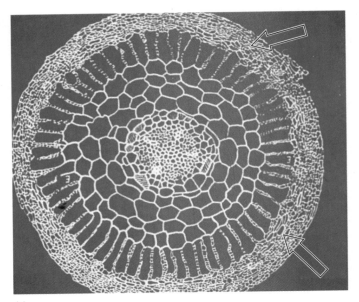

(a)

(b)

FIGURE 32.19
Ectomycorrhizae. (a) Transection of a pine root, showing thick mantle (*arrows*) of ectomycorrhizae. (b) Ectomycorrhizae on roots of pines. From left to right are yellow-brown mycorrhizae formed by *Pisolithus*, white mycorrhizae formed by *Rhizopagon*, and pine roots not associated with a fungus.

physiological characteristics of the plant and its ability to survive under different environmental conditions. At least 5000 species of fungi are involved in ectomycorrhizal relationships, and many of them are restricted to a single species of plant.

> **Mycorrhizae are symbiotic associations between plants and fungi. There are two main types: (1) endomycorrhizae, in which the fungal partner is a zygomycete, and (2) ectomycorrhizae, in which the fungal partner usually is a basidiomycete.**

32.1 Fungi are unlike any other kind of organism.

- The fungi are a distinct kingdom of eukaryotic organisms characterized by a filamentous growth form, lack of chlorophyll and motile cells, chitin-rich cell walls, and external digestion of food by the secretion of enzymes. Together with the bacteria, they are the decomposers of the biosphere.

- Fungal filaments, called hyphae, collectively make up the fungus body, which is called the mycelium. A hypha may contain genetically uniform nuclei and thus be homokaryotic, or it may contain two or more genetically different kinds of nuclei and be heterokaryotic. Mitosis in fungi occurs without centrioles and within the nuclear envelope.

- In many fungi, the two kinds of nuclei that will eventually undergo syngamy occur together in hyphae for a long period before they fuse. Most hyphae of basidiomycetes found in nature, for example, are dikaryotic.

- Meiosis occurs immediately after the formation of the zygote in all fungi; the zygote, therefore, is the only diploid nucleus of the entire life cycle in these organisms.

32.2 Fungi are classified by their reproductive structures.

- There are three phyla of fungi: Zygomycota, the zygomycetes; Ascomycota, the ascomycetes; and Basidiomycota, the basidiomycetes.

- Zygomycetes form septa only when gametangia or sporangia are cut off at the ends of their hyphae; otherwise, their hyphae are multinucleate. Most hyphae of ascomycetes and basidiomycetes have perforated septa through which the cytoplasm, but not necessarily the nuclei, flows freely.

- Cells within the heterokaryotic hyphae of ascomycetes are multinucleate; those within the heterokaryotic hyphae of the basidiomycetes are dikaryotic. Zygotes in ascomycetes form within sac-like structures known as asci, and those in basidiomycetes form within structures known as basidia. Asci and basidia are formed within or on the surfaces of large structures, like mushrooms, that are formed of interwoven hyphae.

- Asexual reproduction in zygomycetes takes place by means of spores from multinucleate sporangia; in ascomycetes, it takes place by means of conidia. Asexual reproduction in basidiomycetes is relatively rare and may involve either hyphal fragmentation or the formation of conidia.

- The yeasts are a group of unicellular fungi, mostly ascomycetes. They are commercially important because of their roles in baking and fermentation.

- The imperfect fungi are a large artificial group of fungi in which sexual reproduction does not occur or is not known; most are ascomycetes. They are classified according to the characteristics of their spore-forming structures and spores. Many are of great commercial importance.

32.3 Fungi form two key symbiotic associations.

- Symbiotic systems involving fungi include lichens and mycorrhizae. The fungal partners in lichens are almost entirely ascomycetes, which derive their nutrients from green algae, cyanobacteria, or both.

- Mycorrhizae are symbiotic associations between plants and fungi that are characteristic of the great majority of plants. Endomycorrhizae, which are more common, have zygomycetes as the fungal partner, while ectomycorrhizae have mainly basidiomycetes.

1. **Fungi** The fungi are eukaryotic, typically multicellular, nonmotile, heterotrophic organisms that digest their food externally rather than ingesting it.

2. **Mycelium** All fungi but yeasts are composed of filaments called hyphae that make up the multicellular body (mycelium) of the organism. The nuclei in these filaments often move from cell to cell.

3. **Hyphal fusion** Reproduction in the fungi usually occurs when cells of two hyphae fuse. The two genetically different nuclei in a hypha may remain separate for a period of time, or they may fuse.

4. **Asexual reproduction** Asexual reproduction occurs almost exclusively in the zygomycetes and the ascomycetes, and it involves structures called sporangia and conidiophores, respectively.

5. **Yeasts** Yeasts are single-celled fungi that can be from any of the fungal classes but are usually ascomycetes. The yeasts are commercially important in fermentation and baking.

6. **Imperfect fungi** The imperfect fungi is a rough category that includes members from all three phyla that have not yet been observed to reproduce sexually.

7. **Symbiosis** In some cases, fungi form symbiotic relationships with other organisms. Lichens involve the symbiotic relationship of algae, cyanobacteria, or both with fungi, while mycorrhizae are symbiotic relationships of fungi and the root cells of vascular plants.

Review Questions

1. What is a hypha? What is a mass of hyphae called? What is a septum as it applies to the fungi? Under what circumstances are septa complete? What is the advantage to having incomplete septa?

2. What is the composition of the fungal cell wall? Why is this composition an advantage to the fungi?

3. Which fungal nuclei are diploid? Which are haploid? What event occurs during sexual reproduction in the fungi? How does this affect the nuclei within the majority of the fungi? To what do the following terms refer: *heterokaryotic, homokaryotic, dikaryotic,* and *monokaryotic?*

4. What are the three reproductive structures that occur in fungi? How do they differ?

5. Fungi are nonmotile. How are they dispersed to new areas?

6. What are the ascomycete asexual spores called? What are the structures that bear these spores called? Do the nonreproductive hyphae of this division have septa?

7. To what phyla do the yeasts belong? How do they differ from other fungi? Is it more likely that this characteristic is primitive or degenerate? What is the yeast's primary mode of reproduction? Why are the yeasts gaining popularity in scientific research?

8. What are the imperfect fungi? Which phylum seems to be best represented in this group? By what means can individuals in this group be classified?

9. What are lichens? Which fungal phylum is best represented in the lichens? What is the function of each partner in this association?

10. What are mycorrhizae? How do endomycorrhizae and ectomycorrhizae differ? Which fungi are most prominent in each? What is the function of each partner in these associations?

Thought Questions

1. Why do we say there are three phyla of fungi when we have mentioned five major groups: Ascomycota, Zygomycota, Basidiomycota, yeasts, and imperfect fungi? What are the major distinguishing characteristics of these groups? Why are only three considered phyla?

2. Many antibiotics, including penicillin, are derived from fungi. Why do you think the fungi produce these substances? Of what use are they to the fungi? In addition, fungi often secrete substances into the foods that they are attacking that make these foods unpalatable or even poisonous to other creatures. What kind of advantage would these substances provide the fungi?

Internet Links

Fungi on the Internet
http://www.keil.ukans.edu/~fungi/funinfo.html
A wonderfully rich library of information about fungi from the University of Kansas, with links to collections, resources, and teaching aids.

Fun Facts About Fungi
http://www.herb.lsa.umich.edu/kidpage/factindx.htm
Explore the world of fungi in this University of Michigan site, which makes mycology fun by focusing on interesting facts and stories.

Mushrooms Galore
http://www.pip.dknet.dk/%7Efvl/mushimage/imageframe.htm
FUNGI IMAGES ON THE NET is the very best directory of fungal photos on the internet.

A Short 'shroom Primer
http://www.lclark.edu/%7Ewstone/stuff/primer.html
A short, easy-to-follow primer on mushrooms.

Mushroom Mania
http://www.aa.net/~reo/mushroom.html
All you ever wanted to know about mushrooms, a site rich with information and links to other sources.

Phear of Phungi
http://149.152.32.5/Plant_Biology/Phungphear.html
A tongue-in-cheek but very clearly illustrated guide to the life cycles of the fungi, intended for these "fungophobes" who cannot stomach mushrooms or figure out their complex life cycles.

For Further Reading

Kosikowski, F. V.: "Cheese," *Scientific American*, May 1985, pages 88–99. A fascinating account how more than 2000 varieties of cheese are made and how bacteria and fungi participate in the process.

Newhouse, J. R.: "Chestnut Blight," *Scientific American*, July 1990, pages 106–11. Biological control is now being used in an effort to stop the ravages of this fungus.

Simon, L., J. Bousquet, R. C. Lévesque, and M. LaLonde: "Origin and Diversification of Endomycorrhizal Fungi and Coincidence with Vascular Land Plants," *Nature*, May 6, 1993, pages 67–69. The critical relationship between fungi and plant roots is discussed in this very technical article.

Sommer, R.: "Why I Will Continue to Eat Corn Smut," *Natural History*, January 1995, pages 18–21. A treatise on the fungus that causes "corn smut," the bane of farmers, but balm to the gastronauts.

Sternberg, S.: "The Emerging Fungal Threat," *Science*, December 9, 1994, pages 1632–34. Gone are the days when fungal infections meant just ringworm or athlete's foot. In today's increasing population of immunocompromised individuals, fungal infections can mean death.

33

Evolutionary History of Plants

Concept Outline

33.1 **Plants are terrestrial, with life cycles that alternate between haploid and diploid.**

The Evolutionary Origins of Plants.   Plants are descendants of multicellular green algae, with key adaptations for meeting the challenges of living on land.

Plant Life Cycles.   Plants alternate between haploid gametophyte and diploid sporophyte generations.

33.2 **Nonvascular plants like mosses are relatively unspecialized.**

Mosses, Liverworts, and Hornworts.   The most conspicuous part of a nonvascular plant is the green photosynthetic gametophyte, which supports the smaller sporophyte nutritionally.

33.3 **Seedless vascular plants like ferns have well-developed conducting tissues.**

Features of Vascular Plants.   Vascular plants have specialized vessels called xylem for conducting water and dissolved minerals up the stem from the roots, and other vessels called phloem for conducting carbohydrates in solution down the stem from the leaves.

Seedless Vascular Plants.   Seedless vascular plants have a much more conspicuous sporophyte than nonvascular plants do, with well-developed conducting systems in stem, roots, and leaves.

33.4 **Seed plants like pine trees and roses have protected embryos specialized for dispersal.**

Seed Plants.   In seed plants, the sporophyte is dominant. Male and female gametophytes develop within the sporophyte and depend on it for food.

Gymnosperms.   In gymnosperms, the female gametophyte (ovule) is not completely enclosed by sporophyte tissue at the time of pollination by male gametophytes (pollen).

Angiosperms.   In angiosperms, the ovule is completely enclosed by sporophyte tissue at the time of pollination. Angiosperms produce flowers and are by far the most successful plant group.

FIGURE 33.1
A forest meadow community. The plant kingdom is astonishingly diverse, often displaying remarkable adaptations to even the most extreme terrestrial biomes.

Plants dominate all terrestrial communities, from forests (figure 33.1) to pastures, from agricultural lands to deserts. They provide most of our food, directly and indirectly, as well as much of our shelter, clothing, and medicines. These multicellular, eukaryotic organisms have cellulose-rich cell walls and chloroplasts that contain chlorophylls *a* and *b*, as well as various carotenoids. Most of the nearly 300,000 species of existing plants are photosynthetic, capturing energy from the sun and transforming it into chemical bonds. Photosynthesis also produces the oxygen essential to the respiration of every living organism. In this chapter, we will discuss characteristics of the plant phyla, focusing on differences in their life cycles. In chapters 34 to 40, we will treat specific aspects of plant biology in more detail.

645

The Evolutionary Origins of Plants

The first terrestrial plants evolved from an organism that, if it existed today, would be classified as a multicellular green alga (see chapter 31). Extensive biochemical and morphological similarities between green algae and plants make this conclusion inescapable. Green algae and plants share chlorophylls *a* and *b*, and carotenoids; they also have cellulose in their cell walls and starch as their primary food reserve. Moreover, the details of cell division are similar in the two groups. After mitosis is complete, animal cells pinch inward toward the center, while plant cells form an interior *cell plate*, which grows out from the center between the two daughter cells. Aside from plants, the only organisms that form a cell plate during cytokinesis are a few groups of green algae.

Plant phyla have traditionally been called *divisions*, but the International Botanical Congress of 1993 voted to accept the term *phylum* as an alternative. In this book, we will call all such groups *phyla* because they are comparable in rank and significance to groups called phyla in animals and other groups. Living plants are grouped into 12 phyla.

Members of 9 of the 12 plant phyla are collectively called **vascular plants,** and include, among others, the ferns (figure 33.2), conifers, and flowering plants. Vascular plants have water-conducting xylem and food-conducting phloem strands of cells in their stems, roots, and leaves. Members of the other three phyla either lack such strands or have poorly developed ones. These three phyla are mosses (phylum Bryophyta), liverworts (phylum Hepaticophyta); and hornworts (phylum Anthocerotophyta).

The Green Invasion of the Land

Plants and fungi are the only major groups of organisms that are primarily terrestrial. There are some 267,000 known species of living plants. They dominate almost every part of the terrestrial landscape, except for extreme deserts, polar regions, and the tops of the highest mountains. Sev-

FIGURE 33.2
Five-finger fern. *Adiantum aleuticum,* a maidenhair fern (phylum Pterophyta).

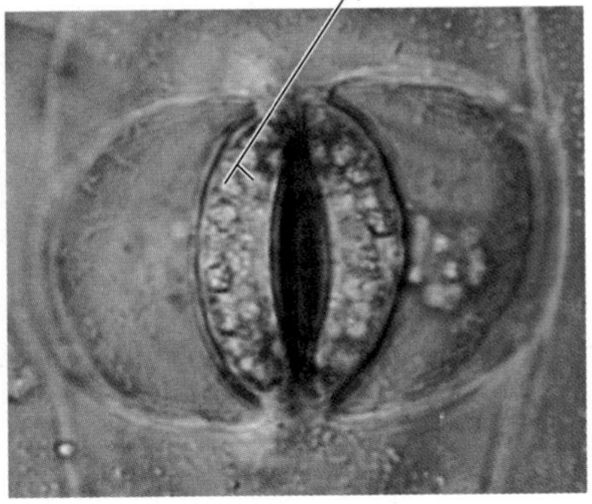

Chloroplasts

FIGURE 33.3
A stoma. Stomata allow water vapor and gases to diffuse into and out of a plant.

eral other groups, including arthropods, chordates, and mollusks, are also well represented on land. Plants probably came first, their cell walls helping to protect their propagules during dispersal. Most plants are protected from **desiccation**—the tendency of organisms to lose water to the air—by a waxy **cuticle** secreted onto their exposed surfaces. The cuticle is relatively impermeable and provides an effective barrier to water loss.

Special structures in this barrier allow the entry of carbon dioxide, which is necessary for photosynthesis, and the exit of oxygen by diffusion. These tiny slit-like or eye-shaped pores are called **stomata** (singular **stoma;** figure 33.3) and make up about 1% of most leaf and green stem surfaces. Water and gas diffusion into and out of a plant is regulated by the opening and closing of stomata, which typically number into the thousands per square centimeter.

At an early stage of their evolution, plants evidently also developed a special kind of relationship with fungi that has been a key to their successful occupation of terrestrial habitats. *Mycorrhizae* (a symbiotic association of fungi with plant roots) are found in about 80% of all plants and are frequently seen in fossils of the earliest plants. Mycorrhizae play an important role among today's plants in the assimilation of phosphorus and probably other ions. We can surmise that mycorrhizae also did so in the raw, unaltered soils that existed before the earliest plants.

Many other features, such as efficient water- and food-conducting systems, developed and contributed to the success of plants on land. Roots, shoots, and leaves—expanded areas of photosynthetically active tissue—also evolved and diversified in relationship to the varied habitats that existed on land (see chapter 34). Lastly, specializations in key reproductive features improved the chances for survival of embryonic tissues within seeds as they were dispersed.

Plants evolved from multicellular green algae and eventually developed cuticles and stomata, mycorrhizae, conducting systems, and other features that adapt them well for life on land.

Plant Life Cycles

Plants exhibit an **alternation of generations,** which also occurs in brown, red, and green algae (see chapter 31). The diploid generation, or **sporophyte,** alternates with the haploid generation, or **gametophyte.** Sporophyte means "spore plant," and gametophyte means "gamete plant." These terms indicate the kinds of reproductive cells the respective generations produce.

Most adult animals are diploid, and in this respect they resemble the sporophyte generation of a plant. Animals, through meiosis, produce gametes—eggs and sperm, which fuse directly to form a zygote. In contrast, the sporophyte generation of a plant does not produce gametes as a result of meiosis. Instead, meiosis takes place in structures call **sporangia,** where diploid **spore mother cells (sporocytes)** undergo meiosis, each producing four haploid **spores.** Spores divide by mitosis, producing a multicellular haploid individual, the gametophyte. Spores are the first cells of the gametophyte generation.

In turn, the gametophyte, by mitosis, produces *gametes.* The gametes, like the gametophyte that produces them, are haploid. When the gametes fuse, the zygote they form is diploid and is the first cell of the next sporophyte generation. The zygote grows into a diploid sporophyte that produces sporangia in which meiosis ultimately occurs. The basic cycle involved in the alternation of generations is summarized in figure 33.4.

Such a cycle is characteristic of all plants. In some, including mosses, liverworts, and ferns, the gametophyte is green and free-living; in others it is not green and is either nutritionally dependent on the sporophyte, or **saprobic** (deriving its energy directly from nonliving organic matter). When you look at a moss or liverwort, what you see is largely gametophyte tissue (figure 33.5*a*); their sporophytes are usually smaller, brownish or yellowish structures attached to or enclosed within tissues of the gametophyte. In vascular plants, the gametophytes are always much smaller than the sporophytes (figure 33.5*b*). In seed plants, the gametophytes are nutritionally dependent on the sporophytes and are enclosed within their tissues. When you look at a vascular plant what you see, with rare exceptions, is a sporophyte.

Having completed our overview of plant life cycles, we will consider the major plant groups. As we do, we will see a progressive reduction of the gametophyte from group to group, a loss of multicellular **gametangia** (structures in which sex cells are produced), and increasing specialization

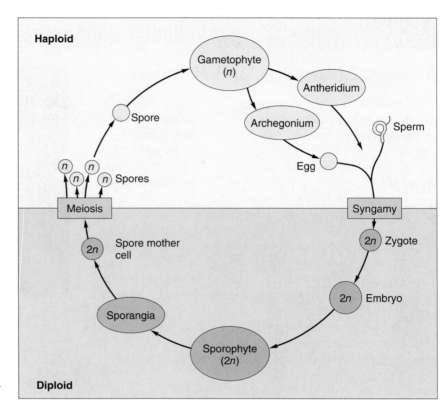

FIGURE 33.4
A generalized plant life cycle.

(a)   (b)

FIGURE 33.5
Two types of gametophytes. (a) This gametophyte of a moss (a nonvascular plant) is green and free-living. (b) Gametophytes of a pine (a vascular plant) are just large enough to be visible to the naked eye.

for life on the land, culminating with the remarkable structural adaptations of the flowering plants, the dominant plants today. Most of the plant groups living today did not directly give rise to one another, and they do not reflect the evolutionary stages that led from simpler to more complex organisms. Nonetheless, similar trends must have characterized the progression to seed plants over the hundreds of millions of years of their history.

Plant life cycles involve an alternation of generations of diploid sporophytes with haploid gametophytes.

Mosses, Liverworts, and Hornworts

The **bryophytes**—mosses, liverworts, and hornworts—were in the past grouped together because of various similarities in reproduction, form, and habitat, and it was assumed they were all related to one another. Scientists now agree that bryophytes consist of three quite distinct phyla of relatively unspecialized plants that did not arise from a common ancestor. Their gametophytes are green and nutritionally independent of the sporophytes. Their sporophytes are attached to the gametophytes and depend on them nutritionally to varying degrees. Bryophytes, like ferns and certain other vascular plants, require water external to the plant (e.g., rainwater) to reproduce sexually. It is not surprising, therefore, that they are especially common in moist places, both in the tropics and temperate regions.

Most bryophytes are small; few exceed 7 centimeters in height. The gametophytes are more conspicuous than the sporophytes, which obtain their food directly from the gametophytes. Some of the sporophytes are completely enclosed within gametophyte tissue; others are not enclosed and usually turn brownish or straw-colored at maturity.

Bryophyta: Mosses

The gametophytes of mosses typically consist of small leaves arranged spirally or alternately around a stem-like axis (figure 33.6); the axis is anchored to its substrate by means of root-like **rhizoids.** Each rhizoid consists of several cells that absorb water, but nothing like the volume of water absorbed by a vascular plant root. Moss leaves have little in common with true leaves, except for the superficial appearance of the green, flattened blade and slightly thickened midrib that runs lengthwise down the middle. Moss leaves, unlike true leaves, are only one cell thick (except at the midrib); there are no vascular strands or stomata, and all the cells are haploid.

Water may rise up a strand of specialized cells in the center of a moss gametophyte axis, but most water used by the plant travels up the outside of the plant by capillarity. Some mosses also have specialized food-conducting cells surrounding those that conduct water.

Multicellular gametangia are formed at the tips of the leafy gametophytes (figure 33.7). Female **archegonia** may develop

**FIGURE 33.6
A hair-cup moss, *Polytrichum* (phylum Bryophyta).** The leaves belong to the gametophyte. Each of the yellowish brown stalks, with the capsule, or sporangium, at its summit, is a sporophyte.

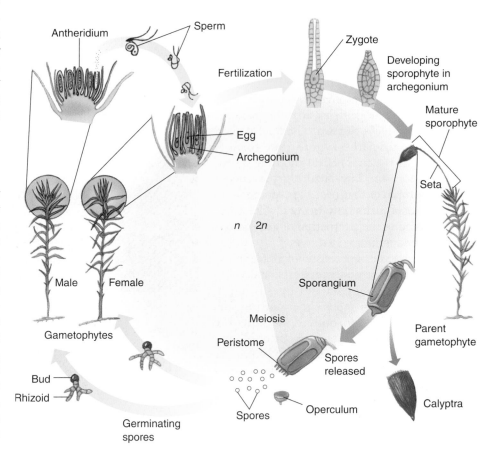

**FIGURE 33.7
Life cycle of a typical moss.**

either on the same gametophyte as the male **antheridia** or on separate plants. A single egg is produced in the swollen lower part of an archegonium while numerous sperm are produced in an antheridium. When sperm are released from an antheridium, they swim with the aid of flagella through a film of dew or rainwater to the archegonia. One sperm (which is haploid) unites with an egg (also haploid), forming a diploid zygote. The zygote divides by mitosis and develops into the sporophyte, a slender, basal stalk, or *seta*, with a swollen capsule, the *sporangium*, at its apex. As the sporophyte develops, its base becomes embedded in gametophyte tissue, from which it derives its energy. The sporangium is often cylindrical or club-shaped, with a row of teeth (the peristome) at the top end. The peristome teeth are protected by a lid-like cap (the operculum), which is itself covered by another hood-like cap (the calyptra). While all the cells of the sporophyte are diploid, those of the calyptra, which grows from the gametophyte, are haploid. The calyptra is lifted upward as the sporophyte grows in length.

Spore mother cells within the sporangium undergo meiosis, each becoming four haploid spores. At maturity, the operculum (and the calyptra, if still present) pops off, and the spores are released. A spore that lands in a suitable damp location may germinate and grow into a thread-like *protonema*; each protonema cell has several chloroplasts. Certain cells of the protonema develop rhizoids and "buds." Each bud develops into a new gametophyte plant consisting of an axis with leaves, thereby completing the life cycle.

In the Arctic and the Antarctic, mosses are the most abundant plants, boasting not only the largest number of individuals in these harsh regions, but also the largest number of species. Many mosses are able to withstand prolonged periods of drought, although they are not common in deserts. Most are remarkably sensitive to air pollution and are rarely found in abundance in or near cities or other areas with high levels of air pollution. Some mosses, such as the peat mosses (*Sphagnum*), can absorb up to 25 times their weight in water and are valuable commercially as a soil conditioner, or as a fuel when dry. Peat mosses, which are naturally acidic, have also been used in emergencies as a cotton substitute in poultices for wounds.

Hepaticophyta: Liverworts

The old English word *wyrt* means "plant" or "herb." Some common liverworts have flattened bodies called *thalli* (singular *thallus*) with lobes resembling those of liver—hence the combination "liverwort." Although the thalloid liverworts are the best known representatives of this phylum, they constitute only about 20% of the species (figure 33.8). The other 80% are leafy and superficially resemble mosses. All liverworts are less complex than mosses and differ from them in several respects: their gametophytes develop almost directly from spores, growth is prostrate instead of erect, and the rhizoids are one-celled.

Thalloid liverworts have air chambers containing upright, branching rows of photosynthetic cells, each chamber having a pore at the top. Unlike stomata, the pores are fixed open and cannot close. Leafy liverworts have two rows of partially overlapping leaves whose cells contain oil bodies.

FIGURE 33.8
A common thalloid liverwort, *Marchantia* **(phylum Hepaticophyta).** The sporophytes are borne within the tissues of the umbrella-shaped structures that arise from the surface of the flat, green, creeping gametophyte.

Sexual reproduction in liverworts is similar to that in mosses. Thalloid liverworts produce gametangia on upright, umbrella-like structures that develop from the thalli. These liverworts can also reproduce asexually by means of *gemmae*, small lens-shaped pieces of tissue produced in shallow cups. When rainwater splashes the gemmae out of the cups, each can potentially grow into a new thallus.

Anthocerotophyta: Hornworts

In hornworts, the sporophytes resemble tiny green broom handles rising from filmy thalloid gametophytes usually less than 2 centimeters in diameter (figure 33.9). The base of the sporophyte is embedded in gametophyte tissue, from which it derives some of its nutrition. However, the sporophyte has stomata, is photosynthetic, and provides much of the energy needed for growth and reproduction.

Hornwort cells usually have a single chloroplast with a pyrenoid like those of green algae. This suggests that hornworts may be more closely related to green algae than to any other group of plants. Hornwort gametophytes have internal cavities that are often filled with mucilage instead of air. It is common for the cyanobacterium *Nostoc* (which is capable of nitrogen fixation) to form colonies in the mucilage. Under favorable conditions, hornwort sporophytes elongate and split at the top into two or three ribbon-like segments. Spores are released as the segments curl back.

Among nonvascular plants, the conspicuous photosynthetic individuals are haploid gametophytes. The three major phyla are all relatively unspecialized.

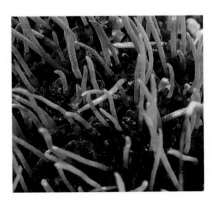

FIGURE 33.9
Hornworts (phylum Anthocerotophyta). Hornwort sporophytes are seen in this photo.

Features of Vascular Plants

The first vascular plants for which we have a relatively complete record belonged to the phylum Rhyniophyta; they flourished some 410 million years ago but are now extinct. We are not certain what the very earliest of these vascular plants looked like, but fossils of *Cooksonia* provide some insight into their characteristics (figure 33.10). *Cooksonia*, the first known vascular land plant, appeared in the late Silurian period about 420 million years ago. It was successful partly because it encountered little competition as it spread out over vast tracts of land. The plants were only a few centimeters tall and had no roots or leaves. They consisted of little more than a branching axis, the branches forking evenly and expanding slightly toward the tips. They were **homosporous** (producing only one type of spore), their spores being produced in sporangia at the ends of the branches. Other ancient vascular plants that followed evolved more complex arrangements of sporangia and leaves, which first appeared as simple emergences from the stems.

Cooksonia and the other early plants that followed it became successful colonizers of the land through the development of efficient water- and food-conducting systems known as **vascular tissues.** The word *vascular* comes from the Latin *vasculum*, meaning a "vessel or duct" (figure 33.11). These tissues consist of strands of specialized cylindrical or elongated cells that form a network throughout a plant, extending from near the tips of the roots, through the stems, and into the leaves. One type of vascular tissue, the *xylem*, conducts water and dissolved minerals upward from the roots; another type of tissue, *phloem*, conducts carbohydrates in solution away from their sources in the green parts of the plant. The carbohydrates provide energy for the synthesis of various cell components. The presence of a cuticle and stomata are also characteristic of vascular plants. (See the discussion of vascular tissue structure in chapter 34.)

A Very Successful Group: The Vascular Plants

The nine phyla of vascular plants (table 33.1), which originated in connection with their invasion of the land, have come to dominate terrestrial habitats everywhere, except for the highest mountains and the polar regions. All plants have an alternation of generations, but this alternation has become modified by reduction of the gametophyte during evolution of the various phyla of vascular plants.

Accompanying this reduction in size and complexity of the gametophytes has been the appearance of the seed. Seeds are highly resistant structures well suited to protect a plant embryo from drought and to some extent from predators. In addition, most seeds contain a supply of food for the young plant. Seeds occur only in **heterosporous** plants (plants that produce two types of spores), and

Sporangia

FIGURE 33.10
Cooksonia, **the first known vascular land plant.** Its upright, branched stems, which were no more than a few centimeters tall, terminated in sporangia, as seen here. It probably lived in moist environments such as mudflats, had a resistant cuticle, and produced spores typical of vascular plants. This fossil represents a plant that lived some 410 million years ago, in what is now New York State. *Cooksonia* belongs to phylum Rhyniophyta, consisting entirely of extinct plants.

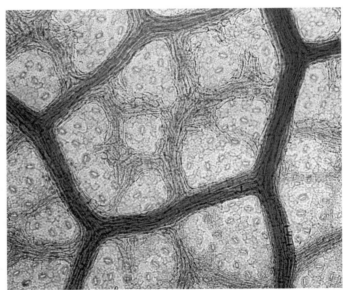

FIGURE 33.11
Vascular tissues. The veins of a vascular plant leaf; each vein contains strands of specialized cells for conducting water and dissolved minerals, and carbohydrates through the roots, shoots and leaves.

heterospory occurs only in vascular plants. Flowers, which evolved among the angiosperms, are primarily a device that induces insects and animals to spread the pollen from the flowers of one individual plant to those of another. They are, in effect, a mechanism whereby plants can overcome their rooted, immobile nature and secure the benefits of wide outcrossing in promoting genetic diversity.

Most vascular plants have well-developed conducting tissues, specialized stems, leaves, and roots, cuticles, stomata, and, in many, seeds.

Table 33.1 The Nine Phyla of Extant Vascular Plants

Phylum	Examples		Key Characteristics	Approximate Number of Living Species
Anthophyta	Flowering plants (angiosperms)		Heterosporous seed plants. Sperm not motile; conducted to egg by a pollen tube. Seeds enclosed within a fruit. Leaves greatly varied in size and form. Herbs, vines, shrubs, trees. About 14,000 genera.	240,000
Pterophyta	Ferns		Primarily homosporous (a few heterosporous) vascular plants. Sperm motile. External water necessary for fertilization. Leaves are megaphylls that uncoil as they mature. Sporophytes and virtually all gametophytes photosynthetic. About 365 genera.	12,000
Lycophyta	Club mosses		Homosporous or heterosporous vascular plants. Sperm motile. External water necessary for fertilization. Leaves are microphylls. About 12–13 genera.	1,000
Coniferophyta	Conifers (including pines, spruces, firs, yews, redwoods, and others)		Heterosporous seed plants. Sperm not motile; conducted to egg by a pollen tube. Leaves mostly needle-like or scale-like. Trees, shrubs. About 50 genera.	550
Cycadophyta	Cycads		Heterosporous vascular seed plants. Sperm flagellated and motile but confined within a pollen tube that grows to the vicinity of the egg. Palm-like plants with pinnate leaves. Secondary growth slow compared with that of the conifers. Ten genera.	100
Gnetophyta	Gnetophytes		Heterosporous vascular seed plants. Sperm not motile; conducted to egg by a pollen tube. The only gymnosperms with vessels. Trees, shrubs, vine. Three very diverse genera (*Ephedra, Gnetum, Welwitschia*).	70
Arthrophyta	Horsetails		Homosporous vascular plants. Sperm motile. External water necessary for fertilization. Stems ribbed, jointed, either photosynthetic or nonphotosynthetic. Leaves scale-like, in whorls, nonphotosynthetic at maturity. One genus.	15
Psilophyta	Whisk ferns		Homosporous vascular plants. Sperm motile. External water necessary for fertilization. No differentiation between root and shoot. No leaves; one of the two genera has scale-like enations and the other leaf-like appendages.	6
Ginkgophyta	*Ginkgo*		Heterosporous vascular seed plants. Sperm flagellated and motile but conducted to the vicinity of the egg by a pollen tube. Deciduous tree with fan-shaped leaves that have evenly forking veins. Seeds resemble a small plum with fleshy, ill-scented outer covering.	1

Seedless Vascular Plants

The earliest vascular plants lacked seeds, which, as we previously indicated, are characteristic of the more advanced plant phyla. Members of four phyla of living vascular plants lack seeds, as do at least three other phyla known only from fossils. We begin our discussion of seedless vascular plants with the most familiar phylum, the ferns.

Pterophyta: The Ferns

Ferns are the most abundant group of seedless vascular plants, with about 12,000 living species. The fossil record indicates that they originated during the Devonian period about 350 million years ago and became abundant and varied in form during the next 50 million years. Their apparent ancestors, known as *preferns*, had no broad leaves and were established on land as much as 375 million years ago.

Today, ferns flourish in a wide range of habitats throughout the world; about three-fourths of the species, however, occur in the tropics. The conspicuous sporophytes, with which most of us are acquainted, may be less than a centimeter in diameter—as seen in small aquatic ferns such as *Azolla*. Most ferns are bigger, with the largest tree ferns having trunks more than 24 meters tall and leaves up to 5 meters or more long (figure 33.12). The sporophytes and the smaller gametophytes, which seldom are more than 6 millimeters in diameter, are both photosynthetic. The fern life cycle differs from that of a moss primarily in the much greater development, independence, and dominance of the fern's sporophyte. The fern's sporophyte is much more complex than that of the moss's; the fern sporophyte has vascular tissue and well-differentiated roots, stems, and leaves.

FIGURE 33.12
A tree fern (phylum Pterophyta) in the forests of Malaysia. The ferns are by far the largest group of seedless vascular plants.

Fern sporophytes typically have a horizontal underground stem called a *rhizome*, with roots emerging from the sides. The leaves, referred to as *fronds*, usually develop at the tip of the rhizome as tightly rolled up coils ("fiddleheads") that unroll and expand. Many fronds are highly dissected and feathery, making the ferns that produce them prized as ornamentals. Some ferns, such as *Marsilea*, have fronds that resemble a four-leaf clover, but *Marsilea* fronds still begin as coiled fiddleheads. Other ferns produce a mixture of photosynthetic fronds and nonphotosynthetic reproductive fronds that tend to be brownish in color.

Nearly all ferns are homosporous, producing distinctive, stalked *sporangia*, usually in clusters called *sori*, typically on the backs of the fronds. Most sori are protected during their development by a transparent, umbrella-like covering called an *indusium*, which later shrivels, exposing the sporangia. Diploid *spore mother cells* in each sporangium undergo meiosis, producing haploid spores. At maturity, the spores are catapulted from the sporangium by a snapping action, and those that land in suitable damp locations may germinate, producing gametophytes known as *prothalli* (singular: *prothallus*). The prothalli, which are often heart-shaped, are only one cell thick (except in the center) and have rhizoids that anchor them to their substrate. Flask-shaped *archegonia* and globular *antheridia* are produced on either the same prothallus or on different prothalli. The sperm formed in the antheridia have flagella, with which they swim toward the archegonia when water is present. One sperm unites with the single egg toward the base of an archegonium, forming a *zygote*. The zygote then develops into a new sporophyte, completing the life cycle (figure 33.13).

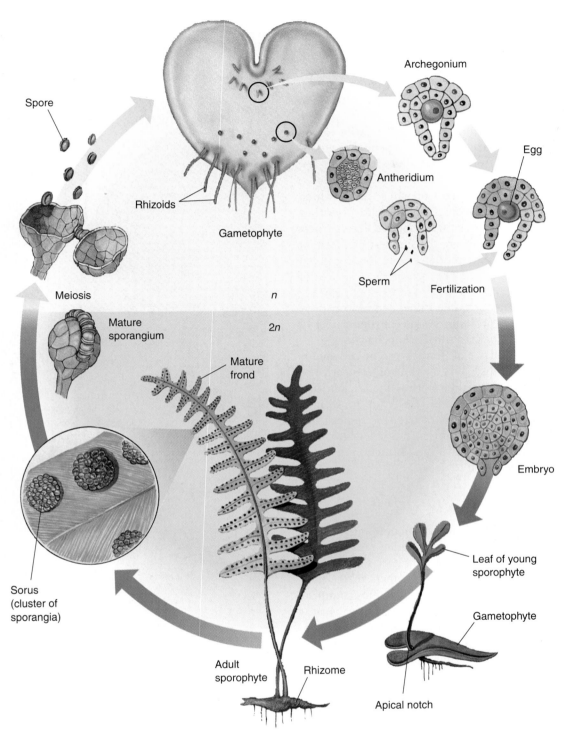

Spore

Archegonium

Rhizoids

Gametophyte

Antheridium

Egg

Sperm

Fertilization

Meiosis

n

2*n*

Mature
sporangium

Mature
frond

Embryo

Sorus
(cluster of
sporangia)

Leaf of young
sporophyte

Gametophyte

Adult
sporophyte

Rhizome

Apical notch

FIGURE 33.13
Life cycle of a typical fern.

Psilophyta: Whisk Ferns

The three other phyla of seedless vascular plants, the Psilophyta, (whisk ferns), Lycophyta (club mosses), and Arthrophyta (horsetails), have many features in common with ferns. For example, they all form antheridia and archegonia. Free water is required for the process of fertilization, during which the sperm, which have flagella, swim to and unite with the eggs. In contrast, most seed plants have non-flagellated sperm; none form antheridia, although a few form archegonia.

The origins of the two genera of whisk ferns, which occur in the tropics and subtropics, are not clear, but they are considered to be living remnants of the very earliest vascular plants. Certainly they are the simplest of all extant vascular plants, consisting merely of evenly forking green stems without roots or leaves. The two or three species of the genus *Psilotum* do, however, have tiny, green, spirally arranged, scale-like flaps of tissue called *enations* on their stems; the enations have no veins or stomata (figure 33.14). Another genus, *Tmespiteris*, has leaf-like appendages.

The gametophytes of whisk ferns were unknown for many years until their discovery in the soil beneath the sporophytes. They are essentially colorless and are less than 2 millimeters in diameter, but they can be up to 18 millimeters long. They form saprobic or parasitic associations with fungi, which furnish their nutrients. Some develop elements of vascular tissue and have the distinction of being the only gametophytes known to do so.

Lycophyta: Club Mosses

The club mosses are worldwide in distribution but are most abundant in the tropics and moist temperate regions. Several genera of club mosses, some of them tree-like, became extinct about 270 million years ago. Members of the four genera and nearly 1,000 living species of club mosses superficially resemble true mosses, but once their internal structure and reproductive processes became known it was clear that these vascular plants are quite unrelated to mosses.

Modern club mosses are either homosporous or heterosporous. The sporophytes have leafy stems that are seldom more than 30 centimeters long. In the homosporous genus *Lycopodium*, sporangia are produced either in cone-like clusters at the tips of upright stems, or on some of the upper leaves. The leaves (*microphylls*, see chapter 34) are short, linear, and arranged in whorls or tight spirals (figure 33.15). In the heterosporous genus *Selaginella*, the leaves tend to branch freely and may be flattened, forming attractive sprays in some species.

The gametophytes of *Lycopodium*, which develop in the ground, are shaped like tiny carrots with an exposed area at the top where archegonia and antheridia are produced. The male gametophytes of *Selaginella* develop within a small **microspore** and consist of little more than an antheridium. Female gametophytes of *Selaginella* develop within larger

FIGURE 33.14
A whisk fern (phylum Psilophyta). Whisk ferns have no roots or leaves.

megaspores. The female gametophytes are also relatively simple in structure, but are multicellular and usually contain several archegonia.

Many club mosses are attractive plants that were in the past used as Christmas ornaments or in floral wreaths. Several species of *Lycopodium* are now threatened with extinction and have been placed on the endangered species list. A few of the 700 species of *Selaginella* have become weeds in greenhouses. A couple of species called "resurrection plants" shrivel up when dry, but quickly unfold and turn green again when sprinkled with water.

Another group of heterosporous plants called *quillworts*, all in the genus *Isoetes*, are included in the phylum. The 60 living species of quillworts are small plants, seldom growing more than 20 centimeters tall. Their leaves are elongated microphylls that resemble small porcupine quills emerging from a relatively hard base to which roots are attached. Most occur in areas that are underwater for part of the year. They first appeared in the Cretaceous period about 130 million years ago. Modern quillworts are virtually indistinguishable in form from their ancient relatives. Their life cycle is similar to that of *Selaginella*.

Arthrophyta: Horsetails

The 15 living species of *horsetails*, also called *scouring rushes*, are all heterosporous and herbaceous. They constitute a single genus, *Equisetum*. Fossil forms of *Equisetum* extend back 300 million years to an era when some of their relatives were tree-like. Today, they are widely scattered around the world, mostly in damp places. Some that grow among the coastal redwoods of California may reach a height of 3 meters, but most are less than a meter tall (figure 33.16).

Horsetail sporophytes consist of ribbed, jointed, photosynthetic stems that arise from branching underground rhizomes with roots at their nodes. A whorl of nonphotosynthetic, scale-like leaves emerges at each node. The stems, which are hollow in the center, have silica deposits in the epidermal cells of the ribs, and the interior parts of the stems have two sets of vertical, somewhat tubular canals. The larger outer canals, which alternate with the ribs, contain air, while the smaller inner canals opposite the ribs contain water.

Horsetail species are divided into two groups. The sporophytes of one group are more or less unbranched and develop cone-shaped *strobili* at the tops of the stems. The strobili consist of spore-bearing *sporangiophores* that are produced around a central axis. The sporophytes of the other group also have terminal strobili, but they develop only on nonphotosynthetic stems; the photosynthetic stems are highly branched, and slightly resemble a horse's tail—hence, the common name "horsetail."

Horsetail spores have two ribbon-like *elaters* that uncoil and aid spore dispersal from the sporangium. The elaters curl round the spore when a pocket of humidity is encountered, thereby promoting their deposit in a damp area suitable for their germination. The gametophytes that develop from the spores vary from about 1 millimeter to 3 centimeters in diameter and are either male or bisexual; archegonia develop before the antheridia in bisexual gametophytes. The numerous flagellated sperm swim to the archegonia, where, after fertilization, the zygote develops into a new sporophyte.

Ferns and other seedless vascular plants have a much larger and more conspicuous sporophyte, with vascular tissue and well-differentiated roots, stem, and leaves.

FIGURE 33.15
The club moss *Lycopodium lucidulum* (phylum Lycophyta). Although superficially similar to the gametophytes of mosses, the conspicuous club moss plants shown here are sporophytes.

FIGURE 33.16
A horsetail, *Equisetum telmateia*, a representative of the only living genus of the phylum Arthrophyta. This species forms two kinds of erect stems; one is green and photosynthetic, and the other, which terminates in a spore-producing "cone," is mostly light brown.

Seed plants like pine trees and roses have protected embryos specialized for dispersal.

Seed Plants

Fossil evidence indicates that seeds first evolved during the Devonian period some 360 million years ago. The evidence also suggests that today's seed plants were derived from a common ancestor. The ancestors of seed plants appear to have been spore-bearing plants known as progymnosperms. Progymnosperms shared several features with modern gymnosperms, including secondary xylem and phloem, and some species had leaves. Their reproduction was very simple, however, and it is not certain which particular group of progymnosperms gave rise to seed plants.

From an evolutionary and ecological perspective, the seed represents an important advance. Developing from an ovule, it protects the embryo with specialized structures derived from the parent sporophyte. As we discussed earlier, having a seed is of great benefit to a plant that grows on land: its embryos are protected from drought and, to some extent, from predators; dispersal is enhanced; and there is no immediate need for external water prior to germination.

Seed plants produce two kinds of gametophytes—male and female, each of which consists of just a few cells. Both kinds of gametophytes develop separately within the sporophyte and are completely dependent on it for their nutrition. Male gametophytes arise from microspores, which develop into pollen grains, and the male gametophytes become mature when sperm are produced. The sperm are conveyed to the egg in the female gametophyte without using free water. A female gametophyte develops from a megaspore produced within an ovule. In flowering plants (angiosperms), the ovules are completely enclosed within diploid sporophyte tissue (ovaries) at the time pollen is transferred from its source to the vicinity of the female gametophytes. The transfer of pollen by insects, wind, or other agents is referred to as pollination. In gymnosperms (mostly cone-bearing seed plants), the ovules are not completely enclosed by sporophyte tissue at the time of pollination.

A common ancestor with seeds gave rise to the gymnosperms and the angiosperms, or flowering plants.

A Vocabulary of Plant Terms

androecium The stamens of a flower.

anther The pollen-producing portion of a stamen.

antheridium The male sperm-producing structure found in the gametophytes of seedless plants and certain fungi.

archegonium The multicellular egg-producing structure in the gametophytes of seedless plants and gymnosperms.

carpel A leaf-like organ in angiosperms that encloses one or more ovules; a unit of a gynoecium.

double fertilization The process, unique to angiosperms, in which one sperm fuses with the egg, forming a zygote, and the other sperm fuses with the two polar nuclei, forming the primary endosperm nucleus.

endosperm The usually triploid food supply of some angiosperm seeds.

filament The stalk-like structure that supports the anther of a stamen.

gametophyte The multicellular, haploid phase of a plant life cycle in which gametes are produced.

gynoecium The pistil(s) of a flower.

heterosporous Refers to a vascular plant that produces two types of spores: microspores and megaspores.

homosporous Refers to a vascular plant that produces only one type of spore.

integument The outer layer of an ovule; integuments become the seed coat of a seed.

micropyle The opening in the ovule integument through which the pollen tube grows.

nucellus The tissue of an ovule in which an embryo sac develops.

ovary The basal, swollen portion of a pistil (gynoecium); it contains the ovules and develops into the fruit.

ovule A seed plant structure within an ovary; it contains a female gametophyte surrounded by the nucellus and one or two integuments. At maturity, an ovule becomes a seed.

pistil An organ in the flower consisting of one or more fused carpels.

pollen grain A binucleate seed plant structure produced from a microspore in a microsporangium.

pollination The transfer of a pollen grain from an anther to a stigma in angiosperms, or to the vicinity of the ovule in gymnosperms.

primary endosperm nucleus The triploid nucleus resulting from the fusion of a single sperm with the two polar nuclei.

seed A reproductive structure that develops from an ovule in seed plants. It consists of an embryo and a food supply surrounded by a seed coat.

seed coat The protective layer of a seed; it develops from the integument or integuments.

spore A haploid reproductive cell, produced when a diploid spore mother cell undergoes meiosis; it gives rise by meiosis to a gametophyte.

sporophyte The multicellular, diploid phase of a plant life cycle; it is the generation that ultimately produces spores.

stamen A unit of an androecium; it consists of a pollen-bearing anther and usually a stalk-like filament.

stigma The uppermost pollen-receptive portion of a gynoecium.

Gymnosperms

In all of the gymnosperm groups recognized in this book (conifers, cycads, gnetophytes, and *Ginkgo;* figure 33.17) the **ovule,** which becomes a seed, rests exposed on a scale and is not completely enclosed by sporophyte tissues at the time of pollination. The name gymnosperm combines the Greek root *gymnos,* or "naked," with *sperma,* or "seed." In other words, gymnosperms are naked-seeded plants. However, although the ovules are naked at the time of pollination, the seeds of gymnosperms are sometimes enclosed by other sporophyte tissues by the time they are mature. There are several groups of living gymnosperms, none of which are directly related to one another, but all of which lack the special characteristics of angiosperms.

Gymnosperm stems may be green and pencil-thin, vine-like, bowl-shaped, or trunks that range from slender shoots to huge specimens up to 7 meters in diameter. Gymnosperm leaves may be needle-like, fan-shaped, scale-like, strap-like, or in the form of a large frond. Details of reproduction vary somewhat in gymnosperms, and their forms vary greatly. For example, cycads and *Ginkgo* have motile sperm, even though the sperm are borne within a pollen tube, while many others have sperm with no flagella. The female cones range from tiny woody structures weighing less than 25 grams with a diameter of a few millimeters, to massive structures weighing more than 45 kilograms and growing to a length of more than a meter.

(a)

(b)

(c)

(d)

FIGURE 33.17
Representatives of the four phyla of gymnosperms with living members. (a) Slash pines, *Pinus palustris,* in Florida, representative of the Coniferophyta, the largest phylum of gymnosperms.
(b) An African cycad, *Encephalartos ferox.*
(c) *Welwitschia mirabilis,* one of the three genera of Gnetophytes. (d) Maidenhair tree, *Ginkgo biloba,* the only living representative of the phylum Ginkgophyta.

Coniferophyta: The Conifers

The most familiar gymnosperms are conifers (phylum Coniferophyta), which include pines (see figure 33.17*a*), spruces, firs, cedars, hemlocks, yews, larches, cypresses, and others. The Coastal Redwood (*Sequoia sempervirens*), a conifer native to northwestern California and southwestern Oregon, is the tallest living vascular plant; it may attain nearly 100 meters (300 feet) in height. Another conifer, the bristlecone pine (*Pinus longaeva*) of the White Mountains of California is the oldest living tree; one is 4900 years of age. Conifers are found in the colder temperate and sometimes drier regions of the world, especially in the northern hemisphere. They are sources of timber, paper, resin, turpentine, and other economically important products.

Pines. More than 100 species of pines exist today, all native to the northern hemisphere, although the range of one species does extend a little south of the equator. Pines and spruces are members of the vast coniferous forests that lie between the arctic tundra and the temperate deciduous forests and prairies to their south. During the past century, various pines have also been extensively planted in the southern hemisphere.

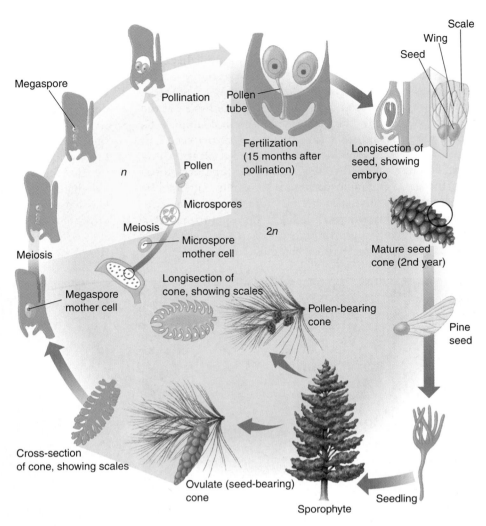

FIGURE 33.18
Life cycle of a pine.

Pines have tough, needle-like leaves produced mostly in clusters of two to five. The leaves, which have a thick cuticle and recessed stomata, represent an evolutionary adaptation for retarding water loss. This is important because many of the trees grow in areas where the topsoil is frozen for part of the year, making it difficult for the roots to obtain water. The leaves and other parts of the sporophyte also have canals into which surrounding cells secrete resin. The resin, which appears to be secreted in response to wounding, deters insect and fungal attacks. The resin of certain pines is harvested commercially for its volatile liquid portion, called *turpentine*, and for the solid *rosin*, which is used with stringed instruments and in baseball. The wood of pines consists primarily of tracheids and differs from that of woody flowering plants in having neither vessel members nor fibers. The absence of fibers causes the wood to be "soft," as opposed to that of broad-leaved trees, which is said to be "hard." The thick bark of pines represents another adaptation for surviving fires and subzero temperatures.

As mentioned earlier, all seed plants are heterosporous (figure 33.18). The male gametophytes of pines develop from pollen grains, which are produced in male cones that develop in clusters of 30 to 70, typically at the tips of the lower branches; there may be hundreds of such clusters on any single tree. The male cones generally are 1 to 4 centimeters long and consist of small, papery scales arranged in a spiral or in whorls. A pair of sacs, called microsporangia, form within each scale. Numerous microspore mother cells in the microsporangia undergo meiosis, each becoming four microspores. The microspores develop into four-celled pollen grains with a pair of air sacs that gives them added buoyancy after they are released into the air. A single cluster of male pine cones may produce more than 1 million pollen grains.

Female cones typically are produced on the upper branches of the same tree that produces male cones. Female cones are larger than male cones, and their scales become woody. Two ovules develop toward the base of each scale. Each ovule contains a megasporangium embedded in nutritive tissue called the **nucellus.** The nucellus itself is completely surrounded by a thick integument that has a small opening (the **micropyle**) toward one end. One of the layers of the integument later becomes the seed coat. A single megaspore mother cell within each megasporangium undergoes meiosis, becoming a row of four megaspores. Three of the megaspores break down, but the remaining one, over the better part of a year, slowly develops into a female gametophyte. The female gametophyte at maturity may consist of thousands of cells, with two to six archegonia formed at the micropylar end. Each archegonium contains an egg so large it can be seen without a microscope.

Female cones usually take two or more seasons to mature. At first they may be reddish or purplish in color, but they soon turn green, and during the first spring, the scales spread apart. While the scales are open, pollen grains carried by the wind drift down between them, some catching in sticky fluid oozing out of the micropyle. As the sticky fluid evaporates, the pollen grains are slowly drawn down through the micropyle to the top of the nucellus, and the scales close shortly thereafter. The archegonia and the rest of the female gametophyte are not mature until about a year later. While the female gametophyte is developing, a pollen tube emerges from a pollen grain at the bottom of the micropyle and slowly digests its way through the nucellus to the archegonia. While the pollen tube is growing, one of the pollen grain's four cells, the generative cell, divides by mitosis, with one of the resulting two cells dividing once more. These last two cells function as sperm. The germinated pollen grain with its pollen tube and two sperm is the mature male gametophyte.

About 15 months after pollination the pollen tube reaches an archegonium, and discharges its contents into it. One sperm unites with the egg, forming a zygote. The other sperm and cells of the pollen grain degenerate. The zygote develops into an embryo within a seed. After dispersal and germination of the seed, the young sporophyte of the next generation grows into a tree.

Cycadophyta: Cycads

Cycads are slow-growing gymnosperms of tropical and subtropical regions. The sporophytes of most of the 100 known species resemble palm trees (see figure 33.17b) with trunks that can attain heights of 15 meters or more. Unlike palm trees—which are flowering plants—cycads produce cones and have a life cycle similar to that of pines. The female cones, which develop upright among the leaf bases, are huge in some species and can weigh up to 45 kilograms. The sperm of cycads, although conveyed to an archegonium by a pollen tube, have thousands of spirally arranged flagella. Several species are facing extinction in the wild and soon may exist only in botanical gardens. Florida arrowroot starch used in infant cookies was obtained from the stems of a cycad, but the slow growth of the plant eventually made this commercial use unprofitable.

Gnetophyta: Gnetophytes

There are three genera and about 70 living species of Gnetophyta. Gnetophytes are the closest living relatives of angiosperms and probably share a common ancestor with that group. They are the only gymnosperms with vessels in their xylem—a common feature in angiosperms. The members of the three genera differ greatly from one another in form. One of the most bizarre of all plants is *Welwitschia*, which occurs in the Namib and Mossamedes deserts of southwestern Africa (see figure 33.17c). The stem is shaped like a large, shallow cup that tapers into a taproot below the surface. It has two strap-shaped, leathery leaves that grow continuously from their base, splitting as they flap in the wind. The reproductive structures of *Welwitschia* are cone-like, appear toward the bases of the leaves around the rims of the stems, and are produced on separate male and female plants.

More than half of the gnetophyte species are in the genus *Ephedra*, which is common in arid regions of the western United States and Mexico. The plants are shrubby, with stems that superficially resemble those of horsetails as they are jointed and have tiny, scale-like leaves at each node. Male and female reproductive structures may be produced on the same or different plants. The drug ephedrine, widely used in the treatment of respiratory problems, was in the past extracted from Chinese species of *Ephedra*, but it has now been largely replaced with synthetic preparations. Mormon tea is brewed from *Ephedra* stems in the southwestern United States.

The best known species of *Gnetum* is a tropical tree, but most species are vine-like. All species have broad leaves similar to those of angiosperms. One *Gnetum* species is cultivated in Java for its tender shoots, which are cooked as a vegetable.

Ginkgophyta: Ginkgo

The fossil record indicates that members of the Ginkgo family were once widely distributed, particularly in the Northern Hemisphere; today only one living species, the maidenhair tree (*Ginkgo biloba*), remains. The tree, which sheds its leaves in the fall, was first encountered by Europeans in cultivation in Japan and China; it apparently no longer exists in the wild (see figure 33.17d). The common name comes from the resemblance of its fan-shaped leaves to the leaflets of maidenhair ferns. Like the sperm of cycads, those of *Ginkgo* have flagella. The *Ginkgo* is diecious, that is, the male and female reproductive structures of *Ginkgo* are produced on separate trees. The fleshy outer coverings of the seeds of female *Ginkgo* plants exude the foul smell of rancid butter caused by butyric and isobutyric acids. In the Orient, however, the seeds are considered a delicacy. In western countries, because of the seed odor, male plants vegetatively propagated from shoots are preferred for cultivation. Because it is resistant to air pollution, *Ginkgo* is commonly planted along city streets.

Gymnosperms are mostly cone-bearing seed plants. In gymnosperms, the ovules are not completely enclosed by sporophyte tissue at pollination.

Angiosperms

The 240,000 known species of flowering plants (figure 33.19) are called angiosperms because their ovules, unlike those of gymnosperms, are enclosed within other tissues at the time of pollination. The name *angiosperm* derives from the Greek words *angeion*, "vessel," and *sperma*, "seed." The "vessel" in this instance refers to the carpel, which is like an inrolled leaf with seeds along its margins. A seed develops from an ovule within a carpel, which is part of an ovary (gynoecium) that develops into a fruit.

Flowering plants (Anthophyta) exhibit an almost infinite variety of shapes, sizes, and textures. They vary, for example, from the huge Tasmanian *Eucalyptus* trees, which have nearly as much mass as the giant redwoods, to the tiniest duckweeds, which are less than 1 millimeter long. In addition to the typical flattened green leaves with which everyone is familiar, flowering plant leaves may be succulent, floating, submerged, cup-shaped, spine-like, scale-like, feathery, papery, hairy, or insect-trapping, and of almost any color. Some are so tiny one needs a microscope to examine them, while others, such as those of the Seychelles Island palm, can be up to six meters long. Their flowers vary from the simple blossoms of buttercups to the extraordinarily complex flowers of some orchids, which may lure their pollinators with drugs, forcibly attach bags of pollen to their bodies, or dunk them in fluid they secrete. The flowers may weigh less than 1 gram and remain functional for only a few minutes, or they can weigh up to 9 kilograms and be functional for months. Plants of several families are parasitic or partially parasitic (e.g., dodder, or mistletoe) on other plants, or *mycotrophic* (deriving their nutrients from fungi that form a mutualism with plant roots). Others, such as many orchids, are *epiphytic* (attached to other plants, with no roots in the ground, and not in any way parasitic).

The Structure of Flowers

Flowers are considered to be modified stems bearing modified leaves. Regardless of their size and shape, they all share certain features (figure 33.20). Each flower originates as a **primordium** that develops into a bud at the end of a stalk called a **pedicel.** The pedicel expands slightly at the tip into a base, the **receptacle,** to which the remaining flower parts are attached. The other flower parts typically are at-

FIGURE 33.19
A flowering plant. Wild geranium, *Geranium maculatum.*

tached in circles called *whorls.* The outermost whorl, called the **calyx,** is composed of **sepals.** In most flowers there are three to five sepals, which are green and somewhat leaf-like; they often function in protecting the immature flower and in some species may drop off as the flower opens. The next whorl, called the **corolla,** consists of **petals** that more often than not are colored and attract pollinators such as insects and birds. The petals, which commonly number three to five, may be separate, fused together, or missing altogether—particularly in wind-pollinated flowers. The calyx and corolla may both be petal-like in some flowers; these two outer whorls together are called a **perianth.**

The third whorl consists of **stamens,** collectively called the **androecium,** a term derived from the Greek words *andros,* "male," and *oikos,* "house." Each stamen consists of a pollen-bearing **anther** and a stalk called a **filament,** which may be missing in some flowers. The **gynoecium,** consisting of one or more **carpels**—traditionally called **pistils**—is at the center of the flower. The term *gynoecium* derives from the Greek words *gynos,* which means "female," and *oikos,* or "house." The first pistil is believed to have been formed from a leaf-like structure with ovules along its margins. The edges of the blade then rolled inward and fused together, forming a carpel. The name *pistil* is derived from its resemblance in shape to the pestles used to grind powders in mortars. Primitive flowers can have several to many separate pistils (carpels), but in most flowers, two to several carpels are fused together into a compound pistil. Such a pistil can be seen in a tomato sliced in half; each segment represents one carpel of the compound pistil.

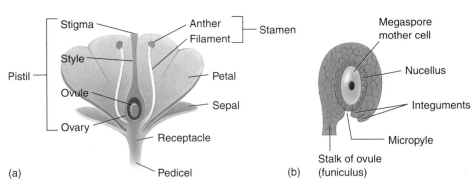

(a)   (b)

FIGURE 33.20
Diagram of an angiosperm flower. (a) The main structures of the flower are labeled. (b) Details of an ovule. The ovary as it matures will become a fruit; as the ovule's outer layers (integuments) mature, they will become a seed coat.

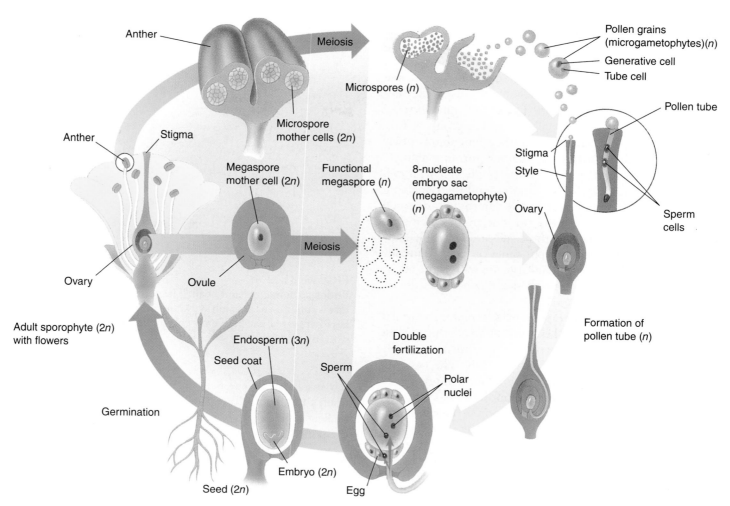

FIGURE 33.21
Life cycle of an angiosperm.

A pistil has three major regions. The **ovary** is the swollen base, which contains from one to hundreds of **ovules;** the ovary later develops into a **fruit.** The tip of the pistil, which may be pad-like, come to a point, or be divided, is called a **stigma.** Most stigmas are sticky or feathery, causing pollen grains that land on them to adhere usually. Typically there is a neck or stalk called a **style** connecting the stigma and the ovary; in some flowers, the style may be very short or even missing. Many flowers have nectar-secreting glands called *nectaries,* often located toward the base of the ovary. Nectar is a fluid containing sugars, amino acids, and other molecules; it plays an important role in attracting insects, birds, and other animals to flowers.

At the start of the twentieth century, many botanists believed that evolution proceeded only from the simple to the more complex, and that angiosperms evolved from gymnosperms. Plants whose flowers are missing petals and sepals, like the wind-pollinated flowers of oaks, were considered primitive, while more complex flower structures, like daisies, were considered advanced. As knowledge increased and investigative techniques became more sophisticated it

was realized that floral simplification can also be an evolutionary advancement; some flowers with missing parts now are no longer considered primitive. Botanists also concluded that angiosperms, although they do have seeds, could not have evolved directly from any of the living gymnosperm phyla because they differ in too many important ways. If we knew exactly what the ancestor of the angiosperm was, however, it probably would be classified as a gymnosperm anyway, because it must have had naked ovules and formed seeds. Enclosed seeds define an angiosperm, and seeds presumably must originally have been present—and naked—for this form of enclosure to have evolved.

The Angiosperm Life Cycle

While a flower bud is developing, a single megaspore mother cell in the ovule undergoes meiosis, producing four megaspores (figure 33.21). In most flowering plants, three of the megaspores soon disappear while the nucleus of the remaining one divides mitotically, and the cell slowly expands until it becomes many times its original size. While this expansion is occurring, each of the daughter nuclei divide again, and then once more, resulting in eight haploid nuclei arranged in two groups of four. At the same time, two layers of the ovule, the **integuments,** differentiate and become the *seed coat* of a seed. The integuments, as they develop, leave a small gap or pore at one end—the *micropyle* (see figure 33.20). One nucleus from each group of four migrates toward the center, where they function as **polar nuclei.** Polar nuclei may fuse together, forming a single diploid nucleus, or they may form a single cell with two haploid nuclei. Cell walls also form around the remaining nuclei. In the group closest to the micropyle, one cell functions as the **egg;** the other two nuclei are called **synergids.** At the other end, the three cells are now called **antipodals;** they have no apparent function and eventually break down and disappear. The large sac with eight nuclei in seven cells is called an **embryo sac;** it constitutes the female gametophyte.

While the female gametophyte is developing, a similar but less complex process takes place in the anthers. Most anthers have patches of tissue (usually four) that eventually become chambers lined with nutritive cells. The tissue in each patch is composed of many diploid microspore mother cells that undergo meiosis more or less simultaneously, each producing four microspores. The four microspores at first remain together as a quartet or tetrad, and the nucleus of each microspore divides once; in most species the microspores of each quartet then separate. At the same time, a two-layered wall develops around each microspore. As the anther matures, the wall between adjacent pairs of chambers breaks down, leaving two larger sacs. At this point, the binucleate microspores have become **pollen grains.** The outer pollen grain wall layer, called the *exine,* often becomes beautifully sculptured, and it contains chemicals that may react with others in a stigma to signal whether or not development of the male gametophyte should proceed to completion. The exine also has areas called *apertures,* through which a pollen tube may later emerge. Most pollen grains have three apertures, but others may have one to many.

Pollination is simply the mechanical transfer of pollen from its source (an anther) to a receptive area (the stigma of a flowering plant). Most pollination takes place between flowers of different plants and is brought about by insects, wind, water, gravity, bats, and other animals. In a few instances, however, a pollen grain may be deposited directly on the stigma of its own flower, and self-pollination occurs. Pollination may or may not be followed by *fertilization,* depending on the genetic compatibility of the pollen grain and the flower on whose stigma it has landed. If the stigma is receptive, the pollen grain's dense cytoplasm absorbs substances from the stigma and bulges through an aperture. The bulge develops into a *pollen tube* that responds to chemicals released by the embryo sac. It follows a diffusion gradient of the chemicals and grows down through the style and into the micropyle. The pollen tube usually takes several hours to two days to reach the micropyle, but in a

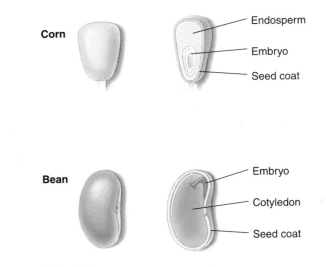

FIGURE 33.22
Endosperm in corn and bean. The corn kernel has endosperm that is still present at maturity, while the endosperm in the bean has disappeared; the bean embryo's cotyledons take over food storage functions.

few instances, it may take up to a year. One of the pollen grain's two nuclei, the *generative nucleus,* lags behind. This nucleus divides, either in the pollen grain or, more often, in the pollen tube, producing two sperm nuclei. Unlike sperm in mosses, ferns, and some gymnosperms, the sperm of flowering plants have no flagella. At this point, the pollen grain with its tube and sperm has become a mature male gametophyte.

As the pollen tube enters the embryo sac, it destroys a synergid in the process and then discharges its contents. Both sperm are functional, and an event called **double fertilization,** unique to angiosperms, follows. One sperm unites with the egg and forms a zygote, which develops into an embryo sporophyte plant. The other sperm and the two polar nuclei unite, forming a triploid primary endosperm nucleus. The primary endosperm nucleus begins dividing rapidly and repeatedly, becoming triploid **endosperm** tissue that may soon consist of thousands of cells. Endosperm tissue can become an extensive part of the seed in grasses such as corn. But in most flowering plants, it provides nutrients for the embryo that develops from the zygote; in many species, such as peas and beans, it disappears completely by the time the seed is mature (figure 33.22). Following double fertilization, the integuments harden and become the seed coat of a seed. The haploid cells remaining in the embryo sac (antipodals, synergid, tube nucleus) degenerate.

A mature angiosperm male gametophyte consists of a germinated pollen grain with two sperm and a tube nucleus.

Angiosperms are characterized by ovules that at pollination are enclosed within an ovary at the base of a pistil—a structure unique to the phylum; a fruit develops from the ovary.

33.1 **Plants are terrestrial, with life cycles that alternate between haploid and diploid.**

- Plants are derived from an aquatic ancestor, a multicellular green algae, but the evolution of their conducting tissues, cuticle, stomata, and seeds has made them progressively less dependent on external water for reproduction. The oldest plant fossils date from the Silurian period, some 330 million years ago.

- The evidence that plants are derived from green algae lies in their common photosynthetic pigments (chlorophylls *a* and *b*, carotenoids); chief storage product (starch); cellulose-rich cell walls (in some green algae only); and cell division by means of a cell plate (in certain green algae only).

- All plants have a life cycle involving an alternation of generations in which haploid gametophytes alternate with diploid sporophytes. The spores that sporophytes form as a result of meiosis grow into gametophytes, which produce gametes—sperm and eggs—by mitosis.

33.2 **Nonvascular plants like mosses are relatively unspecialized.**

- The three phyla of plants that lack well-developed vascular tissue are the simplest in structure and have been grouped as bryophytes. This grouping does not reflect a common ancestry or close relationship between them.

- Sporophytes of mosses, liverworts, and hornworts are usually nutritionally dependent on the gametophytes. They are mostly brown or straw-colored at maturity.

The gametophytes, which are more conspicuous, are green and nutritionally independent of the sporophytes.

33.3 **Seedless vascular plants like ferns have well-developed conducting tissues.**

- Nine of the 12 plant phyla contain vascular plants, which have two kinds of well-defined conducting tissues: xylem, which is specialized to conduct water and dissolved minerals; and phloem, which is specialized to conduct the food molecules the plants manufacture.

- In ferns and other seedless vascular plants, the sporophyte generation is dominant. The fern sporophyte has vascular tissue and well-differentiated roots, stems, and leaves.

33.4 **Seed plants like pine trees and roses have protected embryos specialized for dispersal.**

- Seeds were an important evolutionary advance. In seed plants, the gametophytes are nutritionally dependent on the sporophytes.

- In gymnosperms, ovules are exposed directly to pollen at the time of pollination; in angiosperms, ovules are enclosed within an ovary, and a pollen tube grows from the stigma to the ovule.

- The pollen of gymnosperms is usually disseminated by wind; although some angiosperms are wind-pollinated, in many the pollen is transported by insects and other animals. The ripened ovaries of angiosperms develop into fruits. Both flowers and fruits are found only in angiosperms.

Discussing Key Terms

1. **Plants** Plants, a kingdom of predominately terrestrial organisms, arose from green algae, a predominately aquatic phylum. Living plants are grouped into 12 phyla.

2. **The transition to land** The development of features such as a cuticle, stomata, vascular tissue, seeds, and pollen made plants less dependent on water. These adaptations facilitated the transition of plants to terrestrial habitats, allowing them to flourish and expand their ranges.

3. **Alternation of generations** Plant life cycles exhibit an alternation of generations in which a haploid gametophyte generation alternates with a diploid sporophyte generation. In some phyla, such as mosses, the gametophyte is green and nutritionally independent; in others, such as seed plants, the gametophyte is enclosed within the tissues of, and nutritionally dependent on, the sporophyte.

4. **Bryophytes** The simplest plants, the liverworts and hornworts lack vascular tissue and have reduced sporophytes. Mosses also have simple structures, but many do possess simple vascular tissues.

5. **Vascular plants** Vascular plants have conducting tissues that allow water, dissolved minerals, and food molecules to move between leaves, stems, and roots.

6. **Gymnosperms and angiosperms** Seed plants consist of four phyla of gymnosperms that have living representatives, together with the angiosperms, or flowering plants. Within the seed, an efficient unit of dispersal, the young embryo is nourished and protected. Angiosperms are distinguished from the other seed plants by their possession of flowers and fruits, and by their specialized reproductive structures.

Review Questions

1. What characteristics are shared by green algae and plants? Which detail of cell division in some green algae and plants is unknown in the cells of other organisms?

2. What does the alternation of generations refer to in plants? Distinguish between sporophyte and gametophyte. With which stage is an adult animal comparable?

3. How does the appearance and nutritional status of the gametophyte differ between seedless plants and seed plants? How does the sporophyte of the mosses compare with that of vascular plants in general?

4. Distinguish between male gametophytes and female gametophytes. Which specific haploid spores give rise to each of these? What term is applied to plants that produce two morphologically different kinds of spores? What is one that produces only one type of spore called? Where are the spores produced?

5. What is the relationship between sporophyte and gametophyte in mosses? Which is more conspicuous? Which one is dependent upon the other nutritionally?

6. There are two types of conducting tissues in vascular plants; what are they called? What do they each transport?

7. What is a seed? Why is the seed a crucial adaptation to terrestrial life?

8. Where are the archegonia and antheridia of a fern located? Which generation(s) of the fern are nutritionally independent? What is a rhizome? Where are the sporangia located on a fern?

9. Do the early stages of the familiar, woody, large pine cones produce megaspores or microspores? Are the tissues of the mature pine seed all diploid? What tissues does the seed coat surround?

10. What is the principal difference between gymnosperms and angiosperms? From what part of the flower does an angiosperm fruit develop?

11. What is meant by "double fertilization"? What two distinctly different tissues arise from this process? What is the genetic complement and function of each?

Thought Questions

1. Compare and contrast the alternation of gametophyte and sporophyte generations in mosses, ferns, pines, and flowering plants. Which generation is dominant in each case, if any? Is it haploid or diploid?

2. Why do mosses and ferns both require free water to complete their life cycles? At what stage of the life cycle is the water required? Do angiosperms also require water to complete their life cycles? What are the reasons for the difference, if any?

Internet Links

A Fern Foray
http://www.visuallink.net/fern/
This delightful site provides a great introduction to ferns, with tips on how to grow them, and a link to "FERNET>"

Meeting the Mosses
http://www.science.siu.edu/bryophytes/index.html
BRYOPHYTES is a site devoted to mosses, liverworts, and hornworts, the bryophytes. It provides infromation on the classification, structural features, natural history, ecology, and evolutionary relationships of these plants.

Cycads of the World
http://www.nd.edu/~fboze/cycads.htm
An on-line monograph telling all you ever wanted to know about cycads, ancient gymnosperms which flourished at the time of the dinosaurs.

A Plant Index
http://www.perspective.com/nature/plantae/index.html
From NATURAL PERSPECTIVE comes this well-organized overview of the plant kingdom.

For Further Reading

Gensel, P. G. and H. N. Andrews: "The Evolution of Early Land Plants," *American Scientist*, 1987, vol. 75, pages 478–89. Excellent, well-illustrated account of the earliest plants.

Kenrick, P. and P. R. Crane: "The Origin and Early Evolution of Plants on Land, *Nature*, September 4, 1997, vol. 389, pages 33–39. A review of the evolution of the early terrestrial plants based on new paleobotanical discoveries and advances in the systematics of plants.

Niklas, K. J.: "Aerodynamics of Wind Pollination," *Scientific American*, July 1987, pages 90–95. The complex aerodynamics of pollen grains and the structures on which they land exert a high degree of influence on the apparently random process of pollen dispersal.

Norstog, K.: "Cycads and the Origin of Insect Pollination," *American Scientist*, 1987, vol. 75, pages 270–79. Studies of one of the most ancient lines of living seed plants suggest that insect pollination was established before the origin of the angiosperms.

Raven, P. H. R. F. Evert, and S. E. Eichhorn: *Biology of Plants*, 5th ed., Worth, New York, 1992. A widely-used comprehensive general botany text by one of this text's authors.

Richardson, D. H. S.: *The Biology of Mosses*, John Wiley and Sons, Inc., New York, 1981. Excellent, concise account of the mosses.

Scagel, R. F. and others: *Plants: An Evolutionary Survey*, Wadsworth, Belmont, CA, 1984. Comprehensive account of all groups of plants; highly recommended.

Stewart, W. N. and G. W. Rothwell: *Paleobotany and the Evolution of Plants*, 2d ed., Cambridge University Press, New York, 1993. A survey of fossil plants.

Vidakovic, M., transl. by M. Soljan: *Conifers*, Graficki Zavod Hrvatske, 1991. A comprehensive, well-illustrated worldwide treatment of the conifers. (U.S. distributor of the English translation: University of Arizona Press).

34

Vascular Plant Structure

Concept Outline

FIGURE 34.1
All vascular plants share certain characteristics. Vascular plants such as this tree require an elaborate system of support and fluid transport to grow this large. Smaller plants have similar (though simpler) structures.

Although the similarities between a cactus, an orchid plant, and a tree might not be obvious at first sight, most plants have a basic unity of structure (figure 34.1). This unity is reflected in how the plants are constructed; in the way they grow, manufacture, and transport their food; and in how their development is regulated. The cells and tissues of vascular plants are the major focus of this chapter. We will deal, in particular, with the structure of angiosperms, or flowering plants. Both anatomy, which focuses on structural details, and morphology, which relates to a plant's form, will concern us here. We will discuss the fundamental differences between roots and shoots—as well as the functional and structural relationships between them—and then we will elaborate on the leaves associated with shoots.

Organization of the Plant Body

A vascular plant consists of a **root system** and a **shoot system** (figure 34.2). The root system anchors the plant and penetrates the soil, from which it absorbs water and ions crucial to the plant's nutrition. The shoot system consists of the **stems** and their **leaves**. The stem serves as a framework for positioning the leaves, the principal sites of photosynthesis. The arrangement, size, and other features of the leaves are of critical importance in the plant's production of food. Flowers, other reproductive organs, and, ultimately, fruits and seeds are also formed on the shoot (see chapters 33 and 37).

Four basic types of tissues exist in plants: *meristems, ground tissue, epidermis,* and *vascular tissue*. Each of the four basic tissues has its own distinctive, functionally related cell types. Some of these cell types will be discussed later in this chapter. **Meristems** give rise to all other cells and tissues in the plant. **Ground tissue** consists primarily of thin-walled **parenchyma** cells that are initially (but briefly) more or less spherical. However, the cells, which have living protoplasts, push up against each other shortly after they are produced and assume other shapes, often ending up with 11 to 17 sides. Parenchyma cells may live for many years; they function in storage, photosynthesis, and secretion.

Epidermis, which is one cell thick in most plants, forms the outer protective covering of the plant. In young exposed parts of the plant, the epidermis is covered with a fatty **cutin** layer constituting the **cuticle;** in plants such as the desert succulents, a layer of wax may be added outside the cuticle.

Vascular tissue includes two kinds of conducting tissues: (1) **xylem,** which conducts water and dissolved minerals; and (2) **phloem,** which conducts carbohydrates—mainly sucrose—used by plants for food. The phloem also transports hormones, amino acids, and other substances that are necessary for plant growth. Xylem and phloem differ in structure as well as in function.

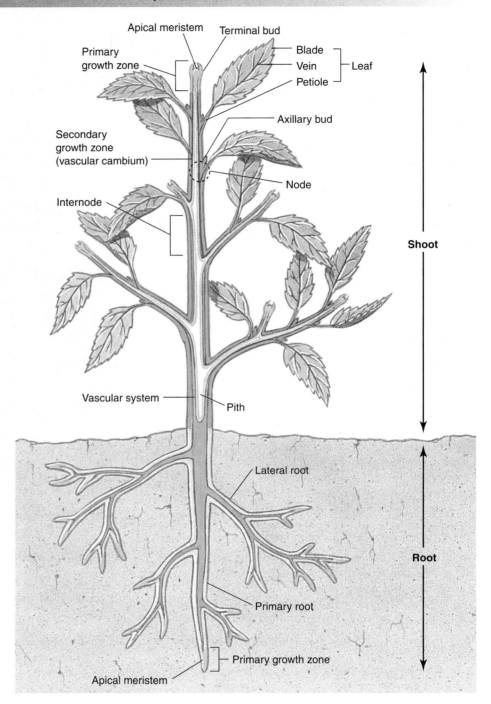

FIGURE 34.2
Diagram of a plant body. The lime green areas are zones of active elongation; secondary growth occurs in the lavender areas.

The four basic types of tissues in plants are
(1) meristems, (2) ground tissue, (3) epidermis, and
(4) vascular tissue. Vascular tissue includes xylem and
phloem.

Primary and Secondary Growth

Early vascular plants exhibited *primary growth* and later *secondary growth* from **meristems,** groups of cells that give rise to all other cell and tissue types found in plants. **Primary growth** results from cell division at the **apical meristems,** the tips of the plant body, and increases the plant's length. **Secondary growth,** which increases a plant's girth, comes from the **lateral meristems,** internal meristematic cylinders within the plant axis. In the earliest vascular plants, the vascular tissues produced by these meristems played the same conducting roles they do in contemporary vascular plants. There was no differentiation of the plant body into stems, leaves, and roots. The presence of these three kinds of organs is a property of most modern plants. It reflects increasing specialization in relation to the demands of a terrestrial existence.

With the evolution of secondary growth, vascular plants could develop thick trunks and become tree-like (figure 34.3). This evolutionary advance made possible the development of forests and the domination of the land by plants. Judging from the fossil record, secondary growth evolved independently in several groups of vascular plants by the middle of the Devonian period, some 380 million years ago.

There were two types of conducting systems in the earliest plants—systems that have become characteristic of vascular plants as a group. *Sieve-tube members* conduct carbohydrates away from areas where they are manufactured. *Vessel members* and *tracheids* are hard-walled cells that transport water and dissolved minerals up from the roots. Both kinds of cells are elongated and occur in strands. Sieve-tube members are characteristic of phloem tissue; vessel members and tracheids are characteristic of xylem tissue. In primary tissues, which result from primary growth, these two types of tissue are typically associated with each other in the same vascular strands.

Plants grow from the division of meristematic tissue. Primary growth results from cell division at the apical meristem at the tip of the plant, making the shoot longer. Secondary growth results from cell division at the lateral meristem in a cylinder encasing the shoot, and increases the shoot's girth.

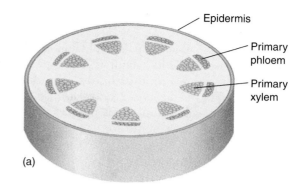

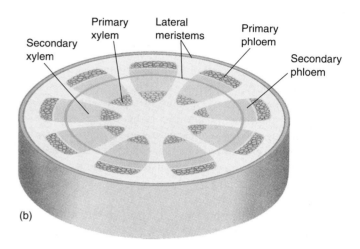

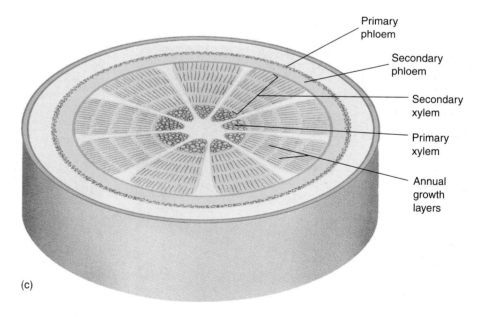

FIGURE 34.3
Secondary growth. (a) Before secondary growth begins, primary tissues continue to elongate as the apical meristems produce primary growth. (b) As secondary growth begins, the lateral meristems produce secondary tissues, and the stem's girth increases. (c) In this three-year-old stem, the secondary tissues continue to widen, and the trunk has become thick and woody.

Monocots and Dicots

There are two classes of angiosperms, phylum Anthophyta: the Dicotyledonae, or **dicots** (about 175,000 species), and the Monocotyledonae, or **monocots** (about 65,000 species). Included in the dicots are the great majority of familiar angiosperms—almost all kinds of trees and shrubs, snapdragons, mints, peas, sunflowers, and other plants. Monocots include the lilies, grasses, cattails, palms, agaves, yuccas, pondweeds, orchids, and irises.

Dicots and monocots differ from one another in a number of features. Aside from the differences listed in table 34.1, monocots and dicots differ fundamentally in other ways. For example, about a sixth of all dicot species are **annuals** (plants that complete their entire growth cycle within a year); there are, however, very few annual monocots. Underground swollen storage organs, such as bulbs, occur much more frequently in monocots than they do in dicots. There are many species of woody dicots (mostly trees or shrubs), but no monocots have true wood; however, a few monocots, such as palms and bamboos, produce extra bundles of conducting tissues that give them a woody texture. Endosperm, which is usually present in mature monocot seeds, is largely absent in mature dicot seeds. Other specific differences between monocots and dicots will be mentioned throughout the remainder of this chapter and the next (figure 34.4).

Monocots and dicots are believed to have been distinct from each other since early in the history of the phylum.

Table 34.1   A Comparison of Monocots and Dicots
Monocots
1. Seed with one cotyledon ("seed leaf").
2. Leaves with more or less parallel veins.
3. Secondary meristems (cambia) present.
4. Groups of vascular tissues scattered.
5. Pollen grains mostly with one aperture.
6. Flower parts mostly in threes or multiples of three.
Dicots
1. Seed with two cotyledons ("seed leaves").
2. Leaves with a network of veins.
3. Secondary meristems (cambia) present.
4. Groups of vascular tissues in a ring.
5. Pollen grains mostly with three or more apertures.
6. Flower parts mostly in fours or fives or multiples of four or five.

Dicots are the more primitive of the two classes, with monocots apparently having derived from early dicots, possibly before the middle of the Cretaceous period.

> **Two-thirds of flowering plants are dicots, including most trees and annual plants. Monocots, which include all the grasses and grains, evolved from them.**

(a)

(b)

FIGURE 34.4
A typical dicot and a typical monocot. (a) The bright red flowers of fire-pink, *Silene virginica*, with their flower parts in fives, are typical of the dicots. (b) Flowers of *Trillium pusillum*, with flower parts in threes, are typical of the monocots. Fire-pink and trillium occur side by side in the understory of the woods throughout the temperate parts of eastern and central Canada and the United States. Fire-pink is pollinated primarily by hummingbirds. The flowers of this trillium, like those of many angiosperms, change their color after pollination, thus signaling the pollinator which flowers are still unvisited and have a supply of food.

Meristems

Both apical and lateral meristems consist of unspecialized cells that are mostly box-like in form. After a meristematic cell divides, one of the two cells produced becomes part of the plant body; the other remains meristematic. The cell that becomes part of the plant body may divide again before differentiating and developing into a specific type of cell.

Apical Meristems Produce Primary Growth

The apical meristem gives rise to three types of embryonic tissue systems called **primary meristems.** Cell division continues in these partly differentiated tissues as they develop into the primary tissues of the plant body (figure 34.5). The three primary meristems are the **protoderm,** which becomes the epidermis; the **procambium,** which produces primary vascular tissues (primary xylem and primary phloem); and the **ground meristem,** which differentiates further into ground tissue, which is composed of parenchyma cells. In some plants, such as horsetails, **intercalary meristems** arise in stem internodes, adding to the internode lengths.

Apical meristems are located at the tips of stems (figure 34.6) and at the tips of roots, just behind the root cap. The plant tissues that result from primary growth are called **primary tissues.** During periods of growth, the cells of apical meristems divide and continually add more cells to the tips of a seedling's body. Thus, the seedling lengthens.

Lateral Meristems Produce Secondary Growth

Secondary growth causes a plant shoot to increase in width. The tissues that develop as a result of secondary growth are called **secondary tissues.** Although secondary growth occurs in many nonwoody plants, its effects are most dramatic in woody plants.

Woody plants have two cylinders of actively dividing cells called **lateral meristems.** Within the bark of a woody stem is the **cork cambium,** a lateral meristem that produces the cork cells of the outer bark. Just beneath the bark is the vascular cambium, a lateral meristem that produces secondary vascular tissue. The vascular cambium forms between the xylem and phloem in vascular bundles, adding secondary vascular tissue on opposite sides of the vascular cambium. *Secondary xylem* is the main component of wood. *Secondary phloem* is very close to the outer surface of a woody stem. Removing the bark of a tree damages the phloem and may eventually kill the tree.

Meristems are actively dividing tissues responsible for both primary and secondary growth.

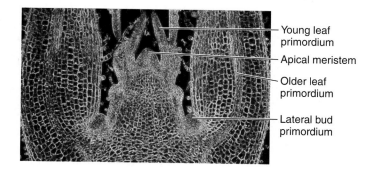

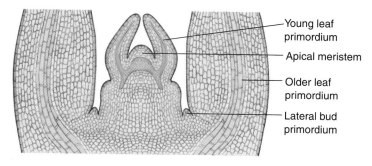

FIGURE 34.5
An apical shoot meristem. This longitudinal section through a shoot apex in *Coleus* shows the tip of a stem. Between the young leaf primordia is the apical meristem.

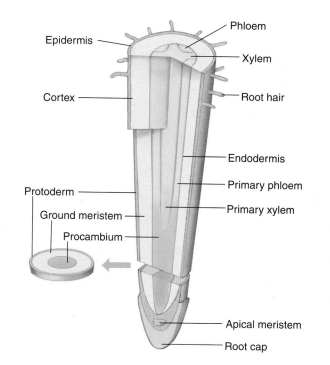

FIGURE 34.6
An apical root meristem. This diagram of meristems in the root shows their relation to the root tip.

Ground Tissue

Parenchyma

Parenchyma cells, which have large vacuoles, thin walls, and an average of 14 sides at maturity, are the most common type of plant cell. They are the most abundant cells of primary tissues and may also occur, to a much lesser extent, in secondary tissues (figure 34.7*a*). Most parenchyma cells have only primary walls, which are walls laid down while the cells are still maturing. Parenchyma are less specialized than other plant cells, although there are many variations that do have special functions such as nectar and resin secretion, or storage of latex, proteins, and metabolic wastes.

Parenchyma cells, which have functional nuclei and are capable of dividing, commonly also store food and water, and usually remain alive after they mature; in some plants (e.g., cacti), they may live to be over 100 years old. The majority of cells in fruits such as apples are parenchyma. Some parenchyma contain chloroplasts, especially in leaves and in the outer parts of herbaceous stems. Such photosynthetic parenchyma tissue is called *chlorenchyma.*

Collenchyma

Collenchyma cells, like parenchyma cells, have living protoplasts and may live for many years. The cells, which are usually a little longer than wide, have walls that vary in thickness (figure 34.7*b*). Collenchyma cells, which are relatively flexible, provide support for plant organs, allowing them to bend without breaking. They often form strands or continuous cylinders beneath the epidermis of stems or leaf petioles (stalks) and along the veins in leaves. Strands of collenchyma provide much of the support for stems in which secondary growth has not taken place. The parts of celery that we eat (petioles, or leaf stalks), have "strings" that consist mainly of collenchyma and vascular bundles (conducting tissues).

Sclerenchyma

Sclerenchyma cells have tough, thick walls; they usually lack living protoplasts when they are mature. Their secondary cell walls are often impregnated with **lignin,** a highly branched polymer that makes cell walls more rigid. Cell walls containing lignin are said to be **lignified.** Lignin is common in the walls of plant cells that have a supporting or mechanical function. Some kinds of cells have lignin deposited in primary as well as secondary cell walls.

There are two types of sclerenchyma: fibers and sclereids. **Fibers** are long, slender cells that are usually grouped together in strands. Linen, for example, is woven from strands of sclerenchyma fibers that occur in the phloem of flax. **Sclereids** are variable in shape but often branched. They may occur singly or in groups; they are not elongated, but may have various forms, including that of a star. The gritty texture of a pear is caused by groups of sclereids that occur throughout the soft flesh of the fruit (figure 34.7*c*). Both of these tough, thick-walled cell types serve to strengthen the tissues in which they occur.

Parenchyma cells are the most common type of plant cells and have various functions. Collenchyma cells provide much of the support in young stems and leaves. Sclerenchyma cells strengthen plant tissues and may be nonliving at maturity.

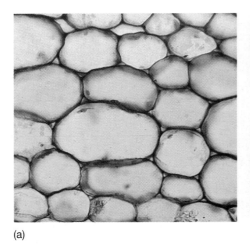
(a)

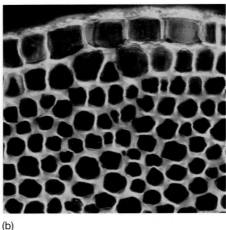
(b)

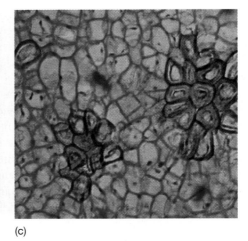
(c)

FIGURE 34.7
The three types of ground tissue. (a) Parenchyma cells. Only primary cell walls are seen in this cross-section of parenchyma cells from grass. (b) Collenchyma cells. Thickened side walls are seen in this cross-section of collenchyma cells from a young branch of elderberry (*Sambucus*). In other kinds of collenchyma cells, the thickened areas may occur at the corners of the cells or in other kinds of strips. (c) Sclereids. Clusters of sclereids ("stone cells"), stained red in this preparation, in the pulp of a pear. The surrounding thin-walled cells, stained light blue, are *parenchyma.* These sclereid clusters give pears their gritty texture.

Epidermis

Epidermal cells, which originate from the protoderm, cover all parts of the primary plant body. The exposed outer walls have a cuticle that varies in thickness, depending on the species and environmental conditions. A number of types of specialized cells occur in the epidermis, including **guard cells, trichomes,** and **root hairs.**

Guard cells are paired sausage- or dumbbell-shaped cells flanking a **stoma** (plural **stomata**), a slit-like or eye-shaped epidermal opening. Guard cells, unlike other epidermal cells, contain chloroplasts. Stomata occur in the epidermis of leaves (figure 34.8), and sometimes on other parts of the plant, such as stems or fruits. The passage of oxygen and carbon dioxide, as well as diffusion of water in vapor form, takes place almost exclusively through the stomata. There are between 1000 to more than 1 million stomata per square centimeter of leaf surface. In most plants, stomata are more numerous on the lower epidermis than on the upper epidermis of the leaf. Some plants have stomata only on the lower epidermis, and a few, such as water lilies, have them only on the upper epidermis. The stomata open and shut in response to external factors such as light, temperature, and availability of water. During periods of active photosynthesis, the stomata are open, allowing the free passage of carbon dioxide into and oxygen out of the leaf. We will consider the mechanism that governs such movements in chapter 35.

Trichomes are hair-like outgrowths of the epidermis (figure 34.9). They occur frequently on stems, leaves, and reproductive organs. A "fuzzy" or "woolly" leaf is covered with trichomes that can be seen clearly with a microscope under low magnification. Trichomes play an important role in keeping the leaf surface cool and in reducing the rate of evaporation. Trichomes vary greatly in form in different kinds of plants; some consist of a single cell, while others may consist of several cells. Some are glandular, often secreting sticky or toxic substances to deter herbivory.

Root hairs, which are tubular extensions of individual epidermal cells, occur in a zone just behind the tips of young, growing roots (see figure 34.6). Because a root hair is simply an extension and not a separate cell, there is no crosswall isolating it from the epidermal cell. Root hairs keep the root in intimate contact with the surrounding soil particles and greatly increase the root's surface area and the efficiency of absorption. As the root grows, the extent of the root hair zone remains roughly constant as root hairs at the older end slough off while new ones are produced at the other end. Most of the absorption of water and minerals occurs through root hairs, especially in herbaceous plants. In plants that have *ectomycorrhizae* (see chapter 32), root hairs are often greatly reduced in number, with fungal filaments mostly replacing them in their function or absorbing water and minerals.

Some epidermal cells are specialized for protection, others for absorption.

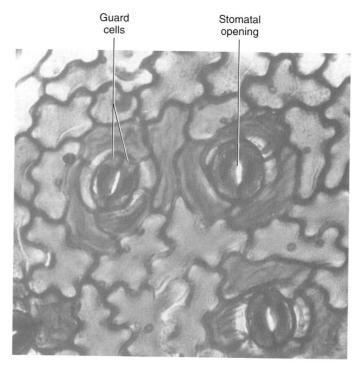

Guard cells   Stomatal opening

FIGURE 34.8
Epidermis of a leaf (250×). Stomata occur frequently among the epidermal cells of this member of the aralia family (Araliaceae). The epidermis has been peeled off the leaf and stained with a red dye.

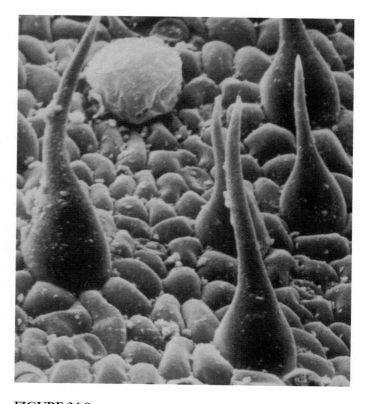

FIGURE 34.9
Trichomes. A covering of trichomes, teardrop-shaped structures above, creates a layer of more humid air near the leaf surface, enabling the plant to conserve available water supplies.

Vascular Tissue

Xylem

Xylem, the principal water-conducting tissues of plants, usually contains a combination of **vessels,** which are continuous tubes formed from dead, hollow, cylindrical cells (**vessel members**) arranged end to end, and **tracheids,** which are dead cells that taper at the ends and overlap one another (figure 34.10). In some plants, such as gymnosperms, tracheids are the only water-conducting cells present; water passes in an unbroken stream through the xylem from the roots up through the shoot and into the leaves. When the water reaches the leaves, much of it passes into a film of water on the outside of the parenchyma cells, and then it diffuses in the form of water vapor into the intercellular spaces and out of the leaves into the surrounding air, mainly through the stomata. This diffusion of water vapor from a plant is known as **transpiration.** In addition to conducting water, dissolved minerals, and inorganic ions such as nitrates and phosphates throughout the plant, xylem supplies support for the plant body.

Primary xylem is derived from the procambium, which comes from the apical meristem. *Secondary xylem* is formed by the vascular cambium, a lateral meristem that develops later. Wood consists of accumulated secondary xylem.

Vessel members are found almost exclusively in angiosperms. In primitive angiosperms, vessel members tend to resemble fibers and are relatively long. In more advanced angiosperms, vessel members tend to be shorter and wider, resembling microscopic, squat coffee cans with both ends removed. Both vessel members and tracheids have thick, lignified secondary walls and no living protoplasts at maturity. When the continuous stream of water in a plant flows through tracheids, it passes through **pits,** which are small, mostly rounded-to-elliptical areas where no secondary wall material has been deposited. The pits of adjacent cells occur opposite one another and are separated from one another by a flexible, porous *pit membrane* with a central thickened area. The pit membrane can swing from one side of the pit pair to the other and close either opening, thereby controlling the water flow. In contrast to tracheids, which have only pit-pairs in their walls, vessel members, which are joined end to end, may be almost completely open or may have bars or strips of wall material across the open ends.

Vessels appear to conduct water more efficiently than do the overlapping strands of tracheids. We know this partly because vessel members have evolved from tracheids independently in several groups of plants, suggesting that they are favored by natural selection. It is also probable that some types of fibers have evolved from tracheids, becoming specialized

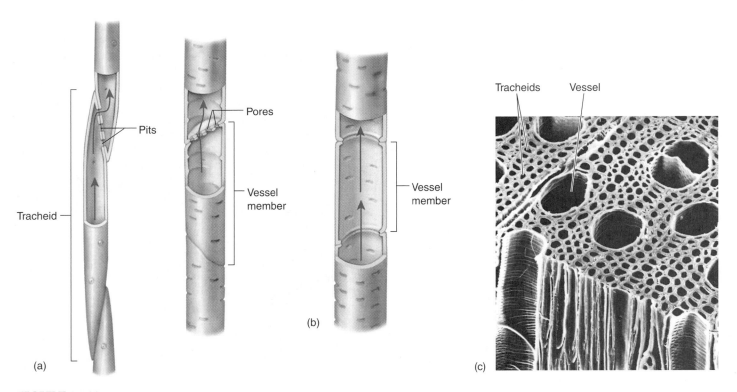

FIGURE 34.10

Comparison between vessel members and tracheids. (a) In tracheids, the water passes from cell to cell by means of pits, (b) while in vessel members, it moves by way of perforation plates or between bars of wall material. In gymnosperm wood, tracheids both conduct water and provide support; in most kinds of angiosperms, vessels are present in addition to tracheids, or present exclusively. These two types of cells conduct the water, and fibers provide additional support. (c) Scanning micrograph of the wood of red maple, *Acer rubrum* (350×).

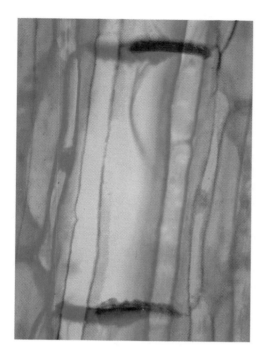
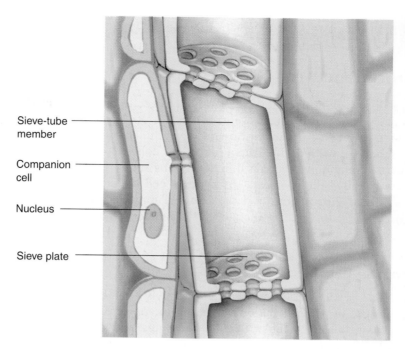

Sieve-tube
member

Companion
cell

Nucleus

Sieve plate

FIGURE 34.11

A sieve-tube member. A sieve-tube member from the phloem of squash (*Cucurbita*), connected with the cells above and below to form a sieve tube. Note the thickened end walls, which are at right angles to the sieve tube. The narrow cell with the nucleus at the left of the sieve-tube member is a companion cell.

for strengthening rather than conducting. Some ancient flowering plants have only tracheids, but virtually all modern angiosperms have vessels.

In addition to conducting cells, xylem typically includes fibers and parenchyma cells. The parenchyma cells, which are usually produced in horizontal rows called **rays** by special *ray initials* of the vascular cambium, function in lateral conduction and food storage. In cross-sections of woody stems and roots, the rays can be seen radiating out from the center of the xylem like the spokes of a wheel (see figure 34.22). Fibers are abundant in some kinds of wood, such as oak (*Quercus*), and the wood is correspondingly dense and heavy. The arrangements of these and other kinds of cells in the xylem make it possible to identify most plant genera and many species from their wood alone.

Phloem

Phloem, which is located toward the outer part of roots and stems, is the principal food-conducting tissue in vascular plants. If a plant is *girdled* (by removing a substantial strip of bark down to the vascular cambium), the plant eventually dies from starvation of the roots.

Food conduction in phloem is carried out through two kinds of elongated cells: **sieve cells** and **sieve-tube members** (figure 34.11). Seedless vascular plants and gymnosperms have only sieve cells; most angiosperms have sieve-tube members, but at least one primitive angiosperm only has sieve cells. Both types of cells have clusters of

pores known as **sieve areas.** Sieve areas are more abundant on the overlapping ends of the cells and connect the protoplasts of adjoining sieve cells and sieve-tube members. Both of these types of cells are living, but although some sieve cells in primitive plants have nuclei, most sieve cells and all sieve-tube members lack a nucleus at maturity.

In sieve-tube members, some sieve areas have larger pores and are called **sieve plates.** Sieve-tube members occur end to end, forming longitudinal series called **sieve tubes.** Sieve cells are less specialized than sieve-tube members, and the pores in all of their sieve areas are roughly of the same diameter. In an evolutionary sense, sieve-tube members are more advanced, more specialized, and, presumably, more efficient.

Each sieve-tube member is associated with an adjacent specialized parenchyma cell known as a **companion cell.** Companion cells apparently carry out some of the metabolic functions that are needed to maintain the associated sieve-tube member. Companion cells have all of the components of normal parenchyma cells, including nuclei, and their numerous **plasmodesmata** (minute holes in primary cell walls) connect their cytoplasm with that of the associated sieve-tube members. Fibers and parenchyma cells are often abundant in phloem.

Xylem conducts water and dissolved minerals from the roots to the shoots to the leaves. Phloem carries organic materials from one part of the plant to another.

Root Structure

We will now consider the three kinds of vegetative organs in plants: **roots, stems,** and **leaves.** Roots have a simpler pattern of organization and development than stems, and we will consider them first. Four zones or regions are commonly recognized in developing roots. The zones are called the **root cap,** the **zone of cell division,** the **zone of elongation,** and the **zone of maturation** (figure 34.12). In three of the zones, the boundaries are not clearly defined.

The Root Cap

The root cap, which has no equivalent in stems, is a thimble-shaped mass of parenchyma cells covering the tip of each young root. In some plants with larger roots it is quite obvious. Its most obvious function is to protect the delicate tissues behind it as growth extends the root through mostly abrasive soil particles. Golgi bodies in the outer root cap cells secrete and release a slimy substance that passes through the cell walls to the outside. The cells, which have an average life of less than a week, are constantly being replaced from the inside, forming a mucilaginous lubricant that eases the root through the soil. The slimy mass also provides a medium for the growth of beneficial nitrogen-fixing bacteria in the roots of some plants such as legumes.

A new root cap is produced when an existing one is artificially or accidentally removed from a root. Until the new root cap is mature, however, the root cap also functions in the *perception of gravity.* Amyloplasts (plastids with starch grains) have been observed to collect on the sides of cells facing the pull of gravity. When a potted plant is placed on its side, the amyloplasts drift or tumble down to the side nearest the source of gravity, and the root bends in that direction. The precise nature of the gravitational response is not known, but some evidence indicates that calcium ions in the amyloplasts influence the distribution of growth hormones in the cells.

The Zone of Cell Division

The apical meristem is shaped like an inverted, concave dome of cells and is located in the center of the root tip in the area protected by the root cap. Most of the activity in this *zone of cell division* takes place toward the edges of the dome, where the cells divide every 12 to 36 hours, often rhythmically, reaching a peak of division once or twice a day. Most of the cells are essentially cuboidal with small vacuoles and proportionately large, centrally located nuclei. The apical meristem soon subdivides into the three primary meristems previously discussed: *protoderm, procambium,* and *ground meristem.*

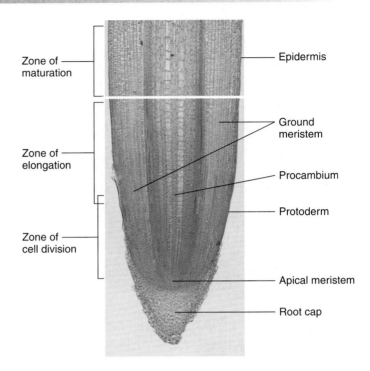

FIGURE 34.12
Root structure. A root tip in corn, *Zea mays.* This longitudinal section of a root shows the root cap, apical meristem, procambium, protoderm, epidermis, and ground meristem.

The Zone of Elongation

In the *zone of elongation,* the cells produced by the primary meristems become several times longer than wide, and their width also increases slightly. The small vacuoles present merge and grow until they occupy 90% or more of the volume of each cell. No further increase in cell size occurs above the zone of elongation, and the mature parts of the root, except for an increase in girth, remain stationary for the life of the plant.

The Zone of Maturation

The cells that have elongated in the zone of elongation become differentiated into specific cell types in the *zone of maturation.* The cells of the root surface cylinder mature into *epidermal cells,* which have a very thin cuticle. Many of the epidermal cells each develop a **root hair;** the protuberance is not separated by a crosswall from the main part of the cell and the nucleus may move into it. Root hairs, which can number over 35,000 per square centimeter of root surface and many billions per plant, greatly increase the surface area and therefore the absorptive capacity of the root. The root hairs usually are alive and functional for only a few days before they are sloughed off at the older part of the zone of maturation, while new ones are being produced toward the zone of elongation.

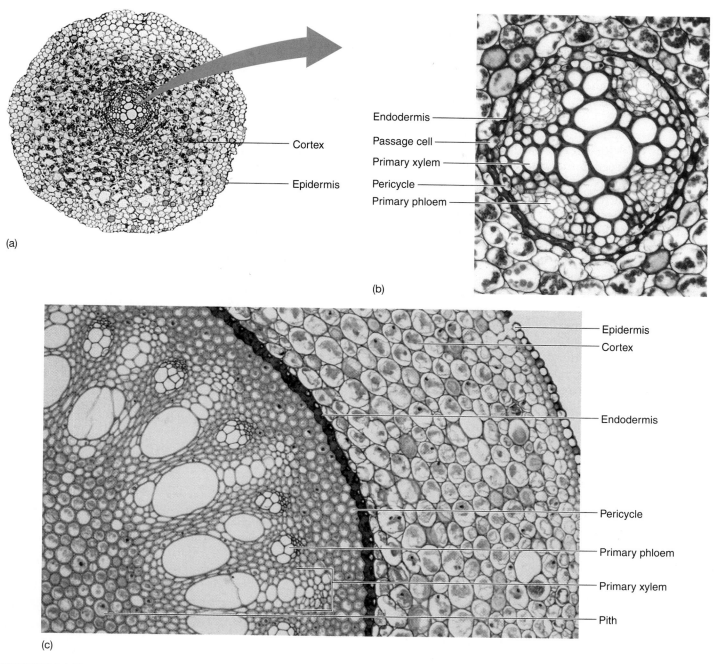

FIGURE 34.13
Cross-section of the zone of maturation of young roots. (a) Buttercup (*Ranunculus*), a dicot (40×). (b) The enlargement shows the various tissues present (600×). (c) Greenbrier (*Smilax*), a monocot (100×).

Parenchyma cells are produced by the ground meristem immediately to the interior of the epidermis. The parenchymatous tissue, known as the **cortex,** may be many cells wide and functions in food storage. The inner boundary of the cortex differentiates into a single-layered cylinder of **endodermis,** whose primary walls are impregnated with **suberin,** a fatty substance that is impervious to water. The suberin is produced in bands, called **Casparian strips** that surround each adjacent endodermal cell wall perpendicular to the root's surface. The Casparian strips are fused to the plasma membranes of the endodermal cells and prevent water and dissolved minerals from passing through them. All water leaving or entering the tissues is forced to pass through either intercellular spaces or plasmodesmata. This has a regulatory effect; it can largely exclude harmful substances from the center of the root, where conducting tissues develop. As a root expands in girth, the epidermis, cortex, and endodermis may be sloughed off in some roots, but in others, substances such as waxes, cellulose, and lignin may be added to the endodermal cell walls (figure 34.13).

All the tissues interior to the endodermis are collectively referred to as the **stele.** Immediately adjacent and interior to the endodermis is a cylinder of parenchyma cells known as the **pericycle.** Pericycle cells can divide, even after they mature, and they ultimately may give rise to *lateral* (branch) *roots* (figure 34.14) or, in dicots, to part of the *vascular cambium.*

The water-conducting cells of the *primary xylem* are differentiated as a solid core in the center of young dicot roots. In cross-section, the central core of primary xylem often is somewhat star-shaped, with one or two to several radiating arms that point toward the pericycle. In monocot and a few dicot roots, the primary xylem is in discrete **vascular bundles** arranged in a ring, which surrounds parenchyma cells, called **pith,** at the very center of the root. Branch or lateral roots normally arise in the pericycle opposite the xylem arms. **Primary phloem,** composed of cells involved in food conduction, is differentiated in discrete groups of cells between the arms of the xylem in both dicot and monocot roots.

In dicots and other plants with **secondary growth,** part of the pericycle and the parenchyma cells between the phloem patches and the xylem arms become the root vascular cambium; which starts producing **secondary xylem** to the inside and **secondary phloem** to the outside (figure 34.15). Eventually, the secondary tissues acquire the form of concentric cylinders. The primary phloem, cortex, and epidermis become crushed and are sloughed off as more secondary tissues are added. In the pericycle of woody plants, the cork cambium produces **cork cells** to the outside and **phelloderm** parenchyma to the inside. The tissue associated with the cork cambium is known as **periderm** (outer bark), which will be discussed further when we take up stems (see figure 34.23).

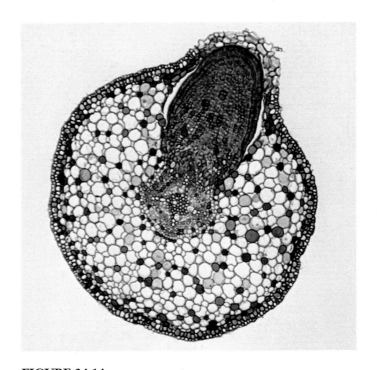

FIGURE 34.14
A lateral root. A lateral root growing out through the cortex of black willow, *Salix nigra.* Lateral roots arise in the pericycle beneath the surface of the main root, while lateral stems arise from buds located in the axils of the leaves.

Roots grow behind the root cap, the cells maturing as the root cap grows away from them. Some of them will develop into an efficient conducting system.

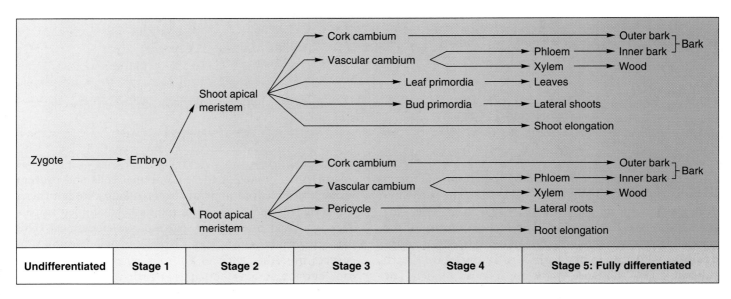

FIGURE 34.15
Stages in the differentiation of plant tissues.

Modified Roots

Most plants produce either a **taproot system** in which there is a single large root with several to many branch roots, or a **fibrous root system** in which there are many smaller roots of similar diameter. Some plants, however, have modified roots that have specific functions in addition to those of anchorage and absorption. We will now briefly discuss some of the modifications.

Aerial roots. Some plants, such as epiphytic orchids (orchids that are attached to tree branches and grow unconnected to the ground without being parasitic in any way) have roots that extend out into the air. Some aerial roots, known as *velamen roots*, have an epidermis that is several cells thick; the velamen apparently reduces water loss. These aerial roots may also be green and photosynthetic, as in the vanilla orchid. Some monocots, such as corn, produce thick roots from the lower parts of the stem; these *prop roots* grow down to the ground and brace the plants against wind. Climbing plants such as ivies also produce roots from their stems; these anchor the stems to the substrate, which may be a tree trunk or a brick wall. Any root that arises along a stem or in some place other than at the base of the plant is called an *adventitious root.* Prop roots and ivy stem anchorage roots are therefore examples of adventitious roots.

Pneumatophores. Some plants that grow in swamps and other wet places may produce spongy outgrowths called *pneumatophores* from their underwater roots (figure 34.16*a*). The pneumatophores commonly extend several centimeters above water, facilitating the oxygen supply to the roots beneath. Although pneumatophores are not produced by the common red mangrove of the tropics—which is known for its prop roots—several other species of mangrove, like the black mangrove, do. The "knees" of bald cypress (*Taxodium distichum*) were thought to function as pneumatophores, but evidence for this is inconclusive.

Contractile roots. The roots from the bulbs of lilies and of several other plants such as dandelions contract by spiralling and pull the plant a little deeper into the soil each year until they reach an area of relatively stable temperatures. The roots may contract to a third of their original length as they spiral like a corkscrew due to cellular thickening and constricting.

(a)

(b)

(c)

FIGURE 34.16
Three types of modified roots. (a) Pneumatophores (foreground) are spongy outgrowths from the roots below. (b) A water storage root weighing over 25 kilograms (60 pounds). (c) Buttress roots of a tropical fig tree.

Parasitic roots. The stems of certain plants that lack chlorophyll, such as dodder (*Cuscuta*), produce peg-like roots called *haustoria* that penetrate the host plants around which they are twined. The haustoria establish contact with the conducting tissues of the host and effectively parasitize them.

Food storage roots. The xylem of branch roots of sweet potatoes and similar plants produce at intervals many extra parenchyma cells that store large quantities of carbohydrates. Carrots, beets, parsnips, radishes, and turnips have combinations of stem and root that also function in food storage.

Water storage roots. Some members of the pumpkin family (Cucurbitaceae), especially those that grow in arid regions, may produce water-storage roots weighing 50 or more kilograms (figure 34.16*b*).

Buttress roots. Certain species of fig and other tropical trees produce huge buttress roots toward the base of the trunk, which provide considerable stability (figure 34.16*c*).

Some plants have modified roots that carry out photosynthesis, gather oxygen, parasitize other plants, store food or water, or stabilize the stem.

Stem Structure

External Form

A woody twig consists of an axis to which leaves are attached. The leaves may be arranged in a spiral around the stem, or they may be in pairs opposite one another; they also may occur in *whorls* (circles) of three or more. The *region* or *area* (no structure is involved) of leaf attachment is called a **node**; the area of stem between two nodes is called an **internode**. A dicot leaf usually has two main parts—a flattened **blade**, and a **petiole** (stalk). Sometimes the petiole is missing, and the leaf is then said to be **sessile**. Note that the word *sessile* as applied to plants has a different meaning than it does when applied to animals; in animals, it means *immobile* or *attached to a substrate*. Each leaf forms an *angle* called an **axil** between its petiole (or blade) and the stem. An **axillary bud** is produced in each axil, and a **terminal bud** is often present at the tip of the twig. Axillary buds frequently develop into branches or may contain tissues that will develop into flowers; terminal buds usually extend the length of the twig during the growing season. Some buds, such as those of geraniums, are unprotected, but most buds of woody plants have protective winter *bud scales* that drop off, leaving tiny *bud scars* as the buds expand. Some twigs have tiny scars of a different origin. A pair of appendages called **stipules** may occur toward the base of a leaf's petiole. In some plants, the stipules fall off and leave *stipule scars*. The leaves of deciduous trees leave *leaf scars* with tiny *bundle scars* within when they drop in the fall. The

FIGURE 34.18
A shoot apex (50×). Scanning electron micrograph of shoot apex of silver maple, *Acer saccharinum*, showing a developing shoot during spring or early summer, the season of active growth. The apical meristem, leaf primordia, and trichomes are plainly visible at this stage.

shapes, sizes, and other features of leaf scars can be distinctive enough to identify the plants in winter (figure 34.17).

Herbaceous stems do not produce a cork cambium and therefore lack a periderm. The stems are usually green and photosynthetic, with at least the outer cells of the cortex containing chloroplasts. Herbaceous stems commonly have stomata, and may have various types of trichomes (hairs).

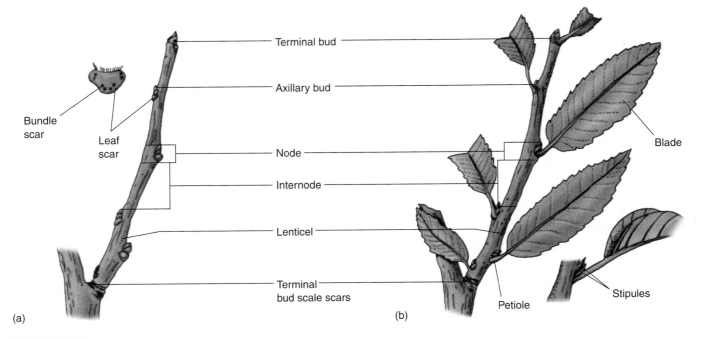

FIGURE 34.17
A woody twig. (a) In winter. (b) In summer.

Internal Form

As in roots, there is an *apical meristem* at the tip of each stem, which produces *primary tissues* that contribute to the stem's increases in length. The apical meristem is dormant until the growing season, at which time the bud scales fall off, revealing the **leaf primordia** (embryonic leaves) and **bud primordia** in their axils (figure 34.18). Three primary meristems develop from the apical meristem. As in roots, the *protoderm* gives rise to the *epidermis;* a cylinder of **procambial strands,** which give rise to cylinders of *primary xylem* and *primary phloem,* is produced interior to the protoderm. The *ground meristem* produces parenchyma cells. Parenchymatous tissue in the center of the stem constitutes the *pith;* parenchymatous tissue away from the center constitutes the *cortex.*

A strand of xylem and phloem, called a *trace,* branches off from the main cylinder of xylem and phloem and enters the developing leaf or bud, leaving a thumbnail-shaped space. These spaces in the main cylinder of conducting tissues are called **leaf gaps** and **bud gaps.** In dicots, a **vascular cambium** develops between the primary xylem and primary phloem (figure 34.19). The cells of the vascular cambium divide indefinitely, producing **secondary tissues** (mainly *secondary xylem* and *secondary phloem*), which contribute to the girth of the stem. In woody dicots, a second cambium, the *cork cambium,* arises in the outer cortex (occasionally in the epidermis or phloem) and produces box-like *cork cells* to the outside and also may produce parenchyma-like *phelloderm* cells to the inside. Cork tissues, whose cells become impregnated with *suberin* shortly after they are formed and then die, constitute the **outer bark.** The cork tissue, whose suberin is impervious to moisture, cuts off water and food to the epidermis, which dies and sloughs off. In young stems, gas exchange between stem tissues and the air takes place through stomata, but as the cork cambium produces cork, it also produces patches of unsuberized cells beneath the stomata. These unsuberized cells, which permit gas exchange to continue, are called **lenticels** (figure 34.20).

The stem consists of an axis to which leaves are attached. Buds that develop into branches or flowers develop at the junction of leaf and stem.

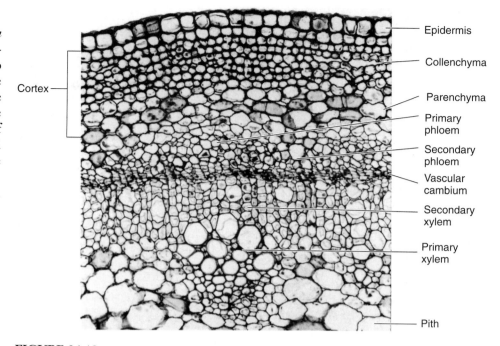

Cortex

Epidermis

Collenchyma

Parenchyma

Primary phloem

Secondary phloem

Vascular cambium

Secondary xylem

Primary xylem

Pith

FIGURE 34.19
Early stage in differentiation of vascular cambium in an elderberry stem, *Sambucus canadensis* (250×). The outer part of the cortex consists of collenchyma, and the inner part of parenchyma.

FIGURE 34.20
Lenticels. (a) Lenticels, the numerous, small, pale, raised areas shown here on cherry tree bark (*Prunus cerasifera*), allow gas exchange between the external atmosphere and the living tissues immediately beneath the bark of woody plants. Highly variable in form in different species, lenticels are an aid to the identification of deciduous trees and shrubs in winter. (b) Transverse section through a lenticel (extruding area) in a stem of elderberry, *Sambucus canadensis* (30×).

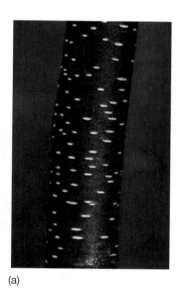

(a)

(b)

Stem Tissue Patterns

In young stems (and roots), the central core of tissues, composed of primary xylem, primary phloem, and pith (if present), is called the *stele*.

Herbaceous Dicot Stems

Most herbaceous dicots have vascular bundles arranged in a ring, although a few plants, such as foxgloves, have the xylem and phloem in continuous rings instead of in discrete bundles. The ring of vascular bundles separates the *pith* to its interior from the cortex to its exterior. When a vascular cambium arises between the patches of primary xylem and primary phloem, it may remain confined to each vascular bundle, or it may extend between the vascular bundles as well. The vascular cambium produces secondary xylem and phloem within the vascular bundles and parenchyma cells between the bundles (figure 34.21).

Woody Dicot Stems

During the early stages of growth, woody dicot stem tissues appear similar to those of herbaceous dicots, but differences become apparent once the vascular cambium and the cork cambium start producing secondary tissues. The most obvious difference involves the secondary xylem, or **wood.**

Annual Rings. Except in tropical rainforests, where growth may be continuous the year round, most growth in woody dicots is confined to the spring and summer. When the vascular cambium of a typical angiosperm tree first produces xylem in the spring, the xylem consists of relatively large vessel members in what is known as *spring*, or *early*, *wood.* As the season progresses, the vessel members usually become progressively smaller, or more tracheids than vessel members are produced until only tracheids (or occasionally fibers) are being differentiated in what is called *summer*, or *late*, *wood.* The switch from the summer wood of the previous year and the spring wood of the following year results in a band of light wood alternating with a band of darker wood, with one year's growth of xylem appearing as an annual ring. Annual rings are also discernible in conifers, even though only tracheids are produced, because the spring tracheids are significantly larger than those of summer wood. Xylem cells are produced in greater numbers and have stronger walls than phloem cells, which usually become crushed and are sloughed off with the older bark. As a result, much of a woody dicot's stem is wood.

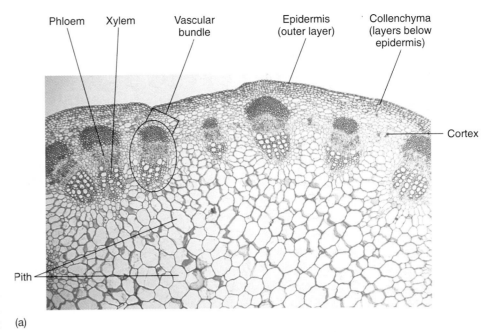

(a)

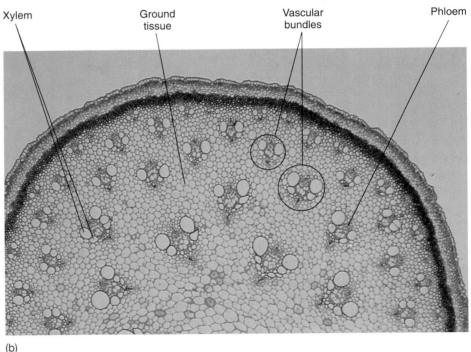

(b)

FIGURE 34.21
Stems. Transverse sections of a young stem in (a) a dicot, the common sunflower, *Helianthus annuus*, in which the vascular bundles are arranged around the outside of the stem (10×); and (b) a monocot, asparagus, *Asparagus officinalis*, with the scattered vascular bundles characteristic of the class (32×).

A science known as *dendrochronology* has arisen around the examination of tree rings and matching them to past events. The age of a tree can be determined by counting the annual rings, which give more historical information than just age alone. For example, if there has been more than the usual amount of rainfall during one season, the annual ring will be wider. If a fire sweeps through a forest but does not kill the tree, the year of the fire may be determined by counting the rings back to the fire scar. Such methods have been used to date wood from European oaks used in various types of construction back more than 9000 years, and to determine the ages of frames used to mount old paintings. Researchers have found that red spruce (*Picea rubra*) trees in the eastern United States have produced unusually thin rings during the past 30 years, indicating a growth decline. Some scientists attribute this to the effects of acid precipitation, while others think that the condition arises as a result of a variety of environmental effects, including herbicides, soot, and other forms of industrial pollution.

Wood. In cross-sections of trees, lines of lighter cells can be seen radiating out from the center. These lines, known as **rays,** are tiers of parenchyma cells produced by the vascular cambium. They may remain alive for several years; they function primarily in lateral conduction of water and dissolved minerals from the stele to the cortex. Within the xylem, the ray is called a *xylem ray*. Rays extend across the phloem, where they are called *phloem rays* (figure 34.22).

As wood toward the center grows older, the protoplasts of the parenchyma cells surrounding the vessels and tracheids may grow through the pits and occupy the cavities. This effectively blocks conduction of water, and darker colored waste materials begin to accumulate. The older, now nonfunctional wood is called **heartwood,** while the lighter-colored, functioning wood is called **sapwood.** Since the heartwood can no longer conduct water, it is of little use to the tree—except, perhaps, in a strengthening capacity—and its removal does not affect its health. This is why it has been possible to cut holes large enough for a car to drive through in a few of California's

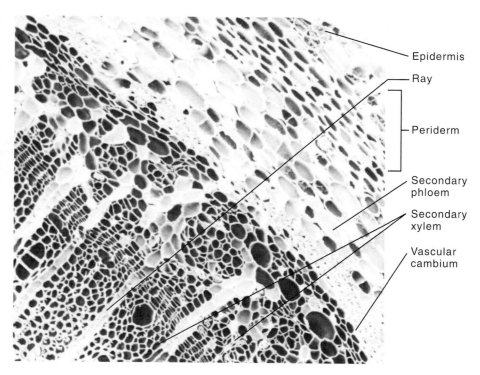

FIGURE 34.22
Scanning electron micrograph of a diagonal view of a cross-section of the outer portion of a trunk of silver maple, *Acer saccharinum* (100×). Note especially the secondary xylem (wood), composed of tracheids, vessel elements, and rays, the vascular cambium, secondary phloem cells, and the periderm.

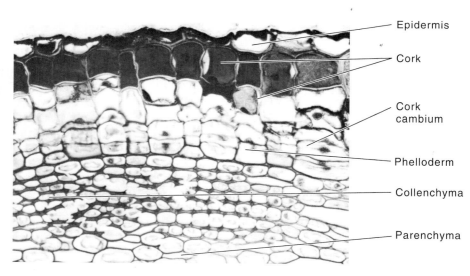

FIGURE 34.23
Section of periderm (250×). An early stage in the development of periderm in elderberry (*Sambucus canadensis*).

huge redwoods, without killing the tree or noticeably affecting their health.

The cork cambium and the tissues it produces constitute the *outer bark*, or *periderm* (figure 34.23). Bark also includes the phloem, which is referred to as the *inner bark*.

Commercial Uses of Wood. Wood is one of the most beautiful and economically important products we obtain from plants. It is used for building materials (figure 34.24), furniture and, as pulpwood, to produce paper. The wood we use commercially is usually obtained from stems rather than roots.

Wood is divided into *hardwoods* and *softwoods*. Hardwoods are the woods of dicots, regardless of how hard or soft they actually may be; softwoods are the woods of conifers. Many hardwoods used commercially come from the tropics, while almost all softwoods come from the great forests of the north temperate zone.

Individual woods differ widely in their microscopic characteristics, including such features as the width and height of rays, type of pits in the conducting elements, abundance of fibers, and nature of the conducting elements and other cells. They can be identified readily by experts, and often provide valuable clues to the evolutionary position of the plants that form them. In addition, these same features contribute to the rich textures and different durability properties of various kinds of wood.

Monocot Stems

Because grasses, greenbriers, and all other monocots lack both a vascular cambium and a cork cambium and the tissues they produce, they are mostly herbaceous. They are similar to herbaceous dicots in having a surface layer of epidermis, but the arrangement of tissues on the inside is quite different. The vascular bundles of primary xylem and primary phloem are not arranged in a cylinder, but are instead scattered throughout the parenchyma cells of ground tissues. The ground tissue is not separated into cortex and pith, and it makes up the remaining volume of the stem. Although the vascular bundles are scattered, each is oriented so that the xylem points toward the center and the phloem points toward the surface. The vascular bundles also tend to be a little more numerous toward the epidermis than they are in the center of the stem. The xylem of a typical monocot vascular bundle consists of tracheids and usually two large vessels; the phloem consists entirely of sieve tubes and companion cells. The bundle is surrounded by a sheath of sclerenchyma cells that provide support.

Although there are no *cambia* (plural of *cambium*) present in monocots, grasses (corn, sugarcane) have, in addition to apical meristems, *intercalary meristems*, present in the vicinity of nodes (areas of leaf attachment). The intercalary meristems produce tissues that increase the length of the stem but not its girth. This is why even tall grasses have stems that are of roughly the same diameter toward the bottom as they are toward the top. Palm trees have parenchyma that continue to divide after they are produced, and as a result, they grow considerably taller than most other monocots.

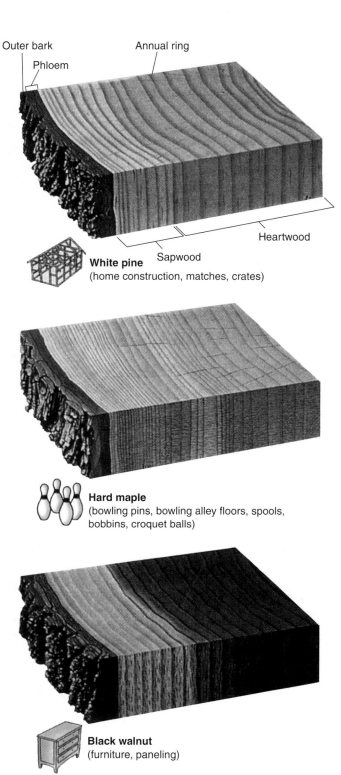

FIGURE 34.24
Sections of mature white pine, hard maple, and black walnut stems. Each type of wood has specific features that make it ideal for certain commercial uses.
Source: St. Regis Paper Company.

Wood is accumulated secondary xylem; it is marked by annual rings when it is formed in regions with a seasonal climate. The woods of dicots are called hardwoods; those of conifers are called softwoods.

Modified Stems

Although most stems grow erect, there are some modifications that serve special purposes, including that of natural *vegetative propagation*. In fact, the widespread artificial vegetative propagation of plants, both commercial and private, frequently involves the cutting of modified stems into segments, which are then planted and grow into new plants. As you become acquainted with the following modified stems, keep in mind that stems have *leaves* at *nodes*, with *internodes* between the nodes, and *buds* in the *axils* of the leaves, while roots have no leaves, nodes, or axillary buds.

Bulbs. Onions, lilies, and tulips have swollen underground stems that are really large buds with adventitious roots at the base. Most of a *bulb* consists of fleshy leaves attached to a small, knob-like stem. In onions, the fleshy leaves are surrounded by papery, scale-like leaf bases of the long, green aboveground leaves.

Corms. Crocuses, gladioluses, and other popular garden plants produce **corms** that superficially resemble bulbs. Cutting a corm in half, however, reveals no fleshy leaves. Instead, almost all of a corm consists of stem, with a few papery, brown nonfunctional leaves on the outside, and adventitious roots below.

Rhizomes. Perennial grasses, ferns, irises, and many other plants produce *rhizomes*, which typically are horizontal stems that grow underground, often close to the surface. Each node has an inconspicuous scale-like leaf with an axillary bud; much larger photosynthetic leaves may be produced at the rhizome tip. Adventitious roots are produced throughout the length of the rhizome, mainly on the lower surface.

Runners and stolons. Strawberry plants produce horizontal stems with long internodes, which, unlike rhizomes, usually grow along the surface of the ground. Several *runners* may radiate out from a single plant. Some botanists use the term *stolon* synonymously with runner; others reserve the term stolon for a stem with long internodes that grows underground, as seen in Irish (white) potato plants. An Irish potato itself, however, is another type of modified stem—a *tuber*.

Tubers. In Irish potato plants, carbohydrates may accumulate at the tips of stolons, which swell, becoming *tubers;* the stolons die after the tubers mature (figure 34.25*a*). The "eyes" of a potato are axillary buds formed in the axils of scale-like leaves. The scale-like leaves, which are present when the potato is starting to form, soon drop off; the tiny ridge adjacent to each "eye" of a mature potato is a leaf scar.

Tendrils. Many climbing plants, such as grapes and Boston ivy, produce modified stems knows as *tendrils*, which twine around supports and aid in climbing. Some tendrils, such as those of peas and pumpkins, are actually modified leaves or leaflets (figure 34.25*b*).

(a)

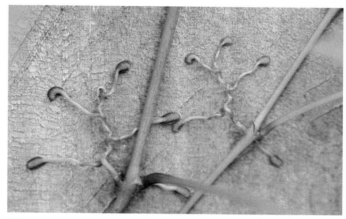

(b)

(c)

FIGURE 34.25
Types of modified stems. (a) White potato tubers. (b) Tendrils of Boston ivy. (c) These young koa tree branches look like leaves, but they are actually cladophylls.

Cladophylls. Cacti and several other plants produce flattened, photosynthetic stems called *cladophylls* that resemble leaves (figure 34.25*c*). In cacti, the real leaves are modified as spines.

Some plants possess modified stems that serve special purposes such as food storage or adventitious growth.

Leaf External Structure

Leaves, which are initiated as *primordia* by the apical meristems (see figures 34.5 and 34.18), are vital to life as we know it. They are the principal sites of photosynthesis on land. Leaves expand primarily by cell enlargement and some cell division and are not capable of further growth once they are mature. Since leaves are crucial to a plant, features such as their arrangement, form, size, and internal structure are highly significant and can differ greatly.

Most dicot leaves have a flattened *blade*, and a slender stalk, the *petiole*. In addition, a pair of *stipules* may be present at the base of the petiole. The stipules, which may be leaf-like or modified as *spines* (as in the black locust—*Robinia pseudo-acacia*) or *glands* (as in cherry trees—*Prunus cerasifera*), vary considerably in size from microscopic to almost half the size of the leaf blade. Grass and other monocot leaves usually lack a petiole and tend to sheathe the stem toward the base. **Veins** (a term used for the vascular bundles in leaves), consisting of both xylem and phloem, are distributed throughout the leaf blades. The main veins are parallel in most monocot leaves; the veins of dicots, on the other hand, form an often intricate network (figure 34.26).

All leaves have an *axillary bud* in the axil of the leaf (the angle between their petiole or base and the stem to which they are attached). In **simple leaves,** such as those of lilacs or birch trees, the blades are undivided, but simple leaves may have teeth, indentations, or lobes of various sizes, as in the leaves of maples and oaks. In compound leaves, such as those of ashes, box elders, and walnuts, the blade is divided into **leaflets.** Leaflets, unlike the whole leaf, do not have an axillary bud at their base. If the leaflets are arranged in pairs along a common axis (the axis is called a *rachis*—the equivalent of the main central vein, or *midrib*, in simple leaves), the leaf is **pinnately compound.** If, however, the leaflets radiate out from a common point at the blade end of the petiole, the leaf is **palmately compound.** Palmately compound leaves occur in buckeyes (*Aesculus* spp.) and Virginia creeper (*Parthenocissus quinquefolia*) (figures 34.27 and 34.28). The leaf blades themselves may have similar arrangements of their veins, and are said to be **pinnately** or **palmately** veined.

(a)

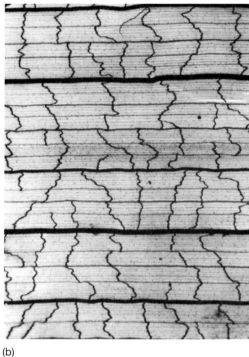
(b)

FIGURE 34.26
Dicot and monocot leaves. The leaves of dicots, such as this (a) African violet relative from Sri Lanka, have netted, or reticulate, veins; (b) those of monocots, like this Latin American palm, have parallel veins. Both leaves have been cleared with chemicals, and the dicot leaf has been stained with a red dye to make its veins show up more clearly.

FIGURE 34.27
Virginia creeper, *Parthenocissus quinquefolia.* Palmately compound leaves of this familiar vine.

(a)

(b)

(c)

(d)

(e)

FIGURE 34.28

Angiosperm leaves are stunningly variable: they are the primary factor in controlling the specific ways plants capture light and regulate water loss. (a) Diverse leaves in the herb layer of a Costa Rican rainforest. (b) A palmately compound leaf of marijuana (*Cannabis sativa*). A compound leaf is associated with a single lateral bud, located where the petiole is attached to the stem. (c) A simple leaf, its margin deeply lobed, from the tulip tree (*Liriodendron tulipifera*). (d) A pinnately compound leaf, from a member of the legume family in the lowland forest of Peru. (e) Many unusual arrangements of leaves occur in different kinds of plants. For example, in miner's lettuce (*Claytonia perfoliata*), an herb from the Pacific states, two leaves are completely fused below each of the clusters of flowers, which seem, therefore, to arise from the center of a single leaf.

Leaves, regardless of whether they are simple or compound, may be **alternately** arranged (alternate leaves usually spiral around a shoot) or they may be in **opposite** pairs. Less often, three or more leaves may be in a **whorl,** a circle of leaves at the same level at a node (figure 34.29).

A *microphyll* is a leaf with one vein that does not leave a gap when it branches from the vascular cylinder of the stem; microphylls are mostly small and are associated primarily with the phylum Lycophyta (see chapter 33). Most plants have leaves called *megaphylls*, which have several to many veins; a megaphyll's conducting tissue leaves a gap in the stem's vascular cylinder as it branches from it.

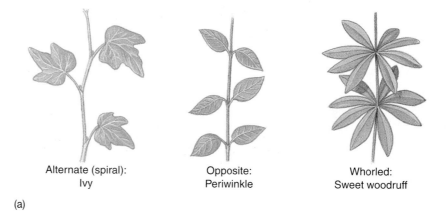

Alternate (spiral): Ivy

Opposite: Periwinkle

Whorled: Sweet woodruff

(a)

(b)

FIGURE 34.29

Types of leaf arrangements. (a) The three common types of leaf arrangements are alternate, opposite, and whorled. (b) The often colorful leaves of *Coleus* are a familiar sight in greenhouses and garden beds. They are produced in pairs of opposite leaves.

Leaves are the principle sites of plant photosynthesis. Their blades may be arranged in a variety of ways. In simple leaves the blades are undivided, while in compound leaves the leaf is composed of two or more leaflets.

Leaf Internal Structure

The entire surface of a leaf is covered by a transparent epidermis, most of whose cells have no chloroplasts. The epidermis itself has a waxy *cuticle* of variable thickness, and may have different types of glands and trichomes (hairs) present. The lower epidermis (and occasionally the upper epidermis) of most leaves contains numerous slit-like or eye-shaped *stomata*. Stomata, as discussed on page 697, are flanked by *guard cells* and function in gas exchange and regulation of water movement through the plant (figure 34.30).

The tissue between the upper and lower epidermis is called **mesophyll.** Mesophyll is interspersed with veins (vascular bundles) of various sizes. In most dicot leaves there are two distinct types of mesophyll. Closest to the upper epidermis are one to several (usually two) rows of tightly packed, barrel-shaped to cylindrical *chlorenchyma* cells (parenchyma with chloroplasts) that constitute the **palisade mesophyll.** Some plants, including species of *Eucalyptus*, have leaves that hang down, rather than extending horizontally. They have palisade parenchyma on both sides of the leaf, and there is, in effect, no upper side. In nearly all leaves there are loosely arranged **spongy mesophyll** cells between the palisade mesophyll and the lower epidermis, with many air spaces throughout the tissue. The inter-

connected intercellular spaces, along with the stomata, function in gas exchange and the passage of water vapor from the leaves (figure 34.31). The mesophyll of monocot leaves is not differentiated into palisade and spongy layers and there is often little distinction between the upper and lower epidermis.

Leaves are basically flattened bags of epidermis containing tightly packed palisade mesophyll rich in chloroplasts and loosely packed spongy mesophyll with many interconnected air spaces that function in gas and water vapor exchange.

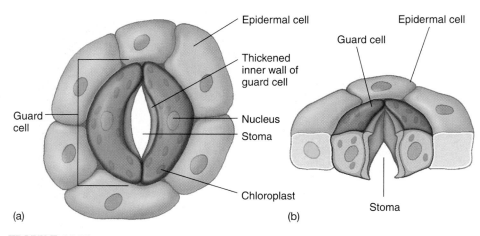

FIGURE 34.30
A stoma. (a) Surface view. (b) View in cross-section.

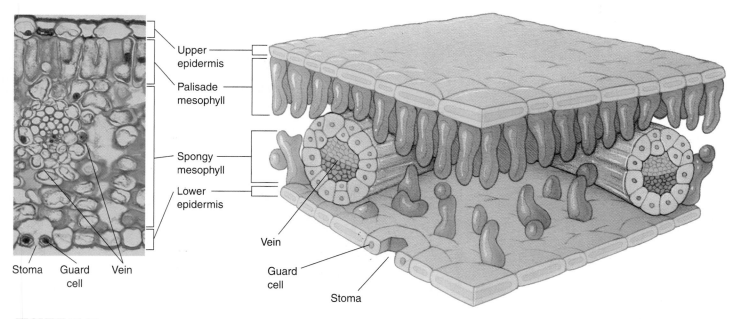

FIGURE 34.31
A leaf in cross-section. Transection of a leaf showing the arrangement of palisade and spongy mesophyll, a vascular bundle or vein, and the epidermis with paired guard cells flanking the stoma.

Leaf Abscission

Deciduous plants in temperate areas produce new leaves in the spring and then lose them in the fall. In the tropics, however, the production and subsequent loss of leaves in some species is correlated with wet and dry seasons. Evergreen plants, such as most conifers, usually have a complete change of leaves every two to seven years, periodically losing some but not all of their leaves. The process by which leaves are shed is called **abscission** (figure 34.32).

Abscission involves changes that take place in an *abscission zone* at the base of the petiole (figure 34.33). Young leaves produce hormones that inhibit the development of specialized layers of cells in the abscission zone. Hormonal changes take place as the leaf ages, however, and two layers of cells become differentiated. A *protective layer*, which may be several cells wide, develops on the stem side of the petiole base. These cells become impregnated with *suberin*, which, as you will recall, is a fatty substance that is impervious to moisture. A *separation layer* develops on the side of the leaf blade; the cells of the separation layer sometimes divide, swell, and become gelatinous. When temperatures drop, the duration and intensity of light diminishes as the days grow shorter, or other environmental changes occur, enzymes break down the pectins in the middle lamellae of the separation cells. Wind and rain can then easily separate the leaf from the stem. Left behind is a sealed leaf scar that is protected from bacteria and other disease organisms.

As the abscission zone develops, the green chlorophyll pigments present in the leaf break down, revealing the yellows and oranges of other pigments, such as carotenoids, that previously had been masked by the intense green colors. At the same time, water-soluble red or blue pigments called *anthocyanins* and *betacyanins* may also accumulate in the vacuoles of the leaf cells—all contributing to an array of fall colors in leaves.

Abscission occurs when a plant sheds its leaves. Two layers of cells in the abscission zone differentiate into a protective layer and a separation layer. As pectins in the separation layer break down, wind and rain can easily separate the leaf from the stem.

FIGURE 34.32
Leaf abscission in a temperate deciduous plant. This type of plant produces leaves in the spring and sheds them in the fall.

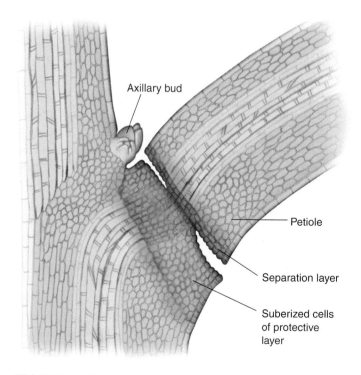

FIGURE 34.33
The abscission zone of a leaf. Hormonal changes in this zone cause abscission.

Modified Leaves

With plants established in a wide variety of environments, from deserts to lakes to tropical rainforests, one would expect to see modifications of plant organs that would adapt the plants to their specific habitats. Leaves, in particular, have evolved some remarkable modifications that adapt them to their surroundings. A brief discussion of a few of these modifications follows.

Floral leaves (bracts). Poinsettias and dogwoods have relatively inconspicuous, small, greenish yellow flowers. However, both plants produce large modified leaves, called **bracts,** (mostly colored red in poinsettias and white or pink in dogwoods). These bracts surround the true flowers and perform the same function as showy petals (figure 34.34*a*). It should be noted, however, that bracts can also be quite small and not as conspicuous as those of the examples mentioned.

Spines. The leaves of many cacti, barberries, and other plants are modified as **spines** (figure 34.34*b*). In the case of cacti, the reduction of leaf surface reduces water loss and also may deter predators. Spines should not be confused with *thorns*, such as those on honey locust (*Gleditsia triacanthos*), which are modified stems, or with the *prickles* on raspberries and rose bushes, which are simply outgrowths from the epidermis or the cortex just beneath it.

Reproductive leaves. Several plants, notably *Kalanchoë*, produce tiny but complete plantlets along their margins. Each plantlet, when separated from the leaf, is capable of growing independently into a full-sized plant. The walking fern (*Asplenium rhizophyllum*) produces new plantlets at the tips of its fronds.

Window leaves. Several genera of plants growing in arid regions produce succulent, cone-shaped leaves with transparent tips. The leaves often become mostly buried in sand blown by the wind, but the transparent tips, which have a thick epidermis and cuticle, admit light to the hollow interiors. This allows photosynthesis to take place beneath the surface of the ground.

Shade leaves. Leaves produced where they receive significant amounts of shade tend to be larger in surface area, but thinner and with less mesophyll than leaves on the same tree receiving more direct light.

Insectivorous leaves. Almost 200 species of flowering plants are known to have leaves that trap insects, with some digesting their soft parts (figure 34.34*c*). Plants

(a)   (b)   (c)

FIGURE 34.34
Three types of modified leaves. (a) Floral leaves (bracts). In this poinsettia "flower," the red colored bracts (modified leaves) surround the several true flowers without petals in the center. (b) Spines. The spines of this barberry are modified leaves. (c) Insectivorous leaves. The leaves of these pitcher plants are modified to trap insects.

with insectivorous leaves often grow in acid swamps deficient in needed elements, or containing elements in forms not readily available to the plants; this inhibits the plants' capacities to maintain metabolic processes sufficient to meet their growth requirements. Their needs are, however, met by the supplementary absorption of nutrients from the animal kingdom.

Pitcher plants (for example, *Sarracenia*, *Darlingtonia*, *Nepenthes*) have cone-shaped leaves in which rainwater can accumulate. The insides of the leaves are very smooth, but there are stiff, downward-pointing hairs at the rim. An insect falling into such a leaf finds it very difficult to escape and eventually drowns. The nutrients released when bacteria, and in most species digestive enzymes, decompose the insect bodies are absorbed into the leaf. Other plants, such as sundews (*Drosera*), have glands that secrete sticky mucilage that trap insects, which are then digested by enzymes. The Venus flytrap (*Dionaea muscipula*) produces leaves that look hinged at the midrib. When tiny trigger hairs on the leaf blade are stimulated by a moving insect, the two halves of the leaf snap shut, and digestive enzymes break down the soft parts of the trapped insect into nutrients that can be absorbed through the leaf surface.

The two remaining chapters of this section are devoted to a review of our knowledge concerning the way in which external and internal factors interact to produce a mature, functioning plant.

The leaves of plants exhibit a variety of adaptations, including spines, reproduction, and even leaves that are carnivorous.

Summary of Concepts

34.1 Vascular plants have root and shoot systems.

- A plant body is basically an axis that includes two parts: root and shoot—with associated leaves. There are four basic types of tissues in plants: meristems, ground tissue, epidermis, and vascular tissue.

- Monocots and dicots, the two classes of angiosperms, differ in many respects. Dicots generally have flower parts in fours or fives, net venation, and two seedling leaves. Monocots generally have their flower parts in threes, parallel venation, and a single seedling leaf. Monocots were derived from early dicots.

34.2 Plants have four basic tissues, each exhibiting several cell types.

- Meristems are actively dividing tissues that occur at shoot and root tips and in sheaths around the stem. They are responsible for primary and secondary growth.

- Ground tissue supports the plant and stores food and water. Parenchyma, collenchyma and sclerenchyma cells make up this tissue.

- Epidermis forms an outer protective covering for the plant. Stomata, trichomes, and root hairs occur in the epidermis, serving specialized functions such as protection and absorption.

- Vascular tissue conducts water, carbohydrates, and dissolved minerals to different parts of the plant. Xylem conducts water and minerals from the roots to shoots and leaves, and phloem conducts food molecules from sources to all parts of the plant.

34.3 Roots have four growth zones.

- Roots have four growth zones: the root cap, zone of cell division, zone of elongation, and zone of maturation.

- Primary growth occurs just behind the root cap. Secondary growth in both roots and stems takes place after the formation of lateral meristems known as cambia.

- Some plants have roots modified for photosynthesis, food or water storage, or other functions.

34.4 Stems conduct liquids and support the shoot.

- Stems branch by means of buds that form externally at the points where the leaves join the stem.

- The vascular cambium is a cylinder of dividing cells, which become secondary xylem internally and secondary phloem externally. As a result of their activity, the girth of a plant increases.

- Cork cambium forms in the cortex of both roots and stems during the initial stages of secondary growth. It produces cork externally and phelloderm, a tissue composed mainly of parenchyma-like cells, internally. The cork, cork cambium, and phelloderm collectively are called the periderm. The outermost layer of periderm, or stem bark, is perforated by groups of loosely organized unsuberized cork cells called lenticels.

34.5 Leaves are organized to transport the products of photosynthesis to the stem.

- Most leaves consist of a flattened blade and a stalk called a petiole. Leaf arrangements on the stem exhibit considerable variation.

- The interior of the leaf, called mesophyll, has veins traversing it.

Discussing Key Terms

1. **Primary and secondary growth**   The tips of plant stems and roots are capable of perpetual growth in length, called primary growth; many stems and roots are also capable of indefinite growth in girth, called secondary growth.

2. **Meristems**   Meristems are groups of cells capable of producing new cells through repeated divisions. Apical meristems are established at root and shoot tips early in the differentiation of a plant embryo; they add to root and shoot length (primary growth). Lateral meristems within woody and many herbaceous plants, add to girth (secondary growth). Intercalary meristems are localized between the base and the apex of stems and leaves of some plants.

3. **Cambia**   There are two lateral meristems. 1. The vascular cambium gives rise to the secondary xylem internally and the secondary phloem externally. 2. The cork cambium gives rise to cork externally and phelloderm internally. Bark may break away from a stem or root at the vascular cambium.

4. **Modified roots, stems, and leaves**   Modifications of roots, stems, and leaves allow plants to survive in a wide variety of environments.

Review Questions

1. What are the four major tissue systems in plants? What are their functions? What are the three primary meristems? Into what type of tissues will each develop?

2. What are the characteristics of a parenchyma cell? of a collenchyma cell? What are the two kinds of sclerenchyma cells? Where are these three cell types found? Are they alive at maturity?

3. What is the function of xylem? How do primary and secondary xylem differ in origin? What are the two types of conducting cells within xylem? How do they compare with each other? What other cell types are present in xylem?

4. What is the function of phloem? What two types of conducting cells are present in phloem? How do they differ?

5. Compare monocot and dicot roots. Which tissues do they have in common? Are any of the tissues unique to dicot roots? How

does the arrangement of the tissues differ? With which layer are Casparian strips associated? How are lateral branches of roots formed?

6. What is the location and function of the vascular cambium? From what tissue(s) is it derived? What types of cells are produced when the vascular cambium divides outwardly, inwardly, or laterally?

7. What three tissues make up the periderm? What is the function of cork cambium? What fatty substance is contained within some of these cells, and what is its function?

8. What are the basic components of most leaves? How do simple and compound leaves differ from each other? Name and describe the three types of growth pattern exhibited by most leaves.

Thought Questions

1. If you hammer a nail into the trunk of a tree 2 meters above the ground when the tree is 6 meters tall, how far above the ground will the nail be when the tree is 12 meters tall?

2. Can you suggest a reason why branches are not formed the same way in roots that they are in stems?

Internet Links

Botany Online
http://herb.biol.uregina.ca/liu/bio/idb.shtml
The result of an international effort, the INTERNET DIRECTORY FOR BOTANY is an immense collection of links and information about all aspects of the plant kingdom, arranged in an easily navigable and searchable form.

What Flower Is That?
http://elib.cs.berkeley.edu/flowers/
Search for wildflowers by common or scientific name, color, and family; very pretty pictures and links to USDA and California sites.

Everything You Wanted to Know About Flowering Plants
http://biodiversity.uno.edu/cgi-bin/hl?botany
An immense collection of descriptions and illustrations of flowering plants, with an impressive collection of links to every conceivable botanical resource.

Visit the Missouri Botanical Garden
http://www.mobot.org/welcome.html
Ever wonder how a botanical garden works? The site of the Missouri Botanical Garden provides a rich introduction to the activities and research at one of the world's great "plant zoos."

For Further Reading

Addicott, F. T.: *Abscission*, University of California Press, Berkeley, 1982. A detailed look at how plants lose their leaves.

Dale, J. E.: "How Do Leaves Grow?" *BioScience*, 1992, vol. 42, pages 423–32. Advances in cell and molecular biology that are unraveling some of the mysteries of leaf development.

Epstein, E.: "Roots," *Scientific American*, 1973, vol. 228, pages 48–58. A very straightforward, well-illustrated account of the structure of plant roots.

Fahn, A.: *Plant Anatomy*, 4th ed., Pergamon Press, Elmsford, NY, 1990. An up-to-date general text on plant structure.

Feininger, A.: *Leaves: 177 Photographs*, Dover, Mineola, NY, 1984. A visual exploration of the incredible diversity of plant leaves.

Flynn, J. H., Jr. (editor): *A Guide to Useful Woods of the World*, King Philip Publishing, Portland, ME, 1994. An interesting account of a diverse substance we often take for granted.

Foskett, D. E.: *Plant Growth and Development*, Academic Press, San Diego, 1994. A recent review of what is known about the mechanisms of plant growth and development.

Raven, P. H., R. F. Evert, and S. E. Eichhorn: *Biology of Plants*, 5th ed., Worth Publishers, New York, 1992. A comprehensive treatment of general botany, emphasizing structural botany.

Romberger, J. A.: *Plant Structure, Function and Development*, Springer-Verlag, New York, 1993. A good general treatment of plant structure and function.

Sandved, K. B. and G. T. Prance: *Leaves*, Crown Publishers, New York, 1984. This book provides an incredibly beautiful introduction to the diversity of leaves.

Zimmermann, M. H.: *Xylem Structure and the Ascent of Sap*, Springer-Verlag, New York, 1983. An outstanding monograph on the structure of xylem and how it functions.

35

Transport in Plants

FIGURE 35.1
Root hairs. Delicate hairs protrude from root epidermal cells, increasing the surface area for the absorption of water and nutrients from the soil.

Concept Outline

35.1 Evaporation of water from leaves draws water up the stem.

Overview of Water Movement through Plants. Water is drawn up through the stem of a plant by suction as evaporation removes water from the leaves.

35.2 Plants absorb water into their roots and lose it from their leaves.

The Absorption of Water by Roots. Water and minerals enter the plant through the roots.

Transpiration of Water from Leaves. Water leaves the plant through openings in the leaves called stomata.

35.3 Nutrients move up the stem, carbohydrates down.

Nutrient Movement. Ions and other minerals are actively transported into the roots and travel up the xylem; carbohydrates enter phloem passively and travel down the phloem.

35.4 Plants, although terrestrial, can adapt to life in water.

Flooding: Too Much of a Good Thing. Too much water is harmful to a plant, although many plants have adaptations that make them tolerant of flooding.

For vascular plants to function normally, two kinds of internal transport are essential. First, carbohydrates produced in leaves by photosynthesis must be carried, in solution, to the other living cells throughout the plant. Second, minerals and water in the ground must be absorbed by the roots (figure 35.1) and transported up to the leaves and other parts of the plant. As you saw in chapter 34, the two types of transport take place in specialized cells, xylem and phloem. Perhaps you are wondering why xylem and phloem are microscopically sized when larger-diameter "pipes" like our blood vessels would make more rapid movement of larger volumes of water possible. The answer is that our hearts actively pump our blood, while plants rely on passive forces to drive liquids through their bodies. As you will see, these passive forces depend heavily on the existence of very narrow transport tubes.

Overview of Water Movement through Plants

It is not unusual for a large tree to have leaves more than 10 stories off the ground (figure 35.2). Did you ever wonder how water gets from the roots to the top of a tree that high? We know that there are interconnected, water-conducting xylem elements extending throughout a plant. We also know that water first enters the roots and then moves to the xylem by osmosis. After that, however, water rises through the xylem because of a combination of factors.

The Puzzle: How Does Water Get to the Top of a Tree?

Water in a dish will rise in upright narrow tubes in proportion to the diameter of the tube; the narrower the tube, the higher the water will rise (figure 35.3). The height limit is set by the pressure exerted by the atmosphere on the surface of the water. At sea level, this pressure is sufficient to raise water in a very narrow tube to a height of about 10.4 meters. The weight of the water in a tube taller than about 10.4 meters pulls the water down. But even though water can, indeed, rise by capillarity in a narrow tube, air must be present above the water column for it to do so, which is not the case within a plant. Also, the diameters of the tubes in plants are not narrow enough to raise water by capillarity more than about a meter. How, then, does water get to the top of a tall tree?

The Solution: Evaporation Sucks It Up

The answer was suggested by Otto Renner in Germany in 1911. To understand his suggestion, imagine a plant as a cylinder partially embedded in moist soil. Imagine that water fills the cylinder, which has small pores at the top, and, for simplicity, further assume that the water is held at that height by capillary action—although we will later identify the true cause as something else. Now if air is blown across the top of the cylinder it will cause water molecules to evaporate from the pores. Does the level of water in the cylinder fall? No. As water molecules are drawn from the top, they are replenished by new ones that are taken in at the bottom. This, in essence, is what Renner proposed actually happens in plants. Passage of air across leaf surfaces results in loss of water by evaporation, creating a pull at the open upper end of the "tube."

In a plant, water rises beyond the point supported by atmospheric pressure (10.4 m at sea level) because evaporation from the leaves produces a tension on the entire water column that extends all the way down to the roots. Water has an inherent tensile strength that arises from the cohesion of its molecules, their tendency to form hydrogen

FIGURE 35.2
How does water get to the tops of these trees? We would expect gravity to make such a tall column of water too heavy to be maintained by capillary action. What pulls the water up?

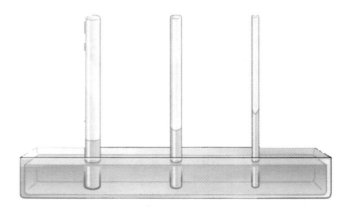

FIGURE 35.3
Capillarity in narrow tubes. The smaller the diameter of the tube, the higher the fluid rises.

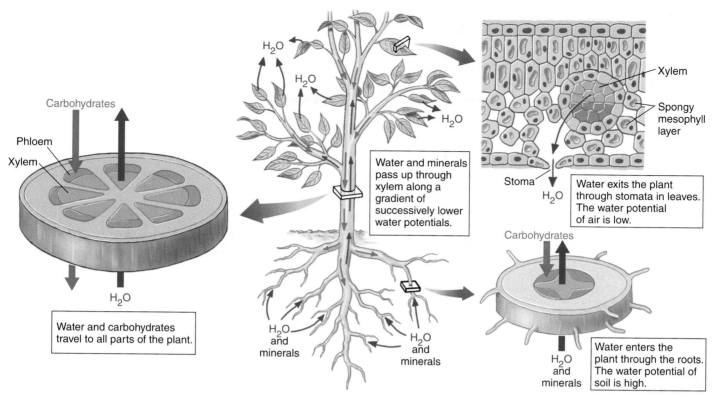

FIGURE 35.4
Water movement through a plant. This diagram illustrates the path of water and inorganic materials as they move into, through, and out of the plant body.

Labels within figure:

Carbohydrates

Phloem
Xylem

H_2O

Water and carbohydrates travel to all parts of the plant.

H_2O

H_2O

H_2O

Water and minerals pass up through xylem along a gradient of successively lower water potentials.

H_2O and minerals

H_2O and minerals

Xylem

Spongy mesophyll layer

Stoma

H_2O

Water exits the plant through stomata in leaves. The water potential of air is low.

Carbohydrates

H_2O and minerals

Water enters the plant through the roots. The water potential of soil is high.

---

bonds with one another. The tensile strength of a column of water varies inversely with the diameter of the column; that is, the smaller the diameter of the column, the greater the tensile strength. Because plants have transporting vessels of very narrow diameter, the cohesive forces in them are strong.

Deformed cells or freezing can cause small bubbles of air to form within xylem cells. Recall that individual tracheids and vessel members are connected by one of more *pits* (cavities) in their walls. Air bubbles are generally larger than the openings, so they cannot pass through them. Furthermore, the cohesive force of water is so great that the bubbles are forced into rigid spheres that have no plasticity and therefore cannot squeeze through the openings. Any bubbles that do form are limited to the xylem elements where they originate, and water may continue to rise in parallel columns.

Water Potential

Plant biologists often discuss the forces that act on water within a plant in terms of **potentials.** By applying pressure, it is possible to prevent osmosis from taking place. The smallest amount of pressure needed to stop fluid from moving by osmosis is referred to as the **solute** (or **osmotic**) **potential** of the solution; to put it another way, solute potential, which is based on concentration gradients of solute and solvent across a membrane, is the pressure needed to stop osmosis from occurring. Water will enter a cell by os-

mosis until the solute potential is offset by the resistance of the cell wall to expansion. The *turgor pressure*, which is a physical pressure that results as water enters the cell vacuoles, is referred to as **pressure potential.** Water coming through a garden hose is an example of physical pressure. The **water potential** of a plant cell is, in essence, the combination of its pressure potential and solute potential; it represents the total potential energy of the water in a plant. If two adjacent cells have different water potentials, water will move from the cell with the higher water potential to the cell with the lower water potential. Water in a plant moves along a gradient between the relatively high water potential in the soil to successively lower water potentials in the roots, stems, leaves, and atmosphere.

Water potential in a plant, and thus movement of water through the plant, depends on its osmotic absorption by the roots and the negative pressures created by water loss from the leaves and other plant surfaces. This process is defined as **transpiration,** the evaporation of water from a living plant (figure 35.4). As you will see, the negative pressure generated by transpiration is largely responsible for the upward movement of water in xylem.

The loss of water from the leaf surface, called transpiration, literally sucks water up the stem from the roots. Water rises up the stem because the water potential in the roots is greater than that in the leaves.

The Absorption of Water by Roots

Most of the water absorbed by the plant comes in through root hairs, which collectively have an enormous surface area. Root hairs are always turgid because their solute potential is greater than that of the surrounding soil due to mineral ions being actively pumped into the cells. Because the mineral ion concentration in the soil water is usually much lower than it is in the plant, an expenditure of energy (supplied by ATP) is required for the accumulation of such ions in root cells. The plasma membranes of root hair cells contain a variety of protein transport channels, through which *proton pumps* (see page 122) transport specific ions against even large concentration gradients. Once in the roots, the ions, which are plant nutrients, are transported via the xylem throughout the plant.

The ions may follow the cell walls and the spaces between them or more often go directly through the plasma membranes and the protoplasm of adjacent cells (figure 35.5). When mineral ions pass between the cell walls, they do so nonselectively. Eventually, on their journey inward, they reach the endodermis and any further passage through the cell walls is blocked by the Casparian strips. Water and ions must pass through the plasma membranes and protoplasts of the endodermal cells to reach the xylem. However, transport through the cells of the endodermis is selective. The endodermis, with its unique structure, along with the cortex and epidermis, controls which ions reach the xylem.

Transpiration from the leaves (figure 35.6), which creates a pull on the water columns, indirectly plays a role in helping water, with its dissolved ions, enter the root cells. However, at night, when the relative humidity may approach 100%, there may be no transpiration. Under these circumstances, the negative pressure component of water potential becomes small or nonexistent.

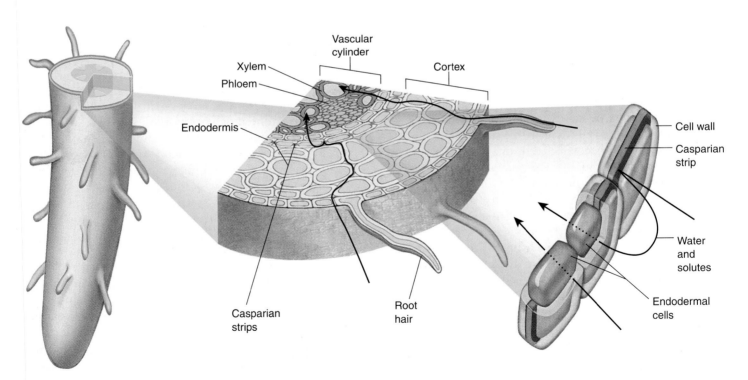

FIGURE 35.5
The pathways of mineral transport in roots. Minerals are absorbed at the surface of the root, mainly by the root hairs. In passing through the cortex, they must either follow the cell walls and the spaces between them or go directly through the plasma membranes and the protoplasts of the cells, passing from one cell to the next by way of the plasmodesmata. When they reach the endodermis, however, their further passage through the cell walls is blocked by the Casparian strips, and they must pass through the membrane and protoplast of an endodermal cell before they can reach the xylem.

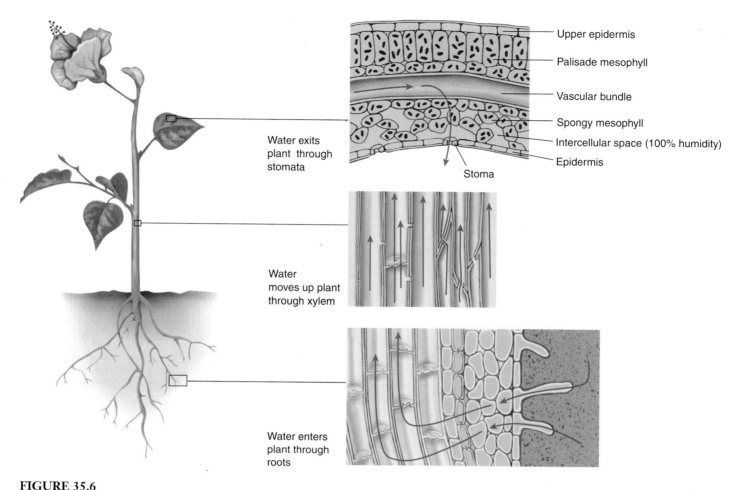

FIGURE 35.6

Transpiration. Water evaporating from the leaves through the stomata causes the movement of water upward in the xylem and the entrance of water through the roots.

Active transport of ions into the roots still continues to take place under these circumstances. This results in an increasingly high ion concentration with the cells, which causes more water to enter the root hair cells by osmosis. In terms of water potential, we say that active transport increases the solute potential of the roots. The result is movement of water into the plant and up the xylem columns despite the absence of transpiration. This phenomenon is called **root pressure,** which in reality is an osmotic phenomenon.

Under certain circumstances, root pressure is so strong that water will ooze out of a cut plant stem for hours or even days. When root pressure is very high, it can force water up to the leaves, where it may be lost in a liquid form through a process known as **guttation** (figure 35.7). Guttation does not take place through the stomata, but instead occurs through special groups of cells located near the ends of small veins that function only in this process. Root pressure is never sufficient to push water up great distances.

> **Water enters the plant by osmosis. Root pressure, which often occurs at night, is caused by the continued, active accumulation of ions in the roots at times when transpiration from the leaves is very low or absent.**

FIGURE 35.7

Guttation. In herbaceous plants, water passes through specialized groups of cells at the edges of the leaves; it is visible here as small droplets around the edge of the leaf in this strawberry plant (*Fragaria ananassa*).

Transpiration of Water from Leaves

More than 90% of the water that is taken in by the roots of a plant is ultimately lost to the atmosphere through transpiration from the leaves. It passes out in the form of water vapor through the stomata. Water passes into the pockets of air within the leaf from the moist surfaces of the walls of the mesophyll cells. As you saw in chapter 34, these intercellular spaces are in contact with the air outside of the leaf by way of the stomata. Water that evaporates from the surfaces of the mesophyll cells is continuously replenished from the tips of the veinlets in the leaves; the supply of water vapor in the intercellular spaces is therefore also continually replenished.

Conflicting Requirements of Photosynthesis and Water Retention

Growing plants depend on a continuous stream of water entering and leaving because they are consistently losing water to the atmosphere, and water is essential to their metabolic activities. Plants have two somewhat conflicting requirements: they need to minimize the loss of water to the atmosphere, and they need to admit carbon dioxide. Structural features such as stomata, the cuticle, and the substomatal chambers have evolved in response to one or both of these requirements. Before considering how this problem has been resolved in plants, however, we must consider absorption of water by roots.

Because of the interactions of the forces that we have just considered, water can be transported to the tops of tall trees and to the uppermost branches of vines. The tension that arises within the xylem of trees during periods of active transpiration may create such a strong pull on the conducting cells that their walls may be drawn closer together, and the diameter of the tree trunk as a whole may be millimeters less during the day than it is at night. Negative water potential can also occur in columns of older vessel members or tracheids as water is drawn laterally from newer xylem elements. This negative water potential created in a column of older xylem cells becomes better established as the season progresses from spring to autumn.

Each morning, the sun illuminates and warms the leaves and small branches of a tree, leading to an increased evaporation of water. The increased evaporation creates a pull in the veinlets of the leaves at the upper end of the water column. This pull extends to the water column in the small branches and on through those in the main trunk, ultimately creating a force that draws water in through the roots. After the sun sets and transpiration from the leaves decreases, the water potential becomes less negative, first in the upper portions of the plant. In the movement of water through the plant, the sun is the ultimate source of potential energy. The water potential that is responsible for water movement is largely the product of negative pressure generated by transpiration, which is driven by the warming effects of sunlight.

The Regulation of Transpiration Rate

On a short-term basis, water loss in plants may be controlled by the closing of the stomata. This occurs in many plants when they are subjected to water stress. However, the stomata must be open at least part of the time so that CO_2, which is necessary for photosynthesis, can enter the plant. A plant must respond both to the need to conserve water and to the need to admit CO_2.

As CO_2 enters the intercellular spaces, it must dissolve in water before it can enter the plant's cells. The gas dissolves mainly in water on the walls of the intercellular spaces below the stomata. These walls are kept moist by the continuous stream of water that reaches the leaves from the roots.

Stomata open and close because of changes in the turgor pressure of their guard cells. The sausage- or dumbbell-shaped guard cells stand out from other epidermal cells not only because of their shape, but also because they are the only epidermal cells containing chloroplasts. Their distinctive wall construction, which is thicker on the inside and thinner elsewhere, results in their bulging out and bowing when they become turgid (figure 35.8a).

Loss of turgor in guard cells causes the uptake of potassium (K^+) ions through ATP-powered ion transport channels in their plasma membranes. This creates a solute potential within the guard cells that causes water to enter osmotically. As a result, these cells accumulate water and become turgid, opening the stomata (figure 35.8b). Keeping the stomata open requires a constant expenditure of ATP, and the guard cells remain turgid only as long as ions are pumped into the cells. When active transport of ions into the guard cells ceases, the higher concentration of ions within the guard cells causes ions to diffuse out of the cells, which reduces the solute potential. Water then leaves the guard cells, which lose turgor, and the stomata close (figure 35.8c).

Potassium ions (K^+) are most important in controlling the turgidity of guard cells. Large numbers of K^+ are held in the cells surrounding the guard cells. The gradient of K^+ between the guard cells and the surrounding cells may change rapidly under certain circumstances, with K^+ moving from one to the other. When such a change occurs, the stomata may open or close rapidly. Photosynthesis in the guard cells apparently provides an immediate source of ATP, which drives the active transport of K^+ by way of a specific K^+ channel; this K^+ channel has now been isolated and studied. In some species, Cl^- accompany the K^+ in and out of the guard cells, thus maintaining electrical neutrality. In other species, H^+ (protons) move opposite K^+ movement.

When a whole plant wilts because there is insufficient water available, the guard cells may also lose turgor, and as a result, the stomata may close. The guard cells of many plant species regularly become turgid in the morning, when photosynthesis occurs, and lose turgor in the evening, regardless of the availability of water. When they are turgid, the stomata open, and CO_2 enters freely; when they are flaccid, CO_2 is largely excluded, but water loss is also retarded.

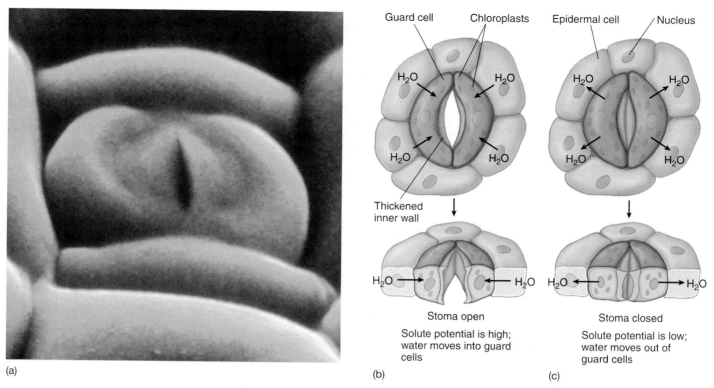

(a)

(b)

(c)

FIGURE 35.8

How a stoma opens and closes. (a) When potassium ions from surrounding cells are pumped into guard cells, the guard cell turgor pressure increases as water enters by osmosis. (b) The increased turgor pressure causes the guard cells to bulge, with the thick walls on the inner side of each guard cell bowing outward, thereby opening the stoma. (c) When the potassium ions leave the guard cells and their solute potential becomes low, they lose water and turgor, and the stoma closes.

Abscisic acid, a plant hormone discussed in chapter 39, plays a primary role in allowing K+ to pass rapidly out of guard cells, causing the stomata to close. This hormone, produced in leaves under water stress, binds to specific receptor sites in the plasma membranes of guard cells, where it brings about a direct response. Applying abscisic acid to crops grown in dry regions might prove useful in conserving water, provided that enough CO_2 could still reach the mesophyll of the plants whose stomata had thus been closed. Plants likely control the duration of their stomatal opening through the integration of several stimuli.

Other Factors Regulating Transpiration

Factors such as CO_2 concentration, light, and temperature can also affect stomatal opening. When CO_2 concentrations are high, guard cells of many plant species lose turgor, and their stomata close. Additional CO_2 is not needed at such times, and water is conserved when the guard cells are closed. The stomata also close when the temperature exceeds 30° C to 34° C. In the dark, stomata will open at low concentrations of CO_2. In chapter 10, we mentioned CAM photosynthesis, which occurs in some kinds of succulent plants such as cacti. In this process, CO_2 is taken in at night and fixed during the day. CAM photosynthesis conserves water in dry environments where succulent plants grow.

Many mechanisms to regulate the rate of water loss have evolved in plants. One involves dormancy during dry times of the year; another involves loss of leaves. Deciduous plants are common in areas that periodically experience a severe drought. Plants are often deciduous in regions with severe winters, when water is locked up in ice and snow and therefore unavailable to them. In a general sense, annual plants conserve water simply by not being present, except as seeds, when conditions are unfavorable.

Thick, hard leaves, often with relatively few stomata—and frequently with stomata only on the lower side of the leaf—lose water much more slowly than large, pliable leaves with abundant stomata. Temperatures are significantly reduced in leaves covered with masses of woolly-looking trichomes, and an increased humidity of the air at their surfaces is also maintained; both factors retard the loss of water vapor to the air. Plants in arid or semiarid habitats often have their stomata in crypts or pits in the leaf surface. Within these depressions the water vapor content of the air may be high, thereby reducing the rate of water loss.

Transpiration from leaves pulls water up the xylem. Stomata open when their guard cells become turgid. To maintain guard cell turgor, a constant expenditure of ATP is required to pump potassium ions into the guard cells, causing water to enter by osmosis.

Nutrient Movement

Ion Transport through the Xylem

Plants acquire ions and other essential nutrients through their roots. Although the details of the process are not well understood, the majority of ions entering a plant evidently, because of the Casparian strips, come in through the protoplasts of cells rather than between their walls. Ion passage through cell membranes appears to be an active, carrier-mediated, and selective process. We know that the movement of nutrients into roots is an active process that requires energy. As a result, specific ions can be maintained within the plant at much higher concentrations than those in the surrounding soil. When roots are deprived of oxygen, they lose their ability to absorb ions, an indication that the absorption process requires energy. A "starving" plant—one deprived of light—will eventually exhaust the available nutrient supply, and replacement nutrients in the soil will be, in effect, no longer available.

Ultimately, ions are removed from the roots and relocated through the xylem to other metabolically active parts of the plant. Phosphorus, potassium, nitrogen, and sometimes iron may be abundant in the xylem during certain seasons. In many plants, such a pattern of ionic concentration helps to conserve these essential nutrients, which may move from mature deciduous parts such as leaves and twigs to areas of active growth. Calcium, one of the essential ions, is not, however, transported elsewhere once it has been deposited in plant parts.

Carbohydrate Translocation through the Phloem

Most carbohydrates manufactured in leaves and other green parts are distributed through the phloem to the rest of the plant. This process, known as **translocation,** is responsible for the availability of suitable carbohydrate building blocks in roots and other actively growing regions of the plant. Carbohydrates concentrated in storage organs such as tubers, often in the form of starch, are also converted into transportable molecules, such as sucrose, and moved through the phloem. The pathway that sugars and other substances travel within the plant has been demonstrated precisely by using radioactive tracers, despite the fact that living phloem is delicate and the process of transport within it is easily disturbed.

Aphids, a groups of insects that extract plant sap for food, have been valuable tools in understanding translocation. Aphids thrust their *stylets* (piercing mouthparts) into phloem cells of leaves and stems to obtain abundant sugars there. When a feeding aphid is removed by cutting its stylet, the liquid from the phloem continues to flow through the detached mouthpart and is thus available in pure form for analysis (figure 35.9). The liquid in the phloem contains

(a)

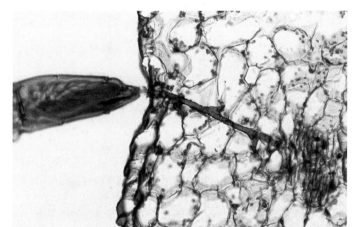

(b)

FIGURE 35.9
Feeding on phloem. (a) Aphids, like this individual of *Macrosiphon rosae* shown here on the edge of a rose leaf, feed on the food-rich contents of the phloem, which they extract through their piercing mouthparts (b), called stylets. When an aphid is separated from its stylet and the cut stylet is left in the plant, the phloem fluid oozes out of it and can then be collected and analyzed.

10% to 25% dry matter, almost all of which is sucrose. Using aphids to obtain the critical samples and radioactive tracers to mark them, it has been demonstrated that movement of substances in phloem can be remarkably fast; rates of 50 to 100 centimeters per hour have been measured.

The most widely accepted theory of how carbohydrates in solution move through the phloem has been called the **mass-flow hypothesis, pressure flow hypothesis,** or **bulk**

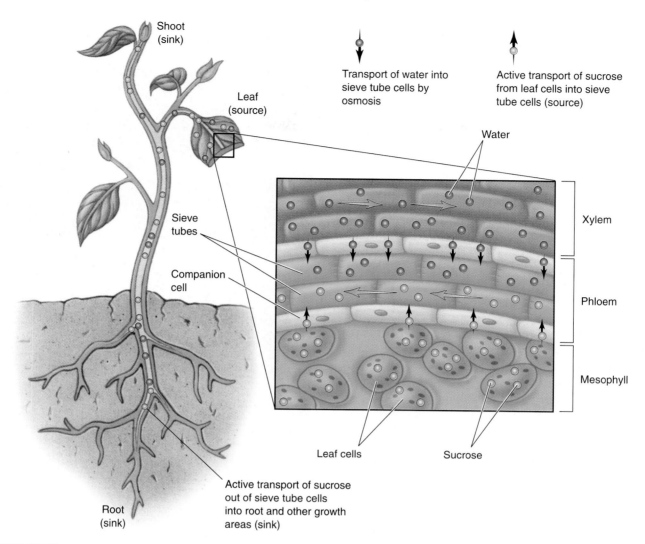

Shoot
(sink)

Leaf
(source)

Transport of water into
sieve tube cells by
osmosis

Active transport of sucrose
from leaf cells into sieve
tube cells (source)

Water

Sieve
tubes

Xylem

Companion
cell

Phloem

Mesophyll

Leaf cells

Sucrose

Active transport of sucrose
out of sieve tube cells
into root and other growth
areas (sink)

Root
(sink)

FIGURE 35.10
Diagram of mass flow. In this diagram, red dots represent sucrose molecules, and blue dots symbolize water molecules. Moving from the mesophyll cells of a leaf or another part of the plant into the conducting cells of the phloem, the sucrose molecules are then transported to other parts of the plant by mass flow and unloaded where they are required.

flow hypothesis. According to this theory, much of which has been experimentally demonstrated, dissolved carbohydrates flow from a **source**, where water is taken up by osmosis, and are released at a **sink**—a place where they are utilized. Carbohydrate sources include tissues where photosynthesis occurs, such as the mesophyll of leaves, and food-storage tissues, such as cortex of roots. Sinks occur primarily at the growing tips of roots and stems and in developing fruits.

In a process known as *phloem loading*, carbohydrates (mostly sucrose) enter the sieve tubes of the smallest veinlets at the source through active transport, thereby decreasing the sieve tubes' water potential. Companion cells and parenchyma cells adjacent to the sieve tubes provide the ATP energy to drive this transport. Then, because of the difference between the water potentials in the sieve tubes and in the nearby xylem cells, water flows into the sieve

tubes by osmosis. As the osmosis occurs, turgor pressure in the sieve tubes increases; this increased turgor pressure drives the fluid throughout the plant's system of sieve tubes. At the sink, water also leaves as the carbohydrates are actively removed, and the turgor pressure there drops, causing a mass flow from the higher pressure at the source to the lower pressure (figure 35.10). Most of the water diffuses then back into the xylem, where it may either be recirculated or lost through transpiration.

Accumulation of ions by plants is an active process that requires the expenditure of energy and usually takes place against a concentration gradient. Transport of sucrose and other carbohydrates through sieve tubes does not require energy. The loading and unloading of these substances from the sieve tubes does.

Flooding: Too Much of a Good Thing

The continued existence and growth of plants depends on an adequate supply of water. Plants lack a closed circulation system, and only the continuous stream of water that flows through them keeps them healthy.

Plant Responses to Flooding

Even plants, however, can receive too much water. This occurs when the soil is flooded, a condition that can arise when rivers or streams overflow their banks or when rainfall is heavy, irrigation is excessive, or drainage is poor. Flooding rapidly depletes the available oxygen in the soil and blocks the normal reactions that take place in roots and make possible the transport of minerals and carbohydrates. Abnormal growth patterns may result, and the plants may ultimately "drown." Hormone levels change in flooded plants—ethylene, for example, often increases, while gibberellins and cytokinins usually decrease—and these changes may also contribute to the abnormal growth patterns. Flooding involving moving water, which brings in new supplies of oxygen, is much less harmful than flooding involving standing water, which does not. Flooding that occurs when a plant is dormant is much less harmful than flooding when it is growing actively.

Physical changes that occur in the roots as a result of oxygen deprivation may halt the flow of water through the plant, paradoxically drying out the leaves—even though the roots of the same plant may be standing in water. Because of such stresses, the stomata of flooded plants often close. In some plants, the closing of the stomata maintains leaf turgor.

FIGURE 35.11
Adaptation to flooded conditions. The "knees" of the bald cypress (*Taxodium*) form whenever it grows in wet conditions, increasing its ability to take in oxygen.

Adapting to Life in Fresh Water

Many plants, of course, grow in places that are often or always flooded naturally; they have adapted to these conditions during the course of their evolution (figure 35.11). One of the most frequent adaptations among such plants is the formation of **aerenchyma,** loose parenchymal tissue with large air spaces in it. Aerenchyma is very prominent in water lilies and many other aquatic plants. Oxygen may be transported from the parts of the plant above water to those below by way of passages in the aerenchyma. This supply of oxygen allows oxidative respiration to take place even in the submerged portions of the plant.

Some plants normally form aerenchyma, whereas others, subject to periodic flooding, can form it when necessary. In corn, ethylene, which becomes abundant under the anaerobic conditions of flooding, induces aerenchyma formation. Plants also respond to flooded conditions by forming larger lenticels (which facilitate gas exchange) and additional adventitious roots.

Adapting to Life in Salt Water

Plants such as mangroves that are normally flooded with salt water must not only provide a supply of oxygen for their submerged parts, but also control their salt balance. The salt must be excluded, actively secreted, or diluted as it enters. The arching silt roots of mangroves are connected to long, spongy, air-filled roots that emerge above the mud. These spongy air roots have large lenticels on their above-water portions through which oxygen enters; it is then transported to the submerged roots. In addition, the succulent leaves of mangroves contain large quantities of water, which dilutes the salt that reaches them. Many plants that grow in such conditions secrete large quantities of salt through more or less specialized glands.

Plants are harmed by excess water. However, plants can survive flooded conditions, and even thrive in them, if they can deliver oxygen to their submerged parts.

35.1 Evaporation of water from leaves draws water up the stem.

- Water flows through plants in continuous columns, with transpiration from the stomata exerting a pull on the columns.

35.2 Plants absorb water into their roots and lose it from their leaves.

- Water enters plant roots by osmosis. The cohesion of water molecules and their adhesion to the walls of the very narrow cell columns through which they pass are additional important factors in maintaining the flow of water to the tops of plants.

- Water loss is controlled primarily by the closing of a plant's stomata. Root pressure develops when the stomata are closed, and a high osmotic concentration builds up in the root's cells. This pressure forces water up the xylem when transpiration is absent. Root pressure is mostly only a minor factor in the flow of water through plants.

- Stomata open when their guard cells are turgid and bulge, causing the thickened inner walls of these cells to bow away from the opening. Abscisic acid, which is produced in the leaf under conditions of water stress, allows potassium ions to move rapidly out of the guard cells; these cells then lose their turgidity and the stomata close.

35.3 Nutrients move up the stem, carbohydrates down.

- Nutrients move through plants primarily as solutes in the water columns of the xylem. Their selective admission into the plant and their subsequent movement between living cells require the expenditure of energy.

- The movement of water containing dissolved sucrose and other substances in the phloem requires energy. Sucrose is loaded into the phloem near sites of synthesis, or sources, using energy supplied by the companion cells or other nearby parenchyma cells. The sucrose is unloaded in sinks (places where it is required). The water potential drops where the sucrose is loaded into the sieve tubes, and the turgor pressure drops where it is unloaded.

35.4 Plants, although terrestrial, can adapt to life in water.

- Flooding can kill plants by preventing oxygen from reaching submerged portions of the plant.

- Plants can tolerate long submersion in water, however, if they can deliver oxygen to their submerged tissues.

1. **Phloem and xylem** Structurally, a typical plant may be likened to a cylinder filled with cells and tubes, partially embedded in soil, with roots below, and with leaf-bearing shoots above. Sugars produced by photosynthesis in the shoot are transported in solution through specialized conducting cells of phloem, and water with dissolved minerals that enter through the roots is transported through specialized conducting cells of xylem.

2. **Water** Most of a plant's water enters through the roots. Water rises in the plant because transpiration from the leaves produces a tension on the interior columns of water that extends all the way down to the roots. More than 90% of the water that enters a plant is lost through the leaves by transpiration.

3. **Root pressure** Root pressure is caused by the active accumulation of ions in a plant's roots, often at night, when transpiration from the leaves is relatively low.

4. **Stomata** Stomata are specialized pores in the leaves, and sometimes in stems and other above-ground plant parts; they are each flanked by a pair of guard cells that cause the stomata to open when they become turgid after potassium ions accumulate in them.

5. **Carbohydrate transport** Plants must transport carbohydrates from source sites to sites of use. Sugars produced by photosynthesis in the leaf cells enter phloem sieve tubes at their source by means of active transport. The increased sugar concentration causes water from the xylem to diffuse into the sieve tubes, thereby increasing turgor pressure. The higher turgor pressure drives the fluid in the sieve tubes toward a sink (cells where the sugars are removed); this causes the pressure in the sieve tubes at the sink to drop, which, in turn, causes a mass flow from the higher pressure at the source to the lower pressure at the sink. Most of the water then diffuses back into the xylem.

Review Questions

1. What is pressure potential? How does it differ from solute potential? How do these pressures cause water to rise in a plant? Are they positive pressures or negative pressures?

2. What proportion of water that enters a plant leaves it via transpiration? What leaf structures help regulate this activity? Why must plants be exposed to a constant source of water?

3. Why are root hairs usually turgid? Does the accumulation of minerals within a plant root require the expenditure of energy? Why or why not? Is mineral passage into most plant cells a selective or a nonselective process? What is an exception to this generalization? Why?

4. Under what environmental condition is water transport through the xylem reduced to near zero? How much transpiration occurs under these circumstances? Does active transport at the roots continue under these conditions? What is the result? What term is associated with this phenomenon?

5. What is guttation? Does it occur through the leaf stomata? Is it more likely to occur in very tall or very short plants? Why?

6. Why is a tree's diameter often smaller during the day than it is at night? What is the ultimate source of a plant's water potential?

7. Does stomatal control require energy? Explain. Which ion is an integral part of this control mechanism? Which epidermal cells possess chloroplasts? Why? When is abscisic acid produced in leaf tissue? What is its effect on leaf cells?

8. What are some mechanisms that have evolved in plants to regulate their rate of water loss? What types of leaves lose water more slowly? How do trichomes help to reduce water loss?

9. What is translocation? In what form are plant carbohydrates generally stored? In what form are they transported? What is the driving force behind translocation?

10. Describe the movement of carbohydrates through a plant, beginning with the source and ending with the sink. Is this process active or passive? How are the companion cells involved in this process?

Thought Questions

1. Why do gardeners often remove many of a plant's leaves after transplanting it?

2. When plant roots are deprived of oxygen, they lose their ability to absorb ions. Why is this? What does this say about the ion absorption process?

Internet Links

Angiosperm Anatomy
http://www.botany.uwc.ac.za/sci_ed/std8/anatomy/
A very useful introduction to plant anatomy, from Western Cape University, South Africa.

The Virtual Plant
http://www.ctpm.uq.edu.au/Programs/IPI/ipivp.html
Computer simulations of the structural development and growth of individual plants in 3-D space. A fun site.

Is It a Monocot or a Dicot?
http://www.ucmp.berkeley.edu/glossary/gloss8/monocotdicot.html
This site explores a distinction that is not as clear-cut as simple definitions imply.

Test Yourself
http://www.colorado.edu/itrc/quiz/halloran/EPOB1220888.6.html
An interactive quiz on plant structure and growth from the University of Colorado; for a quiz on transport in plants, try EPOB1220888.7.html; for one on plant nutrition, try EPOB1220888.8.html.

For Further Reading

Borghetti, M., and others (editors): *Analysis of Water Transport in Plants*, Cambridge University Press, New York, 1993. An up-to-date survey of current research in water transport.

Canny, M. J.: "Transporting Water in Plants," *American Scientist*, vol. 86, pages 152–159, March-April 1998. An up-to-date review of what is known about how evaporation from leaves pulls water to the tops of trees, aided by living cells that protect the stretched water.

Chapin, F. S., III: "Integrated Responses of Plants to Stress," *BioScience*, vol. 41, pages 29–36, 1991. Describes the centralized system of physiological responses in plants.

Galston, A. W.: *Life Processes of Plants: Mechanisms for Survival*, W. H. Freeman, New York, 1993. A good overview of transpiration and translocation in plants.

Loughman, B. C., et. al.: *Structural and Functional Aspects of Transport in Roots*, Kluwer Academic Publications, Norwell, MA, 1990. A detailed look at the movement of water and solutes through roots.

Marschner, H.: *Mineral Nutrition of Higher Plants*, Academic Press, New York, 1986. A comprehensive account of the subject.

Mooney, H. A., and others: "Plant Physiological Ecology Today," *BioScience*, vol. 37, pages 18–67, 1988. Nearly an entire issue of *BioScience* devoted to the ways in which plants adapt to their environments and how scientists study them. Highly recommended.

Salisbury, F. B. and C. W. Ross: *Plant Physiology*, 4th ed., Wadsworth, Belmont, CA, 1992. An outstanding textbook of plant physiology.

Zimmermann, M. H.: *Xylem Structure and the Ascent of Sap*, Springer-Verlag, New York, 1983. A delightfully written "idea" book on xylem structure and function by one who has contributed a great deal to our understanding of functional xylem anatomy and sap movement in plants.

36

Plant Nutrition

Concept Outline

36.1 Plant growth is significantly influenced by the nature of the soil.

The Soil. Soils, produced by the weathering of rocks, also contain organic matter and bound water.

Cultivation. Soil fertility can be preserved by rotating crops, allowing fields to lie fallow, or plowing under crops.

Fertilizers. Fertility of agricultural soils can be maintained with fertilizers that supply nitrogen, phosphorus, and potassium.

36.2 Plants require quite a lot of some nutrients, but only trace amounts of others.

Plant Nutrients. Plants require a few macronutrients in large amounts and several micronutrients in lesser amounts.

36.3 Some plants obtain nutrients by capturing and digesting insects.

Carnivorous Plants. Venus flytraps and other carnivorous plants lure and capture insects and then digest them to obtain energy and nutrients.

FIGURE 36.1
A carnivorous plant. Most plants absorb water and essential nutrients from the soil, but carnivorous plants are able to obtain some nutrients directly from small animals.

Just as animals need certain nutrients, such as carbohydrates, amino acids, and vitamins, to survive, plants also need various nutrients to remain alive and healthy. Lack of an important nutrient may slow a plant's growth or make the plant more susceptible to disease or even death. Plants acquire these nutrients primarily from soil through their roots, although some take a more direct approach (figure 36.1). In this chapter, we will discuss the essential plant nutrients and how these nutrients cycle through the plant's environment. But first, we will examine soil, the substrate in which nearly all plants grow.

The Soil

Soil is the highly weathered outer layer of the earth's crust. It is composed of a mixture of ingredients, which may include sand, rocks of various sizes, clay, silt, humus, and various other forms of mineral and organic matter; pore spaces containing water and air occur between the particles. The mineral fraction of soils varies according to the composition of the rocks. The crust includes about 92 naturally occurring elements; table 2.1 in Chapter 2 lists the most common of these elements and their percentage of the earth's crust by weight. Most elements are combined as inorganic compounds called **minerals;** most rocks consist of several different minerals.

Composition of Soil

Millions of years of weathering of rocks by physical and biological processes have produced the soil's mineral particles. The thousands of soils on earth show unmistakable signs of biological processes: these signs are both direct (evident in the structure of soils and their organic content) and indirect. For example, the atmosphere itself, which plays the major role in weathering rocks on earth, has changed profoundly over time by the effects of biological evolution. Biological effects are obvious not only in the upper layers of soil, where organisms and organic debris are abundant and diverse (figure 36.2), but also lower down, where roots penetrate and organic material is also deposited. The ways in which organic debris is broken down and recycled (mainly as a result of the activities of organisms living in the soil) have a great deal to do with the fertility of the soil and play a key role in controlling the cycling of substances through it.

Most roots occur in **topsoil** (figure 36.3), which is a mixture of mineral particles of varying size (most less than 2 mm thick), living organisms, and **humus.** Humus consists of partly decayed organic material. When topsoil is lost

FIGURE 36.2
Soil organisms. About 5 metric tons of carbon is tied up in the organisms that are present in the soil under a hectare (0.06 mile^2) of wheat land in England, an amount that approximately equals the weight of 100 sheep!

FIGURE 36.3
Most roots occur in topsoil. The extensive root systems of these grasses help to anchor topsoil and prevent soil erosion.

because of erosion or poor landscaping, both the water-holding capacity and the nutrient relationships of the soil are adversely affected.

Soil can contain particles of varying sizes, from coarse sand, with particles 200 to 2000 micrometers in diameter and visible to the naked eye, to clay, with microscopic particles that are less than 2 micrometers in diameter. The size of particles in fine sand and silt lies between these two extremes. Most soils are a mixture of particles of different sizes, but may also contain larger rocks (figure 36.4).

About half of the total soil volume is occupied by spaces or pores, which may be filled with air or water, depending on moisture conditions. Some of the soil water, because of its properties described below, is unavailable to plants.

Water in Soil

You may recall from chapter 2 that the chemical and physical behavior of water is dictated largely by its tendency to form hydrogen bonds with itself and other materials. Two aspects of the behavior of water are particularly important here. First, water is tightly bound to objects that bear electrostatic charges, as well as to those that are able to form hydrogen bonds. Many soils contain minerals that have exactly these properties. Clays, for example, are often negatively charged, and water binds tightly to clay particles by hydrogen bonds.

Second, the more particles there are in a given quantity of soil, the greater the available surface area to which water can adhere. In other words, more water will adhere to soils with very small particles than to soils with large particles. Accordingly, water drains rapidly through sandy soils, which consist of relatively coarse particles. Clay soils hold a great deal more water, but much of this water may be held so tightly because of the electrostatic charges and hydrogen bonds that it is unavailable to plants. For these reasons, a soil composed of a balanced mixture of coarse and fine particles—of sand, silt, and clay—is ideal for plant growth. Such soils are known as **loams.**

Due to gravity, some of the water that reaches a given soil will drain through it immediately. Another fraction of the water is held in small soil pores, which are generally less than about 50 micrometers in diameter. This water, which is readily available to plants, is called **capillary water.** The **field capacity** of the soil is the amount of water held in a given soil after the excess has been removed by gravity. When evaporation and removal by plants has eliminated the capillary water from the soil, leaving only water that is not available to plants, that soil has reached its **permanent wilting point.** This is defined as the point at which plants will wilt permanently in a given soil unless additional water is added; it is measured as moisture content.

FIGURE 36.4

Plants grow in a variety of substrates. Soils contain rocks, but usually not ones this big. In this rare example, a live oak grew from an acorn lodged in a small crack in a large rock. The acorn imbibed runoff water, and a root is now widening the split.

Soils, which are produced by the weathering of their parent rocks, differ in composition. The amount of organic matter in soils affects their fertility and other properties. Water is tightly bound to objects that bear electrostatic charges and to those that form hydrogen bonds. The more soil particle surface area in a given soil, the more water it will bind. For these reasons, water runs through sandy soils but is so tightly bound to clay soils that it is often unavailable to plants.

Cultivation

Natural processes, such as decomposition and nitrogen fixation, continually add nutrients to the soil. However, plants can remove available nutrients from the soil around the roots more quickly than the nutrients can be replaced. The nutrients that plants take from the soil become part of the plants. These nutrients must be replaced if the soil is to maintain its **fertility,** its ability to supply plants with nutrients. Many of the nutrients taken from soil by plants are returned to the soil when the plants die and decay. However, few nutrients are returned to the soil when a crop is harvested. When another crop is planted in the same soil, there may not be enough nutrients in the soil for the crop to grow.

Over the centuries, farmers have developed many methods to help the soil maintain its fertility. For example, a farmer might grow corn in a field one year and soybeans the next year. Both crops remove nutrients from the soil, but the plants have different nutritional requirements, and therefore the soil does not lose the same nutrients two years in a row. Soybean plants even add nitrogen compounds to the soil, released by nitrogen-fixing bacteria growing in nodules on their roots. Alternating two or more crops in the same field from year to year is called **crop rotation.** Sometimes farmers allow a field to lie fallow—that is, they do not grow a crop in the field for a year or two. This allows natural processes to rebuild the field's store of nutrients.

Other farming practices that help maintain soil fertility involve plowing under plant material left in fields (figure 36.5). You can do the same thing in a lawn or garden by leaving grass clippings and dead leaves. Farmers often plow crop residues such as stems, roots, husks, and leaves back into the soil. Occasionally farmers grow a crop not for harvest, but strictly for plowing back into the field. Plowing under plant material in this manner adds organic matter to the soil. Decomposers in the soil do the rest, turning the plant material into humus.

Farming practices like crop rotation and plowing crops under can significantly increase the level of nutrients available to crop plants.

**FIGURE 36.5
Plowing under plant material.** This farmer is mowing a crop of alfalfa, which is a legume. The plants will be plowed into the soil as "green manure," which adds humus to the soil.

Fertilizers

In natural communities, nutrients are recycled and made available to organisms on a continuous basis. When these communities are replaced by cultivated crops, the situation changes drastically. First, the soil is much more exposed to erosion and the loss of nutrients, which may be removed with the crops themselves or with the animals to which the crops are fed. For this reason, cultivated crops and garden plants usually must be supplied with additional mineral nutrients. In the tropics, soils are often particularly poor in nutrients, and fertilizers significantly increase their yield.

The most important mineral nutrients that need to be added to soils are nitrogen (N), phosphorus (P), and potassium (K). All of these elements are needed in large quantities (see table 36.1) and are the most likely to become deficient in the soil. But because they are often added in large quantities, they can also be significant sources of pollution in certain situations (see chapter 25). Commercial fertilizers are normally assigned a "grade" that reflects the percentages of N, P, and K they contain by dry weight; thus, 6-6-6 means 6% of each of these elements (N, P, and K) in the fertilizer (figure 36.6). The optimal proportions are best determined by testing the fertility of the soil and knowing the requirements of the particular crop or plant to be grown. Nitrogen is often supplied to crops in very large amounts, since it seems to have a major effect on crop yield. Nutrients other than these three may also be scarce in specific soils, but such situations are rare and must be dealt with individually.

Organic fertilizers were widely used long before chemical fertilizers were available. Substances such as manure or the remains of dead animals have traditionally been applied to crops, and plants are often plowed under to increase the soil's fertility. There is no basis for believing that organic fertilizers supply any element to plants that inorganic fertilizers cannot provide. However, organic fertilizers build up the humus content of the soil, which often enhances its water- and nutrient-retaining properties. For this reason nutrient availability to plants at different times of the year may be improved, under certain circumstances, with organic fertilizers.

The most important mineral nutrients added to soils in fertilizers are nitrogen, phosphorus, and potassium.

GUARANTEED ANALYSIS	
Total Nitrogen (N)	12%
4.7% Ammoniacal Nitrogen	
7.3% Urea Nitrogen	
Available Phosphoric Acid (P_2O_2)	12%
Soluble Potash (K_2O)	12%

FIGURE 36.6
Nutrients in a commercial fertilizer. The three large numbers on the label of this fertilizer bag show its NPK rating, which is the percentage of available nitrogen, phosphorus, and potassium in the fertilizer.

Plant Nutrients

Plants require a number of inorganic nutrients (table 36.1). Some of these are macronutrients, which the plants need in relatively large amounts, and others are micronutrients, which are required in trace amounts. There are nine macronutrients: carbon, hydrogen, and oxygen—the three elements found in all organic compounds—as well as nitrogen, potassium, calcium, phosphorus, magnesium, and sulfur. Each of these nutrients approaches or, as in the case with carbon, may greatly exceed 1% of the dry weight of a healthy plant. The seven micronutrient elements—iron, chlorine, copper, manganese, zinc, molybdenum, and boron—constitute from less than one to several hundred parts per million in most plants.

The macronutrients were generally discovered in the last century, but the micronutrients have been detected much more recently because they are required in such small quantities. When they study nutritional requirements, scientists usually grow plants in hydroponic cultures; the plants roots are suspended in aerated water containing nutrients.

The solutions in which test plants are grown contain all the necessary nutrients in the right proportions but with certain known or suspected nutrients left out. The plants are then allowed to grow and are studied for the presence of abnormal symptoms that might indicate a need for the missing element (figure 36.7). However, the water or vessels used often contain enough micronutrients to allow the plants to grow normally, even though these substances were not added deliberately to the solutions. To give an idea of how small the quantities of micronutrients may be, the standard dose of molybdenum added to seriously deficient soils in Australia amounts to about 34 grams (about one handful) per hectare, once every 10 years! Most plants grow satisfactorily in hydroponic culture, and the method, although expensive, is occasionally practical for commercial purposes.

The plant macronutrients carbon, oxygen, and hydrogen constitute about 94% of a plant's dry weight; the other macronutrients nitrogen—potassium, calcium, phosphorus, magnesium, and sulfur—each approach or exceed 1% of a plant's dry weight.

Elements	Principal Form in which Element Is Absorbed	Approximate Percent of Dry Weight	Examples of Important Functions
Table 36.1	**Essential Nutrients in Plants**		
MACRONUTRIENTS			
Carbon	(CO_2)	44	Major component of organic molecules
Oxygen	(O_2, H_2O)	44	Major component of organic molecules
Hydrogen	(H_2O)	6	Major component of organic molecules
Nitrogen	(NO_3^-, NH_4^+)	1–4	Component of amino acids, proteins, nucleotides, nucleic acids, chlorophyll, coenzymes, enzymes
Potassium	(K^+)	0.5–6	Protein synthesis, operation of stomata
Calcium	(Ca^{++})	0.2–3.5	Component of cell walls, maintenance of membrane structure and permeability, activates some enzymes
Magnesium	(Mg^{++})	0.1–0.8	Component of chlorophyll molecule, activates many enzymes
Phosphorus	$(H_2PO_4^-, HPO_4^=)$	0.1–0.8	Component of ADP and ATP, nucleic acids, phospholipids, several coenzymes
Sulfur	$(SO_4^=)$	0.05–1	Components of some amino acids and proteins, coenzyme A
MICRONUTRIENTS (CONCENTRATIONS IN PPM)			
Chlorine	(Cl^-)	100–10,000	Osmosis and ionic balance
Iron	(Fe^{++}, Fe^{+++})	25–300	Chlorophyll synthesis, cytochromes, nitrogenase
Manganese	(Mn^{++})	15–800	Activator of certain enzymes
Zinc	(Zn^{++})	15–100	Activator of many enzymes, active in formation of chlorophyll
Boron	$(BO_3^- \text{ or } B_4O_7^=)$	5–75	Possibly involved in carbohydrate transport, nucleic acid synthesis
Copper	(Cu^{++})	4–30	Activator or component of certain enzymes
Molybdenum	$(MoO_4^=)$	0.1–5	Nitrogen fixation, nitrate reduction

FIGURE 36.7

Mineral deficiencies in plants. Individuals of Marglobe tomatoes (*Lycopersicon esculentum*) are grown in hydroponic solution (water with selected nutrients added). (a) Control (healthy) plant, grown in a complete nutrient solution (one that contains all of the essential minerals), with a plant showing the symptoms of molybdenum deficiency. Molybdenum was not added to the nutrient solution of the plant at right, resulting in its unhealthy appearance. (b) Leaves of a healthy control plant. (c) Chlorine-deficient plant with necrotic leaves (leaves with patches of dead tissue). (d) Copper-deficient plant with blue-green, curled leaves. (e) Zinc-deficient plant with small, necrotic leaves. (f) Manganese-deficient plant with chlorosis (yellowing) between the veins. This chlorosis is caused by a drastic reduction in chlorophyll production. The agricultural implications of deficiencies such as these are obvious; a trained observer can determine the nutrient deficiencies that are affecting a plant simply by inspecting it.

Carnivorous Plants

Some plants are able to obtain nitrogen directly from other organisms, just as animals do. These carnivorous plants often grow in acidic soils, such as bogs—habitats not favorable for the growth of most legumes or of nitrifying bacteria. By capturing and digesting small animals directly, such plants obtain adequate nitrogen supplies and thus are able to grow in these seemingly unfavorable environments.

Carnivorous plants have adaptations used to lure and trap insects and other small animals (figure 36.8). The plants digest their prey with enzymes secreted from various types of glands.

The Venus flytrap *(Dionaea muscipula)*, which grows in the bogs of coastal North and South Carolina, has three sensitive hairs on each side of each leaf, which, when touched, trigger the two halves of the leaf to snap together. Once the Venus flytrap enfolds a prey item within a leaf, enzymes secreted from the leaf surfaces digest the prey.

Pitcher plants attract insects by the bright, flower-like colors within their pitcher-shaped leaves and perhaps also by sugar-rich secretions. Once inside the pitchers, insects slide down into the cavity of the leaf, which is filled with water and digestive enzymes.

Bladderworts, *Utricularia*, are aquatic. They sweep small animals into their bladder-like leaves by the rapid action of a spring-like trapdoor, and then they digest these animals. In the sundews, the glandular trichomes secrete both sticky mucilage, which traps small animals, and digestive enzymes.

Carnivorous plants obtain nutrients directly by capturing and digesting insects and other organisms.

(a)

(b)

(c)

(d)

(e)

FIGURE 36.8
Carnivorous plants. (a) The Venus flytrap *(Dionaea muscipula)* inhabits low, boggy ground in North and South Carolina. (b) A Venus flytrap has snapped together, imprisoning a fly. (c) A tropical Asian pitcher plant, *Nepenthes*. Insects enter the pitchers and are trapped and digested. Complex communities of invertebrate animals and protists inhabit the pitchers. (d) The yellow pitcher plant *(Sarracenia flava)* grows in bogs in the southeastern United States. Its pitchers, with their yellow borders, resemble flowers and secrete a sweet-smelling nectar that aids the plants in trapping insects. The flowers of this species, with their hanging petals, are also evident in this photograph. (e) Sundew *(Drosera)*. A small fly has been trapped by the glandular hairs.

36.1 **Plant growth is significantly influenced by the nature of the soil.**

- Soils are formed by the weathering of rocks and the accumulation of organic matter. Most soils have been deeply modified by the action of biological processes.
- Clay soils, which are composed of very fine particles, hold a great deal of water, but much of it adheres so tightly to the particles that it is unavailable for plant growth. Sandy soils, in contrast, hold water poorly, but the water present is readily available to plants.
- Agricultural practices such as crop rotation and plowing under crops can help preserve soil fertility.
- Fertilizers supply nitrogen, phosphorus, and potassium to depleted soils.

36.2 **Plants require quite a lot of some nutrients, but only trace amounts of others.**

- The nine macronutrients, substances that are each present at concentrations of 1% or more of a plant's dry weight, are carbon, hydrogen, oxygen, nitrogen, potassium, calcium, phosphorus, magnesium, and sulfur.
- The seven micronutrients, each present at concentrations from one to several hundred parts per million of dry weight, are iron, chlorine, copper, manganese, zinc, molybdenum, and boron.
- Macronutrients and micronutrients play different roles in plant metabolism.

36.3 **Some plants obtain nutrients by capturing and digesting insects.**

- Carnivorous plants have adaptations to lure and trap insects and other small animals.
- Carnivorous plants often grow in unfavorable environments but are able to obtain nitrogen and other nutrients from the animals they capture and digest.

1. **Soil** The soil, the highly weathered outer layer of the earth's crust, is composed of minute rock fragments mixed with organic matter. Soil textures differ greatly, depending on the sizes of their weathered mineral particle components and organic content.

2. **Field capacity** The field capacity of a soil is the amount of water held in the soil after the excess has been removed by gravity.

3. **Topsoil** Topsoil is a rich surface mixture of mineral particles and humus, which is partly decayed organic material.

4. **Fertilizer** The most important mineral nutrients that need to be replenished in large amounts to maintain soil fertility are nitrogen, phosphorus, and potassium. Commercial fertilizers are mixtures of these three nutrients.

5. **Plant nutrients** The plant macronutrients—carbon, hydrogen, oxygen, nitrogen, potassium, calcium, phosphorus, magnesium, and sulfur—make up roughly 98% of a plant's dry weight, while the micronutrients—iron, chlorine, copper, manganese, zinc, molybdenum, and boron—are present only in minute amounts. Each of these elements is necessary for growth.

6. **Macronutrients** A macronutrient is a major constituent of the plant body. There are nine plant macronutrients (*macro*, large), each constituting at least 1% of the dry weight of healthy plants.

7. **Micronutrients** A micronutrient is a necessary but trace constituent of the plant body. There are seven principle micronutrients, each present in a few to several hundred parts per million.

Review Questions

1. How are sand particles different from clay particles? Why does water pass quickly through sandy soils but is held tightly by clay soils? Why is the water in sandy soils more available to plants than the water in clay soils?

2. What is capillary water? What is meant by a soil's field capacity? When is a soil said to have reached its permanent wilting point?

3. What are macronutrients? List them. In what ways are they important to a plant's survival? What are micronutrients? What are the primary micronutrients? What are the most common ways in which they function?

4. Why are fertilizers generally required in cultivated environments when they are not needed in natural communities? What three mineral nutrients are most commonly found in commercial fertilizers? For what reason(s) may organic fertilizers be better than chemical fertilizers?

Thought Questions

1. Is a legume likely to become a carnivorous plant? Why or why not? If you did find a carnivorous legume, what else would you expect to be true about it?

2. If you grew a potted plant that initially weighed 200 grams, but eventually weighed 50 kilograms, would you expect the soil in the pot to change weight? If so, how much, and why?

Internet Links

Say Hello to a Tomato
http://moby.ucdavis.edu:80/Coursepages/PLB105/students/Tomato/Intro/Intro/PLB.htm
This neat INTRODUCTION TO THE TOMATO site from UC Davis teaches plant anatomy by focusing carefully on one typical angiosperm, the tomato.

Plant Tissues
http://www.botany.uwc.ac.za/sci_ed/std8/plant_tissues/
Part of an on-line, very nicely diagrammed text on plant structure and function.

Careers in Botany
http://www.ou.edu/cas/botany-micro/careers/
This is an online version of the brochure "Careers in Botany" distributed by the Botanical Society of America.

Library of Plant Photographs
gopher://wiscinfo.wisc.edu:70/11/.data/.bot/.130new
From the University of Wisconsin, a large collection of excellent photographs, slides, and illustrations concerning plant physiology and anatomy, indexed usefully.

For Further Reading

Bem, R.: *Everyone's Guide to Home Composting*, Van Nostrand Reinhold, New York, 1978. A comprehensive, easy-to-read manual.

Campbell, S.: *The Gardener's Guide to Composting*, revised edition, Storey Publishing, Pownal, VT, 1990. A concise presentation of all you need to know about composting.

Campbell, S.: *The Mulch Book: A Complete Guide for Gardeners*, Storey Publishing, Pownal, VT, 1991. A concise presentation of all you need to know about mulching.

Epstein, E.: *Mineral Nutrition of Plants: Principles and Perspectives.* John Wiley & Sons, New York, 1972. A thorough coverage of the whole field of mineral nutrition, well-illustrated and written by one of the leading students of the subject.

Gresshoff, P. (editor): *Molecular Biology of Symbiotic Nitrogen Fixation*, CRC Press, Boca Raton, FL, 1990. A multiauthored book summarizing current knowledge, which is undergoing rapid expansion.

Lewis, O.: *Plants and Nitrogen*, Studies in Biology No. 166, Edward Arnold Publishers, London, 1986. A concise introduction to the association that exists between plants and nitrogen. Ecological, physiological, and biochemical aspects are combined to provide the reader with an overall view of nitrogen in the biosphere.

Lipske, M.: "Forget Hollywood: These Bloodthirsty Beauties Are For Real," *Smithsonian*, vol. 23, pages 49–59, 1992. Lively description of carnivorous plants and the biological communities that live and function within and on their leaves.

Marschner, H.: *Mineral Nutrition of Higher Plants*, Academic Press, New York, 1986. A comprehensive account of mineral nutrition in plants.

Mengel, K. and E. A. Kirkby: *Principles of Plant Nutrition*, 3rd ed., International Potash Institute, Worblaufen-Bern, Switzerland, 1982. A textbook for students interested in plant science and crop production; integrates soil science, plant physiology, and biochemistry.

37

Plant Reproduction

Concept Outline

37.1 Many plants can clone themselves by asexual reproduction.

Asexual Reproduction. Some plants do without sexual reproduction, instead cloning new individuals from parts of themselves.

37.2 Angiosperms utilize temporary reproductive structures called flowers.

Rise of the Flowering Plants. Angiosperms, the flowering plants, are the most successful of all plants, with almost a quarter million species.

Evolution of the Flower. A complete flower has four whorls containing protective sepals, attractive petals, male stamens, and female ovules.

Formation of Angiosperm Gametes. In most flowers, pollen and egg formation occur in the same individual flowers.

37.3 Flowering plants use animals or wind to transfer pollen between flowers.

Pollination. Insects and wind are the two most common pollinators of angiosperm flowers.

Self-Pollination versus Outcrossing. Self-pollination is favored in environments where recombination is not.

37.4 Dispersal in angiosperms is aided by seeds and fruits.

Fertilization. Angiosperms use two sperm cells, one to fertilize the egg, the other to produce a nutrient tissue called endosperm.

Seeds. Seeds are growth-arrested embryos in drought-resistant packages.

Fruits. Fruits are mature ovaries that aid in seed dispersal.

FIGURE 37.1
Flowering plants. Because of their unique reproductive systems, flowering plants have secondarily become a key aesthetic phenomenon in human culture.

Flowering plants, or angiosperms, are the dominant photosynthetic organisms in most temperate and tropical terrestrial ecosystems (figure 37.1). Among the features that have contributed to their success are their unique reproductive structures, which include the flower and the fruit. As discussed in chapter 33, flowers allow the precise transfer of pollen by insects and other animals. Thus, plants can exchange gametes with one another while being rooted in one place. Fruits play an important role in the dispersal of angiosperms. Not only were both flowers and fruit key elements in the early success of angiosperms, but their evolution produced most of the striking differences we see among different angiosperms today. After examining how some plants reproduce asexually, this chapter explores sexual reproduction among flowering plants.

Asexual Reproduction

Plant reproduction usually involves a marked alternation of generations, with diploid sporophyte alternating with haploid gametophyte in a process called sexual reproduction. Some plants, however, reproduce without going through an alternation of generations in a process called **asexual reproduction.**

Vegetative Reproduction

In a very common form of asexual reproduction called vegetative reproduction, new plant individuals are simply cloned from parts of adults (figure 37.2). The forms of vegetative reproduction in plants are many and varied.

Stolons.   Some plants reproduce by means of runners, or stolons—long, slender stems that grow along the surface of the soil. In the cultivated strawberry (*Fragaria ananassa*), for example, leaves, flowers, and roots are produced at every other node on the runner. Just beyond each second node, the tip of the runner turns up and becomes thickened. This thickened portion first produces adventitious roots and then a new shoot that continues the runner.

Rhizomes.   Underground stems, or rhizomes, are also important reproductive structures, particularly in grasses and sedges. Rhizomes invade areas near the parent plant, and each node can give rise to a new flowering shoot. The noxious character of many weeds results from this type of growth pattern, and many garden plants, such as irises, are propagated almost entirely from rhizomes. Corms, bulbs, and tubers are rhizomes specialized for storage and reproduction. White potatoes are propagated artificially from tuber segments, each with one or more "eyes." The eyes, or "seed pieces," of potato give rise to the new plant.

Suckers.   The roots of some plants—for example, cherry, apple, raspberry, and blackberry—produce "suckers," or sprouts, which give rise to new plants. Commercial varieties of banana do not produce seeds and are propagated by suckers that develop from buds on underground stems. When the root of a dandelion is broken, as it may be if one attempts to pull it from the ground, each root fragment may give rise to a new plant.

Adventitious Leaves.   In a few species, even the leaves are reproductive. One example is the house plant *Kalanchoë daigremontiana*, familiar to many people as the "maternity plant," or "mother of thousands." The common names of this plant are based on the fact that numerous plantlets arise from meristematic tissue located in notches along the leaves. The maternity plant is ordinarily propagated by

FIGURE 37.2
Vegetative reproduction. Small plants arise from notches along the leaves of the house plant *Kalanchoë daigremontiana*.

means of these small plants, which, when they mature, drop to the soil and take root. The walking fern *(Asplenium rhizophyllum)* is another plant with adventitious leaves; young plants form where the leaf tips touch the ground.

Apomixis

In certain plants, including some citruses, certain grasses (such as Kentucky bluegrass, *Poa pratensis*), and dandelions, the embryos in the seeds may be produced asexually from the parent plant. This kind of asexual reproduction is known as *apomixis*. The seeds produced in this way give rise to individuals that are genetically identical to their parents. Thus, although these plants reproduce asexually by cloning diploid cells in the ovule, they also gain the advantage of seed dispersal, an adaptation usually associated with sexual reproduction.

In general, vegetative reproduction, apomixis, and other forms of asexual reproduction promote the exact reproduction of individuals that are particularly well suited to a certain environment or habitat. Asexual reproduction among plants is far more common in harsh or marginal environments, where there is little margin for variation. There is a greater proportion of asexual plants in the arctic, for example, than in temperate regions.

Plants that reproduce asexually clone new individuals from portions of the root, stem, leaves, or ovules of adult individuals. The asexually produced progeny are genetically identical to the parent individual.

Rise of the Flowering Plants

Most of the plants we see daily are angiosperms (phylum Anthophyta). The 240,000 species of flowering plants range in size from almost microscopic herbs to giant *Eucalyptus* trees, and their form varies from cacti, grasses, and daisies to aquatic pondweeds. Most shrubs and trees (other than conifers and *Ginkgo*) are also in this phylum. Virtually all of our food is derived, directly or indirectly, from flowering plants; in fact, more than 90% of the calories we consume come from just over 100 species. The remarkable evolutionary success of angiosperms, and their central role in our lives as sources of food, medicine, clothing, and building materials, makes this phylum merit special attention.

Why Were the Angiosperms Successful?

When flowering plants originated, Africa and South America were still connected to each other, as well as to Antarctica and India, and, via Antarctica, to Australia and New Zealand (figure 37.3). These landmasses formed the great continent known as Gondwanaland. In the north, Eurasia and North America were united, forming another supercontinent called Laurasia. The huge landmass formed by the union of South America and Africa spanned the equator and probably had a climate characterized by extreme temperatures and aridity in its interior. Similar climates occur in the interiors of major continents at present. Much of the early evolution of angiosperms probably took place in patches of drier and less favorable habitat found in the interior of Gondwanaland. Many features of flowering plants seem to correlate with successful growth under arid and semiarid conditions.

The transfer of pollen between flowers of separate plants, sometimes over long distances, ensures *outcrossing* (cross-pollination between individuals of the same species) and may have been important in the early success of angiosperms. The various means of effective fruit dispersal that evolved in the group would also have been significant in the success of angiosperms. The leaves of angiosperms, with their protective surface waxes and hairs, their stomatal mechanisms, their specialized conducting tissues, and their different modes of photosynthesis, all would have been important in survival and growth under somewhat arid conditions, just as they are today. Angiosperms produce many natural insecticides (discussed in chapter 24), which would have added to their ability to survive. As early angiosperms evolved, all of these advantageous features became further elaborated and developed, and the pace of their evolution accelerated.

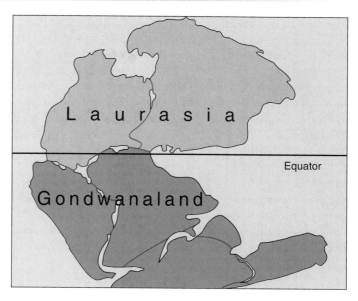

FIGURE 37.3
The alignment of the continents when the angiosperms first appeared in the fossil record about 130 million years ago. Africa, Madagascar, South America, India, Australia, and Antarctica were all connected and part of the huge continent of Gondwanaland, which eventually separated into the discrete landmasses we have today.

The Rise to Dominance

Angiosperms began to dominate temperate and tropical terrestrial communities about 80 to 90 million years ago, during the second half of the Cretaceous Period. We can document the relative abundance of different groups of plants by studying fossils that occur at the same time and place. In rocks more than 80 million years old, the fossil remains of plant phyla other than angiosperms, including lycopods, horsetails, ferns, and gymnosperms, are most common. Angiosperms arose in temperate and tropical terrestrial communities in a relatively short time.

At about the time that angiosperms became abundant in the fossil record, pollen, leaves, flowers, and fruits of some families that still survive began to appear. For example, representatives of the magnolia, beech, and legume families, which were in existence before the end of the Cretaceous Period (65 million years ago), are alive and flourishing today.

A number of insect orders that are particularly associated with flowers, such as Lepidoptera (butterflies and moths) and Diptera (flies), appeared or became more abundant during the rise of angiosperms. Plants and insects have clearly played a major role in each other's patterns of evolution, and their interactions continue to be of fundamental importance.

By 80 to 90 million years ago, angiosperms were dominant in terrestrial habitats throughout the world.

Evolution of the Flower

Pollination in angiosperms does not involve direct contact between the pollen grain and the ovule. Pollen matures within the anthers and is transported, often by insects, birds, or other animals, to the stigma of another flower. When pollen reaches the stigma, it germinates, and a pollen tube grows down, carrying the sperm nuclei to the embryo sac. After double fertilization takes place, development of the embryo and endosperm begins. The seed matures within the ripening fruit; the germination of the seed initiates another life cycle.

Successful pollination in many angiosperms depends on the regular attraction of **pollinators** such as insects, birds, and other animals, so that pollen is transferred between plants of the same species. When animals disperse pollen, they perform the same functions for flowering plants that they do for themselves when they actively search out mates.

Characteristics of Floral Evolution

By studying fossil angiosperm features, and correlating floral features with known primitive features—such as those of wood and pollen—scientists have reached general agreement on which angiosperm groups are primitive and which are advanced. Comparisons between sequences of bases in certain regions of DNA have revealed additional information about relationships within this phylum. Fossil deposits of angiosperms are found mostly in lowland basins where conditions favored fossil preservation. Paleobotanists (botanists who study fossil plants) have suggested that the early evolution of angiosperms could have occurred in tropical uplands where fossils were much less likely to form. Later, about 125 million years ago, when angiosperms invaded the lowlands during the Cretaceous Period, flowering plants were already quite diverse. One group of Cretaceous angiosperms resembled magnolias (figure 37.4), and another resembled flowers of the pepper and sycamore families (Piperaceae and Platanaceae, respectively), which have few flower parts. Fossils of both groups show considerable variation in number and arrangement of flower parts. This is in contrast to several modern families of angiosperms with highly specialized flowers such as honeysuckles (Caprifoliaceae) and heaths (Ericaceae), whose members show little variation.

Using these different lines of evidence, botanists generally deduced that the first angiosperm flowers probably had numerous, spirally arranged sepals, petals, stamens, and carpels. Petals and sepals may have been similar in color and form, the colors and shapes grading into each other throughout the spiral arrangement. All parts of primitive flowers tended to be free (not fused together or with other flower parts). In more advanced flowers, the parts are not spirally arranged but are, instead, arranged in definite whorls (attached in a circle). However, recent paleobotanical data and studies of floral form and development now challenge the view that the earliest flowers were similar to those of magnolias; they suggest the flowers may have been more like those of the pepper and sycamore families, which have few flower parts.

Like leaves, but unlike stems, flowers have **determinate** growth because new cells are not added by the **apical meristems** (the active tissues that produce new cells at the tips of stems and roots), which become incorporated into the tissues that differentiate into the young flower bud. In contrast, most leafy shoots themselves are **indeterminate;** when conditions are favorable, they keep growing, adding new leaves as growth continues.

(a)

(b)

(c)

FIGURE 37.4

A primitive flower. Species of *Magnolia* have large, relatively primitive flowers. (a) A flower from *M. virginiana*, the southern bay magnolia, illustrates the numerous free, spirally arranged pistils and stamens, and the whorled petals. The stamens have fallen into the bowl-shaped flower, and a longhorn beetle is eating pollen from them. (b) The sepal and petals have been removed from this young flower from *M. grandiflora*, the evergreen magnolia, showing the spirally arranged pistils (the curly structures at the top of the floral axis are their styles) and stamens (packed around the lower two-thirds of the floral axis). (c) This fruiting axis of *M. grandiflora* shows the numerous, spirally arranged mature ovaries. The bright red seeds of this species protrude from the individual fruits and hang by slender threads. These seeds are attractive to and dispersed by birds, which eat them.

Calyx

A *complete flower* has four whorls of parts (calyx, corolla, androecium, and gynoecium), while an *incomplete flower* lacks one or more of the whorls. In both complete and incomplete flowers, the **calyx** usually constitutes the outermost whorl; it consists of flattened appendages called **sepals,** which protect the flower in the bud (figure 37.5). Leaves and sepals may have had a common evolutionary origin because they have similar venation patterns and are usually similar in coloration and form. Certain genes that do not influence petal features are known to affect both sepals and leaves. In many monocots, the sepals may become petaloid and take on the form and color of the petals.

Corolla

The evolution of petals, which collectively make up the **corolla,** is more complex and difficult to interpret than that of sepals. Petals appear to have had two different evolutionary origins. In most plants, petals and stamens have a similar basic anatomy and might be homologous structures, with petals evidently having become flattened and leaf-like in the course of their evolution. Petals are also affected by genes that do not affect the carpels or sepals. In view of this relationship, the petals and stamens of most flowering plants might be homologous structures that share a common evolutionary origin. In other plants, such as water lilies, petals seem to have originated as modified sepals that became larger, colored, and attractive to insects during the course of their evolution. In such plants there are often structures that are obviously transitional between sepals and petals.

Androecium

Androecium (from the Greek *andros*, man, and *oikos*, house) is a collective term for all the **stamens** (male structures) of a flower. Stamens are specialized structures that bear the angiosperm microsporangia. There are similar structures bearing the microsporangia in the pollen cones of gymnosperms. In both cases, the structures probably evolved from leaves that bore the microsporangia. These structures have progressively become reduced and modified over the course of evolutionary history. Most living angiosperms have stamens whose **filaments** ("stalks") are slender and often thread-like, and whose four microsporangia are evident at the apex in a swollen portion, the **anther.** Some of the more primitive angiosperms have stamens that are flattened and leaf-like, with the sporangia producing from the upper or lower surface.

Gynoecium

The word **gynoecium** (from the Greek *gyne*, woman, and *oikos*, house) is a collective term for all the female parts of a flower. In most flowers, the gynoecium, which is unique to angiosperms, consists of a single **pistil,** which may be compound (composed of two or more fused *carpels*). Most flowers with which we are familiar—for example, those of tomatoes and oranges—have a single compound pistil. In other mostly primitive flowers—for example, buttercups and stonecups—there may be several to many separate pistils, each formed from a single carpel. **Ovules** (which develop into seeds) are produced in the pistil's swollen lower portion, the **ovary,** which usually narrows at the top into a slender, neck-like **style** with a pollen-receptive **stigma** at its apex. Sometimes the stigma is divided, with the number of stigma branches indicating how many carpels are in the particular pistil.

You will recall that carpels are essentially inrolled floral leaves with ovules along the margins. Some who have studied floral evolution believe that the first carpels were leaf blades that folded longitudinally; the margins, which had hairs, did not actually fuse until the fruit developed, but the hairs interlocked and were receptive to pollen. In the course of evolution there is evidence the hairs became localized into a stigma, a style was formed, and the fusing of the carpel margins ultimately resulted in a pistil. In many modern flowering plants, the carpels have become highly modified and are not visually distinguishable from one another unless the pistil is cut open.

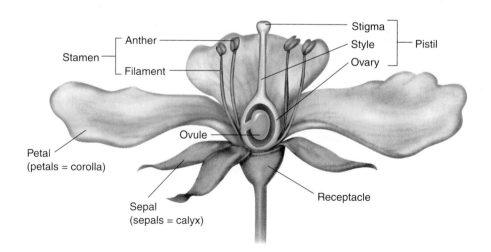

FIGURE 37.5
Structure of an angiosperm flower.

Trends of Floral Specialization

Two major evolutionary trends led to the wide diversity of modern flowering plants: (1) separate floral parts have grouped together, or fused, and (2) there has been a reduction or loss of floral parts (figure 37.6). In the more advanced angiosperms, the number of parts in each whorl has often been reduced from many to few. The spiral patterns of attachment of all floral parts in primitive angiosperms have in the course of evolution given way to a single whorl at each level. The central axis of many flowers has shortened, and the whorls are close to one another. In some evolutionary lines, the members of one or more whorls have fused with one another, sometimes joining into a tube. In other kinds of flowering plants, different whorls may be fused together. Whole whorls may even be lost from the flower, which may lack sepals, petals, stamens, carpels, or various combinations of these structures.

Trends in Floral Symmetry

Other trends in floral evolution have affected the symmetry of the flower (figure 37.7). Primitive flowers such as those of buttercups are *radically symmetrical*; that is, one could draw a line anywhere through the center and have two roughly equal halves. Flowers of many advanced groups are *bilaterally symmetrical*; that is, they are divisible into two equal parts along only a single plane. Examples of such flowers are snapdragons, mints, and orchids. Such bilaterally symmetrical flowers are also common among violets and peas. In these groups, they are often associated with advanced and highly precise pollination systems.

The first angiosperms likely had numerous free, spirally arranged flower parts. Sepals appear to be modified leaves. Petals of most kinds of plants seem to be homologous with stamens. In other plants, petals evolved as enlarged, colored sepals.

(a)

(b)

FIGURE 37.6
Trends in floral specialization.
(a) Baobab, *Adansonia digitata*. In the hanging flowers of this species, the numerous stamens are united into a tube around the style. (b) Wild geranium, *Geranium maculatum*. The petals are reduced to five each, the stamens to ten.

(a)

(b)

FIGURE 37.7
Trends in floral symmetry. (a) In most members of the very large sunflower or daisy family (Asteraceae), the flowers are crowded together in radially symmetrical heads; part of the corolla of each ray flower is greatly enlarged and extends out into a petal-like ray. (b) In members of the orchid family (Orchidaceae), the flowers are bilaterally symmetrical.

Formation of Angiosperm Gametes

As mentioned previously, plant sexual life cycles are characterized by an alternation of generations, in which a diploid sporophyte generation gives rise to a haploid gametophyte generation. In angiosperms, the gametophyte generation is very small and is completely enclosed within the tissues of the parent sporophyte. The male gametophytes, or microgametophytes, are **pollen grains**. The female gametophyte, or megagametophyte, is the **embryo sac**. Pollen grains and the embryo sac both are produced in separate, specialized structures of the angiosperm flower.

Like animals, angiosperms have separate structures for producing male and female gametes, but the reproductive organs of angiosperms are different from those of animals in two ways. First, in angiosperms, both male and female structures usually occur together in the same individual flower. Second, angiosperm reproductive structures are not permanent parts of the adult individual. Angiosperm flowers and reproductive organs develop seasonally, at times of the year most favorable for pollination.

Pollen Formation

Pollen grains form in the two pollen sacs located in the anther. Each pollen sac contains specialized chambers in which the *microspore mother cells* are enclosed and protected. The microspore mother cells undergo meiosis to form four haploid microspores. Subsequently, mitotic divisions form four pollen grains.

Pollen grain shapes are specialized for specific flower species. As discussed in more detail later in the chapter, fertilization requires that the pollen grain grow a tube that penetrates the style until it encounters the ovary. Most pollen grains have a furrow from which this pollen tube emerges; some grains have three furrows (figure 37.8).

Egg Formation

Eggs develop in the ovules of the angiosperm flower. Within each ovule is a megaspore mother cell. Each megaspore mother cell undergoes meiosis to produce four haploid megaspores. In most plants, only one of these megaspores, however, survives; the rest are absorbed by the ovule. The lone remaining megaspore undergoes repeated mitotic divisions to produce eight haploid nuclei that are enclosed within a seven-celled embryo sac. Within the embryo sac, the eight nuclei are arranged in precise positions. One nucleus is located near the opening of the embryo sac in the egg cell. Two are located in a single cell in the middle of the embryo sac and are called polar nuclei; two nuclei are contained in cells called synergids that flank the egg cell; and the other three nuclei reside in cells called the antipodals, located at the end of the sac, opposite the egg cell (figure 37.9).

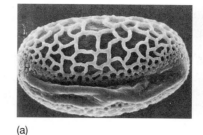
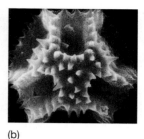

(a)   (b)

FIGURE 37.8
Pollen grains. (a) In the Easter lily, *Lilium candidum*, the pollen tube emerges from the pollen grain through the groove or furrow that occurs on one side of the grain. (b) In a plant of the sunflower family, *Hyoseris longiloba*, three pores are hidden among the ornamentation of the pollen grain. The pollen tube may grow out through any one of them.

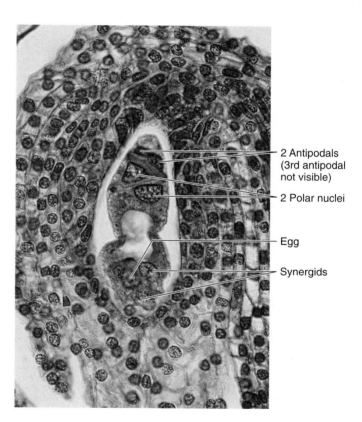

FIGURE 37.9
A mature embryo sac of a lily. The eight haploid nuclei produced by mitotic divisions of the haploid megaspore are labeled.

In angiosperms, both male and female structures usually occur together in the same individual flower. These reproductive structures are not a permanent part of the adult individual.

Pollination

Pollination is the process by which pollen is placed on the stigma. The pollen may be carried to the flower by wind or by animals, or it may originate within the individual flower itself. When pollen from a flower's anther pollinates the same flower's stigma, the process is called *self-pollination*.

Pollination in Early Seed Plants

Early seed plants were pollinated passively, by the action of the wind. As in present-day conifers, great quantities of pollen were shed and blown about, occasionally reaching the vicinity of the ovules of the same species. Individual plants of any given species must grow relatively close to one another for such a system to operate efficiently. Otherwise, the chance that any pollen will arrive at the appropriate destination is very small. The vast majority of windblown pollen travels less than 100 meters. This short distance is significant compared with the long distances pollen is routinely carried by certain insects, birds, and other animals.

A few kinds of living gymnosperms (some gnetophytes and cycads) are insect-pollinated. It is possible, therefore, that insect pollination may already have been established in the angiosperm ancestors.

Pollination by Animals

The spreading of pollen from plant to plant by pollinators visiting flowers of specific angiosperm species has played an important role in the evolutionary success of the group. It now seems clear that the earliest angiosperms, and perhaps their ancestors also, were insect-pollinated, and the coevolution of insects and plants has been important for both groups for over 100 million years. Such interactions have also been important in bringing about increased floral specialization. As flowers become increasingly specialized, so do their relationships with particular groups of insects and other animals.

Bees. Among insect-pollinated angiosperms, the most numerous groups are those pollinated by bees (figure 37.10). Like most insects, bees initially locate sources of food by odor, then orient themselves on the flower or group of flowers by its shape, color, and texture. Flowers that bees characteristically visit are often blue or yellow. Many have stripes or lines of dots that indicate the location

FIGURE 37.10
Pollination by a bumblebee. As this bumblebee, *Bombus*, squeezes into the bilaterally symmetrical, advanced flower of a member of the mint family, the stigma contacts its back and picks up any pollen that the bee may have acquired during a visit to a previous flower.

of the nectaries, which often occur within the throats of specialized flowers. Some bees collect nectar, which is used as a source of food for adult bees and occasionally for larvae. Most of the approximately 20,000 species of bees visit flowers to obtain pollen. Pollen is used to provide food in cells where bee larvae complete their development.

Only a few hundred species of bees are social or semisocial in their nesting habits. These bees live in colonies, as do the familiar honeybee, *Apis mellifera*, and the bumblebee, *Bombus*. Such bees produce several generations a year and must shift their attention to different kinds of flowers as the season progresses. To maintain large colonies, they also must use more than one kind of flower as a food source at any given time.

Except for these social and semisocial bees and about 1000 species that are parasitic in the nests of other bees, the great majority of bees—at least 18,000 species—are solitary. Solitary bees in temperate regions characteristically have only a single generation in the course of a year. Often they are active as adults for as little as a few weeks a year.

Solitary bees often use the flowers of a given group of plants almost exclusively as sources of their larval food. The highly constant relationships of such bees with those flowers may lead to modifications, over time, in both the flowers and the bees. For example, the time of day when the flowers open may correlate with the time when the bees appear; the mouthparts of the bees may become elongated in relation to tubular flowers; or the bees' pollen-collecting apparatuses may be adapted to the pollen of the plants that they normally visit. When such relationships are established, they provide both an efficient mechanism of pollination for the flowers and a constant source of food for the bees that "specialize" on them.

Insects Other Than Bees. Among flower-visiting insects other than bees, a few groups are especially prominent. Flowers such as phlox, which are visited regularly by butterflies, often have flat "landing platforms" on which butterflies perch. They also tend to have long, slender floral tubes filled with nectar that is accessible to the long, coiled proboscis characteristic of Lepidoptera, the order of insects that includes butterflies and moths. Flowers like jimsonweed, evening primrose, and others visited regularly by moths are often white, yellow, or some other pale color; they also tend to be heavily scented, thus serving to make the flowers easy to locate at night.

Birds. Several interesting groups of plants are regularly visited and pollinated by birds, especially the hummingbirds of North and South America and the sunbirds of Africa (figure 37.11). Such plants must produce large amounts of nectar because if the birds do not find enough food to maintain themselves, they will not continue to visit flowers of that plant. Flowers producing large amounts of nectar have no advantage in being visited by insects because an insect could obtain its energy requirements at a single flower and would not cross-pollinate the flower. How are these different selective forces balanced in flowers that are "specialized" for hummingbirds and sunbirds?

To insects, ultraviolet light is highly visible. Carotenoids, yellow or orange pigments frequently found in plants, are responsible for the colors of many flowers, such as sunflowers and mustard. Carotenoids reflect both in the yellow range and in the ultraviolet range, the mixture resulting in a distinctive color called "bee's purple." Such yellow flowers may also be marked in distinctive ways normally invisible to us, but highly visible to bees and other insects (figure 37.12).

Red does not stand out as a distinct color to most insects, but it is a very conspicuous color to birds. To most insects, the red upper leaves of poinsettias look just like the other leaves of the plant. Consequently, even though the flowers produce abundant supplies of nectar and attract hummingbirds, insects tend to bypass them. Thus, the red color both signals to birds the presence of abundant nectar and makes that nectar as inconspicuous as possible to insects.

As hummingbirds migrate northward through North America every summer, they encounter a progression of red flowers that provide an abundant source of food for them along the way. Even if the birds are unfamiliar with these flowers, the red color of the flowers signals the birds that they are suitable sources of food. Hummingbird-visited flowers (such as those on cacti) are also specialized in other, analogous ways. Such flowers generally are odorless, because birds normally do not have well-developed olfactory senses. The nectar in hummingbird-visited flowers is often held within strong tubes or otherwise protected and can be obtained only by the beak of the bird and not, in general, by insects.

Wind-Pollinated Angiosperms

Many angiosperms, representing a number of different groups, are wind-pollinated—a characteristic of early seed plants. Among them are such familiar plants as oaks, birches, cottonwoods, grasses, sedges, and nettles. The flowers of these plants are small, greenish, and odorless; their corollas are reduced or absent (see figures 37.13 and 37.14). Such flowers often are grouped together in fairly large numbers and may hang down in tassels that wave about in the wind and shed pollen freely. Many wind-pollinated plants have stamen- and pistil-containing flowers

FIGURE 37.11
Hummingbirds and flowers. A long-tailed hermit hummingbird extracts nectar from the flowers of *Heliconia imbricata* in the forests of Costa Rica. Note the pollen on the bird's beak. Hummingbirds of this group obtain nectar primarily from long, curved flowers that more or less match the length and shape of their beaks.

(a)

(b)

FIGURE 37.12
How a bee sees a flower. (a) The yellow flower of *Ludwigia peruviana* (Onagraceae) photographed in normal light and (b) with a filter that selectively transmits ultraviolet light. The outer sections of the petals reflect both yellow and ultraviolet, a mixture of colors called "bee's purple"; the inner portions of the petals reflect yellow only and therefore appear dark in the photograph that emphasizes ultraviolet reflection. To a bee, this flower appears as if it has a conspicuous central bull's-eye.

separated among individuals or on a single individual. If the pollen-producing and ovule-bearing flowers are separated, it is certain that pollen released to the wind will reach a flower other than the one that sheds it, a strategy that greatly promotes outcrossing. Some wind-pollinated plants, especially trees and shrubs, flower in the spring, before the development of their leaves can interfere with the wind-borne pollen.

Bees are the most frequent and characteristic pollinators of flowers. Insects often are attracted by the odors of flowers. Bird-pollinated flowers are characteristically odorless and red, with the nectar not readily accessed by insects.

Self-Pollination versus Outcrossing

Self-Pollination

All of the modes of pollination that we have considered thus far tend to lead to outcrossing, which is as highly advantageous for plants as it is for eukaryotic organisms generally. Nevertheless, self-pollination also occurs among angiosperms, particularly in temperate regions. Most of the self-pollinating plants have small, relatively inconspicuous flowers that shed pollen directly onto the stigma, sometimes even before the bud opens. You might logically ask why there are many self-pollinated plant species if outcrossing is just as important genetically for plants as it is for animals. There are two basic reasons for the frequent occurrence of self-pollinated angiosperms:

1. Self-pollination obviously is ecologically advantageous under certain circumstances because self-pollinators do not need to be visited by animals to produce seed. As a result, self-pollinated plants expend less energy in the production of pollinator attractants and can grow in areas where the kinds of insects or other animals that might visit them are absent or very scarce—as in the Arctic or at high elevations.

2. In genetic terms, self-pollination produces progenies that are more uniform than those that result from outcrossing. Such progenies may contain high proportions of individuals well-adapted to particular habitats. Self-pollination in normally outcrossing species tends to produce large numbers of ill-adapted individuals because it brings together deleterious recessive genes; but some of these combinations may be highly advantageous in particular habitats. In such habitats, it may be advantageous for the plant to continue self-pollinating indefinitely. This is the main reason many self-pollinating plant species are weeds—not only have humans made weed habitats uniform, but they have also spread the weeds all over the world.

Factors That Promote Outcrossing

Outcrossing, as we have stressed, is of critical importance for the adaptation and evolution of all eukaryotic organisms. In contrast to most animals, in which males and females are distinct, flowers of most kinds of plants have both stamens and one or more pistils, with self-fertilization being possible. In some kinds of plants, certain flowers may lack either stamens or pistils; in these, the missing organs have been lost in the course of evolution. Flowers lacking stamens and having only a pistil or pistils are called **pistillate** flowers; those with stamens but no pistils are called **staminate** flowers.

In various species of flowering plants—for example, willows and some mulberries—staminate and pistillate flowers may occur on separate plants. Such plants, which produce

FIGURE 37.13
Staminate and pistillate flowers of a birch, *Betula*. Birches are monoecious; their staminate flowers hang down in long, yellowish tassels, while their pistillate flowers mature into clusters of small, brownish, conelike structures.

only ovules or only pollen, are called **dioecious,** from the Greek words for "two houses." Obviously, they cannot self-pollinate and must rely exclusively on outcrossing. In other kinds of plants, such as oaks, birches, corn (maize), and pumpkins, separate male and female flowers may both be produced on the same plant. Such plants are called **monoecious,** meaning "one house" (figure 37.13). In monoecious plants, the separation of pistillate and staminate flowers, which may mature at different times, greatly enhances the probability of outcrossing.

Even if, as usually is the case, functional stamens and pistils are both present in each flower of a particular plant species, these organs may reach maturity at different times. Plants in which this occurs are called **dichogamous.** If the stamens mature first, shedding their pollen before the stigmas are receptive, the flower is effectively staminate at that time. Once the stamens have finished shedding pollen, the stigma or stigmas may then become receptive, and the flower may become essentially pistillate (figures 37.14 and 37.15). This has the same effect as if the flower completely lacked either functional stamens or functional pistils; its outcrossing rate is thereby significantly increased.

FIGURE 37.14
Wind-pollinated flowers. Most of the parts of the flowers of grasses and other wind-pollinated plants are greenish; they are therefore not particularly attractive to insects. At this stage, the large yellow anthers, dangling on very slender filaments, are hanging out, about to shed their pollen to the wind; later, these flowers will become pistillate, with long, feathery stigmas—well suited for trapping windblown pollen—sticking far out of them. Many grasses, like this one, are therefore dichogamous.

Many flowers are constructed such that the stamens and stigmas do not come in contact with each other. With such an arrangement, there is a natural tendency for the pollen to be transferred to the stigma of another flower rather than to the stigma of its own flower, thereby promoting outcrossing.

Genetic self-incompatibility, which is widespread in flowering plants, increases outcrossing even when a flower's stamens and stigma mature at the same time. In self-incompatible plants, a given flower's pollen may germinate on its own stigma, but growth of the pollen tube is arrested or germination may not occur at all. In other self-incompatible plants embryos resulting from self-fertilization abort soon after fertilization occurs.

(a)

(b)

FIGURE 37.15
Dichogamy, as illustrated by the flowers of fireweed, *Epilobium angustifolium.* More than 200 years ago (in the 1790s) fireweed, which is outcrossing, was one of the first plant species to have its process of pollination described. First, the anthers shed pollen, and then the style elongates above the stamens while the four lobes of the stigma curl back and become receptive. Consequently, the flowers are functionally staminate at first, becoming pistillate about two days later. The flowers open progressively up the stem, so that the lowest are visited first. Working up the stem, the bees encounter pollen-shedding, staminate-phase flowers and become covered with pollen, which they then carry to the lower, functionally pistillate flowers of another plant. Shown here are flowers in (a) the staminate phase and (b) the pistillate phase.

Self-pollinated angiosperms are frequent where there is a strong selective pressure to produce large numbers of genetically uniform individuals adapted to specific, relatively uniform habitats. Outcrossing in plants may be promoted through dioecism, monoecism, self-incompatibility, or the physical separation or different maturation times of the stamens and pistils.

Fertilization

Fertilization in angiosperms is a complex, somewhat unusual process in which two sperm cells are utilized in a unique process called **double fertilization** (figure 37.16). Double fertilization results in two key developments: (1) the fertilization of the egg, and (2) the formation of a nutrient substance called endosperm that nourishes the embryo. Once fertilization is complete, the embryo develops by dividing numerous times. Meanwhile, protective tissues enclose the embryo, resulting in the formation of the seed. The seed, in turn, is enclosed in another structure called the fruit. These typical angiosperm structures evolved in response to the need for seeds to be dispersed over long distances to ensure genetic variability.

Once a pollen grain has been spread by wind, by animals, or through self-pollination, it adheres to the sticky, sugary substance that covers the stigma and begins to grow a **pollen tube** that pierces the style. The pollen tube, nourished by the sugary substance, grows until it reaches the ovule in the ovary. Meanwhile, one of the cells within the pollen grain inside the tube divides to form two sperm cells.

The pollen tube eventually reaches the embryo sac in the ovule. At the entry to the embryo sac, the tip of the pollen tube bursts and releases the two sperm cells. Simultaneously, the two nuclei that flank the egg cell disintegrate, and one of the sperm cells fertilizes the egg cell, forming a zygote. The other sperm cell fuses with the two polar nuclei located at the center of the embryo sac, forming the triploid (3*n*) primary endosperm nucleus. The primary endosperm nucleus eventually develops into the endosperm.

In double fertilization, angiosperms utilize two sperm cells. One fertilizes the egg, while the other helps form a substance called endosperm that nourishes the embryo.

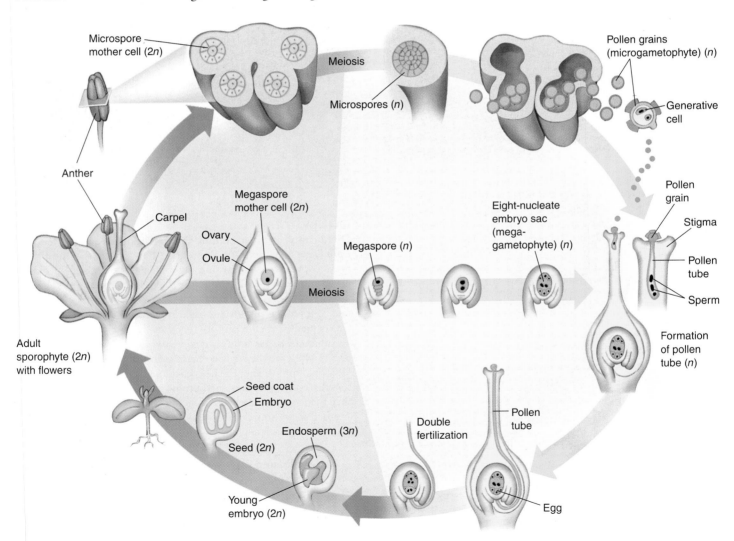

FIGURE 37.16

Angiosperm life cycle. Eggs form within the embryo sac inside the ovules, which, in turn, are enclosed in the carpels. The pollen grains, meanwhile, are formed within the sporangia of the anthers and are shed. Fertilization is a double process. A sperm and an egg come together, producing a zygote; at the same time, another sperm fuses with the polar nuclei to produce the endosperm. The endosperm is the tissue, unique to angiosperms, that nourishes the embryo and young plant.

Seeds

A seed is a compact, drought-resistant package consisting of an *embryo*—the young sporophyte whose embryonic development has been temporarily arrested—stored food, and a usually tough protective covering, the *seed coat* (figure 37.17). Seeds provide a means of dispersal of progeny to new places for mature plants that are anchored in the ground. The dispersal of seeds is often aided by wind, animals, or water (figure 37.18). Pine seeds, for example, have wings, as do the fruits of maples, elms, and ashes, and dandelion fruitlets have plumes to facilitate wind dispersal. Birds carry seeds to new locations, and seeds with hooks catch in the fur of animals. Coconuts are conveyed great distances by ocean currents, and the air sacs in sedge fruits allow them to float down watercourses. In addition, seeds can remain dormant under unfavorable conditions such as drought and germinate when favorable circumstances return.

The seed is a crucial adaptation to life on land because it protects the embryonic plant from drying out during its most vulnerable stage. The seed coat also provides protection for the embryo and its stored food material from being eaten by predators, attacked by fungi, or otherwise destroyed. Most kinds of seeds have abundant food stored in them, either within the embryo or in specialized storage tissue. This food, which plays the same role as the yolk of an egg, is used as a ready source of energy for the rapidly growing young plant.

Seeds evolved in advanced vascular plants. They characterize the cone-bearing gymnosperms, which produce seeds exposed to the air, and the flowering plants, or angiosperms, which have seeds enclosed in a fruit. The first seed-producing plants appeared about 360 million years ago. In seed plants there are two kinds of gametophytes, one producing male gametes and the other female gametes. Both are greatly reduced, and seed plants experience a distinctive kind of life cycle. Their life cycles represent the culmination of a major trend of evolution among vascular plants: there has been a progressive reduction of size of the gametophyte, and the gametophyte has become completely dependent on the sporophyte.

Seed Formation

During their development, plant tissues become progressively more specialized. The first stage in the development of a plant zygote is active cell division to form an organized mass of cells, the embryo. In angiosperms, the differentiation of cell types within the embryo begins almost immediately after fertilization. By the fifth day, the principal tissue systems can be detected within the embryo mass, and within another day, the root and shoot apical meristems can be detected.

Early in the development of an angiosperm embryo, a profoundly significant event occurs: the embryo simply stops developing and becomes dormant. In many plants, embryo development is arrested soon after apical meri-

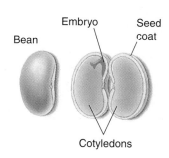

FIGURE 37.17
Diagram of a seed. The hard seed coat, formed from ovule tissues called integuments, protects the embryo—the dormant young plant of the next sporophyte generation—within. The first leaves, specialized in a bean for food storage, are the cotyledons. Most seeds are drought-resistant and readily dispersed.

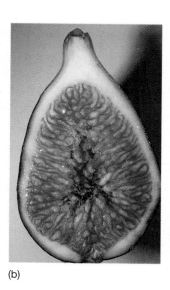

(a)   (b)

FIGURE 37.18
Wind- and animal-dispersed seeds. (a) The fruits and seeds of maples, *Acer*, when mature, are blown considerable distances from their parent trees. (b) When birds and mammals consume mature figs, the seeds are scattered about.

stems and the first leaves, or **cotyledons,** are differentiated. The integuments—the coats surrounding the embryo—develop into a relatively impermeable seed coat, which encloses the embryo within the seed, together with a source of stored food.

Once the seed coat fully develops around the embryo, most of the embryo's metabolic activities cease; a mature seed contains only about 10 percent water. Under these conditions, the seed and the young plant within it are very stable. **Germination,** or the resumption of metabolic activities that lead to the growth of a mature plant, cannot take place until water and oxygen reach the embryo, a process that sometimes involves cracking the seed. Seeds of some plants have been known to remain viable for hundreds of years. Environmental factors help ensure that the plant will germinate only under appropriate conditions.

A seed is a sporophyte embryo whose development has been temporarily arrested. Because the seed can be carried far from the parent plant before it germinates and resumes growth, it provides a powerful mechanism for dispersal.

Fruits

Paralleling the evolution of angiosperm flowers, and nearly as spectacular, has been the evolution of their **fruits,** which are defined simply as mature ovaries. During seed formation, the flower ovary begins to develop into fruit. Fruits form in many ways and exhibit a wide array of specializations in relation to their dispersal.

The Formation of Fruits

The differences among most of the fruit types seen today are shown in table 37.1 and in figure 37.19. Examples of the fruit types are as follows:

Pomes. Apples, pears, quinces.

Drupes. Peaches, apricots, plums, cherries. The coconuts sold in stores are the seeds of drupes that have had the fibrous flesh removed.

True berries. Blueberries, cranberries, tomatoes, grapes, eggplants, peppers. Many fruits that include "berry" in their name (mulberry, blackberry, raspberry, strawberry) are not true berries because they originate from more than one pistil.

Hesperidiums. Oranges, tangerines, lemons, limes, kumquats.

Pepos. Pumpkins, squash, cantaloupe and other melons, cucumbers, gourds.

Aggregate fruits. Strawberries, raspberries, blackberries.

Multiple fruits. Mulberries, pineapples, osage oranges. Figs are a unique type of multiple fruit called a syconium that develops from an "outside-in" inflorescence of tiny flowers on a common receptacle.

Follicles. Milkweed, larkspur. The individual fruitlets of Magnolia are follicles.

Legumes. Peas, beans, soybeans. Peanuts are atypical legumes that usually fail to split; in nature, the fruit wall is broken down by bacteria.

Siliques/Silicles. These fruits are produced by members of the large Mustard family (Brassicaceae); they

Pomes

Drupes

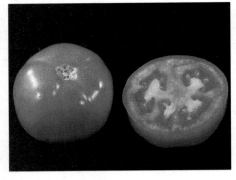
True berries

Aggregate fruits

Follicles

Silicles

Capsules

Samaras

FIGURE 37.19

Examples of some kinds of fruits. Pomes (apples), drupes (peach), true berries (tomato), aggregate fruits (blackberry), follicles (milkweed), silicles (dollar plant, *Lunaria*), capsules (orchid), and samaras (big-leaf maple).

Table 37.1 Key to the Fruits

Characteristic	Type of Fruit
FLESHY FRUITS	
1. Fleshy fruits (fruits that are more or less fleshy at maturity)	
2. Simple fleshy fruits (fleshy fruits that develop from a flower with a single pistil)	
3. Fruits whose flesh is derived primarily from the receptacle	Pomes
3. Fruits whose flesh is derived primarily from the ovary	
4. Fruits with a single seed enclosed in a hard pit	Drupes
4. Fruits with more than one seed, the seed not enclosed in a hard pit (berries)	
5. Fruits with a thin skin	True berries
5. Fruits with a leathery skin containing oils, or a rind	
6. Fruits with a leathery skin	Hesperidiums
6. Fruits with a rind	Pepos
2. Compound fleshy fruits (fleshy fruits that develop from flowers with more than one ovary or from more than one flower).	
7. Fruits derived from a single flower with more than one ovary	Aggregate fruits
7. Fruits that develop from a cluster of flowers	Multiple fruits
DRY FRUITS	
1. Dry fruits (fruits that are dry at maturity)	
8. Dry fruits that split at maturity	
9. Fruits that are split along one carpel edge only	Follicles
10. Fruits that are split along two carpel edges	
11. Fruits whose seeds are attached to the carpel edges	Legumes
11. Fruits whose seeds are attached to a central partition	Siliques; Silicles
10. Fruits that split variously but not along two carpel edges	Capsules
8. Dry fruits that do not split at maturity	
12. Fruits without wings; seed and fruit walls tightly fused	Caryopses
12. Fruits with wings; or fruits with seeds loosely attached to the fruit wall	
13. Fruits with a wing formed from the outer tissues	Samaras
13. Fruits without wings	
14. Fruits with a (sometimes cup-like) cluster of bracts at the base	Nuts
14. Fruits without a cluster of bracts at the base	Achenes

*Note: Duplicate numbers indicate branch points in the key, where an "either/or" statement can be made.

include cabbages, broccoli, radishes, and shepherd's purse. Siliques are distinguished from silicles by length. Siliques are at least four times as long as they are wide; silicles are shorter.

Capsules. Irises, lilies, orchids, snapdragons. Some poppies have unique capsules that release their seeds via pores. Capsules are the most commonly produced dry fruits that split at maturity.

Caryopses (singular caryopsis). Corn, wheat, barley, rye, oats, rice, sugarcane, bamboo. Also known as grains, caryopses are produced only by members of the large Grass family (Poaceae).

Nuts. Chestnuts, filberts (hazelnuts), acorns. Note that walnuts, almonds, coconuts, and peanuts are not nuts, botanically speaking.

Achenes. Sunflowers, buttercups, buckwheat.

Samaras. Maples, elms, ashes.

Schizocarps. A fruit called a schizocarp is produced exclusively by all members of the Parsley family (Apiaceae). Schizocarps consist of two one-seeded mericarps that separate at maturity. The Parsley family includes parsley, dill, carrots, fennel, caraway, celery, and anise.

The Dispersal of Fruits

Aside from the many ways fruits can form, they also exhibit a wide array of specialized dispersal methods. Fruits with fleshy coverings, often shiny black or bright blue or red, normally are dispersed by birds or other vertebrates (see figures 37.4c and 37.20a). Like red flowers, red fruits signal an abundant food supply. By feeding on these fruits, birds and other animals may carry seeds from place to place and thus transfer plants from one suitable habitat to another.

Fruits with hooked spines, like those of burgrass (figure 37.20b), are typical of several genera of plants that occur in the northern deciduous woods. Such fruits are often disseminated by mammals, including humans. Squirrels and similar mammals disperse and bury fruits such as acorns and other nuts. Other fruits, such as those of maples, elms, and ashes, have wings which aid in their distribution by the wind. The dandelion provides another familiar example of a fruit type that is dispersed by wind (figure 37.21), and the dispersal of seeds from plants such as milkweeds, willows, and cottonwoods is similar. Orchids have minute, dust-like seeds, which are likewise blown away by the wind.

Coconuts and other plants that characteristically occur on or near beaches are regularly spread throughout a region by water (figure 37.22). This sort of dispersal is especially important in the colonization of distant island groups, such as the Hawaiian Islands. It has been calculated that seeds of about 175 original angiosperms, nearly a third from North America, must have reached Hawaii to have evolved into the roughly 970 species found there today. Some of these seeds blew through the air, others were transported on the feathers or in the guts of birds, and still others drifted across the Pacific. Although the distances are rarely as great as the distance between Hawaii and the mainland, dispersal is just as important for mainland plant species that have discontinuous habitats, such as mountaintops, marshes, or north-facing cliffs. Dispersal is one of the most fascinating aspects of plant biology.

Fruits, which are characteristic of angiosperms, are extremely diverse. The evolution of specialized structures allows fruits to be dispersed by animals, wind, and water.

(a)　　　(b)

FIGURE 37.20
Animal-dispersed fruits. (a) The bright red berries of this honeysuckle, *Lonicera hispidula*, are highly attractive to birds, just as are red flowers. After eating the fruits, birds may carry the seeds they contain for great distances either internally or, because of their sticky pulp, stuck to their feet or other body parts. (b) The spiny fruits of this burgrass, *Cenchrus incertus*, adhere readily to any passing animal, as you will know if you have ever stepped on them.

FIGURE 37.21
Wind-dispersed fruits. False dandelion, *Pyrrhopappus carolinanus*. The "parachutes" disperse the fruits of both false and true dandelions widely in the wind, much to the gardener's despair.

FIGURE 37.22
A water-dispersed fruit. This fruit of the coconut, *Cocos nucifers*, is sprouting on a sandy beach. Coconuts, one of the most useful plants for humans in the tropics, have become established on even the most distant islands by drifting on the waves.

Summary of Concepts

37.1 Many plants can clone themselves by asexual reproduction.

- In asexual reproduction, plants clone new individuals from portions of adult roots, stems, leaves, or ovules.
- The progeny produced by asexual reproduction are all genetically identical to the parent individual.

37.2 Angiosperms utilize temporary reproductive structures called flowers.

- The ancestor of angiosperms was a seed-bearing plant that was probably already pollinated by insects to some degree. No living group of plants has the right combination of characteristics to be this ancestor, but if it were known, it would be considered a gymnosperm.
- By 80 to 90 million years ago, angiosperms were more common worldwide than other plant groups. They became more abundant and diverse as drier habitats became widespread.
- One of the reasons angiosperms have been successful is their relatively drought-resistant vegetative features, including vascular systems, cuticles, and stomata. Most important, however, are their flowers and fruits. Flowers make possible the precise transfer of pollen and, therefore, outcrossing, even when the stationary individual plants are widely separated. Fruits, with their complex adaptations, facilitate the wide dispersal of angiosperms.
- The flowers of primitive angiosperms had numerous separate, spirally arranged flower parts. Sepals are homologous with leaves. Petals of most angiosperms appear to be homologous with stamens, although some appear to have originated from sepals.

37.3 Flowering plants use animals or wind to transfer pollen between flowers.

- Bees are the most frequent and constant pollinators of flowers. Solitary bees, which constitute about 90% of the estimated 20,000 bee species, often have morphological and physiological specializations that link them to the flowers of a particular plant species.
- Insects often are attracted by the odors of flowers and do not perceive red as a distinct color. Birds are not attracted by odors and do perceive red. Bird-pollinated flowers, which produce large quantities of nectar to support the birds that visit them, are characteristically odorless and red, with the nectar well-protected.
- Outcrossing in different angiosperms is promoted by the separation of the pollen- and ovule-producing structures into different flowers, or even onto different individuals; by genetic self-incompatibility; and by the separation of the pollen and the stigmas within a given flower with respect to space or time of maturation.

37.4 Dispersal in angiosperms is aided by seeds and fruits.

- The nutritive tissue in gymnosperm seeds derives from the expanded, food-rich gametophyte. In angiosperm seeds, the unique nutritive tissue, endosperm, forms from a cell that results from fusion of the polar nuclei of the embryo sac with a sperm cell.
- Fruits and seeds are highly diverse in terms of their dispersal mechanisms, often displaying wings, barbs, or other structures that aid in their transport from place to place. Fruit dispersal methods are especially important in the colonization of islands or other distant patches of suitable habitat.

Discussing Key Terms

1. **Angiosperms** Angiosperms or flowering plants (phylum Magnoliophyta or Anthophyta), dominate much of the temperate and tropical regions of the earth's landmasses. These plants probably originated about 150 million years ago and achieved world dominance only 80 to 90 million years ago; they apparently have originated more recently than any other phylum.

2. **Adaptive traits** Flowers and fruits are the unique reproductive features of flowering plants. Other traits that have contributed to the success of angiosperms include the efficient conducting systems throughout their stems, roots and leaves and their wide array of chemical compounds, which play a key role in defending them against pests and diseases.

3. **Flowers and fruits** Flowers play the central role in the precise transfer of pollen by insects, birds, and other animals—an important adaptation in bringing about outcrossing in stationary organisms. Fruits play a role of major importance in the dispersal of angiosperms from place to place.

4. **Coevolutionary interactions** Interactions between angiosperms and their pollinators have been important for both groups since the origin of the angiosperm. These coevolutionary interactions have played a key role in bringing about increased specialization, through morphological and physiological modifications in both plants and their pollinators.

5. **Angiosperm development** Angiosperms arose from another group of seed plants that was probably already pollinated by insects to some degree. All seed plants that are not angiosperms are called gymnosperms.

Review Questions

1. Why is it highly probable that all angiosperms were derived from a single ancestor? Which of the four living groups of gymnosperms is most closely related to the angiosperms?

2. List nine characteristics of early angiosperms that are thought to contribute to their success.

3. What are the characteristics of a primitive angiosperm flower? How can scientists determine whether a given flower is primitive or advanced?

4. Explain the difference between determinate and indeterminate growth in plants. What parts of the plant exhibit determinate growth? What parts exhibit indeterminate growth?

5. What flower whorl is collectively made up of petals? With which other flower parts are the petals of most flowers homologous? In what ways? What is the origin of petals in small number of flowers?

6. What is an androecium? Of which flower parts is it composed? What is the probable evolution of these flower parts? In what structure are the pollen grains produced?

7. What are pistillate flowers and staminate flowers? What is a dioecious versus a monoecious plant? What is the advantage of each of these characteristics?

8. What does it mean if a plant is dichogamous? Explain genetic self-incompatibility. Of what advantage to the plant are these characteristics?

9. How does the reproductivity and colony size of social versus solitary bees relate to the kinds of flowers that they pollinate? Is it more likely that a flower visited by a social or a solitary bee will become highly specialized toward that bee? Why? How is such coevolution advantageous to both parties?

10. Why must bird-pollinated flowers produce greater amounts of nectar than those pollinated by most kinds of insects? What deterrent to the wasting of nectar by insects is exhibited by these plants?

11. What are the primary characteristics of wind-pollinated flowers? Do most wind-pollinated plants have separate pistillate and staminate flowers, or do they have flowers with both parts functional? Why?

12. Discuss three ways seeds have reached the Hawaiian Islands. What types of structures aided these seeds in their dispersal?

Thought Questions

1. Why was the development of seeds and fruits so important to the success of angiosperms? How did these characteristics help angiosperms become the dominant photosynthetic organisms on land? In what terrestrial communities are angiosperms not dominant? What plant groups are predominant in those areas? Why?

2. Angiosperms usually produce both pollen and ovules within a single flower, or at least on a single plant. Why aren't all angiosperms self-pollinated? What are the advantages to self-pollination?

Internet Links

Botanical Diversity
http://www.euronet.nl/users/mbleeker
A rich site of botanical exploration, including an interactive trip into the rainforests of Suriname, and a "botanical sorting machine" where the user can click on hundreds of alphabetized plant names to find out that species place in the taxonomic world, with pictures and links to specialized databases.

Flowers for Kids
http://www.calpoly.edu/~dchippin/kids1.html
The FOR KIDS site of the California Native Plant Society provides a simple, fun look at how and why flowers work.

Name the Fruit
http://arnica.csustan.edu/maps/fruit.html
Click on the image of the fruit to learn what type it is.

For Further Reading

Gibson, A. C. and P. S. Nobel: *The Cactus Primer*, Harvard University Press, Cambridge, MA, 1986. Outstanding account of all aspects of the biology of this fascinating plant family.

Heywood, V. H. (editor): *Flowering Plants of the World*, Prentice-Hall, New York, 1985. An outstanding guide to the families of flowering plants, with excellent illustrations and diagrams.

Hobson, G.: "How the Tomato Lost Its Taste," *New Scientist*, September 1988, vol. 119, pages 46–50. No one seems to know exactly when the flavor fled the tomato, only that it did. Scientists are searching for the lost ingredient.

Johnson, W. C. and C. S. Adkisson: "Airlifting the Oaks," *Natural History*, October 1986, pages 40–46. Bluejays spread acorns widely to new habitats; an example of animal dispersal of plant fruits and seeds.

Jones, S. B., Jr. and A. E. Luchsinger: *Plant Systematics*, 2d. ed., McGraw-Hill, New York, 1986. A comprehensive account of plant systematics, including a useful review of plant families.

Proctor, M. and P. Yeo: *The Pollination of Flowering Plants*, Taplinger, New York, 1973. An excellent introduction to pollination biology, clearly and interestingly presented.

Richards, A. J.: *Plant Breeding Systems*, George Allen & Unwin, London, 1986. An excellent account of all aspects of the subject.

38
Plant Development

Concept Outline

FIGURE 38.1
This plant has just emerged from its seed. It is extending its shoot and leaves up into the air, toward light.

The development of plants is not as well understood as that of other multicellular organisms. Among the simplest multicellular organisms are the fungi. Fungal maturation is primarily a process of growth rather than specialization. Animal development is far more complex, but it has been extensively researched and is relatively well understood. Scientists are only beginning to seriously study the mechanisms underlying plant development. The adult plant contains a variety of specialized cells organized into tissues and organs, but the way in which the plant body develops is quite different from animals (figure 38.1). As you will learn in this chapter, a hallmark of plant development is flexibility. As a plant develops, its precise array of tissues is greatly influenced by its environment.

An Overview of Plant Development

Plant development is notably different from animal development in that plant cells do not move. Instead, special groups of self-renewing cells called *apical meristems*, located at the tips of stems and roots, produce new cells that will differentiate in place. As a result, these apical meristems are continually pushed ahead of the columns of xylem and other cells that become differentiated. Woody plants have additional meristems that develop in cylindrical fashion within the stems and roots. The new cells the meristems produce become encased in rigid cellulose walls and differentiate into various plant tissues where they are produced.

While the cells of other organisms are mobile and can move away when their environment becomes unfavorable, plant cells are anchored in position. Because they cannot move, plants are often able to adapt to environmental change in various ways, such as losing their leaves or going dormant. However, if the environmental change is too drastic, such as fire, flooding, or drought, plants may die.

Plant bodies do not have a fixed size. Parts such as leaves, roots, branches, and flowers all vary in size and number from plant to plant—even within a species. The development of the form and structure of plant parts may be relatively rigidly controlled, but some aspects of leaf, stem, and root development (and utilization) are quite flexible. As a plant grows, the number, location, size, and even structure of leaves and roots are largely influenced by the environment.

Early Cell Division.   The first division of the fertilized egg in a flowering plant, which takes place in an *embryo sac* within an *ovule*, is off-center, so that one of the daughter cells is small, with dense cytoplasm (figure 38.2*a*). That cell, which will become the embryo, begins to divide repeatedly, forming a ball of cells. The other daughter cell also divides repeatedly, forming an elongated structure called a *suspensor*, which links the embryo to the nutrient tissue of the seed. The suspensor also provides a route for nutrients to reach the developing embryo. The root-shoot axis of the plant embryo forms at this time. Cells near the suspensor are destined to form a root, while those at the other end of the axis ultimately become a shoot.

Tissue Formation.   Three basic tissues differentiate while the plant embryo is still a ball of cells (figure 38.2*b*), although no cell movements are involved. The outermost cells in a plant embryo become *epidermal cells*. The bulk of the embryonic interior consists of *ground tissue* cells that eventually function in food and water storage. Lastly, cells at the core of the embryo are destined to form the future *vascular tissue*.

Seed Formation.   Soon after the three basic tissues form, a flowering plant embryo develops one or two seed leaves called *cotyledons* and then ceases development. A protective seed coat forms from the outer layers of ovule cells, and the embryo within is now either surrounded by nutritive tissue or has amassed stored food in its cotyledons (figure 38.2*c*). The resulting package, known as a *seed*, is resistant to drought and other unfavorable conditions; in its dormant state, it is a vehicle for dispersing the embryo to distant sites and allows a plant embryo to survive in environments that might kill a mature plant.

Germination.   A seed germinates in response to changes in its environment brought about by water, temperature, or other factors. The embryo within the seed resumes development and grows rapidly, its roots extending downward and its leaf-bearing shoots extending upward (figure 38.2*d*).

Meristematic Development.   Plants exhibit great flexibility as their organs develop into a mature body. Apical meristems at the root and shoot tips generate the large numbers of cells needed to form leaves, flowers, and all other components of the mature plant (figure 38.2*e*). Apical meristem cells of corn, for example, divide every 12 hours, producing half a million cells a day in an actively growing corn plant. At the same time, secondary meristems (the vascular cambium and the cork cambium) within the stems and roots of many plants produce wood and other tissues that increase their girth. A variety of hormones produced by plant tissues influence meristem activity and, thus, the development of the plant body. Plant hormones (see chapter 39) are the tools that allow plant development to adjust to the environment.

Morphogenesis.   Because plant cells cannot move, the form of a plant body is largely determined by the plane in which cells divide and by controlled changes in cell shape as they expand osmotically after they form (see figure 38.2*e*). Many of the details of how a plant's body shape develops in response to its environment are not yet known. However, it appears that plant growth-regulating hormones and other factors influence the orientation of bundles of microtubules on the interior of the plasma membrane. These microtubules seem to guide cellulose deposition as the cell wall forms around the outside of a new cell. The orientation of the cellulose fibers, in turn, determines how the cell will elongate as it increases in volume, and so determines the cell's final shape.

In a developing plant, leaves, flowers, and branches are added to the growing body in ways that are strongly influenced by the environment.

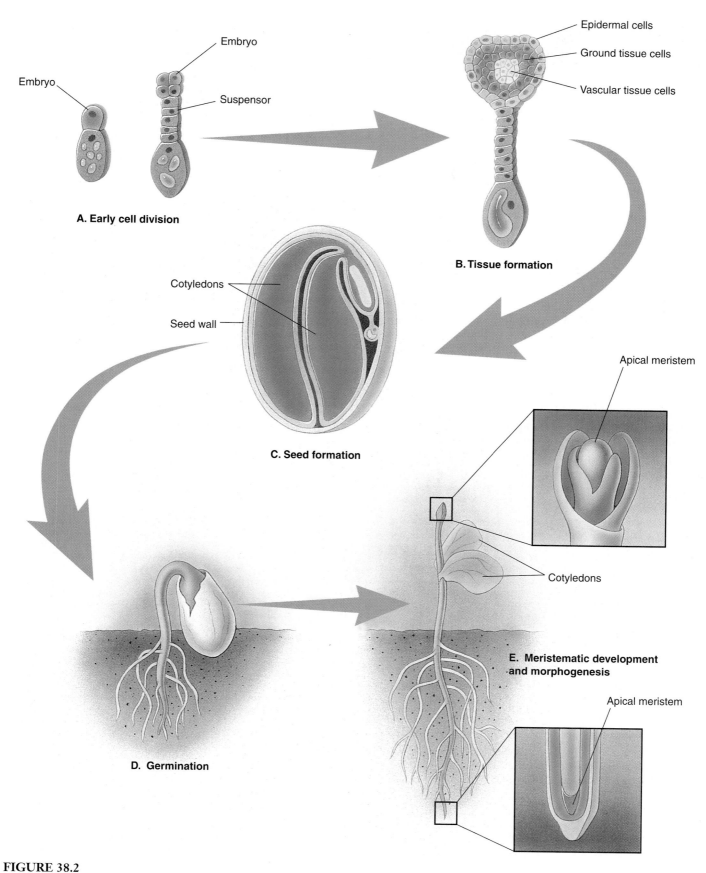

FIGURE 38.2

The path of plant development. An illustration of the developmental stages of *Arabidopsis thaliana*. (a) Early cell division. (b) Tissue formation. (c) Seed formation. (d) Germination. (e) Meristematic development and morphogenesis.

Molecular Mechanisms of Development

Scientists are only beginning to unravel the molecular biology of plant development, largely through intensive recent study of a small weedy relative of the mustard plant, the wall cress *Arabidopsis thaliana*. Easy to grow and cross, and with a short generation time, *Arabidopsis* makes an ideal model for investigating plant development. It is able to self-fertilize, like Mendel's pea plants, making genetic analysis convenient. *Arabidopsis* can be grown indoors in test tubes, a single plant producing thousands of offspring after only two months. Its genome is approximately the same size as those of the nematode *Caenorhabditis elegans* and the fruit fly *Drosophila melanogaster*. An ordered library of *Arabidopsis* gene clones is now available to researchers, and a complete genome sequence was anticipated by 1999.

Pattern Formation

Much of the current work investigating *Arabidopsis* development has centered on obtaining and studying mutations that alter the plant's development. Many different sorts of mutations have been identified. Some of the most interesting of them alter the basic architecture of the embryo, the pattern of tissues laid down as the embryo first forms. Mutations in over 50 different genes that alter pattern formation in *Arabidopsis* embryos are now known, affecting every stage of development (figure 38.3). While work in this area is still very preliminary, it appears that the mechanisms that establish patterns in the early *Arabidopsis* embryo are broadly similar to those known to function in animal development, described in chapter 57.

Organ Formation

Importantly, the subsequent development of organs in *Arabidopsis* also seems to parallel organ development in animals, and a similar set of regulatory genes control development in *Arabidopsis*, *Drosophila*, and mice. In both plants and animals, these genes are called **homeotic genes**. *Arabidopsis* flowers, for example, are modified leaves formed as four whorls in a specific order, and homeotic mutations have been identified that convert one part of the pattern to another, just as the homeotic mutations described in chapter 57 do to the body segments of a fly.

> Scientists are only beginning to understand the molecular biology of plant development. In broad outline, it appears quite similar to the development in animals. The genes that determine pattern formation and organ development, for example, operate in the same way in plants and animals.

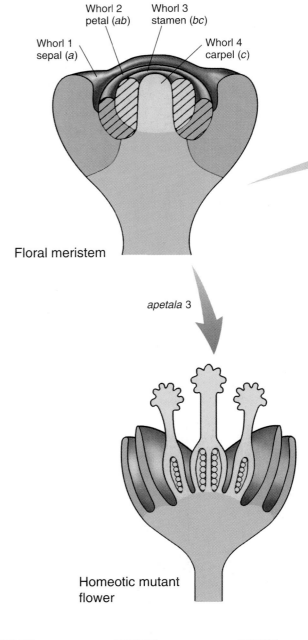

Gene *a* (*apetala* 2) expressed in meristem

Gene *b* (*apetala* 3) expressed in meristem

Gene *c* (*agamous*) expressed in meristem

FIGURE 38.3
Studying development in a flowering plant. Mutations in genes that regulate development alter pattern and organ formation. In this flower, *apetala 3* caused abnormal carpel growth and reduced or absent petals and stamens.

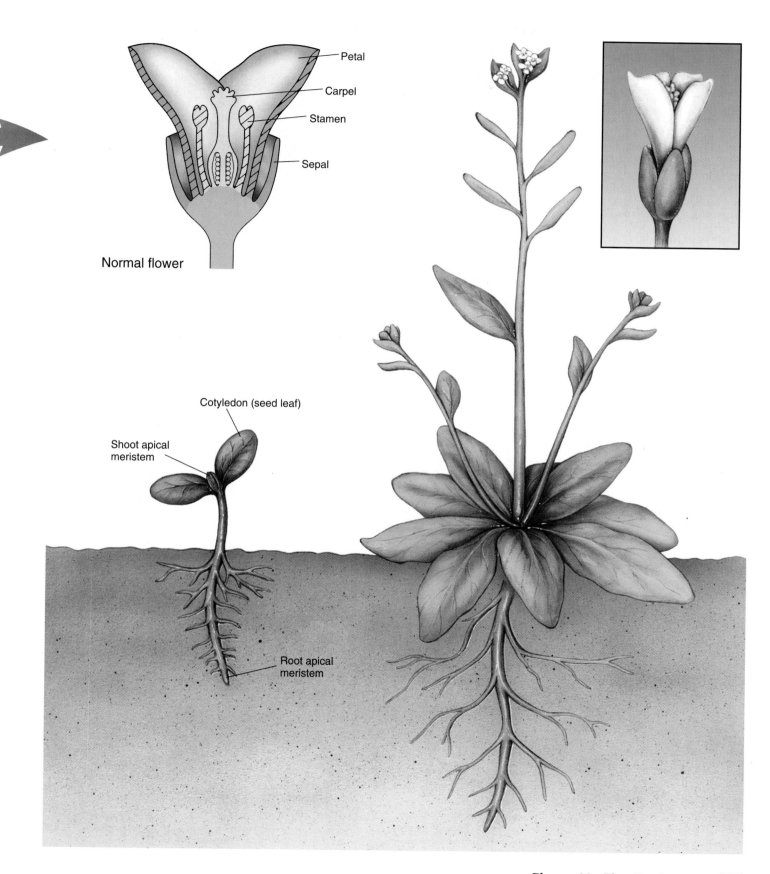

Petal

Carpel

Stamen

Sepal

Normal flower

Cotyledon (seed leaf)

Shoot apical meristem

Root apical meristem

Embryonic Development

We will now consider the way in which a zygote, the first diploid cell of the plant life cycle, develops into a plant. At first the zygote undergoes mitosis repeatedly, and the cells divide; the resulting mass of cells develops rapidly into an embryo sporophyte. As the embryo develops, one can observe the origin of the organs and tissues that make up the plant body. **Meristems** (tissues composed of cells capable of repeated division) are established early at the root and shoot apices, setting the stage for further development of these organs.

Angiosperm Embryos

In angiosperms, differentiation of cell types in the embryo begins almost immediately after fertilization (figure 38.4). Although there are many variations in the details of embryo development, the first division of the zygote is typically transverse to its long axis, establishing the polarity of the embryo. When the lower cell divides, it produces a stalk-like structure called the **suspensor,** while the upper cell develops into a more or less spherical **embryo,** which is called a **proembryo** before it becomes spherical. The suspensor is actively involved in absorbing nutrients from the endosperm (triploid food storage tissue) and extends the embryo into the food source.

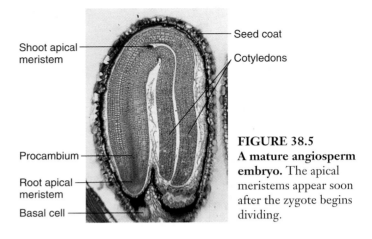

**FIGURE 38.5
A mature angiosperm embryo.** The apical meristems appear soon after the zygote begins dividing.

Meanwhile, as the nearly spherical embryo develops, the **protoderm** (future epidermis), the **procambium** (which produces primary xylem and phloem), and the **ground meristem** (which produces thin-walled parenchyma cells) differentiate. The apical meristems can be detected only a few days after the first division of the zygote (figure 38.5). Leaf and branch **primordia** differentiate as the shoot meristem grows upward; the root meristem grows downward at the same time. Both meristems may continue to function throughout the life of the plant; however, if a shoot meristem becomes reproductive (as, for example, in annuals) that part of the plant loses its capacity for further growth.

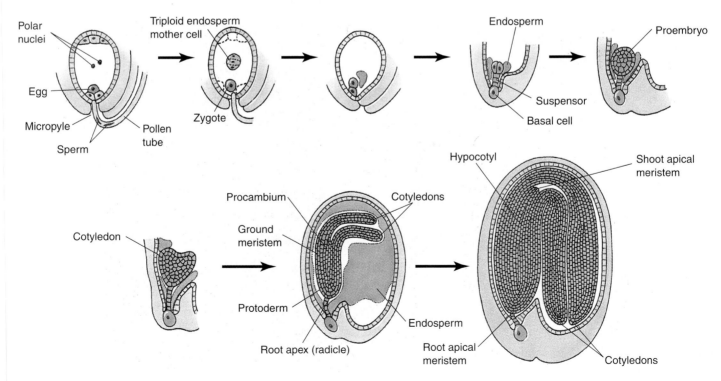

**FIGURE 38.4
Stages of development in an angiosperm embryo.** Differentiation begins almost immediately after fertilization.

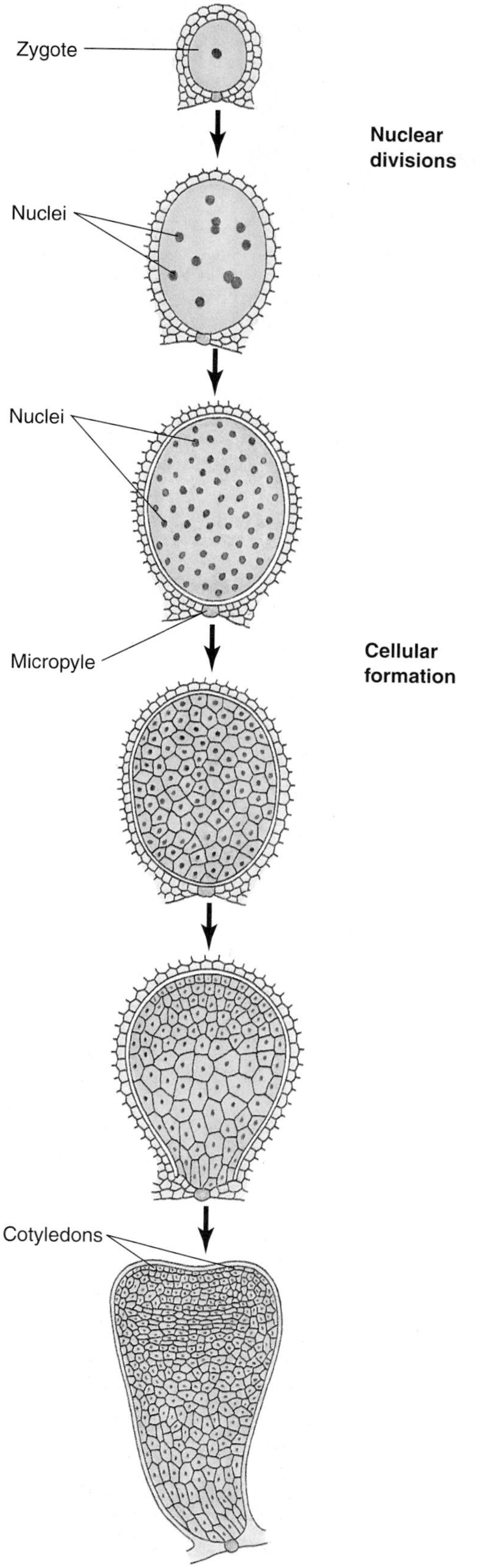

Zygote
Nuclear divisions
Nuclei
Nuclei
Micropyle
Cellular formation
Cotyledons

FIGURE 38.6
Diagram of a developing embryo in a gymnosperm. Cell walls often form only after several rounds of cell division.

Gymnosperm Embryos

In gymnosperms (figure 38.6), the zygote nucleus divides repeatedly after fertilization, but in many gymnosperms, no cell walls form between the daughter nuclei. After eight rounds of cell division, a single large embryonic cell usually contains 256 nuclei. Cell walls then form, producing a mass of 256 cells of equal size, and differentiation begins. The cells farthest from the micropyle divide faster than those nearer to it. This results in a marked gradient in cell size, since the rapidly dividing cells are smaller than the others. The larger cells near the micropyle develop into a suspensor. At the same time, the smaller cells on the opposite pole of the embryo, which constitute about a third of the total cell mass, give rise first to the apical meristem of the root and then to that of the shoot (figure 38.7).

Although the details are different, the basic pattern outlined in the preceding paragraph is apparent in the embryogenesis of all seed plants. An important feature, and one that clearly separates plants from animals, is that cell movement does not occur in plants during the course of their embryonic development: plant cells differentiate where they are formed. Their position relative to other cells is important in determining how they differentiate; their specific course of development probably is determined in part by chemical gradients, but information localized on cell walls also seems to play an important role. The first hint that cell walls regulate development came when Ralph Quatrano of Washington University in St. Louis and colleagues found that enzymatic removal of the cell wall of the brown algae *Fucus* destroyed the ability of the embryo to "remember" where its rhizoid is supposed to grow. *Fucus* cells already specialized to form either rhizoids or fronds, when liberated from their cell walls, could give rise to both. Similar results are reported for the plant *Arabidopsis.* Clearly the embryo cells don't contain the information to specifiy their own fates—but their cell walls do. Cell walls contain a wide variety of carbohydrates and proteins attached to the wall's structural fibers. Attempting to pin down the identities of these suspected developmental signals is an area of active research.

Early in the development of a plant, cells differentiate into meristems near the root and shoot apices.

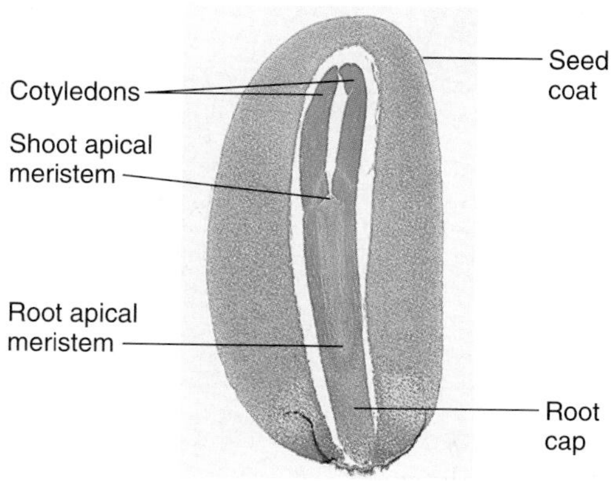

Cotyledons
Shoot apical meristem
Root apical meristem
Seed coat
Root cap

FIGURE 38.7
A mature gymnosperm embryo. The apical meristems arise from cells opposite the micropyle.

Determining Orientation of the Shoot

Patterns of growth and development typical of plants emerge in the early stages of embryo development. Within the growing seed, the mature angiosperm embryo consists of an axis with either one or two **cotyledons** ("seed leaves"). Monocot embryos have only one cotyledon, and dicot embryos generally have two. In monocots such as corn and cereals, food may be stored as endosperm in the mature seed or, in dicots such as peas and beans, the food may be absorbed from the endosperm into the cotyledons, which become thick and fleshy. In monocots, the single cotyledon (scutellum) functions mainly as a food-absorbing organ, transferring food from the endosperm to the young embryo during seed germination. In either case, food that is initially concentrated in the endosperm during the early development of the embryo is used during germination and early establishment of the young plant. Starches, fats, and oils are converted into sugars and used to nourish and sustain the young plant before it becomes a photosynthetically active, independent entity.

The apical meristems differentiate early in embryonic development (figure 38.8). The shoot apical meristem is located either between the cotyledons as a mound of tissue or at the tips of the **epicotyl**, the part of the stem axis that extends above the cotyledons in some seeds (figure 38.9*a*). The epicotyl may be short and relatively undifferentiated, or it may be long and include one or more seed leaves. The epicotyl, together with its young leaves, is called a **plumule**. The stem axis of the embryo below the cotyledons is called the **hypocotyl**. A distinct embryonic root at the lower end of the hypocotyl is called the **radicle**; it later

(a)   (b)

FIGURE 38.8
Developing seedlings. Apical meristems differentiate early in development, as seen in this dicot, (a) a soybean, and in this monocot, (b) corn.

develops into the primary root. If a radicle is lacking, the axis below the cotyledons is referred to as the hypocotyl root axis, the root being represented by only an apical meristem and a root cap. In grasses, such as corn (figure 38.9*b*), there is a single cotyledon called a **scutellum** that absorbs endosperm. The plumule is enclosed within a capped tubular sheath called the **coleoptile,** and the radicle within a similar sheath is called the **coleorhiza.** We will encounter coleoptiles again in chapter 39, in connection with important experiments in plant physiology.

Development of the shoot into stem and leaves occurs soon after germination.

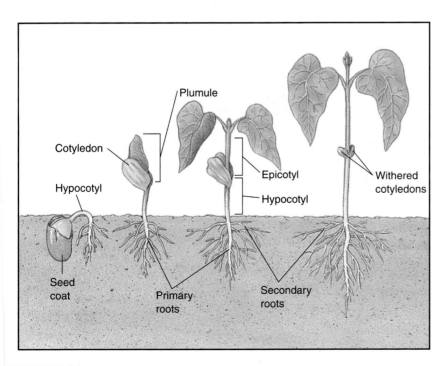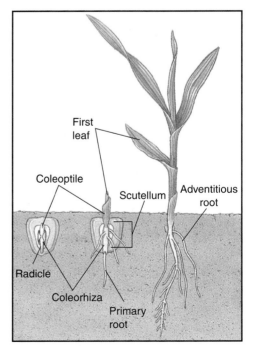

FIGURE 38.9
Shoot development. The stages shown are for a dicot, the common bean, (a) *Phaseolus vulgaris,* and a monocot, corn, (b) *Zea mays.*

Adaptive Importance of Seeds

Early in the development of an angiosperm embryo, a profoundly significant event occurs: the embryo stops developing. In many plants, development of the embryo is arrested soon after the meristems and cotyledons differentiate. The *integuments*—the outer cell layers of the ovule—develop into a relatively impermeable seed coat, which encloses the seed with its dormant embryo and stored food. Seeds are important adaptively in at least five aspects:

1. They maintain dormancy under unfavorable conditions and postpone development until better conditions arise. If conditions are marginal, a plant can "afford" to have some seeds germinate, because others will remain dormant.
2. Seed embryo development is influenced by critical aspects of the plant's habitat, including environmental factors such as temperature and moisture.
3. The seed affords maximum protection to the young plant at its most vulnerable stage of development.
4. The seed contains stored food that permits development of a young plant prior to the availability of an adequate food supply from photosynthetic activity.
5. Perhaps most important, the dispersal of seeds facilitates the migration and dispersal of plant genotypes into new habitats.

Once a seed coat forms, most of the embryo's metabolic activities cease. A mature seed contains only about 5% to 20% water. Under these conditions, the seed and the young plant within it are very stable; it is primarily the progressive and severe desiccation of the embryo and the associated reduction in metabolic activity that are responsible for its arrested growth. Germination cannot take place until water and oxygen reach the embryo, a process that sometimes involves cracking the seed coat through abrasion or alternate freezing and thawing. Seeds of some plants have been known to remain viable for hundreds and, in rare instances, thousands of years (figure 38.10).

Specific adaptations often help ensure that the plant will germinate only under appropriate conditions. Sometimes, seeds lie within tough cones that do not open until they are exposed to the heat of a fire (figure 38.11). This causes the plant to germinate in an open, fire-cleared habitat; nutrients will be relatively abundant, having been released from plants burned in the fire. Seeds of other plants will germinate only when inhibitory chemicals have been leached from their seed coats, thus guaranteeing their germination when sufficient water is available. Still other plants will germinate only after they pass through the intestines of birds or mammals or are regurgitated by them, which both weakens the seed coats and ensures the dispersal of the plants involved. Sometimes seeds of plants thought to be extinct in a particular area may germinate under unique or improved environmental circumstances, and the plants may then reappear.

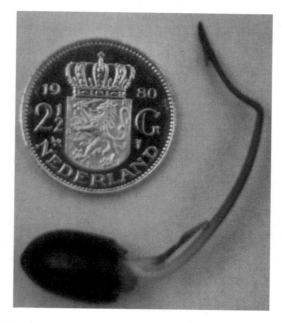

FIGURE 38.10
Seeds can remain dormant for long periods. This seedling was grown from seeds of lotus recovered from the mud of a dry lake bed in Manchuria, northern China. The radiocarbon age of this seed indicates that it was formed around the year 1515. The coin is included in the photo to give some idea of the size.

FIGURE 38.11
Fire induces seed germination in some chaparral plants. Manzanita (*Arctostaphylos viscida*) seeds only germinate when fire or other agents break down the seed coat.

> **Seed dormancy is an important evolutionary factor in plants, ensuring their survival in unfavorable conditions and allowing them to germinate when the chances of survival for the young plants are the greatest.**

Germination

Animals become adults after growing rapidly during a juvenile period; they then maintain a fairly constant size during adulthood. Many plants, after germination, keep on growing—some indefinitely. They add new cells and tissues, both at the ends of their shoots and roots and also internally; as a result, they continue to increase in size. Some patches of prairie grasses, several to many meters in diameter, are believed to be proliferations of a single plant that has been growing in one place since the glaciers receded more than 10,000 years ago. Trees attain great ages; potatoes, bananas, and many other crops are simply propagated over and over again as parts of a single cloned plant, producing generation after generation of genetically identical individuals.

Mechanisms of Germination

Germination is the first step in the development of the plant outside of its seed coat. Germination occurs when a seed absorbs water and its metabolism resumes. Initially, metabolism may be anaerobic, but when the seed coat ruptures, oxidative metabolism takes over. At this point, it is important that oxygen be available to the developing embryo because plants, like animals, require oxygen for active growth. Few plants produce seeds that germinate successfully under water, although some, such as rice, have evolved a tolerance to anaerobic conditions.

A dormant seed, although it may have imbibed a full supply of water and may be respiring, synthesizing proteins and RNA, and apparently carrying on normal metabolism, may nonetheless fail to germinate without an additional signal from the environment. This signal may be light of the correct wavelengths and intensity, a series of cold days, or simply the passage of time at temperatures appropriate for germination. Seeds of many plants will not germinate unless they have been **stratified**—held for periods of time at low temperatures. This phenomenon prevents seeds of plants that grow in cold areas from germinating until they have passed the winter, thus protecting their seedlings from cold conditions.

Germination can occur over a wide temperature range (5 °C to 30 °C), although certain species and specific habitats may have relatively narrow optimum ranges. Some seeds will not germinate even under the best conditions. In some species, a significant fraction of a season's seeds remain dormant, providing a genetic reservoir, or gene pool, of great evolutionary significance to the future plant population.

The Utilization of Reserves

Germination occurs when all internal and external requirements are met. Germination and early seedling growth require the utilization of metabolic reserves; these reserves are stored in the starch grains of **amyloplasts** (colorless plastids that store starch) and protein bodies. Fats and oils also are important food reserves in some kinds of seeds. They can readily be digested during germination, producing glycerol and fatty acids, which yield energy through oxidative respiration; they can also be converted to glucose. Depending on the kind of plant, any of these reserves may be stored in the embryo itself or in the endosperm.

In the kernels of cereal grains, the cotyledon is modified into a relatively massive structure called the **scutellum** (see figure 38.9b), from the Latin word meaning "shield." The abundant food stored in the scutellum is used up first because these plants do not need to use the endosperm during germination. Later, while the seedling is becoming established, the scutellum, in a two-stage process, absorbs the additional food that is stored in the endosperm:

1. The initial utilization of starch in the endosperm is facilitated by hydrolase enzymes, which are secreted by the epidermal layer of the scutellum.
2. The later and more extensive utilization of endosperm starches is achieved by the secretion of amylase and other hydrolytic enzymes from the **aleurone layer.** This layer of specialized endosperm cells lies just inside the seed coat, which is fused to the pericarp. The synthesis and secretion of these aleurone hydrolases are controlled by a class of hormones called **gibberellins,** which we will consider further in the next chapter.

The emergence of the embryonic root and shoot from the seed during germination varies widely from species to species. In most plants, the root emerges before the shoot appears and anchors the young seedling in the soil (see figure 38.10). In plants such as peas and corn, the cotyledons may be held below ground; in other plants, such as beans, radishes, and sunflowers, the cotyledons are held above ground. The cotyledons may or may not become green and contribute to the nutrition of the seedling as it becomes established. The period from the germination of the seed to the establishment of the young plant is a very critical one for the plant's survival; the seedling is unusually susceptible to disease and drought during this period.

During germination and early seedling establishment, the utilization of food reserves stored in the embryo or the endosperm is mediated by hormones, which, in some cases, are gibberellins.

Establishment of the Meristems

Once a seed has germinated, the plant's further development depends on the activities of the meristematic tissues. Clumps of small cells with dense cytoplasm and proportionately large nuclei at the shoot and root apices are the apical meristems. Elongation of both root and shoot takes place as a result of repeated cell divisions and subsequent elongation of the cells produced by the **apical meristems.**

Apical Meristems

Primary growth in plants is brought about by the apical meristems. The elongation of the root and stem forms what is known as the **primary plant body,** which is made up of **primary tissues.** The primary plant body comprises the young, soft shoots and roots of a tree or shrub, or the entire plant body in some herbaceous plants.

Lateral Meristems

Many herbaceous plants exhibit only primary growth, but trees, shrubs, and some herbs also exhibit **secondary growth.** Secondary growth involves the activity of **lateral meristems,** which are cylinders of meristematic tissue within the stems and roots (figure 38.12). The continued division of cells in lateral meristems primarily increases the girth of the plant body. There are two kinds of lateral meristems: (1) **vascular cambium,** which may ultimately give rise to thick accumulations of secondary xylem (wood) and also to secondary phloem, the outermost layers of which become crushed as new layers of xylem are added; and (2) **cork cambium,** which produces the outer layers of bark in both older roots and older stems. Tissues formed from lateral meristems, which comprise most of the trunk, branches, and older roots of trees and shrubs, are known as **secondary tissues** and are collectively called the secondary plant body.

Apical meristems are responsible for a plant's primary growth, principally involving growth in length of the root or shoot. The primary plant body, which includes the young, soft shoots and roots, arises from the apical meristems. Once the lateral meristems begin to function, they produce the secondary plant body, which is often characterized by thick accumulations of conducting tissue and the other cell types associated with it.

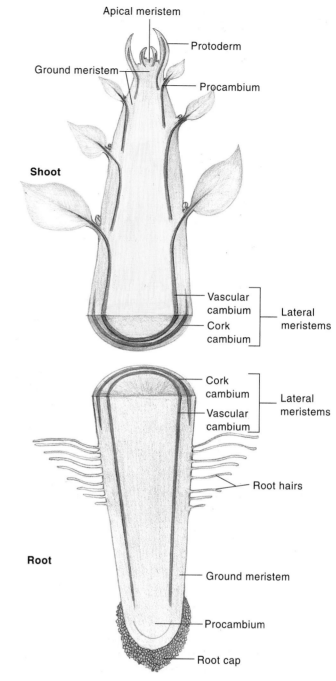

FIGURE 38.12
Apical and lateral meristems. Apical meristems produce primary growth, the elongation of the root and stem. In some plants, the lateral meristems, which form the vascular cambium and the cork cambium, produce secondary growth, an increase in the girth of a plant.

How Long Do Individual Plants Live?

Once established, plants live for highly variable periods of time, depending on the species. Woody plants, which have extensive secondary growth, nearly always live longer than herbaceous plants, which have limited or no secondary growth. Some herbaceous plants send new stems above the ground every year, producing them from woody underground structures. Others germinate and grow, flowering just once before they die. Shorter-lived plants rarely become very woody, since there is not enough time for the accumulation of secondary tissues. Depending on the length of their life cycles, herbaceous plants may be annual, biennial, or perennial, while woody plants are always perennial (figure 38.13).

Annual plants grow, flower, and form fruits and seeds within one growing season; they then die when the process is complete. Many crop plants are annuals, including corn, wheat, and soybeans. Annuals generally grow rapidly under favorable conditions and in proportion to the availability of water or nutrients. The lateral meristems of some annuals, like sunflowers or giant ragweed, do produce poorly developed secondary tissues, but most are entirely herbaceous.

Biennial plants, which are much less common than annuals, have life cycles that take two years to complete. During the first year, biennials generally form a **rosette** (symmetrical tuft) of leaves on the ground. During the second year of growth, in a process called **bolting**, flowering stems are produced using energy stored in the rosette and in the underground parts of the plant. Certain crop plants, including carrots, cabbage, and beets, are biennials, but these plants generally are harvested for food during their first season, before they flower. They are grown for their leaves or roots, not for their fruits or seeds. Wild biennials include evening primroses, Queen Anne's lace, and mullein. Many plants that are considered biennials actually do not flower until they are three or more years of age, but all biennial plants flower only once before they die.

Perennial plants continue to grow year after year and may be herbaceous, as are many woodland, wetland, and prairie wildflowers, or woody, as are trees and shrubs. The majority of vascular plant species are perennials. Herbaceous perennials rarely experience any secondary growth in their stems; the stems die each year after a period of relatively rapid growth and food accumulation. Food is often stored in the plants' roots or underground stems, which can become quite large in comparison to their less substantial aboveground counterparts.

Trees and shrubs are either deciduous, with all the leaves falling at one particular time of year and the plants remaining bare for a period, or evergreen, with the leaves dropping throughout the year and the plants never appearing completely bare. In northern temperate regions, conifers are the most familiar evergreens; but in tropical and subtropical regions, most angiosperms are evergreen, except where there is a severe seasonal drought. In these areas, many angiosperms are deciduous, losing their leaves during the drought and thus conserving water.

(a)

(b)

FIGURE 38.13
Annual and perennial plants. Plants live for very different lengths of time. (a) Desert annuals complete their entire lifespan in a few weeks. (b) Some trees, such as the giant redwood (*Sequoiadendron giganteum*), which occurs in scattered groves along the western slopes of the Sierra Nevada in California, live 2000 years or more.

Annual plants complete their whole growth cycle within a single year. Biennial plants flower only once, normally after two seasons of growth. Perennials flower repeatedly and live for many years.

38.1 The course of plant development is strongly influenced by the environment.

- Plants develop in a linear fashion, their cells prevented from moving by their rigid cell walls.

- Meristems continuously produce new tissues, which then differentiate into body parts. This differentiation is significantly influenced by the environment.

38.2 Cellular mechanisms of development in a flowering plant are broadly similar to those in animals.

- Scientists have researched the cellular mechanisms of plant development mainly in the flowering plant *Arabidopsis thaliana*.

38.3 Differentiation of cell types occurs early in plant development.

- Embryo development in animals involves extensive movements of cells in relation to one another, but the same process in plants consists of an orderly production of cells, rigidly bound by their cell walls.

- In angiosperms, cell walls form with each division of the zygote; in other seed plants (gymnosperms), free nuclei form within a common cytoplasm before cell walls are formed and differentiation begins.

- Seed dormancy is important because it allows plants to survive unfavorable conditions and to germinate when chances of survival are higher.

38.4 Development of the adult plant body involves active meristems.

- In a seed, the embryo with its food supply is encased within a sometimes rigid, relatively impermeable seed coat.

- At germination, the mobilization of the food reserves is critical. Hormones control this process.

- Apical meristems at the tips of roots and shoots produce cells that elongate and increase a plant's length.

- The apical meristem gives rise to three types of primary meristems, partly differentiated tissues in which active cell division continues to take place. These are the protoderm, which gives rise to the epidermis; the procambium, which gives rise to the primary vascular tissues; and the ground meristem, which becomes ground tissue. This is primary growth.

- Secondary meristems consist of the vascular cambium and the cork cambium.

- Growth of the secondary meristems produces secondary (woody) tissues that increase the plant's girth. This is secondary growth.

Discussing Key Terms

1. **Homeotic genes**   The location of specific organs is controlled by these genes, which work much the same way in animals and plants.

2. **Seed plants**   The zygote—the first cell of the sporophyte generation of seed plants—develops into an embryo within a seed, which is protected by a relatively waterproof seed coat. The seed contains a food supply consisting of carbohydrates, often in the form of endosperm, which will be used by the young, germinating seedling.

3. **Germination**   A seed's resumption of growth and development.

4. **Primary growth**   Growth that originates in the apical meristems of the shoot and root and causes an increase in length.

5. **Secondary growth**   Growth that originates in the lateral meristems and causes an increase in root and shoot diameter.

Review Questions

1. How does early development in a gymnosperm differ from that in an angiosperm? When and how does cell differentiation occur in gymnosperms?

2. In seed plants, when does development of the embryo come to a temporary halt? Why are seeds adaptively important? Why may a seed showing proper respiration and synthesis of proteins and nucleic acids and all other normal metabolic activities still fail to germinate?

3. Where in a plant does primary growth occur? How does primary growth change the body of a plant? What kinds of plants exhibit primary growth? In what kind of tissue does secondary growth occur? Where specifically is this tissue found? How does it change the body of a plant?

4. What are the differences among annual, biennial, and perennial plants? What is the difference between deciduous and evergreen trees and shrubs? In what types of climates are deciduous trees found? Why?

Thought Questions

1. How do the developmental control systems of animals and plants compare? How is each affected by changes in the environment? Why is plant development so much more closely linked with environmental cues than animal development?

2. Many plants have seeds that require certain conditions for germination. Besides moisture and proper temperatures, what other types of conditions may be required? How do these requirements increase the survival of the young plants?

Internet Links

The Naked Seeds of *Pinus*
http://koning.ecsu.ctstateu.edu/Plant_Biology/Pinus.html
A well-illustrated guide to the life cycle of gymnosperms from Conneticut State University.

Seeds of Life
http://versicolores.ca/SeedsOfLife/home.html
From Quebec, a beautiful site devoted to the anatomy and biology of seeds.

For Further Reading

Coen, E. and R. Carpenter: "The Power Behind the Flower," *New Scientist*, vol. 134, April 25, 1992, pages 24–27. Clear and interesting discussion of the factors involved in making a plant flower.

Esau, K.: *Anatomy of Seed Plants*, 2d. ed., John Wiley & Sons, New York, 1977. Short but outstanding textbook on plant anatomy.

Finkelstein, R., M. Estelle, J. Martinez-Zapater, and C. Somerville: "*Arabidopsis* as a Tool for the Identification of Genes Involved in Plant Development," In *Temporal and Spatial Regulation of Plant Genes*, pages 1–25, Springer-Verlag, New York, 1988. An account of how genetic procedures are used to find genes involved in key aspects of plant development.

Hardwick, R.: "Construction Kits for Modular Plants," *New Scientist*, April 10, 1986, pages 39–42. Much attention is properly being paid to the ways in which plant growth follows repetitive patterns, and this short article provides an introduction to the topic.

Hutchings, M. J. and I. K. Bradbury: "Ecological Perspectives on Clonal Perennial Herbs," *BioScience*, 1986, vol. 36, pages 178–82. Does a clonal plant have an advantage in nature if it is functionally integrated or if the parts are independent?

Jurgens, G.: "Pattern Formation in the Flowering Plant Embryo," *Curr. Opin. Genet. Dev.*, 1992, vol. 2, pages 567–70. A lucid example of research into the early stages of plant development.

Meyerowitz, E. M.: "The Genetics of Flower Development," *Scientific American*, 1994, vol. 271, pages 56–65. Flower cells "learn" which organs to become from genes that convey positional information.

Steeves, T. and I. Sussex: *Patterns in Plant Development*, 2d. ed., Prentice-Hall, Englewood Cliffs, N.J., 1989. A classic text presenting a structural approach to plant development, with emphasis on experimental data.

Strauss, E.: "When Walls Can Talk, Plant Biologists Listen," *Science*, vol. 282, pages 28–29, October 2, 1998. Plant cell walls, once thought to be inert boxes, are being shown by current research to be platforms for powerful molecular signals that determine the fate of plant cells during development.

Various Authors: "Development," *Science*, vol. 266, October 1994. An entire issue devoted to recent advances in developmental biology.

39

Plant Growth and Regulation

Concept Outline

39.1 Every plant cell contains all the information needed to grow into an adult plant.

Differentiation in Plants: Experimental Evidence. By isolating a single plant cell, researchers were able to demonstrate that plants are totipotent.

39.2 Hormones are the tools plants use to regulate their growth.

Plant Hormones. Five hormones interact to determine the pattern of growth in the plant body.
Auxin. Auxin is involved in the elongation of stems.
Cytokinins. In conjunction with auxin, cytokinins stimulate cell division.
Gibberellins. Gibberellins interact with auxin to control stem elongation.
Ethylene. Ethylene controls the abscission of leaves, flowers, and fruits.
Abscisic Acid. Abscisic acid suppresses growth of buds and promotes leaf senescence.

39.3 Plant growth is often guided by environmental cues.

Tropisms. Plant growth is often influenced by light, gravity, and contact with other plants.
Turgor Movement. Reversible changes in turgor pressure of particular cells can alter the orientations of leaves and so improve photosynthetic efficiency.

39.4 Plant flowering is often keyed to day length.

Flowering Responses. Many plants flower only when they experience certain day lengths.
The Chemical Basis of the Photoperiodic Response. Plants contain a special pigment responsive to red light to regulate photoperiodism.
Dormancy. The ability to cease growth altogether when conditions are not favorable is a key factor in the survival of many plants.

FIGURE 39.1
Tropisms often guide plant growth. The branches of this fallen tree are growing straight up because they are gravitropic.

Plants grow or die partly in response to their environment. Unlike animals, whole plants do not move from place to place, although some of their reproductive cells (for example, sperm cells) may actively move, and they may be spread to new locations by the dispersal of their seeds. Also, unlike animals, plants may develop intermittently for indefinite and sometimes lengthy time periods. Plants, as well as all other organisms, develop according to a genetic blueprint, but the way in which their particular blueprint is expressed is greatly influenced by external factors (figure 39.1). The differentiation of specific tissues in plants is controlled by chemical substances called hormones. The actions of these hormones and external environmental factors on growth is the subject of this chapter.

Differentiation in Plants: Experimental Evidence

The Totipotency of Single Cells

Significant numbers of living plant cells—those that retain a living protoplast and nucleus at maturity—are capable of *dedifferentiation;* that is, the cells can cease to remain in their current differentiated form and can develop into another form. Gene expression is reactivated, triggering alternative modes of differentiation or even the growth of a complete plant. In 1902, German botanist Gottlieb Haberlandt proposed that all living plant cells are **totipotent,** that is, that each cell possesses the full genetic potential of the organism. Before this hypothesis could be confirmed, however, it was necessary to learn how to grow plant cells in culture.

It was relatively easy to isolate and maintain a single plant cell in culture, but it was more difficult to stimulate the cell to divide. For differentiation to occur, the cell might first have to divide repeatedly. It turned out that the substances needed for cell division diffuse out of individual cells when they are isolated in an agitated medium, so the cells will not divide. This problem was finally solved in several different ways in the 1950s. One solution, devised by William Muir, a graduate student working at the University of Wisconsin, involved placing a single cell, isolated within a tiny drop of culture medium, on a piece of filter paper resting on an established cell culture. The single cell was thus isolated from the other cells by filter paper, but at the same time it could be influenced by them. Any substances necessary for the promotion of cell growth and division in the isolated cell could diffuse to it through the filter paper from the medium that it shared with the established cell culture. Under these conditions, the isolated cell grew and divided rapidly, establishing a mass of dividing, undifferentiated cells called a **callus.** This callus could then be grown indefinitely in culture.

Using such methods, it has been possible to determine the nutritional requirements of plants by growing them in defined media, a subject that we will discuss further in the next chapter. For most plants, it was necessary to add coconut "milk," the liquid endosperm produced within coconuts, to the medium. As we will see, additional factors

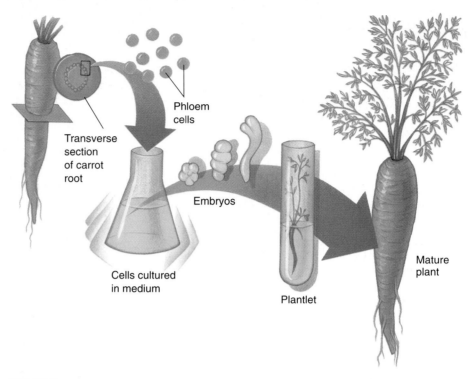

FIGURE 39.2
Growing a new plant from a bit of mature tissue. This technique of isolating phloem tissue from carrots *(Daucus carota)* was carried out in the laboratory of F. C. Steward at Cornell University. The discs of tissue were grown in a flask in which the medium was constantly agitated, bringing a fresh supply of nutrients to the masses of callus being formed.

present in coconut "milk" proved to be of great interest in understanding plant growth and development.

Tissue Culture

In 1958, Cornell University plant physiologist F. C. Steward and his coworkers expanded previously developed techniques of culturing plant cells. Their methods also depended on supplying differentiated cells with substances obtained from dividing cells. Steward isolated small bits of secondary phloem tissue from carrot root and placed them in a flask of liquid growth medium (figure 39.2). Such growth media contained sucrose and minerals essential for plant growth, as well as certain vitamins—organic molecules that the isolated plant parts cannot manufacture for themselves.

In such a medium, many of the new cell clumps, each of which had originated from a single phloem cell, differentiated roots (figure 39.3*a*). When placed on *agar* (a gelatin-like substance obtained from seaweeds), the clumps also

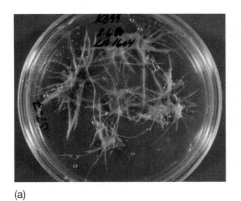

(a)

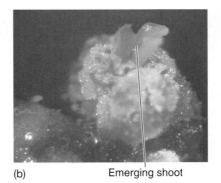

(b)   Emerging shoot

(c)

FIGURE 39.3

Tomato *(Lycopersicon esculentum)* plants regenerated from single parenchyma cells. (a) Roots differentiating from masses of callus tissue in culture. (b) A mass of callus tissue anchored on agar, a firm medium, with a shoot emerging. (c) A regenerated tomato plant, which will mature, producing flowers and fruits, if it is transferred to soil. F. C. Steward obtained similar results with carrots, thus confirming Haberlandt's totipotency hypothesis.

developed shoots (figure 39.3*b*) and eventually grew into whole plants, thus confirming Haberlandt's hypothesis (figure 39.3*c*). Steward's results clearly demonstrated that the original differentiated phloem tissue still contained all of the genetic potential needed for the differentiation of a whole plant. Indeed, the embryonic growth of the isolated cells closely resembled that of normal zygotes, dividing and differentiating into embryos. These "embryoids" had shoot and root apices from which further development proceeded, as well as the same distinct tissue systems that ordinary embryos have at comparable stages of development.

Regeneration in Nature

If the cues that trigger plant development come from the external environment, they ought to be familiar stimuli we see around us. Are everyday plants regenerating whole individuals from differentiated tissue out in the real world? Yes. The common practice of using *cuttings* (segments of stems or roots) to produce mature plants demonstrates this property. Adventitious roots can arise from root primordia that are produced at the base of a cut stem or from phloem tissues after the cut has been made. These primordia can then differentiate into root meristems.

Adventitious roots form easily in many plants, while adventitious shoots form less frequently and apparently with difficulty. Cuttings from plants such as beans, garden geraniums *(Pelargonium)*, and a host of other plants will form roots if they are simply left with their lower ends in water or in sand that is watered; an example is shown in figure 39.4. In other plants, root

FIGURE 39.4

Adventitious roots. A leaf of African violet, *Saintpaulia*, forming adventitious roots and new plantlets at the base of the petiole.

formation seldom occurs, if at all. Stems with leaves will form roots much more readily than stems whose leaves have been removed. The leaves are the primary source of *auxin*, a plant hormone that stimulates root production. Buds at the bases of the leaves are a secondary source of auxin.

Whole plants may be regenerated from bits of leaves of some kinds of plants, such as succulents (for example, jade plants, *Crassula argentea;* rex begonias, *Begonia rex;* and others). A naturally occurring type of vegetative propagation is seen in the familiar houseplant *Kalanchoë* (aptly called "mother of thousands"), which produces up to 30 or more tiny plantlets along its leaf margins; each plantlet can grow independently when it falls to a suitable spot on the ground. As you saw in the previous chapter, many plants can differentiate whole new individuals from cuttings, and from segments of modified stems, such as rhizomes, runners, tubers, bulbs, and corms. Both normal and modified roots also can undergo such vegetative propagation. Sometimes modifications of normal plant reproductive structures function in plant propagation. For example, in some century plants *(Agave)*, plantlets may form among the flowers. Seeds with embryos arising from unfertilized eggs can also be produced, a process that is genetically equivalent to the other kinds of asexual reproduction discussed here.

Plant stems regularly produce adventitious roots under the appropriate circumstances; roots produce adventitious shoots less frequently.

All living plant cells are totipotent, able to regenerate an entire plant.

Plant Hormones

The tissue regeneration experiments of F. C. Steward and many others have led to the general conclusion that differentiated plant tissue is capable of expressing its hidden genetic complement when provided with suitable environmental signals. What halts the expression of genetic potential when the same kinds of cells are incorporated into normal, growing plants? As we will see, the expression of some of these genes is controlled by plant hormones—factors that occur, for example, in coconut "milk," the liquid endosperm of coconuts.

Hormones are chemical substances produced in small, often minute, quantities in one part of an organism and then transported to another part of the organism, where they bring about physiological responses. The activity of hormones results from their capacity to stimulate certain physiological processes and to inhibit others. How they act in a particular instance is influenced both by the hormone and by how it affects the tissue that receives the message.

In animals, hormones are usually produced at definite sites, normally in organs that are solely concerned with hormone production. In plants, hormones are not produced in specialized tissues but, instead, in tissues that also carry out other, usually more obvious, functions. There are five major kinds of plant hormones: auxin, cytokinins, gibberellins, ethylene, and abscisic acid (table 39.1). Other less well-understood kinds of plant hormones also exist. The study of plant hormones, especially attempts to understand how they produce their effects (figure 39.5), is an active and important field of current research.

There are five major kinds of plant hormones: auxin, cytokinins, gibberellins, ethylene, and abscisic acid.

Table 39.1 Functions of the Major Plant Hormones

Hormone		Major Functions	Where Produced or Found in Plant
Auxin (IAA)		Promotion of stem elongation and growth; formation of adventitious roots; inhibition of leaf abscission; promotion of cell division (with cytokinins); inducement of ethylene production; promotion of lateral bud dormancy	Apical meristems; other immature parts of plants
Cytokinins		Stimulation of cell division, but only in the presence of auxin; promotion of chloroplast development; delay of leaf aging; promotion of bud formation	Root apical meristems; immature fruits
Gibberellins		Promotion of stem elongation; stimulation of enzyme production in germinating seeds	Roots and shoot tips; young leaves; seeds
Ethylene		Control of leaf, flower, and fruit abscission; promotion of fruit ripening	Roots, shoot apical meristems; leaf nodes; aging flowers; ripening fruits
Abscisic acid		Inhibition of bud growth; control of stomatal closure; some control of seed dormancy; inhibition of effects of other hormones	Leaves, fruits, root caps, seeds

FIGURE 39.5
Effects of plant hormones. Plant hormones, often acting together, influence many aspects of plant growth and development, including (a) fruit ripening, (b) leaf abscission, (c) shoot elongation, (d) and the formation of mature fruit.

Auxin

More than a century ago, an organic substance known as **auxin** became the first plant hormone to be discovered. Auxin increases the plasticity of plant cell walls and is involved in elongation of stems, although **gibberellins,** another class of plant hormones, have a more pronounced role in affecting stem elongation.

Discovery of Auxin

In his later years, the great evolutionist, Charles Darwin, became increasingly devoted to the study of plants. In 1881, he and his son Francis published a book called *The Power of Movement of Plants*. In this book, the Darwins reported their systematic experiments on the response of growing plants to light—responses that came to be known as **phototropisms.** They used germinating oat *(Avena sativa)* and canary grass *(Phalaris canariensis)* seedlings in their experiments and made many observations in this field.

Charles and Francis Darwin knew that if light came primarily from one direction, the seedlings would bend strongly toward it. If they covered the tip of the shoot with a thin glass tube, the shoot would bend as if it were not covered. However, if they used a metal foil cap to exclude light from the plant tip, the shoot would not bend (figure 39.6). They also found that using an opaque collar to exclude light from the stem below the tip did not keep the area above the collar from bending.

In explaining these unexpected findings, the Darwins hypothesized that when the shoots were illuminated from one side, they bent toward the light in response to an "influence" that was transmitted downward from its source at the tip of the shoot. For some 30 years, the Darwins' perceptive experiments remained the sole source of information about this interesting phenomenon. Then, several other botanists performed a series of related experiments. Some of the most significant were those conducted independently by the Danish plant physiologist Peter Boysen-Jensen and the Hungarian plant physiologist Arpad Paal, who demonstrated that the substance that caused the shoots to bend was a chemical. They showed that if the tip of a germinating grass seedling was cut off and then replaced with a small block of agar separating it from the rest of the seedling, the seedling would grow as if there had been no change. Something evidently was passing from the tip of the seedling through the agar into the region where the bending occurred. On the basis of these observations under conditions of uniform illumination or of darkness, Paal suggested that an unknown substance continually moves down from the tips of grass seedlings and promotes growth on all sides. Such a light pattern would not, of course, cause the shoot to bend.

The next telling results in this area, published in 1926, were obtained by a Dutch plant physiologist, Frits Went, while working on his doctoral dissertation. Carrying Paal's

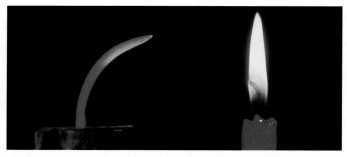

(a)

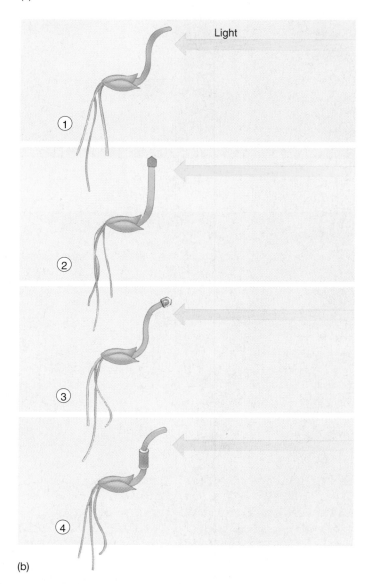

(b)

FIGURE 39.6
The Darwins' experiment. (a) Young grass seedlings normally bend toward the light. (b) The bending (*1*) did not occur when the tip of a seedling was covered with a lightproof cap (*2*), but did occur when it was covered with a transparent one (*3*). When a collar was placed below the tip (*4*), the characteristic light response took place. From these experiments, the Darwins concluded that, in response to light, an "influence" that caused bending was transmitted from the tip of the seedling to the area below, where bending normally occurs.

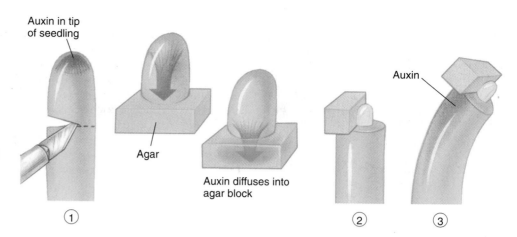

FIGURE 39.7
Frits Went's experiment. (*1*) Went removed the tips of oat seedlings and put them in agar, an inert, gelatinous substance. (*2*) Blocks of agar were then placed off-center on the ends of other oat seedlings from which the tips had been removed. (*3*) The seedlings bent away from the side on which the agar block was placed. Went concluded that the substance that he named *auxin* promoted the elongation of the cells and that it accumulated on the side of an oat seedling away from the light.

experiments an important step further, Went cut off the tips of oat seedlings that had been illuminated normally and set these tips on agar. He then took oat seedlings that had been grown in the dark and cut off their tips in a similar way. Finally, Went cut tiny blocks from the agar on which the tips of the light-grown seedlings had been placed and placed them off-center on the tops of the decapitated dark-grown seedlings (figure 39.7). Even though these seedlings had not been exposed to the light themselves, they bent away from the side on which the agar blocks were placed.

Went then put blocks of pure agar on the decapitated stem tips and noted either no effect or a slight bending toward the side where the agar blocks were placed. Finally, Went cut sections out of the lower portions of the light-grown seedlings to see whether the active principle was present in them. He placed these sections on the tips of decapitated, dark-green oat seedlings and again observed no effect.

As a result of his experiments, Went was able to show that the substance that had diffused into the agar from the tips of light-grown oat seedlings could make seedlings curve when they otherwise would have remained straight. He also showed that this chemical messenger caused the cells on the side of the seedling into which it flowed to grow more than those on the opposite side (figure 39.8). In other words, it enhanced rather than retarded cell elongation. He named the substance that he had discovered **auxin**, from the Greek word *auxein*, which means "to increase."

Went's experiments provided a basis for understanding the responses that the Darwins had obtained some 45 years earlier. The grass seedlings bent toward the light because of differences in the auxin concentrations on the two sides of the shoot. The side of the shoot that was in the shade had more auxin, and its cells therefore elongated more than those on the lighted side, bending the plant toward the light.

Auxin acts to adapt the plant to its environment in a highly advantageous way. It promotes growth and elongation and facilitates the plant's response to its environment. Environmental signals directly influence the distribution

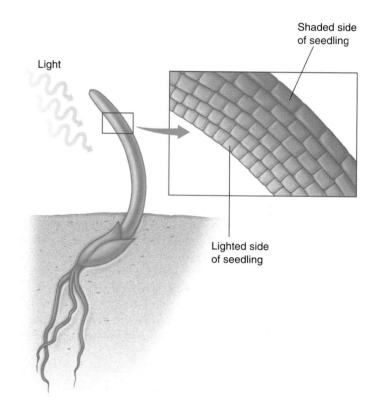

FIGURE 39.8
Auxin causes cells on the dark side to elongate. Went determined that a substance called auxin enhanced cell elongation. Plant cells that are in the shade have more auxin and grow faster than cells on the lighted side, causing the plant to bend toward light. Further experiments showed exactly why there is more auxin on the shaded side of a plant.

of auxin in the plant. How does the environment—specifically, light—exert this influence? Theoretically, it might destroy the auxin, decrease the cells' sensitivity to auxin, or cause the auxin molecules to migrate away from the light into the shaded portion of the shoot. This last possibility has proved to be the case.

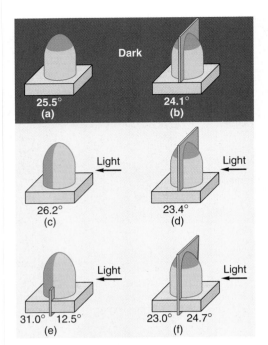

FIGURE 39.9

Phototropism and auxin: diagrams of experiments originally performed by Winslow Briggs. Experiments (a) and (b), performed in the dark, showed that splitting the tip of a seedling leaf and inserting a barrier did not significantly affect the total amount of auxin that diffused from the tip. The numbers below the agar blocks represent the degrees of curvature produced when the agar blocks were later placed on decapitated oat seedlings. Experiments (c) to (f) were performed with light coming from the right side, as indicated. A comparison of (c) and (d) with (a) and (b) shows that auxin production does not depend on light. The slight differences in curvature shown are not significant. If a barrier was inserted in the agar block (e), light caused the displacement of the auxin away from the light. Finally, experiment (f) showed that it was displacement that had occurred, and not different rates of auxin production on the dark and light sides, because when displacement was prevented with a barrier, auxin production did not significantly differ in the two sides. The exact nature of the auxin migration process is not known, but it is thought to involve a light-sensitive pigment that perhaps alters membrane permeability to auxin. Only light of wavelengths of less than 500 nanometers (blue light) can promote the lateral migration of auxin.

How Auxins Work

In a simple but effective experiment, Winslow Briggs, while working at Stanford University, inserted a thin sheet of mica vertically between the half of the shoot oriented toward the light and the half of the shoot oriented away from it (figure 39.9). He found that light from one side does not cause a shoot with such a barrier to bend. When Briggs examined the illuminated plant, he found equal auxin levels on both the light and dark sides of the barrier. He concluded that a normal plant's response to light from one direction involves auxin migrating from the light side to the dark side, and that the mica barrier prevented a response by blocking the migration of auxin.

In the past, botanists believed that plants produce several natural auxins, but the only active one seemed to be indoleacetic acid (IAA). The chemical structure of IAA resembles that of the amino acid tryptophan, from which it is probably synthesized by plants (figure 39.10). Scientists also believed that other organic acids that act like auxin were converted to IAA in plants. However, at least three other natural, auxin-like substances that are not converted to IAA are now known to occur. Phenylacetic acid (PAA) is actually more abundant but less active than IAA; current evidence suggests PAA is unimportant. Indolebutyric acid (IBA), previously known only in synthetic form, has been discovered in the leaves of corn and various dicots. The germinating seeds of legumes produce 4-chloroindoleacetic acid (4-chloroIAA). All three of these "growth regulators" bring about responses similar to those produced by IAA.

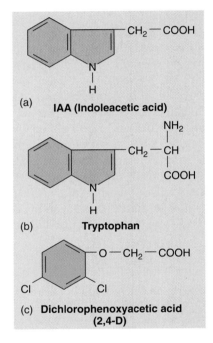

(a) **IAA (Indoleacetic acid)**

(b) **Tryptophan**

(c) **Dichlorophenoxyacetic acid (2,4-D)**

FIGURE 39.10

Auxins. (a) Indoleacetic acid (IAA), the principal naturally occurring auxin. (b) Tryptophan, the amino acid from which plants probably synthesize IAA. (c) Dichlorophenoxyacetic acid (2,4-D), a synthetic auxin, is a widely used herbicide.

Numerous experiments have also yielded a great deal of information about how auxins act in plants. The term *auxin* is now used to refer both to the naturally occurring substance and those related synthetic molecules that produce similar effects.

Auxin and Plant Growth

Auxin increases the plasticity of the plant cell wall. A more plastic wall will stretch more while its protoplast is swelling during active cell growth. Cell wall plasticity is best promoted by very low concentrations of auxin, and thus the hormone must be broken down rapidly to prevent its accumulation. Using methods involving monoclonal antibodies (see chapter 7), scientists have been able to locate the transport sites involved in the movement of auxin downward from the shoot apex through the plant. These sites are in the plasma membranes of the cells that are polarly transporting auxins. Further application of such methods will undoubtedly help to clarify other properties of the auxin-mediated process of cell elongation.

The speed with which auxin increases cell plasticity—grass seedlings may begin to bend significantly within ten minutes of its application—has made it difficult to determine the chemical basis of these reactions. It seems highly unlikely that these rapid-response effects could be caused by changes in the rates of transcription or translation of genes. To act so rapidly, auxin must affect some system that already exists. Extensive changes have been reported in the polysaccharides of plant cell walls, involving the formation and breaking of covalent bonds, in response to treatment with auxin. An increase in the concentration of H^+ ions in the wall is also likely to occur. In addition, auxin appears to mediate the stimulation of mRNA transcription, which would lead to long-term growth changes.

Auxin promotes the activity of the vascular cambium and the vascular tissues. Also, auxins are present in pollen in large quantities and play a key role in the development of fruits. Synthetic auxins are used commercially for the same purpose. Fruits will normally not develop if fertilization has not occurred and seeds are not present, but frequently they will if auxins are applied.

Synthetic Auxins. Synthetic auxins such as NAA (naphthalene acetic acid) and IBA (indolebutyric acid) have many uses in agriculture and horticulture. One of their most important uses is based on their prevention of abscission, the process that causes a leaf or other organ to fall from a plant. Synthetic auxins are used to prevent fruit drop in apples before they are ripe and to hold berries on holly that is being prepared for shipping. Synthetic auxins are also used to promote flowering and fruiting in pineapples and to induce the formation of roots in cuttings (figure 39.11).

Furthermore, synthetic auxins are routinely used to control weeds. When used as herbicides, they are applied in higher concentrations than IAA would normally occur in plants. One of the most important synthetic auxin herbicides is 2,4-dichlorophenoxyacetic acid, usually known as 2,4-D (see figure 39.10c). It kills weeds in grass lawns by selectively eliminating broad-leaved dicots. The stems of the dicot weeds cease all axial growth.

FIGURE 39.11
Synthetic auxins are used to promote the growth of roots in cuttings. The African violet cutting on the left was placed in a solution containing NAA (naphthalene acetic acid). The cutting on the right was placed in pure water.

The herbicide 2,4,5-trichlorophenoxyacetic acid, better known as 2,4,5-T, is closely related to 2,4-D. 2,4,5-T was widely used as a broad-spectrum herbicide to kill weeds and seedlings of woody plants. It became notorious during the Vietnam War as a component of a jungle defoliant known as Agent Orange and was banned in 1979 for most uses in the United States. When 2,4,5-T is manufactured, it is unavoidably contaminated with minute amounts of a product called dioxin (2,3,7,8-tetrachlorodibenzoparadioxin). Dioxin, in doses as low as a few parts per billion, has produced liver and lung diseases, leukemia, miscarriages, birth defects, and even death in laboratory animals. Vietnam veterans and children of Vietnam veterans exposed to Agent Orange have been among the victims.

> Auxin makes young cell walls more plastic and stimulates the elongation of cells. One auxin, indoleacetic acid (IAA), is synthesized in apical meristems of shoots. It causes young stems to bend toward light when it migrates toward the darker side, where it promotes cell elongation. By interacting with other hormones, auxin also stimulates the activity of the vascular cambium, promoting an increase in girth.

Cytokinins

Cytokinins comprise another group of naturally occurring growth hormones in plants. Studies by Gottlieb Haberlandt of Austria around 1913 demonstrated the existence of an unknown chemical in various tissues of vascular plants that, in cut potato tubers, would cause parenchyma cells to become meristematic, and would induce the differentiation of a cork cambium. The role of cytokinins, active components of coconut milk, in promoting the differentiation of organs in masses of plant tissue growing in culture later led to their discovery. Subsequent studies have focused on the role cytokinins play in the differentiation of tissues from callus.

A cytokinin is a plant hormone that, in combination with auxin, stimulates cell division in plants and determines the course of differentiation. Substances with these properties are widespread; in vascular plants, most cytokinins are produced in the root apical meristems and transported throughout the plant. Developing fruits are also important sites of cytokinin synthesis. In mosses, cytokinins cause the formation of vegetative buds on the protonemata. In all plants, cytokinins, working with other hormones, seem to regulate growth patterns.

All naturally occurring cytokinins are purines that appear to be derivatives of, or have molecule side chains similar to, those of adenine (figure 39.12). Other chemically diverse molecules, not known to occur naturally, have effects similar to those of cytokinins. Cytokinins promote growth of lateral buds into branches (figure 39.13); though, along with auxin and ethylene, they also play a role in apical dominance (the suppression of lateral bud growth). Conversely, cytokinins inhibit formation of lateral roots, while auxins promote their formation. As a consequence of these relationships, the balance between cytokinins and auxin, along with other factors, determines the appearance of a mature plant. In addition, the application of cytokinins to leaves detached from a plant retards their yellowing.

The action of cytokinins, like that of other hormones, has been studied in terms of its effects on growth and differentiation of masses of tissue growing in defined media. In the early cell-growth experiments reported in the first pages of this chapter, coconut "milk" was an essential factor. Eventually it was discovered that coconut "milk" is not only rich in amino acids and other reduced nitrogen compounds required for growth, but it also contains cytokinins. Cytokinins seem to be essential for mitosis and cell division. They apparently promote the synthesis or activation of proteins that are specifically required for mitosis.

Cytokinins are plant hormones that, in combination with auxin, stimulate cell division and, along with a number of other factors, determine the course of differentiation. In contrast to auxins, cytokinins are purines that are related to or derived from adenine.

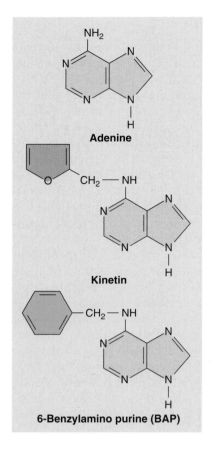

FIGURE 39.12
Some cytokinins. Two commonly used synthetic cytokinins: kinetin and 6-benzylamino purine. Note their resemblance to the purine base adenine.

(a)   (b)

FIGURE 39.13
Cytokinins stimulate lateral bud growth. (a) When the apical meristem of a plant is intact, auxin from the apical bud will inhibit the growth of lateral buds. (b) When the apical bud is removed, cytokinins are able to produce the growth of lateral buds into branches.

Gibberellins

Gibberellins are named after the fungus *Gibberella fujikuroi*, which causes rice plants, on which it is parasitic, to grow abnormally tall. This disease of rice, called "foolish seedling," was investigated in the 1920s by Japanese plant pathologist Eiichi Kurosawa. He grew *Gibberella* in culture and obtained a substance from the fungal filtrate that, when applied to rice plants, would produce in them the "foolish seedling" disease. This substance was isolated and the structural formula identified by Japanese chemists in 1939. British chemists reconfirmed the formula in 1954. Although such chemicals were first thought to be only a curiosity, they have since turned out to belong to a large class of more than 100 naturally occurring plant hormones called gibberellins. All are acidic and are usually abbreviated to **GA** (for gibberellic acid), with a different subscript (GA_1, GA_2, and so forth) to distinguish each one.

Gibberellins, which are synthesized in the apical portions of stems and roots, have important effects on stem elongation. They play the leading role in controlling this process in the growing stems of mature trees and shrubs, where the application of gibberellins characteristically promotes internode elongation. The elongation effect is enhanced if auxin is also present. The application of gibberellins to dwarf mutants is known to restore the normal growth and development in many plants (figure 39.14). Dwarf mutants are known to lack naturally occurring gibberellins in sufficient quantities to promote normal growth.

In the last chapter, we noted the role gibberellins play in stimulating the production of α-amylase and other hydrolytic enzymes that are used in the utilization of food resources during germination and establishment of cereal seedlings. How are the genes encoding these enzymes transcribed? Experimental studies in the aleurone layer surrounding the endosperms of cereal grains have shown that transcription occurs when the gibberellins initiate a burst of messenger RNA (mRNA) and protein synthesis. GA somehow enhances DNA binding proteins, which in turn allow DNA transcription of a gene. Synthesis of DNA does not seem to occur during the early stages of seed germination but becomes important when the radicle has grown through the seed coats.

Gibberellins also affect a number of other aspects of plant growth and development. The application of gibberellins can often induce some plants to grow out of the rosette stage and flower early during their first year of growth (figure 39.15). These hormones also hasten seed germination, apparently because they can substitute for the effects of cold or light requirements in this process.

According to analyses carried out by B. O. Phinney of the University of California, Los Angeles, there is only one kind of gibberellin, GA_1, that is active in the control of shoot elongation in most plants. The natural functions of other gibberellins (for examples, GA_3 and GA_7) that occur widely in eukaryotes other than flowering plants are not known,

FIGURE 39.14
Stem elongation in a dwarf variety of the kidney bean *(Phaseolus vulgaris)*. When treated with gibberellins, the plants on the right attained normal stature.

FIGURE 39.15
Effects of gibberellins. This mustard family plant *(Brassica rapa)* will "bolt" and flower when it is treated with gibberellins.

although a number of them function like GA_1 if they are applied to flowering plants. GA_1 seems to have acquired the properties of a plant hormone in the earliest vascular plants and is probably characteristic of all vascular plants.

> **Gibberellins are an important class of plant hormones that are produced in the apical regions of shoots and roots. They play the major role in controlling stem elongation for most plants, acting in concert with auxin and other hormones.**

Ethylene

Long before its role as a plant hormone was appreciated, the simple, gaseous hydrocarbon ethylene ($H_2C{=}CH_2$) was known to defoliate plants when it leaked from gaslights in streetlamps. Ethylene is, however, a natural product of plant metabolism that, in minute amounts, interacts with other plant hormones. We have already mentioned the way auxin, transported down from the apical meristem of the stem, stimulates the production of ethylene in the tissues around the lateral buds and thus retards their growth. Ethylene also suppresses stem and root elongation, probably in a similar way.

Ethylene plays a major role in fruit ripening. At first, auxin, which is produced in significant amounts in pollinated flowers and developing fruits, stimulates ethylene production; this, in turn, hastens fruit ripening. As ripening continues, the higher concentrations of ethylene eventually counter the effects of auxin, which, as mentioned in the previous chapter, is the chief factor in the formation of the separation layer at the base of leaf petioles (and fruit peduncles) during the process of abscission. Abscission will not occur if auxin is present or if it is applied to the flowers or fruits.

Just before fruits that respond to ethylene ripen, there is often a major increase in respiration, called a **climateric,** which is accompanied by a dramatic increase in ethylene production. The ethylene production in climacteric fruits may be as much as 100 times greater than it was a day or two earlier. At this phase, when the respiration of fruits is proceeding at its most rapid rate, complex carbohydrates are broken down into simple sugars, chlorophylls are broken down, cell walls become soft, and the volatile compounds associated with flavor and scent in ripe fruits are produced. When ethylene is applied to fruits, it hastens their ripening.

One of the first observations that led to the recognition of ethylene as a plant hormone was the premature ripening in bananas produced by gases coming from oranges. Such relationships have led to major commercial uses of ethylene. For example, tomatoes are often picked green and artificially ripened later by the application of ethylene. Ethylene is widely used to speed the ripening of lemons and oranges as well. Carbon dioxide produces an opposite effect; to arrest ripening, fruits are often shipped in an atmosphere of carbon dioxide.

Studies have shown that ethylene plays an important ecological role. Ethylene production increases rapidly when a plant is exposed to ozone and other toxic chemicals, temperature extremes, drought, attack by pathogens or herbivores, and other stresses. The increased production of ethylene that occurs can accelerate the abscission of leaves (figure 39.16) or fruits that have been damaged by these stresses. It now appears that some of the damage associated with exposure to ozone is due to the ethylene produced by the plants. Some studies suggest that the production of ethylene by plants subjected to attack by herbivores or infected with diseases may be a signal to activate the defense mechanisms of the plants. Such mechanisms may include the production of molecules toxic to the animals or pests attacking them. A full understanding of these relationships is obviously important for agriculture and forestry.

Ethylene, a simple gaseous hydrocarbon, is a naturally occurring plant hormone. It controls the abscission of leaves, flowers, and fruits from the plants on which they form; their abscission is counteracted by auxin.

FIGURE 39.16
The effects of ethylene. A holly twig was placed under the glass jar on the *left* for a week. Under the jar on the *right*, a holly twig spent a week with a ripe apple. Ethylene produced by the apple caused abscission of the holly leaves.

Abscisic Acid

Abscisic acid, a naturally occurring plant hormone, appears to be synthesized mainly in mature green leaves, fruits, and root caps. The hormone earned its name because applications of it appear to stimulate leaf senescence (aging) and abscission, but there is little evidence that it plays an important role in this process. In fact, it is believed that abscisic acid may cause ethylene synthesis, and that it is actually the ethylene that promotes senescence and abscission. When abscisic acid is applied to a green leaf, the areas of contact turn yellow. Thus, abscisic acid has the exact opposite effect on a leaf from that of the cytokinins; a yellowing leaf will remain green in an area where cytokinins are applied.

Abscisic acid probably induces the formation of winter buds—dormant buds that remain through the winter—by suppressing growth. The conversion of leaf primordia into bud scales follows (figure 39.17a). Like ethylene, it may also suppress growth of dormant lateral buds. It appears that abscisic acid, by suppressing growth and elongation of buds, can counteract some of the effects of gibberellins (which stimulate growth and elongation of buds); it also promotes senescence by counteracting auxin (which tends to retard senescence). Abscisic acid plays a role in causing the dormancy of many seeds and is also important in controlling the opening and closing of stomata (figure 39.17b), as discussed in chapter 35.

Abscisic acid occurs in all groups of plants and apparently has been functioning as a growth-regulating substance since early in the evolution of the plant kingdom. Relatively little is known about the exact nature of its physiological and biochemical effects. These effects are very rapid—often taking place within a minute or two—and therefore they must be at least partly independent of gene expression. Some longer term effects of abscisic acid involve the regulation of gene expression, but the way this occurs is poorly understood. Abscisic acid levels become greatly elevated when the plant is subject to stress, especially drought. Like other plant hormones, abscisic acid probably will prove to have valuable commercial applications when its mode of action is better understood.

Abscisic acid, produced chiefly in mature green leaves and in fruits, suppresses growth of buds and promotes leaf senescence. It also plays an important role in controlling the opening and closing of stomata.

(a)

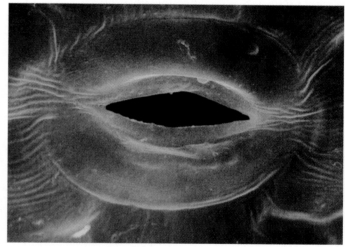

(b)

FIGURE 39.17
Effects of abscisic acid. (a) Abscisic acid plays a role in the formation of these winter buds of an American basswood. These buds will remain dormant for the winter, and bud scales—modified leaves—will protect the buds from desiccation.
(b) Abscisic acid also affects the closing of stomata by influencing the movement of potassium ions out of guard cells.

Tropisms

Tropisms (from *trope*, the Greek word for "turn") are positive or negative growth responses of plants to external stimuli that usually come from one direction. Because tropisms influence the growth patterns of plants, they also influence their appearance. Growth responses produce adjustments of plants to the conditions of their environment. Here we will consider three major classes of plant tropisms: phototropism, gravitropism, and thigmotropism.

Phototropism

In our discussion of auxin in plants, we mentioned phototropic responses, which involve the bending of growing stems and other plant parts toward sources of light (figure 39.18). In general, stems are positively phototropic, growing toward a light source, while most roots do not respond to light or, in exceptional cases, exhibit only a weak negative phototropic response. The phototropic reactions of stems are clearly of adaptive value, giving plants greater exposure to available light. They are also important in determining the development of plant organs and, therefore, the appearance of the plant. Individual leaves may display phototropic responses, in addition to other movements discussed in the next section. The position of leaves is important to the photosynthetic efficiency of the plant. Auxin is probably involved in most, if not all, of the phototropic growth responses of plants.

Gravitropism

Gravitropism, the response of a plant organ to gravity, was formerly known as *geotropism*. This kind of tropism causes stems to grow upward and roots downward. Both of these responses are clearly of adaptive significance; stems that grow upward are apt to receive more light than those that do not; roots that grow downward are more apt to arrive in areas that enhance their functions than those that do not. The phenomenon is now called gravitropism because it is clearly a response to gravity and not to the earth, which the term geotropism implies.

If a potted plant with normal vertical shoots is tipped over so that the shoots are then horizontal, auxin becomes more concentrated on the lower side of the stem/root axis than on the upper side. The increased auxin concentration on the lower side causes the cells in that area to grow more than the cells on the upper side. The result is a bending upward of the stem against the force of gravity—in other words there is a *negative gravitropic response* (figure 39.19). Such differences in hormone concentration have not been as well documented in roots. Nevertheless, the upper sides of roots oriented horizontally grow more rapidly than the

FIGURE 39.18
Phototropism. Corn shoots growing toward light.

FIGURE 39.19
Gravitropism. The stem of this potted plant, which has been tipped over, is curving upward because it is negatively gravitropic.

lower sides, causing the root ultimately to grow downward; this phenomenon is known as *positive gravitropism*. Some evidence suggests the perception of gravity by roots is related to stimulation of root cap cells by amyloplasts (plastids with starch grains) in the cytoplasm. The amyloplasts rest on the "bottom" of the cells, and when a plant is tipped on its side, the amyloplasts tumble or drift down in the direction of the gravity force, with new cell growth then causing the root to turn down. The roots of mutant plants without such amyloplasts grow randomly.

It may surprise you to learn that in tropical rainforests, roots of some plants may grow up the stems of neighboring plants, instead of exhibiting the normal positive gravitropic responses typical of other roots. The rainwater dissolves nutrients, both while passing through the lush upper canopy of the forest, and also subsequently as it trickles down tree trunks. Such water functions as a more reliable source of nutrients for the roots than the nutrient-poor soils in which the plants are anchored.

(a)

(b)

FIGURE 39.20
Thigmotropism. (a) The thigmotropic response of these twining stems causes them to coil around the object with which they have come in contact. (b) The direction of coiling of a manroot tendril reverses near the midpoint.

Thigmotropism

Thigmotropism is a name derived from the Greek root *thigma*, meaning "touch." A thigmotropism is a response of a plant or plant part to contact with a solid object (figure 39.20*a*). When a tendril makes contact with an object, specialized epidermal cells, whose action is not clearly understood, perceive the contact and promote uneven growth, causing the tendril to curl around the object, sometimes within as little as 3 to 10 minutes. Tendrils of many plants will coil first in the direction of contact and then later reverse the coiling (figure 39.20*b*), but those of other plants always coil in the same direction, regardless of the initial point of contact. Both auxin and ethylene appear to be involved in tendril movements, and they can induce coiling in the absence of any contact stimulus. In other plants, such as clematis, bindweed, and dodder, leaf petioles or unmodified stems twine around other stems or solid objects.

Other Tropisms

The tropisms just discussed are among the best known, but others have been recognized. They include *electrotropism* (responses to electricity); *chemotropism* (response to chemicals); *traumotropism* (response to wounding); *thermotropism* (response to temperature); *aerotropism* (response to oxygen); *skototropism* (response to dark); and *geomagnetotropism* (response to magnetic fields). Roots will often follow a diffusion gradient of water coming from a cracked pipe and enter the crack. Some call such growth movement *hydrotropism*, but most plant physiologists doubt that responses to water and several other "stimuli" are true tropisms.

> **Phototropisms are growth responses of plants to a unidirectional source of light. Gravitropism, the response of a plant to gravity, generally causes shoots to grow up (negative gravitropism) and roots to grow down (positive gravitropism). Thigmotropisms are growth responses of plants to contact.**

Turgor Movement

Unlike tropisms, some plant movements are not based on growth responses, but instead result from reversible changes in the turgor pressure of specific cells. *Turgor* is pressure within a living cell resulting from diffusion of water into it. If water leaves turgid cells (ones with turgor pressure), the cells may collapse, causing plant movement; conversely, water entering a limp cell may also cause movement as the cell once more becomes turgid. Some of the most familiar of these reversible changes are seen in leaves and flowers that "open" during the day and "close" at night. The blades of plant leaves that exhibit such a daily shift in position may not actually fold; instead, their orientation may be changed as a result of **turgor movements.** For example, the attractively spotted leaves of the prayer plant (*Maranta*) are spread horizontally during the day, but become more or less vertical at night (figure 39.21).

Many other plants, especially those of the legume family (Fabaceae), exhibit leaf movements not necessarily associated with daily rhythms, like those of the prayer plant. After exposure to a stimulus, the changes in leaf orientation are mostly associated with rapid turgor pressure changes in **pulvini** (singular: pulvinus), which are multicellular swellings located at the base of each leaf or leaflet. When leaves with pulvini, such as those of the sensitive plant (*Mimosa pudica*), are stimulated by wind, heat, touch, or, in some instances, intense light, an electrical signal is generated. The electrical signal is translated into a chemical signal, with potassium ions, followed by water, migrating from the cells in one half of a pulvinus to the intercellular spaces in the other half. The loss of turgor in half of the pulvinus causes the leaf to "fold." The movements of the leaves and leaflets of the sensitive plant are especially rapid; the folding occurs within a second or two after the leaves are touched (figure 39.22). Over a span of about 15 to 30 minutes after the leaves and leaflets have folded, water usually diffuses back into the same cells from which it left, and the leaf returns to its original position. The leaves of some plants with similar mechanisms may track the sun, with their blades oriented at right angles to it; how their orientation is directed is, however, poorly understood. Such leaves can move quite rapidly (as much as 15° an hour). Contrary to widespread popular belief, when sunflowers open, they do not follow the sun, but instead generally face east.

The rapid movement of a leaf of the carnivorous Venus flytrap (*Dionaea muscipula*) is even more spectacular. Mechanisms other than changes in turgor pressure in pulvini are involved in the movement. There are groups of motor cells beneath six tiny trigger hairs—three on the surface of each

(a)   (b)

FIGURE 39.21
Turgor pressure. In the prayer plant, (*Maranta*), leaves are (a) oriented horizontally during the day and (b) vertically at night.

FIGURE 39.22
Sensitive plant (*Mimosa pudica*). The blades of Mimosa leaves are divided into numerous leaflets. When leaves are touched (center two leaves above), they fold due to loss of turgor.

half of the leaf blade. When an insect or anything else touches two trigger hairs simultaneously, or one trigger hair twice, an electrical signal travels to the motor cells, and the trap snaps shut in less than a second. The process involves irreversible cell enlargement, which is initiated by a sudden lowering of pH in the cell walls. The leaves close most rapidly at a pH of 3 to 4, a level at which the walls of the motor cells are very flexible. When the leaf halves are snapping shut, ATP powers the rapid release of H^+, and the walls of the motor cells expand rapidly. About 29% of the ATP in the cells is lost as the snapping action occurs. H^+ from the motor cells travel to the outer leaf cells, where the ions acidify the cell walls. This rapid transfer of H^+ is, in effect, the electrical signal. During closure, the outer surface of the leaf also expands. If a closed trap is empty, reopening may take about 10 hours, or if digestion of an insect has taken place, up to several days.

Turgor movements of plants are reversible and involve changes in the turgor pressure of specific cells.

Eukaryotic organisms, with very few exceptions, are affected by the alternation between light and dark in a 24-hour day, and by shifts in the daily duration of light and dark throughout the year. Features in the growth and development of most plants are keyed to changes in the proportion of light to dark in the daily 24-hour cycle. Responses to these changes are called **photoperiodism,** a mechanism by which organisms respond to seasonal changes in the relative length of day and night. One of the most obvious photoperiodic responses concerns the initiation of flowering in angiosperms.

Flowering Responses

Day length changes with the seasons; the farther from the equator, the greater the variation. Flowering responses of plants to day length fall into several basic categories. When the daylight becomes shorter than a critical length, flowering is initiated in **short-day plants** (figure 39.23). When the daylight becomes longer than a critical length, flowering is initiated in **long-day plants.** Other plants, such as snapdragons, roses, and many native to the tropics (for example, tomatoes), will flower when mature regardless of day length, as long as they have received enough light for normal growth. These are referred to as **day-neutral plants.** Several grasses (for example, Indian grass, *Sorghastrum nutans*) have two critical photoperiods; they will not flower if the days are too long, and they also will not flower if the days are too short. There are several categories of plants that respond to various segments of day length. Such plants are called **intermediate-day plants.** In all of these plants, it is actually the length of darkness (night), not the length of day, that is significant.

Long-day plants usually flower only if the day length is between 12 and 16 hours. Short-day plants tend to flower only if the day length is less than 14 hours. At middle latitudes, most long-day plants flower in the spring and early summer; examples of such plants include clover, irises, lettuce, spinach, and hollyhocks. Short-day plants usually flower in late summer and fall, and include chrysanthemums, goldenrods, poinsettias, soybeans, and many weeds. Commercial plant growers use these responses to day length to bring plants into flower at specific times. The geographic distribution of certain plants may be determined by flowering responses to day length. For example, in the tropics, day lengths may be neither long enough or short enough to initiate flowering in certain species of both long-day and short-day plants.

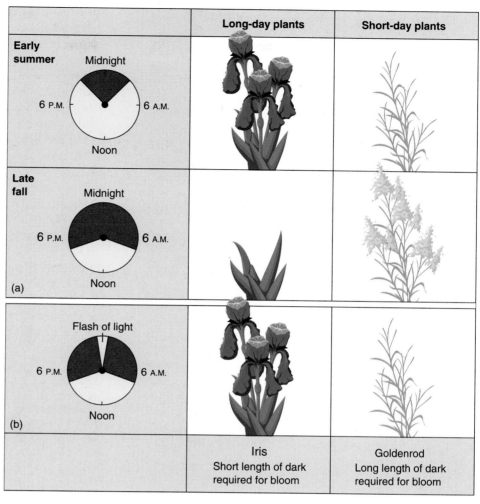

FIGURE 39.23

How flowering responds to day length. (a) This iris is a long-day plant that is stimulated by short nights to flower in the spring. The goldenrod is a short-day plant that, throughout its natural distribution in the Northern Hemisphere, is stimulated by long nights to flower in the fall. (b) If the long night of winter is artificially interrupted by a flash of light, the goldenrod will not flower, and the iris will. In each case, it is the duration of uninterrupted darkness that determines when flowering will occur.

Flowering is initiated in short-day plants when the days become shorter than a certain critical length and in long-day plants when the days become longer than a critical length. Day-neutral plants do not have specific day-length requirements for flowering, while intermediate-day plants have two critical photoperiods.

The Chemical Basis of the Photoperiodic Response

Students of flowering responses made an early discovery that was to prove extremely valuable in their later studies: short-day plants require a certain amount of darkness in each 24-hour cycle to initiate flowering; however, if the period of darkness is interrupted by a brief period of light, often less than a minute, the plants will not flower. Once this discovery was made, scientists proceeded to determine the wavelength of light that was most effective in inhibiting flowering. They found it was red light with a wavelength of about 660 nanometers. Curiously, if plants were exposed to red light at 660 nanometers and then to far-red light at a wavelength of about 730 nanometers, the effect was canceled, and the plants flowered. What is the chemical basis for this strange observation?

Plants contain a pigment, **phytochrome,** which exists in two interconvertible forms, P_r and P_{fr}. In the first form, phytochrome absorbs red light; in the second, it absorbs far-red light. When a molecule of P_r absorbs a photon of red light (660 nm), it is instantly converted into a molecule of P_{fr}, and when a molecule of P_{fr} absorbs a photon of far-red light (730 nm), it is instantly converted to P_r. P_{fr} is biologically active and P_r is biologically inactive. In other words, when P_{fr} is present, a given biological reaction that is affected by phytochrome will occur. When most of the P_{fr} has been replaced by P_r, the reaction will not occur.

Phytochrome is a light receptor, but it does not act directly to bring about reactions to light. In short-day plants, the presence of P_{fr} leads to a biological reaction that suppresses flowering. The amount of P_{fr} steadily declines in darkness, the molecules converting to P_r. When the period of darkness is long enough, the suppression reaction ceases and the flowering response is triggered. However, a single flash of red light at a wavelength of about 660 nanometers will convert most of the molecules of P_r to P_{fr}, and the flowering reaction will be blocked (figure 39.24). Still, since most of the P_{fr} is converted to P_r within the first three to four hours of darkness, the conversion of P_r to P_{fr} cannot fully explain the flowering responses of short-day plants; other factors, still not understood, must also be involved.

The existence of phytochrome was conclusively demonstrated in 1959 by Harry A. Borthwick and his collaborators at the U.S. Department of Agriculture Research Center at Beltsville, Maryland. It has since been shown that the molecule consists of two parts: a smaller one that is sensitive to light and a larger portion that is a protein. The phytochrome

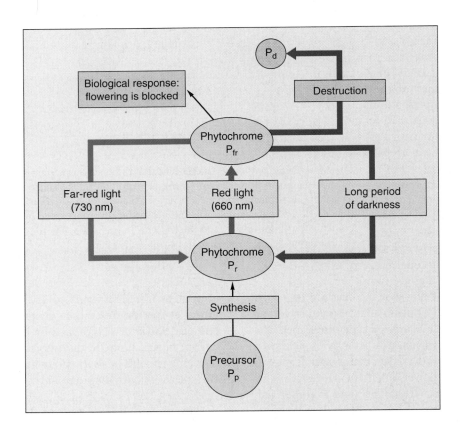

FIGURE 39.24

How phytochrome works. Phytochrome is synthesized in the P_r form from amino acids, designated P_p for phytochrome precursor. When exposed to red light, P_r changes to P_{fr}, which is the active form that elicits a response in plants. P_{fr} is converted to P_r when exposed to far-red light, and it also converts to P_r or is destroyed in darkness. The destruction product is designated P_d.

pigment is blue, and its light-sensitive portion is similar in structure to the phycobilins that occur in cyanobacteria and red algae. Phytochrome is present in all groups of plants and in a few genera of green algae, but not in bacteria, fungi, or protists (other than the few green algae). Therefore, it is provisionally assumed that phytochrome systems for measuring light evolved among the green algae and were present in the common ancestor of the plants.

Phytochrome is also involved in many other plant growth responses. For example, seed germination is inhibited by far-red light and stimulated by red light in many plants. Since chlorophyll absorbs red light strongly but does not absorb far-red light, light passing through green leaves inhibits seed germination. Consequently, seeds on the ground under deciduous plants that lose their leaves in winter are more apt to germinate in the spring after the leaves have decomposed and the seedlings are exposed to direct sunlight.

FIGURE 39.25
Etiolated corn plant grown in darkness and normal corn plant grown in light. Seedlings are naturally etiolated when they are growing through the ground. Etiolation may promote the utilization of the plant's energy in such a way that it accelerates the plant's speed in reaching the surface and light, where normal growth can begin.

This greatly improves the chances the seedlings will become established. A second example of these relationships is the elongation of the shoot in an *etiolated* seedling (one that is pale and slender from having been kept in the dark). Such plants become normal when exposed to light, especially red light, but the effects of such exposure are canceled by far-red light. This indicates a relationship similar to that observed in seed germination (figure 39.25).

In 1983, Dina Mandoli and Winslow Briggs, working at the Department of Plant Biology at the Carnegie Institution of Washington, reported findings that added to our understanding of the possible ways in which a plant could receive light and react to it. Mandoli and Briggs found that an etiolated seedling acts like a bundle of fiber-optic strands, guiding light over distances as great as 4.5 centimeters through the interiors of dozens, even hundreds, of cells, as well as through the junctions between them. Such a system may enable light to directly and immediately affect the responses of plant portions below ground. These findings have opened a promising avenue for research into plant growth.

The Flowering Hormone: Does It Exist?

Working with long-day and short-day plants, some investigators have gathered evidence for the existence of a flowering hormone. It has been shown that plants will not flower in response to day-length stimuli if their leaves have been removed before exposure to the light. However, the presence of a single leaf or exposure of a single leaf to the appropriate stimuli will usually initiate flowering. If the leaf is removed immediately after exposure, the plant will not produce flowers; but if it is left on the plant for a few hours and then removed, flowering will occur normally. These results indicate that a substance passes from the leaves to the apices of the plant, where it induces flowering. Other experiments have shown that, unlike auxin, the substance cannot be transmitted through agar but actually requires a connection through living plant parts.

Scientists have searched for a flowering hormone for more than 50 years, but their quest has been unsuccessful. A considerable amount of evidence demonstrates the existence of substances that promote flowering and substances that inhibit it. These poorly understood substances appear to interact in complex ways. The complexity of their interactions, as well as the fact that multiple chemical messengers are evidently involved, has made this scientifically and commercially interesting search very difficult, and to this day, the existence of a flowering hormone remains strictly hypothetical.

Phytochrome, a plant pigment, comes in two forms. Each form can convert to the other in response to exposure to red light of the proper wavelength. Phytochrome plays an important role in controlling the initiation of flowering in short-day plants.

Dormancy

Plants respond to their external environments mostly by changes in growth rate. As you might imagine, the ability to cease growing altogether when conditions are not favorable is a critical factor in their survival.

In temperate regions, we generally associate dormancy with winter, when freezing temperatures and the accompanying unavailability of water make it impossible for plants to grow. During this season, buds of deciduous trees and shrubs remain dormant, and apical meristems remain well protected inside enfolding scales. Perennial herbs spend the winter underground as stout stems or roots packed with stored food. Many other kinds of plants, including most annuals, pass the winter as seeds.

In some seasonally dry climates, seed dormancy occurs primarily during the dry season, which may occur in the summer. Plants that remain dormant in dry conditions do so for reasons similar to the reasons plants of other areas are dormant in winter.

Annual plants occur frequently in areas of seasonal drought. Seeds are ideal for allowing annual plants to bypass the dry season, when there is insufficient water for growth. When it rains, they can germinate and the plants can grow rapidly, having adapted to the relatively short periods when water is available. Chapter 38 covered some of the mechanisms involved in breaking seed dormancy and allowing germination under favorable circumstances. These include the leaching from the seed coats of chemicals that inhibit germination, or mechanically cracking the seed coats, a procedure that is particularly suitable for promoting growth in seasonally dry areas. Whenever rains occur, they will leach out the chemicals from the seed coats, and the hard coats of other seeds may be cracked when they are being washed down along temporarily flooded arroyos (figure 39.26).

Seeds may remain dormant for a surprisingly long time. Many legumes (plants of the pea and bean family, Fabaceae—also known as Leguminosae) have tough seeds that are virtually impermeable to water and oxygen. These seeds often last decades and even longer without special care; they will eventually germinate when their seed coats have been cracked and water is available.

A period of cold is necessary before some kinds of seeds will germinate, as we mentioned in chapter 38. The seeds of other plants will germinate only when adequate water is available and the temperatures are relatively high. For this reason, certain weeds germinate and grow in the cooler part of the year and others in the warmer part of the year. Similarly, a period of cold is needed before the buds of some trees and shrubs will break dormancy and develop normally. For this reason, many plants that normally grow in temperate regions do not thrive in warmer regions near the equator, because even at high elevations in the tropics it still does not get cold enough, and the day-length relationships are different from those of temperate regions.

Mature plants may become dormant in dry or cold seasons that are unfavorable for growth. Dormant plants usually lose their leaves and drought-resistant winter buds are produced. Long unfavorable periods may be bypassed through the production of seeds, which themselves can remain dormant for long periods.

(a)

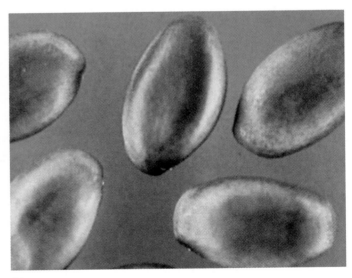

(b)

FIGURE 39.26
Palo verde *(Cercidium floridum)*. This desert tree (a) has tough seeds (b) that germinate only after they are cracked.

39.1 Every plant cell contains all the information needed to grow into an adult plant.

- Development of living plant cells is, in many instances, fully reversible. Plant cells are totipotent as long as they retain a living protoplast and a nucleus, and whole plants can be regenerated from cultures of single cells under the appropriate conditions.

39.2 Hormones are the tools plants use to regulate their growth.

- There are five major classes of naturally occurring plant hormones: auxins, cytokinins, gibberellins, ethylene, and abscisic acid. They often interact with one another in bringing about growth responses.

- Auxin is produced at the tips of shoots and is transported downward. Auxin also migrates away from light and promotes the elongation of plant cells on the dark side, causing stems to bend in the direction of light.

- Cytokinins are necessary for mitosis and cell division in plants. They promote growth of lateral buds and inhibit formation of lateral roots.

- Gibberellins, along with auxin, play a major role in stem elongation in most plants. They also tend to hasten the germination of seeds, to break dormancy in buds, and to cause stems in the rosettes of biennials to elongate.

- Ethylene is a gas that functions as a plant hormone. Ethylene influences leaf abscission and is widely used to hasten fruit ripening.

- Abscisic acid promotes the formation of winter buds and plays a key role in closing stomata.

39.3 Plant growth is often guided by environmental cues.

- Tropisms in plants are growth responses to external stimuli. A phototropism is a response to light, a gravitropism is a response to gravity, and a thigmotropism is a response to contact.

- Turgor movements are reversible and are important elements in the adaptation of plants to their environments. Turgor movements in leaves, flowers, and other structures of plants bring about orientations favorable to more efficient functioning of the organs involved.

39.4 Plant flowering is often keyed to day length.

- Short-day plants initiate flowering when the days become shorter than a given critical length; long-day plants do so when the days become longer than a critical length; intermediate-day plants have two critical photoperiods. The period of dark is actually the critical factor in all three categories of plants. Day-neutral plants have no critical photoperiods.

- A blue pigment known as phytochrome occurs in two interconvertible forms, each converted to the other by red light of different wavelengths. Phytochrome plays a role in determining the flowering response and in mediating other plant responses.

- Dormancy is a plant adaptation that carries a plant through unfavorable seasons, such as winter, or periods of drought. Dormancy facilitates survival of plants in many areas where they would otherwise be unable to grow.

Discussing Key Terms

1. **Auxins** Auxins, produced in the immature parts of plants, have many effects, including increasing the plasticity of plant cell walls and promoting apical dominance.

2. **Cytokinins and gibberellins** Cytokinins are plant hormones that, in combination with auxins, stimulate cell division and determine the course of differentiation. Gibberellins, produced in the apical regions of shoots and roots, immature seeds, fungi, and other organisms, play a major role, along with auxins, in promoting stem elongation.

3. **Ethylene and abscisic acid** Ethylene, a gas, controls the abscission of leaves, flowers, and fruits. Abscisic acid, produced chiefly in mature green leaves and in fruits, suppresses the growth of buds, influences the opening and closing of stomata, promotes leaf senescence, and hastens fruit ripening.

4. **Tropisms** Tropisms, or orientations of growth in response to external stimuli, control the developmental patterns of plants and, thus, their appearances. Among the major tropisms are phototropism (response to light), gravitropism (response to gravity), and thigmotropism (response to contact).

5. **Turgor pressure** Turgor movements in plants are reversible and involve changes in the turgor pressures of specific cells, which cause movements such as leaf folding in diurnal rhythms or in response to touch—enhancing orientation to light.

6. **Photoperiods** Photoperiodic reactions in plants include flowering responses, which are mediated by phytochrome, a pigment that exists in two interconvertible forms.

Review Questions

1. How reversible is the differentiation of most plant cells? Explain. What are the limitations of this capacity? What term defines this capacity? How did Steward experimentally confirm this concept?

2. In what ways are plant and animal hormones similar to one another? In what ways are plant hormones significantly different from animal hormones? What are the five major kinds of plant hormones?

3. What is the principal naturally occurring auxin? From what biomolecule is it probably synthesized? Where in the mature plant is it produced, and to where is it transported? Does this occur via the plant's vascular tissue? Explain.

4. How does auxin affect the plasticity of the plant cell walls? What is the function of indoleacetic acid oxidase? What is the relative rate of speed with which auxin acts in response to light?

5. What are five ways synthetic auxins are used in agriculture and horticulture? What is 2,4-D? What is it used for? Explain how it works.

6. What is the primary role of cytokinins in plant development? How do their actions compare with those of auxins? Where are most cytokinins produced? From what biomolecule do cytokinins appear to be derived?

7. What plant hormone is usually lacking in genetically dwarfed plants? How can one determine if this hormone is indeed absent? What are three secondary effects of this hormone? What are three commercial uses of this hormone?

8. What is the role of ethylene in leaf and fruit abscission? What effect does it have upon the ripening of fruit? What common gas counteracts this effect of ethylene? What are some of the commercial applications of ethylene? Why is this compound a valuable component in weed control products?

9. Where is abscisic acid produced in a mature plant? What is the effect of spotting this hormone on a green leaf? What hormone reverses this? What are the associations of abscisic acid with leaf buds?

10. In general, which part of a plant is positively phototropic? Are any parts negatively phototropic? Explain. What is the adaptive significance of each reaction? What hormone is most involved in mediating this reaction?

11. What is gravitropism? Why did the term *gravitropism* replace an earlier term for the same phenomenon? Which part of the plant is positively gravitropic and which is negatively gravitropic?

12. What is a pulvinus? What is its function? How are motor cells involved in the function of the pulvinus? What happens in the motor cells of the sensitive plant *(Mimosa pudica)* when its leaves are touched?

13. A burst of which wavelength of light is most effective in inhibiting flowering? How can this effect be cancelled? What are the two forms of the compound that mediates these reactions? How do they affect the flowering response in plants?

Thought Questions

1. Why is plant development so much more closely linked with environmental cues than animal development?

2. What experiments would you suggest for studying differentiation of individual plant cells? How could you begin to understand the factors involved in their assumption of mature forms?

3. If day length in a particular place at a particular time of year were 10 hours, which would produce flowers: a short-day plant, a long-day plant, both, or neither? Why? Do you think there are any short-day plants in the tropics? Why?

Internet Links

Grow Your Own Plants
http://www.boldweb.com/greenweb.htm
THE GREENWEB provides information on growing and maintaining a thriving garden, from backyard gardens to in-house planters.

What Does Its Flower Look Like?
http://www.flowerbase.com/main.asp
FLOWERWEB is a comprehensive Dutch database with descriptions and beautiful color images of more than 7000 flowers.

The Virtual Garden
http://www.pathfinder.com/vg/
A marvelous home gardening site for the amateur gardener. You can look up a particular plant and learn how best to grow it, talk to a professional gardener, check your growing season, and search the Virtual Garden for topics of interest to you.

For Further Reading

Chapin, F. S., III: "Integrated Responses of Plants to Stress," *Bioscience*, vol. 41, 1991, pages 29–36. All plants respond to stress of many types in basically the same way, which we are beginning to understand more completely.

Evans, M. L., R. Moore, and K. H. Hasenstein: "How Roots Respond to Gravity," *Scientific American*, December, 1986, pages 112–19. Modern studies of this fascinating phenomenon.

Simons, P.: "The Secret Feelings of Plants," *New Scientist*, vol. 136, pages 29–32, 1992. A fascinating account of recent investigations into the ways in which plants respond to touch.

Sisler, E. C. and S. F. Yang: "Ethylene, the Gaseous Plant Hormone," *BioScience*, vol. 33, 1994, page 238. A review of this important plant hormone.

40

Plant Molecular Biology

Concept Outline

40.1 Genomic organization is much more varied in plants than in animals.

Traditional Ways to Study Plant Genomes. Until relatively recently, plant biologists focused their research efforts on variation in chromosomes, but work is now shifting increasingly to the molecular level.

Organization of Plant Genomes. Plant genomes are more complex than those of other eukaryotic organisms due to the presence of multiple chromosome copies and extensive amounts of DNA with repetitive sequences.

Comparative Genome Mapping and Model Systems. RFLP and AFLP techniques are useful for mapping traits in plant genomes, and the genome is being sequenced for model systems such as *Arabidopsis thaliana* and rice.

40.2 Advances in plant tissue culture are revolutionizing agriculture.

Overview of Plant Tissue Culture. Because plants are totipotent, bits of tissue can be used to regenerate whole plants.

Types of Plant Tissue Cultures. Plant cells, tissues, and organs can be grown in an artificial culture medium, and some cells can be directed to generate whole plants.

Applications of Plant Tissue Culture. Plant tissue cultures can be used for the industrial production of plant products, propagation of horticultural plants, and crop improvement.

40.3 Plant genetic engineering and biotechnology now affect every aspect of agriculture.

World Population in Relation to Advances Made in Crop Production. It is uncertain whether advances made in crop production by improved farming practices and crop breeding can provide for an increasing world population.

Plant Biotechnology for Agricultural Improvement. The genetic engineering of plants will have a significant impact on crop production and is based upon introduction of foreign DNA into plant cells.

Useful Traits That Can Be Introduced into Plants. Plants can be genetically engineered to have altered levels of oils and amino acids and to provide vaccines against human diseases.

FIGURE 40.1
Genetically engineered herbicide resistance. The plants on the right have been sprayed with an herbicide, but they look as healthy as unsprayed plants (left) because they are engineered to be resistant to the herbicide, which will now just kill the weeds.

The organization of plant genomes can be more complex than that observed for most other eukaryotic organisms. This complexity is not only observed at the DNA sequence level itself but also at the chromosome level. In plants, duplication of entire sets of chromosomes can result in variable ploidy numbers even within a given genus. Plants also differ because their cells can display *totipotency*; under appropriate conditions, the developmental program of even differentiated (mature) cells can be altered to allow them to produce organs (roots, stems, leaves), embryos, and even whole plants. As we will see in this chapter, the unique characteristics of plant cell genomes have been an important factor in developing genetic engineering as a tool for producing improved food crops and novel plant products (figure 40.1).

Traditional Ways to Study Plant Genomes

Plant genomes are more complex than other eukaryotic genomes, and analysis reveals many evolutionary flips and turns of the DNA sequences over time. Plants show different chromosome numbers and varied ploidy levels that have conferred selective advantages under different environmental conditions (figure 40.2).

As plants adapt to heterogeneous environments, the amount of genetic variation may contribute significantly to their survival and proliferation. Overall, the size of plant genomes (both number of chromosomes and total number of nucleotide base-pairs) exhibits the greatest variation of any kingdom in the biological world. For example, tulips contain over 170 times as much DNA as a small weed called *Arabidopsis thaliana* (table 40.1). The DNA of plants can also contain regions of sequence repeats, sequence inversions, or transposable element insertions which further modify their genetic content. Traditionally, biologists have studied variation in chromosome inversions and ploidy to build up a picture of how plant species have evolved (figure 40.3). Increasingly, researchers are turning to studying the organization of plant DNA sequences to obtain important information about the evolutionary history of a plant species.

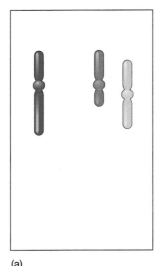 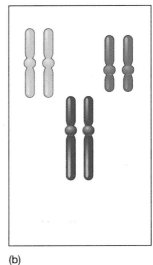 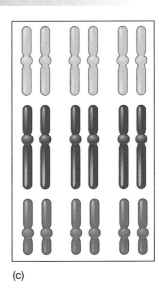

(a) (b) (c)

FIGURE 40.2
Chromosome numbers possible in plant genomes. (a) Haploid: a set of chromosomes without their pairs; for example, the chromosome number present in a gamete. (b) Diploid: a single set of chromosome pairs. (c) Polyploid: multiple sets of chromosome pairs; for example, bananas have a triple set of chromosomes and are therefore polyploid.

> Traditionally, biologists have examined variation among plants at the chromosome level; today, researchers are focusing more of their efforts at the DNA sequence level.

	Table 40.1 Genome Size of Plants	
Scientific Name	**Common Name**	**Genome Size (Millions of Base-Pairs)**
Arabidopsis thaliana	Arabidopsis	145
Prunus persica	Peach	262
Ricinus communis	Castor bean	323
Citrus sinensis	Orange	367
Oryza sativa spp. *javanica*	Rice	424
Petunia parodii	Petunia	1,221
Pisum sativum	Garden pea	3,947
Avena sativa	Oats	11,315
Tulipa spp.	Garden tulip	24,704

From *Plant Biochemistry and Molecular Biology*, by P. J. Lea and R. C. Leegods, eds. Copyright © 1993 John Wiley & Sons, Limited. Reproduced with permission.

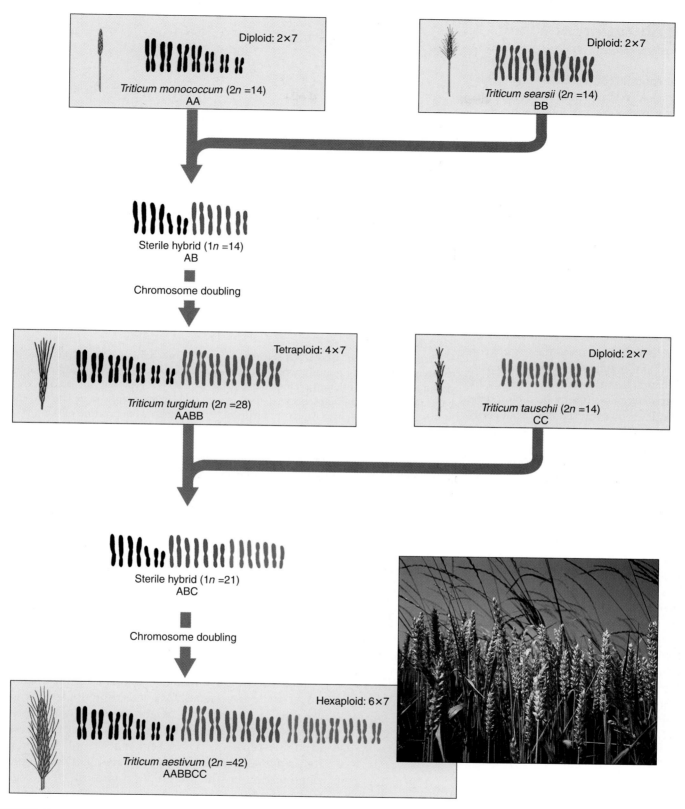

FIGURE 40.3

Evolutionary history of wheat. Domestic wheat arose in southwestern Asia in the hilly country of what is now Iraq. In this region there is a rich assembly of grasses of the genus *Triticum*. Domestic wheat *(T. aestivum)* is a polyploid species of *Triticum* that arose through two so-called "allopolyploid" events. (1) Two different diploid species, AA and BB, hybridized to form an AB polyploid; the species were so different that A and B chromosomes could not pair in meiosis, so the AB polyploid was sterile. However, in some plants the chromosome number spontaneously doubled due to a failure of chromosomes to separate in meiosis, producing a fertile tetraploid species AABB. (2) In a similar fashion, the tetraploid species AABB hybridized with another diploid species CC to produce the hexaploid *T. aestivum*, AABBCC.

Organization of Plant Genomes

By looking at conserved arrangements of genes in related species, a technique called **synteny,** we can tell which plants are closely related and which are more divergent. Plant genomes can vary by the number of repeated sequences and sequence inversions or by the effects of transposable elements. However, the large amount of DNA present in most plant genomes can mask the effects of most of these sequence alterations when they occur in noncoding regions.

Low-, Medium-, and High-Copy-Number DNA

Most higher plants contain quantities of DNA that greatly exceed their needs for coding and regulatory function. Hence, for plants, a very small percentage of the genome may actually be used to encode genes involved in the production of protein. This portion of the genome which encodes most of the transcribed genes is often referred to as "low-copy-number DNA," because the DNA sequences comprising these genes are present in single or small numbers of copies. "Medium-copy-number DNA" includes DNA sequences encoding ribosomal RNA (rRNA), which represents the cellular machinery involved in the translation of messenger RNA into protein. In plant genomes, rRNA genes may be repeated several hundred to several thousand times. This is in contrast to animal cells, where only 100 to 200 rRNA genes are normally present. The extent of variability in plant genomes with respect to the number of rRNA genes and mutations in them has provided a useful tool for analyzing the evolutionary patterns of plant species. Plant cells may also contain excess DNA in their genomes in the form of highly repetitive sequences, or "high-copy-number DNA." At present, the function of this high-copy-number DNA in plant genomes is unknown.

Sequence Replication and Inversion

High-copy-number DNA sequences in the plant genome may be short, such as the nucleotide sequence "GAA," or much longer, involving up to several hundred nucleotides. Moreover, the number of copies of an individual high-copy repetitive DNA sequence can total from 10,000 to 100,000. There are several possibilities for how high-copy repetitive DNA sequences may be organized within a plant genome (figure 40.4*a*). Several copies of a single repetitive

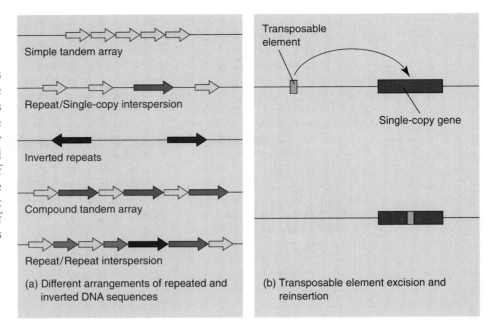

FIGURE 40.4

Organization of repeated DNA sequences and the mechanism of transposable elements in altering gene function. (a) Repeated DNA sequences can occur in plant genomes in several different arrangements. The thick arrows represent repeated DNA sequences, while thin lines represent single-copy DNA. Arrows of the same size and color represent DNA sequences which are identical to each other. The direction of the arrowhead indicates the orientation of the DNA sequence. (b) Transposable elements can be a source of repetitive DNA that alters gene function. Following excision from its original location, a transposable element may reinsert in the single-copy DNA sequence comprising a gene and alter the gene's function.

DNA sequence may be present together in the same orientation, in a pattern called "simple tandem array." Alternatively, repetitive DNA sequences can be dispersed among single-copy DNA in the same orientation ("repeat/single-copy interspersion") or the opposite orientation ("inverted repeats"). In addition, groups of repetitive DNA sequences can also occur together in plant genomes in a variety of possible arrangements, such as a "compound tandem array" or a "repeat/repeat interspersion." The presence of repetitive DNA can vastly increase the size of a plant genome, making it difficult to find and characterize individual single-copy genes. Characterizing single-copy genes can thus become a sort of "needle-in-the-haystack" hunt.

A variety of mechanisms can account for the presence of highly repetitive DNA sequences in plant genomes. Repetitive sequences can be generated by DNA sequence amplification in which multiple rounds of DNA replication occur for specific chromosomal regions. Repetitive sequences can also be generated by unequal crossing over of the chromosomes during meiosis or mitosis (translocation) or the action of transposable elements (see next section). On the other hand, sequence inversions could occur by chromosomal breakage and subsequent reinsertion of the segment in a reverse orientation to the original chromosome.

Transposable Elements

Transposable elements are special sequences of DNA which are characterized by their ability to move from place to place in the genome. They can excise from one site at unpredictable times and reinsert in another site. For this reason, transposable elements have been called "jumping genes." Transposable elements can insert into coding regions or regulatory regions of a gene and so affect expression of that gene, resulting in a mutation that may or may not be detectable (figure 40.4b). A detectable transposable element mutation may "revert" back to the wild-type phenotype if the transposable element disassociates from the gene responsible for that phenotype. Barbara McClintock won the Nobel Prize in 1983 for her work describing transposable elements in corn (see figure 17.23). The importance of McClintock's "jumping genes" was recognized only after subsequent discovery and analysis of transposable elements in bacteria.

Due to their capacity to replicate independently and move through the genome, transposable elements can also be involved in generating repetitive DNA sequences. Retention of the repetitive DNA sequence at a particular site in the genome would involve a mutation in the transposable element itself which removes its capacity to transpose.

Chloroplast Genome and Its Evolution

The chloroplast is a plant organelle that functions in photosynthesis, and it can independently replicate in the plant cell. Plant chloroplasts have their own specific DNA, which is separate from that present in the nucleus. This DNA is maternally inherited and encodes unique chloroplast proteins. Many of the proteins encoded by chloroplast DNA are involved in photosynthesis. Chloroplasts are thought to have originated from a photosynthetic prokaryote that became part of a plant cell by endosymbiosis. In support of this concept, research has shown that chloroplast DNA has many prokaryote-like features. Chloroplast DNA is present as circular loops of double-stranded DNA similar to prokaryotic chromosomal DNA. Moreover, chloroplast DNA contains genes for ribosomes which are very similar to those present in prokaryotes.

The DNA in chloroplasts of all land plants have about the same number of genes (~100), and they are present in about the same order (figure 40.5). In contrast to the evolution of the DNA in the plant cell nucleus, chloroplast DNA has evolved at a more conservative pace, and therefore shows a more interpretable evolutionary pattern when scientists study DNA sequence similarities. Chloroplast DNA is also not subject to modification caused by transposable elements and mutations due to recombination.

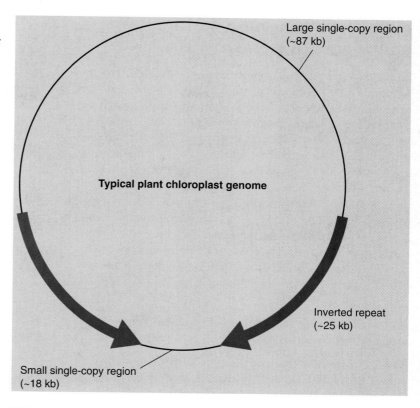

FIGURE 40.5
Chloroplast genome. A schematic drawing of a typical plant chloroplast genome indicates two regions containing single-copy genes, one containing about 87,000 nucleotides (87 kb) and another about 18 kb, and two symmetrical inverted repeats, each containing about 25 kb. Chloroplast DNA does not show recombination events that are common in the nuclear genome. It is thus a good subject for DNA phylogenetic analysis.

A characteristic feature of the chloroplast genome is the presence of two identical inverted repeats in the DNA sequence. Other DNA sequence inversions or deletions occur rarely, but when they do occur, they provide a character or a tool to analyze evolutionary relationships between plants. For instance, a large inversion in chloroplast DNA is found in the Asteraceae, or sunflower family, and not in other plant families. While previous work on the evolutionary relationships between plants has emphasized the comparative analysis of plant anatomy or morphology, there is increasing use of plant molecular data such as chloroplast DNA sequences. When considered together, morphological and molecular information can provide a clearer understanding of the evolutionary processes that govern biological diversity.

> **Plant nuclear genomes may contain large amounts of DNA in comparison to other eukaryotic organisms, but only a small amount of this DNA represents functional genes. Excess DNA in plant genomes can result from increased chromosome copy number (polyploidy), and DNA sequence repeats. Chloroplast genomes evolve more slowly than nuclear genomes and can provide important evolutionary information.**

Comparative Genome Mapping and Model Systems

Knowledge of plant genomes has been growing with the advent of new techniques to study DNA sequences, such as gene mapping and chromosome synteny. An increased understanding of plant genomes can lead to better manipulation of genetic traits such as crop yield, disease resistance, growth abilities, nutritive qualities, or drought tolerance. Each of these traits could be encoded by multiple genes. If genomes between related species are found to be similar by gene mapping, then finding important quantitative trait loci (QTLs) in one of the species means that the loci could likely be found in related species. By genome mapping model plants, plant biologists can lay a foundation for future plant breeding and for an understanding of plant evolution at the genetic level.

RFLP and AFLP as Tools to Map Genomes and Detect Polymorphisms

RFLP (restriction fragment length polymorphism) mapping is a tool used to identify characteristic DNA sequence markers in plant genomes. This approach, described in detail in chapter 18 (see figures 18.2, 18.4, 18.9, and 18.10), involves analysis of the RFLP map, or the pattern of DNA fragments, produced when DNA is treated with restriction endonucleases that cleave at specific sites. RFLP mapping can identify important regions of the genome at a glance, while sequence data require sophisticated computer-based searching and matching systems. A comparison of the RFLP maps of parents and progeny can give an indication of the heritability of gene traits and of heritable loci that are characteristic of traits. Moreover, after full genomes are sequenced at the nucleotide level, the genetic identification of RFLP markers in regions of interest will be facilitated.

Another tool that utilizes sequence variability is AFLP, or amplified fragment length polymorphisms. AFLP maps are generated by hybridizing DNA primers with genomic DNA fragments that have been cut with restriction endonucleases, usually EcoR1 and Mse1, and then subsequently amplified using the polymerase chain reaction (PCR). The resulting PCR products, which represent each piece of DNA cut by a restriction endonuclease, are separated by size via gel electrophoresis. The band sizes on an AFLP gel tend to show more polymorphisms than those found with RFLP mapping because the entire genome is visible on the gel (figure 40.6). Both RFLPs and AFLPs (among many other tools for genomes analysis) can provide markers of traits which are inherited from parents to progeny through crosses.

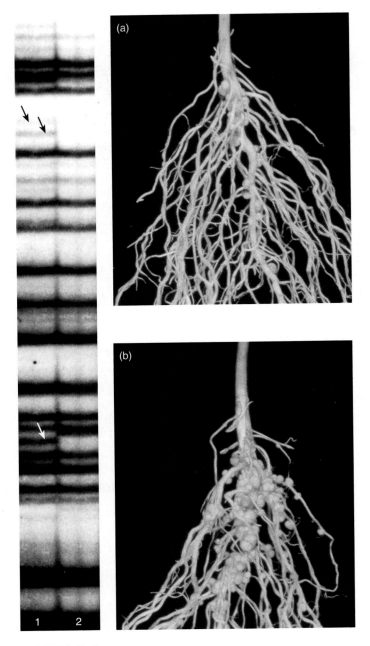

FIGURE 40.6

A portion of an AFLP fingerprint pattern from the DNA of two soybeans, one normal, and the other a mutant "hypernodulator" with a large number of root nodules.
The root nodules are able to form a favorable environment for the soil bacteria *Rhizobium japonicum*, which "fix" atmospheric nitrogen for the plant to use. It is still not known what determines the nodule number on the "hypernodulating" soybean mutant (a), Compared to a normal root (b). The slight genetic differences between these plants can be evaluated by AFLP. The banding pattern changes indicate what genetic markers are linked to the "hypernodulation" mutation. Lane 1: normal soybean DNA; Lane 2: "hypernodulating" soybean DNA.

Arabidopsis thaliana as a Model System for Plant Genome Analysis

Arabidopsis thaliana is a small dicotyledonous weed plant, related to the mustards, which has become a model system for plant genome analysis for a number of practical reasons. First, its life cycle is quite rapid—it can germinate, flower, and reproduce over a period of about five weeks. This short life cycle is very useful for genetic experiments involving crosses and the production of mutant plants. In addition, due to its small size, *Arabidopsis thaliana* can grow in small containers convenient for laboratory experiments and can also grow in tissue culture (see next section). Most importantly, it has a relatively small genome, with repetitive sequences comprising only about 20% of the DNA. Therefore, single-copy gene sequences are easier to find.

There is currently an international effort to sequence the entire *Arabidopsis thaliana* genome to determine the location and function of every gene. The full sequence was expected in 1999. Information obtained from a completely sequenced plant genome will have far-reaching uses in agricultural breeding and evolutionary analysis. This information can be expected to help plant breeders in the future because the localization of genes in one plant species can help indicate where that gene might also be located in another species. In plant genomes, local gene order seems to be more conserved than the nucleotide sequences of homologous genes. Thus, the complete genomic sequence of *Arabidopsis thaliana* will facilitate gene cloning from many plant species, using information on relative genomic location as well as similarity of sequences.

Due to its rapid life cycle and the ease of growing large numbers of plants in the laboratory, *Arabidopsis thaliana* has proven extremely useful in the production and selection of mutant plants for a variety of genes. Mutations in the *Arabidopsis thaliana* genome have been created by transposable elements, radiation, UV-light, or chemical mutagens. A number of mutants have been produced involving plant development and metabolic pathways, and this has provided a means for pinpointing particular genes responsible for particular phenotypes (figure 40.7).

(a)

(b)

FIGURE 40.7

Flowering mutants of *Arabidopsis*. *Arabidopsis* floral mutants help identify genes involved in floral organ development. (a) A normal flower of *Arabidopsis thaliana* contains four green sepals, four white petals, six stamens (male reproductive organs) and one pistil (female reproductive organ). (b) Mutant flowers show altered floral organization due to mutation of developmental genes.

Genome Sequencing of Rice and Other Grains

A sequencing effort similar to the one underway for *Arabidopsis thaliana* is also being conducted for rice, which would serve as a model for a small monocot genome. Grain plants such as rice, corn, barley, and wheat are monocots that represent a major food source for humans. By understanding the rice genome at the level of its DNA sequence, it would be much easier to identify and isolate genes from grains with larger genomes. Even though these plants diverged more than 50 million years ago, the chromosomes of rice, corn, barley, wheat, and other grass crops show extensive conserved arrangements of segments (figure 40.8). DNA sequence analysis of cereal grains will be important for identifying genes associated with disease resistance,

crop yield, nutritional quality, and growth capacity. It will also be possible to construct an approximate map of the ancestral cereal genome.

Restriction fragment length polymorphisms (RFLPs) and amplified fragment length polymorphisms (AFLPs) represent important tools for mapping genetic traits in plant genomes. Due to its short life cycle, small size, and small genome the mustard relative, *Arabidopsis thaliana*, is being used as a model plant for genetic studies, generation of interesting mutants, and genome sequencing. The genome of rice is also being sequenced and this would serve as a valuable model for other monocot cereal grains such as wheat, barley, oats and corn.

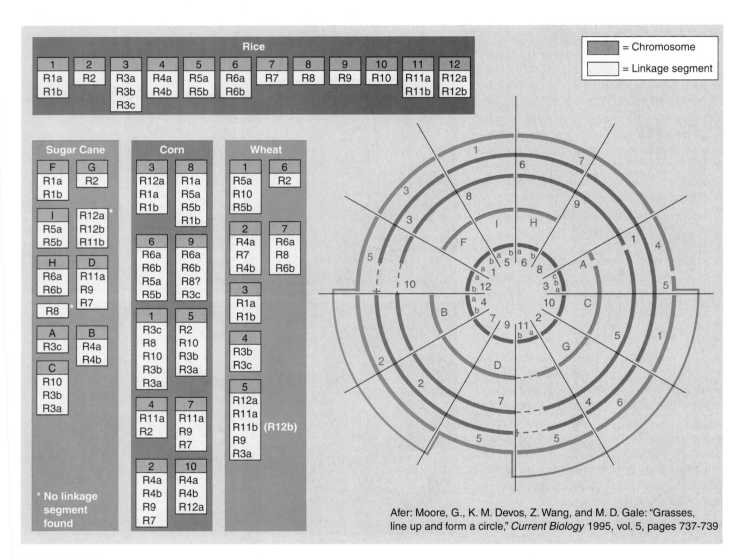

FIGURE 40.8
Cereal genome evolution. The genomes of major grass species can be aligned by splitting the individual chromosomes into segments and rearranging these linkage segments into highly similar structures. This implies that the order of the linkage segments in the ancestral grass genome has been rearranged by recombination as the grasses have evolved. The radiating lines indicate linkage segments in rice and how these linkage segments appear in the genomes of the other grass species. Linkage segments forming parts of chromosome 5 in wheat are shown as a series of segments connected by colored lines. Dashed lines indicate duplicated segments.

Overview of Plant Tissue Culture

An unusual characteristic of plant cells is their ability to dedifferentiate and regenerate into whole plants. When a human cell is removed or dies, it is rarely replaced with identical tissue. Most animal cells are not easily able to dedifferentiate, or alter their developmental pattern after they have matured, although recent advances in cloning sheep and other animals seems to be changing this generalization. Plant cells are able to express portions of their previously unexpressed genetic potential under appropriate conditions, which can even lead to development of a whole plant from a single somatic cell. This attribute of plant cells, as we learned in chapter 39, is called totipotency. The totipotency of plant cells forms the basis for plant tissue culture where plant cells, tissues, or organs can be grown in an artificial growth medium, as is done with more simple cultures of bacteria and fungi.

The successful culture of plant cells, tissues, or organs requires utilizing the proper plant starting material, appropriate nutrient medium, and timing of hormonal treatments to maximize growth potential and drive differentiation (figure 40.9). Most plant tissue cultures are initiated from *explants*, or small sections of tissue removed from an intact plant under sterile conditions. After being placed on a sterile growth medium containing nutrients, vitamins, and combinations of plant growth regulators, cells present in the explant will begin to divide and proliferate. Under appropriate culture conditions, plant cells can multiply and form organs (roots, shoots, embryos, leaf primordia, and so on) and can even regenerate a whole plant. The regeneration of a whole plant from tissue-cultured plant cells represents an important step in the production of genetically engineered plants. Using plant tissue cultures, genetic manipulation can be conducted at the level of single cells in culture, and whole plants can then be produced bearing the introduced genetic trait.

Plant cells growing in culture can also be used for the mass production of genetically identical plants (clones) with valuable inheritable traits. For example, this approach of clonal propagation using plant tissue culture is commonly used in the commercial production of many ornamental plants such as chrysanthemums and ferns. As we will describe next, different types of cultures can be generated based on the initial type of plant tissue used for the explant and on the composition of the growth medium.

It is often possible to regenerate an entire plant from one or a few cells. Depending on the plant tissue used and the growth medium selected, different types of cultures may be produced.

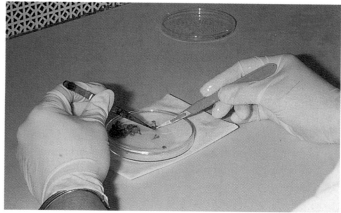
(a)

(b)

(c)

FIGURE 40.9
Culture of orchid plants. While the natural maturation of a single orchid plant may take up to seven years, commercial growers can produce thousands of cultured orchards in a relatively short time. (a) The apical meristem is removed from an orchid plant. (b) The meristematic tissue is grown in flasks containing hormone-enhanced media, and roots and shoots begin to form. (c) The plantlets are then separated and grown to maturity.

Types of Plant Tissue Cultures

Depending on the type of plant tissue used as the explant and the composition of the growth medium, a variety of different types of plant tissue cultures can be generated. These different types of plant tissue cultures have applications both in basic plant research and commercial plant production.

Callus Culture

Callus culture refers to the growth of unorganized masses of plant cells in culture. To generate a callus culture, an explant, usually containing a region of meristematic cells, is incubated on a growth medium containing certain plant growth regulators such as auxin and cytokinin (figure 40.10). The cells grow from the explant and divide to form an undifferentiated mass of cells called a callus. This unorganized mass of growing cells is analogous to a plant tumor. Cells can proliferate indefinitely if they are periodically transferred to fresh growth media. However, if the callus cells are transferred to a growth medium containing a different combination of plant growth regulators, the cells can be directed to differentiate into roots and/or shoots. This process of converting unorganized growth into the production of shoots and roots is called **organogenesis,** and it represents one means by which a whole plant can be regenerated from tissue culture cells. When a plantlet produced by organogenesis is large enough, it can be transferred to a large container with nutrients or soil and grown to maturity.

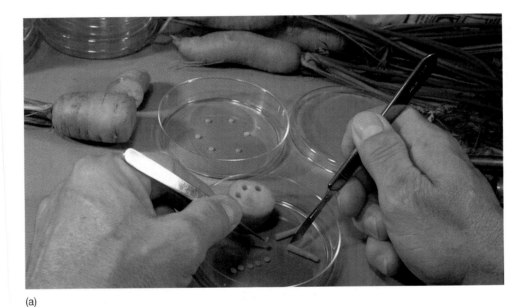

(a)

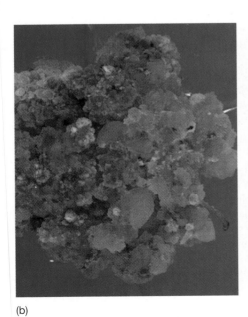

(b)

(c)

(d)

FIGURE 40.10
Callus culture. (a) An explant is incubated on growth media. (b) The cells grow and divide and form a callus. (c) The callus cells, grown on media containing new plant growth regulators, differentiate into plant parts. (d) After the plantlet is large enough, it is grown to maturity in soil.

Cell Suspension Culture

Plant cell suspension culture involves the growth of single or small groups of plant cells in a liquid growth medium. Cell suspension cultures are usually initiated by the transfer of plant callus cells into a liquid medium containing a combination of plant growth regulators and chemicals that promote the disaggregation of the cells into single cells or small clumps of cells (figure 40.11). Continued cell growth requires that the liquid cultures be shaken at low speed to promote aeration and chemical exchange with the medium. Suspension cell cultures are often used in research applications where access to single cells is important. The suspension bath can provide an efficient means for selecting out cells with desirable traits such as herbicide tolerance or salt tolerance, since the bath is in uniform contact with all the cells at once. This differs from callus culture, where only those cells in contact with the solid medium can be selected by chemical additions to the medium. Suspension cultures can also provide a convenient means for producing and collecting the plant chemicals cells secrete. These can include important plant metabolites, such as food products, oils, and medicinal chemicals. In addition, plant suspension cell cultures can often be used to produce whole plants via a process known as **somatic cell embryogenesis** (figure 40.12). For some plants, this provides a more convenient means of regenerating a whole plant after genetic engineering takes place at the single-cell level. In somatic cell embryogenesis, plant suspension culture cells are transferred to a medium containing

FIGURE 40.11
Cell suspension culture. Plant cells can be grown as individual cells or small groups of cells in a liquid culture medium. Liquid suspension culture of plant cells ensures that most cells are in contact with the growth medium.

a combination of growth regulators that drive differentiation and organization of the cells to form individual embryos. Under a dissection microscope, these embryos can be isolated and transferred to a new growth medium, where they grow into individual plants.

(a)

(b)

(c)

(d)

(e)

FIGURE 40.12
Somatic cell embryogenesis. A large number of plants can be cloned from a single soybean seed via somatic cell embryogenesis. (a) Immature soybean seeds placed on culture medium. (b) Embryos appear on the seeds after two weeks in culture. (c) Four embroys at different stages of development (globular, heart, torpedo, and plantlet). (d) Seedlings with shoots and roots. (e) Mature soybean plants.

Protoplast Isolation and Culture

Protoplasts are plant cells that have had their thick cell walls removed by an enzymatic process, leaving behind a plant cell enclosed only by the plasma membrane. Plant protoplasts have been extremely useful in research on the plant plasma membrane, a structure normally inaccessible due to its close association with the cell wall. Within hours of their isolation, plant protoplasts usually begin to resynthesize cell walls, so this process has also been useful in studies on cell wall production in plants. Plant protoplasts are also more easily transformed with foreign DNA using approaches such as electroporation (see the subsequent section). In addition, protoplasts isolated from different plants can be forced to fuse together to form a hybrid. If they are regenerated into whole plants, these hybrids formed from protoplast fusion can represent genetic combinations that would never occur in nature. Hence, protoplast fusion can provide an additional means of genetic engineering, allowing beneficial traits from one plant to be incorporated into another plant despite broad differences between the species. When either single or fused protoplasts are transferred to a culture growth medium, cell wall regeneration takes place, followed by cell division to form a callus (figure 40.13). Once a callus is formed, whole plants can be produced either by organogenesis or by somatic cell embryogenesis in culture.

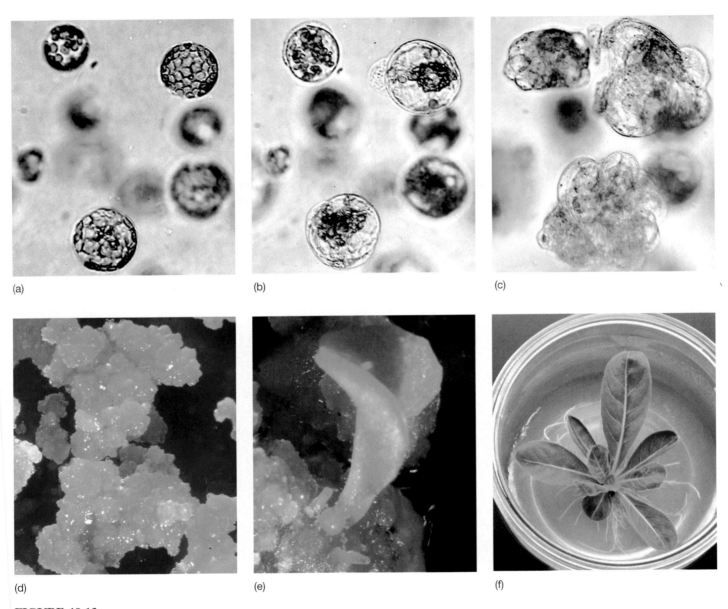

FIGURE 40.13
Protoplast regeneration. Different stages in the recovery of intact plants from single plant protoplasts of nightshade *(Atropa belladona)*. (a) Individual plant protoplasts. (b) Regeneration of the cell wall and the beginning of cell division. (c, d) Aggregates of plant cells resulting from cell division which can form a callus. (e) Production of somatic cell embryos from the callus. (f) Recovery of a plantlet from the somatic cell embryo through the process described in figure 40.12.

Anther/Pollen Culture

In flowers, the anthers are the anatomical structures that contain the pollen. In normal flower development, the anthers mature and open to allow pollen dispersal. In anther culture, anthers are excised from the flowers of a plant and then transferred to an appropriate growth medium. After a short period of time, pollen cells can be manipulated to form individual plantlets, which can be grown in culture and used to produce mature plants. The development of these plantlets usually proceeds through the formation of embryos (figure 40.14). Plants produced by anther/pollen culture can be haploid because they were originally derived from pollen cells which have undergone meiosis. However, these plants may be sterile and thus not useful for breeding or genetic manipulation. On the other hand, plants derived from anther/pollen culture can be treated at an early stage with chemical agents such as colchicine, which allows chromosome duplication. Chromosome duplication results in the conversion of sterile haploid plants into a fertile diploid organism. Under these conditions, plants can be produced that are homozygous for every single trait, even those which tend to be recessive traits. These homozygous plants can become a very useful tool, allowing breeders to introduce a normally recessive trait.

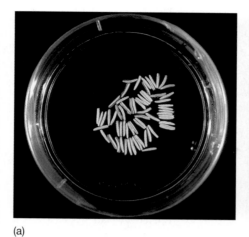

(a)

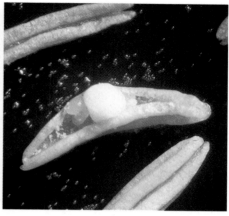

(b)

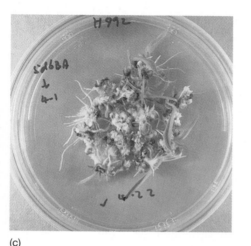

(c)

FIGURE 40.14
Anther culture. Anthers containing pollen can be regenerated on tissue culture medium. The pollen in the anthers contain a haploid set of chromosomes, which can be doubled to form a homozygous diploid cell. Regenerated homozygous diploid plants are important for plant breeding purposes. (a) Maize anthers in culture medium. (b) Callus formation from pollen. (c) Callus and shoot formation.

Plant Organ Culture

Plant organs can also be grown under culture conditions, and this has provided a useful tool in the study of plant organ development. For example, pollinated flowers of a plant such as a tomato can be excised and transferred to a culture flask containing an appropriate medium. Over time, the ovular portion of the plant will develop into a tomato fruit that will eventually turn red and ripen. Sections of plant roots can also be excised and transferred to a liquid growth medium. In this medium, the roots can proliferate extensively, forming both primary and secondary root branches (figure 40.15).

Because plant cells are totipotent, a whole plant may be regenerated from a single plant cell. Depending upon the explant type, culture medium, and combinations of plant growth regulators, it is possible to grow plant cells, tissues, or organs in sterile cultures.

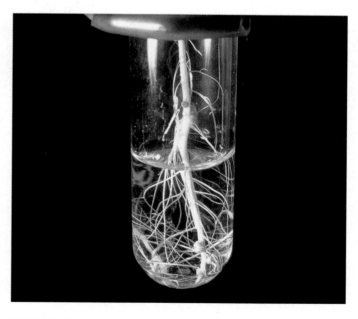

FIGURE 40.15
Plant organ culture. Plant roots growing in a liquid culture medium. From small excised sections of plant roots, the roots will grow and proliferate with extensive lateral root formation (branching).

Applications of Plant Tissue Culture

In addition to the applications already described, plant tissue cultures have a variety of uses both in agriculture and in industry.

Suspension Cultures as Biological Factories

An important industrial application of plant tissue culture involves the use of plant cells as biological factories. Large-scale suspension cultures can be grown to produce antimicrobial compounds, antitumor alkaloids, vitamins, insecticides, and food flavors. Plant roots can also be grown in liquid culture, creating a mesh of roots that can produce a number of useful plant compounds.

Horticultural Uses

Plants with valuable traits can be mass propagated through tissue-culture cloning. In this application of plant tissue culture, hundreds or even thousands of genetically identical plants can be produced by vegetative asexual propagation from one plant source. This has been extensively used in the flower industry where genetically identical plants can be produced from a superior parent plant. Propagation of plant tissue in the sterile environment of the growth medium can also help in the production disease-free plants, such as those cultured from the meristematic (apical dome) tissue untouched by viruses or other diseases because it is new growth. This approach has been particularly useful in the culture of disease-free orchids.

Somaclonal Variation

Plant tissue culture also has a problematic side effect that can be used as an asset under certain conditions. During periods of extended growth of plant cells in callus or suspension cell culture, various parts of the plant genome may become more or less "active" due to a release of control over gene expression. Transposable elements may also become more active, and chromosomal rearrangements may occur. This altered control provides a new source of genetic diversity that can result in novel traits which were not even present in the original plant material used as the ex-

(a)

(b)

(c)

FIGURE 40.16
Somaclonal variation. Regeneration of plants from tissue culture can produce plants that are not similar to their parents due to chromosomal alterations. This variability can be used to select plants with altered traits. These maize plants show evidence of somaclonal variation.
(a) Yellow leaf stripe.
(b) Dwarf maize.
(c) Yellow leaf tip.

plant to start the cultures (figure 40.16). This increased genetic diversity following extended time in tissue culture is called **somaclonal variation.** It can be problematic if the desired goal is the propagation or production of identical plant clones. However, somaclonal variation, induced by intentionally growing plant cells in tissue over a longer time period, can be very useful to generate novel plants with traits not currently present in a given gene pool. These traits can be identified either at the tissue culture stage (for example, disease resistance or heat tolerance) or following the regeneration of whole plants by either organogenesis or embryogenesis (plant size, photosynthetic rates, and so forth).

Plant cell, tissue, and organ cultures have important applications in agriculture and industry.

Plant biotechnology provides an efficient means to produce an array of novel products and tools for use by our global society. Agricultural biotechnology has the potential to increase farming revenue, lower the cost of raw materials, and improve environmental quality. Plant genetic engineering is becoming a key tool for improving crop production.

World Population in Relation to Advances Made in Crop Production

Due in a large part to scientific advances in crop breeding and farming techniques, world food production has doubled since 1960. Moreover, productivity from existing agricultural land and water usage has tripled over this time period. While major genetic improvements have been made in crops through crop breeding, this can be a slow process. Furthermore, most crops grown in the United States produce less than 50% of their genetic potential. These shortfalls in yield are due in large part to the inability of crops to tolerate or adapt to environmental stresses, pests, and disease (figure 40.17).

The world now farms an area the size of South America, but without the scientific advances of the past 30 years, farmland equalling the entire Western hemisphere would be required to feed the world. Nevertheless, the world population is expected to double to 12 billion by the year 2030, and it is not clear whether current levels of food production can keep pace with this rate of population growth. Many believe the exploitation of conventional crop breeding programs may have reached their limit. The question is how best to feed billions of additional people without destroying much of the planet in the process. In this respect, the disappearance of tropical rainforests, wetlands, and other vital habitats will accelerate unless agriculture becomes more productive and less taxing to the environment.

> Although improved farming practices and crop breeding have increased crop yields, it is uncertain whether these approaches can keep pace with the food demands of an ever-increasing world population.

FIGURE 40.17
Corn crop productivity well below its genetic potential due to drought stress. Corn production can be limited by water deficiencies due to drought during the growing season in dry climates.

Plant Biotechnology for Agricultural Improvement

It seems certain that plant genetic engineering will play a major role in resolving the problem of feeding an increasing world population. Biotechnology can now be employed to improve the quality of seed grains, increase protein levels in forage crops, and transform plants to improve their resistance to disease, insects, herbicides, and viruses. Other stresses on plants, such as heat or salt, can be improved by engineering higher tolerance levels. The nutritional quality of plants can also be improved by increasing the levels of nutrients they contain, such as beta-carotene and vitamins C and E, which may protect people from health problems such as cancer and heart disease.

Compared with approaches that rely on plant breeding, genetic engineering can compress the time frame required for the development of improved crop varieties. Moreover, in genetic engineering, genetic barriers, such as pollen compatibility with the pistil, no longer limit the introduction of advantageous traits from diverse species into crop plants. An ideal trait to transfer would be encoded by a single gene. Multigene traits, such as nitrogen fixation, would provide many technical difficulties. Once a useful trait has been identified at the level of individual genes and their DNA sequences, the incorporation of this trait into a crop plant requires only the introduction of the DNA bearing these genes into the crop plant genome. The process of incorporating foreign DNA into an existing plant genome is called **plant transformation**. At present, there are several approaches for plant transformation; the use of *Agrobacterium tumefaciens* in this process was described in chapter 18. This approach works best if the plant being transformed is a dicot. However, many food crops, such as the cereal grains (rice, wheat, corn, barley, oats, and so on) are monocots. This means that the use of *Agrobacterium tumefaciens* in plant transformation may be limited. Two additional methods for plant transformation that are useful in agricultural biotechnology can be used with both dicots and monocots.

Plant Transformation Using the Particle Gun

Using a "gun" to blast plant cells does not seem like a suitable method for introducing foreign DNA into a plant genome. However, it works, and many whole plants have been regenerated after foreign DNA is shot through the

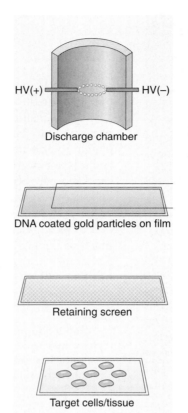

Discharge chamber

DNA coated gold particles on film

Retaining screen

Target cells/tissue

(a)

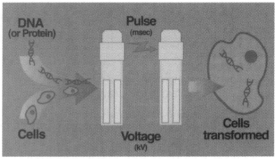

Contributed by Bio-Rad Laboratories

(b)

FIGURE 40.18

Methods for plant transformation. (a) The particle gun is one method for introducing foreign DNA into plant cells. Here an electrical discharge propels DNA-coated gold particles into plant cells or tissue. A retaining screen reduces cellular damage associated with bombardment by only allowing the DNA-coated particles to pass and retaining fragments of the mounting film. (b) Foreign DNA can also be introduced into plant protoplasts by electroporation. A brief pulse of electricity generates pores in the plasma membrane, allowing DNA to enter the cells.

cell wall and then integrated into the plant genome. The *particle gun* utilizes microscopic gold particles coated with the foreign DNA, shooting these particles into plant cells at high velocity. Acceleration of the particles to a sufficient velocity to pass through the plant cell wall can be achieved by a burst of high-pressure helium gas or an electrical discharge (figure 40.18a). Only a few cells actually receive the foreign DNA and survive this treatment. These cells are identified with the help of a **selectable marker** also present on the foreign DNA. The selectable marker allows only those cells receiving the foreign DNA to survive on a particular growth medium (figure 40.19). The selectable markers include genes for resistance to a herbicide or antibiotic. Plant cells which survive growth in the selection medium are then tested for the presence of the foreign gene(s) of interest.

Plant Transformation Using Electroporation

Foreign DNA can also be "shocked" into cells that lack a cell wall, such as the plant protoplasts described earlier. A pulse of high-voltage electricity in a solution containing plant protoplasts and DNA briefly opens up small pores in the protoplasts' plasma membranes, allowing the foreign DNA to enter the cell (figure 40.18b). Following electroporation, the protoplasts are transferred to a growth

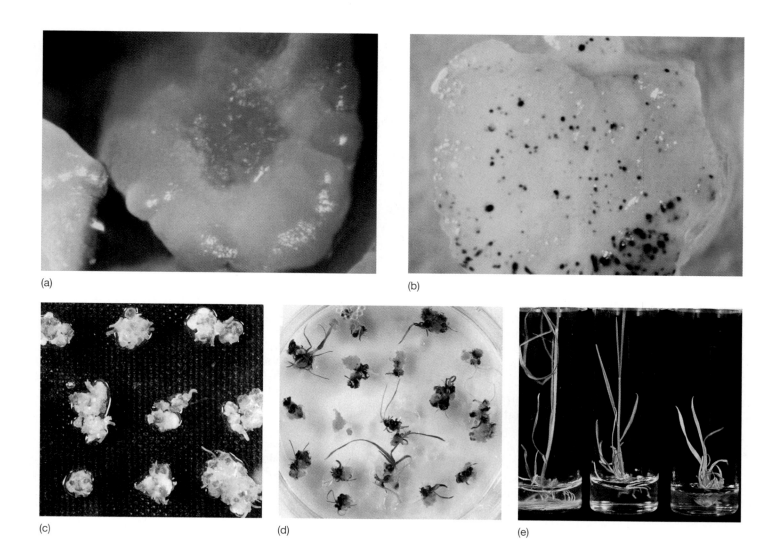

(a)

(b)

(c)

(d)

(e)

(f)

FIGURE 40.19

Regeneration after transformation with the use of a selectable marker. Stages in the recovery of a plant containing foreign DNA introduced by the "particle gun" method for plant cell transformation. A selectable marker, in this case a gene for resistance to herbicide, aids in the identification and recovery of plants containing the DNA insert. (a) Embryonic callus just prior to particle gun bombardment. (b) Following bombardment, callus cells containing the foreign DNA are indicated by color from the *gus* gene used as a tag or label on the foreign DNA. (c) Shoot formation in the transformed plants growing on a selective medium. Here, the gene for herbicide resistance in the transformed plants allows growth on the selective medium containing the herbicide. Nontransformed plants do not contain the herbicide resistance gene and do not grow well. (d) Production of plantlets from transformed plants growing on the selective medium. (e) Comparison of growth on the selection medium for transformed plants (*left*) bearing the herbicide resistance gene and a nontransformed plant (*right*). (f) Mature transgenic plants resulting from this process.

medium for cell wall regeneration, cell division, and, eventually, the regeneration of whole plants. As with the use of the particle gun, a selectable marker is typically present in the foreign DNA, and protoplasts containing foreign DNA are selected based upon their ability to survive and proliferate in a growth medium containing the selection treatment (antibiotic or herbicide). Once regenerated from electroporated protoplasts, whole plants can then be evaluated for the presence of the beneficial trait.

Plant biotechnology may play an important role in the further improvement of crop plants. The particle gun and electroporation are useful methods for introducing foreign DNA into plants.

Useful Traits That Can Be Introduced into Plants

Although plant transformation represents a relatively new technology, extensive efforts are underway to utilize this approach to develop plants and food products with beneficial characteristics. We discussed a variety of biotechnological applications for crop improvement in chapter 18. Two further applications of this approach involve modifications of nutritional quality of foods and using plants as "edible vaccines."

Improved Nutritional Quality of Food Crops

Approximately 75% of the world's production of oils and fats come from plant sources. For medical and dietary reasons, there is a trend away from the use of animal fats and toward the use of high-quality vegetable oils. Genetic engineering has allowed researchers to modify seed oil biochemistry to produce "designer oils" for edible and nonedible products (figure 40.20). One technique modifies canola oil to replace cocoa butter as a source of saturated fatty acids; others modify the enzyme ACP desaturase for the creation of monounsaturated fatty acids in transgenic plants.

There is also great interest in modifying the amino acid contents of various plant seeds to present a more complete nutritional diet to the consumer. A high-lysine corn seed is being developed; this would cut down on the need for lysine supplements that are currently added to livestock feed. Biotechnology has the potential to make plant foods healthier and more nutritious for human consumption. Fruits and vegetables, such as tomatoes, may be engineered to contain increased levels of vitamins A and C and beta-carotene, which, when included in the human diet, may help protect against chronic diseases. Genetic engineering is also used in the development of inedible products, including biodegradable plastics, that can be produced at a lower cost in plants than in transgenic bacteria.

Plants Bearing Vaccines for Human Diseases

Another very interesting application of plant genetic engineering includes the introduction of "vaccine genes" into edible plants. Here, genes encoding the antigen (for example, a viral coat protein) for a particular human pathogen would be introduced into the genome of an edible plant such as a banana, tomato, or apple via plant transformation. This antigen protein would then be present in the cells of the edible plant, and a human individual that consumed the plant would develop antibodies against the pathogenic organism. Currently, researchers are trying to develop such edible vaccines for a coat protein of hepatitis B, an enterotoxin B of *E. coli*, and a viral capsid protein of the Norwalk virus.

Genetic engineering of crop plants has allowed researchers to alter the oil content, amino acid composition, and vitamin content of food crops. Genetic engineering may also allow the production of food crops bearing "edible vaccines" against human pathogens.

FIGURE 40.20
Future directions in the genetic engineering of vegetable oils?

Summary of Concepts

40.1 Genomic organization is much more varied in plants than in animals.

- Plant genomes are very large in comparison to other eukaryotes, mainly due to a high amount of repetitive DNA. The bulk of the DNA is high-copy-number DNA—repetitive sequences or inversions—and also ribosomal DNA and transposable elements.

- Transposable elements are "jumping genes" that can alter the plant genome by exiting or inserting into the DNA.

- Plant genomes can be compared with one another by mapping the locations of certain genes or gene traits in various plants. RFLPs and AFLPs can be used to map plant DNA.

- *Arabidopsis thaliana* has a small genome, for a plant. This complete genome is being sequenced, so all genes and their positions will be known in the next few years.

- The molecular maps of the genomes of rice and other grains demonstrate remarkable similarity. Hence, sequencing the rice genome will give us an understanding of the genomic organization of other cereal crops such as wheat, oats, corn and barley.

40.2 Advances in plant tissue culture are revolutionizing agriculture.

- Most plant cells are totipotent, allowing them to dedifferentiate at maturity and restart growth and development. Therefore a single plant cell can regenerate into a whole plant under the appropriate growth conditions.

- With the addition of appropriate combinations of plant growth regulators (auxin, cytokinin), plant cells in culture can be directed to form organs, embryos or whole plants.

- Plant cells can be treated with enzymes to remove the cell wall, producing protoplasts. These plant cells, devoid of the cell wall, can provide a useful system for introducing foreign DNA into the cell (transformation).

- Anther cultures can produce haploid plants or plants that are homozygous for all traits.

- Plant tissue culture has a number of practical applications, including the industrial production of plant chemicals, clonal propagation of horticultural plants, and the generation of disease-free plants.

- Growth of plant cells in tissue culture over an extended time results in an increase in genetic variation called somaclonal variation. This variation can extend beyond the traits present in the gene pool and can generate novel genetic variations in breeding studies.

40.3 Plant genetic engineering and biotechnology now affect every aspect of agriculture.

- Genetic engineering and biotechnology can be utilized to improve the quality of food crops, increase disease resistance, and improve the tolerance of crops to environmental stress.

- A key aspect of plant genetic engineering is the introduction of foreign DNA into plant cells. This can be achieved using a particle gun or electroporation.

- Using genetic engineering, scientists have modified the oil content and amino acid composition in crop plants. Current work is underway to engineer plants to contain "edible vaccines" against human pathogens.

Discussing Key Terms

1. **Synteny**   Synteny, the analysis of conserved sequence arrangements of genes in related species, can indicate the degree to which species are closely related or divergent.

2. **AFLPs**   Amplified fragment length polymorphisms are generated by hybridizing DNA primers to genomic DNA which has been cleaved with restriction endonuclease and then amplified using the polymerase chain reaction (PCR). The resulting PCR products are then separated by electrophoresis and generally show a greater degree of polymorphism than RFLPs do (see chapter 18).

3. **Callus**   A callus is an unorganized mass of plant cells, resembling a tumor, grown in cell culture. The cells originate from an explant which is a section of plant tissue used to initiate the cell culture.

4. **Somatic cell embryogenesis**   This process allows plant cells in suspension cultures to be directed to form individual embryos. These embryos can grow into a whole plant under appropriate culture conditions.

5. **Plant protoplasts**   Protoplasts are plant cells that have been treated with cell-wall-degrading enzymes to leave only the cell bounded by the plasma membrane. Within several hours after their isolation, protoplasts will regenerate their cell wall and divide to form a callus if transferred to a growth medium.

6. **Somaclonal variation**   If plant cells are grown in culture over extended periods, they generate increased genetic variation called somaclonal variation.

Review Questions

1. Why are plant genomes more complex than the genomes of other eukaryotic cells? What types of arrangements are possible for highly repetitive DNA in plant genomes? Describe mechanisms for the generation of highly repetitive DNA in plants.

2. What are transposable elements, and how do they affect genes located at their site of incorporation?

3. What characteristics of *Arabidopsis thaliana* make it useful as a model system in genetic studies and for the sequencing of its entire genome? Why is rice useful as a model system for the analysis of the genome of a monocot plant?

4. What does the term *totipotency* mean? Why is this property of plant cells important for genetic engineering of plants?

5. Describe how whole plants can be regenerated from tissue-cultured plant cells using either organogenesis or somatic cell embryogenesis. Which approach requires the use of suspension cell cultures?

6. How are plant protoplasts generated, and what is protoplast fusion? How can plant protoplasts be used to generate hybrid plants that would not occur in nature?

7. Describe how the particle gun and electroporation can be used to introduce foreign DNA into plant cells. Which approach requires the use of plant protoplasts? Why?

8. How can a plant be "engineered" to produce an edible vaccine for a human disease?

Thought Questions

1. How does a transposable element generate a mutation for a particular trait in plants? If the sequence of the transposable element is known, how can it be used to identify the location of the gene involved in the particular trait being mutated?

2. In soybeans, a disease known as "brown stem rot" causes substantial losses in yield. The disease is caused by a toxin produced by a fungus that infects the plant. Describe how plant tissue culture could be used to generate a soybean plant resistant to brown stem rot.

3. In the horticultural industry, chrysanthemums are often mass produced using plant tissue culture. Why is it important to only grow cells in culture for a limited time prior to plant regeneration if the objective is to generate clonal copies of a superior chrysanthemum "parent" plant? What effect might it have on the quality of chrysanthemum flowers if the cell culture is grown over a longer period?

Internet Links

Basic Botany Images
http://ike.engr.washington.edu/software/eduimg/botany.html
A collection of plant images from Education Interactive Imaging.

Plants That Eat Animals
http://www.novaproj.org/~sambo/carnivor.html
A NOVA project, this site devoted to CARNIVOROUS PLANTS provides a wealth of information, links, and a carnivorous plant FAQ.

Plant Biotech
http://biotech.chem.indiana.edu
This rich BIOTECH site has a thousand pages crammed with primers and links to diverse topics in biotechnology, including CYBERBOTANICA and an 8200-entry, hyperlinked life sciences dictionary.

Meet the Biomolecules
http://www.kenyon.edu/depts/bmb/chime/gallery.htm
Bring biomolecules to life at the ONLINE MACROMOLECULAR MUSEUM, which features tutorials on such subjects as how the "DNA clamp" of the DNA polymerase molecule allows it to slide along double-helical DNA. You can grab and turn biomolecules or zoom in. A fun learning site.

For Further Reading

Helentjaris, T.: "Implications for Conserved Genomic Structure Among Plant Species," *Proceedings of the National Academy of Sciences*, vol. 90, 1993, pages 8308–9. This article considers the importance and usefulness of synteny in mapping the genomes of related crop species.

Meinke, J., and others: "*Arabidopsis thaliana:* A model Plant for Genome Analysis," *Science*, vol. 282, October 1998, pages 662–681. A good overview of progress in sequencing the 20,000 genes of this widely-studied plant, accompanied by a detailed genomic map.

Raubeson, L. A. and R. K. Jansen: "Chloroplast DNA Evidence on the Ancient Evolutionary Split in Vascular Land Plants," *Science*, vol. 255, 1992, pages 1697–99. These authors seek to answer the question regarding which group of plants represents the living representations of the earliest diverging lineage in vascular plant evolution.

Snow, A. A. and P. M. Palma: "Commercialization of Transgenic Plants: Potential Ecological Risks—Will Evolutionary Effects of Engineered Crops Exacerbate Weed and Pest Problems?" *BioScience*, vol. 47, no. 2, 1997, pages 86–96. Discusses the possible risks associated with key species used in genetically engineered crop production, such as hybridization with populations of free-living relatives.

Various authors: *Trends in Biotechnology*, vol. 13, September, 1995. This volume contains many review articles describing new and anticipated uses of plant genetic engineering, such as "Plants as Bioreactors," "Floriculture: Genetic Engineering of Commercial Traits," and "Transgenic Plants as Vaccine Production Systems."

41

The Noncoelomate Animals

Concept Outline

41.1 Animals are multicellular heterotrophs without cell walls.

Some General Features of Animals.   Animals lack cell walls and move more rapidly and in more complex ways than other organisms.

The Classification of Animals.   The animal kingdom is divided into animals without symmetry and tissues, and animals with symmetry and tissues.

Four Key Transitions in Body Plan.   Over the course of their evolution, animals have undergone significant changes in symmetry, body cavity, degree of segmentation, and mode of development.

41.2 The simplest animals are not bilaterally symmetrical.

The Sponges.   Sponges are the most primitive animals, without either tissues or, for the most part, symmetry.

Radially Symmetrical Animals.   Cnidarians and ctenophorans have radial symmetry and distinct tissues.

41.3 Acoelomates are solid worms that lack a body cavity.

Solid Worms.   Flatworms are the simplest bilaterally symmetrical animals; they lack a body cavity, but possess true organs.

41.4 Pseudocoelomates have a simple body cavity.

The Pseudocoelomates.   Nematodes and rotifers possess a simple body cavity.

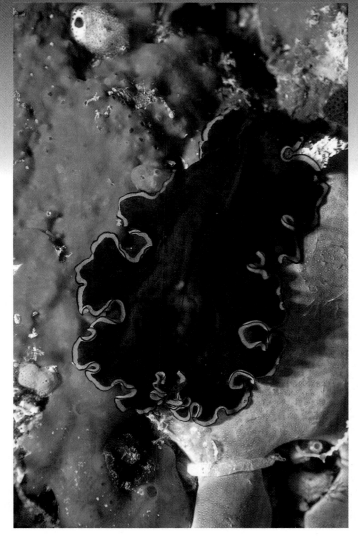

FIGURE 41.1
A noncoelomate: a marine flatworm. The simpler invertebrates, the first to evolve, lack the mesoderm-encased body cavity we call a coelom.

We will now explore the great diversity of animals, the result of a long evolutionary history. Animals, constituting millions of species, are among the most abundant living things. Found in every conceivable habitat, they bewilder us with their diversity. We will start with the simplest members of the animal kingdom—sponges, jellyfish and simple worms. These animals lack a body cavity called a coelom, and are thus called noncoelomates (figure 41.1). The major organization of the animal body first evolved in these animals, a basic body plan upon which all the rest of animal evolution has depended. In chapters 42 through 45, we will consider the more complex animals. Despite their great diversity, you will see that all animals have much in common.

Some General Features of Animals

Animals are the eaters of the earth. All are heterotrophs and depend directly or indirectly for nourishment on plants, photosynthetic protists (algae), or autotrophic bacteria. Many animals are able to move from place to place in search of food (figure 41.2). In most, ingestion of food is followed by digestion in an internal cavity.

Multicellular Heterotrophs. All animals are multicellular heterotrophs. Several decades ago, a biologist would not have made that statement. In recent years, the generally accepted scientific definition of the kingdom Animalia, which comprises animals, has been changed in one respect. The unicellular, heterotrophic organisms called Protozoa, which were at one time regarded as simple animals, are now considered to be members of the kingdom Protista, the large and diverse group we discussed in chapter 31.

Diverse in Form. Almost all animals (99%) are **invertebrates,** animal species that lack a backbone. Of the estimated 10 million living species, only 42,500 have a backbone and are referred to as **vertebrates.** Animals are very diverse in form, ranging in size from ones too small to see with the naked eye to enormous whales and giant squids. The animal kingdom includes about 35 phyla, most of which occur in the sea. Far fewer phyla occur in fresh water and fewer still occur on land. Members of three phyla, Arthropoda (spiders and insects), Mollusca (snails), and Chordata (vertebrates) dominate animal life on land.

No Cell Walls. Animal cells are very diverse in structure and function. They are distinct among multicellular organisms because they lack rigid cell walls and are usually quite flexible. As we shall learn, the cells of all animals but sponges are organized into structural and functional units called **tissues.** A tissue is a collection of cells that have joined together and are specialized to perform a specific function (such as epithelial tissue or muscle tissue).

Active Movement. The ability of animals to move more rapidly and in more complex ways than members of other kingdoms is perhaps their most striking characteristic and one that is directly related to the flexibility of their cells. A remarkable form of movement unique to animals is flying, an ability that is well developed among both insects and vertebrates. Among vertebrates, birds, bats, and pterosaurs (now-extinct flying reptiles) were or are all strong fliers. The only terrestrial vertebrate group never to have had flying representatives is amphibians.

FIGURE 41.2
Animals on the move. These large antelopes, known as wildebeests, move about in large groups consuming grass and finding new areas to feed.

Sexual Reproduction. Most animals reproduce sexually. Animal eggs, which are nonmotile, are much larger than the small, usually flagellated sperm. In animals, cells formed in meiosis function directly as gametes. The haploid cells do not divide by mitosis first, as they do in plants and fungi, but rather fuse directly with each other to form the zygote. Consequently, with a few exceptions, there is no counterpart among animals to the alternation of haploid (gametophyte) and diploid (sporophyte) generations characteristic of plants (see chapter 33).

Embryonic Development. The complex form of a given animal develops from a zygote by a characteristic process of embryonic development. The zygote first undergoes a series of mitotic divisions and becomes a solid ball of cells, the **morula,** then a hollow ball of cells, the **blastula,** a developmental stage that occurs in all animals. In most, the blastula folds inward at one point to form a hollow sac with an opening at one end called the **blastopore.** An embryo at this stage is called a **gastrula.** The subsequent growth and movement of the cells of the gastrula produce the digestive system, also called the gut or intestine. The details of embryonic development differ widely from one phylum of animals to another and often provide important clues to the evolutionary relationships among them. Taken as a whole, this pattern of embryology is characteristic of the animal kingdom.

Animals are complex multicellular organisms typically characterized by mobility and heterotrophy. Most animals also exhibit sexual reproduction and tissue-level organization.

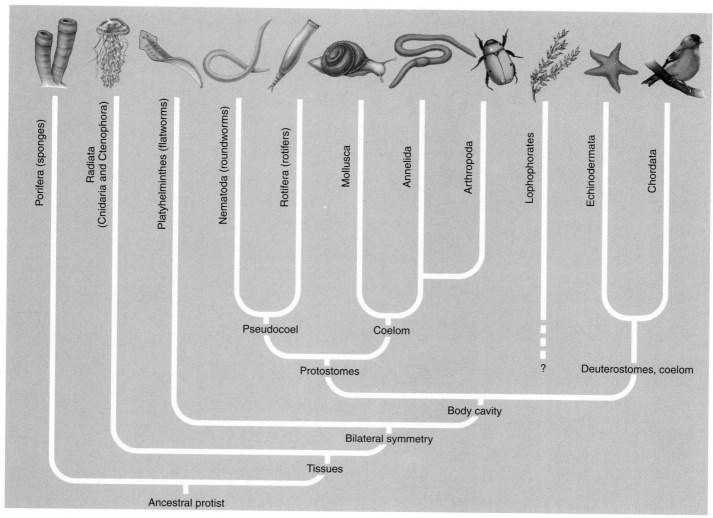

FIGURE 41.3
A possible phylogeny of the major groups of the kingdom Animalia.

The Classification of Animals

Two subkingdoms are generally recognized within the kingdom Animalia: **Parazoa**—animals that for the most part lack a definite symmetry and possess neither tissues nor organs; and **Eumetazoa**—animals that have a definite shape and symmetry and, in most cases, tissues organized into organs and organ systems. The subkingdom Parazoa consists primarily of the sponges, phylum Porifera. The other animals, composing about 35 phyla, belong to the subkingdom Eumetazoa. Because sponges and Eumetazoa have so many features in common, they are grouped as a single kingdom (figure 41.3).

The structure of eumetazoans is much more complex than that of sponges. All eumetazoans form distinct embryonic layers that differentiate into the tissues of the adult animal. The radially symmetrical eumetazoans have two layers, an outer **ectoderm** and an inner **endoderm,** and thus are called **diploblastic.** All other eumetazoans, those that are bilaterally symmetrical, are **triploblastic** and produce a third layer, the **mesoderm,** between the ectoderm and endoderm. No such layers are present in sponges.

The major phyla of animals are listed in table 41.1. The so-called "lower," or "primitive," invertebrates, which have a less complex tissue organization than the "higher" invertebrates, make up about 14 phyla. In this chapter, we will discuss 8 of these 14 phyla and focus in detail on 4 major phyla: phylum Porifera (sponges), which lacks any tissue organization; phylum Cnidaria (radially symmetrical jellyfish, hydroids, sea anemones, and corals); phylum Platyhelminthes (bilaterally symmetrical flatworms); and phylum Nematoda (nematodes), a phylum that includes both free-living and parasitic roundworms.

The animal kingdom is traditionally divided into two subkingdoms. In the simpler subkingdom, Parazoa, the animals lack symmetry and possess neither tissues nor organs. In the more advanced subkingdom, Eumetazoa, the animals are symmetrical and possess tissues.

Table 41.1 The Major Animal Phyla

Phylum	Typical Examples		Key Characteristics	Approximate Number of Named Species
Arthropoda (arthropods)	Beetles, other insects, crabs, spiders		Most successful of all animal phyla; chitinous exoskeleton covering segmented bodies with paired, jointed appendages; many insect groups have wings	1,000,000
Mollusca (mollusks)	Snails, oysters, octopuses, nudibranchs		Soft-bodied coelomates whose bodies are divided into three parts: head-foot, visceral mass, and mantle; many have shells; almost all possess a unique rasping tongue, called a radula; 35,000 species are terrestrial	110,000
Chordata (chordates)	Mammals, fish, reptiles, birds, amphibians		Segmented coelomates with a notochord; possess a dorsal nerve cord, pharyngeal slits, and a tail at some stage of life; in vertebrates, the notochord is replaced during development by the spinal column; 20,000 species are terrestrial	42,500
Platyhelminthes (flatworms)	*Planaria*, tapeworms, liver flukes		Solid, unsegmented, bilaterally symmetrical worms; no body cavity; digestive cavity, if present, has only one opening	20,000
Nematoda (roundworms)	*Ascaris*, pinworms, hookworms, *Filaria*		Pseudocoelomate, unsegmented, bilaterally symmetrical worms; tubular digestive tract passing from mouth to anus; tiny; without cilia; live in great numbers in soil and aquatic sediments; some are important animal parasites	12,000+
Annelida (segmented worms)	Earthworms, polychaetes, beach tube worms, leeches		Coelomate, serially segmented, bilaterally symmetrical worms; complete digestive tract; most have bristles called setae on each segment that anchor them during crawling	12,000

Table 41.1 The Major Animal Phyla (continued)

Phylum	Typical Examples		Key Characteristics	Approximate Number of Named Species
Cnidaria (cnidarians)	Jellyfish, hydra, corals, sea anemones		Soft, gelatinous, radially symmetrical bodies whose digestive cavity has a single opening; possess tentacles armed with stinging cells called cnidocytes that shoot sharp harpoons called nematocysts; almost entirely marine	10,000
Echinodermata (echinoderms)	Sea stars, sea urchins, sand dollars, sea cucumbers		Deuterostomes with radially symmetrical adult bodies; endoskeleton of calcium plates; five-part body plan and unique water vascular system with tube feet; able to regenerate lost body parts; marine	6,000
Porifera (sponges)	Barrel sponges, boring sponges, basket sponges, vase sponges		Asymmetrical bodies without distinct tissues or organs; sac-like body consists of two layers breached by many pores; internal cavity lined with food-filtering cells called choanocytes; most marine (150 species live in fresh water)	5,150
Bryozoa (moss animals)	*Bowerbankia*, *Plumatella*, sea mats, sea moss		Microscopic, aquatic deuterostomes that form branching colonies, possess circular or U-shaped row of ciliated tentacles for feeding called a lophophore that usually protrudes through pores in a hard exoskeleton; also called Ectoprocta because the anus or proct is external to the lophophore; marine or freshwater	4,000
Rotifera (wheel animals)	Rotifers		Small, aquatic pseudocoelomates with a crown of cilia around the mouth resembling a wheel; almost all live in fresh water	2,000

Table 41.1   The Major Animal Phyla (continued)

Phylum	Typical Examples	Key Characteristics	Approximate Number of Named Species
Five phyla of minor worms	Velvet worms, acorn worms, arrow worms, giant tube worms	**Chaetognatha** (arrow worms): coelomate deuterostomes; bilaterally symmetrical; large eyes (some) and powerful jaws **Hemichordata** (acorn worms): marine worms with dorsal *and* ventral nerve cords **Onychophora** (velvet worms): protostomes with a chitinous exoskeleton; evolutionary relics **Pogonophora** (tube worms): sessile deep-sea worms with long tentacles; live within chitinous tubes attached to the ocean floor **Nemertea** (ribbon worms): acoelomate, bilaterally symmetrical marine worms with long extendable proboscis	980
Brachiopoda (lamp shells)	*Lingula*	Like bryozoans, possess a lophophore, but within two clam-like shells; more than 30,000 species known as fossils	250
Ctenophora (sea walnuts)	Comb jellies, sea walnuts	Gelatinous, almost transparent, often bioluminescent marine animals; eight bands of cilia; largest animals that use cilia for locomotion; complete digestive tract with anal pore	100
Phoronida (phoronids)	*Phoronis*	Lophophorate tube worms; often live in dense populations; unique U-shaped gut, instead of the straight digestive tube of other tube worms	12
Loricifera (loriciferans)	*Nanaloricus mysticus*	Tiny, bilaterally symmetrical, marine pseudocoelomates that live in spaces between grains of sand; mouthparts include a unique flexible tube; a recently discovered animal phylum (1983)	6

Four Key Transitions in Body Plan

Radial Versus Bilateral Symmetry

The simplest animals, sponges, mostly lack any definite symmetry, growing asymmetrically as irregular masses. Virtually all other animals have a definite shape and symmetry.

Radial Symmetry. Symmetrical bodies first evolved in marine animals belonging to two primitive phyla whose tissues are not organized into organs, Cnidaria (jellyfish, sea anemones, and corals) and Ctenophora (comb jellies). The bodies of members of these two phyla exhibit **radial symmetry,** a body design in which the parts of the body are arranged around a central axis in such a way that any plane passing through the central axis divides the organism into halves that are approximate mirror images (figure 41.4*a*).

Bilateral Symmetry. The bodies of all other animals are marked by a fundamental **bilateral symmetry,** a body design in which the body has a right and a left half that are mirror images of each other (figure 41.4*b*). A bilaterally symmetrical body plan has a top and a bottom, better known respectively as the *dorsal* and *ventral* portions of the body. It also has a front, or *anterior* end, and a back, or *posterior* end.

Bilateral symmetry constitutes a major evolutionary advance. This unique form of organization allows parts of the body to evolve in different ways, permitting different organs to be located in different parts of the body. In some higher animals like echinoderms (starfish), the adults are radially symmetrical, but even in them the larvae are bilaterally symmetrical.

Bilaterally symmetrical animals move from place to place more efficiently than radially symmetrical ones, which, in general, lead a sessile or passively floating existence. Bilaterally symmetrical animals are therefore efficient in seeking food and mates and avoiding predators.

During the early evolution of bilaterally symmetrical animals, structures that were important to the organism in monitoring its environment, and thereby capturing prey or avoiding enemies, came to be grouped at the anterior end. Other functions tended to be located farther back in the body. The number and complexity of sense organs are much greater in bilaterally symmetrical animals than they are in radially symmetrical ones.

Much of the nervous system in bilaterally symmetrical animals is in the form of major longitudinal nerve cords. In a very early evolutionary advance, nerve cells became grouped around the anterior end of the body. These nerve cells probably first functioned mainly to transmit impulses from the anterior sense organs to the rest of the nervous system. This trend ultimately led to the evolution of a definite head and brain area, a process called **cephalization,** as well as to the increasing dominance and specialization of these organs in the more advanced animal phyla.

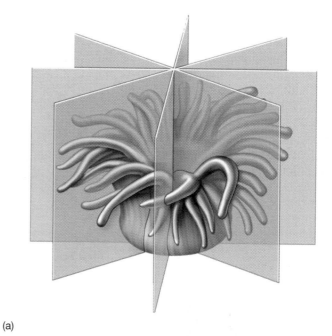

(a)

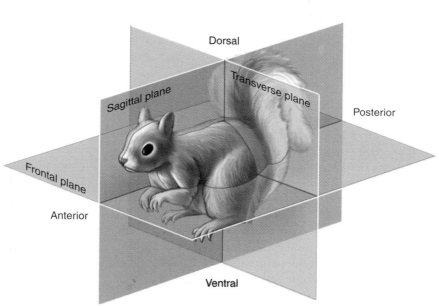

(b)

FIGURE 41.4

A comparison of radial and bilateral symmetry. (a) Radially symmetrical animals, such as this sea anemone, can be bisected into equal halves in any two-dimensional plane. (b) Bilaterally symmetrical animals, such as this squirrel, can only be bisected into equal halves in one plane (the sagittal plane).

Pseudocoelom versus Coelom

A second key transition in the evolution of the animal body plan was the evolution of the body cavity. The evolution of efficient organ systems within the animal body depended critically upon a body cavity for supporting organs, distributing materials, and fostering complex developmental interactions.

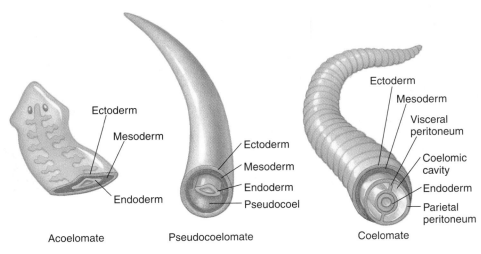

Ectoderm

Mesoderm

Endoderm

Acoelomate

Ectoderm

Mesoderm

Endoderm

Pseudocoel

Pseudocoelomate

Ectoderm

Mesoderm

Visceral peritoneum

Coelomic cavity

Endoderm

Parietal peritoneum

Coelomate

FIGURE 41.5
Three body plans for bilaterally symmetrical animals.

Importance of a Body Cavity. The presence of a body cavity allows the dramatic expansion and sometimes lengthening of portions of the digestive tract. This longer passage allows for storage of undigested food, longer exposure to enzymes for more complete digestion, and even storage and final processing of food remnants. Such an arrangement allows an animal to eat a great deal when it is safe to do so and then to hide during the digestive process, thus limiting the animal's exposure to predators. The tube within the body cavity architecture is also more flexible, thus allowing the animal greater freedom to move.

An internal body cavity also provides space within which the gonads (ovaries and testes) can expand, allowing the accumulation of large numbers of eggs and sperm. Such accumulation helps to make possible all of the diverse modifications of breeding strategy that characterize the more advanced phyla of animals. Furthermore, large numbers of gametes can be released when the conditions are as favorable as possible for the survival of the young animals.

Kinds of Body Cavities. Three basic kinds of body plans are found in bilaterally symmetrical animals. **Acoelomates** have no body cavity. **Pseudocoelomates** have a body cavity called the **pseudocoel** located between the mesoderm and endoderm. A third way of organizing the body is one in which the fluid-filled body cavity develops not between endoderm and mesoderm, but rather entirely within the mesoderm. Such a body cavity is called a **coelom,** and animals that possess such a cavity are called **coelomates.** In coelomates, the gut is suspended, along with other organ systems of the animal, within the coelom; the coelom, in turn, is surrounded by a layer of epithelial cells entirely derived from the mesoderm. The portion of the epithelium that lines the outer wall of the coelom is called the **parietal peritoneum,** and the portion that covers the internal organs suspended within the cavity is called the **visceral peritoneum** (figure 41.5).

The development of the coelom poses a problem—circulation—solved in pseudocoelomates by churning the fluid within the body cavity. In coelomates, the gut is again surrounded by tissue that presents a barrier to diffusion, just as it was in solid worms. This problem is solved among coelomates by the development of a **circulatory system,** a network of vessels that carry fluids to parts of the body. The circulating fluid, or blood, carries nutrients and oxygen to the tissues and removes wastes and carbon dioxide. Blood is usually pushed through the circulatory system by contraction of one or more muscular hearts. In an **open circulatory system,** the blood passes from vessels into sinuses, mixes with body fluid, and then re-enters the vessels later in another location. In a **closed circulatory system,** the blood is separate from the body fluid and can be separately controlled. Also, blood moves through a closed circulatory system faster and more efficiently than it does through an open system.

The evolutionary relationship among coelomates, pseudocoelomates, and acoelomates is not clear. Acoelomates, for example, could have given rise to coelomates, but scientists also cannot rule out the possibility that acoelomates were derived from coelomates. The different phyla of pseudocoelomates form two groups that do not appear to be closely related.

Advantages of a Coelom. What is the functional difference between a pseudocoel and a coelom? The answer has to do with the nature of animal embryonic development. In animals, development of specialized tissues involves a process called **primary induction** in which one of the three primary tissues (endoderm, mesoderm, and ectoderm) interacts with another. The interaction requires physical contact. A major advantage of the coelomate body plan is that it allows contact between mesoderm and endoderm, so that primary induction can occur during development. For example, contact between mesoderm and endoderm permits localized portions of the digestive tract to develop into complex, highly specialized regions like the stomach. In pseudocoelomates, mesoderm and endoderm are separated by the body cavity, limiting developmental interactions between these tissues.

Nonsegmented versus Segmented Bodies

The third key transition in animal body plan involved the subdivision of the body into **segments.** Just as it is efficient for workers to construct a tunnel from a series of identical prefabricated parts, so segmented animals are "assembled" from a succession of identical segments. During the animal's early development, these segments become most obvious in the mesoderm but later are reflected in the ectoderm and endoderm as well. Two advantages result from early embryonic segmentation:

1. In annelids and other highly segmented animals, each segment may go on to develop a more or less complete set of adult organ systems. Damage to any one segment need not be fatal to the individual, since the other segments duplicate that segment's functions.

2. Locomotion is far more effective when individual segments can move independently because the animal as a whole has more flexibility of movement. Because the separations isolate each segment into an individual hydrostatic skeletal unit, each is able to contract or expand autonomously. Therefore, a long body can move in ways that are often quite complex. When an earthworm crawls on a flat surface, it lengthens some parts of its body while shortening others. The elaborate and obvious segmentation of the annelid body clearly represents an evolutionary specialization. Many scientists believe that segmentation is an adaptation for burrowing; it makes possible the production of strong peristaltic waves along the length of the worm, thus enabling vigorous digging.

Segmentation underlies the organization of all advanced animals. In some adult arthropods the segments are fused, but segmentation is usually apparent in their embryological development. In vertebrates, the backbone and muscular areas are segmented.

Protostomes versus Deuterostomes

Two outwardly dissimilar large phyla, Echinodermata (starfish) and Chordata (vertebrates), together with two smaller phyla, have a series of key embryological features different from those shared by the other animal phyla. Because it is extremely unlikely that these features evolved more than once, it is believed that these four phyla share a common ancestry. They are the members of a group called the **deuterostomes.** Members of the other coelomate animal phyla are called **protostomes.** Deuterostomes evolved from protostomes more than 630 million years ago.

Deuterostomes, like protostomes, are coelomates. They differ fundamentally from protostomes, however, in the way in which the embryo grows. Early in embryonic growth, when the embryo is a hollow ball of cells, a portion invaginates inward to form an opening called the blastopore. The blastopore of a protostome becomes the animal's mouth, and the anus develops at the other end. In a deuterostome, by contrast, the blastopore becomes the animal's anus, and the mouth develops at the other end (figure 41.6).

Deuterostomes differ in many other aspects of embryo growth, including the plane in which the cells divide. Perhaps most importantly, the cells that make up an embryonic protostome each contain a different portion of the regulatory signals present in the egg, so no one cell of the embryo (or adult) can develop into a complete organism. In marked contrast, any of the cells of a deuterostome can develop into a complete organism.

> **Four key transitions in body design are responsible for most of the differences we see among the major animal phyla: (1) radial to bilateral body symmetry; (2) no body cavity to possessing a body cavity; (3) unsegmented to segmented bodies; and (4) protostome to deuterostome development.**

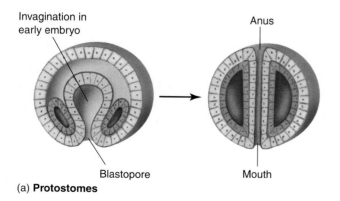

(a) **Protostomes**

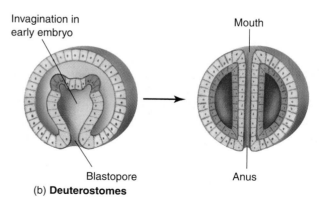

(b) **Deuterostomes**

FIGURE 41.6
The fate of the blastopore. (a) In protostomes, the blastopore becomes the animal's mouth. (b) In deuterostomes, the blastopore becomes the animal's anus.

The Sponges

There are perhaps 5000 species of marine sponges, phylum Porifera (figure 41.7) and about 150 species that live in fresh water. In the sea, sponges are abundant at all depths. Although some sponges are tiny, no more than a few millimeters across, some, like the loggerhead sponges, may reach 2 meters or more in diameter. A few small ones are radially symmetrical, but most members of this phylum completely lack symmetry. Many sponges are colonial. Some have a low and encrusting form, while others may be erect and lobed, sometimes in complex patterns. Although larval sponges are free-swimming, adults are **sessile,** or anchored in place. Individual adults remain attached to submerged objects once they settle down.

There is relatively little coordination among sponge cells. A sponge seems to be little more than a mass of cells embedded in a gelatinous matrix, but these cells recognize one another with a high degree of fidelity and are specialized for different functions of the body.

Sponges feed in a unique way. The beating of flagella that line the inside of the sponge draw water in through numerous small pores; the name of the phylum, Porifera, refers to this system of pores. Plankton and other small organisms are filtered from the water, which flows through passageways and eventually is forced out through an **osculum,** a specialized, larger pore (figure 41.8).

Sponges lack specialized tissues and organs (figure 41.9). The basic structure of a sponge can best be understood by examining the form of a young individual. A small, anatomically simple sponge first attaches to a substrate and then grows into a vase-like shape. The walls of the "vase" have three functional layers:

1. Facing into the internal cavity are specialized flagellated cells called **choanocytes,** or collar cells. These cells line either the entire body cavity or, in many large and more complex sponges, specialized chambers.
2. The bodies of sponges are bounded by an outer epithelial layer consisting of flattened cells somewhat like those that make up the epithelia, or outer layers,

FIGURE 41.7
Barrel sponge. This large sponge has a somewhat organized form.

FIGURE 41.8
Aplysina longissima. This beautiful, bright blue and yellow elongated sponge is found on deep regions of coral reefs. The oscula are ringed with yellow.

of other animal phyla. Some portions of this layer contract when touched or exposed to appropriate chemical stimuli, and this contraction may cause some of the pores to close.
3. Between these two layers, sponges consist mainly of a gelatinous, protein-rich matrix called the **mesohyl,** within which various types of amoeboid cells occur. In addition, many kinds of sponges have minute needles of calcium carbonate or silica known as **spicules,** or fibers of a tough protein called **spongin,** or both, within this matrix. Spicules and spongin strengthen the bodies of the sponges in which they occur. A spongin skeleton is the model for the bathtub sponge, once the skeleton of a real animal, but now largely known from its cellulose and plastic mimics.

The Choanocyte

Each choanocyte closely resembles a protist with a single flagellum, a similarity that reflects its evolutionary derivation. The beating of the flagella of the many choanocytes that line the body cavity draws water in through the pores and through the sponge, thus bringing in food and oxygen and expelling wastes. Each choanocyte flagellum beats independently, and the pressure they create collectively in the cavity forces water out of the osculum. In some sponges, the inner wall of the body cavity is highly convoluted, increasing the surface area and, therefore, the number of flagella that can drive the water. In such a sponge, 1 cubic centimeter of sponge can propel more than 20 liters of water a day.

Microscopic examination of individual choanocytes reveals an important substructure: the base of each flagellum is surrounded by a collar of small, cytoplasmic projections, analogous to a picket fence. The strands that make up this collar are connected to one another by delicate microfibrils. The beating flagellum of the choanocyte draws water, and any food particles in it, through the openings between these microfibrils. The flagellum then forces the water out through the open top of the collar. Any food particles that enter the collar are trapped in the microfibrils at the base of the flagellum. The food of a sponge is about four-fifths particulate organic matter and one-fifth small organisms. Once trapped

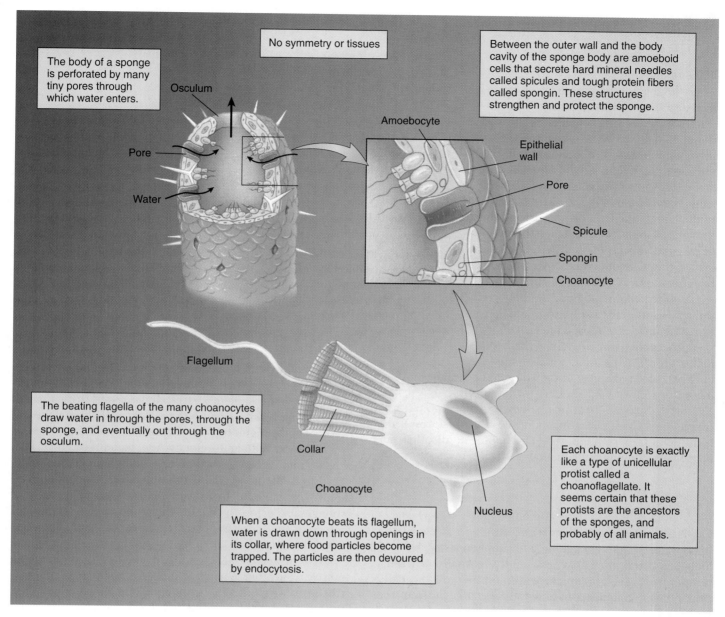

No symmetry or tissues

The body of a sponge is perforated by many tiny pores through which water enters.

Between the outer wall and the body cavity of the sponge body are amoeboid cells that secrete hard mineral needles called spicules and tough protein fibers called spongin. These structures strengthen and protect the sponge.

Osculum

Pore

Water

Amoebocyte

Epithelial wall

Pore

Spicule

Spongin

Choanocyte

Flagellum

The beating flagella of the many choanocytes draw water in through the pores, through the sponge, and eventually out through the osculum.

Collar

Choanocyte

Nucleus

Each choanocyte is exactly like a type of unicellular protist called a choanoflagellate. It seems certain that these protists are the ancestors of the sponges, and probably of all animals.

When a choanocyte beats its flagellum, water is drawn down through openings in its collar, where food particles become trapped. The particles are then devoured by endocytosis.

FIGURE 41.9
The simplest multicellular animals lack tissues. The body of a sponge is multicellular, containing several distinctly different types of cells whose activities are loosely coordinated with one another. The sponge body is not symmetrical and has no organized tissues.

at the collar, this food is digested either by the choanocyte itself or by a neighboring amoeboid cell.

Reproduction in Sponges

Some sponges will re-form themselves once they have passed through a silk mesh. Thus, as you might suspect, sponges frequently reproduce by simply breaking into fragments. If a sponge breaks up, the resulting fragments usually are able to reconstitute whole new individuals. Sexual reproduction is also exhibited by sponges, with some mature individuals producing eggs and sperm. Larval sponges may undergo their initial stages of development within the

parent. They have numerous external, flagellated cells and are free-swimming. After a short planktonic stage, they settle down on a suitable substrate, where they begin their transformation into adults.

Sponges probably represent the most primitive animals, possessing neither tissue-level development nor body symmetry. Their cellular organization hints at the evolutionary ties between the unicellular protists and the multicellular animals. Sponges are unique in the animal kingdom in possessing choanocytes, special flagellated cells whose beating drives water through the body cavity.

Radially Symmetrical Animals

The subkingdom Eumetazoa contains animals that have a definite shape and symmetry and have distinct tissues. Two distinct cell layers form in the embryos of these animals: an outer ectoderm and an inner endoderm. These embryonic tissues give rise to the basic body plan, differentiating into the many tissues of the adult body. In general, the outer covering of the body (called the epidermis) and the nervous system develop from the ectoderm, and the layer of digestive tissue (called the **gastrodermis**) develops from the endoderm. A layer of gelatinous material, called the **mesoglea,** lies between the epidermis and gastrodermis and contains the muscles in most eumetazoans.

Eumetazoans are divided into two major groups. The first, which includes only two phyla, consists of radially symmetrical organisms. These two phyla are Cnidaria (pronounced ni-DAH-ree-ah), or cnidarians—hydroids, jellyfish, sea anemones, and corals—and Ctenophora (pronounced tea-NO-fo-rah), the comb jellies, or ctenophores. The bodies of all other eumetazoans exhibit a fundamental bilateral symmetry.

The Cnidarians

Cnidarians are nearly all marine, although a few live in fresh water. These fascinating and simply constructed animals are often abundant, especially in shallow, warm-temperature or subtropical waters. They are basically gelatinous in composition. They differ markedly from the sponges in organization; their bodies are made up of distinct tissues, although they lack organs. These animals are carnivores. For the most part, they do not actively move from place to place, but rather capture their prey (which includes fishes, crustaceans, and many other kinds of animals) with the tentacles that ring their mouths.

Cnidarians may have two basic body forms, polyps and medusae (figure 41.10). Polyps are cylindrical and are usually found attached to a firm substrate. They may be solitary or colonial. In a polyp, the mouth faces away from the substrate on which the animal is growing, and, therefore, often faces upward. Many polyps build up a chitinous or calcareous (made up of calcium carbonate) external or internal skeleton, or both. Only a few polyps are free-floating. In contrast, most medusae are free-floating and are often umbrella-shaped. Their mouths usually point downward, and the tentacles hang down around them. Medusae, particularly those of the class Scyphozoa, are commonly known as jellyfish because their mesoglea is thick and jelly-like.

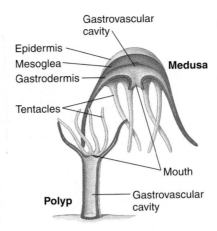

FIGURE 41.10
Two body forms of cnidarians, the medusa (*above*) and the polyp (*below*). These two phases alternate in the life cycles of many cnidarians, but a number—including the corals and sea anemones, for example—exist only as polyps.

Many cnidarians occur only as polyps, while others exist only as medusae; still others alternate between these two phases during the course of their life cycles. Both phases consist of diploid individuals. Polyps may reproduce asexually by budding; if they do, they may produce either new polyps or medusae. Medusae are produced in specialized buds.

Medusae reproduce sexually, and in most cnidarians, fertilized eggs give rise to free-swimming, multicellular, ciliated larvae known as **planulae.** Planulae are common in the plankton at times and may be dispersed widely in the currents.

A major evolutionary innovation in cnidarians, compared with sponges, is the internal extracellular digestion of food (figure 41.11). Digestion takes place within a gut cavity, rather than only within individual cells. Digestive enzymes (primarily proteases) are released from cells lining the walls of the cavity and partially break down food. Cnidarian digestion is not completely extracellular. Food is fragmented extracellularly into small bits that the cells lining the gut subsequently engulf by phagocytosis.

The extracellular fragmentation that precedes phagocytosis and intracellular digestion allows cnidarians to digest animals larger than individual cells, an important improvement over the strictly intracellular digestion of sponges. Any food particles that cannot be digested are released out the same opening through which they entered the animal.

Nets of nerve cells coordinate contraction of cnidarian muscles. Such nerve nets apparently have little central control in most cnidarians, although some, such as jellyfish, display a degree of coordination. Cnidarians have no blood vessels, no respiratory system, and no excretory organs.

On their tentacles and sometimes on their body surface, cnidarians bear specialized cells called **cnidocytes.** The name of the phylum Cnidaria refers to these cells, which are highly distinctive and occur in no other group of organisms. Within each cnidocyte is a **nematocyst,** a small but powerful "harpoon." Each nematocyst features a coiled, thread-like tube. Lining the inner wall of the tube is a series of barbed spines. Cnidarians use the thread-like tube to spear their prey and then draw the harpooned prey back with the tentacle containing the cnidocyte. Nematocysts may also serve a defensive purpose. To propel the harpoon, the cnidocyte uses water pressure. Before firing, the cnidocyte builds up a very high internal osmotic pressure. This is done by using active transport to build a high concentration of ions inside, while keeping its wall impermeable to water. Within the undischarged nematocyst, osmotic pressure reaches about 140 atmospheres.

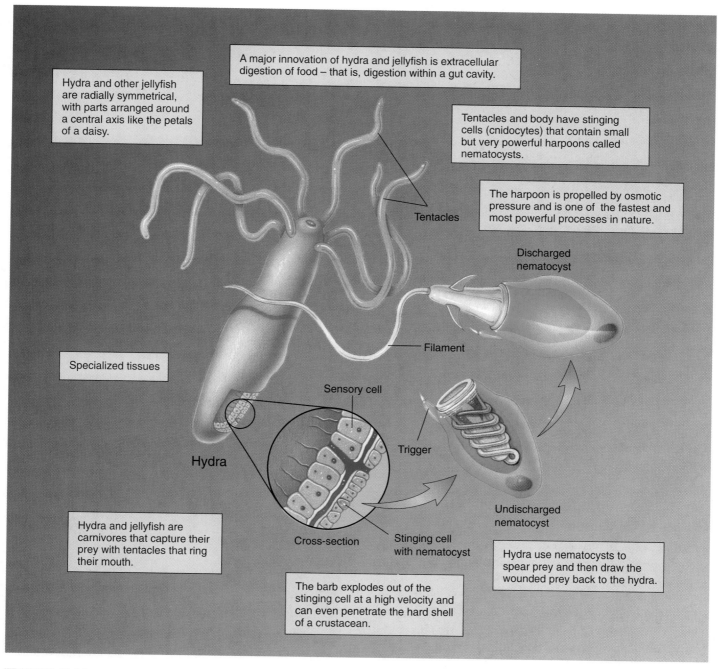

Hydra and other jellyfish are radially symmetrical, with parts arranged around a central axis like the petals of a daisy.

A major innovation of hydra and jellyfish is extracellular digestion of food – that is, digestion within a gut cavity.

Tentacles and body have stinging cells (cnidocytes) that contain small but very powerful harpoons called nematocysts.

The harpoon is propelled by osmotic pressure and is one of the fastest and most powerful processes in nature.

Tentacles

Discharged nematocyst

Filament

Specialized tissues

Sensory cell

Trigger

Hydra

Cross-section

Stinging cell with nematocyst

Undischarged nematocyst

Hydra and jellyfish are carnivores that capture their prey with tentacles that ring their mouth.

Hydra use nematocysts to spear prey and then draw the wounded prey back to the hydra.

The barb explodes out of the stinging cell at a high velocity and can even penetrate the hard shell of a crustacean.

FIGURE 41.11

Eumetazoans all have symmetry and tissues. The cnidarian body plan is more complex than that of sponges. The cells of a cnidarian like this *Hydra* are organized into specialized tissues. The interior gut cavity is specialized for extracellular digestion—that is, digestion within a gut cavity rather than within individual cells. Unlike sponges, cnidarians are radially symmetrical, with parts arranged around a central axis like the petals of a daisy.

If a flagellum-like trigger on the cnidocyte is touched (and certain chemical stimuli are also present), the nematocyst is stimulated to discharge. Its walls become permeable to water, which rushes inside and violently pushes out the barbed filament. As the filament flies outward, it turns inside out, exposing the barbs. Nematocyst discharge is one of the fastest cellular processes in nature. The nematocyst is pushed outward so explosively that the barb can penetrate even the hard shell of a crab. The stinging sensation associated with the discharge of the nematocysts results from the injection of a toxic protein. Some cnidarians are called "stinging nettles" because of their cnidocytes.

The powerful neurotoxins (nerve poisons) that are secreted by the nematocysts of the Portuguese man-of-war, a large, floating, colonial, marine cnidarian, are dangerous and occasionally even fatal to humans. A number of species of jellyfish also have exceedingly painful stings; the box jelly (*Chironex fleckeri*) has been responsible for the deaths of at least 60 people in northern Australia alone.

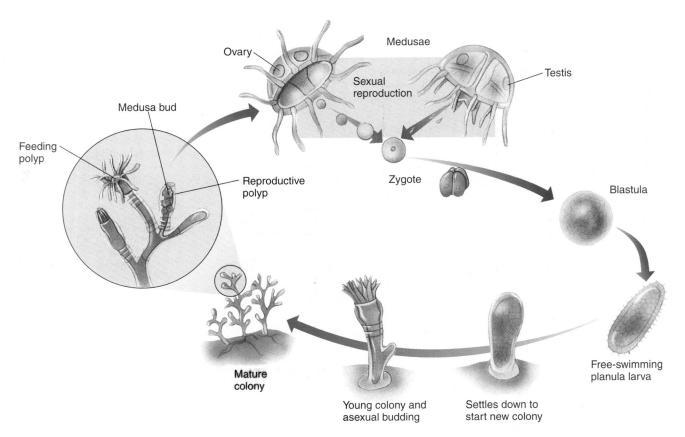

FIGURE 41.12
The life cycle of *Obelia*, a marine colonial hydroid. Polyps reproduce by asexually budding, forming colonies. They may also give rise to medusae, which produce gametes. These gametes fuse, producing zygotes that develop into planulae, which, in turn, settle down to produce polyps.

Classes of Cnidarians

There are four classes of cnidarians: Hydrozoa (hydroids), Scyphozoa (jellyfish), Cubozoa (box jellyfish), and Anthozoa (anemones and corals).

Class Hydrozoa: The Hydroids. Most of the approximately 2700 species of hydroids (class Hydrozoa) have both polyp and medusa stages in their life cycle (figure 41.12). Most of these animals are marine and colonial, such as Obelia and the already mentioned, very unusual Portuguese man-of-war. Some of the marine hydroids are bioluminescent.

A well-known hydroid is the abundant freshwater genus *Hydra*, which is exceptional in that it has no medusa stage and exists as a solitary polyp. Each polyp sits on a basal disk, which it can use to glide around, aided by mucous secretions. It can also move by somersaulting, bending over and attaching itself to the substrate by its tentacles, and then looping over to a new location. If the polyp detaches itself from the substrate, it can float to the surface.

Class Scyphozoa: The Jellyfish. The approximately 200 species of jellyfish (class Scyphozoa) are transparent or translucent marine organisms, some of a striking orange, blue, or pink color (figure 41.13). These animals spend

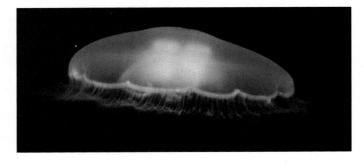

FIGURE 41.13
Class Scyphozoa. Jellyfish, *Aurelia aurita*.

most of their time floating near the surface of the sea. In all of them, the medusa stage is dominant—much larger and more complex than the polyps. The medusae are bell-shaped, with hanging tentacles around their margins. The polyp stage is small, inconspicuous, and simple in structure.

The outer layer, or epithelium, of a jellyfish contains a number of specialized epitheliomuscular cells, each of which can contract individually. Together, the cells form a muscular ring around the margin of the bell that pulses rhythmically and propels the animal through the water. Jellyfish have separate male and female individuals. After fertilization, planulae form, which then attach and develop into polyps. The polyps can reproduce asexually as well as budding off medusae. In some jellyfish that live in the open ocean, the polyp stage is suppressed, and planulae develop directly into medusae.

Class Cubozoa: The Box Jellyfish. Until recently the cubozoa were considered an order of Scyphozoa. As their name implies, they are box-shaped medusa (the polyp stage is inconspicuous and in many cases not known). Most are only a few cm in height, although some are 25 cm tall. A tentacle or group of tentacles is found at each corner of the box (figure 41.14). Box jellies are strong swimmers and voracious predators of fish. Stings of some species can be fatal to humans.

Class Anthozoa: The Sea Anemones and Corals. By far the largest class of cnidarians is Anthozoa, the "flower animals" (from the Greek *anthos*, meaning "flower"). The approximately 6200 species of this group are solitary or colonial marine animals. They include stone-like corals, soft-bodied sea anemones, and other groups known by such fanciful names as sea pens, sea pansies, sea fans, and sea whips (figure 41.15). All of these names reflect a plant-like body topped by a tuft or crown of hollow tentacles. Like other cnidarians, anthozoans use these tentacles in feeding. Nearly all members of this class that live in shallow waters harbor symbiotic algae, which supplement the nutrition of their hosts through photosynthesis. Fertilized eggs of anthozoans usually develop into planulae that settle and develop into polyps; no medusae are formed.

Sea anemones are a large group of soft-bodied anthozoans that live in coastal waters all over the world and are especially abundant in the tropics. When touched, most sea anemones retract their tentacles into their bodies and fold up. Sea anemones are highly muscular and relatively complex organisms, with greatly divided internal cavities. These animals range from a few millimeters to about 10 centimeters in diameter and are perhaps twice that high.

FIGURE 41.14
Class Cubozoa. Box jelly, *Chironex fleckeri.*

FIGURE 41.15
Class Anthozoa. The sessile soft-bodied sea anemone.

Corals are another major group of anthozoans. Many of them secrete tough outer skeletons, or exoskeletons, of calcium carbonate and are thus stony in texture. Others, including the gorgonians, or soft corals, do not secrete exoskeletons. Some of the hard corals help form coral reefs, which are shallow-water limestone ridges that occur in warm seas. Although the waters where coral reefs develop are often nutrient-poor, the coral animals are able to grow actively because of the abundant algae found within them.

The Ctenophorans (Comb Jellies)

The members of this small phylum range from spherical to ribbon-like and are known as comb jellies or sea walnuts. Traditionally, the roughly 90 marine species of ctenophores (phylum Ctenophora) were considered closely related to the cnidarians. However, ctenophores are structurally more complex than cnidarians. They have anal pores, so that water and other substances pass completely through the animal. Comb jellies, abundant in the open ocean, are transparent and usually only a few centimeters long. The members of one group have two long, retractable tentacles that they use to capture their prey.

Ctenophores propel themselves through water with eight comb-like plates of fused cilia that beat in a coordinated fashion (figure 41.16). They are the largest animals that use cilia for locomotion. Many ctenophores are bioluminescent, giving off bright flashes of light particularly evident in the open ocean at night.

Cnidarians and ctenophores are radially symmetrical animals. Cnidarians have a specialized kind of cell called a cnidocyte. Ctenophores propel themselves through the water by means of eight comb-like plates of fused cilia.

FIGURE 41.16
A comb jelly (phylum Ctenophora).
Note the comb-like plates and two tentacles.

Solid Worms

Phylum Platyhelminthes: The Flatworms

All of the higher animals are fundamentally bilaterally symmetrical. Among bilaterally symmetrical animals, different parts of the body are specialized in relation to different functions. Among bilaterally symmetrical animals, those with the simplest body plan are the acoelomates; they lack any internal cavity other than the digestive tract. We will focus our discussion of the acoelomates on the largest phylum of the group, the flatworms.

Phylum Platyhelminthes consists of some 20,000 species. These ribbon-shaped, soft-bodied animals are flattened dorsoventrally, from top to bottom. Flatworms are among the simplest of bilaterally symmetrical animals, but they do have a definite head at the anterior end and they do possess organs. Their bodies are solid: the only internal space consists of the digestive cavity (figure 41.17).

Flatworms range in size from a millimeter or less to many meters long, as in some tapeworms. Most species of flatworms are parasitic, occurring within the bodies of many other kinds of animals (figure 41.18). Other flatworms are free-living, occurring in a wide variety of marine and freshwater habitats, as well as moist places on land. Free-living flatworms are carnivores and scavengers; they eat various small animals and bits of organic debris. They move from place to place by means of ciliated epithelial cells, which are particularly concentrated on their ventral surfaces.

Those flatworms that have a digestive cavity have an incomplete gut, one with only one opening. As a result, they cannot feed, digest, and eliminate undigested particles of food simultaneously, and thus, flatworms cannot feed continuously, as more advanced animals can. Muscular contractions in the upper end of the gut cause a strong sucking force allowing flatworms to ingest their food and tear it into small bits. The gut is branched and extends throughout the body, functioning in both digestion and transport of food. Cells that line the gut engulf most of the food particles by phagocytosis and digest them; but, as in the cnidarians, some of these particles are partly digested extracellularly. Tapeworms, which are parasitic flatworms, lack digestive systems. They absorb their food directly through their body walls.

Unlike cnidarians, flatworms have an excretory system, which consists of a network of fine tubules (little tubes) that runs throughout the body. Cilia line the hollow centers of bulb-like **flame cells,** which are located on the side branches of the tubules. Cilia in the flame cells move water and excretory substances into the tubules and then to exit pores located between the epidermal cells. Flame cells were named because of the flickering movements of the tuft of cilia within them. They primarily regulate the water balance of the organism. The excretory function of flame cells appears to be a secondary one. A large proportion of the metabolic wastes excreted by flatworms probably diffuses directly into the gut and is eliminated through the mouth.

Like sponges, cnidarians, and ctenophorans, flatworms lack circulatory systems for the transport of oxygen and food molecules. Consequently, all flatworm cells must be within diffusion distance of oxygen and food. Flatworms have thin bodies and highly branched digestive cavities that make such a relationship possible.

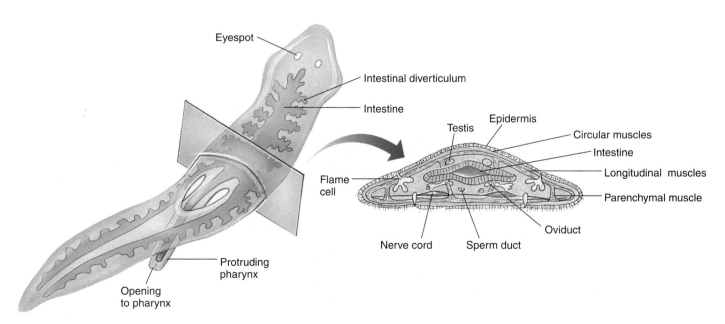

FIGURE 41.17

Architecture of a solid worm. This organism is *Dugesia*, the familiar freshwater "planaria" of many biology laboratories.

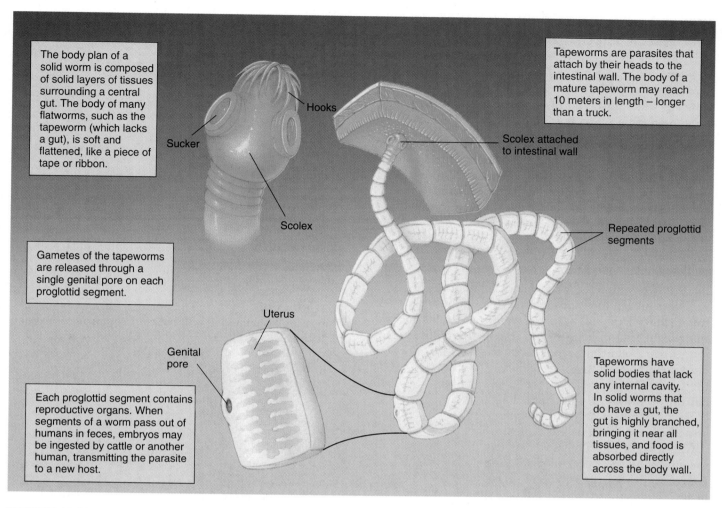

The body plan of a solid worm is composed of solid layers of tissues surrounding a central gut. The body of many flatworms, such as the tapeworm (which lacks a gut), is soft and flattened, like a piece of tape or ribbon.

Tapeworms are parasites that attach by their heads to the intestinal wall. The body of a mature tapeworm may reach 10 meters in length – longer than a truck.

Gametes of the tapeworms are released through a single genital pore on each proglottid segment.

Each proglottid segment contains reproductive organs. When segments of a worm pass out of humans in feces, embryos may be ingested by cattle or another human, transmitting the parasite to a new host.

Tapeworms have solid bodies that lack any internal cavity. In solid worms that do have a gut, the gut is highly branched, bringing it near all tissues, and food is absorbed directly across the body wall.

Hooks

Sucker

Scolex

Scolex attached to intestinal wall

Repeated proglottid segments

Uterus

Genital pore

FIGURE 41.18

The evolution of internal organs and bilateral symmetry. Acoelomate solid worms like this beef tapeworm, *Taenia saginata,* have three embryonic layers. Flatworms were the first bilaterally symmetrical animals with a distinct head.

The nervous system of flatworms is very simple. Like cnidarians, some primitive flatworms have only a nerve net. However, most members of this phylum have longitudinal nerve cords that constitute a simple central nervous system. Between the longitudinal cords are cross-connections, so that the flatworm's nervous system resembles a ladder.

Free-living flatworms use sensory pits or tentacles along the sides of their heads to detect food, chemicals, or movements of the fluid in which they are living. Once they have sensed food, they move directly toward it and begin to feed. Free-living members of this phylum also have eyespots on their heads. These are inverted, pigmented cups containing light-sensitive cells connected to the nervous system. These eyespots enable the worms to distinguish light from dark; worms move away from strong light.

Flatworms are far more active than cnidarians or ctenophores. Such activity is characteristic of bilaterally symmetrical animals. In flatworms, this activity seems to be related to the greater concentration of sensory organs and, to some degree, the nervous system elements in the heads of these animals.

The reproductive systems of flatworms are complex. Most flatworms are **hermaphroditic,** with each individual containing both male and female sexual structures. In many of them, fertilization is internal. When they mate, each partner deposits sperm in the copulatory sac of the other. The sperm travel along special tubes to reach the eggs. In most free-living flatworms, fertilized eggs are laid in cocoons strung in ribbons and hatch into miniature adults. In some parasitic flatworms, there is a complex succession of distinct larval forms. Flatworms are also capable of asexual regeneration. In some genera, when a single individual is divided into two or more parts, each part can regenerate an entirely new flatworm.

Class Turbellaria: The Turbellarians. Only one of the three classes of flatworms, the turbellarians (class Turbellaria) are free-living (figure 41.19*a*). One of the most familiar is the freshwater genus *Dugesia*, the common planaria used in biology laboratory exercises. Other members of this class are widespread and often abundant in lakes, ponds, and the sea. Some also occur in moist places on land.

Class Trematoda: The Flukes. Two classes of parasitic flatworms live within the bodies of other animals: flukes (class Trematoda) and tapeworms (class Cestoda). Both groups of worms have epithelial layers resistant to the digestive enzymes and immune defenses produced by their hosts—an important feature in their parasitic way of life. However, they lack certain features of the free-living flatworms, such as cilia in the adult stage, eyespots, and other sensory organs that lack adaptive significance for an organism that lives within the body of another animal.

Flukes take in food through their mouth, just like their free-living relatives. There are more than 10,000 named species, ranging in length from less than 1 millimeter to more than 8 centimeters. Flukes attach themselves within the bodies of their hosts by means of suckers, anchors, or hooks. Some have a life cycle that involves only one host, usually a fish. Most have life cycles involving two or more hosts. Their larvae almost always occur in snails, and there may be other intermediate hosts. The final host of these flukes is almost always a vertebrate.

To human beings, one of the most important flatworms is the human liver fluke, *Clonorchis sinensis* (figure 41.19*b*). It lives in the bile passages of the liver of humans, cats, dogs, and pigs. It is especially common in Asia. The worms are 1 to 2 centimeters long and have a complex life cycle. Although they are hermaphroditic, cross-fertilization usually occurs between different individuals. Eggs, each containing a complete, ciliated first-stage larva, or **miracidium,** are passed in the feces (figure 41.20). If they reach water, they may be ingested by a snail. Within the snail an egg transforms into a *sporocyst*—a bag-like structure with embryonic germ cells. Within the sporocysts are produced **rediae,** which are elongated, nonciliated larvae. These larvae continue growing within the snail, giving rise to several individuals of the tadpole-like next larval stage, **cercariae.**

Cercariae escape into the water, where they swim about freely. If they encounter a fish of the family Cyprinidae—the family that includes carp and goldfish—they bore into the muscles or under the scales, lose their tails, and transform into **metacercariae** within cysts in the muscle tissue. If a human being or other mammal eats raw infected fish, the cysts dissolve in the intestine, and the young flukes migrate to the bile duct, where they mature. An individual fluke may live for 15 to 30 years in the liver. In humans, a heavy infestation of liver flukes may cause cirrhosis of the liver and death.

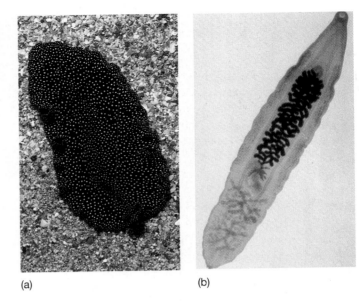

FIGURE 41.19
Flatworms (phylum Platyhelminthes). (a) A marine, free-living turbellarian. (b) The human liver fluke, *Clonorchis sinensis*.

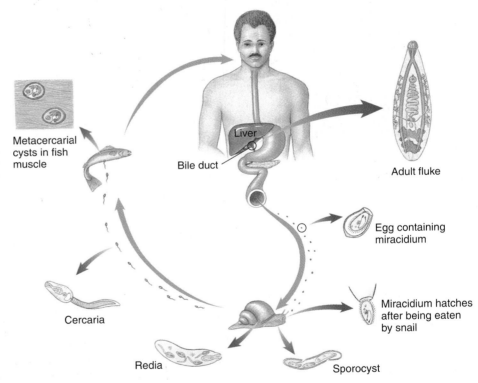

FIGURE 41.20
Life cycle of the human liver fluke, *Clonorchis sinensis*.

Other very important flukes are the blood flukes of genus *Schistosoma*. They afflict about 1 in 20 of the world's population, more than 200 million people throughout tropical Asia, Africa, Latin America, and the Middle East. Three species of *Schistosoma* cause the disease called schistosomiasis, or bilharzia. Some 800,000 people die each year from this disease.

Recently, there has been a great deal of effort to control schistosomiasis. The worms protect themselves in part from the body's immune system by coating themselves with a variety of the host's own antigens that effectively render the worm immunologically invisible (see chapter 54). Despite this difficulty, the search is on for a vaccine that would cause the host to develop antibodies to one of the antigens of the young worms before they protect themselves with host antigens. This vaccine would prevent humans from infection. The disease can be cured with drugs after infection.

Class Cestoda: The Tapeworms.   Class Cestoda is the third class of flatworms; like flukes, they live as parasites within the bodies of other animals. In contrast to flukes, tapeworms simply hang on to the inner walls of their hosts by means of specialized terminal attachment organs and absorb food through their skins. Tapeworms lack digestive cavities as well as digestive enzymes. They are extremely specialized in relation to their parasitic way of life. Most species of tapeworms occur in the intestines of vertebrates, about a dozen of them regularly in humans.

The long, flat bodies of tapeworms are divided into three zones: the **scolex,** or attachment organ; the unsegmented **neck;** and a series of repetitive segments, the **proglottids** (see figure 41.18). The scolex usually bears several suckers and may also have hooks. Each proglottid is a complete hermaphroditic unit, containing both male and female reproductive organs. Proglottids are formed continuously in an actively growing zone at the base of the neck, with maturing ones moving farther back as new ones are formed in front of them. Ultimately the proglottids near the end of the body form mature eggs. As these eggs are fertilized, the zygotes in the very last segments begin to differentiate, and these segments fill with embryos, break off, and leave their host with the host's feces. Embryos, each surrounded by a shell, emerge from the proglottid through a pore or the ruptured body wall. They are deposited on leaves, in water, or in other places where they may be picked up by another animal.

The beef tapeworm *Taenia saginata* occurs as a juvenile in the intermuscular tissue of cattle but as an adult in the intestines of human beings. A mature adult beef tapeworm may reach a length of 10 meters or more. These worms attach themselves to the intestinal wall of their host by a scolex with four suckers. The segments that are shed from the end of the worm pass from the human in the feces and may crawl onto vegetation. The segments ultimately rupture and scatter the embryos. Embryos may remain viable for up to five months. If they are ingested by cattle, they burrow through the wall of the intestine and ultimately reach muscle tissues through the blood or lymph vessels. About 1% of the cattle in the United States are infected, and some 20% of the beef consumed is not federally inspected. When infected beef is eaten rare, infection of humans by these tapeworms is likely. As a result, the beef tapeworm is a frequent parasite of humans.

Phylum Nemertea: The Ribbon Worms

The members of phylum Nemertea (figure 41.21) share many characteristics with free-living flatworms but differ from them in several significant ways. These acoelomate aquatic worms, the majority of which are marine, may be thread-shaped or ribbon-shaped. There are about 900 species. They are characterized by their proboscis, a long muscular tube that can be thrust out quickly from a sheath to capture prey. Ribbon worms are large, often 10 to 20 centimeters and sometimes many meters in length. They are the simplest animals that possess a **complete digestive system,** one that has two separate openings, a mouth and an anus. Ribbon worms also exhibit a circulatory system in which blood flows in vessels. Ribbon worms provide early indications of important evolutionary trends that become fully developed in more advanced animals.

The acoelomates, typified by flatworms, are the most primitive bilaterally symmetrical animals and the simplest animals in which organs occur.

FIGURE 41.21
A ribbon worm, *Lineus* **(phylum Nemertea).** This is the simplest animal with a complete digestive system.

The Pseudocoelomates

All bilaterally symmetrical animals except solid worms possess an internal body cavity. Among them, seven phyla are characterized by their possession of a pseudocoel (see figure 41.5). Only one of the seven, phylum Nematoda, includes a large number of species. In all pseudocoelomates, the pseudocoel serves as a hydrostatic skeleton—one that gains its rigidity from being filled with fluid under pressure. The animals' muscles can work against this "skeleton," thus making the movements of pseudocoelomates far more efficient than those of the acoelomates.

Pseudocoelomates lack a defined circulatory system; this role is performed by the fluids that move within the pseudocoel. Most pseudocoelomates have a complete, one-way digestive tract that acts like an assembly line. Food is first broken down, then absorbed, and then treated and stored.

Phylum Nematoda: The Roundworms

Nematodes, eelworms, and other roundworms constitute a large phylum, Nematoda, with some 12,000 recognized species. Scientists estimate that the actual number might approach 100 times that many. Members of this phylum are found everywhere. Nematodes are abundant and diverse in marine and freshwater habitats, and many members of this phylum are parasites of animals and plants (figure 41.22). Many nematodes are microscopic and live in soil. It has been estimated that a spadeful of fertile soil may contain, on the average, a million nematodes.

Nematodes are bilaterally symmetrical, cylindrical, unsegmented worms (figure 41.23). They are covered by a flexible, thick cuticle, which is molted as they grow. Their muscles constitute a layer beneath the epidermis and extend along the length of the worm, rather than encircling its body. These longitudinal muscles pull both against the cuticle and the pseudocoel, which forms a hydrostatic skeleton. When nematodes move, their bodies whip about from side to side.

Near the mouth of a nematode, at its anterior end, are usually 16 raised, hair-like, sensory organs. The mouth is often equipped with piercing organs called **stylets.** Food

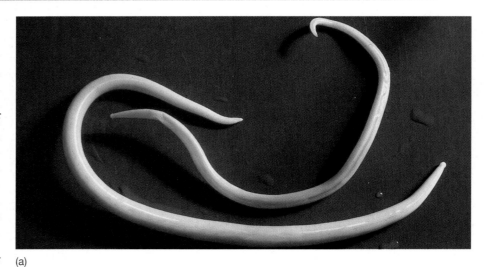

(a)

(b)

(c)

FIGURE 41.22
Nematodes. (a) *Ascaris lumbricoides,* an intestinal roundworm that infects humans and some other animals. Its fertilized eggs pass out with feces and can remain viable in soil for years. In unsanitary conditions, the eggs may be ingested because of improper hand washing. The young worms bore through the wall of the gut, pass through the heart and lungs, and pass out the breathing passage and are swallowed again. Adult females can be up to 30 centimeters long. (b) A pig's intestine opened up to show that it is filled to capacity with *Ascaris.* (c) Soil nematodes (42×). Although most are similar in form, they range from about 0.2 mm to about 6 mm long.

passes through the mouth as a result of the sucking action of a muscular chamber called the **pharynx.** After passing through a short corridor into the pharynx, food continues through the other portions of the digestive tract, where it is broken down and then digested. Some of the water with which the food has been mixed is reabsorbed near the end of the digestive tract, and material that has not been digested is eliminated through the anus.

Nematodes completely lack flagella or cilia, even on sperm cells. Reproduction in nematodes is sexual, with sexes usually separate. Their development is simple, and the adults consist of very few cells. For this reason, nematodes have become extremely important subjects for genetic and developmental studies (see chapter 57). The 1-millimeter-long *Caenorhabditis elegans* matures in only three days, its body is transparent, and it has only 959 cells. It is the only animal whose complete developmental cellular anatomy is known.

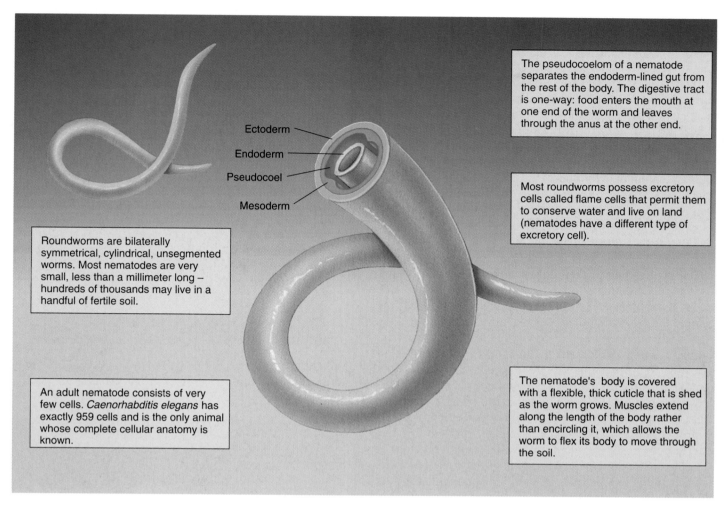

Ectoderm
Endoderm
Pseudocoel
Mesoderm

The pseudocoelom of a nematode separates the endoderm-lined gut from the rest of the body. The digestive tract is one-way: food enters the mouth at one end of the worm and leaves through the anus at the other end.

Most roundworms possess excretory cells called flame cells that permit them to conserve water and live on land (nematodes have a different type of excretory cell).

Roundworms are bilaterally symmetrical, cylindrical, unsegmented worms. Most nematodes are very small, less than a millimeter long – hundreds of thousands may live in a handful of fertile soil.

An adult nematode consists of very few cells. *Caenorhabditis elegans* has exactly 959 cells and is the only animal whose complete cellular anatomy is known.

The nematode's body is covered with a flexible, thick cuticle that is shed as the worm grows. Muscles extend along the length of the body rather than encircling it, which allows the worm to flex its body to move through the soil.

FIGURE 41.23
The evolution of a simple body cavity. The major innovation in body design in roundworms (phylum Nematoda) is a body cavity between the gut and the body wall. This cavity is the pseudocoel. It allows chemicals to circulate throughout the body and prevents organs from being deformed by muscle movements.

About 50 species of nematodes, including several that are rather common in the United States, regularly parasitize human beings. The most serious common nematode-caused disease in temperate regions is trichinosis, caused by worms of the genus *Trichinella*. These worms live in the small intestine of pigs, where fertilized female worms burrow into the intestinal wall. Once it has penetrated these tissues, each female produces about 1500 live young. The young enter the lymph channels and travel to muscle tissue throughout the body, where they mature and form highly resistant, calcified cysts (figure 41.24). Infection in human beings or other animals arises from eating undercooked or raw pork in which the cysts of *Trichinella* are present. If the worms are abundant, a fatal infection can result, but such infections are rare. It is thought that about 2.4% of the people in the United States are infected with trichinosis, but only about 20 deaths have been attributed to this disease during the past decade. The same parasite occurs in bears, and when humans eat improperly cooked bear meat, there seems to be an even greater chance of infection than with pork.

FIGURE 41.24
Trichinella **encysted in pork.** The serious disease trichinosis results from eating undercooked pork or bear meat containing such cysts.

Phylum Rotifera: Rotifers

Phylum Rotifera includes common, small, bilaterally symmetrical, basically aquatic animals that have a crown of cilia at their heads (figure 41.25). Rotifers are pseudocoelomates but are very unlike nematodes. They have several features that suggest their ancestors may have resembled flatworms. There are about 2000 species of this phylum. Some of them are marine, and a few live in soil or in the capillary water in cushions of mosses. Most occur in fresh water, and they are common everywhere. Very few rotifers are marine. Most rotifers are between 50 and 500 micrometers in length, smaller than many protists.

Rotifers have a well-developed food-processing apparatus. A conspicuous organ on the tip of the head called the corona gathers food. It is composed of a circle of cilia which sweeps their food into their mouths, as well as being used for locomotion. Rotifers are often called "wheel animals" because the cilia, when they are beating together, resemble the movement of spokes radiating from a wheel.

Free-living rotifers feed on organic detritus, bacteria, protists, and small animals. A rotifer has a muscular pharynx with complex jaws inside it, which grind rapidly. They also have flame cells like those of flatworms, and they regulate their osmotic pressure by excreting water from these cells. Some rotifers are sessile, while others creep, float, or swim. The sexes of rotifers are separate. In many species, **parthenogenesis** (in which unfertilized eggs develop) is the rule; males do not occur in some of these species. Each female rotifer lays between 8 and 20 large eggs during its lifetime, which is short. A typical lifespan is between one and two weeks.

A New Phylum: Cycliophora

In December 1995, two Danish biologists reported the discovery of a strange new kind of creature, smaller than a period on a printed page. The tiny organism had a striking circular mouth surrounded by a ring of fine, hair-like cilia and was so unusual that they assigned it to an entirely new phylum, Cycliophora (Greek for "carrying a small wheel"). Found adhering to the mouthparts of a lobster like a suction cup, cycliophorans apparently suck up stray food particles as the lobster eats (figure 41.26). There are only about 35 known animal phyla, so finding a new one is extremely rare!

So far there have been five species discovered in the phylum. The scientists gave one species the name *Symbion pandora*; the genus name refers to the animal's life as a symbiont, the species name alludes to its intricate life cycle. While adhering to the lobster, the tiny animal reproduces asexually from buds, but when the lobster starts to molt, the symbiont begins a bizarre form of sexual reproduction. Dwarf males emerge, with nothing but brains and reproductive organs. Each dwarf male seeks out another female symbiont on the molting lobster and fertilizes its eggs, generating free-swimming individuals that can seek out another lobster and renew the life cycle. It is this highly un-

FIGURE 41.25
Rotifers (phylum Rotifera). These are common, small aquatic animals (150×).

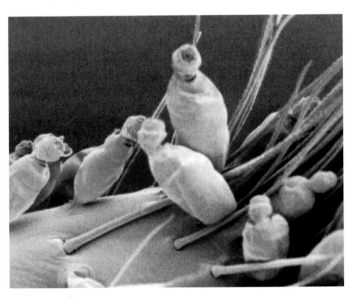

FIGURE 41.26
Phylum Cycliophora. Many individuals of the species *Symbion pandora* are shown attached to a cilia-covered mouthpart of a lobster (50×).

usual life cycle, with both asexual and sexual stages, that has prompted scientists to assign *S. pandora* to a unique phylum. The relationship of cycliophorans to other phyla is not clear. They may be related to rotifers or to the lophophorates discussed in the next chapter.

The pseudocoelomates, including nematodes and rotifers, all have fluid-filled pseudocoels.

41.1 **Animals are multicellular heterotrophs without cell walls.**

- Animals are heterotrophic, multicellular, and usually have the ability to move. Almost all animals reproduce sexually. Animal cells lack rigid cell walls and digest their food internally.

- The kingdom Animalia is divided into two subkingdoms: Parazoa, which includes only the asymmetrical phylum Porifera, and Eumetazoa.

- Most animals are characterized by a symmetrical appearance. In a few instances, the symmetry is radial: a midbody axis can be drawn in any two-dimensional plane. Most animals exhibit bilateral symmetry; the body can be bisected in only one plane.

- Bilaterally symmetrical animals include three major evolutionary lines: acoelomates, which lack a body cavity; pseudocoelomates, which develop a body cavity (pseudocoel) between the mesoderm and the endoderm; and coelomates, which develop a body cavity (coelom) within the mesoderm.

41.2 **The simplest animals are not bilaterally symmetrical.**

- The sponges (phylum Porifera) are characterized by specialized, flagellated cells called choanocytes. They do not possess tissues or organs, and most species lack symmetry in their body organization.

- Animals with radially symmetrical body plans and distinct tissues include the phyla Cnidaria and Ctenophora.

- Cnidarians (phylum Cnidaria) are predominantly marine animals with unique stinging cells called cnidocytes, each of which contains a specialized harpoon-like apparatus, or nematocyst.

- The comb jellies (phylum Ctenophora) are small marine animals that propel themselves through the water by means of eight comb-like plates of fused cilia.

41.3 **Acoelomates are solid worms that lack a body cavity.**

- Acoelomates are the most primitive bilaterally symmetrical animals. They lack an internal cavity, except for the digestive system, and are the simplest animals that have organs—structures made up of two or more tissues.

- The most prominent phylum of acoelomates, Platyhelminthes, includes the free-living flatworms and the parasitic flukes and tapeworms.

- Ribbon worms (phylum Nemertea) are similar to free-living flatworms, but have a complete digestive system and a circulatory system in which the blood flows in vessels.

41.4 **Pseudocoelomates have a simple body cavity.**

- Pseudocoelomates, exemplified by the nematodes (phylum Nematoda), have a body cavity that develops between the mesoderm and the endoderm. This pseudocoel gives these worms enhanced powers of movement.

- Rotifers (phylum Rotifera), or wheel animals, are another phylum of pseudocoelomates.

1. **Animals** Animals are multicellular heterotrophs that lack cell walls and ingest their food. With few exceptions, animals are diploid and reproduce sexually. Their ability to move is their most striking characteristic.

2. **Tissues** Tissues are specialized groups of cells that carry out particular functions in the body. All animals but sponges have tissues.

3. **Symmetry** All animals but sponges have body parts arranged along an axis. In radially symmetrical animals, parts are arranged around a central axis, while bilaterally symmetrical animals have a right and left half that are mirror images.

4. **Parazoa and Eumetazoa** The animal kingdom is typically partitioned into two groups: the Parazoa, or sponges, which possess no tissue level of organization, and the Eumetazoa, which possess tissue-level organization and include the rest of the animal phyla.

5. **Body cavity** An internal body cavity, whether a pseudocoel or a coelom, not only makes an animal rigid but also allows circulatory, digestive, and other systems to function free of the pressure of surrounding muscles.

6. **Pseudocoelomate** Animals that develop a body cavity between embryonic mesoderm and endoderm are referred to as *pseudocoelomates*. While there are many smaller animal groups in this category, the phylum Nematoda is enormous, both in numbers of species and individuals.

Review Questions

1. What are the characteristics that distinguish animals from other living organisms? What is the structural nature of animal cells that differentiates them from plant cells?

2. What are the two subkingdoms of animals? How do they differ in terms of symmetry and body organization?

3. Why are sponges considered one of the most primitive groups of animals? From what kind of ancestor did sponges probably evolve? Why?

4. How does a sponge capture food? What are the specialized cells used for this purpose called? How do they work? Where and how is food digested in a sponge?

5. What are the two ways sponges reproduce? What do larval sponges look like? How do they become adult sponges?

6. What is the basic composition of a cnidarian body? What are the two body forms present in this group? Describe each, including the ploidy of each form. What is a planula?

7. What body plan do members of the phylum Platyhelminthes possess? What is their general appearance? Are these animals parasitic or free-living? How do they move from place to place? What type of digestive system do they possess?

8. How are tapeworms different from flukes? What are the three parts of a tapeworm? How do tapeworms reproduce?

9. What type of body plan does a nematode possess? Briefly describe the digestive tract of a typical nematode. Why are nematodes structurally unique in the animal world? How are these animals frequently damaging to humans?

10. What type of body plan do rotifers have? How do they capture food? What are the two ways they can reproduce?

Thought Questions

1. How might you investigate whether or not sponges shared a common ancestor with the other animals?

2. Coral reefs are often found in nutrient-poor waters. What type of symbiotic relationship helps these coral animals to grow actively? What are the advantages for each member in the relationship? What would happen to the coral reefs if the water they are growing in became cloudy with pollution?

3. Parasites, especially those that require two or more hosts to complete their life cycles, often produce very large numbers of offspring. What advantage would this present to them?

Internet Links

Meet the Invertebrates
http://www.ucmp.berkeley.edu/porifera/porifera.html
The University of California Museum of Paleontology maintains a network of sites that present a wealth of information on animal groups. The above address is for porifera *(sponges). Others include* cnidaria, platyhelminthes *(flatworms),* annelida, molluska, echinodermata, *and* arthropoda.

A Tragic Film from the Microscopic World
http://www.microscopy-uk.org.uk/mag/art97/mmpond.html
The story of birth, life, and death of a female rotifer and her offspring, illustrated with extraordinary video microscope footage. For a remarkable microscopic video of a live Rotifer birth, check out as well http://micscape.simplenet.com/avis/walker1.htm

For Further Reading

Alexander, R. M.: *Animals*, Cambridge University Press, Cambridge, 1990. An exciting and lucid account of the ways in which the various groups of animals have adjusted to their modes of life.

Erwin, D., J. Valentine, and D. Jablonski: "The Origin of Animal Body Plans," *American Scientist*, vol. 85, March 1997, pages 126–37. Recent fossil finds and new insights into animal development are producing fresh perspectives on the origin of animal body plans in the Cambrian.

Funch, P. and R. M. Kristensen: "Cycliophora Is a New Phylum with Affinities to Entoprocta and Ectoprocta," *Nature*, vol. 378, December 14, 1995, pages 711–14. This paper reports the exciting zoological discovery.

Levinton, J. S.: "The Big Bang of Animal Evolution," *Scientific American*, November 1992, pages 84–91. An interesting discussion of the sudden appearance of large numbers of new species in the fossil record.

Raff, R.: *The Shape of Life: Genes, Development, and the Evolution of Animal Form*, University of Chicago Press, Chicago, 1996. An account of how new molecular information is reshaping our ideas about the origins of the animal phyla.

Sebens, K. P.: "Biodiversity of the Coral Reefs: What Are We Losing and Why?" *American Zoologist*, vol. 34, 1994, pages 115–33. The extraordinary diversity of species in the coral reef ecosystem is threatened by people, either directly (through pollution) or indirectly (through global warming).

Vacelet, J. and N. Boury-Esnault: "Carnivorous Sponges," *Nature*, vol. 373, January 26, 1995, pages 33–35. A shallow-water cave in the Mediterranean yields a microcrustacean-eating sponge previously only encountered in very deep water. A fascinating discussion of an unusual mode of feeding in sponges.

42

Mollusks and Annelids

Concept Outline

42.1 Mollusks were among the first coelomates.

The Mollusks.   Snails, clams, octopuses, and their relatives are one of the most successful phyla, second only to arthropods in numbers of species. There are more terrestrial mollusk species than terrestrial vertebrates!

Body Plan of the Mollusks.   The mollusk body plan is characterized by three distinct sections, a unique rasping tongue, and distinctive free-swimming larvae also found in annelid worms.

The Classes of Mollusks.   The three major classes of mollusks are the gastropods (snails and slugs), the bivalves (oysters and clams), and the cephalopods (octopuses and squids). While they seem very different at first glance, on closer inspection, they all have the same basic mollusk body plan.

42.2 Annelids were the first segmented animals.

The Annelids.   Annelids are segmented coelomate worms, most of which live in the sea.

Body Plan of the Annelids.   The annelid body is composed of numerous similar segments.

Classes of Annelids.   The three major classes of annelids are the polychaetes (marine worms), the oligochaetes (earthworms), and the hirudines (leeches).

42.3 Lophophorates appear to be a transitional group.

Lophophorates.   The three phyla of lophophorates share a unique ciliated feeding structure, but differ in many other ways.

FIGURE 42.1
An annelid, the Christmas-tree worm, *Spirobranchus giganteus.* Mollusks and annelids inhabit both terrestrial and aquatic habitats. They are large and successful groups, with some of their most spectacular members represented in marine environments.

Even though acoelomates and pseudocoelomates have proven very successful, a third way of organizing the animal body has also evolved, one that occurs in the bulk of the animal kingdom. We will begin our discussion of these animals with mollusks and annelids. Mollusks include such animals as clams, snails, slugs, and octopuses; annelids include earthworms and clamworms (figure 42.1). The lophophorate phyla, a fascinating group of primarily marine animals with features that seem intermediate between those of protostomes and deuterostomes, will also be discussed in this chapter. The remaining groups of coelomate animals will be discussed in chapters 43, 44, and 45.

The Mollusks

Mollusks (phylum Mollusca) include snails, slugs, clams, scallops, oysters, cuttlefish, octopuses, and many other familiar animals (figure 42.2). The durable shells of some mollusks are often beautiful and elegant; they have long been favorite objects for professional scientists and amateurs alike to collect, preserve, and study. Mollusks are characterized by a coelom (figure 42.3) and are one of the most successful of all phyla. There are at least 110,000 named species and probably at least that many more still to be discovered. In terms of named species, they are the largest animal phylum except for arthropods. Mollusks are widespread and often abundant in marine, freshwater, and terrestrial habitats.

A number of mollusks have invaded the land, including the snails and slugs that live in your garden. Terrestrial mollusks are often abundant in places that are at least seasonally moist. Some of these places, such as the crevices of desert rocks, may appear very dry, but even these habitats have at least a temporary supply of water at certain times. There are so many terrestrial mollusks, that only the arthropods have more species adapted to a terrestrial way of life. The 35,000 species of terrestrial mollusks far outnumber the roughly 20,000 species of terrestrial vertebrates.

As a group, mollusks are an important source of food for humans. Oysters, clams, scallops, mussels, octopuses, and squids are among the culinary delicacies that belong to this large phylum. Mollusks are also of economic significance to us in many other ways—for example, the production of pearls and of the shell material that is used as mother-of-pearl in jewelry and other decorative objects. Mollusks are not wholly

(b)

(c)

(a)

FIGURE 42.2

Molluscan diversity. Second only to the arthropods in number of described species, members of the phylum Mollusca occupy almost every habitat on earth. The blue-ringed octopus (a) is one of the few mollusks dangerous to humans. Strikingly beautiful, it is equipped with a sharp beak and poison glands—divers give it a wide berth! Gastropod mollusks such as (b) the garden snail are ubiquitous in marine, freshwater, and terrestrial environments. (c) A giant clam, *Tridacna maxima*. The green color is caused by the presence of symbiotic dinoflagellates (zooxanthellae), which probably contribute most of the food supply of the clam, although it remains a filter feeder like most bivalves. Some individual giant clams may be nearly 1.5 meters long and weigh up to 270 kilograms.

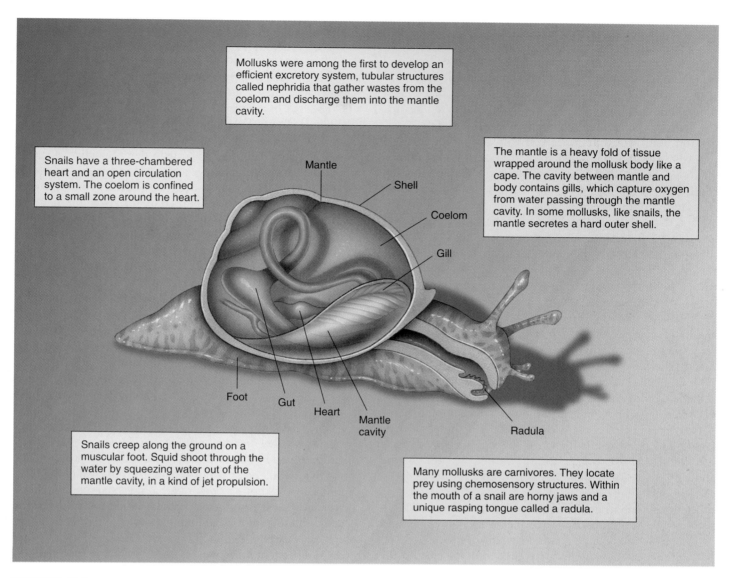

Mollusks were among the first to develop an efficient excretory system, tubular structures called nephridia that gather wastes from the coelom and discharge them into the mantle cavity.

Snails have a three-chambered heart and an open circulation system. The coelom is confined to a small zone around the heart.

The mantle is a heavy fold of tissue wrapped around the mollusk body like a cape. The cavity between mantle and body contains gills, which capture oxygen from water passing through the mantle cavity. In some mollusks, like snails, the mantle secretes a hard outer shell.

Snails creep along the ground on a muscular foot. Squid shoot through the water by squeezing water out of the mantle cavity, in a kind of jet propulsion.

Many mollusks are carnivores. They locate prey using chemosensory structures. Within the mouth of a snail are horny jaws and a unique rasping tongue called a radula.

Mantle · Shell · Coelom · Gill · Foot · Gut · Heart · Mantle cavity · Radula

FIGURE 42.3
Evolution of the coelom. The body cavity of a mollusk like this snail (phylum Mollusca) is a coelom, completely enclosed within the mesoderm. This allows physical contact between the mesoderm and the endoderm, permitting interactions that lead to development of highly specialized organs such as a stomach.

beneficial to humans, however. Bivalve mollusks called 'shipworms' destroy timbers submerged in the sea, including those used to build boats, docks, and pilings. Other bivalves such as the zebra mussel have recently invaded North American ecosystems via the ballast water of cargo ships from Europe, wreaking havoc in many aquatic ecosystems. Slugs and terrestrial snails often cause extensive damage to garden flowers, vegetables, and crops. Other mollusks serve as hosts to the intermediate stages of many serious parasites, including several nematodes and flatworms, which we discussed in chapter 41.

Most mollusks measure several centimeters in their largest dimension. Some, however, are minute, while others reach formidable sizes. The giant squid, which is occasionally cast ashore but has rarely been observed in its natural environment, may grow up to 21 meters long! Weighing up to 250 kilograms, the giant squid is the largest invertebrate and, along with the giant clam, the heaviest. Millions of giant squid probably inhabit the ocean, even though they are seldom caught. Another large mollusk is the bivalve *Tridacna maxima*, the giant clam, which may have a pair of shells as long as 1.5 meters and may weigh as much as 270 kilograms.

Mollusks are the second-largest phylum of animals in terms of named species. They were among the first coelomate animals and are one of the most successful phyla.

Body Plan of the Mollusks

In their basic body plan (figure 42.4), mollusks have distinct bilateral symmetry. They have a **visceral mass** covered with a soft epithelium and a muscular foot that is used in locomotion. They may also have a differentiated head at the anterior end of the body. The organs of digestion, excretion, and reproduction are found within the visceral mass. Folds (often two) arise from the dorsal body wall and enclose a cavity between themselves and the visceral mass; these folds constitute the **mantle.** In some mollusks the mantle cavity acts as a lung; in others it contains gills. **Gills** are specialized portions of the mantle that usually consist of a system of filamentous projections rich in blood vessels. These projections greatly increase the surface area available for gas exchange and, therefore, the animal's overall respiratory potential. Oxygen moves inward, carbon dioxide outward. Mollusk gills are very efficient, and many gilled mollusks extract 50% or more of the dissolved oxygen from the water that passes through the mantle cavity. Finally, in most members of this phylum, the outer surface of the mantle also secretes a protective shell.

A mollusk shell consists of a horny outer layer, rich in protein, which protects the two underlying calcium-rich layers from erosion. The middle layer consists of densely packed crystals of calcium carbonate. The inner layer is pearly and increases in thickness throughout the animal's life. Pearls are occasionally produced between the mantle and this pearly inner shell layer of **bivalve mollusks** (two-shelled), including clams and oysters, as a result of secretions of shell material around foreign bodies. Many mollusks can withdraw for protection into their shell if they have one.

In aquatic mollusks, a continuous stream of water passes into and out of the mantle cavity, drawn by the cilia on the gills. This water brings in oxygen and, in the case of the bivalves, food; it also carries out waste materials. When the gametes are being produced, they are frequently carried out in the same stream. In the squids and octopuses, the mantle cavity has been modified to create the jet-propulsion system that enables these animals to move rapidly through the water.

The foot of a mollusk is muscular and may be adapted for locomotion, attachment, food capture (in squids and octopuses), or various combinations of these functions. Some mollusks secrete mucus, forming a path that they glide along on their foot. In cephalopods—squids and octopuses—the foot is divided into arms, also called tentacles. In some **pelagic** forms—mollusks that are perpetually free-swimming—the foot is modified into wing-like projections or thin fins.

One of the most characteristic features of all the mollusks except the bivalves is the **radula,** a rasping, tongue-like organ used for feeding. The radula consists primarily of dozens to thousands of microscopic, chitinous teeth arranged in rows (figure 42.5). Gastropods (snails and their relatives) use their radula to scrape algae and other food materials off their substrates and then to convey this food to the digestive tract. Other gastropods are active predators, some using a modified radula to drill through the shells of prey and extract the food. The small holes often seen in oyster shells are produced by gastropods that have bored holes to kill the oyster and extract its body for food.

The circulatory system of all mollusks except cephalopods consists of a heart and an open system in which blood circulates freely. The mollusk heart usually has three chambers, two that collect aerated blood from the gills, while the third pumps it to the other body tissues. In cephalopods, a closed circulatory system of vessels carries blood to and from the heart. These dexterous and rapidly moving animals also have auxiliary hearts, which increase the efficiency of their overall pumping system. In mollusks, the coelom is primarily represented by a small cavity around the heart.

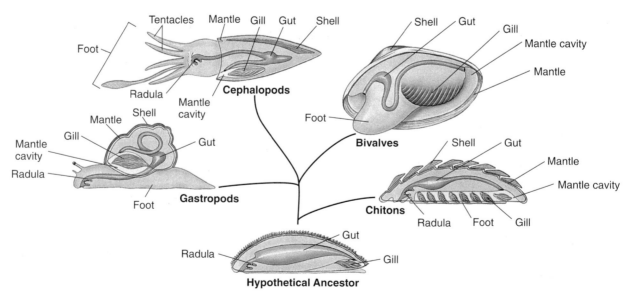

FIGURE 42.4
Body plans among the mollusks.

Nitrogen-rich wastes are removed from the mollusk by one or two tubular structures called **nephridia.** A typical nephridium has an open funnel, the **nephrostome,** which is lined with cilia. From the nephrostome, a coiled tubule runs into an enlarged bladder, which in turn connects to an excretory pore. Wastes are gathered by the nephridia from the coelom, which is located around the heart only, and discharges into the mantle cavity. The wastes are then expelled from the mantle cavity by the continuous pumping of the gills. Sugars, salts, water, and other materials are reabsorbed by the walls of the nephridia and returned to the animal's body as needed to achieve an appropriate osmotic balance.

In animals with a closed circulatory system, such as annelids, some mollusks, and vertebrates, the coiled tubule of a nephridium is surrounded by a network of capillaries. Wastes are extracted from the circulatory system through these capillaries and are transferred into the nephridium, then subsequently discharged. Salts, water, and other associated materials may also be reabsorbed from the tubule of the nephridium back into the capillaries. For this reason, the excretory systems of these coelomates are much more efficient than the flame cells of the acoelomates, which pick up substances only from the body fluids. Mollusks were one of the earliest evolutionary lines to develop an efficient excretory system. Other than chordates, coelomates with closed circulation have similar excretory systems.

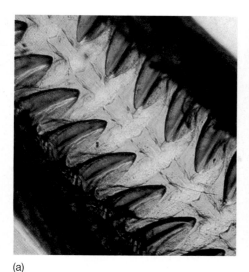
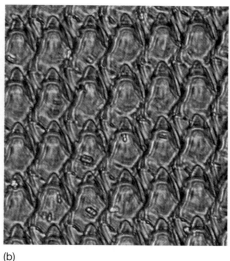

(a)   (b)

FIGURE 42.5
Structure of the radula in a snail. (a) The radula consists of chitin and is covered with rows of teeth. (b) Enlargement of the rasping teeth on a radula.

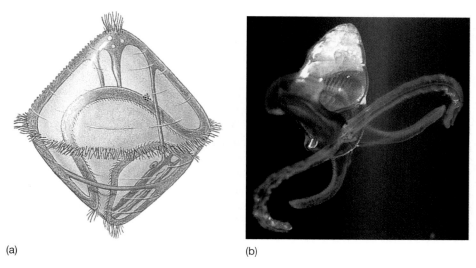

(a)   (b)

FIGURE 42.6
Stages in the molluscan life cycle. (a) The trochophore larva of a mollusk. Similar larvae, as you will see, are characteristic of some annelid worms as well as a few other phyla. (b) Veliger stage of a mollusk.

Reproduction in Mollusks

Most mollusks have distinct male and female individuals, although a few bivalves and many gastropods are hermaphroditic. Even in hermaphroditic mollusks, cross-fertilization, rather than self-fertilization, is most common. Remarkably, some sea slugs and oysters are able to change from one sex to the other several times during a single season.

Many marine mollusks have free-swimming larvae called **trochophores** (figure 42.6a). Trochophores are distinctive in structure; they are propelled through the water by a row of cilia that encircles the middle of their body. In most ma-

rine snails and in bivalves, a second free-swimming stage follows the trochophore stage. In this **veliger** stage, the beginnings of a foot, shell, and mantle can be seen (figure 42.6b). Mollusks are dispersed from place to place largely as trochophores and veligers. Trochophores and veligers drift widely in the ocean currents to new areas.

Mollusks were among the earliest animals to evolve an efficient excretory system. They characteristically have bodies with three distinct sections: head, visceral mass, and foot. All mollusks, except the bivalves, possess a unique rasping tongue called a radula.

The Classes of Mollusks

There are seven classes of mollusks. Some of the smaller classes have greatly helped us to understand the limits of variation in the phylum and the nature of its probable ancestor. By comparative study of these animals, some scientists have concluded that the ancestral mollusk was probably a dorsoventrally flattened, unsegmented, worm-like animal that glided on its ventral surface. This animal may also have had a moderate amount of chitinous cuticle and overlapping calcareous scales. Alternatively, other scientists believe that mollusks are secondarily unsegmented and that their ancestors were segmented.

A contemporary class of mollusks in which many of these characteristics still exist is the Polyplacophora, the chitons. These marine mollusks have oval bodies with eight overlapping calcareous plates (figure 42.7). Some scientists believe these plates are evidence of segmentation in molluscan ancestors. However, underneath the plates, the body is not segmented. Chitons creep along using a broad, flat foot surrounded by a groove or mantle cavity in which the gills are arranged. Most chitons are grazing herbivores that live in shallow marine habitats, but some live at depths of more than 7000 meters.

We will examine three classes of mollusks in some detail as representatives of the phylum: (1) Gastropoda—snails, slugs, limpets, and their relatives; (2) Bivalvia—clams, oysters, scallops, and their relatives; and (3) Cephalopoda—squids, octopuses, cuttlefishes, and nautilus.

FIGURE 42.7
A chiton, *Tonicella lineata*, class Polyplacophora. The shells of chitons, unlike those of other mollusks, consist of overlapping calcareous plates.

Class Gastropoda: The Snails and Slugs

About 40,000 named species of snails and slugs constitute the gastropods (class Gastropoda). This class is primarily a marine group that is also abundant in freshwater and terrestrial habitats (figure 42.8). Gastropods have a shell, lost in some groups, and a body divisible into a head, foot, and visceral mass. Slugs and nudibranchs (marine slugs) are descended from ancestors that had shells. These animals generally creep along on a foot, which may be modified for swimming. The shells of most marine gastropods can be closed by a plate, the operculum, that the animal can pull into place. Most adult land gastropods lack an operculum.

The heads of most gastropods have a pair of tentacles with eyes. These tentacles have been lost in some of the more advanced forms of the class. The mouth opening in gastropods may be simple or modified as a proboscis. Within the mouth cavity of many members of this class are horny jaws and a radula.

The visceral mass of gastropods has become asymmetrical during the course of evolution. This **torsion** occurs during the embryological development of the gastropod when disproportionate growth of the lateral muscles results in a 120-degree relocation of the anus. In some groups of gastropods, varying degrees of detorsion have taken place. The **twisting** of the shell is a separate process and results from the fact that one side of the larva grows much more rapidly than the other. This evolutionary process, which is considerably advanced in certain members of this class, is sometimes associated with other changes. For example, many gastropods have lost their right gill and sometimes also their right nephridium.

Gastropods display extremely varied feeding habits. Some are predatory, others scrape algae off rocks (or aquarium glass), and others are scavengers. Many are herbivores, and some terrestrial ones are serious garden and agricultural pests. The radula of oyster drills is used to bore holes in the shells of other mollusks, through which the contents of the prey can be removed. In cone shells (figure 42.8a), the radula has been modified into a kind of poisonous harpoon, which is shot with great speed into the prey. Sea slugs, or nudibranchs (figure 42.8d), are active predators; a few species of nudibranchs protect themselves with nematocysts they obtain from the cnidarian polyps they eat.

In terrestrial gastropods, an area under the mantle that is extremely rich in blood vessels serves, in effect, as a lung. This lung evolved in animals living in environments with plentiful oxygen. It absorbs oxygen more efficiently than a gill does for an animal living on land and breathing air. The lung is essentially an empty mantle cavity in the space that gills occupy in the aquatic ancestors of terrestrial mollusks.

(a)

(b)

(c)

(d)

FIGURE 42.8
Gastropod mollusks. (a) The textile cone, *Conus textilis*. The radulas of the cone shells, active hunters, are modified into harpoons. The animals of this genus, some of which hunt fishes, secrete powerful neurotoxins; they include some of the most dangerous marine animals. (b) Hawaiian limpet, *Helcioniscus exaratus*, viewed from below. (c) A slug, *Arion ater*, one of the approximately 500 species of terrestrial mollusks that lack shells. (d) Nudibranchs, *Hermissenda crassicornis*. (e) The terrestrial snail *Allogona townsendiana*.

(e)

Class Bivalvia: The Bivalves

Members of the class Bivalvia, which includes clams, scallops, mussels, and oysters, have two lateral (left and right) shells hinged together dorsally and a wedge-shaped foot (figure 42.9). A ligament hinges the shells together and causes them to gape open. Pulling against this ligament are one or two large adductor muscles that can draw the shells together. The mantle secretes the shells and ligament and envelops the internal organs within the pair of shells. Additionally, the mantle is frequently drawn out to form two siphons, one for the incoming and one for the outgoing stream of water. A complex folded gill lies on each side of the visceral mass. These gills consist of pairs of filaments that contain many blood vessels. Within the gills is a rather elaborate pattern of water circulation.

Bivalves do not have distinct heads or radulas, differing from gastropods in this respect (see figure 42.4). However, most have a muscular foot. It may be adapted, in different species, for creeping, burrowing, cleansing the animal, or anchoring it in its burrow. Some species of clams can dig into sand or mud very rapidly by means of muscular contractions of their foot.

Most bivalves are sessile filter-feeders. They extract small organisms from water that they filter through their mantle cavity by ciliary action of their gills, squirting the water back out through a siphon. The food particles that bivalves ingest are entangled in masses of mucus secreted by glands, mainly located on the gills. They then use cilia to convey the food particles to their mouth. On either side of the mouth is usually a pair of organs, the **palps,** which aid in handling particles of food.

Bivalves disperse from place to place largely as larvae. Most adults are fixed and relatively immobile or confined to a burrowing way of life. Some adult bivalves can move about readily. Many genera of scallops can move swiftly through the water by using their large adductor muscles to clap their shells together. These muscles are what we usually eat as "scallops." One of a scallop's shells is larger than the other, and the edge of its body is lined with tentacle-like projections. Their complex eyes (figure 42.9*a*) can differentiate light and darkness, so they can probably see the shadows that potential predators, such as starfish, cast. Scallops can also detect predators by means of chemical signals, which prompt them to flee.

There are about 10,000 species of bivalves. Most species are marine, although many also live in fresh water. One of the more interesting freshwater groups is the pearly freshwater mussels, or naiads. There are nearly 1200 species of this group distributed worldwide. One of its families, Unionidae, includes more than 500 species that occur in the rivers and lakes of North America (figure 42.10). The larvae of this family of mollusks are brooded in a special pouch within a gill before they are released to parasitize fishes; they thus exhibiting a very unusual life cycle for a mollusk.

(a)

(b)

FIGURE 42.9
Bivalves. (a) Scallop, *Chlamys hericia*. Note the eyes (light-colored spots) around the margin of the body. (b) File shell, *Lima scabra*, swimming.

FIGURE 42.10
Pearly freshwater mussels, members of the family Unionidae, from the United States. More than 50 of the estimated 500 species in North America have become extinct, and many of the remainder are threatened with extinction by the pollution of the waters in which they occur.

Class Cephalopoda: The Octopuses, Squids, and Nautilus

The more than 600 species of the class Cephalopoda—octopuses, squids, and nautilus—are the most intelligent of the invertebrates. They are active marine predators that swim, often swiftly, and compete successfully with fish. The giant squid, mentioned earlier, plays an ecological role similar to that of the large marine mammals such as killer whales, and large, predaceous fishes. Cephalopods feed primarily on fishes, other mollusks, crustaceans, and worms. The foot has evolved into a series of tentacles equipped with suction cups, adhesive structures, or hooks that seize prey efficiently. Squids have 10 tentacles; octopuses, as indicated by their name, have eight; and the nautilus, about 80 to 90 (figure 42.11). Once the tentacles have snared the prey, it is bitten with strong, beak-like paired jaws and pulled into the mouth by the tongue-like action of the radula.

Cephalopods have highly developed nervous systems, and their brains are unique among mollusks. Their rapid responses are made possible by a bundle of giant nerve fibers attached to the muscles of the mantle. Their eyes are very elaborate, and the retina has a structure much like that of vertebrate eyes, although there are fundamental differences and they evolved separately (see chapter 52). Squid eyes can also be quite large; the eyes of a giant squid that washed up on a beach in New Zealand in 1933 were 40 centimeters across, the largest eyes known in any animal. Many cephalopods exhibit complex patterns of behavior and a high level of intelligence; octopuses can be easily trained to distinguish among classes of objects. Most members of this class have closed circulatory systems and are the only mollusks that do.

Although they evolved from shelled ancestors, living cephalopods, except for the few species of nautilus, lack an external shell. Squids and cuttlefish retain an internal remnant of their ancestral shells, which serves as a stiffening support. Octopuses have no trace of a shell. Cuttlefish "bones" are used in bird cages to provide calcium for pet birds, such as canaries.

Like other mollusks, cephalopods take water into the mantle cavity and expel it through a siphon. Cephalopods have modified this system into a means of jet propulsion. When threatened, they eject water violently and shoot themselves through the water. Squids and octopuses can

(a)

(b)

(c)

FIGURE 42.11
Cephalopod diversity. (*a*) An octopus. Octopuses generally move slowly along the bottom of the sea. (*b*) Pearly nautilus, *Nautilus pompilius*. (*c*) A squid. Squids are active predators, competing effectively with fish for prey.

swim in various directions by turning their siphon. Both can release a dark fluid, once used for ink, that clouds the water and helps to disguise the direction of their escape.

In cephalopods, the sexes are separate. Sperm are stored in spermatophores in a sac that opens into the mantle cavity of the male. During copulation, the male uses a specialized arm, or tentacle, to transmit a **spermatophore**—a mass enclosing many sperm—from its own mantle cavity into the female's. The female then fertilizes the eggs as they leave the **oviduct,** the passageway through which they leave the female, attaching them to stones and other objects.

The three major classes of mollusks are gastropods, bivalves, and cephalopods. Gastropods typically live in a hard shell. Bivalves have hinged shells but do not have a distinct head area. Cephalopods possess well-developed brains and are the most intelligent invertebrates.

The Annelids

One of the key innovations in body plan to arise among the coelomates was **segmentation,** the building of a body from a series of similar segments. Just as it is efficient for workers to construct a tunnel from a series of identical prefabricated parts, so these advanced coelomates are "assembled" from a succession of identical segments. The great advantage of segmentation is the evolutionary flexibility it offers—a small change in an existing segment can produce a new segment with a different function. Segments can be modified for reproduction, feeding, or eliminating wastes.

The first segmented animals to evolve were most likely **annelid worms** (figure 42.12). Two-thirds of all annelids (phylum Annelida) live in the sea (about 8000 species), and most of the rest, some 3100 species, are earthworms. Annelids are characterized by three principal features:

1. **Repeated segments.** The body segments of an annelid are visible as a series of ring-like structures running the length of the body, looking like a stack of donuts (figure 42.13). Internally, the segments are divided from one another by partitions called **septa,** just as bulkheads separate segments of a submarine. In each of the cylindrical segments, the excretory and locomotor organs are repeated. The fluid within the coelom of each segment creates a hydrostatic (liquid-supported) skeleton that gives the segment rigidity, like an inflated balloon. Muscles within each segment push against the fluid in the coelom. Because each segment is separate, each can expand or contract independently. This lets the worm move in complex ways.

2. **Specialized segments.** The anterior (front) segments of annelids have become modified to contain specialized sensory organs. Some are sensitive to light, and elaborate eyes with lenses and retinas have evolved in some annelids. A well-developed cerebral ganglion, or brain, is contained in one anterior segment.

3. **Connections.** Although partitions separate the segments, materials and information do pass between segments. Annelids have a closed circulatory system that carries blood from one segment to another. Ventral nerve cords connect the nerve centers or ganglia in each segment with one another and the brain, so the brain can coordinate the worm's activities.

Annelids are a diverse group of coelomate animals characterized by serial segmentation.

Body Plan of the Annelids

The basic annelid body plan is a tube within a tube, with the internal digestive tract—a tube running from mouth to anus—suspended within the coelom. In most annelids groups, each segment typically possesses **setae,** bristles of chitin that help anchor the worms during locomotion or in their burrows. A crawling worm retracts the setae in the expanding segments and extends them in the contracted segments to make contact with the surface. Because of the setae, annelids are sometimes called "bristleworms."

Unlike the arthropods and most mollusks, the annelids have a closed circulatory system. Annelids exchange oxygen

(a)

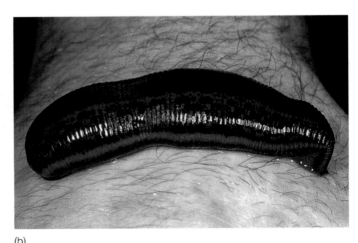
(b)

FIGURE 42.12

Annelids. (*a*) *Nereis virens* is a wide-ranging predatory marine polychaete worm equipped with feathery parapodia for movement and respiration, as well as jaws for hunting. You may have purchased *Nereis* as fishing bait! (*b*) *Hirudo medicinalis*, the medicinal leech, is seen here feeding on a human arm. Leeches uses chitinous, blade-like jaws to make an incision to access blood and secrete an anticoagulant to keep the blood from clotting. Both the anticoagulant *and* the leech itself have made important contributions to modern medicine.

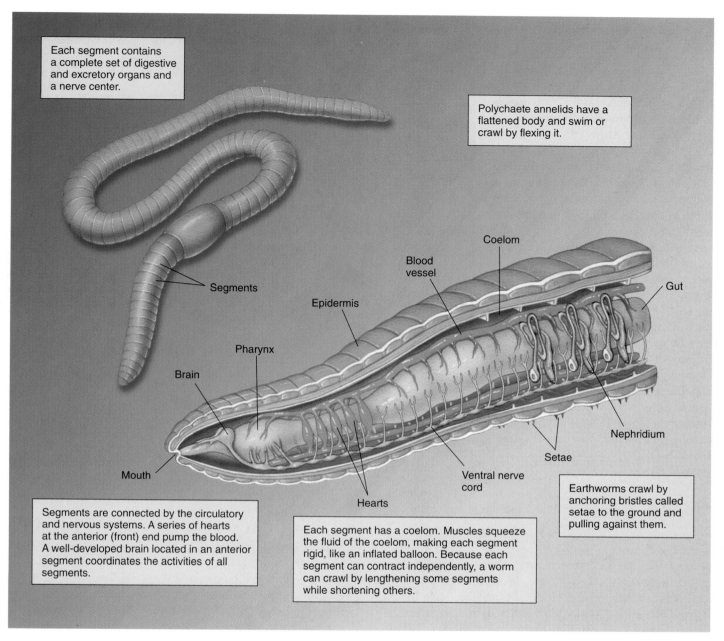

Each segment contains a complete set of digestive and excretory organs and a nerve center.

Polychaete annelids have a flattened body and swim or crawl by flexing it.

Coelom

Blood vessel

Gut

Segments

Epidermis

Pharynx

Brain

Nephridium

Mouth

Setae

Ventral nerve cord

Hearts

Segments are connected by the circulatory and nervous systems. A series of hearts at the anterior (front) end pump the blood. A well-developed brain located in an anterior segment coordinates the activities of all segments.

Each segment has a coelom. Muscles squeeze the fluid of the coelom, making each segment rigid, like an inflated balloon. Because each segment can contract independently, a worm can crawl by lengthening some segments while shortening others.

Earthworms crawl by anchoring bristles called setae to the ground and pulling against them.

FIGURE 42.13

The evolution of segmentation. Marine polychaetes and earthworms (phylum Annelida) were most likely the first organisms to evolve a body plan based on partly repeated body segments. Segments are separated internally from each other by partitions.

and carbon dioxide with the environment through their body surfaces; most lack gills or lungs. However, much of their oxygen supply reaches the different parts of their bodies through their blood vessels. Some of these vessels are enlarged and heavily muscular, serving as hearts that pump the blood. Earthworms have five pulsating blood vessels on each side that serve as hearts, helping to pump blood from the main dorsal vessel, which is their major pumping structure, to the main ventral vessel.

The excretory system of annelids consists of ciliated, funnel-shaped nephridia generally similar to those of mollusks. These nephridia—each segment has a pair—collect waste products and transport them out of the body through the coelom by way of specialized excretory tubes.

The annelid body is composed of numerous similar segments, each with its own circulatory, excretory, and neural elements and each with its own array of setae.

Classes of Annelids

The roughly 12,000 described species of annelids occur in many different habitats. They range in length from as little as 0.5 millimeter to the more than 3-meter length of some polychaetes and giant Australian earthworms. There are three classes of annelids: (1) Polychaeta, which are free-living, almost entirely marine bristleworms, comprising some 8000 species; (2) Oligochaeta, terrestrial earthworms and related marine and freshwater worms, with some 3100 species; and (3) Hirudinea, leeches, mainly freshwater predators or bloodsuckers, with about 500 species. The annelids are believed to have evolved in the sea, with polychaetes being the most primitive class. Oligochaetes seem to have evolved from polychaetes, perhaps by way of brackish water to estuaries and then to streams. Leeches share with oligochaetes an organ called a **clitellum,** which secretes a cocoon specialized to receive the eggs. It is generally agreed that leeches evolved from oligochaetes, specializing in their bloodsucking lifestyle as external parasites.

(a)

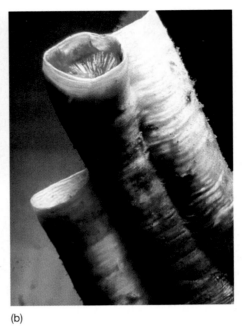
(b)

(c)

FIGURE 42.14
Polychaete annelids. (*a*) Fan worm, *Sabella melanostigma.* (*b*) Tube worms, *Eudistylia polymorpha,* withdrawn into their tubes. (*c*) Shiny bristleworm, *Oenone fulgida.*

Class Polychaeta: The Polychaetes

Polychaetes (class Polychaeta) include clamworms, plumed worms, scaleworms, lugworms, twin-fan worms, sea mice, peacock worms, and many others. These worms are often surprisingly beautiful, with unusual forms and sometimes iridescent colors (figure 42.14). Polychaetes live in burrows, under rocks, in tubes of hardened mud, sand, mucuslike secretions, or calcium carbonate, and inside shells. They are often a crucial part of marine food chains, since they are extremely abundant in certain habitats. A number of polychaetes are commensal; they live inside sponges, in the shells of mollusks, on echinoderms or crustaceans, and on other animals, eating the food particles left over by these organisms. A few of these worms are parasites; some are active predators.

Polychaetes have a well-developed head with specialized sense organs; they differ from other annelids in this respect. Their bodies are often highly organized into distinct regions formed by groups of segments related in function and structure. Their sense organs include eyes, which range from simple eyespots to quite large and conspicuous stalked eyes. All of these eyes serve primarily to concentrate light and indicate the direction from which it is coming.

Another distinctive characteristic of polychaetes is the paired, fleshy, paddle-like flaps, called **parapodia,** on most of their segments. These parapodia, which bear bristle-like setae, are used in swimming, burrowing, or crawling. They also play an important role in gas exchange because they greatly increase the surface area of the body. Some polychaetes that live in burrows or tubes may have parapodia featuring hooks to help anchor the worm. Slow crawling is carried out by means of the parapodia, and rapid crawling is aided by undulating motions of the body. In addition, the polychaete epidermis often includes ciliated cells, which aid in respiration and food procurement.

The sexes of polychaetes are usually separate, and fertilization is often external, occurring in the water and away from both parents. Unlike other annelids, polychaetes usually lack permanent **gonads,** the sex organs that produce gametes. These polychaetes produce their gametes directly from cells in the lining of the coelom or in their septa. The eggs of polychaetes may be laid in gelatinous masses and attached to the substrate, or incubated in a tube or brood chamber in or on the female's body. Fertilization results in the production of ciliated, mobile *trochophore larvae* similar to the larvae of mollusks. The trochophores develop for long periods in the plankton before beginning to add segments and thus changing to a juvenile form that more closely resembles the adult form.

Class Oligochaeta: The Earthworms

Earthworms (class Oligochaeta) literally eat their way through the soil. They suck in organic and other material by expanding their strong pharynx. Everything that they ingest passes through their long, straight digestive tracts. In one region of this tract, the gizzard, muscles are concentrated; the worm uses the gizzard to grind up the organic material with the help of the soil particles that it also takes in. The body of an earthworm consists of 100 to 175 similar segments, with a mouth on the first and an anus on the last.

FIGURE 42.15
Earthworms mating. The anterior ends are pointing in opposite directions.

The material that passes through an earthworm is deposited outside of its burrow in the form of castings, familiar objects that look as if they were extruded from a toothpaste tube. In this way, earthworms aerate and enrich the soil, a subject that fascinated Charles Darwin, particularly in his later years. In his book, *The Formation of Vegetable Mould Through the Action of Worms*, Darwin pointed out that a worm can eat its own weight in soil every day and that the equivalent of 22 to 40 metric tons of soil per hectare pass through their intestines every year. This indicates a population of well over 16,000 worms per hectare!

In view of the underground lifestyle that earthworms have evolved, it is not surprising that they have no eyes. However, earthworms do have light-sensitive and touch-sensitive organs. The light-sensitive organs of the oligochaetes are concentrated in the segments near each end of the body—those regions most likely to be exposed to light. Earthworms have fewer setae than polychaetes and have no parapodia. In addition, earthworms lack the distinct head regions that characterize polychaetes.

Earthworms are hermaphroditic, another way in which they differ from most polychaetes. When they mate (figure 42.15), their anterior ends point in opposite directions, and their ventral surfaces touch. The clitellum is obvious as a thickened band on an earthworm's body. The mucus it secretes holds the worms together during copulation. Sperm cells are released from pores in specialized segments of one partner into the sperm receptacles of the other, the process going in both directions simultaneously.

Two or three days after the worms separate, the clitellum of each worm secretes a mucous cocoon, surrounded by a protective layer of chitin. As this sheath passes over the female pores of the body—a process that takes place as the worm moves—it receives eggs. As it subsequently passes along the body, it incorporates the sperm that were deposited during copulation. When the mucous sheath finally passes over the end of the worm, its ends pinch together. The sheath then encloses the fertilized eggs in a cocoon from which the young worms ultimately hatch.

Class Hirudinea: The Leeches

Leeches (class Hirudinea) occur mostly in fresh water, although a few are marine and some tropical leeches occupy terrestrial habitats. Most leeches are 2 to 6 centimeters long, but one tropical species reaches up to 30 centimeters. Leeches are usually flattened dorsoventrally, like flatworms. They are hermaphroditic, and develop a clitellum during the breeding season; cross-fertilization is obligatory. Leech eggs are enclosed in cocoons and laid on damp earth, attached to objects in water, or occasionally attached to the host or to the parent leech itself.

A leech's coelom is reduced and continuous throughout the body, not divided into individual segments as in the polychaetes and oligochaetes. Leeches have evolved suckers at one or both ends of the body. Those that have suckers at both ends move by attaching first one and then the other end to the substrate looping along. Except for one species, leeches have no setae.

Many leeches have evolved the ability to suck blood from animals, including humans, other vertebrates, and crustaceans. Many freshwater leeches live as external parasites. They remain on their hosts for long periods and suck their blood from time to time.

The best-known leech is the medicinal leech, *Hirudo medicinalis*. Individuals of *Hirudo* are 10 to 12 centimeters long and have blade-like, chitinous jaws that rasp through the skin of the victim. The leech secretes an anticoagulant into the wound to prevent the blood from clotting as it flows out, and its powerful sucking muscles pump the blood out quickly once the hole has been opened. European pharmaceutical companies still collect leeches to remove excess blood after surgery. Following surgery, blood may accumulate because veins may function improperly and fail to circulate the blood. The accumulating blood "turns off" the arterial supply of fresh blood, and the tissue often dies. When leeches remove the excess blood, new capillaries form in about a week, and the tissues remain healthy.

Segmented annelids evolved in the sea. Earthworms are their descendents, as are parasitic leeches.

Lophophorates

Three phyla of marine animals—Ectoprocta, Brachiopoda, and Phoronida—are characterized by a **lophophore**, a circular or U-shaped ridge around the mouth bearing one or two rows of ciliated, hollow tentacles (figure 42.16). Because of this unusual feature, they are thought to be related to one another. The lophophore presumably arose in a common ancestor. The coelomic cavity of lophophorates extends into the lophophore and its tentacles. The lophophore functions as a surface for gas exchange and as a food-collection organ. Lophophorates use the cilia of their lophophore to capture the organic detritus and plankton on which they feed. Lophophorates are attached to their substrate or move slowly.

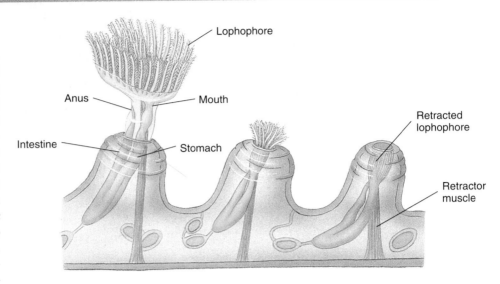

FIGURE 42.16
A lophophorate. A small portion of a colony of the freshwater ectoproct *Plumatella* (phylum Ectoprocta), which grows on the underside of rocks. The individual at the left has a fully extended lophophore, the structure characteristic of the three lophophorate phyla. The tiny individuals of *Plumatella* disappear into their shells when disturbed.

Lophophorates share some features with mollusks, annelids and arthropods (all protostomes) and share others with deuterostomes. Cleavage is mostly radial, as in deuterostomes. The formation of the coelom varies; some lophophorates resemble protostomes in this respect, others deuterostomes. In the Phoronida, the mouth forms from the blastopore, while in the other two phyla, it forms from the end of the embryo opposite the blastopore. Evidence has been supplied recently that shows the ribosomes of all lophophorates are decidedly protostome-like, lending strength to placing them within the protostome phyla. Despite the differences among the three phyla, the unique structure of the lophophore seems to indicate that the members share a common ancestor. Their relationships continue to present a fascinating puzzle.

Phylum Phoronida: The Phoronids

Phoronids (phylum Phoronida) superficially resemble common tube worms seen on dock pilings. Each phoronid secretes a chitinous tube and lives out its life within it. They also extend tentacles to feed and quickly withdraw them when disturbed, but there the resemblance to the tube worm ends. Instead of a straight tube-within-a-tube body plan, phoronids have a U-shaped gut. Only about 10 phoronid species are known, ranging in length from a few millimeters to 30 centimeters. Some species lie buried in sand, others are attached to rocks either singly or in groups. Phoronids develop as protostomes, with radial cleavage and the anus developing secondarily.

Phylum Ectoprocta: The Bryozoans

Ectoprocts (phylum Ectoprocta) look like tiny, short versions of phoronids. Because they are small—usually less than 0.5 millimeter long—and mostly colonial, they were called moss animals, or Bryozoa. The name Ectoprocta refers to the location of the anus (proct), which is external to the lophophore. The 4000 species include both marine and freshwater forms—the only nonmarine lophophorates. Individual ectoprocts secrete a tiny chitinous chamber, or **zooecium**, attached to rocks and other members of the colony. Individuals communicate chemically through pores between chambers. Ectoprocts develop as deuterostomes, with the mouth developing secondarily; cleavage is radial.

Phylum Brachiopoda: The Brachiopods

Brachiopods, or lamp shells, superficially resemble clams, with two calcified shells. Many species attach to rocks or sand by a stalk that protrudes through an opening in one shell. The lophophore lies within the shell and functions when the brachiopod's shells are opened slightly. Although slightly more than 300 species of brachiopods (phylum Brachiopoda) exist today, more than 30,000 species of this phylum are known as fossils. Brachiopods develop as deuterostomes and show radial cleavage.

The three phyla of lophophorates probably share a common ancestor, and they show a mixture of protostome and deuterostome characteristics.

42.1 **Mollusks were among the first coelomates.**

- Mollusks contain a true body cavity, or coelom, within the embryonic mesoderm and were among the first coelomate animals.

- The mollusks constitute the second largest phylum of animals in terms of named species. Their body plan consists of distinct parts: a head, a visceral mass, and a foot.

- Of the seven classes of mollusks, the gastropods (snails and slugs), bivalves (clams and scallops), and cephalopods (octopuses, squids, and nautilus), are best known.

- Gastropods typically live in a hard shell. During development, one side of the embryo grows more rapidly than the other, producing a characteristic twisting of the visceral mass.

- Members of the class Bivalvia have two shells hinged together dorsally and a wedge-shaped foot. They lack distinct heads and radulas. Most bivalves are filter-feeders.

- Octopuses and other cephalopods are efficient and often large predators. They possess well-developed brains and are the most intelligent invertebrates.

42.2 **Annelids were the first segmented animals.**

- Segmentation is a characteristic seen only in coelomate animals at the annelid evolutionary level and above. Segmentation, or the repetition of body regions, greatly facilitates the development of specialized regions of the body.

- Annelids are worms with bodies composed of numerous similar segments, each with its own circulatory, excretory, neural elements, and array of setae. There are three classes of annelids, the largely marine Polychaeta, the largely terrestrial Oligochaeta, and the largely freshwater Hirudinea.

- Polychaetes have a well-developed head with specialized sense organs, and distinctive paired, fleshy, paddle-like parapodia on most of their segments. Parapodia aid in locomotion and gas exchange.

- Oligochaetes (earthworms) have a body consisting of 100 to 175 similar segments, with a mouth on the first of these and an anus on the last. Earthworms are hermaphroditic with light- and touch-sensitive organs and sensitive moisture-detecting cells.

- Leeches (class Hirudinea) are usually flattened dorsoventrally and occur mostly in fresh water. Most leeches have evolved the ability to suck blood from vertebrates and other animals.

42.3 **Lophophorates appear to be a transitional group.**

- The lophophorates consist of three phyla of marine animals, Phoronida, Ectoprocta, and Brachiopoda, characterized by a circular or U-shaped ridge, the lophophore, around the mouth.

- Some lophophorates have characteristics like protostomes, others like deuterostomes. All are characterized by a lophophore and are thought to share a common ancestor.

1. **Coelom**   The coelom is a body cavity between layers of embryonic mesoderm. In animals possessing a coelom, organs are typically packaged within the body cavity. In animals that do not possess such a cavity, organs are in contact with one another.

2. **Segmentation**   The body plan of animals was fundamentally changed with the advent of segmentation, in which the body is built of multiple similar units, some of which become specialized to carry out particular functions.

3. **Lophophore**   The lophophore is a unique ring or U-shaped ridge of tentacles, into which the coelom extends, used for feeding and gas exchange. Lophophores are found only among the three lophophorate phyla, strongly implying that they have a common ancestor.

4. **Trochophore**   A trochophore is a free-swimming larval stage seen in the mollusks and annelids. Their common possession of trochophores argues that mollusks and annelids had a common ancestor.

5. **Radula**   A radula is a unique rasping tongue found in most mollusks and no other phylum.

1. What is the difference among an acoelomate, a pseudocoelomate, and a coelomate animal? What advantages does a coelomate animal have over acoelomate or pseudocoelomate animals?

2. What is the basic body plan of a mollusk? Where is the mantle located? Why is it important in the mollusks? What occurs in the mantle cavity of aquatic mollusks?

3. What is a radula? Do all classes of mollusks possess this structure? How is it used in different types of mollusks?

4. What part of the mollusk body is represented by the coelom?

5. How does the mollusk excretory structure work? Why is it better than the flame cells of acoelomates?

6. What are trochophores? What is a veliger?

7. Do bivalves generally disperse as larvae or adults? Explain.

8. What evolutionary advantages does segmentation confer upon an organism?

9. What are annelid setae? What function do they serve?

10. How are annelids developmentally similar to mollusks? What is the likely evolutionary connection between these phyla?

11. What are parapodia? What class of annelids possess them? What is their function?

12. How do earthworms obtain their nutrients? What sensory structures do earthworms possess? How do these animals reproduce?

13. What adaptations enable leeches to suck blood?

14. What prominent feature characterizes the lophophorate animals? What are the functions of this feature?

Thought Questions

1. Scientists believe the ancestral mollusk had a very limited shell, consisting mainly of calcareous plates; many contemporary mollusks now seem to be in the process of losing their shells. Most mollusks, however, still have well-developed shells. What is the evolutionary advantage of having a shell? of not having one?

Internet Links

Squid: The Inside Story
http://seawifs.gsfc.nasa.gov/OCEAN_PLANET/HTML/
squid_inside_story.html
An unusual and interesting site that lets you explore a squid's body from the inside, learning how it eats, moves, and reproduces.

What Seashell Is That?
http://museum.nhm.uga.edu/GSC/shellpic.htm
A gallery of seashell photographs with brief species descriptions.

Sluggish Beauties
http://home.mem.net/~zipper/
A stunning collection of 135 photographs of Philippine nudibranchs.

Seaslug Central
http://www.slugsite.tierranet.com
For those who want to learn more about nudibranchs, this link offers a fine collection of sea slug web sites.

For Further Reading

Baverdam, F.: "Even for Ethereal Phantasms, It's a Dog-Eat-Dog World," *Smithsonian*, August 1989, pages 94–101. A fascinating account of the ways in which nudibranchs (marine slugs) deter predators.

Campbell, D. G.: "The Bottom of the Bottom of the World," *Natural History*, vol. 101, November 1992, pages 46–53. An interesting look at deep-sea polychaete annelids.

Clark, R. B.: *Dynamics in Metazoan Evolution; The Origin of the Coelom and Segments*, Clarendon Press, Oxford, 1964. Dated but classic treatment of the impact of the coelom on animal evolution.

Gorman, J.: "Worms Stand Tall: A Lowly Creature Gains New Status in the Coming Age of Compost," *Audubon*, vol. 95, November–December 1993, pages 106–7. *Eisenia foetida*, despite its unattractive name, is a little earthworm with a big contribution to composting.

Gosline, J. M. and M. E. DeMont: "Jet-Propelled Swimming in Squids," *Scientific American*, January 1985, pages 96–103. Describes how a squid jets through the water as rapidly as a fish for short distances by contracting muscles in its boneless mantle wall.

Halanych, K. M., and others: "Evidence from 18S Ribosomal DNA that the Lophophorates Are Protostome Animals," *Science*, vol. 267, March 17, 1995, pages 1641–43. Molecular biological data is used to help place the lophophorates among the Metazoa phylogenetically.

Hedeen, R. A.: *The Oyster: The Life and Lore of the Celebrated Bivalve*, Tidewater Publishers, Centreville, MD, 1986. The natural history of oysters, with emphasis on their role in Chesapeake Bay.

Lent, C. M. and M. H. Dickinson: "The Neurobiology of Feeding in Leeches," *Scientific American*, June 1988, pages 98–103. Illustrates beautifully the use of leeches as experimental animals.

Richardson, J. R.: "Brachiopods," *Scientific American*, September 1986, pages 100–106. Fascinating discussion of the ecology and diversity of this ancient phylum.

Sawyer, R. T.: "In Search of the Giant Amazon Leech," *Natural History*, December 1990, pages 66–67. In order to feed, leeches secrete an anticoagulant. Leech anticoagulants continue to be important life-savers in medicine today.

43

Arthropods

FIGURE 43.1
An arthropod. One of the major arthropod groups is represented here by *Polistes*, the common paper wasp (class Insecta).

The evolution of segmentation among annelids marked the first major innovation in body structure among coelomates. An even more profound innovation followed. It marks the origin of the body plan characteristic of the most successful of all animal groups. This innovation was the development of jointed appendages in arthropods, a phylum that almost certainly evolved from an annelid ancestor. Arthropod bodies are segmented like those of annelids, but the individual segments often exist only during early development and fuse into functional groups as adults. In arthropods like the wasp above (figure 43.1) jointed appendages include legs, antennae, and a complex array of mouthparts. The functional flexibility provided by such a broad array of appendages has made arthropods the most successful of animal groups.

Arthropods

With the evolution of the first annelids, many of the major innovations of animal structure had already appeared: the division of tissues into three primary types (endoderm, mesoderm, and ectoderm), bilateral symmetry, a coelom, and segmentation. Another innovation yet remained—the development of jointed appendages. The most successful of all animal groups, the arthropods (phylum Arthropoda) have jointed appendages.

Jointed Appendages

The name "arthropod" comes from two Greek words, *arthros*, jointed, and *podes*, feet. All arthropods have jointed appendages. The numbers of these appendages are reduced in the more advanced members of the phylum. Individual appendages may be modified into antennae, mouthparts of various kinds, or legs. Some appendages, such as the wings of certain insects, are not homologous to the other appendages; insect wings evolved separately.

To gain some idea of the importance of jointed appendages, imagine yourself without them—no hips, knees, ankles, shoulders, elbows, wrists, or knuckles. Without jointed appendages, you could not walk or grasp any object. Arthropods use jointed appendages such as legs for walking, antennae to sense their environment, and mouthparts for feeding.

The arthropod body plan has a second major innovation: a rigid external skeleton, or **exoskeleton,** made of chitin. In any animal, the skeleton functions to provide places for muscle attachment. In arthropods, the muscles attach to the interior surface of their hard chitin shell, which also protects the animal from predators and impedes water loss. However, there is a limitation. Chitin is tough but brittle and cannot support great weight. As a result, the exoskeleton must be much thicker to bear the pull of the muscles in large insects than in small ones. That is why you

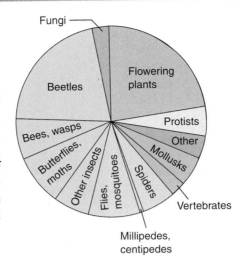

FIGURE 43.2
Arthropods are a successful group.
About two-thirds of all named species are arthropods. About 90% of all arthropods are insects, and about half of the named species of insects are beetles.

FIGURE 43.3
An ancient arthropod. Trilobites, early members of the arthropod phylum, were abundant over 250 million years ago.

FIGURE 43.4
Peripatus, a velvet worm (phylum Onychophora). These animals have features in common with both annelids and arthropods, but most scientists believe they are most closely related to arthropods.

don't see beetles as big as birds, or crabs the size of a cow—the exoskeleton would be so thick the animal couldn't move its great weight. Because this size limitation is inherent in the body design of arthropods, there are no large arthropods—few are larger than your thumb.

The Most Successful of All Animals

Arthropods, especially the largest class—insects—are by far the most successful of all animals. Approximately 1,000,000 species—about two-thirds of all the named species on earth—are members of this phylum (figure 43.2). One scientist recently estimated, based on the number and diversity of insects in tropical forests, that there might be as many as 30 million species in this one class alone. About 200 million insects are alive at any one time for each human! Insects and other arthropods abound in every habitat on the planet, but they especially dominate the land, along with flowering plants and vertebrates.

The majority of arthropod species consist of small animals, mostly about a millimeter in length. Members of the phylum range in adult size from about 80 micrometers long (some parasitic mites) to 3.6 meters across (a gigantic crab found in the sea off Japan).

Fossil records indicate that arthropods existed with the earliest well-preserved multicellular animals, over 630 million years ago. The trilobites were an important group of early arthropods that have been extinct for 250 million years (figure 43.3). Trilobites, which lived in the early seas, were the first animals whose eyes were capable of a high degree of resolution. A group of living animals called the velvet worms (phylum Onychophora) have several features in common with arthropods and appear to be closely related to them (figure 43.4). Fossils similar to Onychophora go back some 500 million years. The earliest known insect fossils are more than 300 million years old. Some groups of insects that date back

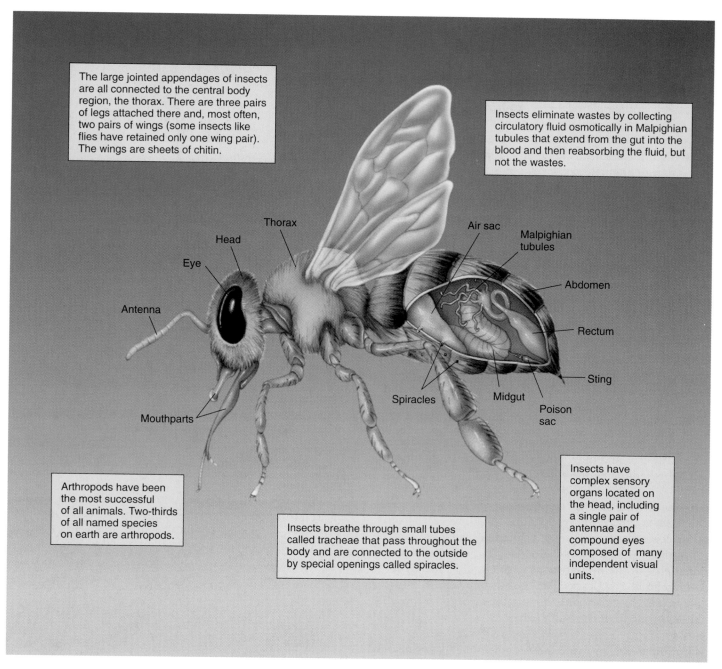

The large jointed appendages of insects are all connected to the central body region, the thorax. There are three pairs of legs attached there and, most often, two pairs of wings (some insects like flies have retained only one wing pair). The wings are sheets of chitin.

Insects eliminate wastes by collecting circulatory fluid osmotically in Malpighian tubules that extend from the gut into the blood and then reabsorbing the fluid, but not the wastes.

Arthropods have been the most successful of all animals. Two-thirds of all named species on earth are arthropods.

Insects breathe through small tubes called tracheae that pass throughout the body and are connected to the outside by special openings called spiracles.

Insects have complex sensory organs located on the head, including a single pair of antennae and compound eyes composed of many independent visual units.

Thorax — Head — Eye — Antenna — Mouthparts — Spiracles — Midgut — Poison sac — Sting — Rectum — Abdomen — Malpighian tubules — Air sac

FIGURE 43.5

The evolution of jointed appendages and an exoskeleton. Insects and other arthropods (phylum Arthropoda) have a coelom, segmented bodies, and jointed appendages. The three body regions of an insect (head, thorax, and abdomen) are each actually composed of a number of segments that fuse during development. All arthropods have a strong exoskeleton made of chitin. One class, the insects, has evolved wings that permit them to fly rapidly through the air.

that far in the fossil record still have living representatives! For example, during the Carboniferous Period, giant dragonflies lived that had wingspans of 60 centimeters.

Arthropods, especially insects (figure 43.5), are of enormous economic importance and affect all aspects of human life. They compete with humans for food of every kind, play a key role in the pollination of certain crops, and cause billions of dollars of damage to crops, before and after harvest. They are by far the most important her-

bivores in all terrestrial ecosystems. Virtually every kind of plant is eaten by one or more species of insect. Diseases spread by insects cause enormous financial damage each year and strike every kind of domesticated animal and plant, as well as human beings.

Arthropods are segmented protostomes with jointed appendages. Arthropods are the most successful of all animal groups.

General Characteristics of Arthropods

Arthropod bodies are segmented like annelids, a phylum to which at least some arthropods are clearly related. Members of some classes of arthropods have many body segments. In others, the segments have become fused together into functional groups, or **tagmata** (singular, **tagma**), such as the head or thorax of an insect (figure 43.6). This fusing process, known as **tagmatization,** is of central importance in the evolution of arthropods. In most arthropods, the original segments can be distinguished during larval development. All arthropods have a distinct head, sometimes fused with the thorax to form a tagma called the **cephalothorax.**

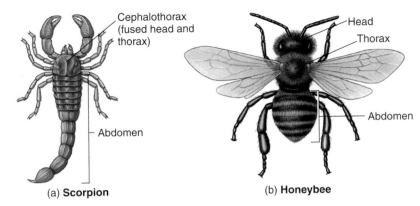

(a) **Scorpion** (b) **Honeybee**

FIGURE 43.6
Arthropod evolution from many to few body segments. The (a) scorpion and the (b) honeybee are arthropods with different numbers of body segments.

Head Appendages Define Key Groups

Living members of the phylum Arthropoda are divided into three subphyla: Chelicerata (spiders, horseshoe crabs, and sea spiders), Crustacea (lobsters, sowbugs, and barnacles), and Uniramia (insects, centipedes, and millipedes). Arthropods that lack jaws are called the **chelicerates** and include the Chelicerata. Their mouthparts are called **chelicerae.** The Crustacea and the Uniramia have jaws, or **mandibles,** and are called the **mandibulates.**

Chelicerates. Among chelicerates, the first pair of appendages are the chelicerae (figure 43.7a), the second pair of appendages is usually pincer-like or feeler-like, and the remaining pairs of appendages are legs. The fundamental differences in the derivation and structure of their mouthparts indicate that chelicerates and mandibulates represent different evolutionary lines among the arthropods and that neither group gave rise to the other.

Mandibulates. In mandibulates, the appendages nearest the anterior end are one or more pairs of sensory antennae, and the next appendages are the mandibles (figure 43.7b). All crustacean appendages are basically **biramous,** or "two-branched" (figure 43.8), although some of these appendages have become single-branched by reduction in the course of their evolution. Members of the Uniramia have **uniramous,** or single-branched, mandibles and other appendages.

> Chelicerates (subphylum Chelicerata) have mouthparts that are usually pincers or fangs; mandibulates have biting jaws. The appendages of mandibulates may either be biramous (subphylum Crustacea) or uniramous (subphylum Uniramia).

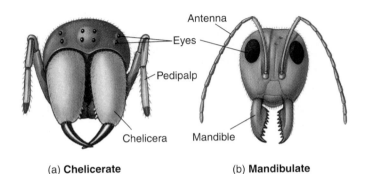

(a) **Chelicerate** (b) **Mandibulate**

FIGURE 43.7
Chelicerates and mandibulates. In the chelicerates, such as a spider (a), the chelicerae are the foremost appendages of the body. In contrast, the foremost appendages in the mandibulates, such as an ant (b), are the antennae, followed by the mandibles.

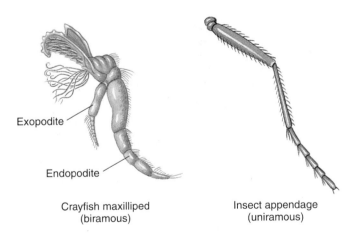

Crayfish maxilliped (biramous)   Insect appendage (uniramous)

FIGURE 43.8
Mandibulate appendages. A biramous leg in a crustacean (crayfish) and a uniramous leg in an insect.

External Features

Aside from the segmentation and appendages of arthropods, two characteristic external features of the phylum deserve special discussion: the exoskeleton and the compound eye.

Exoskeleton

The bodies of all arthropods are covered by a chitinous exoskeleton, or cuticle. This tough outer covering, against which the muscles work, is secreted by the epidermis and fused with it. The exoskeleton may be fairly soft and flexible or it can be thick and very hard, as in some large insects, horseshoe crabs, and other groups. In most crustaceans, the exoskeleton is impregnated with calcium carbonate and is relatively inflexible. The exoskeleton protects arthropods from water loss and helps to protect them from predators, parasites, and injury.

Molting.   Arthropods periodically undergo **ecdysis,** or molting, the shedding of the outer cuticular layer. When they outgrow their exoskeleton, they form a new one underneath. This process is controlled by hormones. When the new exoskeleton is complete, it becomes separated from the old one by fluid. This fluid dissolves the chitin and, if it is present, calcium carbonate, from the old exoskeleton. The fluid increases in volume until, finally, the original exoskeleton cracks open, usually along the back, and is shed. The arthropod emerges, clothed in a new, pale, and still somewhat soft exoskeleton. The arthropod then "puffs itself up," ultimately expanding to full size. The blood circulation to all parts of the body aids them in this expansion, and many insects and spiders take in air to assist them. The expanded exoskeleton subsequently hardens. While the exoskeleton is soft, the animal is especially vulnerable. At this stage, arthropods often hide under stones, leaves, or branches.

Compound Eye

Another important structure in many arthropods is the **compound eye** (figure 43.9a). Compound eyes are composed of many independent visual units, often thousands of them, called **ommatidia.** Each ommatidium is covered with a lens and linked to a complex of eight retinula cells and a light-sensitive central core, or **rhabdom.** Compound eyes among insects are of two main types: **apposition eyes** and **superposition eyes.** Apposition eyes are like those of bees, in which each ommatidium acts in isolation, collecting light from one sector of the external world and throwing an inverted image of that scene on the rhabdom of the ommatidium. In such an eye, the individual ommatidia are surrounded by pigment cells that keep the light that reaches each ommatidium separate. The individual images are combined in the insect's brain to form its visual image of the external world. In superposition eyes, such as those found in moths, the images from a series of ommatidia are combined on the retina at the back of the compound eye; there are no screening pigment cells. Such an image is right-side-up.

Simple eyes, or **ocelli,** with single lenses are found in the other arthropod groups and sometimes occur together with compound eyes, as is often the case in insects (figure 43.9b). Ocelli function in distinguishing light from darkness. The ocelli of some flying insects, namely locusts and dragonflies, function as horizon detectors and help the insect visually stabilize its course in flight.

All arthropods have a rigid, chitinous exoskeleton that provides places for muscle attachment, protects the animal from predators and injury, and, most important, impedes water loss. Many arthropods have compound eyes.

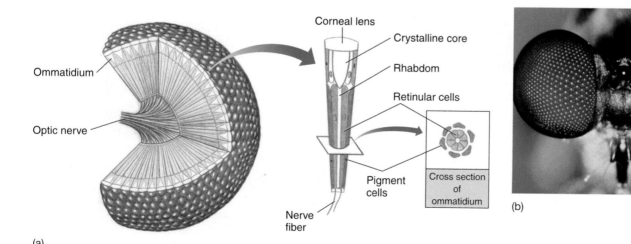

(a)

Ommatidium

Optic nerve

Corneal lens

Crystalline core

Rhabdom

Retinular cells

Pigment cells

Nerve fiber

Cross section of ommatidium

(b)

FIGURE 43.9
The compound eye. (a) The compound eyes found in insects are complex structures. (b) Three ocelli are visible between the compound eyes of the robberfly (order Diptera).

Internal Features

In the course of arthropod evolution, the coelom has become greatly reduced, consisting only of cavities that house the reproductive organs and some glands. Arthropods completely lack cilia, both on the external surfaces of the body and on the internal organs. Like annelids, arthropods have a tubular gut that extends from the mouth to the anus. In the next paragraphs we will discuss the circulatory, respiratory, excretory, and nervous systems of the arthropods (figure 43.10).

Circulatory System

The circulatory system of arthropods is open; their blood flows through cavities between the internal organs and not through closed vessels. The principal component of an insect's circulatory system is a longitudinal vessel called the heart. This vessel runs near the dorsal surface of the thorax and abdomen. When it contracts, blood flows into the head region of the insect.

When an insect's heart relaxes, blood returns to it through a series of valves. These valves are located in the posterior region of the heart and allow the blood to flow inward only. Thus, blood from the head and other anterior portions of the insect gradually flows through the spaces between the tissues toward the posterior end and then back through the one-way valves into the heart. Blood flows most rapidly when the insect is running, flying, or otherwise active. At such times, the blood efficiently delivers nutrients to the tissues and removes wastes from them.

Respiratory System

Insects and other members of subphylum Uniramia, which are fundamentally terrestrial, depend on their respiratory rather than their circulatory system to carry oxygen to their tissues. In vertebrates, blood moves within a closed circulatory system to all parts of the body, carrying the oxygen with it. This is a much more efficient arrangement than exists in arthropods, in which all parts of the body need to be near a respiratory passage to obtain oxygen. As a result, the size of the arthropod body is much more limited than that of the vertebrates.

Unlike most animals, arthropods have no single major respiratory organ. The respiratory system of most terrestrial arthropods consists of small, branched, cuticle-lined air ducts called **tracheae** (figure 43.11). These tracheae, which ultimately branch into very small **tracheoles,** are a series of tubes that transmit oxygen throughout the body. Tracheoles are in direct contact with individual cells, and oxygen diffuses directly across the cell membranes. Air passes into the tracheae by way of specialized openings in the exoskeleton called **spiracles,** which, in most insects, can be opened and closed by valves. The ability to prevent water loss by closing the spiracles was a key adaptation that facilitated the invasion of the land by arthropods. In many insects, especially larger ones, muscle contraction helps to increase the flow of gases in and out of the tracheae. In other terrestrial arthropods, the flow of gases is essentially a passive process.

Many spiders and some other chelicerates have a unique respiratory system that involves **book lungs,** a series of leaf-like plates within a chamber. Air is drawn in and expelled out of this chamber by muscular contraction. Book lungs may exist alongside tracheae, or they may function instead of tracheae. One small class of marine chelicerates, the horseshoe crabs, have book gills, which are analogous to book lungs but function in water. Tracheae, book lungs, and book gills are all structures found only in arthropods and in the phylum Onychophora, which have tracheae. Crustaceans lack such structures and have gills.

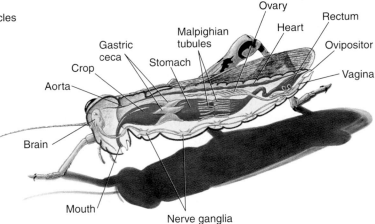

FIGURE 43.10
A grasshopper (order Orthoptera). This grasshopper illustrates the major structural features of the insects, the most numerous group of arthropods. (a) External anatomy. (b) Internal anatomy.

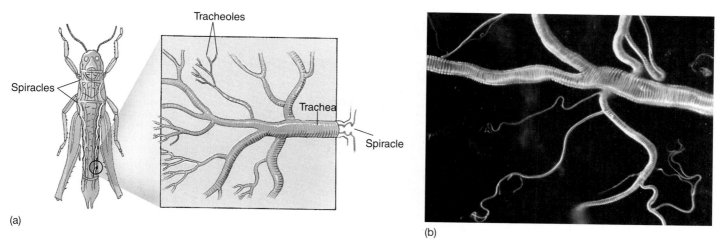

(a)

(b)

FIGURE 43.11

Tracheae and tracheoles. Tracheae and tracheoles are connected to the exterior by specialized openings called spiracles and carry oxygen to all parts of a terrestrial insect's body. (a) The tracheal system of a grasshopper. (b) A portion of the tracheal system of a cockroach.

Excretory System

Though there are various kinds of excretory systems in different groups of arthropods, we will focus here on the unique excretory system consisting of **Malpighian tubules** that evolved in terrestrial uniramians. Malpighian tubules are slender projections from the digestive tract that are attached at the junction of the midgut and hindgut (figure 43.12). Fluid passes through the walls of the Malpighian tubules to and from the blood in which the tubules are bathed. As this fluid passes through the tubules toward the hindgut, nitrogenous wastes are precipitated as concentrated uric acid or guanine. These substances are then emptied into the hindgut and eliminated. Most of the water and salts in the fluid are reabsorbed by the hindgut and rectum and returned to the arthropod's body. Malpighian tubules are an efficient mechanism for water conservation and were another key adaptation facilitating invasion of the land by arthropods.

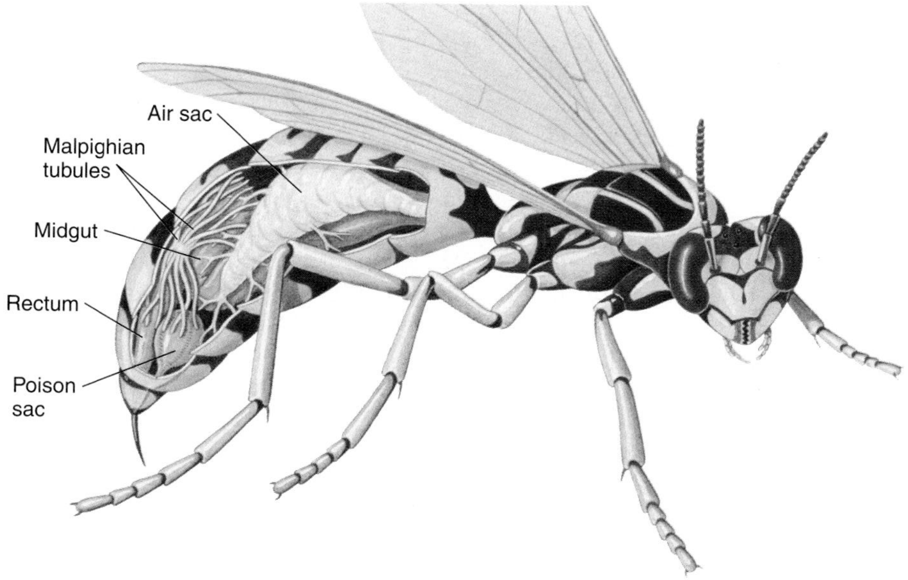

FIGURE 43.12

Malpighian tubules. The Malpighian tubules of insects are extensions of the digestive tract that collect water and wastes from the body's circulatory system. Water is later reabsorbed, while wastes are eliminated.

Nervous System

The central feature of the arthropod nervous system is a double chain of segmented ganglia running along the animal's ventral surface. At the anterior end of the animal are three fused pairs of dorsal ganglia, which constitute the brain. However, much of the control of an arthropod's activities is relegated to ventral ganglia. Therefore, the animal can carry out many functions, including eating, movement, and copulation, even if the brain has been removed. The brain of arthropods seems to be a control point, or inhibitor, for various actions, rather than a stimulator, as it is in vertebrates.

Arthropods have no cilia and only a limited coelom. Their circulatory system is open. Most terrestrial insects have a network of tubes called tracheae that transmit oxygen from the outside to the organs. Many arthropods eliminate metabolic wastes by a unique system of Malpighian tubules.

Chelicerates (subphylum Chelicerata) are a distinct evolutionary line of arthropods in which the appendages nearest the anterior end of the body have been modified into chelicerae. These chelicerae, which often function as fangs or pincers, had a different evolutionary origin from the mandibles that originated separately in the crustaceans and in the Uniramia. There are three classes of chelicerates: arachnids, horseshoe crabs, and sea spiders.

Class Arachnida: The Arachnids

By far the largest of the three classes of chelicerates is the largely terrestrial Arachnida, with some 57,000 named species; it includes spiders, ticks, mites, scorpions, and daddy longlegs. Arachnids have a pair of chelicerae, a pair of pedipalps, and four pairs of walking legs. The chelicerae are the foremost appendages; they consist of a stout basal portion and a movable fang often connected to a poison gland. **Pedipalps,** the next pair of appendages, may resemble legs, but they have one less segment. In male spiders, pedipalps are specialized copulatory organs; in other groups of arthropods, they may have a specialized sensory function. In scorpions, the pedipalps are large and pinching. Despite their appearance, pedipalps of carnivorous arachnids, such as spiders, are often used for catching and handling prey. Some spiders also chew with the basal portions of their pedipalps. Pedipalps are rarely used for locomotion.

Most arachnids are carnivorous. The main exception is mites, which are largely herbivorous. Most arachnids can ingest only preliquified food, which they often digest externally by secreting enzymes into their prey. They can then suck up the digested material with their muscular, pumping pharynx. Arachnids are primarily, but not exclusively, terrestrial. Some 4000 known species of mites and one species of spider live in fresh water, and a few mites live in the sea. Arachnids breathe by means of tracheae, book lungs, or both.

There are eleven orders of arachnids that include living species. Of these, we will briefly discuss four of the most familiar: Opiliones (harvestmen, or daddy longlegs), Scorpiones (scorpions), Araneae (spiders), and Acari (mites and ticks).

FIGURE 43.13
A harvestman, or daddy longlegs.

Order Opiliones: The Daddy Longlegs

A familiar group of arachnids consists of the daddy longlegs, or harvestmen (order Opiliones). Members of this order are easily recognized by their oval, compact bodies and extremely long, slender legs (figure 43.13). They respire by means of a primary pair of tracheae and are unusual among the arachnids in that they engage in direct copulation. The males have a penis, and the females an **ovipositor,** or egg-laying organ which deposits their eggs in cracks and crevices. Most daddy longlegs are predators of insects and other arachnids, but some live on plant juices and many scavenge dead animal matter. The order includes about 5000 species.

Order Scorpiones: The Scorpions

Scorpions (order Scorpiones) are arachnids whose pedipalps are modified into pincers. Scorpions use these pincers to handle and tear apart their food (figure 43.14). The venomous stings of scorpions are used mainly to stun their prey and less commonly in self-defense. The stinging apparatus is located in the terminal segment of the abdomen. A scorpion holds its abdomen folded forward over its body when it is moving about. The elongated, jointed abdomens of scorpions are distinctive; in most chelicerates, the abdominal segments are more or less fused together and appear as a single unit.

Scorpions are probably the most ancient group of terrestrial arthropods; they are known from the Silurian Period, some 425 million years ago. Adults of this order of arachnids range in size from 1 to 18 centimeters. There are some 1200 species of scorpions, all terrestrial, which occur throughout the world. They are most common in tropical, subtropical, and desert regions. The young are born alive, with 1 to 95 in a given brood.

FIGURE 43.14
The scorpion *Uroctonus mordax.* This photograph shows the characteristic pincers and segmented abdomen, ending in a stinging apparatus, raised over the animal's back. The white mass is the scorpion's young.

Order Araneae: The Spiders

There are about 35,000 named species of spiders (order Araneae). These animals play a major role in virtually all terrestrial ecosystems. They are particularly important as predators of insects and other small animals. Spiders hunt their prey or catch it in silk webs of remarkable diversity. The silk is formed from a fluid protein that is forced out of spinnerets (figure 43.15), which are modified appendages on the posterior portion of the spider's abdomen. There may be up to six pairs of these silk glands. Different kinds of spiders produce many adaptive modifications of webs. Some spiders can spin gossamer floats that allow them to drift away in the breeze to a new site. The webs and habits of individual kinds of spiders are often distinctive.

Many kinds of spiders, like the familiar wolf spiders and tarantulas, do not spin webs but instead hunt their prey actively (figure 43.16). Others, called trap-door spiders, construct silk-lined burrows with lids, seizing their prey as it passes by. One species of spider, *Argyroneta aquatica*, lives in fresh water, spending most of its time below the surface. Its body is surrounded by a bubble of air, while its legs, which are used both for underwater walking and for swimming, are not. Several other kinds of spiders walk about freely on the surface of water.

Spiders have poison glands leading through their chelicerae, which are pointed and used to bite and paralyze prey. Some members of this order, such as the black widow and brown recluse (figure 43.17), have bites that are poisonous to humans and other large mammals.

Male spiders make a sperm web from special silk glands in the anterior portion of their abdomen. A drop of sperm deposited on the sperm web is picked up by the male spider with his pedipalps. Many spiders have an elaborate courtship, following which the male spider fits his pedipalps into a special plate on the lower side of the female's abdomen, permitting the sperm to enter specific receptacles. In many species, the female eats the male once fertilization is completed. The eggs are enclosed in a silken egg sac, which is abandoned, guarded, or carried about, depending on the species of spider. Young spiders resemble adults and become mature after 3 to 15 molts, depending on the species.

FIGURE 43.15
Spinnerets of a Peruvian orb-weaving spider in action. This spider is producing fine strands of silk.

FIGURE 43.16
Not all spiders spin webs. Tarantulas are large hunting spiders, powerful enough to prey on small lizards, mammals, and birds, as well as insects. They do not spin webs, but instead actively hunt their prey.

(a)

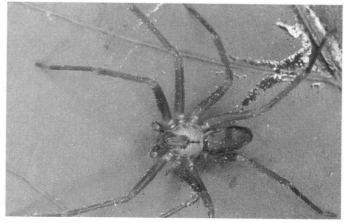

(b)

FIGURE 43.17
Two common poisonous spiders. (a) The black widow spider, *Latrodectus mactans*. (b) The brown recluse spider, *Loxosceles reclusa*. Both species are common throughout temperate and subtropical North America, but bites are rare in humans.

(a)

(b)

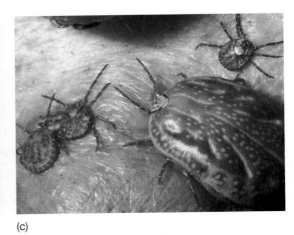
(c)

FIGURE 43.18
Mites and ticks (order Acari). (a) The tiny red flecks on this flower of creamcups, *Platystemon californicus*, are mites of the genus *Balaustium*; they have eaten all the pollen. (b) A water mite, perhaps of the genus *Hydrachna*. Note the feathery legs, modified for swimming. (c) Ticks (the large one is engorged) on the hide of a tapir in Peru. Many ticks spread diseases in humans and other vertebrates.

Order Acari: The Mites

The order Acari, the mites, is the largest in terms of number of species and the most diverse of the arachnids (figure 43.18). Although only about 30,000 species of mites have been named, scientists that study the group estimate that there may be a million or more members of this order in existence. In many regions of the world, the members of this group are poorly known and few scientists study them.

Most mites are small, less than 1 millimeter long, but adults of different species range from 100 nanometers to 2 centimeters. In most mites, the cephalothorax and abdomen are fused into an unsegmented ovoid body. Respiration occurs either by means of tracheae or directly through the exoskeleton. Many mites pass through several distinct stages during their life cycle. In most, an inactive eight-legged prelarva gives rise to an active six-legged larva, which in turn produces a succession of three eight-legged stages and, finally, the adult males and females. In a number of different mites, however, juvenile stages have become reproductive, and the development of those particular species then stops at that stage. The evolutionary process whereby juvenile stages become reproductive is known as **paedomorphosis.**

Mites are diverse, not only in their structure but also in their habits. They are found in virtually every terrestrial, freshwater, and marine habitat known and feed on fungi, plants, and animals. They act as predators and as internal and external parasites of both invertebrates and vertebrates. Many plants—at least 1000 species—have pits, pores, or crypts on their leaves that are inhabited by predaceous mites, which protect the plants from herbivores.

Many mites are well known to human beings because of their irritating bites and the diseases they transmit. Follicle mites live in the hair follicles and wax glands of the human forehead, nose, and chin, but usually cause no symptoms (figure 43.19).

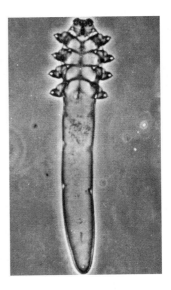

FIGURE 43.19
A follicle mite *(Demodex folliculorum)*. Extensively modified in structure in relation to their symbiotic habits, follicle mites occur in the glands at the base of the hairs of virtually every mammal (including humans), feeding on host cells and fluids. Usually a single mite occurs in each follicle, although there may be as many as three.

Ticks, which are also members of this order, are blood-feeding **ectoparasites,** parasites that occur on the surface of their host (see figure 43.18c). They are larger than most other mites and cause discomfort by sucking the blood of humans and other animals. Ticks can carry many diseases, including some caused by viruses, bacteria, and protozoa. The spotted fevers (Rocky Mountain spotted fever is a familiar example) are caused by bacteria carried by ticks. Lyme disease is apparently caused by spirochaetes transmitted by ticks. Red-water fever, or Texas fever, is an important tick-borne protozoan disease of cattle, horses, sheep, and dogs. In addition to diseases in humans and other animals, mites cause extensive and often severe damage to plants. Spider mites, also called red spiders, are often the most serious pests of houseplants. Mites of this group also damage many crops.

Scorpions, spiders, and mites are all arachnids, the largest class of chelicerates.

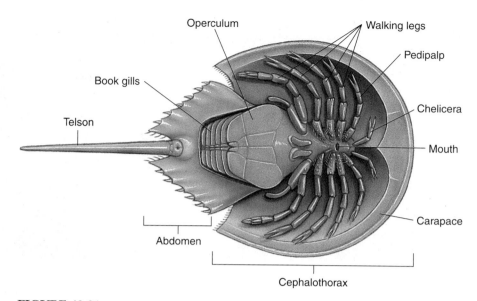

Operculum
Walking legs
Pedipalp
Book gills
Chelicera
Telson
Mouth
Carapace
Abdomen
Cephalothorax

FIGURE 43.20
Limulus. Horseshoe crabs, emerging from the sea to mate at the edge of Delaware Bay, New Jersey, in early May.

FIGURE 43.21
Diagram of a horseshoe crab, *Limulus*, from below. This diagram illustrates the principal features of this archaic animal.

Class Merostomata: Horseshoe Crabs

A second class of chelicerates is the horseshoe crabs (class Merostomata). There are three genera of horseshoe crabs. One, *Limulus* (figure 43.20), is common along the East Coast of North America. The other two genera live in the Asian tropics. Horseshoe crabs are an ancient group, with fossils virtually identical to *Limulus* dating back 220 million years to the Triassic Period. Other members of the class, the now-extinct eurypterans, are known from 400 million years ago. Horseshoe crabs may have been derived from trilobites, a relationship suggested by the appearance of their larvae. Individuals of *Limulus* grow up to 60 centimeters long. They mature in 9 to 12 years and have a lifespan of 14 to 19 years. *Limulus* individuals live in deep water, but they migrate to shallow coastal waters every spring, emerging from the sea to mate on moonlit nights when the tide is high.

Horseshoe crabs feed at night, primarily on mollusks and annelids. They swim on their backs by moving their abdominal plates. They can also walk on their four pairs of legs, which are protected, along with their chelicerae and pedipalps, by their shell (figure 43.21). Each of the five pairs of book gills of the horseshoe crab is located under a pair of covers, or opercula, posterior to the legs.

Horseshoe crabs are a very ancient group.

FIGURE 43.22
A marine pycnogonid. The sea spider *Pycnogonum littorale* (yellow animal) crawling over a sea anemone.

Class Pycnogonida: The Sea Spiders

The third class of chelicerates is the sea spiders (class Pycnogonida). Sea spiders are relatively common, especially in coastal waters, and more than 1000 species are in the class. These animals are not often observed because many are small, only about 1 to 3 centimeters long, and rather inconspicuous. They are found in oceans throughout the world but are most abundant in the far north and far south. Adult sea spiders are mostly external parasites or predators of other animals like sea anemones (figure 43.22).

Sea spiders have a sucking proboscis, with the mouth located at its end. Their abdomen is much reduced, and their body appears to consist almost entirely of the cephalothorax, with no well-defined head. Sea spiders usually have four, or less commonly five or six, pairs of legs. Male sea spiders carry the eggs on their legs until they hatch, thus providing a measure of parental care. Sea spiders completely lack excretory and respiratory systems. They appear to carry out these functions by direct diffusion, with waste products flowing outward through the cells and oxygen flowing inward through them. Sea spiders are not closely related to either of the other two classes of chelicerates.

Sea spiders are very common in marine habitats. They are not closely related to terrestrial spiders.

Crustaceans

The crustaceans (subphylum Crustacea) are a large group of primarily aquatic organisms, consisting of some 35,000 species of crabs, shrimps, lobsters, crayfish, barnacles, water fleas, pillbugs, and related groups. Most crustaceans have two pairs of antennae, three types of chewing appendages, and various numbers of pairs of legs. All crustacean appendages, with the possible exception of the first pair of antennae, are basically biramous. In some crustaceans, appendages appear to have only a single branch; in those cases, one of the branches has been lost during the course of evolutionary specialization. The **nauplius** larva stage through which all crustaceans pass (figure 43.23) provides evidence that all members of this diverse group are descended from a common ancestor. The nauplius hatches with three pairs of appendages and metamorphoses through several stages before reaching maturity. In many groups, this nauplius stage is passed in the egg, and development of the hatchling to the adult form is direct.

Crustaceans differ from insects but resemble centipedes and millipedes in that they have appendages on their abdomen as well as on their thorax. They are the only arthropods with two pairs of antennae. Their mandibles likely originated from a pair of limbs that took on a chewing function during the course of evolution, a process that apparently occurred independently in the common ancestor of the terrestrial mandibulates. Many crustaceans have compound eyes. In addition, they have delicate tactile hairs that project from the cuticle all over the body. Larger crustaceans have feathery gills near the bases of their legs. In smaller members of this class, gas exchange takes place directly through the thinner areas of the cuticle or the entire body. Most crustaceans have separate sexes. Many different kinds of specialized copulation occur among the crustaceans, and the members of some orders carry their eggs with them, either singly or in egg pouches, until they hatch.

Decapod Crustaceans

Large, primarily marine crustaceans such as shrimps, lobsters, and crabs, along with their freshwater relatives, the crayfish, are collectively called *decapod crustaceans* (figure 43.24). The term

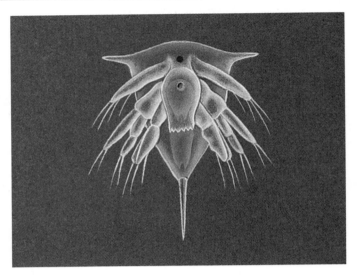

FIGURE 43.23
Although crustaceans are diverse, they have fundamentally similar larvae. The nauplius larva of a crustacean is an important unifying feature found in all members of this group.

(a)

(b)

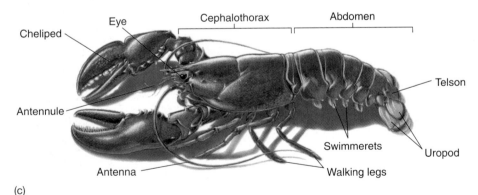

(c)

FIGURE 43.24
Decapod crustaceans. Many of the decapod crustaceans are important sources of food for humans. (a) Montague's shrimp. (b) Edible crab, *Cancer pagurus*. (c) Lobster, *Homarus americanus*. The principal features are labeled.

decapod means "ten-footed." In these animals, the exoskeleton is usually reinforced with calcium carbonate. Most of their body segments are fused into a cephalothorax covered by a dorsal shield, or **carapace,** which arises from the head. The crushing pincers common in many decapod crustaceans are used in obtaining food, for example, by crushing mollusk shells.

In lobsters and crayfish, appendages called **swimmerets** occur in lines along the ventral surface of the abdomen and are used in reproduction and swimming. In addition, flattened appendages known as **uropods** form a kind of compound "paddle" at the end of the abdomen. These animals may also have a **telson,** or tail spine. By snapping its abdomen, the animal propels itself through the water rapidly and forcefully. Crabs (figure 43.24*b*) differ from lobsters and crayfish in proportion; their carapace is much larger and broader and the abdomen is tucked under it. Shrimps (figure 43.24*a*) and their relatives have a carapace that is proportionately smaller than that of lobsters or crabs.

Terrestrial and Freshwater Crustaceans

Although most crustaceans are marine, many occur in fresh water and a few have become terrestrial. These include pillbugs and sowbugs (figure 43.25*a*), the terrestrial members of a large order of crustaceans known as the isopods (order Isopoda). About half of the estimated 4500 species of this order are terrestrial and live primarily in places that are moist, at least seasonally. Sand fleas or beach fleas (order Amphipoda) are other familiar crustaceans, many of which are semiterrestrial (intertidal) species.

Along with the larvae of larger species, minute crustaceans are abundant in the plankton. Especially significant are the tiny copepods (order Copepoda; figure 43.25*b*), which are among the most abundant multicellular organisms on earth. Other orders of minute aquatic crustaceans include the water fleas (Cladocera), ostracods (Ostracoda), and fairy shrimps (Anostraca). One of the fairy shrimps, the brine shrimp, is a standard food for aquarium fish.

Sessile Crustaceans

Barnacles (order Cirripedia; figure 43.26) are a group of crustaceans that are sessile as adults. Barnacles have free-swimming larvae, which ultimately attach their heads to a piling, rock, or other submerged object and then stir food into their mouth with their feathery legs. Calcareous plates protect the barnacle's body, and these plates are usually attached directly and solidly to the substrate. Although most crustaceans have separate sexes, barnacles are hermaphroditic, but they generally cross-fertilize.

Crustaceans include marine, freshwater, and terrestrial forms. All possess a nauplius larval stage and branched appendages.

(a)

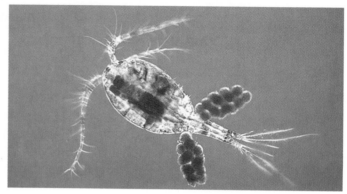

(b)

FIGURE 43.25
Terrestrial and freshwater crustaceans. (a) Sowbugs, *Porcellio scaber,* representatives of the terrestrial isopods (order Isopoda). (b) A copepod with attached eggs, a member of an abundant group of marine and freshwater crustaceans (order Copepoda), most of which are a few millimeters long. Copepods are important components of the plankton.

FIGURE 43.26
Gooseneck barnacles, *Lepas anatifera,* feeding. These are stalked barnacles; many others lack a stalk.

The subphylum Uniramia is an enormous group of mandibulates that includes millipedes, centipedes, and insects, three distinct but clearly related classes. The Uniramia were certainly derived from annelids, probably ones similar to the oligochaetes, which they resemble in their embryology. All Uniramia respire by means of tracheae and excrete their waste products through Malpighian tubules.

Classes Chilopoda and Diplopoda: The Centipedes and Millipedes

The centipedes (class Chilopoda) and millipedes (class Diplopoda) both have bodies that consist of a head region followed by numerous segments, all more or less similar and nearly all bearing paired appendages. Although the name *centipede* would imply an animal with a hundred legs and the name *millipede* one with a thousand, adult centipedes usually have 30 or more legs, adult millipedes 60 or more. Centipedes have one pair of legs on each body segment (figure 43.27), millipedes two (figure 43.28). Each segment of a millipede is a tagma that originated during the group's evolution when two ancestral segments fused. This explains why millipedes have twice as many legs per segment as centipedes.

In both centipedes and millipedes, fertilization is internal and takes place by direct transfer of sperm. The sexes are separate, and all species lay eggs. Young millipedes usually hatch with three pairs of legs; they experience a number of growth stages, adding segments and legs as they mature, but do not change in general appearance.

Centipedes, of which some 2500 species are known, are all carnivorous and feed mainly on insects. The appendages of the first trunk segment are modified into a pair of poison fangs. The poison is often quite toxic to human beings, and many centipede bites are extremely painful, sometimes even dangerous.

In contrast, most millipedes are herbivores, feeding mainly on decaying vegetation. A few millipedes are carnivorous. Many millipedes can roll their bodies into a flat coil or sphere because the dorsal area of each of their body segments is much longer than the ventral one. More than 10,000 species of millipedes have been named, but this is

FIGURE 43.27
A centipede. Centipedes, like this member of the genus *Scolopendra*, are active predators.

estimated to be no more than one-sixth of the actual number of species that exists. In each segment of their body, most millipedes have a pair of complex glands that produces a bad-smelling fluid. This fluid is exuded for defensive purposes through openings along the sides of the body. The chemistry of the secretions of different millipedes has become a subject of considerable interest because of the diversity of the compounds involved and their effectiveness in protecting millipedes from attack. Some produce cyanide gas from segments near their head end. Millipedes live primarily in damp, protected places, such as under leaf litter, in rotting logs, under bark or stones, or in the soil.

Centipedes are segmented hunters with one pair of legs on each segment. Millipedes are segmented herbivores with two pairs of legs on each segment.

FIGURE 43.28
A millipede. Millipedes, such as this *Sigmoria* individual, are herbivores.

Class Insecta: The Insects

The insects, class Insecta, are by far the largest group of organisms on earth, whether measured in terms of numbers of species or numbers of individuals. Insects live in every conceivable habitat on land and in fresh water, and a few have even invaded the sea. More than half of all the named animal species are insects, and the actual proportion is doubtless much higher because millions of additional forms await detection, classification, and naming. Approximately 90,000 described species occur in the United States and Canada, and the actual number of species in this area probably approaches 125,000. A hectare of lowland tropical forest is estimated to be inhabited by as many as 41,000 species of insects, and many suburban gardens may have 1500 or more species. It has been estimated that approximately a billion billion (10^{18}) individual insects are alive at any one time. A glimpse at the enormous diversity of insects is presented in figure 43.29 and later in table 43.1.

(a)

(b)

(c)

(d)

(e)

(f)

FIGURE 43.29

Insect diversity. (a) Luna moth, *Actias luna*. Luna moths and their relatives are among the most spectacular insects (order Lepidoptera). (b) Soldier fly, *Ptecticus trivittatus* (order Diptera). (c) Boll weevil, *Anthonomus grandis*. Weevils are one of the largest groups of beetles (order Coleoptera). (d) A thorn-shaped leafhopper, *Umbonica crassicornis* (order Hemiptera). (e) Copulating grasshoppers (order Orthoptera). (f) Termite, *Macrotermes bellicosus* (order Isoptera). The large, sausage-shaped individual is a queen, specialized for laying eggs; most of the smaller individuals around the queen are nonreproductive workers, but the larger individual at the lower left is a reproductive male.

External Features

Insects are primarily a terrestrial group, and most, if not all, of the aquatic insects probably had terrestrial ancestors. Most insects are relatively small, ranging in size from 0.1 millimeter to about 30 centimeters in length or wingspan. Insects have three body sections, the head, thorax, and abdomen; three pairs of legs, all attached to the thorax; and one pair of antennae. In addition, they may have one or two pairs of wings. Insect mouthparts all have the same basic structure but are modified in different groups in relation to their feeding habits (figure 43.30). Most insects have compound eyes, and many have ocelli as well.

The insect thorax consists of three segments, each with a pair of legs. Occasionally, one or more of these pairs of legs is absent. Legs are completely absent in the larvae of certain groups, for example, in most members of the order Diptera (flies) (figure 43.31). If two pairs of wings are present, they attach to the middle and posterior segments of the thorax. If only one pair of wings is present, it usually attaches to the middle segment. The thorax is almost entirely filled with muscles that operate the legs and wings.

The wings of insects arise as sac-like outgrowths of the body wall. In adult insects, the wings are solid except for the veins. Insect wings are not homologous to the other appendages. Basically, insects have two pairs of wings, but in some groups, like flies, the second set has been reduced to a pair of balancing knobs called halteres during the course of evolution. Most insects can fold their wings over their abdomen when they are at rest; but a few, such as the dragonflies and damselflies (order Odonata), keep their wings erect or outstretched at all times.

Insect forewings may be tough and hard, as in beetles. If they are, they form a cover for the hindwings and usually open during flight. The tough forewings also serve a protective function in the order Orthoptera, which includes grasshoppers and crickets. The wings of many insects are made of sheets of chitin; their strengthening veins are tubules of chitin. Moths and butterflies have wings that are covered with detachable scales that provide most of their bright colors (figure 43.32). In some wingless insects, such as the springtails or silverfish, wings never evolved. Other wingless groups, such as fleas and lice, are derived from ancestral groups of insects that had wings.

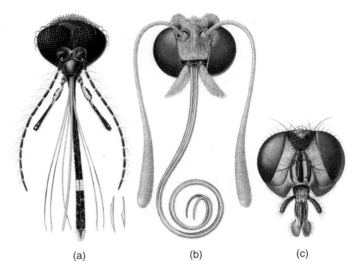

FIGURE 43.30
Modified mouthparts in three kinds of insects. Mouthparts are modified for (a) piercing in the mosquito, *Culex*, (b) sucking nectar from flowers in the Alfalfa butterfly, *Colias*, and (c) sopping up liquids in the housefly, *Musca domestica*.

FIGURE 43.31
Larvae of a mosquito, *Culex pipiens*. The aquatic larvae of mosquitoes are quite active. They breathe through tubes from the surface of the water, as shown here. Covering the water with a thin film of oil, a method used widely to control mosquitoes, causes them to drown.

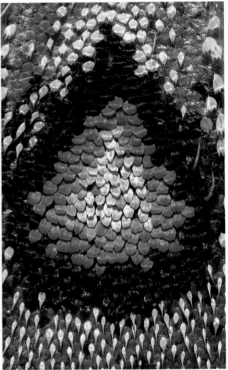

FIGURE 43.32
Scales on the wing of *Parnassius imperator*, a butterfly from China. Scales of this sort account for most of the colored patterns on the wings of butterflies and moths.

Internal Organization

The internal features of insects resemble those of the other arthropods in many ways. The digestive tract is a tube, usually somewhat coiled. It is often about the same length as the body. However, in the order Hemiptera, which consists of the leafhoppers, cicadas, and related groups, and in many flies (order Diptera), the digestive tube may be greatly coiled and several times longer than the body. Such long digestive tracts are generally found in insects that have sucking mouthparts and feed on juices rather than on protein-rich solid foods because they offer a greater opportunity to absorb fluids and their dissolved nutrients. The digestive enzymes of the insect also are more dilute and thus less effective in a highly liquid medium than in a more solid one. Longer digestive tracts give these enzymes more time to work while food is passing through.

The anterior and posterior regions of an insect's digestive tract are lined with cuticle. Digestion takes place primarily in the stomach, or midgut; and excretion takes place through Malpighian tubules. Digestive enzymes are mainly secreted from the cells that line the midgut, although some are contributed by the salivary glands near the mouth.

The tracheae of insects extend throughout the body and permeate its different tissues. In many winged insects, the tracheae are dilated in various parts of the body, forming air sacs. These air sacs are surrounded by muscles and form a kind of bellows system to force air deep into the tracheal system. The spiracles, a maximum of 10 on each side of the insect, are paired and located on or between the segments along the sides of the thorax and abdomen. In most insects, the spiracles can be opened by muscular action. Closing the spiracles at times may be important in retarding water loss. In some parasitic and aquatic groups of insects, the spiracles are permanently closed. In these groups, the tracheae run just below the surface of the insect, and gas exchange takes place by diffusion.

The **fat body** is a group of cells located in the insect body cavity. This structure may be quite large in relation to the size of the insect, and it serves as a food-storage reservoir, also having some of the functions of a vertebrate liver. It is often more prominent in immature insects than in adults, and it may be completely depleted when metamorphosis is finished. Insects that do not feed as adults retain their fat bodies and live on the food stored in them throughout their adult lives (which may be very short).

FIGURE 43.33
Sensory hairs. This honeybee has sensory hairs all over its body.

Sense Receptors

In addition to their eyes, insects have several characteristic kinds of sense receptors. These include **sensory hairs,** which are usually widely distributed over their bodies (figure 43.33). The sensory hairs are linked to nerve cells and are sensitive to mechanical stimulation. They are particularly abundant on the antennae and legs—the parts of the insect most likely to come into contact with other objects.

Sound, which is of vital importance to insects, is detected by tympanal organs in groups such as grasshoppers and crickets, cicadas, and some moths. These organs are paired structures composed of a thin membrane, the **tympanum,** associated with the tracheal air sacs. In many other groups of insects, sound waves are detected by sensory hairs. Male mosquitoes use thousands of sensory hairs on their antennae to detect the sounds made by the vibrating wings of female mosquitoes.

Sound detection in insects is important not only for protection but also for communication. Many insects communicate by making sounds, most of which are quite soft, very high-pitched, or both, and thus inaudible to humans. Only a few groups of insects, especially grasshoppers, crickets, and cicadas, make sounds that people can hear. Male crickets and longhorned grasshoppers produce sounds by rubbing their two front wings together. Shorthorned grasshoppers do so by rubbing their hind legs over specialized areas on their wings. Male cicadas vibrate the membranes of air sacs located on the lower side of the most anterior abdominal segment.

In addition to using sound, nearly all insects communicate by means of chemicals or mixtures of chemicals known as **pheromones.** These compounds, extremely diverse in their chemical structure, are sent forth into the environment, where they are active in very small amounts and convey a variety of messages to other individuals. These messages not only convey the attraction and recognition of members of the same species for mating, but they also mark trails for members of the same species, as in the ants.

All insects possess three body segments (tagmata): the head, the thorax, and the abdomen. The three pairs of legs are attached to the thorax. Most insects have compound eyes, and many have one or two pairs of wings. Insects possess sophisticated means of sensing their environment, including sensory hairs, tympanal organs, and chemoreceptors.

Insect Life Histories

Most young insects hatch from fertilized eggs laid outside their mother's body. The zygote develops within the egg into a young insect, which escapes by chewing through or bursting the shell. Some immature insects have specialized projections on the head that assist in this process. In a few insects, eggs hatch within the mother's body.

During the course of their development, young insects undergo ecdysis a number of times before they become adults. Most insects molt four to eight times during the course of their development, but some may molt as many as 30 times. The stages between the molts are called **instars.** When an insect first emerges following ecdysis, it is pale, soft, and especially susceptible to predators. Its exoskeleton generally hardens in an hour or two. It must grow to its new size, usually by taking in air or water, during this brief period. The wings are expanded by forcing blood into their veins.

There are two principal kinds of metamorphosis in insects: **simple metamorphosis** and **complete metamorphosis** (figure 43.34). In simple metamorphosis, the wings, if present, develop externally during the juvenile stages, and ordinarily no "resting" stage exists before the last molt. In complete metamorphosis, the wings develop internally during the juvenile stages and appear externally only during the resting stage that immediately precedes the final molt (figure 43.35). During this stage, the insect is called a **pupa** or **chrysalis,** depending on the group to which it belongs. A pupa or chrysalis does not normally move around much, although mosquito pupae do move around freely. A large amount of internal body reorganization takes place while the insect is a pupa or chrysalis.

In insects with simple metamorphosis, immature stages are often called **nymphs.** Nymphs are usually quite similar to adults, differing mainly in their smaller size, less well-developed wings, and sometimes color. The most primitive orders of insects, such as the springtails and silverfish, descended from ancestors that never had wings during their evolutionary history. In them, development is direct, and the nymphs only change in size. In other insect orders with simple metamorphosis, such as mayflies and dragonflies, nymphs are aquatic and extract oxygen from the water through gills. The adult stages are terrestrial and look very different from the nymphs. In still other groups, such as grasshoppers and their relatives, nymphs and adults live in the same habitat. Such insects usually change gradually during their life cycles with respect to wing development, body proportions, the appearance of ocelli, and other features.

More than 90% of the insects, including the members of all of the largest and most successful orders, display complete metamorphosis. The juvenile stages and adults often live in distinct habitats, have different habits, and are usually extremely different in form. In these insects, development is indirect. The immature stages, called **larvae,** often worm-like, differing greatly in appearance from the

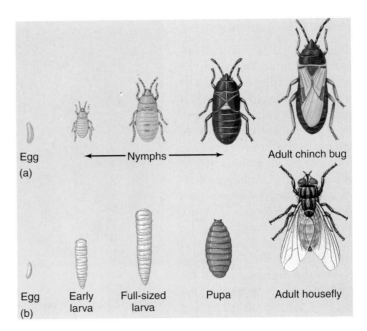

FIGURE 43.34
Metamorphosis. (a) Simple metamorphosis in a chinch bug (order Hemiptera), and (b) complete metamorphosis in a housefly, *Musca domestica* (order Diptera).

adults. Larvae do not have compound eyes. They may be legless or have legs or leg-like appendages on the abdomen. They usually have chewing mouthparts, even in those orders in which the adults have sucking mouthparts; chewing mouthparts are the primitive condition in these groups. When larvae and adults play different ecological roles, they do not compete directly for the same resources, an advantage to the species.

Pupae do not feed and are usually relatively inactive. As pupae, insects are extremely vulnerable to predators and parasites, but they are often covered by a cocoon or some other protective structure. Groups of insects with complete metamorphosis include moths and butterflies; beetles; bees, wasps, and ants; flies; and fleas. Fleas (order Siphonaptera) are among the orders of wingless insects descended from winged ancestors; their winglessness is an advantage in their specialized way of life.

Hormones control both ecdysis and metamorphosis. **Molting hormone,** or **ecdysone,** is released from a gland in the thorax when that gland has been stimulated by brain hormone, which in turn is produced by neurosecretory cells and released into the blood. The effects of the molting induced by ecdysone are determined by juvenile hormone, which is present during the immature stages but declines in quantity as the insect passes through successive molts. When the level of juvenile hormone is relatively high, the molt produces another larva; when it is lower, it produces the pupa and then the final development of the adult.

Insects undergo either simple or complete metamorphosis.

FIGURE 43.35
The life cycle of the monarch butterfly, *Danaus plexippus* (order Lepidoptera), illustrates complete metamorphosis. (a) Egg.
(b) Larva (caterpillar) feeding on a leaf. The pseudopods, or false legs, on the abdomen of larval butterflies and moths are not related to the true legs in the adults. (c) Larva preparing to shed its skin and become a chrysalis. (d) Chrysalis. (e) The colors of the wings of the adult butterfly are visible through the thin outer skin of the chrysalis; the adult is nearly ready to emerge. (f)–(h) Adult butterfly emerging.
(i) Adult monarch butterfly. Monarch butterflies feed on poisonous plants of the milkweed family (Asclepiadaceae). They are protected from their predators, primarily birds, by the chemicals they obtain from these plants. Their conspicuous warning coloration advertises their toxic nature, as we learned in chapter 24.

Table 43.1   Major Orders of Insects

Order	Typical Examples		Key Characteristics	Approximate Number of Named Species
Coleoptera	Beetles		The most diverse animal order; two pairs of wings; front pair of wings is a hard cover that partially protects the transparent rear pair of flying wings; heavily armored exoskeleton; biting and chewing mouthparts; complete metamorphosis	350,000
Diptera	Flies		Some that bite people and other mammals are considered pests; front flying wings are transparent; hind wings are reduced to knobby balancing organs; sucking, piercing, and lapping mouthparts; complete metamorphosis	120,000
Lepidoptera	Butterflies, moths		Often collected for their beauty; two pairs of broad, scaly, flying wings, often brightly colored; hairy body; tube-like, sucking mouthparts; complete metamorphosis	120,000
Hymenoptera	Bees, wasps, ants		Often social, known to many by their sting; two pairs of transparent flying wings; mobile head and well-developed eyes; often possess stingers; chewing and sucking mouthparts; complete metamorphosis	100,000
Hemiptera	True bugs, bedbugs, leafhoppers		Often live on blood; two pairs of wings, or wingless; piercing, sucking mouthparts; simple metamorphosis	60,000
Orthoptera	Grasshoppers, crickets, roaches		Known for their jumping; two pairs of wings or wingless; among the largest insects; biting and chewing mouthparts in adults; simple metamorphosis.	20,000
Odonata	Dragonflies		Among the most primitive of the insect order; two pairs of transparent flying wings; large, long, and slender body; chewing mouthparts; simple metamorphosis	5,000
Isoptera	Termites		One of the few types of animals able to eat wood; two pairs of wings, but some stages are wingless; social insects; there are several body types with division of labor; chewing mouthparts; simple metamorphosis	2,000
Siphonaptera	Fleas		Small, known for their irritating bites; wingless; small flattened body with jumping legs; piercing and sucking mouthparts; complete metamorphosis	1,200

Summary of Concepts

43.1 The evolution of jointed appendages has made arthropods very successful.

- Segmentation in the annelids gave rise to jointed appendages in the arthropods. Jointed appendages greatly expand locomotive and manipulative capabilities for the arthropod and chordate phyla.

- Arthropods are the most successful of all animals in terms of numbers of individuals and numbers of species, as well as in terms of ecological diversification.

- Like annelids, arthropods have segmented bodies, but some of their segments have become fused into tagmata during the course of evolution.

- Arthropods consist of three subphyla: chelicerates (subphylum Chelicerata); crustaceans (subphylum Crustacea); and insects, centipedes, and millipedes (subphylum Uniramia).

- The arthropod subphyla probably evolved from different ancestors, as suggested by the structure and derivation of their mouthparts and other appendages and by the important embryological differences among them.

- All arthropods possess a rigid external skeleton, or exoskeleton, which provides a surface for muscular attachment, a shield against predators and diseases, and a means of conserving water.

- Arthropods have an open circulatory system that is not as important in conducting oxygen to the tissues as the respiratory system.

- The respiratory system in most terrestrial arthropods consists of tracheae, a network of tubes that open to the outside through spiracles. Many chelicerates have book lungs, and crustaceans have gills.

- Many arthropods excrete wastes through Malpighian tubules, which extend from the digestive tract into the blood.

43.2 Scorpions, spiders, and mites all have fangs or pincers.

- Chelicerates consist of three classes: Arachnida (spiders, ticks, mites, and scorpions); Merostomata (horseshoe crabs); and Pycnogonida (sea spiders).

- Spiders, the best known arachnids, have a pair of chelicerae, a pair of pedipalps, and four pairs of walking legs.

- Spiders secrete digestive enzymes into their prey then suck the contents out.

43.3 Crustaceans have branched appendages.

- Crustaceans comprise some 35,000 species of crabs, shrimps, lobsters, barnacles, sowbugs, beach fleas, and many other groups. Their appendages are basically biramous, and their embryology is distinctive.

43.4 Insects are the most diverse of all animal groups.

- Uniramia—insects, centipedes, and millipedes—share a unique pattern of embryological development with oligochaete annelids. It seems clear they evolved from ancestors similar to oligochaete annelids.

- Centipedes and millipedes are segmented uniramia. Centipedes are hunters with one pair of legs per segment, and millipedes are herbivores with two pair of legs per segment.

- Insects have three body segments, three pairs of legs, and often one or two pairs of wings. Many have complex eyes and other specialized sensory structures.

- Insects exhibit either simple metamorphosis, moving through a succession of forms relatively similar to the adult, or complete metamorphosis, in which an often worm-like larva becomes a usually sedentary pupa, and then an adult.

Discussing Key Terms

1. **Jointed appendages** Appendages such as legs, mouthparts, and antennae that could swivel or bend marked a great advance first seen among arthropods.

2. **Exoskeleton** An exoskeleton serves as a muscle attachment site and provides protection for the organism. However, it places a limitation on overall size: exoskeletons cannot grow. To increase in size, arthropods must shed their old exoskeletons and "puff themselves up" while their new exoskeleton is soft.

3. **Spiracles** Most terrestrial arthropods respire through spiracles, holes in the exoskeleton that lead into interior respiratory branches called tracheae. Oxygen diffuses directly from air into the tissues of the arthropod—gases are not circulated by a circulatory system. This is another factor that limits arthropod size.

4. **Malpighian tubules** Many arthropods have a unique and highly effective excretory apparatus made up of several spaghetti-like tubes called Malpighian tubules. These tubules pouch off of the midgut and are exposed to the open circulatory system. Wastes in the body cavity move into the tubules and are excreted through the digestive tract.

Review Questions

1. Into what two groups are arthropods traditionally divided? Describe each group in terms of its mouthparts and appendages, and give several examples of each.

2. How do biramous and uniramous appendages differ? Which subgroup of arthropods has fundamentally biramous appendages? Which subgroup has uniramous appendages?

3. What are the advantages of an exoskeleton? What occurs during ecdysis? What controls this process?

4. What type of circulatory system do arthropods have? Describe the direction of blood flow. What helps to maintain this one-way flow?

5. What are Malpighian tubules? How do they work? What other system are they connected to? How does this system process wastes? How does it regulate water loss?

6. What is the typical arrangement of appendages in crustaceans? On which parts of the body do they possess legs?

7. How are millipedes and centipedes similar to each other? How do they differ?

8. What type of digestive system do most insects possess? What digestive adaptations occur in insects that feed on juices low in protein? Why? What respiratory adaptations have occurred in some insects?

9. What is an instar as it relates to insect metamorphosis? What are the two different kinds of metamorphosis in insects? How do they differ? What are the immature forms of each type called?

Thought Questions

1. What are the main factors that limit the size of terrestrial arthropods? What effect has this had on the evolution of intelligence in the group?

2. What are the key adaptations that facilitated the invasion of the land by arthropods? Why is the phylum so successful?

3. What evidence indicates that the arthropods consist of groups that evolved from different ancestors at different times in the past? Why have the arthropods been treated as a single phylum if this evidence is so convincing?

Internet Links

The Bug Page
http://www.ex.ac.uk/~gjlramel/welcome.html
A wonderful ENTOMOLOGICAL HOME PAGE featuring excellent sections on all kinds of insects and spiders, including amazing facts, pictures, and essays. Fun to explore.

Forensic Entomology
http://www.uio.no/~mostarke/forens_ent/forensic_entomology.html
Forensic entomology is the use of insect knowledge in the investigation of crimes. While some of the contents of this site may disturb you, you will be amazed at how much insects can tell us about a crime scene.

For Further Reading

Berenbaum, M.: *Bugs in the System: Insects and Their Impact on Human Affairs,* Addison-Wesley, Reading, MA, 1995. An informative and delightfully written book, discussing both well-known and hardly known insects that are pests or are beneficial. Includes recipes!

Grimaldi, D.: "Captured in Amber," *Scientific American,* April 1996, 84–91. The exquisitely preserved tissues of insects in 25-million-year-old amber is revealing some genetic secrets of evolution.

Hadley, N. F.: "The Arthropod Cuticle," *Scientific American,* July 1986, pages 104–12. This complex covering accounts for much of the adaptive success of the arthropods.

Heinrich, B.: "Thermoregulation in Winter Moths," *Scientific American,* March 1987, pages 104–11. How can certain moth species fly, feed, and mate at near-freezing temperatures?

Holldobler, B. and E. O. Wilson: *The Ants,* Harvard University Press, Cambridge, MA, 1990. A wonderful book, filled with exciting and informative insights into this incredibly diverse group of insects. Exquisite photographs.

Rinderer, T.: "Africanized Bees in the U.S." *Scientific American,* December 1993, pages 84–90. Africanized honeybees, nicknamed "killer bees" for their aggressive behavior, have now reached the United States.

Shear, W. A.: "One Small Step for an Arthropod: In a Giant Leap in the History of Life, Invertebrates Set Foot on Land More than 400 Million Years Ago," *Natural History,* March 1993, pages 46–51. Who were the first terrestrial animals? New evidence uncovered in New York, Scotland, and Wales points to tiny carnivorous arachnids and myriapods.

White, W.: "The Bees from Rio Claro." *New Yorker,* September 16, 1991, pages 36–60. A wonderful account of how "killer bees" were created and accidentally released. Highly recommended.

Wooton, R. J.: "The Mechanical Design of Insect Wings," *Scientific American,* November 1990, pages 114–20. Subtle details of engineering and design reveal how insect wings are remarkably adapted to the acrobatics of flight.

44

Echinoderms

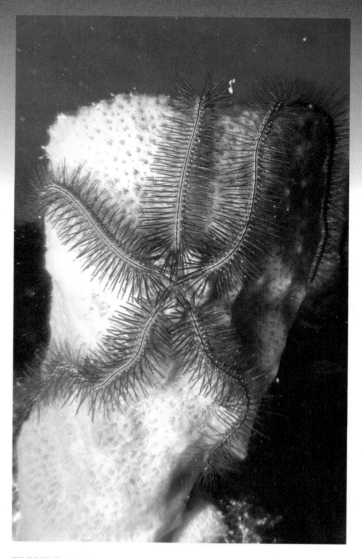

FIGURE 44.1
An echinoderm. Brittle star, *Ophiothrix*, a member of the largest group of echinoderms.

Concept Outline

44.1 The embryos of deuterostomes develop quite differently from those of protostomes.

Protostomes and Deuterostomes. Deuterostomes—the echinoderms, chordates, and a few other groups—share a mode of development that is quite different from other animals.

44.2 Echinoderms are deuterostomes with an endoskeleton.

The Echinoderms. Echinoderms are bilaterally symmetrical as larvae but metamorphose to radially symmetrical adults.

Echinoderm Body Plan. Echinoderms have an endoskeleton and a unique water vascular system seen in no other phylum.

44.3 The six classes of echinoderms are all radially symmetrical as adults.

Class Crinoidea: The Sea Lilies and Feather Stars. Crinoids are the only echinoderms that are attached to the sea bottom for much of their lives.

Class Asteroidea: The Sea Stars. Sea stars, also called starfish, are five-armed mobile predators.

Class Ophiuroidea: The Brittle Stars. Brittle stars are quite different from the sea stars for whom they are sometimes mistaken.

Class Echinoidea: The Sea Urchins and Sand Dollars. Sea urchins and sand dollars have five-part radial symmetry but lack arms.

Class Holothuroidea: The Sea Cucumbers. Sea cucumbers are soft-bodied echinoderms without arms.

Class Concentricycloidea: The Sea Daisies. The most recently discovered class of echinoderms, sea daisies are tiny, primitive echinoderms that live at great depths.

44.4 Arrow worms and acorn worms are two minor phyla.

Arrow Worms. Animals in this phylum are deuterostomes, but their relationship to other phyla is unclear.

Acorn Worms. Hemichordates, or acorn worms, are deuterostomate animals that share features with both echinoderms and chordates.

Echinoderms, which include the familiar starfish, have been described as a "noble group especially designed to puzzle the zoologist." They are bilaterally symmetrical as larvae, but undergo a bizarre metamorphosis to a radial adult (figure 44.1). A compartment of the coelom is transformed into a unique water-vascular system that uses hydraulic power to operate a multitude of tiny tube feet that move the body over the sea bottom. Some echinoderms have an endoskeleton of dermal plates beneath the skin, fused together like body armor. Many have miniature jaw-like pincers scattered over their body surface, often on stalks and sometimes bearing poison glands. This collection of characteristics is unique in the animal kingdom.

Protostomes and Deuterostomes

Two major kinds of coelomate animals represent two distinct evolutionary lines. All the coelomates we have met so far have essentially the same kind of embryo, starting as a hollow ball of cells, a blastula, which indents to form a two-layer-thick ball with a blastopore opening to the outside. In mollusks, annelids, and arthropods, the mouth (stoma) develops from or near the blastopore (figure 44.2*a*). This same pattern of development, in a general sense, is seen in all noncoelomate animals. An animal whose mouth develops in this way is called a **protostome** (from the Greek words *protos*, first, and *stoma*, mouth). If such an animal has a distinct anus or anal pore, it develops later in another region of the embryo. The fact that this kind of developmental pattern is so widespread in diverse phyla makes it likely that it is the original pattern for animals as a whole and that it was characteristic of the common ancestor of all eumetazoan animals.

A second distinct pattern of embryological development occurs in the echinoderms, the chordates, and a few other small related phyla; it doubtless arose in the common ancestor of this group. In these animals, the anus forms from or near the blastopore, and the mouth forms subsequently on another part of the blastula (figure 44.2*b*). This group of phyla consists of animals that are called the **deuterostomes** (Greek, *deuteros*, second, and *stoma*, mouth). They are clearly related to one another by their shared pattern of embryonic development, which differs radically from the protostome pattern and was most likely derived from it.

Deuterostomes represent a revolution in embryonic development. In addition to the pattern of blastopore formation, deuterostomes differ from protostomes in a number of other fundamental embryological features:

1. The progressive division of cells during embryonic growth is called *cleavage*. The cleavage pattern relative to the embryo's polar axis determines how the cells will array. In nearly all protostomes, each new cell buds off at an angle oblique to the polar axis. As a result, a new cell nestles into the space between the older ones in a closely packed array. This pattern is called **spiral cleavage** because a line drawn through a sequence of dividing cells spirals outward from the polar axis.

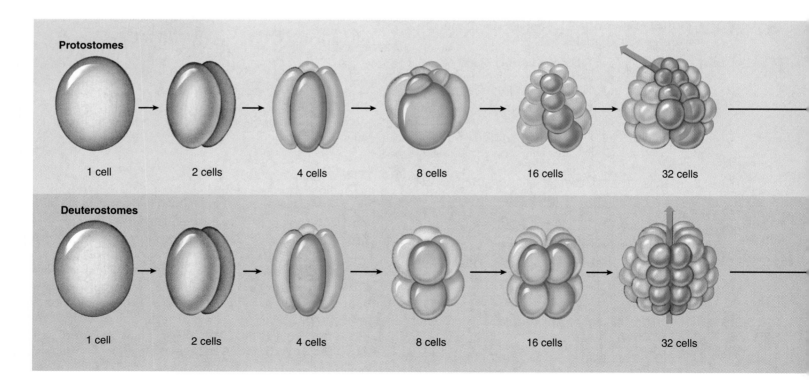

FIGURE 44.2
Embryonic development in protostomes and deuterostomes. Cleavage of the egg produces a hollow ball of cells called the blastula. Invagination of the blastula produces the blastopore. In protostomes, embryonic cells cleave in a spiral pattern and become tightly packed. The blastopore becomes the animal's mouth, and the coelom originates from a mesodermal split. In deuterostomes,

In deuterostomes, the cells divide parallel to and at right angles to the polar axis. As a result, the pairs of cells from each division are positioned directly above and below one another; this process gives rise to a loosely packed array of cells. This pattern is called **radial cleavage** because a line drawn through a sequence of dividing cells describes a radius outward from the polar axis.

2. In protostomes, the developmental fate of each cell in the embryo is fixed when that cell first appears. Even at the four-celled stage, each cell is different, and no one cell, if separated from the others, can develop into a complete animal because the chemicals that act as developmental signals have already been localized in different parts of the egg. Consequently, the cleavage divisions that occur after fertilization separate different signals into different daughter cells. In deuterostomes, on the other hand, the first cleavage divisions of the fertilized embryo produce identical daughter cells, and any single cell, if separated, can develop into a complete organism. The commitment to prescribed developmental pathways occurs later.

3. In all coelomates, the coelom originates from mesoderm. In protostomes, this occurs simply and directly: the cells simply move away from one another as the coelomic cavity expands within the mesoderm. However, in deuterostomes, whole groups of cells usually move around to form new tissue associations. The coelom is normally produced by an evagination of the **archenteron**—the main cavity within the gastrula, also called the primitive gut. This cavity, lined with endoderm, opens to the outside via the blastopore and eventually becomes the gut cavity.

The first abundant and well-preserved animal fossils are about 630 million years old; they occur in the Ediacara series of Australia and similar formations elsewhere. Among these fossils, many represent groups of animals that no longer exist. In addition, these ancient rocks bear evidence of the coelomates, the most advanced evolutionary line of animals, and it is remarkable that their two major subdivisions were differentiated so early. In the coelomates, it is clear that deuterostomes are derived from protostomes. The event, however, occurred very long ago and presumably did not involve groups of organisms that closely resemble any that are living now.

In deuterostomes, the egg cleaves radially, and the blastopore becomes the animal's anus. In protostomes, the egg cleaves spirally, and the blastopore becomes the mouth.

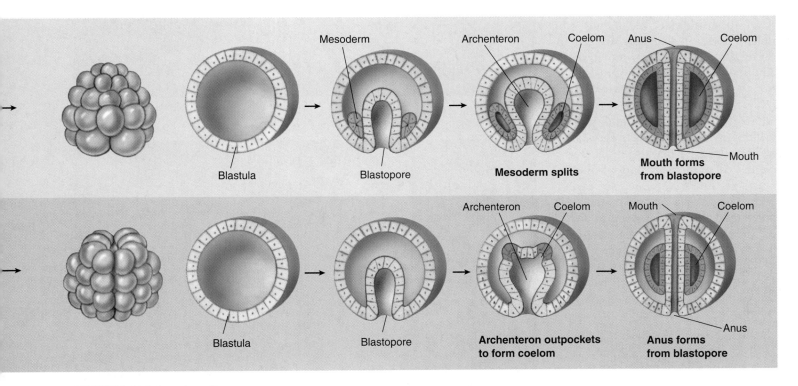

FIGURE 44.2 (continued)
embryonic cells cleave radially and form a loosely packed array. The blastopore becomes the animal's anus, and the mouth develops at the other end. The coelom originates from an evagination, or outpouching, of the archenteron in deuterostomes.

The Echinoderms

Deuterostomate marine animals called **echinoderms** appeared more than 650 million years ago (figure 44.3). Echinoderms (phylum Echinodermata) are an ancient group of marine animals consisting of about 6000 living species and are also well represented in the fossil record. The term *echinoderm* means "spiny skin" and refers to an **endoskeleton** composed of hard calcium-rich plates just beneath the delicate skin (figure 44.4). When they first form, the plates are enclosed in living tissue and so are truly an endoskeleton, although in adults they frequently fuse, forming a hard shell. Another key evolutionary innovation in echinoderms is the development of a hydraulic system to aid in movement or feeding. Called a **water vascular system,** this fluid-filled system is composed of a central ring canal from which five radial canals extend out into the arms.

Many of the most familiar animals seen along the seashore, sea stars (starfish), brittle stars, sea urchins, sand dollars, and sea cucumbers, are echinoderms. All are radially symmetrical as adults. While some other kinds of animals are radially symmetrical, none have the complex organ systems of adult echinoderms. Echinoderms are well represented not only in the shallow waters of the sea but also in its abyssal depths. All of them are bottom-dwellers except for a few swimming sea cucumbers. The adults range from a few millimeters to more than a meter in diameter (for one species of sea star) or in length (for a species of sea cucumber).

There is an excellent fossil record of the echinoderms, extending back into the Cambrian. However, despite this

(a)

(b)

(c)

FIGURE 44.3
Diversity in echinoderms. (a) Sea star, *Oreaster occidentalis* (class Asteroidea), in the Gulf of California, Mexico. (b) Warty sea cucumber, *Parastichopus parvimensis* (class Holothuroidea), Philippines. (c) Sea urchin (class Echinoidea).

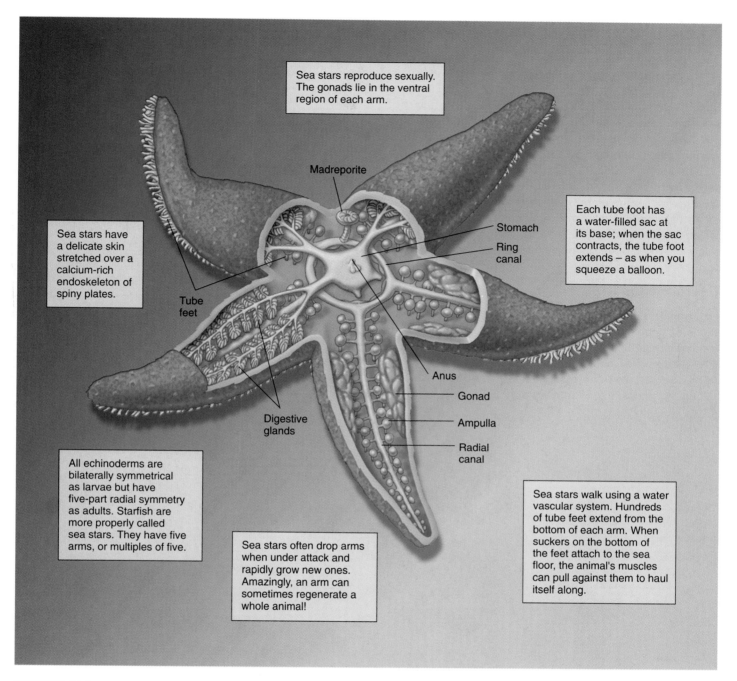

Sea stars reproduce sexually. The gonads lie in the ventral region of each arm.

Madreporite

Stomach

Ring canal

Each tube foot has a water-filled sac at its base; when the sac contracts, the tube foot extends – as when you squeeze a balloon.

Sea stars have a delicate skin stretched over a calcium-rich endoskeleton of spiny plates.

Tube feet

Anus

Gonad

Ampulla

Radial canal

Digestive glands

All echinoderms are bilaterally symmetrical as larvae but have five-part radial symmetry as adults. Starfish are more properly called sea stars. They have five arms, or multiples of five.

Sea stars often drop arms when under attack and rapidly grow new ones. Amazingly, an arm can sometimes regenerate a whole animal!

Sea stars walk using a water vascular system. Hundreds of tube feet extend from the bottom of each arm. When suckers on the bottom of the feet attach to the sea floor, the animal's muscles can pull against them to haul itself along.

FIGURE 44.4

The evolution of deuterostome development and an endoskeleton. Echinoderms, such as sea stars (phylum Echinodermata), are coelomates with a deuterostome pattern of development. A delicate skin stretches over an endoskeleton made of calcium-rich plates, often fused into a continuous, tough, spiny layer.

wealth of information, the origin of echinoderms remains unclear. They are thought to have evolved from bilaterally symmetrical ancestors because echinoderm larvae are bilateral. The radial symmetry that is the hallmark of echinoderms develops later, in the adult body. Many biologists believe that early echinoderms were sessile and evolved radiality as an adaptation to the sessile existence. Bilaterality is of adaptive value to an animal that travels through its en-

vironment, while radiality is of value to an animal whose environment meets it on all sides. Echinoderms attached to the sea bottom by a central stalk were once common, but only about 80 such species survive today.

Echinoderms are a unique, exclusively marine group of organisms in which deuterostome development and an endoskeleton are seen for the first time.

Echinoderm Body Plan

The body plan of echinoderms undergoes a fundamental shift during development. All echinoderms have **secondary radial symmetry,** that is, they are bilaterally symmetrical during larval development but become radially symmetrical as adults. In sea cucumbers and some other echinoderms, the animal's axis lies horizontally; radial symmetry is not as obvious as it is in other members of the phylum.

Echinoderms have a five-part body plan corresponding to the arms of a sea star or the design on the "shell" of a sand dollar. As adults, these animals have no head or brain. Their nervous systems consist of a central **nerve ring** from which branches arise. The animals are capable of complex response patterns, but there is no centralization of function.

Endoskeleton

Echinoderms have a delicate epidermis, containing thousands of neurosensory cells, stretched over an endoskeleton composed of either movable or fixed calcium-rich (calcite) plates called ossicles. The animals grow more or less continuously, but their growth slows down with age. When the plates first form, they are enclosed in living tissue. In many cases, these plates bear spines. The plates in certain portions of the body of some echinoderms are perforated. Through these perforations extend **tube feet,** part of the water vascular system that is a unique feature of this phylum.

The Water Vascular System

The water vascular system of an echinoderm radiates from a **ring canal** that encircles the animal's esophagus. Five **radial canals,** their positions determined early in the development of the embryo, extend into each of the five parts of the body and determine its basic symmetry (figure 44.5). Water enters the water vascular system through a **madreporite,** a sieve-like plate on the animal's surface, and flows to the ring canal through a tube, or stone canal, so named because of the surrounding rings of calcium carbonate. The five radial canals in turn extend out through short side branches into the hollow tube feet (figure 44.6). In some echinoderms, each tube foot has a sucker at its end; in others, suckers are absent. At the base of each tube foot is a muscular sac, the **ampulla,** which contains fluid. When the ampulla contracts, the fluid is prevented from entering the radial canal by a one-way valve and is forced into the tube

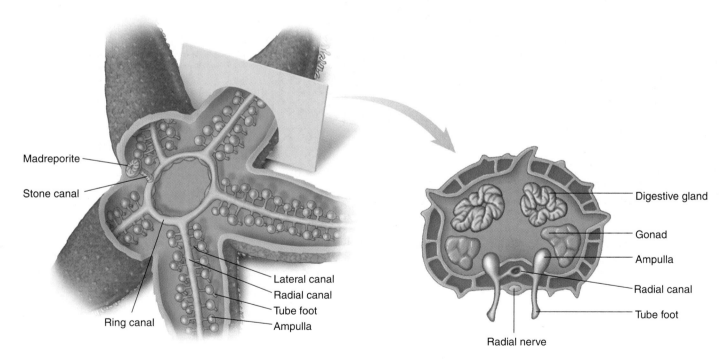

FIGURE 44.5
The water vascular system of an echinoderm. Radial canals allow water to flow into the tube feet. As the ampulla in each tube foot contracts, the tube extends and can attach to the substrate. Subsequently, muscles in the tube feet contract, and the tube foot shortens, pulling the animal forward.

FIGURE 44.6
Tube feet. The tube feet of the sea star *Ludia magnifica* are extended.

foot, thus extending it. When extended, the foot can attach itself to the substrate. Longitudinal muscles in the tube foot wall then contract, shortening it, and the animal is pulled forward to the new position as the water in the tube foot is forced back into the ampulla. By such actions, repeated in many tube feet, sea stars, sand dollars, and sea urchins move along.

Sea cucumbers (see figure 44.3*b*) usually have five rows of tube feet on the body surface that are used in locomotion. They also have modified tube feet around their mouth cavity that are used in feeding. In sea lilies, the tube feet arise from the branches of the arms, which extend from the margins of an upward-directed cup. With these tube feet, the animals take food from the surrounding water. In brittle stars (see figure 44.1), the tube feet are pointed and specialized for feeding.

Body Cavity

In echinoderms, the coelom, which is proportionately large, connects with a complicated system of tubes and helps provide circulation and respiration. In many echinoderms, respiration and waste removal take place by means of **skin gills,** which are small, finger-like projections that occur near the spines. The digestive system is simple but usually complete, consisting of a mouth, gut, and anus.

Reproduction

Many echinoderms are able to regenerate lost parts, and some, especially sea stars and brittle stars, drop various parts when under attack. In a few echinoderms, asexual reproduction takes place by splitting, and the broken parts of sea stars can sometimes regenerate whole animals. Some of the smaller brittle stars, especially tropical species, regu-

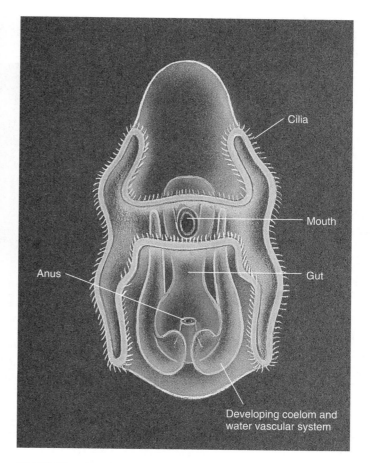

FIGURE 44.7
The free-swimming larva of an echinoderm. The bands of cilia by which the larva moves are prominent in this drawing. Such bilaterally symmetrical larvae demonstrate convincingly that the ancestors of the echinoderms were not radially symmetrical, like the living members of the phylum.

larly reproduce by breaking into two equal parts; each half then regenerates a whole animal.

Despite the ability of many echinoderms to break into parts and regenerate new animals from them, most reproduction in the phylum is sexual and external. The sexes in most echinoderms are separate, although there are few external differences. Fertilized eggs of echinoderms usually develop into free-swimming, bilaterally symmetrical larvae (figure 44.7), which differ from the trochophore larvae of mollusks and annelids. These larvae form a part of the plankton until they metamorphose through a series of stages into the more sedentary adults.

Echinoderms are characterized by a secondary radial symmetry and a five-part body plan. They have characteristic calcium-rich plates called ossicles and a unique water vascular system that includes hollow tube feet.

There are more than 20 extinct classes of echinoderms and an additional 6 with living members: (1) Crinoidea, sea lilies and feather stars; (2) Asteroidea, sea stars, or starfish; (3) Ophiuroidea, brittle stars; (4) Echinoidea, sea urchins and sand dollars; (5) Holothuroidea, sea cucumbers, and (6) Concentricycloidea, sea daisies. Sea daisies were recently discovered living on submerged wood in the deep sea.

Class Crinoidea: The Sea Lilies and Feather Stars

Sea lilies and feather stars, or crinoids (class Crinoidea) differ from all other living echinoderms in that the mouth and anus are located on their upper surface in an open disc. The two structures are connected by a simple gut. These animals have simple excretory and reproductive systems and an extensive water vascular system. The arms, which are the food-gathering structures of crinoids, are located around the margins of the disk. Different species of crinoids may have from 5 to more than 200 arms extending upward from their bodies, with smaller structures called pinnules branching from the arms. In all crinoids, the number of arms is initially small. Species with more than 10 arms add additional arms progressively during growth. Crinoids are filter feeders, capturing the microscopic organisms on which they feed by means of the mucus that coats their tube feet, which are abundant on the animals' pinnules.

Scientists that study echinoderms believe that the common ancestors of this phylum were sessile, sedentary, radially symmetrical animals that resembled crinoids. Such animals were abundant in ancient seas, and crinoids were present when the Burgess Shale was deposited about 530 million years ago. More than 6000 fossil species of this class are known, in comparison with the approximately 600 living species.

FIGURE 44.8
Sea lilies, *Cenocrinus asterius.* Two specimens showing a typical parabola of arms forming a "feeding net." The water current is flowing from right to left, carrying small organisms to the stalked crinoid's arms. Prey, when captured, are passed down the arms to the central mouth. This photograph was taken at a depth of about 400 meters in the Bahamas from the Johnson-Sea-Link Submersible of the Harbor Branch Foundation, Inc.

FIGURE 44.9
Feather star. This feather star is on the Great Barrier Reef in Australia.

Sea Lilies

There are two basic crinoid body plans. In sea lilies, the flower-shaped body is attached to its substrate by a stalk that is from 15 to 30 cm long, although in some species the stalk may be as much as a meter long (figure 44.8). Some fossil species had stalks up to 20 meters long. If they are detached from the substrate, some sea lilies can move slowly by means of their feather-like arms. All of the approximately 80 living species of sea lilies are found below a depth of 100 meters in the ocean. Sea lilies are the only living echinoderms that are fully sessile.

Feather Stars

In the second group of crinoids, the 520 or so species of feather stars, the disc detaches from the stalk at an early stage of development (figure 44.9). Adult feather stars have long, many-branched arms and usually anchor themselves to their substrate by claw-like structures. However, some feather stars are able to swim for short distances, and many of them can move along the substrate. Feather stars range into shallower water than do sea lilies, and only a few species of either group are found at depths greater than 500 meters. Along with sea cucumbers, crinoids are the most abundant and conspicuous large invertebrates in the warm waters and among the coral reefs of the western Pacific Ocean. They have separate sexes, with the sex organs simple masses of cells in special cavities of the arms and pinnules. Fertilization is usually external, with the male and female gametes shed into the water, but brooding—in which the female shelters the young—occurs occasionally.

Crinoids, the sea lilies and feather stars, were once far more numerous. Crinoids are the only echinoderms attached for much of their lives to the sea bottom.

Class Asteroidea: The Sea Stars

Sea stars, or starfish (class Asteroidea; figure 44.10), are perhaps the most familiar echinoderms. These attractive animals are among the most important predators in many marine ecosystems. They are often brightly colored and range in size from a centimeter to a meter across. They are abundant in the intertidal zone, but they also occur at depths as great as 10,000 meters. Around 1500 species of sea stars occur throughout the world.

The body of a sea star is composed of a central disc that merges gradually with five tapering arms. The arms are prominent and sharply set off from the disc. Although many sea stars have five arms, conforming to the basic symmetry of the phylum, members of some families have many more, typically in multiples of five. The body is somewhat flattened, flexible, and covered with a pigmented epidermis.

Endoskeleton

Beneath the epidermis is an endoskeleton of small calcium-rich plates called ossicles, bound together with connective tissue. From these ossicles project spines that make up the spiny upper surface. Around the base of the spines are minute, pincer-like *pedicellariae*, bearing tiny jaws manipulated by muscles. These keep the body surface free of debris and may aid in food capture.

The Water Vascular System

A deep groove runs along each arm from the central mouth out to the tip of the arm. This groove is bordered by rows of tube feet, which the animal uses to move about. Within each arm, there is a radial canal that connects the tube feet to a ring canal in the central body. This system of piping is used by sea stars to power a unique hydraulic system. Contraction of small chambers called ampullae attached to the tube feet forces water into the podium of the feet, extending them. Conversely, contraction of muscles in the tube foot retracts the podium, forcing fluid back into the ampulla. Small muscles at the end of each tube foot can raise the center of the disc-like end, creating suction when the foot is pressed against a substrate. Hundreds of tube feet, moving in unison, pull the arm along the surface.

Feeding

The mouth of a sea star is located in the center of its lower surface. Some sea stars have an extraordinary way of feeding on bivalve mollusks. They grasp both parts of the bivalve shell with their tube feet (figure 44.11) and extrude their stomach through their mouth into the bivalve. A sea star can push its stomach into a bivalve through an opening as little as 0.1 millimeter wide, corresponding to the natural openings in many irregular shells. Within the mollusk, the sea star secretes its digestive enzymes and digests

FIGURE 44.10
Class Asteroidea. This class includes the familiar starfish, or sea stars.

FIGURE 44.11
A sea star attacking a clam. The tube feet, each of which ends in a suction cup, are located along grooves on the underside of the arms.

the soft tissues of its prey, retracting its stomach when the process is complete.

Reproduction

Most sea stars have separate sexes, with a pair of gonads lying in the ventral region inside each arm. Eggs and sperm are shed into the water so that fertilization is external. In some species, fertilized eggs are brooded in special cavities or simply under the animal. They mature into larvae that swim by means of conspicuous bands of cilia.

> Sea stars, also called starfish, are five-armed, mobile predators.

Class Ophiuroidea: The Brittle Stars

Brittle stars (class Ophiuroidea; figure 44.12) are the largest class of echinoderms in numbers of species (about 2000) and they are probably the most abundant also. In shallow water, brittle stars occur largely on hard substrates, and in the deep ocean are one of the most abundant organisms. They tend to be secretive, living on the sea bottom where little or no light penetrates. They avoid light and are more active at night.

Brittle stars have slender, branched arms. These arms are more sharply set off from their central disc than in sea stars. The most mobile of echinoderms, brittle stars move by pulling themselves along, "rowing" over the substrate by moving their arms, often in pairs or groups, from side to side. In a number of species, the arms are covered with spines, which also assist in their movements. Some brittle stars use their arms to swim, a very unusual habit among echinoderms.

Brittle stars feed by capturing suspended microplankton and organic detritus with their tube feet, long arm spines, or branching arms, or by climbing over objects on the ocean floor in search of small animals. In addition, the tube feet are important sensory organs and assist in directing food into the mouth once the animal has captured it. As implied by their common name, the arms of brittle stars detach easily, a characteristic that helps to protect the brittle stars from their predators.

Like sea stars, brittle stars have five arms. More closely related to the sea stars than to the other classes of the phylum, they are sometimes grouped with them in a single class. However, on closer inspection they are surprisingly different. They have no pedicellariae, as sea stars have, and the groove running down the length of each arm is closed over and covered with ossicles. Their tube feet lack ampullae, have no suckers, and are used for feeding, not locomotion.

Brittle stars usually have separate sexes, with the male and female gametes in most species being released into the water and fusing there. Some brood their young in special cavities and release the larvae. These larvae are free-swimming, using their conspicuous bands of cilia to move.

Brittle stars are dwellers of the deep ocean floor and shallow waters, where they pull themselves along two arms at a time, like oars rowing a boat.

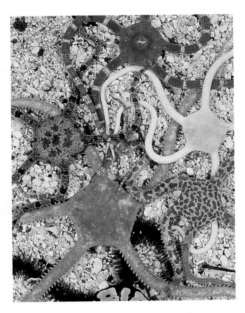

FIGURE 44.12
Class Ophiuroidea. Brittle stars use their arms to crawl actively across their marine substrates.

Class Echinoidea: The Sea Urchins and Sand Dollars

The members of the class Echinoidea, sand dollars and sea urchins, lack distinct arms but have the same five-part body plan as all other echinoderms (figure 44.13). Five rows of tube feet protrude through the plates of the calcareous skeleton, and there are also openings for the mouth and anus. These different openings can be seen in the globular skeletons of sea urchins and in the flat skeletons of sand dollars. Both types of endoskeleton, often common along the seashore, consist of fused calcareous plates. About 950 living species constitute the class Echinoidea.

Echinoids walk by means of their tube feet or their movable spines, which are hinged to the skeleton by a joint that makes free rotation possible. Sea urchins and sand dollars move along the sea bottom, feeding on algae and small fragments of organic material. They scrape these off the substrate with the large, triangular teeth that ring their mouths. The gonads of sea urchins are considered a great delicacy by people in different parts of the world. Because of their calcareous plates, sea urchins and sand dollars are well preserved in the fossil record, with more than 5000 additional species described.

As with most other echinoderms, the sexes of sea urchins and sand dollars are separate. The eggs and sperm are shed separately into the water, where they fuse. Some brood their young, and others have free-swimming larvae, with bands of cilia extending onto their long, graceful arms—an unusual feature of this class.

Sand dollars and sea urchins lack arms but have a five-part symmetry.

(a)

(b)

FIGURE 44.13
Class Echinoidea. (a) Sand dollar, *Echinarachnius parma*, New England. (b) Giant red sea urchin, *Strongylocentrotus franciscanus*.

Class Holothuroidea: The Sea Cucumbers

Sea cucumbers (class Holothuroidea) are shaped somewhat like their plant namesakes. They differ from the preceding classes in that they are soft, slug-like organisms, often with a tough, leathery outside skin (figure 44.14). The class consists of about 1500 species found worldwide. Except for a few forms that swim, sea cucumbers lie on their sides at the bottom of the ocean. Their mouth is located at one end and is surrounded by eight to 30 modified tube feet called tentacles; the anus is at the other end. The tentacles around the mouth may secrete mucus, used to capture the small planktonic organisms on which the animals feed. Each tentacle is periodically wiped off within the esophagus and then brought out again, covered with a new supply of mucus.

Sea cucumbers are soft because their calcareous skeletons are reduced to widely separated microscopic plates. These animals have extensive internal branching systems, called respiratory trees, which arise from the **cloaca,** or anal cavity. Water is pulled into and expelled from the respiratory tree by contractions of the cloaca; gas exchange takes place as this process occurs. The sexes of most cucumbers are separate, but some of them are hermaphroditic.

Most kinds of sea cucumbers have tube feet on the body in addition to tentacles. These additional tube feet, which might be restricted to five radial grooves or scattered over the surface of the body, may enable the animals to move about slowly. On the other hand, sea cucumbers may simply wriggle along whether or not they have additional tube feet. Most sea cucumbers are quite sluggish, but some, especially among the deep-sea forms, swim actively. Sea cucumbers, when irritated, sometimes eject a portion of their intestines by a strong muscular contraction that may send the intestinal fragments through the anus or even rupture the body wall.

Sea cucumbers are soft-bodied, slug-like animals without arms.

Class Concentricycloidea: The Sea Daisies

The most recently described class of echinoderms (1986), sea daisies are strange little disc-shaped animals, less than 1 cm in diameter, discovered in waters over 1000 m deep

FIGURE 44.14
Class Holothuroidea. Sea cucumber.

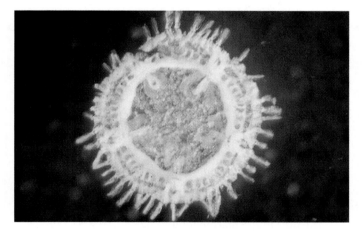

FIGURE 44.15
Class Concentricycloidea. Sea daisy.

off New Zealand (figure 44.15). Only two species are known so far. They have five-part radial symmetry, but no arms. Their tube feet are located around the periphery of the disc, rather than along radial lines, as in other echinoderms. One species has a shallow, sac-like stomach but no intestine or anus; the other species has no digestive tract at all—the surface of its mouth is covered by a membrane through which it apparently absorbs nutrients.

The newly discovered sea daisies are the most mysterious echinoderms. Tiny and primitive, they live at great depths, absorbing food from their surroundings.

Arrow Worms

There are roughly 70 species of arrow worms (phylum Chaetognatha). Given their English name because of their shape (figure 44.16*a*), arrow worms are abundant predators in the marine plankton. Arrow worms occur all over the world but they are most common in warm, shallow seas. In most marine planktonic environments, their numbers are exceeded only by the copepod crustaceans. They can dart rapidly forward or backward through the water to capture their prey, which includes mainly copepods but also cnidarian medusae, fish larvae, and other small animals. Arrow worms that feed on fish larvae may be harmful predators in areas where commercially important fishes breed.

Arrow worms, which range from 0.6 to 7 centimeters long as adults, are translucent and bilaterally symmetrical coelomates. Transverse septa divide arrow worms into head, trunk, and tail segments, foreshadowing the kind of specialization that occurs in chordates. The arrow worms have powerful jaws and a head ringed with numerous sharp movable hooks (figure 44.16*b*), key factors in their predatory role. These animals, which are linked to the other deuterostomes only in the details of their embryonic development, seem to lack close relatives. Arrow worms date back in the fossil record at least 500 million years and appear to have remained essentially unchanged throughout this long history.

Arrow worms are among the most abundant primary carnivores in the marine food web.

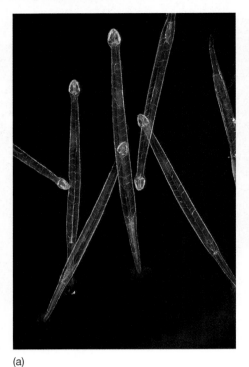
(a)

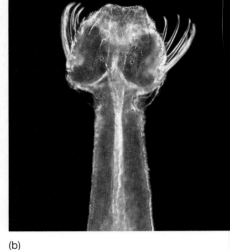
(b)

FIGURE 44.16
Arrow worms (phylum Chaetognatha).
(a) Several individuals of *Sagitta elegans*.
(b) The head of an arrow worm, *Sagitta setosa*, showing the eyes and the hooks ringing the jaws.

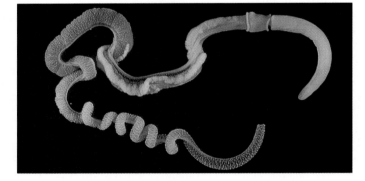

FIGURE 44.17
Acorn worm (phylum Hemichordata). *Saccoglossus kowaleskii*, extracted from its burrow.

Acorn Worms

The exclusively marine acorn worms consist of about 90 species comprised by the phylum Hemichordata, meaning "half chordates" (figure 44.17). In addition to being deuterostomes, the acorn worms share a number of features with both the echinoderms and the chordates. For example, their ciliated larvae closely resemble those of sea stars. A dorsal nerve cord is present in hemichordates, in addition to a ventral cord. In some hemichordates, part of the dorsal nerve cord is hollow, a feature otherwise found only among the chordates. Also found in both acorn worms and in chordates is the pharynx, or throat region, perforated with holes known as pharyngeal slits; these structures are found in no other animals. The acorn worms are soft-bodied animals that live in burrows in sand or mud in the sea. Most range between about 2.5 centimeters and 2.5 meters in length. Their bodies are fleshy and contractile and consist of a proboscis, collar, and trunk.

Acorn worms live in burrows in the sea. Both arrow worms and acorn worms are deuterostomes whose evolutionary relationships to the other animal phyla are unclear.

Summary of Concepts

44.1 The embryos of deuterostomes develop quite differently from those of protostomes.

- The two major evolutionary lines of coelomate animals—the protostomes and the deuterostomes—are both represented among the oldest known fossils of multicellular animals, dating back some 650 million years.

- In the protostomes, the mouth develops from or near the blastopore, and the early divisions of the embryo are spiral. At early stages of development, the fate of the individual cells is already determined, and they cannot develop individually into a whole animal.

- In the deuterostomes, the anus develops from or near the blastopore, and the mouth forms subsequently on another part of the gastrula. The early divisions of the embryo are radial. At early stages of development, each cell of the embryo can differentiate into a whole animal.

44.2 Echinoderms are deuterostomes with an endoskeleton.

- Echinoderms are exclusively marine deuterostomes that are radially symmetrical as adults.

- The epidermis of an echinoderm stretches over an endoskeleton made of separate or fused calcium-rich plates.

- Echinoderms use a unique water vascular system that includes tube feet for locomotion and feeding.

44.3 The six classes of echinoderms are all radially symmetrical as adults.

- The six classes of echinoderms with living members are Crinoidea, sea lilies and feather stars; Asteroidea, sea stars or starfish; Ophiuroidea, brittle stars; Echinoidea, sea urchins and sand dollars; Holothuroidea, sea cucumbers; and Concentricycloidea, sea daisies.

- Crinoids are sessile for some or all of their lives and have a mouth and anus located on the upper surface of the animal.

- Sea stars are active predators that move about on their tube feet.

- Brittle stars use their tube feet for feeding and move about using two arms at a time.

- The endoskeletons of sea urchins and sand dollars consist of fused calcareous plates that have been well preserved in the fossil record.

- The endoskeletons of sea cucumbers are drastically reduced and separated, making them soft-bodied.

- Sea daisies are a newly described class of echinoderms with disc-shaped bodies.

44.4 Arrow worms and acorn worms are two minor phyla.

- Arrow worms and acorn worms are animals that share features with both the echinoderms and the chordates.

Discussing Key Terms

1. **Protostomes** During embryological development, protostomes form a mouth from or near the blastopore. This group of animals typically exhibits spiral determinate cleavage. The largest, best-known phyla are mollusks, annelids, and arthropods.

2. **Deuterostomes** Deuterostomes, in contrast to protostomes, develop an anus from the blastopore during embryological development. These organisms typically exhibit radial, indeterminate cleavage.

The largest, best-known phyla are echinoderms and chordates.

3. **Endoskeleton** Echinoderms are characterized by the presence of hard inclusions constituting an endoskeleton beneath the skin. This is the first group of animals that used an internal arrangement of hard material for support and locomotion.

4. **Water vascular system** Echinoderms use a fluid-filled hydraulic system to aid movement.

Review Questions

1. What patterns of embryonic development related to cleavage and the blastopore occur in protostome coelomates? What patterns occur in deuterostome coelomates?

2. Which major coelomate phyla are protostomes and which are deuterostomes? How does the early developmental fate of cells differ between the two groups? How is the development of the coelom from mesodermal tissue different between them?

3. What type of symmetry and body plan do adult echinoderms exhibit?

4. What is the composition and location of the echinoderm skeleton? How centralized is their nervous system? What is the nature of their water vascular system? How are tube feet specialized for functions other than locomotion?

5. How do echinoderms respire? How developed is their digestive system?

6. In what two ways do members of the phylum Echinodermata reproduce? What type of larva do they possess?

7. How do sea cucumbers superficially differ from other echinoderms? How are some of their tube feet specially modified? What is the extent of their skeleton? What is the function of their unique respiratory tree? How is their reproduction different from that of other echinoderms?

Thought Questions

1. Would you expect the pattern of cell division in the early embryos of the acoelomate and pseudocoelomate animals to be spiral? Why?

2. Why do we believe the ancestor of the deuterostomes is a protostome?

3. Why is it believed that echinoderms and chordates, which are so dissimilar, are members of the same evolutionary line?

Internet Links

Life As Art
http://www.pbrc.hawaii.edu/bemf/microangela/
Dozens of wonderful scanning electron micrographs, colored in fanciful hues — a feast for screen-strained eyes.

Coral Reef Biology
http://www.uvi.edu/coral.reefer/index.html
This site presents a good general introduction to coral reef biology, with information on feeding, reproduction, & bleaching.

For Further Reading

Baker, A., W. Rowe, and H. Clark: "A New Class of Echinodermata from New Zealand," *Nature*, 1986, vol. 321, pages 862–64. Discovery of the strange sea daisies on sunken wood.

Billet, D.: "The Rise and Rise of the Sea Cucumber," *New Scientist*, March 20, 1986, pages 48–51. Sea cucumbers often swim, and there are many more than once thought; beautiful illustrations.

Birkeland, C.: "The Faustian Traits of the Crown-of-Thorns Starfish," *American Scientist*, 1989, vol. 77, pages 154–63. Very rapid growth during its early years causes *Acanthaster planci* to lose its body integrity later in life.

Davidson, E., B. Hough-Evans, and R. Britten: "Molecular Biology of the Sea Urchin Embryo," *Science*, 1982, vol. 217, pages 17–26. Many fundamental insights into the molecular biology of animal development have been gained by studying sea urchins.

Hickman, C. P., Jr., L. S. Roberts, and A. Larson: *Integrated Principles of Zoology*, 10th ed., WCB/McGraw-Hill, 1997. One of the most widely used general zoology texts, with a fine treatment of echinoderms in chapter 24.

Kozloff, E.: *Invertebrates*, Saunders, Philadelphia, 1990. A good general text on invertebrates, with a fine treatment of echinoderms.

Lawrence, J.: *A Functional Biology of Echinoderms*, Johns Hopkins University Press, Baltimore, MD, 1987. A comprehensive, modern perspective on echinoderm biology. Clearly written and well illustrated, with emphasis on feeding, maintenance, and reproduction.

Pearse, V. and J., and Buchsbaum, M. and R.: *Living Invertebrates*, Blackwell, Palo Alto, CA, 1987. A brief look at starfish and other echinoderms.

Pechenik, J.: *Biology of the Invertebrates*, Prindle, Weber & Schmidt, Boston, 1985. An excellent brief overview of invertebrates, with a good treatment of echinoderms.

Williamson, D. I.: *Larvae and Evolution: Toward a New Zoology*, Chapman and Hall, New York, 1992. Often the critical information about an organism that places it within its correct taxon comes from larval anatomy—echinoderms were considered radially symmetrical until their larval forms were studied! With a foreword by Lynn Margulis.

45

Vertebrates

Concept Outline

45.1 Attaching muscles to an internal framework greatly improves movement.

The Chordates. Chordates have an internal flexible rod, the first stage in the evolution of a truly internal endoskeleton.

45.2 Nonvertebrate chordates have a notochord but no backbone.

The Nonvertebrate Chordates. Lancelets are thought to be the ancestors of vertebrates.

45.3 The vertebrates have an interior framework of bone.

Characteristics of Vertebrate Chordates. Vertebrates have a true, usually bony endoskeleton, with a backbone encasing the spinal column, and a skull-encased brain.

45.4 The evolution of vertebrates involves successful invasions of sea, land, and air.

Fishes. Over half of all vertebrate species are fishes, the group from which all other vertebrates evolved.
History of the Fishes. Swim bladders have made bony fishes a particularly successful group.
Amphibians. The key innovation that made life on land possible for vertebrates was the pulmonary vein.
History of the Amphibians. In the past, amphibians were far more diverse, and included many large, armored terrestrial forms.
Reptiles. Reptiles were the first vertebrates to completely master the challenge of living on dry land.
The Rise and Fall of Dominant Reptile Groups. Now-extinct forms of reptiles dominated life on land for 250 million years. Four orders survive today.
Birds. Birds are much like reptiles, but with feathers.
History of the Birds. Birds are thought to have evolved from dinosaurs.
Mammals. Mammals are the only vertebrates that possess hair and milk glands.
History of the Mammals. Mammals evolved at the same time as dinosaurs, but only became common when dinosaurs disappeared.

FIGURE 45.1
A typical vertebrate. This snow leopard, *Panthera uncia*, shares many characteristics with humans and all other mammals. Today mammals dominate vertebrate life on land, but for over 200 million years in the past they were a minor group in a world dominated by reptiles.

Members of the phylum Chordata exhibit great improvements in the endoskeleton over what is seen in echinoderms. As we saw in the previous chapter, the endoskeleton of echinoderms is functionally similar to the exoskeleton of arthropods; it is a hard shell that encases the body, with muscles attached to its inner surface. Chordates employ a very different kind of endoskeleton, one that is truly internal. Members of the phylum Chordata are characterized by a flexible rod that develops along the back of the embryo. Muscles attached to this rod allowed early chordates to swing their backs from side to side, swimming through the water. This key evolutionary advance, attaching muscles to an internal element, started chordates along an evolutionary path that led to the vertebrates (figure 45.1)—and, for the first time, to truly large animals.

The Chordates

Chordates (phylum Chordata) are deuterostome coelomates whose nearest relations in the animal kingdom are the echinoderms, also deuterostomes. The chordates are the best known and most familiar of all animal groups. There are some 43,000 species of chordates, a phylum that includes birds, reptiles, amphibians, fishes, and mammals.

Four principal features characterize the chordates and have played an important role in the evolution of the phylum (figure 45.2):

1. A single, hollow **nerve cord** runs just beneath the dorsal surface of the animal. In vertebrates, the dorsal nerve cord differentiates into the brain and spinal cord.
2. A flexible rod, the **notochord,** forms on the dorsal side of the primitive gut in the early embryo and is present at some stage in the life cycle in all chordates. The notochord is located just below the nerve cord. The notochord, which persists throughout the life cycle of some invertebrate chordates, becomes surrounded and then replaced during embryological development in most vertebrates by the vertebral column that forms around the nerve cord.
3. **Pharyngeal slits** connect the **pharynx,** a muscular tube that links the mouth cavity and the esophagus, with the outside. In most vertebrates, the slits do not actually connect to the outside and are better termed pharyngeal pouches. Pharyngeal pouches are present in the embryos of all vertebrates but are lost later in the development of terrestrial vertebrates. The presence of these structures in all vertebrate embryos provides a clue to the aquatic ancestry of the group.
4. Chordates have a **postanal tail** that extends beyond the anus, at least during their embryonic development. Nearly all other animals have a terminal anus.

All chordates have all four of these characteristics at some time in their lives. For example, humans have pharyngeal slits, a dorsal nerve cord, and a notochord as embryos. As

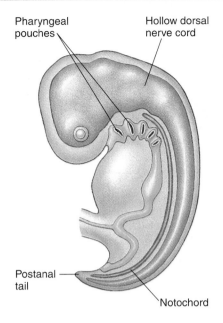

FIGURE 45.2
Some of the principal features of the chordates, as shown in a generalized embryo.

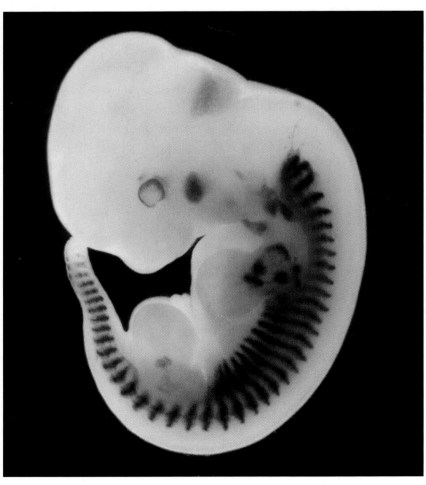

FIGURE 45.3
A mouse embryo. At 11.5 days of development, the muscle is already divided into segments called somites (stained dark in this photo), reflecting the fundamentally segmented nature of all chordates.

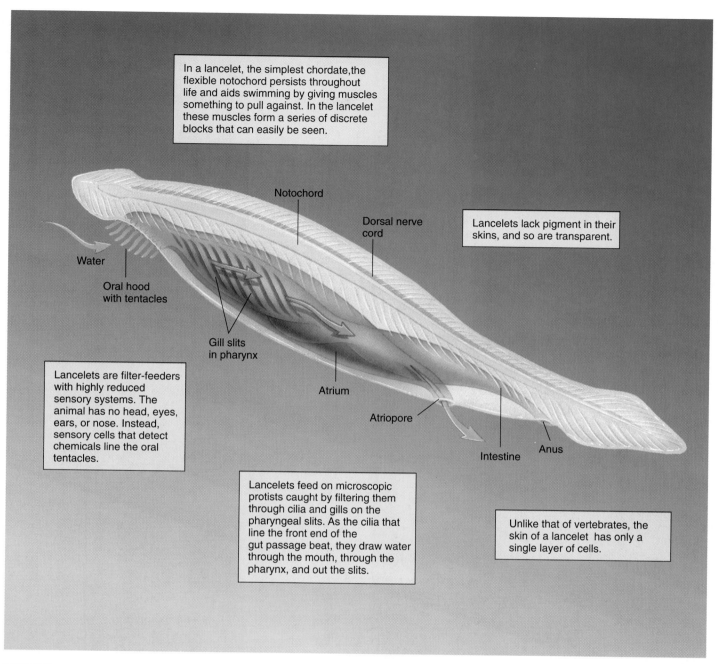

In a lancelet, the simplest chordate, the flexible notochord persists throughout life and aids swimming by giving muscles something to pull against. In the lancelet these muscles form a series of discrete blocks that can easily be seen.

Notochord

Dorsal nerve cord

Lancelets lack pigment in their skins, and so are transparent.

Water

Oral hood with tentacles

Gill slits in pharynx

Atrium

Atriopore

Lancelets are filter-feeders with highly reduced sensory systems. The animal has no head, eyes, ears, or nose. Instead, sensory cells that detect chemicals line the oral tentacles.

Intestine

Anus

Lancelets feed on microscopic protists caught by filtering them through cilia and gills on the pharyngeal slits. As the cilia that line the front end of the gut passage beat, they draw water through the mouth, through the pharynx, and out the slits.

Unlike that of vertebrates, the skin of a lancelet has only a single layer of cells.

FIGURE 45.4

Evolution of a notochord. Vertebrates, tunicates, and lancelets are chordates (phylum Chordata), coelomate animals with a flexible rod, the notochord, that anchors internal muscles and permits rapid body movements. Chordates also possess pharyngeal slits (reflecting their aquatic ancestry and present habitat in some) and a hollow dorsal nerve cord. In vertebrates, the notochord is replaced during embryonic development by the vertebral column.

adults, humans retain only the nerve cord and one pair of pharyngeal slits, which are the Eustachian tubes that connect the throat to the middle ear.

In addition to these four principal features, a number of other characteristics distinguish the chordates fundamentally from other animals. Chordates have a more or less segmented body plan, and distinct blocks of muscles can often be clearly seen in embryos of this phylum (figure 45.3). Most chordates have an internal skeleton against which the muscles work. Either this internal skeleton or the notochord (figure 45.4) makes possible the extraordinary powers of locomotion that characterize the members of this group.

Chordates are characterized by a hollow dorsal nerve cord, a notochord, pharyngeal gill slits, and a postanal tail at some point in their development. The flexible notochord anchors internal muscles and allows rapid, versatile movement.

The Nonvertebrate Chordates

Tunicates

The tunicates (subphylum Urochordata) are a group of about 1250 species of marine animals. Most of them are sessile as adults (figure 45.5*a,b*), with only the larvae having a notochord and nerve cord. As adults, they lack a major body cavity and visible signs of segmentation. Most species occur in shallow waters, but some are found at great depths. In some tunicates, adults are colonial, living in masses on the ocean floor. The pharynx is lined with numerous cilia, and the animals obtain their food by ciliary action. The cilia beat, drawing a stream of water into the pharynx, where microscopic food particles are trapped in a mucous sheet secreted from a structure called an *endostyle*.

The tadpole-like larvae of tunicates plainly exhibit all of the basic characteristics of chordates and mark the tunicates as having the most primitive combination of features found in any chordate (figure 45.5*c*). The larvae do not feed and have a poorly developed gut. They remain free-swimming for only a few days before settling to the bottom and attaching themselves to a suitable substrate by means of a sucker.

Tunicates change so much as they mature and adjust developmentally to a sessile, filter-feeding existence that it would be difficult to discern their evolutionary relationships by examining an adult. Many adult tunicates secrete a **tunic,** a tough sac composed mainly of cellulose. The tunic surrounds the animal and gives the subphylum its name. Cellulose is a substance frequently found in the cell walls of plants and algae but is rarely found in animals. In colonial tunicates, there may be a common sac and a common opening to the outside. Many tunicates are inhabited by photosynthetic bacteria, which provide their principal nutrition.

(a)

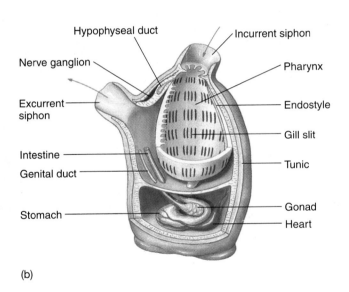

(b)

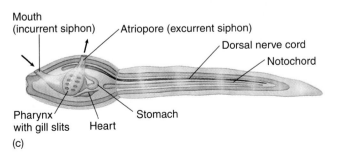

(c)

FIGURE 45.5
Tunicates (phylum Chordata, subphylum Tunicata). (a) The sea peach, *Halocynthia auranthium.* (b) Diagram of the structure of an adult tunicate. (c) Diagram of the structure of a larval tunicate, showing the characteristic tadpole-like form. Larval tunicates resemble the postulated common ancestor of the chordates.

Lancelets

Lancelets are scaleless, fish-like marine chordates a few centimeters long that occur widely in shallow water throughout the oceans of the world. Lancelets (subphylum Cephalochordata) were given their English name because of their similarity in appearance to a lancet—a small, two-edged surgical knife. There are about 23 species of this subphylum. Most of them belong to the genus *Branchiostoma*, formerly called *Amphioxus*, a name still used widely. In lancelets, the notochord runs the entire length of the dorsal nerve cord and persists throughout the animal's life.

Lancelets spend most of their time partly buried in sandy or muddy substrates, with only their anterior ends protruding (figure 45.6). They can swim efficiently, although they rarely do so. Their muscles can easily be seen as a series of discrete blocks. Lancelets have many more pharyngeal gill slits than fishes, which they resemble in overall shape. They lack pigment in their skin, which has only a single layer, unlike the multilayered skin of vertebrates. The lancelet body is pointed at both ends. There is no distinguishable head or separate eyes, nose, or ears, although there are pigmented light receptors.

Lancelets feed on microscopic members of the plankton, filtering them through a current created by beating cilia that line the gill slits at the anterior end of their alimentary canal (figure 45.7). They have an oral hood that projects beyond the mouth and bears sensory tentacles, which also ring the mouth. Males and females are separate, but no obvious external differences exist between them.

Biologists are not sure whether lancelets are primitive chordates that have survived with a number of ancestral features or are actually degenerate fishes. In the latter case, they would be vertebrates whose structural features have been reduced and simplified during the course of evolution. The fact that lancelets feed by means of cilia and have a single-layered skin, coupled with distinctive features of their excretory systems, seem to make it unlikely that they are degenerate fishes. The recent discovery of fossil forms similar to living lancelets in rocks 550 million years old—well before the appearance of any fishes—also argues for the antiquity of this group. Recent studies by molecular systematists further support the hypothesis that lancelets are vertebrates' closest ancestors.

FIGURE 45.6
Lancelets. Two lancelets, *Branchiostoma lanceolatum* (phylum Chordata, subphylum Cephalochordata), partly buried in shell gravel, with their anterior ends protruding. The muscle segments are clearly visible; the square objects along the side of the body are gonads, indicating that these are male lancelets.

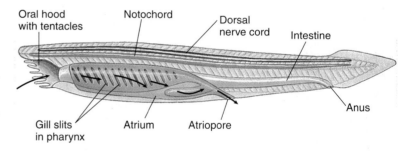

FIGURE 45.7
The structure of a lancelet. This diagram shows the path through which the lancelet's cilia pull water.

Nonvertebrate chordates, including tunicates and lancelets, have notochords but not vertebrae. They are the closest relatives of vertebrates.

Characteristics of Vertebrate Chordates

Vertebrates (subphylum Vertebrata) are chordates with a spinal column. The name *vertebrate* comes from the individual bony segments that make up the spine, which are called vertebrae. Vertebrates differ from the tunicates and lancelets in two important respects:

1. **Vertebral column.** In vertebrates, the notochord becomes surrounded, and then replaced during the course of embryonic development by a bony vertebral column. The column is a tube of hollow bones called vertebrae that encloses and protects the dorsal nerve cord like a sleeve (figure 45.8).
2. **Head.** In all vertebrates but the earliest fishes, there is a distinct and well-differentiated head, with a skull and brain. For this reason, the vertebrates are sometimes called the **craniate chordates** (Greek *kranion,* skull).

In addition to these two key characteristics, vertebrates differ from other chordates in other important respects:

1. **Neural crest.** A unique group of embryonic cells called the neural crest contributes to the development of many vertebrate structures. These cells develop on the crest of the neural tube as it forms by an invagination and pinching together of the neural plate (see chapter 58 for a detailed account). Neural crest cells then migrate to various locations in the developing embryo, where they participate in the development of a variety of structures.
2. **Internal organs.** Among the internal organs of vertebrates, livers, kidneys, and endocrine organs are characteristic of the group. Endocrine organs are ductless glands that secrete hormones that play a critical role in controlling the functions of the vertebrate body. All vertebrates have a heart and a closed circulatory system. In both their circulatory and their excretory functions, vertebrates differ markedly from other animals.
3. **Endoskeleton.** The endoskeleton of most vertebrates is made of bone (in a few, like sharks, there is flexible cartilage instead). Bone is a special form of tissue containing fibers of the protein collagen compacted together with crystals of a calcium phosphate salt. Bone forms in two stages. First, collagen is laid down in a matrix of fibers along stress lines to provide flexibility, and then calcium minerals infiltrate the fibers, providing rigidity. The great advantage of bone over chitin as a structural material is that bone is strong without being brittle. The vertebrate endoskeleton makes possible the great size and extraordinary powers of movement that characterize this group.

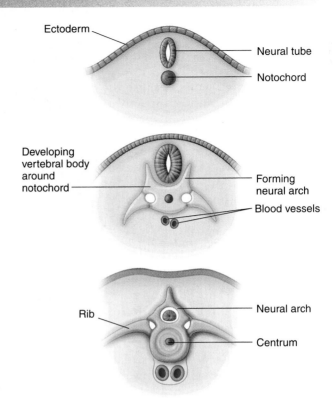

FIGURE 45.8
Embryonic development of a vertebra. During the course of evolution, or of development, the flexible notochord is surrounded and eventually replaced by a cartilaginous or bony covering, the centrum. The neural tube is protected by an arch above the centrum, and the vertebra may also have a hemal arch, which protects major blood vessels below the centrum. The vertebral column functions as a strong, flexible rod that the muscles pull against when the animal swims or moves.

Overview of the Evolution of Vertebrates

The first vertebrates evolved about 470 million years ago in the oceans as fishes without jaws or paired fins. Many of them looked like a flat hot dog, with a hole at one end and a fin at the other. Jawed fishes became the dominant creatures in the sea, some of them growing larger than cars. Their descendants, the amphibians, invaded the land. Salamander-like amphibians and other, much larger now-extinct amphibians were the first vertebrates to live successfully on land. Amphibians, in turn, gave rise to the first reptiles about 300 million years ago. Within 50 million years, reptiles, better suited than amphibians to living out of water, replaced them as the dominant land animal on earth.

With the success of reptiles, vertebrates truly came to dominate the surface of the earth. Many kinds of reptiles evolved, ranging in size from smaller than a chicken to bigger than a truck. Some flew, and others swam. Among them evolved reptiles that gave rise to the two great lines of terrestrial vertebrates, dinosaurs (and their bird descendants) and mammals. Dinosaurs and mammals appear at about the same time in the fossil record, 220 million years ago. For over 150 million years, dinosaurs dominated the face of the earth. Over all these centuries (think of it—over *a million centuries!*) the largest mammal was no bigger than a cat. Then, about 65 million years ago, the dinosaurs abruptly disappeared, for reasons that are still hotly debated. Mammals quickly took their place, becoming in turn abundant and diverse.

The history of vertebrates has been a series of evolutionary advances that have allowed vertebrates to first invade the sea and then the land. In this chapter, we will examine the key evolutionary advances that permitted vertebrates to invade the land successfully. As you will see, this invasion was a staggering evolutionary achievement, involving fundamental changes in many body systems.

Vertebrates are a diverse group, containing members adapted to life in aquatic habitats, on land, and in the air. There are eight principal classes of living vertebrates (figure 45.9). Four of the classes are fishes that live in the water, and four are land-dwelling **tetrapods,** animals with four limbs. (The name *tetrapod* comes from two Greek words meaning "four-footed.") The extant classes of fishes are the superclass Agnatha (the jawless fishes), which includes the class Myxini, the hagfish, and the class Cephalaspidomorphi, the lampreys; Chondrichthyes, the cartilaginous fishes, sharks, skates, and rays; and Osteichthyes, the bony fishes that are dominant today. The four classes of tetrapods are Amphibia, the amphibians; Reptilia, the reptiles; Aves, the birds; and Mammalia, the mammals.

Vertebrates, the principal chordate group, are characterized by a vertebral column and a distinct head.

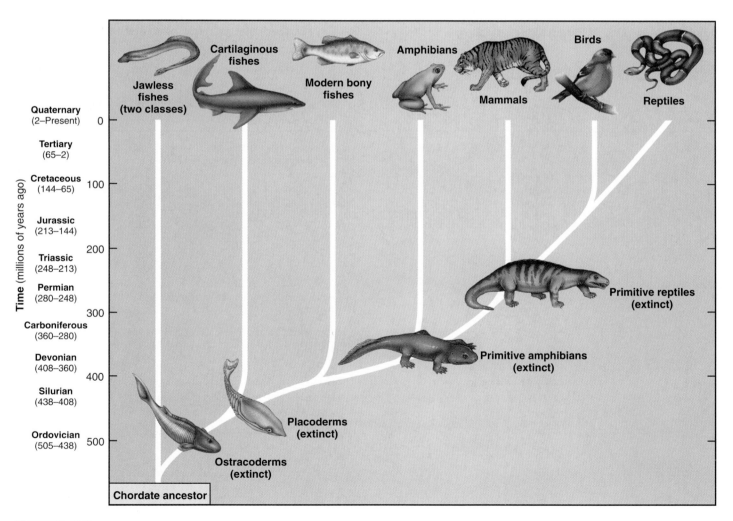

FIGURE 45.9
Vertebrate family tree. Two classes of vertebrates comprise the Agnatha, or jawless fishes. Primitive amphibians arose from fish. Primitive reptiles arose from amphibians and gave rise to mammals and to dinosaurs, which survive today as birds.

Fishes

Over half of all vertebrates are fishes. The most diverse and successful vertebrate group (figure 45.10), they provided the evolutionary base for invasion of land by amphibians. In many ways, amphibians, the first terrestrial vertebrates, can be viewed as transitional—fish out of water. In fact, fishes and amphibians share many similar features, among the host of obvious differences. First, let us look at the fishes (table 45.1).

The story of vertebrate evolution started in the ancient seas of the Cambrian Period (570–505 million years ago), when the first backboned animals appeared (figure 45.11). Wriggling through the water, jawless and toothless, these first fishes sucked up small food particles from the ocean floor like miniature vacuum cleaners. Most less than a foot long, they respired with gills but had no fins—just a primitive tail to push them through the water. For 50 million years, during the Ordovician Period (505–438 million years ago), these simple fishes were the only vertebrates. By the end of this pe-

FIGURE 45.10
Fish are diverse and include more species than all other kinds of vertebrates combined.

riod, they had developed primitive fins to help them swim and massive shields of bone to protect them from attack by the sea scorpions that were fearsome predators then.

	Table 45.1	Major Classes of Fishes		
Class	**Typical Examples**		**Key Characteristics**	**Approximate Number of Living Species**
Acanthodii	Spiny fishes		Fishes with jaws; all now extinct; paired fins supported by sharp spines	Extinct
Placodermi	Armored fishes		Jawed fishes with heavily armored heads; often quite large	Extinct
Osteichthyes	Ray-finned fishes		Most diverse group of vertebrates; swim bladders and bony skeletons; paired fins supported by bony rays	20,000
	Lobe-finned fishes		Largely extinct group of bony fishes; ancestral to amphibians; paired lobed fins	7
Chondrichthyes	Sharks, skates, rays		Streamlined hunters; cartilaginous skeletons; no swim bladders; internal fertilization	850
Myxini	Hagfishes		Jawless fishes with no paired appendages; scavengers; mostly blind, but a well-developed sense of smell	43
Cephalaspidomorphi	Lampreys		Largely extinct group of jawless fishes with no paired appendages; parasitic and nonparasitic types; all breed in fresh water	17

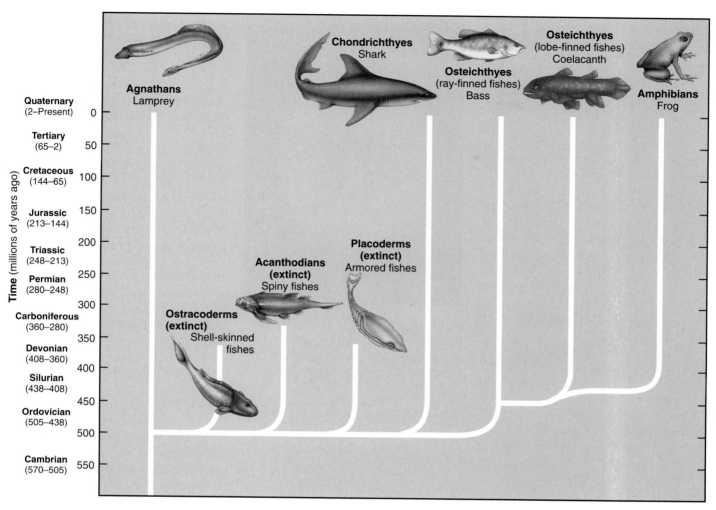

FIGURE 45.11
Evolution of the fishes. The evolutionary relationships among the different groups of fishes as well as between fishes and amphibians is shown. The spiny and armored fishes that dominated the early seas are now extinct.

Characteristics of Fishes

From whale sharks that are 18 meters long to tiny cichlids no larger than your fingernail, fishes vary considerably in size, shape, color, and appearance. Some live in freezing Arctic seas, others in warm freshwater lakes, and still others spend a lot of time out of water entirely. However varied, all fishes have important characteristics in common:

1. **Gills.** Fishes are water-dwelling creatures and must extract oxygen for their metabolism from oxygen gas dissolved in the water around them. They do this by gulping a great deal of water. The water passes over the gills, fine filaments of tissue rich in blood vessels, in the back of the mouth, and then washes back out of the body through slits in the side of the throat. The blood moves opposite the flow of water.
2. **Vertebral column.** All fishes have an internal skeleton with a spine surrounding the dorsal nerve cord, although it may not necessarily be made of bone. The brain is fully encased within a protective box, the skull or cranium, made of bone or cartilage.
3. **Single-loop blood circulation.** Blood is pumped from the heart to the gills. From the gills, the oxygenated blood passes to the rest of the body, then returns to the heart. The heart is a muscular tube-pump made of four chambers that contract in sequence.
4. **Nutritional deficiencies.** Fishes are unable to synthesize the aromatic amino acids and must consume them in their diet. This inability has been inherited by all their vertebrate descendants.

Fishes were the first vertebrates to make their appearance, and today they are still the largest vertebrate class. They are the vertebrate group from which all other vertebrates evolved.

History of the Fishes

The First Fishes

The first fishes were members of the five Ostracoderm orders (the word means "shell-skinned"). Only their head-shields were made of bone; their elaborate internal skeletons were constructed of cartilage. Many ostracoderms were bottom dwellers, their mouth underneath their flat head, their eyes on its upper surface. Ostracoderms thrived in the Ordovician Period and in the period which followed, the Silurian Period (438–408 million years ago), only to become almost completely extinct at the close of the following Devonian Period (408–360 million years ago). One group, the jawless Agnatha, survive today as hagfish and parasitic lampreys (figure 45.12).

A fundamentally important evolutionary advance was achieved in the late Silurian Period, 410 million years ago—the development of jaws. Jaws evolved from the frontmost of a series of arch-supports made of cartilage that were used to reinforce the tissue between gill slits, holding the slits open (figure 45.13). This transformation was not as radical as it might at first appear. Each gill arch was formed by a series of several cartilages (later to become bones) arranged somewhat in the shape of a V turned on its side, with the point directed outward. Imagine the fusion of the front pair of arches at top and bottom, with hinges at the points, and you have the primitive vertebrate jaw. The top half of the jaw is not attached to the skull directly except at the rear. Teeth developed on the jaws from skin that lined the mouth.

The first fishes to develop jaws were called acanthodians, or spiny fishes. Spiny fishes were very common during the early Devonian, largely replacing ostracoderms, but became extinct themselves at the close of the Devonian. Like ostracoderms, they had internal skeletons made of cartilage, but their skin scales contained small plates of bone, foreshadowing the much larger role bone would play in the future of vertebrates. Spiny fishes were predators and far better swimmers than ostracoderms, with as many as seven paired fins to aid their swimming. All of these fins were reinforced with strong spines, giving these fishes their name. No spiny fishes survive today.

By the mid-Devonian, spiny fishes were being replaced by fishes with better jaws, the heavily armored placoderms. A very diverse and successful group, seven orders of placoderms dominated the seas of the late Devonian, only to become extinct at the end of that period. The front of the placoderm body was more heavily armored than the rear. The placoderm jaw was much improved from the primitive jaw of spiny fishes, with the upper jaw fused to the skull and the skull hinged on the shoulder. Many of the placoderms grew to enormous sizes, some over 30 feet long, with two-foot skulls that had an enormous bite. No placoderms survive today.

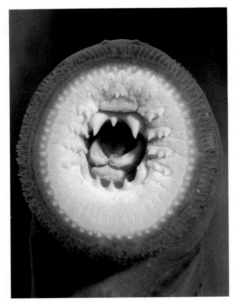

FIGURE 45.12
Specialized mouth of a lamprey.
Lampreys use their sucker-like mouths to attach themselves to the fishes on which they prey. When they have done so, they bore a hole in the fish with their teeth and feed on its blood.

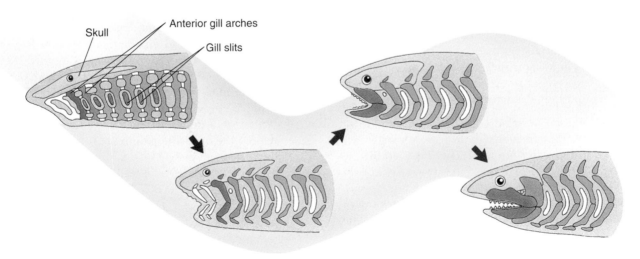

FIGURE 45.13
Evolution of the jaw. Jaws evolved from the anterior gill arches of ancient, jawless fishes.

The Rise of Active Swimmers

At the end of the Devonian, essentially all of these pioneer vertebrates disappeared, replaced by sharks and bony fishes. Sharks and bony fishes first evolved in the early Devonian, 400 million years ago. In these fishes, the jaw was improved even further, with the first gill arch behind the jaws being transformed into a supporting strut or prop, joining the rear of the lower jaw to the rear of the skull. This allowed the mouth to open very wide, into almost a full circle. In a great white shark, this wide-open mouth can be a very efficient weapon.

The major factor responsible for the replacement of primitive fishes by sharks and bony fishes was that they had a superior design for swimming. The typical shark and bony fish is streamlined. The head of the fish acts as a wedge to cleave through the water, and the body tapers back to the tail, allowing the fish to slip through the water with a minimum amount of turbulence.

In addition, sharks and bony fishes have an array of mobile fins that greatly aid swimming. First, there is a propulsion fin: a large and efficient tail (caudal) fin that helps drive the fish through the water with a back-and-forth waving motion, pushing against the water and thrusting the fish forward. Second, there are two stabilizing fins: one (or sometimes two) dorsal fins on the back that act as a stabilizer to prevent rolling as the fish swims through the water, while another ventral fin acts as a keel to prevent side-slip. Third, there are the paired fins at shoulder and hip ("A fin at each corner"), consisting of a front (pectoral) pair and a rear (pelvic) pair. These fins act like the elevator flaps of an airplane to assist the fish in going up or down through the water, as rudders to help it turn sharply left or right, and as brakes to help it stop quickly.

Sharks Become Top Predators

In the period following the Devonian, the Carboniferous Period (360–280 million years ago), sharks became the dominant vertebrates in the sea. Sharks (class Chondrichthyes) have a skeleton made of cartilage, like primitive fishes, but it is "calcified," strengthened by granules of calcium carbonate deposited in the outer layers of cartilage. The result is a very light and strong skeleton. Streamlined, with paired fins and a light, flexible skeleton, sharks are superior swimmers (figure 45.14). Their pectoral fins are particularly large, jutting out stiffly like airplane wings—and that is how they function, adding lift to compensate for the downward thrust of the tail fin. Very aggressive predators, some sharks reached enormous size.

Sharks were among the first vertebrates to develop teeth. These teeth are not set into the jaw, as yours are, but rather sit atop it. The teeth are not firmly anchored and are easily lost. In a shark's mouth, the teeth are arrayed in 6 to 20 rows, the teeth in front doing the biting and cutting, while behind them other teeth grow and await their turn.

FIGURE 45.14
Chondrichthyes. Members of the class Chondrichthyes, such as this bull shark, are mainly predators or scavengers and spend most of their time in graceful motion. As they move, they create a flow of water past their gills, extracting oxygen from the water.

When a tooth breaks or is worn down, a replacement moves forward. One shark may eventually use more than 20,000 teeth. This programmed loss of teeth offers a great advantage: the teeth in use are always new and sharp. The skin is covered with tiny teeth-like scales, giving it a rough "sandpaper" texture. Like the teeth, which are themselves modified scales, these scales are constantly replaced throughout the shark's life.

Reproduction among the Chondrichthyes is the most advanced of any fishes. Shark eggs are fertilized internally. During mating, the male grasps the female with modified fins called claspers. Sperm run from the male into the female through grooves in the claspers. Although a few species lay fertilized eggs, the eggs of most species develop within the female's body, and the pups are born alive.

Many of the early evolutionary lines of sharks died out during the great extinction at the end of the Permian Period (280–248 million years ago). The survivors thrived and underwent a burst of evolution in the age of dinosaurs that followed, when most of the modern groups of sharks appeared. Skates and rays (flattened sharks that are bottom-dwellers) evolved then, some 200 million years after the sharks first appeared. Sharks competed successfully with the marine reptiles of that time and are still the dominant predators of the sea. Today there are 275 species of sharks, more kinds than existed in the Carboniferous.

Bony Fishes Dominate the Sea

Bony fishes (members of the class Osteichthyes) evolved at the same time as sharks, some 400 million years ago, but took quite a different evolutionary road. Instead of gaining speed through lightness, as sharks did, bony fishes adopted a heavy internal skeleton made completely of bone. Such an internal skeleton is very strong, providing a base against which very strong muscles could pull. The process of *ossification* (the evolutionary replacement of cartilage by bone) happened suddenly in evolutionary terms, completing a process started by sharks, who lay down a thin film of bone over their cartilage. Not only is the internal skeleton ossified, but also the external skeleton, the outer covering of plates and scales. Many scientists believe bony fishes evolved from spiny sharks, which also had bony plates set in their skin. Bony fishes are the most successful of all fishes, indeed of all vertebrates. There are several dozen orders containing more than 20,000 living species (figure 45.15).

Unlike sharks, bony fishes evolved in fresh water. The most ancient fossils of bony fishes are found in freshwater lake beds from the middle Devonian. These first bony fishes were small and possessed paired air sacs connected to the back of the throat. These sacs could be inflated with air to buoy the fish up or deflated to sink it down in the water.

Most bony fishes have highly mobile fins, very thin scales, and completely symmetrical tails (which keep the fish on a straight course as it swims through the water). This is a very successful design for a fish. Almost all the fishes that live today (21,000 species) are bony fishes. Two great groups arose from these pioneers: the lobe-finned fishes, ancestors of the first tetrapods, and the ray-finned fishes, which include the vast majority of today's fishes.

The characteristic feature of all ray-finned fishes is an internal skeleton of parallel bony rays that support and stiffen each fin. There are no muscles within the fins; they are moved by muscles within the body. In ray-finned fishes, the primitive air sacs are transformed into an air pouch, which provides a remarkable degree of control over buoyancy.

Important Adaptations of Bony Fishes

The remarkable success of the bony fishes has resulted from a series of significant adaptations that have enabled them to dominate life in the water. These include the swim bladder, lateral line system, and gill cover.

Swim Bladder. Although bones are heavier than cartilaginous skeletons, bony fishes are still buoyant because they possess a swim bladder, a gas-filled sac that allows them to regulate their buoyant density and so remain suspended at any depth in the water effortlessly (figure 45.16). Sharks, by contrast, must move through the water or sink, as their bodies are denser than water. In primitive bony fishes, the swim bladder is a ventral outpocketing of the pharynx behind the throat, and these species fill the swim

(a)

(b)

FIGURE 45.15
The bony fishes. The bony fishes (class Osteichthyes) are extremely diverse. (a) This perch is one of the most advanced: highly maneuverable and very adaptable. (b) This Korean angelfish in Fiji is one of the many striking fishes that live around coral reefs in tropical seas.

bladder by simply gulping air at the surface of the water. In most of today's bony fishes, the swim bladder is an independent organ that is filled and drained of gases, mostly nitrogen and oxygen, internally. How do bony fishes manage this remarkable trick? It turns out that the gases are released from their blood. The gas flow is regulated by lactic acid, the acidity of which drives nitrogen out of the blood. The lower pH also alters the shape of hemoglobin, an oxygen-binding molecule in the blood, so that it is less able to bind oxygen.

Lateral Line System. Although precursors are found in sharks, bony fishes possess a fully developed lateral line system. The lateral line system consists of a series of sensory organs that project into a canal beneath the surface of the skin. The canal runs the length of the fish's body and is open

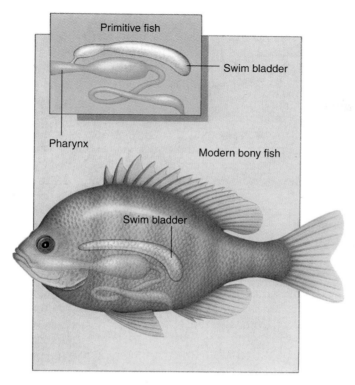

FIGURE 45.16
Diagram of a swim bladder. The bony fishes use this structure, which evolved as a ventral outpocketing of the pharynx, to control their buoyancy in water.

FIGURE 45.17
The living coelacanth, *Latimeria chalumnae*. Discovered in the western Indian Ocean in 1938, this coelacanth represents a group of fishes that had been thought to be extinct for about 70 million years. Scientists who studied living individuals in their natural habitat at depths of 100 to 200 meters observed them drifting in the current and hunting other fishes at night. Some individuals are nearly 3 meters long; they have a slender, fat-filled swim bladder. *Latimeria* is a strange animal, and its discovery was a complete surprise.

to the exterior through a series of sunken pits. Movement of water past the fish forces water through the canal. The sensory organs are deflected by the slightest movement of water over them. The pits are oriented so that some are stimulated no matter what direction the water moves. Nerve impulses from these sensory organs permit the fish to assess its rate of movement through water, sensing the movement as pressure waves against its lateral line. This is how a trout orients itself with its head upstream.

The lateral line system also enables a fish to detect motionless objects at a distance by the movement of water reflected off the object. In a very real sense, this is the fish equivalent of hearing. The basic mechanism of cilia deflection by pressure waves does not differ from what happens in human ears. You detect a pattern of pressure waves in the medium around you. The sound receptors within the ears of terrestrial vertebrates are thought to have evolved through modification of lateral line receptors.

Gill Cover. Most bony fishes have a hard plate called the operculum that covers the gills on each side of the head. Flexing the operculum permits bony fishes to pump water over their gills. The gills are suspended between the mouth and the exit behind the mouth. When the operculum is closed, it seals off the exit. When the mouth is open, closing the operculum increases the volume of the mouth cavity, so

that water is drawn into the mouth. When the mouth is closed, opening the operculum decreases the volume of the mouth cavity, forcing water past the gills to the outside. Using this very efficient bellows, bony fishes can pass water over the gills while stationary in the water. That is what a goldfish is doing when it seems to be gulping in a fish tank.

The Path to Land

Lobe-finned fishes (figure 45.17) evolved 390 million years ago, shortly after the first bony fishes appeared. Only seven species survive today, a single species of coelacanth and six species of lungfish. Lobe-finned fishes have paired fins that consist of a long fleshy muscular lobe (hence their name), supported by a central core of bones that form fully articulated joints with one another. There are bony rays only at the tips of each lobed fin. Muscles within each lobe can move the fin rays independently of one another, a feat no ray-finned fish could match. Although rare today, lobe-finned fishes played an important part in the evolutionary story of vertebrates. Amphibians almost certainly evolved from the lobe-finned fishes.

Fishes are characterized by gills and a simple, single-loop circulatory system. Sharks are fast swimmers, while the very successful bony fishes have unique characteristics such as swim bladders and lateral line systems.

Amphibians

Frogs, salamanders, and caecilians, the damp-skinned vertebrates, are direct descendants of fishes. They are the sole survivors of a very successful group, the amphibians, the first vertebrates to walk on land. Most present-day amphibians are small and live largely unnoticed by humans. When amphibians were first described by Carl Linnaeus in 1758, he said

"These foul and loathsome animals are abhorrent because of their cold body, pale color, filthy skin, fierce aspect, calculating eye, offensive smell, harsh voice, squalid habitation, and terrible venom; and so their Creator has not exerted his powers to make many of them."

In fact, Linnaeus was wrong. Amphibians are among the most numerous of terrestrial animals; there are more species of amphibians than of mammals. Throughout the world amphibians play key roles in terrestrial food chains.

Characteristics of Living Amphibians

Biologists have classified living species of amphibians into three orders (table 45.2): 3680 species of frogs and toads in 22 families make up the order Anura ("without a tail"); 369 species of salamanders and newts in 9 families make up the order Urodela or Caudata ("visible tail"); and 168 species (6 families) of worm-like, nearly blind organisms called caecilians that live in the tropics make up the order Apoda or Gymnophiona ("without legs"). While the three orders arose at different times, they have key characteristics in common:

1. **Legs.** Frogs and salamanders have four legs and can move about on land quite well. Legs were one of the key adaptations to life on land. Caecilians have lost their legs during the course of adapting to a burrowing existence.
2. **Cutaneous respiration.** Frogs, salamanders, and caecilians all supplement the use of lungs by respiring directly across their skin, which is kept moist and provides an extensive surface area. This mode of respiration limits the body size of amphibians, as it is only efficient for a high surface-to-volume ratio.
3. **Lungs.** Most amphibians possess a pair of lungs, although the internal surfaces are poorly developed, with much less surface area than reptilian or mammalian lungs. Amphibians still breathe by lowering the floor of the mouth to suck air in, then raising it back to force the air down into the lungs.
4. **Pulmonary veins.** After blood is pumped through the lungs, two large veins called pulmonary veins return the aerated blood to the heart for repumping. This allows the aerated blood to be pumped to the tissues at a much higher pressure than when it leaves the lungs.
5. **Partially divided heart.** The initial chamber of the fish heart is absent in amphibians, and the second and last chambers are separated by a dividing wall that helps prevent aerated blood from the lungs from mixing with nonaerated blood being returned to the heart from the rest of the body. This separates the blood circulation into two separate paths, pulmonary and systemic. The separation is imperfect; the third chamber has no dividing wall.

Several other specialized characteristics are shared by all present-day amphibians. In all three orders, there is a zone of weakness between the base and the crown of the teeth. Caecilians have greatly reduced eyes and ears, while frogs and salamanders possess two bones in the middle ear (rather than the one seen in reptiles). They also have a peculiar type of sensory rod cell in the retina of the eye called a "green rod." The exact function of this rod is unknown.

Amphibians, with legs and more efficient blood circulation than fishes, were the first vertebrates to walk on land.

	Table 45.2	Orders of Amphibians		
Order	**Typical Examples**		**Key Characteristics**	**Approximate Number of Living Species**
Anura	Frogs, toads		Compact tailless body; large head fused to the trunk; rear limbs specialized for jumping	3680
Caudata	Salamanders, newts		Slender body; long tail and limbs set out at right angles to the body	369
Apoda (Gymnophiona)	Caecilians		Tropical group with a snake-like body; no limbs; little or no tail	168

History of the Amphibians

The word *amphibia* (a Greek word meaning "both lives") nicely describes the essential quality of today's amphibians, referring to their ability to live in two worlds: the aquatic world of their fish ancestors and in the terrestrial world that they first invaded. In this section we will review the checkered history of this group, almost all of whose members have been extinct for the last 200 million years. Then, in the following section, we will examine in more detail what the few kinds of surviving amphibians are like.

Origin of Amphibians

Paleontologists (biologists who study fossils) agree that amphibians must have evolved from the lobe-finned fishes, although for some years there was considerable disagreement about whether the direct ancestors were coelacanths, lungfish, or the extinct rhipidistian fishes. Good arguments can be made for each. Many details of amphibian internal anatomy resemble those of the coelacanth. Lungfish and rhipidistians have openings in the tops of their mouths similar to the internal nostrils of amphibians. In addition,

lungfish have paired lungs, very like those of amphibians. Recent DNA analysis indicates lungfish are in fact far more closely related to amphibians than are coelacanths. Most paleontologists consider that amphibians evolved from rhipidistian fishes, rather than lungfish, because the pattern of bones in the amphibian skull and limbs bears a remarkable resemblance to the rhipidistians.

The successful invasion of land by amphibians involved a number of major innovations:

1. Legs were necessary to support the body's weight as well as to allow movement from place to place (figure 45.18).
2. Lungs were necessary, even though there is far more oxygen available to gills in air than in water, because the delicate structure of fish gills requires the buoyancy of water to support them.
3. The heart had to be redesigned to deliver the greater amounts of oxygen required by walking muscles.
4. Reproduction had to be carried out in water until methods evolved to prevent eggs from drying out.
5. Most importantly, a system had to be developed to prevent the body itself from drying out.

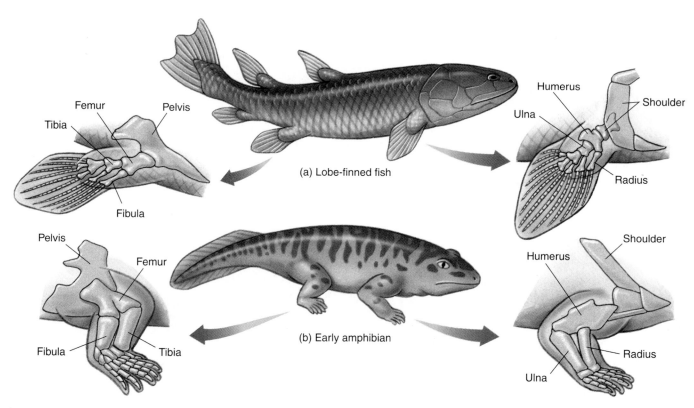

FIGURE 45.18
A comparison between the limbs of a lobe-finned fish and those of a primitive amphibian. (a) A lobe-finned fish. Some of these animals could probably move onto land. (b) A primitive amphibian. As illustrated by their skeletal structure, the legs of such an animal could clearly function on land much better than the fins of the lobe-finned fish.

The First Amphibian

Amphibians solved these problems only partially, but their solutions worked well enough that amphibians have survived for 350 million years. Evolution does not insist on perfect solutions, only workable ones.

Ichthyostega, the earliest amphibian fossil (figure 45.19), was found in a 370-million-year-old rock in Greenland. At that time, Greenland was part of the North American continent and lay near the equator. For the next 100 million years, all amphibian fossils are found in North America. Only when Asia and the southern continents all merged with North America to form the supercontinent Pangaea did amphibians spread throughout the world.

Ichthyostega was a strongly built animal, with four sturdy legs well supported by hip and shoulder bones, braced against a backbone much more substantial than that of a fish. To strengthen the backbone further, long, broad ribs that overlap each other formed a solid cage for the lungs and heart. The rib cage was so solid that it probably couldn't expand and contract for breathing. Instead, *Ichthyostega* obtained oxygen somewhat as a fish does, by lowering the floor of the mouth to draw air in, then raising it to push air down the windpipe into the lungs.

The Rise and Fall of Amphibians

Amphibians first became common during the Carboniferous Period (360–280 million years ago). Fourteen families of amphibians are known from the early Carboniferous, nearly all aquatic or semiaquatic, like Ichthyostega. By the late Carboniferous, much of North America was covered by low-lying tropical swamplands, and 34 families of amphibians thrived in this wet terrestrial environment, sharing it with pelycosaurs and other early reptiles. In the early Permian Period that followed (280–248 million years ago), a remarkable change occurred among amphibians—they began to leave the marshes for dry uplands. Many of these terrestrial amphibians had bony plates and armor covering their bodies and grew to be very large, some as big as a pony (figure 45.20). Both their large size and the complete covering of their bodies indicate that these amphibians did not use the skin respiratory system of present-day amphibians, but rather had an impermeable leathery skin to prevent water loss. By the mid-Permian, there were 40 families of amphibians. Only 25% of them were still semiaquatic like Ichthyostega; 60% of the amphibians were fully terrestrial, 15% were semiterrestrial.

This was the peak of amphibian success. By the end of the Permian, a reptile called a therapsid had become common, ousting the amphibians from their newly acquired niche on land. Over half of all amphibians were aquatic at the end of the Permian. This trend continued in the following Triassic Period (248–213 million years ago), which saw the virtual extinction of amphibians from land. By the end of the Triassic, there were only 15 families of amphibians (including the first frog), and almost without exception they were aquatic. Some of these grew to great size; one was three meters long. Only two groups of amphibians are known from the following Jurassic Period (213–144 million years ago), the anurans (frogs and toads) and the urodeles (salamanders and newts). The Age of Amphibians was over.

FIGURE 45.19
Amphibians were the first vertebrates to walk on land. Reconstruction of *Ichthyostega*, one of the first amphibians with efficient limbs for crawling on land, an improved olfactory sense associated with a lengthened snout, and a relatively advanced ear structure for picking up airborne sounds. Despite these features, *Ichthyostega*, which lived about 350 million years ago, was still quite fish-like in overall appearance and represents a very early amphibian.

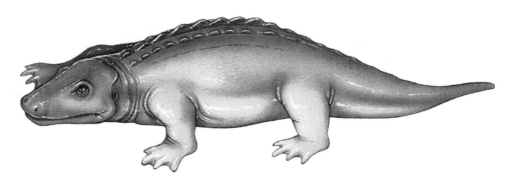

FIGURE 45.20
A terrestrial amphibian of the Permian. *Cacops*, a large, extinct amphibian, had extensive body armor.

Amphibians Today

All of today's amphibians descended from the two groups of amphibians that survived the Age of the Dinosaurs. During the Tertiary Period (65–2 million years ago), these moist-skinned amphibians underwent a highly successful invasion of wet habitats all over the world, and today there are over 4200 species of amphibians in 37 different families.

Anura. Frogs and toads, amphibians without tails, live in a variety of environments from deserts and mountains to ponds and puddles (figure 45.21*a*). Frogs have smooth, moist skin, a broad body, and long hind legs that make them excellent jumpers. Most frogs live in or near water, although some tropical species live in trees. Unlike frogs, toads have a dry, bumpy skin, short legs, and are well adapted to dry environments. All adult anurans are carnivores, eating a wide variety of invertebrates.

Most frogs and toads return to water to reproduce, laying their eggs directly in water. Their eggs lack water-tight external membranes and would dry out quickly out of the water. Eggs are fertilized externally and hatch into swimming larval forms called tadpoles. Tadpoles live in the water, where they generally feed on minute algae. After considerable growth, the body of the tadpole gradually changes into that of an adult frog. Tail, gills, and lateral line system disappear, and legs grow from the body. The mouth broadens, developing teeth and jaws. A sac-like bladder in the throat (the position of the air sacs in the anurans' fish ancestors) divides into two sacs that become lungs. The pulmonary vein appears, and the heart develops its internal wall. This process of abrupt change in body form is called **metamorphosis.**

Urodela (Caudata). Salamanders have elongated bodies, long tails, and smooth moist skin (figure 45.21*b*). They typically range in length from a few inches to a foot, although giant Asiatic salamanders of the genus *Andrias* are as much as 1.5 meters long and weigh up to 33 kilograms. Salamanders are generally unable to remain away from water for long periods, although some manage to live in dry areas by remaining inactive during the day. Most salamanders live in moist places, such as under stones or logs, or among the leaves of tropical plants. Some salamanders live entirely in water.

Salamanders lay their eggs in water or in moist places. Fertilization is usually external, although a few species practice a type of internal fertilization in which the female picks up sperm packets deposited by the male. Unlike anurans, the young that hatch from salamander eggs do not undergo profound metamorphosis, but are born looking like small adults and are carnivorous.

Apoda (Gymnophiona). Caecilians, members of the order Apoda (Gymnophiona), are a highly specialized group of tropical burrowing amphibians (figure 45.21*c*). These legless, worm-like creatures average about 30 cen-

(a)

(b)

(c)

FIGURE 45.21
Class Amphibia. (a) Red-eyed tree frog, *Agalychnis callidryas* (order Anura). (b) An adult barred tiger salamander, *Ambystoma tigrinum* (order Caudata). (c) A Mexican caecilian, *Dermophis mexicanus* (order Gymnophiona).

timeters long, but can be up to 1.3 meters long. They have very small eyes and are often blind. They eat worms and other soil invertebrates. The caecilian male deposits sperm directly into the female, and the female bears live young. Mud eels, small amphibians with tiny forelimbs and no hind limbs that live in the eastern United States, are not apodans, but highly specialized urodelians.

Amphibians ventured onto land some 370 million years ago. They are characterized by moist skin, legs (secondarily lost in some species), lungs (usually), and a more complex and divided circulatory system. They are still tied to water for reproduction.

Reptiles

If one thinks of amphibians as a first draft of a manuscript about survival on land, then reptiles are the finished book. For each of the five key challenges of living on land, reptiles improved on the innovations first seen in amphibians. Legs were arranged to support the body's weight more effectively, allowing reptile bodies to be bigger and to *run*. Lungs and heart were altered to make them more efficient. The skin was covered with dry scales to minimize water loss (figure 45.22), and eggs were encased in watertight covers. Reptiles were the first truly *terrestrial* vertebrates.

Over 7000 species of reptiles (class Reptilia) now live on earth (table 45.3). They are a highly successful group in today's world, more common than mammals. There are three reptile species for every two mammal species. While it is traditional to think of reptiles as more primitive than mammals, the great majority of reptiles that live today evolved from lines that appeared after therapsids did (the line that leads directly to mammals).

FIGURE 45.22

A common desert-dweller. Lizards are among the most common vertebrates in desert environments. They are superbly adapted to retain water and require far less food than mammals.

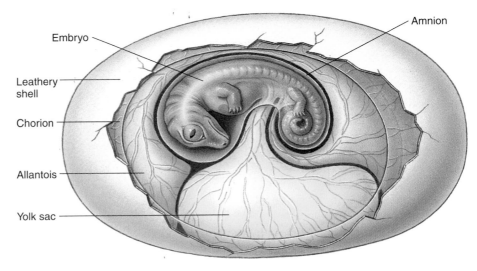

FIGURE 45.23

The watertight egg. The amniotic egg is perhaps the most important feature that allows reptiles to live in a wide variety of terrestrial habitats.

Key Characteristics of Reptiles

All living reptiles share certain fundamental characteristics, features they retain from the time when they replaced amphibians as the dominant terrestrial vertebrates. Among the most important are:

1. **Amniotic egg.** Amphibians never succeeded in becoming fully terrestrial because amphibian eggs must be laid in water to avoid drying out. Most reptiles lay watertight eggs that contain a food source (the yolk) and a series of four membranes—the yolk sac, the amnion, the allantois, and the chorion (figure 45.23). Each membrane plays a role in making the egg an independent life-support system. The outermost membrane of the egg is the **chorion,** which lies just beneath the porous shell. It allows oxygen to enter, but retains water within the egg. Within, the **amnion** encases the developing embryo within a fluid-filled cavity. The **yolk sac** provides food from the yolk for the embryo via blood vessels connecting to the embryo's gut. The **allantois** surrounds a cavity into which waste products from the embryo are excreted. All modern reptiles (as well as birds and mammals) show exactly this same pattern of membranes within the egg.

2. **Dry skin.** Living amphibians have a moist skin and must remain in moist places to avoid drying out. Like earlier amphibians, reptiles have dry skin. A layer of scales or armor covers their bodies, preventing water loss. These scales develop as surface cells fill with keratin, the same protein that forms human fingernails and bird feathers.

3. **Thoracic breathing.** Amphibians breathe by squeezing their throat to pump air into their lungs; this limits their breathing capacity to the volume of their mouth. Reptiles developed pulmonary breathing, expanding and contracting the rib cage to suck air into the lungs and then force it out. The capacity of this system is limited only by the volume of the lungs.

Reptiles were the first vertebrates to completely master the challenge of living on dry land.

Table 45.3   Orders of Reptiles

Order	Typical Examples		Key Characteristics	Approximate Number of Living Species
Ornithischia	Stegosaur		Dinosaurs with two pelvic bones facing backward, like a bird's pelvis; herbivores, with turtle-like upper beak; legs under body	Extinct
Saurischia	Tyrannosaur		Dinosaurs with one pelvic bone facing forward, the other back, like a lizard's pelvis; both plant- and flesh-eaters; legs under body	Extinct
Pterosauria	Pterosaur		Flying reptiles; wings were made of skin stretched between fourth fingers and body; wingspans of early forms typically 60 centimeters, later forms nearly 8 meters	Extinct
Plesiosaura	Plesiosaur		Barrel-shaped marine reptiles with sharp teeth and large, paddle-shaped fins; some had snake-like necks twice as long as their bodies	Extinct
Ichthyosauria	Ichthyosaur		Streamlined marine reptiles with many body similarities to sharks and modern fishes	Extinct
Squamata, suborder Sauria	Lizards		Lizards; limbs set at right angles to body; anus is in transverse (sideways) slit; most are terrestrial	3800
Squamata, suborder Serpentes	Snakes		Snakes; no legs; move by slithering; scaly skin is shed periodically; most are terrestrial	3000
Chelonia	Turtles, tortoises, sea turtles		Ancient armored reptiles with shell of bony plates to which vertebrae and ribs are fused; sharp, horny beak without teeth	250
Crocodylia	Crocodiles, alligators, gavials, caimans		Advanced reptiles with four-chambered heart and socketed teeth; anus is a longitudinal (lengthwise) slit; closest living relatives to birds	25
Rhynchocephalia	Tuataras		Sole survivors of a once successful group that largely disappeared before dinosaurs; fused, wedge-like, socketless teeth; primitive third eye under skin of forehead	2

Carboniferous　　　　　　　　　　Permian　　　　　　　　　　Triassic

FIGURE 45.24
Dinosaurs. Some of the remarkable diversity of dinosaurs, as shown in a reconstruction from the Peabody Museum at Yale University. This painting covers a span of approximately 160 million years during the later Mesozoic Era, ending 65 million years ago at the end of the Cretaceous Period. Throughout this vast period of time, the remarkable increase in the structural complexity and overall diversity of

The Rise and Fall of Dominant Reptile Groups

During the 250 million years that reptiles were the dominant large land vertebrates, a parade of changes occurred. Four major forms of reptiles took turns as the dominant type: pelycosaurs, therapsids, thecodonts, and dinosaurs (figure 45.24).

Pelycosaurs: Becoming a Better Predator

Early reptiles like *pelycosaurs* were better adapted to life on dry land than amphibians were because they evolved water-tight eggs. They had powerful jaws because of an innovation in skull design and muscle arrangement. Pelycosaurs (see figure 45.27) were **synapsids,** meaning that their skulls had a pair of temporal holes in addition to the openings for the eyes. Their jaw muscles were anchored to these holes, which allowed them to bite more powerfully. An individual pelycosaur weighed in the neighborhood of 200 kilograms. With long, sharp, "steak knife" teeth, pelycosaurs were the first land vertebrates to kill beasts their own size. Dominant for 50 million years, pelycosaurs once made up 70% of all land vertebrates. They died out about 250 million years ago, replaced by their direct descendants—the therapsids.

Therapsids: Speeding Up Metabolism

Therapsids (figure 45.25) ate ten times more frequently than their pelycosaur ancestors, burning the extra food to

FIGURE 45.25
Therapsids. The direct descendants of pelycosaurs, these synapsids are the immediate ancestors of mammals. Many grew to be quite large. Therapsids walked on all fours, but adopted a more upright stance then pelycosaurs. Therapsids had complex teeth and a secondary palate so they could breathe and chew at the same time.

produce body heat. They were endotherms, able to regulate their own body temperature. This permitted therapsids to be far more active than other vertebrates of that time, when winters were cold and long. For 20 million years, therapsids (also called "mammal-like reptiles") were the dominant land vertebrate, until largely replaced 230 million years ago by a cold-blooded, or ectothermic, reptile line—the thecodonts. Therapsids became extinct 170 million years ago, but not before giving rise to their descendants—the mammals.

Jurassic

Cretaceous

FIGURE 45.24 (continued)
the dinosaurs is evident, until they abruptly become extinct, giving way to the dominant mammals of the Cenozoic Era. Flowering plants can be seen for the first time at the right-hand side of the illustration. The names of the geological periods run along the bottom of the painting.

Thecodonts: Wasting Less Energy

Thecodonts were **diapsids,** their skulls having two pairs of temporal holes, and like amphibians and early reptiles, they were ectotherms (figure 45.26). Thecodonts largely replaced therapsids when the world's climate warmed 230 million years ago. The therapsid's endothermy was no longer advantageous in the warmer climate, and ectothermic thecodonts needed only a tenth as much food. Thecodonts were the first land vertebrates to be bipedal—to stand and walk on two feet. They were dominant for 15 million years, until replaced by their direct descendants—the dinosaurs.

Dinosaurs: Learning to Run Upright

Dinosaurs evolved from thecodonts as bipedal predators about 220 million years ago. Their legs were positioned directly underneath their bodies, a significant improvement in body design. This placed the weight of the body directly over the legs, which allowed dinosaurs to run with great speed and agility. A dinosaur fossil can be distinguished from a thecodont fossil by the presence of a hole in the side of the hip socket. Because the dinosaur leg is positioned underneath the socket, the force is directed upward, not inward, so there was no need for bone on the side of the

FIGURE 45.26
Thecodonts. These diapsid reptiles were the first bipedal land vertebrates. Most early thecodonts resembled crocodiles, but later forms stood upright. They were dominant for 15 million years until they were replaced by their direct descendants, the dinosaurs.

socket. Dinosaurs went on to become the most successful of all land vertebrates, dominating for 150 million years. All dinosaurs became extinct rather abruptly 65 million years ago, apparently as a result of an asteroid's impact.

Figures 45.27 and 45.28 (see next pages) summarize the evolutionary relationships among the extinct and living groups of reptiles, birds, and mammals (the amniotes).

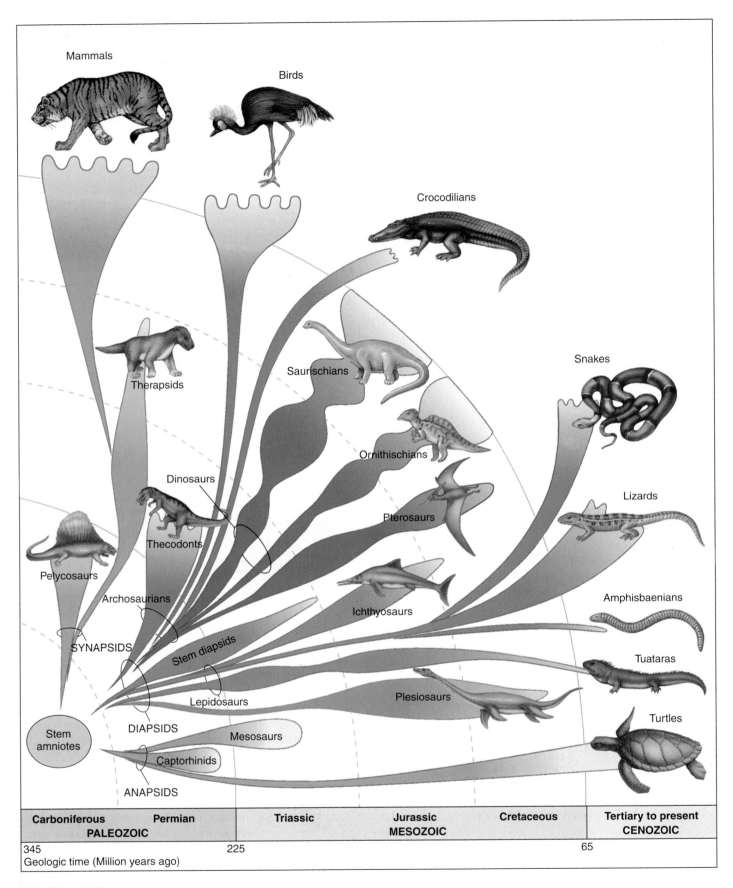

Mammals

Birds

Crocodilians

Snakes

Saurischians

Therapsids

Ornithischians

Lizards

Dinosaurs

Pterosaurs

Thecodonts

Pelycosaurs

Ichthyosaurs

Amphisbaenians

Archosaurians

Stem diapsids

SYNAPSIDS

Tuataras

Lepidosaurs

Plesiosaurs

Stem
amniotes

DIAPSIDS

Turtles

Mesosaurs

Captorhinids

ANAPSIDS

Carboniferous	Permian	Triassic	Jurassic	Cretaceous	Tertiary to present
PALEOZOIC		MESOZOIC			CENOZOIC

345   225   65
Geologic time (Million years ago)

FIGURE 45.27
Evolution of the amniotes (reptiles, birds, and mammals).

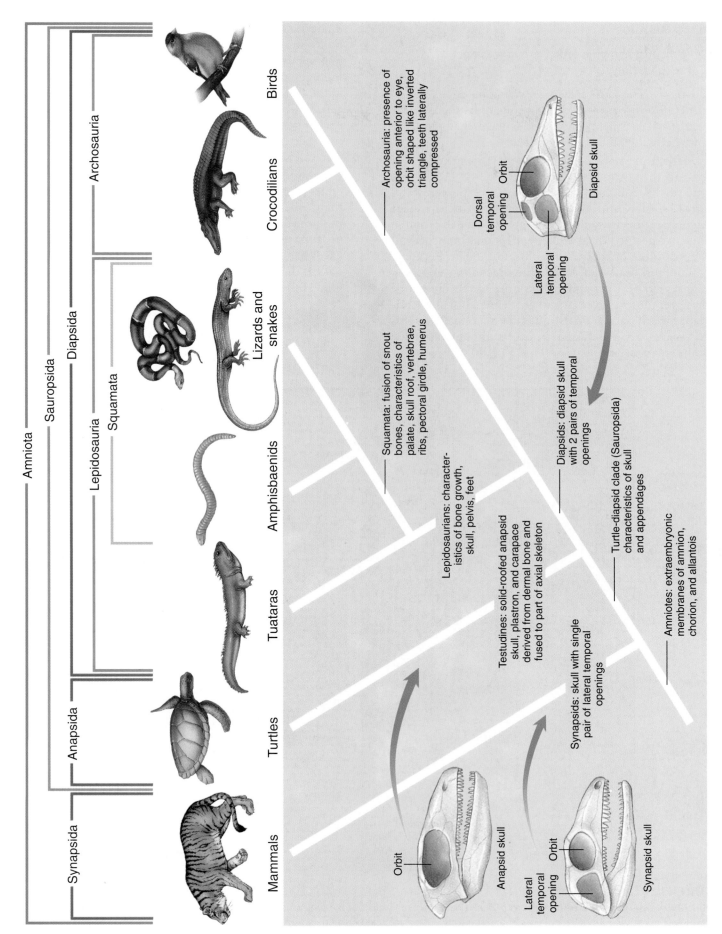

FIGURE 45.28
Cladogram of the living groups of amniotes.

Birds

Crocodilians

Lizards and snakes

Amphisbaenids

Tuataras

Turtles

Mammals

Amniota

Sauropsida

Synapsida

Diapsida

Archosauria

Anapsida

Lepidosauria

Squamata

Archosauria: presence of opening anterior to eye, orbit shaped like inverted triangle, teeth laterally compressed

Squamata: fusion of snout bones, characteristics of palate, skull roof, vertebrae, ribs, pectoral girdle, humerus

Lepidosaurians: characteristics of bone growth, skull, pelvis, feet

Testudines: solid-roofed anapsid skull, plastron, and carapace derived from dermal bone and fused to part of axial skeleton

Synapsids: skull with single pair of lateral temporal openings

Diapsids: diapsid skull with 2 pairs of temporal openings

Turtle-diapsid clade (Sauropsida) characteristics of skull and appendages

Amniotes: extraembryonic membranes of amnion, chorion, and allantois

Dorsal temporal opening

Orbit

Lateral temporal opening

Diapsid skull

Orbit

Anapsid skull

Lateral temporal opening

Orbit

Synapsid skull

Today's Reptiles

Most of the major reptile orders are now extinct. Of 16 orders of reptiles, only 4 survive.

Turtles. The most ancient surviving lineage is that of turtles. Turtles have solid skulls much like those of the first reptiles. The earliest turtles have changed little since before the time of the dinosaurs.

Lizards and snakes. Most reptiles living today belong to the second lineage to evolve, the lizards and snakes. As you read this, some 6000 species walk, crawl, slither, and climb on every continent except Antarctica. Lizards and snakes are descended from an ancient lineage of lizard-like reptiles that branched off the main line of reptile evolution in the late Permian, 250 million years ago, before the thecodonts appeared (figure 45.29). Throughout the time of the dinosaurs, these reptiles survived as minor elements of the landscape, much as mammals did. Like mammals, lizards and snakes became diverse and common only after the dinosaurs disappeared.

Tuataras. The third lineage of surviving reptiles to evolve were the Rhynchocephalonts, small diapsid reptiles that appeared shortly before the dinosaurs. They lived throughout the time of the dinosaurs and were common in the Jurassic. They began to decline in the Cretaceous, apparently unable to compete with lizards, and were already rare by the time dinosaurs disappeared. Today only two species of the order Rhynchocephalia survive, both tuataras living on small islands near New Zealand.

Crocodiles. The fourth lineage of living reptile, crocodiles, appeared on the evolutionary scene much later than other living reptiles. Crocodiles are descended from the same line of thecodonts that gave rise to the dinosaurs and resemble dinosaurs in many ways. They have changed very little in over 200 million years. Crocodiles, thecodonts, and dinosaurs together make up a group called archosaurs ("ruling reptiles").

Crocodiles resemble birds far more than they do other living reptiles. Alone among living reptiles, crocodiles care for their young and have a four-chambered heart, as birds

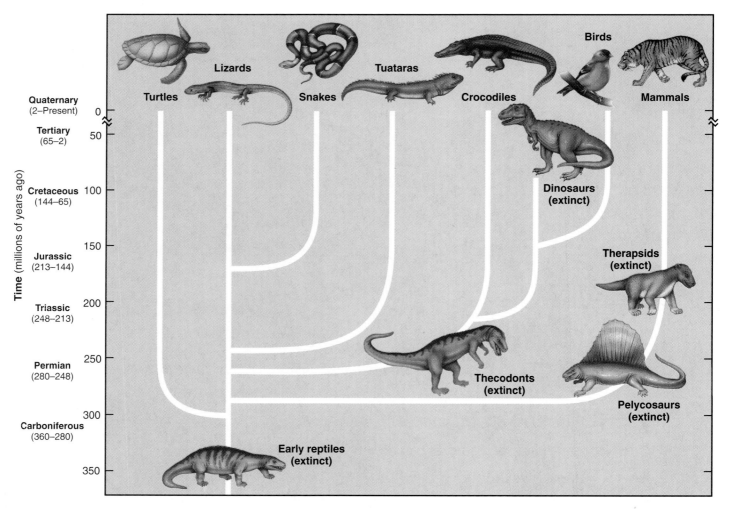

FIGURE 45.29
Evolutionary relationships among the reptiles. There are four orders of living reptiles: turtles, lizards and snakes, tuataras, and crocodiles. This phylogenetic tree shows how these four orders are related to one another and to dinosaurs, birds, and mammals.

do. There are also many other points of anatomy in which crocodiles differ from all living reptiles and resemble birds. Why are crocodiles more like birds than living reptiles? Most biologists now believe that birds are in fact the direct descendants of dinosaurs. Both crocodiles and birds are more closely related to dinosaurs, and each other, than they are related to lizards and snakes.

Other Important Characteristics

As you might imagine from the structure of the amniotic egg, reptiles do not practice external fertilization as amphibians do. There would be no way for a sperm to penetrate the membrane barriers protecting the egg. Instead, the male places sperm inside the female, where they fertilize the egg before the membranes are formed. This is called internal fertilization.

The circulatory system of reptiles is improved over that of fish and amphibians, providing more oxygen to the body (figure 45.30). The improvement is achieved by extending the septum within the heart from the atrium partway across the ventricle. This septum creates a wall that tends to lessen

mixing of oxygen-poor blood with oxygen-rich blood within the ventricle. In crocodiles, the septum completely divides the ventricle, creating a four-chambered heart, just as it does in birds (and almost certainly did in dinosaurs).

All living reptiles are **ectothermic,** obtaining their heat from external sources. In contrast, **endothermic** animals are able to generate their heat internally. In addition, **homeothermic** animals have a constant body temperature, and **poikilothermic** animals have a body temperature that fluctuates with ambient temperature. Thus, a deep-sea fish may be an ectothermic homeotherm because its heat comes from an external source, but its body temperature is constant. Reptiles are largely ectothermic poikilotherms; their body temperature is largely determined by their surroundings. Reptiles also regulate their temperature through behavior. They may bask in the sun to warm up or seek shade to prevent overheating. The thecodont ancestors of crocodiles were ectothermic, as crocodiles are today. The later dinosaurs from which birds evolved were endothermic, which is why crocodiles and birds differ in this one important respect—and a principal reason why crocodiles have been grouped among the reptiles.

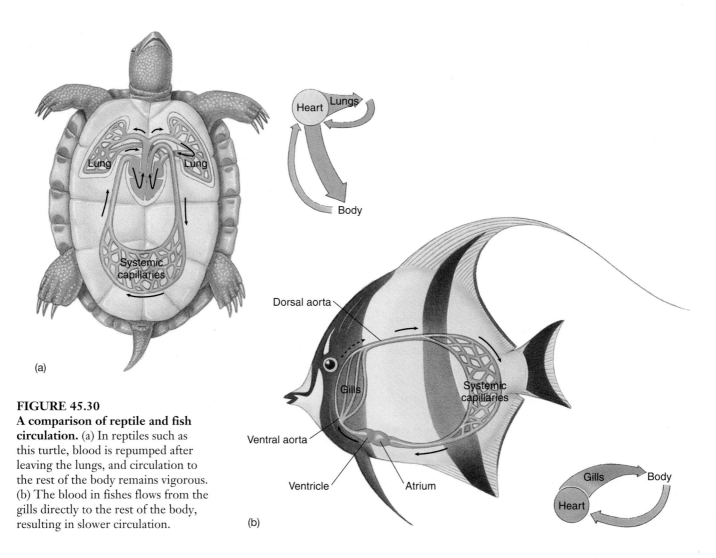

(a)

(b)

FIGURE 45.30
A comparison of reptile and fish circulation. (a) In reptiles such as this turtle, blood is repumped after leaving the lungs, and circulation to the rest of the body remains vigorous. (b) The blood in fishes flows from the gills directly to the rest of the body, resulting in slower circulation.

Kinds of Living Reptiles

Biologists have classified reptiles into 16 orders, 12 of which are extinct. The four surviving orders contain about 7000 species. Reptiles occur worldwide except in the coldest regions, where it is impossible for ectotherms to survive. In recent times, humans have had a powerful negative impact on the number and distribution of reptiles. Many species of turtles are prized for food and have been hunted almost to extinction. Despite hunting and the destruction of many of their natural habitats, reptiles remain among the most numerous and diverse of terrestrial vertebrates. The four living orders of the class Reptilia are Chelonia, Rhynchocephalia, Squamata, and Crocodilia.

FIGURE 45.31
Red-bellied turtles, *Pseudemys rubriventris.* This turtle is common in the northeastern United States.

Order Chelonia: Turtles and Tortoises. The order Chelonia consists of about 250 species of turtles (most of which live in water; figure 45.31) and tortoises (which live on land). They differ from all other reptiles because their bodies are encased within a protective shell. Many of them can pull their head and legs into the shell as well, for total protection from predators. Turtles and tortoises lack teeth but have sharp beaks.

Today's turtles and tortoises have changed very little since the first turtles appeared 200 million years ago. Turtles are **anapsid**—they lack the temporal openings in the skull characteristic of other reptiles, which are diapsid. This evolutionary stability of turtles may reflect the continuous benefit of their basic design—a body covered with a shell. In some species, the shell is made of hard plates; in other species, it is a covering of tough, leathery skin. In either case, the shell consists of two basic parts. The carapace is the dorsal covering, while the plastron is the ventral portion. In a fundamental commitment to this shell architecture, the vertebrae and ribs of most turtle and tortoise species are fused to the inside of the carapace. All of the support for muscle attachment comes from the shell.

While most tortoises have a domed-shaped shell into which they can retract their head and limbs, water-dwelling turtles have a streamlined, disc-shaped shell that permits rapid turning in water. Freshwater turtles have webbed toes, and in marine turtles, the forelimbs have evolved into flippers. Although marine turtles spend their life at sea, they must return to land to lay their eggs. Many species migrate long distances to do this. Atlantic green turtles migrate from their feeding grounds off the coast of Brazil to Ascension Island in the middle of the South Atlantic—a distance of more than 2000 kilometers—to lay their eggs on the same beaches where they hatched.

Chelonians are so different in behavior from other reptiles, and their anatomy so specialized, that many biologists place them in an entirely separate subclass (Testudines).

Order Rhynchocephalia: Tuatara. The order Rhynchocephalia contains only two species today, the tuataras, large, lizard-like animals about half a meter long. The only place in the world where tuataras are found is on a cluster of small islands off the coast of New Zealand. The native Maoris of New Zealand named the tuatara for the conspicuous spiny crest running down its back.

An unusual feature of the tuatara (and some lizards) is the inconspicuous "third eye" on the top of its head, called a parietal eye. Concealed under a thin layer of scales, the eye has a lens and retina and is connected by nerves to the brain. Why have an eye, if it is covered up? The parietal eye may function to alert the tuatara when it has been exposed to too much sun, protecting it against overheating. Unlike most reptiles, tuataras are most active at low temperatures. They burrow during the day and feed at night on insects, worms, and small animals.

Order Squamata: Lizards and Snakes. The order Squamata (figure 45.32) consists of two suborders: Sauria, some 3800 species of lizards, and Serpentes, about 3000 species of snakes. The distinguishing characteristics of this order are the presence of paired copulatory organs in the male and a lower jaw that is not joined directly to the skull. A movable hinge with five joints (your jaw has only one) allows great flexibility in the movements of the jaw. In addition, the loss of the lower arch of bone below the lower opening in the skull of lizards makes room for large muscles to operate their jaws. Most lizards and snakes are carnivores, preying on insects and small animals, and these improvements in jaw design have made a major contribution to their evolutionary success.

The chief difference between lizards and snakes is that most lizards have limbs and snakes do not. Snakes also lack movable eyelids and external ears. Lizards are a more ancient group than modern snakes, which evolved only 20 million years ago. Common lizards include iguanas, chameleons,

(a)

(b)

FIGURE 45.32
Representatives from the order Squamata. (a) An Australian skink, *Sphenomorophus.* Some burrowing lizards lack legs, and the snakes evolved from one line of legless lizards. (b) A smooth green snake, *Orpheodrys vernalis.*

FIGURE 45.33
River crocodile, *Crocodilus acutus.* Most crocodiles resemble birds and mammals in having four-chambered hearts; all other living reptiles have three-chambered hearts. Crocodiles, like birds, are more closely related to dinosaurs than to any of the other living reptiles.

Order Crocodilia: Crocodiles and Alligators. The order Crocodilia is composed of 25 species of large, primitive-looking reptiles (figure 45.33). In addition to crocodiles and alligators, the order includes two less familiar animals: the caimans and gavials. Crocodilians have remained relatively unchanged since they first evolved from thecodonts over 200 million years ago.

Crocodiles are largely nocturnal animals that live in or near water in tropical or subtropical regions of Africa, Asia, and South America. Nile crocodiles can grow to enormous size and are responsible for many human fatalities each year. There are only two species of alligators: one living in the southern United States and the other a rare endangered species living in China. Caimans, which resemble alligators, are native to Central America. Gavials are a group of fish-eating crocodilians with long, slender snouts that live only in India and Burma.

All crocodilians are carnivores. They generally hunt by stealth, waiting in ambush for prey, then attacking ferociously. Their bodies are well adapted for this form of hunting: their eyes are on top of their heads and their nostrils on top of their snouts, so they can see and breathe while lying quietly submerged in water. They have enormous mouths, studded with sharp teeth, and very strong necks. A valve in the back of the mouth prevents water from entering the air passage when a crocodilian feeds underwater.

Alone among living reptiles, crocodilians care for their young (a trait they share with at least some dinosaurs). Among Nile crocodiles, both parents carry the young in their jaws from the nest to a special nursery area where they care for them as they develop.

Many major reptile groups that dominated life on land for 250 million years are now extinct. The four living orders of reptiles include the turtles, lizards and snakes, tuataras, and crocodiles.

---

geckos, and anoles. Most are small, measuring less than a foot in length. The largest lizards belong to the monitor family. The largest of all monitors is the Komodo dragon of Indonesia, which reaches 3 meters in length and weighs up to 100 kilograms.

Lizards and snakes rely on agility and speed to catch prey and elude predators. Only two species of lizard are venomous, the Gila monster of the southwestern United States and the beaded lizard of western Mexico. Similarly, most species of snakes are nonvenomous. Of the 13 families of snakes, only 4 are venomous: the elapids (cobras, kraits, and coral snakes); the sea snakes; the vipers (adders, bushmasters, rattlesnakes, water moccasins, and copperheads); and some colubrids (African boomslang and twig snake).

Many lizards, including skinks and geckos, have the ability to lose their tails and then regenerate a new one. This apparently allows these lizards to escape from predators.

Birds

The air has been successfully conquered only four times—by insects, pterosaurs, birds, and bats. Pterosaurs, flying reptiles, evolved from gliding reptiles and flew for 130 million years before becoming extinct with the dinosaurs. There are startling similarities in how these very different animals meet the challenges of flight. Like water running downhill through similar gullies, evolution tends to seek out similar adaptations. There are major differences as well. The success of birds lies in the development of a structure unique in the animal world—the feather. Developed from reptilian scales, feathers are the ideal adaptation for flight—lightweight airfoils that are easily replaced if damaged (unlike the vulnerable skin wings of pterosaurs and bats). Today, birds (class Aves) are the most successful of all terrestrial vertebrates (figure 45.34), with 28 orders containing a total of 166 families and about 8800 species (table 45.4).

FIGURE 45.34
Bee-eater in Africa. The birds contain more species than any other class of vertebrates except the bony fishes. This bee-eater, *Merops bullockoides*, nests communally, with many adult individuals participating in the care of the young.

Key Characteristics of Birds

Modern birds lack teeth and have only vestigial tails, but they still retain many reptilian characteristics. For instance, birds lay amniotic eggs, although the shells of bird eggs are hard rather than leathery. Also, reptilian scales are present on the feet and lower legs of birds. What makes birds unique? What distinguishes them from living reptiles?

1. **Feathers.** Feathers are modified reptilian scales that serve two functions: providing lift for flight and conserving heat. The structure of feathers combines maximum flexibility and strength with minimum weight (figure 45.35). Feathers develop from tiny pits in the skin called follicles. A shaft emerges from the follicle, and pairs of vanes develop from its opposite sides. At maturity, each vane has many branches called barbs. The barbs, in turn, have many projections called barbules that are equipped with microscopic hooks. These hooks link the barbs to one another, giving the feather a continuous surface and a sturdy but flexible shape. Like scales, feathers can be replaced. Feathers are unique—no reptile or other animal has evolved them.

2. **Flight skeleton.** The bones of birds are thin and hollow. Many of the bones are fused, making the bird skeleton more rigid than a reptilian skeleton. The fused sections of backbone and of the shoulder and hip girdles form a sturdy frame that anchors muscles during flight. The power for active flight comes from large breast muscles that can make up 30% of a bird's total body weight. They stretch down from the wing and attach to the breastbone, which is greatly enlarged and bears a prominent keel for muscle attachment. They also attach to the fused collarbones that form the so-called "wishbone." No other living vertebrates have a fused collarbone or a keeled breastbone.

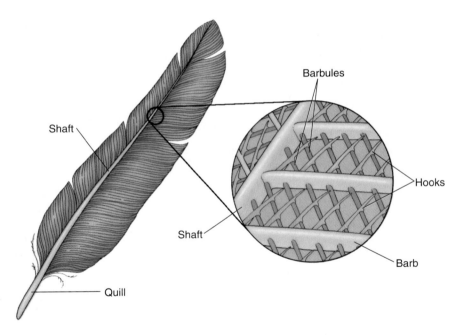

FIGURE 45.35
A feather. This enlargement shows how the vanes, secondary branches and barbs, are linked together by microscopic barbules.

Birds are the most diverse of all terrestrial vertebrates. They are closely related to reptiles, but unlike reptiles or any other animals, birds have feathers.

Table 45.4 Major Orders of Birds

Order	Typical Examples	Key Characteristics	Approximate Number of Living Species
Passeriformes	Crows, mockingbirds, robins, sparrows, starlings, warblers	*Songbirds* Well-developed vocal organs; perching feet; dependent young	5276 (largest of all bird orders; contains over 60% of all species)
Apodiformes	Hummingbirds, swifts	*Fast fliers* Short legs; small bodies; rapid wing beat	428
Piciformes	Honeyguides, toucans, woodpeckers	*Woodpeckers or toucans* Grasping feet; chisel-like, sharp bills can break down wood	383
Psittaciformes	Cockatoos, parrots	*Parrots* Large, powerful bills for crushing seeds; well-developed vocal organs	340
Charadriiformes	Auks, gulls, plovers, sandpipers, terns	*Shorebirds* Long, stilt-like legs; slender probing bills	331
Columbiformes	Doves, pigeons	*Pigeons* Perching feet; rounded, stout bodies	303
Falconiformes	Eagles, falcons, hawks, vultures	*Birds of prey* Carnivorous; keen vision; sharp, pointed beaks for tearing flesh; active during the day	288
Galliformes	Chickens, grouse, pheasants, quail	*Gamebirds* Often limited flying ability; rounded bodies	268
Gruiformes	Bitterns, coots, cranes, rails	*Marsh birds* Long, stilt-like legs; diverse body shapes; marsh-dwellers	209
Anseriformes	Ducks, geese, swans	*Waterfowl* Webbed toes; broad bill with filtering ridges	150
Strigiformes	Barn owls, screech owls	*Owls* Nocturnal birds of prey; strong beaks; powerful feet	146
Ciconiiformes	Herons, ibises, storks	*Waders* Long-legged; large bodies	114
Procellariiformes	Albatrosses, petrels	*Seabirds* Tube-shaped bills; capable of flying for long periods of time	104
Sphenisciformes	Emperor penguins, crested penguins	*Penguins* Marine; modified wings for swimming; flightless; found only in southern hemisphere; thick coats of insulating feathers	18
Dinornithiformes	Kiwis	*Kiwis* Flightless; small; primitive; confined to New Zealand	2
Struthioniformes	Ostriches	*Ostriches* Powerful running legs; flightless; only two toes; very large	1

History of the Birds

A fossil of the first bird, *Archaeopteryx* (figure 45.36), was found in a limestone quarry in Bavaria. In the late Jurassic, 150 million years ago in a quiet lagoon, the shells of microscopic marine organisms formed a very fine-grained rock that preserved and fossilized the delicate structures of animals that died there. In 1862, a slab of limestone split open to reveal the almost complete skeleton of what appeared to be a bird, the impression of its feathers stamped into the rocks. The miners gave the fossil to the mine's medical officer to pay off old fees, and he sold it (at a very high price) to the British Museum. A second and even finer *Archaeopteryx* fossil was discovered in 1877, and four others have been found since.

The skeleton of *Archaeopteryx* shares many features with small theropod dinosaurs. About the size of a crow, its skull has teeth, and very few of its bones are fused to one another—dinosaurian features, not avian. Its bones are solid, not hollow like a bird's. Also, it has a long reptilian tail, and no breastbone such as modern birds use to anchor flight muscles. Finally, it has the forelimbs of a dinosaur. Because of its many dinosaur features, several *Archaeopteryx* fossils were originally classified as the coelurosaur *Compsognathus*, a small theropod dinosaur of similar size—until feathers were discovered on the fossils. What makes *Archaeopteryx* distinctly avian is the presence of feathers on its wings and tail. It also has other bird-like features, notably the presence of a wishbone. Dinosaurs lack a wishbone, although thecodonts had them.

Today almost all paleontologists agree that *Archaeopteryx* is in fact very closely related to *Compsognathus*. Indeed, some paleontologists go so far as to classify *Archaeopteryx* as a "feathered dinosaur," not a bird, and speak flippantly of "carving the dinosaur" at Thanksgiving dinner. However, biologists continue to classify birds as Aves, a separate class, because of the key evolutionary novelties of birds: feathers, hollow bones, and physiological mechanisms such as superefficient lungs that permit sustained, powered flight. It is because of their unique adaptations and great diversity that birds are assigned to a separate class. This practical judgment should not conceal the basic agreement among almost all biologists that birds are the direct descendants of theropod dinosaurs, as closely related to coelurosaurs as are other theropods.

By the early Cretaceous, only a few million years after *Archaeopteryx*, a diverse array of birds had evolved, with many of the features of modern birds. Fossils in Mongolia, Spain, and China discovered within the last few years reveal a diverse collection of toothed birds with the hollow bones and breastbones necessary for sustained flight. Other fossils reveal highly specialized, flightless diving birds. The diverse birds of the Cretaceous shared the skies with pterosaurs for 70 million years.

Since the impression of feathers is rarely fossilized and modern birds have hollow, delicate bones, the fossil record of birds is incomplete. Relationships among the 166 families of modern birds are mostly inferred from studies of the degree of DNA similarity among living birds. These studies suggest that the most ancient living birds are the flightless birds, like the ostrich. Many others, like the elephant bird and the Moa, only recently became extinct. Ducks, geese, and other waterfowl evolved next, in the early Cretaceous, followed by a diverse group of woodpeckers, parrots, swifts, and owls. The largest of the bird orders, Passeriformes, or songbirds (60% of all species of birds today), evolved in the mid-Cretaceous. The more specialized orders of birds, such as shorebirds, birds of prey, flamingos, and penguins, did not appear until the late Cretaceous. All but a few of the modern orders of toothless birds are thought to have arisen before the disappearance of the pterosaurs and dinosaurs at the end of the Cretaceous 65 million years ago.

FIGURE 45.36

Archaeopteryx. An artist's reconstruction of *Archaeopteryx*, an early bird about the size of a crow. Closely related to its ancestors among the bipedal dinosaurs, *Archaeopteryx* lived in the forests of central Europe 150 million years ago. The teeth and long, jointed tail are features not found in modern birds. Discovered in 1862, *Archaeopteryx* was cited by Darwin in support of his theory of evolution. The true feather colors of *Archaeopteryx* are not known.

Birds Today

You can tell a great deal about the habits and food of a bird by examining its beak and feet. For instance, carnivorous birds such as hawks have curved talons for seizing prey and sharp beaks for tearing apart their meal. The beaks of ducks are flat for shoveling through mud, while the beaks of finches are short, thick seed-crushers. Figure 45.37 presents four of the 28 orders of birds.

Flight is an energy-demanding activity. Many internal changes were necessary in a bird's physiology to cope with the heavy energy demands of flight:

1. **Efficient respiration.** Flight muscles consume an enormous amount of oxygen during active flight. The reptilian lung has a limited internal surface area, not nearly enough to absorb all the oxygen needed. Mammalian lungs have a greater surface area, but as we will see in chapter 50, bird lungs satisfy this challenge with a radical redesign. When a bird inhales, the air goes past the lungs to a series of air sacs located near and within the hollow bones of the back; from there the air travels to the lungs and then a set of anterior air sacs and is exhaled. Because air always passes through the lungs in the same direction, and blood circulates past the lung in the opposite direction, the bird lung extracts oxygen from air much more efficiently.

(a)

(b)

(c)

(d)

FIGURE 45.37
Class Aves. (a) Great-horned owl, *Bubo virginianus* (order Strigiformes). (b) Ostrich, *Struthio camelus* (order Struthioniformes). (c) A pair of wood ducks, *Aix sponsa* (order Anseriformes). (d) Western tanager, *Piranga ludoviciana* (order Passeriformes).

2. **Efficient circulation.** The revved-up metabolism needed to power active flight also requires very efficient blood circulation, so that the oxygen captured by the lungs can be delivered to the flight muscles quickly. In the heart of most living reptiles, oxygen-rich blood coming from the lungs mixes with oxygen-poor blood returning from the body because the wall dividing the ventricle into two chambers is not complete. In birds, the wall dividing the ventricle is complete, and the two blood circulations do not mix, so flight muscles receive fully oxygenated blood.

In comparison with reptiles and most other vertebrates, birds have a rapid heartbeat. The hummingbird's heart beats about 600 times a minute. An active chickadee's heart beats 1000 times a minute. In contrast, the heart of the large, flightless ostrich averages 70 beats per minute—the same rate as the human heart.

3. **Endothermy.** Birds, like mammals, are endothermic. Many paleontologists believe the dinosaurs that birds evolved from were endothermic as well. Birds maintain body temperatures significantly higher than most mammals, ranging from 40 °C to 42 °C (your body temperature is 37 °C). The high temperatures maintained by endothermy permit metabolism in the bird's flight muscles to proceed at a rapid pace, to metabolize food molecules into the ATP necessary to drive rapid muscle contraction. Feathers provide excellent insulation, helping to conserve body heat.

The class Aves probably debuted 150 million years ago with *Archaeopteryx*. Modern birds are characterized by feathers, scales, a thin, hollow skeleton, auxiliary air sacs, and a four-chambered heart. Birds lay amniotic eggs and are endothermic.

Mammals

There are about 4100 living species of mammals (class Mammalia), the smallest number of species in any of the five classes of vertebrates. Almost all large, land-dwelling vertebrates are mammals, and they tend to dominate terrestrial communities, as did the dinosaurs that they replaced. When you look out over an African plain, you see the big mammals, the lions, zebras, gazelles, and antelope. Your eye does not as readily pick out the many birds, lizards, and frogs that live in the grassland community with them. Nor is the typical mammal all that large. Of the 4100 species of mammals, 3200 are rodents, bats, shrews, or moles.

Key Mammalian Characteristics

All the really large land animals are mammals. Size is what sets mammals apart from all other kinds of land animals today. But large body size is not a unique mammalian characteristic; 100 million years ago, dinosaurs were the large land animals, and mammals occupied a small-body ecological niche similar to the one frogs occupy today.

Mammals are distinguished from all other classes of vertebrates by two fundamental characteristics that all mammals possess and no other living vertebrates do:

1. **Hair.** All mammals have hair. Even apparently naked whales and dolphins grow sensitive bristles on their snouts. Evolution of fur and the ability to regulate body temperature enabled mammals to invade colder climates than ectothermic reptiles could inhabit, and the insulation fur provided may have ensured the survival of mammals when the dinosaurs perished.

   Unlike feathers, which evolved from modified reptilian scales, mammalian hair is a completely new form of skin structure. An individual mammalian hair is a long, protein-rich filament that extends like a stiff thread from a bulb-like foundation beneath the skin known as a hair follicle. The filament is composed mainly of dead cells filled with the fibrous protein keratin.

   One of the most important functions of hair is insulation against heat loss. Mammals are endothermic animals, and typically maintain body temperatures higher than the temperature of their surroundings. The dense undercoat of many mammals reduces the amount of body heat that escapes.

   Another function of hair is camouflage. The coloration and pattern of a mammal's coat usually matches its background. A little brown mouse is practically invisible against the brown leaf litter of a forest floor, while the orange and black stripes of a Bengal tiger disappear against the orange-brown color of the tall grass in which it hunts. Hairs also function as sensory structures. The whiskers of cats and dogs are

FIGURE 45.38
All mammals have hair. Some mammals, like this lemur, a nocturnal primate, also have sensory whiskers that, in addition to the large eyes, receive sensory information in the dark.

stiff hairs that are very sensitive to touch. Mammals that are active at night or live underground often rely on their whiskers to locate prey or to avoid colliding with objects (figure 45.38). Hair can also serve as a defense weapon. Porcupines and hedgehogs protect themselves with long, sharp, stiff hairs called quills.

2. **Milk-producing glands.** All female mammals possess mammary glands that secrete milk. Newborn mammals, without teeth, suckle this milk. Even baby whales are nursed by their mother's milk. Milk is a fluid rich in fat, sugar, and protein. A liter of human milk contains 11 grams of protein, 45 grams of fat, 70 grams of carbohydrate (chiefly the sugar lactose), and 2 grams of minerals critical to early growth, such as calcium. About 95% of the volume is water, critical to avoid dehydration. Milk is a very high calorie food (human milk has 750 kcal per liter), important because of the high energy needs of a rapidly growing newborn mammal. About 50% of the energy in the milk comes from fat.

Mammals first appeared 220 million years ago, evolving to their present position of dominance in modern terrestrial ecosystems. Mammals are the only vertebrates that possess hair and milk glands.

History of the Mammals

Mammals have been around since the time of the dinosaurs, although they were never common until dinosaurs disappeared. We have learned a lot about the evolutionary history of mammals from their fossils.

Origin of Mammals

The first mammals arose from therapsids in the mid-Triassic about 220 million years ago, just as the first dinosaurs evolved from thecodonts. Tiny, shrew-like creatures that lived in trees chasing insects, mammals were only a minor element in a land that quickly came to be dominated by dinosaurs. Fossils reveal that these early mammals had large eye sockets, evidence that they may have been active at night. Early mammals improved on the endothermic advances of therapsids in many ways. Instead of attaching jaw muscles to a jaw with five bones, the mammal jaw was reduced to a single massive bone, removing potentially weak junctions. Two of the bones forming the therapsid jaw joint retreated into the middle ear of mammals, linking with a bone already there to form a chain of three bones that greatly improved hearing.

Table 45.5 Some Groups of Extinct Mammals

Group	Description
Cave bears	Numerous in the ice ages; this enormous vegetarian bear slept through the winter in large groups.
Irish elk	Neither Irish nor an elk (it is a kind of deer), *Megaloceros* was the largest deer that ever lived, with horns spanning 12 feet. Seen in French cave paintings, they became extinct about 2500 years ago.
Mammoths	Although only two species of elephants survive today, the elephant family was far more diverse during the late Tertiary. Many were cold-adapted mammoths with fur.
Giant ground sloths	*Megatherium* was a giant 20-foot ground sloth that weighed three tons and was as large as a modern elephant.
Sabertooth cats	The jaws of these large, lion-like cats opened an incredible 120 degrees to allow the animal to drive its huge upper pair of saber teeth into prey.

Early Mammals

For 155 million years, all the time the dinosaurs flourished, mammals were a minor group that changed little. Only five orders of mammals arose in that time, and their fossils are scarce, indicating that mammals were not abundant. However, the two groups to which present-day mammals belong did appear. The most primitive mammals, direct descendents of therapsids, were members of the subclass Prototheria. Most prototherians were small and resembled modern shrews. All prototherians laid eggs, as did their therapsid ancestors. The only prototherians surviving today are the monotremes—the duckbill platypus and the echidnas, or spiny anteaters. The other major mammalian group is the subclass Theria. All of the mammals you are familiar with, including humans, are therians. Therians are viviparous (that is, their young are born alive). The two major living therian groups are marsupials, or pouched mammals, and placental mammals. Kangaroos, opossums, and koalas are marsupials. Dogs, cats, humans, horses, and most other mammals are placentals.

The Age of Mammals

At the end of the Cretaceous Period 65 million years ago, the dinosaurs and numerous other land and marine animals became extinct, but mammals survived, possibly because of the insulation their fur provided. In the Tertiary Period (lasting from 65 million years to 2 million years ago), mammals rapidly diversified, taking over many of the ecological roles once dominated by dinosaurs (table 45.5). Mammals reached their maximum diversity late in the Tertiary Period, about 15 million years ago. At that time, tropical conditions existed over much of the world. During the last 15 million years, world climates have deteriorated, and the area covered by tropical habitats has decreased, causing a decline in the total number of mammalian species. There are now 19 orders of mammals.

Characteristics of Modern Mammals

Endothermy. Mammals, like their therapsid ancestors, are endothermic, a crucial adaptation that has allowed mammals to be active at any time of the day or night and to colonize severe environments, from deserts to ice fields. Many characteristics, such as hair that provides insulation, played important roles in making endothermy possible. Also, the more efficient blood circulation provided by the four-chambered heart and the more efficient breathing provided by the *diaphragm* (a special sheet of muscles below the rib cage that aids breathing) make possible the higher metabolic rate upon which endothermy depends.

Placenta. In most mammal species, females carry their young in the uterus during development, nourishing them by a placenta, and give birth to live young. The placenta is a specialized organ within the womb of the mother that brings the bloodstream of the fetus into close contact with the bloodstream of the mother (figure 45.39). Food, water, and oxygen can pass across from mother to child, and wastes can pass over to the mother's blood and be carried away.

Teeth. Reptiles have homodont dentition: their teeth are all the same. However, mammals have heterodont dentition, with different types of teeth that are highly specialized to match particular eating habits (figure 45.40). It is usually possible to determine a mammal's diet simply by examining its teeth. Compare the skull of a dog (a carnivore) and a deer (an herbivore). The dog's long canine teeth are well suited for biting and holding prey, and its premolar and molar teeth are triangular and sharp for ripping off chunks of flesh. In contrast, canine teeth are absent in deer; instead the deer clips off mouthfuls of plants with its flat, chisel-like incisors. The deer's molars are large and covered with ridges to effectively grind and break up tough plant tissues. Rodents, such as a beaver, are gnawers and have long incisors for chewing through branches or stems. These incisors are ever-growing; that is, the ends may become sharp and wear down, but new incisor growth maintains the length.

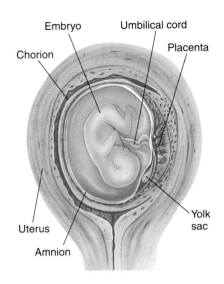

FIGURE 45.39
The placenta. The placenta is characteristic of the largest group of mammals, the placental mammals. It evolved from membranes in the amniotic egg. The umbilical cord evolved from the allantois. The chorion, or outermost part of the amniotic egg, forms most of the placenta itself. The placenta serves as the provisional lungs, intestine, and kidneys of the embryo, without ever mixing maternal and fetal blood.

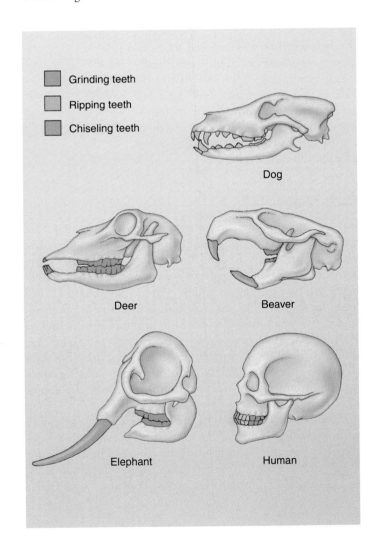

FIGURE 45.40
Mammals have different types of specialized teeth. While reptiles have all the same kind of teeth, mammals have different types of teeth specialized for different feeding habits. In carnivores such as the dog, *canine* teeth are able to rip food. Herbivores, such as deer, have *incisors* to chisel off vegetation and *molars* to grind up the plant material. In the beaver, the chiseling incisors dominate. In the elephant, the incisors have become specialized weapons, and molars grind up vegetation. Humans are omnivores; we have ripping, chiseling, and grinding teeth.

Digestive Systems for Eating Plants. Most mammals are herbivores, eating mostly or only plants. Cellulose, the major component of plant cell walls, forms the bulk of a plant's body and is a major source of food for mammalian herbivores. The cellulose molecule has the structure of a pearl necklace, with each pearl a glucose sugar molecule. Mammals do not have enzymes that can break the links between the pearls to release the glucose elements for use as food. Herbivorous mammals rely on a mutualistic partnership with bacteria that have the necessary cellulose-splitting enzymes to digest cellulose into sugar for them.

Mammals such as cows, buffalo, antelopes, goats, deer, and giraffes have huge, four-chambered stomachs that function as storage and fermentation vats. The first chamber is the largest and holds a dense population of cellulose-digesting bacteria. When the animal swallows, chewed plant material passes into this chamber, where the bacteria attack the cellulose. The material is then regurgitated and chewed again. After another thorough grinding, the cud is swallowed and digested further in the rest of the stomach.

Rodents, horses, rabbits, and elephants are herbivores that employ mutualistic bacteria to digest cellulose in a different way. They have relatively small stomachs and do not chew a cud. Instead, they digest plant material in their large intestine, like a termite. The bacteria that actually carry out the digestion of the cellulose live in a pouch called the cecum that branches from the large intestine.

Even with these complex adaptations for breaking down cellulose, a mouthful of plant is less nutritious than a mouthful of flesh. Herbivores must consume large amounts of plant material to gain sufficient nutrition. An elephant eats 135 to 150 kg (300–400 pounds) of food each day.

Horns and Hooves. Keratin, the protein of hair, is a versatile protein that is also a component of many other mammalian structures. For example, keratin is the structural building material in claws, fingernails, and hooves. Hooves are specialized keratin pads on the toes of horses, cows, sheep, antelopes, and other running mammals. The pads are hard and horny, protecting the toe and cushioning it from impact.

The horns of cattle and sheep are composed of a core of bone surrounded by a sheath of keratin. The bony core is attached to the skull, and the horn is not shed. The horn that you see is the outer sheath, made of hair-like fibers of keratin that are compacted into a very hard structure. Deer antlers are made not of keratin but of bone. Male deer grow and shed a set of antlers each year. While growing during the summer, antlers are covered by a thin layer of skin known as velvet. The velvet dies and is scraped off when the antlers are fully grown. Antlers are used by male deer during the fall and winter to attract females and to combat other males. In the spring, after the breeding season is over, the male sheds his antlers and begins the process of growing another set. A third type of horn, the rhinoceros horn, is composed only of keratinized hair.

FIGURE 45.41
Greater horseshoe bat, *Rhinolophus ferrumequinum.* The bat is the only mammal capable of true flight.

Flying Mammals. Bats are the only mammals capable of powered flight (figure 45.41). Like the wings of birds, bat wings are modified forelimbs. The bat wing is a leathery membrane of skin and muscle stretched over the bones of four fingers. The edges of the membrane attach to the side of the body and to the hind leg. When resting, most bats prefer to hang upside down from their legs. Bats are the second largest order of mammals, after rodents. They have been a particularly successful group because they have been able to utilize a food resource that most birds do not have access to—night-flying insects.

How do bats navigate in the dark? Late in the eighteenth century, the Italian biologist Lazzaro Spallanzani showed that a blinded bat could fly without crashing into things and still capture insects. Clearly another sense other than vision was being used by bats to navigate in the dark. When Spallanzani plugged the ears of a bat, it was unable to navigate and collided with objects. Spallanzani concluded that bats "hear" their way through the night world.

We now know that bats have evolved a sonar system that functions much like the sonar devices used by ships and submarines to locate underwater objects. As a bat flies, it emits a very rapid series of extremely high-pitched "clicking" sounds well above our range of human hearing. The high-frequency pulses are emitted either through the mouth or, in some cases, through the nose. The soundwaves bounce off obstacles or flying insects, and the bat hears the echo. Through sophisticated processing of this echo within its brain, a bat can determine not only the direction of an object but also the distance to the object.

The Orders of Mammals

There are 19 orders of mammals. Seventeen of them (containing 94% of the species) are placental (table 45.6). The other two are the primitive monotremes and the marsupials.

Monotremes: Egg-laying Mammals. The duck-billed platypus and two species of echidna, or spiny anteater, are the only living monotremes (figure 45.42). Among living mammals, only monotremes lay shelled eggs. The structure of their shoulder and pelvis is more similar to that of the early reptiles than to any other living mammal. Also like reptiles, monotremes have a single opening through which feces, urine, and reproductive products leave the body. Monotremes are more closely related to early mammals than are any other living mammal.

In addition to many reptilian features, monotremes have both defining mammalian features: fur and functioning mammary glands. Young monotremes drink their mother's milk after they hatch from eggs. Because of their strange mouths and the absence of well-developed nipples on the females, the babies cannot suckle. Instead, the milk oozes onto the mother's fur, and the babies lap it off with their tongues.

The platypus lives much of its life in the water and is a good swimmer. It uses its bill much as a duck does, rooting in the mud for worms and other soft-bodied animals. Echidnas have very strong, sharp claws, which they use for burrowing and digging. The echidna probes with its long, beak-like snout for insects, especially ants and termites.

Marsupials: Pouched Mammals. The major difference between marsupials (figure 45.43) and other mammals is their pattern of embryonic development. In marsupials, a fertilized egg is surrounded by chorion and amniotic membranes, but no shell forms around the egg as it does in monotremes. During most of its early development, the marsupial embryo is nourished by an abundant yolk within the egg. Shortly before birth, a short-lived placenta forms from the chorion membrane. Soon after, sometimes within eight days of fertilization, the embryonic marsupial is born. It emerges tiny and hairless, and crawls into the marsupial pouch, where it latches onto a nipple and continues to develop.

Marsupials evolved shortly before placental mammals, about 100 million years ago. Today, most species of marsupials live in Australia and South America, areas that have been historically isolated. Marsupials in Australia and New Guinea have diversified to fill ecological positions occupied by placental mammals elsewhere in the world. For example, kangaroos are the Australian grazers, playing the role antelope, horses, and buffalo perform elsewhere. The placental mammals in Australia and New Guinea today arrived relatively recently and include some introduced by humans. The only marsupial found in the United States is the Virginia opossum.

(a)

(b)

FIGURE 45.42
Monotremes (class Mammalia). (a) Echidna, *Tachyglossus aculeatus*. (b) Duck-billed platypus, *Ornithorhynchus anatinus*, at the edge of a stream in Australia.

FIGURE 45.43
Marsupials. A kangaroo with young in its pouch.

Table 45.6 Major Orders of Mammals

Order	Typical Examples		Key Characteristics	Approximate Number of Living Species
Rodentia	Beavers, mice, porcupines, rats		*Small plant-eaters* Chisel-like incisor teeth	1814
Chiroptera	Bats		*Flying mammals* Primarily fruit- or insect-eaters; elongated fingers; thin wing membrane; nocturnal; navigate by sonar	986
Insectivora	Moles, shrews		*Small, burrowing mammals* Insect-eaters; most primitive placental mammals; spend most of their time underground	390
Marsupialia	Kangaroos, koalas		*Pouched mammals* Young develop in abdominal pouch	280
Carnivora	Bears, cats, raccoons, weasels, dogs		*Carnivorous predators* Teeth adapted for shearing flesh; no native families in Australia	240
Primates	Apes, humans, lemurs, monkeys		*Tree-dwellers* Large brain size; binocular vision; opposable thumb; end product of a line that branched off early from other mammals	233
Artiodactyla	Cattle, deer, giraffes, pigs		*Hoofed mammals* With two or four toes; mostly herbivores	211
Cetacea	Dolphins, porpoises, whales		*Fully marine mammals* Streamlined bodies; front limbs modified into flippers; no hind limbs; blowholes on top of head; no hair except on muzzle	79
Lagomorpha	Rabbits, hares, pikas		*Rodent-like jumpers* Four upper incisors (rather than the two seen in rodents); hind legs often longer than forelegs; an adaptation for jumping	69
Pinnipedia	Sea lions, seals, walruses		*Marine carnivores* Feed mainly on fish; limbs modified for swimming	34
Edentata	Anteaters, armadillos, sloths		*Toothless insect-eaters* Many are toothless, but some have degenerate, peg-like teeth	30
Perissodactyla	Horses, rhinoceroses, zebras		*Hoofed mammals with one or three toes* Herbivorous teeth adapted for chewing	17
Proboscidea	Elephants		*Long-trunked herbivores* Two upper incisors elongated as tusks; largest living land animal	2

FIGURE 45.44
Placental mammals. (a) Grey wolves, *Canis lupus*, during breeding season (order Carnivora). (b) African elephants, *Loxodonta africana*, at a water hole (order Proboscidea). (c) A golden lion tamarin monkey, *Leontopithecus rosalia* (order Primates). (d) A female African lion, *Panthera leo* (order Carnivora). (e) A killer whale, *Orcinus orca*, breaching (order Cetacea).

Placental Mammals. Mammals that produce a true placenta that nourishes the embryo throughout its entire development are called placental mammals. Most species of mammals living today, including humans, are placental mammals. Of the 19 orders of living mammals, 17 are placental mammals. They are a very diverse group, ranging in size from 1.5 g pygmy shrews to 100,000 kg whales (figure 45.44).

Early in the course of embryonic development, the placenta forms. Both fetal and maternal blood vessels are abundant in the placenta, and substances can be exchanged efficiently between the bloodstreams of mother and offspring. The fetal placenta is formed from the membranes of the chorion and allantois. The maternal side of the placenta is part of the wall of the uterus, the organ in which the young develop. In placental mammals, unlike marsupials, the young undergo a considerable period of development before they are born.

A New Mammal

A recent addition to the mammal family is a tree-dwelling kangaroo discovered in New Guinea in 1994 (figure 45.45). This was an exciting find, as new macroscopic (particularly *mammalian*) discoveries are rare.

FIGURE 45.45
The newest member of the mammalian club. This tree kangaroo is a newly discovered mammal, inhabiting an area so remote that scientists had never seen it before.

Mammals were not a major group until the dinosaurs disappeared. Mammal specializations include the placenta, a tooth design suited to diet, and specialized sensory systems. There are 2 orders of nonplacental mammals (monotremes and marsupials) and 17 orders of placental mammals.

45.1 Attaching muscles to an internal framework greatly improves movement.

- Complex animals that contain muscular systems must have something to which the muscles can attach. In the case of the arthropods, a chitinous exoskeleton serves this function. In the vertebrates, a bony endoskeleton provides attachment sites for skeletal muscle.

- The chordates are characterized by a single, hollow dorsal nerve cord and by the presence, at least early in development, of a notochord, pharyngeal slits, and a postanal tail.

45.2 Nonvertebrate chordates have a notochord but no backbone.

- The approximately 43,000 species of chordates are classified into three subphyla. Two are small, exclusively marine subphyla, the tunicates and the lancelets, which seem to represent ancient evolutionary offshoots from the group.

45.3 The vertebrates have an interior framework of bone.

- The third subphylum is the vertebrates, whose members include the fishes as well as four classes that have radiated extensively in terrestrial habitats.

- Vertebrates differ from other chordates in that they possess a vertebral column, a distinct and well-differentiated head, and a bony skeleton. There are four classes of fishes and four classes of tetrapods (four-footed vertebrates).

45.4 The evolution of vertebrates involves successful invasions of sea, land, and air.

- Members of the group Agnatha differ from other vertebrates because they lack jaws. Once abundant and diverse, they are now represented only by the hagfish, which are scavengers, and the lampreys, which are parasites of other fishes.

- Jawed fishes constitute more than half of the estimated 42,500 species of vertebrates and are dominant in fresh and salt water everywhere. The Chondrichthyes, or cartilaginous fishes, consist of about 850 species of sharks, rays, and skates; the Osteichthyes, or bony fishes, comprise about 18,000 species.

- The first land vertebrates were the amphibians. Amphibians are dependent on water and lay their eggs in moist places.

- Reptiles were the first vertebrates fully adapted to terrestrial habitats. Scales, and amniotic eggs represented significant adaptations to the dry conditions on land.

- Birds and mammals were derived from reptiles and are now among the dominant groups of animals on land. The members of these two classes have independently become endothermic, capable of regulating their own body temperatures; all other living animals are ectothermic, their temperatures set by external conditions.

- The living mammals are divided into three major groups: (1) the monotremes, or egg-laying mammals, consisting only of the echidnas and the duck-billed platypus; (2) the marsupials, in which the young are born at a very early stage of development and complete their development in a pouch; and (3) the placental mammals, which lack pouches and suckle their young.

Discussing Key Terms

1. **Notochord** This flexible rod, located dorsally, is present in at least some part of the life cycle in all chordates. In vertebrate chordates, the notochord is surrounded and replaced by the vertebral column during embryonic development.

2. **Endothermy** The first terrestrial vertebrates were ectothermic, their bodies assuming the temperature of their surroundings. Mammals and birds are endothermic, expending up to 90% of the energy they consume to regulate body temperature.

3. **Respiratory adaptations** Fish acquire oxygen from water by passing the water over a thin film of tissue rich in blood vessels, the blood moving opposite the flow of the water. Amphibians use lungs to extract oxygen from air by passing air in and out of a sac with walls rich in blood vessels. Reptiles and mammals have increased surface area inside the lungs, improving respiratory efficiency. Birds capture oxygen even more efficiently with reengineered lungs in which blood moves opposite the flow of air, much as blood moves opposite the flow of water in a fish gill.

4. **Pulmonary circulation** The fish heart pumps blood to the gills, where it loses much of its force as it passes through narrow blood capillaries; the blood then circulates to the rest of the body with less pressure. In terrestrial vertebrates, a pulmonary vein returns blood to the heart for repumping after it passes through the narrow capillaries of the lungs; the blood then circulates to the rest of the body vigorously.

5. **Amniotic egg** Reptiles were the first vertebrates to reproduce on land using watertight, fluid-filled eggs.

Review Questions

1. What are the four primary characteristics of the chordates? What are the three subphyla of the chordates? Give an example of each.

2. What is the relationship between the notochord and the vertebral column in vertebrates? How is the latter structure associated with the nerve cord?

3. What are the eight classes of vertebrates? Give examples of each.

4. What is one advantage of possessing jaws? From what existing structures did jaws evolve? What is the evolutionary derivation of teeth as found in jawed animals?

5. What is the primary disadvantage of a bony skeleton compared to one made of cartilage? What special structure do most bony fishes have to counter this? What is the evolutionary derivation of this structure?

6. What is the lateral line system in fishes? How does it function?

7. The successful invasion of land by amphibians involved five major innovations. What were they, and why was each important?

8. What are the five key characteristics of amphibians? What is the differentiating characteristic of each order of amphibians?

9. What are three fundamental characteristics of reptiles? What are three other important characteristics? What are the four living orders of reptiles? Give an example of each.

10. What type of egg was first exhibited by the reptiles? What are its evolutionary advantages? How does the embryo obtain nutrients and excrete wastes while contained within the egg? How does the developing embryo respire?

11. How do birds differ from other tetrapods? Why are they more efficient flyers than reptiles or mammals? From what reptilian structure are feathers derived?

12. How do amphibian, reptile, and mammal legs differ? What are the advantages of mammal legs over the other two?

13. What are the two fundamental characteristics that distinguish mammals from all other classes of vertebrates? What are five other key characteristics of mammals, and why are they important?

Thought Questions

1. What characteristics allowed vertebrates to attain great sizes?

2. What limits the ability of amphibians to occupy the full range of terrestrial habitats and allows other terrestrial vertebrates to occupy them successfully?

3. List some of the advantages that the early birds, in which flight was not nearly as efficient as it is in most of their modern descendants, might have had as a result of their feathered wings.

Internet Links

Biology of the Past
http://www.ucmp.berkeley.edu/Paleonet
The PALEONET site of the Natural History Museum in London is a rich directory of links and information on fossil life, from the latest dinosaur find to ancient mammals, with hundreds of links and much more.

Find That Fish
http://www.birminghamzoo.com/ao/fishes.htm
The FISHES website allows the visitor to search for a particular species, and to browse through a large collection of "fishy" links.

SHARK!
http://www.brunel.ac.uk/admin/alumni/sharks/home.html
A comprehensive look at sharks, their anatomy, physiology, behavior, and conservation, from Brunel University, UK.

Frogland
http://www.teleport.com/~dstroy/index.html
A site to delight frog lovers everywhere, FROGLAND has lots of information and links, instructions on keeping a frog as a pet, frog conservation, and much more. A fun site.

Dinosaurs Today
http://www.ucmp.berkeley.edu/diapsids/dinobuzz.html
This wonderful UC Berkeley site, DINOBUZZ, is devoted to indepth analysis of current interest topics concerning dinosaurs. Excellent site!

Guide to the Mammals
http://www.ucmp.berkeley.edu/mammal/mammal.html
The UC Museum of Palentology's HALL OF MAMMALS is a great guide for exploring the mammals.

For Further Reading

Colbert, E. H.: *Evolution of the Vertebrates: A History of the Back-boned Animals Through Time*, Wiley-Liss, New York, 1991. A complete, up-to-date discussion of vertebrate phylogeny, including fossil vertebrates and evolution of the group as a whole.

Droser, M., R. Fortey, and X. Li: "The Ordovician Radiation," *American Scientist*, March/April 1996, pages 122–31. Changes in world climates played a pivotal role in biological diversification.

Hanken, J.: "Development and Evolution in Amphibians," *American Scientist*, vol. 77, 1989, pages 336–43. The evolution of morphological diversity in amphibians has been achieved by modifications in development.

Norman, D.: *Prehistoric Life: The Rise of the Vertebrates*, Macmillan, New York, 1994. A very readable account of the history of terrestrial vertebrates.

Rismiller, P. D. and R. S. Seymour: "The Echidna," *Scientific American*, February 1991, pages 96–103. The secrets of the natural history and reproductive behavior of this spiny mammal from Australia and New Guinea are now being explored.

Tuttle, R.: "Apes of the World," *American Scientist*, vol. 78, March 1990, pages 115–25. A survey of our closest living relatives reveals rich prospects for further study and an urgent need for conservation.

46

Organization of the Animal Body

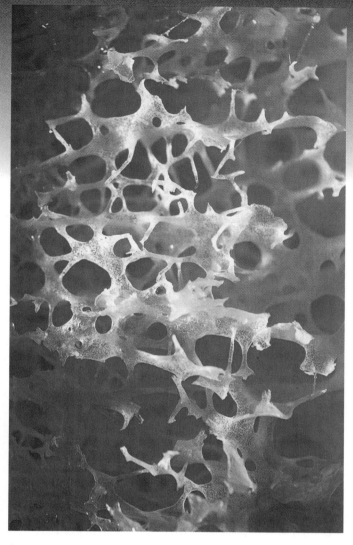

FIGURE 46.1
Bone. Like most of the tissues in the vertebrate body, bone is a dynamic structure, constantly renewing itself.

Concept Outline

46.1 The bodies of vertebrates are organized into functional systems.

Organization of the Body. Cells are organized into tissues, and tissues are organized into organs. Several organs can cooperate to form organ systems.

46.2 Epithelial tissue forms membranes and glands.

Characteristics of Epithelial Tissue. Epithelial membranes cover all body surfaces, and thus can serve for protection or for transport of materials. Glands are also epithelial tissue. Epithelial membranes may be composed of one layer or many.

46.3 Connective tissues contain abundant extracellular material.

Connective Tissue Proper. Connective tissues have abundant extracellular material. In connective tissue proper, this material consists of protein fibers within an amorphous ground substance.

Special Connective Tissues. These tissues include cartilage, bone and blood, each with their own unique form of extracellular material.

46.4 Muscle tissue provides for movement, and nerve tissue provides for control.

Muscle Tissue. Muscle tissue contains the filaments actin and myosin, which enable the muscles to contract. There are three types of muscle: smooth, cardiac, and skeletal.

Nerve Tissue. Nerve cells, or neurons, have specialized regions and different neurons are specialized for different functions, such as sensory, motor, and associative activities.

W hen most people think of animals, they think of their pet dogs and cats and the animals that they've seen in a zoo, on a farm, in an aquarium, or out in the wild. When they think about the diversity of animals, they may think of the differences between the predatory lions and tigers and the herbivorous deer and antelope, between a ferocious-looking shark and a playful dolphin. Despite the differences among these animals, they are all vertebrates. All vertebrates share the same basic body plan, with the same sorts of organs operating in much the same way. In this chapter, we will begin a detailed consideration of the biology of vertebrates and of the fascinating structure and function of their bodies (figure 46.1).

Organization of the Body

The bodies of all mammals have the same general architecture (figure 46.2), and are very similar to the general body plan of other vertebrate groups. This body plan is basically a tube suspended within a tube. Starting from the inside, it is composed of the digestive tract, a long tube that travels from one end of the body to the other (mouth to anus). This tube is suspended within an internal body cavity, the *coelom* that was mentioned in chapter 41. In humans and other mammals, the coelom is divided by a sheet of muscle, the *diaphragm*, into two cavities. The **thoracic cavity** contains the heart and lungs, and the **abdominal cavity** contains the stomach, intestines, and liver. All vertebrate bodies are supported by an internal **skeleton** made of jointed bones that grow as the body grows. A bony *skull* surrounds the brain, and a column of separate bones, the *vertebrae*, surrounds the dorsal nerve cord, or *spinal cord*.

There are four levels of organization in the vertebrate body: (1) cells, (2) tissues, (3) organs, and (4) organ systems. Like those of all animals, the bodies of vertebrates are composed of different cell types. In adult vertebrates, there are between 50 and several hundred different kinds of cells, depending upon which vertebrate is being considered and how finely one distinguishes between cell types.

Tissues

Groups of cells similar in structure and function are organized into **tissues.** Early in development, the cells of the growing embryo differentiate (specialize) into three fundamental embryonic tissues, called *germ layers.* From innermost to outermost layers, these are the **endoderm, mesoderm,** and **ectoderm.** These germ layers, in turn, differentiate into the scores of different cell types and tissues that are characteristic of the vertebrate body. In adult vertebrates, there are four principal kinds of tissues, or *primary tissues*: epithelial, connective, muscle, and nerve (figure 46.3). Each of these will be discussed in separate sections of this chapter.

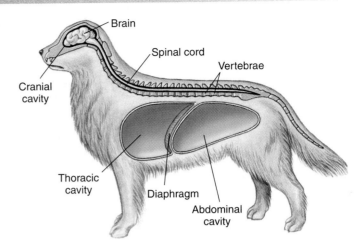

FIGURE 46.2
Architecture of the vertebrate body. Mammals, like all vertebrates, have a dorsal central nervous system, consisting of a spinal cord and brain enclosed in vertebrae and the skull. In mammals, a muscular diaphragm divides the coelom into the thoracic cavity and the abdominal cavity.

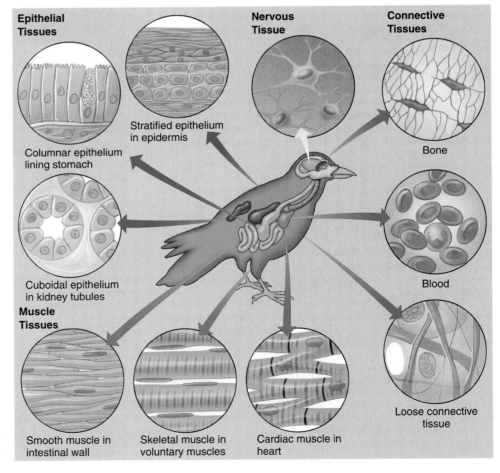

FIGURE 46.3
Vertebrate tissue types. Epithelial tissues are indicated by blue arrows, connective tissues by green arrows, muscle tissues by red arrows, and nervous tissue by a yellow arrow.

Organs and Organ Systems

Organs are body structures composed of several different tissues that form a structural and functional unit (figure 46.4). One example is the heart, which contains cardiac muscle, connective tissue, and epithelial tissue and is laced with nerve tissue that helps regulate the heart beat. An **organ system** is a group of organs that function together to carry out the major activities of the body. For example, the digestive system is composed of the digestive tract, liver, gallbladder, and pancreas. These organs cooperate in the digestion of food and the absorption of the digestion products into body. The vertebrate body contains 11 principal organ systems (table 46.1 and figure 46.5).

> The bodies of humans and other mammals contain a cavity divided by the diaphragm into thoracic and abdominal cavities. The bodies, cells are organized into tissues, which are, in turn, organized into organs and organ systems.

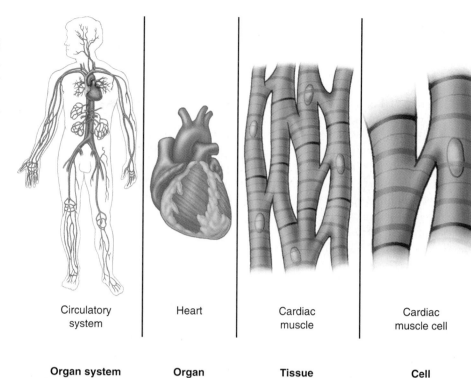

Circulatory system | Heart | Cardiac muscle | Cardiac muscle cell

Organ system | **Organ** | **Tissue** | **Cell**

FIGURE 46.4
Levels of organization within the body. Similar cell types operate together and form tissues. Tissues functioning together form organs. Several organs working together to carry out a function for the body are called an organ system. The circulatory system is an example of an organ system.

Table 46.1	The Major Vertebrate Organ Systems		
System	**Functions**	**Components**	**Detailed Treatment**
Circulatory	Transports cells, respiratory gases, and chemical compounds throughout the body	Heart, blood vessels, lymph, and lymph structures	Chapter 49
Digestive	Captures soluble nutrients from ingested food	Mouth, esophagus, stomach, intestines, liver, and pancreas	Chapter 48
Endocrine	Coordinates and integrates the activities of the body	Pituitary, adrenal, thyroid, and other ductless glands	Chapter 53
Integumentary	Covers and protects the body	Skin, hair, nails, scales, feathers, and sweat glands	Chapter 54
Lymphatic/ Immune	Vessels transport extracellular fluid and fat to circulatory system; lymph nodes and lymphatic organs provide defenses to microbial infection and cancer	Lymphatic vessels, lymph nodes, thymus, tonsils, spleen	Chapter 54
Muscular	Produces body movement	Skeletal muscle, cardiac muscle, and smooth muscle	Chapter 47
Nervous	Receives stimuli, integrates information, and directs the body	Nerves, sense organs, brain, and spinal cord	Chapters 51, 52
Reproductive	Carries out reproduction	Testes, ovaries, and associated reproductive structures	Chapter 56
Respiratory	Captures oxygen and exchanges gases	Lungs, trachea, gills, and other air passageways	Chapter 50
Skeletal	Protects the body and provides support for locomotion and movement	Bones, cartilage, and ligaments	Chapter 47
Urinary	Removes metabolic wastes from the bloodstream	Kidney, bladder, and associated ducts	Chapter 55

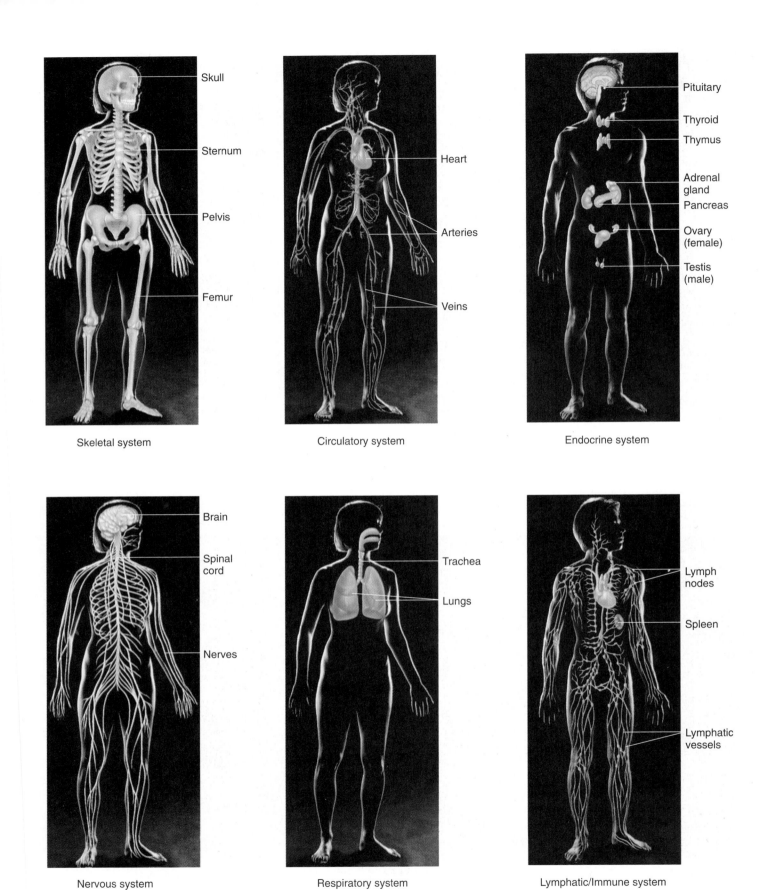

FIGURE 46.5
Vertebrate organ systems. The eleven principal organ systems of the human body are shown, including both male and female reproductive systems.

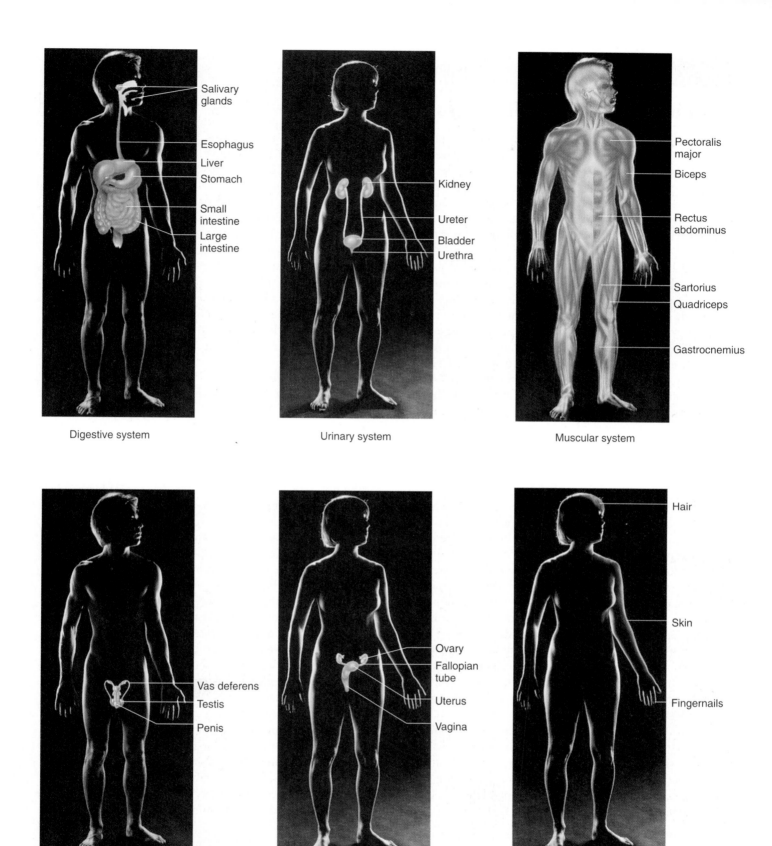

Digestive system

Urinary system

Muscular system

Salivary glands

Esophagus
Liver
Stomach

Small intestine

Large intestine

Kidney

Ureter

Bladder
Urethra

Pectoralis major

Biceps

Rectus abdominus

Sartorius
Quadriceps

Gastrocnemius

Reproductive system
(male)

Reproductive system
(female)

Integumentary system

Vas deferens

Testis

Penis

Ovary
Fallopian tube

Uterus

Vagina

Hair

Skin

Fingernails

FIGURE 46.5 (continued)

Characteristics of Epithelial Tissue

An epithelial membrane, or **epithelium,** covers every surface of the vertebrate body. Epithelial membranes are derived from all three germ layers. The epidermis, derived from ectoderm, constitutes the outer portion of the skin. The inner surface of the digestive tract is lined by an epithelium derived from endoderm, and the inner surfaces of the body cavities are lined with an epithelium derived from mesoderm.

Since all body surfaces are covered by epithelial membranes, a substance must pass through an epithelium in order to enter or leave the body. Epithelial membranes thus provide a barrier that can impede the passage of some substances while facilitating the passage of others. For land-dwelling vertebrates, the relative impermeability of the surface epithelium (the epidermis) to water offers essential protection from dehydration and from airborne pathogens (disease-causing organisms). On the other hand, the epithelial lining of the digestive tract must allow selective entry of the products of digestion while providing a barrier to toxic substances, and the epithelium of the lungs must allow for the rapid diffusion of gases.

Some epithelia become modified in the course of embryonic development into glands, which are specialized for secretion. A characteristic of all epithelia is that the cells are tightly bound together, with very little space between them. As a consequence, blood vessels cannot be interposed between adjacent epithelial cells. Therefore, nutrients and oxygen must diffuse to the epithelial cells from blood vessels in nearby tissues. This places a limit on the thickness of epithelial membranes; most are only one or a few cell layers thick.

Epithelium possesses remarkable regenerative powers, constantly replacing its cells throughout the life of the animal. For example, the liver, a gland formed from epithelial tissue, can readily regenerate after substantial portions of it have been surgically removed. The epidermis is renewed every two weeks, and the epithelium inside the stomach is replaced every two to three days.

There are two general classes of epithelial membranes: simple and stratified. These classes are further subdivided into squamous, cuboidal, and columnar, based upon the shape of the cells (table 46.2). Squamous cells are flat, cuboidal cells are about as thick as they are tall, and columnar cells are taller than they are wide.

Types of Epithelial Tissues

Simple epithelial membranes are one cell layer thick. A *simple, squamous epithelium* is composed of squamous epithelial cells that have an irregular, flattened shape with tapered edges. Such membranes line the lungs and blood capillaries, for example, where the thin, delicate nature of these membranes permits the rapid movement of molecules (such as the diffusion of gases). A *simple cuboidal epithelium* lines the small ducts of some glands, and a *simple columnar epithelium* is found in the airways of the respiratory tract and in the gastrointestinal tract, among other locations. Interspersed among the columnar epithelial cells are numerous *goblet cells,* specialized to secrete mucus. The columnar epithelial cells of the respiratory airways contain cilia on their apical surface (the surface facing the lumen, or cavity), which move mucus toward the throat. In the small intestine, the apical surface of the columnar epithelial cells form finger-like projections called *microvilli,* that increase the surface area for the absorption of food.

Stratified epithelial membranes are several cell layers thick and are named according to the features of their uppermost layers. For example, the epidermis is a *stratified squamous epithelium.* In terrestrial vertebrates it is further characterized as a *keratinized epithelium,* because its upper layer consists of dead squamous cells and filled with a water-resistant protein called *keratin.* The deposition of keratin in the skin can be increased in response to abrasion, producing calluses. The water-resistant property of keratin is evident when the skin is compared with the red portion of the lips, which can easily become dried and chapped because it is covered by a nonkeratinized, stratified squamous epithelium.

The glands of vertebrates are derived from invaginated epithelium. In **exocrine glands,** the connection between the gland and the epithelial membrane is maintained as a duct. The duct channels the product of the gland to the surface of the epithelial membrane and thus to the external environment (or to an interior compartment that opens to the exterior, such as the digestive tract). Examples of exocrine glands include sweat and sebaceous (oil) glands, which secrete to the external surface of the skin, and accessory digestive glands such as the salivary glands, liver, and pancreas, which secrete to the surface of the epithelium lining the digestive tract.

Endocrine glands are ductless glands; their connections with the epithelium from which they were derived are lost during development. Therefore, their secretions, called *hormones,* are not channeled onto an epithelial membrane. Instead, hormones enter blood capillaries and thus stay within the body. Endocrine glands are discussed in more detail in chapter 53.

Epithelial tissues include membranes that cover all body surfaces and glands. The epidermis of the skin is an epithelial membrane specialized for protection, whereas membranes that cover the surfaces of hollow organs are often specialized for transport.

Table 46.2 Epithelial Tissue

Simple Epithelium

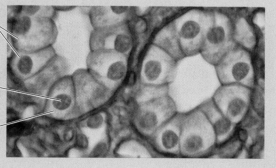

Red blood cell

Simple squamous epithelium

Nucleus

SQUAMOUS

Typical Location
Lining of lungs, capillary walls, and blood vessels

Function
Cells very thin; provides thin layer across which diffusion can readily occur

Characteristic Cell Types
Epithelial cells

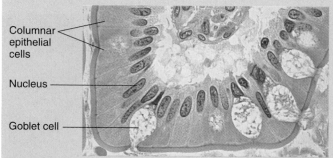

Cuboidal epithelial cells

Nucleus

Cytoplasm

CUBOIDAL

Typical Location
Lining of some glands and kidney tubules; covering of ovaries

Function
Cells rich in specific transport channels; functions in secretion and absorption

Characteristic Cell Types
Gland cells

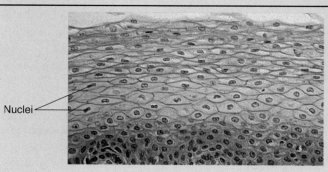

Columnar epithelial cells

Nucleus

Goblet cell

COLUMNAR

Typical Location
Surface lining of stomach, intestines, and parts of respiratory tract

Function
Thicker cell layer; provides protection and functions in secretion and absorption

Characteristic Cell Types
Epithelial cells

Stratified Epithelium

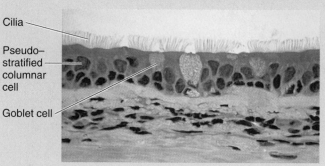

Nuclei

SQUAMOUS

Typical Location
Outer layer of skin; lining of mouth

Function
Tough layer of cells; provides protection

Characteristic Cell Types
Epithelial cells

Cilia

Pseudo–stratified columnar cell

Goblet cell

PSEUDOSTRATIFIED COLUMNAR

Typical Location
Lining parts of the respiratory tract

Function
Secretes mucus; dense with cilia that aid in movement of mucus; provides protection

Characteristic Cell Types
Gland cells; ciliated epithelial cells

Connective Tissue Proper

Connective tissues are derived from embryonic mesoderm and occur in many different forms (table 46.3). These various forms are divided into two major classes: **connective tissue proper**, which is further divided into loose and dense connective tissues; and **special connective tissues** that include cartilage, bone, and blood. At first glance, it may seem odd that such diverse tissues are placed in the same category. Yet all connective tissues do share a common structural feature: they all have abundant extracellular material because their cells are spaced widely apart. This extracellular material is generically known as the **matrix** of the tissue. In bone, the extracellular matrix contains crystals that make the bones hard; in blood, the extracellular matrix is plasma, the fluid portion of the blood.

Loose connective tissue consists of cells scattered within an amorphous mass of proteins that form a **ground substance**. This gelatinous material is strengthened by a loose scattering of protein fibers such as *collagen* (figure 46.6), *elastin*, which makes the tissue elastic, and *reticulin*, which supports the tissue by forming a meshwork. The flavored gelatin we eat for dessert consists of the extracellular material from loose connective tissues. The cells that secrete collagen and other fibrous proteins are known as *fibroblasts*.

Loose connective tissue contains other cells as well, including *mast cells* that produce histamine (a blood vessel dilator) and heparin (an anticoagulant) and *macrophages*, the immune system's first defense against invading organisms, as will be described in detail in chapter 54.

Adipose cells are found in loose connective tissue, usually in large groups that form what is referred to as *adipose tissue* (figure 46.7). Each adipose cell contains a droplet of fat (triglycerides) within a storage vesicle. When that fat is needed for energy, the adipose cell hydrolyzes its stored triglyceride and secretes fatty acids into the blood for oxidation by the cells of the muscles, liver, and other organs. The number of adipose cells in an adult is generally fixed. When a person gains weight, the cells become larger, and when weight is lost, the cells shrink.

Dense connective tissue contains tightly packed collagen fibers, making it stronger than loose connective tissue. It consists of two types: regular and irregular. The collagen fibers of **dense regular connective tissue** are lined up in parallel, like the strands of a rope. This is the structure of *tendons*, which bind muscle to bone, and *ligaments*, which bind bone to bone. In contrast, the collagen fibers of **dense irregular connective tissue** have many different orientations. This type of connective tissue produces the tough coverings that package organs, such as

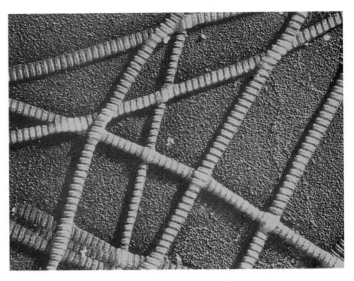

FIGURE 46.6
Collagen fibers. Each fiber is composed of many individual collagen strands and can be very strong.

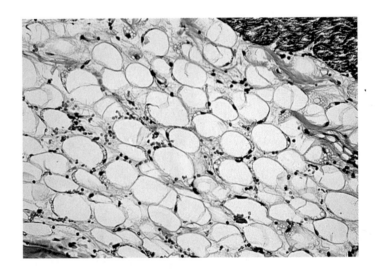

FIGURE 46.7
Adipose tissue. Fat is stored in globules of adipose tissue, a type of loose connective tissue. As a person gains or loses weight, the size of the fat globules increases or decreases. A person cannot decrease the number of fat cells by losing weight.

the *capsules* of the kidneys and adrenal glands. It also covers muscle as *perimysium*, nerves as *perineurium*, and bones as *periosteum*.

Connective tissues are characterized by abundant extracellular materials in the matrix between cells. Connective tissue proper may be either loose or dense.

Table 46.3   Connective Tissue

LOOSE CONNECTIVE TISSUE

Typical Location
Beneath skin; between organs

Function
Provides support, insulation, food storage, and nourishment for epithelium

Characteristic Cell Types
Fibroblasts, macrophages, mast cells, fat cells

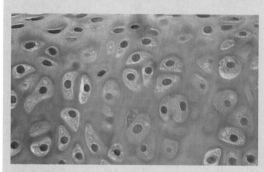

DENSE CONNECTIVE TISSUE

Typical Location
Tendons; sheath around muscles; kidney; liver; dermis of skin

Function
Provides flexible, strong connections

Characteristic Cell Types
Fibroblasts

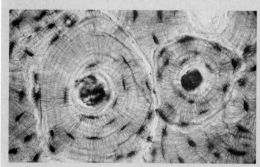

CARTILAGE

Typical Location
Spinal discs; knees and other joints; ear; nose; tracheal rings

Function
Provides flexible support, shock absorption, and reduction of friction on load-bearing surfaces

Characteristic Cell Types
Chondrocytes

BONE

Typical Location
Most of skeleton

Function
Protects internal organs; provides rigid support for muscle attachment

Characteristic Cell Types
Osteocytes

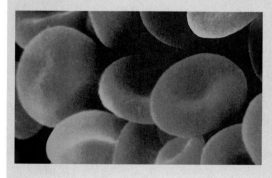

BLOOD

Typical Location
Circulatory system

Function
Functions as highway of immune system and primary means of communication between organs

Characteristic Cell Types
Erythrocytes, leukocytes

Special Connective Tissues

The special connective tissues—cartilage, bone, and blood—each have a unique extracellular matrix that allows them to perform their specialized functions.

Cartilage

Cartilage (figure 46.8) is a specialized connective tissue in which the ground substance is formed from a characteristic type of glycoprotein, and the collagen fibers are laid down along the lines of stress in long, parallel arrays. The result is a firm and flexible tissue that does not stretch, is far tougher than loose or dense connective tissue, and has great tensile strength. Cartilage makes up the entire skeletal system of the modern agnathans and cartilaginous fishes (see chapter 45), replacing the bony skeletons that were characteristic of the ancestors of these vertebrate groups. In most adult vertebrates, however, cartilage is restricted to the articular (joint) surfaces of bones that form freely movable joints and to other specific locations. In humans, for example, the tip of the nose, the pinna (outer ear flap), the intervertebral discs of the backbone, the larynx (voice box) and a few other structures are composed of cartilage.

Chondrocytes, the cells of the cartilage, live within spaces called *lacunae* within the cartilage ground substance. These cells remain alive, even though there are no blood vessels within the cartilage matrix, because they receive oxygen and nutrients by diffusion through the cartilage ground substance from surrounding blood vessels. This diffusion can only occur because the cartilage matrix is not calcified, as is bone.

Bone

In the course of fetal development, the bones of vertebrate fins, arms, and legs, among others, are first "modeled" in cartilage. The cartilage matrix then calcifies at particular locations, so that the chondrocytes are no longer able to obtain oxygen and nutrients by diffusion through the matrix. The dying and degenerating cartilage is then replaced by living bone. Bone cells, or osteocytes, can remain alive even though the extracellular matrix becomes hardened with crystals of calcium phosphate. This is because blood

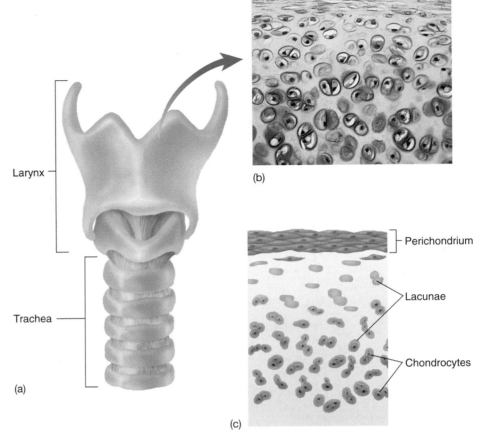

FIGURE 46.8
Cartilage is a strong, flexible tissue that makes up the larynx (voice box) and several other structures in the human body. The larynx (a) is seen under the light microscope in (b), where the cartilage cells, or chondrocytes, are visible within cavities, or lacunae, in the matrix (extracellular material) of the cartilage. This is diagrammed in (c).

vessels travel through canals into the bone, and the osteocytes can communicate with those blood vessels through tiny canals, or *canaliculi* (figure 46.9).

It should be noted here that some bones, such as those of the cranium, are not formed first as cartilage models. These bones instead develop within a membrane of dense, irregular connective tissue. The structure and formation of bone are discussed in chapter 47.

Blood

Blood is classified as a connective tissue because it contains abundant extracellular material, the fluid plasma. The cells of blood are erythrocytes, or red blood cells, and leukocytes, or white blood cells (figure 46.10). Blood also contains platelets, or *thrombocytes*, which are fragments of a type of bone marrow cell.

Erythrocytes are the most common blood cells; there are about 5 billion in every milliliter of blood. During their maturation in mammals, they lose their nucleus, mitochondria, and endoplasmic reticulum. As a result, mammalian erythrocytes are relatively inactive metabolically. Each erythrocyte contains about 300 million molecules of the iron-containing protein *hemoglobin*, the principal carrier of oxygen in vertebrates and in many other groups of animals.

There are several types of leukocytes, but together they are only one-thousandth as numerous as erythrocytes. Unlike erythrocytes, leukocytes have nuclei and mitochondria but lack the red pigment hemoglobin. These cells are therefore hard to see under a microscope without special staining. The names *neutrophils, eosinophils,* and *basophils* distinguish three types of leukocytes on the basis of their staining properties; other leukocytes include *lymphocytes* and *monocytes*. These different types of leukocytes play critical roles in immunity, as will be described in chapter 54.

The blood plasma is the "commons" of the body; it (or a derivative of it) travels to and from every cell in the body. As the plasma circulates, it carries nourishment, waste products, and regulatory molecules. Practically every substance used by cells, including sugars, lipids, and amino acids, is delivered by the plasma to the body cells. Waste products from the cells are carried by the plasma to the kidneys, liver, and lungs or gills for disposal, and regulatory molecules (hormones) that endocrine gland cells secrete are carried by the plasma to regulate the activities of most organs of the body. The plasma also contains sodium, calcium, and other inorganic ions that all cells need, as well as numerous proteins. Plasma proteins include *fibrinogen*, produced by the liver, which helps blood to clot; *albumin*, also produced by the liver, which exerts an osmotic force needed for fluid balance; and *antibodies* produced by lymphocytes and needed for immunity.

Special connective tissues each have a unique extracellular matrix between cells. The matrix of cartilage is composed of organic material, whereas that of bone is impregnated with calcium phosphate crystals. The matrix of blood is fluid, the plasma.

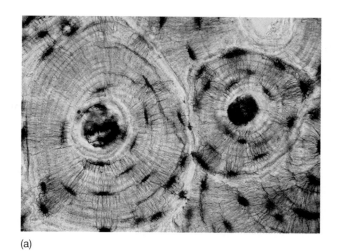

(a)

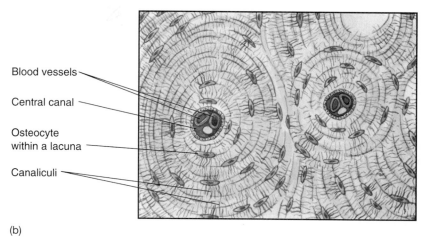

Blood vessels

Central canal

Osteocyte within a lacuna

Canaliculi

(b)

FIGURE 46.9
The structure of bone. A photomicrograph (a) and diagram (b) of the structure of bone, showing the bone cells, or osteocytes, within their lacunae (cavities) in the bone matrix. Though the bone matrix is calcified, the osteocytes remain alive because they can be nourished by blood vessels that enter the bone. Nourishment is brought to the osteocytes through tiny canals, or canaliculi.

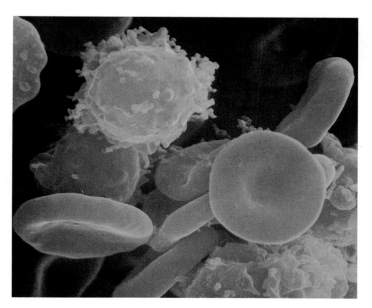

FIGURE 46.10
White and red blood cells (500×). White blood cells, or leukocytes, are roughly spherical and have irregular surfaces with numerous extending pili. Red blood cells, or erythrocytes, are flattened spheres, typically with a depressed center, forming biconcave discs.

Muscle Tissue

Muscle cells are the motors of the vertebrate body. The characteristic that makes them unique is the relative abundance and organization of actin and myosin filaments within them. Although these filaments form a fine network in all eukaryotic cells, where they contribute to cellular movements, they are far more common in muscle cells, which are specialized for contraction. Vertebrates possess three kinds of muscle: smooth, skeletal, and cardiac (table 46.4). Skeletal and cardiac muscles are also known as **striated muscles** because their cells have transverse stripes when viewed in longitudinal section under the microscope. The contraction of each skeletal muscle is under voluntary control, whereas the contraction of cardiac and smooth muscles is generally involuntary. Muscles are described in more detail in chapter 47.

Smooth Muscle

Smooth muscle was the earliest form of muscle to evolve, and it is found throughout the animal kingdom. In vertebrates, smooth muscle is found in the organs of the internal environment, or *viscera*, and is sometimes known as visceral muscle. Smooth muscle tissue is organized into sheets of long, spindle-shaped cells, each cell containing a single nucleus. In some tissues, the cells contract only when they are stimulated by a nerve, and then all of the cells in the sheet contract as a unit. In vertebrates, muscles of this type line the walls of many blood vessels and make up the iris of the eye. In other smooth muscle tissues, such as those in the wall of the gut, the muscle cells themselves may spontaneously initiate electric impulses and contract, leading to a slow, steady contraction of the tissue. Nerves regulate, rather than cause, this activity.

Skeletal Muscle

Skeletal muscles are usually attached by tendons to bones, so that, when the muscles contract, they cause the bones to move at their joints. A skeletal muscle is made up of numerous, very long muscle cells, called *muscle fibers*, which lie parallel to each other within the muscle and insert into the tendons on the ends of the muscle. Each skeletal muscle fiber is stimulated to contract by a nerve fiber; therefore, a stronger muscle contraction will result when more of the muscle fibers are stimulated by nerve fibers to contract. In this way, the nervous system can vary the strength of skeletal muscle contraction. Each muscle fiber contracts by means of substructures called *myofibrils* (figure 46.11) that contain highly ordered arrays of actin and myosin filaments, that give the muscle fiber its striated appearance. Skeletal muscle fibers are produced during development by the fusion of several cells, end to end. This embryological development explains why a mature muscle fiber contains

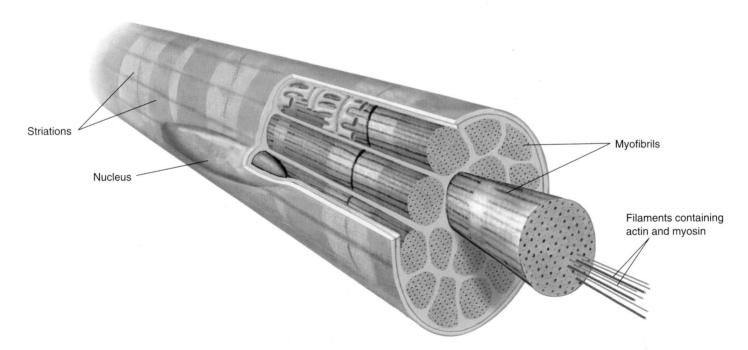

Striations

Nucleus

Myofibrils

Filaments containing actin and myosin

FIGURE 46.11
A muscle fiber, or muscle cell. Each muscle fiber is composed of numerous myofibrils, which, in turn, are composed of actin and myosin filaments. Each muscle fiber is multinucleate as a result of its embryological development from the fusion of smaller cells.

Table 46.4 Muscle Tissue

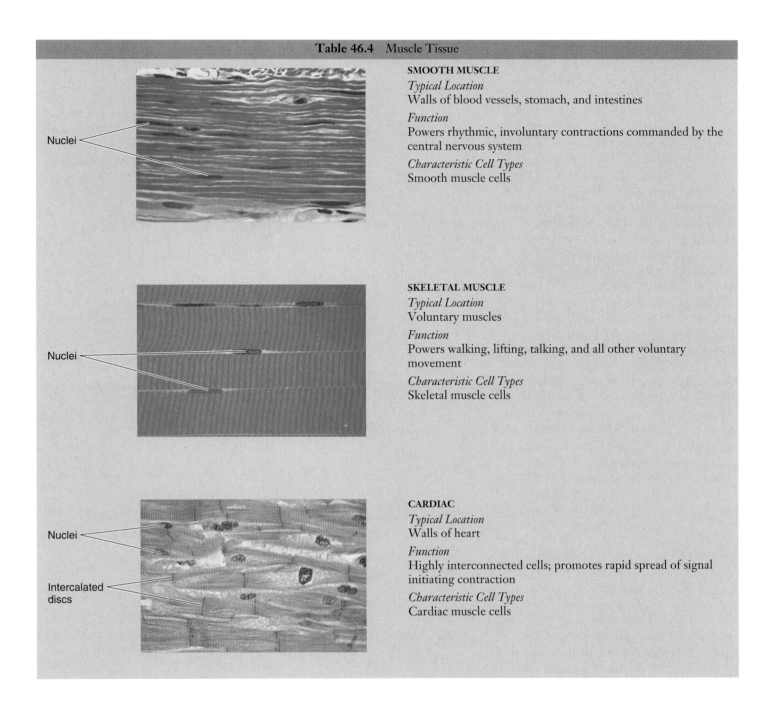

SMOOTH MUSCLE

Typical Location
Walls of blood vessels, stomach, and intestines

Function
Powers rhythmic, involuntary contractions commanded by the central nervous system

Characteristic Cell Types
Smooth muscle cells

SKELETAL MUSCLE

Typical Location
Voluntary muscles

Function
Powers walking, lifting, talking, and all other voluntary movement

Characteristic Cell Types
Skeletal muscle cells

CARDIAC

Typical Location
Walls of heart

Function
Highly interconnected cells; promotes rapid spread of signal initiating contraction

Characteristic Cell Types
Cardiac muscle cells

many nuclei. The structure and function of skeletal muscle is explained in more detail in chapter 47.

Cardiac Muscle

The hearts of vertebrates are composed of striated muscle cells arranged very differently from the fibers of skeletal muscle. Instead of having very long, multinucleate cells running the length of the muscle, cardiac muscle is composed of smaller, interconnected cells, each with a single nucleus. The interconnections between adjacent cells appear under the microscope as dark lines called *intercalated discs*. In reality, these lines are regions where adjacent cells are linked by *gap junctions*. As we noted in chapter 7, gap junctions have openings that permit the movement of small substances and electric charges from one cell to another. These interconnections enable the cardiac muscle cells to form a single, functioning unit known as a myocardium. Certain cardiac muscle cells generate electric impulses spontaneously, and these impulses spread across the gap junctions from cell to cell, causing all of the cells in the myocardium to contract. We will describe this process more fully in chapter 49.

Skeletal muscles enable the vertebrate body to move. Cardiac muscle powers the heartbeat, while smooth muscles provide a variety of visceral functions.

Nerve Tissue

The fourth major class of vertebrate tissue is nerve tissue (table 46.5). Its cells include neurons and supporting cells. Neurons are specialized to produce and conduct electrochemical events, or "impulses." Each neuron consists of three parts: cell body, dendrites, and axon (figure 46.12). The cell body of a neuron contains the nucleus. Dendrites are thin, highly branched extensions that receive incoming stimulation and conduct electric events to the cell body. As a result of this stimulation and the electric events produced in the cell body, outgoing impulses may be produced at the origin of the axon. The axon is a single extension of cytoplasm that conducts impulses away from the cell body. Some axons can be quite long. The cell bodies of neurons that control the muscles in your feet, for example, lie in the spinal cord, and their axons may extend over a meter to your feet.

Many axons are covered with an insulating layer called a *myelin sheath*. There are periodic interruptions in the sheath, known as *nodes of Ranvier*, that serve as the sites for impulse production in these axons. The myelin sheaths are formed from specific supporting cells, as described in chapter 51.

The nervous system is divided into the central nervous system (CNS), which includes the brain and spinal cord, and the peripheral nervous system (PNS), which includes *nerves* and *ganglia*. Nerves consist of axons in the PNS that are bundled together in much the same way as wires are bundled together in a cable. Ganglia are collections of neuron cell bodies.

Neurons are specialized to receive, produce and conduct electrical signals. Some neurons convey sensory information to the brain and spinal cord, some carry signals from the brain and spinal cord to muscles or glands, and still others associate and integrate information.

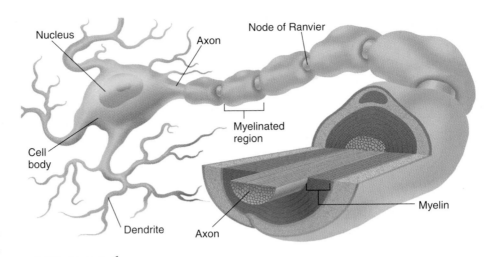

FIGURE 46.12
Idealized structure of a vertebrate neuron. Many dendrites arise from the cell body, as does a single long axon. In some neurons specialized for rapid signal conduction, the axon is encased in a myelin sheath that is interrupted at intervals. At its far end, the axon may branch to terminate on more than one cell.

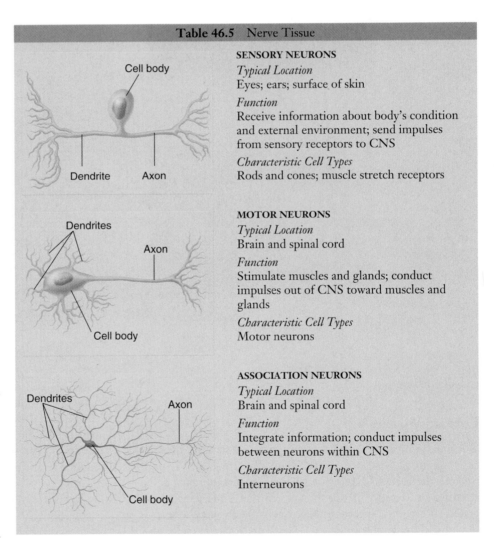

Table 46.5	Nerve Tissue
	SENSORY NEURONS
	Typical Location
	Eyes; ears; surface of skin
	Function
	Receive information about body's condition and external environment; send impulses from sensory receptors to CNS
	Characteristic Cell Types
	Rods and cones; muscle stretch receptors
	MOTOR NEURONS
	Typical Location
	Brain and spinal cord
	Function
	Stimulate muscles and glands; conduct impulses out of CNS toward muscles and glands
	Characteristic Cell Types
	Motor neurons
	ASSOCIATION NEURONS
	Typical Location
	Brain and spinal cord
	Function
	Integrate information; conduct impulses between neurons within CNS
	Characteristic Cell Types
	Interneurons

46.1 The bodies of vertebrates are organized into functional systems.

- The vertebrate body is organized into cells, tissues, organs, and organ systems, which are specialized for different functions.
- The four primary tissues of the vertebrate adult body—epithelial, connective, muscle, and nerve—are derived from three embryonic germ layers.

46.2 Epithelial tissue forms membranes and glands.

- Epithelial membranes cover all body surfaces and are divided into those that are simple and those that are stratified.
- Stratified membranes, particularly the keratinized epithelium of the epidermis, provides protection, whereas simple membranes are more adapted for secretion and transport.
- Exocrine glands secrete into ducts that conduct the secretion to the outside of an epithelial membrane; endocrine glands secrete hormones into the blood.

46.3 Connective tissues contain abundant extracellular material.

- Connective tissues are characterized by abundant extracellular matrix, which is composed of fibrous proteins and a gel-like ground substance in connective tissue proper.
- Loose connective tissues contain many cell types such as adipose cells and mast cells; dense regular connective tissues form tendons and ligaments.
- Special connective tissues include cartilage, bone, and blood. Nutrients can diffuse through the cartilage matrix but not through the calcified matrix of bone, which contains canaliculi for that purpose.

46.4 Muscle tissue provides for movement, and nerve tissue provides for control.

- Smooth muscles are composed of spindle-shaped cells and are found in the organs of the internal environment and in the walls of blood vessels.
- Skeletal and cardiac muscles are striated; skeletal muscles, however, are under voluntary control whereas cardiac muscle is involuntary.
- A neuron consists of a cell body with one or more dendrites and one axon.
- Neuron cell bodies form ganglia, and their axons form nerves in the peripheral nervous system.
- Neurons may be classified as sensory, motor, or association neurons depending on their function.

1. **Epithelial membrane**   Composed of cells joined tightly together, epithelial membranes cover all body surfaces. The shape and number of cell layers varies in different epithelial membranes.

2. **Connective tissues**   Connective tissues are the primary tissue in which there is abundant extracellular material. The extracellular material is called matrix, and its nature varies with the particular type of connective tissue.

3. **Muscle tissue**   Muscles are specialized for contraction, and can be categorized as striated or smooth on the basis of their appearance, or as voluntary or involuntary on the basis of their regulation by the nervous system.

4. **Nerve tissue**   Nerve tissue consists of neurons, that produce electrochemical impulses, and supporting cells. The nervous system is subdivided into the peripheral nervous system and central nervous system.

Review Questions

1. What is a tissue? What is an organ? What is an organ system?

2. What are the different types of epithelial membranes, and how do they differ in structure and function?

3. What are the two types of glands, and how do they differ in structure and function?

4. What feature do all connective tissues share? What are the different categories of connective tissue? Give an example of each.

5. What is the structure of a ligament? How do cartilage and bone differ? Why is blood considered to be a connective tissue?

6. From what embryonic tissue is muscle derived? What two contractile proteins are abundant in muscle? What are the three categories of muscle tissue? Which two are striated?

7. Why are skeletal muscle fibers multinucleated? What is the functional significance of intercalated discs in heart muscle?

8. What are the parts of a neuron and the function of each? What are the two major divisions of the nervous system and what structures do they include? What is a nerve?

Thought Questions

1. Chemotherapy is frequently used as a cancer treatment designed to kill the rapidly dividing malignant cells. Why do you think this treatment typically causes patients to lose their hair and often interferes with the function of their gastrointestinal tract?

2. Land was successfully invaded four times—by plants, fungi, arthropods, and vertebrates. Each of these four groups evolved a characteristic hard substance to lend mechanical support, since bodies are far less buoyant in air than in water. Describe and compare these four substances, discussing their advantages and disadvantages. Can you imagine any other substance (plastic, perhaps) that would have been superior to these four?

3. Devise an experiment to test whether a fever is beneficial in fighting an infection, using lizards as your experimental animals. Why might you be justified in extrapolating your findings from lizards to humans?

Internet Links

The Virtual Body
http://www.medtropolis.com/vbody/
This multimedia supermarket explores the major human body systems (skeleton, heart, digestive system, brain), providing narrative, animations, interactives, and links for each.

Looking at the Body in 3D
http://www.nlm.nih.gov/research/visible/visible_human.html
This remarkable site, THE VISIBLE HUMAN project from the National Library of Medicine, aims to create a complete, anatomically detailed, three-dimensional representation of the female and male human body, using CAT and MRI scans.

Human Anatomy On-line
http://www.innerbody.com/indexbody.html
An award-winning interactive site that explores the human body in depth, containing over one hundred illustrations of the human body with animations, and thousands of descriptive links.

Medical Student Tutorial
http://www–sci.lib.uci.edu/HSG/HSGuide.html
An on-line HEALTH SCIENCE CENTER maintained by UC Irvine for medical students, containing over 49,000 teaching files, over 124,000 medical cases, 864 multimedia coursebooks, 1350 multimedia tutorials, 3350 databases, and 10,300 movies. Whew!

For Further Reading

Caplan, A.: "Cartilage," *Scientific American*, October 1984, pages 84–97. An interesting account of the many roles played by cartilage in the vertebrate body.

Currey, J.: *The Mechanical Adaptations of Bones*, Princeton University Press, Princeton, NJ, 1984. A functional analysis of why different bones are constructed the way they are, with an unusually well-integrated evolutionary perspective.

Kent, G.: *Comparative Anatomy of the Vertebrates*, 7th ed., Times Mirror/Mosby College Publishing, St. Louis, 1992. A very readable introduction to the comparative anatomy of the vertebrates.

National Geographic Society: *The Incredible Machine*, National Geographic Society, Washington, DC, 1986. A series of outstanding articles on the human body, focusing on its major organ systems. Beautifully illustrated and fun to read.

Rosenfeld, A.: "There's More to Skin than Meets the Eye," *Smithsonian*, May 1988, pages 159–80. An entertaining and informative account of the many tasks carried out by the body's largest organ.

Schmidt-Nielsen, K.: *Animal Physiology: Adaptation and Environment*, Cambridge University Press, Cambridge, England, 5th ed., 1996. A good standard reference on the subject by one of the leading research scientists in this field.

47

Locomotion

Concept Outline

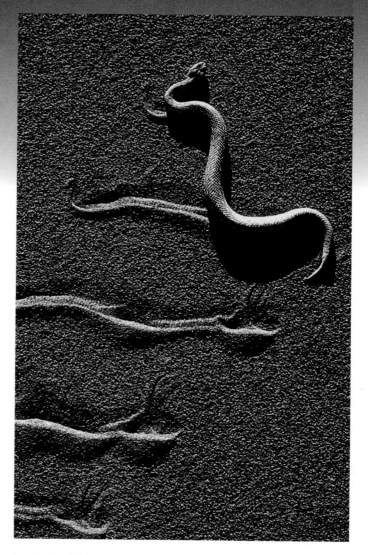

FIGURE 47.1
On the move. The movement of this sidewinder rattlesnake is the result of strong muscle contractions.

Plants and fungi move only by growing, or as the passive passengers of wind and water. Of the three multicellular kingdoms that evolved from protists, only animals explore their environment in an active way, through **locomotion**. In this chapter, we will examine how vertebrates use muscles connected to bones to achieve movement. The rattlesnake in figure 47.1 slithers across the sand by a rhythmic contraction of the muscles sheathing its body. Humans walk by contracting muscles in their legs. Although our focus in this chapter will be on vertebrates, it is important to realize that *all* animals employ muscles. When a mosquito flies, its wings are moved rapidly through the air by quickly contracting flight muscles. When an earthworm burrows through the soil, its movement is driven by strong muscles pushing its body past the surrounding dirt. A starfish uses muscles to creep along the ocean floor. Locomotion driven by muscle contraction is a fundamental property of animals.

Types of Skeletons

Animal locomotion is accomplished through the force of muscles acting on a rigid skeletal system. There are three types of skeletal systems in the animal kingdom: hydraulic skeletons, exoskeletons, and endoskeletons.

Hydraulic skeletons are found in soft-bodied invertebrates such as earthworms and jellyfish. In this case, a fluid-filled cavity is encircled by muscle fibers that raise the pressure of the fluid when they contract. In an earthworm, for example, a wave of contractions of circular muscles begins anteriorly and compresses the body, so that the fluid pressure pushes it forward. Contractions of longitudinal muscles then pull the rest of the body.

Exoskeletons surround the body as a rigid hard case in most animals. Arthropods, such as crustaceans and insects, have exoskeletons made of the polysaccharide *chitin*. An animal with an exoskeleton cannot get too large because its exoskeleton would have to become thicker and heavier, in order to prevent collapse, as the animal grew larger. If an insect were the size of a human being, its exoskeleton would have to be so thick and heavy it would hardly be able to move.

Endoskeletons, found in vertebrates and echinoderms, are rigid internal skeletons to which muscles are attached. Vertebrates have a soft, flexible exterior that stretches to accommodate the movements of their skeleton. The endoskeleton of vertebrates is composed of bone. Unlike chitin, bone is a cellular, living tissue capable of growth, self-repair, and remodeling in response to physical stresses.

The Human Skeleton

The human skeleton (figure 47.2), a typical vertebrate endoskeleton, is divided into an axial and an appendicular skeleton. The axial skeleton's 80 bones form the axis of the body and support and protect the organs of the head, neck, and chest. The appendicular skeleton's 126 bones include the bones of the limbs, and the pectoral and pelvic girdles that attach them to the axial skeleton.

The bones of the skeletal system support and protect the body, and serve as levers for the forces produced by contraction of skeletal muscles. The skeletal system also serves other functions; blood cells form within the bone marrow, and the calcified matrix of bones acts as a reservoir for calcium and phosphate ions.

There are three types of animal skeletons: a hydraulic skeleton, an endoskeleton, and an exoskeleton. Vertebrates have an endoskeleton composed of bone. The vertebrate skeleton consists of axial and appendicular portions. The skeletal system serves a variety of functions, including support and protection, levers for the force of skeletal muscle contraction, blood cell formation, and calcium homeostasis.

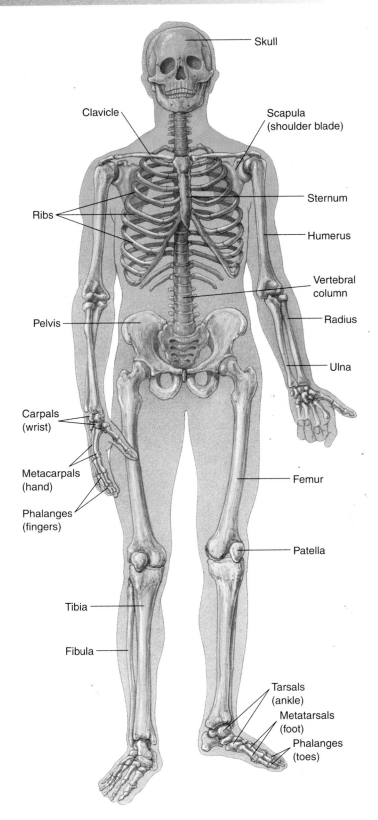

FIGURE 47.2
The human skeleton. The axial skeleton is shown in the purplish color, the appendicular skeleton in the peach color.

The Structure of Bone

Bone, the building material of the vertebrate skeleton, is a special form of connective tissue (see chapter 46). In bone, an organic extracellular matrix containing collagen fibers is impregnated with small, needle-shaped crystals of calcium phosphate in the form of hydroxyapatite crystals. Hydroxyapatite is brittle but rigid, giving bone great strength. Collagen, on the other hand, is flexible but weak. As a result, bone is both strong and flexible. The collagen acts to spread the stress over many crystals, making bone more resistant to fracture than hydroxyapatite is by itself.

Bone is a dynamic, living tissue that is constantly reconstructed throughout the life of an individual. New bone is formed by *osteoblasts*, which secrete the collagen-containing organic matrix in which calcium phosphate is later deposited. After the calcium phosphate is deposited, the cells, now known as osteocytes, are encased within lacunae in the calcified matrix. Yet another type of bone cells, called osteoclasts, act to dissolve bone and thereby aid in the remodeling of bone in response to physical stress.

Bone is constructed in thin, concentric layers, or *lamellae*, which are laid down around narrow channels called *Haversian canals* that run parallel to the length of the bone. Haversian canals contain nerve fibers and blood vessels, which keep the osteocytes alive even though they are entombed in a calcified matrix. The concentric lamellae of bone, with their entrapped osteocytes, that surround a Haversian canal form the basic unit of bone structure, called a Haversian system.

Bone formation occurs in two ways, as described in chapter 46. In flat bones, such as those of the skull, osteoblasts located in a web of dense connective tissue produce bone within that tissue. In long bones, the bone is first "modeled" in cartilage. Calcification then occurs, and bone is formed as the cartilage degenerates. At the end of this process, cartilage remains only at the articular (joint) surfaces of the bones and at the growth plates located in the necks of the long bones. A child grows taller as the cartilage thickens in the growth plates and then is partly replaced with bone. A person stops growing (usually by the late teenage years) when the entire cartilage growth plate

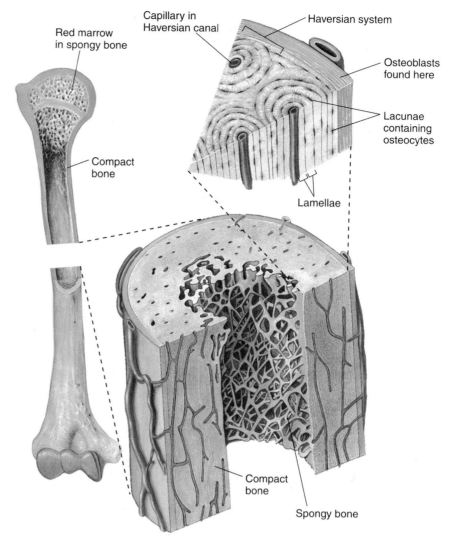

FIGURE 47.3

The organization of bone, shown at three levels of detail. Some parts of bone are dense and compact, giving the bone strength. Other parts are spongy, with a more open lattice; it is there that most blood cells are formed.

becomes replaced by bone. At this point, only the articular cartilage at the ends of the bone remains.

The ends and interiors of long bones are composed of an open lattice of bone called *spongy bone*. The spaces within contain marrow, where most blood cells are formed (figure 47.3). Surrounding the spongy bone tissue are concentric layers of *compact bone*, where the bone is much denser. Compact bone tissue gives bone the strength to withstand mechanical stress.

Bone consists of cells and an extracellular matrix that contains collagen fibers, which provide flexibility, and hydroxyapatite crystals, which provide strength. Bone contains blood vessels and nerves and is capable of growth and remodeling.

Types of Joints

The skeletal movements of the body are produced by contraction and shortening of muscles. Skeletal muscles are generally attached by tendons to bones, so when the muscles shorten, the attached bones move. These movements of the skeleton occur at joints, or articulations, where one bone meets another. There are three kinds of joints (figure 47.4):

1. **Immovable joints** include the *sutures* that join the bones of the skull. In a fetus, the skull bones are not fully formed, and there are open areas of dense connective tissue ("soft spots," or *fontanels*) between the bones. These areas allow the bones to flex slightly as the fetus moves through the birth canal during child-birth. Later, bone replaces most of this connective tissue.

2. **Slightly movable joints** include those in which the bones are bridged by cartilage. The vertebral bones of the spine are separated by pads of cartilage called *intervertebral discs*. These *cartilaginous joints* allow some movement while acting as efficient shock absorbers.

3. **Freely movable joints** are also called synovial joints, because the articulating ends of the bones are located within a *synovial capsule* filled with a lubricating fluid. The ends of the bones are capped with cartilage, and the synovial capsule is strengthened by ligaments.

Synovial joints allow the bones to move in directions dictated by the structure of the joint. For example, the joint between the thigh bone (femur) and pelvis has a ball-and-socket structure that permits a variety of different movements, while the knee joint allows only a hinge-like movement.

Joints confer flexibility to a rigid skeleton, allowing a range of motions determined by the type of joint.

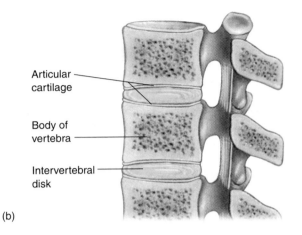

Articular cartilage

Body of vertebra

Intervertebral disk

(b)

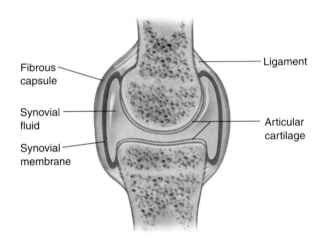

Fibrous capsule

Ligament

Synovial fluid

Articular cartilage

Synovial membrane

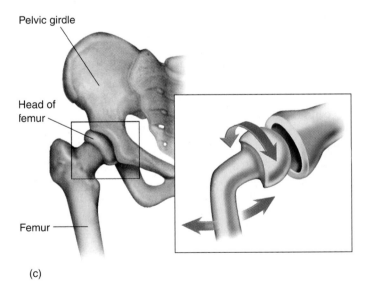

Pelvic girdle

Head of femur

Femur

(c)

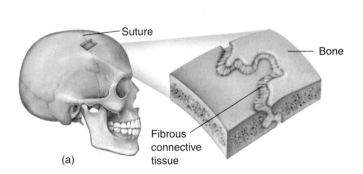

Suture

Bone

Fibrous connective tissue

(a)

FIGURE 47.4
Three types of joints. (a) Immovable joints include the sutures of the skull; (b) slightly movable joints include the cartilaginous joints between the vertebrae; and (c) freely movable joints are the synovial joints, such as the knee joint and hip joints.

Actions of Skeletal Muscles

Skeletal muscles produce movement of the skeleton when they contract. Usually, the two ends of a skeletal muscle are attached to different bones (although in some cases, one or both ends may be connected to some other kind of structure, such as skin). The attachments to bone are made by means of dense connective tissue straps called *tendons.* One attachment of the muscle, the **origin,** remains relatively stationary during a contraction. The other end of the muscle, the **insertion,** is attached to the bone that moves when the muscle contracts. For example, contraction of the biceps muscles in the upper arm causes the forearm (the insertion of the muscle) to move toward the shoulder (the origin of the muscle).

Muscles that cause the same action at a joint are **synergists,** while those that produce opposing actions are **antagonists.** For example, the various muscles of the quadriceps group are synergists: they all act to extend the knee joint (figure 47.5). When the hamstring muscles contract, in contrast, they cause flexion of the knee joint. Therefore, the quadriceps and hamstrings are antagonists to each other. In general, the muscles that antagonize a given movement are relaxed when that movement is performed. Thus, when the hamstrings flex the knee joint, the quadriceps muscles relax.

Muscle Fiber Twitches

An isolated skeletal muscle can be studied by stimulating it artificially with electric shocks. If a muscle is stimulated with a single electric shock, it will quickly contract and relax in a response called a **twitch.** Increasing the stimulus voltage increases the strength of the twitch up to a maximum. The strength of a muscle contraction can thus be *graded,* or varied—an obvious requirement for the proper control of skeletal movements. If a second electric shock is delivered immediately after the first, it will produce a second twitch that may partially "ride piggyback" on the first. This cumulative response is called **summation** (figure 47.6).

If the stimulator is set to deliver an increasing frequency of electric shocks automatically, the relaxation time between successive twitches will get shorter and shorter, as the strength of contraction increases. Finally, at a particular frequency of stimulation, there is no visible relaxation between successive twitches. Contraction is smooth and sustained, as it is during normal muscle contraction in the body. This smooth, sustained contraction is called **tetanus.** (The term *tetanus* should not be confused with the disease of the same name, which is accompanied by a painful state of muscle contracture, or *tetany.*)

Isotonic and Isometric Contractions

In order for muscle fibers to shorten when they contract, they must generate a force that is greater than the opposing

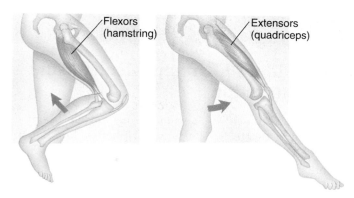

FIGURE 47.5

Flexor and extensor muscles of the leg. The hamstrings, a group of three muscles, produce flexion of the knee joint, whereas the quadriceps, a group of four muscles, produce extension.

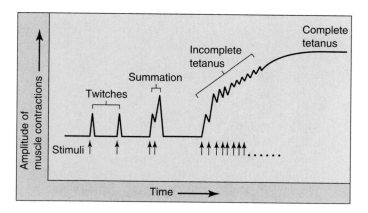

FIGURE 47.6

Muscle twitches summate to produce a sustained, tetanized contraction. This pattern is produced when the muscle is stimulated electrically. Tetanus, a smooth, sustained contraction, is the normal type of muscle contraction in the body.

forces that act to prevent movement of the muscle's insertion. When you lift a weight by contracting muscles in your biceps, for example, the force produced by the muscle is greater than the force of gravity on the object you are lifting. In this case, the muscle and all of its fibers shorten in length.

Contraction that results in muscle shortening is called **isotonic contraction,** because the force of contraction remains relatively constant throughout the shortening process (*iso* = same; *tonic* = strength). If the opposing forces are too great or the number of muscle fibers activated is too few to shorten the muscle, however, the contraction is called an **isometric** (literally, "same length") **contraction.**

> Synergistic muscles have the same action, whereas antagonistic muscles have opposite actions. Muscles contract through summation of the contractions of their fibers, producing tension that may result in shortening of the muscle.

The Sliding Filament Mechanism of Contraction

Each skeletal muscle contains numerous **muscle fibers,** as described in chapter 46. Each muscle fiber encloses a bundle of 4 to 20 elongated structures called **myofibrils.** Each myofibril, in turn, is composed of **thick** and **thin myofilaments** (figure 47.7). The muscle fiber is striated (has cross-striping) because its myofibrils are striated, with dark and light bands. The banding pattern results from the organization of the myofilaments within the myofibril. The thick myofilaments are stacked together to produce the dark bands, called *A bands*; the thin filaments alone are found in the light bands, or *I bands*.

Each I band in a myofibril is divided in half by a disc of protein, called a *Z line* because of its appearance in electron micrographs. The thin filaments are anchored to these discs of proteins that form the Z lines. If you look at an electron micrograph of a myofibril (figure 47.8), you will see that the structure of the myofibril repeats from Z line to Z line. This repeating structure, called a **sarcomere,** is the smallest subunit of muscle contraction.

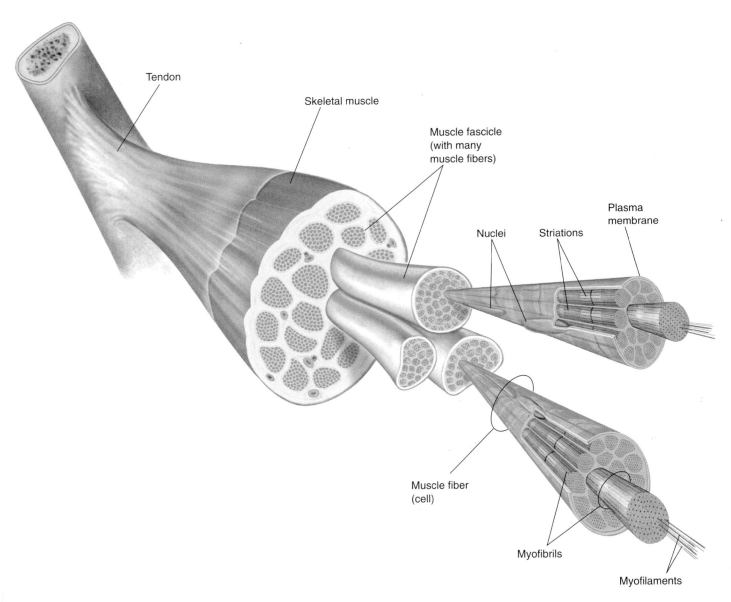

FIGURE 47.7

The organization of skeletal muscle. Each muscle is composed of many fascicles, which are bundles of muscle cells, or fibers. Each fiber is composed of many myofibrils, which are each, in turn, composed of myofilaments.

The thin filaments stick partway into the stack of thick filaments on each side of an A band, but, in a resting muscle, do not project all the way to the center of the A band. As a result, the center of an A band (called an *H band*) is lighter than each side, with its interdigitating thick and thin filaments. This appearance of the sarcomeres changes when the muscle contracts.

A muscle contracts and shortens because its myofibrils contract and shorten. When this occurs, the myofilaments do *not* shorten; instead, the thin filaments slide deeper into the A bands (figure 47.9). This makes the H bands narrower until, at maximal shortening, they disappear entirely. It also makes the I bands narrower, because the dark A bands are brought closer together. This is the **sliding filament** mechanism of contraction.

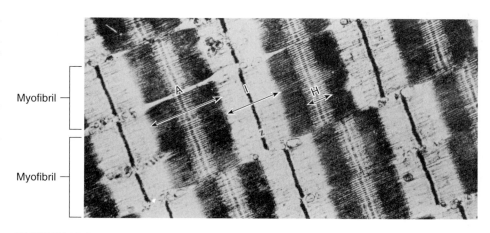

FIGURE 47.8
An electron micrograph of a skeletal muscle fiber. The Z lines that serve as the borders of the sarcomeres are clearly seen within each myofibril. The thick filaments comprise the A bands; the thin filaments are within the I bands and stick partway into the A bands, overlapping with the thick filaments. There is no overlap of thick and thin filaments at the central region of an A band, which is therefore lighter in appearance. This is the H band.

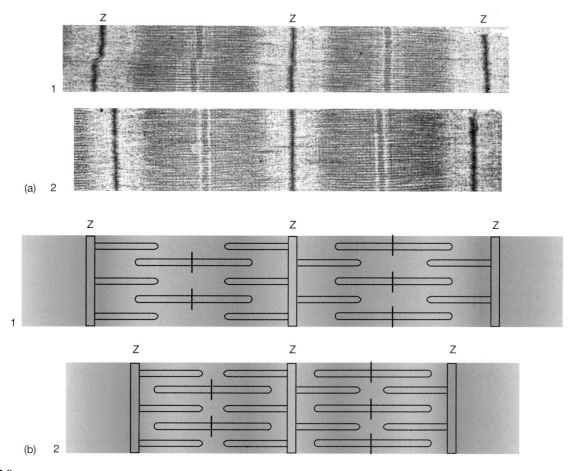

FIGURE 47.9
Electron micrograph (a) and diagram (b) of the sliding filament mechanism of contraction. As the thin filaments slide deeper into the centers of the sarcomeres, the Z lines are brought closer together. (*1*) Relaxed muscle; (*2*) partially contracted muscle.

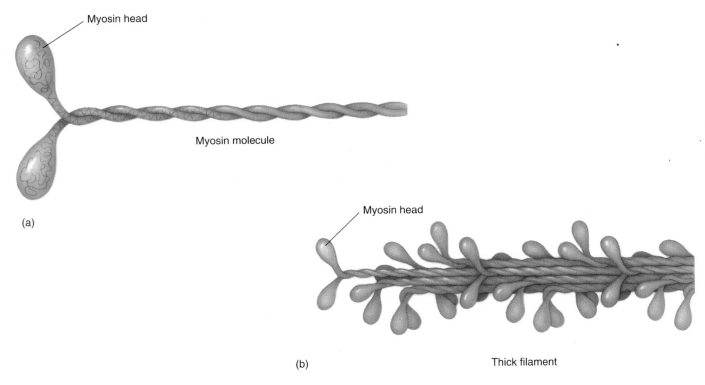

Myosin head

Myosin molecule

(a)

Myosin head

(b)

Thick filament

FIGURE 47.10

Thick filaments are composed of myosin. (a) Each myosin molecule consists of two polypeptide chains wrapped around each other; at the end of each chain is a globular region referred to as the "head." (b) Thick filaments consist of myosin molecules combined into bundles from which the heads protrude at regular intervals.

Electron micrographs reveal **cross-bridges** that extend from the thick to the thin filaments, suggesting a mechanism that might cause the filaments to slide. In order to understand how this is accomplished, we have to examine the thick and thin filaments at a molecular level. Biochemical studies show that each thick filament is composed of many **myosin** proteins packed together, and every myosin molecule has a "head" region that protrudes from the thick filaments (figure 47.10). These myosin heads form the cross-bridges seen in electron micrographs. Biochemical studies also show that each thin filament consists primarily of many globular **actin** proteins twisted into a double helix (figure 47.11). Therefore, if we were able to see a sarcomere at a molecular level, it would have the structure depicted in figure 47.12a.

Before the myosin heads bind to the actin of the thin filaments, they act as ATPase enzymes, splitting ATP into ADP and Pi. This activates the heads, "cocking" them so that they can bind to actin and form cross-bridges. Once a myosin head binds to actin, it undergoes a conformational

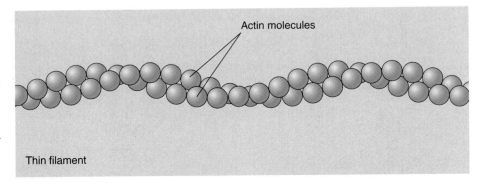

Actin molecules

Thin filament

FIGURE 47.11

Thin filaments are composed of globular actin proteins. Two rows of actin proteins are twisted together in a helix to produce the thin filaments.

(shape) change, pulling the thin filament toward the center of the sarcomere (figure 47.12b) in a *power stroke*. At the end of the power stroke, the myosin head binds to a new molecule of ATP. This allows the head to detach from actin and continue the **cross-bridge cycle** (figure 47.13), which repeats as long as the muscle continues to contract.

In death, the cell can no longer produce ATP and therefore the cross-bridges cannot be broken—this causes the muscle stiffness of death, or *rigor mortis*. A living cell,

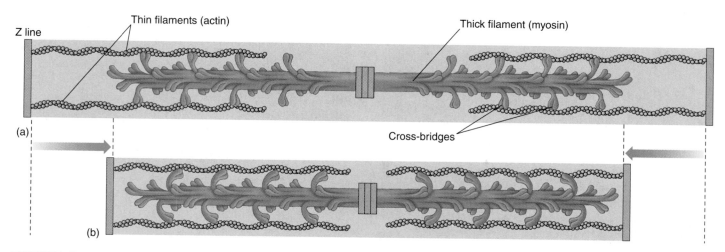

FIGURE 47.12

The interaction of thick and thin filaments in striated muscle sarcomeres. The heads on the two ends of the thick filaments are oriented in opposite directions (a), so that the cross-bridges pull the thin filaments and the Z lines on each side of the sarcomere toward the center. (b) This sliding of the filaments produces muscle contraction.

FIGURE 47.13

The cross-bridge cycle in muscle contraction. (a) With ADP and P_i attached to the myosin head, (b) the head is in a conformation that can bind to actin and form a cross-bridge. (c) Binding causes the myosin head to assume a more bent conformation, moving the thin filament along the thick filament (to the left in this diagram) and releasing ADP and P_i. (d) Binding of ATP to the head detaches the cross-bridge; cleavage of ATP into ADP and P_i puts the head into its original conformation, allowing the cycle to begin again.

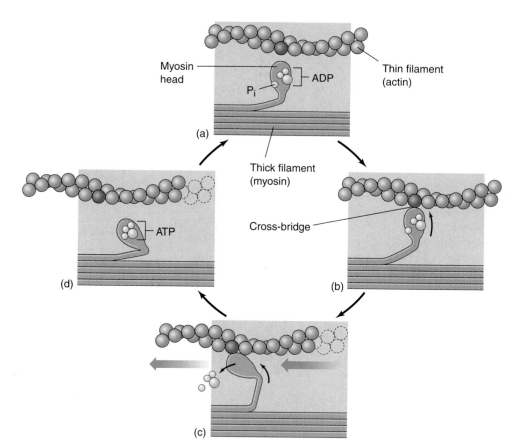

however, always has enough ATP to allow the myosin heads to detach from actin. How, then, is the cross-bridge cycle arrested so that the muscle can relax? The regulation of muscle contraction and relaxation requires additional factors that we will discuss in the next section.

Thick and thin filaments are arranged to form sarcomeres within the myofibrils. Myosin proteins comprise the thick filaments, and the heads of the myosin form cross-bridges with the actin proteins of the thin filaments. ATP provides the energy for the cross-bridge cycle and muscle contraction.

The Role of Ca⁺⁺ in Contraction

When a muscle is relaxed, its myosin heads are "cocked" and ready, through the splitting of ATP, but are unable to bind to actin. This is because the attachment sites for the myosin heads on the actin are physically blocked by another protein, known as **tropomyosin,** in the thin filaments. Cross-bridges therefore cannot form in the relaxed muscle, and the filaments cannot slide.

In order to contract a muscle, the tropomyosin must be moved out of the way so that the myosin heads can bind to actin. This requires the function of **troponin,** a regulatory protein that binds to the tropomyosin. The troponin and tropomyosin form a complex that is regulated by the calcium ion (Ca⁺⁺) concentration of the muscle cell cytoplasm.

When the Ca⁺⁺ concentration of the muscle cell cytoplasm is low, tropomyosin inhibits cross-bridge formation and the muscle is relaxed (figure 47.14). When the Ca⁺⁺ concentration is raised, Ca⁺⁺ binds to troponin. This causes the troponin-tropomyosin complex to be shifted away from the attachment sites for the myosin heads on the actin. Cross-bridges can thus form, undergo power strokes, and produce muscle contraction.

Where does the Ca⁺⁺ come from? Muscle fibers store Ca⁺⁺ in a modified endoplasmic reticulum called a **sarcoplasmic reticulum,** or **SR** (figure 47.15). When a muscle fiber is stimulated to contract, Ca⁺⁺ is released from the SR and diffuses into the myofibrils, where it binds to troponin and causes contraction. The contraction of muscles is regulated by nerve activity, and so nerves must influence the distribution of Ca⁺⁺ in the muscle fiber.

The cross-bridges are prevented from binding to actin by tropomyosin in a relaxed muscle. In order to contract a muscle, Ca⁺⁺ must be released from the sarcoplasmic reticulum, where it is stored, so that it can bind to troponin and cause the tropomyosin to shift its position in the thin filaments.

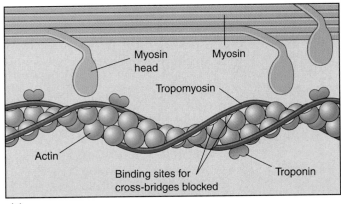

(a)

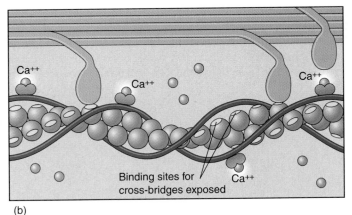

(b)

FIGURE 47.14
How calcium controls striated muscle contraction. (a) When the muscle is at rest, a long filament of the protein tropomyosin blocks the myosin-binding sites on the actin molecule. Since myosin is unable to form cross-bridges with actin at these sites, muscle contraction cannot occur. (b) When Ca⁺⁺ binds to another protein, troponin, the Ca⁺⁺-troponin complex displaces tropomyosin and exposes the myosin-binding sites on actin, permitting cross-bridges to form and contraction to occur.

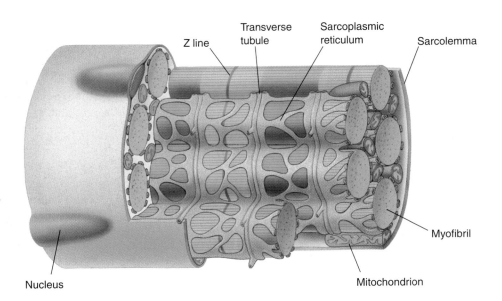

FIGURE 47.15
The relationship between the myofibrils, transverse tubules, and sarcoplasmic reticulum. Impulses conducted along the transverse tubules stimulate the release of Ca⁺⁺ from the sarcoplasmic reticulum into the cytoplasm. Ca⁺⁺ diffuses toward the myofibrils and causes contraction.

Nerves Stimulate Contraction

Muscles are stimulated to contract by motor neurons. The particular motor neurons that stimulate skeletal muscles, as opposed to cardiac and smooth muscles, are called *somatic motor neurons*. The axon (see chapter 46) of a somatic motor neuron branches to make functional connections, or *synapses*, with a number of muscle fibers. (Synapses are discussed in more detail in chapter 51.) Although one axon can stimulate many muscle fibers, each muscle fiber only has a single synapse with a branch of one axon.

When a somatic motor neuron produces electrochemical impulses, it stimulates contraction of the muscle fibers it innervates (makes synapses with) through the following events:

1. The motor neuron, at its synapse with the muscle fibers, releases a chemical known as a *neurotransmitter*. The specific neurotransmitter chemical released by somatic motor neurons is **acetylcholine (ACh)**. ACh acts on the muscle fiber membrane to stimulate the muscle fiber to produce its own electrochemical impulses.
2. The impulses spread along the membrane of the muscle fiber. In addition, they are carried more deeply into the muscle fiber by invaginations of the cell membrane called **transverse, or T, tubules.**
3. The T tubules conduct the impulses toward the sarcoplasmic reticulum (SR). Through a mechanism not yet completely understood, this stimulates the SR to release Ca^{++}. The Ca^{++} diffuses into the cytoplasm, binds to troponin, and causes the troponin-tropomyosin complex to shift position in the thin filaments, thereby stimulating contraction.

When impulses along the nerve stop, the nerve stops releasing ACh. This stops the production of impulses in the muscle fiber. When the T tubules no longer produce impulses, Ca^{++} is brought back into the SR by active transport. Troponin is no longer bound to Ca^{++}, so tropomyosin returns to its inhibitory position, allowing the muscle to relax.

Notice that neurons produce electric excitation of the muscle fiber. Electric excitation then causes the release of Ca^{++} from the SR, and the Ca^{++} causes muscle contraction. A shorthand way of describing this is to say that Ca^{++} is responsible for **excitation-contraction coupling.**

Motor Units and Recruitment

The set of muscle fibers innervated by all axonal branches of a given motor neuron is defined as a **motor unit** (figure 47.16). Every time the motor neuron produces impulses, all muscle fibers in that motor unit contract together. The division of the muscle into motor units allows the muscle's strength of contraction to be finely graded, a requirement for coordinated movements of the skeleton. Muscles that require a finer degree of control have smaller motor units (fewer muscle fibers per neuron) than muscles that require

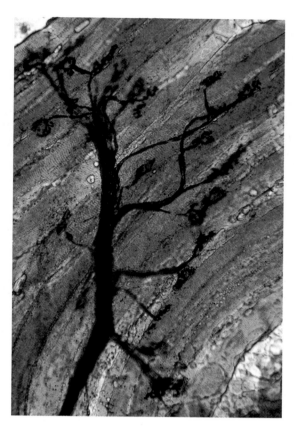

FIGURE 47.16
A motor unit. This micrograph shows the physical connections between a single motor neuron's axon and a group of skeletal muscle fibers. Small branches arise from the thick, main part of the axon and terminate on individual muscle fibers, forming synapses called neuromuscular junctions.

less precise control but must exert more force. For example, there are only a few muscle fibers per motor neuron in the muscles that move the eyes, while there are several hundred per motor neuron in the large muscles of the legs.

Most muscles contain motor units in a variety of sizes, which can be selectively activated by the nervous system. The weakest contractions of a muscle involve the activation of a few small motor units. If a slightly stronger contraction is necessary, additional small motor units are also activated. The initial increments to the total force generated by the muscle are therefore relatively small. As ever greater forces are required, more and larger motor units are brought into action, and the force increments become larger. The nervous system's use of increased numbers and sizes of motor units to produce a stronger contraction is termed **recruitment.**

Muscle contraction is stimulated by the production of electrochemical impulses in the muscle fiber. This causes Ca^{++} to be released from the sarcoplasmic reticulum, so that cross-bridges can form between the thin and thick filaments. Muscle relaxation occurs when impulses cease and Ca^{++} is transported back into the sarcoplasmic reticulum.

Types of Muscle Fibers

Skeletal muscle fibers can be divided on the basis of their contraction speed into **slow-twitch, or type I, fibers,** and **fast-twitch, or type II, fibers.** The muscles that move the eyes, for example, have a high proportion of fast-twitch fibers and reach maximum tension in about 7.3 milliseconds; the soleus muscle in the leg, by contrast, has a high proportion of slow-twitch fibers and requires about 100 milliseconds to reach maximum tension (figure 47.17).

Muscles like the soleus must be able to sustain a contraction for a long period of time without fatigue. The resistance to fatigue demonstrated by these muscles is aided by other characteristics of slow-twitch (type I) fibers that endow them with a high capacity for aerobic respiration. Slow-twitch fibers have a rich capillary supply, numerous mitochondria and aerobic respiratory enzymes, and a high concentration of **myoglobin** pigment. Myoglobin is a red pigment, similar to the hemoglobin in red blood cells, that improves the delivery of oxygen to the slow-twitch fibers. Because of their high myoglobin content, slow-twitch fibers are also called *red fibers.*

The thicker, fast-twitch (type II) fibers have fewer capillaries and mitochondria than slow-twitch fibers and not as much myoglobin; hence, these fibers are also called *white fibers.* Fast-twitch fibers are adapted to respire anaerobically by using a large store of glycogen and high concentrations of glycolytic enzymes. Fast-twitch fibers are adapted for the rapid generation of power and can grow thicker and stronger in response to weight training.

In addition to the type I (slow-twitch) and type II (fast-twitch) fibers, human muscles also have an intermediate form of fibers that are fast-twitch but also have a high oxidative capacity, and so are more resistant to fatigue. Endurance training increases the proportion of these fibers in muscles.

Muscle Metabolism during Rest and Exercise

Skeletal muscles at rest obtain most of their energy from the aerobic respiration of fatty acids. During exercise, muscle glycogen and blood glucose are also used as energy sources. The energy obtained by cell respiration is used to make ATP, which is needed for (1) the movement of the cross-bridges during muscle contraction and (2) the pumping of Ca^{++} into the sarcoplasmic reticulum for muscle relaxation. ATP can be obtained by skeletal muscles quickly by combining ADP with phosphate derived from creatine phosphate. This compound was produced previously in the resting muscle by combining creatine with phosphate derived from the ATP generated in cell respiration.

Skeletal muscles respire anaerobically for the first 45 to 90 seconds of moderate-to-heavy exercise, because the cardiopulmonary system requires this amount of time to sufficiently increase the oxygen supply to the exercising muscles. If exercise is moderate, aerobic respiration contributes the major portion of the skeletal muscle energy requirements following the first 2 minutes of exercise.

Whether exercise is light, moderate, or intense for a given person depends upon that person's maximal capacity for aerobic exercise. The maximum rate of oxygen consumption in the body (by aerobic respiration) is called the maximal oxygen uptake, or the aerobic capacity. The intensity of exercise can also be defined by the lactate threshold. This is the percentage of the maximal oxygen uptake at which a significant rise in blood lactate levels occurs as a result of anaerobic respiration. For average, healthy people, for example, a significant amount of blood lactate appears when exercise is performed at about 50 to 70% of the maximal oxygen uptake.

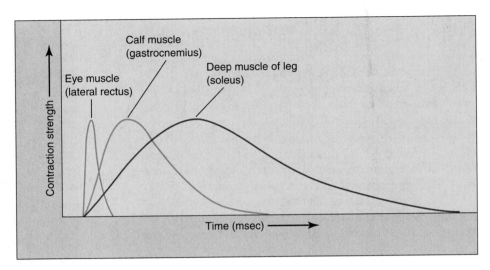

FIGURE 47.17
Skeletal muscles have different proportions of fast-twitch and slow-twitch fibers. The muscles that move the eye contain mostly fast-twitch fibers, whereas the deep muscle of the leg (the soleus) contains mostly slow-twitch fibers. The calf muscle (gastrocnemius) is intermediate in its composition.

Skeletal Muscle Fiber Types and Athletic Training

The various muscles of the body have different functions. Finger and eye muscles must contract rapidly and they fatigue with continued use. The muscles that control posture are slow to contract and slow to fatigue. These functional differences occur because there are three distinct muscle fiber types: **slow oxidative, fast oxidative,** and **fast glycolytic.**

Muscle fibers are divided into fast and slow types on the basis of their speed of contraction, which is related to the form of ATPase possessed by their myosin heads. The myosin in fast oxidative and fast glycolytic fibers has a high ATPase activity, while that in slow oxidative fibers has a low ATPase activity. The higher rate of ATP hydrolysis, the faster the cross-bridges can cycle and the more rapidly a muscle can shorten. The price of faster shortening is a higher rate of ATP use.

Muscle fibers are divided into oxidative and glycolytic types on the basis of the metabolic pathway they use to provide energy. Oxidative muscle fibers rely on aerobic respiration to synthesize ATP. They have a high density of mitochondria, and the muscles in which they occur have a profuse capillary blood supply. They appear red because they are rich in **myoglobin,** an oxygen-binding protein similar to hemoglobin, the red pigment of blood. Myoglobin stores oxygen within the muscle fibers for use in respiration. In contrast, glycolytic fibers are limited to anaerobic glycolysis for ATP production. They have large stores of glycogen but little myoglobin and hence appear pale. While oxidative fibers resist fatigue, glycolytic fibers fatigue quickly.

Individual muscles contain both glycolytic and oxidative fibers, but in different muscles the relative proportions of each vary according to the function of the muscle. For example, the postural muscles of the neck and the back, which are used almost constantly, contain predominantly oxidative fibers; the muscles of the arms, which receive more occasional use, have mostly glycolytic fibers. In each individual, the numbers of the various fiber types in each skeletal muscle are constant after maturity. Athletic training does not affect their relative proportions.

However, training can result in changes in the muscle fibers themselves. Muscle fibers that are used frequently respond in ways that serve to increase their performance, and the changes that occur are different for oxidative and glycolytic fiber types. Oxidative fibers respond with increases in myoglobin content and number of mitochondria, and muscles with large numbers of oxidative fibers develop more capillaries. These changes have little effect on the mass of the muscle. Endurance exercises that call for sustained effort, such as distance swimming, running, and skiing, favor the development of oxidative fibers.

The situation is quite different for brief, high-intensity training such as weight lifting, which promotes the development of glycolytic fibers. These fibers respond by increasing the number of their myofibrils and the size of their glycogen stores, changes that necessitate an increase in fiber diameter and thus an increase in muscle mass, termed **hypertrophy.** The muscle hypertrophy experienced by bodybuilders also results from the addition of collagen and other connective tissue required to sustain the passive tension of heavy loads. The rate and amount of tension developed during training exercises is the most important factor contributing to an increased incorporation of contractile protein into muscle. The short-term demands of maximum force development do not cause the circulatory and metabolic changes associated with endurance training.

Muscle Fatigue and Physical Training

Muscle fatigue refers to the use-dependant decrease in the ability of a muscle to generate force. The reasons for fatigue are not entirely understood. In most cases, however, muscle fatigue is correlated with the production of lactic acid by the exercising muscles. Lactic acid is produced by the anaerobic respiration of glucose, and glucose is obtained from muscle glycogen and from the blood. Lactate production and muscle fatigue are therefore also related to the depletion of muscle glycogen.

Because the depletion of muscle glycogen places a limit on exercise, any adaptation that spares muscle glycogen will improve physical endurance. Trained athletes have an increased proportion of energy derived from the aerobic respiration of fatty acids, resulting in a slower depletion of their muscle glycogen reserve. The greater the level of physical training, the higher the proportion of energy derived from the aerobic respiration of fatty acids. Since the aerobic capacity of endurance-trained athletes is higher than that of untrained people, athletes can perform more exercise before lactic acid production and glycogen depletion cause muscle fatigue.

Endurance training does not increase muscle size. Muscle enlargement is produced only by frequent periods of high-intensity exercise in which muscles work against high resistance, as in weight lifting. As a result of resistance training, type II (fast-twitch) muscle fibers become thicker as a result of the increased size and number of their myofibrils. Weight training, therefore, causes skeletal muscles to grow by **hypertrophy** (increased cell size) rather than by cell division and an increased number of cells.

The Oxygen Debt

When a person stops exercising, the rate of oxygen uptake does not immediately return to preexercise levels; it returns slowly (the person continues to breathe heavily for some time afterward). This extra oxygen is used to repay the **oxygen debt** incurred during exercise. The oxygen debt includes: (1) oxygen withdrawn from savings deposits—hemoglobin in blood and myoglobin in muscle; (2) extra oxygen required for metabolism by tissues warmed during exercise; and (3) oxygen needed for the metabolism of the lactic acid produced during anaerobic respiration.

Slow-twitch skeletal muscle fibers are adapted for aerobic respiration and are slower to fatigue than fast-twitch fibers, which are more adapted for the rapid generation of power.

Comparing Cardiac and Smooth Muscles

Cardiac and smooth muscle are similar in that both are found within internal organs and both are generally not under conscious control. Cardiac muscle, however, is like skeletal muscle in that it is striated and contracts by means of a sliding filament mechanism. Smooth muscle (as its name implies) is not striated. Smooth muscle does contain actin and myosin filaments, but they are arranged less regularly within the cell.

Cardiac Muscle

Cardiac muscle in the vertebrate heart is composed of striated muscle cells that are arranged differently from the fibers in a skeletal muscle. Instead of the long, multinucleate cells that form skeletal muscle, cardiac muscle is composed of shorter, branched cells, each with its own nucleus, that interconnect with one another at intercalated discs (figure 47.18). Intercalated discs are regions where the membranes of two cells fuse together, and the fused membranes are pierced by **gap junctions** (chapter 7). The gap junctions permit the diffusion of ions, and thus the spread of electric excitation, from one cell to the next. The mass of interconnected cardiac muscle cells forms a single, functioning unit called a *myocardium*. Electric impulses begin spontaneously in a specific region of the myocardium known as the *pacemaker*. These impulses are *not* initiated by impulses in motor neurons, as they are in skeletal muscle, but rather are produced by the cardiac muscle cells themselves. From the pacemaker, the impulses spread throughout the myocardium via gap junctions, causing contraction.

The heart has two myocardia, one that receives blood from the body and one that ejects blood into the body. Since all of the cells in a myocardium are stimulated as a unit, cardiac muscle cannot produce summated contractions or tetanus. This would interfere with the alternation between contraction and relaxation that is necessary for pumping.

Smooth Muscle

Smooth muscle surrounds hollow internal organs, including the stomach, intestines, bladder, and uterus, as well as all blood vessels except capillaries. Smooth muscle cells are long and spindle-shaped, and each contains a single nucleus.

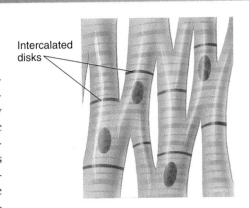

Intercalated disks

FIGURE 47.18
Cardiac muscle. Cells are organized into long branching chains that interconnect, forming a lattice; neighboring cells are linked by structures called intercalated discs.

They also contain actin and myosin, but these contractile proteins are not organized into sarcomeres. Parallel arrangements of thick and thin filaments cross diagonally from one side of the cell to the other. The thick filaments are attached either to structures called dense bodies, the functional equivalents of Z lines, or to the plasma membrane. Most smooth muscle cells have 10 to 15 thin filaments per thick filament, compared to 3 per thick filament in striated muscle fibers.

Smooth muscle cells do not have a sarcoplasmic reticulum; during a contraction, Ca^{++} enters from the extracellular fluid. In the cytoplasm, Ca^{++} binds to calmodulin, a protein that is structurally similar to troponin. The Ca^{++}-calmodulin complex activates an enzyme that phosphorylates (adds a phosphate group to) the myosin heads. Unlike the case with striated muscles, this phosphorylation is required for the myosin heads to form cross-bridges with actin.

This mechanism allows gradations in the strength of contraction in a smooth muscle cell, increasing contraction strength as more Ca^{++} enters the cytoplasm. Heart patients sometimes take drugs that block Ca^{++} entry into smooth muscle cells, reducing the cells' ability to contract. This treatment causes vascular smooth muscle to relax, dilating the blood vessels and reducing the amount of work the heart must do to pump blood through them.

In some smooth muscle tissues, the cells contract only when they are stimulated by the nervous system. These muscles line the walls of many blood vessels and make up the iris of the eye. Other smooth muscle tissues, like those in the wall of the gut, contains cells that produce electric impulses spontaneously. These impulses spread to adjoining cells through gap junctions, leading to a slow, steady contraction of the tissue.

Neither skeletal nor cardiac muscle can be greatly stretched because if the thick and thin filaments no longer overlay in the sarcomere, cross-bridges cannot form. Unlike these striated muscles, smooth muscle can contract even when it is greatly stretched. If one considers the degree to which some internal organs may be stretched—a uterus during pregnancy, for example—it is no wonder that these organs contain smooth muscle instead of striated muscle.

Cardiac muscle cells interconnect physically and electrically to form a single, functioning unit called a myocardium, which produces its own impulses at a pacemaker region.

Summary of Concepts

47.1 The vertebrate skeleton is composed of living bone.

- The skeleton of vertebrates is an endoskeleton composed of bone.
- Bone is formed by the secretion of an organic matrix by osteoblasts; this organic matrix becomes calcified.
- The human skeleton is divided into the axial and appendicular skeletons, the latter consisting of the limbs and the pectoral and pelvic girdles.

47.2 Skeletal muscles contract to produce movements at joints.

- Freely movable joints surround the articulating bones with a synovial capsule filled with a lubricating fluid.
- Skeletal muscles can work together as synergists, or oppose each other as antagonists.

47.3 Sliding of the myofilaments produces muscle contraction.

- A muscle fiber, or cell, contains numerous myofibrils within it. The thick filaments are composed of myosin, the thin filaments of actin.
- There are small cross-bridges of myosin that extend out toward the actin; the cross-bridges are activated by the hydrolysis of ATP so that it can bind to actin and undergo a power stroke that causes the sliding of the myofilaments.

- During contraction, the thin filaments are moved toward the center of the dark bands, thus shortening the sarcomeres and the myofibrils.
- In a resting muscle, the cross-bridges are prevented from attaching to actin by tropomyosin, a protein in the thin filaments.
- When Ca^{++} binds to troponin, the tropomyosin shifts position in the thin filament, allowing the cross-bridges to bind to actin and undergo a power stroke.
- The release of Ca^{++} from the sarcoplasmic reticulum is stimulated by impulses in the muscle fiber produced by neural stimulation.
- Slow-twitch fibers are adapted for aerobic respiration and are resistant to fatigue; fast-twitch fibers can provide power quickly but produce lactic acid and fatigue quickly.
- Endurance-trained athletes have an increased proportion of their muscle energy derived from the aerobic respiration of fatty acids, and thus a slower depletion of their muscle glycogen reserve.

47.4 Cardiac and smooth muscles are involuntary.

- Cardiac muscle cells have gap junctions that permit the spread of electric impulses from one cell to the next; like skeletal muscle.
- Cardiac and smooth muscles are involuntary and regulated by autonomic nerves; the contractions are automatically produced in cardiac muscle and some smooth muscles.

Discussing Key Terms

1. **Haversian system**   A Haversian system is the microscopic unit of bone structure in which osteocytes are arranged in concentric circles around a central canal containing blood vessels.

2. **Growth plate**   Growth plates are discs of cartilage at each end of a growing long bone. Since the growth plates are composed of cartilage, their cells can divide to make the plates thicker. Part of the plates are then converted to bone so that the bone grows in length.

3. **Synovial joint**   Synovial joints permit movement of the articulating bones at the joints. Each bone is capped with articular cartilage and the articulating ends of the bones are surrounded by a synovial capsule and are lubricated with synovial fluid.

4. **Sarcomere**   A sarcomere is the smallest structural unit in a striated muscle. It extends from Z line to Z line and contains a stack of thick filaments at its center. When a muscle cell contracts, the thin filaments slide over the thick filaments and the sarcomere shortens.

5. **Cross-bridge cycle**   The myosin cross-bridge binds to actin and undergoes a power stroke that produces sliding of the filaments. Afterward, a cross-bridge binds to ATP and releases from the actin. Hydrolysis of the ATP causes the cross-bridge to resume its resting orientation to begin the next cycle.

6. **Sarcoplasmic reticulum**   The sarcoplasmic reticulum is a modified smooth endoplasmic reticulum that stores Ca^{++} by active transport when a muscle is relaxed. It releases its stored Ca^{++} when the muscle fiber is stimulated by a nerve.

7. **Gap junction**   A gap junction is a region of contact between adjacent heart muscle cells where their membranes fuse together but are perforated by proteins that form channels through which ions can pass. This allows electric impulses to spread from one cell to the next.

Review Questions

1. What are the two major components of the extracellular matrix in bone? What structural properties does each component have? How do the two components combine to make bone resistant to fracture?

2. What are the three types of joints in a vertebrate skeleton? Give an example of where each type is found in the body.

3. What is the difference between a skeletal muscle's origin and its insertion? If two muscles are synergists, what effect do they have on the movement of bones at a joint? What effect do antagonistic muscles have?

4. Define each of the following: myofibril, sarcomere, Z line, myofilament.

5. Of what proteins are thick and thin filaments composed? What are the two regulatory proteins, where are they found, and what are their functions?

6. Describe the steps involved in the cross-bridge cycle. What functions does ATP perform in the cycle?

7. Describe the steps involved in excitation-contraction coupling. What functions do acetylcholine and Ca^{++} perform in this process?

8. How does a somatic motor neuron stimulate a muscle fiber to contract? Explain the sequence of steps involved. How does a nerve signal a muscle fiber to relax, and how is this accomplished?

9. What is the difference between an isometric and an isotonic contraction? Between a muscle twitch and tetanus? How is tetanus produced?

10. How does cardiac muscle differ from skeletal muscle in terms of tissue organization and initiation of contraction? Why can't a myocardium produce a sustained contraction?

11. How does smooth muscle differ from skeletal muscle in terms of thick and thin filament organization, the role of Ca^{++} in contraction, and the effect of stretching on the muscle's ability to contract?

Thought Questions

1. When myofibrils shorten due to sliding of the myofilaments, they pull inward on the membranes attached to their two ends, shortening the muscle cell. However, myofilaments never actively move in the opposite direction, pushing the membranes attached to their two ends apart and forcefully expanding the cell. Why is it that myofilaments pull but don't push? What causes a muscle fiber to lengthen when it relaxes?

2. When you begin to lift a heavy weight your muscles "tighten" before the weight begins to move. How can you use the terms *isometric* and *isotonic* to describe this process? How do you convert an isometric to an isotonic contraction? Use the concepts of motor units and recruitment in your answer.

Internet Links

An On-line Histology Course
http://www.meddean.luc.edu/lumen/MedEd/Histo/frames/histo_frames.html
A huge on-line collection of well-described histology slides of all major body tissues from Loyola University Medical School, with associated practical exams to test your knowledge.

Know Your Tissues
http://www.mc.vanderbilt.edu/histo/
Interactive lessons on the structure of blood, epithelium, nerves, and glands, from Vanderbilt University Medical Center.

Muscle Physiology
http://ortho84-13.ucsd.edu/MusIntro/
A comprehensive interactive index of topics in muscle physiology from UC San Diego.

How a Muscle Works
http://web.indstate.edu/thcme/mwking/muscle.html
A detailed look from Iowa State University at how the sarcomere is organized and how its components interact to bring about muscle contraction.

Know Your Bones
http://www.leeds.ac.uk/chb/lectures/anatomy3.html
An in-depth lecture on bones.

For Further Reading

Carafoli, E. and J. Penniston: "The Calcium Signal," *Scientific American*, November 1985, pages 70–80. The release of calcium ion is the only known way in which the electricity of the nervous system is able to produce changes in the body. Nerves regulate all muscle contractions and hormone secretions by controlling the level of Ca^{++}.

Cohen, C.: "The Protein Switch of Muscle Contraction," *Scientific American*, November 1975, pages 36–45. How proteins associated with myofilaments interact with Ca^{++} to trigger contraction.

Gordon, K.: "Adaptive Nature of Skeleton Design," *BioScience*, December 1989, pages 784–90. Plasticity allows changes in strength and locomotion.

Rogers, M.: "The Nature Of Muscles," *National Wildlife*, October/November 1990, pages 34–41. How efficient are the muscles of larger animals (including humans) compared to smaller ones? You may be surprised!

48

Fueling Body Activities: Digestion

FIGURE 48.1
Animals are heterotrophs. All animals must consume plant material or other animals in order to live. The nuts in this chipmunk's cheeks will be consumed and converted to body tissue, energy, and refuse.

Concept Outline

Plants and other photosynthetic organisms can produce the organic molecules they need from inorganic components. Therefore, they are autotrophs, or self-sustaining. Animals are heterotrophs: they must consume organic molecules other organisms have already produced (figure 48.1). The molecules heterotrophs eat must be digested into smaller molecules in order to be absorbed into the animal's body. Once these products of digestion enter the body, the animal can use them for energy in cell respiration or for the construction of the larger molecules that make up its tissues. The process of animal digestion is the focus of this chapter.

935

Types of Digestive Systems

Heterotrophs are divided into three groups on the basis of their food sources. Animals that eat plants exclusively are classified as **herbivores;** common examples include cows and horses, rabbits and sparrows. Animals that are meat-eaters, such as cats and eagles, trout and frogs, are **carnivores. Omnivores** are animals that eat both plants and other animals. We humans are omnivores, as are pigs, bears, and crows.

Single-celled organisms (as well as sponges) digest their food intracellularly. Other animals digest their food extracellularly, within a digestive cavity. In this case, the digestive enzymes are released into a cavity that is continuous with the animal's external environment. In coelenterates and flatworms (such as *Planaria*), the digestive cavity has only one opening that serves as both mouth and anus. There can be no specialization within this type of digestive system, called a *gastrovascular cavity*, because every cell is exposed to all stages of food digestion (figure 48.2).

Specialization occurs when the digestive tract, or alimentary canal, has a separate mouth and anus, so that transport of food is one-way. The most primitive digestive tract is seen in nematodes (phylum Nematoda), where it is simply a tubular *gut* lined by an epithelial membrane. Earthworms (phylum Annelida) have a digestive tract specialized in different regions for the ingestion, storage, fragmentation, digestion, and absorption of food. All higher animal groups, including all vertebrates, show similar specializations (figure 48.3).

The ingested food may be stored in a specialized region of the digestive tract or may first be subjected to physical fragmentation. This fragmentation may occur through the chewing action of teeth (in the mouth of many vertebrates), or the grinding action of pebbles (in the gizzard of earthworms and birds). Chemical digestion then occurs, breaking down the larger food molecules of polysaccharides and disaccharides, fats, and proteins into their smallest subunits. Chemical digestion involves hydrolysis reactions that liberate the subunit molecules—primarily monosaccharides, amino acids, and fatty acids—from the food. These products of chemical digestion pass through the epithelial lining of the gut into the blood, in a process known as absorption. Any molecules in the food that are not absorbed cannot be used by the animal. These waste products are excreted, or defecated, from the anus.

Most animals digest their food extracellularly. The digestive tract, with a one-way transport of food and specialization of regions for different functions, allows food to be ingested, physically fragmented, chemically digested, and absorbed.

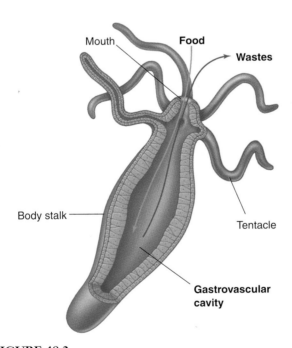

FIGURE 48.2
The gastrovascular cavity of *Hydra*, a coelenterate. Since there is only one opening, the mouth is also the anus, and no specialization is possible in the different regions that participate in extracellular digestion.

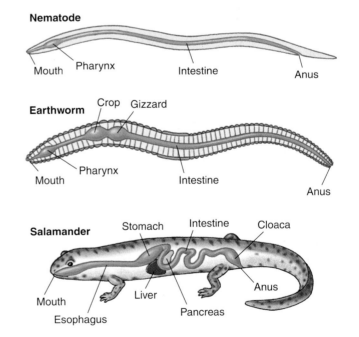

FIGURE 48.3
The one-way digestive tract of nematodes, earthworms, and vertebrates. One-way movement through the digestive tract allows different regions of the digestive system to become specialized for different functions.

Vertebrate Digestive Systems

In humans and other vertebrates, the digestive system consists of a tubular gastrointestinal tract and accessory digestive organs (figure 48.4). The initial components of the gastrointestinal tract are the mouth and the pharynx, which is the common passage of the oral and nasal cavities. The pharynx leads to the esophagus, a muscular tube that delivers food to the stomach, where some preliminary digestion occurs. From the stomach, food passes to the first part of the small intestine, where a battery of digestive enzymes continues the digestive process. The products of digestion then pass across the wall of the small intestine into the bloodstream. The small intestine empties what remains into the large intestine, where water and minerals are absorbed. In most vertebrates other than mammals, the waste products emerge from the large intestine into a cavity called the cloaca (see figure 48.3), which also receives the products of the urinary and reproductive systems. In mammals, the urogenital products are separated from the fecal material in the large intestine; the fecal material enters the rectum and is expelled through the anus.

In general, carnivores have shorter intestines for their size than do herbivores. A short intestine is adequate for a carnivore, but herbivores ingest a large amount of plant cellulose, which resists digestion. These animals have a long, convoluted small intestine. In addition, mammals called *ruminants* (such as cows) that consume grass and other vegetation have stomachs with multiple chambers, where bacteria aid in the digestion of cellulose. Other herbivores, including rabbits and horses, digest cellulose (with the aid of bacteria) in a blind pouch called the **cecum** located at the beginning of the large intestine.

The accessory digestive organs (described in detail later in the chapter) include the liver, which produces *bile* (a green solution that emulsifies fat), the gallbladder, which stores and concentrates the bile, and the pancreas. The pancreas produces *pancreatic juice*, which contains digestive enzymes and bicarbonate. Both bile and pancreatic juice are secreted into the first region of the small intestine and aid digestion.

The tubular gastrointestinal tract of a vertebrate has a characteristic layered structure (figure 48.5). The innermost layer is the mucosa, an epithelium that lines the interior of the tract (the lumen). The next major tissue layer, made of connective tissue, is called the submucosa. Just outside the submucosa is the muscularis, which consists of a double layer of smooth muscles. The muscles in the inner layer have a circular orientation, and those in the outer layer are arranged longitudinally. Another connective tissue layer, the serosa, covers the external surface of the tract. Nerves, intertwined in regions called *plexuses*, are located in the submucosa and help regulate the gastrointestinal activities.

The vertebrate digestive system consists of a tubular gastrointestinal tract, which is modified in different animals, composed of a series of tissue layers.

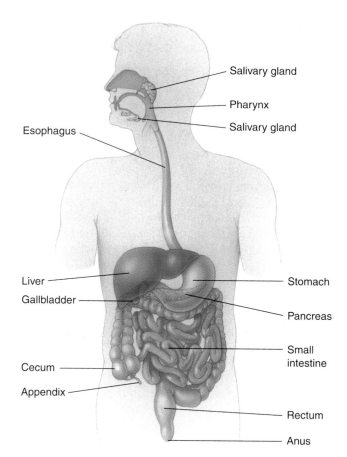

FIGURE 48.4
The human digestive system. Humans, like all placental mammals, lack a cloaca and have a separate exit from the digestive tract through the rectum and anus.

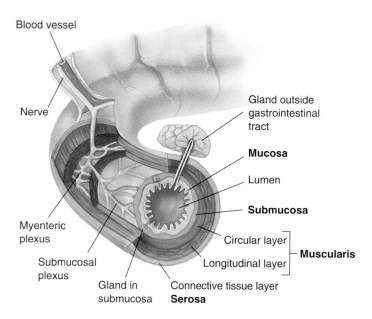

FIGURE 48.5
The layers of the gastrointestinal tract. The mucosa contains a lining epithelium; the submucosa is composed of connective tissue (as is the serosa), and the muscularis consists of smooth muscles.

The Mouth and Teeth

Specializations of the digestive systems in different kinds of vertebrates reflect differences in the way these animals live. Fishes have a large pharynx with gill slits, while air-breathing vertebrates have a greatly reduced pharynx. Many vertebrates have teeth (figure 48.6), and chewing (*mastication*) breaks up food into small particles and mixes it with fluid secretions. Birds, which lack teeth, break up food in their two-chambered stomachs (figure 48.7). In one of these chambers, the gizzard, small pebbles ingested by the bird are churned together with the food by muscular action. This churning grinds up the seeds and other hard plant material into smaller chunks that can be digested more easily.

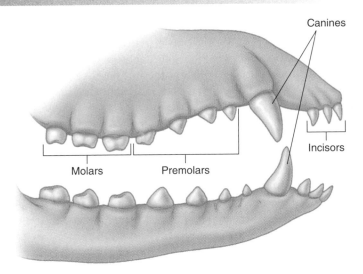

FIGURE 48.6
Diagram of generalized vertebrate dentition. Different vertebrates will have specific variations from this generalized pattern, depending on whether the vertebrate is an herbivore, carnivore, or omnivore.

Vertebrate Teeth

Carnivorous mammals have pointed teeth that lack flat grinding surfaces. Such teeth are adapted for cutting and shearing. Carnivores often tear off pieces of their prey but have little need to chew them, since digestive enzymes can act directly on animal cells. (Recall how a cat or dog gulps down its food.) By contrast, grass-eating herbivores, such as cows and horses, must pulverize the cellulose cell walls of plant tissue before digesting it. These animals have large, flat teeth with complex ridges well-suited to grinding.

Humans are omnivores, and human teeth are specialized for eating both plant and animal food. Viewed simply, humans are carnivores in the front of the mouth and herbivores in the back (figure 48.8). The four front teeth in the upper and lower jaws are sharp, chisel-shaped incisors used for biting. On each side of the incisors are sharp, pointed teeth called cuspids (sometimes referred to as "canine" teeth), which are used for tearing food. Behind the canines are two premolars and three molars, all with flattened, ridged surfaces for grinding and crushing food. Children have only 20 teeth, but these deciduous teeth are lost during childhood and are replaced by 32 adult teeth.

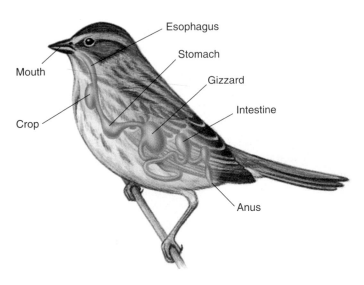

FIGURE 48.7
Birds grind up food in the crop and gizzard. Birds, which lack teeth, have two chambers, the crop and the gizzard, where food is pulverized and crushed before passing into the small intestine.

The Mouth

Inside the mouth, the tongue mixes food with a mucous solution, saliva. In humans, three pairs of salivary glands secrete saliva into the mouth through ducts in the mouth's mucosal lining. Saliva moistens and lubricates the food so that it is easier to swallow and does not abrade the tissue it passes on its way through the esophagus. Saliva also contains the hydrolytic enzyme salivary amylase, which initiates the breakdown of the polysaccharide starch into the disaccharide maltose. This digestion is usually minimal in humans, however, because most people don't chew their food very long.

The secretions of the salivary glands are controlled by the nervous system, which in humans maintains a constant flow of about half a milliliter per minute when the mouth is empty of food. This continuous secretion keeps the mouth moist. The presence of food in the mouth triggers an increased rate of secretion, as taste-sensitive neurons in the mouth send impulses to the brain, which responds by stimulating the salivary glands. The most potent stimuli are acidic solutions; lemon juice, for example, can increase the rate of salivation eightfold. The sight, sound, or smell of food can stimulate salivation markedly in dogs, but in humans, these stimuli are much less effective than thinking or talking about food.

When food is ready to be swallowed, the tongue moves it to the back of the mouth. In mammals, the process of swallowing begins when the soft palate elevates, pushing against the back wall of the pharynx (figure 48.9). Elevation of the soft palate seals off the nasal cavity and prevents food from entering it. Pressure against the pharynx stimulates neurons within its walls, which send impulses to the swallowing center in the brain. In response, muscles are stimulated to contract and raise the *larynx* (voice box). This pushes the *glottis*, the opening from the larynx into the trachea (windpipe), against a flap of tissue called the *epiglottis*. These actions keep food out of the respiratory tract, directing it instead into the esophagus.

> In many vertebrates ingested food is fragmented through the tearing or grinding action of specialized teeth. In birds, this is accomplished through the grinding action of pebbles in the gizzard. Food mixed with saliva is swallowed and enters the esophagus.

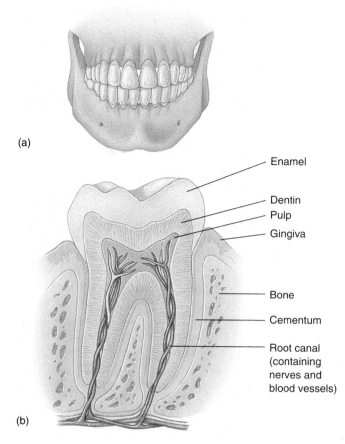

(a)

Enamel
Dentin
Pulp
Gingiva
Bone
Cementum
Root canal (containing nerves and blood vessels)

(b)

FIGURE 48.8
Human teeth. (a) The front six teeth on the upper and lower jaws are cuspids and incisors. The remaining teeth, running along the sides of the mouth, are grinders called premolars and molars. Hence, humans are carnivores in the front of their mouth and herbivores in the back. (b) Each tooth is alive, with a central pulp containing nerves and blood vessels. The actual chewing surface is a hard enamel layered over the softer dentin, which forms the body of the tooth.

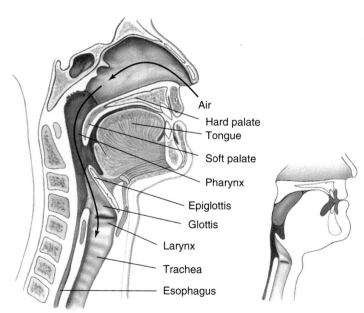

Air
Hard palate
Tongue
Soft palate
Pharynx
Epiglottis
Glottis
Larynx
Trachea
Esophagus

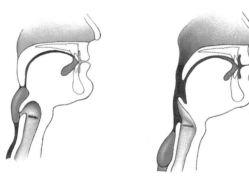

FIGURE 48.9
The human pharynx, palate, and larynx. Food that enters the pharynx is prevented from entering the nasal cavity by elevation of the soft palate, and is prevented from entering the larynx and trachea (the airways of the respiratory system) by elevation of the larynx against the epiglottis.

Esophagus and Stomach

Structure and Function of the Esophagus

Swallowed food enters a muscular tube called the esophagus, which connects the pharynx to the stomach. In adult humans, the esophagus is about 25 centimeters long; the upper third is enveloped in skeletal muscle, for voluntary control of swallowing, while the lower two-thirds is surrounded by involuntary smooth muscle. The swallowing center stimulates successive waves of contraction in these muscles that move food along the esophagus to the stomach. These rhythmic waves of muscular contraction are called peristalsis (figure 48.10); they enable humans and other vertebrates to swallow even if they are upside down.

In many vertebrates, the movement of food from the esophagus into the stomach is controlled by a ring of circular smooth muscle, or a *sphincter*, that opens in response to the pressure exerted by the food. Contraction of this sphincter prevents food in the stomach from moving back into the esophagus. Rodents and horses have a true sphincter at this site and thus cannot regurgitate, while humans lack a true sphincter and so are able to regurgitate. Normally, however, the human esophagus is closed off except during swallowing.

Structure and Function of the Stomach

The stomach (figure 48.11) is a sac-like portion of the digestive tract. Its inner surface is highly convoluted, enabling it to fold up when empty and open out like an expanding balloon as it fills with food. Thus, while the human stomach has a volume of only about 50 milliliters when empty, it may expand to contain 2 to 4 liters of food when full. Carnivores that engage in sporadic gorging as an important survival strategy possess stomachs that are able to distend much more than that.

Secretory Systems

The stomach contains an extra layer of smooth muscle for churning food and mixing it with *gastric juice*, an acidic secretion of the tubular gastric glands of the mucosa (figure 48.12). These exocrine glands contain two kinds of secretory cells: parietal cells, which secrete hydrochloric acid (HCl); and chief cells, which secrete pepsinogen, a weak protease (protein-digesting enzyme) that requires a very low pH to be active. This low pH is provided by the HCl. Activated pepsinogen molecules then cleave one another at specific sites, producing a much more active protease, pepsin. This process of secreting a relatively inactive enzyme that is then converted

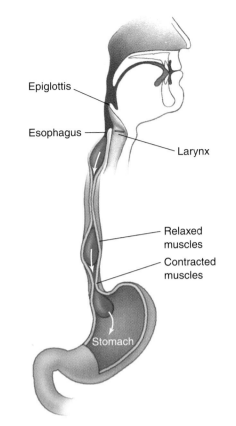

FIGURE 48.10
The esophagus and peristalsis. After food has entered the esophagus, rhythmic waves of muscular contraction, called peristalsis, move the food down to the stomach.

Epiglottis

Esophagus

Larynx

Relaxed muscles

Contracted muscles

Stomach

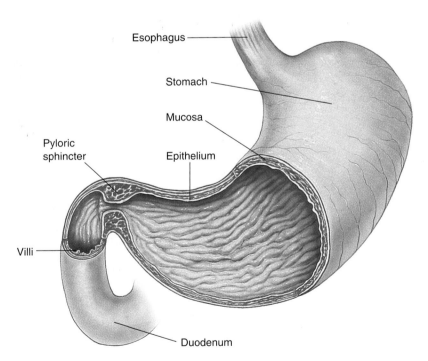

Esophagus

Stomach

Mucosa

Pyloric sphincter

Epithelium

Villi

Duodenum

FIGURE 48.11
The stomach and duodenum. Food enters the stomach from the esophagus. The epithelial walls of the stomach are dotted with gastric pits, which contain glands that secrete hydrochloric acid and the enzyme pepsinogen. A band of smooth muscle called the pyloric sphincter controls the entrance to the duodenum, the upper part of the small intestine.

into a more active enzyme outside the cell prevents the chief cells from digesting themselves. It should be noted that only proteins are partially digested in the stomach—there is no significant digestion of carbohydrates or fats.

Action of Acid

The human stomach produces about 2 liters of HCl and other gastric secretions every day, creating a very acidic solution inside the stomach. The concentration of HCl in this solution is about 10 millimolar, corresponding to a pH of 2. Thus, gastric juice is about 250,000 times more acidic than blood, whose normal pH is 7.4. The low pH in the stomach helps denature food proteins, making them easier to digest, and keeps pepsin maximally active. Active pepsin hydrolyzes food proteins into shorter chains of polypeptides that are not fully digested until the mixture enters the small intestine. The mixture of partially digested food and gastric juice is called chyme.

The acidic solution within the stomach also kills most of the bacteria that are ingested with the food. The few bacteria that survive the stomach and enter the intestine intact are able to grow and multiply there, particularly in the large intestine. In fact, most vertebrates harbor thriving colonies of bacteria within their intestines, and bacteria are a major component of feces. As we will discuss later, bacteria that live within the digestive tract of cows and other ruminants play a key role in the ability of these mammals to digest cellulose.

Ulcers

Overproduction of gastric acid can occasionally eat a hole through the wall of the stomach. Such *gastric ulcers* are rare, however, because epithelial cells in the mucosa of the stomach are protected somewhat by a layer of alkaline mucus, and because those cells are rapidly replaced by cell division if they become damaged (gastric epithelial cells are replaced every 2 to 3 days). Over 90% of gastrointestinal ulcers are *duodenal ulcers*. These may be produced when excessive amounts of acidic chyme are delivered into the duodenum, so that the acid cannot be properly neutralized through the action of alkaline pancreatic juice (described later). Susceptibility to ulcers is increased when the mucosal barriers to self-digestion are weakened by an infection of the bacterium *Helicobacter pylori*. Indeed, modern antibiotic treatments of this infection can reduce symptoms and often even cure the ulcer.

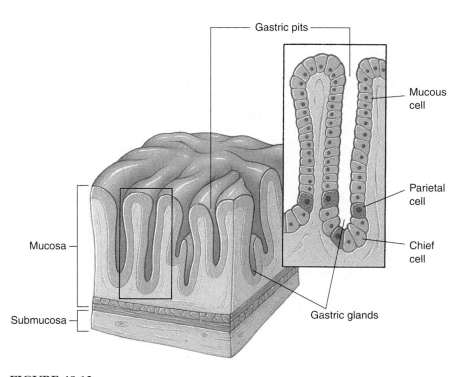

FIGURE 48.12
Gastric pits and gastric glands of the stomach mucosa. Gastric glands consist of mucous cells, chief cells that secrete pepsinogen, and parietal cells that secrete HCl. Gastric pits are the openings of the gastric glands.

In addition to producing HCl, the parietal cells of the stomach also secrete intrinsic factor, a polypeptide needed for the intestinal absorption of vitamin B_{12}. Because this vitamin is required for the production of red blood cells, persons who lack sufficient intrinsic factor develop a type of anemia (low red blood cell count) called *pernicious anemia*.

Leaving the Stomach

Chyme leaves the stomach through the *pyloric sphincter* (see figure 48.11) to enter the small intestine. This is where all terminal digestion of carbohydrates, lipids, and proteins occurs, and where the products of digestions—amino acids, glucose, and so on—are absorbed into the blood. Only some of the water in chyme and a few substances such as aspirin and alcohol are absorbed through the wall of the stomach.

Peristaltic waves of contraction propel food along the esophagus to the stomach. Gastric juice contains strong hydrochloric acid and the protein-digesting enzyme pepsin, which begins the digestion of proteins into shorter polypeptides. The acidic chyme is then transferred through the pyloric sphincter to the small intestine.

The Small Intestine

Digestion in the Small Intestine

The capacity of the small intestine is limited, and its digestive processes take time. Consequently, efficient digestion requires that only relatively small amounts of chyme be introduced from the stomach into the small intestine at any one time. Coordination between gastric and intestinal activities is regulated by neural and hormonal signals, which we will describe in a later section.

The small intestine is approximately 4.5 meters long in a living person, but is 6 meters long at autopsy when the muscles relax. The first 25 centimeters is the **duodenum;** the remainder of the small intestine is divided into the **jejunum** and the **ileum.** The duodenum receives acidic chyme from the stomach, digestive enzymes and bicarbonate from the pancreas, and bile from the liver and gallbladder. The pancreatic juice enzymes digest larger food molecules into smaller fragments. This occurs primarily in the duodenum and jejunum.

The epithelial wall of the small intestine is covered with tiny, finger-like projections called villi (singular, villus; figure 48.13). In turn, each of the epithelial cells lining the villi is covered on its apical surface (the side facing the lumen) by many foldings of the plasma membrane that form cytoplasmic extensions called *microvilli.* These are quite tiny and can be seen clearly only with an electron microscope (figure 48.14). In a light micrograph, the microvilli resemble the bristles of a brush, and for that reason the epithelial wall of the small intestine is also called a brush border.

The villi and microvilli greatly increase the surface area of the small intestine; in humans, this surface area is 300 square meters! It is over this vast surface that the products of digestion are absorbed. The microvilli also participate in

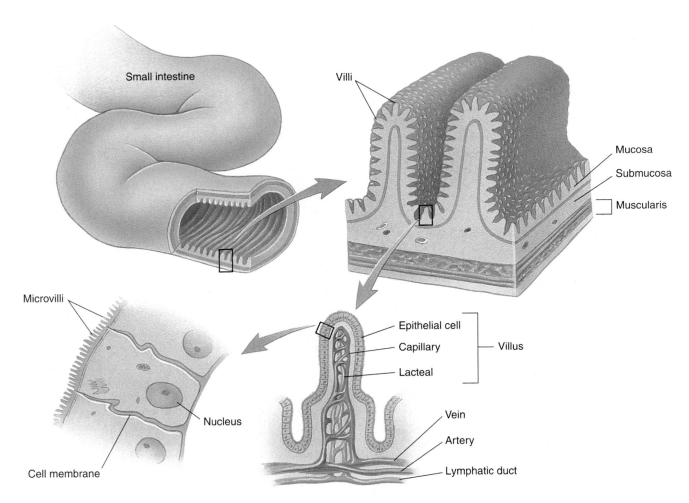

FIGURE 48.13
The small intestine. Cross-section of the small intestine; the enlargements show villi and an epithelial cell with numerous microvilli.

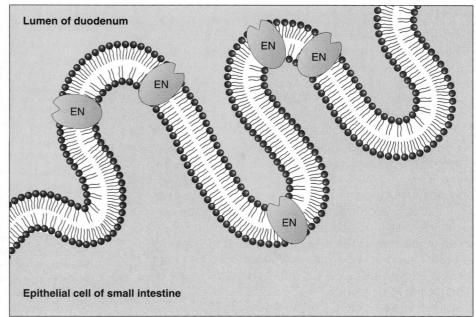

Lumen of duodenum

Epithelial cell of small intestine

FIGURE 48.15
Brush border enzymes. These enzymes, which are labeled "EN" in this diagram, are part of the plasma membrane of the microvilli in the small intestine. They catalyze many of the terminal steps in digestion.

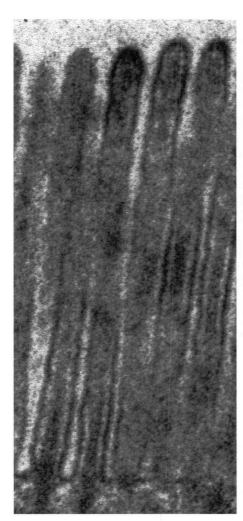

FIGURE 48.14
Intestinal microvilli. Microvilli, shown in an electron micrograph, are very densely clustered, giving the small intestine an enormous surface area important in efficient absorption of the digestion products.

digestion because a number of digestive enzymes are embedded within the epithelial cells' plasma membranes, with their active sites exposed to the chyme (figure 48.15). These brush border enzymes include those that hydrolyze the disaccharides lactose and sucrose, among others (table 48.1). Many adult humans lose the ability to produce the brush border enzyme *lactase* and therefore cannot digest lactose (milk sugar), a rather common condition called *lactose intolerance*. The brush border enzymes complete the digestive process that started with the action of the pancreatic enzymes released into the duodenum.

Table 48.1	Digestive Enzymes		
Location	**Enzymes**	**Substrates**	**Digestion Products**
Salivary glands	Amylase	Starch, glycogen	Disaccharides
Stomach	Pepsin	Proteins	Short peptides
Small intestine	Peptidases	Short peptides	Amino acids
(brush border)	Nucleases	DNA, RNA	Sugars, nucleic acid bases
	Lactase, maltase, sucrase	Disaccharides	Monosaccharides
Pancreas	Lipase	Triglycerides	Fatty acids, glycerol
	Trypsin, chymotrypsin	Proteins	Peptides
	DNase	DNA	Nucleotides
	RNase	RNA	Nucleotides

Absorption in the Small Intestine

The amino acids and monosaccharides resulting from the digestion of proteins and carbohydrates, respectively, are transported across the brush border into the epithelial cells that line the intestine (figure 48.16a). They then move to the other side of the epithelial cells, and from there are transported across the membrane and into the blood capillaries within the villi. The blood carries these products of digestion from the intestine to the liver via the hepatic portal vein. The term *portal* here refers to a special arrangement of vessels, seen only in a couple of instances, where one organ (the liver, in this case) is located "downstream" from another organ (the intestine). As a result, the second organ receives blood-borne molecules from the first. Because of the hepatic portal vein, the liver is the first organ to receive most of the products of digestion. This arrangement is important for the functions of the liver, as will be described in a later section.

The products of fat digestion are absorbed by a different mechanism (figure 48.16b). Fats (triglycerides) are hydrolyzed into fatty acids and monoglycerides, which are absorbed into the intestinal epithelial cells and reassembled into triglycerides. The triglycerides then combine with proteins to form small particles called chylomicrons. Instead of entering the hepatic portal circulation, the chylomicrons are absorbed into lymphatic capillaries (see chapter 49), which empty their contents into the blood in veins near the neck. Chylomicrons can make the blood plasma appear cloudy if a sample of blood is drawn after a fatty meal.

The amount of fluid passing through the small intestine in a day is startlingly large: approximately 9 liters. However, almost all of this fluid is absorbed into the body rather than eliminated in the feces. About 8.5 liters are absorbed in the small intestine and an additional 350 milliliters in the large intestine. Only about 50 grams of solid and 100 milliliters of liquid leave the body as feces. The normal fluid absorption efficiency of the human digestive tract thus approaches 99%, which is very high indeed.

Digestions occurs primarily in the duodenum, which receives the pancreatic juice enzymes. The small intestine provides a large surface area for absorption. Glucose and amino acids from food are absorbed through the small intestine and enter the blood via the hepatic portal vein, going to the liver. Fat from food enters the lymphatic system, which eventually drains the fat into veins near the neck.

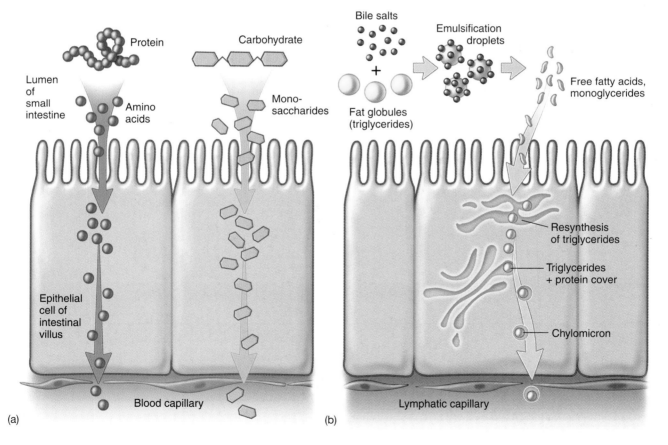

FIGURE 48.16
Absorption of the products of digestion. (a) Monosaccharides and amino acids are transported into blood capillaries. (b) Fatty acids and monoglycerides within the intestinal lumen are absorbed and converted within the intestinal epithelial cells into triglycerides. These are then coated with proteins to form tiny structures called chylomicrons, which enter lymphatic capillaries.

The Large Intestine

The large intestine, or colon, is much shorter than the small intestine, occupying approximately the last meter of the digestive tract; it is called "large" only because of its larger diameter. The small intestine empties directly into the large intestine at a junction where two vestigial structures, the cecum and the appendix, remain (figure 48.17). No digestion takes place within the large intestine, and only about 4% of the absorption of fluids by the intestine occurs there. The large intestine is not as convoluted as the small intestine, and its inner surface has no villi. Consequently, the large intestine has less than one-thirtieth the absorptive surface area of the small intestine. Although sodium, vitamin K, and some products of bacterial metabolism are absorbed across its wall, the primary function of the large intestine is to concentrate waste material. Within it, undigested material, primarily bacterial fragments and cellulose, is compacted and stored. Many bacteria live and reproduce within the large intestine, and the excess bacteria are incorporated into the refuse material, called *feces*. Bacterial fermentation produces gas within the colon at a rate of about 500 milliliters per day. This rate increases greatly after the consumption of beans or other vegetable matter because the passage of undigested plant material (fiber) into the large intestine provides substrates for fermentation.

The human colon has evolved to process food with a relatively high fiber content. Diets that are low in fiber, which are common in the United States, result in a slower passage of food through the colon. Low dietary fiber content is thought to be associated with the level of colon cancer in the United States, which is among the highest in the world.

Compacted feces, driven by peristaltic contractions of the large intestine, pass from the large intestine into a short tube called the rectum. From the rectum, the feces exit the body through the anus. Two sphincters control passage through the anus. The first is composed of smooth muscle and opens involuntarily in response to pressure inside the rectum. The second, composed of striated muscle, can be controlled voluntarily by the brain, thus permitting a conscious decision to delay defecation.

In all vertebrates except placental mammals, the reproductive and urinary tracts empty together with the digestive tract into a common cavity, the cloaca. In some reptiles and birds, additional water from either the feces or urine may be absorbed in the cloaca before the products are expelled from the body.

The large intestine concentrates wastes for excretion by absorbing water. Some ions and vitamin K are also absorbed by the large intestine.

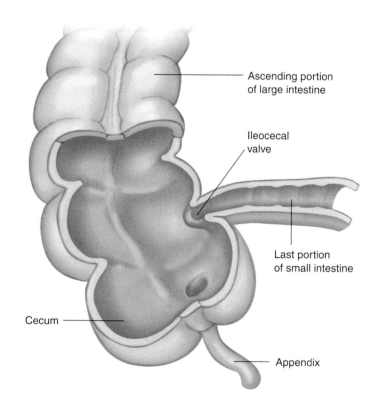

FIGURE 48.17
The junction of the small and large intestines in humans. The large intestine, or colon, starts with the cecum, which is relatively small in humans compared with that in other mammals. A vestigial structure called the appendix extends from the cecum.

Variations in Vertebrate Digestive Systems

Most animals lack the enzymes necessary to digest cellulose, the carbohydrate that functions as the chief structural component of plants. The digestive tracts of some animals, however, contain bacteria and protists that convert cellulose into substances the host can digest. Although digestion by gastrointestinal microorganisms plays a relatively small role in human nutrition, it is an essential element in the nutrition of many other kinds of animals, including insects like termites and cockroaches, and a few groups of herbivorous mammals. The relationships between these microorganisms and their animal hosts are mutually beneficial and provide an excellent example of symbiosis.

Cows, deer, and other ruminants have large, divided stomachs (figure 48.18). The first portion consists of the rumen and a smaller chamber, the reticulum; the second portion consists of two additional chambers: the omasum and abomasum. The rumen which may hold up to 50 gallons, serves as a fermentation vat in which bacteria and protozoa convert cellulose and other molecules into a variety of simpler compounds. The location of the rumen at the front of the four chambers is important because it allows the animal to regurgitate and rechew the contents of the rumen, an activity called *rumination*, or "chewing the cud." The cud is then swallowed and enters the reticulum, from which it passes to the omasum and then the abomasum, where it is finally mixed with gastric juice. Hence, only the abomasum is equivalent to the human stomach in its function. This process leads to a far more efficient digestion of cellulose in ruminants than in mammals that lack a rumen, such as horses.

In horses, rodents, and lagomorphs (rabbits and hares), the digestion of cellulose by microorganisms takes place in the cecum, which is greatly enlarged (figure 48.19). Since the cecum is located beyond the stomach, regurgitation of its contents is impossible. However, rodents and lagomorphs have evolved another way to digest cellulose that achieves a degree of efficiency similar to that of ruminant digestion. They do this by eating their feces, thus passing the food through their digestive tract a second time. The second passage makes it possible for the animal to absorb the nutrients produced by the microorganisms in its cecum. Animals that engage in this practice of **coprophagy** (from the Greek words *copros*, excrement, and *phagein*, eat) cannot remain healthy if they are prevented from eating their feces.

Cellulose is not the only plant product that vertebrates can use as a food source because of the digestive activities

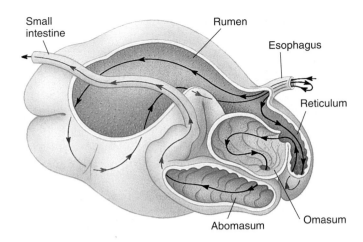

FIGURE 48.18
Four-chambered stomach of a ruminant. The grass and other plants that a ruminant, such as a cow, eats enter the rumen, where they are partially digested. Before moving into a second chamber, the reticulum, the food may be regurgitated and rechewed. The food is then transferred to the rear two chambers: the omasum and abomasum. Only the abomasum is equivalent to the human stomach in its function of secreting gastric juice.

of intestinal microorganisms. Wax, a substance indigestible by most terrestrial animals, is digested by symbiotic bacteria living in the gut of honey guides, African birds that eat the wax in bee nests. In the marine food chain, wax is a major constituent of copepods (crustaceans in the plankton), and many marine fish and birds appear to be able to digest wax with the aid of symbiotic microorganisms.

Another example of the way intestinal microorganisms function in the metabolism of their animal hosts is provided by the synthesis of vitamin K. All mammals rely on intestinal bacteria to synthesize this vitamin, which is necessary for the clotting of blood. Birds, which lack these bacteria, must consume the required quantities of vitamin K in their food. In humans, prolonged treatment with antibiotics greatly reduces the populations of bacteria in the intestine; under such circumstances, it may be necessary to provide supplementary vitamin K.

Much of the food value of plants is tied up in cellulose, and the digestive tract of many animals harbors colonies of cellulose-digesting microorganisms. Intestinal microorganisms also produce molecules such as vitamin K that are important to the well-being of their vertebrate hosts.

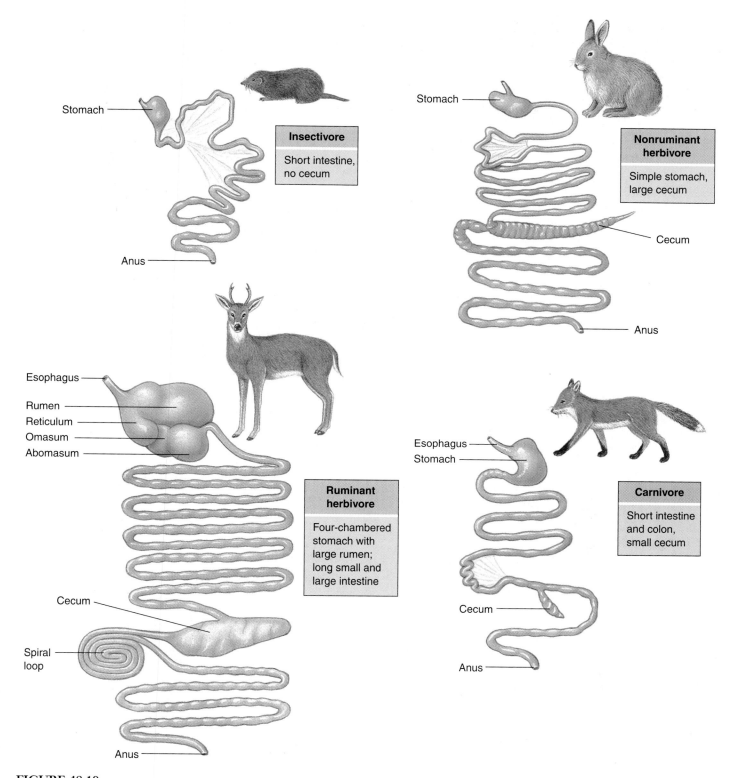

Stomach

Insectivore

Short intestine, no cecum

Anus

Stomach

Nonruminant herbivore

Simple stomach, large cecum

Cecum

Anus

Esophagus

Rumen

Reticulum

Omasum

Abomasum

Ruminant herbivore

Four-chambered stomach with large rumen; long small and large intestine

Cecum

Spiral loop

Anus

Esophagus

Stomach

Carnivore

Short intestine and colon, small cecum

Cecum

Anus

FIGURE 48.19
The digestive systems of different mammals reflect their diets. Herbivores require long digestive tracts with specialized compartments for the breakdown of plant matter. Protein diets are more easily digested; thus, insectivorous and carnivorous mammals have short digestive tracts with few specialized pouches.

Accessory Organs

Secretions of the Pancreas

The pancreas (figure 48.20), a large gland situated near the junction of the stomach and the small intestine, is one of the accessory organs that contribute secretions to the digestive tract. Fluid from the pancreas is secreted into the duodenum through the *pancreatic duct*; thus, the pancreas functions as an exocrine organ. This fluid contains a host of enzymes, including trypsin and chymotrypsin, which digest proteins; pancreatic amylase, which digests starch; and lipase, which digests fat. These enzymes are released into the duodenum primarily as inactive zymogens and are then activated by the brush border enzymes of the intestine. Pancreatic enzymes digest proteins into smaller polypeptides, polysaccharides into shorter chains of sugars, and fat into free fatty acids and other products. The complete digestion of these molecules is then completed by the brush border enzymes, as previously described.

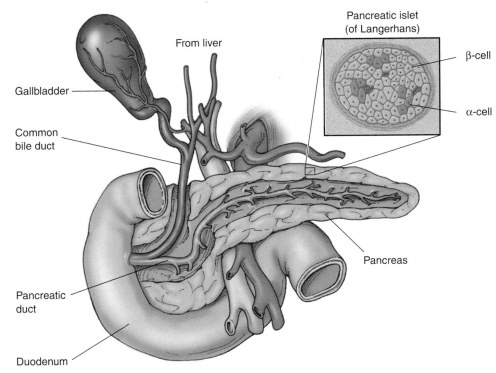

FIGURE 48.20
The pancreas and bile duct empty into the duodenum. The pancreas secretes pancreatic juice into the pancreatic duct. The pancreatic islets of Langerhans secrete hormones into the blood; α-cells secrete glucagon and β-cells secrete insulin.

Pancreatic fluid also contains bicarbonate, which neutralizes the HCl from the stomach and gives the chyme in the duodenum a slightly alkaline pH. The digestive enzymes and bicarbonate are produced by clusters of secretory cells known as *acini*.

In addition to its exocrine role in digestion, the pancreas also functions as an endocrine gland, secreting several hormones into the blood that control the blood levels of glucose and other nutrients. These hormones are produced in the **islets of Langerhans,** clusters of endocrine cells scattered throughout the pancreas. The two most important pancreatic hormones, insulin and glucagon, are discussed later in this chapter and in chapter 53.

The Liver and Gallbladder

The liver is the largest internal organ of the body (see figure 48.4). In an adult human, the liver weighs about 1.5 kilograms and is the size of a football. The main exocrine secretion of the liver is bile, a fluid mixture consisting of *bile pigments* and *bile salts* that is delivered into the duodenum during the digestion of a meal. The bile pigments do not participate in digestion; they are waste products resulting from the liver's destruction of old red blood cells and ultimately are eliminated with the feces. If the excretion of bile pigments by the liver is blocked, the pigments can accumulate in the blood and cause a yellow staining of the tissues known as *jaundice*.

In contrast, the bile salts play a very important role in the digestion of fats. Because fats are insoluble in water, they enter the intestine as drops within the watery chyme. The bile salts, which are partly lipid-soluble and partly water-soluble, work like detergents, dispersing the large drops of fat into a fine suspension of smaller droplets. This emulsification process produces a greater surface area of fat upon which the lipase enzymes can act, and thus allows the digestion of fat to proceed more rapidly.

After it is produced in the liver, bile is stored and concentrated in the gallbladder. The arrival of fatty food in the duodenum triggers a neural and endocrine reflex (discussed later) that stimulates the gallbladder to contract, causing bile to be transported through the common bile duct and injected into the duodenum. If the bile duct is blocked by a *gallstone* (formed from a hardened precipitate of cholesterol), contraction of the gallbladder will cause pain that is generally felt under the right scapula (shoulder blade).

Regulatory Functions of the Liver

Because the hepatic portal vein carries blood from the stomach and intestine directly to the liver, the liver is in a position to chemically modify the substances absorbed in the gastrointestinal tract before they reach the rest of the body. For example, ingested alcohol and other drugs are taken into liver cells and metabolized; this is why the liver is often damaged as a result of alcohol and drug abuse. The liver also removes toxins, pesticides, carcinogens, and other poisons, converting them into less toxic forms. An important example of this is the liver's conversion of the toxic ammonia produced by intestinal bacteria into urea, a compound that can be contained safely and carried by the blood at higher concentrations.

Similarly, the liver regulates the levels of many compounds produced within the body. Steroid hormones, for instance, are converted into less active and more water-soluble forms by the liver. These molecules are then included in the bile and eliminated from the body in the feces, or carried by the blood to the kidneys and excreted in the urine.

The liver also produces most of the proteins found in blood plasma. The total concentration of plasma proteins is significant because it must be kept within normal limits in order to maintain osmotic balance between blood and interstitial (tissue) fluid. If the concentration of plasma proteins drops too low, as can happen as a result of liver disease such as cirrhosis, fluid accumulates in the tissues, a condition called *edema*.

Regulation of Blood Glucose Concentration

The neurons in the brain obtain their energy primarily from the aerobic respiration of glucose obtained from the blood plasma. It is therefore extremely important that the blood glucose concentration not fall too low, as might happen during fasting or prolonged exercise. It is also important that the blood glucose concentration not stay at too high a level, as it does in people with uncorrected *diabetes mellitus*, because this can lead to tissue damage.

After a carbohydrate-rich meal, the liver and skeletal muscles remove excess glucose from the blood and store it as the polysaccharide glycogen. This process is stimulated by the hormone insulin, secreted by the β- *(beta) cells* in the islets of Langerhans of the pancreas. When blood glucose levels decrease, as they do between meals, during periods of fasting, and during exercise, the liver secretes glucose into the blood. This glucose is obtained in part from the breakdown of liver glycogen to glucose-6-phosphate, a process called glycogenolysis. The phosphate group is then removed, and free glucose is secreted into the blood. Skeletal muscles lack the enzyme needed to remove the phosphate group, and so, even though they have glycogen stores, they cannot secrete glucose into the blood. The breakdown of liver glycogen is stimulated by another hormone, glucagon, which is secreted by the α- *(alpha) cells* of the islets of Langerhans in the pancreas (figure 48.21).

If fasting or exercise continues, the liver begins to convert other molecules, such as amino acids and lactic acid, into glucose. We mentioned this process, called **gluconeogenesis** ("new formation of glucose") in relation to exercise in chapter 47. The amino acids used for gluconeogenesis are obtained from muscle protein, which explains the severe muscle wasting that occurs during prolonged fasting.

The pancreas secretes digestive enzymes and bicarbonate into the pancreatic duct. The liver produces bile, which is stored and concentrated in the gallbladder. The liver and the pancreatic hormones regulate blood glucose concentration.

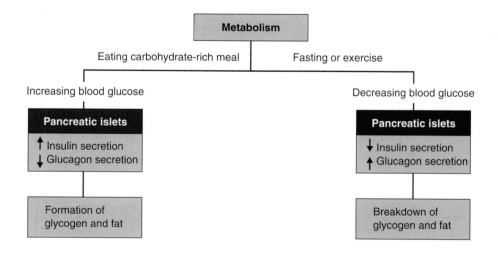

FIGURE 48.21
The actions of insulin and glucagon. An increased secretion of insulin after a meal promotes the deposition of glycogen and fat; a decreased insulin and increased glucagon secretion during fasting or exercising promotes the breakdown (through hydrolysis reactions) of glycogen and fat.

Neural and Hormonal Regulation of Digestion

The activities of the gastrointestinal tract are coordinated by the nervous system and the endocrine system. The nervous system, for example, stimulates salivary and gastric secretions in response to the sight and smell of food. When food arrives in the stomach, proteins in the food stimulate the secretion of a stomach hormone called gastrin (table 48.2), which in turn stimulates the secretion of pepsinogen and HCl from the gastric glands (figure 48.22). The secreted HCl then lowers the pH of the gastric juice, which acts to inhibit further secretion of gastrin. Since inhibition of gastrin secretion will reduce the amount of HCl released into the gastric juice, a negative feedback loop is completed. In this way, the secretion of gastric acid is kept under tight control.

The passage of chyme from the stomach into the duodenum inhibits the contractions of the stomach, so that no additional chyme can enter the duodenum until the previous amount has been processed. This inhibition is mediated by a neural reflex and by a hormone secreted by the small intestine that inhibits gastric emptying. The hormone is known generically as an enterogastrone (*entero* refers to the intestine; *gastro* to the stomach). The chemical identity of the enterogastrone is currently controversial. A hormone known as gastric inhibitory peptide (GIP), released by the duodenum, was named for this function but may not be the only, or even the major, enterogastrone. The secretion of enterogastrone is stimulated most strongly by the presence of fat in the chyme. Fatty meals therefore remain in the stomach longer than meals low in fat.

The duodenum secretes two additional hormones. Cholecystokinin (CCK), like enterogastrone, is secreted in response to the presence of fat in the chyme. CCK stimulates the contractions of the gallbladder, injecting bile into the duodenum so that fat can be emulsified and more efficiently digested. The other duodenal hormone is secretin. Released in response to the acidity of the chyme that arrives in the duodenum, secretin stimulates the pancreas to release bicarbonate, which then neutralizes some of the acidity. Secretin has the distinction of being the first hormone ever discovered.

Neural and hormonal reflexes regulate the activity of the digestive system. The stomach's secretions are regulated by food and by the hormone gastrin. Other hormones, secreted by the duodenum, inhibit stomach emptying and promote the release of bile from the gallbladder and the secretion of bicarbonate in pancreatic juice.

Table 48.2	Hormones of Digestion				
Hormone	**Class**	**Source**	**Stimulus**	**Action**	**Note**
Gastrin	Polypeptide	Pyloric portion of stomach	Entry of food into stomach	Stimulates secretion of HCl and pepsinogen by stomach	Unusual in that it acts on same organ that secretes it
Cholecystokinin	Polypeptide	Duodenum	Fatty chyme in duodenum	Stimulates gallbladder contraction and secretion of digestive enzymes by pancreas	Structurally similar to gastrin
Gastric inhibitory peptide	Polypeptide	Duodenum	Fatty chyme in duodenum	Inhibits stomach emptying	Also stimulates insulin secretion
Secretin	Polypeptide	Duodenum	Acidic chyme in duodenum	Stimulates secretion of bicarbonate by pancreas	The first hormone to be discovered (1902)

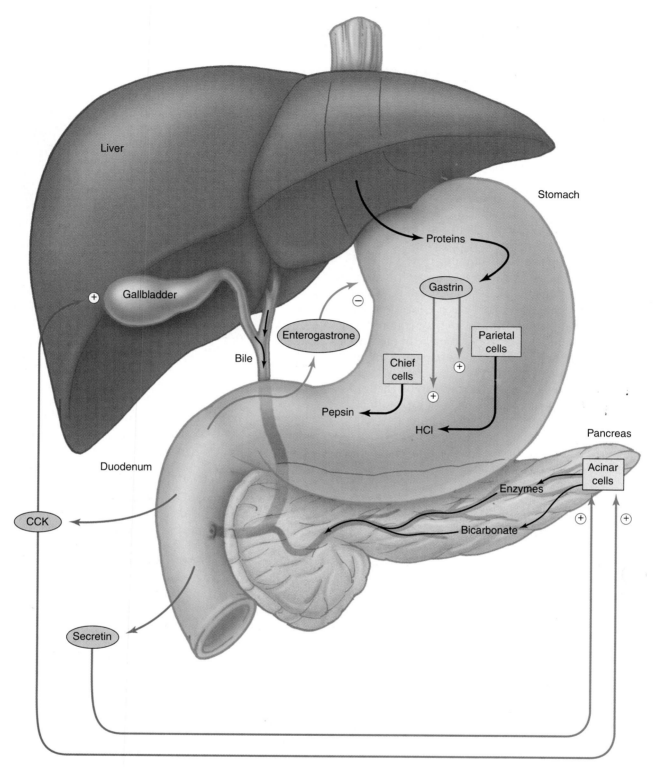

FIGURE 48.22
Hormonal control of the gastrointestinal tract. Gastrin is secreted by the mucosa of the stomach and stimulates the secretion of pepsinogen (which is converted into pepsin) and HCl. The duodenum secretes three hormones: cholecystokinin (CCK), which stimulates contraction of the gallbladder and secretion of pancreatic enzymes; secretin, which stimulates secretion of pancreatic bicarbonate; and an enterogastrone, which inhibits stomach emptying.

Food Energy and Energy Expenditure

The ingestion of food serves two primary functions: it provides a source of energy, and it provides raw materials the animal is unable to manufacture for itself. Even an animal that is completely at rest requires energy to support its metabolism; the minimum rate of energy consumption under defined resting conditions is called the basal metabolic rate (BMR). The BMR is relatively constant for a given individual, depending primarily on the person's age, sex, and body size.

Exercise raises the metabolic rate above the basal levels, so the amount of energy that the body consumes per day is determined not only by the BMR but also by the level of physical activity. If food energy taken in is greater than the energy consumed per day, the excess energy will be stored in glycogen and fat. Since glycogen reserves are limited, however, continued ingestion of excess food energy results primarily in the accumulation of fat. The intake of food energy, measured in Calories (where 1 Calorie with a capital C = 1 kilocalorie), can be altered by the choice of diet, and the amount of energy expended in exercise can be changed by the choice of lifestyle. The daily energy expenditures (metabolic rates) of people vary between 1300 and 5000 kilocalories per day, depending on the person's BMR and level of physical activity. If the food Calories ingested exceed the metabolic rate for a sustained period, the person will accumulate an amount of fat that is deleterious to health, a condition called **obesity.**

In adults, obesity is due mainly to an increase in adipose cell size, although the number of cells may also increase in extreme weight gains. When weight is lost, the size of the adipose cells decreases, but their number does not. It is therefore easy for formerly obese individuals to gain weight again. In the United States, about 30% of middle-aged women and 15% of middle-aged men are classified as obese, which means they weigh at least 20% more than the average weight for their height.

Regulation of Food Intake

Scientists have for years suspected that adipose tissue secretes a hormonal *satiety factor* (a circulating chemical that decreases appetite), because genetically obese mice lose weight when their circulatory systems are surgically joined with those of normal mice. Apparently, some weight-loss hormone was passing into the obese mice! The satiety factor secreted by adipose tissue has recently been identified. It is the product of a gene first observed in a strain of mice known as *ob/ob* (*ob* stands for "obese"; the double symbols indicate that the mice are homozygous for this gene—they inherit it from both parents). The *ob* gene has been cloned in mice, and more recently in humans, and has been found to be expressed (that is, to produce mRNA) only in

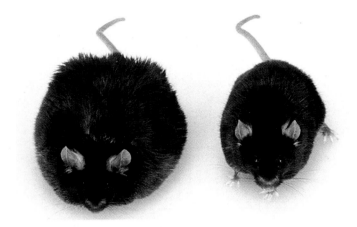

FIGURE 48.23
Injection of the hormone leptin causes genetically obese mice to lose weight. These two mice are identical twins, both members of a mutant strain of obese mice. The mouse on the right has been injected with the hormone leptin. It lost 30% of its body weight in just two weeks, with no apparent side effects.

adipocytes. The protein product of this gene, the presumed satiety factor, is called leptin. The *ob* mice produce a mutated and ineffective form of leptin, and it is this defect that causes their obesity. When injected with normal leptin, they stop eating and lose weight (figure 48.23).

More recent studies in humans show that the activity of the *ob* gene and the blood concentrations of leptin are actually higher in obese than in lean people, and that the leptin produced by obese people appears to be normal. It has therefore been suggested that most cases of human obesity may result from a reduced sensitivity to the actions of leptin in the brain rather than from reduced leptin production by adipose cells. Aggressive research is ongoing, as might be expected from the possible medical and commercial applications of these findings.

In the United States, serious eating disorders have become much more common since the mid-1970s. The most common of these disorders are *anorexia nervosa*, a condition in which the afflicted individuals literally starve themselves, and *bulimia*, in which individuals gorge themselves and then vomit, so that their weight stays constant. For reasons that we do not understand, 90 to 95% of those suffering from these disorders are female; researchers estimate that 2 to 5% of the adolescent girls and young women in the United States have eating disorders.

> The amount of caloric energy expended by the body depends on the basal metabolic rate and the additional calories consumed by exercise. Obesity results if the ingested food energy exceeds the energy expenditure by the body over a prolonged period.

Essential Nutrients

Over the course of their evolution, many animals have lost the ability to synthesize specific substances that nevertheless continue to play critical roles in their metabolism. Substances that an animal cannot manufacture for itself but which are necessary for its health must be obtained in the diet and are referred to as **essential nutrients.**

Included among the essential nutrients are *vitamins*, certain organic substances required in trace amounts. Humans, apes, monkeys, and guinea pigs, for example, have lost the ability to synthesize ascorbic acid (vitamin C). If vitamin C is not supplied in sufficient quantities in their diets, they will develop scurvy, a potentially fatal disease characterized by weakness, spongy gums, and bleeding of the skin and mucous membranes. Vitamin K, as we noted earlier, is obtained by mammals from their symbiotic intestinal bacteria, but must be consumed by birds with their food. Humans require at least 13 different vitamins (table 48.3).

Some essential nutrients are required in more than trace amounts. Many vertebrates, for example, are unable to synthesize 1 or more of the 20 amino acids used in making proteins. These *essential amino acids* must be obtained from proteins in the food they eat (figure 48.24). There are nine essential amino acids for humans: lysine, tryptophan, threonine, methionine, phenylalanine, leucine, isoleucine, valine, and (in children) histidine. People who are vegetarians must choose their foods so that the essential amino acids in one food complement those in another. Corn (with tryptophan, methionine, isoleucine, and lysine) and beans (with valine, threonine, phenylalanine, and leucine), for example, supply all of the essential amino acids when eaten together.

In addition, all vertebrates have lost the ability to synthesize certain unsaturated fatty acids and therefore must obtain them in food. On the other hand, some essential nutrients that vertebrates can synthesize cannot be manufactured by the members of other animal groups. For example, vertebrates can synthesize cholesterol, a key component of steroid hormones, but some carnivorous insects cannot.

Food also supplies **essential minerals** such as calcium, phosphorus, and other inorganic substances, including a wide variety of *trace elements*, which are required in very small amounts (see table 2.1). Among the trace elements are iodine (a component of thyroid hormone), cobalt (a component of vitamin B$_{12}$), zinc and molybdenum (compo-

nents of enzymes), manganese, and selenium. Animals obtain trace elements either directly from plants or from animals that have eaten plants.

The body requires vitamins and minerals obtained in food. Also, food must provide particular essential amino acids and fatty acids that the body cannot manufacture by itself.

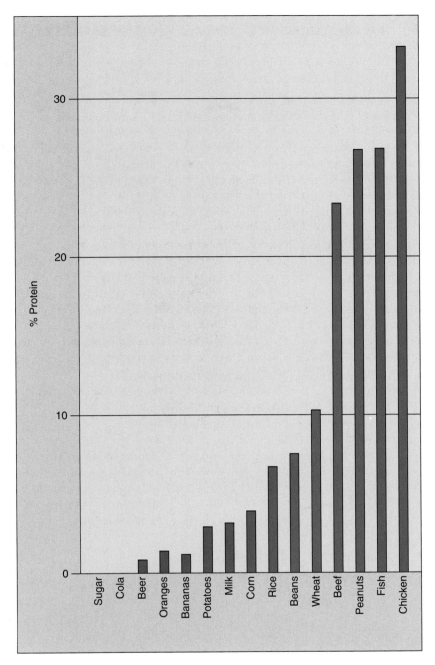

FIGURE 48.24
The protein content of common foods. The percent of protein in each of the foods is given as a proportion of total weight.

Table 48.3 Major Vitamins

Vitamin	Function	Source	Recommended Daily Allowance (milligrams)	Deficiency Symptoms	Solubility
Vitamin A (retinol)	Used in making visual pigments, maintenance of epithelial tissues	Green vegetables, milk products, liver	1	Night blindness, flaky skin	Fat
B-complex vitamins					
B$_1$	Coenzyme in CO_2 removal during cellular respiration	Meat, grains, legumes	1.5	Beriberi, weakening of heart, edema	Water
B$_2$ (riboflavin)	Part of coenzymes FAD and FMN, which play metabolic roles	Many different kinds of foods	1.8	Inflammation and breakdown of skin, eye irritation	Water
B$_3$ (niacin)	Part of coenzymes NAD$^+$ and NADP$^+$	Liver, lean meats, grains	20	Pellagra, inflammation of nerves, mental disorders	Water
B$_5$ (pantothenic acid)	Part of coenzyme-A, a key connection between carbohydrate and fat metabolism	Many different kinds of foods	5–10	Rare: fatigue, loss of coordination	Water
B$_6$ (pyridoxine)	Coenzyme in many phases of amino acid metabolism	Cereals, vegetables, meats	2	Anemia, convulsions, irritability	Water
B$_{12}$ (cyano-cobalamin)	Coenzyme in the production of nucleic acids	Red meats, dairy products	0.003	Pernicious anemia	Water
Biotin	Coenzyme in fat synthesis and amino acid metabolism	Meat, vegetables	Minute	Rare: depression, nausea	Water
Folic acid	Coenzyme in amino acid and nucleic acid metabolism	Green vegetables	0.4	Anemia, diarrhea	Water
Vitamin C	Important in forming collagen, cement of bone, teeth, connective tissue of blood vessels; may help maintain resistance to infection	Fruit, green leafy vegetables	45	Scurvy, breakdown of skin, blood vessels	Water
Vitamin D (calciferol)	Increases absorption of calcium and promotes bone formation	Dairy products, cod liver oil	0.01	Rickets, bone deformities	Fat
Vitamin E (tocopherol)	Protects fatty acids and cell membranes from oxidation	Margarine, seeds, green leafy vegetables	15	Rare	Fat
Vitamin K	Essential to blood clotting	Green leafy vegetables	0.03	Severe bleeding	Fat

48.1 Animals employ a digestive system to prepare food for assimilation by cells.

- The digestive system of vertebrates consists of a gastrointestinal tract and accessory digestive organs.
- Different regions of the digestive tract display specializations of structure and function; the entire tubular tract, however, consists of the same tissue layers of mucosa, submucosa, muscularis, and serosa.

48.2 Food is ingested, swallowed, and transported to the stomach.

- The teeth of carnivores are different from those of herbivores, and omnivores such as humans have dental characteristics of both.
- The esophagus contracts in peristaltic waves to drive the swallowed food to the stomach.
- Parietal cells of the gastric mucosa secrete hydrochloric acid, which activates pepsin, an enzyme that promotes the partial hydrolysis of ingested proteins.

48.3 The small and large intestines have very different functions.

- The duodenum receives pancreatic juice and bile, which help digest the chyme that arrives from the stomach through the pyloric valve.
- Digestive enzymes called brush border enzymes are located in the microvilli of the small intestine and help finish the chemical breakdown of food into small molecules that can be absorbed by the small intestine.
- Glucose and amino acids are absorbed by the small intestine and carried to the liver in the hepatic portal vein; absorbed fat travels through the lymphatic system before eventually emptying into veins.

- The large intestine absorbs water and ions, as well as certain organic molecules such as vitamin K; the remaining material passes out of the anus.
- Ruminants have a divided stomach which includes the rumen, a large fermentation chamber containing microorganisms that digest cellulose.

48.4 Accessory organs, neural stimulation, and endocrine secretions assist in digestion.

- Pancreatic juice contains bicarbonate to neutralize the acid chyme from the stomach that arrives in the duodenum; it also has a variety of digestive enzymes.
- Bile contains bile pigment and bile salts, which emulsify fat; bile is produced by the liver and stored in the gallbladder, so that it can be delivered to the duodenum.
- The liver metabolizes toxins and hormones that are delivered to it in the hepatic portal vein; the liver also helps to regulate the blood glucose concentration and produces most of the proteins in the blood plasma.
- The stomach secretes the hormone gastrin, and the small intestine secretes various hormones that help to regulate the digestive system.

48.5 All animals require food energy and essential nutrients.

- The basal metabolic rate (BMR) is the lowest level of energy consumption of the body; metabolism is raised by exercise and other activities.
- Food energy, measured as Calories, as well as vitamins, minerals, and the essential amino acids and fatty acids must be supplied in the diet.
- Hunger is controlled by a part of the brain called the hypothalamus and is regulated in part by the hormone leptin secreted by the adipose tissue.

1. **Omnivore** An omnivore is a vertebrate capable of eating both plant and animal foods, as compared to more specialized animals that eat only meat (carnivores) or plants (herbivores).

2. **Absorption** After food molecules are digested into their smallest component parts, these parts are absorbed or transported through the lining of the digestive tract so they can enter the body.

3. **Cloaca** Most vertebrates other than mammals have a cloaca, a common chamber into which the large intestine, urinary system, and reproductive system empty.

4. **Cecum** The cecum, a blind pouch at the beginning of the large intestine, is the site of cellulose digestion in some mammals, such as rabbits and horses.

5. **Mucosa** The mucosa is the innermost layer of the digestive tract, consisting of a lining epithelium, a thin layer of connective tissue, and a very thin border of smooth muscle.

6. **Rumen** Mammals that chew their cud are called ruminants because they have a rumen where microorganisms digest cellulose; they regurgitate this material so that they can rechew it and then swallow it again.

Review Questions

1. What are the layers that make up the wall of the vertebrate gastrointestinal tract? What type of tissue is found in each layer?

2. How does tooth structure vary among carnivores, herbivores, and omnivores? What are the four types of teeth present in adult humans, and what are their functions? What class of vertebrates lacks teeth, and what do they use instead?

3. What is the mechanism by which food is moved along the esophagus? What normally prevents regurgitation in humans, and why can't horses regurgitate?

4. What inorganic substance is secreted by parietal cells? What substance is secreted by chief cells? How do these two substances interact to promote digestion of proteins in the stomach?

5. How are the products of protein and carbohydrate digestion absorbed across the intestinal wall, and where do they go after they are absorbed? How are the products of fat digestion absorbed across the intestinal wall, and where do they go after they are absorbed?

6. What are the main exocrine secretions of the pancreas, and what are their functions? What are the main endocrine secretions of the pancreas, and what are their functions?

7. Where are bile pigment and bile salts produced and where are they stored? From what source is bile pigment derived? What is the function of bile salts in digestion?

8. Describe the role of each of the following hormones in the regulation of digestion: gastrin, enterogastrone, cholecystokinin, and secretin. What is the major stimulus for the secretion of each hormone?

9. What anatomical and behavioral specializations do ruminants have for making use of microorganisms? What anatomical and behavioral specializations do rodents and rabbits have for making use of microorganisms?

10. What is a vitamin? What is the difference between an essential amino acid and any other amino acid?

Thought Questions

1. Many birds possess crops, although few mammals do. Suggest a reason for this difference between birds and mammals.

2. Humans obtain vitamin K from symbiotic bacteria living in their gastrointestinal tract. Many bacteria also produce vitamin C

(ascorbic acid). Why do you think humans have not evolved a symbiotic relationship with bacteria that would result in humans obtaining bacterial vitamin C?

Internet Links

How to Get an Ulcer
http://www.unsw.edu.au/clients/microbiology/learn.html
From the University of New South Wales in Australia, a homepage for Helicobacter pylori *that describes its role in peptic ulcers and provides many links to other sources of information.*

An Interactive Food Pyramid
http://www.ganesa.com/food/foodpyramid.html
The FOOD PYRAMID GUIDE allows you to click on items in each food group to jump to a discussion of its caloric content and proper role in your diet.

Food Bytes
http://ificinfo.health.org/
The INTERNATION FOOD INFORMATION COUNCIL SITE provides information for educators and reporters about food safety and nutrition aspects of late-breaking news.

Learning About Diabetes
http://diabetes.sciweb.com/
The DIABETES INFORMATION CENTER provides a collection of basic resources for learning about diabetes, with a real-time discussion board and many links.

Shape Up America
http://www.shapeup.org/sua/general/whatis.htm
Part of a national initiative founded by former Surgeon General C. Everett Koop, this useful site helps you determine your body mass index and learn healthy ways to maintain your ideal weight.

How a Cow Digests Grass
http://hammock.ifas.vfl.edu/txt/fairs/2842
This interesting site, from the Florida State Agricultural Extension Service, explains how the ruminant digestive system works, using a dairy cow.

For Further Reading

Achord, J. L. "Alcohol and the Liver," *Scientific American Science and Medicine*, March/April, 1995, pages 16–25. Before excessive alcohol consumption produces cirrhosis of the liver, it affects the liver in.more subtle ways.

Blaser, M. J.: "The Bacteria Behind Ulcers," *Scientific American*, February 1996, page 104. This article presents a clear description of the association between gastrointestinal bacteria and peptic ulcers.

Brown, L. and E. Pollitt: "Malnutrition, Poverty, and Intellectual Development," *Scientific American*, February 1996, pages 38–43. A poor diet influences mental development in many unexpected ways.

Cohen, L. A.: "Diet and Cancer," *Scientific American*, vol. 257, November, 1987, page 42. Summary of evidence for an association between high-fat, low-fiber diets and certain cancers.

Gibbs, W.: "More Fun than a Root Canal," *Scientific American*, vol. 269, November 1993, page 106. Well, what isn't? A new synthetic molecule may stimulate regrowth of dentin in infected teeth.

Service, R.: "Stalking the Start of Colon Cancer," *Science*, vol. 263, March 1994, pages 1559–60. Colon cancer is one of the deadliest varieties there is—and it's a disease that's strongly heritable.

49

Circulation

Concept Outline

49.1 The circulatory systems of animals may be open or closed.

Open and Closed Circulatory Systems. All vertebrates have a closed circulation, while many smaller animals have open circulatory systems.

49.2 A network of vessels transports blood through the body.

The Blood Plasma. The blood plasma transports a variety of solutes, including ions, metabolites, proteins, and hormones.

The Blood Cells. The blood cells include erythrocytes, which transport oxygen, leukocytes, which provide defenses for the body, and platelets, which function in blood clotting.

Characteristics of Blood Vessels. Blood leaves the heart in arteries and returns in veins; in between, the blood passes through capillaries, where all exchanges with tissues occur.

The Lymphatic System. The lymphatic system returns interstitial fluid to veins.

49.3 The vertebrate heart has undergone progressive evolutionary change.

The Fish Heart. The fish heart consists of a row of four chambers that receive blood from the body at one end and pump blood to the gills at the other end.

Amphibian and Reptile Circulation. Land vertebrates have a double circulation, where blood from the lungs returns to the heart to be pumped to the rest of the body.

Mammalian and Bird Hearts. Mammals and birds have a complete separation between the two sides of the heart.

49.4 The cardiac cycle drives the cardiovascular system.

The Cardiac Cycle. The right and left sides of the heart rest and receive blood at the same time, then pump the blood into arteries at the same time.

Electrical Excitation and Contraction of the Heart. The impulse begins in one area of the heart and is conducted to the rest of the heart.

Blood Flow and Blood Pressure. Blood flow and blood pressure depend on the diameter of the arterial vessels and on the amount of blood pumped by the heart.

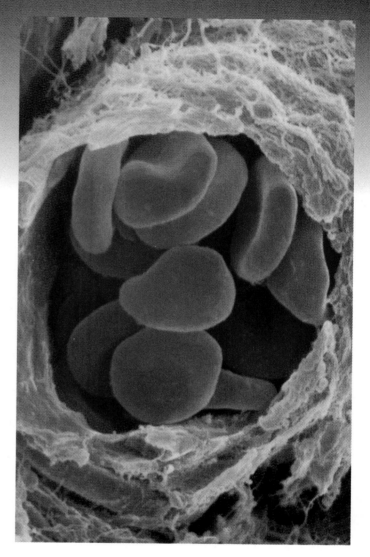

FIGURE 49.1
Red blood cells. This ruptured blood vessel, seen in a scanning electron micrograph, is full of red blood cells, which move through vessels transporting oxygen from one place to another in the body.

Every cell in the animal body must acquire the energy it needs for living from other molecules. Like residents of a city whose food is imported from farms in the countryside, cells in the body need trucks to carry the food, highways for the trucks to travel on, and a way to cook the food when it arrives. In animals, the organ system that provides the trucks and highways, or blood and blood vessels (figure 49.1), is called the circulatory system and is discussed in this chapter. The organ system that provides the fuel (oxygen) to process the food (glucose) is called the respiratory system and will be discussed in the following chapter.

Open and Closed Circulatory Systems

Among the unicellular protists, oxygen and nutrients are obtained directly by simple diffusion from the aqueous external environment. Cnidaria, such as *Hydra*, and flatworms, such as *Planaria*, have cells that are directly exposed to either the external environment or to the gastrovascular cavity (figure 49.2*a*). The gastrovascular cavity of *Hydra* (see chapter 48) extends even into the tentacles, and that of *Planaria* branches extensively to supply every cell with oxygen and the nourishment obtained by digestion. Larger animals, however, have tissues that are several cell layers thick, so that many cells are too far away from the body surface or digestive cavity to exchange materials directly with the environment. Instead, oxygen and nutrients are transported from the environment and digestive cavity to the body cells by an internal fluid within a *circulatory system.*

There are two main types of circulatory systems: *open* or *closed*. In an **open circulatory system,** such as that found in mollusks and arthropods (figure 49.2*b*), there is no distinction between the circulating fluid (blood) and the extracellular fluid of the body tissues (interstitial fluid or lymph). This fluid is thus called **hemolymph.** Insects have a muscular tube that serves as a heart to pump the hemolymph through a network of channels and cavities in the body. The fluid then drains back into the central cavity.

In a **closed circulatory system,** the circulating fluid, or blood, is always enclosed within blood vessels that transport blood away from and back to a pump, the **heart.** Annelids (see chapter 42) and all vertebrates have a closed circulatory system. In annelids such as an earthworm, for example, a dorsal artery contracts rhythmically to function as a pump. Blood is pumped through five small connecting arteries which also function as pumps, to a ventral artery, which transports the blood posteriorly until it eventually reenters the dorsal artery. Smaller vessels branch from each artery to supply the tissues of the earthworm with oxygen and nutrients and to transport waste products (figure 49.2*c*).

Blood vessels form a tubular network that permits blood to flow from the heart to all the cells of the body and then back to the heart. *Arteries* carry blood away from the heart, whereas *veins* return blood to the heart. Blood passes from the arterial to the venous system in *capillaries*, which are the thinnest and most numerous of the blood vessels.

As blood plasma passes through capillaries, the pressure of the blood forces some of this fluid out of the capillary walls. Fluid derived from plasma that passes out of capillary walls into the surrounding tissues is called **interstitial fluid.** Some of this fluid returns directly to capillaries, and some enters into **lymph vessels,** located in the connective tissues around the blood vessels. This fluid, now called *lymph*, is returned to the venous blood at specific sites. The lymphatic system is considered a part of the circulatory system and is discussed later in this chapter.

The Functions of Vertebrate Circulatory Systems

The vertebrate circulatory system has three principal functions: transportation, regulation, and protection.

1. **Transportation.** All of the substances essential for cellular metabolism are transported by the circulatory system. These substances can be categorized as follows:
   a. Respiratory. Red blood cells, or *erythrocytes*, transport oxygen to the tissue cells. In the capillaries of the lungs or gills, oxygen attaches to hemoglobin molecules within the erythrocytes and is transported to the cells for aerobic respiration. Carbon dioxide produced by cell respiration is carried by the blood to the lungs or gills for elimination.

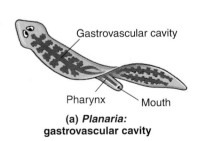

(a) *Planaria:*
gastrovascular cavity

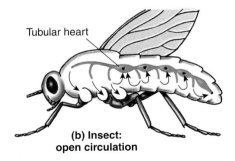

(b) Insect:
open circulation

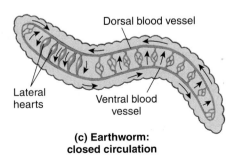

(c) Earthworm:
closed circulation

FIGURE 49.2
Circulatory systems of the animal kingdom. (*a*) The gastrovascular cavity of *Planaria* serves as both a digestive and circulatory system, delivering nutrients directly to the tissue cells by diffusion from the digestive cavity. (b) In the open circulation of an insect, hemolymph is pumped from a tubular heart into cavities in the insect's body; the hemolymph then returns to the blood vessels so that it can be recirculated. (c) In the closed circulation of the earthworm, blood pumped from the hearts remains within a system of vessels that returns it to the hearts. All vertebrates also have closed circulatory systems.

b. *Nutritive.* The digestive system is responsible for the break-down of food so that nutrients can be absorbed through the intestinal wall and into the blood vessels of the circulatory system. The blood then carries these absorbed products of digestion through the liver and to the cells of the body.

c. *Excretory.* Metabolic wastes, excessive water and ions, and other molecules in the fluid portion of blood are filtered through the capillaries of the kidneys and excreted in urine.

2. **Regulation.** The cardiovascular system transports hormones and participates in temperature regulation.

a. *Hormone transport.* The blood carries hormones from the endocrine glands, where they are secreted, to the distant target organs they regulate.

b. *Temperature regulation.* In warm-blooded vertebrates, or **homeotherms,** a constant body temperature is maintained regardless of the ambient temperature. This is accomplished in part by blood vessels located just under the epidermis. When the ambient temperature is cold, the superficial vessels constrict to divert the warm blood to deeper vessels. When the ambient temperature is warm, the superficial vessels dilate so that the warmth of the blood can be lost by radiation (figure 49.3).

Some vertebrates also retain heat in a cold environment by using a **countercurrent heat exchange.** In this process, a vessel carrying warm blood from deep within the body passes next to a vessel carrying cold blood from the surface of the body (figure 49.4). The warm blood going out heats the cold blood returning from the body surface, so that this blood is no longer cold when it reaches the interior of the body.

3. **Protection.** The circulatory system protects against injury and foreign microbes or toxins introduced into the body.

a. *Blood clotting.* The clotting mechanism protects against blood loss when vessels are damaged. This clotting mechanism involves both proteins from the blood plasma and cell structures called platelets (discussed in the next section).

b. *Immune defense.* The blood contains white blood cells, or leukocytes, that provide immunity against many disease-causing agents. Some white blood cells are phagocytic, some produce antibodies, and some act by other mechanisms to protect the body.

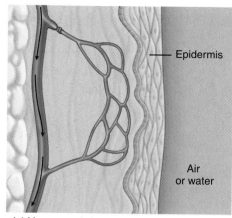 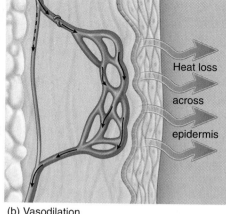

(a) Vasoconstriction (b) Vasodilation

FIGURE 49.3
Regulation of heat loss. The amount of heat lost at the body's surface can be regulated by controlling the flow of blood to the surface. (a) Constriction of the surface blood vessels limits flow and heat loss; (b) dilation of these vessels increases flow and heat loss.

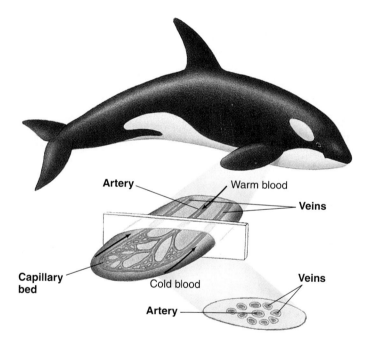

FIGURE 49.4
Countercurrent heat exchange. Many marine animals, such as this killer whale, limit heat loss in cold water by countercurrent flow. The warm blood pumped from within the body in arteries warms the cold blood returning from the skin in veins, so that the core body temperature can remain constant in cold water.

Circulatory systems may be open or closed. All vertebrates have a closed circulatory system, in which blood circulates away from the heart in arteries and back to the heart in veins. The circulatory system serves a variety of functions, including transportation, regulation, and protection.

The Blood Plasma

Blood is composed of a fluid **plasma** and several different kinds of cells that circulate within that fluid (figure 49.5). Blood platelets, although included in figure 49.5, are not complete cells; rather, they are fragments of cells that reside in the bone marrow. Blood plasma is the extracellular environment of the blood cells and platelets, and is used to produce the other fluids of the body that provide the extracellular environment of most body cells.

Plasma contains the following solutes:

1. **Metabolites, wastes, and hormones.** Dissolved within the plasma are all of the metabolites used by cells, including glucose, amino acids, and vitamins. Also dissolved in the plasma are wastes such as nitrogen compounds and CO_2 produced by metabolizing cells, and the hormones that regulate cellular activities.

2. **Ions.** Like the water of the seas in which life arose, blood plasma is a dilute salt solution. The chief plasma ions are sodium, chloride, and bicarbonate ions. In addition, there are trace amounts of other ions such as calcium, magnesium, copper, potassium, and zinc. The composition of the plasma, therefore, is similar to seawater, but plasma has a lower total ion concentration than that of present-day seawater.

3. **Proteins.** The liver produces most of the plasma proteins, including **albumin,** which comprises most of the plasma protein; the alpha (α) and beta (β) **globulins,** which serve as carriers of lipids and steroid hormones; and *fibrinogen*, which is required for blood clotting. When blood in a test tube clots, the fibrinogen is converted into insoluble threads of *fibrin* that become part of the clot. The fluid that's left, which lacks fibrinogen and so cannot clot, is called **serum.**

Plasma, the liquid portion of the blood, contains different types of proteins, ions, metabolites, wastes, and hormones. This liquid, and fluids derived from it, provide the extracellular environment of most of the cells of the body.

Blood cell	Life span in blood	Function
 Erythrocyte	120 days	O_2 and CO_2 transport
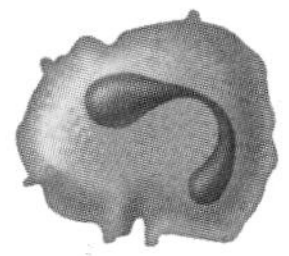 Neutrophil	7 hours	Immune defenses
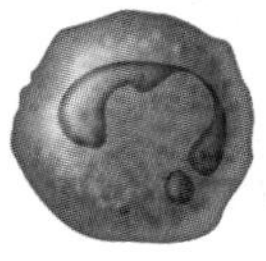 Eosinophil	Unknown	Defense against parasites
 Basophil	Unknown	Inflammatory response
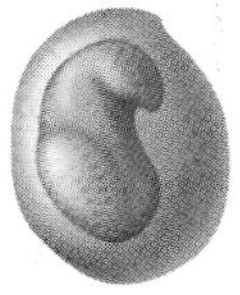 Monocyte	3 days	Immune surveillance (precursor of tissue macrophage)
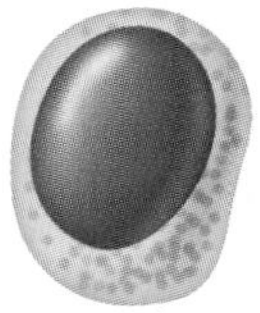 B - lymphocyte	Unknown	Antibody production (precursor of plasma cells)
 T - lymphocyte	Unknown	Cellular immune response
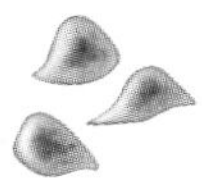 Platelets	7-8 days	Blood clotting

FIGURE 49.5
Types of blood cells. Erythrocytes are red blood cells, platelets are fragments of a bone marrow cell, and all the other cells are different types of leukocytes, or white blood cells.

The Blood Cells

The red blood cells function in oxygen transport, the white blood cells in immunological defenses, and the platelets in blood clotting.

Erythrocytes and Oxygen Transport

Each cubic millimeter of blood contains about 5 million **erythrocytes,** or **red blood cells.** The fraction of the total blood volume that is occupied by erythrocytes is called the blood's *hematocrit*; in humans, it is typically around 45%. A flat disc with a central depression, each erythrocyte resembles a doughnut with a hole that does not go all the way through. As we've already seen, the erythrocytes of vertebrates contain hemoglobin and function in oxygen transport. Unlike the situation in many invertebrates, the oxygen-carrying pigment (hemoglobin) of vertebrates is found exclusively in the erythrocytes; none is normally in the plasma.

In mammals, erythrocytes lose their nuclei and protein-synthesizing machinery when they mature. This is different from the mature erythrocytes of all other vertebrates, which remain nucleated. As mammalian erythrocytes age, they are removed from the blood by the phagocytic cells of the spleen, bone marrow, and liver. Balancing this loss, new erythrocytes are constantly being formed in the bone marrow. Each erythrocyte develops from an unspecialized cell, or *stem cell*, that produces hemoglobin and gradually loses its nucleus in a developmental process known as *erythropoiesis*. This process is stimulated by the hormone *erythropoietin*, which is secreted by the kidneys in response to a fall in the plasma oxygen concentration.

Leukocytes Defend the Body

Less than 1% of the cells in human blood are **leukocytes,** or **white blood cells;** there are only 1 or 2 leukocytes for every 1000 erythrocytes. Leukocytes are larger than erythrocytes and have nuclei. Furthermore, leukocytes are not confined to the blood as erythrocytes are, but can migrate out of capillaries into the interstitial (tissue) fluid.

There are several kinds of leukocytes (see figure 49.5), each of which plays a specific role in defending the body against invading microorganisms and other foreign substances. Since leukocytes lack hemoglobin, they are barely visible under a microscope unless they are stained. **Granular leukocytes** include **neutrophils, eosinophils,** and **basophils,** which are named according to the staining properties of granules in their cytoplasm. **Nongranular leukocytes** include **monocytes** and **lymphocytes.** Neutrophils are the most numerous of the leukocytes, followed in order by lymphocytes, monocytes, eosinophils, and basophils.

When tissue is damaged, neutrophils leave the blood capillaries and accumulate at the site of injury. They are soon joined by monocytes, which change into *macrophages*

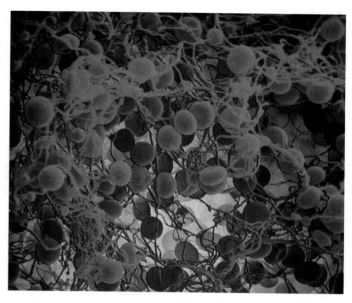

FIGURE 49.6
A scanning electron micrograph showing threads of fibrin (1430×). Fibrin is formed from a soluble protein, fibrinogen, in the plasma. This reaction is catalyzed by the enzyme thrombin, which is formed from an inactive enzyme called prothrombin. The activation of thrombin is the last step in a cascade of enzymatic reactions that produces a blood clot when a blood vessel is damaged.

within the loose connective tissues (see chapter 46). Phagocytosis by the neutrophils and macrophages helps to eliminate many disease-causing organisms that may have entered the body through the wound. The lymphocytes provide a more specific immune defense, as you'll learn in chapter 54. Eosinophils aid in defense against parasitic infections and play a role in allergic responses.

Platelets Help Blood to Clot

Certain large cells within the bone marrow, called *megakaryocytes*, regularly pinch off bits of their cytoplasm. These membrane-bound cell fragments, known as **platelets,** play an important role in blood clotting. When a blood vessel is broken, smooth muscle in the vessel walls contract, causing the vessel to constrict. Platelets then accumulate at the injured site and form a plug by sticking to each other and to the surrounding tissues. This plug is reinforced by threads of the protein **fibrin** (figure 49.6), which contract to form a tighter mass. The tightening plug of platelets, fibrin, and often trapped erythrocytes constitutes a blood clot.

Erythrocytes contain hemoglobin and serve in oxygen transport. The different types of leukocytes have specialized functions that serve to protect the body from invading pathogens, and the platelets participate in blood clotting.

Characteristics of Blood Vessels

Blood leaves the heart through vessels known as **arteries.** These continually branch, forming a hollow "tree" that enters each of the organs of the body. The finest, microscopically-sized branches of the arterial trees are the **arterioles.** Blood from the arterioles enters the **capillaries** (from the Latin *capillus,* "a hair"), an elaborate latticework of very narrow, thin-walled tubes. After traversing the capillaries, the blood is collected into the **venules;** the venules lead to larger vessels called **veins,** which carry the blood back to the heart.

Arteries, arterioles, veins, and venules all have the same basic structure (figure 49.7). The innermost layer is an epithelial sheet called the *endothelium.* Covering the endothelium is a thin layer of elastic fibers, a smooth muscle layer, and a connective tissue layer. The walls of these vessels are thus too thick to permit any exchange of materials between the blood and the tissues outside the vessels. The walls of capillaries, however, contain only the endothelium, so molecules and ions can leave the blood plasma by filtration through pores in the capillary walls and by transport through the endothelial cells. Therefore, it is while blood is in the capillaries that gases and metabolites are exchanged with the cells of the body.

Arteries and Arterioles

The larger arteries contain extra elastic fibers in their walls, allowing them to recoil each time they receive a volume of blood pumped by the heart. Smaller arteries and arterioles are less elastic, but their disproportionately thicker smooth muscle layer enables them to resist bursting.

The vast tree of arteries presents a frictional resistance to blood flow. The narrower the vessel, the greater the frictional resistance to flow. In fact, a vessel that is half the diameter of another has 16 times the frictional resistance! This is because the resistance to blood flow is inversely proportional to the fourth power of the radius of the vessel. Therefore, within the arterial tree, it is the small arteries and arterioles that provide the greatest resistance to blood flow. Contraction of the smooth muscle layer of the arterioles produces **vasoconstriction,** which greatly increases resistance and decreases flow. Relaxation of the smooth muscle layer causes **vasodilation,** decreasing resistance and increasing blood flow to an organ (see figure 49.3).

In addition, blood flow through some organs is regulated by rings of smooth muscle around arterioles near the region where they empty into capillaries. These **precapillary sphincters** (figure 49.8) can close off specific capillary beds completely. For example, the closure of precapillary sphincters in the skin contributes to the vasoconstriction that limits heat loss in cold environments.

Exchange in the Capillaries

Each time the heart contracts, it must produce sufficient pressure to pump blood against the resistance of the arterial tree and into the capillaries. The vast number and extensive branching of the capillaries ensure that *every cell in the body is within 100 μm of a capillary.* On the average, capillaries are about 1 mm long and 8 μm in diameter, only slightly larger than a red blood cell (5–7 μm in diameter). Despite the close fit, red blood cells can squeeze through capillaries without difficulty (figure 49.9).

Although each capillary is very narrow, there are so many of them that the capillaries have the greatest *total*

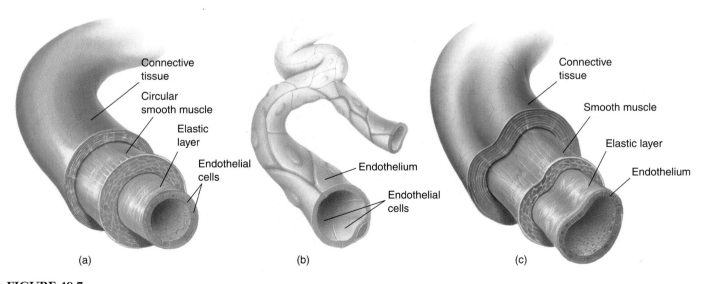

Connective tissue
Circular smooth muscle
Elastic layer
Endothelial cells

(a)

Endothelium
Endothelial cells

(b)

Connective tissue
Smooth muscle
Elastic layer
Endothelium

(c)

FIGURE 49.7
The structure of blood vessels. (a) Arteries and (c) veins have the same tissue layers. (b) Capillaries are composed of only a single layer of endothelial cells.

cross-sectional area of any other type of vessel. Consequently, the blood decreases in velocity as it passes through the capillary beds, allowing more time for it to exchange materials with the surrounding extracellular fluid. By the time the blood reaches the end of a capillary, it has released some of its oxygen and nutrients and picked up carbon dioxide and other waste products. Blood also loses most of its pressure in passing through the vast capillary networks, and so is under very low pressure when it enters the veins.

Venules and Veins

Blood flows from the venules to ever larger veins, and ultimately back to the heart. Venules and veins have the same tissue layers as arteries, but they have a thinner layer of smooth muscle. Less muscle is needed because the pressure in the veins is only about one-tenth that in the arteries. Most of the blood in the cardiovascular system is contained within veins, which can expand when needed to hold additional amounts of blood. You can see the expanded veins in your feet when you stand for a long time.

Since the blood pressure in the veins is so low, how does the blood return to the heart from the feet and legs? The venous pressure alone is not sufficient, but several sources provide help. Most significantly, skeletal muscles surrounding the veins can contract to move blood by squeezing the veins. **Venous valves** ensure that blood moves through the veins in only one direction, back to the heart. When a person's veins expand too much with blood, the venous valves may no longer work and the blood may pool in the veins. Veins in this condition are known as *varicose veins.*

Blood is pumped from the heart into the arterial system, which branches into fine arterioles. This blood is delivered into the thinnest and most numerous of vessels, the capillaries, where exchanges with the tissues occur. Blood returns to the heart through veins.

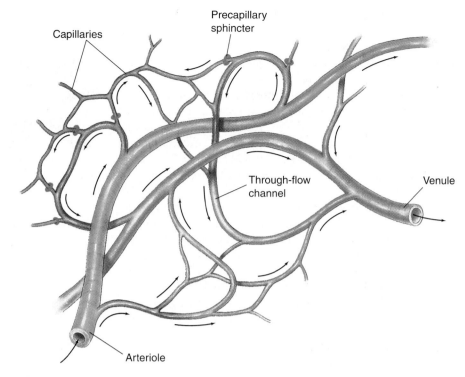

FIGURE 49.8
The capillary network connects arteries with veins. Most of the exchange between the blood and the extracellular fluid occurs while the blood is in the capillaries. Entrance to the capillaries is controlled by bands of muscle called precapillary sphincters at the entrance to each capillary. When a sphincter contracts, it closes off the capillary. By contracting these sphincters, the body can limit the amount of blood in the capillary network of a particular tissue, and thus control the rate of exchange in that tissue.

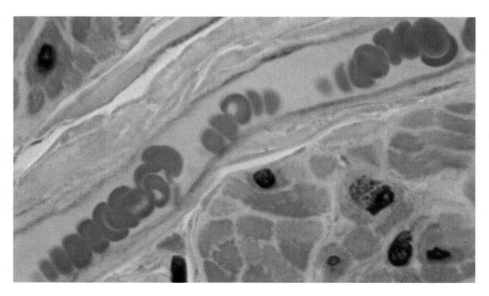

FIGURE 49.9
The red blood cells in this capillary are passing along in single file (500×). Many capillaries are even narrower than that shown here, which is in the bladder of a monkey. However, red blood cells can even pass through capillaries narrower than the cells' own diameters, pushed along by the pressure generated by the heart.

The Lymphatic System

The cardiovascular system is considered to be a closed system because all of its vessels are connected with one another—none are simply open-ended. However, some water and solutes in the blood plasma do filter through the walls of the capillaries to form the interstitial (tissue) fluid. This filtration is driven by the pressure of the blood, and it helps supply the tissue cells with oxygen and nutrients. Most of the fluid is filtered from the capillaries near their arteriolar ends, where the blood pressure is higher, and returned to the capillaries near their venular ends. This return of fluid occurs by osmosis, which is driven by a higher solute concentration within the capillaries. Since most of the plasma proteins cannot escape through the capillary pores because of their large size, the concentration of proteins in the plasma is greater than the protein concentration in the interstitial fluid. The difference in protein concentration produces an osmotic pressure, called the *oncotic pressure*, that causes osmosis of water into the capillaries (figure 49.10).

Since interstitial fluid is produced because of the blood pressure, high capillary blood pressure could cause too much interstitial fluid to be produced. A common example of this occurs in pregnant women, when the fetus compresses veins and thereby increases the capillary blood pressure in the mother's lower limbs. The increased interstitial fluid can cause swelling of the tissues, or *edema*, of the feet. Edema may also result if the plasma protein concentration (and thus the oncotic pressure) is too low, so that the osmotic return of fluid to the blood is inadequate. This may be caused by liver disease, because the liver produces most of the plasma proteins, or by protein malnutrition (*kwashiorkor*).

Even under normal conditions, the amount of fluid filtered out of the capillaries is greater than the amount that returns to the capillaries by osmosis. The remainder does eventually return to the cardiovascular system, however, by way of an *open* circulatory system called the **lymphatic system.** The lymphatic system consists of lymphatic capillaries, lymphatic vessels, lymph nodes, and lymphatic organs, including the spleen and thymus. Excess fluid in the tissues drains into blind-ended lymph capillaries with highly permeable walls. This fluid, now called **lymph,** passes into progressively larger lymphatic vessels, which resemble veins and have one-way valves (figure 49.11). The lymph eventually enters two major lymphatic vessels, which drain into veins on each side of the neck.

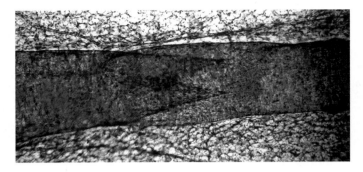

FIGURE 49.11
A lymphatic vessel valve (25×). Valves allow lymph to flow in one direction (from left to right in this figure) but not in the reverse direction.

Movement of lymph in mammals is accomplished by the squeezing action of skeletal muscles against the lymphatic vessels, a mechanism similar to the one that moves blood through veins. In some cases, the lymphatic vessels also contract rhythmically. In many fishes, all amphibians and reptiles, bird embryos, and some adult birds, movement of lymph is propelled by **lymph hearts.**

As the lymph moves through lymph nodes and lymphatic organs, it is modified by phagocytic cells that line the channels of those organs. In addition, the lymph nodes and lymphatic organs contain *germinal centers* for the production of lymphocytes, a type of white blood cell critically important in immunity.

> Lymphatic vessels carry excess interstitial fluid back to the vascular system. This fluid, called lymph, travels through lymph nodes and lymphatic organs where it encounters the immune cells called lymphocytes that are produced in these organs.

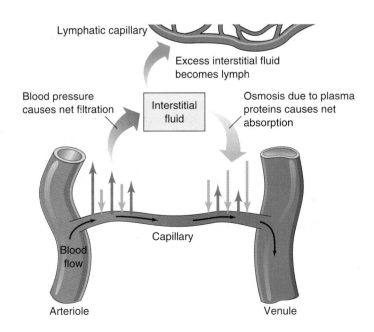

FIGURE 49.10
Plasma fluid, minus proteins, is filtered out of capillaries. This forms interstitial fluid, which bathes the tissues. Much of the interstitial fluid is returned to the capillaries by the osmotic pressure generated by the higher protein concentration in plasma. The excess interstitial fluid is drained into open-ended lymphatic capillaries, which ultimately return the fluid to the cardiovascular system.

The Fish Heart

The chordates that were ancestral to the vertebrates are thought to have had simple tubular hearts, similar to those now seen in lancelets (see chapter 45). The heart was little more than a specialized zone of the ventral artery, more heavily muscled than the rest of the arteries, which contracted in simple peristaltic waves. A pumping action results because the uncontracted portions of the vessel have a larger diameter than the contracted portion, and thus present less resistance to blood flow.

The development of gills by fishes required a more efficient pump, and in fishes we see the evolution of a true chamber-pump heart. The fish heart is, in essence, a tube with four chambers arrayed one after the other (figure 49.12*a*). The first two chambers—the **sinus venosus** and **atrium**—are collection chambers, while the second two, the **ventricle** and **conus arteriosus,** are pumping chambers.

As might be expected from the early chordate hearts from which the fish heart evolved, the sequence of the heartbeat in fishes is a peristaltic sequence, starting at the rear and moving to the front. The first of the four chambers to contract is the sinus venosus, followed by the atrium, the ventricle, and finally the conus arteriosus. Despite shifts in the relative positions of the chambers in the vertebrates that evolved later, this heartbeat sequence is maintained in all vertebrates. In fish, the electrical impulse that produces the contraction is initiated in the sinus venosus; in other vertebrates, the electrical impulse is initiated by their equivalent of the sinus venosus.

The fish heart is remarkably well suited to the gill respiratory apparatus and represents one of the major evolutionary innovations in the vertebrates. Perhaps its greatest advantage is that the blood it delivers to the tissues of the body is fully oxygenated. Blood is pumped first through the gills, where it becomes oxygenated; from the gills, it flows through a network of arteries to the rest of the body; then it returns to the heart through the veins (figures 49.12*b*). This arrangement has one great limitation, however. In passing through the capillaries in the gills, the blood loses much of the pressure developed by the contraction of the heart, so the circulation from the gills through the rest of the body is sluggish. This feature limits the rate of oxygen delivery to the rest of the body.

The fish heart is a modified tube, consisting of a series of four chambers. Blood first enters the heart at the sinus venosus, where the wave-like contraction of the heart begins.

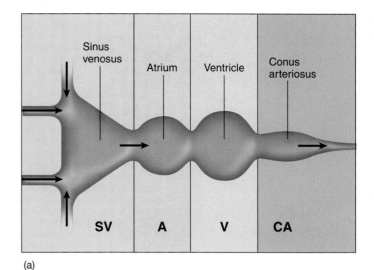

(a)

(b)

FIGURE 49.12
The heart and circulation of a fish. (a) Diagram of a fish heart, showing the chambers in series with each other. (b) Diagram of fish circulation, showing that blood is pumped by the ventricle through the gills and then to the body. Blood rich in oxygen (oxygenated) is shown in red; blood low in oxygen (deoxygenated) is shown in blue.

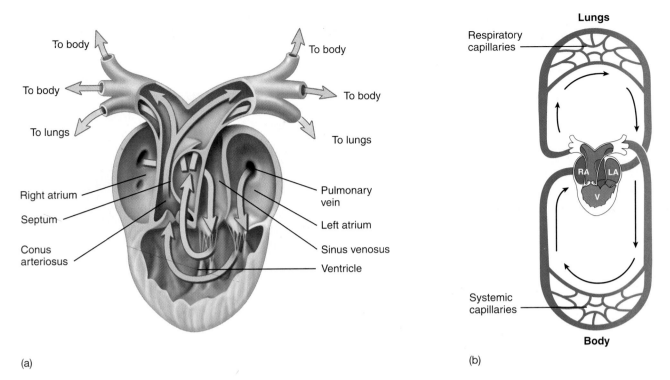

(a)

(b)

FIGURE 49.13
The heart and circulation of an amphibian. (a) The frog heart has two atria but only one ventricle, which pumps blood both to the lungs and to the body. (b) Despite the potential for mixing, the oxygenated and deoxygenated bloods (red and blue, respectively) mix very little as they are pumped to the body and lungs. The slight mixing is shown in purple. RA = right atrium; LA = left atrium; V = ventricle.

Amphibian and Reptile Circulation

The advent of lungs involved a major change in the pattern of circulation. After blood is pumped by the heart through the *pulmonary arteries* to the lungs, it does not go directly to the tissues of the body but is instead returned via the *pulmonary veins* to the heart. This results in two circulations: one that goes from the heart to the lungs and back, called the **pulmonary circulation,** and one that goes from the heart to the rest of the body and back, called the **systemic circulation.**

If no changes had occurred in the structure of the heart, the oxygenated blood from the lungs would be mixed in the heart with the deoxygenated blood returning from the rest of the body. Consequently, the heart would pump a mixture of oxygenated and deoxygenated blood rather than fully oxygenated blood. The amphibian heart has two structural features that help reduce this mixing (figure 49.13). First, the atrium is divided into two chambers: the right atrium receives deoxygenated blood from the systemic circulation, and the left atrium receives oxygenated blood from the lungs. These two stores of blood therefore do not mix in the atria, but some mixing might be expected when the contents of each atrium enter the single, common ventricle. Surprisingly, however, little mixing actually occurs. Second, the conus arteriosus is partially separated by a *septum*, or dividing wall, which directs deoxygenated blood into the pulmonary

arteries to the lungs and oxygenated blood into the *aorta*, the major artery of the systemic circulation to the body.

Since there is only one ventricle in an amphibian heart, the separation of the pulmonary and systemic circulations is incomplete. Amphibians in water, however, can obtain additional oxygen by diffusion through their skin. This process, called **cutaneous respiration,** helps to supplement the oxygenation of the blood in these vertebrates.

Among reptiles, additional modifications have reduced the mixing of blood in the heart still further. In addition to having two separate atria, reptiles have a septum that partially subdivides the ventricle. This results in an even greater separation of oxygenated and deoxygenated blood within the heart. The separation is complete in one order of reptiles, the crocodiles, which have two separate ventricles divided by a complete septum. Crocodiles therefore have a completely divided pulmonary and systemic circulation. Another change in the circulation of reptiles is that the conus arteriosus has become incorporated into the trunks of the large arteries leaving the heart.

Amphibians and reptiles have two circulations, pulmonary and systemic, that deliver blood to the lungs and rest of the body, respectively. The oxygenated blood from the lungs is kept relatively separate from the deoxygenated blood from the rest of the body by incomplete divisions within the heart.

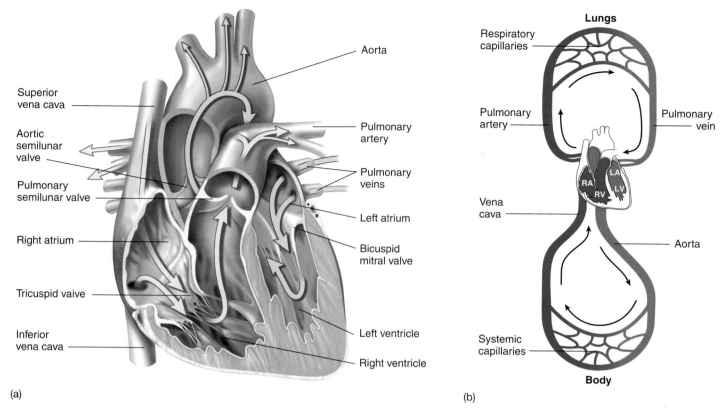

Superior vena cava

Aortic semilunar valve

Pulmonary semilunar valve

Right atrium

Tricuspid valve

Inferior vena cava

Aorta

Pulmonary artery

Pulmonary veins

Left atrium

Bicuspid mitral valve

Left ventricle

Right ventricle

(a)

Lungs

Respiratory capillaries

Pulmonary artery

Pulmonary vein

Vena cava

Aorta

RA LA
RV LV

Systemic capillaries

Body

(b)

FIGURE 49.14

The heart and circulation of mammals and birds. (a) The path of blood through the four-chambered heart. (b) The right side of the heart receives deoxygenated blood and pumps it to the lungs; the left side of the heart receives oxygenated blood and pumps it to the body. In this way, the pulmonary and systemic circulations are kept completely separate. RA = right atrium; LA = left atrium; RV = right ventricle; LV = left ventricle.

Mammalian and Bird Hearts

Mammals, birds, and crocodiles have a four-chambered heart with two separate atria and two separate ventricles (figure 49.14). The right atrium receives deoxygenated blood from the body and delivers it to the right ventricle, which pumps the blood to the lungs. The left atrium receives oxygenated blood from the lungs and delivers it to the left ventricle, which pumps the oxygenated blood to the rest of the body. This completely double circulation is powered by a two-cycle pump. Both atria fill with blood and simultaneously contract, emptying their blood into the ventricles. Both ventricles contract at the same time, pushing blood simultaneously into the pulmonary and systemic circulations. The increased efficiency of the double circulatory system in mammals and birds is thought to have been important in the evolution of endothermy (warm-bloodedness), because a more efficient circulation is necessary to support the high metabolic rate required.

Because the overall circulatory system is closed, the same volume of blood must move through the pulmonary circulation as through the much larger systemic circulation with each heartbeat. Therefore, the right and left ventricles must pump the same amount of blood each time they contract. If the output of one ventricle did not match that

of the other, fluid would accumulate and pressure would increase in one of the circuits. The result would be increased filtration out of the capillaries and edema (as occurs in congestive heart failure, for example.) Although the volume of blood pumped by the two ventricles is the same, the pressure they generate is not. The left ventricle, which pumps blood through the higher-resistance systemic pathway, is more muscular and generates more pressure than does the right ventricle.

Throughout the evolutionary history of the vertebrate heart, the sinus venosus has served as a pacemaker, the site where the impulses that produce the heartbeat originate. Although it constitutes a major chamber in the fish heart, it is reduced in size in amphibians and further reduced in reptiles. In mammals and birds, the sinus venosus is no longer evident as a separate chamber, but its disappearance is not really complete. Some of its tissue remains in the wall of the right atrium, near the point where the systemic veins empty into the atrium. This tissue, which is called the *sinoatrial (SA) node*, is still the site where each heartbeat originates.

The oxygenated blood from the lungs returns to the left atrium and is pumped out the left ventricle. The deoxygenated blood from the body returns to the right atrium and is pumped out the right ventricle to the lungs.

The Cardiac Cycle

The human heart, like that of all mammals and birds, is really two separate pumping systems operating within a single organ. The right pump sends blood to the lungs, and the left pump sends blood to the rest of the body.

The heart has two pairs of valves. One pair, the **atrioventricular (AV) valves,** guards the opening between the atria and ventricles. The AV valve on the right side is the **tricuspid valve,** and the AV valve on the left is the **bicuspid,** or **mitral, valve.** Another pair of valves, together called the **semilunar valves,** guard the exits of the ventricles to the arterial system; the **pulmonary valve** is located at the exit of the right ventricle, and the **aortic valve** is located at the exit of the left ventricle. These valves open and close as the heart goes through its **cardiac cycle** of rest (*diastole*) and contraction (*systole*).

Blood returns to the resting heart through veins that empty into the right and left atria. As the atria fill and the pressure in them rises, the AV valves open to admit the blood into the ventricles. The ventricles become about 80% filled during this time. Contraction of the atria wrings out the final 20% of the 80 milliliters of blood the ventricles will receive, on average, in a resting person. These events occur while the ventricles are relaxing, a period called ventricular **diastole.**

After a slight delay, the ventricles contract; this period of contraction is known as ventricular **systole.** Contraction of each ventricle increases the pressure within each chamber, causing the AV valves to forcefully close and thereby preventing blood from backing up into the atria. Immediately after the AV valves close, the pressure in the ventricles forces the semilunar valves open so that blood can be pushed out into the arterial system.

The right and left **pulmonary arteries** deliver oxygen-depleted blood to the right and left lungs. As previously mentioned, these return blood to the left atrium of the heart via the **pulmonary veins.** The **aorta** and all its branches are systemic arteries (figure 49.15), carrying oxygen-rich blood from the left ventricle to all parts of the body. The **coronary arteries** are the first branches off the aorta; these supply the heart muscle itself. Other systemic arteries branch from the aorta as it makes an arch above the heart, and as it descends and traverses the thoracic and abdominal cavities. These branches provide all body organs with oxygenated blood. The blood from the body organs, now lower in oxygen, returns to the heart in the systemic veins. These eventually empty into two major veins: the **superior vena cava,** which drains the upper body, and the **inferior vena cava,** which drains the lower body. These veins empty into the right atrium and thereby complete the systemic circulation.

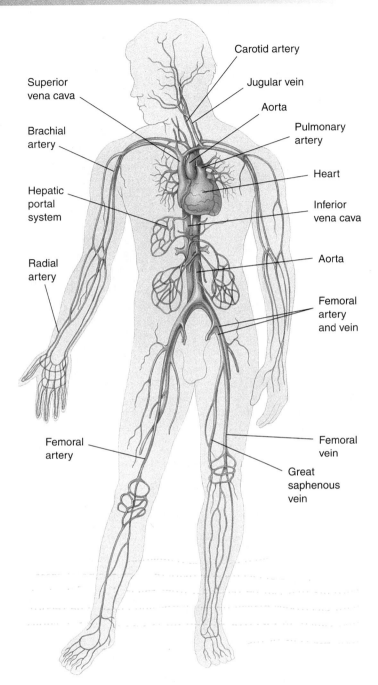

FIGURE 49.15
The human circulatory system. Some of the major arteries and veins are shown.

The cardiac cycle consists of systole and diastole; the ventricles contract at systole and relax at diastole.

Electrical Excitation and Contraction of the Heart

As in other types of muscle, contraction of heart muscle is stimulated by membrane **depolarization,** a reversal of the electrical polarity that normally exists across the plasma membrane (see chapter 51). In skeletal muscles, the nervous system initiates depolarization. However, in the heart, the depolarization is triggered by the **sinoatrial (SA) node** (figure 49.16), the small cluster of cardiac muscle cells derived from the sinus venosus. The SA node acts as a **pacemaker** for the rest of the heart by producing depolarization impulses spontaneously at a particular rate. Each depolarization initiated within this pacemaker region passes quickly from one cardiac muscle cell to another in a wave that envelops the right and left atria nearly simultaneously. The spread of depolarization is possible because the cardiac muscle cells are electrically coupled by gap junctions.

After a delay of almost 0.1 second, the wave of depolarization spreads to the ventricles. The reason for this delay is that connective tissue separates the atria from the ventricles, and connective tissue cannot transmit depolarization. The depolarization would not pass to the ventricles at all, were it not for a group of specialized cardiac muscle cells known as the **atrioventricular (AV) node.** The cells in the AV node transmit the depolarization slowly, causing the delay. This delay permits the atria to finish contracting and emptying their blood into the ventricles before the ventricles contract.

From the AV node, the wave of depolarization is conducted rapidly over both ventricles by a network of fibers called the **atrioventricular bundle** or **bundle of His.** It is then transmitted by **Purkinje fibers,** which directly stimulate the myocardial cells of the ventricles. The rapid conduction of the depolarization along the bundle of His and the Purkinje fibers causes the almost simultaneous contraction of the left and right ventricles. The rate can be increased or decreased by neural regulation.

The spread of electrical activity through the heart creates currents that can be recorded from the surface of the body with electrodes placed on the limbs and chest. The recording, called an **electrocardiogram** (**ECG** or **EKG**), shows how the cells of the heart depolarize and repolarize during the cardiac cycle. As was explained in chapter 47, depolarization causes contraction of a muscle (including the heart), while repolarization causes relaxation. The first peak in the recording is produced by the depolarization of the atria, and thus is associated with atrial systole. The second, larger peak is produced by ventricular depolarization; during this time, the ventricles contract (ventricular systole) and eject blood into the arteries. The last peak is produced by ventricular repolarization; at this time the ventricles begin diastole.

Cardiac Output

Cardiac output is the volume of blood pumped by each ventricle per minute. Since humans (like all vertebrates)

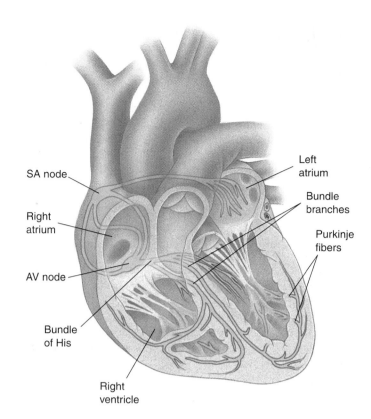

FIGURE 49.16
The path of electrical excitation in the heart. A wave of depolarization begins at the sinoatrial node. After passing over the atria and causing them to contract, the depolarization reaches the atrioventricular node, from which it passes to the ventricles along the septum by the bundle of His. Finer Purkinje fibers carry the depolarization into the right and left ventricular muscles.

have a closed circulation, the cardiac output is the same as the volume of blood that traverses the systemic or pulmonary circulations per minute. It is calculated by multiplying the heart rate by the *stroke volume*, which is the volume of blood ejected by each ventricle per beat. For example, if the heart rate is 72 beats per minute and the stroke volume is 70 milliliters, the cardiac output is 5 liters per minute, which is about average in a resting person.

Cardiac output increases during exercise because of an increase in heart rate and stroke volume. When exercise begins, the heart rate increases up to about 100 beats per minute. As exercise becomes more intense, skeletal muscles squeeze on veins more vigorously, returning blood to the heart more rapidly. In addition, the ventricles contract more strongly, so they empty more completely with each beat.

The SA node in the right atrium serves as the pacemaker, initiating waves of depolarization that are passed by the conduction tissue of the heart and stimulate first the atria and then the ventricles to contract. The cardiac output depends on the rate of the heart and how much blood is ejected per beat.

Blood Flow and Blood Pressure

During exercise, the cardiac output increases to a maximum of about 25 liters per minute in an average young adult. Although the cardiac output has increased five times, not all organs receive five times the blood flow. This is because the arterioles in some organs, such as in the digestive system, constrict, while the arterioles in the exercising muscles and heart dilate. As previously mentioned, the resistance to flow decreases as the fourth power of the radius of the vessel. As a consequence, vasodilation greatly increases and vasoconstriction greatly decreases blood flow.

Blood Pressure and the Baroreceptor Reflex

The arterial blood pressure depends on two factors: how much blood the ventricles pump (the cardiac output) and how great a resistance to flow the blood encounters in the entire arterial system. An increased blood pressure, therefore, could be produced by an increased heart rate or an increased blood volume (because both increase the cardiac output) or by vasoconstriction, which increases the resistance to blood flow. Conversely, blood pressure will fall if the heart rate slows or if the blood volume is reduced, for example by dehydration or excessive bleeding (hemorrhage).

Changes in the arterial blood pressure are detected by **baroreceptors** located in the arch of the aorta and in the carotid arteries. These receptors activate sensory neurons that relay information to *cardiovascular control centers* in the medulla oblongata, a region of the brain stem. When the baroreceptors detect a fall in blood pressure, they stimulate neurons that go to blood vessels in the skin and viscera, causing arterioles in these organs to constrict and raise the blood pressure. This baroreceptor reflex therefore completes a negative feedback loop that acts to correct the fall in blood pressure and restore homeostasis.

Blood Volume Reflexes

Blood pressure depends in part on the total blood volume. A decrease in blood volume, therefore, will decrease blood pressure, if all else remains equal. Blood volume regulation involves the effects of four hormones: (1) antidiuretic hormone; (2) aldosterone; (3) atrial natriuretic hormone; and (4) nitric oxide.

Antidiuretic Hormone. *Antidiuretic Hormone* (ADH), also called *vasopressin*, is secreted by the posterior pituitary gland in response to an increase in the osmotic concentration of the blood plasma. Dehydration, for example, causes the blood volume to decrease while the remaining plasma becomes more concentrated. This stimulates *osmoreceptors* in the hypothalamus of the brain, a region located immediately above the pituitary. The osmoreceptors promote thirst and stimulate ADH secretion from the posterior pituitary gland. ADH, in turn, stimulates the

kidneys to retain more water in the blood, excreting less in the urine (urine is derived from blood plasma—see chapter 55). A dehydrated person thus drinks more and urinates less, helping to raise the blood volume and restore homeostasis.

Aldosterone. If a person's blood volume is lowered (by dehydration, for example), the flow of blood through the organs will be reduced if no compensation occurs. Whenever the kidneys experience a decreased blood flow, a group of kidney cells initiate the release of a short polypeptide known as *angiotensin II*. This is a very powerful molecule: it stimulates vasoconstriction throughout the body while it also stimulates the adrenal cortex (the outer region of the adrenal glands) to secrete the hormone *aldosterone*. This important steroid hormone is necessary for life; it acts on the kidneys to promote the retention of Na^+ and water in the blood. An animal that lacks aldosterone will die if untreated, because so much of the blood volume is lost in urine that the blood pressure falls too low to sustain life.

Atrial Matriuretic Hormone. When the body needs to eliminate excessive Na^+, less aldosterone is secreted by the adrenals, so that less Na^+ is retained by the kidneys. In recent years, scientists have learned that Na^+ excretion in the urine is promoted by another hormone. Surprisingly, this hormone is secreted by the right atrium of the heart—the heart is an endocrine gland! The right atrium secretes *atrial natriuretic hormone* in response to stretching of the atrium by an increased blood volume. The action of atrial natriuretic hormone completes a negative feedback loop, because the elimination of Na^+ and water that it promotes will lower the blood volume and pressure.

Nitric Oxide. *Nitric oxide* (NO) is a gas that acts as a hormone in vertebrates, regulating blood pressure and blood flow. As described in Chapter 7, nitric oxide gas is a paracrine hormone, is produced by one cell, penetrates through membranes, and alters the activities of other neighboring cells. In 1998 the Nobel Prize for Medicine was awarded for the discovery of this signal transmission activity. How does NO regulate blood pressure? Nitric oxide gas produced by the surface endothelial cells of blood vessels passes inward through the cell layers of the vessel, causing the smooth muscles that encase it to relax and the blood vessel to dilate (become wider). For over a century, heart patients have been prescribed nitroglycerin to relieve chest pain, but only now has it become clear that nitroglycerin acts by releasing nitric oxide gas.

Blood flow is regulated by the degree of constriction of the arteries, which affects the resistance to flow. Blood pressure is influenced by blood volume. The volume of water retained in the vascular system is regulated by hormones that act on the kidneys and blood vessels.

Summary of Concepts

49.1 The circulatory systems of animals may be open or closed.

- Vertebrates have a closed circulation, where the blood stays within vessels as it travels away from and back to the heart.
- The circulatory system serves a variety of functions, including transport, regulation, and protection.

49.2 A network of vessels transports blood through the body.

- Plasma is the liquid portion of the blood, and it contains a variety of plasma proteins, ions, metabolites, wastes, and hormones dissolved in the plasma water.
- Erythrocytes, or red blood cells, contain hemoglobin and function to transport oxygen; the leukocytes, or white blood cells, function in immunological defenses.
- The heart pumps blood into arteries, which branch into smaller arterioles; the smooth muscle in the walls of arterioles can constrict or dilate these vessels to vary the resistance to flow.
- Blood from the arterial system empties into capillaries, the most numerous of vessels and those with the thinnest walls; all exchanges between the blood and tissues occurs across the walls of capillaries.
- Blood returns to the heart in veins, which have one-way valves to ensure that blood travels toward the heart only.
- Lymphatic vessels return interstitial fluid to the venous system; the lymphatic system also includes lymph nodes and lymphatic organs, such as the spleen and thymus.

49.3 The vertebrate heart has undergone progressive evolutionary change.

- The fish heart consists of four chambers in a row; the beat originates in the sinus venosus and spreads to the atrium, ventricle, and conus arteriosus.
- In the circulation of fishes, blood from the heart goes to the gills and then to the rest of the body before returning to the heart; in terrestrial vertebrates, blood returns from the lungs to the heart before it is pumped to the body.
- Amphibians have a three-chambered heart, as do reptiles other than the crocodiles; in crocodiles, birds, and mammals, there is a complete division between left and right ventricles as well as left and right atria, so blood from the lungs does not mix with blood from the body.

49.4 The cardiac cycle drives the cardiovascular system.

- Electrical excitation of the heart is initiated by the SA (sinoatrial) node, spreads through gap junctions between myocardial cells in the atria, and then is conducted into the ventricles by specialized conducting tissue.
- The cardiac output is regulated by nerves that influence the cardiac rate and by factors that influence the stroke volume.
- A fall in blood pressure activates the baroreceptor reflex, which causes increased heart rate and constriction of vessels in the viscera and skin, raising the blood pressure and helping to maintain homeostasis.

Discussing Key Terms

1. **Interstitial fluid**   The pressure of the blood in capillaries causes fluid derived from plasma to be filtered through the walls of capillaries into the surrounding tissues; this fluid, called interstitial fluid, is eventually returned to the vascular system.

2. **Vasoconstriction and vasodilation**   The smooth muscle walls of smaller arteries and arterioles can contract to constrict these vessels and reduce blood flow, or they relax to dilate these vessels and increase blood flow.

3. **Sinus venosus**   The sinus venosus is the first chamber of a fish heart to receive blood from the circulation; the electrical stimulation of the heart originates here or in its derivative, the sinoatrial node, which is the pacemaker of mammalian and bird hearts.

4. **Cardiac cycle**   The cardiac cycle refers to the alternating phases of contraction, or systole, and relaxation, or diastole, of the ventricles. It can be monitored by the ECG, by listening to the heart sounds, by taking blood pressure, and by feeling the pulse.

5. **Baroreceptor reflex**   A fall in blood pressure is sensed by stretch receptors called baroreceptors in the aorta and carotid arteries that activate a reflex response; this reflex sets about compensatory changes that include an increase in heart rate and constriction of arterioles.

6. **Cardiac output**   The cardiac output is a measure of the amount of blood pumped per minute by each ventricle, and is equal to the product of the cardiac rate and the stroke volume. The cardiac output is adjusted according to the needs of the body.

Review Questions

1. What is the difference between a closed circulatory system and an open circulatory system? In what types of animals would you find each?

2. Describe the structure of arteries and veins, explaining their similarities and differences. Why do arteries differ in structure from veins?

3. How is interstitial fluid formed? How is it returned to the capillaries?

4. What is the relationship between vessel diameter and the resistance to blood flow? How do the arterial trees adjust their resistance to flow?

5. How does the lymphatic system participate in the function of the cardiovascular system? What drives the flow of fluid within this system, and in what direction does the fluid flow?

6. What are the major components of blood plasma? In general terms, what are the functions of erythrocytes, leukocytes, and platelets?

7. Describe the pattern of circulation through a fish and an amphibian, and compare the structure of their hearts. What new circulatory pattern accompanies the evolution of lungs?

8. How does the baroreceptor reflex help to maintain blood pressure? How do ADH and aldosterone maintain blood volume and pressure? What causes their secretion?

Thought Questions

1. Explain why the capillaries collectively have a lower flow resistance than the arterioles, even though the diameter of a capillary is smaller than that of an arteriole.

2. Mean blood pressure continuously declines from the aorta to the right atrium. Where does the largest pressure drop occur, and why does it occur there?

3. What aspect of the lifestyle of a crocodile might warrant it having a four-chambered heart, as compared with the hearts of other reptiles?

4. Using two arguments, one from function and one from evolution, explain why the atria contract before the ventricles.

Internet Links

Coming to Understand the Heart
http://outcast.gene.com/ae/AE/AEC/CC/heart_background.html
A historical account of how we came to understand the circulation of the blood, with a comparison of circulation in different animals, from the AC-CESS EXCELLENCE CLASSIC COLLECTION.

The Virtual Heart
http://sln.fi.edu/biosci/heart.html
From the Franklin Institute, THE HEART: AN ONLINE EXPLORATION allows you to investigate the structure and functioning of the mammalian heart.

American Heart Association
http://www.amhrt.org/
The homepage of the American Heart Association, providing a wealth of information about heart attack and stroke risk factors, and many links to other up-to-date information sources.

For Further Reading

Cantin, M. and J. Genest: "The Heart as an Endocrine Gland," *Scientific American*, February 1986, pages 76–81. An account of the discovery of atrial peptides, and interesting speculation on how they work in regulating blood pressure and volume.

Eisenberg, M., and others: "Sudden Cardiac Death," *Scientific American*, May 1986, pages 37–93. An account of the role of rapid medical intervention in coping with serious heart attacks.

Goldstein, G. and A. L. Betz: "The Blood-Brain Barrier," *Scientific American*, September 1986, pages 74–83. A description of brain capillaries and the special ways in which they guard what enters and leaves the brain.

Hajjar, D. P. and A. C. Nicholson: "Atherosclerosis," *American Scientist*, vol. 83, September–October, 1995, page 460. An excellent review of the cellular and molecular mechanisms that normally protect blood vessels against damage and of how abnormal responses produce atherosclerosis.

Perutz, M. F.: "Hemoglobin Structure and Respiratory Transport," *Scientific American*, December 1978, pages 92–125. An account of how hemoglobin changes its shape to facilitate oxygen binding and unloading, by the man who won the 1962 Nobel Prize in medicine for unraveling the structure of hemoglobin.

Robinson, T., S. Factor, and E. Sonneblink: "The Heart as a Suction Pump," *Scientific American*, June 1986, pages 84–91. Between birth and death our hearts beat millions of times. These authors argue that the heart is aided greatly in this Herculean task by a very clever trick: contraction compresses elastic elements within the heart muscles, which then bounce back to expand the ventricles.

Zucker, M. B.: "The Functioning of the Blood Platelets," *Scientific American*, June 1980, pages 86–103. A description of the many roles of platelets, with emphasis on their role in blood clotting.

50

Respiration

Concept Outline

FIGURE 50.1
Elephant seals are respiratory champions. Diving to depths greater than those of all other marine animals, including sperm whales and sea turtles, elephant seals can hold their breath for over two hours, descend and ascend rapidly in the water, and endure repeated dives without suffering any apparent respiratory distress.

Animals pry energy out of food molecules using the biochemical process called cellular respiration. While the term *cellular respiration* pertains to the use of oxygen and production of carbon dioxide at the cellular level, the general term **respiration** describes the uptake of oxygen from the environment and the disposal of carbon dioxide at the body system level. Respiration at the body system level involves a host of processes not found at the cellular level, like the mechanics of breathing and the exchange of oxygen and carbon dioxide in the arteries and veins. These processes, one of the principal physiological challenges facing all animals (figure 50.1), are the subject of this chapter.

Fick's Law of Diffusion

Respiration involves the diffusion of gases across plasma membranes. Because plasma membranes must be surrounded by water to be stable, the external environment in gas exchange is always aqueous. This is true even in terrestrial animals; in these cases, oxygen from air dissolves in a thin layer of fluid that covers the respiratory surfaces, such as the alveoli in lungs.

In vertebrates, the gases diffuse into the aqueous layer covering the epithelial cells that line the respiratory organs. The diffusion process is passive, driven only by the difference in O_2 and CO_2 concentrations on the two sides of the membranes. In general, the rate of diffusion between two regions is governed by a relationship known as **Fick's law of diffusion:**

$$R = D \times A \frac{\Delta p}{d}$$

In this equation,

R = the rate of diffusion; the amount of oxygen or carbon dioxide diffusing per unit of time;
D = the diffusion constant;
A = the area over which diffusion takes place;
Δp = the difference in concentration (for gases, the difference in their partial pressures) between the interior of the organism and the external environment; and
d = the distance across which diffusion takes place.

Major changes in the mechanism of respiration have occurred during the evolution of animals (figure 50.2) that have tended to optimize the rate of diffusion R. By inspecting Fick's law, you can see that natural selection can optimize R by favoring changes that (1) increase the surface area A; (2) decrease the distance d; or (3) increase the concentration difference, as indicated by Δp. The evolution of respiratory systems has involved changes in all of these factors.

Fick's law of diffusion states that the rate of diffusion across a membrane depends on surface area, concentration difference, and distance.

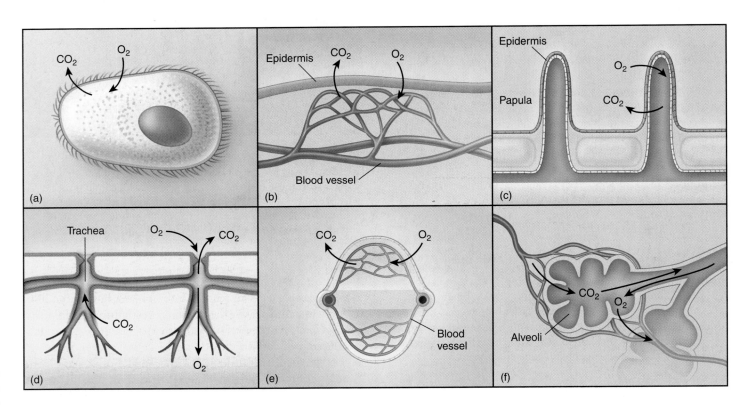

FIGURE 50.2
Gas exchange may take place in a variety of ways. (a) Gases diffuse directly into single-celled organisms. (b) Amphibians and many other animals respire across their skin. (c) Echinoderms have protruding papulae, which provide an increased respiratory surface area. (d) Insects respire through an extensive tracheal system. (e) The gills of fishes provide a very large respiratory surface area and countercurrent exchange. (f) The alveoli in mammalian lungs provide a large respiratory surface area but do not permit countercurrent exchange.

How Animals Maximize the Rate of Diffusion

The levels of oxygen required by oxidative metabolism cannot be obtained by diffusion alone over distances greater than about 0.5 millimeter. This restriction severely limits the size of organisms that obtain their oxygen entirely by diffusion directly from the environment. Protists are small enough that such diffusion can be adequate (see figure 50.2a), but most multicellular animals are much too large.

Most of the more primitive phyla of invertebrates lack special respiratory organs. The sponges (phylum Porifera), cnidarians (phylum Cnidaria), many flatworms (phylum Platyhelminthes), roundworms (phylum Nematoda), and some annelids (phylum Annelida) all obtain their oxygen by diffusion directly from the surrounding water. In a number of different ways, many of which involve beating cilia, these organisms create a *water current* that continuously replaces the water over the respiratory surfaces. Because of this continuous replenishment with water containing fresh oxygen, *the external oxygen concentration does not decrease as diffusion proceeds.* Although each oxygen molecule that passes into the organism has been removed from the surrounding water, new water continuously replaces the oxygen-depleted water. This increases the rate of diffusion by maximizing the concentration difference—the Δp of the Fick equation.

All of the more advanced invertebrates (mollusks, arthropods, echinoderms), as well as vertebrates, possess respiratory organs that increase the surface area available for diffusion and bring the external environment (either water or air) close to the internal fluid, which is usually circulated throughout the body. The respiratory organs thus increase the rate of diffusion by maximizing surface area and decreasing the distance the diffusing gases must travel (the A and d factors, respectively, in the Fick equation).

Atmospheric Pressure and Partial Pressures

Dry air contains 78.09% nitrogen (N_2), 20.95% oxygen, 0.93% argon and other inert gases, and 0.03% carbon dioxide. Convection currents cause air to maintain a constant composition to altitudes of at least 100 kilometers, although the *amount* of air that is present decreases with altitude (figure 50.3).

Imagine a column of air extending from the ground to the limits of the atmosphere. All of the gas molecules in this column experience the force of gravity, so they have weight and can exert pressure. If this column were on top of one end of a U-shaped tube of mercury at sea level, it would exert enough pressure to raise the other end of the tube 760 millimeters under a set of specified, standard conditions (see figure 50.3). An apparatus that measures air pressure is called a barometer, and 760 mm Hg (millimeters of mercury) is the barometric pressure of the air at sea level. A pressure of 760 mm Hg is also defined as **one atmosphere** of pressure.

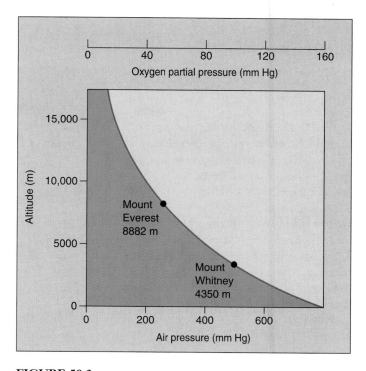

FIGURE 50.3
The relationship between air pressure and altitude above sea level. At the high altitudes characteristic of mountaintops, air pressure is much less than at sea level. At the top of Mount Everest, the world's highest mountain, the air pressure is only one-third that at sea level.

Each type of gas in the atmosphere exerts a fraction of the total atmospheric pressure. That fraction is called a **partial pressure** and indicated by P_{N_2}, P_{O_2}, P_{CO_2}, and so on. For dry air, the partial pressures are calculated simply by multiplying the fractional composition of each gas in the air by the atmospheric pressure. Thus, at sea level, the partial pressures of N_2^+ inert gases, O_2, and CO_2 are:

$$P_{N_2} = 760 \times 79.02\% = 600.6 \text{ mm Hg,}$$
$$P_{O_2} = 760 \times 20.95\% = 159.2 \text{ mm Hg, and}$$
$$P_{CO_2} = 760 \times 0.03\% = 0.2 \text{ mm Hg.}$$

Humans do not survive long at altitudes above 6000 meters. Although the air at these altitudes still contains 20.95% oxygen, the atmospheric pressure is only about 380 mm Hg (see figure 50.3), so its P_{O_2} is only 80 mm Hg (380 × 20.95%). This figure is only half of the amount of oxygen that is available at sea level.

Respiration involves the exchange of oxygen and carbon dioxide between an organism and its environment. This exchange occurs by diffusion of dissolved gases across plasma membranes and is maximized by increasing the concentration gradient and the surface area and by decreasing the distance that the diffusing gases must travel.

The Gill as a Respiratory Structure

Aquatic respiratory organs increase the diffusion surface area by extensions of tissue, called *gills*, that project out into the water. Gills can be simple, as in the papulae of echinoderms (see figure 50.2*c*), or complex, as in the highly convoluted gills of fish (see figure 50.2*e*). The great increase in diffusion surface area provided by gills enables aquatic organisms to extract far more oxygen from water than would be possible from their body surface alone.

External gills (gills that are not enclosed within body structures) provide a greatly increased surface area for gas exchange. Examples of vertebrates with external gills are the larvae of many fish and amphibians, as well as developmentally arrested *(neotenic)* amphibian larvae that remain permanently aquatic, such as the axolotl (figure 50.4). The disadvantage of external gills is that they must constantly be moved or the surrounding water becomes depleted in oxygen as the oxygen diffuses from the water to the blood of the gills. The highly branched gills, however, offer significant resistance to movement, making this form of respiration ineffective except in smaller animals.

Other types of aquatic animals evolved specialized *branchial chambers*, which provide a means of pumping water past stationary gills. Mollusks, for example, have an internal *mantle cavity* that opens to the outside and contains the gills. Contraction of the muscular walls of the mantle cavity draws water in and then expels it. In crustaceans, the branchial chamber lies between the bulk of the body and the hard exoskeleton of the animal. This chamber contains gills and opens to the surface beneath a limb. Movement of the limb draws water through the branchial chamber, thus creating currents over the gills.

FIGURE 50.4
The gills of the axolotl are external. By sweeping its gills through the water, the axolotl moves water over the diffusion surface.

The Gills of Bony Fishes

The gills of bony fishes are located between the *buccal* (mouth) *cavity* and the *opercular cavities* (figure 50.5). The buccal cavity can be opened and closed by opening and closing the mouth, and the opercular cavity can be opened and closed by movements of the **operculum,** or gill cover. The two sets of cavities function as pumps that expand alternately to move water into the mouth, through the gills, and out of the fish through the open operculum. Water is brought into the buccal cavity by lowering the jaw and floor of the mouth, and then is moved through the gills into the opercular cavity by the opening of the operculum. The lower pressure in the opercular cavity causes water to move in the correct direction across the gills, and tissue that acts as valves ensures that the movement is one-way.

Some fishes that swim continuously, such as tuna, have practically immobile opercula. These fishes swim with their mouths partly open, constantly forcing water over the gills in a form of **ram ventilation.** Most bony fishes, however, have flexible gill covers that permit a pumping action. The remora, a fish that rides "piggyback" on sharks, for example, uses ram ventilation while the shark swims and the pumping action of its opercula when the shark stops swimming.

There are four **gill arches** on each side of the fish head. Each gill arch is composed of two rows of *gill filaments*, and each gill filament contains thin membranous plates, or *lamellae*, that project out into the flow of water (figure 50.6). Water flows past the lamellae in one direction only. Within each lamella, blood flows in a direction that is *opposite* the direction of water movement. This arrangement is called **countercurrent flow,** and it acts to maximize the oxygenation of the blood.

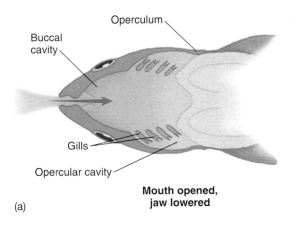

Mouth opened, jaw lowered

(a)

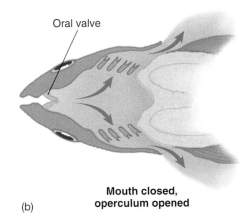

Mouth closed, operculum opened

(b)

FIGURE 50.5
How most bony fishes respire. The gills are suspended between the buccal (mouth) cavity and the opercular cavity. Respiration occurs in two stages. (a) The oral valve in the mouth is opened and the jaw is depressed, drawing water into the buccal cavity while the opercular cavity is closed. (b) The oral valve is closed and the operculum is opened, drawing water through the gills to the outside.

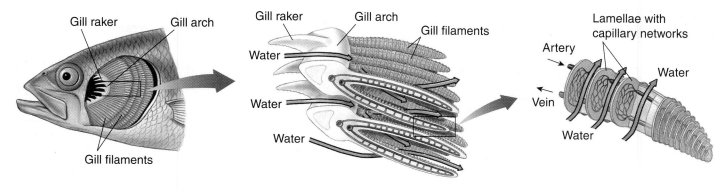

FIGURE 50.6
Structure of a fish gill. Water passes from the gill arch over the filaments (from left to right in the diagram). Water always passes the lamellae in a direction that is opposite to the direction of blood flow through the lamellae. The success of the gill's operation critically depends on this countercurrent flow of water and blood.

To see the advantage of countercurrent flow, examine figure 50.7*a.* Blood low in oxygen enters the back of the lamella, where it comes in close proximity to water that has already had most of its oxygen removed as it flowed through the lamella in the opposite direction. The water still has a higher oxygen concentration than the blood at this point, however, so oxygen diffuses from the water to the blood. As the blood flows toward the front of the lamella, it runs next to water that has a still higher oxygen content, so oxygen continuously diffuses from the water to the blood. Thus, countercurrent flow ensures that a concentration gradient remains between blood and water throughout the flow. This permits oxygen to continue to diffuse all along the lamellae, so that the blood leaving the gills has nearly as high an oxygen concentration as the water entering the gills.

This concept is easier to understand if we look at what would happen if blood and water flowed in the same direction, that is, had a *concurrent flow.* The difference in oxygen concentration would be very high at the front of each lamella, where oxygen-depleted blood would meet oxygen-rich water entering the gill (figure 50.7*b*). The concentration difference would fall rapidly, however, as the water lost oxygen to the blood. Net diffusion of oxygen would cease when the oxygen concentration of blood matched that of the water. At this point, much less oxygen would have been transferred to the blood than is the case with countercurrent flow. The flow of blood and water in a fish gill is in fact countercurrent, and because of the countercurrent exchange of gases, fish gills are the most efficient of all respiratory organs.

> **In bony fishes, water is forced past gills by the pumping action of the buccal and opercular cavities, or by active swimming in ram ventilation. In the gills, blood flows in an opposite direction to the flow of water. This countercurrent flow maximizes gas exchange, making the fish's gill an efficient respiratory organ.**

FIGURE 50.7
Countercurrent exchange. This process allows for the most efficient blood oxygenation known in nature.

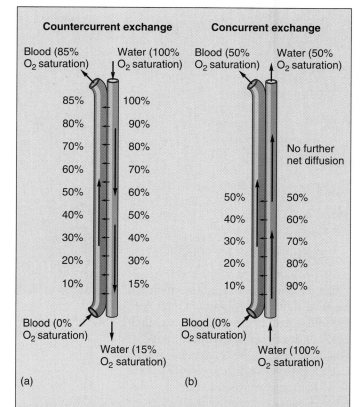

When blood and water flow in *opposite* directions (*a*), the initial oxygen concentration difference between water and blood is not large, but is sufficient for oxygen to diffuse from water to blood. As more oxygen diffuses into the blood, raising the blood's oxygen concentration, the blood encounters water with ever higher oxygen concentrations. At every point, the oxygen concentration is higher in the water, so that diffusion continues. In this example, blood attains an oxygen concentration of 85%. When blood and water flow in the *same* direction (*b*), oxygen can diffuse from the water into the blood rapidly at first, but the diffusion rate slows as more oxygen diffuses from the water into the blood, until finally the concentrations of oxygen in water and blood are equal. In this example, blood's oxygen concentration cannot exceed 50%.

Respiration in Air-Breathing Animals

Despite the high efficiency of gills as respiratory organs in aquatic environments, gills were replaced in terrestrial animals for two principal reasons:

1. **Air is less buoyant than water.** The fine membranous lamellae of gills lack structural strength and rely on water for their support. A fish out of water, although awash in oxygen (water contains only 5–10 mL O_2/L, compared with air with 210 mL O_2/L), soon suffocates because its gills collapse into a mass of tissue. This collapse greatly reduces the diffusion surface area of the gills. Unlike gills, internal air passages can remain open, since the body itself provides the necessary structural support.

2. **Water diffuses into air through evaporation.** Atmospheric air is rarely saturated with water vapor, except immediately after a rainstorm. Consequently, terrestrial organisms that are surrounded by air constantly lose water to the atmosphere. Gills would provide an enormous surface area for water loss.

Two main types of respiratory organs are used by terrestrial animals, and both sacrifice respiratory efficiency to some extent in exchange for reduced evaporation. The first are the **tracheae** of insects (see chapter 43 and figure 50.2*d*). Tracheae comprise an extensive series of air-filled passages connecting the surface of an insect to all portions of its body. Oxygen diffuses from these passages directly into cells, without the intervention of a circulatory system. Piping air directly from the external environment to the cells works very well in insects because their small bodies give them a high surface area-to-volume ratio. Insects prevent excessive water loss by closing the external openings of the tracheae whenever their internal CO_2 levels fall below a certain point.

The other main type of terrestrial respiratory organ is the **lung** (figure 50.8), and respiration involving lungs is known as **ventilation.** A lung minimizes evaporation by moving air through a branched tubular passage; the air becomes saturated with water vapor before reaching the

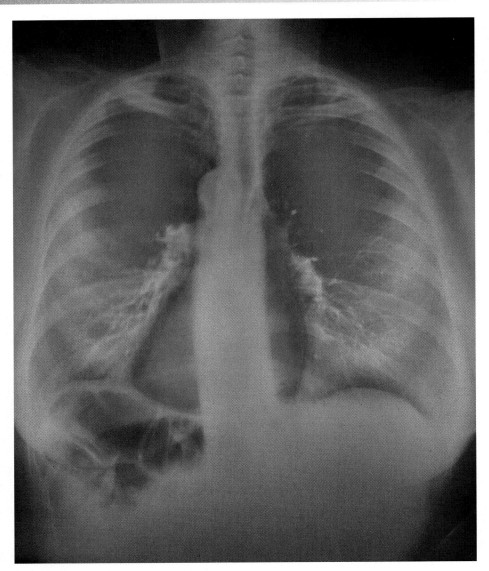

FIGURE 50.8
Human lungs. This chest X ray was color-enhanced to show the lungs clearly. The heart is the pear-shaped object behind the vertical white column that is the esophagus.

portion of the lung where a thin, wet membrane permits gas exchange. In the lungs of all terrestrial vertebrates except birds, the air moves back out again by way of the same airway passages. This two-way flow of air replaces the one-way flow of water that is so effective in the respiratory function of gills. Let us now examine the structure and function of lungs in the four classes of terrestrial vertebrates.

Air is piped directly to the body cells of insects, but the cells of terrestrial vertebrates obtain oxygen from the blood. The blood obtains its oxygen by diffusion across the wet membranes of the lungs, which are filled with air in the process of ventilation.

Respiration in Amphibians and Reptiles

The lungs of amphibians are formed as sac-like outpouchings of the gut (figure 50.9). Although the internal surface area of these sacs is increased by folds, much less surface area is available for gas exchange in amphibian lungs than in the lungs of other terrestrial vertebrates. Each amphibian lung is connected to the rear of the oral cavity, or pharynx, and the opening to each lung is controlled by a valve, the glottis.

Amphibians do not breathe the same way other terrestrial vertebrates do. Amphibians force air into their lungs by creating a greater-than-atmospheric pressure (positive pressure) in the air outside their lungs. They do this by filling their pharynx with air, closing their mouth and nostrils, and then elevating the floor of their oral cavity. This pushes air into their lungs in the same way that a pressurized tank of air is used to fill balloons. This is called positive pressure breathing; in humans, it would be analogous to forcing air into a victim's lungs by performing mouth-to-mouth resuscitation.

All other terrestrial vertebrates breathe by expanding their lungs and thereby creating a lower-than-atmospheric pressure (a negative pressure) within the lungs. This is called negative pressure breathing and is analogous to taking air into an accordion by pulling the accordion out to a greater volume. In reptiles, birds, and mammals, this is accomplished by expanding the thoracic (chest) cavity through muscular contractions, as will be described in a later section.

The oxygenation of amphibian blood by the lungs is supplemented by cutaneous respiration—the exchange of gases across the skin, which is wet and well vascularized in amphibians. Cutaneous respiration is actually more significant than pulmonary (lung) ventilation in frogs during winter, when their metabolisms are slow. Lung function becomes more important during the summer as the frog's metabolism increases.

Reptiles expand their rib cages by muscular contraction, and thereby take air into their lungs through negative pressure breathing. Their lungs have somewhat more surface area than the lungs of amphibians and so are more efficient at gas exchange. Terrestrial reptiles have dry, tough, scaly skins that prevent desiccation, and so cannot have cutaneous respiration. Cutaneous respiration, however, has been demonstrated in marine sea snakes.

Amphibians force air into their lungs by positive pressure breathing, whereas reptiles and all other terrestrial vertebrates take air into their lungs by expanding their lungs when they increase rib cage volume through muscular contractions. This creates a subatmospheric pressure in the lungs.

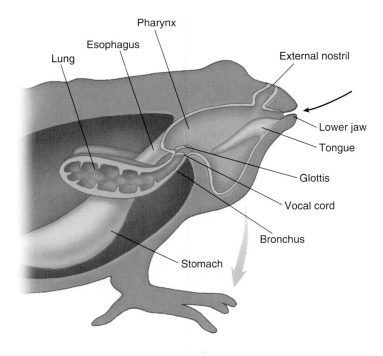
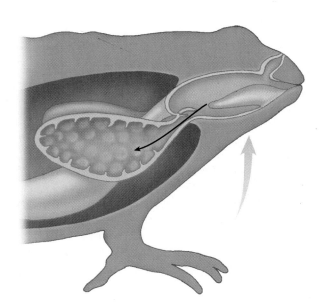

FIGURE 50.9
Amphibian lungs. Each lung of this frog is an outpouching of the gut and is filled with air by the creation of a positive pressure in the pharynx. The amphibian lung lacks the structures present in the lungs of other terrestrial vertebrates that provide an enormous surface area for gas exchange, and so are not as efficient as the lungs of other vertebrates.

Respiration in Mammals

The metabolic rate, and therefore the demand for oxygen, is much greater in birds and mammals, which are endothermic and thus require a more efficient respiratory system.

The lungs of mammals are packed with millions of *alveoli*, tiny sacs clustered like grapes (figure 50.10). This provides each lung with an enormous surface area for gas exchange. Air is brought to the alveoli through a system of air passages. Inhaled air is taken in through the **larynx** (voice box), where it passes through an opening in the vocal cords, the *glottis*, into a tube supported by C-shaped rings of cartilage, the **trachea** (windpipe). The trachea bifurcates into right and left **bronchi** (singular, *bronchus*), which enter each lung and further subdivide into **bronchioles** that deliver the air into blind-ended sacs called **alveoli**. The alveoli are surrounded by an extremely extensive capillary network. All gas exchange between the air and blood takes place across the walls of the alveoli.

The branching of bronchioles and the vast number of alveoli combine to increase the respiratory surface area far above that of amphibians or reptiles. In humans, there are about 300 million alveoli in each of the two lungs, and the total surface area available for diffusion can be as much as 80 square meters, or about 42 times the surface area of the body. Respiration in mammals will be considered in more detail in a separate section later.

Mammalian lungs are composed of millions of alveoli that provide a huge surface area for gas exchange. Air enters and leaves these alveoli through the same system of airways.

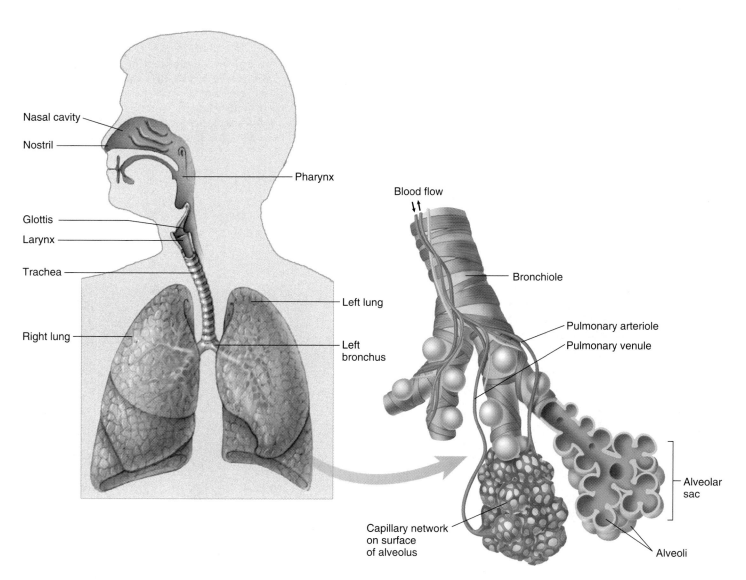

FIGURE 50.10
The human respiratory system and the structure of the mammalian lung. The lungs of mammals have an enormous surface area because of the millions of alveoli that cluster at the ends of the bronchioles. This provides for efficient gas exchange with the blood.

Respiration in Birds

The avian respiratory system has a unique structure that affords birds the most efficient respiration of all terrestrial vertebrates. Unlike the blind-ended alveoli in the lungs of mammals, the bird lung channels air through tiny air vessels called parabronchi, where gas exchange occurs (figure 50.11a). Air flows through the parabronchi in one direction only; this is similar to the unidirectional flow of water through a fish gill, but markedly different from the two-way flow of air through the airways of other terrestrial vertebrates. In other terrestrial vertebrates, the inhaled fresh air is mixed with "old" oxygen-depleted air that was not exhaled from the previous breathing cycle. In birds, only fresh air enters the parabronchi of the lung, and the old air exits the lung by a different route.

The unidirectional flow of air through the parabronchi of an avian lung is achieved through the action of air sacs, which are unique to birds (figure 50.11b). There are two groups of air sacs, anterior and posterior. When they are expanded during *inspiration* they take in air, and when they are compressed during *expiration* they push air into and through the lungs.

If we follow the path of air through the avian respiratory system, we will see that respiration occurs in *two cycles.*

Each cycle has an inspiration and expiration phase—but the air inhaled in one cycle is not exhaled until the second cycle. Upon inspiration, both anterior and posterior air sacs expand and take in air. The inhaled air, however, only enters the posterior air sacs; the anterior air sacs fill with air from the lungs (figure 50.11c). Upon expiration, the air forced out of the anterior air sacs is exhaled, but the air forced out of the posterior air sacs enters the lungs. This process is repeated in the second cycle, so that air flows through the lungs in one direction and is exhaled at the end of the second cycle.

The unidirectional flow of air also permits a second respiratory efficiency: the flow of blood through the avian lung runs at a 90° angle to the air flow. This cross-current flow is not as efficient as the 180° countercurrent flow in fish gills, but it still enables the avian lung to extract more oxygen from the air than a mammalian lung can. Because of the unidirectional air flow in the parabronchi and cross-current blood flow, a sparrow can be active at an altitude of 6000 meters while a mouse, which has a similar body mass and metabolic rate, cannot respire successfully at that elevation.

> **The avian respiration system is the most efficient among terrestrial vertebrates because it has unidirectional air flow and cross-current blood flow through the lungs.**

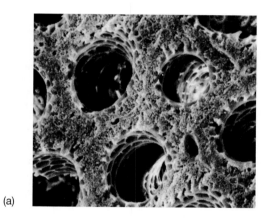

(a)

(b)

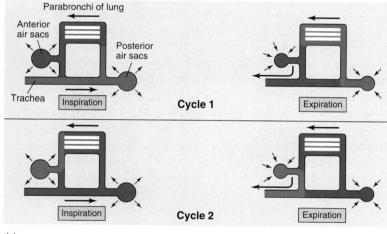

(c)

FIGURE 50.11

How a bird breathes. (a) Cross section of lung of a domestic chicken (75×). Air travels through tiny tunnels in the lungs, called parabronchi, while blood circulates within the fine lattice at right angles to the air flow. This crosscurrent flow makes the bird lung very efficient at extracting oxygen. (b) Birds have a system of air sacs, divided into an anterior group and posterior group, that extend between the internal organs and into the bones. (c) Breathing occurs in two cycles. *Cycle 1:* Inhaled air (shown in red) is drawn from the trachea into the posterior air sacs and then is exhaled into the lungs. *Cycle 2:* Air is drawn from the lungs into the anterior air sacs and then is exhaled through the trachea. Passage of air through the lungs is always in the same direction, from posterior to anterior (right to left in this diagram).

Structures and Mechanisms of Breathing

In mammals, inspired air travels through the trachea, bronchi, and bronchioles to reach the alveoli, where gas exchange occurs. Each alveolus is composed of an epithelium only one cell thick, and is surrounded by blood capillaries with walls that are also only one cell layer thick. There are about 30 billion capillaries in both lungs, or about 100 capillaries per alveolus. Thus, an alveolus can be visualized as a microscopic air bubble whose entire surface is bathed by blood. Since the alveolar air and the capillary blood are separated by only two cell layers, the distance between the air and blood is only 0.5 to 1.5 micrometers, allowing for the rapid exchange of gases by diffusion.

The blood leaving the lungs, as a result of this gas exchange, normally contains a partial oxygen pressure (P_{O_2}) of about 100 millimeters of mercury. As previously discussed, the P_{O_2} is a measure of the concentration of dissolved oxygen—you can think of it as indicating the plasma oxygen. Since the P_{O_2} of the blood leaving the lungs is close to the P_{O_2} of the air in the alveoli (about 105 mm Hg), the lungs do a very effective, but not perfect, job of oxygenating the blood. After gas exchange in the systemic capillaries, the blood that returns to the right side of the heart is depleted in oxygen, with a P_{O_2} of about 40 millimeters of mercury. These changes in the P_{O_2} of the blood, as well as the changes in plasma carbon dioxide (indicated as the P_{CO_2}), are shown in figure 50.12.

The outside of each lung is covered by a thin membrane called the **visceral pleural membrane.** A second, **parietal pleural membrane** lines the inner wall of the thoracic cavity. The space between these two membranes, the **pleural cavity,** is normally very small and filled with fluid. This fluid links the two membranes in the same way a thin film of water can hold two plates of glass together, effectively coupling the lungs to the thoracic cavity. The pleural membranes package each lung separately—if one collapses due to a perforation of the membranes, the other lung can still function.

Mechanics of Breathing

As in all other terrestrial vertebrates except amphibians, air is drawn into the lungs by the creation of a negative, or subatmospheric, pressure. In accordance with *Boyle's Law,* when the volume of a given quantity of gas increases its pressure decreases. This occurs because the volume of the thorax is increased during inspiration (inhalation), and the lungs likewise expand because of the adherence of the visceral and parietal pleural membranes. When the pressure within the lungs is lower than the atmospheric pressure, air enters the lungs.

The thoracic volume is increased through contraction of two sets of muscles: the *external intercostals* and the *diaphragm.* During inspiration, contraction of the external intercostal muscles between the ribs raises the ribs and expands the rib cage. Contraction of the diaphragm, a convex sheet of striated muscle separating the thoracic cavity from the abdominal cavity, causes the diaphragm to lower and assume a more flattened shape. This expands the volume of the thorax and lungs while it increases the pressure on the abdomen (causing the belly to protrude). You can force a deeper inspiration by contracting other muscles that insert on the sternum or rib cage and expand the thoracic cavity and lungs to a greater extent (figure 50.13a).

The thorax and lungs have a degree of *elasticity*—they tend to resist distension and they recoil when the distending force subsides. Expansion of the thorax and lungs during inspiration places these structures under elastic tension. It is the relaxation of the external intercostal muscles and diaphragm

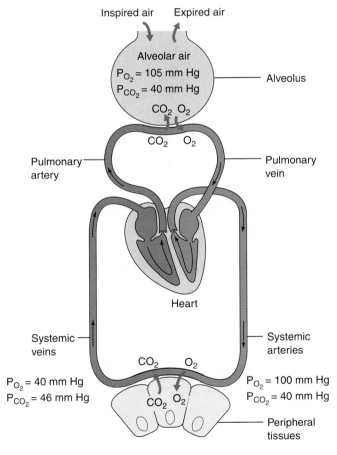

FIGURE 50.12

Gas exchange in the blood capillaries of the lungs and systemic circulation. As a result of gas exchange in the lungs, the systemic arteries carry oxygenated blood with a relatively low carbon dioxide concentration. After the oxygen is unloaded to the tissues, the blood in the systemic veins has a lowered oxygen content and an increased carbon dioxide concentration.

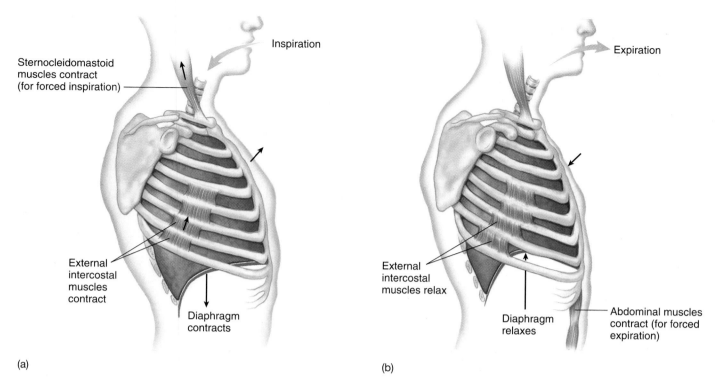

(a)

(b)

FIGURE 50.13

How a human breathes. (a) Inspiration. The diaphragm contracts and the walls of the chest cavity expand, increasing the volume of the chest cavity and lungs. As a result of the larger volume, air is drawn into the lungs. (b) Expiration. The diaphragm and chest walls return to their normal positions as a result of elastic recoil, reducing the volume of the chest cavity and forcing air out of the lungs through the trachea. Note that inspiration can be forced by contracting accessory respiratory muscles (such as the sternocleidomastoid), and expiration can be forced by contracting abdominal muscles.

that produces unforced expiration, because it relieves that elastic tension and allows the thorax and lungs to recoil. You can force a greater expiration by contracting your abdominal muscles and thereby pressing the abdominal organs up against the diaphragm (figure 50.13*b*).

Breathing Measurements

A variety of terms are used to describe the volume changes of the lung during breathing. At rest, each breath moves a **tidal volume** of about 500 milliliters of air into and out of the lungs. About 150 milliliters of the tidal volume is contained in the tubular passages (trachea, bronchi, and bronchioles), where no gas exchange occurs. The air in this *anatomical dead space* mixes with fresh air during inspiration. This is one of the reasons why respiration in mammals is not as efficient as in birds, where air flow through the lungs is one-way.

The maximum amount of air that can be expired after a forceful, maximum inspiration is called the **vital capacity.** This measurement, which averages 4.6 liters in young men and 3.1 liters in young women, can be clinically important, since an abnormally low vital capacity may indicate damage to the alveoli in various pulmonary disorders. For example,

in **emphysema,** a potentially fatal condition usually caused by cigarette smoking, vital capacity is reduced as the alveoli are progressively destroyed.

A person normally breathes at a rate and depth that properly oxygenate the blood and remove carbon dioxide, keeping the blood P_{O_2} and P_{CO_2} within a normal range. If breathing is insufficient to maintain normal blood gas measurements (a rise in the blood P_{CO_2} is the best indicator), the person is **hypoventilating.** If breathing is excessive for a particular metabolic rate, so that the blood P_{CO_2} is abnormally lowered, the person is said to be **hyperventilating.** Perhaps surprisingly, the increased breathing that occurs during moderate exercise is not necessarily hyperventilation, because the faster breathing is matched to the faster metabolic rate, and blood gas measurements remain normal. The next section describes how breathing is regulated to keep pace with metabolism.

Humans inspire by contracting muscles that insert on the rib cage and by contracting the diaphragm. Expiration is produced primarily by muscle relaxation and elastic recoil. As a result, the blood oxygen and carbon dioxide levels are maintained in a normal range through adjustments in the depth and rate of breathing.

Mechanisms That Regulate Breathing

Each breath is initiated by neurons in a *respiratory control center* located in the medulla oblongata, a part of the brain stem (see chapter 51). These neurons send impulses to the diaphragm and external intercostal muscles, stimulating them to contract, and contractions of these muscles expand the chest cavity, causing inspiration. When these neurons stop producing impulses, the inspiratory muscles relax and expiration occurs. Although the muscles of breathing are skeletal muscles, they are usually controlled automatically. This control can be voluntarily overridden, however, as in hypoventilation (breath holding) or hyperventilation.

A proper rate and depth of breathing is required to maintain the blood oxygen and carbon dioxide levels in the normal range. Thus, although the automatic breathing cycle is driven by neurons in the brain stem, these neurons must be responsive to changes in blood P_{O_2} and P_{CO_2} in order to maintain homeostasis. You can demonstrate this mechanism, a negative feedback loop, by simply holding your breath. When you do this, your blood carbon dioxide immediately rises and your blood oxygen falls. After a short time, the urge to breathe induced by the changes in blood gases becomes overpowering. This is due primarily to the rise in blood carbon dioxide, as indicated by a rise in P_{CO_2}, rather than to the fall in oxygen levels.

A rise in P_{CO_2} causes an increased production of carbonic acid (H_2CO_3), which is formed from carbon dioxide and water and acts to lower the blood pH (increase its H^+ concentration). A fall in blood pH stimulates neurons in the **aortic** and **carotid bodies,** which are sensory structures known as *peripheral chemoreceptors* in the aorta and the carotid artery. These receptors send impulses to the respiratory control center in the medulla oblongata, which then stimulates increased breathing. The brain also contains chemoreceptors, but they cannot be stimulated by blood H^+ because the blood is unable to enter the brain. After a brief delay, however, the increased blood P_{CO_2} also causes a decrease in the pH of the cerebrospinal fluid (CSF) bathing the brain. This stimulates the **central chemoreceptors** in the brain (figure 50.14).

The peripheral chemoreceptors are responsible for the immediate stimulation of breathing when the blood P_{CO_2} rises, and the central chemoreceptors are responsible for

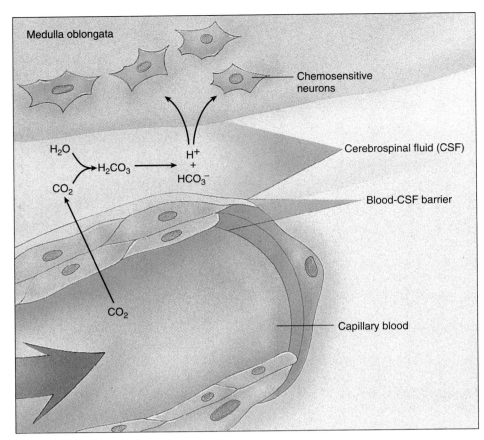

FIGURE 50.14
The effect of blood CO_2 on cerebrospinal fluid (CSF). Changes in the pH of the CSF are detected by chemosensitive neurons in the brain that help regulate breathing.

the sustained increase if P_{CO_2} remains elevated. The increased respiratory rate then acts to eliminate the extra CO_2, bringing the blood pH back to normal (figure 50.15).

A person cannot voluntarily hyperventilate for too long. The decrease in plasma P_{CO_2} and increase in pH of plasma and CSF caused by hyperventilation extinguish the reflex drive to breathe. They also lead to constriction of cerebral blood vessels, causing dizziness. People can hold their breath longer if they hyperventilate first, because it takes longer for the CO_2 levels to build back up, not because hyperventilation increases the P_{O_2} of the blood. Actually, in people with normal lungs, P_{O_2} becomes a significant stimulus for breathing only at high altitudes, where the P_{O_2} is low. Low P_{O_2} can also stimulate breathing in patients with emphysema, where the lungs are so damaged that blood CO_2 can never be adequately eliminated.

Breathing serves to keep the blood gases and pH in the normal range and is under the reflex control of peripheral and central chemoreceptors. These chemoreceptors sense the pH of the blood and cerebrospinal fluid, and they regulate the respiratory control center in the medulla oblongata of the brain.

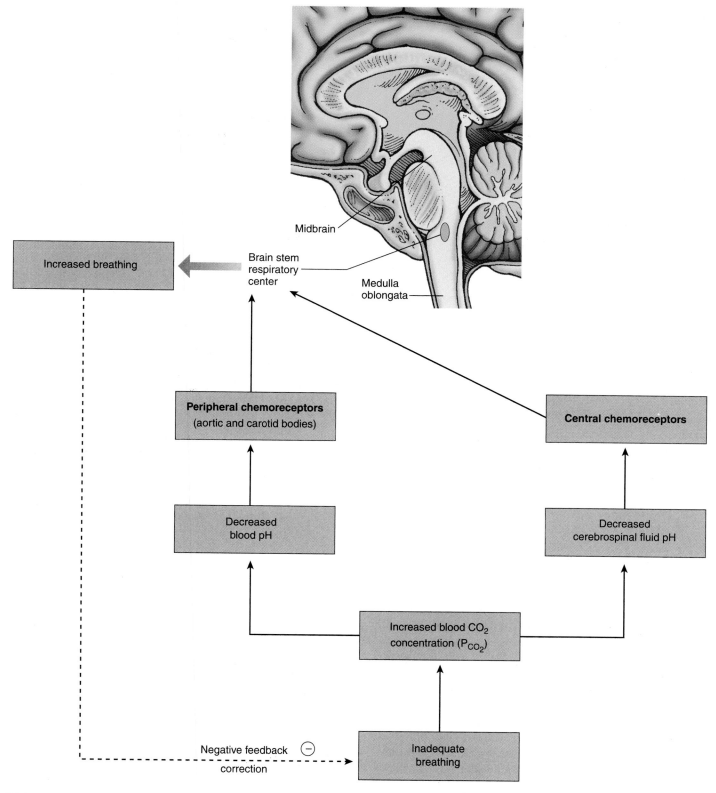

FIGURE 50.15

The regulation of breathing by chemoreceptors. Peripheral and central chemoreceptors sense a fall in the pH of blood and cerebrospinal fluid, respectively, when the blood carbon dioxide levels rise as a result of inadequate breathing. In response, they stimulate the respiratory control center in the medulla oblongata, which directs an increase in breathing. As a result, the blood carbon dioxide concentration is returned to normal, completing the negative feedback loop.

Hemoglobin and Oxygen Transport

When oxygen diffuses from the alveoli into the blood, its journey is just beginning. As described in chapter 49, the circulatory system delivers oxygen to all tissues for cell respiration and carries away carbon dioxide. The transport of oxygen and carbon dioxide by the blood is itself an interesting and physiologically important process.

The amount of oxygen that can be dissolved in the blood plasma depends directly on the P_{O_2} of the air in the alveoli, as we explained earlier. When the lungs are functioning normally, the blood plasma leaving the lungs has almost as much dissolved oxygen as is theoretically possible, given the P_{O_2} of the air. Because of oxygen's low solubility in water, however, blood plasma can contain a maximum of only about 3 milliliters O_2 per liter. Yet whole blood carries almost 200 milliliters O_2 per liter! The reason for the difference between these numbers is that most of the oxygen is bound to molecules of hemoglobin inside the red blood cells.

Hemoglobin is a protein composed of four polypeptide chains and four organic compounds called *heme groups.* At the center of each heme group is an atom of iron, which can bind to a molecule of oxygen (figure 50.16). Thus, each hemoglobin molecule can carry up to four molecules of oxygen. Hemoglobin loads up with oxygen in the lungs, forming **oxyhemoglobin.** This molecule has a bright red, tomato juice color. As blood passes through capillaries in the rest of the body, some of the oxyhemoglobin releases oxygen and becomes **deoxyhemoglobin.** Deoxyhemoglobin has a dark red color (the color of blood that is collected from the veins of blood donors), but it imparts a bluish tinge to tissues. Because of these color changes, vessels that carry oxygenated blood are always shown in artwork with a red color, and vessels that carry oxygen-depleted blood are indicated with a blue color.

Hemoglobin is an ancient protein that is not only the oxygen-carrying molecule in all vertebrates, but is also used as an oxygen carrier by many invertebrates, including annelids, mollusks, echinoderms, flatworms, and even some protists. Many other invertebrates, however, employ different oxygen carriers, such as *hemocyanin.* In hemocyanin, the oxygen-binding atom is copper instead of iron. Hemocyanin is not found in blood cells, but is instead dissolved in the circulating fluid (hemolymph) of invertebrates.

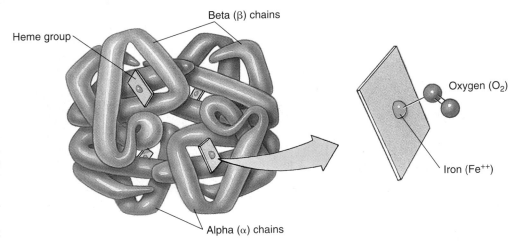

FIGURE 50.16
Hemoglobin consists of four polypeptide chains—two alpha (α) chains and two beta (β) chains. Each chain is associated with a heme group, and each heme group has a central iron atom, which can bind to a molecule of O_2.

Oxygen Transport

The P_{O_2} of the air within alveoli at sea level is approximately 105 millimeters of mercury (mm Hg), which is less than the P_{O_2} of the atmosphere because of the mixing of freshly inspired air with "old" air in the anatomical dead space of the respiratory system. The P_{O_2} of the blood leaving the alveoli is slightly less than this, about 100 mm Hg, because the blood plasma is not completely saturated with oxygen due to slight inefficiencies in lung function. At a blood P_{O_2} of 100 mm Hg, approximately 97% of the hemoglobin within red blood cells is in the form of oxyhemoglobin—indicated as a percent oxyhemoglobin saturation of 97%.

As the blood travels through the systemic blood capillaries, oxygen leaves the blood and diffuses into the tissues. Consequently, the blood that leaves the tissue in the veins has a P_{O_2} that is decreased (in a resting person) to about 40 mm Hg. At this lower P_{O_2}, the percent saturation of hemoglobin is only 75%. A graphic representation of these changes is called an oxyhemoglobin dissociation curve (figure 50.17a). In a person at rest, therefore, 22% (97% minus 75%) of the oxyhemoglobin has released its oxygen to the tissues. Put another way, roughly one-fifth of the oxygen is unloaded in the tissues, leaving four-fifths of the oxygen in the blood as a reserve.

This large reserve of oxygen serves two important functions. First, it enables the blood to supply the body's oxygen needs during exercise as well as at rest. During exercise, the muscles' accelerated metabolism uses more oxygen from the capillary blood and thus decreases the venous blood P_{O_2}. For example, the P_{O_2} of the venous blood could drop to 20 mm Hg; in this

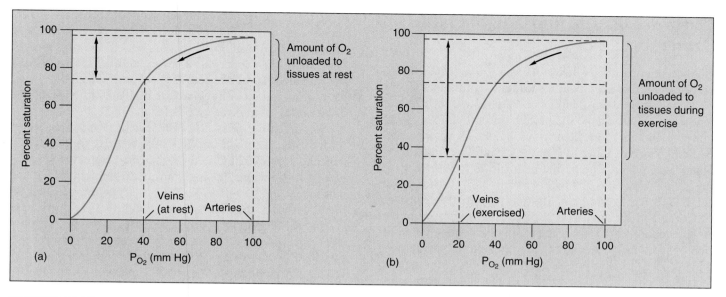

FIGURE 50.17

The oxyhemoglobin dissociation curve. Hemoglobin combines with O_2 in the lungs, and this oxygenated blood is carried by arteries to the body cells. After oxygen is removed from the blood to support cell respiration, the blood entering the veins contains less oxygen. The difference in O_2 content between arteries and veins during (a) rest and (b) exercise shows how much O_2 was unloaded to the tissues.

case, the percent saturation of hemoglobin will be only 35% (figure 50.17*b*). Since arterial blood still contains 97% oxyhemoglobin (ventilation increases proportionately with exercise), the amount of oxygen unloaded is now 62% (97% minus 35%), instead of the 22% at rest. The second function of the oxygen reserve is that it ensures that the blood contains enough oxygen to maintain life for four to five minutes if breathing is interrupted or if the heart stops pumping.

The CO_2 produced by metabolizing tissues as a product of aerobic respiration combines with H_2O to form carbonic acid (H_2CO_3), lowering the pH of the blood. This reaction occurs primarily inside red blood cells, where the lowered pH reduces hemoglobin's affinity for oxygen and thus causes it to release oxygen more readily. The effect of pH on hemoglobin's affinity for oxygen is known as the Bohr effect and is shown graphically by a shift of the oxyhemoglobin dissociation curve to the right (figure 50.18). Since skeletal muscles produce carbon dioxide more rapidly during exercise, the blood unloads a higher percentage of the oxygen it carries during exercise as a result of the Bohr effect.

Deoxyhemoglobin combines with oxygen in the lungs to form oxyhemoglobin, which dissociates in the tissue capillaries to release its oxygen. The degree to which the loading reaction occurs depends on ventilation; the degree of unloading is influenced by such factors as temperature and pH.

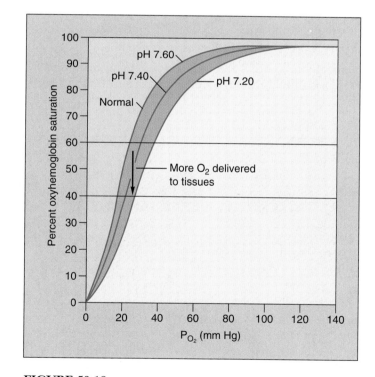

FIGURE 50.18

The effect of pH on the oxyhemoglobin dissociation curve. Lower blood pH shifts the oxyhemoglobin dissociation curve to the right, facilitating oxygen unloading. This can be seen as a lowering of the oxyhemoglobin percent saturation from 60% to 40% in the example shown, indicating that the difference of 20% more oxygen is unloaded to the tissues.

Carbon Dioxide and Nitric Oxide Transport

The systemic capillaries deliver oxygen to the tissues and remove carbon dioxide. About 8% of the CO_2 in blood is simply dissolved in plasma; another 20% is bound to hemoglobin. (Since CO_2 binds to the protein portion of hemoglobin, however, and not to the heme irons, it does not compete with oxygen.) The remaining 72% of the CO_2 diffuses into the red blood cells, where the enzyme carbonic anhydrase catalyzes the combination of CO_2 with water to form carbonic acid (H_2CO_3). Carbonic acid dissociates into bicarbonate (HCO_3^-) and hydrogen (H^+) ions. This reaction removes large amounts of CO_2 from the plasma, facilitating the diffusion of additional CO_2 into the plasma from the surrounding tissues. The formation of carbonic acid is also important in maintaining the acid-base balance of the blood, because bicarbonate serves as the major buffer of the blood plasma.

The blood carries CO_2 in these forms to the lungs. The lower P_{CO_2} of the air inside the alveoli causes the carbonic anhydrase reaction to proceed in the reverse direction, converting H_2CO_3 into H_2O and CO_2. The CO_2 diffuses out of the red blood cells and into the alveoli, so that it can leave the body in the next exhalation (figure 50.19).

Nitric Oxide Transport

Hemoglobin also has the ability to hold and release nitric oxide gas (NO). Although a noxious gas in the atmosphere, nitric oxide has an important physiological role in the body and acts on many kinds of cells to change their shapes and functions. For example, in blood vessels the presence of NO causes the blood vessels to expand because it relaxes the surrounding muscle cells (see chapters 7 and 49). Thus, blood flow and blood pressure are regulated by the amount of NO released into the bloodstream.

Hemoglobin carries NO in a special form called super nitric oxide. In this form, NO has acquired an extra electron and is able to bind to the amino acid cysteine in hemoglobin. In the lungs, hemoglobin that is dumping CO_2 and picking up O_2 also picks up NO as super nitric oxide. In blood vessels at the tissues, hemoglobin that is releasing its O_2 and picking up CO_2 can do one of two things with nitric oxide. To increase blood flow, hemoglobin can release the super nitric oxide as NO into the blood, making blood vessels expand because NO acts as a relaxing agent. Or, hemoglobin can trap any excess of NO on its iron atoms left vacant by the release of oxygen, causing blood vessels to constrict. When the red blood cells return to the lungs, hemoglobin dumps its CO_2 and the regular form of NO bound to the iron atoms. It is then ready to pick up O_2 and super nitric oxide and continue the cycle.

Carbon dioxide combines with water in the tissue capillaries to form carbonic acid, which is primarily formed within the red blood cells as a result of an enzymatic reaction. The reverse reaction occurs in the lungs so that CO_2 gas can be exhaled.

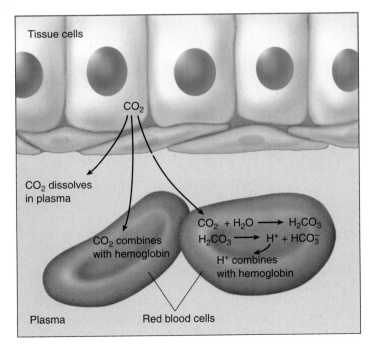
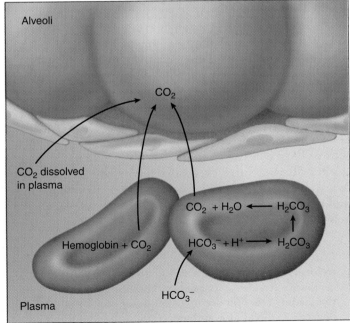

FIGURE 50.19
The transport of carbon dioxide by the blood. CO_2 is transported in three ways: dissolved in plasma, bound to the protein portion of hemoglobin, and as carbonic acid and bicarbonate, which form in the red blood cells. When the blood passes through the pulmonary capillaries, these reactions are reversed so that CO_2 gas is formed, which is exhaled.

50.1 Respiration involves the diffusion of gases.

- The factors that influence the rate of diffusion are described by Fick's Law, which states that the rate of diffusion varies directly with the surface area and the concentration gradient, and inversely with diffusion distance.

- Animals have evolved to maximize the diffusion rate across respiratory membranes by increasing the respiratory surface area, increasing the concentration gradient across the membrane, or decreasing the diffusion distance.

50.2 Gills are used for respiration by aquatic vertebrates.

- Water is forced past the surface of the gills of bony fish by either a pumping action or by ram ventilation.

- As water flows past the gill lamellae, it comes close to blood flowing in an opposite, or countercurrent, direction; this maximizes the concentration difference between the two fluids, thereby maximizing the diffusion of gases.

50.3 Lungs are used for respiration by terrestrial vertebrates.

- Amphibians force air into their lungs by positive pressure breathing; they also obtain a significant amount of oxygen through cutaneous respiration.

- Reptiles, birds, and mammals use negative pressure breathing; air is taken into the lungs when the lung volume is expanded to create a partial vacuum.

- Mammals have lungs composed of millions of alveoli, where gas exchange occurs; this is very efficient, but new air going into the lungs is mixed with some old air inspiration and expiration occur through the same airways.

- Birds have the most efficient respiratory system of terrestrial vertebrates because the flow of air in their respiratory system is unidirectional and the flow of blood is at a 90° angle to the flow of air.

50.4 Mammalian breathing is a dynamic process.

- The lungs are covered with a wet membrane that sticks to the wet membrane lining the thoracic cavity, so the lungs can expand as the chest expands through muscular contractions.

- A resting breath is called a tidal volume; the maximum amount of air that can be forcefully expired after a maximum inspiration is the vital capacity.

- Breathing is controlled by centers in the medulla oblongata of the brain; breathing is stimulated by a rise in blood CO_2, and consequent fall in blood pH, as sensed by chemoreceptors located in the aorta and carotid artery.

50.5 Blood transports oxygen and carbon dioxide.

- Hemoglobin loads with oxygen in the lungs to produce a normal percent oxyhemoglobin saturation of 97% in the arterial blood; this oxyhemoglobin then unloads a part of the oxygen it carries as the blood goes through the systemic capillaries.

- As the bond between hemoglobin and oxygen gets weaker because of increased temperature or decreased pH, the hemoglobin unloads a higher proportion of the oxygen it carries.

- Carbon dioxide combines with water to form carbonic acid, which dissociates to bicarbonate and H^+; this helps to regulate the acid-base balance of the blood as the carbon dioxide is transported to the lungs for exhalation.

Discussing Key Terms

1. **Partial pressure** The partial pressure of a gas is the product of its percent contribution to the gas mixture and the total pressure of the gas mixture; the amount of the gas that can be dissolved in a fluid is directly proportional to its partial pressure.

2. **Countercurrent flow** The flow of two fluids in opposite directions past each other is countercurrent flow. In the gills of fishes, water flows in a direction opposite to the direction of blood, maximizing gas exchange by diffusion.

3. **Negative/positive pressure breathing** In negative pressure breathing, inhalation is produced by creating a partial vacuum, or negative pressure, in the lungs. This is different from positive pressure breathing, in which air is forced into the lungs.

4. **Alveoli** The tiny air sacs of mammalian lungs are called alveoli; their large number contributes to an enormous surface area, thereby maximizing the rate of diffusion of gases.

5. **Hyperventilation** Hyperventilation refers to excessive breathing relative to the rate of metabolism such that the blood levels of carbon dioxide are decreased below normal.

Review Questions

1. What is the most common gas in dry air? Approximately what percentage of dry air is oxygen, and what percentage is carbon dioxide?

2. Why is it that only very small organisms can satisfy their respiratory requirements by direct diffusion to all cells from the body surface?

3. What is countercurrent flow, and how does it help make the fish gill the most efficient respiratory organ?

4. How do amphibians get air into their lungs? How does this differ from the way other terrestrial vertebrates get air into their lungs?

5. What two features of the respiratory system in birds make it the most efficient of all terrestrial respiratory systems?

6. How are the lungs connected to and supported within the thoracic cavity? Describe the pathway of air flow in the lungs.

7. How does the brain control inspiration and expiration? How do peripheral and central chemoreceptors influence the brain's control of breathing?

8. In what form does most of the carbon dioxide travel in the blood? How and where is this molecule produced? What is the physiological significance of the formation of this molecule?

Thought Questions

1. People who appear to have drowned can often be revived, in some cases after being underwater for as long as half an hour. In every case of full recovery after extended submergence, however, the person had been submerged in very cold water. Can you explain this observation?

2. Why can't fish obtain oxygen from air? Why can't humans or other mammals obtain oxygen from water?

3. In what ways is the amphibian respiratory system inefficient, and how do these animals compensate for that fact?

4. Frogs store fat in internal fat bodies; they lack subcutaneous fat. Why can't they have subcutaneous fat the way mammals do?

5. Why should birds have the most efficient respiratory system of all terrestrial vertebrates? In addition to their function in avian respiration, what other benefits might the air sacs of birds provide?

6. Horror movies sometimes show insects the size of elephants terrorizing a town. Can insects theoretically get that big? Why?

Internet Links

The Link Between Cigarette Smoking and Lung Cancer
http://www.oncolink.upenn.edu/disease/lung1/
A site providing the latest information on lung cancer and its relation to cigarette smoking, ONCOLINK is maintained by the University of Pennsylvania.

Lung Cancer Information Library
http://www.meds.com/lung/lunginfo.html
This site provides an electronic library of comprehensive educational materials on lung cancer for doctors and patients.

How Birds Breath
http://www.sci.sdsu.edu/multimedia/birdlungs/
From San Diego State University's HOW ANIMALS WORK collection of shockwave animations, a very clear animation of this multiphase process.

For Further Reading

Feder, M. and W. W. Burggren: "Skin Breathing in Vertebrates," *Scientific American*, November 1985, pages 126–42. Many terrestrial vertebrates breathe though their skin. Indeed, some amphibians have dispensed with lungs altogether.

Perutz, M. F.: "Hemoglobin Structure and Respiratory Transport," *Scientific American*, December 1978, pages 92–125. An account of how hemoglobin changes its shape to facilitate oxygen binding and unloading, by the man who won a Nobel Prize in chemistry for unraveling the structure of hemoglobin.

Randall, D. J., W. W. Burggren, A. P. Farrell, and M. S. Haswell: *The Evolution of Air Breathing in Vertebrates*, Cambridge University Press, Cambridge, England, 1981. An account of the physiological changes in respiration that occurred during the evolution of the terrestrial vertebrates.

Schmidt-Nielsen, K.: "How Birds Breathe," *Scientific American*, December 1971, pages 73–79. A fascinating account of the discovery of unidirectional flow in avian lungs, by a great comparative physiologist.

Storey, K. and J. Storey: "Frozen and Alive," *Scientific American*, December 1990, pages 92–97. Some animals survive the winter by freezing solid, using special proteins and antifreeze to avoid ice damage.

Uterman, G.: "The Mysteries of Lipoprotein *a*," *Science*, vol. 246, November 1989, pages 904–10. Lipoprotein *a*, when present in high concentrations in the bloodstream, signals the likelihood of heart attack and stroke.

51

The Nervous System

Concept Outline

51.1 The nervous system consists of neurons and supporting cells.

Neuron Organization. Neurons are organized into the central nervous system (the brain and spinal cord) and the peripheral nervous system (sensory and motor neurons).

51.2 Nerve impulses are produced on the axon membrane.

The Membrane Potential. The inside of the membrane is electrically negative in comparison with the outside.
Generation of Action Potentials. In response to a stimulus that reduces the membrane voltage, membrane channels open, producing a nerve impulse.
Propagation of Action Potentials. One action potential stimulates the production of the next along the axon.

51.3 Neurons form junctions called synapses with other cells.

Structure of Synapses. Neurotransmitters diffuse across to the postsynaptic cell and combine with receptor proteins.
Neurotransmitters and Their Functions. Some neurotransmitters cause a depolarization in the postsynaptic membrane; others produce inhibition by hyperpolarization.

51.4 The central nervous system consists of the brain and spinal cord.

The Evolution of the Vertebrate Brain. Vertebrate brains include a forebrain, midbrain, and hindbrain.
The Human Forebrain. The cerebral cortex contains areas specialized for different functions.
The Spinal Cord. Reflex responses and messages to and from the brain are coordinated by the spinal cord.

51.5 The peripheral nervous system consists of sensory and motor neurons.

Components of the Peripheral Nervous System. A spinal nerve contains sensory and motor neurons.
The Autonomic Nervous System. Sympathetic motor neurons arouse the body for fight or flight; parasympathetic motor neurons have antagonistic actions.

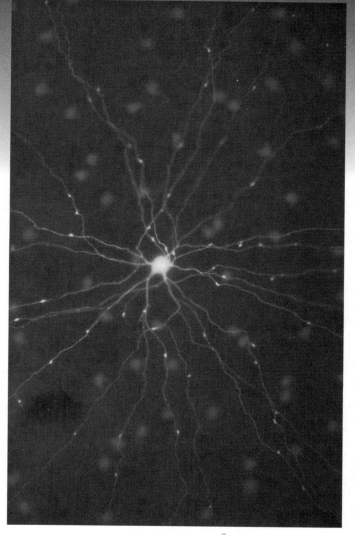

FIGURE 51.1
A neuron in the retina of the eye (500×). This neuron has been injected with a fluorescent dye, making its cell body and long dendrites readily apparent.

All animals except sponges use a network of nerve cells to gather information about the body's condition and the external environment, to process and integrate that information, and to issue commands to the body's muscles and glands. Just as telephone cables run from every compartment of a submarine to the conning tower, where the captain controls the ship, so bundles of nerve cells called neurons connect every part of an animal's body to its command and control center, the brain and spinal cord (figure 51.1). The animal body is run just like a submarine, with status information about what is happening in organs and outside the body flowing into the command center, which analyzes the data and issues commands to glands and muscles.

Neuron Organization

An animal must be able to respond to environmental stimuli. To do this, it must have sensory receptors that can detect the stimulus and motor *effectors* that can respond to it. In most invertebrate phyla and in all vertebrate classes, sensory receptors and motor effectors are linked by way of the nervous system. As described in chapter 46, the nervous system consists of neurons and supporting cells. **Sensory (or afferent) neurons** carry impulses from sensory receptors; **motor (or efferent) neurons** carry impulses to effectors—muscles and glands (figure 51.2).

In addition to sensory and motor neurons, a third type of neuron is present in the nervous systems of most invertebrates and all vertebrates: **association neurons (or interneurons)**. These neurons are located in the brain and spinal cord of vertebrates, together called the **central nervous system (CNS)**, where they help provide more complex reflexes and higher associative functions, including learning and memory. Sensory neurons carry impulses into the CNS, and motor neurons carry impulses away from the CNS. Together, sensory and motor neurons constitute the **peripheral nervous system (PNS)** of vertebrates. Motor neurons that stimulate skeletal muscles to contract are **somatic motor neurons,** and those that regulate the activity of the smooth muscles, cardiac muscle, and glands are **autonomic motor**

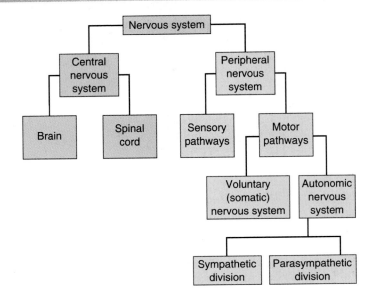

FIGURE 51.3
The divisions of the vertebrate nervous system. The major divisions are the central and peripheral nervous systems.

neurons. The autonomic motor neurons are further subdivided into the **sympathetic** and **parasympathetic** systems, which act to counterbalance each other (figure 51.3).

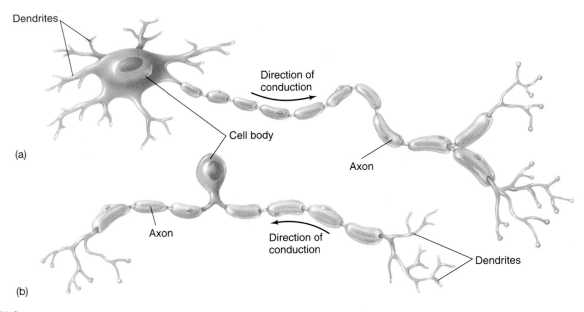

FIGURE 51.2
A motor and a sensory neuron. (a) The motor neuron carries impulses or "commands" to muscles and glands (effectors). (b) The sensory neuron carries information about the environment to the brain and spinal cord.

Despite their varied appearances, all neurons have the same functional architecture (figure 51.4). The cell body is an enlarged region containing the nucleus. Extending from the cell body are one or more cytoplasmic extensions called **dendrites.** Motor and association neurons possess a profusion of highly branched dendrites, enabling those cells to receive information from many different sources simultaneously. The surface of the cell body integrates the information arriving at its dendrites. If the resulting membrane excitation is sufficient, it triggers impulses that are conducted away from the cell body along an **axon.** Each neuron has a single axon leaving its cell body, although an axon may produce small terminal branches to stimulate a number of cells. An axon can be quite long: the axons controlling the muscles in your feet are more than a meter long, and the axons that extend from the skull to the pelvis in a giraffe are about three meters long!

Neurons are supported both structurally and functionally by **supporting cells,** which are called **neuroglia** in the CNS. These cells are ten times more numerous than neurons and serve a variety of functions. Two of the most important kinds of supporting neurons are the **Schwann cells** and **oligodendrocytes,** which produce **myelin sheaths** that surround the axons of many neurons. Schwann cells produce myelin in the PNS, while oligodendrocytes produce myelin in the CNS. During development, these cells wrap themselves around each axon several times to form the myelin sheath, an insulating covering consisting of multiple layers of membrane (figure 51.5). Axons that have myelin sheaths are said to be myelinated, and those that don't are unmyelinated. In the CNS, myelinated axons form the **white matter,** and the unmyelinated dendrites and cell bodies form the **gray matter.** In the PNS, both myelinated and unmyelinated axons are bundled together, much like wires in a cable, to form **nerves.**

The myelin sheath is interrupted at intervals of 1 to 2 mm by small gaps known as **nodes of Ranvier** (see figure 51.4). The role of the myelin sheath in impulse conduction will be discussed later in this chapter.

Neurons and supporting cells make up the central and peripheral nervous systems in vertebrates. Sensory, motor, and association neurons play different roles in the nervous system, and the supporting cells aid their function, in part by producing myelin sheaths.

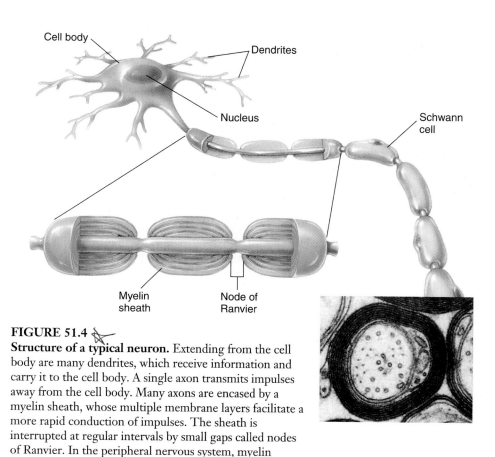

FIGURE 51.4
Structure of a typical neuron. Extending from the cell body are many dendrites, which receive information and carry it to the cell body. A single axon transmits impulses away from the cell body. Many axons are encased by a myelin sheath, whose multiple membrane layers facilitate a more rapid conduction of impulses. The sheath is interrupted at regular intervals by small gaps called nodes of Ranvier. In the peripheral nervous system, myelin sheaths are formed by supporting Schwann cells.

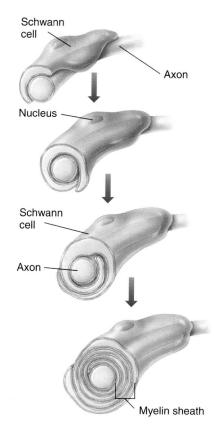

FIGURE 51.5
The formation of the myelin sheath around a peripheral axon. The myelin sheath is formed by successive wrappings of Schwann cell membranes, leaving most of the Schwann cell cytoplasm outside the myelin.

The Membrane Potential

Nerve impulses are electrical signals produced on the plasma membrane of a neuron. To understand how these signals are generated and transmitted within the nervous system, we must first examine some of the electrical properties of plasma membranes.

The battery in a car or a flashlight separates electrical charges between its two poles. There is said to be a *potential difference*, or voltage, between the poles, with one pole being positive and the other negative. Similarly, a potential difference exists across every cell's plasma membrane. The side of the membrane exposed to the cytoplasm is the negative pole, and the side exposed to the extracellular fluid is the positive pole. This potential difference is called the membrane potential; in a neuron that is "at rest" (not producing impulses), it is referred to as the *resting* membrane potential. The magnitude of the potential difference across a battery (or a plasma membrane) is expressed in units called volts. In a car battery it is typically 12 volts, but in a cell it is much smaller.

For example, the resting membrane potential of a neuron is often about 70 millivolts (70 mV, or 0.07 volt). To indicate that the inside of the membrane is negative with respect to the outside, the membrane potential is written as −70 mV.

A cell's membrane potential is a result of three factors: (1) the action of the sodium-potassium pumps; (2) the different degrees of permeability of the plasma membrane to different ions; and (3) the presence of *fixed anions* (negatively charged ions)—proteins and organic phosphates that are negatively charged and unable to leave the cell. Since these conditions prevail in all cells, all cells have a membrane potential.

The action of the sodium-potassium pumps is of lesser overall significance in the generation of a membrane potential than the other two factors. The sodium-potassium pumps do contribute to the membrane potential, however, because they pump out three sodium ions (Na⁺) for every two potassium ions (K⁺) that they pump into the cell (figure 51.6). This primary active transport (see chapter 6) results in an unequal distribution of charges, with the inside of the cell having fewer positive charges than the outside.

FIGURE 51.6
The sodium-potassium pump. This pump transports three Na⁺ to the outside of the cell and simultaneously transports two K⁺ to the inside of the cell. This is an active transport carrier requiring the breakdown of ATP.

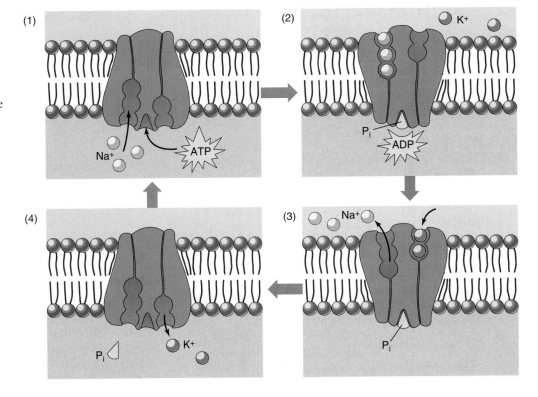

The action of the sodium-potassium pumps complements the more important effect of the fixed anions within the cell. These anions, which cannot leave the cell, attract *cations* (positively charged ions) from the extracellular environment. The plasma membrane, however, is differentially permeable to cations. The plasma membrane of a resting neuron is much more permeable to K⁺ than to any other cation, and so K⁺ enters the cell because of its electrical attraction for the fixed anions. Other cations can also enter to a lesser degree, but we can best understand the membrane potential if we first examine what would happen if K⁺ were the only cation able to enter the cell.

In that case, the electrical attraction of the fixed anions would cause a buildup in the K⁺ concentration inside the cell. The higher concentration of K⁺, in turn, would produce a diffusion gradient out of the cell. There are thus two opposing forces: an electrical gradient attracting K⁺ into the cell, and a diffusion gradient causing K⁺ to leave the cell. An equilibrium state would be reached when these two forces are balanced, and under these conditions, the membrane potential would be –90 mV. Put a different way, the inside of the cell would have to be 90 mV negative just to prevent the net diffusion of K⁺ out of the cell. This voltage is therefore called the **equilibrium potential** for potassium (table 51.1).

The concentration of Na⁺ is ten times higher outside the cell than inside; consequently, the membrane potential would have to be quite positive on the inside (+60 mV, as shown in table 51.1) to balance the tendency of Na⁺ to diffuse inward. Since the resting membrane potential of a neuron is –70 mV, it is clearly closer to the equilibrium potential of K⁺ (figure 51.7), as expected from the fact that the membrane is more permeable to K⁺. The resting membrane potential is a little less negative than –90 mV because some Na⁺ does diffuse into the cell. Since –70 mV is less negative than the –90 mV equilibrium potential for K⁺, some K⁺ also diffuses outward.

The resting membrane potential of an axon can be seen using an oscilloscope and a pair of electrodes, one outside and one inside the axon. A cell's membrane potential can change in response to appropriate stimulation. A shift in membrane potential in the positive direction, from –70 mV to –50 mV, for example, is called a depolarization. Conversely, a change in membrane potential that makes the inside of the membrane even more negative than it is at rest, such as a shift from –70 mV to –85 mV, is called a hyperpolarization.

The resting axon membrane maintains a potential difference as a result of the uneven distribution of charges, where the inside of the membrane is negatively charged in comparison with the outside (–70 mV). The magnitude, measured in millivolts, of this potential difference primarily reflects the difference in K⁺ concentration.

	Table 51.1 The Ionic Composition of Cytoplasm and Extracellular Fluid (ECF)			
Ion	Concentration in Cytoplasm (mM)	Concentration in ECF (mM)	Ratio	Equilibrium Potential (mV)
Na⁺	15	150	10:1	+60
K⁺	150	5	1:30	–90
Cl⁻	7	110	15:1	–70

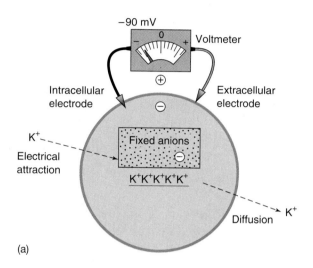

(a)

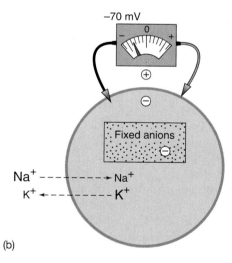

(b)

FIGURE 51.7

The establishment of the resting membrane potential. The fixed anions (primarily proteins and organic phosphates) attract cations from the extracellular fluid. (a) If the membrane were only permeable to K⁺, an equilibrium would be established and the membrane potential would be –90 mV. (b) A true resting membrane potential is about –70 mV, because the membrane does allow a low rate of Na⁺ diffusion into the cell. This is not quite sufficiently negative to prevent the outward diffusion of K⁺, so the cell is not at equilibrium and the action of the sodium-potassium pumps is required to maintain stability.

Generation of Action Potentials

Na$^+$ and K$^+$ channels in the plasma membrane have *gates*, portions of the channel protein that open or close the channel's pore. In the axons of neurons and in muscle fibers, the gates are closed or open depending on the membrane potential. Such channels are therefore known as *voltage-gated ion channels* (figure 51.8). In a resting cell, the Na$^+$ gates are closed, but many of the K$^+$ gates are open. As a result, the cell is more permeable to K$^+$ than to Na$^+$.

If the plasma membrane is depolarized slightly, an oscilloscope will show a small upward deflection of the line that soon decays back to the resting membrane potential. Once a particular level of depolarization is reached (about –55 mV in mammalian axons), however, a nerve impulse, or **action potential,** is produced. The level of depolarization needed to produce an action potential is called the *threshold*.

A depolarization that reaches or exceeds the threshold opens both the Na$^+$ and the K$^+$ channels, but the Na$^+$ channels open first. The rapid diffusion of Na$^+$ into the cell shifts the membrane potential toward the equilibrium potential for Na$^+$ (+60 mV—the positive sign indicates that the membrane reverses polarity as Na$^+$ rushes in). When the action potential is recorded on an oscilloscope, this part of the action potential appears as the *rising phase* of a spike (figure 51.9). The membrane potential never quite reaches +60 mV because the Na$^+$ channels close and, at about the same time, the K$^+$ channels that were previously closed begin to open. The action potential thus peaks at about +30 mV. Opening the K$^+$ channels allows K$^+$ to diffuse out of the cell, restoring the resting membrane potential. On an oscilloscope, this repolarization of the membrane appears as the *falling phase* of the action potential. In many cases, the repolarization carries the membrane potential to a value slightly more negative than the resting potential for a brief period, resulting in an *undershoot*. The entire sequence of events in an action potential is over in a few milliseconds.

Action potentials have two distinguishing characteristics. First, they follow an all-or-none law: each depolarization stimulus produces either a full action potential (if the stimulus is above threshold) or none at all. Secondly, action potentials are always separate events; they cannot add together or interfere with one another in any way because the membrane enters a refractory period for a short time after it generates an action potential.

The production of an action potential results entirely from the passive diffusion of ions. However, at the end of each action potential, the cytoplasm has a little more Na$^+$ and a little less K$^+$ than it did at rest. The constant activity of the sodium-potassium pumps compensates for these changes. Thus, although active transport is not required to produce action potentials, it is needed to maintain the ion gradients.

> The rapid inward diffusion of Na$^+$ followed by the outward diffusion of K$^+$ produces a rapid change in the membrane potential called an action potential. Action potentials are all-or-none events and cannot summate.

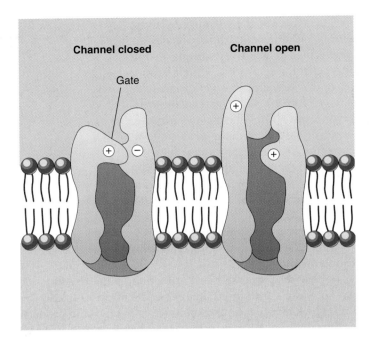

FIGURE 51.8
Voltage-gated ion channels. In neurons and muscle cells, the channels for Na$^+$ and K$^+$ have gates that are closed at the resting membrane potential but open when a threshold level of depolarization is attained.

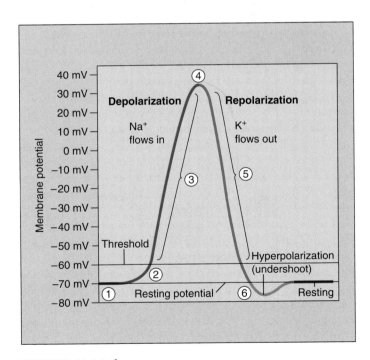

FIGURE 51.9
The action potential. (1) The resting membrane potential. (2) The depolarization stimulus reaches threshold. (3) Rapid depolarization occurs because sodium channels open, allowing Na$^+$ to diffuse into the axon. (4) Sodium channels close; potassium channels begin to open. (5) Repolarization occurs because of the diffusion of K$^+$ out of the axon. (6) Undershoot occurs, and then the membrane returns to its original resting potential.

Propagation of Action Potentials

Although we often speak of axons as conducting action potentials (impulses), action potentials do not really travel along an axon—they are events that are reproduced at different points along the axon membrane. This can occur for two reasons: action potentials are stimulated by depolarization, and an action potential can serve as a depolarization stimulus. Each action potential, during its rising phase, reflects a reversal in membrane polarity (from –70 mV to +30 mV) as Na$^+$ diffuses rapidly into the axon. The positive charges can depolarize the next region of membrane to threshold, so that the next region produces its own action potential (figure 51.10). Meanwhile, the previous region of membrane repolarizes back to the resting membrane potential. This is analogous to people in a stadium performing the "wave": individuals stay in place as they stand up (depolarize), raise their hands (peak of the action potential), and sit down again (repolarize).

Action potentials are conducted without decrement (without decreasing in amplitude); thus, the last action potential at the end of the axon is just as "new" as the first action potential. The velocity of conduction is greater if the diameter of the axon is large or if the axon is myelinated (table 51.2). Myelinated axons conduct impulses more rapidly than nonmyelinated axons because the action potentials in myelinated axons are only produced at the nodes of Ranvier. One action potential still serves as the depolarization stimulus for the next, but the depolarization at one node must spread to the next before the voltage-gated channels can be opened. The impulses therefore seem to jump from node to node (figure 51.11) in a process called **saltatory conduction** (Latin *saltare*, "to jump").

Action potentials are regenerated along an axon as one action potential serves as the depolarization stimulus for the next action potential.

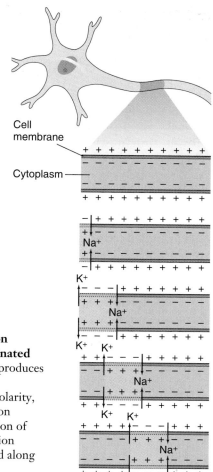

FIGURE 51.10
Propagation of an action potential in an unmyelinated axon. When one region produces an action potential and undergoes a reversal of polarity, it serves as a depolarization stimulus for the next region of the axon. In this way, action potentials are regenerated along each small region of the unmyelinated axon membrane.

Table 51.2	Conduction Velocities of Some Axons		
	Axon Diameter (mm)	**Myelin**	**Conduction Velocity (m/s)**
Squid giant axon	500	No	25
Large motor axon to human leg muscle	20	Yes	120
Axon from human skin pressure receptor	10	Yes	50
Axon from human skin temperature receptor	5	Yes	20
Motor axon to human internal organ	1	No	2

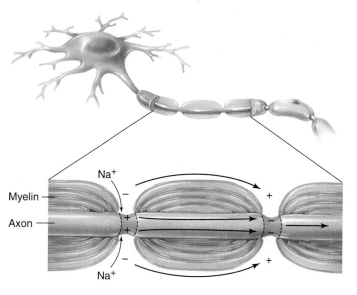

FIGURE 51.11
Saltatory conduction in a myelinated axon. Action potentials are only produced at the nodes of Ranvier in a myelinated axon. One node depolarizes the next node so that the action potentials can skip between nodes. As a result, saltatory ("leaping") conduction in a myelinated axon is more rapid than conduction in an unmyelinated axon.

Structure of Synapses

An action potential passing down an axon eventually reaches the end of the axon and all of its branches. These branches may form junctions with the dendrites of other neurons, with muscle cells, or with gland cells. Such intercellular junctions are called **synapses.** The neuron whose axon transmits action potentials to the synapse is the *presynaptic cell*, while the cell on the other side of the synapse is the *postsynaptic cell*. Although the presynaptic and postsynaptic cells may appear to touch when the synapse is seen under a light microscope, examination with an electron microscope reveals that most synapses have a **synaptic cleft,** a narrow space that separates these two cells (figure 51.12).

The end of the presynaptic axon is swollen and contains numerous **synaptic vesicles,** which are each packed with chemicals called **neurotransmitters.** When action potentials arrive at the end of the axon, they stimulate the opening of voltage-gated Ca^{++} channels, causing a rapid inward diffusion of Ca^{++}. This serves as the stimulus for the fusion of the synaptic vesicles membrane with the plasma membrane of the axon, so that the contents of the vesicles can be released by exocytosis (figure 51.13). The higher the frequency of action potentials in the presynaptic axon, the more vesicles will release their contents of neurotransmitters. The neurotransmitters diffuse rapidly to the other side of the cleft and bind to **receptor proteins** in the membrane of the postsynaptic cell. There are different types of neurotransmitters, and different ones act in different ways. We will next consider the action of a few of the important neurotransmitter chemicals.

The presynaptic axon is separated from the postsynaptic cell by a narrow synaptic cleft. Neurotransmitters diffuse across it to transmit a nerve impulse.

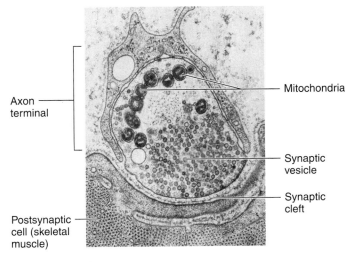

FIGURE 51.12
A synaptic cleft. An electron micrograph showing a neuromuscular synapse.

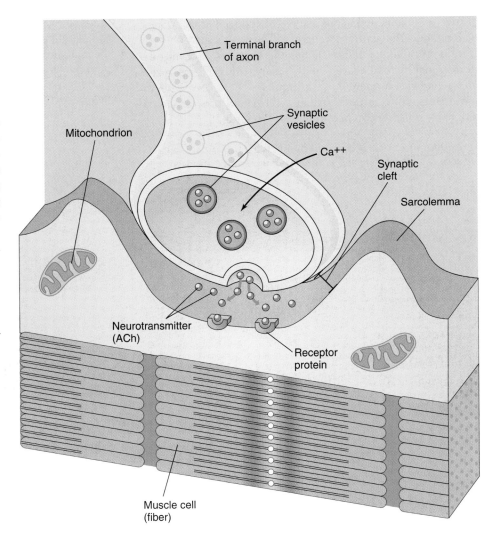

FIGURE 51.13 ⭐

The release of neurotransmitter. Action potentials arriving at the end of an axon trigger the uptake of Ca^{++}, which causes synaptic vesicles to fuse with the plasma membrane and release their neurotransmitters (acetylcholine [ACh] in this case), which diffuse across the synaptic gap and bind to receptors in the postsynaptic membrane.

Neurotransmitters and Their Functions

Acetylcholine was the first neurotransmitter chemical to be discovered and is widely used in the nervous system. Many other neurotransmitter chemicals have been shown to play important roles, however, and ongoing research continues to produce new information about neurotransmitter function.

Acetylcholine

The neurotransmitter that crosses the synapse between a neuron and a skeletal muscle cell, called a *neuromuscular junction* (see figures 51.12 and 51.13), is **acetylcholine (ACh)**. Acetylcholine binds to its receptor proteins in the postsynaptic membrane and thereby causes ion channels within these proteins to open (figure 51.14). The gates to these ion channels are said to be *chemically regulated* because they open in response to ACh, rather than in response to depolarization. The opening of the chemically regulated channels permits Na^+ to diffuse into the postsynaptic cell and K^+ to diffuse out. Although both ions move at the same time, the inward diffusion of Na^+ occurs at a faster rate and has the predominant effect. As a result, that site on the postsynaptic membrane produces a depolarization called an **excitatory postsynaptic potential (EPSP)**. The EPSP can now open the voltage-regulated channels for Na^+ and K^+

that are responsible for action potentials. Since the postsynaptic cell we are discussing is a skeletal muscle cell, the action potentials it produces stimulate muscle contraction through the mechanisms discussed in chapter 47.

If ACh stimulates muscle contraction, we must be able to eliminate ACh from the synaptic cleft in order to relax our muscles. This illustrates a general principle: molecules such as neurotransmitters and hormones must be quickly eliminated after secretion if they are to be effective regulators. In the case of ACh, the elimination is achieved by an enzyme in the postsynaptic membrane called **acetylcholinesterase (AChE)**. This enzyme is one of the fastest known, cleaving ACh into inactive fragments. Nerve gas and the agricultural insecticide parathion are potent inhibitors of AChE and can produce severe spastic paralysis and even death if the respiratory muscles become paralyzed.

Although ACh acts as a neurotransmitter between motor neurons and skeletal muscle cells, many neurons also use ACh as a neurotransmitter at their synapses with other neurons; in these cases, the postsynaptic membrane is generally on the dendrites or cell body of the second neuron. The EPSPs produced must then travel through the dendrites and cell body to the initial segment of the axon, where the first voltage-regulated channels needed for action potentials are located. This is where the first action potentials will be produced, providing that the EPSP depolarization is above the threshold needed to trigger action potentials.

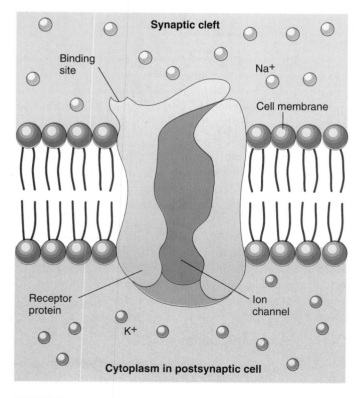

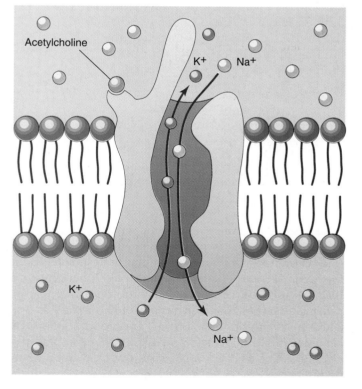

FIGURE 51.14

The binding of ACh to its receptor opens ion channels. The chemically regulated gates to these channels open when the neurotransmitter ACh binds to the receptor.

Glutamate, Glycine, and GABA

Glutamate is the major excitatory neurotransmitter in the vertebrate CNS, producing EPSPs and stimulating action potentials in the postsynaptic neurons. Although normal amounts produce physiological stimulation, excessive stimulation by glutamate has been shown to cause neurodegeneration, as in Huntington's chorea.

Glycine and **GABA** (an acronym for gamma-aminobutyric acid) are inhibitory neurotransmitters. If you remember that action potentials are triggered by a threshold level of depolarization, you will understand why hyperpolarization of the membrane would cause inhibition. These neurotransmitters cause the opening of chemically regulated gated channels for Cl^-, which has a concentration gradient favoring its diffusion into the neuron. Since Cl^- is negatively charged, it makes the inside of the membrane even more negative than it is at rest—from –70 mV to –85 mV, for example (figure 51.15). This hyperpolarization is called an **inhibitory postsynaptic potential (IPSP),** and is very important for neural control of body movements and other brain functions. Interestingly, the drug diazepam (Valium) causes its sedative and other effects by enhancing the binding of GABA to its receptors and thereby increasing the effectiveness of GABA at the synapse.

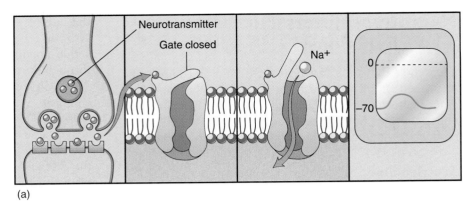

(a)

(b)

FIGURE 51.15
Different neurotransmitters can have different effects. (a) An excitatory neurotransmitter promotes a depolarization, or excitatory postsynaptic potential (EPSP). (b) An inhibitory neurotransmitter promotes a hyperpolarization, or inhibitory postsynaptic potential (IPSP).

Biogenic Amines

The **biogenic amines** include the hormone epinephrine (adrenaline), together with the neurotransmitters dopamine, norepinephrine, and serotonin. Epinephrine, norepinephrine, and dopamine are derived from the amino acid tyrosine and are included in the subcategory of *catecholamines.* Serotonin is a biogenic amine derived from a different amino acid, tryptophan.

Dopamine is a very important neurotransmitter used in the brain to control body movements and other functions. Degeneration of particular dopamine-releasing neurons produces the resting muscle tremors of Parkinson's disease, and people with this condition are treated with L-dopa (an acronym for dihydroxyphenylalanine), a precursor of dopamine. Additionally, studies suggest that excessive activity of dopamine-releasing neurons in other areas of the brain is associated with schizophrenia. As a result, patients with schizophrenia are sometimes helped by drugs that block the production of dopamine.

Norepinephrine is used by neurons in the brain and also by particular autonomic neurons, where its action as a neurotransmitter complements the action of the hormone epinephrine, secreted by the adrenal gland. The autonomic nervous system will be discussed in a later section of this chapter.

Serotonin is a neurotransmitter involved in the regulation of sleep and is also implicated in various emotional states. Insufficient activity of neurons that release serotonin may be one cause of clinical depression; this is suggested by the fact that antidepressant drugs, particularly fluoxetine (Prozac) and related compounds, specifically block the elimination of serotonin from the synaptic cleft (figure 51.16). The drug lysergic acid diethylamide (LSD) specifically blocks serotonin receptors in a region of the brain stem known as the raphe nuclei.

Other Neurotransmitters

Axons also release various polypeptides, called **neuropeptides,** at synapses. These neuropeptides may have a neurotransmitter function or they may have more subtle, long-term action on the postsynaptic neurons. In the latter case, they are often referred to as **neuromodulators.** A given axon generally releases only one kind of neurotransmitter, but many can release both a neurotransmitter and a neuromodulator.

One important neuropeptide is **substance P,** which is released at synapses in the CNS by sensory neurons activated by painful stimuli. The perception of pain, however, can vary depending on circumstances; an injured football player may not feel the full extent of his trauma, for example, until he's taken out of the game. The intensity with which pain is perceived partly depends on the effects of neuropeptides called **enkephalins** and **endorphins.** Enkephalins are released by axons descending from the brain and inhibit the passage of pain information to the brain. Endorphins are released by neurons in the brain stem and also block the perception of pain. Opium and its derivatives, morphine and heroin, have an analgesic (pain-reducing) effect because they are similar enough in chemical structure to bind to the receptors normally utilized by enkephalins and endorphins. For this reason, the enkephalins and the endorphins are referred to as *endogenous opiates.*

Nitric oxide (NO) is the first gas known to act as a regulatory molecule in the body. Produced from the amino acid arginine, nitric oxide's actions are very different from those of the more familiar nitrous oxide (N_2O), or laughing gas, sometimes used by dentists. Nitric oxide diffuses out of the presynaptic axon and into neighboring cells by simply passing through the lipid portions of the cell membranes. In the PNS, nitric oxide is released by some neurons that innervate the gastrointestinal tract, penis, respiratory passages, and cerebral blood vessels. These are autonomic neurons that cause smooth muscle relaxation in their target organs. This can produce, for example, the engorgement of the spongy tissue of the penis with blood, causing erection. The drug *VIAGRA* increases the release of nitric oxide in the penis, prolonging erection. Nitric oxide is also released as a neurotransmitter in the brain, and has been implicated in the processes of learning and memory.

Synaptic Integration

The activity of a postsynaptic neuron in the brain and spinal cord of vertebrates is influenced by different types of input from a number of presynaptic neurons. For example, a single motor neuron in the spinal cord can receive as many as 50,000 synapses from presynaptic axons! Each postsynaptic neuron may receive both excitatory and inhibitory synapses. The EPSPs (depolarizations) and IPSPs (hyperpolarizations) from these synapses interact with each other when they reach the cell body of the neuron. Small EPSPs add together to bring the membrane potential closer to the threshold, while IPSPs subtract from the depolarizing effect of the EPSPs, keeping the membrane potential below the threshold (figure 51.17). This process is called **synaptic integration.**

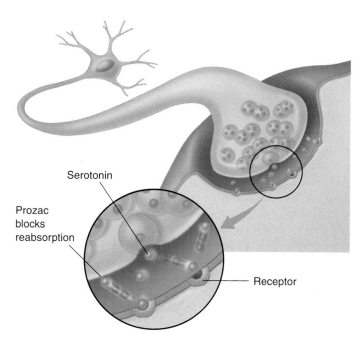

FIGURE 51.16
Serotonin and depression. Depression can result from a shortage of the neurotransmitter serotonin. The antidepressant drug Prozac works by blocking reabsorption of serotonin in the synapse, making up for the shortage.

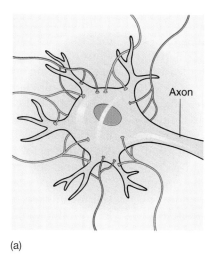
(a)

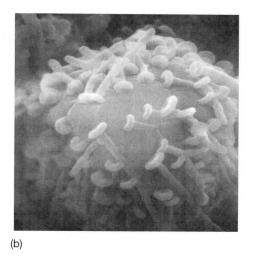
(b)

FIGURE 51.17
Integration of EPSPs and IPSPs takes place on the neuronal cell body. (a) The synapses made by some axons are excitatory *(blue);* the synapses made by other axons are inhibitory *(red).* The summed influence of all of these inputs determines whether the axonal membrane of the postsynaptic cell will be sufficiently depolarized to produce an action potential. (b) Micrograph of a neuronal cell body with numerous synapses (15,000×).

Neurotransmitters and Drug Addiction

When a cell of your body is exposed to a chemical signal for a prolonged period, it tends to lose its ability to respond to the stimulus with its original intensity. (You are familiar with this loss of sensitivity—when you sit in a chair, how long are you aware of the chair?) Nerve cells are particularly prone to this loss of sensitivity. If receptor proteins within synapses are exposed to high levels of neurotransmitter molecules for prolonged periods, that nerve cell often responds by inserting fewer receptor proteins into the membrane. This feedback is a normal function in all neurons, a simple mechanism that has evolved to make the cell more efficient by adjusting the number of "tools" (receptor proteins) in the membrane "workshop" to suit the workload.

Cocaine. The drug cocaine is a neuromodulator that causes abnormally large amounts of neurotransmitter to remain in the synapses for long periods of time. Cocaine affects nerve cells in the brain's pleasure pathways (the so-called limbic system). These cells transmit pleasure messages using the neurotransmitter dopamine. Using radioactively labeled cocaine molecules, investigators found that cocaine binds tightly to the transporter proteins in the gaps between nerves. These proteins normally remove the neurotransmitter dopamine after it has acted. Like a game of musical chairs in which all the chairs become occupied, there are no unoccupied carrier proteins available to the dopamine molecules, so the dopamine stays in the gap, firing the receptors again and again. As new signals arrive, more and more dopamine is added, firing the pleasure pathway more and more often (figure 51.18).

When receptor proteins on limbic system nerve cells are exposed to high levels of dopamine neurotransmitter molecules for prolonged periods of time, the nerve cells "turn down the volume" of the signal by lowering the number of receptor proteins on their surfaces. They respond to the greater number of neurotransmitter molecules by simply reducing the number of targets available for these molecules to hit. The cocaine user is now addicted (figure 51.19). With so few receptors, the user needs the drug to maintain even normal levels of limbic activity.

Is Nicotine an Addictive Drug? Investigators attempting to explore the habit-forming nature of nicotine used what had been learned about cocaine to carry out what seems a reasonable experiment—they introduced radioactively labeled nicotine into the brain and looked to see what sort of carrier protein it attached itself to. To their great surprise, the nicotine ignored proteins in the between-cell gaps and instead bound directly to a specific receptor on the receiving nerve cell surface! This was totally unexpected, as nicotine does not normally occur in the brain—why should it have a receptor there?

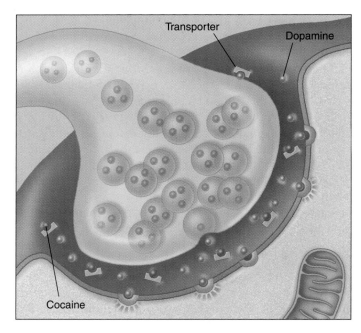

FIGURE 51.18
How cocaine alters events at the synapse. When cocaine binds to the dopamine transporters, the neurotransmitter survives longer in the synapse and continues to stimulate the postsynaptic cell. Cocaine thus acts to intensify pleasurable sensations.

Intensive research followed, and researchers soon learned that the "nicotine receptors" normally served to bind the neurotransmitter acetylcholine. It was just an accident of nature that nicotine, an obscure chemical from a tobacco plant, was also able to bind to them. What, then, is the normal function of these receptors? The target of considerable research, these receptors turned out to be one of the brain's most important tools. The brain uses them to coordinate the activities of many other kinds of receptors, acting to "fine tune" the sensitivity of a wide variety of behaviors.

When neurobiologists compare the limbic system nerve cells of smokers to those of nonsmokers, they find changes in both the number of nicotine receptors and in the levels of RNA used to make the receptors. They have found that the brain adjusts to prolonged exposure to nicotine by "turning down the volume" in two ways: (1) by making fewer receptor proteins to which nicotine can bind; and (2) by altering the pattern of *activation* of the nicotine receptors (that is, their sensitivity to neurotransmitter).

It is this second adjustment that is responsible for the profound effect smoking has on the brain's activities. By overriding the normal system used by the brain to coordinate its many activities, nicotine alters the pattern of release into gaps between nerve cells of many neurotransmitters, including acetylcholine, dopamine, serotonin, and many others. As a result, changes in level of activity occur in a wide variety of nerve pathways within the brain.

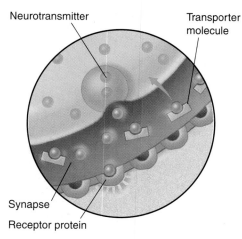

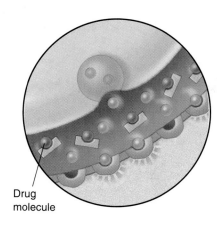

(a) Neurotransmitter is reabsorbed at a normal synapse.

(b) Drug molecules prevent reabsorption and cause overstimulation of postsynaptic membrane.

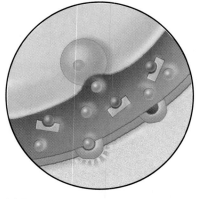

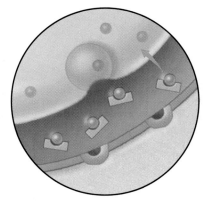

(c) The number of receptors decreases.

(d) The synapse is less sensitive when the drug is removed.

FIGURE 51.19

Drug addiction. (a) In a normal synapse, the neurotransmitter binds to a transporter molecule and is rapidly reabsorbed after it has acted. (b) When a drug molecule binds to the transporters, reabsorption of the neurotransmitter is blocked, and the postsynaptic cell is over-stimulated by the increased amount of neurotransmitter left in the synapse. (c) The central nervous system adjusts to the increased firing by producing fewer receptors in the postsynaptic membrane. The result is addiction. (d) When the drug is removed, normal absorption of the neurotransmitter resumes, and the decreased number of receptors creates a less-sensitive nerve pathway. Physiologically, the only way a person can then maintain normal functioning is to continue to take the drug. Only if the drug is removed permanently will the nervous system eventually adjust again and restore the original amount of receptors.

Addiction occurs when chronic exposure to nicotine induces the nervous system to adapt physiologically. The brain compensates for the many changes nicotine induces by making other changes. Adjustments are made to the numbers and sensitivities of many kinds of receptors within the brain, restoring an appropriate balance of activity.

Now what happens if you stop smoking? Everything is out of whack! The newly coordinated system *requires* nicotine to achieve an appropriate balance of nerve pathway activities. This is addiction in any sensible use of the term.

The body's physiological response is profound and unavoidable. There is no way to prevent addiction to nicotine with willpower, any more than willpower can stop a bullet when playing Russian roulette with a loaded gun. If you smoke cigarettes for a prolonged period, you will become addicted.

What do you do if you are addicted to smoking cigarettes and you want to stop? When use of an addictive drug like nicotine is stopped, the level of signaling will change to levels far from normal. If the drug is not reintroduced, the altered level of signaling will eventually induce the nerve cells to once again make compensatory changes that restore an appropriate balance of activities within the brain. Over time, receptor numbers, their sensitivity, and patterns of release of neurotransmitters all revert to normal, once again producing normal levels of signaling along the pathways. There is no way to avoid the down side of addiction. The pleasure pathways will not function at normal levels until the number of receptors on the affected nerve cells have time to readjust.

Many people attempt to quit smoking by using patches containing nicotine; the idea is that providing gradually lesser doses of nicotine allows the smoker to be weaned of his or her craving for cigarettes. The patches do reduce the craving for cigarettes—so long as you keep using the patches! Actually, using such patches simply substitutes one (admittedly less dangerous) nicotine source for another. If you are going to quit smoking, there is no way to avoid the necessity of eliminating the drug to which you are addicted. Hard as it is to hear the bad news, there is no easy way out. The only way to quit is to quit.

Acetylcholine stimulates the opening of chemically regulated ion channels, causing a depolarization called an excitatory postsynaptic potential (EPSP). Glycine and GABA are inhibitory neurotransmitters that produce hyperpolarization of the postsynaptic membrane. There are also many other neurotransmitters, including the biogenic amines: dopamine, norepinephrine, and serotonin. The effects of different neurotransmitters are integrated through summation of depolarizations and hyperpolarizations.

The Evolution of the Vertebrate Brain

Sponges are the only major phylum of multicellular animals that lack nerves. The simplest nervous systems occur among cnidarians (figure 51.20): all neurons are similar and are linked to one another in a web, or *nerve net*. There is no associative activity, no control of complex actions, and little coordination The first associative activity in nervous systems is seen in the free-living flatworms, phylum Platyhelminthes. Running down the bodies of these flatworms are two nerve cords; peripheral nerves extend outward to the muscles of the body. The two nerve cords converge at the front end of the body, forming an enlarged mass of nervous tissue that also contains associative neurons with synapses connecting neurons to one another. This primitive "brain" is a rudimentary central nervous system and permits a far more complex control of muscular responses than is possible in cnidarians.

All of the subsequent evolutionary changes in nervous systems can be viewed as a series of elaborations on the characteristics already present in flatworms. For example, earthworms exhibit a central nervous system that is connected to all other parts of the body by peripheral nerves. And, in arthropods, the central coordination of complex response is increasingly localized in the front end of the nerve cord. As this region evolved, it came to contain a progressively larger number of associative interneurons, and to develop tracts, which are highways within the brain that connect associative elements.

Casts of the interior braincases of fossil agnathans, fishes that swam 500 million years ago, have revealed much about the early evolutionary stages of the vertebrate brain. Although small, these brains already had the three divisions that characterize the brains of all contemporary vertebrates: (1) the hindbrain, or rhombencephalon; (2) the midbrain, or mesencephalon; and (3) the forebrain, or prosencephalon (figure 51.21).

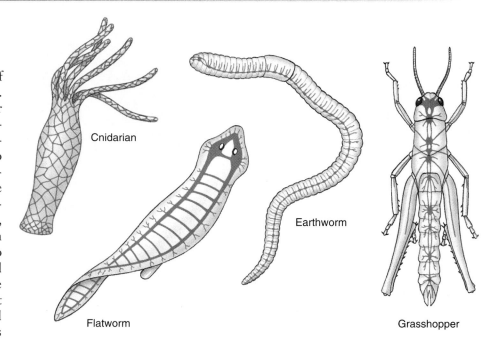

FIGURE 51.20
Evolution of the nervous system. Invertebrates exhibit a progressive elaboration of organized nerve cords and the centralization of complex responses in the front end of the nerve cord.

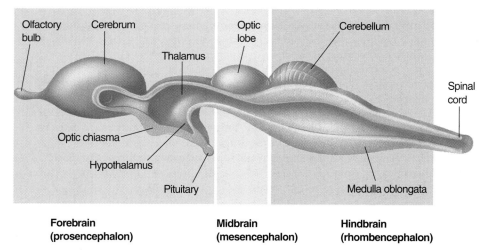

FIGURE 51.21
The basic organization of the vertebrate brain can be seen in the brains of primitive fishes. The brain is divided into three regions that are found in differing proportions in all vertebrates: the hindbrain, which is the largest portion of the brain in fishes; the midbrain, which in fishes is devoted primarily to processing visual information; and the forebrain, which is concerned mainly with olfaction (the sense of smell) in fishes. In terrestrial vertebrates, the forebrain plays a far more dominant role in neural processing than it does in fishes.

The hindbrain was the major component of these early brains, as it still is in fishes today. Composed of the *cerebellum*, *pons*, and *medulla oblongata*, the hindbrain may be considered an extension of the spinal cord devoted primarily to coordinating motor reflexes. Tracts containing large numbers of axons run like cables up and down the spinal cord to the hindbrain. The hindbrain, in turn, integrates the many sensory signals coming from the muscles and coordinates the pattern of motor responses.

Much of this coordination is carried on within a small extension of the hindbrain called the cerebellum ("little cerebrum"). In more advanced vertebrates, the cerebellum plays an increasingly important role as a coordinating center and is correspondingly larger than it is in the fishes. In all vertebrates, the cerebellum processes data on the current position and movement of each limb, the state of relaxation or contraction of the muscles involved, and the general position of the body and its relation to the outside world. These data are gathered in the cerebellum, synthesized, and the resulting commands issued to efferent pathways.

In fishes, the remainder of the brain is devoted to the reception and processing of sensory information. The midbrain is composed primarily of the **optic lobes,** which receive and process visual information, while the forebrain is devoted to the processing of *olfactory* (smell) information. The brains of fishes continue growing throughout their lives. This continued growth is in marked contrast to the brains of other classes of vertebrates, which generally complete their development by infancy. The human brain continues to develop through early childhood, but no new neurons are produced once development has ceased, except in the tiny hippocampus, which controls which experiences are filed away into long-term memory and which are forgotten.

The Dominant Forebrain

Starting with the amphibians and continuing more prominently in the reptiles, sensory information is increasingly centered in the forebrain. This pattern was the dominant evolutionary trend in the further development of the vertebrate brain (figure 51.22).

The forebrain in reptiles, amphibians, birds, and mammals is composed of two elements that have distinct functions. The *diencephalon* (Greek *dia*, "between") consists of the thalamus and hypothalamus. The **thalamus** is an integrating and relay center between incoming sensory information and the cerebrum. The **hypothalamus** participates in basic drives and emotions and controls the secretions of the pituitary gland. The *telencephalon*, or "end brain" (Greek *telos*, "end"), is located at the front of the forebrain and is devoted largely to associative activity. In mammals, the telencephalon is called the **cerebrum.**

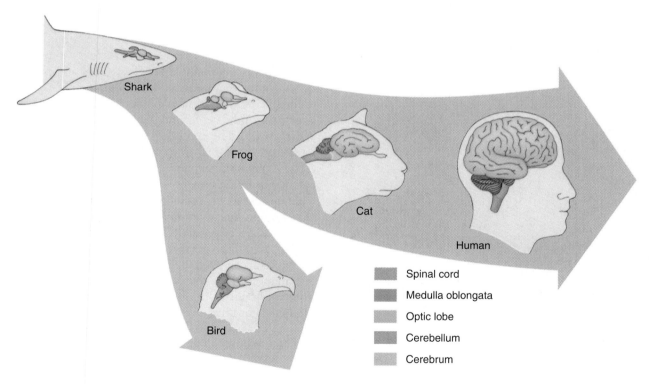

- Spinal cord
- Medulla oblongata
- Optic lobe
- Cerebellum
- Cerebrum

FIGURE 51.22
The evolution of the vertebrate brain involved changes in the relative sizes of different brain regions. In sharks and other fishes, the hindbrain is predominant, and the rest of the brain serves primarily to process sensory information. In amphibians and reptiles, the forebrain is far larger, and it contains a larger cerebrum devoted to associative activity. In birds, which evolved from reptiles, the cerebrum is even more pronounced. In mammals, the cerebrum is the largest portion of the brain. The dominance of the cerebrum is greatest in humans, in whom it envelops much of the rest of the brain.

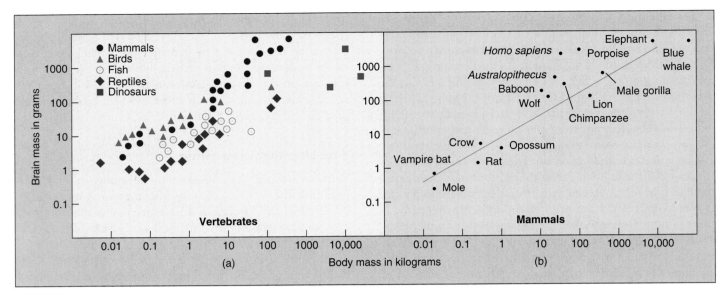

FIGURE 51.23

Brain mass versus body mass. Among most vertebrates, brain mass is a relatively constant proportion of body mass, so that a plot of brain mass versus body mass gives a straight line. (a) However, the proportion of brain mass to body mass is much greater in birds than in reptiles, and it is greater still in mammals. (b) Among mammals, humans have the greatest brain mass per unit of body mass (that is, the farthest perpendicular distance from the plotted line). In second place are the porpoises.

Table 51.3 Subdivisions of the Central Nervous System	
Major Subdivision	**Function**
Spinal cord	Spinal reflexes; relays sensory information
Medulla oblongata	Sensory nuclei; reticular activating system; visceral control
Pons	Reticular activating system; visceral control
Midbrain	Reflexes involving eyes and ears
Cerebellum	Coordination of movements; balance
Diencephalon	
Thalamus	Relay station for ascending sensory and descending tracts; visceral control
Hypothalamus	Visceral control; neuroendocrine control
Telencephalon (cerebrum)	
Basal ganglia	Motor control
Corpus callosum	Connects the two hemispheres
Hippocampus (limbic system)	Memory; emotion
Cerebral cortex	Higher functions

The Expansion of the Cerebrum

In examining the relationship between brain mass and body mass among the vertebrates (figure 51.23), you can see a remarkable difference between fishes and reptiles, on the one hand, and birds and mammals, on the other. Mammals have brains that are particularly large relative to their body mass. This is especially true of porpoises and humans; the human brain weighs about 1.4 kilograms. The increase in brain size in the mammals largely reflects the great enlargement of the cerebrum, the dominant part of the mammalian brain. The cerebrum is the center for correlation, association, and learning in the mammalian brain. It receives sensory data from the thalamus and issues motor commands to the spinal cord via descending tracts of axons.

In vertebrates, the central nervous system is composed of the brain and the spinal cord (table 51.3). These two structures are responsible for most of the information processing within the nervous system and consist primarily of interneurons and neuroglia. Ascending tracts carry sensory information to the brain. Descending tracts carry impulses from the brain to the motor neurons and interneurons in the spinal cord that control the muscles of the body.

The vertebrate brain consists of three primary regions: the forebrain, midbrain, and hindbrain. The hindbrain was the principal component of the brain of early vertebrates; it was devoted to the control of motor activity. In vertebrates more advanced than fishes, the processing of information is increasingly centered in the forebrain.

The Human Forebrain

The human cerebrum is so large that it appears to envelop the rest of the brain (figure 51.24). It is split into two cerebral hemispheres, which are connected by a tract called the *corpus callosum*. The hemispheres are further divided into the *frontal, parietal, temporal,* and *occipital lobes.*

Each hemisphere receives sensory input from the opposite, or contralateral, side of the body and exerts motor control primarily over that side. Therefore, a touch on the right hand, for example, is relayed primarily to the left hemisphere, which may then initiate movement of that hand in response to the touch. Damage to one hemisphere due to a stroke often results in a loss of sensation and paralysis on the contralateral side of the body.

Cerebral Cortex

Much of the neural activity of the cerebrum occurs within a layer of gray matter only a few millimeters thick on its outer surface. This layer, called the **cerebral cortex,** is densely packed with nerve cells. In humans, it contains over 10 billion nerve cells, amounting to roughly 10% of all the neurons in the brain. The surface of the cerebral cortex is highly convoluted; this is particularly true in the human brain, where the convolutions increase the surface area of the cortex threefold.

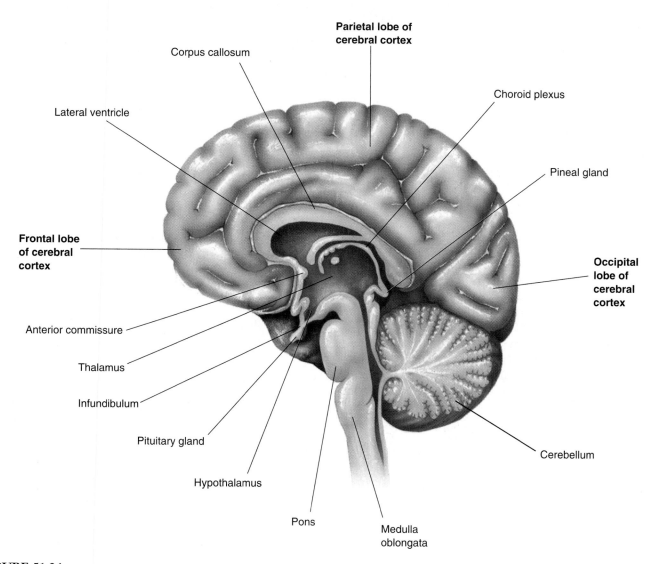

FIGURE 51.24

A section through the human brain. In this sagittal section, the corpus callosum, a fiber tract connecting the two cerebral hemispheres, can be clearly seen.

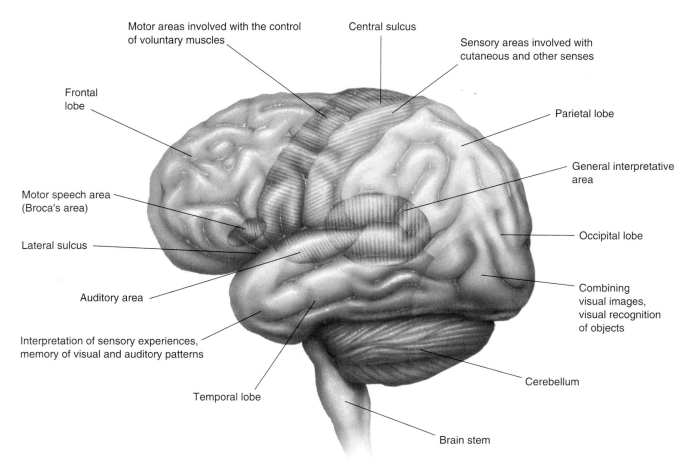

FIGURE 51.25
The lobes of the cerebrum. Some of the known regions of specialization are indicated in this diagram.

The activities of the cerebral cortex fall into one of three general categories: motor, sensory, and associative. The primary motor cortex lies along the *gyrus* (convolution) on the posterior border of the frontal lobe, just in front of the central *sulcus* (crease) (figure 51.25). Each point on its surface is associated with the movement of a different part of the body (figure 51.26). Just behind the central sulcus, on the anterior edge of the parietal lobe, lies the primary somatosensory cortex. Each point in this area receives input from sensory neurons serving cutaneous and muscle senses in a particular part of the body. Large areas of the motor cortex and primary somatosensory cortex are devoted to the fingers, lips, and tongue because of the need for manual dexterity and speech. The auditory cortex lies within the temporal lobe, and different regions of this cortex deal with different sound frequencies. The visual cortex lies on the occipital lobe, with different sites processing information from different positions on the retina, equivalent to particular points in the visual fields of the eyes.

The portion of the cerebral cortex that is not occupied by these motor and sensory cortices is referred to as *association cortex*. The site of higher mental activities, the association cortex reaches its greatest extent in primates, especially humans, where it makes up 95% of the surface of the cerebral cortex.

Basal Ganglia

Buried deep within the white matter of the cerebrum are several collections of cell bodies and dendrites that produce islands of gray matter. These aggregates of neuron cell bodies, which are collectively termed the basal ganglia, receive sensory information from ascending tracts and motor commands from the cerebral cortex and cerebellum. Outputs from the basal ganglia are sent down the spinal cord, where they participate in the control of body movements. Damage to specific regions of the basal ganglia can produce the resting tremor of muscles that is characteristic of people with Parkinson's disease.

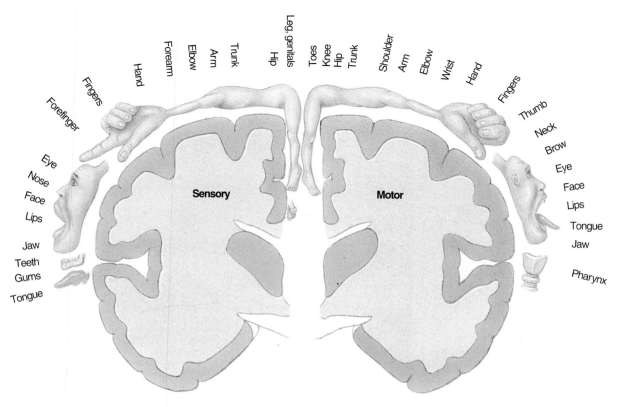

FIGURE 51.26
The primary somatosensory cortex *(left)* and the primary motor cortex *(right)*. Each region in these parts of the cerebral cortex is associated with a different region of the body, as indicated in this stylized map.

Thalamus and Hypothalamus

The thalamus is a primary site of sensory integration in the brain. Visual, auditory, and somatosensory information is sent to the thalamus, where the sensory tracts synapse with association neurons. The sensory information is then relayed via the thalamus to the occipital, temporal, and parietal lobes of the cerebral cortex, respectively. The transfer of each of these types of sensory information is handled by specific aggregations of neuron cell bodies within the thalamus.

The hypothalamus integrates the visceral activities. It helps regulate body temperature, hunger and satiety, thirst, and—along with the limbic system—various emotional states. The hypothalamus also controls the pituitary gland, which in turn regulates many of the other endocrine glands of the body. By means of its interconnections with the cerebral cortex and with control centers in the brain stem (a term used to refer collectively to the midbrain, pons, and medulla oblongata), the hypothalamus helps coordinate the neural and hormonal responses to many internal stimuli and emotions.

The *hippocampus* and *amygdala* are, together with the hypothalamus, the major components of the **limbic system** (figure 51.27). This is an evolutionarily ancient group of linked structures deep within the cerebrum that are respon-

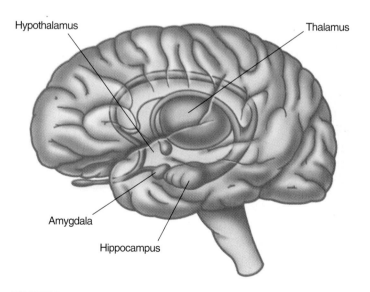

FIGURE 51.27
The limbic system. The hippocampus and the amygdala are the major components of the limbic system, which controls our most deep-seated drives and emotions.

sible for emotional responses. The hippocampus is also believed to be important in the formation and recall of memories, a topic we will discuss later.

Language and Other Higher Functions

Arousal and Sleep. The brain stem contains a diffuse collection of neurons referred to as the *reticular formation*. One part of this formation, the reticular activating system, controls consciousness and alertness. All of the sensory pathways feed into this system, which monitors the information coming into the brain and identifies important stimuli. When the reticular activating system has been stimulated to arousal, it increases the level of activity in many parts of the brain. Neural pathways from the reticular formation to the cortex and other brain regions are depressed by anesthetics and barbiturates.

The reticular activating system controls both sleep and the waking state. It is easier to sleep in a dark room than in a lighted one because there are fewer visual stimuli to stimulate the reticular activating system. In addition, activity in this system is reduced by serotonin, a neurotransmitter we previously discussed. Serotonin causes the level of brain activity to fall, bringing on sleep.

Sleep is not the loss of consciousness. Rather, it is an active process whose multiple states can be revealed by recording the electrical activity of the brain in an electroencephalogram (EEG). In a relaxed but awake individual whose eyes are shut, the EEG consists primarily of large, slow waves that occur at a frequency of 8 to 13 hertz (cycles per second). These waves are referred to as *alpha waves*. In an alert subject whose eyes are open, the EEG waves are more rapid (*beta waves* are seen at frequencies of 13–30 hertz) and is more desynchronized because multiple sensory inputs are being received, processed, and translated into motor activities.

Theta waves (4–7 hertz) and *delta waves* (0.5–4 hertz) are seen in various stages of sleep. The first change seen in the EEG with the onset of drowsiness is a slowing and reduction in the overall amplitude of the waves. This slow-wave sleep has several stages but is generally characterized by decreases in arousability, skeletal muscle tone, heart rate, blood pressure, and respiratory rate. During REM sleep (named for the rapid eye movements that occur during this stage), the EEG resembles that of a relaxed, awake individual, and the heart rate, blood pressure, and respiratory rate

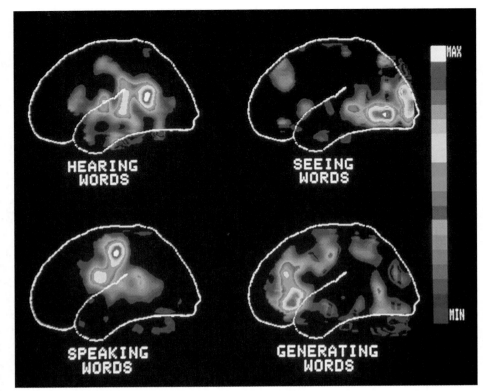

FIGURE 51.28
Different brain regions control various activities. This illustration shows how the brain reacts in human subjects asked to listen to a spoken word, to read that same word silently, to repeat the word out loud, and then to speak a word related to the first. Regions of white, red, and yellow show the greatest activity. Compare this with figure 51.25 to see how regions of the brain are mapped.

are all increased. Paradoxically, individuals in REM sleep are difficult to arouse and are more likely to awaken spontaneously. Dreaming occurs during REM sleep, and the rapid eye movements resemble the tracking movements made by the eyes when awake, suggesting that dreamers "watch" their dreams.

Language and Spatial Recognition. Although the two cerebral hemispheres seem structurally similar, they are responsible for different activities. The most thoroughly investigated example of this lateralization of function is language. The left hemisphere is the "dominant" hemisphere for language—the hemisphere in which most neural processing related to language is performed—in 90% of right-handed people and nearly two-thirds of left-handed people. There are two language areas in the dominant hemisphere. Wernicke's area, located in the parietal lobe between the primary auditory and visual areas, is important for language comprehension and the formulation of thoughts into speech (figure 51.28). Broca's area, found near the part of the motor cortex controlling the face, is responsible for the generation of motor output needed for language communication. Damage to these brain areas can cause language disorders known as *aphasias*. For example, if Wernicke's area is damaged, the person's speech is rapid and fluid but lacks meaning; words are tossed together as in a "word salad."

While the dominant hemisphere for language is adept at sequential reasoning, like that needed to formulate a sentence, the nondominant hemisphere (the right hemisphere in most people) is adept at spatial reasoning, the type of reasoning needed to assemble a puzzle or draw a picture. It is also the hemisphere primarily involved in musical ability—a person with damage to Broca's speech area in the left hemisphere may not be able to speak but may retain the ability to sing! Damage to the nondominant hemisphere may lead to an inability to appreciate spatial relationships and may impair musical activities such as singing. Even more specifically, damage to the inferior temporal cortex in that hemisphere eliminates the capacity to recall faces. Reading, writing, and oral comprehension remain normal, and patients with this disability can still recognize acquaintances by their voices. The nondominant hemisphere is also important for the consolidation of memories of nonverbal experiences.

Memory and Learning. One of the great mysteries of the brain is the basis of memory and learning. There is no one part of the brain in which all aspects of a memory appear to reside. Specific cortical sites cannot be identified for particular memories because relatively extensive cortical damage does not selectively remove memories. Although memory is impaired if portions of the brain, particularly the temporal lobes, are removed, it is not lost entirely. Many memories persist in spite of the damage, and the ability to access them gradually recovers with time. Therefore, investigators who have tried to probe the physical mechanisms underlying memory often have felt that they were grasping at a shadow. Although we still do not have a complete understanding of these mechanisms, we have learned a good deal about the basic processes in which memories are formed.

There appear to be fundamental differences between short-term and long-term memory. Short-term memory is transient, lasting only a few moments. Such memories can readily be erased by the application of an electrical shock, leaving previously stored long-term memories intact. This result suggests that short-term memories are stored electrically in the form of a transient neural excitation. Long-term memory, in contrast, appears to involve structural changes in certain neural connections within the brain. Two parts of the temporal lobes, the hippocampus and the amygdala, are involved in both short-term memory and its consolidation into long-term memory. Damage to these structures impairs the ability to process recent events into long-term memories.

Synapses that are used intensively for a short period of time display more effective synaptic transmission upon subsequent use. This phenomenon is called long-term potentiation (LTP). During LTP, the presynaptic neuron may release increased amounts of neurotransmitter with each action potential frequency, and the postsynaptic neuron may become increasingly sensitive to the neurotransmitter. It is believed that these changes in synaptic transmission may be responsible for some aspects of memory storage.

Mechanism of Alzheimer's Disease Still a Mystery

In the past, little was known about *Alzheimer's disease*, a condition in which the memory and thought processes of the brain become dysfunctional. Drug companies are eager to develop new products for the treatment of Alzheimer's, but they have little concrete evidence to go on. Scientists disagree about the biological nature of the disease and its cause. Two hypotheses have been proposed: one that nerve cells in the brain are killed from the outside in, and the other that the cells are killed from the inside out.

In the first hypothesis, external proteins called β-amyloid peptides kill nerve cells. A mistake in protein processing produces an abnormal form of the peptide, which then forms aggregates, or plaques. The plaques begin to fill in the brain and then damage and kill nerve cells. However, these amyloid plaques have been found in autopsies of people that did not have Alzheimer's disease.

The second hypothesis maintains that the nerve cells are killed by an abnormal form of an internal protein. This protein, called tau (τ), normally functions to maintain protein transport microtubules. Abnormal forms of τ assemble into helical segments that form tangles, which interfere with the normal functioning of the nerve cells. Researchers continue to study whether tangles and plaques are causes or effects of Alzheimer's disease.

Progress has been made in identifying genes that increase the likelihood of developing Alzheimer's and genes that, when mutated, can cause Alzheimer's disease. However, the genes may not reveal much about Alzheimer's since they do not show up in most Alzheimer's patients, and they cause symptoms that start much earlier than when most Alzheimer's patients show symptoms.

The cerebrum is composed of two cerebral hemispheres. Each hemisphere consists of the gray matter of the cerebral cortex overlying white matter and islands of gray matter (nuclei) called the basal ganglia. These areas are involved in the integration of sensory information, control of body movements, and such associative functions as learning and memory.

The Spinal Cord

The spinal cord is a cable of neurons extending from the brain down through the backbone (figure 51.29). It is enclosed and protected by the vertebral column and layers of membranes called *meninges*. Inside the spinal cord there are two zones. The inner zone, called gray matter, consists of interneurons and the cell bodies of motor neurons. The outer zone, called white matter, contains the axons and dendrites of nerve cells. Messages from the body and the brain run up and down the spinal cord, an "information highway."

In addition to relaying messages, the spinal cord also functions in reflexes, the sudden, involuntary movement of muscles. A reflex produces a rapid motor response to a stimulus because the sensory neuron passes its information to a motor neuron in the spine, without higher level processing. The escape reaction of a fly about to be swatted is a reflex. One of the most frequently used reflexes in your body is blinking, a reflex that protects your eyes. If anything, such as an insect or a cloud of dust, approaches your eye, the eyelid blinks before you realize what has happened. The reflex occurs before the cerebrum is aware the eye is in danger.

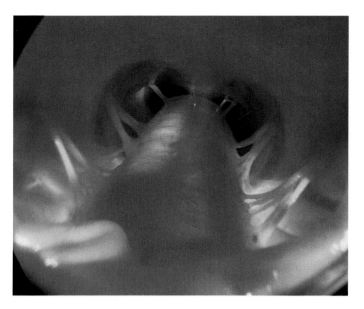

FIGURE 51.29
A view down the human spinal cord. Pairs of spinal nerves can be seen extending from the spinal cord. It is along these nerves, as well as the cranial nerves that arise from the brain, that the central nervous system communicates with the rest of the body.

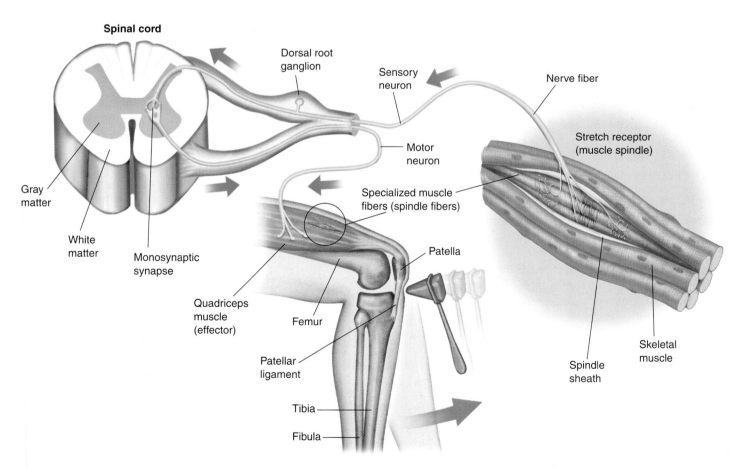

FIGURE 51.30
The knee-jerk reflex. This is the simplest reflex, involving only sensory and motor neurons.

Because they pass information along only a few neurons, reflexes are very fast. Many reflexes never reach the brain. The nerve impulse travels only as far as the spinal cord and then comes right back as a motor response. A few reflexes, like the knee-jerk reflex (see figure 51.30), are monosynaptic reflex arcs. In these, the sensory nerve cell makes synaptic contact directly with a motor neuron in the spinal cord whose axon travels directly back to the muscle. The knee-jerk reflex is also an example of a *muscle stretch reflex*. When the muscle is briefly stretched by tapping the patellar ligament with a rubber mallet, the *muscle spindle apparatus* is also stretched. The spindle apparatus is embedded within the muscle, and, like the muscle fibers outside the spindle, is stretched along with the muscle. Stretching of the spindle activates sensory neurons that synapse directly with somatic motor neurons within the spinal cord. As a result, the somatic motor neurons conduct action potentials to the skeletal muscle fibers and stimulate the muscle to contract. This reflex is the simplest in the vertebrate body because only one synapse is crossed in the reflex arc.

Most reflexes in vertebrates, however, involve a single connecting interneuron between the sensory and the motor neuron. The withdrawal of a hand from a hot stove or the blinking of an eye in response to a puff of air involve a relay of information from a sensory neuron through one or more interneurons to a motor neuron. The motor neuron then stimulates the appropriate muscle to contract (figure 51.31).

Spinal Cord Regeneration

In the past, scientists have tried to repair severed spinal cords by installing nerves from another part of the body to bridge the gap and act as guides for the spinal cord to regenerate. But most of these experiments have failed because the nerve bridges did not go from white matter to gray matter. Also, there is a factor that inhibits nerve growth in the spinal cord. After discovering that fibroblast growth factor stimulates nerve growth, neurobiologists tried gluing on the nerves, from white to gray matter, with fibrin that had been mixed with the fibroblast growth factor.

Three months later, rats with the nerve bridges began to show movement in their lower bodies. In further analyses of the experimental animals, dye tests indicated that the spinal cord nerves had regrown from both sides of the gap. Many scientists are encouraged by the potential to use a similar treatment in human medicine. However, most spinal cord injuries in humans do not involve a completely severed spinal cord; often, nerves are crushed. Also, while the rats with nerve bridges did regain some locomotory ability, tests indicated that they were barely able to walk or stand.

The spinal cord relays messages to and from the brain and processes some sensory information directly.

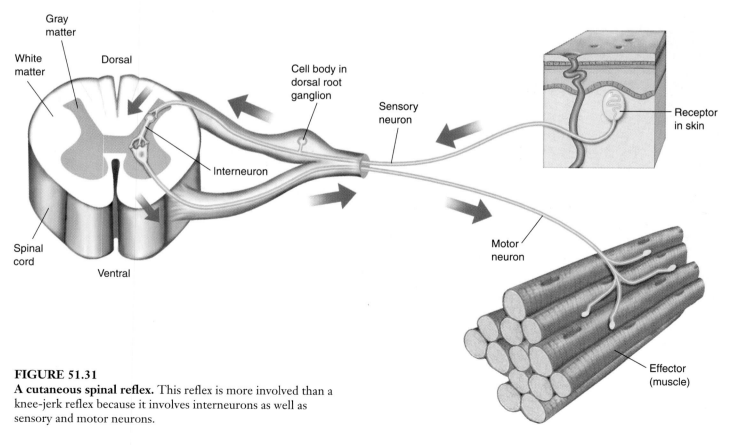

FIGURE 51.31
A cutaneous spinal reflex. This reflex is more involved than a knee-jerk reflex because it involves interneurons as well as sensory and motor neurons.

Components of the Peripheral Nervous System

The peripheral nervous system consists of nerves and ganglia (figure 51.32). Nerves are cable-like collections of axons, usually containing both sensory and motor neurons. Ganglia are aggregations of neuron cell bodies located outside the central nervous system.

At its origin, a spinal nerve separates into sensory and motor components. The axons of sensory neurons enter the dorsal surface of the spinal cord and form the **dorsal root** of the spinal nerve, whereas motor axons leave from the ventral surface of the spinal nerve and form the **ventral root** of the spinal nerve. The cell bodies of sensory neurons are grouped together outside each level of the spinal cord in the **dorsal root ganglia.** The cell bodies of somatic motor neurons, on the other hand, are located within the spinal cord and so are not located in ganglia.

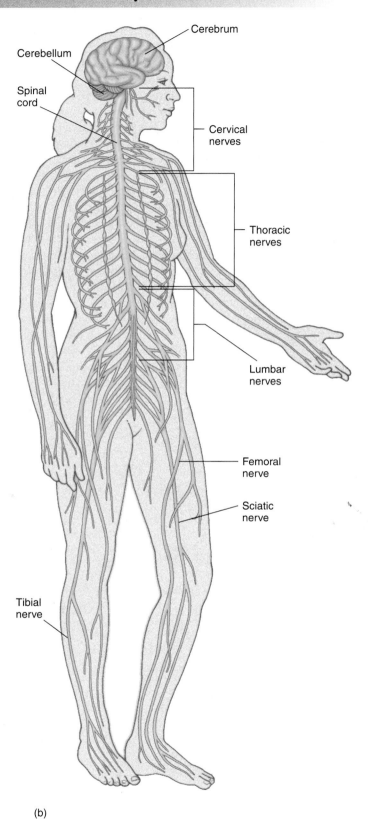

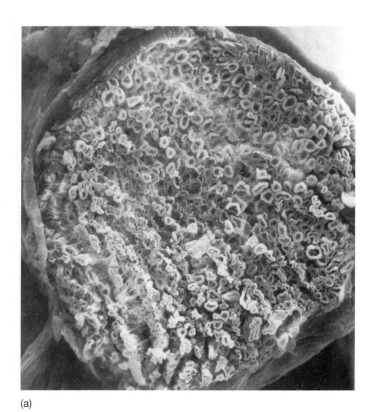

(a)

FIGURE 51.32
Nerves in the peripheral nervous system. (a) Photomicrograph (1600×) showing a cross section of a bullfrog nerve. The nerve is a bundle of axons bound together by connective tissue. Many myelinated axons are visible, each looking somewhat like a doughnut. (b) Some of the major nerves in humans are shown.

(b)

Somatic motor neurons stimulate skeletal muscles to contract, and autonomic motor neurons innervate involuntary effectors—smooth muscles, cardiac muscle, and glands. A comparison of the somatic and autonomic nervous systems is provided in table 51.4 and each will be discussed in turn. Somatic motor neurons stimulate the skeletal muscles of the body to contract in response to conscious commands and as part of reflexes that do not require conscious control. Conscious control of skeletal muscles is achieved by activation of tracts of axons that descend from the cerebrum to the appropriate level of the spinal cord. Some of these descending axons will stimulate spinal cord motor neurons directly, while others will activate interneurons that in turn stimulate the spinal motor neurons. When a particular muscle is stimulated to contract, however, its antagonist must be inhibited. In order to flex the arm, for example, the flexor muscles must be stimulated while the antagonistic extensor muscle is inhibited (figure 51.33). Descending motor axons produce this necessary inhibition by causing hyperpolarizations (IPSPs) of the spinal motor neurons that innervate the antagonistic muscles.

A spinal nerve contains sensory neurons that enter the dorsal root and motor neurons that enter the ventral root of the nerve. Somatic motor neurons innervate skeletal muscles and stimulate the muscles to contract.

Table 51.4	Comparison of the Somatic and Autonomic Nervous Systems	
Characteristic	**Somatic**	**Autonomic**
Effectors	Skeletal muscle	Cardiac muscle
		Smooth muscle
		Gastrointestinal tract
		Blood vessels
		Airways
		Exocrine glands
Effect on motor nerves	Excitation	Excitation or inhibition
Innervation of effector cells	Always single	Typically dual
Number of neurons in path to effector	One	Two
Neurotransmitter	Acetylcholine	Acetylcholine
		Norepinephrine

(a)

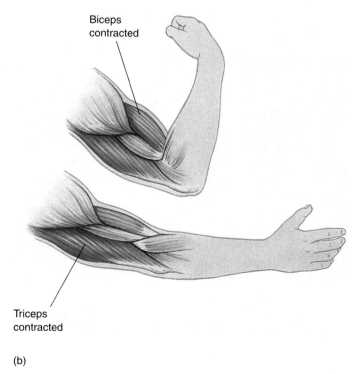

Biceps contracted

Triceps contracted

(b)

FIGURE 51.33
Conscious control of skeletal muscles. (a) A light micrograph shows axons branching to make contact with several individual muscle fibers. (b) When you flex your arm, one set of muscles, the flexors or biceps, contracts, and another set, the extensors or triceps, relaxes. When you extend your arm, the extensors contract and the flexors relax.

The Autonomic Nervous System

The autonomic nervous system is composed of the sympathetic and parasympathetic divisions and the medulla oblongata of the hindbrain, which coordinates this system. Though they differ, the sympathetic and parasympathetic divisions share several features. In both, the efferent motor pathway involves two neurons: the first has its cell body in the CNS and sends an axon to an autonomic ganglion, while the second has its cell body in the autonomic ganglion and sends its axon to synapse with a smooth muscle, cardiac muscle, or gland cell (figure 51.34). The first neuron is called a *preganglionic neuron*, and it always releases ACh at its synapse. The second neuron is a *postganglionic neuron*; those in the parasympathetic division release ACh, while those in the sympathetic division release norepinephrine.

In the sympathetic division, the preganglionic neurons originate in the thoracic and lumbar regions of the spinal cord (figure 51.35). Most of the axons from these neurons synapse in two parallel chains of ganglia immediately outside the spinal cord. These structures are usually called the *sympathetic chain* of ganglia. The sympathetic chain contains the cell bodies of postganglionic neurons, and it is the axons from these neurons that innervate the different visceral organs. There are some exceptions to this general pattern, however. Most importantly, the axons of some preganglionic sympathetic neurons pass through the sympathetic chain without synapsing and, instead, terminate

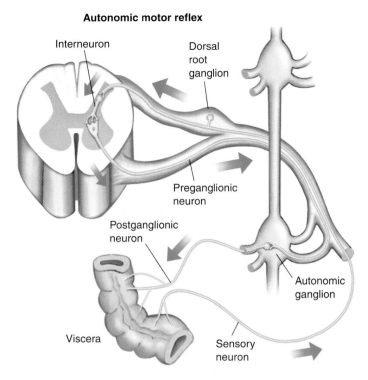

Autonomic motor reflex

FIGURE 51.34
An autonomic reflex. There are two motor neurons in the efferent pathway. The first, or preganglionic neuron, exits the CNS and synapses at an autonomic ganglion. The second, or postganglionic neuron, exits the ganglion and regulates the visceral effectors (smooth muscle, cardiac muscle, or glands).

Table 51.5 Autonomic Innervation of Target Tissues		
Target Tissue	**Sympathetic Stimulation**	**Parasympathetic Stimulation**
Pupil of eye	Dilation	Constriction
Glands		
Salivary	Vasoconstriction; slight secretion	Vasodilation; copious secretion
Gastric	Inhibition of secretion	Stimulation of gastric activity
Liver	Stimulation of glucose secretion	Inhibition of glucose secretion
Sweat	Sweating	None
Gastrointestinal tract		
Sphincters	Increased tone	Decreased tone
Wall	Decreased tone	Increased motility
Gallbladder	Relaxation	Contraction
Urinary bladder		
Muscle	Relaxation	Contraction
Sphincter	Contraction	Relaxation
Heart muscle	Increased rate and strength	Decreased rate
Lungs	Dilation of bronchioles	Constriction of bronchioles
Blood vessels		
In muscles	Dilation	None
In skin	Constriction	None
In viscera	Constriction	Dilation

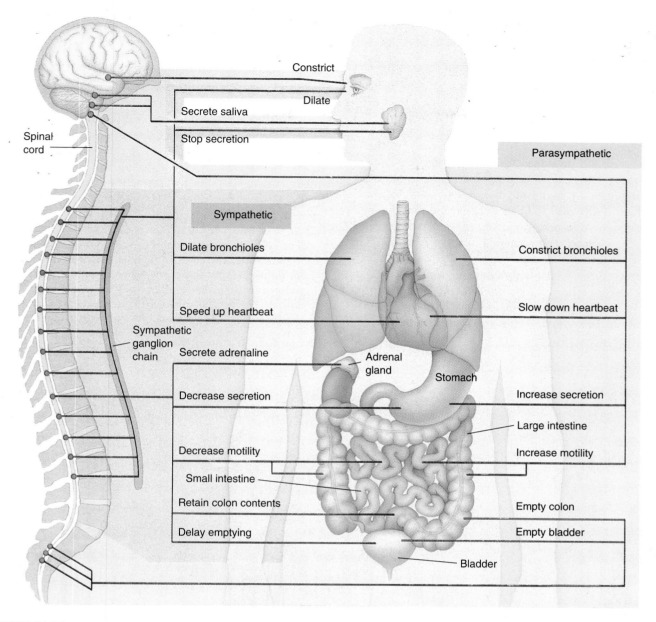

FIGURE 51.35

The sympathetic and parasympathetic divisions of the autonomic nervous system. The preganglionic neurons of the sympathetic division exit the thoracic and lumbar regions of the spinal cord, while those of the parasympathetic division exit the brain and sacral region of the spinal cord. The ganglia of the sympathetic division are located near the spinal cord, while those of the parasympathetic division are located near the organs they innervate. Most of the internal organs are innervated by both divisions.

within the adrenal gland. The adrenal gland consists of an outer part, or cortex, and an inner part, or medulla. The adrenal medulla receives sympathetic nerve innervation and secretes the hormone epinephrine (adrenaline) in response.

When the sympathetic division becomes activated, epinephrine is released into the blood as a hormonal secretion, and norepinephrine is released at the synapses of the postganglionic neurons. Epinephrine and norepinephrine act to prepare the body for fight or flight. The heart beats faster and stronger, blood glucose concentration increases, blood flow is diverted to the muscles and heart, and the bronchioles dilate (table 51.5).

These responses are antagonized by the parasympathetic division. Preganglionic parasympathetic neurons originate in the brain and sacral regions of the spinal cord. Because of this origin, there cannot be a chain of parasympathetic ganglia analogous to the sympathetic chain. Instead, the preganglionic axons, many of which travel in the vagus (the tenth cranial) nerve, terminate in ganglia located near or even within the internal organs. The postganglionic neurons then regulate the internal organs by releasing ACh at their synapses. Parasympathetic nerve effects include a slowing of the heart, increased secretions and activities of digestive organs, and so on.

G Proteins Mediate Cell Responses to Autonomic Nerves

You might wonder how ACh can slow the heart rate—an inhibitory effect—when it has excitatory effects elsewhere. This inhibitory effect in the pacemaker cells of the heart is produced because ACh causes the opening of potassium channels, leading to the outward diffusion of potassium and thus to hyperpolarization. This and other parasympathetic effects of ACh are produced indirectly, using a group of membrane proteins called **G proteins** (so-called because they are regulated by guanosine diphosphate and guanosine triphosphate [GDP and GTP]). Because the ion channels are located some distance away from the receptor proteins for ACh, the G proteins are needed to serve as connecting links between them.

There are three G protein subunits, designated α, β, and γ, bound together and attached to the receptor protein for ACh. When ACh, released by parasympathetic endings, binds to its receptor, the G protein subunits dissociate (figure 51.36). Specific G protein components (the β-γ complex in the figure) move within the membrane to the potassium channel and cause it to open, producing hyperpolarization and a slowing of the heart. In other organs, the G proteins have different effects that lead to excitation. In this way, for example, the parasympathetic nerves that innervate the stomach can cause increased gastric secretions and contractions.

The sympathetic nerve effects also involve the action of G proteins. Stimulation by norepinephrine from sympathetic nerve endings and epinephrine from the adrenal medulla requires G proteins to activate the target cells. We will describe this in more detail, together with hormone action, in chapter 53.

The sympathetic division of the autonomic system, together with the adrenal medulla, activates the body for fight-or-flight responses, whereas the parasympathetic division generally has antagonistic effects. The actions of parasympathetic nerves are produced by ACh, whereas the actions of sympathetic nerves are produced by norepinephrine.

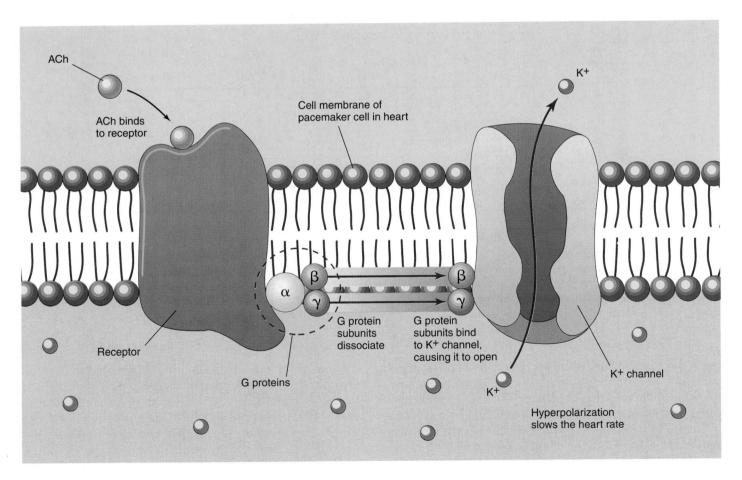

FIGURE 51.36

The parasympathetic effects of ACh require the action of G proteins. The binding of ACh to its receptor causes dissociation of a G protein complex, releasing some components of this complex to move within the membrane and bind to other proteins that form ion channels. The example shown here is the effects of ACh on the heart, where the G protein components cause the opening of potassium channels. This leads to outward diffusion of potassium and hyperpolarization, slowing the heart rate.

51.1 **The nervous system consists of neurons and supporting cells.**

- The nervous system is subdivided into the central nervous system (CNS) and peripheral nervous system (PNS).
- Schwann cells produce myelin sheaths in the PNS and oligodendrocytes produce myelin in the CNS; the gray matter in the CNS consists of cell bodies and dendrites, which are not myelinated, whereas the white matter consists of myelinated axons.

51.2 **Nerve impulses are produced on the axon membrane.**

- The resting axon has a membrane potential of -70 mV; the magnitude of this voltage is produced primarily by the distribution of K^+.
- A depolarization stimulus opens voltage-regulated Na^+ channels and then K^+ channels, producing first the upward phase and then the repolarization phase of the action potential as first Na^+ enters the axon and then K^+ leaves.
- Action potentials are all or none, cannot be summated, and are conducted without decrease in amplitude because each action potential serves as the stimulus for the production of the next action potential along the axon.

51.3 **Neurons form junctions called synapses with other cells.**

- The presynaptic axon releases neurotransmitter chemicals that diffuse across the synapse and stimulate the production of either a depolarization or a hyperpolarization in the postsynaptic membrane.
- Depolarizations and hyperpolarization can summate in the dendrites and cell bodies of the postsynaptic neuron, allowing integration of information.

51.4 **The central nervous system consists of the brain and spinal cord.**

- The vertebrate brain is divided into a forebrain, midbrain and hindbrain, and these are further subdivided into other brain regions.
- The forebrain is divided into the cerebrum and diencephalon; the former is the structurally dominant part of the mammalian brain, while the latter, which includes the thalamus and hypothalamus, has important functions subservient to the cerebrum.
- The cerebral cortex has a primary motor area and a primary somatosensory area, as well as areas devoted to the analysis of vision and hearing and the integration and association of information.
- The basal ganglia are areas of gray matter within the cerebrum that aid in motor control.
- Short-term memories are converted into long-term memories by the function of the hippocampus and amygdala.
- The spinal cord carries information to and from the brain and coordinates many reflex movements.

51.5 **The peripheral nervous system consists of sensory and motor neurons.**

- The sympathetic division, together with the adrenal medulla, are activated during fight-or-flight responses when they stimulate their effector organs through the release of norepinephrine and epinephrine.
- The parasympathetic division opposes the action of the sympathetic division in most activities through the release of ACh as a neurotransmitter.

Discussing Key Terms

1. **Action potential**   Also known as a nerve impulse, an action potential is an all-or-none change in the membrane potential at one region of an axon membrane. Action potentials are produced by the opening of voltage-gated Na^+ and K^+ channels in response to depolarization.

2. **Refractory period**   A refractory period is the period of time during an action potential and slightly afterward when a membrane cannot be stimulated again.

3. **Saltatory conduction**   In a myelinated axon, the action potentials are only produced at the nodes of Ranvier where the myelin sheath is interrupted. This is called saltatory conduction because the impulse seems to jump from node to node.

4. **Cerebral cortex**   The cerebral cortex is the outer region of gray matter in the cerebrum; this region contains centers that are responsible for the analysis of sensory information, control of motor activity, and higher associative functions.

5. **Autonomic nervous system**   The autonomic nervous system controls the involuntary effectors—smooth muscles, cardiac muscles, and glands—through its two divisions: the sympathetic and parasympathetic.

Review Questions

1. Which cation is most concentrated in the cytoplasm of a cell, and which is most concentrated in the extracellular fluid? How are these concentration differences maintained? In a resting neuron, to which of these cations is the plasma membrane most permeable?

2. What is a voltage-gated ion channel? During an action potential, which type of channel opens first, and what effect does this opening have on the membrane potential? Which type of channel opens second, and what effect does this opening have on the membrane potential?

3. What happens to the size of an action potential as it is propagated? How does conduction of action potentials differ in myelinated and unmyelinated axons?

4. Where is the reticular activating system located and what is its function? Where are the basal ganglia located, and what is their function?

5. How are short-term and long-term memory thought to differ in terms of their basic underlying mechanisms? What is long-term potentiation, and how is it thought to be related to memory?

6. How do the sympathetic and parasympathetic divisions of the autonomic nervous system differ in the locations of the preganglionic neurons, the locations of the postganglionic neurons, and the neurotransmitter released by the postganglionic neurons? What are the functions of each division?

Thought Questions

1. Tetraethylammonium (TEA) is a drug that blocks voltage-gated K^+ channels. What effect would TEA have on the action potentials produced by a neuron? If TEA could be applied selectively to a presynaptic neuron that released an excitatory neurotransmitter, how would it alter the synaptic effect of that neuron on the postsynaptic cell?

2. If a nerve impulse can jump from node to node along a myelinated axon, why can't it jump from the presynaptic cell to the postsynaptic cell across a synaptic cleft?

3. Atropine blocks the parasympathetic effects of ACh. What do you think atropine would do to the heart, pupils, salivary glands, mucous glands, and intestine? How might these actions be medically useful?

4. A monkey will pick up a chair and move it under a shelf so that it can climb up and get food stored on the shelf. Is the monkey "thinking"? In your opinion, do other animals, such as dogs and fish, think, or has thinking evolved only in humans?

Internet Links

Neurobiology Today
http://ivory.lm.com/!nab
An immense collection of neuroscience links (over 2000 sites) in an attractive, easily navigated format.

Nonhuman Nervous Systems
http://www./m.com:80/~nab/andp.html
The TOUR OF NEUROSCIENCE presents a comprehensive list of neuroscience research labs, and a wonderful NON HUMAN SYSTEMS index of what is being done in flies, mice, fish, and even plants.

Neuroscience Tutorial
http://thalamus.wustl.edu/course/
An illustrated tutorial to the essential basics of neuroscience created in conjunction with the first year course for medical students at Washington University School of Medicine in St Louis.

Drug Interactions
http://www.drugfreeamerica.org/drug_info.html
A detailed database on drugs and how they affect your nervous system. Click on the name of the drug, or its slang name, to find detailed FAQs and pictures. Tobacco and alcohol are included.

For Further Reading

Alkon, D.: "Memory Storage and Neural Systems," *Scientific American*, July 1989, pages 42–50. Changes in the molecular and electrical properties of nerve cells take place during learning.

Allport, S.: *Explorers of the Black Box: The Search for the Cellular Basis of Memory*, W. W. Norton & Company, New York, 1986. A vivid account of the pioneering studies of Eric Kandel and others in their efforts to demonstrate how we remember. Easy to read, this book shows scientists in action, gathering data and disputing among themselves about what the data mean.

Crick, F.: *The Astonishing Hypothesis*. Sir Francis Crick, codiscoverer of the DNA double-helix molecule, uses visual perception to explore the nature of consciousness.

Eccles, J.: *Evolution of the Brain*, Routledge, London, 1989. A deeply stimulating and thoughtful look at how our brains differ from those of apes, by a Nobel-prize winning neurobiologist.

Holloway, M.: "Rx for Addiction," *Scientific American*, March 1991, pages 94–104. An up-to-date look at the molecular mechanisms underlying drug addiction.

Musto, D.: "Opium, Cocaine, and Marijuana in American History," *Scientific American*, July 1991, pages 40–47. A brief history of drug use by the general public over the last 200 years.

52

Sensory Systems

Concept Outline

52.1 Animals employ a wide variety of sensory receptors.

Categories of Sensory Receptors and Their Actions. Sensory receptors can be classified according to the type of stimuli to which they can respond.

52.2 Mechanical and chemical receptors sense the body's condition.

Detecting Temperature and Pressure. Receptors within the skin respond to touch, pressure, pain, heat and cold.

Sensing Muscle Contraction and Blood Pressure. A muscle spindle responds to stretching of the muscle; receptors in arteries monitor changes in blood pressure.

Sensing Taste, Smell, and Body Position. Receptors that respond to chemicals produce sensations of taste and smell. Hair cells send nerve impulses when they are bent.

52.3 Auditory receptors detect pressure waves in the air.

The Ears and Hearing. Sound causes vibrations in the ear that bend hair cell processes, initiating a nerve impulse.

Sonar. Bats orient themselves in space by emitting sounds and detecting the time required for the sounds to bounce off objects and return to their ears.

52.4 Optic receptors detect light over a broad range of wavelengths.

Evolution of the Eye. True image-forming eyes evolved independently in several phyla.

Vertebrate Photoreceptors and the Retina. Light causes a pigment molecule in a rod or cone cell to dissociate; this "bleaching" reaction activates the photoreceptor.

Visual Processing in the Vertebrate Retina. Action potentials travel from the retina of the eyes to the brain for visual perception.

52.5 Some vertebrates use heat, electricity, or magnetism for orientation.

Diversity of Sensory Experiences. Special receptors can detect heat, electrical currents, and magnetic fields.

FIGURE 52.1
Photoreceptors in the vertebrate eye. Rods, the broad, tubular cells, allow black-and-white vision, while cones, the short, tapered cells, are responsible for color vision. Not all vertebrates have both types of receptors.

All input from sensory neurons to the central nervous system arrives in the same form, as action potentials propagated by afferent (inward-conducting) sensory neurons. Different sensory neurons project to different brain regions, and so are associated with different sensory modalities (figure 52.1). The intensity of the sensation depends on the frequency of action potentials conducted by the sensory neuron. A sunset, a symphony, and a searing pain are distinguished by the brain only in terms of the identity of the sensory neuron carrying the action potentials and the frequency of these impulses. Thus, if the auditory nerve is artificially stimulated, the brain perceives the stimulation as sound. But if the optic nerve is artificially stimulated in exactly the same manner and degree, the brain perceives a flash of light.

Categories of Sensory Receptors and Their Actions

Sensory information is conveyed to the CNS and perceived in a four-step process (figure 52.2): (1) *stimulation*—a physical stimulus impinges on a sensory neuron or an accessory structure; (2) *transduction*—the stimulus energy is used to produce electrochemical nerve impulses in the dendrites of the sensory neuron; (3) *transmission*—the axon of the sensory neuron conducts action potentials along an afferent pathway to the CNS; and (4) *interpretation*—the brain creates a sensory perception from the electrochemical events produced by afferent stimulation. We actually see (as well as hear, touch, taste, and smell) with our brains, not with our sense organs.

Sensory receptors differ with respect to the nature of the environmental stimulus that best activates their sensory dendrites. Broadly speaking, we can recognize three classes of environmental stimuli: (1) mechanical forces, which stimulate **mechanoreceptors;** (2) chemicals, which stimulate **chemoreceptors;** and (3) electromagnetic and thermal energy, which stimulate a variety of receptors, including the **photoreceptors** of the eyes (table 52.1).

The simplest sensory receptors are *free nerve endings* that respond to bending or stretching of the sensory neuron membrane, to changes in temperature, or to chemicals like oxygen in the extracellular fluid. Other sensory receptors are more complex, involving the association of the sensory neurons with specialized epithelial cells.

Sensing the External and Internal Environments

Exteroceptors are receptors that sense stimuli that arise in the external environment. Almost all of a vertebrate's exterior senses evolved in water before vertebrates invaded the land. Consequently, many senses of terrestrial vertebrates emphasize stimuli that travel well in water, using receptors that have been retained in the transition from the sea to the land. Mammalian hearing, for example, converts an airborne stimulus into a waterborne one, using receptors similar to those that originally evolved in the water. A few vertebrate sensory systems that function well in the water, such as the electrical organs of fish, cannot function in the air and are not found among terrestrial vertebrates. On the other hand, some land-dwellers have sensory systems, such as infrared receptors, that could not function in the sea.

Sensory systems can provide several levels of information about the external environment. Some sensory systems provide only enough information to determine that an object is present; they call the animal's attention to the object but give little or no indication of where it is located. Other sensory systems provide information about the location of an ob-

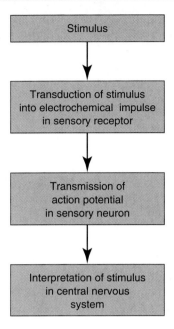

FIGURE 52.2
The path of sensory information. Sensory stimuli must be transduced into electrochemical nerve impulses that are conducted to the brain for interpretation.

Table 52.1	Classes of Environmental Stimuli	
Mechanical Forces	**Chemicals**	**Electromagnetic Energy**
Pressure	Taste	Light
Gravity	Smell	Heat
Inertia	Humidity	Electricity
Sound		Magnetism
Touch		
Vibration		

ject, permitting the animal to move toward it. Still other sensory systems enable the brain to construct a three-dimensional image of an object and its surroundings.

Interoceptors sense stimuli that arise from within the body. These internal receptors detect stimuli related to muscle length and tension, limb position, pain, blood chemistry, blood volume and pressure, and body temperature. Many of these receptors are simpler than those that monitor the external environment and are believed to bear a closer resemblance to primitive sensory receptors. In the rest of this chapter, we will consider the different types of interoceptors and exteroceptors according to the kind of stimulus each is specialized to detect (table 52.2).

Table 52.2 Sensory Transduction Among the Vertebrates

Stimulus	Receptor	Location	Structure	Transduction Process
INTEROCEPTION				
Temperature	Heat receptors and cold receptors	Skin, hypothalamus	Free nerve ending	Temperature change opens/closes ion channels in membrane
Touch	Meissner's corpuscles, Merkel cells	Surface of skin	Nerve ending within elastic capsule	Rapid or extended change in pressure deforms membrane
Vibration	Pacinian corpuscles	Deep within skin	Nerve ending within elastic capsule	Severe change in pressure deforms membrane
Pain	Nociceptors	Throughout body	Free nerve ending	Chemicals or changes in pressure or temperature open/close ion channels in membrane
Muscle stretch	Stretch receptors	Within muscles	Spiral nerve endings wrapped around muscle spindle	Stretch of spindle deforms membrane
Blood pressure	Baroreceptors	Arterial branches	Nerve endings over thin part of arterial wall	Stretch of arterial wall deforms membrane
EXTEROCEPTION				
Gravity	Statocysts	Outer chambers of inner ear	Otoliths and cilia	Otoliths deform cilia
Motion	Cupula	Semicircular canals of inner ear	Collection of cilia	Fluid movement deforms cilia
	Lateral line organ	Within grooves on body surface of fish	Collection of cilia	Fluid movement deforms cilia
Taste	Taste bud cells	Mouth; skin of fish	Chemoreceptors	Chemicals bind to membrane receptors
Smell	Olfactory neurons	Nasal passages	Chemoreceptors	Chemicals bind to membrane receptors
Hearing	Organ of Corti	Cochlea of inner ear	Cilia between basilar and tectorial membranes	Sound waves in fluid deform membranes
Vision	Rod and cone cells	Retina of eye	Array of photosensitive pigments	Light initiates process that closes ion channels
Heat	Pit organ	Face of snake	Temperature receptors in two chambers	Receptors compare temperatures of surface and interior chambers
Electricity	Ampullae of Lorenzini	Within skin of fishes	Closed vesicles with asymmetrical ion channel distribution	Electrical field alters ion distribution on membranes
Magnetism	Unknown	Unknown	Unknown	Deflection at magnetic field initiates nerve impulses?

Sensory Transduction

Sensory cells respond to stimuli because they possess *stimulus-gated ion channels* in their membranes. The sensory stimulus causes these ion channels to open or close, depending on the sensory system involved. In doing so, a sensory stimulus produces a change in the membrane potential of the receptor cell. In most cases, the sensory stimulus produces a depolarization of the receptor cell, analogous to the excitatory postsynaptic potential (EPSP, described in chapter 51) produced in a postsynaptic cell in response to neurotransmitter. A depolarization that occurs in a sensory receptor upon stimulation is referred to as a **receptor potential** (figure 52.3*a*).

Like an EPSP, a receptor potential is graded: the larger the sensory stimulus, the greater the degree of depolarization. Once a threshold level of depolarization is reached, the receptor potential stimulates the production of action potentials that are conducted by a sensory axon into the CNS (figure 52.3*b*). The greater the sensory stimulus, the greater the depolarization of the receptor potential and the higher the frequency of action potentials. There is generally a logarithmic relationship between stimulus intensity and action potential frequency—a sensory stimulus that is ten times greater than another stimulus will produce action potentials at twice the frequency of the other stimulus. This allows the brain to interpret the incoming signals as indicating a sensory stimulus of a particular strength.

Sensory receptors transduce stimuli in the internal or external environment into graded depolarizations, which stimulates the production of action potentials. Sensory receptors may be classified on the basis of the type of stimulus energy to which they respond.

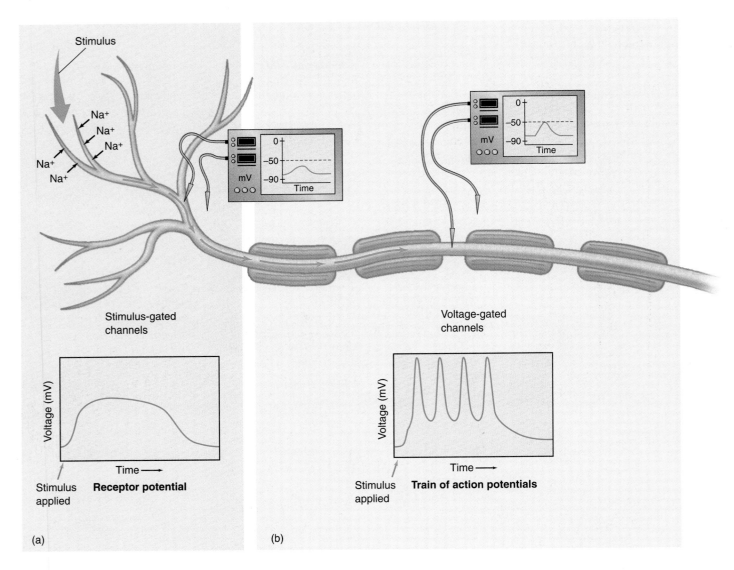

FIGURE 52.3
Events in sensory transduction. (a) Depolarization of a free nerve ending leads to a receptor potential that spreads by local current flow to the axon. (b) Action potentials are produced in the axon in response to a sufficiently large receptor potential.

Detecting Temperature and Pressure

While the receptors of the skin, called the **cutaneous receptors,** are classified as interoceptors, they in fact respond to stimuli at the border between the external and internal environments. These receptors serve as good examples of the specialization of receptor structure and function, responding to heat, cold, pain, touch, and pressure.

The skin contains two populations of **thermoreceptors,** which are naked dendritic endings of sensory neurons that are sensitive to changes in temperature. *Cold receptors* are stimulated by a fall in temperature and inhibited by warming, while *warm receptors* are stimulated by a rise in temperature and inhibited by cooling. Cold receptors are located immediately below the epidermis, while warm receptors are located slightly deeper, in the dermis. Thermoreceptors are also found within the hypothalamus of the brain, where they monitor the temperature of the circulating blood and thus provide the CNS with information on the body's internal (core) temperature.

A stimulus that causes or is about to cause tissue damage is perceived as pain. The receptors that transmit impulses that are perceived by the brain as pain are called **nociceptors.** They consist of free nerve endings located throughout the body, especially near surfaces where damage is most likely to occur. Different nociceptors may respond to extremes in temperature, very intense mechanical stimulation, or specific chemicals in the extracellular fluid, including some that are released by injured cells. The thresholds of these sensory cells vary; some nociceptors are sensitive only to actual tissue damage, while others respond before damage has occurred.

Several types of mechanoreceptors are present in the skin, some in the dermis and others in the underlying subcutaneous tissue (figure 52.4). Morphologically specialized receptors that respond to fine touch are most concentrated on areas such as the fingertips and face. They are used to localize cutaneous stimuli very precisely and can be either phasic (intermittently activated) or tonic (continuously activated). The phasic receptors include *hair follicle receptors* and *Meissner's corpuscles,* which are present on body surfaces that do not contain hair, such as the fingers, palms, and nipples. The tonic receptors consist of *Ruffini endings* in the dermis and *touch dome endings (Merkel cells)* located near the surface of the skin. These receptors monitor the duration of a touch and the extent to which it is applied.

Deep below the skin in the subcutaneous tissue lie phasic, pressure-sensitive receptors called **Pacinian corpuscles.** Each of these receptors consists of the end of an afferent axon, surrounded by a capsule of alternating layers of connective tissue cells and extracellular fluid. When sustained pressure is applied to the corpuscle, the elastic capsule absorbs much of the pressure and the axon ceases to produce impulses. Pacinian corpuscles thus monitor only the onset and removal of pressure, as may occur repeatedly when something that vibrates is placed against the skin.

Different cutaneous receptors respond to touch, pressure, heat, cold, and pain. Some of these receptors are naked dendrites of sensory neurons, while others have supporting cells that modify the activities of their sensory dendrites.

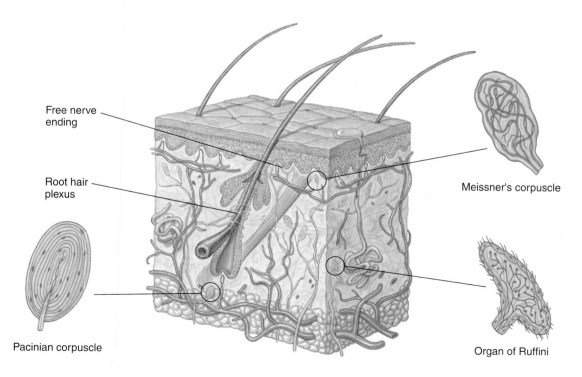

Free nerve ending

Root hair plexus

Meissner's corpuscle

Pacinian corpuscle

Organ of Ruffini

FIGURE 52.4
Sensory receptors in human skin. Cutaneous receptors may be free nerve endings or sensory dendrites in association with other supporting structures.

Sensing Muscle Contraction and Blood Pressure

Mechanoreceptors contain sensory cells with ion channels that are sensitive to a mechanical force applied to the membrane. These channels open in response to mechanical distortion of the membrane, initiating a depolarization (receptor potential) that causes the sensory neuron to generate action potentials.

Muscle Length and Tension

Buried within the skeletal muscles of all vertebrates except the bony fishes are muscle spindles, sensory stretch receptors that lie in parallel with the rest of the fibers in the muscle (figure 52.5). Each spindle consists of several thin muscle fibers wrapped together and innervated by a sensory neuron, which becomes activated when the muscle, and therefore the spindle, is stretched. Muscle spindles, together with other receptors in tendons and joints, are known as **proprioceptors.** The sensory neurons conduct action potentials into the spinal cord, where they synapse with somatic motor neurons that innervate the muscle. This pathway constitutes the muscle stretch reflex, including the knee-jerk reflex, previously discussed in chapter 51.

When a muscle contracts, it exerts tension on the tendons attached to it. The Golgi tendon organs, another type of proprioceptor, monitor this tension; if it becomes too high, they elicit a reflex that inhibits the motor neurons innervating the muscle. This reflex helps to ensure that muscles do not contract so strongly that they damage the tendons to which they are attached.

Blood Pressure

Blood pressure is monitored at two main sites in the body. One is the *carotid sinus,* an enlargement of the left and right internal carotid arteries, which supply blood to the brain. The other is the *aortic arch,* the portion of the aorta very close to its emergence from the heart. The walls of the blood vessels at both sites contain a highly branched network of afferent neurons called **baroreceptors,** which detect tension in the walls. When the blood pressure decreases, the frequency of impulses produced by the baroreceptors decreases. The CNS responds to this reduced input by stimulating the sympathetic division of the autonomic nervous system, causing an increase in heart rate and vasoconstriction. Both effects help to raise the blood pressure, thus maintaining homeostasis. A rise in blood pressure, conversely, reduces sympathetic activity and stimulates the parasympathetic division, slowing the heart and lowering the blood pressure.

Mechanical distortion of the plasma membrane of mechanoreceptors produces nerve impulses that serve to monitor muscle length from skeletal muscle spindles and to monitor blood pressure from baroreceptors within arteries.

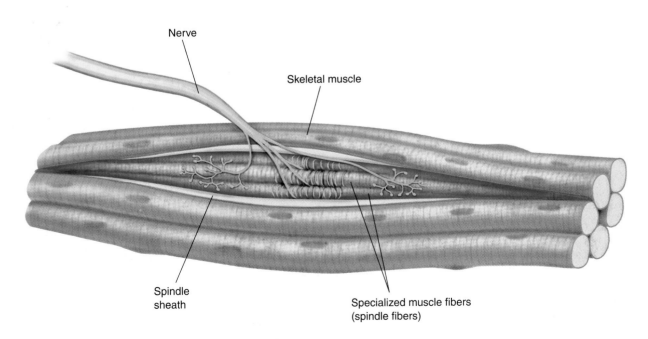

FIGURE 52.5
A muscle spindle is a stretch receptor embedded within skeletal muscle. Stretching of the muscle elongates the spindle fibers and stimulates the sensory dendritic endings wrapped around them, causing the sensory neuron to send impulses to the CNS.

Sensing Taste, Smell, and Body Position

Some sensory cells, called **chemoreceptors,** contain membrane proteins that can bind to particular chemicals in the extracellular fluid. In response to this chemical interaction, the membrane of the sensory neuron becomes depolarized, leading to the production of action potentials. Chemoreceptors are used in the senses of taste and smell and are also important in monitoring the chemical composition of the blood and cerebrospinal fluid.

Taste

Taste buds—collections of chemosensitive cells associated with afferent neurons—mediate the sense of taste in vertebrates. In a fish, the taste buds are scattered over the surface of the body. These are the most sensitive vertebrate chemoreceptors known. They are particularly sensitive to amino acids; a catfish, for example, can distinguish between two different amino acids at a concentration of less than 100 parts per billion (1 g in 10,000 L of water)! The ability to taste the surrounding water is very important to bottom-feeding fish, enabling them to sense the presence of food in an often murky environment.

The taste buds of all terrestrial vertebrates are located in the epithelium of the tongue, within raised areas called *papillae* (figure 52.6). Humans have four kinds of taste buds—salty, sweet, sour, and bitter. The salty taste is produced by the effects of sodium (Na^+) and the sour taste by the effects of hydrogen (H^+). Organic molecules that produce the sweet and bitter tastes, such as sugars and quinine, respectively, are varied in structure. Taste buds that respond best to specific tastes are concentrated in specific regions of the tongue: sweet at the tip, sour at the sides, bitter at the back, and salty over most of the tongue's surface. Our complex perception of taste is the result of different combinations of impulses in the sensory neurons from these four kinds of taste buds, together with information related to smell. The effect of smell on the sense of taste can easily be demonstrated by eating an onion with the nose open and then eating it with the nose plugged.

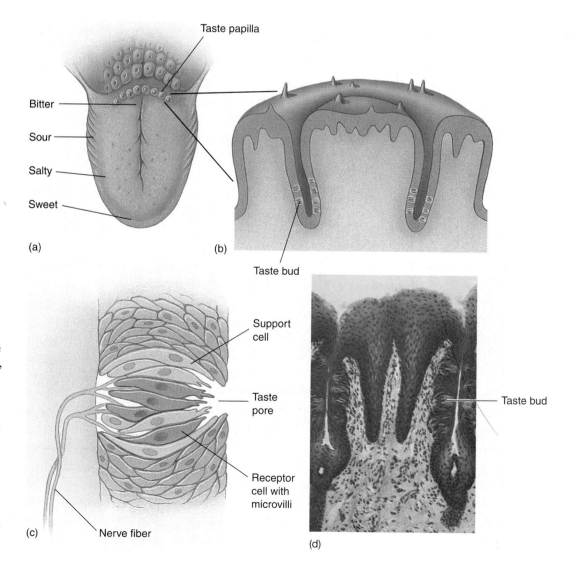

FIGURE 52.6

Taste. (a) Human beings have four kinds of taste buds (bitter, sour, salty, and sweet), located on different regions of the tongue. (b) Groups of taste buds are typically organized in sensory projections called papillae. (c) Individual taste buds are bulb-shaped collections of chemosensitive receptors that open out into the mouth through a pore. (d) Photomicrograph of a taste buds in papillae.

Smell

In terrestrial vertebrates, the sense of smell, or olfaction, involves chemoreceptors located in the upper portion of the nasal passages (figure 52.7). These receptors are bipolar neurons whose dendrites end in tassels of cilia that project into the nasal mucosa, and whose axon projects directly into the cerebral cortex. A terrestrial vertebrate uses its sense of smell in much the same way that a fish uses its sense of taste—to sample the chemical environment around it. Because terrestrial vertebrates are surrounded by air rather than water, their sense of smell has become specialized to detect airborne particles (but these particles must first dissolve in extracellular fluid before they can activate the olfactory receptors). The sense of smell can be extremely acute in many mammals, so much so that a single odorant molecule may be all that is needed to excite a given receptor.

Although humans can detect only four modalities of taste, they can discern thousands of different smells. New research suggests that there may be as many as a thousand different genes coding for different receptor proteins for smell. The particular set of olfactory neurons that respond to a given odor might serve as a "fingerprint" the brain can use to identify the odor.

Internal Chemoreceptors

Sensory receptors within the body detect a variety of chemical characteristics of the blood or fluids derived from the blood, including cerebrospinal fluid. Included among these receptors are the *peripheral chemoreceptors* of the aortic and carotid bodies, which are sensitive primarily to plasma pH, and the *central chemoreceptors* in the medulla oblongata of the brain, which are sensitive to the pH of cerebrospinal fluid. These receptors were discussed together with the regulation of breathing in chapter 50. When the breathing rate is too low, the concentration of plasma CO_2 increases, producing more carbonic acid and causing a fall in the blood pH. The carbon dioxide can also enter the cerebrospinal fluid and cause a lowering of the pH, thereby stimulating the central chemoreceptors. This chemoreceptor stimulation indirectly affects the respiratory control center of the brain stem, which increases the breathing rate. The aortic bodies can also respond to a lowering of blood oxygen concentrations, but this effect is normally not significant unless a person goes to a high altitude.

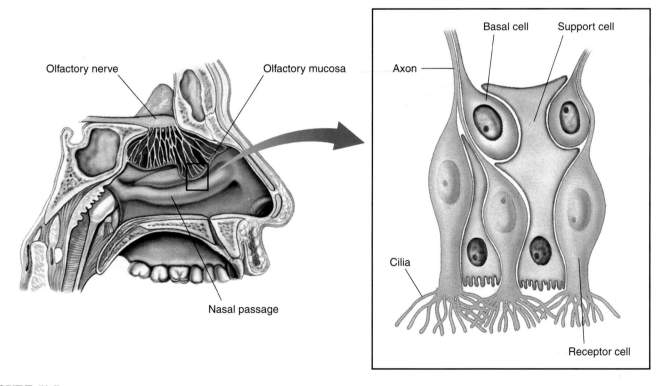

FIGURE 52.7
Smell. Humans detect smells by means of olfactory neurons located in the lining of the nasal passages. The axons of these neurons transmit impulses directly to the brain via the olfactory nerve.

The Lateral Line System

The lateral line system provides fish with a sense of "distant touch," enabling them to sense objects that reflect pressure waves and low-frequency vibrations. This enables a fish to detect prey, for example, and to swim in synchrony with the rest of its school. It also enables a blind cave fish to sense its environment by monitoring changes in the patterns of water flow past the lateral line receptors. The lateral line system is found in amphibian larvae, but is lost at metamorphosis and is not present in any terrestrial vertebrate. The sense provided by the lateral line system supplements the fish's sense of hearing, which is performed by a different sensory structure. The structures and mechanisms involved in hearing will be described in a later section.

The lateral line system consists of sensory structures within a longitudinal canal in the fish's skin that extends along each side of the body and within several canals in the head (figure 52.8a). The sensory structures are known as hair cells because they have hair-like processes at their surface that project into a gelatinous membrane called a *cupula* (Latin, "little cup"). The hair cells are innervated by sensory neurons that transmit impulses to the brain.

Hair cells have several hair-like processes of the same length, called *stereocilia*, and one longer process called a *kinocilium* (figure 52.8b). Vibrations carried through the fish's environment produce movements of the cupula, which cause the hairs to bend. When the stereocilia bend in the direction of the kinocilium, the associated sensory neurons are stimulated and generate a receptor potential. As a result, the frequency of action potentials produced by the sensory neuron is increased. If the stereocilia are bent in the opposite direction, on the other hand, the activity of the sensory neuron is inhibited.

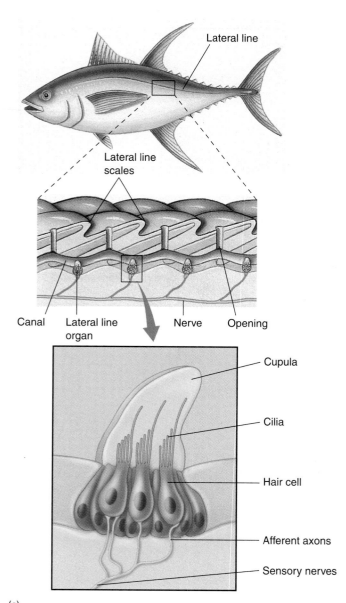

(a)

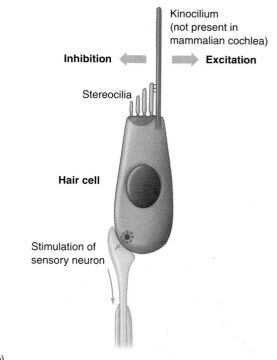

(b)

FIGURE 52.8

The lateral line system. (a) This system consists of canals running the length of the fish's body beneath the surface of the skin. Within these canals are sensory structures containing hair cells with cilia that project into a gelatinous cupula. Pressure waves traveling through the water in the canals deflect the cilia and depolarize the sensory neurons associated with the hair cells. (b) Hair cells are mechanoreceptors with hair-like cilia that project into a gelatinous membrane. The hair cells of the lateral line system (and the membranous labyrinth of the vertebrate inner ear) have a number of smaller cilia called stereocilia and one larger kinocilium. When the cilia bend in the direction of the kinocilium, the hair cell releases a chemical transmitter that depolarizes the associated sensory neuron. Bending of the cilia in the opposite direction has an inhibitory effect.

Gravity and Angular Acceleration

Most invertebrates can orient themselves with respect to gravity due to a sensory structure called a *statocyst*. Statocysts generally consist of ciliated hair cells with the cilia embedded in a gelatinous membrane containing crystals of calcium carbonate. These "stones," or *statoliths*, increase the mass of the gelatinous membrane so that it can bend the cilia when the animal's position changes. If the animal tilts to the right, for example, the statolith membrane will bend the cilia on the right side and activate associated sensory neurons.

A similar structure is found in the membranous labyrinth of the inner ear of vertebrates. The labyrinth is a system of fluid-filled membranous chambers and tubes that constitute the organs of equilibrium and hearing in vertebrates. This membranous labyrinth is filled with fluid (endolymph) and is surrounded by a different fluid (perilymph) and bone. Though intricate, the entire structure is very small; in a human, it is about the size of a pea.

The receptors for gravity in vertebrates consist of two chambers of the membranous labyrinth called the utricle and saccule (figure 52.9). Within these structures are hair cells with stereocilia and a kinocilium, similar to those in the lateral line system of fish. The hair-like processes are embedded within a gelatinous membrane containing calcium carbonate crystals; this is known as an *otolith membrane*, because of its location in the inner ear (*oto* is derived from the Greek word for ear). Since the otolith organ is oriented differently in the utricle and saccule, the utricle is more sensitive to horizontal acceleration (as in a moving car) and the saccule to vertical acceleration (as in an elevator). In both cases, the acceleration causes the stereocilia to bend and consequently produces action potentials in an associated sensory neuron.

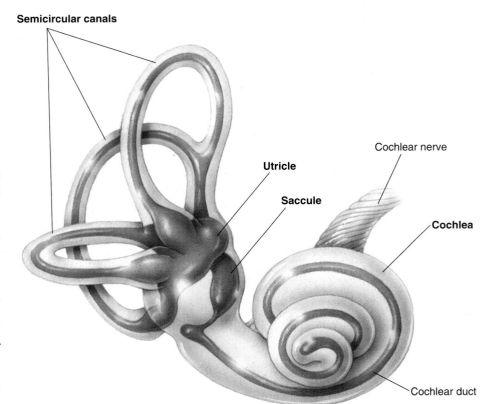

(a)

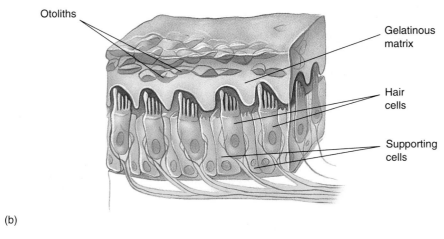

(b)

FIGURE 52.9

The structure of the utricle and saccule. (a) The relative positions of the utricle and saccule within the membranous labyrinth of the human inner ear. (b) Enlargement of a section of the utricle or saccule showing the otoliths embedded in the gelatinous matrix that covers the hair cells.

The membranous labyrinth of the utricle and saccule is continuous with three semicircular canals, oriented in different planes so that angular acceleration in any direction can be detected (figure 52.10). At the ends of the canals are swollen chambers called *ampullae*, into which protrude the cilia of another group of hair cells. The tips of the cilia are embedded within a sail-like wedge of gelatinous material called a *cupula* (similar to the culpa of the fish lateral line system) that protrudes into the endolymph fluid of each semicircular canal.

When the head rotates, the fluid inside the semicircular canals pushes against the cupula and causes the cilia to bend. This bending either depolarizes or hyperpolarizes the hair cells, depending on the direction in which the cilia are bent. This is similar to the way the lateral line system works in a fish: if the stereocilia are bent in the direction of the kinocilium, a depolarization (receptor potential) is pro-

duced, which stimulates the production of action potentials in associated sensory neurons.

The saccule, utricle, and semicircular canals are collectively referred to as the *vestibular apparatus*. While the saccule and utricle provide a sense of linear acceleration, the semicircular canals provide a sense of angular acceleration. The brain uses information that comes from the vestibular apparatus about the body's position in space to maintain balance and equilibrium.

Receptors that sense chemicals originating outside the body are responsible for the senses of odor, smell, and taste. Internal chemoreceptors help to monitor chemicals produced within the body and are needed for the regulation of breathing. Hair cells in the lateral line organ of fishes detect water movements, and hair cells in the vestibular apparatus of terrestrial vertebrates provide a sense of acceleration.

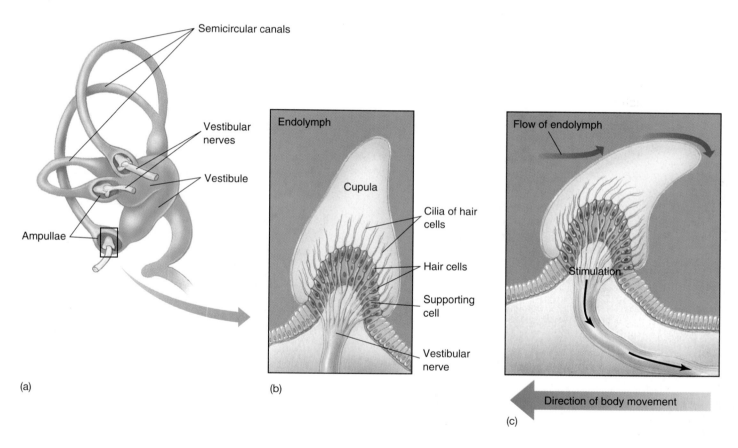

(a)   (b)   (c)

FIGURE 52.10
The structure of the semicircular canals. (a) The position of the semicircular canals in relation to the rest of the inner ear. (b) Enlargement of a section of one ampulla, showing how hair cell cilia insert into the cupula. (c) Angular acceleration in the plane of the semicircular canal causes bending of the cupula, thereby stimulating the hair cells.

The Ears and Hearing

Fish detect vibrational pressure waves in water by means of their lateral line system. Terrestrial vertebrates detect similar vibrational pressure waves in air by means of similar hair cell mechanoreceptors in the inner ear. Hearing actually works better in water than in air because water transmits pressure waves more efficiently. Despite this limitation, hearing is widely used by terrestrial vertebrates to monitor their environments, communicate with other members of the same species, and to detect possible sources of danger (figure 52.11). Auditory stimuli travel farther and more quickly than chemical ones, and auditory receptors provide better directional information than do chemoreceptors. Auditory stimuli alone, however, provide little information about distance.

Structure of the Ear

Fish use their lateral line system to detect water movements and vibrations emanating from relatively nearby objects, and their hearing system to detect vibrations that originate from a greater distance. The hearing system of fish consists of the otolith organs in the membranous labyrinth (utricle and saccule) previously described, together with a very small outpouching of the membranous labyrinth called the lagena. Sound waves travel through the body of the fish as easily as through the surrounding water, since the body is also composed primarily of water. Therefore, an object of different density is needed in order for the sound to be detected. This function is served by the otolith (calcium carbonate crystals) in many fish. In catfish, minnows, and suckers, however, this function is served by an air-filled swim bladder that vibrates with the sound. A chain of small bones, Weberian ossicles, then transmits the vibrations to the saccule in some of these fish.

In the ears of terrestrial vertebrates, vibrations in air may be channeled through an ear canal to the eardrum, or tympanic membrane. These structures are part of the *outer ear*. Vibrations of the tympanic membrane cause movement of three small bones (ossicles)—the *malleus* (hammer), *incus* (anvil), and *stapes* (stirrup)—that are located in a bony cavity known as the *middle ear* (figure 52.12). These middle ear ossicles are analogous to the Weberian ossicles in fish. The

FIGURE 52.11
Kangaroo rats have specialized ears. Kangaroo rats *(Dipodomys)* are unique in having an enlarged tympanic membrane (eardrum), a lengthened and freely rotating malleus (ear bone), and an increased volume of air-filled chambers in the middle ear. These and other specializations result in increased sensitivity to sound, especially to low-frequency sounds. Experiments have shown that the kangaroo rat's ears are adapted to nocturnal life and allow them to hear the low-frequency sounds of their predators, such as an owl's wingbeats or a sidewinder rattlesnake's scales rubbing against the ground. Also, the ears seem to be adapted to the poor sound-carrying quality of dry, desert air.

middle ear is connected to the throat by the *Eustachian tube*, which equalizes the air pressure between the middle ear and the external environment. The "ear popping" you may have experienced when flying in an airplane or driving on a mountain is caused by pressure equalization between the two sides of the eardrum.

The stapes vibrates against a flexible membrane, the *oval window*, which leads into the *inner ear*. Since the oval window is smaller in diameter than the tympanic membrane, vibrations against it produce more force per unit area, transmitted into the inner ear. The inner ear consists of the cochlea (Latin for "snail"), a bony structure containing part of the membranous labyrinth called the cochlear duct. The cochlear duct is located in the center of the cochlea; the area above the cochlear duct is the *vestibular canal*, and the area below is the *tympanic canal*. All three chambers are filled with fluid, as previously described. The oval window opens to the upper vestibular canal, so that when the stapes causes it to vibrate, it produces pressure waves of fluid. These pressure waves travel down to the tympanic canal, pushing another flexible membrane, the *round window*, that transmits the pressure back into the middle ear cavity.

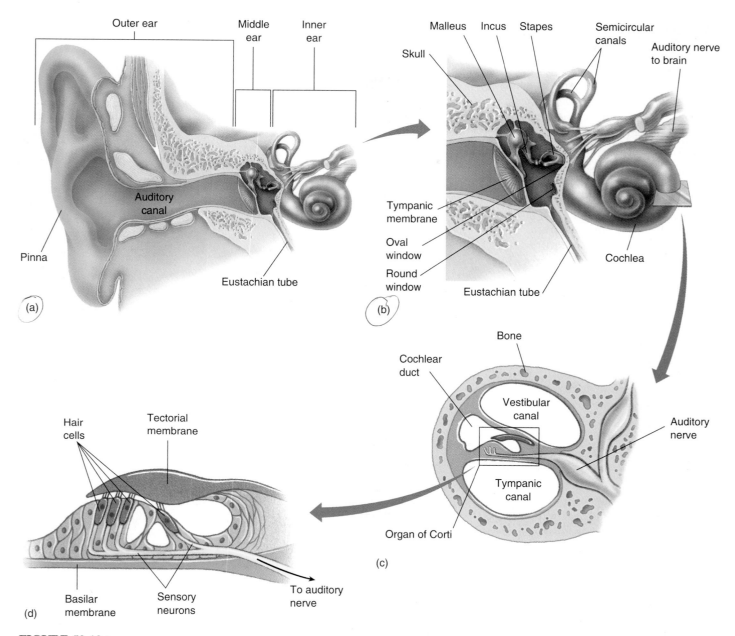

FIGURE 52.12
Structure of the human ear. (a) Sound waves passing through the ear canal produce vibrations of the tympanic membrane, which cause movement of the (b) middle ear ossicles (the malleus, incus, and stapes) against an inner membrane called the oval window. Vibration of the oval window sets up pressure waves that (c and d) travel through the fluid in the vestibular and tympanic canals of the cochlea.

Transduction in the Cochlea

As the pressure waves produced by vibrations of the oval window are transmitted through the cochlea to the round window, they cause the cochlear duct to vibrate. The bottom of the cochlear duct, called the *basilar membrane*, is quite flexible and vibrates in response to these pressure waves. The surface of the basilar membrane contains sensory hair cells, similar to those of the vestibular apparatus and lateral line system but lacking a kinocilium. The cilia from the hair cells project into an overhanging gelatinous membrane, the *tectorial membrane*. This sensory apparatus, consisting of the basilar membrane, hair cells with associated sensory neurons, and tectorial membrane, is known as the **Organ of Corti.**

As the basilar membrane vibrates, the cilia of the hair cells bend in response to the movement of the basilar membrane relative to the tectorial membrane. As in the lateral line organs and the vestibular apparatus, the bending of these cilia depolarizes the hair cells. The hair cells, in turn, stimulate the production of action potentials in sensory neurons that project to the brain, where they are interpreted as sound.

Frequency Localization in the Cochlea

The basilar membrane of the cochlea consists of elastic fibers of varying length and stiffness, like the strings of a musical instrument, embedded in a gelatinous material. At the base of the cochlea (near the oval window), the fibers of the basilar membrane are short and stiff. At the far end of the cochlea (the apex), the fibers are 5 times longer and 100 times more flexible. Therefore, the resonant frequency of the basilar membrane is higher at the base than the apex; the base responds to higher pitches, the apex to lower.

When a wave of sound energy enters the cochlea from the oval window, it initiates a traveling up-and-down motion of the basilar membrane. However, this wave imparts most of its energy to that part of the basilar membrane with a resonant frequency near the frequency of the sound wave, resulting in a maximum deflection of the basilar membrane at that point (figure 52.13). As a result, the hair cell depolarization is greatest in that region, and the afferent axons from that region are stimulated to produce action potentials more than those from other regions. When these action potentials arrive in the brain, they are interpreted as representing a sound of that particular frequency, or pitch.

The flexibility of the basilar membrane limits the frequency range of human hearing to between approximately 20 and 20,000 cycles per second (hertz) in children. Our ability to hear high-pitched sounds decays progressively throughout middle age. Other vertebrates can detect sounds at frequencies lower than 20 hertz and much higher than 20,000 hertz. Dogs, for example, can detect sounds at 40,000 hertz, enabling them to hear high-pitched dog whistles that seem silent to a human listener.

Hair cells are also innervated by efferent axons from the brain, and impulses in those axons can make hair cells less sensitive. This central control of receptor sensitivity can increase an individual's ability to concentrate on a particular auditory signal (for example, a single voice) in the midst of background noise, which is effectively "tuned out" by the efferent axons.

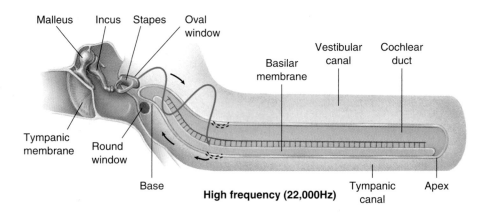

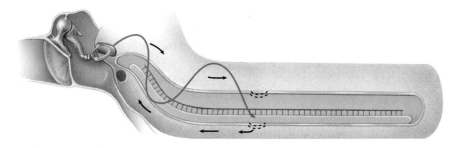

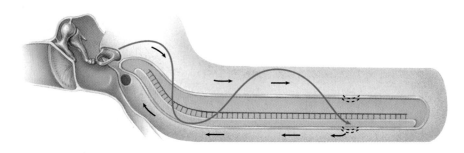

FIGURE 52.13

Frequency localization in the cochlea. The cochlea is shown unwound, so that the length of the basilar membrane can be seen. The fibers within the basilar membrane vibrate in response to different frequencies of sound, related to the pitch of the sound. Thus, regions of the basilar membrane show maximum vibrations in response to different sound frequencies. Notice that low-frequency (pitch) sounds vibrate the basilar membrane more toward the apex, while high frequencies cause vibrations more toward the base.

The middle ear ossicles vibrate in response to sound waves, creating fluid vibrations within the inner ear. This causes the hair cells to bend, transducing the sound into action potentials. The pitch of a sound is determined by which hair cells (and thus which sensory neurons), are activated by the vibration of the basilar membrane.

Sonar

Since terrestrial vertebrates have two ears located on opposite sides of the head, the information provided by hearing can be used by the CNS to determine *direction* of a sound source with some precision. Sound sources vary in strength, however, and sounds are attenuated (weakened) to varying degrees by the presence of objects in the environment. For these reasons, auditory sensors do not provide a reliable measure of *distance*.

A few groups of mammals that live and obtain their food in dark environments have circumvented the limitations of darkness. A bat flying in a completely dark room easily avoids objects that are placed in its path—even a wire less than a millimeter in diameter (figure 52.14). Shrews use a similar form of "lightless vision" beneath the ground, as do whales and dolphins beneath the sea. All of these mammals perceive distance by means of sonar. They emit sounds and then determine the time it takes these sounds to reach an object and return to the animal. This process is called echolocation. A bat, for example, produces clicks that last 2 to 3 milliseconds and are repeated several hundred times per second. The three-dimensional imaging achieved with such an auditory sonar system is quite sophisticated.

Being able to "see in the dark" has opened a new ecological niche to bats, one largely closed to birds because birds must rely on vision. There are no truly nocturnal birds; even owls rely on vision to hunt, and do not fly on dark nights. Because bats are able to be active and efficient in total darkness, they are one of the most numerous and widespread of all orders of mammals.

Some mammals emit sounds and then determine the time it takes for the sound to return, using the method of sonar to locate themselves and other objects in a totally dark environment by the characteristics of the echo. Bats are the most adept at this echolocation.

FIGURE 52.14

Sonar. As it flies, a bat emits high-frequency "chirps" and listens for the return of the chirps after they are reflected by objects such as moths. By timing how long it takes for a chirp to return, the bat can locate its prey and catch it even in total darkness.

Evolution of the Eye

Vision begins with the capture of light energy by photoreceptors. Because light travels in a straight line and arrives virtually instantaneously, visual information can be used to determine both the direction and the distance of an object. No other stimulus provides as much detailed information.

Many invertebrates have simple visual systems with photoreceptors clustered in an eyespot. Although an eyespot can perceive the direction of light, it cannot be used to construct a visual image. The members of four phyla—annelids, mollusks, arthropods, and vertebrates—have evolved well-developed, image-forming eyes. True image-forming eyes in these phyla, though strikingly similar in structure, are believed to have evolved independently (figure 52.15). Interestingly, the photoreceptors in all of them use the same light-capturing molecule, suggesting that not many alternative molecules are able to play this role.

Structure of the Vertebrate Eye

The eye of a human is typical of the vertebrate eye (figure 52.16). The "white of the eye" is the sclera, formed of tough connective tissue. Light enters the eye through a transparent cornea, which begins to focus the light. This occurs because light is refracted (bent) when it travels into a medium of different density. The colored portion of the eye is the iris; contraction of the iris muscles in bright light decreases the size of its opening, the pupil. Light passes through the pupil to the lens, a transparent structure that completes the focusing of the light onto the retina at the back of the eye. The lens is attached by the *suspensory ligament* to the ciliary muscles.

The shape of the lens is influenced by the amount of tension in the suspensory ligament, which surrounds the

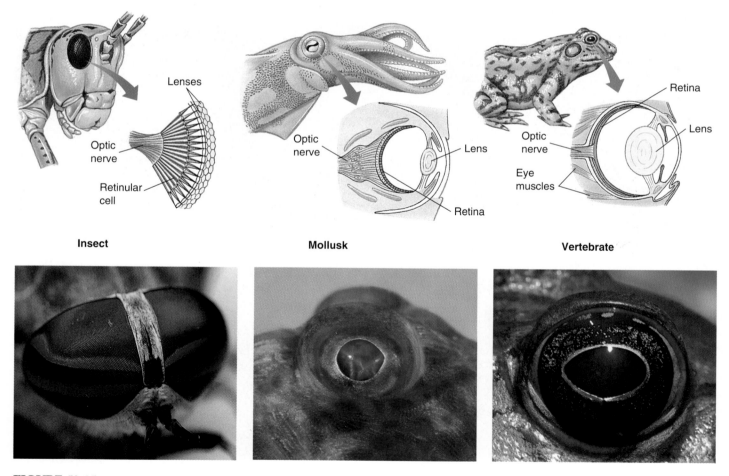

Insect **Mollusk** **Vertebrate**

FIGURE 52.15
Eyes in three phyla of animals. Although they are superficially similar, these eyes differ greatly in structure and are not homologous. Each has evolved separately and, despite the apparent structural complexity, has done so from simpler structures.

lens and attaches it to the circular ciliary muscle. When the ciliary muscle contracts, it puts slack in the suspensory ligament and the lens becomes more rounded and powerful. This is required for close vision; in far vision, the ciliary muscles relax, moving away from the lens and tightening the suspensory ligament. The lens thus becomes more flattened and less powerful, keeping the image focused on the retina. People who are nearsighted or farsighted do not properly focus the image on the retina (figure 52.17). Interestingly, the lens of an amphibian or a fish does not change shape; these animals instead focus images by moving their lens in and out, just as you would do to focus a camera.

Annelids, mollusks, arthropods, and vertebrates have independently evolved image-forming eyes. The vertebrate eye admits light through a pupil and then focuses this light by means of an adjustable lens onto the retina at the back of the eye.

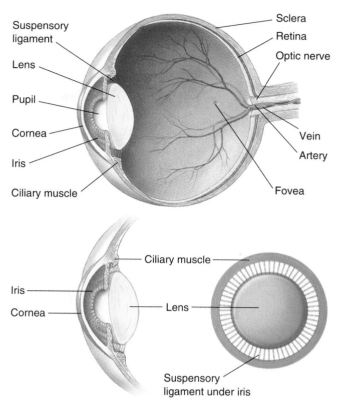

FIGURE 52.16
Structure of the human eye. The transparent cornea and lens focus light onto the retina at the back of the eye, which contains the rods and cones. The center of each eye's visual field is focused on the fovea. Focusing is accomplished by contraction and relaxation of the ciliary muscle, which adjusts the curvature of the lens.

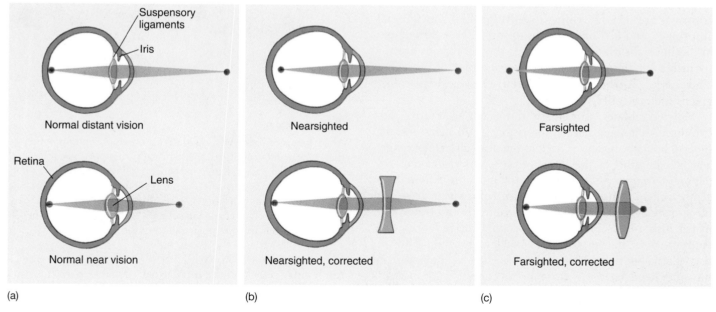

FIGURE 52.17
Focusing the human eye. (a) In people with normal vision, the image remains focused on the retina in both near and far vision because of changes produced in the curvature of the lens. When a person with normal vision stands 20 feet or more from an object, the lens is in its least convex form and the image is focused on the retina. (b) In nearsighted people, the image comes to a focus in front of the retina and the image thus appears blurred. (c) In farsighted people, the focus of the image would be behind the retina because the distance from the lens to the retina is too short.

Vertebrate Photoreceptors and the Retina

The vertebrate retina contains two kinds of photoreceptors, called rods and cones (figure 52.18). Rods are responsible for black-and-white vision when the illumination is dim, while cones are responsible for high visual acuity (sharpness) and color vision. Humans have about 100 million rods and 3 million cones in each retina. Most of the cones are located in the central region of the retina known as the fovea, where the eye forms its sharpest image. Rods are almost completely absent from the fovea.

Rods and cones have the same basic cellular structure. An inner segment rich in mitochondria contains numerous vesicles filled with neurotransmitter molecules. It is connected by a narrow stalk to the outer segment, which is packed with hundreds of flattened discs stacked on top of one another. The light-capturing molecules, or photopigments, are located on the membranes of these discs.

In rods, the photopigment is called rhodopsin. It consists of the protein opsin bound to a molecule of *cis*-retinal (figure 52.19), which is derived from carotene, a photosynthetic pigment in plants. The photopigments of cones, called photopsins, are structurally very similar to rhodopsin. Humans have three kinds of cones, each of which possesses a photopsin consisting of *cis*-retinal bound to a protein with a slightly different amino acid sequence. These differences shift the *absorption maximum*—the region of the electromagnetic spectrum that is best absorbed by the pigment—(figure 52.20). The absorption maximum of the *cis*-retinal in rhodopsin is 500 nanometers (nm); the absorption maxima of the three kinds of cone photopsins, in contrast, are 455 nm (blue-absorbing), 530 nm (green-absorbing), and 625 nm (red-absorbing). These differences in the light-absorbing properties of the photopsins are responsible for the different color sensitivities of the three kinds of cones, which are often referred to as simply blue, green, and red cones.

Most vertebrates, particularly those that are diurnal (active during the day), have color vision, as do many insects. Indeed, honeybees can see light in the near-ultraviolet range, which is invisible to the human eye. Color vision requires the presence of more than one photopigment in different receptor cells, but not all animals with color vision have the three-cone system characteristic of humans and other primates. Fish, turtles, and birds, for example, have four or five kinds of cones; the "extra" cones enable these animals to see near-ultraviolet light. Many mammals (such as squirrels), on the other hand, have only two types of cones.

The retina is made up of three layers of cells (figure 52.21): the layer closest to the external surface of the eyeball consists of the rods and cones, the next layer contains bipolar cells, and the layer closest to the cavity of the eye is composed of ganglion cells. Thus, light must first pass through the ganglion cells and bipolar cells in order to reach the photoreceptors! The rods and cones synapse with

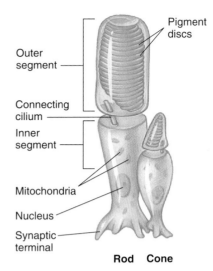

FIGURE 52.18
Rods and cones. The pigment-containing outer segment in each of these cells is separated from the rest of the cell by a partition through which there is only a narrow passage, the connective cilium.

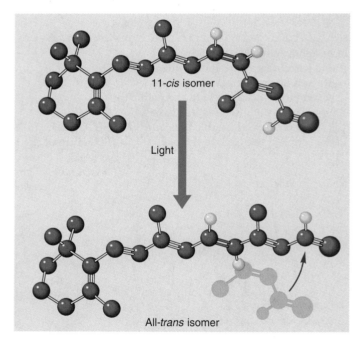

FIGURE 52.19
Absorption of light. When light is absorbed by a photopigment, the 11-*cis* isomer of retinal, the light-capturing portion of the pigment undergoes a change in shape: the linear end of the molecule (at the right in this diagram) rotates about a double bond (indicated here in *red*). The resulting isomer is referred to as all-*trans* retinal. This change in retinal's shape initiates a chain of events that leads to hyperpolarization of the photoreceptor.

the bipolar cells, and the bipolar cells synapse with the ganglion cells, which transmit impulses to the brain via the optic nerve. The flow of sensory information in the retina is therefore opposite to the path of light through the retina. It should also be noted that the retina contains two additional types of neurons, called horizontal cells and amacrine cells, that synapse with other neurons within a given layer of the retina.

Sensory Transduction in Photoreceptors

The transduction of light energy into nerve impulses follows a sequence that is the inverse of the usual way that sensory stimuli are detected. This is because, in the dark, the photoreceptors release an inhibitory neurotransmitter that hyperpolarizes the bipolar neurons. Thus inhibited, the bipolar neurons do not release excitatory neurotransmitter to the ganglion cells. Light *inhibits* the photoreceptors from releasing their inhibitory neurotransmitter, and by this means, *stimulates* the bipolar cells and thus the ganglion cells, which transmit action potentials to the brain.

A rod or cone contains many Na^+ channels in the plasma membrane of its outer segment, and in the dark, many of these channels are open. As a consequence, Na^+ continuously diffuses into the outer segment and across the narrow stalk to the inner segment. This flow of Na^+ that occurs in the absence of light is called the dark current, and it causes the membrane of a photoreceptor to be somewhat depolarized in the dark. In the light, the Na^+ channels in the outer segment rapidly close, reducing the dark current and causing the photoreceptor to hyperpolarize.

Researchers have discovered that cyclic guanosine monophosphate (cGMP) is required to keep the Na^+ channels open, and that the channels will close if the cGMP is converted into GMP. How does light cause this conversion and consequent closing of the Na^+ channels? When a photopigment absorbs light, *cis*-retinal isomerizes and dissociates from opsin in what is known as the bleaching reaction. As a result of this dissociation, the opsin protein changes shape. Each opsin is associated with over a hundred regulatory *G proteins* (see chapters 7 and 51). When the opsin changes shape, the G proteins dissociate, releasing subunits that activate hundreds of molecules of the enzyme *phosphodiesterase*. This enzyme converts cGMP to GMP, thus closing the Na^+ channels at a rate of about 1000 per second and inhibiting the dark current. The absorption of a single photon of light can block the entry of more than a million sodium ions, thereby causing the photoreceptor to hyperpolarize and release less inhibitory neurotransmitters. Freed from inhibition, the bipolar cells activate ganglion cells, which transmit action potentials to the brain.

Photoreceptor rods and cones contain the photopigment *cis*-retinal bound to protein. The photopigment dissociates in response to light and indirectly activates bipolar neurons and then ganglion cells, which transmit action potentials to the brain.

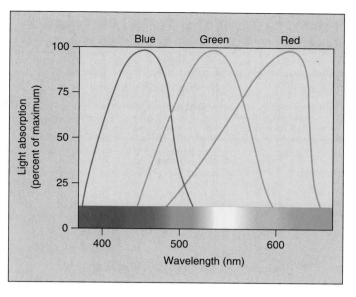

FIGURE 52.20
Color vision. The absorption maximum of *cis*-retinal in the rhodopsin of rods is 500 nanometers (nm). However, the "blue cones" have their maximum light absorption at 455 nm; the "green cones" at 530 nm, and the red cones at 625 nm. The brain perceives all other colors from the combined activities of these three cones' systems.

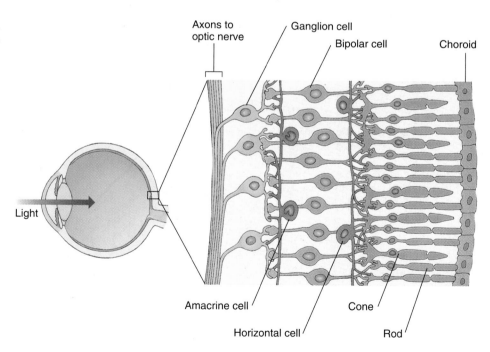

FIGURE 52.21
Structure of the retina. Note that the rods and cones are at the rear of the retina, not the front. Light passes through four other types of cells in the retina before it reaches the rods and cones. Once the photoreceptors are activated, they stimulate bipolar cells, which in turn stimulate ganglion cells. The direction of nerve impulses in the retina is thus opposite to the direction of light.

Visual Processing in the Vertebrate Retina

Action potentials propagated along the axons of ganglion cells are relayed through structures called the lateral geniculate nuclei of the thalamus and projected to the occipital lobe of the cerebral cortex (figure 52.22). There the brain interprets this information as light in a specific region of the eye's receptive field. The pattern of activity among the ganglion cells across the retina encodes a point-to-point map of the receptive field, allowing the retina and brain to image objects in visual space. In addition, the frequency of impulses in each ganglion cell provides information about the light intensity at each point, while the relative activity of ganglion cells connected (through bipolar cells) with the three types of cones provides color information.

The relationship between receptors, bipolar cells, and ganglion cells varies in different parts of the retina. In the

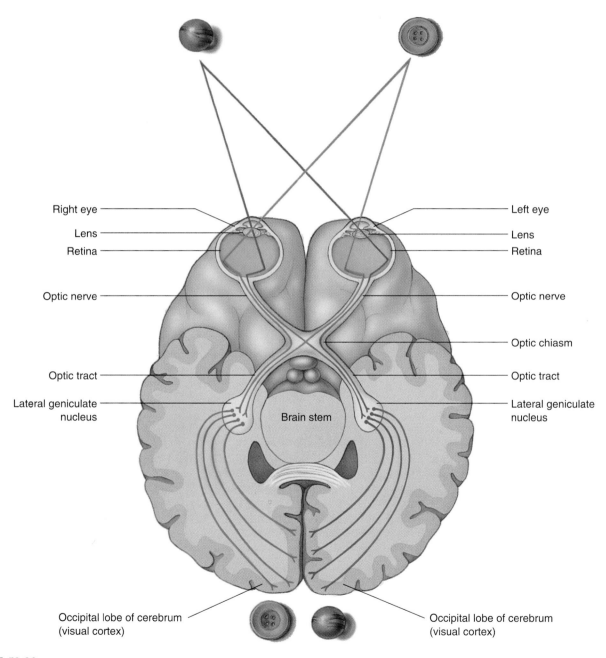

FIGURE 52.22
The pathway of visual information. Action potentials in the optic nerves are relayed from the retina to the lateral geniculate nuclei, and from there to the visual cortex of the occipital lobes. Notice that the medial fibers of the optic nerves cross to the other side at the optic chiasm, so that each hemisphere of the cerebrum receives input from both eyes.

fovea, each cone makes a one-to-one connection with a bipolar cell, and each bipolar cell synapses, in turn, with one ganglion cell. This point-to-point relationship is responsible for the high acuity of foveal vision. Outside the fovea, many rods can converge on a single bipolar cell, and many bipolar cells can converge on a single ganglion cell. This convergence permits the summation of neural activity, making the area of the retina outside of the fovea more sensitive to dim light than the fovea, but at the expense of acuity and color vision. This is why dim objects, such as faint stars at night, are best seen when you don't look directly at them. It has been said that we use the periphery of the eye as a detector and the fovea as an inspector.

Color blindness is due to an inherited lack of one or more types of cones. People with normal color vision are *trichromats*; those with only two types of cones are *dichromats*. People with this condition may lack red cones (have *protanopia*), for example, and have difficulty distinguishing red from green. Men are far more likely to be color blind than women, because the trait for color blindness is carried on the X chromosome; men have only one X chromosome per cell, whereas women have two X chromosomes and so can carry the trait in a recessive state.

Binocular Vision

Primates (including humans) and most predators have two eyes, one located on each side of the face. When both eyes are trained on the same object, the image that each sees is slightly different because each eye views the object from a different angle. This slight displacement of the images (an effect called parallax) permits **binocular vision,** the ability to perceive three-dimensional images and to sense depth. Having their eyes facing forward maximizes the field of overlap in which this stereoscopic vision occurs (figure 52.23).

In contrast, prey animals generally have eyes located to the sides of the head, preventing binocular vision but enlarging the overall receptive field. Depth perception is less important to prey than detection of potential enemies from any quarter. The eyes of the American Woodcock, for example, are located at exactly opposite sides of its skull so that it has a 360-degree field of view without turning its head! Most birds have laterally placed eyes and, as an adaptation, have two foveas in each retina. One fovea provides sharp frontal vision, like the single fovea in the retina of mammals, and the other fovea provides sharper lateral vision.

The axons of ganglion cells transmit action potentials to the thalamus, which in turn relays visual information to the occipital lobe of the brain. The fovea provides high visual acuity, whereas the retina outside the fovea provides high sensitivity to dim light. Binocular vision with overlapping visual fields provides depth perception.

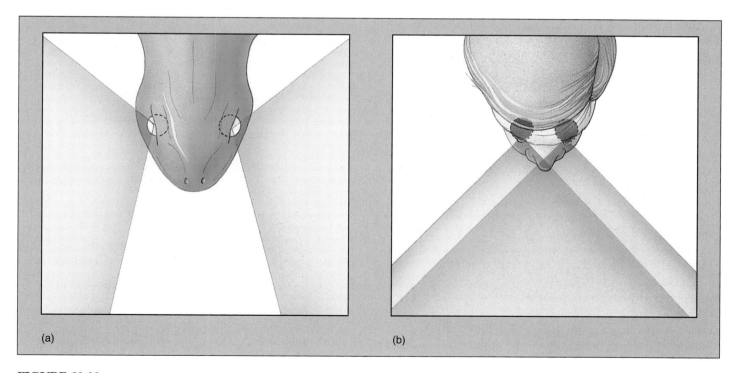

FIGURE 52.23
Stereoscopic vision. (a) When the eyes are located on the sides of the head, their fields of vision do not overlap, and stereoscopic vision cannot occur. (b) Placing the eyes at the front of the head allows their fields of vision to overlap and enables the visual system to form three-dimensional images.

Diversity of Sensory Experiences

Vision is the primary sense used by all vertebrates that live in a light-filled environment, but visible light is by no means the only part of the electromagnetic spectrum that vertebrates use to sense their environment.

Heat

Electromagnetic radiation with wavelengths longer than those of visible light is too low in energy to be detected by photoreceptors. Radiation from this *infrared* ("below red") portion of the spectrum is what we normally think of as radiant heat. Heat is an extremely poor environmental stimulus in water because water has a high thermal capacity and readily absorbs heat. Air, in contrast, has a low thermal capacity, so heat in air is a potentially useful stimulus. However, the only vertebrates known to have the ability to sense infrared radiation are the snakes known as pit vipers.

The pit vipers possess a pair of heat-detecting pit organs located on either side of the head between the eye and the nostril (figure 52.24). The pit organs permit a blindfolded rattlesnake to accurately strike at a warm, dead rat. Each pit organ is composed of two chambers separated by a membrane, and the organ apparently operates by comparing the temperatures of the two chambers. The nature of the pit organ's thermal receptor is not known; it probably consists of temperature-sensitive neurons innervating the two chambers. The two pit organs appear to provide stereoscopic information, in much the same way that two eyes do. Indeed, the information transmitted from the pit organs is processed by the visual center of the snake brain.

Electricity

While air does not readily conduct an electrical current, water is a good conductor. All aquatic animals generate electrical currents from contractions of their muscles. A number of different groups of fishes can detect these electrical currents. The *electrical fish* even have the ability to produce electrical discharges from specialized electrical organs. Electrical fish use these weak discharges to locate their prey and mates and to construct a three-dimensional image of their environment even in murky water.

The elasmobranchs (sharks, rays, and skates) have electroreceptors called the ampullae of Lorenzini. The receptor cells are located in sacs that open through jelly-filled canals to pores on the body surface. The jelly is a very good conductor, so a negative charge in the opening of the canal can depolarize the receptor at the base, causing the release of neurotransmitter and increased activity of sensory neurons. This allows sharks, for example, to detect the electrical fields generated by the muscle contractions of their

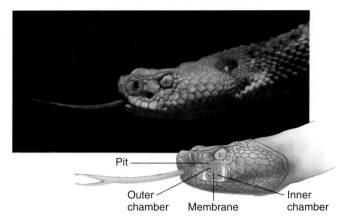

FIGURE 52.24
"Seeing" heat. The depression between the nostril and the eye of this rattlesnake opens into the pit organ. In the cutaway portion of the diagram, you can see that the organ is composed of two chambers separated by a membrane. Snakes known as pit vipers have this unique ability to sense infrared radiation (heat).

prey. Although the ampullae of Lorenzini were lost in the evolution of teleost fish (most of the bony fish), electroreception reappeared in some groups of teleost fish that use sensory structures analogous to the ampullae of Lorenzini. Electroreceptors evolved yet another time, independently, in the duck-billed platypus, an egg-laying mammal. The receptors in its bill can detect the electrical currents created by the contracting muscles of shrimp and fish, enabling the mammal to detect its prey at night and in muddy water.

Magnetism

Eels, sharks, and many birds appear to navigate along the magnetic field lines of the earth. Even some bacteria use such forces to orient themselves. Birds kept in blind cages, with no visual cues to guide them, will peck and attempt to move in the direction in which they would normally migrate at the appropriate time of the year. They will not do so, however, if the cage is shielded from magnetic fields by steel. Indeed, if the magnetic field of a blind cage is deflected 120° clockwise by an artificial magnet, a bird that normally orients to the north will orient toward the east-southeast. There has been much speculation about the nature of the magnetic receptors in these vertebrates, but the mechanism is still very poorly understood.

Pit vipers can locate warm prey by infrared radiation (heat), and many aquatic vertebrates can locate prey and ascertain the contours of their environment by means of electroreceptors. Some vertebrates can even orient themselves with respect to the earth's magnetic field.

52.1 Animals employ a wide variety of sensory receptors.

- Mechanoreceptors, chemoreceptors, and photoreceptors are responsive to different categories of sensory stimuli; interoceptors and exteroceptors respond to stimuli that originate in the internal and external environments, respectively.

- In response to a stimulus, a sensory receptor produces a depolarization called a receptor potential.

52.2 Mechanical and chemical receptors sense the body's condition.

- Muscle spindles respond to stretching of the skeletal muscle, and Golgi tendon organs respond to tension exerted by a muscle on a tendon.

- The sensory organs of taste are taste buds, scattered over the surface of a fish's body but located on the papillae of the tongue in terrestrial vertebrates.

- The sense of smell is mediated by bipolar neurons that project into the nasal mucosa.

- Chemoreceptors in the aortic and carotid bodies sense the blood pH and oxygen levels, helping to regulate breathing.

- Hair cells in the lateral line system of fish are responsive to water movements and vibrations, giving fish a sense of "distant touch."

- Hair cells in the membranous labyrinth of the inner ear are the receptors of the vestibular apparatus; this sense organ includes the utricle, saccule, and semicircular canals, which provide a sense of acceleration.

52.3 Auditory receptors detect pressure waves in the air.

- In terrestrial vertebrates, sound waves cause vibrations of the tympanic membrane, which cause movements of the middle ear ossicles; these, in turn, cause vibrations of the oval window of the cochlea.

- Different pitches of sounds vibrate different regions of the basilar membrane, and therefore stimulate different hair cells.

- Bats and some other vertebrates use sonar to provide a sense of "lightless vision."

52.4 Optic receptors detect light over a broad range of wavelengths.

- A flexible lens focuses light onto the retina, which contains the photoreceptors; the amount of light admitted into the eye through the pupil is adjusted by the iris.

- Light causes the photodissociation of the visual pigment, thereby blocking the dark current and hyperpolarizing the photoreceptor; this inverse effect stops the inhibitory effect of the photoreceptor and thereby activates the bipolar cells.

52.5 Some vertebrates use heat, electricity, or magnetism for orientation.

- The pit organs of certain snakes allows them to detect the position and movements of warm-bodied prey.

- Many aquatic vertebrates can detect the electrical currents produced by muscular contraction of their prey.

- Although the mechanisms are poorly understood, it seems clear that some vertebrates can orient themselves using the magnetic field of the earth.

Discussing Key Terms

1. **Hair cells** Mechanoreceptor cells with hair-like cilia are called hair cells; bending of these cilia causes a receptor potential, providing a method of sensory transduction in the lateral line system and membranous labyrinth.

2. **Lateral line system** The lateral line system has been described as providing a sense of "distant touch," because movements and vibrations of water can be detected by their effects on hair cells in the lateral line system of fish.

3. **Membranous labyrinth** The membranous labyrinth is a small but complex system of membranous tubes and chambers filled with fluid; it is located in the inner ear and contains the vestibular apparatus responsible for a sense of acceleration and the cochlea for hearing.

4. **Cochlea** The snail-shaped cochlea is the organ of hearing; it contains sensory hair cells on a vibrating membrane; it is in the cochlea that sound energy is transduced into action potentials.

5. **Echolocation** Bats emit sounds that bounce off objects and return to the ears; this method of orienting in space is called echolocation.

6. **Dark current** In the dark, there is a constant inward diffusion of Na^+ into photoreceptors, called a dark current; in response to light, this dark current is blocked and the photoreceptor becomes hyperpolarized.

7. **Rhodopsin** Rhodopsin is the visual pigment of rods and is composed of *cis*-retinal and a protein called opsin; in response to light, the *cis*-retinal converts to *trans*-retinal and dissociates from the opsin.

Review Questions

1. How are sound waves transmitted and amplified through the middle ear? How is the pitch of the sound determined?

2. How does focusing in fishes and amphibians differ from that in other vertebrates? What structure controls the amount of light entering the eye?

3. When a photoreceptor absorbs light, what happens to the Na^+ channels in its outer segment?

4. What is the advantage of having two eyes instead of just one? How does the position of the eyes on the head in most predators differ from that in most prey? What perceptual differences result from this difference in positioning?

5. How do rattlesnakes sense their prey? Why would they strike a moving lightbulb but not a dead animal that was no longer warmer than its environment?

6. How do sharks detect their prey? How is this sense used by some fish to navigate? Why don't terrestrial vertebrates have this sense?

7. What mechanoreceptors detect muscle stretch and the tension on a tendon? What type of mechanoreceptors detect water movements in the lateral line system of fish? How do these work?

8. What structures in the vertebrate ear detect changes in the body's position with respect to gravity? What structures detect angular motion? What do these sensory structures have in common with the lateral line system of fish?

Thought Questions

1. Owls cannot fly on a completely dark night, but bats can navigate without trouble. How?

2. How would the otolith organs of an astronaut in zero gravity behave? Would the astronaut still have a subjective impression of motion? Would the semicircular canals detect angular acceleration equally well at zero gravity?

3. How can the *hyperpolarization* of a photoreceptor by light lead to the production of action potentials by retinal ganglion cells, when action potential generation requires membrane *depolarization*?

Internet Links

Seeing, Hearing, and Smelling
http://www.hhmi.org/senses/
A wonderful exploration of the primary senses by the Howard Hughes Medical Institute. Very nice!

Optical Illusions
http://www.illusionworks.com/
An extensive collection of illusions and puzzles that illustrate how the eye can be tricked, with links to educational projects and vision research sites. Also consult http://kahuna.psych.uiuc.edu/ipl/weblinks.html for many other sites.

More Optical Illusions
http://valley.uml.edu/psychology/DEPTH.html
Wonder why ILLUSIONWORKS fools your eye? This site, the ILLUSION GALLERY, provides information on the science behind a host of classic optical illusions.

Beye
http://cvs.anu.edu.au/andy/beye/beyehome.html
See the world through the eyes of a honey bee.

For Further Reading

Barlow, R.: "What the Brain Tells the Eye," *Scientific American*, April 1990, pages 90–95. The brain may exercise substantial control over just what the eye can detect.

Gilbert, A. and C. Wysocki: "Smell: The Survey Results," *National Geographic*, October 1987, pages 514–25. A delightful and interesting account of how well we smell, based on a national survey. Among many surprising findings: women smell more accurately than men.

Hudspeth, A.: "How the Ear's Works Work," *Nature*, vol. 341, October 1989, pages 397–404. A review of how sensory receptors in the ear activate ion channels to initiate nerve impulses.

Masland, R.: "The Functional Architecture of the Retina," *Scientific American*, December 1986, pages 102–11. An up-to-date description of the interconnections of the many cells beneath the retina.

Nathans, J.: "The Genes for Color Vision," *Scientific American*, February 1989, pages 42–49. An interesting account of how genetic engineers isolated the genes encoding the color-detecting proteins.

Schnapf, J. and D. Baylor: "How Photoreceptor Cells Respond to Light," *Scientific American*, April 1987, pages 40–47. A detailed account of how an individual photoreceptor in the eye is able to detect a single photon of light.

Stryer, L.: "The Molecules of Visual Excitation," *Scientific American*, July 1987, pages 42–50. A lucid description of the cascade that a photon initiates in a rod cell.

Suga, N.: "Biosonar and Neural Computation in Bats," *Scientific American*, June 1990, pages 60–68. How the bat brain is organized to extract information from biosonar signals. Highly recommended.

Waldvogel, J.: "The Bird's Eye View," *American Scientist*, vol. 78, July–August, 1990, pages 342–53. Bird vision shows some interesting and surprising variations from the visual system of mammals. This is a clearly written and well-illustrated article summarizing the bird's-eye view of the world.

53

The Endocrine System

FIGURE 53.1
The endocrine system controls when animals breed. These Japanese macaques live in a close-knit community whose members cooperate to ensure successful breeding and raising of offspring. Not everybody breeds at the same time because hormone levels vary among individuals.

Concept Outline

53.1 Regulation is often accomplished by chemical messengers.

Types of Regulatory Molecules. Regulatory molecules may function as neurotransmitters, hormones, or as organ-specific regulators.

Endocrine Glands and Hormones. Endocrine glands secrete molecules called hormones into the blood.

Autocrine and Paracrine Regulation. Autocrine and paracrine regulators act within organs that produce them.

53.2 Lipophilic and polar hormones regulate their target cells by different means.

Hormones That Enter Cells. Steroid and thyroid hormones act by entering target cells and stimulating specific genes.

Hormones That Do Not Enter Cells. All other hormones bind to receptors on the cell surface and activate second-messenger molecules within the target cells.

53.3 The hypothalamus controls the secretions of the pituitary gland.

The Posterior Pituitary Gland. The posterior pituitary receives and releases hormones from the hypothalamus.

The Anterior Pituitary Gland. The anterior pituitary produces a variety of hormones under stimulation from hypothalamic releasing hormones.

53.4 Endocrine glands secrete hormones that regulate many body functions.

The Thyroid and Parathyroid Glands. The thyroid hormones regulate metabolism; the parathyroid glands regulate calcium balance.

The Adrenal Glands. The adrenal medulla secretes epinephrine during the fight-or-flight reaction, while the adrenal cortex secretes steroid hormones that regulate glucose and mineral balance.

The Pancreas. The islets of Langerhans in the pancreas secrete insulin, which acts to lower blood glucose, and glucagon, which acts to raise blood glucose.

Other Endocrine Glands. The gonads, pineal gland, thymus, kidneys, and other organs secrete important hormones that have a variety of functions.

The tissues and organs of the vertebrate body cooperate to maintain homeostasis of the body's internal environment. Homeostasis is achieved through the actions of many regulatory mechanisms that involve all the organs of the body. Two systems, however, are devoted exclusively to the regulation of the body organs: the nervous system and the endocrine system (figure 53.1). Both release regulatory molecules that control the body organs by first binding to receptor proteins in the cells of those organs. In this chapter we will examine these regulatory molecules, the cells and glands that produce them, and how they function to regulate the body's activities.

Types of Regulatory Molecules

As we discussed in chapter 51, the axons of neurons secrete chemical messengers called *neurotransmitters* into the synaptic cleft. These chemicals diffuse only a short distance to the postsynaptic membrane, where they bind to their receptor proteins and stimulate the postsynaptic cell (another neuron, or a muscle or gland cell). Synaptic transmission generally affects only the one postsynaptic cell that receives the neurotransmitter.

A **hormone** is a regulatory chemical that is secreted into the blood by an **endocrine gland.** The blood carries the hormone to every cell in the body, but only the **target cells** for a given hormone can respond to it. Thus, the difference between a neurotransmitter and a hormone is not a difference in the chemical nature of the regulatory molecule, but rather a difference in the way it is transported to its target cells. A chemical regulator called norepinephrine, for example, is released as a neurotransmitter by sympathetic nerve endings (chapter 51) and is also secreted by the adrenal gland as a hormone.

Some specialized neurons secrete chemical messengers into the blood rather than into a narrow synaptic cleft. In these cases, the chemical that the neurons secrete is some-times called a **neurohormone.** The distinction between the nervous system and endocrine system blurs when it comes to such molecules. Indeed, because some neurons in the brain secrete hormones, the brain can be considered an endocrine gland!

In addition to the chemical messengers released as neurotransmitters and as hormones, other chemical regulatory molecules are released and act *within* an organ. In this way, the cells of an organ regulate one another. This type of regulation is not endocrine, because the regulatory molecules work without being transported by the blood, but is otherwise similar to the way that hormones regulate their target cells. Such regulation is called **autocrine,** if similar cells regulate one another, or **paracrine** if the regulatory molecules are produced and act in different tissues of the same organ. A comparison of the different types of chemical messengers used for regulation is given in figure 53.2.

Regulatory molecules released by axons at a synapse are neurotransmitters and those released by endocrine glands into the blood are hormones, while those that act within the organ in which they are produced are autocrine or paracrine regulators.

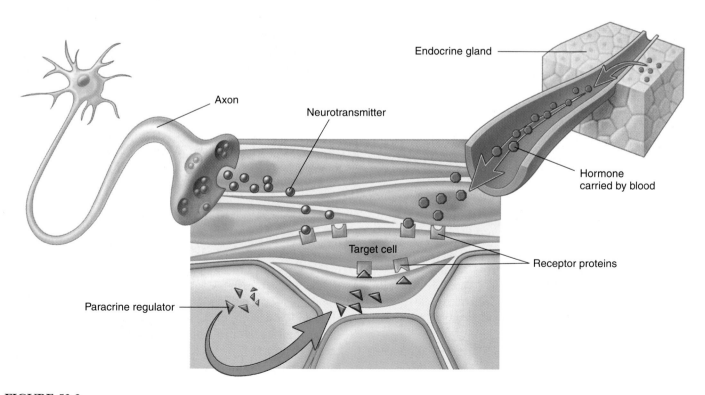

FIGURE 53.2
The functions of organs are influenced by neural, autocrine, paracrine, and endocrine regulators. Each type of chemical regulator binds in a specific fashion to receptor proteins on the surface of or within the cells of target organs.

Endocrine Glands and Hormones

The endocrine system (figure 53.3) includes all of the organs that function exclusively as endocrine glands—such as the thyroid gland, pituitary gland, adrenal glands, and so on (table 53.1)—as well as organs that secrete hormones in addition to other functions. Endocrine glands lack ducts and thus must secrete into surrounding blood capillaries, unlike exocrine glands, which secrete their products into a duct. The pancreas, for example, is both an exocrine and endocrine gland. The exocrine function of the pancreas is to produce pancreatic juice, which is secreted via the pancreatic duct into the intestine (chapter 48). The endocrine portion of the pancreas secretes the hormones insulin and glucagon into the blood.

Hormones belong to four different chemical categories:

1. **Polypeptides.** These hormones are composed of chains of amino acids that are shorter than about 100 amino acids. Some important examples include insulin and antidiuretic hormone (ADH).
2. **Glycoproteins.** These are composed of a polypeptide significantly longer than 100 amino acids to which is attached a carbohydrate. Examples include follicle-stimulating hormone (FSH) and luteinizing hormone (LH).
3. **Amines.** Derived from the amino acids tyrosine and tryptophan, they include hormones secreted by the adrenal medulla, thyroid, and pineal glands.
4. **Steroids.** These hormones are lipids derived from cholesterol, and include the hormones testosterone, estradiol, progesterone, and cortisol.

Steroid hormones can be subdivided into **sex steroids,** secreted by the testes, ovaries, placenta, and adrenal cortex, and **corticosteroids,** secreted only by the adrenal cortex (the outer portion of the adrenal gland). The corticosteroids include cortisol, which regulates glucose balance, and aldosterone, which regulates salt balance.

The amine hormones secreted by the adrenal medulla (the inner portion of the adrenal gland), known as **catecholamines,** include epinephrine (adrenaline) and norepinephrine. These are derived from the amino acid tyrosine. Another hormone derived from tyrosine is thyroxine, secreted by the thyroid gland. The pineal gland secretes a different amine hormone, melatonin, derived from tryptophan.

All hormones may be categorized as lipophilic (fat-soluble) or water-soluble. The **lipophilic hormones** include the steroid hormones and thyroxine; all other hormones are water-soluble. This distinction will be useful later in understanding how these hormones regulate their target cells.

Neural and Endocrine Interactions

The endocrine system is an extremely important regulatory system in its own right, but it also interacts and cooperates with the nervous system to regulate the activities of the other organ systems of the body. The secretory activity of

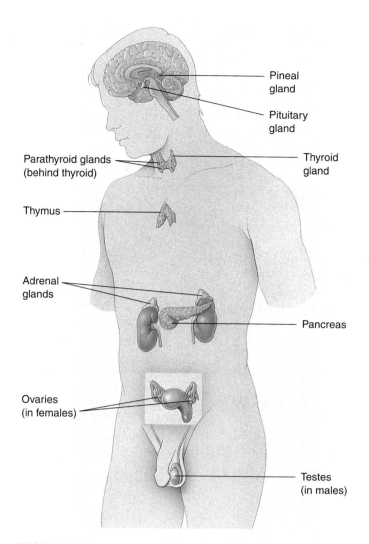

FIGURE 53.3
The human endocrine system. The major endocrine glands are shown, but many other organs secrete hormones in addition to their primary functions.

many endocrine glands is controlled by the nervous system. Among such glands are the adrenal medulla, posterior pituitary, and pineal gland. These three glands are derived from the neural ectoderm (to be discussed in chapter 57), the same embryonic tissue layer that forms the nervous system. The major site for neural regulation of the endocrine system, however, is the brain's regulation of the anterior pituitary gland. As we'll see, the hypothalamus controls the hormonal secretions of the anterior pituitary, which in turn regulates other endocrine glands. On the other hand, the secretion of a number of hormones is largely independent of neural control. The release of insulin by the pancreas and aldosterone by the adrenal cortex, for example, are stimulated primarily by increases in the blood concentrations of glucose and potassium (K^+), respectively.

Any organ that secretes a hormone from a ductless gland is part of the endocrine system. Hormones may be any of a variety of different chemicals.

Table 53.1   Principal Endocrine Glands and Their Hormones

Endocrine Gland and Hormone	Target Tissue	Principal Actions	Chemical Nature
POSTERIOR LOBE OF PITUITARY			
Antidiuretic hormone (ADH)	Kidneys	Stimulates reabsorption of water; conserves water	Peptide (9 amino acids)
Oxytocin	Uterus	Stimulates contraction	Peptide (9 amino acids)
	Mammary glands	Stimulates milk ejection	
ANTERIOR LOBE OF PITUITARY			
Growth hormone (GH)	Many organs	Stimulates growth by promoting protein synthesis and fat breakdown	Protein
Adrenocorticotropic hormone (ACTH)	Adrenal cortex	Stimulates secretion of adrenal cortical hormones such as cortisol	Peptide (39 amino acids)
Thyroid-stimulating hormone (TSH)	Thyroid gland	Stimulates thyroxine secretion	Glycoprotein
Luteinizing hormone (LH)	Gonads	Stimulates ovulation and corpus luteum formation in females; stimulates secretion of testosterone in males	Glycoprotein
Follicle-stimulating hormone (FSH)	Gonads	Stimulates spermatogenesis in males; stimulates development of ovarian follicles in females	Glycoprotein
Prolactin (PRL)	Mammary glands	Stimulates milk production	Protein
Melanocyte-stimulating hormone (MSH)	Skin	Stimulates color change in reptiles and amphibians; unknown function in mammals	Peptide (two forms; 13 and 22 amino acids)
THYROID GLAND			
Thyroxine (thyroid hormone)	Liver	Stimulates metabolic rate; essential to normal growth and development	Iodinated amino acid
Calcitonin	Bone	Lowers blood calcium level by inhibiting loss of calcium from bone	Peptide (32 amino acids)
PARATHYROID GLANDS			
Parathyroid hormone	Bone, kidneys, digestive tract	Raises blood calcium level by stimulating bone breakdown; stimulates calcium reabsorption in kidneys; activates vitamin D	Peptide (34 amino acids)

Autocrine and Paracrine Regulation

Autocrine and paracrine regulation occurs in many organs and among the cells of the immune system. Many autocrine regulatory molecules are known as **cytokines,** particularly if they regulate different cells of the immune system. Cytokines produced by lymphocytes (the type of white blood cell involved in specific immunity—see chapter 54) are also known as **lymphokines.** Many paracrine regulators are called **growth factors,** because they promote growth and cell division in specific organs. Examples include *platelet-derived growth factor, epidermal growth factor,* and the *insulin-like growth factors* that stimulate cell division and proliferation of their target cells. *Nerve growth factor* is a regulatory molecule that belongs to a family of paracrine regulators of the nervous system called **neurotrophins.**

Nitric oxide, which can function as a neurotransmitter (chapter 51), is also produced by the endothelium of blood vessels. In this context, it is a paracrine regulator because it diffuses to the smooth muscle layer of the blood vessel and promotes vasodilation. The endothelium of blood vessels also produces other paracrine regulators, including *endothelin,* which stimulates vasoconstriction, and *bradykinin,* which promotes vasodilation. This paracrine regulation supplements the regulation of blood vessels by autonomic nerves.

The most diverse group of paracrine regulators are the **prostaglandins.** A prostaglandin is a 20-carbon-long fatty acid that contains a five-member carbon ring. This molecule is derived from the precursor molecule *arachidonic acid,* which can be released from phospholipids in the cell membrane under hormonal or other stimulation. Prostaglandins are produced in almost every organ and participate in a variety of regulatory functions, including:

1. **Immune system.** Prostaglandins promote many aspects of inflammation, including pain and fever. Drugs that inhibit prostaglandin synthesis help to alleviate these symptoms.

Table 53.1   Principal Endocrine Glands and Their Hormones *(continued.)*

Endocrine Gland and Hormone	Target Tissue	Principal Actions	Chemical Nature
ADRENAL MEDULLA			
Epinephrine (adrenaline) and norepinephrine (noradrenaline)	Skeletal muscle, cardiac muscle, blood vessels	Initiate stress responses; raise heart rate, blood pressure, metabolic rate; dilate blood vessels; mobilize fat; raise blood glucose level	Amino acid derivatives
ADRENAL CORTEX			
Aldosterone	Kidney tubules	Maintains proper balance of Na^+ and K^+ ions	Steroid
Cortisol	Many organs	Adaptation to long-term stress; raises blood glucose level; mobilizes fat	Steroid
PANCREAS			
Insulin	Liver, skeletal muscles, adipose tissue	Lowers blood glucose level; stimulates storage of glycogen in liver	Peptide (51 amino acids)
Glucagon	Liver, adipose tissue	Raises blood glucose level; stimulates breakdown of glycogen in liver	Peptide (29 amino acids)
OVARY			
Estradiol	General	Stimulates development of secondary sex characteristics in females	Steroid
	Female reproductive structures	Stimulates growth of sex organs at puberty and monthly preparation of uterus for pregnancy	
Progesterone	Uterus	Completes preparation for pregnancy	Steroid
	Breasts	Stimulates development	
TESTIS			
Testosterone	Many organs	Stimulates development of secondary sex characteristics in males and growth spurt at puberty	Steroid
	Male reproductive structures	Stimulates development of sex organs; stimulates spermatogenesis	
PINEAL GLAND			
Melatonin	Gonads, pigment cells	Function not well understood; influences pigmentation in some vertebrates; may control biorhythms in some animals; may influence onset of puberty in humans	Amino acid derivative

2. **Reproductive system.** Prostaglandins may play a role in ovulation. Excessive prostaglandin production may be involved in premature labor, endometriosis, or dysmenorrhea (painful menstrual cramps).

3. **Digestive system.** Prostaglandins produced by the stomach and intestines may inhibit gastric secretions and influence intestinal motility and fluid absorption.

4. **Respiratory system.** Some prostaglandins cause constriction, whereas others cause dilation of blood vessels in the lungs and of bronchiolar smooth muscle.

5. **Circulatory system.** Prostaglandins are needed for proper function of blood platelets in the process of blood clotting.

6. **Urinary system.** Prostaglandins produced in the renal medulla cause vasodilation, resulting in increased renal blood flow and increased excretion of urine.

The functions of prostaglandins are inhibited by aspirin. Aspirin is the most widely used of the **nonsteroidal anti-inflammatory drugs (NSAIDs),** a class of drugs that also includes indomethacin and ibuprofen. These drugs produce their effects because they specifically inhibit the enzyme cyclooxygenase-2 (cox-2), needed to produce prostaglandins from arachidonic acid. Through this action, the NSAIDs inhibit inflammation and associated pain. Unfortunately, NSAIDs also inhibit another similar enzyme, cox-1, which helps maintain the wall of the digestive tract, and in so doing can produce severe unwanted side effects, including gastric bleeding and prolonged clotting time. A new kind of pain reliever, celebrex, inhibits cox-2 but not cox-1, a potentially great benefit to arthritis sufferers and others who must use pain relievers regularly.

The neural and endocrine control systems are supplemented by the autocrine and paracrine regulators, including the prostaglandins, which perform many diverse functions. Other such regulators include lymphokines, growth factors, and neurotrophins.

Hormones That Enter Cells

As we mentioned previously, hormones can be divided into those that are lipophilic (lipid-soluble) and those that are polar (water-soluble). The lipophilic hormones—all of the steroid hormones (figure 53.4) and thyroxine—as well as other lipophilic regulatory molecules (including the retinoids, or vitamin A) can easily enter cells. This is because the lipid portion of the cell membrane does not present a barrier to the entry of lipophilic regulators. Therefore, all lipophilic regulatory molecules have a similar mechanism of action. Water-soluble hormones, in contrast, cannot pass through cell membranes. They must regulate their target cells through different mechanisms.

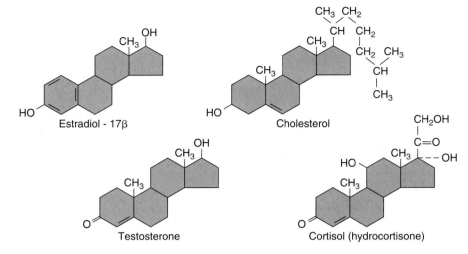

FIGURE 53.4

Chemical structures of some steroid hormones. Steroid hormones are derived from the blood lipid cholesterol. The hormones shown, estradiol, testosterone, and cortisol, differ only slightly in chemical structure yet have widely different effects on the body. Steroid hormones are secreted by the testes, ovaries, and placenta.

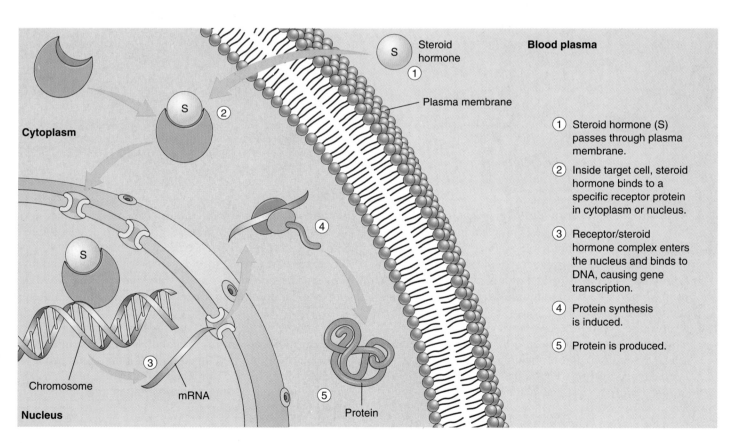

FIGURE 53.5

The mechanism of steroid hormone action. Steroid hormones are lipid-soluble and thus readily diffuse through the plasma membrane of cells. They bind to receptor proteins in either the cytoplasm or nucleus (not shown). If the steroid binds to a receptor in the cytoplasm, the steroid/receptor complex moves into the nucleus. The steroid/receptor complex then binds to specific regions of the DNA, stimulating the production of messenger RNA (mRNA).

Steroid hormones are lipids themselves and thus lipophilic; thyroxine is lipophilic because it is derived from a nonpolar amino acid. Since these hormones are not water-soluble, they don't dissolve in plasma but rather travel in the blood attached to protein carriers. When the hormones arrive at their target cells, they dissociate from their carriers and pass through the plasma membrane of the cell (figure 53.5). Some steroid hormones then bind to very specific receptor proteins in the cytoplasm, and then move as a hormone-receptor complex into the nucleus. Other steroids travel directly into the nucleus before encountering their receptor proteins. Whether the steroid finds its receptor in the nucleus or translocates with its receptor to the nucleus from the cytoplasm, the rest of the story is the same.

The hormone receptor, activated by binding to the lipophilic hormone, is now also able to bind to specific regions of the DNA. These DNA regions are known as the hormone response elements; when bound by the receptor-hormone complex, they are activated to begin genetic transcription. This produces messenger RNA (mRNA), which then codes for the production of specific proteins. These proteins often have enzymatic activity that changes the metabolism of the target cell in a specific fashion.

The thyroid hormone's mechanism of action resembles that of the steroid hormones. Thyroxine contains four iodines and so is often abbreviated T_4 (for *tetraiodothyronine*). The thyroid gland also secretes smaller amounts of a similar molecule that has only three iodines, called *triiodothyronine* (and abbreviated T_3). Both hormones enter target cells, but all of the T_4 that enters is changed into T_3 (figure 53.6). Thus, only the T_3 form of the hormone enters the nucleus and binds to nuclear receptor proteins. The hormone/receptor complex, in turn, binds to the appropriate hormone response elements in DNA.

The lipophilic hormones pass through the target cell's plasma membrane and bind to intracellular receptor proteins. The hormone/receptor complex then binds to specific regions of DNA, thereby activating genes and regulating the target cells.

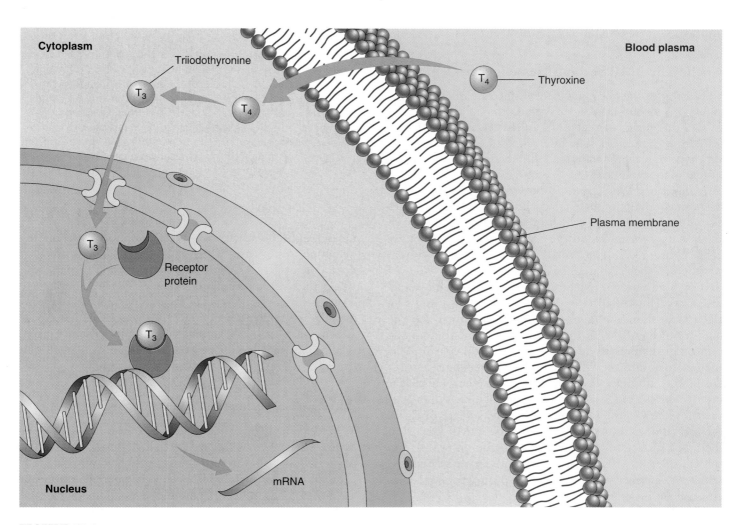

FIGURE 53.6
The mechanism of thyroxine action. Thyroxine contains four iodines. When it enters the target cell, thyroxine is changed into triiodothyronine, with three iodines. This hormone moves into the nucleus and binds to nuclear receptors. The hormone-receptor complex then binds to regions of the DNA and stimulates gene transcription.

Hormones That Do Not Enter Cells

Hormones that are too polar to cross the plasma membranes of their target cells include all of the peptide and glycoprotein hormones, as well as the catecholamine hormones epinephrine and norepinephrine. These hormones bind to receptor proteins located on the outer surface of the plasma membrane—the hormones do *not* enter the cell. If you think of the hormone as a messenger sent from an endocrine gland to the target cell, it is evident that a second messenger is needed within the target cell to produce the effects of the hormone. Second messengers are particular organic molecules and Ca^{++}. The interaction between the hormone and its receptor activates mechanisms in the plasma membrane that increase the concentration of the second messengers within the target cell cytoplasm.

The binding of a water-soluble hormone to its receptor is reversible and usually very brief. After the hormone binds to its receptor and activates a second-messenger system, it dissociates from the receptor and may travel in the blood to another target cell somewhere else in the body. Eventually, enzymes (primarily in the liver) degrade the hormone by converting it into inactive derivatives.

The Cyclic AMP Second-Messenger System

The action of the hormone epinephrine can serve as an example of a second-messenger system. Epinephrine can bind to two categories of receptors, called *alpha* (α)- and *beta* (β)-*adrenergic receptors.* The interaction of epinephrine with each type of receptor activates a different second-messenger system in the target cell.

In the early 1960s, Earl Sutherland showed that cyclic adenosine monophosphate, or **cyclic AMP (cAMP),** serves as a second messenger when epinephrine binds to β-adrenergic receptors on the plasma membranes of liver cells (figure 53.7). The cAMP second-messenger system was the first such system to be described.

The β-adrenergic receptors are associated with membrane proteins called *G proteins* (see chapters 7 and 51). Each G protein is composed of three subunits, and the binding of epinephrine to its receptor causes one of the G protein subunits to dissociate from the other two. This subunit then diffuses within the plasma membrane until it encounters **adenylyl cyclase,** a membrane enzyme that is inactive until it binds to the G protein subunit. When activated by the G protein subunit, adenylyl cyclase catalyzes the formation of cAMP from ATP. The cAMP formed at the inner surface of the plasma membrane diffuses within the cytoplasm, where it binds to and activates **protein kinase-A,** an enzyme that adds phosphate groups to specific cellular proteins.

The identities of the proteins that are phosphorylated by protein kinase-A varies from one cell type to the next, and this variation is one of the reasons epinephrine has such diverse effects on different tissues. In liver cells, protein

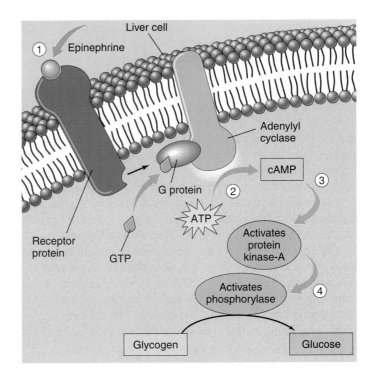

FIGURE 53.7
The action of epinephrine on a liver cell. *(1)* Epinephrine binds to specific receptor proteins on the cell surface. *(2)* Acting through intermediary G proteins, the hormone-bound receptor activates the enzyme adenylyl cyclase, which converts ATP into cyclic AMP (cAMP). *(3)* Cyclic AMP performs as a second messenger and activates protein kinase-A, an enzyme that was previously present in an inactive form. *(4)* Protein kinase-A phosphorylates and thereby activates the enzyme phosphorylase, which catalyzes the hydrolysis of glycogen into glucose.

kinase-A phosphorylates and thereby activates another enzyme, phosphorylase, which converts glycogen into glucose. Through this multistep mechanism, epinephrine causes the liver to secrete glucose into the blood during the fight-or-flight reaction, when the adrenal medulla is stimulated by the sympathetic division of the autonomic nervous system (chapter 51). In cardiac muscle cells, protein kinase-A phosphorylates a different set of cellular proteins, which cause the heart to beat faster and more forcefully.

The IP₃/Ca⁺⁺ Second-Messenger System

When epinephrine binds to α-adrenergic receptors, it doesn't activate adenylyl cyclase and cause the production of cAMP. Instead, through a different type of G protein, it activates another membrane-bound enzyme, phospholipase C (figure 53.8). This enzyme cleaves certain membrane phospholipids to produce the second messenger, inositol trisphosphate (IP_3). IP_3 diffuses into the cytoplasm from the plasma membrane and binds to receptors located on the surface of the endoplasmic reticulum.

Recall from chapter 5 that the endoplasmic reticulum is a system of membranous sacs and tubes that serves a variety of functions in different cells. One of its functions is to accumulate Ca^{++} by actively transporting Ca^{++} out of the cytoplasm. Other pumps transport Ca^{++} from the cytoplasm through the plasma membrane to the extracellular fluid. These two mechanisms keep the concentration of Ca^{++} in the cytoplasm very low. Consequently, there is an extremely steep concentration gradient for Ca^{++} between the cytoplasm and the inside of the endoplasmic reticulum, and between the cytoplasm and the extracellular fluid.

When IP_3 binds to its receptors on the endoplasmic reticulum, it stimulates the endoplasmic reticulum to release its stored Ca^{++}. Calcium channels in the plasma membrane may also open, allowing Ca^{++} to diffuse into the cell from the extracellular fluid. Some of the Ca^{++} that has suddenly entered the cytoplasm then binds to a protein called calmodulin, which has regulatory functions analogous to those of cyclic AMP. One of the actions of calmodulin is to activate another type of protein kinase, resulting in the phosphorylation of a different set of cellular proteins.

What is the advantage of having multiple second-messenger systems? Consider the antagonistic actions of epinephrine and insulin on liver cells. Epinephrine uses cAMP as a second messenger to promote the hydrolysis of glycogen to glucose, while insulin stimulates the synthesis of glycogen from glucose. Clearly, insulin cannot use cAMP as a second messenger. Although the exact mechanism of insulin's action is still not well understood, insulin may act in part through the IP_3/Ca^{++} second-messenger system.

The water-soluble hormones cannot pass through the plasma membrane; they must rely on second messengers within the target cells to mediate their actions. Such second messengers include cyclic AMP (cAMP), inositol trisphosphate (IP_3), and Ca^{++}. In many cases, the second messengers activate previously inactive enzymes.

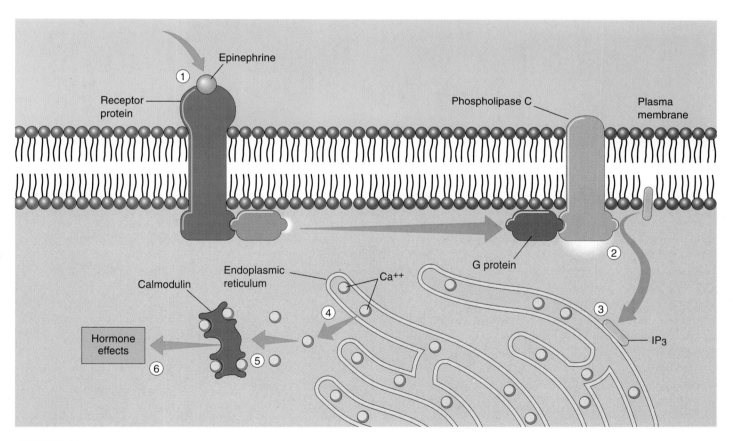

FIGURE 53.8

The IP_3/Ca^{++} second-messenger system. *(1)* The hormone epinephrine binds to specific receptor proteins on the cell surface. *(2)* Acting through G proteins, the hormone-bound receptor activates the enzyme phospholipase C, which converts membrane phospholipids into inositol trisphosphate (IP_3). *(3)* IP_3 diffuses through the cytoplasm and binds to receptors on the endoplasmic reticulum. *(4)* The binding of IP_3 to its receptors stimulates the endoplasmic reticulum to release Ca^{++} into the cytoplasm. *(5)* Some of the released Ca^{++} binds to a regulatory protein called calmodulin. *(6)* The Ca^{++}/calmodulin complex activates other intracellular proteins, ultimately producing the effects of the hormone.

The Pituitary Gland: Posterior

The **pituitary gland** hangs by a stalk from the hypothalamus of the brain (figure 53.9) posterior to the optic chiasma (chapter 52). A microscopic view reveals that the gland consists of two parts. The anterior portion appears glandular and is called the **anterior pituitary;** the posterior portion appears fibrous and is the **posterior pituitary.** These two portions of the pituitary gland have different embryonic origins, secrete different hormones, and are regulated by different control systems.

The Posterior Pituitary Gland

The posterior pituitary appears fibrous because it contains axons that originate in cell bodies within the hypothalamus and extend along the stalk of the pituitary as a tract of fibers. This anatomical relationship results from the way

that the posterior pituitary is formed in embryonic development. As the floor of the third ventricle of the brain forms the hypothalamus, part of this neural tissue grows downward to produce the posterior pituitary. The hypothalamus and posterior pituitary thus become interconnected by a tract of axons.

The endocrine role of the posterior pituitary gland first became evident in 1912, when a remarkable medical case was reported: a man who had been shot in the head developed the need to urinate every 30 minutes or so, 24 hours a day. The bullet had lodged in his pituitary gland. Subsequent research demonstrated that removal of this gland produces the same symptoms. Pituitary extracts were found to contain a substance that makes the kidneys conserve water, and in the early 1950s investigators isolated a peptide from the posterior pituitary, **antidiuretic hormone (ADH,** also known as **vasopressin),** that stimulates water retention by

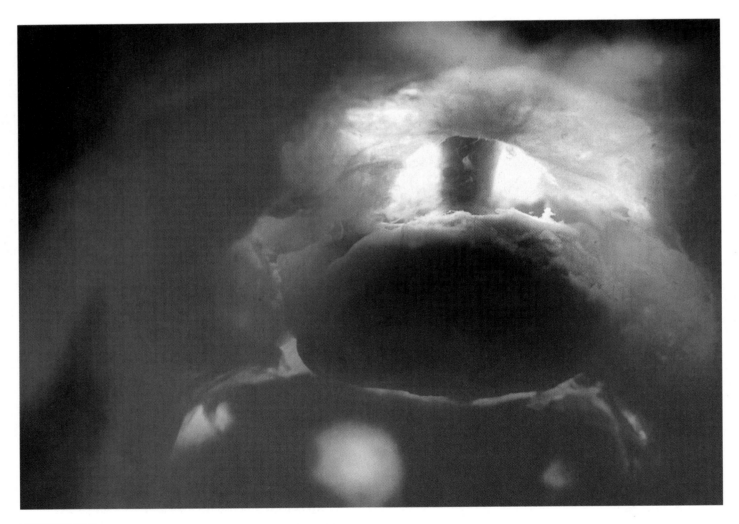

FIGURE 53.9
The pituitary gland hangs by a short stalk from the hypothalamus. The pituitary gland (the oval structure hanging from the stalk), shown here enlarged 15 times, regulates hormone production in many of the body's endocrine glands.

FIGURE 53.10

The effects of antidiuretic hormone (ADH). An increase in the osmotic concentration of the blood stimulates the posterior pituitary gland to secrete ADH, which promotes water retention by the kidneys. This works as a negative feedback loop to correct the initial disturbance of homeostasis.

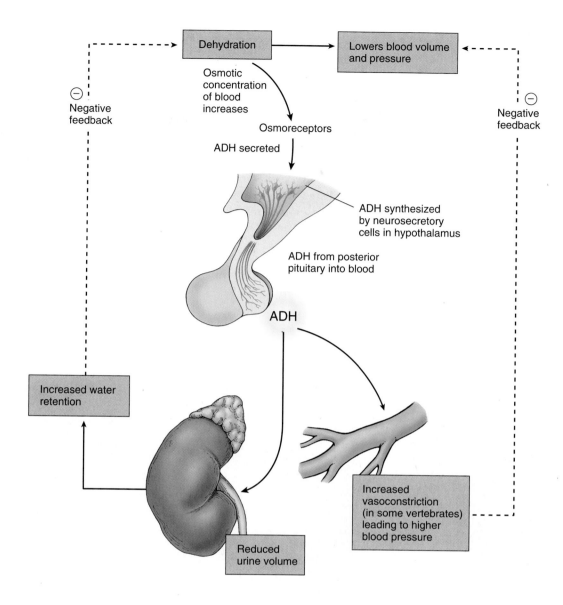

the kidneys (figure 53.10). When ADH is missing, the kidneys do not retain water and excessive quantities of urine are produced. This is why the consumption of alcohol, which inhibits ADH secretion, stimulates urination.

The posterior pituitary also secretes **oxytocin,** a second peptide hormone which, like ADH, is composed of nine amino acids. Oxytocin stimulates the milk-ejection reflex, so that contraction of the smooth muscles around the mammary glands and ducts causes milk to be ejected from the ducts through the nipple. During suckling, sensory receptors in the nipples send impulses to the hypothalamus, which triggers the release of oxytocin. Oxytocin is also needed to stimulate uterine contractions in women during childbirth. Oxytocin secretion continues after childbirth in a woman who is breast-feeding, which is why the uterus of a nursing mother returns to its normal size after pregnancy more quickly than does the uterus of a mother who does not breast-feed.

ADH and oxytocin are actually *produced* by neuron cell bodies located in the hypothalamus. These two hormones are transported along the axon tract that runs from the hypothalamus to the posterior pituitary and are stored in the posterior pituitary. In response to the appropriate stimulation—increased blood plasma concentration in the case of ADH, the suckling of a baby in the case of oxytocin—the hormones are released by the posterior pituitary into the blood. Since this reflex control involves both the nervous and endocrine systems, the secretion of ADH and oxytocin are **neuroendocrine reflexes.**

The posterior pituitary gland contains axons originating from neurons in the hypothalamus. These neurons produce ADH and oxytocin, which are stored in and released from the posterior pituitary gland in response to neural stimulation from the hypothalamus.

The Anterior Pituitary Gland

The anterior pituitary, unlike the posterior pituitary, does not develop from a downgrowth of the brain; instead, it develops from a pouch of epithelial tissue that pinches off from the roof of the embryo's mouth. Since it is epithelial tissue, the anterior pituitary is a complete gland—it produces the hormones it secretes. Many of these hormones stimulate growth in their target organs, including other endocrine glands. Therefore, the hormones of the anterior pituitary gland are collectively termed *tropic hormones* (Greek *trophe*, "nourishment"), or *tropins*. When the target organ of a tropic hormone is another endocrine gland, that gland is stimulated by the tropic hormone to secrete its own hormones.

The hormones produced and secreted by the anterior pituitary gland (figure 53.11) include the following:

1. **Growth hormone (GH,** or *somatotropin*) stimulates the growth of muscle, bone, and other tissues and is also essential for proper metabolic regulation.
2. **Adrenocorticotropic hormone (ACTH,** or *corticotropin*) stimulates the adrenal cortex to produce corticosteroid hormones, including cortisol (in humans) and corticosterone (in many other vertebrates), which regulate glucose homeostasis.
3. **Thyroid-stimulating hormone (TSH,** or *thyrotropin*) stimulates the thyroid gland to produce thyroxine, which in turn stimulates oxidative respiration.
4. **Luteinizing hormone (LH)** is needed for ovulation and the formation of a corpus luteum in the female menstrual cycle (see chapter 56). It also stimulates the testes to produce testosterone, which is needed for sperm production and for the development of male secondary sexual characteristics.

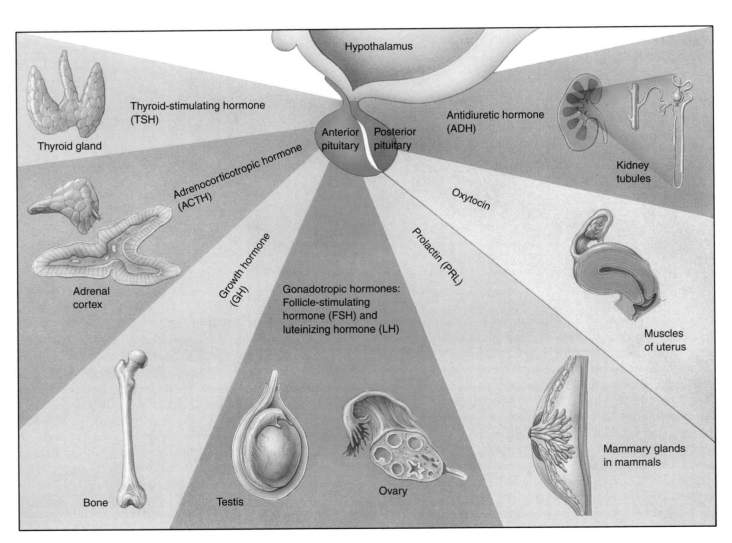

FIGURE 53.11
The major hormones of the anterior and posterior pituitary glands. Only a few of the actions of these hormones are shown.

5. **Follicle-stimulating hormone (FSH)** is required for the development of ovarian follicles in females. In males, it is required for the development of sperm. FSH and LH are both referred to as *gonadotropins*.

6. **Prolactin (PRL)** stimulates the breasts to produce milk in mammals. It also helps regulate kidney function in vertebrates, the production of "crop milk" (fed to chicks by regurgitation) in some birds, and acts on the gills of fish that travel from salt to fresh water to promote sodium retention.

7. **Melanocyte-stimulating hormone (MSH)** stimulates the synthesis and dispersion of melanin pigment, which darken the epidermis of some fish, amphibians, and reptiles. MSH has no known specific function in mammals, but abnormally high amounts of ACTH can cause skin darkening because it contains the amino acid sequence of MSH within its structure.

Growth Hormone

The importance of the anterior pituitary gland first became understood in 1909, when a 38-year-old South Dakota farmer was cured of the growth disorder acromegaly by the surgical removal of a pituitary tumor. Acromegaly is a form of gigantism in which the jaw begins to protrude and other facial features thicken. It was discovered that gigantism is almost always associated with pituitary tumors. Robert Wadlow, born in 1928 in Alton, Illinois, stood 8 feet, 11 inches tall and weighed 485 pounds before he died from infection at age 22 (figure 53.12). He was the tallest human being ever recorded, and he was still growing the year he died.

We now know that gigantism is caused by the excessive secretion of growth hormone (GH) by the anterior pituitary gland in a growing child. GH stimulates protein synthesis and growth of muscles and connective tissues; it also promotes the elongation of bones by stimulating cell division in the cartilaginous epiphyseal growth plates of bones. Researchers found that this stimulation does not occur in the absence of blood plasma, suggesting that the stimulation by GH was indirect. We now know that GH stimulates the production of **insulin-like growth factors,** which are produced by the liver and secreted into the blood in response to stimulation by GH. The insulin-like growth factors then stimulate growth of the epiphyseal growth plates and thus elongation of the bones.

When a person's skeletal growth plates have converted from cartilage into bone, however, GH can no longer cause an increase in height. Therefore, excessive GH secretion in an adult produces bone and soft tissue deformities in the condition called acromegaly. A deficiency in GH secretion during childhood results in pituitary dwarfism, a failure to achieve normal growth.

FIGURE 53.12

The Alton giant. This photograph of Robert Wadlow of Alton, Illinois, taken on his 21st birthday, shows him at home with his father and mother and four siblings. Born normal size, he developed a growth hormone-secreting pituitary tumor as a young child and never stopped growing during his 22 years of life.

Other Anterior Pituitary Hormones

Prolactin is like growth hormone in that it acts on organs that are not endocrine glands. In addition to its stimulation of milk production in mammals and "crop milk" production in birds, prolactin has varied effects on electrolyte balance by acting on the kidneys, the gills of fish, and the salt glands of marine birds (discussed in chapter 55). Unlike growth hormone and prolactin, the other anterior pituitary hormones act on specific glands.

Some of the anterior pituitary hormones that act on specific glands have common names, such as thyroid-stimulating hormone (TSH), and alternative names that emphasize the tropic nature of the hormone, such as thyrotropin. TSH stimulates only the thyroid gland, and adrenocorticotropic hormone (ACTH) stimulates only the adrenal cortex (outer portion of the adrenal glands). Follicle-stimulating hormone (FSH) and luteinizing hormone (LH) act only on the gonads (testes and ovaries); hence, they are collectively called *gonadotropic hormones.* Although both FSH and LH act on the gonads, they each act on different target cells in the gonads of both females and males.

Hypothalamic Control of Anterior Pituitary Gland Secretion

The anterior pituitary gland, unlike the posterior pituitary, is not derived from the brain and does not receive an axon tract from the hypothalamus. Nevertheless, the hypothalamus controls production and secretion of the anterior pituitary hormones. This control is exerted hormonally rather than by means of nerve axons. Neurons in the hypothalamus secrete releasing hormones and inhibiting hormones into blood capillaries at the base of the hypothalamus (figure 53.13). These capillaries drain into small veins that run within the stalk of the pituitary to a second bed of capillaries in the anterior pituitary. This unusual system of vessels is known as the hypothalamohypophyseal portal system. It is called a portal system because it has a second capillary bed downstream from the first; the only other body location with a similar system is the liver, where capillaries receive blood drained from the gastrointestinal tract (via the hepatic portal vein—see chapter 49). Since the second bed of capillaries receives little oxygen from such vessels, the vessels must be delivering something else of importance.

Each releasing hormone delivered by the hypothalamohypophyseal system regulates the secretion of a specific anterior pituitary hormone. For example, *thyrotropin-releasing hormone (TRH)* stimulates the release of TSH, *corticotropin-releasing hormone (CRH)* stimulates the release of ACTH, and *gonadotropin-releasing hormone (GnRH)* stimulates the release of FSH and LH. A releasing hormone for growth hormone, called *growth hormone-releasing hormone (GHRH)* has also been discovered, and a releasing hormone for prolactin has been postulated but has thus far not been identified.

The hypothalamus also secretes hormones that *inhibit* the release of certain anterior pituitary hormones. To date, three such hormones have been discovered: *somatostatin* inhibits the secretion of GH; *prolactin-inhibiting hormone (PIH)* inhibits the secretion of prolactin; and *melanotropin-inhibiting hormone (MIH)* inhibits the secretion of MSH.

Negative Feedback Control of Anterior Pituitary Gland Secretion

Since hypothalamic hormones control the secretions of the anterior pituitary gland, and the anterior pituitary hormones control the secretions of some other endocrine glands, it may seem that the hypothalamus functions as a "master gland," in charge of hormonal secretion in the body. This idea is not valid, however, for two reasons. First, a number of endocrine organs, such as the adrenal medulla and the pancreas, are not directly regulated by this control system. Second, the hypothalamus and the anterior pituitary gland are themselves controlled by the very hormones whose secretion they stimulate! In most cases this is an inhibitory control, where the target gland hormones inhibit the secretions of the hypothalamus and anterior pituitary (figure 53.14). This type of control system is called negative feedback inhibition.

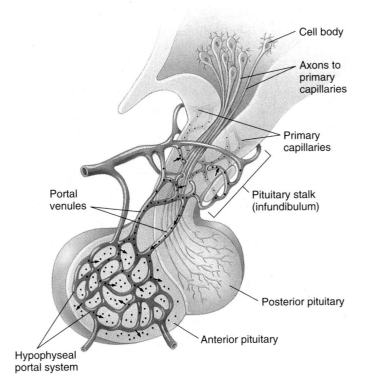

FIGURE 53.13
Hormonal control of the anterior pituitary gland by the hypothalamus. Neurons in the hypothalamus secrete hormones that are carried by short blood vessels directly to the anterior pituitary gland, where they either stimulate or inhibit the secretion of anterior pituitary hormones.

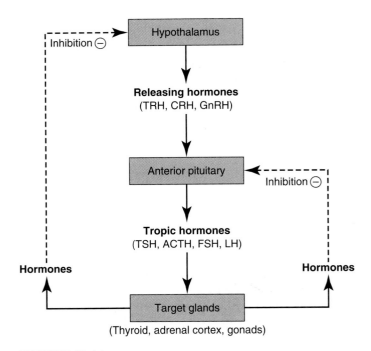

FIGURE 53.14
Negative feedback inhibition. The hormones secreted by some endocrine glands feed back to inhibit the secretion of hypothalamic releasing hormones and anterior pituitary tropic hormones.

Let's consider the hormonal control of the thyroid gland. The hypothalamus secretes TRH into the hypothalamohypophyseal portal system, which stimulates the anterior pituitary gland to secrete TSH, which in turn stimulates the thyroid gland to release thyroxine. Among thyroxine's many target organs are the hypothalamus and the anterior pituitary gland themselves. Thyroxine acts upon these organs to inhibit their secretion of TRH and TSH, respectively (figure 53.15). This negative feedback inhibition is essential for homeostasis.

We can see the importance of negative feedback inhibition by examining a person who lacks sufficient iodine in the diet. Without iodine, the thyroid gland cannot produce thyroxine (which contains four iodines per molecule). As a result, thyroxine levels in the blood fall drastically, and the hypothalamus and anterior pituitary receive far less negative feedback inhibition than is normal. This reduced inhibition causes an elevated secretion of TRH and TSH. The high levels of TSH stimulate the thyroid gland to grow, but it still could not produce thyroxine without iodine. The consequence of this interruption of the normal inhibition by thyroxine is an enlarged thyroid gland, a condition known as a goiter (figure 53.16).

There is only one example of positive feedback in the control of the hypothalamus and anterior pituitary by the target glands, because positive feedback cannot maintain constancy of the internal environment (homeostasis). Positive feedback accentuates change, driving the change in the same direction. The one example of positive control involves the control of ovulation, an explosive event that culminates in the expulsion of the egg cell from the ovary. In that case, an ovarian hormone, estradiol, actually stimulates the secretion of an anterior pituitary hormone, LH. This will be discussed in detail in chapter 56.

The hypothalamus controls the anterior pituitary gland by means of hormones, and the anterior pituitary gland controls some other glands through the hormones it secretes. However, both the hypothalamus and the anterior pituitary gland are controlled by other glands through negative feedback inhibition.

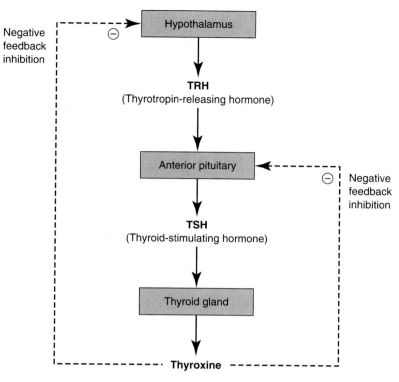

FIGURE 53.15
Regulation of thyroxine secretion. The hypothalamus secretes TRH, which stimulates the anterior pituitary to secrete TSH. The TSH then stimulates the thyroid to secrete thyroxine, which exerts negative feedback control of the hypothalamus and anterior pituitary.

FIGURE 53.16
A person with a goiter. This condition is caused by a lack of iodine in the diet. As a result, thyroxine secretion is low, so there is less negative feedback inhibition of TSH. The elevated TSH secretion, in turn, stimulates the thyroid to grow and produce the goiter.

The Thyroid and Parathyroid Glands

The endocrine glands that are regulated by the anterior pituitary, and those endocrine glands that are regulated by other means, help to control metabolism, electrolyte balance, and reproductive functions. Some of the major endocrine glands will be considered in this section.

The Thyroid Gland

The thyroid gland (Greek *thyros*, "shield") is shaped like a shield and lies just below the Adam's apple in the front of the neck. We have already mentioned that the thyroid gland secretes thyroxine and smaller amounts of triiodothyronine (T_3), which stimulate oxidative respiration in most cells in the body and, in so doing, help set the body's basal metabolic rate (chapter 48). In children, these thyroid hormones also promote growth and stimulate maturation of the central nervous system. Children with underactive thyroid glands are therefore stunted in their growth and suffer severe mental retardation, a condition called cretinism. This differs from pituitary dwarfism, which results from inadequate GH and is not associated with abnormal intellectual development.

People who are hypothyroid (whose secretion of thyroxine is too low) can take thyroxine orally, as pills. Only thyroxine and the steroid hormones (as in contraceptive pills), can be taken orally because they are nonpolar and can pass through the plasma membranes of intestinal epithelial cells without being digested.

There is an additional function of the thyroid gland that is unique to amphibians—thyroid hormones are needed for the metamorphosis of the larvae into adults (figure 53.17). If the thyroid gland is removed from a tadpole, it will not change into a frog. Conversely, if an immature tadpole is fed pieces of a thyroid gland, it will undergo premature metamorphosis and become a miniature frog!

The thyroid gland also secretes calcitonin, a peptide hormone that plays a role in maintaining proper levels of calcium (Ca^{++}) in the blood. When the blood Ca^{++} concentration rises too high, calcitonin stimulates the uptake of Ca^{++} into bones, thus lowering its level in the blood. Although calcitonin may be important in the physiology of

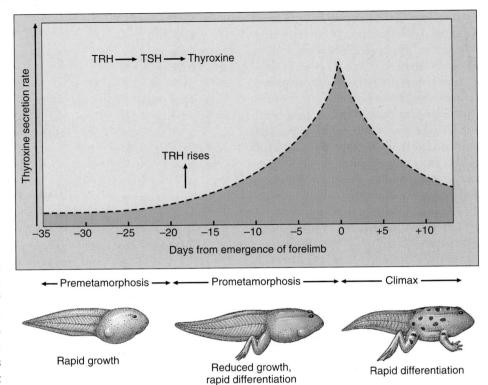

FIGURE 53.17

Thyroxine triggers metamorphosis in amphibians. In tadpoles at the premetamorphic stage, the hypothalamus releases TRH (thyrotropin-releasing hormone), which causes the anterior pituitary to secrete TSH (thyroid-stimulating hormone). TSH then acts on the thyroid gland, which secretes thyroxine. The hindlimbs then begin to form. As metamorphosis proceeds, thyroxine reaches its maximal level, after which the forelimbs begin to form.

some vertebrates, its significance in normal human physiology is controversial. A hormone that plays an even more important role in Ca^{++} homeostasis is secreted by the parathyroid glands, described in the next section.

The Parathyroid Glands and Calcium Homeostasis

The parathyroid glands are four small glands attached to the thyroid. Because of their size, researchers ignored them until well into this century. The first suggestion that these organs have an endocrine function came from experiments on dogs: if their parathyroid glands were removed, the Ca^{++} concentration in the dogs' blood plummeted to less than half the normal value. The Ca^{++} concentration returned to normal when an extract of parathyroid gland was administered. However, if too much of the extract was administered, the dogs' Ca^{++} levels rose far above normal as the calcium phosphate crystals in their bones was dissolved. It was clear that the parathyroid glands produce a hormone that stimulates the release of Ca^{++} from bone.

The hormone produced by the parathyroid glands is a peptide called parathyroid hormone (PTH). It is one of only

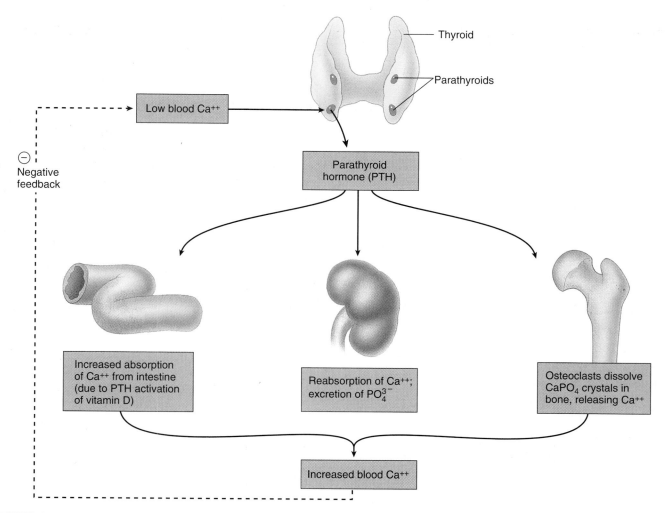

FIGURE 53.18

Regulation of blood Ca⁺⁺ levels by parathyroid hormone (PTH). When blood Ca^{++} levels are low, parathyroid hormone (PTH) is released by the parathyroid glands. PTH directly stimulates the dissolution of bone and the reabsorption of Ca^{++} by the kidneys. PTH indirectly promotes the intestinal absorption of Ca^{++} by stimulating the production of the active form of vitamin D.

two hormones in humans that are absolutely essential for survival (the other is aldosterone, which will be discussed in the next section). PTH is synthesized and released in response to falling levels of Ca^{++} in the blood. This cannot be allowed to continue uncorrected, because a significant fall in the blood Ca^{++} level can cause severe muscle spasms. A normal blood Ca^{++} is important for the functioning of muscles, including the heart, and for proper functioning of the nervous and endocrine systems.

PTH stimulates the osteoclasts in bone to dissolve the hydroxyapatite crystals of the bone matrix and release Ca^{++} into the blood (figure 53.18). PTH also stimulates the kidneys to reabsorb Ca^{++} from the urine and leads to the activation of vitamin D, needed for the absorption of Ca^{++} from food in the intestine.

Vitamin D is produced in the skin from a cholesterol derivative in response to ultraviolet light. It is called a vitamin because a dietary source is needed to supplement the amount that the skin produces. Secreted into the blood from the skin, vitamin D is actually an inactive form of a hormone. In order to become activated, the molecule must gain two hydroxyl groups (—OH); one of these is added by an enzyme in the liver, the other by an enzyme in the kidneys. The enzyme needed for this final step is stimulated by PTH, thereby producing the active form of vitamin D known as 1, 25-dihydroxyvitamin D. This hormone stimulates the intestinal absorption of Ca^{++} and thereby helps to raise blood Ca^{++} levels so that bone can become properly mineralized. A diet deficient in vitamin D thus leads to poor bone formation, a condition called rickets.

Thyroxine helps to set the basal metabolic rate by stimulating the rate of cell respiration throughout the body; this hormone is also needed for amphibian metamorphosis. Parathyroid hormone acts to raise the blood Ca⁺⁺ levels, in part by stimulating osteoclasts to dissolve bone.

The Adrenal Glands

The adrenal glands are located just above each kidney (figure 53.19). Each gland is composed of an inner portion, the adrenal medulla, and an outer layer, the adrenal cortex.

The Adrenal Medulla

The adrenal medulla receives neural input from axons of the sympathetic division of the autonomic nervous system, and it secretes epinephrine and norepinephrine in response to stimulation by these axons. The actions of these hormones trigger "alarm" responses similar to those elicited by the sympathetic division, helping to prepare the body for "fight or flight." Among the effects of these hormones are an increased heart rate, increased blood pressure, dilation of the bronchioles, elevation in blood glucose, and reduced blood flow to the skin and digestive organs. The actions of epinephrine released as a hormone supplement those of norepinephrine released as a sympathetic nerve neurotransmitter.

The Adrenal Cortex: Homeostasis of Glucose and Na⁺

Cortisol (also called hydrocortisone) and related steroids secreted by the adrenal cortex act on various cells in the body to maintain glucose homeostasis. These hormones are therefore referred to as glucocorticoids. The glucocorticoids stimulate the breakdown of muscle protein into amino acids, which are carried by the blood to the liver. They also stimulate the liver to produce the enzymes needed for gluconeogenesis, the conversion of amino acids into glucose. This creation of glucose from protein is particularly important during very long periods of fasting or exercise, when blood glucose levels might otherwise become dangerously low.

In addition to regulating glucose metabolism, the glucocorticoids modulate some aspects of the immune response. Glucocorticoids are given medically to suppress the immune system in persons with immune disorders, such as rheumatoid arthritis. Derivatives of cortisol, such as prednisone, have widespread medical use as antiinflammatory agents.

Aldosterone, the other major corticosteroid, is classified as a mineralocorticoid because it helps regulate mineral balance through two functions. One of the functions of aldosterone is to stimulate the kidneys to reabsorb Na^+ from the urine. (Urine is formed by filtration of blood plasma, so the blood levels of Na^+ will decrease if Na^+ is not reabsorbed from the urine.) Sodium is the major extracellular solute and is needed for the maintenance of normal blood volume and pressure. Without aldosterone, the kidneys would lose excessive amounts of blood Na^+ in the urine, followed by Cl^- and water; this would cause the blood volume and pressure to fall. By stimulating the kidneys to reabsorb salt and water, aldosterone thus maintains the normal blood volume and pressure essential to life.

The other function of aldosterone is to stimulate the kidneys to secrete K^+ into the urine. Thus, when aldosterone levels are too low, the concentration of K^+ in the blood may rise to dangerous levels. Because of these essential functions performed by aldosterone, removal of the adrenal glands, or diseases that prevent aldosterone secretion, are invariably fatal without hormone therapy.

The adrenal medulla is stimulated by sympathetic neurons to secrete epinephrine and norepinephrine during the fight-or-flight reaction. The adrenal cortex is stimulated to secrete its steroid hormones by ACTH from the anterior pituitary. Cortisol helps to regulate blood glucose and aldosterone acts to regulate blood Na^+ and K^+ levels.

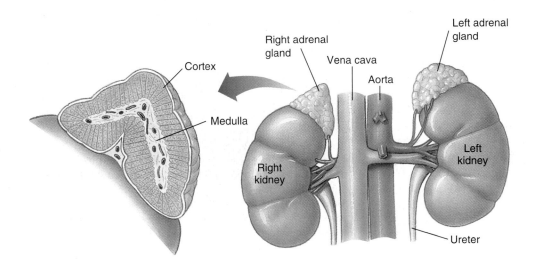

FIGURE 53.19
The adrenal glands. The inner portion of the gland, the adrenal medulla, produces epinephrine and norepinephrine, which initiate a response to stress. The outer portion of the gland, the adrenal cortex, produces steroid hormones that influence blood glucose levels.

The Pancreas

The pancreas is located adjacent to the stomach and is connected to the duodenum of the small intestine by the pancreatic duct. It secretes bicarbonate ions and a variety of digestive enzymes into the small intestine through this duct (chapter 48), and for a long time the pancreas was thought to be solely an exocrine gland. In 1869, however, a German medical student named Paul Langerhans described some unusual clusters of cells scattered throughout the pancreas; these clusters came to be called islets of Langerhans. Laboratory workers later observed that the surgical removal of the pancreas caused glucose to appear in the urine, the hallmark of the disease diabetes mellitus. This suggested that the pancreas, specifically the islets of Langerhans, might be producing a hormone that prevents this disease.

That hormone is insulin, secreted by the beta (β)-cells of the islets. Insulin was not isolated until 1922, when two young doctors working in a Toronto hospital succeeded where many others had not. On January 11, 1922, they injected an extract purified from beef pancreas into a 13-year-old diabetic boy, whose weight had fallen to 65 pounds and who was not expected to survive. With that single injection, the glucose level in the boy's blood fell 25%. A more potent extract soon brought the level down to near normal. The doctors had achieved the first instance of successful insulin therapy.

Two forms of diabetes mellitus are now recognized. People with *type I*, or insulin-dependent diabetes mellitus, lack the insulin-secreting β-cells. Treatment for these patients therefore consists of insulin injections. (Since insulin is a peptide hormone, it would be digested if taken orally and must instead be injected into the blood.) In the past, only insulin extracted from the pancreas of pigs or cattle was available, but today people with insulin-dependent diabetes can inject themselves with human insulin produced by genetically engineered bacteria. Active research on the possibility of transplanting islets of Langerhans holds much promise of a lasting treatment for these patients. Most diabetic patients, however, have *type II*, or non-insulin-dependent diabetes mellitus. They generally have normal or even above-normal levels of insulin in their blood, but their cells have a reduced sensitivity to insulin. These peo-

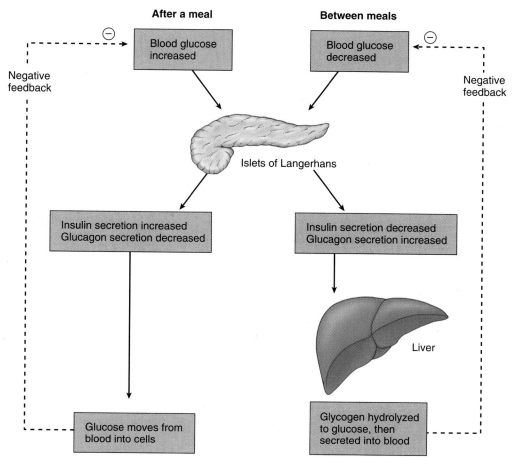

FIGURE 53.20
The antagonistic actions of insulin and glucagon on blood glucose. Insulin stimulates the cellular uptake of blood glucose into skeletal muscles and the liver after a meal. Glucagon stimulates the hydrolysis of liver glycogen between meals, so that the liver can secrete glucose into the blood. These antagonistic effects help to maintain homeostasis of the blood glucose concentration.

ple do not require insulin injections and can usually control their diabetes through diet and exercise.

The islets of Langerhans produce another hormone; the alpha (α)-cells of the islets secrete glucagon, which acts antagonistically to insulin (figure 53.20). When a person eats carbohydrates, the blood glucose concentration rises. This stimulates the secretion of insulin by β-cells and inhibits the secretion of glucagon by the α-cells. Insulin promotes the cellular uptake of glucose into liver and muscle cells, where it is stored as glycogen, and into adipose cells, where it is stored as fat. Between meals, when the concentration of blood glucose falls, insulin secretion decreases and glucagon secretion increases. Glucagon promotes the hydrolysis of stored glycogen in the liver and fat in adipose tissue. As a result, glucose and fatty acids are released into the blood and can be taken up by cells and used for energy.

The β-cells of the islets of Langerhans secrete insulin, and the α-cells secrete glucagon. These two hormones have antagonistic actions on the blood glucose concentration; insulin lowers and glucagon raises blood glucose.

Other Endocrine Glands

Sexual Development, Biological Clocks, and Immune Regulation in Vertebrates

The ovaries and testes are important endocrine glands, producing the steroid sex hormones (including estrogens, progesterone, and testosterone), to be described in detail in chapter 56. Estrogens and progesterone regulate the menstrual cycle and are secreted by the placenta during pregnancy as well as by the ovaries. Testosterone and other androgens promote protein synthesis in many organs and are responsible for the generally larger male body size compared with females. Some bodybuilders illegally take androgens to increase muscle mass, but this practice can cause liver disorders as well as a number of other serious side effects.

Another major endocrine gland is the pineal gland, located in the roof of the third ventricle of the brain in most vertebrates (see figure 51.24). It is about the size of a pea and is shaped like a pine-cone (hence its name). The pineal gland evolved from a median light-sensitive eye (sometimes called a "third eye," although it could not form images) at the top of the skull in primitive vertebrates. This pineal eye is still present in primitive fish (cyclostomes) and some reptiles. In other vertebrates, however, the pineal gland is buried deep in the brain and functions as an endocrine gland by secreting the hormone melatonin. One of the actions of melatonin is to cause blanching of the skin of lower vertebrates by reducing the dispersal of melanin granules.

The secretion of melatonin is stimulated by activity of the *suprachiasmatic nucleus (SCN)* of the hypothalamus. The SCN is known to function as the major biological clock in vertebrates, entraining (synchronizing) various body processes to a circadian rhythm (one that repeats every 24 hours). Through regulation by the SCN, the secretion of melatonin by the pineal gland is entrained to cycles of light and dark, decreasing during the day and increasing at night. In some vertebrates, this activity helps to regulate reproductive physiology, but the role of melatonin in human reproduction is controversial.

There are a variety of hormones secreted by nonendocrine organs. The thymus is the site of production of particular lymphocytes called T cells, and it secretes a number of hormones that function in the regulation of the immune system. The right atrium of the heart secretes atrial natriuretic hormone, which stimulates the kidneys to excrete salt and water in the urine. This hormone, therefore, acts antagonistically to aldosterone, which promotes salt and water retention. The kidneys secrete erythropoietin, a hormone that stimulates the bone marrow to produce red blood cells. Even the skin has an endocrine

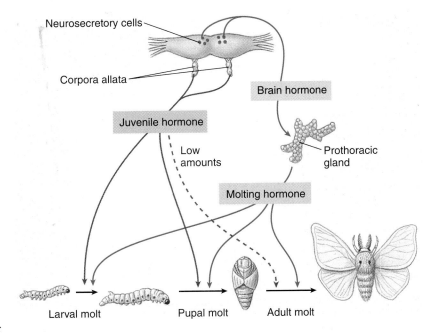

FIGURE 53.21
The hormonal control of metamorphosis in the silkworm moth, *Bombyx mori*. While molting hormone (ecdysone), produced by the prothoracic gland, triggers when molting will occur, juvenile hormone, produced by bodies near the brain called the corpora allata, determines the result of a particular molt. High levels of juvenile hormone inhibit the formation of the pupa and adult forms. At the late stages of metamorphosis, therefore, it is important that the corpora allata not produce large amounts of juvenile hormone.

function, since it secretes vitamin D. The gas nitric oxide, made by many different cells, controls blood pressure by dilating arteries. The drug *Viagra* counters impotence by causing nitric oxide to dialate the blood vessels of the penis.

Molting and Metamorphosis in Insects

In insects, hormonal secretions influence both metamorphosis and molting. Prior to molting, neurosecretory cells on the surface of the brain secrete brain hormone, which in turn stimulates a gland in the thorax called the prothoracic gland to produce *molting hormone*, or *ecdysone* (figure 53.21). Another hormone, called juvenile hormone, is produced by bodies near the brain called the corpora allata. For insects to molt, both ecdysone and juvenile hormone must be present, but the level of juvenile hormone determines the result of a particular molt. When juvenile hormone levels are high, the molt produces another larvae. At the late stages of metamorphosis, juvenile hormone levels decrease, and the molt produces a pupa and eventually an adult insect.

Sex steroid hormones from the gonads regulate reproduction, melatonin secreted by the pineal gland helps regulate circadian rhythms, and thymus hormones help regulate the vertebrate immune system. Molting hormone, or ecdysone, and juvenile hormone regulate metamorphosis and molting in insects.

53.1 Regulation is often accomplished by chemical messengers.

- Endocrine glands secrete hormones into the blood, which are then transported to target cells.

- Hormones may be lipophilic, such as the steroid hormones and thyroxine, or polar, such as amine, polypeptide, and glycoprotein hormones.

- Prostaglandins and other paracrine regulatory molecules are produced by one tissue and regulate a different tissue within the same organ.

53.2 Lipophilic and polar hormones regulate their target cells by different means.

- The lipid-soluble hormones enter their target cells, bind to intracellular receptor proteins, and the complex then binds to hormone response elements on the DNA, activating specific genes.

- The polar hormones do not enter their target cells, but instead bind to receptor proteins on the cell membrane and activate second-messenger systems.

- Cyclic AMP (cAMP) is a second messenger produced in response to some hormones; it activates a previously inactive enzyme, a protein kinase, that then activates other enzymes.

53.3 The hypothalamus controls the secretions of the pituitary gland.

- Axons from neurons in the hypothalamus enter the posterior pituitary, carrying ADH and oxytocin; the posterior pituitary stores these hormones and secretes them in response to neural activity.

- The anterior pituitary produces and secretes a variety of hormones, many of which control other endocrine glands; the anterior pituitary, however, is itself controlled by the hypothalamus via releasing and inhibiting hormones secreted by the hypothalamus.

53.4 Endocrine glands secrete hormones that regulate many body functions.

- The thyroid secretes thyroxine and triiodothyronine, which set the basal metabolic rate by stimulating the rate of cell respiration in most cells of the body.

- The adrenal medulla secrets epinephrine; the adrenal cortex secretes cortisol, which regulates glucose balance, and aldosterone, which regulates Na^+ and K^+ balance.

- The β-cells of the islets of Langerhans in the pancreas secrete insulin, which lowers the blood glucose by promoting the movement of blood glucose into tissue cells; glucagon, secreted by the α-cells, raises the blood glucose by promoting the hydrolysis of liver glycogen.

Discussing Key Terms

1. **Hormone**   A hormone is a regulatory molecule secreted by an endocrine gland into the blood. The blood then distributes the hormone molecules to their target cells.

2. **Cytokine**   An autocrine or paracrine regulatory molecule, a cytokine is a molecule that is produced by one cell type and acts on other cells within a given organ.

3. **Prostaglandin**   Prostaglandins are a specific class of paracrine regulators that have a wide variety of functions; they are derived from the membrane lipid arachidonic acid.

4. **Cyclic AMP**   A second messenger in the action of many polar hormones, cyclic AMP is produced by the activation of the enzyme adenylyl cyclase and in turn activates the enzyme protein kinase. This begins a cascade of activation steps that leads to the effects of the hormone.

5. **Thyroxine**   Thyroxine, also called tetraiodothyronine, is the major hormone secreted by the thyroid gland; once in its target cells, it is changed into triiodothyronine, which activates specific genes in the target cells.

6. **Parathyroid Hormone**   Parathyroid hormone directly stimulates the loss of calcium from bone and the reabsorption of Ca^{++} by the kidneys; it also promotes the activation of vitamin D, which stimulates the absorption of Ca^{++} in the small intestine.

Review Questions

1. What is the definition of a hormone? How do hormones reach their target cells? Why are only certain cells capable of being target cells for a particular hormone?

2. How do hormones, paracrine regulators, and autocrine regulators differ from one another?

3. How does epinephrine result in the production of cAMP in its target cells? How does cAMP bring about specific changes inside target cells?

4. What hormones are secreted by the posterior pituitary gland? What function does each hormone serve? Where are these hormones actually *produced*, and how are they transported to the posterior pituitary gland?

5. Why are the hormones of the anterior pituitary gland called tropic hormones? What effect do they have on other endocrine

glands, such as the thyroid gland and the adrenal cortex? What happens to these other endocrine glands if the pituitary is surgically removed?

6. How does the hypothalamus regulate the secretion of the anterior pituitary? What anatomical structures are involved? How are the hypothalamus and anterior pituitary regulated by their target endocrine glands?

7. What hormones are produced by the adrenal cortex? What functions do these hormones serve? What stimulates the secretion of these hormones?

8. What pancreatic hormone is produced when the body's blood glucose level becomes elevated? How does this hormone act to return the level to normal?

Thought Questions

1. Why do you suppose steroid hormones do not employ second messengers as peptide hormones do? Why are there different kinds of second messengers?

2. Two different organs, such as the liver and heart, are sensitive to a particular hormone (such as epinephrine). The cells in both organs have identical receptors for the hormone, and hormone-receptor binding produces the same intracellular second

messenger in both organs. However, the hormone produces different effects in the two organs. Explain how this can happen.

3. Many physiological parameters, such as blood Ca^{++} concentration and blood glucose levels, are controlled by two hormones that have opposite effects. What is the advantage of achieving regulation in this manner instead of by using a single hormone that changes the parameter in one direction only?

Internet Links

Neuroendrocrinology
http://www.neuroguide.com/
A comprehensive index of neurosciences links, with many devoted to interactions between the endocrine system and the nervous system—or play NEUROROULETTE and go to a link selected at random from the index.

How Intracellular Hormone Receptors Work
http://www.ks.uiuc.edu/Research/pro_DNA/ster_horm_rec/
Study the interaction of intracellular hormone receptors with DNA as well as their interaction with hormones on this site from the Theoretical Biophysics Group at the University of Illinois.

Endocrine Disorders
http://www.niddk.nih.gov/health/endo/endo.htm
Explanations of endocrine disorders and diseases, and a host of links, provided on the home page of the NATIONAL INSTITUTE OF DIABETES AND DIGESTIVE AND KIDNEY DISEASES.

Getting a Job in Endocrinology
http://www.faseb.org/aps/
The AMERICAN PHYSIOLOGICAL SOCIETY provides a good overview of career options available in endocrinology and other aspects of physiology.

For Further Reading

Atkinson, M. and N. MacLaren: "What Causes Diabetes?" *Scientific American*, July 1990, pages 62–71. For insulin-dependent diabetic patients, the answer is an autoimmune ambush of the body's insulin-producing cells.

Davis, D. L. and H. L. Bradlow: "Can Environmental Estrogens Cause Breast Cancer?" *Scientific American*, October 1995, pages 166–72. Non-naturally occurring estrogens ("xenoestrogens") in the environment may be contributing to the high rates of breast cancer in women.

Hoberman, J. and C. Yesalis: "The History of Synthetic Testosterone," *Scientific American*, February 1995, pages 76–81. Long banned in sports, these anabolic steroids greatly enhance athletic performance.

Kalin, N.: "The Neurobiology of Fear," *Scientific American*, vol. 268, May 1993, pages 94–99. The hormonal response to anxiety, fear, and stress is examined, along with the localized regions of the nervous system in which the activity occurs.

Lacy, P: "Treating Diabetes with Transplanted Cells," *Scientific American*, July 1995, pages 50–58. Implanting islet cells of the pancreas can potentially cure diabetes—if a way can be found to avoid immune attacks.

Rosen, O.: "After Insulin Binds," *Science*, September 1987, pages 1452–58. A review of how the insulin receptor works—it is much like growth factor receptors and some cancer-inducing genes.

54

The Immune System

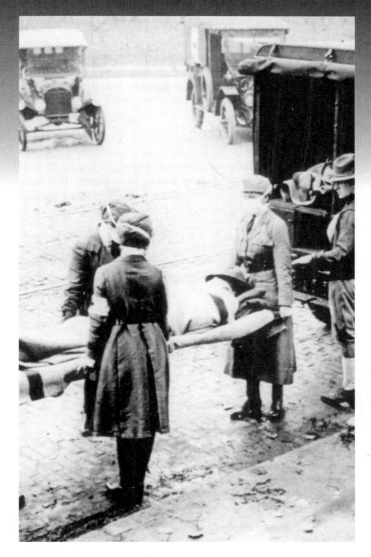

FIGURE 54.1
The influenza epidemic of 1918–1919 killed 22 million people in 18 months. With 25 million Americans infected, the Red Cross often worked around the clock.

Concept Outline

When you consider how animals defend themselves, it is natural to think of dinosaurs, turtles, armadillos, and other animals covered like tanks with heavy plates of armor. However, armor offers no protection against the greatest dangers vertebrates face—microorganisms and viruses. We live in a world awash with attackers too tiny to see with the naked eye, and no vertebrate could long withstand their onslaught unprotected. We survive because we have evolved a variety of very effective defenses against this constant attack. As we review these defenses, it is important to keep in mind that they are far from perfect. Some 22 million Americans and Europeans died from influenza over an 18-month period in 1918–1919 (figure 54.1), and more than 3 million people will die of malaria *this year*. Attempts to improve our defenses against infection are among the most active areas of scientific research today.

Nonspecific Defense by the Skin

Vertebrates employ a multilevel defense that resembles the way medieval cities were defended against attack by outsiders. Like walls and moats around a city, the skin and mucous membranes constitute a very effective first line of defense against attack by viruses and microorganisms.

If an invader gains entrance to the vertebrate body, a second line of defense comes into play. Within the body, circulating cells function as roaming patrols that kill foreign cells with a battery of **nonspecific immune defenses,** including chemicals. As we will see, these defensive cells employ a negative test similar to the one invertebrates use, but this one cannot be foiled by "copycat" foreign cells. The surfaces of most vertebrate cells possess glycoproteins produced by a group of genes called the **major histocompatibility complex (MHC).** These glycoproteins are called **MHC proteins** or **human leukocyte antigens (HLA).** Clinical tissue typing is usually performed on leukocytes. The genes encoding the MHC proteins are highly polymorphic (have many forms), so that very few individuals in a population possess the same set of alleles. As a result, the MHC proteins on the tissue cells serve as self markers that enable the individual's immune system to distinguish its cells from foreign cells, an ability called **self-versus-nonself recognition.**

Vertebrates employ a third, very powerful defensive strategy that relies on a *positive* test to identify foreign cells as well as to detect cancer cells. This defense involves cells with surface receptor proteins that recognize specific "nonself" molecules, and thus is termed the **specific immune defense.** The characteristics of the specific immune response will be described in later sections.

The skin (figure 54.2) is the largest organ of the vertebrate body, accounting for 15% of an adult human's total weight. Packed among the skin cells are many other kinds of cells: one square centimeter of human skin, about the area covered by a dime, contains 10 hairs and muscles, 100 sweat glands, 15 oil glands, 3 blood vessels, and 200 nerve endings, including 25 pressure sensors, 12 heat sensors, and 2 cold sensors.

The Skin as a Barrier to Infection

The skin not only defends the body by providing a nearly impenetrable barrier, but also reinforces this defense with chemical weapons on the surface. Oil and sweat glands give the skin's surface a pH of 3 to 5, acidic enough to inhibit the growth of many microorganisms. Sweat also contains the enzyme **lysozyme,** which digests bacterial cell walls. In addition to defending the body against invasion by viruses and microorganisms, the skin prevents excessive loss of water to the air through evaporation.

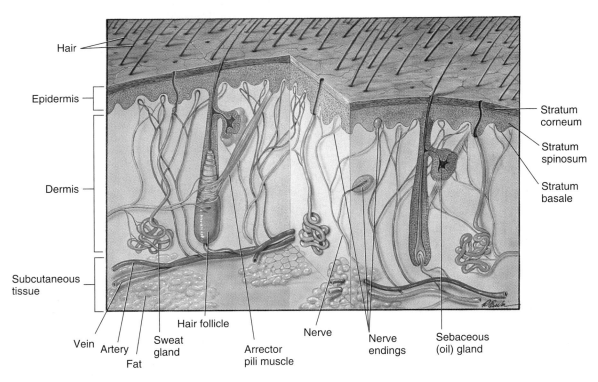

FIGURE 54.2
Human skin. The skin serves as the first barrier to invasion by microorganisms.

The epidermis of skin is approximately 10 to 30 cells thick, about as thick as this page. The outer layer, called the stratum corneum, contains cells that are continuously abraded, injured, and worn by friction and stress during the body's many activities. The body deals with this damage not by repairing the cells, but by replacing them. Cells are shed continuously from the stratum corneum and are replaced by new cells produced in the innermost layer of the epidermis, the stratum basale, which contains some of the most actively dividing cells in the vertebrate body. The cells formed in this layer migrate upward and enter a broad intermediate stratum spinosum layer. As they move upward they form the protein keratin, which makes skin tough and water-resistant. These new cells eventually arrive at the stratum corneum, where they normally remain for about a month before they are shed and replaced by newer cells from below. Psoriasis, which afflicts some 4 million Americans, is a chronic skin disorder in which epidermal cells are replaced every 3 to 4 days, about eight times faster than normal.

The dermis of skin is 15 to 40 times thicker than the epidermis. It provides structural support for the epidermis and a matrix for the many blood vessels, nerve endings, muscles, and other structures situated within skin. The wrinkling that occurs as we grow older takes place in the dermis, and the leather used to manufacture belts and shoes is derived from very thick animal dermis.

The layer of subcutaneous tissue below the dermis contains primarily adipose cells. These cells act as shock absorbers and provide insulation, conserving body heat. Subcutaneous tissue varies greatly in thickness in different parts of the body. It is nonexistent in the eyelids, is a half-centimeter thick or more on the soles of the feet, and may be much thicker in other areas of the body, such as the buttocks and thighs.

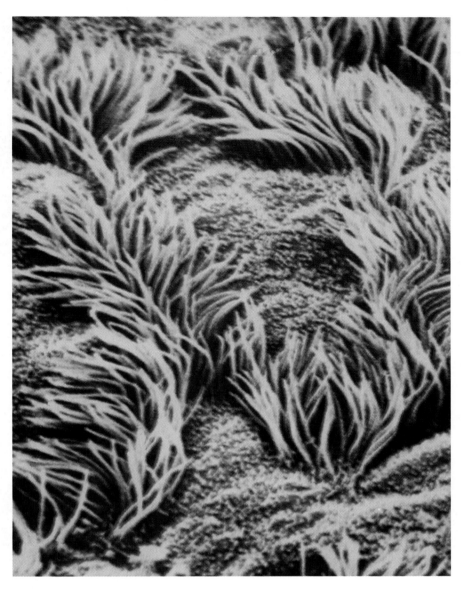

FIGURE 54.3
Cilia line the respiratory tract. Cilia (3500×) cover the epithelial linings of the nasal cavity, trachea, and bronchi. These cilia filter foreign bodies out of inhaled air and sweep them toward the glottis for swallowing.

Other External Surfaces

In addition to the skin, two other potential routes of entry by viruses and microorganisms must be guarded: the *digestive tract* and the *respiratory tract*. Microbes are present in food, but many are killed by saliva (which also contains lysozyme), by the very acidic environment of the stomach, and by digestive enzymes in the intestine. Microorganisms are also present in inhaled air. The cells lining the smaller bronchi and bronchioles secrete a layer of sticky mucus that traps most microorganisms before they can reach the warm, moist lungs, which would provide ideal breeding grounds for them. Other cells lining these passages have cilia (figure 54.3) that continually sweep the mucus toward the glottis. There it can be swallowed, carrying potential invaders out of the lungs and into the digestive tract.

> The surface defenses of the body consist of the skin and the mucous membranes lining the digestive and respiratory tracts, which eliminate many microorganisms before they can invade the body tissues.

Cells and Proteins That Kill Invading Microbes

The surface defenses of the vertebrate body are very effective but are occasionally breached, allowing invaders to enter the body. At this point, the body uses a host of non-specific cellular and chemical devices to defend itself. These devices all have one property in common: they respond to *any* microbial infection without pausing to determine the invader's identity.

Cells That Kill Invading Microbes

Perhaps the most important of the vertebrate body's nonspecific defenses are cells that circulate through the body and attack invading microbes within tissues. There are three basic kinds of these cells, and each kills invading microorganisms differently.

Macrophages ("big eaters") are large, irregularly shaped cells that kill microbes one at a time by ingesting them through *phagocytosis*, much as an amoeba ingests a food particle (figure 54.4). Within the macrophage, the membrane-bound vacuole containing the bacterium fuses with a lysosome. Fusion activates lysosomal enzymes that kill the microbe by liberating large quantities of oxygen free-radicals. Macrophages also engulf viruses, cellular debris, and dust particles in the lungs. Macrophages circulate continuously in the extracellular fluid, and their phagocytic actions supplement those of the specialized phagocytic cells that are part of the structure of the liver, spleen, and bone marrow. In response to an infection, monocytes (a type of leukocyte) from the blood squeeze through capillaries to enter the connective tissues. There, at the site of the infection, the monocytes are transformed into additional macrophages.

Neutrophils are leukocytes that, like macrophages, ingest and kill bacteria by phagocytosis. However, while macrophages kill only one invading cell at a time but may repeat the process many times, neutrophils release chemicals (some of which are identical to household bleach) that kill other bacteria in the neighborhood as well as neutrophils themselves.

Natural killer cells do not attack invading microbes directly. Instead, they kill cells of the body that have been infected with viruses. They kill not by phagocytosis, but rather by creating a pore in the plasma membrane of the target cell

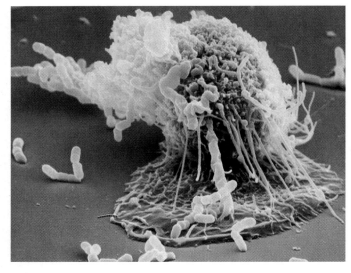

FIGURE 54.4

A macrophage in action (1800×). In this scanning electron micrograph, a macrophage is "fishing" with long, sticky cytoplasmic extensions. Bacterial cells that come in contact with the extensions are drawn toward the macrophage and engulfed.

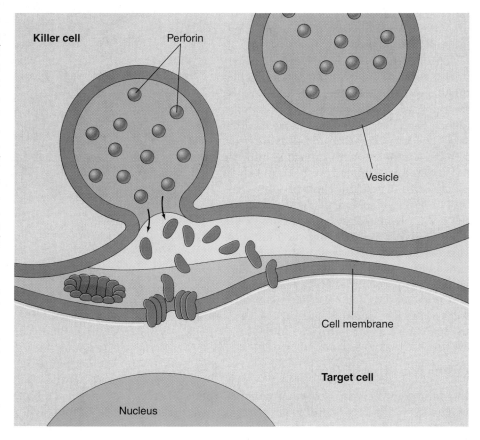

FIGURE 54.5

How natural killer cells kill target cells. The initial event is the tight binding of the killer cell to the target cell. Binding initiates a chain of events within the killer cell in which vesicles loaded with *perforin* molecules move to the plasma membrane and disgorge their contents into the intercellular space over the target cell. The perforin molecules insert into the plasma membrane of the target cell like the staves of a barrel, forming a pore that admits water and ruptures the cell.

(figure 54.5). This pore allows water to rush into the target cell, which then swells and bursts (figure 54.6). Natural killer cells also attack cancer cells, often before the cancer cells have had a chance to develop into a detectable tumor. The vigilant surveillance by natural killer cells is one of the body's most potent defenses against cancer.

These three kinds of cells can distinguish the body's cells (self) from foreign cells (nonself) because the body's cells contain self-identifying MHC proteins. When the body's defensive cells fail to make the self-versus-nonself distinction correctly, they may attack the body's own tissues. Diseases resulting from this failure are known as *autoimmune diseases*.

Proteins That Kill Invading Microbes

The cellular defenses of vertebrates are complemented by a very effective chemical defense called the complement system. This system consists of approximately 20 different proteins that circulate freely in the blood plasma. When they encounter a bacterial or fungal cell wall, these proteins aggregate to form a *membrane attack complex* that inserts itself into the foreign cell's plasma membrane, forming a pore like that produced by natural killer cells (figure 54.7). Water enters the foreign cell through this pore, causing the cell to swell and burst. Aggregation of the complement proteins is also triggered by the binding of antibodies to invading microbes, as we will see in a later section.

The proteins of the complement system can augment the effects of other body defenses. Some amplify the inflammatory response (discussed next) by stimulating histamine release; others attract phagocytes to the area of infection; and still others coat invading microbes, roughening the microbes' surfaces so that phagocytes may attach to them more readily.

Another class of proteins that play a key role in body defense are interferons. There are three major categories of interferons: *alpha*, *beta*, and *gamma*. Almost all cells in the body make alpha and beta interferons. These polypeptides act as messengers that protect other cells in the vicinity from viral infection. The viruses are still able to penetrate the other cells, but the ability of the viruses to replicate and assemble new virus particles is inhibited. Gamma interferon is produced only by particular lymphocytes and natural killer cells. The secretion of gamma interferon by these cells is part of the immunological defense against infection and cancer, as we will describe later.

> A patrolling army of macrophages, neutrophils, and natural killer cells attack and destroy invading viruses and bacteria and eliminate infected cells. In addition, a system of proteins called complement may be activated to destroy foreign cells, and body cells infected with a virus secrete proteins called interferons that protect other cells.

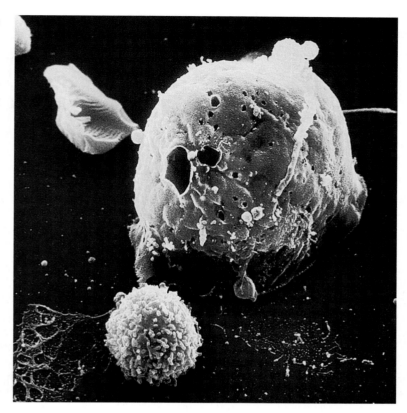

FIGURE 54.6
Death of a tumor cell. A natural killer cell has attacked this cancer cell (larger cell in the background), punching a hole in its plasma membrane. Water has rushed in, making the cell swell. Soon it will burst.

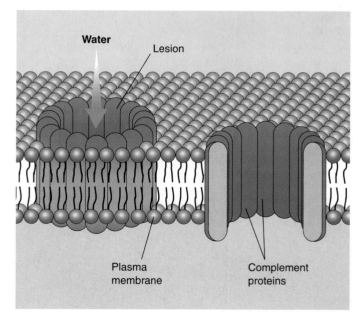

FIGURE 54.7
How complement creates a hole in a cell membrane. As the diagram shows, the complement proteins form a complex transmembrane pore resembling the perforin-lined pores formed by natural killer cells.

The Inflammatory Response

The inflammatory response is a localized, nonspecific response to infection. Infected or injured cells release chemical alarm signals, most notably histamine and prostaglandins. These chemicals promote the dilation of local blood vessels, which increases the flow of blood to the site of infection or injury and causes the area to become red and warm. They also increase the permeability of capillaries in the area, producing the edema (tissue swelling) so often associated with infection. The more permeable capillaries allow phagocytes (monocytes and neutrophils) to migrate from the blood to the extracellular fluid, where they can attack bacteria. Neutrophils arrive first, spilling out chemicals that kill the bacteria in the vicinity (as well as tissue cells and themselves); the *pus* associated with some infections is a mixture of dead or dying pathogens, tissue cells, and neutrophils. Monocytes follow, become macrophages and engulf pathogens and the remains of the dead cells (figure 54.8). In some autoimmune diseases such as rheumatoid arthritis, inflammation occurs in the absence of infection.

The Temperature Response

Macrophages that encounter invading microbes release a regulatory molecule called interleukin-1, which is carried by the blood to the brain. Interleukin-1 and other pyrogens (Greek *pyr*, "fire") such as bacterial endotoxins cause neurons in the hypothalamus to raise the body's temperature several degrees above the normal value of 37 °C (98.6 °F). The elevated temperature that results is called a fever.

Experiments with lizards, which regulate their body temperature by moving to warmer or colder locations, demonstrate that infected lizards choose a warmer environment—they give themselves a fever! Further, if lizards are prevented from elevating their body temperature, they have a slower recovery from their infection. Fever contributes to the body's defense by stimulating phagocytosis and causing the body to reduce blood levels of iron, which bacteria need in large amounts to grow. However, very high fevers are hazardous because excessive heat may inactivate critical enzymes. In general, temperatures greater than 39.4 °C (103 °F) are considered dangerous for humans, and those greater than 40.6 °C (105 °F) are often fatal.

Histamine and other chemicals are released during an inflammation, producing redness, warmth, and edema. Neutrophils, monocytes, and macrophages attack the invading microbes and contribute to the pus. The infection may also cause the release of pyrogens, producing a fever.

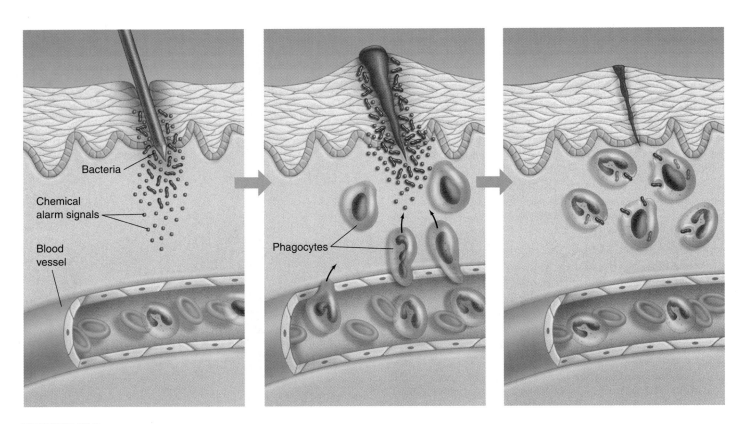

FIGURE 54.8
The events in a local inflammation. When an invading microbe has penetrated the skin, chemicals, such as histamine and prostaglandins, cause nearby blood vessels to dilate. Increased blood flow brings a wave of phagocytic cells, which attack and engulf invading bacteria.

Characteristics of the Immune Response

Few of us pass through childhood without contracting some sort of infection. Measles, for example, is an illness that many of us experience before we reach our teens. It is a disease of childhood, because most of us contract it as children and *never catch it again*. Once you have had the measles, you are usually immune to the disease. Specific immune defense mechanisms provide this immunity.

Terms Used in Describing Specific Immunity

An **antigen** is a molecule that provokes a specific immune response. Antigens are large, complex molecules such as proteins; they are generally foreign to the body, usually belonging to bacteria and viruses. A large antigen may have several parts, and each stimulate a different specific immune response. In this case, the different parts are known as **antigenic determinant sites,** and each serves as a different antigen.

Particular lymphocytes have receptor proteins on their surfaces that recognize an antigen and direct a specific immune response against either the antigen or the cell that carries the antigen. Sometimes the immune system fails in self-nonself recognition; if that occurs the immune system regards its own molecules as antigens and produces an autoimmune disease, as previously mentioned.

Particular lymphocytes respond to antigens by producing proteins called **antibodies.** Antibody proteins are secreted into the blood and other body fluids and thus provide **humoral immunity.** (The term *humor* here is used in its ancient sense, referring to a body fluid). Other lymphocytes do not secrete antibodies but instead directly attack the cells that carry the specific antigens. These cells are thus described as producing **cell-mediated immunity.**

The specific immune responses protect the body in two ways. First, an individual can gain immunity by being exposed to a *pathogen* (disease-causing agent) and perhaps getting the disease. This is *acquired immunity*, such as the resistance to the measles that you acquire after having the disease in childhood. Another term for this process is **active immunity.** Second, an individual can gain immunity by obtaining the antibodies from another individual. This happened to you before you were born, with antibodies made by your mother being transferred to you across the placenta. Immunity gained in this way is called **passive immunity.**

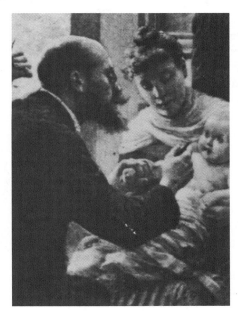

FIGURE 54.9
The birth of immunology. This famous painting shows Edward Jenner inoculating patients with cowpox in the 1790s and thus protecting them from smallpox.

Discovery of the Immune Response

In 1796, an English country doctor named Edward Jenner carried out an experiment that marks the beginning of the study of immunology. Smallpox was a common and deadly disease in those days. Jenner observed, however, that milkmaids who had caught a much milder form of "the pox" called cowpox (presumably from cows) rarely caught smallpox. Jenner set out to test the idea that cowpox conferred protection against smallpox. He infected people with the mild cowpox (figure 54.9), and as he had predicted, many of them became immune to smallpox.

We now know that smallpox and cowpox are caused by two different but similar viruses. Jenner's patients who were injected with the cowpox virus mounted a defense that was also effective against a later infection of the smallpox virus. Jenner's procedure of injecting a harmless microbe in order to confer resistance to a dangerous one is called **vaccination.** Modern attempts to develop resistance to malaria, herpes, and other diseases often involve delivering antigens with a harmless vaccinia virus that is related to the cowpox virus.

Many years passed before anyone learned how exposure to one disease can confer resistance to another. A key step toward answering this question was taken more than a half-century later by the famous French scientist Louis Pasteur. Pasteur was studying fowl cholera, and he isolated a culture of bacteria from diseased chickens that would produce the disease if injected into healthy birds. Before departing on a two-week vacation, he accidentally left his bacterial culture out on a shelf. When he returned, he injected this old culture into healthy birds and found that it had been weakened; the injected birds became only slightly ill and then recovered. Surprisingly, however, those birds did not get sick when subsequently infected with fresh fowl cholera. They remained healthy even if given massive doses of active fowl cholera bacteria that did produce the disease in control chickens. Clearly, something about the bacteria could elicit immunity as long as the bacteria did not kill the animals first. We now know that molecules protruding from the surfaces of the bacterial cells evoked active immunity in the chickens.

Antigens are molecules, usually foreign, that provoke a specific immune attack. This immune attack may involve secreted proteins called antibodies, or it may invoke a cell-mediated attack.

Evolution of the Immune System

All organisms possess mechanisms to protect themselves from the onslaught of smaller organisms and viruses. Bacteria defend against viral invasion by means of *restriction endonucleases*, enzymes that degrade any foreign DNA lacking the specific pattern of DNA methylation characteristic of that bacterium. Multicellular organisms face a more difficult problem in defense because their bodies often take up whole viruses, bacteria, or fungi instead of naked DNA.

Invertebrates

Invertebrate animals solve this problem by marking the surfaces of their cells with proteins that serve as "self" labels. Special amoeboid cells in the invertebrate attack and engulf any invading cells that lack such labels. By looking for the absence of specific markers, invertebrates employ a *negative* test to recognize foreign cells and viruses. This method provides invertebrates with a very effective surveillance system, although it has one great weakness: any microorganism or virus with a surface protein resembling the invertebrate self marker will not be recognized as foreign. An invertebrate has no defense against such a "copycat" invader.

In 1882, Russian zoologist Elie Metchnikoff became the first to recognize that invertebrate animals possess immune defenses. On a beach in Sicily, he collected the tiny transparent larva of a common starfish. Carefully he pierced it with a rose thorn. When he looked at the larva the next morning, he saw a host of tiny cells covering the surface of the thorn as if trying to engulf it (figure 54.10). The cells were attempting to defend the larva by ingesting the invader by phagocytosis (described in chapter 6). For this discovery of what came to be known as the **cellular immune response,** Metchnikoff was awarded the 1908 Nobel Prize in medicine, along with Paul Ehrlich for his work on the other major part of the immune defense, the antibody or **humoral immune response.** The invertebrate immune response shares several elements with the vertebrate immune response.

Phagocytes. All animals possess phagocytic cells that attack invading microbes. These phagocytic cells travel through the animal's circulatory system or circulate within the fluid-filled body cavity. In simple animals like sponges that lack either a circulatory system or a body cavity, the phagocytic cells circulate among the spaces between cells.

Distinguishing Self from Nonself. The ability to recognize the difference between cells of one's own body and those of another individual appears to have evolved early in the history of life. Sponges, thought to be the oldest animals, attack grafts from other sponges, as do insects and starfish. None of these invertebrates, however, exhibit any evidence of immunological memory; apparently, the antibody-based humoral immune defense did not evolve until the vertebrates.

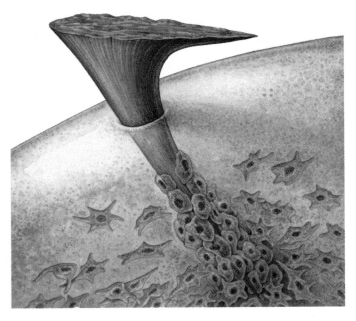

FIGURE 54.10
Discovering the cellular immune response in invertebrates. In a Nobel-Prize-winning experiment, the Russian zoologist Metchnikoff pierced the larva of a starfish with a rose thorn and the next day found tiny phagocytic cells covering the thorn.

Complement. While invertebrates lack complement, many arthropods (including crabs and a variety of insects) possess an analogous nonspecific defense called the prophenyloxidase (proPO) system. Like the vertebrate complement defense, the proPO defense is activated as a cascade of enzyme reactions, the last of which converts the inactive protein prophenyloxidase into the active enzyme phenyloxidase. Phenyloxidase both kills microbes and aids in encapsulating foreign objects.

Lymphocytes. Invertebrates also lack lymphocytes, but annelid earthworms and other invertebrates do possess lymphocyte-like cells that may be evolutionary precursors of lymphocytes.

Antibodies. All invertebrates possess proteins called lectins that may be the evolutionary forerunners of antibodies. Lectins bind to sugar molecules on cells, making the cells stick to one another. Lectins isolated from sea urchins, mollusks, annelids, and insects appear to tag invading microorganisms, enhancing phagocytosis. The genes encoding vertebrate antibodies are part of a very ancient gene family, the immunoglobulin superfamily. Proteins in this group all have a characteristic recognition structure called the Ig fold. The fold probably evolved as a self-recognition molecule in early metazoans. Insect immunoglobulins have been described in moths, grasshoppers, and flies that bind to

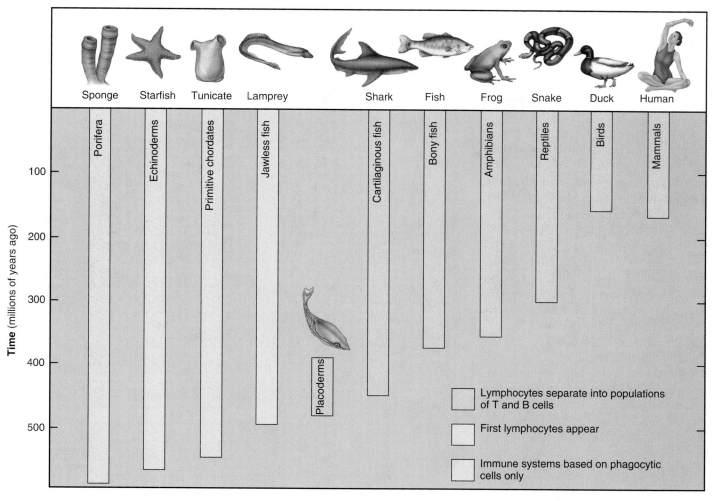

FIGURE 54.11

How immune systems evolved in vertebrates. Lampreys were the first vertebrates to possess an immune system based on lymphocytes, although distinct B and T cells did not appear until the jawed fishes evolved. By the time sharks and other cartilaginous fish appeared, the vertebrate immune response was fully formed.

microbial surfaces and promote their destruction by phagocytes. The antibody immune response appears to have evolved from these earlier, less complex systems.

Vertebrates

The earliest vertebrates of which we have any clear information, the jawless lampreys that first evolved some 500 million years ago, possess an immune system based on lymphocytes. At this early stage of vertebrate evolution, however, lampreys lack distinct populations of B and T cells such as found in all higher vertebrates (figure 54.11).

With the evolution of fish with jaws, the modern vertebrate immune system first arose. These early fishes, including placoderms and spiny fishes, are now extinct. The oldest surviving group of jawed fishes are the sharks, which evolved some 450 million years ago. By then, the vertebrate immune defense had fully evolved. Sharks have an immune response much like that seen in mammals, with a cellular response

carried out by T-cell lymphocytes and an antibody-mediated humoral response carried out by B cells. The similarities of the cellular and humoral immune defenses are far more striking than the differences. Both sharks and mammals possess a thymus that produces T cells and a spleen that is a rich source of B cells. Four hundred fifty million years of evolution did little to change the antibody molecule—the amino acid sequences of shark and human antibody molecules are very similar. The most notable difference between sharks and mammals is that their antibody-encoding genes are arrayed somewhat differently.

The sophisticated two-part immune defense of mammals evolved about the time jawed fishes appeared. Before then, animals utilized a simpler immune defense based on mobile phagocytic cells. Many of the elements of today's mammalian immune response can be recognized in analogous systems in invertebrates.

Cells of the Vertebrate Immune System

The immune defense mechanisms of the body involve the actions of white blood cells, or leukocytes. Leukocytes include neutrophils, eosinophils, basophils, and monocytes, all of which are phagocytic, as well as two types of lymphocytes (*T cells* and *B cells*), which are not phagocytic but are critical to the specific immune response (table 54.1). T cells direct the cell-mediated response, B cells the humoral response.

After their origin in the bone marrow, T cells migrate to the thymus (hence the designation "T"), a gland just above the heart. There they develop the ability to identify microorganisms and viruses by the antigens exposed on their surfaces. Tens of millions of different T cells are made, each specializing in the recognition of one particular antigen. No invader can escape being recognized by at least a few T cells. There are four principal kinds of T cells: inducer T cells oversee the development of T cells in the thymus; helper T cells (often symbolized T_H) initiate the immune response; cytotoxic ("cell-poisoning") T cells (often symbolized T_C) lyse cells that have been infected by viruses; and suppressor T cells terminate the immune response.

Unlike T cells, B cells do not travel to the thymus; they complete their maturation in the bone marrow. (B cells are so named because they were originally characterized in a region of chickens called the bursa.) From the bone marrow, B cells are released to circulate in the blood and lymph. Individual B cells, like T cells, are specialized to recognize particular foreign antigens. When a B cell encounters the antigen to which it is targeted, it begins to divide rapidly, and its progeny differentiate into plasma cells and memory cells. Each plasma cell is a miniature factory producing antibodies that stick like flags to that antigen wherever it occurs in the body, marking any cell bearing the antigen for destruction. The immunity that Pasteur observed resulted from such antibodies and from the continued presence of the B cells that produced them.

Table 54.1   Cells of the Immune System

	Cell Type	Function
T cell	Helper T cell	Commander of the immune response; detects infection and sounds the alarm, initiating both T-cell and B-cell responses
	Inducer T cell	Not involved in the immediate response to infection; mediates the maturation of other T cells in the thymus
	Cytotoxic T cell	Detects and kills infected body cells; recruited by helper T cells
	Suppressor T cell	Dampens the activity of T and B cells, scaling back the defense after the infection has been checked
B cell	B cell	Precursor of plasma cell; specialized to recognize specific foreign antigens
Plasma cell	Plasma cell	Biochemical factory devoted to the production of antibodies directed against specific foreign antigens
Mast cell	Mast cell	Initiator of the inflammatory response, which aids the arrival of leukocytes at a site of infection; secretes histamine and is important in allergic responses
Monocyte	Monocyte	Precursor of macrophage
Macrophage	Macrophage	The body's first cellular line of defense; also serves as antigen-presenting cell to B and T cells and engulfs antibody-covered cells
Killer cell	Killer cell	Recognizes and kills infected body cells; natural killer (NK) cell detects and kills cells infected by a broad range of invaders; killer (K) cell attacks only antibody-coated cells

Active Immunity through Clonal Selection

As we will discuss in later sections, B and T cells have receptors on their cell surfaces that recognize and bind to specific antigens. When a particular antigen enters the body, it must, by chance, encounter the specific lymphocyte with the appropriate receptor in order to provoke an immune response. The first time a pathogen invades the body, there are only a few B or T cells that may have the receptors that can recognize the invader's antigens. Binding of the antigen to its receptor on the lymphocyte surface, however, stimulates cell division and produces a *clone* (a population of genetically identical cells). This process is known as **clonal selection.** In this first encounter, there are only a few cells that can mount an immune response and the response is relatively weak. This is called a **primary immune response** (figure 54.12).

If the primary immune response involves B cells, some become plasma cells that secrete antibodies, and some become memory cells. Because a clone of memory cells specific for that antigen develops after the primary response, the immune response to a second infection by the same pathogen is swifter and stronger. The next time the body is invaded by the same pathogen, the immune system is ready. As a result of the first infection, there is now a large clone of lymphocytes that can recognize that pathogen (figure 54.13). This more effective response, elicited by subsequent exposures to an antigen, is called a **secondary immune response.**

Memory cells can survive for several decades, which is why people rarely contract measles a second time after they have had it once. Memory cells are also the reason that vaccinations are effective. The viruses causing childhood diseases have surface antigens that change little from year to year, so the same antibody is effective for decades. Other diseases, such as influenza, are caused by viruses whose protein coat-specifying genes mutate rapidly. This rapid genetic change causes new strains to appear every year or so that are not recognized by memory cells from previous infections.

T cells develop in the thymus, while B cells develop in the bone marrow. When a B cell encounters a specific antigen, it gives rise to plasma cells that produce antibodies. Active immunity is produced by clonal selection and expansion. This occurs because interaction of an antigen with its receptor on the lymphocyte surface stimulates cell division, so that more lymphocytes are available to combat subsequent exposures to the same antigen.

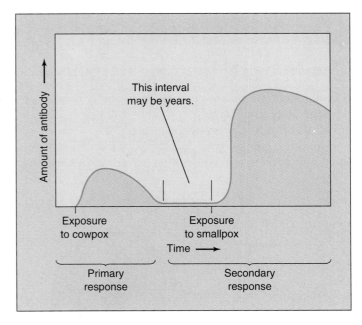

FIGURE 54.12
The development of active immunity. Immunity to smallpox in Jenner's patients occurred because their inoculation with cowpox stimulated the development of lymphocyte clones with receptors that could bind not only to cowpox but also to smallpox antigens. As a result of clonal selection, a second exposure, this time to smallpox, stimulates the immune system to produce large amounts of the antibody more rapidly than before.

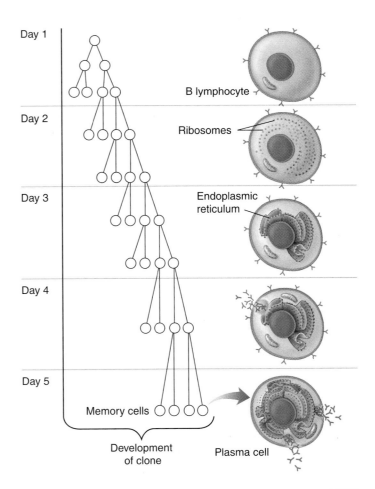

FIGURE 54.13
The clonal selection theory of active immunity. In response to interaction with an antigen that binds specifically to its surface receptors, a B cell divides many times to produce a clone of B cells. Some of these become plasma cells that secrete antibodies for the primary response, while others become memory cells that await subsequent exposures to the antigen for the mounting of a secondary immune response.

The Humoral Immune Response

The humoral immune response, carried out by B cells, protects the body from bacteria and other invading cells, by labeling these cells for destruction.

The B cell lymphocytes that carry out this response produce and secrete antibodies that circulate in the blood plasma, lymph, and other extracellular fluids—hence the term *humoral* (relating to fluid) is used in describing their immune response. In response to antigen exposure, a B cell divides to produce plasma cells that serve as short-lived antibody factories, and to produce long-lived memory cells, as previously described (figure 54.14).

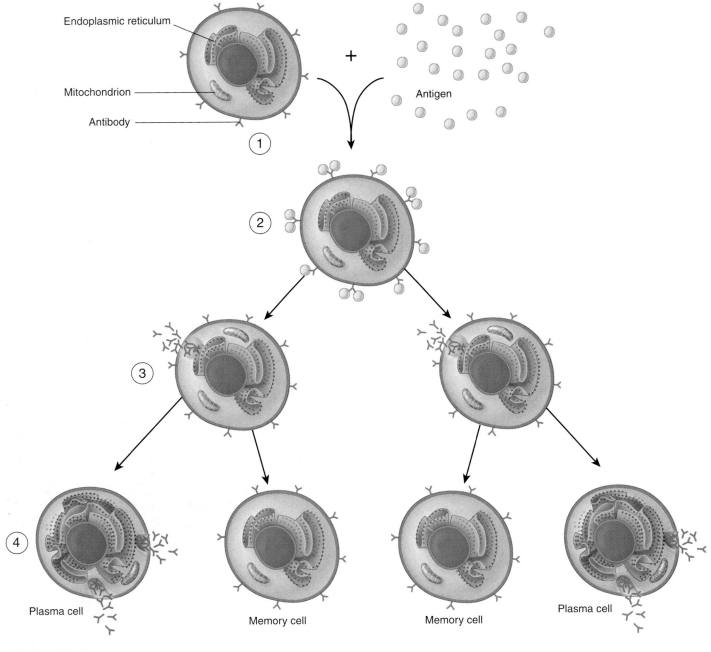

FIGURE 54.14
The formation of plasma cells and memory cells. In response to an antigen, B cells divide to produce plasma cells and memory cells. Plasma cells have an extensive rough endoplasmic reticulum and are specialized for the production of antibody proteins.

Antibodies are proteins in a class called **immunoglobu-lins** (abbreviated Ig), which is divided into subclasses based on the structures and functions of the antibodies. The different immunoglobulin subclasses are as follows:

1. **IgM.** This is the first type of antibody to be secreted during the primary response and they serve as receptors on the lymphocyte surface. These antibodies also promote agglutination reactions (causing antigen-containing particles to stick together, or agglutinate).
2. **IgG.** This is the major form of antibody secreted in a secondary response and the major one in the blood plasma.
3. **IgD.** These antibodies serve as receptors for antigens on the B-cell surface. Their other functions are unknown.
4. **IgA.** This is the form of antibody in external secretions, such as saliva and mother's milk.
5. **IgE.** This form of antibodies promotes the release of histamine and other agents that produce allergic symptoms, such as those of hay fever.

Each B cell has on its surface about 100,000 IgM or IgD receptors. Unlike the receptors on T cells, which bind only to antigens presented by certain cells (described later), B receptors can bind to *free* antigens. This provokes a primary response in which antibodies of the IgM class are secreted, and also stimulates cell division and clonal expansion. Upon subsequent exposure, the plasma cells secrete large amounts of antibodies that are generally of the IgG class. Although plasma cells live only a few days, they produce a vast number of antibodies. In fact, antibodies constitute about 20% by weight of the total protein in blood plasma. Production of IgG antibodies peaks after about three weeks (figure 54.15).

When IgM (and to a lesser extent IgG) antibodies bind to antigens on a cell, they cause the aggregation of complement proteins. As we mentioned earlier, these proteins form a pore that pierces the plasma membrane of the infected cell (figure 54.16), allowing water to enter and causing the cell to burst. In contrast, when IgG antibodies bind to antigens on a cell, they serve as markers that stimulate phagocytosis by macrophages. Since certain complement proteins attract phagocytic cells, activation of complement is generally accompanied by increased phagocytosis. Notice that antibodies don't kill invading pathogens directly; rather, they cause destruction of the pathogens by activating the complement system and by targeting the pathogen for attack by phagocytic cells.

In the humoral immune response, B cells recognize antigens and divide to produce plasma cells, producing large numbers of circulating antibodies directed against those antigens. IgM antibodies are produced first, and they activate the complement system. Thereafter, IgG antibodies are produced and promote phagocytosis.

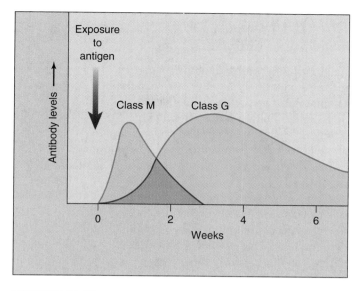

FIGURE 54.15
Class M and class G antibodies. The first antibodies produced in the humoral immune response are class M antibodies, which are very effective at activating the complement system. This initial wave of antibody production peaks after about one week and is followed by a far more extended production of class G antibodies.

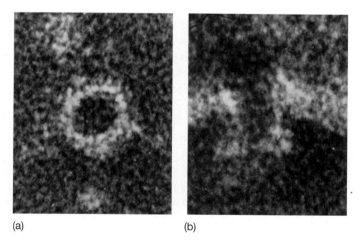

(a)   (b)

FIGURE 54.16
Electron micrographs of complement lesions in the plasma membrane of a cell (375,000×). The lesions appear as transmembrane channels when viewed (a) from above and (b) in cross-section. The micrographs are negatively stained, so that the complement proteins appear white and the pore through the channel appears black.

Antibodies

Structure of Antibodies

Each antibody molecule consists of two identical short polypeptides, called **light chains,** and two identical long polypeptides, called **heavy chains** (figure 54.17). The four chains in an antibody molecule are held together by disulfide (—S—S—) bonds, forming a Y-shaped molecule (figure 54.18).

Comparing the amino acid sequences of different antibody molecules shows that the specificity of antibodies for antigens resides in the two arms of the Y, which have a variable amino acid sequence. The amino acid sequence of the polypeptides in the stem of the Y is constant within a given class of immunoglobulins. Most of the sequence variation between antibodies of different specificity is found in the terminal half of each arm. Here, six small *hypervariable segments* (three in the light chain and three in the heavy chain) form a cleft that acts as the binding site for the antigen. Both arms always have exactly the same cleft and so bind to the same antigen.

Antibodies with the same hypervariable segments have identical clefts and therefore recognize the same antigen, but they may differ in the stem portions of the antibody molecule. The stem is formed by the so-called "constant" regions of the heavy chains. In mammals there are five different classes of heavy chain that form five classes of immunoglobulins: IgM, IgG, IgA, IgD, and IgE. We have already discussed the roles of IgM and IgG antibodies in the humoral immune response.

IgE antibodies bind to **mast cells.** The heavy-chain stems of the IgE antibody molecules insert into receptors on the mast cell plasma membrane, in effect creating B receptors on the mast cell surface. When these cells encounter the specific antigen recognized by the arms of the antibody, they initiate the inflammatory response by releasing histamine. The resulting vasodilation and increased capillary permeability enable lymphocytes, macrophages, and complement proteins to more easily reach the site where the mast cell encountered the antigen. Allergy will be discussed in more detail in a later section.

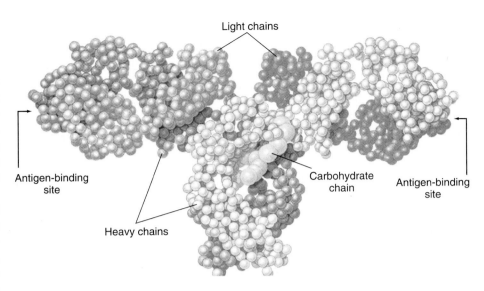

FIGURE 54.17

The structure of an antibody molecule. A molecular model, in which each amino acid is represented by a small sphere. The heavy chains are colored blue; the light chains are red. The four chains wind about one another to form a Y shape, with two identical antigen-binding sites at the arms of the Y and a stem region that directs the antibody to a particular portion of the immune response.

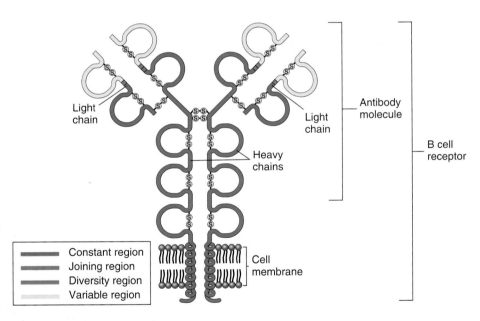

FIGURE 54.18

Structure of the B-cell receptor. The receptor molecules are characterized by domains of about 100 amino acids (represented as loops) joined by —S—S— covalent bonds. Each receptor has a constant region (*green*) and a variable region (*yellow*), joined by a diversity region (*blue*) and a joining region (*red*). The receptor binds to antigens at its two variable regions.

IgA antibodies are present in secretions such as milk, mucus, and saliva. In milk, these antibodies are thought to provide immune protection to nursing infants, whose own immune systems are not yet fully developed.

Antibody Diversity

The vertebrate immune system is capable of recognizing as foreign practically any nonself molecule presented to it—literally millions of different antigens. Although vertebrate chromosomes contain only a few hundred receptor-encoding genes, it is estimated that human B cells can make between 10^6 and 10^9 different antibody molecules. How do vertebrates generate millions of different antigen receptors when their chromosomes contain only a few hundred copies of the genes encoding those receptors?

The answer to this question is that the millions of immune receptor genes do not have to be inherited at conception because they do not exist as single sequences of nucleotides. Rather, they are assembled by stitching together three or four DNA segments that code for different parts of the receptor molecule. When an antibody is assembled, the different sequences of DNA are brought together to form a composite gene (figure 54.19). This process is called **somatic rearrangement.** For example, combining DNA in different ways can produce 16,000 different heavy chains and about 1200 different light chains (in mouse antibodies).

Two other processes generate even more sequences. First, the DNA segments are often joined together with one or two nucleotides off-register, shifting the reading frame during gene transcription and so generating a totally different sequence of amino acids in the protein. Second, random mistakes occur during successive DNA replications as the lymphocytes divide during clonal expansion. Both mutational processes produce changes in amino acid sequences, a phenomenon known as **somatic mutation** because it takes place in a somatic cell rather than in a gamete.

Because a cell may end up with any heavy-chain gene and any light-chain gene during its maturation, the total number of different antibodies possible is staggering: 16,000 heavy-chain combinations × 1200 light-chain combinations = 19 million different possible antibodies. If one also takes into account the changes induced by somatic mutation, the total can exceed 200 million! It should be understood that, although this discussion has centered on B cell and their receptors, the receptors on T cells are as diverse as those on B cells because they also are subject to similar somatic rearrangements and mutations.

Immunological Tolerance

A mature animal's immune system normally does not respond to that animal's own tissue. This acceptance of self cells is known as **immunological tolerance.** The immune system of an embryo, on the other hand, is able to respond to both foreign and self molecules, but it loses the ability to respond to self molecules as its development proceeds. Indeed, if foreign tissue is introduced into an embryo before its immune system has developed, the mature animal that results will not recognize that tissue as foreign and will accept grafts of similar tissue without rejection.

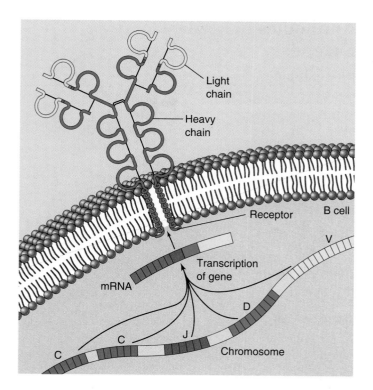

FIGURE 54.19

The lymphocyte receptor molecule is produced by a composite gene. Different regions of the DNA code for different regions of the receptor structure (*C*, constant regions; *J*, joining regions; *D*, diversity regions; and *V*, variable regions) and are brought together to make a composite gene that codes for the receptor. Through different somatic rearrangements of these DNA segments, an enormous number of different receptor molecules can be produced.

There are two general explanations for immunological tolerance: clonal deletion and clonal suppression. During the normal maturation of hemopoietic stem cells in an embryo, fetus, or newborn, most lymphocyte clones that have receptors for self antigens are either eliminated (clonal deletion) or suppressed (clonal suppression). Thus, the only clones that survive this phase of development are those that are directed against foreign rather than self molecules.

Immunological tolerance sometimes breaks down, causing either B cells or T cells (or both) to recognize their own tissue antigens. This loss of immune tolerance results in autoimmune disease. Myasthenia gravis, for example, is an autoimmune disease in which individuals produce antibodies directed against acetylcholine receptors on their own skeletal muscle cells, causing paralysis. Autoimmunity will be discussed in more detail later in this chapter.

An antibody molecule is composed of constant and variable regions. The variable regions recognize a specific antigen because they possess clefts into which the antigen can fit. Lymphocyte receptors are encoded by genes that are assembled by somatic rearrangement and mutation of the DNA.

Antibodies in Medical Diagnosis

Blood Typing

The blood type denotes the class of antigens found on the red blood cell surface. Red blood cell antigens are clinically important because their types must be matched between donors and recipients for blood transfusions. There are several groups of red blood cell antigens, but the major group is known as the **ABO system.** In terms of the antigens present on the red blood cell surface, a person may be *type A* (with only A antigens), *type B* (with only B antigens), *type AB* (with both A and B antigens), or *type O* (with neither A nor B antigens).

The immune system is tolerant to its own red blood cell antigens. A person who is type A, for example, does not produce anti-A antibodies. Surprisingly, however, people with type A blood do make antibodies against the B antigen, and conversely, people with blood type B make antibodies against the A antigen. This is believed to result from the fact that antibodies made in response to some common bacteria cross-react with the A or B antigens. A person who is type A, therefore, acquires antibodies that can react with B antigens by exposure to these bacteria but does not develop antibodies that can react with A antigens. People who are type AB develop tolerance to both antigens and thus do not produce either anti-A or anti-B antibodies. Those who are type O, in contrast, do not develop tolerance to either antigen and, therefore, have both anti-A and anti-B antibodies in their plasma.

If type A blood is mixed on a glass slide with serum from a person with type B blood, the anti-A antibodies in the serum will cause the type A red blood cells to clump together, or **agglutinate** (figure 54.20). These tests allow the blood types to be matched prior to transfusions, so that agglutination will not occur in the blood vessels, where it could lead to inflammation and organ damage.

Rh Factor. Another group of antigens found in most red blood cells is the *Rh factor* (Rh stands for rhesus monkey, in which these antigens were first discovered). People who have these antigens are said to be **Rh-positive,** whereas those who do not are **Rh-negative.** There are fewer Rh-negative people because this condition is recessive to Rh-positive. The Rh factor is of particular significance when Rh-negative mothers give birth to Rh-positive babies.

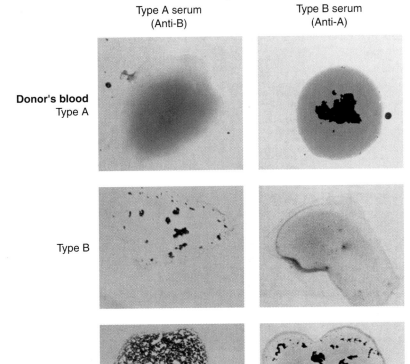

Recipient's blood

Type A serum (Anti-B) Type B serum (Anti-A)

Donor's blood Type A

Type B

Type AB

FIGURE 54.20
Blood typing. Agglutination of the red blood cells is seen when blood types are mixed with sera containing antibodies against the ABO antigens. Note that no agglutination would be seen if type O blood (not shown) were used.

Since the fetal and maternal blood are normally kept separate across the placenta (chapter 56), the Rh-negative mother is not usually exposed to the Rh antigen of the fetus during the pregnancy. At the time of birth, however, a variable degree of exposure may occur, and the mother's immune system may become sensitized and produce antibodies against the Rh antigen. If the woman does produce antibodies against the Rh factor, these antibodies can cross the placenta in subsequent pregnancies and cause hemolysis of the Rh-positive red blood cells of the fetus. The baby is therefore born anemic, with a condition called *erythroblastosis fetalis,* or *hemolytic disease of the newborn.*

Erythroblastosis fetalis can be prevented by injecting the Rh-negative mother with an antibody preparation against the Rh factor within 72 hours after the birth of each Rh-positive baby. This is a type of passive immunization in which the injected antibodies inactivate the Rh antigens and thus prevent the mother from becoming actively immunized to them.

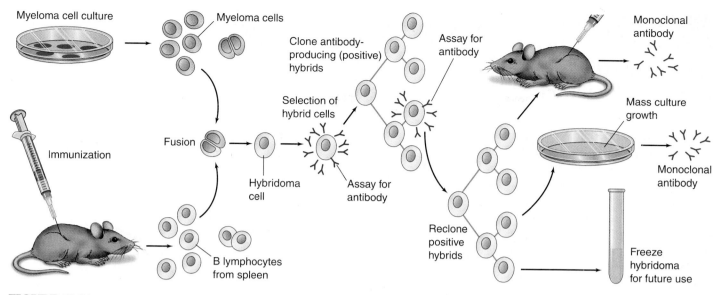

FIGURE 54.21

The production of monoclonal antibodies. These antibodies are produced by cells that arise from successive divisions of a single B cell, and hence all of the antibodies target a single antigenic determinant site. Such antibodies are used for a variety of medical applications, including pregnancy testing.

Monoclonal Antibodies

Antibodies are commercially prepared for use in medical diagnosis and research. In the past, antibodies were obtained by chemically purifying a specific antigen and then injecting this antigen into animals. Since an antigen typically has many different antigenic determinant sites, however, the antibodies obtained by this method were *polyclonal*; they stimulated the development of different B-cell clones with different specificities. This decreased their sensitivity to a particular antigenic site and resulted in some degree of cross-reaction with closely related antigen molecules.

Monoclonal antibodies, by contrast, exhibit specificity for one antigenic determinant only. In the preparation of monoclonal antibodies, an animal (frequently, a mouse) is injected with an antigen and subsequently killed. B lymphocytes are then obtained from the animal's spleen and placed in thousands of different in vitro incubation vessels. These cells soon die, however, unless they are hybridized with cancerous multiple myeloma cells. The fusion of a B lymphocyte with a cancerous cell produces a hybrid that undergoes cell division and produces a clone called a *hybridoma*. Each hybridoma secretes large amounts of identical, monoclonal antibodies. From among the thousands of hybridomas produced in this way, the one that produces the desired antibody is cultured for large-scale production, and the rest are discarded (figure 54.21).

The availability of large quantities of pure monoclonal antibodies has resulted in the development of much more sensitive clinical laboratory tests. Modern pregnancy tests, for example, use particles (latex rubber or red blood cells) that are covered with monoclonal antibodies produced against a pregnancy hormone (abbreviated hCG—see

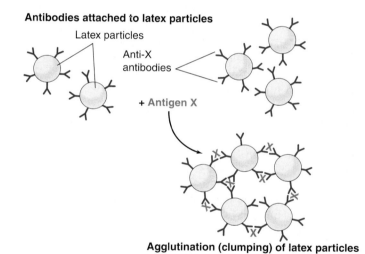

FIGURE 54.22

Using monoclonal antibodies to detect an antigen. In many clinical tests (such as pregnancy testing), the monoclonal antibodies are bound to particles of latex which agglutinate in the presence of the antigen.

chapter 56) as the antigen. When these particles are mixed with a sample that contains this hormone antigen from a pregnant woman, the antigen-antibody reaction causes a visible agglutination of the particles (figure 54.22).

An agglutination reaction will occur if different types of blood are mixed together because different antibodies exist for the ABO and Rh factor antigens on the surface of red blood cells. Monoclonal antibodies are commercially produced antibodies that react against one specific antigen.

Role of MHC Proteins in T-Cell Function

The cell surfaces of leukocytes contain a variety of proteins that play roles in the immune response (table 54.2). We have already mentioned the MHC proteins that serve as *self labels*, identifying leukocytes and every other body cell as belonging to a specific individual. There are two classes of MHC proteins. MHC-I is present on every nucleated cell of the body. MHC-II, however, is found only on macrophages, B cells, and a subtype of T cells called CD4+ T cells. These three cell types work together in one form of the immune response, and their MHC-II markers permit them to recognize one another.

The human MHC proteins are specified by genes that are the most polymorphic known, with as many as 50 alleles each. Only rarely will two individuals have the same combination of alleles and, therefore, the same mix of MHC proteins. These MHC proteins can thus distinguish self cells from nonself cells, a distinction critical for the function of the T cells. This is because, unlike B cells, T cells cannot bind to free antigens—they can only bind to antigens that are presented to them on the surfaces of cells! Further, the T cells can only recognize antigens presented to them associated with the MHC proteins. The cells that perform this function are known as antigen-presenting cells.

Cytotoxic T lymphocytes, which act to destroy infected cells as previously described, can only interact with antigens presented to them with MHC-I proteins. Helper T lymphocytes, whose functions will soon be described,

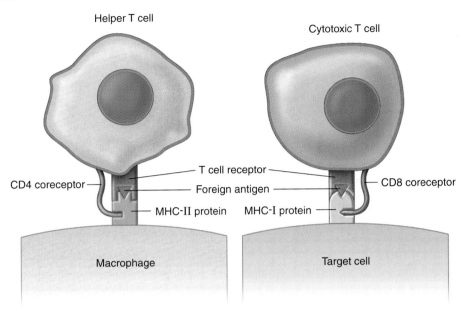

FIGURE 54.23
T cells bind to foreign antigens in conjunction with MHC proteins. The CD4 coreceptor on helper T cells requires that these cells interact with class-2 MHC (or MHC-II) proteins. The CD8 coreceptor on cytotoxic T cells requires that these cells interact only with cells bearing class-1 MHC (or MHC-I) proteins.

can interact only with antigens presented with MHC-II proteins. These restrictions result from the presence of coreceptors, which are proteins associated with the T-cell receptors. The coreceptor known as CD8 is associated with the cytotoxic T-cell receptor (these cells can therefore be indicated as CD8+). The CD8 coreceptor can interact only with the MHC-I proteins of an infected cell. The coreceptor known as CD4 is associated with the helper T-cell receptor (these cells can thus be indicated as CD4+) and interacts only with the MHC-II proteins of another cell (figure 54.23).

	Immune Receptors		MHC Proteins	
Cell Type	**T Receptor**	**B Receptor**	**MHC-I**	**MHC-II**
B cell	–	+	+	+
CD4+ T cell	+	–	+	+
CD8+ T cell	+	–	+	–
Macrophage	–	–	+	+

Table 54.2   Key Cell Surface Proteins of the Immune System

Note: CD4+ T cells include inducer T cells and helper T cells; CD8+ T cells include cytotoxic T cells and suppressor T cells. + means present; – means absent.

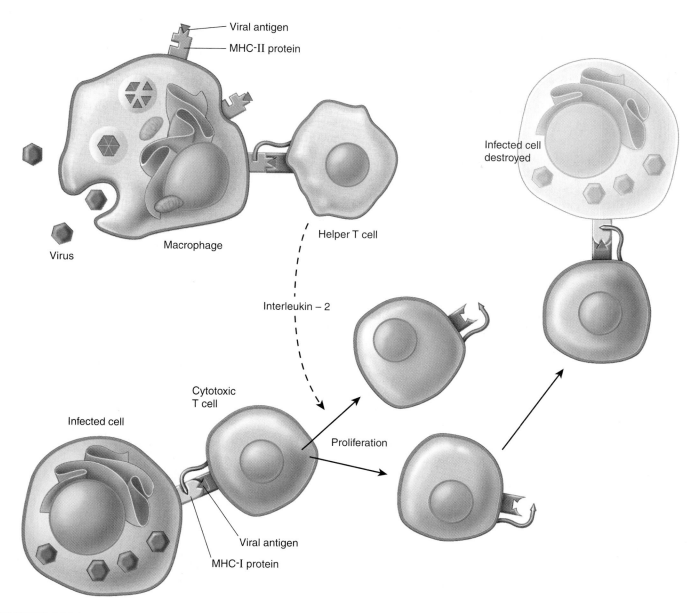

FIGURE 54.24
Macrophages present foreign antigens to helper T cells. The helper T cells then secrete interleukin-2, a lymphokine that stimulates the proliferation of sensitized cytotoxic T cells.

When a foreign particle, such as a virus, infects the body, it is taken into macrophages by phagocytosis and partially digested. Within the macrophage, the partially digested virus particles provide foreign antigens that move to the surface of the cell membrane. At the membrane, these foreign antigens form a complex with the MHC-II proteins. This combination of MHC proteins and foreign antigens is required for interaction with the receptors on the surface of helper T cells. The macrophages thus "present" the antigens to the helper T cells and, in this way, begin the activation of the T cells (figure 54.24).

Antigen-presenting cells must present foreign antigens together with MHC-II proteins in order to activate helper T cells, which have the CD4 coreceptor. Cytotoxic T cells (with the CD8 coreceptor) must interact with both MHC-I proteins and the foreign antigen.

The Cell-Mediated Immune Response

The cell-mediated immune response, carried out by T cells, protects the body from virus infection and cancer, killing abnormal or virus-infected body cells.

Once this helper T cell that initiates this response is presented with foreign antigen together with MHC proteins by a macrophage or other antigen-presenting cell, a complex series of steps is initiated. An integral part of this process is the secretion of autocrine regulatory molecules (chapter 53) known generally as **cytokines,** or more specifically as **lymphokines** if they are secreted by lymphocytes.

Cytokines

When a cytokine is first discovered, it is named according to its biological activity (such as *B-cell-stimulating factor*). Since each cytokine has many different actions, however, such names can be misleading. Scientists have thus agreed to use the name **interleukin,** followed by a number, to indicate a cytokine whose amino acid sequence has been determined.

Interleukin-1, for example, is secreted by macrophages and can activate the T-cell system. B-cell-stimulating factor, now called interleukin-4, is secreted by T cells and is required for the proliferation and clone development of B cells. *Interleukin-2* is released by helper T cells and, among its effects, is required for the activation of cytotoxic T lympho-

cytes. We will consider the actions of the cytokines as we describe the development of the T-cell immune response.

Cell Interactions in the T-Cell Response

When macrophages process the foreign antigens, they secrete interleukin-1, which stimulates cell division and proliferation of T cells. Once the helper T cells have been activated by the antigens presented to them by the macrophages, they secrete the cytokines known as macrophage colony-stimulating factor and gamma interferon, which promote the activity of macrophages. In addition, the helper T cells secrete interleukin-2, which stimulates the proliferation of cytotoxic T cells and B cells that are specific for the antigen. Cytotoxic T cells can destroy infected cells only if those cells display the foreign antigen together with their MHC-I proteins (see figure 54.24).

The network of interaction among different cell types of the immune system now spreads outward. Helper T cells also promote the humoral immune response of B cells. This occurs when the foreign antigen attaches to the immunoglobulin receptors on the B cells, so that the B cells can present this antigen together with their MHC-II proteins to the helper T cells. This interaction stimulates proliferation of the B cells, their conversion to plasma cells, and their secretion of antibodies against the foreign antigens. Helper T cells are thus essential in both the cell-mediated and humoral immune responses (figure 54.25).

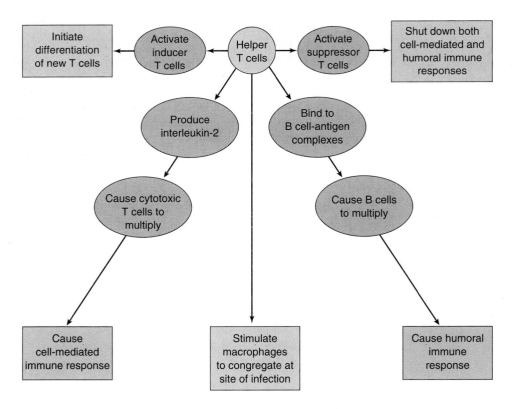

FIGURE 54.25
The many roles of helper T cells.
Helper T cells, through their secretion of lymphokines and interaction with other cells of the immune system, participate in every aspect of the immune response.

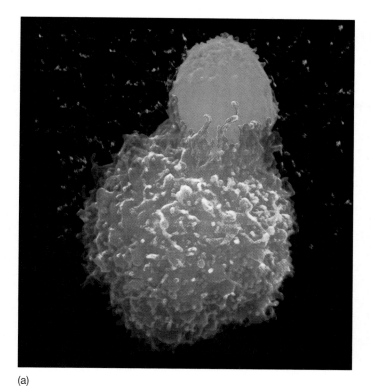

(a)

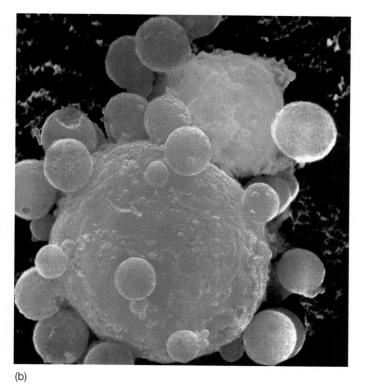

(b)

FIGURE 54.26
Cytotoxic T cells destroy cancer cells. (a) The cytotoxic T cell *(orange)* comes into contact with a cancer cell *(pink)*. (b) The T cell recognizes that the cancer cell is "nonself" and causes the destruction of the cancer.

T Cells in Transplant Rejection and Surveillance against Cancer

Cytotoxic T cells will also attack any foreign version of MHC-I as if it signaled a virus-infected cell. Therefore, even though vertebrates did not evolve the immune system as a defense against tissue transplants, their immune systems will attack transplanted tissue and cause graft rejection. The closer two individuals are related, the less variance in their MHC proteins and the more likely they will tolerate each other's tissues—this is why relatives are often sought for kidney transplants. The drug cyclosporin inhibits graft rejection by inactivating cytotoxic T cells.

As tumors develop, they reveal surface antigens that can stimulate the immune destruction of the tumor cells. Tumor antigens activate the immune system, initiating an attack primarily by cytotoxic T cells (figure 54.26) and natural killer cells. The concept of **immunological surveillance** against cancer was introduced in the early 1970s to describe the proposed role of the immune system in fighting cancer.

The production of human interferons by genetically engineered bacteria has made large amounts of these substances available for the experimental treatment of cancer. Thus far, interferons have proven to be a useful addition to

the treatment of particular forms of cancer, including some types of lymphomas, renal carcinoma, melanoma, Kaposi's sarcoma, and breast cancer.

Interleukin-2 (IL-2), which activates both cytotoxic T cells and B cells, is now also available for therapeutic use through genetic-engineering techniques. Particular lymphocytes from cancer patients have been removed, treated with IL-2, and given back to the patients together with IL-2 and gamma interferon. Scientists are also attempting to identify specific antigens and their genes that may become uniquely expressed in cancer cells, in an effort to help the immune system to better target cancer cells for destruction.

Helper T cells are only activated when a foreign antigen is presented together with MHC antigens by a macrophage or other antigen-presenting cells. The helper T cells are also stimulated by interleukin-1 secreted by the macrophages, and, when activated, secrete a number of lymphokines. Interleukin-2, secreted by helper T cells, activates both cytotoxic T cells and B cells. Cytotoxic T cells destroy infected cells, transplanted cells, and cancer cells by cell-mediated attack.

T-Cell Destruction: AIDS

One mechanism for defeating the vertebrate immune system is to attack the immune mechanism itself. Helper T cells and inducer T cells are CD4+ T cells. Therefore, any pathogen that inactivates CD4+ T cells leaves the immune system unable to mount a response to *any* foreign antigen. Acquired immune deficiency syndrome (AIDS) is a deadly disease for just this reason. The AIDS retrovirus, called human immunodeficiency virus (HIV), mounts a direct attack on CD4+ T cells because it recognizes the CD4 coreceptors associated with these cells.

HIV's attack on CD4+ T cells cripples the immune system in at least three ways. First, HIV-infected cells die only after releasing replicated viruses that infect other CD4+T cells, until the entire population of CD4+ T cells is destroyed (figure 54.27). In a normal individual, CD4+ T cells make up 60% to 80% of circulating T cells; in AIDS patients, CD4+ T cells often become too rare to detect (figure 54.28). Second, HIV causes infected CD4+ T cells to secrete a soluble suppressing factor that blocks other T cells from responding to the HIV antigen. Finally, HIV may block transcription of MHC genes, hindering the recognition and destruction of infected CD4+ T cells and thus protecting those cells from any remaining vestiges of the immune system.

The combined effect of these responses to HIV infection is to wipe out the human immune defense. With no defense against infection, any of a variety of otherwise commonplace infections proves fatal. With no ability to recognize and destroy cancer cells when they arise, death by cancer becomes far more likely. Indeed, AIDS was first recognized as a disease because of a cluster of cases of an unusually rare form of cancer. More AIDS victims die of cancer than from any other cause.

Although HIV became a human disease vector only recently, possibly through transmission to humans from chimpanzees in central Africa, it is already clear that AIDS is one of the most serious diseases in human history (figure 54.29). The fatality rate of AIDS is 100%; no patient exhibiting the symptoms of AIDS has ever been known to survive more than a few years. However, the disease is *not* highly contagious, since it is transmitted from one individual to another through the transfer of internal body fluids, typically in semen and in blood during transfusions. Not all individuals exposed to HIV (as judged by anti-HIV antibodies in their blood) have yet acquired the disease. However, most will die within two years of onset of the symptoms unless additional strategies for the treatment of AIDS are discovered first.

Until recently, the only effective treatment for slowing the progression of the disease involved treatment with drugs such as AZT that inhibit the activity of reverse transcriptase, the enzyme needed by the virus to produce DNA from RNA. Recently, however, a new type of drug has become available

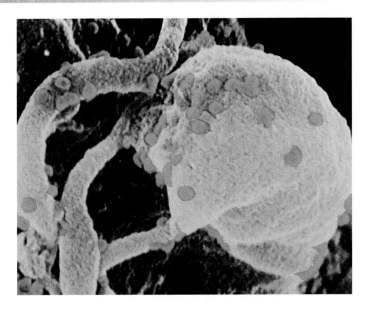

FIGURE 54.27
HIV, the virus that causes AIDS. Viruses released from infected CD4+ T cells soon spread to neighboring CD4+ T cells, infecting them in turn. The individual viruses, colored blue in this scanning electron micrograph, are extremely small; over 200 million would fit on the period at the end of this sentence.

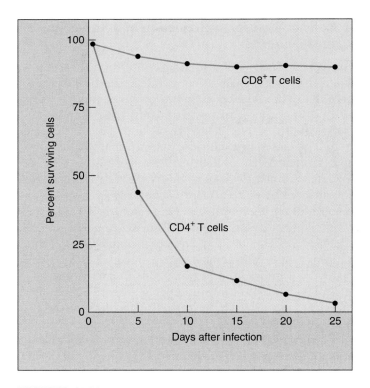

FIGURE 54.28
Survival of T cells in culture after exposure to HIV. The virus has little effect on the number of CD8+ T cells, but it causes the number of CD4+ T cells (this group includes helper T cells) to decline dramatically.

that acts to inhibit protease, an enzyme needed for viral assembly. Treatments that include a combination of reverse transcriptase inhibitors and protease inhibitors (p. 582) appear to lower levels of HIV, though they are very costly. Efforts to develop a vaccine against AIDS continue, both by splicing portions of the HIV surface protein gene into vaccinia virus and by attempting to develop a harmless strain of HIV. These approaches, while promising, have not yet proved successful and are limited by the fact that different strains of HIV seem to possess different surface antigens. Like the influenza virus, HIV engages in some form of antigen shifting, making it difficult to develop an effective vaccine.

AIDS destroys the ability of the immune system to mount a defense against any infection. HIV, the virus that causes AIDS, induces a state of immune deficiency by attacking and destroying CD4+T cells.

FIGURE 54.29

The AIDS epidemic in the United States: new cases. The U.S. Centers for Disease Control (CDC) reports that 54,656 new AIDS cases were reported in 1996 and 31,153 new cases in 1997, with a total of 641,086 cases and 390,692 deaths in the United States. Over 1.5 million other individuals are thought to be infected with the HIV virus in the United States, and 14 million worldwide. The 100,000th AIDS case was reported in August 1989, eight years into the epidemic; the next 100,000 cases took just 26 months; the third 100,000 cases took barely 19 months (May 1993), and the fourth 100,000 took only 13 months (June 1994). The extraordinarily high numbers seen in 1993 reflect an expansion of the definition of what constitutes an AIDS case.

Source: Data from U.S. Centers for Disease Control and Prevention, Atlanta, GA.

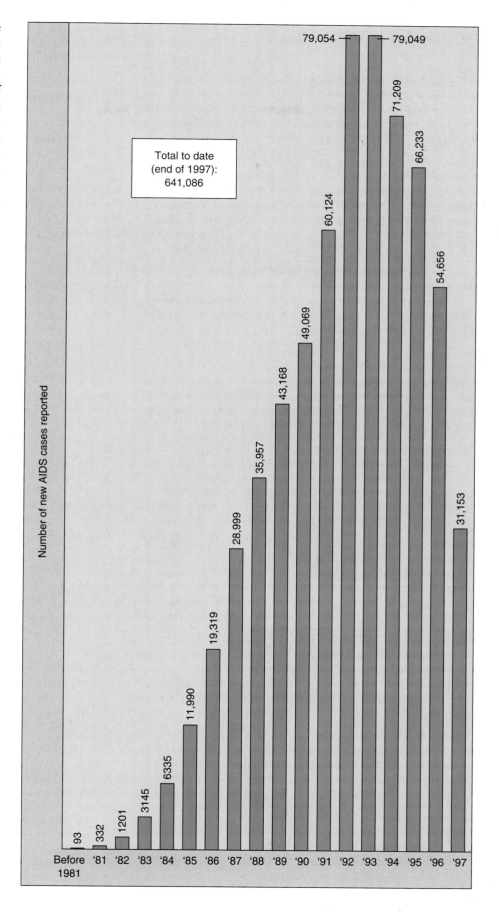

Total to date (end of 1997): 641,086

Antigen Shifting

A second way that a pathogen may defeat the immune system is to mutate frequently so that it varies the nature of its surface antigens. The virus which causes influenza uses this mechanism, and so we have to be immunized against a different strain of this virus periodically. This way of escaping immune attack is known as antigen shifting, and is practiced very effectively by trypanosomes, the protists responsible for sleeping sickness (see chapter 31). Trypanosomes possess several thousand different versions of the genes encoding their surface protein, but the cluster containing these genes has no promoter and so is not transcribed as a unit. The necessary promoter is located within a transposable element that jumps at random from one position to another within the cluster, transcribing a different surface protein gene with every move. Because such moves occur in at least one cell of an infective trypanosome population every few weeks, the vertebrate immune system is unable to mount an effective defense against trypanosome infection. By the time a significant number of antibodies have been generated against one form of trypanosome surface protein, another form is already present in the trypanosome population that survives immunological attack, and the infection cycle is renewed. People with sleeping sickness rarely rid themselves of the infection.

How Malaria Hides from the Immune System

Every year, about a half-million people become infected with the protozoan parasite *Plasmodium falciparum*, which multiplies in their bodies to cause the disease malaria. About one million die of the disease, most of them children. The plasmodium parasites enter the red blood cells and consume the hemoglobin of their hosts. Normally this sort of damage to a red blood cell would cause the damaged cell to be transported to the spleen for disassembly, destroying the plasmodium as well. The plasmodium avoids this fate, however, by secreting knob-like proteins that extend through the surface of the red blood cell and anchor the cell to the inner surface of the blood vessel.

Over the course of several days, the immune system of the infected person slowly brings the infection under control. During this time, however, a small proportion of the plasmodium parasites change their knob proteins to a form different from those that sensitized the immune system. Cells infected with these individuals survive the immune response, only to start a new wave of infection.

Scientists have recently discovered how the malarial parasite carries out this antigen-shifting defense. About 6% of the total DNA of the plasmodium is devoted to encoding a block of some 150 *var* genes, which are shifted on and off in multiple combinations. Each time a plasmodium divides, it alters the pattern of *var* gene expression about 2%, an incredibly rapid rate of antigen shifting. The exact means by which this is done is not yet completely understood.

DNA Vaccines May Get around Antigen Shifting

Vaccination against diseases such as smallpox, measles, and polio involves introducing into your body a dead or disabled pathogen, or a harmless microbe with pathogen proteins displayed on its surface. The vaccination triggers an immune response against the pathogen, and the bloodstream of the vaccinated person contains B cells which will remember and quickly destroy the pathogen in future infections. However, for some diseases, vaccination is nearly impossible because of antigen shifting; the pathogens change over time, and the B cells no longer recognize them. Influenza, as we have discussed, presents different surface proteins yearly. The trypanosomes responsible for sleeping sickness change their surface proteins every few weeks.

A new type of vaccine, based on DNA, may prove to be effective against almost any disease. The vaccine makes use of the killer T cells instead of the B cells of the immune system. Pathogens floating around in the bloodstream are recognized by B cells of the immune system, which then signal the production of antibodies against the pathogen. Once inside a cell, however, pathogens are protected from antibodies. Only when the cell is able to display on its outside surface some of the proteins from the pathogen it contains will killer T cells come and destroy that cell. The problem is that the cell will not stick bits of the pathogen's proteins to its surface unless it contains a live version of the disease-causing agent. This means that any vaccine will have to be alive and may cause an actual infection.

DNA vaccines avoid this risk. They consist of a plasmid, a harmless circle of bacterial DNA, that contains a gene from the pathogen that encodes an internal protein, one which is critical to the function of the pathogen and does not change. When this plasmid is injected into cells, the gene they carry is transcribed into protein but is not incorporated into the DNA of the cell's nucleus. Fragments of the pathogen protein are then stuck on the cell's membrane, marking it for destruction. In actual infections later, the immune system will be able to respond immediately. Studies are now underway to isolate the critical, unchanging proteins of pathogens and to investigate fully the use of the vaccines in humans.

Antigen shifting refers to the way a pathogen may defeat the immune system by changing its surface antigens and thereby escaping immune recognition. Pathogens that employ this mechanism include flu viruses, trypanosomes, and the protozoans that cause malaria.

Autoimmunity and Allergy

The previous section described ways that pathogens can elude the immune system to cause diseases. There is another way the immune system can fail; it can itself be the agent of disease. Such is the case with autoimmune diseases and allergies—the immune system is the cause of the problem, not the cure.

Autoimmune Diseases

Autoimmune diseases are produced by failure of the immune system to recognize and tolerate self antigens. This failure results in the activation of autoreactive T cells and the production of autoantibodies by B cells, causing inflammation and organ damage. There are over 40 known or suspected autoimmune diseases that affect 5% to 7% of the population. For reasons that are not well understood, two-thirds of the people with autoimmune diseases are women.

Autoimmune diseases can result from a variety of mechanisms. The self antigen may normally be hidden from the immune system, for example, so that the immune system treats it as foreign if exposure later occurs. This occurs when a protein normally trapped in the thyroid follicles triggers autoimmune destruction of the thyroid (Hashimoto's thyroiditis). It also occurs in systemic lupus erythematosus, in which antibodies are made to nucleoproteins. Since the immune attack triggers inflammation, and inflammation causes organ damage, the immune system must be suppressed to alleviate the symptoms of autoimmune diseases. Immune suppression is generally accomplished with corticosteroids (including hydrocortisone) and by nonsteroidal antiinflammatory drugs, including aspirin.

Allergy

The term *allergy*, often used interchangeably with *hypersensitivity*, refers to particular types of abnormal immune responses to antigens, which are called *allergens* in these cases. There are two major forms of allergy: (1) **immediate hypersensitivity,** which is due to an abnormal B-cell response to an allergen that produces symptoms within seconds or minutes, and (2) **delayed hypersensitivity,** which is an abnormal T-cell response that produces symptoms within about 48 hours after exposure to an allergen.

Immediate hypersensitivity can produce allergic rhinitis (chronic runny or stuffy nose); conjunctivitis (red eyes); allergic asthma; atopic dermatitis (urticaria, or hives); and other symptoms. These symptoms result from the production of antibodies of the IgE subclass instead of the normal IgG antibodies. Unlike IgG antibodies, IgE antibodies do not circulate in the blood. Instead, they attach to tissue mast cells and basophils, which have membrane receptors for these antibodies. When the person is again exposed to the same allergen, the allergen binds to the antibodies attached to the mast cells and basophils. This stimulates these cells to secrete various chemicals, including histamine, which produce the symptom of the allergy (figure 54.30).

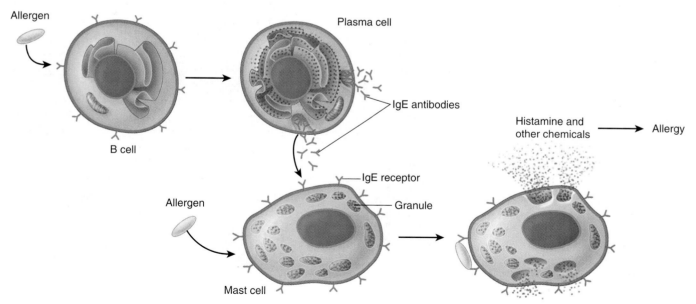

FIGURE 54.30

An allergic reaction. This is an immediate hypersensitivity response, in which B cells secrete antibodies of the IgE class. These antibodies attach to the plasma membranes of mast cells, which secrete histamine in response to antigen-antibody binding.

Allergens that provoke immediate hypersensitivity include various foods, bee stings, and pollen grains. The most common allergy of this type is seasonal hay fever, which may be provoked by ragweed *(Ambrosia)* pollen grains (figure 54.31a). People with chronic allergic rhinitis and asthma due to an allergy to dust or feathers are usually allergic to a tiny mite (figure 54.31b) that lives in dust and eats the scales of skin that are constantly shed from the body. Actually, most of the antigens from the dust mite are not in its body but rather in its feces, which are tiny particles that can enter the nasal mucosa like pollen grains. These allergic reactions are generally mild, but in some allergies (as to penicillin or peanuts in susceptible people) the widespread and excessive release of histamine may cause **anaphylactic shock,** an uncontrolled fall in blood pressure.

In delayed hypersensitivity, symptoms take a longer time (hours to days) to develop than in immediate hypersensitivity. This may be due to the fact that immediate hypersensitivity is mediated by antibodies, whereas delayed hypersensitivity is a T-cell response. One of the best-known examples of delayed hypersensitivity is **contact dermatitis,** caused by poison ivy, poison oak, and poison sumac. Since the symptoms are caused by the secretion of lymphokines rather than by the secretion of histamine, treatment with antihistamines provides little benefit. At present, corticosteroids are the only drugs that can effectively treat delayed hypersensitivity.

Autoimmune diseases are produced by inflammation that results when the immune system fails to tolerate self antigens. Allergies are produced by abnormal responses of the immune system and involve IgE antibodies in immediate hypersensitivity or an abnormal T-cell response in delayed hypersensitivity.

(a)

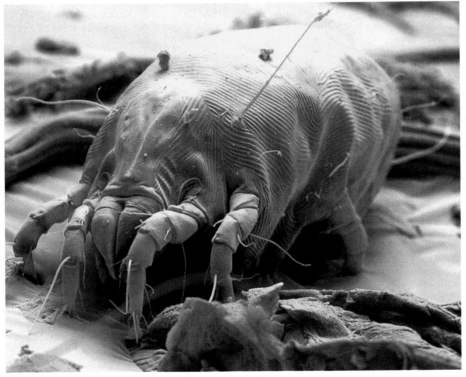

(b)

FIGURE 54.31
Common causes of immediate hypersensitivity. (a) A scanning electron micrograph of ragweed pollen, responsible for hay fever. (b) A scanning electron micrograph of the house dust mite, often responsible for allergic rhinitis and asthma.

Summary of Concepts

54.1 **Many of the body's most effective defenses are nonspecific.**

- Nonspecific defenses include physical barriers such as the skin and mucous membranes.
- Cells of the immune system make the self-versus-nonself recognition by means of surface proteins on cells called MHC proteins.
- Neutrophils and monocytes are phagocytic.
- Natural killer cells produce pores in the plasma membrane of their victims; complement proteins similarly destroy microbial cells.

54.2 **Specific immune defenses require the recognition of antigens.**

- Lymphocytes called B cells secrete antibodies and produce the humoral response; lymphocytes called T cells are responsible for cell-mediated immunity.
- Active immunity occurs when an individual gains immunity by prior exposure to a pathogen; passive immunity is produced by the transfer of antibodies from one individual to another.

54.3 **B cells label specific cells for destruction.**

- The antibody molecules consist of two heavy and two light polypeptide regions arranged like a "Y"; the two arms contain variable regions that bind to antigens, while the stalk has a constant amino acid sequence within each immunoglobulin class.
- An individual can produce a tremendous variety of different antibodies because the genes which produce those antibodies don't all have to be inherited.

- When serum from a person with one blood type is mixed with blood from another, the red blood cells will likely agglutinate; the exceptions are if the serum is from type AB or type O blood.
- Monoclonal antibodies are prepared commercially from a clone of cells that produces antibodies against one antigenic determinant site; such monoclonal antibodies are used for pregnancy and other diagnostic tests.

54.4 **T cells organize attacks against invading microbes.**

- T cells only respond to antigens presented to them by macrophages or other antigen-presenting cells together with MHC proteins.
- Cytotoxic T cells kill cells that have foreign antigens presented together with MHC-I proteins.
- T cells are also responsible for graft rejection and the immunological surveillance against cancer cells.

54.5 **The immune system can be defeated.**

- The HIV virus causes destruction of CD4+ cells, thereby weakening the immune system so much that the person dies of cancer or infections.
- Flu viruses, the trypanosomes that cause sleeping sickness, and the protozoan that causes malaria are able to evade the immune system by mutating the genes that produce their surface antigens.
- In autoimmune diseases, the immune system targets the body's own antigens, producing inflammation that can cause organ damage.

Discussing Key Terms

1. **MHC proteins** MHC stands for major histocompatibility complex and refers to the products of genes that code for cell surface proteins needed for self-nonself recognition and for activation of T cells.

2. **Lymphokines** Lymphokines are cytokines, or autocrine regulatory molecules, released from lymphocytes, and include interleukin-2, which promotes clonal expansion of cytotoxic T cells and B cells.

3. **Immunoglobulin** Antibodies are immunoglobulins, and are divided into classes based on different constant regions of the four polypeptide chains that constitute each antibody molecule.

4. **Plasma cells** When an antigen binds to its receptor on the B cell surface, the B cell is stimulated to divide.

Some of the progeny become plasma cells, which have an extensive rough endoplasmic reticulum devoted to the production of antibody proteins.

5. **Clonal selection** The interaction of an antigen with a receptor protein on a lymphocyte surface stimulates the rapid cell division of that lymphocyte, producing a clone of lymphocytes which all have the same receptor and all target the same antigen. This is how the secondary response of active immunity is produced.

6. **Hypersensitivity** Allergy is divided into immediate hypersensitivity and delayed hypersensitivity. Immediate hypersensitivity is a B-cell response involving IgE antibodies and histamine; delayed hypersensitivity is a T-cell response.

Review Questions

1. How do macrophages destroy foreign cells? How do natural killer cells destroy foreign cells?

2. What is the complement system, and how does it participate in defense against infection?

3. On what types of cells are the two classes of MHC proteins found? What roles do these proteins play in the response to infection?

4. In what two ways do macrophages activate helper T cells? How do helper T cells stimulate the proliferation of cytotoxic T cells?

5. What type of cell produces antibodies? How do IgM antibodies trigger the destruction of infected cells? How do IgG antibodies trigger the destruction of infected cells?

6. Describe the clonal selection model and how this helps to explain active immunity.

7. How are lymphocytes able to produce millions of different types of immune receptors from only a few hundred copies of immune receptor genes?

8. How does the immune system perform self-versus-nonself recognition? How is tolerance to self antigens achieved? What might cause an immune attack of self antigens, and what are some of the consequences of this attack?

9. How do trypanosomes defeat vertebrate immune defenses? How does HIV defeat human immune defenses? How do the effects of HIV on the immune system cause death in patients with AIDS?

Thought Questions

1. Some zoos have deliberately sought to produce highly inbred lines of very rare animals such as the Speke's gazelle, with individuals as genetically alike as possible. Other zoos start with highly inbred lines, an unavoidable consequence of the shortage of wild-captured individuals. Although these inbred animals possess perfectly functional immune systems, they are sometimes subject to disastrous outbreaks of disease. Why?

2. Do you think a virus or other pathogen that invades a vertebrate host, avoids the host's immune defenses entirely, reproduces quickly within the host, and leads to the rapid death of the host is well adapted or poorly adapted to that host? Explain.

Internet Links

The Immune System: Friend and Foe
http://www.hhmi.org/grants/lectures/1996/
A super series of lectures on basic aspects of the immune system sponsored by the Howard Hughes Medical Institute.

Ask the CDC
http://www.cdc.gov.cdc.gov
The home page of the CENTER FOR DISEASE CONTROL provides a wide range of up-to-date information about communicable disease outbreaks, with monthly updates on the status of the AIDS epidemic.

Poison Ivy: An Exaggerated Immune Response to Nothing Much
http://www.bio.umass.edu/immunology/poisoniv.htm
The operation of the immune system is taught in this Amherst site by examining in detail how we respond to poison ivy.

Anatomy of a Splinter
http://www.cellsalive.com/ouch.htm
The operation of the immune system is taught in this CELLS ALIVE multimedia site by using animations to illustrate how the immune system responds to a splinter.

For Further Reading

Ada, G. L. and G. Nossal: "The Clonal Selection Theory," *Scientific American*, August 1987, page 62. Lucid description of one of the most important concepts in immunology.

Beck, G. and G. S. Habicht: "Immunity and the Invertebrates," *Scientific American*, November 1996, pages 60–66. Invertebrates can distinguish self versus nonself, and even have the equivalent of IL-1 and other cytokines. Some important medical advances have resulted from the study of the invertebrate immune system.

Boon, T.: "Teaching the Immune System to Fight Cancer," *Scientific American*, vol. 268, March 1993, pages 82–89. Can antigens on the surfaces of cancerous tumors be used to sensitize the immune system to the development of subsequent tumors?

Caldwell, J. and P. Caldwell: "The African AIDS Epidemic," *Scientific American*, March 1996, pages 62–68. In parts of Africa, nearly 25% of the population is infected with the HIV virus as a result of heterosexual transmission of the virus.

Donelson, J. and M. Turner: "How the Trypanosome Changes Its Coat," *Scientific American*, February 1985, pages 44–51. An easily understood discussion of how parasites responsible for sleeping sickness use transposition to avoid immune surveillance.

Hoffman, M.: "AIDS: Solving the Molecular Puzzle," *American Scientist*, vol. 82, March/April 1994, pages 171–77. Despite more than a decade of solid, intensive, well-funded research, researchers are still not all that close to understanding how exactly the HIV virus actually *causes* AIDS.

Jaret, P.: "Our Immune System: The Wars Within," *National Geographic*, June 1986, pages 702–36. A very readable account of current progress in the study of the human immune system, with striking photographs by Lennart Nilsson.

Litman, G. W.: "Sharks and the Origins of Vertebrate Immunity," *Scientific American*, November, 1996, pages 67–71. Sharks have a very effective immune system that shares similarities with the human immune system.

55

Maintaining Homeostasis

FIGURE 55.1
Regulating the body with water. One of the ways an elephant can regulate its temperature is to spray water on its body. Water also cycles through the elephant's body in enormous quantities each day and helps to regulate its internal environment.

The first vertebrates evolved in seawater, and the physiology of all vertebrates reflects this origin. Approximately two-thirds of every vertebrate's body is water. If the amount of water in the body of a vertebrate falls much lower than this, the animal will die. In this chapter, we discuss the various mechanisms by which animals avoid gaining or losing too much water. As we shall see, these mechanism are closely tied to the way animals exploit the varied environments in which they live and to the regulatory systems of the body (figure 55.1).

The Need to Maintain Homeostasis

As the animal body has evolved, specialization has increased. Each cell is a sophisticated machine, finely tuned to carry out a precise role within the body. Such specialization of cell function is possible only when extracellular conditions are kept within narrow limits. Temperature, pH, the concentrations of glucose and oxygen, and many other factors must be held fairly constant for cells to function efficiently and interact properly with one another. These extracellular conditions constitute the internal environment of the animal.

Homeostasis may be defined as the dynamic constancy of the internal environment. The term *dynamic* is used because conditions are never absolutely constant, but fluctuate continuously within narrow limits. Homeostasis is essential for life, and most of the regulatory mechanisms of the vertebrate body that are not devoted to reproduction are concerned with maintaining homeostasis. As long as homeostasis is maintained, an animal is healthy. This is why a physician tests body temperature, blood composition, and other measures of a patient's internal environment as part of a routine physical examination.

Negative Feedback Loops

To maintain internal constancy, the vertebrate body must have *sensors* that are able to detect deviations from a *set point* (figure 55.2). The set point is analogous to the temperature set on a house thermostat. In a similar manner, there are set points for body temperature, blood glucose concentration, the tension on a tendon, and so on. When a sensor detects a deviation from a particular set point, it must relay this information to an *integrating center*, which usually receives information from many different sensors. The integrating center is often a particular region of the brain or spinal cord, but in some cases it can also be cells of endocrine glands. The relative strengths of different sensory inputs are weighed in the integrating center, and, in response, the integrating center either increases or decreases the activity of particular **effectors**, generally muscles or glands.

To return to the idea of a home thermostat, suppose you set the thermostat at a set point of 70 °F. If the temperature of the house rises sufficiently above the set point, a sensor within the thermostat detects the deviation. This sensor will then, via the thermostat's equivalent of an integrating center, activate the effector. The effector in this case may be an air conditioner, which acts to reverse the deviation from the set point.

In a human, if the body temperature exceeds the set point of 37 °C, sensors in a part of the brain detect this deviation. Acting via an integrating center (also in the brain), these sensors stimulate effectors (including sweat glands) that lower the temperature (figure 55.3). One can think of the effectors as "defending" the set points of the body against deviations. Since the activity of the effectors is influenced by the effects they produce, and since this regulation is in a negative, or reverse, direction, this type of control system is known as a *negative feedback loop.*

The nature of the negative feedback loop becomes clear when we again refer to the analogy of the thermostat and air conditioner. After the air conditioner has been on for some time, the room temperature may fall significantly below the set point of the thermostat. When this occurs, the air conditioner will be turned off. The effector (air conditioner) is turned on by a high temperature; and, when activated, it produces a negative change (lowering of the temperature) that ultimately causes the effector to be turned off. In this way, constancy is maintained.

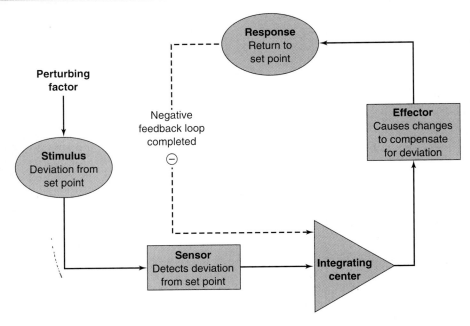

FIGURE 55.2
A generalized diagram of a negative feedback loop. Negative feedback loops maintain a state of homeostasis, or dynamic constancy of the internal environment, by correcting deviations from a set point.

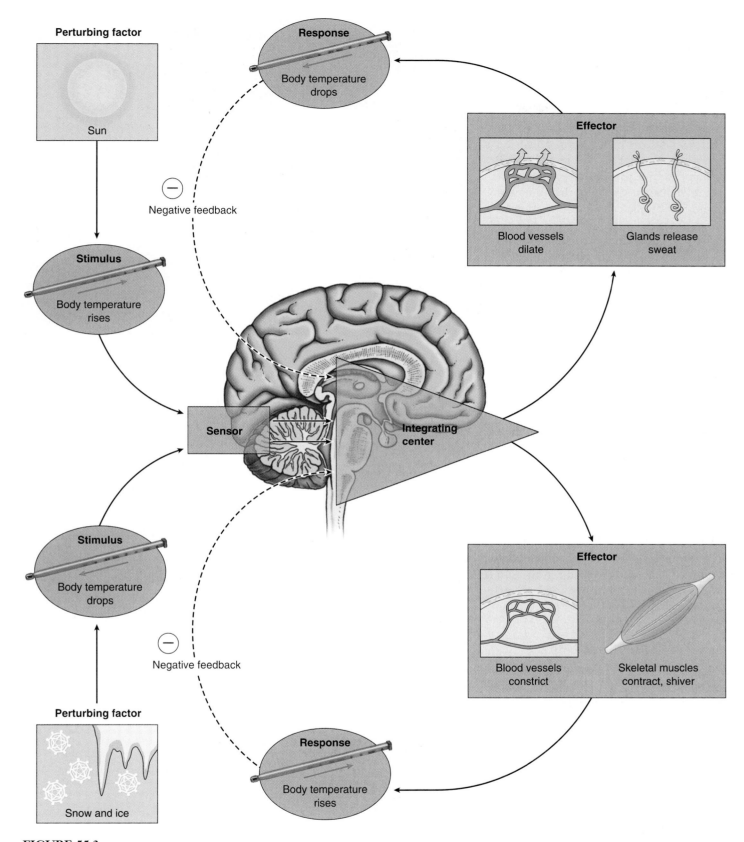

FIGURE 55.3
Negative feedback loops keep the body temperature within a normal range. An increase (*top*) or decrease (*bottom*) in body temperature is sensed by the brain. The integrating center in the brain then processes the information and activates effectors, such as surface blood vessels, sweat glands, and skeletal muscles. When the body temperature responds by returning to normal, negative feedback prevents further stimulation of the effectors by the integrating center.

Regulating Body Temperature

Humans, together with other mammals and with birds, are *endothermic*; they can maintain relatively constant body temperatures independent of the environmental temperature. When the temperature of your blood exceeds 37 °C (98.6 °F), neurons in a part of the brain called the hypothalamus detect the temperature change. Acting through the control of motor neurons, the hypothalamus responds by promoting the dissipation of heat through sweating, dilation of blood vessels in the skin, and other mechanisms. These responses tend to counteract the rise in body temperature. When body temperature falls, the hypothalamus coordinates a different set of responses, such as shivering and the constriction of blood vessels in the skin, which help to raise body temperature and correct the initial challenge to homeostasis.

Vertebrates other than mammals and birds are *ectothermic*; their body temperatures are more or less dependent on the environmental temperature. However, to the extent that it is possible, many ectothermic vertebrates attempt to maintain some degree of temperature homeostasis. Certain large fish, including tuna, swordfish, and some sharks, for example, can maintain parts of their body at a significantly higher temperature than that of the water. Reptiles attempt to maintain a constant body temperature through behavioral means—by placing themselves in varying locations of sun and shade. That's why you frequently see lizards basking in the sun (figure 55.4). Sick lizards even give themselves a "fever" by seeking warmer locations (see chapter 54)!

Regulating Blood Glucose

When you digest a carbohydrate-containing meal, you absorb glucose into your blood. This causes a temporary rise in the blood glucose concentration, which is brought back down in a few hours. What counteracts the rise in blood glucose following a meal?

When glucose levels within the blood exceed normal values, the rise is detected by a sensor, the islets of Langerhans in the pancreas. In response, the islets secrete the hormone *insulin*, which stimulates the uptake of blood glucose into muscles, liver, and adipose tissue. The islets are, in this case, the sensor, integrating center, and effector. The muscles and liver can convert the glucose into the polysaccharide glycogen; adipose cells can convert glucose into fat. These actions lower the blood glucose (figure 55.5) and help to store energy in forms that can be used later by the body.

FIGURE 55.4
Behavioral regulation of body temperature. This African lizard regulates its body temperature by basking in the sun. Its dark skin helps it absorb heat.

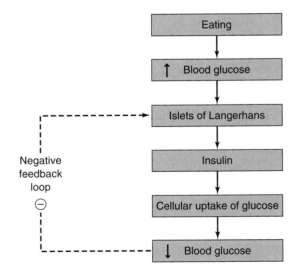

FIGURE 55.5
The negative feedback control of blood glucose. The rise in blood glucose concentration following a meal stimulates the secretion of insulin from the islets of Langerhans in the pancreas. Insulin is a hormone that promotes the entry of glucose in skeletal muscle and other tissue, thereby lowering the blood glucose and compensating for the initial rise.

Negative feedback mechanisms correct deviations from a set point for different internal variables. In this way, body temperature and blood glucose, for example, are kept within normal limits.

Antagonistic Effectors and Positive Feedback

The negative feedback mechanisms that maintain homeostasis often oppose each other to produce a finer degree of control. In a few cases positive feedback mechanisms, which push a change further in the same direction, are used by the body.

Antagonistic Effectors

Most factors in the internal environment are controlled by several effectors, which often have antagonistic actions. Control by antagonistic effectors is sometimes described as "push-pull," in which the increasing activity of one effector is accompanied by decreasing activity of an antagonistic effector. This affords a finer degree of control than could be achieved by simply switching one effector on and off.

Room temperature can be maintained, for example, by simply turning an air conditioner on and off, or by just turning a heater on and off. A much more stable temperature, however, can be achieved if the air conditioner and heater are both controlled by a thermostat (figure 55.6). Then the heater is turned on when the air conditioner shuts off, and vice versa. Antagonistic effectors are similarly involved in the control of body temperature and blood glucose. Whereas insulin, for example, lowers blood glucose following a meal, other hormones act to raise the blood glucose concentration between meals, especially when a person is exercising. The heart rate is similarly controlled by antagonistic effectors. Stimulation of one group of nerve fibers increases the heart rate, while stimulation of another group slows the heart rate.

Positive Feedback Loops

Feedback loops that accentuate a disturbance are called positive feedback loops. In a positive feedback loop, perturbations cause the effector to drive the value of the controlled variable even *farther* from the set point. Hence, systems in which there is positive feedback are highly unstable, analogous to a spark that ignites an explosion. Nevertheless, such systems are important components of some physiological mechanisms. For example, positive feedback occurs in blood clotting, where one clotting factor activates another in a cascade that leads quickly to the formation of a clot. Positive feedback also plays a role in the contractions of the uterus during childbirth (figure 55.7). In this case, stretching of the uterus by the fetus stimulates contraction, and contraction stimulates further stretching; the cycle continues until the fetus is expelled from the uterus. In the body, most positive feedback systems act as part of some larger mechanism that maintains homeostasis. In the examples we've described, formation of a blood clot stops bleeding and hence tends to keep blood volume constant, and expulsion of the fetus reduces the contractions of the uterus.

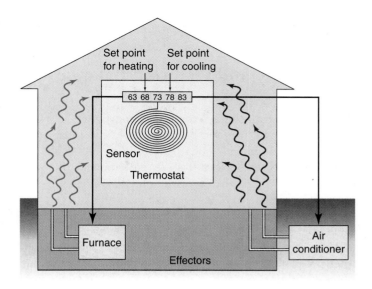

FIGURE 55.6
Room temperature is maintained by antagonistic effectors. If a thermostat senses a low temperature, the heater is turned on and the air conditioner is turned off. If the temperature is too high, the air conditioner is activated, and the heater is turned off.

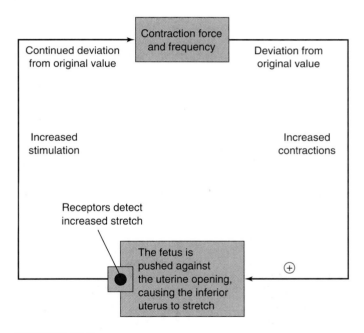

FIGURE 55.7
An example of positive feedback during childbirth. This is one of the few examples of positive feedback that operate in the vertebrate body.

Antagonistic effectors that act antagonistically to each other are more effective than effectors that act alone. Positive feedback mechanisms accentuate changes and have limited functions in the body.

Osmolality and Osmotic Balance

Water in an animal's body is distributed between the intracellular and extracellular compartments (figure 55.8). In order to maintain osmotic balance, the extracellular compartment of an animal's body (including its blood plasma) must be able to take water from its environment or to excrete excess water into its environment. Inorganic ions must also be exchanged between the extracellular body fluids and the external environment to maintain homeostasis. Such exchanges of water and electrolytes between the body and the external environment occur across specialized epithelial cells and, in most vertebrates, through a filtration process in the kidneys.

Most vertebrates maintain homeostasis in regard to the total solute concentration of their extracellular fluids and in regard to the concentration of specific inorganic ions. Sodium (Na^+) is the major cation in extracellular fluids, and chloride (Cl^-) is the major anion. The divalent cations, calcium (Ca^{++}) and magnesium (Mg^{++}), as well as other ions, also have important functions and must be maintained at their proper concentrations.

Osmolality and Osmotic Pressure

Osmosis is the diffusion of water across a membrane, and it always occurs from a more dilute solution (with a lower solute concentration) to a less dilute solution (with a higher solute concentration). Since the total solute concentration of a solution determines its osmotic behavior, the total moles of solute per kilogram of water is expressed as the **osmolality** of the solution. Solutions that have the same osmolality are *isosmotic*. A solution with a lower or higher osmolality than another is called *hypoosmotic* or *hyperosmotic*, respectively.

If one solution is hyperosmotic compared with another, and if the two solutions are separated by a semipermeable membrane, water may move by osmosis from the more dilute solution to the hyperosmotic one. In this case, the hyperosmotic solution is also **hypertonic** ("higher strength") compared with the other solution, and it has a higher osmotic pressure. The **osmotic pressure** of a solution is a measure of its tendency to take in water by osmosis. A cell placed in a hypertonic solution, which has a higher osmotic pressure than the cell cytoplasm, will lose water to the surrounding solution and shrink. A cell placed in a **hypotonic** solution, in contrast, will gain water and expand.

If a cell is placed in an isosmotic solution, there may be no net water movement. In this case, the isosmotic solution can also be said to be **isotonic.** Isotonic solutions such as normal saline and 5% dextrose are used in medical care to bathe exposed tissues and to be given as intravenous fluids.

Osmoconformers and Osmoregulators

Most marine invertebrates are **osmoconformers;** the osmolality of their body fluids is the same as that of seawater (although the concentrations of particular solutes, such as magnesium ion, are not equal). Because the extracellular fluids are isotonic to seawater, there is no osmotic gradient and no tendency for water to leave or enter the body. Therefore, osmoconformers are in osmotic equilibrium with their environment. Among the vertebrates, only the primitive hagfish are strict osmoconformers. The sharks and their relatives in the class Chondrichthyes (cartilaginous fish) are also isotonic to seawater, even though their blood level of NaCl is lower than that of seawater; the difference in total osmolality is made up by retaining urea at a high concentration in their blood plasma.

All other vertebrates are osmoregulators—that is, animals that maintain a relatively constant blood osmolality despite the different concentration in the surrounding environment. The maintenance of a relatively constant body fluid osmolality has permitted vertebrates to exploit a wide variety of ecological niches. Achieving this constancy, however, requires continuous regulation.

Freshwater vertebrates have a much higher solute concentration in their body fluids than that of the surrounding water. In other words, they are hypertonic to their environment. Because of their higher osmotic pressure, water tends to enter their bodies. Consequently, they must prevent water from entering their bodies as much as possible and eliminate the excess water that does enter. In addition, they tend to lose inorganic ions to their environment and so must actively transport these ions back into their bodies.

In contrast, marine vertebrates are hypotonic to their environment; their body fluids have only about one-third the osmolality of the surrounding seawater. These animals are therefore in danger of losing water by osmosis and must retain water to prevent dehydration. They do this by drinking seawater and eliminating the excess ions through their kidneys and gills.

The body fluids of terrestrial vertebrates have a higher concentration of water than does the air surrounding them. Therefore, they tend to lose water to the air by evaporation from the skin and lungs. All reptiles, birds, and mammals, as well as amphibians during the time when they live on land, face this problem. These vertebrates have evolved excretory systems that help them retain water.

Marine invertebrates are isotonic with their environment, but most vertebrates are either hypertonic or hypotonic to their environment and thus tend to gain or lose water. Physiological mechanisms help most vertebrates to maintain a constant blood osmolality and constant concentrations of individual ions.

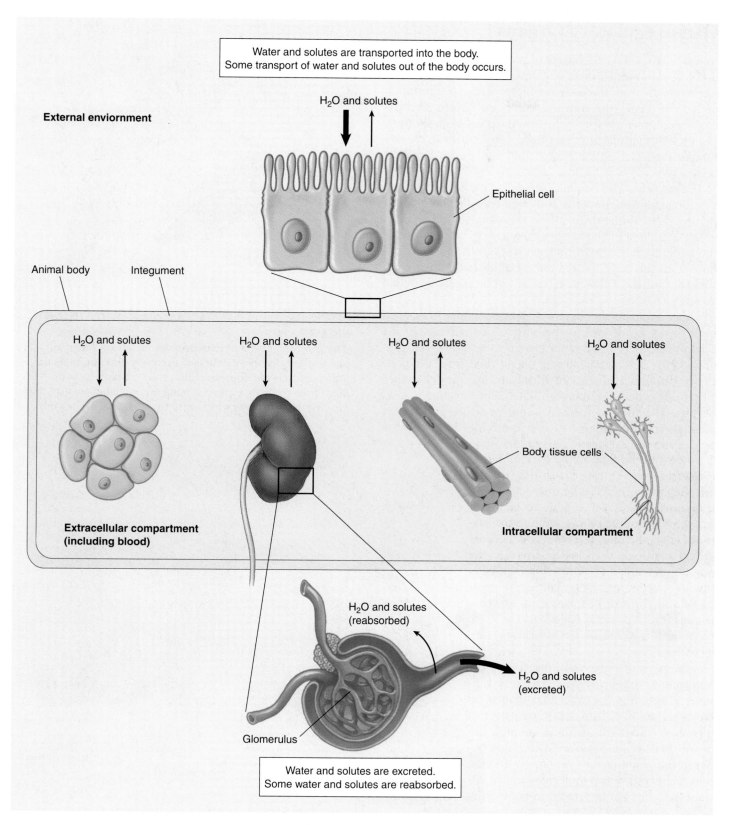

FIGURE 55.8

The interaction between intracellular and extracellular compartments of the body and the external environment. Water can be taken in from the environment or lost to the environment. Exchanges of water and solutes between the extracellular fluids of the body and the environment occur across transport epithelia, and water and solutes can be filtered out of the blood by the kidneys.

Osmoregulatory Organs

Animals have evolved a variety of mechanisms to cope with problems of water balance. In many animals, the removal of water or salts from the body is coupled with the removal of metabolic wastes through the excretory system. Protists employ contractile vacuoles for this purpose, as do sponges. Other multicellular animals have a system of excretory tubules (little tubes) that expel fluid and wastes from the body.

In flatworms, these tubules are called *protonephridia*, and they branch throughout the body into bulb-like **flame cells** (figure 55.9). While these simple excretory structures open to the outside of the body, they do not open to the inside of the body. Rather, cilia within the flame cells must draw in fluid from the body. Water and metabolites are then reabsorbed, and the substances to be excreted are expelled through excretory pores.

Other invertebrates have a system of tubules that open both to the inside and to the outside of the body. In the earthworm, these tubules are known as *metanephridia* (figure 55.10). The metanephridia obtain fluid from the body cavity through a process of filtration into funnel-shaped structures called *nephrostomes*. The term *filtration* is used because the fluid is formed under pressure and passes through small openings, so that molecules larger than a certain size are excluded. This filtered fluid is isotonic to the fluid in the coelom, but as it passes through the tubules of the metanephridia, NaCl is removed by active transport processes. A general term for transport out of the tubule and into the surrounding body fluids is *reabsorption*. Because salt is reabsorbed from the filtrate, the urine excreted is more dilute than the body fluids (is hypotonic). The kidneys of mollusks and the excretory organs of crustaceans (called *antennal glands*) also produce urine by filtration and reclaim certain ions by reabsorption.

The excretory organs in insects are the Malpighian tubules (figure 55.11), extensions of the digestive tract that branch off anterior to the hindgut. Urine is not formed by filtration in these tubules, because there is no pressure difference between the blood in the body cavity and the tubule. Instead, waste molecules and potassium (K^+) ions are secreted into the tubules by active transport. Secretion is the opposite of reabsorption—ions or molecules are transported from the body fluid into the tubule. The secretion of K^+ creates an

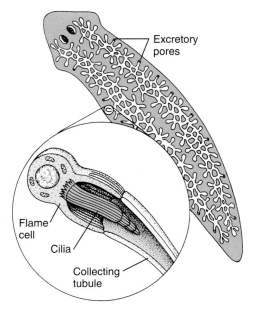

FIGURE 55.9

The protonephridia of flatworms. A branching system of tubules, bulb-like flame cells, and excretory pores make up the protonephridia of flatworms. Cilia inside the flame cells draw in fluids from the body by their beating action. Substances are then expelled through pores which open to the outside of the body.

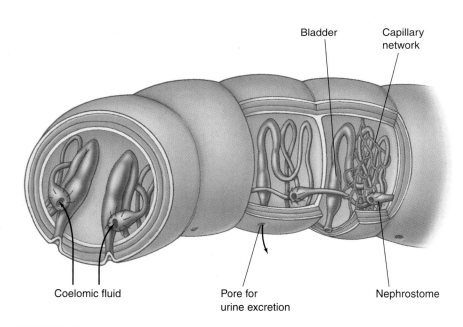

FIGURE 55.10

The metanephridia of annelids. Most invertebrates, such as the annelid shown here, have metanephridia. These consist of tubules that receive a filtrate of coelomic fluid, which enters the funnel-like nephrostomes. Salt can be reabsorbed from these tubules, and the fluid that remains, urine, is released from pores into the external environment.

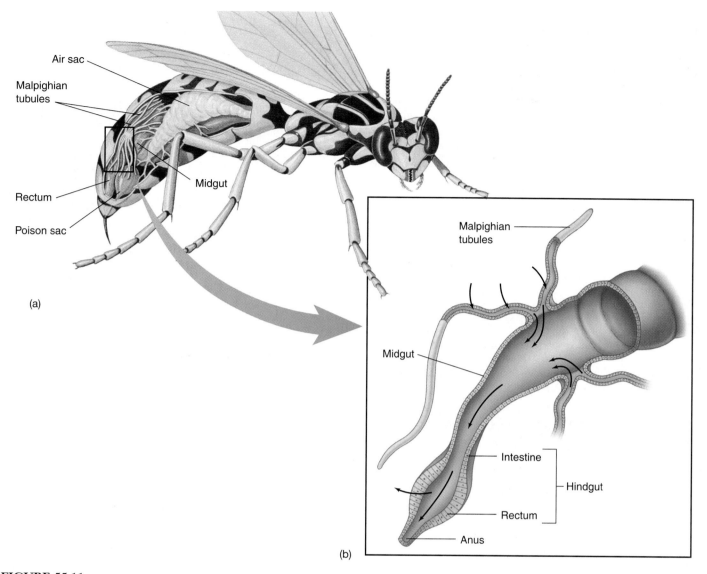

FIGURE 55.11
The Malpighian tubules of insects. (a) The Malpighian tubules of insects are extensions of the digestive tract that collect water and wastes from the body's circulatory system. (b) K$^+$ is secreted into these tubules, drawing water with it osmotically. Much of this water (see arrows) is reabsorbed across the wall of the hindgut.

osmotic gradient that causes water to enter the tubules by osmosis from the body's open circulatory system. Most of the water and K$^+$ is then reabsorbed into the circulatory system through the epithelium of the hindgut, leaving only small molecules and waste products to be excreted from the rectum along with feces. Malpighian tubules thus provide a very efficient means of water conservation.

The kidneys of vertebrates, unlike the Malpighian tubules of insects, create a tubular fluid by filtration of the blood under pressure. In addition to containing waste products and water, the filtrate contains many small molecules that are of value to the animal, including glucose, amino acids, and vitamins. These molecules and most of the water are reabsorbed from the tubules into the blood, while wastes remain in the filtrate. Additional wastes may be secreted by the tubules and added to the filtrate, and the final waste product, urine, is eliminated from the body.

It may seem odd that the vertebrate kidney should filter out almost everything from blood plasma (except proteins, which are too large to be filtered) and then spend energy to take back or reabsorb what the body needs. But selective reabsorption provides great flexibility, since various vertebrate groups have evolved the ability to reabsorb different molecules that are especially valuable in particular habitats. This flexibility is a key factor underlying the successful colonization of many diverse environments by the vertebrates.

Many invertebrates filter fluid into a system of tubules and then reabsorb ions and water, leaving waste products for excretion. Insects create an excretory fluid by secreting K$^+$ into tubules, which draws water osmotically. The vertebrate kidney produces a filtrate that enters tubules and is modified to become urine.

The Mammalian Kidney

In humans, the kidneys are fist-sized organs located in the region of the lower back. Each kidney receives blood from a renal artery, and it is from this blood that urine is produced. Urine drains from each kidney through a **ureter,** which carries the urine to a **urinary bladder.** Within the kidney, the mouth of the ureter flares open to form a funnel-like structure, the *renal pelvis.* The renal pelvis, in turn, has cup-shaped extensions that receive urine from the renal tissue. This tissue is divided into an outer **renal cortex** and an inner **renal medulla** (figure 55.12). Together, these structures perform filtration, reabsorption, secretion, and excretion.

Nephron Structure and Filtration

On a microscopic level, each kidney contains about one million functioning units known as **nephrons.** Mammalian kidneys contain a mixture of juxtamedullary nephrons, which have long loops which dip deeply into the medulla, and cortical nephrons with shorter loops (figure 55.13). The significance of the length of the loops will be explained a little later.

Each nephron consists of a system of tubules and associated small blood vessels. First, blood is carried by an *afferent arteriole* to a tuft of capillaries in the renal cortex, the **glomerulus** (figure 55.14). Here the blood is filtered as the blood pressure forces fluid through the porous capillary walls. Blood

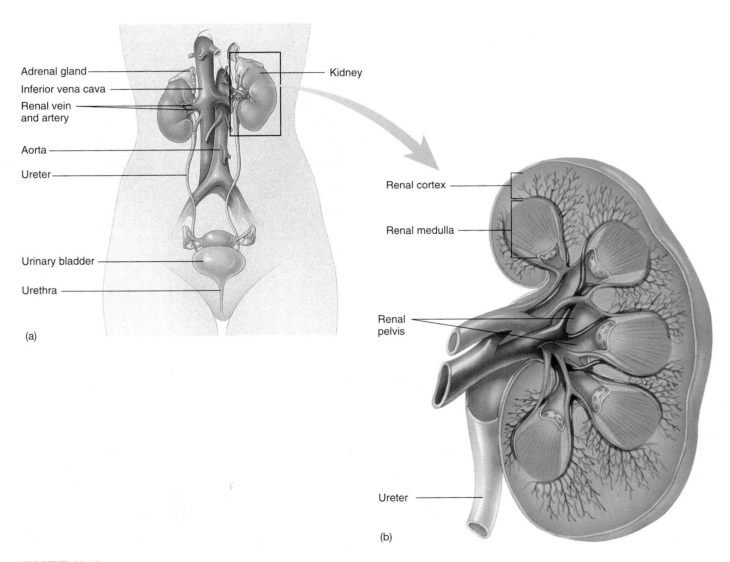

(a)

Adrenal gland
Inferior vena cava
Renal vein and artery
Aorta
Ureter
Urinary bladder
Urethra
Kidney

Renal cortex
Renal medulla
Renal pelvis
Ureter

(b)

FIGURE 55.12
The urinary system of a human female. (a) The positions of the organs of the urinary system. (b) A sectioned kidney, revealing the internal structure.

cells and plasma proteins are too large to enter this **glomerular filtrate,** but large amounts of water and dissolved molecules leave the vascular system at this step. The filtrate immediately enters the first region of the nephron tubules. This region, *Bowman's capsule*, envelops the glomerulus much as a large, soft balloon surrounds your fist if you press your fist into it. The capsule has slit openings so that the glomerular filtrate can enter the system of nephron tubules.

After the filtrate enters Bowman's capsule it goes into a portion of the nephron called the **proximal convoluted tubule,** located in the cortex. The fluid then moves down into the medulla and back up again into the cortex in a **loop of Henle.** Only the kidneys of mammals and birds have loops of Henle, and this is why only birds and mammals have the ability to concentrate their urine. After leaving the loop, the fluid is delivered to a **distal convoluted tubule** in the cortex that next drains into a **collecting duct.** The collecting duct again descends into the medulla, where it merges with other collecting ducts to empty its contents, now urine, into the renal pelvis.

Blood components that were not filtered out of the glomerulus drain into an *efferent arteriole,* which then empties into a second bed of capillaries called *peritubular capillaries* that surround the tubules. This is the only location in the body where a capillary bed (the glomerulus) is drained by an arteriole rather than by a venule, and the only location where this second arteriole delivers blood to a second capillary bed, the peritubular capillaries. As described later, the peritubular capillaries are needed for the processes of reabsorption and secretion.

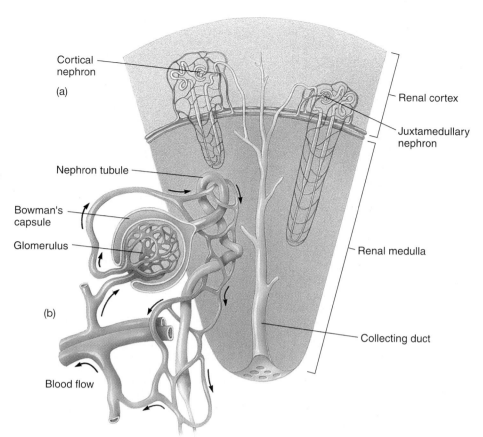

FIGURE 55.13
Nephrons are the structural components of the kidney. (a) The position of nephrons in the mammalian kidney. Cortical nephrons are located predominantly in the renal cortex, while juxtamedullary nephrons have long loops that extend deep into the renal medulla. (b) An expanded view of the relationship between the glomerular capillaries and Bowman's capsule of a nephron tubule.

FIGURE 55.14
Glomerulus in the kidney (300×). The red spherical structure in this micrograph is a glomerulus, a fine network of capillaries associated with a nephron tubule and connected to a pair of arterioles. The glomerulus is surrounded by the Bowman's capsule, partly visible as a white-brown membrane at center right.

From R. G. Kessel and R. H. Kardon, *Tissues and Organs: A Text Atlas of Scanning Electron Microscopy,* 1979, W. H. Freeman Co.

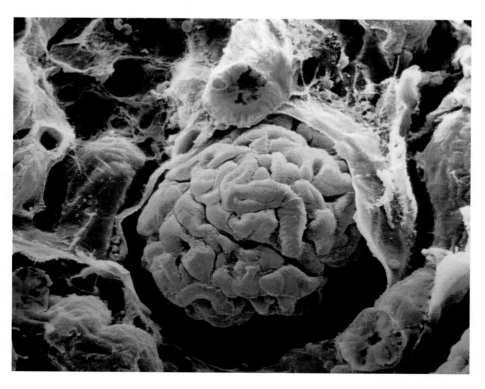

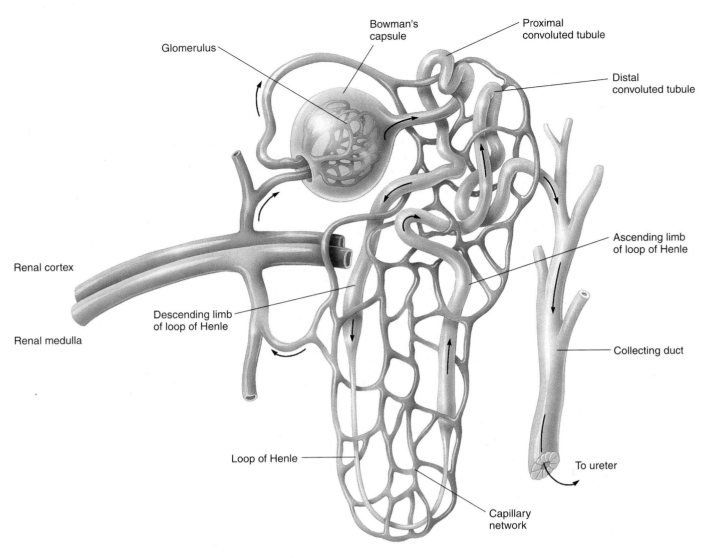

FIGURE 55.15
A nephron in a mammalian kidney. The nephron tubules are surrounded by blood capillaries, which carry away molecules and ions that are reabsorbed from the filtrate.

Reabsorption and Secretion

Most of the water and dissolved solutes that enter the glomerular filtrate must be returned to the blood (figure 55.15), or the animal would literally urinate to death. In a human, for example, approximately 2000 liters of blood passes through the kidneys each day, and 180 liters of water leaves the blood and enters the glomerular filtrate. Since we only have a total blood volume of about 5 liters and only produce 1 to 2 liters of urine per day, it is obvious that most of the filtered water is reabsorbed. The reabsorption of water occurs as a consequence of salt (NaCl) reabsorption through mechanisms that will be described shortly.

The reabsorption of glucose, amino acids, and many other molecules needed by the body is driven by active transport carriers (figure 55.16). As in all carrier-mediated transport, a maximum rate of transport is reached whenever the carriers are saturated (chapter 6). For the renal glucose carriers, saturation occurs when the concentration of glucose in the blood (and thus in the glomerular filtrate) is about 180 milligrams per 100 milliliters of blood. If a person has a blood glucose concentration in excess of this amount, as happens in untreated diabetes mellitus, the glucose left untransported in the filtrate is expelled in the urine. Indeed, the presence of glucose in the urine is diagnostic of diabetes mellitus.

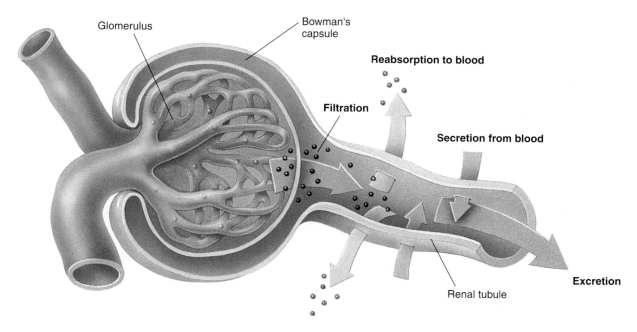

Glomerulus

Bowman's capsule

Reabsorption to blood

Filtration

Secretion from blood

Renal tubule

Excretion

FIGURE 55.16

Four functions of the kidney. Molecules enter the urine by *filtration* out of the glomerulus and by *secretion* into the tubules from surrounding blood vessels. Molecules that entered the filtrate can be returned to the blood by *reabsorption* from the tubules into surrounding blood vessels, or they may be eliminated from the body by *excretion* through the tubule to a ureter, then to the bladder.

The secretion of foreign molecules and particular waste products of the body involves the transport of these molecules across the membranes of the blood capillaries and kidney tubules into the filtrate. This process is similar to reabsorption, but it proceeds in the opposite direction. Some secreted molecules are eliminated in the urine so rapidly that they may be cleared from the blood in a single pass through the kidneys. This rapid elimination explains why penicillin, which is secreted by the tubules, must be administered in very high doses.

Excretion

A major function of the kidney is the elimination of a variety of potentially harmful substances that animals eat and drink. In addition, urine contains nitrogenous wastes, such as urea and uric acid, that are products of the catabolism of amino acids and nucleic acids. Urine may also contain excess K^+, H^+, and other ions that are removed from the blood. Urine's generally high H^+ concentration (pH 5–7)

helps maintain the acid-base balance of the blood within a narrow range (pH 7.35–7.45). Moreover, the excretion of water in urine contributes to the maintenance of blood volume and pressure; the larger the volume of urine excreted, the lower the blood volume.

The purpose of kidney function is therefore homeostasis—the kidneys are critically involved in maintaining the constancy of the internal environment. When disease interferes with kidney function, it causes a rise in the blood concentration of nitrogenous waste products, disturbances in electrolyte and acid-base balance, and a failure in blood pressure regulation. Such potentially fatal changes highlight the central importance of the kidneys in normal body physiology.

The mammalian kidney is divided into a cortex and medulla and contains microscopic functioning units called nephrons. The nephron tubules receive a blood filtrate from the glomeruli and modify this filtrate to produce urine, which empties into the renal pelvis and is expelled from the kidney through the ureter.

Evolution of the Vertebrate Kidney

In the following sections, we will discuss the structure and function of the kidney in the major groups of vertebrates. Although the same basic design has been retained in all vertebrate kidneys, there have been a few modifications. Since the original glomerular filtrate is isotonic to blood, all vertebrates can produce a urine that is isotonic to (by reabsorbing ions) or hypotonic to (more dilute than) blood. Only birds and mammals can reabsorb water from their glomerular filtrate to produce a urine that is hypertonic to (more concentrated than) blood.

Freshwater Fish

Kidneys are thought to have evolved first among the freshwater teleosts, or bony fish (figure 55.17). Because the body fluids of a freshwater fish have a greater osmotic concentration than the surrounding water, these animals face two serious problems: (1) water tends to enter the body from the environment; and (2) solutes tend to leave the body and enter the environment. Freshwater fish address the first problem by *not* drinking water and by excreting a large volume of dilute urine, which is hypotonic to their body fluids. They address the second problem by reabsorbing ions across the nephron tubules, from the glomerular filtrate back into the blood. In addition, they actively transport ions across their gills from the surrounding water into the blood.

Marine Bony Fish

Although most groups of animals seem to have evolved first in the sea, marine bony fish (teleosts) probably evolved from freshwater ancestors, as was mentioned in chapter 45. They faced significant new problems in making the transition to the sea because their body fluids are hypotonic to the surrounding seawater. Consequently, water tends to leave their bodies by osmosis across their gills, and they also lose water in their urine. To compensate for this continuous water loss, marine fish drink large amounts of seawater (figure 55.18).

Many of the divalent cations (principally Ca^{++} and Mg^{++}) in the seawater that a marine fish drinks remain in the digestive tract and are eliminated through the anus. Some, however, are absorbed into the blood, as are the monovalent ions K^+, Na^+, and Cl^-. Most of the monovalent ions are actively transported out of the blood across the gills, while the divalent ions that enter the blood are secreted into the nephron tubules and excreted in the urine. In these two ways, marine bony fish eliminate the ions they get from the seawater they drink. The urine they excrete is isotonic to their body fluids. It is more concentrated than the urine of freshwater fish, but not as concentrated as that of birds and mammals.

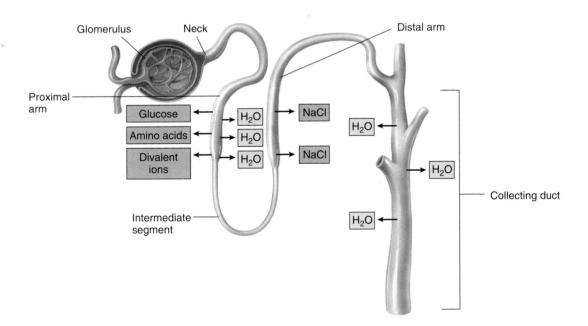

FIGURE 55.17
The basic organization of the vertebrate nephron. The nephron tubule of the freshwater fish is a basic design that has been retained in the kidneys of marine fish and terrestrial vertebrates that evolved later. Sugars, amino acids, and divalent ions such as Ca^{++} are recovered in the proximal arm; monovalent ions such as Na^+ and Cl^- are recovered in the distal arm; and water is recovered in the collecting duct.

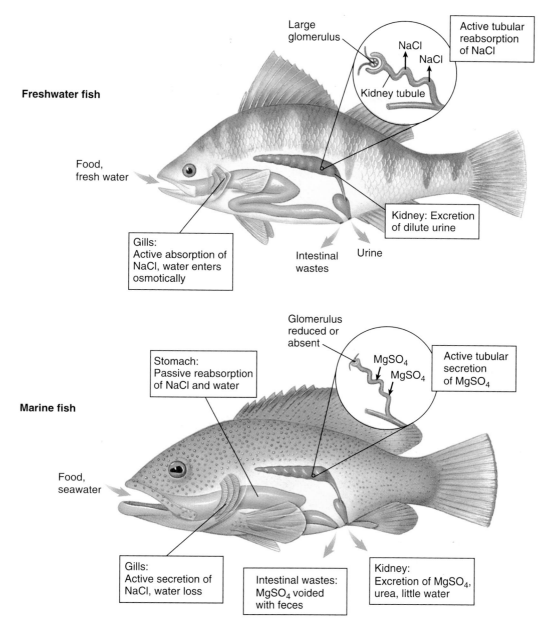

FIGURE 55.18
Freshwater and marine teleosts (bony fish) face different osmotic problems. Whereas the freshwater teleost is hypertonic to its environment, the marine teleost is hypotonic to seawater. To compensate for its tendency to take in water and lose ions, a freshwater fish excretes dilute urine, avoids drinking water, and reabsorbs ions across the nephron tubules. To compensate for its osmotic loss of water, the marine teleost drinks seawater and eliminates the excess ions through transport across epithelia.

Cartilaginous Fish

The elasmobranchs, including sharks and rays, are by far the most common subclass in the class Chondrichthyes (cartilaginous fish). Elasmobranchs have solved the osmotic problem posed by their seawater environment in a different way than have the bony fish. Instead of having body fluids that are hypotonic to seawater, so that they have to continuously drink seawater and actively pump out ions, the elasmobranchs reabsorb urea from the nephron tubules and maintain a blood urea concentration that is 100 times higher than that of mammals. This added urea makes their blood approximately isotonic to the surrounding sea. Since there is no net water movement between isotonic solutions, water loss is prevented. Hence, these fishes do not need to drink seawater for osmotic balance, and their kidneys and gills do not have to remove large amounts of ions from their bodies. The enzymes and tissues of the cartilaginous fish have evolved to tolerate the high urea concentrations.

Amphibians and Reptiles

The first terrestrial vertebrates were the amphibians, and the amphibian kidney is identical to that of freshwater fish. This is not surprising, since amphibians spend a significant portion of their time in fresh water, and when on land, they generally stay in wet places. Amphibians produce a very dilute urine and compensate for their loss of Na^+ by actively transporting Na^+ across their skin from the surrounding water.

Reptiles, on the other hand, live in diverse habitats. Those living mainly in fresh water, like some of the crocodilians, occupy a habitat similar to that of the freshwater fish and amphibians and thus have similar kidneys. Marine reptiles, which consist of other crocodilians, turtles, sea snakes, and one lizard, possess kidneys similar to those of their freshwater relatives but face opposite problems; they tend to lose water and take in salts. Like marine teleosts (bony fish), they drink the seawater and excrete an isotonic urine. Marine teleosts eliminate the excess salt by transport across their gills, while marine reptiles eliminate excess salt through salt glands located near the nose or the eye (figure 55.19).

The kidneys of terrestrial reptiles also reabsorb much of the salt and water in the nephron tubules, helping somewhat to conserve blood volume in dry environments. Like fish and amphibians, they cannot produce urine that is more concentrated than the blood plasma. However, when their urine enters their cloaca (the common exit of the digestive and urinary tracts), additional water can be reabsorbed.

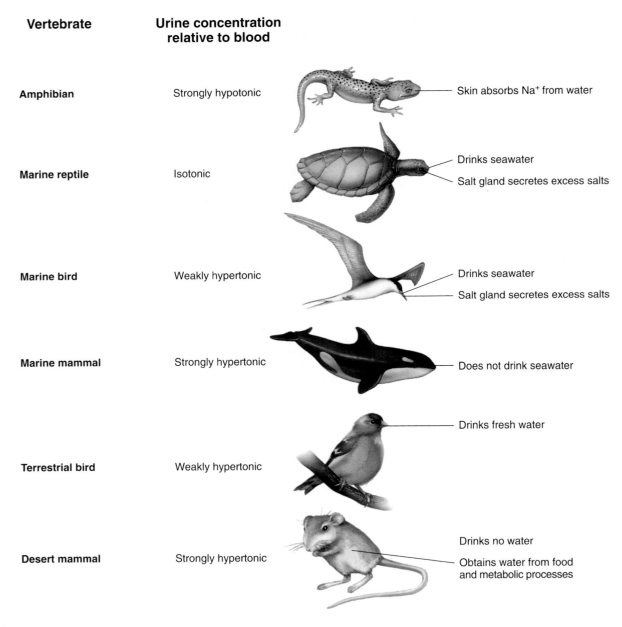

Vertebrate	Urine concentration relative to blood	
Amphibian	Strongly hypotonic	Skin absorbs Na+ from water
Marine reptile	Isotonic	Drinks seawater / Salt gland secretes excess salts
Marine bird	Weakly hypertonic	Drinks seawater / Salt gland secretes excess salts
Marine mammal	Strongly hypertonic	Does not drink seawater
Terrestrial bird	Weakly hypertonic	Drinks fresh water
Desert mammal	Strongly hypertonic	Drinks no water / Obtains water from food and metabolic processes

FIGURE 55.19

Osmoregulation by some vertebrates. Only birds and mammals can produce a hypertonic urine and thereby retain water efficiently, but marine reptiles and birds can drink seawater and excrete the excess salt through salt glands.

Mammals and Birds

Mammals and birds are the only vertebrates able to produce urine with a higher osmotic concentration than their body fluids. This allows these vertebrates to excrete their waste products in a small volume of water, so that more water can be retained in the body. Human kidneys can produce urine that is as much as 4.2 times as concentrated as blood plasma, but the kidneys of some other mammals are even more efficient at conserving water. For example, the camel, gerbil, and pocket mouse *Perognathus* can excrete urine 8, 14, and 22 times as concentrated as their blood plasma, respectively. The kidneys of the kangaroo rat (figure 55.20) are so efficient it never has to drink water; it can obtain all the water it needs from its food and from water produced in aerobic cell respiration!

The production of hypertonic urine is accomplished by the loop of Henle portion of the nephron (see figure 55.15), found only in mammals and birds. A nephron with a long loop of Henle extends deeper into the renal medulla, where the hypertonic osmotic environment draws out more water, and so can produce more concentrated urine. Most mammals have some nephrons with short loops and other nephrons with loops that are much longer (see figure 55.13). Birds, however, have relatively few or no nephrons with long loops, so they cannot produce urine that is as concentrated as that of mammals. At most, they can only reabsorb enough water to produce a urine that is about twice the concentration of their blood. Marine birds solve the problem of water loss by drinking salt water and then excreting the excess salt from salt glands near the eyes (figure 55.21).

The moderately hypertonic urine of a bird is delivered to its cloaca, along with the fecal material from its digestive tract. If needed, additional water can be absorbed across the wall of the cloaca to produce a semisolid white paste or pellet, which is excreted.

The kidneys of freshwater fish must excrete copious amounts of very dilute urine, while marine teleosts drink seawater and excrete an isotonic urine. The basic design and function of the nephron of freshwater fishes have been retained in the terrestrial vertebrates. Modifications, particularly the presence of a loop of Henle, allow mammals and birds to reabsorb water and produce a hypertonic urine.

FIGURE 55.20
The kangaroo rat, *Dipodomys panamintensis*. This mammal has very efficient kidneys that can concentrate urine to a high degree by reabsorbing water, thereby minimizing water loss from the body. This feature is extremely important to the kangaroo rat's survival in dry or desert habitats.

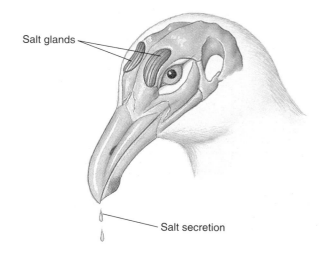

FIGURE 55.21
Marine birds drink seawater and then excrete the salt through salt glands. The salty fluid excreted by these glands can then dribble down the beak.

Transport Processes in the Mammalian Nephron

As previously described, approximately 180 liters (in a human) of isotonic glomerular filtrate enters the Bowman's capsules each day. After passing through the remainder of the nephron tubules, this volume of fluid would be lost as urine if it were not reabsorbed back into the blood. It is clearly impossible to produce this much urine, yet water is only able to pass through a cell membrane by osmosis, and osmosis is not possible between two isotonic solutions. Therefore, some mechanism is needed to create an osmotic gradient between the glomerular filtrate and the blood, allowing reabsorption.

Proximal Tubule

Approximately two-thirds of the NaCl and water filtered into Bowman's capsule is immediately reabsorbed across the walls of the proximal convoluted tubule. This reabsorption is driven by the active transport of Na^+ out of the filtrate and into surrounding blood vessels. Cl^- follows Na^+ passively because of electrical attraction, and water follows them both because of osmosis. Since NaCl and water are removed from the filtrate in proportionate amounts, the filtrate that remains in the tubule is still isotonic to the blood plasma.

Although only one-third of the initial volume of filtrate remains in the nephron tubule after the initial reabsorption of NaCl and water, it still represents a large volume (60 L out of the original 180 L of filtrate produced per day by both human kidneys). Obviously, no animal can excrete that much urine, so most of this water must be reabsorbed. It is reabsorbed primarily across the wall of the collecting duct because the interstitial fluid of the renal medulla surrounding the collecting ducts is hypertonic. The hypertonic renal medulla draws water out of the collecting duct by osmosis, leaving behind a hypertonic urine for excretion.

Loop of Henle

The reabsorption of much of the water in the tubular filtrate thus depends on the creation of a hypertonic renal medulla; the more hypertonic the medulla is, the steeper the osmotic gradient will be and the more water will leave the collecting ducts. It is the loops of Henle that create the hypertonic renal medulla in the following manner (figure 55.22):

1. The ascending limb of the loop actively extrudes Na^+, and Cl^- follows. The mechanism that extrudes NaCl from the ascending limb of the loop differs from that which extrudes NaCl from the proximal tubule, but the most important difference is that the ascending limb

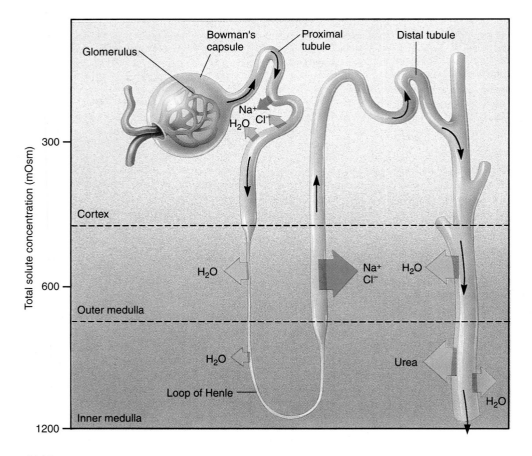

FIGURE 55.22
The reabsorption of salt and water in the mammalian kidney.
Active transport of Na^+ out of the proximal tubules is followed by the passive movement of Cl^- and water. Active extrusion of NaCl from the ascending limb of the loop of Henle creates the osmotic gradient required for the reabsorption of water from the collecting duct. The changes in osmolality from the cortex to the medulla is indicated to the left of the figure.

is *not permeable to water*. As Na⁺ exits, the fluid within the ascending limb becomes increasingly dilute (hypotonic) as it enters the cortex, while the surrounding tissue becomes increasingly concentrated (hypertonic).

2. The NaCl pumped out of the ascending limb of the loop is trapped within the surrounding interstitial fluid. This is because the blood vessels in the medulla also have loops, called *vasa recta*, so that NaCl can diffuse from the blood leaving the medulla to the blood entering the medulla. Thus, the vasa recta functions in a countercurrent exchange, similar to that described for oxygen in the countercurrent flow of water and blood in the gills of fish (chapter 50). In the case of the renal medulla, the diffusion of NaCl between the blood vessels keeps much of the NaCl within the interstitial fluid, making it hypertonic.

3. The descending limb is permeable to water, so water leaves by osmosis as the fluid descends into the hypertonic renal medulla. This water enters the blood vessels of the vasa recta and is carried away in the general circulation.

4. The loss of water from the descending limb multiplies the concentration that can be achieved at each level of the loop through the active extrusion of NaCl by the ascending limb. The longer the loop of Henle, the longer the region of interaction between the descending and ascending limbs, and the greater the total concentration that can be achieved. In a human kidney, the concentration of filtrate entering the loop is 300 milliosmolal, and this concentration is multiplied to more than 1200 milliosmolal at the bottom of the longest loops of Henle in the renal medulla.

Because fluid flows in opposite directions in the two limbs of the loop, the action of the loop of Henle in creating a hypertonic renal medulla is known as the *countercurrent multiplier system*. The high solute concentration of the renal medulla is primarily the result of NaCl accumulation by the countercurrent multiplier system, but urea also contributes to the total osmolality of the medulla. This is because the descending limb of the loop of Henle and the collecting duct are permeable to urea, which leaves these regions of the nephron by diffusion.

Distal Tubule and Collecting Duct

Because NaCl was pumped out of the ascending limb, the filtrate that arrives at the distal convoluted tubule and enters the collecting duct in the renal cortex is hypotonic (with a concentration of only 100 mOsm). The collecting duct carrying this dilute fluid now plunges into the medulla. As a result of the hypertonic interstitial fluid of the renal medulla, there is a strong osmotic gradient that pulls water out of the collecting duct and into surrounding blood vessels.

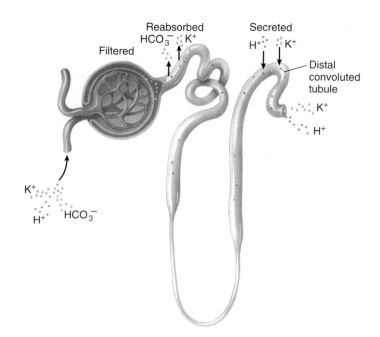

FIGURE 55.23
The nephron controls the amounts of K⁺, H⁺, and HCO₃⁻ excreted in the urine. K⁺ is completely reabsorbed in the proximal tubule and then secreted in varying amounts into the distal tubule. HCO₃⁻ is filtered but normally completely reabsorbed. H⁺ is filtered and also secreted into the distal tubule, so that the final urine has an acidic pH.

The osmotic gradient is normally constant, but the permeability of the collecting duct to water is adjusted by a hormone, **antidiuretic hormone** (**ADH,** also called **vasopressin**), discussed in chapter 53. When an animal needs to conserve water, the posterior pituitary gland secretes more ADH, and this hormone increases the number of water channels in the plasma membranes of the collecting duct cells. This increases the permeability of the collecting ducts to water so that more water is reabsorbed and less is excreted in the urine. The animal thus excretes a hypertonic urine.

In addition to the regulation of water balance, the kidneys regulate the balance of electrolytes in the blood by reabsorption and secretion. For example, the kidneys reabsorb K⁺ in the proximal tubule and then secrete an amount of K⁺ needed to maintain homeostasis into the distal convoluted tubule (figure 55.23). The kidneys also maintain acid-base balance by excreting H⁺ into the urine and reabsorbing bicarbonate (HCO₃⁻), as previously described.

The loop of Henle creates a hypertonic renal medulla as a result of the active extrusion of NaCl from the ascending limb and the interaction with the descending limb. The hypertonic medulla then draws water osmotically from the collecting duct, which is permeable to water under the influence of antidiuretic hormone.

Ammonia, Urea, and Uric Acid

Amino acids and nucleic acids are nitrogen-containing molecules. When animals catabolize these molecules for energy or convert them into carbohydrates or lipids, they produce nitrogen-containing by-products called **nitrogenous wastes** (figure 55.24) that must be eliminated from the body.

The first step in the metabolism of amino acids and nucleic acids is the removal of the amino (—NH₂) group and its combination with H⁺ to form **ammonia** (NH₃) in the liver. Ammonia is quite toxic to cells and therefore is safe only in very dilute solutions. The excretion of ammonia is not a problem for the bony fish and tadpoles, which eliminate most of it by diffusion through the gills and less by excretion in very dilute urine. In elasmobranchs, adult amphibians, and mammals, the nitrogenous wastes are eliminated in the far less toxic form of **urea**. Urea is water-soluble and so can be excreted in large amounts in the urine. Reptiles, birds, and insects excrete nitrogenous wastes in the form of **uric acid,** which is only slightly soluble in water. As a result of its low solubility, uric acid precipitates and thus can be excreted using very little water. Uric acid forms the pasty white material in bird droppings.

Thus, animals have evolved three general solutions to the problem of nitrogenous waste removal:

1. **Flushing.** Bony fish remove the amino portion of amino acids and form ammonia. The ammonia can be carried away by the water that passes over the gills.
2. **Detoxification.** All mammals and many other land animals incorporate ammonia into urea, which is far less toxic than ammonia and so can be transported and excreted at far higher concentrations. Urea is carried by the bloodstream from its place of synthesis in the liver to the kidneys, where it is excreted in the urine.
3. **Insolubilization.** Birds and terrestrial reptiles face a special problem: because their eggs are encased within shells, metabolic wastes build up as the embryo grows within the egg. The solution these animals have evolved is to incorporate ammonia into uric acid, which is insoluble in water. This metabolic process, while lengthy and requiring considerable energy, produces a compound that crystallizes and precipitates. In this way, the toxic nitrogenous waste product is removed from solution. Adult birds and terrestrial reptiles carry out the same process, permitting nitrogenous wastes to be eliminated in the urine with less water than is required for the excretion of urea. The semisolid paste produced is called **guano.**

Mammals also produce some uric acid, but it is a waste product of the degradation of purine nucleotides (see chapter 3), not of amino acids. Most mammals have an enzyme called *uricase*, which converts uric acid into a more soluble derivative, **allantoin.** Only humans, apes, and the dalmatian dog lack this enzyme and so must excrete the uric acid. In humans, excessive accumulation of uric acid in the joints produces a condition known as *gout.*

The metabolic breakdown of amino acids and nucleic acids produces ammonia as a by-product. Ammonia is excreted by bony fish, but other vertebrates convert nitrogenous wastes into urea and uric acid, which are less toxic nitrogenous wastes.

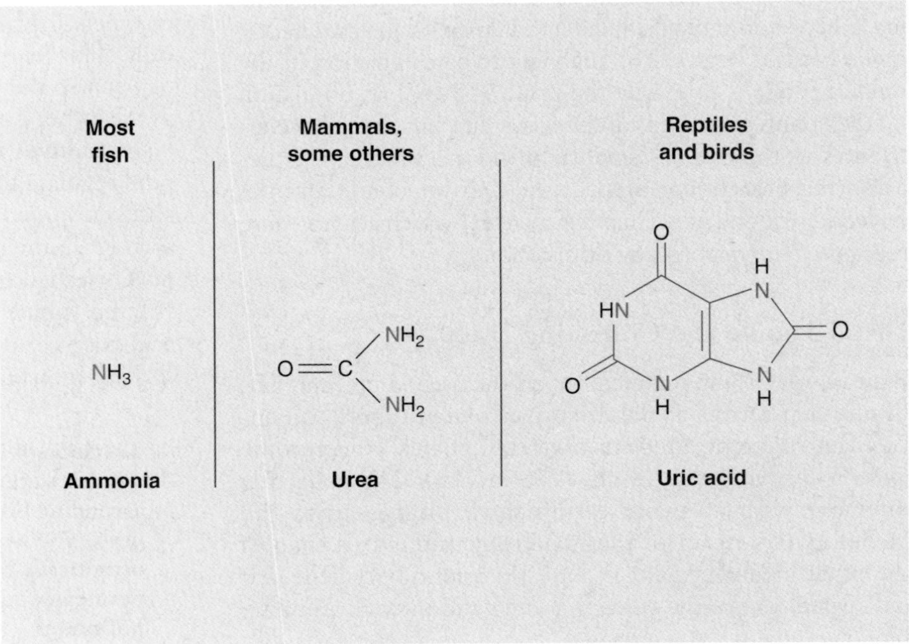

FIGURE 55.24
Nitrogenous wastes. When amino acids and nucleic acids are metabolized, the immediate by-product is ammonia, which is quite toxic but which can be eliminated through the gills of teleost fish. Mammals convert ammonia into urea, which is less toxic. Birds and terrestrial reptiles convert it instead into uric acid, which is insoluble in water.

Antidiuretic Hormone and Aldosterone

In mammals and birds, the amount of water excreted in the urine, and thus the concentration of the urine, varies according to the changing needs of the body. Acting through the mechanisms described next, the kidneys will excrete a hypertonic urine when the body needs to conserve water. If an animal drinks too much water, the kidneys will excrete a hypotonic urine. As a result, the volume of blood, the blood pressure, and the osmolality of blood plasma are maintained relatively constant by the kidneys, no matter how much water you drink. The kidneys also regulate the plasma K^+ and Na^+ concentrations and blood pH within very narrow limits. These homeostatic functions of the kidneys are coordinated primarily by two hormones, antidiuretic hormone and aldosterone (chapter 53).

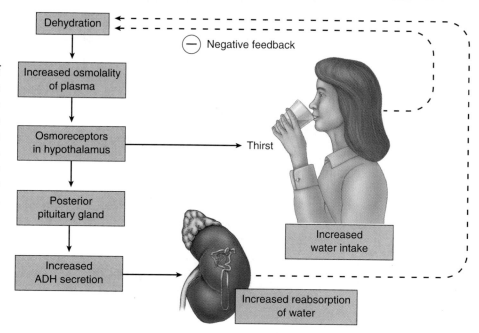

FIGURE 55.25
Antidiuretic hormone stimulates the reabsorption of water by the kidneys. This action completes a negative feedback loop and helps to maintain homeostasis of blood volume and osmolality.

Antidiuretic Hormone

Antidiuretic hormone (ADH) is produced by the hypothalamus and secreted by the posterior pituitary gland. The primary stimulus for ADH secretion is an increase in the osmolality of the blood plasma. The osmolality of plasma increases when a person is dehydrated or when a person eats salty food. Osmoreceptors in the hypothalamus respond to the elevated blood osmolality by triggering a sensation of thirst and an increase in the secretion of ADH (figure 55.25).

ADH causes the walls of the collecting ducts in the kidney to become more permeable to water. This occurs because water channels are contained within the membranes of intracellular vesicles in the epithelium of the collecting ducts, and ADH stimulates the fusion of the vesicle membrane with the plasma membrane, similar to the process of exocytosis. When the secretion of ADH is reduced, the plasma membrane pinches in to form new vesicles that contain the water channels, so that the plasma membrane becomes less permeable to water.

Since the extracellular fluid in the renal medulla is hypertonic to the filtrate in the collecting ducts, water leaves the filtrate by osmosis and is reabsorbed into the blood. Under conditions of maximal ADH secretion, a person excretes only 600 milliliters of highly concentrated urine per day. A person who lacks ADH due to pituitary damage has the disorder known as *diabetes insipidus* and constantly excretes a large volume of dilute urine. Such a person is in danger of becoming severely dehydrated and succumbing to dangerously low blood pressure.

Aldosterone and Atrial Natriuretic Hormone

Sodium ion is the major solute in the blood plasma. When the blood concentration of Na^+ falls, therefore, the blood osmolality also falls. This drop in osmolality inhibits ADH secretion, causing more water to remain in the collecting duct for excretion in the urine. As a result, the blood volume and blood pressure decrease. If Na^+ deprivation is severe, the blood volume may fall so low that there is insufficient blood pressure to sustain life. For this reason, salt is necessary for life. Many animals have a "salt hunger" and actively seek salt, such as the deer at "salt licks."

A drop in blood Na^+ concentration is normally compensated by the kidneys under the influence of the hormone aldosterone, which is secreted by the adrenal cortex. Aldosterone stimulates the distal convoluted tubules to reabsorb Na^+, decreasing the excretion of Na^+ in the urine. Indeed, under conditions of maximal aldosterone secretion, Na^+ may be completely absent from the urine. The reabsorption of Na^+ is followed by Cl^- and by water, so aldosterone has the net effect of promoting the retention of both salt and water. It thereby helps to maintain blood volume and pressure.

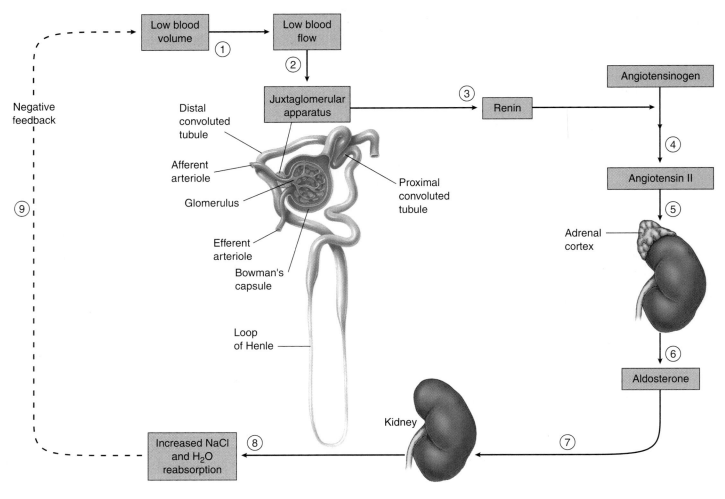

FIGURE 55.26
A lowering of blood volume activates the renin-angiotensin-aldosterone system. Aldosterone stimulates the reabsorption of NaCl and water from the nephron, serving as a negative feedback control on the blood volume.

The secretion of aldosterone in response to a decreased blood level of Na⁺ is indirect. Since a fall in blood Na⁺ is accompanied by a decreased blood volume, there is a reduced flow of blood past a group of cells called the juxtaglomerular apparatus, located in the region of the kidney between the distal convoluted tubule and the afferent arteriole (figure 55.26). The juxtaglomerular apparatus responds by secreting the enzyme *renin* into the blood, which catalyzes the production of the polypeptide angiotensin I from the protein angiotensinogen (see chapter 49). Angiotensin I is then converted by another enzyme into angiotensin II, which stimulates blood vessels to constrict and the adrenal cortex to secrete aldosterone. Thus, homeostasis of blood volume and pressure can be maintained by the activation of this renin-angiotensin-aldosterone system.

In addition to stimulating Na⁺ reabsorption, aldosterone also promotes the secretion of K⁺ into the distal convoluted tubules. Consequently, aldosterone lowers the blood K⁺ concentration, helping to maintain constant blood K⁺ levels in the face of changing amounts of K⁺ in the diet. People who lack the ability to produce aldosterone will die if untreated because of the excessive loss of salt and water in the urine and the buildup of K⁺ in the blood.

The action of aldosterone in promoting salt and water retention is opposed by another hormone, atrial natriuretic hormone (see chapter 49). This hormone is secreted by the right atrium of the heart in response to an increased blood volume, which stretches the atrium. Under these conditions, aldosterone secretion from the adrenal cortex will decrease and atrial natriuretic hormone secretion will increase, thus promoting the excretion of salt and water in the urine and lowering the blood volume.

ADH stimulates the insertion of water channels into the cells of the collecting duct, making the collecting duct more permeable to water. Thus, ADH stimulates the reabsorption of water and the excretion of a hypertonic urine. Aldosterone promotes the reabsorption of NaCl and water across the distal convoluted tubule, as well as the secretion of K⁺ into the tubule.

55.1 The regulatory systems of the body maintain homeostasis.

- Negative feedback loops maintain nearly constant extracellular conditions in the internal environment of the body, a condition called homeostasis.
- Antagonistic effectors afford an even finer degree of control.

55.2 The extracellular fluid concentration is constant in most vertebrates.

- Osmoconformers maintain a tissue fluid osmolality equal to that of their environment, whereas osmoregulators maintain a constant blood osmolality that is different from that of their environment.
- The kidneys of most vertebrates eliminate water by filtering blood into kidney tubules, whereas insects eliminate water by secreting K^+ into Malpighian tubules and the water follows the K^+ by osmosis.

55.3 The functions of the vertebrate kidney are performed by nephrons.

- The glomerular filtrate enters Bowman's capsule as an isotonic fluid; blood cells and plasma proteins, however, are excluded from the filtrate.
- The primary function of the kidneys is homeostasis of blood volume, pressure, and composition, including the concentration of particular solutes in the blood and the blood pH.
- Freshwater bony fish are hypertonic to their environment, and saltwater bony fish are hypotonic to their environment; these conditions place different demands upon their kidneys and other regulatory systems.

- Birds and mammals are the only vertebrates that have loops of Henle and thus are capable of producing a hypertonic urine.
- The ascending limb of the loop of Henle actively extrudes NaCl but is impermeable to water; the filtrate entering the distal convoluted tubule is thus hypotonic.
- The descending limb of the loop of Henle is permeable to water, so water leaves by osmosis; the interaction between the two limbs of the loop create a hypertonic renal medulla.
- The wall of the collecting duct is permeable to water under stimulation by ADH, and so water is drawn out, and a hypertonic urine is excreted.
- Bony fish remove the amine portions of amino acids and excrete them as ammonia across the gills.
- Elasmobranchs, adult amphibians, and mammals produce urea, which is quite soluble but much less toxic than ammonia, and excrete the urea in the urine.
- Insects, reptiles, and birds produce uric acid from the amino groups in amino acids; the precipitates so that little water is required for its excretion.

55.4 The kidney is regulated by hormones.

- Antidiuretic hormone is secreted by the posterior pituitary gland in response to an increase in blood osmolality, and acts to increase the number of water channels in the wall of the collecting duct.
- Aldosterone stimulates the distal convoluted tubule to reabsorb Na^+ and secrete K^+ into the filtrate.

1. **Homeostasis** Homeostasis refers to the dynamic constancy of the internal environment. Most vertebrate regulatory systems function to maintain homeostasis.

2. **Negative feedback** Mechanisms that maintain homeostasis operate through negative feedback, in which a deviation from constancy stimulates the activity of effectors that act to counteract the deviation.

3. **Osmolality** The osmolality is the total moles of solute per kilogram water in a solution.

4. **Osmoregulator** An osmoregulator is an organism that maintains a constant plasma osmolality despite differences in concentration between the body fluids and the external environment. The opposite of an osmoregulator is an osmoconformer, which maintains the same body fluid osmolality as the external environment.

5. **Nephron** The nephron is a microscopic structure in the kidney, consisting of tubules and associated blood vessels. It is the functional unit of the kidney, wherein a blood filtrate is modified and converted into urine.

6. **Reabsorption and secretion** Reabsorption refers to the transport of molecules out of the filtrate, through the tubule wall, and into the surrounding blood vessels. Secretion refers to the transport in the opposite direction, enabling the tubule to excrete molecules that were not originally present in the glomerular filtrate.

7. **Glomerulus** A glomerulus is a tuft of capillaries in the renal cortex that filters the blood to produce a filtrate that enters the renal tubules and is converted by reabsorption and secretion processes into urine.

Review Questions

1. What is homeostasis? What is a negative feedback loop? Give an example of how homeostasis is maintained by a negative feedback loop.

2. What is the difference between an osmoconformer and an osmoregulator? What are examples of each?

3. How does the body fluid osmolality of a freshwater vertebrate compare with that of its environment? Does water tend to enter or exit its body? What must it do to maintain proper body water levels?

4. In what type of animal are Malpighian tubules found? By what mechanism is fluid caused to flow into these tubules? How is this fluid further modified before it is excreted?

5. What drives the movement of fluid from the blood to the inside of the nephron tubule at Bowman's capsule? What substances move into the tubule, and what remains in the blood?

6. In what portion of the nephron is most of the NaCl and water reabsorbed from the filtrate? How does the osmolality of the filtrate remaining in this portion compare with that of the blood plasma?

7. In what part of the loop of Henle does the active transport of Na^+ take place? How do the descending and ascending limbs interact, and what is the purpose of this interaction?

8. What causes water reabsorption from the collecting duct? How is this influenced by antidiuretic hormone?

9. What effects does aldosterone have on kidney function? How is the secretion of aldosterone stimulated?

Thought Questions

1. The glomeruli filter out a tremendous amount of water and molecules needed by the body, which must later be reclaimed by the energy-requiring process of reabsorption. The Malpighian tubules of the insects might seem to function more logically, secreting molecules and ions that need to be excreted. What advantages might a filtration-reabsorption process provide over a strictly secretion process of elimination?

2. In the mammalian kidney, salt that is extruded from the ascending limbs of the loops of Henle is trapped in the renal medulla, making it hypertonic, but water that is removed from the descending limb is reabsorbed back into the blood and not used to dilute the salty medulla. How do the arrangement of blood vessels in the medulla permit this to happen?

Internet Links

Digestive and Renal Disorders
http://www.niddk.nih.gov/kidney/kidney.htm
The home page of the NATIONAL INSTITUTE OF DIABETES AND DIGESTIVE AND KIDNEY DISEASES, the single largest institute of the NIH, with links to a wealth of information about diabetes and about the digestive and renal systems. Fun to explore.

How Your Kidney Works
http://www.renalnet.org/Renalnet/renalnet.cfm
RENALNET is a kidney information clearinghouse for information on kidney function and disease.

For Further Reading

Beauchamp, G. K.: "The Human Preference for Excess Salt," *American Scientist*, vol. 75, page 27, 1987. Discusses salt appetite and the possible relationship of salt consumption to hypertension.

Marshall, E.: "Testing Urine for Drugs," *Science*, vol. 241, July 1988, pages 150–52. Urinalysis now provides fast and accurate tests for the presence of many drugs.

McArthur, A. and J. Clark: "Body Temperature and Heat and Water Balance," *Nature*, vol. 326, pages 647–48, April 16, 1987. The role of water in the body in thermoregulation is discussed.

Murray, J.: "Human Organ Transplantation: Background and Consequences," *Science*, vol. 256, June 5, 1992, pages 1411–16. The kidney is one of the most frequently transplanted human organs, and discussed as the primary example in this paper on tissue and organ transplantation.

O'Brien, C.: "Lucky Break for Kidney Disease Gene," *Science*, vol. 264, June 24, 1994, page 1844. Identification of the gene apparently responsible for the development of kidney disease offers a wealth of possible treatments—including gene therapy.

Palfrey, C. and A. Cossins: "Fishy Tales of Kidney Function," *Nature*, vol. 371, September 29, 1994, pages 377–78. From an evolutionary standpoint, understanding the fish kidney is critical to understanding osmoregulation. Fishes also have the added challenges of living in environments high in (obviously) water content, and, in the case of marine fishes, salt content.

Schmidt-Nielsen, K.: *Animal Physiology: Adaptation and Environment*, 4th ed., Cambridge University Press, Cambridge, 1990. This general animal physiology text has an excellent chapter on excretion by one of the pioneers in comparative vertebrate physiology.

56

Sex and Reproduction

Concept Outline

56.1 Animals employ both sexual and asexual reproductive strategies.

Asexual and Sexual Reproduction.   Some animals reproduce asexually, but most reproduce sexually; male and female are usually different individuals, but not always.

56.2 The evolution of reproduction among the vertebrates has led to internalization of fertilization and development.

Fertilization and Development.   Among vertebrates that have internal fertilization, the young are nourished by egg yolk or from their mother's blood.

Fish and Amphibians.   Most bony fish and amphibians have external fertilization, while most cartilaginous fish have internal fertilization.

Reptiles and Birds.   Most reptiles and all birds lay eggs externally, and the young develop inside the egg.

Mammals.   Monotremes lay eggs, marsupials have pouches where their young develop, and placental mammals have placentas that nourish the young within the uterus.

56.3 Male and female reproductive systems are specialized for different functions.

Structure and Function of the Male Reproductive System.   The testes produce sperm and secrete the male sex hormone, testosterone.

Structure and Function of the Female Reproductive System.   An egg cell within an ovarian follicle develops and is released from the ovary; the egg cell travels into the female reproductive tract, which undergoes cyclic changes due to hormone secretion.

56.4 The physiology of human sexual intercourse is becoming better known.

Physiology of Human Sexual Intercourse.   The human sexual response can be divided into four phases: excitement, plateau, orgasm, and resolution.

Birth Control.   Various methods of birth control are employed, including barriers to fertilization, prevention of ovulation, and prevention of the implantation.

FIGURE 56.1
The bright color of male golden toads serves to attract mates. The rare golden toads, found in the Monteverde Cloud Forest Reserve of Costa Rica, are nearly voiceless and so use bright colors to attract mates.

The cry of a cat in heat, insects chirping outside the window, frogs croaking in swamps, and wolves howling in a frozen northern forest are all sounds of evolution's essential act, reproduction. These distinct vocalizations, as well as the bright coloration characteristic of some animals like the tropical golden toads of figure 56.1, function to attract mates. Few subjects pervade our everyday thinking more than sex, and few urges are more insistent. This chapter deals with sex and reproduction among the vertebrates, including humans.

Asexual and Sexual Reproduction

Asexual reproduction is the primary means of reproduction among the protists, cnidaria, and tunicates, but it may also occur in some of the more complex animals. Indeed, the formation of identical twins (by the separation of two identical cells of a very early embryo) is a form of asexual reproduction.

Through mitosis, genetically identical cells are produced from a single parent cell. This permits asexual reproduction to occur in protists by division of the organism, or **fission.** Cnidaria commonly reproduce by **budding,** where a part of the parent's body becomes separated from the rest and differentiates into a new individual. The new individual may become an independent animal or may remain attached to the parent, forming a colony.

Sexual reproduction occurs when a new individual is formed by the union of two sex cells, or **gametes,** a term that includes **sperm** and **eggs** (or ova). The union of sperm and egg cells produces a fertilized egg, or **zygote,** that develops by mitotic division into a new multicellular organism. The zygote and the cells it forms by mitosis are diploid; they contain both members of each homologous pair of chromosomes. The gametes, formed by meiosis in the sex organs, or **gonads**—the **testes** and **ovaries**—are haploid (chapter 12). The process of spermatogenesis (sperm formation) and oogenesis (egg formation) will be described in later sections.

(a)

(b)

FIGURE 56.2

Hermaphroditism and protogyny. (a) The hamlet bass (genus *Hypoplectrus*) is a deep-sea fish that is a hermaphrodite—both male and female at the same time. In the course of a single pair-mating, one fish may switch sexual roles as many as four times, alternately offering eggs to be fertilized and fertilizing its partner's eggs. Here the fish acting as a male curves around its motionless partner, fertilizing the upward-floating eggs. (b) The bluehead wrasse, *Thalassoma bifasciatium*, is protogynous—females sometimes turn into males. Here a large male, or sex-changed female, is seen among females, typically much smaller.

Different Approaches to Sex

Parthenogenesis (virgin birth) is common in many species of arthropods; some species are exclusively parthenogenic (and all female), while others switch between sexual reproduction and parthenogenesis in different generations. In honeybees, for example, a queen bee mates only once and stores the sperm. She then can control the release of sperm. If no sperm are released, the eggs develop parthenogenetically into drones, which are males; if sperm are allowed to fertilize the eggs, the fertilized eggs develop into other queens or worker bees, which are female.

The Russian biologist Ilya Darevsky reported in 1958 one of the first cases of unusual modes of reproduction among vertebrates. He observed that some populations of small lizards of the genus *Lacerta* were exclusively female, and suggested that these lizards could lay eggs that were viable even if they were not fertilized. In other words, they were capable of asexual reproduction in the absence of sperm, a type of parthenogenesis. Further work has shown that parthenogenesis also occurs among populations of other lizard genera.

Another variation in reproductive strategies is **hermaphroditism,** when one individual has both testes and ovaries, and so can produce both sperm and eggs (figure 56.2*a*). A tapeworm is hermaphroditic and can fertilize itself, a useful strategy since it is unlikely to encounter another tapeworm. Most hermaphroditic animals, however, require another individual to reproduce. Two earthworms, for example, are required for reproduction—each functions as both male and female, and each leaves the encounter with fertilized eggs.

There are some deep-sea fish that are hermaphrodites—both male and female at the same time. Numerous fish genera include species in which individuals can change their sex, a process called *sequential hermaphroditism*. Among coral reef fish, for example, both **protogyny** ("first female," a change from female to male) and **protandry** ("first male," a change from male to female) occur. In fish that practice protogyny (figure 56.2b), the sex change appears to be under social control. These fish commonly live in large groups, or schools, where successful reproduction is typically limited to one or a few large, dominant males. If those males are removed, the largest female rapidly changes sex and becomes a dominant male.

Sex Determination

Among the fish just described, and in some species of reptiles, environmental changes can cause changes in the sex of the animal. In mammals, the sex is determined early in embryonic development. The reproductive systems of human males and females appear similar for the first 40 days after conception. During this time, the cells that will give rise to ova or sperm migrate from the yolk sac to the embryonic gonads, which have the potential to become either ovaries in females or testes in males. For this reason, the embry-onic gonads are said to be "indifferent." If the embryo is a male, it will have a Y chromosome with a gene whose product converts the indifferent gonads into testes. In females, which lack a Y chromosome, this gene and the protein it encodes are absent, and the gonads become ovaries. Recent evidence suggests that the sex-determining gene may be one known as *SRY* (for "sex-determining region of the Y chromosome") (figure 56.3). The *SRY* gene appears to have been highly conserved during the evolution of different vertebrate groups.

Once testes form in the embryo, the testes secrete testosterone and other hormones that promote the development of the male external genitalia and accessory reproductive organs. If the embryo lacks testes (the ovaries are nonfunctional at this stage), the embryo develops female external genitalia and sex accessory organs. In other words, all mammalian embryos will develop female sex accessory organs and external genitalia unless they are masculinized by the secretions of the testes.

Sexual reproduction is most common among animals, but many reproduce asexually by fission, budding, or parthenogenesis. Sexual reproduction generally involves the fusion of gametes derived from different individuals of a species, but some species are hermaphroditic.

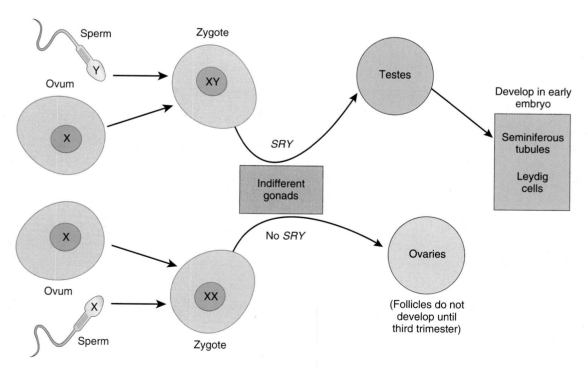

FIGURE 56.3
Sex determination in mammals is made by a region of the Y chromosome designated *SRY*. Testes are formed when the Y chromosome and *SRY* are present; ovaries are formed when they are absent.

Fertilization and Development

Vertebrate sexual reproduction evolved in the ocean before vertebrates colonized the land. The females of most species of marine bony fish produce eggs or ova in batches and release them into the water. The males generally release their sperm into the water containing the eggs, where the union of the free gametes occurs. This process is known as **external fertilization.**

Although seawater is not a hostile environment for gametes, it does cause the gametes to disperse rapidly, so their release by females and males must be almost simultaneous. Thus, most marine fish restrict the release of their eggs and sperm to a few brief and well-defined periods. Some reproduce just once a year, while others do so more frequently. There are few seasonal cues in the ocean that organisms can use as signals for synchronizing reproduction, but one all-pervasive signal is the cycle of the moon. Once each month, the moon approaches closer to the earth than usual, and when it does, its increased gravitational attraction causes somewhat higher tides. Many marine organisms sense the tidal changes and entrain the production and release of their gametes to the lunar cycle.

The invasion of land posed the new danger of desiccation, a problem that was especially severe for the small and vulnerable gametes. On land, the gametes could not simply be released near each other, as they would soon dry up and perish. Consequently, there was intense selective pressure for terrestrial vertebrates (as well as some groups of fish) to evolve **internal fertilization,** that is, the introduction of male gametes into the female reproductive tract. By this means, fertilization still occurs in a nondesiccating environment, even when the adult animals are fully terrestrial. The vertebrates that practice internal fertilization have three strategies for embryonic and fetal development:

1. **Oviparity.** This is found in some bony fish, most reptiles, some cartilaginous fish, some amphibians, a few mammals, and all birds. The eggs, after being fertilized internally, are deposited outside the mother's body to complete their development.

2. **Ovoviviparity.** This is found in some bony fish (including mollies, guppies, and mosquito fish), some cartilaginous fish, and many reptiles. The fertilized eggs are retained within the mother to complete their development, but the embryos still obtain all of their nourishment from the egg yolk. The young are fully developed when they are hatched and released from the mother.

3. **Viviparity.** This is found in most cartilaginous fish, some amphibians, a few reptiles, and almost all mammals. The young develop within the mother and obtain nourishment directly from their mother's blood, rather than from the egg yolk (figure 56.4).

Fertilization is external in most fish but internal in most other vertebrates. Depending upon the relationship of the developing embryo to the mother and egg, those vertebrates with internal fertilization may be classified as oviparous, ovoviviparous, or viviparous.

FIGURE 56.4
Viviparous fish carry live, mobile young within their bodies. The young complete their development within the body of the mother and are then released as small but competent adults. Here a lemon shark has just given birth to a young shark, which is still attached by the umbilical cord.

Fish and Amphibians

Most fish and amphibians, unlike other vertebrates, reproduce by means of external fertilization.

Fish

Fertilization in most species of bony fish (teleosts) is external, and the eggs contain only enough yolk to sustain the developing embryo for a short time. After the initial supply of yolk has been exhausted, the young fish must seek its food from the waters around it. Development is speedy, and the young that survive mature rapidly. Although thousands of eggs are fertilized in a single mating, many of the resulting individuals succumb to microbial infection or predation, and few grow to maturity.

In marked contrast to the bony fish, fertilization in most cartilaginous fish is internal. The male introduces sperm into the female through a modified pelvic fin. Development of the young in these vertebrates is generally viviparous.

Amphibians

The amphibians invaded the land without fully adapting to the terrestrial environment, and their life cycle is still tied to the water. Fertilization is external in most amphibians, just as it is in most species of bony fish. Gametes from both males and females are released through the cloaca. Among the frogs and toads, the male grasps the female and discharges fluid containing the sperm onto the eggs as they are released into the water (figure 56.5). Although the eggs of most amphibians develop in the water, there are some interesting exceptions. In two species of frogs, for example, the eggs develop in the vocal sacs and the stomach, and the young frogs leave through their mother's mouth!

The time required for development of amphibians is much longer than that for fish, but amphibian eggs do not include a significantly greater amount of yolk. Instead, the process of development in most amphibians is divided into embryonic, larval, and adult stages, in a way reminiscent of the life cycles found in some insects. The embryo develops within the egg, obtaining nutrients from the yolk. After hatching from the egg, the aquatic larva then functions as a free-swimming, food-gathering machine, often for a con-

FIGURE 56.5
The eggs of frogs are fertilized externally. When frogs mate, as these two are doing, the clasp of the male induces the female to release a large mass of mature eggs, over which the male discharges his sperm.

siderable period of time. The larvae may increase in size rapidly; some tadpoles, which are the larvae of frogs and toads, grow in a matter of weeks from creatures no bigger than the tip of a pencil into individuals as big as a goldfish. When the larva has grown to a sufficient size, it undergoes a developmental transition, or metamorphosis, into the terrestrial adult form.

The eggs of most bony fish and amphibians are fertilized externally. In amphibians, however, the eggs develop into a larval stage that undergoes metamorphosis. Most cartilaginous fish are fertilized internally, and the young develop within the mother.

Reptiles and Birds

Most reptiles and all birds are oviparous—after the eggs are fertilized internally, they are deposited outside of the mother's body to complete their development. Like most vertebrates that fertilize internally, male reptiles utilize a tubular organ, the penis, to inject sperm into the female (figure 56.6). The penis, containing erectile tissue, can become quite rigid and penetrate far into the female reproductive tract. Most reptiles are oviparous, laying eggs and then abandoning them. These eggs are surrounded by a leathery shell that is deposited as the egg passes through the oviduct, the part of the female reproductive tract leading from the ovary. A few species of reptiles are ovoviviparous or viviparous, forming eggs that develop into embryos within the body of the mother.

All birds practice internal fertilization, though most male birds lack a penis. In some of the larger birds (including swans, geese, and ostriches), however, the male cloaca extends to form a false penis. As the egg passes along the oviduct, glands secrete albumin proteins (the egg white) and the hard, calcareous shell that distinguishes bird eggs from reptilian eggs. While modern reptiles are poikilotherms (animals whose body temperature varies with the temperature of their environment), birds are homeotherms (animals that maintain a relatively constant body temperature independent of environmental temperatures). Hence, most birds incubate their eggs after laying them to keep them warm (figure 56.7). The young that hatch from the eggs of most bird species are unable to survive unaided, since their development is still incomplete. These young birds are fed and nurtured by their parents, and they grow to maturity gradually.

The shelled eggs of reptiles and birds constitute one of the most important adaptations of these vertebrates to life on land, since shelled eggs can be laid in dry places. Such eggs are known as amniotic eggs because the embryo develops within a fluid-filled cavity surrounded by a membrane called the amnion. The amnion is an extraembryonic membrane—that is, a membrane formed from embryonic cells but located outside the body of the embryo. Other extraembryonic membranes in amniotic eggs include the chorion, which lines the inside of the eggshell, the yolk sac, and the allantois. In contrast, the eggs of fish and amphibians contain only one extraembryonic membrane, the yolk sac. The viviparous mammals, including humans, also have extraembryonic membranes that will be described in chapter 58.

FIGURE 56.6
The introduction of sperm by the male into the female's body is called copulation. Reptiles such as these turtles were the first terrestrial vertebrates to develop this form of reproduction, which is particularly suited to a terrestrial environment.

FIGURE 56.7
Crested penguins incubating their egg. This nesting pair is changing the parental guard in a stylized ritual.

Most reptiles and all birds are oviparous, laying amniotic eggs that are protected by watertight membranes from desiccation. Birds, being homeotherms, must keep the eggs warm by incubation.

Mammals

Some mammals are seasonal breeders, reproducing only once a year, while others have shorter reproductive cycles. Among the latter, the females generally undergo the reproductive cycles, while the males are more constant in their reproductive activity. Cycling in females involves the periodic release of a mature ovum from the ovary in a process known as ovulation. Most female mammals are "in heat," or sexually receptive to males, only around the time of ovulation. This period of sexual receptivity is called **estrus,** and the reproductive cycle is therefore called an **estrous cycle.** The females continue to cycle until they become pregnant.

In the estrous cycle of most mammals, changes in the secretion of follicle-stimulating hormone (FSH) and luteinizing hormone (LH) by the anterior pituitary gland cause changes in egg cell development and hormone secretion in the ovaries. Humans and apes have menstrual cycles that are similar to the estrous cycles of other mammals in their cyclic pattern of hormone secretion and ovulation. Unlike mammals with estrous cycles, however, human and ape females bleed when they shed the inner lining of their uterus, a process called menstruation, and may engage in copulation at any time during the cycle.

Rabbits and cats differ from most other mammals in that they are induced ovulators. Instead of ovulating in a cyclic fashion regardless of sexual activity, the females ovulate only after copulation as a result of a reflex stimulation of LH secretion (described later). This makes these animals extremely fertile.

The most primitive mammals, the **monotremes** (consisting solely of the duck-billed platypus and the echidna), are oviparous, like the reptiles from which they evolved. They incubate their eggs in a nest (figure 56.8a) or specialized pouch, and the young hatchlings obtain milk from their mother's mammary glands by licking her skin, since monotremes lack nipples. All other mammals are viviparous, and are divided into two subcategories based on how they nourish their young. The **marsupials,** a group that includes opossums and kangaroos, give birth to fetuses that are incompletely developed. The fetuses complete their development in a pouch of their mother's skin, where they can obtain nourishment from nipples of the mammary glands (figure 56.8b). The **placental mammals** (figure 56.8c) retain their young for a much longer period of development within the mother's uterus. The fetuses are nourished by a structure known as the placenta, which is derived from both an extraembryonic membrane (the chorion) and the mother's uterine lining. Because the fetal and maternal blood vessels are in very close proximity in the placenta, the fetus can obtain nutrients by diffusion from the mother's blood. The functioning of the placenta is discussed in more detail in chapter 58.

Among mammals that are not seasonal breeders, the females undergo shorter cyclic variations in ovarian function. These are estrous cycles in most mammals and menstrual cycles in humans and apes. Some mammals are induced ovulators, ovulating in response to copulation.

(a) (b) (c)

FIGURE 56.8
Reproduction in mammals. (a) Monotremes, like the duck-billed platypus shown here, lays eggs in a nest. (b) Marsupials, such as this kangaroo, give birth to small fetuses which complete their development in a pouch. (c) In placental mammals, like this domestic cat, the young remain inside the mother's uterus for a longer period of time and are born relatively more developed.

Structure and Function of the Male Reproductive System

The structures of the human male reproductive system, typical of mammals, are illustrated in figure 56.9. If testes form in the human embryo, they develop seminiferous tubules beginning at around 43 to 50 days after conception. The seminiferous tubules are the sites of sperm production. At about 9 to 10 weeks, the Leydig cells, located in the interstitial tissue between the seminiferous tubules, begin to secrete testosterone (the major male sex hormone, or androgen). Testosterone secretion during embryonic development converts indifferent structures into the male external genitalia, the penis and the scrotum, a sac that contains the testes. In the absence of testosterone, these structures develop into the female external genitalia.

In an adult, each testis is composed primarily of the highly convoluted seminiferous tubules (figure 56.10). Although the testes are actually formed within the abdominal cavity, shortly before birth they descend through an opening called the inguinal canal into the scrotum, which suspends them outside the abdominal cavity. The scrotum maintains the testes at around 34 °C, slightly lower than the core body temperature (37 °C). This lower temperature is required for normal sperm development in humans.

Production of Sperm

The wall of the seminiferous tubule consists of *germinal cells*, which become sperm by meiosis, and supporting *Ser-*

toli cells. The germinal cells near the outer surface of the seminiferous tubule are diploid (with 46 chromosomes in humans), while those located closer to the lumen of the tubule are haploid (with 23 chromosomes each). Each parent cell duplicates by mitosis, and one of the two daughter cells then undergoes meiosis to form sperm; the other remains as a parent cell. In that way, the male never runs out of parent cells to produce sperm. Adult males produce an average of 100 to 200 million sperm each day and can continue to do so throughout most of the rest of their lives.

The diploid daughter cell that begins meiosis is called a primary spermatocyte. It has 23 pairs of homologous chromosomes (in humans) and each chromosome is duplicated, with two chromatids. The first meiotic division separates the homologous chromosomes, producing two haploid secondary spermatocytes. However, each chromosome still consists of two duplicate chromatids. Each of these cells then undergoes the second meiotic division to separate the chromatids and produce two haploid cells, the spermatids. Therefore, a

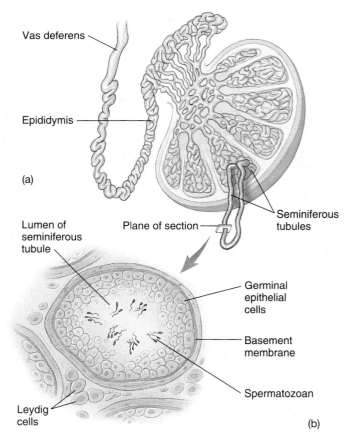

(a)

(b)

FIGURE 56.10
The testis. (a) A section through a testis and associated structures. (b) A cross-section of a seminiferous tubule.

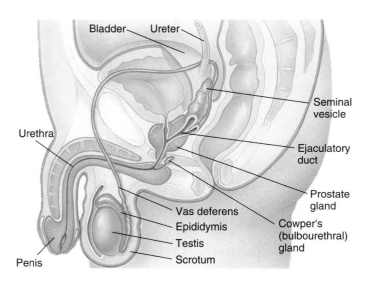

FIGURE 56.9
Organization of the human male reproductive system. The penis and scrotum are the external genitalia, the testes are the gonads, and the other organs are sex accessory organs, aiding the production and ejaculation of semen.

total of four haploid spermatids are produced by each primary spermatocyte. All of these cells constitute the germinal epithelium of the seminiferous tubules because they "germinate" the gametes.

In addition to the germinal epithelium, the walls of the seminiferous tubules contain nongerminal cells known as Sertoli cells. The Sertoli cells nurse the developing sperm and secrete products required for spermatogenesis (sperm production). They also help convert the spermatids into spermatozoa by engulfing their extra cytoplasm (figure 56.11).

Spermatozoa are relatively simple cells, consisting of a head, body, and tail (figure 56.12). The head encloses a compact nucleus and is capped by a vesicle called an acrosome, which is derived from the Golgi complex. The acrosome contains enzymes that aid in the penetration of the protective layers surrounding the egg. The body and tail provide a propulsive mechanism: within the tail is a flagellum, while inside the body are a centriole, which acts as a basal body for the flagellum, and mitochondria, which generate the energy needed for flagellar movement.

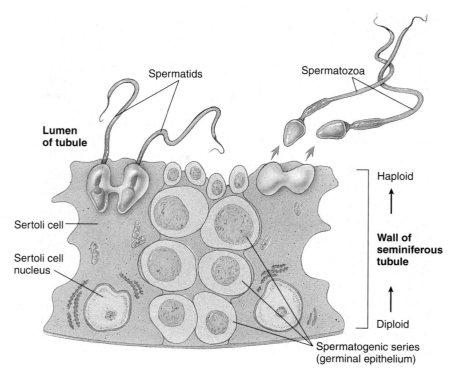

FIGURE 56.11
Sertoli cells are nongerminal cells within the walls of the seminiferous tubules. They assist spermatogenesis in several ways, such as helping to convert spermatids into spermatozoa.

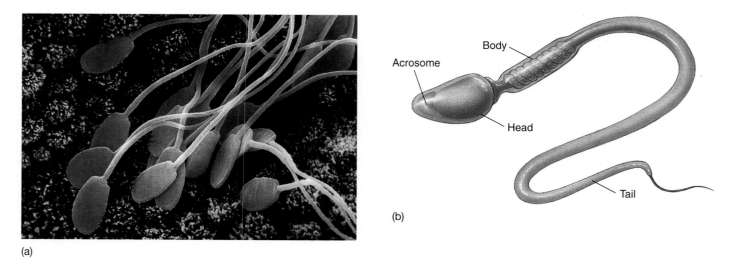

(a)

(b)

FIGURE 56.12
Human sperm. (a) A scanning electron micrograph. (b) A diagram of the main components of a sperm cell.

Male Accessory Sex Organs

After the sperm are produced within the seminiferous tubules, they are delivered into a long, coiled tube called the epididymis (figure 56.13). The sperm are not motile when they arrive in the epididymis, and they must remain there for at least 18 hours before their motility develops. From the epididymis, the sperm enter another long tube, the vas deferens, which passes into the abdominal cavity via the inguinal canal.

The vas deferens from each testis joins with one of the ducts from a pair of glands called the seminal vesicles (see figure 56.9), which produce a fructose-rich fluid. From this point, the vas deferens continues as the ejaculatory duct and enters the prostate gland at the base of the urinary bladder. In humans, the prostate gland is about the size of a golf ball and is spongy in texture. It contributes about 60% of the bulk of the semen, the fluid that contains the products of the testes, fluid from the seminal vesicles, and the products of the prostate gland. Within the prostate gland, the ejaculatory duct merges with the urethra from the urinary bladder. The urethra carries the semen out of the body through the tip of the penis. A pair of pea-sized bulbourethral glands secrete a fluid that lines the urethra and lubricates the tip of the penis prior to coitus (sexual intercourse).

In addition to the urethra, there are two columns of erectile tissue, the corpora cavernosa, along the dorsal side of the penis and one column, the corpus spongiosum, along the ventral side (figure 56.14). Penile erection is produced by neurons in the parasympathetic division of the autonomic nervous system. As a result of the release of nitric oxide by these neurons, arterioles in the penis dilate, causing the erectile tissue to become engorged with blood and turgid. This increased pressure in the erectile tissue compresses the veins, so blood flows into the penis but cannot flow out. The drug **viagra** prolongs erection by stimulating release of nitric oxide in the penis. Some mammals, such as the walrus, have a bone in the penis that contributes to its stiffness during erection, but humans do not.

The result of erection and continued sexual stimulation is ejaculation, the ejection from the penis of about 5 milliliters of semen containing an average of 300 million sperm. Successful fertilization requires such a high sperm count because the odds against any one sperm cell successfully completing the journey to the egg and fertilizing it are extraordinarily high, and the acrosomes of several sperm need to interact with the egg before a single sperm can penetrate the egg. Males with fewer than 20 million sperm per milliliter are generally considered sterile. Despite their large numbers, sperm constitute only about 1% of the volume of the semen ejaculated.

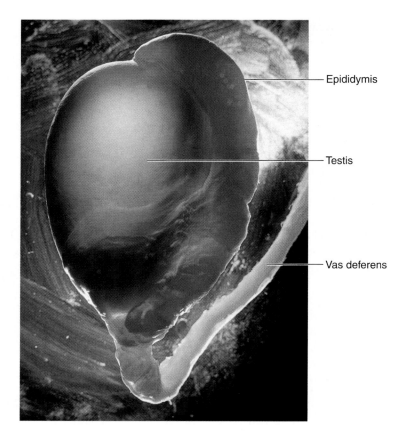

FIGURE 56.13
Photograph of the human testis. The dark, round object in the center of the photograph is a testis, within which sperm are formed. Cupped around it is the epididymis, a highly coiled passageway in which sperm complete their maturation. Mature sperm are stored in the vas deferens, a long tube that extends from the epididymis.

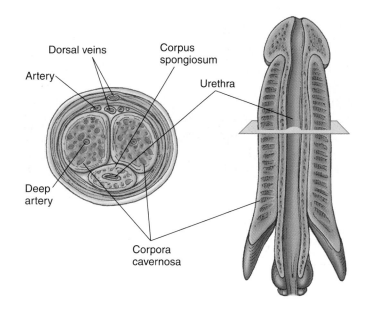

FIGURE 56.14
A penis in cross-section (*left*) and longitudinal section (*right*). Note that the urethra runs through the corpus spongiosum.

Table 56.1 Mammalian Reproductive Hormones

MALE

Follicle-stimulating hormone (FSH)	Stimulates spermatogenesis
Luteinizing hormone (LH)	Stimulates secretion of testosterone by Leydig cells
Testosterone	Stimulates development and maintenance of male secondary sexual characteristics and accessory sex organs

FEMALE

Follicle-stimulating hormone (FSH)	Stimulates growth of ovarian follicles and secretion of estradiol
Luteinizing hormone (LH)	Stimulates ovulation, conversion of ovarian follicles into corpus luteum, and secretion of estradiol and progesterone by corpus luteum
Estradiol	Stimulates development and maintenance of female secondary sexual characteristics; prompts monthly preparation of uterus for pregnancy
Progesterone	Completes preparation of uterus for pregnancy; helps maintain female secondary sexual characteristics
Oxytocin	Stimulates contraction of uterus and milk-ejection reflex
Prolactin	Stimulates milk production

Hormonal Control of Male Reproduction

As we saw in chapter 53, the anterior pituitary gland se-cretes two gonadotropic hormones: FSH and LH. Al-though these hormones are named for their actions in the female, they are also involved in regulating male reproduc-tive function (table 56.1). In males, FSH stimulates the Ser-toli cells to facilitate sperm development, and LH stimu-lates the Leydig cells to secrete testosterone.

The principle of negative feedback inhibition discussed in chapter 53 applies to the control of FSH and LH secre-tion (figure 56.15). The hypothalamic hormone, gonadotropin-releasing hormone (GnRH), stimulates the anterior pituitary gland to secrete both FSH and LH. FSH causes the Sertoli cells to release a peptide hormone called inhibin that specifically inhibits FSH secretion. Similarly, LH stimulates testosterone secretion, and testosterone feeds back to inhibit the release of LH, both directly at the anterior pituitary gland and indirectly by reducing GnRH release. The importance of negative feedback inhibition can be demonstrated by removing the testes; in the absence of testosterone and inhibin, the secretion of FSH and LH from the anterior pituitary is greatly increased.

An adult male produces sperm continuously by meiotic division of the germinal cells lining the seminiferous tubules. Semen consists of sperm from the testes and fluid contributed by the seminal vesicles and prostate gland. Production of sperm and secretion of testosterone from the testes are controlled by FSH and LH from the anterior pituitary.

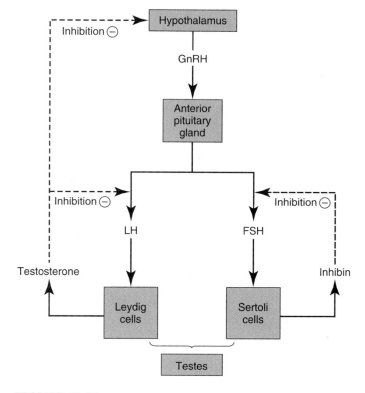

FIGURE 56.15

Hormonal interactions between the testes and anterior pituitary. LH stimulates the Leydig cells to secrete testosterone, and FSH stimulates the Sertoli cells of the seminiferous tubules to secrete inhibin. Testosterone and inhibin, in turn, exert negative feedback inhibition on the secretion of LH and FSH, respectively.

Structure and Function of the Female Reproductive System

The structures of the reproductive system in a human female are shown in figure 56.16. In contrast to the testes, the ovaries develop much more slowly. In the absence of testosterone, the female embryo develops a **clitoris** and **labia majora** from the same embryonic structures that produce a penis and scrotum in males. Thus clitoris and penis, and the labia majora and scrotum, are said to be *homologous structures*. The clitoris, like the penis, contains corpora cavernosa and is therefore erectile. The ovaries contain microscopic structures called **ovarian follicles,** which each contain an egg cell and smaller **granulosa cells.** The ovarian follicles are the functional units of the ovary.

At puberty, the granulosa cells begin to secrete the major female sex hormone estradiol (also called estrogen), triggering **menarche,** the onset of menstrual cycling. Estradiol also stimulates the formation of the **female secondary sexual characteristics,** including breast development and the production of pubic hair. In addition, estradiol and another steroid hormone, progesterone, help to maintain the female accessory sex organs: the Fallopian tubes, uterus, and vagina.

Female Accessory Sex Organs

The Fallopian tubes (also called uterine tubes or oviducts) transport ova from the ovaries to the uterus. In humans, the uterus is a muscular, pear-shaped organ that narrows to form a neck, the cervix, which leads to the vagina (figure 56.17*a*). The uterus is lined with a stratified epithelial membrane called the endometrium. The surface of the endometrium is shed during menstruation, while the underlying portion remains to generate a new surface during the next cycle.

Mammals other than primates have more complex female reproductive tracts, where part of the uterus divides to form uterine "horns," each of which leads to an oviduct (figure 56.17*b,c*). In cats, dogs, and cows, for example, there is one cervix but two uterine horns separated by a septum, or wall. Marsupials, such as opossums, carry the split even further, with two unconnected uterine horns, two cervices, and two vaginas. A male marsupial has a forked penis that can enter both vaginas simultaneously.

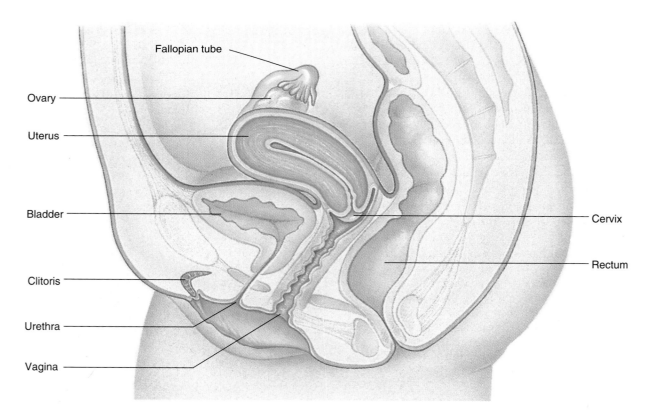

FIGURE 56.16

Organization of the human female reproductive system. The ovaries are the gonads, the Fallopian tubes receive the ovulated ova, and the uterus is the womb, the site of development of an embryo if the egg cell becomes fertilized.

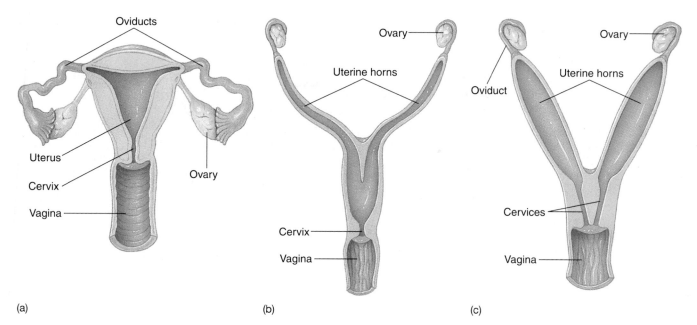

(a)

(b)

(c)

FIGURE 56.17
A comparison of mammalian uteruses. (a) Humans and other primates; (b) cats, dogs, and cows; and (c) rats, mice, and rabbits.

Menstrual and Estrous Cycles

At birth, a female's ovaries contain some 2 million follicles, each with an ovum that has begun meiosis but which is arrested in prophase of the first meiotic division. At this stage, the ova are called primary oocytes. Some of these primary-oocyte-containing follicles are stimulated to develop during each cycle. The human menstrual (Latin *mens*, "month") cycle lasts approximately one month (28 days on the average) and can be divided in terms of ovarian activity into a follicular phase and luteal phase, with the two phases separated by the event of ovulation.

Follicular Phase

During the follicular phase, a few follicles are stimulated to grow under FSH stimulation, but only one achieves full maturity as a tertiary, or Graafian, follicle (figure 56.18). This follicle forms a thin-walled blister on the surface of the ovary. The primary oocyte within the Graafian follicle completes the first meiotic division during the follicular phase. Instead of forming two equally large daughter cells, however, it produces one large daughter cell, the secondary oocyte, and one tiny daughter cell, called a polar body. Thus, the secondary oocyte acquires almost all of the cytoplasm from the primary oocyte, increasing its chances of sustaining the early embryo should the oocyte be fertilized. The polar body, on the other hand, often disintegrates. The secondary oocyte then begins the second meiotic division, but its progress is arrested at metaphase II. It is in this form that the egg cell is discharged from the ovary at ovulation, and it does not complete the second meiotic division unless it becomes fertilized in the Fallopian tube.

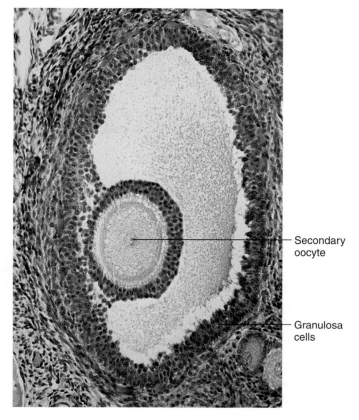

FIGURE 56.18
A mature Graafian follicle in a cat ovary (50×). Note the ring of granulosa cells that surrounds the secondary oocyte. This ring will remain around the egg cell when it is ovulated, and sperm must tunnel through the ring in order to reach the plasma membrane of the egg cell.

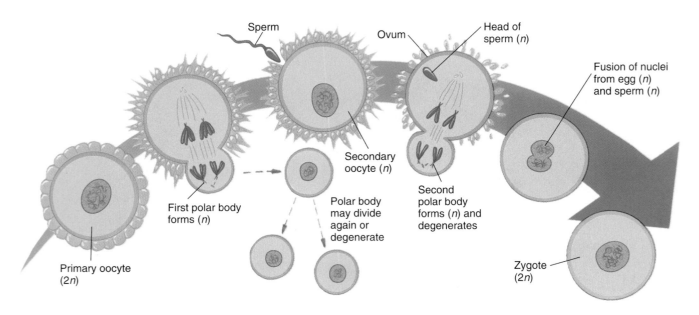

FIGURE 56.19
The meiotic events of oogenesis. A primary oocyte is diploid. At the completion of the first meiotic division, one division product is eliminated as a polar body, while the other, the secondary oocyte, is released during ovulation. The secondary oocyte does not complete the second meiotic division until after fertilization; that division yields a second polar body and a single haploid egg. Fusion of the haploid egg with a haploid sperm produces a diploid zygote, from which an embryo subsequently forms.

Ovulation

The increasing level of estradiol in the blood during the follicular phase stimulates the anterior pituitary gland to secrete LH about midcycle. This sudden secretion of LH causes the fully developed Graafian follicle to burst in the process of ovulation, releasing its secondary oocyte. The released oocyte enters the abdominal cavity near the fimbriae, the feathery projections surrounding the opening to the Fallopian tube. The ciliated epithelial cells lining the Fallopian tube propel the oocyte through the Fallopian tube toward the uterus. If it is not fertilized, the oocyte will disintegrate within a day following ovulation. If it is fertilized, the stimulus of fertilization allows it to complete the second meiotic division, forming a fully mature ovum and a second polar body. Fusion of the two nuclei from the ovum and the sperm produces a diploid zygote (figure 56.19). Fertilization normally occurs in the upper one-third of the Fallopian tube, and in a human the zygote takes approximately three days to reach the uterus, then another two to three days to implant in the endometrium (figure 56.20).

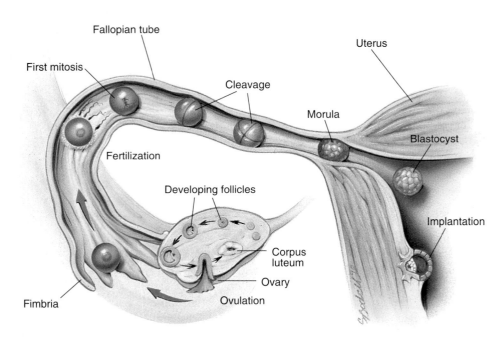

FIGURE 56.20
The journey of an egg. Produced within a follicle and released at ovulation, an egg is swept into a Fallopian tube and carried along by waves of ciliary motion in the tube walls. Sperm journeying upward from the vagina fertilize the egg within the Fallopian tube. The resulting zygote undergoes several mitotic divisions while still in the tube, so that by the time it enters the uterus, it is a hollow sphere of cells called a blastocyst. The blastocyst implants within the wall of the uterus, where it continues its development. (The egg and its subsequent stages have been enlarged for clarification.)

Luteal Phase

After ovulation, LH stimulates the empty Graafian follicle to develop into a structure called the corpus luteum (Latin, "yellow body"). For this reason, the second half of the menstrual cycle is referred to as the **luteal phase** of the cycle. The corpus luteum secretes both estradiol and another steroid hormone, progesterone. The high blood levels of estradiol and progesterone during the luteal phase now exert negative feedback inhibition of FSH and LH secretion by the anterior pituitary gland. This inhibition during the luteal phase is in contrast to the stimulation exerted by estradiol on LH secretion at midcycle, which caused ovulation. The inhibitory effect of estradiol and progesterone on FSH and LH secretion after ovulation acts as a natural contraceptive mechanism, preventing both the development of additional follicles and continued ovulation.

During the follicular phase the granulosa cells secrete increasing amounts of estradiol, which stimulates the growth of the endometrium. Hence, this portion of the cycle is also referred to as the **proliferative phase** of the endometrium. During the luteal phase of the cycle, the combination of estradiol and progesterone cause the endometrium to become more vascular, glandular, and enriched with glycogen deposits. Because of the endometrium's glandular appearance, this portion of the cycle is known as the **secretory phase** of the endometrium (figure 56.21).

In the absence of fertilization, the corpus luteum triggers its own atrophy, or regression, toward the end of the luteal phase. It does this by secreting hormones (estradiol and progesterone) that inhibit the secretion of LH, the hormone needed for its survival. In many mammals, atrophy of the corpus luteum is assisted by luteolysin, a paracrine regulator believed to be a prostaglandin. The disappearance of the corpus luteum results in an abrupt decline in the blood concentration of estradiol and progesterone at the end of the luteal phase, causing the built-up endometrium to be sloughed off with accompanying bleeding. This process is called menstruation, and the portion of the cycle in which it occurs is known as the **menstrual phase** of the endometrium.

If the ovulated oocyte is fertilized, however, regression of the corpus luteum and subsequent menstruation is averted by the tiny embryo! It does this by secreting human chorionic gonadotropin (hCG), an LH-like hormone produced by the chorionic membrane of the embryo. By maintaining the corpus luteum, hCG keeps the levels of estradiol and progesterone high and thereby prevents menstruation, which would terminate the pregnancy. Since hCG comes from the embryonic chorion and not the mother, it is the hormone that is tested for in all pregnancy tests.

Menstruation is absent in mammals with an estrous cycle. Although such mammals do cyclically shed cells from the endometrium, they don't bleed in the process. The estrous cycle is divided into four phases: proestrus, estrus, metestrus, and diestrus, which correspond to the proliferative, mid-cycle, secretory, and menstrual phases of the endometrium in the menstrual cycle.

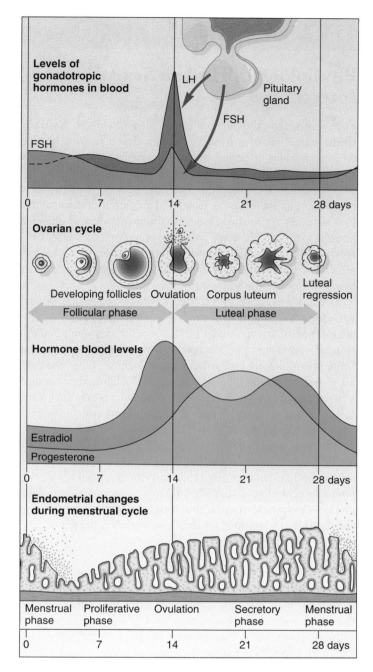

FIGURE 56.21

The human menstrual cycle. The growth and thickening of the endometrial (uterine) lining is stimulated by estradiol and progesterone. The decline in the levels of these two hormones triggers menstruation, the sloughing off of built-up endometrial tissue.

The ovarian follicles develop under FSH stimulation, and one follicle ovulates under LH stimulation. During the follicular and luteal phases, the hormones secreted by the ovaries stimulate the development of the endometrium, so an embryo can implant there if fertilization has occurred. A secondary oocyte is released from an ovary at ovulation, and it only completes meiosis if it is fertilized.

Physiology of Human Sexual Intercourse

Few physical activities are more pleasurable to humans than sexual intercourse. The sex drive is one of the strongest drives directing human behavior, and as such, it is circumscribed by many rules and customs. Sexual intercourse acts as a channel for the strongest of human emotions such as love, tenderness, and personal commitment. Few subjects are at the same time more private and of more general interest. Here we will limit ourselves to a very narrow aspect of sexual behavior, its immediate physiological effects. The emotional consequences are no less real, but they are beyond the scope of this book.

Until relatively recently, the physiology of human sexual activity was largely unknown. Perhaps because of the prevalence of strong social taboos against the open discussion of sexual matters, no research was carried out on the subject, and detailed information was lacking. Over the past 30 years, however, investigations by William Masters and Virginia Johnson, as well as an army of researchers who followed them, have revealed much about the biological nature of human sexual activity.

The sexual act is referred to by a variety of names, including sexual intercourse, copulation, and coitus, as well as a host of informal terms. It is common to partition the physiological events that accompany intercourse into four phases—**excitement, plateau, orgasm,** and **resolution**—although there are no clear divisions between these phases.

Excitement

The sexual response is initiated by the nervous system. In both males and females, commands from the brain increase the respiratory rate, heart rate, and blood pressure. The nipples commonly harden and become more sensitive. Other changes increase the diameter of blood vessels, leading to increased circulation. In some people, these changes may produce a reddening of the skin around the face, breasts, and genitals (the sex flush). Increased circulation also leads to vasocongestion, producing erection of the male's penis and similar swelling of the female's clitoris. The female experiences changes that prepare the vagina for sexual intercourse: the labia majora and labia minora, lips of tissue that cover the opening to the vagina, swell and separate due to the increased circulation; the vaginal walls become moist; and the muscles encasing the vagina relax.

Plateau

The penetration of the vagina by the thrusting penis continuously stimulates nerve endings both in the tip of the penis and in the clitoris. The clitoris, which is now swollen, becomes very sensitive and withdraws up into a sheath or "hood." Once it has withdrawn, the clitoris is stimulated indirectly when the thrusting movements of the penis rub the clitoral hood against the clitoris. The nervous stimulation produced by the repeated movements of the penis within the vagina elicits a continuous response in the autonomic nervous system, greatly intensifying the physiological changes initiated during the excitement phase. In the female, pelvic thrusts may begin, while in the male the penis reaches its greatest length and rigidity.

Orgasm

The climax of intercourse is reached when the stimulation is sufficient to initiate a series of reflexive muscular contractions. The nerve impulses producing these contractions are associated with other activity within the central nervous system, activity that we experience as intense pleasure. In females, the contractions are initiated by impulses in the hypothalamus, which causes the posterior pituitary gland to release large amounts of oxytocin. This hormone, in turn, causes the muscles in the uterus and around the vaginal opening to contract and the cervix to be pulled upward. Contractions occur at intervals of about one per second. There may be one to several intense peaks of contractions (orgasms), or the peaks may be more numerous but less intense.

Analogous contractions take place in the male. The first contractions, which occur in the vas deferens and prostate gland, cause *emission*, the peristaltic movement of sperm and seminal fluid into a collecting zone of the urethra located at the base of the penis. Shortly thereafter, violent contractions of the muscles at the base of the penis result in *ejaculation* of the collected semen through the penis. As in the female, the contractions are spaced about one second apart, although in the male they continue for only a few seconds and are almost invariably restricted to a single intense wave.

Resolution

After ejaculation, males rapidly lose their erection and enter a refractory period lasting 20 minutes or longer, in which sexual arousal is difficult to achieve and ejaculation is almost impossible. By contrast, many women can be aroused again almost immediately. After intercourse, the bodies of both men and women return over a period of several minutes to their normal physiological state.

Sexual intercourse is a physiological series of events leading to the ultimate deposition of sperm within the female reproductive tract. The phases are similar in males and females.

Birth Control

In most vertebrates, copulation is associated solely with reproduction. Reflexive behavior that is deeply ingrained in the female limits sexual receptivity to those periods of the sexual cycle when she is fertile. In humans and a few species of apes, the female can be sexually receptive throughout her reproductive cycle, and this extended receptivity to sexual intercourse serves a second important function—it reinforces pair-bonding, the emotional relationship between two individuals living together. The evolution of strong pair-bonding is certainly not unique to humans and apes, but it was probably a necessary precondition for the subsequent evolution of increased mental capacity. The associative activities that constitute human thinking are largely based on learning, and learning takes time. Human children are very vulnerable during the extended period of learning that follows their birth, and they require parental nurturing. Perhaps for this reason, human pair-bonding is a continuous process, not restricted to short episodes coinciding with ovulation.

Not all human couples want to initiate a pregnancy every time they have sexual intercourse, yet sexual intercourse may be a necessary and important part of their emotional lives together. The solution to this dilemma is to find a way to avoid reproduction without avoiding sexual intercourse; this approach is commonly called birth control or contraception. A variety of approaches differing in effectiveness and in their acceptability to different couples are commonly taken to achieve birth control (figure 56.22 and table 56.2).

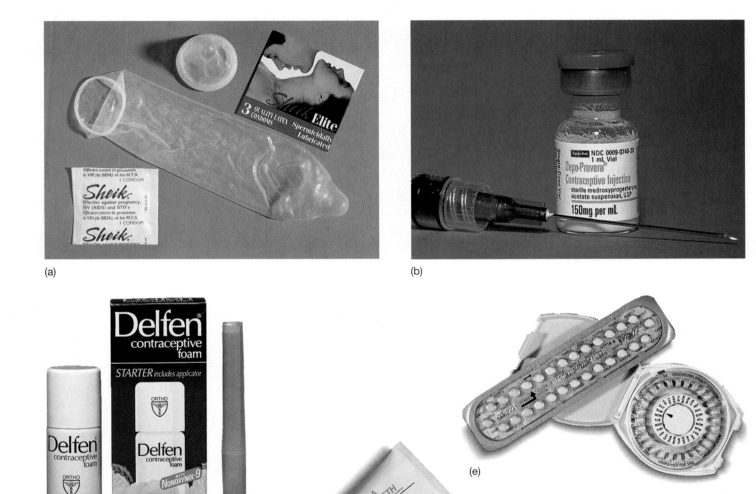

(a)

(b)

(c)

(d)

(e)

FIGURE 56.22
Five common methods of birth control. (a) Condom; (b) Depo-Provera; (c) foam; (d) diaphragm and spermicidal jelly; (e) oral contraceptives.

Table 56.2 Methods of Birth Control

Device	Action	Failure Rate*	Advantages	Disadvantages
Intrauterine device (IUD)	Small plastic or metal device placed in the uterus; prevents implantation; some contain copper, others release hormones	1–5	Convenient; highly effective; infrequent replacement	Can cause excess menstrual bleeding and pain; risk of perforation, infection, expulsion, pelvic inflammatory disease, and infertility; not recommended for those who eventually intend to conceive or are not monogamous; dangerous in pregnancy
Oral contraceptive	Hormones (progesterone analogue alone or in combination with other hormones) primarily prevent ovulation	1–5, depending on type	Convenient; highly effective; provides significant noncontraceptive health benefits, such as protection against ovarian and endometrial cancers	Must be taken regularly; possible minor side effects which new formulations have reduced; not for women with cardiovascular risks (mostly smokers over age 35)
Condom	Thin sheath for penis that collects semen; "female condoms" sheath vaginal walls	3–15	Easy to use; effective; inexpensive; protects against some sexually transmitted diseases	Requires male cooperation; may diminish spontaneity; may deteriorate on the shelf
Diaphragm	Soft rubber cup covers entrance to uterus, prevents sperm from reaching egg, holds spermicide	4–25	No dangerous side effects; reliable if used properly; provides some protection against sexually transmitted diseases and cervical cancer	Requires careful fitting; some inconvenience associated with insertion and removal; may be dislodged during intercourse
Cervical cap	Miniature diaphragm covers cervix closely, prevents sperm from reaching egg, holds spermicide	Probably similar to that of diaphragm	No dangerous side effects; fairly effective; can remain in place longer than diaphragm	Problems with fitting and insertion; comes in limited number of sizes
Foams, creams, jellies, vaginal suppositories	Chemical spermicides inserted in vagina before intercourse that prevent sperm from entering uterus	10–25	Can be used by anyone who is not allergic; protect against some sexually transmitted diseases; no known side effects	Relatively unreliable; sometimes messy; must be used 5–10 minutes before each act of intercourse
Implant (levonorgestrel; Norplant)	Capsules surgically implanted under skin slowly release hormone that blocks ovulation	.03	Very safe, convenient, and effective; very long-lasting (5 years); may have nonreproductive health benefits like those of oral contraceptives	Irregular or absent periods; minor surgical procedure needed for insertion and removal; some scarring may occur
Injectable contraceptive (medroxy-progesterone; Depo-Provera)	Injection every 3 months of a hormone that is slowly released and prevents ovulation	1	Convenient and highly effective; no serious side effects other than occasional heavy menstrual bleeding	Animal studies suggest it may cause cancer, though new studies in humans are mostly encouraging; occasional heavy menstrual bleeding

*Failure rate is expressed as pregnancies per 100 actual users per year.

Source: Data from American College of Obstetricians and Gynecologists: Contraception, Patient Education Pamphlet No. AP005.ACOG, Washington, D.C., 1990.

Abstinence

The simplest and most reliable way to avoid pregnancy is not to have sexual intercourse at all. Of all methods of birth control, this is the most certain. It is also the most limiting, since it denies a couple the emotional support of a sexual relationship.

A variant of complete abstinence is to avoid sexual relations around the time of ovulation, when successful fertil-

ization is most likely to occur. Since sperm can live for only about two days in the female reproductive tract, and an unfertilized ovum can survive for only about one day after ovulation, this approach involves abstinence for two days preceding and a day or two following ovulation. The rest of the menstrual cycle, approximately three out of the four weeks, is relatively "safe" for intercourse. This approach, called the rhythm method, is satisfactory in principle but difficult in application, since ovulation is not easy to predict and may occur unexpectedly. Consequently, the rhythm method is not very effective; its failure rate is estimated to be 13% to 21% (13–21 pregnancies per 100 women practicing the method per year).

Another variant of abstinence is to have incomplete sex, that is, to withdraw the penis before ejaculation, a procedure known as coitus interruptus. This method requires considerable willpower and often destroys the emotional bonding of sexual intercourse. It also is not as reliable as it might seem. The penis can eject sperm within its lubricating fluid prior to ejaculation, and a second sexual act may transfer sperm ejaculated earlier. The failure rate of this approach is estimated at 9% to 25%, no better than that of the rhythm method.

Sperm Blockage

If sperm cannot reach the uterus, fertilization cannot occur. One way to prevent the delivery of sperm is to encase the penis within a thin sheath, or condom. Many males do not favor the use of condoms, which tend to decrease their sensory pleasure during intercourse. In principle, this method is easy to apply and foolproof, but in practice it has a failure rate of 3% to 15%. Nevertheless, it is the most commonly employed form of birth control in the United States. Condoms are also widely used to prevent the transmission of AIDS and other sexually transmitted diseases (STDs). Over a billion condoms were sold in the United States last year.

A second way to prevent the entry of sperm into the uterus is to place a cover over the cervix. The cover may be a relatively tight-fitting cervical cap, which is worn for days at a time, or a rubber dome called a diaphragm, which is inserted immediately before intercourse. Because the dimensions of individual cervices vary, a cervical cap or diaphragm must be fitted by a physician. Failure rates average 4% to 25% for diaphragms, perhaps because of the propensity to insert them carelessly when in a hurry. Failure rates for cervical caps are somewhat lower.

Sperm Destruction

A third general approach to birth control is to eliminate the sperm after ejaculation. This can be achieved in principle by washing out the vagina immediately after intercourse, before the sperm have a chance to enter the uterus. Such a procedure is called a douche (French, "wash"). The douche method is difficult to apply well, since it involves a rapid dash to the bathroom immediately after ejaculation and a very thorough washing. Its failure rate is as high as 40%. Alternatively, sperm delivered to the vagina can be destroyed there with spermicidal jellies or foams. These treatments generally require application immediately before intercourse. Their failure rates vary from 10% to 30%.

Prevention of Ovulation

Since about 1960, a widespread form of birth control in the United States has been the daily ingestion of birth control pills, or oral contraceptives, by women. These pills contain analogues of progesterone, sometimes in combination with estrogens. As described earlier, progesterone and estradiol act by negative feedback to inhibit the secretion of FSH and LH during the luteal phase of the menstrual cycle, thereby preventing follicle development and ovulation. They also cause a buildup of the endometrium. The hormones in birth control pills have the same effects. Because the pills block ovulation, no ovum is available to be fertilized. A woman generally takes the hormone-containing pills for three weeks; during the fourth week, she takes pills without hormones (placebos), allowing the levels of those hormones in her blood to fall, which causes menstruation. Oral contraceptives provide a very effective means of birth control, with a failure rate of only 1% to 5%. In a variation of the oral contraceptive, hormone-containing capsules are implanted beneath the skin. These implanted capsules have failure rates below 1%.

A small number of women using birth control pills or implants experience undesirable side effects, such as blood clotting and nausea. These side effects have been reduced in newer generations of birth control pills, which contain less estrogen and different analogues of progesterone. Moreover, these new oral contraceptives provide a number of benefits, including reduced risks of endometrial and ovarian cancer, cardiovascular disease, and osteoporosis (for older women). However, they may increase the risk of contracting breast cancer and cervical cancer. The current consensus is that the health benefits of oral contraceptives outweigh their risks for most women, although a physician must help each woman determine the relative risks and benefits.

Prevention of Embryo Implantation

The insertion of a coil or other irregularly shaped object into the uterus is an effective means of birth control, since the irritation it produces in the uterus prevents the implantation of an embryo within the uterine wall. Such intrauterine devices (IUDs) have a failure rate of only 1% to 5%. Their high degree of effectiveness probably reflects their convenience; once they are inserted, they can be forgotten. The great disadvantage of this method is that almost a third of the women who attempt to use IUDs experience cramps, pain, and sometimes bleeding and therefore must discontinue using them.

Another method of preventing embryo implantation is the "morning after pill," which contains 50 times the dose of estrogen present in birth control pills. Its failure rate is 1% to 10%, but many women are uneasy about taking such high hormone doses, as side effects can be severe.

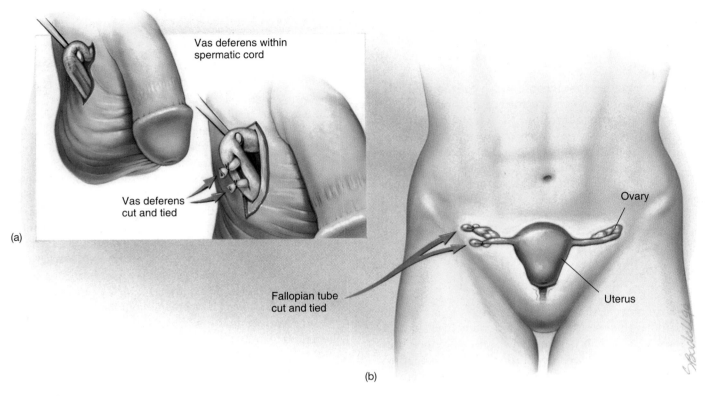

FIGURE 56.23
Birth control through sterilization. (a) Vasectomy; (b) tubal ligation.

Sterilization

A completely effective means of birth control is sterilization, the surgical removal of portions of the tubes that transport the gametes from the gonads (figure 56.23). Sterilization may be performed on either males or females, preventing sperm from entering the semen in males and preventing an ovulated oocyte from reaching the uterus in females. These procedures do not affect other aspects of sexual function.

In males, sterilization involves a vasectomy, the removal of a portion of the vas deferens from each testis. The procedure is simple and can be carried out in a physician's office. However, reversing a vasectomy is difficult and sometimes impossible. Reversal is particularly difficult if the man has developed antibodies against his own sperm, as often occurs in men who have had a vasectomy for a long period of time.

In females, the comparable operation involves the removal of a section of each Fallopian tube. Since these tubes are located within the abdomen, the operation, called a tubal ligation, is more difficult to perform than a vasectomy and is even more difficult to reverse. The surgical removal of the entire uterus, an operation that is usually performed for health reasons rather than birth control, is called a hysterectomy.

Abortion

Reproduction can be prevented after fertilization by abortion, removal of the developing embryo or fetus before birth. During the first three months of pregnancy, abortion is performed by vacuum suction or by dilation and curettage, in which the cervix is dilated and the uterine wall is scraped with a spoon-shaped surgical knife called a curette. Drugs are also being developed that cause abortion early in the first trimester, apparently with complete safety to the woman. One such drug, RU 486, is already in use in France and China but has not been approved for use in the United States. When administered within 49 days of the woman's last menstrual period and followed by prostaglandins that induce uterine contractions, RU 486 is almost 100% effective in causing abortion. Similar results are obtained by injecting the anticancer drug methotrexate (which blocks DNA replication, stopping cell proliferation), followed a week later by prostaglandins.

In the second trimester, the fetus can be removed by injecting a 20% saline solution into the uterus, which induces labor and delivery. In general, the more advanced the pregnancy, the more difficult and dangerous abortion is to the woman.

Fertilization can be prevented by a variety of birth control methods, including barrier contraceptives, hormonal inhibition, surgery, and abstinence. Postfertilization birth control involves removal of a fertilized egg. Efficacy rates vary from method to method and only latex condoms (and abstinence) afford protection from STDs such as AIDS.

Summary of Concepts

56.1 Animals employ both sexual and asexual reproductive strategies.

- Parthenogenesis is a form of asexual reproduction that is practiced by many insects and some lizards.
- Sequential hermaphroditism, in which the sex of an individual changes in response to environmental cues, is seen in some fish.
- Among mammals, the sex is determined by the presence of a Y chromosome in males and its absence in females.

56.2 The evolution of reproduction among the vertebrates has led to internalization of fertilization and development.

- Most bony fish practice external fertilization, releasing eggs and sperm into the water where fertilization occurs.
- Amphibians have external fertilization and the young go through a larval stage before metamorphosis occurs.
- Reptiles and birds are oviparous, the young developing in eggs that are deposited externally.
- Most mammals are viviparous, the young developing within the mother.

56.3 Male and female reproductive systems are specialized for different functions.

- The testes contain seminiferous tubules with germinal epithelium that produce sperm by meiosis.

- Sperm leave the testes and pass through the epididymis and vas deferens; the ejaculatory duct merges with the urethra, which empties at the tip of the penis.
- The clitoris is homologous to the penis and contains erectile tissue, and the labia major are homologous to the scrotum.
- An egg cell released from the ovary in ovulation is drawn by fimbria into the Fallopian tube, which conducts the egg cell to the lining of the uterus, or endometrium, where it implants if fertilized.
- As the follicles grow during the follicular phase of the cycle, they secrete increasing amounts of estradiol; this stimulates a surge of LH secretion, which causes the follicle to rupture in the process of ovulation.
- If fertilization does not occur, the corpus luteum regresses at the end of the cycle and the resulting fall in estradiol and progesterone secretion cause menstruation to occur in humans and apes.

56.4 The physiology of human sexual intercourse is becoming better known.

- The physiological events that occur in the human sexual response are grouped into four phases: excitement, plateau, orgasm, and resolution.
- Males and females have similar phases, but males enter a refractory period following orgasm that is absent in many women.
- There are a variety of methods of birth control available that range in ease of use, effectiveness, and permanence.

Discussing Key Terms

1. **Hermaphrodite**   A hermaphroditic animal has both testes and ovaries, though few hermaphroditic species are capable of self-fertilization. A sequential hermaphrodite is capable of changing sexes during its lifespan in response to environmental cues.

2. **Parthenogenesis**   This is "virgin birth," in which a species produces offspring from egg cells that are not fertilized.

3. **Oviparity**   Oviparous species are fertilized internally and lay eggs covered with a leathery skin (reptiles) or a calcareous shell (birds) in which the young develop.

4. **Ovoviviparity**   Ovoviviparous species have young that develop within eggs inside the mother.

5. **Viviparity**   Viviparous species have young that develop within the body of the mother and obtain their

nourishment from their mother's blood rather than from egg yolk.

6. **Ovulation**   Ovulation is the process whereby an egg cell, in the form of a secondary oocyte, is released from the ovary.

7. **Amnion**   The amnion is an extraembryonic membrane, formed from cells derived from the embryo, that surrounds the embryo and encases it within amniotic fluid.

8. **Endometrium**   The endometrium is the inner epithelial lining of the uterus. The more superficial layers are shed once per cycle, and in those species that undergo menstruation (humans and apes), this shedding is accompanied by bleeding.

Review Questions

1. How are oviparity, ovoviviparity, and viviparity different?

2. How does fetal development differ in the monotremes, marsupials, and placental mammals?

3. Briefly describe the function of each of the following male accessory sex organs: seminal vesicles, prostate gland, and bulbourethral glands.

4. When do the ova in a female mammal begin meiosis? When do they complete the first meiotic division? When do they complete the second meiotic division?

5. What hormone is secreted by the granulosa cells in a Graafian follicle? What effect does this hormone have on the endometrium, and what effect does it have on hormonal secretion by the anterior pituitary gland near the midpoint of the menstrual cycle?

6. What triggers ovulation in a female mammal? After ovulation, what happens to the empty follicle, and what hormones does it secrete? What effect do these hormones have on the endometrium, and what effect do they have on hormonal secretion by the anterior pituitary gland?

7. What are the four phases in the physiological events of sexual intercourse in humans? During the first phase, what events occur specifically in males, and what events occur specifically in females?

8. How do birth control pills prevent pregnancy?

Thought Questions

1. Relatively few kinds of animals have both male and female sex organs in the same animal, while most plants do. Propose an explanation for this.

2. In mammals, female sexual development is under *negative* control; that is, the absence of the *SRY* gene product and the absence of testosterone result in the development of female structures. Negative control makes it possible for embryos of either sex to develop inside female parents, which have no *SRY* gene product and very low levels of testosterone. If female sexual development were triggered by *positive* controls (the presence of certain hormones) and male sexual development were under negative controls, how might the development of embryos inside female parents be affected?

Internet Links

The Embryonic Zoo
http://www.ucalgary.ca/UofC/eduweb/virtualembryo/zoo.html
Video clips of developing embryos of Dictyostelium *(a fungus),* Caenorhabditis elegans *(a flatworm),* Drosophila melanogaster *(a fruitfly), sea urchin (an echinoderm), fish, chick, pig, and human.*

AIDS Information Center
http://www.ama-assn.org/special/hiv/hivhome.htm
The American Medical Association's AIDS INFORMATION CLEARING HOUSE, with extensive information resources, a library of literature, and reviews of links. A super-reliable comprehensive resource.

For Further Reading

Aral, S. and K. Holmes: "Sexually Transmitted Diseases in the AIDS Era," *Scientific American,* February 1991, pages 52–59. Gonorrhea, syphilis, and other infections still exact a terrible toll.

Culotta, E.: "Birth Tale Gets a New Twist," *Science,* vol. 268, page 365, April 21, 1995. The phenomenon of mammalian birth is reexamined in this short but interesting paper.

Fischman, J. and L. Ray: "Insights into Reproduction," *Science,* vol. 266, page 1459, December 2, 1994. Reproduction in humans is discussed.

Greene, W.: "AIDS and the Immune System," *Scientific American,* vol. 269, pages 98–105, September 1993. The feature article of this issue discusses how the virus is transmitted and how it "programs" the immune system to replicate the virus—with the subsequent consequences.

Holden, C.: "The Hazards of Estrogen," *Science,* vol. 260, pages 1238–39, May 28, 1993. Estrogen is essential for a properly functioning reproductive system in females, but what effect does estrogen, either naturally produced or supplied as a supplement, have on individuals, including developing fetuses?

Lagercrantz, H. and A. Slotkin: "The 'Stress' of Being Born," *Scientific American,* April 1986, pages 100–107. Passage through the narrow birth canal triggers the release of hormones important to the newborn's future survival.

Paabo, S.: "The Y Chromosome and the Origin of All of Us (Men)," *Science,* vol. 268, pages 1141–42, May 26, 1995. Why are there Y chromosomes? Are men really necessary? The origin of man and population genetics are explored under these blanket questions.

57

Cellular Mechanisms of Development

Concept Outline

57.1 Development is a regulated process.

Studying Development in Animals. Studies of cellular mechanisms have focused on mice, fruit flies, and nematodes.

Vertebrate Development. Vertebrates develop in a highly orchestrated fashion.

Insect Development. Insect development is highly specialized, many key events occurring in a fused mass of cells.

57.2 All animals employ the same basic mechanisms of development.

Cell Movement. Animal cells move by extending protein cables that they use to pull themselves past surrounding cells.

Induction. Transcription within cells is influenced by signal molecules from other cells.

Determination. Cells become reversibly committed to particular developmental paths.

Pattern Formation. Diffusion of chemical inducers governs pattern formation in fly embryos.

Expression of Homeotic Genes. Master genes determine the form body segments will take.

Programmed Cell Death. Some genes, when activated, kill their cells.

57.3 Three model developmental systems of animals have been extensively researched.

The Mouse. *Mus musculus.*
The Fruit Fly. *Drosophila melanogaster.*
The Nematode. *Caenorhabditis elegans.*

57.4 Aging can be considered a developmental process.

Theories of Aging. While there are many ideas about why cells age, no one theory of aging is widely accepted.

FIGURE 57.1
A collection of future fish undergo embryonic development.
Inside a transparent fish egg, a single cell becomes millions of cells that form eyes, fins, gills, and other body parts.

In the previous chapter, we explored how sex and reproduction lead to the formation of a new individual, haploid sperm fertilizing haploid egg to form a diploid zygote. Now we will broaden our perspective and look at the unique challenge posed by the development of that single diploid cell into a multicellular organism (figure 57.1). In the course of this developmental journey, a pattern of decisions about transcription are made that isolate particular lines of cells and cause them to proceed along different paths, spinning an incredibly complex web of cause and effect. Yet, for all its complexity, this developmental program works with impressive precision. In this chapter, we will explore the mechanisms used by animals to control their development and achieve this precision.

Studying Development in Animals

Organisms in all three multicellular kingdoms—fungi, plants, and animals—realize cell specialization by orchestrating gene expression (see chapter 38 for plant development). That is, different cells express different genes at different times. To understand development, we need to focus on how cells determine which genes to activate, and when.

In animals, development is complex and rigidly controlled, producing a bewildering array of specialized cell types. The subject of intensive study, animal development has in the last decades become relatively well understood. Here we will focus our attention on three developmental systems which researchers have studied intensively: (1) an animal with a very complexly arranged body, a mammal; (2) a less complex animal with an intricate developmental cycle, an insect; and (3) a very simple animal, a nematode.

To begin our investigation of development, we will first examine the overall process of development in two quite different animals, so we can sort through differences in the gross process to uncover basic similarities in underlying mechanisms. We will start by describing the overall process in vertebrates, because it is the best understood. Then we will examine the very different developmental process carried out by insects, in which genetics has allowed us to gain detailed knowledge of many aspects of the process (figure 57.2).

Almost all multicellular organisms undergo development. The process has been well studied in animals, especially in mammals, insects, and nematodes.

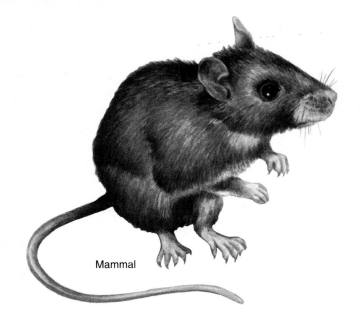

Mammal

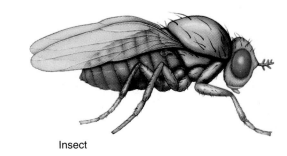

Insect

Nematode

FIGURE 57.2
Three developmental systems. Researchers studying the cellular mechanisms of development in animals have focused on these three organisms.

Vertebrate Development

Vertebrate development is a dynamic process in which cells divide rapidly and move over each other as they first establish the basic geometry of the body (figure 57.3). At different sites, particular cells then proceed to form the body's organs, and then the body grows to a size and shape that will allow it to survive after birth. The entire process, described more fully in chapter 58, is traditionally divided into phases. As in mitosis, however, the boundaries between phases are somewhat artificial, and the phases, in fact, grade into one another.

Cleavage

Vertebrates begin development as a single fertilized egg, the zygote. Within an hour after fertilization, the zygote begins to divide rapidly into a larger and larger number of smaller and smaller cells called **blastomeres,** until a solid ball of cells is produced (figure 57.4). This initial period of cell division, termed **cleavage,** is not accompanied by any increase in the overall size of the embryo; rather, the contents of the zygote are simply partitioned into the daughter cells. The two ends of the zygote are traditionally referred to as the **animal** and **vegetal poles.** In general, the blastomeres of the animal pole will go on to form the external tissues of the body, while those of the vegetal pole will form the internal tissues. The initial top-bottom (dorsal-ventral) orientation of the embryo is determined at fertilization by the location where the sperm nucleus enters the egg, a point that corresponds roughly to the future belly. After about 12 divisions, the burst of cleavage divisions slows, and transcription of key genes begins within the embryo cells.

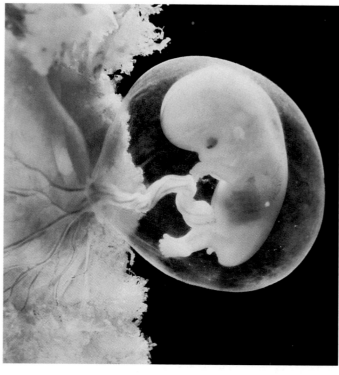

FIGURE 57.3
The miracle of development. This nine-week-old human fetus started out as a single cell: a fertilized egg, or zygote. The zygote's daughter cells have been repeatedly dividing and specializing to produce the distinguishable features of a fetus.

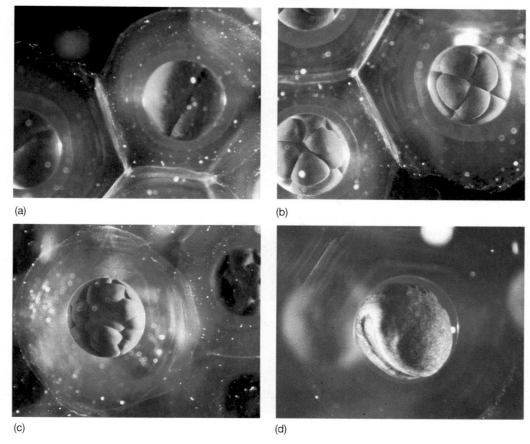

FIGURE 57.4
Cleavage divisions producing a frog embryo. (a) The initial divisions are, in this case, on the side of the embryo facing you, producing (b) a cluster of cells on this side of the embryo, which soon expands to become a (c) compact mass of cells. (d) This mass eventually invaginates into the interior of the embryo, forming a gastrula.

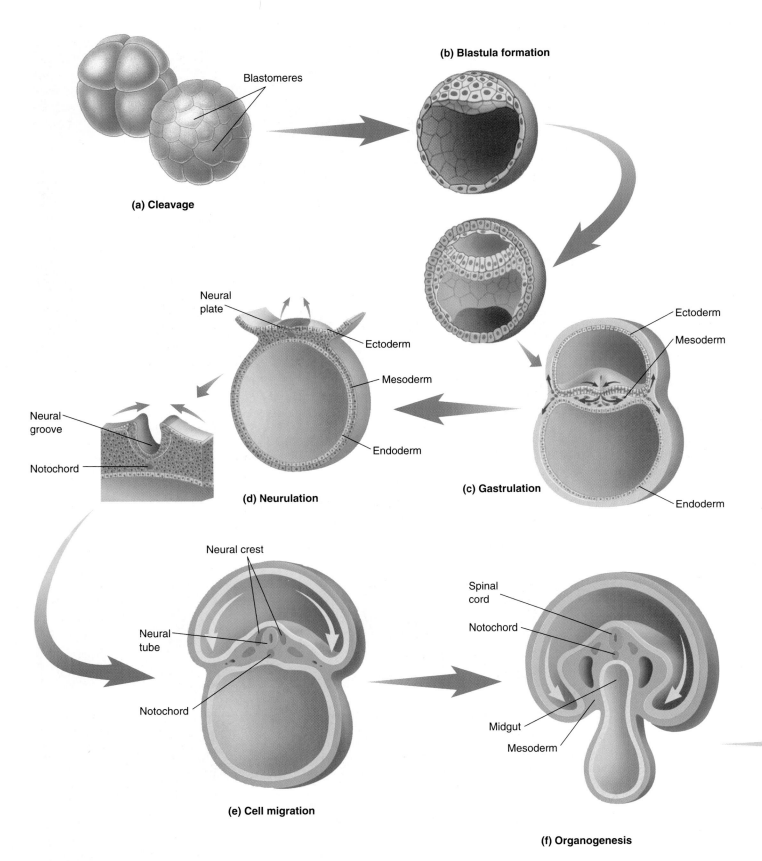

(a) Cleavage

Blastomeres

(b) Blastula formation

(c) Gastrulation

Ectoderm

Mesoderm

Endoderm

(d) Neurulation

Neural plate

Ectoderm

Mesoderm

Endoderm

Neural groove

Notochord

(e) Cell migration

Neural crest

Neural tube

Notochord

(f) Organogenesis

Spinal cord

Notochord

Midgut

Mesoderm

FIGURE 57.5

The path of vertebrate development. An illustration of the major events in the development of *Mus musculus,* the house mouse. (a) Cleavage. (b) Formation of blastula. (c) Gastrulation. (d) Neurulation. (e) Cell migration. (f) Organogenesis. (g) Growth.

Formation of the Blastula

The outermost blastomeres (figure 57.5a) in the ball of cells produced during cleavage are joined to one another by tight junctions, which, as you may recall from chapter 7, are belts of protein that encircle a cell and weld it firmly to its neighbors. These tight junctions create a seal that isolates the interior of the cell mass from the surrounding medium. At about the 16-cell stage, the cells in the interior of the mass begin to pump Na+ from their cytoplasm into the spaces between cells. The resulting osmotic gradient causes water to be drawn into the center of the cell mass, enlarging the intercellular spaces. Eventually, the spaces coalesce to form a single large cavity within the cell mass. The resulting hollow ball of cells is called a **blastula** (figure 57.5b).

Gastrulation

One wall of the blastula then pushes inward at the vegetal pole, forming a **gastrula** that is invaginated like a collapsed tennis ball. Cells at the vegetal pole move by using extensions called lamellipodia to crawl over neighboring cells, which respond by forming lamellipodia of their own. Soon the sheet of cells at the vegetal pole contracts on itself and shoves inward, starting the invagination. Called **gastrulation** (figure 57.5c), this process creates the main axis of the vertebrate body, converting the blastula into a bilaterally symmetrical embryo with a central gut tube that opens to the outside. From this point on, the embryo has three **germ layers** whose organization foreshadows the future organization of the adult body. The cells that invaginate and form the tube of the primitive gut are endoderm; they give rise to the stomach, lungs, liver, and most of the other internal organs. The cells that remain on the exterior are ectoderm, and their derivatives include the skin on the outside of the body and the nervous system. The cells that break away from the invaginating cells and invade the space between the gut and the exterior wall are mesoderm; they eventually form the notochord, bones, blood vessels, connective tissues, and muscles.

Neurulation

Soon after gastrulation is complete, a broad zone of ectoderm begins to thicken on the dorsal surface of the embryo, an event triggered by the presence of the notochord beneath it. The thickening is produced by the elongation of certain ectodermal cells. Those cells then assume a wedge shape by contracting bundles of actin filaments at one end. This change in shape causes the cells to roll up into a tube, which eventually pinches off from the rest of the ectoderm and gives rise to the brain and spinal cord. This tube is called the **neural tube,** and the process by which it forms is termed **neurulation** (figure 57.5d).

Cell Migration

During the next stage of vertebrate development, a variety of cells migrate to form distant tissues, following specific paths through the embryo to particular locations (figure 57.5e). These migrating cells include those of the **neural crest,** which pinch off from the neural tube and form a number of structures, including some of the body's sense organs; cells that migrate from central blocks of muscle tissue called **somites** and form the skeletal muscles of the body; and the precursors of blood cells and gametes. When a migrating cell reaches its destination, receptor proteins on its surface interact with proteins on the surfaces of cells in the destination tissue, triggering changes in the cy-toskeleton of the migrating cell that cause it to cease moving.

Organogenesis and Growth

At the end of this wave of cell migration and colonization, the basic vertebrate body plan has been established, although the embryo is only a few millimeters long and has only about 10^5 cells. Over the course of subsequent development, tissues will develop into organs (figure 57.5f), and the embryo will grow to be a hundred times larger, with a million times as many cells (figure 57.5g).

(g) Growth

Vertebrates develop in a highly orchestrated fashion. The zygote divides rapidly, forming a hollow ball of cells that then pushes inward, forming the main axis of an embryo that goes on to develop tissues, and after a process of cell migration, organs.

Insect Development

Like all animals, insects develop through an orchestrated series of cell changes, but the path of development is quite different from that of a vertebrate. Many insects produce two different kinds of bodies during their development, the first a tubular eating machine called a **larva**, and the second a flying machine with legs and wings. The passage from one body form to the other is called **metamorphosis** and involves a radical shift in development. Here we will describe development in the fruit fly *Drosophila* (figure 57.6), which is the subject of much genetic research.

FIGURE 57.6
The fruit fly, *Drosophila melanogaster*. A dorsal view of *Drosophila*, one of the most intensively studied animals in development.

Maternal Genes

The development of an insect like *Drosophila* begins before fertilization, with the construction of the egg. Specialized nurse cells that help the egg to grow move some of their own mRNA into the end of the egg nearest them (figure 57.7*a*). As a result, mRNAs produced by maternal genes are positioned in particular locations in the egg, so that after repeated divisions subdivide the fertilized egg, different daughter cells will contain different maternal products. Thus, the action of maternal (rather than zygotic) genes determines the initial course of development.

Syncytial Blastoderm

After fertilization, 12 rounds of nuclear division without cytokinesis produce about 4000 nuclei, all within a single cytoplasm. All of the nuclei within this **syncytial blastoderm** (figure 57.7*b*) can freely communicate with one another, but nuclei located in different sectors of the egg experience different maternal products. The nuclei space themselves evenly along the surface of the blastoderm, and membranes grow between them, producing a hollow ball of cells. Folding of the embryo and primary tissue development soon follow, in a process fundamentally similar to that seen in vertebrate development. The tubular body that results within a day of fertilization is a larva.

Larval Instars

The larva begins to feed immediately, and as it does so, it grows. Its chitinous exoskeleton cannot stretch much, however, and within a day it sheds the exoskeleton. Before the new exoskeleton has had a chance to harden, the larva expands in size. A total of three larval stages, or **instars,** are produced over a period of four days (figure 57.7*c*).

Imaginal Discs

During embryonic growth, about a dozen groups of cells called **imaginal discs** are set aside in the body of the larva (figure 57.7*d*). Imaginal disks play no role in the life of the larva, but are committed to form key parts of the adult fly's body.

Metamorphosis

After the last larval stage, a hard outer shell forms, and the larva is transformed into a **pupa** (figure 57.7*e*). Within the pupa, the larval cells break down and release their nutrients, which are used in the growth and development of the various imaginal discs (eye discs, wing discs, leg discs, and so on). The imaginal discs then associate with one another, assembling themselves into the body of the adult fly (figure 57.7*f*). The metamorphosis of a *Drosophila* larva into a pupa and then into adult fly takes about four days, after which the pupal shell splits and the fly emerges.

Drosophila development proceeds through two discrete phases, the first a larval phase that gathers food, then an adult phase that is capable of flight and reproduction.

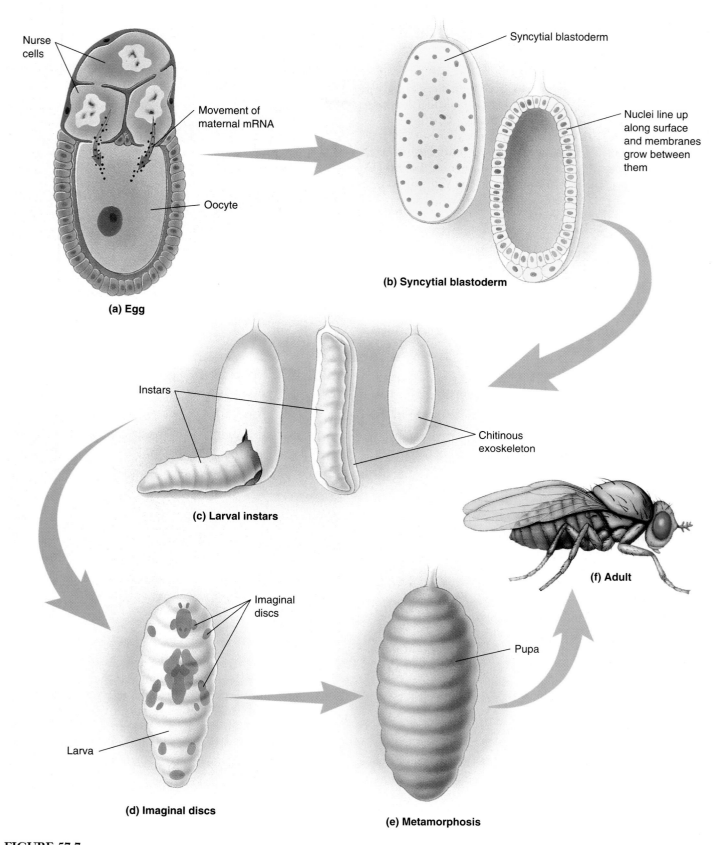

FIGURE 57.7
The path of insect development. An illustration of the major events in the development of *Drosophila melanogaster*. (a) Egg. (b) Syncytial blastoderm. (c) Larval instars. (d) Imaginal discs. (e) Metamorphosis. (f) Adult.

Despite the many differences in the two developmental paths we have just discussed, it is becoming increasingly clear that all animals develop according to molecular mechanisms that are fundamentally very similar. This observation suggests that these mechanisms evolved very early in the history of multicellular life. Here, we will focus on six mechanisms that seem to be of particular importance in the development of a wide variety of organisms. We will consider them in roughly the order in which they first become important during development.

Cell Movement

Cells migrate during many stages in animal development, sometimes traveling great distances before reaching the site where they are destined to develop. By the time vertebrate development is complete, most tissues contain cells that originated from quite different parts of the early embryo. One way cells move is by pulling themselves along using cell adhesion molecules, such as the cadherin proteins you read about in chapter 7. Cadherins span the plasma membrane, protruding into the cytoplasm and extending out from the cell surface. The cytoplasmic portion of the molecule is attached to actin or intermediate filaments of the cytoskeleton, while the extracellular portion has five 100-amino acid segments linked end-to-end; three or more of these segments have Ca++ binding sites that play a critical role in the attachment of the cadherin to other cells (figure 57.8). Over a dozen different cadherins have been discovered to date. Each type of cadherin attaches to others of its own type at its terminal segments, forming a two-cadherin link between the cytoskeletons of adjacent cells. As a cell migrates to a different tissue, the nature of the cadherin it expresses changes, and if cells expressing two different cadherins are mixed, they quickly sort themselves out, aggregating into two separate masses. This is how the different imaginal discs of a *Drosophila* larva assemble into an adult. Other calcium-independent cell adhesion molecules, such as the neural cell adhesion molecules (N-CAMs) expressed by migrating nerve cells, reinforce the associations made by cadherins, but cadherins play the major role in holding aggregating cells together.

In some tissues, such as connective tissue, much of the volume of the tissue is taken up by the spaces *between* cells. These spaces are not vacant, however. Rather, they are filled with a network of molecules secreted by surrounding cells, principally, a matrix of long polysaccharide chains covalently linked to proteins (proteoglycans), within which are embedded strands of fibrous protein (collagen, elastin, and fibronectin). Migrating cells traverse this matrix by binding to it with cell surface proteins called integrins (figure 57.9), which was also described in chapter 7. Integrins are attached to actin filaments of the cytoskeleton and protrude out from the cell surface in pairs, like two hands. The

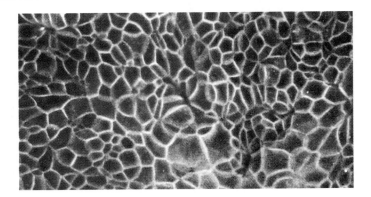

FIGURE 57.8
Cadherins and cell-cell contact. In this culture of epithelial cells, which has been specially stained white for cadherins, notice that the concentration of cadherins is high in regions where cells come into contact with one another, visible here in the intercellular spaces.

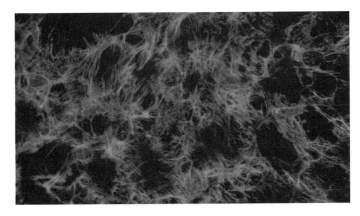

FIGURE 57.9
Integrins connect the cytoskeleton of cells to the extracellular matrix. Here, cells have been stained to show only a fibronectin extracellular matrix. The fibronectin is adhered to the cells by integrins.

"hands" grasp a specific component of the matrix such as collagen or fibronectin, thus linking the cytoskeleton to the fibers of the matrix. In addition to providing an anchor, this binding can initiate changes within the cell, alter the growth of the cytoskeleton, and change the way in which the cell secretes materials into the matrix.

Thus, cell migration is largely a matter of changing patterns of cell adhesion. As a migrating cell travels, it continually extends projections that probe the nature of its environment. Tugged this way and that by different tentative attachments, the cell literally feels its way toward its ultimate target site.

Cells migrate by extending probes to neighboring cells which they use to pull themselves along.

Induction

In *Drosophila* the initial cells created by cleavage divisions contain different developmental signals (called **determinants**) from the egg, setting individual cells off on different developmental paths. This pattern of development is called **mosaic development**. In mammals, by contrast, all of the blastomeres receive equivalent sets of determinants; body form is determined by cell-cell interactions, a pattern called **regulative development**.

We can demonstrate the importance of cell-cell interactions in development by separating the cells of an early blastula and allowing them to develop independently. Under these conditions, animal pole blastomeres develop features of ectoderm and vegetal pole blastomeres develop features of endoderm, but none of the cells ever develop features characteristic of mesoderm. However, if animal pole and vegetal pole cells are placed next to each other, some of the animal pole cells will develop as mesoderm. The interaction between the two cell types triggers a switch in the developmental path of the cells! When a cell switches from one path to another as a result of interaction with an adjacent cell, **induction** has taken place (figure 57.10).

How do cells induce developmental changes in neighboring cells? Apparently, the inducing cells secrete proteins that act as intercellular signals. Signal molecules, which we discussed in detail in chapter 7, are capable of producing abrupt changes in the patterns of gene transcription.

In some cases, particular groups of cells called **organizers** produce diffusible signal molecules that convey positional information to other cells. Organizers can have a profound influence on the development of surrounding tissues. Working as signal beacons, they inform surrounding cells of their distance from the organizer. The closer a particular cell is to an organizer, the higher the concentration of the signal molecule, or **morphogen**, it experiences (figure 57.11). Although only a few morphogens have been isolated, they are thought to be part of a widespread mechanism for determining relative position during development.

A single morphogen can have different effects, depending upon how far away from the organizer the affected cell is located. Thus, low levels of the morphogen activin will cause cells of the animal pole of an early *Xenopus* embryo to develop into epidermis, while slightly higher levels will induce the cells to develop into muscles, and levels a little higher than that will induce them to form notochord (figure 57.12).

Interactions between cells strongly influence the developmental paths they take. Signal molecules from an inducing cell alter patterns of transcription in cells which come in contact with it.

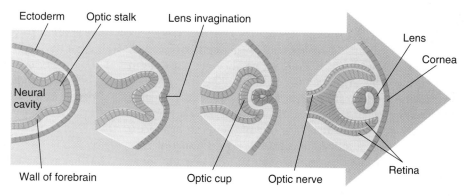

FIGURE 57.10
Development of the vertebrate eye proceeds by induction. The eye develops as an extension of the forebrain called the optic stalk that grows out until it contacts the ectoderm. This contact induces the formation of a lens from the ectoderm.

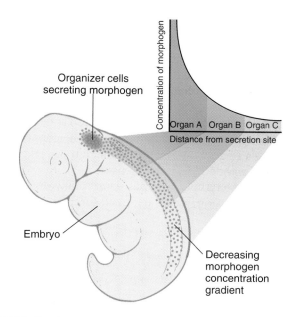

FIGURE 57.11
An organizer creates a morphogen gradient. As a morphogen diffuses from the organizer site, it becomes less concentrated. Different concentrations of the morphogen stimulate the development of different organs.

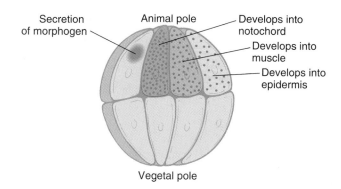

FIGURE 57.12
Fate of cells in an early *Xenopus* embryo. The fates of the individual cells are determined by the concentration of morphogen washing over them.

Determination

The mammalian egg is symmetrical in its contents as well as its shape, so that all of the cells of an early blastoderm are equivalent up to the eight-cell stage. The cells are said to be **totipotent,** meaning that they are potentially capable of expressing all of the genes of their genome. If they are separated from one another, any one of them can produce a completely normal individual. Indeed, just this sort of procedure has been used to produce sets of four or eight identical offspring in the commercial breeding of particularly valuable lines of cattle. The reverse process works, too; if cells from two different eight-cell-stage embryos are combined, a single normal individual results. Such an individual is called a **chimera,** because it contains cells from different genetic lines (figure 57.13).

Mammalian cells start to become different after the eight-cell stage as a result of cell-cell interactions like those we just discussed. At this point, the pathway that will influence the future developmental fate of the cells is determined. The commitment of a particular cell to a specialized developmental path is called **determination.** A cell in the prospective brain region of an amphibian embryo at the early gastrula stage has not yet been determined; if transplanted elsewhere in the embryo, it will develop like its new neighbors. By the late gastrula stage, however, determination has taken place, and the cell will develop as neural tissue no matter where it is transplanted. Determination must be carefully distinguished from **differentiation,** which is the cell specialization that occurs at the end of the developmental path. Cells may become *determined* to give rise to particular tissues long before they actually *differentiate* into those tissues. The cells of a *Drosophila* eye imaginal disc, for example, are fully determined to produce an eye, but they remain totally undifferentiated during most of the course of larval development.

The Mechanism of Determination

What is the molecular mechanism of determination? The gene regulatory proteins discussed in detail in chapter 16 are the tools used by cells to initiate developmental changes. When genes encoding these proteins are activated, one of their effects is to reinforce their own activation. This makes the developmental switch deterministic, initiating a chain of events that leads down a particular developmental pathway. Cells in which a set of regulatory genes have been activated may not actually undergo differentiation until some time later, when other factors interact with the regulatory protein and cause it to activate still other genes. Nevertheless, once the initial "switch" is thrown, the cell is fully committed to its future developmental path.

Often, before a cell becomes fully committed to a particular developmental path, it first becomes partially committed, acquiring **positional labels** that reflect its location in the embryo. These labels can have a great influence on how the pattern of the body subsequently develops. In a chicken embryo, if tissue at the base of the leg bud (which would normally give rise to the thigh) is transplanted to the tip of the identical-looking wing bud (which would normally give rise to the wing tip), that tissue will develop into a toe rather than a thigh! The tissue has already been determined as leg but is not yet committed to being a particular part of the leg. Therefore, it can be influenced by the positional signaling at the tip of the wing bud to form a tip (in this case a tip of leg).

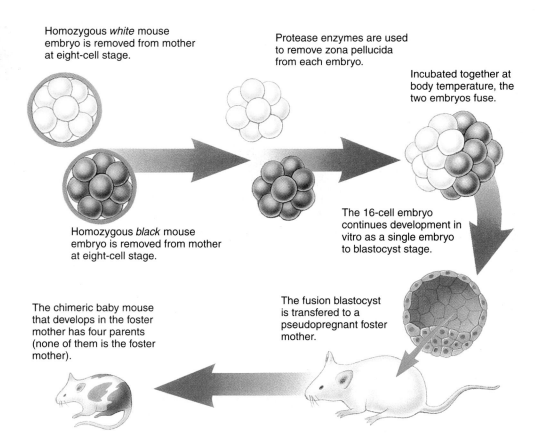

Homozygous *white* mouse embryo is removed from mother at eight-cell stage.

Protease enzymes are used to remove zona pellucida from each embryo.

Incubated together at body temperature, the two embryos fuse.

Homozygous *black* mouse embryo is removed from mother at eight-cell stage.

The 16-cell embryo continues development in vitro as a single embryo to blastocyst stage.

The fusion blastocyst is transfered to a pseudopregnant foster mother.

The chimeric baby mouse that develops in the foster mother has four parents (none of them is the foster mother).

FIGURE 57.13
Constructing a chimeric mouse.
Cells from two eight-cell individuals fuse to form a single individual.

Is Determination Irreversible?

Until very recently, biologists thought determination was irreversible. Experiments carried out in the 1950s and 1960s by John Gurdon and others (described in chapter 14) made what seemed a convincing case: using very fine pipettes (hollow glass tubes) to suck the nucleus out of a frog or toad egg, these researchers replaced the egg nucleus with a nucleus sucked out of a body cell taken from another individual. If the transplanted nucleus was obtained from an advanced embryo, the egg went on to develop into a tadpole, but died before becoming an adult.

Nuclear transplant experiments were attempted without success by many investigators, until finally, in 1984, Steen Willadsen, a Danish embryologist working in Texas, succeeded in cloning a sheep using the nucleus from a cell of an early embryo. The key to his success was in picking a cell very early in development. This exciting result was soon replicated by others in a host of other organisms, including pigs and monkeys.

Only early embryo cells seemed to work, however. Researchers became convinced, after many attempts to transfer older nuclei, that animal cells become irreversibly committed after the first few cell divisions of the developing embryo.

We now know this conclusion to have been unwarranted. The key advance unraveling this puzzle was made in Scotland by geneticists Keith Campbell and Ian Wilmut, who reasoned that perhaps the egg and the donated nucleus needed to be at the same stage in the cell cycle (figure 57.14). They removed mammary cells from the udder of a six-year-old sheep. The origin of these cells gave the clone its name, "Dolly," after the country singer Dolly Parton. The cells were grown in tissue culture; then, in preparation for cloning, the researchers substantially reduced for five days the concentration of serum nutrients on which the sheep mammary cells were subsisting. Starving the cells caused them to pause at the beginning of the cell cycle. In parallel preparation, eggs obtained from a ewe were enucleated, the nucleus of each egg carefully removed with a micropipette.

Mammary cells and egg cells were then surgically combined in January of 1996, inserting the mammary cells inside the covering around the egg cell. The researchers then applied a brief electrical shock. This caused the plasma membranes surrounding the two cells to become leaky, so that the nucleus of the mammary cell passed into the egg cell—a neat trick. The shock also kick-started the cell cycle, causing the cell to begin to divide.

After six days, in 30 of 277 tries, the dividing embryo reached the hollow-ball "blastula" stage, and 29 of these were transplanted into surrogate mother sheep. A little over five months later, on July 5, 1996, one sheep gave birth to a lamb, Dolly, the first clone generated from a fully differentiated animal cell. Dolly established beyond all dispute that determination is reversible, that with the right techniques the fate of a fully differentiated cell *can* be altered.

The commitment of particular cells to certain developmental fates is fully reversible.

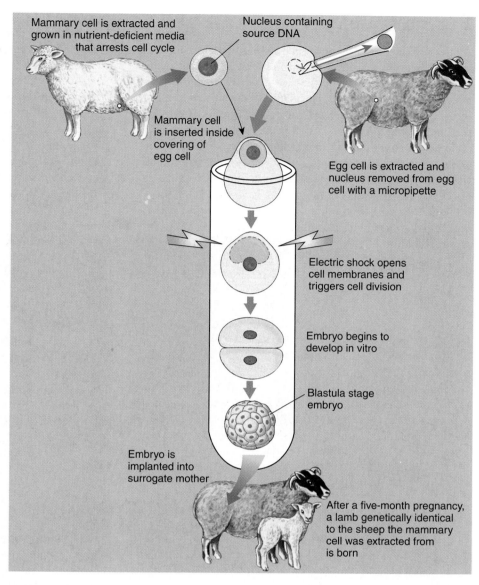

Mammary cell is extracted and grown in nutrient-deficient media that arrests cell cycle

Nucleus containing source DNA

Mammary cell is inserted inside covering of egg cell

Egg cell is extracted and nucleus removed from egg cell with a micropipette

Electric shock opens cell membranes and triggers cell division

Embryo begins to develop in vitro

Blastula stage embryo

Embryo is implanted into surrogate mother

After a five-month pregnancy, a lamb genetically identical to the sheep the mammary cell was extracted from is born

FIGURE 57.14
Proof that determination is reversible. This experiment by Campbell and Wilmut was the first successful cloning of an adult animal.

Pattern Formation

All animals seem to use positional information to determine the basic pattern of body compartments and, thus, the overall architecture of the adult body. How is positional information encoded in labels and read by cells? To answer this question, let us consider how positional labels are used in pattern formation in *Drosophila*. The Nobel Prize in physiology or medicine was awarded in 1995 for the unraveling of this puzzle.

As we noted previously, a *Drosophila* egg acquires an initial asymmetry long before fertilization as a result of maternal mRNA molecules that are deposited in one end of the egg by nurse cells. Part of this maternal mRNA, from a gene called *bicoid*, remains near its point of entry, marking what will become the embryo's front end. Fertilization causes this mRNA to be translated into bicoid protein, which diffuses throughout the syncytial blastoderm, forming a morphogen gradient. Mothers unable to make bicoid protein produce embryos without a head or thorax (in effect, these embryos are two-tailed, or bicaudal—hence the name "bicoid"). Bicoid protein establishes the anterior (front) end of the embryo. If bicoid protein is injected into the anterior end of mutant

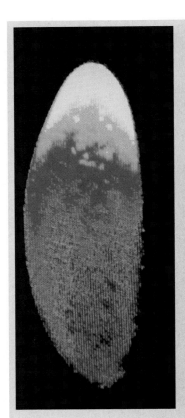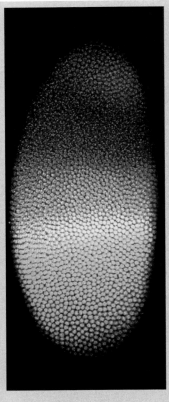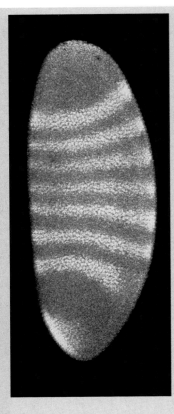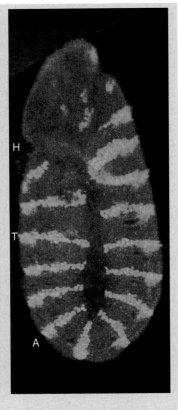

Establishing polarity of the embryo: Fertilization of the egg triggers the production of bicoid protein from maternal RNA in the egg. The bicoid protein diffuses through the egg, forming a gradient. This gradient determines the polarity of the embryo, with the head and thorax developing in the zone of high concentration (yellow through red).

Setting the stage for segmentation: About 2½ hours after fertilization, bicoid protein turns on a series of brief signals from so-called gap genes. The gap proteins act to divide the embryo into large blocks. In this photo, fluorescent dyes in antibodies that bind to the gap proteins Krüppel *(red)* and hunchback *(green)* make the blocks visible; the region of overlap is yellow.

Laying down the fundamental regions: About ½ hour later, the *gap* genes switch on a so-called "pair-rule" gene called *hairy. Hairy* produces a series of boundaries within each block, dividing the embryo into seven fundamental regions.

Forming the segments: The final stage of segmentation occurs when a "segment-polarity" gene called *engrailed* divides each of the seven regions into halves, producing 14 narrow compartments. Each compartment corresponds to one segment of the future body. There are three head segments (H, top left), three thoracic segments (T, lower left), and eight abdominal segments (A, from bottom left to upper right).

FIGURE 57.15

Body organization in an early *Drosophila* embryo. In these images by 1995 Nobel laureate, Christiane Nüsslein-Volhard, and Sean Carroll, we watch a *Drosophila* egg pass through the early stages of development, in which the basic segmentation pattern of the embryo is established.

embryos unable to make it, the embryos will develop normally. If it is injected into the opposite (posterior) end of normal embryos, a head and thorax will develop at that end.

Bicoid protein exerts this profound effect on the organization of the embryo by activating genes that encode the first mRNAs to be transcribed after fertilization. Within the first two hours, before cellularization of the syncytial blastoderm, a group of six genes called the **gap genes** begins to be transcribed. These genes map out the coarsest subdivision of the embryo (figure 57.15). One of them is a gene called *hunchback* (since an embryo without *hunchback* lacks a thorax and so, takes on a hunched shape). Although *hunchback* mRNA is distributed throughout the embryo, its translation is controlled by the protein product of another maternal mRNA called *nanos* (named after the Greek word for "dwarf," as mutants without *nanos* genes lack abdominal segments and hence, are small). The **nanos** protein binds to *hunchback* mRNA, preventing it from being translated. The only place in the embryo where there is too little nanos protein to block translation of *hunchback* mRNA is the far anterior end. Consequently, hunchback protein is made primarily at the anterior end of the embryo. As it diffuses back toward the posterior end, it sets up a second morphogen gradient responsible for establishing the thoracic and abdominal segments.

Other gap genes act in more posterior regions of the embryo. They, in turn, activate 11 or more **pair-rule genes.** (When mutated, each of these genes alters every other body segment.) One of the pair-rule genes, named *hairy*, produces seven bands of protein, which look like stripes when visualized with fluorescent markers. These bands establish boundaries that divide the embryo into seven zones. Finally, a group of 16 or more **segment polarity genes** subdivide these zones. The *engrailed* gene, for example, divides each of the seven zones established by *hairy* into anterior and posterior compartments. The 14 compartments that result correspond to the three head segments, three thoracic segments, and eight abdominal segments of the embryo.

Thus, within three hours after fertilization, a highly orchestrated cascade of segmentation gene activity produces the fly embryo's basic body plan. The activation of these and other developmentally important genes (figure 57.16) depends upon the free diffusion of morphogens that is possible within a syncytial blastoderm. In mammalian embryos with cell partitions, other mechanisms must operate.

In *Drosophila* diffusion of chemical inducers produces the embryo's basic body plan, a cascade of genes dividing it into 14 compartments.

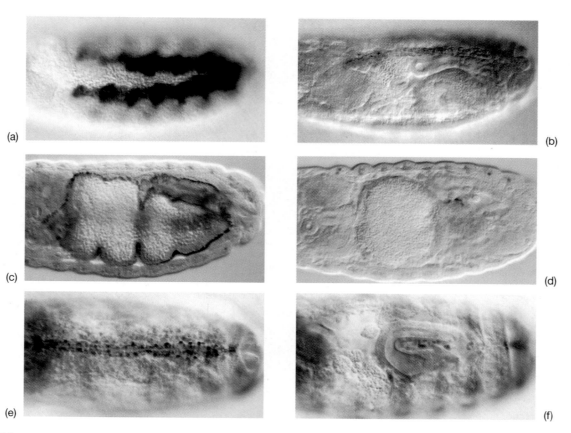

FIGURE 57.16
A gene controlling organ formation in *Drosophila*. Called *tinman*, this gene is responsible for the formation of gut musculature and the heart. The dye shows expression of the *tinman* in five-hour (a) and seventeen-hour (b) *Drosophila* embryos. The gut musculature then appears along the edges of normal embryos (c) but is not present in embyros in which the gene has been mutated (d). The heart tissue develops along the center of normal embryos (e) but is missing in *tinman* mutant embryos (f).

Expression of Homeotic Genes

After pattern formation has successfully established the number of body segments in *Drosophila*, a series of **homeotic genes** act as master switches to determine the forms these segments will assume. Homeotic genes code for proteins that function as transcription factors. Each homeotic gene activates a particular module of the genetic program, initiating the production of specific body parts within each of the 14 compartments.

Homeotic Mutations

Mutations in homeotic genes lead to the appearance of perfectly normal body parts in unusual places. Mutations in *bithorax* (figure 57.17), for example, cause a fly to grow an extra pair of wings, as if it had a double thoracic segment, and mutations in *Antennapedia* cause legs to grow out of the head in place of antennae! In the early 1950s, geneticist Edward Lewis discovered that several homeotic genes, including *bithorax*, map together on the third chromosome of *Drosophila*, in a tight cluster called the **bithorax complex.** Mutations in these genes all affect body parts of the thoracic and abdominal segments, and Lewis concluded that the genes of the bithorax complex control the development of body parts in the rear half of the thorax and all of the abdomen. Most interestingly, the order of the genes in the bithorax complex mirrors the order of the body parts they control, as if the genes are activated serially! Genes at the beginning of the cluster switch on development of the

FIGURE 57.17
Mutations in homeotic genes. Three separate mutations in the *bithorax* gene caused this fruit fly to develop an extra thoracic segment, with accompanying wings. Compare this photograph with that of the normal fruit fly in figure 57.6.

FIGURE 57.18

FIGURE 57.18
***Drosophila* homeotic genes.**
Called the homeotic gene complex, or HOM complex, the genes are grouped into two clusters, the Antennapedia complex (anterior) and the bithorax complex (posterior).

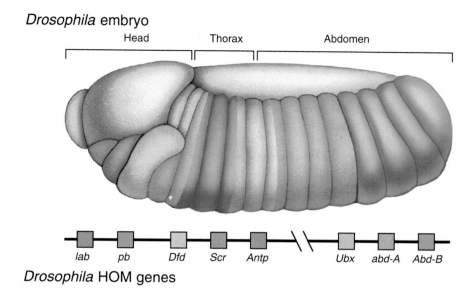

Drosophila embryo

Head | Thorax | Abdomen

lab | *pb* | *Dfd* | *Scr* | *Antp* | *Ubx* | *abd-A* | *Abd-B*

Drosophila HOM genes

thorax, those in the middle control the anterior part of the abdomen, and those at the end affect the tip of the abdomen. A second cluster of homeotic genes, the **Antennapedia complex,** was discovered in 1980 by Thomas Kaufmann. The Antennapedia complex governs the anterior end of the fly, and the order of genes in it also corresponds to the order of segments they control (figure 57.18).

The Homeobox

Drosophila homeotic genes typically contain the **homeobox,** a sequence of 180 nucleotides that codes for a 60-amino acid DNA-binding peptide domain called the homeodomain (figure 57.19). As we saw in chapter 16, proteins that contain the homeodomain function as transcription factors, ensuring that developmentally related genes are transcribed at the appropriate time. Segmentation genes such as *bicoid* and *engrailed* also contain the homeobox sequence. Clearly, the homeobox distinguishes those portions of the genome devoted to pattern formation.

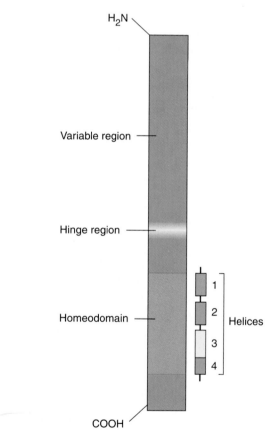

H₂N

Variable region

Hinge region

Homeodomain

1
2 } Helices
3
4

COOH

FIGURE 57.19
Homeodomain protein. This protein plays an important regulatory role when it binds to DNA and regulates expression of specific genes. The variable region of the protein determines the specific activity of the protein. Also included in this protein is a small hinge region and the homeodomain, a 60-amino-acid sequence common to all proteins of this type. The homeodomain region of the protein is coded for by the homeobox region of genes and is composed of four α-helices. One of the helices recognizes and binds to a specific DNA sequence in target genes.

Evolution of Homeobox Genes

Since their initial discovery in *Drosophila*, homeotic genes have also been found in mice and humans, which are separated from insects by over 600 million years of evolution. Their presence in mammals and insects indicates that homeotic genes governing the positioning of body parts must have arisen very early in the evolutionary history of animals. Similar genes also appear to operate in flowering plants. Gene probes made using the homeobox sequence of *Drosophila* have been used to identify very similar sequences in a wide variety of other organisms, including frogs, mice, humans, cows, chickens, beetles, and even earthworms. Mice and humans have four clusters of homeobox-containing genes, called *Hox* genes in mice. Just as in flies, the homeotic genes of mammals appear to be lined up in the same order as the segments they control (figure 57.20). Thus, the ordered nature of homeotic gene clusters is highly conserved in evolution (figure 57.21). There is a total of 38 *Hox* genes in the four homeotic clusters of a mouse, and we are only beginning to understand how they interact.

Homeotic genes encode transcription factors that activate blocks of genes specifying particular body parts.

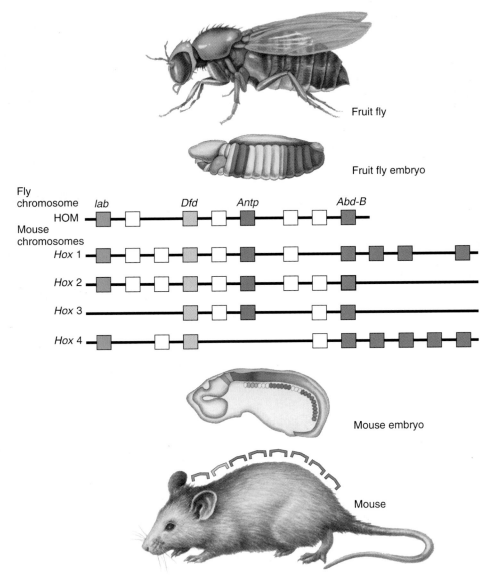

FIGURE 57.20
A comparison of homeotic gene clusters in the fruit fly *Drosophila melanogaster* and the mouse *Mus musculus*. Similar genes, the *Drosophila* HOM genes and the mouse *Hox* genes, control the development of front and back parts of the body. These genes are located on a single chromosome in the fly, and on four separate chromosomes in mammals. The genes are color-coded to match the parts of the body in which they are expressed.

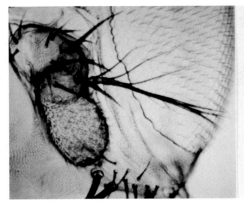
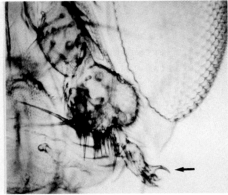

FIGURE 57.21

The remarkably conserved homeobox series. By inserting a mouse homeobox-containing gene into a fruit fly, a mutant fly (*right*) can be manufactured with a leg (*arrow*) growing from where its antenna would be in a normal fly (*left*).

Programmed Cell Death

Not every cell that is produced during development is destined to survive. For example, the cells between your fingers and toes die; if they did not, you would have paddles rather than digits. Vertebrate embryos produce a very large number of neurons, ensuring that there are enough neurons to make all of the necessary synaptic connections, but over half of these neurons never make connections and die in an orderly way as the nervous system develops. Unlike accidental cell deaths due to injury, these cell deaths are planned for and indeed required for proper development. Cells that die due to injury typically swell and burst, releasing their contents into the extracellular fluid. This form of cell death is called **necrosis**. In contrast, cells programmed to die shrivel and shrink in a process called **apoptosis** (from the Greek word meaning shedding of leaves in autumn), and their remains are taken up by surrounding cells (figure 57.22).

Gene Control of Apoptosis

This sort of developmentally regulated cell suicide occurs when a "death program" is activated. All animal cells appear to possess such programs. In the nematode worm, for example, the same 131 cells always die during development in a predictable and reproducible pattern of apoptosis. Three genes govern this process. Two (*ced-3* and *ced-4*) constitute the death program itself; if either is mutant, those 131 cells do not die, and go on instead to form nervous and other tissue. The third gene (*ced-9*) represses the death program encoded by the other two. The same sorts of apoptosis programs occur in human cells: the *bax* gene encodes the cell death program, and another, an oncogene called *bcl-2*, represses it. The mechanism of apoptosis appears to have been highly conserved during the course of animal evolution. The protein made by *bcl-2* is 25% identical in amino acid sequence to that made by *ced-9*. If a copy of the human *bcl-2* gene is transferred into a nematode with a defective *ced-9* gene, *bcl-2* suppresses the cell death program of *ced-3* and *ced-4!*

How does *bax* kill a cell? The bax protein seems to induce apoptosis by binding to the permeability pore of the cell's mitochondria, increasing its permeability and in doing so triggering cell death. How does *bcl-2* prevent cell death? One suggestion is that it prevents damage from free radicals, highly reactive fragments of atoms that can damage cells severely. Proteins or other molecules that destroy free radicals are called **antioxidants**. Antioxidants are almost as effective as *bcl-2* in blocking apoptosis.

Animal development involves programmed cell death (apoptosis), in which particular genes, when activated, kill their cells.

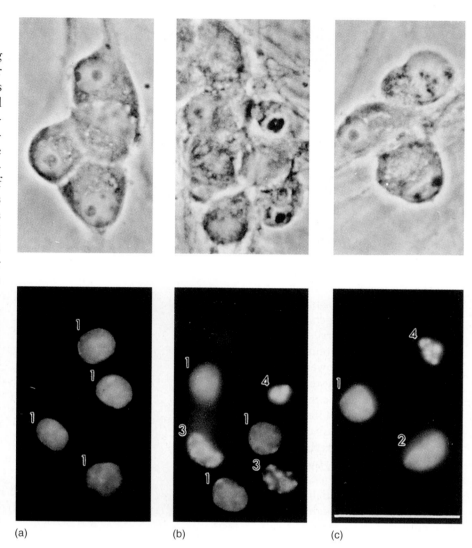

(a)　　　　　(b)　　　　　(c)

FIGURE 57.22
Programmed cell death. In the natural progression of apoptosis, normal cell nuclei (column *a*) shrink and fragment (columns *b* and *c*) concomitantly with the fragmentation of the nuclear DNA. In the bottom row, fluorescence microscopy has been used to visualize the nuclear shape and distribution of the nuclear chromatin in the clusters of cell bodies (shown in the top row by phase contrast microscopy). The nuclei change from a normal shape (labeled 1) to shrunken and abnormal (2 and 3) and finally fragmented (4). Condensation of the nuclear chromatin is apparent in nuclei labeled 3 and 4.

The Mouse

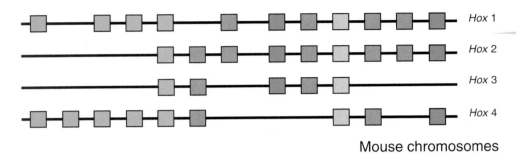

Mouse chromosomes

Some of the most elegant investigations of the cellular mechanisms of development are being done with mammals, particularly the mouse *Mus musculus*. Mice have a battery of homeotic genes, the *Hox* genes (figure 57.23), which seem to be closely related to the homeotic genes of *Drosophila*. Very interestingly, not only do the same genes occur, but they also seem to operate in the same order! Clearly, the homeotic gene system has been highly conserved during the course of animal evolution.

What lends great power to this developmental model system is the ability to create chimeric mice containing cells from two different genetic lines. Mammalian embryos are unusual among vertebrates in that they arise from symmetrical eggs; there are no chemical gradients, and during the initial cleavage divisions, all of the daughter cells are identical. Up to the eight-cell stage, any one of the cells, if isolated, will form a normal adult. Moreover, two different eight-cell-stage embryos can be fused to form a single embryo that will go on to form a normal adult. The resulting adult is a chimera, containing cells from both embryos. In a very real sense, these chimeric mice each have four parents!

The *Hox* genes control body part development in mice.

FIGURE 57.23
Studying development in the mouse.

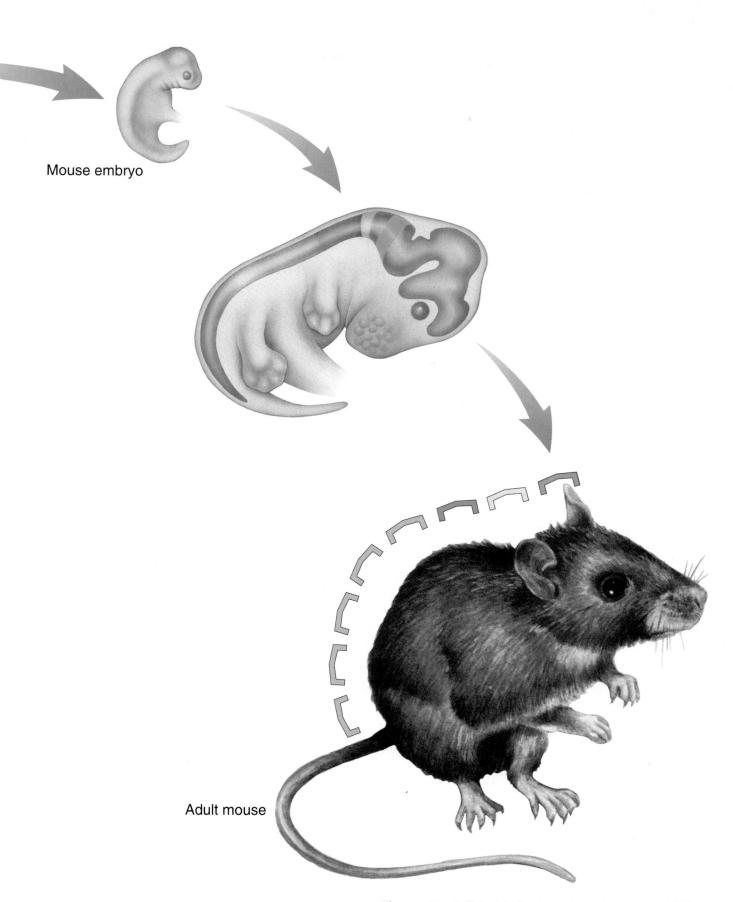

Mouse embryo

Adult mouse

The Fruit Fly

The tiny fruit fly *Drosophila melanogaster* has been a favorite of geneticists for over 90 years and is now playing a key role in our growing understanding of the cellular mechanisms of development. Over the last 10 years, researchers have pieced together a fairly complete picture of how genes expressed early in fruit fly development determine the pattern of the adult body (figure 57.24). The major parts of the adult body are determined as patches of tissue called imaginal discs that float within the body of the larva; during the pupal stage, these discs grow, develop, and associate to form the adult body.

The adult *Drosophila* body is divided into 17 segments, some bearing jointed appendages such as wings or legs. These segments are established during very early development, before the many nuclei of the blastoderm are fully separated from one another. Chemical gradients, established within the egg by material from the mother, create a polarity that directs embryonic development. Reacting to this gradient, a series of segmentation genes progressively subdivide the embryo, first into four broad stripes, and then into 7, 14, and finally 17 segments.

Within each segment, the development of key body parts is under the control of homeotic genes that determine where the body part will form. As we have seen, there are two clusters of homeotic genes, one called Antennapedia that governs the front (anterior) end of the body, and another called bithorax that governs the rear (posterior) end. The organization of genes within each cluster corresponds nicely with the order of the segments they affect. A very similar set of homeotic genes governs body architecture in mice and humans.

A series of segmentation genes divides a *Drosophila* embryo into parts; Antennapedia genes control anterior development, and bithorax genes control the development of the posterior.

Drosophila egg

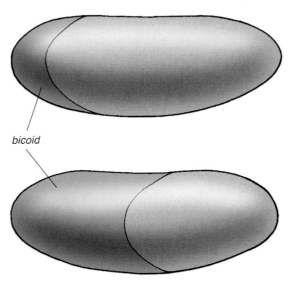

bicoid

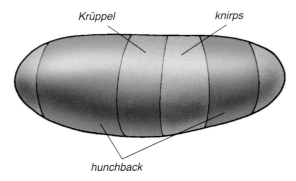

Krüppel   knirps

hunchback

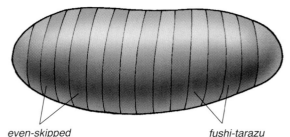

even-skipped   fushi-tarazu

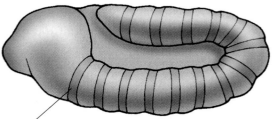

engrailed

FIGURE 57.24
Studying development in the fruit fly.

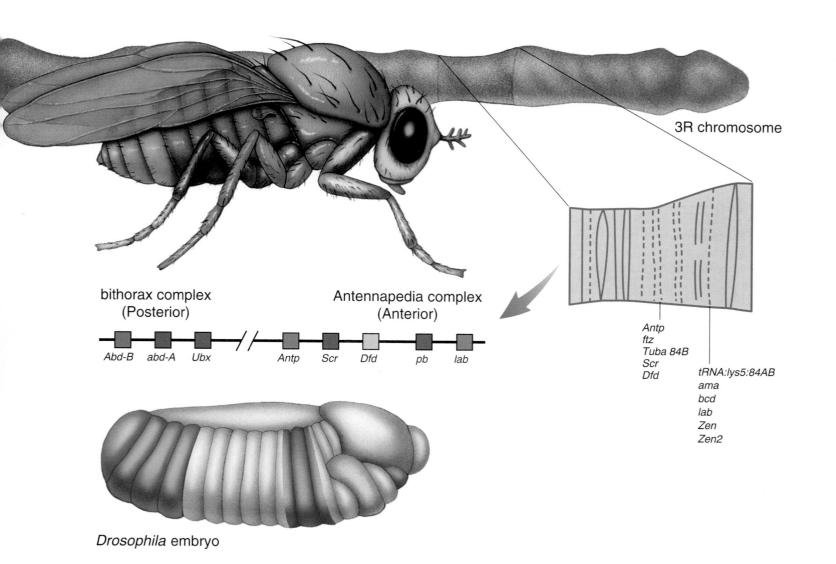

3R chromosome

bithorax complex
(Posterior)

Antennapedia complex
(Anterior)

Abd-B abd-A Ubx   Antp Scr Dfd pb lab

Antp
ftz
Tuba 84B
Scr
Dfd

tRNA:lys5:84AB
ama
bcd
lab
Zen
Zen2

Drosophila embryo

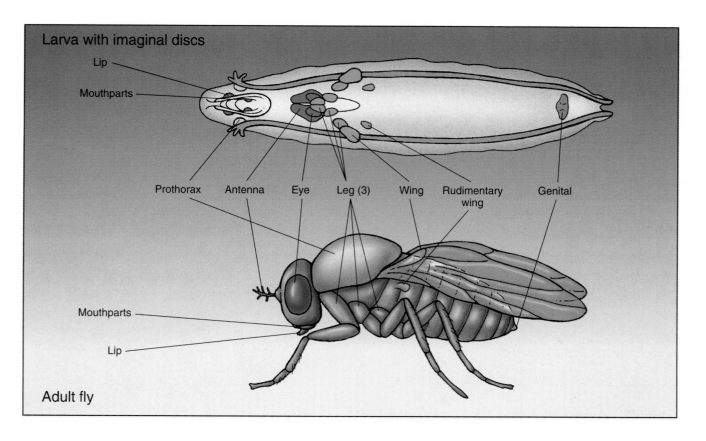

Larva with imaginal discs

Lip
Mouthparts

Prothorax Antenna Eye Leg (3) Wing Rudimentary wing Genital

Mouthparts

Lip

Adult fly

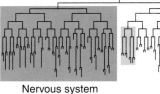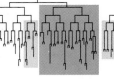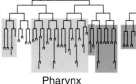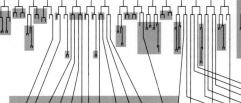

Nervous system

Pharynx

The Nematode

One of the most powerful models of animal development is the tiny nematode *Caenorhabditis elegans*. Only about 1 mm long, it consists of 959 somatic cells and has about the same amount of DNA as *Drosophila*. The entire genome has been mapped as a series of overlapping fragments, and a serious effort is underway to determine the complete DNA sequence of the genome.

Because *C. elegans* is transparent, individual cells can be followed as they divide. By observing them, researchers have learned how each of the cells that make up the adult worm is derived from the fertilized egg. As shown on this lineage map (figure 57.25), the egg divides into two, and then its daughter cells continue to divide. Each horizontal line on the map represents one round of cell division. The length of each vertical line represents the time between cell divisions, and the end of each vertical line represents one fully differentiated cell. In figure 57.25, the major organs of the worm are color-coded to match the colors of the corresponding groups of cells on the lineage map.

Some of these differentiated cells, such as some of the cells that generate the worm's external cuticle, are "born" after only 8 rounds of cell division; other cuticle cells require as many as 14 rounds. The cells that make up the worm's pharynx, or feeding organ, are born after 9 to 11 rounds of division, while cells in the gonads require up to 17 divisions.

Exactly 302 nerve cells are destined for the worm's nervous system. Exactly 131 cells are programmed to die, mostly within minutes of their birth. The fate of each cell is the same in every *C. elegans* individual, except for the cells that will become eggs and sperm.

> The nematode develops 959 somatic cells from a single fertilized egg in a carefully orchestrated series of cell divisions which have been carefully mapped by researchers.

Cuticle-making cells

Cuticle

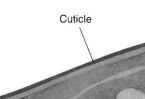

FIGURE 57.25
Studying development in the nematode.

Gonad

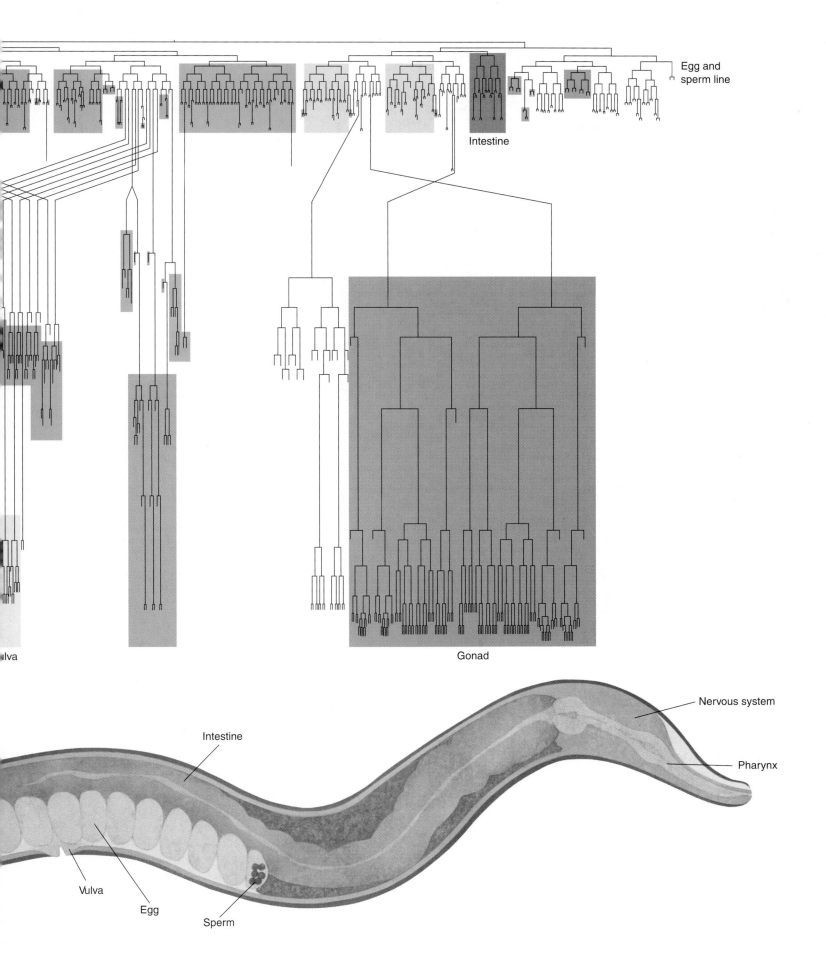

Egg and sperm line

Intestine

Gonad

Vulva

Nervous system

Pharynx

Intestine

Vulva

Egg

Sperm

Theories of Aging

All humans die. The oldest documented person, Jeanne Louise Calment of Arles, France, reached the age of 122 years before her death in 1997 (figure 57.26). The "safest" age is around puberty. As you can see in figure 57.27, 10- to 15-year-olds have the lowest risk of dying. The death rate begins to increase rapidly after puberty; the rate of mortality then begins to increase as an exponential function of increasing age. Plotted on a log scale as in figure 57.27 (in a so-called Gompertz plot), the mortality rate increases as a straight line from about 15 to 90 years, doubling about every eight years (the "Gompertz number"). By the time we reach 100, age has taken such a toll that the risk of dying reaches 50% per year.

A wide variety of theories have been advanced to explain why humans and other animals age. No one theory has gained general acceptance, but the following five are being intensively investigated:

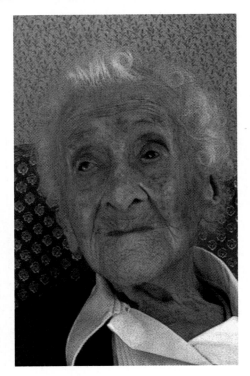

FIGURE 57.26
Jeanne Louise Calment (1875–1997).
She is the oldest documented human.

An explanation of the "Hayflick limit" was suggested in 1986 when Howard Cooke first glimpsed an extra length of DNA at the ends of chromosomes. These **telomeric regions** (figure 57.29), repeats of the sequence TTAGGG, were found to be substantially shorter in older somatic tissue, and Cooke speculated that a 100 base-pair portion of the telomere cap was lost by a chromosome during each cycle of DNA replication. Eventually, after some 50 replication cycles, the protective telomeric cap would be used up, and the cell line would then enter senescence, no longer able to proliferate. Cancer cells appear to avoid telomeric shortening.

Research reported in 1998 has confirmed Cooke's hypothesis, providing direct evidence for a causal relation between telomeric shortening and cell senescence. Using genetic engineering, researchers transferred into human primary cell cultures a gene that leads to expression of telomerase, an enzyme that builds TTAGGG telomeric caps. The result was unequivocal. New telomeric caps were added to the chromosomes of the cells, and the cells with the artificially elongated telomeres did not senesce at the Hayflick limit, continuing to divide in a healthy and vigorous manner for more than 20 additional generations.

Accumulated Mutation Hypothesis

The oldest general theory of aging is that cells accumulate mutations as they age, leading eventually to lethal damage. Careful studies have shown that somatic mutations do indeed accumulate during aging. As cells age, for example, they tend to accumulate the modified base 8-hydroxyguanine, in which an —OH group is added to the base guanine. There is little direct evidence, however, that these mutations *cause* aging. No acceleration in aging occurred among survivors of Hiroshima and Nagasaki despite their enormous added mutation load, arguing against any general relationship between mutation and aging.

Telomere Depletion Hypothesis

In a seminal experiment carried out in 1961, Leonard Hayflick demonstrated that fibroblast cells growing in tissue culture will divide only a certain number of times (figure 57.28). After about 50 population doublings, cell division stops, the cell cycle blocked just before DNA replication. If a cell sample is taken after 20 doublings and frozen, when thawed it resumes growth for 30 more doublings, then stops.

Wear-and-Tear Hypothesis

Numerous theories of aging focus in one way or another on the general idea that cells wear out over time, accumulating damage until they are no longer able to function. Loosely dubbed the "wear-and-tear" hypothesis, this idea implies that there is no inherent designed-in limit to aging, just a statistical one—over time, disruption, wear, and damage eventually erode a cell's ability to function properly.

There is considerable evidence that aging cells do accumulate damage. Some of the most interesting evidence concerns free radicals, fragments of molecules or atoms that contain an unpaired electron. Free radicals are very reactive chemically and can be quite destructive in a cell. Free radicals are produced as natural by-products of oxidative metabolism, but most are mopped up by special enzymes that function to sweep the cell interior free of their destructive effects.

One of the most damaging free radical reactions that occurs in cells causes glucose to become linked to proteins, a nonenzymatic process called glycation. Two of the most

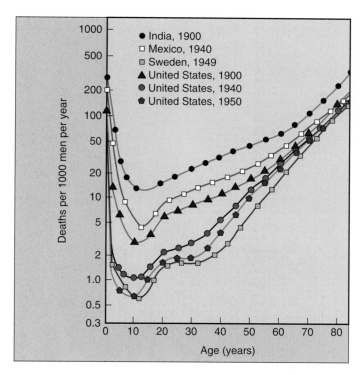

FIGURE 57.27
Gompertz curves. While human populations may differ 25-fold in their mortality rates before puberty, the slopes of their Gompertz curves are about the same in later years.

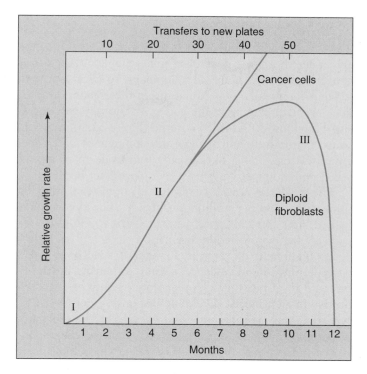

FIGURE 57.28
Hayflick's experiment. Fibroblast cells stop growing after about 50 doublings. Growth is rapid in phases I and II, but slows in phase III, as the culture becomes senescent, until the final doubling. Cancer cells, by contrast, do not "age."

commonly glycated proteins are collagen and elastin, key components of the connective tissues in our joints. Glycated collagen and elastin are not replaced, and individual molecules may be as old as the individual.

Glycation of collagen, elastin, and a diverse collection of other proteins within the cell produces a complex mixture of glucose-linked proteins called advanced glycosylation end products (AGEs). AGEs can cross-link to one another, reducing the flexibility of connective tissues in the joints and producing many of the other characteristic symptoms of aging.

Immunological Exhaustion Hypothesis

With age, T cells respond more slowly to stimuli. As we become old, we become more susceptible to infectious diseases like flu and are more likely to die from them. The body's stock of "virgin" T cells dwindles over time, as more and more of them become memory T cells committed to responding to one particular antigen. Some investigators have suggested that this depletion is responsible for the diminished immune response in older individuals. Other investigators blame lower levels of interleukin-2, a T-cell growth factor. As we age, it appears we produce progressively less interleukin-2. Perhaps not surprisingly, "IL-2 cocktails" are being investigated by pharmaceutical companies as anti-aging remedies!

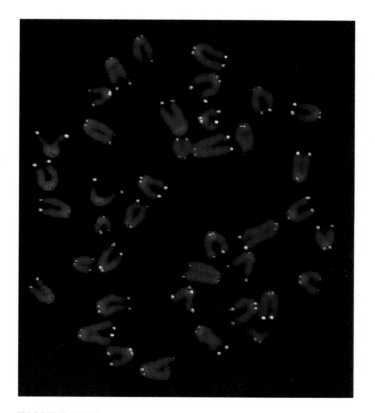

FIGURE 57.29
Telomeres. The bright orange stain marks the telomeres of these mouse chromosomes.

Chapter 57 Cellular Mechanisms of Development **1165**

Gene Clock Hypothesis

There is very little doubt that at least some aspects of aging are under the direct control of genes (table 57.1). Just as genes regulate the body's development, so they appear to regulate its rate of aging. Mutations in these genes can produce premature aging in the young. In the very rare recessive Hutchinson-Gilford syndrome (figure 57.30), growth, sexual maturation, and skeletal development are retarded; atherosclerosis and strokes usually lead to death by age 12 years. Only some 20 cases have ever been described.

The similar Werner's syndrome is not as rare, affecting some 10 people per million worldwide. The syndrome is named after Otto Werner, who in Germany in 1904 reported a family affected by premature aging and said a genetic component was at work. Werner's syndrome makes its appearance in adolescence, usually producing death before age 50 of heart attack or one of a variety of rare connective tissue cancers. The gene responsible for Werner's syndrome was identified in 1996. Located on the short arm of chromosome 8, it seems to affect a helicase enzyme involved in the repair of DNA. The gene, which codes for a 1432-amino-acid protein, has been fully sequenced, and four mutant alleles identified. Helicase enzymes are needed to unwind the DNA double helix whenever DNA has to be replicated, repaired, or transcribed. The high incidence of certain cancers among Werner's syndrome patients leads investigators to speculate that the mutant helicase may fail to activate critical tumor suppressor genes. The potential role of helicases in aging is the subject of heated research.

Research on aging in other animals strongly supports the hypothesis that genes regulate the rate of aging. Particularly impressive results have been obtained in the nematode *Caenorhabditis elegans*, where genes discovered in 1996 seem to affect an intrinsic genetic clock. A combination of mutations can increase the worm's lifespan fivefold, the largest increase in lifespan seen in any organism! Mutations in the clock gene *clk-1* cause individual cells to divide more slowly, and the animal spends more time in each phase of its life cycle. Mutations in two other clock genes, *clk-2* and *clk-3*, have similar effects. Nematodes with mutations in two of the clock genes lived three to four times longer than normal. It seems that slowing life down in nematodes extends it. Perhaps, as the "wear-and-tear" theory suggests, aging results from damage to cells and their DNA by highly reactive oxidative by-products of metabolism. Living more slowly, destructive by-products may be produced less frequently, accumulate more slowly, and their damage be repaired more efficiently. Similar genes have been reported in yeasts, and attempts are now underway to isolate and clone these genes.

Among the many theories advanced to explain aging, many involve the progressive accumulation of damage to DNA. When genes affecting aging have been isolated, they affect DNA repair processes.

FIGURE 57.30

A child with Hutchinson-Gilford syndrome. Such children have normal intelligence for their age, but never reach adult size, typically growing only to 30 pounds and 40 inches height. They have many symptoms associated with aging and have frequent strokes and heart attacks. Few survive beyond 15 years.

Table 57.1 Theories of Aging	
Theory	**Description**
Accumulated mutation	Cells accumulate deleterious mutations as they age, leading eventually to lethal effects.
Telomere deletion	As cells divide, telomeres get progressively shorter, till they run out.
Wear-and-tear	Over time, free radicals damage cells by linking proteins together.
Immunological exhaustion	With age, the body's stock of "virgin" uncommitted T cells becomes depleted.
Gene clock hypothesis	An intrinsic genetic clock governs aging, perhaps by controlling access to DNA.

57.1 Development is a regulated process.

- Vertebrate development is initiated by a rapid cleavage of the fertilized egg into a hollow ball of cells, the blastula. Cell movements then form primary germ layers and organize the structure of the embryo.

- Insect cleavage divisions produce a syncytial blastoderm in which nuclei are not separated by cell membranes. Development within this blastoderm is guided by mRNA contributed by maternal genes. Cells determined in the embryo are carried within the body of larvae as imaginal discs, which are assembled into the adult body during pupation.

57.2 All animals employ the same basic mechanisms of development.

- Cell movement in animal development is carried out by altering a cell's complement of surface adhesion molecules, which it uses to pull itself over other cells.

- A key to animal development is the ability of cells to alter the developmental paths of adjacent cells, a process called induction. Induction is achieved by diffusible chemicals called morphogens.

- Determination of a cell's ultimate developmental fate often involves the addition to it of positional labels that reflect its location in the embryo. In *Drosophila*, the attachment of the positional labels that determine segmentation is guided by chemical gradients set up in the egg by maternal genes.

- The location of structures within body segments is dictated by a spatially organized assembly of homeotic genes, first discovered in *Drosophila* but now known to occur in all animals. All of these genes contain a homeobox domain.

- Many cells are genetically programmed to die, usually soon after they are formed during development, in a process called apoptosis.

57.3 Three model developmental systems of animals have been extensively researched.

- The three most intensively studied model systems of animal development are the mouse *Mus musculus*, the fruit fly *Drosophila melanogaster*, and the nematode *Caenorhabditis elegans*.

57.4 Aging can be considered a developmental process.

- Aging is not well understood, although not for want of theories, most of which involve progressive damage to DNA.

Discussing Key Terms

1. **Mosaic development**   In many animals (but not mammals), developmental determinants are positioned asymmetrically within the egg. As a result, cleavage divisions produce cells with different developmental instructions, destined from their formation to follow different developmental paths.

2. **Induction**   Very often in animal development, the developmental path of a cell is changed by interactions with adjacent cells, a process called induction. These interactions play a key role in organizing tissues.

3. **Organizers**   Some tissues in the embryo act as organizers, signal markers that other cells can use to assess their own position within the embryo.

4. **Morphogens**   Organizers release chemicals called morphogens that establish diffusion gradients within the embryo. The higher the morphogen concentration a cell encounters, the closer the cell is to the organizer.

5. **Differentiation**   A fully determined cell may not yet express the entire complement of genes it is capable of expressing. Thus, a *Drosophila* cell determined to become a leg cell may exist as part of an imaginal disc through several larval stages before finally becoming a leg cell in the adult.

6. **Segmentation genes**   In *Drosophila*, three sets of genes act in sequence to determine the pattern of segmentation in the embryo. These genes amplify and make permanent a chemical gradient set up in the egg by the mother. It is not clear whether similar gradients function in vertebrate embryos.

7. **Homeotic genes**   In animals as diverse as the fruit fly and the mouse, the same set of genes seems to organize the location of major body parts. Each gene contains a homeobox domain, and the gene's position relative to other homeotic genes mirrors the position of the body segment it affects.

Review Questions

1. What is cleavage? How does the type of cleavage influence subsequent embryonic development?

2. What is a blastula? How does it form and what does it turn into?

3. What is a gastrula? Where are the germ layers in a gastrula? What is each region of a gastrula destined to become?

4. What is neurulation? How and when does it occur?

5. What are the major differences between vertebrate and insect developmental pathways? What are the similarities?

6. Why is cell movement important in animal development? What role do cadherins and integrins play in cell movement?

7. What is the difference between mosaic development and regulative development?

8. What is induction? How do organizers and morphogens participate in induction?

9. What is determination? Why must it occur for development to proceed normally? How is it distinguished from differentiation?

10. What role does maternal mRNA play in the development of a *Drosophila* embryo? What roles do the gap genes, pair-rule genes, and segment polarity genes play in pattern formation in a Drosophila embryo?

11. What are homeotic genes and what do they do? In what types of organisms are they found? What is the homeobox?

12. How is apoptosis different from necrosis? Which is more important in development?

Thought Questions

1. If you had the tools and the know-how, how would you construct chimeras using the information supplied about the organisms in this chapter?

2. What would happen if you put mouse *Hox* genes into a human embryo?

Internet Links

The Interactive Fly
http://sdb.bio.purdue.edu/fly/aimain/1aahome.htm
A site that links 350 (and counting) genes in Drosophila to their roles in development. Loaded with information on protein function, mutation effects, and links to similar genes in other organisms.

How Plants Develop: *Arabidopsis*
genome-www.ARABIdopsis.com
A compilation of links to databases, resources, conferences, and lab procedures concerning this key model of flowering plant development.

FlyBase
http://flybase.bio.indiana.edu
A central storehouse for Drosophila *information funded by the NIH, with chromosome maps, gene and protein data, and many links.*

For Further Reading

Beardsley, T.: "Smart Genes," *Scientific American*, August 1991, pages 86–95. An overview of how transcription complexes choreograph development.

Derobertis, E., and others: "Homeobox Genes and the Vertebrate Body Plan," *Scientific American*, July 1990, pages 46–52. An account of the discovery of the genes that subdivide animal embryos along the head-to-tail axis into fields of cells that eventually develop into antennae, limbs, and other structures.

Douglas, K.: "Making Friends with Death Wish Genes," *New Scientist*, vol. 143, July 1994, pages 31–34. A hot area of cancer research involves attempts to activate genes triggering programmed cell death within cancer cells.

Gould, S. J.: "Geoffroy and the Homeobox," *Natural History*, November 1985, pages 12–23. An entertaining account of the French zoologist Etienne Geoffroy Saint-Hilaire's attempt in the 1830s to establish a connection between insect segments and mammalian vertebrae, just the relationship we now believe to be dictated by homeotic genes.

Hart, S.: "The Drama of Cell Death," *BioScience*, July 1994, pages 451–55. An engaging account of research on the genes that govern programmed cell death, genes that play key roles in development and cancer.

McGinnis, W. and M. Kuziora: "The Molecular Architects of Body Design," *Scientific American*, February 1994, pages 58–66. An account of research on homeotic genes, which govern pattern formation in much the same way in all animals.

Patel, N.: "Developmental Evolution: Insights from Studies of Insect Segmentation," *Science*, vol. 266, October 1994, pages 581–90. A review of what we know about how the genes governing pattern formation have been conserved during evolution.

Steller, H.: "Mechanisms and Genes of Cellular Suicide," *Science*, March 1995, pages 1445–49. A review of the fast-emerging field of apoptosis, programmed cell death, which is now known to play a major role in many diseases, including cancer and AIDS.

58

Vertebrate Development

Concept Outline

58.1 Fertilization is the initial event in development.

Stages of Development.   Fertilization of an egg cell by a sperm occurs in three stages: penetration, activation of the egg cell, and fusion of the two haploid nuclei.

58.2 Cell cleavage and the formation of a blastula set the stage for later development.

Cell Cleavage Patterns.   The cytoplasm of the zygote is divided into smaller cells by a mitotic cell division in a process called cleavage.

58.3 Gastrulation forms the three germ layers of the embryo.

The Process of Gastrulation.   Cells of the blastula invaginate and involute to produce an outer ectoderm, an inner endoderm layer, and a third layer, the mesoderm.

58.4 Body architecture is determined during the next stages of embryonic development.

Developmental Processes during Neurulation.   The mesoderm of chordates forms a notochord, and the overlying ectoderm rolls to produce a neural tube.
Embryonic Development and Vertebrate Evolution.   The embryonic development of a mammal includes stages that are characteristic of more primitive vertebrates.
Extraembryonic Membranes.   Embryonic cells form several membranes outside of the embryo that provide protection, nourishment, and gas exchange for the embryo.

58.5 Human development is divided into trimesters.

First Trimester.   A blastocyst implants into the mother's endometrium, and the formation of body organs begins during the fourth week.
Second and Third Trimesters.   Further growth and development take place during this time.
Birth and Postnatal Development.   Birth occurs as a result of uterine contractions; the human brain continues to grow significantly after birth.

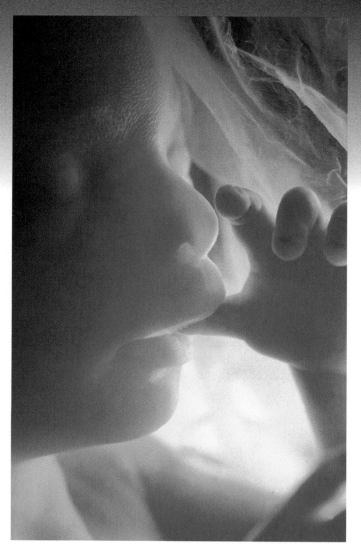

FIGURE 58.1
Development is the process that determines an organism's form and function. A human fetus at 18 weeks is not yet halfway through the 38 weeks—about 9 months—it will spend within its mother, but it has already developed many distinct behaviors, such as the sucking reflex that is so important to survival after birth.

Reproduction in all but a few vertebrates unites two haploid gametes to form a single diploid cell called a zygote. The zygote develops by a process of cell division and differentiation into a complex multicellular organism, composed of many different tissues and organs (figure 58.1). Although some of the details vary among different vertebrate groups, development in all vertebrates occurs in six stages. In this chapter, we will consider the mechanisms governing development and conclude with a description of the events that occur during human development.

Stages of Development

In vertebrates, as in all sexual animals, the first step in development is the union of male and female gametes, a process called **fertilization.** Fertilization is typically external in fish and amphibians, which reproduce in water, and internal in all other vertebrates. In internal fertilization, small, motile sperm are introduced into the female reproductive tract during mating. The sperm swim up the reproductive tract until they encounter a mature egg or oocyte in an oviduct, where fertilization occurs. Fertilization consists of three stages: penetration, activation, and fusion.

Penetration

As described in chapter 56, the secondary oocyte is released from a fully developed Graafian follicle at ovulation. It is surrounded by the same layer of small granulosa cells that surrounded it within the follicle (figure 58.2). Between the granulosa cells and the egg's plasma membrane is a glycoprotein layer called the zona pellucida. The head of each sperm is capped by an organelle called the acrosome, which contains glycoprotein-digesting enzymes. These enzymes become exposed as the sperm begin to work their way into the layer of granulosa cells, and the activity of the enzymes enables the sperm to tunnel their way through the zona pellucida to the egg's plasma membrane. Egg cytoplasm bulges out at this point, engulfing the head of the sperm and permitting the sperm nucleus to enter the cytoplasm of the egg (figure 58.3).

Activation

The series of events initiated by sperm penetration are collectively called egg activation. In some frogs, reptiles, and birds, more than one sperm may penetrate the egg, but only the first one that enters is successful in fertilizing it. In mammals, by contrast, the penetration of the first sperm initiates changes in the egg membrane that prevent the entry of other sperm. In addition to these changes, sperm penetration has three other effects on the egg. First, it stimulates the chromosomes in the egg nucleus to complete

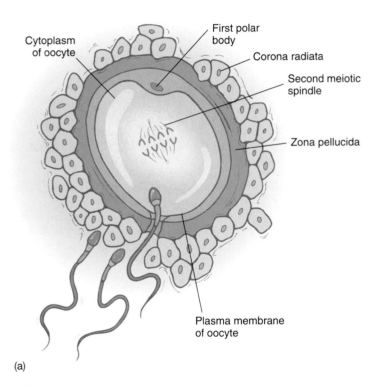

Cytoplasm of oocyte

First polar body

Corona radiata

Second meiotic spindle

Zona pellucida

Plasma membrane of oocyte

(a)

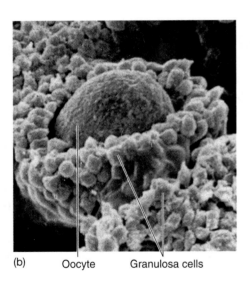

(b) Oocyte Granulosa cells

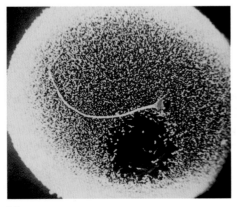

(c)

FIGURE 58.2
Mammalian reproductive cells. (a) A sperm must penetrate a layer of granulosa cells called the corona radiata and then a layer of glycoprotein called the zona pellucida, before it reaches the oocyte membrane. This penetration is aided by digestive enzymes in the acrosome of the sperm. These scanning electron micrographs show (b) a human oocyte (90×) surrounded by numerous granulosa cells, and (c) a human sperm on an egg (3000×).

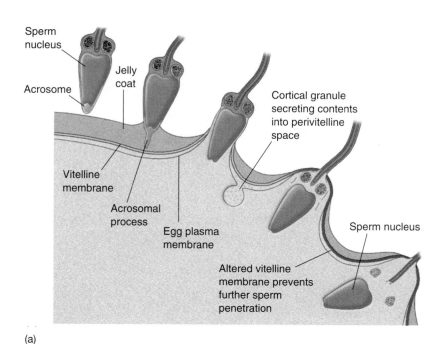

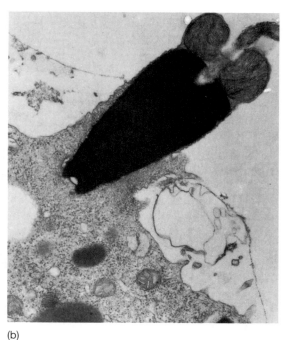

(a)

(b)

FIGURE 58.3

Sperm penetration of a sea urchin egg. (a) The stages of penetration. (b) An electron micrograph (50,000×) of penetration. Penetration in both invertebrate and vertebrate eggs is similar.

the second meiotic division, producing two egg nuclei. One of these nuclei is extruded from the egg as a second polar body (chapter 56), leaving a single haploid egg nucleus within the egg.

Second, sperm penetration triggers movements of the egg cytoplasm around the point of sperm entry. These movements ultimately establish the bilateral symmetry of the developing animal. In frogs, for example, sperm penetration causes an outer pigmented cap of egg cytoplasm to rotate toward the point of entry, uncovering a gray crescent of interior cytoplasm opposite the point of penetration (figure 58.4). The position of the gray crescent determines the orientation of the first cell division. A line drawn between the point of sperm entry and the gray crescent would bisect the right and left halves of the future adult. Third, activation is characterized by a sharp increase in protein synthesis and an increase in metabolic activity in general. Experiments demonstrate that the protein synthesis in the activated oocyte is coded by mRNA that was previously produced and already present in the cytoplasm of the unfertilized egg cell.

In some vertebrates, it is possible to activate an egg without the entry of a sperm, simply by pricking the egg membrane. An egg that is activated in this way may go on to develop parthenogenetically. A few kinds of amphibians, fish, and reptiles rely entirely on parthenogenetic reproduction in nature, as we mentioned in chapter 56.

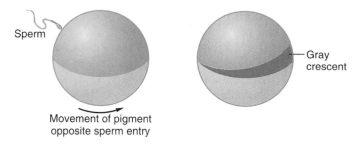

FIGURE 58.4

Gray crescent formation in frog eggs. The gray crescent appears opposite the point of penetration by the sperm.

Fusion

The third stage of fertilization is the fusion of the entering sperm nucleus with the haploid egg nucleus to form the diploid nucleus of the zygote. This fusion is triggered by the activation of the egg. If a sperm nucleus is microinjected into an egg without activating the egg, the two nuclei will not fuse. The nature of the signals that are exchanged between the two nuclei, or sent from one to the other, is not known.

The three stages of fertilization are penetration, activation, and fusion. Penetration initiates a complex series of developmental events, including major movements of cytoplasm, which eventually lead to the fusion of the egg and sperm nuclei.

Cell Cleavage Patterns

Following fertilization, the second major event in vertebrate reproduction is the rapid division of the zygote into a larger and larger number of smaller and smaller cells (table 58.1). This period of division, called cleavage, is not accompanied by an increase in the overall size of the embryo. The resulting tightly packed mass of about 32 cells is called a **morula,** and each individual cell in the morula is referred to as a **blastomere.** As the blastomeres continue to divide, they secrete a fluid into the center of the morula. Eventually, a hollow ball of 500 to 2000 cells, the **blastula,** is formed. The fluid-filled cavity within the blastula is known as the *blastocoel.*

The pattern of cleavage division is influenced by the presence and location of yolk, which is abundant in the eggs of many vertebrates (figure 58.5). As we discussed in the previous chapter, vertebrates have embraced a variety of reproductive strategies involving different patterns of yolk utilization.

Primitive Chordates

When eggs contain little or no yolk, cleavage occurs throughout the whole egg (figure 58.6). This pattern of cleavage, called holoblastic cleavage, was characteristic of the ancestors of the vertebrates and is still seen in groups such as the lancelets and agnathans. In these animals, holoblastic cleavage results in the formation of a symmetrical blastula composed of cells of approximately equal size.

Amphibians and Advanced Fish

The eggs of bony fish and frogs contain much more yolk in one hemisphere than the other. Because yolk-rich cells divide much more slowly than those that have little yolk, holoblastic cleavage in these eggs results in a very asymmetrical blastula (figure 58.7), with large cells containing a lot of yolk at one pole and a concentrated mass of small cells containing very little yolk at the other. In these blastulas, the pole that is rich in yolk is called the vegetal pole, while the pole that is relatively poor in yolk is called the animal pole.

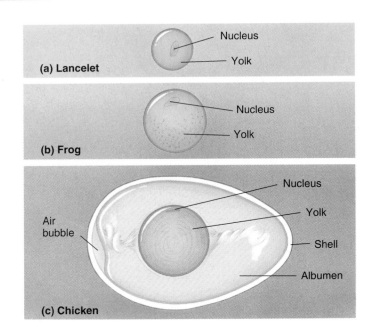

FIGURE 58.5
Yolk distribution in three kinds of eggs. (a) In the lancelet, a primitive chordate, the egg consists of a central nucleus surrounded by a small amount of yolk. (b) In a frog egg there is much more yolk, and the nucleus is displaced toward one pole. (c) Bird eggs are complexly organized, with the nucleus just under the surface of a large, central yolk.

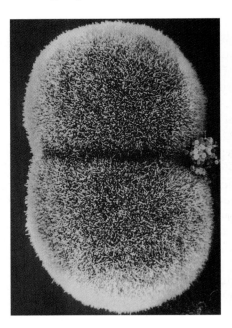

FIGURE 58.6
Holoblastic cleavage (3000×). In this type of cleavage, cell division occurs throughout the entire egg.

FIGURE 58.7
Dividing frog eggs. The closest cells in this photo (those near the animal pole) divide faster and are smaller than those near the vegetal pole.

Table 58.1 Stages of Vertebrate Development

	Fertilization	The haploid male and female gametes fuse to form a diploid zygote.
	Cleavage	The zygote rapidly divides into many cells, with no overall increase in size. These divisions affect future development, since different cells receive different portions of the egg cytoplasm and, hence, different regulatory signals.
Ectoderm — Mesoderm — Endoderm	Gastrulation	The cells of the embryo move, forming three primary cell layers: ectoderm, mesoderm, and endoderm.
Neural groove — Notochord	Neurulation	In all chordates, the first organ to form is the notochord; second is the dorsal nerve cord.
Neural crest — Neural tube — Notochord	Neural crest	During neurulation, the neural crest is produced as the neural tube is formed. The neural crest gives rise to several uniquely vertebrate structures.
	Organogenesis	Cells from the three primary layers combine in various ways to produce the organs of the body.

Reptiles and Birds

The eggs produced by reptiles, birds, and some fish are composed almost entirely of yolk, with a small amount of cytoplasm concentrated at one pole. Cleavage in these eggs occurs only in the tiny disc of polar cytoplasm, called the blastodisc, that lies astride the large ball of yolk material. This type of cleavage pattern is called meroblastic cleavage (figure 58.8). The resulting embryo is not spherical, but rather has the form of a thin cap perched on the yolk.

Mammals

Mammalian eggs are in many ways similar to the reptilian eggs from which they evolved, except that they contain very little yolk. Because cleavage is not impeded by yolk in mammalian eggs, it is holoblastic, forming a ball of cells surrounding a blastocoel. However, an inner cell mass is concentrated at one pole (figure 58.9). This inner cell mass is analogous to the blastodisc of reptiles and birds, and it goes on to form the developing embryo. The outer sphere of cells, called a trophoblast, is analogous to the cells that form the watertight membrane covering the reptilian egg. These cells have changed during the course of mammalian evolution to carry out a very different function: part of the trophoblast enters the endometrium (the epithelial lining of the uterus) and contributes to the placenta, the organ that permits exchanges between the fetal and maternal bloods. While part of the placenta is composed of fetal tissue (the trophoblast), part is composed of the modified endometrial tissue (called the *decidua basalis*) of the mother's uterus. The placenta will be discussed in more detail in a later section.

The Blastula

Viewed from the outside, the blastula looks like a simple ball of cells all resembling each other. In many animals, this appearance is misleading, as unequal distribution of developmental signals from the egg produces some degree of mosaic development, so that the cells are already committed to different developmental paths. In mammals, however, all of the blastomeres receive equivalent sets of signals, and body form is determined by cell-cell interactions. In a mammalian blastula, called a blastocyst, each cell is in contact with a different set of neighboring cells, and these interactions with neighboring cells are a major factor influencing the developmental fate of each cell. This positional information is of particular importance in the orientation of mammalian embryos, setting up different patterns of development along three embryonic axes: anterior-posterior, dorsal-ventral, and proximal-distal.

For a short period of time, just before they implant in the uterus, the cells of the mammalian blastocyst have the power to develop into many of the 210 different types of cells in the body—and probably all of them. Biologist have

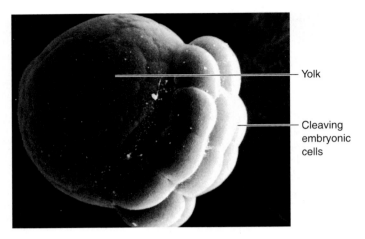

FIGURE 58.8
Meroblastic cleavage (400×). In this type of cleavage, only a portion of the egg actively divides to form a mass of cells.

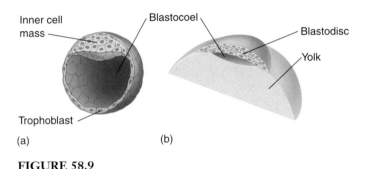

FIGURE 58.9
The embryos of mammals and birds are more similar than they seem. (a) A mammalian blastula is composed of a sphere of cells, the trophoblast, surrounding a cavity, the blastocoel, and an inner cell mass. (b) An avian (bird) blastula consists of a cap of cells, the blastodisc, resting atop a large yolk mass; the blastocoel is a small cavity between the blastodisc and the yolk.

long sought to grow these cells, called *embryonic stem cells*, in tissue culture, as such stem cells might in principle be used to produce tissues for human transplant operations. Injected into a patient, for example, they might be able to respond to local signals and produce new tissue. The first success in growing stem cells in culture was reported in 1998, when researchers isolated cells from the inner cell mass of human blastocysts and successfully grew them in tissue culture. These stem cells continue to grow and divide in culture indefinitely, unlike ordinary body cells, which divide only 50 or so times and then die.

A series of rapid cell divisions called cleavage transforms the zygote into a hollow ball of cells, the blastula. The cleavage pattern is influenced by the amount of yolk and its distribution in the egg.

The Process of Gastrulation

The first visible results of prepatterning and cell position within the blastula can be seen immediately after the completion of cleavage. Certain groups of cells **invaginate** (dent inward) and **involute** (roll inward) from the surface of the blastula in a carefully orchestrated activity called **gastrulation.** The events of gastrulation determine the basic developmental pattern of the vertebrate embryo. By the end of gastrulation, the cells of the embryo have rearranged into three primary cell layers, or **germ layers: ectoderm, mesoderm,** and **endoderm.** The cells in each layer have very different developmental fates. The ectoderm is destined to form the epidermis and neural tissue; the mesoderm gives rise to connective tissue, muscle, and vascular elements; and the endoderm forms the lining of the gut and its derivatives (table 58.2).

How is cell movement during gastrulation brought about? Apparently, migrating cells creep over stationary cells by means of actin filament contractions that change the shapes of the migrating cells affecting an invagination of blastula tissue. Each cell that moves possesses particular cell surface polysaccharides, which adhere to similar polysaccharides on the surfaces of the other moving cells. This interaction between cell surface molecules enables the migrating cells to adhere to one another and move as a single mass (see chapter 7).

Just as the pattern of cleavage divisions in different groups of vertebrates depends heavily on the amount and distribution of yolk in the egg, so the pattern of gastrulation among vertebrate groups depends on the shape of the blastulas produced during cleavage.

Table 58.2	Developmental Fates of the Primary Cell Layers
Ectoderm	Epidermis, central nervous system, sense organs, neural crest
Mesoderm	Skeleton, muscles, blood vessels, heart, gonads
Endoderm	Lining of digestive and respiratory tracts; liver, pancreas

Gastrulation in Primitive Chordates

In primitive chordates such as lancelets, which develop from symmetrical blastulas, gastrulation begins as the surface of the blastula invaginates into the blastocoel. About half of the blastula's cells move into the interior of the blastula, forming a structure that looks something like an indented tennis ball. Eventually, the inward-moving wall of cells pushes up against the opposite side of the blastula and then stops moving. The resulting two-layered, cup-shaped embryo is the gastrula (figure 58.10). The hollow structure resulting from the invagination is called the archenteron, and it is the progenitor of the gut. The opening of the archenteron, the future anus, is known as the blastopore.

This process produces an embryo with two cell layers: an outer ectoderm and an inner endoderm. Soon afterward, a third cell layer, the mesoderm, forms between the ectoderm and endoderm. In lancelets, the mesoderm forms from pouches that pinch off the endoderm. The appearance of these three primary cell layers sets the stage for all subsequent tissue and organ differentiation.

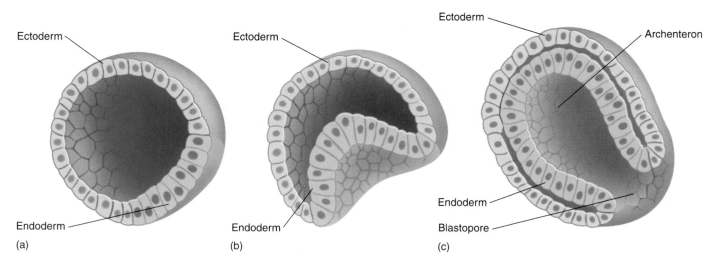

FIGURE 58.10
Gastrulation in a lancelet. In these chordates, the endoderm is formed by invagination of surface cells (a, b). This produces the primitive gut, or archenteron (c). Mesoderm will later be formed from pouches off the endoderm.

Gastrulation in Most Aquatic Vertebrates

In the blastulas of amphibians and those aquatic vertebrates with asymmetrical yolk distribution, the yolk-laden cells of the vegetal pole are fewer and much larger than the yolk-free cells of the animal pole. Consequently, gastrulation is more complex than it is in the lancelets. First, a layer of surface cells invaginates to form a small, crescent-shaped slit where the blastopore will soon be located. Next, cells from the animal pole involute over the dorsal lip of the blastopore (figure 58.11), at the same location as the gray crescent of the fertilized egg (see figure 58.4). As in the lancelets, the involuting cell layer eventually presses against the inner surface of the opposite side of the embryo, eliminating the blastocoel and producing an archenteron with a blastopore. In this case, however, the blastopore gradually fills with yolk-rich cells, forming the yolk plug. The outer layer of cells resulting from these movements is the ectoderm, and the inner layer is the endoderm. Other cells that involute over the dorsal lip and ventral lip (the two lips of the blastopore that are separated by the yolk plug) migrate between the ectoderm and endoderm to form the third germ layer, the mesoderm (figure 58.11).

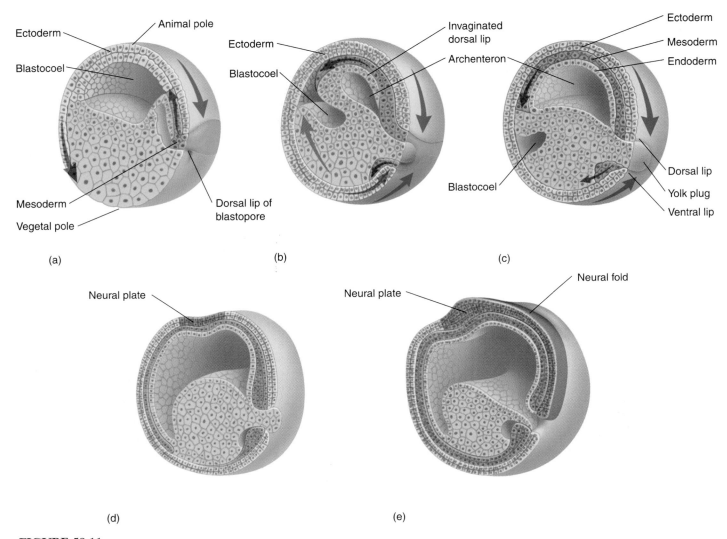

FIGURE 58.11
Frog gastrulation. (a) A layer of cells from the animal pole moves toward the yolk cells ultimately involuting through the dorsal lip of the blastopore. (b) Cells in the dorsal lip zone then involute into the hollow interior, or blastocoel, eventually pressing against the far wall. The three primary tissues (ectoderm, mesoderm, and endoderm) become distinguished. Ectoderm is shown in blue, mesoderm in red, and endoderm in yellow. (c) The movement of cells in the dorsal lip creates a new internal cavity, the archenteron, which opens to the outside through the plug of yolk remaining at the point of invagination. (d) The neural plate later forms from ectoderm. (e) This will next form a neural groove and then a neural tube as the embryo begins the process of neurulation. The cells of the neural ectoderm are shown in green.

Gastrulation in Reptiles, Birds, and Mammals

In the blastodisc of a bird or reptile and the inner cell mass of a mammal, the developing embryo is a small cap of cells rather than a sphere. No yolk separates the two sides of the embryo, and, as a result, the lower cell layer is able to differentiate into endoderm and the upper layer into ectoderm without cell movement. Just after these two primary cell layers form, the mesoderm arises by invagination and involution of cells from the upper layer. The surface cells begin moving to the midline where they involute and migrate laterally to form a mesodermal layer between the ectoderm and endoderm. A furrow along the longitudinal midline marks the site of this involution (figures 58.12 and 58.13). This furrow, analogous to an elongated blastopore, is called the **primitive streak**.

Gastrulation in lancelets involves the formation of ectoderm and endoderm by the invagination of the blastula, and the mesoderm layer forms from pouches pinched from the endoderm. In those vertebrates with extensive amounts of yolk, gastrulation requires the involution of surface cells into a blastopore or primitive streak, and the mesoderm is derived from some of these involuted cells.

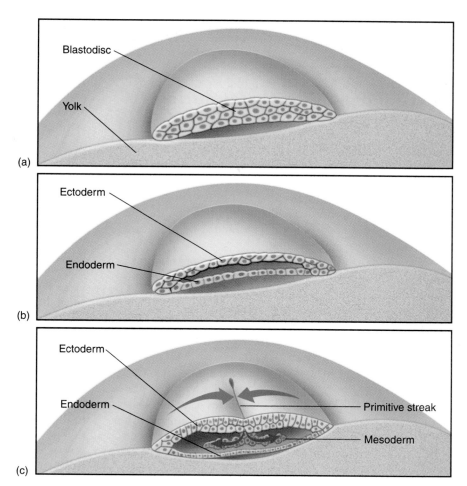

FIGURE 58.12
Gastrulation in birds. The upper layer of the blastodisc (a) differentiates into ectoderm, the lower layer into endoderm (b). Among the cells that migrate into the interior through the dorsal primitive streak are future mesodermal cells (c).

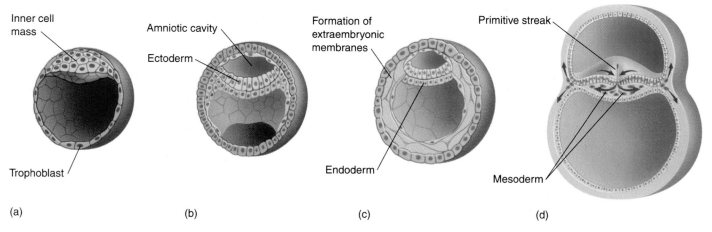

FIGURE 58.13
Mammalian gastrulation. (a) The amniotic cavity forms within the inner cell mass and its base. Layers of ectoderm and endoderm differentiate (b and c) as in the avian blastodisc. (d) A primitive streak develops, through which cells destined to become mesoderm migrate into the interior, again reminiscent of gastrulation in birds. The trophoblast has now moved further away from the embryo and begins to play a role in forming the placenta.

Developmental Processes during Neurulation

During the next step in vertebrate development, the three primary cell layers begin their transformation into the body's tissues and organs. The process of tissue differentiation begins with the formation of two morphological features found only in chordates, the **notochord** and the hollow dorsal nerve cord. This development of the dorsal nerve cord is called **neurulation.**

The notochord is first visible soon after gastrulation is complete, forming from mesoderm along the dorsal midline of the embryo. It is a flexible rod located along the dorsal midline in the embryos of all chordates, although its function is replaced by the vertebral column when it develops from mesoderm in the vertebrates. After the notochord has been laid down, a layer of ectodermal cells situated above the notochord invaginates, forming a long crease, the **neural groove,** down the long axis of the embryo. The edges of the neural groove then move toward each other and fuse, creating a long hollow cylinder, the **neural tube** (figure 58.14), which runs beneath the surface of the embryo's back. The neural tube later differentiates into the spinal cord and brain.

The process of induction, when one embryonic region influences the development of an adjacent region, was discussed in chapter 57. The dorsal lip of the blastopore induces the formation of a notochord, and the presence of the notochord induces the overlying ectoderm to differentiate into the neural tube.

While the neural tube is forming from ectoderm, the rest of the basic architecture of the body is being determined rapidly by changes in the mesoderm. On either side of the developing notochord, segmented blocks of tissue called **somites** form; more somites are added as development continues. Ultimately, the somites give rise to the muscles, vertebrae, and connective tissues. Some body organs, including the kidneys, adrenal glands, and gonads, develop within another strip of mesoderm that runs alongside the somites. The remainder of the mesoderm moves out and around the endoderm and eventually surrounds it completely. As a result of this movement, the mesoderm becomes separated into two layers. The outer layer is associated with the body wall and the inner layer is associated with the gut. Between these two layers of mesoderm is the coelom (chapter 41), which becomes the body cavity of the adult.

The Neural Crest

Neurulation occurs in all chordates, and the process in a lancelet is much the same as it is in a human. The next stage in the development of vertebrates, however, is unique to that group and is largely responsible for the characteristic body architecture of its members. Just before the neural groove closes to form the neural tube, its edges develop a small strip of cells, the **neural crest,** which becomes incorporated into the roof of the neural tube (see figure 58.14*d* and *e*). The cells of the neural crest later move to the sides of the developing embryo. The appearance of the neural crest was a key event in the evolution of the vertebrates because neural crest cells, after migrating to different parts of the embryo, ultimately develop into the structures characteristic of (though not necessarily unique to) the vertebrate body.

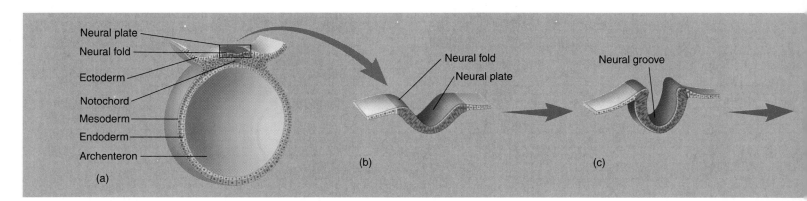

FIGURE 58.14
Neural tube formation. (a) The neural tube forms above the notochord when (b) cells of the neural plate fold together to form the (c) neural groove, which (d) eventually closes to form a hollow tube. (e) As the tube closes, some of the cells from the dorsal margin of the neural tube differentiate into the neural crest, which is characteristic of vertebrates.

The differentiation of neural crest cells depends on their location. At the anterior end of the embryo, they merge with the anterior portion of the brain, the forebrain. Nearby clusters of ectodermal cells associated with the neural crest cells thicken into **placodes,** structures that subsequently develop into parts of the sense organs in the head. The neural crest and associated placodes exist in two lateral strips, which is why the vertebrate sense organs that develop from them are paired.

Neural crest cells located in more posterior positions have very different developmental fates. These cells migrate away from the neural tube to other locations in the head and trunk, where they form connections between the neural tube and the surrounding tissues. At these new locations, they dictate the development of a variety of structures that are particularly characteristic of the vertebrates. The migration of neural crest cells is unique in that it is not simply a change in the relative positions of cells, such as that seen in gastrulation. Instead, neural crest cells actually pass through other tissues.

The Gill Chamber

Primitive chordates such as lancelets are filter-feeders, using the rapid beating of cilia to draw water into their bodies through slits in their pharynx. These pharyngeal slits evolved into the vertebrate gill chamber, a structure that provides a greatly improved means of respiration. The evolution of the gill chamber was certainly a key event in the transition from filter-feeding to active predation.

In the development of the gill chamber, some of the neural crest cells form cartilaginous bars between the embryonic pharyngeal slits. Other neural crest cells induce portions of the mesoderm to form muscles along the cartilage, while still others form neurons that carry impulses between the nerve cord and these muscles. A major blood vessel called the aortic arch passes through each of the bars. Lined by still more neural crest cells, these bars, with their internal blood supply, become highly branched and form the gills.

Because the stiff bars of the gill chamber can be bent inward by powerful muscles controlled by nerves, the whole structure is a very efficient pump that drives water past the gills. The gills themselves act as highly efficient oxygen exchangers, greatly increasing the respiratory capacity of the animals that possess them. The structure and operation of the gill chamber were described in more detail in chapter 50.

Elaboration of the Nervous System

Some neural crest cells migrate ventrally toward the notochord and form sensory neurons in the dorsal root ganglia (chapter 51). Others become specialized as Schwann cells, which insulate nerve fibers and permit the rapid conduction of nerve impulses. Still others form the autonomic ganglia and the adrenal medulla. Cells in the adrenal medulla secrete epinephrine when stimulated by the sympathetic division of the autonomic nervous system during the fight-or-flight reaction. The similarity in the chemical nature of the hormone epinephrine and the neurotransmitter norepinephrine, released by sympathetic neurons, is understandable—both adrenal medullary cells and sympathetic neurons derive from the neural crest.

Sensory Organs and Skull

A variety of sense organs develop from the placodes whose formation is induced by the anterior neural crest. Included among them are the olfactory (smell) and lateral line (primitive hearing) organs discussed in chapter 52. The teeth also develop from neural crest cells, as do the cranial bones that protect the brain.

The appearance of the neural crest in the developing embryo marked the beginning of the first truly vertebrate phase of development, since many of the structures characteristic of vertebrates derive directly or indirectly from neural crest cells.

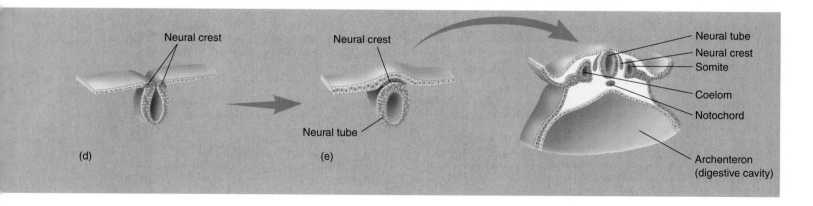

Embryonic Development and Vertebrate Evolution

The primitive chordates that gave rise to vertebrates were initially slow-moving, filter-feeding animals with relatively low metabolic rates. Adaptations derived from the neural crest allowed the vertebrates to assume a very different ecological role. The vertebrates became fast-swimming predators with much higher metabolic rates. This accelerated metabolism permitted a greater level of activity than was possible among the more primitive chordates. Other evolutionary changes associated with the derivatives of the neural crest provided better detection of prey, a greatly improved ability to orient spatially during prey capture, and the means to respond quickly to sensory information. The evolution of the neural crest and of the structures derived from it were thus crucial steps in the evolution of the vertebrates (figure 58.15).

Ontogeny Recapitulates Phylogeny

The patterns of development in the vertebrate groups that evolved most recently reflect in many ways the simpler patterns occurring among earlier forms. Thus, mammalian development and bird development are elaborations of reptile development, which is an elaboration of amphibian development, and so forth (figure 58.16). During the development of a mammalian embryo, traces can be seen of appendages and organs that are apparently relics of more primitive chordates. For example, at certain stages a human embryo possesses pharyngeal slits, which occur in all chordates and are homologous to the gill slits of fish. At later stages, a human embryo also has a tail.

In a sense, the patterns of development in chordate groups has built up in incremental steps over the evolutionary history of those groups. The developmental instructions for each new form seem to have been layered on top of the previous instructions, contributing additional steps in the developmental journey. This hypothesis, proposed in the nineteenth century by Ernst Haeckel, is referred to as the "biogenic law." It is usually stated as an aphorism: *ontogeny recapitulates phylogeny*; that is, embryological development (ontogeny) involves the same progression of changes that have occurred during evolution (phylogeny). However, the biogenic law is not literally true when stated in this way because embryonic stages are not reflections of *adult* ancestors. Instead, the embryonic stages of a particular vertebrate often reflect the *embryonic* stages of that vertebrate's ancestors. Thus, the pharyngeal slits of

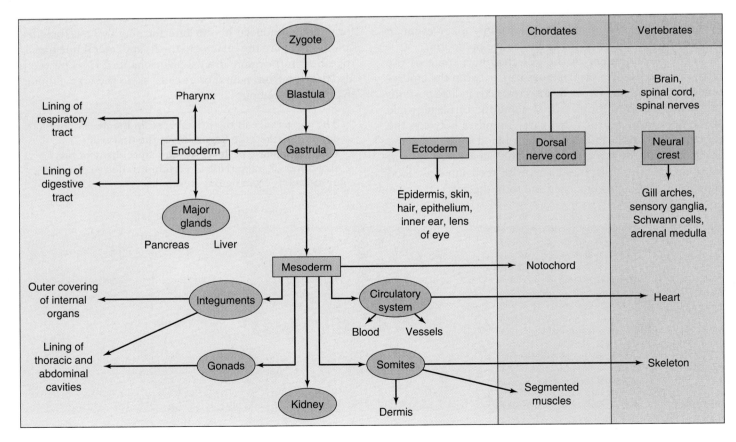

FIGURE 58.15
Derivation of the major tissue types. The key role of the neural crest is evident from the many characteristically vertebrate features that derive from it.

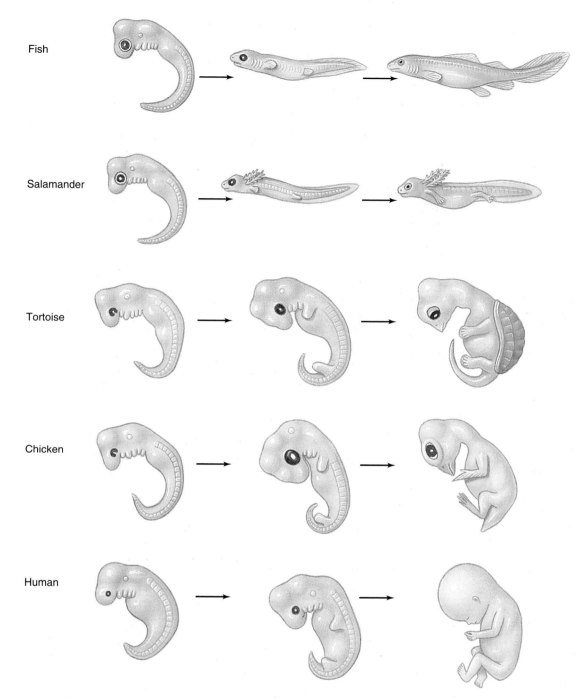

FIGURE 58.16
Embryonic development of vertebrates. Notice that the early embryonic stages of these vertebrates bear a striking resemblance to each other, even though the individuals are from different classes (fish, amphibians, reptiles, birds, and mammals). All vertebrates start out with an enlarged head region, gill slits, and a tail regardless of whether these characteristics are retained in the adults.

a mammalian embryo are *not* like the gill slits its ancestors had *when they were adults.* Rather, they are like the pharyngeal slits its ancestors had *when they were embryos.*

Vertebrates seem to have evolved largely by the addition of new instructions to the developmental program. Development of a mammal thus proceeds through a series of stages, and the earlier stages are unchanged from those that occur in the development of more primitive vertebrates.

Extraembryonic Membranes

As an adaptation to terrestrial life, the embryos of reptiles, birds, and mammals develop within a fluid-filled **amniotic membrane** (chapter 56). The amniotic membrane and several other membranes form from embryonic cells but are located outside of the body of the embryo. For this reason, they are known as **extraembryonic membranes.** The extraembryonic membranes, later to become the **fetal membranes,** include the amnion, chorion, yolk sac, and allantois.

In birds, the **amnion** and **chorion** arise from two folds that grow to completely surround the embryo (figure 58.17). The amnion is the inner membrane that surrounds the embryo and suspends it in *amniotic fluid*, thereby mimicking the aquatic environments of fish and amphibian embryos. The chorion is located next to the eggshell and is separated from the other membranes by a cavity, the *extraembryonic coelom.* The **yolk sac** plays a critical role in the nutrition of bird and reptile embryos; it is also present in mammals, though it does not nourish the embryo. The **allantois** is derived as an outpouching of the gut and serves to store the uric acid excreted in the urine of birds. During development, the allantois of a bird embryo expands to form a sac that eventually fuses with the overlying chorion, just under the eggshell. The fusion of the allantois and chorion form a functioning unit in which embryonic blood vessels, carried in the allantois, are brought close to the porous eggshell for gas exchange. The allantois is thus the functioning "lung" of a bird embryo.

In mammals, the embryonic cells form an inner cell mass that will become the body of the embryo and a layer of surrounding cells called the trophoblast (see figure 58.9). The trophoblast implants into the endometrial lining of its mother's uterus and becomes the chorionic membrane (figure 58.18). The part of the chorion in contact with endometrial tissue contributes to the placenta, as is described in more detail in the next section. The allantois in mammals contributes blood vessels to the structure that will become the umbilical cord, so that fetal blood can be delivered to the placenta for gas exchange.

The extraembryonic membranes include the yolk sac, amnion, chorion, and allantois. These are derived from embryonic cells and function in a variety of ways to support embryonic development.

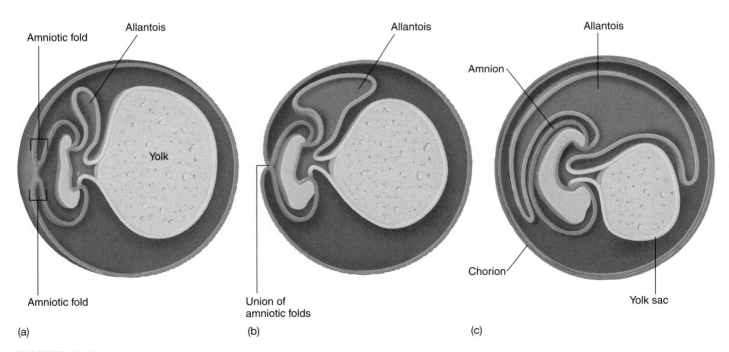

(a)   (b)   (c)

FIGURE 58.17
The extraembryonic membranes of a chick embryo. The membranes begin as amniotic folds from the embryo (a) that unite (b) to form a separate amnion and chorion (c). The allantois continues to grow until it will eventually unite with the chorion just under the eggshell.

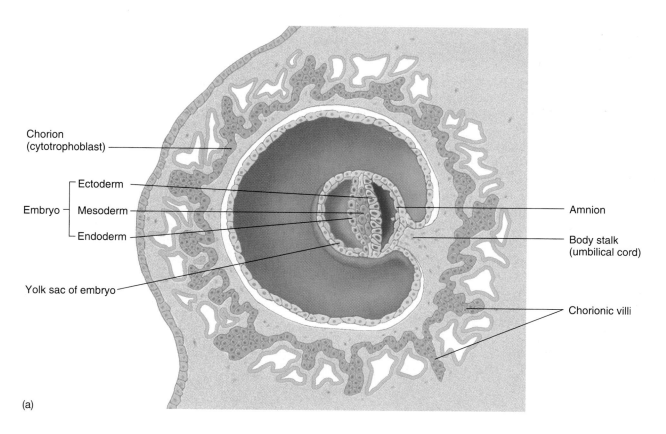

Chorion
(cytotrophoblast)

Ectoderm

Embryo — Mesoderm

Endoderm

Yolk sac of embryo

Amnion

Body stalk
(umbilical cord)

Chorionic villi

(a)

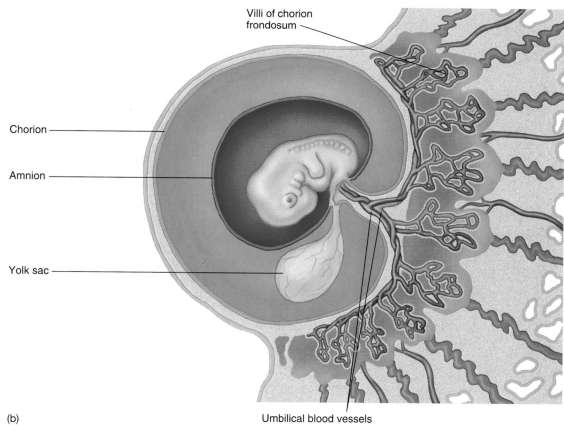

Villi of chorion
frondosum

Chorion

Amnion

Yolk sac

Umbilical blood vessels

(b)

FIGURE 58.18
The extraembryonic membranes of a mammalian embryo. (a) After the embryo implants into the mother's endometrium, the trophoblast becomes the chorion, and the yolk sac and amnion are produced. (b) The chorion develops extensions, called villi, that interdigitate with surrounding endometrial tissue. The embryo is encased within an amniotic sac.

First Trimester

The development of the human embryo shows its evolutionary origins. Without an evolutionary perspective, we would be unable to account for the fact that human development proceeds in much the same way as development in a bird. In both animals, the embryo develops from a flattened collection of cells—the blastodisc in birds or the inner cell mass in humans. While the blastodisc of a bird is flattened because it is pressed against a mass of yolk, the inner cell mass of a human is flat despite the absence of a yolk mass. In humans as well as in birds, a primitive streak forms and gives rise to the three primary germ layers. Human development, from fertilization to birth, takes an average of 266 days. This time is commonly divided into three periods, called trimesters.

The First Month

About 30 hours after fertilization, the zygote undergoes its first cleavage; the second cleavage occurs about 30 hours after that. By the time the embryo reaches the uterus, it is a blastula, which in mammals is referred to as a blastocyst. As we mentioned earlier, the blastocyst consists of an inner

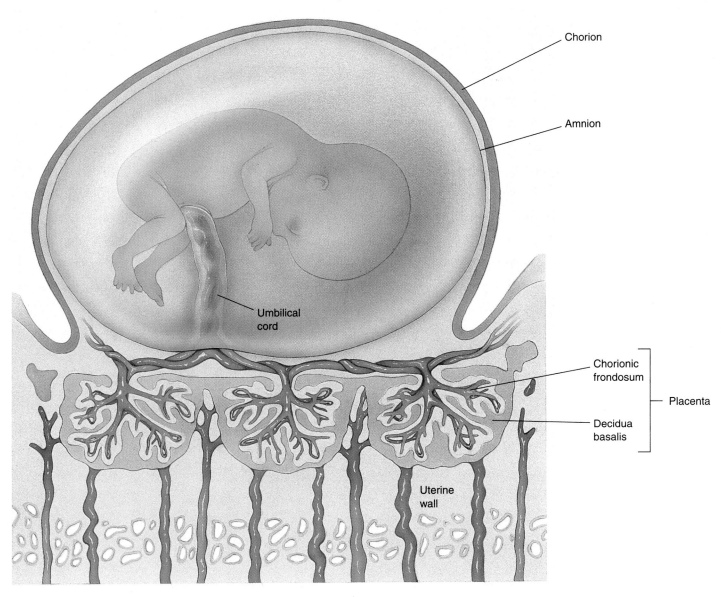

FIGURE 58.19
Structure of the placenta. The placenta contains a fetal component, the chorionic frondosum, and a maternal component, the decidua basalis.

cell mass, which will become the body of the embryo, and a surrounding layer of trophoblast cells (see figure 58.9). The blastocyst begins to grow rapidly and initiates the formation of the amnion and the chorion. The blastocyst digests its way into the endometrial lining of the uterus in the process known as implantation.

During the second week after fertilization, the developing chorion forms branched extensions, the *chorionic frondosum* (fetal placenta) that protrude into the endometrium (figure 58.19). These extensions induce the surrounding endometrial tissue to undergo changes and become the *decidua basalis* (maternal placenta). Together, the chorionic frondosum and decidua basalis form a single functioning unit, the placenta (figure 58.20). Within the placenta, the mother's blood and the blood of the embryo come into close proximity but do not mix (figure 58.21). Oxygen can thus diffuse from the mother to the embryo, and carbon dioxide can diffuse in the opposite direction. In addition to exchanging gases, the placenta provides nourishment for the embryo, detoxifies certain molecules that may pass into the embryonic circulation, and secretes hormones.

One of the hormones released by the placenta is human chorionic gonadotropin (hCG), which was discussed in chapter 56. This hormone is secreted by the trophoblast cells even before they become the chorion, and is the hormone assayed in pregnancy tests. Because its action is almost identical to that of luteinizing hormone (LH), hCG maintains the mother's corpus luteum. The corpus luteum, in turn, continues to secrete estradiol and progesterone, thereby preventing menstruation and further ovulations (as described in chapter 56).

Gastrulation also takes place in the second week after fertilization. The primitive streak can be seen on the surface of the embryo, and the three germ layers (ectoderm, mesoderm, and endoderm) are differentiated.

In the third week, neurulation occurs. This stage is marked by the formation of the neural tube along the dorsal axis of the embryo, as well as by the appearance of the first somites, which give rise to the muscles, vertebrae, and connective tissues. By the end of the week, over a dozen somites are evident, and the blood vessels and gut have begun to develop. At this point, the embryo is about 2 millimeters long.

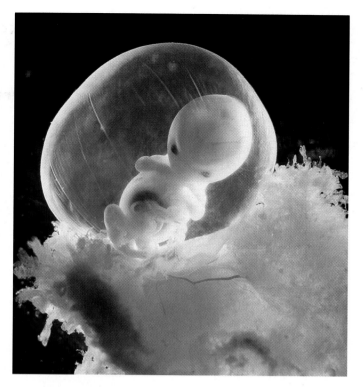

FIGURE 58.20
Placenta and fetus at seven weeks.

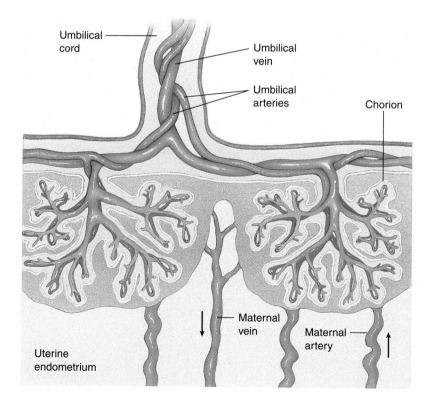

FIGURE 58.21
A close-up view of the placenta. Deoxygenated fetal blood enters the placenta via the umbilical arteries (shown in blue), where it picks up oxygen and nutrients from the mother's blood. Oxygenated fetal blood returns to the fetus through the umbilical vein (shown in red).

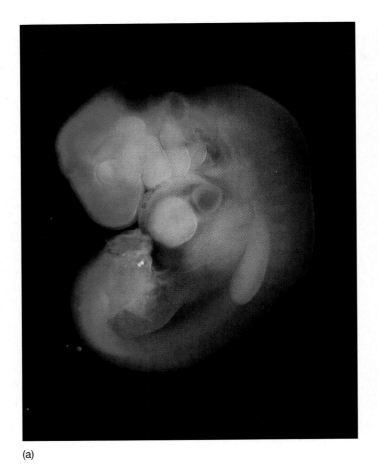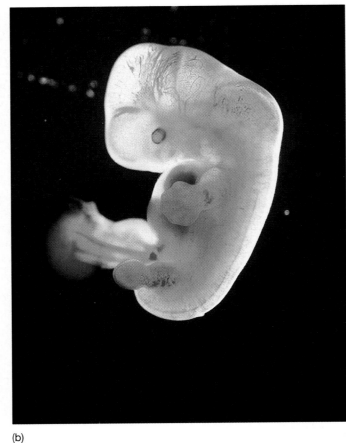

(a)

(b)

FIGURE 58.22
The developing human. (a) Four weeks, (b) seven weeks, (c) three months, and (d) four months.

Organogenesis (the formation of body organs) begins during the fourth week (figure 58.22*a*). The eyes form. The tubular heart develops its four chambers and starts to pulsate rhythmically, as it will for the rest of the individual's life. At 70 beats per minute, the heart is destined to beat more than 2.5 billion times during a lifetime of 70 years. Over 30 pairs of somites are visible by the end of the fourth week, and the arm and leg buds have begun to form. The embryo has increased in length to about 5 millimeters. Although the developmental scenario is now far advanced, many women are unaware they are pregnant at this stage.

Early pregnancy is a very critical time in development, since the proper course of events can be interrupted easily. In the 1960s, for example, many pregnant women took the tranquilizer thalidomide to minimize the discomforts associated with early pregnancy. Unfortunately, this drug had not been adequately tested. It interferes with limb bud development, and its widespread use resulted in many deformed babies. Organogenesis may also be disrupted during the first and second months of pregnancy if the mother contracts rubella (German measles). Most spontaneous abortions occur during this period.

The Second Month

Morphogenesis (the formation of shape) takes place during the second month (figure 58.22*b*). The miniature limbs of the embryo assume their adult shapes. The arms, legs, knees, elbows, fingers, and toes can all be seen—as well as a short bony tail! The bones of the embryonic tail, an evolutionary reminder of our past, later fuse to form the coccyx. Within the abdominal cavity, the major organs, including the liver, pancreas, and gallbladder, become evident. By the end of the second month, the embryo has grown to about 25 millimeters in length, weighs about one gram, and begins to look distinctly human.

The Third Month

The nervous system and sense organs develop during the third month, and the arms and legs start to move (figure 58.22*c*). The embryo begins to show facial expressions and carries out primitive reflexes such as the startle reflex and sucking. The eighth week marks the transition from embryo to fetus. At this time, all of the major organs of the

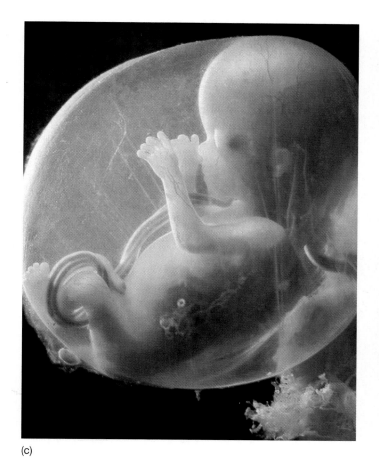

(c)

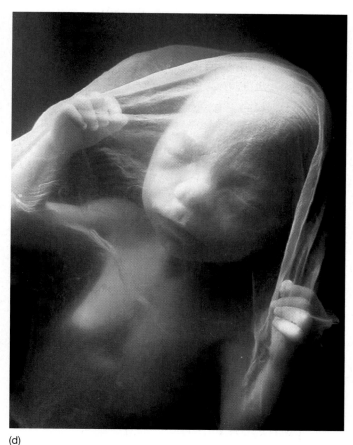

(d)

body have been established. What remains of development is essentially growth.

At around 10 weeks, the secretion of human chorionic gonadotropin (hCG) by the placenta declines, and the corpus luteum regresses as a result. However, menstruation does not occur because the placenta itself secretes estradiol and progesterone (figure 58.23). In fact, the amounts of these two hormones secreted by the placenta far exceed the amounts that are ever secreted by the ovaries. The high levels of estradiol and progesterone in the blood during pregnancy continue to inhibit the release of FSH and LH, thereby preventing ovulation. They also help maintain the uterus and eventually prepare it for labor and delivery, and they stimulate the development of the mammary glands in preparation for lactation after delivery.

The embryo implants into the endometrium, differentiates the germ layers, forms the extraembryonic membranes, and undergoes organogenesis during the first month and morphogenesis during the second month. After eight weeks it becomes a fetus with all of the major body organs.

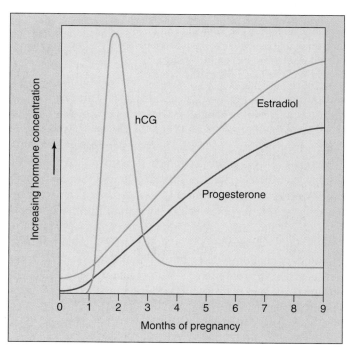

FIGURE 58.23
Hormonal secretion by the placenta. The placenta secretes chorionic gonadotropin (hCG) for 10 weeks. Thereafter, it secretes increasing amounts of estradiol and progesterone.

Second and Third Trimesters

The second and third trimesters are characterized by the tremendous growth and development required for the viability of the baby after its birth.

Second Trimester

Bones actively enlarge during the fourth month (see figure 58.22d), and by the end of the month, the mother can feel the baby kicking. During the fifth month, the head and body grow a covering of fine hair. This hair, called **lanugo,** is another evolutionary relict but is lost later in development. By the end of the fifth month, the rapid heartbeat of the fetus can be heard with a stethoscope. The fetus has grown to about 175 millimeters in length and attained a weight of about 225 grams. Growth begins in earnest in the sixth month; by the end of that month, the baby weighs 600 grams (1.3 lbs) and is over 300 millimeters (1 ft) long. However, most of its prebirth growth is still to come. The baby cannot yet survive outside the uterus without special medical intervention.

Third Trimester

The third trimester is predominantly a period of growth rather than development. The weight of the fetus doubles several times, but this increase in bulk is not the only kind of growth that occurs. Most of the major nerve tracts in the brain, as well as many new neurons (nerve cells), are formed during this period. The developing brain produces neurons at an average rate estimated at more than 250,000 per minute! Neurological growth is far from complete at the end of the third trimester, when birth takes place. If the fetus remained in the uterus until its neurological development was complete, it would grow too large for safe delivery through the pelvis (figure 58.24). Instead, the infant is born as soon as the probability of its survival is high, and its brain continues to develop and produce new neurons for months after birth.

The mother's blood fuels all of the growth of the fetus with the nutrients it provides. Undernourishment of the fetus by a malnourished mother can adversely affect the growth of the fetus and result in severe retardation of the infant. This form of retardation is a serious problem in many developing countries and regions of some developed nations where poverty is common.

The critical stages of human development take place quite early, and the following six months are essentially a period of growth. The growth of the brain is not yet complete, however, by the end of the third trimester, and must be completed postnatally.

FIGURE 58.24

Position of the fetus just before birth. A developing fetus is a major addition to a woman's anatomy. The stomach and intestines are pushed far up, and there is often considerable discomfort from pressure on the lower back. In a natural delivery, the fetus exits through the vagina, which must dilate (expand) considerably to permit passage.

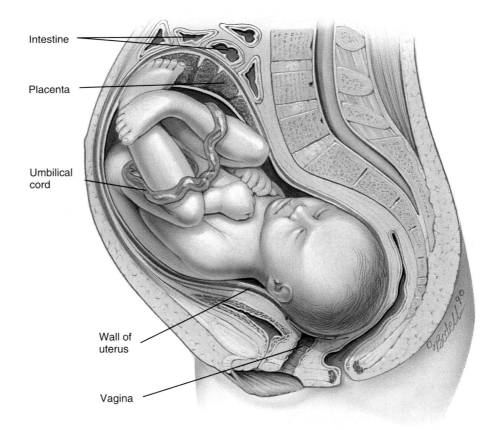

Intestine

Placenta

Umbilical cord

Wall of uterus

Vagina

Birth and Postnatal Development

In some mammals, changing hormone levels in the developing fetus initiate the process of birth. The fetuses of these mammals have an extra layer of cells in their adrenal cortex, called a fetal zone. Before birth, the fetal pituitary gland secretes corticotropin, which stimulates the fetal zone to secrete steroid hormones. These corticosteroids then induce the uterus of the mother to manufacture prostaglandins, which trigger powerful contractions of the uterine smooth muscles.

The adrenal glands of human fetuses lack a fetal zone, and human birth does not seem to be initiated by this mechanism. In a human, the uterus releases prostaglandins, possibly as a result of the high levels of estradiol secreted by the placenta. Estradiol also stimulates the uterus to produce more oxytocin receptors, and as a result, the uterus becomes increasingly sensitive to oxytocin. Prostaglandins begin the uterine contractions, but then sensory feedback from the uterus stimulates the release of oxytocin from the mother's posterior pituitary gland. Working together, oxytocin and prostaglandins further stimulate uterine contractions, forcing the fetus downward. Initially, only a few contractions occur each hour, but the rate eventually increases to one contraction every two to three minutes. Finally, strong contractions, aided by the mother's pushing, expel the fetus, which is now a newborn baby.

After birth, continuing uterine contractions expel the placenta and associated membranes, collectively called the afterbirth. The umbilical cord is still attached to the baby, and to free the newborn a doctor or midwife ties and cuts the cord. Blood clotting and contraction of muscles in the cord prevent excessive bleeding.

Nursing

Milk production, or lactation (figure 58.25), occurs in the alveoli of mammary glands when they are stimulated by the anterior pituitary hormone, prolactin. Milk from the alveoli is secreted into a series of alveolar ducts, which are surrounded by smooth muscle and lead to the nipple (figure 58.26). During pregnancy, high levels of progesterone stimulate the development of the mammary alveoli, and high levels of estradiol stimulate the development of the alveolar ducts. However, estradiol blocks the actions of prolactin on the mammary glands, and it inhibits prolactin secretion by promoting the release of prolactin-inhibiting hormone from the hypothalamus. During pregnancy, therefore, the mammary glands are prepared for lactation but prevented from lactating.

When the placenta is discharged after birth, the concentrations of estradiol and progesterone in the mother's blood decline rapidly. This decline allows the anterior pituitary gland to secrete prolactin, which stimulates the mammary alveoli to produce milk. Sensory impulses associated with the baby's suckling trigger the posterior pituitary gland to

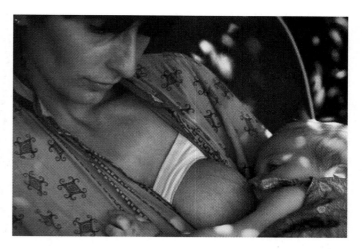

FIGURE 58.25
Nursing is an important bonding time between mother and child. It also provides the child with nourishment and the protection of its mother's immune system while its own develops.

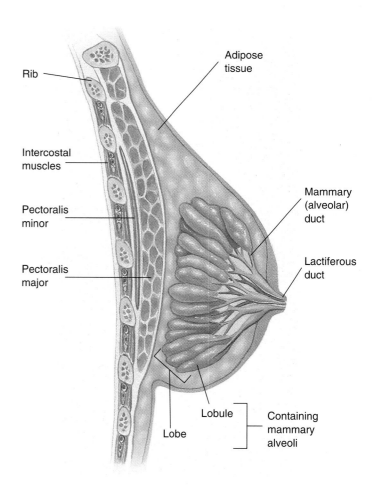

FIGURE 58.26
A sagittal section of a mammary gland. The mammary alveoli produce milk in response to stimulation by prolactin, and milk is ejected through the lactiferous duct in response to stimulation by oxytocin.

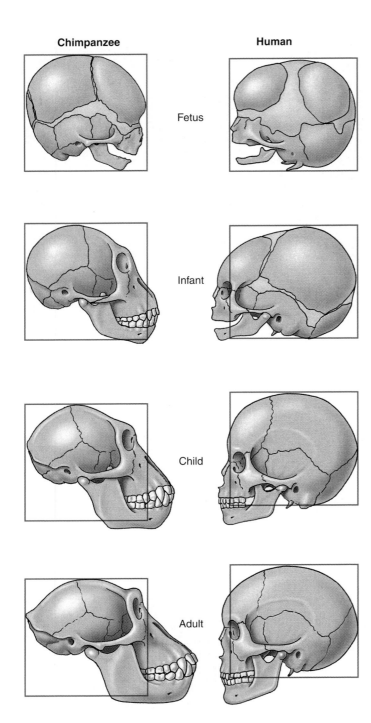

Chimpanzee **Human**

Fetus

Infant

Child

Adult

FIGURE 58.27

Allometric growth. In young chimpanzees, the jaw grows at a faster rate than the rest of the head. As a result, the adult chimpanzee head shape differs greatly from its head shape as a newborn. In humans, the difference in growth between the jaw and the rest of the head is much smaller, and the adult head shape is similar to that of the newborn.

release oxytocin. Oxytocin stimulates contraction of the smooth muscle surrounding the alveolar ducts, thus causing milk to be ejected by the breast. This pathway is known as the milk-ejection reflex. The secretion of oxytocin during lactation also causes some uterine contractions, as it did during labor. These contractions help to restore the tone of uterine muscles in mothers who are breastfeeding.

The first milk produced after birth is a yellowish fluid called colostrum, which is both nutritious and rich in maternal antibodies. Many mothers nurse for a year or longer. During this period, important pair-bonding occurs between the mother and child. When nursing stops, the accumulation of milk in the breasts signals the brain to stop prolactin secretion, and milk production ceases.

Postnatal Development

Growth of the infant continues rapidly after birth. Babies typically double their birth weight within two months. However, different organs grow at different rates and cease growing at different times. Therefore, the body proportions of infants are different from those of adults. The head, for example, is disproportionately large in newborns, but after birth it grows more slowly than the rest of the body. Such a pattern of growth, in which different components grow at different rates, is referred to as allometric growth.

In most mammals, brain growth is mainly a fetal phenomenon. In chimpanzees, for instance, the brain and the cerebral portion of the skull grow very little after birth, while the bones of the jaw continue to grow. As a result, the head of an adult chimpanzee looks very different from that of a fetal chimpanzee (figure 58.27). In human infants, on the other hand, the brain and cerebral skull grow at the same rate as the jaw. Therefore, the jaw-skull proportions do not change after birth, and the head of a human adult looks very similar to that of a human fetus. It is primarily for this reason that an early human fetus seems so remarkably adult-like. The fact that the human brain continues to grow significantly for the first few years of postnatal life means that adequate nutrition and a rich, safe environment are particularly crucial during this period for the full development of the person's intellectual potential.

Birth occurs in response to uterine contractions stimulated by oxytocin and prostaglandins. Lactation is stimulated by prostaglandin, but the milk ejection reflex requires the action of oxytocin.

58.1 Fertilization is the initial event in development.

- Fertilization is the union of an egg and a sperm to form a zygote. It is accomplished externally in most fish and amphibians, and internally in all other vertebrates.

- The three stages of fertilization are (1) penetration, in which the sperm moves through the cells and other materials surrounding the egg and penetrates the egg membrane; (2) activation, which consists of cytoplasmic movements and other changes within the egg; and (3) fusion, in which the sperm and egg nuclei fuse.

58.2 Cell cleavage and the formation of a blastula set the stage for later development.

- Cleavage is the rapid division of the newly formed zygote into a mass of cells, without any increase in overall size.

- The cleavages produce a hollow ball of cells, called blastomeres, which forms a disc-shaped mass in reptiles and birds over a large accumulation of yolk, and an inner cell mass in mammalian embryos.

58.3 Gastrulation forms the three germ layers of the embryo.

- During gastrulation, cells in the blastula change their relative positions, forming the three primary cell layers: ectoderm, mesoderm, and endoderm.

- In eggs with moderate or large amounts of yolk, cells involute down and around the yolk, through a blastopore or primitive streak to form the three germ layers.

58.4 Body architecture is determined during the next stages of embryonic development.

- Neurulation involves the formation of a structure found only in chordates, the notochord and dorsal hollow nerve cord.

- The formation of the neural crest is the first developmental event unique to vertebrates. Most of the distinctive structures associated with vertebrates are derived from the neural crest.

- The embryos of more advanced vertebrates repeat the developmental stages of the embryos of less advanced vertebrates; this is summarized by the phrase "ontogeny recapitulates phylogeny."

- The embryos of amniotic vertebrates are surrounded by an amniotic membrane and a chorionic membrane; together with the yolk sac and allantois, these are known as the extraembryonic membranes.

58.5 Human development is divided into trimesters.

- Most of the critical events in human development occur in the first month of pregnancy. Cleavage occurs during the first week, gastrulation during the second week, neurulation during the third week, and organogenesis during the fourth week.

- The second and third months of human development are devoted to morphogenesis and to the elaboration of the nervous system and sensory organs.

- During the last six months before birth, the human fetus grows considerably, and the brain produces large numbers of neurons and establishes major nerve tracts.

Discussing Key Terms

1. **Holoblastic cleavage** This is the type of cleavage in which the entire cytoplasm of the egg cell is divided into blastomeres by cell division. This occurs in vertebrates that have no or relatively little amounts of yolk.

2. **Meroblastic cleavage** This is the type of cleavage that occurs in reptiles and birds, in which large amounts of yolk cause only a disc of cytoplasm over the yolk to be divided by cleavages into a blastodisc.

3. **Gastrulation** Gastrulation is the developmental stage following the formation of a blastula, in which cell movements result in the formation of an archenteron lined with endoderm, surface ectoderm, and a middle germ layer of mesoderm.

4. **Archenteron** The archenteron is the primitive gut, lined with endoderm, that is formed by invagination and involution during gastrulation; it opens by means of a blastopore that will become the anus of the animal.

5. **Blastopore** The blastopore is the opening created by invagination of surface cells, through which cells involute to form the endoderm and mesoderm of the gastrula; the dorsal lip of the blastopore induces the formation of a notochord and neural tube in chordates.

6. **Neural crest** The neural crest is unique to the vertebrates and forms along the dorsal and lateral sides of the neural tube; it contributes to many structures characteristic of the vertebrate body.

Review Questions

1. What are the three stages of fertilization, and what happens during each stage?

2. What is the difference between holoblastic cleavage and meroblastic cleavage? What characteristic of the egg is responsible for an embryo undergoing one or the other type of cleavage?

3. What is an archenteron, and during what developmental stage does it form? What is the opening of the archenteron called? What is the future fate of this opening in vertebrates?

4. How is gastrulation in amphibians different from that in lancelets? How does the mesoderm form in an amphibian embryo?

5. What structure unique to chordates forms during neurulation? From what type(s) of tissue does it arise? From what tissue does the neural crest arise, and what structures does it ultimately produce?

6. What are the functions of the amnion, chorion, and allantois in birds and mammals?

7. From what structures does the placenta form? How does the placenta prevent menstruation during the first two months of pregnancy? How does it prevent menstruation during the remainder of pregnancy?

8. At what times during human pregnancy do gastrulation, neurulation, and organogenesis occur? Why is this stage of pregnancy critical?

9. What major processes occur during the second and third trimesters of human pregnancy? Is neurological growth complete at birth?

10. What hormone stimulates lactation (milk production)? What hormone stimulates milk ejection from the breast? What stimulus normally initiates the milk-ejection reflex?

Thought Questions

1. Female armadillos always give birth to four offspring of the same sex. Can you suggest a mechanism that would account for this phenomenon?

2. During animal development, cells gradually lose the ability to express portions of their genome, and at some point they become irreversibly committed to a specific fate. Design a transplantation experiment that would identify the time in development when a particular primary tissue becomes committed to differentiate along a specific path (e.g., when ectoderm is committed to become spinal cord).

3. In what sense is the statement "ontogeny recapitulates phylogeny" true, and in what sense is it false? Is this good evidence for evolution? Explain.

Internet Links

STD's
http://med-www.bu.edu/people/sycamore/std/std.htm
The STD HOMEPAGE provides extensive information and links for all of the common sexually transmitted diseases. Also try http://www.stdservices.on.net/.

Having a Baby?
http://www.parentsoup.com/
STORKNET provides tons of helpful information and advice for the expectant mother. A very nice site.

For Further Reading

Cooke, J.: "Vertebrate Development: Induction and All That," *Nature*, vol. 358, pages 111–12, July 9, 1992. A look at the interactions among cells that would turn them ultimately into an amphibian.

Cross, J., Z. Werb, and S. Fisher: "Implantation and the Placenta: Key Pieces of the Development Puzzle," *Science*, vol. 266, pages 1508–18, December 2, 1994. What kinds of lethal mutations will affect implantation of a mammalian embryo in the placenta?

DeRobertis, E.: "Dismantling the Organizer," *Nature*, vol. 374, pages 407–8, March 30, 1995. A discussion of mutation and organogenesis of the head in mammals, the "organizer" being, of course, the homeobox.

Holden C.: "Perils of Freezing Embryos," *Science*, vol. 267, pages 618–19, February 3, 1995. How do we save fertilized embryos in vitro for later implantation? Are the processes currently available good enough?

Holmes, B.: "Genetic Engineering in the Womb," *New Scientist*, vol. 145, page 16, March 4, 1995. Why wait until a baby is born to try to correct genetic defects? Scientists have already met with some success in in utero gene transfer.

Munson, P., D. Counts, and G. Cutler: "Patterns of Human Growth," *Science*, vol. 268, pages 442–47, April 21, 1995. Human infant development is explored in detail.

Nilsson, L. and J. Lindberg: *Behold Man*, Little, Brown & Co., Boston, 1974. A wonderful collection of color photographs of the developing human fetus.

Tickle, C.: "Vertebrate Development: A Tool for Transgenesis," *Nature*, vol. 358, pages 188–89, July 16, 1992. A discussion of homeobox genes and the development of vertebrate extremities.

59

Animal Behavior

Concept Outline

59.1 Ethology focuses on how neural units influence behavior.

Approaches to the Study of Behavior. Field biologists focus on evolutionary aspects of behavior.

The Neural Basis of Behavior. Many behaviors reflect neural "hard-wiring."

Behavioral Genetics. At least some behaviors are genetically determined.

59.2 Comparative physiology focuses on how learning influences behavior.

Learning. Association plays a major role in learning.

The Development of Behavior. Parent-offspring interactions play a key role in the development of behavior.

The Physiology of Behavior. Hormones influence many behaviors, particularly reproductive ones.

Behavioral Rhythms. Many behaviors are governed by innate biological clocks.

59.3 Communication is a key element of many animal behaviors.

Courtship. Animals use many kinds of signals to court one another.

Communication in Social Groups. Bees and other social animals communicate in complex ways.

59.4 Migratory behavior presents many puzzles.

Orientation and Migration. Animals use many cues from the environment to navigate during migrations.

59.5 To what degree animals "think" is a subject of lively dispute.

Animal Cognition. It is not clear to what degree animals "think."

FIGURE 59.1
Rearing offspring involves complex behaviors. Living in groups called prides makes lions better mothers. Females share the responsibilities of nursing and protecting the pride's young, increasing the probability that the youngsters will survive into adulthood.

Why do animals behave the way they do? What influences how an animal behaves? These are very simple questions with profound answers and implications. In this chapter, we will consider the mechanisms that allow animals to adaptively respond to their environment (figure 59.1), as well as the ways in which behavior develops in an individual. We will discuss different approaches to the study of behavior, some focusing on instinct, some emphasizing learning, and still others combining the elements of both approaches. We will also examine how the environment influences the neurobiological, hormonal, and developmental processes that mediate animal behavior.

Approaches to the Study of Behavior

During the past two decades, the study of animal behavior has emerged as an important and diversified science that bridges several disciplines within biology. Evolution, ecology, physiology, genetics, and psychology all have natural and logical linkages with the study of behavior, each discipline adding a different perspective and addressing different questions.

Research in animal behavior has made major contributions to our understanding of nervous system organization, child development, and human communication, as well as the process of speciation, community organization, and the mechanism of natural selection itself. The study of the behavior of nonhuman animals has led to the identification of general principles of behavior, which have been applied, often controversially, to humans. This has changed the way we think about the origins of human behavior and the way we perceive ourselves.

Behavior can be defined as the way an organism responds to stimuli in its environment. These stimuli might be as simple as the odor of food. In this sense, a bacterial cell "behaves" by moving toward higher concentrations of sugar. This behavior is very simple and well-suited to the life of bacteria, allowing these organisms to live and reproduce. As animals evolved, they occupied different environments and faced diverse problems that affected their survival and reproduction. Their nervous systems and behavior concomitantly became more complex. Nervous systems perceive and process information concerning environmental stimuli and trigger adaptive motor responses, which we see as patterns of behavior.

When we observe animal behavior, we can explain it in two different ways. First, we might ask *how* it all works, that is, how the animal's senses, nerve networks, or internal state provide a physiological basis for the behavior. In this way, we would be asking a question of **proximate causation.** To analyze the proximate cause of behavior, we might measure hormone levels or record the impulse activity of neurons in the animal. We could also ask *why* the behavior evolved, that is, what is its adaptive value? This is a question concerning **ultimate causation.** To study the ultimate cause of a behavior, we would attempt to determine how it influenced the animal's survival or reproductive success. Thus, a male songbird may sing during the breeding season because it has a level of the steroid sex hormone, testosterone, which binds to hormone receptors in the brain and triggers the production of song; this would be the proximate cause of the male bird's song. But the male sings to defend a territory from other males and to attract a female to reproduce; this is the ultimate, or evolutionary, explanation for the male's vocalization.

The study of behavior has had a long history of controversy. One source of controversy has been the question of whether behavior is determined more by an individual's genes or its learning and experience. In other words, is behavior the result of nature (instinct) or nurture (experience)? In the past, this question has been considered an "either/or" proposition, but we now know that instinct and experience both play significant roles, often interacting in complex ways to produce the final behavior. The scientific study of instinct and learning, as well as their interrelationship, has led to the growth of several scientific disciplines, including ethology, behavioral genetics, behavioral neuroscience, and psychology.

Ethology

Ethology is the study of the natural history of behavior. Early ethologists (figure 59.2) were trained in zoology and evolutionary biology, fields that emphasize the study of animal behavior under natural conditions. As a result of this training, they believed that behavior is largely instinctive, or innate—the product of natural selection. Because behavior is often **stereotyped** (appearing in the same way in different individuals of a species), they argued that it must be based on preset neural circuits. In their view, these circuits are structured from genetic blueprints and cause animals to show a relatively complete behavior the first time it is produced.

The early ethologists based their opinions on behaviors such as egg retrieval by geese. Geese incubate their eggs in a nest. If a goose notices that an egg has been knocked out of the nest, it will extend its neck toward the egg, get up, and roll the egg back into the nest with a side-to-side motion of its neck while the egg is tucked beneath its bill. Even if the egg is removed during retrieval, the goose completes the behavior, as if driven by a program released by the initial sight of the egg outside the nest. According to ethologists, egg retrieval behavior is triggered by a **sign stimulus** (also called a **key stimulus**), the appearance of an egg out of the nest; a component of the goose's nervous system, the **innate releasing mechanism,** provides the neural instructions for the motor program, or **fixed action pattern** (figure 59.3). More generally, the sign stimulus is a "signal" in the environment that triggers a behavior. The innate releasing mechanism is the sensory mechanism that detects the signal, and the fixed action pattern is the stereotyped act. Similarly, a frog unfolds its long, sticky tongue at the sight of an insect, and a male stickleback fish will attack another male showing a bright red underside. Such responses certainly appear to be programmed and instinctive, but what evidence supports the ethological view that behavior has an underlying neural basis?

FIGURE 59.2
The founding fathers of ethology: Karl von Frisch, Konrad Lorenz, and Niko Tinbergen pioneered the study of behavioral science. In 1973, they received the Nobel Prize in physiology or medicine for their path-making contributions. Von Frisch led the study of honeybee communication and sensory biology. Lorenz focused on social development (imprinting) and the natural history of aggression. Tinbergen examined the functional significance of behavior and was the first behavioral ecologist.

Behavior as a Response to Stimuli in the Environment.

In the example of egg retrieval behavior in geese, a goose must first perceive that an egg is outside of the nest. To respond to this stimulus, it must convert one form of energy—the energy of photons of light—which is an input to its visual system into a form of energy its nervous system can understand and use to respond: the electrical energy of a nerve impulse. Animals need to respond to other stimuli in the environment as well. For an animal to orient from a food source back to its nest, it might rely on the position of the sun. To find a mate, an animal might use a particular chemical scent. The electromagnetic energy of light and the chemical energy of an odor must be converted to the electrical energy of a nerve impulse. This is done through **transduction**, the conversion of energy in the environment to an action potential, and the first step in the processing of stimuli perceived by the senses. For example, rhodopsin is responsible for the transduction of visual stimuli. Rhodopsin is made of *cis*-retinal and the protein opsin. Light is absorbed by the visual pigment *cis*-retinal causing it to change its shape to *trans*-retinal (see chapter 52). This in turn changes the shape of

FIGURE 59.3
Lizard prey capture. The complex series of movements of the tongue this chameleon uses to capture an insect represents a fixed action pattern.

the companion protein opsin, and induces the first step in a cascade of molecular events that finally triggers a nerve impulse. Sound, odor, and tastes are transduced to nerve impulses by similar processes.

Ethologists study behavior from an evolutionary perspective, focusing on the neural basis of behaviors.

The Neural Basis of Behavior

The transduction of stimuli in the environment is the first in a series of events that govern behavioral responses to environmental stimuli. The way an animal perceives its world depends upon the structure of its sensory systems. Hearing is often limited to certain frequencies of sound and vision to certain wavelengths of light. The sense of smell decodes information about specific chemicals in the air or water. These and other senses monitor the environment by means of sensory neurons specialized for the detection of particular stimuli. Research on the neurobiology of stimulus detection has revealed how animal senses and other aspects of the nervous system are organized to govern perception and behavior, thus providing support for the ethological concept of the innate releasing mechanism. The study of the neural basis of behavior is called **neuroethology.**

Neuroethology in Action

One behavior studied at the neuroethological level is prey capture in frogs and toads. These animals prey on insects by capturing them with their sticky tongues. To do so, a toad must be able to detect and follow prey in the environment. When an insect appears in front of a toad, its image enters the lens of the toad's eye and is focused on the retina. Light-sensitive photoreceptor cells in the retina (chapter 52) relay information about the position of the insect via other retinal cells to the brain (figure 59.4). As the insect moves, its image passes over different groups of retinal cells, which ultimately inform the brain of the changing position of the insect. The size and movement of the retinal image enable neurons in the toad's brain to identify the insect as prey. These "insect detector" cells respond to the sign stimulus (the insect) by triggering a releasing mechanism in the brain, which in turn initiates the fixed action pattern of flipping out the tongue at the insect.

A number of animals have been extremely useful as "models" for research on the neural basis of behavior because they have relatively simple nervous systems and show behaviors that are easy to record. Included among these animals are a variety of marine mollusks, whose nervous systems contain relatively few neurons, many of which are quite large. One simple but important behavior shown by the nudibranch mollusk *Tritonia* is its **escape response,** which occurs when the animal detects the presence of a predator such as a sea star. The response consists of a series of alternating dorsal and ventral body flexions that move *Tritonia* through the water. These flexions result from impulses in three groups of neurons in the brain. One group, the dorsal flexion neurons, stimulate the muscles that cause the animal to flex dorsally, while a second group, the ventral flexion neurons, stimulate the muscles that cause ventral flexion. The third group of neurons causes impulse activity to alternate in the two sets of flexion neurons.

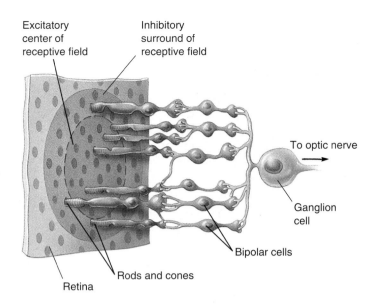

FIGURE 59.4
Prey detectors in the eye of the toad. Light-sensitive photoreceptor cells (rods and cones) embedded in the retina monitor prey motion. Signals are relayed to bipolar cells, which in turn communicate with ganglion cells. Each ganglion cell has a receptive field, a portion of the retina's surface in which it monitors photoreceptor activity. A small object (a potential prey) whose image lies in the center of a ganglion cell's receptive field excites the ganglion cell to generate action potentials that are transmitted via the optic nerve to the brain. The image of a large object may extend into the surrounding area of the receptive field, inhibiting the ganglion cell from producing action potentials.

The neurobiology of escape behavior has also been examined in the cockroach. In this animal, sensory hairs located on the cerci (paired appendages at the end of the abdomen) are moved by microcurrents of air, such as those produced by an approaching predator. Sensory neurons associated with the hairs generate impulses when the hairs move. These impulses are conducted across only one synapse, where they excite a "giant" interneuron. The giant interneuron, in turn, transmits its own impulses rapidly along its axis to a ganglion (a cluster of neural tissue) in the thorax, where they excite motor neurons that initiate escape locomotion. The response is extremely fast—only about 60 milliseconds elapse between the detection of air currents and the beginning of the escape behavior. Other neuroethological studies describe the neural structures and circuits that underscore behaviors such as navigation and communication.

Studies of nerve networks underlying behavior support the concept of the innate nature of behavior. Research on the neuroethology of simple behaviors helps us understand the neural units from which more complex nervous systems and behaviors are assembled.

Behavioral Genetics

In a famous experiment carried out in the 1940s, Robert Tryon studied the ability of rats to find their way through a maze with many blind alleys and only one exit, where a reward of food awaited. Some rats quickly learned to zip right through the maze to the food, making few incorrect turns, while other rats took much longer to learn the correct path (figure 59.5). Tryon bred the fast learners with one another to establish a "maze-bright" colony, and he similarly bred the slow learners with one another to establish a "maze-dull" colony. He then tested the offspring in each colony to see how quickly they learned the maze. The offspring of maze-bright rats learned even more quickly than their parents had, while the offspring of maze-dull parents were even poorer at maze learning. After repeating this procedure over several generations, Tryon was able to produce two behaviorally distinct types of rat with very different maze-learning abilities. Clearly the ability to learn the maze was to some degree hereditary, governed by genes passed from parent to offspring. Furthermore, those genes were specific to this behavior, since the two groups of rats did not differ in their ability to perform other behavioral tasks, such as running a completely different kind of maze. Tryon's research demonstrates how a study can reveal that behavior has a heritable component.

Further support for the genetic basis of behavior has come from studies of hybrids. William Dilger of Cornell University has examined two species of lovebird (genus *Agapornis*), which differ in the way they carry twigs, paper, and other materials used to build a nest. *A. personata* holds nest material in its beak, while *A. roseicollis* carries material tucked under its flank feathers (figure 59.6). When Dilger crossed the two species to produce hybrids, he found that the hybrids carry nest material in a way that seems intermediate between that of the parents: they repeatedly shift material between the bill and the flank feathers. Other studies conducted on courtship songs in crickets and tree frogs also demonstrate the intermediate nature of hybrid behavior.

The maze-learning and hybrid studies just described suggest genes play a role in behavior, but recent research has provided much greater detail on the genetic basis of behavior. In the fruit fly *Drosophila*, and in mice, many mutations have been associated with particular behavioral abnormalities. In mice, for example, three genes serve as markers for "personality." Also, male mice genetically engineered (as "knock-outs") to lack the ability to synthesize nitric oxide, a brain neurotransmitter, show aggressive behavior. In 1996, scientists using the knock-out technique discovered a new gene, *fosB*, that seems to determine whether or not female mice will nurture their young.

The genetic basis of behavior is supported by artificial selection experiments, hybridization studies, and studies on the behavior of mutants. Research has also identified specific genes that control behavior.

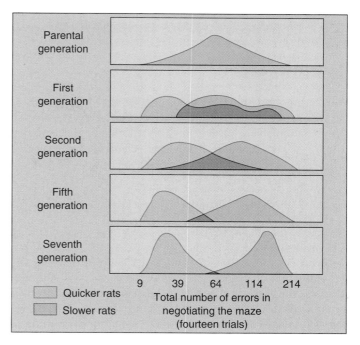

FIGURE 59.5
The genetics of learning. Tryon selected rats for their ability to learn to run a maze and demonstrated that this ability is influenced by genes. He tested a large group of rats, selected those that ran the maze in the shortest time, and let them breed with one another. He then tested their progeny and again selected those with the quickest maze-running times for breeding. After seven generations, he had succeeded in halving the average time an inexperienced rat required to negotiate the maze. Parallel "artificial selection" for slow running time more than doubled the average running time.

FIGURE 59.6
Genetics of lovebird behavior. Lovebirds inherit the tendency to carry nest material, such as these paper strips, under their flank feathers.

Learning

While ethologists were attempting to explain behavior as an instinctive process, **comparative psychologists** focused almost exclusively on learning as the major element that shapes behavior. These behavioral scientists, working primarily on rats in laboratory settings, identified the ways in which animals learn. Learning is any modification of behavior that results from experience rather than maturation.

The simplest type of learning, **nonassociative learning,** does not require an animal to form an association between two stimuli or between a stimulus and a response. One form of nonassociative learning is **habituation,** which can be defined as a decrease in response to a repeated stimulus that has no positive or negative consequences (i.e., no *reinforcement*). In many cases, the stimulus evokes a strong response when it is first encountered, but the magnitude of the response gradually declines with repeated exposure. For example, young birds see many types of objects moving overhead. At first, they may respond by crouching down and remaining still. Some of the objects, like falling leaves or members of their own species flying by, are seen very frequently and have no positive or negative consequence to the nestlings. Over time, the young birds may habituate to such stimuli and stop responding. Thus, habituation can be thought of as learning *not* to respond to a stimulus. Being able to ignore unimportant stimuli is critical to an animal confronting a barrage of stimuli in a complex environment. Another form of nonassociative learning is **sensitization,** characterized by an increased responsiveness to a stimulus. Sensitization is essentially the opposite of habituation.

A change in behavior that involves an association between two stimuli or between a stimulus and a response is termed **associative learning** (figure 59.7). The behavior is modified, or **conditioned,** through the association. This form of learning is more complex than habituation or sensitization. The two major types of associative learning are called **classical conditioning** and **operant conditioning;** they differ in the way the associations are established.

In classical conditioning, the paired presentation of two different kinds of stimuli causes the animal to form an association between the stimuli. Classical conditioning is also called **Pavlovian conditioning,** after Russian psychologist Ivan Pavlov, who first described it. Pavlov presented meat powder, an *unconditioned stimulus*, to a dog and noted that the dog responded by salivating, an *unconditioned response*. If an unrelated stimulus, such as the ringing of a bell, was presented at the same time as the meat powder, over repeated trials the dog would salivate in response to the sound of the bell *alone*. The dog had learned to associate the unrelated sound stimulus with the meat powder stimulus. Its response to the sound stimulus was, therefore, conditioned, and the sound of the bell is referred to as a *conditioned stimulus*.

(a)

(b)

(c)

FIGURE 59.7

Learning what is edible. Associative learning is involved in predator-prey interactions. (a) A naive toad is offered a bumblebee as food. (b) The toad is stung, and (c) subsequently avoids feeding on bumblebees or any other insects having their black and yellow coloration. The toad has associated the appearance of the insect with pain, and modifies its behavior.

In operant conditioning, an animal learns to associate its behavioral response with a reward or punishment. American psychologist B. F. Skinner studied operant conditioning in rats by placing them in an apparatus that came to be called a "Skinner box" (figure 59.8). As the rat explored the box, it would occasionally press a lever by accident, causing a pellet of food to appear. At first, the rat would ignore the lever, eat the food pellet, and continue to move about. Soon, however, it learned to associate pressing the lever (the behavioral response) with obtaining food (the reward). When it was hungry, it would spend all its time pressing the lever. This sort of trial-and-error learning is of major importance to most vertebrates.

Comparative psychologists used to believe that any two stimuli could be linked in classical conditioning and that animals could be conditioned to perform any learnable behavior in response to any stimulus by operant conditioning. As you will see below, this view has changed. Today, it is thought that instinct guides learning by determining what type of information can be learned through conditioning.

Instinct

It is now clear that some animals have innate predispositions toward forming certain associations. For example, if a rat is offered a food pellet at the same time it is exposed to X rays (which *later* produces nausea), the rat will remember the taste of the food pellet but not its size. Conversely, if a rat is given a food pellet at the same time an electric shock is delivered (which *immediately* causes pain), the rat will remember the size of the pellet but not its taste. Similarly, pigeons can learn to associate *food* with colors but not with sounds; on the other hand, they can associate *danger* with sounds but not with colors.

These examples of **learning preparedness** demonstrate that what an animal can learn is biologically influenced—that is, learning is possible only within the boundaries set by instinct. Innate programs have evolved because they underscore adaptive responses. Rats, which forage at night and have a highly developed sense of smell, are better able to identify dangerous food by its odor than by its size or color. The seed a pigeon eats may have a distinctive color that the pigeon can see, but it makes no sound the pigeon can hear. The study of learning has expanded to include its ecological significance, so that we are now able to consider the "evolution of learning." An animal's ecology, of course, is key to understanding what an animal is capable of learning. Some species of birds, like Clark's nutcracker, feed on seeds. Birds store seeds in caches they bury when seeds are abundant so they will have food during the winter. Thousands of seed caches may be buried and then later recovered. One would expect the birds to have an extraordinary spatial memory, and this is indeed what has been found (figure 59.9). Clark's nutcracker, and other seed-hoarding birds, have an unusually large hippocampus, the center for memory storage in the brain (see chapter 51).

FIGURE 59.8
A Skinner box. The rat rapidly learns that pressing the lever results in the appearance of a food pellet. This kind of learning, consisting of trial and error with a reward for success, can also work for far more complex tasks.

FIGURE 59.9
The Clark's nutcracker has an extraordinary memory. A Clark's nutcracker can remember the locations of up to 2000 seed caches months after hiding them. After conducting experiments, scientists have concluded that the birds use features of the landscape and other surrounding objects as spatial references to memorize the locations of the caches.

Habituation and sensitization are simple forms of learning in which there is no association between stimuli and responses. In contrast, associative learning (classical and operant conditioning) involves the formation of an association between two stimuli or between a stimulus and a response.

The Development of Behavior

Behavioral biologists now recognize that behavior has both genetic and learned components, and the schools of ethology and psychology are less polarized than they once were. Thus far in this chapter we have discussed the influence of genes and learning separately. As we will see, these factors interact during development to shape behavior.

Parent-Offspring Interactions

As an animal matures, it may form social attachments to other individuals or form preferences that will influence behavior later in life. This process, called **imprinting,** is sometimes considered a type of learning. In **filial imprinting,** social attachments form between parents and offspring. For example, young birds begin to follow their mother within a few hours after hatching, and their following response results in a bond between mother and young. However, the young birds' initial experience determines how this imprint is established. The German ethologist Konrad Lorenz showed that birds will follow the first object they see after hatching and direct their social behavior toward that object. Lorenz raised geese from eggs, and when he offered himself as a model for imprinting, the goslings treated him as if he were their parent, following him dutifully (figure 59.10). Black boxes, flashing lights, and watering cans can also be effective imprinting objects (figure 59.11). Imprinting occurs during a **sensitive phase,** or a **critical period** (roughly 13–16 hours after hatching in geese), when the success of imprinting is highest.

Several studies demonstrate that the social interactions that occur between parents and offspring during the critical period are key to the normal development of behavior. The psychologist Harry Harlow gave orphaned rhesus monkey infants the opportunity to form social attachments with two surrogate "mothers," one made of soft cloth covering a wire frame and the other made only of wire. The infants chose to spend time with the cloth mother, even if only the wire mother provided food, indicating that texture and tactile contact, rather than providing food, may be among the key qualities in a mother that promote infant social attachment. If infants are deprived of normal social contact, their development is abnormal. Greater degrees of deprivation lead to greater abnormalities in social behavior during childhood and adulthood. Studies on orphaned human infants suggest that a constant "mother figure" is required for normal growth and psychological development.

Recent research has revealed a biological need for the stimulation that occurs during parent-offspring interactions early in life. Female rats lick their pups after birth, and this stimulation inhibits the release of an endorphin (see chapter 51) that can block normal growth. Pups that receive normal tactile stimulation also have more brain receptors for glucocorticoid hormones, longer lived brain neurons, and a greater tolerance for stress. Premature human infants who are massaged gain weight rapidly. These studies indicate that the need for normal social interaction is based in the brain and that touch and other aspects of contact between parents and offspring are important for physical as well as behavioral development.

Sexual imprinting is a process in which an individual learns to direct its sexual behavior at members of its own species. **Cross-fostering** studies, in which individuals of one species are raised by parents of another species, reveal that this form of imprinting also occurs early in life. In most species of birds, these studies have shown that the fostered bird will attempt to mate with members of its foster species when it is sexually mature.

FIGURE 59.10
An unlikely parent. The eager goslings following Konrad Lorenz think he is their mother. He is the first object they saw when they hatched, and they have used him as a model for imprinting.

(a)

(b)

FIGURE 59.11
How imprinting is studied. Ducklings will imprint on the first object they see, even (a) a black box or (b) a white sphere.

Interaction between Instinct and Learning

The work of Peter Marler and his colleagues on the acquisition of courtship song by white-crowned sparrows provides an excellent example of the interaction between instinct and learning in the development of behavior. The song is sung by mature males and is species-specific. By rearing male birds in soundproof incubators provided with speakers and microphones, Marler could control what a bird heard as it matured and record the song it produced as an adult. He found that birds that heard no song at all during development, or that heard only the song of a different species, the song sparrow, sang a poorly developed song as adults (figure 59.12). But birds that heard the song of their own species, or that heard the songs of *both* the white-crowned sparrow and the song sparrow, sang a fully developed, white-crowned sparrow song as adults. These results suggest that these birds have a **genetic template,** or instinctive program, that guides them to learn the appropriate song. During a critical period in development, the template will accept the correct song as a model. Thus, song acquisition depends on learning, but only the song of the correct species can be learned. The genetic template for learning is *selective.* However, learning plays a prominent role as well. If a young white-crowned sparrow is surgically deafened *after* it hears its species' song during the critical period, it will also sing a poorly developed song as an adult. Therefore, the bird must "practice" listening to himself sing, matching what he hears to the model his template has accepted.

Although this explanation of song development stood unchallenged for many years, recent research has shown that white-crowned sparrow males *can* learn another species' song under certain conditions. If a live male strawberry finch is placed in a cage next to a young male sparrow, the young sparrow will learn to sing the strawberry finch's song! This finding indicates that social stimuli may be more effective than a tape-recorded song in overriding the innate program that guides song development. Furthermore, the males of some bird species have no opportunity to hear the song of their own species. In such cases, it appears that the males instinctively "know" their own species' song. For example, cuckoos are brood parasites; females lay their eggs in the nest of another species of bird, and the young that hatch are reared by the foster parents (figure 59.13). When the cuckoos become adults, they sing the song of their own species rather than that of their foster parents. Since male brood parasites would most likely hear the song of their host species during development, it is adaptive for them to ignore such "incorrect" stimuli. They hear no adult males of their own species singing, so no correct song models are available. In these species, natural selection has programmed the male with a genetically guided song. In all bird species, singing depends upon the physiological maturation of certain centers of the brain and other organs that control song production, as we will see in the next section.

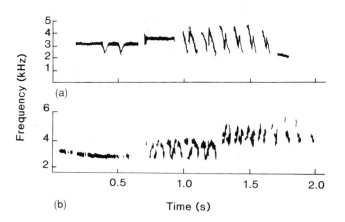

FIGURE 59.12

Song development in birds. (a) The sonograms of songs produced by male white-crowned sparrows that had been exposed to their own species' song during development are different from (b) those of male sparrows that heard no song during rearing. This difference indicates that the genetic program itself is insufficient to produce a normal song.

FIGURE 59.13

Brood parasite. Cuckoos lay their eggs in the nests of other species of birds. Since the young cuckoos (large bird to the right) are raised by a different species (like this meadow pipit, smaller bird to the left), they have no opportunity to *learn* the cuckoo song; the cuckoo song they later sing is innate.

Interactions that occur during sensitive phases of imprinting are critical to normal behavioral development. Physical contact plays an important role in the development of psychological well-being and growth.

The Physiology of Behavior

The early ethologists' emphasis on instinct sometimes overlooked the internal factors that control behavior. If asked why a male bird defends a territory and sings only during the breeding season, they would answer that a bird sings when it is in the right *motivational state* or *mood* and has the appropriate *drive*. But what do these phrases mean in terms of physiological control mechanisms?

Part of our understanding of the physiological control of behavior has come from the study of reproductive behavior. Animals show reproductive behaviors such as courtship only during the breeding season. Research on lizards, birds, rats, and other animals has revealed that hormones play an important role in these behaviors. Changes in day length trigger the secretion of gonadotropin-releasing hormone by the hypothalamus, which stimulates the release of the gonadotropins, follicle-stimulating hormone (FSH) and luteinizing hormone, by the anterior pituitary gland. These hormones cause the development of reproductive tissues to ready the animal for breeding. The gonadotropins, in turn, stimulate the secretion of the steroid sex hormones, estrogens and progesterone in females and testosterone in males. The sex hormones act on the brain to trigger behaviors associated with reproduction. For example, bird song and territorial behavior depend upon the level of testosterone in the male, and the receptivity of females to male courtship depends upon estrogen levels.

Hormones have both organizational and activational effects. In the example of bird song given above, estrogen in the male causes the development of the song system, which is composed of neural tissue in the forebrain and its connections to the spinal cord and the syrinx (a structure like our larynx that allows the bird to sing). Early in a male's development, the gonads produce estrogen, which stimulates neuron growth in the brain. In the mature male, the testes produce testosterone, which activates song. Thus the development of the neural systems that are responsible for behavior is first organized, then activated by hormones.

FIGURE 59.14
Hormonal control of reproductive behavior. Reproduction in the ring dove involves a sequence of behaviors regulated by hormones: (1) courtship and copulation; (2) nest building; (3) egg laying; (4) incubation; and (5) feeding crop milk to the young after they hatch.

Research on the physiology of reproductive behavior shows that there are important interactions among hormones, behavior, and stimuli in both the physical and social environments of an individual. Daniel Lehrman's work on reproduction in ring doves provides an excellent example of how these factors interact (figure 59.14). Male courtship behavior is controlled by testosterone and related steroid hormones. The male's behavior causes the release of FSH in the female, and FSH promotes the growth of the ovarian follicles (see chapter 56). The developing follicles release estrogens, which affect other reproductive tissues. Nest construction follows after one or two days. The presence of the nest then triggers the secretion of progesterone in the female, initiating incubation behavior after the egg is laid. Feeding occurs once the eggs hatch, and this behavior is also hormonally controlled.

The research of Lehrman and his colleagues paved the way for many additional investigations in **behavioral endocrinology,** the study of the hormonal regulation of behavior. For example, male *Anolis* lizards begin courtship after a seasonal rise in temperature, and the male's courtship is needed to stimulate the growth of ovarian follicles in the female. These and other studies demonstrate the interactive effects of the physical environment (e.g., temperature and day length) and the social environment (e.g., the presence of a nest and the courtship display of a mate) on the hormonal condition of an animal. Hormones are, therefore, a proximate cause of behavior. To control reproductive behavior, they must be released when the conditions are most favorable for the growth of young. Other behaviors, such as territoriality and dominance behavior, also have hormonal correlates.

Hormones may interact with neurotransmitters to alter behavior. Estrogen affects serotonin in female mice, and may be in part responsible for the "mood swings" experienced by some human females during the menstrual cycle.

Hormones have important influences on reproductive and social behavior.

Behavioral Rhythms

Many animals exhibit behaviors that vary at regular intervals of time. Geese migrate south in the fall, birds sing in the early morning, bats fly at night rather than during the day, and most humans sleep at night and are active in the daytime. Some behaviors are timed to occur in concert with lunar or tidal cycles (figure 59.15). Why do regular repeating patterns of behavior occur, and what determines when they occur? The study of questions like these has revealed that rhythmic animal behaviors are based on both **endogenous** (internal) **rhythms** and **exogenous** (external) **timers.**

Most studies of behavioral rhythms have focused on behaviors that appear to be keyed to a daily cycle, such as sleeping. Rhythms with a period of about 24 hours are called **circadian** ("about a day") **rhythms.** Many of these behaviors have a strong endogenous component, as if they were driven by a **biological clock.** Such behaviors are said to be *free-running*, continuing on a regular cycle even in the absence of any cues from the environment. Almost all fruit fly pupae hatch in the early morning, for example, even if they are kept in total darkness throughout their week-long development. They keep track of time with an internal clock whose pattern is determined by a single gene, called the *per* (for *period*) gene. Different mutations of the *per* gene shorten or lengthen the daily rhythm. The *per* gene produces a protein in a regular 24-hour cycle in the brain, serving as the fly's pacemaker of activity. The protein probably affects the expression of other genes that ultimately regulate activity. As the per protein accumulates, it seems to turn off the gene. In mice, the *clock* gene is responsible for regulating the animal's daily rhythm.

Most biological clocks do not exactly match the rhythms of the environment. Therefore, the behavioral rhythm of an individual deprived of external cues gradually drifts out of phase with the environment. Exposure to an environmental cue resets the biological clock and keeps the behavior properly synchronized with the environment. Light is the most common cue for resetting circadian rhythms.

The most obvious circadian rhythm in humans is the sleep-activity cycle. In controlled experiments, humans have lived for months in underground apartments, where all light is artificial and there are no external cues whatsoever indicating day length. Left to set their own schedules, most of these people adopt daily activity patterns (one phase of activity plus one phase of sleep) of about 25 hours, although there is considerable variation. Some individuals exhibit 50-hour clocks, active for as long as 36 hours during each period! Under normal circumstances, the day-night cycle resets an individual's free-running clock every day to a cycle period of 24 hours.

What constitutes an animal's biological clock? In some insects, the clock is thought to be located in the optic lobes of the brain, and timekeeping appears to be based on hormones. In mammals, including humans, the biological

FIGURE 59.15
Tidal rhythm. Oysters open their shells for feeding when the tide is in and close them when the tide is out.

clock lies in a specific region of the hypothalamus called the **suprachiasmatic nucleus (SCN).** The SCN is a self-sustaining oscillator, which means it undergoes spontaneous, cyclical changes in activity. This oscillatory activity helps the SCN to act as a pacemaker for circadian rhythms, but in order for the rhythms to be entrained to external light-dark cycles, the SCN must be influenced by light. In fact, there are both direct and indirect neural projections from the retina to the SCN.

The SCN controls circadian rhythms by regulating the secretion of the hormone **melatonin** by the **pineal gland.** During the daytime, the SCN suppresses melatonin secretion. Consequently, more melatonin is secreted over a 24-hour period during short days than during long days. Variations in melatonin secretion thus serve as an indicator of seasonal changes in day length, and these variations participate in timing the seasonal reproductive behavior of many mammals. Disturbances in melatonin secretion may be partially responsible for the "jet-lag" people experience when air travel suddenly throws their internal clocks out of register with the day-night cycle.

Many important behavioral rhythms have cycle periods longer than 24 hours. For example, **circannual behaviors** such as breeding, hibernation, and migration occur on a yearly cycle. These behaviors seem to be largely timed by hormonal and other physiological changes keyed to exogenous factors such as day length. The degree to which endogenous biological clocks underlie circannual rhythms is not known, since it is very difficult to perform constant-environment experiments of several years' duration. The mechanism of the biological clock remains one of the most tantalizing puzzles in biology today.

Endogenous circadian rhythms have free-running cycle periods of approximately 24 hours; they are entrained to a more exact 24-hour cycle period by environmental cues.

Much of the research in animal behavior is devoted to analyzing the nature of communication signals, determining how they are perceived, and identifying the ecological roles they play and their evolutionary origins.

Courtship

During courtship, animals produce signals to communicate with potential mates and with other members of their own sex. A **stimulus-response chain** sometimes occurs, in which the behavior of one individual in turn releases a behavior by another individual (figure 59.16).

Courtship Signaling

A male stickleback fish will defend the nest it builds on the bottom of a pond or stream by attacking *conspecific* males (i.e., males of the same species) that approach the nest. Niko Tinbergen studied the social releasers responsible for this behavior by making simple clay models. He found that a model's shape and degree of resemblance to a fish were unimportant; any model with a red underside (like the underside of a male stickleback) could release the attack behavior. Tinbergen also used a series of clay models to demonstrate that a male stickleback recognizes a female by her abdomen, swollen with eggs.

Courtship signals are often *species-specific*, limiting communication to members of the same species and thus playing a key role in reproductive isolation. The flashes of fireflies (which are actually beetles) are such species-specific signals. Females recognize conspecific males by their flash pattern (figure 59.17), and males recognize conspecific females by their flash response. This series of reciprocal responses provides a continuous "check" on the species identity of potential mates.

Visual courtship displays sometimes have more than one component. The male *Anolis* lizard extends and retracts his fleshy and often colorful dewlap while perched on a branch in his territory (figure 59.18). The display thus involves both color and movement (the extension of the dewlap as well as a series of lizard "push-ups"). To which component of the display does the female respond? Experiments in which the dewlap color is altered with ink show that color is unimportant for some species; that is, a female can be courted successfully by a male with an atypically colored dewlap.

Pheromones

Chemical signals also mediate interactions between males and females. **Pheromones,** chemical messengers used for communication between individuals of the same species, serve as sex attractants among other functions in many animals. Even the human egg produces a chemical attractant to communicate with sperm! Female silk moths *(Bombyx*

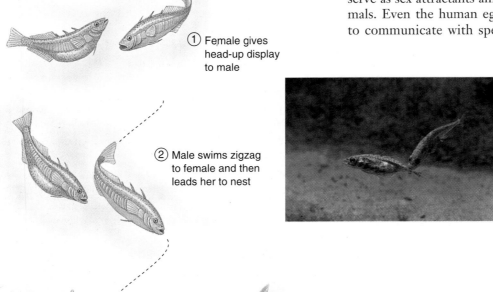

① Female gives head-up display to male

② Male swims zigzag to female and then leads her to nest

③ Male shows female entrance to nest

④ Female enters nest and spawns while male stimulates tail

⑤ Male enters nest and fertilizes eggs

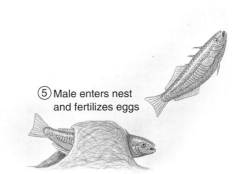

FIGURE 59.16
A stimulus-response chain.
Stickleback courtship involves a sequence of behaviors leading to the fertilization of eggs.

mori) produce a sex pheromone called *bombykol* in a gland associated with the reproductive system. Neurophysiological studies show that the male's antennae contain numerous sensory receptors specific for bombykol. These receptors are extraordinarily sensitive, enabling the male to respond behaviorally to concentrations of bombykol as low as one molecule in 10^{17} molecules of oxygen in the air!

Many insects, amphibians, and birds produce species-specific acoustic signals to attract mates. Bullfrog males call to females by inflating and discharging air from their vocal sacs, located beneath the lower jaw. The female can distinguish a conspecific male's call from the call of other frogs that may be in the same habitat and mating at the same time. Male birds produce songs, complex sounds composed of notes and phrases, to advertise their presence and to attract females. In many bird species, variations in the males' songs identify *particular* males in a population. In these species, the song is individually specific as well as species-specific.

Level of Specificity

Why should different signals have different levels of specificity? The **level of specificity** relates to the function of the signal. Many courtship signals are species-specific to help animals avoid making errors in mating that would produce inviable hybrids or otherwise waste reproductive effort. A male bird's song is individually specific because it allows his presence (as opposed to simply the presence of an unidentifiable member of the species) to be recognized by neighboring birds. When territories are being established, males may sing and aggressively confront neighboring conspecifics to defend their space. Aggression carries the risk of injury, and it is energetically costly to sing. After territorial borders have been established, intrusions by neighbors are few because the outcome of the contests have already been determined. Each male then "knows" his neighbor by the song he sings, and also "knows" that male does not constitute a threat because they have already settled their territorial contests. So, all birds in the population can lower their energy costs by identifying their neighbors through their individualistic songs. In a similar way, mammals mark their territories with pheromones that signal individual identity, which may be encoded as a blend of a number of chemicals. Other signals, such as the mobbing and alarm calls of birds, are anonymous, conveying no information about the identity of the sender. These signals may permit communication about the presence of a predator common to several bird species.

Courtship behaviors are keyed to species-specific visual, chemical, and acoustic signals.

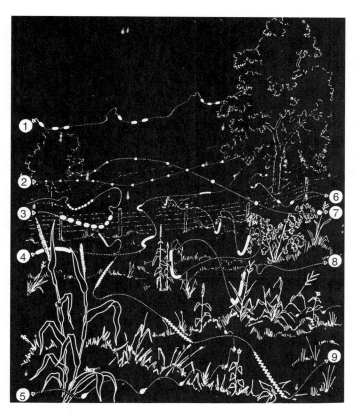

FIGURE 59.17
Firefly fireworks. The bioluminescent displays of these lampyrid beetles are species-specific and serve as behavioral mechanisms of reproductive isolation. Each number represents the flash pattern of a male of a different species.

FIGURE 59.18
Dewlap display of a male *Anolis* lizard. Under hormonal stimulation, males extend their fleshy, colored dewlaps to court females. This behavior also stimulates hormone release and egg-laying in the female.

Communication in Social Groups

Many insects, fish, birds, and mammals live in social groups in which information is communicated between group members. For example, some individuals in mammalian societies serve as "guards." When a predator appears, the guards give an **alarm call,** and group members respond by seeking shelter. Social insects, such as ants and honeybees, produce **alarm pheromones** that trigger attack behavior. Ants also deposit **trail pheromones** between the nest and a food source to induce cooperation during foraging (figure 59.19). Honeybees have an extremely complex **dance language** that directs nestmates to rich nectar sources.

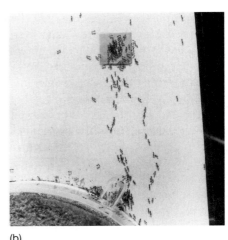
(a)

(b)

FIGURE 59.19

The chemical control of fire ant foraging. Trial pheromones, produced in an accessory gland near the fire ant's sting, organize cooperative foraging. The trails taken by the first ants to travel to a food source (a) are soon followed by most of the other ants (b).

The Dance Language of the Honeybee

The European honeybee, *Apis mellifera,* lives in hives consisting of 30,000 to 40,000 individuals whose behaviors are integrated into a complex colony. Worker bees may forage for miles from the hive, collecting nectar and pollen from a variety of plants and switching between plant species and populations on the basis of how energetically rewarding their food is. The food sources used by bees tend to occur in patches, and each patch offers much more food than a single bee can transport to the hive. A colony is able to exploit the resources of a patch because of the behavior of scout bees, which locate patches and communicate their location to hivemates through a *dance language.* Over many years, Nobel laureate Karl von Frisch was able to unravel the details of this communication system.

After a successful scout bee returns to the hive, she performs a remarkable behavior pattern called a *waggle dance* on a vertical comb (figure 59.20). The path of the bee during the dance resembles a figure-eight. On the straight part of the path, the bee vibrates or waggles her abdomen while producing bursts of sound. She may stop periodically to give her hivemates a sample of the nectar she has carried back to the hive in her crop. As she dances, she is followed closely by other bees, which soon appear as foragers at the new food source.

(a)
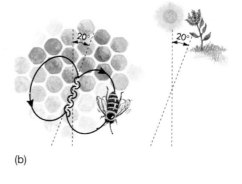
(b)

FIGURE 59.20

The waggle dance of honeybees. (a) A scout bee dances on a comb in the hive. (b) The angle between the food source and the nest is represented by a dancing bee as the angle between the straight part of the dance and vertical. The food is 45° to the left of the sun in (a), 20° to the right of the sun in (b).

Von Frisch and his colleagues claimed that the other bees use information in the waggle dance to locate the food source. According to their explanation, the scout bee indicates the *direction* of the food source by representing the angle between the food source and the hive in reference to the sun as the angle between the straight part of the dance and vertical in the hive. The *distance* to the food source is indicated by the tempo, or degree of vigor, of the dance.

Adrian Wenner, a scientist at the University of California, did not believe that the dance language communicated anything about the location of food, and he challenged von Frisch's explanation. Wenner maintained that flower odor was the most important cue allowing recruited bees to

(a)

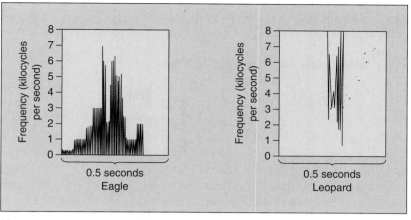

(b)

FIGURE 59.21
Primate semantics. (a) Predators, like this leopard, attack and feed on vervet monkeys. (b) The monkeys give different alarm calls when eagles, leopards, and snakes are sighted by troupe members. Each distinctive call elicits a different and adaptive escape behavior.

arrive at a new food source. A heated controversy ensued as the two groups of researchers published articles supporting their positions.

The "dance language controversy" was resolved (in the minds of most scientists) in the mid-1970s by the creative research of James L. Gould. Gould devised an experiment in which a scout bee gave incorrect information. Gould caused the scout bee to direct her hivemates to a new location—not where she had been, but rather where automated traps would capture the "fooled" bees. This confirmed von Frisch's ideas.

Recently, researchers have extended the study of the honeybee dance language by building robot bees whose dances can be completely controlled. Their dances are programmed by a computer and perfectly reproduce the natural honeybee dance; the robots even stop to give food samples! The use of robot bees has allowed scientists to determine precisely which cues direct hivemates to food sources.

Primate Language

Some primates have a "vocabulary" that allows individuals to communicate the identity of specific predators. The vocalizations of African vervet monkeys, for example, distinguish eagles, leopards, and snakes (figure 59.21). Chimpanzees and gorillas can learn to recognize a large number of symbols and use them to communicate abstract concepts. The complexity of human language would at first appear to defy biological explanation, but closer examination suggests that the differences are in fact superficial—all languages share many basic structural similarities. All of the roughly 3000 languages draw from the same set of 40 consonant sounds (English uses two dozen of them), and any human can learn them. Researchers believe these similari-

ties reflect the way our brains handle abstract information, a genetically determined characteristic of all humans.

Language develops at an early age in humans. Human infants are capable of recognizing the 40 consonant sounds characteristic of speech, including those not present in the particular language they will learn, while they ignore other sounds. In contrast, individuals who have not heard certain consonant sounds as infants can only rarely distinguish or produce them as adults. That is why English speakers have difficulty mastering the throaty French "r," French speakers typically replace the English "th" with "z," and native Japanese often substitute "l" for the unfamiliar English "r." Children go through a "babbling" phase, in which they learn by trial and error how to make the sounds of language. Even deaf children go through a babbling phase using sign language. Next, children quickly and easily learn a vocabulary of thousands of words. Like babbling, this phase of rapid learning seems to be genetically programmed. It is followed by a stage in which children form simple sentences which, though they may be grammatically incorrect, can convey information. Learning the rules of grammar constitutes the final step in language acquisition.

While language is the primary channel of human communication, odor and other nonverbal signals (such as "body language") may also convey information. However, it is difficult to determine the relative importance of these other communication channels in humans.

The study of animal communication involves analysis of the specificity of signals, their information content, and the methods used to produce and receive them. Communication plays an important ecological role in maintaining genetic isolation between species.

Orientation and Migration

Animals may travel to and from a nest to feed or move regularly from one place to another. To do so, they must orient themselves by tracking stimuli in the environment.

Movement toward or away from some stimulus is called **taxis.** The attraction of flying insects to outdoor lights is an example of *positive phototaxis.* Insects that avoid light, such as the common cockroach, exhibit *negative phototaxis.* Other stimuli may be used as orienting cues. For example, trout orient themselves in a stream so as to face against the current. However, not all responses involve a specific orientation; in some cases, individuals simply become more active under certain conditions. If an animal moves randomly but is active under unfavorable conditions and is inactive under favorable ones, it will tend to move into favorable areas and remain there. Such changes in activity that are dependent on stimulus intensity are called **kineses.**

Long-range, two-way movements are known as **migrations.** In many animals, migrations occur circannually. Ducks and geese migrate along flyways from Canada across the United States each fall and return each spring. Monarch butterflies migrate each fall from central and eastern North America to several small, geographically isolated areas of coniferous forest in the mountains of central Mexico (figure 59.22). Each August, the butterflies begin a flight southward to their overwintering sites. At the end of winter, the monarchs begin the return flight to their summer breeding ranges. What is amazing about the migration of the monarch, however, is that two to five generations may be produced as the butterflies fly north. The butterflies that migrate in the autumn to the precisely located overwintering grounds in Mexico have never been there before.

When colonies of bobolinks became established in the western United States, far from their normal range in the Midwest and East, they did not migrate directly to their winter range in South America. Instead, they migrated east to their ancestral range and then south along the original flyway (figure 59.23). Rather than changing the original migration pattern, they simply added a new pattern.

How Migrating Animals Navigate

Biologists have studied migration with great interest, and we now have a good understanding of how these feats of navigation are achieved. It is important to understand the distinction between **orientation** (the ability to follow a bearing) and **navigation** (the ability to set or adjust a bearing, and then follow it). The former is analogous to using a compass, while the latter is like using a compass in conjunction with a map. Experiments on starlings indicate that inexperienced birds migrate by orientation, while older birds that have migrated previously use true navigation (figure 59.24).

Birds and other animals navigate by looking at the sun and the stars. The indigo bunting, which flies during the day and uses the sun as a guide, compensates for the movement of the sun in the sky as the day progresses by reference to the north star, which does not move in the sky. Buntings also use the positions of the constellations and the position

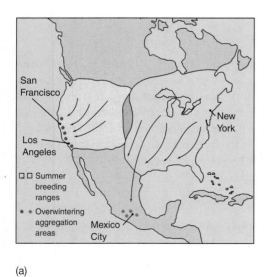

(a)

(b)

(c)

FIGURE 59.22

Migration of monarch butterflies. (a) Monarchs from western North America overwinter in areas of mild climate along the Pacific Coast. Those from the eastern United States and southeastern Canada migrate to Mexico, a journey of over 3000 kilometers that takes from two to five generations to complete. (b) Monarch butterflies arriving at the remote fir forests of the overwintering grounds and (c) forming aggregations on the tree trunks.

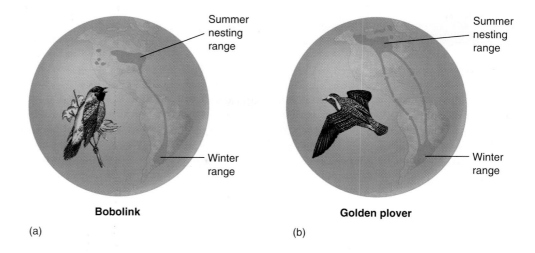

FIGURE 59.23

Birds on the move. (a) The summer range of bobolinks recently extended to the far West from their more established range in the Midwest. When they migrate to South America in the winter, bobolinks that nested in the West do not fly directly to the winter range; instead, they fly to the Midwest first and then use the ancestral flyway. (b) The golden plover has an even longer migration route that is circular. These birds fly from Arctic breeding grounds to wintering areas in southeastern South America, a distance of some 13,000 kilometers.

of the pole star in the night sky, cues they learn as young birds. Starlings and certain other birds compensate for the sun's apparent movement in the sky by using an internal clock. If such birds are shown an experimental sun in a fixed position while in captivity, they will change their orientation to it at a constant rate of about 15° per hour.

Many migrating birds also have the ability to detect the earth's magnetic field and to orient themselves with respect to it. In a closed indoor cage, they will attempt to move in the correct geographical direction, even though there are no visible external cues. However, the placement of a powerful magnet near the cage can alter the direction in which the birds attempt to move. Magnetite, a magnetized iron ore, has been found in the heads of some birds, but the sensory receptors birds employ to detect magnetic fields have not been identified.

It appears that the first migration of a bird is innately guided by both celestial cues (the birds fly mainly at night) and the earth's magnetic field. These cues give the same information about the general direction of the migration, but the information about direction provided by the stars seems to dominate over the magnetic information when the two cues are experimentally manipulated to give conflicting directions. Recent studies, however, indicate that celestial cues tell northern hemisphere birds to move south when they begin their migration, while magnetic cues give them the direction for the specific migratory path (perhaps a southeast turn the bird must make midroute). In short, these new data suggest that celestial and magnetic cues interact during development to fine-tune the bird's navigation.

We know relatively little about how other migrating animals navigate. For instance, green sea turtles migrate from Brazil halfway across the Atlantic Ocean to Ascension Island, where the females lay their eggs. How do they find this tiny island in the middle of the ocean, which they haven't seen for perhaps 30 years? How do the young that hatch on the island know how to find their way to Brazil? Recent studies suggest that wave action is an important cue.

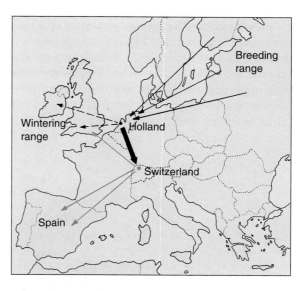

FIGURE 59.24

Migratory behavior of starlings. The navigational abilities of inexperienced birds differ from those of adults who have made the migratory journey before. Starlings were captured in Holland, halfway along their full migratory route from Baltic breeding grounds to wintering grounds in the British Isles; these birds were transported to Switzerland and released. Experienced older birds compensated for the displacement and flew toward the normal wintering grounds (*blue arrow*). Inexperienced young birds kept flying in the same direction, on a course that took them toward Spain (*red arrow*). These observations imply that inexperienced birds fly by orientation, while experienced birds learn true navigation.

Many animals migrate in predictable ways, navigating by looking at the sun and stars, and in some cases by detecting magnetic fields.

Animal Cognition

It is likely each of us could tell an anecdotal story about the behavior of a pet cat or dog that would seem to suggest that the animal had a degree of reasoning ability or was capable of thinking. For many decades, however, students of animal behavior flatly rejected the notion that nonhuman animals can think. In fact, behaviorist Lloyd Morgan stated that one should never assume a behavior represents conscious thought if there is any other explanation that precludes the assumption of consciousness. The prevailing approach was to treat animals as though they responded to the environment through reflex-like behaviors.

In recent years, serious attention has been given to the topic of animal awareness. The central question is whether animals show **cognitive behavior**—that is, do they process information and respond in a manner that suggests thinking (figure 59.25)? What kinds of behavior would demonstrate cognition? Some birds in urban areas remove the foil caps from nonhomogenized milk bottles to get at the cream beneath, and this behavior is known to have spread within a population to other birds. Japanese macaques learned to wash potatoes and float grain to separate it from sand. A chimpanzee pulls the leaves off of a tree branch and uses the stick to probe the entrance to a termite nest and gather termites. As we saw earlier, vervet monkeys have a vocabulary that identifies specific predators.

Only a few experiments have tested the thinking ability of nonhuman animals. Some of these studies suggest that animals may give false information (i.e., they "lie"). Currently, researchers are trying to determine if some primates deceive others to manipulate the behavior of the other members of their troop. There are many anecdotal accounts that appear to support the idea that deception occurs in some nonhuman primate species such as baboons and chimpanzees, but it has been difficult to devise field-based experiments to test this idea. Much of this type of research on animal cognition is in its infancy, but it is sure to grow and to raise controversy. In any case, there is nothing to be gained by a dogmatic denial of the *possibility* of animal consciousness.

Research into animal cognition is in its infancy.

(a)

(b)

FIGURE 59.25
Animal thinking? (a) This chimpanzee is stripping the leaves from a twig, which it will then use to probe a termite nest. This behavior strongly suggests that the chimpanzee is consciously planning ahead, with full knowledge of what it intends to do. (b) This sea otter is using a rock as an "anvil," against which it bashes a clam to break it open. A sea otter will often keep a favorite rock for a long time, as though it has a clear idea of what it is going to use the rock for. Behaviors such as these suggest that animals have cognitive abilities.

Summary of Concepts

59.1 Ethology focuses on how neural units influence behavior.

- Behavior is an adaptive response to stimuli in the environment. An animal's sensory systems detect and process information about these stimuli.

59.2 Comparative physiology focuses on how learning influences behavior.

- Behavior is both instinctive (influenced by genes) and learned through experience. Genes are thought to limit the extent to which behavior can be modified and the types of associations that can be made.

- The simplest forms of learning involve habituation and sensitization. More complex associative learning, such as classical and operant conditioning, may be due to the strengthening or weakening of existing synapses as well as the formation of entirely new synapses.

- An animal's internal state influences when and how a behavior will occur. Hormones can change an animal's behavior and perception of stimuli in a way that facilitates reproduction.

59.3 Communication is a key element of many animal behaviors.

- Animals communicate by producing visual, acoustic, and chemical signals. These signals are involved in mating, finding food, defense against predators, and other social situations.

59.4 Migratory behavior presents many puzzles.

- Animals use cues such as the position of the sun and stars to orient during daily activities and to navigate during long-range migrations.

59.5 To what degree animals "think" is a subject of lively dispute.

- Many anecdotal accounts point to animal cognition, but research is in its infancy.

Discussing Key Terms

1. **Genetic basis of behavior** Many behaviors are instinctive, strongly influenced or even programmed by an individual's genes. Genes may also place limitations on the ability of an animal to learn behaviors that are different from those that are related to its own natural history.

2. **Learning** Simple nonassociative learning involves habituation and sensitization. More complex associative learning occurs in classical and operant conditioning.

3. **Hormones and behavior** Animal behavior is often affected by variations in hormone secretion, which may be tied to daily and seasonal changes in environmental stimuli, as well as changes in an animal's social environment.

4. **Animal communication** Animals communicate through the use of visual, acoustic, chemical, tactile, and electrical signals.

5. **Animal navigation** Animals use the position of the sun and stars to navigate over long distances during their migrations.

Review Questions

1. What is the difference between the proximate cause and the ultimate cause of a behavior?

2. In the egg retrieval behavior of geese, what is the sign stimulus? What is the innate releasing mechanism in this behavior? What is the fixed action pattern?

3. How does a hybrid lovebird's method of carrying nest materials compare with that of its parents? What does this comparison suggest about whether the behavior is instinctive or learned?

4. How does associative learning differ from nonassociative learning? How does classical conditioning differ from operant conditioning?

5. What is filial imprinting? What is sexual imprinting? Why do some young animals imprint on objects like a moving box?

6. How does Marler's work on song development in white-crowned sparrows indicate that behavior is shaped by learning? How does it indicate that behavior is shaped by instinct?

7. In what part of the mammalian body is the "biological clock" located? What property does it have that makes it suitable as a pacemaker for circadian rhythms? What other structure does it regulate to control circadian rhythms, and what substance does that structure secrete?

8. Define the following terms: social releaser; stimulus-response chain; pheromone.

9. How do communication signals participate in reproductive isolation? Give one example each of a visual, acoustic, and chemical signal that is species-specific. Why are some signals individually specific?

10. What is the definition of taxis? What are kineses? What cues do migrating birds use to orient and navigate during their migrations?

Thought Questions

1. If a young animal exhibits a behavior immediately after being born or hatched, is it reasonable to conclude that the behavior is instinctive rather than learned? If not, what sorts of experiments might be performed to distinguish between these two possibilities?

2. A species of bird is found on an island, where it is the only bird species present. Do you think that song learning would occur in the same manner in this species as it does in the white-crowned sparrow? Why or why not?

Internet Links

Dancing Spiders
http://spiders.arizona.edu/salticidae/++salticidae/movies/spiderdances.html
This site from THE TREE OF LIFE presents quicktime movies of the mating dances and behaviors of many kinds of spiders. Amazing.

How a Songbird Learns to Sing
http://cogsci.ucsd.edu/~cfry/bird/modeltour.html
A guided tour of a behavioral model of how a song sparrow learns to sing its species-specific song.

The Song Post
http://www.princeton.edu/~shackles/song.post.html
A comprehensive site from Princeton University devoted to the academic study of bird song.

Neurobiology of Birdsong
http://etho.caltech.edu/~schmidt/
An interesting look at research activity into the neural basis of birdsong, with sample projects.

For Further Reading

Alcock, J.: *Animal Behavior: An Evolutionary Approach*, ed. 5, Sinauer Associates, Sunderland, MA, 1993. Now in its fifth edition, Alcock's book is a standard text in the field.

Cheney, D. L. and Seyfarth, R. M.: "How Monkeys See the World: Inside the Mind of Another Species," University of Chicago Press, 1990. A good reference on animal awareness and the complexity of animal communication.

Crews, D.: "The Annotated Anole: Studies on the Control of Lizard Reproduction," *American Scientist*, vol. 65, pages 428–34, 1977. A fascinating account of how seasonal and social changes trigger the hormonal events that lead to reproduction.

Gould, J. and P. Marler: "Learning by Instinct," *Scientific American*, January 1987, pages 74–85. A clear and interesting account of the relative roles of instinct and learning in behavior. The authors, prominent behaviorists, argue that learning is often limited or controlled by instinct.

Greenspan, R. J.: "Understanding the Genetic Control of Behavior", *Scientific American*, 1995, vol. 272, pages 72–105. Thorough and interesting review of the genetic dissection of courtship behavior in the fruit fly.

Huber, F. and J. Thorson: "Cricket Auditory Communication," *Scientific American*, December 1985, pages 60–68. An unusually clear example of how nervous system activity underlies animal behavior.

Packer, C. and A. E. Pusey: "Divided We Fall: Cooperation among Lions," *Scientific American*, May 1997, pages 52–59. Lions share a complex social organization, all of which can be understood in terms of promoting individual survival.

Rose, K.: *The Body in Time*, New York, Wiley & Sons, 1988. A fascinating journey through the human clock—how our body keeps time.

60

Behavioral Ecology

Concept Outline

FIGURE 60.1
A snake in the throes of death—or is it? When threatened, many organisms feign death, as this snake is doing—foaming at the mouth and going limp or looking paralyzed.

Animal behavior can be investigated in a variety of ways. An investigator can ask, how did the behavior develop? What is the physiology behind the behavior? Or what is the function of the behavior (figure 60.1), and does it confer an advantage to the animal? The field of behavioral ecology deals with the last two questions. Specifically, *behavioral ecologists* study the ways in which behavior serves as adaptation and allows an animal to increase or even maximize its reproductive success. This chapter examines both of these aspects of behavioral ecology.

Behavioral Ecology

In an important essay, Nobel laureate Niko Tinbergen outlined the different types of questions biologists can ask about animal behavior. In essence, he divided the investigation of behavior into the study of its development, physiological basis, and function (evolutionary significance). One type of evolutionary analysis pioneered by Tinbergen himself was the study of the **survival value** of behavior. That is, how does an animal's behavior allow it to stay alive or keep its offspring alive? For example, Tinbergen observed that after gull nestlings hatch, the parents remove the eggshells from the nest. To understand *why* this behavior occurs, he camouflaged chicken eggs by painting them to resemble the natural background where they would lie and distributed them throughout the area in which the gulls were nesting (figure 60.2). He placed broken eggshells next to some of the eggs, and as a control, he left other camouflaged eggs alone without eggshells. He then noted which eggs were found more easily by crows. Because the crows could use the white interior of a broken eggshell as a cue, they ate more of the camouflaged eggs that were near eggshells. Thus, Tinbergen concluded that eggshell removal behavior is *adaptive:* it reduces predation and thus increases the offspring's chances of survival.

Tinbergen is credited with being one of the founders of the field of **behavioral ecology,** the study of how natural selection shapes behavior. This branch of ecology examines the **adaptive significance** of behavior, or how behavior may increase survival and reproduction. Current research in behavioral ecology focuses on the contribution behavior makes to an animal's reproductive success, or **fitness.** To study the relation between behavior and fitness is to study the process of adaptation itself.

Natural selection acts on behavioral differences that have underlying genetic components. Behaviors that favor higher reproductive success become more prevalent in a population over evolutionary time (figure 60.3). Behavioral ecologists focus on behavioral differences among individuals that lead to reproductive advantage. Therefore, testing hypotheses on the evolution and adaptiveness of behavior requires measuring fitness and demonstrating its correlation with behavior. In place of measuring reproductive success directly, behavioral ecologists may measure other factors associated with reproduction, such as the rate at which an animal acquires energy.

Behavioral ecology is the study of how natural selection shapes behavior.

FIGURE 60.2
The adaptive value of egg coloration. Niko Tinbergen painted chicken eggs to resemble the mottled brown camouflage of gull eggs. The eggs were used to test the hypothesis that camouflaged eggs are more difficult for predators to find and thus increase the young's chances of survival.

FIGURE 60.3
Foraging and predator avoidance. A meerkat sentinel on duty. Meerkats, *Suricata suricata,* are a species of highly social mongoose living in the semiarid sands of the Kalahari Desert. This meerkat is taking its turn to act as a lookout for predators. Under the security of its vigilance, the other group members can focus their attention on foraging. David Macdonald's long-term study in the Kalahari Desert has revealed great cooperation among group members.

Foraging Behavior

The best introduction to behavioral ecology is the examination of one well-defined behavior in detail. While many behaviors might be chosen, we will focus on foraging behavior. Animals can be divided into two broad groups based on their foraging behavior—that is, on the range of food items they consume. **Specialists** feed primarily or exclusively on one kind of food. Some species of ants, for example, eat only spider eggs. Oystercatchers are shorebirds that feed mainly on bivalve mollusks; they use their long bills to open the shells of their prey (figure 60.4). **Generalists,** in contrast, consume many different kinds of food. For instance, some insects eat the leaves of a wide variety of plants. While generalists do not appear to be as efficient as specialists at feeding on any one type of food, they have the advantage of collecting more than one kind.

For many predators, foods come in a variety of sizes. Larger foods may contain more energy but are harder to capture and may be less abundant. Hence, foraging for these predators involves a trade-off between a food's energy content and its availability. The *net energy* (in calories or Joules) gained by feeding on each size prey is simply the energy content of the prey minus the energy costs of pursuing and handling it. According to **optimal foraging theory,** natural selection makes foraging behavior as energetically efficient as possible. In other words, animals tend to feed on prey that maximize their net energy intake per unit of foraging time (figure 60.5).

Behavioral ecologists use optimal foraging theory to predict aspects of foraging behavior, such as where an animal will search for food and how long it will stay in one area before moving to another. However, research has shown that the foraging choices of animals are not based solely on maximizing net energy intake. Animals also consume foods that are lower in net energy but that supply important nutrients. Moreover, predation often forces an animal to compromise its energy intake in order to reduce its risk of being eaten itself.

Predators and their prey are in a type of selectional "arms race" in which predators evolve toward greater effectiveness at capturing prey and prey evolve toward greater effectiveness at deterring predation. The defense mechanisms of prey species include camouflage, mimicry, and active defense (see chapter 24). Prey defenses often exploit the behavior of the predator. Batesian mimics are protected by their convergent coloration because a predator, such as a blue jay, will learn to avoid attacking prey that resemble the mimic's model.

Predators can increase their ability to detect cryptic prey over time through learning. In one series of studies, a blue jay was trained through operant conditioning (see chapter 59) to find the image of a cryptically colored moth in a photographic slide flashed in front of the bird. If the bird saw the moth, which was on a background that camou-

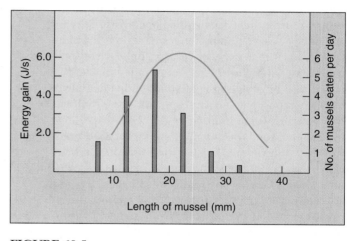

FIGURE 60.4 Foraging behavior in oystercatchers. An oystercatcher forages by stabbing the sand in search of buried mollusks.

**FIGURE 60.5
Optimal diet.** The shore crab selects a diet of energetically profitable prey. The curve describes the energy gain derived from feeding on different sizes of mussels. The bar graph shows the number of mussels of each size in the diet. Shore crabs most often feed on those mussels that provide the most energy.

flaged it, the bird would peck at a key and receive a food reward. If the image of only one species of moth was shown, the jay improved its performance, learning to find the cryptic prey. When two different species of cryptic moths were randomly shown in the slides, the bird was unable to increase its ability to find the prey. This suggests that a specialist predator may be able to learn to perceive cryptic prey, but a generalist predator cannot.

Natural selection has favored the evolution of foraging behaviors that maximize the amount of energy gained per unit time spent foraging. Animals that acquire energy efficiently during foraging will increase their fitness by having more energy available for reproduction.

Searching for a place to nest, finding a mate, and rearing young involve a collection of behaviors loosely referred to as reproductive behavior. These behaviors often involve seeking and defending a particular territory, making choices about mates and about the amount of energy to devote to the rearing of young. Mate selection, in particular, often involves intense natural selection. We will look briefly at each of these components of reproductive behavior.

Territorial Behavior

Animals often move over a large area, or **home range,** during their daily course of activity. In many animal species, the home ranges of several individuals may overlap in time or in space, but each individual defends a *portion* of its home range and uses it *exclusively*. This behavior, in which individual members of a species maintain exclusive use of an area that contains some limiting resource, such as foraging ground, food, or potential mates, is called **territoriality** (figure 60.6). The critical aspect of territorial behavior is *defense* against intrusion by other individuals. Territories are defended by displays that advertise that the territories are occupied and by overt aggression. A bird sings from its perch within a territory to prevent a takeover by a neighboring bird. If an intruder is not deterred by the song, it may be attacked. However, territorial defense has its costs. Singing is energetically expensive, and attacks can lead to injury. In addition, advertisement through song or visual display can reveal one's position to a predator.

Why does an animal bear the costs of territorial defense? Over the past two decades, it has become increasingly clear that an *economic* approach can be useful in answering this question. Although there are costs to defending a territory, there are also benefits, and both costs and benefits can be measured in units of energy. Studies of nectar-feeding birds like hummingbirds and sunbirds demonstrate this point (figure 60.7). A bird benefits from having the exclusive use of a patch of flowers because it can efficiently harvest the nectar they produce. In order to maintain exclusive use, however, the bird must actively defend the flowers. The benefits of exclusive use outweigh the costs of defense only under certain conditions. Sunbirds, for example, expend 3000 calories per hour chasing intruders from a territory. Whether or not the benefit of defending a territory will exceed this cost depends upon the amount of nectar in the flowers and how efficiently the bird can collect it. If flowers are very scarce or nectar levels are very low, for example, a nectar-feeding bird may not gain enough energy to balance the energy used in defense. Under this circumstance, it is not advantageous to be territorial. Similarly, if flowers are

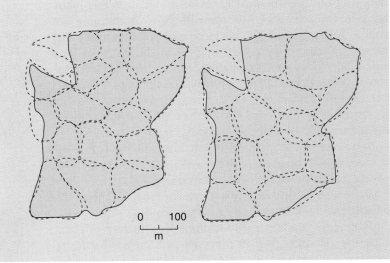

FIGURE 60.6
Competition for space. Territory size in birds is adjusted according to the number of competitors. When six pairs of great tits *(Parus major)* were removed from their territories (the shaded areas in the left figure), their territories were taken over by other birds in the area and by four new pairs (right figure).

FIGURE 60.7
The benefit of territoriality. Sunbirds increase nectar availability by defending flowers.

very abundant, a bird can efficiently meet its daily energy requirements without behaving territorially and adding the costs of defense. From an energy standpoint, defending abundant resources isn't worth the cost. Only for intermediate levels of flower availability and higher levels of nectar production, where the benefits of defense outweigh the costs, will territorial behavior be favored.

An economic approach can be used to explain the evolution and ecology of reproductive behaviors such as territoriality. This approach assumes that animals that gain more energy from a behavior than they expend will have an advantage in survival and reproduction over animals that behave in less efficient ways.

The Ecology of Reproduction

During the breeding season, animals make several important "decisions" concerning their choice of mates, how many mates to have, and how much time and energy to devote to rearing offspring. These decisions are all aspects of an animal's **reproductive strategy,** a set of behaviors that have evolved to maximize reproductive success, as discussed in Chapter 23. Reproductive strategies have evolved partly in response to the energy costs of reproduction and the way food resources, nest sites, and members of the opposite sex are spatially distributed in the environment.

Males and females usually differ in their reproductive strategies. Darwin was the first to observe that females do not simply mate with the first male they encounter, but instead seem to evaluate a male's quality and then decide whether to mate (figure 60.8). This behavior, called **mate choice,** has since been described in many invertebrate and vertebrate species. Natural selection favors mate choice because individuals that select superior quality mates will produce more viable offspring. Mate quality can be measured in terms of the "good genes" that will be passed to offspring and in terms of material gains such as food resources or nesting sites.

Parental Investment and Mate Choice

Which sex should show mate choice? Robert Trivers proposed that patterns of mate choice could be understood by comparing the parental investment made by males and females. **Parental investment** refers to the contributions each sex makes in rearing offspring; it is, in effect, an estimate of the energy expended by males and females in parental care. Trivers predicted that the sex having the higher parental investment would be the sex that exercises mate choice.

Many studies have shown that parental investment is high in females. One reason is that eggs are much larger than sperm—195,000 times larger in humans! Eggs contain proteins and lipids in the yolk and other nutrients for the developing embryo, but sperm are little more than motile DNA. Furthermore, in some groups of animals, females are responsible for gestation and lactation, costly reproductive functions only they can carry out.

Males may show mate choice as well, if their parental investment is high. Male mormon crickets transfer a protein-containing spermatophore to females during mating. Almost 30% of a male's body weight is made up by the spermatophore, which provides nutrition for the female, and helps her develop her eggs. Females of this species compete with each other for access to males, and males choose females based on the female's body weight. Since heavier females have more eggs, males that choose larger females will leave more offspring (figure 60.9). In other cricket species, males make a significant parental investment by defending and feeding the young.

FIGURE 60.8
Feathered display. The Australian male superb lyrebird shows off before a female.

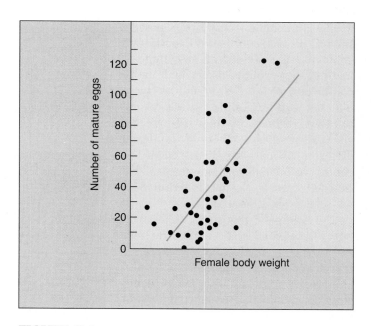

FIGURE 60.9
The advantage of male mate choice. Male mormon crickets choose heavier females as mates, and larger females have more eggs. Thus, male mate selection increases fitness.

Mate choice is strongly influenced by the degree of parental investment.

Reproductive Competition and Sexual Selection

Females that choose their mates often select the largest male, whose size advantage may enable it to compete successfully with other males. Male deer lock antlers while trying to gain access to females for mating. Older males with large body size and multipoint antlers most often win such disputes, enabling them to mate with more females and have a higher reproductive success. Likewise, large male elephant seals are more successful at defending the territories females use for breeding. As a result, a relatively small number of males in the breeding population mate, and they sire the majority of the next generation's offspring (figure 60.10). Similar examples of competition over access to mates occur in many species.

Males also compete in ways that do not involve aggression. For example, male birds of paradise display their elaborate feathers and accompany their mating ritual with complex vocalizations. The females stand by and observe. As Darwin asked, "Is it not possible that the female prefers the most gallant, beautiful, or melodious male?" In fact, recent research has shown that females do select males with the brightest or longest plumage or the most complex songs. But how did this ornamentation in the male evolve?

Darwin thought that long, bright feathers or complex vocalizations posed a survival problem for males. Even if such characteristics yielded an advantage in mating, they would make a male more conspicuous to a predator and thus place him at risk (figure 60.11). Since natural selection favors traits that increase the chances of survival, Darwin thought a different process was involved in the evolution of male ornamentation. He called this process **sexual selection.** Sexual selection occurs when individuals of one sex compete for mates. It involves both **intrasexual selection,** or interactions between members of one sex ("the power to conquer other males in battle," as Darwin put it), and **intersexual selection,** essentially mate choice ("the power to charm"). Sexual selection thus leads to the evolution of structures used in combat with other males, such as a deer's antlers and a ram's horns, as well as ornamentation used to advertise a male's quality as a potential mate and to "persuade" females to mate, such as long tail feathers and bright plumage (figure 60.12). These traits are called **secondary sexual characteristics.**

FIGURE 60.10
Intrasexual competition. Male elephant seals compete aggressively for breeding territories. A small proportion of the males in the population sire the majority of the offspring in the next generation.

(a)

(b)

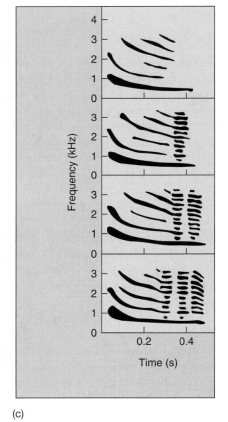

(c)

FIGURE 60.11
The benefits and costs of vocalizing. (a) The male tungara frog, *Physalaemus pustulosus.* (b) The males' calls attract females as well as predatory bats. Calls of greater complexity are represented from top to bottom in (c). Females prefer more complex calls, but so do bats. Males compromise by giving simpler calls, thus lowering their reproductive success but also decreasing their risk of predation.

While it is understandable how antlers would evolve if they aided in combat, it is not as evident why bright color, large flashy tails, and complex songs would evolve. The famous population geneticist R. A. Fisher proposed that once mate choice began, it could lead to the exaggeration of a trait if the trait was preferred by the opposite sex. According to this theory, if there was an initial preference by females for males with longer-than-average tails, tail length would continue to increase as long as females used tail length in mate selection. Such "runaway selection" would stop when natural selection acted as a "brake;" that is, at some point, long tails would reduce locomotion or increase predation. Other explanations suggest that secondary sexual characteristics are true indicators of a mate's quality; they advertise superior genes that can be passed on to the female's offspring. Brightly colored feathers, for example, may indicate that a male is resistant to parasites. Indeed, when males of some species are parasitized, their coloration becomes less vibrant and they are less able to acquire mates.

The longer the tail feathers, the greater the tendency for those feathers to be bilaterally symmetrical. Symmetry is thought to be an indicator of an individual's ability to resist stress during development and is thus a good indicator of mate quality.

Other courtship displays appear to have evolved from a female predisposition to be stimulated by a certain type of signal, such as a color, vocalization, or body ornament. Sensory exploitation involves a male's ability to evolve an attractive signal by "exploiting" a female's predisposition to respond to certain colors, sounds or shapes. Recall the example of the Tungara frog's *(Physalaemus pustulosus)* vocalizations described in figure 60.11. Males add "chuck" notes in complex calls to make their signals more attractive to females (see figure 60.11). In an ancestral species of *Physalaemus*, males do not use chuck notes at all, but females of this species still find the notes attractive! A female sensory system that found such notes stimulating fostered the evolution of the "attractive" notes a male adds to his call. Thus, neuroethology (chapter 59) and behavioral ecology can be bridged to help understand social evolution.

The Benefits of Mate Choice

What does a male or female gain by selecting a mate? Behavioral ecologists recognize two types of potential benefits. First, individuals may ensure that their offspring receive "good genes," which might promote the survival and fitness of the offspring. The progeny of fruit fly females that have been allowed to select their mates survive in larger numbers than the progeny of females whose mates were chosen for them at random. Female spadefoot toads preferentially select large males for mates; the sperm from a large male will produce tadpoles that mature rapidly, giving those tadpoles a greater chance to metamorphose before an ephemeral habitat such as a small pond dries up.

(a)   (b)

FIGURE 60.12
Products of sexual selection. Attracting mates with long feathers is common in bird species such as the African paradise whydah (a), which shows pronounced sexual dimorphism. Bowerbird males try to impress females by adorning their "nests" with rare flowers, or in the case of this satin bowerbird nest (b), pieces of blue plastic. Males of some species of bowerbirds steal long, colorful feathers of bird-of-paradise males to use as ornaments.

Both of these examples suggest that genes coding for traits that affect survival are passed from males to the offspring. A second way an individual may benefit by selecting a mate is to choose one that defends a territory containing high-quality resources necessary for reproduction. These resources may include nutritional materials like nectar-producing flowers or nesting sites that favor the survival of young.

There are a few excellent long-term studies of mating behavior in which the reproductive success of each male and female in a population has been followed for his or her lifetime. Research on red deer has shown that size, age, territory defense ability and the number of females in a male's territory are critical to male breeding success. Female breeding success is largely determined by the quality of grasses upon which she feeds (found in a male's territory), the time of the season when the female gives birth, and her longevity.

Natural selection has favored the evolution of behaviors that maximize the reproductive success of males and females. By evaluating and selecting mates with superior qualities, an animal can increase its reproductive success.

Mating Systems

The number of individuals with which an animal mates during the breeding season varies throughout the animal kingdom. Mating systems such as monogamy (one male mates with one female; figure 60.13), polygyny (one male mates with more than one female; figure 60.14), and polyandry (one female mates with more than one male) are aspects of male and female reproductive strategy that concern how many mates an individual has during the breeding season. Like mate choice, mating systems have evolved to maximize reproductive fitness. Much research has shown that mating systems are strongly influenced by ecology. For instance, a male may defend a territory that holds nest sites or food sources necessary for a female to reproduce, and the territory might have resources sufficient for more than one female. If males differ in the quality of the territories they hold, a female's fitness will be maximized if she mates with a male holding a high-quality territory. Such a male may already have a mate, but it is still more advantageous for the female to breed with that male than with an unmated male that defends a low-quality territory. In this way, natural selection would favor the evolution of polygyny.

Mating systems are also constrained by the needs of offspring. If the presence of both parents is necessary for young to be reared successfully, then monogamy may be favored. This is generally the case in birds, in which over 90% of all species are monogamous. The evolution of mating systems, therefore, depends not only on ecology but also on phylogeny, or a species' evolutionary history. For example, a male may either remain with his mate and provide care for the offspring or desert that mate to search for others; both strategies may increase his fitness. The strategy that evolution will favor depends upon the requirement for male assistance in feeding or defending the offspring. In some species, offspring are **altricial**—they require prolonged and extensive care. In these species, the need for care by two parents will reduce the tendency for the male to desert his mate and seek other matings. In species where the young are **precocial** (requiring little parental care), males may be more likely to be polygynous.

The timing of female reproduction also influences the ability of individuals to have access to more than one mate. If the females in a population become sexually receptive at different times during the breeding season, males will be unable to secure a large number of mates at any given time. This phylogenetic trait may thus dictate whether males can take advantage of ecological opportunities, such as defendable resources, that might favor polygyny.

> **Mating systems represent reproductive adaptations to ecological conditions. The need for parental care, the ability of both sexes to provide it, and the timing of female reproduction are important influences on the evolution of monogamy, polygyny, and polyandry.**

FIGURE 60.13
Pink flamingos. These birds are monogamous and show few differences in size or plumage.

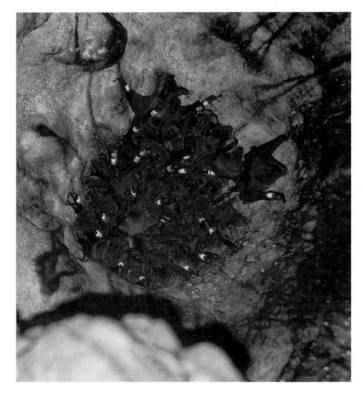

FIGURE 60.14
Female defense polygyny in bats. The male at the right is guarding a group of females.

The Evolution of Animal Societies

Society is a common word; we tend to think of it in terms of the groups in which humans live and the human cultural environment. But organisms as diverse as bacteria, cnidarians, insects, fish, birds, prairie dogs, lions, whales, and chimpanzees also exist in social groups. To encompass the wide variety of social phenomena, we can broadly define a society as *a group of organisms of the same species that are organized in a cooperative manner.*

In the early 1970s, E. O. Wilson of Harvard University, Richard Alexander of the University of Michigan, and Robert Trivers, now at the University of California at Santa Barbara, initiated what has become a major movement within biology. They attempted to understand animal social behavior as a biological process with a genetic basis that is shaped by evolution. An evolutionary view of social behavior predicts that the behavioral characteristics of animals are suited to their mode of living. In other words, behavior is adaptive in a Darwinian sense. This is the approach of the disciplines of **behavioral ecology** and **sociobiology**. It has provided many insights into the origin of behavior, including social behavior.

Why have individuals in some species given up a solitary existence to become members of a group? Recent research has focused on the advantages and disadvantages of group living, and much attention has been given to the evolution of cooperation, the trait that defines social life.

Group Living

Living as a member of a group is actually selfish behavior. A bird that joins a flock may receive greater protection from predators. As flock size increases, the risk of predation decreases because there are more individuals to scan the environment for predators (figure 60.15). A member of a flock may also increase its feeding success if it can acquire information from other flock members about the location of new, rich food sources. However, there are also costs associated with group living. For example, parasites and disease spread more easily within groups. At some point, the disadvantages of increased group size are balanced by the advantages. Cliff swallows offer an excellent illustration of this principle. Swallows in larger groups have increased feeding rates, but they also lose more young to blood-sucking, ectoparasitic insects called "swallow bugs."

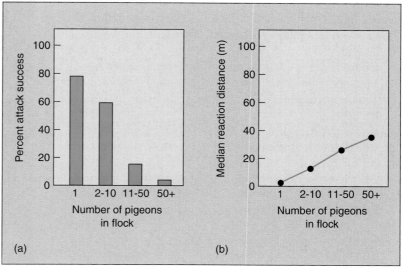

FIGURE 60.15
Flocking behavior decreases predation. (a) As the size of a pigeon flock increases, hawks are less successful at capturing pigeons. (b) When more pigeons are present in a flock, they can detect hawks at greater distances, thus allowing more time for the pigeons to escape.

The Evolution of Altruism

In many species, altruism is an important aspect of cooperation. Worker honeybees, for example, sting intruders in defense of the hive; as the stinging bees fly off, their barbed stingers remain in the skin of the intruder, causing the bee to be eviscerated and die. How did such altruism evolve? One of the great misconceptions about social behavior is that altruism has evolved because it benefits a group or even an entire species. This **group selection** argument has been used (incorrectly) to explain how animals regulate the size of their populations. For example, in some bird species, males display and compete with one another on the mating grounds to secure centrally located territories. Some males are able to hold territories, while others are not. Females choose males with territories as mates, so males unable to hold territories never mate. V. C. Wynne-Edwards, an animal ecologist, explained that nonterritorial males were sacrificing their own reproductive success to limit population growth. He suggested that those males did not reproduce because it was "good for the species." That is, a population that became too large might exhaust limited resources, causing the population or the entire species to become extinct.

There is a very important flaw in the group selection explanation for the evolution of altruism: If males that have the trait of altruism never reproduce and leave offspring, the trait cannot be passed from one generation to the next! We must look for other evolutionary explanations of altruistic behavior.

Reciprocity

Robert Trivers proposed that individuals may form "partnerships" in which mutual exchanges of altruistic acts occur, because it benefits both participants to do so. Trivers' model, called **reciprocal altruism,** requires that the individuals of an altruistic pair be unrelated, that is, that they share no genes in common. Studies of alliances among male baboons show that a male usually solicits the help of a particular unrelated male, and that "favorite partners" occur.

In the evolution of altruism through reciprocity, "cheaters" (nonreciprocators) are discriminated against and are cut off from receiving future aid. According to Trivers, if the altruistic act is relatively inexpensive, the small benefit a cheater receives by not reciprocating is far outweighed by the potential cost of not receiving future aid. Under these conditions, cheating should not occur. Vampire bats roost in hollow trees in groups of 8 to 12 individuals (figure 60.16). Because these bats have a high metabolic rate, individuals that have not fed recently may die. Bats that have found a host imbibe a great deal of blood; giving up a small amount presents no great energy cost to the donor, and it can keep a roostmate from starvation. Vampire bats tend to share blood with past reciprocators. If an individual fails to give blood to a bat from which it had received blood in the past, it will be excluded from future bloodsharing.

Kin Selection

The most influential explanation for the origin of altruism was presented by William D. Hamilton in 1964. It is perhaps best introduced by quoting a passing remark made in a pub in 1932 by the great population geneticist J. B. S. Haldane. Haldane said that he would willingly lay down his life for *two brothers* or *eight first cousins.* Evolutionarily speaking, Haldane's statement makes sense, because for each allele Haldane received from his parents, his brothers each had a 50% chance of receiving the same allele. Consequently, it is statistically expected that two of his brothers would pass on as many of Haldane's particular combination of alleles to the next generation as Haldane himself would. Similarly, Haldane and a first cousin would share an eighth of their alleles. Their sibling parents would each share half their alleles, and each of their children would receive half of these, of which half on the average would be in common: $\frac{1}{2} \times \frac{1}{2} \times \frac{1}{2} = \frac{1}{8}$. Eight first cousins would therefore pass on as many of those alleles to the next generation as Haldane himself would. Hamilton saw Haldane's point clearly: *evolution will favor any strategy that increases the net flow of an adaptive combination of alleles to the next generation.*

Hamilton showed that by directing aid toward kin, or close genetic relatives, an altruist may increase the reproductive success of its relatives enough to compensate for the reduction in its own fitness. Since the altruist's behavior increases the propagation of alleles in relatives, it will be fa-

FIGURE 60.16
A vampire bat roost.

vored by natural selection. Selection that favors altruism directed toward relatives is called **kin selection.** Although the behaviors being favored are cooperative, the genes are actually "behaving selfishly," because they encourage the organism to support copies of themselves in other individuals.

In explaining the evolution of altruism by kin selection, Hamilton developed the concept of **inclusive fitness,** which describes the effect an individual has on propagating its alleles through its own reproduction and through kin selection. It is important to note that inclusive fitness does not simply result from adding the number of alleles passed on directly by an individual and the number passed on by relatives other than offspring. Instead, inclusive fitness is the sum of the number of alleles passed on directly by an individual and those alleles passed on by kin (other than offspring) *whose reproduction benefits from that individual's altruism.* Simply stated, inclusive fitness has both *personal* and *kin-selected* components.

Hamilton's kin selection model predicts that altruism is likely to be directed toward close relatives. The more closely related two individuals are, the greater the potential genetic payoff in inclusive fitness. This relationship is described by Hamilton's rule: $b/c > 1/r$. In this expression, b and c are the benefits and costs of the altruistic act, respectively, and r is the coefficient of relatedness, the proportion of alleles shared by two individuals through common descent.

Altruism is best explained as having evolved to aid relatives, and so improve the chance of an animal's genes being passed on to the next generation.

Insect Societies

A hive of honeybees consists of a single queen, who is the sole egg-layer, and up to 50,000 of her offspring, nearly all of whom are female workers with nonfunctional ovaries (figure 60.17). The sterility of the workers is altruistic: during the course of evolution, these offspring gave up their personal reproduction to help their mother rear more of their sisters.

Sociality has chiefly evolved in two insect orders, the Hymenoptera (ants, bees, and wasps) and the Isoptera (termites), although a few other insect groups include social species. All ants, some bees, some wasps, and all termites are **eusocial** (truly social): they have a division of labor in reproduction (a fertile queen and sterile workers), cooperative care of brood and an overlap of generations so that the queen lives alongside her offspring. Social insect colonies are composed of highly integrated groups called **castes.**

Hamilton explained the origin of altruism in insect societies with his kin selection model. In bees, wasps, and ants, males are haploid and females are diploid. This unusual system of sex determination, called **haplodiploidy,** causes workers to share a very high proportion of alleles (theoretically as high as 75%). Because of this close genetic relatedness, *workers propagate more alleles by giving up their own reproduction to assist their mother in rearing their sisters, some of whom will be new queens and start new colonies and reproduce.* Thus, altruism allows workers to maximize their inclusive fitness.

In honeybees, the queen maintains her dominance in the hive by secreting a pheromone, called "queen substance," that suppresses development of the ovaries in other females, turning them into sterile workers. Drones (male bees) are produced only for purposes of mating. When the colony grows larger in the spring, some members do not receive a sufficient quantity of queen substance, and the colony begins preparations for swarming. Workers make several new queen chambers, in which new queens begin to develop. Scout workers look for a new nest site and communicate its location to the colony. The old queen and a swarm of female workers then move to the new site. Left behind, a new queen emerges, kills the other potential queens, flies out to mate, and returns to assume "rule" of the hive.

The leafcutter ants provide another fascinating example of the remarkable lifestyles of social insects. Leafcutters live in colonies of up to several million individuals, growing crops of fungi beneath the ground. Their mound-like nests are underground "cities" covering more than 100 square meters, with hundreds of entrances and chambers as deep as 5 meters beneath the ground. The division of labor among the worker ants is related to their size (figure 60.18). Every day, workers travel along trails from the nest to a tree or a bush, cut its leaves into small pieces, and carry the pieces back to the nest. Smaller workers chew the leaf fragments into a mulch, which they spread like a carpet in the underground fungus chambers. Even smaller workers implant fungal hyphae in the mulch. Soon a luxuriant

FIGURE 60.17
Reproductive division of labor in honeybees. The queen (shown here with a red spot painted on her thorax) is the sole egg-layer. Her daughters are sterile workers.

FIGURE 60.18
Leafcutter ants carrying leaf sections back to their nest. Behavioral differences among workers of the leafcutter ant are associated with worker size. These workers, which have a head width of roughly 2.4 mm, cut leaves. The largest workers defend the nest, while the smallest tend the fungal gardens.

garden of fungi is growing. While other workers weed out undesirable kinds of fungi, nurse ants carry the larvae of the nest to choice spots in the garden, where the larvae graze. This elaborate social system has evolved to produce reproductive queens that will disperse from the parent nest and start new colonies, repeating the cycle.

A few other invertebrates, including a species of shrimp that lives in a sponge, some thrips (an insect) and a species of weevil are also eusocial. These findings support the theory of cooperation proposed by kin selection.

Social insect workers share a large fraction of their genomes, so they propagate more of their own alleles by helping their mother reproduce than by reproducing themselves.

Vertebrate Societies

In contrast to the highly structured and integrated insect societies and their remarkable forms of altruism, vertebrate social groups are usually less rigidly organized and cohesive. It seems paradoxical that vertebrates, which have larger brains and are capable of more complex behaviors, are generally less altruistic than insects. Apparently, the lower degree of altruism in vertebrates is due to the lower frequency of shared alleles among group members (maximally 50% in nearly all vertebrates). Nevertheless, in some complex vertebrate social systems individuals may be exhibiting both reciprocity and kin-selected altruism. But vertebrate societies also display more conflict and aggression among group members than do insect societies. Conflict in vertebrate societies generally centers on access to food and mates.

Altruism in Vertebrates

Some species of birds, like the African white-fronted bee-eater and the Florida scrubjay, have evolved cooperative breeding systems. A pair of scrubjays may have other birds present in their territory that serve as helpers at the nest. The helpers, the offspring of a breeding pair, assist in feeding their siblings, watching for predators, and defending the territory. Nests with helpers have more offspring than those that do not (table 60.1). Although they are fully capable of breeding on their own, the helpers remain as nonreproductive altruists for a period of time. Because helpers are most often the fledged offspring of the pair they assist, the situation resembles a family.

White-fronted bee-eaters form their nests in the face of a cliff; as many as 25 families are a part of the colony. As in the case of the Florida scrubjay, helpers assist breeding pairs in raising young; the higher the genetic relatedness, the more likely a bird is to help. In African kingfishers, helpers may be related or unrelated to the breeding pair they assist and pursue different reproductive strategies. The evolution of these types of cooperative breeding in birds has been explained by using the inclusive fitness concept we discussed earlier.

From the previous examples, it seems that some aspects of vertebrate behavior are self-sacrificing and, therefore, present a puzzle to evolutionists. Particularly puzzling is the fact that vertebrates are often organized into social groups so that the activities of certain individuals benefit the group at the potential expense of those individuals themselves. The meerkat in figure 60.3, for instance, is exposed to predators and the full glare of the sun while it acts as a sentry atop a termite mound. It draws attention to itself and thus exposes itself to greater danger than if it were not a sentry. Behaviors such as this seem contrary to an individual's self-interest.

Table 60.1 Helpers	Average Number of Offspring Reared With:	
	No Helpers	**Helpers**
Inexperienced pairs	1.24	2.20
Experienced pairs	1.80	2.38

Florida scrubjay helpers increase the reproductive success of breeding pairs whether or not those pairs are experienced breeders. Usually one to two birds serve as helpers.

Meerkats and other social vertebrates may also give an **alarm call** to communicate the presence of a predator when one is sighted. An alarm call causes other members of the social group to seek shelter. Why would an individual place itself in jeopardy by calling and thereby revealing its own location? *Who benefits from the alarm call?*

Paul Sherman of Cornell University has answered this question through years of field observations of alarm calling in Belding's ground squirrel. These animals give alarm calls when they spot a predator such as a coyote or a badger. Such predators may attack a calling squirrel, so giving a signal places the caller at risk. The social unit of a ground squirrel colony consists of a female and her daughters, sisters, aunts, and nieces. Males in the colony are not genetically related to these females. By marking all squirrels in a colony with an individual dye pattern on their fur and by recording which individuals gave calls and the social circumstances of their calling, Sherman found that females who have relatives living nearby are more likely to give alarm calls than females with no kin nearby. Males tend to call much less frequently. Alarm calling therefore seems to represent **nepotism**, that is, it favors relatives. Again, the kin selection model best explains the evolution of alarm calling, even when many other competing hypotheses of the origin of this trait are evaluated.

Organization of Vertebrate Societies

Vertebrate societies, like insect societies, have particular types of organization. Each social group of vertebrates has a certain size, stability of members, number of breeding males and females, and type of mating system (figure 60.19). Behavioral ecologists have learned that the way a group is organized is influenced most often by ecological factors such as food type and predation.

African weaver birds provide an excellent example to illustrate the relationship between ecology and social organization. J. H. Crook, in 1964, showed that these finch-like birds construct nests from vegetation. Their roughly 90 species can be divided according to the type of social group they form. One set of species lives in the forest and builds camouflaged,

(a)

(b)

(c)

FIGURE 60.19

Social diversity among primates. (a) Prosimian bush babies are solitary. (b) Golden lion tamarins form family groups. (c) Gorillas exist in troops headed by a dominant silver-backed male.

solitary nests. Males and females are monogamous; they forage for insects to feed their young. The second group of species nests in colonies in trees on the savanna (figure 60.20). They are polygynous and feed in flocks on seeds. The feeding and nesting habits of these two sets of species are correlated with their mating systems. In the forest, insects are hard to find, and both parents must cooperate in feeding the young. The camouflaged nests do not call the attention of predators to their brood. On the open savanna, building a hidden nest is not an option. Rather, savanna-dwelling weaver birds protect their young from predators by nesting in spiny trees, which are not very abundant. This shortage of safe nest sites means that birds must nest together in colonies. Because seeds occur abundantly, a female can acquire all the food needed to rear young without a male's help. The male, free from the duties of parenting, spends his time competing with other males for the best nest sites in the tree and courting many females, producing a polygynous mating system.

One exception to the general rule that vertebrate societies are not organized like those of insects is the naked mole rat, a small, hairless rodent that lives in and near East Africa. Unlike other kinds of mole rats, which live alone or in small family groups, naked mole rats form large underground colonies with a far-ranging system of tunnels and a central nesting area. It is not unusual for a colony to contain 80 individuals.

Naked mole rats feed on bulbs, roots and tubers, which they locate by constant tunneling. As in insect societies, there is a division of labor among the colony members, with some mole rats working as tunnelers while others perform different tasks, depending upon the size of their body. Large mole rats defend the colony and dig tunnels.

Naked mole rat colonies have a *reproductive* division of labor similar to the one normally associated with the eusocial insects. All of the breeding is done by a single female or

FIGURE 60.20

Weaver bird socioecology. Monogamous pairs of weaver birds build camouflaged nests in the forest. Polygynous, savanna-dwelling species form colonial nests, as shown here.

"queen," who has one or two male consorts. The workers, consisting of both sexes, keep the tunnels clear and forage for food. Recent DNA fingerprinting studies show that colony members may share 80% of their alleles, apparently due to inbreeding. Kin selection may have been important in the evolution of this eusocial mammalian society.

Social behavior in vertebrates is often characterized by kin-selected altruism. Altruistic behavior is involved in cooperative breeding in birds and alarm-calling in mammals.

Human Sociobiology

Sociobiology is a comparative science: the same theoretical concepts can be applied to the evolution of social behavior in very different species. As we saw in our discussion of insect and vertebrate societies, altruism in bees, birds, and mammals may be explained by the same concept of kin selection. We can all agree that social behavior in these animals has a biological basis, and that one unifying theory of evolution can explain it. But are humans just another social animal whose behavior can be explained and fully understood with Darwinian concepts? Scientists who study human sociobiology believe the answer to this question is, to a large extent, yes.

As a social species, humans have an unparalleled complexity. Indeed, we are the only species with the intelligence to contemplate the social behavior of other animals. Intelligence is just one human trait. If an ethologist were to take an inventory of human behavior, he or she would list kin-selected altruism; reciprocity and other elaborate social contracts; extensive parental care; conflicts between parents and offspring; violence and warfare; infanticide; a variety of mating systems, including monogamy, polygyny, and polyandry; along with sexual behaviors such as extra-pair copulation ("adultery") and homosexuality; and behaviors like adoption that appear to defy evolutionary explanation. This incredible variety of behaviors occurs *in one species*, and any trait can change within *any individual*. Are these behaviors rooted in human biology?

Biological and Cultural Evolution

During the course of human evolution and the emergence of civilization, two processes have led to adaptive change. One is **biological evolution.** We have a primate heritage, reflected in the extensive amount of genetic material we share with our closest relatives, the chimpanzees. Our upright posture, bipedal locomotion, and powerful, precise hand grips are adaptations whose origins are traceable through our primate ancestors. Kin-selected and reciprocal altruism, as well as other shared traits like aggression and different types of mating systems, can also be seen in non-human primates, in whom we can demonstrate that these social traits are adaptive. We may speculate, based on various lines of evidence, that similar traits evolved in early humans. If individuals with certain social traits had an advantage in reproduction over other individuals that lacked the traits, and if these traits had a genetic basis, then the alleles for their expression would now be expected to be part of the human genome and to influence our behavior.

The second process that has underscored the emergence of civilization and led to adaptive change is **cultural evolution,** the transfer across generations of information necessary for survival. This is a nongenetic mode of adaptation, although some scientists feel that genes influence culture. Many adaptations—the use of tools, the formation of cooperative hunting groups, the construction of shelters, and marriage practices—do not follow Mendelian rules of inheritance and are passed from generation to generation by *tradition*. To anthropologists interested in the origins of human behavior, cultural inheritance is as valid a way to convey adaptations across generations as genetic inheritance. Human cultures are also extraordinarily diverse. The ways in which children are socialized among Trobriand Islanders, Pygmies, and Yanomamo Indians are very different. Again, we must remember that this fantastic variation occurs within one species, and that individual behavior is very flexible.

Identifying the Biological Components of Human Behavior

Given this great flexibility, how can the biological components of human behavior be identified? One way is to look for common patterns that appear in a wide variety of cultures, that is, to study behaviors that are cross-cultural. In spite of cultural variation, there are some traits that characterize all human societies. For example, all cultures have an incest taboo, forbidding marriages between close relatives. Incestuous matings lead to a greater chance of exposing disorders such as mental retardation and hemophilia. Natural selection may have acted to create a behavioral disposition against incest, and that disposition is now a cultural norm. Genes responsible for guiding this behavior may have become fixed in human populations because of their adaptive effects. Genes thus guide the direction of culture.

Although human mating systems vary, polygyny is found to be the most common among all cultures. Because most mammalian species are polygynous, the human pattern seems to reflect our mammalian evolutionary heritage and thus is a part of our biology. This conclusion is drawn from using the comparative approach, common in evolutionary science. Nonverbal communication patterns, like smiling and raising the hand in a greeting, also occur in many cultures. Perhaps these behaviors represent a common human heritage.

The explanations sociobiology offers to understand human behavior have been and continue to be controversial. For example, the new discipline of **evolutionary psychology** seeks to understand the origins of the human mind. Human behaviors are viewed as being extensions of our genes. The diversity of human cultures are thought to have a common core of characteristics that are generated by our psychology, which evolved as an adaptation to the lifestyle of our hunter-gatherer ancestors during the Pleistocene. Much of human behavior is seen as reflecting ancient, adaptive traits, now expressed in the context of modern civilization. In this controversial view, human behaviors such as jealousy and infidelity are viewed as adaptations; these behaviors increased the fitness of our ancestors, and thus are now part of the human psyche.

Sociobiology offers general explanations of human behavior that are controversial, but are becoming more generally accepted than in the past.

Summary of Concepts

60.1 Evolutionary forces shape behavior.

- Many behaviors are ecologically important and serve as adaptations.
- Foraging and territorial behaviors have evolved because they allow animals to use resources efficiently.

60.2 Reproductive behavior involves many choices influenced by natural selection.

- Male and female animals maximize their fitness with different reproductive behaviors. The differences relate to the extent to which each sex provides care for offspring.
- Usually, males are competitive and females show mate choice because females have higher reproductive costs.
- A species' mating system is related to its ecology.

60.3 There is considerable controversy about the evolution of social behavior.

- Many animals show altruistic, or self-sacrificing, behavior. Altruism may evolve through reciprocity or be directed toward relatives. Cooperative behavior often increases an individual's inclusive fitness.
- Individuals form social groups because it is advantageous for them to do so.
- The benefits of living in a group, such as enhanced feeding success, are often balanced by the cost of increased incidence of disease and parasitism.
- Animal societies are characterized by cooperation and conflict. The organization of a society is related to the ecology of a species.

60.4 Vertebrates exhibit a broad range of social behaviors.

- Human behavior is extremely rich and varied and may result from both biology and culture.
- Evolutionary theory can give us important insight into human nature, but such an approach to the study of human behavior may have political consequences.

Discussing Key Terms

1. **Foraging and territorial behavior**   Foraging and territorial behaviors are shaped by natural selection to maximize the energy gained per unit time. This maximization provides more energy for reproduction, thus increasing fitness.

2. **Mate choice**   There is usually a higher energy cost in reproduction for females than for males, and therefore mate choice is usually exercised by females.

3. **Altruism**   Altruism has selective benefits if it is directed toward genetic relatives, or if it is practiced reciprocally by nonrelatives. Cooperative behavior can increase an individual's inclusive fitness.

4. **Vertebrate social groups**   Vertebrate social groups are formed because the selective advantages outweigh the disadvantages to the individual members of the group.

5. **Human behavior**   Sociobiology can tell us a great deal about the origin of human behavior, which in many ways can be considered to be adaptive. Nonadaptive behavioral variation in human individuals and across human cultures must also be considered. Because most human behaviors cannot be shown to have a clear fitness correlate, the study of the adaptiveness of human behavior is both difficult and controversial.

Review Questions

1. What is behavioral ecology? How do behavioral ecologists assess the adaptive significance of behavior?

2. What does optimal foraging theory predict about an animal's foraging behavior? What factors unrelated to this theory may also influence an animal's foraging choices?

3. What are the benefits of territorial behavior, and what are its costs? Under what circumstances is territorial behavior disadvantageous?

4. Why does natural selection favor mate choice? According to Trivers, what factor is most important in determining which sex exhibits mate choice?

5. What is sexual selection? How does it differ from natural selection?

6. Define the following three types of mating systems: monogamy, polygyny, and polyandry. In birds, how does the amount of parental care required by the offspring affect the evolution of a species' mating system?

7. What is reciprocal altruism? What is kin selection? How does kin selection increase an individual's inclusive fitness?

8. In vertebrate societies, what are the costs to an individual who makes an alarm call? Based on research in ground squirrels, which individuals are most likely to make alarm calls, and what benefits do they receive by doing so?

9. What is evolutionary psychology? How does this discipline view the origin of the human mind?

Thought Questions

1. Swallows often hunt in groups, while hawks and other predatory birds usually are solitary hunters. What do you think is the basis for this difference?

2. Can you suggest an evolutionary reason why polygyny is common in most vertebrate species but polyandry is rare?

3. Imagine an animal society in which the individuals share an average of one-third of their alleles with one another. If the benefits to an individual of performing an altruistic act are twice the costs, does the kin selection model predict that these individuals are likely to be altruistic? Explain your reasoning.

Internet Links

Social Behavior in Mammals
http://www.si.edu/organiza/museums/zoo/hilights/webcams/molerat1/nmcam.htm
The NAKED MOLE-RAT COLONY CAM at the National Zoo in Washington DC allows you to keep an eye on a real colony of these social mammals.

Male Parental Care
http://www.pbs.org/wgbh/pages/nova/seahorse/superdads.html
This ANIMAL SUPER DADS site from NOVA provides detailed information about seven animals that practice male parental care.

For Further Reading

Clutton-Brock, T. and G. Parker: "Punishment in Animal Societies," *Nature*, vol. 373, pages 209–16, January 19, 1995. Aggression, evolutionary fitness, and altruism are examined in animal social groups.

Dawkins, R.: *The Selfish Gene*, Oxford University Press, New York, 1976. An accurate and entertaining account of the behavioral ecologist's view of behavior.

Godfray, H.: "Evolutionary Theory of Parent-Offspring Conflict," *Nature*, vol. 376, pages 133–38, July 13, 1995. Early conflicts between parents and offspring apparently serve as substantial influence on the development of subsequent animal behavior. Authorities are not in agreement about the "evolutionary theory of this conflict," and this paper does a good job of presenting a summary of the prevailing theories.

Halliday, T.: *Sexual Strategy*, University of Chicago Press, 1982. A colorful, broad-based treatment of animal reproduction.

Barkow, J. H., L. Cosmides, and J. Tooby. *The Adapted Mind: Evolutionary Psychology and the Generation of Culture*, Oxford University Press, 1992. An excellent introduction to this new hybrid field.

Holldobler, B. and E. O. Wilson: *The Ants*, Harvard University Press, Cambridge, MA, 1990. A Pulitzer Prize winning tour-de-force of the evolution, ecology, and social biology of the most diverse groups of social species.

Horgan, J. "The New Social Darwinists," *Scientific American*, October, 1995, pages 174–181. A discussion of the advantages and disadvantages (political and scientific) of applying evolutionary theory to humans.

Krebs, J. and N. B. Davies: *An Introduction to Behavioral Ecology*, ed. 3, Sinauer, Sunderland, MA, 1993. Perhaps the best source for entering the study of the adaptiveness of behavior.

deWaal, F.: *Good Natured: The Origin of Right and Wrong in Humans and Other Animals*, Harvard University Press, Cambridge, MA, 1996. Can we trace our moral values to our primate ancestors? Provocative answers to this question.

Appendix
Classification of Organisms

The classification used in this book is presented here and explained in chapter 28. Not all phyla are included. Responding to a wealth of recent molecular data, it recognizes two separate kingdoms for the prokaryotes (bacteria), Archaebacteria and Eubacteria, and divides the eukaryotes into four kingdoms, Protista, Fungi, Plantae, and Animalia. There is no attempt to reorganize the kingdoms into domains, as no consensus has been reached about how this should be done. Viruses, which are considered nonliving, are not included in this appendix, but are treated in chapter 30. Plant and fungal divisions are the taxonomic equivalents of phyla, and are considered to be phyla in this text. For each entry in this classification, the page in the text where the organism is treated is indicated in parentheses. A more complete description of each entry can be found on the Raven & Johnson *BIOLOGY* web site, www.mhhe.com/biosci/genbio/rjbiology

Kingdom Archaebacteria

Prokaryotic bacteria; single-celled, cell walls lack muramic acid. Like all bacteria, they lack a membrane-bound nucleus, sexual recombination, and internal cell compartments. Archaebacterial cells have distinctive membranes, and unique rRNA and metabolic cofactors. Many are capable of living in extreme or anaerobic environments.

Methanogens (page 567)
Extremophiles (page 567)
Nonextreme archaebacteria (page 567)

Kingdom Eubacteria

Single-celled prokaryotes with cell walls containing muramic acid, sometimes forming filaments or other forms of colonies. Eubacteria lack a membrane-bound nucleus, sexual recombination, and internal cell compartments. Their flagella are simple, composed of a single fiber of protein. They are much more diverse metabolically than the eukaryotes. About 4800 species are currently recognized, but that is almost certainly only a small fraction of the actual number.

Actinomycetes (page 597)
Chemoautotrophs (page 597)
Cyanobacteria (page 597)
Enterobacteria (page 597)
Gliding and budding bacteria (page 597)
Pseudomonads (page 597)
Spirochaetes (page 597)

Kingdom Protista

Eukaryotic, primarily single-celled organisms. Eukaryotes have a membrane-bound nucleus containing chromosomes, sexual recombination, and extensive internal compartmentalization of cells. Their flagella are complex, with a 9 + 2 internal microtubular organization. Some protists have flagella, some have pseudopods, and some are sessile (immobile). They are diverse metabolically, but much less so than bacteria. Protists may capture prey, absorb their food, or photosynthesize. Reproduction among protists can be sexual, involving meiosis, but is usually asexual. The kingdom Protista is a catch-all classification, including all eukaryotes which are not plants, animals, or fungi.

Sarcodina—*heterotrophs with no permanent locomotor apparatus*
Phylum Rhizopoda (page 610)
Phylum Foraminifera (page 611)
Phylum Actinopoda (page 610)

Algae—*photosynthetic protists that are multicellular or largely multicellular*
Phylum Rhodophyta (page 613)
Phylum Phaeophyta (page 614)
Phylum Chlorophyta (page 612)

Diatoms—*photosynthetic protists that are unicellular, many with a double shell of silica*
Phylum Chrysophyta (page 615)

Flagellates—*protists with locomotor flagella*
Phylum Pyrrhophyta (page 616)
Phylum Euglenophyta (page 617)
Phylum Zoomastigophora (page 618)

Sporozoans—*nonmotile, spore-forming unicellular parasites*
Phylum Apicomplexa (page 620)

Ciliates—*heterotrophic unicellular protists with cells of fixed shape possessing two nuclei and many cilia*
Phylum Ciliophora (page 622)

Molds—*heterotrophs with restricted mobility that have cell walls made of carbohydrate*
Phylum Acrasiomycota (page 624)
Phylum Myxomycota (page 625)
Phylum Oomycota (page 626)

Kingdom Fungi

Filamentous, multinucleate, heterotrophic eukaryotes with cell walls rich in chitin; no flagellated cells present. Mitosis in fungi takes place within the nuclei, the nuclear envelope never breaking down. Filaments of fungal cells called hyphae grow through the substrate, secreting enzymes and absorbing the products of their digestion. Septa between the nuclei in the hyphae normally complete only at the borders of reproductive structures. Asexual reproduction frequent in some groups. The nuclei of fungi are haploid, with the zygote the only diploid stage of the life cycle. About 77,000 named species.

Fungi with a known sexual cycle

Phylum Zygomycota (page 635)

Phylum Ascomycota (page 636)

Phylum Basidiomycota (page 638)

Fungi with an unknown sexual cycle

Fungi Imperfecti (page 639)

Symbiotic Associations—*fungi and a photosynthetic partner.*

Lichens (page 640)

Mycorrhizae (page 642)

Kingdom Plantae

Multicellular, photosynthetic, terrestrial eukaryotes derived from the green algae, and like them containing chlorophylls *a* and *b*. Pigments are present in chloroplasts which may also store starch. The cell walls of plants have cellulose matrix and sometimes become lignified; cell division is by means of a cell plate that forms across the mitotic spindle. The vascular plants have an elaborate system of xylem and phloem conducting cells that move water through the plant body. Plants have a waxy cuticle that helps them retain water; most have stomata. All plants have an alternation of generations with reduced gametophyte and multicellular gametangia. About 270,000 species.

Nonvascular Plants—*mosses, liverworts, and hornworts*

Phylum Hepaticophyta (page 649)

Phylum Antheroceratophyta (page 649)

Phylum Byrophyta (page 648)

Seedless Vascular Plants—*ferns*

Phylum Psilophyta (page 654)

Phylum Lycophyta (page 654)

Phylum Arthrophyta (page 655)

Phylum Pterophyta (page 652)

Gymnosperms—*conifers and cycads*

Phylum Coniferophyta (page 658)

Phylum Cycadophyta (page 659)

Phylum Ginkgophyta (page 659)

Phylum Gnetophyta (page 659)

Angiosperms—*flowering plants*

Phylum Antherophyta (page 660)

Kingdom Animalia

Animals are mulitcellular eukaryotes that characteristically injest their food. Their cells are usually flexible, without a cell wall. In all of the approximately 35 phyla except sponges, body cells are organized into structural and functional units called tissues. Cells move extensively during the development of the animal embryo; the blastula, a hollow ball of cells, forms early in this process and is characteristic of the group. Most animals reproduce sexually; their non-motile eggs are much larger than their small flagellated sperm. The gametes fuse directly to produce a zygote and do not divide by mitosis as in plants. More than a million species of animals have been described.

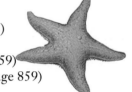

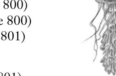

Glossary

A

ABO blood group A set of four phenotypes produced by different combinations of three alleles at a single locus; blood types are A, B, AB, and O, depending on which alleles are expressed as antigens on the red blood cell surface.

abscission (L. *ab*, away, off + *scissio*, dividing) In vascular plants, the dropping of leaves, flowers, fruits, or stems at the end of growing season, as the result of formation of a layer of specialized cells (the abscission zone) and the action of a hormone (ethylene).

acquired immune deficiency syndrome (AIDS) An infectious and usually fatal human disease caused by a retrovirus, HIV (human immunodeficiency virus), which attacks T cells. The affected individual is helpless in the face of microbial infections.

actin (Gr. *actis*, a ray) One of the two major proteins that make up vertebrate muscle (the other is myosin).

action potential A transient, all-or-none reversal of the electric potential across a membrane; in neurons, an action potential initiates transmission of a nerve impulse.

activation energy The energy that must be processed by a molecule in order for it to undergo a specific chemical reaction.

active site The region of an enzyme surface to which a specific set of substrates binds, lowering the activation energy required for a particular chemical reaction and so facilitating it.

active transport The pumping of individual ions or other molecules across a cellular membrane from a region of lower concentration to one of higher concentration (that is, against a concentration gradient); this transport process requires energy, which is typically supplied by the expenditure of ATP.

adaptation (L. *adaptare*, to fit) A peculiarity of structure, physiology, or behavior that promotes the likelihood of an organism's survival and reproduction in a particular environment.

adaptive radiation The evolution of several divergent forms from a primitive and unspecialized ancestor.

adenosine diphosphate (ADP) A nucleotide consisting of adenine, ribose sugar, and two phosphate groups; formed by the removal of one phosphate from an ATP molecule.

adenosine triphosphate (ATP) A nucleotide consisting of adenine, ribose sugar, and three phosphate groups; ATP is the energy currency of cellular metabolism in all organisms.

adenylyl cyclase An enzyme that produces large amounts of cAMP from ATP; the cAMP acts as a second messenger in a target cell.

adipose cells Fat cells, found in loose connective tissue, usually in large groups that form adipose tissue. Each adipose cell can store a droplet of fat (triglycerides).

adventitious (L. *adventicius*, not properly belonging to) Referring to a structure arising from an unusual place, such as stems from roots or roots from stems.

aerobic (Gr. *aer*, air + *bios*, life) Requiring free oxygen; any biological process that can occur in the presence of gaseous oxygen (O_2).

alga, pl. **algae** A unicellular or simple multicellular photosynthetic organism lacking multicellular sex organs.

allantois (Gr. *allas*, sausage + *eidos*, form) A membrane of the amniotic egg that functions in respiration and excretion in birds and reptiles and plays an important role in the development of the placenta in most mammals.

allele (Gr. *allelon*, of one another) One of two or more alternative states of a gene.

allometric growth (Gr. *allos*, other + *ergon*, action) A pattern of growth in which different components grow at different rates.

allopatric speciation The differentiation of geographically isolated populations into distinct species.

allosteric site A part of an enzyme, away from its active site, that serves as an on/off switch for the function of the enzyme.

alternation of generations A reproductive cycle in which a haploid (l*n*) phase, the gametophyte, gives rise to gametes, which, after fusion to form a zygote, germinate to produce a diploid (2*n*) phase, the

sporophyte. Spores produced by meiotic division from the sporophyte give rise to new gametophytes, completing the cycle.

altruism Self-sacrifice for the benefit of others; in formal terms, the behavior that increases the fitness of the recipient while reducing the fitness of the altruistic individual.

alveolus, pl. **alveoli** (L., a small cavity) One of many small, thin-walled air sacs within the lungs in which the bronchioles terminate.

amino acid The subunit structure from which proteins are produced, consisting of a central carbon atom with a carboxyl group (—COOH), an amino group (—NH_2), a hydrogen, and a side group (R group); only the side group differs from one amino acid to another.

amniocentesis (Gr. *amnion*, membrane around the fetus + *centes*, puncture) Examination of a fetus indirectly, by tests on cell cultures grown from fetal cells obtained from a sample of the amniotic fluid surrounding the developing embryo or tests on the fluid itself.

amnion (Gr., membrane around the fetus) The innermost of the extraembryonic membranes; the amnion forms a fluid-filled sac around the embryo in amniotic eggs.

amniotic egg An egg that is isolated and protected from the environment by a more or less impervious shell during the period of its development and that is completely self-sufficient, requiring only oxygen.

amyloplast (Gr. *amylon*, starch + *plastos*, formed) A plant organelle called a plastid that specializes in storing starch.

anabolism (Gr. *ana*, up + *bolein*, to throw) The biosynthetic or constructive part of metabolism; those chemical reactions involved in biosynthesis.

anaerobic (Gr. *an*, without + *aer*, air + *bios*, life) Any process that can occur without oxygen, such as anaerobic fermentation or H_2S photosynthesis.

analogous (Gr. *analogos*, proportionate) Structures that are similar in function but different in evolutionary origin, such as the wing of a bat and a wing of a butterfly.

anaphase In mitosis and meiosis II, the stage initiated by the separation of sister chromatids, during which the daughter chromosomes move to opposite poles of the cell; in meiosis I, marked by separation of replicated homologous chromosomes.

androecium (Gr. *andros*, man + *oilos*, house) The flora whorl that comprises the stamens.

aneuploidy (Gr. *an*, without + *eu*, good + *ploid*, multiple of) The condition in an organism whose cells have lost or gained a chromosome; Down syndrome, which results from an extra copy of human chromosome 21, is an example of aneuploidy in humans.

angiosperms The flowering plants, one of five phyla of seed plants. In angiosperms, the ovules at the time of pollination are completely enclosed by tissues.

animal pole In fish and other aquatic vertebrates with asymmetrical yolk distribution in their eggs, the hemisphere of the blastula, comprising cells relatively poor in yolk.

anion (Gr. *anienae*, to go up) A negatively charged ion.

anther (Gr. *anthos*, flower) In angiosperm flowers, the pollen-bearing portion of a stamen.

antheridium, pl. **antheridia** A sperm-producing organ.

antibody (Gr. *anti*, against) A protein called immunoglobulin that is produced by lymphocytes in response to a foreign substance (antigen) and released into the bloodstream.

anticodon The three-nucleotide sequence at the end of a transfer RNA molecule that is complementary to, and base-pairs with, an amino-acid specifying codon in messenger RNA.

antigen (Gr. *anti*, against + *genos*, origin) A foreign substance, usually a protein or polysaccharide, that stimulates an immune response.

anus The terminal opening of the gut; the solid residues of digestion are eliminated through the anus.

aorta (Gr. *aeirein*, to lift) The major artery of vertebrate systemic blood circulation; in mammals, carries oxygenated blood away from the heart to all regions of the body except the lungs.

apical meristem (L. *apex*, top + Gr. *meristos*, divided) In vascular plants, the growing point at the tip of the root or stem.

apoptosis A process of programmed cell death, in which dying cells shrivel and shrink; used in all animal cell development to produce planned and orderly death of cells not destined to be present in the final tissue.

aposematic coloration An ecological strategy of some organisms that "advertise" their poisonous nature by the use of bright colors.

aquifers Permeable, saturated, underground layers of rock, sand, and gravel, which serve as reservoirs for groundwater.

archaebacteria A group of bacteria that are among the most primitive still in existence, characterized by the absence of peptidoglycan in their cell walls, a feature that distinguishes them from all other bacteria.

archegonium, pl. archegonia (Gr. *archegonos*, first of a race) The multicellular egg-producing organ in bryophytes and some vascular plants.

archenteron (Gr. *arche*, beginning + *enteron*, gut) The principal cavity of a vertebrate embryo in the gastrula stage; lined with endoderm, it opens up to the outside and represents the future digestive cavity.

arteriole A smaller artery, leading from the arteries to the capillaries.

arteriosclerosis Hardening and thickening of the wall of an artery.

ascomycetes A large group comprising part of the "true fungi." They are characterized by separate hyphae, asexually produced conidiospores and sexually produced ascospores within asci.

ascospore A fungal spore produced within an ascus.

ascus, pl. asci (Gr. *askos*, wineskin, bladder) A specialized cell, characteristic of the ascomycetes, in which two haploid nuclei fuse to produce a zygote that divides immediately by meiosis; at maturity, an ascus contains ascospores.

asexual reproduction The process by which an individual inherits all of its chromosomes from a single parent, thus being genetically identical to that parent; cell division is by mitosis only.

aster In animal cell mitosis, a radial array of microtubules extending from the centrioles toward the plasma membrane, possibly serving to brace the centrioles for retraction of the spindle.

atrioventricular (AV) node A slender connection of cardiac muscle cells that receives the heartbeat impulses from the sinoatrial node and conducts them by way of the bundle of His.

atrium (L., vestibule or courtyard) An antechamber; in the heart, a thin-walled chamber that receives venous blood and passes it on to the thick-walled ventricle; in the ear, the tympanic cavity.

autonomic nervous system (Gr. *autos*, self + *nomos*, law) The involuntary neurons and ganglia of the peripheral nervous system of vertebrates; regulates the heart, glands, visceral organs, and smooth muscle.

autosome (Gr. *autos*, self + *soma*, body) Any eukaryotic chromosome that is not a sex chromosome; autosomes are present in the same number and kind in both males and females of the species.

autotroph (Gr. *autos*, self + *trophos*, feeder) An organism able to build all the complex organic molecules that it requires as its own food source, using only simple inorganic compounds.

auxin (Gr. *auxein*, to increase) A plant hormone that controls cell elongation, among other effects.

axon (Gr., axle) A process extending out from a neuron that conducts impulses away from the cell body.

B

bacteriophage (Gr. *bakterion*, little rod + *phagein*, to eat) A virus that infects bacterial cells; also called a *phage*.

Barr body A deeply staining structure seen in the interphase nucleus of a cell of an individual with more than one X chromosome, condensed and inactivated X. Only one X remains active in each cell after early embryogenesis.

basal body A self-reproducing, cylinder-shaped cytoplasmic organelle composed of nine triplets of microtubules from which the flagella or cilia arise.

base-pair A complementary pair of nucleotide bases, consisting of a purine and a pyrimidine.

basidiospore A spore of the basidiomycetes, produced within and borne on a basidium after nuclear fusion and meiosis.

basidium, pl. basidia (L., a little pedestal) A specialized reproductive cell of the basidiomycetes, often club shaped, in which nuclear fusion and meiosis occur.

basophil A leukocyte containing granules that rupture and release chemicals that enhance the inflammatory response. Important in causing allergic responses.

Batesian mimicry A situation in which a palatable or nontoxic organism resembles another kind of organism that is distasteful or toxic. Both species exhibit warning coloration.

B cell A type of lymphocyte that, when confronted with a suitable antigen, is capable of secreting a specific antibody protein.

behavioral ecology The study of how natural selection shapes behavior.

biennial A plant that normally requires two growing seasons to complete its life cycle. Biennials flower in the second year of their lives.

bile salts A solution of organic salts that is secreted by the vertebrate liver and temporarily stored in the gallbladder; emulsifies fats in the small intestine.

binary fission (L. *binarius*, consisting of two things or part + *fissus*, split) Asexual reproduction by division of one cell or body into two equal or nearly equal parts.

binomial distribution The distribution of phenotypes seen among the progeny of a cross in which there are only two alternative alleles.

biomass (Gr. *bios*, life + *maza*, lump or mass) The weight of all the living organisms in a given population, area, or other unit being measured.

biome One of the major terrestrial ecosystems, characterized by climatic and soil conditions; the largest ecological unit.

bipedal Able to walk upright on two feet.

bipolar cell A specialized type of neuron connecting cone cells to ganglion cells in the visual system. Bipolar cells receive a hyperpolarized stimulus from the cone cell and then transmit a depolarization stimulus to the ganglion cell.

blade The broad, expanded part of a leaf; also called the lamina.

blastocoel (Gr. *blastos*, a sprout + *koilos*, hollow) The central cavity of the blastula stage of vertebrate embryos.

blastodisc (Gr. *blastos*, sprout + *discos*, a round plate) In the development of birds, a disclike area on the surface of a large, yolky egg that undergoes cleavage and gives rise to the embryo.

blastomere (Gr. *blastos*, sprout + *meros*, part) One of the cells of a blastula.

blastopore (Gr. *blastos*, sprout + *poros*, a path or passage) In vertebrate development, the opening that connects the archenteron cavity of a gastrula stage embryo with the outside.

blastula (Gr., a little sprout) In vertebrates, an early embryonic stage consisting of a hollow, fluid-filled ball of cells one layer thick; a vertebrate embryo after cleavage and before gastrulation.

Bohr effect The release of oxygen by hemoglobin molecules in response to elevated ambient levels of CO_2.

Bowman's capsule In the vertebrate kidney, the bulbous unit of the nephron, which surrounds the glomerulus.

B-oxidation In the cellular respiration of fats, the process by which two-carbon acetyl groups are removed from a fatty acid and combined with coenzyme A to form acetyl-CoA, until the entire fatty acid has been broken down.

bronchus, pl. bronchi (Gr. *bronchos*, windpipe) One of a pair of respiratory tubes branching from the lower end of the trachea (windpipe) into either lung.

bryophytes Nonvascular plants including mosses, hornworts, and liverworts.

bud An asexually produced outgrowth that develops into a new individual. In plants, an embryonic shoot, often protected by young leaves; buds may give rise to branch shoots.

C

C₃ photosynthesis The main cycle of the dark reactions of photosynthesis, in which CO_2 binds to ribulose 1,5-bisphosphate (RuBP) to form two three-carbon phosphoglycerate (PGA) molecules.

C₄ photosynthesis A process of CO_2 fixation in photosynthesis by which the first product is the four-carbon oxaloacetate molecule.

callus (L. *callos*, hard skin) Undifferentiated tissue; a term used in tissue culture, grafting, and wound healing.

Calvin cycle The dark reactions of C₃ photosynthesis; Also called the Calvin-Benson cycle.

calyx (Gr. *kalyx*, a husk, cup) The sepals collectively; the outermost flower whorl.

capillary (L. *capillaris*, hairlike) The smallest of the blood vessels; the very thin walls of capillaries are permeable to many molecules, and exchanges between blood and the tissues occur across them; the vessels that connect arteries with veins.

capsid The outermost protein covering of a virus.

carapace (Fr. from Sp. *carapacho*, shell) Shieldlike plate covering the cephalothorax of decapod crustaceans; the dorsal part of the shell of a turtle.

carbohydrate (L. *carbo*, charcoal + *hydro*, water) An organic compound consisting of a chain or ring of carbon atoms to which hydrogen and oxygen atoms are attached in a ratio of approximately 2:1; having the generalized formula $(CH_2O)_n$; carbohydrates include sugars, starch, glycogen, and cellulose.

carbon fixation The conversion of CO_2 into organic compounds during photosynthesis; the first stage of the dark reactions of photosynthesis, in which carbon dioxide from the air is combined with ribulose 1,5-bisphosphate.

carpel (Gr. *karpos*, fruit) A leaflike organ in angiosperms that encloses one or more ovules.

carrying capacity The maximum population size that a habitat can support.

cartilage (L. *cartilago*, gristle) A connective tissue in skeletons of vertebrates. Cartilage forms much of the skeleton of embryos, very young vertebrates, and some adult vertebrates, such as sharks and their relatives.

catabolism (Gr. *ketabole*, throwing down) In a cell, those metabolic reactions that result in the breakdown of complex molecules into simpler compounds, often with the release of energy.

catalysis The process by which chemical subunits of larger organic molecules are held and positioned by enzymes that stress their chemical bonds, leading to the disassembly of the larger molecule into its subunits, often with the release of energy.

cation (Gr. *katienai*, to go down) A positively charged ion.

cecum, also caecum (L. *caecus*, blind) In vertebrates, a blind pouch at the beginning of the large intestine.

cell cycle The repeating sequence of growth and division through which cells pass each generation.

cell plate The structure that forms at the equator of the spindle during early telophase in the dividing cells of plants and a few green algae.

cell surface receptor A cell surface protein that binds a signal molecule and converts the extracellular signal into an intracellular one.

cellular respiration The metabolic harvesting of energy by oxidation, ultimately dependent on molecular oxygen; carried out

by the Krebs cycle and oxidative phosphorylation.

cellulose (L. *cellula*, a little cell) The chief constituent of the cell wall in all green plants, some algae, and a few other organisms; an insoluble complex carbohydrate $(C_6H_1OO_5)$ formed of microfibrils of glucose molecules.

cell wall The rigid, outermost layer of the cells of plants, some protists, and most bacteria; the cell wall surrounds the cell (plasma) membrane.

central nervous system (CNS) That portion of the nervous system where most association occurs; in vertebrates, it is composed of the brain and spinal cord; in invertebrates it usually consists of one or more cords of nervous tissue, together with their associated ganglia.

centriole (Gr. *kentron*, center of a circle + L. *olus*, little one) A cytoplasmic organelle located outside the nuclear membrane, identical in structure to a basal body; found in animal cells and in the flagellated cells of other groups; divides and organizes spindle fibers during mitosis and meiosis.

centromere (Gr. *kentron*, center + *meros*, a part) Condensed region on a eukaryotic chromosome where sister chromatids are attached to each other after replication. *See* kinetochore.

cerebellum (L., little brain) The hindbrain region of the vertebrate brain that lies above the medulla (brain stem) and behind the forebrain; it integrates information about body position and motion, coordinates muscular activities, and maintains equilibrium.

cerebral cortex The thin surface layer of neurons and glial cells covering the cerebrum; well developed only in mammals, and particularly prominent in humans. The cerebral cortex is the seat of conscious sensations and voluntary muscular activity.

cerebrum (L., brain) The portion of the vertebrate brain (the forebrain) that occupies the upper part of the skull, consisting of two cerebral hemispheres united by the corpus callosum. It is the primary association center of the brain. It coordinates and processes sensory input and coordinates motor responses.

chelicera, pl. chelicerae (Gr. *chele*, claw + *keras*, horn) The first pair of appendages in horseshoe crabs, sea spiders, and arachnids—the chelicerates, a group of arthropods. Chelicerae usually take the form of pincers or fangs.

chemiosmosis The mechanism by which ATP is generated in mitochondria and chloroplasts; energetic electrons excited by light (in chloroplasts) or extracted by oxidation in the Krebs cycle (in mitochondria) are used to drive proton pumps, creating a proton concentration gradient; when protons subsequently flow back across the membrane, they pass through channels that couple their movement to the synthesis of ATP.

chiasma An X-shaped figure that can be seen in the light microscope during meiosis; evidence of crossing over, where two chromatids have exchanged parts; chiasmata move to the ends of the chromosome arms as the homologues separate.

chitin (Gr. *chiton*, tunic) A tough, resistant, nitrogen-containing polysaccharide that forms the cell walls of certain fungi, the exoskeleton of arthropods, and the epidermal cuticle of other surface structures of certain other invertebrates.

chloroplast (Gr. *chloros*, green + *plastos*, molded) A cell-like organelle present in algae and plants that contains chlorophyll (and usually other pigments) and carries out photosynthesis.

chorion (Gr., skin) The outer member of the double membrane that surrounds the embryo of reptiles, birds, and mammals; in placental mammals, it contributes to the structure of the placenta.

chromatid (Gr. *chroma*, color + L. *-id*, daughters of) One of the two daughter strands of a duplicated chromosome that is joined by a single centromere.

chromatin (Gr. *chroma*, color) The complex of DNA and proteins of which eukaryotic chromosomes are composed; chromatin is highly uncoiled and diffuse in interphase nuclei, condensing to form the visible chromosomes in prophase.

chromosome (Gr. *chroma*, color + *soma*, body) The vehicle by which hereditary information is physically transmitted from one generation to the next; in a bacterium, the chromosome consists of a single naked circle of DNA; in eukaryotes, each chromosome consists of a single linear DNA molecule and associated proteins.

cilium A short cellular projection from the surface of a eukaryotic cell, has the same internal structure of microtubules in a 9 + 2 arrangement as seen in a flagellum.

circadian rhythm An endogenous cyclical rhythm that oscillates on a daily (24 hour) basis.

cisterna A small collecting vessel that pinches off from the end of a Golgi body to form a transport vesicle that moves materials through the cytoplasm.

cladistics A taxonomic technique used for creating hierarchies of organisms that represent true phylogenetic relationship and descent.

class A taxonomic category between phyla (divisions) and orders. A class contains one or more orders, and belongs to a particular phylum or division.

classical conditioning The repeated presentation of a stimulus in association with a response that causes the brain to from an association between the stimulus and the response, even if they have never been associated before.

clathrin A protein located just inside the plasma membrane in eukaryotic cells, in indentations called clathrin-coated pits.

cleavage In vertebrates, a rapid series of successive cell divisions of a fertilized egg, forming a hollow sphere of cells, the blastula.

climax vegetation Vegetation encountered in a self-perpetuating community of plants that has proceeded through all the stages of succession and stabilized.

cloaca (L., sewer) In some animals, the common exit chamber from the digestive, reproductive, and urinary system; in others, the cloaca may also serve as a respiratory duct.

cloning Producing a cell line or culture all of whose members contain identical copies of a particular nucleotide sequence; an essential element in genetic engineering.

coacervate (L. *coacervatus*, heaped up) A spherical aggregation of lipid molecules in water, held together by hydrophobic forces.

cochlea (Gr. *kochlios*, a snail) In terrestrial vertebrates, a tubular cavity of the inner ear containing the essential organs for hearing.

codon (L., code) The basic unit of the genetic code; a sequence of three adjacent nucleotides in DNA or mRNA that code for one amino acid, or for polypeptide chain.

coenzyme (L. *co-*, together + Gr. *en*, in + zyme, leaven) A nonprotein organic molecule such as NAD^+ that plays an accessory role in enzyme-catalyzed processes, often by acting as a donor or acceptor of electrons.

coevolution (L. *co-*, together + *e-*, out + *volvere*, to fill) The simultaneous development of adaptations, in two or more populations, species, or other categories that interact so closely that each is a strong selective force on the other.

cofactor One or more nonprotein components required by enzymes in order to function; many cofactors are metal ions, others are organic coenzymes.

collenchyma (Gr. *kolla*, glue + *en*, in + *chymein*, to pour) In plants, a supporting tissue composed of collenchyma cells; often found in regions of primary growth in stems and in some leaves.

commensalism (L. *cum*, together with + *mensa*, table) A relationship in which one individual lives close to or on another and benefits, and the host is unaffected; a kind of symbiosis.

community (L. *communitas*, community, fellowship) All of the organisms inhabiting a common environment and interacting with one another.

compeptitve exclusion The hypothesis that two species with identical ecological requirements cannot exist in the same locality indefinitely, and that the more efficient of the two in utilizing the available scarce resources will exclude the other; also known as Gause's principle.

complementarity The basis for copying the genetic information, where each nucleotide base has a complementary partner with which it forms a base-pair.

complement system The chemical defense of a vertebrate body that consists of a battery of proteins that become activated by the walls of bacteria and fungi.

concentration gradient The concentration difference of a substance across a distance; in a cell, a greater concentration of its molecules in one region than in another.

cone (1) In plants, the reproductive structure of a conifer. (2) In vertebrates, a type of light-sensitive neuron in the retina, concerned with the perception of color and with the most acute discrimination of detail.

conidia An asexually produced fungal spore.

conjugation (L. *conjugare*, to yolk together) Temporary union of two unicellular organisms, during which genetic material is transferred from one cell to the other; occurs in bacteria, protists, and certain algae and fungi.

contractile vacuole In protists and some animals, a clear fluid-filled vacuole that takes up water from within the cell and then contracts, releasing it to the outside through a pore in a

cyclical manner; functions primarily in osmoregulation and excretion.

conus arteriosus The anteriormost chamber of the embryonic heart in vertebrate animals.

convergent evolution The independent development of similar structures in organisms that are not directly related; often found in organisms living in similar environments.

cork cambium The lateral meristem that forms the periderm, producing cork (phellem) toward the surface (outside) of the plant and phelloderm toward the inside.

cornea (L. *corneus*, horny) The transparent outer layer of the vertebrate eye.

corolla (L. *cornea*, crown) The petals, collectively; usually the conspicuously colored flower whorl.

corpus callosum The band of nerve fibers that connect the two hemispheres of the cerebrum in humans and other primates.

corpus luteum (L., yellowish body) A structure that develops from a ruptured follicle in the ovary after ovulation.

cortex (L., bark) The outer layer of a structure; in animals, the outer, as opposed to the inner, part of an organ; in vascular plants, the primary ground tissue of a stem or root.

cotyledon (Gr. *kotyledon*, a cup-shaped hollow) A seed leaf that generally stores food in dicots or absorbs it in monocots, providing nourishment used during seed germination.

crassulacean acid metabolism (CAM) A mode of carbon dioxide fixation by which CO_2 enters open leaf stomata at night and is used in photosynthesis during the day, when stomata are closed to prevent water loss.

cross-current flow In bird lungs, the latticework of capillaries arranged across the air flow, at a 90° angle.

crossing over In meiosis, the exchange of corresponding chromatid segments between homologous chromosomes; responsible for genetic recombination between homologous chromosomes.

cumulus cells Cells that surround an egg cell and nuture it.

cuticle (L. *cuticula*, little skin) A waxy or fatty, noncellular layer [formed of a substance called cutin] on the outer wall of epidermal cells.

cyanobacteria A group of photosynthetic eubacteria, sometimes called the "blue-green algae," that contain the chlorophyll pigments most

abundant in plants and algae, as well as other pigments.

cyclic AMP (cAMP) A form of adenosine monophosphate (AMP) in which the atoms of the phosphate group form a ring; found in almost all organisms, cAMP functions as an intracellular second messenger that regulates a diverse array of metabolic activities.

cystic fibrosis An autosomal disorder that produces the most common fatal genetic disease in Caucasians, characterized by secretion of thick mucus that clogs passageways in the lungs, liver, and pancreas.

cytochrome (Gr. *kytos*, hollow vessel + *chroma*, color) Any of several iron-containing protein pigments that serve as electron carriers in transport chains of photosynthesis and cellular respiration.

cytokinesis (Gr. *kytos*, hollow vessel + *kinesis*, movement) Division of the cytoplasm of a cell after nuclear division.

cytoplasm (Gr. *ketos*, hollow vessel + *plasma*, anything molded) The material within a cell, excluding the nucleus; the protoplasm.

cytoskeleton A network of protein microfilaments and microtubules within the cytoplasm of a eukaryotic cell that maintains the shape of the cell, anchors its organelles, and is involved in animal cell motility.

cytotoxic T cell (Gr. *kytos*, hollow vessel + toxin) A special T cell activated during cell-mediated immune response that recognizes and destroys infected body cells.

D

dehydration reaction (L. *co-*, together + *densare*, to make dense) A type of chemical reaction in which two molecules join to form one larger molecule, simultaneously splitting out a molecule of water; one molecule is stripped of a hydrogen atom and another is stripped of a hydroxyl group (—OH), resulting in the joining of the two molecules, while the H+ and —OH released may combine to form a water molecule.

demography (Gr. *demos*, people + *graphein*, to draw) The properties of the rate of growth and the age structure of populations.

denaturation The loss of the native configuration of a protein or nucleic acid as a result of excessive heat, extremes of pH, chemical modification, or changes in solvent ionic strength

or polarity that disrupt hydrophobic interactions; usually accompanied by loss of biological activity.

dendrite (Gr. *dendron*, tree) A process extending from the cell body of a neuron, typically branched, that conducts impulses toward the cell body.

deoxyribonucleic acid (DNA) The genetic material of all organisms; composed of two complementary chains of nucleotides wound in a double helix.

depolarization The movement of ions across a cell membrane that wipes out locally an electrical potential difference.

derived character A characteristic used in taxonomic analysis representing a departure from the primitive form.

desmosome A type of anchoring junction that links adjacent cells by connecting their cytoskeletons with cadherin proteins.

diaphragm (Gr. *diaphrassein*, to barricade) (1) In mammals, a sheet of muscle tissue that separates the abdominal and thoracic cavities and functions in breathing. (2) A contraceptive device used to block the entrance to the uterus temporarily and thus prevent sperm from entering during sexual intercourse.

dicot Short for dicotyledon; a class of flowering plants generally characterized as having two cotyledons, net-veined leaves, and flower parts usually in fours or fives.

differentiation A developmental process by which a relatively unspecialized cell undergoes a progressive change to a more specialized form or function.

diffusion (L. *diffundere*, to pour out) The net movement of dissolved molecules or other particles from a region where they are more concentrated to a region where they are less concentrated.

digestion (L. *digestio*, separating out, dividing) The breakdown of complex, usually insoluble foods into molecules that can be absorbed into cells, there to be degraded to yield energy and the raw materials for synthetic processes.

dihybrid (Gr. *dis*, twice + *hibridia*, mixed offspring) An individual heterozygous at two different loci; for example A/a B/b.

dikaryotic (Gr. *di*, two + *karyon*, kernel) In fungi, having pairs of nuclei within each cell.

dioecious (Gr. *di*, two + *oikos*, house) Having the male and female elements on different individuals.

diploid (Gr. *diploos*, double + *eidos*, form) Having two sets of chromosomes (2*n*); in animals, twice the number characteristic of gametes; in plants, the chromosome number characteristic of the sporophyte generation; in contrast to haploid (1*n*).

disaccharide A carbohydrate formed of two simple sugar molecules bonded covalently.

diurnal (L. *diurnalis*, day) Active during the day.

division A major taxonomic group; kingdoms are divided into division (or phyla, which are equivalent), and divisions are divided into classes.

DNA ligase The enzyme that links together Okazaki fragments in DNA replication of the lagging strand; it also links other broken areas of the DNA backbone.

domain A distinct modular region of a protein that serves a particular function in the action of the protein, such as a regulatory domain or DNA-binding domain.

dominant An allele that is expressed when present in either the heterozygous or the homozygous condition.

double fertilization The fusion of the egg and sperm (resulting in a 2*n* fertilized egg, the zygote) and the simultaneous fusion of the second male gamete with the polar nuclei (resulting in a primary endosperm nucleus, which is often triploid, 3*n*); a unique characteristic of all angiosperms.

Down syndrome A congenital syndrome caused by the presence of an extra copy of chromosome 21.

duodenum (L. *duodeni*, 12 each—from its length, about 12 fingers' breadth) In vertebrates, the upper portion of the small intestine.

E

ecdysis (Gr. *ekdysis*, stripping off) Shedding of outer, cuticular layer; molting, as in insects or crustaceans.

ecdysone (Gr. *ekdysis*, stripping off) Molting hormone of arthropods, which stimulates growth of ecdysis.

ecology (Gr. *oikos*, house + *logos*, word) The study of interactions of organisms with one another and with their physical environment.

ecosystem (Gr. *oikos*, house + *systems*, that which is put together) A major interacting system that involves both organisms and their nonliving environment.

ecotype (Gr. *oikos*, house + L. *typus*, image) A locally adapted variant of an organism; differing genetically from other ecotypes.

ectoderm (Gr. *ecto*, outside + *derma*, skin) One of the three embryonic germ layers of early vertebrate embryos; ectoderm gives rise to the outer epithelium of the body (skin, hair, nails) and to the nerve tissue, including the sense organs, brain, and spinal cord.

ectomycorrhizae Externally developing mycorrhizae that do not penetrate the cells they surround.

ectotherms Cold-blooded animal, such as reptiles, fish, or amphibians, whose body temperature is regulated by their behavior or by their surroundings.

electronegativity A property of atomic nuclei that refers to the affinity of the nuclei for valence electrons; a nucleus that is more electronegative has a greater pull on electrons than one that is less electronegative.

electron transport chain The passage of energetic electrons through a series of membrane-associated electron-carrier molecules to proton pumps embedded within mitochondrial or chloroplast membranes. *See* chemiosmosis.

endergonic reaction A chemical reaction in which the products contain more energy than the reactants, so that free energy must be put into the reaction from an outside source to allow it to proceed.

endocrine gland (Gr. *endon*, within + *krinein*, to separate) Ductless gland that secretes hormones into the extracellular spaces, from which they diffuse into the circulatory system.

endocytosis (Gr. *endon*, within + *kytos*, hollow vessel) The uptake of material into cells by inclusion within an invagination of the plasma membrane; the uptake of solid material is phagocytosis, while that of dissolved material is pinocytosis.

endoderm (Gr. *endon*, within + *derma*, skin) One of the three embryonic germ layers of early vertebrate embryos, destined to give rise to the epithelium that lines internal structures and most of the digestive and respiratory tracts.

endodermis (Gr. *endon*, within + *derma*, skin) In vascular plants, a layer of cells forming the innermost layer of the cortex in roots and some stems

endometrium (Gr. *endon*, within + *metrios*, of the womb) The lining of the uterus in mammals; thickens in response to secretion of estrogens and progesterone and is sloughed off in menstruation.

endomycorrhizae Mycorrhizae that develop within cells.

endoplasmic reticulum (ER) An internal membrane system that forms a netlike array of channels and interconnections of organelles within the cytoplasm of eukaryotic cells.

endorphin One of a group of small neuropeptides produced by the vertebrate brain; like morphine, endorphins modulate pain perception.

endosperm (Gr. *endon*, within + *sperma*, seed) A storage tissue characteristic of the seeds of angiosperms, which develops from the union of a male nucleus and the polar nuclei of the embryo sac. The endosperm is digested by the growing sporophyte either before maturation of the seed or during its germination.

endosymbiotic theory Proposes that eukaryotic cells evolved from a symbiosis between different species of prokaryotes.

endothermy The ability of animals to maintain a constant body temperature.

energy The capacity to do work.

enhancer A site of regulatory protein binding on the DNA molecule distant from the promoter and start site for a gene's transcription.

enthalpy In a chemical reaction, the energy contained in the chemical bonds of the molecule, symbolized as H; in a cellular reaction, the free energy is equal to the enthalpy of the reactant molecules in the reaction.

entropy (Gr. *en*, in + *tropos*, change in manner) A measure of the randomness or disorder of a system; a measure of how much energy in a system has become so dispersed (usually as evenly distributed heat) that it is no longer available to do work.

enzyme (Gr. *enzymes*, leavened, from *en*, in + *zyme*, leaven) A protein that is capable of speeding up specific chemical reactions by lowering the required activation energy.

epicotyl The region just above where the cotyledons are attached.

epidermis (Gr. *epi*, on or over + *derma*, skin) The outermost layers of cells; in plants, the exterior primary tissue of leaves, young stems, and roots; in vertebrates, the nonvascular external layer of skin, or ectodermal origin; in invertebrates, a single layer of ectodermal epithelium.

epididymis (Gr. *epi*, on + *didymos*, testicle) A sperm storage vessel; a coiled part of the sperm duct that lies near the testis.

epistasis (Gr. *epi*, on + *stasis*, a standing still) Interaction between two nonallelic genes in which one of them modifies the phenotypic expression of the other.

epithelium (Gr. *epi*, on + *thele*, nipple) In animals, a type of tissue that covers an exposed surface or lines a tube or cavity.

equilibrium (L. *acequus*, equal + *libra*, balance) A stable condition; the point at which a chemical reaction proceeds as rapidly in the reverse direction as it does in the forward direction, so that there is no further net change in the concentrations of products or reactants.

erythrocyte (Gr. *erythros*, red + *kytos*, hollow vessel) Red blood cell, the carrier of hemoglobin.

erythropoiesis The manufacture of blood cells in the bone marrow.

estrous cycle The periodic cycle in which periods of estrus correspond to ovulation events.

estrus (L. *oestrus*, frenzy) The period of maximum female sexual receptivity, associated with ovulation of the egg.

ethology (Gr. *ethos*, habit or custom + *logos*, discourse) The study of patterns of animal behavior in nature.

eubacteria A major group of bacteria, including most of those living today, which have very strong cell walls and a simpler gene architecture than that of the archaebacteria. *See* archaebacteria.

euchromatin (Gr. *eu*, good + *chroma*, color) That portion of a eukaryotic chromosome that is transcribed into mRNA; contains active genes that are not tightly condensed during interphase.

eukaryote (Gr. *eu*, good + *karyon*, kernel) A cell characterized by membrane-bound organelles, most notably the nucleus, and one that possesses chromosomes whose DNA is associated with proteins; an organisms composed of such cells.

eutrophic (Gr. *eutrophos*, thriving) Refers to a lake in which an abundant supply of minerals and organic matter exists.

evolution (L. *evolvere*, to unfold) Genetic change in a population of organisms; in general, evolution leads to progressive change from simple complex.

exergonic reaction A chemical reaction in which the products contain less free energy than the reactants, so that free energy is released in the reaction.

exocrine gland (Gr. *ex*, out of + *krinein*, to separate) A type of gland that releases its secretion through a duct, such as digestive glands and sweat glands.

exocytosis (Gr. *ex*, out + *kytos*, vessel) A type of bulk transport out of cells where a vacuole fuses with the cell membrane, discharging the vacuole's contents to the outside.

exon (Gr. *exo*, outside) A segment of DNA that is both transcribed into RNA and translated into protein.

exoskeleton (Gr. *exo*, outside + *skeletos*, hand) An external skeleton, as in arthropods.

exteroception (L. *exter*, outward + *capere*, to take) The condition when a sense organ is excited by stimuli from the external world.

F

facilitated diffusion Carrier-assisted diffusion of molecules across a cellular membrane through specific channels from a region of higher concentration to one of lower concentration; the process is driven by the concentration gradient and does not require cellular energy from ATP.

fat A molecule composed of glycerol and three fatty acid molecules.

feedback inhibition Control mechanism whereby an increase in the concentration of some molecules inhibits the synthesis of that molecule.

fermentation (L. *fermentum*, ferment) The enzyme-catalyzed extraction of energy from organic compounds without the involvement of oxygen.

fertilization The fusion of two haploid gamete nuclei to form a diploid zygote nucleus.

fibroblast (L. *fibra*, fiber + Gr. *blastos*, sprout) A flat, irregularly branching cell of connective tissue that secretes structurally stong proteins into the matrix between the cells.

first law of thermodynamics Energy cannot be created or destroyed, but can only undergo conversion from one form to another; thus, the amount of energy in the universe is unchangeable.

fitness The genetic contribution of an individual to succeeding generations, relative to the contributions of other individuals in the population.

fixed action pattern A stereotyped animal behavior response, thought by ethologists to be based on programmed neural circuits.

foraging behavior A collective term for the many complex, evolved behaviors that influence what an animal eats and how the food is obtained.

founder principle The effect by which rare alleles and combinations of alleles may be enhanced in new populations.

fovea (L., a small pit) A small depression in the center of the retina with a high concentration of cones; the area of sharpest vision.

free energy Energy available to do work.

free radical An ionized atom with one or more unpaired electrons, resulting from electrons that have been energized by ionizing radiation being ejected from the atom; free radicals react violently with other molecules, such as DNA, causing damage by mutation.

fruit In angiosperms, a mature, ripened ovary (or group of ovaries), containing the seeds.

fundamental niche Also referred to as the hypothetical niche, this is the entire niche an organism could fill if there were no other interacting factors (such as competition or predation).

G

gametangium, pl. **gametangia** (Gr. *gamein*, to marry + L. *tangere*, to touch) A cell or organ in which gametes are formed.

gamete (G., wife) A haploid reproductive cell.

gametocytes Cells in the malarial sporozoite life cycle capable of giving rise to gametes when in the correct host.

gametophyte In plants, the haploid (1*n*), gamete-producing generation, which alternates with the diploid (2*n*) sporophyte.

ganglion, pl. **ganglia** (Gr., a swelling) An aggregation of nerve cell bodies; in invertebrates, ganglia are the integrative centers; in vertebrates, the term is restricted to aggregations of nerve cell bodies located outside the central nervous system.

gap junction A junction between adjacent animal cells that allows the passage of materials between the cells.

gastrula (Gr., little stomach) In vertebrates, the embryonic stage in which the blastula with its single layer of cells turns into a three-layered embryo made up of ectoderm, mesoderm, and endoderm.

gene (Gr. *genos*, birth, race) The basic unit of heredity; a sequence of DNA nucleotides on a chromosome that encodes a protein, tRNA, or rRNA molecule, or regulates the transcription of such a sequence.

gene conversion Alteration of one homologous chromosome by the cell's error-detection and repair system to make it resemble the sequence on the other homologue.

genetic drift Random fluctuation in allele frequencies over time by chance.

genotype (Gr. *genos*, offspring + *typos*, form) The genetic constitution underlying a single trait or set of traits.

genus, pl. **genera** (L., race) A taxonomic group that ranks below a family and above a species.

germination (L. *germinare*, to sprout) The resumption of growth and development by a spore or seed.

germ layers The three cell layers formed at gastrulation of the embryo that foreshadow the future organization of tissues; the layers, from the outside inward, are the ectoderm, the mesoderm, and the endoderm.

gill A respiratory organ of aquatic animals, usually a thin-walled projection from some part of the external body surface, endowed with a rich capillary bed and having a large surface area; basidiomycete fungi, the plates on the underside of the cap.

glomerular filtrate The fluid that passes out of the capillaries of each glomerulus.

glomerulus (L., a little ball) A cluster of capillaries enclosed by Bowman's capsule.

glucagon (Gr. *glukus*, sweet + *ago*, to lead forward) A vertebrate hormone produced in the pancreas that acts to initiate the breakdown of glycogen to glucose subunits.

gluconeogenesis The synthesis of glucose from noncarbohydrates (such as proteins or fats).

glucose A common six-carbon sugar ($C_6H_{12}O_6$); the most common monosaccharide in most organisms.

glycocalyx A "sugar coating" on the surface of a cell resulting from the presence of polysaccharides on glycolipids and glycoproteins embedded in the outer layer of the plasma membrane.

glycogen (Gr. *glykys*, sweet + *gen*, of a kind) Animal starch; a complex branched polysaccharide that serves as a food reserve in animals, bacteria, and fungi.

glycolipid Lipid molecule modified within the Golgi complex by having a short sugar chain (polysaccharide) attached.

glycolysis (Gr. *glykys*, sweet + *lyein*, to loosen) The anaerobic breakdown of glucose; this enzyme-catalyzed process yields two molecules of pyruvate with a net of two molecules of ATP.

glycoprotein Protein molecule modified within the Golgi complex by having a short sugar chain (polysaccharide) attached.

glyoxysome A small cellular organelle or microbody containing enzymes necessary for conversion of fats into carbohydrates.

Golgi complex A collection of flattened stacks of membranes (each called a Golgi body) in the cytoplasm of eukaryotic cells; functions in collection, packaging, and distribution of molecules synthesized in the cell.

G protein A protein that binds guanosine triphosphate (GTP) and assists in the function of cell surface receptors. When the receptor binds its signal molecule, the G protein binds GTP and is activated to start a chain of events within the cell.

gravitropism (L. *gravis*, heavy + *tropes*, turning) Growth response to gravity in plants; formerly called geotropism.

ground meristem (Gr. *meristos*, divisible) The primary meristem, or meristematic tissue, that gives rise to the plant body (except for the epidermis and vascular tissues).

guttation (L. *gutta*, a drop) The exudation of liquid water from leaves due to root pressure.

gymnosperm (Gr. *gymnos*, naked + *sperma*, seed) A seed plant with seed not enclosed in an ovary; conifers are gymnosperms.

gynoecium (Gr. *gyne*, woman + *oikos*, house) The aggregate of carpels in the flower of a seed plant.

H

habitat (L. *hibitare*, to inhabit) The environment of an organism; the place where it is usually found.

habituation (L. *habitus*, condition) A form of learning; a diminishing response to a repeated stimulus.

haplodiploidy A phenomenon encountered in certain organisms such as wasps, wherein both haploid (male) and diploid (female) individuals are encountered.

haploid (Gr. *haploos*, single + *ploion*, vessel) Having only one set of chromosomes (*n*), in contrast to diploid (2*n*).

Hardy-Weinberg equilibrium A mathematical description of the fact that allele and genotype frequencies remain constant in a random-mating population in the absence of inbreeding, selection, or other evolutionary forces; usually stated: if the frequency of allele *a* is *p* and the frequency of allele *b* is *q*, then the genotype frequencies after one generation of random mating will always be $p^2(a) = 2pq(ab) + q^2(b)$.

Haversian canal After Clopton Havers, English anatomist. Narrow channels that run parallel to the length of a bone and contain blood vessels and nerve cells.

helper T cell A class of white blood cells that initiates both the cell-mediated immune response and the humoral immune response; helper T cells are the targets of the AIDS virus (HIV).

hemoglobin (Gr. *haima*, blood + L. *globus*, a ball) A globular protein in vertebrate red blood cells and in the plasma of many invertebrates that carries oxygen and carbon dioxide.

hemophilia (Gr. *haima*, blood + *philios*, friendly) A group of hereditary diseases characterized by failure of the blood to clot.

hemopoietic stem cell (Gr. *haimatos*, blood + *poiesis*, a making) The cells in bone marrow where blood cells are formed.

hermaphrodite (Gr. *hermaphroditos*, containing both sexes) An organism with both male and female functional reproductive organs.

heterochromatin (Gr. *heteros*, different + *chroma*, color) The portion of a eukaryotic chromosome that is not transcribed into RNA; remains condensed in interphase and stains intensely in histological preparations.

heterokaryotic (Gr. *heteros*. other + *karyon*, kernel) In fungi, having two or more genetically distinct types of nuclei within the same mycelium.

heterosporous (Gr. *heteros*, other + *sporos*, seed) In vascular plants, having spores of two kinds, namely, microspores and megaspores.

hetrotroph (Gr. *heteros*, other + *trophos*, feeder) An organism that cannot derive energy from photosynthesis or inorganic chemicals, and so must feed on other plants and animals obtaining chemical energy by degrading their organic molecules.

heterozygous Having two different alleles of the same gene; the term is usually applied to one or more specific loci, as in "heterozygous with respect to the *W* locus" (that is, the genotype is *W/w*).

histone (Gr. *histos*, tissue) One of a group of relatively small, very basic polypeptides, rich in arginine and lysine, forming the core of nucleosomes around which DNA is wrapped in the first stage of chromosome condensation.

holoblastic cleavage (Gr. *holos*, whole + *blastos*, germ) Process in vertebrate embryos in which the cleavage divisions all occur at the same rate, yielding a uniform cell size in the blastula.

homeobox A sequence of 180 nucleotides located in homeotic genes that produces a 60-amino acid peptide sequence (the homeodomain) active in transcription factors.

homeostasis (Gr. *homeos*, similar + *stasis*, standing) The maintenance of a relatively stable internal physiological environment in an organism, usually involves some form of feedback self-regulation.

homeotherm (Gr. *homoios*, same or similar + *therme*, heat) An organism, such as a bird or mammal, capable of maintaining a stable body temperature independent of the environmental temperature. *See* endotherm.

homeotic gene One of a series of "master switch" genes that determine the form of segments developing in the embryo.

hominid (L. *homo*, man) Any primate in the human family, Hominidae. *Homo sapiens* is the only living representative.

hominoid (L. *homo*, man) Collectively, hominids and apes; together with the monkeys, hominoids constitute the anthropoid primates.

homokaryotic (Gr. *homos*, same + *karyon*, kernel) In fungi, having a nuclei with the same genetic makeup within a mycelium.

homologous chromosome (homologue) One of a pair of chromosomes of the same kind located in a diploid cell; one copy of each pair of homologues comes from each gamete that formed the zygote.

homology (Gr. *homologia*, agreement) A condition in which the similarity between two structures or functions is indicative of a common evolutionary origin.

homosporous (Gr. *homos*, same or similar + *sporos*, seed) In some plants, production of only one type of spore rather than differentiated types. Compare heterosporous.

homozygous Being a homozygote, having two identical alleles of the same gene; the term is usually applied to one or more specific loci, as in "homozygous

with respect to the *W* locus" (that is, the genotype is *W/W* or *w/w*).

hormone (Gr. *hormaein*, to excite) A molecule, usually a peptide or steroid, that is produced in one part of an organism and triggers a specific cellular reaction in target tissues and organs some distance away.

human immunodeficiency virus (HIV) The virus responsible for AIDS, a deadly disease that destroys the human immune system. HIV is a retrovirus (its genetic material is RNA) that is thought to have been introduced to humans from African green monkeys.

hybridization The mating of unlike parents.

hydroskeleton (Gr. *hydro*, water + *skeletos*, hard) The skeleton of most soft-bodied invertebrates that have neither an internal nor an external skeleton. They use the relative incompressibility of the water within their bodies as a kind of skeleton.

hyperosmotic The condition in which a (hyperosmotic) solution has a higher osmotic concentration than that of a second solution.

hyperpolarization An above normal negativity of a cell membrane during its resting potential.

hypha, pl. **hyphae** (Gr. *hyphe*, web) A filament of a fungus or oomycete; collectively, the hyphae comprise the mycelium.

hypocotyl The region immediately below where the cotyledons are attached.

hypoosmotic The condition in which a (hypoosmotic) solution has a lower osmotic concentration than that of a second solution.

hypothalamus (Gr. *hypo*, under + *thalamos*, inner room) A region of the vertebrate brain just below the cerebral hemispheres, under the thalamus; a center of the autonomic nervous system, responsible for the integration and correlation of many neural and endocrine functions.

I

imaginal disk One of about a dozen groups of cells set aside in the abdomen of a larval insect and committed to forming key parts of the adult insect's body.

immune response In vertebrates, a defensive reaction of the body to invasion by a foreign substance or organism. *See* antibody and B cell.

immunoglobulin (L. *immunis*, free + *globus*, globe) An antibody.

inbreeding The breeding of genetically related plants or

animals; inbreeding tends to increase homozygosity.

inclusive fitness Describes the sum of the number of genes directly passed on in an individual's offspring and those genes passed on indirectly by kin (other than offspring) whose existence results from the benefit of the individual's altruism.

induction Processes by which one group of embryonic cells affects an adjacent group, thereby inducing those cells to differentiate in a manner they otherwise would not have.

industrial melanism (Gr. *melas*, black) Phrase used to describe the evolutionary process in which initially light-colored organisms become dark as a result of natural selection.

inflammatory response (L. *inflammare*, to flame) A generalized nonspecific response to infection that acts to clear an infected area of infecting microbes and dead tissue cells so that tissue repair can begin.

initiation factor One of several proteins involved in the formation of an initiation complex in prokaryote polypeptide synthesis.

insertional inactivation Destruction of a gene's function by the insertion of a transposon.

instar A larval developmental stage in insects.

interferon In vertebrates, a protein produced in virus-infected cells that inhibit viral multiplication.

interneuron (association neuron) A nerve cell found only in the middle of the spinal cord that acts as a functional link between sensory neurons and motor neurons.

internode In plants, the region of a stem between two successive nodes.

interoception The sensing of information that relates to the body itself, its internal condition, and its position.

interphase The period between two mitotic or meiotic divisions in which a cell grows and its DNA replicates; includes G1, S, and G2 phases.

intron (L. *intra*, within) Portion of mRNA as transcribed from eukaryotic DNA that is removed by enzymes before the matue mRNA is translated into protein. *See* exon.

inversion (L. *invertere*, to turn upside down) A reversal in order of a segment of a chromosome; also, to turn inside out, as in embryogenesis of sponges or discharge of a nematocyst.

ion Any atom or molecule containing an unequal number of electrons and protons and therefore carrying a net positive or net negative charge.

ionizing radiation High-energy radiation that is highly mutagenic, producing free radicals that react with DNA; includes X rays and gamma rays.

isomer (Gr. *isos*, equal + *meros*, part) One of a group of molecules identical in atomic composition but differing in structural arrangement; for example, glucose and fructose.

isosmotic solutions The conditions in which the osmotic concentrations of two solutions are equal, so that no net water movement occurs between them by osmosis.

K

karyotype (Gr. *karyon*, kernel + *typos*, stamp or print) The morphology of the chromosomes of an organism as viewed with a light microscope.

keratin (Gr. *kera*, horn + *in*, suffix used for proteins) A tough, fibrous protein formed in epidermal tissues and modified into skin, feathers, hair and hard structures such as horns and nails.

kidney In vertebrates, the organ that filters the blood to remove nitrogenous wastes and regulates the balance of water and solutes in blood plasma.

kilocalorie Unit describing the amount of heat required to raise the temperature of a kilogram of water by one degree Celsius (°C); sometimes called a Calorie, equivalent to 1000 calories.

kinesis (Gr. *kinesis*, motion) Changes in activity level in an animal that are dependent on stimulus intensity. *See* kinetic energy.

kinetic energy The energy of motion.

kinetochore (Gr. *kinetikos*, putting in motion + *choros*, chorus) Disc-shaped protein structure within the centromere to which the spindle fibers attach during mitosis or meiosis. *See* centromere,

kingdom The highest commonly used taxonomic category.

kin selection Selection favoring relatives; an increase in the frequency of related individuals (kin) in a population, leading to an increase in the relative frequency in the population of those alleles shared by members of the kin group.

Krebs cycle Another name for the citric acid cycle; also called the tricarboxylic acid (TCA) cycle.

L

labrum (L., a lip) The upper lip of insects and crustaceans situated above or in front of the mandibles.

larynx The voice box; a cartilaginous organ that lies between the pharynx and trachea and is responsible for sound production in vertebrates.

lateral line system A sensory system encountered in fish, through which mechanoreceptors in a line down the side of the fish are sensitive to motion.

lateral meristems (L. *latus*, side + Gr. *meristos*, divided) In vascular plants, the meristems that give rise to secondary tissue; the vascular cambium and cork cambium.

Law of independent assortment Mendel's second law of heredity, stating that genes located on nonhomologous chromosomes assort independently of one another.

Law of segregation Mendel's first law of heredity, stating that alternative alleles for the same gene segregate from each other in production of gametes.

leaf primordium, pl. **primordia** (L. *primordium*, beginning) A lateral outgrowth from the apical meristem that will eventually become a leaf.

lenticels (L. *lenticella*, a small window) Spongy areas in the cork surfaces of stem, roots, and other plant parts that allow interchange of gases between internal tissues and the atmosphere through the periderm.

leucoplast (Gr. *leukos*, white + *plasein*, to form) In plant cells, a colorless plastic in which starch grains are stored; usually found in cells not exposed to light.

leukocyte (Gr. *leukos*, a white + *kytos*, hollow vessel) A white blood cell; a diverse array of nonhemoglobin-containing blood cells, including phagocytic macrophages and antibody-producing lymphocytes.

lichen Symbionic association between a fungus and a photosynthetic organism such as a green alga or cyanobacteria.

limbic system The hypothalamus, together with the network of neurons that link the hypothalamus to some areas of the cerebral cortex. Responsible for many of the most deep-seated drives and emotions of vertebrates, including pain, anger, sex, hunger, thirst, and pleasure.

lipase (Gr. *lipos*, fat + *-ase*, enzyme suffix) An enzyme that catalyzes the hydrolysis of fats.

lipid (Gr. *lipos*, fat) A nonpolar hydrophobic organic molecule that is insoluble in water (which is polar) but that dissolves readily in nonpolar organic solvents; includes fats, oils, waxes, steroids, phospholipids, and carotenoids.

lipid bilayer The structure of a cellular membrane, in which two layers of phospholipids spontaneously align so that the hydrophilic head groups are exposed to water, while the hydrophobic fatty acid tails are pointed toward the center of the membrane.

locus The position on a chromosome where a gene is located.

loop of Henle In the kidney of birds and mammals, a hairpin-shaped portion of the renal tubule in which water and salt are reabsorbed from the glomerular filtrate by diffusion.

lophophore A ring or U-shaped arrangement of tentacles, often ciliated, into which the coelom of the animal extends. The organ functions in feeding and gas exchange and is encountered in only a few phyla of small marine invertebrates.

luteal phase The second phase of the reproductive cycle, during which the mature eggs are released into the Fallopian tubes, a process called ovulation.

lymph (L. *lympha*, clear water) In animals, a colorless fluid derived from blood by filtration through capillary walls in the tissues.

lymphatic system In animals, an open vascular system that reclaims water that has entered interstitial regions from the bloodstream (lymph);includes the lymph nodes, spleen, thymus, and tonsils.

lymphocyte (L. *lympha*, water + Gr. *kytos*, hollow vessel) A type of white blood cell. Lymphocytes are responsible for the immune response; there are two principal class—B cells and T cells.

lymphokine A regulatory molecule that is secreted by lymphocytes. In the immune response, lymphokines secreted by helper T cells unleash the cell-mediated immune response.

lysis (Gr., a loosening) Disintegration of a cell by rupture of its cell membrane.

M

macroevolution (Gr. *makros*, large + L. *evolvere*, to unfold) The creation of new species and the extinction of old ones.

macromolecule (Gr. *makros*, large + L. *moleculus*, a little mass) An extremely large biological molecule; refers specifically to proteins, nucleic acids, polysaccharides, lipids, and complexes of these.

macronutrients (Gr. *makros*, large + L. *nutrire*, to nourish) Inorganic chemical elements required in large amounts for plant growth, such as nitrogen, potassium, calcium, phosphorus, magnesium, and sulfur.

macrophage A large phagocytic cell that is able to engulf and digest cellular debris and invading bacteria.

major histocompatibility complex (MHC) A set of protein cell-surface markers anchored in plasma membrane, which the immune system uses to identify "self." All the cells of a given individual have the same "self" marker, called an MHC protein.

Malpighian tubules Blind tubules opening into the hindgut of terrestrial arthropods; they function as excretory organs.

mandibles (L. *mandibula*, jaw) In crustaceans, insects, and myriapods, the appendages immediately posterior to the antennae; used to seize, hold, bite, or chew food.

mantle The soft, outermost layer of the body wall in mollusks; the mantle secretes the shell.

marsupial (L. *marsupium*, pouch) A mammal in which the young are born early in their development, sometimes as soon as eight days after fertilization, and are retained in a pouch.

mass flow The overall process by which materials move in the phloem of plants.

matrix (L. *mater*, mother) In mitochondria, the solution in the interior space surrounded by the cristae that contains the enzymes and other molecules involved in oxidative respiration; more generally, that part of a tissue within which an organ or process is embedded.

menstrual cycle (L. *mens*, month) Periodic sloughing off of the blood-enriched lining of the uterus when pregnancy does not occur.

menstruation Periodic sloughing off of the blood-enriched lining of the uterus when pregnancy does not occur.

meristem (Gr. *merizein*, to divide) Undifferentiated plant tissue from which new cells arise.

meroblastic cleavage (Gr. *meros*, part + *blastos*, sprout) A type of cleavage in the eggs of reptiles, birds, and some fish. Occurs only on the blastodisc.

mesoderm (Gr. *mesos*, middle + *derma*, skin) One of the three embryonic germ layers that form in the gastrula; gives rise to muscle, bone and other connective tissue, the peritoneum, the circulatory system, and most of the excretory and reproductive systems.

mesophyll (Gr. *mesos*, middle + *phyllon*, leaf) The photosynthetic parenchyma of a leaf, located within the epidermis.

messenger RNA (mRNA) The RNA transcribed from structural genes; RNA molecules complementary to a portion of one strand of DNA, which are translated by the ribosomes to form protein.

metabolism (Gr. *metabole*, change) The sum of all chemical processes occurring within a living cell or organism.

metamorphosis (Gr. *meta*, after + *morphe*, form + *osis*, state of) Process in which there is a marked change in form during postembryonic development as, for example, in tadpole to frog.

metaphase (Gr. *meta*, middle + *phasis*, form) The stage of mitosis or meiosis during which microtubules become organized into a spindle and the chromosomes come to lie in the spindle's equatorial plane.

metastasis The process by which cancer cells move from their point of origin to other locations in the body; also a population of cancer cells in a secondary location, the result of movement from the primary tumor.

methanogens Obligate, anaerobic archaebacteria that produce methane.

microbody A cellular organelle bounded by a single membrane and containing a variety of enzymes; generally derived from endoplasmic reticulum; includes peroxisomes and glyoxysomes.

microevolution (Gr. *mikros*, small + L. *evolvere*, to unfold) Refers to the evolutionary process itself. Evolution within a species. Also called adaptation.

micronutrient (Gr. *mikros*, small + L. *nutrire*, to nourish) A mineral required in only minute amounts for plant growth, such as iron, chlorine, copper, manganese, zinc, molybdenum, and boron.

micropyle In the ovules of seed plants, an opening in the integuments through which the pollen tube usually enters.

microtubule (Gr. *mikros*, small + L. *tubulus*, little pipe) In eukaryotic cells, a long, hollow protein cylinder, composed of the protein tubulin; these influence cell shape, move the chromosomes in cell division, and provide the functional internal structure of cilia and flagella.

microvillus (Gr. *mikros*, small + L. *villus*, shaggy hair) Cytoplasmic projection from epithelial cells, microvilli greatly increase the surface area of the small intestine.

middle lamella The layer of intercellular material, rich in pectic compounds, that cements together the primary walls of adjacent plant cells.

mimicry (Gr. *mimos*, mime) The resemblance in form, color, or behavior of certain organisms (mimics) to other more powerful or more protected ones (models).

mitosis (Gr. *mitos*, thread) Somatic cell division; nuclear division in which the duplicated chromosomes separate to form two genetically identical daughter nuclei.

monocot Short for monocotyledon; flowering plant in which the embryos have only one cotyledon, the floral parts are generally in threes, and the leaves typically are parallel-veined.

monocyte (Gr. *monos*, single + *kytos*, hollow vessel) A type of leukocyte that becomes a phagocytic cell (macrophage) after moving into tissues.

monoecious (Gr. *monos*, single + *oecos*, house) A plant in which the staminate and pistillate flowers are separate, but borne on the same individual.

monosaccharide (Gr. *monos*, one + L. *saccharum*, sugar) A simple sugar that cannot be decomposed into smaller sugar molecules.

monotreme (Gr. *mono*, single + *treme*, hole) An egg-laying mammal.

morphogen A signal molecule produced by an embryonic organizer region that informs surrounding cells of their distance from the organizer, thus determining relative position of cells during development.

morula (L., a little mulberry) Solid balls of cells in the early stage of embryonic development.

mosaic development A pattern of embryonic development in which initial cells produced by cleavage divisions contain different developmental signals (determinants) from the egg, setting the individual cells on different developmental paths.

motor (efferent) neuron Neuron that transmits nerve impulses from the central nervous system to an effector, which is typically a muscle or gland.

Müellerian mimicry After Fritz Müeller, German biologist. A phenomenon in which two or more unrelated but protected species resemble one another, thus achieving a kind of group defense.

multigene family A collection of related genes on a single chromosome or on different chromosomes.

muscle fiber A long, cylindrical, multinucleated cell containing numerous myofibrils, which is capable of contraction when stimulated.

mutagen (L. *mutare*, to change + Gr. *genaio*, to produce) An agent that induces changes in DNA (mutations): includes physical agents that damage DNA and chemicals that alter DNA bases.

mutant (L. *mutare*, to change) A mutated gene; alternatively, an organism carrying a gene that has undergone a mutation.

mutation A permanent change in a cell's DNA; includes changes in nucleotide sequence, alteration of gene position, gene loss or duplication, and insertion of foreign sequences.

mutualism (L. *mutuus*, lent, borrowed) The living together of two or more organisms in a symbiotic association in which both members benefit.

mycelium, pl. **mycelia** (Gr. *mykes*, fungus) In fungi, a mass of hyphae.

mycorrhiza, pl. **mycorrhizae** (Gr. *mykes*, fungus + *rhiza*, root) A symbiotic association between fungi and the roots of a plant.

myelin sheath (Gr. *myelinos*, full of marrow) A fatty layer surrounding the long axons of motor neurons in the peripheral nervous system of vertebrates.

myofilament (Gr. *myos*, muscle + L. *filare*, to spin) A contractile microfilament, composed largely of actin and myosin, within muscle. Sometimes called myofibril.

myosin (Gr. *mys*, muscle + *in*, belonging to) One of the two protein components of microfilaments (the other is actin); a principal component of vertebrate muscle.

N

natural killer cell A cell that does not kill invading microbes, but rather, the cells infected by them.

natural selection The differential reproduction of genotypes; caused by factors in the environment; leads to evolutionary change.

negative feedback A homeostatic control mechanism whereby an increase in some substance or activity inhibits the process leading to the increase; also known as feedback inhibition.

nephridium, pl. **nephridia** (Gr. *nephros*, kidney) In invertebrates, a tubular excretory structure.

nephrid organ A filtration system of many freshwater invertebrates in which water and waste pass from the body across the membrane into a collecting organ, from which they are expelled to the outside through a pore.

nephron (Gr. *nephros*, kidney) Functional unit of the vertebrate kidney; one of numerous tubules involved in filtration and selective reabsorption of blood; each nephron consists of a Bowman's capsule, an enclosed glomerulus, and a long attached tubule; in humans, called a renal tubule.

nerve A group or bundle of nerve fibers (axons) with accompanying neurological cells, held together by connective tissue; located in the peripheral nervous system.

neural crest A special strip of cells that develops just before the neural groove closes over to form the neural tube.

neural groove The long groove formed along the long axis of the embryo by a layer of ectodermal cells.

neural tube The dorsal tube, formed from the neural plate, that differentiates into the brain and spinal cord.

neuroethology (Gr. *neuron*, nerve + *ethos*, habit or custom + *logos*, discourse) The study of the neural basis of behavior.

neuroglia (Gr. *neuron*, nerve + *glia*, glue) Nonconducting nerve cells that are intimately associated with neurons and appear to provide nutritional support.

neuromuscular junction The structure formed when the tips of axons contact (innervate) a muscle fiber.

neuron (Gr., nerve) A nerve cell specialized for signal transmission; includes cell body; dendrites, and axon.

neurotransmitter (Gr. *neuron*, nerve + L. *trans*, across + *mittere*, to send) A chemical released at the axon terminal of a neuron that travels across the synaptic cleft, binds a specific receptor on the far side, and, depending on the nature of the receptor, depolarizes or hyperpolarizes a second neuron or a muscle or gland cell.

neurulation A process in early embryonic development by which a dorsal band of ectoderm thickens and rolls into the neural tube.

neutrophil An abundant type of granulocyte capable of engulfing microorganisms and other foreign particles; neutrophils comprise about 50% to 70% of the total number of white blood cells.

niche The role played by a particular species in its environment.

nicotinamide adenine dinucleotide (NAD⁺) A molecule that becomes reduced (to NADH) as it carries high-energy electrons from oxidized molecules and delivers them to ATP-producing pathways in the cell.

nociceptor A naked dendrite that acts as a receptor in response to a pain stimulus.

nocturnal (L. *nocturnus*, night) Active primarily at night.

node (L. *nodus*, knot) The part of a plant stem where one or more leaves are attached. *See* internode.

node of Ranvier After L. A. Ranvier, French histologist. A gap formed at the point where two Schwann cells meet and where the axon is in direct contact with the surrounding intercellular fluid.

nonassociative learning A learned behavior that does not require an animal to form an association between two stimuli, or between a stimulus and a response.

nonsense codon One of three codons (UAA, UAG, and UGA) that are not recognized by tRNAs, thus serving as "stop" signals in the mRNA message and terminating translation.

notochord (Gr. *noto*, back + L. *chorda*, cord) In chordates, a dorsal rod of cartilage that runs the length of the body and forms the primitive axial skeleton in the embryos of all chordates.

nucellus (L. *nucella* a small nut) Tissue composing the chief pair of young ovules, in which the embryo sac develops; equivalent to a megasporangium.

nucleic acid A nucleotide polymer; chief types are deoxyribonucleic acid (DNA), which is double-stranded, and ribonucleic acid (RNA), which is typically single-stranded.

nucleolus (L., a small nucleus) In eukaryotes, the site of rRNA synthesis; a spherical body composed chiefly of rRNA in the process of being transcribed from multiple copies of rRNA genes.

nucleosome (L. *nucleus*, kernel + *soma*, body) The fundamental packaging unit of eukaryotic chromosomes; a complex of DNA and histone proteins in which the double-helical DNA winds around eight molecules of histone; chromatin is composed of long sequences of nucleosomes.

nucleotide A single unit of nucleic acid, composed of a phosphate, a five-carbon sugar (either ribose or deoxyribose), and a purine or a pyrimidine.

nucleus In atoms, the central core, containing positively charged protons and (in all but

hydrogen) electrically neutral neurons; in eukaryotic cells, the membranous organelle that houses the chromosomal DNA; in the central nervous system, a cluster of nerve cell bodies.

O

ocellus, pl. ocelli (L., little eye) A simple light receptor common among invertebrates.

Okazaki fragment A short segment of DNA produced by discontinuous replication elongating in the 5′–> 3′ direction away from the replication.

olfaction (L. *olfactum*, smell) The process or function of smelling.

ommatidium, pl. ommatidia (Gr., little eye) The visual unit in the compound eye of arthropods; contains light-sensitive cells and a lens able to form an image.

oncogene (Gr. *oncos*, cancer + *genos*, birth) A mutant form of a growth-regulating gene that is inappropriately "on," causing unrestrained cell growth and division.

oocyst The zygote in a sporozoan life cycle. It is surrounded by a tough cyst to prevent dehydration or other damage.

oogonium The female gametangium of the Oomycota (fungi).

oospore A thick-walled zygote resulting from the fusion of Oomycote gametes.

operant conditioning A learning mechanism in which the reward follows only after the correct behavioral response.

operator A site of gene regulation; a sequence of nucleotides overlapping the promoter site and recognized by a repressor protein; binding of the repressor prevents binding of the polymerase to the promoter site and so blocks transcription of the structural gene.

operculum A flat, external, bony protective covering over the gill chamber in fish.

operon (L. *operis* work) A cluster of adjacent structural genes transcribed as a unit into a single mRNA molecule.

order A category of classification above the level of family and below that of class.

organ (Gr. *organon*, tool) A body structure composed of several different tissues grouped in a structural and functional unit.

organelle (Gr. *organella*, little tool) Specialized part of a cell; literally a small cytoplasmic organ.

osmoconformer An animal that maintains the osmotic concentration of its body fluids at

about the same level as that of the medium in which it is living.

osmosis (Gr. *osmos*, act of pushing, thrust) The diffusion of water across a selectively permeable membrane (a membrane that permits the free passage of water but prevents or retards the passage of a solute); in the absence of differences in pressure or volume, the net movement of water is from the side containing a lower concentration of solute to the side containing a higher concentration.

osmotic pressure The potential pressure developed by a solution separated from pure water by a differentially permeable membrane. The higher the solute concentration, the greater the osmotic potential of the solution; also called *osmotic potential.*

osteoblast (Gr. *osteon*, bone + *blastos*, bud) A bone-forming cell.

osteocyte (Gr. *osteon*, bone + *kytos*, hollow vessel) A mature osteoblast.

outcrossing Breeding with individuals other than oneself or one's close relatives.

ovary (L. *ovum*, egg) (1) In animals, the organ in which eggs are produced. (2) In flowering plants, the enlarged basal portion of a carpel, which contains the ovule(s); the ovary matures to become the fruit.

oviduct (L. *ovum*, egg + *ductus*, duct) In vertebrates, the passageway through which ova (eggs) travel from the ovary to the uterus.

oviparity (L. *ovum*, egg + *parere*, to bring forth) Refers to a type of reproduction in which the eggs are developed after leaving the body of the mother, as in reptiles.

ovoviviparity Refers to a type of reproduction in which young hatch from eggs that are retained in the mother's uterus.

ovulation In animals, the release of an egg or eggs from the ovary.

ovum, pl. ova (L., egg) The egg cell; female gamete.

oxidation (Fr. *oxider*, to oxidize) Loss of an electron by an atom or molecule; in metabolism, often associated with a gain of oxygen or loss of hydrogen.

oxidative respiration Process of cellular activity in which glucose or other molecules are broken down to water and carbon dioxide with the release of energy.

oxygen debt The amount of oxygen required to convert the lactic acid generated in the muscles during exercise back into glucose.

oxytocin (Gr. *oxys*, sharp + *tokos* birth) A hormone of the posterior pituitary gland that affects uterine

contractions during childbirth and stimulates lactation.

ozone O₃, a stratospheric layer of the earth's atmosphere responsible for filtering out ultraviolet radiation supplied by the sun.

P

pacemaker A patch of excitatory tissue in the vertebrate heart that initiates the heartbeat.

palisade parenchyma (L. *palus*, stake) In plant leaves, the columnar, chloroplast-containing parenchyma cells of the mesophyll. Also called *palisade cells.*

papilla A small projection of tissue.

paracrine A type of chemical signaling between cells where the effects are local and short-lived.

parapodia (Gr. *para*, beside + *pous*, foot) One of the paired lateral processes on each side of most segments in polychaete annelids.

parasexuality In certain fungi, the fusion and segregation of heterokaryotic haploid nuclei to produce recombinant nuclei.

parasitism (Gr. *para*, beside + *sitos*, food) A living arrangement in which an organism lives on or in an organism of a different species and derives nutrients from it.

parthenogenesis (Gr. *parthenos*, virgin + *genesis*, birth) The development of an egg without fertilization, as in aphids, bees, ants, and some lizards.

partial pressure The components of each individual gas—nitrogen, oxygen, and carbon—that together comprise the total air pressure.

pelagic Free-swimming, usually in open water.

pellicle A tough, flexible covering in ciliates and euglenoids.

peptide bond (Gr. *peptein*, to soften, digest) The type of bond that links amino acids together in proteins through a dehydration reaction.

perianth (Gr. *peri*, around + *anthos*, flower) In flowering plants, the petals and sepals taken together.

pericycle (Gr. *peri*, around + *kykos*, circle) In vascular plants, one or more cell layers surrounding the vascular tissues of the root, bounded externally by the endodermis and internally by the phloem.

periderm (Gr. *peri*, around + *derma*, skin) Outer protective tissue in vascular plants that is produced by the cork cambium and functionally replaces epidermis when it is destroyed during secondary growth; the periderm includes the cork, cork cambium, and phelloderm.

peristalsis (Gr. *peristaltikos*, compressing around) In animals, a series of alternating contracting and relaxing muscle movements along the length of a tube such as the oviduct or alimentary canal that tend to force material such as an egg cell or food through the tube.

peroxisome A microbody that plays an important role in the breakdown of highly oxidative hydrogen peroxide by catalase.

petal A flower part, usually conspicuously colored; one of the units of the corolla.

petiole (L. *petiolus*, a little foot) The stalk of a leaf.

phagocyte (Gr. *phagein*, to eat + *kytos*, hollow vessel) Any cell that engulfs and devours microorganisms or other particles.

phagocytosis (Gr., cell-eating) Endocytosis of solid particle; the cell membrane folds inward around the particle (which may be another cell) and engulfs it to form a vacuole.

pharynx (Gr., gullet) In vertebrates, a muscular tube that connects the mouth cavity and the esophagus; it serves as the gateway to the digestive tract and to the trachea.

phenetics Also called numerical taxonomy, this taxonomic technique seeks to classify organisms based on overall morphological similarity and does not take into account historical relationships between organisms.

phenotype (Gr. *phainein*, to show + *typos*, stamp or print) The realized expression of the genotype; the physical appearance or functional expression of a trait.

pheromone (Gr. *pherein*, to carry + *hormonos*, exciting, stirring up) Chemical substance released by one organism that influences the behavior or physiological processes of another organism of the same species. Some pheromones serve as sex attractants, as trail markers, and as alarm signals.

phloem (Gr. *phloos*, bark) In vascular plants, a food-conducting tissue basically composed of sieve elements, various kinds of parenchyma cells, fibers, and sclereids.

phosphodiester bond The type of bond that links nucleotides in a nucleic acid; formed when the phosphate group of one nucleotide binds to the 3′ hydroxyl group of the sugar of another.

phospholipid Similar in structure to a fat, but having only two fatty acids attached to the glycerol backbone, with the third space linked to a phosphorylated molecule; contains a polar hydrophilic "head" end (phosphate group) and a nonpolar hydrophobic "tail" end (fatty acids).

photoperiodism (Gr. *photos*, light + *periodos*, a period) The tendency of biological reactions to respond to the duration and timing of day and night; a mechanism for measuring seasonal time.

photoreceptor (Gr. *photos*, light) A light-sensitive sensory cell.

photosystem An organized collection of chlorophyll and other pigment molecules embedded in the thylakoid of chloroplasts; traps photon energy and channels it as energetic electrons to the thylakoid membrane.

phototropism (Gr. *photos*, light + *trope*, turning) In plants, a growth response to a light stimulus.

phycologist (Gr. *phykos*, seaweed) The scientist who studies algae.

phylum, pl. phyla (Gr. *phylon*, race, tribe) A major category, between kingdom and class, of taxonomic classifications.

phytochrome (Gr. *phyton*, plant + *chroma*, color) A plant pigment that is associated with the absorption of light; photoreceptor for red to far-red light.

pigment (L. *pigmentum*, paint) A molecule that absorbs light.

pilus, pl. pili Extensions of a bacterial cell enabling it to transfer genetic materials from one individual to another or to adhere to substrates.

pinocytosis (Gr. *pinein*, to drink + *kytos*, hollow vessel + *osis*, condition) The process of fluid uptake by endocytosis in a cell.

pistil (L. *pistillum*, pestle) Central organ of flowers, typically consisting of ovary, style, and stigma; a pistil may consist of one or more fused carpels and is more technically and better known as the gynoecium.

pith The ground tissue occupying the center of the stem or root within the vascular cylinder.

placenta, pl. placentae (L., a flat cake) (1) In flowering plants, the part of the ovary wall to which the ovules or seeds are attached. (2) In mammals, a tissue formed in part from the inner lining of the uterus and in part from other membranes, through which the embryo (later the fetus) is nourished while in the uterus and wastes are carried away.

plankton (Gr. *planktos*, wandering) Free-floating, mostly microscopic, aquatic organisms.

plasma (Gr., form) The fluid of vertebrate blood; contains dissolved salts, metabolic wastes, hormones, and a variety of proteins, including antibodies and albumin; blood minus the blood cells.

plasma cell An antibody-producing cell resulting from the multiplication and differentiation of a B lymphocyte that has interacted with an antigen.

plasma membrane The membrane surrounding the cytoplasm of a cell; consists of a single phospholipid bilayer with embedded protein.

plasmid (Gr. *plasma*, a form or mold) A small fragment of extrachromosomal DNA, usually circular, that replicates independently of the main chromosome, although it may have been derived from it.

plasmodium (Gr. *plasma*, a form, mold + *eidos*, form) Stage in the life cycle of myxomycetes (plasmodial slime molds); a multinucleate mass of protoplasm surrounded by a membrane.

plastid (Gr. *plastos*, formed or molded) An organelle in the cells of photosynthetic eukaryotes that is the site of photosynthesis and, in plants and green algae, of starch storage.

platelet (Gr. dim. of *plattus*, flat) In mammals, a fragment of a white blood cell that circulates in the blood and functions in the formation of blood clots at sites of injury.

pleiotropy Condition in which an individual allele has more than one effect on production of the phenotype.

plumule The epicotyl of a plant with its two young leaves.

point mutation An alteration of one nucleotide in a chromosomal DNA molecule.

polar body Minute, nonfunctioning cell produced during the meiotic divisions leading to gamete formation in vertebrates.

pollen tube A tube formed after germination of the pollen grain; carries the male gametes into the ovule.

pollination The transfer of pollen from an anther to a stigma.

polyandry The condition in which a female mates with more than one male.

polyclonal antibody An antibody response in which an antigen elicits many different antibodies, each fitting a different portion of the antigen surface.

polygyny (Gr. *poly* many + *gyne*, woman, wife) A mating choice in which a male mates with more than one female.

polymer (Gr. *polus*, many + *meris*, part) A molecule composed of many similar or identical molecular subunits; starch is a polymer of glucose.

polymerase chain reaction (PCR) A process by which DNA polymerase is used to copy a sequence of interest repeatedly, making millions of copies of the same DNA.

polymorphism (Gr. *polys*, many + *morphe*, form) The presence in a population of more than one allele of a gene at a frequency greater than that of newly arising mutations.

polypeptide (Gr. *polys*, many + *peptein*, to digest) A molecule consisting of many joined amino acids; not usually as complex as a protein.

polyploidy Condition in which one or more entire sets of chromosomes is added to the diploid genome.

polysaccharide (Gr. *polys*, many + *sakcharon*, sugar, from Latin *sakara*, gravel, sugar) A carbohydrate composed of many monosaccharide sugar subunits linked together in a long chain; examples are glycogen, starch, and cellulose.

polyunsaturated fat A fat molecule in which there are at least two double bonds between adjacent carbons, in one or more of the fatty acid chains.

population (L. *populus*, the people) Any group of individuals, usually of a single species, occupying a given area at the same time.

posttranscriptional control A mechanism of control over gene expression that operates after the transcription of mRNA is complete.

potential energy Energy that is not being used, but could be; energy in a potentially usable form; often called "energy of position."

precapillary sphincter A ring of muscle that guards each capillary loop and that, when closed, blocks flow through the capillary.

primary endosperm nucleus In flowering plants, the result of the fusion of a sperm nucleus and the (usually) two polar nuclei.

primary growth In vascular plants, growth originating in the apical meristems of shoots and roots; results in an increase in length.

primary immune response The first response of an immune system to a foreign antigen. If the system is challenged again with the same antigen, the memory cells created during the primary response will respond more quickly.

primary induction Inductions between the three primary tissue types—ectoderm, mesoderm, and endoderm.

primary nondisjunction Failure of chromosomes to separate properly at meiosis I.

primary productivity The amount of energy produced by photosynthetic organisms in a community.

primary structure The specific amino acid sequence of a protein.

primary tissues Tissues that comprise the primary plant body.

primary transcript The initial mRNA molecule copied from a gene by RNA polymerase, containing a faithful copy of the entire gene, including introns as well as exons.

primary wall In plants, the wall layer deposited during the period of cell expansion.

primate Monkeys and apes (including humans).

primitive streak (L. *primus*, first) In the early embryos of birds, reptiles, and mammals, a dorsal, longitudinal strip of ectoderm and mosoderm that is equivalent to the blastopore in other forms.

prions Infectious proteinaceous particles.

procambium (L. *pro*, before + *cambiare*, to exchange) In vascular plants, a primary meristematic tissue that gives rise to primary vascular tissues.

prokaryote (Gr. *pro*, before + *karyon*, kernel) A bacterium; a cell lacking a membrane-bounded nucleus or membrane-bounded organelles.

promoter A specific nucleotide sequence to which RNA polymerase attaches to initiate transcription of mRNA from a gene.

prophase (Gr. *pro*, before + *phasis*, form) An early stage in nuclear division, characterized by the formation of a microtubule spindle along the future axis of division, the shortening and thickening of the chromosomes, and their movement toward the equator of the spindle (the "metaphase plate").

proprioceptor (L. *proprius*, one's own) In vertebrates, a sensory receptor that senses the body's position and movements.

prostaglandins (Gr. *prostas*, a porch or vestibule + *glans*, acorn) A group of modified fatty acids, that function as chemical messengers.

prostate gland (Gr. *prostas*, a porch or vestibule) In male mammals, a mass of glandular tissue at the base of the urethra that secretes an alkaline fluid that has a stimulating effect on the sperm as they are released.

protein (Gr. *proteios*, primary) A chain of amino acids joined by peptide bonds.

protein kinase An enzyme that adds phosphate groups to proteins, changing their activity.

proton pump A protein channel in a membrane of the cell that expends energy to transport protons against a concentration gradient; involved in the chemiosmotic generation of ATP.

proto-oncogene A normal gene that promotes cell division, so called because mutations that cause these genes to become overexpressed convert them into oncogenes that produce excessive cellular proliferation—cancer.

protozoa (Gr. *protos*, first + *zoon*, animal) The traditional name given to heterotrophic protists.

pseudogene (Gr. *pseudos*, false + *genos*, birth) A copy of a gene that is not transcribed.

pseudopod, or pseudopodium (Gr. *pseudes*, false + *pous*, foot) A nonpermanent cytoplasmic extension of the cell body.

punctured equilibrium A hypothesis of the mechanism of evolutionary change that proposes that long periods of little or no change are punctuated by periods of rapid evolution.

pupa (L., girl, doll) A developmental stage of some insects in which the organism is nonfeeding, immotile, and sometimes encapsulated or in a cocoon; the pupal stage occurs between the larval and adult phases.

purine (Gr. *purinos*, fiery, sparkling) The larger of the two general kinds of nucleotide base found in DNA and RNA; a nitrogenous base with a double-ring structure, such as adenine or guanine.

pyrimidine (alt. of pyridine, from G. *pyr*, fire) The smaller of two general kinds of nucleotide base found in DNA and RNA; a nitrogenous base with a single-ring structure, such as cytosine, thymine, or uracil.

Q

quaternary structure The structural level of a protein composed of more than one polypeptide chain, each of which has its own teriary structure; the individual chains are called subunits.

R

radicle (L. *radicula*, root) The part of the plant embryo that develops into the root.

radioactivity The emission of nuclear particles and rays by unstable atoms as they decay into more stable forms.

radula (L., scraper) Rasping tongue found in most mollusks.

realized niche The actual niche occupied by an organism when all biotic and abiotic interactions are taken into account.

receptor-mediated endocytosis Process by which specific macromolecules are transported into eukaryotic cells at clathrin-coated pits, after binding to specific cell-surface receptors.

receptor protein A highly specific cell-surface receptor embedded in a cell membrane that responds only to a specific messenger molecule.

recessive An allele that is only expressed when present in the homozygous condition, while being "hidden" by the expression of a dominant allele in the heterozygous condition.

reciprocal altruism Performance of an altruistic act with the expectation that the favor will be returned. A key and very controversial assumption of many theories dealing with the evolution of social behavior. *See* altruism.

reciprocal recombination A mechanism of genetic recombination that occurs only in eukaryotic organisms, in which two chromosomes trade segments; can occur between nonhomologous chromosomes as well as the more usual exchange between homologous chromosomes in meiosis.

recombinant DNA Fragments of DNA from two different species, such as a bacterium and mammal, spliced together in the laboratory into a single molecule.

reduction (L. *reductio*, a bringing back; originally "bringing back" a metal from its oxide) The gain of an electron by an atom often with an associated proton.

reflex (L. *reflectere*, to bend back) In the nervous system, a motor response subject to little associative modification; among the simplest neural pathways, involving only a sensory neuron, sometimes (but not always) an interneuron, and one or more motor neurons.

reflex arc The nerve path in the body that leads from stimulus to reflex action.

refractory period The recovery period after membrane depolarization during which the membrane is unable to respond to additional stimulation.

replication fork A Y-shaped end of a growing replication bubble in a DNA molecule undergoing replication.

repolarization Return of the ions in a nerve to their resting potential distribution following depolarization.

repressor (L. *reprimere*, to press back, keep back) A protein that regulates DNA transcription by preventing RNA polymerase from attaching to the promoter and transcribing the structural gene. *See* operator.

residual volume The amount of air remaining in the lungs after the maximum amount of air has been exhaled.

resting membrane potential The charge difference (difference in electric potential) that exists across a neuron at rest (about 70 millivolts).

restriction endonuclease An enzyme that cleaves a DNA duplex molecule at a particular base sequence, usually within or near a palindromic sequence; also called restriction enzyme.

retina (L., a small net) The photosensitive layer of the vertebrate eye; contains several layers of neurons and light receptors (rods and cones); receives the image formed by the lens and transmits it to the brain via the optic nerve.

retrovirus (L. *retro*, turning back) An RNA virus. When a retrovirus enters a cell, a viral enzyme (reverse transcriptase) transcribes viral RNA into duplex DNA, which the cell's machinery then replicates and transcribes as if it were its own.

Rh blood group A set of cell surface markers (antigens) on the surface of red blood cells in humans and rhesus monkeys (for which it is named); although there are several alleles, they are grouped into two main types called Rh-positive and Rh-negative.

rhizome (Gr. *rhizoma*, mass of roots) In vascular plants, a usually more or less horizontal underground stem; may be enlarged for storage or may function in vegetative reproduction.

ribonucleic acid (RNA) A class of nucleic acids characterized by the presence of the sugar ribose and the pyrimidine uracil; includes mRNA, tRNA, and rRNA.

ribosomal RNA (rRNA) A class of RNA molecules found, together with characteristic proteins, in ribosomes; transcribed from the DNA of the nucleolus.

ribosome The molecular machine that carries out protein synthesis; the most complicated aggregation of proteins in a cell, also containing three different rRNA molecules.

ribozyme An RNA molecule that can behave as an enzyme, sometimes catalyzing its own assembly; rRNA also acts as a ribozyme in the polymerization of amino acids to form protein.

RNA polymerase An enzyme that catalzyes the assembly of an mRNA molecule, the sequence of which is complementary to a DNA molecule used as a template. *See* transcription.

RNA primer In DNA replication, a sequence of about 10 RNA mucleotides complementary to unwound DNA that attaches at a replication fork; the DNA polymerase uses the RNA primer as a starting point for addition of DNA nucleotides to form the new DNA strand; the RNA primer is later removed and replaced by DNA nucleotides.

RNA slicing A nuclear process by which intron sequences of a primary mRNA transcript are cut out and the exon sequences are spliced together to give the correct linkages of genetic information that will be used in protein construction.

rod Light-sensitive nerve cell found in the vertebrate retina; sensitive to very dim light; responsible for "night vision."

root The usually descending axis of a plant, normally below ground, which anchors the plant and serves as the major point of entry for water and minerals.

rumen An "extra stomach" in cows and related mammals wherein digestion of cellulose occurs and from which partially digested material can be ejected back into the mouth.

S

saltatory conduction A very fast form of nerve impulse conduction in which the impulses leap from node to node over insulation portions.

saprobes Heterotrophic organisms that digest their food externally (such as most fungi).

sarcolemma The specialized cell membrane in a muscle cell.

sarcoma A cancerous tumor arising from cells of connective tissue, bone, or muscle.

sarcomere (Gr. *sarx*, flexh + *meris*, part of) Fundamental unit of contraction in skeletal muscle; repeating bands of actin and myosin that appear between two Z lines.

sarcoplasmic reticulum (Gr. *sarx*, flesh + *plassein*, to form, mold; L. *reticulum*, network) The endoplasmic reticulum of a muscle cell. A sleeve of membrane that wraps around each myofilament.

satellite DNA A nontranscribed region of the chromosome with a distinctive base composition; a short nucleotide sequence repeated tandemly many thousands of times.

saturated fat A fat composed of fatty acids in which all the internal carbon atoms contain the maximum possible number of hydrogen atoms.

Schwann cells The supporting cells associated with projecting axons, along with all the other nerve cells that make up the peripheral nervous system.

sclereid (Gr. *skleros*, hard) In vascular plants, a sclerenchyma cell with a thick, lignified, secondary wall having many pits; not elongate like a fiber.

sclerenchyma (Gr. *skleros*, hard + *en*, in + *chymein*, to pour) A type of tissue made up of sclerenchyma cells.

scrotum (L., bag) The pouch that contains the testes in most mammals.

scuttellum The modified cotyledon in cereal grains.

secondary cell wall In plants, the innermost layer of the cell wall. Secondary walls have a highly organized microfibrillar structure and are often impregnated with lignin.

secondary growth In vascular plants, an increase in stem and root diameter made possible by cell division of the lateral meristems.

secondary immune response The swifter response of the body the second time it is invaded by the same pathogen because of the presence of memory cells, which quickly become antibody-producing plasma cells.

secondary induction An induction between tissues that have already differentiated.

secondary structure In a protein, hydrogen-bonding interactions between CO and NH groups of the primary structure.

second law of thermodynamics A statement concerning the transformation of potential energy into heat; it says that disorder (entropy) is continually increasing in the universe as energy changes occur, so disorder is more likely than order.

second messenger A small molecule or ion that carries the message from a receptor on the target cell surface into the cytoplasm.

segregation The process by which alternative forms of traits are expressed in offspring rather than blending each trait of the parents in the offspring.

selection The process by which some organisms leave more offspring than competing ones, and their genetic traits tend to appear in greater proportions among members of succeeding generations than the traits of those individuals that leave fewer offspring.

selective permeability Condition in which a membrane is permeable to some substances but not to others.

self-fertilization The union of egg and sperm produced by a single hermaphroditic organism.

semen (L., seed) In reptiles and mammals, sperm-bearing fluid expelled from the penis during male orgasm.

semicircular canal Any of three fluid-filled canals in the inner ear that help to maintain balance.

semiconservative replication DNA replication, in which each strand of the original duplex serves as the template for construction of a totally new complementary strand, so the original duplex is partially conserved in each of the two new DNA molecules.

senescent Aged, or in the process of aging.

sensory neuron A neuron that transmits nerve impulses from a sensory receptor to the central nervous system or central ganglion; an afferent neuron.

sepal (L. *sepalum*, a covering) A member of the outermost floral whorl of a flowering plant.

septum, pl. septa (L., fence) A wall between two cavities.

seta, pl. setae (L., bristle) In an annelid, bristles of chitin that help to anchor the worm during locomotion or when it is in its burrow.

sex-linked A trait determined by a gene carried on the X chromosome and absent on the Y chromosome.

sexual reproduction The process of producing offspring through an alternation of fertilization (producing diploid cells) and meiotic reduction in chromosome number (producing haploid cells).

sexual selection A type of differential reproduction that results from variable success in obtaining mates.

shoot In vascular plants, the aboveground portions, such as the stem and leaves.

sieve cell In the phloem of vascular plants, a long, slender sieve element with relatively unspecialized sieve areas and with tapering end walls that lack sieve plates.

sinoatrial (SA) node *See* pacemaker.

sinus (L., curve) A cavity or space in tissues or in bone.

sister chromatid One of two identical copies of each chromosome, still linked at the centromere, produced as the chromosomes duplicate for mitotic division; similarly, one of two identical copies of each homologous chromosome present in a tetrad at meiosis.

sodium-potassium pump Transmembrane channels engaged in the active (ATP-driven) transport of sodium ions, exchanging them for potassium ions, where both ions are being moved against their respective concentration gradients; maintains the resting membrane potential of neurons and other cells.

solute A molecule dissolved in some solution; as a general rule, solutes dissolve only in solutions of similar polarity; for example, glucose (polar) dissolves in (forms hydrogen bonds with) water (also polar), but not in vegetable oil (nonpolar).

solvent The medium in which one or more solutes is dissolved.

somatic cell Any of the cells of a multicellular organism except those that are destined to form gametes (germline cells).

somatic mutation A change in genetic information (mutation) occurring in one of the somatic cells of a multicellular organism, not passed from one generation to the next.

somatic nervous system (Gr. *soma*, body) In vertebrates, the neurons of the peripheral nervous system that control skeletal muscle.

somite (Gr, *soma*, body) One of the blocks, or segments, of tissue into which the mesoderm is divided during differentiation of the vertebrate embryo.

Southern blot A procedure used for identifying a specific gene, in which DNA from the source being tested is cut into fragments with restriction enzymes and separated by gel electrophoresis, then blotted onto a sheet of nitorocellulose and probed with purified, labeled, single-stranded DNA corresponding to a specific gene; if the DNA matching the specific probe is present in the source DNA, it is visible as a band of radioactive label on the sheet.

species, pl. species (L., kind, sort) A kind of organism; species are designated by binomial names written in italics.

spectrin A scaffold of proteins that links plasma membrane proteins to actin filaments in the cytoplasm of red blood cells, producing their characteristic biconcave shape.

sperm (Gr. *sperma*, seed) A mature male gamete, usually motile and smaller than the female gamete.

spermatid (Gr. *sperma*, seed) In animals, each of four haploid (1*n*) cells that result from the meiotic divisions of a spermatocyte; each spermatid differentiates into a sperm cell.

spermatozoa The male gamete, usually smaller than the female gamete, and usually motile.

sphincter (Gr. *sphinkter*, band, from *sphingein*, to bind tight) In vertebrate animals, a ring-shaped muscle capable of closing a tubular opening by constriction (such as the one between stomach and small intestine, or between anus and exterior).

spindle apparatus The assembly that carries out the separation of chromosomes during cell division; composed of microtubules (spindle fibers) and assembled during prophase at the equator of the dividing cell.

spiracle (L. *spiraculum*, from *spirare*, to breathe) External opening of a trachea in arthropods.

spongy parenchyma A leaf tissue composed of loosely arranged, chloroplast-bearing cell. *See* palisade parenchyma.

sporangium, pl. sporangia (Gr. *spora*, seed + *angeion*, a vessel) A structure in which spores are produced.

spore A haploid reproductive cell, usually unicellular, capable of developing into an adult without fusion with another cell.

sporophyte (Gr. *spora*, seed + *phyton*, plant) The spore producing, diploid (2*n*) phase in the life cycle of a plant having alternation of generations.

stabilizing section A form of selection in which selection acts to eliminate both extremes from a range of phenotypes.

stamen (L., thread) The organ of a flower that produces the pollen; usually consists of anther and filament; collectively, the stamens make up the androecium.

starch (Mid. Eng. *sterchen*, to stiffen) An insoluble polymer of glucose; the chief food storage substance of plants.

statocyst (Gr. *statos*, standing + *kystis*, sac) A sensory receptor sensitive to gravity and motion.

stele The central vascular cylinder of stems and roots.

stem cell A relatively undifferentiated cell in animal tissue that can divide to produce more differentiated tissue cell.

stereoscopic vision (Gr. *stereos*, solid + *opitkos*, pertaining to the eye) Ability to perceive a single, three-dimensional image from the simultaneous but slightly divergent two-dimensional images delivered to the brain by each eye.

stigma (Gr., mark, tatoo mark) (1) In angiosperm flowers, the region of a carpel that serves as a receptive surface for pollen grains. (2) Light-sensitive eyespot of some algae.

stipules Leaflike appendages that occur at the base of some flowering plant leaves or stems.

stolon (L. *stolo*, shoot) (1) A stem that grows horizontally along the ground surface and may form adventitious roots, such as runners of the strawberry plant.

stoma, pl. stomata (Gr., mouth) In plants, a minute opening bordered by guard cells in the epidermis of leaves and stems; water passes out of a plant mainly through the stomata.

stratum corneum The outer layer of the epidermis of the skin of the vertebrate body.

striated muscle (L., from *striare*, to groove) Skeletal voluntary muscle and cardiac muscle.

stromatolite A fossilized mat of ancient bacteria formed as long as 2 or 3 billion years ago, in which the bacterial remains individually resemble some modern-day bacteria.

style (Gr. *stylos*, column) In flowers, the slender column of tissue that arises from the top of the ovary and through which the pollen tube grows.

substrate (L. *substratus*, strewn under) The foundation to which an organism is attached; a molecule upon which an enzyme acts.

succession In ecology, the slow, orderly progression of changes in community composition that takes place through time.

summation Repetitive activation of the motor neuron resulting in maximum sustained contraction of a muscle.

surface tension A tautness of the surface of a liquid, caused by the cohesion of the molecules of liquid. Water has an extremely high surface tension.

surface-to-volume ratio Relationship of the surface area of a structure, such as a cell, to the volume it contains.

swim bladder An organ encountered only in the bony fish that helps the fish regulate its buoyancy through increasing or decreasing the amount of gas in the bladder via the esophagus or a specialized network of capillaries.

symbiosis (Gr. *syn*, together with + *bios*, life) The condition in which two or more dissimilar organisms live together in close association; includes parasitism (harmful to one of the organisms), commensalism (beneficial to one, of no significance to the other), and mutualism (advantageous to both).

sympatric speciation The differentiation of populations within a common geographic area into species.

synapse (Gr. *synapsis*, a union) A junction between a neuron and another neuron or muscle cell; the two cells do not touch, the gap being bridged by neurotransmitter molecules.

synapsis (Gr., contact, union) The point-by-point alignment (pairing) of homologous chromosomes that occurs before the first meiotic division; crossing over takes place during synapsis.

synaptic cleft The space between two adjacent neurons.

synaptic vesicle A vesicle of a neurotransmitter produced by the axon terminal of a nerve. The filled vesicle migrates to the presynaptic membrane, fuses with it, and releases the neurotransmitter into the synaptic cleft.

synaptonemal complex A protein lattice that forms between two homologous chromosomes in prophase I of meiosis, holding the replicated chromosomes in precise register with each other so that base-pairs can form between nonsister chromatids for crossing over that is usually exact within a gene sequence.

syncytial blastoderm A structure composed of a single large cytoplasm containing about 4000 nuclei in embryonic development of insects such as *Drosophila*.

syngamy (Gr. *syn*, together with + *gamos*, marriage) The process by which two haploid cells (gametes) fuse to form a diploid zygote; fertilization.

systolic pressure A measurement of how hard the heart is contracting. When measured during a blood pressure reading, ventricular systole (contraction) is what is being monitored.

T

tagma, pl. tagmata (Gr., arrangement, order, row) A compound body section of an arthropod resulting from embryonic fusion of two or more segments; for example, head, thorax, abdomen.

taxis, pl. taxes (Gr., arrangement) An orientation movement by a

(usually) simple organism in response to an environmental stimulus.

T cell A type of lymphocyte involved in cell-mediated immunity and interactions with B cells; the "T" refers to the fact that T cells are produced in the thymus.

telencephalon (Gr. *telos*, end + *encephalon*, brain) The most anterior portion of the brain, including the cerebrum and associated structures.

telomere A specialized nontranscribed structure that caps each end of a chromosome.

tendon (Gr. *tendon*, stretch) A strap of cartilage that attaches muscle to bone.

tertiary structure The folded shape of a protein, produced by hydrophobic interactions with water, ionic and covalent bonding between side chains of different amino acids, and van der Waal's forces; may be changed by denaturation so that the protein becomes inactive.

testcross A mating between a phenotypically dominant individual of unknown genotype and a homozygous "tester," done to determine whether the phenotypically dominant individual is homozygous or heterozygous for the relevant gene.

testis, pl. testes (L., witness) In mammals, the sperm-producing organ.

tetanus Sustained forceful muscle contraction with no relaxation.

thalamus (Gr. *thalamos*, chamber) That part of the vertebrate forebrain just posterior to the cerebrum; governs the flow of information from all other parts of the nervous system to the cerebrum.

thermodynamics (Gr. *therme*, heat + *dynamis*, power) The study of transformations of energy, using heat as the most convenient form of measurement of energy.

thigmotropism In plants, unequal growth in some structure that comes about as a result of physical contact with an object.

threshold The minimum amount of stimulus required for a nerve to fire (depolarize).

thylakoid (Gr. *thylakos*, sac + *-oides*, like) A saclike membranous structure containing chlorophyll in cyanobacteria and the chloroplasts of eukaryotic organisms.

tight junction Region of actual fusion of cell membranes between two adjacent animal cells that prevents materials from leaking through the tissue.

tissue (L. *texere*, to weave) A group of similar cells organized into a structural and functional unit.

totipotent A cell that possesses the full genetic potential of the organism.

trachea, pl. tracheae (L., windpipe) A tube for breathing; in terrestrial vertebrates, the windpipe that carries air between the larynx and bronchi (which leads to the lungs); in insects and some other terrestrial arthropods, a system of chitin-lined air ducts.

transcription (L. *trans*, across + *scribere*, to write) The enzyme-catalyzed assembly of an RNA molecule complementary to a strand of DNA.

transcription factor One of a set of proteins required for RNA polymerase to bind to a eukaryotic promoter region, become stabilized, and begin the transcription process.

transfection The transformation of eukaryotic cells in culture.

transfer RNA (tRNA) (L. *trans*, across + *ferre*, to bear or carry) A class of small RNAs (about 80 nucleotides) with two functional sites; at one site an "activating enzyme" adds a specific amino acid, while the other site carries the nucleotide triplet (anticodon) specific for that amino acid.

translation (L. *trans*, across + *latus*, that which is carried) The assembly of a protein on the ribosomes, using mRNA to specify the order of amino acids.

translocation (L. *trans*, across + *locare*, to put or place) In plants, the long-distance transport of soluble food molecules (mostly sucrose), which occurs primarily in the sieve tubes of phloem tissue; in genetics, the interchange of chromosome segments between nonhomologous chromosomes.

transpiration (L. *trans*, across + *spirare*, to breathe) The loss of water vapor by plant parts; most transpiration occurs through the stomata.

transposition Type of genetic recombination in which transposable elements (transposons) move from one site in the DNA sequence to another, apparently randomly.

transposon (L. *transponere*, to change the position of) A DNA sequence capable of transposition.

triglyceride An individual fat molecule, composed of a glycerol and three fatty acids.

triploid Possessing three sets of chromosomes.

trochophore A free-swimming larval stage unique to the mollusks and annelids.

trophic level (Gr. *trophos*, feeder) A step in the movement of energy through an ecosystem.

trophoblast (Gr. *trephein*, to nourish + *blastos*, germ) In vertebrate embryos, the outer ectodermal layer of the blastodermic vesicle; in mammals it is part of the chorion and attaches to the uterine wall.

tropism (Gr. *trope*, a turning) A response to an external stimulus.

tropomyosin (Gr. *tropos*, turn + *myos*, muscle) Low-molecular-weight protein surrounding the actin filaments of striated muscle.

troponin Complex of globular proteins positioned at intervals along the actin filament of skeletal muscle; thought to serve as a calcium-dependent "switch" in muscle contraction.

tubulin (L. *tubulus*, small tube + *in*, belonging to) Globular protein subunit forming the hollow cylinder of microtubules.

tumor-suppressor gene A gene that normally functions to inhibit cell division; mutated forms can lead to the unrestrained cell division of cancer, but only when both copies of the gene are mutant.

turgor pressure (L. *turgor*, a swelling) The pressure within a cell resulting from the movement of water into the cell; a cell with high turgor pressure is said to be turgid. *See* osmotic pressure.

U

unequal crossing over A process by which a crossover in a small region of misalignment at synapsis causes two homologous chromosomes to exchange segments of unequal length.

unsaturated fat A fat molecule in which one or more of the fatty acids contain fewer than the maximum number of hydrogens attached to their carbons.

urea (Gr. *ouron*, urine) An organic molecule formed in the vertebrate liver; the principal form of disposal of nitrogenous wastes by mammals.

urethra (Gr. from *ourein*, to urinate) The tube carrying urine from the bladder to the exterior of mammals.

uric acid Insoluble nitrogenous waste products produced largely by reptiles, birds, and insects.

urine (Gr. *ouron*, urine) The liquid waste filtered from the blood by the kidney and stored in the bladder pending elimination through the urethra.

uterus (L., womb) In mammals, a chamber in which the developing embryo is contained and nurtured during pregnancy.

V

vacuole A membrane-bounded sac in the cytoplasm of some cells, used for storage or digestion purposes in different kinds of cells; plant cells often contain a large central vacuole that stores water, proteins, and waste materials.

vascular cambium In vascular plants, a cylindrical sheath of meristematic cells, the division of which produces secondary phloem outwardly and secondary xylem inwardly; the activity of the vascular cambium increases stem or root diameter.

vascular tissue (L. *vasculum*, a small vessel) Containing or concerning vessels that conduct fluid.

vas deferens (L. *vas*, a vessel + *ferre*, to carry down) In mammals, the tube carrying sperm from the testes to the urethra.

vasopressin A posterior pituitary hormone that regulates the kidney's retention of water.

vegetal pole The hemisphere of the zygote, comprising cells rich in yolk.

vein (L. *vena*, a blood vessel) (1) In plants, a vascular bundle forming a part of the framework of the conducting and supporting tissue of a stem or leaf. (2) in animals, a blood vessel carrying blood from the tissues to the heart.

veliger The second larval stage of mollusks following the trochophore stage, during which the beginning of a foot, shell, and mantle can be seen.

ventricle A muscular chamber of the heart that receives blood from an atrium and pumps blood out to either the lungs or the body tissues.

vesicle (L. *vesicula*, a little bladder) A small intracellular, membrane-bounded sac in which various substances are transported or stored.

vessel element In vascular plants, a typically elongated cell, dead at maturity, which conducts water and solutes in the xylem.

vestibular apparatus The complicated sensory apparatus of the inner ear that provides for balance and orientation of the head in vertebrates.

villus, pl. villi (L., a tuft of hair) In vertebrates, one of the minute, fingerlike projections lining the small intestine that serve to increase the absorptive surface area of the intestine.

visceral mass (L., internal organs) Internal organs in the body cavity of an animal.

vitamin (L. *vita*, life + *amine*, of chemical origin) An organic substance that cannot be synthesized by a particular organism but is required in small amounts for normal metabolic function.

viviparity (L. *vivus*, alive + *parere*, to bring forth) Refers to reproduction in which eggs develop within the mother's body and young are born freeliving.

voltage-gated ion channel A transmembrane pathway for an ion that is opened or closed by a change in the voltage, or charge difference, across the cell membrane.

W

water potential The potential energy of water molecules. Regardless of the reason (e.g., gravity, pressure, concentration of solute particles) for the water potential, water moves from a region where water potential is greater to a region where water potential is lower.

wild type In genetics, the phenotype or genotype that is characteristic of the majority of individuals of a species in a natural environment.

X

xylem (Gr. *xylon*, wood) In vascular plants, a specialized tissue, composed primarily of elongate thick-walled conducting cells, which transports water and solutes through the plant body.

Y

yolk plug A plug occurring in the blastopore of amphibians during formation of the archenteron in embryological development.

Z

zona pellucida An outer membrane that encases a mammalian egg.

zoospore A motile spore.

zooxanthellae Symbiotic photosynthetic protists in the tissues of corals.

zygomycetes (Gr. *zygon*, yoke + *mykes*, fungus) A type of fungus whose chief characteristic is the production of sexual structures call zygosporangia, which result from the fusion of two of its simple reproductive organs.

zygote (Gr. *zygotos*, paired together) The diploid (2*n*) cell resulting from the fusion of male and female gametes (fertilization).

Credits

15.9b, 15.11, 15.12, 15.13, 15.15, 15.16, 15.18, 16.11, 16.12b, 16.13a-b, 16.14, 16.15, 16.17–1,16.18, 16.21, 16.22, 17.22, 18.2, 18.3, 18.5, 18.6, 18.8, 18.10, 18.11, 18.26, 19.10, 20.3, 21.2, 21.17, 24.6, 24.7, 27.7, 29.7, 29.9, 29.10, 29.13, 35.4, 40.3, 43.6, 43.7, 45.18, 46.2, 48.3, 49.2, 50.6, 57.14, and 58.27

Photos

Chapter 1

Opener: F1.01: ©Christopher Ralling;TA1.A: © John Gerard; TA1.01(left) © E.R. Degginger/ Animals Animals/Earth Scenes; F1.02H: ©Tom J. Ulrich/Visuals Unlimited; F1.02I: ©Maslowski/ Visuals Unlimited; F1.02J: ©Arthur Morris/Visuals Unlimited; F1.02C: From C.P. Morgan & R.A. Jersid, Anatomical Record, 166: 575-586, 1970 © John Wiley & Sons; F1.02D: © Lennart Nilsson; F1.02J(right): © Alan Nelson/ Animals Animals/Earth Scenes; F1.02K: © John D. Cunningham/ Visuals Unlimited; F1.02L: © Robert & Jean Pollock; TA1.02 (right): Courtesy of Stephen Jay Gould; F1.02E: © Ed Reschke; F1.05: Mohr/ Darwin Collection; Herbert Rose Barrand, photographer. Reproduced by permission of the Huntington Library, San Marino, California; F1.07: Courtesy James Moore, from Darwin, 1992/ Warner Books; F1.09A: © Christian Crzimek/Photo Researchers; F1.09B: © George Holton/Photo Researchers; F1.10B: C.H.Greenewalt ©VIREO/Academy of Natural Sciences of Philadelphia; F1.10A: © Frank B. Gill/Vireo; F1.13: © Mary Evans Picture Library/ Photo Researchers; F1.15F: © Stephen J. Krasemann /DRK Photo;F1.15A: © T.J. Beveridge/ Visuals Unlimited; F1.15B: © J.J. Cardamone, Jr. & B.K. Pugashetti/BPS/Tom Stack & Associates; F1.15C: © Marty Snyderman

Chapter 2

Opener: F2.01: © Irving Geis/Photo Researchers; F2.02: Courtesy of O. Marti, B. Drake, P.K. Hansma, Univ. of California-Santa Barbara; F2.11B: © Archive Photos; F2.12A: © Jurgen Schmitt/The Image Bank; F2.12B: © Nicholas Foster/The Image Bank; F2.12C: © John Kelly/The Image Bank; F2.15: © George I. Bernard/Animals Animals/Earth Scenes; F2.23: © Roy Pfortner/ Peter Arnold, Inc.

Chapter 3

Opener: F3.01: © Science Photo Library/Photo Researchers; F3.09A: © J.D. Litvay/Visuals Unlimited; F3.10: © Scott Johnson/Animals Animals/Earth Scenes; F3.15A: © Michael Pasdzior/The Image Bank; F3.15B: © Manfred Kage/Peter Arnold, Inc.; F3.15C: © George Bernard/Animals Animals/Earth Scenes; F3.15D: © OSF/Animals Animals/Earth Scenes; F3.15E: © Scott Blackman/Tom Stack & Associates; F3.22: Courtesy of Lawrence Berkeley National Laboratory

Chapter 4

Opener: F4.01: © Visuals Unlimited; F4.02: © Edward S. Ross; F4.03: © Y. Arthur Bertrand/Peter Arnold, Inc.; F4.04: © T.E. Adams/Visuals Unlimited; F4.05: © John D.Cunningham/Visuals Unlimited; F4.06: © Bob McKeever/Tom Stack & Associates; F4.11: Courtesy of William J. Schopf, UCLA; F4.12: Reproduced with permission from Dr. J. William Schopf, 1992-93, CSEOL/UCLA; F4.13: © Dwight R. Kuhn; F4.14: ©NASA; F4.15: © NASA; F4.16: NASA; F4.17: Courtesy of Andrew H. Knoll

Chapter 5

Opener: F5.01: © Manfred Kage/Peter Arnold; F5.02: © Manfred Kage/Peter Arnold, Inc.; F5.04A: © John Walsh/ Science Photo Library/Photo Researchers; F5.04B: © David M. Phillips/Visuals Unlimited; F5.04C: © David M. Phillips/ Visuals Unlimited; F5.05: © Manfred Kage/Peter Arnold, Inc. F5.07C: © Ed Reschke; F5.07A: © J.J. Cardamore/BPS/ Tom Stack & Associates; F5.07B: © David M. Phillips/Visuals Unlimited; F5.09: Courtesy of E.H. Newcomb & T.D. Pugh, U. of Wisconsin; F5.10A: © L.M. Pope/BPS/Tom Stack & Associates; F5.11: © Biophoto Associates/Science Source/Photo Researchers; F5.12B: © Biophoto Associates/Photo Researchers; F5.13B: Courtesy of Dr. Thomas Tillack; F5.13C: Photo by J. David Robertson, from Charles Flickinger, Medical Cellular Biology, W.B. Saunders, 1979; F5.15: © Ed Reschke; F5.16B: © Alfred Pasiexa/Science Photo Library/Photo Researchers; F5.18: Courtesy of Dr. Charles Flickinger, Medical Cellular Biology, W.B. Saunders,1979; F5.22: From C.P. Morgan & R.A. Jersid, *Anatomical Record* 166:575-586, 1970 © John Wiley & Sons F5.23B: © Don W. Fawcett/Visuals Unlimited; F5.24: Courtesy of Kenneth Miller, Brown University; F5.25A: © Don W. Fawcett/Visuals Unlimited; F5.27: Courtesy of Drs. J.V. Small & G. Rinnerthaler; F5.29A: Courtesy of James Spudich F5.29B: Photo by Susama Ito, from Charles Flickinger, Medical Cellular Biology, W.B. Saunders, 1979

Chapter 6

Opener: F6.01:Image Courtesy Drs. S.A. Porcelli, P.J. Peters, E. van Donselaar & M. Sugita. Reprinted with permission from AAAS/ Science Vol 273-7/1996-Cover; F6.05: Photo by J. David Robertson from Charles Flickinger, Medical Cellular Biology, W.B. Saunders, 1979; F6.07 (left): © W. Omerod/ Visuals Unlimited; F6.07 (right): © Keith Porter/Photo Researchers; F6.16: All: © David M. Phillips/Visuals Unlimited; F6.18C: © Biophoto Associates/ Photo Researchers; F6.18D: © Biophoto Associates/Photo Researchers; F6.19A: Courtesy of M.M. Perry & A.B. Gilbert, Cell Science, 39:257-272,1979; F6.19B: Courtesy of M.M. Perry & A.B. Gilbert, Cell Science, 39:257-272, 1979; F6.19C: Courtesy of M.M. Perry & A.B. Gilbert, Cell Science, 39:257-272, 1979; F6.19D: Courtesy of M.M. Perry & A.B. Gilbert, Cell Science, 39:257-272,1979; F6.20B: Courtesy of Dr. Birgit H. Satir; F6.25: Courtesy of Dr. Paul Quinton

Chapter 7

Opener: F7.01: © Cabisco/Visuals Unlimited; F7.14: © Audrey M. Glauert & G.M.E. Cook; F7.19A: © Don Fawcett/Visuals Unlimited

Chapter 8

Opener: F8.01: © Robert A. Caputo/Aurora; F8.03 (both): © Spencer Grant/ Photo Edit; F8.11B: Courtesy Dr. Lester J. Reed, University of Texas, Austin

Chapter 9

Opener: F9.01: © Jane Buron/ Bruce Coleman Inc.; F9.02: © Steve Kaufman/Peter Arnold, Inc.; F9.09: © Grant Heilman/ Grant Heilman Photography; F9.B-1(All): Edward S. Ross; F9.14: Nature V.376/Aug 24'95/p661-So Iwata, Christian Ostermeler, Bernd Ludwig & Hartmut Michel / Photo Courtesy H. Michel; F9.B-2 (top right): © Edward S. Ross; F9.B-3 (top center): © Edward S. Ross; F9.B4 (bottom center): © Edward S. Ross; F9.B5 (bottom right) © Edward S. Ross

Chapter 10

Opener: F10.01: © Grant Heilman/Grant Heilman Photography; F10.02A: Courtesy of Kenneth Miller, Brown University; F10.02B: © Manfred Kage/Peter Arnold,Inc.; F10.03A: Courtesy of Hunt Institute for Botanical Documentation, Carnegie Mellon University, Pittsburgh, PA; F10.07: © Eric V. Grave/ Photo Researchers; F10.08A (both): © Eric Soder/ Tom Stack & Associates; F10.09B (center): Courtesy Dr. Linda E. Graham & Bruce Iverson; F10.09A(left): © SPL/Photo Researchers; F10.09C (right): © J. Robert Waaland/Visuals Unlimited; F10.13: © Alfred Pasieka/Science Photo Library/ Photo Researchers; F10.20: from G. Schneider, Y. Lindquist, C.-I. Branden, and G. Lorimer-EMBO J. 5(1986): 3411 Photo courtesy Dr. Gunter Schneider; F10.21: © Valorie Hodgson/Visuals Unlimited; F10.22: Ray P. Evert

Chapter 11

Opener: F11.01: © George Musil/Visuals Unlimited; F11.03: © Lee D. Simon/Photo Researchers; F11.04: © Biophoto Associates/Photo Researchers; F11.06: © Science Photo Library/Photo Researchers; F11.09A-F (all): © Dr. Andrew J. Bajer; F11.10: © Jeremy Pickett-Heaps/Photo Researchers; F11.11B: Dr. William Brinkley/Baylor College of Medicine; F11.13: © Dr. Andrew S. Bajer; F11.14: Dr. Jeremy Pickett-Heaps; F11.15: Dr. C.L. Rieder; F11.16A: © David M. Phillips/Visuals Unlimited; F11.16B: © Dr. Guenter Albrecht-Buehler; F11.17A (left): © B.A. Palevitz & E.H. Newcomb/BPS/Tom Stack & Associates; F11.22: Dr. Potu N. Rao; F11.25: © David M. Phillips/Visuals Unlimited

Chapter 12

Opener: F12.01: © L. Maziarski/ Visuals Unlimited; F12.02: © David Cavagnaro/Peter Arnold, Inc.; F12.08: Diter von Wettstein; Reproduced with permission from Annual Review of Genetics Vol. 6, 1972, by Annual Review, Inc.; F12.10: © B. John Cabisco/ Visuals Unlimited; F12.14: © C.A.Hasenkampf/Biological Photo Service; F12.15A (above): Courtesy Sheldon Wolff & Judy Bodycote; F12.15 (below): Courtesy of Bill Morgan,

U. of California, San Francisco; F12.16: Charles J. Cole, Dept. of Herpetology, American Museum of Natural History

Chapter 13

Opener: F13.01: © Alan Carey/The Image Works; F13.02: George Johnson; F13.03: © Visuals Unlimited; F13.04: © Richard Gross/Biological Photography; F13.05: Courtesy of V. Orel, Mendelianum Musei Moraviae, Brno; F13.06: Courtesy of the Mendelianum, Moravian Museum, Brno, Czech Republic; F13.10: Courtesy of R.W. Van Norman; F13.11: Courtesy of V. Orel, Mendelianum Musei Moraviae, Brno; F13.12: © Globe Photos, Inc.; F13.18A: From Albert & Blakeslee, Corn and Man *Journal of Heredity*, Vol. 5, pg. 511, 1914, Oxford University Press; F13.20A © Fred Bruemmer; F13.20B (right): © Fred Bruemmer; F13.21: © Cabisco/Phototake; F13.28: © Leonard Lessin/Peter Arnold; F13.30A: Courtesy of Loris McGavaran, Denver Children's Hospital; F13.30B: © Richard Hutchings/Photo Researchers; F13.34: © Corbis-Bettmann; F13.36: © SPL/Photo Researchers; F13.39: © Hattie Young/Science Photo Library/Photo Researchers; F13.41: © Yoav Levy/Phototake

Chapter 14

Opener: F14.01: © Visuals Unlimited; F14.02: © Dennis Kunkel/Phototake; F14.05: O.T. Avery, C.M. Macleod & M. McCarty, "Studies on the Chemical Nature of the Substance Inducing Transformation of Pneumococcal Types." Reproduced from the Journal of Experimental Medicine 79 (1944): 137-158. fig. 1 by copyright permission of the Rockefeller University Press. Reproduced by permission. Photograph made by Mr. Joseph B. Haulenbeek; F14.06B: © Lee D. Simon/Photo Researchers; F14.06D: Courtesy of Prof. A.K. Kleinschmidt; F14.11: © Cold Spring Harbor Laboratory; F14.12: From J.D. Watson, *The Double Helix*, Atheneum, New York 1968: F14.13: A.C. Barrington Brown from J.D. Watson, *The Double Helix*, Atheneum, New York 1968. Cold Springs Harbor Lab; F14.16: From M. Meselson and F.W. Stahl/Proceedings of the Nat. Acad. of Sci. 44(1958):671; F14.18: From: *Biochemistry 4/e* by Stryer © 1995 by Lubert Stryer. Used with permission of W.H. Freeman and Company; F14.21:

© E. Kiselva & D. Fawcett/Visuals Unlimited; F14.22: © Prof. Ulrich Laemmli/Photo Researchers; F14.23A: Courtesy of David Wolstenholme

Chapter 15

Opener: F15.01: © K.G. Murti/Visuals Unlimited; F15.06: From R.C. Williams; Proc. Nat. Acad. of Sci. 74(1977): 2313; F15.10: Courtesy of Dr. Oscar L. Miller; F15.14: Courtesy of Dr. Oscar L. Miller; F15.17B: Courtesy of Bert O'Malley, Baylor College of Medicine

Chapter 16

Opener: F16.01: Courtesy of Claus Pelling; F16.08A: © R. Feldman/Visuals Unlimited; F16.08B: Courtesy of Brian W. Matthews; F16.13C: © Jack Griffith; F16.17: Harrison Echols; F16.19A: Courtesy of Victoria Foe; F16.23: © Jack Griffith

Chapter 17

Opener: F17.01: © David M. Phillips/Visuals Unlimited; F17.02: © David Scharp; F17.08: © Custom Medical Stock; F17.10: © SPL/Photo Researchers; F17.16: (all): Dr. Kathleen Cho; F17.18: Courtesy of American Cancer Society; F17.23: © Georgiana Silk/Outline Press Syndicate; F17.25: Courtesy of Charles Brinton; F17.27A: Courtesy of Stanley Cohen & J. Shapiro, Scientific American, Feb. 1980, pgs 40-49; F17.34: Courtesy of M.L. Pardue and J.G. Gall

Chapter 18

Opener: F18.01: © Stanley Cohen/Science Photo Library/Photo Researchers; TA18.01 (top left): Alger Meekma; TA18.02: (bottom left): Robert H. Delvin/Fisheries & Oceans Canada; TA18.03 (top right): David Umberger/Purdue News Service; TA18.04 (bottom right): Richard Shade; F18.04B: Courtesy of Bio-Rad Laboratories; F18.13: Nancy Federspiel, Stanford DNA Sequence & Tech. Center; F18.14: Courtesy of Lifecodes Corp., Stamford, CT; F18.15: AP/Wide World Photos; F18.16: R.L. Brinster, U. of Pennsylvania Sch. of Vet. Med.; F18.19: © Calgene/Peter Arnold; F18.20: Courtesy of Monsanto; F18.21: © Biological Photo Service; F18.22: Courtesy of Monsanto; F18.23: Courtesy of Monsanto; F18.25: © Roger Minkoff/Animals Animals/Earth Scenes; F18.26: AP/Wide World Photos

Chapter 19

Opener: F19.01: © Dick Poe/Visuals Unlimited; F19.02: © Renee Lynn/Photo Researchers; F19.03: Biological Photo Service; F19.05: © Edward S. Ross; F19.07A (left): © Joe McDonald/Animals Animals/Earth Scenes; F19.07B (center): © Kenneth W. Fink/The National Audubon Society Collection/Photo Researchers; F19.07C (right): © Tom McHugh/The National Audubon Society Collection/Photo Researchers; F19.09: © Nigel Cattlin/Holt Studios International/Photo Researchers

Chapter 20

Opener: F20.01: © John Cancalosi/Peter Arnold Inc.; F20.02: Courtesy of the University of Chicago Library/Dept. of Special Collections & Todd L. Savitt; F20.05 (both): © Breck P. Kent/Animals Animals/Earth Scenes; F20.07: The Smithsonian Institution; F20.08: J. Ulrich/Academy of Natural Sciences Philadelphia © VIREO; F20.21: © Cleveland P. Hickman

Chapter 21

Opener: F21.01: © John Conway/Visuals Unlimited; F21.03A: Courtesy of Paul R. Ehrlich; F21.03B: Courtesy of Paul R. Ehrlich; F21.07A: © Kjell B. Sandved/Visuals Unlimited; F21.07B: © Fateh Singh Rathore/Peter Arnold; F21.07C: © Porterfield/Chickering/Photo Researchers; F21.08: © Barbara Gerlach/Visuals Unlimited; F21.09 (all): Courtesy of Dr. Kenneth Y. Kaneshiro; F21.10A: © Breck P. Kent/Animals Animals/Earth Scenes; F21.10B: © Ken Lucas/Biological Photo service; F21.12: © Frank Oberle/Photographic Resources; F21.14A (all): © William P. Mull; F21.16: © G. R. Roberts; F21.18B: © Gregory G. Dimijian/Photo Researchers; F21.18A: William J. Weber/Visuals Unlimited

Chapter 22

Opener: F22.01: © National Museum of Kenya/Visuals Unlimited; F22.02B: © Larry Tackett/Tom Stack & Associates; F22.02A: © Alan Nelson/Animals Animals/Earth Scenes; F22.03A: © Denise Tackett/Tom Stack & Associates; F22.03B: © Edward S. Ross; F22.04 (all): A: © Russell A. Mittermeier; F22.06 (all): © David L. Brill; F22.07: © John Reader/Science Photo Library/Photo Researchers; F22.09:

© Dan Dreyfuss & Associates Photography; F22.10 (both): Courtesy of National Museums of Kenya, photo © 1985 David L. Brill; F22.11: © 1994 Tim D. White/Brill Atlanta, housed in National Museum of Ethiopia; F22.15: Courtesy of National Museums of Kenya, photo © 1985 David L. Brill; F22.20: AP/Wide World Photos

Chapter 23

Opener: F23.01: © Dick Poe/Visuals Unlimited; F23.03B: © Manfred Danegger/Peter Arnold, Inc.; F23.10B: © Arthur Morris/Visuals Unlimited; F23.10A: © John Shaw/Tom Stack & Associates; F23.13: Courtesy of National Museum of Natural History, Smithsonian Institution; F23.14A: © Jean Vie/Gamma-Liaison

Chapter 24

Opener: F24.01: © Dave Fleetham/Tom Stack & Associates; F24.02: Courtesy of Save-The-Redwoods League; F24.03A (all): © Edward S. Ross; F24.06A (all): Courtesy of J.B. Losos; F24.08: Merlin D. Tuttle, Bat Conservation International; F24.09: © Jim Harvey/Visuals Unlimited; F24.10: © Gunter Ziesler/Peter Arnold; F24.12 (both): © Edward S. Ross; F24.13A (both): © Edward S. Ross; F24.14A: © Lincoln P. Brower; F24.14B: © Lincoln P. Brower; F24.15: © James L. Castner; F24.16: James L. Castner; F24.17: Courtesy of Rolf O. Peterson; F24.18A: © Biophoto Associates/Science Source/Photo Researchers; F24.19A: © Tom J. Ulrich/Visuals Unlimited; F24.20A: © Edward S. Ross; F24.20B: Courtesy of Dr. Paul A. Opler; F24.21 (all): © Edward S. Ross; F24.22A: Courtesy of G. David Tilman; F24.27: Courtesy of E.O. Wilson; F24.29 (all): Tom Bean; F24.30A: © E.R. Degginger/Animals Animals/Earth Scenes; F24.30B: © Anne Wertheim/Animals Animals/Earth Scenes

Chapter 25

Opener: F25.01: David Cavagnaro/Peter Arnold, Inc.; F25.03: © Walt Anderson/Visuals Unlimited; F25.07A: U.S. Forest Service

Chapter 26

Opener: F26.01: NASA; F26.02: Courtesy of William J. Hamilton III; F 26.08: © Edward S. Ross; F26.10: © Edward S. Ross; F26.16: © Michael Graybill & Jan Hodder/Biological Photo Service;

Chapter 44

Opener: F44.01: © William C. Ober; F44.03A: © Alex Kerstitch/Visuals Unlimited; F44.03B: Randy Morse/Tom Stack & Associates; F44.03C: © Daniel W. Gotshall/Visuals Unlimited; F44.06: Kjell Sandved/Butterfly Alphabet; F44.08: Courtesy of David L. Pawson, Harbor Branch Oceanographic Institute, Inc./National Museum of Natural History; F44.09: © Carl Roessler/Tom Stack & Associates; F44.10: Daniel W. Gotshall/Visuals Unlimited; F44.11: © Gary Milburn/Tom Stack & Associates; F44.12: © Kjell Sandved/Visuals Unlimited; F44.13A: © Jeff Rotman; F44.13B: © Daniel W. Gotshall; F44.14: © Alex Kerstitch/Visuals Unlimited; F44.15: © Alan Baker/Visuals Unlimited; F44.16 (both): © D.P. Wilson/FLPA-Images of Nature; F44.17: © C.R. Wyttenbach, U. of Kansas/Biological Photo Service

Chapter 45

Opener: F45.01: © Brian Parker/Tom Stack & Associates; F45.03: © Eric N. Olson, PhD/The University of Texas MD Anderson Cancer Center; F45.05A: © Rick Harbo Marine Images; F45.06: © Heather Angel; F45.10: © Denise Tackett/Tom Stack & Associates; F45.12: © Tom McHugh/Tom Stack & Associates; F45.14: © 1993 Norbert Wu; F45.15A: © William H. Mullins/Photo Researchers; F45.15B: © John D. Cunningham/Visuals Unlimited; F45.17: © 1979 Peter Scoones/Contact Press Images/Woodfin Camp & Associates Inc.; F45.21C: Courtesy of Prof. M.H. Wake, U. of California; F45.21A: © John Shaw/Tom Stack & Associates; F45.21B: © Suzanne L. Collins & Joseph T. Collins/Photo Researchers; F45.22: © John Gerlach/Tom Stack & Associates; F45.31: © William J. Weber/Visuals Unlimited; F45.32A: © John Cancalosi/Tom Stack & Associates; F45.32B: © Rod Planck/Tom Stack & Associates; F45.33: © Brian Parker/Tom Stack & Associates; F45.34: © Gil Lopez-Espina/Visuals Unlimited; F45.37B: © Bill Kamin Visuals Unlimited; F45.37C: © Jim Davis/National Audubon Society Collection/Photo Researchers; F45.37D: © Art Wolfe; F45.37A: © Thomas Kitchin/Tom Stack & Associates; F45.38: © Denise Tackett/Tom Stack & Associates; F45.41: © Stephen Dalton/National Audubon Society Collection/Photo Researchers; F45.42A: © B.J. Alcock/Visuals Unlimited; F45.42B: © J.E. Wapstra/Australian Museum; F45.43: © Fritz Prenzel/Animals Animals/Earth Scenes; F45.44A: © Carl R. Sams II/Peter Arnold Inc.; F45.44B: © Joe McDonald/Visuals Unlimited; F45.44C: © Denise Tackett/Tom Stack & Associates; F45.44D: © Stephen J. Krasemann/DRK Photo; F45.44E: © Mike Bacon/Tom Stack & Associates; F45.45: © Tim Flannery

Chapter 46

Opener: F46.01: Photo Lennart Nilsson/Albert Bonniers Forlag AB, *Behold Man*, Little Brown & Co.; TA46.01A,B,C: © Ed Reschke; TA46.01D: © Fred Hossler/Visuals Unlimited; TA46.01E: © Ed Reschke; TA46.02A: © Biophoto Associates/Photo Researchers; TA46.02B: © Cleveland P. Hickman; TA46.02C: © Chuck Brown/Photo Researchers; TA46.02D: © Ed Reschke; TA46.02E: © Ken Edward/Science Source/Photo Researchers; TA46.03A,B,C: © Ed Reschke; F46.06: © J. Gross/Science Photo Library/Photo Researchers; F46.07: © Biophoto Associates/Photo Researchers; F46.08B: © Ed Reschke; F46.09A: © Ed Reschke; F46.10: © David M. Phillips/Visuals Unlimited

Chapter 47

Opener: F47.01: © Anthony Bannister/Animals Animals/Earth Scenes; F47.08, F47.09A: © Dr. H.E. Huxley; F47.16: © John D. Cunningham/Visuals Unlimited

Chapter 48

Opener: F48.01: © John Gerlach/Animals Animals/Earth Scenes; F48.14: © Biophoto Associates/Science Source/Photo Researchers; F48.23: © John Sholtis/The Rockefeller University

Chapter 49

Opener: F49.01: © Professors P.M. Motta & S. Correr/Science Photo Library/Photo Researchers; F49.06: © Manfred Kage/Peter Arnold Inc.; F49.09: © Ed Reschke; F49.11: © Ed Reschke

Chapter 50

Opener: F50.01: © Frans Lanting/Minden Pictures; F50.04: © Russ Kinne/Photo Researchers; F50.08: © Chris Bjornberg/Photo Researchers; F50.11A: Courtesy of H.R. Dunker

Chapter 51

Opener: F51.01: Courtesy of David I. Vaney, University of Queensland, Australia; F51.04: © C.S. Raines/Visuals Unlimited; F51.12: © John Heuser, Washington University, School of Medicine, St. Louis, MO.; F51.17B: © E.R. Lewis, YY Zeevi, T.E. Everhart, U. of California/Biological Photo Service; F51.28: Dr. Marcus E. Rachle, Washington University, McDonnell Center for High Brain Function/Peter Arnold, Inc.; F51.29: photo Lennart Nilsson/Albert Bonniers Forlag AB, *Behold Man*, Little Brown & Co.; F51.32A: © E.R. Lewis/Biological Photo Service; F51.33A: © Ed Reschke

Chapter 52

Opener: F52.01: © Omikron/Photo Researchers; F52.06: © Ed Reschke; F52.11: © Wendy Shatil/Bob Rozinski/Tom Stack & Associates; F52.14: © Dalton, S. OSF/Animals Animals/Earth Scenes; F52.15 (left): © David M. Dennis/Tom Stack & Associates; F52.15 (center) © Kjell Sandved/Butterfly Alphabet; F52.15 (right): Kjell Sandved/Butterfly Alphabet; F52.24: © Leonard L. Rue, III

Chapter 53

Opener: F53.01: © Francois Gohier/Science Source/Photo Researchers; F53.09: Photo Lennart Nilsson/Albert Bonniers Forlag AB, *Behold Man*, Little Brown & Co.; F53.12: © UPI/Corbis-Bettmann; F53.16: © John Paul Kay/Peter Arnold, Inc.

Chapter 54

Opener: F54.01: Courtesy of the National Library of Medicine; F54.03: © Ellen Dirkson/Visuals Unlimited; F54.04: © Manfred Kage/Peter Arnold, Inc.; F54.06: © Lennart Nilsson, from *The Body Victorious*; F54.09: © Visuals Unlimited; F54.16: © Dr. Robert Dourmashkin; F54.20: © Stuart Fox; F54.26A,B: © Dr. Andrejs Liepins/Science Photo Library/Photo Researchers; F54.31A: © David Scharf/Peter Arnold Inc.; F54.31B: © Oliver Meckes/Photo Researchers

Chapter 55

Opener: F55.01: © Belinda Wright/DRK Photo; F55.04: © Tim Davis/The National Audubon Society Collection/Photo Researchers; F55.14: © Prof. P. Motta/Dept. of Anatomoy/University "La Sapienza," Rome/Science Photo Library/Photo Researchers; F55.20: © Larry Brock/Tom Stack & Associates; F55.4: © Tim Davis/Photo Researchers

Chapter 56

Opener: F56.01: © Gregory Dimijian, M.D. 1984/The National Audubon Society Collection/Photo Researchers; F56.02A: © Chuck Wise/Animals Animals/Earth Scenes; F56.02B: © Fred McConnaughey/The National Audubon Society Collection/Photo Researchers; F56.04: © David Doubilet; F56.05: © Hans Pfletschinger/Peter Arnold Inc.; F56.06: © Cleveland P. Hickman; F56.07: © Frans Lanting/Minden Pictures; F56.08A: © Jean Phillippe Varin/Jacana/Photo Researchers; F56.08B: © Tom McHugh/The National Audubon Society Collection/Photo Researchers F56.08C: © Fritz Prenzel/Animals Animals/Earth Scenes; F56.12A: © David M. Phillips/Photo Researchers; F56.13: Photo Lennart Nilsson/Bonniers Forlag AB, *A Child is Born*, Dell Publishing Co.; F56.18: © Ed Reschke; F56.22 (all): Bob Coyle photo, © Times Mirror Higher Education Group, Inc.

Chapter 57

Opener: F57.01: © Cabisco/Visuals Unlimited; F57.B: © Ed Lewis; F57.03: photo Lennart Nilsson/Albert Bonniers Forlag AB, *A Child is Born*, Dell Publishing Company; F57.04A (all): © Cabisco/Phototake; F57.08: Dr. Barry Gumbiner; F57.10: Richard O. Hynes; F57.16 (left): Dr. Christiane Nusslein-Volhard/Max Planck Institute as published in From Egg to Adult, © 1992 HHMI; F57.16 (right 3 images): James Langeland, Stephen Paddock & Sean Carroll as published in *From Egg to Adult*, © 1992 HHMI; F57.17: Courtesy of Manfred Frasch; F57.18: Courtesy of E.B. Lewis; F57.22: William McGinnis/Yale University; F57.23: From *Journal of Cell Biology*, by T.L. Deckworth & E.M. Johnson (1993) 123:1207-1222. Copyright permission of the Rockefeller University Press; F57.24: © A. Tsiaras/Photo Researchers; F57.28: © Robert Ricci/Gamma Liaison; F57.31: Courtesy of Dr. P.M. Lansdorp; F57.32: © Mark Wexler/Woodfin Camp & Associates

Chapter 58

Opener: F58.01: Photo Lennart Nilsson/Bonniers Forlag AB, *A Child is Born*, Dell Publishing Co.; F58.02B: © P. Bagavandoss/

Index

Mutation rate, 394
Mutualism, 485–86, 486f
Myasthenia gravis, 1081
Mycelium, **631,** 631f
  primary, **638,** 638f
  secondary, **638,** 638f
myc gene, 344, 346t
Mycobacterium leprae, 601t
Mycobacterium tuberculosis, 376t, 600, 600f, 601t
Mycologist, **630**
Mycoplasma, 601t
Mycoplasma genitalium, 376t
Mycoplasma pneumoniae, 376t
Mycorrhizae, 484, **633,** 642, 642f, 646, 677
Mycotrophic plant, 660
Myc protein, 220f
myc proto-oncogene, 223
Myelinated axon, 997, 997f
Myelin sheath, 916, **993,** 993f
Myocardium, 932
Myofibril, 914, 914f, **924,** 924–25f
Myoglobin, 48, 48t, **930–31**
Myosin, 48t, 49, 914f, **926,** 926f, 931
Myotonic dystrophy, 357, 357t
Myxini (class), 870t
Myxobacteria, 597t
Myxomycota (phylum), 608–9t, 625, 625f

N

NAD+, 57, 153
  as electron carrier, 173, 173f
  reduction to NADH, 153
  regeneration of, 168, 168f, 180, 180f
  structure of, 153f, 173f
NADH, 164, 173
  contributing electrons to electron transport chain, 174–76, 174f, 176f
  from glycolysis, 166, 167f
  from Krebs cycle, 170, 171f
  recycling of, 169, 169f
  structure of, 173f
NADH dehydrogenase, **174**
NADP+, 185, 185f, 195–97, 195–96f
NADPH, 185, 185f, 197–202
NADP reductase, 195–96f, 197
Naked mole rat, 1225
Naked-seeded plant. *See* Gymnosperm
Nanos protein, **1153**
Naphthalene acetic acid, 753, 753f
Nasal passage, 1028
Nashua River, 554
Natural killer cells, **1070**–71, 1070–71f, 1076f
Natural selection, 8, 12–**13,** 390, **396,** 406–11
  invention of theory of, 12–13
Nauplius larva, **838,** 838f
Nautilus, 819, 819f
Navigation, 897, **1208**–9
Neanderthals, 460
Nearly neutral model, 398
Nearsightedness, 1037, 1037f
Neck, of tapeworm, **805**
Necrosis, **1157**
Nectar, 661, 720–21
Nectar-feeding birds, 1216, 1216f
Nectary, 661, 720
Needle, pine, 658
Nef gene, 582, 583f
Negative feedback control, 1058–59, 1058–59f, 1096, 1096–97f
Negative gravitropism, 758, 758f

Negative phototaxis, 1208
Negative pressure breathing, 979
Neisseria gonorrhoeae, 601t, 602
Nekton, 536
Nematocyst, **798**–99, 816
Nematoda (phylum), 789, 790t, 806–7, 806f
Nematode
  development in, 1142f, 1162, 1162–63f
  fungus feeding on, 632, 632f
  in soil, 806, 806f
Nemertea (phylum), 805, 805f
Neotenic larva, 976
Nephridia, **815,** 821
Nephron, **1104**
  filtration in, 1104–5, 1104–5f
  structure of, 1104–5, 1104–5f
  transport processes in, 1112–13, 1112–13f
Nephrostome, **815,** 1102, 1102f
Nepotism, **1224**
Neries virens, 820f
Neritic zone, **535,** 535f
Nerve, 916, **993,** 1014, 1014f
  stimulation of muscle contraction, 929, 929f
Nerve cord, 793, **864,** 1004
Nerve gas, 999
Nerve growth factor, 221, 221t, 1048
Nerve net, 798, 1004
Nerve ring, **854**
Nerve tissue, 904, 904f, 916, 916f, 916t
Nervous system, 905t, 991–1018
  of arthropods, 833
  of cephalopods, 819
  elaboration of, 1179
  evolution of, 569t
  of flatworms, 803
  interaction with endocrine system, 1047
  of invertebrates, 1004, 1004f
Nesting, 1124, 1124f
Net energy, 1215
Net primary productivity, **512**–13, 540t
Neural basis, of behavior, 1196, 1196f
Neural crest, 868, **1145,** 1173t, **1178**–80, 1178f, 1180f
Neural groove, 1176f, **1178,** 1178f
Neural plate, 868, 1176f, 1178f
Neural regulation, of digestion, 950
Neural tube, 868f, **1145,** 1176f, **1178,** 1178f
Neuraminidase, **584**
Neuroblastoma, 346t
Neuroendocrine reflex, **1055**
Neuroethology, **1196**
Neurofibroma, 346t
Neurofilament, 99
Neuroglia, **993**
Neurohormone, **1046,** 1046f
Neuromodulator, **1000**
Neuromuscular junction, 929, 929f
Neuron, 916, 916f, 991–93f, 992–93.
  See also specific types of neurons
  membrane potential, 994–95, 994–95f
Neuropeptide, **1000**
Neurospora crassa, Beadle and Tatum experiment with, 294–95, 294–95f
Neurotoxin, 799
Neurotransmitter, **129,** 129f, 929, **998**–1003, 998f, 1046
  drug addiction and, 1002–3, 1002–3f

Neurotrophin, **1048**
Neurulation, 1144f, **1145,** 1173t, 1176f, **1178**–79, 1178–79f, 1185
Neutral theory, **398**
Neutron, **20**
Neutrophils, 913, **961, 1070,** 1072, 1076
Newt, 876, 876t, 878
Newton, Isaac, 4
New World monkey, **445,** 445f, 449f
New Zealand alpine buttercup, 439, 439f
NF genes, 346t
Niacin. *See* Vitamin B₃
Niche, **425, 481**–82
  competition and, 481, 481f
  fundamental, **481**
  realized, **481**
Niche overlap, 482
Nicolson, G., 108
Nicotinamide adenine dinucleotide. *See* NAD+
Nicotine, 1002–3
Nicotine patch, 1003
Nicotine receptor, 1002
Nierenberg, Marshall, 302
Nif genes, **381**
Night blindness, 954t
Nightshade (*Atropa belladona*), 778f
Nile perch, 438
Nine + two structure, **101,** 101f, 593
Nitella, 613
Nitrates, 507, 507f
Nitric oxide, **131,** 988, **1000,** 1048, 1128
Nitrification, 507, 507f
Nitrites, 507
Nitrobacter, 597t
Nitrogen
  on earth, 25t
  electron energy levels for, 23f
  in human body, 25t
  in plants, 708, 708t
Nitrogenase, 506
Nitrogen-containing base, 57, 57f
Nitrogen cycle, **507,** 507f
  in forest ecosystem, 509, 509f
Nitrogen fertilizer, 381, 707, 707f
Nitrogen fixation, 484, 592
  evolution of, 158
  genetically-engineered plants, 381, 381f
  nitrogen cycle, 507, 507f
Nitrogenous waste, **1114,** 1114f
  detoxification of, 1114
  excretion in urine, 1107
  flushing of, 1114
  insolubilization of, 1115
Nitrogen reductase, 506
Nitrosomonas, 597t
Nitrous oxide, 507, 549
Nociceptor, 1023t, **1025**
Noctiluca, 616f
Nocturnal mammal, **444**
Node (plant stem), **678,** 678f, 683
Nodes of Ranvier, 916, **993,** 993f, 997
Nonassociative learning, **1198**
Noncoelomate animals, 787–808
Noncompetitive inhibitor, **152,** 153f
Noncyclic photophosphorylation, **196**
Nondisjunction
  involving autosomes, 264
  involving sex chromosomes, 265, 265f
  primary, **264**–65
Nonfattening sweets, 43
Nongranular leukocytes, **961**

Nonmyelinated axon, 997, 997f
Nonrandom mating, 393–94, 393t
Nonsegmented body, 795
Nonsense codon, **307,** 330
Nonspecific immune defense, 1068–69
Nonsteroidal anti-inflammatory drugs (NSAIDs), **1049**
Nonvertebrate chordate, 866–67, 866–67f
Norepinephrine, **1000,** 1016–18, 1046–47, 1049t, 1052, 1062, 1179
Nori, 614
Nostoc, 192f, 597t, 649
Notochord, **864,** 865f, 868f, **1178,** 1178f
N-ras gene, 346t
NSAIDs. *See* Nonsteroidal anti-inflammatory drugs
NtrC protein, 326f
Nucellus, 656, **658**–59, 658f
Nuclear envelope, 86–88f, **89,** 215, 568t
Nuclear mitosis, 630
Nuclear pore, 88f, **89**
Nuclear power, 548
Nuclear weapons, 548, 641
Nuclease, 943t
Nucleic acids, 38, 39f, 40t, 56, **57**–58, 284
  catabolism of, 178f
  evolution of, 58
  structure of, 57–58, 57f, 284, 284f
Nuclein, 284
Nucleoid, **80,** 593
Nucleoid region, **595**
Nucleolar organizer region, **359**
Nucleolus, 86–87f, 88, 90t, 95, 95f, 215
Nucleosome, **89,** 89f, **208**–9, 208f, 224, 293, 293f
  promoter blocking by, 327, 327f
Nucleotide, 56f, **57,** 284
Nucleotide substitution, 335t
Nucleus, atomic, 22
Nucleus, cellular, **70,** 74, **80,** 86–89f, **88**–89, 90t, 102t
  hereditary information stored in, 278–79
  movement of molecules into and out of, 89
  transplantation of, 279, 384–85, 1151
Nudibranch, 816, 817f
Numerical taxonomy, **564**–65
Nursing, 1080, 1189–90, 1189f
Nüsslein-Volhard, Christiane, 1152f
Nut, 727–28, 727f
Nutcracker man, 450
Nutrient
  essential, 953, 953f, 954t
  in fish, 871
  plant, 703–10, 708t, 709f
  transport in blood, 959
Nutritional mutant, in *Neurospora crassa,* 294–95, 294–95f
Nutritional quality, of crop plants, 784
Nymph, **844**

O

Oak (*Quercus*), 430–41, 430f, 560, 560f, 673
Oat (*Avena sativa*), 750–51, 768t
Obelia, 800, 800f
Obesity, **952,** 952f